World Checklist of

Myrtaceae

World Checklist of Myrtaceae

Rafaël Govaerts, Marcos Sobral, Peter Ashton, Fred Barrie, Bruce K. Holst, Leslie L. Landrum, Kazue Matsumoto, Fiorella Fernanda Mazine, Eimear Nic Lughadha, Carolyn Proença, Lucia Helena Soares-Silva, Peter G. Wilson & Eve Lucas

Kew Publishing
Royal Botanic Gardens, Kew

PLANTS PEOPLE
POSSIBILITIES

First published in 2008 by
Royal Botanic Gardens, Kew
Richmond, Surrey, TW9 3AB, UK
www.kew.org

ISBN 978-1-84246-391-8

British Library Cataloguing in Publication Data
A catalogue record for this book is available from the British Library

Typesetting and page layout: Christine Beard
Design by Media Resources,
Royal Botanic Gardens, Kew

Printed in the United Kingdom by Lightning Source

For information or to purchase all Kew titles please visit
www.kewbooks.com or email publishing@kew.org

All proceeds go to support Kew's work in saving the world's plants for life

Contents

Introduction

Myrtaceae is generally considered a well-delimited family of two subfamilies comprising 17 tribes. The checklist includes 132 genera and 5671 species. Centres of diversity for the family are in the wet tropics, notably South America, Australia and Tropical Asia. The family is often characterized by having stringy bark, mostly having an inferior ovary, opposite leaves filled with aromatic essential oils in translucent 'gland dots', and often with a secondary marginal vein. Myrtaceae genera are commonly cultivated for timber and shade (*Eucalyptus*), as spices (*Pimenta, Syzygium*), for their essential oils (e.g. *Eucalyptus, Callistemon, Metrosideros*), and for their comestible fruit (e.g. *Eugenia, Plinia, Syzygium*). *Eucalyptus* is commonly found naturalised and invasive throughout the tropics. *Psidium* species are also invasive in certain tropical localities.

Review process

In a large family such as Myrtaceae, with distribution centres in both the old and new worlds, the cooperation of experts from diverse institutions was sought to review a basic checklist produced mainly from published literature. In February 2005 a workshop was held at the Royal Botanic Gardens, Kew, during which a manuscript of the World Checklist of Myrtaceae was reviewed by an international team of Myrtaceae researchers to develop a strategy for checking and editing the status of each name. Contributors (Authors and institutions: Rafaël Govaerts (Royal Botanic Gardens, Kew), Marcos Sobral (Universidade Federal de Minas Gerais), Peter Ashton (Arnold Arboretum), Fred Barrie (Missouri Botanical Garden), Bruce Holst (Marie Selby Botanical Gardens), Leslie Landrum (Arizona State University), Kazue Matsumoto (Universidade Estadual de Campinas), Fiorella Fernanda Mazine (Universidade de São Paulo), Eimear Nic Lughadha (Royal Botanic Gardens, Kew), Carolyn Proença (Universidade de Brasília), Lucia Helena Soares-Silva (Universidade de Brasília), Peter Wilson (Royal Botanic Gardens, Sydney) & Eve Lucas (Royal Botanic Gardens, Kew)) were assigned a taxonomic or geographically defined portion of the list to edit. This resulted in c. 700 corrections and updates being made to the manuscript and a product reflecting consensus among key Myrtaceae specialists.

The World Checklist of Myrtaceae is now online. (http://apps.kew.org/wcsp/myrtaceae). The searchable list provides information on accepted scientific names, synonyms and geographical distribution. It allows for searches of all scientific names of a particular genus, or the distribution of a taxon by country and TDWG code. Corrections made by the team of Myrtaceae researchers and other contributers are listed under the reference tabs on the species pages. The checklist continues to be edited and maintained by screening new literature as well as through continued collaboration with international Myrtaceae researchers and according to the curation of the Kew Myrtaceae collection.

Progress and ongoing research

Dedicated molecular and morphological study of Myrtaceae phylogenetics by local researchers has resulted in impressive increases in our understanding of the systematics of some, particularly Australasian groups. In tropical Asia and the new world however, the morphological similarity of species and the existing unstable, at times turbulent taxonomy have created a situation familiar to taxonomists of large and complex taxa (three genera contain more than 700 species); the most challenging groups have not been treated taxonomically since the 19th century and to do so remains daunting. For groups such as *Syzygium*, certain sections of *Eugenia* and *Myrcia s.l.* the nomenclatural review of the checklist must remain incomplete until '*a botanist shall be found courageous enough to undertake so tedious a task*' (Bentham 1869).

Fortunately, systematic and taxonomic study in Myrtaceae continues. Research slowly fills the gaps in our knowledge so that as this checklist emerges in printed form, it is out of date. Studies in preparation that will generate significant numbers of nomenclatural changes are revisions of *Psdium* prepared independantly by Landrum (ASU) and Proença (UB), of newly defined monophyletic clades of *Eugenia* and *Myrcia s.l.* by Mazine (ESA) and Lucas (K) respectively, of *Neomitranthes* and associated genera by Souza (RJ), of sections of *Syzygium* by Biffin and Craven (CSIRO) and of those taxa occurring in areas of particular Myrtaceae species richness such as the Brazilian state of Espírito Santo, by Sobral (UFMG), central America, by Barrie (MO) and Amazonia, by Holst (SELBY).

As a result of this continuous growth in knowledge, the checklist should be used in conjunction with the on-line version of the database available from the Kew website.

How to use this Checklist

Structure

The checklist is derived from a database encompassing 24 fields and complying with the data standards proposed by the Taxonomic Databases Working Group (TDWG). Its compilation was effected using Microsoft Access 2002 for Windows, within which editing was also carried out. The arrangement of genera being alphabetical with accepted and synonymous genera intercalated. For more information on the process see Govaerts (2005).

Names

Names of accepted genera and their species and infraspecific taxa are listed alphabetically. Synonymised genera (and species) are intercalated. For each accepted taxon, associated synonyms are listed chronologically if heterotypic, with any homotypic synonyms following in a given lead; in addition, all synonyms in an accepted genus are listed alphabetically at the end of that genus. Place and date of publication of all names are given. Citation of authors follows *Authors of Plant Names* (Brummitt & Powell 1992; updated and available electronically at http://www.ipni.org/ipni/query_author.html); for book abbreviations, the standard is *Taxonomic Literature*, 2nd edn. (Stafleu & Cowan 1976–88; supplements, 1992–2000) also available electronically (http://tl2.idcpublishers.info/); and periodicals are abbreviated according to *Botanico-Periodicum-Huntianum/ 2* (Bridson 2004). A question mark (?) following a name and author indicates that a place of publication has yet to be established. Names of hybrids are preceded by a multiplication sign (?), with the place of publication being followed by the names of the parents if known. Basionyms or replaced synonyms of accepted names are designated by an asterisk (*). For genera, the number of accepted species and the geographical distribution are furnished together with general comments and the suprageneric taxa as used at Kew.

Acceptance of taxa

Acceptance of species and infraspecific taxa is based not only on assessments of literature but also, by reference to specialist advice as explained above and (where necessary) to the herbarium or living collections. In principle, the last published synonymy is followed unless the author was obviously not aware of earlier published data (indicated by the references cited) or by advise received from the taxonomic or regional experts on the groups involved. The generic deliminations follow the arrangement in the Kew herbarium.

New names and combinations

A few necessary new names and combinations have been made in the text for nomenclatural reasons and to bring known taxa into line with current practice.

Geographical Distribution

Distributions of **species** and taxa of lower rank are furnished in two ways: firstly by a generalised statement in narrative form, and secondly as TDWG geographical codes (Brummitt, 2001) also available electronically (http://www.nhm.ac.uk/hosted_sites/tdwg/geogrphy.html) expressed to that system's third level. Examples of the former include:

E. & C. U.S.A.
Texas to C. America
Mexico (Veracruz)
Europe to Iran
E. Himalaya, Tibet, China (W. Yunnan)
Philippines (Luzon)
S. Trop. America

When the presence of a taxon in a given region or location is not certainly known, a question mark is used, e.g. New Ireland ?; when an exact location within a country is not known, a question mark within brackets is used, e.g. Mexico (?). Distributions of **genera** are furnished in a relatively simplified form, any special features being given within brackets.

With respect to the TDWG codes, the **region** is indicated by the two-digit number (representative of the first two levels), the first digit also indicating the continent. The letter codes following the digits, when given, represent the third-level **unit** (a country, state or other comparable area). They usually are the first three letters of a given unit's name, but sometimes are contractions. If the country code is not known, '+' is used. For taxa that are known or appear to be extinct in a given region, '†' is used after the country code. Naturalisation is expressed by putting the third-level codes in lower case and, if in a second-level region all occurrences are the result of naturalisation, the code number for the region is placed in brackets. The application of question marks is as indicated above for geographical regions. Examples include:

12 SPA	[SW. Europe: Spain]
32 +	[C. Asia (more exact distribution not known)]
36 CHN? 38 JAP KOR	[Doubtful in China and Eastern Asia; China, Japan, Korea]
51 NZN NZS	[New Zealand: North and South Islands]
76 ARI 77 NWM TEX	[SW. & SC. U.S.A., Mexico & C. America:
79 ALL 80 GUA HON	Arizona, New Mexico, Texas, Mexico, Guatemala and Honduras]
77 TEX†	[SC. U.S.A.: Texas, where extinct]

Life-forms

The terminology for *life-forms*, definitions of which follow, is based on the system of Raunkiær (1934, especially chapters 1 and 2) with modifications derived from *Flora van België, het Groothertogdom Luxemburg, Noord-Frankrijk en de aangrenzende gebieden* (De Langhe *et al.* 1983: pp. xvii–xviii, 869 (fig. 16)).

Main Categories:

nanophanerophyte (*nanophan.*)
stems: woody and persisting for several years
buds: above soil level but normally below 3 m
commonly known as shrubs

chamerophyte (*cham.*)
stems: herbaceous and/or woody and persisting for several years
buds: on or just above soil level, never above 50 cm
commonly known as subshrubs

hemicryptophyte (*hemicr.*)
stems: herbaceous, often dying back after the growing season, with shoots at soil level surviving
buds: just on or below soil level
commonly known as perennials

geophyte
hemicryptophytes that survive unfavourable seasons in the form of a rhizome, bulb, tuber or rootbud

helophyte (*hel.*)
hemicryptophytes that grow in soil saturated with water or in water. leaf and flower bearing shoots rise above water

hydrophyte
stems: vegetative shoots sunk in water
buds: permanently or temporarily on the bottom of the water

therophyte (*ther.*)
plants that survive unfavourable seasons in the form of seeds and complete their life-history during the favourable season.
commonly known as annuals

Additional information:

climbing (*cl.*)

Abbreviations

A	alpine/arctic
Agg.	aggregate
al.	alii: others
Arch.	archipelago
app.	approaching, close to
auct.	of author
C.	Central
cham.	chamaephyte
cit.	citatus: cited
cl.	climbing
Co.	county
comb.	combinatio: combination
cons.	conservandus: to be conserved
cppo	centre page pull-out
cult.	cultus: cultivated
cv.	cultivarietas: cultivar
descr.	description
Distr.	district
DT	dry tropical (desert/steppe)
E.	East(ern)
etc.	et cetera: and the rest
e.g.	exampli gratia: for example
G	temperate
hel.	helophyte
hort.	hortorum: of gardens
I./Is	island(s)
ICBN	International Code of Botanical Nomenclature
i.e.	id est: that is
ign.	ignotus: unknown
in litt.	in litteris: in correspondence
ined.	ineditus: unpublished, provisional name
inq.	inquilinus: naturalised
i.q.	idem quod: the same as
Kep.	kepulauan (islands)
Medit.	mediterranean
MT	monsoon tropical (savanna)
Mt./Mts	mountain(s)
N.	North(ern)
nanophan.	nanophanerophyte
No.	numero: number
noh	new orchid hybrids
nom. cons.	nomen conservandum: name conserved in ICBN

nom. illeg. .... nomen illegitimum: illegitimate name
nom. inval. ... invalid name
nom. nud. ..... nomen nudum: name without a description
nom. rejic. .... nomen rejiciendum: name rejected in ICBN
nom. superfl. .. nomen superfluum: name superfluous when published
nov. .... novus: new
orth. var. ..... orthographic variant
par. .... parasitic
Pen. .... peninsula(r)
phan. .... phanerophyte
p.p. .... pro parte: partly
Prov. .... province
q.e. .... quod est: which is
q.v. .... quod vide: which see
Reg. .... region
Rep. .... republic
S .... Subtropical
S. .... South(ern)
seq. .... sequens: following
s.l. .... sensu lato: in the broad sense
sp. .... species
s.p. .... without page number
sphalm. ...... sphalmate: by mistake
s.s. .... sensu stricto: in the narrow sense
st. .... status
subtrop. ...... subtropical
syn. .... synonymon: synonym
T .... tropical
temp. .... temperate
ther. .... therophyte
trop. .... tropical
vol. .... volume: volume
viz. .... videlicet: namely
W. .... West(ern)
WT .... wet tropical
? .... not known, doubtful
† .... extinct

Acknowledgements

We would particularly like to thank Nick Black and Antonella Linguanti for writing and maintaining the computer programs and reports used to create the checklist. Thanks are also due to Simon Owens, Keeper of the Herbarium at Kew for providing the facilities to allow the work to come to fruition, the staff at the Kew library, R.K.Brummitt, K.Challis and C.Barker for nomenclatural advice and Eve Lucas for initiating and susbequently managing the review process of the checklist. We are also grateful to correspondents who have commented on entries in the online version of the checklist.

References

Bentham, G. (1869). Notes on Myrtaceae. J. Linn. Soc. London, Bot. 10: 101–166.

Bridson, G., comp. & ed. (2004). Botanico-Periodicum Huntianum/Second Edition. 2 vol. Pittsburgh: Hunt Institute for Botanical Documentation

Brummitt, R. K. (2001). World Geographical Scheme for Recording Plant Distributions, ed 2. xvi, 138 pp. Hunt Institute for Botanical Documentation, Carnegie-Mellon University, Pittsburgh, Penna. (for the International Working Group on Taxonomic Databases for Plant Sciences). (Plant Taxonomic Database Standards, 2: version 1.0.) [http://www.tdwg.org/geo2.htm]

Brummitt, R. K. & Powell, C. E., (1992). Authors of Plant Names. 732 pp. Kew: Royal Botanic Gardens. [http://www.ipni.org/ipni/query_author.html]

De Langhe, J. E. et al. (1983). Flora van België, het Groothertogdom Luxemburg, Noord-Frankrijk en de aangrenzende gebieden. civ, 970 pp., illus., map. Patrimonium, Nationale Plantentuin van België, Meise.

Govaerts, R. (2005). The Monocot Checklist Project: a global effort. Acta Univ. Upsal., Symb. Bot. Upsal. 33(3): 175–177.

Greuter, W. et al. (1993). Names in Current Use for Extant Plant Genera (Names in current use, 3). xxvii, 1464 pp. Koeltz, Koenigstein. (Regnum Vegetabile.)

Raunkiær, C. (1934). The Life Forms of Plants and Statistical Plant Geography. xvi, 632 pp., illus. Oxford University Press, London.

Stafleu, F. & Cowan, R. S. (1976–88). Taxonomic Literature: A Selective Guide to Botanical Publications and Collections with Dates, Commentaries and Types. 2nd edn. 7 vols. Utrecht: Bohn, Scheltema & Holkema. (Regnum Vegetabilie 94, 98, 105, 110, 112, 115, 116.) Continued as Stafleu, F. et al.. (1992–2000). Taxonomic Literature, Supplement. Vols. 1–6. Koenigstein, Germany: Koeltz. (Regnum Vegetabile 125, passim. As of 2000 six volumes published.) [http://tl2.idcpublishers.info/]

Myrtaceae

Careya [Lecythidaceae]

Careya jambosoides Lauterb. = ***Syzygium suborbiculare*** (Benth.) T.G.Hartley & L.M.Perry

Abbevillea

Abbevillea O.Berg = ***Campomanesia*** Ruiz & Pav.
Abbevillea bullata Barb.Rodr. = ***Campomanesia sessiliflora*** var. ***bullata*** (Barb.Rodr.) Landrum
Abbevillea cerasoides (Cambess.) O.Berg = ***Campomanesia guaviroba*** (DC.) Kiaersk.
Abbevillea chrysophylla O.Berg = ***Campomanesia eugenioides*** (Cambess.) D.Legrand
Abbevillea eugenioides (Cambess.) O.Berg = ***Campomanesia eugenioides*** (Cambess.) D.Legrand
Abbevillea fenzliana O.Berg = ***Campomanesia guaviroba*** (DC.) Kiaersk.
Abbevillea fenzliana var. *brevipes* O.Berg = ***Campomanesia guaviroba*** (DC.) Kiaersk.
Abbevillea fenzliana var. *intermedia* O.Berg = ***Campomanesia guaviroba*** (DC.) Kiaersk.
Abbevillea fenzliana var. *longipes* O.Berg = ***Campomanesia guaviroba*** (DC.) Kiaersk.
Abbevillea fenzliana var. *obversa* O.Berg = ***Campomanesia guaviroba*** (DC.) Kiaersk.
Abbevillea gardneriana O.Berg = ***Campomanesia viatoris*** Landrum
Abbevillea guaviroba (DC.) O.Berg = ***Campomanesia guaviroba*** (DC.) Kiaersk.
Abbevillea klotzschiana O.Berg = ***Campomanesia guaviroba*** (DC.) Kiaersk.
Abbevillea klotzschiana var. *intermedia* O.Berg = ***Campomanesia guaviroba*** (DC.) Kiaersk.
Abbevillea klotzschiana var. *laurifolia* O.Berg = ***Campomanesia guaviroba*** (DC.) Kiaersk.
Abbevillea klotzschiana var. *longipes* O.Berg = ***Campomanesia guaviroba*** (DC.) Kiaersk.
Abbevillea langsdorffiana O.Berg = ***Campomanesia sessiliflora*** (O.Berg) Mattos var. ***sessiliflora***
Abbevillea martiana (DC.) O.Berg = ***Campomanesia aromatica*** (Aubl.) Griseb.
Abbevillea maschalantha O.Berg = ***Campomanesia guaviroba*** (DC.) Kiaersk.
Abbevillea maschalantha var. *oblongata* O.Berg = ***Campomanesia guaviroba*** (DC.) Kiaersk.
Abbevillea maschalantha var. *ovata* O.Berg = ***Campomanesia guaviroba*** (DC.) Kiaersk.
Abbevillea maschalantha var. *recurvata* O.Berg = ***Campomanesia guaviroba*** (DC.) Kiaersk.
Abbevillea neriiflora O.Berg = ***Campomanesia neriifolia*** (O.Berg) Nied.
Abbevillea phaea O.Berg = ***Campomanesia phaea*** (O.Berg) Landrum
Abbevillea punctulata (DC.) O.Berg = ***Campomanesia guaviroba*** (DC.) Kiaersk.
Abbevillea recurvata O.Berg = ***Psidium rufum*** Mart. ex DC.
Abbevillea regeliana O.Berg = ***Psidium rufum*** Mart. ex DC.
Abbevillea rufa O.Berg = ***Campomanesia rufa*** (O.Berg) Nied.
Abbevillea rugosa O.Berg = ***Campomanesia guazumifolia*** (Cambess.) O.Berg
Abbevillea schlechtendahliana O.Berg = ***Campomanesia schlechtendaliana*** (O.Berg) Nied.
Abbevillea sellowiana O.Berg = ***Campomanesia guaviroba*** (DC.) Kiaersk.
Abbevillea widgreniana (O.Berg) O.Berg = ***Campomanesia pubescens*** (Mart. ex DC.) O.Berg

Acca

Acca O.Berg, Linnaea 27: 138 (1856).
Peru to Uruguay. (25) ken tan uga (51) nzs 83 BOL PER 84 BZL BZS (85) jnf URU. [Myrtaceae]
3 Species
Orthostemon O.Berg, Linnaea 27: 440 (1856).
Feijoa O.Berg, Linnaea 29: 258 (1858).

Acca lanuginosa (Ruiz & Pav. ex G.Don) McVaugh, Taxon 5: 136 (1956).
C. & SC. Peru. 83 PER. Nanophan. or phan.
**Psidium lanuginosum* Ruiz & Pav. ex G.Don, Gen. Hist. 1: 421 (1831).
Acca peruviana O.Berg, Linnaea 27: 139 (1856).
Acca velutina Burret, Repert. Spec. Nov. Regni Veg. 50: 58 (1941).

Acca macrostema (Ruiz & Pav. ex G.Don) McVaugh, Taxon 5: 136 (1956).
C. Peru to Bolivia. 83 BOL PER. Nanophan.
Eugenia acka DC., Prodr. 3: 277 (1828).
**Psidium macrostemum* Ruiz & Pav. ex G.Don, Gen. Hist. 2: 833 (1832).
Acca domingensis O.Berg, Linnaea 27: 138 (1856).
Acca domingensis var. *angustifolia* O.Berg, Linnaea 27: 138 (1856).
Acca domingensis var. *latifolia* O.Berg, Linnaea 27: 138 (1856).
Myrtus eugeniiflora Kunth ex O.Berg, Linnaea 27: 139 (1856), pro syn.

Acca sellowiana (O.Berg) Burret, Repert. Spec. Nov. Regni Veg. 50: 59 (1941).
Brazil (S. Minas Gerais) to Uruguay. (25) ken tan uga (51) nzs 84 BZL BZS (85) jnf URU. Nanophan. or phan. Widely cultivated for its edible fruits (pineapple guava).
Orthostemon obovatus O.Berg in C.F.P.von Martius & auct. suc. (eds.), Fl. Bras. 14(1): 468 (1857). *Feijoa obovata* (O.Berg) O.Berg in C.F.P.von Martius & auct. suc. (eds.), Fl. Bras. 14(1): 616 (1859).
**Orthostemon sellowianus* O.Berg in C.F.P.von Martius & auct. suc. (eds.), Fl. Bras. 14(1): 467 (1857). *Feijoa sellowiana* (O.Berg) O.Berg in C.F.P.von Martius & auct. suc. (eds.), Fl. Bras. 14(1): 615 (1859).
Feijoa schenckiana Kiaersk., Enum. Myrt. Bras.: 186 (1891).
Feijoa sellowiana var. *rugosa* Mattos, Loefgrenia 70: 4 (1976). *Acca sellowiana* var. *rugosa* (Mattos) Mattos, Loefgrenia 99: 3 (1990).
Feijoa sellowiana f. *elongata* Voronova, Sborn. Nauchn. Trudov Prikl. Bot. Genet. Selekts. 139: 79 (1991).

***Synonyms*:**
Acca domingensis O.Berg = ***Acca macrostema*** (Ruiz & Pav. ex G.Don) McVaugh
Acca domingensis var. *angustifolia* O.Berg = ***Acca macrostema*** (Ruiz & Pav. ex G.Don) McVaugh
Acca domingensis var. *latifolia* O.Berg = ***Acca macrostema*** (Ruiz & Pav. ex G.Don) McVaugh
Acca glazioviana Kiaersk. = ***Eugenia moschata*** (Aubl.) Nied. ex T.Durand & B.D.Jacks.
Acca peruviana O.Berg = ***Acca lanuginosa*** (Ruiz & Pav. ex G.Don) McVaugh
Acca sellowiana var. *rugosa* (Mattos) Mattos = ***Acca sellowiana*** (O.Berg) Burret
Acca velutina Burret = ***Acca lanuginosa*** (Ruiz & Pav. ex G.Don) McVaugh

Accara

Accara Landrum, Syst. Bot. 15: 221 (1990).
SE. Brazil. 84 BZL. [Myrtaceae]
1 Species

Accara elegans (DC.) Landrum, Syst. Bot. 15: 221 (1990).
Brazil (Minas Gerais). 84 BZL. Nanophan.
**Myrtus elegans* DC., Prodr. 3: 240 (1828). *Psidium elegans* (DC.) Mart. ex O.Berg in C.F.P.von Martius & auct. suc. (eds.), Fl. Bras. 14(1): 391 (1857), nom. illeg. *Guajava elegans* (DC.) Kuntze, Revis. Gen. Pl. 1: 239 (1891).
Psidium elegans var. *angustifolium* O.Berg in C.F.P.von Martius & auct. suc. (eds.), Fl. Bras. 14(1): 391 (1857).
Psidium elegans var. *latifolium* O.Berg in C.F.P.von Martius & auct. suc. (eds.), Fl. Bras. 14(1): 391 (1857).
Myrtus stictophylla Kiaersk., Enum. Myrt. Bras.: 20 (1893). *Psidium stictophyllum* (Kiaersk.) Mattos, Loefgrenia 64: 1 (1975).

Acicalyptus

Acicalyptus A.Gray = ***Syzygium*** Gaertn.
Acicalyptus elliptica A.C.Sm. = ***Syzygium seemannii*** (A.Gray) Biffin & Craven
Acicalyptus eugenioides F.Muell. = ***Syzygium eugenioides*** (Merr. & L.M.Perry) Biffin & Craven
Acicalyptus fullagarii F.Muell. = ***Syzygium fullagarii*** (F.Muell.) Craven
Acicalyptus longiflora A.C.Sm. = ***Syzygium seemannii*** (A.Gray) Biffin & Craven
Acicalyptus myrtoides A.Gray = ***Syzygium myrtoides*** (A.Gray) R.Schmid
Acicalyptus myrtoides Seem. = ***Syzygium seemannii*** (A.Gray) Biffin & Craven
Acicalyptus nitida Brongn. & Gris = ***Syzygium brongniartii*** (Merr. & L.M.Perry) J.W.Dawson
Acicalyptus pennelii Guillaumin = ***Syzygium pennelii*** (Guillaumin) J.W.Dawson
Acicalyptus seemannii A.Gray = ***Syzygium seemannii*** (A.Gray) Biffin & Craven

Acmena

Acmena DC. = ***Syzygium*** Gaertn.
Acmena acuminatissima (Blume) Merr. & L.M.Perry = ***Syzygium acuminatissimum*** (Blume) DC.
Acmena australis L.A.S.Johnson = ***Syzygium ingens*** (F.Muell. ex C.Moore) Craven & Biffin
Acmena brachyandra (Maiden & Betche) Merr. & L.M.Perry = ***Syzygium ingens*** (F.Muell. ex C.Moore) Craven & Biffin
Acmena bracteolata (Wight) Walp. = ***Syzygium fastigiatum*** (Blume) Merr. & L.M.Perry
Acmena brassii Craven = ***Syzygium lenbrassii*** Craven & Biffin
Acmena caudata Merr. & L.M.Perry = ***Syzygium piluliferum*** Craven & Biffin
Acmena championii Benth. = ***Syzygium championii*** (Benth.) Merr. & L.M.Perry
Acmena chinensis Planch. = [36 CHS]
Acmena claviflora (Roxb.) Walp. = ***Syzygium claviflorum*** (Roxb.) Wall. ex A.M.Cowan & Cowan
Acmena dielsii Merr. & L.M.Perry = ***Syzygium acuminatissimum*** (Blume) DC.
Acmena dispansa (Ridl.) Merr. & L.M.Perry = ***Syzygium dispansum*** (Ridl.) Craven & Biffin
Acmena divaricata Merr. & L.M.Perry = ***Syzygium divaricatum*** (Merr. & L.M.Perry) Craven & Biffin
Acmena elliptica G.Don ex Steud. = ***Syzygium smithii*** (Poir.) Nied.
Acmena elliptica G.Benn. = ***Syzygium smithii*** (Poir.) Nied.
Acmena floribunda (Sm.) A.Cunn. ex DC. = ***Angophora floribunda*** (Sm.) Sweet
Acmena floribunda DC. = ***Syzygium smithii*** (Poir.) Nied.
Acmena floribunda var. *elliptica* A.Cunn. ex DC. = ***Syzygium smithii*** (Poir.) Nied.
Acmena fluvicola T.G.Hartley & Craven = ***Syzygium fluvicola*** (T.G.Hartley & Craven) Craven & Biffin
Acmena gerrardii Harv. ex Hook.f. = ***Syzygium gerrardii*** (Harv. ex Hook.f.) Burtt Davy
Acmena grata (Wight) Walp. = ***Syzygium antisepticum*** (Blume) Merr. & L.M.Perry
Acmena graveolens (F.M.Bailey) L.S.Sm. = ***Syzygium graveolens*** (F.M.Bailey) Craven & Biffin
Acmena hemilampra (F.Muell.) Merr. & L.M.Perry = ***Syzygium hemilamprum*** (F.Muell.) Craven & Biffin
Acmena hemilampra subsp. *orophila* B.Hyland = ***Syzygium hemilamprum*** subsp. ***orophilum*** (B.Hyland) Craven & Biffin
Acmena ingens (F.Muell. ex C.Moore) Guymer & B.Hyland = ***Syzygium ingens*** (F.Muell. ex C.Moore) Craven & Biffin
Acmena inophylla (DC.) Wight = ***Syzygium inophyllum*** DC.
Acmena kingii G.Don = ***Syzygium smithii*** (Poir.) Nied.
Acmena laevifolia (Ridl.) Merr. & L.M.Perry = ***Syzygium acuminatissimum*** (Blume) DC.
Acmena lanceolata (Lam.) Thwaites = ***Syzygium lanceolatum*** (Lam.) Wight & Arn.
Acmena leptantha Walp. = ***Syzygium claviflorum*** (Roxb.) Wall. ex A.M.Cowan & Cowan
Acmena mackinnoniana B.Hyland = ***Syzygium mackinnonianum*** (B.Hyland) Craven & Biffin
Acmena macrocarpa C.T.White = ***Syzygium graveolens*** (F.M.Bailey) Craven & Biffin
Acmena melanosticta (Miq.) Merr. & L.M.Perry = ***Syzygium melanostictum*** (Miq.) Craven & Biffin
Acmena montana T.G.Hartley & Craven = ***Syzygium zhenghei*** Craven & Biffin
Acmena oblata (Roxb.) Walp. = ***Syzygium oblatum*** (Roxb.) Wall. ex A.M.Cowan & Cowan
Acmena parviflora DC. = ***Syzygium zeylanicum*** (L.) DC.
Acmena pendula G.Benn. = ***Syzygium smithii*** (Poir.) Nied.
Acmena polyantha (K.Schum. & Lauterb.) Merr. & L.M.Perry = ***Syzygium acuminatissimum*** (Blume) DC.
Acmena resa B.Hyland = ***Syzygium resa*** (B.Hyland) Craven & Biffin

Acmena smithii (Poir.) Merr. & L.M.Perry = ***Syzygium smithii*** (Poir.) Nied.
Acmena smithii var. *minor* (Maiden) Merr. & L.M. Perry = ***Syzygium smithii*** (Poir.) Nied.
Acmena sorongensis T.G.Hartley & Craven = ***Syzygium sorongense*** (T.G.Hartley & Craven) Craven & Biffin
Acmena triphlebia (Diels) T.G.Hartley & Craven = ***Syzygium triphlebium*** Diels
Acmena wightiana (Wight) Walp. = ***Syzygium lanceolatum*** (Lam.) Wight & Arn.
Acmena zeylanica (L.) Thwaites = ***Syzygium zeylanicum*** (L.) DC.

***Unplaced Names*:**
Acmena chinensis Planch., Hort. Donat.: 84 (1858). = [36 CHS]

Acmenosperma

Acmenosperma Kausel = ***Syzygium*** Gaertn.
Acmenosperma claviflorum (Roxb.) Kausel = ***Syzygium claviflorum*** (Roxb.) Wall. ex A.M.Cowan & Cowan
Acmenosperma pringlei B.Hyland = ***Syzygium pringlei*** (B.Hyland) Craven & Biffin

Acrandra

Acrandra O.Berg = ***Campomanesia*** Ruiz & Pav.
Acrandra guianensis Amshoff = ***Eugenia amshoffiae*** McVaugh
Acrandra laurifolia (Gardner) O.Berg = ***Campomanesia laurifolia*** Gardner
Acrandra sellowiana O.Berg = ***Campomanesia laurifolia*** Gardner
Acrandra verrucosa O.Berg = ***Campomanesia neriifolia*** (O.Berg) Nied.

Acreugenia

Acreugenia Kausel = ***Myrcianthes*** O.Berg
Acreugenia mato (Griseb.) Kausel = ***Myrcianthes mato*** (Griseb.) McVaugh
Acreugenia pungens (O.Berg) Kausel = ***Myrcianthes pungens*** (O.Berg) D.Legrand

Actinodium

Actinodium Schauer ex Schltdl., Linnaea 10: 311 (1836).
SW. Australia. 50 WAU. [Myrtaceae]
1 Species
Triphelia R.Br. ex Endl. in S.L.Endlicher & al., Enum. Pl.: 48 (1837).

Actinodium cunninghamii Schauer ex Lindl., Intr. Nat. Syst. Bot., ed. 2: 440 (1836).
SW. Western Australia. 50 WAU. Cham.
Triphelia brunioides R.Br. ex Endl. in S.L.Endlicher & al., Enum. Pl.: 48 (1837).
Actinodium proliferum Turcz., Bull. Soc. Imp. Naturalistes Moscou 22(2): 17 (1849).

***Synonyms*:**
Actinodium proliferum Turcz. = ***Actinodium cunninghamii*** Schauer ex Lindl.

Agalmanthus

Agalmanthus (Endl.) Hombr. & Jacquinot = ***Metrosideros*** Banks ex Gaertn.
Agalmanthus umbellata (Cav.) Hombr. & Jacquinot = ***Metrosideros umbellata*** Cav.

Agonis

Agonis (DC.) Sweet, Hort. Brit., ed. 2: 209 (1830).
SW. Australia. 50 WAU. [Myrtaceae]
14 Species
Billottia R.Br. ex G.Don, Gen. Hist. 2: 810 (1832).

Agonis angustifolia Schauer in J.G.C.Lehmann, Pl. Preiss. 1: 118 (1844).
SW. Western Australia. 50 WAU. Nanophan.
Agonis glabra Turcz., Bull. Acad. Imp. Sci. Saint-Pétersbourg 10: 334 (1862).

Agonis conspicua Schauer in J.G.C.Lehmann, Pl. Preiss. 1: 118 (1844). *Agonis linearifolia* var. *conspicua* (Schauer) Domin, Vestn. Král. Ceské Spolecn. Nauk, Tr. Mat.-Prír. 2(2): 85 (1923).
SW. Western Australia. 50 WAU. Nanophan.
Agonis spathulata var. *angustifolia* Benth., Fl. Austral. 3: 97 (1867).

Agonis flexuosa (Muhl. ex Willd.) Sweet, Hort. Brit., ed. 2: 209 (1830).
SW. Western Australia. 50 WAU. Nanophan.
**Metrosideros flexuosa* Muhl. ex Willd., Enum. Pl.: 514 (1809). *Leptospermum flexuosum* (Muhl. ex Willd.) Spreng., Novi Provent.: 25 (1818). *Billottia flexuosa* (Muhl. ex Willd.) R.Br., J. Roy. Geogr. Soc. 1: 19 (1832).
Leptospermum resiniferum Bertol., Opusc. Sci. 1: 148 (1817).
Leptospermum glomeratum H.L.Wendl., Flora 2: 678 (1819).
Billottia flexuosa var. *angustifolia* Otto & A.Dietr., Allg. Gartenzeitung 9: 218 (1841).
Billottia flexuosa var. *latifolia* Otto & A.Dietr., Allg. Gartenzeitung 9: 218 (1841).
Agonis flexuosa var. *angustifolia* Schauer in J.G.C.Lehmann, Pl. Preiss. 1: 117 (1845).
Agonis flexuosa var. *latifolia* Schauer in J.G.C.Lehmann, Pl. Preiss. 1: 117 (1845).

Agonis floribunda Turcz., Bull. Soc. Imp. Naturalistes Moscou 22(2): 20 (1849).
SW. Western Australia. 50 WAU. Nanophan.

Agonis fragrans J.R.Wheeler & N.G.Marchant, Nuytsia 13: 567 (2001).
SW. Australia. 50 WAU. Nanophan.

Agonis grandiflora Benth., Fl. Austral. 3: 100 (1867).
SW. Western Australia. 50 WAU. Nanophan.

Agonis hypericifolia (Otto & A.Dietr.) Schauer in J.G.C.Lehmann, Pl. Preiss. 1: 117 (1844).
SW. Western Australia. 50 WAU. Nanophan.
**Leptospermum hypericifolium* Otto & A.Dietr., Allg. Gartenzeitung 10: 243 (1841).
Leptospermum theiforme A.Cunn. ex Schauer in W.G.Walpers, Repert. Bot. Syst. 5: 739 (1846).
Agonis theiformis Schauer in J.G.C.Lehmann, Pl. Preiss. 2: 223 (1848). *Billottia theiformis* (Schauer) G.Don in J.C. Loudon, Encycl. Pl., Suppl. 2: 1380 (1855).

Agonis juniperina Schauer in J.G.C.Lehmann, Pl. Preiss. 1: 118 (1844).
SW. Western Australia. 50 WAU. Nanophan.

Agonis linearifolia (DC.) Sweet, Hort. Brit., ed. 2: 209 (1830).
SW. Australia. 50 WAU. Nanophan.
**Leptospermum linearifolium* DC., Prodr. 3: 277 (1828). *Billottia linearifolia* (DC.) R.Br., J. Roy. Geogr. Soc. 1: 19 (1832).

Agonis marginata (Labill.) Sweet, Hort. Brit., ed. 2: 209 (1830).
SW. Western Australia. 50 WAU. Nanophan.
**Leptospermum marginatum* Labill., Nov. Holl. Pl. 2: 9 (1806). *Billottia marginata* (Labill.) R.Br., J. Roy. Geogr. Soc. 1: 19 (1832).
Fabricia stricta Lodd., Bot. Cab. 13: t. 1219 (1827), nom. nud.
Fabricia angustifolia Otto ex DC., Prodr. 3: 226 (1828).
Leptospermum marginatum var. *glabratum* A.Cunn. ex DC., Prodr. 3: 226 (1828).
Billottia marginata var. *glabrata* (DC.) G.Don, Gen. Hist. 2: 827 (1832).
Fabricia myrtifolia Heynh., Nom. Bot. Hort. 2: 70 (1841).
Billottia marginata var. *angustifolia* (Otto) Heynh., Nom. Bot. Hort. 2: 70 (1846).

Agonis obtusissima F.Muell., Fragm. 11: 119 (1881).
SW. Western Australia. 50 WAU. Nanophan.

Agonis parviceps Schauer in J.G.C.Lehmann, Pl. Preiss. 1: 119 (1844).
SW. Western Australia. 50 WAU. Nanophan.

Agonis spathulata Schauer in J.G.C.Lehmann, Pl. Preiss. 1: 117 (1844).
SW. Western Australia. 50 WAU. Nanophan.

Agonis undulata Benth., Fl. Austral. 3: 100 (1867).
SW. Western Australia. 50 WAU. Nanophan. Provisionally accepted.

***Synonyms*:**
Agonis abnormis (F.Muell. ex Benth.) C.T.White = ***Leptospermum brachyandrum*** (F.Muell.) Druce
Agonis elliptica C.T.White & W.D.Francis = ***Leptospermum whitei*** Cheel
Agonis elliptica var. *angustifolia* C.T.White & W.D.Francis = ***Leptospermum whitei*** Cheel
Agonis ericoides F.M.Bailey = ***Kunzea muelleri*** Benth.
Agonis flexuosa var. *angustifolia* Schauer = ***Agonis flexuosa*** (Muhl. ex Willd.) Sweet
Agonis flexuosa var. *latifolia* Schauer = ***Agonis flexuosa*** (Muhl. ex Willd.) Sweet
Agonis glabra Turcz. = ***Agonis angustifolia*** Schauer
Agonis linearifolia var. *conspicua* (Schauer) Domin = ***Agonis conspicua*** Schauer
Agonis longifolia C.T.White & W.D.Francis = ***Leptospermum madidum*** A.R.Bean subsp. ***madidum***
Agonis luehmannii (F.M.Bailey) C.T.White & W.D.Francis = ***Leptospermum luehmannii*** F.M.Bailey
Agonis lysicephala (F.Muell. & F.M.Bailey) F.M.Bailey = ***Asteromyrtus lysicephala*** (F.Muell. & F.M.Bailey) Craven
Agonis scortechiniana F.Muell. = ***Leptospermum speciosum*** Schauer
Agonis spathulata var. *angustifolia* Benth. = ***Agonis conspicua*** Schauer
Agonis speciosa (Schauer) C.T.White = ***Leptospermum speciosum*** Schauer
Agonis theiformis Schauer = ***Agonis hypericifolia*** (Otto & A.Dietr.) Schauer

Agonomyrtus

Agonomyrtus Schauer ex Rchb. = ***Leptospermum*** J.R.Forst. & G.Forst.

Aguava

Aguava Raf. = ***Myrcia*** DC. ex Guill.
Aguava guianensis (Aubl.) Raf. = ***Myrcia guianensis*** (Aubl.) DC.
Aguava tomentosa (Aubl.) Raf. = ***Myrcia tomentosa*** (Aubl.) DC.

Algrizea

Algrizea Proença & NicLugh., Syst. Bot. 31: 320 (2006).
NE. Brazil. 84 BZE. [Myrtaceae]
1 Species

Algrizea macrochlamys (DC.) Proença & NicLugh., Syst. Bot. 31: 321 (2006).
Brazil (Bahia). 84 BZE. Nanophan.
**Myrcia macrochlamys* DC., Prodr. 3: 247 (1828). *Myrtus macrochlamys* (DC.) Mart. ex O.Berg in C.F.P.von Martius & auct. suc. (eds.), Fl. Bras. 14(1): 419 (1857). *Psidium macrochlamys* (DC.) Mattos, Loefgrenia 64: 1 (1975).

Allostis

Allostis Raf. = ***Baeckea*** L.

Allosyncarpia

Allosyncarpia S.T.Blake, Austrobaileya 1: 43 (1977).
N. Australia. 50 NTA. [Myrtaceae]
1 Species

Allosyncarpia ternata S.T.Blake, Austrobaileya 1: 44 (1977).
Northern Territory. 50 NTA. Nanophan.

Aluta

Aluta Rye & Trudgen, Nuytsia 13: 347 (2000).
W. & C. Australia. 50 NTA SOA WAU. [Myrtaceae]
5 Species

Aluta appressa (C.R.P.Andrews) Rye & Trudgen, Nuytsia 13: 351 (2000).
SC. & S. Western Australia. 50 WAU. Nanophan.
**Thryptomene appressa* C.R.P.Andrews, J. West Austral. Nat. Hist. Soc. 1: 41 (1904).

Aluta aspera (E.Pritz.) Rye & Trudgen, Nuytsia 13: 352 (2000).
W. & SC. Western Australia. 50 WAU. Cham. or nanophan.
**Thryptomene aspera* E.Pritz., Bot. Jahrb. Syst. 35: 413 (1904).

subsp. ***aspera***
SC. Western Australia. 50 WAU. Cham. or nanophan.

subsp. ***hesperia*** Rye & Trudgen, Nuytsia 13: 355 (2000).
SC. Western Australia. 50 WAU. Cham. or nanophan.

subsp. ***localis*** Rye & Trudgen, Nuytsia 13: 357 (2000).
WC. & W. Western Australia. 50 WAU. Cham. or nanophan.

Aluta maisonneuvei (F.Muell.) Rye & Trudgen, Nuytsia 13: 358 (2000).
C. Australia. 50 NTA SOA WAU. Cham. or nanophan.
**Thryptomene maisonneuvei* F.Muell., Fragm. 4: 64 (1864).

subsp. ***auriculata*** (F.Muell.) Rye & Trudgen, Nuytsia 13: 360 (2000).
SC. Western Australia to South Australia. 50 SOA WAU. Cham. or nanophan.
**Thryptomene auriculata* F.Muell., Fragm. 10: 24 (1876).

subsp. ***maisonneuvei***.
C. Australia. 50 NTA SOA WAU. Cham. or nanophan.

Aluta quadrata Rye & Trudgen, Nuytsia 13: 361 (2000).
NW. Western Australia. 50 WAU. Nanophan.

Aluta teres Rye & Trudgen, Nuytsia 13: 364 (2000).
C. Western Australia. 50 WAU. Cham.

Amomis

Amomis O.Berg = ***Pimenta*** Lindl.
Amomis acris (Sw.) O.Berg = ***Pimenta racemosa*** (Mill.) J.W.Moore var. ***racemosa***
Amomis acris var. *grandifolia* O.Berg = ***Pimenta racemosa*** (Mill.) J.W.Moore var. ***racemosa***
Amomis acris var. *obtusata* O.Berg = ***Pimenta racemosa*** (Mill.) J.W.Moore var. ***racemosa***
Amomis acris var. *parvifolia* O.Berg = ***Pimenta racemosa*** (Mill.) J.W.Moore var. ***racemosa***
Amomis anisomera Urb. & Ekman = ***Pimenta racemosa*** var. ***ozua*** (Urb. & Ekman) Landrum
Amomis caryophyllata Krug & Urb. = ***Pimenta racemosa*** (Mill.) J.W.Moore var. ***racemosa***
Amomis caryophyllata var. *grisea* (Kiaersk.) Kiaersk. = ***Pimenta racemosa*** var. ***grisea*** (Kiaersk.) Fosberg
Amomis grisea (Kiaersk.) Britton = ***Pimenta racemosa*** var. ***grisea*** (Kiaersk.) Fosberg
Amomis hispaniolensis Urb. = ***Pimenta racemosa*** var. ***hispaniolensis*** (Urb.) Landrum
Amomis jamaicensis Britton & Harris = ***Pimenta jamaicensis*** (Britton & Harris) Proctor
Amomis oblongata O.Berg = ***Pimenta racemosa*** (Mill.) J.W.Moore var. ***racemosa***
Amomis oblongata var. *occidentalis* O.Berg = ***Pimenta racemosa*** (Mill.) J.W.Moore var. ***racemosa***
Amomis odiolens Urb. = ***Pimenta odiolens*** (Urb.) Burret
Amomis ozua Urb. & Ekman = ***Pimenta racemosa*** var. ***ozua*** (Urb. & Ekman) Landrum
Amomis pauciflora Urb. = ***Pimenta racemosa*** var. ***hispaniolensis*** (Urb.) Landrum
Amomis pilotoana Urb. = ***Pimenta cainitoides*** (Urb.) Burret
Amomis pimento O.Berg = ***Pimenta racemosa*** (Mill.) J.W.Moore var. ***racemosa***
Amomis pimento var. *jamaicensis* O.Berg = ***Pimenta racemosa*** (Mill.) J.W.Moore var. ***racemosa***
Amomis pimento var. *surinamensis* O.Berg = ***Pimenta racemosa*** (Mill.) J.W.Moore var. ***racemosa***
Amomis pimentoides O.Berg = ***Pimenta racemosa*** (Mill.) J.W.Moore var. ***racemosa***

Amomyrtella

Amomyrtella Kausel, Ark. Bot., a.s., 3: 514 (1956).
Bolivia to NW. Argentina. 83 BOL 85 AGW. [Myrtaceae]
1 Species

Amomyrtella guilii (Speg.) Kausel, Ark. Bot., a.s., 3: 515 (1956).
Bolivia to NW. Argentina. 83 BOL 85 AGW. Phan.
**Eugenia guilii* Speg., Anales Soc. Rural Argent. 1910: 388 (1910). *Myrtus guilii* (Speg.) D.Legrand, Darwiniana 5: 469 (1941). *Pseudocaryophyllus guilii* (Speg.) Burret, Notizbl. Bot. Gart. Berlin-Dahlem 15: 515 (1941).

Amomyrtus

Amomyrtus (Burret) D.Legrand & Kausel, Lilloa 13: 145 (1947).
Chile to SW. Argentina. 85 AGS CLC CLS. [Myrtaceae]
2 Species

Amomyrtus luma (Molina) D.Legrand & Kausel, Lilloa 13: 146 (1947).
C. & S. Chile to SW. Argentina. 85 AGS CLC CLS. Nanophan. or phan.
**Myrtus luma* Molina, Sag. Stor. Nat. Chili: 173 (1782). *Myrceugenia luma* (Molina) I.M.Johnst., Contr. Gray Herb., n.s., 61: 92 (1924), nom. illeg.
Myrtus multiflora Juss. ex J.St.-Hil. in H.L.Duhamel du Monceau, Traité Arbr. Arbust., nouv. ed. 1: 208 (1803). *Pseudocaryophyllus multiflorus* (Juss. ex J.St.-Hil.) Burret, Notizbl. Bot. Gart. Berlin-Dahlem 15: 515 (1941).
Eugenia darwinii Hook.f., Fl. Antarct.: 277 (1846). *Myrtus darwinii* (Hook.f.) Barnéoud in C.Gay, Fl. Chil. 2: 383 (1847). *Pseudocaryophyllus darwinii* (Hook.f.) Burret, Notizbl. Bot. Gart. Berlin-Dahlem 15: 514 (1941).
Myrtus luma Barnéoud in C.Gay, Fl. Chil. 2: 384 (1847), nom. illeg. *Myrtus reloncavi* Barnéoud ex F.Phil., Cat. Pl. Vasc. Chil.: 79 (1881).
Myrcia lechleriana Miq., Linnaea 25: 651 (1853). *Myrtus lechleriana* (Miq.) Sealy, Bot. Mag. 161: t. 9523 (1938).
Myrtus valdiviana Phil., Linnaea 28: 688 (1858).

Amomyrtus meli (Phil.) D.Legrand & Kausel, Lilloa 13: 146 (1947).
SC. Chile. 85 CLC CLS. Phan.
**Myrtus meli* Phil., Linnaea 28: 638 (1857). *Pseudocaryophyllus meli* (Phil.) Burret, Notizbl. Bot. Gart. Berlin-Dahlem 15: 514 (1941).

Amyrsia

Amyrsia Raf. = ***Myrteola*** O.Berg
Amyrsia compressa (Kunth) Raf. = ***Myrcianthes fragrans*** (Sw.) McVaugh
Amyrsia discolor (Kunth) Raf. = ***Myrcianthes discolor*** (Kunth) McVaugh
Amyrsia foliosa (Kunth) Raf. = ***Myrcianthes leucoxyla*** (Ortega) McVaugh
Amyrsia hallii (O.Berg) Kausel = ***Myrcianthes hallii*** (O.Berg) McVaugh
Amyrsia limbata (Kunth) Kausel = ***Myrcianthes fragrans*** (Sw.) McVaugh
Amyrsia microphyla Raf. = ***Myrteola phylicoides*** (Benth.) Landrum var. ***phylicoides***

Anamomis

Anamomis Griseb. = ***Myrcianthes*** O.Berg
Anamomis bahamensis (Kiaersk.) Britton ex Small = ***Mosiera longipes*** (O.Berg) Small
Anamomis dichotoma (Poir.) Sarg. = ***Myrcianthes fragrans*** (Sw.) McVaugh
Anamomis dicrana (O.Berg) Britton = ***Myrcianthes fragrans*** (Sw.) McVaugh

Anamomis esculenta (O.Berg) Griseb. = ***Pseudanamomis umbellulifera*** (Kunth) Kausel
Anamomis ferruginea Griseb. = ***Pimenta ferruginea*** (Griseb.) Burret
Anamomis fragrans (Sw.) Griseb. = ***Myrcianthes fragrans*** (Sw.) McVaugh
Anamomis fragrans var. *cuneata* Griseb. = ***Myrcianthes fragrans*** (Sw.) McVaugh
Anamomis grandis Britton = ***Myrcianthes fragrans*** (Sw.) McVaugh
Anamomis guayabillo (A.Rich.) Griseb. = ***Myrcianthes fragrans*** (Sw.) McVaugh
Anamomis longipes (O.Berg) Britton ex Small = ***Mosiera longipes*** (O.Berg) Small
Anamomis lucayana Britton = ***Myrcianthes fragrans*** (Sw.) McVaugh
Anamomis punctata Griseb. = ***Myrcianthes fragrans*** (Sw.) McVaugh
Anamomis punctata var. *dichotoma* (Poir.) Kiaersk. = ***Myrcianthes fragrans*** (Sw.) McVaugh
Anamomis reticulata Britton & P.Wilson = ***Myrcia manacalensis*** Urb.
Anamomis simpsonii Small = ***Myrcianthes fragrans*** (Sw.) McVaugh
Anamomis umbellulifera (Kunth) Britton = ***Pseudanamomis umbellulifera*** (Kunth) Kausel

Anetholea

Anetholea Peter G.Wilson = ***Syzygium*** Gaertn.
Anetholea anisata (Vickery) Peter G.Wilson = ***Syzygium anisatum*** (Vickery) Craven & Biffin

Angasomyrtus

Angasomyrtus Trudgen & Keighery, Nuytsia 4: 435 (1983).
SW. Australia. 50 WAU. [Myrtaceae]
1 Species

Angasomyrtus salina Trudgen & Keighery, Nuytsia 4: 435 (1983).
SW. Australia. 50 WAU. Cham. or nanophan.

Angophora

Angophora Cav., Icon. 4: 21 (1797).
E. & SE. Australia. 50 NSW QLD VIC. [Myrtaceae]
14 Species

Angophora bakeri C.C.Hall, J. Proc. Roy. Soc. New S. Wales 47: 101 (1913). *Eucalyptus angustata* Brooker, Austral. Syst. Bot. 13: 135 (2000).
New South Wales. 50 NSW. Nanophan.

subsp. ***bakeri***.
EC. New South Wales. 50 NSW. Nanophan.
Angophora lanceolata var. *angustifolia* A.Gray, U.S. Expl. Exped., Phan. 1: 556 (1854).

subsp. ***paludosa*** G.J.Leach, Telopea 2: 764 (1986).
NE. New South Wales. 50 NSW. Nanophan.

Angophora* × *clelandii Maiden, J. Proc. Roy. Soc. New S. Wales 54: 175 (1920). *A. bakeri* × *A. hispida*.
New South Wales. 50 NSW. Nanophan.

Angophora costata (Gaertn.) Hochr. ex Britten, J. Bot. 54: 62 (1916).
E. New South Wales. 50 NSW. Phan.
**Metrosideros costata* Gaertn., Fruct. Sem. Pl. 1: 171 (1788). *Angophora lanceolata* Cav., Icon. 4: 22 (1797), nom. illeg. *Metrosideros lanceolata* Pers., Syn. Pl. 2: 25 (1806), nom. illeg. *Melaleuca costata* (Gaertn.) Raeusch., Nomencl. Bot., ed. 3: 142 (1797).
Metrosideros apocynifolia Salisb., Prodr. Stirp. Chap. Allerton: 351 (1796). *Eucalyptus apocynifolia* (Salisb.) Brooker, Austral. Syst. Bot. 13: 135 (2000).
Angophora lanceolata var. *hispida* A.Cunn. ex A.Gray, U.S. Expl. Exped., Phan. 1: 555 (1854).

Angophora crassifolia (G.J.Leach) L.A.S.Johnson & K.D.Hill, Telopea 4: 38 (1990).
CE. New South Wales. 50 NSW. Nanophan.
**Angophora bakeri* subsp. *crassifolia* G.J.Leach, Telopea 2: 766 (1986). *Eucalyptus crassifolia* (G.J.Leach) Brooker, Austral. Syst. Bot. 13: 136 (2000).

Angophora* × *dichromophloia Blakely, Contr. New South Wales Natl. Herb. 1: 34 (1939). *A. costata* × *A. hispida*.
New South Wales. 50 NSW. Phan.

Angophora euryphylla (L.A.S.Johnson ex G.J.Leach) L.A.S.Johnson & K.D.Hill, Telopea 4: 38 (1990).
New South Wales. 50 NSW. Nanophan. or phan.
**Angophora costata* subsp. *euryphylla* L.A.S.Johnson ex G.J.Leach, Telopea 2: 759 (1986). *Eucalyptus euryphylla* (L.A.S.Johnson ex G.J.Leach) Brooker, Austral. Syst. Bot. 13: 136 (2000).

Angophora exul K.D.Hill, Telopea 7: 100 (1997). *Eucalyptus exul* (K.D.Hill) Brooker, Austral. Syst. Bot. 13: 136 (2000).
New South Wales. 50 NSW. Nanophan. or phan.

Angophora floribunda (Sm.) Sweet, Hort. Brit., ed. 2: 209 (1830).
E. & SE. Australia. 50 NSW QLD VIC. Nanophan. or phan.
**Metrosideros floribunda* Sm., Trans. Linn. Soc. London 3: 267 (1797). *Acmena floribunda* (Sm.) A.Cunn. ex DC., Prodr. 3: 262 (1828), nom. illeg. *Eucalyptus florida* Brooker, Austral. Syst. Bot. 13: 136 (2000).
Angophora intermedia A.Cunn. ex DC., Prodr. 3: 222 (1828).
Angophora ochrophylla R.T.Baker, Proc. Linn. Soc. New South Wales 38: 601 (1913 publ. 1914).

Angophora hispida (Sm.) Blaxell, Kew Bull. 31: 272 (1976).
EC. New South Wales. 50 NSW. Phan.
Angophora cordifolia Cav., Icon. 4: 21 (1797). *Metrosideros cordifolia* (Cav.) Pers., Syn. Pl. 2: 25 (1806).
**Metrosideros hispida* Sm., Trans. Linn. Soc. London 3: 267 (1797). *Eucalyptus hispida* (Sm.) Brooker, Austral. Syst. Bot. 13: 136 (2000).
Metrosideros anomala Vent., Jard. Malmaison: 5 (1803).
Metrosideros hirsuta Andrews, Bot. Repos. 4: t. 281 (1803).
Eucalyptus hirsuta Link, Enum. Hort. Berol. Alt. 2: 31 (1822).
Angophora cordata Vis., Orto Bot. Padova: 133 (1842).

Angophora inopina K.D.Hill, Telopea 7: 97 (1997). *Eucalyptus inopina* (K.D.Hill) Brooker, Austral. Syst. Bot. 13: 136 (2000).
New South Wales. 50 NSW. Nanophan. or phan.

Angophora leiocarpa (L.A.S.Johnson ex G.J.Leach) K.R.Thiele & Ladiges, Cladistics 4: 41 (1988).
E. Australia. 50 NSW QLD. Nanophan.
**Angophora costata* subsp. *leiocarpa* L.A.S.Johnson ex G.J.Leach, Telopea 2: 760 (1986). *Eucalyptus leiocarpa* (L.A.S.Johnson ex G.J.Leach) Brooker, Austral. Syst. Bot. 13: 137 (2000).

Angophora melanoxylon F.Muell. ex R.T.Baker, Proc. Linn. Soc. New South Wales 25: 85 (1900). *Angophora intermedia* var. *melanoxylon* (F.Muell. ex R.T.Baker) Maiden & Betche, Census N.S.W. Pl.: 145 (1916). *Eucalyptus melana* Brooker, Austral. Syst. Bot. 13: 137 (2000).
SC. Queensland to NC. New South Wales. 50 NSW QLD. Phan.

Angophora paludosa (G.J.Leach) K.R.Thiele & Ladiges, Cladistics 4: 41 (1988).
NE. New South Wales. 50 NSW. Nanophan.

Angophora robur L.A.S.Johnson & K.D.Hill, Telopea 4: 40 (1990). *Eucalyptus robur* (L.A.S.Johnson & K.D.Hill) Brooker, Austral. Syst. Bot. 13: 137 (2000).
New South Wales. 50 NSW. Nanophan. or phan.

Angophora subvelutina F.Muell., Fragm. 1: 31 (1858). *Eucalyptus subvelutina* (F.Muell.) Brooker, Austral. Syst. Bot. 13: 137 (2000).
SE. Queensland to E. New South Wales. 50 NSW QLD. Phan.
Angophora velutina F.Muell., Fragm. 4: 170 (1864), orth. var.

Angophora woodsiana F.M.Bailey, Proc. Linn. Soc. New South Wales 6: 143 (1881). *Angophora intermedia* var. *woodsiana* (F.M.Bailey) F.M.Bailey, Queensl. Fl. 2: 605 (1900). *Angophora lanceolata* var. *woodsiana* (F.M.Bailey) Maiden, Forest Fl. N.S.W. 2(11): 16 (1904). *Angophora floribunda* var. *woodsiana* (F.M.Bailey) Domin, Biblioth. Bot. 89: 1013 (1928). *Eucalyptus woodsiana* (F.M.Bailey) Brooker, Austral. Syst. Bot. 13: 137 (2000).
SE. Queensland to NE. New South Wales. 50 NSW QLD. Phan.

Synonyms:
Angophora bakeri subsp. *crassifolia* G.J.Leach = ***Angophora crassifolia*** (G.J.Leach) L.A.S.Johnson & K.D.Hill
Angophora cordata Vis. = ***Angophora hispida*** (Sm.) Blaxell
Angophora cordifolia Cav. = ***Angophora hispida*** (Sm.) Blaxell
Angophora costata subsp. *euryphylla* L.A.S.Johnson ex G.J.Leach = ***Angophora euryphylla*** (L.A.S.Johnson ex G.J.Leach) L.A.S.Johnson & K.D.Hill
Angophora costata subsp. *leiocarpa* L.A.S.Johnson ex G.J.Leach = ***Angophora leiocarpa*** (L.A.S.Johnson ex G.J.Leach) K.R.Thiele & Ladiges
Angophora floribunda var. *woodsiana* (F.M.Bailey) Domin = ***Angophora woodsiana*** F.M.Bailey
Angophora intermedia A.Cunn. ex DC. = ***Angophora floribunda*** (Sm.) Sweet
Angophora intermedia var. *melanoxylon* (F.Muell. ex R.T.Baker) Maiden & Betche = ***Angophora melanoxylon*** F.Muell. ex R.T.Baker
Angophora intermedia var. *woodsiana* (F.M.Bailey) F.M.Bailey = ***Angophora woodsiana*** F.M.Bailey
Angophora lanceolata Cav. = ***Angophora costata*** (Gaertn.) Hochr. ex Britten
Angophora lanceolata var. *angustifolia* A.Gray = ***Angophora bakeri*** C.C.Hall subsp. ***bakeri***
Angophora lanceolata var. *hispida* A.Cunn. ex A.Gray = ***Angophora costata*** (Gaertn.) Hochr. ex Britten
Angophora lanceolata var. *woodsiana* (F.M.Bailey) Maiden = ***Angophora woodsiana*** F.M.Bailey
Angophora ochrophylla R.T.Baker = ***Angophora floribunda*** (Sm.) Sweet
Angophora velutina F.Muell. = ***Angophora subvelutina*** F.Muell.

Anticoryne

Anticoryne Turcz. = ***Baeckea*** L.
Anticoryne diosmoides Turcz. = ***Baeckea ovalifolia*** (F.Muell.) F.Muell.

Aphanomyrtus

Aphanomyrtus Miq. = ***Syzygium*** Gaertn.
Aphanomyrtus alata Lauterb. = ***Syzygium alatum*** (Lauterb.) Diels
Aphanomyrtus camphorata Valeton = ***Syzygium skiophilum*** (Duthie) Airy Shaw
Aphanomyrtus forbesii Greves = ***Syzygium minimum*** (Blume) Airy Shaw
Aphanomyrtus minima (Blume) Merr. = ***Syzygium minimum*** (Blume) Airy Shaw
Aphanomyrtus octandra Koord. & Valeton = ***Syzygium minimum*** (Blume) Airy Shaw
Aphanomyrtus rostrata Miq. = ***Syzygium singaporense*** (King) Airy Shaw
Aphanomyrtus skiophila (Duthie) Valeton = ***Syzygium skiophilum*** (Duthie) Airy Shaw
Aphanomyrtus tenuifiolia (Ridl.) Merr. = ***Syzygium tenuifolium*** (Ridl.) Airy Shaw
Aphanomyrtus tetraquetra (Miq.) Valeton = ***Syzygium minimum*** (Blume) Airy Shaw

Archirhodomyrtus

Archirhodomyrtus (Nied.) Burret, Repert. Spec. Nov. Regni Veg. 50: 59 (1941).
E. Australia, New Caledonia. 50 NSW QLD 60 NWC. [Myrtaceae]
5 Species

Archirhodomyrtus baladensis (Brongn. & Gris) Burret, Repert. Spec. Nov. Regni Veg. 50: 60 (1941).
New Caledonia. 60 NWC. Phan.
**Myrtus baladensis* Brongn. & Gris, Bull. Soc. Bot. France 12: 177 (1865). *Rhodomyrtus baladensis* (Brongn. & Gris) Nied. in H.G.A.Engler & K.A.E.Prantl, Nat. Pflanzenfam. 3(7): 70 (1893).

Archirhodomyrtus beckleri (F.Muell.) A.J.Scott, Kew Bull. 33: 326 (1978).
Queensland to NE. New South Wales. 50 NSW QLD. Phan.
**Myrtus beckleri* F.Muell., Fragm. 2: 85 (1860). *Rhodomyrtus beckleri* (F.Muell.) L.S.Sm., Proc. Roy. Soc. Queensland 67: 37 (1956).
Myrtus cymiflora F.Muell., Fragm. 5: 12 (1865). *Rhodomyrtus cymiflora* (F.Muell.) Benth., Fl. Austral. 3: 273 (1867).
Pilidiostigma cuneatum Burret, Notizbl. Bot. Gart. Berlin-Dahlem 15: 549 (1941).
Pilidiostigma parviflorum Burret, Notizbl. Bot. Gart. Berlin-Dahlem 15: 549 (1941).

Archirhodomyrtus paitensis (Schltr.) Burret, Repert. Spec. Nov. Regni Veg. 50: 60 (1941).
New Caledonia. 60 NWC. Phan.

Myrtus paitensis Schltr., Bot. Jahrb. Syst. 39: 203 (1906). *Rhodomyrtus paitensis* (Schltr.) Burret, Notizbl. Bot. Gart. Berlin-Dahlem 15: 496 (1941).

Archirhodomyrtus turbinata (Schltr.) Burret, Repert. Spec. Nov. Regni Veg. 50: 60 (1941).
New Caledonia. 60 NWC. Phan.
Myrtus turbinata Schltr., Bot. Jahrb. Syst. 40(92): 31 (1908). *Rhodomyrtus turbinata* (Schltr.) Burret, Notizbl. Bot. Gart. Berlin-Dahlem 15: 496 (1941).

Archirhodomyrtus vieillardii (Brongn. & Gris) Burret, Repert. Spec. Nov. Regni Veg. 50: 60 (1941).
New Caledonia. 60 NWC. Nanophan.
Myrtus vieillardii Brongn. & Gris, Bull. Soc. Bot. France 12: 177 (1865). *Rhodomyrtus vieillardii* (Brongn. & Gris) Burret, Notizbl. Bot. Gart. Berlin-Dahlem 15: 496 (1941).

Arillastrum

Arillastrum Planch. ex Baill, Hist. Pl. 6: 363 (1877).
New Caledonia. 60 NWC. [Myrtaceae]
1 Species
Spermolepis Brongn. & Gris, Bull. Soc. Bot. France 10: 577 (1863), nom. illeg. *Myrtomera* B.C.Stone, Pacific Sci. 16: 241 (1962).

Arillastrum gummiferum (Brongn. & Gris) Planch. ex Baill., Hist. Pl. 6: 364 (1877).
C. & SE. New Caledonia. 60 NWC. Phan.
Spermolepis gummifera Brongn. & Gris, Bull. Soc. Bot. France 10: 578 (1863). *Nania gummifera* (Brongn. & Gris) Kuntze, Revis. Gen. Pl. 1: 242 (1891). *Spermolepis tannifera* Heckel, Compt. Rend. Hebd. Séances Acad. Sci. 153: 322 (1911), nom. illeg. *Myrtomera gummifera* (Brongn. & Gris) B.C.Stone, Pacific Sci. 16: 241 (1962).
Spermolepis rubra Vieill. ex Guillanmin, Ann. Inst. Bot.-Géol. Colon. Marseille, II, 9: 146 (1911), nom. inval.

Aromadendron

Aromadendron Andrews ex Steud. = ***Eucalyptus*** L'Hér.

Aromadendrum

Aromadendrum W.Anderson ex R.Br. = ***Eucalyptus*** L'Hér.

Aspidogenia

Aspidogenia Burret = ***Myrcianthes*** O.Berg
Aspidogenia coquimbensis (Barnéoud) Burret = ***Myrcianthes coquimbensis*** (Barnéoud) Landrum & Grifo

Astartea

Astartea DC. in J.B.G. Bory de Saint-Vincent, Dict. Class. Hist. Nat. 11: 400 (1827).
Australia. 50 NTA WAU. [Myrtaceae]
10 Species

Astartea affinis (Endl.) Rye, Nuytsia 16: 150 (2006).
SW. Western Australia. 50 WAU. Nanophan.
Leptospermum dubium Spreng., Syst. Veg. 2: 492 (1825).
Baeckea affinis Endl. in S.L.Endlicher & al., Enum. Pl.: 51 (1837). *Astartea endlicheriana* Schauer, Linnaea 17: 242 (1843), nom. superfl.

Astartea ambigua F.Muell., Fragm. 2: 32 (1861).
Baeckea ambigua (F.Muell.) Nied. in H.G.A.Engler & K.A.E.Prantl, Nat. Pflanzenfam. 3(7): 99 (1892).
S. Western Australia. 50 WAU. Nanophan.

Astartea arbuscula (R.Br. ex Benth.) Rye, Nuytsia 16: 152 (2006).
SW. Western Australia. 50 WAU. Nanophan.
Baeckea arbuscula R.Br. ex Benth., Fl. Austral. 3: 79 (1867).

Astartea astarteoides (Benth.) Rye, Nuytsia 16: 153 (2006).
SW. Western Australia. 50 WAU. Nanophan.
Baeckea astarteoides Benth., Fl. Austral. 3: 80 (1867).

Astartea clavifolia C.A.Gardner, J. Roy. Soc. Western Australia 27: 188 (1942).
SW. Australia. 50 WAU. Nanophan.

Astartea clavulata Turcz., Bull. Cl. Phys.-Math. Acad. Imp. Sci. Saint-Pétersbourg 10: 333 (1852).
SW. Australia. 50 WAU. Nanophan.

Astartea fascicularis (Labill.) A.Cunn. ex DC., Prodr. 3: 210 (1828).
SW. & S. Western Australia. 50 WAU. Nanophan.
Melaleuca fascicularis Labill., Nov. Holl. Pl. 2: 29 (1806). *Baeckea fascicularis* (Labill.) Nied. in H.G.A. Engler & K.A.E.Prantl, Nat. Pflanzenfam. 3(7): 99 (1892).
Astartea aspera Schauer, Linnaea 17: 242 (1843).
Astartea glomerulosa Schauer, Linnaea 17: 242 (1843).
Astartea leptophylla Schauer, Linnaea 17: 242 (1843).
Astartea scoparia Schauer, Linnaea 17: 242 (1843).
Astartea corniculata Schauer in J.G.C.Lehmann, Pl. Preiss. 1: 113 (1844).
Astartea muricata Turcz., Bull. Cl. Phys.-Math. Acad. Imp. Sci. Saint-Pétersbourg 10: 334 (1852).
Astartea fascicularis var. *brachyphylla* Domin, Vestn. Král. Ceské Spolecn. Nauk, Tr. Mat.-Prír. 2(2): 84 (1923).

Astartea heteranthera C.A.Gardner, J. Roy. Soc. Western Australia 13: 65 (1927).
SW. Australia. 50 WAU. Nanophan.

Astartea intratropica F.Muell., Fragm. 1: 83 (1859).
Baeckea intratropica (F.Muell.) Nied. in H.G.A. Engler & K.A.E.Prantl, Nat. Pflanzenfam. 3(7): 99 (1892).
Northern Territory. 50 NTA. Nanophan.

Astartea laricifolia Schauer, Linnaea 17: 242 (1843).
SW. Australia. 50 WAU. Nanophan.
Leptospermum laricifolium A.Cunn. ex Schauer in W.G.Walpers, Repert. Bot. Syst. 2: 922 (1843), nom. inval.

Synonyms:
Astartea aspera Schauer = ***Astartea fascicularis*** (Labill.) A.Cunn. ex DC.
Astartea corniculata Schauer = ***Astartea fascicularis*** (Labill.) A.Cunn. ex DC.
Astartea endlicheriana Schauer = ***Astartea affinis*** (Endl.) Rye
Astartea fascicularis var. *brachyphylla* Domin = ***Astartea fascicularis*** (Labill.) A.Cunn. ex DC.
Astartea glomerulosa Schauer = ***Astartea fascicularis*** (Labill.) A.Cunn. ex DC.
Astartea leptophylla Schauer = ***Astartea fascicularis*** (Labill.) A.Cunn. ex DC.

Astartea muricata Turcz. = ***Astartea fascicularis*** (Labill.) A.Cunn. ex DC.
Astartea scoparia Schauer = ***Astartea fascicularis*** (Labill.) A.Cunn. ex DC.

Asteromyrtus

Asteromyrtus Schauer, Linnaea 17: 242 (1843).
S. New Guinea to N. Australia. 43 NWG 50 NTA QLD. [Myrtaceae]
7 Species
Sinoga S.T.Blake, Proc. Roy. Soc. Queensland 69: 80 (1958).

Asteromyrtus angustifolia (Gaertn.) Craven, Austral. Syst. Bot. 1: 377 (1988 publ. 1989).
N. Queensland. 50 QLD. Nanophan. or phan.
**Melaleuca angustifolia* Gaertn., Fruct. Sem. Pl. 1: 172 (1788). *Asteromyrtus gaertneri* Schauer, Linnaea 17: 243 (1843), nom. illeg. *Austromyrtus gaertneri* Schauer, Linnaea 17: 234 (1843), nom. illeg. *Myrtoleucodendron angustifolium* (Gaertn.) Kuntze, Revis. Gen. Pl. 1: 241 (1891).

Asteromyrtus arnhemica (Byrnes) Craven, Austral. Syst. Bot. 1: 378 (1988 publ. 1989).
N. Northern Territory. 50 NTA. Nanophan. or phan.
**Melaleuca arnhemica* Byrnes, Austrobaileya 2: 74 (1984).

Asteromyrtus brassii (Byrnes) Craven, Austral. Syst. Bot. 1: 379 (1988 publ. 1989).
S. New Guinea to N. Queensland. 43 NWG 50 QLD. Nanophan. or phan.
**Melaleuca brassii* Byrnes, Austrobaileya 2: 74 (1984).

Asteromyrtus lysicephala (F.Muell. & F.M.Bailey) Craven, Austral. Syst. Bot. 1: 380 (1988 publ. 1989).
S. New Guinea to N. Queensland. 43 NWG 50 QLD. Nanophan.
**Melaleuca lysicephala* F.Muell. & F.M.Bailey, Occas. Pap. Queensland Fl. 1: 2 (1886). *Agonis lysicephala* (F.Muell. & F.M.Bailey) F.M.Bailey, Syn. Queensl. Fl., Suppl. 2: 24 (1888). *Sinoga lysicephala* (F.Muell. & F.M.Bailey) S.T.Blake, Proc. Roy. Soc. Queensland 69: 80 (1958).

Asteromyrtus magnifica (Specht) Craven, Austral. Syst. Bot. 1: 382 (1988 publ. 1989).
N. Northern Territory (incl. Groote Eylandt). 50 NTA. Nanophan.
**Melaleuca magnifica* Specht, Rec. Amer.-Austral. Sci. Exped. Arnhem Land, Bot. Pl. Ecol. 3: 270 (1958).

Asteromyrtus symphyocarpa (F.Muell.) Craven, Austral. Syst. Bot. 1: 383 (1988 publ. 1989).
S. New Guinea, N. Northern Territory, N. Queensland. 43 NWG 50 NTA QLD. Phan.
**Melaleuca symphyocarpa* F.Muell., Trans. & Proc. Philos. Inst. Victoria 3: 44 (1859). *Myrtoleucodendron symphyocarpum* (F.Muell.) Kuntze, Revis. Gen. Pl. 1: 241 (1891).

Asteromyrtus tranganensis Craven, Austral. Syst. Bot. 1: 384 (1988 publ. 1989)".
New Guinea (Kep. Aru). 43 NWG. Nanophan. or phan.

***Synonyms*:**
Asteromyrtus gaertneri Schauer = ***Asteromyrtus angustifolia*** (Gaertn.) Craven

Astraea

Astraea Schauer = ***Thryptomene*** Endl.
Astraea saxicola (A.Cunn. ex Hook.) Schauer = ***Thryptomene saxicola*** (A.Cunn. ex Hook.) Schauer

Astus

Astus Trudgen & Rye, Nuytsia 15: 502 (2005).
SW. Australia. 50 WAU. [Myrtaceae]
4 Species

Astus duomilius Trudgen & Rye, Nuytsia 15: 504 (2005).
SW. Australia. 50 WAU. Nanophan.

Astus subroseus Trudgen & Rye, Nuytsia 15: 505 (2005).
SW. Australia. 50 WAU. Nanophan.

Astus tetragonus (F.Muell. ex Benth.) Trudgen & Rye, Nuytsia 15: 508 (2005).
Western Australia. 50 WAU. Nanophan.

Astus wittweri Trudgen & Rye, Nuytsia 15: 509 (2005).
SW. Australia. 50 WAU. Nanophan.

Aulomyrcia

Aulomyrcia O.Berg = ***Myrcia*** DC. ex Guill.
Aulomyrcia abrantea O.Berg = [84]
Aulomyrcia acetosans (Poir.) O.Berg = ***Myrcia citrifolia*** (Aubl.) Urb.
Aulomyrcia acrantha O.Berg = [84 BZS]
Aulomyrcia acutata O.Berg = [84 BZL]
Aulomyrcia acutifolia O.Berg = ***Myrcia laruotteana*** Cambess.
Aulomyrcia acutifolia var. *pubescens* O.Berg = ***Myrcia laruotteana*** Cambess.
Aulomyrcia acutifolia var. *villosa* O.Berg = ***Myrcia laruotteana*** Cambess.
Aulomyrcia aethusa O.Berg = ***Myrcia aethusa*** (O.Berg) N.Silveira
Aulomyrcia alagoensis O.Berg = [84 BZE]
Aulomyrcia albidotomentosa Amshoff = ***Myrcia albidotomentosa*** (Amshoff) McVaugh
Aulomyrcia albotomentosa (DC.) O.Berg = ***Myrcia albotomentosa*** DC.
Aulomyrcia albotomentosa var. *lutescens* O.Berg = ***Myrcia albotomentosa*** DC.
Aulomyrcia albotomentosa var. *rubrescens* O.Berg = ***Myrcia albotomentosa*** DC.
Aulomyrcia albotomentosa var. *subcordata* O.Berg = ***Myrcia albotomentosa*** DC.
Aulomyrcia alloiota O.Berg = ***Myrcia tomentosa*** (Aubl.) DC.
Aulomyrcia alloiota var. *cuneata* O.Berg = ***Myrcia tomentosa*** (Aubl.) DC.
Aulomyrcia alloiota var. *obovata* O.Berg = ***Myrcia tomentosa*** (Aubl.) DC.
Aulomyrcia alloiota var. *ovalis* O.Berg = ***Myrcia tomentosa*** (Aubl.) DC.
Aulomyrcia alloiota var. *pyramidalis* O.Berg = ***Myrcia tomentosa*** (Aubl.) DC.
Aulomyrcia alloiota var. *subcordata* O.Berg = ***Myrcia tomentosa*** (Aubl.) DC.
Aulomyrcia alternifolia (Miq.) O.Berg = ***Myrcia guianensis*** (Aubl.) DC.
Aulomyrcia amazonica (DC.) O.Berg = ***Myrcia amazonica*** DC.
Aulomyrcia amethystina O.Berg = ***Myrcia amethystina*** (O.Berg) Kiaersk.

Aulomyrcia amethystina var. *dealbata* O.Berg = ***Myrcia amethystina*** (O.Berg) Kiaersk.
Aulomyrcia amethystina var. *pulchra* O.Berg = ***Myrcia amethystina*** (O.Berg) Kiaersk.
Aulomyrcia amshoffiana Steyerm. = ***Myrcia albidotomentosa*** (Amshoff) McVaugh
Aulomyrcia andromedoides (Cambess.) O.Berg = ***Myrcia venulosa*** DC.
Aulomyrcia androsaemoides O.Berg = ***Myrcia guianensis*** (Aubl.) DC.
Aulomyrcia angustifolia O.Berg = ***Myrcia angustifolia*** (O.Berg) Nied.
Aulomyrcia anomala O.Berg = ***Myrcia aegiphylloides*** Mattos
Aulomyrcia apiocarpa O.Berg = ***Myrcia apiocarpa*** (O.Berg) N.Silveira
Aulomyrcia assumptionis (Morong) Kausel = ***Myrcia laruotteana*** Cambess.
Aulomyrcia atropilosa O.Berg = ***Myrcia atropilosa*** (O.Berg) N.Silveira
Aulomyrcia atrovirens O.Berg = ***Myrcia obovata*** (O.Berg) Nied.
Aulomyrcia aureolanata O.Berg = ***Myrcia tomentosa*** (Aubl.) DC.
Aulomyrcia badia O.Berg = ***Myrcia badia*** (O.Berg) N.Silveira
Aulomyrcia bahiensis O.Berg = [84 BZE]
Aulomyrcia bella (Cambess.) O.Berg = ***Myrcia bella*** Cambess.
Aulomyrcia bicarinata O.Berg = ***Myrcia bicarinata*** (O.Berg) D.Legrand
Aulomyrcia bicudoensis O.Berg = [84 BZC]
Aulomyrcia biformis O.Berg = [84 BZL]
Aulomyrcia billardiana (Kunth) O.Berg = ***Myrcia billardiana*** (Kunth) DC.
Aulomyrcia billardiana var. *grandifolia* O.Berg = ***Myrcia billardiana*** (Kunth) DC.
Aulomyrcia billardiana var. *parvifolia* O.Berg = ***Myrcia billardiana*** (Kunth) DC.
Aulomyrcia bimarginata O.Berg = [84 BZL]
Aulomyrcia blanchetiana O.Berg = ***Myrcia blanchetiana*** (O.Berg) Mattos
Aulomyrcia bolivarensis Steyerm. = ***Myrcia bolivarensis*** (Steyerm.) McVaugh
Aulomyrcia bombycina O.Berg = ***Myrcia bombycina*** (O.Berg) Nied.
Aulomyrcia botrys O.Berg = ***Myrcia guianensis*** (Aubl.) DC.
Aulomyrcia bracteata O.Berg = ***Myrcia didrichseniana*** Kiaersk.
Aulomyrcia breviramis O.Berg = ***Myrcia pulchra*** (O.Berg) Kiaersk.
Aulomyrcia bullata O.Berg = [84 BZE]
Aulomyrcia bullata var. *oblongata* O.Berg = ?
Aulomyrcia buxifolia O.Berg = ***Myrcia guianensis*** (Aubl.) DC.
Aulomyrcia buxifolia var. *elliptica* O.Berg = ***Myrcia guianensis*** (Aubl.) DC.
Aulomyrcia buxifolia var. *ovalis* O.Berg = ***Myrcia guianensis*** (Aubl.) DC.
Aulomyrcia buxizans O.Berg = ***Myrcia guianensis*** (Aubl.) DC.
Aulomyrcia caerulescens O.Berg = ***Myrcia multiflora*** (Lam.) DC.
Aulomyrcia caesia O.Berg = ***Myrcia multiflora***
Aulomyrcia calyptranthoides O.Berg = ***Myrcia calyptranthoides*** (O.Berg) Mattos
Aulomyrcia camareana (DC.) O.Berg = ***Myrcia multiflora*** (Lam.) DC.
Aulomyrcia cambessedeana O.Berg = [84 BZL]
Aulomyrcia campestris (DC.) O.Berg = ***Myrcia campestris*** DC.
Aulomyrcia campestris var. *brunnea* O.Berg = ***Myrcia campestris*** DC.
Aulomyrcia campestris var. *rufa* O.Berg = ***Myrcia campestris*** DC.
Aulomyrcia capitata O.Berg = ***Myrcia camapuanensis*** N.Silveira
Aulomyrcia capivarhyensis (Cambess.) O.Berg = ***Myrcia tomentosa*** (Aubl.) DC.
Aulomyrcia caracasana O.Berg = ***Myrcia caracasana*** (O.Berg) Klotzsch ex O.Berg
Aulomyrcia cardiophylla O.Berg = ***Myrcia cardiophylla*** (O.Berg) Reichardt
Aulomyrcia cardiophylla var. *densiflora* O.Berg = ***Myrcia cardiophylla*** (O.Berg) Reichardt
Aulomyrcia cardiophylla var. *laxiflora* O.Berg = ***Myrcia cardiophylla*** (O.Berg) Reichardt
Aulomyrcia cassinioides (DC.) O.Berg = ***Myrcia guianensis*** (Aubl.) DC.
Aulomyrcia cassinioides var. *glabrata* O.Berg = ***Myrcia guianensis*** (Aubl.) DC.
Aulomyrcia cassinioides var. *velutina* O.Berg = ***Myrcia guianensis*** (Aubl.) DC.
Aulomyrcia castrensis O.Berg = ***Myrcia venulosa*** DC.
Aulomyrcia chapadensis O.Berg = [84 BZC]
Aulomyrcia chilensis O.Berg = ***Myrcia aliena*** McVaugh
Aulomyrcia chrysophylla O.Berg = ***Myrcia cuprea*** (O.Berg) Kiaersk.
Aulomyrcia citrifolia (Aubl.) Amshoff = ***Myrcia citrifolia*** (Aubl.) Urb.
Aulomyrcia clausseniana O.Berg = [84 BZL]
Aulomyrcia collina (DC.) O.Berg = ***Myrcia pubiflora*** DC.
Aulomyrcia comosa O.Berg = [84 BZL]
Aulomyrcia conduplicata O.Berg = ***Myrcia guianensis*** (Aubl.) DC.
Aulomyrcia confusa O.Berg = ***Myrcia tomentosa*** (Aubl.) DC.
Aulomyrcia coriacea (DC.) O.Berg = ***Myrcia citrifolia*** (Aubl.) Urb.
Aulomyrcia coriacea var. *parvifolia* O.Berg = ***Myrcia citrifolia*** (Aubl.) Urb.
Aulomyrcia corymbiflora O.Berg = ***Myrcia vestita*** DC.
Aulomyrcia corymbosa (DC.) O.Berg = ***Myrcia amazonica*** DC.
Aulomyrcia costata (DC.) O.Berg = ***Myrcia splendens*** (Sw.) DC.
Aulomyrcia coumeta (Aubl.) O.Berg = ***Myrcia coumeta*** (Aubl.) DC.
Aulomyrcia crassicaulis (Cambess.) O.Berg = ***Myrcia guianensis*** (Aubl.) DC.
Aulomyrcia crassifolia O.Berg = ***Myrcia irwinii*** Mattos & D.Legrand
Aulomyrcia crenulata O.Berg = [84 BZL]
Aulomyrcia crenulata var. *glabrata* O.Berg = ?
Aulomyrcia crenulata var. *hirta* O.Berg = ?
Aulomyrcia cuneata O.Berg = ***Myrcia guianensis*** (Aubl.) DC.
Aulomyrcia cuprea O.Berg = ***Myrcia cuprea*** (O.Berg) Kiaersk.
Aulomyrcia curassavica Amshoff = ***Myrcia curassavica*** (Amshoff) Stoffers
Aulomyrcia curassavica var. *acutata* Amshoff = ***Myrcia curassavica*** (Amshoff) Stoffers
Aulomyrcia curatellifolia (DC.) O.Berg = ***Myrcia tomentosa*** (Aubl.) DC.
Aulomyrcia curatellifolia var. *australis* O.Berg = ***Myrcia tomentosa*** (Aubl.) DC.

Aulomyrcia curatellifolia var. *grandifolia* O.Berg = **Myrcia tomentosa** (Aubl.) DC.
Aulomyrcia curatellifolia var. *parvifolia* O.Berg = **Myrcia tomentosa** (Aubl.) DC.
Aulomyrcia cymosa O.Berg = **Myrcia cymosa** (O.Berg) Nied.
Aulomyrcia daphnoides (DC.) O.Berg = **Myrcia guianensis** (Aubl.) DC.
Aulomyrcia daphnoides var. *ochracea* O.Berg = **Myrcia guianensis** (Aubl.) DC.
Aulomyrcia dealbata (DC.) O.Berg = **Myrcia dealbata** DC.
Aulomyrcia dealbata var. *glaucescens* O.Berg = **Myrcia dealbata** DC.
Aulomyrcia dealbata var. *pallida* O.Berg = **Myrcia dealbata** DC.
Aulomyrcia debilis (Cambess.) O.Berg = **Myrcia debilis** Cambess.
Aulomyrcia decrescens O.Berg = [84 BZC BZL]
Aulomyrcia densa (DC.) O.Berg = **Myrcia densa** (DC.) Sobral
Aulomyrcia desertorum O.Berg = **Myrcia desertorum** (O.Berg) N.Silveira
Aulomyrcia detergens (Miq.) O.Berg = **Myrcia amazonica** DC.
Aulomyrcia detergens var. *depauperata* O.Berg = **Myrcia amazonica** DC.
Aulomyrcia detergens var. *dives* O.Berg = **Myrcia amazonica** DC.
Aulomyrcia diaphana O.Berg = **Myrcia diaphana** (O.Berg) N.Silveira
Aulomyrcia dichroma O.Berg = **Myrcia guianensis** (Aubl.) DC.
Aulomyrcia dictyophleba O.Berg = **Myrcia venulosa** DC.
Aulomyrcia dictyophylla O.Berg = **Myrcia dictyophylla** (O.Berg) Mattos & D.Legrand
Aulomyrcia dictyophylla var. *robustior* O.Berg = **Myrcia dictyophylla** (O.Berg) Mattos & D.Legrand
Aulomyrcia dictyophylla var. *subcordata* O.Berg = **Myrcia dictyophylla** (O.Berg) Mattos & D.Legrand
Aulomyrcia dimorpha O.Berg = **Myrcia dimorpha** (O.Berg) N.Silveira
Aulomyrcia divaricata O.Berg = **Myrcia rufipila** McVaugh
Aulomyrcia doniana O.Berg = **Myrcia doniana** O.Berg
Aulomyrcia dumosa O.Berg = **Myrcia platyclada** DC.
Aulomyrcia edulis O.Berg = **Myrcia platyclada** DC.
Aulomyrcia egensis O.Berg = **Myrcia egensis** (O.Berg) McVaugh
Aulomyrcia ehrenbergiana (O.Berg) Amshoff = **Myrcia ehrenbergiana** (O.Berg) McVaugh
Aulomyrcia elaeodendra (DC.) O.Berg = **Myrcia guianensis** (Aubl.) DC.
Aulomyrcia emarginata O.Berg = **Myrcia emarginata** (O.Berg) Nied.
Aulomyrcia eumecephylla O.Berg = **Myrcia eumecephylla** (O.Berg) Nied.
Aulomyrcia exsucca (DC.) O.Berg = **Myrcia guianensis** (Aubl.) DC.
Aulomyrcia fenzliana O.Berg = **Myrcia oligantha** O.Berg
Aulomyrcia ferruginea O.Berg = **Myrcia lapensis** N.Silveira
Aulomyrcia fragilis O.Berg = **Myrcia guianensis** (Aubl.) DC.
Aulomyrcia gardneriana O.Berg = **Myrcia guianensis** (Aubl.) DC.
Aulomyrcia gardneriana var. *caerulescens* O.Berg = **Myrcia guianensis** (Aubl.) DC.
Aulomyrcia gardneriana var. *virescens* O.Berg = **Myrcia guianensis** (Aubl.) DC.
Aulomyrcia garopabensis (Cambess.) O.Berg = **Myrcia palustris** DC.
Aulomyrcia gaudichaudiana O.Berg = **Myrcia racemosa** (O.Berg) Kiaersk.
Aulomyrcia gigantea O.Berg = **Myrcia gigantea** (O.Berg) Nied.
Aulomyrcia glabra O.Berg = **Myrcia glabra** (O.Berg) D.Legrand
Aulomyrcia glandulosa O.Berg = **Myrcia guianensis** (Aubl.) DC.
Aulomyrcia glandulosa var. *elliptica* O.Berg = **Myrcia guianensis** (Aubl.) DC.
Aulomyrcia glandulosa var. *longifolia* O.Berg = **Myrcia guianensis** (Aubl.) DC.
Aulomyrcia glandulosa var. *obovata* O.Berg = **Myrcia guianensis** (Aubl.) DC.
Aulomyrcia glauca (Cambess.) O.Berg = **Myrcia glauca** Cambess.
Aulomyrcia glaucescens O.Berg = **Myrcia multiflora** (Lam.) DC.
Aulomyrcia glaucescens var. *grandifolia* O.Berg = **Myrcia multiflora** (Lam.) DC.
Aulomyrcia glaucescens var. *parvifolia* O.Berg = **Myrcia multiflora** (Lam.) DC.
Aulomyrcia goyazensis O.Berg = [84 BZC]
Aulomyrcia grandiflora O.Berg = **Myrcia pubipetala** Miq.
Aulomyrcia grandifolia O.Berg = [84 BZN]
Aulomyrcia hepatica O.Berg = **Myrcia guianensis** (Aubl.) DC.
Aulomyrcia hirtiflora (DC.) O.Berg = **Myrcia tomentosa** (Aubl.) DC.
Aulomyrcia holosericea O.Berg = **Myrcia neesiana** DC.
Aulomyrcia hostmanniana O.Berg = **Myrcia amazonica** DC.
Aulomyrcia hostmanniana var. *gracilior* O.Berg = **Myrcia amazonica** DC.
Aulomyrcia hypericoides (Cambess.) O.Berg = **Myrcia hypericoides** Cambess.
Aulomyrcia imbricata (Gardner) O.Berg = **Myrcia racemulosa** DC.
Aulomyrcia imbricata var. *intermedia* O.Berg = **Myrcia racemulosa** DC.
Aulomyrcia imbricata var. *microphylla* O.Berg = **Myrcia racemulosa** DC.
Aulomyrcia imbricata var. *rotundifolia* O.Berg = **Myrcia racemulosa** DC.
Aulomyrcia imbricata var. *scutulifera* (DC.) O.Berg = **Myrcia racemulosa** DC.
Aulomyrcia inaequiloba (DC.) Amshoff = **Myrcia inaequiloba** (DC.) Lemée
Aulomyrcia inaequiloba var. *nitida* (Benth.) Amshoff = **Myrcia inaequiloba** (DC.) Lemée
Aulomyrcia inaequiloba var. *paniculata* (O.Berg) Amshoff = **Myrcia inaequiloba** (DC.) Lemée
Aulomyrcia insularis (Gardner) O.Berg = **Myrcia insularis** Gardner
Aulomyrcia insularis var. *opaca* O.Berg = **Myrcia insularis** Gardner
Aulomyrcia insularis var. *punctata* O.Berg = **Myrcia insularis** Gardner
Aulomyrcia intermedia O.Berg = **Myrcia guianensis** (Aubl.) DC.
Aulomyrcia jacquiniana O.Berg = **Myrcia citrifolia** (Aubl.) Urb.
Aulomyrcia jequitinhonhensis O.Berg = **Myrcia torta** DC.

Aulomyrcia jequitinhonhensis var. *glauca* O.Berg = ***Myrcia torta*** DC.
Aulomyrcia jequitinhonhensis var. *grandifolia* O.Berg = ***Myrcia torta*** DC.
Aulomyrcia jequitinhonhensis var. *parvifolia* O.Berg = ***Myrcia torta*** DC.
Aulomyrcia karuaiensis Steyerm. = ***Marlierea karuaiensis*** (Steyerm.) McVaugh
Aulomyrcia klotzschiana O.Berg = [84 BZL]
Aulomyrcia klotzschiana var. *depauperata* O.Berg = ?
Aulomyrcia klotzschiana var. *itacolumensis* O.Berg = ?
Aulomyrcia klotzschiana var. *multiflora* O.Berg = ?
Aulomyrcia lacunosa O.Berg = ***Myrcia lacunosa*** (O.Berg) N.Silveira
Aulomyrcia lancea O.Berg = ***Myrcia tomentosa*** (Aubl.) DC.
Aulomyrcia lancifolia O.Berg = ***Myrcia inaequiloba*** (DC.) Lemée
Aulomyrcia langsdorffii O.Berg = [84 BZL]
Aulomyrcia lanuginosa O.Berg = ***Myrcia tomentosa*** (Aubl.) DC.
Aulomyrcia lanuginosa var. *pyramidata* O.Berg = ***Myrcia tomentosa*** (Aubl.) DC.
Aulomyrcia laricina O.Berg = ***Myrcia laricina*** (O.Berg) Burret ex Luetzelb.
Aulomyrcia laruotteana (Cambess.) O.Berg = ***Myrcia laruotteana*** Cambess.
Aulomyrcia laruotteana var. *membranacea* O.Berg = ***Myrcia laruotteana*** Cambess.
Aulomyrcia laruotteana var. *opaca* O.Berg = ***Myrcia laruotteana*** Cambess.
Aulomyrcia laruotteana var. *paraguayensis* O.Berg = ***Myrcia laruotteana*** Cambess.
Aulomyrcia laruotteana var. *peruviana* O.Berg = ***Myrcia multiflora*** (Lam.) DC.
Aulomyrcia laruotteana var. *punctata* O.Berg = ***Myrcia laruotteana*** Cambess.
Aulomyrcia laureola O.Berg = ***Myrcia venulosa*** DC.
Aulomyrcia lauriflora (DC.) O.Berg = ***Myrcia guianensis*** (Aubl.) DC.
Aulomyrcia laurifolia O.Berg = ***Myrcia laurifolia*** Cambess.
Aulomyrcia laxiflora (Cambess.) O.Berg = ***Myrcia laxiflora*** Cambess.
Aulomyrcia laxiflora var. *angustifolia* O.Berg = ***Myrcia laxiflora*** Cambess.
Aulomyrcia laxiflora var. *latifolia* O.Berg = ***Myrcia laxiflora*** Cambess.
Aulomyrcia leptoclada (DC.) O.Berg = ***Myrcia amazonica*** DC.
Aulomyrcia leucadendron (DC.) O.Berg = ***Myrcia leucadendron*** DC.
Aulomyrcia leucantha O.Berg = ***Myrcia tomentosa*** (Aubl.) DC.
Aulomyrcia linearifolia (Cambess.) O.Berg = ***Myrcia linearifolia*** Cambess.
Aulomyrcia lineata O.Berg = ***Myrcia lineata*** (O.Berg) Nied.
Aulomyrcia lingua O.Berg = ***Myrcia guianensis*** (Aubl.) DC.
Aulomyrcia lingua var. *glabrata* O.Berg = ***Myrcia guianensis*** (Aubl.) DC.
Aulomyrcia lingua var. *rufa* O.Berg = ***Myrcia guianensis*** (Aubl.) DC.
Aulomyrcia linguiformis O.Berg = ***Myrcia vestita*** DC.
Aulomyrcia longipes O.Berg = ***Myrcia tomentosa*** (Aubl.) DC.
Aulomyrcia longipes var. *latifolia* O.Berg = ***Myrcia tomentosa*** (Aubl.) DC.
Aulomyrcia longipes var. *obovata* O.Berg = ***Myrcia tomentosa*** (Aubl.) DC.
Aulomyrcia longipes var. *spathulata* O.Berg = ***Myrcia tomentosa*** (Aubl.) DC.
Aulomyrcia lucida O.Berg = ***Myrcia dilucida*** G.M.Barroso
Aulomyrcia lucida var. *grandifolia* O.Berg = ***Myrcia dilucida*** G.M.Barroso
Aulomyrcia lucida var. *parvifolia* O.Berg = ***Myrcia dilucida*** G.M.Barroso
Aulomyrcia macrocarpa (DC.) O.Berg = ***Myrcia macrocarpa*** DC.
Aulomyrcia macrophylla O.Berg = ***Myrcia egensis*** (O.Berg) McVaugh
Aulomyrcia mansonii O.Berg = ***Myrcia mansonii*** (O.Berg) N.Silveira
Aulomyrcia maraguana (DC.) O.Berg = ***Myrcia maraguana*** DC.
Aulomyrcia maritima O.Berg = ***Myrcia guianensis*** (Aubl.) DC.
Aulomyrcia martiana O.Berg = ***Myrcia torta*** DC.
Aulomyrcia mathewsiana O.Berg = ***Myrcia mathewsiana*** (O.Berg) McVaugh
Aulomyrcia micrantha O.Berg = ***Myrcia tomentosa*** (Aubl.) DC.
Aulomyrcia microcarpa (Cambess.) O.Berg = ***Myrcia guianensis*** (Aubl.) DC.
Aulomyrcia micropetala (Mart.) O.Berg = ***Myrcia micropetala*** (Mart.) Nied.
Aulomyrcia minutiflora (Sagot) Amshoff = ***Myrcia minutiflora*** Sagot
Aulomyrcia montana (Cambess.) O.Berg = ***Myrcia montana*** Cambess.
Aulomyrcia mugiensis (Cambess.) O.Berg = ***Blepharocalyx salicifolius*** (Kunth) O.Berg
Aulomyrcia multiflora (Lam.) O.Berg = ***Myrcia multiflora*** (Lam.) DC.
Aulomyrcia multiflora var. *grandifolia* O.Berg = ***Myrcia multiflora*** (Lam.) DC.
Aulomyrcia mutabilis O.Berg = ***Myrcia mutabilis*** (O.Berg) N.Silveira
Aulomyrcia myrtillifolia (DC.) O.Berg = ***Myrcia myrtillifolia*** DC.
Aulomyrcia neesiana (DC.) O.Berg = ***Myrcia neesiana*** DC.
Aulomyrcia nigrescens (DC.) O.Berg = ***Marlierea umbraticola*** (Kunth) O.Berg
Aulomyrcia nigrescens var. *angustifolia* O.Berg = ***Marlierea umbraticola*** (Kunth) O.Berg
Aulomyrcia nigrescens var. *dives* O.Berg = ***Marlierea umbraticola*** (Kunth) O.Berg
Aulomyrcia nigrescens var. *latifolia* O.Berg = ***Marlierea umbraticola*** (Kunth) O.Berg
Aulomyrcia nigricans O.Berg = ***Myrcia nigricans*** (O.Berg) N.Silveira
Aulomyrcia nigropunctata O.Berg = ***Myrcia nigropunctata*** (O.Berg) N.Silveira
Aulomyrcia nivea (Cambess.) O.Berg = ***Myrcia nivea*** Cambess.
Aulomyrcia nivea var. *andromedifolia* O.Berg = ***Myrcia nivea*** Cambess.
Aulomyrcia nivea var. *rosmarinifolia* O.Berg = ***Myrcia nivea*** Cambess.
Aulomyrcia oblongata (DC.) O.Berg = ***Myrcia oblongata*** DC.
Aulomyrcia obovata O.Berg = ***Myrcia obovata*** (O.Berg) Nied.
Aulomyrcia obscura O.Berg = ***Myrcia guianensis*** (Aubl.) DC.

Aulomyrcia obscura var. *longipes* O.Berg = ***Myrcia guianensis*** (Aubl.) DC.
Aulomyrcia obtecta O.Berg = ***Myrcia guianensis*** (Aubl.) DC.
Aulomyrcia obtusa (Schauer) O.Berg = ***Myrcia guianensis*** (Aubl.) DC.
Aulomyrcia obtusa var. *grandifolia* O.Berg = ***Myrcia guianensis*** (Aubl.) DC.
Aulomyrcia obtusa var. *longipes* O.Berg = ***Myrcia guianensis*** (Aubl.) DC.
Aulomyrcia obtusa var. *panicularis* O.Berg = ***Myrcia guianensis*** (Aubl.) DC.
Aulomyrcia obtusa var. *pauciflora* O.Berg = ***Myrcia guianensis*** (Aubl.) DC.
Aulomyrcia obtusa var. *surinamensis* (Miq.) Amshoff = ***Myrcia guianensis*** (Aubl.) DC.
Aulomyrcia obtusa var. *tenuifolia* O.Berg = ***Myrcia guianensis*** (Aubl.) DC.
Aulomyrcia obtusata O.Berg = [82 GUY]
Aulomyrcia orthophylla O.Berg = ***Myrcia orthophylla*** (O.Berg) Kiaersk.
Aulomyrcia ottonis O.Berg = ***Myrcia tomentosa*** (Aubl.) DC.
Aulomyrcia ovalifolia O.Berg = ***Myrcia multiflora*** (Lam.) DC.
Aulomyrcia ovalis O.Berg = ***Myrcia multiflora*** (Lam.) DC.
Aulomyrcia ovata O.Berg = ***Myrcia pyrifolia*** (Desv.) Nied.
Aulomyrcia pachyclada O.Berg = ***Myrcia vestita*** DC.
Aulomyrcia pachyclada var. *elliptica* O.Berg = ***Myrcia vestita*** DC.
Aulomyrcia pachyclada var. *prolifera* O.Berg = ***Myrcia vestita*** DC.
Aulomyrcia pachyclada var. *spathulata* O.Berg = ***Myrcia vestita*** DC.
Aulomyrcia pallens (DC.) O.Berg = ***Myrcia guianensis*** (Aubl.) DC.
Aulomyrcia pallens var. *ovalis* O.Berg = ***Myrcia guianensis*** (Aubl.) DC.
Aulomyrcia pallens var. *ovata* O.Berg = ***Myrcia guianensis*** (Aubl.) DC.
Aulomyrcia pallens var. *petiolaris* O.Berg = ***Myrcia guianensis*** (Aubl.) DC.
Aulomyrcia pallens var. *subcordata* O.Berg = ***Myrcia guianensis*** (Aubl.) DC.
Aulomyrcia pallida O.Berg = ***Myrcia guianensis*** (Aubl.) DC.
Aulomyrcia panicularis O.Berg = ***Myrcia panicularis*** (O.Berg) N.Silveira
Aulomyrcia paniculata O.Berg = ***Myrcia inaequiloba*** (DC.) Lemée
Aulomyrcia paraensis O.Berg = ***Myrcia amazonica*** DC.
Aulomyrcia parnahibensis O.Berg = ***Myrcia parnahibensis*** (O.Berg) Kiaersk.
Aulomyrcia parnahibensis var. *acutifolia* O.Berg = ***Myrcia parnahibensis*** (O.Berg) Kiaersk.
Aulomyrcia parnahibensis var. *imbricata* O.Berg = ***Myrcia parnahibensis*** (O.Berg) Kiaersk.
Aulomyrcia parnahibensis var. *verticillata* O.Berg = ***Myrcia parnahibensis*** (O.Berg) Kiaersk.
Aulomyrcia parvifolia Steyerm. = ***Myrcia rotundata***
Aulomyrcia pauciflora O.Berg = [84 BZL]
Aulomyrcia perforata O.Berg = ***Myrcia multiflora*** (Lam.) DC.
Aulomyrcia pertusa (DC.) O.Berg = ***Myrcia pertusa*** DC.
Aulomyrcia piauhiensis O.Berg = [84 BZE]
Aulomyrcia pilantha O.Berg = ***Myrcia rufipes*** DC.
Aulomyrcia pilantha var. *latifolia* O.Berg = ***Myrcia rufipes*** DC.
Aulomyrcia pilantha var. *longifolia* O.Berg = ***Myrcia rufipes*** DC.
Aulomyrcia pilantha var. *parvifolia* O.Berg = ***Myrcia rufipes*** DC.
Aulomyrcia pinifolia (Cambess.) O.Berg = ***Myrcia pinifolia*** Cambess.
Aulomyrcia pirarensis O.Berg = ***Myrcia inaequiloba*** (DC.) Lemée
Aulomyrcia platyclada O.Berg = [84 BZL]
Aulomyrcia platyclada (DC.) Amshoff = ***Myrcia platyclada*** DC.
Aulomyrcia platyclada var. *kaieteurensis* Amshoff = ***Myrcia platyclada*** DC.
Aulomyrcia plumbea O.Berg = [84 BZL]
Aulomyrcia poeppigiana O.Berg = ***Myrcia guianensis*** (Aubl.) DC.
Aulomyrcia pohliana O.Berg = ***Myrcia pubiflora*** DC.
Aulomyrcia polyantha (DC.) O.Berg = ***Myrcia polyantha*** DC.
Aulomyrcia polyantha var. *coriacea* O.Berg = ***Myrcia polyantha*** DC.
Aulomyrcia polyantha var. *membranacea* O.Berg = ***Myrcia polyantha*** DC.
Aulomyrcia polyantha var. *parviflora* O.Berg = ***Myrcia polyantha*** DC.
Aulomyrcia polymorpha O.Berg = ***Myrcia decorticans*** DC.
Aulomyrcia polymorpha var. *decorticans* (DC.) O.Berg = ***Myrcia decorticans*** DC.
Aulomyrcia polymorpha var. *duriuscula* (Mart. ex DC.) O.Berg = ***Myrcia decorticans*** DC.
Aulomyrcia polymorpha var. *lasiopus* (DC.) O.Berg = ***Myrcia decorticans*** DC.
Aulomyrcia polymorpha var. *leucophloea* (DC.) O.Berg = ***Myrcia decorticans*** DC.
Aulomyrcia pruinosa O.Berg = ***Myrcia guianensis*** (Aubl.) DC.
Aulomyrcia prunifolia (DC.) O.Berg = ***Myrcia tomentosa*** (Aubl.) DC.
Aulomyrcia prunifolia var. *brevipes* O.Berg = ***Myrcia tomentosa*** (Aubl.) DC.
Aulomyrcia prunifolia var. *longipes* O.Berg = ***Myrcia tomentosa*** (Aubl.) DC.
Aulomyrcia ptariensis Steyerm. = ***Myrcia ptariensis*** (Steyerm.) McVaugh
Aulomyrcia puberula (Cambess.) O.Berg = ***Myrcia tomentosa*** (Aubl.) DC.
Aulomyrcia pubiflora (DC.) O.Berg = ***Myrcia pubiflora*** DC.
Aulomyrcia pulchra O.Berg = ***Myrcia pulchra*** (O.Berg) Kiaersk.
Aulomyrcia punctata O.Berg = [84 BZL]
Aulomyrcia pusilla O.Berg = [84 BZL]
Aulomyrcia pyrifolia (Desv.) O.Berg = ***Myrcia pyrifolia*** (Desv.) Nied.
Aulomyrcia pyrifolia var. *gracilis* O.Berg = ***Myrcia pyrifolia*** (Desv.) Nied.
Aulomyrcia pyrifolia var. *robusta* O.Berg = ***Myrcia pyrifolia*** (Desv.) Nied.
Aulomyrcia racemosa O.Berg = ***Myrcia racemosa*** (O.Berg) Kiaersk.
Aulomyrcia ramuliflora O.Berg = ***Myrcia ramuliflora*** (O.Berg) N.Silveira
Aulomyrcia ramulosa (DC.) O.Berg = ***Myrcia selloi*** (Spreng.) N.Silveira
Aulomyrcia ramulosa var. *acutata* O.Berg = ***Myrcia selloi*** (Spreng.) N.Silveira
Aulomyrcia ramulosa var. *australis* O.Berg = ***Myrcia selloi*** (Spreng.) N.Silveira

Aulomyrcia ramulosa var. *colorata* O.Berg = ***Myrcia selloi*** (Spreng.) N.Silveira
Aulomyrcia ramulosa var. *panicularis* O.Berg = ***Myrcia selloi*** (Spreng.) N.Silveira
Aulomyrcia ramulosa var. *pauciflora* O.Berg = ***Myrcia selloi*** (Spreng.) N.Silveira
Aulomyrcia ramulosa var. *subcordata* O.Berg = ***Myrcia selloi*** (Spreng.) N.Silveira
Aulomyrcia ramulosa var. *triflora* (Cambess.) O.Berg = ***Myrcia selloi*** (Spreng.) N.Silveira
Aulomyrcia regeliana O.Berg = [84 BZC BZL]
Aulomyrcia regeliana var. *oppositifolia* O.Berg = ?
Aulomyrcia regeliana var. *sparsifolia* O.Berg = ?
Aulomyrcia reticulata O.Berg = ***Myrcia venulosa*** DC.
Aulomyrcia reticulata var. *dives* O.Berg = ***Myrcia venulosa*** DC. var. ***dives***
Aulomyrcia reticulata var. *pauciflora* O.Berg = ***Myrcia venulosa*** DC. var. ***dives***
Aulomyrcia reticulosa (Miq.) O.Berg = ***Myrcia reticulosa*** Miq.
Aulomyrcia retusa O.Berg = ***Myrcia guianensis*** (Aubl.) DC.
Aulomyrcia richardiana O.Berg = ***Myrcia aethusa*** (O.Berg) N.Silveira
Aulomyrcia riedeliana O.Berg = ***Myrcia venulosa***
Aulomyrcia roraimae (Oliv.) Steyerm. = ***Myrcia guianensis*** (Aubl.) DC.
Aulomyrcia roraimensis O.Berg = ***Myrcia guianensis*** (Aubl.) DC.
Aulomyrcia rorida O.Berg = ***Myrcia guianensis*** (Aubl.) DC.
Aulomyrcia rostrata O.Berg = ***Myrcia neorostrata*** Sobral
Aulomyrcia rosulans O.Berg = ***Myrcia tomentosa*** (Aubl.) DC.
Aulomyrcia rotundata Amshoff = ***Myrcia rotundata*** (Amshoff) McVaugh
Aulomyrcia rotundifolia O.Berg = ***Myrcia rotundifolia*** (O.Berg) Kiaersk.
Aulomyrcia rubella (Cambess.) O.Berg = ***Myrcia rubella*** Cambess.
Aulomyrcia rubella var. *glabra* O.Berg = ***Myrcia rubella*** Cambess.
Aulomyrcia rubella var. *puberula* O.Berg = ***Myrcia rubella*** Cambess.
Aulomyrcia rufa O.Berg = ***Myrcia pulchra*** (O.Berg) Kiaersk.
Aulomyrcia rufipes (DC.) O.Berg = ***Myrcia rufipes*** DC.
Aulomyrcia rufipes var. *angustifolia* O.Berg = ***Myrcia rufipes*** DC.
Aulomyrcia rufipes var. *bracteata* O.Berg = ***Myrcia rufipes*** DC.
Aulomyrcia rufipes var. *dives* O.Berg = ***Myrcia rufipes*** DC.
Aulomyrcia rufipes var. *grandiflora* O.Berg = ***Myrcia rufipes*** DC.
Aulomyrcia rufipes var. *latifolia* O.Berg = ***Myrcia rufipes*** DC.
Aulomyrcia rugosa O.Berg = ***Myrcia venulosa*** DC.
Aulomyrcia sagraea O.Berg = ***Myrcia valenzuelana*** (A.Rich.) Griseb.
Aulomyrcia salicifolia (DC.) O.Berg = ***Myrcia salicifolia*** DC.
Aulomyrcia salticola Steyerm. = ***Myrcia salticola*** (Steyerm.) McVaugh
Aulomyrcia salzmannii O.Berg = [84 BZE]
Aulomyrcia saxatilis Amshoff = ***Myrcia saxatilis*** (Amshoff) McVaugh
Aulomyrcia schaueriana (Miq.) Amshoff = ***Myrcia graciliflora*** Sagot
Aulomyrcia schomburgkiana O.Berg = ***Myrcia guianensis*** (Aubl.) DC.
Aulomyrcia schrankiana (DC.) O.Berg = ***Myrcia guianensis*** (Aubl.) DC.
Aulomyrcia scrobiculata O.Berg = ***Myrcia scrobiculata*** (O.Berg) O.Berg
Aulomyrcia selloi (Spreng.) Kausel = ***Myrcia selloi*** (Spreng.) N.Silveira
Aulomyrcia sonderiana O.Berg = ***Myrcia venulosa*** DC.
Aulomyrcia spathulata O.Berg = ***Myrcia spathulata*** (O.Berg) Kiaersk.
Aulomyrcia speciosa Amshoff = ***Myrcia speciosa*** (Amshoff) McVaugh
Aulomyrcia sphaerocarpa (DC.) O.Berg = ***Myrcia multiflora*** (Lam.) DC.
Aulomyrcia sphaerocarpa var. *arborescens* O.Berg = ***Myrcia multiflora*** (Lam.) DC.
Aulomyrcia sphaerocarpa var. *complicata* O.Berg = ***Myrcia multiflora*** (Lam.) DC.
Aulomyrcia sphaerocarpa var. *gracilis* O.Berg = ***Myrcia multiflora*** (Lam.) DC.
Aulomyrcia sphaerocarpa var. *intermedia* O.Berg = ***Myrcia multiflora*** (Lam.) DC.
Aulomyrcia sphaerocarpa var. *obtusata* O.Berg = ***Myrcia multiflora*** (Lam.) DC.
Aulomyrcia sphaerocarpa var. *ovata* O.Berg = ***Myrcia multiflora*** (Lam.) DC.
Aulomyrcia sphaerocarpa var. *pauciflora* O.Berg = ***Myrcia multiflora*** (Lam.) DC.
Aulomyrcia sphenoides O.Berg = ***Myrcia ramuliflora*** (O.Berg) N.Silveira
Aulomyrcia spruceana O.Berg = ***Myrcia amazonica*** DC.
Aulomyrcia stenophylla O.Berg = [85 PAR]
Aulomyrcia stictophylla O.Berg = ***Myrcia stictophylla*** (O.Berg) N.Silveira
Aulomyrcia stricta O.Berg = ***Myrcia stricta*** (O.Berg) Kiaersk.
Aulomyrcia suaveolens (Cambess.) O.Berg = ***Myrcia guianensis*** (Aubl.) DC.
Aulomyrcia subalpestris (DC.) O.Berg = ***Myrcia subalpestris*** DC.
Aulomyrcia subavenia O.Berg = ***Myrcia subavenia*** (O.Berg) N.Silveira
Aulomyrcia subcordata (DC.) O.Berg = ***Myrcia subcordata*** DC.
Aulomyrcia subobliqua O.Berg = ***Myrcia subobliqua*** (O.Berg) Nied.
Aulomyrcia subverticillaris O.Berg = ***Myrcia subverticillaris*** (O.Berg) Kiaersk.
Aulomyrcia subverticillaris var. *angustifolia* O.Berg = ***Myrcia subverticillaris*** (O.Berg) Kiaersk.
Aulomyrcia subverticillaris var. *incanescens* O.Berg = ***Myrcia subverticillaris*** (O.Berg) Kiaersk.
Aulomyrcia subverticillaris var. *rufa* O.Berg = ***Myrcia subverticillaris*** (O.Berg) Kiaersk.
Aulomyrcia suffruticosa O.Berg = ***Myrcia paracatuensis*** Kiaersk.
Aulomyrcia suffruticosa var. *arenaria* O.Berg = ***Myrcia paracatuensis*** Kiaersk.
Aulomyrcia suffruticosa var. *venosa* O.Berg = ***Myrcia paracatuensis*** Kiaersk.
Aulomyrcia supraaxillaris O.Berg = ***Myrcia jacobinensis*** Mattos
Aulomyrcia surinamensis (Miq.) O.Berg = ***Myrcia guianensis*** (Aubl.) DC.
Aulomyrcia tenuifolia O.Berg = ***Myrcia tenuifolia*** (O.Berg) Sobral
Aulomyrcia ternifolia O.Berg = [84 BZL]
Aulomyrcia tetramera Amshoff = ***Myrcia decorticans*** DC.

Aulomyrcia thyrsiflora O.Berg = ***Myrcia vestita*** DC.
Aulomyrcia thyrsiflora var. *lateriflora* O.Berg = ***Myrcia vestita*** DC.
Aulomyrcia thyrsiflora var. *obtusifolia* (DC.) O.Berg = ***Myrcia vestita*** DC.
Aulomyrcia thyrsiflora var. *petiolaris* O.Berg = ***Myrcia vestita*** DC.
Aulomyrcia tobagensis (Krug & Urb.) Amshoff = ***Myrcia decorticans*** DC.
Aulomyrcia tomentosa (Aubl.) Amshoff = ***Myrcia tomentosa*** (Aubl.) DC.
Aulomyrcia torta (DC.) O.Berg = ***Myrcia torta*** DC.
Aulomyrcia tortuosa O.Berg = ***Myrcia tortuosa*** (O.Berg) N.Silveira
Aulomyrcia trichantha Wawra = [84]
Aulomyrcia triflora O.Berg = ***Myrcia graciliflora*** Sagot
Aulomyrcia trifolia O.Berg = [84 BZL]
Aulomyrcia uaupensis O.Berg = ***Myrcia guianensis*** (Aubl.) DC.
Aulomyrcia undulata O.Berg = ***Myrcia lajeana*** D.Legrand
Aulomyrcia vacciniifolia O.Berg = ***Myrcia vacciniifolia*** (O.Berg) Nied.
Aulomyrcia variabilis (Mart. ex DC.) O.Berg = ***Myrcia variabilis*** Mart. ex DC.
Aulomyrcia variabilis var. *intermedia* Mart. ex O.Berg = ***Myrcia variabilis*** Mart. ex DC.
Aulomyrcia variabilis var. *nummulaira* Mart. ex O.Berg = ***Myrcia variabilis*** Mart. ex DC.
Aulomyrcia variabilis var. *ovatifolia* Mart. ex O.Berg = ***Myrcia variabilis*** Mart. ex DC.
Aulomyrcia variabilis var. *suffruticosa* O.Berg = ***Myrcia variabilis*** Mart. ex DC.
Aulomyrcia vautheriana O.Berg = ***Myrcia lundiana*** Kiaersk.
Aulomyrcia velhensis O.Berg = ***Myrcia velhensis*** (O.Berg) N.Silveira
Aulomyrcia venulosa (DC.) O.Berg = ***Myrcia venulosa*** DC.
Aulomyrcia venulosa var. *capoeirensis* (DC.) O.Berg = ***Myrcia venulosa*** DC.
Aulomyrcia venulosa var. *ochracea* O.Berg = ***Myrcia venulosa*** DC.
Aulomyrcia venulosa var. *parvifolia* O.Berg = ***Myrcia venulosa*** DC.
Aulomyrcia venulosa var. *rufa* O.Berg = ***Myrcia venulosa*** DC.
Aulomyrcia vestita (DC.) O.Berg = ***Myrcia vestita*** DC.
Aulomyrcia vestita var. *grandifolia* O.Berg = ***Myrcia vestita*** DC.
Aulomyrcia vestita var. *parvifolia* O.Berg = ***Myrcia vestita*** DC.
Aulomyrcia vinacea Steyerm. = ***Myrcia multiflora*** (Lam.) DC.
Aulomyrcia virgata (Cambess.) O.Berg = ***Myrcia virgata*** Cambess.
Aulomyrcia widgreniana O.Berg = ***Myrcia pulchra*** (O.Berg) Kiaersk.
Aulomyrcia wullschlaegeliana O.Berg = ***Myrcia splendens*** (Sw.) DC.
Aulomyrcia zetekiana (Standl.) Amshoff = ***Myrcia zetekiana*** (Standl.) B.Holst

***Unplaced Names*:**
Aulomyrcia abrantea O.Berg, Linnaea 30: 659 (1861). = [84]
Aulomyrcia acrantha O.Berg in C.F.P.von Martius & auct. suc. (eds.), Fl. Bras. 14(1): 71 (1857). = [84 BZS]
Aulomyrcia acutata O.Berg in C.F.P.von Martius & auct. suc. (eds.), Fl. Bras. 14(1): 71 (1857). = [84 BZL]
Aulomyrcia alagoensis O.Berg in C.F.P.von Martius & auct. suc. (eds.), Fl. Bras. 14(1): 120 (1857). = [84 BZE]
Aulomyrcia bahiensis O.Berg in C.F.P.von Martius & auct. suc. (eds.), Fl. Bras. 14(1): 113 (1857). = [84 BZE]
Aulomyrcia bicudoensis O.Berg in C.F.P.von Martius & auct. suc. (eds.), Fl. Bras. 14(1): 557 (1859). = [84 BZC]
Aulomyrcia biformis O.Berg in C.F.P.von Martius & auct. suc. (eds.), Fl. Bras. 14(1): 141 (1857). = [84 BZL]
Aulomyrcia bimarginata O.Berg in C.F.P.von Martius & auct. suc. (eds.), Fl. Bras. 14(1): 115 (1857). = [84 BZL]
Aulomyrcia bullata O.Berg in C.F.P.von Martius & auct. suc. (eds.), Fl. Bras. 14(1): 96 (1857). = [84 BZE]
Aulomyrcia caesia O.Berg in C.F.P.von Martius & auct. suc. (eds.), Fl. Bras. 14(1): 83 (1857). = [84 BZE]
Aulomyrcia cambessedeana O.Berg, Linnaea 27: 40 (1855). = [84 BZL]
Aulomyrcia chapadensis O.Berg in C.F.P.von Martius & auct. suc. (eds.), Fl. Bras. 14(1): 554 (1859). = [84 BZC]
Aulomyrcia clausseniana O.Berg in C.F.P.von Martius & auct. suc. (eds.), Fl. Bras. 14(1): 118 (1857). = [84 BZL]
Aulomyrcia comosa O.Berg in C.F.P.von Martius & auct. suc. (eds.), Fl. Bras. 14(1): 133 (1857). = [84 BZL]
Aulomyrcia crenulata O.Berg in C.F.P.von Martius & auct. suc. (eds.), Fl. Bras. 14(1): 141 (1857). = [84 BZL]
Aulomyrcia decrescens O.Berg in C.F.P.von Martius & auct. suc. (eds.), Fl. Bras. 14(1): 135 (1857). = [84 BZC BZL]
Aulomyrcia goyazensis O.Berg in C.F.P.von Martius & auct. suc. (eds.), Fl. Bras. 14(1): 85 (1857). = [84 BZC]
Aulomyrcia grandifolia O.Berg in C.F.P.von Martius & auct. suc. (eds.), Fl. Bras. 14(1): 97 (1857). = [84 BZN]
Aulomyrcia klotzschiana O.Berg in C.F.P.von Martius & auct. suc. (eds.), Fl. Bras. 14(1): 111 (1857). = [84 BZL]
Aulomyrcia langsdorffii O.Berg in C.F.P.von Martius & auct. suc. (eds.), Fl. Bras. 14(1): 550 (1859). = [84 BZL]
Aulomyrcia obtusata O.Berg, Linnaea 27: 39 (1855). = [82 GUY]
Aulomyrcia parvifolia Steyerm., Fieldiana, Bot. 28: 1006 (1957). = [82 VEN]
Aulomyrcia pauciflora O.Berg in C.F.P.von Martius & auct. suc. (eds.), Fl. Bras. 14(1): 560 (1859). = [84 BZL]
Aulomyrcia piauhiensis O.Berg in C.F.P.von Martius & auct. suc. (eds.), Fl. Bras. 14(1): 87 (1857). = [84 BZE]
Aulomyrcia platyclada O.Berg in C.F.P.von Martius & auct. suc. (eds.), Fl. Bras. 14(1): 545 (1859). = [84 BZL]
Aulomyrcia plumbea O.Berg in C.F.P.von Martius & auct. suc. (eds.), Fl. Bras. 14(1): 142 (1857). = [84 BZL]
Aulomyrcia punctata O.Berg in C.F.P.von Martius & auct. suc. (eds.), Fl. Bras. 14(1): 546 (1859). = [84 BZL]
Aulomyrcia pusilla O.Berg in C.F.P.von Martius & auct. suc. (eds.), Fl. Bras. 14(1): 140 (1857). = [84 BZL]
Aulomyrcia regeliana O.Berg in C.F.P.von Martius & auct. suc. (eds.), Fl. Bras. 14(1): 557 (1859). = [84 BZC BZL]

Aulomyrcia riedeliana O.Berg in C.F.P.von Martius & auct. suc. (eds.), Fl. Bras. 14(1): 551 (1859). = [84 BZL]
Aulomyrcia salzmannii O.Berg in C.F.P.von Martius & auct. suc. (eds.), Fl. Bras. 14(1): 116 (1857). = [84 BZE]
Aulomyrcia stenophylla O.Berg, Linnaea 30: 654 (1861). = [85 PAR]
Aulomyrcia ternifolia O.Berg in C.F.P.von Martius & auct. suc. (eds.), Fl. Bras. 14(1): 134 (1857). = [84 BZL]
Aulomyrcia trichantha Wawra, Oesterr. Bot. Z. 29: 215 (1879). = [84]
Aulomyrcia trifolia O.Berg in C.F.P.von Martius & auct. suc. (eds.), Fl. Bras. 14(1): 107 (1857). = [84 BZL]

Austromyrtus

Austromyrtus (Nied.) Burret, Notizbl. Bot. Gart. Berlin-Dahlem 15: 500 (1941).
E. Australia, SW. Pacific. 50 NSW QLD 60 NWC. [Myrtaceae]
21 Species

Austromyrtus alaternoides (Brongn. & Gris) Burret, Notizbl. Bot. Gart. Berlin-Dahlem 15: 504 (1941).
New Caledonia (Balade). 60 NWC. Nanophan. or phan.
**Myrtus alaternoides* Brongn. & Gris, Bull. Soc. Bot. France 12: 177 (1865).

Austromyrtus aphthosa (Vieill. ex Brongn. & Gris) Burret, Notizbl. Bot. Gart. Berlin-Dahlem 15: 504 (1941).
New Caledonia (Wagap). 60 NWC. Nanophan. or phan.
**Eugenia aphthosa* Vieill. ex Brongn. & Gris, Bull. Soc. Bot. France 13: 469 (1866).

Austromyrtus cataractarum (Guillaumin) Burret, Notizbl. Bot. Gart. Berlin-Dahlem 15: 503 (1941).
New Caledonia. 60 NWC. Nanophan. or phan.
**Eugenia cataractarum* Guillaumin, Bull. Soc. Bot. France 85: 636 (1938 publ. 1939).

Austromyrtus clusioides (Brongn. & Gris) Burret, Notizbl. Bot. Gart. Berlin-Dahlem 15: 503 (1941).
New Caledonia. 60 NWC. Nanophan. or phan.
**Eugenia clusioides* Brongn. & Gris, Bull. Soc. Bot. France 12: 180 (1865).

Austromyrtus conspicua (Vieill. ex Guillaumin) Burret, Notizbl. Bot. Gart. Berlin-Dahlem 15: 505 (1941).
New Caledonia (Gatope). 60 NWC. Nanophan. or phan.
**Myrtus conspicua* Vieill. ex Guillaumin, Bull. Soc. Bot. France 85: 631 (1938 publ. 1939).

Austromyrtus diversifolia (Brongn. & Gris) Burret, Notizbl. Bot. Gart. Berlin-Dahlem 15: 503 (1941).
New Caledonia. 60 NWC. Nanophan. or phan.
**Eugenia diversifolia* Brongn. & Gris, Bull. Soc. Bot. France 12: 180 (1865). *Myrtus diversifolia* (Brongn. & Gris) Guillaumin, Bull. Soc. Bot. France 85: 631 (1938 publ. 1938).

Austromyrtus dulcis (C.T.White) L.S.Sm., Proc. Roy. Soc. Queensland 67: 35 (1956).
SE. Queensland to NE. New South Wales. 50 NSW QLD. Cham. or nanophan.
Myrtus tenuifolia var. *latifolia* Maiden & Betche, Census N.S.W. Pl.: 143 (1916).
**Myrtus dulcis* C.T.White, Proc. Roy. Soc. Queensland 50: 76 (1938 publ. 1939).

Austromyrtus glabra N.Snow & Guymer, Syst. Bot. Monogr. 65: 23 (2003).
SE. Queensland. 50 QLD. Cham. or nanophan.

Austromyrtus kanalaensis (Hochr.) Burret, Notizbl. Bot. Gart. Berlin-Dahlem 15: 504 (1941).
New Caledonia (Kanala). 60 NWC. Nanophan. or phan.
**Eugenia kanalaensis* Hochr., Bull. New York Bot. Gard. 6: 280 (1910).

Austromyrtus kuakuensis (Baker f.) Burret, Notizbl. Bot. Gart. Berlin-Dahlem 15: 503 (1941).
New Caledonia. 60 NWC. Nanophan. or phan.
**Psidium kuakuense* Baker f., J. Linn. Soc., Bot. 45: 318 (1921).

Austromyrtus lotoides (Guillaumin) Burret, Notizbl. Bot. Gart. Berlin-Dahlem 15: 504 (1941).
New Caledonia (Gatope). 60 NWC. Nanophan. or phan.
**Myrtus lotoides* Guillaumin, Bull. Soc. Bot. France 85: 632 (1938 publ. 1939).

Austromyrtus luteoviridis (Baker f.) Burret, Notizbl. Bot. Gart. Berlin-Dahlem 15: 505 (1941).
New Caledonia (Ngoye River). 60 NWC. Nanophan. or phan.
**Myrtus luteoviridis* Baker f., J. Linn. Soc., Bot. 45: 312 (1921).

Austromyrtus mendute (Guillaumin) Burret, Notizbl. Bot. Gart. Berlin-Dahlem 15: 505 (1941).
New Caledonia. 60 NWC. Nanophan. or phan.
**Eugenia mendute* Guillaumin, Bull. Soc. Bot. France 85: 638 (1938 publ. 1939).

Austromyrtus nigripes (Guillaumin) Burret, Notizbl. Bot. Gart. Berlin-Dahlem 15: 505 (1941).
New Caledonia (Kanala). 60 NWC. Nanophan. or phan.
**Myrtus nigripes* Guillaumin, Bull. Soc. Bot. France 85: 632 (1938 publ. 1939).

Austromyrtus pancheri (Brongn. & Gris) Burret, Notizbl. Bot. Gart. Berlin-Dahlem 15: 503 (1941).
New Caledonia. 60 NWC. Nanophan. or phan.
**Eugenia pancheri* Brongn. & Gris, Bull. Soc. Bot. France 12: 180 (1865).

Austromyrtus ploumensis (Däniker) Burret, Notizbl. Bot. Gart. Berlin-Dahlem 15: 504 (1941).
New Caledonia (near Ploume). 60 NWC. Nanophan. or phan.
**Eugenia ploumensis* Däniker, Vierteljahrsschr. Naturf. Ges. Zürich 78(19): 298 (1933).

Austromyrtus poimbailensis (Guillaumin) Burret, Notizbl. Bot. Gart. Berlin-Dahlem 15: 505 (1941).
New Caledonia (Poimbail). 60 NWC. Nanophan. or phan.
**Myrtus poimbailensis* Guillaumin, Bull. Soc. Bot. France 85: 633 (1938 publ. 1939).

Austromyrtus prolixa (Baker f.) Burret, Notizbl. Bot. Gart. Berlin-Dahlem 15: 505 (1941).
New Caledonia (Pleine des Sacs). 60 NWC. Nanophan. or phan.
**Myrtus prolixa* Baker f., J. Linn. Soc., Bot. 45: 311 (1921).

Austromyrtus styphelioides (Schltr.) Burret, Notizbl. Bot. Gart. Berlin-Dahlem 15: 504 (1941).
New Caledonia. 60 NWC. Nanophan. or phan.
**Myrtus styphelioides* Schltr., Bot. Jahrb. Syst. 40(29): 31 (1908).

Austromyrtus tenuifolia (Sm.) Burret, Notizbl. Bot. Gart. Berlin-Dahlem 15: 501 (1941).
CE. New South Wales. 50 NSW. Nanophan.
**Myrtus tenuifolia* Sm., Trans. Linn. Soc. London 3: 280 (1797).

Austromyrtus vieillardii (Brongn. & Gris) Burret, Notizbl. Bot. Gart. Berlin-Dahlem 15: 503 (1941).
New Caledonia. 60 NWC. Nanophan. or phan.
**Eugenia vieillardii* Brongn. & Gris, Bull. Soc. Bot. France 12: 180 (1865).
Eugenia heckelii Pancher & Sebert, Not. Bois Nouv. Caléd.: 259 (1874).
Eugenia angustibracteolata Baker f., J. Linn. Soc., Bot. 45: 313 (1921).

Synonyms:
Austromyrtus acmenoides (F.Muell.) Burret = ***Gossia acmenoides*** (F.Muell.) N.Snow & Guymer
Austromyrtus acutiuscula Burret = ***Gossia acmenoides*** (F.Muell.) N.Snow & Guymer
Austromyrtus aneityensis (Guillaumin) Burret = ***Gossia aneityensis*** (Guillaumin) N.Snow
Austromyrtus bidwillii (Benth.) Burret = ***Gossia bidwillii*** (Benth.) N.Snow & Guymer
Austromyrtus dallachyana (F.Muell. ex Benth.) L.S.Sm. = ***Gossia dallachyana*** (F.Muell. ex Benth.) N.Snow & Guymer
Austromyrtus decapermoides (Domin) Burret = ***Gossia inophloia*** (J.F.Bailey & C.T.White) N.Snow & Guymer
Austromyrtus exaltata (F.M.Bailey) Burret = ***Syzygium luehmannii*** (F.Muell.) L.A.S.Johnson
Austromyrtus floribunda (A.J.Scott) Guymer = ***Gossia floribunda*** (A.J.Scott) N.Snow & Guymer
Austromyrtus fragrantissima (F.Muell. ex Benth.) Burret = ***Gossia fragrantissima*** (F.Muell. ex Benth.) N.Snow & Guymer
Austromyrtus gaertneri Schauer = ***Asteromyrtus angustifolia*** (Gaertn.) Craven
Austromyrtus gatopensis (Guillaumin) Burret = ***Eugenia gatopensis*** Guillaumin
Austromyrtus gomonenensis (Guillaumin) Burret = ***Uromyrtus gomonenensis*** (Guillaumin) Burret
Austromyrtus gonoclada (F.Muell. ex Benth.) Burret = ***Gossia gonoclada*** (F.Muell. ex Benth.) N.Snow & Guymer
Austromyrtus gyrosepala (Baker f.) Burret = ***Eugenia gyrosepala*** Baker f.
Austromyrtus hillii (Benth.) Burret = ***Gossia hillii*** (Benth.) N.Snow & Guymer
Austromyrtus horizontalis (Pancher ex Brongn. & Gris) Burret = ***Eugenia horizontalis*** Pancher ex Brongn. & Gris
Austromyrtus inophloia (J.F.Bailey & C.T.White) Burret = ***Gossia inophloia*** (J.F.Bailey & C.T.White) N.Snow & Guymer
Austromyrtus lasioclada (F.Muell.) L.S.Sm. = ***Lenwebbia lasioclada*** (F.Muell.) N.Snow & Guymer
Austromyrtus lucida (Gaertn.) L.S.Sm. = ***Gossia lucida*** (Gaertn.) N.Snow & Guymer
Austromyrtus metrosideros (F.M.Bailey) Burret = ***Uromyrtus metrosideros*** (F.M.Bailey) A.J.Scott
Austromyrtus minutiflora Burret = ***Gossia myrsinocarpa*** (F.Muell.) N.Snow & Guymer
Austromyrtus opaca (C.T.White) L.S.Sm. = ***Gossia hillii*** (Benth.) N.Snow & Guymer
Austromyrtus pubiflora (C.T.White) L.S.Sm. = ***Gossia pubiflora*** (C.T.White) N.Snow & Guymer
Austromyrtus racemulosa Burret = ***Gossia bidwillii*** (Benth.) N.Snow & Guymer
Austromyrtus shepherdii (F.Muell.) L.S.Sm. = ***Gossia shepherdii*** (F.Muell.) N.Snow & Guymer
Austromyrtus stricta (Pancher ex Brongn. & Gris) Burret = ***Eugenia stricta*** Pancher ex Brongn. & Gris

Babingtonia

Babingtonia Lindl., Edwards's Bot. Reg. 28: t. 10 (1842).
Borneo, SW. & E. Australia, New Caledonia. 42 BOR 50 NSW QLD VIC WAU 60 NWC. [Myrtaceae]
25 Species
Camphoromyrtus Schauer, Linnaea 17: 240 (1843).
Harmogia Schauer, Linnaea 17: 238 (1843).
Tetrapora Schauer, Linnaea 17: 238 (1843).

Babingtonia angusta A.R.Bean, Austrobaileya 5: 163 (1999).
E. Australia. 50 NSW QLD. Nanophan.

Babingtonia behrii (Schltdl.) A.R.Bean, Austrobaileya 4: 637 (1997).
NW. Victoria. 50 VIC. Nanophan.
**Camphoromyrtus behrii* Schltdl., Linnaea 20: 651 (1847). *Baeckea behrii* (Schltdl.) F.Muell., Fragm. 4: 68 (1864).
Baeckea behrii var. *brevifolia* F.Muell., J. Bot. 15: 280 (1877).

Babingtonia bidwillii A.R.Bean, Austrobaileya 5: 161 (1999).
E. Queensland. 50 QLD. Nanophan. or phan.
Baeckea virgata var. *parvula* F.M.Bailey, Queensl. Fl. 2: 585 (1900).

Babingtonia brachypoda A.R.Bean, Austrobaileya 5: 168 (1999).
SE. Queensland. 50 QLD. Nanophan. or phan.

Babingtonia camphorosmae (Endl.) Lindl., Edwards's Bot. Reg. 28: t. 10 (1842).
W. Western Australia. 50 WAU. Nanophan.
**Baeckea camphorosmae* Endl. in S.L.Endlicher & al., Enum. Pl.: 51 (1837).

Babingtonia collina A.R.Bean, Austrobaileya 5: 167 (1999).
New South Wales, Queensland. 50 NSW QLD. Nanophan.

Babingtonia crassa A.R.Bean, Austrobaileya 5: 166 (1999).
E. New South Wales. 50 NSW. Nanophan.

Babingtonia crenulata (F.Muell.) A.R.Bean, Austrobaileya 4: 635 (1997).
NE. Victoria. 50 VIC. Nanophan.
**Camphoromyrtus crenulata* F.Muell., Defin. Austral. Pl.: 43 (1855). *Baeckea crenatifolia* F.Muell., Fragm. 4: 70 (1864). *Baeckea crenulata* (F.Muell.) Druce, Bot. Soc. Exch. Club Brit. Isles 1916: 608 (1917), nom. illeg.
Harmogia crenulata Miq., Ned. Kruidk. Arch. 4: 148 (1856).

Babingtonia cunninghamii (Schauer) A.R.Bean, Austrobaileya 4: 638 (1997).
New South Wales. 50 NSW. Nanophan.
Baeckea microphylla A.Cunn. ex Schauer, Linnaea 17: 238 (1843), pro syn.
**Harmogia cunninghamii* Schauer, Linnaea 17: 238 (1843). *Baeckea cunninghamii* (Schauer) Benth., Fl. Austral. 2: 82 (1864).

Babingtonia densifolia (Sm.) F.Muell., Fragm. 4: 74 (1864).
E. Australia. 50 NSW QLD. Nanophan.
**Baeckea densifolia* Sm., Trans. Linn. Soc. London 3: 260 (1797). *Harmogia densifolia* (Sm.) Schauer, Linnaea 17: 238 (1843).
Baeckea fasciculata Sieber ex Spreng., Syst. Veg. 4(2): 149 (1827).
Harmogia baueriana Schauer, Linnaea 17: 238 (1843).
Harmogia propinqua Schauer, Linnaea 17: 238 (1843).
Baeckea nova-anglica F.Muell., Fragm. 4: 71 (1864).

Babingtonia granitica A.R.Bean, Austrobaileya 4: 642 (1997).
SE. Queensland. 50 QLD. Nanophan.

Babingtonia jucunda (S.T.Blake) A.R.Bean, Austrobaileya 4: 639 (1997).
Queensland to N. New South Wales. 50 NSW QLD. Nanophan.
**Baeckea jucunda* S.T.Blake, Proc. Roy. Soc. Queensland 69: 75 (1958).

Babingtonia leratii (Schltr.) A.R.Bean, Austrobaileya 4: 633 (1997).
SE. New Caledonia. 60 NWC. Nanophan.
Baeckea ericoides Brongn. & Gris, Bull. Soc. Bot. France 11: 184 (1864), nom. illeg.
**Baeckea leratii* Schltr., Bot. Jahrb. Syst. 40(92): 32 (1908).
Baeckea ericoides f. *tetrasticha* Dunker, Vierteljahrsschr. Naturf. Ges. Zürich 75(19): 320 (1933).

Babingtonia odontocalyx A.R.Bean, Austrobaileya 4: 644 (1997).
NE. New South Wales. 50 NSW. Nanophan.

Babingtonia papillosa A.R.Bean, Austrobaileya 5: 169 (1999).
NE. Queensland. 50 QLD. Nanophan.

Babingtonia pinifolia (Labill.) A.R.Bean, Austrobaileya 4: 633 (1997).
New Caledonia. 60 NWC. Nanophan. or phan.
**Leptospermum pinifolium* Labill., Sert. Austro-Caledon.: 63 (1825). *Baeckea pinifolia* (Labill.) DC., Prodr. 3: 229 (1828). *Harmogia pinifolia* (Labill.) Schauer, Linnaea 17: 238 (1843).

Babingtonia pluriflora (F.Muell.) A.R.Bean, Austrobaileya 5: 165 (1999), without basionym page.
New South Wales to Victoria. 50 NSW VIC. Nanophan. or phan.
**Camphoromyrtus pluriflora* F.Muell., Defin. Austral. Pl.: 43 (1855).

Babingtonia procera (J.W.Dawson) A.R.Bean, Austrobaileya 4: 633 (1997).
NW. New Caledonia. 60 NWC. Nanophan.
**Baeckea procera* J.W.Dawson, in Fl. Nouv.-Caléd. 18: 22 (1992).

Babingtonia prominens A.R.Bean, Austrobaileya 4: 639 (1997).
NE. New South Wales. 50 NSW. Nanophan.

Babingtonia silvestris A.R.Bean, Austrobaileya 4: 641 (1997).
SE. Queensland to NE. New South Wales. 50 NSW QLD. Nanophan.

Babingtonia similis A.R.Bean, Austrobaileya 5: 162 (1999).
SE. Queensland to E. New South Wales. 50 NSW QLD. Nanophan.

Babingtonia squarrulosa (Domin) A.R.Bean, Austrobaileya 4: 642 (1997).
NE. Queensland. 50 QLD. Nanophan.
**Baeckea squarrulosa* Domin, Biblioth. Bot. 89: 450 (1928).

Babingtonia taxifolia (Merr.) A.R.Bean, Austrobaileya 4: 632 (1997).
Borneo (N. Sarawak). 42 BOR. Nanophan.
**Baeckea taxifolia* Merr., Sarawak Mus. J. 3: 534 (1928).

Babingtonia tozerensis A.R.Bean, Austrobaileya 4: 634 (1997).
Queensland (Cook). 50 QLD. Nanophan.

Babingtonia virgata (J.R.Forst. & G.Forst.) F.Muell., Fragm. 4: 74 (1864).
New Caledonia. 60 NWC. Nanophan.
**Leptospermum virgatum* J.R.Forst. & G.Forst., Char. Gen. Pl.: 36 (1775). *Melaleuca virgata* (J.R.Forst. & G.Forst.) L.f., Suppl. Pl.: 343 (1782). *Baeckea virgata* (J.R.Forst. & G.Forst.) Andrews, Bot. Repos. 10: t. 598 (1810). *Harmogia virgata* (J.R.Forst. & G.Forst.) Schauer, Linnaea 17: 238 (1843).
Leptospermum parvulum Labill., Sert. Austro-Caledon.: 62 (1825). *Baeckea parvula* (Labill.) DC., Prodr. 3: 229 (1828).
Harmogia parvula Schauer, Linnaea 17: 238 (1843).
Harmogia umbellata F.Muell., Fragm. 2: 31 (1860).
Baeckea obtusifolia Brongn. & Gris, Bull. Soc. Bot. France 11: 185 (1864).
Baeckea parvula var. *latifolia* Brongn. & Gris, Bull. Soc. Bot. France 11: 184 (1864).
Baeckea neglecta Vieill. ex Guillaumin, Ann. Inst. Bot.-Géol. Colon. Marseille, II, 9: 143 (1911), nom. nud.

***Synonyms*:**
Babingtonia corynophylla (F.Muell.) F.Muell. = ***Baeckea corynophylla*** (F.Muell.) F.Muell.
Babingtonia crispiflora (F.Muell.) F.Muell. = ***Baeckea crispiflora*** (F.Muell.) F.Muell.
Babingtonia gracilis (Schauer) F.Muell. = ***Baeckea polyandra*** F.Muell.
Babingtonia nova-anglica (F.Muell.) F.Muell. = ***Babingtonia densifolia*** (Sm.) F.Muell.
Babingtonia ovalifolia (F.Muell.) F.Muell. = ***Baeckea ovalifolia*** (F.Muell.) F.Muell.
Babingtonia pentagonantha (F.Muell.) F.Muell. = ***Baeckea pentagonantha*** F.Muell.
Babingtonia pentandra (F.Muell.) F.Muell. = ***Baeckea pentandra*** (F.Muell.) F.Muell.
Babingtonia preissiana (Schauer) F.Muell. = ***Baeckea preissiana*** (Schauer) Druce
Babingtonia robusta (F.Muell.) F.Muell. = ***Baeckea robusta*** F.Muell.
Babingtonia subcuneata (F.Muell.) F.Muell. = ***Baeckea subcuneata*** F.Muell.

Backhousia

Backhousia Hook. & Harv., Bot. Mag. 71: t. 4133 (1845).
E. Australia. 50 NSW QLD. [Myrtaceae]
9 Species

Backhousia angustifolia F.Muell., Fragm. 1: 79 (1859).
Queensland. 50 QLD. Nanophan.

Backhousia bancroftii F.M.Bailey & F.Muell. in F.M.Bailey, Syn. Queensl. Fl., Suppl. 1: 24 (1886).
Queensland (Cook, South Kennedy). 50 QLD. Nanophan. or phan.

Backhousia citriodora F.Muell., Fragm. 1: 78 (1859).
Queensland. 50 QLD. Nanophan.

Backhousia enata A.J.Ford, Craven & J.Holmes, Austrobaileya 7: 121 (2005).
NE. Queensland. 50 QLD. Nanophan.

Backhousia hughesii C.T.White, Proc. Roy. Soc. Queensland 47: 61 (1935 publ. 1936).
Queensland (Cook). 50 QLD. Nanophan. or phan.

Backhousia kingii Guymer, Austrobaileya 2: 567 (1988).
Queensland. 50 QLD. Nanophan. or phan.

Backhousia myrtifolia Hook. & Harv., Bot. Mag. 71: t. 4133 (1845).
SE. Queensland to New South Wales. 50 NSW QLD. Nanophan. or phan.
Backhousia riparia Hook. & Harv., Bot. Mag. 71: t. 4133 (1845).
Backhousia australis G.Benn., Nat. Austr.: 284 (1860), nom. nud.

Backhousia oligantha A.R.Bean, Austrobaileya 6: 533 (2003).
Queensland. 50 QLD.

Backhousia sciadophora F.Muell., Fragm. 2: 26, 171 (1860).
SE. Queensland to NE. New South Wales. 50 NSW QLD. Nanophan. or phan.

***Synonyms*:**
Backhousia anisata Vickery = ***Syzygium anisatum*** (Vickery) Craven & Biffin
Backhousia arfakensis Gibbs = ***Kania hirsutula*** (F.Muell.) A.J.Scott
Backhousia aurea Ridl. = ***Kania eugenioides*** Schltr.
Backhousia australis G.Benn. = ***Backhousia myrtifolia*** Hook. & Harv.
Backhousia floribunda A.J.Scott = ***Gossia floribunda*** (A.J.Scott) N.Snow & Guymer
Backhousia riparia Hook. & Harv. = ***Backhousia myrtifolia*** Hook. & Harv.

Baeckea

Baeckea L., Sp. Pl.: 358 (1753).
Trop. & Subtrop. Asia to Australia. 36 CHS 41 LAO MYA THA VIE 42 BOR jaw MLY PHI SUL SUM 43 NWG 50 NSW NTA QLD SOA TAS VIC WAU. [Myrtaceae]
52 Species
Tjongina Adans., Fam. Pl. 2: 611 (1763).
Jungia Gaertn., Fruct. Sem. Pl. 1: 175 (1788), nom. illeg.
Mollia J.F.Gmel., Syst. Nat.: 420 (1791), nom. illeg.
Imbricaria Sm., Trans. Linn. Soc. London 3: 257 (1797), nom. illeg.
Neuhofia Stokes, Bot. Mat. Med. 1: 439 (1812).
Allostis Raf., Sylva Tellur.: 104 (1838).
Murrinea Raf., Sylva Tellur.: 104 (1838).
Oxymyrrhine Schauer, Linnaea 17: 240 (1843).
Schidiomyrtus Schauer, Linnaea 17: 237 (1843).
Drosodendron M.Roem., Fam. Nat. Syn. Monogr. 1: 140 (1846).
Ericomyrtus Turcz., Bull. Soc. Imp. Naturalistes Moscou 20(1): 154 (1847).
Anticoryne Turcz., Bull. Cl. Phys.-Math. Acad. Imp. Sci. Saint-Pétersbourg 10: 332 (1852).
Cyathostemon Turcz., Bull. Cl. Phys.-Math. Acad. Imp. Sci. Saint-Pétersbourg 10: 331 (1852).
Tetraspora Miq., Ned. Kruidk. Arch. 4: 150 (1856).

Baeckea baileyana C.A.Gardner, J. Roy. Soc. Western Australia 19: 89 (1934).
Western Australia (Avon). 50 WAU. Nanophan.

Baeckea blackettii F.Muell., Fragm. 8: 181 (1874).
SW. Western Australia. 50 WAU. Nanophan.

Baeckea brevifolia (Rudge) DC., Prodr. 3: 230 (1828).
SE. New South Wales. 50 NSW. Nanophan.
**Leptospermum brevifolium* Rudge, Trans. Linn. Soc. London 8: 299 (1807).
Baeckea carnosula Sieber ex Spreng., Syst. Veg. 4(2): 149 (1827).

Baeckea clavifolia S.Moore, J. Linn. Soc., Bot. 45: 176 (1920).
Western Australia (Austin, Coolgardie). 50 WAU. Nanophan.

Baeckea corymbulosa Benth., Fl. Austral. 3: 87 (1867).
Western Australia (?). 50 WAU. Nanophan.

Baeckea corynophylla (F.Muell.) F.Muell., Fragm. 4: 72 (1864).
Western Australia (Eyre). 50 WAU. Nanophan.
**Harmogia corynophylla* F.Muell., Fragm. 2: 30 (1860). *Babingtonia corynophylla* (F.Muell.) F.Muell., Fragm. 4: 74 (1864).

Baeckea crassifolia Lindl. in T.L.Mitchell, Three Exped. Australia 2: 114 (1838).
S. & SE. Australia. 50 NSW SOA VIC WAU.
Baeckea crassifolia var. *icosandra* F.Muell. ex Benth., Fl. Austral. 3: 76 (1867).
Baeckea crassifolia var. *pentamera* J.M.Black, Trans. & Proc. Roy. Soc. South Australia 59: 259 (1935).

Baeckea crispiflora (F.Muell.) F.Muell., Fragm. 4: 72 (1864).
Western Australia. 50 WAU. Nanophan.
**Harmogia crispiflora* F.Muell., Fragm. 2: 31 (1860). *Babingtonia crispiflora* (F.Muell.) F.Muell., Fragm. 4: 74 (1864).
Baeckea crispiflora var. *tenuior* Ewart, Proc. Roy. Soc. Victoria, n.s. 20: 76 (1907).
Baeckea crispiflora var. *tenuior* Ewart, Proc. Roy. Soc. Victoria, n.s. 20: 76 (1907).

Baeckea cryptandroides F.Muell., Fragm. 10: 29 (1876).
Western Australia. 50 WAU. Nanophan.

Baeckea decipiens W.Fitzg., J. West Austral. Nat. Hist. Soc. 1: 20 (1904).
Western Australia (Irwin, Avon). 50 WAU. Nanophan.

Baeckea diosmifolia Rudge, Trans. Linn. Soc. London 8: 198 (1807). *Schidiomyrtus diosmifolia* (Rudge) Schauer, Linnaea 17: 237 (1843).
SE. Queensland to New South Wales. 50 NSW QLD. Cham. or nanophan.

Baeckea elderiana Pritz., Bot. Jahrb. Syst. 35: 418 (1905).
Western Australia. 50 WAU. Nanophan.

Baeckea ericaea F.Muell., Fragm. 1: 31 (1858).
SE. Australia. 50 SOA VIC.

Baeckea exserta S.Moore, J. Linn. Soc., Bot. 45: 177 (1920).
Western Australia. 50 WAU. Nanophan.

Baeckea floribunda Benth., Fl. Austral. 3: 87 (1867).
Western Australia. 50 WAU. Nanophan.

Baeckea frutescens L., Sp. Pl.: 358 (1753).
SE. China to E. Australia. 36 CHS 41 LAO MYA THA VIE 42 BOR jaw MLY PHI SUL SUM 43 NWG 50 NSW QLD. Nanophan.
Baeckea chinensis Gaertn., Fruct. Sem. Pl. 1: 157 (1788).
Baeckea sinensis Gaertn., Fruct. Sem. Pl. 1: 31 (1788).
Neuhofia rosmarinifolia Stokes, Bot. Mat. Med. 1: 439 (1812).
Baeckea cumingiana Schauer in W.G.Walpers, Repert. Bot. Syst. 2: 920 (1843).
Baeckea ericoides Schltdl., Bot. Zeitung (Berlin) 4: 724 (1846).
Drosodendron rosmarinus (Lour.) M.Roem., Fam. Nat. Syn. Monogr. 1: 140 (1846).
Baeckea cochinchinensis Blume, Mus. Bot. 1: 69 (1850), nom. illeg.
Baeckea sumatrana Blume, Mus. Bot. 1: 69 (1850).
Baeckea stenophylla F.Muell., Fragm. 1: 13 (1858).

Baeckea grandibracteata Pritz., Bot. Jahrb. Syst. 35: 417 (1905).
SW. Western Australia. 50 WAU. Nanophan.

Baeckea grandiflora Benth., Fl. Austral. 3: 89 (1867).
Western Australia. 50 WAU. Nanophan.

Baeckea grandis Pritz., Bot. Jahrb. Syst. 35: 417 (1905).
Western Australia. 50 WAU. Nanophan.

Baeckea gunniana Schauer ex Walp., Repert. Bot. Syst. 2: 920 (1843). *Tetrapora gunniana* (Schauer ex Walp.) Miq., Ned. Kruidk. Arch. 4: 150 (1856). *Tetraspora gunniana* (Schauer ex Walp.) Miq., Ned. Kruidk. Arch. 4: 150 (1856).
New South Wales to Tasmania. 50 NSW TAS VIC. Nanophan.
Baeckea micrantha Hook.f., Hooker's Icon. Pl. 4: t. 309 (1841), nom. illeg.

Baeckea imbricata (Gaertn.) Druce, Bot. Soc. Exch. Club Brit. Isles 1916: 608 (1917).
Queensland to New South Wales. 50 NSW QLD. Cham. or nanophan.
**Jungia imbricata* Gaertn., Fruct. Sem. Pl. 1: 175 (1788). *Mollia imbricata* (Gaertn.) J.F.Gmel., Syst. Nat.: 420 (1791). *Imbricaria crenulata* Sm., Trans. Linn. Soc. London 3: 259 (1797), nom. illeg. *Baeckea crenulata* (Sm.) DC., Prodr. 3: 239 (1828), nom. illeg. *Schidiomyrtus crenulata* (Sm.) Schauer, Linnaea 17: 237 (1843). *Baeckea imbricata* var. *typica* Domin, Biblioth. Bot. 89: 450 (1928), nom. inval.
Jungia tenella Gaertn., Fruct. Sem. Pl. 1: 175 (1788). *Imbricaria tenella* (Gaertn.) Dryand., Ann. Bot. (König & Sims) 2: 16 (1805). *Baeckea crenulata* var. *tenella* (Gaertn.) Benth., Fl. Austral. 3: 78 (1867). *Baeckea crenulata* var. *tenella* (Gaertn.) Benth., Fl. Austral. 3: 78 (1867). *Baeckea imbricata* var. *tenella* (Gaertn.) Hochr., Candollea 2: 467 (1925).
Baeckea diosmoides Sieber ex DC., Prodr. 3: 230 (1828).
Schidiomyrtus sieberi Schauer, Linnaea 17: 237 (1843).

Baeckea kandos A.R.Bean, Telopea 7: 260 (1997).
New South Wales (CW. Slopes). 50 NSW. Nanophan.

Baeckea latens C.R.P.Andrews, J. West Austral. Nat. Hist. Soc. 1: 41 (1904).
SW. Western Australia. 50 WAU. Nanophan.

Baeckea latifolia (Benth.) A.R.Bean, Telopea 7: 263 (1997).
SE. New South Wales to E. Victoria. 50 NSW VIC. Nanophan.
**Baeckea gunniana* var. *latifolia* Benth., Fl. Austral. 3: 79 (1867). *Baeckea utilis* var. *latifolia* (Benth.) J.H.Willis, Muelleria 1: 139 (1967).

Baeckea leptocaulis Hook.f., Hooker's Icon. Pl. 3: t. 298 (1840).
W. & SW. Tasmania. 50 TAS. Cham. or nanophan.

Baeckea leptophylla (Turcz.) Domin, Vestn. Král. Ceské Spolecn. Nauk, Tr. Mat.-Prír. 2: 82 (1923).
Western Australia. 50 WAU. Nanophan.
**Harmogia leptophylla* Turcz., Bull. Cl. Phys.-Math. Acad. Imp. Sci. Saint-Pétersbourg 10: 330 (1852).

Baeckea linifolia Rudge, Trans. Linn. Soc. London 8: 297 (1807).
Queensland to Victoria. 50 NSW QLD VIC. Nanophan.
Baeckea trichophylla Sieber ex Spreng., Syst. Veg. 4(2): 149 (1827). *Baeckea linifolia* f. *trichophylla* (Sieber ex Spreng.) Domin, Biblioth. Bot. 89: 450 (1928). *Baeckea linifolia* f. *trichophylla* (Sieber ex Spreng.) Domin, Biblioth. Bot. 89: 450 (1928).
Baeckea linifolia var. *brevifolia* Ewart, Fl. Victoria: 871 (1931).
Baeckea linifolia var. *brevifolia* Ewart, Fl. Victoria: 871 (1931).

Baeckea muricata C.A.Gardner, J. Roy. Soc. Western Australia 13: 64 (1927).
Western Australia. 50 WAU. Nanophan.

Baeckea ochropetala F.Muell., Fragm. 10: 29 (1876).
Western Australia. 50 WAU. Nanophan.

Baeckea omissa A.R.Bean, Telopea 7: 256 (1997).
SE. Queensland to NE. New South Wales. 50 NSW QLD. Cham. or nanophan.

Baeckea ovalifolia (F.Muell.) F.Muell., Fragm. 4: 72 (1864).
Western Australia (Eyre). 50 WAU. Nanophan.
Anticoryne diosmoides Turcz., Bull. Cl. Phys.-Math. Acad. Imp. Sci. Saint-Pétersbourg 10: 332 (1852). *Baeckea diosmoides* (Turcz.) Domin, Vestn. Král. Ceské Spolecn. Nauk, Tr. Mat.-Prír. 2: 83 (1923), nom. illeg.
**Harmogia ovalifolia* F.Muell., Fragm. 2: 32 (1860). *Babingtonia ovalifolia* (F.Muell.) F.Muell., Fragm. 4: 74 (1864).

Baeckea pachyphylla Benth., Fl. Austral. 3: 85 (1867).
SW. Western Australia. 50 WAU. Nanophan.

Baeckea pentagonantha F.Muell., Fragm. 4: 73 (1864). *Babingtonia pentagonantha* (F.Muell.) F.Muell., Fragm. 4: 74 (1864).
Western Australia (Irwin). 50 WAU. Nanophan.

Baeckea pentandra (F.Muell.) F.Muell., Fragm. 4: 72 (1864).
Northern Territory. 50 NTA. Nanophan.
Tetrapora glomerata Turcz., Bull. Cl. Phys.-Math. Acad. Imp. Sci. Saint-Pétersbourg 10: 329 (1852).
**Harmogia pentandra* F.Muell., Fragm. 2: 31 (1860). *Babingtonia pentandra* (F.Muell.) F.Muell., Fragm. 4: 74 (1864).

Baeckea platycephala Pritz., Bot. Jahrb. Syst. 35: 419 (1905).
Western Australia. 50 WAU. Nanophan.

Baeckea polyandra F.Muell., Fragm. 4: 72 (1864).
Western Australia (?). 50 WAU. Nanophan.

Oxymyrrhine gracilis Schauer, Linnaea 17: 240 (1843). *Babingtonia gracilis* (Schauer) F.Muell., Fragm. 4: 74 (1864). *Baeckea gracilis* (Schauer) C.A.Gardner, Enum. Pl. Austr. Occ.: 94 (1931), nom. illeg.

Baeckea polystemonea F.Muell., Fragm. 2: 124 (1860).
C. Australia. 50 NTA WAU.

Baeckea preissiana (Schauer) Druce, Bot. Soc. Exch. Club Brit. Isles 1916: 608 (1917).
Western Australia. 50 WAU. Nanophan.
**Tetrapora preissiana* Schauer, Linnaea 17: 238 (1843). *Babingtonia preissiana* (Schauer) F.Muell., Fragm. 4: 74 (1864).
Tetrapora verrucosa Turcz., Bull. Cl. Phys.-Math. Acad. Imp. Sci. Saint-Pétersbourg 10: 329 (1852).

Baeckea pulchella A.Cunn. ex DC., Prodr. 3: 230 (1828).
Western Australia. 50 WAU. Nanophan.
Ericomyrtus drummondii Turcz., Bull. Soc. Imp. Naturalistes Moscou 20(1): 155 (1847).

Baeckea pygmaea R.Br. ex Benth., Fl. Austral. 3: 86 (1867).
SW. Western Australia. 50 WAU. Nanophan.

Baeckea robusta F.Muell., Fragm. 4: 72 (1864). *Babingtonia robusta* (F.Muell.) F.Muell., Fragm. 4: 74 (1864).
Western Australia (Irwin, Avon). 50 WAU. Nanophan.

Baeckea serpyllifolia (Turcz.) F.Muell., Fragm. 10: 30 (1876).
Australia (?). 50+. Nanophan. Provisionally accepted.
**Harmogia serpyllifolia* Turcz., Bull. Cl. Phys.-Math. Acad. Imp. Sci. Saint-Pétersbourg 10: 330 (1852).

Baeckea staminosa Pritz., Bot. Jahrb. Syst. 35: 417 (1905).
Western Australia. 50 WAU. Nanophan.

Baeckea stowardii S.Moore, J. Linn. Soc., Bot. 45: 176 (1920).
Western Australia. 50 WAU. Nanophan.

Baeckea subcuneata F.Muell., Fragm. 4: 73 (1864). *Babingtonia subcuneata* (F.Muell.) F.Muell., Fragm. 4: 74 (1864).
Western Australia (Irwin). 50 WAU. Nanophan.

Baeckea tenuifolia (Turcz.) Domin, Vestn. Král. Ceské Spolecn. Nauk, Tr. Mat.-Prír. 2: 82 (1923).
SW. Western Australia. 50 WAU. Nanophan.
**Cyathostemon tenuifolius* Turcz., Bull. Cl. Phys.-Math. Acad. Imp. Sci. Saint-Pétersbourg 10: 332 (1852).

Baeckea tenuiramea S.Moore, J. Linn. Soc., Bot. 45: 177 (1920).
Western Australia (?). 50 WAU. Nanophan.

Baeckea thymoides S.Moore, J. Linn. Soc., Bot. 45: 177 (1920).
SW. Western Australia. 50 WAU. Nanophan.

Baeckea trapeza A.R.Bean, Telopea 7: 255 (1997).
E. Queensland. 50 QLD. Nanophan.

Baeckea tuberculata Trudgen, Nuytsia 5: 441 (1986).
NW. South Australia. 50 SOA. Nanophan.

Baeckea uncinella Benth., Fl. Austral. 3: 84 (1867).
SW. Western Australia. 50 WAU. Nanophan.

Baeckea utilis F.Muell. ex Miq., Ned. Kruidk. Arch. 4: 150 (1856).
New South Wales, Victoria. 50 NSW VIC. Nanophan.

Synonyms:
Baeckea affinis Hook. = ***Euryomyrtus ramosissima*** (A.Cunn.) M.E.Trudgen subsp. ***ramosissima***
Baeckea affinis Endl. = ***Astartea affinis*** (Endl.) Rye
Baeckea alpina Lindl. = ***Euryomyrtus ramosissima*** (A.Cunn.) M.E.Trudgen subsp. ***ramosissima***
Baeckea ambigua (F.Muell.) Nied. = ***Astartea ambigua*** F.Muell.
Baeckea arbuscula R.Br. ex Benth. = ***Astartea arbuscula*** (R.Br. ex Benth.) Rye
Baeckea astarteoides Benth. = ***Astartea astarteoides*** (Benth.) Rye
Baeckea behrii (Schltdl.) F.Muell. = ***Babingtonia behrii*** (Schltdl.) A.R.Bean
Baeckea behrii var. *brevifolia* F.Muell. = ***Babingtonia behrii*** (Schltdl.) A.R.Bean
Baeckea calycina Lindl. = ***Thryptomene calycina*** (Lindl.) Stapf
Baeckea camphorata R.Br. ex Sims = ***Triplarina imbricata*** (Sm.) A.R.Bean
Baeckea camphorosmae Endl. = ***Babingtonia camphorosmae*** (Endl.) Lindl.
Baeckea capitata (F.Muell. ex Benth.) F.Muell. = ***Scholtzia capitata*** F.Muell. ex Benth.
Baeckea carnosa S.Moore = ***Rinzia carnosa*** (S.Moore) Trudgen
Baeckea carnosula Sieber ex Spreng. = ***Baeckea brevifolia*** (Rudge) DC.
Baeckea chinensis Gaertn. = ***Baeckea frutescens*** L.
Baeckea ciliata (F.Muell.) F.Muell. = ***Scholtzia ciliata*** F.Muell.
Baeckea citriodora Penfold & J.L.Willis = ***Ochrosperma citriodorum*** (Penfold & J.L.Willis) Trudgen
Baeckea cochinchinensis Blume = ***Baeckea frutescens*** L.
Baeckea crassifolia var. *icosandra* F.Muell. ex Benth. = ***Baeckea crassifolia*** Lindl.
Baeckea crassifolia var. *pentamera* J.M.Black = ***Baeckea crassifolia*** Lindl.
Baeckea crenatifolia F.Muell. = ***Babingtonia crenulata*** (F.Muell.) A.R.Bean
Baeckea crenulata (Sm.) DC. = ***Baeckea imbricata*** (Gaertn.) Druce
Baeckea crenulata (F.Muell.) Druce = ***Babingtonia crenulata*** (F.Muell.) A.R.Bean
Baeckea crenulata var. *tenella* (Gaertn.) Benth. = ***Baeckea imbricata*** (Gaertn.) Druce
Baeckea crenulata var. *tenella* (Gaertn.) Benth. = ***Baeckea imbricata*** (Gaertn.) Druce
Baeckea crispiflora var. *tenuior* Ewart = ***Baeckea crispiflora*** (F.Muell.) F.Muell.
Baeckea crispiflora var. *tenuior* Ewart = ***Baeckea crispiflora*** (F.Muell.) F.Muell.
Baeckea cumingiana Schauer = ***Baeckea frutescens*** L.
Baeckea cunninghamii (Schauer) Benth. = ***Babingtonia cunninghamii*** (Schauer) A.R.Bean
Baeckea densifolia Sm. = ***Babingtonia densifolia*** (Sm.) F.Muell.
Baeckea denticulata Maiden & Betche = ***Euryomyrtus denticulata*** (Maiden & Betche) Trudgen
Baeckea diffusa Sieber ex DC. = ***Euryomyrtus ramosissima*** (A.Cunn.) M.E.Trudgen subsp. ***ramosissima***
Baeckea diffusa var. *striata* DC. = ***Euryomyrtus ramosissima*** (A.Cunn.) M.E.Trudgen subsp. ***ramosissima***

Baeckea dimorphandra F.Muell. ex Benth. = ***Rinzia dimorphandra*** (F.Muell. ex Benth.) Trudgen
Baeckea diosmoides Sieber ex DC. = ***Baeckea imbricata*** (Gaertn.) Druce
Baeckea diosmoides (Turcz.) Domin = ***Baeckea ovalifolia*** (F.Muell.) F.Muell.
Baeckea drummondii Benth. = *Rinzia* sp. ?
Baeckea eatoniana Ewart & Jean White = ***Scholtzia eatoniana*** (Ewart & Jean White) C.A.Gardner
Baeckea ericoides Schltdl. = ***Baeckea frutescens*** L.
Baeckea ericoides Brongn. & Gris = ***Babingtonia leratii*** (Schltr.) A.R.Bean
Baeckea ericoides f. *tetrasticha* Dunker = ***Babingtonia leratii*** (Schltr.) A.R.Bean
Baeckea fascicularis (Labill.) Nied. = ***Astartea fascicularis*** (Labill.) A.Cunn. ex DC.
Baeckea fasciculata Sieber ex Spreng. = ***Babingtonia densifolia*** (Sm.) F.Muell.
Baeckea fumana (Schauer) F.Muell. = ***Rinzia fumana*** Schauer
Baeckea gracilis (Schauer) C.A.Gardner = ***Baeckea polyandra*** F.Muell.
Baeckea gracilis A.Cunn. = ***Micromyrtus ciliata*** (Sm.) Druce
Baeckea gunniana var. *latifolia* Benth. = ***Baeckea latifolia*** (Benth.) A.R.Bean
Baeckea imbricata S.Moore = ?
Baeckea imbricata var. *tenella* (Gaertn.) Hochr. = ***Baeckea imbricata*** (Gaertn.) Druce
Baeckea imbricata var. *typica* Domin = ***Baeckea imbricata*** (Gaertn.) Druce
Baeckea intratropica (F.Muell.) Nied. = ***Astartea intratropica*** F.Muell.
Baeckea involucrata Endl. = ***Scholtzia obovata*** (DC.) Schauer
Baeckea jucunda S.T.Blake = ***Babingtonia jucunda*** (S.T.Blake) A.R.Bean
Baeckea laxiflora (Benth.) F.Muell. = ***Scholtzia laxiflora*** Benth.
Baeckea leptantha (Benth.) F.Muell. = ***Scholtzia leptantha*** Benth.
Baeckea leptocalyx F.Muell. = ***Micromyrtus leptocalyx*** (F.Muell.) Benth.
Baeckea leptospermoides C.A.Gardner = ***Euryomyrtus leptospermoides*** (C.A.Gardner) Trudgen
Baeckea leratii Schltr. = ***Babingtonia leratii*** (Schltr.) A.R.Bean
Baeckea linearis C.T.White = ***Ochrosperma lineare*** (C.T.White) Trudgen
Baeckea linifolia var. *brevifolia* Ewart = ***Baeckea linifolia*** Rudge
Baeckea linifolia var. *brevifolia* Ewart = ***Baeckea linifolia*** Rudge
Baeckea linifolia f. *trichophylla* (Sieber ex Spreng.) Domin = ***Baeckea linifolia*** Rudge
Baeckea linifolia f. *trichophylla* (Sieber ex Spreng.) Domin = ***Baeckea linifolia*** Rudge
Baeckea maidenii Ewart & Jean White = ***Euryomyrtus maidenii*** (Ewart & Jean White) Trudgen
Baeckea micrantha Hook.f. = ***Baeckea gunniana*** Schauer ex Walp.
Baeckea micrantha DC. = ***Thryptomene baeckeacea*** F.Muell.
Baeckea microphylla A.Cunn. ex Schauer = ***Babingtonia cunninghamii*** (Schauer) A.R.Bean
Baeckea microphylla Sieber ex Spreng. = ***Micromyrtus ciliata*** (Sm.) Druce
Baeckea minutifolia Cheel = ***Rinzia carnosa*** (S.Moore) Trudgen
Baeckea neglecta Vieill. ex Guillaumin = ***Babingtonia virgata*** (J.R.Forst. & G.Forst.) F.Muell.
Baeckea nelitrioides Seem. = ***Cloezia floribunda*** Brongn. & Gris
Baeckea nova-anglica F.Muell. = ***Babingtonia densifolia*** (Sm.) F.Muell.
Baeckea obovata DC. = ***Scholtzia obovata*** (DC.) Schauer
Baeckea obtusifolia Brongn. & Gris = ***Babingtonia virgata*** (J.R.Forst. & G.Forst.) F.Muell.
Baeckea oligandra (F.Muell. ex Benth.) F.Muell. = ***Scholtzia oligandra*** F.Muell. ex Benth.
Baeckea oligomera Radlk. = ***Ochrosperma oligomerum*** (Radlk.) A.R.Bean
Baeckea oxycoccoides (Turcz.) Benth. = ***Rinzia oxycoccoides*** Turcz.
Baeckea parviflora (F.Muell.) F.Muell. = ***Scholtzia parviflora*** F.Muell.
Baeckea parvula (Labill.) DC. = ***Babingtonia virgata*** (J.R.Forst. & G.Forst.) F.Muell.
Baeckea parvula var. *latifolia* Brongn. & Gris = ***Babingtonia virgata*** (J.R.Forst. & G.Forst.) F.Muell.
Baeckea phylicoides A.Cunn. ex Schauer = ***Kunzea ericoides*** (A.Rich.) Joy Thomps.
Baeckea pinifolia (Labill.) DC. = ***Babingtonia pinifolia*** (Labill.) A.R.Bean
Baeckea platystemona Benth. = ***Rinzia crassifolia*** Turcz.
Baeckea plicata F.Muell. = ***Micromyrtus ciliata*** (Sm.) Druce
Baeckea procera J.W.Dawson = ***Babingtonia procera*** (J.W.Dawson) A.R.Bean
Baeckea prostrata Hook.f. = ***Euryomyrtus ramosissima*** subsp. ***prostrata*** (Hook.f.) M.E.Trudgen
Baeckea ramosissima A.Cunn. = ***Euryomyrtus ramosissima*** (A.Cunn.) M.E.Trudgen
Baeckea ramosissima subsp. *prostrata* (Hook.f.) G.W.Carr = ***Euryomyrtus ramosissima*** subsp. ***prostrata*** (Hook.f.) M.E.Trudgen
Baeckea saxicola A.Cunn. ex Hook. = ***Thryptomene saxicola*** (A.Cunn. ex Hook.) Schauer
Baeckea schollerifolia Lehm. = ***Rinzia schollerifolia*** (Lehm.) Trudgen
Baeckea sinensis Gaertn. = ***Baeckea frutescens*** L.
Baeckea spatulata (Turcz.) F.Muell. = ***Scholtzia spathulata*** (Turcz.) Benth.
Baeckea speciosa A.Cunn. = ***Hypocalymma strictum*** subsp. ***strictum***
Baeckea spinosa Sieber ex Spreng. = [50+]
Baeckea squarrulosa Domin = ***Babingtonia squarrulosa*** (Domin) A.R.Bean
Baeckea stenophylla F.Muell. = ***Baeckea frutescens*** L.
Baeckea stuartiana F.Muell. ex Miq. = ***Euryomyrtus ramosissima*** (A.Cunn.) M.E.Trudgen subsp. ***ramosissima***
Baeckea sumatrana Blume = ***Baeckea frutescens*** L.
Baeckea taxifolia Merr. = ***Babingtonia taxifolia*** (Merr.) A.R.Bean
Baeckea teretifolia (Benth.) F.Muell. = ***Scholtzia teretifolia*** Benth.
Baeckea tetragona F.Muell. ex Benth. = [50 WAU]
Baeckea thymifolia Hook.f. = ***Euryomyrtus ramosissima*** (A.Cunn.) M.E.Trudgen subsp. ***ramosissima***
Baeckea trichophylla Sieber ex Spreng. = ***Baeckea linifolia*** Rudge
Baeckea uberiflora (F.Muell.) F.Muell. = ***Scholtzia uberiflora*** F.Muell.
Baeckea umbellata F.Muell. = ***Stenostegia congesta*** A.R.Bean
Baeckea umbellifera (F.Muell.) F.Muell. = ***Scholtzia umbellifera*** F.Muell.

Baeckea utilis var. *latifolia* (Benth.) J.H.Willis = ***Baeckea utilis*** F.Muell. ex Miq.
Baeckea virgata (J.R.Forst. & G.Forst.) Andrews = ***Babingtonia virgata*** (J.R.Forst. & G.Forst.) F.Muell.
Baeckea virgata var. *parvula* F.M.Bailey = ***Babingtonia bidwillii*** A.R.Bean
Baeckea virgata var. *polyandra* Maiden & Betche = ***Kunzea ericoides*** (A.Rich.) Joy Thomps.

***Unplaced Names*:**
Baeckea drummondii Benth., Fl. Austral. 3: 75 (1867). = *Rinzia* sp. ?
Baeckea imbricata S.Moore, J. Linn. Soc., Bot. 45: 178 (1920), nom. illeg. = ?
Baeckea spinosa Sieber ex Spreng., Syst. Veg. 4(2): 149 (1827). = [50+]
Baeckea tetragona F.Muell. ex Benth., Fl. Austral. 3: 77 (1867). = [50 WAU]

Balaustion

Balaustion Hook., Hooker's Icon. Pl. 9: t. 852 (1851).
SW. Australia. 50 WAU. [Myrtaceae]
2 Species
Punicella Turcz., Bull. Cl. Phys.-Math. Acad. Imp. Sci. Saint-Pétersbourg 10: 333 (1852).
Cheynia J.Drumm. ex Harv., Hooker's J. Bot. Kew Gard. Misc. 7: 56 (1855).

Balaustion microphyllum C.A.Gardner, J. Roy. Soc. Western Australia 13: 66 (1927).
Western Australia (Avon). 50 WAU. Cham. or nanophan.

Balaustion pulcherrimum Hook., Hooker's Icon. Pl. 9: t. 852 (1851).
SW. Western Australia. 50 WAU. Cham. or nanophan.
Punicella carinata Turcz., Bull. Cl. Phys.-Math. Acad. Imp. Sci. Saint-Pétersbourg 10: 333 (1852).
Cheynia pulchella Harv., Hooker's J. Bot. Kew Gard. Misc. 7: 56 (1855).

Ballardia

Ballardia Montrouz. = ***Metrosideros*** Banks ex Gaertn.
Ballardia elegans Montrouz. = ***Metrosideros elegans*** (Montrouz.) Beauvis.

Barongia

Barongia Peter G.Wilson & B.Hyland, Telopea 3: 257 (1988).
Queensland. 50 QLD. [Myrtaceae]
1 Species

Barongia lophandra Peter G.Wilson & B.Hyland, Telopea 3: 258 (1988).
Queensland (Cook). 50 QLD. Phan.

Basisperma

Basisperma C.T.White, J. Arnold Arbor. 23: 84 (1942).
New Guinea. 43 NWG. [Myrtaceae]
1 Species

Basisperma lanceolata C.T.White, J. Arnold Arbor. 23: 84 (1942).
Papua New Guinea. 43 NWG. Phan.

Baudinia

Baudinia Lesch. ex DC. = ***Calothamnus*** Labill.

Beaufortia

Beaufortia R.Br. in W.T.Aiton, Hortus Kew. 4: 418 (1812).
SW. Australia. 50 WAU. [Myrtaceae]
20 Species
Manglesia Lindl., Sketch Veg. Swan R.: t. 3 (1839).
Schizopleura Endl., Gen. Pl.: 1228 (1840).

Beaufortia aestiva K.J.Brooks, Nuytsia 12: 164 (1998).
SW. Australia. 50 WAU. Nanophan.

Beaufortia anisandra Schauer, Nov. Actorum Acad. Caes. Leop.-Carol. Nat. Cur. 21: 13 (1844).
SW. Australia. 50 WAU. Nanophan.

Beaufortia bicolor Strid, Pl. Syst. Evol. 155: 339 (1987).
Western Australia (Irwin). 50 WAU. Nanophan.

Beaufortia bracteosa Diels, Bot. Jahrb. Syst. 35: 431 (1905).
SW. Australia. 50 WAU. Nanophan.

Beaufortia cyrtodonta (Turcz.) Benth., Fl. Austral. 3: 167 (1867).
SW. Australia. 50 WAU. Nanophan.
**Melaleuca cyrtodonta* Turcz., Bull. Soc. Imp. Naturalistes Moscou 22(2): 24 (1849).

Beaufortia decussata R.Br. in W.T.Aiton, Hortus Kew. 4: 418 (1812).
SW. Australia. 50 WAU. Nanophan.

Beaufortia elegans Schauer, Nova Acta Phys.-Med. Acad. Caes. Leop.-Carol. Nat. Cur. 21: 16 (1844).
SW. Australia. 50 WAU. Nanophan.
Beaufortia microphylla Turcz., Bull. Soc. Imp. Naturalistes Moscou 22(2): 24 (1849).
Beaufortia elegans var. *minor* Benth., Fl. Austral. 3: 169 (1867).

Beaufortia empetrifolia (Rchb.) Schauer, Nova Acta Phys.-Med. Acad. Caes. Leop.-Carol. Nat. Cur. 21: 17 (1844).
SW. Australia. 50 WAU. Nanophan.
**Melaleuca empetrifolia* Rchb., Iconogr. Bot. Exot. 2: 1 (1829). *Beaufortia micrantha* var. *empetrifolia* (Rchb.) Benth., Fl. Austral. 3: 170 (1867).

Beaufortia eriocephala W.Fitzg., J. West Austral. Nat. Hist. Soc. 2: 23 (1905).
SW. Australia. 50 WAU. Nanophan.

Beaufortia heterophylla Turcz., Bull. Cl. Phys.-Math. Acad. Imp. Sci. Saint-Pétersbourg 10: 345 (1852).
SW. Australia. 50 WAU. Nanophan.

Beaufortia incana (Benth.) A.S.George, Nuytsia 1: 290 (1972).
SW. Australia. 50 WAU. Nanophan.
**Beaufortia macrostemon* var. *incana* Benth., Fl. Austral. 3: 167 (1867).

Beaufortia interstans F.Muell., Fragm. 10: 30 (1876).
SW. Australia. 50 WAU. Nanophan.

Beaufortia macrostemon Lindl., Sketch Veg. Swan R.: 31 (1839). *Schizopleura macrostemon* (Lindl.) Walp., Repert. Bot. Syst. 2: 162 (1843).
SW. Australia. 50 WAU. Nanophan.

Beaufortia micrantha Schauer, Nov. Actorum Acad. Caes. Leop.-Carol. Nat. Cur. 21: 18 (1844).
SW. Australia. 50 WAU. Nanophan.

var. ***micrantha***
SW. Australia. 50 WAU. Nanophan.

Regelia adpressa Turcz., Bull. Soc. Imp. Naturalistes Moscou 22(1): 25 (1849).
Melaleuca regelii Planch., Hort. Donat.: 88, App. 227 (1858).

var. ***puberula*** Benth., Fl. Austral. 3: 170 (1867).
SW. Australia. 50 WAU. Nanophan.

Beaufortia orbifolia F.Muell., Fragm. 3: 110 (1862).
SW. Australia. 50 WAU. Nanophan.

Beaufortia purpurea Lindl., Sketch Veg. Swan R.: 30 (1839). *Schizopleura purpurea* (Lindl.) Walp., Repert. Bot. Syst. 2: 162 (1843).
SW. Australia. 50 WAU. Nanophan.
Manglesia purpurea Lindl., Sketch Veg. Swan R.: t. 3A (1839).

Beaufortia schaueri Preissler ex Schauer, Nov. Actorum Acad. Caes. Leop.-Carol. Nat. Cur. 21: 14 (1844).
Western Australia (Eyre, Roe). 50 WAU. Nanophan.
Beaufortia schaueri var. *atrorubens* Benth., Fl. Austral. 3: 168 (1867).

Beaufortia sparsa R.Br. in W.T.Aiton, Hortus Kew. 4: 418 (1812).
SW. Australia. 50 WAU. Nanophan.
Beaufortia splendens Baxter ex A.Dietr., Allg. Gartenzeitung 2: 274 (1834).

Beaufortia sprengelioides (DC.) Craven, Taxon 48: 54 (1999).
SW. Australia. 50 WAU. Nanophan.
**Melaleuca sprengelioides* DC., Prodr. 3: 215 (1828).
Beaufortia dampieri A.Cunn., Bot. Mag. 60: t. 3272 (1833). *Schizopleura dampieri* (A.Cunn.) Walp., Repert. Bot. Syst. 2: 162 (1843).

Beaufortia squarrosa Schauer, Nov. Actorum Acad. Caes. Leop.-Carol. Nat. Cur. 21: 11 (1844).
SW. Australia. 50 WAU. Nanophan.

***Synonyms*:**
Beaufortia cymbifolia Diels = ***Regelia cymbifolia*** (Diels) C.A.Gardner
Beaufortia dampieri A.Cunn. = ***Beaufortia sprengelioides*** (DC.) Craven
Beaufortia elegans var. *minor* Benth. = ***Beaufortia elegans*** Schauer
Beaufortia inops Schauer = ***Regelia inops*** (Schauer) Schauer
Beaufortia macrostemon var. *incana* Benth. = ***Beaufortia incana*** (Benth.) A.S.George
Beaufortia micrantha var. *empetrifolia* (Rchb.) Benth. = ***Beaufortia empetrifolia*** (Rchb.) Schauer
Beaufortia microphylla Turcz. = ***Beaufortia elegans*** Schauer
Beaufortia pinifolia Cels = ***Calothamnus quadrifidus*** R.Br. ex W.T.Aiton
Beaufortia puberula Turcz. = [50 WAU]
Beaufortia schaueri var. *atrorubens* Benth. = ***Beaufortia schaueri*** Preissler ex Schauer
Beaufortia splendens Baxter ex A.Dietr. = ***Beaufortia sparsa*** R.Br.
Beaufortia velutina Turcz. = ***Regelia velutina*** (Turcz.) C.A.Gardner

***Unplaced Names*:**
Beaufortia puberula Turcz., Bull. Cl. Phys.-Math. Acad. Imp. Sci. Saint-Pétersbourg 10: 345 (1852). = [50 WAU]

Billottia

Billottia R.Br. ex G.Don = ***Agonis*** (DC.) Sweet
Billottia Colla = ***Calothamnus*** Labill.
Billottia acerosa Colla = ***Calothamnus quadrifidus*** R.Br. ex W.T.Aiton
Billottia elliptica (Endl.) Sweet = ***Pericalymma ellipticum*** (Endl.) Schauer
Billottia flexuosa (Muhl. ex Willd.) R.Br. = ***Agonis flexuosa*** (Muhl. ex Willd.) Sweet
Billottia flexuosa var. *angustifolia* Otto & A.Dietr. = ***Agonis flexuosa*** (Muhl. ex Willd.) Sweet
Billottia flexuosa var. *latifolia* Otto & A.Dietr. = ***Agonis flexuosa*** (Muhl. ex Willd.) Sweet
Billottia hypericifolia (Otto & A.Dietr.) Heynh. = ***Agonis hypericifolia*** (Otto & A.Dietr.) Schauer
Billottia linearifolia (DC.) R.Br. = ***Agonis linearifolia*** (DC.) Sweet
Billottia marginata (Labill.) R.Br. = ***Agonis marginata*** (Labill.) Sweet
Billottia marginata var. *angustifolia* (Otto) Heynh. = ***Agonis marginata*** (Labill.) Sweet
Billottia marginata var. *glabrata* (DC.) G.Don = ***Agonis marginata*** (Labill.) Sweet
Billottia theiformis (Schauer) G.Don = ***Agonis hypericifolia*** (Otto & A.Dietr.) Schauer

Blepharocalyx

Blepharocalyx O.Berg, Linnaea 27: 412 (1856).
Trop. & S. South America. 81 LEE VNA WIN 82 GUY VEN 83 BOL ECU PER 84 BZC BZE BZL BZN BZS 85 AGE AGW CLC CLS PAR URU. [Myrtaceae]
4 Species
Heteromyrtus Blume, Mus. Bot. 1: 76 (1850), nom. illeg.
Temu O.Berg, Linnaea 30: 710 (1861).
Marlieriopsis Kiaersk., Bot. Tidsskr. 17: 281 (1890).

Blepharocalyx cruckshanksii (Hook. & Arn.) Nied. in H.G.A.Engler & K.A.E.Prantl, Nat. Pflanzenfam. 3(7): 71 (1892).
C. & SC. Chile. 85 CLC CLS. Nanophan. or phan.
**Eugenia cruckshanksii* Hook. & Arn., Bot. Misc. 3: 321 (1833). *Luma cruckshanksii* (Hook. & Arn.) A.Gray, U.S. Expl. Exped., Phan. 1: 540 (1854). *Temu cruckshanksii* (Hook. & Arn.) O.Berg, Linnaea 30: 711 (1861). *Myrtus cruckshanksii* (Hook. & Arn.) Kuntze, Revis. Gen. Pl. 3(2): 91 (1898).
Eugenia divaricata O.Berg, Linnaea 27: 257 (1856), nom. illeg. *Temu divaricatum* O.Berg, Linnaea 30: 711 (1861). *Blepharocalyx divaricatus* (O.Berg) Nied. in H.G.A.Engler & K.A.E.Prantl, Nat. Pflanzenfam. 3(7): 71 (1892).
Eugenia divaricata var. *obovata* O.Berg, Linnaea 27: 257 (1856). *Blepharocalyx divaricatus* var. *obovatus* (O.Berg) Reiche, Anales Univ. Chile 98: 707 (1897).
Eugenia divaricata var. *ovalis* O.Berg, Linnaea 27: 258 (1856). *Blepharocalyx divaricatus* var. *ovalis* (O.Berg) Reiche, Anales Univ. Chile 89: 707 (1897).
Eugenia divaricata var. *pauciflora* O.Berg, Linnaea 27: 258 (1856). *Blepharocalyx divaricatus* var. *pauciflorus* (O.Berg) Reiche, Anales Univ. Chile 89: 707 (1897).
Eugenia elliptica Phil., Anales Univ. Chile 1865(2): 315 (1865), nom. illeg.

Blepharocalyx eggersii (Kiaersk.) Landrum, Syst. Bot. 11: 159 (1986).
Lesser Antilles to S. Trop. America. 81 LEE VNA WIN 82 GUY VEN 83 PER 84 BZN. Phan.

Marlieriopsis eggersii Kiaersk., Bot. Tidsskr. 17: 282 (1890). *Mitranthes eggersii* (Kiaersk.) Nied. in H.G.A.Engler & K.A.E.Prantl, Nat. Pflanzenfam. 3(7): 72 (1893).

Blepharocalyx myriophyllus Morais & Sobral, Lundiana 1: 5 (2006).
Brazil (Minas Gerais). 84 BZL. Nanophan.
Eugenia myriophylla Casar., Nov. Stirp. Bras.: 77 (1842). *Myrciaria myriophylla* (Casar.) O.Berg in C.F.P.von Martius & auct. suc. (eds.), Fl. Bras. 14(1): 375 (1857).
Myrcia pinaster Mart. ex O.Berg, Linnaea 27: 338 (1856).

Blepharocalyx salicifolius (Kunth) O.Berg, Linnaea 27: 413 (1856).
Brazil to N. Argentina. 83 BOL ECU PER 84 BZC BZE BZL BZS 85 AGE AGW PAR URU. Nanophan. or phan.
**Myrtus salicifolia* Kunth in F.W.H.von Humboldt, A.J.A.Bonpland & C.S.Kunth, Nov. Gen. Sp. 6: 136 (1823). *Eugenia salicifolia* (Kunth) DC., Prodr. 3: 278 (1828).
Myrtus dumosa Spreng., Syst. Veg. 2: 483 (1825).
Eugenia adamantium Cambess. in A.F.C.de Saint-Hilaire, Fl. Bras. Merid. 2: 342 (1832). *Pseudomyrcianthes adamantium* (Cambess.) Kausel, Ark. Bot., a.s., 3: 505 (1956). *Blepharocalyx acuminatus* var. *adamantium* (Cambess.) Mattos, Loefgrenia 64: 2 (1975).
Eugenia decumbens Cambess. in A.F.C. de Saint-Hilaire, Fl. Bras. Merid. 2: 339 (1832).
Eugenia deserti Cambess. in A.F.C.de Saint-Hilaire, Fl. Bras. Merid. 2: 343 (1832). *Myrciaria deserti* (Cambess.) O.Berg in C.F.P.von Martius & auct. suc. (eds.), Fl. Bras. 14(1): 359 (1857). *Blepharocalyx deserti* (Cambess.) Burret, Notizbl. Bot. Gart. Berlin-Dahlem 15: 535 (1941).
Eugenia suaveolens Cambess. in A.F.C.de Saint-Hilaire, Fl. Bras. Merid. 2: 339 (1832). *Blepharocalyx suaveolens* (Cambess.) Burret, Notizbl. Bot. Gart. Berlin-Dahlem 15: 535 (1941).
Myrcia mugiensis Cambess. in A.F.C.de Saint-Hilaire, Fl. Bras. Merid. 2: 327 (1832). *Aulomyrcia mugiensis* (Cambess.) O.Berg, Linnaea 27: 36 (1855). *Blepharocalyx mugiensis* (Cambess.) Burret, Notizbl. Bot. Gart. Berlin-Dahlem 15: 535 (1941). *Blepharocalyx umbilicatus* var. *mugiensis* (Cambess.) Mattos, Loefgrenia 64: 2 (1975).
Myrtus umbilicata Cambess. in A.F.C.de Saint-Hilaire, Fl. Bras. Merid. 2: 296 (1832). *Heteromyrtus umbilicata* (Cambess.) Blume, Mus. Bot. 1: 76 (1850). *Blepharocalyx umbilicatus* (Cambess.) Burret, Notizbl. Bot. Gart. Berlin-Dahlem 15: 535 (1941). *Blepharocalyx suaveolens* var. *umbilicatus* (Cambess.) D.Legrand, Notul. Syst. (Paris) 15: 272 (1958).
Eugenia depauperata Cambess. in A.F.C.de Saint-Hilaire, Fl. Bras. Merid. 2: 366 (1833). *Blepharocalyx depauperatus* (Cambess.) O.Berg, Linnaea 27: 415 (1856).
Eugenia tweediei Hook. & Arn., Bot. Misc. 3: 323 (1833). *Blepharocalyx tweediei* (Hook. & Arn.) O.Berg, Linnaea 27: 415 (1856). *Blepharocalyx salicifolius* var. *tweediei* (Hook. & Arn.) D.Legrand, in Fl. Ilustr. Catar., Mirt.: 784 (1978).
Eugenia acuminatissima Miq., Linnaea 19: 440 (1846). *Blepharocalyx acuminatissimus* (Miq.) O.Berg, Linnaea 27: 413 (1856).
Blepharocalyx acuminatus O.Berg, Linnaea 27: 413 (1856).
Blepharocalyx angustifolius O.Berg, Linnaea 27: 421 (1856). *Myrtus angustifolia* (O.Berg) Arechav., Anales Mus. Nac. Montevideo 5: 37 (1905), nom. illeg.
Blepharocalyx amarus O.Berg in C.F.P.von Martius & auct. suc. (eds.), Fl. Bras. 14(1): 422 (1857). *Myrtus amara* (O.Berg) Arechav., Anales Mus. Nac. Montevideo 5: 36 (1905).
Blepharocalyx angustifolius O.Berg in C.F.P.von Martius & auct. suc. (eds.), Fl. Bras. 14(1): 421 (1857), nom. illeg.
Blepharocalyx angustissimus O.Berg in C.F.P.von Martius & auct. suc. (eds.), Fl. Bras. 14(1): 421 (1857). *Myrtus angustissima* (O.Berg) Arechav., Anales Mus. Nac. Montevideo 5: 37 (1905).
Blepharocalyx apiculatus O.Berg in C.F.P.von Martius & auct. suc. (eds.), Fl. Bras. 14(1): 425 (1857). *Myrtus apiculata* (O.Berg) Kiaersk., Enum. Myrt. Bras.: 17 (1893).
Blepharocalyx apiculatus var. *rubellus* O.Berg in C.F.P.von Martius & auct. suc. (eds.), Fl. Bras. 14(1): 425 (1857).
Blepharocalyx apiculatus var. *strictus* O.Berg in C.F.P.von Martius & auct. suc. (eds.), Fl. Bras. 14(1): 425 (1857).
Blepharocalyx brunneus O.Berg in C.F.P.von Martius & auct. suc. (eds.), Fl. Bras. 14(1): 427 (1857). *Myrtus brunnea* (O.Berg) Kiaersk., Enum. Myrt. Bras.: 17 (1893).
Blepharocalyx canescens O.Berg in C.F.P.von Martius & auct. suc. (eds.), Fl. Bras. 14(1): 426 (1857). *Myrtus reinhardtiana* Kiaersk., Enum. Myrt. Bras.: 18 (1893).
Blepharocalyx cuspidatus O.Berg in C.F.P.von Martius & auct. suc. (eds.), Fl. Bras. 14(1): 424 (1857).
Blepharocalyx lanceolatus O.Berg in C.F.P.von Martius & auct. suc. (eds.), Fl. Bras. 14(1): 422 (1857). *Myrtus lanceolata* (O.Berg) Arechav., Anales Mus. Nac. Montevideo 5: 39 (1905), nom. illeg.
Blepharocalyx lanceolatus var. *arborescens* O.Berg in C.F.P.von Martius & auct. suc. (eds.), Fl. Bras. 14(1): 423 (1857).
Blepharocalyx lanceolatus var. *frutescens* O.Berg in C.F.P.von Martius & auct. suc. (eds.), Fl. Bras. 14(1): 423 (1857).
Blepharocalyx longipes O.Berg in C.F.P.von Martius & auct. suc. (eds.), Fl. Bras. 14(1): 423 (1857). *Myrtus longipes* (O.Berg) Kiaersk., Enum. Myrt. Bras.: 18 (1893). *Blepharocalyx tweediei* var. *longipes* (O.Berg) Mattos, Loefgrenia 64: 2 (1975). *Blepharocalyx salicifolius* var. *longipes* (O.Berg) D.Legrand, in Fl. Ilustr. Catar., Mirt.: 785 (1978).
Blepharocalyx picrocarpus O.Berg in C.F.P.von Martius & auct. suc. (eds.), Fl. Bras. 14(1): 427 (1857).
Blepharocalyx pilosus O.Berg in C.F.P.von Martius & auct. suc. (eds.), Fl. Bras. 14(1): 422 (1857).
Blepharocalyx ramosissimus O.Berg in C.F.P.von Martius & auct. suc. (eds.), Fl. Bras. 14(1): 424 (1857).
Blepharocalyx sessilifolius O.Berg in C.F.P.von Martius & auct. suc. (eds.), Fl. Bras. 14(1): 428 (1857).
Blepharocalyx sessilifolius var. *pauciflorus* O.Berg in C.F.P.von Martius & auct. suc. (eds.), Fl. Bras. 14(1): 428 (1857).

Blepharocalyx sessilifolius var. *pluriflorus* O.Berg in C.F.P.von Martius & auct. suc. (eds.), Fl. Bras. 14(1): 428 (1857).
Blepharocalyx strictus O.Berg in C.F.P.von Martius & auct. suc. (eds.), Fl. Bras. 14(1): 423 (1857). *Myrtus stricta* (O.Berg) Arechav., Anales Mus. Nac. Montevideo 5: 38 (1905).
Blepharocalyx villosus O.Berg in C.F.P.von Martius & auct. suc. (eds.), Fl. Bras. 14(1): 425 (1857).
Blepharocalyx villosus var. *triflorus* O.Berg in C.F.P.von Martius & auct. suc. (eds.), Fl. Bras. 14(1): 425 (1857).
Blepharocalyx villosus var. *uniflorus* O.Berg in C.F.P. von Martius & auct. suc. (eds.), Fl. Bras. 14(1): 426 (1857).
Blepharocalyx widgrenii O.Berg in C.F.P.von Martius & auct. suc. (eds.), Fl. Bras. 14(1): 427 (1857). *Myrtus widgrenii* (O.Berg) Kiaersk., Enum. Myrt. Bras.: 18 (1893).
Myrcianthes cisplatensis var. *angustifolia* O.Berg in C.F.P.von Martius & auct. suc. (eds.), Fl. Bras. 14(1): 353 (1857). *Myrciaria angustifolia* (O.Berg) Mattos, Loefgrenia 64: 3 (1975).
Myrcianthes cisplatensis var. *brevipedunculata* O.Berg in C.F.P.von Martius & auct. suc. (eds.), Fl. Bras. 14(1): 353 (1857). *Myrciaria brevipedunculata* (O.Berg) Mattos, Loefgrenia 78: 3 (1983).
Blepharocalyx affinis O.Berg in C.F.P.von Martius & auct. suc. (eds.), Fl. Bras. 14(1): 605 (1859).
Blepharocalyx parvifolius O.Berg in C.F.P.von Martius & auct. suc. (eds.), Fl. Bras. 14(1): 606 (1859).
Blepharocalyx ramosissimus var. *latifolius* O.Berg in C.F.P.von Martius & auct. suc. (eds.), Fl. Bras. 14(1): 606 (1859).
Blepharocalyx ramosissimus var. *nanus* O.Berg in C.F.P.von Martius & auct. suc. (eds.), Fl. Bras. 14(1): 606 (1859).
Blepharocalyx ramosissimus var. *obovatus* O.Berg in C.F.P.von Martius & auct. suc. (eds.), Fl. Bras. 14(1): 606 (1859).
Blepharocalyx serra O.Berg in C.F.P.von Martius & auct. suc. (eds.), Fl. Bras. 14(1): 605 (1859).
Blepharocalyx cisplatensis Griseb., Abh. Königl. Ges. Wiss. Göttingen 24: 126 (1879).
Eugenia piedadensis Kiaersk., Enum. Myrt. Bras.: 176 (1893). *Myrciaria piedadensis* (Kiaersk.) Mattos & D.Legrand, Loefgrenia 67: 5 (1975).
Myrtus apiculata var. *microphylla* Kiaersk., Enum. Myrt. Bras.: 17 (1893).
Myrtus reinhardtiana f. *subcordata* Kiaersk., Enum. Myrt. Bras.: 18 (1893).
Eugenia ipehuensis Barb.Rodr. ex Chodat & Hassl., Bull. Herb. Boissier, II, 7: 804 (1907), nom. inval.
Blepharocalyx montanus Lillo, Contr. Conoc. Arb. Argent.: 67 (1910). *Blepharocalyx giganteus* var. *montanus* (Lillo) Lillo, Anales Soc. Ci. Argent. 72: 174 (1911).
Blepharocalyx giganteus Lillo, Anales Soc. Ci. Argent. 72: 175 (1911).
Myrciaria dichotoma D.Legrand, Sellowia 13: 331 (1961).
Blepharocalyx minutiflorus Mattos & D.Legrand, Loefgrenia 67: 31 (1975).
Blepharocalyx myrcianthoides Mattos, Loefgrenia 64: 2 (1975).
Blepharocalyx tweediei var. *septentrionalis* Mattos, Loefgrenia 64: 2 (1975).
Blepharocalyx umbilicatus var. *paranaensis* Mattos, Loefgrenia 66: 6 (1975).
Blepharocalyx salicifolius f. *catharinae* D.Legrand, in Fl. Ilustr. Catar. 1(Mirt.): 787 (1978). *Blepharocalyx tweediei* f. *catharinae* (D.Legrand) Mattos, Loefgrenia 76: 2 (1981).
Blepharocalyx suaveolens var. *abrupticulmeus* D.Legrand, in Fl. Ilustr. Catar., Mirt.: 795 (1978).
Blepharocalyx suaveolens var. *cuneatus* D.Legrand, in Fl. Ilustr. Catar., Mirt.: 797 (1978).

Synonyms:
Blepharocalyx acuminatissimus (Miq.) O.Berg = ***Blepharocalyx salicifolius*** (Kunth) O.Berg
Blepharocalyx acuminatus O.Berg = ***Blepharocalyx salicifolius*** (Kunth) O.Berg
Blepharocalyx acuminatus var. *adamantium* (Cambess.) Mattos = ***Blepharocalyx salicifolius*** (Kunth) O.Berg
Blepharocalyx affinis O.Berg = ***Blepharocalyx salicifolius*** (Kunth) O.Berg
Blepharocalyx amarus O.Berg = ***Blepharocalyx salicifolius*** (Kunth) O.Berg
Blepharocalyx angustifolius O.Berg = ***Blepharocalyx salicifolius*** (Kunth) O.Berg
Blepharocalyx angustifolius O.Berg = ***Blepharocalyx salicifolius*** (Kunth) O.Berg
Blepharocalyx angustissimus O.Berg = ***Blepharocalyx salicifolius*** (Kunth) O.Berg
Blepharocalyx apiculatus O.Berg = ***Blepharocalyx salicifolius*** (Kunth) O.Berg
Blepharocalyx apiculatus var. *rubellus* O.Berg = ***Blepharocalyx salicifolius*** (Kunth) O.Berg
Blepharocalyx apiculatus var. *strictus* O.Berg = ***Blepharocalyx salicifolius*** (Kunth) O.Berg
Blepharocalyx aromatica (O.Berg) Nied. ex B.D.Jacks. = ?
Blepharocalyx brunneus O.Berg = ***Blepharocalyx salicifolius*** (Kunth) O.Berg
Blepharocalyx canescens O.Berg = ***Blepharocalyx salicifolius*** (Kunth) O.Berg
Blepharocalyx cisplatensis Griseb. = ***Blepharocalyx salicifolius*** (Kunth) O.Berg
Blepharocalyx cuspidatus O.Berg = ***Blepharocalyx salicifolius*** (Kunth) O.Berg
Blepharocalyx depauperatus (Cambess.) O.Berg = ***Blepharocalyx salicifolius*** (Kunth) O.Berg
Blepharocalyx deserti (Cambess.) Burret = ***Blepharocalyx salicifolius*** (Kunth) O.Berg
Blepharocalyx divaricatus (O.Berg) Nied. = ***Blepharocalyx cruckshanksii*** (Hook. & Arn.) Nied.
Blepharocalyx divaricatus var. *obovatus* (O.Berg) Reiche = ***Blepharocalyx cruckshanksii*** (Hook. & Arn.) Nied.
Blepharocalyx divaricatus var. *ovalis* (O.Berg) Reiche = ***Blepharocalyx cruckshanksii*** (Hook. & Arn.) Nied.
Blepharocalyx divaricatus var. *pauciflorus* (O.Berg) Reiche = ***Blepharocalyx cruckshanksii*** (Hook. & Arn.) Nied.
Blepharocalyx ellipticus (Hook. & Arn.) O.Berg = ***Myrcianthes cisplatensis*** (Cambess.) O.Berg
Blepharocalyx giganteus Lillo = ***Blepharocalyx salicifolius*** (Kunth) O.Berg
Blepharocalyx giganteus var. *montanus* (Lillo) Lillo = ***Blepharocalyx salicifolius*** (Kunth) O.Berg
Blepharocalyx lanceolatus O.Berg = ***Blepharocalyx salicifolius*** (Kunth) O.Berg
Blepharocalyx lanceolatus var. *arborescens* O.Berg = ***Blepharocalyx salicifolius*** (Kunth) O.Berg
Blepharocalyx lanceolatus var. *frutescens* O.Berg = ***Blepharocalyx salicifolius*** (Kunth) O.Berg
Blepharocalyx longipes O.Berg = ***Blepharocalyx salicifolius*** (Kunth) O.Berg

Blepharocalyx minutiflorus Mattos & D.Legrand = ***Blepharocalyx salicifolius*** (Kunth) O.Berg
Blepharocalyx montanus Lillo = ***Blepharocalyx salicifolius*** (Kunth) O.Berg
Blepharocalyx mugiensis (Cambess.) Burret = ***Blepharocalyx salicifolius*** (Kunth) O.Berg
Blepharocalyx myrcianthoides Mattos = ***Blepharocalyx salicifolius*** (Kunth) O.Berg
Blepharocalyx parvifolius O.Berg = ***Blepharocalyx salicifolius*** (Kunth) O.Berg
Blepharocalyx picrocarpus O.Berg = ***Blepharocalyx salicifolius*** (Kunth) O.Berg
Blepharocalyx pilosus O.Berg = ***Blepharocalyx salicifolius*** (Kunth) O.Berg
Blepharocalyx ramosissimus O.Berg = ***Blepharocalyx salicifolius*** (Kunth) O.Berg
Blepharocalyx ramosissimus var. *latifolius* O.Berg = ***Blepharocalyx salicifolius*** (Kunth) O.Berg
Blepharocalyx ramosissimus var. *nanus* O.Berg = ***Blepharocalyx salicifolius*** (Kunth) O.Berg
Blepharocalyx ramosissimus var. *obovatus* O.Berg = ***Blepharocalyx salicifolius*** (Kunth) O.Berg
Blepharocalyx salicifolius f. *catharinae* D.Legrand = ***Blepharocalyx salicifolius*** (Kunth) O.Berg
Blepharocalyx salicifolius var. *longipes* (O.Berg) D.Legrand = ***Blepharocalyx salicifolius*** (Kunth) O.Berg
Blepharocalyx salicifolius var. *tweediei* (Hook. & Arn.) D.Legrand = ***Blepharocalyx salicifolius*** (Kunth) O.Berg
Blepharocalyx serra O.Berg = ***Blepharocalyx salicifolius*** (Kunth) O.Berg
Blepharocalyx sessilifolius O.Berg = ***Blepharocalyx salicifolius*** (Kunth) O.Berg
Blepharocalyx sessilifolius var. *pauciflorus* O.Berg = ***Blepharocalyx salicifolius*** (Kunth) O.Berg
Blepharocalyx sessilifolius var. *pluriflorus* O.Berg = ***Blepharocalyx salicifolius*** (Kunth) O.Berg
Blepharocalyx spirioides Stapf = [84]
Blepharocalyx strictus O.Berg = ***Blepharocalyx salicifolius*** (Kunth) O.Berg
Blepharocalyx suaveolens (Cambess.) Burret = ***Blepharocalyx salicifolius*** (Kunth) O.Berg
Blepharocalyx suaveolens var. *abrupticulmeus* D.Legrand = ***Blepharocalyx salicifolius*** (Kunth) O.Berg
Blepharocalyx suaveolens var. *cuneatus* D.Legrand = ***Blepharocalyx salicifolius*** (Kunth) O.Berg
Blepharocalyx suaveolens var. *umbilicatus* (Cambess.) D.Legrand = ***Blepharocalyx salicifolius*** (Kunth) O.Berg
Blepharocalyx tweediei (Hook. & Arn.) O.Berg = ***Blepharocalyx salicifolius*** (Kunth) O.Berg
Blepharocalyx tweediei f. *catharinae* (D.Legrand) Mattos = ***Blepharocalyx salicifolius*** (Kunth) O.Berg
Blepharocalyx tweediei var. *longipes* (O.Berg) Mattos = ***Blepharocalyx salicifolius*** (Kunth) O.Berg
Blepharocalyx tweediei var. *septentrionalis* Mattos = ***Blepharocalyx salicifolius*** (Kunth) O.Berg
Blepharocalyx umbilicatus (Cambess.) Burret = ***Blepharocalyx salicifolius*** (Kunth) O.Berg
Blepharocalyx umbilicatus var. *mugiensis* (Cambess.) Mattos = ***Blepharocalyx salicifolius*** (Kunth) O.Berg
Blepharocalyx umbilicatus var. *paranaensis* Mattos = ***Blepharocalyx salicifolius*** (Kunth) O.Berg
Blepharocalyx villosus O.Berg = ***Blepharocalyx salicifolius*** (Kunth) O.Berg
Blepharocalyx villosus var. *triflorus* O.Berg = ***Blepharocalyx salicifolius*** (Kunth) O.Berg
Blepharocalyx villosus var. *uniflorus* O.Berg = ***Blepharocalyx salicifolius*** (Kunth) O.Berg
Blepharocalyx widgrenii O.Berg = ***Blepharocalyx salicifolius*** (Kunth) O.Berg

Unplaced Names:
Blepharocalyx spirioides Stapf, Bot. Mag. 133: t. 8123 (1907). = [84]

Bostrychode

Bostrychode (Miq.) O.Berg = ***Syzygium*** Gaertn.

Britoa

Britoa O.Berg = ***Campomanesia*** Ruiz & Pav.
Britoa acida (DC.) O.Berg = ***Psidium acutangulum*** Mart. ex DC.
Britoa dichotoma O.Berg = ***Campomanesia dichotoma*** (O.Berg) Mattos
Britoa eriantha (Cambess.) O.Berg = ***Campomanesia pubescens*** (Mart. ex DC.) O.Berg
Britoa glazioviana Kiaersk. = ***Campomanesia schlechtendaliana*** (O.Berg) Nied. var. ***schlechtendaliana***
Britoa guazumifolia (Cambess.) D.Legrand = ***Campomanesia guazumifolia*** (Cambess.) O.Berg
Britoa hassleriana Barb.Rodr. ex Chodat & Hassl. = ***Campomanesia guazumifolia*** (Cambess.) O.Berg
Britoa lundiana Kiaersk. = ***Campomanesia lundiana*** (Kiaersk.) Mattos
Britoa psidioides O.Berg = ***Campomanesia dichotoma*** (O.Berg) Mattos
Britoa rugosa O.Berg = ***Campomanesia schlechtendaliana*** var. ***rugosa*** (O.Berg) Landrum
Britoa sellowiana O.Berg = ***Campomanesia guazumifolia*** (Cambess.) O.Berg
Britoa sessiliflora O.Berg = ***Campomanesia sessiliflora*** (O.Berg) Mattos
Britoa triflora O.Berg = ***Campomanesia dichotoma*** (O.Berg) Mattos
Britoa triflora var. *ilhensis* O.Berg = ***Campomanesia dichotoma*** (O.Berg) Mattos

Bucheria

Bucheria Heynh. = ***Thryptomene*** Endl.
Bucheria saxicola (A.Cunn. ex Hook.) Heynh. = ***Thryptomene saxicola*** (A.Cunn. ex Hook.) Schauer

Burchardia

Burchardia Neck. = ***Campomanesia*** Ruiz & Pav.
Burchardia aromatica (Aubl.) Raf. = ***Campomanesia aromatica*** (Aubl.) Griseb.
Burchardia grandiflora (Aubl.) Raf. = ***Campomanesia grandiflora*** (Aubl.) Sagot

Caju-puti

Caju-puti Adans. = ***Melaleuca*** L.

Cajuputi

Cajuputi cuticularis (Labill.) Skeels = ***Melaleuca cuticularis*** Labill.
Cajuputi ericifolia (Sm.) A.Lyons = ***Melaleuca ericifolia*** Sm.
Cajuputi hypericifolia (Sm.) Skeels = ***Melaleuca hypericifolia*** Sm.
Cajuputi leucadendron (L.) A.Lyons = ***Melaleuca leucadendra*** (L.) L.

Cajuputi pubescens (Schauer) Skeels = ***Melaleuca preissiana*** Schauer
Cajuputi pulchella (R.Br.) Skeels = ***Melaleuca pulchella*** R.Br.
Cajuputi viridiflora (Sol. ex Gaertn.) A.Lyons = ***Melaleuca viridiflora*** Sol. ex Gaertn.
Cajuputi wilsonii (F.Muell.) Skeels = ***Melaleuca wilsonii*** F.Muell.

Callistemon

Callistemon R.Br., Voy. Terra Austral. 2: 547 (1814).
Australia. (25) ken tan uga (27) cpp 50 NSW NTA QLD SOA TAS VIC WAU (51) nzn 79 MXG. [Myrtaceae]
36 Species

Callistemon acuminatus Cheel in J.H.Maiden, Ill. N.S.W. Pl.: 63 (1911). *Melaleuca flammea* Craven, Novon 16: 471 (2006).
New South Wales (SW. Plains). 50 NSW. Nanophan.

Callistemon brachyandrus Lindl., J. Hort. Soc. London 4: 112 (1849). *Melaleuca brachyandra* (Lindl.) Craven, Novon 16: 471 (2006).
SE. Australia. 50 NSW SOA VIC. Nanophan.
Callistemon arborescens F.Muell., Linnaea 25: 388 (1853).
Callistemon acerosus Miq., Ned. Kruidk. Arch. 4: 141 (1856), nom. illeg.

Callistemon chisholmii Cheel, Proc. Linn. Soc. New South Wales 50: 260 (1925). *Melaleuca chisholmii* (Cheel) Craven, Novon 16: 471 (2006).
Queensland (Mitchell). 50 QLD. Nanophan.

Callistemon citrinus (Curtis) Skeels, Bull. Bur. Pl. Industr. U.S.D.A. 282: 49 (1913).
E. & SE. Australia. (25) ken tan 50 NSW QLD VIC (79) mxg. Nanophan. or phan.
**Metrosideros citrina* Curtis, Bot. Mag. 8: t. 260 (1794). *Melaleuca citrina* (Curtis) Dum.Cours., Bot. Cult. 3: 282 (1802).
Callistemon citrinus var. *splendens* Stapf, Bot. Mag. 150: t. 9050 (1925).
Callistemon laevis Stapf, Bot. Mag. 150: t. 9050 (1925), pro syn.

Callistemon coccineus F.Muell., Fragm. 1: 13 (1858).
Queensland (Darling Downs). 50 QLD. Nanophan.
Callistemon rugulosus Miq., Ned. Kruidk. Arch. 4: 141 (1856), nom. illeg.
Callistemon coccineus var. *laevifolius* F.Muell., Fragm. 1: 14 (1858).

Callistemon comboynensis Cheel, Proc. Linn. Soc. New South Wales 68: 184 (1943). *Melaleuca comboynensis* (Cheel) Craven, Novon 16: 471 (2006).
SE. Queensland to New South Wales. 50 NSW QLD. Nanophan.

Callistemon flavovirens (Cheel) Cheel, Proc. Linn. Soc. New South Wales 50: 263 (1925).
SE. Queensland to New South Wales. 50 NSW QLD. Nanophan.
**Callistemon rugulosus* var. *flavovirens* Cheel, Ill. N.S.W. Pl. 3: iv (1911). *Melaleuca flavovirens* (Cheel) Craven, Novon 16: 471 (2006).

Callistemon formosus S.T.Blake, Proc. Roy. Soc. Queensland 69: 83 (1958). *Melaleuca formosa* (S.T.Blake) Craven, Novon 16: 472 (2006).
Queensland. 50 QLD. Nanophan.

Callistemon forresterae Molyneux, Muelleria 8: 61 (1993).
Victoria. 50 VIC.

Callistemon genofluvialis Molyneux, Muelleria 10: 57 (1997).
Victoria. 50 VIC. Nanophan.

Callistemon kenmorrisonii Molyneux, Muelleria 8: 319 (1995).
Victoria. 50 VIC. Nanophan.

Callistemon lanceolatus (Sm.) Sweet, Hort. Brit.: 155 (1826).
E. Queensland. 50 QLD. Nanophan.
**Metrosideros lanceolata* Sm., Trans. Linn. Soc. London 3: 272 (1797). *Callistemon lanceolatus* var. *sparsus* Regel, Index Seminum (LE) 1856: 39 (1856), nom. inval.
Metrosideros marginata Cav., Icon. 4: 18 (1797). *Callistemon marginatus* (Cav.) Sweet, Hort. Brit.: 155 (1826).
Metrosideros angustifolia Dum.Cours., Bot. Cult., ed. 2, 5: 378 (1811), nom. illeg.
Metrosideros falcata Dum.Cours., Bot. Cult., ed. 2, 5: 378 (1811).
Metrosideros latifolia Dum.Cours., Bot. Cult., ed. 2, 5: 379 (1811).
Metrosideros longifolia Dum.Cours., Bot. Cult., ed. 2, 5: 378 (1811). *Callistemon longifolius* (Dum. Cours.) auct., Index Seminum (KRA) 1867: 17 (1868).
Callistemon semperflorens Lodd., Bot. Cab. 6: t. 523 (1821).
Metrosideros semperflorens Lodd., Bot. Cab. 6: t. 523 (1821), nom. nud. *Callistemon lanceolatus* var. *semperflorens* (Lodd.) Heynh., Nom. Bot. Hort. 2: 107 (1841). *Callistemon lanceolatus* f. *semperflorens* (Lodd.) Siebert & Voss, Vilm. Blumengärtn. ed. 3, 1: 312 (1896).
Metrosideros myrtifolia Hoffmanns., Verz. Pfl.-Kult.: 80 (1824).
Callistemon pendulus Regel, Gartenflora 8: 289 (1859). *Callistemon pendulus* var. *genuinus* Regel, Index Seminum (LE) 1866: 105 (1867), nom. inval.
Callistemon flavescens Regel, Gartenflora 10: 51 (1861). *Callistemon pendulus* var. *rigidus* Regel, Index Seminum (LE) 1866: 106 (1867).
Callistemon lanceolatus var. *pendulus* Regel, Index Seminum (LE) 1865: 53 (1866).
Callistemon pendulus var. *confertus* Regel, Index Seminum (LE) 1866: 106 (1867).
Callistemon lilacinus Cheel, Proc. Linn. Soc. New South Wales 50: 263 (1925).
Callistemon lilacinus f. *albus* Cheel, Proc. Linn. Soc. New South Wales 50: 262 (1925).
Callistemon lilacinus f. *carminus* Cheel, Proc. Linn. Soc. New South Wales 50: 263 (1925).

Callistemon linearifolius (Link) DC., Prodr. 3: 223 (1828).
New South Wales (C. Coast). 50 NSW. Nanophan.
**Metrosideros linearifolia* Link, Enum. Hort. Berol. Alt. 2: 26 (1822). *Melaleuca linearifolia* (Link) Craven, Novon 16: 472 (2006).

Callistemon linearis (Schrad. & J.C.Wendl.) Colv. ex Sweet, Hort. Brit.: 155 (1826).
SE. Queensland to New South Wales. 50 NSW QLD. Nanophan.
**Melaleuca linearis* Schrad. & J.C.Wendl., Sert. Hannov.: 19 (1796). *Metrosideros linearis* (Schrad.

& J.C.Wendl.) Sm., Trans. Linn. Soc. London 3: 271 (1797).
Metrosideros calyculatus Sieber ex DC., Prodr. 3: 223 (1828), nom. inval.

Callistemon macropunctatus (Dum.Cours.) Court, Victorian Naturalist 83: 175 (1957).
SE. Australia. 50 SOA VIC. Nanophan.
**Metrosideros macropunctata* Dum.Cours., Bot. Cult., ed. 2, 7: 277 (1814).

var. ***laevifolius*** (F.Muell. ex Miq.) H.Eichler, Taxon 12: 295 (1963).
South Australia. 50 SOA. Nanophan.
**Callistemon rugulosus* var. *laevifolius* F.Muell. ex Miq., Ned. Kruidk. Arch. 4: 141 (1856). *Callistemon laevifolius* (F.Muell. ex Miq.) Cheel, Proc. Linn. Soc. New South Wales 50: 262 (1925).

var. ***macropunctatus***.
Victoria. 50 VIC. Nanophan.
Metrosideros rugulosa Sieber ex Link, Enum. Hort. Berol. Alt.: 27 (1822). *Callistemon rugulosus* (Sieber ex Link) DC., Prodr. 3: 223 (1828). *Melaleuca rugulosa* (Schltdl. ex Link) Craven, Novon 16: 473 (2006).
Metrosideros scabra Colla, Hortus Ripul.: 91 (1824).
Callistemon scaber Lodd., Bot. Cab. 13: t. 1288 (1828). *Callistemon rugulosus* var. *scaber* (Lodd.) Heynh., Nom. Bot. Hort. 1: 148 (1840). *Callistemon rugulosus* var. *scaber* (Lodd.) Heynh., Nom. Bot. Hort. 1: 148 (1840).

Callistemon montanus C.T.White ex S.T.Blake, Proc. Roy. Soc. Queensland 69: 84 (1958). *Melaleuca montana* (C.T.White ex S.T.Blake) Craven, Novon 16: 472 (2006).
SE. Queensland to NE. New South Wales. 50 NSW QLD. Nanophan.

Callistemon nyallingensis Molyneux, Muelleria 21: 101 (2005).
Victoria. 50 VIC. Nanophan.

Callistemon pachyphyllus Cheel in J.H.Maiden, Ill. N.S.W. Pl.: 61 (1911). *Melaleuca pachyphylla* (Cheel) Craven, Novon 16: 472 (2006).
E. Queensland to NE. New South Wales. 50 NSW QLD. Nanophan.

var. ***pachyphyllus***.
NE. New South Wales. 50 NSW. Nanophan.
Callistemon pachyphyllus var. *angustifolius* Cheel, Proc. Linn. Soc. New South Wales 50: 261 (1925).

var. ***rubrolilacinus*** Cheel, Proc. Linn. Soc. New South Wales 50: 261 (1925).
Queensland (Moreton). 50 QLD. Nanophan.

var. ***viridis*** Cheel, Proc. Linn. Soc. New South Wales 50: 261 (1925).
Queensland (Wide Bay). 50 QLD. Nanophan.

Callistemon pallidus (Bonpl.) DC., Prodr. 3: 223 (1828).
E. & SE. Australia. 50 NSW QLD TAS VIC. Nanophan.
**Metrosideros pallida* Bonpl., Descr. Pl. Malmaison: 101 (1816). *Melaleuca pallida* (Bonpl.) Craven, Novon 16: 472 (2006).

Callistemon pauciflorus R.D.Spencer & Lumley, Muelleria 6: 295 (1986). *Melaleuca faucicola* Craven, Novon 16: 471 (2006).
S. Northern Territory. 50 NTA. Nanophan.

Callistemon pearsonii R.D.Spencer & Lumley, Muelleria 6: 293 (1986). *Melaleuca pearsonii* (R.D. Spencer & Lumley) Craven, Novon 16: 473 (2006).
Queensland (Leichhardt). 50 QLD. Nanophan.

Callistemon phoeniceus Lindl., Sketch Veg. Swan R.: 10 (1839). *Melaleuca phoenicea* (Lindl.) Craven, Novon 16: 473 (2006).
Western Australia. 50 WAU. Nanophan.

Callistemon pinifolius (J.C.Wendl.) Sweet, Hort. Brit.: 155 (1826).
New South Wales. 50 NSW. Nanophan.
**Metrosideros pinifolia* J.C.Wendl., Coll. Pl. 1: 53 (1806). *Melaleuca linearis* var. *pinifolia* (J.C. Wendl.) Craven, Novon 16: 472 (2006).
Callistemon acerosus Tausch, Flora 19: 411 (1836). *Callistemon brachyandrus* var. *acerosus* (Tausch) Cheel in J.H.Maiden, Forest Fl. N.S.W. 7: 62 (1917).

Callistemon pityoides F.Muell., Chem. & Druggist, Austr. Suppl. 5: 94 (1883). *Callistemon sieberi* var. *pityoides* (F.Muell.) Cheel in J.H.Maiden, Forest Fl. N.S.W. 7: 59 (1917). *Melaleuca pityoides* (F.Muell.) Craven, Novon 16: 473 (2006).
SE. Queensland to NE. Victoria. 50 QLD VIC. Nanophan.

Callistemon polandii F.M.Bailey, Queensl. Fl.: 2003 (1902). *Melaleuca polandii* (F.M.Bailey) Craven, Novon 16: 473 (2006).
Queensland. 50 QLD. Nanophan.

Callistemon pungens Lumley & R.D.Spencer, Muelleria 7: 253 (1990). *Melaleuca williamsii* Craven, Novon 16: 474 (2006).
Queensland to New South Wales. 50 NSW QLD. Nanophan.

Callistemon recurvus R.D.Spencer & Lumley, Muelleria 7: 255 (1990). *Melaleuca recurva* (R.D.Spencer & Lumley) Craven, Novon 16: 473 (2006).
E. Australia. 50 NSW QLD. Nanophan.

Callistemon rigidus R.Br., Bot. Reg. 5: t. 393 (1819). *Metrosideros rigida* (R.Br.) Dum.Cours., Bot. Cult., ed. 2, 5: 381 (1811).
SE. Queensland to NE. New South Wales. (25) tan (27) cpp 50 NSW QLD (51) nzn. Nanophan.
Metrosideros linearis Muhl. ex Willd., Enum. Pl.: 513 (1809), nom. illeg.
Metrosideros linifolia Dum.Cours., Bot. Cult., ed. 2, 7: 277 (1814), pro syn.
Metrosideros aspera Hoffmanns., Verz. Pfl.-Kult.: 80 (1824).
Metrosideros rigidifolia Hoffmanns., Verz. Pfl.-Kult.: 80 (1824).
Metrosideros glandulosa Desf., Tabl. École Bot., ed. 3: 407 (1829).

Callistemon salignus (Sm.) Colv. ex Sweet, Hort. Brit.: 155 (1826).
Queensland to New South Wales. (25) ken uga 50 NSW QLD. Nanophan.
**Metrosideros saligna* Sm., Trans. Linn. Soc. London 3: 272 (1797). *Callistemon salignus* var. *typicus* F.Muell., Fragm. 4: 55 (1864), nom. inval. *Melaleuca salicina* Craven, Novon 16: 473 (2006).

var. ***roseus*** Guilf., Austral. Pl. Parks: 93 (1911).
Queensland (Port Curtis). 50 QLD. Nanophan.

var. ***salignus***
Queensland to New South Wales. 50 NSW QLD. Nanophan.
Metrosideros lophantha Vent., Choix Pl.: t. 69 (1808). *Callistemon lophanthus* (Vent.) Sweet, Fl. Australas.: t. 29 (1828). *Callistemon lanceolatus* var. *lophanthus* (Vent.) Heynh., Nom. Bot. Hort. 2: 107 (1841). *Callistemon salignus* var. *hebestachyus* Benth., Fl. Austral. 3: 121 (1867). *Callistemon salignus* f. *hebestachys* (Benth.) Siebert & Voss, Vilm. Blumengärtn. ed. 3, 1: 312 (1896).
Metrosideros australis R.Br. ex Bonpl., Descr. Pl. Malmaison: 102 (1816).
Callistemon leptostachyus Sweet, Fl. Australas.: t. 29 (1827).
Melaleuca paludosa Schltdl., Linnaea 20: 653 (1847), nom. illeg.
Callistemon salignus var. *angustifolius* Miq., Ned. Kruidk. Arch. 4: 141 (1856).
Melaleuca pithyoides F.Muell. ex Benth., Fl. Austral. 3: 121 (1867).
Callistemon salignus f. *angustifolius* Siebert & Voss, Vilm. Blumengärtn. ed. 3, 1: 312 (1896), nom. illeg.

Callistemon shiressii Blakely, Austral. Naturalist 10: 257 (1941). *Melaleuca shiressii* (Blakely) Craven, Novon 16: 473 (2006).
New South Wales. 50 NSW. Nanophan.

Callistemon sieberi DC., Prodr. 3: 223 (1828). *Callistemon salignus* var. *sieberi* (DC.) F.Muell., Fragm. 4: 55 (1864). *Callistemon salignus* f. *sieberi* (DC.) Siebert & Voss, Vilm. Blumengärtn. ed. 3, 1: 312 (1896). *Melaleuca paludicola* Craven, Novon 16: 472 (2006).
E. & SE. Australia. 50 NSW QLD SOA TAS VIC. Nanophan.
Callistemon paludosus F.Muell., Fragm. 1: 14 (1858). *Callistemon salignus* var. *australis* Benth., Fl. Austral. 3: 121 (1867).

Callistemon speciosus (Sims) Sweet, Hort. Brit.: 155 (1826).
SW. Western Australia. (27) cpp 50 WAU. Nanophan.
Metrosideros crassifolia Dum.Cours., Bot. Cult., ed. 2, 5: 379 (1811), non C. crassifolius Ettingsh. (1861), fossil.
Metrosideros glauca Dum.Cours., Bot. Cult., ed. 2, 5: 380 (1811).
Melaleuca paludosa R.Br. in W.T.Aiton, Hortus Kew. 4: 410 (1812). *Callistemon salignus* var. *paludosus* (R.Br.) F.Muell., Fragm. 4: 55 (1864).
Metrosideros glauca Bonpl., Descr. Pl. Malmaison: 86 (1815), nom. illeg. *Callistemon speciosus* var. *glaucus* DC., Prodr. 3: 224 (1828). Callistemon glaucus (DC.) Sweet, Hort. Brit. , ed. 2: 209 (1830).
**Metrosideros speciosa* Sims, Bot. Mag. 42: t. 1761 (1815).
Callistemon heleogiton Domin, Biblioth. Bot. 89: 454 (1928).

Callistemon subulatus Cheel, Proc. Linn. Soc. New South Wales 1: 259 (1925). *Melaleuca subulata* (Cheel) Craven, Novon 16: 473 (2006).
New South Wales, Victoria. 50 NSW VIC. Nanophan.

Callistemon teretifolius F.Muell., Linnaea 25: 387 (1853). *Melaleuca orophila* Craven, Novon 16: 472 (2006).
South Australia. 50 SOA. Nanophan.

Callistemon viminalis (Sol. ex Gaertn.) G.Don ex Loudon, Hort. Brit. 1: 197 (1830).
Queensland to New South Wales. (25) ken tan 50 NSW QLD. Nanophan.
**Metrosideros viminalis* Sol. ex Gaertn., Fruct. Sem. Pl. 1: 171 (1788). *Melaleuca viminalis* (Sol. ex Gaertn.) Byrnes, Austrobaileya 2: 75 (1984).
Melaleuca viminalis var. *minor* Byrnes, Austrobaileya 2: 75 (1984).
Melaleuca viminalis var. *minor* Byrnes, Austrobaileya 2: 75 (1984).

Callistemon viridiflorus (Sieber ex Sims) Sweet, Hort. Brit.: 155 (1826).
Tasmania. 50 TAS. Nanophan.
**Metrosideros viridiflora* Sieber ex Sims, Bot. Mag. 52: t. 2602 (1825). *Callistemon salignus* var. *viridiflorus* (Sieber ex Sims) F.Muell., Fragm. 4: 55 (1864). *Callistemon salignus* f. *viridiflorus* (Sieber ex Sims) Siebert & Voss, Vilm. Blumengärtn. ed. 3, 1: 312 (1896). *Melaleuca virens* Craven, Novon 16: 473 (2006).
Callistemon viridiflorus var. *violaceus* Kunze, Linnaea 20: 58 (1847).

Synonyms:
Callistemon acerosus Miq. = ***Callistemon brachyandrus*** Lindl.
Callistemon acerosus Tausch = ***Callistemon pinifolius*** (J.C.Wendl.) Sweet
Callistemon amoenus Lem. = ?
Callistemon arborescens F.Muell. = ***Callistemon brachyandrus*** Lindl.
Callistemon brachyandrus var. *acerosus* (Tausch) Cheel = ***Callistemon pinifolius*** (J.C.Wendl.) Sweet
Callistemon brevisepalus J.W.Dawson = ***Melaleuca brevisepala*** (J.W.Dawson) Craven & J.W.Dawson
Callistemon buseanus Guillaumin = ***Melaleuca buseana*** (Guillaumin) Craven & J.W.Dawson
Callistemon buseanus var. *longifolius* Guillaumin = ***Melaleuca buseana*** (Guillaumin) Craven & J.W. Dawson
Callistemon capitatus (Sm.) Rchb. = ***Kunzea capitata*** (Sm.) Heynh.
Callistemon citrinus var. *splendens* Stapf = ***Callistemon citrinus*** (Curtis) Skeels
Callistemon coccineus var. *laevifolius* F.Muell. = ***Callistemon coccineus*** F.Muell.
Callistemon cunninghamii K.Koch = ?
Callistemon flavescens Regel = ***Callistemon lanceolatus*** (Sm.) Sweet
Callistemon gnidioides Guillaumin = ***Melaleuca sphaerodendra*** Craven & J.W.Dawson
Callistemon gnidioides var. *microphyllus* Virot = ***Melaleuca sphaerodendra*** var. ***microphylla*** (Virot) Craven & J.W.Dawson
Callistemon hainesii F.Muell. = ***Kunzea baxteri*** (Klotzsch) Schauer
Callistemon heleogiton Domin = ***Callistemon speciosus*** (Sims) Sweet
Callistemon hortensis Cheel = ?
Callistemon hybridus DC. = ?
Callistemon juniperinus (Rchb. ex Spreng.) Heynh. = ***Melaleuca nodosa*** (Sol. ex Gaertn.) Sm.
Callistemon laevifolius (F.Muell. ex Miq.) Cheel = ***Callistemon macropunctatus*** var. ***laevifolius*** (F.Muell. ex Miq.) H.Eichler
Callistemon laevis Stapf = ***Callistemon citrinus*** (Curtis) Skeels

Callistemon lanceolatus var. *lophanthus* (Vent.) Heynh. = ***Callistemon salignus*** (Sm.) Colv. ex Sweet var. ***salignus***
Callistemon lanceolatus var. *pendulus* Regel = ***Callistemon lanceolatus*** (Sm.) Sweet
Callistemon lanceolatus var. *semperflorens* (Lodd.) Heynh. = ***Callistemon lanceolatus*** (Sm.) Sweet
Callistemon lanceolatus f. *semperflorens* (Lodd.) Siebert & Voss = ***Callistemon lanceolatus*** (Sm.) Sweet
Callistemon lanceolatus var. *sparsus* Regel = ***Callistemon lanceolatus*** (Sm.) Sweet
Callistemon leptostachyus Sweet = ***Callistemon salignus*** (Sm.) Colv. ex Sweet var. ***salignus***
Callistemon lilacinus Cheel = ***Callistemon lanceolatus*** (Sm.) Sweet
Callistemon lilacinus f. *albus* Cheel = ***Callistemon lanceolatus*** (Sm.) Sweet
Callistemon lilacinus f. *carminus* Cheel = ***Callistemon lanceolatus*** (Sm.) Sweet
Callistemon longifolius (Dum.Cours.) auct. = ***Callistemon lanceolatus*** (Sm.) Sweet
Callistemon lophanthus (Vent.) Sweet = ***Callistemon salignus*** (Sm.) Colv. ex Sweet var. ***salignus***
Callistemon marginatus (Cav.) Sweet = ***Callistemon lanceolatus*** (Sm.) Sweet
Callistemon microstachyus Lindl. = ***Kunzea baxteri*** (Klotzsch) Schauer
Callistemon nervosus Lindl. = ***Melaleuca nervosa*** (Lindl.) Cheel
Callistemon pachyphyllus var. *angustifolius* Cheel = ***Callistemon pachyphyllus*** Cheel var. ***pachyphyllus***
Callistemon paludosus F.Muell. = ***Callistemon sieberi*** DC.
Callistemon pancheri Brongn. & Gris = ***Melaleuca pancheri*** (Brongn. & Gris) Craven & J.W.Dawson
Callistemon pendulus Regel = ***Callistemon lanceolatus*** (Sm.) Sweet
Callistemon pendulus var. *confertus* Regel = ***Callistemon lanceolatus*** (Sm.) Sweet
Callistemon pendulus var. *genuinus* Regel = ***Callistemon lanceolatus*** (Sm.) Sweet
Callistemon pendulus var. *rigidus* Regel = ***Callistemon lanceolatus*** (Sm.) Sweet
Callistemon roseus (Guilf.) Cheel = ***Callistemon salignus*** (Sm.) Colv. ex Sweet var. ***salignus***
Callistemon rugulosus Miq. = ***Callistemon coccineus*** F.Muell.
Callistemon rugulosus (Sieber ex Link) DC. = ***Callistemon macropunctatus*** (Dum.Cours.) Court var. ***macropunctatus***
Callistemon rugulosus var. *flavovirens* Cheel = ***Callistemon flavovirens*** (Cheel) Cheel
Callistemon rugulosus var. *laevifolius* F.Muell. ex Miq. = ***Callistemon macropunctatus*** var. ***laevifolius*** (F.Muell. ex Miq.) H.Eichler
Callistemon rugulosus var. *scaber* (Lodd.) Heynh. = ***Callistemon macropunctatus*** (Dum.Cours.) Court var. ***macropunctatus***
Callistemon rugulosus var. *scaber* (Lodd.) Heynh. = ***Callistemon macropunctatus*** (Dum.Cours.) Court var. ***macropunctatus***
Callistemon salignus var. *angustifolius* Miq. = ***Callistemon salignus*** (Sm.) Colv. ex Sweet var. ***salignus***
Callistemon salignus f. *angustifolius* Siebert & Voss = ***Callistemon salignus*** (Sm.) Colv. ex Sweet var. ***salignus***
Callistemon salignus var. *australis* Benth. = ***Callistemon sieberi*** DC.
Callistemon salignus f. *hebestachys* (Benth.) Siebert & Voss = ***Callistemon salignus*** (Sm.) Colv. ex Sweet var. ***salignus***
Callistemon salignus var. *hebestachyus* Benth. = ***Callistemon salignus*** (Sm.) Colv. ex Sweet var. ***salignus***
Callistemon salignus var. *paludosus* (R.Br.) F.Muell. = ***Callistemon speciosus*** (Sims) Sweet
Callistemon salignus var. *sieberi* (DC.) F.Muell. = ***Callistemon sieberi*** DC.
Callistemon salignus f. *sieberi* (DC.) Siebert & Voss = ***Callistemon sieberi*** DC.
Callistemon salignus var. *typicus* F.Muell. = ***Callistemon salignus*** (Sm.) Colv. ex Sweet
Callistemon salignus var. *viridiflorus* (Sieber ex Sims) F.Muell. = ***Callistemon viridiflorus*** (Sieber ex Sims) Sweet
Callistemon salignus f. *viridiflorus* (Sieber ex Sims) Siebert & Voss = ***Callistemon viridiflorus*** (Sieber ex Sims) Sweet
Callistemon scaber Lodd. = ***Callistemon macropunctatus*** (Dum.Cours.) Court var. ***macropunctatus***
Callistemon semperflorens Lodd. = ***Callistemon lanceolatus*** (Sm.) Sweet
Callistemon sieberi var. *pityoides* (F.Muell.) Cheel = ***Callistemon pityoides*** F.Muell.
Callistemon speciosus (Sims) Sweet = ***Callistemon speciosus*** (Sims) Sweet
Callistemon speciosus var. *glaucus* (Bonpl.) DC. = ***Callistemon speciosus*** (Sims) Sweet
Callistemon suberosus Pancher ex Brongn. & Gris = ***Melaleuca dawsonii*** Craven
Callistemon suberosus var. *microphyllus* Guillaumin = ***Melaleuca pancheri*** (Brongn. & Gris) Craven & J.W.Dawson
Callistemon viridiflorus var. *violaceus* Kunze = ***Callistemon viridiflorus*** (Sieber ex Sims) Sweet

***Unplaced Names*:**
Callistemon amoenus Lem., Ill. Hort. 7: t. 247 (1860). = ?
Callistemon cunninghamii K.Koch, Wochenschr. Gärtnerei Pflanzenk. 6: 294 (1863). = ?
Callistemon hortensis Cheel, Proc. Linn. Soc. New South Wales 68: 184 (1943). = ?
Callistemon hybridus DC., Prodr. 3: 224 (1828). = ?

Callobuxus

Callobuxus Planch. ex Brongn. & Gris = ***Tristania*** R.Br.

Calomyrtus

Calomyrtus Blume = ***Hexachlamys*** O.Berg
Calomyrtus excelsa (Cambess.) Blume = ***Hexachlamys edulis*** (O.Berg) Kausel & D.Legrand
Calomyrtus ovalifolia (Cambess.) Blume = ?

Calophylloides

Calophylloides Smeathman ex DC. = ***Eugenia*** P.Micheli ex L.
Calophylloides lucida Smeathman ex DC. = ***Eugenia calophylloides*** DC.

Calothamnus

Calothamnus Labill., Nov. Holl. Pl. 2: 25 (1806).
SW. Australia. 50 WAU. [Myrtaceae]
39 Species
Billottia Colla, Hortus Ripul.: 20 (1824).
Baudinia Lesch. ex DC., Prodr. 3: 211 (1828).

Calothamnus accedens Hawkeswood, Nuytsia 5: 305 (1984 publ. 1985).
Western Australia (Avon). 50 WAU.

Calothamnus affinis Turcz., Bull. Cl. Phys.-Math. Acad. Imp. Sci. Saint-Pétersbourg 10: 346 (1852).
SW. Australia. 50 WAU.
Calothamnus microcarpus var. *teres* Benth., Fl. Austral. 3: 177 (1867).

Calothamnus aridus Hawkeswood, Nuytsia 5: 145 (1984).
Western Australia (Austin). 50 WAU. Nanophan.

Calothamnus asper Turcz., Bull. Soc. Imp. Naturalistes Moscou 22(1): 25 (1849).
Western Australia (Avon). 50 WAU.

Calothamnus blepharospermus F.Muell., Fragm. 3: 111 (1862).
Western Australia (Irwin, Drummond). 50 WAU.
Calothamnus blepharospermus var. *glaber* Benth., Fl. Austral. 3: 176 (1867).

Calothamnus borealis Hawkeswood, Nuytsia 5: 150 (1984).
Western Australia (Carnarvon). 50 WAU. Cham. or nanophan.

Calothamnus brevifolius Hawkeswood, Nuytsia 5: 141 (1984).
Western Australia (Avon). 50 WAU. Cham.

Calothamnus chrysanthereus F.Muell., Fragm. 3: 112 (1862).
SW. Australia. 50 WAU.

Calothamnus crassus (Benth.) Hawkeswood, Nuytsia 6: 86 (1987).
Western Australia (Eyre). 50 WAU.
**Calothamnus lateralis* f. *crassus* Benth., Fl. Austral. 3: 177 (1867).

Calothamnus formosus Hawkeswood, Nuytsia 5: 135 (1984).
Western Australia (Irwin). 50 WAU. Nanophan.

subsp. ***formosus***.
Western Australia (Irwin). 50 WAU. Nanophan.

subsp. ***rigidus*** Hawkeswood, Nuytsia 5: 137 (1984).
Western Australia (Irwin). 50 WAU. Nanophan.

Calothamnus gibbosus Benth., Fl. Austral. 3: 175 (1867).
SW. Australia. 50 WAU.

Calothamnus gilesii F.Muell., Fragm. 10: 31 (1876).
SW. Australia. 50 WAU.

Calothamnus gracilis R.Br. ex W.T.Aiton, Hortus Kew. 4: 418 (1812).
SW. Australia. 50 WAU.

Calothamnus graniticus Hawkeswood, Nuytsia 5: 127 (1984).
Western Australia. 50 WAU. Nanophan.

subsp. ***graniticus***.
Western Australia (Cape Naturaliste Pen. to Yallingup). 50 WAU. Nanophan.

subsp. ***leptophyllus*** (Benth.) Hawkeswood, Nuytsia 5: 129 (1984).
Western Australia (Dwellingup Reg., Wellington Dam). 50 WAU. Nanophan.
**Calothamnus torulosus* var. *leptophyllus* Benth., Fl. Austral. 3: 175 (1867).

Calothamnus hirsutus Hawkeswood, Nuytsia 5: 139 (1984).
Western Australia (Irwin, Drummond, Dale). 50 WAU. Nanophan.
Calothamnus villosus var. *ericifolius* Benth., Fl. Austral. 3: 179 (1867).

Calothamnus homalophyllus F.Muell., Fragm. 3: 111 (1862).
Western Australia (Irwin). 50 WAU. Nanophan.

Calothamnus huegelii Schauer, Nov. Actorum Acad. Caes. Leop.-Carol. Nat. Cur. 21: 30 (1845).
Western Australia (Eyre, Roe). 50 WAU. Nanophan.
Calothamnus nodosus Turcz., Bull. Soc. Imp. Naturalistes Moscou 20(1): 168 (1847).

Calothamnus kalbarriensis Hawkeswood, Nuytsia 5: 147 (1984).
Western Australia (Irwin). 50 WAU. Nanophan.

Calothamnus lateralis Lindl., Sketch Veg. Swan R.: 9 (1839).
SW. Australia. 50 WAU. Nanophan.
Calothamnus longifolius Lehm., Index Seminum (HBG) 1842: 7 (1842). *Calothamnus lateralis* f. *longifolius* (Lehm.) Benth., Fl. Austral. 3: 177 (1867).
Calothamnus lateralis f. *rigidus* Benth., Fl. Austral. 3: 177 (1867).

Calothamnus lehmannii Schauer, Nov. Actorum Acad. Caes. Leop.-Carol. Nat. Cur. 21: 27 (1845).
SW. Australia. 50 WAU. Nanophan.
Calothamnus plumosus Turcz., Bull. Soc. Imp. Naturalistes Moscou 20(1): 168 (1847).

Calothamnus longissimus F.Muell., Fragm. 3: 112 (1862).
SW. Australia. 50 WAU. Nanophan.

Calothamnus macrocarpus Hawkeswood, Nuytsia 5: 146 (1984).
Western Australia (Eyre). 50 WAU. Nanophan.

Calothamnus microcarpus F.Muell., Fragm. 3: 113 (1862).
Western Australia (Austin). 50 WAU. Nanophan.

Calothamnus oldfieldii F.Muell., Fragm. 3: 113 (1862).
Western Australia (Irwin). 50 WAU. Nanophan.

Calothamnus pachystachyus Benth., Fl. Austral. 3: 173 (1867).
Western Australia (Dale). 50 WAU. Nanophan.

Calothamnus pallidifolius (Benth.) Hawkeswood, Nuytsia 6: 100 (1987).
Western Australia (Menzies). 50 WAU. Nanophan.
**Calothamnus planifolius* var. *pallidifolius* Benth., Fl. Austral. 3: 177 (1867).

Calothamnus pinifolius F.Muell., Fragm. 3: 153 (1863).
Western Australia (Eyre). 50 WAU. Nanophan.

Calothamnus planifolius Lehm., Index Seminum (HBG) 1842: 7 (1842). *Calothamnus planifolius* f. *angustifolius* Schauer, Regelia, Beaufortia et Calothamno: 35 (1843), nom. inval.
SW. Australia. 50 WAU. Nanophan.
Calothamnus planifolius f. *latifolius* Schauer, Regelia, Beaufortia et Calothamno: 36 (1843).
Calothamnus planifolius var. *latifolius* Schauer in J.G.C.Lehmann, Pl. Preiss. 1: 155 (1845).

Calothamnus preissii Schauer, Nov. Actorum Acad. Caes. Leop.-Carol. Nat. Cur. 21: 27 (1845).

Western Australia (Eyre, Avon). 50 WAU. Nanophan.
Calothamnus laxus Kunze, Linnaea 20: 58 (1847).

Calothamnus quadrifidus R.Br. ex W.T.Aiton, Hortus Kew. 4: 418 (1812).
SW. Australia. 50 WAU. Nanophan.
Beaufortia pinifolia Cels, Cat.: 8 (1817), nom. nud.
Billottia acerosa Colla, Hortus Ripul.: 20 (1824). *Melaleuca acerosa* (Colla) Sweet ex G.Don, Gen. Hist. 2: 815 (1832). *Calothamnus quadrifidus* f. *acerosus* (Colla) Benth., Fl. Austral. 3: 180 (1867).
Calothamnus purpureus Endl. in S.L.Endlicher & al., Enum. Pl.: 48 (1837).
Calothamnus clavatus J.Mackay ex Schauer, Nov. Actorum Acad. Caes. Leop.-Carol. Nat. Cur. 21: 24 (1845). *Calothamnus quadrifidus* f. *normalis* Benth., Fl. Austral. 3: 180 (1867), nom. inval.
Calothamnus laevigatus Schauer, Nov. Actorum Acad. Caes. Leop.-Carol. Nat. Cur. 21: 26 (1845).
Calothamnus quadrifidus var. *hirsutus* Regel, Index Seminum (LE) 1856: 39 (1856).
Calothamnus quadrifidus f. *obtusus* Benth., Fl. Austral. 3: 180 (1867).
Calothamnus homalophyllus var. *angustifolius* Ewart, Proc. Roy. Soc. Victoria, n.s., 24: 63 (1911).

Calothamnus robustus Schauer, Nov. Actorum Acad. Caes. Leop.-Carol. Nat. Cur. 21: 22 (1845).
Western Australia (Eyre). 50 WAU. Nanophan.
Calothamnus knightii Schauer, Nov. Actorum Acad. Caes. Leop.-Carol. Nat. Cur. 21: 23 (1845), nom. inval.

Calothamnus rupestris Schauer, Nov. Actorum Acad. Caes. Leop.-Carol. Nat. Cur. 21: 22 (1845).
Western Australia (Avon, Dale). 50 WAU. Nanophan.

Calothamnus sanguineus Labill., Nov. Holl. Pl. 2: 25 (1806).
SW. Australia. 50 WAU. Nanophan.
Calothamnus eriocarpus Lindl., Sketch Veg. Swan R.: 9 (1839).
Calothamnus blepharantherus F.Muell., Fragm. 3: 112 (1862).

Calothamnus schaueri Lehm., Index Seminum (HBG) 1842: 7 (1842).
SW. Australia. 50 WAU. Nanophan.
Calothamnus schoenophyllus Schauer, Regelia, Beaufortia, Caloth.: 33 (1843).

Calothamnus superbus Hawkeswood & Mollemans, Nuytsia 8: 311 (1993).
SW. Western Australia. 50 WAU. Nanophan.

Calothamnus torulosus Schauer, Nov. Actorum Acad. Caes. Leop.-Carol. Nat. Cur. 21: 21 (1845).
SW. Australia. 50 WAU.

Calothamnus tuberosus Hawkeswood, Nuytsia 5: 133 (1984).
Western Australia (Avon). 50 WAU. Nanophan.

Calothamnus validus S.Moore, J. Linn. Soc., Bot. 45: 205 (1920).
Western Australia (Eyre). 50 WAU. Nanophan.

Calothamnus villosus R.Br. ex W.T.Aiton, Hortus Kew. 4: 418 (1812).
SW. Australia. 50 WAU. Nanophan.

Synonyms:
Calothamnus baxteri Klotzsch = ***Kunzea baxteri*** (Klotzsch) Schauer
Calothamnus blepharantherus F.Muell. = ***Calothamnus sanguineus*** Labill.
Calothamnus blepharospermus var. *glaber* Benth. = ***Calothamnus blepharospermus*** F.Muell.
Calothamnus clavatus J.Mackay ex Schauer = ***Calothamnus quadrifidus*** R.Br. ex W.T.Aiton
Calothamnus eriocarpus Lindl. = ***Calothamnus sanguineus*** Labill.
Calothamnus homalophyllus var. *angustifolius* Ewart = ***Calothamnus quadrifidus*** R.Br. ex W.T.Aiton
Calothamnus knightii Schauer = ***Calothamnus robustus*** Schauer
Calothamnus laevigatus Schauer = ***Calothamnus quadrifidus*** R.Br. ex W.T.Aiton
Calothamnus lateralis f. *crassus* Benth. = ***Calothamnus crassus*** (Benth.) Hawkeswood
Calothamnus lateralis f. *longifolius* (Lehm.) Benth. = ***Calothamnus lateralis*** Lindl.
Calothamnus lateralis f. *rigidus* Benth. = ***Calothamnus lateralis*** Lindl.
Calothamnus laxus Kunze = ***Calothamnus preissii*** Schauer
Calothamnus longifolius Lehm. = ***Calothamnus lateralis*** Lindl.
Calothamnus microcarpus var. *teres* Benth. = ***Calothamnus affinis*** Turcz.
Calothamnus nodosus Turcz. = ***Calothamnus huegelii*** Schauer
Calothamnus planifolius f. *angustifolius* Schauer = ***Calothamnus planifolius*** Lehm.
Calothamnus planifolius f. *latifolius* Schauer = ***Calothamnus planifolius*** Lehm.
Calothamnus planifolius var. *latifolius* Schauer = ***Calothamnus planifolius*** Lehm.
Calothamnus planifolius var. *pallidifolius* Benth. = ***Calothamnus pallidifolius*** (Benth.) Hawkeswood
Calothamnus plumosus Turcz. = ***Calothamnus lehmannii*** Schauer
Calothamnus purpureus Endl. = ***Calothamnus quadrifidus*** R.Br. ex W.T.Aiton
Calothamnus quadrifidus f. *acerosus* (Colla) Benth. = ***Calothamnus quadrifidus*** R.Br. ex W.T.Aiton
Calothamnus quadrifidus var. *hirsutus* Regel = ***Calothamnus quadrifidus*** R.Br. ex W.T.Aiton
Calothamnus quadrifidus f. *normalis* Benth. = ***Calothamnus quadrifidus*** R.Br. ex W.T.Aiton
Calothamnus quadrifidus f. *obtusus* Benth. = ***Calothamnus quadrifidus*** R.Br. ex W.T.Aiton
Calothamnus schoenophyllus Schauer = ***Calothamnus schaueri*** Lehm.
Calothamnus spathulatus Steud. = ***Kunzea baxteri*** (Klotzsch) Schauer
Calothamnus suberosus Schauer = ***Melaleuca suberosa*** (Schauer) C.A.Gardner
Calothamnus torulosus var. *leptophyllus* Benth. = ***Calothamnus graniticus*** subsp. ***leptophyllus*** (Benth.) Hawkeswood
Calothamnus villosus var. *ericifolius* Benth. = ***Calothamnus hirsutus*** Hawkeswood

Calycampe

Calycampe O.Berg = ***Myrcia*** DC. ex Guill.
Calycampe angustifolia O.Berg = ***Myrcia calycampa*** Amshoff
Calycampe latifolia O.Berg = ***Myrcia calycampae*** Amshoff

Calycolpus

Calycolpus O.Berg, Linnaea 27: 378 (1856).
Trop. America. 80 COS NIC PAN 81 CUB TRT 82 FRG GUY SUR VEN 83 CLM ECU PER 84 BZE BZL BZN. [Myrtaceae]
13 Species
Psidiopsis O.Berg, Linnaea 27: 350 (1856).

Calycolpus aequatorialis Landrum, Novon 15: 442 (2005).
Ecuador (Sucumbíos). 83 ECU. Phan.

Calycolpus alternifolius (Gleason) Landrum, Madroño 36: 10 (1989).
S. Venezuela to N. Brazil. 82 VEN. 84 BZN Nanophan.
**Myrtus alternifolia* Gleason, Brittonia 3: 173 (1939).

Calycolpus australis Landrum, Novon 8: 244 (1998).
Brazil (Minas Gerais). 84 BZL. Nanophan.

Calycolpus bolivarensis Landrum, Ann. Missouri Bot. Gard. 76: 930 (1989).
S. Venezuela. 82 VEN. Nanophan. or phan.

Calycolpus calophyllus (Kunth) O.Berg in C.F.P.von Martius & auct. suc. (eds.), Fl. Bras. 14(1): 412 (1857).
S. Trop. America. 82 VEN 83 CLM PER 84 BZN. Nanophan. or phan.
**Myrtus calophylla* Kunth in F.W.H.von Humboldt, A.J.A.Bonpland & C.S.Kunth, Nov. Gen. Sp. 6: 133 (1823).

var. ***angustifolius*** Steyerm., Ann. Missouri Bot. Gard. 71: 330 (1984).
S. Venezuela. 82 VEN. Nanophan. or phan.

var. ***calophyllus***
S. Trop. America. 82 VEN 83 CLM PER 84 BZN. Nanophan. or phan.
Myrtus citrifolia Willd. ex O.Berg, Linnaea 27: 381 (1856).
Psidium campomanesioides O.Berg in C.F.P.von Martius & auct. suc. (eds.), Fl. Bras. 14(1): 520 (1858). *Guajava campomanesiodes* (O.Berg) Kuntze, Revis. Gen. Pl. 1: 240 (1891). *Calycolpus campomanesioides* (O.Berg) Mattos, Arq. Bot. Estado São Paulo, n.s., f.m., 4: 56 (1966).

Calycolpus cochleatus McVaugh, Mem. New York Bot. Gard. 18(2): 241 (1969).
Venezuela to Guyana. 82 GUY VEN. Nanophan. or phan.

Calycolpus goetheanus (Mart. ex DC.) O.Berg in C.F.P. von Martius & auct. suc. (eds.), Fl. Bras. 14(1): 412 (1857).
Trinidad to S. Trop. America. 81 TRT 82 FRG GUY SUR VEN 84 BZN. Nanophan. or phan.
**Myrtus goetheana* Mart. ex DC., Prodr. 3: 240 (1828).
Campomanesia glabra Benth., J. Bot. (Hooker) 2: 319 (1843). *Calycolpus glaber* (Benth.) O.Berg in C.F.P.von Martius & auct. suc. (eds.), Fl. Bras. 14(1): 411 (1857).
Calycolpus ovalifolius O.Berg, Linnaea 27: 379 (1856).
Calycolpus schomburgkianus O.Berg, Linnaea 27: 380 (1856).
Calycolpus schomburgkianus var. *recurvatus* O.Berg, Linnaea 27: 380 (1856).
Calycolpus schomburgkianus var. *speciosus* O.Berg, Linnaea 27: 380 (1856).
Campomanesia goetheana Hemsl., Biol. Cent.-Amer., Bot. 1: 407 (1880).
Calycolpus angustifolius L.Riley, Bull. Misc. Inform. Kew 1926: 151 (1926).
Calycolpus cordatus L.Riley, Bull. Misc. Inform. Kew 1926: 152 (1926).
Calycolpus glaber var. *angustilobus* L.Riley, Bull. Misc. Inform. Kew 1926: 153 (1926).

Calycolpus legrandii Mattos, Loefgrenia 5: 1 (1962).
Brazil (Bahia). 84 BZE. Nanophan. or phan.

Calycolpus moritzianus (O.Berg) Burret, Repert. Spec. Nov. Regni Veg. 50: 57 (1941).
S. Trop. America. 82 VEN 83 CLM ECU. Nanophan. or phan.
Myrtus ambrosiaca Moritzi ex O.Berg, Linnaea 27: 351 (1856).
**Psidiopsis moritzianum* O.Berg, Linnaea 27: 351 (1856). *Psidium caudatum* McVaugh, Fieldiana, Bot. 29: 226 (1956).

Calycolpus revolutus (Schauer) O.Berg, Linnaea 27: 383 (1856).
N. South America. 82 FRG GUY SUR. Nanophan. or phan.
**Myrtus revoluta* Schauer, Linnaea 21: 272 (1848).
Calycolpus kegelianus O.Berg, Linnaea 27: 381 (1856).
Calycolpus kegelianus var. *gracilis* O.Berg, Linnaea 27: 381 (1856). *Calycolpus gracilis* (O.Berg) L.Riley, Bull. Misc. Inform. Kew 1926: 150 (1926).
Calycolpus kegelianus var. *robustus* O.Berg, Linnaea 27: 381 (1856).
Calycolpus chnoiophyllus L.Riley, Bull. Misc. Inform. Kew 1926: 148 (1926).
Calycolpus megalodon L.Riley, Bull. Misc. Inform. Kew 1926: 149 (1926).
Calycolpus pyrifer L.Riley, Bull. Misc. Inform. Kew 1926: 149 (1926).

Calycolpus roraimensis Steyerm., Fieldiana, Bot. 28: 1009 (1957). *Calycolpus piresianus* Mattos, Loefgrenia 116: 1 (2001), nom. superfl.
S. Trop. America. 82 GUY VEN 83 PER. Nanophan. or phan.
Calycolpus roraimensis Kausel, Lilloa 33: 98 (1971 publ. 1972), nom. illeg. *Psidium roraimense* Mattos, Loefgrenia 99: 4 (1990).

Calycolpus surinamensis McVaugh, Mem. New York Bot. Gard. 18(2): 245 (1969).
Suriname, Guyana. 82 GUY SUR. Phan.

Calycolpus warscewiczianus O.Berg, Linnaea 27: 382 (1856).
C. America. 80 COS NIC PAN. Nanophan. or phan.

Synonyms:
Calycolpus angustifolius L.Riley = ***Calycolpus goetheanus*** (Mart. ex DC.) O.Berg
Calycolpus campomanesioides (O.Berg) Mattos = ***Calycolpus calophyllus*** (Kunth) O.Berg var. ***calophyllus***
Calycolpus chnoiophyllus L.Riley = ***Calycolpus revolutus*** (Schauer) O.Berg
Calycolpus cordatus L.Riley = ***Calycolpus goetheanus*** (Mart. ex DC.) O.Berg
Calycolpus cristalense (Urb.) Bisse = ***Eugenia cristalensis*** Urb.
Calycolpus excisus (Urb.) Bisse = ***Eugenia excisa*** Urb.
Calycolpus glaber (Benth.) O.Berg = ***Calycolpus goetheanus*** (Mart. ex DC.) O.Berg
Calycolpus glaber var. *angustilobus* L.Riley = ***Calycolpus goetheanus*** (Mart. ex DC.) O.Berg

Calycolpus gracilis (O.Berg) L.Riley = ***Calycolpus revolutus*** (Schauer) O.Berg
Calycolpus kegelianus O.Berg = ***Calycolpus revolutus*** (Schauer) O.Berg
Calycolpus kegelianus var. *gracilis* O.Berg = ***Calycolpus revolutus*** (Schauer) O.Berg
Calycolpus kegelianus var. *robustus* O.Berg = ***Calycolpus revolutus*** (Schauer) O.Berg
Calycolpus lucens (Alain) Bisse = ***Eugenia lucens*** Alain
Calycolpus megalodon L.Riley = ***Calycolpus revolutus*** (Schauer) O.Berg
Calycolpus nipense (Urb.) Bisse = ***Eugenia nipensis*** Urb.
Calycolpus ovalifolius O.Berg = ***Calycolpus goetheanus*** (Mart. ex DC.) O.Berg
Calycolpus parviflorus Sagot = ***Psidium salutare*** var. ***salutare***
Calycolpus piresianus Mattos = ***Calycolpus roraimensis*** Steyerm.
Calycolpus pyrifer L.Riley = ***Calycolpus revolutus*** (Schauer) O.Berg
Calycolpus reversus (Urb.) Bisse = ***Psidium reversum*** Urb.
Calycolpus roraimensis Kausel = ***Calycolpus roraimensis*** Steyerm.
Calycolpus schomburgkianus O.Berg = ***Calycolpus goetheanus*** (Mart. ex DC.) O.Berg
Calycolpus schomburgkianus var. *recurvatus* O.Berg = ***Calycolpus goetheanus*** (Mart. ex DC.) O.Berg
Calycolpus schomburgkianus var. *speciosus* O.Berg = ***Calycolpus goetheanus*** (Mart. ex DC.) O.Berg

Calycorectes

Calycorectes O.Berg, Linnaea 27: 317 (1856).
Mexico to Trop. America. 79 MXG MXS 81 DOM 82 FRG SUR VEN 83 CLM PER 84 BZC BZE BZL BZN BZS 85 PAR. [Myrtaceae]
30 Species
Schizocalyx O.Berg, Linnaea 27: 319 (1856), nom. illeg.
Schizocalomyrtus Kausel, Lilloa 32: 367 (1967).

Calycorectes acutatus (Miq.) Toledo, Relat. Anual Inst. Bot. 1945: 40 (1945).
SE. Brazil. 84 BZL. Phan.
Eugenia laurifolia Spring ex Mart., Flora 20(2 Beibl.): 82 (1837), nom. illeg. *Eugenia springiana* O.Berg in C.F.P.von Martius & auct. suc. (eds.), Fl. Bras. 14(1): 299 (1857).
**Eugenia acutata* Miq., Linnaea 22: 535 (1849).

Calycorectes batavorum McVaugh, Mem. New York Bot. Gard. 18(2): 22 (1969). *Eugenia batavorum* (McVaugh) Mattos, Loefgrenia 120: 4 (2005).
Suriname. 82 SUR. Nanophan. or phan.

Calycorectes belemii Mattos, Loefgrenia 106: 1 (1995). *Eugenia belemii* (Mattos) Mattos, Loefgrenia 120: 4 (2005), without basionym page.
Brazil (Bahia). 84 BZE. Phan.

Calycorectes bergii Sandwith, Bull. Misc. Inform. Kew 1932: 212 (1932).
Guianas. 82 FRG GUY SUR. Phan.

Calycorectes beruttii Mattos, Loefgrenia 94: 9 (1989). *Eugenia beruttii* (Mattos) Mattos, Loefgrenia 120: 4 (2005).
Brazil (Bahia). 84 BZE. Nanophan. or phan.

Calycorectes costatus Mattos & D.Legrand, Loefgrenia 67: 8 (1975). *Eugenia manausensis* Mattos, Loefgrenia 120: 6 (2005).
Brazil (Amazonas). 84 BZN. Phan.

Calycorectes cucullatus Mattos, Loefgrenia 94: 10 (1989). *Eugenia ademireana* Mattos, Loefgrenia 120: 7 (2005).
Brazil (São Paulo). 84 BZL. Phan.

Calycorectes dardanoi Mattos, Loefgrenia 99: 1 (1990).
Brazil (Pernambuco). 84 BZE. Phan.

Calycorectes dominicanus Mattos, Loefgrenia 112: 1 (1998).
Dominican Rep. 81 DOM. Nanophan. or phan.

Calycorectes duarteanus D.Legrand, Comun. Bot. Mus. Hist. Nat. Montevideo 4(51): 2 (1970). *Eugenia falkenbergiana* Mattos, Loefgrenia 120: 4 (2005).
S. Brazil. 84 BZS. Phan.

Calycorectes enormis McVaugh, Mem. New York Bot. Gard. 18(2): 223 (1969). *Eugenia enormis* (McVaugh) Mattos, Loefgrenia 120: 5 (2005).
S. Venezuela. 82 VEN. Phan.

Calycorectes ferrugineus Mattos, Loefgrenia 106: 2 (1995). *Eugenia ferruginosa* Mattos, Loefgrenia 120: 5 (2005).
Brazil (Goiás). 84 BZC. Nanophan. or phan.

Calycorectes fluminensis Mattos, Loefgrenia 62: 1 (1974). *Eugenia fontellae* Mattos, Loefgrenia 120: 5 (2005).
Brazil (Rio de Janeiro). 84 BZL. Phan.

Calycorectes grandifolius O.Berg, Linnaea 27: 317 (1856). *Catinga oblongifolia* Sagot, Ann. Sci. Nat., Bot., VI, 20: 197 (1885), nom. superfl. *Eugenia oblongifolia* (Sagot) Nied. ex T.Durand & B.D.Jacks., Index Kew., Suppl. 1: 164 (1902), nom. illeg. *Eugenia neograndifolia* Mattos, Loefgrenia 120: 10 (2005).
Suriname, French Guiana. 82 FRG SUR. Phan.

Calycorectes heringerianus Mattos, Loefgrenia 101: 1 (1992). *Eugenia ezechiasii* Mattos, Loefgrenia 120: 6 (2005).
Brazil (Minas Gerais). 84 BZL. Phan.

Calycorectes langsdorffii O.Berg in C.F.P.von Martius & auct. suc. (eds.), Fl. Bras. 14(1): 597 (1859). *Eugenia fissurata* Mattos, Loefgrenia 120: 5 (2005).
Brazil (Bahia). 84 BZE. Phan.

Calycorectes lourteigiae Mattos & D.Legrand, Loefgrenia 67: 8 (1975). *Eugenia lourteigiae* (Mattos & D.Legrand) Mattos, Loefgrenia 120: 6 (2005).
Brazil (Paraná). 84 BZS. Nanophan.

Calycorectes martianus O.Berg in C.F.P.von Martius & auct. suc. (eds.), Fl. Bras. 14(1): 596 (1859). *Eugenia martiana* (O.Berg) Mattos, Loefgrenia 120: 6 (2005).
Brazil (Rio de Janeiro). 84 BZL. Nanophan. or phan.

Calycorectes maximus McVaugh, Mem. New York Bot. Gard. 18(2): 224 (1969). *Eugenia maxima* (McVaugh) Mattos, Loefgrenia 120: 7 (2005).
Suriname. 82 SUR. Phan.

Calycorectes mexicanus O.Berg, Linnaea 27: 318 (1856). *Eugenia rinconiensis* Mattos, Loefgrenia 120: 9 (2005).
Mexico. 79 MXG MXS. Nanophan. or phan.
Eugenia calycorectoides O.Berg, Linnaea 29: 236 (1858).

Calycorectes orlandoi Mattos, Loefgrenia 102: 1 (1993). *Eugenia orlandoi* (Mattos) Mattos, Loefgrenia 120: 7 (2005).
Brazil (Bahia). 84 BZE. Nanophan. or phan.

Calycorectes paraguayensis Mattos, Loefgrenia 62: 2 (1974).
Paraguay, Brazil (São Paulo). 84 BZL 85 PAR. Phan.

Calycorectes paranaensis Mattos, Loefgrenia 102: 2 (1993). *Eugenia paranaensis* (Mattos) Mattos, Loefgrenia 120: 7 (2005).
Brazil (Paraná). 84 BZS. Phan.

Calycorectes pirataquinensis Mattos, Loefgrenia 107: 1 (1995). *Eugenia pirataquinensis* (Mattos) Mattos, Loefgrenia 120: 8 (2005).
Brazil (Bahia). 84 BZE. Nanophan.

Calycorectes pohlianus (O.Berg) Kiaersk., Enum. Myrt. Bras.: 117 (1893).
Brazil (Rio de Janeiro to Santa Catarina). 84 BZL BZS. Phan.
**Schizocalyx pohliana* O.Berg, Linnaea 27: 320 (1856). *Schizocalomyrtus pohliana* (O.Berg) Kausel, Lilloa 32: 367 (1967).
Schizocalyx pohliana var. *panicularis* O.Berg in C.F.P. von Martius & auct. suc. (eds.), Fl. Bras. 14(1): 358 (1857).
Schizocalyx pohliana var. *triflora* O.Berg in C.F.P.von Martius & auct. suc. (eds.), Fl. Bras. 14(1): 358 (1857).

Calycorectes rodriguesii Mattos & D.Legrand, Loefgrenia 67: 9 (1975). *Eugenia rodriguesii* (Mattos & D.Legrand) Mattos, Loefgrenia 120: 9 (2005).
Brazil (Amazonas). 84 BZN. Phan.

Calycorectes rogersianus Mattos, Loefgrenia 112: 2 (1998).
Colombia. 83 CLM. Nanophan. or phan.

Calycorectes schottianus O.Berg in C.F.P.von Martius & auct. suc. (eds.), Fl. Bras. 14(1): 357 (1857). *Eugenia leonorae* Mattos, Loefgrenia 120: 6 (2005).
Brazil (Rio de Janeiro to Paraná). 84 BZL BZS. Phan.

Calycorectes teixeireanus Mattos, Loefgrenia 109: 1 (1996). *Eugenia teixeireana* (Mattos) Mattos, Loefgrenia 120: 9 (2005).
Brazil (Rio de Janeiro). 84 BZL. Nanophan. or phan.

Calycorectes wurdackii McVaugh, Mem. New York Bot. Gard. 18(2): 225 (1969). *Eugenia wurdackii* (McVaugh) Mattos, Loefgrenia 120: 10 (2005).
NE. Peru. 83 PER. Phan.

***Synonyms*:**
Calycorectes ambivalens Sobral = ***Eugenia lagoensis*** Kiaersk.
Calycorectes australis D.Legrand = ***Eugenia neoaustralis*** Sobral
Calycorectes australis var. *impressovenosus* D.Legrand = ***Eugenia neoaustralis*** Sobral
Calycorectes cubensis Griseb. = ***Myrciaria floribunda*** (H.West ex Willd.) O.Berg
Calycorectes densiflorus (O.Berg) Nied. = ***Siphoneugena densiflora*** O.Berg
Calycorectes ekmanii Urb. = ***Hottea ekmanii*** (Urb.) Borhidi
Calycorectes guyanensis Mattos = **Eugenia galbaoensis**
Calycorectes latifolius (Aubl.) O.Berg = ***Eugenia latifolia*** Aubl.
Calycorectes legrandii Mattos = ***Eugenia psidiiflora***
Calycorectes macrocalyx Rusby = ***Eugenia wentii*** Amshoff
Calycorectes minutifolius Mattos = ***Eugenia neibensis*** Mattos
Calycorectes moana Borhidi & O.Muñiz = ***Hottea moana*** (Borhidi & O.Muñiz) Borhidi
Calycorectes ovigerus (Brongn. & Gris) Guillaumin = ***Stereocaryum ovigerum*** (Brongn. & Gris) Burret
Calycorectes protractus Griseb. = ***Psidium sartorianum*** (O.Berg) Nied.
Calycorectes psidiiflorus (O.Berg) Sobral = ***Eugenia psidiiflora***
Calycorectes psidiiflorus var. *legrandii* (Mattos) Mattos = ***Eugenia psidiiflora***
Calycorectes psidiiflorus var. *schultzianus* (Mattos) Mattos = ***Eugenia psidiiflora***
Calycorectes psidiiflorus var. *triflorus* Mattos = ***Eugenia psidiiflora***
Calycorectes riedelianus O.Berg = ***Eugenia psidiiflora***
Calycorectes rubiginosa (Brongn. & Gris) Guillaumin = ***Stereocaryum rubiginosum*** (Brongn. & Gris) Burret
Calycorectes schultzianus Mattos = ***Eugenia psidiiflora***
Calycorectes sellowianus O.Berg = ***Eugenia neoaustralis*** Sobral
Calycorectes striatulus Mattos = ***Eugenia psidiiflora***
Calycorectes widgrenianus (O.Berg) Nied. = ***Siphoneugena widgreniana*** O.Berg
Calycorectes yatuae McVaugh = ***Eugenia yatuae*** (McVaugh) B.Holst

Calycothrix

Calycothrix Meisn. = ***Calytrix*** Labill.
Calycothrix achaeta F.Muell. = ***Calytrix achaeta*** (F.Muell.) Benth.
Calycothrix angulata (Lindl.) Schauer = ***Calytrix angulata*** Lindl.
Calycothrix arborescens F.Muell. = ***Calytrix arborescens*** (F.Muell.) Benth.
Calycothrix asperula Schauer = ***Calytrix asperula*** (Schauer) Benth.
Calycothrix aurea (Lindl.) Schauer = ***Calytrix aurea*** Lindl.
Calycothrix baueri Schauer = ***Calytrix exstipulata*** DC.
Calycothrix behriana Schltdl. = ***Calytrix tetragona*** Labill.
Calycothrix billardierei Schauer = ***Calytrix tetragona*** Labill.
Calycothrix birdii F.Muell. = ***Calytrix birdii*** (F.Muell.) B.D.Jacks.
Calycothrix brachychaeta F.Muell. = ***Calytrix brachychaeta*** (F.Muell.) Benth.
Calycothrix brachyphylla Turcz. = ***Calytrix leschenaultii*** (Schauer) Benth.
Calycothrix brevifolia Meisn. = ***Calytrix brevifolia*** (Meisn.) Benth.
Calycothrix breviseta (Lindl.) Schauer = ***Calytrix breviseta*** Lindl.
Calycothrix brownii Schauer = ***Calytrix brownii*** (Schauer) Craven
Calycothrix brunioides (A.Cunn.) Schauer = ***Calytrix tetragona*** Labill.
Calycothrix candolleana Schauer = ***Calytrix decandra*** DC.
Calycothrix ciliata Turcz. = ***Calytrix sapphirina*** Lindl.
Calycothrix conanthera F.Muell. = ***Calytrix decandra*** DC.
Calycothrix conferta (A.Cunn.) Schauer = ***Calytrix exstipulata*** DC.
Calycothrix creswellii F.Muell. = ***Calytrix creswellii*** (F.Muell.) B.D.Jacks.
Calycothrix cupressifolia (A.Rich.) Stapf = ***Calytrix exstipulata*** DC.

Calycothrix cupressoides (A.Rich.) Schauer = ***Calytrix exstipulata*** DC.
Calycothrix curtophylla (A.Cunn.) Schauer = ***Calytrix tetragona*** Labill.
Calycothrix cuspidata Turcz. = ***Calytrix breviseta*** Lindl. subsp. ***breviseta***
Calycothrix decandra (DC.) F.Muell. = ***Calytrix decandra*** DC.
Calycothrix depressa Turcz. = ***Calytrix depressa*** (Turcz.) Benth.
Calycothrix diversifolia Turcz. = ***Calytrix tetragona*** Labill.
Calycothrix drummondii Meisn. = ***Calytrix drummondii*** (Meisn.) Craven
Calycothrix empetroides Schauer = ***Calytrix sapphirina*** Lindl.
Calycothrix flavescens (A.Cunn.) Schauer = ***Calytrix flavescens*** A.Cunn.
Calycothrix fraseri (A.Cunn.) Schauer = ***Calytrix fraseri*** A.Cunn.
Calycothrix glabra (R.Br.) Hook.f. = ***Calytrix tetragona*** Labill.
Calycothrix glabra var. *ciliata* Hook.f. = ***Calytrix tetragona*** Labill.
Calycothrix glabra var. *glaberrima* Hook.f. = ***Calytrix tetragona*** Labill.
Calycothrix glabra var. *virgata* (A.Cunn.) Hook.f. = ***Calytrix tetragona*** Labill.
Calycothrix glutinosa (Lindl.) Schauer = ***Calytrix glutinosa*** Lindl.
Calycothrix gracilis (Benth.) F.Muell. = ***Calytrix gracilis*** Benth.
Calycothrix granulosa (Benth.) F.Muell. = ***Calytrix fraseri*** A.Cunn.
Calycothrix laricina (R.Br. ex Benth.) F.Muell. = ***Calytrix brachychaeta*** (F.Muell.) Benth.
Calycothrix lasiantha Meisn. = ***Calytrix strigosa*** A.Cunn.
Calycothrix lasiostachya F.Muell. = ***Calytrix sapphirina*** Lindl.
Calycothrix leptophylla (Benth.) F.Muell. = ***Calytrix leptophylla*** Benth.
Calycothrix leschenaultii Schauer = ***Calytrix leschenaultii*** (Schauer) Benth.
Calycothrix leucantha Miq. = ***Calytrix tetragona*** Labill.
Calycothrix longiflora F.Muell. = ***Calytrix longiflora*** (F.Muell.) Benth.
Calycothrix luteola Schauer = ***Calytrix flavescens*** A.Cunn.
Calycothrix megaphylla F.Muell. = ***Calytrix megaphylla*** (F.Muell.) Benth.
Calycothrix microphylla Schauer = ***Calytrix exstipulata*** DC.
Calycothrix monticola Miq. = ***Calytrix tetragona*** Labill.
Calycothrix muelleri Miq. = ***Calytrix tetragona*** Labill.
Calycothrix muricata F.Muell. = ***Calytrix brevifolia*** (Meisn.) Benth.
Calycothrix oldfieldii (Benth.) F.Muell. = ***Calytrix oldfieldii*** Benth.
Calycothrix plumulosa F.Muell. = ***Calytrix plumulosa*** (F.Muell.) B.D.Jacks.
Calycothrix puberula Meisn. = ***Calytrix flavescens*** A.Cunn.
Calycothrix pulchella Turcz. = ***Calytrix pulchella*** (Turcz.) B.D.Jacks.
Calycothrix rosea Meisn. = ***Calytrix depressa*** (Turcz.) Benth.
Calycothrix rosea Miq. = ***Calytrix tetragona*** Labill.
Calycothrix sapphirina (Lindl.) Schauer = ***Calytrix sapphirina*** Lindl.
Calycothrix scabra (DC.) Schauer = ***Calytrix tetragona*** Labill.
Calycothrix scabra var. *minor* Schltdl. = ***Calytrix tetragona*** Labill.
Calycothrix schlechtendalii Miq. = ***Calytrix tetragona*** Labill.
Calycothrix simplex (Lindl.) Schauer = ***Calytrix simplex*** Lindl.
Calycothrix squarrosa Miq. = ***Calytrix tetragona*** Labill.
Calycothrix strigosa (A.Cunn.) Schauer = ***Calytrix strigosa*** A.Cunn.
Calycothrix sullivanii F.Muell. = ***Calytrix tetragona*** Labill.
Calycothrix tenella Meisn. = ***Calytrix flavescens*** A.Cunn.
Calycothrix tenuifolia Meisn. = ***Calytrix depressa*** (Turcz.) Benth.
Calycothrix tenuiramea Turcz. = ***Calytrix tenuiramea*** (Turcz.) Benth.
Calycothrix tetragona (Labill.) F.Muell. = ***Calytrix tetragona*** Labill.
Calycothrix tetragonophylla Meisn. = ***Calytrix flavescens*** A.Cunn.
Calycothrix variabilis (Lindl.) Schauer = ***Calytrix variabilis*** Lindl.
Calycothrix virgata (A.Cunn.) Schauer = ***Calytrix tetragona*** Labill.
Calycothrix watsonii F.Muell. & Tate = ***Calytrix strigosa*** A.Cunn.

Calyptranthes

Calyptranthes Sw., Prodr.: 79 (1788).
Trop. & Subtrop. America. 78 FLA 79 MXC MXE MXG MXN MXS MXT 80 BLZ COS ELS GUA HON NIC PAN 81 BAH CAY CUB DOM HAI JAM LEE PUE SWC TRT WIN 82 FRG GUY SUR VEN 83 BOL CLM ECU PER 84 BZC BZE BZL BZN BZS 85 AGE PAR URU. [Myrtaceae]
264 Species
Chytraculia P.Browne, Civ. Nat. Hist. Jamaica: 239 (1756).
Suzygium P.Browne, Civ. Nat. Hist. Jamaica: 240 (1756).
Chytralia Adans., Fam. Pl. 2: 80 (1763).

Calyptranthes acevedoi Alain, Phytologia 61: 353 (1986).
Puerto Rico. 81 PUE. Nanophan.

Calyptranthes acunae Borhidi & O.Muñiz, Bot. Közlem. 64: 13 (1977).
Cuba (Sierra de Nipe). 81 CUB. Nanophan.

Calyptranthes acutissima Urb., Symb. Antill. 6: 22 (1909).
W. Jamaica. 81 JAM. Nanophan. or phan.

Calyptranthes affinis O.Berg in C.F.P.von Martius & auct. suc. (eds.), Fl. Bras. 14(1): 43 (1857).
Brazil (Rio de Janeiro). 84 BZL. Phan.

Calyptranthes albicans Borhidi, Bot. Közlem. 64: 18 (1977).
Cuba (Sierra del Cristal). 81 CUB. Nanophan.

Calyptranthes amarulenta B.Holst, Novon 9: 517 (1999 publ. 2000).
Nicaragua. 80 NIC. Nanophan. or phan.

Calyptranthes amoena Pilg., Bot. Jahrb. Syst. 30: 178 (1901).
Brazil (Mato Grosso). 84 BZC. Nanophan. or phan.

Calyptranthes amshoffae McVaugh, Mem. New York Bot. Gard. 10(1): 70 (1958).
Guianas, Brazil (Amapá). 82 FRG SUR 84 BZN. Phan.

Calyptranthes anacletoi Borhidi & O.Muñiz, Bot. Közlem. 64: 15 (1977).
Cuba (Sierra de Moa). 81 CUB. Nanophan. or phan.

Calyptranthes anceps O.Berg in C.F.P.von Martius & auct. suc. (eds.), Fl. Bras. 14(1): 40 (1857). *Chytraculia anceps* (O.Berg) Kuntze, Revis. Gen. Pl. 1: 238 (1891).
Brazil (Rio de Janeiro). 84 BZL. Nanophan. or phan.

Calyptranthes angustifolia Kiaersk., Enum. Myrt. Bras.: 42 (1893).
Brazil (Rio de Janeiro). 84 BZL. Nanophan. or phan.

Calyptranthes apicata (C.Wright ex Griseb.) Urb., Symb. Antill. 9: 482 (1928).
W. Cuba. 81 CUB. Nanophan.
**Calyptranthes rigida* var. *apicata* C.Wright ex Griseb., Cat. Pl. Cub.: 86 (1866).

Calyptranthes apoda Urb., Symb. Antill. 9: 96 (1923).
E. Cuba. 81 CUB. Nanophan.

Calyptranthes arborea Urb. & Ekman, Ark. Bot. 21A(5): 27 (1927).
N. Haiti. 81 HAI. Phan.

Calyptranthes arenicola Urb., Symb. Antill. 9: 478 (1928).
W. Cuba. 81 CUB. Nanophan.

Calyptranthes aromatica A.St.-Hil., Pl. Usuel. Bras. 3: t. 14 (1824). *Chytraculia aromatica* (A.St.-Hil.) Kuntze, Revis. Gen. Pl. 1: 238 (1891).
SE. Brazil. 84 BZL. Nanophan.

Calyptranthes axillaris O.Berg in C.F.P.von Martius & auct. suc. (eds.), Fl. Bras. 14(1): 53 (1857). *Chytraculia axillaris* (O.Berg) Kuntze, Revis. Gen. Pl. 1: 238 (1891).
Brazil (Rio de Janeiro). 84 BZL. Nanophan. or phan.

Calyptranthes banilejoana Alain, Phytologia 61: 354 (1986).
Dominican Rep. 81 DOM. Nanophan. or phan.

Calyptranthes baracoensis Borhidi, Bot. Közlem. 64: 14 (1977).
SE. Cuba. 81 CUB. Nanophan.

Calyptranthes barkeri Ekman & Urb., Ark. Bot. 22A(10): 32 (1929).
Hispaniola. 81 DOM HAI. Nanophan. or phan.

Calyptranthes bartlettii Standl., Publ. Field Mus. Nat. Hist., Bot. Ser. 11: 136 (1932).
Belize. 80 BLZ. Nanophan. or phan.

Calyptranthes bergii Krug & Urb., Bot. Jahrb. Syst. 19: 601 (1895).
W. Cuba. 81 CUB. Nanophan.
**Calyptranthes thomasiana* var. *obscura* O.Berg, Linnaea 30: 652 (1861).

Calyptranthes bialata Urb., Symb. Antill. 9: 95 (1923).
E. Cuba. 81 CUB. Nanophan.

Calyptranthes bimarginata O.Berg in C.F.P.von Martius & auct. suc. (eds.), Fl. Bras. 14(1): 543 (1859). *Chytraculia bimarginata* (O.Berg) Kuntze, Revis. Gen. Pl. 1: 238 (1891).
Brazil (Rio de Janeiro). 84 BZL. Nanophan. or phan.

Calyptranthes bipennis O.Berg, Linnaea 31: 248 (1862). *Chytraculia bipennis* (O.Berg) Kuntze, Revis. Gen. Pl. 1: 238 (1891).
W. South America to N. Brazil 83 BOL ECU PER 84 BZN. Phan.

Calyptranthes blanchetiana O.Berg in C.F.P.von Martius & auct. suc. (eds.), Fl. Bras. 14(1): 40 (1857). *Chytraculia blanchetiana* (O.Berg) Kuntze, Revis. Gen. Pl. 1: 238 (1891).
Brazil (Bahia). 84 BZE. Nanophan. or phan.

Calyptranthes boldinghii Urb., Symb. Antill. 6: 22 (1909).
Leeward Is. (St. Martin). 81 LEE. Nanophan.

Calyptranthes bracteata M.L.Kawas. & B.Holst, Brittonia 46: 137 (1994).
French Guiana. 82 FRG. Phan.

Calyptranthes bracteosa Urb. & Ekman, Ark. Bot. 21A(5): 26 (1927).
N. Haiti. 81 HAI. Nanophan. or phan.

Calyptranthes brasiliensis Spreng., Syst. Veg. 2: 499 (1825). *Chytraculia brasiliensis* (Spreng.) Kuntze, Revis. Gen. Pl. 1: 238 (1891).
E. & S. Brazil. 84 BZE BZL BZS. Nanophan.
**Calyptranthes paniculata* Spreng., Neue Entd. 2: 170 (1821), nom. illeg.
Calyptranthes brasiliensis var. *densa* O.Berg in C.F.P.von Martius & auct. suc. (eds.), Fl. Bras. 14(1): 41 (1857).
Calyptranthes brasiliensis var. *laxa* O.Berg in C.F.P.von Martius & auct. suc. (eds.), Fl. Bras. 14(1): 42 (1857).
Calyptranthes mutabilis O.Berg in C.F.P.von Martius & auct. suc. (eds.), Fl. Bras. 14(1): 540 (1859). *Chytraculia mutabilis* (O.Berg) Kuntze, Revis. Gen. Pl. 1: 238 (1891). *Calyptranthes loranthifolia* var. *mutabilis* (O.Berg) D.Legrand, Lilloa 31: 199 (1962). *Calyptranthes brasiliensis* var. *mutabilis* (O.Berg) D.Legrand, in Fl. Ilustr. Catar. 1(Mirtac.): 516 (1971).

Calyptranthes brevispicata McVaugh, Fieldiana, Bot. 29: 181 (1956).
Peru (Loreto). 83 PER. Phan.

Calyptranthes buchenavioides Parra-Os., Novon 14: 210 (2004).
Colombia. 83 CLM.

Calyptranthes bullata (Salisb.) DC., Prodr. 3: 258 (1828).
Honduras. 80 HON. Nanophan. or phan.
**Myrtus bullata* Salisb., Prodr. Stirp. Chap. Allerton: 354 (1796). *Chytraculia bullata* (Salisb.) Kuntze, Revis. Gen. Pl. 1: 238 (1891).

Calyptranthes calderonii Standl., Publ. Field Mus. Nat. Hist., Bot. Ser. 4: 318 (1929).
Belize to El Salvador. 80 BLZ ELS. Nanophan. or phan.

Calyptranthes calophylla Urb. & Ekman, Ark. Bot. 22A(10): 28 (1929).
Haiti (Massif de la Hotte). 81 HAI. Nanophan.

Calyptranthes calyptrata Griseb., Mem. Amer. Acad. Arts, n.s., 8: 181 (1861).
E. Cuba. 81 CUB. Nanophan.

Calyptranthes canaliculata McVaugh, Mem. New York Bot. Gard. 10(1): 71 (1958).
Colombia. 83 CLM. Phan.

Calyptranthes canapuensis Urb., Symb. Antill. 9: 484 (1928).
Cuba (Sierra de Nipe). 81 CUB. Nanophan.

Calyptranthes capitata Proctor, J. Arnold Arbor. 63: 272 (1982).
Jamaica. 81 JAM. Nanophan.

Calyptranthes capitulata C.Wright in F.A.Sauvalle, Fl. Cub.: 39 (1873).
W. Cuba. 81 CUB. Nanophan.

var. ***capitulata***
W. Cuba. 81 CUB. Nanophan.

var. ***caroli*** (Britton & P.Wilson) Borhidi & O.Muñiz, Bot. Közlem. 62: 26 (1975).
W. Cuba. 81 CUB. Nanophan.
**Calyptranthes caroli* Britton & P.Wilson, Mem. Torrey Bot. Club 16: 87 (1920).
Calyptranthes caroli var. *longipedunculata* Moldenke, Phytologia 2: 215 (1947).

Calyptranthes cardiophylla Urb., Symb. Antill. 9: 96 (1923).
E. Cuba (Sierra Azul). 81 CUB. Nanophan.

Calyptranthes caudata Gardner, London J. Bot. 4: 102 (1845). *Chytraculia caudata* (Gardner) Kuntze, Revis. Gen. Pl. 1: 238 (1891).
Colombia, SE. Brazil. 83 CLM 84 BZL. Phan.

Calyptranthes cephalantha O.Berg in C.F.P.von Martius & auct. suc. (eds.), Fl. Bras. 14(1): 46 (1857). *Chytraculia cephalantha* (O.Berg) Kuntze, Revis. Gen. Pl. 1: 238 (1891).
SE. Brazil. 84 BZL. Nanophan. or phan.

Calyptranthes chiapensis Lundell, Contr. Univ. Michigan Herb. 7: 28 (1942).
SE. Mexico (to Veracruz). 79 MXG MXT 80 GUA?. Phan.

Calyptranthes chrysophylloides Urb. & Ekman, Ark. Bot. 22A(10): 31 (1929).
Haiti (Massif de la Hotte). 81 HAI. Nanophan. or phan.
Calyptranthes chrysophylloides var. *minor* Urb. & Ekman, Ark. Bot. 22A(10): 31 (1929).

Calyptranthes chytraculia (L.) Sw., Prodr.: 79 (1788). *Chytraculia arborea* Kuntze, Revis. Gen. Pl. 1: 238 (1891).
Belize to Colombia, Jamaica, Cuba. 80 BLZ COS GUA HON NIC PAN 81 CUB JAM 83 CLM. Nanophan. or phan.
**Myrtus chytraculia* L., Syst. Nat. ed. 10: 1056 (1759).
Chytraculia chytraculia (L.) Millsp., Publ. Field Columb. Mus., Bot. Ser. 2: 80 (1896), nom. inval.
Chytraculia chytraculia var. *genuina* Sudw., Bull. Div. Forest. U.S.D.A. 14: 305 (1897), nom. inval.

var. ***americana*** McVaugh, Fieldiana, Bot. 29: 404 (1963). *Calyptranthes lindeniana* var. *americana* (McVaugh) Lundell, Wrightia 2: 115 (1964).
Belize to Colombia. 80 BLZ COS GUA HON NIC PAN 83 CLM. Phan.

var. ***chytraculia***
Cuba, Jamaica. 81 CUB JAM. Nanophan. or phan.
Calyptranthes urbanii Fawc. & Rendle, J. Bot. 64: 14 (1926).

Calyptranthes clarensis Britton & P.Wilson, Mem. Torrey Bot. Club 16: 87 (1920).
C. Cuba. 81 CUB. Nanophan.

Calyptranthes clementis Britton & P.Wilson, Mem. Torrey Bot. Club 16: 87 (1920).
C. Cuba. 81 CUB. Nanophan.

Calyptranthes clusiifolia O.Berg in C.F.P.von Martius & auct. suc. (eds.), Fl. Bras. 14(1): 39 (1857). *Chytraculia clusiifolia* (O.Berg) Kuntze, Revis. Gen. Pl. 1: 238 (1891).
Brazil. 84 BZC BZE BZL. Phan.
**Myrcia clusiifolia* Miq., Linnaea 22: 533 (1849), nom. illeg.
Calyptranthes clusiifolia var. *cordifolia* D.Legrand, Lilloa 31: 193 (1962).

Calyptranthes collina Urb., Ark. Bot. 20A(5): 22 (1926).
Hispaniola. 81 DOM HAI. Nanophan. or phan.

Calyptranthes compressa Urb., Repert. Spec. Nov. Regni Veg. 18: 367 (1922).
SE. Cuba. 81 CUB. Nanophan.

Calyptranthes concinna DC., Prodr. 3: 258 (1828). *Chytraculia concinna* (DC.) Kuntze, Revis. Gen. Pl. 1: 238 (1891).
Brazil to NE. Argentina (Misiones). 83 BOL 84 BZC BZL BZS 85 AGE PAR URU. Phan.
Calyptranthes glomerata Cambess. in A.F.C.de Saint-Hilaire, Fl. Bras. Merid. 2: 371 (1833). *Chytraculia glomerata* (Cambess.) Kuntze, Revis. Gen. Pl. 1: 238 (1891). *Calyptranthes concinna* var. *glomerata* (Cambess.) D.Legrand, Sellowia 13: 286 (1961). *Calyptranthes concinna* var. *glomerata* (Cambess.) D.Legrand, Lilloa 31: 193 (1962).
Calyptranthes variabilis var. *oblongata* O.Berg in C.F.P.von Martius & auct. suc. (eds.), Fl. Bras. 14(1): 49 (1857).
Calyptranthes variabilis var. *pulchella* O.Berg in C.F.P.von Martius & auct. suc. (eds.), Fl. Bras. 14(1): 49 (1857).
Calyptranthes variabilis var. *riparia* O.Berg in C.F.P.von Martius & auct. suc. (eds.), Fl. Bras. 14(1): 49 (1857).
Calyptranthes variabilis var. *stricta* O.Berg in C.F.P.von Martius & auct. suc. (eds.), Fl. Bras. 14(1): 49 (1857).
Calyptranthes kleinii D.Legrand, Lilloa 31: 197 (1962).
Calyptranthes reitziana D.Legrand, Lilloa 31: 202 (1962).
Calyptranthes concinna var. *paulistana* D.Legrand, in Fl. Ilustr. Catar. 1(Mirtac.): 543 (1971), nom. illeg.

Calyptranthes conduplicata B.Holst, Selbyana 23: 137 (2002).
S. Venezuela. 82 VEN. Phan.

Calyptranthes contrerasii Lundell, Wrightia 2: 205 (1961).
Guatemala. 80 GUA. Phan.

Calyptranthes crebra McVaugh, Fieldiana, Bot. 29: 181 (1956).
Peru (Loreto). 83 PER. Phan.

Calyptranthes cristalensis Borhidi, Bot. Közlem. 64: 16 (1977).
Cuba (Sierra del Cristal). 81 CUB. Nanophan. or phan.

Calyptranthes cubensis O.Berg, Linnaea 29: 215 (1858). *Chytraculia cubensis* (O.Berg) Kuntze, Revis. Gen. Pl. 1: 238 (1891).
Cuba. 81 CUB. Nanophan.

Calyptranthes cuneifolia Lundell, Amer. Midl. Naturalist 29: 478 (1943).
Belize. 80 BLZ. Phan.

Calyptranthes cuprea O.Berg, Linnaea 30: 653 (1861). *Chytraculia cuprea* (O.Berg) Kuntze, Revis. Gen. Pl. 1: 238 (1891).
W. Cuba. 81 CUB. Nanophan.
Eugenia ferruginea A.Rich., Hist. Phys. Cuba, Pl. Vasc.: 591 (1846), nom. illeg. *Calyptranthes ferruginea* Krug & Urb., Bot. Jahrb. Syst. 19: 603 (1895).
Eugenia stuposa A.Rich. ex O.Berg, Linnaea 30: 699 (1861).
Calyptranthes ferruginea var. *cajalbanica* Borhidi & O.Muñiz, Bot. Közlem. 64: 14 (1977).

Calyptranthes cuspidata Mart. ex DC., Prodr. 3: 258 (1828). *Chytraculia cuspidata* (Mart. ex DC.) Kuntze, Revis. Gen. Pl. 1: 238 (1891).
N. Brazil to Peru. 83 PER 84 BZN. Phan.

Calyptranthes dardanoi Mattos, Loefgrenia 99: 1 (1990).
Brazil (Pernambuco). 84 BZE. Nanophan. or phan.

Calyptranthes decandra Griseb., Mem. Amer. Acad. Arts, n.s., 8: 181 (1861).
W. & C. Cuba. 81 CUB. Nanophan. or phan.

Calyptranthes densiflora Poepp. ex O.Berg, Linnaea 27: 30 (1855). *Chytraculia densiflora* (Poepp. ex O.Berg) Kuntze, Revis. Gen. Pl. 1: 238 (1891).
Ecuador to Peru. 83 ECU PER. Nanophan. or phan.

Calyptranthes densifolia Urb. & Ekman, Ark. Bot. 20A(5): 23 (1926).
Haiti (Massif de la Hotte). 81 HAI. Nanophan. or phan.

Calyptranthes depressa Urb., Ark. Bot. 24A(4): 23 (1932).
Dominican Rep. 81 DOM. Nanophan. or phan.

Calyptranthes dichotoma Casar., Nov. Stirp. Bras. 5: 47 (1843). *Chytraculia dichotoma* (Casar.) Kuntze, Revis. Gen. Pl. 1: 238 (1891).
Brazil (São Paulo). 84 BZL. Nanophan. or phan.

Calyptranthes discolor Urb., Symb. Antill. 5: 443 (1908).
NW. Jamaica. 81 JAM. Nanophan. or phan.

Calyptranthes dryadica M.L.Kawas., Novon 8: 386 (1998).
Brazil (São Paulo). 84 BZL. Nanophan. or phan.

Calyptranthes dusenii Kausel, Lilloa 33: 98 (1971 publ. 1972).
Brazil (Paraná). 84 BZS. Nanophan. or phan.

Calyptranthes elegans Krug, Bot. Jahrb. Syst. 19: 599 (1895).
Windward Is. 81 LEE? WIN. Nanophan. or phan.

Calyptranthes elliptica B.Holst & M.L.Kawas., Selbyana 25: 87 (2004).
Panama. 80 PAN.

Calyptranthes elongata Urb., Symb. Antill. 9: 94 (1923).
Cuba (Sierra de Nipe). 81 CUB. Nanophan.

Calyptranthes enneantha C.Wright, Anales Acad. Ci. Méd. Habana 5: 429 (1868).
W. Cuba. 81 CUB. Nanophan.

Calyptranthes eriocephala Urb., Symb. Antill. 7: 297 (1912).
Dominican Rep. 81 DOM. Nanophan.

Calyptranthes eriopoda DC., Prodr. 3: 257 (1828). *Chytraculia eriopoda* (DC.) Kuntze, Revis. Gen. Pl. 1: 238 (1891).
SE. Brazil. 84 BZL. Nanophan. or phan.

Calyptranthes ermitensis Borhidi, Bot. Közlem. 64: 15 (1977).
SE. Cuba. 81 CUB. Nanophan.

Calyptranthes estoraquensis Parra-Os., Novon 14: 211 (2004).
Colombia (Norte de Santander). 83 CLM.

Calyptranthes estremenae Alain, Phytologia 58: 325 (1985).
Puerto Rico. 81 PUE. Phan.

Calyptranthes eugenioides Cambess. in A.F.C.de Saint-Hilaire, Fl. Bras. Merid. 2: 370 (1833). *Mitranthes eugenioides* (Cambess.) O.Berg in C.F.P.von Martius & auct. suc. (eds.), Fl. Bras. 14(1): 355 (1857). *Chytraculia eugenioides* (Cambess.) Kuntze, Revis. Gen. Pl. 1: 238 (1891). *Mitropsidium eugenioides* (Cambess.) Burret, Notizbl. Bot. Gart. Berlin-Dahlem 15: 489 (1941).
E. Brazil. 84 BZE BZL. Nanophan. or phan.
Mitranthes eugenioides var. *oblongifolia* O.Berg in C.F.P.von Martius & auct. suc. (eds.), Fl. Bras. 14(1): 355 (1857).
Mitranthes eugenioides var. *ovata* O.Berg in C.F.P.von Martius & auct. suc. (eds.), Fl. Bras. 14(1): 355 (1857).

Calyptranthes exasperata Borhidi, Bot. Közlem. 64: 17 (1977).
Cuba (Sierra de Nipe). 81 CUB. Nanophan.

Calyptranthes fasciculata O.Berg, Linnaea 27: 31 (1855). *Chytraculia fasciculata* (O.Berg) Kuntze, Revis. Gen. Pl. 1: 238 (1891). *Calyptranthes fasciculata* var. *genuina* Stehlé & Quentin, Fl. Guadeloupe 2(3): 62 (1949), nom. inval.
Windward Is. to N. Brazil. 81 TRT WIN 82 FRG GUY SUR VEN 83 BOL 84 BZN. Phan.
Calyptranthes sericea Griseb., Fl. Brit. W. I.: 233 (1860).
Calyptranthes sericea var. *hahnii* Krug & Urb., Bot. Jahrb. Syst. 19: 602 (1895). *Calyptranthes fasciculata* var. *hahnii* (Krug & Urb.) Stehlé, Fl. Guadeloupe 2(3): 62 (1949).

Calyptranthes flavoviridis Urb., Symb. Antill. 9: 477 (1928).
W. Cuba. 81 CUB. Nanophan. or phan.

Calyptranthes forsteri O.Berg, Linnaea 27: 23 (1855). *Chytraculia forsteri* (O.Berg) Kuntze, Revis. Gen. Pl. 1: 238 (1891).
Guadeloupe, St. Lucia, Venezuela to Ecuador. 81 LEE WIN 82 VEN 83 CLM ECU 84 BZN. Phan.

Calyptranthes fusiformis M.L.Kawas., Brittonia 48: 508 (1996 publ. 1997).
Brazil (São Paulo). 84 BZL. Nanophan. or phan.

Calyptranthes garciae Alain & M.M.Mejia, Moscosoa 9: 8 (1997).
N. Dominican Rep. 81 DOM. Phan.

Calyptranthes glabrescens Krug & Urb., Bot. Jahrb. Syst. 19: 596 (1895).
E. Cuba. 81 CUB. Phan.

Calyptranthes glandulosa M.L.Kawas. & B.Holst, Sida 21: 1955 (2005).
Ecuador. 83 ECU. Phan.

Calyptranthes glazioviana Kiaersk., Enum. Myrt. Bras.: 35 (1893).
SE. Brazil. 84 BZL. Nanophan. or phan.

Calyptranthes gracilipes C.Wright, Anales Acad. Ci. Méd. Habana 5: 428 (1868).
Cuba. 81 CUB. Nanophan.

Calyptranthes grammica (Spreng.) D.Legrand, Lilloa 31: 196 (1962).
SE. & S. Brazil. 84 BZL BZS. Nanophan. or phan.
**Myrtus grammica* Spreng., Syst. Veg. 2: 480 (1825). *Chytraculia grammica* (Spreng.) Kuntze, Revis. Gen. Pl. 1: 238 (1891).
Calyptranthes uniflora Spreng., Flora 20(2): 87 (1837).
Calyptranthes musciflora O.Berg in C.F.P.von Martius & auct. suc. (eds.), Fl. Bras. 14(1): 54 (1857).
Calyptranthes musciflora var. *angustifolia* O.Berg in C.F.P.von Martius & auct. suc. (eds.), Fl. Bras. 14(1): 54 (1857).
Calyptranthes musciflora var. *obscura* O.Berg in C.F.P.von Martius & auct. suc. (eds.), Fl. Bras. 14(1): 54 (1857).
Calyptranthes musciflora var. *spathulata* O.Berg in C.F.P.von Martius & auct. suc. (eds.), Fl. Bras. 14(1): 54 (1857).
Calyptranthes musciflora var. *triflora* O.Berg in C.F.P.von Martius & auct. suc. (eds.), Fl. Bras. 14(1): 54 (1857). *Calyptranthes grammica* var. *triflora* (O.Berg) Mattos, Loefgrenia 32: 2 (1969).
Calyptranthes musciflora var. *venosa* O.Berg in C.F.P.von Martius & auct. suc. (eds.), Fl. Bras. 14(1): 54 (1857).
Calyptranthes musciflora var. *glauca* O.Berg in C.F.P.von Martius & auct. suc. (eds.), Fl. Bras. 14(1): 543 (1859).

Calyptranthes grandiflora O.Berg in C.F.P.von Martius & auct. suc. (eds.), Fl. Bras. 14(1): 542 (1859).
Brazil (Rio de Janeiro). 84 BZL. Nanophan. or phan.

Calyptranthes grandifolia O.Berg in C.F.P.von Martius & auct. suc. (eds.), Fl. Bras. 14(1): 48 (1857). *Chytraculia grandifolia* (O.Berg) Kuntze, Revis. Gen. Pl. 1: 238 (1891).
SE. & S. Brazil. 84 BZL BZS. Phan.
Calyptranthes rufa O.Berg in C.F.P.von Martius & auct. suc. (eds.), Fl. Bras. 14(1): 541 (1859). *Chytraculia rufa* (O.Berg) Kuntze, Revis. Gen. Pl. 1: 238 (1891). *Calyptranthes grandifolia* var. *rufa* (O.Berg) D.Legrand, in Fl. Ilustr. Catar. 1(Mirtac.): 515 (1971).
Calyptranthes grandifolia var. *macrantha* D.Legrand, Lilloa 31: 196 (1962).

Calyptranthes grandis Urb. & Ekman, Ark. Bot. 24A(4): 20 (1932).
Hispaniola. 81 DOM HAI. Phan.

Calyptranthes guayabillo Alain, Phytologia 61: 354 (1986).
Dominican Rep. 81 DOM. Nanophan.

Calyptranthes hatschbachii D.Legrand, Comun. Bot. Mus. Hist. Nat. Montevideo 3(35): 1 (1958).
Brazil (Paraná). 84 BZS. Nanophan. or phan.

Calyptranthes heineriana Kausel, Lilloa 33: 99 (1971 publ. 1972).
Brazil (São Paulo). 84 BZL. Nanophan.

Calyptranthes hernandezii McVaugh, Fieldiana, Bot. 29: 404 (1963).
Mexico (San Luis Potosí). 79 MXE. Nanophan.

Calyptranthes heterochroa Urb., Symb. Antill. 9: 91 (1923).
E. Cuba. 81 CUB. Nanophan.

Calyptranthes heteroclada Urb. & Ekman, Ark. Bot. 24A(4): 21 (1932).
Hispaniola. 81 DOM HAI. Nanophan. or phan.

Calyptranthes hintonii Lundell, Wrightia 2: 166 (1961).
Mexico (México State). 79 MXC. Nanophan.

Calyptranthes hondurensis Standl., J. Arnold Arbor. 11: 36 (1930).
C. America. 80 BLZ ELS GUA HON. Nanophan. or phan.
Eugenia belizensis Standl., Publ. Field Mus. Nat. Hist., Bot. Ser. 11: 137 (1932). *Calyptranthes belizensis* (Standl.) Lundell, Phytologia 1: 218 (1937).
Calyptranthes aguilarii Standl. & Steyerm., Publ. Field Mus. Nat. Hist., Bot. Ser. 23: 128 (1944).

Calyptranthes hotteana Urb. & Ekman, Ark. Bot. 21A(5): 24 (1927).
Haiti (Massif de la Hotte). 81 HAI. Nanophan. or phan.

Calyptranthes hylobates Standl. ex Amshoff, Ann. Missouri Bot. Gard. 45: 169 (1958).
Mexico (Chiapas) to Panama. 79 MXT 80 COS PAN. Nanophan. or phan.

Calyptranthes insularis Bisse, Revista Jard. Bot. Nac. Univ. Habana 4(2): 5 (1983).
Cuba (I. de la Juventud). 81 CUB. Nanophan.

Calyptranthes involucrata Urb. & Ekman, Ark. Bot. 24A(4): 24 (1932).
Haiti (Massif de la Hotte). 81 HAI. Nanophan. or phan.

Calyptranthes ishoaquinicca M.L.Kawas. & B.Holst, Sida 21: 1957 (2005).
Ecuador. 83 ECU. Phan.

Calyptranthes izabalana Lundell, Wrightia 4: 142 (1970).
Guatemala. 80 GUA. Nanophan. or phan.

Calyptranthes jefensis B.Holst & M.L.Kawas., Selbyana 25: 89 (2004).
Panama. 80 PAN.

Calyptranthes jimenoana Alain, Phytologia 61: 355 (1986).
Dominican Rep. 81 DOM. Nanophan.

Calyptranthes johnstonii McVaugh, Fieldiana, Bot. 29: 405 (1963).
Panama. 80 PAN. Nanophan. or phan.

Calyptranthes karlingii Standl., Publ. Field Mus. Nat. Hist., Bot. Ser. 8: 29 (1930).
SE. Mexico to Guatemala. 79 MXT 80 BLZ GUA. Phan.

Calyptranthes karwinskyana O.Berg, Linnaea 29: 214 (1858). *Chytraculia karwinskyana* (O.Berg) Kuntze, Revis. Gen. Pl. 1: 238 (1891).
Mexico (Veracruz). 79 MXG. Nanophan. or phan.

Calyptranthes kiaerskovii Krug & Urb., Symb. Antill. 1: 43 (1898).
British Virgin Is. (Tortola, Virgin Gorda). 81 LEE. Nanophan. or phan.
**Calyptranthes obovata* Krug & Urb., Bot. Jahrb. Syst. 19: 600 (1895), nom. illeg.

Calyptranthes killipii Standl., Publ. Field Mus. Nat. Hist., Bot. Ser. 22: 93 (1940).
Colombia. 83 CLM. Phan.

Calyptranthes krugii Kiaersk., Bot. Tidsskr. 17: 248 (1889).
Puerto Rico. 81 PUE. Nanophan. or phan.

Calyptranthes krugioides McVaugh, Fieldiana, Bot. 29: 182 (1956).
N. Brazil to N. Peru. 83 PER 84 BZN. Nanophan. or phan.

Calyptranthes laevigata Urb. & Ekman, Ark. Bot. 24A(4): 21 (1932).
Dominican Rep. 81 DOM. Nanophan.

Calyptranthes lanceolata O.Berg in C.F.P.von Martius & auct. suc. (eds.), Fl. Bras. 14(1): 51 (1857). *Chytraculia lanceolata* (O.Berg) Kuntze, Revis. Gen. Pl. 1: 238 (1891).
Bolivia to SE. & S. Brazil. 83 BOL 84 BZL BZS. Nanophan. or phan.
Calyptranthes lanceolata var. *catharinensis* O.Berg in C.F.P.von Martius & auct. suc. (eds.), Fl. Bras. 14(1): 51 (1857).
Calyptranthes lanceolata var. *latifolia* O.Berg in C.F.P.von Martius & auct. suc. (eds.), Fl. Bras. 14(1): 540 (1859).

Calyptranthes langsdorffii O.Berg in C.F.P.von Martius & auct. suc. (eds.), Fl. Bras. 14(1): 539 (1859). *Chytraculia langsdorffii* (O.Berg) Kuntze, Revis. Gen. Pl. 1: 238 (1891).
Brazil (Rio de Janeiro). 84 BZL. Phan.
Calyptranthes langsdorffii var. *grandiflora* Kiaersk., Enum. Myrt. Bras.: 43 (1893).
Calyptranthes langsdorffii var. *macrophylla* Kiaersk., Enum. Myrt. Bras.: 43 (1893).

Calyptranthes leonis Borhidi & O.Muñiz, Bot. Közlem. 64: 17 (1977).
Cuba (Sierra de Imias). 81 CUB. Nanophan.

Calyptranthes lepida McVaugh, Mem. New York Bot. Gard. 18(2): 73 (1969).
French Guiana to Brazil (Amapá). 82 FRG 84 BZN. Phan.

Calyptranthes leptoclada Urb., Symb. Antill. 9: 93 (1923).
SE. Cuba. 81 CUB. Nanophan. or phan.
Calyptranthes montana Britton & P.Wilson, Bull. Torrey Bot. Club 50: 144 (1923).

Calyptranthes levisensis Bisse & A.Rodr., Revista Jard. Bot. Nac. Univ. Habana 1(2–3): 43 (1980 publ. 1981).
Cuba. 81 CUB. Nanophan.

Calyptranthes lilloi Speg. in C.L.Spegazzini & C.D.Girola, Cat. Descr. Maderas: 386 (1910).
NE. Argentina. 85 AGE. Nanophan. or phan.

Calyptranthes limoncillo Alain, Phytologia 61: 355 (1986).
Dominican Rep. 81 DOM. Nanophan. or phan.

Calyptranthes lindeniana O.Berg, Linnaea 29: 213 (1858). *Chytraculia lindeniana* (O.Berg) Kuntze, Revis. Gen. Pl. 1: 238 (1891).
Mexico to Guatemala. 79 MXG MXT 80 BLZ GUA. Nanophan. or phan.
Calyptranthes fluviatilis Lundell, Bull. Torrey Bot. Club 64: 553 (1937).

Calyptranthes linearis Alain, Contr. Ocas. Mus. Hist. Nat. Colegio "De Le Salle" 12: 8 (1953).
Cuba (Sierra de Moa). 81 CUB. Nanophan.

Calyptranthes litoralis Urb. & Ekman, Ark. Bot. 21A(5): 24 (1927).
Haiti. 81 HAI. Nanophan.

Calyptranthes lomensis Urb., Symb. Antill. 9: 478 (1928).
SC. Cuba (Sierra de Trinidad). 81 CUB. Nanophan.

Calyptranthes longicalyptrata B.Holst & M.L.Kawas., Selbyana 25: 90 (2004).
Panama. 80 PAN.

Calyptranthes longifolia O.Berg in C.F.P.von Martius & auct. suc. (eds.), Fl. Bras. 14(1): 46 (1857). *Chytraculia longifolia* (O.Berg) Kuntze, Revis. Gen. Pl. 1: 238 (1891).
N. Brazil to Peru. 83 ECU PER 84 BZN. Nanophan. or phan.
Calyptranthes pleophlebia Diels, Verh. Bot. Vereins Prov. Brandenburg 48: 188 (1906 publ. 1907).

Calyptranthes loranthifolia DC., Prodr. 3: 258 (1828). *Chytraculia loranthifolia* (DC.) Kuntze, Revis. Gen. Pl. 1: 238 (1891).
Brazil (Minas Gerais to Paraná). 84 BZL BZS. Nanophan. or phan.

Calyptranthes lozanoi Parra-Os., Caldasia 23: 435 (2001).
Colombia. 83 CLM. Phan.

subsp. ***lozanoi***
Colombia (Sierra Nevada de Santa Marta). 83 CLM. Phan.

subsp. ***tomentosa*** Parra-Os., Novon 14: 213 (2004).
Colombia (Antioquia). 83 CLM.

Calyptranthes lucida Mart. ex DC., Prodr. 3: 258 (1828). *Chytraculia lucida* (Mart. ex DC.) Kuntze, Revis. Gen. Pl. 1: 238 (1891).
C. America to Brazil. 80 BLZ COS HON PAN 82 FRG SUR 84 BZC BZE BZL BZN BZS. Phan.

var. ***lucida***
C. America to Brazil. 80 BLZ COS HON PAN 82 FRG SUR 84 BZC BZE BZL BZN BZS. Phan.
Calyptranthes pohliana O.Berg in C.F.P.von Martius & auct. suc. (eds.), Fl. Bras. 14(1): 42 (1857). *Chytraculia pohliana* (O.Berg) Kuntze, Revis. Gen. Pl. 1: 238 (1891).
Calyptranthes lucida f. *hilariana* D.Legrand, in Fl. Ilustr. Catar. 1(Mirtac.): 530 (1971).

var. ***polyantha*** (O.Berg) D.Legrand, in Fl. Ilustr. Catar. 1(Mirtac.): 531 (1971).
Brazil. 84 BZC BZE BZL BZS. Phan.
**Calyptranthes polyantha* O.Berg in C.F.P.von Martius & auct. suc. (eds.), Fl. Bras. 14(1): 541 (1859). *Chytraculia polyantha* (O.Berg) Kuntze, Revis. Gen. Pl. 1: 238 (1891).

Calyptranthes luetzelburgii Burret ex Luetzelb., Estud. Bot. Nordéste 3: 200 (1926).
Brazil (Paraíba). 84 BZE. Nanophan. or phan.

Calyptranthes luquillensis Alain, Bull. Torrey Bot. Club 90: 189 (1963).
Puerto Rico. 81 PUE. Nanophan. or phan.

Calyptranthes macrantha Standl. & Steyerm., Publ. Field Mus. Nat. Hist., Bot. Ser. 23: 127 (1944).
Mexico (Chiapas) to Guatemala. 79 MXT 80 GUA. Phan.

Calyptranthes macrocarpa B.Holst & M.L.Kawas., Selbyana 25: 90 (2004).
Panama to Colombia. 80 PAN 83 CLM.

Calyptranthes macrophylla O.Berg in C.F.P.von Martius & auct. suc. (eds.), Fl. Bras. 14(1): 45 (1857). *Chytraculia macrophylla* (O.Berg) Kuntze, Revis. Gen. Pl. 1: 238 (1891).
W. South America to N. Brazil. 83 CLM ECU PER 84 BZN. Nanophan. or phan.

Calyptranthes maestrensis Urb., Symb. Antill. 9: 483 (1928).
SE. Cuba. 81 CUB. Nanophan.

Calyptranthes mammosa Lundell, Wrightia 5: 42 (1974).
Guatemala. 80 GUA. Phan.

Calyptranthes manuensis B.Holst & M.L.Kawas., Sida 22: 931 (2006).
Peru. 83 PER. Phan.

Calyptranthes marmeladensis Urb. & Ekman, Ark. Bot. 22A(10): 30 (1929).
Haiti (Massif du Nord). 81 DOM? HAI. Nanophan. or phan.

Calyptranthes martiusiana DC., Prodr. 3: 257 (1828). *Chytraculia martiusiana* (DC.) Kuntze, Revis. Gen. Pl. 1: 238 (1891).
Brazil (Rio de Janeiro). 84 BZL. Nanophan. or phan.

Calyptranthes martorellii Alain, Phytologia 58: 326 (1985).
Puerto Rico. 81 PUE. Nanophan.

Calyptranthes maxima McVaugh, Fieldiana, Bot. 29: 182 (1956).
Colombia to N. Peru. 83 CLM ECU PER. Nanophan. or phan.

Calyptranthes mayana Lundell, Wrightia 5: 149 (1975).
C. America. 80 ELS GUA HON. Phan.

Calyptranthes mayarensis Borhidi, Bot. Közlem. 64: 15 (1977).
Cuba (Sierra de Nipe). 81 CUB. Nanophan. or phan.

Calyptranthes megistophylla Standl., Publ. Carnegie Inst. Wash. 461: 75 (1935).
C. America. 80 BLZ GUA PAN. Phan.

Calyptranthes melanoclada O.Berg in C.F.P.von Martius & auct. suc. (eds.), Fl. Bras. 14(1): 50 (1857). *Chytraculia melanoclada* (O.Berg) Kuntze, Revis. Gen. Pl. 1: 238 (1891).
Brazil (Rio de Janeiro). 84 BZL. Nanophan. or phan.

Calyptranthes meridensis Steyerm., Fieldiana, Bot. 28: 1009 (1957).
Venezuela. 82 VEN. Nanophan. or phan.

Calyptranthes micrantha C.Wright ex Griseb., Cat. Pl. Cub.: 85 (1866).
W. Cuba. 81 CUB. Nanophan.

Calyptranthes microphylla B.Holst & M.L.Kawas., Selbyana 25: 91 (2004).
Panama. 80 PAN.

Calyptranthes millspaughii Urb., Symb. Antill. 7: 294 (1912).
SE. Mexico to Guatemala. 79 MXG MXT 80 BLZ GUA. Phan.
Calyptranthes euryphylla Standl., Contr. U. S. Natl. Herb. 23: 1034 (1924).

Calyptranthes minutiflora Borhidi, Bot. Közlem. 64: 18 (1977).
Cuba (Sierra de Nipe). 81 CUB. Nanophan.

Calyptranthes mirabilis Bisse & A.Rodr., Revista Jard. Bot. Nac. Univ. Habana 1(2–3): 42 (1980 publ. 1981).
Cuba. 81 CUB. Nanophan.

Calyptranthes moaensis Alain, Contr. Ocas. Mus. Hist. Nat. Colegio "De Le Salle" 12: 9 (1953).
Cuba (Sierra de Moa). 81 CUB. Nanophan.

Calyptranthes monocarpa Urb., Symb. Antill. 9: 95 (1923).
Cuba (Sierra de Nipe). 81 CUB. Nanophan.

Calyptranthes monteverdensis P.E.Sanchez, Brenesia 62: 31 (2004).
Costa Rica. 80 COS. Phan.

Calyptranthes mornicola Urb., Ark. Bot. 22A(10): 29 (1929).
Haiti. 81 HAI. Nanophan. or phan.

Calyptranthes multiflora Poepp. ex O.Berg in C.F.P.von Martius & auct. suc. (eds.), Fl. Bras. 14(1): 42 (1857). *Chytraculia multiflora* (Poepp. ex O.Berg) Kuntze, Revis. Gen. Pl. 1: 238 (1891).
S. Trop. America. 82 VEN 83 BOL CLM ECU PER 84 BZE BZN. Phan.
Calyptranthes poeppigiana O.Berg in C.F.P.von Martius & auct. suc. (eds.), Fl. Bras. 14(1): 45 (1857). *Chytraculia poeppigiana* (O.Berg) Kuntze, Revis. Gen. Pl. 1: 238 (1891).
Calyptranthes florifera McVaugh, Mem. New York Bot. Gard. 10(1): 72 (1958).

Calyptranthes munizii Borhidi, Bot. Közlem. 64: 14 (1977).
Cuba (Sierra de Nipe). 81 CUB. Nanophan.

Calyptranthes myrcioides Urb. & Ekman, Ark. Bot. 24A(4): 25 (1932).
Dominican Rep. 81 DOM. Nanophan. or phan.

Calyptranthes nigrescens B.Holst, Selbyana 23: 138 (2002).
SE. Colombia to S. Venezuela. 82 VEN 83 CLM 84 BZN. Phan.

Calyptranthes nigricans DC., Prodr. 3: 258 (1828). *Chytraculia nigricans* (DC.) Kuntze, Revis. Gen. Pl. 1: 238 (1891).
N. Brazil. 84 BZN. Nanophan.

Calyptranthes nipensis Borhidi & O.Muñiz, Bot. Közlem. 64: 14 (1977).
Cuba (Sierra de Nipe). 81 CUB. Nanophan.

Calyptranthes nodosa Urb., Symb. Antill. 5: 444 (1908).
Jamaica. 81 JAM. Nanophan. or phan.

Calyptranthes nummularia O.Berg, Linnaea 30: 652 (1861). *Chytraculia nummularia* (O.Berg) Kuntze, Revis. Gen. Pl. 1: 238 (1891).
Hispaniola. 81 DOM HAI. Nanophan. or phan.

Calyptranthes oblanceolata Urb., Symb. Antill. 9: 91 (1923).
Cuba (Sierra de Nipe). 81 CUB. Nanophan.

Calyptranthes oblongifolia R.A.Howard, J. Arnold Arbor. 28: 122 (1947).
Cuba (Sierra de Moa). 81 CUB. Nanophan.

Calyptranthes obovata Kiaersk., Enum. Myrt. Bras.: 38 (1893).
Brazil (Rio de Janeiro to Santa Catarina). 84 BZL BZS. Phan.

Calyptranthes obtusa Benth., J. Bot. (Hooker) 2: 319 (1840).
French Guiana. 82 FRG. Phan.

Calyptranthes obversa O.Berg in C.F.P.von Martius & auct. suc. (eds.), Fl. Bras. 14(1): 51 (1857). *Chytraculia obversa* (O.Berg) Kuntze, Revis. Gen. Pl. 1: 238 (1891).
Brazil (São Paulo). 84 BZL. Nanophan. or phan.

Calyptranthes oligantha Urb., Symb. Antill. 9: 92 (1923).
E. Cuba. 81 CUB. Nanophan.

Calyptranthes oreophila Speg. in C.L.Spegazzini & C.D.Girola, Cat. Descr. Maderas: 389 (1910).
NE. Argentina. 85 AGE. Nanophan. or phan.

Calyptranthes ovalifolia Cambess. in A.F.C.de Saint-Hilaire, Fl. Bras. Merid. 2: 371 (1833). *Chytraculia ovalifolia* (Cambess.) Kuntze, Revis. Gen. Pl. 1: 238 (1891).
Brazil (Minas Gerais). 84 BZL. Nanophan.

Calyptranthes ovata O.Berg in C.F.P.von Martius & auct. suc. (eds.), Fl. Bras. 14(1): 50 (1857). *Chytraculia ovata* (O.Berg) Kuntze, Revis. Gen. Pl. 1: 238 (1891).
Brazil (Minas Gerais). 84 BZL. Nanophan. or phan.

Calyptranthes ovoidea McVaugh, Mem. New York Bot. Gard. 10(1): 75 (1958).
Guyana. 82 GUY. Phan.

Calyptranthes pachyadenia Urb. & Ekman in I.Urban, Symb. Antill. 9: 481 (1928).
C. Cuba. 81 CUB. Nanophan. or phan.

Calyptranthes pallens Griseb., Abh. Königl. Ges. Wiss. Göttingen 7: 215 (1857). *Chytraculia pallens* (Griseb.) Britton in G.B.S.Shattuck, Bahama Is.: 260 (1905).
Mexico to C. America, Florida (Key West), Caribbean. 78 FLA 79 MXG MXN MXS MXT 80 BLZ COS GUA HON PAN 81 BAH CAY CUB DOM HAI JAM LEE PUE WIN. Nanophan. or phan.
**Calyptranthes chytraculia* var. *pauciflora* O.Berg, Linnaea 27: 27 (1855). *Chytraculia chytraculia* var. *pauciflora* (O.Berg) Sudw., Bull. Div. Forest. U.S.D.A. 14: 305 (1897).

var. ***mexicana*** (Lundell) McVaugh, Fieldiana, Bot. 29: 409 (1963).
W. & S. Mexico. 79 MXN MXS MXT. Nanophan. or phan.
**Calyptranthes mexicana* Lundell, Wrightia 2: 166 (1961).

var. ***pallens***
S. Mexico to C. America, Florida (Key West), Caribbean. 78 FLA 79 MXG MXS MXT 80 BLZ GUA HON 81 BAH CAY CUB DOM HAI JAM LEE PUE WIN. Nanophan. or phan.
Eugenia pallens Poir. in J.B.A.P.M.de Lamarck, Encycl., Suppl. 3: 122 (1813).

var. ***williamsii*** (Standl.) McVaugh, Fieldiana, Bot. 29: 408 (1963).
Costa Rica to Panama. 80 COS PAN. Nanophan. or phan.
Calyptranthes costaricensis O.Berg, Linnaea 27: 20 (1855). *Chytraculia costaricensis* (O.Berg) Kuntze, Revis. Gen. Pl. 1: 238 (1891).
**Calyptranthes williamsii* Standl., Ceiba 1: 156 (1950).

Calyptranthes palustris Urb. & Ekman, Ark. Bot. 24A(4): 22 (1932).
Dominican Rep. 81 DOM. Nanophan. or phan.

Calyptranthes paniculata Ruiz & Pav., Fl. Peruv. 4: t. 424 (1830). *Chytraculia paniculata* (Ruiz & Pav.) Kuntze, Revis. Gen. Pl. 1: 238 (1891).
Ecuador to Peru. 83 ECU PER. Nanophan. or phan.
Calyptranthes fragrans Ruiz ex O.Berg, Linnaea 27: 20 (1855).

Calyptranthes paradoxa Urb., Symb. Antill. 9: 97 (1923).
E. Cuba. 81 CUB. Nanophan.

Calyptranthes pauciflora O.Berg in C.F.P.von Martius & auct. suc. (eds.), Fl. Bras. 14(1): 543 (1859). *Chytraculia pauciflora* (O.Berg) Kuntze, Revis. Gen. Pl. 1: 238 (1891).
SE. Brazil. 84 BZL. Phan.

Calyptranthes paxillata McVaugh, Fieldiana, Bot. 29: 410 (1963).
C. America. 80 ELS GUA NIC. Phan.

Calyptranthes peduncularis Alain, Bull. Torrey Bot. Club 90: 189 (1963).
Puerto Rico. 81 PUE. Nanophan.
Calyptranthes dumetorum Alain, Bull. Torrey Bot. Club 92: 298 (1965).

Calyptranthes pendula O.Berg, Linnaea 27: 21 (1855). *Chytraculia pendula* (O.Berg) Kuntze, Revis. Gen. Pl. 1: 238 (1891).
Mexico (Oaxaca) to Guatemala. 79 MXS 80 GUA. Nanophan. or phan.

Calyptranthes peninsularis Bisse, Revista Jard. Bot. Nac. Univ. Habana 4(2): 3 (1983).
Cuba (Matanzas). 81 CUB. Nanophan.

Calyptranthes pereireana Mattos & D.Legrand, Loefgrenia 67: 19 (1975).
Brazil (Rio de Janeiro). 84 BZL. Nanophan.

Calyptranthes perlaevigata Lundell, Contr. Univ. Michigan Herb. 7: 29 (1942).
SE. Mexico to El Salvador. 79 MXT 80 ELS. Nanophan. or phan.

Calyptranthes petenensis Lundell, Phytologia 16: 443 (1968).
Guatemala. 80 GUA. Phan.

Calyptranthes picachoana Urb. & Ekman, Ark. Bot. 24A(4): 26 (1932).
Dominican Rep. 81 DOM. Nanophan.

Calyptranthes picardae Krug & Urb., Bot. Jahrb. Syst. 19: 595 (1895).
Hispaniola. 81 DOM HAI. Nanophan.

Calyptranthes pileata D.Legrand, Lilloa 31: 199 (1962).
S. Brazil. 84 BZS. Phan.

var. ***pileata***.
Brazil (Santa Catarina). 84 BZS. Phan.

var. ***riograndensis*** D.Legrand, Lilloa 31: 201 (1962).
Brazil (Rio Grande do Sul). 84 BZS. Phan.

Calyptranthes pinetorum Britton & P.Wilson, Bull. Torrey Bot. Club 43: 465 (1916).
W. Cuba. 81 CUB. Nanophan.

Calyptranthes pitoniana Urb. & Ekman, Ark. Bot. 21A(5): 23 (1927).
Haiti. 81 HAI. Nanophan. or phan.

Calyptranthes pittieri Standl., Publ. Field Mus. Nat. Hist., Bot. Ser. 8: 143 (1930).
Costa Rica. 80 COS. Nanophan. or phan.

Calyptranthes platyphylla O.Berg in C.F.P.von Martius & auct. suc. (eds.), Fl. Bras. 14(1): 539 (1859). *Chytraculia platyphylla* (O.Berg) Kuntze, Revis. Gen. Pl. 1: 238 (1891).
Brazil (São Paulo). 84 BZL. Nanophan. or phan.

Calyptranthes plicata McVaugh, Fieldiana, Bot. 29: 182 (1956).
Ecuador to N. Peru. 83 ECU PER. Nanophan. or phan.

Calyptranthes pocsiana Borhidi, Bot. Közlem. 64: 17 (1977).
Cuba (Sierra de Moa). 81 CUB. Nanophan.

Calyptranthes polysticta Urb., Symb. Antill. 7: 298 (1912).
E. Cuba. 81 CUB. Nanophan.

Calyptranthes portoricensis Britton, Bull. Torrey Bot. Club 51: 11 (1924).
Puerto Rico. 81 PUE. Nanophan. or phan.

Calyptranthes pozasiana Urb., Symb. Antill. 9: 479 (1928).
W. Cuba. 81 CUB. Nanophan. or phan.

Calyptranthes proctorii Acev.-Rodr., Brittonia 51: 168 (1999).
Jamaica. 81 JAM. Nanophan. or phan.
**Calyptranthes uniflora* Proctor, J. Arnold Arbor. 63: 272 (1982), nom. illeg.

Calyptranthes protracta Urb., Symb. Antill. 9: 480 (1928).
W. Cuba. 81 CUB. Nanophan. or phan.

Calyptranthes pseudoapoda Bisse & A.Rodr., Revista Jard. Bot. Nac. Univ. Habana 1(2–3): 44 (1980 publ. 1981).
Cuba. 81 CUB. Nanophan.

Calyptranthes pseudobrunneica Parra-Os., Caldasia 24: 96 (2002).
Colombia. 83 CLM.

Calyptranthes pseudomoaensis Borhidi & O.Muñiz, Bot. Közlem. 64: 16 (1977).
E. Cuba. 81 CUB. Nanophan. or phan.

Calyptranthes pteropoda O.Berg in C.F.P.von Martius & auct. suc. (eds.), Fl. Bras. 14(1): 47 (1857). *Chytraculia pteropoda* (O.Berg) Kuntze, Revis. Gen. Pl. 1: 238 (1891).
Brazil (Minas Gerais). 84 BZL. Nanophan. or phan.

Calyptranthes pulchella DC., Prodr. 3: 257 (1828). *Chytraculia pulchella* (DC.) Kuntze, Revis. Gen. Pl. 1: 238 (1891).
S. Trop. America. 82 GUY SUR VEN 83 ECU PER 84 BZE BZL BZN. Nanophan. or phan.
Calyptranthes pulchella var. *cuneata* O.Berg in C.F.P.von Martius & auct. suc. (eds.), Fl. Bras. 14(1): 43 (1857).
Calyptranthes pulchella var. *grandiflora* O.Berg in C.F.P.von Martius & auct. suc. (eds.), Fl. Bras. 14(1): 44 (1857).
Calyptranthes pulchella var. *latifolia* O.Berg in C.F.P.von Martius & auct. suc. (eds.), Fl. Bras. 14(1): 43 (1857).
Calyptranthes pulchella var. *parviflora* O.Berg in C.F.P.von Martius & auct. suc. (eds.), Fl. Bras. 14(1): 516 (1858).

Calyptranthes pullei Burret ex Amshoff, Recueil Trav. Bot. Néerl. 42: 4 (1950).
S. Trop. America. 82 GUY SUR VEN 83 PER 84 BZN. Nanophan. or phan.
Calyptranthes pullei var. *immaculata* McVaugh, Mem. New York Bot. Gard. 10(1): 78 (1958).

Calyptranthes punctata Griseb., Mem. Amer. Acad. Arts, n.s., 8: 181 (1861). *Chytraculia punctata* (Griseb.) Millsp., Publ. Field Columb. Mus., Bot. Ser. 1: 431 (1895).
Cuba. 81 CUB. Nanophan.

Calyptranthes quinoensis Mattos, Loefgrenia 95: 2 (1989).
Brazil (Sale de Quino). 84 BZN?.

Calyptranthes regeliana O.Berg in C.F.P.von Martius & auct. suc. (eds.), Fl. Bras. 14(1): 542 (1859). *Chytraculia regeliana* (O.Berg) Kuntze, Revis. Gen. Pl. 1: 238 (1891).
Brazil (São Paulo). 84 BZL. Phan.

Calyptranthes restingae Sobral, Napaea 4: 11 (1988).
Brazil (Bahia). 84 BZE. Phan.

Calyptranthes rhodophylla Ekman & Urb. in I.Urban, Symb. Antill. 9: 480 (1928).
SE. Cuba. 81 CUB. Nanophan.

Calyptranthes rigida Sw., Prodr.: 80 (1788). *Myrtus rigida* (Sw.) J.F.Gmel., Syst. Nat.: 792 (1791). *Chytraculia rigida* (Sw.) Kuntze, Revis. Gen. Pl. 1: 238 (1891).
E. Cuba, Jamaica. 81 CUB JAM. Nanophan. or phan.
Calyptranthes fawcettii Krug, Bot. Jahrb. Syst. 19: 600 (1895).

Calyptranthes rostellata Mattos & D.Legrand, Loefgrenia 67: 18 (1975).
Brazil (Rio de Janeiro). 84 BZL. Nanophan.

Calyptranthes rotundata Griseb., Mem. Amer. Acad. Arts, n.s., 8: 181 (1861).
C. & E. Cuba. 81 CUB. Nanophan.

Calyptranthes rubella (O.Berg) D.Legrand, in Fl. Ilustr. Catar. 1(Mirtac.): 535 (1971).
Brazil (E. & SE. Santa Catarina, Rio Grande do Sul). 84 BZS. Nanophan. or phan.
**Calyptranthes variabilis* var. *rubella* O.Berg in C.F.P.von Martius & auct. suc. (eds.), Fl. Bras. 14(1): 49 (1857).

Calyptranthes rufescens Mattos & D.Legrand, Loefgrenia 67: 19 (1975).
Brazil (Minas Gerais). 84 BZL. Nanophan.

Calyptranthes rufotomentosa McVaugh, Fieldiana, Bot. 29: 183 (1956).
N. Brazil. 84 BZN. Nanophan. or phan.

Calyptranthes ruiziana O.Berg, Linnaea 27: 22 (1855). *Chytraculia ruiziana* (O.Berg) Kuntze, Revis. Gen. Pl. 1: 238 (1891).
S. Venezuela to N. Peru. 82 VEN 83 ECU PER 84 BZN. Nanophan. or phan.

Calyptranthes rupicola Urb., Symb. Antill. 7: 295 (1912).
E. Cuba. 81 CUB. Nanophan. or phan.

Calyptranthes salamensis Lundell, Wrightia 5: 81 (1975).
Guatemala. 80 GUA. Phan.

Calyptranthes salicifolia Urb. & Ekman, Ark. Bot. 22A(10) 29 (1929).
Haiti (Î. de la Gonâve). 81 HAI. Nanophan. or phan.

Calyptranthes samuelssonii Urb. & Ekman, Ark. Bot. 21A(5): 26 (1927).
Haiti. 81 HAI. Nanophan. or phan.

Calyptranthes schiedeana O.Berg, Linnaea 27: 28 (1855). *Chytraculia schiedeana* (O.Berg) Kuntze, Revis. Gen. Pl. 1: 238 (1891).
Mexico to C. America. 79 MXG MXS MXT 80 COS PAN. Nanophan. or phan.
Myrcia aromatica Schltdl., Linnaea 13: 415 (1839). *Calyptranthes schlechtendaliana* O.Berg, Linnaea 27: 29 (1855). *Chytraculia schlechtendaliana* (O.Berg) Kuntze, Revis. Gen. Pl. 1: 238 (1891).

Calyptranthes scoparia O.Berg, Linnaea 31: 247 (1862). *Chytraculia scoparia* (O.Berg) Kuntze, Revis. Gen. Pl. 1: 238 (1891).
SE. Brazil. 84 BZL. Nanophan. or phan.

Calyptranthes selleana Urb. & Ekman, Ark. Bot. 22A(10): 33 (1929).
Hispaniola. 81 DOM HAI. Nanophan.

Calyptranthes sessilis McVaugh, Fieldiana, Bot. 29: 183 (1956).
N. Peru. 83 PER. Nanophan. or phan.

Calyptranthes simulata McVaugh, Fieldiana, Bot. 29: 184 (1956).
Ecuador to N. Peru. 83 ECU PER. Nanophan. or phan.

Calyptranthes sintenisii Kiaersk., Bot. Tidsskr. 17: 250 (1889).
Hispaniola to Puerto Rico. 81 DOM HAI PUE. Phan.

Calyptranthes smithii McVaugh, Mem. New York Bot. Gard. 10(1): 78 (1958).
Guyana. 82 GUY. Phan.

Calyptranthes sordida Urb. & Ekman, Ark. Bot. 21A(5): 25 (1927).
Hispaniola. 81 DOM HAI. Nanophan. or phan.

Calyptranthes speciosa Sagot, Ann. Sci. Nat., Bot., VI, 20: 187 (1885).
Guianas to N. Peru. 82 FRG SUR 83 ECU PER. Nanophan. or phan.

var. ***gigantifolia*** (McVaugh) McVaugh, Publ. Field Mus. Nat. Hist., Bot. Ser. 13(4/2): 613 (1958).
N. Peru. 83 PER. Nanophan. or phan.
**Calyptranthes gigantifolia* McVaugh, Fieldiana, Bot. 29: 181 (1956).

var. ***speciosa***
Guianas to N. Peru. 82 FRG SUR 83 ECU PER. Nanophan. or phan.

Calyptranthes spicata Amshoff, Recueil Trav. Bot. Néerl. 39: 148 (1942).
Suriname. 82 SUR. Phan.

Calyptranthes spruceana O.Berg in C.F.P.von Martius & auct. suc. (eds.), Fl. Bras. 14(1): 45 (1857). *Chytraculia spruceana* (O.Berg) Kuntze, Revis. Gen. Pl. 1: 238 (1891).
N. Brazil. 84 BZN. Nanophan. or phan.

Calyptranthes strigipes O.Berg in C.F.P.von Martius & auct. suc. (eds.), Fl. Bras. 14(1): 540 (1859). *Chytraculia strigipes* (O.Berg) Kuntze, Revis. Gen. Pl. 1: 238 (1891).
Brazil (Rio de Janeiro to Santa Catarina). 84 BZL BZS. Phan.

Calyptranthes subcapitata Urb., Symb. Antill. 9: 483 (1928).
Cuba (Sierra de Nipe). 81 CUB. Nanophan. or phan.

Calyptranthes tenuipes McVaugh, Fieldiana, Bot. 29: 411 (1963).
Mexico (Veracruz, Puebla). 79 MXC MXG. Nanophan. or phan.

Calyptranthes terniflora Urb. & Ekman, Ark. Bot. 20A(5): 23 (1926).
Hispaniola. 81 DOM HAI. Nanophan.

Calyptranthes tessmannii Burret ex McVaugh, Fieldiana, Bot. 29: 184 (1956).
N. Brazil to N. Peru. 83 ECU? PER 84 BZN. Nanophan. or phan.

Calyptranthes tetraptera O.Berg in C.F.P.von Martius & auct. suc. (eds.), Fl. Bras. 14(1): 53 (1857). *Chytraculia tetraptera* (O.Berg) Kuntze, Revis. Gen. Pl. 1: 238 (1891).
Brazil (Minas Gerais). 84 BZL. Nanophan. or phan.

Calyptranthes thomasiana O.Berg, Linnaea 27: 26 (1855). *Chytraculia thomasiana* (O.Berg) Kuntze, Revis. Gen. Pl. 1: 238 (1891).
Puerto Rico to Leeward Is. 81 LEE PUE. Nanophan. or phan.

Calyptranthes toaensis Borhidi, Bot. Közlem. 64: 16 (1977).
Cuba (Sierra de Frijol). 81 CUB. Nanophan. or phan.

Calyptranthes tonii Lundell, Wrightia 4: 40 (1968).
Mexico (Chiapas) to Guatemala. 79 MXT 80 GUA. Phan.

Calyptranthes tricona D.Legrand, Lilloa 31: 204 (1962).
S. Brazil to Argentina (Misiones). 84 BZS 85 AGE. Phan.
Calyptranthes iraiensis Mattos, Loefgrenia 10: 2 (1963).

Calyptranthes tridymantha Diels, Verh. Bot. Vereins Prov. Brandenburg 48: 188 (1906 publ. 1907).
N. Brazil. 84 BZN. Phan.

Calyptranthes triflora Alain, Bull. Torrey Bot. Club 90: 189 (1963).
Puerto Rico. 81 PUE. Nanophan.

Calyptranthes tumidonodia Schery, Ann. Missouri Bot. Gard. 30: 95 (1943).
Panama. 80 PAN. Nanophan. or phan.
Calyptranthes urophylla Standl. & L.O.Williams, Ceiba 3: 215 (1953).

Calyptranthes tussaceana O.Berg, Linnaea 27: 25 (1855). *Chytraculia tussaceana* (O.Berg) Kuntze, Revis. Gen. Pl. 1: 238 (1891).
Jamaica. 81 JAM. Nanophan.
**Calyptranthes rigida* Tussac, Fl. Antill. 3: t. 26 (1824), nom. illeg.

Calyptranthes umbelliformis Krug & Urb., Bot. Jahrb. Syst. 19: 596 (1895).
E. Jamaica. 81 JAM. Nanophan. or phan.
Calyptranthes impressa Urb., Symb. Antill. 5: 442 (1908).

Calyptranthes ursina Barroso & Peixoto, Acta Bot. Brasil. 10: 80 (1996).
Brazil (Rio de Janeiro). 84 BZL. Nanophan. or phan.

Calyptranthes variabilis O.Berg in C.F.P.von Martius & auct. suc. (eds.), Fl. Bras. 14(1): 49 (1857).

Chytraculia variabilis (O.Berg) Kuntze, Revis. Gen. Pl. 1: 238 (1891).
SE. & S. Brazil. 84 BZL BZS. Nanophan. or phan.

Calyptranthes venulosa Lundell, Wrightia 4: 40 (1968).
Mexico (Chiapas). 79 MXT. Nanophan.

Calyptranthes vexata McVaugh, Fieldiana, Bot. 29: 412 (1963).
Guyana. 82 GUY. Phan.
**Calyptranthes apoda* McVaugh, Mem. New York Bot. Gard. 10(1): 70 (1958), nom. illeg.

Calyptranthes widgreniana O.Berg in C.F.P.von Martius & auct. suc. (eds.), Fl. Bras. 14(1): 39 (1857). *Chytraculia widgreniana* (O.Berg) Kuntze, Revis. Gen. Pl. 1: 238 (1891).
WC. & SE. Brazil. 84 BZC BZL. Nanophan. or phan.

Calyptranthes wilsonii Griseb., Fl. Brit. W. I.: 233 (1860).
E. Jamaica. 81 JAM. Nanophan.

Calyptranthes woodburyi Alain, Phytologia 54: 107 (1983).
Puerto Rico. 81 PUE. Nanophan. or phan.

Calyptranthes yaquensis Urb. & Ekman, Ark. Bot. 24A(4): 24 (1932).
Dominican Rep. (Cordillera Central). 81 DOM. Nanophan. or phan.

Calyptranthes yaraensis Urb., Symb. Antill. 9: 481 (1928).
SE. Cuba. 81 CUB. Nanophan. or phan.

Calyptranthes zanquinensis A.R.Molina, Ceiba 3: 168 (1953).
Honduras. 80 HON. Nanophan. or phan.

Calyptranthes zuzygium (L.) Sw., Fl. Ind. Occid. 2: 919 (1800).
Florida, Caribbean. 78 FLA 81 BAH CAY CUB DOM HAI JAM LEE PUE SWC. Nanophan. or phan.
**Myrtus zuzygium* L., Syst. Nat. ed. 10, 2: 1056 (1759). *Calyptranthes chytraculia* var. *zuzygium* (L.) O.Berg, Linnaea 27: 28 (1855). *Chytraculia zuzygium* (L.) Kuntze, Revis. Gen. Pl. 1: 238 (1891). *Chytraculia chytraculia* var. *zuzygium* (L.) Sudw., Bull. Div. Forest. U.S.D.A. 14: 305 (1897).
Myrtus syzygium L., Amoen. Acad. 5: 398 (1760), orth. var.
Calyptranthes zyzygium Sw., Prodr.: 79 (1788), orth. var.
Myrtus dichotoma Salisb., Prodr. Stirp. Chap. Allerton: 354 (1796), nom. illeg.
Eugenia glomerata Sieber ex C.Presl, Isis (Oken) 21: 274 (1828).
Calyptranthes chytraculia var. *ovalis* O.Berg, Linnaea 27: 27 (1855). *Chytraculia chytraculia* var. *ovalis* (O.Berg) Sudw., Bull. Div. Forest. U.S.D.A. 14: 305 (1897).
Calyptranthes chytraculia var. *trichotoma* O.Berg, Linnaea 27: 27 (1855). *Chytraculia chytraculia* var. *trichotoma* (O.Berg) Sudw., Bull. Div. Forest. U.S.D.A. 14: 305 (1897).
Calyptranthes rigida Macfad. ex Griseb., Fl. Brit. W. I.: 233 (1860), nom. illeg.

Synonyms:
Calyptranthes aguilarii Standl. & Steyerm. = ***Calyptranthes hondurensis*** Standl.
Calyptranthes apoda McVaugh = ***Calyptranthes vexata*** McVaugh
Calyptranthes aromatica Blume = ***Syzygium antisepticum*** (Blume) Merr. & L.M.Perry
Calyptranthes belizensis (Standl.) Lundell = ***Calyptranthes hondurensis*** Standl.
Calyptranthes brasiliensis var. *densa* O.Berg = ***Calyptranthes brasiliensis*** Spreng.
Calyptranthes brasiliensis var. *laxa* O.Berg = ***Calyptranthes brasiliensis*** Spreng.
Calyptranthes brasiliensis var. *mutabilis* (O.Berg) D.Legrand = ***Calyptranthes brasiliensis*** Spreng.
Calyptranthes capitellata Buch.-Ham. ex Wall. = ***Syzygium cumini*** (L.) Skeels
Calyptranthes caroli Britton & P.Wilson = ***Calyptranthes capitulata*** var. ***caroli*** (Britton & P.Wilson) Borhidi & O.Muñiz
Calyptranthes caroli var. *longipedunculata* Moldenke = ***Calyptranthes capitulata*** var. ***caroli*** (Britton & P.Wilson) Borhidi & O.Muñiz
Calyptranthes caryophyllata (L.) Pers. = ***Syzygium caryophyllatum*** (L.) Alston
Calyptranthes caryophyllifolia Willd. = ***Syzygium cumini*** (L.) Skeels
Calyptranthes chrysophylloides var. *minor* Urb. & Ekman = ***Calyptranthes chrysophylloides*** Urb. & Ekman
Calyptranthes chytraculia var. *ovalis* O.Berg = ***Calyptranthes zuzygium*** (L.) Sw.
Calyptranthes chytraculia var. *pauciflora* O.Berg = ***Calyptranthes pallens*** Griseb.
Calyptranthes chytraculia var. *trichotoma* O.Berg = ***Calyptranthes zuzygium*** (L.) Sw.
Calyptranthes chytraculia var. *zuzygium* (L.) O.Berg = ***Calyptranthes zuzygium*** (L.) Sw.
Calyptranthes clarendonensis Proctor = ***Mitranthes clarendonensis*** (Proctor) Proctor
Calyptranthes clusiifolia var. *cordifolia* D.Legrand = ***Calyptranthes clusiifolia*** O.Berg
Calyptranthes concinna var. *glomerata* (Cambess.) D.Legrand = ***Calyptranthes concinna*** DC.
Calyptranthes concinna var. *glomerata* (Cambess.) D.Legrand = ***Calyptranthes concinna*** DC.
Calyptranthes concinna var. *paulistana* D.Legrand = ***Calyptranthes concinna*** DC.
Calyptranthes cordata O.Berg = ***Myrcia subcordata*** DC.
Calyptranthes cordifolia Moon = ***Syzygium cordifolium*** (Wight) Walp. subsp. ***cordifolium***
Calyptranthes corymbosa Blume = ***Syzygium corymbosum*** (Blume) DC.
Calyptranthes costaricensis O.Berg = ***Calyptranthes pallens*** var. ***williamsii*** (Standl.) McVaugh
Calyptranthes costata Buch.-Ham. = ***Syzygium operculatum*** (Roxb.) Nied.
Calyptranthes cumini (L.) Pers. = ***Syzygium cumini*** (L.) Skeels
Calyptranthes cuminodora Stokes = ***Syzygium cumini*** (L.) Skeels
Calyptranthes cuneata Buch.-Ham. ex Wall. = ***Syzygium operculatum*** (Roxb.) Nied.
Calyptranthes danca Buch.-Ham. ex Wall. = ***Syzygium salicifolium*** (Wight) J.Graham
Calyptranthes densa DC. = ***Myrcia densa*** (DC.) Sobral
Calyptranthes dumetorum Alain = ***Calyptranthes peduncularis*** Alain
Calyptranthes ekmanii Urb. = ***Calyptrogenia ekmanii*** (Urb.) Burret
Calyptranthes eugenioides Seem. = ***Syzygium eugenioides*** (Merr. & L.M.Perry) Biffin & Craven
Calyptranthes eugeniopsoides D.Legrand & Kausel = ***Marlierea eugeniopsoides*** (D.Legrand & Kausel) D.Legrand

Calyptranthes euryphylla Standl. = ***Calyptranthes millspaughii*** Urb.
Calyptranthes fasciculata var. *genuina* Stehlé & Quentin = ***Calyptranthes fasciculata*** O.Berg
Calyptranthes fasciculata var. *hahnii* (Krug & Urb.) Stehlé = ***Calyptranthes fasciculata*** O.Berg
Calyptranthes fastigiata O.Berg = [84 BZL]
Calyptranthes fastigiata Blume = ***Syzygium fastigiatum*** (Blume) Merr. & L.M.Perry
Calyptranthes fawcettii Krug = ***Calyptranthes rigida*** Sw.
Calyptranthes ferruginea Krug & Urb. = ***Calyptranthes cuprea*** O.Berg
Calyptranthes ferruginea var. *cajalbanica* Borhidi & O.Muñiz = ***Calyptranthes cuprea*** O.Berg
Calyptranthes floribunda (H.West ex Willd.) Blume = ***Myrciaria floribunda*** (H.West ex Willd.) O.Berg
Calyptranthes florifera McVaugh = ***Calyptranthes multiflora*** Poepp. ex O.Berg
Calyptranthes fluviatilis Lundell = ***Calyptranthes lindeniana*** O.Berg
Calyptranthes fragrans Ruiz ex O.Berg = ***Calyptranthes paniculata*** Ruiz & Pav.
Calyptranthes gigantifolia McVaugh = ***Calyptranthes speciosa*** var. ***gigantifolia*** (McVaugh) McVaugh
Calyptranthes glomerata Cambess. = ***Calyptranthes concinna*** DC.
Calyptranthes grammica var. *triflora* (O.Berg) Mattos = ***Calyptranthes grammica*** (Spreng.) D.Legrand
Calyptranthes grandifolia var. *macrantha* D.Legrand = ***Calyptranthes grandifolia*** O.Berg
Calyptranthes grandifolia var. *rufa* (O.Berg) D.Legrand = ***Calyptranthes grandifolia*** O.Berg
Calyptranthes grandis Buch.-Ham. ex Wall. = ***Syzygium operculatum*** (Roxb.) Nied.
Calyptranthes guineensis Willd. = ***Syzygium guineense*** (Willd.) DC.
Calyptranthes impressa Urb. = ***Calyptranthes umbelliformis*** Krug & Urb.
Calyptranthes iraiensis Mattos = ***Calyptranthes tricona*** D.Legrand
Calyptranthes jambolana (Lam.) Willd. = ***Syzygium cumini*** (L.) Skeels
Calyptranthes jambolifera Stokes = ***Syzygium cumini*** (L.) Skeels
Calyptranthes kleinii D.Legrand = ***Calyptranthes concinna*** DC.
Calyptranthes lanceolata var. *catharinensis* O.Berg = ***Calyptranthes lanceolata*** O.Berg
Calyptranthes lanceolata var. *latifolia* O.Berg = ***Calyptranthes lanceolata*** O.Berg
Calyptranthes langsdorffii var. *grandiflora* Kiaersk. = ***Calyptranthes langsdorffii*** O.Berg
Calyptranthes langsdorffii var. *macrophylla* Kiaersk. = ***Calyptranthes langsdorffii*** O.Berg
Calyptranthes lateriflora DC. = ***Plinia spiciflora*** (Nees & Mart.) Sobral
Calyptranthes latifolia Poir. = ***Syzygium latifolium*** (Poir.) DC.
Calyptranthes laxiflora Blume = ***Syzygium laxiflorum*** (Blume) DC.
Calyptranthes lindeniana var. *americana* (McVaugh) Lundell = ***Calyptranthes chytraculia*** var. ***americana*** McVaugh
Calyptranthes loranthifolia var. *mutabilis* (O.Berg) D.Legrand = ***Calyptranthes brasiliensis*** Spreng.
Calyptranthes lucida f. *hilariana* D.Legrand = ***Calyptranthes lucida*** Mart. ex DC. var. ***lucida***
Calyptranthes makul (Gaertn.) Raeusch. = ***Syzygium makul*** Gaertn.
Calyptranthes makul Blanco = ***Syzygium operculatum*** (Roxb.) Nied.
Calyptranthes malabarica Dennst. = ***Syzygium zeylanicum*** (L.) DC.
Calyptranthes mangiferifolia Hance ex Walp. = ***Syzygium operculatum*** (Roxb.) Nied.
Calyptranthes maschalantha O.Berg = ***Neomitranthes obscura*** (DC.) N.Silveira
Calyptranthes maschalantha var. *rotundifolia* O.Berg = ***Neomitranthes obscura*** (DC.) N.Silveira
Calyptranthes maxonii Britton & Urb. = ***Mitranthes maxonii*** (Britton & Urb.) Proctor
Calyptranthes mexicana Lundell = ***Calyptranthes pallens*** var. ***mexicana*** (Lundell) McVaugh
Calyptranthes montana Britton & P.Wilson = ***Calyptranthes leptoclada*** Urb.
Calyptranthes musciflora O.Berg = ***Calyptranthes grammica*** (Spreng.) D.Legrand
Calyptranthes musciflora var. *angustifolia* O.Berg = ***Calyptranthes grammica*** (Spreng.) D.Legrand
Calyptranthes musciflora var. *glauca* O.Berg = ***Calyptranthes grammica*** (Spreng.) D.Legrand
Calyptranthes musciflora var. *obscura* O.Berg = ***Calyptranthes grammica*** (Spreng.) D.Legrand
Calyptranthes musciflora var. *spathulata* O.Berg = ***Calyptranthes grammica*** (Spreng.) D.Legrand
Calyptranthes musciflora var. *triflora* O.Berg = ***Calyptranthes grammica*** (Spreng.) D.Legrand
Calyptranthes musciflora var. *venosa* O.Berg = ***Calyptranthes grammica*** (Spreng.) D.Legrand
Calyptranthes mutabilis O.Berg = ***Calyptranthes brasiliensis*** Spreng.
Calyptranthes myrtoides (A.Gray) Seem. = ***Syzygium myrtoides*** (A.Gray) R.Schmid
Calyptranthes nitida Raeusch. = ***Gossia lucida*** (Gaertn.) N.Snow & Guymer
Calyptranthes obovata Krug & Urb. = ***Calyptranthes kiaerskovii*** Krug & Urb.
Calyptranthes obscura DC. = ***Neomitranthes obscura*** (DC.) N.Silveira
Calyptranthes obscura Mart. = ***Neomitranthes obscura*** (DC.) N.Silveira
Calyptranthes obscura var. *fluminensis* O.Berg = ***Neomitranthes obscura*** (DC.) N.Silveira
Calyptranthes obtusa Miq. = ***Marlierea montana*** (Aubl.) Amshoff
Calyptranthes obtusifolia Buch.-Ham. ex Wall. = ***Syzygium inophyllum*** DC.
Calyptranthes oneillii Lundell = ***Syzygium cumini*** (L.) Skeels
Calyptranthes ottonis (O.Berg) C.Wright = ***Mitranthes ottonis*** O.Berg
Calyptranthes paniculata Spreng. = ***Calyptranthes brasiliensis*** Spreng.
Calyptranthes paniculata (Lam.) Raeusch. = ***Syzygium borbonicum*** J.Guého & A.J.Scott
Calyptranthes pleophlebia Diels = ***Calyptranthes longifolia*** O.Berg
Calyptranthes poeppigiana O.Berg = ***Calyptranthes multiflora*** Poepp. ex O.Berg
Calyptranthes pohliana O.Berg = ***Calyptranthes lucida*** Mart. ex DC. var. ***lucida***
Calyptranthes pollicina Willemet = ***Syzygium cymosum*** (Lam.) DC. var. ***cymosum***
Calyptranthes polyantha O.Berg = ***Calyptranthes lucida*** var. ***polyantha*** (O.Berg) D.Legrand

Calyptranthes polyneura Urb. = ***Myrcia polyneura*** (Urb.) Borhidi
Calyptranthes pulchella var. *cuneata* O.Berg = ***Calyptranthes pulchella*** DC.
Calyptranthes pulchella var. *grandiflora* O.Berg = ***Calyptranthes pulchella*** DC.
Calyptranthes pulchella var. *latifolia* O.Berg = ***Calyptranthes pulchella*** DC.
Calyptranthes pulchella var. *parviflora* O.Berg = ***Calyptranthes pulchella*** DC.
Calyptranthes pullei var. *immaculata* McVaugh = ***Calyptranthes pullei*** Burret ex Amshoff
Calyptranthes ramiflora Blanco = ***Syzygium simile*** (Merr.) Merr.
Calyptranthes ramosissima Urb. = ***Plinia ramosissima*** (Urb.) Urb.
Calyptranthes ranulphii D.Legrand = ***Neomitranthes glomerata*** (D.Legrand) D.Legrand
Calyptranthes reitziana D.Legrand = ***Calyptranthes concinna*** DC.
Calyptranthes rigida Macfad. ex Griseb. = ***Calyptranthes zuzygium*** (L.) Sw.
Calyptranthes rigida Tussac = ***Calyptranthes tussaceana*** O.Berg
Calyptranthes rigida var. *apicata* C.Wright ex Griseb. = ***Calyptranthes apicata*** (C.Wright ex Griseb.) Urb.
Calyptranthes rostrata Griseb. = [81 CUB]
Calyptranthes rufa O.Berg = ***Calyptranthes grandifolia*** O.Berg
Calyptranthes schlechtendaliana O.Berg = ***Calyptranthes schiedeana*** O.Berg
Calyptranthes seemannii (A.Gray) Seem. = ***Syzygium seemannii*** (A.Gray) Biffin & Craven
Calyptranthes sericea Griseb. = ***Calyptranthes fasciculata*** O.Berg
Calyptranthes sericea var. *hahnii* Krug & Urb. = ***Calyptranthes fasciculata*** O.Berg
Calyptranthes tatna Buch.-Ham. ex Wall. = ***Syzygium operculatum*** (Roxb.) Nied.
Calyptranthes tenuis Buch.-Ham. ex Wall. = ***Syzygium tenue*** (Duthie) N.P.Balakr.
Calyptranthes thomasiana var. *obscura* O.Berg = ***Calyptranthes bergii*** Krug & Urb.
Calyptranthes tobagensis Krug & Urb. = ***Myrcia decorticans*** DC.
Calyptranthes tonduzii Donn.Sm. = ***Psidium sartorianum*** (O.Berg) Nied.
Calyptranthes tovarensis (O.Berg) Steyerm. = ***Myrcia tovarensis*** O.Berg
Calyptranthes tuberculata O.Berg = ***Neomitranthes obscura*** (DC.) N.Silveira
Calyptranthes uniflora Spreng. = ***Calyptranthes grammica*** (Spreng.) D.Legrand
Calyptranthes uniflora Proctor = ***Calyptranthes proctorii*** Acev.-Rodr.
Calyptranthes urbanii Fawc. & Rendle = ***Calyptranthes chytraculia*** (L.) Sw. var. ***chytraculia***
Calyptranthes urophylla Standl. & L.O.Williams = ***Calyptranthes tumidonodia*** Schery
Calyptranthes variabilis var. *oblongata* O.Berg = ***Calyptranthes concinna*** DC.
Calyptranthes variabilis var. *pulchella* O.Berg = ***Calyptranthes concinna*** DC.
Calyptranthes variabilis var. *riparia* O.Berg = ***Calyptranthes concinna*** DC.
Calyptranthes variabilis var. *rubella* O.Berg = ***Calyptranthes rubella*** (O.Berg) D.Legrand
Calyptranthes variabilis var. *stricta* O.Berg = ***Calyptranthes concinna*** DC.
Calyptranthes warmingiana Kiaersk. = ***Neomitranthes warmingiana*** (Kiaersk.) Mattos
Calyptranthes williamsii Standl. = ***Calyptranthes pallens*** var. ***williamsii*** (Standl.) McVaugh
Calyptranthes zuzygium Blanco = ***Syzygium operculatum*** (Roxb.) Nied.
Calyptranthes zyzygium Sw. = ***Calyptranthes zuzygium*** (L.) Sw.

Unplaced Names:
Calyptranthes fastigiata O.Berg in C.F.P.von Martius & auct. suc. (eds.), Fl. Bras. 14(1): 50 (1857), nom. illeg. = [84 BZL]
Calyptranthes rostrata Griseb., Mem. Amer. Acad. Arts, n.s., 8: 181 (1861), nom. illeg. = [81 CUB]

Calyptranthus

Calyptranthus Blume = ***Syzygium*** Gaertn.
Calyptranthus pyrifolia Blume = ***Syzygium pyrifolium*** (Blume) DC.
Calyptranthus racemosa Blume = ***Syzygium racemosum*** (Blume) DC.
Calyptranthus rostrata Blume = ***Syzygium rostratum*** (Blume) DC.

Calyptrogenia

Calyptrogenia Burret, Notizbl. Bot. Gart. Berlin-Dahlem 15: 545 (1941).
Caribbean. 81 DOM HAI JAM. [Myrtaceae]
6 Species

Calyptrogenia biflora Alain, Moscosoa 1(1): 28 (1976).
Dominican Rep. (Sierra Prieta). 81 DOM. Nanophan. or phan.

Calyptrogenia bracteosa (Urb.) Burret, Notizbl. Bot. Gart. Berlin-Dahlem 15: 546 (1941).
Haiti (Massif de la Hotte). 81 HAI. Nanophan.
**Calyptropsidium bracteosum* Urb., Ark. Bot. 21A(5): 20 (1927).

Calyptrogenia cuspidata Alain, Phytologia 61: 356 (1986).
SW. Dominican Rep. 81 DOM. Nanophan. or phan.

Calyptrogenia ekmanii (Urb.) Burret, Notizbl. Bot. Gart. Berlin-Dahlem 15: 545 (1941).
Haiti (Massif de la Hotte), NE. Jamaica. 81 HAI JAM. Nanophan.
**Calyptranthes ekmanii* Urb., Ark. Bot. 22A(10): 32 (1929).

Calyptrogenia grandiflora Burret, Notizbl. Bot. Gart. Berlin-Dahlem 15: 545 (1941).
Hispaniola. 81 DOM HAI. Nanophan.
**Calyptropsidium ekmanii* Urb., Ark. Bot. 24A(4): 17 (1932). *Neomitranthes ekmanii* (Urb.) Mattos, Loefgrenia 99: 6 (1990).

Calyptrogenia jeremiensis (Urb. & Ekman) Burret, Notizbl. Bot. Gart. Berlin-Dahlem 15: 545 (1941).
Haiti (Massif de la Hotte), NW. Jamaica. 81 HAI JAM. Nanophan. or phan.
**Eugenia jeremiensis* Urb. & Ekman, Ark. Bot. 24A(4): 29 (1932).

Synonyms:
Calyptrogenia hatschbachii D.Legrand = ***Myrceugenia gertii*** Landrum
Calyptrogenia riedeliana (O.Berg) Burret = ***Neomitranthes riedeliana*** (O.Berg) Mattos

Calyptromyrcia

Calyptromyrcia O.Berg = ***Myrcia*** DC. ex Guill.
Calyptromyrcia cordata O.Berg = [84 BZE]
Calyptromyrcia costata (DC.) O.Berg = ***Myrcia splendens*** (Sw.) DC.
Calyptromyrcia cymosa O.Berg = ***Myrcianthes cymosa*** (O.Berg) Mattos
Calyptromyrcia cymosa var. *major* O.Berg = ***Myrcianthes cymosa*** (O.Berg) Mattos
Calyptromyrcia cymosa var. *minor* O.Berg = ***Myrcianthes cymosa*** (O.Berg) Mattos
Calyptromyrcia elegans (DC.) O.Berg = ***Myrcia guianensis*** (Aubl.) DC.
Calyptromyrcia eugenioides (Cambess.) O.Berg = ***Marlierea eugenioides*** (Cambess.) D.Legrand
Calyptromyrcia eugenioides var. *glabra* O.Berg = ***Marlierea eugenioides*** (Cambess.) D.Legrand
Calyptromyrcia eugenioides var. *puberula* O.Berg = ***Marlierea eugenioides*** (Cambess.) D.Legrand
Calyptromyrcia paniculata O.Berg = [84 BZL]
Calyptromyrcia paniculata var. *opaca* O.Berg = ?
Calyptromyrcia puberula (Cambess.) O.Berg = ***Myrcia tomentosa*** (Aubl.) DC.
Calyptromyrcia spixiana (DC.) O.Berg = ***Myrcia guianensis*** (Aubl.) DC.
Calyptromyrcia venosa O.Berg = ***Myrcia tomentosa*** (Aubl.) DC.

***Unplaced Names*:**
Calyptromyrcia cordata O.Berg in C.F.P.von Martius & auct. suc. (eds.), Fl. Bras. 14(1): 56 (1857). = [84 BZE]
Calyptromyrcia paniculata O.Berg in C.F.P.von Martius & auct. suc. (eds.), Fl. Bras. 14(1): 57 (1857). = [84 BZL]

Calyptropsidium

Calyptropsidium O.Berg = ***Psidium*** L.
Calyptropsidium bracteosum Urb. = ***Calyptrogenia bracteosa*** (Urb.) Burret
Calyptropsidium ekmanii Urb. = ***Calyptrogenia grandiflora*** Burret
Calyptropsidium friedrichsthalianum O.Berg = ***Psidium friedrichsthalianum*** (O.Berg) Nied.
Calyptropsidium sartorianum (O.Berg) Krug & Urb. = ***Psidium sartorianum*** (O.Berg) Nied.
Calyptropsidium sintenisii Kiaersk. = ***Psidium sintenisii*** (Kiaersk.) Alain

Calythropsis

Calythropsis C.A.Gardner = ***Calytrix*** Labill.
Calythropsis aurea C.A.Gardner = ***Calytrix ecalycata*** Craven

Calytrix

Calytrix Labill., Nov. Holl. Pl. 2: 8 (1806).
Australia. 50 NTA QLD SOA WAU. [Myrtaceae]
78 Species
Lhotskya Schauer, Linnaea 10: 309 (1836).
Calycothrix Meisn., Pl. Vasc. Gen.: 107 (1837).
Trichocalyx Schauer, Monogr. Myrt. Xerocarp.: 86 (1840), nom. illeg.
Calythropsis C.A.Gardner, J. Roy. Soc. Western Australia 27: 188 (1942).

Calytrix achaeta (F.Muell.) Benth., Fl. Austral. 3: 52 (1867).
N. Australia. 50 NTA QLD WAU. Nanophan.
Lhotskya cuspidata F.Muell., Hooker's J. Bot. Kew Gard. Misc. 8: 324 (1856), nom. nud. *Calytrix cuspidata* (F.Muell.) Druce, Bot. Soc. Exch. Club Brit. Isles 1916: 611 (1917), nom. inval.
**Calycothrix achaeta* F.Muell., Trans. Philos. Inst. Victoria 3: 43 (1859).

Calytrix acutifolia (Lindl.) Craven, Brunonia 10: 122 (1987).
SW. Australia. 50 WAU. Nanophan.
Lhotskya ericoides Schauer in J.Lindley, Intr. Nat. Syst. Bot., ed. 2: 439 (1836).
**Lhotskya acutifolia* Lindl., Sketch Veg. Swan R.: vii (1839).
Calytrix truncata A.Cunn., J. Bot. (Hooker) 4: 271 (1841).
Lhotskya scabra Turcz., Bull. Soc. Imp. Naturalistes Moscou 35(2): 324 (1862).
Lhotskya hirta Regel, Gartenflora 12: 337 (1863).

Calytrix alpestris (Lindl.) Court, Victorian Naturalist 73: 176 (1957).
South Australia. 50 SOA. Nanophan.
**Genetyllis alpestris* Lindl. in T.L.Mitchell, Three Exped. Australia 2: 178 (1838). *Lhotskya genethylloides* F.Muell., Trans. Philos. Inst. Victoria 1: 16 (1854), nom. illeg. *Lhotskya alpestris* (Lindl.) Druce, Rep. Bot. Exch. Club Brit. Isles 1916: 633 (1917).
Lhotskya genethylloides var. *glabra* F.Muell., Trans. Philos. Inst. Victoria 1: 16 (1854).
Lhotskya genethylloides var. *bracteosa* Benth., Fl. Austral. 3: 54 (1867). *Lhotskya alpestris* var. *bracteosa* (Benth.) J.M.Black, Fl. S. Austral. 3: 426 (1986).

Calytrix amethystina Craven, Brunonia 10: 108 (1987).
Western Australia (E. of Sandstone). 50 WAU. Nanophan.

Calytrix angulata Lindl., Sketch Veg. Swan R.: 6 (1839).
Calycothrix angulata (Lindl.) Schauer, Nov. Actorum Acad. Caes. Leop.-Carol. Nat. Cur. 19(Suppl. 2): 256 (1841).
SW. Australia. 50 WAU. Nanophan.

Calytrix arborescens (F.Muell.) Benth., Fl. Austral. 3: 51 (1867).
Northern Territory (Darwin). 50 NTA. Nanophan.
**Calycothrix arborescens* F.Muell., Trans. Philos. Inst. Victoria 3: 42 (1859).

Calytrix asperula (Schauer) Benth., Fl. Austral. 3: 424 (1867).
SW. Australia. 50 WAU. Nanophan.
**Calycothrix asperula* Schauer in J.G.C.Lehmann, Pl. Preiss. 1: 106 (1845).
Calytrix asperula var. *gracilis* Benth., Fl. Austral. 3: 42 (1867).

Calytrix aurea Lindl., Sketch Veg. Swan R.: 5 (1839).
Calycothrix aurea (Lindl.) Schauer, Nov. Actorum Acad. Caes. Leop.-Carol. Nat. Cur. 19(Suppl 2): 258 (1841).
Western Australia (Irwin, Drummond). 50 WAU. Nanophan.

Calytrix birdii (F.Muell.) B.D.Jacks., Index Kew. 1: 398 (1893).
Western Australia (Coolgardie). 50 WAU. Nanophan.
**Calycothrix birdii* F.Muell., Fragm. 10: 26 (1877).

Calytrix brachychaeta (F.Muell.) Benth., Fl. Austral. 3: 52 (1867).
N. Australia. 50 NTA QLD. Nanophan.
**Calycothrix brachychaeta* F.Muell., Trans. Philos. Inst. Victoria 3: 43 (1859).
Calytrix brachychaeta var. *tenuifolia* Benth., Fl. Austral. 3: 52 (1867).
Calytrix laricina R.Br. ex Benth., Fl. Austral. 3: 52 (1867). *Calycothrix laricina* (R.Br. ex Benth.) F.Muell., Syst. Census Austral. Pl. 2: 90 (1889).

Calytrix brevifolia (Meisn.) Benth., Fl. Austral. 3: 45 (1867).
SW. Australia. 50 WAU. Nanophan.
**Calycothrix brevifolia* Meisn., J. Proc. Linn. Soc., Bot. 1: 46 (1857).
Calycothrix muricata F.Muell., Fragm. 1: 224 (1858).
Calytrix muricata (F.Muell.) Benth., Fl. Austral. 3: 44 (1867).
Calytrix muricata var. *parvifolia* Benth., Fl. Austral. 3: 45 (1867).

Calytrix breviseta Lindl., Sketch Veg. Swan R.: 5 (1839). *Calycothrix breviseta* (Lindl.) Schauer, Nov. Actorum Acad. Caes. Leop.-Carol. Nat. Cur. 19(Suppl. 2): 251 (1841).
SW. Australia. 50 WAU. Nanophan.

subsp. ***breviseta***.
SW. Australia. 50 WAU. Nanophan.
Calycothrix cuspidata Turcz., Bull. Soc. Imp. Naturalistes Moscou 20(1): 162 (1847).

subsp. ***stipulosa*** (W.Fitzg.) Craven, Brunonia 10: 113 (1987).
SW. Australia. 50 WAU. Nanophan.
**Calytrix stipulosa* W.Fitzg., J. West Austral. Nat. Hist. Soc. 1: 18 (1904).

Calytrix brownii (Schauer) Craven, Brunonia 10: 42 (1987).
N. Western Australia. 50 WAU. Nanophan.
**Calycothrix brownii* Schauer, Nov. Actorum Acad. Caes. Leop.-Carol. Nat. Cur. 19(Suppl. 2): 260 (1841).

Calytrix carinata Craven, Brunonia 10: 71 (1987).
Australia. 50 NTA QLD SOA WAU. Nanophan.

Calytrix chrysantha Craven, Brunonia 10: 90 (1987).
Western Australia (Eneabba). 50 WAU. Nanophan.

Calytrix creswellii (F.Muell.) B.D.Jacks., Index Kew. 1: 398 (1893).
Western Australia (Avon, Coolgardie). 50 WAU. Nanophan.
**Calycothrix creswellii* F.Muell., Fragm. 10: 27 (1877).

Calytrix decandra DC., Prodr. 3: 208 (1828). *Calycothrix decandra* (DC.) F.Muell., Syst. Census Austral. Pl. 1: 52 (1882).
Western Australia (Eyre, Roe). 50 WAU. Nanophan.
Calycothrix candolleana Schauer, Nov. Actorum Acad. Caes. Leop.-Carol. Nat. Cur. 19(Suppl. 2): 244 (1841).
Calycothrix conanthera F.Muell., Fragm. 1: 146 (1858).

Calytrix decussata Craven, Brunonia 3: 237 (1980).
Northern Territory (Darwin). 50 NTA. Nanophan.

Calytrix depressa (Turcz.) Benth., Fl. Austral. 3: 47 (1867).
SW. Australia. 50 WAU. Nanophan.
**Calycothrix depressa* Turcz., Bull. Soc. Imp. Naturalistes Moscou 20(1): 162 (1847).
Calycothrix rosea Meisn., J. Proc. Linn. Soc., Bot. 1: 46 (1857).
Calycothrix tenuifolia Meisn., J. Proc. Linn. Soc., Bot. 1: 46 (1857). *Calytrix tenuifolia* (Meisn.) Benth., Fl. Austral. 3: 47 (1867).
Calytrix tenuifolia var. *rigidior* Benth., Fl. Austral. 3: 47 (1867).
Calytrix stowardii S.Moore, J. Linn. Soc., Bot. 45: 174 (1920).

Calytrix desolata S.Moore, J. Linn. Soc., Bot. 34: 191 (1899).
Western Australia (Austin, Coolgardie). 50 WAU. Nanophan.

Calytrix divergens Craven, Brunonia 10: 61 (1987).
Western Australia (Murchison River). 50 WAU. Nanophan.

Calytrix drummondii (Meisn.) Craven, Brunonia 10: 93 (1987).
Western Australia (Murchison River, Mingenew-Coorow Distr.). 50 WAU. Nanophan.
**Calycothrix drummondii* Meisn., J. Proc. Linn. Soc., Bot. 1: 47 (1857). *Calytrix flavescens* var. *drummondii* (Meisn.) Benth., Fl. Austral. 3: 42 (1867).

Calytrix duplistipulata Craven, Brunonia 10: 48 (1987).
SW. Australia. 50 WAU. Nanophan.

Calytrix ecalycata Craven, Austral. Syst. Bot. 3: 722 (1990).
Western Australia. 50 WAU. Nanophan.
**Calythropsis aurea* C.A.Gardner, J. Roy. Soc. Western Australia 27: 189 (1942).

subsp. ***brevis*** Keighery, Nuytsia 15: 266 (2004).
Western Australia. 50 WAU. Nanophan.

subsp. ***ecalycata***
Western Australia (Irwin). 50 WAU. Nanophan.

subsp. ***pubescens*** Keighery, Nuytsia 15: 267 (2004).
Western Australia. 50 WAU. Nanophan.

Calytrix eneabbensis Craven, Brunonia 10: 55 (1987).
Western Australia (Eneabbon Distr.). 50 WAU. Nanophan.

Calytrix erosipetala Craven, Brunonia 10: 37 (1987).
Western Australia (Meekotharra Distr., Lake Barlee Distr.). 50 WAU. Nanophan.

Calytrix exstipulata DC., Prodr. 3: 208 (1828).
N. Australia. 50 NTA QLD WAU. Nanophan.
Calytrix conferta A.Cunn., Bot. Mag. 61: t. 3323 (1834). *Calycothrix conferta* (A.Cunn.) Schauer, Nov. Actorum Acad. Caes. Leop.-Carol. Nat. Cur. 19(Suppl. 2): 240 (1841).
Calytrix cupressifolia A.Rich. in J.S.C.Dumont d'Urville, Voy. Astrolabe 2: 41 (1834). *Calycothrix cupressifolia* (A.Rich.) Stapf, Index Lond. 1: 538 (1929).
Calytrix cupressoides A.Rich. in J.S.C.Dumont d'Urville, Voy. Astrolabe 2: 43 (1834). *Calycothrix cupressoides* (A.Rich.) Schauer, Nov. Actorum Acad. Caes. Leop.-Carol. Nat. Cur. 19(Suppl. 2): 270 (1841).
Calytrix microphylla A.Cunn., Bot. Mag. 61: t. 3323 (1834), nom. illeg.
Calycothrix baueri Schauer, Nov. Actorum Acad. Caes.

Leop.-Carol. Nat. Cur. 19(Suppl. 2): 261 (1841).
Calytrix baueri (Schauer) Benth., Fl. Austral. 3: 51 (1867), pro syn.
Calycothrix microphylla Schauer, Nov. Actorum Acad. Caes. Leop.-Carol. Nat. Cur. 19(Suppl. 2): 241 (1841).
Calytrix microphylla var. *longifolia* Benth., Fl. Austral. 3: 49 (1867).
Calytrix interstans S.Moore, J. Linn. Soc., Bot. 45: 200 (1920).
Calytrix wickllamiana S.Moore, J. Linn. Soc., Bot. 45: 198 (1920).

Calytrix faucicola Craven, Brunonia 3: 241 (1980).
Northern Territory (Darwin). 50 NTA. Nanophan.

Calytrix flavescens A.Cunn., Bot. Mag. 61: t. 3323 (1834). *Calycothrix flavescens* (A.Cunn.) Schauer, Nov. Actorum Acad. Caes. Leop.-Carol. Nat. Cur. 19(Suppl. 2): 257 (1841).
SW. Australia. 50 WAU. Nanophan.
Calycothrix luteola Schauer in J.G.C.Lehmann, Pl. Preiss. 1: 106 (1845).
Calycothrix puberula Meisn., J. Proc. Linn. Soc., Bot. 1: 48 (1857). *Calytrix puberula* (Meisn.) Benth., Fl. Austral. 3: 42 (1867).
Calycothrix tenella Meisn., J. Proc. Linn. Soc., Bot. 1: 47 (1857). *Calytrix flavescens* var. *tenella* (Meisn.) Benth., Fl. Austral. 3: 42 (1867).
Calycothrix tetragonophylla Meisn., J. Proc. Linn. Soc., Bot. 1: 47 (1857).

Calytrix formosa Craven, Brunonia 10: 105 (1987).
Western Australia (Kalbarri Distr.). 50 WAU. Nanophan.

Calytrix fraseri A.Cunn., Bot. Mag. 61: t. 3323 (1834). *Calycothrix fraseri* (A.Cunn.) Schauer, Nov. Actorum Acad. Caes. Leop.-Carol. Nat. Cur. 19(Suppl. 2): 250 (1841).
SW. Australia. 50 WAU. Nanophan.
Calytrix granulosa Benth., Fl. Austral. 3: 49 (1867). *Calycothrix granulosa* (Benth.) F.Muell., Syst. Census Austral. Pl. 1: 52 (1882).

Calytrix glaberrima (F.Muell.) Craven, Brunonia 10: 26 (1987).
South Australia (S. Mt. Lofty Range, Kangaroo I.). 50 SOA. Nanophan.
**Lhotskya glaberrima* F.Muell., Fragm. 1: 13 (1858).
Lhotskya glaberrima var. *magnisepala* J.M.Black, Fl. S. Austral., ed. 4, 3: 426 (1986).

Calytrix glutinosa Lindl., Sketch Veg. Swan R.: 5 (1839). *Calycothrix glutinosa* (Lindl.) Schauer, Nov. Actorum Acad. Caes. Leop.-Carol. Nat. Cur. 19(Suppl. 2): 243 (1841).
SW. Australia. 50 WAU. Nanophan.

Calytrix gracilis Benth., Fl. Austral. 3: 45 (1867). *Calycothrix gracilis* (Benth.) F.Muell., Syst. Census Austral. Pl. 1: 52 (1882).
SW. Australia. 50 WAU. Nanophan.

Calytrix gurulmundensis Craven, Brunonia 10: 79 (1987).
Queensland (Gulugula Distr.). 50 QLD. Nanophan.

Calytrix gypsophila Craven, Brunonia 10: 36 (1987).
South Australia. 50 SOA. Nanophan.

Calytrix habrantha Craven, Brunonia 10: 96 (1987).
SW. Australia. 50 WAU. Nanophan.

Calytrix harvestiana (F.Muell.) Craven, Brunonia 10: 59 (1987).
Western Australia (Irwin). 50 WAU. Nanophan.
**Lhotskya harvestiana* F.Muell., Fragm. 11: 8 (1878).

Calytrix inopinata Craven, Austral. Syst. Bot. 4: 535 (1991).
Northern Territory. 50 NTA. Nanophan.

Calytrix involucrata J.M.Black, Trans. & Proc. Roy. Soc. South Australia 52: 225 (1928).
South Australia. 50 SOA. Nanophan.

Calytrix islensis Craven, Brunonia 10: 80 (1987).
Queensland (Theodore). 50 QLD. Nanophan.

Calytrix leptophylla Benth., Fl. Austral. 3: 50 (1867). *Calycothrix leptophylla* (Benth.) F.Muell., Syst. Census Austral. Pl. 1: 52 (1882).
Queensland (Cook). 50 QLD. Nanophan.

Calytrix leschenaultii (Schauer) Benth., Fl. Austral. 3: 46 (1867).
Western Australia. 50 WAU. Nanophan.
**Calycothrix leschenaultii* Schauer in J.G.C.Lehmann, Pl. Preiss. 1: 104 (1844).
Calycothrix brachyphylla Turcz., Bull. Soc. Imp. Naturalistes Moscou 20(1): 161 (1847). *Calytrix brachyphylla* (Turcz.) Benth., Fl. Austral. 3: 45 (1867).

Calytrix longiflora (F.Muell.) Benth., Fl. Austral. 3: 49 (1867).
C. Australia. 50 NTA QLD SOA. Nanophan.
**Calycothrix longiflora* F.Muell., Fragm. 1: 12 (1858).

Calytrix megaphylla (F.Muell.) Benth., Fl. Austral. 3: 50 (1867).
Northern Territory (Darwin). 50 NTA. Nanophan.
**Calycothrix megaphylla* F.Muell., Fragm. 1: 13, (1858).

Calytrix merrelliana (F.Muell. & Tate) Craven, Brunonia 10: 118 (1987).
SW. Australia. 50 WAU. Nanophan.
**Lhotskya violacea* var. *merrilliana* F.Muell. & Tate, Trans. & Proc. Roy. Soc. South Australia 16: 355 (1896).

Calytrix micrairoides Craven, Brunonia 3: 236 (1980).
Northern Territory (Darwin). 50 NTA. Nanophan.

Calytrix microcoma Craven, Brunonia 10: 70 (1987).
Queensland. 50 QLD. Nanophan.

Calytrix mimiana Craven, Brunonia 3: 234 (1980).
Northern Territory (Darwin). 50 NTA. Nanophan.

Calytrix nematoclada Craven, Brunonia 10: 121 (1987).
SW. Australia. 50 WAU. Nanophan.
**Lhotskya ciliata* F.Muell. ex Benth., Fl. Austral. 3: 54 (1867).

Calytrix oldfieldii Benth., Fl. Austral. 3: 46 (1867). *Calycothrix oldfieldii* (Benth.) F.Muell., Syst. Census Austral. Pl. 1: 52 (1882).
SW. Australia. 50 WAU. Nanophan.

Calytrix oncophylla Craven, Austral. Syst. Bot. 3: 720 (1990).
Western Australia (Toodyay Distr.). 50 WAU. Nanophan.

Calytrix parvivallis Craven, Brunonia 10: 120 (1987).
Western Australia (Minnivale Distr.). 50 WAU. Nanophan.

Calytrix paucicostata Craven, Brunonia 10: 106 (1987).
Western Australia (Kolborri Distr.). 50 WAU. Nanophan.

Calytrix pimeleoides C.A.Gardner ex Keighery, Nuytsia 15: 262 (2004).
Western Australia. 50 WAU. Nanophan.

Calytrix platycheiridia Craven, Brunonia 10: 62 (1987).
Western Australia (Coorow-Waterloo Distr.). 50 WAU. Nanophan.

Calytrix plumulosa (F.Muell.) B.D.Jacks., Index Kew. 1: 398 (1893).
SW. Australia. 50 WAU. Nanophan.
**Calycothrix plumulosa* F.Muell., Fragm. 10: 27 (1877).

Calytrix praecipua Craven, Brunonia 10: 50 (1987).
Western Australia (Laverton Distr.). 50 WAU. Nanophan.

Calytrix pulchella (Turcz.) B.D.Jacks., Index Kew. 1: 398 (1893).
SW. Australia. 50 WAU. Nanophan.
**Calycothrix pulchella* Turcz., Bull. Cl. Phys.-Math. Acad. Imp. Sci. Saint-Pétersbourg 10: 328 (1852).

Calytrix purpurea (F.Muell.) Craven, Brunonia 10: 58 (1987).
Western Australia (Irwin). 50 WAU. Nanophan.
**Lhotskya purpurea* F.Muell., Fragm. 1: 224 (1859).

Calytrix rupestris Craven, Brunonia 10: 46 (1987).
Northern Territory (Darwin). 50 NTA. Nanophan.

Calytrix sapphirina Lindl., Sketch Veg. Swan R.: 5 (1839). *Calycothrix sapphirina* (Lindl.) Schauer, Nov. Actorum Acad. Caes. Leop.-Carol. Nat. Cur. 19(Suppl. 2): 255 (1841).
SW. Australia. 50 WAU. Nanophan.
Calycothrix empetroides Schauer, Nov. Actorum Acad. Caes. Leop.-Carol. Nat. Cur. 19(Suppl. 2): 254 (1841). *Calytrix empetroides* (Schauer) Benth., Fl. Austral. 3: 44 (1867).
Calycothrix ciliata Turcz., Bull. Soc. Imp. Naturalistes Moscou 20(1): 161 (1847). *Calytrix ciliata* (Turcz.) Benth., Fl. Austral. 3: 44, 161 (1867).
Calycothrix lasiostachya F.Muell., Fragm. 1: 224 (1858).

Calytrix similis Craven, Brunonia 10: 84 (1987).
SW. Australia. 50 WAU. Nanophan.

Calytrix simplex Lindl., Sketch Veg. Swan R.: 5 (1839). *Calycothrix simplex* (Lindl.) Schauer, Nov. Actorum Acad. Caes. Leop.-Carol. Nat. Cur. 19(Suppl. 2): 253 (1841).
SW. Australia. 50 WAU. Nanophan.

subsp. ***simplex***.
SW. Australia. 50 WAU. Nanophan.
Calytrix stenophylla W.Fitzg., J. West Austral. Nat. Hist. Soc. 1: 18 (1904).

subsp. ***suboppositifolia*** Craven, Brunonia 10: 100 (1987).
SW. Australia. 50 WAU. Nanophan.

Calytrix smeatoniana (F.Muell.) Craven, Brunonia 10: 27 (1987).
South Australia (Kangaroo I.). 50 SOA. Nanophan.
**Lhotskya smeatoniana* F.Muell., Australas. J. Pharm. 2: 5 (1887).

Calytrix strigosa A.Cunn., Bot. Mag. 61: t. 3323 (1834). *Calycothrix strigosa* (A.Cunn.) Schauer, Nov. Actorum Acad. Caes. Leop.-Carol. Nat. Cur. 19(Suppl. 2): 260 (1841).
W. Western Australia. 50 WAU. Nanophan.
Calycothrix lasiantha Meisn., J. Proc. Linn. Soc., Bot. 1: 46 (1857).
Calycothrix watsonii F.Muell. & Tate, Trans. & Proc. Roy. Soc. South Australia 16: 355 (1896). *Calytrix watsonii* (F.Muell. & Tate) C.A.Gardner, Enum. Pl. Austr. Occ.: 96 (1931).

Calytrix superba C.A.Gardner & A.S.George, J. Roy. Soc. Western Australia 46: 134 (1963).
Western Australia (Irwin). 50 WAU. Nanophan.

Calytrix surdiviperana Craven, Brunonia 3: 226 (1980).
Northern Territory (Darwin). 50 NTA. Nanophan.

Calytrix sylvana Craven, Brunonia 10: 117 (1987).
Western Australia (Drummond, Dale). 50 WAU. Nanophan.
**Lhotskya brevifolia* Schauer in J.G.C.Lehmann, Pl. Preiss. 1: 103 (1844).

Calytrix tenuiramea (Turcz.) Benth., Fl. Austral. 3: 48 (1867).
SW. Australia. 50 WAU. Nanophan.
**Calycothrix tenuiramea* Turcz., Bull. Soc. Imp. Naturalistes Moscou 22(2): 20 (1849).

Calytrix tetragona Labill., Nov. Holl. Pl. 2: 8 (1806). *Calycothrix tetragona* (Labill.) F.Muell., Fragm. 4: 36 (1864).
Western Australia, South Australia, Queensland. 50 QLD SOA WAU. Nanophan.
Calytrix glabra R.Br., Bot. Reg. 5: t. 409 (1819). *Calycothrix glabra* (R.Br.) Hook.f., Fl. Tasman. 1: 127 (1857).
Calytrix ericoides A.Cunn., Field New South Wales: 350 (1825).
Calytrix scabra DC., Prodr. 3: 208 (1828). *Calycothrix scabra* (DC.) Schauer, Nov. Actorum Acad. Caes. Leop.-Carol. Nat. Cur. 19(Suppl. 2): 248 (1841).
Calytrix pubescens Sweet ex G.Don, Gen. Hist. 2: 811 (1832).
Calytrix brunioides A.Cunn., Bot. Mag. 61: t. 3323 (1834). *Calycothrix brunioides* (A.Cunn.) Schauer, Nov. Actorum Acad. Caes. Leop.-Carol. Nat. Cur. 19(Suppl. 2): 244 (1841).
Calytrix curtophylla A.Cunn., Bot. Mag. 61: t. 3323 (1834). *Calycothrix curtophylla* (A.Cunn.) Schauer, Nov. Actorum Acad. Caes. Leop.-Carol. Nat. Cur. 19(Suppl. 2): 242 (1841). *Calytrix flavescens* var. *curtophylla* (A.Cunn.) Benth., Fl. Austral. 3: 42 (1867).
Calytrix virgata A.Cunn., Bot. Mag. 61: t. 3323 (1834). *Calycothrix virgata* (A.Cunn.) Schauer, Nov. Actorum Acad. Caes. Leop.-Carol. Nat. Cur. 19(Suppl. 2): 247 (1841). *Calycothrix glabra* var. *virgata* (A.Cunn.) Hook.f., Fl. Tasman. 1: 127 (1857).
Calycothrix billardierei Schauer, Nov. Actorum Acad. Caes. Leop.-Carol. Nat. Cur. 19(Suppl. 2): 245 (1841).
Calycothrix behriana Schltdl., Linnaea 20: 650 (1847). *Calytrix behriana* (Schltdl.) Benth., Fl. Austral. 3: 51 (1867), pro syn.
Calycothrix scabra var. *minor* Schltdl., Linnaea 20: 650 (1847).
Calycothrix diversifolia Turcz., Bull. Cl. Phys.-Math. Acad. Imp. Sci. Saint-Pétersbourg 10: 317 (1852). *Calytrix diversifolia* (Turcz.) B.D.Jacks., Index Kew. 1: 398 (1893).
Calycothrix leucantha Miq., Ned. Kruidk. Arch. 4: 117 (1856).
Calycothrix monticola Miq., Ned. Kruidk. Arch. 4: 118 (1856).

Calycothrix muelleri Miq., Ned. Kruidk. Arch. 4: 119 (1856).
Calycothrix rosea Miq., Ned. Kruidk. Arch. 4: 117 (1856).
Calycothrix schlechtendalii Miq., Ned. Kruidk. Arch. 4: 116 (1856).
Calycothrix squarrosa Miq., Ned. Kruidk. Arch. 4: 118 (1856).
Calycothrix glabra var. *ciliata* Hook.f., Fl. Tasman. 1: 127 (1857).
Calycothrix glabra var. *glaberrima* Hook.f., Fl. Tasman. 1: 127 (1857).
Calytrix billardierei Benth., Fl. Austral. 3: 51 (1867), pro syn.
Calycothrix sullivanii F.Muell., Fragm. 9: 1 (1875). *Calytrix sullivanii* (F.Muell.) B.D.Jacks., Index Kew. 1: 399 (1893).
Calytrix mitchellii S.Moore, J. Linn. Soc., Bot. 45: 199 (1920).

Calytrix truncatifolia Craven, Brunonia 10: 103 (1987).
NW. Western Australia. 50 WAU. Nanophan.

Calytrix uncinata Craven, Austral. Syst. Bot. 3: 719 (1990).
WC. Western Australia. 50 WAU. Nanophan.

Calytrix variabilis Lindl., Sketch Veg. Swan R.: 5 (1839). *Calycothrix variabilis* (Lindl.) Schauer, Nov. Actorum Acad. Caes. Leop.-Carol. Nat. Cur. 19(Suppl. 2): 252 (1841).
SW. Australia. 50 WAU. Nanophan.

Calytrix verruculosa Craven, Brunonia 10: 51 (1987).
Western Australia (Meekatharra Distr.). 50 WAU. Nanophan.

Calytrix verticillata Craven, Brunonia 3: 225 (1980).
Northern Territory (Darwin). 50 NTA. Nanophan.

Calytrix violacea (Lindl.) Craven, Brunonia 10: 119 (1987).
SW. Australia. 50 WAU. Nanophan.
**Lhotskya violacea* Lindl., Sketch Veg. Swan R.: vii (1839).

Calytrix warburtonensis Craven, Brunonia 10: 34 (1987).
SC. Western Australia. 50 WAU. Nanophan.

Synonyms:
Calytrix asperula var. *gracilis* Benth. = ***Calytrix asperula*** (Schauer) Benth.
Calytrix baueri (Schauer) Benth. = ***Calytrix exstipulata*** DC.
Calytrix behriana (Schltdl.) Benth. = ***Calytrix tetragona*** Labill.
Calytrix billardierei Benth. = ***Calytrix tetragona*** Labill.
Calytrix brachychaeta var. *tenuifolia* Benth. = ***Calytrix brachychaeta*** (F.Muell.) Benth.
Calytrix brachyphylla (Turcz.) Benth. = ***Calytrix leschenaultii*** (Schauer) Benth.
Calytrix brunioides A.Cunn. = ***Calytrix tetragona*** Labill.
Calytrix ciliata (Turcz.) Benth. = ***Calytrix sapphirina*** Lindl.
Calytrix conferta A.Cunn. = ***Calytrix exstipulata*** DC.
Calytrix cupressifolia A.Rich. = ***Calytrix exstipulata*** DC.
Calytrix cupressoides A.Rich. = ***Calytrix exstipulata*** DC.
Calytrix curtophylla A.Cunn. = ***Calytrix tetragona*** Labill.
Calytrix cuspidata (F.Muell.) Druce = ***Calytrix achaeta*** (F.Muell.) Benth.
Calytrix diversifolia (Turcz.) B.D.Jacks. = ***Calytrix tetragona*** Labill.
Calytrix empetroides (Schauer) Benth. = ***Calytrix sapphirina*** Lindl.
Calytrix ericoides A.Cunn. = ***Calytrix tetragona*** Labill.
Calytrix flavescens var. *curtophylla* (A.Cunn.) Benth. = ***Calytrix tetragona*** Labill.
Calytrix flavescens var. *drummondii* (Meisn.) Benth. = ***Calytrix drummondii*** (Meisn.) Craven
Calytrix flavescens var. *tenella* (Meisn.) Benth. = ***Calytrix flavescens*** A.Cunn.
Calytrix glabra R.Br. = ***Calytrix tetragona*** Labill.
Calytrix granulosa Benth. = ***Calytrix fraseri*** A.Cunn.
Calytrix interstans S.Moore = ***Calytrix exstipulata*** DC.
Calytrix laricina R.Br. ex Benth. = ***Calytrix brachychaeta*** (F.Muell.) Benth.
Calytrix microphylla A.Cunn. = ***Calytrix exstipulata*** DC.
Calytrix microphylla var. *longifolia* Benth. = ***Calytrix exstipulata*** DC.
Calytrix mitchellii S.Moore = ***Calytrix tetragona*** Labill.
Calytrix muricata (F.Muell.) Benth. = ***Calytrix brevifolia*** (Meisn.) Benth.
Calytrix muricata var. *parvifolia* Benth. = ***Calytrix brevifolia*** (Meisn.) Benth.
Calytrix puberula (Meisn.) Benth. = ***Calytrix flavescens*** A.Cunn.
Calytrix pubescens Sweet ex G.Don = ***Calytrix tetragona*** Labill.
Calytrix scabra DC. = ***Calytrix tetragona*** Labill.
Calytrix stenophylla W.Fitzg. = ***Calytrix simplex*** Lindl. subsp. ***simplex***
Calytrix stipulosa W.Fitzg. = ***Calytrix breviseta*** subsp. ***stipulosa*** (W.Fitzg.) Craven
Calytrix stowardii S.Moore = ***Calytrix depressa*** (Turcz.) Benth.
Calytrix sullivanii (F.Muell.) B.D.Jacks. = ***Calytrix tetragona*** Labill.
Calytrix tenuifolia (Meisn.) Benth. = ***Calytrix depressa*** (Turcz.) Benth.
Calytrix tenuifolia var. *rigidior* Benth. = ***Calytrix depressa*** (Turcz.) Benth.
Calytrix truncata A.Cunn. = ***Calytrix acutifolia*** (Lindl.) Craven
Calytrix virgata A.Cunn. = ***Calytrix tetragona*** Labill.
Calytrix watsonii (F.Muell. & Tate) C.A.Gardner = ***Calytrix strigosa*** A.Cunn.
Calytrix wickllamiana S.Moore = ***Calytrix exstipulata*** DC.

Camphoromyrtus

Camphoromyrtus Schauer = ***Babingtonia*** Lindl.
Camphoromyrtus behrii Schltdl. = ***Babingtonia behrii*** (Schltdl.) A.R.Bean
Camphoromyrtus brownii Schauer = ***Triplarina imbricata*** (Sm.) A.R.Bean
Camphoromyrtus crenulata F.Muell. = ***Babingtonia crenulata*** (F.Muell.) A.R.Bean
Camphoromyrtus pluriflora F.Muell. = ***Babingtonia pluriflora*** (F.Muell.) A.R.Bean

Campomanesia

Campomanesia Ruiz & Pav., Fl. Peruv. Prodr.: 72 (1794).
S. America. 81 TRT 82 FRG GUY SUR VEN 83 BOL CLM ECU PER 84 AGE BAL BZC BZE BZL BZN BZS PAR 85 AGE PAR URU. [Myrtaceae]
37 Species
Burchardia Neck., Elem. Bot. 2: 76 (1790), opus utique oppr.

Abbevillea O.Berg, Linnaea 27: 425 (1856).
Acrandra O.Berg, Linnaea 27: 435 (1856).
Britoa O.Berg, Linnaea 27: 435 (1856).
Lacerdaea O.Berg, Linnaea 27: 437 (1856).
Paivaea O.Berg in C.F.P.von Martius & auct. suc. (eds.), Fl. Bras. 14(1): 614 (1859).

Campomanesia adamantium (Cambess.) O.Berg, Linnaea 27: 434 (1856).
Brazil to Paraguay. 84 BZC BZL BZS 85 PAR. Cham. or nanophan.
**Psidium adamantium* Cambess. in A.F.C.de Saint-Hilaire, Fl. Bras. Merid. 2: 292 (1832).
Psidium campestre Cambess. in A.F.C.de Saint-Hilaire, Fl. Bras. Merid. 2: 289 (1832). *Campomanesia cambessedeana* O.Berg in C.F.P.von Martius & auct. suc. (eds.), Fl. Bras. 14(1): 457 (1857). *Campomanesia campestris* (Cambess.) D.Legrand, Notul. Syst. (Paris) 15: 273 (1958), nom. illeg.
Campomanesia caerulea O.Berg in C.F.P.von Martius & auct. suc. (eds.), Fl. Bras. 14(1): 455 (1857).
Campomanesia desertorum O.Berg in C.F.P.von Martius & auct. suc. (eds.), Fl. Bras. 14(1): 450 (1857).
Campomanesia glabra O.Berg in C.F.P.von Martius & auct. suc. (eds.), Fl. Bras. 14(1): 450 (1857), nom. illeg.
Campomanesia microcarpa O.Berg in C.F.P.von Martius & auct. suc. (eds.), Fl. Bras. 14(1): 455 (1857).
Campomanesia obscura O.Berg in C.F.P.von Martius & auct. suc. (eds.), Fl. Bras. 14(1): 451 (1857).
Campomanesia vaccinioides O.Berg in C.F.P.von Martius & auct. suc. (eds.), Fl. Bras. 14(1): 450 (1857).
Campomanesia caerulescens O.Berg in C.F.P.von Martius & auct. suc. (eds.), Fl. Bras. 14(1): 612 (1859).
Campomanesia resinosa Barb.Rodr., Myrt. Paraguay: 20 (1903).
Campomanesia glareophila Barb.Rodr. ex Chodat & Hassl., Bull. Herb. Boissier, II, 7: 800 (1907), nom. nud.
Campomanesia lancifolia Barb.Rodr. ex Chodat & Hassl., Bull. Herb. Boissier, II, 7: 800 (1907), nom. nud.
Campomanesia paraguayensis Barb.Rodr. ex Chodat & Hassl., Bull. Herb. Boissier, II, 7: 800 (1907), nom. nud.
Campomanesia cambessedeana var. *nana* D.Legrand, Bol. Soc. Argent. Bot. 10: 8 (1962). *Campomanesia adamantium* var. *nana* (D.Legrand) Mattos, Loefgrenia 116: 3 (2001).
Campomanesia cambessedeana var. *pyriformis* Mattos, Loefgrenia 32: 1 (1969).

Campomanesia anemonea Landrum, Brittonia 53: 536 (2001 publ. 2002).
Brazil (Bahia). 84 BZE. Phan.

Campomanesia aprica (Vell.) O.Berg in C.F.P.von Martius & auct. suc. (eds.), Fl. Bras. 14(1): 459 (1857).
SE. Brazil. 84 BZL.
**Psidium apricum* Vell., Fl. Flumin. 5: 213, t. 58 (1829).

Campomanesia aromatica (Aubl.) Griseb., Fl. Brit. W. I.: 242 (1860).
Trinidad to S. Trop. America. 81 TRT 82 FRG GUY SUR VEN 83 BOL 84 BZE BZL BZN BZS. Nanophan. or phan.
**Psidium aromaticum* Aubl., Hist. Pl. Guiane 1: 485 (1775). *Burchardia aromatica* (Aubl.) Raf., Sylva Tellur.: 106 (1828).
Myrtus psidioides Desv. in W.Hamilton, Prodr. Pl. Ind. Occid.: 44 (1825).
Eugenia sparsiflora DC., Prodr. 3: 263 (1828). *Abbevillea martiana* (DC.) O.Berg in C.F.P.von Martius & auct. suc. (eds.), Fl. Bras. 14(1): 435 (1857). *Campomanesia sparsiflora* (DC.) J.F.Macbr., Candollea 5: 394 (1934).
Myrtus fascicularis DC., Prodr. 3: 240 (1828).
Psidium tenuifolium Mart. ex DC., Prodr. 3: 236 (1828). *Campomanesia tenuifolia* (Mart. ex DC.) O.Berg in C.F.P.von Martius & auct. suc. (eds.), Fl. Bras. 14(1): 452 (1857).
Eugenia desvauxiana O.Berg, Linnaea 27: 198 (1856).
Campomanesia ciliata O.Berg in C.F.P.von Martius & auct. suc. (eds.), Fl. Bras. 14(1): 453 (1857).
Campomanesia coaetanea O.Berg in C.F.P.von Martius & auct. suc. (eds.), Fl. Bras. 14(1): 444 (1857).
Campomanesia synchrona O.Berg in C.F.P.von Martius & auct. suc. (eds.), Fl. Bras. 14(1): 444 (1857).
Campomanesia beaurepairiana Kiaersk., Enum. Myrt. Bras.: 15 (1893).
Campomanesia glazioviana Kiaersk., Enum. Myrt. Bras.: 16 (1893).

Campomanesia aurea O.Berg in C.F.P.von Martius & auct. suc. (eds.), Fl. Bras. 14(1): 454 (1857).
S. Brazil to NE. Argentina. 84 BZS 85 AGE PAR URU. Cham. or nanophan.

var. ***aurea***
S. Brazil, Paraguay, Uruguay, NE. Argentina (Misiones). 84 BZS 85 AGE PAR URU. Cham. or nanophan.
Campomanesia cyanea O.Berg in C.F.P.von Martius & auct. suc. (eds.), Fl. Bras. 14(1): 454 (1857).
Campomanesia cyanea var. *cordata* O.Berg in C.F.P.von Martius & auct. suc. (eds.), Fl. Bras. 14(1): 454 (1857).
Campomanesia cyanea var. *ovata* O.Berg in C.F.P.von Martius & auct. suc. (eds.), Fl. Bras. 14(1): 454 (1857).
Campomanesia gardneriana O.Berg in C.F.P.von Martius & auct. suc. (eds.), Fl. Bras. 14(1): 455 (1857).
Campomanesia maracayuensis Barb.Rodr. ex Chodat & Hassl., Bull. Herb. Boissier, II, 7: 800 (1907), nom. nud.

var. ***hatschbachii*** (Mattos) D.Legrand, Bol. Fac. Agron. Univ. Montevideo 101: 68 (1968).
S. Brazil to NC. Uruguay. 84 BZS 85 URU. Cham. or nanophan.
**Campomanesia hatschbachii* Mattos, Loefgrenia 7: 1 (1962).
Campomanesia gracilis Kausel, Lilloa 33: 101 (1971 publ. 1972).

Campomanesia dichotoma (O.Berg) Mattos, Loefgrenia 26: 28 (1967).
E. Brazil. 84 BZE BZL. Nanophan. or phan.
**Britoa dichotoma* O.Berg, Linnaea 27: 436 (1856).
Britoa psidioides O.Berg, Linnaea 27: 436 (1856). *Campomanesia psidioides* (O.Berg) Nied. in H.G.A.Engler & K.A.E.Prantl, Nat. Pflanzenfam. 3(7): 73 (1893).

Britoa triflora O.Berg, Linnaea 27: 436 (1856). *Campomanesia triflora* (O.Berg) Baill., Hist. Pl. 6: 342 (1876).
Eugenia triflora Willd. ex O.Berg in C.F.P.von Martius & auct. suc. (eds.), Fl. Bras. 14(1): 462 (1857), nom. inval.
Britoa triflora var. *ilhensis* O.Berg in C.F.P.von Martius & auct. suc. (eds.), Fl. Bras. 14(1): 613 (1859).

Campomanesia espiritosantensis Landrum, Brittonia 39: 245 (1987).
SE. Brazil. 84 BZL.

Campomanesia eugenioides (Cambess.) D.Legrand, Notul. Syst. (Paris) 15: 274 (1958).
Brazil. 84 BZC BZE BZL BZN BZS. Nanophan. or phan.
Psidium desertorum Mart. ex DC., Prodr. 3: 236 (1828). *Campomanesia eugenioides* var. *desertorum* (Mart. ex DC.) Landrum, Brittonia 36: 241 (1984).
**Psidium eugenioides* Cambess. in A.F.C.de Saint-Hilaire, Fl. Bras. Merid. 2: 290 (1832). *Abbevillea eugenioides* (Cambess.) O.Berg in C.F.P.von Martius & auct. suc. (eds.), Fl. Bras. 14(1): 436 (1857).
Abbevillea chrysophylla O.Berg in C.F.P.von Martius & auct. suc. (eds.), Fl. Bras. 14(1): 437 (1857). *Psidium chrysophyllum* (O.Berg) F.Muell., Select Pl.: 188 (1872). *Campomanesia chrysophylla* (O.Berg) Nied. in H.G.A.Engler & K.A.E.Prantl, Nat. Pflanzenfam. 3(7): 73 (1893).
Campomanesia repanda O.Berg in C.F.P.von Martius & auct. suc. (eds.), Fl. Bras. 14(1): 456 (1857).
Campomanesia dardano-limai Mattos & D.Legrand, Loefgrenia 67: 10 (1975).
Myrcianthes campomanesioides Mattos & D.Legrand, Loefgrenia 67: 11 (1975).
Campomanesia montana D.Legrand, in Fl. Ilustr. Catar. 1(Mirtac.): 616 (1977).

Campomanesia fruticosa (Vell.) O.Berg in C.F.P.von Martius & auct. suc. (eds.), Fl. Bras. 14(1): 458 (1857).
SE. Brazil. 84 BZL.
**Psidium fruticosum* Vell., Fl. Flumin. 5: 213, t. 57 (1829).

Campomanesia grandiflora (Aubl.) Sagot, Ann. Sci. Nat., Bot., VI, 20: 182 (1885).
Guianas to N. & NE. Brazil. 82 FRG GUY SUR 84 BZE BZN. Nanophan. or phan.
**Psidium grandiflorum* Aubl., Hist. Pl. Guiane 1: 483 (1775). *Burchardia grandiflora* (Aubl.) Raf., Sylva Tellur.: 106 (1838).
Campomanesia poiteaui O.Berg, Linnaea 27: 432 (1856).

Campomanesia guaviroba (DC.) Kiaersk., Enum. Myrt. Bras.: 8 (1893).
E. & S. Brazil to Argentina (Misiones). 83 BOL 84 BZE BZL BZS 85 AGE PAR. Phan.
**Psidium guaviroba* DC., Prodr. 3: 235 (1828). *Abbevillea guaviroba* (DC.) O.Berg in C.F.P.von Martius & auct. suc. (eds.), Fl. Bras. 14(1): 432 (1857).
Psidium punctulatum DC., Prodr. 3: 233 (1828). *Abbevillea punctulata* (DC.) O.Berg in C.F.P.von Martius & auct. suc. (eds.), Fl. Bras. 14(1): 432 (1857). *Campomanesia punctulata* (DC.) Mattos & D.Legrand, Loefgrenia 67: 11 (1975).
Psidium dulce Vell., Fl. Flumin. 5: 213, t. 56 (1829). *Campomanesia dulcis* (Vell.) J.F.Macbr., Candollea 5: 394 (1934).
Psidium cerasoides Cambess. in A.F.C.de Saint-Hilaire, Fl. Bras. Merid. 2: 290 (1832). *Campomanesia cerasoides* (Cambess.) A.Gray, U.S. Expl. Exped., Phan. 1: 549 (1854). *Abbevillea cerasoides* (Cambess.) O.Berg in C.F.P.von Martius & auct. suc. (eds.), Fl. Bras. 14(1): 436 (1857).
Abbevillea fenzliana O.Berg in C.F.P.von Martius & auct. suc. (eds.), Fl. Bras. 14(1): 433 (1857).
Abbevillea fenzliana var. *brevipes* O.Berg in C.F.P.von Martius & auct. suc. (eds.), Fl. Bras. 14(1): 434 (1857).
Abbevillea fenzliana var. *intermedia* O.Berg in C.F.P.von Martius & auct. suc. (eds.), Fl. Bras. 14(1): 433 (1857).
Abbevillea fenzliana var. *longipes* O.Berg in C.F.P.von Martius & auct. suc. (eds.), Fl. Bras. 14(1): 433 (1857).
Abbevillea fenzliana var. *obversa* O.Berg in C.F.P.von Martius & auct. suc. (eds.), Fl. Bras. 14(1): 433 (1857).
Abbevillea klotzschiana O.Berg in C.F.P.von Martius & auct. suc. (eds.), Fl. Bras. 14(1): 434 (1857). *Abbevillea klotzschiana* var. *laurifolia* O.Berg in C.F.P.von Martius & auct. suc. (eds.), Fl. Bras. 14(1): 434 (1857), nom. inval.
Abbevillea klotzschiana var. *intermedia* O.Berg in C.F.P.von Martius & auct. suc. (eds.), Fl. Bras. 14(1): 434 (1857).
Abbevillea klotzschiana var. *longipes* O.Berg in C.F.P.von Martius & auct. suc. (eds.), Fl. Bras. 14(1): 434 (1857).
Abbevillea maschalantha O.Berg in C.F.P.von Martius & auct. suc. (eds.), Fl. Bras. 14(1): 432 (1857). *Campomanesia maschalantha* (O.Berg) Kiaersk. ex Engl. & Prantl. in H.G.A.Engler & K.A.E.Prantl, Nat. Pflanzenfam. 3(7): 8 (1893).
Abbevillea maschalantha var. *oblongata* O.Berg in C.F.P.von Martius & auct. suc. (eds.), Fl. Bras. 14(1): 433 (1857).
Abbevillea maschalantha var. *ovata* O.Berg in C.F.P.von Martius & auct. suc. (eds.), Fl. Bras. 14(1): 433 (1857).
Abbevillea maschalantha var. *recurvata* O.Berg in C.F.P.von Martius & auct. suc. (eds.), Fl. Bras. 14(1): 433 (1857).
Abbevillea sellowiana O.Berg in C.F.P.von Martius & auct. suc. (eds.), Fl. Bras. 14(1): 431 (1857). *Campomanesia sellowiana* (O.Berg) Mattos, Loefgrenia 90: 4 (1986), nom. illeg.
Campomanesia stictopetala Kiaersk., Enum. Myrt. Bras.: 8 (1893).
Campomanesia guaviroba var. *insulae* D.Legrand, in Fl. Ilustr. Catar. 1(Mirtac.): 595 (1977).
Campomanesia guaviroba var. *itatiaiae* D.Legrand, in Fl. Ilustr. Catar. 1(Mirtac.): 592 (1977).

Campomanesia guazumifolia (Cambess.) O.Berg, Linnaea 27: 434 (1856).
Brazil to Paraguay. 84 BZC BZE BZL BZS 85 AGE PAR URU. Nanophan. or phan.
**Psidium guazumifolium* Cambess. in A.F.C.de Saint-Hilaire, Fl. Bras. Merid. 2: 280 (1832). *Britoa guazumifolia* (Cambess.) D.Legrand, Notul. Syst. (Paris) 15: 274 (1958).
Psidium guazumifolium var. *griseum* Cambess. in A.F.C.de Saint-Hilaire, Fl. Bras. Merid. 2: 280 (1832). *Campomanesia guazumifolia* var. *grisea*

(Cambess.) O.Berg in C.F.P.von Martius & auct. suc. (eds.), Fl. Bras. 14(1): 434 (1857).
Britoa sellowiana O.Berg, Linnaea 27: 437 (1856).
Lacerdaea luschnathiana O.Berg, Linnaea 27: 437 (1856).
Abbevillea rugosa O.Berg in C.F.P.von Martius & auct. suc. (eds.), Fl. Bras. 14(1): 437 (1857). *Campomanesia itanarensis* Kiaersk., Enum. Myrt. Bras.: 8 (1893).
Campomanesia guazumifolia var. *rubiginosa* O.Berg in C.F.P.von Martius & auct. suc. (eds.), Fl. Bras. 14(1): 458 (1857).
Campomanesia albiflora Rojas, Cat. Hist. Nat. Corrientes: 62 (1897).
Britoa hassleriana Barb.Rodr. ex Chodat & Hassl., Bull. Herb. Boissier, II, 7: 801 (1907), nom. nud.

Campomanesia hirsuta Gardner, London J. Bot. 2: 353 (1843).
Brazil (Rio de Janeiro). 84 BZL. Nanophan. or phan.

Campomanesia ilhoensis Mattos, Loefgrenia 66: 3 (1975).
Brazil (Bahia). 84 BZE. Nanophan. or phan. Provisionally accepted.

Campomanesia laurifolia Gardner, London J. Bot. 2: 353 (1843). *Acrandra laurifolia* (Gardner) O.Berg in C.F.P.von Martius & auct. suc. (eds.), Fl. Bras. 14(1): 459 (1857).
SE. Brazil. 84 BZL. Phan.
Acrandra sellowiana O.Berg in C.F.P.von Martius & auct. suc. (eds.), Fl. Bras. 14(1): 459 (1857). *Campomanesia sellowiana* (O.Berg) Mattos, Arq. Bot. Estado São Paulo 4: 260 (1970). *Campomanesia acrandroides* Mattos, Loefgrenia 90: 4 (1986), nom. illeg.

Campomanesia lineatifolia Ruiz & Pav., Syst. Veg. Fl. Peruv. Chil.: 128 (1798). *Psidium lineatifolium* (Ruiz & Pav.) Pers., Syn. Pl. 2: 27 (1806).
W. South America to Brazil. 83 CLM ECU PER 84 BZC BZN. Phan.
Campomanesia cornifolia Kunth in F.W.H.von Humboldt, A.J.A.Bonpland & C.S.Kunth, Nov. Gen. Sp. 6: 150 (1823).
Psidium rivulare Mart. ex DC., Prodr. 3: 233 (1828). *Campomanesia rivularis* (Mart. ex DC.) Nied. in H.G.A.Engler & K.A.E.Prantl, Nat. Pflanzenfam. 3(7): 73 (1893).

Campomanesia lundiana (Kiaersk.) Mattos, Loefgrenia 26: 35 (1967).
Brazil (Rio de Janeiro). 84 BZL. Nanophan. or phan.
**Britoa lundiana* Kiaersk., Enum. Myrt. Bras.: 7 (1893).

Campomanesia macrobracteolata Landrum, Brittonia 53: 534 (2001 publ. 2002).
Brazil (Espírito Santo). 84 BZL. Phan.

Campomanesia mediterranea (Vell.) O.Berg in C.F.P.von Martius & auct. suc. (eds.), Fl. Bras. 14(1): 458 (1857).
SE. Brazil. 84 BZL.
**Psidium mediterraneum* Vell., Fl. Flumin. 5: 212, t. 52 (1829).

Campomanesia neriifolia (O.Berg) Nied. in H.G.A.Engler & K.A.E.Prantl, Nat. Pflanzenfam. 3(7): 73 (1893).
Brazil (São Paulo to Paraná). 84 BZL BZS. Nanophan. or phan.
**Abbevillea neriiflora* O.Berg in C.F.P.von Martius & auct. suc. (eds.), Fl. Bras. 14(1): 431 (1857).

Acrandra verrucosa O.Berg in C.F.P.von Martius & auct. suc. (eds.), Fl. Bras. 14(1): 459 (1857). *Campomanesia verrucosa* (O.Berg) Mattos, Ci. & Cult. 19: 333 (1967).

Campomanesia pabstiana Mattos & D.Legrand, Loefgrenia 67: 11 (1975).
Brazil (Brasília D.F.). 84 BZC. Nanophan.

Campomanesia phaea (O.Berg) Landrum, Brittonia 36: 241 (1984).
SE. Brazil (to Paraná). 84 BZL BZS. Phan.
**Abbevillea phaea* O.Berg in C.F.P.von Martius & auct. suc. (eds.), Fl. Bras. 14(1): 435 (1857). *Paivaea phaea* (O.Berg) Mattos, Loefgrenia 94: 8 (1989).
Paivaea langsdorffii O.Berg in C.F.P.von Martius & auct. suc. (eds.), Fl. Bras. 14(1): 614 (1859).
Paivaea langsdorffii var. *lauroana* Mattos, Revista Ci. Biol. 19(2): 344 (1967). *Paivaea phaea* var. *lauroana* (Mattos) Mattos, Loefgrenia 99: 6 (1990). *Campomanesia phaea* var. *lauroana* (Mattos) Mattos, Loefgrenia 110: 1 (1997).

Campomanesia prosthecesepala Kiaersk., Enum. Myrt. Bras.: 11 (1893).
Brazil (Minas Gerais). 84 BZL. Phan.

Campomanesia pubescens (Mart. ex DC.) O.Berg in C.F.P.von Martius & auct. suc. (eds.), Fl. Bras. 14(1): 443 (1857).
Brazil to Paraguay. 84 BZC BZE BZL BZS 85 PAR. Nanophan.
Psidium hians Mart. ex DC., Prodr. 3: 234 (1828). *Guajava hians* (Mart. ex DC.) Kuntze, Revis. Gen. Pl. 1: 239 (1891).
**Psidium pubescens* Mart. ex DC., Prodr. 3: 234 (1828).
Psidium corymbosum Cambess. in A.F.C.de Saint-Hilaire, Fl. Bras. Merid. 2: 286 (1832). *Campomanesia corymbosa* (Cambess.) O.Berg, Linnaea 27: 429 (1856).
Psidium erianthum Cambess. in A.F.C.de Saint-Hilaire, Fl. Bras. Merid. 2: 279 (1832). *Britoa eriantha* (Cambess.) O.Berg, Linnaea 27: 436 (1856). *Campomanesia eriantha* (Cambess.) Blume ex B.D.Jacks., Index Kew. 1: 408 (1893).
Psidium suaveolens Cambess. in A.F.C.de Saint-Hilaire, Fl. Bras. Merid. 2: 291 (1832). *Campomanesia suaveolens* (Cambess.) O.Berg, Linnaea 27: 434 (1856).
Psidium corymbosum f. *angustifolium* Miq., Linnaea 22: 532 (1849). *Campomanesia salviifolia* var. *angustifolia* (Miq.) O.Berg in C.F.P.von Martius & auct. suc. (eds.), Fl. Bras. 14(1): 442 (1857).
Psidium erosum Miq., Linnaea 22: 532 (1849). *Campomanesia fusca* var. *erosa* (Miq.) O.Berg in C.F.P.von Martius & auct. suc. (eds.), Fl. Bras. 14(1): 447 (1857). *Campomanesia erosa* (Miq.) Govaerts, World Checklist Seed Pl. 3(1): 12 (1999).
Psidium obversum Miq., Linnaea 22: 532 (1849). *Campomanesia obversa* (Miq.) O.Berg, Linnaea 27: 430 (1856).
Campomanesia affinis O.Berg in C.F.P.von Martius & auct. suc. (eds.), Fl. Bras. 14(1): 446 (1857).
Campomanesia australis O.Berg in C.F.P.von Martius & auct. suc. (eds.), Fl. Bras. 14(1): 445 (1857).
Campomanesia campestris O.Berg in C.F.P.von Martius & auct. suc. (eds.), Fl. Bras. 14(1): 440 (1857).
Campomanesia cuneata O.Berg in C.F.P.von Martius & auct. suc. (eds.), Fl. Bras. 14(1): 446 (1857).

Campomanesia discolor O.Berg in C.F.P.von Martius & auct. suc. (eds.), Fl. Bras. 14(1): 439 (1857).
Campomanesia discolor var. *alternifolia* O.Berg in C.F.P.von Martius & auct. suc. (eds.), Fl. Bras. 14(1): 440 (1857).
Campomanesia discolor var. *oppositifolia* O.Berg in C.F.P.von Martius & auct. suc. (eds.), Fl. Bras. 14(1): 440 (1857).
Campomanesia fusca O.Berg in C.F.P.von Martius & auct. suc. (eds.), Fl. Bras. 14(1): 447 (1857).
Campomanesia fusca var. *integra* O.Berg in C.F.P.von Martius & auct. suc. (eds.), Fl. Bras. 14(1): 447 (1857).
Campomanesia fusca var. *stricta* O.Berg in C.F.P.von Martius & auct. suc. (eds.), Fl. Bras. 14(1): 448 (1857).
Campomanesia fusca var. *subtriflora* O.Berg in C.F.P.von Martius & auct. suc. (eds.), Fl. Bras. 14(1): 448 (1857).
Campomanesia heterophylla O.Berg in C.F.P.von Martius & auct. suc. (eds.), Fl. Bras. 14(1): 440 (1857).
Campomanesia houlletii O.Berg in C.F.P.von Martius & auct. suc. (eds.), Fl. Bras. 14(1): 449 (1857).
Campomanesia obversa var. *angustifolia* O.Berg in C.F.P.von Martius & auct. suc. (eds.), Fl. Bras. 14(1): 446 (1857).
Campomanesia obversa var. *latifolia* O.Berg in C.F.P.von Martius & auct. suc. (eds.), Fl. Bras. 14(1): 445 (1857).
Campomanesia obversa var. *perforata* O.Berg in C.F.P.von Martius & auct. suc. (eds.), Fl. Bras. 14(1): 446 (1857).
Campomanesia ovalifolia O.Berg in C.F.P.von Martius & auct. suc. (eds.), Fl. Bras. 14(1): 452 (1857).
Campomanesia pohliana O.Berg in C.F.P.von Martius & auct. suc. (eds.), Fl. Bras. 14(1): 441 (1857).
Campomanesia pubescens var. *coarctata* O.Berg in C.F.P.von Martius & auct. suc. (eds.), Fl. Bras. 14(1): 443 (1857).
Campomanesia pubescens var. *effusa* O.Berg in C.F.P.von Martius & auct. suc. (eds.), Fl. Bras. 14(1): 443 (1857).
Campomanesia reticulata O.Berg in C.F.P.von Martius & auct. suc. (eds.), Fl. Bras. 14(1): 439 (1857).
Campomanesia rugosa O.Berg in C.F.P.von Martius & auct. suc. (eds.), Fl. Bras. 14(1): 439 (1857).
Campomanesia salviifolia O.Berg in C.F.P.von Martius & auct. suc. (eds.), Fl. Bras. 14(1): 442 (1857).
Campomanesia salviifolia var. *latifolia* O.Berg in C.F.P.von Martius & auct. suc. (eds.), Fl. Bras. 14(1): 442 (1857).
Campomanesia widgreniana O.Berg in C.F.P.von Martius & auct. suc. (eds.), Fl. Bras. 14(1): 447 (1857). *Abbevillea widgreniana* (O.Berg) O.Berg, Linnaea 31: 260 (1862).
Psidium hians var. *cuneatum* O.Berg in C.F.P.von Martius & auct. suc. (eds.), Fl. Bras. 14(1): 394 (1857).
Psidium hians var. *truncatum* O.Berg in C.F.P.von Martius & auct. suc. (eds.), Fl. Bras. 14(1): 394 (1857).
Campomanesia dimorpha O.Berg in C.F.P.von Martius & auct. suc. (eds.), Fl. Bras. 14(1): 609 (1859).
Campomanesia lanceolata O.Berg in C.F.P.von Martius & auct. suc. (eds.), Fl. Bras. 14(1): 612 (1859).
Campomanesia ovalifolia var. *venulosa* O.Berg in C.F.P.von Martius & auct. suc. (eds.), Fl. Bras. 14(1): 611 (1859).
Campomanesia rhytidophylla O.Berg in C.F.P.von Martius & auct. suc. (eds.), Fl. Bras. 14(1): 608 (1859).
Campomanesia virescens O.Berg in C.F.P.von Martius & auct. suc. (eds.), Fl. Bras. 14(1): 611 (1859).
Campomanesia bracteolata Kiaersk., Enum. Myrt. Bras.: 10 (1893).
Campomanesia warmingiana Kiaersk., Enum. Myrt. Bras.: 14 (1893).
Campomanesia diversifolia Barb.Rodr., Myrt. Paraguay: 18 (1903).
Campomanesia hassleri Barb.Rodr., Myrt. Paraguay: 19 (1903).
Campomanesia trichosepala Barb.Rodr., Myrt. Paraguay: 19 (1903).
Campomanesia apiculata Barb.Rodr. ex Chodat & Hassl., Bull. Herb. Boissier, II, 7: 800 (1907), nom. nud.
Campomanesia yerutiensis Barb.Rodr. ex Chodat & Hassl., Bull. Herb. Boissier, II, 7: 800 (1907), nom. nud.
Campomanesia paranensis D.Legrand, Bol. Soc. Argent. Bot. 10: 7 (1962). *Campomanesia discolor* var. *paranensis* (D.Legrand) Mattos, Loefgrenia 110: 1 (1997).
Campomanesia gomesiana Handro & Mattos, Loefgrenia 25: 1 (1967).

Campomanesia racemosa (Vell.) O.Berg in C.F.P.von Martius & auct. suc. (eds.), Fl. Bras. 14(1): 458 (1857).
SE. Brazil. 84 BZL.
**Psidium racemosum* Vell., Fl. Flumin. 5: 212, t. 55 (1829).

Campomanesia reitziana D.Legrand, Sellowia 8: 71 (1957).
Brazil (São Paulo to Santa Catarina). 84 BAL BZS. Nanophan. or phan.

Campomanesia rhombea O.Berg in C.F.P.von Martius & auct. suc. (eds.), Fl. Bras. 14(1): 453 (1857).
SE. & S. Brazil. 84 BZL BZS. Nanophan. or phan.

Campomanesia rufa (O.Berg) Nied. in H.G.A.Engler & K.A.E.Prantl, Nat. Pflanzenfam. 3(7): 73 (1893).
Brazil (Minas Gerais, Goiás). 84 BZC BZL. Nanophan. or phan.
**Abbevillea rufa* O.Berg in C.F.P.von Martius & auct. suc. (eds.), Fl. Bras. 14(1): 437 (1857).

Campomanesia schlechtendaliana (O.Berg) Nied. in H.G.A.Engler & K.A.E.Prantl, Nat. Pflanzenfam. 3(7): 73 (1893).
E. & S. Brazil. 84 BZE BZL BZS. Nanophan. or phan.
**Abbevillea schlechtendahliana* O.Berg in C.F.P.von Martius & auct. suc. (eds.), Fl. Bras. 14(1): 435 (1857).

var. ***rugosa*** (O.Berg) Landrum, Brittonia 36: 241 (1984).
Brazil (Rio de Janeiro: Cabo Frio). 84 BZL. Nanophan. or phan.
**Britoa rugosa* O.Berg in C.F.P.von Martius & auct. suc. (eds.), Fl. Bras. 14(1): 613 (1859). *Campomanesia fluminensis* Mattos, Loefgrenia 26: 37 (1967).

var. ***schlechtendaliana***
E. & S. Brazil. 84 BZE BZL BZS. Nanophan. or phan.

Britoa glazioviana Kiaersk., Enum. Myrt. Bras.: 6 (1893). *Campomanesia kiaerskoviana* Mattos, Loefgrenia 26: 40 (1967).
Campomanesia kiaerskoviana var. *impressovenosa* Mattos, Loefgrenia 26: 42 434 (1967).

Campomanesia sessiliflora (O.Berg) Mattos, Loefgrenia 26: 26 (1967).
Brazil to Paraguay. 83 BOL 84 BZC BZE BZL BZS 85 PAR. Nanophan. or phan.
**Britoa sessiliflora* O.Berg, Linnaea 27: 436 (1856).

var. ***bullata*** (Barb.Rodr.) Landrum, Brittonia 36: 241 (1984).
Brazil to Paraguay. 84 BZC BZS 85 PAR. Nanophan. or phan.
**Abbevillea bullata* Barb.Rodr., Myrt. Paraguay: 17 (1903). *Campomanesia bullata* (Barb.Rodr.) Barb.Rodr. ex Chodat & Hassl., Bull. Herb. Boissier, II, 7: 801 (1907).
Campomanesia igatimiensis Barb.Rodr. ex Chodat & Hassl., Bull. Herb. Boissier, II, 7: 800 (1907), nom. nud.
Campomanesia nitidifolia Barb.Rodr. ex Chodat & Hassl., Bull. Herb. Boissier, II, 7: 800 (1907), nom. nud.

var. ***lanuginosa*** (Barb.Rodr. ex Chodat & Hassl.) Landrum, Brittonia 36: 241 (1984).
Brazil to Paraguay. 83 BOL 84 BZC BZE BZL 85 PAR. Nanophan. or phan.
Campomanesia langsdorffii O.Berg in C.F.P.von Martius & auct. suc. (eds.), Fl. Bras. 14(1): 610 (1859).
Campomanesia apaensis Barb.Rodr. ex Chodat & Hassl., Bull. Herb. Boissier, II, 7: 800 (1907), nom. nud.
Campomanesia mollicarpa Barb.Rodr. ex Chodat & Hassl., Bull. Herb. Boissier, II, 7. 800 (1907), nom. nud.
**Campomanesia rugosa* var. *lanuginosa* Barb.Rodr. ex Chodat & Hassl., Bull. Herb. Boissier, II, 7: 800 (1907).

var. ***sessiliflora***
Brazil (Rio de Janeiro, São Paulo). 84 BZL. Nanophan. or phan.
Abbevillea langsdorffiana O.Berg in C.F.P.von Martius & auct. suc. (eds.), Fl. Bras. 14(1): 608 (1859).

Campomanesia simulans M.L.Kawas., Brittonia 52: 188 (2000).
Brazil (E. Minas Gerais, São Paulo). 84 BZL. Nanophan. or phan.

Campomanesia speciosa (Diels) McVaugh, Publ. Field Mus. Nat. Hist., Bot. Ser. 13(4): 813 (1958).
E. Peru to Brazil (Acre). 83 PER 84 BZN. Nanophan. or phan.
**Psidium speciosum* Diels, Verh. Bot. Vereins Prov. Brandenburg 48: 186 (1906 publ. 1907).

Campomanesia terminalis (Vell.) Mattos, Arq. Bot. Estado São Paulo 4: 260 (1970).
Brazil (Rio de Janeiro). 84 BZL. Phan.
**Psidium terminale* Vell., Fl. Flumin. 5: 212, t. 53 (1829).

Campomanesia transalpina (Vell.) O.Berg in C.F.P.von Martius & auct. suc. (eds.), Fl. Bras. 14(1): 459 (1857).
Brazil (Rio de Janeiro). 84 BZL.
**Psidium transalpinum* Vell., Fl. Flumin. 5: 213, t. 59 (1829).

Campomanesia velutina (Cambess.) O.Berg in C.F.P.von Martius & auct. suc. (eds.), Fl. Bras. 14(1): 441 (1857).
Brazil. 84 BZC BZE BZL BZN. Nanophan. or phan.
**Psidium velutinum* Cambess. in A.F.C.de Saint-Hilaire, Fl. Bras. Merid. 2: 288 (1832).
Campomanesia riedeliana O.Berg in C.F.P.von Martius & auct. suc. (eds.), Fl. Bras. 14(1): 610 (1859).
Campomanesia rabeniana Kiaersk., Enum. Myrt. Bras.: 11 (1893).

Campomanesia viatoris Landrum, Brittonia 36: 242 (1984).
Brazil (Alagoas, Bahia, Sergipe). 84 BZE. Nanophan. or phan.
**Abbevillea gardneriana* O.Berg in C.F.P.von Martius & auct. suc. (eds.), Fl. Bras. 14(1): 436 (1857).
Campomanesia schultziana Mattos, Loefgrenia 90: 4 (1986), nom. illeg.

Campomanesia xanthocarpa (Mart.) O.Berg in C.F.P.von Martius & auct. suc. (eds.), Fl. Bras. 14(1): 451 (1857).
SE. & S. Brazil to NE. Argentina. 84 BZL BZS 85 AGE PAR URU. Phan.
**Eugenia xanthocarpa* Mart., Syst. Mat. Med. Bras.: 31 (1843).

var. ***littoralis*** (D.Legrand) Landrum, Brittonia 36: 242 (1984).
Brazil (Santa Catarina, Rio Grande do Sul). 84 BZS. Cham. or nanophan.
**Campomanesia littoralis* D.Legrand, Sellowia 13: 335 (1961).

var. ***xanthocarpa***
SE. & S. Brazil, Paraguay, Uruguay, NE. Argentina. 84 BZL BZS 85 AGE PAR URU. Phan.
Psidium punctulatum Miq., Linnaea 22: 513 (1849).
Campomanesia crenata O.Berg in C.F.P.von Martius & auct. suc. (eds.), Fl. Bras. 14(1): 456 (1857).
Campomanesia malifolia O.Berg in C.F.P.von Martius & auct. suc. (eds.), Fl. Bras. 14(1): 452 (1857). *Campomanesia xanthocarpa* var. *malifolia* (O.Berg) D.Legrand, in Fl. Ilustr. Catar. 1(Mirtac.): 602 (1977).
Campomanesia rhombea var. *grandifolia* O.Berg in C.F.P.von Martius & auct. suc. (eds.), Fl. Bras. 14(1): 453 (1857).
Campomanesia rhombea var. *parvifolia* O.Berg in C.F.P.von Martius & auct. suc. (eds.), Fl. Bras. 14(1): 453 (1857).
Psidium malifolium F.Muell., Select Pl.: 189 (1872).
Campomanesia dusenii Kausel, Lilloa 33: 100 (1971 publ. 1972).
Campomanesia rhombea var. *kleinii* D.Legrand, in Fl. Ilustr. Catar. 1(Mirtac.): 612 (1977).

Synonyms:
Campomanesia acrandroides Mattos = ***Campomanesia terminalis*** (Vell.) Mattos
Campomanesia adamantium var. *nana* (D.Legrand) Mattos = ***Campomanesia adamantium*** (Cambess.) O.Berg
Campomanesia affinis O.Berg = ***Campomanesia pubescens*** (Mart. ex DC.) O.Berg
Campomanesia albiflora Rojas = ***Campomanesia guazumifolia*** (Cambess.) O.Berg

Campomanesia apaensis Barb.Rodr. ex Chodat & Hassl. = ***Campomanesia sessiliflora*** var. ***lanuginosa*** (Barb.Rodr. ex Chodat & Hassl.) Landrum
Campomanesia apiculata Barb.Rodr. ex Chodat & Hassl. = ***Campomanesia pubescens*** (Mart. ex DC.) O.Berg
Campomanesia arenaria O.Berg = [85 URU]
Campomanesia australis O.Berg = ***Campomanesia pubescens*** (Mart. ex DC.) O.Berg
Campomanesia beaurepairiana Kiaersk. = ***Campomanesia aromatica*** (Aubl.) Griseb.
Campomanesia bracteolata Kiaersk. = ***Campomanesia pubescens*** (Mart. ex DC.) O.Berg
Campomanesia bullata (Barb.Rodr.) Barb.Rodr. ex Chodat & Hassl. = ***Campomanesia sessiliflora*** var. ***bullata*** (Barb.Rodr.) Landrum
Campomanesia caerulea O.Berg = ***Campomanesia adamantium*** (Cambess.) O.Berg
Campomanesia caerulescens O.Berg = ***Campomanesia adamantium*** (Cambess.) O.Berg
Campomanesia cagaiteira Kiaersk. = ***Hexachlamys edulis*** (O.Berg) Kausel & D.Legrand
Campomanesia cambessedeana O.Berg = ***Campomanesia adamantium*** (Cambess.) O.Berg
Campomanesia cambessedeana var. *nana* D.Legrand = ***Campomanesia adamantium*** (Cambess.) O.Berg
Campomanesia cambessedeana var. *pyriformis* Mattos = ***Campomanesia adamantium*** (Cambess.) O.Berg
Campomanesia campestris (Cambess.) D.Legrand = ***Campomanesia adamantium*** (Cambess.) O.Berg
Campomanesia campestris O.Berg = ***Campomanesia pubescens*** (Mart. ex DC.) O.Berg
Campomanesia cerasoides (Cambess.) A.Gray = ***Campomanesia guaviroba*** (DC.) Kiaersk.
Campomanesia chrysophylla (O.Berg) Nied. = ***Campomanesia eugenioides*** (Cambess.) D.Legrand
Campomanesia ciliata O.Berg = ***Campomanesia aromatica*** (Aubl.) Griseb.
Campomanesia coaetanea O.Berg = ***Campomanesia aromatica*** (Aubl.) Griseb.
Campomanesia cornifolia Kunth = ***Campomanesia lineatifolia*** Ruiz & Pav.
Campomanesia corymbosa (Cambess.) O.Berg = ***Campomanesia pubescens*** (Mart. ex DC.) O.Berg
Campomanesia crassifolia Benth. = (Melastomataceae)
Campomanesia crenata O.Berg = ***Campomanesia xanthocarpa*** (Mart.) O.Berg var. ***xanthocarpa***
Campomanesia cuneata O.Berg = ***Campomanesia pubescens*** (Mart. ex DC.) O.Berg
Campomanesia cyanea O.Berg = ***Campomanesia aurea*** O.Berg var. ***aurea***
Campomanesia cyanea var. *cordata* O.Berg = ***Campomanesia aurea*** O.Berg var. ***aurea***
Campomanesia cyanea var. *ovata* O.Berg = ***Campomanesia aurea*** O.Berg var. ***aurea***
Campomanesia dardano-limai Mattos & D.Legrand = ***Campomanesia eugenioides*** (Cambess.) D.Legrand
Campomanesia dentata O.Berg = [84 BZS]
Campomanesia desertorum O.Berg = ***Campomanesia adamantium*** (Cambess.) O.Berg
Campomanesia dimorpha O.Berg = ***Campomanesia pubescens*** (Mart. ex DC.) O.Berg
Campomanesia discolor O.Berg = ***Campomanesia pubescens*** (Mart. ex DC.) O.Berg
Campomanesia discolor var. *alternifolia* O.Berg = ***Campomanesia pubescens*** (Mart. ex DC.) O.Berg
Campomanesia discolor var. *oppositifolia* O.Berg = ***Campomanesia pubescens*** (Mart. ex DC.) O.Berg
Campomanesia discolor var. *paranensis* (D.Legrand) Mattos = ***Campomanesia pubescens*** (Mart. ex DC.) O.Berg
Campomanesia diversifolia Barb.Rodr. = ***Campomanesia pubescens*** (Mart. ex DC.) O.Berg
Campomanesia dulcis (Vell.) J.F.Macbr. = ***Campomanesia guaviroba*** (DC.) Kiaersk.
Campomanesia dusenii Kausel = ***Campomanesia xanthocarpa*** (Mart.) O.Berg var. ***xanthocarpa***
Campomanesia eriantha (Cambess.) Blume ex B.D. Jacks. = ***Campomanesia pubescens*** (Mart. ex DC.) O.Berg
Campomanesia erosa (Miq.) Govaerts = ***Campomanesia pubescens*** (Mart. ex DC.) O.Berg
Campomanesia eugenioides var. *desertorum* (Mart. ex DC.) Landrum = ***Campomanesia eugenioides*** (Cambess.) D.Legrand
Campomanesia fluminensis Mattos = ***Campomanesia schlechtendaliana*** var. ***rugosa*** (O.Berg) Landrum
Campomanesia fusca O.Berg = ***Campomanesia pubescens*** (Mart. ex DC.) O.Berg
Campomanesia fusca var. *erosa* (Miq.) O.Berg = ***Campomanesia pubescens*** (Mart. ex DC.) O.Berg
Campomanesia fusca var. *integra* O.Berg = ***Campomanesia pubescens*** (Mart. ex DC.) O.Berg
Campomanesia fusca var. *stricta* O.Berg = ***Campomanesia pubescens*** (Mart. ex DC.) O.Berg
Campomanesia fusca var. *subtriflora* O.Berg = ***Campomanesia pubescens*** (Mart. ex DC.) O.Berg
Campomanesia gardneriana O.Berg = ***Campomanesia aurea*** O.Berg var. ***aurea***
Campomanesia glabra O.Berg = ***Campomanesia adamantium*** (Cambess.) O.Berg
Campomanesia glabra Benth. = ***Calycolpus goetheanus*** (Mart. ex DC.) O.Berg
Campomanesia glareophila Barb.Rodr. ex Chodat & Hassl. = ***Campomanesia adamantium*** (Cambess.) O.Berg
Campomanesia glazioviana Kiaersk. = ***Campomanesia aromatica*** (Aubl.) Griseb.
Campomanesia goetheana Hemsl. = ***Calycolpus goetheanus*** (Mart. ex DC.) O.Berg
Campomanesia gomesiana Handro & Mattos = ***Campomanesia pubescens*** (Mart. ex DC.) O.Berg
Campomanesia gracilis Kausel = ***Campomanesia aurea*** var. ***hatschbachii*** (Mattos) D.Legrand
Campomanesia guaviroba var. *insulae* D.Legrand = ***Campomanesia guaviroba*** (DC.) Kiaersk.
Campomanesia guaviroba var. *itatiaiae* D.Legrand = ***Campomanesia guaviroba*** (DC.) Kiaersk.
Campomanesia guazumifolia var. *grisea* (Cambess.) O.Berg = ***Campomanesia guazumifolia*** (Cambess.) O.Berg
Campomanesia guazumifolia var. *rubiginosa* O.Berg = ***Campomanesia guazumifolia*** (Cambess.) O.Berg
Campomanesia hassleri Barb.Rodr. = ***Campomanesia pubescens*** (Mart. ex DC.) O.Berg
Campomanesia hatschbachii Mattos = ***Campomanesia aurea*** var. ***hatschbachii*** (Mattos) D.Legrand
Campomanesia heterophylla O.Berg = ***Campomanesia pubescens*** (Mart. ex DC.) O.Berg
Campomanesia houlletii O.Berg = ***Campomanesia pubescens*** (Mart. ex DC.) O.Berg
Campomanesia igatimiensis Barb.Rodr. ex Chodat & Hassl. = ***Campomanesia sessiliflora*** var. ***bullata*** (Barb.Rodr.) Landrum
Campomanesia itanarensis Kiaersk. = ***Campomanesia guazumifolia*** (Cambess.) O.Berg
Campomanesia kiaerskoviana Mattos = ***Campomanesia schlechtendaliana*** (O.Berg) Nied. var. ***schlechtendaliana***
Campomanesia kiaerskoviana var. *impressovenosa* Mattos = ***Campomanesia schlechtendaliana*** (O.Berg) Nied. var. ***schlechtendaliana***

Campomanesia lanceolata O.Berg = ***Campomanesia pubescens*** (Mart. ex DC.) O.Berg
Campomanesia lancifolia Barb.Rodr. ex Chodat & Hassl. = ***Campomanesia adamantium*** (Cambess.) O.Berg
Campomanesia langsdorffii O.Berg = ***Campomanesia sessiliflora*** var. ***lanuginosa*** (Barb.Rodr. ex Chodat & Hassl.) Landrum
Campomanesia littoralis D.Legrand = ***Campomanesia xanthocarpa*** var. ***littoralis*** (D.Legrand) Landrum
Campomanesia malifolia O.Berg = ***Campomanesia xanthocarpa*** (Mart.) O.Berg var. ***xanthocarpa***
Campomanesia maracayuensis Barb.Rodr. ex Chodat & Hassl. = ***Campomanesia aurea*** O.Berg var. ***aurea***
Campomanesia martiana O.Berg = ***Psidium rufum*** Mart. ex DC.
Campomanesia maschalantha (O.Berg) Kiaersk. ex Engl. & Prantl. = ***Campomanesia guaviroba*** (DC.) Kiaersk.
Campomanesia microcarpa O.Berg = ***Campomanesia adamantium*** (Cambess.) O.Berg
Campomanesia mollicarpa Barb.Rodr. ex Chodat & Hassl. = ***Campomanesia sessiliflora*** var. ***lanuginosa*** (Barb.Rodr. ex Chodat & Hassl.) Landrum
Campomanesia montana D.Legrand = ***Campomanesia eugenioides*** (Cambess.) D.Legrand
Campomanesia multiflora (Cambess.) O.Berg = ***Psidium guineense*** Sw.
Campomanesia nitidifolia Barb.Rodr. ex Chodat & Hassl. = ***Campomanesia sessiliflora*** var. ***bullata*** (Barb.Rodr.) Landrum
Campomanesia obscura O.Berg = ***Campomanesia adamantium*** (Cambess.) O.Berg
Campomanesia obversa (Miq.) O.Berg = ***Campomanesia pubescens*** (Mart. ex DC.) O.Berg
Campomanesia obversa var. *angustifolia* O.Berg = ***Campomanesia pubescens*** (Mart. ex DC.) O.Berg
Campomanesia obversa var. *latifolia* O.Berg = ***Campomanesia pubescens*** (Mart. ex DC.) O.Berg
Campomanesia obversa var. *perforata* O.Berg = ***Campomanesia pubescens*** (Mart. ex DC.) O.Berg
Campomanesia ovalifolia O.Berg = ***Campomanesia pubescens*** (Mart. ex DC.) O.Berg
Campomanesia ovalifolia var. *venulosa* O.Berg = ***Campomanesia pubescens*** (Mart. ex DC.) O.Berg
Campomanesia paraguayensis Barb.Rodr. ex Chodat & Hassl. = ***Campomanesia adamantium*** (Cambess.) O.Berg
Campomanesia paranensis D.Legrand = ***Campomanesia pubescens*** (Mart. ex DC.) O.Berg
Campomanesia phaea var. *lauroana* (Mattos) Mattos = ***Campomanesia phaea*** (O.Berg) Landrum
Campomanesia pohliana O.Berg = ***Campomanesia pubescens*** (Mart. ex DC.) O.Berg
Campomanesia poiteaui O.Berg = ***Campomanesia grandiflora*** (Aubl.) Sagot
Campomanesia psidioides (O.Berg) Nied. = ***Campomanesia dichotoma*** (O.Berg) Mattos
Campomanesia pubescens var. *coarctata* O.Berg = ***Campomanesia pubescens*** (Mart. ex DC.) O.Berg
Campomanesia pubescens var. *effusa* O.Berg = ***Campomanesia pubescens*** (Mart. ex DC.) O.Berg
Campomanesia punctulata (DC.) Mattos & D.Legrand = ***Campomanesia guaviroba*** (DC.) Kiaersk.
Campomanesia rabeniana Kiaersk. = ***Campomanesia velutina*** (Cambess.) O.Berg
Campomanesia recurvata (O.Berg) Nied. = ***Psidium rufum*** Mart. ex DC.
Campomanesia regeliana (O.Berg) Kiaersk. = ***Psidium rufum*** Mart. ex DC.
Campomanesia repanda O.Berg = ***Campomanesia eugenioides*** (Cambess.) D.Legrand
Campomanesia resinosa Barb.Rodr. = ***Campomanesia adamantium*** (Cambess.) O.Berg
Campomanesia reticulata O.Berg = ***Campomanesia pubescens*** (Mart. ex DC.) O.Berg
Campomanesia rhombea var. *grandifolia* O.Berg = ***Campomanesia xanthocarpa*** (Mart.) O.Berg var. ***xanthocarpa***
Campomanesia rhombea var. *kleinii* D.Legrand = ***Campomanesia xanthocarpa*** (Mart.) O.Berg var. ***xanthocarpa***
Campomanesia rhombea var. *parvifolia* O.Berg = ***Campomanesia xanthocarpa*** (Mart.) O.Berg var. ***xanthocarpa***
Campomanesia rhytidophylla O.Berg = ***Campomanesia pubescens*** (Mart. ex DC.) O.Berg
Campomanesia riedeliana O.Berg = ***Campomanesia velutina*** (Cambess.) O.Berg
Campomanesia rivularis (Mart. ex DC.) Nied. = ***Campomanesia lineatifolia*** Ruiz & Pav.
Campomanesia rugosa O.Berg = ***Campomanesia pubescens*** (Mart. ex DC.) O.Berg
Campomanesia rugosa var. *lanuginosa* Barb.Rodr. ex Chodat & Hassl. = ***Campomanesia sessiliflora*** var. ***lanuginosa*** (Barb.Rodr. ex Chodat & Hassl.) Landrum
Campomanesia salviifolia O.Berg = ***Campomanesia pubescens*** (Mart. ex DC.) O.Berg
Campomanesia salviifolia var. *angustifolia* (Miq.) O.Berg = ***Campomanesia pubescens*** (Mart. ex DC.) O.Berg
Campomanesia salviifolia var. *latifolia* O.Berg = ***Campomanesia pubescens*** (Mart. ex DC.) O.Berg
Campomanesia schultziana Mattos = ***Campomanesia viatoris*** Landrum
Campomanesia sellowiana (O.Berg) Mattos = ***Campomanesia guaviroba*** (DC.) Kiaersk.
Campomanesia sellowiana (O.Berg) Mattos = ***Campomanesia laurifolia*** Gardner
Campomanesia sparsiflora (DC.) J.F.Macbr. = ***Campomanesia aromatica*** (Aubl.) Griseb.
Campomanesia stictopetala Kiaersk. = ***Campomanesia guaviroba*** (DC.) Kiaersk.
Campomanesia suaveolens (Cambess.) O.Berg = ***Campomanesia pubescens*** (Mart. ex DC.) O.Berg
Campomanesia suffruticosa O.Berg = ***Psidium laruotteanum*** Cambess.
Campomanesia synchrona O.Berg = ***Campomanesia aromatica*** (Aubl.) Griseb.
Campomanesia tenuifolia (Mart. ex DC.) O.Berg = ***Campomanesia aromatica*** (Aubl.) Griseb.
Campomanesia tetramera McVaugh = ***Eugenia tetramera*** (McVaugh) M.L.Kawas. & B.K.Holst
Campomanesia thea (Seem.) Gilg & Strauss = ?
Campomanesia tomentosa Kunth = ***Psidium guineense*** Sw.
Campomanesia trichosepala Barb.Rodr. = ***Campomanesia pubescens*** (Mart. ex DC.) O.Berg
Campomanesia triflora (O.Berg) Baill. = ***Campomanesia dichotoma*** (O.Berg) Mattos
Campomanesia vaccinioides O.Berg = ***Campomanesia adamantium*** (Cambess.) O.Berg
Campomanesia verrucosa (O.Berg) Mattos = ***Campomanesia neriifolia*** (O.Berg) Nied.
Campomanesia virescens O.Berg = ***Campomanesia pubescens*** (Mart. ex DC.) O.Berg
Campomanesia warmingiana Kiaersk. = ***Campomanesia pubescens*** (Mart. ex DC.) O.Berg

Campomanesia widgreniana O.Berg = ***Campomanesia pubescens*** (Mart. ex DC.) O.Berg
Campomanesia xanthocarpa var. *malifolia* (O.Berg) D.Legrand = ***Campomanesia xanthocarpa*** (Mart.) O.Berg var. ***xanthocarpa***
Campomanesia yerutiensis Barb.Rodr. ex Chodat & Hassl. = ***Campomanesia pubescens*** (Mart. ex DC.) O.Berg

Unplaced Names:
Campomanesia arenaria O.Berg in C.F.P.von Martius & auct. suc. (eds.), Fl. Bras. 14(1): 448 (1857). = [85 URU]
Campomanesia dentata O.Berg in C.F.P.von Martius & auct. suc. (eds.), Fl. Bras. 14(1): 457 (1857). = [84 BZS]

Carpolepis

Carpolepis (J.W.Dawson) J.W.Dawson = ***Metrosideros*** Banks ex Gaertn.
Carpolepis elegans (Montrouz.) J.W.Dawson = ***Metrosideros elegans*** (Montrouz.) Beauvis.
Carpolepis laurifolia (Brongn. & Gris) J.W.Dawson = ***Metrosideros laurifolia*** Brongn. & Gris
Carpolepis laurifolia var. *demonstrans* (Tison) J.W. Dawson = ***Metrosideros laurifolia*** Brongn. & Gris
Carpolepis tardiflora J.W.Dawson = ***Metrosideros laurifolia*** Brongn. & Gris

Caryophyllus

Caryophyllus L. = ***Syzygium*** Gaertn.
Caryophyllus adenolepis Miq. = [42 JAW]
Caryophyllus amieuensis Guillaumin = ***Syzygium amieuense*** (Guillaumin) J.W.Dawson
Caryophyllus antisepticus Blume = ***Syzygium antisepticum*** (Blume) Merr. & L.M.Perry
Caryophyllus aromaticus L. = ***Syzygium aromaticum*** (L.) Merr. & L.M.Perry
Caryophyllus baladensis Brongn. & Gris = ***Syzygium baladense*** (Brongn. & Gris) J.W.Dawson
Caryophyllus balansae Guillaumin = ***Syzygium balansae*** (Guillaumin) J.W.Dawson
Caryophyllus corticosus Stokes = ***Syzygium cumini*** (L.) Skeels
Caryophyllus cotinifolius Mill. = ***Eugenia cotinifolia*** Jacq.
Caryophyllus deplanchei Guillaumin = ***Syzygium deplanchei*** (Guillaumin) J.W.Dawson
Caryophyllus elegans Brongn. & Gris = ***Syzygium elegans*** (Brongn. & Gris) J.W.Dawson
Caryophyllus ellipticus Labill. = ***Syzygium deplanchei*** (Guillaumin) J.W.Dawson
Caryophyllus fastigiatus (Blume) Blume = ***Syzygium fastigiatum*** (Blume) Merr. & L.M.Perry
Caryophyllus floribundus (H.West ex Willd.) Blume = ***Myrciaria floribunda*** (H.West ex Willd.) O.Berg
Caryophyllus fruticosus Mill. = ***Eugenia biflora*** (L.) DC.
Caryophyllus garciniifolius Guillaumin = ***Syzygium paniense*** (Baker f.) J.W.Dawson
Caryophyllus hortensis Noronha = ***Syzygium aromaticum*** (L.) Merr. & L.M.Perry
Caryophyllus jambos Stokes = ***Syzygium cumini*** (L.) Skeels
Caryophyllus kriegeri Guillaumin = ***Syzygium kriegeri*** (Guillaumin) J.W.Dawson
Caryophyllus laxeracemosus Guillaumin = ***Syzygium laxeracemosum*** (Guillaumin) J.W.Dawson
Caryophyllus malaccensis (L.) Stokes = ***Syzygium malaccense*** (L.) Merr. & L.M.Perry
Caryophyllus multipetalus (Pancher ex Brongn. & Gris) Guillaumin = ***Syzygium multipetalum*** Pancher ex Brongn. & Gris
Caryophyllus pimenta (L.) Mill. = ***Pimenta dioica*** (L.) Merr.
Caryophyllus propinquus Guillaumin = ***Syzygium propinquum*** (Guillaumin) J.W.Dawson
Caryophyllus pterocalyx (Brongn. & Gris) Guillaumin = ***Syzygium pterocalyx*** Brongn. & Gris
Caryophyllus pterocarpus Vieill. ex Pancher & Sebert = ***Syzygium pterocarpum*** (Vieill. ex Pancher & Sebert) Govaerts
Caryophyllus racemosus Mill. = ***Pimenta racemosa*** (Mill.) J.W.Moore
Caryophyllus rhopalanthus (Schltr.) Guillaumin = ***Syzygium rhopalanthum*** Schltr.
Caryophyllus rugosus Blume ex Miq. = ***Syzygium zeylanicum*** (L.) DC.
Caryophyllus silvestris Teijsm. ex Hassk. = ***Syzygium aromaticum*** (L.) Merr. & L.M.Perry
Caryophyllus undulatus Guillaumin = ***Syzygium arboreum*** (Baker f.) J.W.Dawson
Caryophyllus vieillardii Lenorm. ex Guillaumin = ***Syzygium baladense*** (Brongn. & Gris) J.W.Dawson
Caryophyllus xanthostemifolius Guillaumin = ***Syzygium xanthostemifolium*** (Guillaumin) J.W.Dawson

Unplaced Names:
Caryophyllus adenolepis Miq., Fl. Ned. Ind. 1: 466 (1855). = [42 JAW]

Catinga

Catinga Aubl. = ***Eugenia*** P.Micheli ex L.
Catinga aromatica Aubl. = ***Eugenia moschata*** (Aubl.) Nied. ex T.Durand & B.D.Jacks.
Catinga media Sagot = ?
Catinga moschata Aubl. = ***Eugenia moschata*** (Aubl.) Nied. ex T.Durand & B.D.Jacks.
Catinga oblongifolia Sagot = ***Calycorectes grandifolius*** O.Berg

Unplaced Names:
Catinga media Sagot in ?. = ?

Cerocarpus

Cerocarpus Colebr. ex Hassk. = ***Syzygium*** Gaertn.
Cerocarpus aqueus (Burm.f.) Hassk. = ***Syzygium aqueum*** (Burm.f.) Alston

Cerqueiria

Cerqueiria O.Berg = ***Myrcia*** DC. ex Guill.
Cerqueiria sellowiana O.Berg = [84 BZE]

Unplaced Names:
Cerqueiria sellowiana O.Berg, Linnaea 27: 6 (1855). = [84 BZE]

Cetra

Cetra Noronha = ***Syzygium*** Gaertn.

Chamelaucium

Chamelaucium Desf., Mém. Mus. Hist. Nat. 5: 39 (1819). Western Australia. 50 WAU (51) nzn. [Myrtaceae]
13 Species
Decalophium Turcz., Bull. Soc. Imp. Naturalistes Moscou 20(1): 153 (1847).

Chamelaucium axillare F.Muell. ex Benth., Fl. Austral. 3: 38 (1867). *Darwinia axillaris* (F.Muell. ex Benth.) F.Muell., Syst. Census Austral. Pl. 1: 51 (1882).
Western Australia (Eyre, Roe). 50 WAU. Nanophan.

Chamelaucium brevifolium Benth., Fl. Austral. 3: 37 (1867). *Darwinia brevifolia* (Benth.) F.Muell., Syst. Census Austral. Pl. 1: 51 (1882).
Western Australia (Avon). 50 WAU. Nanophan.

Chamelaucium ciliatum Desf., Mém. Mus. Hist. Nat. 5: 40 (1819). *Darwinia ciliata* (Desf.) F.Muell., Syst. Census Austral. Pl. 1: 51 (1882).
SW. Australia. 50 WAU. Nanophan.
Genetyllis pauciflora Turcz., Bull. Soc. Imp. Naturalistes Moscou 22(2): 17 (1849).
Decalophium darwinioides Turcz., Bull. Cl. Phys.-Math. Acad. Imp. Sci. Saint-Pétersbourg 10: 326 (1852).
Decalophium rugulosum Turcz., Bull. Cl. Phys.-Math. Acad. Imp. Sci. Saint-Pétersbourg 10: 326 (1852).

Chamelaucium confertiflorum Domin, Vestn. Král. Ceské Spolecn. Nauk, Tr. Mat.-Prír. 2(2): 80 (1923).
Western Australia (Menzies, Eyre, Avon). 50 WAU. Nanophan.

Chamelaucium drummondii (Turcz.) Meisn., J. Proc. Linn. Soc., Bot. 1: 44 (1857).
Western Australia (Irwin, Drummond, Avon). 50 WAU. Nanophan.
**Genetyllis drummondii* Turcz., Bull. Soc. Imp. Naturalistes Moscou 20(1): 155 (1847). *Darwinia drummondii* (Turcz.) F.Muell., Syst. Census Austral. Pl. 1: 51 (1882).
Chamelaucium hallii Ewart, Proc. Roy. Soc. Victoria, n.s., 20: 77 (1907).

Chamelaucium gracile F.Muell., Fragm. 4: 62 (1864). *Darwinia gracilis* (F.Muell.) F.Muell., Syst. Census Austral. Pl. 1: 51 (1882).
SW. Australia. 50 WAU. Nanophan.

Chamelaucium heterandrum Benth., Fl. Austral. 3: 36 (1867). *Darwinia heterandra* (Benth.) F.Muell., Syst. Census Austral. Pl. 1: 51 (1882).
SW. Australia. 50 WAU. Nanophan.

Chamelaucium marchantii Strid, Pl. Syst. Evol. 155: 341 (1987).
Western Australia (Irwin). 50 WAU. Nanophan.

Chamelaucium megalopetalum F.Muell. ex Benth., Fl. Austral. 3: 38 (1867). *Darwinia megalopetala* (F.Muell. ex Benth.) F.Muell., Syst. Census Austral. Pl. 1: 51 (1882).
SW. Australia. 50 WAU. Nanophan.
Decalophium juniperinum Turcz., Bull. Cl. Phys.-Math. Acad. Imp. Sci. Saint-Pétersbourg 10: 325 (1852).
Decalophium melaleucum Turcz., Bull. Cl. Phys.-Math. Acad. Imp. Sci. Saint-Pétersbourg 10: 325 (1852).

Chamelaucium micranthum (Turcz.) Domin, Vestn. Král. Ceské Spolecn. Nauk, Tr. Mat.-Prír. 2(2): 81 (1923).
Western Australia (Austin, Irwin, Avon). 50 WAU. Nanophan.
**Decalophium micranthum* Turcz., Bull. Cl. Phys.-Math. Acad. Imp. Sci. Saint-Pétersbourg 10: 326 (1852).

Chamelaucium pauciflorum (Turcz.) Benth., Fl. Austral. 3: 38 (1867).
SW. Australia. 50 WAU. Nanophan.
**Decalophium pauciflorum* Turcz., Bull. Soc. Imp. Naturalistes Moscou 20(1): 154 (1847). *Darwinia turczaninowii* F.Muell., Syst. Census Austral. Pl. 1: 51 (1882).
Darwinia thryptomenioides D.A.Herb., J. Proc. Roy. Soc. W. Australia 8: 35 (1922).

Chamelaucium uncinatum Schauer in J.G.C.Lehmann, Pl. Preiss. 1: 97 (1844). *Darwinia uncinata* (Schauer) F.Muell., Syst. Census Austral. Pl. 1: 51 (1882).
Western Australia (Irwin, Drummond). 50 WAU (51) nzn. Nanophan.
Chamelaucium affine Meisn., J. Proc. Linn. Soc., Bot. 1: 45 (1857).
Chamelaucium uncinatum var. *leptophyllum* Benth., Fl. Austral. 3: 38 (1867).

Chamelaucium virgatum Endl., Stirp. Herb. Hügel.: 5 (1838). *Darwinia endlicheri* F.Muell., Syst. Census Austral. Pl. 1: 51 (1882).
Western Australia (Roe). 50 WAU. Nanophan.

Synonyms:
Chamelaucium affine Meisn. = ***Chamelaucium uncinatum*** Schauer
Chamelaucium brownii Desf. = ***Verticordia brownii*** (Desf.) A.Cunn. ex DC.
Chamelaucium dilatata J.Drumm. = ***Pileanthus peduncularis*** Endl. subsp. ***peduncularis***
Chamelaucium hallii Ewart = ***Chamelaucium drummondii*** (Turcz.) Meisn.
Chamelaucium micropetalum (F.Muell.) F.Muell. = ***Darwinia micropetala*** (F.Muell.) Benth.
Chamelaucium plumosum Desf. = ***Verticordia plumosa*** (Desf.) Druce
Chamelaucium schuermannii (F.Muell.) F.Muell. = ***Homoranthus homoranthoides*** (F.Muell.) Craven & S.R.Jones
Chamelaucium thomasii F.Muell. = ***Homoranthus homoranthoides*** (F.Muell.) Craven & S.R.Jones
Chamelaucium uncinatum var. *leptophyllum* Benth. = ***Chamelaucium uncinatum*** Schauer
Chamelaucium verticordinum F.Muell. = ***Verticordia verticordina*** (F.Muell.) A.S.George

Chamguava Landrum, Syst. Bot. 16: 21 (1991).
Mexico to C. America. 79 MXS MXT 80 BLZ GUA HON PAN. [Myrtaceae]
3 Species

Chamguava gentlei (Lundell) Landrum, Syst. Bot. 16: 23 (1991).
Mexico (Chiapas), C. America. 79 MXT 80 BLZ GUA HON. Nanophan. or phan.
**Eugenia gentlei* Lundell, Publ. Carnegie Inst. Wash. 478: 216 (1937). *Psidium biloculare* McVaugh, Fieldiana, Bot. 29: 520 (1963).

var. ***apodantha*** (Standl.) Landrum, Syst. Bot. 16: 24 (1991).
C. America. 80 GUA HON. Nanophan. or phan.
**Eugenia apodantha* Standl., Publ. Field Mus. Nat. Hist., Bot. Ser. 17: 380 (1938). *Psidium apodanthum* (Standl.) McVaugh, Fieldiana, Bot. 29: 520 (1963).

var. ***gentlei***
Mexico (Chiapas), Guatemala, Belize. 79 MXT 80 BLZ GUA. Nanophan. or phan.

Chamguava musarum (Standl. & Steyerm.) Landrum, Syst. Bot. 16: 27 (1991).
Guatemala. 80 GUA. Nanophan. or phan.
**Eugenia musarum* Standl. & Steyerm., Publ. Field Mus. Nat. Hist., Bot. Ser. 22: 358 (1940). *Psidium musarum* (Standl. & Steyerm.) McVaugh, Fieldiana, Bot. 29: 521 (1963).

Chamguava schippii (Standl.) Landrum, Syst. Bot. 16: 27 (1991).
Mexico (Guerrero, Chiapas), C. America. 79 MXS MXT 80 BLZ GUA PAN. Nanophan. or phan.
**Eugenia schippii* Standl., Publ. Field Mus. Nat. Hist., Bot. Ser. 11: 137 (1932).
Eugenia mouririoides Lundell, Amer. Midl. Naturalist 29: 480 (1943). *Psidium mouririoides* (Lundell) McVaugh, Fieldiana, Bot. 29: 521 (1963).
Eugenia anglohondurensis Lundell, Wrightia 2: 123 (1961). *Psidium anglohondurense* (Lundell) McVaugh, Fieldiana, Bot. 29: 521 (1963).

Cheynia

Cheynia J.Drumm. ex Harv. = ***Balaustion*** Hook.
Cheynia pulchella Harv. = ***Balaustion pulcherrimum*** Hook.

Chloromyrtus

Chloromyrtus Pierre = ***Eugenia*** P.Micheli ex L.
Chloromyrtus klaineana Pierre = ***Eugenia klaineana*** (Pierre) Engl.

Choricarpia

Choricarpia Domin, Biblioth. Bot. 89: 472 (1928).
E. Australia. 50 NSW QLD. [Myrtaceae]
2 Species

Choricarpia leptopetala (F.Muell.) Domin, Biblioth. Bot. 89: 472 (1928).
SE. Queensland to NE. New South Wales. 50 NSW QLD. Nanophan.
**Syncarpia leptopetala* F.Muell., Fragm. 1: 79 (1858). *Metrosideros leptopetala* (F.Muell.) F.Muell., Syst. Census Austral. Pl. 1: 59 (1882). *Nania leptopetala* (F.Muell.) Kuntze, Revis. Gen. Pl. 1: 242 (1891).

Choricarpia subargentea (C.T.White) L.A.S.Johnson, Contr. New South Wales Natl. Herb. 3: 100 (1962).
SE. Queensland. 50 QLD. Nanophan.
**Syncarpia subargentea* C.T.White, Bot. Bull. Dept. Agric. Queensland 21: 8 (1919).
Syncarpia subargentea var. *latifolia* C.T.White, Bot. Bull. Dept. Agric. Queensland 21: 8 (1919).

Chrysorhoe

Chrysorhoe Lindl. = ***Verticordia*** DC.
Chrysorhoe nitens Lindl. = ***Verticordia nitens*** (Lindl.) Endl.
Chrysorhoe serrata Lindl. = ***Verticordia serrata*** (Lindl.) Schauer

Chytraculia

Chytraculia P.Browne = ***Calyptranthes*** Sw.
Chytraculia anceps (O.Berg) Kuntze = ***Calyptranthes anceps*** O.Berg
Chytraculia arborea Kuntze = ***Calyptranthes chytraculia*** (L.) Sw.
Chytraculia aromatica (A.St.-Hil.) Kuntze = ***Calyptranthes aromatica*** A.St.-Hil.
Chytraculia axillaris (O.Berg) Kuntze = ***Calyptranthes axillaris*** O.Berg
Chytraculia bergiana Kuntze = ***Neomitranthes langsdorffii*** (O.Berg) Mattos
Chytraculia bimarginata (O.Berg) Kuntze = ***Calyptranthes bimarginata*** O.Berg
Chytraculia bipennis (O.Berg) Kuntze = ***Calyptranthes bipennis*** O.Berg
Chytraculia blanchetiana (O.Berg) Kuntze = ***Calyptranthes blanchetiana*** O.Berg
Chytraculia brasiliensis (Spreng.) Kuntze = ***Calyptranthes brasiliensis*** Spreng.
Chytraculia browniana (Mart. ex DC.) Kuntze = ***Psidium brownianum*** Mart. ex DC.
Chytraculia bullata (Salisb.) Kuntze = ***Calyptranthes bullata*** (Salisb.) DC.
Chytraculia caudata (Gardner) Kuntze = ***Calyptranthes caudata*** Gardner
Chytraculia cephalantha (O.Berg) Kuntze = ***Calyptranthes cephalantha*** O.Berg
Chytraculia chytraculia (L.) Millsp. = ***Calyptranthes chytraculia*** (L.) Sw.
Chytraculia chytraculia var. *genuina* Sudw. = ***Calyptranthes chytraculia*** (L.) Sw.
Chytraculia chytraculia var. *ovalis* (O.Berg) Sudw. = ***Calyptranthes zuzygium*** (L.) Sw.
Chytraculia chytraculia var. *pauciflora* (O.Berg) Sudw. = ***Calyptranthes pallens*** Griseb.
Chytraculia chytraculia var. *trichotoma* (O.Berg) Sudw. = ***Calyptranthes zuzygium*** (L.) Sw.
Chytraculia chytraculia var. *zuzygium* (L.) Sudw. = ***Calyptranthes zuzygium*** (L.) Sw.
Chytraculia clusiifolia (O.Berg) Kuntze = ***Calyptranthes clusiifolia*** O.Berg
Chytraculia concinna (DC.) Kuntze = ***Calyptranthes concinna*** DC.
Chytraculia cordata (O.Berg) Kuntze = ***Myrcia subcordata*** DC.
Chytraculia costaricensis (O.Berg) Kuntze = ***Calyptranthes pallens*** var. ***williamsii*** (Standl.) McVaugh
Chytraculia cubensis (O.Berg) Kuntze = ***Calyptranthes cubensis*** O.Berg
Chytraculia cuprea (O.Berg) Kuntze = ***Calyptranthes cuprea*** O.Berg
Chytraculia cuspidata (Mart. ex DC.) Kuntze = ***Calyptranthes cuspidata*** Mart. ex DC.
Chytraculia densiflora (Poepp. ex O.Berg) Kuntze = ***Calyptranthes densiflora*** Poepp. ex O.Berg
Chytraculia dichotoma (Casar.) Kuntze = ***Calyptranthes dichotoma*** Casar.
Chytraculia eriopoda (DC.) Kuntze = ***Calyptranthes eriopoda*** DC.
Chytraculia eugenioides (Cambess.) Kuntze = ***Calyptranthes eugenioides*** Cambess.
Chytraculia fasciculata (O.Berg) Kuntze = ***Calyptranthes fasciculata*** O.Berg
Chytraculia fastigiata Kuntze = ***Syzygium fastigiatum*** (Blume) Merr. & L.M.Perry
Chytraculia forsteri (O.Berg) Kuntze = ***Calyptranthes forsteri*** O.Berg
Chytraculia gardneriana (O.Berg) Kuntze = ?
Chytraculia glomerata (Cambess.) Kuntze = ***Calyptranthes concinna*** DC.
Chytraculia grammica (Spreng.) Kuntze = ***Calyptranthes grammica*** (Spreng.) D.Legrand
Chytraculia grandifolia (O.Berg) Kuntze = ***Calyptranthes grandifolia*** O.Berg
Chytraculia karwinskyana (O.Berg) Kuntze = ***Calyptranthes karwinskyana*** O.Berg

Chytraculia lanceolata (O.Berg) Kuntze = ***Calyptranthes lanceolata*** O.Berg
Chytraculia langsdorffii (O.Berg) Kuntze = ***Calyptranthes langsdorffii*** O.Berg
Chytraculia lindeniana (O.Berg) Kuntze = ***Calyptranthes lindeniana*** O.Berg
Chytraculia longifolia (O.Berg) Kuntze = ***Calyptranthes longifolia*** O.Berg
Chytraculia loranthifolia (DC.) Kuntze = ***Calyptranthes loranthifolia*** DC.
Chytraculia lucida (Mart. ex DC.) Kuntze = ***Calyptranthes lucida*** Mart. ex DC.
Chytraculia macrophylla (O.Berg) Kuntze = ***Calyptranthes macrophylla*** O.Berg
Chytraculia martiusiana (DC.) Kuntze = ***Calyptranthes martiusiana*** DC.
Chytraculia maschalantha (O.Berg) Kuntze = ***Neomitranthes obscura*** (DC.) N.Silveira
Chytraculia melanoclada (O.Berg) Kuntze = ***Calyptranthes melanoclada*** O.Berg
Chytraculia multiflora (Poepp. ex O.Berg) Kuntze = ***Calyptranthes multiflora*** Poepp. ex O.Berg
Chytraculia mutabilis (O.Berg) Kuntze = ***Calyptranthes brasiliensis*** Spreng.
Chytraculia nigricans (DC.) Kuntze = ***Calyptranthes nigricans*** DC.
Chytraculia nummularia (O.Berg) Kuntze = ***Calyptranthes nummularia*** O.Berg
Chytraculia obscura (DC.) Kuntze = ***Neomitranthes obscura*** (DC.) N.Silveira
Chytraculia obversa (O.Berg) Kuntze = ***Calyptranthes obversa*** O.Berg
Chytraculia ottonis (O.Berg) Kuntze = ***Mitranthes ottonis*** O.Berg
Chytraculia ovalifolia (Cambess.) Kuntze = ***Calyptranthes ovalifolia*** Cambess.
Chytraculia ovata (O.Berg) Kuntze = ***Calyptranthes ovata*** O.Berg
Chytraculia pallens (Griseb.) Britton = ***Calyptranthes pallens*** Griseb.
Chytraculia paniculata (Ruiz & Pav.) Kuntze = ***Calyptranthes paniculata*** Ruiz & Pav.
Chytraculia pauciflora (O.Berg) Kuntze = ***Calyptranthes pauciflora*** O.Berg
Chytraculia pendula (O.Berg) Kuntze = ***Calyptranthes pendula*** O.Berg
Chytraculia platyphylla (O.Berg) Kuntze = ***Calyptranthes platyphylla*** O.Berg
Chytraculia poeppigiana (O.Berg) Kuntze = ***Calyptranthes multiflora*** Poepp. ex O.Berg
Chytraculia pohliana (O.Berg) Kuntze = ***Calyptranthes lucida*** Mart. ex DC. var. ***lucida***
Chytraculia polyantha (O.Berg) Kuntze = ***Calyptranthes lucida*** var. ***polyantha*** (O.Berg) D.Legrand
Chytraculia pteropoda (O.Berg) Kuntze = ***Calyptranthes pteropoda*** O.Berg
Chytraculia pulchella (DC.) Kuntze = ***Calyptranthes pulchella*** DC.
Chytraculia punctata (Griseb.) Millsp. = ***Calyptranthes punctata*** Griseb.
Chytraculia regeliana (O.Berg) Kuntze = ***Calyptranthes regeliana*** O.Berg
Chytraculia riedeliana (O.Berg) Kuntze = ***Neomitranthes riedeliana*** (O.Berg) Mattos
Chytraculia rigida (Sw.) Kuntze = ***Calyptranthes rigida*** Sw.
Chytraculia rufa (O.Berg) Kuntze = ***Calyptranthes grandifolia*** O.Berg
Chytraculia ruiziana (O.Berg) Kuntze = ***Calyptranthes ruiziana*** O.Berg
Chytraculia sartoriana (O.Berg) Kuntze = ***Psidium sartorianum*** (O.Berg) Nied.
Chytraculia schiedeana (O.Berg) Kuntze = ***Calyptranthes schiedeana*** O.Berg
Chytraculia schlechtendaliana (O.Berg) Kuntze = ***Calyptranthes schiedeana*** O.Berg
Chytraculia scoparia (O.Berg) Kuntze = ***Calyptranthes scoparia*** O.Berg
Chytraculia selloana Kuntze = ***Myrceugenia ovalifolia*** (O.Berg) Landrum
Chytraculia spruceana (O.Berg) Kuntze = ***Calyptranthes spruceana*** O.Berg
Chytraculia strigipes (O.Berg) Kuntze = ***Calyptranthes strigipes*** O.Berg
Chytraculia tetraptera (O.Berg) Kuntze = ***Calyptranthes tetraptera*** O.Berg
Chytraculia thomasiana (O.Berg) Kuntze = ***Calyptranthes thomasiana*** O.Berg
Chytraculia tuberculata (O.Berg) Kuntze = ***Neomitranthes obscura*** (DC.) N.Silveira
Chytraculia tussaceana (O.Berg) Kuntze = ***Calyptranthes tussaceana*** O.Berg
Chytraculia variabilis (O.Berg) Kuntze = ***Calyptranthes variabilis*** O.Berg
Chytraculia widgreniana (O.Berg) Kuntze = ***Calyptranthes widgreniana*** O.Berg
Chytraculia zuzygium (L.) Kuntze = ***Calyptranthes zuzygium*** (L.) Sw.

Chytralia

Chytralia Adans. = ***Calyptranthes*** Sw.

Clavimyrtus

Clavimyrtus Blume = ***Syzygium*** Gaertn.
Clavimyrtus claviflora (Roxb.) Blume = ***Syzygium claviflorum*** (Roxb.) Wall. ex A.M.Cowan & Cowan
Clavimyrtus cylindrica (Wight) Blume = ***Syzygium cylindricum*** (Wight) Alston
Clavimyrtus firma Blume = ***Syzygium ampliflorum*** (Koord. & Valeton) Amshoff
Clavimyrtus glabrata (DC.) Blume = ***Syzygium glabratum*** (DC.) Veldkamp
Clavimyrtus latifolia Blume = ***Syzygium lineatum*** (DC.) Merr. & L.M.Perry
Clavimyrtus lineata (DC.) Blume = ***Syzygium lineatum*** (DC.) Merr. & L.M.Perry
Clavimyrtus marginata Blume = ***Syzygium glabratum*** (DC.) Veldkamp
Clavimyrtus pauciflora (Walp.) Blume = ***Syzygium laetum*** (Buch.-Ham.) Gandhi subsp. ***laetum***
Clavimyrtus ramosissima Blume = ***Syzygium ramosissimum*** (Blume) N.P.Balakr.
Clavimyrtus symphytocarpa Blume = ***Syzygium lineatum*** (DC.) Merr. & L.M.Perry
Clavimyrtus virens Blume = ***Syzygium glabratum*** (DC.) Veldkamp

Cleistocalyx

Cleistocalyx Blume = ***Syzygium*** Gaertn.
Cleistocalyx arcuatinervius (Merr.) Merr. & L.M.Perry = ***Syzygium arcuatinervium*** (Merr.) Craven & Biffin
Cleistocalyx baeuerlenii (F.Muell.) Merr. & L.M.Perry = ***Syzygium baeuerlenii*** (F.Muell.) Craven & Biffin
Cleistocalyx barringtonioides (Ridl.) Merr. & L.M.Perry = ***Syzygium barringtonioides*** (Ridl.) Masam.
Cleistocalyx brongniartii Merr. & L.M.Perry = ***Syzygium brongniartii*** (Merr. & L.M.Perry) J.W.Dawson

Cleistocalyx cerasoides (Roxb.) I.M.Turner = ***Syzygium operculatum*** (Roxb.) Nied.
Cleistocalyx cerasoides var. *paniala* (Roxb.) I.M.Turner = ***Syzygium operculatum*** (Roxb.) Nied.
Cleistocalyx circumscissa (Gagnep.) P.H.Hô = ***Syzygium circumscissum*** (Gagnep.) Craven & Biffin
Cleistocalyx conspersipunctatus Merr. & L.M.Perry = ***Syzygium conspersipunctatum*** (Merr. & L.M.Perry) Craven & Biffin
Cleistocalyx decussatus A.C.Sm. = ***Syzygium decussatum*** (A.C.Sm.) Biffin & Craven
Cleistocalyx ellipticus (A.C.Sm.) Merr. & L.M.Perry = ***Syzygium seemannii*** (A.Gray) Biffin & Craven
Cleistocalyx eugenioides (F.Muell.) Merr. & L.M.Perry = ***Syzygium eugenioides*** (Merr. & L.M.Perry) Biffin & Craven
Cleistocalyx fullagarii (F.Muell.) Merr. & L.M.Perry = ***Syzygium fullagarii*** (F.Muell.) Craven
Cleistocalyx gustavioides (F.M.Bailey) Merr. & L.M.Perry = ***Syzygium gustavioides*** (F.M.Bailey) B.Hyland
Cleistocalyx kasiensis A.C.Sm. = ***Syzygium seemannii*** (A.Gray) Biffin & Craven
Cleistocalyx khaoyaiensis Chantaranothai & J.Parn. = ***Syzygium khaoyaiense*** (Chantaranothai & J.Parn.) Craven & Biffin
Cleistocalyx leucocladus Merr. & L.M.Perry = ***Syzygium calyptrocalyx*** P.S.Ashton
Cleistocalyx longiflorus (A.C.Sm.) Merr. & L.M.Perry = ***Syzygium seemannii*** (A.Gray) Biffin & Craven
Cleistocalyx myrtoides (A.Gray) Merr. & L.M.Perry = ***Syzygium myrtoides*** (A.Gray) R.Schmid
Cleistocalyx nervosus (Lour.) Blume = ***Eugenia nervosa*** Lour.
Cleistocalyx nervosus (A.Cunn. ex DC.) Kosterm. = ***Syzygium operculatum*** (Roxb.) Nied.
Cleistocalyx nervosus var. *paniala* (Roxb.) J.Parn. & Chantaranothai = ***Syzygium operculatum*** (Roxb.) Nied.
Cleistocalyx nicobaricus (King) Merr. & L.M.Perry = ***Syzygium nicobaricum*** (King) Rathakr. & N.C.Nair
Cleistocalyx nigrans (Gagnep.) Merr. & L.M.Perry = ***Syzygium nigrans*** (Gagnep.) Craven & Biffin
Cleistocalyx nitidus (Korth.) Blume = ***Syzygium operculatum*** (Roxb.) Nied.
Cleistocalyx operculatus (Roxb.) Merr. & L.M.Perry = ***Syzygium operculatum*** (Roxb.) Nied.
Cleistocalyx operculatus var. *paniala* (Roxb.) Chantaranothai & J.Parn. = ***Syzygium operculatum*** (Roxb.) Nied.
Cleistocalyx paradoxus (Merr.) Merr. & L.M.Perry = ***Syzygium paradoxum*** (Merr.) Masam.
Cleistocalyx paucipunctatus Merr. & L.M.Perry = ***Syzygium pseudocalcicola*** Craven & Biffin
Cleistocalyx pennelii (Guillaumin) Merr. = ***Syzygium pennelii*** (Guillaumin) J.W.Dawson
Cleistocalyx perspicuinervius (Merr.) Merr. & L.M.Perry = ***Syzygium perspicuinervium*** (Merr.) Masam.
Cleistocalyx phengklaii Chantar. & J.Parn. = ***Syzygium phengklaii*** (Chantar. & J.Parn.) Craven & Biffin
Cleistocalyx retinervius Merr. & L.M.Perry = ***Syzygium retinervium*** (Merr. & L.M.Perry) Craven & Biffin
Cleistocalyx seemannii (A.Gray) Merr. & L.M.Perry = ***Syzygium seemannii*** (A.Gray) Biffin & Craven
Cleistocalyx seemannii var. *punctatus* Merr. & L.M.Perry = ***Syzygium seemannii*** (A.Gray) Biffin & Craven

Cloezia

Cloezia Brongn. & Gris, Bull. Soc. Bot. France 10: 576 (1863).
New Caledonia. 60 NWC. [Myrtaceae]
5 Species
Mooria Montrouz., Mém. Acad. Roy. Sci. Lyon, Sect. Sci. 10: 207 (1860).

Cloezia aquarum (Guillaumin) J.W.Dawson, in Fl. Nouv.-Caléd. 18: 28 (1992).
SE. New Caledonia. 60 NWC. Nanophan.
**Mooria aquarum* Guillaumin, Bull. Mus. Natl. Hist. Nat. 27: 123 (1921).
Mooria streptophylla Guillaumin in C.F.Sarasin & J.Roux, Nova Caledonia, Bot. 1: 247 (1921).

Cloezia artensis (Montrouz.) P.S.Green, Kew Bull. 23: 343 (1969).
New Caledonia. 60 NWC. Nanophan. or phan.
**Mooria artensis* Montrouz., Mém. Acad. Roy. Sci. Lyon, Sect. Sci. 10: 207 (1860).

var. ***artensis***
New Caledonia. 60 NWC. Nanophan. or phan.
Cloezia canescens Brongn. & Gris, Bull. Soc. Bot. France 10: 577 (1863). *Nania canescens* (Brongn. & Gris) Kuntze, Revis. Gen. Pl. 1: 242 (1891). *Mooria canescens* (Brongn. & Gris) Beauvis. ex Guillaumin, Notul. Syst. (Paris) 1: 111 (1909).
Cloezia canescens var. *glabrescens* Brongn. & Gris, Bull. Soc. Bot. France 10: 577 (1863).
Cloezia angustifolia Baker f., J. Linn. Soc., Bot. 45: 310 (1921). *Mooria angustifolia* (Baker f.) Guillaumin, Bull. Mus. Natl. Hist. Nat., II, 4: 690 (1932).
Mooria canescens f. *parvifolia* Däniker, Vierteljahrsschr. Naturf. Ges. Zürich 78(19): 312 (1933).
Cloezia morierei Vieill. ex Guillaumin, Bull. Soc. Bot. France 81: 9 (1934).

var. ***basilaris*** J.W.Dawson, in Fl. Nouv.-Caléd. 18: 44 (1992).
SE. New Caledonia. 60 NWC. Nanophan. or phan.

var. ***riparia*** J.W.Dawson, in Fl. Nouv.-Caléd. 18: 45 (1992).
SE. New Caledonia. 60 NWC. Nanophan. or phan.

Cloezia buxifolia Brongn. & Gris, Bull. Soc. Bot. France 10: 577 (1863). *Nania buxifolia* (Brongn. & Gris) Kuntze, Revis. Gen. Pl. 1: 242 (1891). *Mooria buxifolia* (Brongn. & Gris) Guillaumin, Notul. Syst. (Paris) 1: 111 (1909).
SE. New Caledonia. 60 NWC. Nanophan.

Cloezia deplanchei Brongn. & Gris, Bull. Soc. Bot. France 10: 577 (1863). *Mooria deplanchei* (Brongn. & Gris) Guillaumin, Notul. Syst. (Paris) 1: 111 (1909).
SE. New Caledonia. 60 NWC. Nanophan.
Nania grisii Kuntze, Revis. Gen. Pl. 1: 242 (1891).
Cloezia comptonii Baker f., J. Linn. Soc., Bot. 45: 309 (1921).

Cloezia floribunda Brongn. & Gris, Bull. Soc. Bot. France 10: 577 (1863). *Nania floribunda* (Brongn. & Gris) Kuntze, Revis. Gen. Pl. 1: 242 (1891). *Mooria floribunda* (Brongn. & Gris) Guillaumin, Notul. Syst. (Paris) 1: 111 (1909).
C. & SE. New Caledonia. 60 NWC. Nanophan. or phan.

Cloezia ligustrina Brongn. & Gris, Bull. Soc. Bot. France 10: 577 (1863). *Nania ligustrina* (Brongn. & Gris) Kuntze, Revis. Gen. Pl. 1: 242 (1891).
Cloezia ligustrina var. *angustifolia* Brongn. & Gris, Bull. Soc. Bot. France 10: 577 (1863). *Mooria artensis* var. *angustifolia* (Brongn. & Gris) Guillaumin, Ann. Inst. Bot.-Géol. Colon. Marseille, II, 9: 74 (1911).
Cloezia ligustrina var. *latifolia* Brongn. & Gris, Bull. Soc. Bot. France 10: 577 (1863). *Mooria artensis* var. *latifolia* (Brongn. & Gris) Guillaumin, Ann. Inst. Bot.-Géol. Colon. Marseille, II, 9: 71 (1911).
Cloezia sessilifolia Brongn. & Gris, Bull. Soc. Bot. France 10: 577 (1863). *Nania sessilifolia* (Brongn. & Gris) Kuntze, Revis. Gen. Pl. 1: 242 (1891). *Mooria sessilifolia* (Brongn. & Gris) Guillaumin, Bull. Soc. Bot. France 81: 9 (1934).
Baeckea nelitrioides Seem., J. Bot. 2: 74 (1864).

Cloezia* × *glaberrima (Guillaumin) J.W.Dawson, in Fl. Nouv.-Caléd. 18: 46 (1992). *C. buxifolia* × *C. floribunda*.
SE. New Caledonia. 60 NWC. Nanophan. or phan.
**Mooria* × *glaberrima* Guillaumin, Bull. Mus. Natl. Hist. Nat., II, 27: 325 (1955).

***Synonyms*:**
Cloezia angustifolia Baker f. = ***Cloezia artensis*** (Montrouz.) P.S.Green var. ***artensis***
Cloezia canescens Brongn. & Gris = ***Cloezia artensis*** (Montrouz.) P.S.Green var. ***artensis***
Cloezia canescens var. *glabrescens* Brongn. & Gris = ***Cloezia artensis*** (Montrouz.) P.S.Green var. ***artensis***
Cloezia comptonii Baker f. = ***Cloezia deplanchei*** Brongn. & Gris
Cloezia ligustrina Brongn. & Gris = ***Cloezia floribunda*** Brongn. & Gris
Cloezia ligustrina var. *angustifolia* Brongn. & Gris = ***Cloezia floribunda*** Brongn. & Gris
Cloezia ligustrina var. *latifolia* Brongn. & Gris = ***Cloezia floribunda*** Brongn. & Gris
Cloezia microphylla (A.C.Sm.) A.C.Sm. = ***Decaspermum cryptanthum*** A.J.Scott
Cloezia morierei Vieill. ex Guillaumin = ***Cloezia artensis*** (Montrouz.) P.S.Green var. ***artensis***
Cloezia sessilifolia Brongn. & Gris = ***Cloezia floribunda*** Brongn. & Gris
Cloezia urdanetensis (Elmer) Merr. = ***Kania urdanetensis*** (Elmer) Peter G.Wilson

Cluacena

Cluacena Raf. = ***Myrteola*** O.Berg
Cluacena myrsinoides (Kunth) Raf. = ***Myrcianthes myrsinoides*** (Kunth) Grifo
Cluacena vaccinioides (Kunth) Raf. = ***Myrteola nummularia*** (Lam.) O.Berg

Conothamnus

Conothamnus Lindl., Sketch Veg. Swan R.: 9 (1839).
SW. Australia. 50 WAU. [Myrtaceae]
3 Species
Trichobasis Turcz., Bull. Cl. Phys.-Math. Acad. Imp. Sci. Saint-Pétersbourg 10: 336 (1852).

Conothamnus aureus (Turcz.) Domin, Vestn. Král. Ceské Spolecn. Nauk, Tr. Mat.-Prír. 1921–1922(2): 91 (1923).
SW. Australia. 50 WAU. Nanophan.
**Trichobasis aurea* Turcz., Bull. Cl. Phys.-Math. Acad. Imp. Sci. Saint-Pétersbourg 10: 337 (1852).
Conothamnus divaricatus Benth., Fl. Austral. 3: 164 (1867).

Conothamnus neglectus Diels, Bot. Jahrb. Syst. 35: 430 (1904).
SW. Australia. 50 WAU. Nanophan.

Conothamnus trinervis Lindl., Sketch Veg. Swan R.: 9 (1839).
SW. Australia. 50 WAU. Nanophan.
Melaleuca cuspidata Turcz., Bull. Soc. Imp. Naturalistes Moscou 35(2): 327 (1862).

***Synonyms*:**
Conothamnus divaricatus Benth. = ***Conothamnus aureus*** (Turcz.) Domin

Corymbia

Corymbia K.D.Hill & L.A.S.Johnson, Telopea 6: 214 (1995).
S. New Guinea to Australia. (20) mor (22) ben (23) rwa (25) ken tan uga (26) zim (36) chs 43 NWG 50 NSW NTA QLD SOA VIC WAU (61) sci (63) haw (76) cal (81) pue. [Myrtaceae]
113 Species

Corymbia abbreviata (Blakely & Jacobs) K.D.Hill & L.A.S.Johnson, Telopea 6: 344 (1995).
N. Australia. 50 NTA WAU. Phan.
**Eucalyptus abbreviata* Blakely & Jacobs in W.F.Blakely, Key Eucalypts: 77 (1934).

Corymbia abergiana (F.Muell.) K.D.Hill & L.A.S.Johnson, Telopea 6: 244 (1995).
E. Queensland. 50 QLD. Phan.
**Eucalyptus abergiana* F.Muell., Fragm. 11: 41 (1878).

Corymbia aparrerinja K.D.Hill & L.A.S.Johnson, Telopea 6: 453 (1995). *Eucalyptus aparrerinja* (K.D.Hill & L.A.S.Johnson) Brooker, Austral. Syst. Bot. 13: 137 (2000).
C. Australia. 50 NTA QLD WAU. Phan.

Corymbia arafurica K.D.Hill & L.A.S.Johnson, Telopea 6: 409 (1995).
N. Northern Territory. 50 NTA. Phan.

Corymbia arenaria (Blakely) K.D.Hill & L.A.S.Johnson, Telopea 6: 274 (1995).
N. Western Australia. 50 WAU. Phan.
**Eucalyptus arenaria* Blakely, Key Eucalypts: 81 (1934).

Corymbia arnhemensis (D.J.Carr & S.G.M.Carr) K.D.Hill & L.A.S.Johnson, Telopea 6: 271 (1995).
N. Northern Territory. 50 NTA. Phan.
**Eucalyptus arnhemensis* D.J.Carr & S.G.M.Carr, Eucalyptus 1: 78 (1985).

Corymbia aspera (F.Muell.) K.D.Hill & L.A.S.Johnson, Telopea 6: 448 (1995).
N. Australia. 50 NTA QLD WAU. Phan.
**Eucalyptus aspera* F.Muell., J. Proc. Linn. Soc., Bot. 3: 95 (1859).

Corymbia aureola (Brooker & A.R.Bean) K.D.Hill & L.A.S.Johnson, Telopea 6: 374 (1995).
E. Queensland. 50 QLD. Phan.
**Eucalyptus aureola* Brooker & A.R.Bean, Austrobaileya 3: 430 (1991).

Corymbia bella K.D.Hill & L.A.S.Johnson, Telopea 6: 411 (1995). *Eucalyptus bella* (K.D.Hill & L.A.S.Johnson) Brooker, Austral. Syst. Bot. 13: 137 (2000).
N. Australia. 50 NTA QLD WAU. Phan.

Corymbia blakei K.D.Hill & L.A.S.Johnson, Telopea 6: 437 (1995). *Eucalyptus blakei* (K.D.Hill & L.A.S.Johnson) Brooker, Austral. Syst. Bot. 13: 137 (2000).
Queensland. (25) ken tan 50 QLD. Phan.

subsp. ***blakei***
Queensland. (25) ken tan 50 QLD. Phan.

subsp. ***rasilis*** K.D.Hill & L.A.S.Johnson, Telopea 6: 438 (1995).
W. Queensland. 50 QLD. Phan.

Corymbia bleeseri (Blakely) K.D.Hill & L.A.S.Johnson, Telopea 6: 288 (1995).
N. Australia. 50 NTA WAU. Phan.
**Eucalyptus bleeseri* Blakely, J. Proc. Roy. Soc. New S. Wales 61: 175 (1927).
Eucalyptus terminalis var. *longipedata* Maiden & Blakely in J.H.Maiden, Crit. Revis. Eucalyptus 7: 407 (1928).

Corymbia bloxsomei (Maiden) K.D.Hill & L.A.S.Johnson, Telopea 6: 372 (1995).
Queensland. 50 QLD. Phan.
**Eucalyptus bloxsomei* Maiden, J. Proc. Roy. Soc. New S. Wales 59: 156 (1925).

Corymbia brachycarpa (D.J.Carr & S.G.M.Carr) K.D.Hill & L.A.S.Johnson, Telopea 6: 281 (1995).
WC. Queensland. 50 QLD. Phan.
**Eucalyptus brachycarpa* D.J.Carr & S.G.M.Carr, Eucalyptus 2: 223 (1987).

Corymbia bunites (Brooker & A.R.Bean) K.D.Hill & L.A.S.Johnson, Telopea 6: 373 (1995).
Queensland. 50 QLD. Phan.
**Eucalyptus bunites* Brooker & A.R.Bean, Austrobaileya 3: 423 (1991).

Corymbia byrnesii (D.J.Carr & S.G.M.Carr) K.D.Hill & L.A.S.Johnson, Telopea 6: 331 (1995).
NW. Australia. 50 WAU. Phan.
**Eucalyptus byrnesii* D.J.Carr & S.G.M.Carr, Eucalyptus 2: 159 (1987).

Corymbia cadophora K.D.Hill & L.A.S.Johnson, Telopea 6: 345 (1995). *Eucalyptus cadophora* (K.D.Hill & L.A.S.Johnson) Brooker, Austral. Syst. Bot. 13: 138 (2000).
NW. Australia. 50 WAU. Phan.
**Eucalyptus perfoliata* R.Br. ex Benth., Fl. Austral. 3: 253 (1867), nom. illeg.

subsp. ***cadophora***
NW. Australia. 50 WAU. Phan.

subsp. ***pliantha*** K.D.Hill & L.A.S.Johnson, Telopea 6: 348 (1995).
N. Western Australia. 50 WAU. Phan.

Corymbia calophylla (R.Br. ex Lindl.) K.D.Hill & L.A.S.Johnson, Telopea 6: 240 (1995).
W. & SW. Australia. (25) ken tan 50 WAU (63) haw. Phan.
**Eucalyptus calophylla* R.Br. ex Lindl., Edwards's Bot. Reg. 27(Misc.): 72 (1841).
Eucalyptus glaucophylla Hoffmanns., Verz. Pfl.-Kult., Nachtr. 2: 113 (1841).
Eucalyptus splachnicarpa Hook., Bot. Mag. 69: t. 4036 (1843).
Eucalyptus calophylla var. *multiflora* Guilf., Austral. Pl.: 161 (1911).
Eucalyptus calophylla var. *rosea* Guilf., Austral. Pl.: 161 (1911).
Eucalyptus calophylla var. *rubra* Guilf., Austral. Pl.: 161 (1911).
Eucalyptus calophylla var. *maideniana* Hochr., Candollea 2: 463 (1925).
Eucalyptus calophylla var. *parviflora* Blakely, Key Eucalypts: 85 (1934).

Corymbia candida K.D.Hill & L.A.S.Johnson, Telopea 6: 440 (1995). *Eucalyptus candida* (K.D.Hill & L.A.S. Johnson) Brooker, Austral. Syst. Bot. 13: 138 (2000).
Western Australia. 50 WAU. Phan.

subsp. ***candida***
NW. Western Australia. 50 WAU. Phan.

subsp. ***dipsodes*** K.D.Hill & L.A.S.Johnson, Telopea 6: 444 (1995).
Western Australia. 50 WAU. Phan.

subsp. ***lautifolia*** K.D.Hill & L.A.S.Johnson, Telopea 6: 442 (1995).
NW. Western Australia. 50 WAU. Phan.

Corymbia capricornia (D.J.Carr & S.G.M.Carr) K.D.Hill & L.A.S.Johnson, Telopea 6: 299 (1995).
Northern Territory, NW. Queensland. 50 NTA QLD. Phan.
**Eucalyptus capricornia* D.J.Carr & S.G.M.Carr, Eucalyptus 1: 100 (1985).

Corymbia catenaria K.D.Hill & L.A.S.Johnson, Telopea 6: 376 (1995).
Queensland. 50 QLD. Phan.

Corymbia chartacea K.D.Hill & L.A.S.Johnson, Telopea 6: 353 (1995). *Eucalyptus chartacea* (K.D.Hill & L.A.S.Johnson) Brooker, Austral. Syst. Bot. 13: 138 (2000).
NW. Northern Territory. 50 NTA. Phan.

Corymbia chillagoensis K.D.Hill & L.A.S.Johnson, Telopea 6: 433 (1995).
Queensland. 50 QLD. Phan.

Corymbia chippendalei (D.J.Carr & S.G.M.Carr) K.D.Hill & L.A.S.Johnson, Telopea 6: 313 (1995).
C. & E. Western Australia, SW. Northern Territory. 50 NTA WAU. Phan.
**Eucalyptus chippendalei* D.J.Carr & S.G.M.Carr, Eucalyptus 1: 50 (1985).

Corymbia chlorolampra K.D.Hill & L.A.S.Johnson, Telopea 6: 238 (1995).
W. Western Australia. 50 WAU. Phan.

Corymbia citriodora (Hook.) K.D.Hill & L.A.S.Johnson, Telopea 6: 388 (1995).
Queensland to NE. New South Wales. (20) mor (23) rwa (25) ken tan uga (26) zim (36) chc chs 50 NSW QLD (61) sci (63) haw (76) cal (81) pue. Phan.
**Eucalyptus citriodora* Hook. in T.L.Mitchell, J. Exped. Trop. Australia: 235 (1848). *Eucalyptus maculata* var. *citriodora* (Hook.) F.M.Bailey, Syn. Queensl. Fl.: 181 (1883).

subsp. ***citriodora***
Queensland. (20) mor (23) rwa (25) ken tan uga (26) zim (36) chc chs 50 QLD (63) haw (76) cal (81) pue. Phan.
Eucalyptus melissiodora Lindl. in T.L.Mitchell, J. Exped. Trop. Australia: 235 (1848).

subsp. ***variegata*** (F.Muell.) A.R.Bean & M.W. McDonald, Austrobaileya 5: 735 (2000).
SE. Queensland to NE. New South Wales. 50 NSW QLD. Phan.

Eucalyptus variegata F.Muell., J. Proc. Linn. Soc., Bot. 3: 88 (1859). *Corymbia variegata* (F.Muell.) K.D.Hill & L.A.S.Johnson, Telopea 6: 389 (1995).

Corymbia clandestina (A.R.Bean) K.D.Hill & L.A.S.Johnson, Telopea 6: 282 (1995).
EC. Queensland. 50 QLD. Phan.
Eucalyptus clandestina A.R.Bean, Austrobaileya 4: 205 (1994).

Corymbia clarksoniana (D.J.Carr & S.G.M.Carr) K.D.Hill & L.A.S.Johnson, Telopea 6: 259 (1995).
Queensland. 50 QLD. Phan.
Eucalyptus clarksoniana D.J.Carr & S.G.M.Carr, Eucalyptus 2: 209 (1987).

Corymbia clavigera (A.Cunn. ex Schauer) K.D.Hill & L.A.S.Johnson, Telopea 6: 413 (1995).
N. Western Australia. 50 WAU. Phan.
Eucalyptus clavigera A.Cunn. ex Schauer in W.G.Walpers, Repert. Bot. Syst. 2: 926 (1843).

Corymbia cliftoniana (W.Fitzg.) K.D.Hill & L.A.S.Johnson, Telopea 6: 337 (1995).
N. Western Australia to NW. Northern Territory. 50 NTA WAU. Phan.
Eucalyptus cliftoniana W.Fitzg. in J.H.Maiden, Crit. Revis. Eucalyptus 4: 209 (1919).
Eucalyptus pontis D.J.Carr & S.G.M.Carr, Eucalyptus 1: 88 (1985).

Corymbia collina (W.Fitzg.) K.D.Hill & L.A.S.Johnson, Telopea 6: 290 (1995).
N. Western Australia. 50 WAU. Phan.
Eucalyptus collina W.Fitzg. in J.H.Maiden, Crit. Revis. Eucalyptus 6: 419 (1923).
Eucalyptus macropoda Blakely, Key Eucalypts: 87 (1934).

Corymbia confertiflora (Kippist ex F.Muell.) K.D.Hill & L.A.S.Johnson, Telopea 6: 428 (1995).
N. Australia. 50 NTA QLD WAU. Phan.
Eucalyptus confertiflora Kippist ex F.Muell., J. Proc. Linn. Soc., Bot. 3: 96 (1859).
Eucalyptus floribunda F.Muell., J. Proc. Linn. Soc., Bot. 3: 96 (1859), nom. illeg.
Eucalyptus clavigera var. *diffusa* Blakely & Jacobs, Key Eucalypts: 75 (1934).

Corymbia curtipes (D.J.Carr & S.G.M.Carr) K.D.Hill & L.A.S.Johnson, Telopea 6: 332 (1995).
N. Australia. 50 NTA QLD WAU. Phan.
Eucalyptus curtipes D.J.Carr & S.G.M.Carr, Eucalyptus 2: 149 (1978).

Corymbia dallachiana (Benth.) K.D.Hill & L.A.S. Johnson, Telopea 6: 451 (1995).
Queensland. 50 QLD. Phan.
Eucalyptus tessellaris var. *dallachiana* Benth., Fl. Austral. 3: 251 (1867). *Eucalyptus clavigera* var. *dallachiana* (Benth.) Maiden, J. Proc. Roy. Soc. New S. Wales 47: 77 (1913).

Corymbia dampieri (D.J.Carr & S.G.M.Carr) K.D.Hill & L.A.S.Johnson, Telopea 6: 329 (1995).
NW. Western Australia. 50 WAU. Phan.
Eucalyptus blackwelliana D.J.Carr & S.G.M.Carr, Eucalyptus 2: 162 (1987).
Eucalyptus dampieri D.J.Carr & S.G.M.Carr, Eucalyptus 2: 260 (1987).
Eucalyptus durackiana D.J.Carr & S.G.M.Carr, Eucalyptus 2: 168 (1987).

Corymbia dendromerinx K.D.Hill & L.A.S.Johnson, Telopea 6: 431 (1995).
N. Western Australia. 50 WAU. Phan.

Corymbia deserticola (S.G.M.Carr) K.D.Hill & L.A.S.Johnson, Telopea 6: 365 (1995).
N. Australia. 50 NTA WAU. Phan.
Eucalyptus desertorum D.J.Carr & S.G.M.Carr, Eucalyptus 1: 102 (1985), nom. illeg. *Eucalyptus deserticola* S.G.M.Carr, in Fl. Australia 19: 495 (1988).

subsp. ***deserticola***
NW. Western Australia. 50 WAU. Phan.

subsp. ***mesogeotica*** K.D.Hill & L.A.S.Johnson, Telopea 6: 368 (1995).
Northern Territory. 50 NTA. Phan.

Corymbia dichromophloia (F.Muell.) K.D.Hill & L.A.S.Johnson, Telopea 6: 295 (1995).
Northern Territory. 50 NTA. Phan.
Eucalyptus dichromophloia F.Muell., J. Proc. Linn. Soc., Bot. 3: 89 (1859).
Eucalyptus niphophloia Blakely & Jacobs in W.F.Blakely, Key Eucalypts: 79 (1934).
Eucalyptus atrovirens Brooker & Kleinig, Field Guide Eucalypts 3: 371 (1994).

Corymbia dimorpha (Brooker & A.R.Bean) K.D.Hill & L.A.S.Johnson, Telopea 6: 380 (1995).
Queensland. 50 QLD. Phan.
Eucalyptus peltata subsp. *dimorpha* Brooker & A.R.Bean, Austrobaileya 3: 418 (1991).

Corymbia disjuncta K.D.Hill & L.A.S.Johnson, Telopea 6: 423 (1995).
S. New Guinea to N. Australia. 43 NWG 50 NTA QLD WAU. Phan.

Corymbia dolichocarpa (D.J.Carr & S.G.M.Carr) K.D.Hill & L.A.S.Johnson, Telopea 6: 267 (1995).
E. Australia. 50 NSW QLD. Phan.
Eucalyptus dolichocarpa D.J.Carr & S.G.M.Carr, Eucalyptus 2: 216 (1987).

Corymbia drysdalensis (D.J.Carr & S.G.M.Carr) K.D.Hill & L.A.S.Johnson, Telopea 6: 297 (1995).
N. Western Australia to NW. Northern Territory. 50 NTA WAU. Phan.
Eucalyptus drysdalensis D.J.Carr & S.G.M.Carr, Eucalyptus 1: 84 (1985).

Corymbia dunlopiana K.D.Hill & L.A.S.Johnson, Telopea 6: 354 (1995). *Eucalyptus dunlopiana* (K.D.Hill & L.A.S.Johnson) Brooker, Austral. Syst. Bot. 13: 138 (2000).
N. Northern Territory. 50 NTA. Phan.

Corymbia ellipsoidea (D.J.Carr & S.G.M.Carr) K.D.Hill & L.A.S.Johnson, Telopea 6: 306 (1995).
Queensland. 50 QLD. Phan.
Eucalyptus ellipsoidea D.J.Carr & S.G.M.Carr, Eucalyptus 2: 299 (1987).

Corymbia eremaea (D.J.Carr & S.G.M.Carr) K.D.Hill & L.A.S.Johnson, Telopea 6: 309 (1995).
C. Australia. 50 NTA SOA WAU. Phan.
Eucalyptus eremaea D.J.Carr & S.G.M.Carr, Eucalyptus 1: 38 (1985).

subsp. ***eremaea***
SC. Australia. 50 NTA SOA. Phan.
Eucalyptus australis D.J.Carr & S.G.M.Carr, Eucalyptus 1: 41 (1985).

Eucalyptus connerensis D.J.Carr & S.G.M.Carr, Eucalyptus 1: 45 (1985).
Eucalyptus fordeana D.J.Carr & S.G.M.Carr, Eucalyptus 1: 53 (1985).
Eucalyptus nelsonii D.J.Carr & S.G.M.Carr, Eucalyptus 1: 57 (1985).
Eucalyptus symonii D.J.Carr & S.G.M.Carr, Eucalyptus 1: 36 (1985).

subsp. ***oligocarpa*** (Blakely & Jacobs) K.D.Hill & L.A.S.Johnson, Telopea 6: 311 (1995).
C. Australia. 50 NTA SOA WAU. Phan.
**Eucalyptus polycarpa* var. *oligocarpa* Blakely & Jacobs, Key Eucalypts: 86 (19334).

Corymbia erythrophloia (Blakely) K.D.Hill & L.A.S.Johnson, Telopea 6: 304 (1995).
Queensland. 50 QLD. Phan.
**Eucalyptus erythrophloia* Blakely, Key Eucalypts: 80 (1934).

Corymbia eximia (Schauer) K.D.Hill & L.A.S.Johnson, Telopea 6: 383 (1995).
E. New South Wales. 50 NSW. Phan.
Eucalyptus elongata Link, Enum. Hort. Berol. Alt. 2: 30 (1822), provisional synonym.
**Eucalyptus eximia* Schauer in W.G.Walpers, Repert. Bot. Syst. 2: 925 (1843).

Corymbia ferriticola (Brooker & Edgecombe) K.D.Hill & L.A.S.Johnson, Telopea 6: 446 (1995).
Western Australia. 50 WAU. Phan.
**Eucalyptus ferriticola* Brooker & Edgecombe, Nuytsia 5: 373 (1986).

subsp. ***ferriticola***
Western Australia. 50 WAU. Phan.

subsp. ***sitiens*** L.A.S.Johnson & K.D.Hill, Telopea 6: 447 (1995).
C. Western Australia. 50 WAU. Phan.

Corymbia ferruginea (Schauer) K.D.Hill & L.A.S. Johnson, Telopea 6: 339 (1995).
N. Australia. 50 NTA QLD WAU. Phan.
**Eucalyptus ferruginea* Schauer in W.G.Walpers, Repert. Bot. Syst. 2: 926 (1843).

subsp. ***ferruginea***
N. Australia. 50 NTA QLD WAU. Phan.
Eucalyptus undulata Sweet, Hort. Brit.: 157 (1826), nom. nud.

subsp. ***stypophylla*** K.D.Hill & L.A.S.Johnson, Telopea 6: 343 (1995).
N. Australia. 50 NTA WAU. Phan.

Corymbia ficifolia (F.Muell.) K.D.Hill & L.A.S. Johnson, Telopea 6: 245 (1995).
S. Western Australia. (23) rwa (25) ken tan uga (26) zim 50 WAU (63) haw. Phan.
**Eucalyptus ficifolia* F.Muell., Fragm. 2: 85 (1860).
Eucalyptus ficifolia var. *alba* Guilf., Austral. Pl.: 165 (1911).
Eucalyptus ficifolia var. *bakeri* Guilf., Austral. Pl.: 165 (1911).
Eucalyptus ficifolia var. *rosea* Guilf., Austral. Pl.: 165 (1911).
Eucalyptus ficifolia var. *carmina* Blakely, Key Eucalypts: 85 (1934).

Corymbia flavescens K.D.Hill & L.A.S.Johnson, Telopea 6: 455 (1995). *Eucalyptus flavescens* (K.D.Hill & L.A.S.Johnson) Brooker, Austral. Syst. Bot. 13: 138 (2000).
N. Australia. 50 NTA QLD WAU. Phan.

Corymbia foelscheana (F.Muell.) K.D.Hill & L.A.S.Johnson, Telopea 6: 333 (1995).
Northern Territory. 50 NTA. Phan.
**Eucalyptus foelscheana* F.Muell., Melbourne Chem. Druggist 5(Austr. Suppl.): 56 (1882).
Eucalyptus leiophloia Blakely & Jacobs in W.F.Blakely, Key Eucalypts: 82 (1934).
Eucalyptus leiophloia var. *lepidophloia* Blakely & Jacobs in W.F.Blakely, Key Eucalypts: 83 (1934).
Eucalyptus darwinensis D.J.Carr & S.G.M.Carr, Eucalyptus 2: 131 (1987).
Eucalyptus kakadu D.J.Carr & S.G.M.Carr, Eucalyptus 2: 138 (1987).

Corymbia gilbertensis (Maiden & Blakely) K.D.Hill & L.A.S.Johnson, Telopea 6: 435 (1995).
Queensland. 50 QLD. Phan.
**Eucalyptus clavigera* var. *gilbertensis* Maiden & Blakely in J.H.Maiden, Crit. Revis. Eucalyptus 7: 432 (1928). *Eucalyptus gilbertensis* (Maiden & Blakely) S.T.Blake, Austral. J. Bot. 1: 220 (1953).

Corymbia grandifolia (R.Br. ex Benth.) K.D.Hill & L.A.S.Johnson, Telopea 6: 457 (1995).
Australia. 50 NTA QLD WAU. Phan.
**Eucalyptus grandifolia* R.Br. ex Benth., Fl. Austral. 3: 250 (1867).

subsp. ***grandifolia***
Northern Territory, Queensland. 50 NTA QLD. Phan.

subsp. ***lamprocardia*** L.A.S.Johnson, Telopea 6: 461 (1995).
Western Australia, Northern Territory. 50 NTA WAU. Phan.

subsp. ***longa*** L.A.S.Johnson, Telopea 6: 460 (1995).
Western Australia, Northern Territory. 50 NTA WAU. Phan.

Corymbia greeniana (D.J.Carr & S.G.M.Carr) K.D.Hill & L.A.S.Johnson, Telopea 6: 330 (1995).
Western Australia, Northern Territory. 50 NTA WAU. Phan.
**Eucalyptus greeniana* D.J.Carr & S.G.M.Carr, Eucalyptus 2: 123 (1987).

Corymbia gummifera (Gaertn.) K.D.Hill & L.A.S.Johnson, Telopea 6: 233 (1995).
E. & SE. Australia. (23) rwa (25) ken tan 50 NSW QLD VIC (63) haw. Phan.
**Metrosideros gummifera* Gaertn., Fruct. Sem. Pl. 1: 170 (1788). *Eucalyptus gummifera* (Gaertn.) Hochr., Candollea 2: 464 (1925).
Eucalyptus corymbosa Sm., Spec. Bot. New Holland 4: 43 (1795).
Eucalyptus gummifera var. *intermedia* Domin, Biblioth. Bot. 89: 469 (1928).

Corymbia haematoxylon (Maiden) K.D.Hill & L.A.S.Johnson, Telopea 6: 237 (1995).
W. Western Australia. 50 WAU. Phan.
**Eucalyptus haematoxylon* Maiden, J. Proc. Roy. Soc. New S. Wales 47: 218 (1914).

Corymbia hamersleyana (D.J.Carr & S.G.M.Carr) K.D.Hill & L.A.S.Johnson, Telopea 6: 314 (1995).
NW. Western Australia. 50 WAU. Phan.
Eucalyptus bynoeana D.J.Carr & S.G.M.Carr, Eucalyptus 2: 266 (1987).
**Eucalyptus hamersleyana* D.J.Carr & S.G.M.Carr,

Eucalyptus 2: 256 (1987).
Eucalyptus hesperis D.J.Carr & S.G.M.Carr, Eucalyptus 2: 253 (1987).

Corymbia hendersonii K.D.Hill & L.A.S.Johnson, Telopea 6: 279 (1995). *Eucalyptus hendersonii* (K.D. Hill & L.A.S.Johnson) Brooker, Austral. Syst. Bot. 13: 138 (2000). *Eucalyptus hendrsonii* (K.D.Hill & L.A.S. Johnson) Brooker, Austral. Syst. Bot. 13: 138 (2000).
Queensland. 50 QLD. Phan.

Corymbia henryi (S.T.Blake) K.D.Hill & L.A.S.Johnson, Telopea 6: 396 (1995).
SE. Queensland to NE. New South Wales. 50 NSW QLD. Phan.
**Eucalyptus henryi* S.T.Blake, Austrobaileya 1: 4 (1977).

Corymbia hylandii (D.J.Carr & S.G.M.Carr) K.D.Hill & L.A.S.Johnson, Telopea 6: 283 (1995).
Queensland (Cook). 50 QLD. Phan.
**Eucalyptus hylandii* D.J.Carr & S.G.M.Carr, Eucalyptus 2: 199 (1987).
Eucalyptus hylandii var. *campestris* D.J.Carr & S.G.M.Carr, Eucalyptus 2: 203 (1987).

Corymbia inobvia K.D.Hill & L.A.S.Johnson, Telopea 6: 436 (1995).
Queensland. 50 QLD. Phan.

Corymbia intermedia (F.Muell. ex R.T.Baker) K.D.Hill & L.A.S.Johnson, Telopea 6: 247 (1995).
E. Queensland to E. New South Wales. 50 NSW QLD. Phan.
**Eucalyptus intermedia* F.Muell. ex R.T.Baker, Proc. Linn. Soc. New South Wales 25: 674 (1901).

Corymbia jacobsiana (Blakely) K.D.Hill & L.A.S.Johnson, Telopea 6: 225 (1995).
N. Northern Teritory. 50 NTA. Phan.
**Eucalyptus jacobsiana* Blakely, Key Eucalypts: 92 (1934).

Corymbia karelgica K.D.Hill & L.A.S.Johnson, Telopea 6: 430 (1995).
NC. Western Australia. 50 WAU. Phan.

Corymbia kombolgiensis (Brooker & Dunlop) K.D.Hill & L.A.S.Johnson, Telopea 6: 414 (1995).
N. Northern Territory. 50 NTA. Phan.
**Eucalyptus kombolgiensis* Brooker & Dunlop, Austral. Forest. Res. 8: 212 (1978).

Corymbia lamprophylla (Brooker & A.R.Bean) K.D.Hill & L.A.S.Johnson, Telopea 6: 287 (1995).
Queensland. 50 QLD. Phan.
**Eucalyptus lamprophylla* Brooker & A.R.Bean, Brunonia 10: 195 (1987).

Corymbia latifolia (F.Muell.) K.D.Hill & L.A.S.Johnson, Telopea 6: 327 (1995).
S. New Guinea to N. Australia. 43 NWG 50 NTA QLD WAU. Phan.
**Eucalyptus latifolia* F.Muell., J. Proc. Linn. Soc., Bot. 3: 95 (1859).

Corymbia leichhardtii (F.M.Bailey) K.D.Hill & L.A.S.Johnson, Telopea 6: 370 (1995).
Queensland. 50 QLD. Phan.
**Eucalyptus leichhardtii* F.M.Bailey, Queensland Agric. J. 17: 493 (1906). *Eucalyptus eximia* var. *leichhardtii* (F.M.Bailey) Ewart, Victorian Naturalist 24: 56 (1907). *Eucalyptus peltata* subsp. *leichhardtii* (F.M.Bailey) L.A.S.Johnson & Blaxell, Contr. New South Wales Natl. Herb. 4: 453 (1973).

Corymbia lenziana (D.J.Carr & S.G.M.Carr) K.D.Hill & L.A.S.Johnson, Telopea 6: 312 (1995).
WC. Western Australia. 50 WAU. Phan.
**Eucalyptus lenziana* D.J.Carr & S.G.M.Carr, Eucalyptus 1: 47 (1985).

Corymbia leptoloma (Brooker & A.R.Bean) K.D.Hill & L.A.S.Johnson, Telopea 6: 369 (1995).
Queensland (W. Paluma Range). 50 QLD. Phan.
**Eucalyptus leptoloma* Brooker & A.R.Bean, Austrobaileya 3: 432 (1991).

Corymbia ligans K.D.Hill & L.A.S.Johnson, Telopea 6: 263 (1995).
Queensland. 50 QLD. Phan.

subsp. ***burdelinensis*** K.D.Hill & L.A.S.Johnson, Telopea 6: 265 (1995).
Queensland. 50 QLD. Phan.

subsp. ***ligans***
Queensland. 50 QLD. Phan.

subsp. ***novocastrensis*** K.D.Hill & L.A.S.Johnson, Telopea 6: 265 (1995).
Queensland. 50 QLD. Phan.

Corymbia maculata (Hook.) K.D.Hill & L.A.S.Johnson, Telopea 6: 393 (1995).
SE. Australia. (23) rwa (25) ken tan uga (27) cpp (36) chc chs (38) tai 50 NSW VIC. Phan.
**Eucalyptus maculata* Hook., Hooker's Icon. Pl. 7: t. 619 (1844).

Corymbia maritima K.D.Hill & L.A.S.Johnson, Telopea 6: 261 (1995).
Queensland. 50 QLD. Phan.

Corymbia nesophila (Blakely) K.D.Hill & L.A.S. Johnson, Telopea 6: 242 (1995).
N. Australia. 50 NTA QLD WAU. Phan.
**Eucalyptus nesophila* Blakely, Key Eucalypts: 90 (1934).

Corymbia novoguinensis (D.J.Carr & S.G.M.Carr) K.D.Hill & L.A.S.Johnson, Telopea 6: 257 (1995).
S. New Guinea to N. Queensland. 43 NWG 50 QLD. Phan.
**Eucalyptus novoguinensis* D.J.Carr & S.G.M.Carr, Eucalyptus 2: 192 (1987).

Corymbia oocarpa (D.J.Carr & S.G.M.Carr) K.D.Hill & L.A.S.Johnson, Telopea 6: 293 (1995).
N. Northern Territory. 50 NTA. Phan.
**Eucalyptus oocarpa* D.J.Carr & S.G.M.Carr, Eucalyptus 2: 277 (1987).

Corymbia opaca (D.J.Carr & S.G.M.Carr) K.D.Hill & L.A.S.Johnson, Telopea 6: 318 (1995).
Western Australia, Northern Territory. 50 NTA WAU. Phan.
Eucalyptus centralis D.J.Carr & S.G.M.Carr, Eucalyptus 1: 61 (1985).
**Eucalyptus opaca* D.J.Carr & S.G.M.Carr, Eucalyptus 1: 63 (1985).
Eucalyptus orientalis D.J.Carr & S.G.M.Carr, Eucalyptus 1: 66 (1985).

Corymbia opacula L.A.S.Johnson, Telopea 6: 325 (1995).
NE. Western Australia. 50 WAU. Phan.

Corymbia pachycarpa K.D.Hill & L.A.S.Johnson, Telopea 6: 359 (1995). *Eucalyptus pachycarpa* (K.D. Hill & L.A.S.Johnson) Brooker, Austral. Syst. Bot. 13: 138 (2000).
Western Australia, Northern Territory. 50 NTA WAU. Phan.

subsp. ***glabrescens*** K.D.Hill & L.A.S.Johnson, Telopea 6: 360 (1995).
W. Northern Territory. 50 NTA. Phan.

subsp. ***pachycarpa***
Western Australia, Northern Territory. 50 NTA WAU. Phan.

Corymbia papillosa K.D.Hill & L.A.S.Johnson, Telopea 6: 362 (1995). *Eucalyptus papillosa* (K.D.Hill & L.A.S. Johnson) Brooker, Austral. Syst. Bot. 13: 138 (2000).
NC. Western Australia to Northern Territory. 50 NTA WAU. Phan.

subsp. ***globifera*** K.D.Hill & L.A.S.Johnson, Telopea 6: 363 (1995).
NC. Western Australia. 50 WAU. Phan.

subsp. ***papillosa***
Northern Territory. 50 NTA. Phan.

Corymbia papuana (F.Muell.) K.D.Hill & L.A.S. Johnson, Telopea 6: 405 (1995).
S. New Guinea. 43 NWG. Phan.
**Eucalyptus papuana* F.Muell., Descr. Notes Papuan Pl. 1: 8 (1875).
Eucalyptus papuana var. *aparrerinja* Blakely, Trans. & Proc. Roy. Soc. South Australia 60: 154 (1936).

Corymbia paracolpica K.D.Hill & L.A.S.Johnson, Telopea 6: 407 (1995).
N. Queensland. 50 QLD. Phan.

Corymbia paractia K.D.Hill & L.A.S.Johnson, Telopea 6: 462 (1995).
NW. Western Australia. 50 WAU. Phan.

Corymbia pauciseta K.D.Hill & L.A.S.Johnson, Telopea 6: 426 (1995).
Northern Territory. 50 NTA. Phan.

Corymbia pedimontana L.A.S.Johnson, Telopea 6: 335 (1995).
N. Western Australia. 50 WAU. Phan.

Corymbia peltata (Benth.) K.D.Hill & L.A.S.Johnson, Telopea 6: 381 (1995).
Queensland. 50 QLD. Phan.
**Eucalyptus peltata* Benth., Fl. Austral. 3: 254 (1867).

Corymbia petalophylla (Brooker & A.R.Bean) K.D.Hill & L.A.S.Johnson, Telopea 6: 375 (1995).
Queensland. 50 QLD. Phan.
**Eucalyptus petalophylla* Brooker & A.R.Bean, Austrobaileya 3: 428 (1991).

Corymbia plena K.D.Hill & L.A.S.Johnson, Telopea 6: 268 (1995). *Eucalyptus plena* (K.D.Hill & L.A.S. Johnson) Brooker, Austral. Syst. Bot. 13: 138 (2000).
Queensland. 50 QLD. Phan.

Corymbia pocillum (D.J.Carr & S.G.M.Carr) K.D.Hill & L.A.S.Johnson, Telopea 6: 303 (1995).
Queensland. 50 QLD. Phan.
**Eucalyptus pocillum* D.J.Carr & S.G.M.Carr, Eucalyptus 2: 293 (1987).

Corymbia polycarpa (F.Muell.) K.D.Hill & L.A.S. Johnson, Telopea 6: 254 (1995).
N. Australia. 50 NTA QLD WAU. Phan.
**Eucalyptus polycarpa* F.Muell., J. Proc. Linn. Soc., Bot. 3: 88 (1859). *Eucalyptus pyrophora* var. *polycarpa* (F.Muell.) Maiden, Crit. Revis. Eucalyptus 4: 321 (1920).
Eucalyptus terminalis var. *carnosa* F.M.Bailey, Queensland Agric. J. 15: 898 (1905).
Eucalyptus derbyensis D.J.Carr & S.G.M.Carr, Eucalyptus 2: 189 (1987).
Eucalyptus erubescens D.J.Carr & S.G.M.Carr, Eucalyptus 2: 184 (1987).

Corymbia polysciada (F.Muell.) K.D.Hill & L.A.S. Johnson, Telopea 6: 417 (1995).
Northern Territory. 50 NTA. Phan.
**Eucalyptus polysciada* F.Muell., J. Proc. Linn. Soc., Bot. 3: 98 (1859).

Corymbia porphyritica K.D.Hill & L.A.S.Johnson, Telopea 6: 307 (1995).
Queensland. 50 QLD. Phan.

Corymbia porrecta (S.T.Blake) K.D.Hill & L.A.S. Johnson, Telopea 6: 270 (1995).
N. Northern Territory. 50 NTA. Phan.
**Eucalyptus porrecta* S.T.Blake, Austral. J. Bot. 1: 251 (1953).

Corymbia ptychocarpa (F.Muell.) K.D.Hill & L.A.S.Johnson, Telopea 6: 250 (1995).
N. Australia. 50 NTA QLD WAU. Phan.
**Eucalyptus ptychocarpa* F.Muell., J. Proc. Linn. Soc., Bot. 3: 90 (1859).

subsp. ***aptycha*** K.D.Hill & L.A.S.Johnson, Telopea 6: 251 (1995).
N. Northern Territory. 50 NTA. Phan.

subsp. ***ptychocarpa***
N. Australia. 50 NTA QLD WAU. Phan.

Corymbia punkapitiensis K.D.Hill & L.A.S.Johnson, Telopea 6: 449 (1995).
Western Australia (Walter James Ranges). 50 WAU. Phan.

Corymbia rhodops (D.J.Carr & S.G.M.Carr) K.D.Hill & L.A.S.Johnson, Telopea 6: 276 (1995).
Queensland. 50 QLD. Phan.
**Eucalyptus rhodops* D.J.Carr & S.G.M.Carr, Eucalyptus 2: 283 (1987).

Corymbia rubens K.D.Hill & L.A.S.Johnson, Telopea 6: 301 (1995).
NC. Western Australia. 50 WAU. Phan.

Corymbia scabrida (Brooker & A.R.Bean) K.D.Hill & L.A.S.Johnson, Telopea 6: 382 (1995).
Queensland. 50 QLD. Phan.
**Eucalyptus scabrida* Brooker & A.R.Bean, Austrobaileya 3: 420 (1991).

Corymbia semiclara K.D.Hill & L.A.S.Johnson, Telopea 6: 316 (1995).
NW. Western Australia. 50 WAU. Phan.

Corymbia serendipita (Brooker & Kleinig) A.R.Bean, Austrobaileya 6: 345 (2002).
N. Queensland. 50 QLD. Phan.
**Eucalyptus serendipita* Brooker & Kleinig, Field Guide Eucalypts 3: 371 (1994).
Corymbia arnhemensis subsp. *monticola* K.D.Hill & L.A.S.Johnson, Telopea 6: 273 (1995).

Corymbia setosa (Schauer) K.D.Hill & L.A.S.Johnson, Telopea 6: 356 (1995).

Northern Territory, Queensland. 50 NTA QLD. Phan.
**Eucalyptus setosa* Schauer in W.G.Walpers, Repert. Bot. Syst. 2: 926 (1843).

subsp. ***pedicellaris*** K.D.Hill & L.A.S.Johnson, Telopea 6: 358 (1995).
E. Queensland. 50 QLD. Phan.

subsp. ***setosa***
Northern Territory, Queensland. 50 NTA QLD. Phan.

Corymbia sphaerica K.D.Hill & L.A.S.Johnson, Telopea 6: 351 (1995).
C. Northern Territory. 50 NTA. Phan.

Corymbia stockeri (D.J.Carr & S.G.M.Carr) K.D.Hill & L.A.S.Johnson, Telopea 6: 286 (1995).
Queensland. 50 QLD. Phan.
**Eucalyptus stockeri* D.J.Carr & S.G.M.Carr, Eucalyptus 2: 288 (1987).

subsp. ***peninsularis*** (K.D.Hill & L.A.S.Johnson) A.R.Bean, Austrobaileya 6: 345 (2002).
N. Queensland. 50 QLD. Phan.
**Corymbia hylandii* subsp. *peninsularis* K.D.Hill & L.A.S.Johnson, Telopea 6: 284 (1995).

subsp. ***stockeri***
Queensland (Cook). 50 QLD. Phan.

Corymbia terminalis (F.Muell.) K.D.Hill & L.A.S. Johnson, Telopea 6: 323 (1995).
Northern Territory, Queensland. 50 NTA QLD. Phan.
**Eucalyptus terminalis* F.Muell., J. Proc. Linn. Soc., Bot. 3: 89 (1859). *Eucalyptus corymbosa* var. *terminalis* (F.Muell.) F.M.Bailey, Queensl. Woods: 44 (1886).
Eucalyptus pyrophora Benth., Fl. Austral. 3: 257 (1867).
Eucalyptus pyrophora f. *compacta* Domin, Biblioth. Bot. 89: 1024 (1928).
Eucalyptus pyrophora var. *compacta* Domin, Biblioth. Bot. 89: 1024 (1928).

Corymbia tessellaris (F.Muell.) K.D.Hill & L.A.S. Johnson, Telopea 6: 402 (1995).
Queensland to N. New South Wales. 50 NSW QLD. Phan.
Eucalyptus viminalis Hook. in T.L.Mitchell, J. Exped. Trop. Australia: 157 (1848), nom. illeg. *Eucalyptus hookeri* F.Muell., J. Proc. Linn. Soc., Bot. 3: 90 (1859).
**Eucalyptus tessellaris* F.Muell., J. Proc. Linn. Soc., Bot. 3: 88 (1859).

Corymbia torelliana (F.Muell.) K.D.Hill & L.A.S.Johnson, Telopea 6: 385 (1995).
E. Queensland. (22) ben (36) chs (38) tai 50 QLD. Phan.
**Eucalyptus torelliana* F.Muell., Fragm. 10: 106 (1877).

Corymbia torta K.D.Hill & L.A.S.Johnson, Telopea 6: 418 (1995).
N. & NC. Western Australia. 50 WAU. Phan.

subsp. ***allanii*** K.D.Hill & L.A.S.Johnson, Telopea 6: 420 (1995).
N. Western Australia. 50 WAU. Phan.

subsp. ***mixtifolia*** K.D.Hill & L.A.S.Johnson, Telopea 6: 421 (1995).
N. Western Australia. 50 WAU. Phan.

subsp. ***torta***
NC. Western Australia. 50 WAU. Phan.

Corymbia trachyphloia (F.Muell.) K.D.Hill & L.A.S. Johnson, Telopea 6: 227 (1995).
Queensland to New South Wales. 50 NSW QLD. Phan.
**Eucalyptus trachyphloia* F.Muell., J. Proc. Linn. Soc., Bot. 3: 90 (1859).

subsp. ***amphistomatica*** K.D.Hill & L.A.S.Johnson, Telopea 6: 230 (1995).
SE. Queensland to New South Wales. 50 NSW QLD. Phan.

subsp. ***carnarvonica*** K.D.Hill & L.A.S.Johnson, Telopea 6: 231 (1995).
SE. Queensland. 50 QLD. Phan.

subsp. ***trachyphloia***
Queensland to NE. New South Wales. 50 NSW QLD. Phan.
Eucalyptus trachyphloia f. *fruticosa* F.M.Bailey, Queensland Agric. J. 25: 9 (1910).

Corymbia tumescens K.D.Hill & L.A.S.Johnson, Telopea 6: 321 (1995).
Queensland, N. New South Wales, NE. South Australia. 50 NSW QLD SOA. Phan.

Corymbia umbonata (D.J.Carr & S.G.M.Carr) K.D.Hill & L.A.S.Johnson, Telopea 6: 300 (1995).
Northern Territory. 50 NTA. Phan.
Eucalyptus ollaris D.J.Carr & S.G.M.Carr, Eucalyptus 1: 76 (1985).
**Eucalyptus umbonata* D.J.Carr & S.G.M.Carr, Eucalyptus 1: 95 (1985).

Corymbia watsoniana (F.Muell.) K.D.Hill & L.A.S. Johnson, Telopea 6: 378 (1995).
Queensland. 50 QLD. Phan.
**Eucalyptus watsoniana* F.Muell., Fragm. 10: 98 (1877).

subsp. ***capillata*** (Brooker & A.R.Bean) K.D.Hill & L.A.S.Johnson, Telopea 6: 379 (1995).
Queensland. 50 QLD. Phan.
**Eucalyptus watsoniana* subsp. *capillata* Brooker & A.R.Bean, Austrobaileya 3: 428 (1991).

subsp. ***watsoniana***
Queensland. 50 QLD. Phan.

Corymbia xanthope (A.R.Bean & Brooker) K.D.Hill & L.A.S.Johnson, Telopea 6: 277 (1995).
Queensland. 50 QLD. Phan.
**Eucalyptus xanthope* A.R.Bean & Brooker, Austrobaileya 3: 39 (1989).

Corymbia zygophylla (Blakely) K.D.Hill & L.A.S.Johnson, Telopea 6: 349 (1995).
N. & NW. Western Australia. 50 WAU. Phan.
**Eucalyptus zygophylla* Blakely, Key Eucalypts: 88 (1934).

Synonyms:
Corymbia arnhemensis subsp. *monticola* K.D.Hill & L.A.S.Johnson = ***Corymbia serendipita*** (Brooker & Kleinig) A.R.Bean
Corymbia hylandii subsp. *peninsularis* K.D.Hill & L.A.S.Johnson = ***Corymbia stockeri*** subsp. ***peninsularis*** (K.D.Hill & L.A.S.Johnson) A.R.Bean
Corymbia variegata (F.Muell.) K.D.Hill & L.A.S.Johnson = ***Corymbia citriodora*** subsp. ***variegata*** (F.Muell.) A.R.Bean & M.W.McDonald

Corynanthera

Corynanthera J.W.Green, Nuytsia 2: 368 (1979).
WSW. Western Australia. 50 WAU. [Myrtaceae]
1 Species

Corynanthera flava J.W.Green, Nuytsia 2: 371 (1979).
Western Australia (Irwin). 50 WAU. Nanophan.

Corynemyrtus

Corynemyrtus (Kiaersk.) Mattos = ***Psidium*** L.
Corynemyrtus corynantha (Kiaersk.) Mattos = ?

Cryptorhiza

Cryptorhiza Urb. = ***Pimenta*** Lindl.
Cryptorhiza haitiensis Urb. = ***Pimenta haitiensis*** (Urb.) Landrum

Cryptostemon

Cryptostemon F.Muell. ex Miq. = ***Darwinia*** Rudge
Cryptostemon ericaeus F.Muell. ex Miq. = ***Darwinia fascicularis*** Rudge subsp. ***fascicularis***
Cryptostemon fascicularis (Rudge) F.Muell. ex Miq. = ***Darwinia fascicularis*** Rudge

Cuiavus

Cuiavus Trew = ***Psidium*** L.

Cumetea

Cumetea Raf. = ***Myrcia*** DC. ex Guill.
Cumetea alba Raf. = ***Myrcia coumeta*** (Aubl.) DC.
Cumetea angustifolia (Lam.) Raf. = ***Eugenia pomifera*** (Aubl.) Urb.
Cumetea divaricata (Lam.) Raf. = ***Myrcia splendens*** (Sw.) DC.
Cumetea fragrans (Sw.) Raf. = ***Myrcianthes fragrans*** (Sw.) McVaugh
Cumetea microphyla Raf. = ***Eugenia picardiae*** Krug & Urb.
Cumetea mini (Aubl.) Raf. = ***Eugenia biflora*** (L.) DC.
Cumetea montana (Aubl.) Raf. = ***Marlierea montana*** (Aubl.) Amshoff
Cumetea multiflora (Lam.) Raf. = ***Myrcia multiflora*** (Lam.) DC.
Cumetea tomentosa (Aubl.) Raf. = ***Myrcia tomentosa*** (Aubl.) DC.

Cupheanthus

Cupheanthus Seem. = ***Syzygium*** Gaertn.
Cupheanthus austrocaledonicus Guillaumin = ***Syzygium neocaledonicum*** (Seem.) J.W.Dawson
Cupheanthus comptonii (Baker f.) Guillaumin = ***Syzygium neocaledonicum*** (Seem.) J.W.Dawson
Cupheanthus microphyllus Guillaumin = ***Syzygium virotii*** J.W.Dawson
Cupheanthus neocaledonicus Seem. = ***Syzygium neocaledonicum*** (Seem.) J.W.Dawson
Cupheanthus neocaledonicus (Baker f.) Guillaumin = ***Syzygium toninense*** (Baker f.) J.W.Dawson
Cupheanthus paniensis (Baker f.) Guillaumin = ***Syzygium paniense*** (Baker f.) J.W.Dawson
Cupheanthus serpentinus Airy Shaw = ***Syzygium toninense*** (Baker f.) J.W.Dawson
Cupheanthus toninensis (Baker f.) Guillaumin = ***Syzygium toninense*** (Baker f.) J.W.Dawson

Cyathostemon

Cyathostemon Turcz. = ***Baeckea*** L.
Cyathostemon tenuifolius Turcz. = ***Baeckea tenuifolia*** (Turcz.) Domin

Cynomyrtus

Cynomyrtus Scriv. = ***Rhodomyrtus*** (DC.) Rchb.
Cynomyrtus tomentosa (Sol.) Scriv. = ***Rhodomyrtus tomentosa*** (Sol.) Hassk.

Darwinia

Darwinia Rudge, Trans. Linn. Soc. London 11: 299 (1813).
Australia. 50 NSW QLD SOA VIC WAU. [Myrtaceae]
45 Species
Genetyllis DC., Prodr. 3: 209 (1828).
Polyzone Endl., Stirp. Herb. Hügel.: 3 (1838).
Hedaroma Lindl., Sketch Veg. Swan R.: 7 (1839).
Francisia Endl., Gen. Pl.: 1226 (1840).
Cryptostemon F.Muell. ex Miq., Ned. Kruidk. Arch. 4: 114 (1856).

Darwinia acerosa W.Fitzg., J. West Austral. Nat. Hist. Soc. 1: 17 (1904).
SW. Australia. 50 WAU. Nanophan.

Darwinia apiculata N.G.Marchant, Nuytsia 5: 63 (1984).
SW. Australia. 50 WAU. Nanophan.

Darwinia biflora (Cheel) B.G.Briggs, Contr. New South Wales Natl. Herb. 3: 144 (1962).
New South Wales. 50 NSW. Nanophan.
**Darwinia taxifolia* var. *biflora* Cheel, J. Proc. Roy. Soc. New S. Wales 56: 72 (1922).

Darwinia briggsiae Craven & S.R.Jones, Austral. Syst. Bot. 4: 530 (1991).
New South Wales. 50 NSW. Nanophan.

Darwinia camptostylis B.G.Briggs, Contr. New South Wales Natl. Herb. 3: 141 (1962).
SE. Australia. 50 NSW VIC. Nanophan.

Darwinia capitellata Rye, Nuytsia 4: 423 (1983).
SW. Australia. 50 WAU. Nanophan.

Darwinia carnea C.A.Gardner, J. Roy. Soc. Western Australia 14: 80 (1928).
SW. Australia. 50 WAU. Nanophan.

Darwinia citriodora (Endl.) Benth., J. Linn. Soc., Bot. 9: 180 (1867).
SW. Australia. 50 WAU. Nanophan.
**Genetyllis citriodora* Endl. in S.L.Endlicher & al., Enum. Pl.: 47 (1837).
Hedaroma latifolium Lindl., Sketch Veg. Swan R.: vii (1839).
Genetyllis pimeloides F.Muell., Fragm. 2: 169 (1861).

Darwinia collina Gardner, J. Proc. Roy. Soc. W. Australia 9(1): 41 (1923).
SW. Australia. 50 WAU. Nanophan.

Darwinia diminuta B.G.Briggs, Contr. New South Wales Natl. Herb. 3: 140 (1962).
New South Wales. 50 NSW. Nanophan.

Darwinia diosmoides (DC.) Benth., J. Linn. Soc., Bot. 9: 180 (1867).
SW. Australia. 50 WAU. Nanophan.
**Genetyllis diosmoides* DC., Prodr. 3: 209 (1828).

Genetyllis affinis Turcz., Bull. Soc. Imp. Naturalistes Moscou 20(1): 155 (1847).

Darwinia divisa Keighery & N.G.Marchant, Nordic J. Bot. 22: 45 (2002).
Western Australia. 50 WAU.

Darwinia fascicularis Rudge, Trans. Linn. Soc. London 11: 299 (1815). *Cryptostemon fascicularis* (Rudge) F.Muell. ex Miq., Ned. Kruidk. Arch. 4: 115 (1856).
Queensland to New South Wales. 50 NSW QLD. Nanophan.

subsp. ***fascicularis***
New South Wales. 50 NSW. Nanophan.
Cryptostemon ericaeus F.Muell. ex Miq., Ned. Kruidk. Arch. 4: 115 (1856).

subsp. ***oligantha*** B.G.Briggs, Contr. New South Wales Natl. Herb. 3: 149 (1962).
Queensland to New South Wales. 50 NSW QLD. Nanophan.

Darwinia glaucophylla B.G.Briggs, Contr. New South Wales Natl. Herb. 3: 143 (1962).
New South Wales. 50 NSW. Nanophan.

Darwinia grandiflora (Benth.) R.T.Baker & H.G.Sm., J. Proc. Roy. Soc. New S. Wales 50: 181 (1916 publ. 1917).
New South Wales. 50 NSW. Nanophan.
**Darwinia taxifolia* var. *grandiflora* Benth., Fl. Austral. 3: 12 (1867).

Darwinia helichrysoides (Meisn.) Benth., Journ. Linn. Soc., Bot. 9: 179 (1867).
SW. Australia. 50 WAU. Nanophan.
**Genetyllis helichrysoides* Meisn., J. Proc. Linn. Soc., Bot. 1: 37 (1857).

Darwinia hypericifolia (Turcz.) Domin, Vestn. Král. Ceské Spolecn. Nauk, Tr. Mat.-Prír. 1921–1922(2): 78 (1923).
SW. Australia. 50 WAU. Nanophan.
**Genetyllis hypericifolia* Turcz., Bull. Cl. Phys.-Math. Acad. Imp. Sci. Saint-Pétersbourg 10: 323 (1852).
Genetyllis macrostegia Hook., Bot. Mag. 81: t. 4860 (1855), nom. illeg.
Genetyllis hookeriana Meisn., J. Proc. Linn. Soc., Bot. 1: 37 (1857). *Darwinia hookeriana* (Meisn.) Benth., J. Linn. Soc., Bot. 9: 179 (1867).
Hedaroma hookeri Baines, Gard. Chron. 1874(1): 601 (1874).

Darwinia leiostyla (Turcz.) Domin, Vestn. Král. Ceské Spolecn. Nauk, Tr. Mat.-Prír. 1921–1922(2): 78 (1923).
SW. Australia. 50 WAU. Nanophan.
**Genetyllis leiostyla* Turcz., Bull. Cl. Phys.-Math. Acad. Imp. Sci. Saint-Pétersbourg 10: 323 (1852).

Darwinia leptantha B.G.Briggs, Contr. New South Wales Natl. Herb. 3: 142 (1962).
New South Wales. 50 NSW. Nanophan.

Darwinia luehmannii F.Muell. & Tate, Trans. & Proc. Roy. Soc. South Australia 16: 353 (1896).
SW. Australia. 50 WAU. Nanophan.

Darwinia macrostegia (Turcz.) Benth., J. Linn. Soc., Bot. 9: 179 (1867).
SW. Australia. 50 WAU. Nanophan.
**Genetyllis macrostegia* Turcz., Bull. Soc. Imp. Naturalistes Moscou 22(2): 18 (1849).
Hedaroma tulipiferum Lindl., Gard. Chron. 1854: 323 (1854).

Darwinia masonii C.A.Gardner, J. Roy. Soc. Western Australia 47: 62 (1964).
SW. Australia. 50 WAU. Nanophan.

Darwinia meeboldii C.A.Gardner, J. Roy. Soc. Western Australia 27: 189 (1942).
SW. Australia. 50 WAU. Nanophan.

Darwinia micropetala (F.Muell.) Benth., J. Linn. Soc., Bot. 9: 181 (1867).
South Australia to Victoria. 50 SOA VIC. Nanophan.
**Genetyllis micropetala* F.Muell., Fragm. 1: 12 (1858).
Chamelaucium micropetalum (F.Muell.) F.Muell., Fragm. 4: 138 (1864).

Darwinia neildiana F.Muell., Fragm. 9: 177 (1875).
SW. Australia. 50 WAU. Nanophan.

Darwinia oederoides (Turcz.) Benth., J. Linn. Soc., Bot. 9: 179 (1867).
SW. Australia. 50 WAU. Nanophan.
**Genetyllis oederoides* Turcz., Bull. Soc. Imp. Naturalistes Moscou 22(2): 18 (1849).

Darwinia oldfieldii Benth., J. Linn. Soc., Bot. 9: 180 (1867).
SW. Australia. 50 WAU. Nanophan.

Darwinia oxylepis (Turcz.) N.G.Marchant & Keighery, Nuytsia 3: 181 (1980).
SW. Australia. 50 WAU. Nanophan.
**Genetyllis oxylepis* Turcz., Bull. Cl. Phys.-Math. Acad. Imp. Sci. Saint-Pétersbourg 10: 234 (1852).
Genetyllis meisneri Kippist, J. Proc. Linn. Soc., Bot. 1: 49 (1857). *Darwinia meisneri* (Kippist) Benth., J. Linn. Soc., Bot. 9: 179 (1867).

Darwinia pauciflora Benth., J. Linn. Soc., Bot. 9: 180 (1867).
SW. Australia. 50 WAU. Nanophan.

Darwinia peduncularis B.G.Briggs, Contr. New South Wales Natl. Herb. 3: 144 (1962).
New South Wales. 50 NSW. Nanophan.

Darwinia pimelioides Cayzer & F.W.Wakef., J. Proc. Roy. Soc. W. Australia 8: 40 (1922).
SW. Australia. 50 WAU. Nanophan.

Darwinia pinifolia (Lindl.) Benth., J. Linn. Soc., Bot. 9: 181 (1867).
SW. Australia. 50 WAU. Nanophan.
**Hedaroma pinifolium* Lindl., Sketch Veg. Swan R.: vii (1839). *Genetyllis pinifolia* (Lindl.) Schauer, Monogr. Myrt. Xerocarp. 1: 34 (1843).
Darwinia rhadinophylla F.Muell., Fragm. 9: 175 (1875).

Darwinia polycephala Gardner, J. Roy. Soc. Western Australia 11: 19 (1924).
SW. Australia. 50 WAU. Nanophan.

Darwinia procera B.G.Briggs, Contr. New South Wales Natl. Herb. 3: 145 (1962).
New South Wales. 50 NSW. Nanophan.

Darwinia purpurea (Endl.) Benth., J. Linn. Soc., Bot. 9: 180 (1867).
SW. Australia. 50 WAU. Nanophan.
**Polyzone purpurea* Endl., Stirp. Herb. Hügel.: 3 (1838).
Genetyllis purpurea (Endl.) Schauer, Monogr. Myrt. Xerocarp. 1: 27 (1843).

Darwinia repens A.S.George, J. Roy. Soc. Western Australia 50: 99 (1967).
SW. Australia. 50 WAU. Nanophan.

Darwinia salina Craven & S.R.Jones, Austral. Syst. Bot. 4: 531 (1991).
South Australia. 50 SOA. Nanophan.

Darwinia sanguinea (Meisn.) Benth., J. Linn. Soc., Bot. 9: 181 (1867).
SW. Australia. 50 WAU. Nanophan.
**Genetyllis sanguinea* Meisn., J. Proc. Linn. Soc., Bot. 1: 38 (1857).

Darwinia speciosa (Meisn.) Benth., J. Linn. Soc., Bot. 9: 179 (1867).
SW. Australia. 50 WAU. Nanophan.
**Genetyllis speciosa* Meisn., J. Proc. Linn. Soc., Bot. 1: 36 (1857).

Darwinia squarrosa (Turcz.) Domin, Vestn. Král. Ceské Spolecn. Nauk, Tr. Mat.-Prír. 1921–1922(2): 78 (1923).
SW. Australia. 50 WAU. Nanophan.
**Genetyllis squarrosa* Turcz., Bull. Cl. Phys.-Math. Acad. Imp. Sci. Saint-Pétersbourg 10: 328 (1852).
Genetyllis fimbriata Kippist, J. Proc. Linn. Soc., Bot. 1: 49 (1857). *Darwinia fimbriata* (Kippist) Benth., J. Linn. Soc., Bot. 9: 179 (1867). *Hedaroma fimbriatum* (Kippist) Baines, Gard. Chron. 1874(1): 601 (1874).

Darwinia taxifolia A.Cunn., Field New South Wales: 352 (1825). *Darwinia taxifolia* var. *typica* Cheel, J. Proc. Roy. Soc. New South Wales 56: 72 (1922), nom. inval.
New South Wales. 50 NSW. Nanophan.
Darwinia intermedia A.Cunn. ex Schauer, Nov. Actorum Acad. Caes. Leop.-Carol. Nat. Cur. 19(Suppl. 2): 190 (1841), nom. inval. *Darwinia taxifolia* var. *intermedia* Cheel, J. Proc. Roy. Soc. New South Wales 56: 73 (1922).
Darwinia laxifolia Schauer, Nov. Actorum Acad. Caes. Leop.-Carol. Nat. Cur. 19(Suppl. 2): 190 (1841), nom. illeg.
Darwinia taxifolia subsp. *macrolaena* B.G.Briggs, Contr. New South Wales Natl. Herb. 3(3): 138 (1961).

Darwinia thymoides (Lindl.) Benth., J. Linn. Soc., Bot. 9: 180 (1867).
SW. Australia. 50 WAU. Nanophan.
**Hedaroma thymoides* Lindl., Sketch Veg. Swan R.: vii (1839). *Genetyllis thymoides* (Lindl.) Schauer, Monogr. Myrt. Xerocarp. 1: 33 (1843).
Darwinia brevistyla Turcz., Bull. Soc. Imp. Naturalistes Moscou 20(1): 155 (1847).
Darwinia saturejifolia Turcz., Bull. Cl. Phys.-Math. Acad. Imp. Sci. Saint-Pétersbourg 10: 324 (1852).

Darwinia vestita (Endl.) Benth., J. Linn. Soc., Bot. 9: 180 (1867).
SW. Australia. 50 WAU. Nanophan.
**Genetyllis vestita* Endl. in S.L.Endlicher & al., Enum. Pl.: 47 (1837).
Genetyllis agathosmoides Turcz., Bull. Cl. Phys.-Math. Acad. Imp. Sci. Saint-Pétersbourg 10: 322 (1852).

Darwinia virescens (Meisn.) Benth., J. Linn. Soc., Bot. 9: 179 (1867).
SW. Australia. 50 WAU. Nanophan.
**Genetyllis virescens* Meisn., J. Proc. Linn. Soc., Bot. 1: 38 (1857).

Darwinia wittwerorum N.G.Marchant & Keighery, Nuytsia 3: 179 (1980).
SW. Australia. 50 WAU. Nanophan.

Synonyms:
Darwinia axillaris (F.Muell. ex Benth.) F.Muell. = ***Chamelaucium axillare*** F.Muell. ex Benth.
Darwinia brevifolia (Benth.) F.Muell. = ***Chamelaucium brevifolium*** Benth.
Darwinia brevistyla Turcz. = ***Darwinia thymoides*** (Lindl.) Benth.
Darwinia ciliata (Desf.) F.Muell. = ***Chamelaucium ciliatum*** Desf.
Darwinia decumbens Byrnes = ***Homoranthus decumbens*** (Byrnes) Craven & S.R.Jones
Darwinia drummondii (Turcz.) F.Muell. = ***Chamelaucium drummondii*** (Turcz.) Meisn.
Darwinia endlicheri F.Muell. = ***Chamelaucium virgatum*** Endl.
Darwinia fimbriata (Kippist) Benth. = ***Darwinia squarrosa*** (Turcz.) Domin
Darwinia flavescens A.Cunn. ex Schauer = ***Homoranthus flavescens*** Schauer
Darwinia forrestii F.Muell. = *Chamelaucium* sp. ?
Darwinia gracilis (F.Muell.) F.Muell. = ***Chamelaucium gracile*** F.Muell.
Darwinia heterandra (Benth.) F.Muell. = ***Chamelaucium heterandrum*** Benth.
Darwinia homoranthoides (F.Muell.) J.M.Black = ***Homoranthus homoranthoides*** (F.Muell.) Craven & S.R.Jones
Darwinia hookeriana (Meisn.) Benth. = ***Darwinia hypericifolia*** (Turcz.) Domin
Darwinia intermedia A.Cunn. ex Schauer = ***Darwinia taxifolia*** A.Cunn.
Darwinia laxifolia Schauer = ***Darwinia taxifolia*** A.Cunn.
Darwinia megalopetala (F.Muell. ex Benth.) F.Muell. = ***Chamelaucium megalopetalum*** F.Muell. ex Benth.
Darwinia meisneri (Kippist) Benth. = ***Darwinia oxylepis*** (Turcz.) N.G.Marchant & Keighery
Darwinia porteri C.T.White = ***Homoranthus porteri*** (C.T.White) Craven & S.R.Jones
Darwinia rhadinophylla F.Muell. = ***Darwinia pinifolia*** (Lindl.) Benth.
Darwinia saturejifolia Turcz. = ***Darwinia thymoides*** (Lindl.) Benth.
Darwinia schuermannii (F.Muell.) Benth. = ***Homoranthus homoranthoides*** (F.Muell.) Craven & S.R.Jones
Darwinia taxifolia var. *biflora* Cheel = ***Darwinia biflora*** (Cheel) B.G.Briggs
Darwinia taxifolia var. *grandiflora* Benth. = ***Darwinia grandiflora*** (Benth.) R.T.Baker & H.G.Sm.
Darwinia taxifolia var. *intermedia* Cheel = ***Darwinia taxifolia*** A.Cunn.
Darwinia taxifolia subsp. *macrolaena* B.G.Briggs = ***Darwinia taxifolia*** A.Cunn.
Darwinia taxifolia var. *typica* Cheel = ***Darwinia taxifolia*** A.Cunn.
Darwinia thomasii (F.Muell.) Benth. = ***Homoranthus thomasii*** (F.Muell.) Craven & S.R.Jones
Darwinia thryptomenioides D.A.Herb. = ***Chamelaucium pauciflorum*** (Turcz.) Benth.
Darwinia turczaninowii F.Muell. = ***Chamelaucium pauciflorum*** (Turcz.) Benth.
Darwinia uncinata (Schauer) F.Muell. = ***Chamelaucium uncinatum*** Schauer
Darwinia verticordina (F.Muell.) Benth. = ***Verticordia verticordina*** (F.Muell.) A.S.George
Darwinia virgata (A.Cunn. ex Schauer) F.Muell. = ***Homoranthus virgatus*** A.Cunn. ex Schauer

***Unplaced Names*:**
Darwinia forrestii F.Muell., Fragm. 11: 9 (1878). = *Chamelaucium* sp. ?

Decalophium

Decalophium Turcz. = ***Chamelaucium*** Desf.
Decalophium darwinioides Turcz. = ***Chamelaucium ciliatum*** Desf.
Decalophium juniperinum Turcz. = ***Chamelaucium megalopetalum*** F.Muell. ex Benth.
Decalophium melaleucum Turcz. = ***Chamelaucium megalopetalum*** F.Muell. ex Benth.
Decalophium micranthum Turcz. = ***Chamelaucium micranthum*** (Turcz.) Domin
Decalophium pauciflorum Turcz. = ***Chamelaucium pauciflorum*** (Turcz.) Benth.
Decalophium rugulosum Turcz. = ***Chamelaucium ciliatum*** Desf.

Decaspermum

Decaspermum J.R.Forst. & G.Forst., Char. Gen. Pl.: 37 (1775).
Trop. & Subtrop. Asia to Pacific. 36 CHC CHH CHS 38 TAI 40 ASS 41 AND CBD MYA THA VIE 42 BOR JAW LSI MLY MOL PHI SUL SUM 43 BIS NWG SOL 50 QLD 60 FIJ SAM TON VAN WAL 61 SCI 62 CRL. [Myrtaceae]
34 Species
Dodecaspermum J.R.Forst. ex Scop., Intr. Hist. Nat.: 219 (1777).
Nelitris Gaertn., Fruct. Sem. Pl. 1: 134 (1788).
Pyrenocarpa H.T.Chang & R.H.Miao, Acta Sci. Nat. Univ. Sunyatseni 1975(1): 62 (1975).

Decaspermum albociliatum Merr. & L.M.Perry, J. Arnold Arbor. 19: 202 (1938).
Hainan. 36 CHH. Nanophan. or phan.

Decaspermum alpinum P.Royen, Alp. Fl. New Guinea 4: 2587 (1983).
New Guinea. 43 NWG. Phan.

Decaspermum arfakense Diels, Bot. Jahrb. Syst. 57: 369 (1922).
W. New Guinea. 43 NWG. Nanophan. or phan.

Decaspermum austrohainanicum H.T.Chang & R.H. Miao, Acta Bot. Yunnan. 4: 24 (1982).
Hainan. 36 CHH. Nanophan. or phan.

Decaspermum belense Merr. & L.M.Perry, J. Arnold Arbor. 23: 234 (1942).
W. New Guinea. 43 NWG. Nanophan. or phan.

Decaspermum blancoi Vidal, Phan. Cuming. Philipp.: 172 (1885).
Philippines. 42 PHI. Phan.
**Myrtus communis* Blanco, Fl. Filip.: 422 (1837), nom. illeg.
Nelitris rubra Vidal, Sin. Gen. Pl. Leños. Filip.: t. 50, fig. C (1883).
Decaspermum grandiflorum Elmer, Leafl. Philipp. Bot. 4: 1481 (1912).

Decaspermum bracteatum (Roxb.) A.J.Scott, Kew Bull. 40: 151 (1985).
Sulawesi to New Guinea. 42 MOL SUL 43 BIS NWG. Nanophan. or phan.
**Fabricia bracteata* Roxb., Fl. Ind. ed. 1832, 2: 476 (1832).

var. ***bracteatum***
Sulawesi to New Guinea. 42 MOL SUL 43 BIS NWG. Nanophan. or phan.
Myrtus coriandriformis Zipp. ex Blume, Mus. Bot. 1: 74 (1850).
Nelitris coriandri Blume, Mus. Bot. 1: 74 (1850). *Decaspermum coriandri* (Blume) Diels, Bot. Jahrb. Syst. 57: 372 (1922).
Nelitris laxiflora Blume, Mus. Bot. 1: 74 (1850). *Decaspermum laxiflorum* (Blume) Diels, Bot. Jahrb. Syst. 57: 370 (1922).
Nelitris rubra Blume, Mus. Bot. 1: 73 (1850). *Eugenia rubra* (Blume) Reinw. ex de Vriese, Pl. Ind. Bat. Orient.: 76 (1856). *Decaspermum rubrum* (Blume) Baill., Hist. Pl. 6: 341 (1876).
Decaspermum nitidum Lauterb. in K.M.Schumann & C.A.G.Lauterbach, Fl. Schutzgeb. Südsee, Nachtr.: 325 (1905).
Decaspermum leptanthelium Diels, Bot. Jahrb. Syst. 57: 369 (1922).
Decaspermum petraeum Diels, Bot. Jahrb. Syst. 57: 371 (1922).

var. ***glabrum*** A.J.Scott, Kew Bull. 40: 154 (1985).
New Guinea. 43 NWG. Nanophan. or phan.

Decaspermum cryptanthum A.J.Scott, Kew Bull. 34: 63 (1979).
Fiji (Vanua Levu). 60 FIJ. Nanophan.
**Mooria microphylla* A.C.Sm., Bernice P. Bishop Mus. Bull. 141: 110 (1936). *Cloezia microphylla* (A.C.Sm.) A.C.Sm., J. Arnold Arbor. 36: 286 (1955). *Decaspermum microphyllum* (A.C.Sm.) J.W.Dawson, Allertonia 1: 404 (1978), nom. illeg.

Decaspermum exiguum Merr. & L.M.Perry, J. Arnold Arbor. 23: 235 (1942).
Papua New Guinea. 43 NWG. Phan.

Decaspermum forbesii Baker f., J. Bot. 61(Suppl.): 14 (1923).
Papua New Guinea. 43 NWG. Phan.
Decaspermum simile Merr. & L.M.Perry, J. Arnold Arbor. 23: 236 (1942).

Decaspermum fruticosum J.R.Forst. & G.Forst., Char. Gen. Pl.: 37 (1775). *Psidium decaspermum* L.f., Suppl. Pl.: 252 (1782), nom. illeg. *Nelitris jambosella* Gaertn., Fruct. Sem. Pl. 1: 134 (1788), nom. illeg. *Nelitris fruticosa* (J.R.Forst. & G.Forst.) A.Gray, U.S. Expl. Exped., Phan. 1: 547 (1854). *Nelitris forsteri* Seem., Fl. Vit.: 81 (1866), nom. illeg.
S. Pacific. 42 SUL? 60 SAM TON WAL 61 SCI. Nanophan. or phan.

Decaspermum glabrum H.T.Chang & R.H.Miao, Acta Bot. Yunnan. 4: 25 (1982).
SE. China. 36 CHS. Nanophan. or phan.

Decaspermum gracilentum (Hance) Merr. & L.M.Perry, J. Arnold Arbor. 19: 202 (1938).
Taiwan to Vietnam. 36 CHC CHH CHS 38 TAI 41 VIE. Nanophan. or phan.
**Eugenia gracilenta* Hance, J. Bot. 23: 7 (1885). *Syzygium gracilentum* (Hance) Hu, J. Arnold Arbor. 5: 232 (1924).
Eugenia esquirolii H.Lév., Repert. Spec. Nov. Regni Veg. 9: 459 (1911). *Decaspermum esquirolii* (H.Lév.) H.T.Chang & R.H.Miao, Acta Bot. Yunnan. 4: 25 (1982).

Decaspermum hainanense (Merr.) Merr., Lingnan Sci. J. 14: 42 (1935).
Hainan. 36 CHH. Phan.

Eugenia hainanensis Merr., Philipp. J. Sci. 23: 255 (1923). *Pyrenocarpa hainanensis* (Merr.) H.T.Chang & R.H.Miao, Acta Sci. Nat. Univ. Sunyatseni 1975: 63 (1975).
Pyrenocarpa teretis H.T.Chang & R.H.Miao, Acta Sci. Nat. Univ. Sunyatseni 1975: 64 (1975).

Decaspermum humile (Sweet ex G.Don) A.J.Scott, Kew Bull. 34: 64 (1979).
New Guinea, N. Queensland. 43 NWG 50 QLD. Nanophan. or phan.
Nelitris humilis Sweet ex G.Don, Gen. Hist. 2: 829 (1832).
Nelitris paniculata var. *laxiflora* Benth., Fl. Austral. 3: 280 (1867). *Decaspermum paniculatum* var. *laxiflorum* (Benth.) F.M.Bailey, Syn. Queensl. Fl.: 188 (1883). *Decaspermum laxiflorum* (Benth.) Domin, Biblioth. Bot. 89: 475 (1928), nom. illeg.
Myrtus sericocalyx C.T.White, Proc. Roy. Soc. Queensland 57: 27 (1947).

Decaspermum lanceolatum J.W.Moore, Bernice P. Bishop Mus. Bull. 102: 32 (1933).
Society Is. (Raiatea). 61 SCI. Nanophan.

Decaspermum lorentzii Lauterb., Nova Guinea 8: 849 (1912).
W. New Guinea. 43 NWG. Nanophan.

Decaspermum montanum Ridl., J. Straits Branch Roy. Asiat. Soc. 61: 6 (1912).
Hainan, Indo-China to Pen. Malaysia. 36 CHH 41 CBD THA VIE 42 MLY. Phan.
Decaspermum cambodianum Gagnep., Bull. Mus. Natl. Hist. Nat. 26: 73 (1920).
Eugenia multipunctata Merr., J. Arnold Arbor. 6: 138 (1925).
Eugenia ciliaris Ridl., Bull. Misc. Inform. Kew 1928: 74 (1928).

Decaspermum neoebudicum Guillaumin, J. Arnold Arbor. 12: 254 (1931).
Vanuatu. 60 VAN. Phan.

Decaspermum neurophyllum K.Schum. & Lauterb., Fl. Schutzgeb. Südsee: 468 (1900).
New Guinea. 43 NWG. Phan.
Decaspermum iodochnoum Diels, Nova Guinea 14: 89 (1924).

Decaspermum nitentifolium Merr. & L.M.Perry, J. Arnold Arbor. 23: 234 (1942).
New Guinea. 43 NWG. Nanophan. or phan.

Decaspermum nivale (Ridl.) Merr. & L.M.Perry, J. Arnold Arbor. 23: 236 (1942).
New Guinea. 43 NWG. Nanophan. or phan.
Myrtus nivalis Ridl., Trans. Linn. Soc. London, Bot. 9: 42 (1916).
Decaspermum humifusum Diels, Bot. Jahrb. Syst. 57: 371 (1922).
Decaspermum lamii Diels, Nova Guinea 14: 88 (1924).
Decaspermum prostratum Diels, Nova Guinea 14: 88 (1924).

Decaspermum parviflorum (Lam.) A.J.Scott, Kew Bull. 34: 66 (1979).
Assam to W. Pacific. 36 CHC CHH CHS 40 ASS 41 AND CBD MYA THA VIE 42 BOR JAW MLY MOL PHI SUL SUM 43 NWG 60 FIJ SAM 62 CRL. Nanophan. or phan.
Eugenia parviflora Lam., Encycl. 3: 200 (1789). *Myrtus parviflora* (Lam.) Spreng., Syst. Veg. 2: 486 (1825). *Nelitris parviflora* (Lam.) Blume, Mus. Bot. 1: 75 (1850).

subsp. ***parviflorum***
Assam to W. Pacific. 36 CHC CHH CHS 40 ASS 41 AND CBD MYA THA VIE 42 BOR JAW MLY MOL PHI SUL SUM 43 NWG 60 FIJ SAM 62 CRL. Nanophan. or phan.
Nelitris paniculata Lindl., Coll. Bot.: t. 16 (1821). *Decaspermum paniculatum* (Lindl.) Kurz, J. Asiat. Soc. Bengal, Pt. 2, Nat. Hist. 46(2): 61 (1877).
Nelitris polygama Spreng., Syst. Veg. 2: 488 (1825).
Eugenia polygama Roxb., Fl. Ind. ed. 1832, 2: 491 (1832).
Myrtus longifolia Reinw. ex Blume, Mus. Bot. 1: 75 (1850).
Nelitris alba Blume, Mus. Bot. 1: 74 (1850).
Nelitris bracteata Blume, Mus. Bot. 1: 74 (1850).
Nelitris confinis Blume, Mus. Bot. 1: 74 (1850). *Decaspermum confinis* (Blume) Nied. in H.G.A. Engler & K.A.E.Prantl, Nat. Pflanzenfam. 3(7): 70 (1893).
Nelitris lucida Blume, Mus. Bot. 1: 74 (1850).
Nelitris myrsinoides Blume, Mus. Bot. 1: 75 (1850). *Decaspermum myrsinoides* (Blume) Nied. in H.G.A.Engler & K.A.E.Prantl, Nat. Pflanzenfam. 3(7): 69 (1893).
Nelitris polymorpha Blume, Mus. Bot. 1: 75 (1850).
Nelitris alternifolia Miq., Fl. Ned. Ind. 1(1): 476 (1855). *Decaspermum alternifolium* (Miq.) Nied. in H.G.A.Engler & K.A.E.Prantl, Nat. Pflanzenfam. 3(7): 70 (1893).
Nelitris leucocoma Miq., Fl. Ned. Ind. 1(1): 474 (1855).
Nelitris pubescens Miq., Fl. Ned. Ind. 1(1): 475 (1855).
Nelitris pyrifolia Miq., Fl. Ned. Ind. 1(1): 475 (1855). *Decaspermum pyrifolium* (Miq.) Koord. & Valeton, Meded. Lands Plantentuin 40: 40 (1900).
Nelitris pallescens Miq., Fl. Ned. Ind., Eerste Bijv.: 314 (1861).
Myrtus elachantha F.Muell., Fragm. 4: 56 (1864).
Decaspermum sericeum Hance, J. Bot. 15: 333 (1877).
Decaspermum parviflorum var. *caudatum* A.J.Scott, Kew Bull. 35: 411 (1980).

subsp. ***quadripartitum*** J.Parn. & NicLughadha, Kew Bull. 47: 704 (1992).
Indo-China to Pen. Malaysia. 41 THA VIE 42 MLY. Nanophan. or phan.

Decaspermum parvifolium (Ridl.) A.J.Scott, Kew Bull. 40: 157 (1985).
W. New Guinea. 43 NWG. Nanophan. or phan.
Rhodamnia parvifolia Ridl., Trans. Linn. Soc. London, Bot. 9: 43 (1916).

Decaspermum philippinum A.J.Scott, Kew Bull. 35: 408 (1980).
Philippines (Palawan). 42 PHI. Nanophan. or phan.

Decaspermum prunoides Diels, Bot. Jahrb. Syst. 57: 370 (1922).
Papua New Guinea. 43 NWG. Phan.

Decaspermum raymundi Diels, Bot. Jahrb. Syst. 56: 530 (1921).
Caroline Is. 62 CRL. Nanophan.

Decaspermum salomonense A.J.Scott, Kew Bull. 34: 65 (1979).

Solomon Is. 43 SOL. Phan.

Decaspermum salomonense var. *glabrum* A.J.Scott, Kew Bull. 34: 66 (1979).

Decaspermum struckoilicum N.Snow & Guymer, Novon 11: 475 (2001).
EC. Queensland. 50 QLD. Nanophan. or phan.

Decaspermum teretis Craven, Harvard Pap. Bot. 11: 27 (2006).
Hainan. 36 CHH. Nanophan. or phan.

Decaspermum triflorum A.J.Scott, Kew Bull. 35: 407 (1980).
Lesser Sunda Is. 42 LSI. Phan.

Decaspermum urvillei (DC.) A.J.Scott, Kew Bull. 34: 59 (1979).
Bismarck Arch. 43 BIS. Phan.

**Nelitris urvillei* DC., Prodr. 3: 231 (1828).
Decaspermum rhodoleucum Diels, Bot. Jahrb. Syst. 57: 372 (1922).

Decaspermum vitiense (A.Gray) Nied. in H.G.A.Engler & K.A.E.Prantl, Nat. Pflanzenfam. 3(7): 69 (1893).
Fiji. 60 FIJ. Phan.

**Nelitris vitiensis* A.Gray, U.S. Expl. Exped., Phan. 1: 548 (1854). *Myrtus vitiensis* (A.Gray) F.Muell., Pap. & Proc. Roy. Soc. Tasmania 1875: 165 (1876).

Decaspermum vitis-idaea Stapf, Trans. Linn. Soc. London, Bot. 4: 150 (1894).
Borneo to Sulawesi. 42 BOR PHI SUL. Phan.

Decaspermum microphyllum Merr., Philipp. J. Sci. 18: 289 (1921).

Synonyms:

Decaspermum alternifolium (Miq.) Nied. = ***Decaspermum parviflorum*** (Lam.) A.J.Scott subsp. ***parviflorum***
Decaspermum cambodianum Gagnep. = ***Decaspermum montanum*** Ridl.
Decaspermum confinis (Blume) Nied. = ***Decaspermum parviflorum*** (Lam.) A.J.Scott subsp. ***parviflorum***
Decaspermum coriandri (Blume) Diels = ***Decaspermum bracteatum*** (Roxb.) A.J.Scott var. ***bracteatum***
Decaspermum esquirolii (H.Lév.) H.T.Chang & R.H.Miao = ***Decaspermum gracilentum*** (Hance) Merr. & L.M.Perry
Decaspermum grandiflorum Elmer = ***Decaspermum blancoi*** Vidal
Decaspermum humifusum Diels = ***Decaspermum nivale*** (Ridl.) Merr. & L.M.Perry
Decaspermum iodochnoum Diels = ***Decaspermum neurophyllum*** K.Schum. & Lauterb.
Decaspermum lamii Diels = ***Decaspermum nivale*** (Ridl.) Merr. & L.M.Perry
Decaspermum laxiflorum (Blume) Diels = ***Decaspermum bracteatum*** (Roxb.) A.J.Scott var. ***bracteatum***
Decaspermum laxiflorum (Benth.) Domin = ***Decaspermum humile*** (Sweet ex G.Don) A.J.Scott
Decaspermum leptanthelium Diels = ***Decaspermum bracteatum*** (Roxb.) A.J.Scott var. ***bracteatum***
Decaspermum microphyllum (A.C.Sm.) J.W.Dawson = ***Decaspermum cryptanthum*** A.J.Scott
Decaspermum microphyllum Merr. = ***Decaspermum vitis-idaea*** Stapf
Decaspermum myrsinoides (Blume) Nied. = ***Decaspermum parviflorum*** (Lam.) A.J.Scott subsp. ***parviflorum***
Decaspermum nitidum Lauterb. = ***Decaspermum bracteatum*** (Roxb.) A.J.Scott var. ***bracteatum***
Decaspermum paniculatum (Lindl.) Kurz = ***Decaspermum parviflorum*** (Lam.) A.J.Scott subsp. ***parviflorum***
Decaspermum paniculatum var. *laxiflorum* (Benth.) F.M. Bailey = ***Decaspermum humile*** (Sweet ex G.Don) A.J.Scott
Decaspermum papuanum Lauterb. = ***Pilidiostigma papuanum*** (Lauterb.) A.J.Scott
Decaspermum parviflorum var. *caudatum* A.J.Scott = ***Decaspermum parviflorum*** (Lam.) A.J.Scott subsp. ***parviflorum***
Decaspermum petraeum Diels = ***Decaspermum bracteatum*** (Roxb.) A.J.Scott var. ***bracteatum***
Decaspermum prostratum Diels = ***Decaspermum nivale*** (Ridl.) Merr. & L.M.Perry
Decaspermum pyrifolium (Miq.) Koord. & Valeton = ***Decaspermum parviflorum*** (Lam.) A.J.Scott subsp. ***parviflorum***
Decaspermum rhodoleucum Diels = ***Decaspermum urvillei*** (DC.) A.J.Scott
Decaspermum rubrum (Blume) Baill. = ***Decaspermum bracteatum*** (Roxb.) A.J.Scott var. ***bracteatum***
Decaspermum salomonense var. *glabrum* A.J.Scott = ***Decaspermum salomonense*** A.J.Scott
Decaspermum sericeum Hance = ***Decaspermum parviflorum*** (Lam.) A.J.Scott subsp. ***parviflorum***
Decaspermum simile Merr. & L.M.Perry = ***Decaspermum forbesii*** Baker f.

Diplachne

Diplachne R.Br. ex. Desf. = ***Verticordia*** DC.
Diplachne baueri R.Br. ex Desf. = ***Verticordia plumosa*** (Desf.) Druce var. ***plumosa***

Dodecaspermum

Dodecaspermum J.R.Forst. ex Scop. = ***Decaspermum*** J.R.Forst. & G.Forst.

Draparnaudia

Draparnaudia Montrouz. = ***Xanthostemon*** F.Muell.
Draparnaudia multiflora Montrouz. = ***Xanthostemon multiflorus*** (Montrouz.) Beauvis.

Drosodendron

Drosodendron M.Roem. = ***Baeckea*** L.
Drosodendron rosmarinus (Lour.) M.Roem. = ***Baeckea frutescens*** L.

Emurtia

Emurtia Raf. = ***Eugenia*** P.Micheli ex L.
Emurtia emarginata (Kunth) Raf. = ***Eugenia emarginata*** (Kunth) DC.
Emurtia guayaquilensis (Kunth) Raf. = ***Eugenia guayaquilensis*** (Kunth) DC.
Emurtia micrantha (Kunth) Raf. = ***Eugenia monticola*** (Sw.) DC.
Emurtia punicifolia (Kunth) Raf. = ***Eugenia punicifolia*** (Kunth) DC.

Enosanthes

Enosanthes A.Cunn. ex Schauer = ***Homoranthus*** A.Cunn. ex Schauer
Enosanthes flavescens A.Cunn. ex Schauer = ***Homoranthus flavescens*** Schauer
Enosanthes virgata A.Cunn. ex Schauer = ***Homoranthus virgatus*** A.Cunn. ex Schauer

Episyzygium

Episyzygium Suess. & A.Ludw. = ***Eugenia*** P.Micheli ex L.
Episyzygium cahuense Suess. & A.Ludw. = ***Eugenia reinwardtiana*** (Blume) A.Cunn. ex DC.

Epleienda

Epleienda Raf. = ***Eugenia*** P.Micheli ex L.
Epleienda sinemariensis (Aubl.) Raf. = ***Eugenia coffeifolia*** DC.

Eremaea

Eremaea Lindl., Sketch Veg. Swan R.: 11 (1839).
Western Australia. 50 WAU. [Myrtaceae]
14 Species
Eremaeopsis Kuntze in T.E.von Post & C.E.O.Kuntze, Lex. Gen. Phan., Prosp.: 201 (1903).

Eremaea acutifolia F.Muell., Fragm. 2: 30 (1860).
Western Australia (Irwin). 50 WAU. Cham. or nanophan.

Eremaea asterocarpa Hnatiuk, Nuytsia 9: 208 (1993).
W. Western Australia. 50 WAU. Nanophan.

subsp. ***asterocarpa***
W. Western Australia. 50 WAU. Nanophan.

subsp. ***brachyclada*** Hnatiuk, Nuytsia 9: 213 (1993).
Western Australia (Darling). 50 WAU. Cham. or nanophan.

subsp. ***histoclada*** Hnatiuk, Nuytsia 9: 212 (1993).
Western Australia (Irwin). 50 WAU. Nanophan.

Eremaea atala Hnatiuk, Nuytsia 9: 167 (1993).
Western Australia (Irwin). 50 WAU. Nanophan.

Eremaea beaufortioides Benth., Fl. Austral. 3: 182 (1867).
W. Western Australia. 50 WAU. Nanophan.

var. ***beaufortioides***
Western Australia (Irwin). 50 WAU. Nanophan.

var. ***lachnosanthe*** Hnatiuk, Nuytsia 9: 188 (1993).
Western Australia. 50 WAU. Nanophan.

var. ***microphylla*** Hnatiuk, Nuytsia 9: 186 (1993).
W. Western Australia. 50 WAU. Nanophan.

Eremaea blackwelliana Hnatiuk, Nuytsia 9: 180 (1993).
Western Australia (Avon). 50 WAU. Nanophan.

Eremaea brevifolia (Benth.) Domin, Vestn. Král. Ceské Spolecn. Nauk, Tr. Mat.-Prír. 1921–1922(2): 93 (1923).
Western Australia (Irwin). 50 WAU. Nanophan.
**Eremaea fimbriata* var. *brevifolia* Benth., Fl. Austral. 3: 181 (1867).

Eremaea* × *codonocarpa Hnatiuk, Nuytsia 9: 215 (1993). *E. asterocarpa* × *E. violacea* subsp. *raphiophylla*.
W. Western Australia. 50 WAU. Nanophan.

Eremaea dendroidea Hnatiuk, Nuytsia 9: 194 (1993).
Western Australia (N. Irwin). 50 WAU. Nanophan.

Eremaea ebracteata F.Muell., Fragm. 2: 29 (1860).
Western Australia (Irwin). 50 WAU. Nanophan.

var. ***brachyphylla*** Hnatiuk, Nuytsia 9: 193 (1993).
Western Australia (NC. Irwin). 50 WAU. Cham.

var. ***ebracteata***
Western Australia (N. Irwin). 50 WAU. Nanophan.

Eremaea ectadioclada Hnatiuk, Nuytsia 9: 205 (1993).
Western Australia (Irwin). 50 WAU. Cham. or nanophan.

Eremaea fimbriata Lindl., Sketch Veg. Swan R.: xi (1839).
WSW. Western Australia. 50 WAU. Cham.
Eremaea rosea C.A.Gardner & A.S.George, J. Roy. Soc. Western Australia 46: 134 (1963).

Eremaea hadra Hnatiuk, Nuytsia 9: 170 (1993).
Western Australia (Irwin). 50 WAU. Nanophan.

Eremaea pauciflora (Endl.) Druce, Rep. Bot. Exch. Club Brit. Isles 1916: 622 (1917).
SW. Australia. 50 WAU. Nanophan.
**Metrosideros pauciflora* Endl. in S.L.Endlicher & al., Enum. Pl.: 50 (1837).

var. ***calyptra*** Hnatiuk, Nuytsia 9: 179 (1993).
W. Western Australia. 50 WAU. Nanophan.

var. ***lonchophylla*** Hnatiuk, Nuytsia 9: 176 (1993).
W. Western Australia. 50 WAU. Nanophan.

var. ***pauciflora***
SW. Australia. 50 WAU. Nanophan.
Eremaea ericifolia Lindl., Sketch Veg. Swan R.: xi (1839).
Eremaea pilosa Lindl., Sketch Veg. Swan R.: xi (1839).

Eremaea* × *phoenicea Hnatiuk, Nuytsia 9: 218 (1993). *E. beaufortioides* × *E. violacea* subsp. *raphiophylla*.
W. Western Australia. 50 WAU. Nanophan.

Eremaea purpurea C.A.Gardner, J. Roy. Soc. Western Australia 47: 61 (1964).
Western Australia (Avon, Darling). 50 WAU. Nanophan.

Eremaea violacea F.Muell., Fragm. 11: 10 (1879).
Western Australia (Irwin). 50 WAU. Nanophan.

subsp. ***raphiophylla*** Hnatiuk, Nuytsia 9: 164 (1993).
Western Australia (Irwin). 50 WAU. Cham.

subsp. ***violacea***
Western Australia (N. Irwin). 50 WAU. Nanophan.

***Synonyms*:**
Eremaea ericifolia Lindl. = ***Eremaea pauciflora*** (Endl.) Druce var. ***pauciflora***
Eremaea fimbriata var. *brevifolia* Benth. = ***Eremaea brevifolia*** (Benth.) Domin
Eremaea pilosa Lindl. = ***Eremaea pauciflora*** (Endl.) Druce var. ***pauciflora***
Eremaea rosea C.A.Gardner & A.S.George = ***Eremaea fimbriata*** Lindl.

Eremaeopsis

Eremaeopsis Kuntze = ***Eremaea*** Lindl.

Eremopyxis

Eremopyxis Baill. = ***Triplarina*** Raf.
Eremopyxis camphorata (R.Br. ex Sims) Baill. = ***Triplarina imbricata*** (Sm.) A.R.Bean

Ericomyrtus

Ericomyrtus Turcz. = ***Baeckea*** L.
Ericomyrtus drummondii Turcz. = ***Baeckea pulchella*** A.Cunn. ex DC.

Eucalypton

Eucalypton St.-Lag. = ***Eucalyptus*** L'Hér.

Eucalyptopsis

Eucalyptopsis C.T.White, J. Arnold Arbor. 32: 139 (1951).
Maluku to New Guinea. 42 MOL 43 NWG. [Myrtaceae]
2 Species

Eucalyptopsis alauda Craven, Austral. Syst. Bot. 3: 729 (1990).
New Guinea (Louisiade Arch.). 43 NWG. Phan.

Eucalyptopsis papuana C.T.White, J. Arnold Arbor. 32: 139 (1951).
Maluku to New Guinea. 42 MOL 43 NWG. Phan.

Eucalyptus

Eucalyptus L'Hér., Sert. Angl.: 18 (1789).
Philippines to Australia. (10) ire (12) bal cor fra por sar spa (13) alb grc ita sic (20) mor (21) azo cny mdr (22) ben (23) rwa (25) ken tan uga (26) zim (27) cpp ofs (33) tcs (36) chc chh chs 38 tai (40) ind 42 LSI MOL PHI SUL 43 BIS NWG 50 NSW NTA QLD SAO SOA TAS VIC WAU (51) nzn nzs (61) sci 63 HAW (76) cal (78) fla (80) cos els gua pan 81 DOM (83) bol clm ecu per (85) agw jnf par. [Myrtaceae]
742 Species
Aromadendrum W.Anderson ex R.Br., Prodr.: 553 (1810).
Eudesmia R.Br., Voy. Terra Austral. 2: 599 (1814).
Aromadendron Andrews ex Steud., Nomencl. Bot., ed. 2, 1: 134 (1840), nom. illeg.
Symphyomyrtus Schauer in J.G.C.Lehmann, Pl. Preiss. 1: 126 (1844).
Eucalypton St.-Lag., Ann. Soc. Bot. Lyon 7: 125 (1880).

Eucalyptus abdita Brooker & Hopper, Nuytsia 8: 61 (1991).
Western Australia. 50 WAU. Nanophan. or phan.

Eucalyptus absita Grayling & Brooker, Nuytsia 8: 210 (1992).
Western Australia. 50 WAU. Nanophan. or phan.

Eucalyptus acaciiformis H.Deane & Maiden, Proc. Linn. Soc. New South Wales 24: 454 (1899).
NE. New South Wales. 50 NSW. Nanophan. or phan.

Eucalyptus accedens W.Fitzg., J. West Austral. Nat. Hist. Soc. 1: 21 (1904).
W. Western Australia. 50 WAU. Phan.

Eucalyptus acies Brooker, Nuytsia 1: 245 (1972).
SSW. Western Australia. 50 WAU. Phan.

Eucalyptus acmenoides Schauer in W.G.Walpers, Repert. Bot. Syst. 2: 924 (1843). *Eucalyptus pilularis* var. *acmenoides* (Schauer) Benth., Fl. Austral. 3: 208 (1867).
E. Australia. 50 NSW QLD. Phan.
Eucalyptus acmenoides var. *carnea* Maiden, Crit. Revis. Eucalyptus 1: 266 (1907).

Eucalyptus acroleuca L.A.S.Johnson & K.D.Hill, Telopea 5: 748 (1994).
Queensland. 50 QLD. Phan.

Eucalyptus* × *adjuncta Maiden, J. Proc. Roy. Soc. New S. Wales 54: 167 (1920). *E. longifolia* × ?.
New South Wales. 50 NSW. Phan.

Eucalyptus aenea K.D.Hill, Telopea 7: 101 (1997).
New South Wales. 50 NSW. Phan.

Eucalyptus* × *aequans Blakely, J. Proc. Roy. Soc. New S. Wales 61: 154 (1927). *E. ligustrina* × *E. moorei*.
New South Wales. 50 NSW. Phan.

Eucalyptus aequioperta Brooker & Hopper, Nuytsia 9: 58 (1993).
Western Australia. 50 WAU. Phan.

Eucalyptus* × *affinis H.Deane & Maiden, Proc. Linn. Soc. New South Wales 25: 104 (1900). *E. albens* × *E. sideroxylon*.
New South Wales. 50 NSW. Phan.

Eucalyptus agglomerata Maiden, J. Proc. Roy. Soc. New S. Wales 55: 266 (1921 publ. 1922).
New South Wales to NE. Victoria. 50 NSW VIC. Phan.

Eucalyptus aggregata H.Deane & Maiden, Proc. Linn. Soc. New South Wales 24: 614 (1900).
New South Wales to Victoria. 50 NSW VIC. Phan.
Eucalyptus rydalensis R.T.Baker & H.F.Sm., Res. Eucalypts, ed. 2: 48 (1920).

Eucalyptus alaticaulis R.J.Watson & Ladiges, Brunonia 10: 175 (1987).
Victoria. 50 VIC. Phan.

Eucalyptus alba Reinw. ex Blume, Bijdr.: 1101 (1827).
Lesser Sunda Is. to N. Australia, New Guinea. (36) chs (40) ind 42 LSI 43 NWG 50 NTA QLD WAU. Phan.
Eucalyptus leucadendron Reinw. ex de Vriese, Pl. Ind. Bat. Orient.: 63 (1856).
Eucalyptus alba var. *australasica* Blakely & Jacobs, Key Eucalypts: 137 (1934).

Eucalyptus albens Benth., Fl. Austral. 3: 219 (1867). *Eucalyptus hemiphloia* var. *albens* (Benth.) Maiden, Forest Fl. N.S.W. 1: 131 (1904), nom. illeg.
SE. Queensland to South Australia. (25) ken uga 50 NSW QLD SOA VIC (63) haw. Phan.
Eucalyptus hemiphloia var. *albens* C.Moore & Betche, Handb. Fl. N.S.W.: 201 (1893).
Eucalyptus albens var. *elongata* Blakely, Key Eucalypts: 237 (1934).

Eucalyptus albida Maiden & Blakely, J. Proc. Roy. Soc. New S. Wales 59: 175 (1925).
SW. Australia. 50 WAU. Nanophan. or phan.

Eucalyptus albopurpurea (Boomsma) D.Nicolle, J. Adelaide Bot. Gard. 19: 90 (2000).
S. South Australia. 50 SOA. Phan.
Eucalyptus behriana var. *purpurascens* F.Muell. ex Benth., Fl. Austral. 3: 214 (1867). *Eucalyptus hemiphloia* var. *purpurascens* (F.Muell. ex Benth.) Maiden, Trans. & Proc. Roy. Soc. South Australia 26: 12 (1902). *Eucalyptus odorata* var. *purpurascens* (F.Muell. ex Benth.) Maiden, Trans. & Proc. Roy. Soc. South Australia 32: 283 (1908).
Eucalyptus lansdowneana var. *leucantha* Blakely, Key Eucalypts: 224 (1934).
**Eucalyptus lansdowneana* subsp. *albopurpurea* Boomsma, S. Austral. Naturalist 48: 55 (1974).

Eucalyptus alipes (L.A.S.Johnson & K.D.Hill) D.Nicolle & Brooker, Nuytsia 15: 412 (2005).
Western Australia. 50 WAU. Phan.
**Eucalyptus suggrandis* subsp. *alipes* L.A.S.Johnson & K.D.Hill, Telopea 4: 581 (1992).

Eucalyptus alligatrix L.A.S.Johnson & K.D.Hill, Telopea 4: 237 (1991).
New South Wales to Victoria. 50 NSW VIC. Phan.

subsp. ***alligatrix***
Victoria. 50 VIC. Phan.

subsp. ***limaensis*** Brooker, Slee & J.D.Briggs, Austral. Syst. Bot. 8: 510 (1995).
Victoria. 50 VIC. Phan.

subsp. ***miscella*** Brooker, Slee & J.D.Briggs, Austral. Syst. Bot. 8: 512 (1995).
New South Wales. 50 NSW. Phan.

Eucalyptus alpina Lindl. in T.L.Mitchellin T.L.Mitchell, Three Exped. Australia 2: 175 (1838). *Eucalyptus stellulata* var. *alpina* (Lindl.) Ewart, Fl. Victoria: 835 (1931).
SW. Victoria. 50 VIC. Nanophan. or phan.

Eucalyptus ammophila Brooker & Slee, Austrobaileya 4: 265 (1994).
Queensland. 50 QLD. Phan.

Eucalyptus amplifolia Naudin, Descr. Emploi Eucalypt.: 28 (1891).
New South Wales. (36) chc chs 50 NSW. Phan.

subsp. ***amplifolia***
New South Wales. (36) chc chs 50 NSW. Phan.

subsp. ***sessiliflora*** (Blakely) L.A.S.Johnson & K.D.Hill, Telopea 4: 52 (1990).
New South Wales. 50 NSW. Phan.
**Eucalyptus amplifolia* var. *sessiliflora* Blakely, Key Eucalypts: 131 (1934).

Eucalyptus amygdalina Labill., Nov. Holl. Pl. 2: 14 (1806).
Tasmania. 50 TAS (63) haw. Phan.
Eucalyptus salicifolia Cav., Icon. 4: 24 (1797).
Eucalyptus glandulosa Desf., Tabl. École Bot., ed. 3: 408 (1829).
Eucalyptus amygdalina var. *alpina* Maiden, Rep. Meetings Australas. Assoc. Advancem. Sci. 9: 365 (1903).
Eucalyptus amygdalina var. *numerosa* Maiden, Proc. Linn. Soc. New South Wales 29: 752 (1905).
Eucalyptus calyculata Link ex Maiden, Crit. Revis. Eucalyptus 1: 156 (1905).
Eucalyptus numerosa Maiden, Crit. Revis. Eucalyptus 1: 155 (1905).

Eucalyptus anceps (Maiden) Blakely, Key Eucalypts: 118 (1934).
Western Australia to South Australia. 50 SOA WAU. Phan.
**Eucalyptus conglobata* var. *anceps* Maiden, Crit. Revis. Eucalyptus 6: 275 (1922).

Eucalyptus ancophila L.A.S.Johnson & K.D.Hill, Telopea 4: 82 (1990).
New South Wales. 50 NSW. Phan.

Eucalyptus andrewsii Maiden, Proc. Linn. Soc. New South Wales 29: 472 (1904).
Queensland to NE. New South Wales. 50 NSW QLD. Phan.

subsp. ***andrewsii***
Queensland to NE. New South Wales. 50 NSW QLD. Phan.
Eucalyptus haemastoma var. *inophloia* C.T.White, Queensland Agric. J., II, 14: 70 (1920).

subsp. ***campanulata*** (R.T.Baker & H.G.Sm.) L.A.S. Johnson & Blaxell, Contr. New South Wales Natl. Herb. 4: 381 (1973).
SE. Queensland to NE. New South Wales. 50 NSW QLD. Phan.
**Eucalyptus campanulata* R.T.Baker & H.G.Sm., J. Proc. Roy. Soc. New S. Wales 45: 288 (1911 publ. 1912).

Eucalyptus angophoroides R.T.Baker, Proc. Linn. Soc. New South Wales 25: 676 (1901).
SE. New South Wales to E. Victoria. 50 NSW VIC. Phan.
Eucalyptus stuartiana var. *parviflora* H.Deane & Maiden, Proc. Linn. Soc. New South Wales 24: 109 (1900).

Eucalyptus angularis Brooker & Hopper, Nuytsia 9: 8 (1993).
Western Australia. 50 WAU. Phan.

Eucalyptus angulosa Schauer in W.G.Walpers, Repert. Bot. Syst. 2: 925 (1843). *Eucalyptus incrassata* var. *angulosa* (Schauer) Benth., Fl. Austral. 3: 231 (1867). *Eucalyptus incrassata* subsp. *angulosa* (Schauer) F.C.Johnstone & Hallam, Proc. Roy. Soc. Queensland 91: 204 (1980).
S. Western Australia to South Australia. 50 SOA WAU. Phan.
Eucalyptus cuspidata Turcz., Bull. Soc. Imp. Naturalistes Moscou 22(2): 21 (1849).

Eucalyptus angustissima F.Muell., Fragm. 4: 25 (1863).
S. Western Australia. 50 WAU. Nanophan. or phan.

Eucalyptus annulata Benth., Fl. Austral. 3: 234 (1867).
S. Western Australia. 50 WAU. Phan.

Eucalyptus annuliformis Grayling & Brooker, Nuytsia 8: 213 (1992).
Western Australia. 50 WAU. Phan.

Eucalyptus × anomala Blakely, J. Proc. Roy. Soc. New S. Wales 62: 209 (1928). *E. racemosa × E. umbra*.
New South Wales. 50 NSW. Phan.

Eucalyptus apiculata R.T.Baker & H.G.Sm., Res. Eucalypts: 198 (1902).
EC. New South Wales. 50 NSW. Nanophan. or phan.

Eucalyptus apodophylla Blakely & Jacobs in W.F. Blakely, Key Eucalypts: 165 (1934).
N. Western Australia to NW. Northern Territory. 50 NTA WAU. Phan.

subsp. ***apodophylla***
N. Western Australia to NW. Northern Territory. 50 NTA WAU. Phan.

subsp. ***provecta*** Brooker & Kleinig, Field Guide Eucalypts 3: 371 (1994).
N. Western Australia. 50 WAU. Phan.

Eucalyptus apothalassica L.A.S.Johnson & K.D.Hill, Telopea 4: 84 (1990).
E. Australia. 50 NSW QLD. Phan.

Eucalyptus approximans Maiden, J. Proc. Roy. Soc. New S. Wales 53: 65 (1919).
SE. Queensland to NE. New South Wales. 50 NSW QLD. Phan.

subsp. ***approximans***
SE. Queensland to NE. New South Wales. 50 NSW QLD. Phan.

subsp. ***codonocarpa*** (Blakely & McKie) L.A.S. Johnson & Blaxell, Contr. New South Wales Natl. Herb. 4: 453 (1973).
SE. Queensland to NE. New South Wales. 50 NSW QLD. Phan.
**Eucalyptus codonocarpa* Blakely & McKie, Proc. Linn. Soc. New South Wales 55: 589 (1930).

Eucalyptus aquilina Brooker, Nuytsia 1: 297 (1974).
S. Western Australia. 50 WAU. Phan.

Eucalyptus arachnaea Brooker & Hopper, Nuytsia 8: 68 (1991).
Western Australia. 50 WAU. Phan.
**Eucalyptus redunca* var. *melanophloia* Benth., Fl. Austral. 3: 253 (1867).

subsp. ***arachnaea***
Western Australia. 50 WAU. Phan.

subsp. ***arrecta*** Brooker & Hopper, Nuytsia 8: 70 (1991).
Western Australia. 50 WAU. Phan.

Eucalyptus arborella Brooker & Hopper, Nuytsia 14: 336 (2002).
Western Australia. 50 WAU. Phan.

Eucalyptus archeri Maiden & Blakely in J.H.Maiden, Crit. Revis. Eucalyptus 8: 58 (1929).
Tasmania. 50 TAS. Phan.

Eucalyptus arenacea J.C.Marginson & Ladiges, Austral. Syst. Bot. 1: 163 (1988).
South Australia to Victoria. 50 SOA VIC. Phan.

Eucalyptus argillacea W.Fitzg. ex Maiden, Crit. Revis. Eucalyptus 4: 132 (1918).
N. Australia. 50 NTA QLD WAU. Phan.
Eucalyptus leucophylla Domin, Biblioth. Bot. 89: 464 (1928).
Eucalyptus tropica Cambage ex Maiden, Crit. Revis. Eucalyptus 8: 6 (1929).

Eucalyptus argophloia Blakely, Key Eucalypts: 256 (1934).
SE. Queensland. 50 QLD. Phan.
Eucalyptus corticosa L.A.S.Johnson, Contr. New South Wales Natl. Herb. 3: 108 (1962).

Eucalyptus argutifolia Grayling & Brooker, Nuytsia 8: 215 (1992).
Western Australia. 50 WAU. Phan.

Eucalyptus argyphea L.A.S.Johnson & K.D.Hill, Telopea 4: 603 (1992).
Western Australia. 50 WAU. Phan.

Eucalyptus aromaphloia Pryor & J.H.Willis, Victorian Naturalist 71: 125 (1954).
SE. New South Wales to Victoria. 50 NSW VIC. Phan.

Eucalyptus articulata Brooker & Hopper, Nuytsia 9: 14 (1993).
Western Australia. 50 WAU. Phan.

Eucalyptus aspersa Brooker & Hopper, Nuytsia 9: 28 (1993).
Western Australia. 50 WAU. Phan.

Eucalyptus aspratilis L.A.S.Johnson & K.D.Hill, Telopea 4: 572 (1992).
SW. Australia. 50 WAU. Phan.
**Eucalyptus occidentalis* var. *stenantha* Blakely in J.H.Maiden, Crit. Revis. Eucalyptus 4: 142 (1919).

Eucalyptus assimilans L.A.S.Johnson & K.D.Hill, Telopea 9: 308 (2001).
Western Australia. 50 WAU. Phan.

Eucalyptus astringens (Maiden) Maiden, Crit. Revis. Eucalyptus 7: 55 (1924).
SW. Australia. 50 WAU. Phan.
**Eucalyptus occidentalis* var. *astringens* Maiden, J. Nat. Hist. Sci. Soc. Western Australia 3: 186 (1911).

subsp. ***astringens***
SW. Australia. 50 WAU. Phan.

subsp. ***redacta*** Brooker & Hopper, Nuytsia 14: 353 (2002).
SW. Australia. 50 WAU. Phan.

Eucalyptus atrata L.A.S.Johnson & K.D.Hill, Telopea 4: 334 (1991).
Queensland. 50 QLD. Phan.

Eucalyptus* × *auburnensis Maiden, Crit. Revis. Eucalyptus 6: 116 (1922). *E. melanophloia* × *E. melliodora*.
New South Wales. 50 NSW. Phan.

Eucalyptus badjensis Beuzev. & Welch, J. Proc. Roy. Soc. New S. Wales 58: 177 (1924).
SE. New South Wales. 50 NSW. Phan.

Eucalyptus baeuerlenii F.Muell., Victorian Naturalist 7: 76 (1890). *Eucalyptus viminalis* var. *baeuerlenii* (F. Muell.) H.Deane & Maiden, Proc. Linn. Soc. New South Wales 26: 142 (1901).
CE. & SE. New South Wales. 50 NSW. Phan.

Eucalyptus baileyana F.Muell., Fragm. 11: 37 (1878).
SE. Queensland to NE. New South Wales. 50 NSW QLD. Phan.

Eucalyptus bakeri Maiden, J. Proc. Roy. Soc. New S. Wales 47: 87 (1913).
Queensland to NE. New South Wales. 50 NSW QLD. Phan.

Eucalyptus balanites Grayling & Brooker, Nuytsia 8: 216 (1992).
Western Australia. 50 WAU. Phan.

Eucalyptus balanopelex L.A.S.Johnson & K.D.Hill, Telopea 4: 605 (1992).
Western Australia. 50 WAU. Nanophan. or phan.

Eucalyptus balladoniensis Brooker, Nuytsia 2: 103 (1976).
Western Australia. 50 WAU. Nanophan. or phan.

subsp. ***balladoniensis***
CS. Western Australia. 50 WAU. Nanophan. or phan.

subsp. ***sedens*** L.A.S.Johnson & K.D.Hill, Telopea 4: 617 (1992).
Western Australia. 50 WAU. Nanophan. or phan.

Eucalyptus bancroftii (Maiden) Maiden, Crit. Revis. Eucalyptus 4: 14 (1917).
SE. Queensland to NE. New South Wales. 50 NSW QLD. Phan.
Eucalyptus tereticornis var. *brevifolia* Benth., Fl. Austral. 3: 242 (1867).
**Eucalyptus tereticornis* var. *bancroftii* Maiden, Forest Fl. N.S.W. 2 9 (1904).
Eucalyptus seeana var. *constricta* Blakely, Key Eucalypts: 134 (1934).

Eucalyptus banksii Maiden, Proc. Linn. Soc. New South Wales 29: 774 (1905).
SE. Queensland to NE. New South Wales. 50 NSW QLD. Phan.

Eucalyptus barberi L.A.S.Johnson & Blaxell, Contr. New South Wales Natl. Herb. 4: 288 (1972).
E. Tasmania. 50 TAS. Phan.

Eucalyptus barklyensis L.A.S.Johnson & K.D.Hill, Telopea 5: 753 (1994).
N. Australia. 50 NTA QLD WAU. Phan.

Eucalyptus* × *barmedmanensis Maiden, Crit. Revis. Eucalyptus 6: 108 (1922). *E. sideroxylon* × *E. woollsiana*.
New South Wales. 50 NSW. Phan.

Eucalyptus baueriana Schauer in W.G.Walpers, Repert. Bot. Syst. 2: 924 (1843).
SE. Queensland to Victoria. 50 NSW QLD VIC. Phan.
Eucalyptus fletcheri F.Muell. ex R.T.Baker, Proc. Linn. Soc. New South Wales 25: 682 (1901).

Eucalyptus baxteri (Benth.) J.M.Black, Fl. S. Austral. 3: 415 (1926).
SE. New South Wales to SE. South Australia. 50 NSW SOA VIC. Phan.
Eucalyptus capitellata var. *latifolia* Benth., Fl. Austral. 3: 206 (1867).
**Eucalyptus santalifolia* var. *baxteri* Benth., Fl. Austral. 3: 207 (1867).
Eucalyptus baxteri var. *pedicellata* J.M.Black, Fl. S. Austral. 3: 416 (1926).

Eucalyptus beaniana L.A.S.Johnson & K.D.Hill, Telopea 4: 330 (1991).
Queensland. 50 QLD. Phan.

Eucalyptus beardiana Brooker & Blaxell, Nuytsia 2: 220 (1978).
W. Western Australia. 50 WAU. Phan.

Eucalyptus* × *beaselyi Blakely, Key Eucalypts: 234 (1934). *E. melanophloia* × *E. populnea*.
Queensland. 50 QLD. Phan.

Eucalyptus behriana F.Muell., Trans. Philos. Soc. Victoria 1: 34 (1855).
SC. New South Wales to SE. South Australia. 50 NSW SOA VIC. Phan.

Eucalyptus* × *bennettiae D.J.Carr & S.G.M.Carr, Austral. J. Bot. 28: 541 (1980). *E. lehmannii* × *E. occidentalis*.
Western Australia. 50 WAU. Phan.

Eucalyptus bensonii L.A.S.Johnson & K.D.Hill, Telopea 4: 91 (1990).
New South Wales. 50 NSW. Phan.

Eucalyptus benthamii Maiden & Cambage, J. Proc. Roy. Soc. New S. Wales 48: 418 (1914 publ. 1915).
CE. New South Wales. 50 NSW. Phan.

Eucalyptus beyeri F.Muell. ex R.T.Baker, J. Proc. Roy. Soc. New S. Wales 51: t. XX(3) (1917).
New South Wales. 50 NSW. Phan.
Eucalyptus paniculata var. *angustifolia* Benth., Fl. Austral. 3: 212 (1867).
Eucalyptus panda subsp. *illaquens* L.A.S.Johnson, Contr. New South Wales Natl. Herb. 3: 120 (1962).

Eucalyptus beyeriana L.A.S.Johnson & K.D.Hill, Telopea 4: 83 (1990).
New South Wales. 50 NSW. Phan.

Eucalyptus bicolor A.Cunn. ex Hook. in T.L.Mitchell, J. Exped. Trop. Australia: 390 (1848).
E. & SE. Australia. (36) chs 50 NSW QLD SOA VIC. Phan.
Eucalyptus pendula Page ex Steud., Nomencl. Bot., ed. 2, 1: 600 (1840).
Eucalyptus largiflorens F.Muell., Trans. Philos. Soc. Victoria 1: 34 (1855).
Eucalyptus parviflora F.Muell., J. Proc. Linn. Soc., Bot. 3: 90 (1859).

Eucalyptus bigalerita F.Muell., J. Proc. Linn. Soc., Bot. 3: 96 (1859).
N. Western Australia to Northern Territory. 50 NTA WAU. Phan.
Eucalyptus pastoralis S.Moore, J. Bot. 40: 27 (1902).

Eucalyptus binacag (Elmer) Elmer, Leafl. Philipp. Bot. 8: 2776 (1915).
Philippines (Mindanao). 42 PHI. Phan.
**Eugenia binacag* Elmer, Leafl. Philipp. Bot. 7: 2351 (1914).

Eucalyptus* × *bipileata Blakely, Key Eucalypts: 253 (1934). *E. crebra* × *E. melanophloia*.
New South Wales. 50 NSW. Phan.

Eucalyptus biterranea L.A.S.Johnson & K.D.Hill, Telopea 8: 505 (2000).
New Guinea to Queensland. 43 NWG 50 QLD. Phan.

Eucalyptus biturbinata L.A.S.Johnson & K.D.Hill, in Fl. Australia 19: 507 (1988).
SE. Queensland to NE. New South Wales. 50 NSW QLD. Phan.

Eucalyptus* × *blackburniana Maiden ex R.T.Baker & H.G.Sm., Trans. & Proc. Roy. Soc. South Australia 40: 478 (1916). *E. sideroxylon* subsp. *tricarpa* × *E. viridis*.
Victoria. 50 VIC. Phan.

Eucalyptus blakelyi Maiden, Crit. Revis. Eucalyptus 4: 43 (1917).
Queensland to Victoria. (36) chc chs 50 NSW QLD VIC. Phan.
Eucalyptus blakelyi var. *irrorata* Blakely, Key Eucalypts: 131 (1934).
Eucalyptus blakelyi var. *parvifructa* Blakely, Key Eucalypts: 132 (1934).

Eucalyptus blaxellii L.A.S.Johnson & K.D.Hill, Telopea 4: 564 (1992).
Western Australia. 50 WAU. Phan.

Eucalyptus blaxlandi Maiden & Cambage, J. Proc. Roy. Soc. New S. Wales 52: 495 (1918 publ. 1919).
New South Wales. 50 NSW. Phan.

Eucalyptus boliviana J.B.Williams & K.D.Hill, Telopea 9: 409 (2001).
New South Wales. 50 NSW. Phan.

Eucalyptus* × *boormanii H.Deane & Maiden, Proc. Linn. Soc. New South Wales 26: 339 (1901). *E. fibrosa* × *E. moluccana*.
New South Wales. 50 NSW. Phan.

Eucalyptus bosistoana F.Muell., Australas. J. Pharm. 10: 293 (1895).
SE. New South Wales to E. Victoria. (25) ken uga 50 NSW VIC. Phan.

Eucalyptus botryoides Sm., Trans. Linn. Soc. London 3: 286 (1797). *Eucalyptus saligna* var. *botryoides* (Sm.)

Maiden, Proc. Linn. Soc. New South Wales 30: 502 (1906). *Eucalyptus saligna* subsp. *botryoides* (Sm.) Passioura & J.E.Ash, Austral. Syst. Bot. 6: 182 (1993).
SE. New South Wales to E. Victoria. (12) por sar (13) ita sic (23) rwa (25) tan uga (36) chc chs (38) tai 50 NSW VIC (51) nzn (7)+. Phan.
Eucalyptus platypodos Cav., Icon. 4: 23 (1797).
Eucalyptus botryoides var. *platycarpa* Blakely, Key Eucalypts: 97 (1934).

Eucalyptus brachyandra F.Muell., J. Proc. Linn. Soc., Bot. 3: 97 (1859).
N. Western Australia to NW. Northern Territory. 50 NTA WAU. Phan.

Eucalyptus brachycalyx Blakely, Key Eucalypts: 119 (1934).
SE. Western Australia to South Australia. 50 SOA WAU. Phan.
Eucalyptus incrassata var. *protrusa* J.M.Black, Fl. S. Austral. 3: 421 (1926). *Eucalyptus brachycalyx* var. *protrusa* (J.M.Black) H.Eichler in J.M.Black, Fl. S. Austral., ed. 2(Suppl.): 240 (1965).
Eucalyptus brachycalyx var. *chindoo* Blakely, Key Eucalypts: 119 (1934).

Eucalyptus brachycorys Blakely, Key Eucalypts: 119 (1934).
W. Western Australia. 50 WAU. Phan.

Eucalyptus brachyphylla C.A.Gardner, J. Roy. Soc. Western Australia 27: 186 (1942).
SC. Western Australia. 50 WAU. Nanophan. or phan.

Eucalyptus brassiana S.T.Blake, Austrobaileya 1: 1 (1977).
S. New Guinea, N. Queensland. 43 NWG 50 QLD. Phan.

Eucalyptus brevifolia F.Muell., J. Proc. Linn. Soc., Bot. 3: 84 (1859).
N. Western Australia to NW. Northern Territory. 50 NTA WAU. Phan.
Eucalyptus pallidifolia F.Muell., Fragm. 3: 131 (1863).

Eucalyptus brevipes Brooker, Nuytsia 5: 365 (1986).
SW. Australia. 50 WAU. Phan.

Eucalyptus × ***brevirostris*** Blakely, Key Eucalypts: 183 (1934). *E. macrorhyncha* × *E. muelleriana*.
Victoria. 50 VIC. Phan.

Eucalyptus brevistylis Brooker, Nuytsia 1: 310 (1974).
SW. Australia. 50 WAU. Phan.

Eucalyptus bridgesiana F.Muell. ex R.T.Baker, Proc. Linn. Soc. New South Wales 23: 164 (1898).
SE. Queensland to Victoria. 50 NSW QLD VIC (63) haw. Phan.
Eucalyptus stuartiana var. *amblycorys* Blakely, Key Eucalypts: 145 (1934). *Eucalyptus bridgesiana* var. *amblycorys* (Blakely) Cameron, Victorian Naturalist 63: 41 (1946).

Eucalyptus brockwayi C.A.Gardner, J. Roy. Soc. Western Australia 27: 185 (1942).
SC. Western Australia. (27) cpp 50 WAU. Phan.

Eucalyptus brookeriana A.M.Gray, Austral. Forest. Res. 9: 111 (1979).
Victoria to Tasmania. 50 TAS VIC. Phan.

Eucalyptus broviniensis A.R.Bean, Austrobaileya 6: 117 (2001).
Queensland. 50 QLD. Phan.

Eucalyptus brownii Maiden & Cambage, J. Proc. Roy. Soc. New S. Wales 47: 215 (1913).
Queensland. 50 QLD. Phan.
Eucalyptus bicolor var. *parviflora* Benth., Fl. Austral. 3: 215 (1867).

Eucalyptus brunnea L.A.S.Johnson & K.D.Hill, Telopea 4: 40 (1990).
E. Australia. 50 NSW QLD. Phan.

Eucalyptus × ***bucknellii*** Cambage, Proc. Linn. Soc. New South Wales 51: 325 (1926). *E. microtheca* × *E. populnea*.
New South Wales. 50 NSW. Phan.
Eucalyptus populifolia var. *obconica* Blakely, Key Eucalypts: 243 (1934). *Eucalyptus populnea* var. *obconica* (Blakely) Cameron, Victorian Naturalist 63: 42 (1946).

Eucalyptus buprestium F.Muell., Fragm. 3: 57 (1862).
SSW. Western Australia. 50 WAU. Nanophan. or phan.

Eucalyptus burdettiana Blakely & H.Steedman, Contr. New South Wales Natl. Herb. 1: 35 (1939).
SW. Australia. 50 WAU. Phan.

Eucalyptus burgessiana L.A.S.Johnson & Blaxell, Contr. New South Wales Natl. Herb. 4: 286 (1972).
SE. New South Wales. 50 NSW. Nanophan. or phan.

Eucalyptus burracoppinensis Maiden & Blakely, J. Proc. Roy. Soc. New S. Wales 59: 178 (1925).
SW. Australia. 50 WAU. Phan.

Eucalyptus cadens J.D.Briggs & Crisp, Muelleria 7: 7 (1989).
Victoria. 50 VIC. Phan.

Eucalyptus caesia Benth., Fl. Austral. 3: 227 (1867).
Western Australia. 50 WAU. Phan.

subsp. ***caesia***
SW. Australia. 50 WAU. Phan.

subsp. ***magna*** Brooker & Hopper, Nuytsia 4: 117 (1982).
SC. Western Australia. 50 WAU. Phan.

Eucalyptus calcareana Boomsma, J. Adelaide Bot. Gard. 1: 361 (1979).
SE. Western Australia to S. South Australia. 50 SOA WAU. Phan.

Eucalyptus calcicola Brooker, Nuytsia 1: 302 (1974).
SW. Australia. 50 WAU. Nanophan. or phan.

subsp. ***calcicola***
SW. Western Australia. 50 WAU. Nanophan. or phan.

subsp. ***unita*** D.Nicolle, Nuytsia 15: 71 (2002).
SW. Australia. 50 WAU. Nanophan. or phan.

Eucalyptus caleyi Maiden, Proc. Linn. Soc. New South Wales 30: 512 (1905 publ. 1906).
SE. Queensland to New South Wales. 50 NSW QLD. Phan.

subsp. ***caleyi***
SE. Queensland to New South Wales. 50 NSW QLD. Phan.
Eucalyptus leucoxylon var. *pallens* Benth., Fl. Austral. 3: 210 (1867). *Eucalyptus sideroxylon* var. *pallens* (Benth.) Rehder in L.H.Bailey, Cycl. Amer. Hort. 2: 552 (1900).
Eucalyptus coerulea R.T.Baker & H.G.Sm., Res. Eucalypts, ed. 2: 271 (1920).

subsp. ***ovendenii*** L.A.S.Johnson & K.D.Hill, Telopea 4: 246 (1991).
New South Wales. 50 NSW. Phan.

Eucalyptus caliginosa Blakely & McKie in W.F.Blakely, Key Eucalypts: 181 (1934).
SE. Queensland to NE. New South Wales. 50 NSW QLD. Phan.
Eucalyptus cyathiformis Blakely & McKie in W.F.Blakely, Key Eucalypts: 179 (1934).

Eucalyptus* × *callanii Blakely, J. Proc. Roy. Soc. New S. Wales 61: 160 (1927). *E. globoidea* × *E. pauciflora*.
New South Wales. 50 NSW. Phan.

Eucalyptus calycogona Turcz., Bull. Cl. Phys.-Math. Acad. Imp. Sci. Saint-Pétersbourg 10: 338 (1852).
S. Australia. 50 SOA VIC WAU. Phan.

subsp. ***calycogona***
S. Western Australia, SE. South Australia. 50 SOA WAU. Phan.

subsp. ***spaffordii*** D.Nicolle, Nuytsia 13: 310 (2000).
SE. South Australia. 50 SOA. Phan.
Eucalyptus calycogona var. *spaffordii* Blakely, Key Eucalypts: 265 (1934).

subsp. ***trachybasis*** D.Nicolle, Nuytsia 13: 307 (2000).
SE. Australia. 50 NSW SOA VIC. Phan.

Eucalyptus calyerup McQuoid & Hopper, Nuytsia 15: 64 (2002).
SW. Australia. 50 WAU. Phan.

Eucalyptus camaldulensis Dehnh., Cat. Horti Camald., ed. 2: 20 (1832).
Australia, widely introduced elsewere. (12) cor fra por sar spa (13) alb grc ita sic (20) mor (21) cny (22) ben (23) rwa (25) ken tan uga (27) cpp ofs (33) tcs (36) chc chs (38) tai 50 NSW NTA QLD SOA VIC WAU (76) cal. Phan.

subsp. ***camaldulensis***
Australia, widely introduced elsewere. (12) cor fra por sar spa (13) alb grc ita sic (20) mor (21) cny (22) ben (23) rwa (25) ken tan uga (27) cpp ofs (33) tcs (36) chc chs (38) tai 50 NSW NTA QLD SOA VIC WAU (76) cal. Phan.
Eucalyptus rostrata Schltdl., Linnaea 20: 655 (1847), nom. illeg. *Eucalyptus longirostris* F.Muell. ex Miq., Ned. Kruidk. Arch. 4: 125 (1856).
Eucalyptus acuminata Hook. in T.L.Mitchell, J. Exped. Trop. Australia: 390 (1848). *Eucalyptus camaldulensis* var. *acuminata* (Hook.) Blakely, Key Eucalypts: 135 (1934).
Eucalyptus longirostris f. *brevirostris* Miq., Ned. Kruidk. Arch. 4: 125 (1856). *Eucalyptus rostrata* var. *brevirostris* (Miq.) Maiden, Bull. Herb. Boissier, II, 2: 581 (1902). *Eucalyptus camaldulensis* var. *brevirostris* (Miq.) Blakely, Key Eucalypts: 135 (1934).
Eucalyptus rostrata var. *borealis* R.T.Baker & H.G.Sm., Res. Eucalypts: 75 (1902).
Eucalyptus rostrata var. *acuminata* (Hook.) Maiden, Crit. Revis. Eucalyptus 4(33): 67 (1917).
Eucalyptus tereticornis var. *rostrata* Ewart, Handb. Forest Trees Victorian: 301 (1925).
Eucalyptus camaldulensis var. *obtusaa* Blakely, Key Eucalypts: 135 (1934).
Eucalyptus camaldulensis var. *pendula* Blakely & Jacobs in W.F.Blakely, Key Eucalypts: 135 (1934).
Eucalyptus camaldulensis var. *subcinerea* Blakely, Key Eucalypts: 135 (1934).

subsp. ***simulata*** Brooker & Kleinig, Field Guide Eucalypts 3: 371 (1994).
Queensland. 50 QLD. Phan.

Eucalyptus cambageana Maiden, J. Proc. Roy. Soc. New S. Wales 47: 91 (1913).
Queensland. 50 QLD. Phan.

Eucalyptus cameronii Blakely & McKie in W.F.Blakely, Key Eucalypts: 180 (1934).
NE. New South Wales. 50 NSW. Phan.

Eucalyptus camfieldii Maiden, J. Proc. Roy. Soc. New S. Wales 54: 66 (1920).
CE. New South Wales. 50 NSW. Phan.

Eucalyptus* × *campanifructa Blakely, Key Eucalypts: 228 (1934).
New South Wales. 50 NSW. Phan.

Eucalyptus campaspe S.Moore, J. Linn. Soc., Bot. 34: 193 (1899).
SC. Western Australia. 50 WAU. Phan.

Eucalyptus camphora F.Muell. ex R.T.Baker, Proc. Linn. Soc. New South Wales 24: 298 (1899). *Eucalyptus ovata* var. *camphora* (F.Muell. ex R.T. Baker) Maiden, Crit. Revis. Eucalyptus 3: 148 (1916).
Queensland to Victoria. 50 NSW QLD VIC. Phan.

subsp. ***camphora***
SE. Queensland to Victoria. 50 NSW QLD VIC. Phan.
Eucalyptus ovata var. *aquatica* Blakely, Key Eucalypts: 140 (1934). *Eucalyptus aquatica* (Blakely) L.A.S.Johnson & K.D.Hill, Telopea 4: 56 (1990).

subsp. ***humeana*** L.A.S.Johnson & K.D.Hill, Telopea 4: 55 (1990).
New South Wales to Victoria. 50 NSW VIC. Phan.

subsp. ***relicta*** L.A.S.Johnson & K.D.Hill, Telopea 4: 54 (1990).
Queensland to New South Wales. 50 NSW QLD. Phan.

Eucalyptus canaliculata Maiden, J. Proc. Roy. Soc. New S. Wales 54: 171 (1920).
CE. New South Wales. 50 NSW. Phan.
Eucalyptus punctata var. *grandiflora* H.Deane & Maiden, Proc. Linn. Soc. New South Wales 26: 133 (1901).

Eucalyptus canescens Nicolle, Nuytsia 11: 377 (1997).
South Australia. 50 SOA. Phan.

subsp. ***beadellii*** Nicolle, Nuytsia 11: 381 (1997).
South Australia. 50 SOA. Phan.

subsp. ***canescens***
South Australia. 50 SOA. Phan.

Eucalyptus canobolensis (L.A.S.Johnson & K.D.Hill) J.T.Hunter, Telopea 8: 157 (1998).
New South Wales. 50 NSW. Phan.
**Eucalyptus rubida* subsp. *canobolensis* L.A.S. Johnson & K.D.Hill, Telopea 4: 239 (1991).

Eucalyptus capillosa Brooker & Hopper, Nuytsia 8: 41 (1991).
SW. Australia. 50 WAU. Phan.

subsp. ***capillosa***
SW. Australia. 50 WAU. Phan.

subsp. ***polyclada*** Brooker & Hopper, Nuytsia 8: 45 (1991).
SW. Australia. 50 WAU. Phan.

Eucalyptus capitanea L.A.S.Johnson & K.D.Hill, Telopea 9: 270 (2001).
South Australia. 50 SOA. Phan.

Eucalyptus capitellata Sm., Spec. Bot. New Holland 4: 42 (1795).
New South Wales. 50 NSW. Phan.
Eucalyptus triantha Link, Enum. Hort. Berol. Alt. 2: 30 (1822).
Eucalyptus triantha var. *carnea* Domin, Biblioth. Bot. 89: 462 (1928).

Eucalyptus captiosa Brooker & Hopper, Nuytsia 9: 55 (1993).
SW. Australia. 50 WAU. Phan.

Eucalyptus* × *carnabyi Blakely & H.Steedman, Austral. Naturalist 10: 259 (1941). *E. drummondii* × *E. macrocarpa*.
SW. Australia. 50 WAU. Phan.

Eucalyptus carnea F.Muell. ex R.T.Baker, Proc. Linn. Soc. New South Wales 31: 303 (1906). *Eucalyptus umbra* subsp. *carnea* (F.Muell. ex R.T.Baker) L.A.S.Johnson, Contr. New South Wales Natl. Herb. 3: 123 (1962).
SE. Queensland to E. New South Wales. 50 NSW QLD. Phan.

Eucalyptus carnei C.A.Gardner, J. Roy. Soc. Western Australia 14: 82 (1928).
C. Western Australia. 50 WAU. Phan.

Eucalyptus castrensis K.D.Hill, Telopea 9: 773 (2002).
New South Wales. 50 NSW. Phan.

Eucalyptus celastroides Turcz., Bull. Cl. Phys.-Math. Acad. Imp. Sci. Saint-Pétersbourg 10: 338 (1852). *Eucalyptus calycogona* var. *celastroides* (Turcz.) Maiden, Crit. Revis. Eucalyptus 1: 79 (1903).
Western Australia. 50 WAU. Phan.

subsp. ***celastroides***.
SC. Western Australia. 50 WAU. Phan.

subsp. ***virella*** Brooker, Nuytsia 5: 359 (1986).
SW. Australia. 50 WAU. Phan.

Eucalyptus cephalocarpa Blakely, Key Eucalypts: 164 (1934).
SE. New South Wales to Victoria. 50 NSW VIC. Phan.
**Eucalyptus cinerea* var. *multiflora* Maiden, Crit. Revis. Eucalyptus 3: 7 (1914).

Eucalyptus ceracea Brooker & Done, Nuytsia 5: 382 (1986).
N. Western Australia. 50 WAU. Nanophan.

Eucalyptus cerasiformis Brooker & Blaxell, Nuytsia 2: 226 (1978).
S. Western Australia. 50 WAU. Nanophan. or phan.

Eucalyptus ceratocorys (Blakely) L.A.S.Johnson & K.D.Hill, in Fl. Australia 19: 508 (1988).
SC. Western Australia. 50 WAU. Nanophan. or phan.
**Eucalyptus angulosa* var. *ceratocorys* Blakely, Key Eucalypts: 124 (1934).

Eucalyptus cernua Brooker & Hopper, Nuytsia 14: 342 (2002).
Western Australia. 50 WAU. Phan.

Eucalyptus chapmaniana Cameron, Victorian Naturalist 64: 52 (1947).
SE. New South Wales to E. Victoria. 50 NSW VIC. Phan.

Eucalyptus chartaboma D.Nicolle, Nuytsia 13: 324 (2000).
Queensland. 50 QLD. Phan.

Eucalyptus* × *chisholmii Maiden & Blakely in J.H.Maiden, Crit. Revis. Eucalyptus 7: 61 (1924). *E. piperita* × *E. rosii*.
New South Wales. 50 NSW. Phan.

Eucalyptus chloroclada (Blakely) L.A.S.Johnson & K.D.Hill, in Fl. Australia 19: 508 (1988).
S. Queensland to C. New South Wales. 50 NSW QLD. Phan.
**Eucalyptus dealbata* var. *chloroclada* Blakely, Key Eucalypts: 339 (1934).

Eucalyptus chlorophylla Brooker & Done, Nuytsia 5: 389 (1986).
NE. Western Australia. 50 WAU. Phan.

Eucalyptus* × *chrysantha Blakely & H.Steedman, Proc. Linn. Soc. New South Wales 63: 66 (1938). *E. preissiana* × *E. sepulcralis*.
Western Australia. 50 WAU. Phan.
**Eucalyptus sepulcralis* var. *robusta* C.A.Gardner, J. Roy. Soc. Western Australia 19: 89 (1933).

Eucalyptus cinerea F.Muell. ex Benth., Fl. Austral. 3: 239 (1867).
SE. New South Wales to NE. Victoria. (23) rwa (33) tcs 50 NSW VIC (63) haw. Phan.

subsp. ***cinerea***
SE. New South Wales to NE. Victoria. (23) rwa (33) tcs 50 NSW VIC (63) haw. Phan.
Eucalyptus pulverulenta var. *lanceolata* A.W.Howitt, Rep. Meetings Australas. Assoc. Advancem. Sci. 7: 518 (1898). *Eucalyptus stuartiana* var. *cordata* R.T.Baker & H.G.Sm., Res. Eucalypts: 105 (1902), nom. illeg.

subsp. ***triplex*** (L.A.S.Johnson & K.D.Hill) Brooker, Slee & J.D.Briggs, Austral. Syst. Bot. 8: 507 (1995).
SE. New South Wales. 50 NSW. Phan.
**Eucalyptus triplex* L.A.S.Johnson & K.D.Hill, Telopea 4: 61 (1990).

Eucalyptus cladocalyx F.Muell., Linnaea 25: 388 (1853). *Eucalyptus corynocalyx* F.Muell., Fragm. 2: 43 (1860), nom. superfl.
South Australia. (20) mor (25) ken tan (27) cpp 50 SOA (63) haw (76) cal. Phan.
Eucalyptus langii Maiden & Blakely in J.H.Maiden, Crit. Revis. Eucalyptus 8: 72 (1929).

Eucalyptus clelandii (Maiden) Maiden, Crit. Revis. Eucalyptus 2: 189 (1912).
Western Australia. 50 WAU. Phan.
**Eucalyptus gonianthe* var. *clelandii* Maiden, J. Nat. Hist. Sci. Soc. Western Australia 3: 176 (1911).

Eucalyptus clivicola Brooker & Hopper, Nuytsia 8: 92 (1991).
Western Australia. 50 WAU. Phan.

Eucalyptus cloeziana F.Muell., Fragm. 11: 44 (1878).
Queensland. (23) rwa 50 QLD. Phan.
Eucalyptus stannariensis F.M.Bailey, Queensland Agric. J. 21: 293 (1908).

Eucalyptus cneorifolia A.Cunn. ex DC., Prodr. 3: 220 (1828).
SE. South Australia. 50 SOA. Phan.
Eucalyptus gneorifolia Sweet ex G.Don, Gen. Hist. 2: 820 (1832), orth. var.
Eucalyptus enervifolia Walp., Repert. Bot. Syst. 2: 164 (1843).
Eucalyptus myrtiformis Naudin, Descr. Emploi Eucalypt.: 50 (1891).

Eucalyptus coccifera Hook.f., London J. Bot. 6: 477 (1847).
C. Tasmania. 50 TAS. Phan.
Eucalyptus daphnoides Miq., Ned. Kruidk. Arch. 4: 133 (1856).
Eucalyptus coccifera var. *parviflora* Benth., Fl. Austral. 3: 204 (1867).
Eucalyptus alpina R.Br. ex Maiden, Rep. Meetings Australas. Assoc. Advancem. Sci. 9: 365 (1896), nom. illeg.
Eucalyptus coccifera var. *viridifolia* Summerh., Bot. Mag. 160: t. 9511 (1934).

Eucalyptus comitae-vallis Maiden, Crit. Revis. Eucalyptus 6: 431 (1923).
SC. Western Australia. 50 WAU. Phan.

Eucalyptus communalis Brooker & Hopper, Nuytsia 9: 19 (1993).
Western Australia. 50 WAU. Phan.

Eucalyptus concinna Maiden & Blakely in J.H.Maiden, Crit. Revis. Eucalyptus 8: 49 (1929).
Western Australia to South Australia. 50 SOA WAU. Phan.
Eucalyptus ochrophylla Maiden & Blakely in J.H.Maiden, Crit. Revis. Eucalyptus 8: 50 (1929).
Eucalyptus meeboldii Blakely, Key Eucalypts: 121 (1934).

Eucalyptus conferruminata D.J.Carr & S.G.M.Carr, Austral. J. Bot. 28: 535 (1980).
S. Western Australia. 50 WAU. Phan.

Eucalyptus confluens W.Fitzg. ex Maiden, J. Proc. Roy. Soc. New S. Wales 49: 317 (1916).
N. Western Australia. 50 WAU. Phan.

Eucalyptus* × *congener Maiden & Blakely in J.H.Maiden, Crit. Revis. Eucalyptus 8: 36 (1929). *E. piperita* subsp. *urceolaris* × *E. sclerophylla*.
New South Wales. 50 NSW. Phan.

Eucalyptus conglobata (Benth.) Maiden, Crit. Revis. Eucalyptus 6: 273 (1922).
Western Australia to South Australia. 50 SOA WAU. Phan.
**Eucalyptus dumosa* var. *conglobata* Benth., Fl. Austral. 3: 230 (1867). *Eucalyptus incrassata* var. *conglobata* (Benth.) Maiden, Crit. Revis. Eucalyptus 1: 96 (1904).

subsp. ***conglobata***
Western Australia to South Australia. 50 SOA WAU. Phan.

subsp. ***perata*** Brooker & Slee, Nuytsia 15: 157 (2004).
Western Australia. 50 WAU. Phan.

Eucalyptus conglomerata Maiden & Blakely in J.H. Maiden, Crit. Revis. Eucalyptus 8: 5 (1929).
SE. Queensland. 50 QLD. Phan.

Eucalyptus conica H.Deane & Maiden, Proc. Linn. Soc. New South Wales 24: 613 (1900). *Eucalyptus baueriana* var. *conica* (H.Deane & Maiden) Maiden, Proc. Linn. Soc. New South Wales 27: 216 (1902).
S. Queensland to New South Wales. 50 NSW QLD. Phan.

Eucalyptus conjuncta L.A.S.Johnson & K.D.Hill, Telopea 4: 93 (1990).
New South Wales. 50 NSW. Phan.

Eucalyptus consideniana Maiden, Proc. Linn. Soc. New South Wales 29: 475 (1904).
SE. New South Wales to Victoria. 50 NSW VIC. Phan.

Eucalyptus conspicua L.A.S.Johnson & K.D.Hill, Telopea 4: 235 (1991).
New South Wales to Victoria. 50 NSW VIC. Phan.

Eucalyptus contracta L.A.S.Johnson & K.D.Hill, Telopea 8: 228 (1999).
SE. Queensland (Blackdown Tableland). 50 QLD. Phan.

Eucalyptus conveniens L.A.S.Johnson & K.D.Hill, Telopea 7: 393 (1998).
Western Australia. 50 WAU. Phan.

Eucalyptus coolabah Blakely & Jacobs in W.F.Blakely, Key Eucalypts: 245 (1934).
Queensland to New South Wales. 50 NSW QLD. Phan.

subsp. ***arida*** (Blakely) L.A.S.Johnson & K.D.Hill, Telopea 4: 66 (1990).
New South Wales. 50 NSW. Phan.
**Eucalyptus coolabah* var. *arida* Blakely, Key Eucalypts: 246 (1934).

subsp. ***coolabah***
New South Wales. 50 NSW. Phan.
Eucalyptus coolabah var. *rhodoclada* Blakely, Key Eucalypts: 246 (1934).

subsp. ***excerata*** L.A.S.Johnson & K.D.Hill, Telopea 4: 66 (1990).
Queensland to New South Wales. 50 NSW QLD. Phan.

Eucalyptus cooperiana F.Muell., Fragm. 11: 83 (1880).
S. Western Australia. 50 WAU. Nanophan. or phan.

Eucalyptus copulans L.A.S.Johnson & K.D.Hill, Telopea 4: 261 (1991).
New South Wales. 50 NSW. Phan.

Eucalyptus cordata Labill., Nov. Holl. Pl. 2: 13 (1806).
SE. Tasmania. 50 TAS. Nanophan. or phan.

Eucalyptus* × *cordieri Trab., Bull. Stat. Recherch. Forest. Nord Afr. 1: 149 (1917). *E. globulus* × *E. goniantha*.
Western Australia. 50 WAU. Phan.

Eucalyptus cornuta Labill., Voy. Rech. Pérouse 1: 403 (1800).
SW. & S. Western Australia. 50 WAU (63) haw. Phan.
Eucalyptus macrocera Turcz., Bull. Soc. Imp. Naturalistes Moscou 22(2): 20 (1849).

Eucalyptus coronata C.A.Gardner, J. Roy. Soc. Western Australia 19: 86 (1933). *Eucalyptus mitrata* C.A.Gardner, J. Roy. Soc. Western Australia 22: 127 (1936), nom. illeg.
S. Western Australia. 50 WAU. Nanophan.

Eucalyptus corrugata Luehm., Victorian Naturalist 13: 168 (1897).
SC. Western Australia. 50 WAU. Phan.

Eucalyptus corynodes A.R.Bean & Brooker, Austrobaileya 4: 191 (1994).
Queensland. 50 QLD. Phan.

Eucalyptus cosmophylla F.Muell., Trans. Philos. Soc. Victoria 1: 32 (1855).
SE. South Australia. 50 SOA. Nanophan. or phan.
Eucalyptus cosmophylla f. *leprosula* Miq., Ned. Kruidk. Arch. 4: 135 (1856). *Eucalyptus cosmophylla* var. *leprosula* (Miq.) Maiden, Crit. Revis. Eucalyptus 3: 17 (1914).
Eucalyptus cosmophylla var. *rostrigera* Maiden, Crit. Revis. Eucalyptus 3: 17 (1914).

Eucalyptus costuligera L.A.S.Johnson & K.D.Hill, Telopea 8: 527 (2000).
Western Australia. 50 WAU. Phan.

Eucalyptus* × *crawfordii Maiden & Blakely in J.H. Maiden, Crit. Revis. Eucalyptus 8: 54 (1929). *E. acaciiformis* × *E. saligna*.
New South Wales. 50 NSW. Phan.

Eucalyptus crebra F.Muell., J. Proc. Linn. Soc., Bot. 3: 87 (1859).
E. Australia. (25) ken tan uga (36) chs 50 NSW QLD (61) sci (63) haw. Phan.
Metrosideros salicifolia Sol. ex Gaertn., Fruct. Sem. Pl. 1: 171 (1788).
Eucalyptus racemosa var. *longiflora* Blakely, Key Eucalypts: 249 (1934).

Eucalyptus crenulata Blakely & Beuzev., Contr. New South Wales Natl. Herb. 1: 37 (1939).
Victoria. 50 VIC. Phan.

Eucalyptus creta L.A.S.Johnson & K.D.Hill, Telopea 4: 213 (1991).
Western Australia. 50 WAU. Phan.

Eucalyptus cretata P.J.Lang & Brooker, J. Adelaide Bot. Gard. 13: 71 (1990).
South Australia. 50 SOA. Phan.

Eucalyptus crispata Brooker & Hopper, Nuytsia 8: 100 (1991).
Western Australia. 50 WAU. Phan.

Eucalyptus croajingolensis L.A.S.Johnson & K.D.Hill, Telopea 4: 99 (1990).
New South Wales to Victoria. 50 NSW VIC. Phan.

Eucalyptus crucis Maiden, Crit. Revis. Eucalyptus 6: 514 (1923).
Western Australia. 50 WAU. Phan.

subsp. ***crucis***
SC. Western Australia. 50 WAU. Phan.

subsp. ***lanceolata*** Brooker & Hopper, Nuytsia 4: 120 (1982).
WSW. Western Australia. 50 WAU. Phan.

subsp. ***praecipua*** Brooker & Hopper, Nuytsia 9: 40 (1993).
SW. Western Australia. 50 WAU. Phan.

Eucalyptus cullenii Cambage, J. Proc. Roy. Soc. New S. Wales 54: 48 (1920).
N. Queensland. 50 QLD. Phan.
Eucalyptus cullenii var. *trivalvis* Blakely, Key Eucalypts: 248 (1934).

Eucalyptus cunninghamii Sweet, Hort. Brit., ed. 2: 209 (1830).
CE. New South Wales. 50 NSW. Nanophan.
**Eucalyptus microphylla* A.Cunn., Field New South Wales: 350 (1825), nom. illeg.
Eucalyptus rupicola L.A.S.Johnson & Blaxell, Contr. New South Wales Natl. Herb. 4: 287 (1972).

Eucalyptus cuprea Brooker & Hopper, Nuytsia 9: 64 (1993).
Western Australia. 50 WAU. Phan.

Eucalyptus cupularis C.A.Gardner, J. Roy. Soc. Western Australia 47: 60 (1964).
NE. Western Australia to NW. Northern Territory. 50 NTA WAU. Phan.

Eucalyptus* × *currabubula Blakely, Key Eucalypts: 229 (1934). *E. albens* × *E. viridis*.
New South Wales. 50 NSW. Phan.

Eucalyptus curtisii Blakely & C.T.White, Proc. Roy. Soc. Queensland 42: 82 (1930 publ. 1931).
SE. Queensland. 50 QLD. Phan.

Eucalyptus cyanoclada Blakely, Key Eucalypts: 242 (1934).
Northern Territory. 50 NTA. Phan.

Eucalyptus cyanophylla Brooker, Trans. Roy. Soc. South Australia 101: 15 (1977).
SE. Australia. 50 NSW SOA VIC. Phan.

Eucalyptus cyclostoma Brooker, Brunonia 4: 21 (1981).
CS. Western Australia. 50 WAU. Nanophan. or phan.

Eucalyptus cylindriflora Maiden & Blakely, J. Proc. Roy. Soc. New S. Wales 59: 180 (1925).
S. Western Australia. 50 WAU. Nanophan. or phan.

Eucalyptus cylindrocarpa Blakely, Key Eucalypts: 116 (1934).
SW. & S. Western Australia. 50 WAU. Phan.

Eucalyptus cypellocarpa L.A.S.Johnson, Contr. New South Wales Natl. Herb. 3: 114 (1962).
New South Wales to Victoria. 50 NSW VIC. Phan.
Eucalyptus goniocalyx var. *acuminata* Benth., Fl. Austral. 3: 230 (1867).
Eucalyptus goniocalyx var. *parviflora* Blakely & McKie, Proc. Linn. Soc. New South Wales 63: 66 (1938).

Eucalyptus dalrympleana Maiden, Forest Fl. N.S.W. 7: 137 (1920).
New South Wales to Tasmania. 50 NSW TAS VIC. Phan.

subsp. ***dalrympleana***
New South Wales to Tasmania. 50 NSW TAS VIC. Phan.

subsp. ***heptantha*** L.A.S.Johnson, Contr. New South Wales Natl. Herb. 3: 110 (1962).
NE. New South Wales. 50 NSW. Phan.

Eucalyptus dawsonii R.T.Baker, Proc. Linn. Soc. New South Wales 24: 295 (1889).
New South Wales. 50 NSW. Phan.

Eucalyptus dealbata A.Cunn. ex Schauer in W.G.Walpers, Repert. Bot. Syst. 2: 924 (1843). *Eucalyptus viminalis* var. *dealbata* (A.Cunn. ex Schauer) C.Moore & Betche, Handb. Fl. N.S.W.: 202 (1893). *Eucalyptus tereticornis* var. *dealbata* (A.Cunn. ex Schauer) H.Deane & Maiden, Proc. Linn.

Soc. New South Wales 24: 466 (1899). *Eucalyptus umbellata* var. *dealbata* (A.Cunn. ex Schauer) Domin, Biblioth. Bot. 89: 1021 (1928).
E. Australia. 50 NSW QLD. Phan.
Eucalyptus dealbata var. *populnea* Blakely, Key Eucalypts: 339 (1934).

Eucalyptus deanei Maiden, Proc. Linn. Soc. New South Wales 29: 471 (1904).
SE. Queensland to E. New South Wales. 50 NSW QLD (63) haw. Phan.
Eucalyptus saligna var. *parviflora* H.Deane & Maiden, Proc. Linn. Soc. New South Wales 24: 464 (1899).

Eucalyptus decaisneana Blume, Mus. Bot. 1: 83 (1850).
Lesser Sunda Is. (Timor). 42 LSI. Phan.
Eucalyptus obliqua Decne., Nouv. Ann. Mus. Hist. Nat. 3: 454 (1834), nom. illeg.

Eucalyptus decipiens Endl. in S.L.Endlicher & al., Enum. Pl.: 49 (1837).
SW. Australia. 50 WAU. Phan.

subsp. ***adesmophloia*** Brooker & Hopper, Nuytsia 9: 22 (1993).
SW. Australia. 50 WAU. Phan.

subsp. ***chalara*** Brooker & Hopper, Nuytsia 9: 23 (1993).
SW. Australia. 50 WAU. Phan.

subsp. ***decipiens***
SW. Australia. 50 WAU. Phan.
Eucalyptus concolor Schauer in J.G.C.Lehmann, Pl. Preiss. 1: 129 (1844).
Eucalyptus decipiens var. *angustifolia* Schauer in J.G.C.Lehmann, Pl. Preiss. 1: 130 (1844).
Eucalyptus decipiens var. *latifolia* Schauer in J.G.C.Lehmann, Pl. Preiss. 1: 129 (1844).

Eucalyptus decolor A.R.Bean & Brooker, Austrobaileya 3: 41 (1989).
Queensland. 50 QLD. Phan.

Eucalyptus decorticans (F.M.Bailey) Maiden, Crit. Revis. Eucalyptus 5: 231 (1921).
SE. Queensland. 50 QLD. Phan.
**Eucalyptus siderophloia* f. *decorticans* F.M.Bailey, Queensland Agric. J. 26: 127 (1911).

Eucalyptus decurva F.Muell., Fragm. 3: 130 (1863).
SW. Australia. 50 WAU. Nanophan. or phan.

Eucalyptus deflexa Brooker, Nuytsia 2: 106 (1976).
S. Western Australia. 50 WAU. Nanophan. or phan.

Eucalyptus deglupta Blume, Mus. Bot. 1: 83 (1850).
Philippines to New Guinea. 42 LSI MOL PHI SUL 43 NWG (63) haw. Phan.

Eucalyptus delegatensis F.Muell. ex R.T.Baker, Proc. Linn. Soc. New South Wales 25: 305 (1900).
S. New South Wales to Tasmania. 50 NSW TAS VIC (51) nzn nzs. Phan.

subsp. ***delegatensis***
S. New South Wales to Victoria. 50 NSW VIC (51) nzn nzs. Phan.
Eucalyptus obliqua var. *alpina* Maiden, Proc. Austral. Assoc. Advancem. Sci. 9: 369 (1903).

subsp. ***tasmaniensis*** Boland, Austral. Forest. Res. 15: 177 (1985).
Tasmania. 50 TAS. Phan.
Eucalyptus gigantea Hook.f., London J. Bot. 6: 479 (1847), nom. illeg.
Eucalyptus risdonii var. *elata* Benth., Fl. Austral. 3: 203 (1867). *Eucalyptus tasmanica* Blakely, Key Eucalypts: 214 (1934).

Eucalyptus delicata L.A.S.Johnson & K.D.Hill, Telopea 8: 182 (1999).
SC. Western Australia. 50 WAU. Phan.

Eucalyptus dendromorpha (Blakely) L.A.S.Johnson & Blaxell, Contr. New South Wales Natl. Herb. 4: 286 (1972).
New South Wales. 50 NSW. Phan.
**Eucalyptus obtusiflora* var. *dendromorpha* Blakely, Austral. Naturalist 10: 258 (1941).

Eucalyptus densa Brooker & Hopper, Nuytsia 8: 149 (1991).
Western Australia. 50 WAU. Phan.

subsp. ***densa***
Western Australia. 50 WAU. Phan.

subsp. ***improcera*** Brooker & Hopper, Nuytsia 8: 154 (1991).
Western Australia. 50 WAU. Phan.

Eucalyptus denticulata I.O.Cook & Ladiges, Austral. Syst. Bot. 4: 388 (1991).
Victoria. 50 VIC. Phan.

Eucalyptus depauperata L.A.S.Johnson & K.D.Hill, Telopea 4: 587 (1992).
Western Australia. 50 WAU. Phan.

Eucalyptus desmondensis Maiden & Blakely, J. Proc. Roy. Soc. New S. Wales 59: 183 (1925).
S. Western Australia. 50 WAU. Nanophan. or phan.

Eucalyptus deuaensis Boland & P.M.Gilmour, Brunonia 9: 105 (1987).
SE. New South Wales. 50 NSW. Nanophan. or phan.

Eucalyptus dielsii C.A.Gardner, J. Roy. Soc. Western Australia 12: 67 (1927).
S. Western Australia. 50 WAU. Nanophan. or phan.

Eucalyptus diminuta Brooker & Hopper, Nuytsia 14: 358 (2002).
Western Australia. 50 WAU. Phan.

Eucalyptus diptera C.R.P.Andrews, J. West Austral. Nat. Hist. Soc. 1: 42 (1904).
S. Western Australia. 50 WAU. Phan.

Eucalyptus disclusa L.A.S.Johnson & Blaxell, Telopea 4: 226 (1991).
Queensland. 50 QLD. Phan.

Eucalyptus discreta Brooker, Brunonia 2: 148 (1979).
S. Western Australia. 50 WAU. Nanophan. or phan.

Eucalyptus dissimulata Brooker, Nuytsia 6: 332 (1988).
Western Australia. 50 WAU. Phan.

Eucalyptus dissita K.D.Hill, Telopea 7: 104 (1997).
New South Wales. 50 NSW. Phan.

Eucalyptus distans Brooker, Boland & Kleinig, Austral. Forest. Res. 10: 95 (1980).
N. Northern Territory. 50 NTA. Phan.

Eucalyptus diversicolor F.Muell., Fragm. 3: 131 (1863).
SW. Australia. (25) ken tan 50 WAU. Phan.
Eucalyptus colossea F.Muell., Fragm. 7: 42 (1869).

Eucalyptus diversifolia Bonpl., Descr. Pl. Malmaison: 35 (1814).
Western Australia to Victoria. 50 SOA VIC WAU. Nanophan. or phan.

subsp. ***diversifolia***

SE. Western Australia to Victoria. 50 SOA VIC WAU. Nanophan. or phan.
Eucalyptus santalifolia F.Muell., Trans. Philos. Soc. Victoria 1: 35 (1855).

subsp. ***hesperia*** I.J.Wright & Ladiges, Austral. Syst. Bot. 10: 677 (1997).
Western Australia. 50 WAU. Nanophan. or phan.

subsp. ***megacarpa*** I.J.Wright & Ladiges, Austral. Syst. Bot. 10: 678 (1997).
Victoria. 50 VIC. Nanophan. or phan.

Eucalyptus dives Schauer in W.G.Walpers, Repert. Bot. Syst. 2: 926 (1843).
New South Wales to Victoria. 50 NSW VIC. Phan.
Eucalyptus amygdalina var. *latifolia* H.Deane & Maiden, Proc. Linn. Soc. New South Wales 20: 609 (1896).

Eucalyptus* × *dixsonii N.A.Wakef. in J.H.Maiden, Crit. Revis. Eucalyptus 8: 20 (1929). *E. dives* × *E. radiata*.
SE. Australia. 50 NSW VIC. Phan.
Eucalyptus × *radiodives* R.A.Black, Victorian Naturalist 60: 175 (1944).

Eucalyptus dolichocera L.A.S.Johnson & K.D.Hill, Telopea 8: 196 (1999).
W. Western Australia. 50 WAU. Phan.

Eucalyptus dolichorhyncha (Brooker) Brooker & Hopper, Nuytsia 9: 57 (1993).
CS. Western Australia. 50 WAU. Nanophan. or phan.
**Eucalyptus forrestiana* subsp. *dolichorhyncha* Brooker, J. Roy. Soc. Western Australia 56: 74 (1973).

Eucalyptus dolorosa Brooker & Hopper, Nuytsia 9: 6 (1993).
Western Australia. 50 WAU. Phan.

Eucalyptus doratoxylon F.Muell., Fragm. 2: 55 (1860).
S. Western Australia. 50 WAU. Nanophan. or phan.

Eucalyptus* × *dorisiana Blakely, Key Eucalypts: 227 (1934). *E. intertexta* × *E. viridis*.
New South Wales. 50 NSW. Phan.

Eucalyptus dorrigoensis (Blakely) L.A.S.Johnson & K.D.Hill, Telopea 4: 63 (1990).
NE. New South Wales. 50 NSW. Phan.
**Eucalyptus benthamii* var. *dorrigoensis* Blakely, Key Eucalypts: 162 (1934).

Eucalyptus drepanophylla F.Muell. ex Benth., Fl. Austral. 3: 221 (1867).
Queensland. (25) ken tan 50 QLD. Phan.
Eucalyptus crebra var. *macrocarpa* Domin, Biblioth. Bot. 89: 465 (1928). *Eucalyptus racemosa* var. *macrocarpa* (Domin) Blakely, Key Eucalypts: 249 (1934).

Eucalyptus drummondii Benth., Fl. Austral. 3: 237 (1867). *Eucalyptus oldfieldii* var. *drummondii* (Benth.) Maiden, Crit. Revis. Eucalyptus 2: 223 (1913).
W. Western Australia. 50 WAU. Phan.

Eucalyptus dumosa A.Cunn. ex Oxley, J. Exped. N. S. Wales: 63 (1820). *Eucalyptus incrassata* var. *dumosa* (A.Cunn. ex Oxley) Maiden, Crit. Revis. Eucalyptus 1: 96 (1904).
SE. Australia. 50 NSW SOA VIC. Phan.
Eucalyptus lamprocarpa F.Muell. ex Miq., Ned. Kruidk. Arch. 4: 129 (1856).
Eucalyptus muelleri Miq., Ned. Kruidk. Arch. 4: 130 (1856).
Eucalyptus dumosa var. *rhodophloia* Benth., Fl. Austral. 3: 231 (1867). *Eucalyptus rhodophloia* (Benth.) Blakely, Key Eucalypts: 115 (1934).

Eucalyptus dundasii Maiden, J. Proc. Roy. Soc. New S. Wales 49: 309 (1915 publ. 1916).
Western Australia. 50 WAU. Phan.

Eucalyptus dunnii Maiden, Proc. Linn. Soc. New South Wales 30: 336 (1905).
SE. Queensland to NE. New South Wales. 50 NSW QLD. Phan.

Eucalyptus dura L.A.S.Johnson & K.D.Hill, Telopea 4: 343 (1991).
Queensland. 50 QLD. Phan.

Eucalyptus dwyeri Maiden & Blakely, J. Proc. Roy. Soc. New S. Wales 59: 160 (1925).
S. Queensland to NE. Victoria. 50 NSW QLD VIC. Phan.

Eucalyptus ebbanoensis Maiden, Crit. Revis. Eucalyptus 5: 169 (1921).
Western Australia. 50 WAU. Nanophan. or phan.

subsp. ***ebbanoensis***
WC. Western Australia. 50 WAU. Nanophan. or phan.

subsp. ***glauciramula*** L.A.S.Johnson & K.D.Hill, Telopea 7: 402 (1998).
Western Australia. 50 WAU. Nanophan. or phan.

subsp. ***photina*** Brooker & Hopper, Nuytsia 9: 2 (1993).
Western Australia. 50 WAU. Nanophan. or phan.

Eucalyptus* × *ednaeana Blakely, Key Eucalypts: 247 (1934). *E. intertexta* × *E. sideroxylon*.
New South Wales. 50 NSW. Phan.

Eucalyptus educta L.A.S.Johnson & K.D.Hill, Telopea 4: 627 (1992).
Western Australia. 50 WAU. Phan.

Eucalyptus effusa Brooker, Nuytsia 2: 108 (1976).
Western Australia. 50 WAU. Nanophan. or phan.

subsp. ***effusa***
Western Australia. 50 WAU. Nanophan. or phan.

subsp. ***exsul*** L.A.S.Johnson & K.D.Hill, Telopea 4: 216 (1991).
Western Australia. 50 WAU. Phan.

Eucalyptus elaeophloia Chappill, Crisp & Prober, Austral. Syst. Bot. 3: 275 (1990).
Victoria. 50 VIC. Phan.

Eucalyptus elata Dehnh., Cat. Horti Camald.: 26 (1829).
SE. New South Wales to Victoria. 50 NSW VIC. Phan.
Eucalyptus longifolia Lindl., Bot. Reg. 11: t. 947 (1826), nom. illeg. *Eucalyptus lindleyana* A.Cunn. ex DC., Prodr. 3: 219 (1828).
Eucalyptus andreana Naudin, Rev. Hort. 1880: 346 (1880).
Eucalyptus lindleyana var. *stenophylla* Blakely, Key Eucalypts: 209 (1934). *Eucalyptus andreana* var. *stenophylla* (Blakely) Cameron, Victorian Naturalist 63: 41 (1946).

Eucalyptus elegans A.R.Bean, Austrobaileya 7: 113 (2005).
SE. Queensland to New South Wales. 50 NSW QLD.

Eucalyptus elliptica (Blakely & McKie) L.A.S.Johnson & K.D.Hill, Telopea 4: 60 (1990).
SE. Queensland to NE. New South Wales. 50 NSW QLD. Phan.
**Eucalyptus mannifera* var. *elliptica* Blakely & McKie in W.F.Blakely, Key Eucalypts: 148 (1934). *Eucalyptus mannifera* subsp. *elliptica* (Blakely & McKie) L.A.S.Johnson, Contr. New South Wales Natl. Herb. 3: 108 (1962).

Eucalyptus epruinata L.A.S.Johnson & K.D.Hill, Telopea 8: 530 (2000).
Queensland. 50 QLD. Phan.

Eucalyptus erectifolia Brooker & Hopper, Nuytsia 5: 351 (1986).
SW. Australia. 50 WAU. Nanophan. or phan.

Eucalyptus eremicola Boomsma, S. Austral. Naturalist 50: 28 (1975).
South Australia. 50 SOA. Nanophan. or phan.

subsp. ***eremicola***
South Australia. 50 SOA. Nanophan. or phan.

subsp. ***peeneri*** (Blakely) D.Nicolle, Austral. Syst. Bot. 18: 542 (2005).
SW. South Australia. 50 SOA. Nanophan. or phan.
**Eucalyptus oleosa* var. *peeneri* Blakely, Key Eucalypts: 270 (1934). *Eucalyptus peeneri* (Blakely) Prior & Johnson ex Boomsma, J. Adelaide Bot. Gard. 1: 368 (1979).

Eucalyptus eremophila (Diels) Maiden, J. Proc. Roy. Soc. New S. Wales 54: 71 (1920).
Western Australia. 50 WAU. Nanophan. or phan.
**Eucalyptus occidentalis* var. *eremophila* Diels, Bot. Jahrb. Syst. 35: 442 (1904).
Eucalyptus occidentalis var. *grandiflora* Maiden, Crit. Revis. Eucalyptus 4: 149 (1919). *Eucalyptus eremophila* var. *grandiflora* (Maiden) Maiden, Crit. Revis. Eucalyptus 7: 22 (1923).
Eucalyptus eremophila var. *pterocarpa* Blakely & H.Steedman, Contr. New South Wales Natl. Herb. 1: 36 (1939). *Eucalyptus eremophila* subsp. *pterocarpa* (Blakely & H.Steedman) L.A.S.Johnson & Blaxell, Contr. New South Wales Natl. Herb. 4: 454 (1973).

Eucalyptus erosa A.R.Bean, Austrobaileya 7: 141 (2005).
S. Queensland. 50 QLD.

Eucalyptus* × *erythrandra Blakely & H.Steedman, Proc. Linn. Soc. New South Wales 63: 65 (1938). *E. incrassata* × *E. tetraptera*.
S. Western Australia. 50 WAU. Phan.
**Eucalyptus angulosa* var. *robusta* C.A.Gardner, J. Roy. Soc. Western Australia 19: 88 (1933).

Eucalyptus erythrocorys F.Muell., Fragm. 2: 33 (1860).
W. Western Australia. 50 WAU. Phan.
Eudesmia erythrocorys F.Muell., Fragm. 2: 33 (1860), pro syn.

Eucalyptus erythronema Turcz., Bull. Cl. Phys.-Math. Acad. Imp. Sci. Saint-Pétersbourg 10: 337 (1852).
SW. Australia. 50 WAU. Nanophan. or phan.

var. ***erythronema***
SW. Australia. 50 WAU. Nanophan. or phan.
Eucalyptus conoidea Benth., Fl. Austral. 3: 227 (1867).
Eucalyptus erythronema var. *roei* Maiden, Crit. Revis. Eucalyptus 1: 110 (1909).

var. ***marginata*** (Benth.) Domin, Repert. Spec. Nov. Regni Veg. 12: 389 (1913).
WSW. Western Australia. 50 WAU. Phan.
**Eucalyptus conoidea* var. *marginata* Benth., Fl. Austral. 3: 227 (1867).

Eucalyptus eudesmioides F.Muell., Fragm. 2: 35 (1860).
W. Western Australia. 50 WAU. Nanophan.
Eudesmia eucalyptoides F.Muell., Fragm. 2: 35 (1860), pro syn.
Eucalyptus eudesmioides var. *globosa* Blakely, Key Eucalypts: 69 (1934).

Eucalyptus eugenioides Sieber ex Spreng., Syst. Veg. 4(2): 195 (1827).
SE. Queensland to New South Wales. 50 NSW QLD. Phan.
Eucalyptus acervula Sieber ex DC., Prodr. 3: 217 (1828).
Eucalyptus eugenioides var. *nana* H.Deane & Maiden, Proc. Linn. Soc. New South Wales 23: 79 (1889).
Eucalyptus laevopinea var. *minor* F.Muell. ex R.T.Baker, Proc. Linn. Soc. New South Wales 23: 416 (1898).
Eucalyptus wilkinsoniana R.T.Baker, Proc. Linn. Soc. New South Wales 25: 678 (1901).
Eucalyptus wiburdii Blakely, Key Eucalypts: 177 (1934).
Eucalyptus wilkinsoniana var. *crassifructa* Blakely, Key Eucalypts: 179 (1934).

Eucalyptus ewartiana Maiden, J. Proc. Roy. Soc. New S. Wales 53: 111 (1919).
Western Australia. 50 WAU. Phan.

Eucalyptus exigua Brooker & Hopper, Nuytsia 9: 60 (1993).
Western Australia. 50 WAU. Phan.

Eucalyptus exilipes Brooker & A.R.Bean, Brunonia 10: 189 (1987).
Queensland. 50 QLD. Phan.

Eucalyptus exilis Brooker, Nuytsia 1: 305 (1974).
SW. Australia. 50 WAU. Phan.

Eucalyptus exserta F.Muell., J. Proc. Linn. Soc., Bot. 3: 85 (1859).
Queensland to N. New South Wales. (36) chc chh chs 50 NSW QLD. Phan.
Eucalyptus insulana F.M.Bailey, Queensland Agric. J. 17: 103 (1906).
Eucalyptus exserta var. *parvula* Blakely, Key Eucalypts: 129 (1934).

Eucalyptus extensa L.A.S.Johnson & K.D.Hill, Telopea 4: 218 (1991).
Western Australia. 50 WAU. Phan.

Eucalyptus extrica D.Nicolle, Nuytsia 13: 322 (2000).
S. Western Australia. 50 WAU. Phan.

Eucalyptus falcata Turcz., Bull. Soc. Imp. Naturalistes Moscou 20(1): 163 (1847).
SW. Australia. 50 WAU. Phan.
Eucalyptus falcata var. *ecostata* Maiden, J. Nat. Hist. Sci. Soc. Western Australia 3: 173 (1911).
Eucalyptus dorrienii Domin, Repert. Spec. Nov. Regni Veg. 12: 388 (1913).

Eucalyptus famelica Brooker & Hopper, Nuytsia 7: 12 (1989).
Western Australia. 50 WAU. Phan.

Eucalyptus farinosa K.D.Hill, Telopea 7: 191 (1997).
Queensland. 50 QLD. Phan.

Eucalyptus fasciculosa F.Muell., Trans. Philos. Soc. Victoria 1: 34 (1855). *Eucalyptus paniculata* var. *fasciculosa* (F.Muell.) Benth., Fl. Austral. 3: 212 (1867).
SE. South Australia to W. Victoria. 50 SOA VIC. Phan.

Eucalyptus fastigata H.Deane & Maiden, Proc. Linn. Soc. New South Wales 21: 809 (1897).
New South Wales to NE. Victoria. (25) ken tan 50 NSW VIC. Phan.

Eucalyptus fergusonii R.T.Baker, Hardw. Austral.: 269 (1919).
New South Wales. 50 NSW. Phan.

subsp. ***dorsiventralis*** L.A.S.Johnson & K.D.Hill, Telopea 4: 80 (1990).
New South Wales. 50 NSW. Phan.

subsp. ***fergusonii***
New South Wales. 50 NSW. Phan.

Eucalyptus fibrosa F.Muell., J. Proc. Linn. Soc., Bot. 3: 87 (1859).
E. Australia. 50 NSW QLD. Phan.

subsp. ***fibrosa***
E. Australia. 50 NSW QLD. Phan.
Eucalyptus bowmanii F.Muell. ex Benth., Fl. Austral. 3: 219 (1867).
Eucalyptus siderophloia var. *rostrata* Benth., Fl. Austral. 3: 220 (1867).

subsp. ***nubila*** (Maiden & Blakely) L.A.S.Johnson, Contr. New South Wales Natl. Herb. 3: 103 (1962).
SE. Queensland to New South Wales. 50 NSW QLD. Phan.
Eucalyptus siderophloia var. *glauca* H.Deane & Maiden, Proc. Linn. Soc. New South Wales 24: 461 (1899). **Eucalyptus nubila* Maiden & Blakely in J.H.Maiden, Crit. Revis. Eucalyptus 8: 38 (1929).

Eucalyptus filiformis Rule, Muelleria 20: 24 (2004).
Victoria. 50 VIC.

Eucalyptus fitzgeraldii Blakely, Key Eucalypts: 239 (1934).
N. Western Australia. 50 WAU. Phan.

Eucalyptus flavida Brooker & Hopper, Nuytsia 8: 99 (1991).
Western Australia. 50 WAU. Phan.
Eucalyptus redunca var. *oxymitra* Maiden, Crit. Revis. Eucalyptus 4: 98 (1917).

Eucalyptus flindersii Boomsma, J. Adelaide Bot. Gard. 2: 293 (1980).
South Australia. 50 SOA. Phan.

Eucalyptus flocktoniae (Maiden) Maiden, J. Proc. Roy. Soc. New S. Wales 49: 316 (1915 publ. 1916).
SW. Australia. 50 WAU. Phan.
**Eucalyptus oleosa* var. *flocktoniae* Maiden, J. Nat. Hist. Sci. Soc. Western Australia 3: 172 (1911).

subsp. ***flocktoniae***
SW. Australia. 50 WAU. Phan.

subsp. ***hebes*** D.Nicolle, Austral. Syst. Bot. 12: 225 (1999).
SW. Australia. 50 WAU.

Eucalyptus foecunda Schauer in J.G.C.Lehmann, Pl. Preiss. 1: 130 (1844).
S. & SE. Australia. 50 NSW SOA VIC WAU. Phan.
Eucalyptus leptophylla F.Muell. ex Miq., Ned. Kruidk. Arch. 4: 123 (1856).
Eucalyptus desertorum Naudin, Descr. Emploi Eucalypt.: 56 (1891).
Eucalyptus leptophylla var. *densa* Blakely, Key Eucalypts: 222 (1934).
Eucalyptus leptophylla var. *leptorrhyncha* Blakely, Key Eucalypts: 222 (1934).
Eucalyptus leptophylla var. *floribunda* Blakely, Trans. & Proc. Roy. Soc. South Australia 60: 155 (1936).

Eucalyptus foliosa L.A.S.Johnson & K.D.Hill, Telopea 4: 595 (1992).
Western Australia. 50 WAU. Phan.

Eucalyptus formanii C.A.Gardner, J. Roy. Soc. Western Australia 27: 186 (1942).
SC. Western Australia. 50 WAU. Phan.

Eucalyptus forrestiana Diels, Bot. Jahrb. Syst. 35: 439 (1904).
CS. Western Australia. 50 WAU. Nanophan. or phan.

Eucalyptus* × *forthiana Blakely, Contr. New South Wales Natl. Herb. 1: 37 (1939). *E. moluccana* × *E. siderophloia*.
New South Wales. 50 NSW. Phan.

Eucalyptus fracta K.D.Hill, Telopea 7: 106 (1997).
New South Wales. 50 NSW. Phan.

Eucalyptus fraseri (Brooker) Brooker, Austral. Forest. Res. 7: 66 (1976).
CS. Western Australia. 50 WAU. Phan.
**Eucalyptus conglobata* subsp. *fraseri* Brooker, Nuytsia 1: 251 (1972).

subsp. ***fraseri***
CS. Western Australia. 50 WAU. Phan.

subsp. ***melanobasis*** L.A.S.Johnson & K.D.Hill, Telopea 9: 298 (2001).
Western Australia. 50 WAU.

Eucalyptus fraxinoides H.Deane & Maiden, Proc. Linn. Soc. New South Wales 23: 412 (1898). *Eucalyptus virgata* var. *fraxinoides* (H.Deane & Maiden) Maiden, Forest Fl. N.S.W. 3: 87 (1907).
SE. New South Wales to NE. Victoria. 50 NSW VIC. Phan.

Eucalyptus froggattii Blakely, Key Eucalypts: 225 (1934).
Victoria. 50 VIC. Phan.

Eucalyptus fruticosa Brooker, Brunonia 2: 129 (1979).
W. Western Australia. 50 WAU. Nanophan.

Eucalyptus fulgens Rule, Muelleria 9: 136 (1996).
Victoria. 50 VIC. Phan.

Eucalyptus fusiformis Boland & Kleinig, Brunonia 10: 202 (1987).
New South Wales. 50 NSW. Phan.

Eucalyptus gamophylla F.Muell., Fragm. 11: 40 (1879).
Australia. 50 NTA QLD SOA WAU. Nanophan. or phan.

Eucalyptus gardneri Maiden, Crit. Revis. Eucalyptus 7: 53 (1924).
Western Australia. 50 WAU. Phan.

subsp. ***gardneri***
Western Australia. 50 WAU. Phan.

subsp. ***ravensthorpensis*** Brooker & Hopper, Nuytsia 8: 145 (1991).
Western Australia. 50 WAU. Phan.

Eucalyptus georgei Brooker & Blaxell, Nuytsia 2: 224 (1978).
Western Australia. 50 WAU. Nanophan. or phan.

subsp. ***fulgida*** Brooker & Hopper, Nuytsia 9: 47 (1993).
Western Australia. 50 WAU. Nanophan. or phan.

subsp. ***georgei***
S. Western Australia. 50 WAU. Nanophan. or phan.

Eucalyptus gigantangion L.A.S.Johnson & K.D.Hill, Telopea 4: 322 (1991).
Northern Territory. 50 NTA. Phan.

Eucalyptus gillenii Ewart & L.R.Kerr, Proc. Roy. Soc. Victoria, n.s., 39: 7 (1926).
C. Australia. 50 NTA SOA WAU. Phan.
Eucalyptus incurva Boomsma, S. Austral. Naturalist 50: 31 (1975).

Eucalyptus gillii Maiden, Crit. Revis. Eucalyptus 2: 177 (1912).
E. South Australia to W. New South Wales. 50 NSW SOA. Phan.

Eucalyptus gittinsii Brooker & Blaxell, Nuytsia 2: 228 (1978).
W. Western Australia. 50 WAU. Nanophan. or phan.

subsp. ***gittinsii***
W. Western Australia. 50 WAU. Nanophan. or phan.

subsp. ***illucida*** D.Nicolle, Nuytsia 13: 319 (2000).
WSW. Western Australia. 50 WAU. Phan.

Eucalyptus glaucescens Maiden & Blakely in J.H.Maiden, Crit. Revis. Eucalyptus 8: 56 (1929).
SE. New South Wales to E. Victoria. 50 NSW VIC. Phan.
Eucalyptus gunnii var. *glauca* H.Deane & Maiden, Proc. Linn. Soc. New South Wales 24: 464 (1899).

Eucalyptus glaucina (Blakely) L.A.S.Johnson, Contr. New South Wales Natl. Herb. 3: 104 (1962).
New South Wales. 50 NSW. Phan.
**Eucalyptus umbellata* var. *glaucina* Blakely, Key Eucalypts: 130 (1934). *Eucalyptus tereticornis* var. *glaucina* (Blakely) Cameron, Victorian Naturalist 63: 43 (1946).

Eucalyptus globoidea Blakely, J. Proc. Roy. Soc. New S. Wales 61: 157 (1927).
New South Wales to Victoria. 50 NSW VIC. Phan.
Eucalyptus globoidea var. *subsphaerica* Blakely, Key Eucalypts: 189 (1934).
Eucalyptus yangoura Blakely, Key Eucalypts: 180 (1934).

Eucalyptus globulus Labill., Voy. Rech. Pérouse 1: 153 (1800). *Eucalyptus maidenii* subsp. *globulus* (Labill.) J.B.Kirkp., Bot. J. Linn. Soc. 69: 101 (1975).
New South Walews to Tasmania. (10) ire (12) bal fra por spa spa (13) ita sic (20) mor (21) azo cny mdr (23) rwa (25) ken tan uga (26) zim (27) ofs (33) tcs (36) chc chs 50 NSW TAS VIC (51) nzn nzs (76) cal (80) cos els gua pan (83) bol ecu per (85) jnf par. Phan.

subsp. ***bicostata*** (Maiden, Blakely & Simmonds) J.B.Kirkp., Bot. J. Linn. Soc. 69: 101 (1974).
New South Wales to Victoria. (25) ken tan 50 NSW VIC. Phan.
**Eucalyptus bicostata* Maiden, Blakely & Simmonds, Trees Other Lands New Zealand. Eucalypts: 133 (1927). *Eucalyptus globulus* var. *bicostata* (Maiden, Blakely & Simmonds) Ewart, Fl. Victoria: 804 (1931).

subsp. ***globulus***
Victoria, Tasmania. (10) ire (12) bal fra por spa spa (13) ita sic (20) mor (21) azo cny mdr (23) rwa (25) ken tan uga (26) zim (33) tcs (36) chc chs 50 TAS VIC (51) nzn nzs (76) cal (80) cos els gua pan (83) bol ecu per (85) jnf par. Phan.
Eucalyptus pulverulenta Link, Enum. Hort. Berol. Alt. 2: 31 (1822), nom. illeg.
Eucalyptus glauca A.Cunn. ex DC., Prodr. 3: 221 (1828).
Eucalyptus perfoliata Desf., Tabl. École Bot., ed. 3: 408 (1829).
Eucalyptus gigantea Dehnh., Cat. Horti Camald., ed. 2: 20 (1832).
Eucalyptus globulosus St.-Lag., Ann. Soc. Bot. Lyon 7: 125 (1880).

subsp. ***maidenii*** (F.Muell.) J.B.Kirkp., Bot. J. Linn. Soc. 69: 101 (1974).
SE. New South Wales to E. Victoria. (12) por sar spa (13) ita sic (23) rwa (25) ken tan (27) ofs (36) chc chs 50 NSW VIC. Phan.
**Eucalyptus maidenii* F.Muell., Proc. Linn. Soc. New South Wales, II, 3: 1020 (1890).

subsp. ***pseudoglobulus*** (Naudin ex Maiden) J.B. Kirkp., Bot. J. Linn. Soc. 69: 101 (1974).
Victoria. 50 VIC. Phan.
Eucalyptus globulus var. *stjohniis* F.Muell. ex R.T. Baker, Victorian Naturalist 30: 127 (1913). *Eucalyptus stjohnii* (F.Muell. ex R.T.Baker) F. Muell. ex R.T.Baker, Hardw. Austral.: 218 (1919).
**Eucalyptus pseudoglobulus* Naudin ex Maiden, Crit. Revis. Eucalyptus 8: 28 (1929).

Eucalyptus glomericassis L.A.S.Johnson & K.D.Hill, Telopea 8: 516 (2000).
Northern Territory. 50 NTA. Phan.

Eucalyptus glomerosa Brooker & Hopper, Nuytsia 9: 332 (1993).
Western Australia to South Australia. 50 SOA WAU. Phan.

Eucalyptus gomphocephala A.Cunn. ex DC., Prodr. 3: 220 (1828).
WSW. Western Australia. (12) spa (13) ita sic (20) mor (25) ken tan (26) zim (27) cpp 50 WAU (63) haw. Phan.
Eucalyptus gomphocephala var. *rhodoxylon* Blakely & H.Steedman, Contr. New South Wales Natl. Herb. 1: 36 (1939).

Eucalyptus gongylocarpa Blakely, Trans. & Proc. Roy. Soc. South Australia 60: 153 (1936).
Australia. 50 NTA SOA WAU. Phan.

Eucalyptus goniantha Turcz., Bull. Soc. Imp. Naturalistes Moscou 20(1): 163 (1847). *Eucalyptus incrassata* var. *goniantha* (Turcz.) Maiden, Crit. Revis. Eucalyptus 1: 96 (1904).
Western Australia. 50 WAU. Nanophan. or phan.

subsp. ***goniantha***
S. Western Australia. 50 WAU. Nanophan. or phan.

subsp. ***notactites*** L.A.S.Johnson & K.D.Hill, Telopea 4: 610 (1992).
Western Australia. 50 WAU. Nanophan. or phan.

Eucalyptus goniocalyx F.Muell. ex Miq., Ned. Kruidk. Arch. 4: 134 (1856).
SE. Australia. 50 NSW SOA VIC (63) haw. Phan.
Eucalyptus elaeophora F.Muell., Fragm. 4: 52 (1864).
Eucalyptus cambagei H.Deane & Maiden, Proc. Linn. Soc. New South Wales 25: 106 (1900).
Eucalyptus cordieri var. *brachypoma* Blakely, Key Eucalypts: 147 (1934).

subsp. ***exposa*** D.Nicolle, J. Adelaide Bot. Gard. 19: 87 (2000).
South Australia. 50 SOA. Phan.

subsp. ***goniocalyx***
SE. Australia. 50 NSW SOA VIC (63) haw. Phan.

Eucalyptus goniocarpa L.A.S.Johnson & K.D.Hill, Telopea 4: 582 (1992).
Western Australia. 50 WAU. Phan.

Eucalyptus gracilis F.Muell., Trans. Philos. Soc. Victoria 1: 35 (1855). *Eucalyptus calycogona* var. *gracilis* (F.Muell.) Maiden, Crit. Revis. Eucalyptus 1: 79 (1903).
S. & SE. Australia. 50 NSW SOA VIC WAU. Phan.
Eucalyptus gracilis var. *breviflora* Benth., Fl. Austral. 3: 211 (1867).
Eucalyptus gracilis var. *erecta* Blakely, Key Eucalypts: 266 (1934).
Eucalyptus gracilis var. *viminea* Blakely, Key Eucalypts: 266 (1934).

Eucalyptus grandis W.Hill, Cat. Nat. Indust. Prod. Queensland: 25 (1862).
E. Australia. (20) mor (25) ken tan uga (27) cpp (36) chs (38) tai 50 NSW QLD (78) fla (81) dom (83) clm ecu per. Phan.

Eucalyptus granitica L.A.S.Johnson & K.D.Hill, Telopea 4: 332 (1991).
Queensland. 50 QLD. Phan.

Eucalyptus gratiae (Brooker) L.A.S.Johnson & K.D.Hill, Telopea 4: 571 (1992).
SW. Australia. 50 WAU. Phan.
**Eucalyptus loxophleba* subsp. *gratiae* Brooker, Nuytsia 1: 248 (1972).

Eucalyptus gregoriensis N.G.Walsh & Albr., Muelleria 11: 41 (1998).
Northern Territory. 50 NTA. Phan.

Eucalyptus gregsoniana L.A.S.Johnson & Blaxell, Contr. New South Wales Natl. Herb. 4: 380 (1973).
SE. New South Wales. 50 NSW. Phan.
**Eucalyptus pauciflora* var. *nana* Blakely, Key Eucalypts: 339 (1934).

Eucalyptus griffithsii Maiden, J. Nat. Hist. Sci. Soc. Western Australia 3: 177 (1911).
SC. Western Australia. 50 WAU. Phan.
Eucalyptus griffithsii var. *angustiuscula* Blakely, Key Eucalypts: 105 (1934).

Eucalyptus grisea L.A.S.Johnson & K.D.Hill, Telopea 8: 509 (2000).
Queensland. 50 QLD. Phan.

Eucalyptus grossa F.Muell. ex Benth., Fl. Austral. 3: 232 (1867). *Eucalyptus incrassata* var. *grossa* (F.Muell. ex Benth.) Maiden, Crit. Revis. Eucalyptus 1: 96 (1904).
S. Western Australia. 50 WAU. Nanophan. or phan.

Eucalyptus grossifolia L.A.S.Johnson & K.D.Hill, Telopea 9: 271 (2001).
Western Australia. 50 WAU. Phan.

Eucalyptus guilfoylei Maiden, J. Nat. Hist. Sci. Soc. Western Australia 3: 180 (1911).
SW. Australia. 50 WAU. Phan.

Eucalyptus gunnii Hook.f., London J. Bot. 3: 499 (1844).
Tasmania. (33) tcs 50 TAS (51) nzn nzs. Phan.

subsp. ***divaricata*** (McAulay & Brett) B.M.Potts, Pap. & Proc. Roy. Soc. Tasmania 135: 57 (2001).
Tasmania. 50 TAS. Phan.
**Eucalyptus divaricata* McAulay & Brett, Pap. & Proc. Roy. Soc. Tasmania 1937: 94 (1938).

subsp. ***gunnii***
Tasmania. (33) tcs 50 TAS (51) nzn nzs. Phan.
Eucalyptus whittingehameii Landsb., Trans. & Proc. Bot. Soc. Edinburgh 20: 516 (1896).
Eucalyptus gunnii var. *undulata* Rehder in L.H. Bailey, Cycl. Amer. Hort. 2: 555 (1900).
Eucalyptus gunnii var. *montana* Hook.f., Bot. Mag. 127: t. 7808 (1901).
Eucalyptus perriniana R.T.Baker & H.G.Sm., Pap. & Proc. Roy. Soc. Tasmania 1912: 163 (1913), nom. illeg.

Eucalyptus gymnoteles L.A.S.Johnson & K.D.Hill, Telopea 5: 754 (1994).
Western Australia. 50 WAU. Phan.

Eucalyptus gypsophila Nicolle, Nuytsia 11: 373 (1997).
Western Australia to South Australia. 50 SOA WAU. Phan.

Eucalyptus haemastoma Sm., Trans. Linn. Soc. London 3: 286 (1797).
E. New South Wales. 50 NSW. Phan.

Eucalyptus hallii Brooker, Austral. Forest. Res. 7: 11 (1975).
SE. Queensland. 50 QLD. Phan.

Eucalyptus halophila D.J.Carr & S.G.M.Carr, Nuytsia 3: 173 (1980).
S. Western Australia. 50 WAU. Nanophan. or phan.

Eucalyptus hawkeri Rule, Muelleria 20: 17 (2004).
Victoria. 50 VIC.

Eucalyptus hebetifolia Brooker & Hopper, Nuytsia 8: 51 (1991).
Western Australia. 50 WAU. Phan.

Eucalyptus helenae L.A.S.Johnson & K.D.Hill, Telopea 5: 759 (1994).
Northern Territory. 50 NTA. Phan.

Eucalyptus helidonica K.D.Hill, Telopea 8: 225 (1999).
SE. Queensland. 50 QLD. Phan.

Eucalyptus herbertiana Maiden, Crit. Revis. Eucalyptus 6: 429 (1923).
N. Western Australia to NW. Queensland. 50 NTA QLD WAU. Phan.

Eucalyptus histophylla Brooker & Hopper, Nuytsia 8: 89 (1991).
Western Australia. 50 WAU. Phan.

Eucalyptus horistes L.A.S.Johnson & K.D.Hill, in Fl. Australia 19: 509 (1988).
W. Western Australia. 50 WAU. Phan.
Eucalyptus hypochlamydea Brooker, Nuytsia 6: 328 (1988).

Eucalyptus houseana W.Fitzg. ex Maiden, J. Proc. Roy. Soc. New S. Wales 49: 318 (1915 publ. 1916).

N. Western Australia. 50 WAU. Phan.

Eucalyptus howittiana F.Muell., S. Sci. Rec. 2: 171 (1882).
NE. Queensland. 50 QLD. Phan.

Eucalyptus × hybrida Maiden, J. Proc. Roy. Soc. New S. Wales 47: 85 (1913). *E. moluccana × E. paniculata*.
New South Wales. 50 NSW. Phan.

Eucalyptus hypolaena L.A.S.Johnson & K.D.Hill, Telopea 8: 203 (1999).
SW. Australia. 50 WAU. Phan.

Eucalyptus hypostomatica L.A.S.Johnson & K.D.Hill, Telopea 4: 72 (1990).
New South Wales. 50 NSW. Phan.

Eucalyptus ignorabilis L.A.S.Johnson & K.D.Hill, Telopea 4: 242 (1991).
New South Wales to Victoria. 50 NSW VIC. Phan.

Eucalyptus imitans L.A.S.Johnson & K.D.Hill, Telopea 4: 252 (1991).
New South Wales. 50 NSW. Phan.

Eucalyptus imlayensis Crisp & Brooker, Telopea 2: 41 (1980).
SE. New South Wales (Mt. Imlay). 50 NSW. Phan.

Eucalyptus impensa Brooker & Hopper, Nuytsia 9: 35 (1993).
Western Australia. 50 WAU. Phan.

Eucalyptus incerata Brooker & Hopper, Nuytsia 14: 347 (2002).
Western Australia. 50 WAU. Phan.

Eucalyptus incrassata Labill., Nov. Holl. Pl. 2: 12 (1806).
S. & SE. Australia. 50 NSW SOA VIC WAU. Phan.
Eucalyptus costata F.Muell., Trans. Philos. Soc. Victoria 1: 33 (1855). *Eucalyptus incrassata* var. *costata* (F.Muell.) N.T.Burb., Trans. Roy. Soc. South Australia 71: 150 (1947). *Eucalyptus incrassata* subsp. *costata* (F.Muell.) F.C.Johnstone & Hallam, Proc. Roy. Soc. Queensland 91: 204 (1980).
Eucalyptus costata subsp. *murrayana* L.A.S.Johnson & K.D.Hill, Telopea 9: 269 (2001).

Eucalyptus indurata Brooker & Hopper, Nuytsia 9: 26 (1993).
Western Australia. 50 WAU. Phan.

Eucalyptus infera A.R.Bean, Austrobaileya 3: 291 (1990).
Queensland. 50 QLD. Phan.

Eucalyptus infracorticata L.A.S.Johnson & K.D.Hill, Telopea 9: 310 (2001).
Western Australia. 50 WAU. Phan.

Eucalyptus insularis Brooker, Nuytsia 1: 308 (1974).
S. Western Australia. 50 WAU. Nanophan. or phan.

Eucalyptus interstans L.A.S.Johnson & K.D.Hill, Telopea 4: 47 (1990).
E. Australia. 50 NSW QLD. Phan.

Eucalyptus intertexta R.T.Baker, Proc. Linn. Soc. New South Wales 25: 308 (1900).
Australia. 50 NSW NTA QLD SOA WAU. Phan.
Eucalyptus intertexta var. *diminuta* Blakely, Key Eucalypts: 169 (1934).
Eucalyptus intertexta var. *fruticosa* Blakely & Jacobs in W.F.Blakely, Key Eucalypts: 168 (1934).

Eucalyptus intrasilvatica L.A.S.Johnson & K.D.Hill, Telopea 8: 214 (1999).
W. Australia (near North Bannister). 50 WAU. Phan.

Eucalyptus irritans L.A.S.Johnson & K.D.Hill, Telopea 8: 225 (1999).
NE. Queensland. 50 QLD. Phan.

Eucalyptus jacksonii Maiden, J. Proc. Roy. Soc. New S. Wales 47: 219 (1913).
SW. Australia. 50 WAU. Phan.

Eucalyptus jensenii Maiden, Crit. Revis. Eucalyptus 6: 255 (1922).
N. Western Australia to Northern Territory. 50 NTA WAU. Phan.
Eucalyptus perplexa Maiden & Blakely in J.H.Maiden, Crit. Revis. Eucalyptus 8: 3 (1929).

Eucalyptus jimberlanica L.A.S.Johnson & K.D.Hill, Telopea 4: 208 (1991).
Western Australia. 50 WAU. Phan.

Eucalyptus johnsoniana Brooker & Blaxell, Nuytsia 2: 222 (1978).
W. Western Australia. 50 WAU. Nanophan.

Eucalyptus johnstonii Maiden, Crit. Revis. Eucalyptus 6: 280 (1922).
SE. Tasmania. 50 TAS. Phan.
**Eucalyptus muelleri* J.B.Moore, Pap. & Proc. Roy. Soc. Tasmania 1886: 208 (1887), nom. illeg.

Eucalyptus × joyceae Blakely, J. Proc. Roy. Soc. New S. Wales 62: 201 (1928). *E. haemastoma × E. piperita*.
New South Wales. 50 NSW. Phan.

Eucalyptus jucunda C.A.Gardner, J. Roy. Soc. Western Australia 47: 60 (1964).
E. Western Australia. 50 WAU. Nanophan. or phan.

Eucalyptus jutsonii Maiden, J. Proc. Roy. Soc. New S. Wales 53: 61 (1919).
SC. Western Australia. 50 WAU. Phan.

Eucalyptus kabiana L.A.S.Johnson & K.D.Hill, Telopea 4: 328 (1991).
Queensland. 50 QLD. Phan.

Eucalyptus × kalangadooensis Maiden & Blakely, J. Proc. Roy. Soc. New S. Wales 59: 165 (1925). *E. ovata × E. viminalis*.
South Australia. 50 SOA. Phan.

Eucalyptus × kalganensis Maiden, Crit. Revis. Eucalyptus 6: 349 (1922). *E. marginata × E. preissiana*.
Western Australia. 50 WAU. Phan.

Eucalyptus kartzoffiana L.A.S.Johnson & Blaxell, Contr. New South Wales Natl. Herb. 4: 455 (1973).
SE. New South Wales. 50 NSW. Phan.

Eucalyptus kenneallyi K.D.Hill & L.A.S.Johnson, Telopea 8: 518 (2000).
Western Australia. 50 WAU. Phan.

Eucalyptus kessellii Maiden & Blakely, J. Proc. Roy. Soc. New S. Wales 59: 187 (1925).
S. Western Australia. 50 WAU. Phan.

subsp. ***eugnosta*** L.A.S.Johnson & K.D.Hill, Telopea 4: 613 (1992).
S. Western Australia. 50 WAU. Phan.

subsp. ***kessellii***
S. Western Australia. 50 WAU. Phan.

Eucalyptus kingsmillii (Maiden) Maiden & Blakely in J.H.Maiden, Crit. Revis. Eucalyptus 8: 43 (1929).
Western Australia to South Australia. 50 SOA WAU. Phan.
**Eucalyptus pyriformis* var. *kingsmillii* Maiden, J. Proc. Roy. Soc. New S. Wales 52: 508 (1919).

subsp. ***alatissima*** Brooker & Hopper, Nuytsia 9: 38 (1993).
Western Australia to South Australia. 50 SOA WAU. Phan.

subsp. ***kingsmillii***
Western Australia to South Australia. 50 SOA WAU. Phan.

Eucalyptus* × *kirtoniana F.Muell., Eucalyptographia 1: t. 9 (1879). *E. robusta* × *E. tereticornis*. *Eucalyptus resinifera* var. *kirtoniana* (F.Muell.) H.Deane & Maiden, Proc. Linn. Soc. New South Wales 26: 128 (1901).
New South Wales. 50 NSW. Phan.
Eucalyptus × *patentinervis* F.Muell. ex R.T.Baker, Proc. Linn. Soc. New South Wales 24: 602 (1900).

Eucalyptus kitsoniana Maiden, Crit. Revis. Eucalyptus 3: 164 (1916).
Victoria. 50 VIC. Phan.
**Eucalyptus kitsonii* Luehm. ex Maiden, Victorian Naturalist 21: 112 (1904), nom. illeg.

Eucalyptus kochii Maiden & Blakely in J.H.Maiden, Crit. Revis. Eucalyptus 8: 41 (1929). *Eucalyptus oleosa* var. *kochii* (Maiden & Blakely) C.A.Gardner, J. Roy. Soc. Western Australia 34: 78 (1950).
W. Western Australia. 50 WAU. Phan.

subsp. ***amaryssia*** D.Nicolle, Austral. Syst. Bot. 18: 546 (2005).
W. Western Australia. 50 WAU. Phan.
Eucalyptus oleosa f. *lucida* C.A.Gardner, J. Dept. Agric. Western Australia, III, 2: 413 (1953), nom. inval.

subsp. ***borealis*** (C.A.Gardner) D.Nicolle, Austral. Syst. Bot. 18: 545 (2005).
W. Western Australia. 50 WAU. Phan.
**Eucalyptus oleosa* var. *borealis* C.A.Gardner, J. Roy. Soc. Western Australia 34: 77 (1950).

subsp. ***kochii***
W. Western Australia. 50 WAU. Phan.

subsp. ***yellowdinensis*** D.Nicolle, Austral. Syst. Bot. 18: 548 (2005).
W. Western Australia. 50 WAU. Phan.

Eucalyptus kondininensis Maiden & Blakely, J. Proc. Roy. Soc. New S. Wales 59: 189 (1925).
SW. Australia. 50 WAU. Phan.

Eucalyptus koolpinensis Brooker & Dunlop, Austral. Forest. Res. 8: 214 (1978).
N. Northern Territory. 50 NTA. Phan.

Eucalyptus kruseana F.Muell., Australas. J. Pharm. 10: 233 (1895).
SC. Western Australia. 50 WAU. Nanophan. or phan.
Eucalyptus morrisonii Maiden, J. Nat. Hist. Sci. Soc. Western Australia 3: 44 (1910).

Eucalyptus kumarlensis Brooker, Nuytsia 6: 333 (1988).
Western Australia. 50 WAU. Phan.

Eucalyptus kybeanensis Maiden & Cambage, J. Proc. Roy. Soc. New S. Wales 48: 417 (1914 publ. 1915).
SE. New South Wales to E. Victoria. 50 NSW VIC. Nanophan. or phan.

Eucalyptus lacrimans L.A.S.Johnson & K.D.Hill, Telopea 4: 264 (1991).
New South Wales. 50 NSW. Phan.

Eucalyptus laeliae Podger & Chippend., J. Roy. Soc. Western Australia 51: 65 (1968).
WSW. Western Australia. 50 WAU. Phan.

Eucalyptus laevis L.A.S.Johnson & K.D.Hill, Telopea 9: 275 (2001).
Western Australia. 50 WAU. Phan.

Eucalyptus laevopinea F.Muell. ex R.T.Baker, Proc. Linn. Soc. New South Wales 23: 414 (1898).
E. Australia. 50 NSW QLD. Phan.
Eucalyptus macrorhyncha var. *brachycorys* Benth., Fl. Austral. 3: 207 (1867).

Eucalyptus lane-poolei Maiden, J. Proc. Roy. Soc. New S. Wales 53: 107 (1919).
WSW. Western Australia. 50 WAU. Phan.

Eucalyptus langleyi L.A.S.Johnson & Blaxell, Telopea 4: 259 (1991).
New South Wales. 50 NSW. Phan.

Eucalyptus lansdowneana F.Muell. & J.E.Br. in J.E.Brown, For. Fl. S. Australia 9: t. 31 (1890).
S. South Australia. 50 SOA. Phan.

Eucalyptus laophila L.A.S.Johnson & Blaxell, Telopea 4: 256 (1991).
New South Wales. 50 NSW. Phan.

Eucalyptus largeana Blakely & Beuzev. in W.F.Blakely, Key Eucalypts: 232 (1934).
NE. New South Wales. 50 NSW. Phan.

Eucalyptus* × *laseronii R.T.Baker, Proc. Linn. Soc. New South Wales 37: 585 (1912 publ. 1913). *E. caliginosa* × *E. stellulata*.
New South Wales. 50 NSW. Phan.

Eucalyptus latens Brooker, Nuytsia 6: 332 (1988).
Western Australia. 50 WAU. Phan.

Eucalyptus lateritica Brooker & Hopper, Nuytsia 5: 346 (1986).
W. Western Australia. 50 WAU. Nanophan. or phan.

Eucalyptus latisinensis K.D.Hill, Telopea 8: 244 (1999).
E. Queensland. 50 QLD. Phan.

Eucalyptus lehmannii (Schauer) Benth., Fl. Austral. 3: 233 (1867).
S. Western Australia. (27) cpp 50 WAU. Phan.
**Symphyomyrtus lehmannii* Schauer in J.G.C. Lehmann, Pl. Preiss. 1: 127 (1844).

Eucalyptus leprophloia Brooker & Hopper, Nuytsia 9: 12 (1993).
Western Australia. 50 WAU. Phan.

Eucalyptus leptocalyx Blakely, Key Eucalypts: 118 (1934).
Western Australia. 50 WAU. Nanophan. or phan.

subsp. ***leptocalyx***
S. Western Australia. 50 WAU. Nanophan. or phan.

subsp. ***petilipes*** L.A.S.Johnson & K.D.Hill, Telopea 9: 287 (2001).
Western Australia. 50 WAU. Phan.

Eucalyptus* × *leptocarpa Blakely, Key Eucalypts: 230 (1934). *E. crebra* × *E. viridis*.
New South Wales. 50 NSW. Phan.

Eucalyptus leptophleba F.Muell., J. Proc. Linn. Soc., Bot. 3: 86 (1859). *Eucalyptus drepanophylla* var. *leptophleba* (F.Muell.) Luehm. ex Burtt Davey in L.H. Bailey, Cycl. Amer. Hort..
N. Queensland. (36) chs 50 QLD. Phan.
Eucalyptus stoneana F.M.Bailey, Queensland Agric. J. 23: 259 (1909).

Eucalyptus leptopoda Benth., Fl. Austral. 3: 238 (1867).
Western Australia. 50 WAU. Phan.

subsp. ***arctata*** L.A.S.Johnson & K.D.Hill, Telopea 4: 621 (1992).
Western Australia. 50 WAU. Phan.

subsp. ***elevata*** L.A.S.Johnson & K.D.Hill, Telopea 4: 620 (1992).
Western Australia. 50 WAU. Phan.

subsp. ***leptopoda***
Western Australia. 50 WAU. Phan.
Eucalyptus angustifolia Turcz., Bull. Cl. Phys.-Math. Acad. Imp. Sci. Saint-Pétersbourg 10: 337 (1852), nom. illeg.

subsp. ***subluta*** L.A.S.Johnson & K.D.Hill, Telopea 4: 621 (1992).
Western Australia. 50 WAU. Phan.

Eucalyptus lesouefii Maiden, Crit. Revis. Eucalyptus 2: 187 (1912).
Western Australia. 50 WAU. Phan.

Eucalyptus leucophloia Brooker, Nuytsia 2: 112 (1976).
Australia. 50 NTA QLD WAU. Phan.

subsp. ***euroa*** L.A.S.Johnson & K.D.Hill, Telopea 8: 520 (2000).
Northern Territory to Queensland. 50 NTA QLD. Phan.

subsp. ***leucophloia***
Australia. 50 NTA QLD WAU. Phan.

Eucalyptus leucoxylon F.Muell., Trans. Philos. Soc. Victoria 1: 33 (1855).
SE. Australia. (25) ken tan 50 NSW SOA VIC. Phan.

subsp. ***bellarinensis*** Rule, Muelleria 11: 133 (1998).
Victoria. 50 VIC. Phan.

subsp. ***connata*** Rule, Muelleria 7: 394 (1991).
Victoria. 50 VIC. Phan.

subsp. ***leucoxylon***
SE. South Australia to Victoria. (25) ken tan 50 SOA VIC. Phan.
Eucalyptus leucoxylon var. *erythrostema* Miq., Ned. Kruidk. Arch. 4: 127 (1856).
Eucalyptus leucoxylon var. *rostellata* Miq., Ned. Kruidk. Arch. 4: 127 (1856).
Eucalyptus leucoxylon var. *rugulosa* Miq., Ned. Kruidk. Arch. 4: 127 (1856).
Eucalyptus leucoxylon var. *angulata* Benth., Fl. Austral. 3: 210 (1867).
Eucalyptus gracilipes Naudin, Descr. Emploi Eucalypt.: 37 (1891).
Eucalyptus leucoxylon var. *rubra* Guilf., Austral. Pl.: 165 (1911).

subsp. ***megalocarpa*** Boland, Austral. Forest. Res. 9: 68 (1979).
SE. South Australia. 50 SOA. Phan.
Eucalyptus leucoxylon var. *macrocarpa* J.E.Br., Forest Fl. S. Austral. 2: t. 7 (1883).

subsp. ***petiolaris*** Boland, Austral. Forest. Res. 9: 70 (1979).
S. South Australia. 50 SOA. Phan.

subsp. ***pruinosa*** (Miq.) Boland, Austral. Forest. Res. 9: 68 (1979).
S. New South Wales to SE. South Australia. 50 NSW SOA VIC. Phan.
**Eucalyptus leucoxylon* var. *pruinosa* Miq., Ned. Kruidk. Arch. 4: 127 (1856).
Eucalyptus leucoxylon var. *piperita* J.E.Br., Forest Fl. S. Austral. 2: t. 9 (1883).

subsp. ***stephaniae*** Rule, Muelleria 7: 391 (1991).
South Australia to Victoria. 50 SOA VIC. Phan.

Eucalyptus ligulata Brooker, Nuytsia 1: 300 (1974).
SW. Australia. 50 WAU. Nanophan. or phan.

subsp. ***ligulata***
SSW. Western Australia. 50 WAU. Nanophan. or phan.

subsp. ***stirlingica*** D.Nicolle, Nuytsia 15: 75 (2002).
SW. Australia. 50 WAU. Nanophan. or phan.

Eucalyptus ligustrina A.Cunn. ex DC., Prodr. 3: 219 (1828).
New South Wales. 50 NSW. Phan.

Eucalyptus limitaris L.A.S.Johnson & K.D.Hill, Telopea 8: 530 (2000).
Western Australia to Northern Territory. 50 NTA WAU. Phan.

Eucalyptus lirata W.Fitzg. ex Maiden, Crit. Revis. Eucalyptus 5: 111 (1921).
N. Western Australia. 50 WAU. Phan.

Eucalyptus litoralis Rule, Muelleria 20: 11 (2004).
Victoria. 50 VIC.

Eucalyptus litorea Brooker & Hopper, Nuytsia 7: 10 (1989).
Western Australia. 50 WAU. Phan.

Eucalyptus livida Brooker & Hopper, Nuytsia 8: 47 (1991).
Western Australia. 50 WAU. Phan.

Eucalyptus lockyeri Blaxell & K.D.Hill, Telopea 4: 231 (1991).
Queensland. 50 QLD. Phan.

subsp. ***exuta*** Brooker & Kleinig, Field Guide Eucalypts 3: 371 (1994).
Queensland. 50 QLD. Phan.

subsp. ***lockyeri***
Queensland. 50 QLD. Phan.

Eucalyptus longicornis (F.Muell.) Maiden, J. Proc. Roy. Soc. New S. Wales 52: 504 (1919).
Western Australia. 50 WAU. Phan.
**Eucalyptus oleosa* var. *longicornis* F.Muell., Fragm. 11: 14 (1878).
Eucalyptus grasbyi Maiden & Blakely in J.H.Maiden, Crit. Revis. Eucalyptus 8: 40 (1929).

Eucalyptus longifolia Link, Enum. Hort. Berol. Alt. 2: 29 (1822).
CE. & SE. New South Wales. (25) ken tan (27) cpp 50 NSW. Phan.
Eucalyptus woollsii F.Muell., Fragm. 2: 50 (1860).

Eucalyptus longirostrata (Blakely) L.A.S.Johnson & K.D.Hill, in Fl. Australia 19: 509 (1988).
SE. Queensland. 50 QLD. Phan.
**Eucalyptus punctata* var. *longirostrata* Blakely, Key Eucalypts: 101 (1934).

Eucalyptus longissima D.Nicolle, Austral. Syst. Bot. 18: 550 (2005).
Western Australia. 50 WAU. Tuber nanophan. or phan.

Eucalyptus loxophleba Benth., Fl. Austral. 3: 252 (1867). *Eucalyptus foecunda* var. *loxophleba* (Benth.) W.Fitzg., J. Proc. Linn. Soc., Bot. 1(11): 39 (1903).
SW. Australia. 50 WAU. Phan.

subsp. ***lissophloia*** L.A.S.Johnson & K.D.Hill, Telopea 4: 569 (1992).
Western Australia. 50 WAU. Phan.

subsp. ***loxophleba***
SW. Australia. 50 WAU. Phan.
Eucalyptus loxophleba var. *fruticosa* Benth., Fl. Austral. 3: 252 (1867).

subsp. ***supralaevis*** L.A.S.Johnson & K.D.Hill, Telopea 4: 568 (1992).
SW. Australia. 50 WAU. Phan.

Eucalyptus lucasii Blakely, Key Eucalypts: 226 (1934).
Western Australia. 50 WAU. Phan.

Eucalyptus lucens Brooker & Dunlop, Austral. Forest. Res. 8: 209 (1978).
S. Northern Territory. 50 NTA. Nanophan. or phan.

Eucalyptus luculenta L.A.S.Johnson & K.D.Hill, Telopea 8: 208 (1999).
SW. Australia. 50 WAU. Phan.

Eucalyptus luehmanniana F.Muell., Fragm. 11: 38 (1878). *Eucalyptus stellulata* var. *luehmanniana* (F.Muell.) F.Muell., Eucalyptographia 6: in obs. (1880).
E. New South Wales. 50 NSW. Nanophan. or phan.

Eucalyptus luteola Brooker & Hopper, Nuytsia 8: 82 (1991).
Western Australia. 50 WAU. Phan.

Eucalyptus macarthurii H.Deane & Maiden, Proc. Linn. Soc. New South Wales 24: 448 (1899).
New South Wales. (23) rwa 50 NSW. Phan.
Eucalyptus diversifolia Woolls, Contr. Fl. Austral.: 235 (1867), nom. illeg.

Eucalyptus mackintyi Kottek, Austral. Syst. Bot. 3: 685 (1990).
Victoria. 50 VIC. Phan.

Eucalyptus macmahonii Rule, Muelleria 10: 13 (1997).
Victoria. 50 VIC. Phan.

Eucalyptus macrandra F.Muell. ex Benth., Fl. Austral. 3: 235 (1867). *Eucalyptus occidentalis* var. *macrantha* (F.Muell. ex Benth.) Maiden, J. Nat. Hist. Sci. Soc. Western Australia 3: 187 (1911)].
SW. Australia. 50 WAU. Nanophan. or phan.

Eucalyptus macrocarpa Hook., Hooker's Icon. Pl. 5: t. 405 (1842).
Western Australia. 50 WAU. Phan.

subsp. ***elachantha*** Brooker & Hopper, Nuytsia 9: 37 (1993).
Western Australia. 50 WAU. Phan.

subsp. ***macrocarpa***
W. Western Australia. 50 WAU. Phan.

Eucalyptus macrorhyncha F.Muell. ex Benth., Fl. Austral. 3: 207 (1867).
SE. Australia. 50 NSW SOA VIC. Phan.

subsp. ***cannonii*** (R.T.Baker) L.A.S.Johnson & Blaxell, Contr. New South Wales Natl. Herb. 4: 379 (1973).
New South Wales. 50 NSW. Phan.
**Eucalyptus cannonii* R.T.Baker, Hardw. Austral.: 200 (1919).

subsp. ***macrorhyncha***
SE. Australia. 50 NSW SOA VIC. Phan.
Eucalyptus acervula Miq., Ned. Kruidk. Arch. 4: 137 (1856), nom. illeg.
Eucalyptus macrorhyncha f. *grandiflora* Maiden, Forest Fl. N.S.W. 3: 121 (1907).
Eucalyptus macrorhyncha var. *minor* Blakely, Key Eucalypts: 184 (1934).

Eucalyptus macta L.A.S.Johnson & K.D.Hill, Telopea 8: 507 (2000).
Queensland. 50 QLD. Phan.

Eucalyptus magnificata L.A.S.Johnson & K.D.Hill, Telopea 4: 72 (1990).
Queensland to New South Wales. 50 NSW QLD. Phan.

Eucalyptus major (Maiden) Blakely, Key Eucalypts: 100 (1934).
SE. Queensland. 50 QLD. Phan.
**Eucalyptus propinqua* var. *major* Maiden, Crit. Revis. Eucalyptus 6: 504 (1923).

Eucalyptus malacoxylon Blakely, Key Eucalypts: 145 (1934).
NE. New South Wales. 50 NSW. Phan.

Eucalyptus mannensis Boomsma, Trans. Roy. Soc. South Australia 88: 115 (1964).
C. Australia. 50 NTA SOA WAU. Phan.

subsp. ***mannensis***
C. Australia. 50 NTA SOA WAU. Phan.

subsp. ***vespertina*** L.A.S.Johnson & K.D.Hill, Telopea 4: 600 (1992).
Western Australia. 50 WAU. Phan.

Eucalyptus mannifera Mudie, Trans. Med. Bot. Soc. 1832–1833: 24 (1834).
New South Wales to Victoria. 50 NSW VIC. Phan.

subsp. ***gullickii*** (R.T.Baker & H.G.Sm.) L.A.S. Johnson, Contr. New South Wales Natl. Herb. 3: 108 (1962).
New South Wales. 50 NSW. Phan.
**Eucalyptus gullickii* R.T.Baker & H.G.Sm., Res. Eucalypts, ed. 2: 128 (1920).

subsp. ***maculosa*** (R.T.Baker) L.A.S.Johnson, Contr. New South Wales Natl. Herb. 3: 107 (1962).
New South Wales to Victoria. 50 NSW VIC. Phan.
**Eucalyptus maculosa* R.T.Baker, Proc. Linn. Soc. New South Wales 24: 598 (1900). *Eucalyptus gunnii* var. *maculosa* (R.T.Baker) Maiden, Proc. Linn. Soc. New South Wales 26: 561 (1902).

subsp. ***mannifera***
New South Wales to Victoria. 50 NSW VIC. Phan.

subsp. ***praecox*** (Maiden) L.A.S.Johnson, Contr. New South Wales Natl. Herb. 3: 107 (1962).
New South Wales. 50 NSW. Phan.
Eucalyptus lactea R.T.Baker, Proc. Linn. Soc. New South Wales 25: 691 (1901).

Eucalyptus praecox Maiden, J. Proc. Roy. Soc. New S. Wales 48: 123 (1914 publ. 1915).

Eucalyptus marginata Donn ex Sm., Trans. Linn. Soc. London 6: 302 (1802).
SW. Australia. 50 WAU (63) haw. Phan.

subsp. ***elegantella*** Brooker & Hopper, Nuytsia 9: 4 (1993).
Western Australia. 50 WAU. Phan.

subsp. ***marginata***
SW. Australia. 50 WAU (63) haw. Phan.
Eucalyptus floribunda Hügel ex Endl. in S.L.Endlicher & al., Enum. Pl.: 42 (1837).
Eucalyptus hypoleuca Schauer in J.G.C.Lehmann, Pl. Preiss. 1: 131 (1844).
Eucalyptus mahoganii F.Muell., Fragm. 2: 41 (1860).

subsp. ***thalassica*** Brooker & Hopper, Nuytsia 9: 5 (1993).
Western Australia. 50 WAU. Phan.

Eucalyptus* × *marsdenii C.C.Hall, Proc. Linn. Soc. New South Wales 39: 485 (1914).
New South Wales. 50 NSW. Phan.

Eucalyptus mckieana Blakely, Proc. Linn. Soc. New South Wales 55: 594 (1930).
NE. New South Wales. 50 NSW. Phan.

Eucalyptus mcquoidii Brooker & Hopper, Nuytsia 14: 336 (2002).
Western Australia. 50 WAU. Phan.

Eucalyptus medialis Brooker & Hopper, Nuytsia 8: 131 (1991).
Western Australia. 50 WAU. Phan.

Eucalyptus mediocris L.A.S.Johnson & K.D.Hill, Telopea 8: 237 (1999).
E. Queensland. 50 QLD. Phan.

Eucalyptus megacarpa F.Muell., Fragm. 2: 70 (1860).
SW. Australia. 50 WAU. Phan.

Eucalyptus megacornuta C.A.Gardner, J. Roy. Soc. Western Australia 27: 184 (1942).
Western Australia (Ravensthorpe Range). 50 WAU. Phan.

Eucalyptus megasepala A.R.Bean, Austrobaileya 7: 308 (2006).
Queensland. 50 QLD. Phan.

Eucalyptus melanoleuca S.T.Blake, Austrobaileya 1: 6 (1977).
SE. Queensland. 50 QLD. Phan.

Eucalyptus melanophitra Brooker & Hopper, Nuytsia 8: 135 (1991).
Western Australia. 50 WAU. Phan.

Eucalyptus melanophloia F.Muell., J. Proc. Linn. Soc., Bot. 3: 93 (1859).
E. Australia. 50 NSW QLD. Phan.

Eucalyptus melanoxylon Maiden, Crit. Revis. Eucalyptus 6: 351 (1922).
S. Western Australia. 50 WAU. Phan.
Eucalyptus stuartiana var. *grossa* Maiden, Crit. Revis. Eucalyptus 3: 69 (1915).

Eucalyptus melliodora A.Cunn. ex Schauer in W.G.Walpers, Repert. Bot. Syst. 2: 924 (1843).
SE. Queensland to Victoria. (25) ken uga (36) chc chs 50 NSW QLD VIC. Phan.
Eucalyptus patentiflora F.Muell. ex Miq., Ned. Kruidk. Arch. 4: 125 (1856).
Eucalyptus caerulescens Naudin, Descr. Emploi Eucalypt.: 47 (1891).
Eucalyptus forsythii Maiden, Crit. Revis. Eucalyptus 6: 115 (1922).
Eucalyptus melliodora var. *brachycarpa* Blakely, Key Eucalypts: 262 (1934).
Eucalyptus melliodora var. *elliptocarpa* Blakely, Key Eucalypts: 262 (1934).

Eucalyptus mensalis L.A.S.Johnson & K.D.Hill, Telopea 4: 346 (1991).
Queensland. 50 QLD. Phan.

Eucalyptus merrickiae Maiden & Blakely, J. Proc. Roy. Soc. New S. Wales 59: 192 (1925).
S. Western Australia. 50 WAU. Phan.

Eucalyptus michaeliana Blakely, Proc. Linn. Soc. New South Wales 63: 67 (1938).
SE. Queensland to E. New South Wales. 50 NSW QLD. Phan.

Eucalyptus micranthera F.Muell. ex Benth., Fl. Austral. 3: 218 (1867).
S. Western Australia. 50 WAU. Nanophan. or phan.

Eucalyptus microcarpa (Maiden) Maiden, Crit. Revis. Eucalyptus 6: 438 (1923).
SE. Australia to SE. Queensland. 50 NSW QLD SOA VIC. Phan.
Eucalyptus woollsiana F.Muell. ex R.T.Baker, Proc. Linn. Soc. New South Wales 25: 684 (1901).
**Eucalyptus hemiphloia* var. *microcarpa* Maiden, Trans. & Proc. Roy. Soc. South Australia 26: 11 (1902).

Eucalyptus microcodon L.A.S.Johnson & K.D.Hill, Telopea 4: 348 (1991).
Queensland to New South Wales. 50 NSW QLD. Phan.

Eucalyptus microcorys F.Muell., Fragm. 2: 50 (1860).
SE. Queensland to NE. New South Wales. (23) rwa (25) ken tan uga (36) chs (38) tai 50 NSW QLD (63) haw. Phan.

Eucalyptus microneura Maiden & Blakely, J. Proc. Roy. Soc. New S. Wales 59: 168 (1925).
Queensland. 50 QLD. Phan.

Eucalyptus microschema Brooker & Hopper, Nuytsia 8: 121 (1991).
Western Australia. 50 WAU. Phan.

Eucalyptus microtheca F.Muell., J. Proc. Linn. Soc., Bot. 3: 87 (1858).
Australia. (25) ken tan (27) cpp 50 NSW NTA QLD SOA WAU. Phan.
Eucalyptus raveretiana var. *jerichoensis* Domin, Biblioth. Bot. 89: 464 (1928).

Eucalyptus mimica Brooker & Hopper, Nuytsia 14: 344 (2002).
Western Australia. 50 WAU. Phan.

subsp. ***continens*** Brooker & Hopper, Nuytsia 14: 345 (2002).
Western Australia. 50 WAU. Phan.

subsp. ***mimica***
Western Australia. 50 WAU. Phan.

Eucalyptus miniata A.Cunn. ex Schauer in W.G.Walpers, Repert. Bot. Syst. 2: 925 (1843).

N. Australia. 50 NTA QLD WAU. Phan.
Eucalyptus aurantiaca F.Muell., J. Proc. Linn. Soc., Bot. 3: 91 (1859).

Eucalyptus minniritchi D.Nicolle, Nuytsia 13: 489 (2001).
Australia. 50 NTA SOA WAU. Phan.

Eucalyptus misella L.A.S.Johnson & K.D.Hill, Telopea 4: 593 (1992).
Western Australia. 50 WAU. Phan.

Eucalyptus × ***missilis*** Brooker & Hopper, Nuytsia 14: 332 (2002). *E. angulosa* × *E. cornuta*.
SW. Australia. 50 WAU. Phan.

Eucalyptus mitchelliana Cambage, J. Proc. Roy. Soc. New S. Wales 52: 457, add. (1918 publ. 1919).
E. Victoria. 50 VIC. Phan.
**Eucalyptus mitchellii* Cambage, J. Proc. Roy. Soc. New S. Wales 52: 457 (1918 publ. 1919), nom. illeg.

Eucalyptus moderata L.A.S.Johnson & K.D.Hill, Telopea 8: 198 (1999).
SW. Australia. 50 WAU. Phan.

Eucalyptus moluccana Wall. ex Roxb., Fl. Ind. ed. 1832, 2: 298 (1832).
E. Australia. 50 NSW QLD. Phan.
Eucalyptus hemiphloia Benth., Fl. Austral. 3: 216 (1867).
Eucalyptus hemiphloia var. *parviflora* Benth., Fl. Austral. 3: 217 (1867).

Eucalyptus molyneuxii Rule, Muelleria 12: 163 (1999).
Victoria. 50 VIC. Phan.

Eucalyptus × ***montana*** (H.Deane & Maiden) Blakely, Key Eucalypts: 207 (1934). *E. moorei* × *E. sclerophylla*.
New South Wales. 50 NSW. Phan.
**Eucalyptus haemastoma* var. *montana* H.Deane & Maiden, Proc. Linn. Soc. New South Wales 26: 125 (1901).

Eucalyptus montivaga A.R.Bean, Austrobaileya 5: 130 (1997).
Queensland. 50 QLD. Phan.

Eucalyptus mooreana Maiden, J. Proc. Roy. Soc. New S. Wales 47: 221 (1913).
N. Western Australia. 50 WAU. Phan.

Eucalyptus moorei Maiden & Cambage, Proc. Linn. Soc. New South Wales 30: 191 (1905).
New South Wales. 50 NSW. Phan.

var. ***latiuscula*** Blakely, Key Eucalypts: 207 (1934).
Eucalyptus latiuscula (Blakely) L.A.S.Johnson & K.D.Hill, Telopea 4: 102 (1990).
SE. New South Wales. 50 NSW. Nanophan. or phan.

var. ***moorei***
New South Wales. 50 NSW. Phan.
Eucalyptus stellulata var. *angustifolia* Benth., Fl. Austral. 3: 201 (1867).

Eucalyptus morrisbyi Brett, Pap. & Proc. Roy. Soc. Tasmania 1938: 129 (1939).
SE. Tasmania. 50 TAS. Phan.

Eucalyptus morrisii R.T.Baker, Proc. Linn. Soc. New South Wales 25: 312 (1900).
WC. New South Wales. 50 NSW. Phan.

Eucalyptus muelleriana Howitt, Trans. Roy. Soc. Victoria, n.s., 2: 89 (1891).
SE. New South Wales to Victoria. (25) ken tan 50 NSW VIC. Phan.
Eucalyptus dextropinea R.T.Baker, Proc. Linn. Soc. New South Wales 23: 417 (1898).

Eucalyptus multicaulis Blakely, J. Proc. Roy. Soc. New S. Wales 61: 172 (1927).
SE. New South Wales. 50 NSW. Nanophan. or phan.

Eucalyptus × ***mundijongensis*** Maiden, J. Proc. Roy. Soc. New S. Wales 47: 223 (1913). *E. gomphocephala* × *E. wandoo*.
Western Australia. 50 WAU. Phan.

Eucalyptus × ***murphyi*** Maiden & Blakely in J.H.Maiden, Crit. Revis. Eucalyptus 7: 465 (1928). *E. conica* × *E. fibrosa* subsp. *nubila*.
New South Wales. 50 NSW. Phan.

Eucalyptus myriadena Brooker, Brunonia 4: 9 (1981).
SW. Australia. 50 WAU. Phan.

subsp. ***myriadena***
SW. Australia. 50 WAU. Phan.

subsp. ***parviflora*** Brooker & Hopper, Nuytsia 9: 61 (1993).
SW. Australia. 50 WAU. Phan.

Eucalyptus nandewarica L.A.S.Johnson & K.D.Hill, Telopea 4: 52 (1990).
New South Wales. 50 NSW. Phan.

Eucalyptus naudiniana F.Muell., Australas. J. Pharm. 1: 239 (1886).
Bismarck Arch. 43 BIS. Phan.

Eucalyptus neglecta Maiden, Victorian Naturalist 21: 114 (1904).
Victoria. 50 VIC. Phan.

Eucalyptus × ***nepeanensis*** R.T.Baker & H.G.Sm., Res. Eucalypts, ed. 2: 167 (1920).
New South Wales. 50 NSW. Phan.

Eucalyptus neutra D.Nicolle, Austral. Syst. Bot. 12: 232 (1999).
SW. Australia. 50 WAU.

Eucalyptus newbeyi D.J.Carr & S.G.M.Carr, Austral. J. Bot. 28: 541 (1980).
SW. Australia. 50 WAU. Phan.

Eucalyptus nicholii Maiden & Blakely in J.H.Maiden, Crit. Revis. Eucalyptus 8: 52 (1929).
NE. New South Wales. 50 NSW. Phan.
Eucalyptus acaciiformis var. *linearis* H.Deane & Maiden, Proc. Linn. Soc. New South Wales 24: 455 (1899).

Eucalyptus nigra F.Muell. ex R.T.Baker, Proc. Linn. Soc. New South Wales 25: 689 (1901).
Queensland to NE. New South Wales. 50 NSW QLD. Phan.
Eucalyptus phaeotricha Blakely & McKie in W.F. Blakely, Key Eucalypts: 182 (1934).

Eucalyptus nigrifunda Brooker & Hopper, Nuytsia 8: 51 (1991).
Western Australia. 50 WAU. Phan.

Eucalyptus nitens (H.Deane & Maiden) Maiden, Crit. Revis. Eucalyptus 2: 272 (1913).
New South Wales to Victoria. 50 NSW VIC. Phan.
**Eucalyptus goniocalyx* var. *nitens* H.Deane & Maiden, Proc. Linn. Soc. New South Wales 24: 462 (1899).

Eucalyptus nitida Hook.f., Fl. Tasman. 1: 137 (1856). *Eucalyptus amygdlina* var. *nitida* (Hook.f.) Benth., Fl. Austral. 3: 203 (1867). *Eucalyptus australiana* var. *nitida* (Hook.f.) Ewart, Fl. Victoria: 833 (1931).
Tasmania. 50 TAS. Phan.
Eucalyptus simmondsii Maiden, Crit. Revis. Eucalyptus 6: 344 (1922).

Eucalyptus nobilis L.A.S.Johnson & K.D.Hill, Telopea 4: 59 (1990).
Queensland to New South Wales. 50 NSW QLD. Phan.

Eucalyptus normantonensis Maiden & Cambage, J. Proc. Roy. Soc. New S. Wales 52: 489 (1918 publ. 1919).
C. Northern Territory to Queensland. 50 NTA QLD. Phan.
Eucalyptus bicolor var. *xanthophylla* Blakely, Key Eucalypts: 232 (1934). *Eucalyptus largiflorens* var. *xanthophylla* (Blakely) Cameron, Victorian Naturalist 63: 42 (1946).

Eucalyptus nortonii (Blakely) L.A.S.Johnson, Contr. New South Wales Natl. Herb. 3: 112 (1962).
New South Wales to Victoria. 50 NSW VIC. Phan.
Eucalyptus goniocalyx var. *pallens* Benth., Fl. Austral. 3: 230 (1867). *Eucalyptus cambagei* var. *pallens* (Benth.) H.Deane & Maiden, Proc. Linn. Soc. New South Wales 25: 17 (1900).
**Eucalyptus cordieri* var. *nortonii* Blakely, Key Eucalypts: 147 (1934).

Eucalyptus notabilis Maiden, J. Proc. Roy. Soc. New S. Wales 54: 169 (1920).
SE. Queensland to New South Wales. 50 NSW QLD. Phan.

Eucalyptus nova-anglica H.Deane & Maiden, Proc. Linn. Soc. New South Wales 24: 616 (1900). *Eucalyptus cinerea* var. *nova-anglica* (H.Deane & Maiden) Maiden, Crit. Revis. Eucalyptus 3: 9 (1914).
SE. Queensland to NE. New South Wales. 50 NSW QLD. Phan.

Eucalyptus nudicaulis A.R.Bean, Austrobaileya 3: 467 (1991).
Queensland. 50 QLD. Phan.

Eucalyptus nutans F.Muell., Fragm. 3: 152 (1863). *Eucalyptus platypus* var. *nutans* (F.Muell.) Benth., Fl. Austral. 3: 235 (1867).
SSW. Western Australia. 50 WAU. Phan.

Eucalyptus obconica Brooker & Kleinig, Field Guide Eucalypts 3: 372 (1994).
Western Australia. 50 WAU. Phan.

Eucalyptus obesa Brooker & Hopper, Nuytsia 9: 21 (1993).
Western Australia. 50 WAU. Phan.

Eucalyptus obliqua L'Hér., Sert. Angl.: 18 (1789).
SE. Queensland to SE. Australia. (25) ken tan 50 NSW QLD SOA TAS VIC (51) nzn nzs. Phan.
Eucalyptus pallens A.Cunn. ex DC., Prodr. 3: 219 (1828).
Eucalyptus procera Dehnh., Cat. Horti Camald., ed. 2: 20 (1832).
Eucalyptus fabrorum Schltdl., Linnaea 20: 657 (1847).
Eucalyptus falcifolia Miq., Ned. Kruidk. Arch. 4: 136 (1856).
Eucalyptus heterophylla Miq., Ned. Kruidk. Arch. 4: 141 (1856).
Eucalyptus nervosa F.Muell. ex Miq., Ned. Kruidk. Arch. 4: 139 (1856), nom. illeg.
Eucalyptus obliqua var. *degressa* Blakely, Key Eucalypts: 195 (1934).
Eucalyptus obliqua var. *megacarpa* Blakely, Key Eucalypts: 194 (1934).

Eucalyptus oblonga A.Cunn. ex DC., Prodr. 3: 217 (1828).
New South Wales. 50 NSW. Phan.
Eucalyptus deformis Blakely, J. Proc. Roy. Soc. New S. Wales 61: 152 (1927).
Eucalyptus oblonga var. *rugulosa* Blakely, Key Eucalypts: 191 (1934).
Eucalyptus sparsiflora Blakely, Key Eucalypts: 190 (1934), orth. var.
Eucalyptus sparsifolia Blakely, Key Eucalypts: 190 (1934).

Eucalyptus obstans L.A.S.Johnson & K.D.Hill, Telopea 4: 258 (1991).
New South Wales. 50 NSW. Phan.

Eucalyptus obtusiflora A.Cunn. ex DC., Prodr. 3: 220 (1828). *Eucalyptus virgata* var. *obtusiflora* (A.Cunn. ex DC.) Maiden, Forest Fl. N.S.W. 3: 85 (1907).
Western Australia, SE. New South Wales. 50 NSW WAU. Nanophan. or phan.

subsp. ***cowcowensis*** L.A.S.Johnson & K.D.Hill, Telopea 9: 305 (2001).
Western Australia. 50 WAU. Phan.

subsp. ***dongarraensis*** (Maiden & Blakely) L.A.S. Johnson & K.D.Hill, Telopea 9: 305 (2001).
W. Western Australia. 50 WAU. Phan.
**Eucalyptus dongarraensis* Maiden & Blakely, J. Proc. Roy. Soc. New S. Wales 59: 184 (1925).

subsp. ***obtusiflora***
SE. New South Wales. 50 NSW. Nanophan. or phan.
Eucalyptus piperita var. *pauciflora* A.Cunn. ex DC., Prodr. 3: 219 (1828).

Eucalyptus occidentalis Endl. in S.L.Endlicher & al., Enum. Pl.: 49 (1837).
SW. & S. Western Australia. 50 WAU. Phan.
Eucalyptus agnata Domin, Repert. Spec. Nov. Regni Veg. 12: 389 (1913).
Eucalyptus occidentalis var. *oranensis* A.Vilm. ex Trab., Bull. Stat. Rech. For. N. Afr. 1(5): 150 (1917).

Eucalyptus ochrophloia F.Muell., Fragm. 11: 86 (1878).
SW. Queensland to NW. New South Wales. 50 NSW QLD. Phan.

Eucalyptus odontocarpa F.Muell., J. Proc. Linn. Soc., Bot. 3: 98 (1859).
NC. Western Australia to WC. Queensland. 50 NTA QLD WAU. Nanophan. or phan.

Eucalyptus odorata Behr, Linnaea 20: 652 (1847).
South Australia to W. Victoria. 50 SOA VIC. Phan.

var. ***angustifolia*** Blakely, Key Eucalypts: 226 (1934).
South Australia. 50 SOA. Phan.

var. ***odorata***
South Australia to W. Victoria. 50 SOA VIC. Phan.
Eucalyptus cajuputea F.Muell. ex Miq., Ned. Kruidk. Arch. 4: 126 (1856).
Eucalyptus fruticetorum F.Muell. ex Miq., Ned. Kruidk. Arch. 4: 131 (1856).

Eucalyptus odorata var. *erythrandra* Miq., Ned. Kruidk. Arch. 4: 129 (1856).
Eucalyptus odorata var. *floribunda* Benth., Fl. Austral. 3: 216 (1867).
Eucalyptus odorata var. *refracta* Blakely, Key Eucalypts: 226 (1934).

Eucalyptus oldfieldii F.Muell., Fragm. 2: 37 (1860).
Western Australia. 50 WAU. Phan.

Eucalyptus oleosa F.Muell. ex Miq., Ned. Kruidk. Arch. 4: 127 (1856).
S. & SE. Australia. 50 NSW SOA VIC WAU. Phan.

subsp. ***ampliata*** L.A.S.Johnson & K.D.Hill, Telopea 8: 179 (1999).
S. South Australia. 50 SOA. Phan.

subsp. ***corvina*** L.A.S.Johnson & K.D.Hill, Telopea 8: 178 (1999).
S. Western Australia. 50 WAU. Phan.

subsp. ***cylindroidea*** L.A.S.Johnson & K.D.Hill, Telopea 8: 178 (1999).
S. Western Australia. 50 WAU. Phan.

subsp. ***oleosa***
S. & SE. Australia. 50 NSW SOA VIC WAU. Phan.
Eucalyptus turbinata Behr & F.Muell. ex Miq., Ned. Kruidk. Arch. 4: 137 (1856).
Eucalyptus oleosa var. *angustifolia* Maiden, Crit. Revis. Eucalyptus 4: 278 (1920).

subsp. ***repleta*** L.A.S.Johnson & K.D.Hill, Telopea 8: 176 (1999).
SC. Western Australia to SW. South Australia. 50 SOA WAU. Phan.
Eucalyptus oleosa var. *obtusa* C.A.Gardner, J. Roy. Soc. Western Australia 34: 77 (1950).

subsp. ***victima*** L.A.S.Johnson & K.D.Hill, Telopea 8: 179 (1999).
SE. South Australia. 50 SOA. Phan.

subsp. ***wylieana*** L.A.S.Johnson & K.D.Hill, Telopea 8: 180 (1999).
SC. Western Australia to S. South Australia. 50 SOA WAU. Phan.

Eucalyptus olida L.A.S.Johnson & K.D.Hill, Telopea 4: 103 (1990).
New South Wales. 50 NSW. Phan.

Eucalyptus oligantha Schauer in W.G.Walpers, Repert. Bot. Syst. 2: 926 (1843).
N. Western Australia to Northern Territory. 50 NTA WAU. Phan.

subsp. ***modica*** L.A.S.Johnson & K.D.Hill, Telopea 8: 526 (2000).
Western Australia. 50 WAU. Phan.

subsp. ***oligantha***
N. Western Australia to Northern Territory. 50 NTA WAU. Phan.
Eucalyptus hillii Maiden, J. Proc. Roy. Soc. New S. Wales 53: 63 (1919).
Eucalyptus hillii var. *alleniana* Blakely & Jacobs, Key Eucalypts: 239 (1934).

Eucalyptus olivina Brooker & Hopper, Nuytsia 9: 45 (1993).
Western Australia. 50 WAU. Phan.

Eucalyptus olsenii L.A.S.Johnson & Blaxell, Telopea 1: 395 (1980).
SE. New South Wales. 50 NSW. Phan.

Eucalyptus ophitica L.A.S.Johnson & K.D.Hill, Telopea 4: 67 (1990).
New South wales. 50 NSW. Phan.

Eucalyptus optima L.A.S.Johnson & K.D.Hill, Telopea 8: 203 (1999).
SW. Australia. 50 WAU. Phan.

Eucalyptus oraria L.A.S.Johnson, Contr. New South Wales Natl. Herb. 3: 103 (1962).
W. Western Australia. 50 WAU. Phan.
Eucalyptus baudiniana D.J.Carr & S.G.M.Carr, Proc. Roy. Soc. Victoria, n.s., 88: 12 (1976).
Eucalyptus tamala D.J.Carr & S.G.M.Carr, Proc. Roy. Soc. Victoria, n.s., 88: 12 (1976).

Eucalyptus orbifolia F.Muell., Fragm. 5: 50 (1865).
SW. Australia. 50 WAU. Nanophan. or phan.
Eucalyptus lata L.A.S.Johnson & K.D.Hill, Telopea 4: 630 (1992).

Eucalyptus ordiana Dunlop & Done, Nuytsia 8: 197 (1992).
Western Australia. 50 WAU. Phan.

Eucalyptus oreades F.Muell. ex R.T.Baker, Proc. Linn. Soc. New South Wales 24: 596 (1900). *Eucalyptus virgata* var. *altior* H.Deane & Maiden, Proc. Linn. Soc. New South Wales 26: 124 (1901). *Eucalyptus altior* (H.Deane & Maiden) Maiden, Crit. Revis. Eucalyptus 6: 272 (1922), nom. illeg.
SE. Queensland to New South Wales. 50 NSW QLD. Phan.
Eucalyptus luehmanniana var. *altior* H.Deane & Maiden, Proc. Linn. Soc. New South Wales 22: 713 (1898).

Eucalyptus oresbia J.T.Hunter & J.J.Bruhl, Telopea 8: 260 (1999).
N. New South Wales. 50 NSW. Phan.

Eucalyptus orgadophila Maiden & Blakely in J.H.Maiden, Crit. Revis. Eucalyptus 7: 462 (1928).
Queensland. 50 QLD. Phan.
Eucalyptus intertexta var. *magna* Blakely, Key Eucalypts: 168 (1934).

Eucalyptus ornata Crisp, Nuytsia 5: 311 (1984 publ. 1985).
SW. Australia. 50 WAU. Phan.

Eucalyptus orophila L.D.Pryor, Austral. Syst. Bot. 8: 68 (1995).
Lesser Sunda Is. (E. Timor). 42 LSI. Phan.

Eucalyptus orthostemon D.Nicolle & Brooker, Nuytsia 15: 409 (2005).
Western Australia. 50 WAU. Phan.

Eucalyptus ovata Labill., Nov. Holl. Pl. 2: 13 (1806). *Eucalyptus gunnii* var. *ovata* (Labill.) H.Deane & Maiden, Proc. Linn. Soc. New South Wales 26: 136 (1901).
SE. Australia. 50 NSW SOA TAS VIC (51) nzn. Phan.
Eucalyptus acervula Hook.f., Fl. Tasman. 1: 135 (1856).
Eucalyptus stuartiana F.Muell. ex Miq., Ned. Kruidk. Arch. 4: 131 (1856). *Eucalyptus gunnii* var. *elata* Hook.f., Bot. Mag. 127: t. 7808 (1901).
Eucalyptus stuartiana var. *longifolia* Benth., Fl. Austral. 3: 244 (1867). *Eucalyptus paludosa* F.Muell. ex R.T.Baker, Proc. Linn. Soc. New South Wales 23: 167 (1898).
Eucalyptus mulleri Naudin, Descr. Emploi Eucalypt.: 45 (1891).

Eucalyptus gunnii var. *acervula* H.Deane & Maiden, Proc. Linn. Soc. New South Wales 26: 136 (1901).
Eucalyptus ovata var. *grandiflora* Maiden, Crit. Revis. Eucalyptus 3: 146 (1916).

Eucalyptus* × *oviformis Maiden & Blakely in J.H. Maiden, Crit. Revis. Eucalyptus 8: 32 (1929). *E. globulus* subsp. *pseudoglobulus* × *E. tereticornis*.
Victoria. 50 VIC. Phan.
Eucalyptus × *paradoxa* Maiden & Blakely in J.H. Maiden, Crit. Revis. Eucalyptus 8: 30 (1929).

Eucalyptus ovularis Maiden & Blakely, J. Proc. Roy. Soc. New S. Wales 59: 194 (1925).
S. Western Australia. 50 WAU. Phan.

Eucalyptus oxymitra Blakely, Trans. & Proc. Roy. Soc. South Australia 60: 155 (1936).
C. Australia. 50 NTA SOA WAU. Phan.

Eucalyptus* × *oxypoma Blakely, Key Eucalypts: 244 (1934). *E. camaldulensis* × *E. largiflorens*.
New South Wales. 50 NSW. Phan.

Eucalyptus pachycalyx Maiden & Blakely in J.H. Maiden, Crit. Revis. Eucalyptus 8: 15 (1929).
E. Australia. 50 NSW QLD. Phan.

subsp. ***banyabba*** K.D.Hill, Telopea 7: 187 (1997).
New South Wales. 50 NSW. Phan.

subsp. ***pachycalyx***
NE. Queensland. 50 QLD. Phan.

subsp. ***waajensis*** L.A.S.Johnson & K.D.Hill, Telopea 4: 325 (1991).
Queensland. 50 QLD. Phan.

Eucalyptus pachyloma Benth., Fl. Austral. 3: 237 (1867).
SW. Australia. 50 WAU. Nanophan. or phan.

Eucalyptus pachyphylla F.Muell., J. Proc. Linn. Soc., Bot. 3: 98 (1859).
N. Australia. 50 NTA QLD WAU. Phan.
Eucalyptus pyriformis var. *minor* Maiden, Crit. Revis. Eucalyptus 2: 235 (1913).

Eucalyptus paedoglauca L.A.S.Johnson & Blaxell, Telopea 4: 243 (1991).
Queensland. 50 QLD. Phan.

Eucalyptus paliformis L.A.S.Johnson & Blaxell, Contr. New South Wales Natl. Herb. 4: 382 (1973).
SE. New South Wales. 50 NSW. Phan.

Eucalyptus pallida L.A.S.Johnson & K.D.Hill, Telopea 7: 389 (1998).
Western Australia. 50 WAU. Phan.

Eucalyptus paludicola Nicolle, J. Adelaide Bot. Gard. 16: 75 (1995).
South Australia. 50 SOA. Phan.

Eucalyptus panda S.T.Blake, Proc. Roy. Soc. Queensland 69: 87 (1958).
SE. Queensland. 50 QLD. Phan.

Eucalyptus paniculata Sm., Trans. Linn. Soc. London 3: 287 (1797).
SE. Queensland to New South Wales. (23) rwa (25) ken tan uga (26) zim (36) chs 50 NSW QLD (63) haw. Phan.

subsp. ***matutina*** L.A.S.Johnson & K.D.Hill, Telopea 4: 77 (1990).
New South Wales. 50 NSW. Phan.

subsp. ***paniculata***
SE. Queensland to E. New South Wales. (23) rwa (25) ken tan uga (26) zim (36) chs 50 NSW QLD (63) haw. Phan.
Eucalyptus paniculata var. *conferta* Benth., Fl. Austral. 3: 212 (1867).
Eucalyptus nanglei F.Muell. ex R.T.Baker, Hardw. Austral.: 277 (1919).

Eucalyptus pantoleuca L.A.S.Johnson & K.D.Hill, Telopea 8: 513 (2000).
Western Australia. 50 WAU. Phan.

Eucalyptus paralimnetica L.A.S.Johnson & K.D.Hill, Telopea 9: 315 (2001).
Western Australia. 50 WAU. Phan.

Eucalyptus parramattensis C.C.Hall, Proc. Linn. Soc. New South Wales 37: 568 (1912 publ. 1913).
New South Wales. 50 NSW. Phan.

subsp. ***decadens*** L.A.S.Johnson & Blaxell, Telopea 4: 225 (1991).
New South Wales. 50 NSW. Phan.

subsp. ***parramattensis***.
New South Wales. 50 NSW. Phan.
Eucalyptus parramattensis var. *sphaerocalyx* Blakely, Key Eucalypts: 133 (1934).

Eucalyptus parvula L.A.S.Johnson & K.D.Hill, Telopea 4: 233 (1991).
SE. New South Wales. 50 NSW. Phan.
**Eucalyptus parvifolia* Cambage, Proc. Linn. Soc. New South Wales 34: 336 (1909), non Newberry (1895), fossil sp.

Eucalyptus patellaris F.Muell., J. Proc. Linn. Soc., Bot. 3: 84 (1859).
N. Australia. 50 NTA QLD WAU. Phan.

Eucalyptus patens Benth., Fl. Austral. 3: 247 (1867).
SW. Australia. 50 WAU. Phan.

Eucalyptus pauciflora Sieber ex Spreng., Syst. Veg. 4(2): 195 (1827).
SE. Queensland to Tasmania. 50 NSW QLD TAS VIC. Phan.

subsp. ***acerina*** Rule, Muelleria 8: 223 (1994).
Victoria. 50 VIC. Phan.

subsp. ***debeuzevillei*** (Maiden) L.A.S.Johnson & Blaxell, Contr. New South Wales Natl. Herb. 4: 379 (1973).
S. New South Wales to NE. Victoria. 50 NSW VIC. Nanophan. or phan.
**Eucalyptus debeuzevillei* Maiden, J. Proc. Roy. Soc. New S. Wales 54: 68 (1920).

subsp. ***hedraia*** Rule, Muelleria 8: 227 (1994).
Victoria. 50 VIC. Nanophan. or phan.

subsp. ***niphophila*** (Maiden & Blakely) L.A.S. Johnson & Blaxell, Contr. New South Wales Natl. Herb. 4: 379 (1973).
S. New South Wales to Victoria. 50 NSW VIC. Nanophan. or phan.
**Eucalyptus niphophila* Maiden & Blakely in J.H. Maiden, Crit. Revis. Eucalyptus 88: 34 (1929).

subsp. ***parvifructa*** Rule, Muelleria 8: 229 (1994).
Victoria. 50 VIC. Nanophan. or phan.

subsp. ***pauciflora***
SE. Queensland to Tasmania. 50 NSW QLD TAS

VIC. Phan.
Eucalyptus coriacea A.Cunn. ex Schauer in W.G.Walpers, Repert. Bot. Syst. 2: 925 (1843).
Eucalyptus phlebophylla F.Muell. ex Miq., Ned. Kruidk. Arch. 4: 140 (1856).
Eucalyptus submultiplinervis Miq., Ned. Kruidk. Arch. 4: 138 (1856).
Eucalyptus coriacea var. *alpina* Benth., Fl. Austral. 3: 201 (1867).
Eucalyptus sylvicultrix F.Muell. ex Benth., Fl. Austral. 3: 201 (1867), nom. inval.
Eucalyptus pauciflora var. *alpina* Ewart, Fl. Victoria: 837 (1931).

Eucalyptus* × *peacockeana Maiden, Crit. Revis. Eucalyptus 6: 113 (1922). *E. crebra* × *E. melliodora*.
New South Wales. 50 NSW. Phan.
Eucalyptus × *tenandrensis* Maiden, Crit. Revis. Eucalyptus 6: 112 (1922).

Eucalyptus pellita F.Muell., Fragm. 4: 159 (1864).
Eucalyptus resinifera var. *pellita* (F.Muell.) F.M. Bailey, Syn. Queensl. Fl.: 179 (1883).
E. Australia. (25) ken tan (36) chc chs 50 NSW QLD. Phan.
Eucalyptus spectabilis F.Muell., Fragm. 5: 45 (1865). *Eucalyptus resinifera* var. *spectabilis* (F.Muell.) F.M.Bailey, Syn. Queensl. Fl.: 179 (1883).

Eucalyptus pendens Brooker, Nuytsia 1: 243 (1972).
W. Western Australia. 50 WAU. Phan.

Eucalyptus peninsularis D.Nicolle, Austral. Syst. Bot. 12: 230 (1999).
South Australia. 50 SOA.

Eucalyptus* × *penrithensis Maiden, J. Proc. Roy. Soc. New S. Wales 47: 227 (1914). *E. eugenioides* × *E. sclerophylla*.
New South Wales. 50 NSW. Phan.

Eucalyptus perangusta Brooker, Nuytsia 6: 330 (1988).
Western Australia. 50 WAU. Phan.

Eucalyptus percostata Brooker & P.J.Lang, J. Adelaide Bot. Gard. 13: 65 (1990).
South Australia. 50 SOA. Phan.

Eucalyptus perriniana F.Muell. ex Rodway, Pap. & Proc. Roy. Soc. Tasmania 1893: 181 (1860).
S. New South Wales to Tasmania. 50 NSW TAS VIC. Phan.

Eucalyptus persistens L.A.S.Johnson & K.D.Hill, Telopea 4: 336 (1991).
Queensland. 50 QLD. Phan.

Eucalyptus petraea D.J.Carr & S.G.M.Carr, Nuytsia 4: 279 (1983).
SC. Western Australia. 50 WAU. Phan.

Eucalyptus petrensis Brooker & Hopper, Nuytsia 9: 16 (1993).
Western Australia. 50 WAU. Phan.

Eucalyptus* × *petrophila Blakely, Key Eucalypts: 199 (1934). *E. blaxlandii* × *E. sclerophylla*.
New South Wales. 50 NSW. Phan.

Eucalyptus phaenophylla Brooker & Hopper, Nuytsia 8: 76 (1991).
Western Australia. 50 WAU. Phan.

subsp. ***interjacens*** Brooker & Hopper, Nuytsia 8: 81 (1991).
Western Australia. 50 WAU. Phan.

subsp. ***phaenophylla***
Western Australia. 50 WAU. Phan.

Eucalyptus phenax Brooker & Slee, Muelleria 9: 77 (1996).
Western Australia to Victoria. 50 SOA VIC WAU. Phan.

subsp. ***compressa*** D.Nicolle, J. Adelaide Bot. Gard. 19: 84 (2000).
South Australia. 50 SOA. Phan.

subsp. ***phenax***
Western Australia to Victoria. 50 SOA VIC WAU. Phan.

Eucalyptus phoenicea F.Muell., J. Proc. Linn. Soc., Bot. 3: 91 (1859).
N. Australia. 50 NTA QLD WAU. Phan.

Eucalyptus phylacis L.A.S.Johnson & K.D.Hill, Telopea 4: 591 (1992).
Western Australia. 50 WAU. Phan.

Eucalyptus pilbarensis Brooker & Edgecombe, Nuytsia 5: 376 (1986).
NW. Western Australia. 50 WAU. Nanophan. or phan.

Eucalyptus pileata Blakely, Key Eucalypts: 120 (1934).
Western Australia. 50 WAU. Phan.

Eucalyptus pilligaensis Maiden, J. Proc. Roy. Soc. New S. Wales 54: 163 (1920).
S. Queensland to New South Wales. 50 NSW QLD. Phan.
Eucalyptus odorata var. *woollsiana* Maiden, Crit. Revis. Eucalyptus 2: 32 (1910).

Eucalyptus pilularis Sm., Trans. Linn. Soc. London 2: 284 (1797).
SE. Queensland to E. New South Wales. (25) ken tan 50 NSW QLD (51) nzn (63) haw. Phan.
Eucalyptus persicifolia Lodd., Bot. Cab. 6: t. 501 (1821).
Eucalyptus incrassata Sieber ex DC., Prodr. 3: 217 (1828), nom. illeg.
Eucalyptus discolor Desf., Tabl. École Bot., ed. 3: 284 (1829), provisional synonym.
Eucalyptus semicorticata F.Muell., J. Proc. Linn. Soc., Bot. 3: 81 (1859), nom. illeg.
Eucalyptus pilularis var. *muelleriana* Maiden, Crit. Revis. Eucalyptus 1: 31 (1903).

Eucalyptus pimpiniana Maiden, Crit. Revis. Eucalyptus 2: 211 (1912).
South Australia. 50 SOA. Phan.
Eucalyptus isingiana Maiden, Crit. Revis. Eucalyptus 6: 353 (1922).

Eucalyptus piperita Sm. in J.White, J. Voy. N.S.W.: 226 (1790).
New South Wales. 50 NSW. Phan.

subsp. ***piperita***
New South Wales. 50 NSW. Phan.
Metrosideros aromatica K.D.Koenig & Sims in R.A.Salisbury, Prodr. Stirp. Chap. Allerton: 351 (1796). *Eucalyptus aromatica* (K.D. Koenig & Sims) Domin, Biblioth. Bot. 89: 461 (1928).
Eucalyptus piperata Stokes, Bot. Mat. Med. 3: 69 (1812).
Eucalyptus piperita var. *brachycorys* Benth., Fl. Austral. 3: 208 (1867).
Eucalyptus piperita var. *eugenioides* Benth., Fl. Austral. 3: 208 (1867).

Eucalyptus piperita var. *laxiflora* Benth., Fl. Austral. 3: 207 (1867).
Eucalyptus bottii Blakely, J. Proc. Roy. Soc. New S. Wales 61: 163 (1927).

subsp. ***urceolaris*** (Maiden & Blakely) L.A.S.Johnson & Blaxell, Contr. New South Wales Natl. Herb. 4: 381 (1973).
SE. New South Wales. 50 NSW. Phan.
**Eucalyptus urceolaris* Maiden & Blakely in J.H.Maiden, Crit. Revis. Eucalyptus 8: 10 (1929).

Eucalyptus placita L.A.S.Johnson & K.D.Hill, Telopea 4: 80 (1990).
New South Wales. 50 NSW. Phan.

Eucalyptus planchoniana F.Muell., Fragm. 11: 43 (1878).
SE. Queensland to NE. New South Wales. 50 NSW QLD. Phan.

Eucalyptus planipes L.A.S.Johnson & K.D.Hill, Telopea 9: 281 (2001).
Western Australia. 50 WAU. Phan.

Eucalyptus platycorys Maiden & Blakely in J.H.Maiden, Crit. Revis. Eucalyptus 8: 42 (1929).
S. Western Australia. 50 WAU. Phan.
Eucalyptus helmsii Maiden & Blakely in J.H.Maiden, Crit. Revis. Eucalyptus 8: 47 (1929).

Eucalyptus platydisca D.Nicolle & Brooker, Nuytsia 16: 89 (2006).
Western Australia. 50 WAU. Phan.

Eucalyptus platyphylla F.Muell., J. Proc. Linn. Soc., Bot. 3: 93 (1859).
N. & E. Queensland. (36) chs 50 QLD. Phan.

Eucalyptus platypus Hook.f., Hooker's Icon. Pl. 9: t. 849 (1852).
Western Australia. 50 WAU. Nanophan. or phan.

subsp. ***congregata*** Brooker & Hopper, Nuytsia 14: 350 (2002).
Western Australia. 50 WAU. Phan.

var. ***heterophylla*** Blakely, Key Eucalypts: 107 (1934).
SW. & S. Western Australia. 50 WAU. Phan.

subsp. ***platypus***
S. Western Australia. 50 WAU. Nanophan. or phan.
Eucalyptus obcordata Turcz., Bull. Cl. Phys.-Math. Acad. Imp. Sci. Saint-Pétersbourg 10: 337 (1852).

Eucalyptus plenissima (C.A.Gardner) Brooker, Austral. Forest. Res. 7: 65 (1976).
W. Western Australia. 50 WAU. Phan.
**Eucalyptus oleosa* var. *plenissima* C.A.Gardner, J. Roy. Soc. Western Australia 34: 79 (1950). *Eucalyptus kochii* subsp. *plenissima* (C.A.Gardner) Brooker, Austral. J. Bot. 36: 129 (1988).

Eucalyptus pleurocorys L.A.S.Johnson & K.D.Hill, Telopea 9: 277 (2001).
Western Australia. 50 WAU. Phan.

Eucalyptus pluricaulis Brooker & Hopper, Nuytsia 8: 155 (1991).
Western Australia. 50 WAU. Phan.

subsp. ***pluricaulis***
Western Australia. 50 WAU. Phan.

subsp. ***porphyrea*** Brooker & Hopper, Nuytsia 8: 161 (1991).
Western Australia. 50 WAU. Phan.

Eucalyptus polita Brooker & Hopper, Nuytsia 9: 51 (1993).
Western Australia. 50 WAU. Phan.

Eucalyptus polyanthemos Schauer in W.G.Walpers, Repert. Bot. Syst. 2: 924 (1843).
New South Wales to Victoria. (25) ken tan (36) chc chs 50 NSW VIC (76) cal. Phan.
Eucalyptus ovalifolia F.Muell. ex R.T.Baker, Proc. Linn. Soc. New South Wales 25: 680 (1901).
Eucalyptus ovalifolia var. *lanceolata* R.T.Baker & H.G.Sm., Res. Eucalypts: 124 (1902).

subsp. ***longior*** Brooker & Slee, Muelleria 9: 82 (1996).
New South Wales to Victoria. 50 NSW VIC. Phan.

subsp. ***marginalis*** Rule, Muelleria 20: 29 (2004).
Victoria. 50 VIC. Phan.

subsp. ***polyanthemos***
New South Wales to Victoria. (25) ken tan (36) chc chs 50 NSW VIC (76) cal. Phan.

subsp. ***vestita*** L.A.S.Johnson & K.D.Hill, Telopea 4: 75 (1990).
New South Wales to Victoria. 50 NSW VIC. Phan.

Eucalyptus polybractea F.Muell. ex R.T.Baker, Proc. Linn. Soc. New South Wales 25: 692 (1901).
New South Wales to Victoria. 50 NSW VIC. Phan.

Eucalyptus populnea F.Muell., J. Proc. Linn. Soc., Bot. 3: 93 (1859).
E. Australia. 50 NSW QLD. Phan.
**Eucalyptus populifolia* Hook. in T.L.Mitchell, J. Exped. Trop. Australia: 204 (1848), nom. illeg.

subsp. ***bimbil*** L.A.S.Johnson & K.D.Hill, Telopea 4: 71 (1990).
E. Australia. 50 NSW QLD. Phan.

subsp. ***populnea***
E. Australia. 50 NSW QLD. Phan.
Eucalyptus populifolia Hook.f., Hooker's Icon. Pl. 9: t. 879 (1852), nom. illeg.

Eucalyptus porosa Miq., Ned. Kruidk. Arch. 4: 131 (1856). *Eucalyptus calcicultrix* var. *porosa* (Miq.) Blakely, Key Eucalypts: 224 (1934), nom. illeg.
SE. Australia. 50 NSW SOA VIC. Phan.
Eucalyptus odortata var. *calcicultrix* Miq., Ned. Kruidk. Arch. 4: 129 (1856).
Eucalyptus calcicultrix (Miq.) Blakely, Key Eucalypts: 224 (1934).

Eucalyptus portuensis K.D.Hill, Telopea 8: 234 (1999).
NE. Queensland. 50 QLD. Phan.

Eucalyptus praetermissa Brooker & Hopper, Nuytsia 8: 136 (1991).
Western Australia. 50 WAU. Phan.

Eucalyptus prava L.A.S.Johnson & K.D.Hill, Telopea 4: 48 (1990).
Queensland to New South Wales. 50 NSW QLD. Phan.

Eucalyptus preissiana Schauer in J.G.C.Lehmann, Pl. Preiss. 1: 131 (1844).
Western Australia. 50 WAU. Nanophan.

subsp. ***lobata*** Brooker & Slee, Nuytsia 10: 12 (1995).
Western Australia. 50 WAU. Nanophan.

subsp. ***preissiana***
S. Western Australia. 50 WAU. Nanophan.

Eucalyptus preissiana var. *glauca* Regel, Gartenflora 6: 148 (1857).

Eucalyptus prolixa D.Nicolle, Nuytsia 13: 312 (2000).
S. Western Australia. 50 WAU. Phan.

Eucalyptus prominens Brooker, Nuytsia 2: 115 (1976).
NW. Western Australia. 50 WAU. Nanophan. or phan.

Eucalyptus prominula L.A.S.Johnson & K.D.Hill, Telopea 4: 248 (1991).
New South Wales. 50 NSW. Phan.

Eucalyptus propinqua H.Deane & Maiden, Proc. Linn. Soc. New South Wales, II, 10: 541 (1896).
SE. Queensland to New South Wales. 50 NSW QLD. Phan.

Eucalyptus protensa L.A.S.Johnson & K.D.Hill, Telopea 4: 219 (1991).
Western Australia. 50 WAU. Phan.

Eucalyptus provecta A.R.Bean, Austrobaileya 5: 681 (2000).
C. Queensland. 50 QLD. Phan.

Eucalyptus proxima D.Nicolle & Brooker, Nuytsia 15: 423 (2005).
Western Australia. 50 WAU. Phan.

Eucalyptus pruiniramis L.A.S.Johnson & K.D.Hill, Telopea 4: 563 (1992).
Western Australia. 50 WAU. Phan.

Eucalyptus pruinosa Schauer in W.G.Walpers, Repert. Bot. Syst. 2: 926 (1843).
N. Australia. 50 NTA QLD WAU. Phan.

subsp. ***pruinosa***
N. Australia. 50 NTA QLD WAU. Phan.
Eucalyptus spodophylla F.Muell., Fragm. 2: 71 (1860).

subsp. ***tenuata*** L.A.S.Johnson & K.D.Hill, Telopea 8: 538 (2000).
Western Australia to Northern Territory. 50 NTA WAU. Phan.

Eucalyptus psammitica L.A.S.Johnson & K.D.Hill, Telopea 4: 85 (1990).
NE. New South Wales. 50 NSW. Phan.

Eucalyptus* × *pseudopiperita Maiden & Blakely in J.H. Maiden, Crit. Revis. Eucalyptus 8: 8 (1929). *E. capitellata* × *E. piperita*.
New South Wales. 50 NSW. Phan.

Eucalyptus pterocarpa C.A.Gardner ex P.J.Lang, in Fl. Australia 19: 510 (1988).
S. Western Australia. 50 WAU. Phan.

Eucalyptus pulchella Desf., Tabl. École Bot., ed. 3: 408 (1829).
C. & SE. Tasmania. 50 TAS (51) nzn. Phan.
Eucalyptus linearis Dehnh., Cat. Horti Camald., ed. 2: 20 (1832).
Eucalyptus amygdalina var. *angustifolia* F.Muell. ex L.H.Bailey, Cycl. Amer. Hort., ed. 3, 1: 1157 (1947).

Eucalyptus pulverulenta Sims, Bot. Mag. 46: t. 2087 (1818).
SE. New South Wales. 50 NSW (76) cal. Phan.
Eucalyptus cordata Lodd., Bot. Cab. 4: t. 328 (1819), nom. illeg.
Eucalyptus pulvigera A.Cunn., Field New South Wales: 350 (1825).
Eucalyptus pulverulenta var. *ovatifolia* Dehnh., Cat. Horti Camald., ed. 2: 6 (1832).

Eucalyptus pumila Cambage, J. Proc. Roy. Soc. New S. Wales 52: 453 (1918 publ. 1919).
CE. New South Wales. 50 NSW. Phan.

Eucalyptus punctata A.Cunn. ex DC., Prodr. 3: 217 (1828).
New South Wales. (25) ken tan uga (36) chc chs 50 NSW. Phan.
Eucalyptus tereticornis var. *brachycorys* Benth., Fl. Austral. 3: 242 (1867).
Eucalyptus punctata var. *didyma* R.T.Baker & H.G. Sm., Res. Eucalypts: 127 (1902).
Eucalyptus punctata var. *major* R.T.Baker & H.G.Sm., Res. Eucalypts: 128 (1902).
Eucalyptus shiressii Maiden & Blakely in J.H.Maiden, Crit. Revis. Eucalyptus 6: 512 (1923).

Eucalyptus purpurata D.Nicolle, Nuytsia 15: 81 (2002).
SW. Australia. 50 WAU. Phan.

Eucalyptus* × *pygmaea Blakely, J. Proc. Roy. Soc. New S. Wales 61: 149 (1927). *E. camfieldii* × *E. haemastoma*.
New South Wales. 50 NSW. Phan.

Eucalyptus pyrenea Rule, Muelleria 20: 14 (2004).
Victoria. 50 VIC.

Eucalyptus pyriformis Turcz., Bull. Soc. Imp. Naturalistes Moscou 22(2): 22 (1849).
W. Western Australia. 50 WAU. Phan.
Eucalyptus pruinosa Turcz., Bull. Soc. Imp. Naturalistes Moscou 22(2): 23 (1849), nom. illeg.
Eucalyptus macrocalyx Turcz., Bull. Cl. Phys.-Math. Acad. Imp. Sci. Saint-Pétersbourg 10: 339 (1852), nom. illeg.
Eucalyptus erythrocalyx F.Muell., Fragm. 2: 32 (1860).
Eucalyptus pyriformis var. *elongata* Maiden, Crit. Revis. Eucalyptus 2: 235 (1913).

Eucalyptus pyrocarpa L.A.S.Johnson & Blaxell, Contr. New South Wales Natl. Herb. 4: 454 (1973).
NE. New South Wales. 50 NSW. Phan.
**Eucalyptus pilularis* var. *pyriformis* Maiden, J. Proc. Roy. Soc. New S. Wales 47: 94 (1913).

Eucalyptus quadrangulata H.Deane & Maiden, Proc. Linn. Soc. New South Wales 24: 451 (1899).
SE. Queensland to E. New South Wales. 50 NSW QLD (85) agw. Phan.

Eucalyptus quadrans Brooker & Hopper, Nuytsia 9: 43 (1993).
Western Australia. 50 WAU. Phan.

Eucalyptus quadricostata Brooker, Austrobaileya 2: 151 (1985).
EC. Queensland. 50 QLD. Phan.

Eucalyptus quaerenda (L.A.S.Johnson & K.D.Hill) Byrne, Nuytsia 15: 321 (2004).
Western Australia. 50 WAU. Nanophan. or phan.
**Eucalyptus angustissima* subsp. *quaerenda* L.A.S. Johnson & K.D.Hill, Telopea 4: 598 (1992).

Eucalyptus quinniorum J.T.Hunter & J.J.Bruhl, Telopea 8: 257 (1999).
N. New South Wales. 50 NSW. Phan.

Eucalyptus racemosa Cav., Icon. 4: 24 (1797).
SE. Queensland to E. New South Wales. 50 NSW QLD. Phan.

subsp. ***racemosa***
SE. Queensland to E. New South Wales. 50 NSW QLD. Phan.
Eucalyptus micrantha A.Cunn. ex DC., Prodr. 3: 217 (1828). *Eucalyptus haemastoma* var. *micrantha* (A.Cunn. ex DC.) Benth., Fl. Austral. 3: 212 (1867).
Eucalyptus signata F.Muell., J. Proc. Linn. Soc., Bot. 3: 85 (1859). *Eucalyptus micrantha* var. *signata* (F.Muell.) Blakely, Key Eucalypts: 219 (1934). *Eucalyptus racemosa* var. *signata* (F.Muell.) R.D. Johnst. & Marryatt, Commonw. Austral. Forest. Timber Bur. Leafl. 92: 20 (1965).
Eucalyptus haemastoma var. *capitata* Maiden, Crit. Revis. Eucalyptus 1: 319 (1908).
Eucalyptus haemastoma var. *sclerophylla* Blakely, Key Eucalypts: 218 (1934). *Eucalyptus sclerophylla* (Blakely) L.A.S.Johnson & Blaxell, Contr. New South Wales Natl. Herb. 4: 381 (1973).

subsp. ***rossii*** (R.T.Baker & H.G.Sm.) B.E.Pfeil & Henwood, Telopea 10: 721 (2004).
EC. New South Wales. 50 NSW. Phan.
**Eucalyptus rossii* R.T.Baker & H.G.Sm., Res. Eucalypts: 70 (1902).

Eucalyptus radiata A.Cunn. ex DC., Prodr. 3: 218 (1828). *Eucalyptus amygdalina* var. *radiata* (A.Cunn. ex DC.) Benth., Fl. Austral. 3: 203 (1867).
New South Wales to Victoria. 50 NSW VIC. Phan.

subsp. ***radiata***
New South Wales to Victoria. 50 NSW VIC. Phan.
Eucalyptus australiana R.T.Baker & H.G.Sm., J. Proc. Roy. Soc. New S. Wales 49: 514 (1915 publ. 1916). *Eucalyptus radiata* var. *australiana* (R.T.Baker & H.G.Sm.) Blakely, Key Eucalypts: 211 (1934).
Eucalyptus australiana var. *latifolia* R.T.Baker & H.G.Sm., Res. Eucalypts, ed. 2: 174 (1920).
Eucalyptus phellandra R.T.Baker & H.G.Sm., Res. Eucalypts, ed. 2: 280 (1920).
Eucalyptus radiata var. *subexserta* Blakely, Key Eucalypts: 211 (1934).

subsp. ***sejuncta*** L.A.S.Johnson & K.D.Hill, Telopea 4: 98 (1990).
New South Wales. 50 NSW. Phan.
**Eucalyptus radiata* var. *subplatyphylla* Blakely & McKie in W.F.Blakely, Key Eucalypts: 212 (1934).

Eucalyptus ralla L.A.S.Johnson & K.D.Hill, Telopea 4: 250 (1991).
New South Wales. 50 NSW. Phan.

Eucalyptus rameliana F.Muell., Fragm. 10: 84 (1877). *Eucalyptus pyriformis* var. *rameliana* (F.Muell.) Maiden, Crit. Revis. Eucalyptus 2: 232 (1913).
C. Western Australia. 50 WAU. Phan.

Eucalyptus × rariflora F.M.Bailey, Queensland Agric. J., n.s., 1: 62 (1914). *E. crebra × E. populnea*.
Queensland. 50 QLD. Phan.

Eucalyptus raveretiana F.Muell., Fragm. 10: 99 (1877).
E. Queensland. 50 QLD (63) haw. Phan.

Eucalyptus ravida L.A.S.Johnson & K.D.Hill, Telopea 4: 206 (1991).
SC. Western Australia. 50 WAU. Phan.
**Eucalyptus salubris* var. *glauca* Maiden, Crit. Revis. Eucalyptus 4: 158 (1919).

Eucalyptus recta L.A.S.Johnson & K.D.Hill, Telopea 4: 604 (1992).
Western Australia. 50 WAU. Phan.

Eucalyptus recurva Crisp, Telopea 3: 223 (1988).
New South Wales. 50 NSW. Phan.

Eucalyptus redimiculifera L.A.S.Johnson & K.D.Hill, Telopea 9: 316 (2001).
Western Australia. 50 WAU. Phan.

Eucalyptus reducta L.A.S.Johnson & K.D.Hill, Telopea 4: 344 (1991).
Queensland. 50 QLD. Phan.

Eucalyptus redunca Schauer in J.G.C.Lehmann, Pl. Preiss. 1: 127 (1844).
SW. Australia. 50 WAU. Phan.

Eucalyptus regnans F.Muell., Rep. (Annual) Acclim. Soc. Victoria 1870-1871: 20 (1870). *Eucalyptus amygdalina* var. *regnans* (F.Muell.) auct., Gard. Chron., II, 8: 419 (1877).
S. Victoria to Tasmania. (25) ken tan (27) cpp 50 TAS VIC (51) nzn nzs. Phan. Perhaps the tallest tree in the world. The tallest measured specimen officially taken as 114.3 m.
Eucalyptus regnans var. *fastigata* Ewart, Fl. Victoria: 829 (1931).

Eucalyptus relicta Hopper & Ward.-Johnson, Nuytsia 15: 235 (2004).
Western Australia. 50 WAU.

Eucalyptus remota Blakely, Key Eucalypts: 197 (1934).
South Australia (Kangaroo I.). 50 SOA. Nanophan. or phan.

Eucalyptus repullulans Nicolle, Nuytsia 11: 376 (1997).
Western Australia. 50 WAU. Phan.

Eucalyptus resinifera Sm. in J.White, J. Voy. N.S.W.: 231 (1790).
E. Australia. (12) fra por sar spa (13) ita (25) ken tan uga (26) zim 50 NSW QLD (63) haw. Phan.
Melaleuca gummifera Steud., Nomencl. Bot.: 515 (1821).
Eucalyptus hemilampra F.Muell., J. Proc. Linn. Soc., Bot. 3: 85 (1859). *Eucalyptus resinifera* var. *hemilampra* (F.Muell.) Domin, Biblioth. Bot. 89: 468 (1928). *Eucalyptus resinifera* subsp. *hemilampra* (F.Muell.) L.A.S.Johnson & K.D.Hill, Telopea 4: 46 (1990).
Eucalyptus resinifera var. *grandiflora* Benth., Fl. Austral. 3: 246 (1867).

Eucalyptus retinens L.A.S.Johnson & K.D.Hill, Telopea 4: 56 (1990).
New South Wales. 50 NSW. Phan.

Eucalyptus rhodantha Blakely & H.Steedman, Proc. Linn. Soc. New South Wales 63: 68 (1938).
W. Western Australia. 50 WAU. Nanophan. or phan.
Eucalyptus rhodantha var. *petiolaris* Blakely, Austral. Naturalist 10: 261 (1941).

Eucalyptus rhombica A.R.Bean & Brooker, Austrobaileya 4: 187 (1994).
Queensland. 50 QLD. Phan.

Eucalyptus rigens Brooker & Hopper, Nuytsia 7: 8 (1989).
Western Australia. 50 WAU. Phan.

Eucalyptus × rigescens Blakely, Key Eucalypts: 207 (1934). *E. moorei × E. stricta*.
New South Wales. 50 NSW. Phan.

Eucalyptus rigidula Cambage & Blakely in J.H.Maiden, Crit. Revis. Eucalyptus 7: 403 (1928).
Western Australia. 50 WAU. Phan.
Eucalyptus uncinata var. *major* Benth., Fl. Austral. 3: 216 (1867).
**Eucalyptus angusta* Maiden, Crit. Revis. Eucalyptus 6: 265 (1922), nom. illeg.

Eucalyptus risdonii Hook.f., London J. Bot. 6: 477 (1847).
SE. Tasmania. 50 TAS. Nanophan. or phan.
Eucalyptus hypericifolia Dum.Cours., Bot. Cult., ed. 2, 7: 279 (1814).

Eucalyptus × rivularis Blakely, Key Eucalypts: 253 (1934). *E. melanophloia × E. mecrotheca.*
E. Australia. 50 NSW QLD. Phan.

Eucalyptus robertsonii Blakely, J. Proc. Roy. Soc. New S. Wales 61: 167 (1927). *Eucalyptus radiata* subsp. *robertsonii* (Blakely) L.A.S.Johnson & Blaxell, Contr. New South Wales Natl. Herb. 4: 380 (1973).
New South Wales to Victoria. 50 NSW VIC. Phan.

subsp. ***hemisphaerica*** L.A.S.Johnson & K.D.Hill, Telopea 4: 101 (1990).
New South Wales. 50 NSW. Phan.

subsp. ***robertsonii***
New South Wales to Victoria. 50 NSW VIC. Phan.

Eucalyptus × robsonae Blakely & McKie in W.F. Blakely, Key Eucalypts: 236 (1934). *E. albens × E. melliodora.*
New South Wales. 50 NSW. Phan.
Eucalyptus melliodora var. *murrurundi* Blakely, Key Eucalypts: 261 (1934).

Eucalyptus robusta Sm., Spec. Bot. New Holland 4: 39 (1795).
SE. Queensland to E. New South Wales. (12) fra por sar (13) ita (20) mor (23) rwa (25) ken tan uga (36) chc chh chs (38) tai 50 NSW QLD (61) sci (63) haw (78) fla. Phan.
Eucalyptus rostrata Cav., Icon. 4: 23 (1797). *Eucalyptus robusta* var. *rostrata* (Cav.) Pers., Syn. Pl. 2: 32 (1806).
Eucalyptus multiflora Poir. in J.B.A.P.M.de Lamarck, Encycl., Suppl. 2: 594 (1812).
Eucalyptus multiflora var. *bivalva* Blakely, Key Eucalypts: 98 (1934). *Eucalyptus robusta* var. *bivalva* (Blakely) Blakely, Key Eucalypts, ed. 3: 101 (1947).

Eucalyptus rodwayi R.T.Baker & H.G.Sm., Res. Eucalypts Tasman.: 53 (1912).
Tasmania. 50 TAS. Phan.

Eucalyptus rosacea L.A.S.Johnson & K.D.Hill, Telopea 4: 623 (1992).
Western Australia. 50 WAU. Phan.

Eucalyptus roycei S.G.M.Carr, D.J.Carr & A.S.George, Proc. Roy. Soc. Victoria, n.s., 83: 159 (1970).
W. Western Australia. 50 WAU. Nanophan. or phan.

Eucalyptus rubida H.Deane & Maiden, Proc. Linn. Soc. New South Wales 24: 456 (1899). *Eucalyptus gunnii* var. *rubida* (H.Deane & Maiden) Maiden, Proc. Linn. Soc. New South Wales 27: 561 (1902).
SE. Australia. 50 NSW SOA TAS VIC. Phan.

subsp. ***barbigerorum*** L.A.S.Johnson & K.D.Hill, Telopea 4: 240 (1991).
New South Wales. 50 NSW. Phan.

subsp. ***rubida***
SE. Australia. 50 NSW SOA TAS VIC. Phan.
Eucalyptus viminalis var. *microcarpa* Miq., Ned. Kruidk. Arch. 4: 125 (1856).

subsp. ***septemflora*** L.A.S.Johnson & K.D.Hill, Telopea 4: 241 (1991).
New South Wales to Victoria. 50 NSW VIC. Phan.

Eucalyptus rubiginosa Brooker, Austral. Forest. Res. 14: 311 (1984).
SE. Queensland. 50 QLD. Phan.

Eucalyptus rudderi Maiden, Proc. Linn. Soc. New South Wales 29: 779 (1905).
New South Wales. 50 NSW. Phan.

Eucalyptus rudis Endl. in S.L.Endlicher & al., Enum. Pl.: 49 (1837).
SW. Australia. (12) sar (13) ita sic (25) ken tan uga (36) chs 50 WAU (63) haw (7)+. Phan.
Eucalyptus brachypoda Turcz., Bull. Soc. Imp. Naturalistes Moscou 22(2): 21 (1849).

subsp. ***cratyantha*** Brooker & Hopper, Nuytsia 9: 63 (1993).
SW. Australia. 50 WAU. Phan.

subsp. ***rudis***
SW. Australia. (12) sar (13) ita sic (25) ken tan uga (36) chs 50 WAU (63) haw. Phan.

Eucalyptus rugosa R.Br. ex Blakely, Key Eucalypts: 120 (1934).
SE. Western Australia to S. South Australia. 50 SOA WAU. Phan.

Eucalyptus rugulata D.Nicolle, Nuytsia 15: 79 (2002).
SW. Australia. 50 WAU. Phan.

Eucalyptus rummeryi Maiden, Crit. Revis. Eucalyptus 6: 427 (1923).
NE. New South Wales. 50 NSW. Phan.

Eucalyptus rupestris Brooker & Done, Nuytsia 5: 385 (1986).
N. Western Australia. 50 WAU. Phan.

Eucalyptus sabulosa Rule, Muelleria 9: 138 (1996).
Victoria. (20) mor 50 VIC. Phan.

Eucalyptus salicola Brooker, Nuytsia 6: 329 (1988).
Western Australia. 50 WAU. Phan.

Eucalyptus saligna Sm., Trans. Linn. Soc. London 3: 285 (1797).
E. Australia. (20) mor (25) ken tan uga (36) chs (38) tai 50 NSW QLD (51) nzn (63) haw. Phan.
Eucalyptus saligna var. *pallidivalvis* R.T.Baker & H.G. Sm., Res. Eucalypts: 32 (1902).
Eucalyptus saligna var. *protrusa* Blakely & McKie, Key Eucalypts: 96 (1934).

Eucalyptus salmonophloia F.Muell., Fragm. 11: 11 (1878).
Western Australia. (20) mor 50 WAU. Phan.

Eucalyptus salubris F.Muell., Fragm. 11: 54 (1876).
Western Australia. 50 WAU. Phan.

Eucalyptus sarassa Blume, Mus. Bot. 1: 84 (1850).
Maluku. 42 MOL. Phan.

Eucalyptus sargentii Maiden, Crit. Revis. Eucalyptus 7: 58 (1924).
SW. Australia. 50 WAU. Phan.

subsp. ***fallens*** L.A.S.Johnson & K.D.Hill, Telopea 4: 575 (1992).
SW. Australia. 50 WAU. Phan.

subsp. ***onesia*** D.Nicolle, Nuytsia 15: 398 (2005).
SW. Australia. 50 WAU. Phan.

subsp. ***sargentii***
SW. Australia. 50 WAU. Phan.

Eucalyptus saxatilis J.B.Kirkp. & Brooker, Austral. Forest. Res. 7: 209 (1977).
E. Victoria. 50 VIC. Phan.

Eucalyptus saxicola J.T.Hunter, Telopea 9: 403 (2001).
New South Wales. 50 NSW. Phan.

Eucalyptus schlechteri Diels, Bot. Jahrb. Syst. 57: 423 (1922).
New Guinea. 43 NWG. Phan.

Eucalyptus scias L.A.S.Johnson & K.D.Hill, Telopea 4: 42 (1990).
New South Wales. 50 NSW. Phan.

subsp. ***apoda*** L.A.S.Johnson & K.D.Hill, Telopea 4: 44 (1990).
New South Wales. 50 NSW. Phan.

subsp. ***callimastha*** L.A.S.Johnson & K.D.Hill, Telopea 4: 44 (1990).
New South Wales. 50 NSW. Phan.

subsp. ***scias***
New South Wales. 50 NSW. Phan.

Eucalyptus scoparia Maiden, Proc. Linn. Soc. New South Wales 29: 777 (1905).
SE. Queensland. 50 QLD. Phan.

Eucalyptus scopulorum K.D.Hill, Telopea 7: 192 (1997).
New South Wales. 50 NSW. Phan.

Eucalyptus scyphocalyx (Benth.) Maiden & Blakely in J.H.Maiden, Crit. Revis. Eucalyptus 8: 45 (1929).
Western Australia. 50 WAU. Phan.
**Eucalyptus dumosa* var. *scyphocalyx* Benth., Fl. Austral. 3: 230 (1867).

subsp. ***scyphocalyx***
S. Western Australia. 50 WAU. Phan.

subsp. ***triadica*** L.A.S.Johnson & K.D.Hill, Telopea 9: 289 (2001).
Western Australia. 50 WAU. Phan.

Eucalyptus seeana Maiden, Proc. Linn. Soc. New South Wales 29: 469 (1904).
SE. Queensland to NE. New South Wales. 50 NSW QLD. Phan.
Eucalyptus tereticornis var. *linearis* R.T.Baker & H.G.Sm., Res. Eucalypts: 74 (1902).

Eucalyptus selachiana L.A.S.Johnson & K.D.Hill, Telopea 7: 390 (1998).
Western Australia. 50 WAU. Phan.

Eucalyptus semiglobosa (Brooker) L.A.S.Johnson & K.D.Hill, Telopea 4: 608 (1992).
S. Western Australia. 50 WAU. Nanophan. or phan.
**Eucalyptus goniantha* subsp. *semiglobosa* Brooker, Nuytsia 2: 110 (1976).

Eucalyptus semota Macph. & Grayling, Nuytsia 10: 437 (1996).
Western Australia. 50 WAU. Phan.

Eucalyptus sepulcralis F.Muell., Eucalyptographia 8: t. 10 (1882).
SW. Australia. 50 WAU. Nanophan. or phan.

Eucalyptus serpentinicola L.A.S.Johnson & Blaxell, Telopea 4: 262 (1991).
New South Wales. 50 NSW. Phan.

Eucalyptus serraensis Ladiges & Whiffin, Austral. Syst. Bot. 6: 367 (1993).
Victoria. 50 VIC. Phan.

Eucalyptus sessilis (Maiden) Blakely, Key Eucalypts: 276 (1934).
Western Australia to Northern Territory. 50 NTA WAU. Phan.
**Eucalyptus pachyphylla* var. *sessilis* Maiden, Crit. Revis. Eucalyptus 5: 14 (1920).

Eucalyptus sheathiana Maiden, J. Proc. Roy. Soc. New S. Wales 49: 312 (1915 publ. 1916).
Western Australia. 50 WAU. Phan.

Eucalyptus shirleyi Maiden, Crit. Revis. Eucalyptus 6: 435 (1923).
NC. Queensland. 50 QLD. Phan.

Eucalyptus sicilifolia L.A.S.Johnson & K.D.Hill, Telopea 4: 339 (1991).
Queensland. 50 QLD. Phan.

Eucalyptus siderophloia Benth., Fl. Austral. 3: 220 (1867).
E. Australia. (25) ken uga 50 NSW QLD. Phan.
Eucalyptus decepta Blakely, Key Eucalypts: 249 (1934).

Eucalyptus sideroxylon A.Cunn. ex Woolls, Proc. Linn. Soc. New South Wales, II, 1: 859 (1886).
SE. Queensland to N. Victoria. (25) ken tan uga (27) ofs 50 NSW QLD VIC (7)+. Phan.
Eucalyptus leucoxylon var. *minor* Benth., Fl. Austral. 3: 210 (1867). *Eucalyptus sideroxylon* var. *minor* (Benth.) Maiden, Crit. Revis. Eucalyptus 2: 84 (1910).
Eucalyptus sideroxylon var. *rosea* Rehder in L.H.Bailey, Cycl. Amer. Hort. 2: 552 (1900).

Eucalyptus sieberi L.A.S.Johnson, Contr. New South Wales Natl. Herb. 3: 125 (1962).
SE. New South Wales to NE. Tasmania. 50 NSW TAS VIC. Phan.
Eucalyptus sieberiana F.Muell., Eucalyptographia 2: t. 9 (1879), nom. illeg.
Eucalyptus sieberiana var. *oxleyensis* H.Deane & Maiden, Proc. Linn. Soc. New South Wales 23: 794 (1899).

Eucalyptus silvestris Rule, Muelleria 8: 193 (1994).
South Australia to Victoria. 50 SOA VIC. Phan.

Eucalyptus similis Maiden, J. Proc. Roy. Soc. New S. Wales 47: 90 (1913).
Queensland. 50 QLD. Phan.

Eucalyptus singularis L.A.S.Johnson & D.F.Blaxell, Telopea 9: 272 (2001).
Western Australia. 50 WAU. Phan.

Eucalyptus smithii F.Muell. ex R.T.Baker, Proc. Linn. Soc. New South Wales 24: 292 (1899).
SE. New South Wales to E. Victoria. 50 NSW VIC. Phan.
Eucalyptus viminalis var. *pedicellaris* H.Deane & Maiden, Proc. Linn. Soc. New South Wales 26: 141 (1901).

Eucalyptus socialis F.Muell. ex Miq., Ned. Kruidk. Arch. 4: 132 (1856).

Australia. 50 NSW NTA QLD SOA VIC WAU. Phan.

subsp. ***eucentrica*** (L.A.S.Johnson & K.D.Hill) D.Nicolle, Austral. Syst. Bot. 18: 502 (2005).
Australia. 50 NTA QLD SAO WAU. Tuber phan.
**Eucalyptus eucentrica* L.A.S.Johnson & K.D.Hill, Telopea 4: 326 (1991).

subsp. ***socialis***
Australia. 50 NSW NTA SOA VIC WAU. Phan.

subsp. ***victoriensis*** D.Nicolle, Austral. Syst. Bot. 18: 503 (2005).
Western Australia to South Australia. 50 SOA WAU. Tuber nanophan. or phan.

subsp. ***viridans*** D.Nicolle, Austral. Syst. Bot. 18: 505 (2005).
South Australia. 50 SOA. Tuber nanophan. or phan.

Eucalyptus sparsa Boomsma, J. Adelaide Bot. Gard. 1: 363 (1979).
C. Australia. 50 NTA SOA WAU. Phan.

Eucalyptus sparsicoma Brooker & Hopper, Nuytsia 8: 62 (1991).
Western Australia. 50 WAU. Phan.

Eucalyptus spathulata Hook., Hooker's Icon. Pl. 7: t. 611 (1844). *Eucalyptus occidentalis* var. *spathulata* (Hook.) Maiden, J. Nat. Hist. Sci. Soc. Western Australia 3: 188 (1911).
SW. Australia. 50 WAU. Phan.

subsp. ***grandiflora*** (Benth.) L.A.S.Johnson & Blaxell, Contr. New South Wales Natl. Herb. 4: 453 (1973).
SW. Australia. 50 WAU. Phan.
**Eucalyptus spathulata* var. *grandiflora* Benth., Fl. Austral. 3: 236 (1867).

subsp. ***salina*** D.Nicolle & Brooker, Nuytsia 15: 408 (2005).
SW. Australia. 50 WAU. Phan.

subsp. ***spathulata***
SW. Australia. 50 WAU. Phan.

Eucalyptus spectatrix L.A.S.Johnson & Blaxell, Telopea 4: 255 (1991).
New South Wales. 50 NSW. Phan.

Eucalyptus sphaerocarpa L.A.S.Johnson & Blaxell, Contr. New South Wales Natl. Herb. 4: 284 (1972).
EC. Queensland. 50 QLD. Phan.

Eucalyptus splendens Rule, Muelleria 9: 140 (1996).
SE. Australia. 50 SOA VIC. Phan.

subsp. ***arcana*** Nicolle & Brooker, J. Adelaide Bot. Gard. 18: 103 (1998).
South Australia. 50 SOA. Phan.

subsp. ***splendens***.
Victoria. 50 VIC. Phan.

Eucalyptus sporadica Brooker & Hopper, Nuytsia 14: 356 (2002).
Western Australia. 50 WAU. Phan.

Eucalyptus spreta L.A.S.Johnson & K.D.Hill, Telopea 9: 294 (2001).
Western Australia. 50 WAU. Phan.

Eucalyptus squamosa H.Deane & Maiden, Proc. Linn. Soc. New South Wales 22: 561 (1898). *Eucalyptus tereticornis* var. *squamosa* (H.Deane & Maiden) Maiden, Bull. Herb. Boissier, II, 2: 574 (1902).
CE. New South Wales. 50 NSW. Phan.

Eucalyptus staeri (Maiden) Maiden ex Kessell & C.A.Gardner, W. Austral. Forests Dept. Bull., Bot. Notes 34: 110 (1924).
SW. Australia. 50 WAU. Phan.
**Eucalyptus marginata* var. *staeri* Maiden, J. Proc. Roy. Soc. New S. Wales 47: 230 (1914).

Eucalyptus staigeriana F.Muell. ex F.M.Bailey, Syn. Queensl. Fl.: 176 (1883).
N. Queensland. 50 QLD. Phan.
Eucalyptus crebra var. *citrata* F.Muell., Eucalyptographia 5: t. 3 (1880).

Eucalyptus stannicola L.A.S.Johnson & K.D.Hill, Telopea 4: 86 (1990).
New South Wales. 50 NSW. Phan.

Eucalyptus steedmanii C.A.Gardner, J. Roy. Soc. Western Australia 19: 87 (1933).
WSW. Western Australia. 50 WAU. Phan.

Eucalyptus* × *stellaris Blakely, Key Eucalypts: 192 (1934). *E. blaxlandii* × *E. moorei*.
New South Wales. 50 NSW. Phan.

Eucalyptus stellulata Sieber ex DC., Prodr. 3: 217 (1828).
New South Wales to Victoria. 50 NSW VIC. Phan.

Eucalyptus stenostome L.A.S.Johnson & Blaxell, Contr. New South Wales Natl. Herb. 4: 285 (1972).
SE. New South Wales. 50 NSW. Phan.

Eucalyptus* × *stoataptera E.M.Benn., Nuytsia 10: 1 (1995). *E. stoatei* × *E. tetraptera*.
Western Australia. 50 WAU. Phan.

Eucalyptus stoatei C.A.Gardner, J. Roy. Soc. Western Australia 22: 126 (1936). *Eucalyptus forrestiana* subsp. *stoatei* (C.A.Gardner) C.J.Rob., Nuytsia 5: 197 (1984 publ. 1985).
S. Western Australia. 50 WAU. Phan.

Eucalyptus* × *stopfordii Maiden, Crit. Revis. Eucalyptus 6: 114 (1922). *E. melliodora* × *E. sideroxylon*.
New South Wales. 50 NSW. Phan.

Eucalyptus stowardii Maiden, J. Proc. Roy. Soc. New S. Wales 51: 457 (1917).
WSW. Western Australia. 50 WAU. Phan.

Eucalyptus striaticalyx W.Fitzg., J. West Austral. Nat. Hist. Soc. 1: 20 (1904).
Western Australia to South Australia. 50 SOA WAU. Phan.

subsp. ***delicata*** Nicolle & P.J.Lang, Nuytsia 11: 371 (1997).
Western Australia. 50 WAU. Phan.

subsp. ***striaticalyx***
Western Australia to South Australia. 50 SOA WAU. Phan.

Eucalyptus stricklandii Maiden, J. Nat. Hist. Sci. Soc. Western Australia 3: 175 (1911).
SC. Western Australia. 50 WAU. Phan.

Eucalyptus stricta Sieber ex Spreng., Syst. Veg. 4(2): 195 (1827). *Eucalyptus virgata* var. *stricta* (Sieber ex Spreng.) Maiden, Forest Fl. N.S.W. 3: 86 (1907).
SE. New South Wales. 50 NSW. Nanophan. or phan.
Eucalyptus ambigua A.Cunn. ex DC., Prodr. 3: 219 (1828).
Eucalyptus stricta var. *subcampanulata* Blakely, Key Eucalypts: 202 (1934).

Eucalyptus strzeleckii Rule, Muelleria 7: 497 (1992).
Victoria. 50 VIC. Phan.

Eucalyptus* × *studleyensis Maiden, Crit. Revis. Eucalyptus 6: 121 (1922). *E. camaldulensis* × *E. ovata*. *Eucalyptus rostrata* var. *studleyensis* (Maiden) Ewart, Fl. Victoria: 822 (1931).
SE. Australia. 50 SOA VIC. Phan.
Eucalyptus × *mcintyrensis* Maiden, Crit. Revis. Eucalyptus 6: 166 (1922).

Eucalyptus sturgissiana L.A.S.Johnson & Blaxell, Contr. New South Wales Natl. Herb. 4: 289 (1972).
SE. New South Wales. 50 NSW. Phan.

Eucalyptus subangusta (Blakely) Brooker & Hopper, Nuytsia 8: 104 (1991).
Western Australia. 50 WAU. Phan.
**Eucalyptus redunca* var. *subangusta* Blakely, Key Eucalypts: 111 (1934).

subsp. ***cerina*** Brooker & Hopper, Nuytsia 8: 109 (1991).
Western Australia. 50 WAU. Phan.

subsp. ***pusilla*** Brooker & Hopper, Nuytsia 8: 108 (1991).
Western Australia. 50 WAU. Phan.

subsp. ***subangusta***
Western Australia. 50 WAU. Phan.

subsp. ***virescens*** Brooker & Hopper, Nuytsia 8: 115 (1991).
Western Australia. 50 WAU. Phan.

Eucalyptus subcaerulea K.D.Hill, Telopea 7: 195 (1997).
New South Wales. 50 NSW. Phan.

Eucalyptus subcrenulata Maiden & Blakely in J.H.Maiden, Crit. Revis. Eucalyptus 7: 59 (1929).
W. & C. Tasmania. 50 TAS. Phan.

Eucalyptus suberea Brooker & Hopper, Nuytsia 5: 343 (1986).
W. Western Australia. 50 WAU. Nanophan.

Eucalyptus sublucida L.A.S.Johnson & K.D.Hill, Telopea 8: 186 (1999).
C. Western Australia to SW. South Australia. 50 SOA WAU. Phan.

Eucalyptus subtilior L.A.S.Johnson & K.D.Hill, Telopea 4: 89 (1990).
Queensland to New South Wales. 50 NSW QLD. Phan.

Eucalyptus subtilis Brooker & Hopper, Nuytsia 8: 118 (1991).
Western Australia. 50 WAU. Phan.

Eucalyptus* × *subviridis Maiden & Blakely in J.H. Maiden, Crit. Revis. Eucalyptus 8: 16 (1929). *E. blakelyi* × *E. cinerea*.
New South Wales. 50 NSW. Phan.

Eucalyptus suffulgens L.A.S.Johnson & K.D.Hill, Telopea 4: 341 (1991).
Queensland. 50 QLD. Phan.

Eucalyptus suggrandis L.A.S.Johnson & K.D.Hill, Telopea 4: 580 (1992).
SW. Australia. 50 WAU. Phan.

subsp. ***promiscua*** D.Nicolle & Brooker, Nuytsia 15: 419 (2005).
SW. Australia. 50 WAU. Phan.

subsp. ***suggrandis***
SW. Australia. 50 WAU. Phan.

Eucalyptus surgens Brooker & Hopper, Nuytsia 9: 53 (1993).
Western Australia. 50 WAU. Phan.

Eucalyptus synandra Crisp, Nuytsia 4: 129 (1982).
Western Australia. 50 WAU. Phan.

Eucalyptus* × *taeniola R.T.Baker & H.G.Sm., Res. Eucalypts Tasman.: 60 (1912). *E. amygdalina* × *E. sieberi*.
Tasmania. 50 TAS. Phan.

Eucalyptus talyuberlup D.J.Carr & S.G.M.Carr, Austral. J. Bot. 28: 543 (1980).
SW. Australia. 50 WAU. Nanophan. or phan.

Eucalyptus tardecidens (L.A.S.Johnson & K.D.Hill) A.R.Bean, Austrobaileya 5: 683 (2000).
NE. Queensland. 50 QLD. Nanophan. or phan.
**Eucalyptus persistens* subsp. *tardecidens* L.A.S. Johnson & K.D.Hill, Telopea 4: 337 (1991).

Eucalyptus taurina A.R.Bean & Brooker, Austrobaileya 4: 189 (1994).
Queensland. 50 QLD. Phan.

Eucalyptus* × *taylorii Maiden, Crit. Revis. Eucalyptus 7: 63 (1924). *E. conica* × *E. crebra*.
New South Wales. 50 NSW. Phan.

Eucalyptus tectifica F.Muell., J. Proc. Linn. Soc., Bot. 3: 92 (1859).
N. Australia. 50 NTA QLD WAU. Phan.
Eucalyptus spenceriana Maiden in A.J.Ewart & O.B.Davies, Fl. N. Territory: 307 (1917).

Eucalyptus tenella L.A.S.Johnson & K.D.Hill, Telopea 4: 249 (1991).
New South Wales. 50 NSW. Phan.

Eucalyptus tenera L.A.S.Johnson & K.D.Hill, Telopea 4: 586 (1992).
Western Australia. 50 WAU. Phan.

Eucalyptus tenuipes (Maiden & Blakely) Blakely & C.T.White, Proc. Roy. Soc. Queensland 42: 84 (1930 publ. 1931).
Queensland. 50 QLD. Phan.
**Eucalyptus acmenoides* var. *tenuipes* Maiden & Blakely in J.H.Maiden, Crit. Revis. Eucalyptus 7: 464 (1928).

Eucalyptus tenuiramis Miq., Ned. Kruidk. Arch. 4: 128 (1856).
E. & S. Tasmania. 50 TAS (51) nzn. Phan.

Eucalyptus tenuis Brooker & Hopper, Nuytsia 9: 48 (1993).
Western Australia. 50 WAU. Phan.

Eucalyptus tephroclada L.A.S.Johnson & K.D.Hill, Telopea 4: 586 (1992).
Western Australia. 50 WAU. Phan.

Eucalyptus tephrodes L.A.S.Johnson & K.D.Hill, Telopea 8: 533 (2000).
Western Australia. 50 WAU. Phan.

Eucalyptus* × *tephrophloia Blakely, Key Eucalypts: 198 (1934). *E. sclerophylla* × *E. stricta*.
New South Wales. 50 NSW. Phan.

Eucalyptus terebra L.A.S.Johnson & K.D.Hill, Telopea 4: 211 (1991).

Western Australia. 50 WAU. Phan.

Eucalyptus tereticornis Sm., Spec. Bot. New Holland 4: 41 (1795).
New Guinea, E. & SE. Australia. (12) por (13) ita (20) mor (23) rwa (25) ken tan uga (36) chc chs 43 NWG 50 NSW QLD VIC (51) nzn (76) cal. Phan.
Leptospermum umbellatum Gaertn., Fruct. Sem. Pl. 1: 174 (1788). *Eucalyptus umbellata* (Gaertn.) Domin, Biblioth. Bot. 89: 467 (1928), nom. illeg.
Eucalyptus populifolia Desf., Tabl. École Bot., ed. 3: 408 (1829).
Eucalyptus subulata A.Cunn. ex Schauer in W.G.Walpers, Repert. Bot. Syst. 2: 924 (1843).
Eucalyptus tereticornis var. *brevirostris* Benth., Fl. Austral. 3: 241 (1867).
Eucalyptus insignis Naudin, Descr. Emploi Eucalypt.: 30 (1891).
Eucalyptus coronata Tausch ex Maiden, Bull. Herb. Boissier, II, 2: 570 (1902), nom. inval.
Eucalyptus tereticornis var. *cineolifera* R.T.Baker & H.G.Sm., Res. Eucalypts, ed. 2: 181 (1920).
Eucalyptus umbellata var. *pruiniflora* Blakely, Key Eucalypts: 130 (1934). *Eucalyptus tereticornis* var. *pruiniflora* (Blakely) Cameron, Victorian Naturalist 63: 43 (1946).

Eucalyptus terrica A.R.Bean, Austrobaileya 3: 468 (1991).
Queensland. 50 QLD. Phan.

Eucalyptus tetragona (R.Br.) F.Muell., Fragm. 4: 51 (1864).
SW. Australia. 50 WAU. Nanophan.
**Eudesmia tetragona* R.Br., Voy. Terra Austral. 2: 599 (1814).
Eucalyptus pleurocarpa Schauer in J.G.C.Lehmann, Pl. Preiss. 1: 132 (1844).

Eucalyptus tetrapleura L.A.S.Johnson, Contr. New South Wales Natl. Herb. 3: 119 (1962).
NE. New South Wales. 50 NSW. Phan.

Eucalyptus tetraptera Turcz., Bull. Soc. Imp. Naturalistes Moscou 22(2): 22 (1849).
S. Western Australia. 50 WAU. Nanophan. or phan.
Eucalyptus acutangula Turcz., Bull. Cl. Phys.-Math. Acad. Imp. Sci. Saint-Pétersbourg 10: 338 (1852).

Eucalyptus tetrodonta F.Muell., J. Proc. Linn. Soc., Bot. 3: 97 (1859).
N. Australia. 50 NTA QLD WAU. Phan.

Eucalyptus thamnoides Brooker & Hopper, Nuytsia 14: 353 (2002).
Western Australia. 50 WAU. Phan.

subsp. ***megista*** Brooker & Hopper, Nuytsia 14: 354 (2002).
Western Australia. 50 WAU. Phan.

subsp. ***thamnoides***
Western Australia. 50 WAU. Phan.

Eucalyptus tholiformis A.R.Bean & Brooker, Austrobaileya 4: 188 (1994).
Queensland. 50 QLD. Phan.

Eucalyptus thozetiana F.Muell. ex R.T.Baker, Proc. Linn. Soc. New South Wales 31: 305 (1906).
S. Northern Territory to Queensland. 50 NTA QLD. Phan.
Eucalyptus gracilis var. *thozetii* F.M.Bailey, Queensl. Fl. 2: 615 (1900).

Eucalyptus tindaliae Blakely in J.H.Maiden, Crit. Revis. Eucalyptus 8: 61 (1929).
NE. New South Wales. 50 NSW. Phan.

Eucalyptus × tinghaensis Blakely & McKie, Proc. Linn. Soc. New South Wales 55: 591 (1930). *E. caliginosa* × *E. mckieana*.
New South Wales. 50 NSW. Phan.

Eucalyptus tintinnans (Blakely & Jacobs) L.A.S.Johnson & K.D.Hill, in Fl. Australia 19: 509 (1988).
N. Northern Territory. 50 NTA. Phan.
**Eucalyptus platyphylla* var. *tintinnans* Blakely & Jacobs in W.F.Blakely, Key Eucalypts: 138 (1934).

Eucalyptus todtiana F.Muell., S. Sci. Rec. 2: 171 (1882).
W. Western Australia. 50 WAU. Phan.

Eucalyptus torquata Luehm., Victorian Naturalist 13: 147 (1897).
SC. Western Australia. (20) mor 50 WAU. Phan.

Eucalyptus tortilis L.A.S.Johnson & K.D.Hill, Telopea 4: 209 (1991).
Western Australia. 50 WAU. Phan.

Eucalyptus trachybasis L.A.S.Johnson & K.D.Hill, Telopea 9: 281 (2001).
Western Australia. 50 WAU. Phan.

Eucalyptus transcontinentalis Maiden, J. Proc. Roy. Soc. New S. Wales 53: 58 (1919).
SW. Australia. 50 WAU. Phan.
**Eucalyptus oleosa* var. *glauca* Maiden, J. Nat. Hist. Sci. Soc. Western Australia 3: 171 (1911).

subsp. ***semivestita*** L.A.S.Johnson & K.D.Hill, Telopea 8: 201 (1999).
SW. Australia. 50 WAU. Phan.

subsp. ***transcontinentalis***
SW. Australia. 50 WAU. Phan.
Eucalyptus uncinata var. *rostrata* Benth., Fl. Austral. 3: 216 (1867).

Eucalyptus tricarpa (L.A.S.Johnson) L.A.S.Johnson & K.D.Hill, Telopea 4: 247 (1991).
SE. New South Wales to Victoria. 50 NSW VIC. Phan.
**Eucalyptus sideroxylon* subsp. *tricarpa* L.A.S. Johnson, Contr. New South Wales Natl. Herb. 3: 122 (1962).

subsp. ***decora*** Rule, Muelleria 20: 27 (2004).
Victoria. 50 VIC.

subsp. ***tricarpa***
SE. New South Wales to Victoria. 50 NSW VIC. Phan.

Eucalyptus triflora (Maiden) Blakely, Key Eucalypts: 201 (1934).
SE. New South Wales. 50 NSW. Phan.
**Eucalyptus virgata* var. *triflora* Maiden, Forest Fl. N.S.W. 3: 87 (1907). *Eucalyptus fraxinoides* var. *triflora* (Maiden) Maiden, Crit. Revis. Eucalyptus 4: 301 (1920).

Eucalyptus trivalvis Blakely, Trans. & Proc. Roy. Soc. South Australia 60: 155 (1936).
C. Australia. 50 NTA SOA WAU. Phan.

Eucalyptus tumida Brooker & Hopper, Nuytsia 8: 85 (1991).
Western Australia. 50 WAU. Phan.

Eucalyptus ultima L.A.S.Johnson & K.D.Hill, Telopea 8: 188 (1999).
W. Western Australia. 50 WAU. Phan.

Eucalyptus umbra F.Muell. ex R.T.Baker, Proc. Linn. Soc. New South Wales 25: 687 (1901).
CE. New South Wales. 50 NSW. Nanophan. or phan.

Eucalyptus umbrawarrensis Maiden, Crit. Revis. Eucalyptus 6: 257 (1922).
N. Northern Territory. 50 NTA. Phan.

Eucalyptus uncinata Turcz., Bull. Soc. Imp. Naturalistes Moscou 22(2): 23 (1849).
S. Western Australia. 50 WAU. Phan.
Eucalyptus dumosa var. *puncticulata* Benth., Fl. Austral. 3: 230 (1867). *Eucalyptus puncticulata* (Benth.) Blakely, Key Eucalypts: 114 (1934).
Eucalyptus uncinata var. *latifolia* Benth., Fl. Austral. 3: 216 (1867).
Eucalyptus uncinata var. *pedicellata* Hochr., Candollea 2: 463 (1925).

Eucalyptus* × *unialata R.T.Baker & H.G.Sm., Res. Eucalypts Tasman.: 38 (1912). *E. globulus* × *E. viminalis*.
Tasmania. 50 TAS. Phan.
Eucalyptus viminalis var. *macrocarpa* Rodway, Tasm. Fl.: 57 (1903).

Eucalyptus urna D.Nicolle, Austral. Syst. Bot. 12: 227 (1999).
SW. Australia. 50 WAU.

Eucalyptus urnigera Hook.f., London J. Bot. 6: 477 (1847).
Tasmania. (33) tcs 50 TAS. Phan.
Eucalyptus urnigera var. *elongata* Rodway, Tasm. Fl.: 58 (1903).

Eucalyptus urophylla S.T.Blake, Austrobaileya 1: 7 (1977).
Lesser Sunda Is. 42 LSI. Phan.

Eucalyptus utilis Brooker & Hopper, Nuytsia 14: 349 (2002).
Western Australia. 50 WAU. Phan.

Eucalyptus uvida K.D.Hill, Telopea 8: 231 (1999).
NE. Queensland. 50 QLD. Phan.

Eucalyptus valens L.A.S.Johnson & K.D.Hill, Telopea 9: 299 (2001).
Western Australia. 50 WAU. Phan.

Eucalyptus varia Brooker & Hopper, Nuytsia 8: 162 (1991).
Western Australia. 50 WAU. Phan.

subsp. ***salsuginosa*** Brooker & Hopper, Nuytsia 8: 165 (1991).
Western Australia. 50 WAU. Phan.

subsp. ***varia***
Western Australia. 50 WAU. Phan.

Eucalyptus vegrandis L.A.S.Johnson & K.D.Hill, Telopea 4: 577 (1992).
SW. Australia. 50 WAU. Phan.

subsp. ***recondita*** D.Nicolle & Brooker, Nuytsia 15: 423 (2005).
SW. Australia. 50 WAU. Phan.

subsp. ***vegrandis***
SW. Australia. 50 WAU. Phan.

Eucalyptus vernicosa Hook.f., London J. Bot. 6: 478 (1847).
W. Tasmania. 50 TAS. Nanophan.

Eucalyptus verrucata Ladiges & Whiffin, Austral. Syst. Bot. 8: 123 (1995).
Victoria. 50 VIC. Phan.
**Eucalyptus verrucosa* Ladiges & Whiffin, Austral. Syst. Bot. 6: 367 (1993), nom. illeg.

Eucalyptus versicolor Blume, Mus. Bot. 1: 84 (1850).
Maluku. 42 MOL. Phan.

Eucalyptus vesiculosa Brooker & Hopper, Nuytsia 14: 341 (2002).
Western Australia. 50 WAU. Phan.

Eucalyptus vicina L.A.S.Johnson & K.D.Hill, Telopea 4: 229 (1991).
New South Wales. 50 NSW. Phan.

Eucalyptus victoriana Ladiges & Whiffin, Austral. Syst. Bot. 6: 366 (1993).
Victoria. 50 VIC. Phan.

Eucalyptus victrix L.A.S.Johnson & K.D.Hill, Telopea 5: 765 (1994).
Western Australia to Northern Territory. 50 NTA WAU. Phan.

Eucalyptus viminalis Labill., Nov. Holl. Pl. 2: 12 (1806).
SE. Australia to SE. Queensland. (12) fra por spa (13) ita sic (25) ken tan (33) tcs 50 NSW QLD SOA VIC (51) nzn nzs (76) cal. Phan.

subsp. ***cygnetensis*** Boomsma, J. Adelaide Bot. Gard. 2: 295 (1980).
SE. South Australia to W. Victoria. (25) ken tan 50 SOA VIC. Phan.

subsp. ***hentyensis*** Brooker & Slee, J. Adelaide Bot. Gard. 21: 92 (2007).
Tasmania. 50 TAS. Phan.

subsp. ***pryoriana*** (L.A.S.Johnson) Brooker & Slee, Muelleria 9: 80 (1996).
S. Victoria. 50 VIC. Phan.
Eucalyptus viminalis var. *racemosa* Maiden, Forest Fl. N.S.W. 7: 131 (1920). **Eucalyptus pryoriana* L.A.S. Johnson, Contr. New South Wales Natl. Herb. 3: 115 (1962).

subsp. ***viminalis***
SE. Australia to SE. Queensland. (12) fra por spa (13) ita sic (33) tcs 50 NSW QLD SOA VIC (51) nzn nzs (76) cal. Phan.
Eucalyptus angustifolia Desf. ex Link, Enum. Hort. Berol. Alt. 2: 30 (1822).
Eucalyptus gunnii Miq., Ned. Kruidk. Arch. 4: 126 (1856), nom. illeg.
Eucalyptus huberiana Naudin, Descr. Emploi Eucalypt.: 42 (1891). *Eucalyptus viminalis* var. *huberiana* (Naudin) N.T.Burb., Trans. Roy. Soc. South Australia 71: 147 (1947).
Eucalyptus viminalis var. *rhynchocorys* Maiden, Forest Fl. N.S.W. 7: 131 (1920).

Eucalyptus virens Brooker & A.R.Bean, Brunonia 9: 223 (1987).
S. Queensland. 50 QLD. Phan.

Eucalyptus* × *virgata Sieber ex DC., Prodr. 3: 217 (1828). *E. luehmanniana* × *E. obtusiflora*.
New South Wales. 50 NSW. Phan.

Eucalyptus virginea Hopper & Ward.-Johnson, Nuytsia 15: 229 (2004).
Western Australia. 50 WAU.

Eucalyptus viridis F.Muell. ex R.T.Baker, Proc. Linn. Soc. New South Wales 25: 316 (1900).
SE. Australia to SE. Queensland. 50 NSW QLD SOA VIC. Phan.

subsp. ***viridis***
SE. Australia to SE. Queensland. 50 NSW QLD SOA VIC. Phan.
Eucalyptus acacioides A.Cunn. ex Maiden, Crit. Revis. Eucalyptus 3(21): 45 (1914), nom. illeg.
Eucalyptus viridis var. *ovata* Blakely, Key Eucalypts: 229 (1934).

subsp. ***wimmerensis*** (Rule) Brooker & Slee, Muelleria 9: 81 (1996).
South Australia to Victoria. 50 SOA VIC. Phan.
**Eucalyptus wimmerensis* Rule, Muelleria 7: 193 (1990).

Eucalyptus × vitrea F.Muell. ex R.T.Baker, Proc. Linn. Soc. New South Wales 25: 303 (1900). *E. pauciflora × E. radiata*.
New South Wales. 50 NSW. Phan.
Eucalyptus × vitrea var. *thryptomena* Blakely, Key Eucalypts: 204 (1934).

Eucalyptus vokesensis D.Nicolle & L.A.S.Johnson & K.D.Hill, Telopea 8: 194 (1999).
W. South Australia. 50 SOA. Phan.

Eucalyptus volcanica L.A.S.Johnson & K.D.Hill, Telopea 4: 58 (1990).
New South Wales. 50 NSW. Phan.

Eucalyptus walshii Rule, Muelleria 20: 22 (2004).
Victoria. 50 VIC.

Eucalyptus wandoo Blakely, Key Eucalypts: 112 (1934).
SW. Australia. 50 WAU. Phan.
**Eucalyptus redunca* var. *elata* Benth., Fl. Austral. 3: 253 (1867).

subsp. ***pulverea*** Brooker & Hopper, Nuytsia 8: 40 (1991).
SW. Australia. 50 WAU. Phan.

subsp. ***wandoo***
SW. Australia. 50 WAU. Phan.

Eucalyptus × wardii Blakely, J. Proc. Roy. Soc. New S. Wales 62: 212 (1928). *E. oblonga × E. pilularis*.
New South Wales. 50 NSW. Phan.

Eucalyptus websteriana Maiden, J. Proc. Roy. Soc. New S. Wales 49: 313 (1915 publ. 1916).
Western Australia. 50 WAU. Phan.

subsp. ***norsemanica*** L.A.S.Johnson & K.D.Hill, Telopea 4: 627 (1992).
Western Australia. 50 WAU. Phan.

subsp. ***websteriana***
SC. Western Australia. 50 WAU. Phan.

Eucalyptus × westonii Maiden & Blakely, J. Proc. Roy. Soc. New S. Wales 59: 170 (1925). *E. goniocalyx ×* E. *mannifera* subsp. *maculosa*.
New South Wales. 50 NSW. Phan.

Eucalyptus wetarensis L.D.Pryor, Austral. Syst. Bot. 8: 67 (1995).
Lesser Sunda Is. 42 LSI. Phan.

Eucalyptus whitei Maiden & Blakely, J. Proc. Roy. Soc. New S. Wales 59: 172 (1925).
Queensland. 50 QLD. Phan.

Eucalyptus wilcoxii Boland & Kleinig, Brunonia 6: 242 (1984).
SE. New South Wales. 50 NSW. Phan.

Eucalyptus williamsiana L.A.S.Johnson & K.D.Hill, Telopea 4: 87 (1990).
Queensland to New South Wales. 50 NSW QLD. Phan.

Eucalyptus willisii Ladiges, Humphries & Brooker, Austral. J. Bot. 31: 583 (1983).
SE. South Australia to S. Victoria. 50 SOA VIC. Phan.

subsp. ***falciformis*** M.R.Newnham, Ladiges & Whiffin, Austral. J. Bot. 34: 348 (1986).
SW. Victoria. 50 VIC. Phan.

subsp. ***willisii***
SE. South Australia to S. Victoria. 50 SOA VIC. Phan.

Eucalyptus woodwardii Maiden, J. Nat. Hist. Sci. Soc. Western Australia 3: 42 (1910).
SC. Western Australia. 50 WAU. Phan.

Eucalyptus wubinensis L.A.S.Johnson & K.D.Hill, Telopea 9: 307 (2001).
Western Australia. 50 WAU. Phan.

Eucalyptus wyolensis Boomsma, J. Adelaide Bot. Gard. 11: 59 (1988).
South Australia. 50 SOA. Phan.

Eucalyptus xanthoclada Brooker & A.R.Bean, Brunonia 10: 193 (1987).
Queensland. 50 QLD. Phan.

Eucalyptus xanthonema Turcz., Bull. Soc. Imp. Naturalistes Moscou 20(1): 163 (1847).
Western Australia. 50 WAU. Nanophan. or phan.

subsp. ***apposita*** Brooker & Hopper, Nuytsia 8: 129 (1991).
Western Australia. 50 WAU. Phan.

subsp. ***xanthonema***
S. Western Australia. 50 WAU. Nanophan. or phan.
Eucalyptus redunca var. *angustifolia* Benth., Fl. Austral. 3: 253 (1867).

Eucalyptus xerothermica L.A.S.Johnson & K.D.Hill, Telopea 8: 534 (2000).
Western Australia. 50 WAU. Phan.

Eucalyptus × yagobiei Maiden, Crit. Revis. Eucalyptus 6: 118 (1922). *E. albens × E. microtheca*.
New South Wales. 50 NSW. Phan.

Eucalyptus yalatensis Boomsma, S. Austral. Naturalist 50: 29 (1975).
SE. Western Australia to S. South Australia. 50 SOA WAU. Nanophan. or phan.

Eucalyptus yarraensis Maiden & Cambage in J.H.Maiden, Crit. Revis. Eucalyptus 6: 17 (1922).
Victoria. 50 VIC. Phan.

Eucalyptus yilgarnensis (Maiden) Brooker, Nuytsia 5: 366 (1986).
SW. Australia. 50 WAU. Phan.
**Eucalyptus gracilis* var. *yilgarnensis* Maiden, J. Proc. Roy. Soc. New S. Wales 52: 489 (1919).

Eucalyptus youmanii Blakely & McKie, Proc. Linn. Soc. New South Wales 55: 590 (1930).

SE. Queensland to NE. New South Wales. 50 NSW QLD. Phan.
Eucalyptus youmanii var. *sphaerocarpa* Blakely & McKie in W.F.Blakely, Key Eucalypts: 185 (1934).

Eucalyptus youngiana F.Muell., Fragm. 10: 5 (1876).
Eucalyptus pyriformis subsp. *youngiana* (F.Muell.) Boomsma, Trans. Roy. Soc. South Australia 93: 161 (1969).
Western Australia to South Australia. 50 SOA WAU. Phan.

Eucalyptus yumbarrana Boomsma, J. Adelaide Bot. Gard. 1: 366 (1979).
South Australia. 50 SOA. Phan.

Eucalyptus zopherophloia Brooker & Hopper, Nuytsia 9: 10 (1993).
Western Australia. 50 WAU. Phan.

***Synonyms*:**

Eucalyptus abbreviata Blakely & Jacobs = ***Corymbia abbreviata*** (Blakely & Jacobs) K.D.Hill & L.A.S. Johnson
Eucalyptus abergiana F.Muell. = ***Corymbia abergiana*** (F.Muell.) K.D.Hill & L.A.S.Johnson
Eucalyptus acaciiformis var. *linearis* H.Deane & Maiden = ***Eucalyptus nicholii*** Maiden & Blakely
Eucalyptus acacioides A.Cunn. ex Maiden = ***Eucalyptus viridis*** F.Muell. ex R.T.Baker subsp. ***viridis***
Eucalyptus acervula Sieber ex DC. = ***Eucalyptus eugenioides*** Sieber ex Spreng.
Eucalyptus acervula Miq. = ***Eucalyptus macrorhyncha*** F.Muell. ex Benth. subsp. ***macrorhyncha***
Eucalyptus acervula Hook.f. = ***Eucalyptus ovata*** Labill.
Eucalyptus acmenoides var. *carnea* Maiden = ***Eucalyptus acmenoides*** Schauer
Eucalyptus acmenoides var. *tenuipes* Maiden & Blakely = ***Eucalyptus tenuipes*** (Maiden & Blakely) Blakely & C.T.White
Eucalyptus acuminata Hook. = ***Eucalyptus camaldulensis*** Dehnh. subsp. ***camaldulensis***
Eucalyptus acutangula Turcz. = ***Eucalyptus tetraptera*** Turcz.
Eucalyptus agnata Domin = ***Eucalyptus occidentalis*** Endl.
Eucalyptus alata Sweet = ?
Eucalyptus alba var. *australasica* Blakely & Jacobs = ***Eucalyptus alba*** Reinw. ex Blume
Eucalyptus albens var. *elongata* Blakely = ***Eucalyptus albens*** Benth.
Eucalyptus albicans F.Muell. = ?
Eucalyptus albicaulis Sweet = ?
Eucalyptus × *algeriensis* A.Vilm. ex Trab. = *E. camaldulensis* × *E. rudris*
Eucalyptus alpina R.Br. ex Maiden = ***Eucalyptus coccifera*** Hook.f.
Eucalyptus altior (H.Deane & Maiden) Maiden = ***Eucalyptus oreades*** F.Muell. ex R.T.Baker
Eucalyptus ambigua A.Cunn. ex DC. = ***Eucalyptus stricta*** Sieber ex Spreng.
Eucalyptus amplifolia var. *sessiliflora* Blakely = ***Eucalyptus amplifolia*** subsp. ***sessiliflora*** (Blakely) L.A.S.Johnson & K.D.Hill
Eucalyptus amygdalina var. *alpina* Maiden = ***Eucalyptus amygdalina*** Labill.
Eucalyptus amygdalina var. *angustifolia* F.Muell. ex L.H.Bailey = ***Eucalyptus pulchella*** Desf.
Eucalyptus amygdalina var. *hypericifolia* Benth. = [50 TAS] *E. amygdalina* × *E. risdonii*
Eucalyptus amygdalina var. *latifolia* H.Deane & Maiden = ***Eucalyptus dives*** Schauer
Eucalyptus amygdalina var. *numerosa* Maiden = ***Eucalyptus amygdalina*** Labill.
Eucalyptus amygdalina var. *radiata* (A.Cunn. ex DC.) Benth. = ***Eucalyptus radiata*** A.Cunn. ex DC.
Eucalyptus amygdalina var. *regnans* (F.Muell.) auct. = ***Eucalyptus regnans*** F.Muell.
Eucalyptus amygdlina var. *nitida* (Hook.f.) Benth. = ***Eucalyptus nitida*** Hook.f.
Eucalyptus andreana Naudin = ***Eucalyptus elata*** Dehnh.
Eucalyptus andreana var. *stenophylla* (Blakely) Cameron = ***Eucalyptus elata*** Dehnh.
Eucalyptus androsaemifolia Hoffmanns. = ?
Eucalyptus angulosa var. *ceratocorys* Blakely = ***Eucalyptus ceratocorys*** (Blakely) L.A.S.Johnson & K.D.Hill
Eucalyptus angulosa var. *robusta* C.A.Gardner = ***Eucalyptus* × *erythrandra*** Blakely & H.Steedman
Eucalyptus angusta Maiden = ***Eucalyptus rigidula*** Cambage & Blakely
Eucalyptus angustata Brooker = ***Angophora bakeri*** C.C.Hall
Eucalyptus angustifolia Turcz. = ***Eucalyptus leptopoda*** Benth. subsp. ***leptopoda***
Eucalyptus angustifolia Desf. ex Link = ***Eucalyptus viminalis*** Labill. subsp. ***viminalis***
Eucalyptus angustissima subsp. *quaerenda* L.A.S.Johnson & K.D.Hill = ***Eucalyptus quaerenda*** (L.A.S.Johnson & K.D.Hill) Byrne
Eucalyptus × *antipolitensis* Trab. ex Maiden = ?
Eucalyptus aparrerinja (K.D.Hill & L.A.S.Johnson) Brooker = ***Corymbia aparrerinja*** K.D.Hill & L.A.S.Johnson
Eucalyptus apocynifolia (Salisb.) Brooker = ***Angophora costata*** (Gaertn.) Hochr. ex Britten
Eucalyptus apodophylla var. *brachyphylla* Blakely & Jacobs = *E. alba* × *E. apodophylla*
Eucalyptus aquatica (Blakely) L.A.S.Johnson & K.D.Hill = ***Eucalyptus camphora*** F.Muell. ex R.T.Baker subsp. ***camphora***
Eucalyptus arenaria Blakely = ***Corymbia arenaria*** (Blakely) K.D.Hill & L.A.S.Johnson
Eucalyptus arnhemensis D.J.Carr & S.G.M.Carr = ***Corymbia arnhemensis*** (D.J.Carr & S.G.M.Carr) K.D.Hill & L.A.S.Johnson
Eucalyptus aromatica (K.D.Koenig & Sims) Domin = ***Eucalyptus piperita*** Sm. subsp. ***piperita***
Eucalyptus aspera F.Muell. = ***Corymbia aspera*** (F.Muell.) K.D.Hill & L.A.S.Johnson
Eucalyptus atrovirens Brooker & Kleinig = ***Corymbia dichromophloia*** (F.Muell.) K.D.Hill & L.A.S.Johnson
Eucalyptus aurantiaca F.Muell. = ***Eucalyptus miniata*** A.Cunn. ex Schauer
Eucalyptus aureola Brooker & A.R.Bean = ***Corymbia aureola*** (Brooker & A.R.Bean) K.D.Hill & L.A.S.Johnson
Eucalyptus australiana R.T.Baker & H.G.Sm. = ***Eucalyptus radiata*** A.Cunn. ex DC. subsp. ***radiata***
Eucalyptus australiana var. *latifolia* R.T.Baker & H.G.Sm. = ***Eucalyptus radiata*** A.Cunn. ex DC. subsp. ***radiata***
Eucalyptus australiana var. *nitida* (Hook.f.) Ewart = ***Eucalyptus nitida*** Hook.f.
Eucalyptus australis D.J.Carr & S.G.M.Carr = ***Corymbia eremaea*** (D.J.Carr & S.G.M.Carr) K.D.Hill & L.A.S.Johnson subsp. ***eremaea***
Eucalyptus baudiniana D.J.Carr & S.G.M.Carr = ***Eucalyptus oraria*** L.A.S.Johnson

Eucalyptus baueriana var. *conica* (H.Deane & Maiden) Maiden = ***Eucalyptus conica*** H.Deane & Maiden
Eucalyptus bauerlesii F.Muell. = ?
Eucalyptus baxteri var. *pedicellata* J.M.Black = ***Eucalyptus baxteri*** (Benth.) J.M.Black
Eucalyptus beauchampiana Elwes & A.Henry = ?
Eucalyptus behriana var. *purpurascens* F.Muell. ex Benth. = ***Eucalyptus albopurpurea*** (Boomsma) D.Nicolle
Eucalyptus bella (K.D.Hill & L.A.S.Johnson) Brooker = ***Corymbia bella*** K.D.Hill & L.A.S.Johnson
Eucalyptus benthamii var. *dorrigoensis* Blakely = ***Eucalyptus dorrigoensis*** (Blakely) L.A.S.Johnson & K.D.Hill
Eucalyptus × *biangularis* Simmonds = *E. globulus* × *E. urnigera*
Eucalyptus bicolor var. *parviflora* Benth. = ***Eucalyptus brownii*** Maiden & Cambage
Eucalyptus bicolor var. *xanthophylla* Blakely = ***Eucalyptus normantonensis*** Maiden & Cambage
Eucalyptus bicostata Maiden, Blakely & Simmonds = ***Eucalyptus globulus*** subsp. ***bicostata*** (Maiden, Blakely & Simmonds) J.B.Kirkp.
Eucalyptus blackwelliana D.J.Carr & S.G.M.Carr = ***Corymbia dampieri*** (D.J.Carr & S.G.M.Carr) K.D.Hill & L.A.S.Johnson
Eucalyptus blakei (K.D.Hill & L.A.S.Johnson) Brooker = ***Corymbia blakei*** K.D.Hill & L.A.S.Johnson
Eucalyptus blakelyi var. *irrorata* Blakely = ***Eucalyptus blakelyi*** Maiden
Eucalyptus blakelyi var. *parvifructa* Blakely = ***Eucalyptus blakelyi*** Maiden
Eucalyptus bleeseri Blakely = ***Corymbia bleeseri*** (Blakely) K.D.Hill & L.A.S.Johnson
Eucalyptus bloxsomei Maiden = ***Corymbia bloxsomei*** (Maiden) K.D.Hill & L.A.S.Johnson
Eucalyptus botryoides var. *lynei* Blakely = [50 NSW] *E. botryoides* × *E. robusta*
Eucalyptus botryoides var. *platycarpa* Blakely = ***Eucalyptus botryoides*** Sm.
Eucalyptus bottii Blakely = ***Eucalyptus piperita*** Sm. subsp. ***piperita***
Eucalyptus × *bourlieri* Trab. = ?
Eucalyptus bowmanii F.Muell. ex Benth. = ***Eucalyptus fibrosa*** F.Muell. subsp. ***fibrosa***
Eucalyptus brachycalyx var. *chindoo* Blakely = ***Eucalyptus brachycalyx*** Blakely
Eucalyptus brachycalyx var. *protrusa* (J.M.Black) H.Eichler = ***Eucalyptus brachycalyx*** Blakely
Eucalyptus brachycarpa D.J.Carr & S.G.M.Carr = ***Corymbia brachycarpa*** (D.J.Carr & S.G.M.Carr) K.D.Hill & L.A.S.Johnson
Eucalyptus brachypoda Turcz. = ***Eucalyptus rudis*** Endl.
Eucalyptus bridgesiana var. *amblycorys* (Blakely) Cameron = ***Eucalyptus bridgesiana*** F.Muell. ex R.T.Baker
Eucalyptus bunites Brooker & A.R.Bean = ***Corymbia bunites*** (Brooker & A.R.Bean) K.D.Hill & L.A.S.Johnson
Eucalyptus bynoeana D.J.Carr & S.G.M.Carr = ***Corymbia hamersleyana*** (D.J.Carr & S.G.M.Carr) K.D.Hill & L.A.S.Johnson
Eucalyptus byrnesii D.J.Carr & S.G.M.Carr = ***Corymbia byrnesii*** (D.J.Carr & S.G.M.Carr) K.D.Hill & L.A.S.Johnson
Eucalyptus cadophora (K.D.Hill & L.A.S.Johnson) Brooker = ***Corymbia cadophora*** K.D.Hill & L.A.S.Johnson
Eucalyptus caerulescens Naudin = ***Eucalyptus melliodora*** A.Cunn. ex Schauer
Eucalyptus cajuputea F.Muell. ex Miq. = ***Eucalyptus odorata*** Behr var. ***odorata***
Eucalyptus calcicultrix (Miq.) Blakely = ***Eucalyptus porosa*** Miq.
Eucalyptus calcicultrix var. *obscura* Blakely = [50 SOA] *E. albens* × *E. odorata*
Eucalyptus calcicultrix var. *porosa* (Miq.) Blakely = ***Eucalyptus porosa*** Miq.
Eucalyptus × *californica* Kinney = ?
Eucalyptus calophylla R.Br. ex Lindl. = ***Corymbia calophylla*** (R.Br. ex Lindl.) K.D.Hill & L.A.S.Johnson
Eucalyptus calophylla var. *hawkeyi* Blakely = [50 WAU] *Corymbia calophylla* × *Corymbia ficifolia*
Eucalyptus calophylla var. *maideniana* Hochr. = ***Corymbia calophylla*** (R.Br. ex Lindl.) K.D.Hill & L.A.S.Johnson
Eucalyptus calophylla var. *multiflora* Guilf. = ***Corymbia calophylla*** (R.Br. ex Lindl.) K.D.Hill & L.A.S.Johnson
Eucalyptus calophylla var. *parviflora* Blakely = ***Corymbia calophylla*** (R.Br. ex Lindl.) K.D.Hill & L.A.S.Johnson
Eucalyptus calophylla var. *rosea* Guilf. = ***Corymbia calophylla*** (R.Br. ex Lindl.) K.D.Hill & L.A.S.Johnson
Eucalyptus calophylla var. *rubra* Guilf. = ***Corymbia calophylla*** (R.Br. ex Lindl.) K.D.Hill & L.A.S.Johnson
Eucalyptus calycogona var. *celastroides* (Turcz.) Maiden = ***Eucalyptus celastroides*** Turcz.
Eucalyptus calycogona var. *gracilis* (F.Muell.) Maiden = ***Eucalyptus gracilis*** F.Muell.
Eucalyptus calycogona var. *spaffordii* Blakely = ***Eucalyptus calycogona*** subsp. ***spaffordii*** D.Nicolle
Eucalyptus calyculata Link ex Maiden = ***Eucalyptus amygdalina*** Labill.
Eucalyptus camaldulensis var. *acuminata* (Hook.) Blakely = ***Eucalyptus camaldulensis*** Dehnh. subsp. ***camaldulensis***
Eucalyptus camaldulensis var. *brevirostris* (Miq.) Blakely = ***Eucalyptus camaldulensis*** Dehnh. subsp. ***camaldulensis***
Eucalyptus camaldulensis var. *obtusaa* Blakely = ***Eucalyptus camaldulensis*** Dehnh. subsp. ***camaldulensis***
Eucalyptus camaldulensis var. *pendula* Blakely & Jacobs = ***Eucalyptus camaldulensis*** Dehnh. subsp. ***camaldulensis***
Eucalyptus camaldulensis var. *subcinerea* Blakely = ***Eucalyptus camaldulensis*** Dehnh. subsp. ***camaldulensis***
Eucalyptus cambagei H.Deane & Maiden = ***Eucalyptus goniocalyx*** F.Muell. ex Miq.
Eucalyptus cambagei var. *pallens* (Benth.) H.Deane & Maiden = ***Eucalyptus nortonii*** (Blakely) L.A.S.Johnson
Eucalyptus campanulata R.T.Baker & H.G.Sm. = ***Eucalyptus andrewsii*** subsp. ***campanulata*** (R.T.Baker & H.G.Sm.) L.A.S.Johnson & Blaxell
Eucalyptus candida (K.D.Hill & L.A.S.Johnson) Brooker = ***Corymbia candida*** K.D.Hill & L.A.S.Johnson
Eucalyptus cannonii R.T.Baker = ***Eucalyptus macrorhyncha*** subsp. ***cannonii*** (R.T.Baker) L.A.S.Johnson & Blaxell
Eucalyptus capitellata var. *latifolia* Benth. = ***Eucalyptus baxteri*** (Benth.) J.M.Black

Eucalyptus capricornia D.J.Carr & S.G.M.Carr = ***Corymbia capricornia*** (D.J.Carr & S.G.M.Carr) K.D.Hill & L.A.S.Johnson
Eucalyptus carnea C.A.Gardner = ?
Eucalyptus centralis D.J.Carr & S.G.M.Carr = ***Corymbia opaca*** (D.J.Carr & S.G.M.Carr) K.D.Hill & L.A.S.Johnson
Eucalyptus chartacea (K.D.Hill & L.A.S.Johnson) Brooker = ***Corymbia chartacea*** K.D.Hill & L.A.S.Johnson
Eucalyptus chippendalei D.J.Carr & S.G.M.Carr = ***Corymbia chippendalei*** (D.J.Carr & S.G.M.Carr) K.D.Hill & L.A.S.Johnson
Eucalyptus cinerea var. *multiflora* Maiden = ***Eucalyptus cephalocarpa*** Blakely
Eucalyptus cinerea var. *nova-anglica* (H.Deane & Maiden) Maiden = ***Eucalyptus nova-anglica*** H.Deane & Maiden
Eucalyptus citriodora Hook. = ***Corymbia citriodora*** (Hook.) K.D.Hill & L.A.S.Johnson
Eucalyptus clandestina A.R.Bean = ***Corymbia clandestina*** (A.R.Bean) K.D.Hill & L.A.S.Johnson
Eucalyptus clarksoniana D.J.Carr & S.G.M.Carr = ***Corymbia clarksoniana*** (D.J.Carr & S.G.M.Carr) K.D.Hill & L.A.S.Johnson
Eucalyptus clavigera A.Cunn. ex Schauer = ***Corymbia clavigera*** (A.Cunn. ex Schauer) K.D.Hill & L.A.S.Johnson
Eucalyptus clavigera var. *dallachiana* (Benth.) Maiden = ***Corymbia dallachiana*** (Benth.) K.D.Hill & L.A.S.Johnson
Eucalyptus clavigera var. *diffusa* Blakely & Jacobs = ***Corymbia confertiflora*** (Kippist ex F.Muell.) K.D.Hill & L.A.S.Johnson
Eucalyptus clavigera var. *gilbertensis* Maiden & Blakely = ***Corymbia gilbertensis*** (Maiden & Blakely) K.D.Hill & L.A.S.Johnson
Eucalyptus cliftoniana W.Fitzg. = ***Corymbia cliftoniana*** (W.Fitzg.) K.D.Hill & L.A.S.Johnson
Eucalyptus coccifera var. *parviflora* Benth. = ***Eucalyptus coccifera*** Hook.f.
Eucalyptus coccifera var. *viridifolia* Summerh. = ***Eucalyptus coccifera*** Hook.f.
Eucalyptus codonocarpa Blakely & McKie = ***Eucalyptus approximans*** subsp. ***codonocarpa*** (Blakely & McKie) L.A.S.Johnson & Blaxell
Eucalyptus coerulea R.T.Baker & H.G.Sm. = ***Eucalyptus caleyi*** Maiden subsp. ***caleyi***
Eucalyptus collina W.Fitzg. = ***Corymbia collina*** (W.Fitzg.) K.D.Hill & L.A.S.Johnson
Eucalyptus colossea F.Muell. = ***Eucalyptus diversicolor*** F.Muell.
Eucalyptus concolor Schauer = ***Eucalyptus decipiens*** Endl. subsp. ***decipiens***
Eucalyptus confertiflora Kippist ex F.Muell. = ***Corymbia confertiflora*** (Kippist ex F.Muell.) K.D.Hill & L.A.S.Johnson
Eucalyptus conglobata var. *anceps* Maiden = ***Eucalyptus anceps*** (Maiden) Blakely
Eucalyptus conglobata subsp. *fraseri* Brooker = ***Eucalyptus fraseri*** (Brooker) Brooker
Eucalyptus × *coniophloia* D.J.Carr & S.G.M.Carr = [50 NTA WAU] *Corymbia capricornia* × *Corymbia drysdalensis*
Eucalyptus connata Dum.Cours. = ?
Eucalyptus connerensis D.J.Carr & S.G.M.Carr = ***Corymbia eremaea*** (D.J.Carr & S.G.M.Carr) K.D.Hill & L.A.S.Johnson subsp. ***eremaea***
Eucalyptus conoidea Benth. = ***Eucalyptus erythronema*** Turcz. var. ***erythronema***
Eucalyptus conoidea var. *marginata* Benth. = ***Eucalyptus erythronema*** var. ***marginata*** (Benth.) Domin
Eucalyptus coolabah var. *arida* Blakely = ***Eucalyptus coolabah*** subsp. ***arida*** (Blakely) L.A.S.Johnson & K.D.Hill
Eucalyptus coolabah var. *rhodoclada* Blakely = ***Eucalyptus coolabah*** Blakely & Jacobs subsp. ***coolabah***
Eucalyptus cordata Lodd. = ***Eucalyptus pulverulenta*** Sims
Eucalyptus cordieri var. *brachypoma* Blakely = ***Eucalyptus goniocalyx*** F.Muell. ex Miq.
Eucalyptus cordieri var. *nortonii* Blakely = ***Eucalyptus nortonii*** (Blakely) L.A.S.Johnson
Eucalyptus coriacea A.Cunn. ex Schauer = ***Eucalyptus pauciflora*** Sieber ex Spreng. subsp. ***pauciflora***
Eucalyptus coriacea var. *alpina* Benth. = ***Eucalyptus pauciflora*** Sieber ex Spreng. subsp. ***pauciflora***
Eucalyptus coronata Tausch ex Maiden = ***Eucalyptus tereticornis*** Sm.
Eucalyptus corticosa L.A.S.Johnson = ***Eucalyptus argophloia*** Blakely
Eucalyptus corymbosa Sm. = ***Corymbia gummifera*** (Gaertn.) K.D.Hill & L.A.S.Johnson
Eucalyptus corymbosa var. *terminalis* (F.Muell.) F.M.Bailey = ***Corymbia terminalis*** (F.Muell.) K.D.Hill & L.A.S.Johnson
Eucalyptus corynocalyx F.Muell. = ***Eucalyptus cladocalyx*** F.Muell.
Eucalyptus cosmophylla f. *leprosula* Miq. = ***Eucalyptus cosmophylla*** F.Muell.
Eucalyptus cosmophylla var. *leprosula* (Miq.) Maiden = ***Eucalyptus cosmophylla*** F.Muell.
Eucalyptus cosmophylla var. *rostrigera* Maiden = ***Eucalyptus cosmophylla*** F.Muell.
Eucalyptus costata F.Muell. = ***Eucalyptus incrassata*** Labill.
Eucalyptus costata subsp. *murrayana* L.A.S.Johnson & K.D.Hill = ***Eucalyptus incrassata*** Labill.
Eucalyptus cotinifolia Colla = ?
Eucalyptus crassifolia (G.J.Leach) Brooker = ***Angophora crassifolia*** (G.J.Leach) L.A.S.Johnson & K.D.Hill
Eucalyptus crebra var. *citrata* F.Muell. = ***Eucalyptus staigeriana*** F.Muell. ex F.M.Bailey
Eucalyptus crebra var. *macrocarpa* Domin = ***Eucalyptus drepanophylla*** F.Muell. ex Benth.
Eucalyptus cullenii var. *trivalvis* Blakely = ***Eucalyptus cullenii*** Cambage
Eucalyptus × *cultrifolia* Naudin = ?
Eucalyptus curtipes D.J.Carr & S.G.M.Carr = ***Corymbia curtipes*** (D.J.Carr & S.G.M.Carr) K.D.Hill & L.A.S. Johnson
Eucalyptus curvula Sieber ex Spreng. = ?
Eucalyptus cuspidata Turcz. = ***Eucalyptus angulosa*** Schauer
Eucalyptus cyathiformis Blakely & McKie = ***Eucalyptus caliginosa*** Blakely & McKie
Eucalyptus dampieri D.J.Carr & S.G.M.Carr = ***Corymbia dampieri*** (D.J.Carr & S.G.M.Carr) K.D. Hill & L.A.S.Johnson
Eucalyptus daphnoides Miq. = ***Eucalyptus coccifera*** Hook.f.
Eucalyptus darwinensis D.J.Carr & S.G.M.Carr = ***Corymbia foelscheana*** (F.Muell.) K.D.Hill & L.A.S. Johnson
Eucalyptus dealbata var. *chloroclada* Blakely = ***Eucalyptus chloroclada*** (Blakely) L.A.S.Johnson & K.D.Hill

Eucalyptus dealbata var. *populnea* Blakely = ***Eucalyptus dealbata*** A.Cunn. ex Schauer
Eucalyptus debeuzevillei Maiden = ***Eucalyptus pauciflora*** subsp. ***debeuzevillei*** (Maiden) L.A.S.Johnson & Blaxell
Eucalyptus decepta Blakely = ***Eucalyptus siderophloia*** Benth.
Eucalyptus decipiens var. *angustifolia* Schauer = ***Eucalyptus decipiens*** Endl. subsp. ***decipiens***
Eucalyptus decipiens var. *latifolia* Schauer = ***Eucalyptus decipiens*** Endl. subsp. ***decipiens***
Eucalyptus deformis Blakely = ***Eucalyptus oblonga*** A.Cunn. ex DC.
Eucalyptus derbyensis D.J.Carr & S.G.M.Carr = ***Corymbia polycarpa*** (F.Muell.) K.D.Hill & L.A.S. Johnson
Eucalyptus deserticola S.G.M.Carr = ***Corymbia deserticola*** (S.G.M.Carr) K.D.Hill & L.A.S.Johnson
Eucalyptus desertorum Naudin = ***Eucalyptus foecunda*** Schauer
Eucalyptus desertorum D.J.Carr & S.G.M.Carr = ***Corymbia deserticola*** (S.G.M.Carr) K.D.Hill & L.A.S.Johnson
Eucalyptus dextropinea R.T.Baker = ***Eucalyptus muelleriana*** Howitt
Eucalyptus dichromophloia F.Muell. = ***Corymbia dichromophloia*** (F.Muell.) K.D.Hill & L.A.S.Johnson
Eucalyptus discolor Desf. = ***Eucalyptus pilularis*** Sm.
Eucalyptus divaricata McAulay & Brett = ***Eucalyptus gunnii*** subsp. ***divaricata*** (McAulay & Brett) B.M.Potts
Eucalyptus diversifolia Woolls = ***Eucalyptus macarthurii*** H.Deane & Maiden
Eucalyptus dolichocarpa D.J.Carr & S.G.M.Carr = ***Corymbia dolichocarpa*** (D.J.Carr & S.G.M.Carr) K.D.Hill & L.A.S.Johnson
Eucalyptus dongarraensis Maiden & Blakely = ***Eucalyptus obtusiflora*** subsp. ***dongarraensis*** (Maiden & Blakely) L.A.S.Johnson & K.D.Hill
Eucalyptus dorrienii Domin = ***Eucalyptus falcata*** Turcz.
Eucalyptus drepanophylla var. *leptophleba* (F.Muell.) Luehm. ex Burtt Davey = ***Eucalyptus leptophleba*** F.Muell.
Eucalyptus drysdalensis D.J.Carr & S.G.M.Carr = ***Corymbia drysdalensis*** (D.J.Carr & S.G.M.Carr) K.D.Hill & L.A.S.Johnson
Eucalyptus dumosa var. *conglobata* Benth. = ***Eucalyptus conglobata*** (Benth.) Maiden
Eucalyptus dumosa var. *puncticulata* Benth. = ***Eucalyptus uncinata*** Turcz.
Eucalyptus dumosa var. *rhodophloia* Benth. = ***Eucalyptus dumosa*** A.Cunn. ex Oxley
Eucalyptus dumosa var. *scyphocalyx* Benth. = ***Eucalyptus scyphocalyx*** (Benth.) Maiden & Blakely
Eucalyptus dunlopiana (K.D.Hill & L.A.S.Johnson) Brooker = ***Corymbia dunlopiana*** K.D.Hill & L.A.S.Johnson
Eucalyptus durackiana D.J.Carr & S.G.M.Carr = ***Corymbia dampieri*** (D.J.Carr & S.G.M.Carr) K.D.Hill & L.A.S.Johnson
Eucalyptus elaeophora F.Muell. = ***Eucalyptus goniocalyx*** F.Muell. ex Miq.
Eucalyptus ellipsoidea D.J.Carr & S.G.M.Carr = ***Corymbia ellipsoidea*** (D.J.Carr & S.G.M.Carr) K.D.Hill & L.A.S.Johnson
Eucalyptus elongata Link = ***Corymbia eximia*** (Schauer) K.D.Hill & L.A.S.Johnson
Eucalyptus enervifolia Walp. = ***Eucalyptus cneorifolia*** A.Cunn. ex DC.
Eucalyptus eremaea D.J.Carr & S.G.M.Carr = ***Corymbia eremaea*** (D.J.Carr & S.G.M.Carr) K.D.Hill & L.A.S.Johnson
Eucalyptus eremophila var. *grandiflora* (Maiden) Maiden = ***Eucalyptus eremophila*** (Diels) Maiden
Eucalyptus eremophila var. *pterocarpa* Blakely & H.Steedman = ***Eucalyptus eremophila*** (Diels) Maiden
Eucalyptus eremophila subsp. *pterocarpa* (Blakely & H.Steedman) L.A.S.Johnson & Blaxell = ***Eucalyptus eremophila*** (Diels) Maiden
Eucalyptus erubescens D.J.Carr & S.G.M.Carr = ***Corymbia polycarpa*** (F.Muell.) K.D.Hill & L.A.S.Johnson
Eucalyptus erythrocalyx F.Muell. = ***Eucalyptus pyriformis*** Turcz.
Eucalyptus erythronema var. *roei* Maiden = ***Eucalyptus erythronema*** Turcz. var. ***erythronema***
Eucalyptus erythrophloia Blakely = ***Corymbia erythrophloia*** (Blakely) K.D.Hill & L.A.S.Johnson
Eucalyptus eucentrica L.A.S.Johnson & K.D.Hill = ***Eucalyptus socialis*** subsp. ***eucentrica*** (L.A.S.Johnson & K.D.Hill) D.Nicolle
Eucalyptus eudesmioides var. *globosa* Blakely = ***Eucalyptus eudesmioides*** F.Muell.
Eucalyptus eugenioides var. *nana* H.Deane & Maiden = ***Eucalyptus eugenioides*** Sieber ex Spreng.
Eucalyptus euryphylla (L.A.S.Johnson ex G.J.Leach) Brooker = ***Angophora euryphylla*** (L.A.S.Johnson ex G.J.Leach) L.A.S.Johnson & K.D.Hill
Eucalyptus eximia Schauer = ***Corymbia eximia*** (Schauer) K.D.Hill & L.A.S.Johnson
Eucalyptus eximia var. *leichhardtii* (F.M.Bailey) Ewart = ***Corymbia leichhardtii*** (F.M.Bailey) K.D.Hill & L.A.S.Johnson
Eucalyptus exserta var. *parvula* Blakely = ***Eucalyptus exserta*** F.Muell.
Eucalyptus exul (K.D.Hill) Brooker = ***Angophora exul*** K.D.Hill
Eucalyptus fabrorum Schltdl. = ***Eucalyptus obliqua*** L'Hér.
Eucalyptus falcata var. *ecostata* Maiden = ***Eucalyptus falcata*** Turcz.
Eucalyptus falcifolia Miq. = ***Eucalyptus obliqua*** L'Hér.
Eucalyptus ferriticola Brooker & Edgecombe = ***Corymbia ferriticola*** (Brooker & Edgecombe) K.D.Hill & L.A.S.Johnson
Eucalyptus ferruginea Schauer = ***Corymbia ferruginea*** (Schauer) K.D.Hill & L.A.S.Johnson
Eucalyptus ficifolia F.Muell. = ***Corymbia ficifolia*** (F.Muell.) K.D.Hill & L.A.S.Johnson
Eucalyptus ficifolia var. *alba* Blakely = *E. calophylla* × *E. ficifolia*
Eucalyptus ficifolia var. *alba* Guilf. = ***Corymbia ficifolia*** (F.Muell.) K.D.Hill & L.A.S.Johnson
Eucalyptus ficifolia var. *bakeri* Guilf. = ***Corymbia ficifolia*** (F.Muell.) K.D.Hill & L.A.S.Johnson
Eucalyptus ficifolia var. *carmina* Blakely = ***Corymbia ficifolia*** (F.Muell.) K.D.Hill & L.A.S.Johnson
Eucalyptus ficifolia var. *guilfoylei* F.M.Bailey = *Corymbia calophylla* × *Corymbia ficifolia*
Eucalyptus ficifolia var. *rosea* Guilf. = ***Corymbia ficifolia*** (F.Muell.) K.D.Hill & L.A.S.Johnson
Eucalyptus firma F.Muell. ex Miq. = ?
Eucalyptus fissilis Guilf. = ?
Eucalyptus flavescens (K.D.Hill & L.A.S.Johnson) Brooker = ***Corymbia flavescens*** K.D.Hill & L.A.S.Johnson

Eucalyptus flavieri Añon = ?
Eucalyptus fletcheri F.Muell. ex R.T.Baker = ***Eucalyptus baueriana*** Schauer
Eucalyptus flexilis Regel = ?
Eucalyptus floribunda Hügel ex Endl. = ***Eucalyptus marginata*** Donn ex Sm. subsp. ***marginata***
Eucalyptus floribunda F.Muell. = ***Corymbia confertiflora*** (Kippist ex F.Muell.) K.D.Hill & L.A.S.Johnson
Eucalyptus florida Brooker = ***Angophora floribunda*** (Sm.) Sweet
Eucalyptus foecunda var. *loxophleba* (Benth.) W.Fitzg. = ***Eucalyptus loxophleba*** Benth.
Eucalyptus foelscheana F.Muell. = ***Corymbia foelscheana*** (F.Muell.) K.D.Hill & L.A.S.Johnson
Eucalyptus fordeana D.J.Carr & S.G.M.Carr = ***Corymbia eremaea*** (D.J.Carr & S.G.M.Carr) K.D.Hill & L.A.S.Johnson subsp. ***eremaea***
Eucalyptus forrestiana subsp. *dolichorhyncha* Brooker = ***Eucalyptus dolichorhyncha*** (Brooker) Brooker & Hopper
Eucalyptus forrestiana subsp. *stoatei* (C.A.Gardner) C.J.Rob. = ***Eucalyptus stoatei*** C.A.Gardner
Eucalyptus forsythii Maiden = ***Eucalyptus melliodora*** A.Cunn. ex Schauer
Eucalyptus fraxinoides var. *triflora* (Maiden) Maiden = ***Eucalyptus triflora*** (Maiden) Blakely
Eucalyptus fruticetorum F.Muell. ex Miq. = ***Eucalyptus odorata*** Behr var. ***odorata***
Eucalyptus gigantea Hook.f. = ***Eucalyptus delegatensis*** subsp. ***tasmaniensis*** Boland
Eucalyptus gigantea Dehnh. = ***Eucalyptus globulus*** Labill. subsp. ***globulus***
Eucalyptus gilbertensis (Maiden & Blakely) S.T.Blake = ***Corymbia gilbertensis*** (Maiden & Blakely) K.D.Hill & L.A.S.Johnson
Eucalyptus gillii var. *petiolaris* Maiden = [50 SOA] *E. gillii* × *E. socialis*
Eucalyptus glandulosa Desf. = ***Eucalyptus amygdalina*** Labill.
Eucalyptus glauca A.Cunn. ex DC. = ***Eucalyptus globulus*** Labill. subsp. ***globulus***
Eucalyptus glaucophylla Hoffmanns. = ***Corymbia calophylla*** (R.Br. ex Lindl.) K.D.Hill & L.A.S.Johnson
Eucalyptus globoidea var. *largifructa* Blakely = ?
Eucalyptus globoidea var. *subsphaerica* Blakely = ***Eucalyptus globoidea*** Blakely
Eucalyptus globulosus St.-Lag. = ***Eucalyptus globulus*** Labill. subsp. ***globulus***
Eucalyptus globulus var. *bicostata* (Maiden, Blakely & Simmonds) Ewart = ***Eucalyptus globulus*** subsp. ***bicostata*** (Maiden, Blakely & Simmonds) J.B.Kirkp.
Eucalyptus globulus var. *compacta* Maiden = ?
Eucalyptus globulus var. *stjohniis* F.Muell. ex R.T.Baker = ***Eucalyptus globulus*** subsp. ***pseudoglobulus*** (Naudin ex Maiden) J.B.Kirkp.
Eucalyptus gneorifolia Sweet ex G.Don = ***Eucalyptus cneorifolia*** A.Cunn. ex DC.
Eucalyptus gomphocephala var. *rhodoxylon* Blakely & H.Steedman = ***Eucalyptus gomphocephala*** A.Cunn. ex DC.
Eucalyptus × *gomphocornuta* A.Vilm. ex Trab. = *E. cornuta* × *E. gomphocephala*
Eucalyptus goniantha var. *clelandii* Maiden = ***Eucalyptus clelandii*** (Maiden) Maiden
Eucalyptus goniantha subsp. *semiglobosa* Brooker = ***Eucalyptus semiglobosa*** (Brooker) L.A.S.Johnson & K.D.Hill
Eucalyptus goniocalyx var. *acuminata* Benth. = ***Eucalyptus cypellocarpa*** L.A.S.Johnson
Eucalyptus goniocalyx var. *nitens* H.Deane & Maiden = ***Eucalyptus nitens*** (H.Deane & Maiden) Maiden
Eucalyptus goniocalyx var. *pallens* Benth. = ***Eucalyptus nortonii*** (Blakely) L.A.S.Johnson
Eucalyptus goniocalyx var. *parviflora* Blakely & McKie = ***Eucalyptus cypellocarpa*** L.A.S.Johnson
Eucalyptus gracilipes Naudin = ***Eucalyptus leucoxylon*** F.Muell. subsp. ***leucoxylon***
Eucalyptus gracilis var. *breviflora* Benth. = ***Eucalyptus gracilis*** F.Muell.
Eucalyptus gracilis var. *erecta* Blakely = ***Eucalyptus gracilis*** F.Muell.
Eucalyptus gracilis var. *thozetii* F.M.Bailey = ***Eucalyptus thozetiana*** F.Muell. ex R.T.Baker
Eucalyptus gracilis var. *viminea* Blakely = ***Eucalyptus gracilis*** F.Muell.
Eucalyptus gracilis var. *yilgarnensis* Maiden = ***Eucalyptus yilgarnensis*** (Maiden) Brooker
Eucalyptus grandifolia R.Br. ex Benth. = ***Corymbia grandifolia*** (R.Br. ex Benth.) K.D.Hill & L.A.S.Johnson
Eucalyptus grandis var. *grandiflora* Maiden = *E. grandis* × *E. robusta*
Eucalyptus grasbyi Maiden & Blakely = ***Eucalyptus longicornis*** (F.Muell.) Maiden
Eucalyptus greeniana D.J.Carr & S.G.M.Carr = ***Corymbia greeniana*** (D.J.Carr & S.G.M.Carr) K.D.Hill & L.A.S.Johnson
Eucalyptus griffithsii var. *angustiuscula* Blakely = ***Eucalyptus griffithsii*** Maiden
Eucalyptus gullickii R.T.Baker & H.G.Sm. = ***Eucalyptus mannifera*** subsp. ***gullickii*** (R.T.Baker & H.G.Sm.) L.A.S.Johnson
Eucalyptus gummifera (Gaertn.) Hochr. = ***Corymbia gummifera*** (Gaertn.) K.D.Hill & L.A.S.Johnson
Eucalyptus gummifera var. *intermedia* Domin = ***Corymbia gummifera*** (Gaertn.) K.D.Hill & L.A.S.Johnson
Eucalyptus gunnii Miq. = ***Eucalyptus viminalis*** Labill. subsp. ***viminalis***
Eucalyptus gunnii var. *acervula* H.Deane & Maiden = ***Eucalyptus ovata*** Labill.
Eucalyptus gunnii var. *elata* Hook.f. = ***Eucalyptus ovata*** Labill.
Eucalyptus gunnii var. *glauca* H.Deane & Maiden = ***Eucalyptus glaucescens*** Maiden & Blakely
Eucalyptus gunnii var. *maculosa* (R.T.Baker) Maiden = ***Eucalyptus mannifera*** subsp. ***maculosa*** (R.T.Baker) L.A.S.Johnson
Eucalyptus gunnii var. *montana* Hook.f. = ***Eucalyptus gunnii*** subsp. ***gunnii***
Eucalyptus gunnii var. *ovata* (Labill.) H.Deane & Maiden = ***Eucalyptus ovata*** Labill.
Eucalyptus gunnii var. *rubida* (H.Deane & Maiden) Maiden = ***Eucalyptus rubida*** H.Deane & Maiden
Eucalyptus gunnii var. *undulata* Rehder = ***Eucalyptus gunnii*** subsp. ***gunnii***
Eucalyptus haemastoma var. *capitata* Maiden = ***Eucalyptus racemosa*** subsp. ***racemosa***
Eucalyptus haemastoma var. *inophloia* C.T.White = ***Eucalyptus andrewsii*** Maiden subsp. ***andrewsii***
Eucalyptus haemastoma var. *micrantha* (A.Cunn. ex DC.) Benth. = ***Eucalyptus racemosa*** subsp. ***racemosa***
Eucalyptus haemastoma var. *montana* H.Deane & Maiden = ***Eucalyptus*** × ***montana*** (H.Deane & Maiden) Blakely
Eucalyptus haemastoma var. *sclerophylla* Blakely = ***Eucalyptus racemosa*** subsp. ***racemosa***
Eucalyptus haematoxylon Maiden = ***Corymbia***

haematoxylon (Maiden) K.D.Hill & L.A.S.Johnson
Eucalyptus hamersleyana D.J.Carr & S.G.M.Carr = ***Corymbia hamersleyana*** (D.J.Carr & S.G.M.Carr) K.D.Hill & L.A.S.Johnson
Eucalyptus helmsii Maiden & Blakely = ***Eucalyptus platycorys*** Maiden & Blakely
Eucalyptus hemilampra F.Muell. = ***Eucalyptus resinifera*** Sm.
Eucalyptus hemiphloia Benth. = ***Eucalyptus moluccana*** Wall. ex Roxb.
Eucalyptus hemiphloia var. *albens* C.Moore & Betche = ***Eucalyptus albens*** Benth.
Eucalyptus hemiphloia var. *albens* (Benth.) Maiden = ***Eucalyptus albens*** Benth.
Eucalyptus hemiphloia var. *microcarpa* Maiden = ***Eucalyptus microcarpa*** (Maiden) Maiden
Eucalyptus hemiphloia var. *parviflora* Benth. = ***Eucalyptus moluccana*** Wall. ex Roxb.
Eucalyptus hemiphloia var. *purpurascens* (F.Muell. ex Benth.) Maiden = ***Eucalyptus albopurpurea*** (Boomsma) D.Nicolle
Eucalyptus hendersonii (K.D.Hill & L.A.S.Johnson) Brooker = ***Corymbia hendersonii*** K.D.Hill & L.A.S.Johnson
Eucalyptus hendrsonii (K.D.Hill & L.A.S.Johnson) Brooker = ***Corymbia hendersonii*** K.D.Hill & L.A.S.Johnson
Eucalyptus henryi S.T.Blake = ***Corymbia henryi*** (S.T.Blake) K.D.Hill & L.A.S.Johnson
Eucalyptus hesperis D.J.Carr & S.G.M.Carr = ***Corymbia hamersleyana*** (D.J.Carr & S.G.M.Carr) K.D.Hill & L.A.S.Johnson
Eucalyptus heterophylla Miq. = ***Eucalyptus obliqua*** L'Hér.
Eucalyptus hillii Maiden = ***Eucalyptus oligantha*** Schauer subsp. ***oligantha***
Eucalyptus hillii var. *alleniana* Blakely & Jacobs = ***Eucalyptus oligantha*** Schauer subsp. ***oligantha***
Eucalyptus hirsuta Link = ***Angophora hispida*** (Sm.) Blaxell
Eucalyptus hispida (Sm.) Brooker = ***Angophora hispida*** (Sm.) Blaxell
Eucalyptus hookeri F.Muell. = ***Corymbia tessellaris*** (F.Muell.) K.D.Hill & L.A.S.Johnson
Eucalyptus huberiana Naudin = ***Eucalyptus viminalis*** Labill. subsp. ***viminalis***
Eucalyptus hylandii D.J.Carr & S.G.M.Carr = ***Corymbia hylandii*** (D.J.Carr & S.G.M.Carr) K.D.Hill & L.A.S.Johnson
Eucalyptus hylandii var. *campestris* D.J.Carr & S.G.M.Carr = ***Corymbia hylandii*** (D.J.Carr & S.G.M.Carr) K.D.Hill & L.A.S.Johnson
Eucalyptus hypericifolia Dum.Cours. = ***Eucalyptus risdonii*** Hook.f.
Eucalyptus hypochlamydea Brooker = ***Eucalyptus horistes*** L.A.S.Johnson & K.D.Hill
Eucalyptus hypoleuca Schauer = ***Eucalyptus marginata*** Donn ex Sm. subsp. ***marginata***
Eucalyptus incrassata Sieber ex DC. = ***Eucalyptus pilularis*** Sm.
Eucalyptus incrassata var. *angulosa* (Schauer) Benth. = ***Eucalyptus angulosa*** Schauer
Eucalyptus incrassata subsp. *angulosa* (Schauer) F.C.Johnstone & Hallam = ***Eucalyptus angulosa*** Schauer
Eucalyptus incrassata var. *conglobata* (Benth.) Maiden = ***Eucalyptus conglobata*** (Benth.) Maiden
Eucalyptus incrassata var. *costata* (F.Muell.) N.T.Burb. = ***Eucalyptus incrassata*** Labill.
Eucalyptus incrassata subsp. *costata* (F.Muell.) F.C.Johnstone & Hallam = ***Eucalyptus incrassata*** Labill.
Eucalyptus incrassata var. *dumosa* (A.Cunn. ex Oxley) Maiden = ***Eucalyptus dumosa*** A.Cunn. ex Oxley
Eucalyptus incrassata var. *goniantha* (Turcz.) Maiden = ***Eucalyptus goniantha*** Turcz.
Eucalyptus incrassata var. *grossa* (F.Muell. ex Benth.) Maiden = ***Eucalyptus grossa*** F.Muell. ex Benth.
Eucalyptus incrassata var. *protrusa* J.M.Black = ***Eucalyptus brachycalyx*** Blakely
Eucalyptus incurva Boomsma = ***Eucalyptus gillenii*** Ewart & L.R.Kerr
Eucalyptus inophloea Brogli = [50]
Eucalyptus inopina (K.D.Hill) Brooker = ***Angophora inopina*** K.D.Hill
Eucalyptus insignis Naudin = ***Eucalyptus tereticornis*** Sm.
Eucalyptus insingiana Maiden = ?
Eucalyptus × *insizwaensis* Maiden = *E. globulus* × *E. robusta*
Eucalyptus insulana F.M.Bailey = ***Eucalyptus exserta*** F.Muell.
Eucalyptus intermedia F.Muell. ex R.T.Baker = ***Corymbia intermedia*** (F.Muell. ex R.T.Baker) K.D.Hill & L.A.S.Johnson
Eucalyptus intertexta var. *diminuta* Blakely = ***Eucalyptus intertexta*** R.T.Baker
Eucalyptus intertexta var. *fruticosa* Blakely & Jacobs = ***Eucalyptus intertexta*** R.T.Baker
Eucalyptus intertexta var. *magna* Blakely = ***Eucalyptus orgadophila*** Maiden & Blakely
Eucalyptus irbyi R.T.Baker & H.F.Sm. = [50 TAS]
Eucalyptus isingiana Maiden = ***Eucalyptus pimpiniana*** Maiden
Eucalyptus jacobsiana Blakely = ***Corymbia jacobsiana*** (Blakely) K.D.Hill & L.A.S.Johnson
Eucalyptus × *jugalis* Naudin = ?
Eucalyptus kakadu D.J.Carr & S.G.M.Carr = ***Corymbia foelscheana*** (F.Muell.) K.D.Hill & L.A.S.Johnson
Eucalyptus kitsonii Luehm. ex Maiden = ***Eucalyptus kitsoniana*** Maiden
Eucalyptus kitsonii H.Deane = Fossil species.
Eucalyptus kochii subsp. *plenissima* (C.A.Gardner) Brooker = ***Eucalyptus plenissima*** (C.A.Gardner) Brooker
Eucalyptus kombolgiensis Brooker & Dunlop = ***Corymbia kombolgiensis*** (Brooker & Dunlop) K.D.Hill & L.A.S.Johnson
Eucalyptus lactea R.T.Baker = ***Eucalyptus mannifera*** subsp. ***praecox*** (Maiden) L.A.S.Johnson
Eucalyptus laevopinea var. *minor* F.Muell. ex R.T.Baker = ***Eucalyptus eugenioides*** Sieber ex Spreng.
Eucalyptus laevopinea var. *turbinata* Blakely = ?
Eucalyptus × *lamprocalyx* Blakely = [50 WAU] *Corymbia cadophora* × *Corymbia polycarpa*
Eucalyptus lamprocarpa F.Muell. ex Miq. = ***Eucalyptus dumosa*** A.Cunn. ex Oxley
Eucalyptus × *lamprocarpa* Blakely = ?
Eucalyptus lamprophylla Brooker & A.R.Bean = ***Corymbia lamprophylla*** (Brooker & A.R.Bean) K.D.Hill & L.A.S.Johnson
Eucalyptus langii Maiden & Blakely = ***Eucalyptus cladocalyx*** F.Muell.
Eucalyptus lansdowneana subsp. *albopurpurea* Boomsma = ***Eucalyptus albopurpurea*** (Boomsma) D.Nicolle
Eucalyptus lansdowneana var. *leucantha* Blakely = ***Eucalyptus albopurpurea*** (Boomsma) D.Nicolle

Eucalyptus largiflorens F.Muell. = ***Eucalyptus bicolor*** A.Cunn. ex Hook.
Eucalyptus largiflorens var. *xanthophylla* (Blakely) Cameron = ***Eucalyptus normantonensis*** Maiden & Cambage
Eucalyptus laseronii var. *doleiformis* Blakely = ?
Eucalyptus laseronii var. *maxima* Blakely = ?
Eucalyptus lata L.A.S.Johnson & K.D.Hill = ***Eucalyptus orbifolia*** F.Muell.
Eucalyptus latifolia F.Muell. = ***Corymbia latifolia*** (F.Muell.) K.D.Hill & L.A.S.Johnson
Eucalyptus latiuscula (Blakely) L.A.S.Johnson & K.D.Hill = ***Eucalyptus moorei*** var. ***latiuscula*** Blakely
Eucalyptus leichhardtii F.M.Bailey = ***Corymbia leichhardtii*** (F.M.Bailey) K.D.Hill & L.A.S.Johnson
Eucalyptus leiocarpa (L.A.S.Johnson ex G.J.Leach) Brooker = ***Angophora leiocarpa*** (L.A.S.Johnson ex G.J.Leach) K.R.Thiele & Ladiges
Eucalyptus leiophloia Blakely & Jacobs = ***Corymbia foelscheana*** (F.Muell.) K.D.Hill & L.A.S.Johnson
Eucalyptus leiophloia var. *lepidophloia* Blakely & Jacobs = ***Corymbia foelscheana*** (F.Muell.) K.D.Hill & L.A.S.Johnson
Eucalyptus lenziana D.J.Carr & S.G.M.Carr = ***Corymbia lenziana*** (D.J.Carr & S.G.M.Carr) K.D.Hill & L.A.S.Johnson
Eucalyptus leptoloma Brooker & A.R.Bean = ***Corymbia leptoloma*** (Brooker & A.R.Bean) K.D.Hill & L.A.S.Johnson
Eucalyptus leptophylla F.Muell. ex Miq. = ***Eucalyptus foecunda*** Schauer
Eucalyptus leptophylla var. *densa* Blakely = ***Eucalyptus foecunda*** Schauer
Eucalyptus leptophylla var. *floribunda* Blakely = ***Eucalyptus foecunda*** Schauer
Eucalyptus leptophylla var. *leptorrhyncha* Blakely = ***Eucalyptus foecunda*** Schauer
Eucalyptus leucadendron Reinw. ex de Vriese = ***Eucalyptus alba*** Reinw. ex Blume
Eucalyptus leucophylla Domin = ***Eucalyptus argillacea*** W.Fitzg. ex Maiden
Eucalyptus leucoxylon var. *angulata* Benth. = ***Eucalyptus leucoxylon*** F.Muell. subsp. ***leucoxylon***
Eucalyptus leucoxylon var. *erythrostema* Miq. = ***Eucalyptus leucoxylon*** F.Muell. subsp. ***leucoxylon***
Eucalyptus leucoxylon var. *macrocarpa* J.E.Br. = ***Eucalyptus leucoxylon*** subsp. ***megalocarpa*** Boland
Eucalyptus leucoxylon var. *minor* Benth. = ***Eucalyptus sideroxylon*** A.Cunn. ex Woolls
Eucalyptus leucoxylon var. *pallens* Benth. = ***Eucalyptus caleyi*** Maiden subsp. ***caleyi***
Eucalyptus leucoxylon var. *piperita* J.E.Br. = ***Eucalyptus leucoxylon*** subsp. ***pruinosa*** (Miq.) Boland
Eucalyptus leucoxylon var. *pluriflora* Miq. = ?
Eucalyptus leucoxylon var. *pruinosa* Miq. = ***Eucalyptus leucoxylon*** subsp. ***pruinosa*** (Miq.) Boland
Eucalyptus leucoxylon var. *rostellata* Miq. = ***Eucalyptus leucoxylon*** F.Muell. subsp. ***leucoxylon***
Eucalyptus leucoxylon var. *rubra* Guilf. = ***Eucalyptus leucoxylon*** F.Muell. subsp. ***leucoxylon***
Eucalyptus leucoxylon var. *rugulosa* Miq. = ***Eucalyptus leucoxylon*** F.Muell. subsp. ***leucoxylon***
Eucalyptus lindleyana A.Cunn. ex DC. = ***Eucalyptus elata*** Dehnh.
Eucalyptus lindleyana var. *stenophylla* Blakely = ***Eucalyptus elata*** Dehnh.
Eucalyptus linearis Dehnh. = ***Eucalyptus pulchella*** Desf.
Eucalyptus longifolia Lindl. = ***Eucalyptus elata*** Dehnh.
Eucalyptus longifolia var. *multiflora* Maiden = *E. longifolia* × *E. robusta*
Eucalyptus longifolia var. *turbinata* Blakely = [50 NSW] *E. longifolia* × *E. tereticornis*
Eucalyptus longirostris F.Muell. ex Miq. = ***Eucalyptus camaldulensis*** Dehnh. subsp. ***camaldulensis***
Eucalyptus longirostris f. *brevirostris* Miq. = ***Eucalyptus camaldulensis*** Dehnh. subsp. ***camaldulensis***
Eucalyptus loxophleba var. *fruticosa* Benth. = ***Eucalyptus loxophleba*** Benth. subsp. ***loxophleba***
Eucalyptus loxophleba subsp. *gratiae* Brooker = ***Eucalyptus gratiae*** (Brooker) L.A.S.Johnson & K.D.Hill
Eucalyptus luehmanniana var. *altior* H.Deane & Maiden = ***Eucalyptus oreades*** F.Muell. ex R.T.Baker
Eucalyptus macrocalyx Turcz. = ***Eucalyptus pyriformis*** Turcz.
Eucalyptus macrocera Turcz. = ***Eucalyptus cornuta*** Labill.
Eucalyptus macropoda Blakely = ***Corymbia collina*** (W.Fitzg.) K.D.Hill & L.A.S.Johnson
Eucalyptus macrorhyncha var. *brachycorys* Benth. = ***Eucalyptus laevopinea*** F.Muell. ex R.T.Baker
Eucalyptus macrorhyncha f. *grandiflora* Maiden = ***Eucalyptus macrorhyncha*** F.Muell. ex Benth. subsp. ***macrorhyncha***
Eucalyptus macrorhyncha var. *minor* Blakely = ***Eucalyptus macrorhyncha*** F.Muell. ex Benth. subsp. ***macrorhyncha***
Eucalyptus maculata Hook. = ***Corymbia maculata*** (Hook.) K.D.Hill & L.A.S.Johnson
Eucalyptus maculata var. *citriodora* (Hook.) F.M.Bailey = ***Corymbia citriodora*** (Hook.) K.D.Hill & L.A.S.Johnson
Eucalyptus maculosa R.T.Baker = ***Eucalyptus mannifera*** subsp. ***maculosa*** (R.T.Baker) L.A.S.Johnson
Eucalyptus mahoganii F.Muell. = ***Eucalyptus marginata*** Donn ex Sm. subsp. ***marginata***
Eucalyptus maidenii F.Muell. = ***Eucalyptus globulus*** subsp. ***maidenii*** (F.Muell.) J.B.Kirkp.
Eucalyptus maidenii subsp. *globulus* (Labill.) J.B.Kirkp. = ***Eucalyptus globulus*** Labill.
Eucalyptus maidenii var. *williamsonii* Blakely = [50 VIC] *E. botruoides* × *E. globulus* subsp. *pseudoglobulus*
Eucalyptus mannifera var. *elliptica* Blakely & McKie = ***Eucalyptus elliptica*** (Blakely & McKie) L.A.S.Johnson & K.D.Hill
Eucalyptus mannifera subsp. *elliptica* (Blakely & McKie) L.A.S.Johnson = ***Eucalyptus elliptica*** (Blakely & McKie) L.A.S.Johnson & K.D.Hill
Eucalyptus marginata var. *staeri* Maiden = ***Eucalyptus staeri*** (Maiden) Maiden ex Kessell & C.A.Gardner
Eucalyptus mazeliana Naudin = ?
Eucalyptus × *mcclatchiei* Kinney = *E. globulus* × *E. ovata*
Eucalyptus × *mcintyrensis* Maiden = ***Eucalyptus*** × ***studleyensis*** Maiden
Eucalyptus × *media* Link = [50+]
Eucalyptus meeboldii Blakely = ***Eucalyptus concinna*** Maiden & Blakely
Eucalyptus melana Brooker = ***Angophora melanoxylon*** F.Muell. ex R.T.Baker
Eucalyptus melanophloia var. *senta* Blakely = ?
Eucalyptus melissiodora Lindl. = ***Corymbia citriodora*** (Hook.) K.D.Hill & L.A.S.Johnson subsp. ***citriodora***
Eucalyptus melliodora var. *brachycarpa* Blakely = ***Eucalyptus melliodora*** A.Cunn. ex Schauer

Eucalyptus melliodora var. *elliptocarpa* Blakely = ***Eucalyptus melliodora*** A.Cunn. ex Schauer
Eucalyptus melliodora var. *murrurundi* Blakely = ***Eucalyptus*** × ***robsonae*** Blakely & McKie
Eucalyptus micrantha A.Cunn. ex DC. = ***Eucalyptus racemosa*** subsp. ***racemosa***
Eucalyptus micrantha var. *signata* (F.Muell.) Blakely = ***Eucalyptus racemosa*** subsp. ***racemosa***
Eucalyptus microphylla A.Cunn. = ***Eucalyptus cunninghamii*** Sweet
Eucalyptus microphylla Muhl. ex Willd. = not Eucalyptus
Eucalyptus microtheca var. *cymbaliformis* Blakely & Jacobs = [50 NTA] *E. cyanoclada* × *E. microtheca*
Eucalyptus mitchellii Cambage = ***Eucalyptus mitchelliana*** Cambage
Eucalyptus mitchellii Ettingsh. = Fossil species.
Eucalyptus mitrata C.A.Gardner = ***Eucalyptus coronata*** C.A.Gardner
Eucalyptus moorei var. *arborea* Blakely = *E. moorei* × *E. piperita*
Eucalyptus morrisonii Maiden = ***Eucalyptus kruseana*** F.Muell.
Eucalyptus × *mortoniana* Kinney = *E. globulus* × *E. viminalis*
Eucalyptus mucronata Link = ?
Eucalyptus muelleri Miq. = ***Eucalyptus dumosa*** A.Cunn. ex Oxley
Eucalyptus muelleri J.B.Moore = ***Eucalyptus johnstonii*** Maiden
Eucalyptus mulleri Naudin = ***Eucalyptus ovata*** Labill.
Eucalyptus multiflora Poir. = ***Eucalyptus robusta*** Sm.
Eucalyptus multiflora Rich. ex A.Gray = [42 PHI]
Eucalyptus multiflora var. *bivalva* Blakely = ***Eucalyptus robusta*** Sm.
Eucalyptus myrtifolia Link = ?
Eucalyptus myrtiformis Naudin = ***Eucalyptus cneorifolia*** A.Cunn. ex DC.
Eucalyptus nanglei F.Muell. ex R.T.Baker = ***Eucalyptus paniculata*** Sm. subsp. ***paniculata***
Eucalyptus nelsonii D.J.Carr & S.G.M.Carr = ***Corymbia eremaea*** (D.J.Carr & S.G.M.Carr) K.D.Hill & L.A.S.Johnson subsp. ***eremaea***
Eucalyptus nervosa F.Muell. ex Miq. = ***Eucalyptus obliqua*** L'Hér.
Eucalyptus × *nervosa* Hoffmanns. = ?
Eucalyptus nesophila Blakely = ***Corymbia nesophila*** (Blakely) K.D.Hill & L.A.S.Johnson
Eucalyptus niphophila Maiden & Blakely = ***Eucalyptus pauciflora*** subsp. ***niphophila*** (Maiden & Blakely) L.A.S.Johnson & Blaxell
Eucalyptus niphophloia Blakely & Jacobs = ***Corymbia dichromophloia*** (F.Muell.) K.D.Hill & L.A.S.Johnson
Eucalyptus novoguinensis D.J.Carr & S.G.M.Carr = ***Corymbia novoguinensis*** (D.J.Carr & S.G.M.Carr) K.D.Hill & L.A.S.Johnson
Eucalyptus × *nowraensis* Maiden = [50 NSW] *Corymbia gummifera* × *Corymbia maculata*
Eucalyptus nubila Maiden & Blakely = ***Eucalyptus fibrosa*** subsp. ***nubila*** (Maiden & Blakely) L.A.S.Johnson
Eucalyptus numerosa Maiden = ***Eucalyptus amygdalina*** Labill.
Eucalyptus obcordata Turcz. = ***Eucalyptus platypus*** Hook.f. subsp. ***platypus***
Eucalyptus obliqua Decne. = ***Eucalyptus decaisneana*** Blume
Eucalyptus obliqua var. *alpina* Maiden = ***Eucalyptus delegatensis*** F.Muell. ex R.T.Baker subsp. ***delegatensis***
Eucalyptus obliqua var. *degressa* Blakely = ***Eucalyptus obliqua*** L'Hér.
Eucalyptus obliqua var. *discocarpa* Blakely = [50 VIC] *E. muelleriana* × *E. obliqua*
Eucalyptus obliqua var. *megacarpa* Blakely = ***Eucalyptus obliqua*** L'Hér.
Eucalyptus obliqua var. *microstoma* Blakely = ?
Eucalyptus obliqua var. *pilula* Blakely = ?
Eucalyptus oblonga var. *rugulosa* Blakely = ***Eucalyptus oblonga*** A.Cunn. ex DC.
Eucalyptus obtusiflora var. *dendromorpha* Blakely = ***Eucalyptus dendromorpha*** (Blakely) L.A.S.Johnson & Blaxell
Eucalyptus occidentalis var. *astringens* Maiden = ***Eucalyptus astringens*** (Maiden) Maiden
Eucalyptus occidentalis var. *californica* Kinney = ?
Eucalyptus occidentalis var. *eremophila* Diels = ***Eucalyptus eremophila*** (Diels) Maiden
Eucalyptus occidentalis var. *grandiflora* Maiden = ***Eucalyptus eremophila*** (Diels) Maiden
Eucalyptus occidentalis var. *macrantha* (F.Muell. ex Benth.) Maiden = ***Eucalyptus macrandra*** F.Muell. ex Benth.
Eucalyptus occidentalis var. *oranensis* A.Vilm. ex Trab. = ***Eucalyptus occidentalis*** Endl.
Eucalyptus occidentalis var. *spathulata* (Hook.) Maiden = ***Eucalyptus spathulata*** Hook.
Eucalyptus occidentalis var. *stenantha* Blakely = ***Eucalyptus aspratilis*** L.A.S.Johnson & K.D.Hill
Eucalyptus ochrophylla Maiden & Blakely = ***Eucalyptus concinna*** Maiden & Blakely
Eucalyptus odorata var. *erythrandra* Miq. = ***Eucalyptus odorata*** Behr var. ***odorata***
Eucalyptus odorata var. *floribunda* Benth. = ***Eucalyptus odorata*** Behr var. ***odorata***
Eucalyptus odorata var. *macrocarpa* Blakely = [50 SOA] *E. leucoxyla* × *E. odorata*
Eucalyptus odorata var. *purpurascens* (F.Muell. ex Benth.) Maiden = ***Eucalyptus albopurpurea*** (Boomsma) D.Nicolle
Eucalyptus odorata var. *refracta* Blakely = ***Eucalyptus odorata*** Behr var. ***odorata***
Eucalyptus odorata var. *woollsiana* Maiden = ***Eucalyptus pilligaensis*** Maiden
Eucalyptus odortata var. *calcicultrix* Miq. = ***Eucalyptus porosa*** Miq.
Eucalyptus oldfieldii var. *drummondii* (Benth.) Maiden = ***Eucalyptus drummondii*** Benth.
Eucalyptus oleosa var. *angustifolia* Maiden = ***Eucalyptus oleosa*** F.Muell. ex Miq. subsp. ***oleosa***
Eucalyptus oleosa var. *borealis* C.A.Gardner = ***Eucalyptus kochii*** subsp. ***borealis*** (C.A.Gardner) D.Nicolle
Eucalyptus oleosa var. *flocktoniae* Maiden = ***Eucalyptus flocktoniae*** (Maiden) Maiden
Eucalyptus oleosa var. *glauca* Maiden = ***Eucalyptus transcontinentalis*** Maiden
Eucalyptus oleosa var. *kochii* (Maiden & Blakely) C.A.Gardner = ***Eucalyptus kochii*** Maiden & Blakely
Eucalyptus oleosa var. *longicornis* F.Muell. = ***Eucalyptus longicornis*** (F.Muell.) Maiden
Eucalyptus oleosa f. *lucida* C.A.Gardner = ***Eucalyptus kochii*** subsp. ***amaryssia*** D.Nicolle
Eucalyptus oleosa var. *obtusa* C.A.Gardner = ***Eucalyptus oleosa*** subsp. ***repleta*** L.A.S.Johnson & K.D.Hill
Eucalyptus oleosa var. *peeneri* Blakely = ***Eucalyptus eremicola*** subsp. ***peeneri*** (Blakely) D.Nicolle
Eucalyptus oleosa var. *plenissima* C.A.Gardner = ***Eucalyptus plenissima*** (C.A.Gardner) Brooker

Eucalyptus ollaris D.J.Carr & S.G.M.Carr = ***Corymbia umbonata*** (D.J.Carr & S.G.M.Carr) K.D.Hill & L.A.S.Johnson
Eucalyptus oocarpa D.J.Carr & S.G.M.Carr = ***Corymbia oocarpa*** (D.J.Carr & S.G.M.Carr) K.D.Hill & L.A.S.Johnson
Eucalyptus opaca D.J.Carr & S.G.M.Carr = ***Corymbia opaca*** (D.J.Carr & S.G.M.Carr) K.D.Hill & L.A.S.Johnson
Eucalyptus × *oppositifolia* Desf. = ?
Eucalyptus orientalis D.J.Carr & S.G.M.Carr = ***Corymbia opaca*** (D.J.Carr & S.G.M.Carr) K.D.Hill & L.A.S.Johnson
Eucalyptus ovalifolia F.Muell. ex R.T.Baker = ***Eucalyptus polyanthemos*** Schauer
Eucalyptus ovalifolia var. *lanceolata* R.T.Baker & H.G.Sm. = ***Eucalyptus polyanthemos*** Schauer
Eucalyptus ovata var. *aquatica* Blakely = ***Eucalyptus camphora*** F.Muell. ex R.T.Baker subsp. ***camphora***
Eucalyptus ovata var. *camphora* (F.Muell. ex R.T.Baker) Maiden = ***Eucalyptus camphora*** F.Muell. ex R.T.Baker
Eucalyptus ovata var. *grandiflora* Maiden = ***Eucalyptus ovata*** Labill.
Eucalyptus pachycarpa (K.D.Hill & L.A.S.Johnson) Brooker = ***Corymbia pachycarpa*** K.D.Hill & L.A.S.Johnson
Eucalyptus pachyphylla var. *sessilis* Maiden = ***Eucalyptus sessilis*** (Maiden) Blakely
Eucalyptus pachypoda Benth. = ?
Eucalyptus × *pachypoda* F.Muell. = ?
Eucalyptus pallens A.Cunn. ex DC. = ***Eucalyptus obliqua*** L'Hér.
Eucalyptus pallidifolia F.Muell. = ***Eucalyptus brevifolia*** F.Muell.
Eucalyptus paludosa F.Muell. ex R.T.Baker = ***Eucalyptus ovata*** Labill.
Eucalyptus palustris Brooker = ***Angophora paludosa*** (G.J.Leach) K.R.Thiele & Ladiges
Eucalyptus panda subsp. *illaquens* L.A.S.Johnson = ***Eucalyptus beyeri*** F.Muell. ex R.T.Baker
Eucalyptus paniculata var. *angustifolia* Benth. = ***Eucalyptus beyeri*** F.Muell. ex R.T.Baker
Eucalyptus paniculata var. *conferta* Benth. = ***Eucalyptus paniculata*** Sm. subsp. ***paniculata***
Eucalyptus paniculata var. *fasciculosa* (F.Muell.) Benth. = ***Eucalyptus fasciculosa*** F.Muell.
Eucalyptus papillosa (K.D.Hill & L.A.S.Johnson) Brooker = ***Corymbia papillosa*** K.D.Hill & L.A.S.Johnson
Eucalyptus papuana F.Muell. = ***Corymbia papuana*** (F.Muell.) K.D.Hill & L.A.S.Johnson
Eucalyptus papuana var. *aparrerinja* Blakely = ***Corymbia papuana*** (F.Muell.) K.D.Hill & L.A.S.Johnson
Eucalyptus × *paradoxa* Maiden & Blakely = ***Eucalyptus* × *oviformis*** Maiden & Blakely
Eucalyptus parramattensis var. *sphaerocalyx* Blakely = ***Eucalyptus parramattensis*** C.C.Hall subsp. ***parramattensis***
Eucalyptus parviflora F.Muell. = ***Eucalyptus bicolor*** A.Cunn. ex Hook.
Eucalyptus parvifolia Cambage = ***Eucalyptus parvula*** L.A.S.Johnson & K.D.Hill
Eucalyptus pastoralis S.Moore = ***Eucalyptus bigalerita*** F.Muell.
Eucalyptus patentiflora F.Muell. ex Miq. = ***Eucalyptus melliodora*** A.Cunn. ex Schauer
Eucalyptus × *patentinervis* F.Muell. ex R.T.Baker = ***Eucalyptus* × *kirtoniana*** F.Muell.
Eucalyptus pauciflora var. *alpina* Ewart = ***Eucalyptus pauciflora*** Sieber ex Spreng. subsp. ***pauciflora***
Eucalyptus pauciflora var. *cylindrocarpa* Blakely = [50 NSW]
Eucalyptus pauciflora var. *densiflora* Blakely & McKie = [50 NSW]
Eucalyptus pauciflora var. *nana* Blakely = ***Eucalyptus gregsoniana*** L.A.S.Johnson & Blaxell
Eucalyptus pauciflora var. *rusticata* Blakely = ?
Eucalyptus peeneri (Blakely) Prior & Johnson ex Boomsma = ***Eucalyptus eremicola*** subsp. ***peeneri*** (Blakely) D.Nicolle
Eucalyptus peltata Benth. = ***Corymbia peltata*** (Benth.) K.D.Hill & L.A.S.Johnson
Eucalyptus peltata subsp. *dimorpha* Brooker & A.R.Bean = ***Corymbia dimorpha*** (Brooker & A.R.Bean) K.D.Hill & L.A.S.Johnson
Eucalyptus peltata subsp. *leichhardtii* (F.M.Bailey) L.A.S.Johnson & Blaxell = ***Corymbia leichhardtii*** (F.M.Bailey) K.D.Hill & L.A.S.Johnson
Eucalyptus pendula Page ex Steud. = ***Eucalyptus bicolor*** A.Cunn. ex Hook.
Eucalyptus perfoliata Desf. = ***Eucalyptus globulus*** Labill. subsp. ***globulus***
Eucalyptus perfoliata R.Br. ex Benth. = ***Corymbia cadophora*** K.D.Hill & L.A.S.Johnson
Eucalyptus perplexa Maiden & Blakely = ***Eucalyptus jensenii*** Maiden
Eucalyptus perriniana R.T.Baker & H.G.Sm. = ***Eucalyptus gunnii*** subsp. ***gunnii***
Eucalyptus persicifolia Lodd. = ***Eucalyptus pilularis*** Sm.
Eucalyptus persistens subsp. *tardecidens* L.A.S.Johnson & K.D.Hill = ***Eucalyptus tardecidens*** (L.A.S.Johnson & K.D.Hill) A.R.Bean
Eucalyptus petalophylla Brooker & A.R.Bean = ***Corymbia petalophylla*** (Brooker & A.R.Bean) K.D.Hill & L.A.S.Johnson
Eucalyptus phaeotricha Blakely & McKie = ***Eucalyptus nigra*** F.Muell. ex R.T.Baker
Eucalyptus phellandra R.T.Baker & H.G.Sm. = ***Eucalyptus radiata*** A.Cunn. ex DC. subsp. ***radiata***
Eucalyptus phlebophylla F.Muell. ex Miq. = ***Eucalyptus pauciflora*** Sieber ex Spreng. subsp. ***pauciflora***
Eucalyptus pilularis var. *acmenoides* (Schauer) Benth. = ***Eucalyptus acmenoides*** Schauer
Eucalyptus pilularis var. *muelleriana* Maiden = ***Eucalyptus pilularis*** Sm.
Eucalyptus pilularis var. *pyriformis* Maiden = ***Eucalyptus pyrocarpa*** L.A.S.Johnson & Blaxell
Eucalyptus piperata Stokes = ***Eucalyptus piperita*** Sm. subsp. ***piperita***
Eucalyptus piperita var. *brachycorys* Benth. = ***Eucalyptus piperita*** Sm. subsp. ***piperita***
Eucalyptus piperita var. *eugenioides* Benth. = ***Eucalyptus piperita*** Sm. subsp. ***piperita***
Eucalyptus piperita var. *laxiflora* Benth. = ***Eucalyptus piperita*** Sm. subsp. ***piperita***
Eucalyptus piperita var. *orophila* Blakely = [50 NSW] *E. oreodes* × *E. piperita*
Eucalyptus piperita var. *pauciflora* A.Cunn. ex DC. = ***Eucalyptus obtusiflora*** A.Cunn. ex DC. subsp. ***obtusiflora***
Eucalyptus platyphylla var. *tintinnans* Blakely & Jacobs = ***Eucalyptus tintinnans*** (Blakely & Jacobs) L.A.S.Johnson & K.D.Hill
Eucalyptus platypodos Cav. = ***Eucalyptus botryoides*** Sm.
Eucalyptus platypus var. *nutans* (F.Muell.) Benth. = ***Eucalyptus nutans*** F.Muell.

Eucalyptus plena (K.D.Hill & L.A.S.Johnson) Brooker = ***Corymbia plena*** K.D.Hill & L.A.S.Johnson
Eucalyptus pleurocarpa Schauer = ***Eucalyptus tetragona*** (R.Br.) F.Muell.
Eucalyptus × *plurilocularis* F.Muell. = ?
Eucalyptus pocillum D.J.Carr & S.G.M.Carr = ***Corymbia pocillum*** (D.J.Carr & S.G.M.Carr) K.D.Hill & L.A.S.Johnson
Eucalyptus polycarpa F.Muell. = ***Corymbia polycarpa*** (F.Muell.) K.D.Hill & L.A.S.Johnson
Eucalyptus polycarpa var. *oligocarpa* Blakely & Jacobs = ***Corymbia eremaea*** subsp. ***oligocarpa*** (Blakely & Jacobs) K.D.Hill & L.A.S.Johnson
Eucalyptus polysciada F.Muell. = ***Corymbia polysciada*** (F.Muell.) K.D.Hill & L.A.S.Johnson
Eucalyptus pontis D.J.Carr & S.G.M.Carr = ***Corymbia cliftoniana*** (W.Fitzg.) K.D.Hill & L.A.S.Johnson
Eucalyptus populifolia Hook. = ***Eucalyptus populnea*** F.Muell.
Eucalyptus populifolia Hook.f. = ***Eucalyptus populnea*** F.Muell. subsp. ***populnea***
Eucalyptus populifolia Desf. = ***Eucalyptus tereticornis*** Sm.
Eucalyptus populifolia var. *obconica* Blakely = ***Eucalyptus* × *bucknellii*** Cambage
Eucalyptus populnea var. *obconica* (Blakely) Cameron = ***Eucalyptus* × *bucknellii*** Cambage
Eucalyptus porrecta S.T.Blake = ***Corymbia porrecta*** (S.T.Blake) K.D.Hill & L.A.S.Johnson
Eucalyptus praecox Maiden = ***Eucalyptus mannifera*** subsp. ***praecox*** (Maiden) L.A.S.Johnson
Eucalyptus preissiana var. *glauca* Regel = ***Eucalyptus preissiana*** Schauer subsp. ***preissiana***
Eucalyptus procera Dehnh. = ***Eucalyptus obliqua*** L'Hér.
Eucalyptus propinqua var. *major* Maiden = ***Eucalyptus major*** (Maiden) Blakely
Eucalyptus pruinosa Turcz. = ***Eucalyptus pyriformis*** Turcz.
Eucalyptus pryoriana L.A.S.Johnson = ***Eucalyptus viminalis*** subsp. ***pryoriana*** (L.A.S.Johnson) Brooker & Slee
Eucalyptus pseudoglobulus Naudin ex Maiden = ***Eucalyptus globulus*** subsp. ***pseudoglobulus*** (Naudin ex Maiden) J.B.Kirkp.
Eucalyptus ptychocarpa F.Muell. = ***Corymbia ptychocarpa*** (F.Muell.) K.D.Hill & L.A.S.Johnson
Eucalyptus pulverulenta Link = ***Eucalyptus globulus*** Labill. subsp. ***globulus***
Eucalyptus pulverulenta var. *lanceolata* A.W.Howitt = ***Eucalyptus cinerea*** F.Muell. ex Benth. subsp. ***cinerea***
Eucalyptus pulverulenta var. *ovatifolia* Dehnh. = ***Eucalyptus pulverulenta*** Sims
Eucalyptus pulvigera A.Cunn. = ***Eucalyptus pulverulenta*** Sims
Eucalyptus punctata var. *didyma* R.T.Baker & H.G.Sm. = ***Eucalyptus punctata*** A.Cunn. ex DC.
Eucalyptus punctata var. *grandiflora* H.Deane & Maiden = ***Eucalyptus canaliculata*** Maiden
Eucalyptus punctata var. *longirostrata* Blakely = ***Eucalyptus longirostrata*** (Blakely) L.A.S.Johnson & K.D.Hill
Eucalyptus punctata var. *major* R.T.Baker & H.G.Sm. = ***Eucalyptus punctata*** A.Cunn. ex DC.
Eucalyptus puncticulata (Benth.) Blakely = ***Eucalyptus uncinata*** Turcz.
Eucalyptus × *purpurascens* Link = ?
Eucalyptus × *purpurascens* var. *petiolaris* A.Cunn. ex DC. = ?
Eucalyptus × *purpurascens* var. *petiolulata* A.Cunn. ex DC. = ?
Eucalyptus pyriformis var. *elongata* Maiden = ***Eucalyptus pyriformis*** Turcz.
Eucalyptus pyriformis var. *kingsmillii* Maiden = ***Eucalyptus kingsmillii*** (Maiden) Maiden & Blakely
Eucalyptus pyriformis var. *minor* Maiden = ***Eucalyptus pachyphylla*** F.Muell.
Eucalyptus pyriformis var. *rameliana* (F.Muell.) Maiden = ***Eucalyptus rameliana*** F.Muell.
Eucalyptus pyriformis subsp. *youngiana* (F.Muell.) Boomsma = ***Eucalyptus youngiana*** F.Muell.
Eucalyptus pyrophora Benth. = ***Corymbia terminalis*** (F.Muell.) K.D.Hill & L.A.S.Johnson
Eucalyptus pyrophora f. *compacta* Domin = ***Corymbia terminalis*** (F.Muell.) K.D.Hill & L.A.S.Johnson
Eucalyptus pyrophora var. *compacta* Domin = ***Corymbia terminalis*** (F.Muell.) K.D.Hill & L.A.S.Johnson
Eucalyptus pyrophora var. *polycarpa* (F.Muell.) Maiden = ***Corymbia polycarpa*** (F.Muell.) K.D.Hill & L.A.S.Johnson
Eucalyptus racemosa var. *longiflora* Blakely = ***Eucalyptus crebra*** F.Muell.
Eucalyptus racemosa var. *macrocarpa* (Domin) Blakely = ***Eucalyptus drepanophylla*** F.Muell. ex Benth.
Eucalyptus racemosa var. *signata* (F.Muell.) R.D.Johnst. & Marryatt = ***Eucalyptus racemosa*** subsp. ***racemosa***
Eucalyptus radiata var. *australiana* (R.T.Baker & H.G.Sm.) Blakely = ***Eucalyptus radiata*** A.Cunn. ex DC. subsp. ***radiata***
Eucalyptus radiata subsp. *robertsonii* (Blakely) L.A.S.Johnson & Blaxell = ***Eucalyptus robertsonii*** Blakely
Eucalyptus radiata var. *subexserta* Blakely = ***Eucalyptus radiata*** A.Cunn. ex DC. subsp. ***radiata***
Eucalyptus radiata var. *subplatyphylla* Blakely & McKie = ***Eucalyptus radiata*** subsp. ***sejuncta*** L.A.S.Johnson & K.D.Hill
Eucalyptus × *radiodives* R.A.Black = ***Eucalyptus* × *dixsonii*** N.A.Wakef.
Eucalyptus × *rameliana* A.Vilm. ex Trab. = ?
Eucalyptus × *rameliana* var. *trabutii* Vilm. ex Trab. = *E. botryoides* × *E. camaldulensis*
Eucalyptus raveretiana var. *jerichoensis* Domin = ***Eucalyptus microtheca*** F.Muell.
Eucalyptus redunca var. *angustifolia* Benth. = ***Eucalyptus xanthonema*** Turcz. subsp. ***xanthonema***
Eucalyptus redunca var. *elata* Benth. = ***Eucalyptus wandoo*** Blakely
Eucalyptus redunca var. *melanophloia* Benth. = ***Eucalyptus arachnaea*** Brooker & Hopper
Eucalyptus redunca var. *oxymitra* Maiden = ***Eucalyptus flavida*** Brooker & Hopper
Eucalyptus redunca var. *subangusta* Blakely = ***Eucalyptus subangusta*** (Blakely) Brooker & Hopper
Eucalyptus regnans var. *fastigata* Ewart = ***Eucalyptus regnans*** F.Muell.
Eucalyptus resinifera var. *grandiflora* Benth. = ***Eucalyptus resinifera*** Sm.
Eucalyptus resinifera var. *hemilampra* (F.Muell.) Domin = ***Eucalyptus resinifera*** Sm.
Eucalyptus resinifera subsp. *hemilampra* (F.Muell.) L.A.S.Johnson & K.D.Hill = ***Eucalyptus resinifera*** Sm.
Eucalyptus resinifera var. *kirtoniana* (F.Muell.) H.Deane & Maiden = ***Eucalyptus* × *kirtoniana*** F.Muell.
Eucalyptus resinifera var. *pellita* (F.Muell.) F.M.Bailey = ***Eucalyptus pellita*** F.Muell.
Eucalyptus resinifera var. *spectabilis* (F.Muell.) F.M.Bailey = ***Eucalyptus pellita*** F.Muell.

Eucalyptus × *reticulata* Link = ?
Eucalyptus rhodantha var. *petiolaris* Blakely = ***Eucalyptus rhodantha*** Blakely & H.Steedman
Eucalyptus rhodophloia (Benth.) Blakely = ***Eucalyptus dumosa*** A.Cunn. ex Oxley
Eucalyptus rhodops D.J.Carr & S.G.M.Carr = ***Corymbia rhodops*** (D.J.Carr & S.G.M.Carr) K.D.Hill & L.A.S.Johnson
Eucalyptus × *rigida* Hoffmanns. = ?
Eucalyptus × *rigida* var. *luehmanniana* F.Muell. = ?
Eucalyptus risdonii var. *elata* Benth. = ***Eucalyptus delegatensis*** subsp. ***tasmaniensis*** Boland
Eucalyptus risdonii var. *hypericifolia* (Benth.) Rodway = ?
Eucalyptus robur (L.A.S.Johnson & K.D.Hill) Brooker = ***Angophora robur*** L.A.S.Johnson & K.D.Hill
Eucalyptus robusta var. *bivalva* (Blakely) Blakely = ***Eucalyptus robusta*** Sm.
Eucalyptus robusta var. *rostrata* (Cav.) Pers. = ***Eucalyptus robusta*** Sm.
Eucalyptus rossii R.T.Baker & H.G.Sm. = ***Eucalyptus racemosa*** subsp. ***rossii*** (R.T.Baker & H.G.Sm.) B.E.Pfeil & Henwood
Eucalyptus rostrata Schltdl. = ***Eucalyptus camaldulensis*** Dehnh. subsp. ***camaldulensis***
Eucalyptus rostrata Cav. = ***Eucalyptus robusta*** Sm.
Eucalyptus rostrata var. *acuminata* (Hook.) Maiden = ***Eucalyptus camaldulensis*** Dehnh. subsp. ***camaldulensis***
Eucalyptus rostrata var. *borealis* R.T.Baker & H.G.Sm. = ***Eucalyptus camaldulensis*** Dehnh. subsp. ***camaldulensis***
Eucalyptus rostrata var. *brevirostris* (Miq.) Maiden = ***Eucalyptus camaldulensis*** Dehnh. subsp. ***camaldulensis***
Eucalyptus rostrata var. *studleyensis* (Maiden) Ewart = ***Eucalyptus*** × ***studleyensis*** Maiden
Eucalyptus rubida subsp. *canobolensis* L.A.S.Johnson & K.D.Hill = ***Eucalyptus canobolensis*** (L.A.S.Johnson & K.D.Hill) J.T.Hunter
Eucalyptus × *rubricaulis* Desf. = ?
Eucalyptus rupicola L.A.S.Johnson & Blaxell = ***Eucalyptus cunninghamii*** Sweet
Eucalyptus rydalensis R.T.Baker & H.F.Sm. = ***Eucalyptus aggregata*** H.Deane & Maiden
Eucalyptus salicifolia Cav. = ***Eucalyptus amygdalina*** Labill.
Eucalyptus salicifolia var. *hypericifolia* (Benth.) Blakely = ?
Eucalyptus saligna var. *botryoides* (Sm.) Maiden = ***Eucalyptus botryoides*** Sm.
Eucalyptus saligna subsp. *botryoides* (Sm.) Passioura & J.E.Ash = ***Eucalyptus botryoides*** Sm.
Eucalyptus saligna var. *pallidivalvis* R.T.Baker & H.G.Sm. = ***Eucalyptus saligna*** Sm.
Eucalyptus saligna var. *parviflora* H.Deane & Maiden = ***Eucalyptus deanei*** Maiden
Eucalyptus saligna var. *protrusa* Blakely & McKie = ***Eucalyptus saligna*** Sm.
Eucalyptus salubris var. *glauca* Maiden = ***Eucalyptus ravida*** L.A.S.Johnson & K.D.Hill
Eucalyptus santalifolia F.Muell. = ***Eucalyptus diversifolia*** Bonpl. subsp. ***diversifolia***
Eucalyptus santalifolia var. *baxteri* Benth. = ***Eucalyptus baxteri*** (Benth.) J.M.Black
Eucalyptus scabra Dum.Cours. = ***Eucalyptus piperita*** Sm. subsp. ***piperita***
Eucalyptus scabrida Brooker & A.R.Bean = ***Corymbia scabrida*** (Brooker & A.R.Bean) K.D.Hill & L.A.S.Johnson
Eucalyptus sclerophylla (Blakely) L.A.S.Johnson & Blaxell = ***Eucalyptus racemosa*** subsp. ***racemosa***
Eucalyptus scyphoidea Maiden = ?
Eucalyptus seeana var. *constricta* Blakely = ***Eucalyptus bancroftii*** (Maiden) Maiden
Eucalyptus semicorticata F.Muell. = ***Eucalyptus pilularis*** Sm.
Eucalyptus sepulcralis var. *robusta* C.A.Gardner = ***Eucalyptus*** × ***chrysantha*** Blakely & H.Steedman
Eucalyptus serendipita Brooker & Kleinig = ***Corymbia serendipita*** (Brooker & Kleinig) A.R.Bean
Eucalyptus setosa Schauer = ***Corymbia setosa*** (Schauer) K.D.Hill & L.A.S.Johnson
Eucalyptus shiressii Maiden & Blakely = ***Eucalyptus punctata*** A.Cunn. ex DC.
Eucalyptus siderophloia f. *decorticans* F.M.Bailey = ***Eucalyptus decorticans*** (F.M.Bailey) Maiden
Eucalyptus siderophloia var. *glauca* H.Deane & Maiden = ***Eucalyptus fibrosa*** subsp. ***nubila*** (Maiden & Blakely) L.A.S.Johnson
Eucalyptus siderophloia var. *rostrata* Benth. = ***Eucalyptus fibrosa*** F.Muell. subsp. ***fibrosa***
Eucalyptus sideroxylon var. *minor* (Benth.) Maiden = ***Eucalyptus sideroxylon*** A.Cunn. ex Woolls
Eucalyptus sideroxylon var. *pallens* (Benth.) Rehder = ***Eucalyptus caleyi*** Maiden subsp. ***caleyi***
Eucalyptus sideroxylon var. *rosea* Rehder = ***Eucalyptus sideroxylon*** A.Cunn. ex Woolls
Eucalyptus sideroxylon subsp. *tricarpa* L.A.S.Johnson = ***Eucalyptus tricarpa*** (L.A.S.Johnson) L.A.S.Johnson & K.D.Hill
Eucalyptus sieberiana F.Muell. = ***Eucalyptus sieberi*** L.A.S.Johnson
Eucalyptus sieberiana var. *oxleyensis* H.Deane & Maiden = ***Eucalyptus sieberi*** L.A.S.Johnson
Eucalyptus signata F.Muell. = ***Eucalyptus racemosa*** subsp. ***racemosa***
Eucalyptus simmondsii Maiden = ***Eucalyptus nitida*** Hook.f.
Eucalyptus sparsiflora Blakely = ***Eucalyptus oblonga*** A.Cunn. ex DC.
Eucalyptus sparsifolia Blakely = ***Eucalyptus oblonga*** A.Cunn. ex DC.
Eucalyptus spathulata var. *grandiflora* Benth. = ***Eucalyptus spathulata*** subsp. ***grandiflora*** (Benth.) L.A.S.Johnson & Blaxell
Eucalyptus spectabilis F.Muell. = ***Eucalyptus pellita*** F.Muell.
Eucalyptus spenceriana Maiden = ***Eucalyptus tectifica*** F.Muell.
Eucalyptus splachnicarpa Hook. = ***Corymbia calophylla*** (R.Br. ex Lindl.) K.D.Hill & L.A.S.Johnson
Eucalyptus spodophylla F.Muell. = ***Eucalyptus pruinosa*** Schauer subsp. ***pruinosa***
Eucalyptus stannariensis F.M.Bailey = ***Eucalyptus cloeziana*** F.Muell.
Eucalyptus stellulata var. *alpina* (Lindl.) Ewart = ***Eucalyptus alpina*** Lindl.
Eucalyptus stellulata var. *angustifolia* Benth. = ***Eucalyptus moorei*** Maiden & Cambage var. ***moorei***
Eucalyptus stellulata var. *luehmanniana* (F.Muell.) F.Muell. = ***Eucalyptus luehmanniana*** F.Muell.
Eucalyptus stenophylla Link = ?
Eucalyptus stjohnii (F.Muell. ex R.T.Baker) F.Muell. ex R.T.Baker = ***Eucalyptus globulus*** subsp. ***pseudoglobulus*** (Naudin ex Maiden) J.B.Kirkp.
Eucalyptus stockeri D.J.Carr & S.G.M.Carr = ***Corymbia stockeri*** (D.J.Carr & S.G.M.Carr) K.D.Hill & L.A.S.Johnson

Eucalyptus stoneana F.M.Bailey = ***Eucalyptus leptophleba*** F.Muell.
Eucalyptus stricta var. *pyrifera* Blakely = [50 NSW] *E. sieberi* × *E. stricta*
Eucalyptus stricta var. *rigida* H.Deane & Maiden = ?
Eucalyptus stricta var. *subcampanulata* Blakely = ***Eucalyptus stricta*** Sieber ex Spreng.
Eucalyptus stuartiana F.Muell. ex Miq. = ***Eucalyptus ovata*** Labill.
Eucalyptus stuartiana var. *amblycorys* Blakely = ***Eucalyptus bridgesiana*** F.Muell. ex R.T.Baker
Eucalyptus stuartiana var. *cordata* R.T.Baker & H.G.Sm. = ***Eucalyptus cinerea*** F.Muell. ex Benth. subsp. ***cinerea***
Eucalyptus stuartiana var. *grossa* Maiden = ***Eucalyptus melanoxylon*** Maiden
Eucalyptus stuartiana var. *longifolia* Benth. = ***Eucalyptus ovata*** Labill.
Eucalyptus stuartiana var. *parviflora* H.Deane & Maiden = ***Eucalyptus angophoroides*** R.T.Baker
Eucalyptus submultiplinervis Miq. = ***Eucalyptus pauciflora*** Sieber ex Spreng. subsp. ***pauciflora***
Eucalyptus subulata A.Cunn. ex Schauer = ***Eucalyptus tereticornis*** Sm.
Eucalyptus subvelutina (F.Muell.) Brooker = ***Angophora subvelutina*** F.Muell.
Eucalyptus suggrandis subsp. *alipes* L.A.S.Johnson & K.D.Hill = ***Eucalyptus alipes*** (L.A.S.Johnson & K.D.Hill) D.Nicolle & Brooker
Eucalyptus sylvicultrix F.Muell. ex Benth. = ***Eucalyptus pauciflora*** Sieber ex Spreng. subsp. ***pauciflora***
Eucalyptus symonii D.J.Carr & S.G.M.Carr = ***Corymbia eremaea*** (D.J.Carr & S.G.M.Carr) K.D.Hill & L.A.S.Johnson subsp. ***eremaea***
Eucalyptus tamala D.J.Carr & S.G.M.Carr = ***Eucalyptus oraria*** L.A.S.Johnson
Eucalyptus tasmanica Blakely = ***Eucalyptus delegatensis*** subsp. ***tasmaniensis*** Boland
Eucalyptus × *tenandrensis* Maiden = ***Eucalyptus* × *peacockeana*** Maiden
Eucalyptus tereticornis var. *bancroftii* Maiden = ***Eucalyptus bancroftii*** (Maiden) Maiden
Eucalyptus tereticornis var. *brachycorys* Benth. = ***Eucalyptus punctata*** A.Cunn. ex DC.
Eucalyptus tereticornis var. *brevifolia* Benth. = ***Eucalyptus bancroftii*** (Maiden) Maiden
Eucalyptus tereticornis var. *brevirostris* Benth. = ***Eucalyptus tereticornis*** Sm.
Eucalyptus tereticornis var. *cineolifera* R.T.Baker & H.G.Sm. = ***Eucalyptus tereticornis*** Sm.
Eucalyptus tereticornis var. *dealbata* (A.Cunn. ex Schauer) H.Deane & Maiden = ***Eucalyptus dealbata*** A.Cunn. ex Schauer
Eucalyptus tereticornis var. *glaucina* (Blakely) Cameron = ***Eucalyptus glaucina*** (Blakely) L.A.S.Johnson
Eucalyptus tereticornis var. *latifolia* Benth. = [50 QLD] *E. platyphylla* × *E. tereticornis*
Eucalyptus tereticornis var. *linearis* R.T.Baker & H.G.Sm. = ***Eucalyptus seeana*** Maiden
Eucalyptus tereticornis var. *media* (Blakely) Cameron = ?
Eucalyptus tereticornis var. *pruiniflora* (Blakely) Cameron = ***Eucalyptus tereticornis*** Sm.
Eucalyptus tereticornis var. *rostrata* Ewart = ***Eucalyptus camaldulensis*** Dehnh. subsp. ***camaldulensis***
Eucalyptus tereticornis var. *squamosa* (H.Deane & Maiden) Maiden = ***Eucalyptus squamosa*** H.Deane & Maiden
Eucalyptus teretiuscula Ettingsh. = [50]
Eucalyptus terminalis F.Muell. = ***Corymbia terminalis*** (F.Muell.) K.D.Hill & L.A.S.Johnson
Eucalyptus terminalis var. *carnosa* F.M.Bailey = ***Corymbia polycarpa*** (F.Muell.) K.D.Hill & L.A.S.Johnson
Eucalyptus terminalis var. *longipedata* Maiden & Blakely = ***Corymbia bleeseri*** (Blakely) K.D.Hill & L.A.S.Johnson
Eucalyptus tessellaris F.Muell. = ***Corymbia tessellaris*** (F.Muell.) K.D.Hill & L.A.S.Johnson
Eucalyptus tessellaris var. *dallachiana* Benth. = ***Corymbia dallachiana*** (Benth.) K.D.Hill & L.A.S.Johnson
Eucalyptus × *tokwa* D.J.Carr & S.G.M.Carr = [43 NWG 50 NTA QLD WAU] *Corymbia latifolia* × *Corymbia novoguinensis*
Eucalyptus torelliana F.Muell. = ***Corymbia torelliana*** (F.Muell.) K.D.Hill & L.A.S.Johnson
Eucalyptus trachyphloia F.Muell. = ***Corymbia trachyphloia*** (F.Muell.) K.D.Hill & L.A.S.Johnson
Eucalyptus trachyphloia f. *fruticosa* F.M.Bailey = ***Corymbia trachyphloia*** (F.Muell.) K.D.Hill & L.A.S.Johnson subsp. ***trachyphloia***
Eucalyptus triantha Link = ***Eucalyptus capitellata*** Sm.
Eucalyptus triantha var. *carnea* Domin = ***Eucalyptus capitellata*** Sm.
Eucalyptus triartha Domin = [50]
Eucalyptus trinervis Ettingsh. = [50]
Eucalyptus triplex L.A.S.Johnson & K.D.Hill = ***Eucalyptus cinerea*** subsp. ***triplex*** (L.A.S.Johnson & K.D.Hill) Brooker, Slee & J.D.Briggs
Eucalyptus tropica Cambage ex Maiden = ***Eucalyptus argillacea*** W.Fitzg. ex Maiden
Eucalyptus tuberculata Parm. ex DC. = ?
Eucalyptus turbinata Behr & F.Muell. ex Miq. = ***Eucalyptus oleosa*** F.Muell. ex Miq. subsp. ***oleosa***
Eucalyptus umbellata (Gaertn.) Domin = ***Eucalyptus tereticornis*** Sm.
Eucalyptus umbellata Dum.Cours. = [50]
Eucalyptus umbellata var. *dealbata* (A.Cunn. ex Schauer) Domin = ***Eucalyptus dealbata*** A.Cunn. ex Schauer
Eucalyptus umbellata var. *glaucina* Blakely = ***Eucalyptus glaucina*** (Blakely) L.A.S.Johnson
Eucalyptus umbellata var. *latifolia* (Benth.) Blakely = ?
Eucalyptus umbellata var. *media* Blakely = [50 VIC] *E. camaldulensis* × *E. tereticornis*
Eucalyptus umbellata var. *pruiniflora* Blakely = ***Eucalyptus tereticornis*** Sm.
Eucalyptus umbonata D.J.Carr & S.G.M.Carr = ***Corymbia umbonata*** (D.J.Carr & S.G.M.Carr) K.D.Hill & L.A.S.Johnson
Eucalyptus umbra subsp. *carnea* (F.Muell. ex R.T.Baker) L.A.S.Johnson = ***Eucalyptus carnea*** F.Muell. ex R.T.Baker
Eucalyptus uncinata var. *latifolia* Benth. = ***Eucalyptus uncinata*** Turcz.
Eucalyptus uncinata var. *major* Benth. = ***Eucalyptus rigidula*** Cambage & Blakely
Eucalyptus uncinata var. *pedicellata* Hochr. = ***Eucalyptus uncinata*** Turcz.
Eucalyptus uncinata var. *rostrata* Benth. = ***Eucalyptus transcontinentalis*** Maiden subsp. ***transcontinentalis***
Eucalyptus undulata Sweet = ***Corymbia ferruginea*** (Schauer) K.D.Hill & L.A.S.Johnson subsp. ***ferruginea***
Eucalyptus urceolaris Maiden & Blakely = ***Eucalyptus piperita*** subsp. ***urceolaris*** (Maiden & Blakely) L.A.S.Johnson & Blaxell

Eucalyptus urnigera var. *elongata* Rodway = ***Eucalyptus urnigera*** Hook.f.
Eucalyptus × *urnularis* D.J.Carr & S.G.M.Carr = [50 NTA] *Corymbia dichromophloia* × *Corymbia latifolia*
Eucalyptus variegata F.Muell. = ***Corymbia citriodora*** subsp. ***variegata*** (F.Muell.) A.R.Bean & M.W. McDonald
Eucalyptus verrucosa Ladiges & Whiffin = ***Eucalyptus verrucata*** Ladiges & Whiffin
Eucalyptus verrucosa Sweet = ?
Eucalyptus verrucosa Colla = ?
Eucalyptus viminalis Hook. = ***Corymbia tessellaris*** (F.Muell.) K.D.Hill & L.A.S.Johnson
Eucalyptus viminalis var. *baeuerlenii* (F.Muell.) H.Deane & Maiden = ***Eucalyptus baeuerlenii*** F.Muell.
Eucalyptus viminalis var. *dealbata* (A.Cunn. ex Schauer) C.Moore & Betche = ***Eucalyptus dealbata*** A.Cunn. ex Schauer
Eucalyptus viminalis var. *huberiana* (Naudin) N.T.Burb. = ***Eucalyptus viminalis*** Labill. subsp. ***viminalis***
Eucalyptus viminalis var. *macrocarpa* Rodway = ***Eucalyptus*** × ***unialata*** R.T.Baker & H.G.Sm.
Eucalyptus viminalis var. *microcarpa* Miq. = ***Eucalyptus rubida*** H.Deane & Maiden subsp. ***rubida***
Eucalyptus viminalis var. *pedicellaris* H.Deane & Maiden = ***Eucalyptus smithii*** F.Muell. ex R.T.Baker
Eucalyptus viminalis var. *racemosa* Maiden = ***Eucalyptus viminalis*** subsp. ***pryoriana*** (L.A.S. Johnson) Brooker & Slee
Eucalyptus viminalis var. *rhynchocorys* Maiden = ***Eucalyptus viminalis*** Labill. subsp. ***viminalis***
Eucalyptus virgata var. *altior* H.Deane & Maiden = ***Eucalyptus oreades*** F.Muell. ex R.T.Baker
Eucalyptus virgata var. *fraxinoides* (H.Deane & Maiden) Maiden = ***Eucalyptus fraxinoides*** H.Deane & Maiden
Eucalyptus virgata var. *obtusiflora* (A.Cunn. ex DC.) Maiden = ***Eucalyptus obtusiflora*** A.Cunn. ex DC.
Eucalyptus virgata var. *stricta* (Sieber ex Spreng.) Maiden = ***Eucalyptus stricta*** Sieber ex Spreng.
Eucalyptus virgata var. *triflora* Maiden = ***Eucalyptus triflora*** (Maiden) Blakely
Eucalyptus viridis var. *latiuscula* Blakely = [50 QLD] *E. microcarpa* × *E. viridis*
Eucalyptus viridis var. *ovata* Blakely = ***Eucalyptus viridis*** F.Muell. ex R.T.Baker subsp. ***viridis***
Eucalyptus × *vitrea* var. *thryptomena* Blakely = ***Eucalyptus*** × ***vitrea*** F.Muell. ex R.T.Baker
Eucalyptus × *vittellina* Naudin = ?
Eucalyptus watsoniana F.Muell. = ***Corymbia watsoniana*** (F.Muell.) K.D.Hill & L.A.S.Johnson
Eucalyptus watsoniana subsp. *capillata* Brooker & A.R. Bean = ***Corymbia watsoniana*** subsp. ***capillata*** (Brooker & A.R.Bean) K.D.Hill & L.A.S.Johnson
Eucalyptus whittingehameii Landsb. = ***Eucalyptus gunnii*** subsp. ***gunnii***
Eucalyptus whittingehamensis G.Nicholson ex Elwes & Henry = ?
Eucalyptus wiburdii Blakely = ***Eucalyptus eugenioides*** Sieber ex Spreng.
Eucalyptus wilkinsoniana R.T.Baker = ***Eucalyptus eugenioides*** Sieber ex Spreng.
Eucalyptus wilkinsoniana var. *crassifructa* Blakely = ***Eucalyptus eugenioides*** Sieber ex Spreng.
Eucalyptus wimmerensis Rule = ***Eucalyptus viridis*** subsp. ***wimmerensis*** (Rule) Brooker & Slee
Eucalyptus woodsiana (F.M.Bailey) Brooker = ***Angophora woodsiana*** F.M.Bailey
Eucalyptus woollsiana F.Muell. ex R.T.Baker = ***Eucalyptus microcarpa*** (Maiden) Maiden
Eucalyptus woollsii F.Muell. = ***Eucalyptus longifolia*** Link
Eucalyptus xanthope A.R.Bean & Brooker = ***Corymbia xanthope*** (A.R.Bean & Brooker) K.D.Hill & L.A.S. Johnson
Eucalyptus yangoura Blakely = ***Eucalyptus globoidea*** Blakely
Eucalyptus youmanii var. *sphaerocarpa* Blakely & McKie = ***Eucalyptus youmanii*** Blakely & McKie
Eucalyptus yumbarrana subsp. *striata* Boomsma = [50 SOA] *E. canescens* × *E. volkesensis*
Eucalyptus zygophylla Blakely = ***Corymbia zygophylla*** (Blakely) K.D.Hill & L.A.S.Johnson

Unplaced Names:
Eucalyptus alata Sweet, Hort. Brit.: 157 (1826). = ?
Eucalyptus albicans F.Muell., Fragm. 7: 42 (1869). = ?
Eucalyptus albicaulis Sweet, Hort. Brit.: 157 (1826). = ?
Eucalyptus × *algeriensis* A.Vilm. ex Trab., Rev. Hort. Algérie Tunisie Maroc 8: 146 (1904). = E. *camaldulensis* × *E. rudris*
Eucalyptus amygdalina var. *hypericifolia* Benth., Fl. Austral. 3: 203 (1867). = [50 TAS] *E. amygdalina* × *E. risdonii*
Eucalyptus androsaemifolia Hoffmanns., Verz. Pfl.-Kult., Nachtr. 2: 113 (1841). = ?
Eucalyptus × *antipolitensis* Trab. ex Maiden, Crit. Revis. Eucalyptus 6: 70 (1922). = ?
Eucalyptus apodophylla var. *brachyphylla* Blakely & Jacobs, Key Eucalypts: 166 (1934). = *E. alba* × *E. apodophylla*
Eucalyptus bauerlesii F.Muell., Victorian Naturalist 7: 76 (1890). = ?
Eucalyptus beauchampiana Elwes & A.Henry, Trees Great Britain 6: 1615 (1912). = ?
Eucalyptus × *biangularis* Simmonds in J.H.Maiden, Crit. Revis. Eucalyptus 7: 382 (1927). = *E. globulus* × *E. urnigera*
Eucalyptus botryoides var. *lynei* Blakely, Key Eucalypts: 97 (1934). = [50 NSW] *E. botryoides* × *E. robusta*
Eucalyptus ×*bourlieri* Trab., Rev. Hort. Algérie Tunisie Maroc 1903: 327 (1903). = ?
Eucalyptus calcicultrix var. *obscura* Blakely, Key Eucalypts: 224 (1934). = [50 SOA] *E. albens* × *E. odorata*
Eucalyptus × *californica* Kinney, Eucalyptus: 191 (1895). = ?
Eucalyptus calophylla var. *hawkeyi* Blakely, Key Eucalypts: 85 (1934). = [50 WAU] *Corymbia calophylla* × *Corymbia ficifolia*
Eucalyptus carnea C.A.Gardner, J. Roy. Soc. Western Australia 14: 82 (1928), nom. illeg. = ?
Eucalyptus × *coniophloia* D.J.Carr & S.G.M.Carr, Eucalyptus 1: 97 (1985). = [50 NTA WAU] *Corymbia capricornia* × *Corymbia drysdalensis*
Eucalyptus connata Dum.Cours., Bot. Cult., ed. 2, 7: 280 (1814). = ?
Eucalyptus cotinifolia Colla, Herb. Pedem. 2: 423 (1834). = ?
Eucalyptus × *cultrifolia* Naudin, Descr. Emploi Eucalypt.: 64 (1891). = ?
Eucalyptus curvula Sieber ex Spreng., Syst. Veg. 4(2): 195 (1827). = ?
Eucalyptus ficifolia var. *alba* Blakely, Key Eucalypts: 86 (1934), nom. illeg. = *E. calophylla* × *E. ficifolia*
Eucalyptus ficifolia var. *guilfoylei* F.M.Bailey, Proc. Roy. Soc. Queensland 10: 18 (1894). = *Corymbia calophylla* × *Corymbia ficifolia*
Eucalyptus firma F.Muell. ex Miq., Ned. Kruidk. Arch. 4: 133 (1856). = ?

Eucalyptus fissilis Guilf., Rep. (Annual) Director Bot. Gard. Melbourne 1877: 25 (1877). = ?
Eucalyptus flavieri Añon, Rev. Int. Bot. Appl. Agric. Colon. 26: 234 (1946). = ?
Eucalyptus flexilis Regel, Gartenflora 7 7: 284 (1858). = ?
Eucalyptus gillii var. *petiolaris* Maiden, J. Proc. Roy. Soc. New S. Wales 53: 69 (1919). = [50 SOA] *E. gillii* × *E. socialis*
Eucalyptus globoidea var. *largifructa* Blakely, Key Eucalypts: 189 (1934). = ?
Eucalyptus globulus var. *compacta* Maiden, Crit. Revis. Eucalyptus 8: 23 (1929). = ?
Eucalyptus × *gomphocornuta* A.Vilm. ex Trab., Rev. Hort., II, 3: 326 (1903). = *E. cornuta* × *E. gomphocephala*
Eucalyptus grandis var. *grandiflora* Maiden, J. Proc. Roy. Soc. New S. Wales 52: 502 (1919). = *E. grandis* × *E. robusta*
Eucalyptus inophloea Brogli, Beitr. Anat. Myrtac.-Rinden: 20 (1926). = [50]
Eucalyptus insingiana Maiden, Crit. Revis. Eucalyptus 6: 353 (1923). = ?
Eucalyptus × *insizwaensis* Maiden, Crit. Revis. Eucalyptus 6: 82 (1922). = *E. globulus* × *E. robusta*
Eucalyptus irbyi R.T.Baker & H.F.Sm., Res. Eucalypts, ed. 2: 242 (1920). = [50 TAS]
Eucalyptus × *jugalis* Naudin, Descr. Emploi Eucalypt.: 37 (1891). = ?
Eucalyptus kitsonii H.Deane in ?. = Fossil species.
Eucalyptus laevopinea var. *turbinata* Blakely, Key Eucalypts: 183 (1934). = ?
Eucalyptus × *lamprocalyx* Blakely, Key Eucalypts: 87 (1934). = [50 WAU] *Corymbia cadophora* × *Corymbia polycarpa*
Eucalyptus laseronii var. *doleiformis* Blakely, Key Eucalypts: 193 (1934). = ?
Eucalyptus laseronii var. *maxima* Blakely, Key Eucalypts: 193 (1934). = ?
Eucalyptus leucoxylon var. *pluriflora* Miq., Ned. Kruidk. Arch. 4: 127 (1856). = ?
Eucalyptus longifolia var. *multiflora* Maiden, Crit. Revis. Eucalyptus 6: 432 (1923). = *E. longifolia* × *E. robusta*
Eucalyptus longifolia var. *turbinata* Blakely, Key Eucalypts: 102 (1934). = [50 NSW] *E. longifolia* × *E. tereticornis*
Eucalyptus maidenii var. *williamsonii* Blakely, Key Eucalypts: 157 (1934). = [50 VIC] *E. botruoides* × *E. globulus* subsp. *pseudoglobulus*
Eucalyptus mazeliana Naudin, Descr. Emploi Eucalypt.: 41 (1891). = ?
Eucalyptus × *mcclatchiei* Kinney, Eucalyptus: 188 (1895). = *E. globulus* × *E. ovata*
Eucalyptus × *media* Link, Enum. Hort. Berol. Alt. 2: 30 (1822). = [50+]
Eucalyptus melanophloia var. *senta* Blakely, Key Eucalypts: 253 (1934). = ?
Eucalyptus microphylla Muhl. ex Willd., Enum. Pl.: 515 (1809). = not Eucalyptus
Eucalyptus microtheca var. *cymbaliformis* Blakely & Jacobs in W.F.Blakely, Key Eucalypts: 245 (1934). = [50 NTA] *E. cyanoclada* × *E. microtheca*
Eucalyptus mitchellii Ettingsh., Blatt-Skel. Dikot.: 203 (1861). = Fossil species.
Eucalyptus moorei var. *arborea* Blakely, Key Eucalypts: 207 (1934). = *E. moorei* × *E. piperita*
Eucalyptus × *mortoniana* Kinney, Eucalyptus: 193 (1895). = *E. globulus* × *E. viminalis*
Eucalyptus mucronata Link, Enum. Hort. Berol. Alt. 2: 30 (1822). = ?
Eucalyptus multiflora Rich. ex A.Gray, U.S. Expl. Exped., Phan. 1: 554 (1854), nom. illeg. = [42 PHI]
Eucalyptus myrtifolia Link, Enum. Hort. Berol. Alt. 2: 30 (1822). = ?
Eucalyptus × *nervosa* Hoffmanns., Verz. Pfl.-Kult. 1: 134 (1824). = ?
Eucalyptus × *nowraensis* Maiden, Crit. Revis. Eucalyptus 7: 68 (1924). = [50 NSW] *Corymbia gummifera* × *Corymbia maculata*
Eucalyptus obliqua var. *discocarpa* Blakely, Key Eucalypts: 194 (1934). = [50 VIC] *E. muelleriana* × *E. obliqua*
Eucalyptus obliqua var. *microstoma* Blakely, Key Eucalypts: 194 (1934). = ?
Eucalyptus obliqua var. *pilula* Blakely, Key Eucalypts: 194 (1934). = ?
Eucalyptus occidentalis var. *californica* Kinney, Eucalyptus: 92 (1895). = ?
Eucalyptus odorata var. *macrocarpa* Blakely, Key Eucalypts: 226 (1934). = [50 SOA] *E. leucoxyla* × *E. odorata*
Eucalyptus pachypoda Benth., Fl. Austral. 3: 233 (1867). = ?
Eucalyptus × *pachypoda* F.Muell., Fragm. 7: 41 (1870), nom. illeg. = ?
Eucalyptus pauciflora var. *cylindrocarpa* Blakely, Key Eucalypts: 205 (1934). = [50 NSW]
Eucalyptus pauciflora var. *densiflora* Blakely & McKie, Proc. Linn. Soc. New South Wales 63: 66 (1938). = [50 NSW]
Eucalyptus pauciflora var. *rusticata* Blakely, Key Eucalypts: 205 (1934). = ?
Eucalyptus piperita var. *orophila* Blakely, Key Eucalypts: 339 (1934). = [50 NSW] *E. oreodes* × *E. piperita*
Eucalyptus × *plurilocularis* F.Muell., Fragm. 2: 70 (1860). = ?
Eucalyptus × *purpurascens* Link, Enum. Hort. Berol. Alt. 2: 31 (1822). = ?
Eucalyptus × *rameliana* A.Vilm. ex Trab., Assoc. Franç. 'Avancem. Sci. Marseille 1891: 463 (1892), nom. illeg. = ?
Eucalyptus × *rameliana* var. *trabutii* Vilm. ex Trab., Bull. Stat. Recherch. Forest. Nord Afr. 1: 141 (1917). = *E. botryoides* × *E. camaldulensis*
Eucalyptus × *reticulata* Link, Enum. Hort. Berol. Alt. 2: 29 (1822). = ?
Eucalyptus × *rigida* Hoffmanns., Verz. Pfl.-Kult., Nachtr. 2: 114 (1841). = ?
Eucalyptus × *rigida* var. *luehmanniana* F.Muell., Eucalyptographia 4: t. 6 (1879). = ?
Eucalyptus × *rubricaulis* Desf., Tabl. École Bot., ed. 3: 408 (1829). = ?
Eucalyptus scyphoidea Maiden, Proc. Linn. Soc. New South Wales 28: 899 (1904). = ?
Eucalyptus stenophylla Link, Enum. Hort. Berol. Alt. 2: 30 (1822). = ?
Eucalyptus stricta var. *pyrifera* Blakely, Key Eucalypts: 202 (1934). = [50 NSW] *E. sieberi* × *E. stricta*
Eucalyptus stricta var. *rigida* H.Deane & Maiden, Proc. Linn. Soc. New South Wales 22: 710 (1898). = ?
Eucalyptus tereticornis var. *latifolia* Benth., Fl. Austral. 3: 242 (1867). = [50 QLD] *E. platyphylla* × *E. tereticornis*
Eucalyptus teretiuscula Ettingsh., Blatt-Skel. Dikot.: 201 (1861). = [50]
Eucalyptus × *tokwa* D.J.Carr & S.G.M.Carr, Eucalyptus 2: 152 (1987). = [43 NWG 50 NTA QLD WAU] *Corymbia latifolia* × *Corymbia novoguinensis*

Eucalyptus triartha Domin, Biblioth. Bot. 89: 461 (1928). = [50]
Eucalyptus trinervis Ettingsh., Blatt-Skel. Dikot.: 204 (1861). = [50]
Eucalyptus tuberculata Parm. ex DC., Prodr. 3: 221 (1828). = ?
Eucalyptus umbellata Dum.Cours., Bot. Cult., ed. 2, 7: 279 (1814). = [50]
Eucalyptus umbellata var. *media* Blakely, Key Eucalypts: 130 (1934). = [50 VIC] *E. camaldulensis* × *E. tereticornis*
Eucalyptus × *urnularis* D.J.Carr & S.G.M.Carr, Eucalyptus 1: 87 (1985). = [50 NTA] *Corymbia dichromophloia* × *Corymbia latifolia*
Eucalyptus verrucosa Sweet, Hort. Brit.: 157 (1826). = ?
Eucalyptus verrucosa Colla, Herb. Pedem. 2: 422 (1834), nom. illeg. = ?
Eucalyptus viridis var. *latiuscula* Blakely, Key Eucalypts: 229 (1934). = [50 QLD] *E. microcarpa* × *E. viridis*
Eucalyptus × *vittellina* Naudin, Descr. Emploi Eucalypt.: 65 (1891). = ?
Eucalyptus whittingehamensis G.Nicholson ex Elwes & Henry, Trees Great Britain 6: 1642 (1912). = ?
Eucalyptus yumbarrana subsp. *striata* Boomsma, J. Adelaide Bot. Gard. 2: 298 (1980). = [50 SOA] *E. canescens* × *E. volkesensis*

Eudesmia

Eudesmia R.Br. = ***Eucalyptus*** L'Hér.
Eudesmia erythrocorys F.Muell. = ***Eucalyptus erythrocorys*** F.Muell.
Eudesmia eucalyptoides F.Muell. = ***Eucalyptus eudesmioides*** F.Muell.
Eudesmia tetragona R.Br. = ***Eucalyptus tetragona*** (R.Br.) F.Muell.

Eugenia

Eugenia P.Micheli ex L., Sp. Pl.: 470 (1753).
Trop. 22 BEN GHA GUI IVO LBR MLI NGA SIE TOG 23 BUR CAB CAF CMN CON GAB GGI RWA ZAI 24 SOM 25 KEN TAN UGA 26 ANG MLW MOZ ZAM ZIM 27 CPP NAT SWZ TVL 29 ALD COM MAU MDG REU ROD 40 BAN IND SRL 41 MYA THA VIE 42 BOR LSI MOL PHI SUL 43 SOL 50 QLD WAU? 60 FIJ NUE NWC TON 61 MRQ SCI TUA 62 MRN 63 HAW 78 FLA 79 MXC MXE MXG MXN MXS MXT 80 BLZ COS CPI ELS GUA HON NIC PAN 81 BAH BER CAY COM CUB DOM HAI JAM LEE NLA PUE TRT WIN 82 FRG GUY SUR VEN 83 BOL CLM ECU PER 84 BZC BZE BZL BZN BZS 85 AGE AGW PAR URU. [Myrtaceae]
999 Species
Catinga Aubl., Hist. Pl. Guiane 1: 511 (1775).
Greggia Gaertn., Fruct. Sem. Pl. 1: 168 (1788), nom. illeg.
Olynthia Lindl., Coll. Bot.: t. 19 (1821).
Calophylloides Smeathman ex DC., Prodr. 3: 272 (1828).
Jossinia Comm. ex DC., Prodr. 3: 237 (1828).
Emurtia Raf., Sylva Tellur.: 106 (1838).
Epleienda Raf., Sylva Tellur.: 107 (1838).
Phyllocalyx O.Berg, Linnaea 27: 306 (1856).
Stenocalyx O.Berg, Linnaea 27: 309 (1856).
Myrtopsis O.Hoffm., Linnaea 43: 133 (1881).
Psidiastrum Bello, Anales Soc. Esp. Hist. Nat. 10: 272 (1881).
Chloromyrtus Pierre, Bull. Mens. Soc. Linn. Paris, n.s., 1: 71 (1898).
Myrcialeucus Rojas, Bull. Acad. Int. Géogr. Bot. 24: 217 (1914).
Episyzygium Suess. & A.Ludw., Mitt. Bot. Staatssamml. München 1: 18 (1950).
Pilothecium (Kiaersk.) Kausel, Ark. Bot., a.s., 4: 401 (1962).
Pseudeugenia D.Legrand & Mattos, Arq. Bot. Estado São Paulo, n.s., f.m., 4(2): 63 (1966).

Eugenia abbreviata Urb., Symb. Antill. 6: 24 (1909).
Jamaica. 81 JAM. Nanophan.

Eugenia aboukirensis Proctor, J. Arnold Arbor. 63: 273 (1982).
Jamaica. 81 JAM. Phan.

Eugenia acapulcensis Steud., Nomencl. Bot., ed. 2, 1: 601 (1840).
Mexico to Brazil. 79 MXC MXE MXG MXN MXS MXT 80 BLZ COS ELS GUA HON NIC PAN 83 CLM 84 BZE. Nanophan. or phan.
Eugenia carthagenensis Jacq., Enum. Syst. Pl.: 23 (1760), provisional synonym.
**Myrtus maritima* Kunth in F.W.H.von Humboldt, A.J.A.Bonpland & C.S.Kunth, Nov. Gen. Sp. 6: 146 (1823). *Eugenia maritima* (Kunth) DC., Prodr. 3: 282 (1828), nom. illeg. *Eugenia ilhensis* O.Berg in C.F.P.von Martius & auct. suc. (eds.), Fl. Bras. 14(1): 283 (1857), nom. illeg.
Eugenia bonplandiana O.Berg, Linnaea 27: 228 (1856).
Eugenia mosquitensis O.Berg, Linnaea 31: 257 (1862).
Eugenia deltoidea Standl., Contr. U. S. Natl. Herb. 23: 1045 (1924).
Eugenia antiquae L.Riley, Bull. Misc. Inform. Kew 1927: 121 (1927).
Eugenia escuintlensis Lundell, Phytologia 2: 4 (1941).
Eugenia bartlettiana Lundell, Contr. Univ. Michigan Herb. 7: 30 (1942).
Eugenia bracteolosa Lundell, Contr. Univ. Michigan Herb. 7: 31 (1942).
Eugenia campechiana Lundell, Contr. Univ. Michigan Herb. 7: 32 (1942).
Eugenia ovatifolia Lundell, Bull. Torrey Bot. Club 69: 396 (1942).
Eugenia sibunensis Lundell, Contr. Univ. Michigan Herb. 7: 34 (1942).
Eugenia comitanensis Lundell, Wrightia 3: 12 (1961).

Eugenia aceitillo Urb., Symb. Antill. 9: 513 (1928).
E. Cuba. 81 CUB. Nanophan. or phan.

Eugenia acrantha Urb., Symb. Antill. 9: 101 (1923).
E. Cuba. 81 CUB. Nanophan.

Eugenia acrensis McVaugh, Fieldiana, Bot. 29: 202 (1956).
N. Brazil. 84 BZN. Phan.

Eugenia acrisepala Govaerts, World Checklist Myrtaceae: 130 (2008).
C. Jamaica. 81 JAM. Nanophan. or phan.
**Eugenia acutisepala* Proctor, Bull. Inst. Jamaica, Sci. Ser. 16: 29 (1967), nom. illeg.

Eugenia acunae Alain, Revista Soc. Cub. Bot. 10: 30 (1953).
E. Cuba. 81 CUB. Nanophan. or phan.

Eugenia acutissima Urb. & Ekman in I.Urban, Symb. Antill. 9: 502 (1928).
NW. Cuba. 81 CUB. Nanophan. or phan.

Eugenia adenantha O.Berg in C.F.P.von Martius & auct. suc. (eds.), Fl. Bras. 14(1): 578 (1859).

Brazil (Rio de Janeiro). 84 BZL. Nanophan. or phan.

Eugenia adenocarpa O.Berg in C.F.P.von Martius & auct. suc. (eds.), Fl. Bras. 14(1): 569 (1859). *Luma adenocarpa* (O.Berg) Burret, Notizbl. Bot. Gart. Berlin-Dahlem 15: 531 (1941).
Brazil (Rio de Janeiro). 84 BZL. Nanophan. or phan.

Eugenia aerosa McVaugh, Fieldiana, Bot. 29: 203 (1956).
N. Brazil to N. Peru. 83 PER 84 BZN. Phan.

Eugenia aeruginea DC., Prodr. 3: 283 (1828).
E. Cuba, Jamaica, SE. Mexico to C. America. 79 MXT 80 BLZ ELS GUA HON 81 CUB JAM. Nanophan. or phan.
Eugenia fadyenii Krug & Urb., Bot. Jahrb. Syst. 19: 622 (1895).
Eugenia fadyenii var. *glabra* Krug & Urb., Bot. Jahrb. Syst. 19: 623 (1895).

Eugenia afzelii Engl., Notizbl. Königl. Bot. Gart. Berlin 2: 290 (1899). *Myrtus afzelii* (Engl.) Kuntze, Deutsche Bot. Monatsschr. 21: 173 (1903).
Sierra Leone. 22 SIE. Nanophan.

Eugenia agathopoda Diels, Verh. Bot. Vereins Prov. Brandenburg 48: 192 (1906 publ. 1907).
N. Brazil. 84 BZN. Phan.

Eugenia aherniana C.B.Rob., Philipp. J. Sci., C 4: 344 (1909). *Jossinia aherniana* (C.B.Rob.) Merr., Philipp. J. Sci. 79: 358 (1951).
Philippines to Sulawesi. 42 PHI SUL. Phan.
Eugenia melastomoides Elmer, Leafl. Philipp. Bot. 4: 1429 (1912).

Eugenia alagoensis (O.Berg) Mattos, Loefgrenia 94: 3 (1989).
NE. Brazil. 84 BZE. Nanophan. or phan.
**Stenocalyx alagoensis* O.Berg in C.F.P.von Martius & auct. suc. (eds.), Fl. Bras. 14(1): 350 (1857).

Eugenia alainii Borhidi, Acta Bot. Acad. Sci. Hung. 19: 39 (1973).
Cuba. 81 CUB. Nanophan. or phan.

Eugenia alaotrensis H.Perrier, Mém. Inst. Sci. Madagascar, Sér. B, Biol. Vég. 4: 173 (1953).
C. Madagascar. 29 MDG. Nanophan. or phan.

Eugenia albicans (O.Berg) Urb., Bot. Jahrb. Syst. 19: 617 (1895).
Lesser Antilles to Guianas. 81 LEE TRT WIN 82 FRG GUY SUR. Nanophan. or phan.
**Stenocalyx albicans* O.Berg, Linnaea 30: 698 (1861).

Eugenia albimarginata Urb. & Ekman, Ark. Bot. 24A(4): 28 (1932).
Haiti (Massif de la Hotte). 81 HAI. Nanophan. or phan.

Eugenia alnifolia McVaugh, Fieldiana, Bot. 29: 425 (1963).
SW. Mexico. 79 MXS. Nanophan. or phan.

Eugenia aloysii C.J.Saldanha, Fl. Karnataka 2: 23 (1996).
India (Karnataka). 40 IND. Nanophan. or phan.

Eugenia alpina (Sw.) Willd., Sp. Pl. 2: 961 (1799).
Jamaica. 81 JAM. Nanophan. or phan.
**Myrtus alpina* Sw., Prodr.: 77 (1788).

Eugenia amatenangensis Lundell, Wrightia 3: 10 (1961).
SE. Mexico. 79 MXT. Nanophan. or phan.

Eugenia amblyophylla Urb., Symb. Antill. 9: 511 (1928).
SE. Cuba. 81 CUB. Nanophan.

Eugenia amblyosepala McVaugh, Mem. New York Bot. Gard. 18(2): 162 (1969).
Venezuela (Amazonas). 82 VEN. Nanophan. or phan.

Eugenia amoena Thwaites, Enum. Pl. Zeyl.: 114 (1859).
Sri Lanka. 40 SRL. Nanophan. or phan.

Eugenia amplifolia Urb., Symb. Antill. 5: 455 (1908).
Jamaica. 81 JAM. Nanophan. or phan.

Eugenia amshoffiae McVaugh, Mem. New York Bot. Gard. 18(2): 163 (1969).
Guyana. 82 GUY. Nanophan. or phan.
**Acrandra guianensis* Amshoff, Acta Bot. Neerl. 5: 279 (1956).

Eugenia anafensis Urb., Repert. Spec. Nov. Regni Veg. 14: 337 (1916).
W. Cuba. 81 CUB. Nanophan.

Eugenia analamerensis H.Perrier, Mém. Inst. Sci. Madagascar, Sér. B, Biol. Vég. 4: 171 (1953).
NW. Madagascar. 29 MDG. Nanophan. or phan.

Eugenia anastomosans DC., Prodr. 3: 269 (1828).
S. Trop. America. 82 FRG GUY SUR VEN 83 CLM ECU 84 BZN. Phan.
Eugenia ptariensis Steyerm., Fieldiana, Bot. 28: 1013 (1957).

Eugenia ancorifera Amshoff, Adansonia, n.s., 14: 481 (1974).
Cameroon. 23 CMN. Nanophan. or phan.

Eugenia angelyana Mattos, Loefgrenia 94: 2 (1989).
Brazil (Santa Catarina). 84 BZS. Phan.

Eugenia angustissima O.Berg in C.F.P.von Martius & auct. suc. (eds.), Fl. Bras. 14(1): 569 (1859).
Bolivia to Brazil (Goiás to São Paulo). 83 BOL 84 BZC BZL. Nanophan. or phan.

Eugenia ankarensis (H.Perrier) A.J.Scott, Kew Bull. 34: 546 (1980).
W. Madagascar. 29 MDG. Nanophan.
**Myrtus ankarensis* H.Perrier, Mém. Inst. Sci. Madagascar, Sér. B, Biol. Vég. 4: 162 (1953).

Eugenia anthacanthoides Urb. & Ekman in I.Urban, Symb. Antill. 9: 486 (1928).
C. & E. Cuba. 81 CUB. Nanophan.
Eugenia squarrosa Urb. & Ekman in I.Urban, Symb. Antill. 9: 487 (1928).

Eugenia antongilensis H.Perrier, Mém. Inst. Sci. Madagascar, Sér. B, Biol. Vég. 4: 173 (1953).
E. Madagascar. 29 MDG. Nanophan. or phan.

Eugenia apiocarpa O.Berg in C.F.P.von Martius & auct. suc. (eds.), Fl. Bras. 14(1): 284 (1857).
Brazil (Rio de Janeiro). 84 BZL. Nanophan. or phan.

Eugenia arawakorum Sandwith, Bull. Misc. Inform. Kew 1932: 211 (1932).
Guyana. 82 GUY. Phan.

Eugenia arayan Seem., Bot. Voy. Herald: 125 (1854).
Colombia. 83 CLM. Phan.

Eugenia ardisioides Lundell, Amer. Midl. Naturalist 29: 479 (1943).
Belize. 80 BLZ. Nanophan. or phan.
Eugenia oblanceifolia Lundell, Amer. Midl. Naturalist 29: 480 (1943).

Eugenia arenaria Cambess. in A.F.C.de Saint-Hilaire, Fl. Bras. Merid. 2: 349 (1832).

SE. & S. Brazil. 84 BZL BZS. Nanophan. or phan.
Eugenia exechusa O.Berg in C.F.P.von Martius & auct. suc. (eds.), Fl. Bras. 14(1): 269 (1857).
Eugenia exechusa var. *uniflora* Kiaersk., Enum. Myrt. Bras.: 122 (1893).

Eugenia arenicola H.Perrier, Mém. Inst. Sci. Madagascar, Sér. B, Biol. Vég. 4: 181 (1953).
S. Madagascar. 29 MDG. Nanophan.

Eugenia arenosa Mattos, Loefgrenia 42: 2 (1970).
S. Brazil. 84 BZS. Hemicr. or cham.
**Psidium herbaceum* O.Berg in C.F.P.von Martius & auct. suc. (eds.), Fl. Bras. 14(1): 410 (1857). *Guajava herbacea* (O.Berg) Kuntze, Revis. Gen. Pl. 1: 239 (1891).
Eugenia hagelundii Mattos, Loefgrenia 76: 1 (1981).

Eugenia argentea Bedd., Fl. Sylv. S. India: cix (1872).
S. India. 40 IND. Nanophan. or phan.

Eugenia argyrophylla B.Holst & M.L.Kawas., Brittonia 52: 20 (2000).
French Guiana. 82 FRG. Phan.

Eugenia armeniaca Sagot, Ann. Sci. Nat., Bot., VI, 20: 190 (1885).
Guianas. 82 FRG GUY. Phan.

Eugenia arrabidae O.Berg, Linnaea 27: 302 (1856).
Brazil (Rio de Janeiro). 84 BZL. Nanophan. or phan.
**Eugenia campestris* Vell., Fl. Flumin. 5: 208, t. 35 (1829), nom. illeg.

Eugenia arthroopoda Baill. ex Drake, Bull. Mens. Soc. Linn. Paris 2: 1222 (1896).
Madagascar. 29 MDG. Nanophan. or phan.
Eugenia arthroopoda var. *ambalavensis* H.Perrier, Mém. Inst. Sci. Madagascar, Sér. B, Biol. Vég. 4: 177 (1952).
Eugenia arthroopoda var. *angustata* H.Perrier, Mém. Inst. Sci. Madagascar, Sér. B, Biol. Vég. 4: 177 (1952).
Eugenia arthroopoda var. *baroniana* H.Perrier, Mém. Inst. Sci. Madagascar, Sér. B, Biol. Vég. 4: 177 (1952).
Eugenia arthroopoda var. *elatipes* H.Perrier, Mém. Inst. Sci. Madagascar, Sér. B, Biol. Vég. 4: 177 (1952).
Eugenia arthroopoda var. *menabeensis* H.Perrier, Mém. Inst. Sci. Madagascar, Sér. B, Biol. Vég. 4: 177 (1952).

Eugenia arvensis Vell., Fl. Flumin. 5: 209, t. 37 (1829).
Brazil (Rio de Janeiro). 84 BZL. Nanophan. or phan.
Eugenia oxyphylla O.Berg in C.F.P.von Martius & auct. suc. (eds.), Fl. Bras. 14(1): 251 (1857).

Eugenia aschersoniana F.Hoffm., Beitr. Fl. Centr.-Ost-Afr.: 35 (1889). *Eugenia capensis* subsp. *aschersoniana* (F.Hoffm.) F.White, Kirkia 10: 403 (1977).
Somalia, Tanzania (Tabora Distr.), Mozambique. 24 SOM 25 TAN 26 MOZ. Nanophan.
Eugenia mossambicensis Engl., Notizbl. Königl. Bot. Gart. Berlin 2: 289 (1899). *Myrtus mossambicensis* (Engl.) Kuntze, Deutsche Bot. Monatsschr. 21: 173 (1903).
Eugenia somalensis Chiov., Fl. Somala 2: 217 (1932).

Eugenia asperifolia O.Berg, Linnaea 30: 674 (1861).
Cuba. 81 CUB. Nanophan.
Eugenia microphylla A.Rich., Hist. Phys. Cuba, Pl. Vasc.: 584 (1846).

Eugenia atricha Urb., Repert. Spec. Nov. Regni Veg. 19: 305 (1924).
W. Cuba. 81 CUB. Nanophan.

Eugenia atroracemosa McVaugh, Fieldiana, Bot. 29: 203 (1956).
N. Brazil to Peru. 83 PER 84 BZN. Phan.

Eugenia atrosquamata McVaugh, Fieldiana, Bot. 29: 204 (1956).
Colombia to Peru. 83 CLM PER. Phan.

Eugenia augustana Kiaersk., Enum. Myrt. Bras.: 153 (1893).
Brazil (Rio de Janeiro). 84 BZL. Nanophan. or phan.

Eugenia aurata O.Berg in C.F.P.von Martius & auct. suc. (eds.), Fl. Bras. 14(1): 273 (1857).
Brazil to Bolivia. 83 BOL 84 BZC BZE BZL BZS 85 PAR. Nanophan. or phan.
Eugenia chrysantha O.Berg in C.F.P.von Martius & auct. suc. (eds.), Fl. Bras. 14(1): 272 (1857).

Eugenia austin-smithii Standl., Publ. Field Mus. Nat. Hist., Bot. Ser. 18: 1561 (1938).
C. America. 80 COS ELS NIC PAN. Nanophan. or phan.

Eugenia avicenniae Standl., Contr. U. S. Natl. Herb. 23: 1043 (1924).
SW. Mexico. 79 MXS. Nanophan.

Eugenia axillaris (Sw.) Willd., Sp. Pl. 2: 960 (1799).
S. Florida, Mexico to C. America, Caribbean. 78 FLA 79 MXG MXT 80 BLZ ELS GUA HON NIC 81 BAH BER CAY COM CUB HAI JAM LEE PUE WIN. Nanophan. or phan.
**Myrtus axillaris* Sw., Prodr.: 78 (1788)

var. ***axillaris***
S. Florida, Mexico to C. America, Caribbean. 78 FLA 79 MXG MXT 80 BLZ ELS GUA HON NIC 81 BAH BER CAY COM CUB HAI JAM LEE PUE WIN. Nanophan. or phan.
Eugenia carthagenensis var. *baruensis* Jacq., Select. Stirp. Amer. Hist.: 153 (1780). *Eugenia baruensis* (Jacq.) Jacq., Collectanea 3: 183 (1791). *Myrtus baruensis* (Jacq.) Spreng., Syst. Veg. 2: 483 (1825).
Eugenia guadalupensis DC., Prodr. 3: 275 (1828).
Eugenia verrucosa A.Rich., Hist. Phys. Cuba, Pl. Vasc.: 589 (1846).
Eugenia cabanisiana O.Berg, Linnaea 27: 235 (1856).
Eugenia divaricata Willd. ex O.Berg, Linnaea 27: 199 (1856), nom. illeg.
Eugenia matanzensis O.Berg, Linnaea 27: 235 (1856).
Eugenia yumuryensis O.Berg, Linnaea 27: 234 (1856).
Psidiastrum dubium Bello, Anales Soc. Esp. Hist. Nat. 10: 272 (1881).
Eugenia axillaris var. *microcarpa* Krug & Urb., Bot. Jahrb. Syst. 19: 641 (1895).
Eugenia anthera Small, Man. S.E. Fl.: 935 (1933).
Eugenia itzana Lundell, Bull. Torrey Bot. Club 69: 395 (1942).

var. ***cozumelensis*** (Lundell) Lundell, Phytologia 16: 443 (1968).
SE. Mexico to Guatemala. 79 MXT 80 GUA. Nanophan. or phan.
**Eugenia cozumelensis* Lundell, Wrightia 3: 13 (1961).

Eugenia azurensis O.Berg in C.F.P.von Martius & auct. suc. (eds.), Fl. Bras. 14(1): 224 (1857).
NE. Brazil. 84 BZE. Nanophan. or phan.

Eugenia bacopari D.Legrand, Sellowia 13: 317 (1961).
S. Brazil. 84 BZS. Phan.

Eugenia badia O.Berg in C.F.P.von Martius & auct. suc. (eds.), Fl. Bras. 14(1): 266 (1857).
Brazil (Rio de Janeiro). 84 BZL. Nanophan. or phan.
Myrtus glabra Vell., Fl. Flumin. 5: 215, t. 67 (1829).

Eugenia bahiana Mattos, Loefgrenia 94: 4 (1989).
NE. Brazil. 84 BZE. Nanophan. or phan.
**Stenocalyx bahiensis* O.Berg in C.F.P.von Martius & auct. suc. (eds.), Fl. Bras. 14(1): 351 (1857).

Eugenia bahiensis DC., Prodr. 3: 271 (1828).
E. Brazil. 84 BZE BZL. Nanophan. or phan.
Eugenia velutiflora Kiaersk., Enum. Myrt. Bras.: 144 (1893).

Eugenia bahorucana Alain, Phytologia 50: 166 (1982).
Dominican Rep. 81 DOM. Cham. or nanophan.

Eugenia baileyi Britton, Bull. Torrey Bot. Club 48: 334 (1921 publ. 1922).
Trinidad to N. South America. 81 TRT 82 SUR VEN. Phan.
Eugenia peregrina McVaugh, Mem. New York Bot. Gard. 18(2): 201 (1969).

Eugenia bajaverapazana Lundell, Wrightia 5: 166 (1975).
Guatemala. 80 GUA. Nanophan. or phan.

Eugenia balansae Guillaumin, Bull. Soc. Bot. France 85: 636 (1938 publ. 1939).
New Caledonia. 60 NWC. Nanophan. or phan.

Eugenia banderensis Urb., Symb. Antill. 9: 495 (1928).
W. Cuba. 81 CUB. Nanophan.
Eugenia mogotensis Urb. & Ekman in I.Urban, Symb. Antill. 9: 497 (1928).

Eugenia barbata McVaugh, Fieldiana, Bot. 29: 204 (1956).
Peru. 83 PER. Nanophan.

Eugenia barbosae Barb.Rodr. ex Chodat & Hassl., Bull. Herb. Boissier, II, 7: 807 (1907).
Paraguay. 85 PAR. Nanophan. or phan.
**Stenocalyx nanus* Barb.Rodr., Myrt. Paraguay: 8 (1903).

Eugenia barriosana Lundell, Wrightia 5: 82 (1975).
Guatemala. 80 GUA. Nanophan. or phan.

Eugenia basilaris McVaugh, Fieldiana, Bot. 29: 464 (1963).
Costa Rica. 80 COS. Nanophan. or phan.

Eugenia batingabranca Sobral, Napaea 1: 23 (1987).
Brazil (Espírito Santo). 84 BZL. Phan.

Eugenia bayatensis Urb., Symb. Antill. 9: 99 (1923).
E. Cuba. 81 CUB. Nanophan. or phan.

Eugenia belemitana McVaugh, Mem. New York Bot. Gard. 18(2): 166 (1969).
N. Brazil. 84 BZN. Nanophan. or phan.

Eugenia belladerensis Urb. & Ekman, Ark. Bot. 21A(5): 37 (1927).
Haiti. 81 HAI. Nanophan. or phan.

Eugenia belloi Barrie, Novon 15: 4 (2005).
Costa Rica. 80 COS. Nanophan. or phan.

Eugenia bellonis Krug & Urb., Bot. Jahrb. Syst. 19: 611 (1895). *Myrtus bellonis* (Krug & Urb.) Burret, Notizbl. Bot. Gart. Berlin-Dahlem 15: 482 (1941).
SW. Puerto Rico. 81 PUE. Cham.

Eugenia bergii in H.G.A.Engler & K.A.E.Prantl, Nat. Pflanzenfam. 3(7): 81 (1893).
WC. Brazil. 84 BZC. Nanophan. or phan.
**Eugenia uniflora* O.Berg in C.F.P.von Martius & auct. suc. (eds.), Fl. Bras. 14(1): 215 (1857), nom. illeg.
Eugenia lineatifolia Mattos, Loefgrenia 102: 3 (1993), nom. illeg.
Eugenia diaphana Kiaersk., Enum. Myrt. Bras.: 154 (1893).

Eugenia beyeri Urb., Symb. Antill. 9: 502 (1928).
Cuba (Sierra de Nipe). 81 CUB. Nanophan.

Eugenia biflora (L.) DC., Prodr. 3: 276 (1828).
Trop. America. 79 MXT 80 BLZ COS ELS GUA PAN 81 DOM HAI JAM LEE PUE WIN 82 FRG GUY SUR VEN 83 BOL CLM ECU PER 84 BZC BZE BZN. Nanophan. or phan.
**Myrtus biflora* L., Syst. Nat. ed. 10, 2: 1056 (1759).
Caryophyllus fruticosus Mill., Gard. Dict. ed. 8: 3 (1768).
Eugenia mini Aubl., Hist. Pl. Guiane 1: 498 (1775). *Myrtus mini* (Aubl.) Spreng., Syst. Veg. 2: 486 (1825). *Myrcia splendens* var. *mini* (Aubl.) DC., Prodr. 3: 244 (1828). *Myrcia mini* (Aubl.) Sweet, Hort. Brit., ed. 2: 211 (1830). *Cumetea mini* (Aubl.) Raf., Sylva Tellur.: 106 (1838). *Myrcia schomburgkiana* O.Berg, Linnaea 27: 110 (1855), nom. illeg. *Eugenia biflora* var. *mini* (Aubl.) Amshoff, Recueil Trav. Bot. Néerl. 42: 12 (1950).
Myrtus virgultosa Sw., Prodr.: 70 (1788). *Eugenia virgultosa* (Sw.) DC., Prodr. 3: 280 (1828). *Eugenia biflora* var. *virgultosa* (Sw.) Krug & Urb., Bot. Jahrb. Syst. 19: 628 (1895).
Eugenia microcarpos Lam., Encycl. 3: 201 (1789).
Myrtus pallens Vahl, Symb. Bot. 2: 57 (1791). *Eugenia pallens* (Vahl) DC., Prodr. 3: 284 (1828), nom. illeg. *Eugenia biflora* var. *pallens* (Vahl) Krug & Urb., Bot. Jahrb. Syst. 19: 629 (1895).
Eugenia albida Humb. & Bonpl., Pl. Aequinoct. 2: 107 (1813).
Eugenia lancea Poir. in J.B.A.P.M.de Lamarck, Encycl., Suppl. 3: 123 (1813). *Myrtus lancea* (Poir.) Spreng., Syst. Veg. 2: 482 (1825). *Eugenia biflora* var. *lancea* (Poir.) Krug & Urb., Bot. Jahrb. Syst. 19: 631 (1895).
Myrtus berteroana Spreng., Syst. Veg. 2: 487 (1825). *Eugenia berteroana* (Spreng.) DC., Prodr. 3: 285 (1828).
Eugenia australis Colla, Mem. Reale Accad. Sci. Torino 31: 123 (1827), nom. illeg.
Eugenia acutiloba DC., Prodr. 3: 281 (1828).
Eugenia ludibunda Bertero ex DC., Prodr. 3: 280 (1828). *Eugenia biflora* var. *ludibunda* (Bertero ex DC.) Krug & Urb., Bot. Jahrb. Syst. 19: 630 (1895).
Eugenia racemosa DC., Prodr. 3: 381 (1828), nom. illeg.
Eugenia xylopifolia DC., Prodr. 3: 279 (1828).
Eugenia dysantha Benth., Pl. Hartw.: 175 (1845).
Eugenia sericiflora Benth., Bot. Voy. Sulphur: 98 (1845).
Eugenia wallenii Macfad., Fl. Jamaica 2: 118 (1850). *Eugenia biflora* var. *wallenii* (Macfad.) Krug & Urb., Bot. Jahrb. Syst. 19: 629 (1895).
Eugenia jamaicensis O.Berg, Linnaea 27: 237 (1856).
Eugenia macarensis O.Berg, Linnaea 27: 283 (1856).

Eugenia acuminatissima O.Berg in C.F.P.von Martius & auct. suc. (eds.), Fl. Bras. 14(1): 315 (1857), nom. illeg. *Eugenia amanuensis* Steyerm., Fieldiana, Bot. 28: 1010 (1957).
Eugenia freireana O.Berg in C.F.P.von Martius & auct. suc. (eds.), Fl. Bras. 14(1): 305 (1857).
Eugenia freireana var. *angustifolia* O.Berg in C.F.P.von Martius & auct. suc. (eds.), Fl. Bras. 14(1): 305 (1857).
Eugenia freireana var. *latifolia* O.Berg in C.F.P.von Martius & auct. suc. (eds.), Fl. Bras. 14(1): 305 (1857).
Eugenia hoffmannseggii O.Berg in C.F.P.von Martius & auct. suc. (eds.), Fl. Bras. 14(1): 315 (1857). *Eugenia biflora* var. *hoffmannseggii* (O.Berg) Amshoff, in Fl. Surinam 3: 123 (1951).
Eugenia hoffmannseggii var. *grandifolia* O.Berg in C.F.P.von Martius & auct. suc. (eds.), Fl. Bras. 14(1): 315 (1857).
Eugenia hoffmannseggii var. *parvifolia* O.Berg in C.F.P.von Martius & auct. suc. (eds.), Fl. Bras. 14(1): 316 (1857).
Eugenia modesta var. *jamaicensis* O.Berg in C.F.P.von Martius & auct. suc. (eds.), Fl. Bras. 14(1): 314 (1857). *Eugenia virgultosa* var. *jamaicensis* (O.Berg) Proctor, J. Arnold Arbor. 63: 276 (1982).
Eugenia salicifolia O.Berg in C.F.P.von Martius & auct. suc. (eds.), Fl. Bras. 14(1): 312 (1857), nom. illeg. *Myrtus biflora* var. *salicifolia* (O.Berg) Kuntze, Revis. Gen. Pl. 3(2): 89 (1898).
Myrcia erythroxylon var. *virescens* O.Berg in C.F.P.von Martius & auct. suc. (eds.), Fl. Bras. 14(1): 173 (1857).
Eugenia fieldingii O.Berg, Linnaea 29: 242 (1858).
Eugenia inundata var. *acutifolia* O.Berg in C.F.P.von Martius & auct. suc. (eds.), Fl. Bras. 14(1): 520 (1858).
Eugenia meyeriana O.Berg in C.F.P.von Martius & auct. suc. (eds.), Fl. Bras. 14(1): 588 (1859).
Eugenia meyeriana var. *depauperata* O.Berg in C.F.P. von Martius & auct. suc. (eds.), Fl. Bras. 14(1): 588 (1859).
Eugenia meyeriana var. *dives* O.Berg in C.F.P.von Martius & auct. suc. (eds.), Fl. Bras. 14(1): 588 (1859).
Eugenia richardiana O.Berg, Linnaea 30: 694 (1861).
Eugenia xylopifolia var. *brevipes* O.Berg, Linnaea 30: 694 (1861).
Eugenia myriostigma Sagot, Ann. Sci. Nat., Bot., VI, 20: 194 (1885). *Eugenia biflora* var. *myriostigma* (Sagot) Amshoff, in Fl. Surinam 3: 124 (1951).
Eugenia hartii Kiaersk., Bot. Tidsskr. 17: 271 (1890).
Eugenia alexandri Krug & Urb., Bot. Jahrb. Syst. 19: 626 (1895).
Myrtus biflora f. *subsericea* Kuntze, Revis. Gen. Pl. 3(2): 89 (1898).
Myrtus biflora var. *yapacani* Kuntze, Revis. Gen. Pl. 3(2): 89 (1898).
Eugenia loretensis Diels, Bot. Jahrb. Syst. 37: 597 (1906).
Eugenia leptophlebia Diels, Verh. Bot. Vereins Prov. Brandenburg 48: 192 (1906 publ. 1907).
Eugenia alfaroana Standl., J. Wash. Acad. Sci. 14: 240 (1924).
Eugenia nicholsii Fawc. & Rendle, J. Bot. 64: 14 (1926).
Eugenia caurensis Steyerm., Fieldiana, Bot. 28: 1010 (1957).

Eugenia bimarginata DC., Prodr. 3: 271 (1828).
Brazil to NE. Argentina. 84 BZC BZE BZL BZS 85 AGE PAR. Nanophan. or phan.
Eugenia umbellata DC., Prodr. 3: 273 (1828).
Myrtus albida Vell., Fl. Flumin. 5: 216, t. 72 (1829), nom. illeg.
Eugenia bimarginata var. *impunctata* Cambess. in A.F.C.P.de Saint-Hilaire & al., Fl. Bras. Merid. 2: 347 (1832).
Eugenia bimarginata var. *tomentosa* Cambess. in A.F.C.P.de Saint-Hilaire & al., Fl. Bras. Merid. 2: 347 (1832).
Eugenia dicrossa O.Berg in C.F.P.von Martius & auct. suc. (eds.), Fl. Bras. 14(1): 291 (1857).
Eugenia dicrossa var. *latifolia* O.Berg in C.F.P.von Martius & auct. suc. (eds.), Fl. Bras. 14(1): 292 (1857).
Eugenia dicrossa var. *longifolia* O.Berg in C.F.P.von Martius & auct. suc. (eds.), Fl. Bras. 14(1): 292 (1857).
Eugenia dicrossa var. *parvifolia* O.Berg in C.F.P.von Martius & auct. suc. (eds.), Fl. Bras. 14(1): 292 (1857).
Eugenia rubrocincta O.Berg in C.F.P.von Martius & auct. suc. (eds.), Fl. Bras. 14(1): 291 (1857). *Eugenia bimarginata* var. *rubrocincta* (O.Berg) Kiaersk., Enum. Myrt. Bras.: 132 (1893).
Eugenia umbellaris O.Berg in C.F.P.von Martius & auct. suc. (eds.), Fl. Bras. 14(1): 292 (1857). *Eugenia bimarginata* var. *umbellaris* (O.Berg) Kiaersk., Enum. Myrt. Bras.: 132 (1893).
Eugenia warmingiana var. *pubescens* Kiaersk., Enum. Myrt. Bras.: 138 (1893).
Eugenia maracayuensis Barb.Rodr., Myrt. Paraguay: 3 (1903).
Myrciaria atiraensis Barb.Rodr. ex Chodat & Hassl., Bull. Herb. Boissier, II, 7: 808 (1907), nom. nud.

Eugenia blanchetiana O.Berg in C.F.P.von Martius & auct. suc. (eds.), Fl. Bras. 14(1): 273 (1857).
Brazil (Bahia). 84 BZE. Nanophan. or phan.

Eugenia blanda Sobral, Bradea 6: 234 (1993).
Brazil (Minas Gerais). 84 BZL. Nanophan.

Eugenia blastantha (O.Berg) D.Legrand, Sellowia 13: 310 (1961).
Brazil (S. Minas Gerais to Rio Grande do Sul). 84 BZL BZS. Nanophan. or phan.
**Stenocalyx blastanthus* O.Berg in C.F.P.von Martius & auct. suc. (eds.), Fl. Bras. 14(1): 348 (1857).

Eugenia bocainensis Mattos, Loefgrenia 64: 4 (1974).
SE. Brazil. 84 BZL. Nanophan. or phan.

Eugenia bojeri Baker, Fl. Mauritius: 115 (1877).
Mauritius. 29 MAU. (Cl.) nanophan.

Eugenia boliviana (D.Legrand) Mattos, Loefgrenia 105: 2 (1995).
Brazil (Paraná). 84 BZS. Nanophan. or phan.
Hexachlamys boliviensis Kausel, Lilloa 33: 103 (1971 publ. 1972).
**Hexachlamys boliviana* D.Legrand, Loefgrenia 55: 1 (1972).

Eugenia boqueronensis Britton, Bull. Torrey Bot. Club 51: 10 (1924).
Puerto Rico. 81 PUE. Nanophan. or phan.

Eugenia bosseri J.Guého & A.J.Scott, Kew Bull. 34: 475 (1980).
Réunion. 29 REU. Phan.

Eugenia botequimensis Kiaersk., Enum. Myrt. Bras.: 156 (1893).
SE. Brazil. 84 BZL. Nanophan. or phan.

Eugenia brachyclada Urb. & Ekman, Ark. Bot. 21A(5): 32 (1927).
Haiti. 81 HAI. Cham.

Eugenia brachysepala Kiaersk., Enum. Myrt. Bras.: 125 (1893).
SE. Brazil. 84 BZL. Nanophan. or phan.

Eugenia brachythrix Urb., Symb. Antill. 6: 23 (1909).
E. Jamaica. 81 JAM. Nanophan. or phan.

Eugenia brasiliana Aubl., Hist. Pl. Guiane 1: 511 (1775).
N. Brazil. 84 BZN. Phan.

Eugenia brasiliensis Lam., Encycl. 3: 203 (1789). *Stenocalyx brasiliensis* (Lam.) O.Berg in C.F.P.von Martius & auct. suc. (eds.), Fl. Bras. 14(1): 347 (1857).
E. & S. Brazil. 84 BZE BZL BZS. Phan.
Myrtus dombeyi Spreng., Syst. Veg. 2: 485 (1825), nom. illeg. *Eugenia dombeyi* Skeels, Bull. Bur. Pl. Industr. U.S.D.A. 233: 51 (1912), nom. illeg.
Eugenia bracteolaris Lam. ex DC., Prodr. 3: 267 (1828).
Myrtus grumixama Vell., Fl. Flumin. 5: 216, t. 69 (1829).
Eugenia ubensis Cambess. in A.F.C. de Saint-Hilaire, Fl. Bras. Merid. 2: 351 (1832). *Stenocalyx ubensis* (Cambess.) O.Berg in C.F.P.von Martius & auct. suc. (eds.), Fl. Bras. 14(1): 344 (1857).
Stenocalyx brasiliensis var. *erythrocarpa* (Cambess.) O.Berg in C.F.P.von Martius & auct. suc. (eds.), Fl. Bras. 14(1): 348 (1857).
Stenocalyx brasiliensis var. *iocarpa* O.Berg in C.F.P.von Martius & auct. suc. (eds.), Fl. Bras. 14(1): 348 (1857).
Stenocalyx brasiliensis var. *leucocarpa* (Cambess.) O.Berg in C.F.P.von Martius & auct. suc. (eds.), Fl. Bras. 14(1): 348 (1857).
Stenocalyx brasiliensis var. *silvestris* O.Berg in C.F.P.von Martius & auct. suc. (eds.), Fl. Bras. 14(1): 594 (1859).
Eugenia filipes Baill. in A.Grandidier, Hist. Phys. Madagascar, Atlas: t. 328 (1895).

Eugenia breedlovei Barrie, Novon 15: 6 (2005).
Mexico (Chiapas). 79 MXT. Nanophan. or phan.

Eugenia brevipedunculata Kiaersk., Enum. Myrt. Bras.: 139 (1893).
Brazil (Rio de Janeiro to Santa Catarina). 84 BZL BZS. Nanophan. or phan.

Eugenia brevipes A.Rich., Hist. Phys. Cuba, Pl. Vasc.: 584 (1846).
Cuba. 81 CUB. Nanophan.

Eugenia brevistyla D.Legrand, in Fl. Ilustr. Catar. 1(Mirt., Suppl. 1): 18 (1977).
S. Brazil. 84 BZS. Nanophan. or phan.

Eugenia brongniartiana Guillaumin, Notul. Syst. (Paris) 3: 261 (1916).
New Caledonia. 60 NWC. Nanophan. or phan.
**Eugenia crassifolia* Vieill. ex Brongn. & Gris, Bull. Soc. Bot. France 12: 469 (1865), nom. illeg.

Eugenia brownei Urb., Repert. Spec. Nov. Regni Veg. 58: 368 (1922).
SW. Jamaica. 81 JAM. Nanophan. or phan.

Eugenia brownsbergii Amshoff, Recueil Trav. Bot. Néerl. 42: 14 (1950).
Guianas to N. Brazil. 82 FRG SUR 84 BZN. Phan.
Eugenia martini O.Berg, Linnaea 27: 230 (1856), provisional synonym.
Eugenia brownsbergii var. *glauca* Amshoff, Recueil Trav. Bot. Néerl. 42: 14 (1950).

Eugenia brunoi Mattos, Loefgrenia 99: 2 (1990).
Brazil (São Paulo). 84 BZL. Nanophan. or phan.
**Eugenia littoralis* Mattos, Loefgrenia 42: 1 (1970), nom. illeg.

Eugenia bryanii Kaneh., Bot. Mag. (Tokyo) 51: 913 (1937). *Jossinia bryanii* (Kaneh.) Hosok., J. Jap. Bot. 16: 542 (1940).
Marianas. 62 MRN. Nanophan. or phan.

Eugenia buchholzii Engl., Notizbl. Königl. Bot. Gart. Berlin 2: 291 (1899). *Myrtus buchholzii* (Engl.) Kuntze, Deutsche Bot. Monatsschr. 21: 173 (1903).
Cameroon. 23 CMN. Nanophan.

Eugenia bukobensis Engl., Notizbl. Königl. Bot. Gart. Berlin 2: 289 (1899).
E. Trop. Africa. 25 KEN TAN UGA. Nanophan. or phan.

Eugenia bullata Pancher ex Guillaumin, Bull. Soc. Bot. France 85: 636 (1938).
New Caledonia. 60 NWC. Nanophan. or phan.

Eugenia bumelioides Standl., Publ. Carnegie Inst. Wash. 461: 75 (1935).
Belize to Guatemala. 80 BLZ GUA. Nanophan. or phan.
Eugenia brevistipitata Lundell, Wrightia 5: 276 (1976).

Eugenia burkartiana (D.Legrand) D.Legrand, Sellowia 13: 321 (1961).
Paraguay to S. Brazil. 84 BZS 85 AGE PAR. Phan.
**Pseudocaryophyllus burkartianus* D.Legrand, Darwiniana 9: 287 (1950).

Eugenia buxifolia Lam., Encycl. 3: 202 (1789). *Myrtus buxifolia* (Lam.) Poir. in J.B.A.P.M.de Lamarck, Encycl. 4: 413 (1798). *Eugenia coriacea* F.Dietr., Nachtr. Vollst. Lex. Gärtn. 3: 263 (1817), nom. illeg. *Myrtus borbonica* Spreng., Syst. Veg. 2: 481 (1825), nom. illeg. *Jossinia buxifolia* (Lam.) DC., Prodr. 3: 238 (1828).
Réunion. 29 REU. Nanophan. or phan.
Jossinia buxifolia var. *microphylla* Blume, Mus. Bot. 1: 122 (1850).
Eugenia buxifolia var. *minor* Cordem., Fl. Réunion: 428 (1895).
Eugenia heteromorpha Cordem., Fl. Réunion: 428 (1895).

Eugenia byssacea McVaugh, Mem. New York Bot. Gard. 18(2): 168 (1969).
Venezuela. 82 VEN. Nanophan.

Eugenia cachoeirensis O.Berg in C.F.P.von Martius & auct. suc. (eds.), Fl. Bras. 14(1): 265 (1857).
SE. Colombia to N. Brazil. 82 VEN 83 CLM 84 BZN. Phan.
Eugenia lingua O.Berg in C.F.P.von Martius & auct. suc. (eds.), Fl. Bras. 14(1): 266 (1857).
Eugenia cachoeirensis var. *sessilifolia* Kiaersk., Enum. Myrt. Bras.: 132 (1893).

Eugenia cacuminoides Govaerts, World Checklist Myrtaceae: 135 (2008).
N. Puerto Rico. 81 PUE. Nanophan.
**Eugenia cacuminis* Alain, Phytologia 61: 356 (1986), nom. illeg.

Eugenia cacuminum Standl. & Steyerm., Publ. Field Mus. Nat. Hist., Bot. Ser. 22: 357 (1940).
Guatemala to El Salvador. 80 ELS GUA. Nanophan.

or phan.

Eugenia cahosiana Urb. & Ekman, Ark. Bot. 21A(5): 28 (1927).
Haiti (Massif des Cahos). 81 HAI. Nanophan.

Eugenia cajalbanica Borhidi & O.Muñiz, Bot. Közlem. 64: 216 (1977 publ. 1978).
Cuba. 81 CUB. Nanophan. or phan.

Eugenia callichroma McVaugh, Mem. New York Bot. Gard. 18(2): 169 (1969).
Venezuela (Bolívar). 82 VEN. Nanophan. or phan.

Eugenia calophylloides DC., Prodr. 3: 272 (1828).
W. Trop. Africa. 22 GUI IVO LBR SIE. (Cl.) nanophan.
Calophylloides lucida Smeathman ex DC., Prodr. 3: 272 (1828), nom. inval.

Eugenia calumettae Urb. & Ekman, Ark. Bot. 22A(10): 35 (1929).
Haiti. 81 HAI. Nanophan.

Eugenia calva McVaugh, Fieldiana, Bot. 29: 205 (1956).
Ecuador to Peru. 83 ECU PER 84 BZN. Phan.

Eugenia calycina Cambess. in A.F.C.de Saint-Hilaire, Fl. Bras. Merid. 2: 352 (1832). *Phyllocalyx calycinus* (Cambess.) O.Berg, Linnaea 27: 307 (1856).
Brazil. 84 BZC BZL BZS. Nanophan. or phan.

var. ***calycina***
Brazil (Goiás, Minas Gerais). 84 BZC BZL. Nanophan. or phan.

var. ***herbacea*** (O.Berg) Mattos, Loefgrenia 113: 2 (1999).
S. Brazil. 84 BZS. Cham. or nanophan.
**Phyllocalyx herbaceus* O.Berg in C.F.P.von Martius & auct. suc. (eds.), Fl. Bras. 14(1): 328 (1857). *Eugenia suffrutescens* Nied. in H.G.A. Engler & K.A.E.Prantl, Nat. Pflanzenfam. 3(7): 82 (1893). *Eugenia jaguariaivensis* Mattos, Loefgrenia 11: 1 (1963), nom. illeg.
Eugenia jaguariaivensis var. *brevipedunculata* Mattos, Loefgrenia 11: 1 (1963). *Eugenia suffrutescens* var. *brevipedunculata* (Mattos) Mattos, Revista Ci. Cult. 19: 334 (1967).

Eugenia calystegia (O.Berg) Nied. in H.G.A.Engler & K.A.E.Prantl, Nat. Pflanzenfam. 3(7): 82 (1893).
Brazil (São Paulo). 84 BZL. Nanophan. or phan.
**Phyllocalyx calystegius* O.Berg in C.F.P.von Martius & auct. suc. (eds.), Fl. Bras. 14(1): 311 (1857).

Eugenia camarioca C.Wright, Anales Acad. Ci. Méd. Habana 5: 431 (1868).
Cuba. 81 CUB. Nanophan.
Eugenia leptosticta Urb., Symb. Antill. 7: 301 (1912).
Eugenia havanensis Britton & P.Wilson, Mem. Torrey Bot. Club 16: 89 (1920).

Eugenia cana DC., Prodr. 3: 264 (1828).
SE. Brazil. 84 BZL. Nanophan. or phan.

Eugenia canapuensis Urb., Symb. Antill. 9: 491 (1928).
Cuba (Sierra de Nipe). 81 CUB. Nanophan. or phan.

Eugenia candolleana DC., Prodr. 3: 281 (1828).
Brazil. 84 BZC BZE BZL BZS. Nanophan. or phan.
Myrtus nitida Vell., Fl. Flumin. 5: 216, t. 68 (1829).
Eugenia jequitinhonhensis Cambess. in A.F.C.de Saint-Hilaire, Fl. Bras. Merid. 2: 358 (1833).
Eugenia mikanioides O.Berg in C.F.P.von Martius & auct. suc. (eds.), Fl. Bras. 14(1): 298 (1857).
Eugenia recurvata O.Berg in C.F.P.von Martius & auct. suc. (eds.), Fl. Bras. 14(1): 319 (1857).
Eugenia christovana Kiaersk., Enum. Myrt. Bras.: 184 (1893).
Eugenia glandulosissima Kiaersk., Enum. Myrt. Bras.: 147 (1893).
Eugenia friburgensis Glaz., Bull. Soc. Bot. France 54: 231 (1908), nom. nud.

Eugenia cantuana Lundell, Wrightia 3: 122 (1965).
C. Mexico. 79 MXC. Phan.
**Eugenia mirandae* D.Ramírez Cantu, Anales Inst. Biol. Univ. Nac. México 14: 487 (1943), nom. illeg.

Eugenia capensis (Eckl. & Zeyh.) Harv., Gen. S. Afr. Pl.: 416 (1838).
Somalia to S. Africa. 23 ZAI 24 SOM 25 KEN TAN 26 MLW MOZ ZAM ZIM 27 CPP NAT TVL. Nanophan. or phan.
Myrtus capensis (Eckl. & Zeyh.) Harv., Gen. S. Afr. Pl.: 99 (1838), nom. illeg.

subsp. ***albanensis*** (Sond.) F.White, Kirkia 10: 402 (1977).
Mozambique to S. Africa. 26 MOZ 27 CPP NAT TVL. Nanophan.
**Eugenia albanensis* Sond. in W.H.Harvey & auct. suc. (eds.), Fl. Cap. 2: 522 (1862).

subsp. ***capensis***
Mozambique to S. Africa. 26 MOZ 27 CPP NAT. Nanophan. or phan.
Eugenia capensis var. *major* Sond. in W.H.Harvey & auct. suc. (eds.), Fl. Cap. 2: 523 (1862).
Myrtus major Kuntze, Revis. Gen. Pl. 3(2): 92 (1898), nom. illeg.

subsp. ***gracilipes*** F.White, Kirkia 10: 403 (1977).
S. Trop. Africa. 26 MLW MOZ ZIM. Nanophan.

subsp. ***gueinzii*** (Sond.) F.White, Kirkia 10: 403 (1977).
Mozambique to KwaZulu-Natal. 26 MOZ 27 CPP NAT. Nanophan. or phan.
**Eugenia gueinzii* Sond. in W.H.Harvey & auct. suc. (eds.), Fl. Cap. 2: 523 (1862).

subsp. ***multiflora*** Verdc., Kew Bull. 54: 43 (1999).
E. Trop. Africa to Mozambique. 25 KEN TAN 26 MOZ. Nanophan. or phan.

subsp. ***natalitia*** (Sond.) F.White, Kirkia 10: 402 (1977).
S. Mozambique to S. Africa. 26 MOZ 27 CPP NAT SWZ TVL. Nanophan. or phan.
**Eugenia natalitia* Sond. in W.H.Harvey & auct. suc. (eds.), Fl. Cap. 2: 522 (1862).
Eugenia rudatisii Engl. & Brehmer, Bot. Jahrb. Syst. 54: 335 (1917).

subsp. ***nyassensis*** (Engl.) F.White, Kirkia 10: 403 (1977).
Rwanda to S. Trop. Africa. 23 RWA ZAI 25 TAN 26 MLW MOZ ZAM ZIM. Nanophan. or phan.
**Eugenia nyassensis* Engl., Notizbl. Königl. Bot. Gart. Berlin 2: 290 (1899). *Myrtus nyassensis* (Engl.) Kuntze, Deutsche Bot. Monatsschr. 21: 173 (1903).
Eugenia chirindensis Baker f., J. Linn. Soc., Bot. 40: 70 (1911).

subsp. ***simii*** (Dummer) F.White, Kirkia 10: 402 (1977).
Cape Prov. to KwaZulu-Natal. 27 CPP NAT. Nanophan. or phan.
**Eugenia simii* Dummer, Gard. Chron., III, 52: 179 (1912).

subsp. ***zeyheri*** (Harv.) F.White, Kirkia 10: 402 (1977).
Cape Prov. 27 CPP. Nanophan. or phan.
**Myrtus zeyheri* Harv., Gen. S. Afr. Pl.: 99 (1838).

Eugenia capitulifera O.Berg in C.F.P.von Martius & auct. suc. (eds.), Fl. Bras. 14(1): 263 (1857).
Brazil (Rio de Janeiro). 84 BZL. Nanophan. or phan.

Eugenia capparidifolia DC., Prodr. 3: 281 (1828). *Eugenia robusta* var. *capparidifolia* (DC.) O.Berg in C.F.P.von Martius & auct. suc. (eds.), Fl. Bras. 14(1): 320 (1857).
Brazil (Minas Gerais). 84 BZL. Nanophan. or phan.
Eugenia firma DC., Prodr. 3: 282 (1828). *Eugenia robusta* var. *firma* (DC.) O.Berg in C.F.P.von Martius & auct. suc. (eds.), Fl. Bras. 14(1): 320 (1857).
Eugenia robusta O.Berg, Linnaea 27: 298 (1856).

Eugenia capuli (Schltdl. & Cham.) Hook. & Arn., Bot. Beechey Voy.: 291 (1841).
Mexico to C. America. 79 MXE MXG MXN MXT 80 BLZ ELS GUA HON. Nanophan. or phan.
**Myrtus capuli* Schltdl. & Cham., Linnaea 5: 561 (1830).
Eugenia schiedeana Schltdl., Linnaea 13: 418 (1839).
Eugenia capuli var. *macroterantha* O.Berg, Linnaea 27: 239 (1856).
Eugenia capuli var. *micrantha* O.Berg, Linnaea 29: 238 (1858).
Eugenia capuli var. *rigida* O.Berg, Linnaea 29: 238 (1858).
Eugenia lindeniana O.Berg, Linnaea 29: 240 (1858). *Eugenia capuli* var. *lindeniana* (O.Berg) Lundell, Wrightia 3: 123 (1965).
Eugenia contrerasii Lundell, Wrightia 2: 206 (1961).

Eugenia capuliodes Lundell, Contr. Univ. Michigan Herb. 7: 32 (1942).
SE. Mexico. 79 MXT. Nanophan. or phan.

Eugenia cararensis Barrie & Q.Jiménez, Novon 15: 7 (2005).
Costa Rica. 80 COS. Nanophan. or phan.

Eugenia carranzae Lundell, Wrightia 4: 92 (1968).
SE. Mexico. 79 MXT. Nanophan.

Eugenia cartagensis O.Berg, Linnaea 27: 240 (1856).
Costa Rica to Panama. 80 COS PAN. Nanophan. or phan.

Eugenia casearioides (Kunth) DC., Prodr. 3: 275 (1828).
Colombia to Venezuela. 82 VEN 83 CLM. Nanophan. or phan.
**Myrtus casearioides* Kunth in F.W.H.von Humboldt, A.J.A.Bonpland & C.S.Kunth, Nov. Gen. Sp. 6: 145 (1823).

Eugenia cassinoides Lam., Encycl. 3: 205 (1789). *Myrtus cassinoides* (Lam.) Spreng., Syst. Veg. 2: 481 (1825). *Jossinia cassinoides* (Lam.) DC., Prodr. 3: 238 (1828).
SE. Madagascar. 29 MDG. Nanophan. or phan.

Eugenia catharinae O.Berg in C.F.P.von Martius & auct. suc. (eds.), Fl. Bras. 14(1): 259 (1857).
S. Brazil. 84 BZS. Nanophan. or phan.

Eugenia catharinensis D.Legrand, Sellowia 8: 73 (1957).
S. Brazil. 84 BZS. Nanophan.

Eugenia catingiflora Griseb., Cat. Pl. Cub.: 89 (1866). *Pseudanamomis catingiflora* (Griseb.) Bisse, Feddes Repert. 96: 511 (1985).
Cuba. 81 CUB. Nanophan. or phan.
Eugenia cati Britton & P.Wilson, Bull. Torrey Bot. Club 50: 44 (1923). *Pseudanamomis cati* (Britton & P.Wilson) Bisse, Feddes Repert. 96: 511 (1985).

Eugenia cavalcanteana Mattos, Loefgrenia 94: 1 (1989).
Brazil (Minas Gerais). 84 BZL. Nanophan.
**Phyllocalyx riedelianus* O.Berg in C.F.P.von Martius & auct. suc. (eds.), Fl. Bras. 14(1): 590 (1859).

Eugenia cayoana Lundell, Wrightia 4: 143 (1968).
Belize. 80 BLZ. Nanophan.

Eugenia ceibana Urb., Symb. Antill. 9: 497 (1928).
Cuba (I. de la Juventud). 81 CUB. Nanophan.

Eugenia cerasiflora Miq., Linnaea 22: 793 (1850).
E. & S. Brazil. 84 BZE BZL BZS. Phan.
Myrtus aggregata Vell., Fl. Flumin. 5: 215, t. 65 (1829). *Phyllocalyx cerasiflorus* O.Berg, Linnaea 27: 308 (1856), nom. illeg. *Eugenia aggregata* (Vell.) Kiaersk., Enum. Myrt. Bras.: 162 (1893), nom. illeg.
Eugenia lucida Cambess. in A.F.C.de Saint-Hilaire, Fl. Bras. Merid. 2: 356 (1832), nom. illeg. *Luma lucidissima* Herter, Revista Sudamer. Bot. 7: 219 (1943).
Eugenia itacolumensis O.Berg in C.F.P.von Martius & auct. suc. (eds.), Fl. Bras. 14(1): 253 (1857).

Eugenia cereja D.Legrand, Sellowia 13: 316 (1961).
S. Brazil. 84 BZS. Phan.
Eugenia paranaguensis Mattos & D.Legrand, Loefgrenia 67: 25 (1975).

Eugenia cerrocacaoensis Barrie, Novon 15: 7 (2005).
Costa Rica. 80 COS. Nanophan. or phan.

Eugenia cervina Standl. & Steyerm., Publ. Field Mus. Nat. Hist., Bot. Ser. 23: 128 (1944).
Guatemala. 80 GUA. Nanophan. or phan.

Eugenia chacoensis (D.Legrand) Kausel, Lilloa 32: 329 (1967).
NE. Argentina to Paraguay. 85 AGE PAR.
**Eugenia ovalifolia* var. *chacoensis* D.Legrand, Darwiniana 9: 296 (1950).

Eugenia chacueyana Alain, Phytologia 25: 267 (1973).
Dominican Rep. 81 DOM. Nanophan.

Eugenia chahalana Lundell, Wrightia 4: 93 (1968).
Guatemala. 80 GUA. Nanophan. or phan.

Eugenia chartacea McVaugh, Fieldiana, Bot. 29: 206 (1956).
N. Brazil. 84 BZN. Phan.

Eugenia chavarriae Barrie, Novon 15: 9 (2005).
Costa Rica. 80 COS. Nanophan. or phan.

Eugenia chepensis Standl., Publ. Field Mus. Nat. Hist., Bot. Ser. 8: 144 (1930).
C. America. 80 NIC PAN. Nanophan.

Eugenia chiapensis Lundell, Amer. Midl. Naturalist 20: 238 (1938).
SE. Mexico. 79 MXT. Nanophan. or phan.

Eugenia chinajensis Standl. & Steyerm., Publ. Field Mus. Nat. Hist., Bot. Ser. 23: 129 (1944).
Belize to Guatemala. 80 BLZ GUA. Nanophan.

Eugenia chlorocarpa O.Berg in C.F.P.von Martius & auct. suc. (eds.), Fl. Bras. 14(1): 220 (1857).

Brazil (Rio de Janeiro). 84 BZL. Nanophan. or phan.

Eugenia chlorophylla O.Berg in C.F.P.von Martius & auct. suc. (eds.), Fl. Bras. 14(1): 583 (1859).
SE. & S. Brazil. 84 BZL BZS. Nanophan. or phan.
Eugenia multiovulata Mattos & D.Legrand, Loefgrenia 67: 23 (1975).

Eugenia choapamensis Standl., Publ. Field Mus. Nat. Hist., Bot. Ser. 22: 93 (1940).
S. Mexico to Guatemala. 79 MXS MXT 80 GUA. Nanophan. or phan.

Eugenia christii Urb., Symb. Antill. 7: 303 (1912).
Haiti. 81 HAI. Nanophan.
Eugenia bicincta Urb. & Ekman, Ark. Bot. 21A(5): 33 (1927).

Eugenia chrootricha Urb. & Ekman, Ark. Bot. 21A(5): 30 (1927).
Hispaniola. 81 DOM HAI. Nanophan. or phan.

Eugenia chrootrichoides Proctor, Bull. Inst. Jamaica, Sci. Ser. 16: 30 (1967).
Jamaica. 81 JAM. Nanophan.

Eugenia chrysobalanoides DC., Prodr. 3: 285 (1828).
Lesser Antilles. 81 LEE WIN. Phan.

Eugenia chrysophyllum Poir. in J.B.A.P.M.de Lamarck, Encycl., Suppl. 3: 129 (1813).
S. Trop. America. 82 FRG SUR VEN 83 CLM PER 84 BZN. Nanophan. or phan.
Eugenia chrysophylloides DC., Prodr. 3: 276 (1828).
Eugenia llewelynii Steyerm., Fieldiana, Bot. 28: 1011 (1957).

Eugenia churutensis Cornejo, Harvard Pap. Bot. 10: 61 (2005).
Ecuador. 83 ECU.

Eugenia cincta Griseb., Mem. Amer. Acad. Arts, n.s., 8: 182 (1861).
E. Cuba. 81 CUB. Nanophan. or phan.

Eugenia cinerascens Gardner, London J. Bot. 4: 102 (1845).
Brazil (Rio de Janeiro). 84 BZL. Nanophan. or phan.

Eugenia cintalapana Barrie, Novon 15: 11 (2005).
Mexico (Chiapas). 79 MXT. Nanophan. or phan.

Eugenia citrifolia Poir. in J.B.A.P.M.de Lamarck, Encycl., Suppl. 3: 129 (1813).
N. South America. 82 FRG SUR VEN 84 BZE?. Nanophan. or phan.
Myrtus cayennensis Spreng., Syst. Veg. 2: 480 (1825).
Eugenia adenocalyx DC., Prodr. 3: 271 (1828).

Eugenia citroides Lundell, Contr. Univ. Michigan Herb. 7: 33 (1942).
SE. Mexico. 79 MXT. Nanophan.

Eugenia clarendonensis Urb., Symb. Antill. 7: 305 (1912).
Jamaica. 81 JAM. Nanophan.

Eugenia clarensis Britton & P.Wilson, Mem. Torrey Bot. Club 16: 90 (1920).
C. Cuba. 81 CUB. Nanophan. or phan.

Eugenia cloiselii H.Perrier, Mém. Inst. Sci. Madagascar, Sér. B, Biol. Vég. 4: 183 (1953).
SE. Madagascar. 29 MDG. Phan.

Eugenia coaetanea O.Berg in C.F.P.von Martius & auct. suc. (eds.), Fl. Bras. 14(1): 220 (1857). *Myrtus coaetanea* (O.Berg) Kuntze, Revis. Gen. Pl. 3(2): 91 (1898).
Brazil (Rio Grande do Sul). 84 BZS. Nanophan.

Eugenia coccifera O.Berg in C.F.P.von Martius & auct. suc. (eds.), Fl. Bras. 14(1): 278 (1857).
Brazil (Rio de Janeiro). 84 BZL. Nanophan. or phan.

Eugenia cocosensis Barrie, Novon 15: 11 (2005).
C. American Pacific Is. (Cocos I.). 80 CPI. Nanophan. or phan.

Eugenia codyensis Munro ex Wight, Ill. Ind. Bot. 2: 13 (1841). *Syzygium codyensis* (Munro ex Wight) Chandrash., Biol. Membr. Abstr. 2: 57 (1977). *Eugenia cotinifolia* subsp. *codyensis* (Munro ex Wight) P.S.Ashton, in Revised Handb. Fl. Ceyl. 2: 412 (1981).
S. India, Sri Lanka. 40 IND SRL. Nanophan.

Eugenia hypoleuca Thwaites ex Kosterm., Quart. J. Taiwan Mus. 34: 164 (1981).

Eugenia coffeifolia DC., Prodr. 3: 272 (1828).
Lesser Antilles, N. South America to N. Brazil. 81 LEE WIN 82 FRG GUY SUR VEN 84 BZN. Phan.
Eugenia sinemariensis Aubl., Hist. Pl. Guiane 1: 501 (1775), provisional synonym. *Myrtus sinemariensis* (Aubl.) Spreng., Syst. Veg. 2: 483 (1825). *Epleienda sinemariensis* (Aubl.) Raf., Sylva Tellur.: 107 (1838).
Myrciaria ramiflora O.Berg, Linnaea 27: 334 (1856).
Eugenia coffeifolia var. *grandifolia* O.Berg, Linnaea 29: 231 (1858).
Eugenia coffeifolia var. *parvifolia* O.Berg, Linnaea 29: 231 (1858).
Eugenia melinonis Sagot, Ann. Sci. Nat., Bot., VI, 20: 194 (1885).

Eugenia coibensis Barrie, Novon 15: 13 (2005).
Panama. 80 PAN. Nanophan. or phan.

Eugenia colipensis O.Berg, Linnaea 29: 243 (1858).
Mexico (Veracruz). 79 MXG. Nanophan. or phan.

Eugenia colnettiana Guillaumin, Mém. Mus. Natl. Hist. Nat., B, Bot. 8: 292 (1962).
New Caledonia. 60 NWC. Nanophan. or phan.

Eugenia columbiensis O.Berg, Linnaea 27: 469 (1856).
Colombia. 82 VEN? 83 CLM. Nanophan. or phan.

Eugenia commutata O.Berg, Linnaea 27: 174 (1856).
Jamaica. 81 JAM. Nanophan. or phan.

Eugenia comorensis H.Perrier, Mém. Inst. Sci. Madagascar, Sér. B, Biol. Vég. 4: 174 (19553).
Comoros. 29 COM. Phan.

Eugenia complicata O.Berg in C.F.P.von Martius & auct. suc. (eds.), Fl. Bras. 14(1): 313 (1857).
Brazil (Minas Gerais to Goiás). 84 BZC BZL. Nanophan. or phan.
Eugenia barrerensis O.Berg in C.F.P.von Martius & auct. suc. (eds.), Fl. Bras. 14(1): 313 (1857).
Eugenia brasiliae Mattos & D.Legrand, Loefgrenia 67: 27 (1975).

Eugenia conchalensis D.Legrand & Mattos, Loefgrenia 121: 1 (2006).
Brazil (São Paulo). 84 BZL.

Eugenia conchalensis D.Legrand & Mattos, Loefgrenia 121: 1 (2006).
Brazil (São Paulo).

Eugenia concolor Mattos, Loefgrenia 94: 5 (1989).
Brazil (São Paulo). 84 BZL. Nanophan. or phan.

Eugenia conduplicata B.Holst, Selbyana 23: 140 (2002).
S. Venezuela to N. Brazil. 82 VEN 84 BZN. Nanophan. or phan.

Eugenia confusa DC., Prodr. 3: 279 (1828).
S. Florida, Caribbean. 78 FLA 81 BAH CUB DOM HAI JAM LEE PUE TRT WIN. Nanophan. or phan.
Eugenia garberi Sarg., Gard. & Forest 2: 28 (1889).
Eugenia krugii Kiaersk., Bot. Tidsskr. 17: 259 (1890).

Eugenia congolensis De Wild. & T.Durand, Bull. Soc. Roy. Bot. Belgique, Compt.-Rend. 38: 124 (1899).
WC. Trop. Africa to Angola. 23 BUR GAB ZAI 26 ANG. Nanophan.

Eugenia conjuncta Amshoff, Kew Bull. 7: 259 (1952).
Guyana. 82 GUY. Nanophan. or phan.

Eugenia consolatae Chiov., Racc. Bot. Miss. Consol. Kenya: 45 (1935).
Kenya. 25 KEN. Nanophan. or phan.

Eugenia constanzae Alain, Phytologia 61: 357 (1986).
Dominican Rep. 81 DOM. Nanophan.

Eugenia convexinervia D.Legrand, Sellowia 13: 307 (1961).
SE. & S. Brazil. 84 BZL BZS. Phan.

Eugenia corcovadensis Kiaersk., Enum. Myrt. Bras.: 126 (1893).
Brazil (Rio de Janeiro). 84 BZL. Nanophan. or phan.

Eugenia cordata (Sw.) DC., Prodr. 3: 272 (1828).
Puerto Rico to Leeward Is. 81 LEE PUE. Nanophan. or phan.
**Myrtus cordata* Sw., Prodr.: 78 (1788). *Pseudanamomis cordata* (Sw.) Bisse, Feddes Repert. 96: 511 (1985).

var. **cordata**
Puerto Rico to Leeward Is. 81 LEE PUE. Nanophan. or phan.
Eugenia cordata var. *parvifolia* Schltdl., Linnaea 5: 199 (1830).
Myrciaria stirpiflora O.Berg, Linnaea 30: 702 (1861). *Eugenia stirpiflora* (O.Berg) Krug & Urb., Bot. Jahrb. Syst. 19: 672 (1894).
Eugenia cordata var. *ovata* Kiaersk., Bot. Tidsskr. 17: 258 (1890).

var. **sintenisii** (Kiaersk.) Krug & Urb., Bot. Jahrb. Syst. 19: 656 (1895).
Puerto Rico to Lesser Antilles. 81 LEE PUE WIN. Nanophan. or phan.
**Eugenia sintenisii* Kiaersk., Bot. Tidsskr. 17: 263 (1890).
Eugenia sintenisii Krug & Urb., Bot. Jahrb. Syst. 19: 650 (1895).
Eugenia borinquensis Britton, Sci. Surv. Porto Rico & Virgin Islands 6: 38 (1925).

Eugenia coronata Vahl ex DC., Prodr. 3: 271 (1828).
Ivory Coast to Benin. 22 BEN GHA IVO TOG. Nanophan. or phan.
Eugenia myrtoides G.Don, Gen. Hist. 2: 854 (1832), nom. illeg.
Eugenia littorea Engl. & Brehmer, Bot. Jahrb. Syst. 54: 335 (1917).

Eugenia corozalensis Britton, Bull. Torrey Bot. Club 51: 11 (1924).
Puerto Rico. 81 PUE. Nanophan. or phan.

Eugenia corrientina Barb.Rodr., Myrt. Paraguay: 5 (1903).
Paraguay. 85 PAR. Nanophan. or phan.

Eugenia corusca Barrie, Novon 15: 14 (2005).
Costa Rica. 80 COS. Nanophan. or phan.

Eugenia costaricensis O.Berg, Linnaea 27: 213 (1856).
C. America. 80 COS ELS NIC PAN. Nanophan.
Eugenia valerioi Standl., Publ. Field Mus. Nat. Hist., Bot. Ser. 18: 775 (1937).

Eugenia costenoblei Merr., Philipp. J. Sci., C 9: 123 (1914). *Jossinia costenoblei* (Merr.) Diels, Bot. Jahrb. Syst. 56: 531 (1921).
Marianas. 62 MRN. Nanophan. or phan.

Eugenia cotinifolia Jacq., Observ. Bot. 3: 3 (1768). *Myrtus cotinifolia* (Jacq.) Spreng., Syst. Veg. 2: 481 (1825). *Jossinia cotinifolia* (Jacq.) DC., Prodr. 3: 238 (1828).
Venezuela. 82 VEN. Nanophan. or phan.
Caryophyllus cotinifolius Mill., Gard. Dict. ed. 8: 4 (1768).

Eugenia coursiana H.Perrier, Mém. Inst. Sci. Madagascar, Sér. B, Biol. Vég. 4: 170 (1953).
C. Madagascar. 29 MDG. Nanophan.

Eugenia cowanii McVaugh, Mem. New York Bot. Gard. 18(2): 173 (1969).
N. South America. 82 FRG GUY SUR. Nanophan. or phan.
Eugenia cryptadena var. *gracilis* Amshoff, Recueil Trav. Bot. Néerl. 42: 18 (1950).

Eugenia cowellii Britton & P.Wilson, Mem. Torrey Bot. Club 16: 88 (1920).
Cuba. 81 CUB. Nanophan.

Eugenia coyolensis Standl., Publ. Field Mus. Nat. Hist., Bot. Ser. 9: 320 (1940).
Honduras. 80 HON. Nanophan. or phan.

Eugenia crassicaulis Proctor, J. Arnold Arbor. 63: 274 (1982).
Jamaica. 81 JAM. Nanophan. or phan.

Eugenia crassifolia DC., Prodr. 3: 266 (1828).
Brazil (São Paulo). 84 BZL. Nanophan. or phan.

Eugenia crassipetala J.Guého & A.J.Scott, Kew Bull. 34: 476 (1980).
Mauritius. 29 MAU. Nanophan. or phan.

Eugenia crenata Vell., Fl. Flumin. 5: 209, t. 39 (1829). *Eugenia velloziana* O.Berg, Linnaea 27: 302 (1856), nom. illeg.
Brazil (Rio de Janeiro). 84 BZL. Nanophan. or phan.

Eugenia crenularis Lundell, Wrightia 3: 12 (1961).
C. & SW. Mexico. 79 MXC MXS. Nanophan. or phan.
Eugenia hintonii Lundell, Wrightia 3: 14 (1961).

Eugenia crenulata (Sw.) Willd., Sp. Pl. 2: 961 (1799).
E. Cuba to Hispaniola. 81 CUB DOM HAI. Nanophan. or phan.
**Myrtus crenulata* Sw., Prodr.: 78 (1788).
Eugenia crenulata var. *cubensis* O.Berg, Linnaea 27: 244 (1856).
Eugenia crenulata var. *domingensis* O.Berg, Linnaea 27: 244 (1856).

Eugenia cribrata McVaugh, Mem. New York Bot. Gard. 18(2): 174 (1969).
SE. Colombia to Venezuela (Bolívar). 82 VEN 83 CLM. Phan.

Eugenia cricamolensis Standl., Trop. Woods 16: 20 (1928).
Panama. 80 PAN. Nanophan. or phan.

Eugenia cristaensis O.Berg in C.F.P.von Martius & auct. suc. (eds.), Fl. Bras. 14(1): 222 (1857).
Brazil (Goiás). 84 BZC. Hemicr. or cham.

Eugenia cristalensis Urb., Symb. Antill. 9: 493 (1928).
Calycolpus cristalense (Urb.) Bisse, Revista Jard. Bot. Nac. Univ. Habana 4: 5 (1983).
E. Cuba (Sierra del Cristal). 81 CUB. Nanophan. or phan.

Eugenia cristata C.Wright, Anales Acad. Ci. Méd. Habana 5: 430 (1868).
W. & C. Cuba. 81 CUB. Nanophan.
Psidium vicentinum Urb., Symb. Antill. 9: 463 (1928).

Eugenia crossopterygoides A.Chev., Bull. Soc. Bot. France 55(8): 36 (1908).
Central African Rep. 23 CAF. Nanophan. or phan.

Eugenia crucicalyx McVaugh, Fieldiana, Bot. 29: 206 (1956).
Peru. 83 PER. Nanophan.

Eugenia crucigera Däniker, Vierteljahrsschr. Naturf. Ges. Zürich 78(19): 295 (1933).
New Caledonia. 60 NWC. Nanophan. or phan.

Eugenia cuaoensis McVaugh, Mem. New York Bot. Gard. 18(2): 174 (1969).
Venezuela (Amazonas). 82 VEN. Nanophan. or phan.

Eugenia cucullata Amshoff, Recueil Trav. Bot. Néerl. 42: 17 (1950).
Guianas. 82 FRG GUY SUR. Phan.

Eugenia culminicola McVaugh, Fieldiana, Bot. 29: 437 (1963).
C. Mexico. 79 MXC. Nanophan.

Eugenia cuprea (O.Berg) Nied. in H.G.A.Engler & K.A.E.Prantl, Nat. Pflanzenfam. 3(7): 82 (1893).
SE. Brazil (to Paraná). 84 BZL BZS. Phan.
**Phyllocalyx cupreus* O.Berg in C.F.P.von Martius & auct. suc. (eds.), Fl. Bras. 14(1): 592 (1859).
Eugenia grandisepala Mattos, Loefgrenia 66: 7 (1975).

Eugenia cupulata Amshoff, Recueil Trav. Bot. Néerl. 39: 160 (1942).
N. South America to Brazil. 82 FRG SUR VEN 84 BZC BZN. Nanophan. or phan.

var. ***cupulata***
N. South America to Brazil. 82 FRG SUR VEN 84 BZC BZN. Nanophan. or phan.

var. ***macrophylla*** McVaugh, Mem. New York Bot. Gard. 18(2): 176 (1969).
N. Brazil. 84 BZN. Nanophan. or phan.
**Eugenia macrophylla* O.Berg in C.F.P.von Martius & auct. suc. (eds.), Fl. Bras. 14(1): 268 (1857), nom. illeg.

Eugenia cupuligera Urb., Symb. Antill. 9: 109 (1923).
Pseudanamomis cupuligera (Urb.) Bisse, Feddes Repert. 96: 511 (1985).
E. Cuba. 81 CUB. Nanophan. or phan.
Eugenia brunnescens Urb., Symb. Antill. 9: 504 (1928).

Eugenia curvipilosa McVaugh, Fieldiana, Bot. 29: 206 (1956).
Peru. 83 PER. Nanophan.

Eugenia curvivenia McVaugh, Fieldiana, Bot. 29: 207 (1956).
N. Brazil. 84 BZN. Phan.

Eugenia cuspidifolia DC., Prodr. 3: 279 (1828).
S. Trop. America. 83 ECU PER 84 BZN. Phan.

Eugenia cycloidea Urb., Symb. Antill. 9: 490 (1928).
E. Cuba. 81 CUB. Nanophan.

Eugenia cyclophylla O.Berg in C.F.P.von Martius & auct. suc. (eds.), Fl. Bras. 14(1): 287 (1857).
Brazil (Bahia, Espírito Santo). 84 BZE BZL. Nanophan. or phan.

Eugenia cydoniifolia O.Berg in C.F.P.von Martius & auct. suc. (eds.), Fl. Bras. 14(1): 229 (1857).
Bolivia. 83 BOL. Nanophan. or phan.

Eugenia cymatodes O.Berg in C.F.P.von Martius & auct. suc. (eds.), Fl. Bras. 14(1): 578 (1859).
NE. Brazil. 84 BZE. Nanophan. or phan.

Eugenia cyphophloea Griseb., Cat. Pl. Cub.: 87 (1866).
Cuba. 81 CUB. Nanophan.

Eugenia daaouiensis Guillaumin, Bull. Soc. Bot. France 85: 637 (1938 publ. 1939).
New Caledonia. 60 NWC. Nanophan. or phan.
Eugenia daaouiensis var. *glabriflora* Guillaumin, Bull. Soc. Bot. France 85: 637 (1938 publ. 1939).

Eugenia daenikeri Guillaumin, Bull. Soc. Bot. France 85: 637 (1938 publ. 1939).
New Caledonia. 60 NWC. Nanophan. or phan.
**Eugenia quaternifolia* Guillaumin, Notul. Syst. (Paris) 3: 262 (1916), nom. illeg.

Eugenia darcyi Barrie, Monogr. Syst. Bot. Missouri Bot. Gard. 104: 358 (2005).
Panama. 80 PAN. Nanophan. or phan.

Eugenia decussata (Vell.) Mattos, Loefgrenia 102: 3 (1993).
SE. Brazil. 84 BZL. Nanophan. or phan.
**Myrtus decussata* Vell., Fl. Flumin. 5: 27, t. 73 (1829). *Eugenia tenuifolia* O.Berg in C.F.P.von Martius & auct. suc. (eds.), Fl. Bras. 14(1): 249 (1857), nom. illeg.
Eugenia tenuifolia var. *axillaris* O.Berg in C.F.P.von Martius & auct. suc. (eds.), Fl. Bras. 14(1): 249 (1857).
Eugenia tenuifolia var. *laterifolia* O.Berg in C.F.P.von Martius & auct. suc. (eds.), Fl. Bras. 14(1): 250 (1857).

Eugenia delpechiana Urb. & Ekman, Ark. Bot. 22A(10): 37 (1929).
Haiti (Massif des Matheux). 81 HAI. Nanophan.

Eugenia demeusei De Wild., Ann. Mus. Congo Belge, Bot., V, 2: 325 (1908).
Zaïre. 23 ZAI. Nanophan.
Eugenia demeusei f. *lukolelaensis* De Wild., Ann. Mus. Congo Belge, Bot., V, 2: 325 (1908).

Eugenia denigrata McVaugh, Mem. New York Bot. Gard. 18(2): 176 (1969).
French Guiana to Brazil (Amapá). 82 FRG 84 BZN. Phan.

Eugenia dentata (O.Berg) Nied. in H.G.A.Engler & K.A.E.Prantl, Nat. Pflanzenfam. 3(7): 82 (1893).
French Guiana to N. Brazil. 82 FRG 84 BZN. Nanophan. or phan.
**Stenocalyx dentatus* O.Berg in C.F.P.von Martius & auct. suc. (eds.), Fl. Bras. 14(1): 338 (1857).

Eugenia dewevrei De Wild. & T.Durand, Bull. Soc. Roy. Bot. Belgique, Compt.-Rend. 38: 125 (1899).
Zaïre to Angola. 23 ZAI 26 ANG. Nanophan. or phan.

Eugenia dibrachiata McVaugh, Fieldiana, Bot. 29: 207 (1956).
Ecuador to Peru. 83 ECU PER. Phan.

Eugenia dichroma O.Berg in C.F.P.von Martius & auct. suc. (eds.), Fl. Bras. 14(1): 290 (1857).
Brazil (Rio de Janeiro). 84 BZL. Nanophan. or phan.

Eugenia dictyophleba O.Berg in C.F.P.von Martius & auct. suc. (eds.), Fl. Bras. 14(1): 226 (1857).
Brazil (Bahia). 84 BZE. Nanophan. or phan.

Eugenia dictyophylla Urb., Repert. Spec. Nov. Regni Veg. 20: 341 (1924).
Hispaniola. 81 DOM HAI. Nanophan. or phan.

Eugenia diminutiflora Amshoff, Acta Bot. Neerl. 7: 55 (1958).
Congo. 23 CON. Nanophan. or phan.

Eugenia dimorpha O.Berg in C.F.P.von Martius & auct. suc. (eds.), Fl. Bras. 14(1): 264 (1857). *Luma dimorpha* (O.Berg) Herter, Revista Sudamer. Bot. 7: 219 (1943).
Brazil (Rio Grande do Sul). 84 BZS. Nanophan. or phan.
Eugenia oeidocarpa O.Berg in C.F.P.von Martius & auct. suc. (eds.), Fl. Bras. 14(1): 270 (1857). *Eugenia dimorpha* var. *oeidocarpa* (O.Berg) Mattos, Loefgrenia 61: 3 (1974).
Eugenia dimorpha var. *australis* Mattos, Loefgrenia 66: 8 (1965).

Eugenia dinklagei Engl. & Brehmer, Bot. Jahrb. Syst. 54: 338 (1917).
Liberia. 22 LBR. Nanophan. or phan.

Eugenia diospyroides H.Perrier, Mém. Inst. Sci. Madagascar, Sér. B, Biol. Vég. 4: 176 (1953).
N. & NE. Madagascar. 29 MDG. Nanophan.

Eugenia diplocampta Diels, Verh. Bot. Vereins Prov. Brandenburg 48: 191 (1906 publ. 1907).
S. Venezuela to Bolivia. 82 VEN 83 BOL 84 BZN. Nanophan. or phan.

Eugenia discifera Gamble, Bull. Misc. Inform. Kew 1918: 239 (1918).
S. India. 40 IND. Nanophan. or phan.

Eugenia discolorans C.Wright, Anales Acad. Ci. Méd. Habana 5: 480 (1868).
W. Cuba. 81 CUB. Nanophan. or phan.

Eugenia discors McVaugh, Publ. Field Mus. Nat. Hist., Bot. Ser. 13(4): 692 (1958).
N. Peru. 83 PER. Nanophan. or phan.

Eugenia discreta McVaugh, Fieldiana, Bot. 29: 208 (1956).
N. Brazil to Peru. 83 PER 84 BZN. Nanophan. or phan.

Eugenia disperma Vell., Fl. Flumin. 5: 209, t. 38 (1829). *Eugenia phaea* O.Berg in C.F.P.von Martius & auct. suc. (eds.), Fl. Bras. 14(1) 303 (1857), nom. superfl.
SE. Brazil. 84 BZL. Phan.

Eugenia disticha (Sw.) DC., Prodr. 3: 274 (1828).
Jamaica. 81 JAM. Nanophan.
**Myrtus disticha* Sw., Prodr.: 78 (1788).
Myrtus horizontalis Vent., Jard. Malmaison: 60 (1804).
Olynthia disticha (Sw.) Lindl., Coll. Bot.: t. 19 (1821).

Eugenia dittocrepis O.Berg in C.F.P.von Martius & auct. suc. (eds.), Fl. Bras. 14(1): 292 (1857).
S. Trop. America. 82 VEN 83 ECU PER 84 BZN. Nanophan. or phan.
Eugenia congestissima Diels, Verh. Bot. Vereins Prov. Brandenburg 48: 190 (1906 publ. 1907).

Eugenia dodoana Engl. & Brehmer, Bot. Jahrb. Syst. 54: 338 (1917).
Cameroon. 23 CMN. Nanophan. or phan.

Eugenia dodonaeifolia Cambess. in A.F.C.de Saint-Hilaire, Fl. Bras. Merid. 2: 364 (1833).
Brazil. 84 BZC BZL BZS. Nanophan. or phan.
Eugenia ramiflora Miq., Linnaea 22: 536 (1849), nom. illeg.
Eugenia sphenophylla O.Berg in C.F.P.von Martius & auct. suc. (eds.), Fl. Bras. 14(1): 288 (1857).
Eugenia sphenophylla var. *coaetanea* O.Berg in C.F.P. von Martius & auct. suc. (eds.), Fl. Bras. 14(1): 288 (1857).
Eugenia sphenophylla var. *prasina* O.Berg in C.F.P.von Martius & auct. suc. (eds.), Fl. Bras. 14(1): 288 (1857).
Eugenia sphenophylla var. *tristis* O.Berg in C.F.P.von Martius & auct. suc. (eds.), Fl. Bras. 14(1): 288 (1857).
Eugenia dodoneifolia var. *grandifolia* O.Berg in C.F.P.von Martius & auct. suc. (eds.), Fl. Bras. 14(1): 581 (1859).
Eugenia curvatopetiolata Kiaersk., Enum. Myrt. Bras.: 133 (1893).

Eugenia domingensis O.Berg, Linnaea 27: 296 (1856).
Mexico to C. America, Caribbean. 79 MXG MXT 80 BLZ ELS GUA 81 CUB DOM HAI LEE PUE TRT WIN. Phan.
Eugenia domingensis var. *membranacea* O.Berg, Linnaea 29: 245 (1858).
Eugenia calyculata Bello, Anales Soc. Esp. Hist. Nat. 10: 271 (1881).

Eugenia doubledayi Standl., J. Arnold Arbor. 11: 36 (1930).
C. America. 80 COS NIC. Nanophan. or phan.

Eugenia duarteana Cambess. in A.F.C.de Saint-Hilaire, Fl. Bras. Merid. 2: 348 (1832). *Stenocalyx duarteanus* (Cambess.) O.Berg in C.F.P.von Martius & auct. suc. (eds.), Fl. Bras. 14(1): 342 (1857).
Brazil (Minas Gerais). 84 BZL. Nanophan. or phan.

Eugenia duchassaingiana O.Berg, Linnaea 27: 225 (1856).
Lesser Antilles. 81 LEE WIN. Phan.

Eugenia duckeana Mattos, Loefgrenia 28: 1 (1969).
Brazil (Ceará). 84 BZE. Nanophan. or phan.

Eugenia dulcis O.Berg in C.F.P.von Martius & auct. suc. (eds.), Fl. Bras. 14(1): 256 (1857).
Brazil (São Paulo). 84 BZL. Nanophan. or phan.

Eugenia dumosa (Vahl) DC., Prodr. 3: 276 (1828).
Jamaica. 81 JAM. Nanophan. or phan.
**Myrtus dumosa* Vahl, Symb. Bot. 2: 57 (1791).

Eugenia duplicata Britton & P.Wilson in ?.
W. Cuba. 81 CUB. Nanophan.

Eugenia dusenii Engl., Notizbl. Königl. Bot. Gart. Berlin 2: 289 (1899). *Myrtus dusenii* (Engl.) Kuntze, Deutsche Bot. Monatsschr. 21: 173 (1903).
Cameroon. 23 CMN. Nanophan.

Eugenia dysenterica DC., Prodr. 3: 268 (1828). *Stenocalyx dysentericus* (DC.) O.Berg in C.F.P.von Martius & auct. suc. (eds.), Fl. Bras. 14(1): 351 (1857).
Bolivia to Brazil. 83 BOL 84 BZC BZE BZL BZN. Nanophan. or phan.
Myrtus dysenterica Mart., Reise Bras. 2: 571 (1828).

Eugenia earhartii Acev.-Rodr., Brittonia 45: 133 (1993).
Virgin Is. 81 LEE. Nanophan. or phan.

Eugenia earlei Britton & P.Wilson, Mem. Torrey Bot. Club 16: 89 (1920).
W. & C. Cuba. 81 CUB. Nanophan.

Eugenia earthiana P.E.Sanchez, Lankesteriana 4: 179 (2004).
Costa Rica. 80 COS. Nanophan. or phan.

Eugenia egensis DC., Prodr. 3: 281 (1828).
C. & S. Trop. America to NE. Argentina (Formosa). 80 COS PAN 82 FRG GUY SUR VEN 83 BOL CLM ECU PER 84 BZC BZL BZN BZS 85 AGE PAR. Nanophan. or phan.
Eugenia sphaerosperma DC., Prodr. 3: 279 (1828).
Eugenia perforata O.Berg, Linnaea 27: 236 (1856).
Eugenia egensis var. *angustifolia* O.Berg in C.F.P.von Martius & auct. suc. (eds.), Fl. Bras. 14(1): 296 (1857).
Eugenia egensis var. *bimarginata* O.Berg in C.F.P.von Martius & auct. suc. (eds.), Fl. Bras. 14(1): 296 (1857).
Eugenia egensis var. *grandifolia* O.Berg in C.F.P.von Martius & auct. suc. (eds.), Fl. Bras. 14(1): 296 (1857).
Eugenia egensis var. *latifolia* O.Berg in C.F.P.von Martius & auct. suc. (eds.), Fl. Bras. 14(1): 296 (1857).
Eugenia egensis var. *parvifolia* O.Berg in C.F.P.von Martius & auct. suc. (eds.), Fl. Bras. 14(1): 296 (1857).
Eugenia maculata O.Berg in C.F.P.von Martius & auct. suc. (eds.), Fl. Bras. 14(1): 297 (1857).
Eugenia parodiana Morong, Ann. New York Acad. Sci. 7: 107 (1893).

Eugenia eggersii Kiaersk., Bot. Tidsskr. 17: 368 (1890).
Puerto Rico. 81 PUE. Phan.

Eugenia ehrenbergiana O.Berg, Linnaea 27: 153 (1856).
Hispaniola. 81 DOM HAI. Nanophan.
Eugenia blepharidantha Urb., Notizbl. Bot. Gart. Berlin-Dahlem 7: 498 (1921).

Eugenia eliasii Lundell, Phytologia 16: 443 (1968).
Guatemala. 80 GUA. Nanophan. or phan.

Eugenia elliotii Engl. & Brehmer, Bot. Jahrb. Syst. 54: 331 (1917).
Sierra Leone to Guinea. 22 GUI SIE. Nanophan. or phan.
Eugenia djalonensis A.Chev., Explor. Bot. Afrique Occ. Franç. 1: 267 (1920), nom. nud.

Eugenia ellipsoidea Kiaersk., Enum. Myrt. Bras.: 134 (1893).
SE. Brazil. 84 BZL. Nanophan. or phan.

Eugenia elliptica Lam., Encycl. 3: 206 (1789). *Myrtus elliptica* (Lam.) Spreng., Syst. Veg. 2: 483 (1825). *Eugenia cotinifolia* var. *elliptica* (Lam.) Baker ex Engl. & Nied. in H.G.A.Engler, Pflanzenw. Ost-Afrikas, C: 287 (1895).
Mauritius. 29 MAU. Nanophan. or phan.
Jossinia elliptica DC., Prodr. 3: 237 (1828).
Jossinia cordifolia Bojer, Hortus Maurit.: 141 (1837).
Jossinia lanceolata Bojer, Hortus Maurit.: 141 (1837), nom. inval.

Eugenia elongata Nied. in H.G.A.Engler & K.A.E. Prantl, Nat. Pflanzenfam. 3(7): 82 (1893).
Brazil (São Paulo). 84 BZL. Nanophan. or phan.
**Phyllocalyx racemosus* O.Berg in C.F.P.von Martius & auct. suc. (eds.), Fl. Bras. 14(1): 334 (1857).

Eugenia emarginata (Kunth) DC., Prodr. 3: 290 (1828).
Venezuela (Amazonas). 82 VEN. Nanophan. or phan.
**Myrtus emarginata* Kunth in F.W.H.von Humboldt, A.J.A.Bonpland & C.S.Kunth, Nov. Gen. Sp. 6: 142 (1823). *Emurtia emarginata* (Kunth) Raf., Sylva Tellur.: 106 (1838).

Eugenia ependytes McVaugh, Publ. Field Mus. Nat. Hist., Bot. Ser. 13(4): 696 (1958).
Peru to N. Brazil. 83 PER 84 BZN. Nanophan. or phan.

Eugenia eperforata Urb., Symb. Antill. 6: 25 (1909).
Jamaica (St. Ann). 81 JAM. Phan.

Eugenia eriantha Urb., Symb. Antill. 9: 104 (1923).
SE. Cuba. 81 CUB. Nanophan.

Eugenia ericoides Guillaumin, Bull. Soc. Bot. France 85: 637 (1938 publ. 1939).
New Caledonia. 60 NWC. Nanophan. or phan.

Eugenia erythrocarpa (Kunth) DC., Prodr. 3: 270 (1828).
Colombia. 83 CLM. Nanophan. or phan.
Eugenia umbellata Spreng., Neue Entd. 2: 169 (1821).
**Myrtus erythrocarpa* Kunth in F.W.H.von Humboldt, A.J.A.Bonpland & C.S.Kunth, Nov. Gen. Sp. 6: 148 (1823).

Eugenia erythrophylla Strey, Bothalia 10: 569 (1972).
Cape Prov. to KwaZulu-Natal. 27 CPP NAT. Nanophan. or phan.

Eugenia erythroxyloides Mart. ex O.Berg in C.F.P.von Martius & auct. suc. (eds.), Fl. Bras. 14(1): 346 (1857).
Brazil. 84+. Nanophan. or phan.

Eugenia essequiboensis Sandwith, Bull. Misc. Inform. Kew 1932: 211 (1932).
Guyana. 82 GUY. Phan.

Eugenia esteliensis Barrie, Novon 15: 16 (2005).
Nicaragua. 80 NIC. Nanophan. or phan.

Eugenia eurycheila O.Berg, Linnaea 27: 213 (1856).
Guyana to N. Brazil. 82 GUY 84 BZN. Nanophan. or phan.
Eugenia eurycheila var. *grandifolia* O.Berg, Linnaea 27: 214 (1856).
Eugenia eurycheila var. *parvifolia* O.Berg, Linnaea 27: 214 (1856).

Eugenia exaltata A.Rich. ex O.Berg, Linnaea 30: 687 (1861).
French Guiana to N. Brazil. 82 FRG 84 BZN. Phan.

Eugenia excelsa O.Berg in C.F.P.von Martius & auct. suc. (eds.), Fl. Bras. 14(1): 277 (1857).

Guianas to Brazil. 82 FRG GUY SUR 84 BZE BZL BZN BZS. Phan.
Eugenia cuspidata O.Berg in C.F.P.von Martius & auct. suc. (eds.), Fl. Bras. 14(1): 286 (1857).
Eugenia obovata O.Berg in C.F.P.von Martius & auct. suc. (eds.), Fl. Bras. 14(1): 289 (1857), nom. illeg.

Eugenia excisa Urb., Symb. Antill. 9: 100 (1923). *Calycolpus excisus* (Urb.) Bisse, Revista Jard. Bot. Nac. Univ. Habana 4: 6 (1983).
E. Cuba (Bayate). 81 CUB. Nanophan. or phan.

Eugenia excoriata O.Berg in C.F.P.von Martius & auct. suc. (eds.), Fl. Bras. 14(1): 312 (1857).
Brazil (Rio de Janeiro). 84 BZL. Nanophan. or phan.

Eugenia expansa Spring ex Mart., Flora 20(Beibl. 2): 84 (1837).
Brazil (Rio de Janeiro). 84 BZL. Nanophan. or phan.

Eugenia farameoides A.Rich., Hist. Phys. Cuba, Pl. Vasc.: 588 (1846).
Mexico to C. America. 79 MXG MXS MXT 80 BLZ GUA HON NIC. Nanophan.
Eugenia flavifolia Standl., Publ. Field Mus. Nat. Hist., Bot. Ser. 8: 320 (1931).
Eugenia ceibensis Standl., Publ. Field Mus. Nat. Hist., Bot. Ser. 9: 319 (1940).

Eugenia farinacea Barrie, Novon 15: 16 (2005).
Nicaragua. 80 NIC. Nanophan. or phan.

Eugenia fernandopoana Engl. & Brehmer, Bot. Jahrb. Syst. 54: 377 (1917).
Ivory Coast, Bioko, Gabon. 22 IVO 23 GAB GGI. Nanophan. or phan.

Eugenia ferreiraeana O.Berg in C.F.P.von Martius & auct. suc. (eds.), Fl. Bras. 14(1): 285 (1857).
N. South America to Brazil. 82 FRG GUY SUR VEN 84 BZE BZN. Nanophan. or phan.
Eugenia nemoralis DC., Prodr. 3: 267 (1828).
Eugenia ramiflora var. *montana* Amshoff, Bull. Torrey Bot. Club 75: 535 (1948).

Eugenia flamingensis O.Berg in C.F.P.von Martius & auct. suc. (eds.), Fl. Bras. 14(1): 576 (1859).
Brazil (Rio de Janeiro). 84 BZL. Nanophan. or phan.

Eugenia flavescens DC., Prodr. 3: 272 (1828).
S. Trop. America. 82 FRG GUY SUR VEN 83 BOL ECU 84 BZC BZE BZN. Nanophan. or phan.
Eugenia flavescens var. *guyanensis* O.Berg in C.F.P.von Martius & auct. suc. (eds.), Fl. Bras. 14(1): 272 (1857).
Eugenia flavescens var. *longifolia* O.Berg in C.F.P.von Martius & auct. suc. (eds.), Fl. Bras. 14(1): 272 (1857).
Eugenia flavescens var. *parvifolia* O.Berg in C.F.P.von Martius & auct. suc. (eds.), Fl. Bras. 14(1): 272 (1857).

Eugenia flavoviridis Lundell, Amer. Midl. Naturalist 29: 479 (1943).
SE. Mexico to Guatemala. 79 MXT 80 BLZ GUA. Nanophan. or phan.
Eugenia dissitiflora Lundell, Wrightia 2: 207 (1961).
Eugenia flavida Lundell, Wrightia 3: 14 (1961).

Eugenia fleuryi A.Chev., Veg. Ut. Afr. Trop. Franç. 9: 214 (1917).
Gabon. 23 GAB. Nanophan. or phan.

Eugenia floccosa Bedd., Fl. Sylv. S. India 1: t. 200 (1872).
India. 40 IND. Phan.

Eugenia florida DC., Prodr. 3: 283 (1828).
C. & S. Trop. America. 80 BLZ COS PAN 82 FRG GUY SUR VEN 83 BOL CLM ECU PER 84 BZC BZE BZL BZN BZS 85 AGE AGW. Nanophan. or phan.
Eugenia patula DC., Prodr. 3: 284 (1828).
Eugenia sylvatica Cambess. in A.F.C.de Saint-Hilaire, Fl. Bras. Merid. 2: 337 (1832).
Eugenia atropunctata Steud., Flora 26: 762 (1843).
Eugenia atropunctata var. *gracilis* O.Berg, Linnaea 27: 294 (1856).
Eugenia atropunctata var. *robusta* O.Berg, Linnaea 27: 293 (1856).
Eugenia gardneriana O.Berg in C.F.P.von Martius & auct. suc. (eds.), Fl. Bras. 14(1): 316 (1857). *Eugenia moraviana* var. *gardneriana* (O.Berg) Mattos, Loefgrenia 85: 2 (1984).
Eugenia gardneriana var. *depauperata* O.Berg in C.F.P.von Martius & auct. suc. (eds.), Fl. Bras. 14(1): 317 (1857).
Eugenia gardneriana var. *dives* O.Berg in C.F.P.von Martius & auct. suc. (eds.), Fl. Bras. 14(1): 316 (1857).
Eugenia oligoneura O.Berg in C.F.P.von Martius & auct. suc. (eds.), Fl. Bras. 14(1): 321 (1857).
Eugenia gardneriana var. *ovata* O.Berg in C.F.P.von Martius & auct. suc. (eds.), Fl. Bras. 14(1): 589 (1859).
Eugenia membranacea O.Berg in C.F.P.von Martius & auct. suc. (eds.), Fl. Bras. 14(1): 589 (1859).
Eugenia gardneriana var. *rigida* O.Berg, Linnaea 30: 694 (1861).
Eugenia racemifera Sagot, Ann. Sci. Nat., Bot., VI, 20: 195 (1885).
Eugenia seriatoracemosa Kiaersk., Enum. Myrt. Bras.: 151 (1893).
Eugenia tinge-lingua S.Moore, Trans. Linn. Soc. London, Bot. 4: 357 (1895).
Eugenia perorebi Parodi ex Speg. & Girola, Cat. Descr. Maderas: 361 (1910).
Eugenia melanosticta Standl., J. Arnold Arbor. 11: 126 (1930), nom. illeg. *Eugenia coloradoensis* Standl., Trop. Woods 52: 27 (1937).

Eugenia floscellus D.Legrand, in Fl. Ilustr. Catar. 1(Mirt., Suppl. 1): 28 (1977).
S. Brazil. 84 BZS. Nanophan. or phan.

Eugenia fluminensis O.Berg in C.F.P.von Martius & auct. suc. (eds.), Fl. Bras. 14(1): 265 (1857).
Brazil (Rio de Janeiro). 84 BZL. Nanophan. or phan.

Eugenia foetida Pers., Syn. Pl. 2: 29 (1806). *Myrtus foetida* (Pers.) Spreng., Syst. Veg. 2: 481 (1825).
SE. Mexico to Guatemala, S. Florida to Caribbean. 78 FLA 79 MXT 80 BLZ GUA 81 BAH CUB DOM HAI JAM LEE PUE. Nanophan. or phan.
Myrtus buxifolia Sw., Prodr.: 78 (1788). *Eugenia buxifolia* (Sw.) Willd., Sp. Pl. 2: 960 (1799), nom. illeg. *Eugenia triplinervia* O.Berg, Linnaea 27: 190 (1856), nom. illeg. *Eugenia triplinervia* var. *buxifolia* (Sw.) O.Berg, Linnaea 27: 191 (1856).
Myrtus axillaris Poir. in J.B.A.P.M.de Lamarck, Encycl. 4: 412 (1798), nom. illeg. *Myrtus poiretii* Spreng., Syst. Veg. 2: 483 (1825). *Eugenia poiretii* (Spreng.) DC., Prodr. 3: 274 (1828).
Eugenia fetida Poir. in J.B.A.P.M.de Lamarck, Encycl., Suppl. 3: 129 (1813), orth. var.
Eugenia myrtoides Poir. in J.B.A.P.M.de Lamarck, Encycl., Suppl. 3: 125 (1813).
Myrtus acka Juss. ex DC., Prodr. 3: 277 (1828).

Eugenia macrostemon O.Berg, Linnaea 27: 188 (1856).
Eugenia triplinervia var. *angustifolia* O.Berg, Linnaea 27: 192 (1856).
Eugenia triplinervia var. *laevigata* O.Berg, Linnaea 27: 192 (1856).
Eugenia triplinervia var. *latissima* O.Berg, Linnaea 27: 191 (1856).
Eugenia triplinervia var. *oblongata* O.Berg, Linnaea 27: 192 (1856).
Eugenia triplinervia var. *obovata* O.Berg, Linnaea 27: 191 (1856).
Eugenia unedifolia Spreng. ex O.Berg, Linnaea 27: 192 (1856), pro syn.
Eugenia mayana Standl., Contr. U. S. Natl. Herb. 23: 1042 (1924).
Eugenia ossaeana Urb., Symb. Antill. 9: 514 (1928).

Eugenia francavilleana O.Berg, Linnaea 30: 686 (1861).
E. Brazil. 84 BZE BZL. Nanophan. or phan.
Eugenia glazioviana Kiaersk., Enum. Myrt. Bras.: 128 (1893).
Eugenia glazioviana var. *macrophylla* Kiaersk., Enum. Myrt. Bras.: 129 (1893).
Eugenia glazioviana f. *parvifolia* Kiaersk., Enum. Myrt. Bras.: 129 (1893).
Eugenia glazioviana var. *pauciflora* Kiaersk., Enum. Myrt. Bras.: 129 (1893).

Eugenia froesii McVaugh, Mem. New York Bot. Gard. 18(2): 186 (1969).
Brazil (Pará). 84 BZN. Nanophan.

Eugenia fulva Thwaites, Enum. Pl. Zeyl.: 115 (1859).
Sri Lanka. 40 SRL. Nanophan. or phan.
Eugenia floccifera Thwaites, Enum. Pl. Zeyl.: 115 (1859).
Eugenia insignis Thwaites, Enum. Pl. Zeyl.: 416 (1864).
Eugenia haeckeliana Trimen, J. Bot. 23: 207 (1885).

Eugenia funkiana O.Berg, Linnaea 29: 228 (1858).
S. America (?). 8+. Nanophan. or phan.

Eugenia fusca O.Berg in C.F.P.von Martius & auct. suc. (eds.), Fl. Bras. 14(1): 290 (1857).
E. Brazil. 84 BZE BZL. Nanophan. or phan.

Eugenia fuscopunctata Kiaersk., Enum. Myrt. Bras.: 155 (1893).
Brazil (Minas Gerais). 84 BZL. Nanophan. or phan.
**Stenocalyx grandifolius* O.Berg in C.F.P.von Martius & auct. suc. (eds.), Fl. Bras. 14(1): 340 (1857).

Eugenia gabonensis Amshoff, Acta Bot. Neerl. 7: 55 (1958).
Ivory Coast to Zaïre. 22 IVO 23 GAB ZAI. Cham. or nanophan.
Eugenia miegeana Aké Assi, Bull. Jard. Bot. État 30: 15 (1960).

Eugenia gacognei Montrouz., Mém. Acad. Roy. Sci. Lyon, Sect. Sci. 10: 208 (1860).
New Caledonia. 60 NWC. Nanophan. or phan.
Eugenia homei Seem., Fl. Vit.: 76 (1865).

Eugenia galalonensis (C.Wright ex Griseb.) Krug & Urb., Bot. Jahrb. Syst. 19: 641 (1895).
W. & C. Cuba, Mexico to Ecuador. 79 MXG MXS MXT 80 COS GUA HON NIC PAN 81 CUB 83 CLM ECU. Nanophan.
**Eugenia disticha* var. *galalonensis* C.Wright ex Griseb., Cat. Pl. Cub.: 88 (1866).
Eugenia argyrea Lundell, Wrightia 3: 10 (1961).
Eugenia chacteana Lundell, Wrightia 4: 93 (1968).

Eugenia galbaoensis Mattos, Loefgrenia 120: 4 (2005).
French Guiana. 82 FRG. Phan.
**Calycorectes guyanensis* Mattos, Loefgrenia 101: 1 (1992)

Eugenia galeata Urb., Symb. Antill. 7: 307 (1912).
W. Cuba. 81 CUB. Nanophan. or phan.

Eugenia gatopensis Guillaumin, Bull. Soc. Bot. France 85: 638 (1938 publ. 1939). *Austromyrtus gatopensis* (Guillaumin) Burret, Notizbl. Bot. Gart. Berlin-Dahlem 15: 505 (1941).
New Caledonia (Gatope). 60 NWC. Nanophan. or phan.

Eugenia gaudichaudiana O.Berg in C.F.P.von Martius & auct. suc. (eds.), Fl. Bras. 14(1): 228 (1857).
Brazil (Rio de Janeiro). 84 BZL. Nanophan. or phan.

Eugenia gaumeri Standl., Publ. Field Mus. Nat. Hist., Bot. Ser. 8: 28 (1930). *Eugenia laevis* var. *gaumeri* (Standl.) McVaugh, Fieldiana, Bot. 29: 469 (1963).
SE. Mexico to Guatemala. 79 MXT 80 BLZ GUA. Nanophan. or phan.
Eugenia lundellii Standl., Publ. Carnegie Inst. Wash. 461: 7 (1935).

Eugenia gemmiflora O.Berg in C.F.P.von Martius & auct. suc. (eds.), Fl. Bras. 14(1): 223 (1857).
Brazil. 84 BZC BZE. Nanophan, or phan.

Eugenia gibberosa Urb., Symb. Antill. 9: 107 (1923). *Pseudanamomis gibberosa* (Urb.) Bisse, Feddes Repert. 96: 511 (1985).
E. Cuba. 81 CUB. Nanophan. or phan.
Eugenia brevipetiolata Britton & P.Wilson, Bull. Torrey Bot. Club 50: 45 (1923).
Eugenia arnoldsonii Urb., Symb. Antill. 9: 506 (1928).

Eugenia gigas Lundell, Wrightia 4: 41 (1968).
Guatemala. 80 GUA. Nanophan. or phan.
Eugenia brihondurensis Lundell, Wrightia 6: 50 (1979).

Eugenia gilgii Engl. & Brehmer, Bot. Jahrb. Syst. 54: 334 (1917).
Cameroon. 23 CMN. Nanophan. or phan.

Eugenia glabra Alston in H.Trimen, Handb. Fl. Ceylon 6(Suppl.): 120 (1931).
Sri Lanka. 40 SRL. Nanophan. or phan.
**Eugenia decora* Thwaites, Enum. Pl. Zeyl.: 115 (1859), nom. illeg.

Eugenia glabrata (Sw.) DC., Prodr. 3: 274 (1828).
Caribbean. 81 CUB DOM HAI JAM PUE. Nanophan. or phan.
**Myrtus glabrata* Sw., Prodr.: 78 (1788).
Eugenia affinis DC., Prodr. 3: 272 (1828).

Eugenia glandulosa Cambess. in A.F.C.de Saint-Hilaire, Fl. Bras. Merid. 2: 352 (1832). *Phyllocalyx glandulosus* (Cambess.) D.Legrand, Notul. Syst. (Paris) 15: 270 (1958).
Brazil (Minas Gerais). 84 BZL. Nanophan. or phan.

Eugenia glandulosopunctata P.E.Sánchez & Poveda, Brenesia 24: 413 (1985 publ. 1986).
Costa Rica. 80 COS. Nanophan. or phan.

Eugenia gomesiana O.Berg in C.F.P.von Martius & auct. suc. (eds.), Fl. Bras. 14(1): 254 (1857).
S. Trop. America. 82 FRG GUY SUR VEN 83 BOL PER 84 BZN. Nanophan. or phan.
Eugenia prosoneura O.Berg, Linnaea 31: 255 (1862).

Eugenia gomezii Barrie, Novon 15: 18 (2005).
Costa Rica. 80 COS. Nanophan. or phan.

Eugenia gonavensis Urb., Ark. Bot. 22A(10): 38 (1929).
Haiti (Î. de la Gonâve). 81 HAI. Nanophan.

Eugenia gongylocarpa M.L.Kawas. & B.K.Holst, Brittonia 46: 138 (1994).
French Guiana to Brazil (Pará). 82 FRG 84 BZN. Phan.

Eugenia goviala H.Perrier, Mém. Inst. Sci. Madagascar, Sér. B, Biol. Vég. 4: 177 (1953).
EC. Madagascar. 29 MDG. Phan.

Eugenia goyazensis Nied. in H.G.A.Engler & K.A.E.Prantl, Nat. Pflanzenfam. 3(7): 83 (1893).
Brazil (Minas Gerais). 84 BZL. Nanophan.
**Phyllocalyx regelianus* O.Berg in C.F.P.von Martius & auct. suc. (eds.), Fl. Bras. 14(1): 591 (1859). *Eugenia lundiana* Kiaersk., Enum. Myrt. Bras.: 162 (1893).

Eugenia gracilipes Kiaersk., Enum. Myrt. Bras.: 158 (1893).
SE. Brazil. 84 BZL. Nanophan. or phan.

Eugenia gracillima Kiaersk., Enum. Myrt. Bras.: 120 (1893).
SE. Brazil. 84 BZL. Nanophan. or phan.
Eugenia leptomischa Kiaersk., Enum. Myrt. Bras.: 121 (1893).
Eugenia klappenbachiana Mattos & D.Legrand, Loefgrenia 67: 20 (1975).
Eugenia multipunctata Mattos & D.Legrand, Loefgrenia 67: 26 (1975), nom. illeg. *Eugenia neomultipunctata* Sobral, Napaea 11: 36 (1995).

Eugenia grandiflora O.Berg, Linnaea 29: 226 (1858).
Venezuela. 82 VEN. Nanophan. or phan.

Eugenia grandifolia O.Berg in C.F.P.von Martius & auct. suc. (eds.), Fl. Bras. 14(1): 324 (1857).
Brazil (Rio de Janeiro). 84 BZL. Nanophan. or phan.

Eugenia grayumii Barrie, Novon 15: 19 (2005).
Costa Rica. 80 COS. Nanophan. or phan.

Eugenia grazielae Mattos & D.Legrand, Loefgrenia 67: 24 (1975).
Brazil (Rio de Janeiro). 84 BZL. Nanophan. or phan.

Eugenia greggii (Sw.) Poir. in J.B.A.P.M.de Lamarck, Encycl., Suppl. 3: 126 (1813).
Lesser Antilles to Trinidad. 81 LEE TRT WIN. Phan.
Greggia aromatica Gaertn., Fruct. Sem. Pl. 1: 168 (1788).
**Myrtus greggii* Sw., Prodr.: 78 (1788). *Pimentus greggii* (Sw.) Raf., Sylva Tellur.: 105 (1838).
Eugenia sieberiana DC., Prodr. 3: 277 (1828).
Eugenia sieberiana var. *crassifolia* O.Berg, Linnaea 30: 690 (1861).

Eugenia grifensis Urb., Symb. Antill. 9: 491 (1928).
W. Cuba. 81 CUB. Nanophan.

Eugenia grijalvae Barrie, Novon 15: 21 (2005).
Nicaragua. 80 NIC. Nanophan. or phan.

Eugenia grisebachii Krug & Urb., Bot. Jahrb. Syst. 19: 627 (1895).
E. Cuba. 81 CUB. Nanophan.

Eugenia griseiflora McVaugh, Mem. New York Bot. Gard. 18(2): 187 (1969).
Suriname to Peru. 82 SUR 83 PER 84 BZN. Nanophan. or phan.

Eugenia grisiana Guillaumin, Bull. Soc. Bot. France 85: 640 (1938 publ. 1939).
New Caledonia. 60 NWC. Nanophan. or phan.

Eugenia gryposperma Krug & Urb., Bot. Jahrb. Syst. 19: 650 (1895).
Martinique (Mts. du Vauclin). 81 WIN. Nanophan. or phan.

Eugenia guanensis Urb., Symb. Antill. 9: 495 (1928).
W. Cuba. 81 CUB. Nanophan.

Eugenia guatemalensis Donn.Sm., Bot. Gaz. 23: 245 (1897).
Mexico to C. America. 79 MXC MXN MXS MXT 80 COS ELS GUA NIC. Nanophan. or phan.
Eugenia patalensis Standl. & Steyerm., Publ. Field Mus. Nat. Hist., Bot. Ser. 23: 130 (1944).
Eugenia laughlinii Lundell, Phytologia 16: 443 (1968).

Eugenia guayaquilensis (Kunth) DC., Prodr. 3: 275 (1828).
Ecuador. 83 ECU. Nanophan. or phan.
**Myrtus guayaquilensis* Kunth in F.W.H.von Humboldt, A.J.A.Bonpland & C.S.Kunth, Nov. Gen. Sp. 6: 147 (1823). *Emurtia guayaquilensis* (Kunth) Raf., Sylva Tellur.: 106 (1838).

Eugenia guillotii Hochr., Annuaire Conserv. Jard. Bot. Genève 11–12: 76 (1908).
E. Madagascar. 29 MDG. Phan.

Eugenia guttata Lundell, Wrightia 2: 207 (1961).
Guatemala. 80 GUA. Nanophan. or phan.

Eugenia gyrosepala Baker f., J. Linn. Soc., Bot. 45: 313 (1921). *Austromyrtus gyrosepala* (Baker f.) Burret, Notizbl. Bot. Gart. Berlin-Dahlem 15: 506 (1941).
New Caledonia (Ngoye River). 60 NWC. Nanophan. or phan.

Eugenia haberi Barrie, Sida 22: 1071 (2006).
Costa Rica. 80 COS.

Eugenia haematocarpa Alain, Bull. Torrey Bot. Club 90: 190 (1963).
Puerto Rico. 81 PUE. Phan.

Eugenia haitiensis Krug & Urb., Bot. Jahrb. Syst. 19: 616 (1895).
Hispaniola. 81 DOM HAI. Nanophan.

Eugenia hammelii Barrie, Novon 15: 22 (2005).
Costa Rica. 80 COS. Nanophan. or phan.

Eugenia hamoniana Mattos, Loefgrenia 12: 1 (1963).
Brazil (Santa Catarina). 84 BZS. Nanophan. or phan.

Eugenia handroana D.Legrand, Sellowia 13: 320 (1961).
Brazil (S. Minas Gerais to Santa Catarina). 84 BZL BZS. Phan.
Eugenia lepida Mattos & D.Legrand, Loefgrenia 67: 29 (1975).

Eugenia hanoverensis Proctor, J. Arnold Arbor. 63: 274 (1982).
Jamaica. 81 JAM. Nanophan. or phan.

Eugenia haputalense Kosterm., Quart. J. Taiwan Mus. 34: 164 (1981).
Sri Lanka. 40 SRL. Nanophan. or phan.

Eugenia harrisii Krug & Urb., Bot. Jahrb. Syst. 19: 632 (1895).
Jamaica. 81 JAM. Nanophan. or phan.
Eugenia crenata O.Berg, Linnaea 27: 226 (1856), nom. illeg.
Eugenia harrisii var. *grandiflora* Krug & Urb., Bot. Jahrb. Syst. 19: 633 (1895).

Eugenia hartmanniae Mattos, Loefgrenia 94: 6 (1989).
Brazil (Mato Grosso). 84 BZC. Nanophan. or phan.

Eugenia hartshornii Barrie, Novon 15: 24 (2005).
Costa Rica. 80 COS. Nanophan. or phan.

Eugenia hastilis (Blume) J.Guého & A.J.Scott, Kew Bull. 34: 478 (1980).
Mauritius. 29 MAU. Nanophan. or phan.
**Jossinia hastilis* Blume, Mus. Bot. 1: 122 (1850).

Eugenia hazompasika H.Perrier, Mém. Inst. Sci. Madagascar, Sér. B, Biol. Vég. 4: 183 (1953).
EC. Madagascar. 29 MDG. Phan.
Eugenia hazompasika var. *extensa* H.Perrier, in Fl. Madag. 152: 46 (1953), nom. inval.

Eugenia herbacea O.Berg in C.F.P.von Martius & auct. suc. (eds.), Fl. Bras. 14(1): 570 (1859).
SE. Brazil to Paraguay. 84 BZL 85 PAR. Nanophan. or phan.
Eugenia leptophylla Barb.Rodr., Myrt. Paraguay: 6 (1903). *Myrciaria leptophylla* (Barb.Rodr.) Chodat & Hassl., Bull. Herb. Boissier, II, 7: 808 (1907).

Eugenia heringeriana Mattos, Loefgrenia 62: 3 (1974).
WC. Brazil. 84 BZC. Nanophan.

Eugenia hermesiana Mattos, Loefgrenia 94: 1 (1989).
Brazil (São Paulo). 84 BZL. Nanophan.

Eugenia herrerae Barrie, Novon 15: 24 (2005).
Costa Rica. 80 COS. Nanophan. or phan.

Eugenia heterochroa Urb., Symb. Antill. 7: 299 (1912).
SC. Jamaica. 81 JAM. Nanophan.

Eugenia heterochroma Diels, Verh. Bot. Vereins Prov. Brandenburg 48: 190 (1906 publ. 1907).
Amazon Basin. 83 BOL CLM ECU PER 84 BZN. Nanophan. or phan.

Eugenia heterophylla A.Rich., Hist. Phys. Cuba, Pl. Vasc.: 586 (1846).
Cuba. 81 CUB. Nanophan.

Eugenia higueyana Alain, Phytologia 61: 357 (1986).
Dominican Rep. 81 DOM. Nanophan.

Eugenia hilariana DC., Prodr. 3: 269 (1828).
Brazil (Minas Gerais). 84 BZL. Nanophan.

Eugenia hiraeifolia Standl., Publ. Field Mus. Nat. Hist., Bot. Ser. 17: 202 (1937). *Eugenia salamensis* var. *hiraeifolia* (Standl.) McVaugh, Fieldiana, Bot. 29: 457 (1963).
C. America. 80 COS NIC PAN. Nanophan. or phan.

Eugenia hirta O.Berg in C.F.P.von Martius & auct. suc. (eds.), Fl. Bras. 14(1): 574 (1859).
NE. Brazil. 84 BZE. Nanophan. or phan.

Eugenia hodgei McVaugh, J. Arnold Arbor. 54: 320 (1973).
Windward Is. 81 LEE? WIN. Nanophan. or phan.

Eugenia holdridgei Alain, Phytologia 61: 358 (1986).
Haiti. 81 HAI. Nanophan. or phan.

Eugenia hondurensis Ant.Molina, Ceiba 1: 261 (1951).
C. America. 80 HON NIC. Nanophan. or phan.
Eugenia nicaraguensis Amshoff, Acta Bot. Neerl. 5: 278 (1950).
Eugenia crassifolia Ant.Molina, Ceiba 3: 169 (1953), nom. illeg.

Eugenia horizontalis Pancher ex Brongn. & Gris, Bull. Soc. Bot. France 12: 179 (1865). *Austromyrtus horizontalis* (Pancher ex Brongn. & Gris) Burret, Notizbl. Bot. Gart. Berlin-Dahlem 15: 504 (1941).
New Caledonia. 60 NWC. Nanophan. or phan.

Eugenia hovarum H.Perrier, Mém. Inst. Sci. Madagascar, Sér. B, Biol. Vég. 4: 170 (1953).
C. Madagascar. 29 MDG. Nanophan.

Eugenia howardiana Proctor, Bull. Inst. Jamaica, Sci. Ser. 16: 33 (1967).
Jamaica. 81 JAM. Nanophan. or phan.

Eugenia humblotii Engl. & Brehmer, Bot. Jahrb. Syst. 54: 336 (1917). *Eugenia anjouanensis* H.Perrier, Mém. Inst. Sci. Madagascar, Sér. B, Biol. Vég. 4: 172 (1953), nom. illeg.
Comoros. 29 COM. Phan.

Eugenia hurlimannii Guillaumin, Mém. Mus. Natl. Hist. Nat., B, Bot. 8: 293 (1962).
New Caledonia. 60 NWC. Nanophan. or phan.

Eugenia hyemalis Cambess. in A.F.C.de Saint-Hilaire, Fl. Bras. Merid. 2: 360 (1833).
Brazil to NE. Argentina. 84 BZC BZL BZS 85 AGE PAR URU. Nanophan. or phan.
Eugenia multiflora Cambess. in A.F.C.de Saint-Hilaire, Fl. Bras. Merid. 2: 361 (1833), nom. illeg. *Luma multiflora* Herter, Revista Sudamer. Bot. 7: 219 (1943).
Eugenia multiflora var. *lutescens* Cambess. in A.F.C.P.de Saint-Hilaire & al., Fl. Bras. Merid. 2: 361 (1833).
Eugenia multiflora var. *rubiginosa* Cambess. in A.F.C.P.de Saint-Hilaire & al., Fl. Bras. Merid. 2: 361 (1833).
Eugenia bicolor O.Berg in C.F.P.von Martius & auct. suc. (eds.), Fl. Bras. 14(1): 282 (1857).
Eugenia multiflora var. *glabra* O.Berg in C.F.P.von Martius & auct. suc. (eds.), Fl. Bras. 14(1): 271 (1857).
Eugenia polycarpa O.Berg in C.F.P.von Martius & auct. suc. (eds.), Fl. Bras. 14(1): 281 (1857).
Eugenia polycarpa var. *bimarginata* O.Berg in C.F.P.von Martius & auct. suc. (eds.), Fl. Bras. 14(1): 281 (1857).
Eugenia polycarpa var. *marginata* O.Berg in C.F.P.von Martius & auct. suc. (eds.), Fl. Bras. 14(1): 281 (1857). *Eugenia hyemalis* var. *marginata* (O.Berg) D.Legrand, Bol. Fac. Agron. Univ. Montevideo 101: 43 (1968).
Eugenia polycarpa var. *ovata* O.Berg in C.F.P.von Martius & auct. suc. (eds.), Fl. Bras. 14(1): 281 (1857).
Myrciaria itacurubiensis Barb.Rodr. ex Chodat & Hassl., Bull. Herb. Boissier, II, 7: 808 (1907), nom. nud.
Eugenia cycliantha D.Legrand, in Fl. Ilustr. Catar. 1(Mirt., Suppl. 1): 15 (1977).

Eugenia hypargyrea Standl., Contr. U. S. Natl. Herb. 23: 1044 (1924).
Mexico. 79 MXG MXT. Nanophan. or phan.

Eugenia ibarrae Lundell, Wrightia 2: 56 (1960).
Belize to Guatemala. 80 BLZ GUA. Nanophan. or phan.

Eugenia ignambiensis Baker f., J. Linn. Soc., Bot. 45: 315 (1921).
New Caledonia. 60 NWC. Nanophan. or phan.

Eugenia ignota Britton & P.Wilson, Mem. Torrey Bot. Club 16: 90 (1920).
Cuba (I. de la Juventud). 81 CUB. Nanophan.

Eugenia ilalensis Hieron., Bot. Jahrb. Syst. 21: 326 (1895).
Ecuador. 83 ECU. Phan.

Eugenia illepida McVaugh, Fieldiana, Bot. 29: 210 (1956).
N. Brazil. 84 BZN. Phan.

Eugenia imaruiensis D.Legrand, in Fl. Ilustr. Catar. 1(Mirt., Suppl. 1): 22 (1977).
S. Brazil. 84 BZS. Nanophan. or phan.

Eugenia imbricata O.Berg in C.F.P.von Martius & auct. suc. (eds.), Fl. Bras. 14(1): 586 (1859).
Brazil (Minas Gerais). 84 BZL. Nanophan. or phan.

Eugenia imbricatocordata Amshoff, Adansonia, n.s., 8: 353 (1968).
Gabon. 23 GAB. Nanophan. or phan.

Eugenia impressa (O.Berg) Mattos, Loefgrenia 94: 4 (1989).
SE. Brazil. 84 BZL. Nanophan. or phan.
**Stenocalyx impressus* O.Berg in C.F.P.von Martius & auct. suc. (eds.), Fl. Bras. 14(1): 593 (1859).

Eugenia impunctata O.Berg in C.F.P.von Martius & auct. suc. (eds.), Fl. Bras. 14(1): 239 (1857).
Brazil (São Paulo). 84 BZL. Nanophan. or phan.

Eugenia incanescens Benth., J. Bot. (Hooker) 2: 321 (1840).
Guyana. 82 GUY. Phan.
Eugenia incanescens var. *complicata* O.Berg, Linnaea 27: 209 (1856).
Eugenia incanescens var. *elliptica* O.Berg, Linnaea 27: 209 (1856).
Eugenia incanescens var. *parvifolia* O.Berg, Linnaea 27: 209 (1856).

Eugenia inconspicua Standl., Contr. U. S. Natl. Herb. 23: 1043 (1924).
NW. Mexico. 79 MXN. Nanophan. or phan.

Eugenia indica (Wight) Chithra in N.C.Nair & A.N.Henry, Fl. Tamil Nadu 1: 153 (1983).
India. 40 IND. Nanophan. or phan.
**Jossinia indica* Wight, Icon. Pl. Ind. Orient. 2: t. 523 (1842). *Eugenia cuneata* B.Heyne ex Bedd., Fl. Sylv. S. India 2: cxii (1872), nom. superfl. *Eugenia jossinia* Duthie in J.D.Hooker, Fl. Brit. India 2: 500 (1879), nom. illeg.
Myrtus indica Walp., Repert. Bot. Syst. 2: 172 (1843).

Eugenia inirebensis P.E.Sánchez, Phytologia 61: 127 (1986).
Mexico. 79 MXG. Phan.

Eugenia intermedia O.Berg, Linnaea 27: 283 (1856).
Hispaniola. 81 DOM HAI. Nanophan.

Eugenia intibucana Barrie, Novon 15: 26 (2005).
Honduras. 80 HON. Nanophan. or phan.

Eugenia inundata DC., Prodr. 3: 280 (1828). *Eugenia inundata* var. *membranacea* O.Berg in C.F.P.von Martius & auct. suc. (eds.), Fl. Bras. 14(1): 318 (1857), nom. inval.
W. South America to N. Brazil. 83 CLM PER 84 BZN. Nanophan. or phan.
Eugenia leptantha Benth., J. Bot. (Hooker) 2: 321 (1840).
Eugenia inundata var. *coriacea* O.Berg in C.F.P.von Martius & auct. suc. (eds.), Fl. Bras. 14(1): 319 (1857).

Eugenia inversa Sobral, Sida 21: 1465 (2005).
Brazil (Espírito Santo). 84 BZL. Nanophan. or phan.

Eugenia involucrata DC., Prodr. 3: 264 (1828). *Phyllocalyx involucratus* (DC.) O.Berg, Linnaea 27: 307 (1856). *Stenocalyx involucratus* (DC.) Kausel, Lilloa 32: 333 (1967).
Bolivia to Uruguay and Brazil. 83 BOL 84 BZC BZE BZL BZS 85 AGE PAR URU. Nanophan. or phan.
Phyllocalyx laevigatus O.Berg in C.F.P.von Martius & auct. suc. (eds.), Fl. Bras. 14(1): 329 (1857). *Eugenia laevigata* (O.Berg) D.Legrand, Anais Reunião Sul-Amer. Bot. 3: 113 (1938 publ. 1940), nom. illeg.
Eugenia involucrata var. *minutifolia* Mattos & D.Legrand, Loefgrenia 67: 30 (1975). *Eugenia minutifolia* (Mattos & D.Legrand) Mattos, Loefgrenia 85: 2 (1984).

Eugenia irazuensis Hemsl., Biol. Cent.-Amer., Bot. 1: 410 (1880).
Costa Rica. 80 COS. Nanophan. or phan.

Eugenia irirensis O.Berg in C.F.P.von Martius & auct. suc. (eds.), Fl. Bras. 14(1): 586 (1859).
Brazil (Rio de Janeiro). 84 BZL. Nanophan. or phan.
Eugenia yrirensis O.Berg in C.F.P.von Martius & auct. suc. (eds.), Fl. Bras. 14(1): 647 (1859), orth. var.

Eugenia isabeliana Kiaersk., Bot. Tidsskr. 17: 272 (1890).
Hispaniola. 81 DOM HAI. Nanophan. or phan.
Eugenia odorata O.Berg, Linnaea 27: 171 (1856), nom. illeg.
Eugenia odorata var. *longifolia* O.Berg, Linnaea 27: 171 (1856).
Eugenia odorata var. *parvifolia* O.Berg, Linnaea 27: 171 (1856).
Eugenia mornicola Urb., Symb. Antill. 7: 525 (1913).

Eugenia ischnosceles O.Berg in C.F.P.von Martius & auct. suc. (eds.), Fl. Bras. 14(1): 309 (1857).
Brazil (São Paulo). 84 BZL. Nanophan. or phan.

Eugenia isosticta Urb., Symb. Antill. 7: 305 (1912).
Jamaica. 81 JAM. Nanophan. or phan.

Eugenia itacarensis Mattos, Loefgrenia 62: 3 (1974).
Brazil (Bahia). 84 BZE. Nanophan. or phan.

Eugenia itaguahiensis Nied. in H.G.A.Engler & K.A.E.Prantl, Nat. Pflanzenfam. 3(7): 81 (1893).
Brazil (Bahia). 84 BZE.
**Stenocalyx langsdorffii* O.Berg in C.F.P.von Martius & auct. suc. (eds.), Fl. Bras. 14(1): 594 (1859).

Eugenia itahypensis O.Berg in C.F.P.von Martius & auct. suc. (eds.), Fl. Bras. 14(1): 580 (1859).
Brazil (Bahia). 84 BZE. Nanophan. or phan.

Eugenia itajurensis Cambess. in A.F.C.de Saint-Hilaire, Fl. Bras. Merid. 2: 340 (1832). *Pilothecium itajurense* (Cambess.) Kausel, Ark. Bot., a.s., 4: 403 (1962).
SE. Brazil. 84 BZL. Nanophan. or phan.

Eugenia itapemirimensis Cambess. in A.F.C.de Saint-Hilaire, Fl. Bras. Merid. 2: 346 (1832).
SE. Brazil. 84 BZL. Nanophan. or phan.
Eugenia tocaiana O.Berg in C.F.P.von Martius & auct. suc. (eds.), Fl. Bras. 14(1): 269 (1857).

Eugenia iteophylla Krug & Urb., Bot. Jahrb. Syst. 19: 611 (1895).
E. Cuba. 81 CUB. Nanophan.

Eugenia izabalana Lundell, Wrightia 4: 144 (1970).
Guatemala. 80 GUA. Nanophan. or phan.

Eugenia jambosoides C.Wright ex Griseb., Cat. Pl. Cub.: 89 (1866). *Pseudanamomis jambosoides* (C.Wright ex Griseb.) Bisse, Feddes Repert. 96: 511 (1985).
W. Cuba. 81 CUB. Nanophan. or phan.

Eugenia janeirensis O.Berg in C.F.P.von Martius & auct. suc. (eds.), Fl. Bras. 14(1): 578 (1859).
Brazil (Rio de Janeiro). 84 BZL. Nanophan. or phan.

Eugenia jimenezii Alain, Phytologia 61: 358 (1986).
Dominican Rep. 81 DOM. Nanophan. or phan.

Eugenia joenssonii Kausel, Lilloa 33: 102 (1971 publ. 1972).
S. Brazil. 84 BZS. Nanophan.
Eugenia lanosa Mattos & D.Legrand, Loefgrenia 67: 29 (1975).

Eugenia kaalensis Guillaumin, Mém. Mus. Natl. Hist. Nat., B, Bot. 4: 34 (1953).
New Caledonia. 60 NWC. Nanophan. or phan.

Eugenia kaieteurensis Amshoff, Bull. Torrey Bot. Club 75: 536 (1948).
Venezuela to Guyana. 82 GUY VEN. Nanophan. or phan.

Eugenia kalbreyeri Engl. & Brehmer, Bot. Jahrb. Syst. 54: 337 (1917).
W. & WC. Trop. Africa. 22 IVO SIE 23 CMN. Phan.
Eugenia dawei Hutch. & Dalziel, Fl. W. Trop. Afr. 1: 200 (1927).

Eugenia kamelii Merr., Philipp. J. Sci., C 10: 219 (1915). *Jossinia kamelii* (Merr.) Merr., Philipp. J. Sci. 79: 359 (1951).
Philippines. 42 PHI. Nanophan. or phan.

Eugenia kameruniana Engl., Notizbl. Königl. Bot. Gart. Berlin 2: 291 (1899). *Myrtus kameruniana* (Engl.) Kuntze, Deutsche Bot. Monatsschr. 21: 173 (1903).
Cameroon. 23 CMN. Nanophan. or phan.
Eugenia hankeana H.J.P.Winkl., Bot. Jahrb. Syst. 41: 283 (1908).

Eugenia karwinskyana O.Berg, Linnaea 29: 244 (1858).
Mexico to C. America. 79 MXE MXT 80 BLZ GUA HON. Nanophan. or phan.
Eugenia yucatanensis Standl., Publ. Field Mus. Nat. Hist., Bot. Ser. 8: 28 (1930).
Eugenia tabascensis Lundell, Phytologia 1: 482 (1941).
Eugenia tenejapensis Lundell, Wrightia 4: 119 (1969).

Eugenia kellyana Proctor, J. Arnold Arbor. 63: 275 (1982).
Jamaica. 81 JAM. Nanophan. or phan.

Eugenia kerstingii Engl. & Brehmer, Bot. Jahrb. Syst. 54: 333 (1917).
Ivory Coast to Nigeria. 22 IVO NGA TOG. Nanophan.

Eugenia klaineana (Pierre) Engl., Bot. Jahrb. Syst. 54: 339 (1917).
WC. Trop. Africa. 23 CAB GAB ZAI. Nanophan. or phan.
**Chloromyrtus klaineana* Pierre, Bull. Mens. Soc. Linn. Paris, n.s., 1: 72 (1898).
Eugenia soyauxii Engl., Notizbl. Königl. Bot. Gart. Berlin 2: 291 (1899). *Myrtus soyauxii* (Engl.) Kuntze, Deutsche Bot. Monatsschr. 21: 173 (1903).

Eugenia kleinii D.Legrand, Sellowia 13: 313 (1961).
SE. & S. Brazil. 84 BZL BZS. Nanophan. or phan.

var. ***aristata*** D.Legrand, Sellowia 13: 314 (1961).
Brazil (Paraná). 84 BZS. Nanophan. or phan.

var. ***glaziovii*** D.Legrand, Sellowia 13: 313 (1961).
Brazil (Rio de Janeiro). 84 BZL. Nanophan. or phan.

var. ***kleinii***
Brazil (Santa Catarina). 84 BZS. Nanophan. or phan.

Eugenia klotzschiana O.Berg in C.F.P.von Martius & auct. suc. (eds.), Fl. Bras. 14(1): 255 (1857).
WC. Brazil (to Minas Gerais). 84 BZC BZL. Nanophan. or phan.
Eugenia klotzschiana var. *glabrata* O.Berg in C.F.P. von Martius & auct. suc. (eds.), Fl. Bras. 14(1): 575 (1859).

Eugenia koepperi Standl., Publ. Field Mus. Nat. Hist., Bot. Ser. 9: 320 (1940).
Mexico (Chiapas) to Honduras. 79 MXT 80 GUA HON. Nanophan. or phan.

Eugenia koolauensis O.Deg., Fl. Hawaiiensis 273: s.p. (1932).
Hawaiian Is. (Molokai, Oahu). 63 HAW. Nanophan. or phan.
Eugenia molokaiana K.A.Wilson & Rock, Pacific Sci. 11: 175 (1957).

Eugenia kuebuniensis Guillaumin, Mém. Mus. Natl. Hist. Nat., B, Bot. 8: 293 (1962).
New Caledonia. 60 NWC. Nanophan. or phan.

Eugenia kuhlmanniana Mattos & D.Legrand, Loefgrenia 67: 30 (1975).
Brazil (Rio de Janeiro). 84 BZL. Nanophan. or phan.

Eugenia laeteviridis Urb., Symb. Antill. 9: 100 (1923).
SE. Cuba. 81 CUB. Nanophan. or phan.

Eugenia laevis O.Berg, Linnaea 27: 177 (1856).
SE. Mexico to C. America, Caribbean. 79 MXT 80 GUA 81 DOM HAI PUE. Nanophan. or phan.
Eugenia subverticillaris O.Berg, Linnaea 29: 234 (1858).
Eugenia prenleloupii Kiaersk., Bot. Tidsskr. 17: 275 (1890).
Eugenia calciphila Lundell, Wrightia 3: 11 (1961).
Eugenia macancheana Lundell, Wrightia 4: 41 (1968).

Eugenia lagoensis Kiaersk., Enum. Myrt. Bras.: 123 (1893).
Brazil (SC. Minas Gerais). 84 BZL. Nanophan. or phan.
Calycorectes ambivalens Sobral, Daphne 3(4): 5 (1993).

Eugenia lambertiana DC., Prodr. 3: 270 (1828).
Lesser Antilles, S. Trop. America. 81 LEE WIN 82 FRG GUY SUR VEN 83 BOL CLM PER 84 BZE BZN. Nanophan. or phan.
Eugenia schomburgkii Benth., J. Bot. (Hooker) 2: 321 (1840).
Eugenia smaragdina O.Berg, Linnaea 27: 218 (1856).
Eugenia smaragdina var. *angustifolia* O.Berg, Linnaea 27: 218 (1856).
Eugenia smaragdina var. *brevipes* O.Berg, Linnaea 27: 219 (1856).
Eugenia smaragdina var. *rigida* O.Berg, Linnaea 27: 219 (1856).
Eugenia correae O.Berg in C.F.P.von Martius & auct. suc. (eds.), Fl. Bras. 14(1): 277 (1857).
Eugenia flavonigra A.Rich. ex O.Berg, Linnaea 30: 691 (1861).
Eugenia flavonigra var. *guadalupensis* O.Berg, Linnaea 30: 691 (1861).

Eugenia flavonigra var. *martinicensis* O.Berg, Linnaea 30: 692 (1861).
Eugenia oligophylla A.Rich. ex O.Berg, Linnaea 30: 685 (1861).
Eugenia lambertiana var. *hispidula* McVaugh, Mem. New York Bot. Gard. 18(2): 190 (1969).

Eugenia lamprophylla Urb., Symb. Antill. 7: 308 (1912).
Jamaica. 81 JAM. Phan.

Eugenia lancetillae Standl., Publ. Field Mus. Nat. Hist., Bot. Ser. 4: 317 (1929).
Honduras. 80 HON. Nanophan. or phan.

Eugenia langsdorffii O.Berg in C.F.P.von Martius & auct. suc. (eds.), Fl. Bras. 14(1): 568 (1859). *Myrtus langsdorffii* (O.Berg) Kuntze, Revis. Gen. Pl. 3(2): 91 (1898).
Brazil. 84 BZC BZL. Nanophan. or phan.

Eugenia laruotteana Cambess. in A.F.C.de Saint-Hilaire, Fl. Bras. Merid. 2: 50 (1829).
Brazil (NC. Minas Gerais). 84 BZL. Nanophan. or phan.

Eugenia lateriflora Willd., Sp. Pl. 2: 749 (1799).
Leeward Is. 81 LEE. Nanophan. or phan.

Eugenia latifolia Aubl., Hist. Pl. Guiane 1: 502 (1775). *Myrtus latifolia* (Aubl.) Spreng., Syst. Veg. 2: 481 (1825), nom. illeg. *Calycorectes latifolius* (Aubl.) O.Berg, Linnaea 30: 701 (1861). *Eugenia aubletiana* Mattos, Loefgrenia 120: 9 (2005), nom. superfl.
N. South America. 82 FRG GUY SUR VEN. Nanophan. or phan.

Eugenia laurae Proctor, J. Arnold Arbor. 63: 275 (1982).
Jamaica. 81 JAM. Nanophan. or phan.

Eugenia laxa DC., Prodr. 3: 263 (1828). *Stenocalyx laxus* (DC.) O.Berg in C.F.P.von Martius & auct. suc. (eds.), Fl. Bras. 14(1): 349 (1857).
Brazil (Bahia). 84 BZE. Nanophan. or phan.

Eugenia ledermannii Engl. & Brehmer, Bot. Jahrb. Syst. 54: 332 (1917).
Cameroon. 23 CMN. Nanophan. or phan.

Eugenia ledophylla (Standl.) McVaugh, Fieldiana, Bot. 29: 430 (1963).
Mexico (Veracruz). 79 MXG. Nanophan. or phan.
**Myrtus ledophylla* Standl., Contr. U. S. Natl. Herb. 23: 1038 (1924).

Eugenia lempana Barrie, Novon 15: 27 (2005).
Honduras. 80 HON. Nanophan. or phan.

Eugenia leonanii Mattos, Loefgrenia 94: 5 (1989).
Brazil (São Paulo). 84 BZL. Nanophan.

Eugenia leonensis Engl. & Brehmer, Bot. Jahrb. Syst. 54: 329 (1917).
W. Trop. Africa. 22 GHA GUI IVO LBR SIE. Nanophan. or phan.
Eugenia rupestris Engl. & Brehmer, Bot. Jahrb. Syst. 54: 330 (1917), nom. illeg.

Eugenia leonis Borhidi & O.Muñiz, Acta Bot. Acad. Sci. Hung. 19: 39 (1973).
Cuba. 81 CUB. Nanophan. or phan.

Eugenia lepidota O.Berg, Linnaea 27: 226 (1856).
Costa Rica. 80 COS. Nanophan. or phan.
Eugenia lepidota var. *corymbosa* O.Berg, Linnaea 27: 227 (1856).

Eugenia leptoclada O.Berg in C.F.P.von Martius & auct. suc. (eds.), Fl. Bras. 14(1): 248 (1857).
E. Brazil. 84 BZE BZL. Nanophan. or phan.

Eugenia letreroana Lundell, Wrightia 3: 15 (1961).
SE. Mexico to Guatemala. 79 MXT 80 GUA. Nanophan. or phan.
Eugenia coffeoides Lundell, Wrightia 5: 83 (1975).

Eugenia leucadendron O.Berg, Linnaea 27: 202 (1856).
Costa Rica. 80 COS. Nanophan. or phan.

Eugenia levinervis (Fosberg) A.J.Scott, Kew Bull. 34: 496 (1980).
Aldabra. 29 ALD. Nanophan.
**Eugenia elliptica* var. *levinervis* Fosberg, Kew Bull. 33: 134 (1978).

Eugenia lheritieriana DC., Prodr. 3: 272 (1828).
Tobago. 81 TRT. Nanophan. or phan.
Myrtus dumosa L'Hér. ex DC., Prodr. 3: 272 (1828), nom. inval.

Eugenia lhotzkyana O.Berg in C.F.P.von Martius & auct. suc. (eds.), Fl. Bras. 14(1): 305 (1857).
Brazil (Minas Gerais). 84 BZL. Nanophan. or phan.

Eugenia libanensis Urb., Symb. Antill. 9: 503 (1928).
E. Cuba. 81 CUB. Nanophan.

Eugenia liberiana Amshoff, Acta Bot. Neerl. 7: 57 (1958).
Sierra Leone to Liberia. 22 LBR SIE. Nanophan. or phan.
**Eugenia calycina* Benth. in W.J.Hooker, Niger Fl.: 358 (1849), nom. illeg.

Eugenia librevillensis Amshoff, Acta Bot. Neerl. 7: 53 (1958).
NW. Gabon. 23 GAB. Phan.

Eugenia liebmannii Standl., Contr. U. S. Natl. Herb. 23: 1044 (1924).
Mexico. 79 MXE MXG MXS. Nanophan. or phan.

Eugenia liesneri Barrie, Novon 15: 29 (2005).
Honduras. 80 HON. Nanophan. or phan.

Eugenia ligustrina (Sw.) Willd., Sp. Pl. 2: 962 (1799).
Caribbean to Brazil. 81 CUB DOM HAI JAM LEE PUE TRT WIN 82 GUY SUR VEN 83 BOL 84 BZE BZL BZN BZS. Nanophan. or phan.
**Myrtus ligustrina* Sw., Prodr.: 78 (1788). *Stenocalyx ligustrina* (Sw.) O.Berg, Linnaea 27: 312 (1856). *Phyllocalyx ligustrinus* (Sw.) O.Berg in C.F.P.von Martius & auct. suc. (eds.), Fl. Bras. 14(1): 592 (1859).

var. ***hebecarpa*** Amshoff in A.A.Pulle, Fl. Suriname 3(2): 177 (1951).
Guyana to Suriname. 82 GUY SUR. Nanophan. or phan.

var. ***ligustrina***
Caribbean to Brazil. 81 CUB DOM HAI JAM LEE PUE TRT WIN 82 GUY VEN 83 BOL 84 BZE BZL BZN BZS. Nanophan. or phan.
Myrtus cerasina Vahl, Symb. Bot. 2: 57 (1791).
Stenocalyx ligustrina var. *caribaea* O.Berg in C.F.P.von Martius & auct. suc. (eds.), Fl. Bras. 14(1): 343 (1857).
Stenocalyx ligustrinus var. *fluminensis* O.Berg in C.F.P.von Martius & auct. suc. (eds.), Fl. Bras. 14(1): 344 (1857).
Stenocalyx ligustrinus var. *minensis* O.Berg in C.F.P.von Martius & auct. suc. (eds.), Fl. Bras. 14(1): 343 (1857). *Eugenia ligustrina* var.

minensis (O.Berg) Mattos, Loefgrenia 64: 4 (1975). *Eugenia minensis* (O.Berg) Mattos, Loefgrenia 94: 3 (1989), nom. illeg.
Stenocalyx squamiflorus O.Berg in C.F.P.von Martius & auct. suc. (eds.), Fl. Bras. 14(1): 345 (1857).

Eugenia ligustroides Urb., Symb. Antill. 9: 500 (1928).
Cuba (Sierra de Nipe). 81 CUB. Nanophan. or phan.

Eugenia lilloana D.Legrand, Darwiniana 9: 296 (1950).
NE. Argentina (Misiones). 85 AGE. Cham.

Eugenia limbosa O.Berg, Linnaea 27: 294 (1856).
Trinidad-Tobago to S. Trop. America. 81 TRT 82 GUY SUR 83 BOL PER 84 BZN. Nanophan. or phan.
Eugenia cassapensis O.Berg, Linnaea 27: 296 (1856).
Myrtus limbosa Ruiz ex O.Berg, Linnaea 27: 295 (1856), nom. inval.
Eugenia maynensis O.Berg in C.F.P.von Martius & auct. suc. (eds.), Fl. Bras. 14(1): 318 (1857).
Eugenia coriacea O.Berg, Linnaea 30: 694 (1861), nom. illeg. *Eugenia polystachyoides* Amshoff, Recueil Trav. Bot. Néerl. 42: xlii, err. (1950). *Eugenia quitarensis* Amshoff, Recueil Trav. Bot. Néerl. 42: 12 (1950), nom. illeg.
Eugenia cruegeri Krug & Urb., Bot. Jahrb. Syst. 19: 625 (1895).

Eugenia lindahlii Urb. & Ekman, Ark. Bot. 21A(5): 28 (1927).
Hispaniola (incl. I. Gonave). 81 DOM HAI. Nanophan. or phan.
Eugenia orthioneura Urb., Ark. Bot. 21A(5): 29 (1927).

Eugenia lindbergiana O.Berg, Linnaea 30: 685 (1861).
Brazil. 84+. Nanophan. or phan.

Eugenia linearis A.Rich. ex O.Berg, Linnaea 30: 673 (1861).
Hispaniola. 81 DOM HAI. Nanophan. or phan.
Eugenia linearis var. *longifolia* O.Berg, Linnaea 30: 673 (1861).
Eugenia linearis var. *parvifolia* O.Berg, Linnaea 30: 674 (1861).

Eugenia lineata (Sw.) DC., Prodr. 3: 273 (1828).
E. Cuba to Hispaniola. 81 CUB DOM HAI. Nanophan. or phan.
**Myrtus lineata* Sw., Prodr.: 78 (1788).
Eugenia bergiana Griseb., Mem. Amer. Acad. Arts, n.s., 8: 182 (1861). *Eugenia lineata* var. *bergiana* (Griseb.) Krug & Urb., Bot. Jahrb. Syst. 19: 615 (1895).
Eugenia lineata var. *racemosa* O.Berg, Linnaea 30: 683 (1861).

Eugenia lineolata Urb. & Ekman, Ark. Bot. 21A(5): 35 (1927).
Hispaniola. 81 DOM HAI. Nanophan. or phan.

Eugenia lithosperma Barrie, Novon 15: 30 (2005).
Costa Rica. 80 COS. Nanophan. or phan.

Eugenia littoralis Pancher ex Brongn. & Gris, Bull. Soc. Bot. France 12: 178 (1865). *Eugenia oraria* Guillaumin, Bull. Mus. Natl. Hist. Nat. 25: 504 (1919), nom. illeg.
New Caledonia. 60 NWC. Nanophan. or phan.

Eugenia livida O.Berg in C.F.P.von Martius & auct. suc. (eds.), Fl. Bras. 14(1): 584 (1859).
Brazil (São Paulo). 84 BZL. Nanophan. or phan.

Eugenia locuples Barrie, Novon 15: 30 (2005).
Honduras. 80 HON. Nanophan. or phan.

Eugenia loeseneri Urb., Symb. Antill. 9: 512 (1928).
W. & WC. Cuba. 81 CUB. Nanophan.

Eugenia loheri C.B.Rob., Philipp. J. Sci., C 4: 345 (1909). *Jossinia loheri* (C.B.Rob.) Merr., Philipp. J. Sci. 79: 360 (1951).
Philippines. 42 PHI. Nanophan. or phan.

Eugenia lokobensis H.Perrier, Mém. Inst. Sci. Madagascar, Sér. B, Biol. Vég. 4: 183 (1953).
NE. & E. Madagascar. 29 MDG. Phan.

Eugenia lomensis Britton & P.Wilson, Bull. Torrey Bot. Club 50: 44 (1923).
SE. Cuba. 81 CUB. Nanophan.

Eugenia longicuspis McVaugh, Fieldiana, Bot. 29: 211 (1956).
N. Peru. 83 PER. Phan.

Eugenia longifolia DC., Prodr. 3: 269 (1828).
NE. Brazil. 84 BZL. Nanophan. or phan.

Eugenia longipedunculata Nied. in H.G.A.Engler & K.A.E.Prantl, Nat. Pflanzenfam. 3(7): 81 (1893).
Brazil (Minas Gerais). 84 BZL. Nanophan. or phan.
**Stenocalyx longipes* O.Berg in C.F.P.von Martius & auct. suc. (eds.), Fl. Bras. 14(1): 345 (1857).

Eugenia longipetiolata Mattos, Dusenia 8: 162 (1968).
Brazil (Rio de Janeiro). 84 BZL. Nanophan. or phan.
**Stenocalyx mutabilis* O.Berg in C.F.P.von Martius & auct. suc. (eds.), Fl. Bras. 14(1): 347 (1857). *Eugenia mutabilis* (O.Berg) Nied. in H.G.A.Engler & K.A.E. Prantl, Nat. Pflanzenfam. 3(8): 79 (1893), nom. illeg.

Eugenia longiracemosa Kiaersk., Enum. Myrt. Bras.: 149 (1893).
Brazil (Espírito Santo). 84 BZL. Nanophan. or phan.

Eugenia louvelii H.Perrier, Mém. Inst. Sci. Madagascar, Sér. B, Biol. Vég. 4: 178 (1953).
NE. Madagascar. 29 MDG. Phan.
Eugenia louvelii var. *suborbiculata* H.Perrier, Mém. Inst. Sci. Madagascar, Sér. B, Biol. Vég. 4: 178 (1952).

Eugenia lucens Alain, Contr. Ocas. Mus. Hist. Nat. Colegio "De La Salle" 12: 12 (1953). *Calycolpus lucens* (Alain) Bisse, Revista Jard. Bot. Nac. Univ. Habana 4: 6 (1983).
E. Cuba (Sierra de Moa). 81 CUB. Nanophan. or phan.

Eugenia luciae Amshoff, Recueil Trav. Bot. Néerl. 42: 14 (1950).
Guianas. 82 FRG GUY SUR. Phan.

Eugenia lucida Lam., Encycl. 3: 203 (1789). *Myrtus commersonii* Spreng., Syst. Veg. 2: 479 (1825), nom. illeg. *Jossinia lucida* (Lam.) DC., Prodr. 3: 237 (1828).
Mauritius. 29 MAU. Nanophan. or phan.
Jossinia lamarckii Blume, Mus. Bot. 1: 121 (1850).

Eugenia luschnathiana (O.Berg) Klotzsch ex B.D.Jacks., Index Kew. 1: 908 (1893).
Brazil (Bahia). (23) gab 84 BZE. Nanophan. or phan.
**Phyllocalyx luschnathianus* O.Berg in C.F.P.von Martius & auct. suc. (eds.), Fl. Bras. 14(1): 333 (1857). *Eugenia lucescens* Nied. in H.G.A.Engler & K.A.E.Prantl, Nat. Pflanzenfam. 3(7): 82 (1893).

Eugenia lutescens Cambess. in A.F.C.de Saint-Hilaire, Fl. Bras. Merid. 2: 341 (1832). *Pilothecium lutescens* (Cambess.) Kausel, Ark. Bot., a.s., 4: 404 (1962).
Brazil (Mato Grosso, Minas Gerais). 84 BZC BZL. Nanophan. or phan.

Eugenia mabaeoides Wight, Ill. Ind. Bot. 2: 13 (1841).
Sri Lanka. 40 SRL. Nanophan. or phan.

subsp. **mabaeoides**
Sri Lanka. 40 SRL. Nanophan. or phan.

subsp. **pedunculata** (Trimen) P.S.Ashton, in Revised Handb. Fl. Ceyl. 2: 414 (1981).
Sri Lanka. 40 SRL. Nanophan. or phan.
**Eugenia pedunculata* Trimen, J. Bot. 27: 162 (1889).
Eugenia aprica Trimen, Handb. Fl. Ceylon 2: 186 (1894), nom. illeg.

Eugenia macahensis O.Berg in C.F.P.von Martius & auct. suc. (eds.), Fl. Bras. 14(1): 589 (1859).
SE. Brazil. 84 BZL. Nanophan. or phan.

Eugenia macedoi Mattos & D.Legrand, Loefgrenia 67: 24 (1975).
Brazil (Minas Gerais). 84 BZL. Nanophan. or phan.

Eugenia mackeeana Guillaumin, Mém. Mus. Natl. Hist. Nat., B, Bot. 8: 146 (1959).
New Caledonia. 60 NWC. Nanophan. or phan.

Eugenia macnabiana (Krug & Urb.) Urb., Symb. Antill. 6: 104 (1909).
Jamaica. 81 JAM. Nanophan. or phan. Provisionally accepted.
**Eugenia oligandra* var. *macnabiana* Krug & Urb., Bot. Jahrb. Syst. 19: 613 (1895).

Eugenia macradenia Urb. & Ekman, Ark. Bot. 24A(4): 28 (1932).
Dominican Rep. 81 DOM. Nanophan. or phan.

Eugenia macrantha O.Berg in C.F.P.von Martius & auct. suc. (eds.), Fl. Bras. 14(1): 301 (1857).
E. Brazil. 84 BZE BZL. Nanophan. or phan.
Eugenia macrantha var. *glomerata* O.Berg in C.F.P. von Martius & auct. suc. (eds.), Fl. Bras. 14(1): 584 (1859).

Eugenia macrobracteolata Mattos, Loefgrenia 64: 5 (1975).
Brazil (Paraná). 84 BZS. Phan.

Eugenia macrocalyx Mart. ex B.D.Jacks., Index Kew. 1: 908 (1893).
Brazil (Minas Gerais). 84 BZL. Nanophan. or phan.
**Phyllocalyx macrosepalus* O.Berg in C.F.P.von Martius & auct. suc. (eds.), Fl. Bras. 14(1): 332 (1857). *Eugenia macrosepala* (O.Berg) Mattos, Loefgrenia 94: 1 (1989), nom. illeg.

Eugenia macrocarpa Schltdl. & Cham., Linnaea 5: 560 (1830). *Pseudanamomis macrocarpa* (Schltdl. & Cham.) Bisse, Feddes Repert. 96: 511 (1985).
Mexico (Veracruz). 79 MXG. Nanophan. or phan.
Eugenia macrocarpa Cham. & Schltdl., Linnaea 13: 417 (1839), nom. illeg. *Eugenia mexicana* Steud., Nomencl. Bot., ed. 2, 1: 603 (1840).
Eugenia americana Makoy ex E.Morren, Belgique Hort. 10: t. 95 (1860).

Eugenia macrosperma DC., Prodr. 3: 277 (1828).
SE. Brazil (to Bahia). 84 BZE BZL. Nanophan. or phan.
Eugenia velutina O.Berg in C.F.P.von Martius & auct. suc. (eds.), Fl. Bras. 14(1): 317 (1857).

Eugenia madagascariensis (H.Perrier) A.J.Scott, Kew Bull. 34: 546 (1980).
W. Madagascar. 29 MDG. Nanophan.
**Myrtus madagascariensis* H.Perrier, Mém. Inst. Sci. Madagascar, Sér. B, Biol. Vég. 4: 162 (1953).

Eugenia madugodaense Kosterm., Quart. J. Taiwan Mus. 34: 167 (1981).
Sri Lanka. 40 SRL. Nanophan. or phan.

Eugenia maestrensis Urb., Symb. Antill. 9: 107 (1923). *Pseudanamomis maestrensis* (Urb.) Bisse, Feddes Repert. 96: 511 (1985).
Cuba (Sierra Maestra). 81 CUB. Nanophan. or phan.
Eugenia ginoriaefolia Britton & P.Wilson, Bull. Torrey Bot. Club 50: 44 (1923).
Psidium cacuminis Britton & P.Wilson, Bull. Torrey Bot. Club 50: 43 (1923).

Eugenia magna B.Holst, Selbyana 23: 142 (2002).
Brazil (Amazonas), Venezuela, Guyana. 82 GUY VEN 84 BZN. Phan.

Eugenia magnibracteolata Mattos & D.Legrand, Loefgrenia 67: 28 (1975).
Brazil (Paraná). 84 BZS. Nanophan. or phan.

Eugenia magnifica Spring ex Mart., Flora 20(2 Beibl.): 86 (1837).
Brazil (Bahia to Paraná). 84 BZE BZL BZS. Phan.
Myrceugenia mosenii Kausel, Lilloa 33: 105 (1971 publ. 1972). *Eugenia mosenii* (Kausel) Sobral, Napaea 1: 25 (1987).

Eugenia magniflora Barrie, Novon 15: 32 (2005).
Costa Rica. 80 COS. Nanophan. or phan.

Eugenia magnifolia Kiaersk., Enum. Myrt. Bras.: 136 (1893).
SE. Brazil. 84 BZL. Nanophan. or phan.

Eugenia magoana Lundell, Wrightia 4: 117 (1968).
Guatemala. 80 GUA. Nanophan. or phan.

Eugenia malacantha D.Legrand, Sellowia 13: 324 (1961).
S. Brazil. 84 BZS. Nanophan. or phan.

Eugenia malangensis (O.Hoffm.) Nied. in H.G.A.Engler & K.A.E.Prantl, Nat. Pflanzenfam. 3(7): 81 (1893).
Kenya to S. Trop. Africa. 23 BUR RWA ZAI 25 KEN TAN 26 ANG MLW MOZ ZAM ZIM. Cham. or nanophan.
**Myrtopsis malangensis* O.Hoffm., Linnaea 43: 134 (1881).
Eugenia angolensis Engl., Notizbl. Königl. Bot. Gart. Berlin 2: 288 (1899). *Myrtus angolensis* (Engl.) Kuntze, Deutsche Bot. Monatsschr. 21: 173 (1903).
Eugenia laurentii Engl., Notizbl. Königl. Bot. Gart. Berlin 2: 288 (1899). *Myrtus laurentii* (Engl.) Kuntze, Deutsche Bot. Monatsschr. 21: 173 (1903).
Eugenia marquesii Engl., Notizbl. Königl. Bot. Gart. Berlin 2: 290 (1899). *Myrtus marquesii* (Engl.) Kuntze, Deutsche Bot. Monatsschr. 21: 173 (1903).
Eugenia poggei Engl., Notizbl. Königl. Bot. Gart. Berlin 2: 289 (1899). *Myrtus poggei* (Engl.) Kuntze, Deutsche Bot. Monatsschr. 21: 173 (1903).
Eugenia stolzii Engl. & Brehmer, Bot. Jahrb. Syst. 54: 330 (1917).

Eugenia malpighioides (Kunth) DC., Prodr. 3: 275 (1828).
Peru. 83 PER. Nanophan. or phan.
**Myrtus malpighioides* Kunth in F.W.H.von

Humboldt, A.J.A.Bonpland & C.S.Kunth, Nov. Gen. Sp. 6: 146 (1823).

Eugenia mandevillensis Urb., Symb. Antill. 7: 306 (1912).
Jamaica. 81 JAM. Nanophan. or phan.

var. ***mandevillensis***
Jamaica. 81 JAM. Nanophan. or phan.

var. ***perratonii*** (Proctor) Proctor, J. Arnold Arbor. 63: 276 (1982).
Jamaica. 81 JAM. Nanophan. or phan.
**Eugenia perratonii* Proctor, Rhodora 59: 305 (1958).

Eugenia mandioccensis O.Berg in C.F.P.von Martius & auct. suc. (eds.), Fl. Bras. 14(1): 585 (1859).
Brazil (Rio de Janeiro). 84 BZL. Nanophan. or phan.

Eugenia mandonii McVaugh, Fieldiana, Bot. 29: 212 (1956).
Bolivia. 83 BOL. Nanophan. or phan.

Eugenia manickamiana Murugan, J. Econ. Taxon. Bot. 26: 414 (2002).
S. India. 40 IND.

Eugenia mansoi O.Berg in C.F.P.von Martius & auct. suc. (eds.), Fl. Bras. 14(1): 223 (1857).
SE. & S. Brazil to Paraguay. 84 BZL BZS 85 AGE PAR URU. Nanophan. or phan.
Eugenia vincifolia O.Berg in C.F.P.von Martius & auct. suc. (eds.), Fl. Bras. 14(1): 270 (1857).

Eugenia maranhaoensis G.Don, Gen. Hist. 2: 854 (1832).
NE. Brazil. 84 BZE. Nanophan. or phan.

Eugenia maricaensis G.M.Barroso, Bradea 5: 357 (1990).
Brazil (Rio de Janeiro). 84 BZL. Nanophan.

Eugenia maritima DC., Prodr. 3: 271 (1828).
NE. Brazil. 84 BZE. Nanophan. or phan.

Eugenia marlierioides Rusby, Bull. New York Bot. Gard. 8: 108 (1912).
S. Venezuela, Bolivia. 82 VEN 83 BOL. Nanophan. or phan.

Eugenia marowynensis Miq., Stirp. Surinam. Select.: 39 (1851). *Myrciaria marowynensis* (Miq.) O.Berg, Linnaea 27: 335 (1856).
S. Trop. America. 82 FRG SUR 83 ECU. Nanophan. or phan.

Eugenia marshiana Griseb., Fl. Brit. W. I.: 238 (1860).
Jamaica. 81 JAM. Nanophan. or phan.

Eugenia matagalpensis P.E.Sánchez, Brenesia 25–26: 307 (1986 publ. 1988).
Nicaragua. 80 NIC. Nanophan. or phan.

Eugenia mattosii D.Legrand, Sellowia 13: 310 (1961).
Brazil (Santa Catarina). 84 BZS. Nanophan.

Eugenia matudae Lundell, Phytologia 1: 218 (1937).
SE. Mexico. 79 MXT. Phan.

Eugenia mcphersonii Barrie, Novon 15: 34 (2005).
Panama. 80 PAN. Nanophan. or phan.

Eugenia mcvaughii Steyerm. & Lasser, Brittonia 33: 25 (1981).
Venezuela. 82 VEN. Nanophan. or phan.

Eugenia megaflora Govaerts, World Checklist Myrtaceae: 152 (2008).
Brazil (Goiás). 84 BZC. Nanophan. or phan.
**Stenocalyx grandiflorus* O.Berg in C.F.P.von Martius & auct. suc. (eds.), Fl. Bras. 14(1): 347 (1857).
Eugenia grandiflora (O.Berg) Mattos, Loefgrenia 94: 3 (1989), nom. illeg.

Eugenia megalopetala Griseb., Cat. Pl. Cub.: 89 (1866).
W. & WC. Cuba. 81 CUB. Nanophan.

Eugenia melanadenia Krug & Urb., Bot. Jahrb. Syst. 19: 607 (1895).
Cuba to Hispaniola. 81 CUB DOM. Nanophan. or phan.

var. ***melanadenia***
Cuba to Hispaniola. 81 CUB DOM. Nanophan. or phan.

var. ***santayana*** Urb., Symb. Antill. 9: 489 (1928).
EC. Cuba. 81 CUB. Nanophan. or phan.

Eugenia melanogyna (D.Legrand) Sobral, Napaea 11: 35 (1995).
SE. Brazil. 84 BZL. Nanophan. or phan.
**Eugenia stictosepala* var. *melanogyna* D.Legrand, Sellowia 13: 324 (1961).

Eugenia membranifolia Nied. in H.G.A.Engler & K.A.E.Prantl, Nat. Pflanzenfam. 3(7): 82 (1893).
Brazil (Rio de Janeiro). 84 BZL. Nanophan. or phan.
**Phyllocalyx membranaceus* O.Berg in C.F.P.von Martius & auct. suc. (eds.), Fl. Bras. 14(1): 334 (1857).

Eugenia memecylifolia Talbot, J. Bombay Nat. Hist. Soc. 11: 236 (1897).
S. India. 40 IND. Nanophan. or phan.

Eugenia memecyloides Benth. in W.J.Hooker, Niger Fl.: 359 (1849).
Liberia to Ivory Coast. 22 IVO LBR. Nanophan.

Eugenia mensurensis Urb., Symb. Antill. 9: 98 (1923).
Cuba (Sierra de Nipe). 81 CUB. Nanophan.

Eugenia meridensis Steyerm., Fieldiana, Bot. 28: 1012 (1957).
Venezuela. 82 VEN. Nanophan. or phan.

Eugenia mespiloides Lam., Encycl. 3: 205 (1789). *Myrtus mespiloides* (Lam.) Spreng., Syst. Veg. 2: 481 (1825). *Jossinia mespiloides* (Lam.) DC., Prodr. 3: 237 (1828).
Réunion. 29 REU. Nanophan. or phan.
Eugenia hermannii Cordem., Fl. Réunion: 426 (1895).
Eugenia rubiginosa Cordem., Fl. Réunion: 425 (1895), nom. illeg.

Eugenia michoacanensis Lundell, Wrightia 3: 16 (1961).
SW. Mexico. 79 MXS. Nanophan. or phan.

Eugenia micranthoides McVaugh, Fieldiana, Bot. 29: 212 (1956).
Peru. 83 PER. Nanophan. or phan.

Eugenia micropora DC., Prodr. 3: 269 (1828).
N. Brazil. 84 BZN. Nanophan. or phan.

Eugenia mimus McVaugh, Mem. New York Bot. Gard. 18(2): 193 (1969).
S. Trop. America. 82 FRG GUY 83 ECU 84 BZE BZN. Nanophan. or phan.

Eugenia minguetii Urb., Ark. Bot. 20A(5): 24 (1926).
Haiti. 81 HAI. Nanophan. or phan.

Eugenia minimiflora Lundell, Wrightia 3: 124 (1965).
Belize. 80 BLZ. Phan.

Eugenia minuscula McVaugh, Mem. New York Bot. Gard. 18(2): 194 (1969).
Suriname. 82 SUR. Nanophan. or phan.

Eugenia modesta DC., Prodr. 3: 279 (1828).
Brazil to N. Argentina. 83 BOL 84 BZC BZE BZL BZS 85 AGE AGW PAR. Nanophan. or phan.
Eugenia bella Cambess. in A.F.C.de Saint-Hilaire, Fl. Bras. Merid. 2: 362 (1833).
Eugenia modesta var. *brasiliensis* O.Berg in C.F.P.von Martius & auct. suc. (eds.), Fl. Bras. 14(1): 314 (1857).
Eugenia moraviana O.Berg in C.F.P.von Martius & auct. suc. (eds.), Fl. Bras. 14(1): 304 (1857). *Myrtus moraviana* (O.Berg) Kuntze, Revis. Gen. Pl. 3(2): 92 (1898).
Eugenia racemulosa O.Berg in C.F.P.von Martius & auct. suc. (eds.), Fl. Bras. 14(1): 304 (1857).
Eugenia racemulosa var. *grandiflora* O.Berg in C.F.P. von Martius & auct. suc. (eds.), Fl. Bras. 14(1): 304 (1857).
Eugenia racemulosa var. *parviflora* O.Berg in C.F.P. von Martius & auct. suc. (eds.), Fl. Bras. 14(1): 304 (1857).

Eugenia moensis Britton & P.Wilson, Mem. Torrey Bot. Club 16: 88 (1920).
Cuba (Sierra de Moa). 81 CUB. Nanophan. or phan.

Eugenia molinae Barrie, Novon 15: 35 (2005).
Honduras. 80 HON. Nanophan. or phan.

Eugenia mollicoma Mart. ex O.Berg, Linnaea 27: 313 (1856).
NE. Brazil. 84 BZE. Nanophan. or phan.
Stenocalyx mollicoma O.Berg in C.F.P.von Martius & auct. suc. (eds.), Fl. Bras. 14(1): 350 (1857).

Eugenia mollifolia Urb., Symb. Antill. 9: 500 (1928).
W. Cuba. 81 CUB. Nanophan.

Eugenia monosperma Vell., Fl. Flumin. 5: 209, t. 42 (1829).
SE. Brazil. 84 BZL. Nanophan. or phan.
Eugenia compactiflora Spring ex Mart., Flora 20(Beibl. 2): 81 (1837).
Eugenia linguiformis O.Berg in C.F.P.von Martius & auct. suc. (eds.), Fl. Bras. 14(1): 268 (1857).

Eugenia montalbanica Merr., Philipp. J. Sci. 30: 417 (1926). *Jossinia montalbanica* (Merr.) Merr., Philipp. J. Sci. 79: 360 (1951).
Philippines. 42 PHI. Nanophan. or phan.
**Eugenia diospyrifolia* Merr., Philipp. J. Sci. 27: 39 (1925), nom. illeg.

Eugenia monteverdensis Barrie, Novon 15: 35 (2005).
Costa Rica. 80 COS. Nanophan. or phan.

Eugenia monticola (Sw.) DC., Prodr. 3: 275 (1828).
Trop. America. 80 COS GUA HON NIC PAN 81 CUB DOM HAI JAM LEE PUE TRT WIN 82 FRG GUY SUR VEN 83 CLM ECU. Nanophan. or phan.
**Myrtus monticola* Sw., Prodr.: 78 (1788).
Eugenia maleolens Pers., Syn. Pl. 2: 29 (1806).
Myrtus micrantha Kunth in F.W.H.von Humboldt, A.J.A.Bonpland & C.S.Kunth, Nov. Gen. Sp. 6: 144 (1823). *Eugenia micrantha* (Kunth) DC., Prodr. 3: 282 (1828). *Emurtia micrantha* (Kunth) Raf., Sylva Tellur.: 106 (1838).
Eugenia flavovirens O.Berg, Linnaea 27: 184 (1856). *Eugenia monticola* var. *flavovirens* (O.Berg) Stehlé & Quentin, Fl. Guadeloupe 2(3): 65 (1949).
Eugenia flavovirens var. *obscura* O.Berg, Linnaea 27: 185 (1856).
Eugenia flavovirens var. *pallida* O.Berg, Linnaea 27: 184 (1856).
Eugenia foetida var. *parvifolia* O.Berg, Linnaea 27: 212 (1856).
Eugenia obtusata Willd. ex O.Berg, Linnaea 27: 132 (1856).
Eugenia canescens O.Berg, Linnaea 29: 239 (1858).
Eugenia insularis O.Berg, Linnaea 29: 238 (1858).
Eugenia monticola var. *latifolia* Krug & Urb., Bot. Jahrb. Syst. 19: 636 (1895).
Eugenia monticola var. *racemosa* Amshoff in A.A. Pulle, Fl. Suriname 3(2): 121 (1951).

Eugenia mooniana Wight, Ill. Ind. Bot. 2: 13 (1841).
S. India, Sri Lanka. 40 IND SRL. Nanophan. or phan.
Eugenia gracilis Bedd., Madras J. Lit. Sci., III, 1: 46 (1854).
Eugenia concinna Thwaites, Enum. Pl. Zeyl.: 416 (1864), nom. illeg. *Eugenia thwaitesii* Duthie in J.D.Hooker, Fl. Brit. India 2: 506 (1879).

Eugenia moonioides O.Berg in C.F.P.von Martius & auct. suc. (eds.), Fl. Bras. 14(1): 262 (1857).
Brazil (Rio de Janeiro). 84 BZL. Nanophan. or phan.

Eugenia morii B.Holst & M.L.Kawas., Brittonia 52: 21 (2000).
French Guiana. 82 FRG. Phan.

Eugenia moritziana H.Karst., Auswahl Gew. Venez.: 18 (1848).
Venezuela. 82 VEN. Nanophan. or phan.

Eugenia moschata (Aubl.) Nied. ex T.Durand & B.D.Jacks., Index Kew., Suppl. 1: 164 (1902).
S. Trop. America. 82 FRG GUY SUR 83 ECU PER 84 BZE BZL BZN. Cham. or nanophan.
Catinga aromatica Aubl., Hist. Pl. Guiane 1: 511 (1775).
**Catinga moschata* Aubl., Hist. Pl. Guiane 1: 511 (1775). *Eugenia catinga* Baill., Hist. Pl. 6: 344 (1876), nom. illeg.
Eugenia fasciculiflora O.Berg, Linnaea 27: 233 (1856).
Eugenia feijoi O.Berg in C.F.P.von Martius & auct. suc. (eds.), Fl. Bras. 14(1): 283 (1857).
Eugenia leucanthera O.Berg in C.F.P.von Martius & auct. suc. (eds.), Fl. Bras. 14(1): 277 (1857).
Eugenia paraensis O.Berg in C.F.P.von Martius & auct. suc. (eds.), Fl. Bras. 14(1): 301 (1857).
Eugenia pisonis O.Berg in C.F.P.von Martius & auct. suc. (eds.), Fl. Bras. 14(1): 228 (1857).
Eugenia costata O.Berg in C.F.P.von Martius & auct. suc. (eds.), Fl. Bras. 14(1): 577 (1859), nom. illeg. *Luma bergii* Herter, Revista Sudamer. Bot. 7: 219 (1943).
Acca glazioviana Kiaersk., Enum. Myrt. Bras.: 115 (1893). *Eugenia kiaerskoviana* Mattos & D. Legrand, Loefgrenia 67: 30 (1975).
Eugenia pleurosiphonea Diels, Verh. Bot. Vereins Prov. Brandenburg 48: 191 (1907).

Eugenia mouensis Baker f., J. Linn. Soc., Bot. 45: 313 (1921).
New Caledonia. 60 NWC. Nanophan. or phan.
Jambosa canalensis Pancher ex Guillaumin, Bull. Soc. Bot. France 85: 642 (1938 publ. 1939).

Eugenia mozomboensis P.E.Sánchez, Phytologia 61: 126 (1986).
Mexico. 79 MXG. Nanophan.

Eugenia mucronata O.Berg, Linnaea 27: 172 (1856).
Cuba to Hispaniola. 81 CUB DOM HAI. Cham. or nanophan.

Eugenia mufindiensis Verdc., Kew Bull. 54: 50 (1999).
Tanzania (Iringa Distr.). 25 TAN. Nanophan. or phan.

Eugenia multicostata D.Legrand, Sellowia 13: 309 (1961).
Brazil (São Paulo to Santa Catarina). 84 BZL BZS. Phan.
Eugenia multicostata var. *octocostata* D.Legrand, Sellowia 13: 310 (1961).

Eugenia multirimosa McVaugh, Fieldiana, Bot. 29: 213 (1956).
W. South America to N. Brazil. 83 CLM ECU PER 84 BZN. Nanophan. or phan.

Eugenia muscicola H.Perrier, Mém. Inst. Sci. Madagascar, Sér. B, Biol. Vég. 4: 173 (1953).
C. Madagascar. 29 MDG. Nanophan.

Eugenia myrciariifolia Soares-Silva & Sobral, Novon 14: 236 (2004).
Brazil (Paraná). 84 BZS.

Eugenia myrobalana DC., Prodr. 3: 277 (1828).
N. Brazil to Peru. 83 PER 84 BZN. Phan.
Eugenia corymbosa O.Berg, Linnaea 31: 256 (1862).

Eugenia myrtopsidoides Guillaumin, Mém. Mus. Natl. Hist. Nat., B, Bot. 8: 294 (1962).
New Caledonia. 60 NWC. Nanophan. or phan.

Eugenia naguana Urb., Symb. Antill. 9: 509 (1928).
E. Cuba. 81 CUB. Nanophan.

Eugenia nannophylla Urb. & Ekman, Ark. Bot. 21A(5): 27 (1927).
Haiti. 81 HAI. Nanophan.

Eugenia neglecta O.Berg in C.F.P.von Martius & auct. suc. (eds.), Fl. Bras. 14(1): 320 (1857).
Brazil (Minas Gerais). 84 BZL. Nanophan.

Eugenia neibensis Mattos, Loefgrenia 122: 2 (2006).
Dominican Rep. 81 DOM. Nanophan. or phan.
**Calycorectes minutifolius* Mattos, Loefgrenia 113: 1 (1999). *Eugenia minutifolia* (Mattos) Mattos, Loefgrenia 120: 7 (2005), nom. illeg.

Eugenia nematopoda Urb., Symb. Antill. 9: 106 (1923).
Cuba (Sierra Maestra). 81 CUB. Nanophan. or phan.

Eugenia neoaustralis Sobral, Fam. Myrtac. Rio Grande do Sul: 68 (2003).
SE. & S. Brazil. 84 BZL BZS. Phan.
Calycorectes sellowianus O.Berg in C.F.P.von Martius & auct. suc. (eds.), Fl. Bras. 14(1): 356 (1857). *Eugenia vattimoana* Mattos, Loefgrenia 120: 9 (2005).
**Calycorectes australis* D.Legrand, Sellowia 13: 333 (1961).
Calycorectes australis var. *impressovenosus* D.Legrand, Sellowia 13: 334 (1961). *Eugenia neoaustralis* var. *impressovenosa* (D.Legrand) Mattos, Loefgrenia 120: 11 (2005).

Eugenia neofasciculata Bennet, Indian Forester 109: 220 (1983).
Mauritius. 29 MAU. Nanophan. or phan.
**Eugenia fasciculata* J.Guého & A.J.Scott, Kew Bull. 34: 477 (1980), nom. illeg.

Eugenia neoformosa Sobral, Napaea 11: 36 (1995).
Brazil (Minas Gerais). 84 BZL. Cham.
**Eugenia formosa* Cambess. in A.F.C.de Saint-Hilaire, Fl. Bras. Merid. 2: 354 (1832), nom. illeg. *Phyllocalyx formosus* (Cambess.) O.Berg, Linnaea 27: 307 (1856).

Eugenia neoglomerata Sobral, Napaea 11: 35 (1995).
Brazil. 84 BZC BZE BZL BZS. Nanophan. or phan.
**Eugenia glomerata* Spring ex Mart., Flora 20(Beibl. 2): 82 (1837), nom. illeg. *Myrtus subglomerata* (Spring ex Mart.) Kuntze, Revis. Gen. Pl. 3(2): 92 (1898).
Eugenia glomerata var. *grandifolia* O.Berg in C.F.P.von Martius & auct. suc. (eds.), Fl. Bras. 14(1): 267 (1857).
Eugenia glomerata var. *parvifolia* O.Berg in C.F.P.von Martius & auct. suc. (eds.), Fl. Bras. 14(1): 267 (1857).

Eugenia neolaurifolia Sobral, Napaea 11: 36 (1995).
Brazil (São Paulo). 84 BZL. Nanophan. or phan.
**Eugenia laurifolia* Cambess. in A.F.C.de Saint-Hilaire, Fl. Bras. Merid. 2: 357 (1833), nom. illeg.

Eugenia neomyrtifolia Sobral, Napaea 11: 36 (1995).
SE. & S. Brazil. 84 BZL BZS. Nanophan. or phan.
**Eugenia myrtifolia* Cambess. in A.F.C.de Saint-Hilaire, Fl. Bras. Merid. 2: 367 (1833), nom. illeg. *Stenocalyx myrtifolius* O.Berg in C.F.P.von Martius & auct. suc. (eds.), Fl. Bras. 14(1): 342 (1857).

Eugenia neonitida Sobral, Napaea 11: 36 (1995).
SE. Brazil. 84 BZL. Nanophan. or phan.
**Eugenia nitida* Cambess. in A.F.C.de Saint-Hilaire, Fl. Bras. Merid. 2: 349 (1832), nom. illeg.

Eugenia neosericea Morais & Sobral, Lundiana 1: 12 (2006).
Brazil (Minas Gerais). 84 BZL. Nanophan.
**Myrciaria sericea* O.Berg in C.F.P.von Martius & auct. suc. (eds.), Fl. Bras. 14(1): 375 (1857).

Eugenia neosilvestris Sobral, Napaea 11: 36 (1995).
SE. & S. Brazil. 84 BZL BZS. Phan.
**Stenocalyx silvestris* O.Berg in C.F.P.von Martius & auct. suc. (eds.), Fl. Bras. 14(1): 346 (1857). *Eugenia silvestris* (O.Berg) Mattos, Loefgrenia 54: 1 (1971), nom. illeg.

Eugenia neotristis Sobral, Napaea 11: 36 (1995).
Brazil (Santa Catarina). 84 BZS. Nanophan. or phan.
**Eugenia tristis* D.Legrand, Sellowia 12: 318 (1961), nom. illeg.

Eugenia neoverrucosa Sobral, Napaea 11: 36 (1995).
SE. & S. Brazil. 84 BZL BZS. Phan.
**Eugenia verrucosa* D.Legrand, Sellowia 13: 326 (1961), nom. illeg.
Eugenia verrucosa f. *glabrata* D.Legrand, Sellowia 13: 327 (1961).

Eugenia nervosa Lour., Fl. Cochinch.: 308 (1790). *Cleistocalyx nervosus* (Lour.) Blume, Mus. Bot. 1: 85 (1850).
N. Vietnam. 41 VIE. Nanophan. or phan.
Myrtus loureiroi Spreng., Syst. Veg. 2: 488 (1825).

Eugenia nesiotica Standl., Publ. Field Mus. Nat. Hist., Bot. Ser. 17: 203 (1937).
Panama (I. Carro Colorado). 80 PAN. Nanophan.

Eugenia nigerina A.Chev., Explor. Bot. Afr. Occ. Fr.: 268 (1920).
W. Trop. Africa, Kenya. 22 IVO MLI 25 KEN. Nanophan. or phan.

Eugenia nigrita Lundell, Wrightia 3: 16 (1961).
SE. Mexico. 79 MXT. Nanophan. or phan.

Eugenia nipensis Urb., Symb. Antill. 9: 501 (1928). *Calycolpus nipense* (Urb.) Bisse, Revista Jard. Bot. Nac. Univ. Habana 4: 6 (1983).
E. Cuba. 81 CUB. Nanophan. or phan.

Eugenia nodosa Engl., Notizbl. Königl. Bot. Gart. Berlin 2: 290 (1899). *Myrtus nodosa* (Engl.) Kuntze, Deutsche Bot. Monatsschr. 21: 173 (1903).
Gabon. 23 GAB. Nanophan. or phan.

Eugenia nodulosa Urb., Symb. Antill. 9: 102 (1923).
SE. Cuba. 81 CUB. Nanophan.

Eugenia nompa H.Perrier, Mém. Inst. Sci. Madagascar, Sér. B, Biol. Vég. 4: 181 (1953).
N. Madagascar. 29 MDG. Nanophan.

var. ***arborea*** H.Perrier, Mém. Inst. Sci. Madagascar, Sér. B, Biol. Vég. 4: 181 (1952).
N. Madagascar. 29 MDG. Phan.

var. ***nompa***
N. Madagascar. 29 MDG. Nanophan.

Eugenia noumeensis Guillaumin, Bull. Soc. Bot. France 84: 638 (1939).
New Caledonia. 60 NWC. Nanophan. or phan.

Eugenia nutans O.Berg in C.F.P.von Martius & auct. suc. (eds.), Fl. Bras. 14(1): 299 (1857).
E. & S. Brazil. 84 BZE BZL BZS. Nanophan. or phan.
Eugenia nutans var. *pellucida* O.Berg in C.F.P.von Martius & auct. suc. (eds.), Fl. Bras. 14(1): 583 (1859).
Eugenia eurysepala Kiaersk., Enum. Myrt. Bras.: 117 (1893).

Eugenia oaxacana O.Berg, Linnaea 30: 683 (1861).
SW. Mexico. 79 MXS. Nanophan. or phan.

Eugenia obanensis Baker f. in A.B.Rendle & al., Cat. Pl. Oban: 30 (1913).
Ivory Coast to Angola. 22 GHA IVO NGA 23 CAB CMN GAB GGI 26 ANG. Nanophan.

Eugenia oblongata O.Berg in C.F.P.von Martius & auct. suc. (eds.), Fl. Bras. 14(1): 302 (1857).
SE. Brazil (to Paraná). 84 BZL BZS. Nanophan. or phan.
Eugenia oblongata var. *velutina* O.Berg in C.F.P.von Martius & auct. suc. (eds.), Fl. Bras. 14(1): 302 (1857).
Eugenia striata Mattos & D.Legrand, Loefgrenia 67: 20 (1975), nom. illeg.

Eugenia obscura (Lindl.) DC., Prodr. 3: 265 (1828).
NE. Brazil. 84 BZE. Nanophan. or phan.
**Myrtus obscura* Lindl., Bot. Reg. 13: t. 1044 (1827).

Eugenia ocanensis Linden ex Regel, Cat. Pl. Hort. Aksakov.: 57 (1860).
Colombia. 83 CLM. Nanophan. or phan.

Eugenia ochrophloea Diels, Verh. Bot. Vereins Prov. Brandenburg 48: 189 (1906 publ. 1907).
Peru. 83 PER. Nanophan.

Eugenia octopleura Krug & Urb., Bot. Jahrb. Syst. 19: 653 (1895).
C. America, Lesser Antilles. 80 BLZ COS ELS GUA HON NIC PAN 81 LEE WIN. Nanophan. or phan.

Eugenia oerstediana O.Berg, Linnaea 27: 285 (1856).
Mexico to W. South America, Lesser Antilles. 79 MXE MXG MXS MXT 80 BLZ COS ELS GUA HON NIC PAN 81 LEE WIN 83 ECU. Nanophan. or phan.
Eugenia vincentina Krug & Urb., Bot. Jahrb. Syst. 19: 621 (1895).
Eugenia conzattii Standl., Contr. U. S. Natl. Herb. 23: 1041 (1924).
Eugenia cocquericotensis Lundell, Bull. Torrey Bot. Club 64: 554 (1937).
Eugenia balancanensis Lundell, Phytologia 1: 481 (1941).
Eugenia petenensis Lundell, Bull. Torrey Bot. Club 69: 397 (1942).
Eugenia eutenuipes Lundell, Wrightia 2: 106 (1960).

Eugenia ogoouensis Amshoff, in Fl. Gabon 11: 26 (1966).
Gabon. 23 GAB. Nanophan.

Eugenia oligadenia Urb., Symb. Antill. 9: 104 (1923).
E. Cuba. 81 CUB. Nanophan.

Eugenia oligandra Krug & Urb., Bot. Jahrb. Syst. 19: 612 (1895).
W. & C. Cuba. 81 CUB. Nanophan.

Eugenia ombrophila H.Perrier, Mém. Inst. Sci. Madagascar, Sér. B, Biol. Vég. 4: 180 (1953).
EC. Madagascar. 29 MDG. Nanophan.

Eugenia omissa McVaugh, Mem. New York Bot. Gard. 18(2): 197 (1969).
S. Trop. America. 82 FRG GUY SUR VEN 83 BOL PER 84 BZN. Phan.

Eugenia ophthalmantha Kiaersk., Enum. Myrt. Bras.: 159 (1893).
SE. Brazil. 84 BZL. Nanophan. or phan.
**Stenocalyx gemmiflorus* O.Berg in C.F.P.von Martius & auct. suc. (eds.), Fl. Bras. 14(1): 344 (1857).

Eugenia orbiculata Lam., Encycl. 3: 202 (1789). *Myrtus orbiculata* (Lam.) Spreng., Syst. Veg. 2: 480 (1825). *Jossinia orbiculata* (Lam.) DC., Prodr. 3: 237 (1828).
Mauritius. 29 MAU. Nanophan. or phan.
Jossinia terminalis Bojer, Hortus Maurit.: 141 (1837), nom. inval.
Eugenia oleifolia Thouars ex H.Perrier, Mém. Inst. Sci. Madagascar, Sér. B, Biol. Vég. 4: 166 (1953), nom. illeg. *Eugenia oleioides* H.Perrier, in Fl. Madag. 152: 10 (1953).

Eugenia orbignyana O.Berg in C.F.P.von Martius & auct. suc. (eds.), Fl. Bras. 14(1): 314 (1857).
C. Brazil to Bolivia. 83 BOL 84 BZC BZL. Cham.
Eugenia chiquitensis O.Berg in C.F.P.von Martius & auct. suc. (eds.), Fl. Bras. 14(1): 519 (1858).
Eugenia orbignyana var. *grandifolia* O.Berg in C.F.P.von Martius & auct. suc. (eds.), Fl. Bras. 14(1): 587 (1859).
Eugenia miniata S.Moore, Trans. Linn. Soc. London, Bot. 4: 358 (1895).
Eugenia pseudoverticillata S.Moore, Trans. Linn. Soc. London, Bot. 4: 357 (1895), nom. illeg.

Eugenia oreinoma O.Berg, Linnaea 27: 158 (1856).
C. America. 80 COS ELS GUA PAN. Nanophan. or phan.
Eugenia jutiapensis Standl. & Steyerm., Publ. Field Mus. Nat. Hist., Bot. Ser. 23: 129 (1944).

Eugenia ouen-toroensis Guillaumin, Mém. Mus. Natl. Hist. Nat., B, Bot. 4: 34 (1953).
New Caledonia. 60 NWC. Nanophan. or phan.

Eugenia ovalis O.Berg, Linnaea 27: 156 (1856).
Ecuador. 83 ECU. Nanophan. or phan.
Myrtus ovalis O.Berg, Linnaea 27: 156 (1856), nom. inval.

Eugenia ovandensis Lundell, Wrightia 3: 17 (1961).
SE. Mexico. 79 MXT. Nanophan. or phan.

Eugenia oxysepala Urb., Symb. Antill. 9: 102 (1923).
SE. Cuba. 81 CUB. Nanophan.
Eugenia revoluta O.Berg, Linnaea 29: 224 (1858), nom. illeg.

Eugenia pachnantha O.Berg in C.F.P.von Martius & auct. suc. (eds.), Fl. Bras. 14(1): 585 (1859).
Brazil (Bahia). 84 BZE. Nanophan. or phan.

Eugenia pachychlamys Donn.Sm., Bot. Gaz. 27: 333 (1899).
C. America. 80 COS ELS GUA. Nanophan. or phan.
Eugenia skutchii C.V.Morton & Standl., Publ. Field Mus. Nat. Hist., Bot. Ser. 18: 774 (1937).
Eugenia chiquimulana Standl. & Steyerm., Publ. Field Mus. Nat. Hist., Bot. Ser. 22: 358 (1940).

Eugenia pachychremastra Guillaumin, Bull. Soc. Bot. France 85: 639 (1938 publ. 1939).
New Caledonia. 60 NWC. Nanophan. or phan.

Eugenia pachyclada D.Legrand, Sellowia 13: 319 (1961).
Brazil (Santa Catarina). 84 BZS. Nanophan. or phan.

Eugenia pachystachya McVaugh, Mem. New York Bot. Gard. 18(2): 197 (1969).
Venezuela (Amazonas). 82 VEN. Nanophan. or phan.

Eugenia pacifica Benth., Bot. Voy. Sulphur: 98 (1845).
Cocos I. 80 CPI. Nanophan. or phan.

Eugenia padronii Alain, Phytologia 61: 359 (1986).
Puerto Rico. 81 PUE. Phan.

Eugenia pallescens Kiaersk., Enum. Myrt. Bras.: 162 (1893).
SE. Brazil. 84 BZL. Nanophan. or phan.

Eugenia paloverdensis Barrie, Novon 15: 37 (2005).
Costa Rica. 80 COS. Nanophan. or phan.

Eugenia paludosa Pancher ex Brongn. & Gris, Bull. Soc. Bot. France 12: 178 (1865).
New Caledonia. 60 NWC. Nanophan. or phan.

Eugenia palumbis Merr., Philipp. J. Sci., C 9: 122 (1914). *Jossinia palumbis* (Merr.) Diels, Bot. Jahrb. Syst. 56: 531 (1921).
Marianas. 62 MRN. Nanophan.

Eugenia pantagensis O.Berg in C.F.P.von Martius & auct. suc. (eds.), Fl. Bras. 14(1): 246 (1857).
SE. & S. Brazil. 84 BZL BZS. Nanophan. or phan.
Eugenia pantagensis var. *elliptica* O.Berg in C.F.P.von Martius & auct. suc. (eds.), Fl. Bras. 14(1): 573 (1859).

Eugenia papalensis Standl. & Steyerm., Publ. Field Mus. Nat. Hist., Bot. Ser. 23: 130 (1944).
Guatemala. 80 GUA. Nanophan. or phan.

Eugenia papayoensis Urb., Symb. Antill. 9: 103 (1923).
C. & E. Cuba. 81 CUB. Nanophan.

Eugenia paracatuana O.Berg in C.F.P.von Martius & auct. suc. (eds.), Fl. Bras. 14(1): 588 (1859).
SE. Brazil to Paraguay. 84 BZL 85 PAR. Nanophan. or phan.
Eugenia moraviana var. *impunctata* O.Berg in C.F.P. von Martius & auct. suc. (eds.), Fl. Bras. 14(1): 586 (1859).

Eugenia paranahybensis O.Berg in C.F.P.von Martius & auct. suc. (eds.), Fl. Bras. 14(1): 587 (1859).
Brazil (Goiás, Minas Gerais). 84 BZC BZL. Nanophan. or phan.

Eugenia paranapiacabensis Mattos, Loefgrenia 94: 4 (1989).
Brazil (São Paulo). 84 BZL. Nanophan. or phan.

Eugenia pardensis O.Berg in C.F.P.von Martius & auct. suc. (eds.), Fl. Bras. 14(1): 583 (1859).
Brazil (São Paulo). 84 BZL. Nanophan. or phan.

Eugenia pasacaensis C.B.Rob., Philipp. J. Sci., C 4: 346 (1909). *Jossinia pasacaensis* (C.B.Rob.) Merr., Philipp. J. Sci. 79: 360 (1951).
Philippines (Luzon). 42 PHI. Nanophan. or phan.

Eugenia patens Poir. in J.B.A.P.M.de Lamarck, Encycl., Suppl. 3: 124 (1813).
S. Trop. America. 82 FRG GUY SUR 83 BOL CLM ECU PER 84 BZN. Nanophan. or phan.
Eugenia muricata DC., Prodr. 3: 383 (1828).
Eugenia riparia DC., Prodr. 3: 283 (1828).
Eugenia rugosa Ruiz & Pav. ex.DC., Prodr. 3: 280 (1828).
Eugenia rutidocarpa Ruiz & Pav. ex G.Don, Gen. Hist. 2: 865 (1832), nom. nud.
Eugenia lugens O.Berg, Linnaea 27: 299 (1856).
Eugenia amazonica O.Berg in C.F.P.von Martius & auct. suc. (eds.), Fl. Bras. 14(1): 322 (1857).
Eugenia fenzliana O.Berg in C.F.P.von Martius & auct. suc. (eds.), Fl. Bras. 14(1): 323 (1857).
Eugenia racemifera O.Berg in C.F.P.von Martius & auct. suc. (eds.), Fl. Bras. 14(1): 322 (1857).
Eugenia casaretteana O.Berg in C.F.P.von Martius & auct. suc. (eds.), Fl. Bras. 14(1): 520 (1858).
Eugenia muricata var. *guyanensis* O.Berg, Linnaea 30: 695 (1861).
Eugenia calothyrsa Diels, Verh. Bot. Vereins Prov. Brandenburg 48: 189 (1907).
Eugenia ripidocarpa Ruiz & Pav., Anales Inst. Bot. Cavanilles 15: 185 (1957), nom. illeg.

Eugenia patrisii Vahl, Eclog. Amer. 2: 35 (1798). *Myrtus patrisii* (Vahl) Spreng., Syst. Veg. 2: 480 (1825). *Stenocalyx patrisii* (Vahl) O.Berg, Linnaea 29: 247 (1858).
S. Trop. America
Eugenia inocarpa DC., Prodr. 3: 264 (1828).
Eugenia berlynensis O.Berg, Linnaea 27: 468 (1856).
Eugenia baruensis Griseb., Abh. Königl. Ges. Wiss. Göttingen 7: 214 (1857), nom. illeg.
Eugenia teffensis O.Berg in C.F.P.von Martius & auct. suc. (eds.), Fl. Bras. 14(1): 255 (1857).
Eugenia teffensis var. *subcordata* O.Berg in C.F.P.von Martius & auct. suc. (eds.), Fl. Bras. 14(1): 256 (1857).
Eugenia teffensis var. *truncata* O.Berg in C.F.P.von Martius & auct. suc. (eds.), Fl. Bras. 14(1): 256 (1857).
Eugenia vellozii O.Berg in C.F.P.von Martius & auct. suc. (eds.), Fl. Bras. 14(1): 255 (1857).
Stenocalyx patrisii var. *grandifolius* O.Berg, Linnaea 30: 700 (1861).
Stenocalyx patrisii var. *parvifolius* O.Berg, Linnaea 30: 699 (1861).

Eugenia pauciflora DC., Prodr. 3: 269 (1828).
Brazil (Minas Gerais). 84 BZL. Nanophan. or phan.

Eugenia peninsularis Urb., Symb. Antill. 9: 487 (1928).
E. Cuba. 81 CUB. Cham. or nanophan.

Eugenia percincta McVaugh, Fieldiana, Bot. 29: 214 (1956).
N. Brazil. 84 BZN. Nanophan. or phan.

Eugenia percivalii Lundell, Wrightia 2: 124 (1961).
Belize to Guatemala. 80 BLZ GUA. Phan.

Eugenia percrenata McVaugh, Fieldiana, Bot. 29: 215 (1956).
Brazil (Mato Grosso). 84 BZC. Nanophan. or phan.

Eugenia perriniana Urb. & Ekman, Ark. Bot. 21A(5): 31 (1927).
Haiti (Massif de la Hotte). 81 HAI. Nanophan.

Eugenia persicifolia O.Berg in C.F.P.von Martius & auct. suc. (eds.), Fl. Bras. 14(1): 226 (1857).
Brazil (Bahia). 84 BZE. Nanophan. or phan.

Eugenia peruibensis Mattos, Loefgrenia 42: 1 (1970).
Brazil (São Paulo). 84 BZL. Nanophan. or phan.

Eugenia petrophila Urb., Symb. Antill. 9: 489 (1928).
E. Cuba. 81 CUB. Nanophan. or phan.

Eugenia philippioides H.Perrier, Mém. Inst. Sci. Madagascar, Sér. B, Biol. Vég. 4: 166 (1953).
C. Madagascar. 29 MDG. Nanophan.

Eugenia phillyraeoides Trimen, J. Bot. 23: 207 (1885). *Syzygium phillyraeoides* (Trimen) Santapau, Kew Bull. 3: 276 (1948). *Eugenia cotinifolia* subsp. *phyllyraeoides* (Trimen) P.S.Ashton, in Revised Handb. Fl. Ceyl. 2: 413 (1981).
Sri Lanka. 40 SRL. Nanophan.
Eugenia mooniana var. *gracilis* Duthie in J.D.Hooker, Fl. Brit. India 2: 505 (1879). *Eugenia cotinifolia* var. *gracilis* (Duthie) M.R.Almeida, Fl. Maharashtra 2: 265 (1988).

Eugenia phyllocardia Urb., Symb. Antill. 9: 499 (1928).
W. Cuba. 81 CUB. Nanophan.

Eugenia pia DC., Prodr. 3: 263 (1828). *Stenocalyx pius* (DC.) O.Berg in C.F.P.von Martius & auct. suc. (eds.), Fl. Bras. 14(1): 342 (1857).
Brazil (Minas Gerais). 84 BZL. Nanophan. or phan.

Eugenia picardiae Krug & Urb., Bot. Jahrb. Syst. 19: 608 (1895).
Haiti. 81 HAI. Nanophan.
**Eugenia microphylla* A.Rich. ex O.Berg, Linnaea 30: 672 (1861), nom. illeg.
Eugenia pitoniana Urb. & Ekman, Ark. Bot. 21A(5): 30 (1927).
Eugenia formonica Urb. & Ekman, Ark. Bot. 22A(10): 36 (1929).

Eugenia piedraensis Urb., Symb. Antill. 9: 108 (1923).
E. Cuba. 81 CUB. Nanophan.

Eugenia piloesis Cambess. in A.F.C.de Saint-Hilaire, Fl. Bras. Merid. 2: 357 (1833).
Brazil (Goiás to Minas Gerais. 84 BZC BZL. Nanophan. or phan.

Eugenia pilosula Krug & Urb., Bot. Jahrb. Syst. 19: 612 (1895).
Hispaniola. 81 DOM HAI. Nanophan.
**Eugenia angustifolia* O.Berg, Linnaea 27: 159 (1856), nom. illeg.

Eugenia pinariensis Urb., Symb. Antill. 9: 508 (1928).
NW. Cuba. 81 CUB. Nanophan.

Eugenia pinetorum Urb., Symb. Antill. 7: 300 (1912).
E. Cuba. 81 CUB. Nanophan.

Eugenia pinifolia Mart. ex O.Berg in C.F.P.von Martius & auct. suc. (eds.), Fl. Bras. 14(1): 375 (1857).
Brazil. 84+. Nanophan. or phan.

Eugenia piresiana Cambess. in A.F.C.de Saint-Hilaire, Fl. Bras. Merid. 2: 341 (1832).
Brazil (Minas Gerais). 84 BZL. Nanophan. or phan.

Eugenia piresii Mattos, Loefgrenia 94: 3 (1989).
NE. Brazil. 84 BZE. Nanophan. or phan.
**Stenocalyx gardnerianus* O.Berg in C.F.P.von Martius & auct. suc. (eds.), Fl. Bras. 14(1): 350 (1857).

Eugenia pisiformis Cambess. in A.F.C.de Saint-Hilaire, Fl. Bras. Merid. 2: 356 (1832).
Brazil (Espírito Santo). 84 BZL. Nanophan. or phan.

Eugenia pistaciifolia DC., Prodr. 3: 270 (1828). *Stenocalyx pistaciifolius* (DC.) O.Berg in C.F.P.von Martius & auct. suc. (eds.), Fl. Bras. 14(1): 343 (1857).
Brazil (Bahia). 84 BZE. Nanophan. or phan.

Eugenia pitanga (O.Berg) Nied. in H.G.A.Engler & K.A.E.Prantl, Nat. Pflanzenfam. 3(7): 82 (1893).
Paraguay to Brazil. 84 BZC BZL BZS 85 AGE PAR. Cham.
**Stenocalyx pitanga* O.Berg in C.F.P.von Martius & auct. suc. (eds.), Fl. Bras. 14(1): 341 (1857). *Myrtus pitanga* (O.Berg) Kuntze, Revis. Gen. Pl. 3(2): 92 (1898). *Luma pitanga* (O.Berg) Herter, Revista Sudamer. Bot. 7: 219 (1943).
Eugenia camporum Morong, Ann. New York Acad. Sci. 7: 106 (1893). *Eugenia pitanga* var. *camporum* (Morong) Mattos, Loefgrenia 76: 2 (1981).
Eugenia dolichophylla Kiaersk., Enum. Myrt. Bras.: 157 (1893).
Myrtus pitanga var. *angustifolia* Kuntze, Revis. Gen. Pl. 3(2): 92 (1898).
Myrtus pitanga f. *fasciculata* Kuntze, Revis. Gen. Pl. 3(2): 92 (1898).
Myrtus pitanga f. *subsolitaria* Kuntze, Revis. Gen. Pl. 3(2): 92 (1898).
Eugenia montigena Barb.Rodr., Myrt. Paraguay: 5 (1903).
Stenocalyx pitanga var. *nana* Mattos, Reunión (Anual) Soc. Bot. Bras. 5: 100 (1957).
Eugenia pitanga var. *venosa* Mattos, Loefgrenia 61: 2 (1974).

Eugenia pitrensis Urb., Ark. Bot. 22A(10): 35 (1929).
Hispaniola. 81 DOM HAI. Nanophan.

Eugenia pittieri Standl., Publ. Field Mus. Nat. Hist., Bot. Ser. 8: 145 (1930).
C. America. 80 COS GUA NIC PAN. Nanophan. or phan.
Eugenia kellermanii Lundell, Wrightia 3: 116 (1964).

Eugenia platyphylla O.Berg in C.F.P.von Martius & auct. suc. (eds.), Fl. Bras. 14(1): 294 (1857).
E. Brazil. 84 BZE BZL. Nanophan. or phan.

Eugenia platysema O.Berg in C.F.P.von Martius & auct. suc. (eds.), Fl. Bras. 14(1): 276 (1857).
Brazil (Rio Grande do Sul). 84 BZS. Nanophan. or phan.
Eugenia oblongata Mattos & D.Legrand, Loefgrenia 67: 22 (1975), nom. illeg. *Eugenia pinhaesensis* Mattos, Loefgrenia 94: 5 (1989).

Eugenia pleurantha O.Berg in C.F.P.von Martius & auct. suc. (eds.), Fl. Bras. 14(1): 280 (1857).
E. Brazil. 84 BZE BZL. Nanophan. or phan.

Eugenia pleurocarpa Standl., Publ. Field Mus. Nat. Hist., Bot. Ser. 4: 243 (1929).
SW. Mexico. 79 MXS. Nanophan. or phan.

Eugenia plicata Nied. in H.G.A.Engler & K.A.E.Prantl, Nat. Pflanzenfam. 3(7): 91 (1893).
Brazil (Rio de Janeiro). 84 BZL. Nanophan. or phan.
**Stenocalyx riedelianus* O.Berg in C.F.P.von Martius & auct. suc. (eds.), Fl. Bras. 14(1): 349 (1857).

Eugenia plicatocostata O.Berg in C.F.P.von Martius & auct. suc. (eds.), Fl. Bras. 14(1): 575 (1859).
Brazil (Bahia). 84 BZE. Nanophan. or phan.

Eugenia plicatula C.Wright, Anales Acad. Ci. Méd. Habana 5: 439 (1868).
Cuba. 81 CUB. Nanophan.
Eugenia mangasiana Urb., Symb. Antill. 9: 498 (1928).

Eugenia plinioides Urb. & Ekman, Ark. Bot. 21A(5): 34 (1927).
Haiti (Massif de la Selle). 81 HAI. Nanophan.

Eugenia pluricymosa H.Perrier, Mém. Inst. Sci. Madagascar, Sér. B, Biol. Vég. 4: 182 (1953).
NE. Madagascar. 29 MDG. Phan.

Eugenia pluriflora DC., Prodr. 3: 270 (1828).
SE. & S. Brazil to Bolivia. 83 BOL 84 BZL BZS 85 PAR. Nanophan. or phan.
Eugenia pluriflora var. *opaca* O.Berg in C.F.P.von Martius & auct. suc. (eds.), Fl. Bras. 14(1): 282 (1857).
Eugenia pluriflora var. *perforata* O.Berg in C.F.P.von Martius & auct. suc. (eds.), Fl. Bras. 14(1): 282 (1857).
Eugenia osoriana Mattos & D.Legrand, Loefgrenia 67: 21 (1975).
Eugenia bresolinii D.Legrand, in Fl. Ilustr. Catar. 1(Mirt., Suppl. 1): 11 (1977).

Eugenia pobeguinii Aubrév., Fl. Forest. Soudano-Guin.: 87 (1950).
W. Trop. Africa. 22 GUI IVO SIE. Nanophan. or phan.

Eugenia pocsiana Borhidi, Bot. Közlem. 64: 216 (1977 publ. 1978).
Cuba. 81 CUB. Nanophan.

Eugenia pohliana DC., Prodr. 3: 264 (1828).
Brazil (Minas Gerais). 84 BZL. Nanophan. or phan.

Eugenia poiteaui O.Berg, Linnaea 27: 181 (1856).
French Guiana. 82 FRG. Nanophan. or phan. Provisionally accepted.

Eugenia poliensis Aubrév. & Pellegr., Notul. Syst. (Paris) 14: 62 (1950).
Cameroon. 23 CMN. Nanophan. or phan.

Eugenia pollicina J.Guého & A.J.Scott, Kew Bull. 34: 480 (1980).
Mauritius. 29 MAU. Nanophan. or phan.

Eugenia polyadena O.Berg, Linnaea 31: 258 (1862).
Peru. 83 PER. Nanophan. or phan.

Eugenia polyclada Urb. & Ekman, Ark. Bot. 22A(10): 34 (1929).
SW. Dominican Rep. 81 DOM. Cham. or nanophan.

Eugenia polypora Urb., Symb. Antill. 6: 24 (1909).
NW. Jamaica. 81 JAM. Phan.

Eugenia polystachya Rich., Actes Soc. Hist. Nat. Paris 1: 110 (1792).
S. Trop. America. 82 FRG GUY 83 BOL CLM PER 84 BZE BZN. Nanophan. or phan.
Eugenia verruculosa DC., Prodr. 3: 284 (1828).
Eugenia schlechtendaliana O.Berg in C.F.P.von Martius & auct. suc. (eds.), Fl. Bras. 14(1): 321 (1857).
Eugenia forsteri O.Berg, Linnaea 29: 241 (1858).

Eugenia pomifera (Aubl.) Urb., Repert. Spec. Nov. Regni Veg. 16: 150 (1919).
Hispaniola. 81 DOM HAI. Nanophan.
**Myrtus pomifera* Aubl., Hist. Pl. Guiane 1: 513 (1775).
Eugenia angustifolia Lam., Encycl. 3: 203 (1789). *Myrtus angustifolia* (Lam.) Spreng., Syst. Veg. 2: 479 (1825), nom. illeg. *Cumetea angustifolia* (Lam.) Raf., Sylva Tellur.: 107 (1838).
Myrtus linearis J.F.Gmel., Syst. Nat.: 793 (1791).
Eugenia rosmarinifolia Poir. in J.B.A.P.M.de Lamarck, Encycl., Suppl. 3: 129 (1813).
Eugenia rosmarinifolia var. *angustissima* O.Berg, Linnaea 27: 160 (1856). *Eugenia angustifolia* var. *angustissima* (O.Berg) Krug & Urb., Bot. Jahrb. Syst. 19: 614 (1895). *Eugenia pomifera* var. *angustissima* (O.Berg) Urb., Symb. Antill. 8: 482 (1921).
Eugenia rosmarinifolia var. *latifolia* O.Berg, Linnaea 27: 161 (1856). *Eugenia angustifolia* var. *latifolia* (O.Berg) Krug & Urb., Bot. Jahrb. Syst. 19: 614 (1895). *Eugenia pomifera* var. *latifolia* (O.Berg) Urb., Symb. Antill. 8: 483 (1921).
Eugenia rosmarinifolia var. *longifolia* O.Berg, Linnaea 27: 469 (1856).

Eugenia pozasia Urb. & Ekman in I.Urban, Symb. Antill. 9: 490 (1928).
W. Cuba. 81 CUB. Nanophan.

Eugenia praeterita McVaugh, Fieldiana, Bot. 29: 450 (1963).
Mexico. 79 MXG. Nanophan. or phan.

Eugenia prasina O.Berg in C.F.P.von Martius & auct. suc. (eds.), Fl. Bras. 14(1): 255 (1857).
Brazil. 84 BZC BZE BZL BZS. Nanophan. or phan.
Eugenia jurujubensis Kiaersk., Enum. Myrt. Bras.: 143 (1893).
Eugenia stictosepala Kiaersk., Enum. Myrt. Bras.: 119 (1893).
Eugenia prasina var. *grandifolia* Huber, Bull. Herb. Boissier, II, 1: 318 (1901).

Eugenia principium McVaugh, Fieldiana, Bot. 29: 451 (1963).
S. Mexico to C. America. 79 MXS 80 PAN. Nanophan. or phan.

Eugenia procera (Sw.) Poir. in J.B.A.P.M.de Lamarck, Encycl., Suppl. 3: 129 (1811).
Caribbean to Colombia. 81 CUB DOM HAI LEE NLA PUE TRT WIN 83 CLM 84 BZN. Nanophan. or phan.
**Myrtus procera* Sw., Prodr.: 77 (1788).
Eugenia parkeriana DC., Prodr. 3: 271 (1828). *Myrtus brasiliana* var. *parkeriana* (DC.) Kuntze, Revis. Gen. Pl. 3(2): 90 (1898).
Myrtus brachystemon DC., Prodr. 3: 240 (1828).

Eugenia producta DC., Prodr. 3: 266 (1828).
N. South America. 82 FRG VEN. Nanophan. or phan.

Eugenia prolixa S.Moore, Trans. Linn. Soc. London, Bot. 4: 358 (1895).
WC. Brazil. 84 BZC. Nanophan. or phan.
Eugenia prolixa var. *vestita* S.Moore, Trans. Linn. Soc. London, Bot. 4: 359 (1895).

Eugenia pronyensis Guillaumin, Bull. Mus. Natl. Hist. Nat. 28: 545 (1922).
New Caledonia. 60 NWC. Nanophan. or phan.

Eugenia prostrata Mattos, Loefgrenia 94: 4 (1989).
Brazil (Rio de Janeiro). 84 BZL. Nanophan. or phan.
**Stenocalyx protractus* O.Berg in C.F.P.von Martius & auct. suc. (eds.), Fl. Bras. 14(1): 593 (1859).

Eugenia protenta McVaugh, Mem. New York Bot. Gard. 18(2): 204 (1969).
Venezuela (Amazonas) to N. Brazil. 82 VEN 84 BZN. Nanophan. or phan.
**Eugenia protracta* O.Berg, Linnaea 31: 254 (1862), nom. illeg.

Eugenia pruinosa D.Legrand, Sellowia 13: 323 (1961).
Brazil (São Paulo to Santa Catarina). 84 BZL BZS. Phan.

Eugenia pruniformis Cambess. in A.F.C.de Saint-Hilaire, Fl. Bras. Merid. 2: 340 (1832).
Brazil (Espírito Santo to Rio de Janeiro). 84 BZL. Nanophan. or phan.
Myrtus quadrisperma Vell., Fl. Flumin. 5: 215, t. 63 (1829). *Eugenia mikaniana* O.Berg in C.F.P.von Martius & auct. suc. (eds.), Fl. Bras. 14(1): 297 (1857), nom. illeg.
Eugenia olivacea O.Berg in C.F.P.von Martius & auct. suc. (eds.), Fl. Bras. 14(1): 303 (1857).

Eugenia pseudomabaeoides Kosterm., Quart. J. Taiwan Mus. 34: 169 (1981).
Sri Lanka. 40 SRL. Nanophan. or phan.

Eugenia pseudomalacantha D.Legrand, in Fl. Ilustr. Catar. 1(Mirt.): 207 (1969).
Brazil (Santa Catarina). 84 BZS. Phan.

Eugenia pseudopsidium Jacq., Enum. Syst. Pl.: 23 (1760). *Myrtus pseudopsidium* (Jacq.) Spreng., Syst. Veg. 2: 480 (1825). *Stenocalyx pseudopsidium* (Jacq.) O.Berg, Linnaea 27: 314 (1856). *Eugenia willdenowii* Nied. in H.G.A.Engler & K.A.E.Prantl, Nat. Pflanzenfam. 3(7): 81 (1893).
Caribbean, S. Trop. America. 81 DOM LEE PUE WIN 82 FRG GUY SUR VEN 84 BZE BZN. Nanophan. or phan.
Eugenia portoricensis DC., Prodr. 3: 266 (1828). *Stenocalyx portoricensis* (DC.) O.Berg, Linnaea 29: 246 (1858). *Eugenia pseudopsidium* var. *portoricensis* (DC.) Krug & Urb., Bot. Jahrb. Syst. 19: 647 (1895).
Eugenia portoricensis var. *brevipes* DC., Prodr. 3: 266 (1828).
Eugenia psidioides DC., Prodr. 3: 268 (1828).
Eugenia compta A.Rich. ex O.Berg, Linnaea 30: 677 (1861).
Eugenia prieurii O.Berg, Linnaea 30: 681 (1861).
Eugenia prieurii var. *robusta* O.Berg, Linnaea 30: 681 (1861).
Eugenia prieurii var. *tenuiramis* O.Berg, Linnaea 30: 682 (1861).
Eugenia megalocarpa Urb., Symb. Antill. 5: 444 (1908).
Eugenia cryptadena Amshoff, Recueil Trav. Bot. Néerl. 42: 17 (1950).

Eugenia pseudovenosa H.Perrier, Mém. Inst. Sci. Madagascar, Sér. B, Biol. Vég. 4: 180 (1953).
E. Madagascar. 29 MDG. Nanophan.

Eugenia psidiiflora O.Berg in C.F.P.von Martius & auct. suc. (eds.), Fl. Bras. 14(1): 223 (1857). *Calycorectes psidiiflorus* (O.Berg) Sobral, Candollea 40: 636 (1985).
Brazil to NE. Argentina. 84 BZC BZE BZL BZS 85 AGE PAR URU. Nanophan. or phan.
Calycorectes riedelianus O.Berg in C.F.P.von Martius & auct. suc. (eds.), Fl. Bras. 14(1): 596 (1859). *Eugenia psidiiflora* var. *riedelinana* (O.Berg) Mattos, Loefgrenia 120: 8 (2005).
Stenocalyx rufescens Kausel, Lilloa 32: 368 (1967).
Calycorectes legrandii Mattos, Loefgrenia 62: 1 (1974). *Calycorectes psidiiflorus* var. *legrandii* (Mattos) Mattos, Loefgrenia 102: 3 (1993).
Calycorectes striatulus Mattos, Loefgrenia 94: 9 (1989).
Calycorectes schultzianus Mattos, Loefgrenia 101: 2 (1992). *Calycorectes psidiiflorus* var. *schultzianus* (Mattos) Mattos, Loefgrenia 108: 2 (1996). *Eugenia psidiiflora* var. *schultziana* (Mattos) Mattos, Loefgrenia 120: 8 (2005).
Calycorectes psidiiflorus var. *triflorus* Mattos, Loefgrenia 102: 3 (1993). *Eugenia psidiiflora* var. *triflora* (Mattos) Mattos, Loefgrenia 120: 8 (2005).

Eugenia psiloclada Urb., Symb. Antill. 9: 106 (1923).
E. Cuba. 81 CUB. Nanophan.

Eugenia pterocarpa Baill., Hist. Pl. 6: 338 (1876).
New Caledonia. 60 NWC. Nanophan. or phan.

Eugenia pteroclada Urb., Symb. Antill. 9: 493 (1928).
E. Cuba. 81 CUB. Nanophan.

Eugenia puberula Nied. in H.G.A.Engler & K.A.E.Prantl, Nat. Pflanzenfam. 3(7): 82 (1893).
Brazil (Rio de Janeiro). 84 BZL. Nanophan. or phan.
**Phyllocalyx pubescens* O.Berg in C.F.P.von Martius & auct. suc. (eds.), Fl. Bras. 14(1): 592 (1859).

Eugenia pubescens (Kunth) DC., Prodr. 3: 283 (1828).
S. Venezuela to Peru. 82 VEN 83 CLM PER. Nanophan. or phan.
**Myrtus pubescens* Kunth in F.W.H.von Humboldt, A.J.A.Bonpland & C.S.Kunth, Nov. Gen. Sp. 6: 143 (1823).
Eugenia ottonis O.Berg, Linnaea 27: 298 (1856).

Eugenia pubicalyx Alain, Phytologia 25: 268 (1973).
Dominican Rep. 81 DOM. Nanophan. or phan.

Eugenia pueblana Lundell, Wrightia 2: 167 (1961).
Mexico. 79 MXC MXE. Nanophan. or phan.

Eugenia pulcherrima Kiaersk., Enum. Myrt. Bras.: 159 (1893).
SE. Brazil. 84 BZL. Nanophan. or phan.

Eugenia puncticulata Merr., Philipp. J. Sci., C 9: 381 (1914).
Philippines (Leyte). 42 PHI. Nanophan. or phan. Provisionally accepted.

Eugenia punicifolia (Kunth) DC., Prodr. 3: 267 (1828).
W. Cuba, S. Trop. America. 81 CUB 82 FRG GUY SUR VEN 83 BOL CLM PER 84 BZC BZE BZL BZN BZS 85 AGE PAR. Nanophan. or phan.
Myrtus oleifolia Kunth in F.W.H.von Humboldt, A.J.A.Bonpland & C.S.Kunth, Nov. Gen. Sp. 6: 147 (1823). *Eugenia oleifolia* (Kunth) DC., Prodr. 3: 270 (1828).

*Myrtus punicifolia Kunth in F.W.H.von Humboldt, A.J.A.Bonpland & C.S.Kunth, Nov. Gen. Sp. 6: 149 (1823). *Emurtia punicifolia* (Kunth) Raf., Sylva Tellur.: 106 (1838). *Eugenia vaga* var. *punicifolia* (Kunth) O.Berg in C.F.P.von Martius & auct. suc. (eds.), Fl. Bras. 14(1): 238 (1857).
Eugenia clinocarpa DC., Prodr. 3: 267 (1828).
Eugenia coarensis DC., Prodr. 3: 267 (1828).
Eugenia dipoda DC., Prodr. 3: 269 (1828). *Eugenia vaga* var. *dipoda* (DC.) O.Berg in C.F.P.von Martius & auct. suc. (eds.), Fl. Bras. 14(1): 238 (1857).
Eugenia dipoda var. *brachypoda* DC., Prodr. 3: 268 (1828). *Eugenia vaga* var. *brachypoda* (DC.) O.Berg in C.F.P.von Martius & auct. suc. (eds.), Fl. Bras. 14(1): 238 (1857). *Eugenia punicifolia* var. *brachypoda* (DC.) Krug & Urb., Bot. Jahrb. Syst. 19: 617 (1895).
Eugenia fruticulosa DC., Prodr. 3: 270 (1828).
Eugenia kochiana DC., Prodr. 3: 268 (1828). *Pseudomyrcianthes kochiana* (DC.) Kausel, Ark. Bot., a.s., 3: 505 (1956).
Eugenia kunthiana DC., Prodr. 3: 264 (1828).
Eugenia myrtillifolia DC., Prodr. 3: 265 (1828).
Eugenia sancta DC., Prodr. 3: 267 (1828).
Eugenia adstringens Cambess. in A.F.C.de Saint-Hilaire, Fl. Bras. Merid. 2: 260 (1830).
Eugenia ovalifolia Cambess. in A.F.C.de Saint-Hilaire, Fl. Bras. Merid. 2: 350 (1832).
Eugenia sabulosa Cambess. in A.F.C.de Saint-Hilaire, Fl. Bras. Merid. 2: 355 (1832).
Eugenia insipida Cambess. in A.F.C.de Saint-Hilaire, Fl. Bras. Merid. 2: 360 (1833).
Eugenia nhanica Cambess. in A.F.C.de Saint-Hilaire, Fl. Bras. Merid. 2: 362 (1833).
Eugenia obtusifolia Cambess. in A.F.C.de Saint-Hilaire, Fl. Bras. Merid. 2: 368 (1833).
Myrtus psychotrioides Colla, Herb. Pedem. 2: 432 (1834).
Eugenia insipida Mart., Flora 20(2 Beibl.): 85 (1837), nom. illeg.
Eugenia subalterna Benth., J. Bot. (Hooker) 2: 320 (1840). *Eugenia pungens* var. *subalterna* (Benth.) Amshoff, Recueil Trav. Bot. Néerl. 42: 25 (1950).
Eugenia surinamensis Miq., Linnaea 18: 743 (1845).
Eugenia dipoda f. *grandifolia* Miq., Stirp. Surinam. Select.: 38 (1851).
Eugenia benthamii O.Berg, Linnaea 27: 164 (1856).
Eugenia pyrroclada O.Berg, Linnaea 27: 165 (1856).
Eugenia pyrroclada var. *columbiensis* O.Berg, Linnaea 27: 166 (1856).
Eugenia pyrroclada var. *guianensis* O.Berg, Linnaea 27: 165 (1856).
Eugenia surinamensis var. *impunctata* O.Berg, Linnaea 27: 182 (1856).
Eugenia surinamensis var. *punctata* O.Berg, Linnaea 27: 182 (1856).
Eugenia ambigua O.Berg in C.F.P.von Martius & auct. suc. (eds.), Fl. Bras. 14(1): 241 (1857).
Eugenia arbutifolia O.Berg in C.F.P.von Martius & auct. suc. (eds.), Fl. Bras. 14(1): 306 (1857). *Myrtus arbutifolia* (O.Berg) Kuntze, Revis. Gen. Pl. 3(2): 92 (1898).
Eugenia arbutifolia var. *obscura* O.Berg in C.F.P.von Martius & auct. suc. (eds.), Fl. Bras. 14(1): 306 (1857).
Eugenia arbutifolia var. *pallida* O.Berg in C.F.P.von Martius & auct. suc. (eds.), Fl. Bras. 14(1): 306 (1857).
Eugenia arctostaphyloides O.Berg in C.F.P.von Martius & auct. suc. (eds.), Fl. Bras. 14(1): 217 (1857).
Eugenia boliviensis O.Berg in C.F.P.von Martius & auct. suc. (eds.), Fl. Bras. 14(1): 221 (1857).
Eugenia ciarensis O.Berg in C.F.P.von Martius & auct. suc. (eds.), Fl. Bras. 14(1): 235 (1857).
Eugenia decorticans O.Berg in C.F.P.von Martius & auct. suc. (eds.), Fl. Bras. 14(1): 240 (1857).
Eugenia diantha O.Berg in C.F.P.von Martius & auct. suc. (eds.), Fl. Bras. 14(1): 239 (1857).
Eugenia diantha var. *ciliata* O.Berg in C.F.P.von Martius & auct. suc. (eds.), Fl. Bras. 14(1): 240 (1857).
Eugenia diantha var. *glabra* O.Berg in C.F.P.von Martius & auct. suc. (eds.), Fl. Bras. 14(1): 240 (1857).
Eugenia flava O.Berg in C.F.P.von Martius & auct. suc. (eds.), Fl. Bras. 14(1): 273 (1857).
Eugenia kunthiana var. *opaca* O.Berg in C.F.P.von Martius & auct. suc. (eds.), Fl. Bras. 14(1): 236 (1857).
Eugenia kunthiana var. *pellucida* O.Berg in C.F.P.von Martius & auct. suc. (eds.), Fl. Bras. 14(1): 236 (1857).
Eugenia linearifolia O.Berg in C.F.P.von Martius & auct. suc. (eds.), Fl. Bras. 14(1): 230 (1857).
Eugenia linearifolia var. *oppositifolia* O.Berg in C.F.P. von Martius & auct. suc. (eds.), Fl. Bras. 14(1): 230 (1857).
Eugenia linearifolia var. *ternifolia* O.Berg in C.F.P.von Martius & auct. suc. (eds.), Fl. Bras. 14(1): 230 (1857).
Eugenia macroclada O.Berg in C.F.P.von Martius & auct. suc. (eds.), Fl. Bras. 14(1): 243 (1857).
Eugenia maximiliana O.Berg in C.F.P.von Martius & auct. suc. (eds.), Fl. Bras. 14(1): 250 (1857).
Eugenia obversa O.Berg in C.F.P.von Martius & auct. suc. (eds.), Fl. Bras. 14(1): 295 (1857), nom. illeg.
Eugenia phillyraeoides Willd. ex O.Berg in C.F.P.von Martius & auct. suc. (eds.), Fl. Bras. 14(1): 238 (1857), pro syn.
Eugenia platyclada O.Berg in C.F.P.von Martius & auct. suc. (eds.), Fl. Bras. 14(1): 257 (1857).
Eugenia platyclada var. *cuneata* O.Berg in C.F.P.von Martius & auct. suc. (eds.), Fl. Bras. 14(1): 256 (1857).
Eugenia platyclada var. *ovalis* O.Berg in C.F.P.von Martius & auct. suc. (eds.), Fl. Bras. 14(1): 257 (1857).
Eugenia polyphylla O.Berg in C.F.P.von Martius & auct. suc. (eds.), Fl. Bras. 14(1): 240 (1857).
Eugenia prominens O.Berg in C.F.P.von Martius & auct. suc. (eds.), Fl. Bras. 14(1): 235 (1857).
Eugenia psychotrioides Mart. ex O.Berg in C.F.P.von Martius & auct. suc. (eds.), Fl. Bras. 14(1): 281 (1857).
Eugenia pyramidalis O.Berg in C.F.P.von Martius & auct. suc. (eds.), Fl. Bras. 14(1): 236 (1857).
Eugenia rhombocarpa O.Berg in C.F.P.von Martius & auct. suc. (eds.), Fl. Bras. 14(1): 239 (1857).
Eugenia romana O.Berg in C.F.P.von Martius & auct. suc. (eds.), Fl. Bras. 14(1): 264 (1857).
Eugenia spathophylla O.Berg in C.F.P.von Martius & auct. suc. (eds.), Fl. Bras. 14(1): 241 (1857).
Eugenia spathulata O.Berg in C.F.P.von Martius & auct. suc. (eds.), Fl. Bras. 14(1): 240 (1857).
Eugenia subcorymbosa O.Berg in C.F.P.von Martius & auct. suc. (eds.), Fl. Bras. 14(1): 295 (1857).

Eugenia suffruticosa O.Berg in C.F.P.von Martius & auct. suc. (eds.), Fl. Bras. 14(1): 242 (1857).
Eugenia suffruticosa var. *laeta* O.Berg in C.F.P.von Martius & auct. suc. (eds.), Fl. Bras. 14(1): 242 (1857).
Eugenia suffruticosa var. *opaca* O.Berg in C.F.P.von Martius & auct. suc. (eds.), Fl. Bras. 14(1): 242 (1857).
Eugenia triphylla O.Berg in C.F.P.von Martius & auct. suc. (eds.), Fl. Bras. 14(1): 243 (1857).
Eugenia vaga O.Berg in C.F.P.von Martius & auct. suc. (eds.), Fl. Bras. 14(1): 238 (1857).
Eugenia vaga var. *brasiliensis* O.Berg in C.F.P.von Martius & auct. suc. (eds.), Fl. Bras. 14(1): 238 (1857).
Eugenia vaga var. *rigida* O.Berg in C.F.P.von Martius & auct. suc. (eds.), Fl. Bras. 14(1): 239 (1857).
Eugenia arctostaphyloides var. *ovalis* O.Berg in C.F.P. von Martius & auct. suc. (eds.), Fl. Bras. 14(1): 567 (1859).
Eugenia dasyantha O.Berg in C.F.P.von Martius & auct. suc. (eds.), Fl. Bras. 14(1): 572 (1859).
Eugenia glareosa O.Berg in C.F.P.von Martius & auct. suc. (eds.), Fl. Bras. 14(1): 579 (1859).
Eugenia mugiensis O.Berg in C.F.P.von Martius & auct. suc. (eds.), Fl. Bras. 14(1): 571 (1859).
Eugenia platyclada var. *microphylla* O.Berg in C.F.P.von Martius & auct. suc. (eds.), Fl. Bras. 14(1): 576 (1859).
Eugenia platyclada var. *multiflora* O.Berg in C.F.P.von Martius & auct. suc. (eds.), Fl. Bras. 14(1): 576 (1859).
Eugenia polyphylla var. *obovata* O.Berg in C.F.P.von Martius & auct. suc. (eds.), Fl. Bras. 14(1): 572 (1859).
Eugenia suffruticosa var. *latifolia* O.Berg in C.F.P.von Martius & auct. suc. (eds.), Fl. Bras. 14(1): 572 (1859).
Eugenia calycolpoides Griseb., Fl. Brit. W. I.: 238 (1860).
Eugenia diversiflora O.Berg, Linnaea 30: 675 (1861).
Eugenia pyramidalis var. *angustifolia* O.Berg, Linnaea 30: 680 (1861).
Eugenia vaga var. *pumila* O.Berg, Linnaea 30: 681 (1861).
Eugenia psammophila Diels, Bot. Jahrb. Syst. 37: 598 (1906).
Eugenia romana var. *pickeli* Mattos, Loefgrenia 53: 1 (1971).
Eugenia rubrescens Mattos & D.Legrand, Loefgrenia 67: 25 (1975).

Eugenia purpusii Standl., Contr. U. S. Natl. Herb. 23: 1678 (1926).
Mexico (Oaxaca). 79 MXS. Nanophan. or phan.
**Eugenia oaxacana* Standl., Contr. U. S. Natl. Herb. 23: 1043 (1924), nom. illeg.

Eugenia pusilla N.E.Br., Bull. Misc. Inform. Kew 1912: 276 (1912).
Northern Prov. 27 TVL. Nanophan. or phan.

Eugenia pustulescens McVaugh, Fieldiana, Bot. 29: 215 (1956).
Ecaudor. 83 ECU. Nanophan. or phan.

Eugenia pycnantha Benth., Pl. Hartw.: 174 (1845). *Myrcia pycnantha* (Benth.) Steyerm., Fieldiana, Bot. 28: 1018 (1957).
Colombia to Ecuador. 83 CLM ECU. Nanophan. or phan.
Eugenia alaternifolia Benth. ex O.Berg, Linnaea 27: 249 (1856).
Eugenia sodiroi Diels, Bot. Jahrb. Syst. 40(91): 47 (1907).

Eugenia pyriflora O.Berg in C.F.P.von Martius & auct. suc. (eds.), Fl. Bras. 14(1): 299 (1857).
NE. Brazil. 84 BZE. Nanophan. or phan.
Eugenia pyriflora var. *major* O.Berg in C.F.P.von Martius & auct. suc. (eds.), Fl. Bras. 14(1): 299 (1857).
Eugenia pyriflora var. *minor* O.Berg in C.F.P.von Martius & auct. suc. (eds.), Fl. Bras. 14(1): 299 (1857).

Eugenia pyriformis Cambess. in A.F.C.de Saint-Hilaire, Fl. Bras. Merid. 2: 366 (1833). *Myrtus pyriformis* (Cambess.) Parodi, Anales Soc. Ci. Argent. 7: 216 (1879). *Pseudomyrcianthes pyriformis* (Cambess.) Kausel, Ark. Bot., a.s., 3: 504 (1956).
Brazil to NE. Argentina. 84 BZC BZL BZS 85 AGE PAR URU?. Phan.
Eugenia albotomentosa Cambess. in A.F.C.de Saint-Hilaire, Fl. Bras. Merid. 2: 369 (1833).
Eugenia albotomentosa var. *goyazensis* Cambess. in A.F.C.P.de Saint-Hilaire & al., Fl. Bras. Merid. 2: 369 (1833).
Eugenia uvalha Cambess. in A.F.C.de Saint-Hilaire, Fl. Bras. Merid. 2: 367 (1833). *Eugenia pyriformis* var. *uvalha* (Cambess.) D.Legrand, Bol. Fac. Agron. Univ. Montevideo 11: 46 (1968).
Eugenia albotomentosa var. *urussangensis* O.Berg in C.F.P.von Martius & auct. suc. (eds.), Fl. Bras. 14(1): 221 (1857).
Eugenia turbinata O.Berg in C.F.P.von Martius & auct. suc. (eds.), Fl. Bras. 14(1): 218 (1857). *Luma turbinata* (O.Berg) Herter, Revista Sudamer. Bot. 7: 220 (1943).
Eugenia vauthiereana O.Berg in C.F.P.von Martius & auct. suc. (eds.), Fl. Bras. 14(1): 217 (1857).
Stenocalyx lanceolatus O.Berg in C.F.P.von Martius & auct. suc. (eds.), Fl. Bras. 14(1): 341 (1857). *Eugenia phlebotomonides* Kiaersk., Enum. Myrt. Bras.: 159 (1893).
Eugenia viminalis O.Berg in C.F.P.von Martius & auct. suc. (eds.), Fl. Bras. 14(1): 587 (1859).
Myrtus conceptionis Kuntze, Revis. Gen. Pl. 3(2): 91 (1898).
Eugenia dumicola Barb.Rodr., Myrt. Paraguay: 7 (1903). *Myrciaria dumicola* (Barb.Rodr.) Chodat & Hassl., Bull. Herb. Boissier, II, 7: 808 (1907).
Eugenia hassleriana Barb.Rodr., Myrt. Paraguay: 4 (1903).
Eugenia pyriformis f. *ponhi* D.Legrand, Bol. Soc. Argent. Bot. 10: 10 (1962).
Eugenia pyriformis var. *argentea* Mattos & D.Legrand, Loefgrenia 67: 30 (1975).

Eugenia quadrangularis Duch. ex Griseb., Abh. Königl. Ges. Wiss. Göttingen 7: 214 (1857).
Guadeloupe. 81 LEE. Nanophan. or phan.

Eugenia quadriflora H.Perrier, Mém. Inst. Sci. Madagascar, Sér. B, Biol. Vég. 4: 168 (1953).
EC. Madagascar. 29 MDG. Nanophan.

Eugenia quadrijuga McVaugh, Fieldiana, Bot. 29: 216 (1956).
N. Brazil to Peru. 83 PER 84 BZN. Nanophan. or phan.

Eugenia quadriovulata Amshoff, Recueil Trav. Bot. Néerl. 39: 160 (1942).

Guianas. 82 FRG GUY SUR. Nanophan. or phan.

Eugenia quebradensis McVaugh, Fieldiana, Bot. 29: 217 (1956).
Peru. 83 PER. Nanophan. or phan.

Eugenia quercetorum Standl. & L.O.Williams ex Barrie, Novon 15: 39 (2005).
Honduras. 80 HON. Nanophan. or phan.

Eugenia racemiflora O.Berg, Linnaea 27: 232 (1856).
Peru. 83 PER. Nanophan. or phan.
Psidium emarginatum Ruiz & Pav. ex G.Don, Gen. Hist. 2: 833 (1832).

Eugenia radiciflora H.Perrier, Mém. Inst. Sci. Madagascar, Sér. B, Biol. Vég. 4: 176 (1953).
N. Madagascar. 29 MDG. Nanophan.

Eugenia ramboi D.Legrand, Sellowia 13: 321 (1961).
SE. & S. Brazil to Argentina (Misiones). 84 BZL BZS 85 AGE PAR. Phan.

Eugenia ramiflora Desv. in W.Hamilton, Prodr. Pl. Ind. Occid.: 43 (1825).
N. South America to N. Brazil. 82 FRG SUR 84 BZN. Nanophan. or phan.
Eugenia brachypoda DC., Prodr. 3: 274 (1828).
Eugenia fulvipes Sagot, Ann. Sci. Nat., Bot., VI, 20: 189 (1885).

Eugenia ramonae Borhidi & O.Muñiz, Acta Bot. Acad. Sci. Hung. 21: 228 (1975 publ. 1976).
Cuba. 81 CUB. Nanophan. or phan.

Eugenia ramoniana Urb., Symb. Antill. 9: 499 (1928).
W. & C. Cuba. 81 CUB. Nanophan.
**Eugenia heterophylla* Urb., Bot. Jahrb. Syst. 19: 610 (1895), nom. illeg.

Eugenia randrianasoloi J.S.Mill., Adansonia, III, 22: 112 (2000).
Madagascar. 29 MDG. Phan.

Eugenia ravenii Lundell, Wrightia 4: 94 (1968).
SE. Mexico. 79 MXT. Phan.

Eugenia reinwardtiana (Blume) A.Cunn. ex DC., Prodr. 3: 267 (1828).
Malesia to Pacific. 42 BOR LSI MOL 50 QLD WAU? 60 FIJ NUE TON 61 MRQ SCI TUA 63 HAW. Nanophan. or phan.
**Myrtus reinwardtiana* Blume, Bijdr.: 1082 (1826). *Jossinia reinwardtiana* (Blume) Blume, Mus. Bot. 1: 120 (1850).
Eugenia rariflora Benth., London J. Bot. 2: 221 (1843).
Eugenia carissoides F.Muell., Fragm. 3: 130 (1863).
Eugenia hypospodia F.Muell., Fragm. 5: 15 (1865).
Jossinia tahitensis Nadeaud, Énum. Pl. Tahiti: 79 (1873).
Eugenia rariflora var. *parvifolia* Hillebr., Fl. Hawaiian Isl.: 129 (1888). *Eugenia waianensis* O.Deg., Fl. Hawaiiensis 273: s.p. (1932).
Eugenia kangeanensis Valeton, Icon. Bogor.: t. 333 (1912).
Eugenia macrohila C.T.White & W.D.Francis, Proc. Roy. Soc. Queensland 35: 69 (1923).
Eugenia koolauensis var. *glabra* O.Deg., Fl. Hawaiiensis 273: s.p. (1932).
Episyzygium cahuense Suess. & A.Ludw., Mitt. Bot. Staatssamml. München 1: 18 (1950).
Eugenia reinwardtiana f. *lutea* H.St.John, Phytologia 37: 441 (1977).

Eugenia reitziana D.Legrand, Sellowia 13: 315 (1961).
Brazil (Santa Catarina). 84 BZS. Cham.

Eugenia rekoi Standl., Contr. U. S. Natl. Herb. 23: 1044 (1924).
W. Mexico. 79 MXN MXS. Nanophan. or phan.

Eugenia rendlei Urb., Symb. Antill. 7: 302 (1912).
SE. Jamaica. 81 JAM. Nanophan. or phan.

Eugenia repanda O.Berg in C.F.P.von Martius & auct. suc. (eds.), Fl. Bras. 14(1): 304 (1857). *Luma repanda* (O.Berg) Herter, Revista Sudamer. Bot. 7: 220 (1943).
SE. & S. Brazil to N. Argentina. 84 BZL BZS 85 AGE AGW PAR URU. Nanophan. or phan.
Eugenia ypanemensis O.Berg in C.F.P.von Martius & auct. suc. (eds.), Fl. Bras. 14(1): 316 (1857).
Eugenia oxyoentophylla Kiaersk., Enum. Myrt. Bras.: 145 (1893).

Eugenia reticularis O.Berg, Linnaea 27: 170 (1856).
Hispaniola. 81 DOM HAI. Nanophan.

Eugenia retinadenia C.Wright, Anales Acad. Ci. Méd. Habana 5: 432 (1868).
W. Cuba. 81 CUB. Nanophan. or phan.

Eugenia rheophytica Kosterm., Quart. J. Taiwan Mus. 34: 170 (1981).
Sri Lanka. 40 SRL. Nanophan. or phan.

Eugenia rhombea (O.Berg) Krug & Urb., Bot. Jahrb. Syst. 19: 644 (1895).
Florida (Key West) , Caribbean, Mexico to C. America. 78 FLA 79 MXG MXS MXT 80 BLZ GUA HON NIC 81 BAH CUB DOM HAI JAM LEE PUE WIN. Nanophan. or phan.
**Eugenia foetida* var. *rhombea* O.Berg, Linnaea 27: 212 (1856).
Eugenia fiscalensis Donn.Sm., Bot. Gaz. 54: 235 (1912).
Eugenia leptopa Lundell, Bull. Torrey Bot. Club 69: 396 (1942).
Eugenia pusilana Lundell, Wrightia 6: 121 (1980).

Eugenia rigida DC., Prodr. 3: 273 (1828).
NE. Brazil. 84 BZL. Nanophan. or phan.
Eugenia rigida var. *ovalis* O.Berg in C.F.P.von Martius & auct. suc. (eds.), Fl. Bras. 14(1): 258 (1857).
Eugenia rigida var. *ovata* O.Berg in C.F.P.von Martius & auct. suc. (eds.), Fl. Bras. 14(1): 258 (1857).

Eugenia rigidifolia A.Rich., Hist. Phys. Cuba, Pl. Vasc.: 585 (1846). *Eugenia rigidifolia* var. *genuina* Krug & Urb., Bot. Jahrb. Syst. 19: 608 (1895), nom. inval. *Myrtus rigidifolia* (A.Rich.) Kuntze, Revis. Gen. Pl. 3(2): 89 (1898).
W. Cuba. 81 CUB. Nanophan.

subsp. ***rangelensis*** Kitan., God. Sofiisk. Univ. Biol. Fak., 2, Bot. Mikrobiol. Fiziol. Biokhim. Rast. 64(2): 4 (1969–1970 publ. 1972).
W. Cuba. 81 CUB. Nanophan.

subsp. ***rigidifolia***
W. Cuba. 81 CUB. Nanophan.
Eugenia phyllyreifolia A.Rich., Hist. Phys. Cuba, Pl. Vasc.: 585 (1846). *Eugenia rigidifolia* var. *phyllyreifolia* (A.Rich.) Krug & Urb., Bot. Jahrb. Syst. 19: 609 (1895).
Eugenia rigidifolia var. *angustifolia* O.Berg, Linnaea 30: 679 (1861).
Eugenia rigidifolia var. *latifolia* O.Berg, Linnaea 30: 679 (1861).

Eugenia organosia Urb., Symb. Antill. 9: 101 (1923).

subsp. ***tsugifolia*** Kitan., God. Sofiisk. Univ. Biol. Fak., 2, Bot. Mikrobiol. Fiziol. Biokhim. Rast. 64(2): 4 (1969–1970 publ. 1972).
W. Cuba. 81 CUB. Nanophan.

Eugenia rigidula Britton & P.Wilson, Bull. Torrey Bot. Club 50: 45 (1923).
E. Cuba. 81 CUB. Nanophan.

Eugenia rimosa C.Wright, Anales Acad. Ci. Méd. Habana 5: 429 (1868).
Cuba. 81 CUB. Nanophan.

Eugenia riograndis Lundell, Field & Lab. 13: 10 (1945).
Belize. 80 BLZ. Phan.

Eugenia riosae Barrie, Novon 15: 40 (2005).
Costa Rica. 80 COS. Nanophan. or phan.

Eugenia rivulorum Thwaites, Enum. Pl. Zeyl.: 115 (1859).
Sri Lanka. 40 SRL. Nanophan. or phan.

Eugenia rizziniana Mattos, Loefgrenia 117: 1 (2002).
Brazil (Brasília D.F.). 84 BZC.

Eugenia rizzoana Mattos, Loefgrenia 64: 4 (1974).
SE. Brazil. 84 BZL. Nanophan. or phan.
**Myrtus warmingiana* Kiaersk., Enum. Myrt. Bras.: 19 (1893).

Eugenia rocana Britton & P.Wilson, Mem. Torrey Bot. Club 16: 90 (1920).
W. Cuba. 81 CUB. Nanophan.

Eugenia rodriguesensis J.Guého & A.J.Scott, Kew Bull. 34: 481 (1980).
Rodrigues. 29 ROD. Nanophan. or phan.

Eugenia roigii Urb., Symb. Antill. 9: 498 (1928).
W. Cuba. 81 CUB. Nanophan.

Eugenia rosariensis Borhidi, Acta Bot. Hung. 29: 187 (1983).
W. Cuba. 81 CUB. Nanophan.
**Eugenia bakeri* Britton & P.Wilson, Mem. Torrey Bot. Club 16: 91 (1920), nom. illeg.

Eugenia rosea DC., Prodr. 3: 264 (1828). *Pseudomyrcianthes rosea* (DC.) Kausel, Ark. Bot., a.s., 3: 505 (1956).
SE. Brazil. 84 BZL. Nanophan. or phan.

Eugenia roseiflora McVaugh, Mem. New York Bot. Gard. 18(2): 210 (1969).
SE. Colombia to Guyana. 82 GUY VEN 83 CLM 84 BZN. Nanophan. or phan.

Eugenia rostrata O.Berg in C.F.P.von Martius & auct. suc. (eds.), Fl. Bras. 14(1): 282 (1857).
SE. Brazil. 84 BZL. Nanophan. or phan.

Eugenia rostratofalcata Mattos & D.Legrand, Loefgrenia 67: 23 (1975).
Brazil (Paraná). 84 BZS. Nanophan. or phan.

Eugenia rostrifolia D.Legrand, Sellowia 8: 76 (1957).
S. Brazil. 84 BZS. Phan.

Eugenia rottleriana Wight & Arn., Prodr. Fl. Ind. Orient.: 331 (1834).
SW. India. 40 IND. Nanophan. or phan.

Eugenia rotundata (Trimen) Trimen, Handb. Fl. Ceylon 2: 185 (1894).
Sri Lanka. 40 SRL. Nanophan.
**Eugenia amoena* var. *rotundata* Trimen, Cat.: 33 (1885).

Eugenia rotundicosta D.Legrand, in Fl. Ilustr. Catar. 1(Mirt., Suppl. 1): 25 (1977).
S. Brazil. 84 BZS. Nanophan. or phan.

Eugenia roxburghii DC., Prodr. 3: 271 (1828). *Eugenia bracteata* var. *roxburghii* (DC.) Duthie in J.D.Hooker, Fl. Brit. India 2: 502 (1879).
India to Indo-China. 40 BAN IND SRL 41 MYA THA.
Myrtus bracteata Willd., Sp. Pl. 2: 969 (1799). *Eugenia bracteata* (Willd.) Raeusch. ex Roxb., Hort. Bengal.: 37 (1814), nom. illeg. *Syzygium bracteatum* (Willd.) Raizada, Indian Forester 74: 336 (1948).
Myrtus ruscifolia Willd., Sp. Pl. 2: 970 (1799). *Syzygium ruscifolium* (Willd.) Santapau & Wagh, Bull. Bot. Surv. India 5: 109 (1964).
Myrtus latifolia B.Heyne ex Roth, Nov. Pl. Sp.: 232 (1821). *Myrtus heynei* Spreng., Syst. Veg. 2: 482 (1825), nom. illeg. *Eugenia heynei* Rathakr. & N.C.Nair, J. Econ. Taxon. Bot. 5: 232 (1984). *Eugenia rothii* Panigrahi, J. Econ. Taxon. Bot. 5: 994 (1984), nom. illeg.
**Eugenia zeylanica* Roxb., Fl. Ind. ed. 1832, 2: 490 (1832), nom. illeg.
Myrtus coromandelina J.König ex Roxb., Fl. Ind. ed. 1832, 2: 491 (1832), nom. inval.
Myrtus littoralis Roxb. ex Wight & Arn., Prodr. Fl. Ind. Orient.: 331 (1834).
Eugenia willdenowii Wight, Ill. Ind. Bot. 2: 13 (1841), nom. illeg.
Syzygium latifolium Blanco, Fl. Filip., ed. 2: 294 (1845), nom. illeg.
Eugenia fasciculata Wall. ex Blume, Mus. Bot. 1: 87 (1850). *Eugenia bracteata* var. *fasciculata* (Wall. ex Blume) Duthie in J.D.Hooker, Fl. Brit. India 2: 502 (1879). *Eugenia rothii* var. *fasciculata* (Duthie) H.B.Naithani, Fl. Pl. India, Nepal & Bhutan: 165 (1990).
Myrtus quadripartita Royen ex Blume, Mus. Bot. 1: 87 (1850).
Eugenia macrosepala Duthie in J.D.Hooker, Fl. Brit. India 2: 501 (1879).

Eugenia rubella Lundell, Wrightia 3: 18 (1961).
Guatemala. 80 GUA. Nanophan. or phan.

Eugenia rufidula Lundell, Wrightia 2: 209 (1961).
Belize. 80 BLZ. Phan.

Eugenia rufoflavescens Mattos, Loefgrenia 114: 1 (2000).
Brazil (Pará). 84 BZN.

Eugenia rufofulva Thwaites, Enum. Pl. Zeyl.: 416 (1864).
Sri Lanka. 40 SRL. Phan.
Eugenia xanthocarpa Thwaites, Enum. Pl. Zeyl.: 416 (1864), nom. illeg.

Eugenia rugosissima Sobral, Bol. Mus. Biol. Mello Leitão. Nova Sér. 20: 74 (2007).
Brazil (Espírito Santo). 84 BZL. Phan.

Eugenia ruscifolia Poir. in J.B.A.P.M.de Lamarck, Encycl., Suppl. 3: 123 (1813). *Myrtus daphnoides* Spreng., Syst. Veg. 2: 480 (1825).
Caribbean (?). 81+. Nanophan. or phan.

Eugenia sachetae Proctor, Bull. Inst. Jamaica, Sci. Ser. 16: 34 (1967).
NW. Jamaica. 81 JAM. Nanophan.

Eugenia sagraea O.Berg, Linnaea 30: 690 (1861).
Cuba. 81 CUB. Nanophan. or phan.

Eugenia salacioides G.Lawson ex Hutch. & Dalziel, Fl. W. Trop. Afr. 1: 200 (1927).
W. Trop. Africa. 22 BEN IVO LBR SIE. Nanophan. or phan.
Eugenia salicifolia G.Lawson in D.Oliver & auct. suc. (eds.), Fl. Trop. Afr. 2: 437 (1871), nom. illeg.

Eugenia salamensis Donn.Sm., Bot. Gaz. 27: 333 (1899).
W. Mexico to El Salvador. 79 MXN MXS 80 ELS GUA. Nanophan. or phan.
Eugenia tomentulosa Standl., Contr. U. S. Natl. Herb. 23: 1043 (1924).
Psidium rensonianum Standl., J. Wash. Acad. Sci. 14: 241 (1924). *Eugenia salamensis* var. *rensoniana* (Standl.) McVaugh, Fieldiana, Bot. 29: 456 (1963).
Eugenia mexiae Standl., Publ. Field Mus. Nat. Hist., Bot. Ser. 4: 243 (1929).

Eugenia salomonica G.T.White, J. Arnold Arbor. 32: 141 (1951).
Solomon Is. 43 SOL. Nanophan. or phan.

Eugenia samanensis Alain, Phytologia 61: 359 (1986).
Dominican Rep. 81 DOM. Nanophan. or phan.

Eugenia samuelssonii Ekman & Urb. in I.Urban, Symb. Antill. 9: 504 (1928). *Pseudanamomis samuelssonii* (Ekman & Urb.) Bisse, Feddes Repert. 96: 511 (1985).
SE. Cuba. 81 CUB. Nanophan.

Eugenia sancarlosensis Barrie, Novon 15: 42 (2005).
Costa Rica. 80 COS. Nanophan. or phan.

Eugenia sanjuanensis P.E.Sánchez, Phytologia 63: 402 (1987).
Nicaragua, Panama. 80 NIC PAN. Nanophan. or phan.

Eugenia santensis Kiaersk., Enum. Myrt. Bras.: 163 (1893).
SE. Brazil. 84 BZL. Nanophan. or phan.
**Phyllocalyx grandifolius* O.Berg in C.F.P.von Martius & auct. suc. (eds.), Fl. Bras. 14(1): 333 (1857). *Eugenia bunchosiifolia* Nied. in H.G.A.Engler & K.A.E.Prantl, Nat. Pflanzenfam. 3(7): 82 (1893), nom. illeg.
Phyllocalyx grandifolius var. *pyriformis* O.Berg in C.F.P.von Martius & auct. suc. (eds.), Fl. Bras. 14(1): 591 (1859).

Eugenia sarapiquensis P.E.Sánchez, Brenesia 25–26: 317 (1986 publ. 1988).
Costa Rica. 80 COS. Nanophan. or phan.

Eugenia sarasinii Guillaumin in C.F.Sarasin & J.Roux, Nova Caledonia, Bot. 1: 189 (1921). *Xanthomyrtus sarasinii* (Guillaumin) Guillaumin, Bull. Soc. Bot. France 85: 630 (1938 publ. 1939).
New Caledonia. 60 NWC. Nanophan. or phan.
Xanthomyrtus pergracilis Diels, Bot. Jahrb. Syst. 57: 362 (1922).

Eugenia sargentii Merr., Philipp. J. Sci. 18: 290 (1921). *Jossinia sargentii* (Merr.) Merr., Philipp. J. Sci. 79: 360 (1951).
Philippines. 42 PHI. Nanophan. or phan.

Eugenia sasoana Standl. & Steyerm., Publ. Field Mus. Nat. Hist., Bot. Ser. 23: 131 (1944).
Guatemala to El Salvador. 80 ELS GUA. Nanophan. or phan.

Eugenia sauvallei Krug & Urb., Bot. Jahrb. Syst. 19: 661 (1895).
W. Cuba. 81 CUB. Nanophan.

Eugenia savannarum Standl. & Steyerm., Publ. Field Mus. Nat. Hist., Bot. Ser. 23: 132 (1944).
Guatemala. 80 GUA. Nanophan. or phan.
Eugenia perpunctata Lundell, Wrightia 4: 118 (1969).
Eugenia eustipitata Lundell, Wrightia 5: 43 (1974).

Eugenia saviifolia Alain, Bull. Torrey Bot. Club 90: 190 (1963).
Puerto Rico. 81 PUE. Nanophan. or phan.

Eugenia scalariformis McVaugh, Fieldiana, Bot. 29: 218 (1956).
Colombia, N. Peru. 83 CLM PER. Nanophan. or phan.

Eugenia scaphephylla C.Wright, Anales Acad. Ci. Méd. Habana 5: 430 (1868).
Cuba. 81 CUB. Nanophan.
Eugenia sylvicola C.Wright, Anales Acad. Ci. Méd. Habana 5: 430 (1868).

Eugenia schadrackiana D.Legrand, Sellowia 13: 312 (1961).
Brazil (Santa Catarina). 84 BZS. Phan.

Eugenia schatzii J.S.Mill., Adansonia, III, 22: 114 (2000).
Madagascar. 29 MDG. Phan.

Eugenia scheffleri Engl. & Brehmer, Bot. Jahrb. Syst. 54: 331 (1917).
Tanzania (E. Usambara Mts.). 25 TAN. Nanophan. or phan.

Eugenia schottiana O.Berg in C.F.P.von Martius & auct. suc. (eds.), Fl. Bras. 14(1): 286 (1857).
Brazil (Rio de Janeiro). 84 BZL. Nanophan. or phan.

Eugenia schulziana Urb., Symb. Antill. 7: 304 (1912).
Jamaica. 81 JAM. Phan.

Eugenia schunkei McVaugh, Fieldiana, Bot. 29: 218 (1956).
Ecuador to Peru. 83 ECU PER. Phan.

Eugenia sclerocalyx D.Legrand, Sellowia 13: 311 (1961).
Brazil (E. Santa Catarina). 84 BZS. Nanophan. or phan.
Eugenia sclerocalyx var. *cambajuvensis* D.Legrand, Sellowia 13: 312 (1961).

Eugenia scottii H.Perrier, Mém. Inst. Sci. Madagascar, Sér. B, Biol. Vég. 4: 168 (1953).
C. & E. Madagascar. 29 MDG. Nanophan. or phan.

Eugenia sebastiani Urb., Symb. Antill. 9: 494 (1928).
W. Cuba. 81 CUB. Nanophan.

Eugenia seithurensis Gopalan & S.R.Sriniv., J. Bombay Nat. Hist. Soc. 100: 78 (2003).
S. India. 40 IND.

Eugenia selloi B.D.Jacks., Index Kew. 1: 911 (1893).
Brazil (Rio de Janeiro to Paraná). 84 BZL BZS. Nanophan. or phan.
**Phyllocalyx edulis* O.Berg in C.F.P.von Martius & auct. suc. (eds.), Fl. Bras. 14(1): 327 (1857). *Eugenia edulis* (O.Berg) Kiaersk., Enum. Myrt. Bras.: 162 (1893), nom. illeg.
Phyllocalyx edulis var. *depauperata* O.Berg in C.F.P.von Martius & auct. suc. (eds.), Fl. Bras. 14(1): 327 (1857).
Phyllocalyx edulis var. *dives* O.Berg in C.F.P.von Martius & auct. suc. (eds.), Fl. Bras. 14(1): 327 (1857).

Eugenia selvana Barrie, Novon 15: 42 (2005).
Costa Rica. 80 COS. Nanophan. or phan.

Eugenia serrasuela Krug & Urb., Bot. Jahrb. Syst. 19: 652 (1895).
Puerto Rico. 81 PUE. Phan.

Eugenia serrei Urb., Repert. Spec. Nov. Regni Veg. 19: 304 (1924).
W. Cuba. 81 CUB. Nanophan.

Eugenia sessiliflora Vahl in H.West, Bidr. Beskr. Ste Croix: 290 (1793). *Myrtus sessiliflora* (Vahl) Spreng., Syst. Veg. 2: 479 (1825).
Puerto Rico to Leeward Is. 81 LEE PUE. Nanophan. or phan.
Eugenia laterifolia Pers. ex DC., Prodr. 3: 273 (1828).

Eugenia sessilifolia DC., Prodr. 3: 263 (1828). *Stenocalyx sessilifolius* (DC.) O.Berg in C.F.P.von Martius & auct. suc. (eds.), Fl. Bras. 14(1): 344 (1857). *Eugenia amplectens* Nied. in H.G.A.Engler & K.A.E.Prantl, Nat. Pflanzenfam. 3(7): 81 (1893), nom. illeg.
Brazil (Bahia). 84 BZE. Nanophan. or phan.

Eugenia shaferi Urb., Symb. Antill. 7: 298 (1912).
EC. Cuba. 81 CUB. Nanophan.

Eugenia shettyana Murugan & Gopalan, Nordic J. Bot. 23: 625 (2005).
India (Kerala). 40 IND. Nanophan. or phan.

Eugenia shimishito Barrie, Novon 15: 44 (2005).
El Salvador. 80 ELS. Nanophan. or phan.

Eugenia shookii Lundell, Wrightia 2: 209 (1961).
Guatemala. 80 GUA. Nanophan. or phan.

Eugenia sieberi J.Guého & A.J.Scott, Kew Bull. 34: 481 (1980).
Mauritius. 29 MAU. Nanophan. or phan.
**Jossinia sieberiana* Blume, Mus. Bot. 1: 121 (1850).
Eugenia cotinifolia var. *gardneri* Baker, Fl. Mauritius: 114 (1877).

Eugenia siggersii Standl., Publ. Field Mus. Nat. Hist., Bot. Ser. 18: 773 (1937).
Costa Rica. 80 COS. Nanophan. or phan.

Eugenia sigillata McVaugh, Mem. New York Bot. Gard. 18(2): 211 (1969).
Guyana. 82 GUY. Phan.

Eugenia sihanakensis H.Perrier, Mém. Inst. Sci. Madagascar, Sér. B, Biol. Vég. 4: 180 (1953).
C. Madagascar. 29 MDG. Nanophan. or phan.

Eugenia siltepecana Lundell, Amer. Midl. Naturalist 19: 430 (1938).
SE. Mexico. 79 MXT. Phan.

Eugenia simiarum Standl. & Steyerm., Publ. Field Mus. Nat. Hist., Bot. Ser. 22: 359 (1940).
Guatemala. 80 GUA. Nanophan. or phan.

Eugenia sinaloae Standl., Contr. U. S. Natl. Herb. 23: 1042 (1924).
NW. Mexico. 79 MXN. Nanophan. or phan.

Eugenia singampattiana Bedd., Icon. Pl. Ind. Or. 1: 65 (1874).
India. 40 IND. Nanophan. or phan.

Eugenia sloanei Urb., Repert. Spec. Nov. Regni Veg. 14: 338 (1916).
Jamaica. 81 JAM. Nanophan.

Eugenia solimoensis O.Berg in C.F.P.von Martius & auct. suc. (eds.), Fl. Bras. 14(1): 298 (1857).
N. Brazil. 84 BZN. Nanophan. or phan.

Eugenia sonderiana O.Berg in C.F.P.von Martius & auct. suc. (eds.), Fl. Bras. 14(1): 270 (1857).
Brazil (Minas Gerais). 84 BZL. Nanophan.

Eugenia sooiana Borhidi, Acta Bot. Acad. Sci. Hung. 19: 37 (1973).
Cuba. 81 CUB. Nanophan. or phan.

Eugenia sotoesparzae P.E.Sánchez, Phytologia 61: 139 (1986).
Mexico. 79 MXG. Nanophan. or phan.

Eugenia sparsa S.Moore, Trans. Linn. Soc. London, Bot. 4: 358 (1895).
WC. Brazil. 84 BZC. Nanophan. or phan.

Eugenia speciosa Cambess. in A.F.C.de Saint-Hilaire, Fl. Bras. Merid. 2: 351 (1832). *Phyllocalyx speciosus* (Cambess.) O.Berg, Linnaea 27: 307 (1856).
Brazil. 84 BZC BZL BZS. Nanophan. or phan.
Phyllocalyx limbatus O.Berg in C.F.P.von Martius & auct. suc. (eds.), Fl. Bras. 14(1): 332 (1857). *Eugenia caldensis* Kiaersk., Enum. Myrt. Bras.: 162 (1893). *Eugenia undulatifolia* Nied. in H.G.A.Engler & K.A.E.Prantl, Nat. Pflanzenfam. 3(7): 82 (1893).
Phyllocalyx marginatus O.Berg in C.F.P.von Martius & auct. suc. (eds.), Fl. Bras. 14(1): 332 (1857).
Phyllocalyx retusus O.Berg in C.F.P.von Martius & auct. suc. (eds.), Fl. Bras. 14(1): 331 (1857). *Eugenia retusa* (O.Berg) Nied. in H.G.A.Engler & K.A.E.Prantl, Nat. Pflanzenfam. 3(7): 82 (1893).

Eugenia sphenoides O.Berg in C.F.P.von Martius & auct. suc. (eds.), Fl. Bras. 14(1): 259 (1857).
Brazil (São Paulo to Paraná). 84 BZL BZS. Nanophan. or phan.

Eugenia splendens O.Berg in C.F.P.von Martius & auct. suc. (eds.), Fl. Bras. 14(1): 226 (1857).
Brazil (Bahia to Minas Gerais). 84 BZE BZL. Nanophan. or phan.

Eugenia sprengelii DC., Prodr. 3: 276 (1828).
SE. Brazil. 84 BZL. Nanophan. or phan.
Eugenia angustifolia Spreng., Novi Provent.: 18 (1818), nom. illeg.
Myrtus linifolia Spreng., Syst. Veg. 2: 479 (1825).
Myrtus spiciflora Spreng., Syst. Veg. 4(2): 193 (1827).

Eugenia spruceana O.Berg in C.F.P.von Martius & auct. suc. (eds.), Fl. Bras. 14(1): 257 (1857).
N. Brazil to Peru. 83 PER 84 BZN. Phan.

Eugenia squamiflora Mattos, Loefgrenia 28: 1 (1969).
Brazil (São Paulo). 84 BZL. Nanophan. or phan.
**Stenocalyx stipularis* O.Berg in C.F.P.von Martius & auct. suc. (eds.), Fl. Bras. 14(1): 348 (1857). *Eugenia stipularis* (O.Berg) Mattos, Loefgrenia 94: 4 (1989), nom. illeg.

Eugenia sripadaense Kosterm., Misc. Pap. Landbouwhoogeschool 19: 227 (1980).
Sri Lanka. 40 SRL. Nanophan. or phan.

Eugenia stahlii (Kiaersk.) Krug & Urb., Bot. Jahrb. Syst. 19: 650 (1895).
Puerto Rico. 81 PUE. Nanophan. or phan.
**Myrtus stahlii* Kiaersk., Bot. Tidsskr. 17: 286 (1890).

Eugenia standleyi McVaugh, Fieldiana, Bot. 29: 458 (1963).

SW. Mexico. 79 MXT. Nanophan. or phan.
Myrtus oaxacana Standl., Contr. U. S. Natl. Herb. 23: 1038 (1924).

Eugenia staudtii Engl. & Brehmer, Bot. Jahrb. Syst. 54: 336 (1917).
Cameroon. 23 CMN. Nanophan. or phan.

Eugenia stenoptera Urb., Symb. Antill. 7: 301 (1912).
SE. Cuba. 81 CUB. Nanophan.

Eugenia stenosepala Kiaersk., Enum. Myrt. Bras.: 155 (1893).
SE. Brazil. 84 BZL. Nanophan. or phan.

Eugenia stenosepaloides Mattos, Loefgrenia 66: 8 (1975).
Brazil (Rondônia). 84 BZN. Phan.

Eugenia stenoxipha Urb., Symb. Antill. 9: 104 (1923).
E. Cuba. 81 CUB. Nanophan.

Eugenia stephanii O.Berg in C.F.P.von Martius & auct. suc. (eds.), Fl. Bras. 14(1): 242 (1857).
Brazil. 84 BZC BZL. Nanophan. or phan.
Eugenia regeliana O.Berg in C.F.P.von Martius & auct. suc. (eds.), Fl. Bras. 14(1): 573 (1859).
Eugenia stephanii var. *angustifolia* O.Berg in C.F.P.von Martius & auct. suc. (eds.), Fl. Bras. 14(1): 572 (1859).
Eugenia stephanii var. *latifolia* O.Berg in C.F.P.von Martius & auct. suc. (eds.), Fl. Bras. 14(1): 573 (1859).

Eugenia stephanophylla Baker f., J. Linn. Soc., Bot. 45: 314 (1921).
New Caledonia. 60 NWC. Nanophan. or phan.

Eugenia stereophylla Urb., Symb. Antill. 9: 98 (1923).
Cuba (Sierra de Nipe). 81 CUB. Nanophan.

Eugenia stewardsonii Britton, Bull. Torrey Bot. Club 51: 11 (1924).
Puerto Rico. 81 PUE. Phan.

Eugenia stictopetala DC., Prodr. 3: 270 (1828).
Trop. America. 80 PAN 81 LEE? TRT WIN 82 FRG GUY SUR VEN 83 PER 84 BZC BZE BZL BZN. Nanophan. or phan.
Eugenia martiusiana DC., Prodr. 3: 269 (1828).
Eugenia ochra O.Berg, Linnaea 27: 216 (1856).
Eugenia roraimana O.Berg, Linnaea 27: 219 (1856).
Eugenia tapacumensis O.Berg, Linnaea 27: 222 (1856).
Eugenia tapacumensis var. *angustifolia* O.Berg, Linnaea 27: 222 (1856).
Eugenia tapacumensis var. *latifolia* O.Berg, Linnaea 27: 222 (1856).
Eugenia piauhiensis O.Berg in C.F.P.von Martius & auct. suc. (eds.), Fl. Bras. 14(1): 285 (1857).
Eugenia doniana O.Berg in C.F.P.von Martius & auct. suc. (eds.), Fl. Bras. 14(1): 643 (1859).
Eugenia eschholtziana O.Berg in C.F.P.von Martius & auct. suc. (eds.), Fl. Bras. 14(1): 582 (1859).
Eugenia eschholtziana var. *angustifolia* O.Berg in C.F.P.von Martius & auct. suc. (eds.), Fl. Bras. 14(1): 582 (1859).
Eugenia eschholtziana var. *latifolia* O.Berg in C.F.P. von Martius & auct. suc. (eds.), Fl. Bras. 14(1): 582 (1859).

Eugenia stigmatosa DC., Prodr. 3: 268 (1828).
E. & S. Brazil. 84 BZE BZL BZS. Nanophan. or phan.
Eugenia stigmatosa var. *acutata* O.Berg in C.F.P.von Martius & auct. suc. (eds.), Fl. Bras. 14(1): 252 (1857).
Eugenia stigmatosa var. *obtusata* O.Berg in C.F.P.von Martius & auct. suc. (eds.), Fl. Bras. 14(1): 251 (1857).

Eugenia stipitata McVaugh, Fieldiana, Bot. 29: 219 (1956).
N. Brazil to W. South America. 83 BOL CLM ECU PER 84 BZN. Nanophan. or phan.

subsp. ***sororia*** McVaugh, Fieldiana, Bot. 29: 220 (1956).
Ecuador to Peru. 83 ECU PER. Phan.

subsp. ***stipitata***
N. Brazil to W. South America. 83 BOL CLM ECU PER 84 BZN. Nanophan. or phan.
Eugenia stipitata subsp. *sororia* McVaugh, Fieldiana, Bot. 29: 219 (1956).

Eugenia strellensis O.Berg in C.F.P.von Martius & auct. suc. (eds.), Fl. Bras. 14(1): 256 (1857).
Brazil (Rio de Janeiro). 84 BZL. Nanophan. or phan.

Eugenia stricta Pancher ex Brongn. & Gris, Bull. Soc. Bot. France 12: 179 (1865). *Austromyrtus stricta* (Pancher ex Brongn. & Gris) Burret, Notizbl. Bot. Gart. Berlin-Dahlem 15: 505 (1941).
New Caledonia. 60 NWC. Nanophan. or phan.

Eugenia strictissima Govaerts, World Checklist Myrtaceae: 166 (2008).
Brazil (Rio de Janeiro to Paraná). 84 BZL BZS. Nanophan. or phan.
Phyllocalyx strictus O.Berg in C.F.P.von Martius & auct. suc. (eds.), Fl. Bras. 14(1): 330 (1857). *Eugenia stricta* (O.Berg) Kiaersk., Enum. Myrt. Bras.: 163 (1893), nom. illeg.

Eugenia strigipes O.Berg in C.F.P.von Martius & auct. suc. (eds.), Fl. Bras. 14(1): 263 (1857).
Brazil (Rio de Janeiro). 84 BZL.

Eugenia sturrockii R.A.Howard, J. Arnold Arbor. 28: 124 (1947).
Cuba (Sierra de Moa). 81 CUB. Nanophan. or phan.

Eugenia stylaris McVaugh, Mem. New York Bot. Gard. 18(2): 212 (1969).
N. Brazil. 84 BZN. Phan.

Eugenia subamplexicaulis DC., Prodr. 3: 277 (1828).
SE. Brazil. 84 BZL. Nanophan. or phan.

Eugenia subavenia O.Berg in C.F.P.von Martius & auct. suc. (eds.), Fl. Bras. 14(1): 571 (1859).
Brazil (S. Minas Gerais to Santa Catarina). 84 BZL BZS. Phan.

Eugenia subcordata Wight & Arn., Prodr. Fl. Ind. Orient.: 331 (1834).
India. 40 IND.

Eugenia subdisticha Urb., Symb. Antill. 7: 308 (1912).
C. Cuba. 81 CUB. Nanophan.

Eugenia suberosa Cambess. in A.F.C.de Saint-Hilaire, Fl. Bras. Merid. 2: 364 (1833).
Brazil to Paraguay. 84 BZC BZL 85 PAR. Nanophan. or phan.
Eugenia quaternifolia Cambess. in A.F.C.de Saint-Hilaire, Fl. Bras. Merid. 2: 363 (1833).
Eugenia atrata Mattos & D.Legrand, Loefgrenia 67: 26 (1975).

Eugenia subherbacea A.Chev., Bull. Soc. Bot. France 58(8): 170 (1911 publ. 1912).

Ivory Coast to Central African Rep. 22 GHA IVO TOG 23 CAF. Hemicr. or cham.
Eugenia herbacea A.Chev., Bull. Soc. Bot. France 55(8): 36 (1908), nom. illeg.

Eugenia subreticulata Glaz., Bull. Soc. Bot. France 52(3): 233 (1905).
Brazil (Minas Gerais). 84 BZL. Nanophan. or phan.
Eugenia ligustrina Miq., Linnaea 22: 532 (1849), nom. illeg. **Stenocalyx reticulatus* O.Berg in C.F.P.von Martius & auct. suc. (eds.), Fl. Bras. 14(1): 340 (1857). *Eugenia angustisepala* Mattos, Loefgrenia 94: 3 (1989), nom. illeg.

Eugenia subspinulosa Borhidi & O.Muñiz, Acta Bot. Acad. Sci. Hung. 19: 38 (1973).
Cuba. 81 CUB. Nanophan. or phan.

Eugenia subterminalis DC., Prodr. 3: 263 (1828).
W. South America to N. Brazil. 83 BOL CLM ECU PER 84 BZN. Nanophan. or phan.

Eugenia subundulata Kiaersk., Enum. Myrt. Bras.: 129 (1893).
SE. Brazil. 84 BZL. Nanophan. or phan.

var. ***longipetiolata*** Mattos, Loefgrenia 99: 4 (1990).
Brazil (Rio de Janeiro). 84 BZL. Nanophan. or phan.

var. ***subundulata***
SE. Brazil. 84 BZL. Nanophan. or phan.
Eugenia mooniana Gardner, London J. Bot. 2: 352 (1843), nom. illeg. *Eugenia neomooniana* Sobral, Napaea 11: 36 (1995).

Eugenia sulcata Spring ex Mart., Flora 20(2 Beibl.): 85 (1837). *Stenocalyx sulcatus* (Spring ex Mart.) O.Berg in C.F.P.von Martius & auct. suc. (eds.), Fl. Bras. 14(1): 339 (1857).
Brazil (Rio de Janeiro to Santa Catarina). 84 BZL BZS. Nanophan.
Stenocalyx sulcatus var. *latifolius* O.Berg in C.F.P.von Martius & auct. suc. (eds.), Fl. Bras. 14(1): 339 (1857).
Stenocalyx sulcatus var. *ligustrinus* O.Berg in C.F.P.von Martius & auct. suc. (eds.), Fl. Bras. 14(1): 339 (1857).
Stenocalyx sulcatus var. *longifolius* O.Berg in C.F.P. von Martius & auct. suc. (eds.), Fl. Bras. 14(1): 339 (1857).
Stenocalyx sulcatus var. *strictus* O.Berg in C.F.P.von Martius & auct. suc. (eds.), Fl. Bras. 14(1): 339 (1857). *Eugenia sulcata* var. *stricta* (O.Berg) Mattos, Loefgrenia 64: 4 (1975).
Eugenia sulcata var. *latifolia* Kiaersk., Enum. Myrt. Bras.: 156 (1893).
Eugenia sulcata var. *pubescens* Mattos, Loefgrenia 119: 4 (2004).

Eugenia sulcivenia Krug & Urb., Bot. Jahrb. Syst. 19: 632 (1895).
SE. Jamaica. 81 JAM. Nanophan. or phan.

Eugenia sumbensis Greves, J. Bot. 66(Suppl. 1): 175 (1928).
WC. & S. Trop. Africa. 23 CAB ZAI 26 ANG ZIM?. Nanophan. or phan.

Eugenia supraaxillaris Spreng., Flora 20(2 Beibl.)P: 83 (1837).
E. Brazil. 84 BZE BZL. Phan.
**Eugenia axillaris* Vell., Fl. Flumin. 5: 209, t. 41 (1829), nom. illeg.
Eugenia cambucarana Kiaersk., Enum. Myrt. Bras.: 151 (1893).

Eugenia symphoricarpos McVaugh, Fieldiana, Bot. 29: 458 (1963).
Mexico. 79 MXE MXG. Nanophan. or phan.

Eugenia tachirensis Steyerm., Fieldiana, Bot. 28: 1014 (1957).
Venezuela. 82 VEN. Nanophan. or phan.

Eugenia tafelbergica Amshoff, Bull. Torrey Bot. Club 75: 535 (1948).
Guyana to Suriname. 82 GUY SUR. Nanophan. or phan.

Eugenia talbotii Keay, Kew Bull. 8: 288 (1953).
Nigeria. 22 NGA. Nanophan.

Eugenia tamaensis Steyerm., Fieldiana, Bot. 28: 1014 (1957).
Venezuela. 82 VEN. Nanophan. or phan.

Eugenia tanaensis Verdc., Kew Bull. 54: 44 (1999).
E. Trop. Africa. 25 KEN TAN. Nanophan.

Eugenia tapirorum Standl., Publ. Field Mus. Nat. Hist., Bot. Ser. 9: 321 (1940).
Honduras. 80 HON.

Eugenia teapensis McVaugh, Fieldiana, Bot. 29: 459 (1963).
SE. Mexico. 79 MXT. Nanophan. or phan.

Eugenia tenuimarginata McVaugh, Fieldiana, Bot. 29: 220 (1956).
Peru. 83 PER. Nanophan. or phan.

Eugenia tenuipedunculata Kiaersk., Enum. Myrt. Bras.: 131 (1893).
Brazil (S. Minas Gerais, São Paulo). 84 BZL. Nanophan. or phan.

Eugenia tenuissima Lundell, Wrightia 3: 19 (1961).
Mexico (Oaxaca). 79 MXS. Nanophan.

Eugenia tepuiensis Steyerm., Fieldiana, Bot. 28: 1015 (1957).
Venezuela. 82 VEN. Nanophan. or phan.

Eugenia teresae J.F.Morales, Phytologia 81: 361 (1996 publ. 1997).
Costa Rica. 80 COS. Nanophan. or phan.

Eugenia ternatifolia Cambess. in A.F.C.de Saint-Hilaire, Fl. Bras. Merid. 2: 336 (1832). *Pilothecium ternatifolium* (Cambess.) Kausel, Ark. Bot., a.s., 4: 402 (1962).
Brazil to Paraguay. 84 BZC BZE BZL BZS 85 PAR. Nanophan. or phan.
Myrtus beaurepairiana Kiaersk., Enum. Myrt. Bras.: 22 (1893). *Eugenia beaurepairiana* (Kiaersk.) D.Legrand, Sellowia 13: 308 (1961). *Pilothecium beaurepairianum* (Kiaersk.) Kausel, Ark. Bot., a.s., 4: 404 (1962).

Eugenia terpnophylla Thwaites, Enum. Pl. Zeyl.: 114 (1859).
Sri Lanka. 40 SRL. Phan.

Eugenia tetramera (McVaugh) M.L.Kawas. & B.K.Holst, Brittonia 46: 142 (1994).
N. South America. 82 FRG GUY SUR. Nanophan. or phan.
**Campomanesia tetramera* McVaugh, Mem. New York Bot. Gard. 18(2): 239 (1969).

Eugenia tetrasticha Poepp. ex O.Berg in C.F.P.von Martius & auct. suc. (eds.), Fl. Bras. 14(1): 324 (1857).
Peru (Loreto). 83 PER. Nanophan. or phan.

Eugenia theodorae Kiaersk., Enum. Myrt. Bras.: 160 (1893).
SE. Brazil. 84 BZL. Nanophan. or phan.
Eugenia theodorae var. *brevipedunculata* Kiaersk., Enum. Myrt. Bras.: 161 (1893).

Eugenia thikaensis Verdc., Kew Bull. 54: 56 (1999).
Kenya (Fort Hall Distr.). 25 KEN. Nanophan.

Eugenia thollonii Amshoff, Acta Bot. Neerl. 7: 53 (1958).
Gabon. 23 GAB. Nanophan.

Eugenia thomasiana O.Berg, Linnaea 27: 183 (1856).
Leeward Is. (St. Thomas). 81 LEE. Nanophan. or phan.

Eugenia thompsonii Merr., Philipp. J. Sci., C 9: 121 (1914).
Marianas. 62 MRN. Nanophan. or phan.

Eugenia thouvenotiana H.Perrier, Mém. Inst. Sci. Madagascar, Sér. B, Biol. Vég. 4: 182 (1953).
EC. Madagascar. 29 MDG. Phan.

Eugenia tikalana Lundell, Wrightia 2: 210 (1961).
SE. Mexico to El Salvador. 79 MXT 80 ELS GUA HON. Nanophan. or phan.

Eugenia tilarana Barrie, Novon 15: 45 (2005).
Costa Rica. 80 COS. Nanophan. or phan.

Eugenia tinguyensis Cambess. in A.F.C.de Saint-Hilaire, Fl. Bras. Merid. 2: 358 (1833).
E. & S. Brazil. 84 BZE BZL BZS. Nanophan. or phan.
Eugenia sericea O.Berg in C.F.P.von Martius & auct. suc. (eds.), Fl. Bras. 14(1): 280 (1857).
Eugenia vicozensis O.Berg in C.F.P.von Martius & auct. suc. (eds.), Fl. Bras. 14(1): 279 (1857).
Eugenia vicozensis var. *glabrata* O.Berg in C.F.P.von Martius & auct. suc. (eds.), Fl. Bras. 14(1): 280 (1857).
Eugenia vicozensis var. *sericea* O.Berg in C.F.P.von Martius & auct. suc. (eds.), Fl. Bras. 14(1): 279 (1857).
Eugenia microcarpa O.Berg in C.F.P.von Martius & auct. suc. (eds.), Fl. Bras. 14(1): 574 (1859).
Eugenia sericea var. *angustifolia* O.Berg in C.F.P.von Martius & auct. suc. (eds.), Fl. Bras. 14(1): 579 (1859).
Eugenia sericea var. *robusta* O.Berg in C.F.P.von Martius & auct. suc. (eds.), Fl. Bras. 14(1): 579 (1859).

Eugenia tinifolia Lam., Encycl. 3: 202 (1789). *Myrtus tinifolia* (Lam.) Spreng., Syst. Veg. 2: 481 (1825). *Jossinia tinifolia* (Lam.) DC., Prodr. 3: 238 (1828).
Mauritius. 29 MAU. Nanophan. or phan.
Jossinia ferruginea Bojer, Hortus Maurit.: 141 (1837), nom. inval.
Jossinia revoluta Bojer, Hortus Maurit.: 140 (1837), nom. inval.

Eugenia tisserantii Aubrév. & Pellegr., Notul. Syst. (Paris) 14: 61 (1950).
Central African Rep. 23 CAF. Nanophan. or phan.

Eugenia toaensis Borhidi & O.Muñiz, Bot. Közlem. 64: 216 (1977 publ. 1978).
Cuba. 81 CUB. Nanophan. or phan.

Eugenia tocaiana O.Berg in C.F.P.von Martius & auct. suc. (eds.), Fl. Bras. 14(1): 269 (1857).
Brazil. 84+. Nanophan. or phan.

Eugenia togoensis Engl., Notizbl. Königl. Bot. Gart. Berlin 2: 288 (1899). *Myrtus togoensis* (Engl.) Kuntze, Deutsche Bot. Monatsschr. 21: 173 (1903).
Togo. 22 TOG. Nanophan.

Eugenia toledinensis Lundell, Phytologia 1: 307 (1939).
Belize. 80 BLZ. Phan.

Eugenia tomasina Urb., Symb. Antill. 9: 492 (1928).
W. Cuba. 81 CUB. Nanophan.

Eugenia tonii Lundell, Wrightia 4: 95 (1968).
SE. Mexico. 79 MXT. Phan.

Eugenia toxanatolica Verdc., Kew Bull. 54: 49 (1999).
Tanzania. 25 TAN. Nanophan. or phan.

Eugenia trahyra Barb.Rodr., Contr. Jard. Bot. Rio de Janeiro 4: 990 (1907).
SE. Brazil. 84 BZL. Nanophan. or phan.

Eugenia triflora (Jacq.) Ham., Prodr. Pl. Ind. Occid.: 44 (1825).
Colombia. 83 CLM. Nanophan. or phan.
**Myrtus triflora* Jacq., Enum. Syst. Pl.: 23 (1760). *Pimentus triflora* (Jacq.) Raf., Sylva Tellur.: 105 (1838).

Eugenia trikii Lundell, Wrightia 2: 211 (1961).
Belize to Guatemala. 80 BLZ GUA. Nanophan.

Eugenia trinervia Vahl, Eclog. Amer. 2: 36 (1798).
Lesser Antilles to N. Brazil. 81 LEE TRT WIN 82 FRG GUY VEN 84 BZN. Phan.
Myrtus vahlii Spreng., Syst. Veg. 2: 486 (1825). *Eugenia vahlii* DC., Prodr. 3: 279 (1828).
Eugenia brachystachys O.Berg, Linnaea 27: 301 (1856).
Eugenia perplexans R.O.Williams, Fl. Trinidad 1: 350 (1934).

Eugenia trinitatis DC., Prodr. 3: 280 (1828).
Windward Is., Trinidad ?. 81 TRT? WIN. Nanophan. or phan.
Eugenia grenadensis Urb., Repert. Spec. Nov. Regni Veg. 14: 337 (1916).

Eugenia tropophylla H.Perrier, Mém. Inst. Sci. Madagascar, Sér. B, Biol. Vég. 4: 167 (1953).
W. Madagascar. 29 MDG. Nanophan. or phan.
Eugenia tropophylla var. *majoriflora* H.Perrier, Mém. Inst. Sci. Madagascar, Sér. B, Biol. Vég. 4: 167 (1952).
Eugenia tropophylla var. *piriformis* H.Perrier, Mém. Inst. Sci. Madagascar, Sér. B, Biol. Vég. 4: 167 (1952).

Eugenia truncata O.Berg, Linnaea 27: 157 (1856).
Costa Rica. 80 COS. Nanophan. or phan.
Eugenia guanacastensis Standl., Publ. Field Mus. Nat. Hist., Bot. Ser. 8: 144 (1930).

Eugenia trunciflora (Schltdl. & Cham.) G.Don, Gen. Hist. 2: 867 (1832).
Mexico. 79 MXG. Nanophan. or phan.
**Myrtus trunciflora* Schltdl. & Cham., Linnaea 5: 561 (1830). *Pseudanamomis trunciflora* (Schltdl. & Cham.) Bisse, Feddes Repert. 96: 511 (1985).

Eugenia tuberculata (Kunth) DC., Prodr. 3: 273 (1828).
Cuba. 81 CUB. Nanophan.
**Myrtus tuberculata* Kunth in F.W.H.von Humboldt,

A.J.A.Bonpland & C.S.Kunth, Nov. Gen. Sp. 6: 148 (1823).
Eugenia revoluta Willd. ex O.Berg, Linnaea 27: 184 (1856).
Eugenia tuberculata var. *uniflora* O.Berg, Linnaea 30: 682 (1861).

Eugenia tulanan Merr., Philipp. J. Sci., C 11: 201 (1916). *Jossinia tulanan* (Merr.) Merr., Philipp. J. Sci. 79: 360 (1951).
Philippines. 42 PHI. Nanophan. or phan.

Eugenia tumulescens McVaugh, Fieldiana, Bot. 29: 221 (1956).
S. Venezuela to N. Brazil. 82 VEN 84 BZN. Nanophan. or phan.

Eugenia tungo Hiern, Cat. Afr. Pl. 1: 362 (1898).
Angola. 26 ANG. Nanophan. or phan.

Eugenia turneri McVaugh, Fieldiana, Bot. 29: 460 (1963).
SW. Mexico. 79 MXS. Nanophan. or phan.

Eugenia ulei (Diels) McVaugh, Publ. Field Mus. Nat. Hist., Bot. Ser. 13(4): 742 (1958).
Peru. 83 PER. Nanophan. or phan.
**Psidium ulei* Diels, Verh. Bot. Vereins Prov. Brandenburg 48: 187 (1906 publ. 1907).

Eugenia uliginosa Lundell, Wrightia 3: 20 (1961).
SE. Mexico. 79 MXT. Nanophan. or phan.

Eugenia umbelliflora O.Berg in C.F.P.von Martius & auct. suc. (eds.), Fl. Bras. 14(1): 290 (1857).
Brazil (São Paulo to Santa Catarina). 84 BZL BZS. Nanophan. or phan.

Eugenia umbonata McVaugh, Mem. New York Bot. Gard. 18(2): 215 (1969).
Venezuela (Amazonas). 82 VEN. Nanophan. or phan.

Eugenia umbrosa O.Berg in C.F.P.von Martius & auct. suc. (eds.), Fl. Bras. 14(1): 582 (1859).
SE. Brazil. 84 BZL. Nanophan. or phan.
Eugenia crassiflora Kiaersk., Enum. Myrt. Bras.: 132 (1893).
Eugenia robustovenosa Kiaersk., Enum. Myrt. Bras.: 137 (1893).

Eugenia umtamvunensis A.E.van Wyk, S. African J. Bot. 1: 158 (1982).
Cape Prov. to KwaZulu-Natal. 27 CPP NAT. Nanophan. or phan.

Eugenia underwoodii Britton, Bull. Torrey Bot. Club 51: 10 (1924).
Puerto Rico. 81 PUE. Nanophan.

Eugenia undulata Aubl., Hist. Pl. Guiane 1: 508 (1775). *Myrtus undulata* (Aubl.) Spreng., Syst. Veg. 2: 482 (1825).
Guianas. 82 FRG SUR. Nanophan. or phan.
Eugenia arivoa Aubl., Hist. Pl. Guiane 1: 510 (1775).

Eugenia uniflora L., Sp. Pl.: 470 (1753). *Plinia petiolata* L., Amoen. Acad., Schreb. ed. 8: 257 (1785), nom. illeg. *Stenocalyx uniflorus* (L.) Kausel, Lilloa 32: 331 (1967).
E. & S. Brazil to S. South America. (22) ben (23) gab (61) sci (80) blz (81) cub dom lee trt win (82) sur 83 BOL 84 BZE BZL BZS 85 AGE AGW PAR URU. Nanophan. or phan. Widely cultivated for its edible fruits (Surinam cherry).
Myrtus brasiliana L., Sp. Pl.: 701 (1753). *Myrtus brasiliana* var. *normalis* Kuntze, Revis. Gen. Pl. 3(2): 90 (1898), nom. inval.
Plinia rubra L., Mant. Pl. 2: 243 (1771). *Stenocalyx ruber* (L.) Kausel, Ark. Bot., a.s., 3: 495 (1956).
Plinia tetrapetala L., Mant. Pl. 2: 402 (1771).
Plinia pedunculata L.f., Suppl. Pl.: 253 (1782).
Eugenia michelii Lam., Encycl. 3: 203 (1789). *Stenocalyx michelii* (Lam.) O.Berg in C.F.P.von Martius & auct. suc. (eds.), Fl. Bras. 14(1): 337 (1857). *Syzygium michelii* (Lam.) Duthie in J.D.Hooker, Fl. Brit. India 2: 505 (1879).
Eugenia myrtifolia Salisb., Prodr. Stirp. Chap. Allerton: 353 (1796).
Eugenia zeylanica Willd., Sp. Pl. 2: 963 (1799). *Myrtus willdenowii* Spreng., Syst. Veg. 2: 480 (1825). *Eugenia willdenowii* (Spreng.) DC., Prodr. 3: 265 (1828).
Eugenia costata Cambess. in A.F.C.de Saint-Hilaire, Fl. Bras. Merid. 2: 359 (1833). *Stenocalyx costatus* (Cambess.) O.Berg in C.F.P.von Martius & auct. suc. (eds.), Fl. Bras. 14(1): 340 (1857). *Luma costata* (Cambess.) Herter, Revista Sudamer. Bot. 7: 219 (1943).
Stenocalyx affinis O.Berg in C.F.P.von Martius & auct. suc. (eds.), Fl. Bras. 14(1): 336 (1857).
Stenocalyx brunneus O.Berg in C.F.P.von Martius & auct. suc. (eds.), Fl. Bras. 14(1): 335 (1857).
Stenocalyx dasyblastus O.Berg in C.F.P.von Martius & auct. suc. (eds.), Fl. Bras. 14(1): 337 (1857). *Eugenia dasyblasta* (O.Berg) Nied. in H.G.A.Engler & K.A.E.Prantl, Nat. Pflanzenfam. 3(7): 82 (1893). *Luma dasyblasta* (O.Berg) Herter, Revista Sudamer. Bot. 7: 219 (1943).
Stenocalyx glaber O.Berg in C.F.P.von Martius & auct. suc. (eds.), Fl. Bras. 14(1): 337 (1857).
Stenocalyx impunctatus O.Berg in C.F.P.von Martius & auct. suc. (eds.), Fl. Bras. 14(1): 336 (1857).
Stenocalyx lucidus O.Berg in C.F.P.von Martius & auct. suc. (eds.), Fl. Bras. 14(1): 337 (1857). *Myrtus brasiliana* var. *lucida* (O.Berg) Kuntze, Revis. Gen. Pl. 3(2): 90 (1898).
Stenocalyx michelii var. *membranacea* O.Berg in C.F.P.von Martius & auct. suc. (eds.), Fl. Bras. 14(1): 338 (1857).
Stenocalyx michelii var. *rigida* O.Berg in C.F.P.von Martius & auct. suc. (eds.), Fl. Bras. 14(1): 338 (1857).
Stenocalyx oblongifolius O.Berg in C.F.P.von Martius & auct. suc. (eds.), Fl. Bras. 14(1): 339 (1857). *Eugenia oblongifolia* (O.Berg) Nied. in H.G.A. Engler & K.A.E.Prantl, Nat. Pflanzenfam. 3(7): 82 (1893), nom. illeg. *Eugenia oblongifolia* (O.Berg) Arechav., Anales Mus. Nac. Montevideo 5: 62 (1905). *Eugenia arechavaletae* Herter, Estud. Bot. Reg. Urug. 4: 93 (1931). *Luma arechavaletae* (Herter) Herter, Revista Sudamer. Bot. 7: 218 (1943).
Stenocalyx strigosus O.Berg in C.F.P.von Martius & auct. suc. (eds.), Fl. Bras. 14(1): 335 (1857). *Eugenia strigosa* (O.Berg) Arechav., Anales Mus. Nac. Montevideo 5: 60 (1905). *Luma strigosa* (O.Berg) Herter, Revista Sudamer. Bot. 7: 220 (1943).
Myrtus brasiliana var. *diversifolia* Kuntze, Revis. Gen. Pl. 3(2): 90 (1898).
Myrtus brasiliana var. *lanceolata* Kuntze, Revis. Gen. Pl. 3(2): 90 (1898).
Stenocalyx rhampiri Barb.Rodr., Myrt. Paraguay: 8 (1903).

Eugenia decidua Merr., Philipp. J. Sci., C 9: 121 (1914).
Eugenia uniflora var. *atropurpurea* Mattos, Loefgrenia 85: 2 (1984).

Eugenia uninervia Rusby, Mem. Torrey Bot. Club 6: 36 (1895).
Bolivia. 83 BOL. Nanophan. or phan.

Eugenia urophylla Welw. ex Hiern, Cat. Afr. Pl. 1: 360 (1898).
Angola. 26 ANG. Nanophan. or phan.

Eugenia urschiana H.Perrier, Mém. Inst. Sci. Madagascar, Sér. B, Biol. Vég. 4: 171 (1953).
C. Madagascar. 29 MDG. Phan.
Eugenia urschiana var. *subacuminata* H.Perrier, Mém. Inst. Sci. Madagascar, Sér. B, Biol. Vég. 4: 171 (1952).

Eugenia ursina Lundell, Wrightia 3: 116 (1964).
Guatemala. 80 GUA. Nanophan. or phan.

Eugenia uruguayensis Cambess. in A.F.C.de Saint-Hilaire, Fl. Bras. Merid. 2: 362 (1833). *Luma uruguayensis* (Cambess.) Herter, Revista Sudamer. Bot. 7: 220 (1943).
Paraguay to Brazil. 84 BZC BZL BZS 85 AGE PAR URU. Phan.
Eugenia batucaryensis O.Berg, Linnaea 27: 262 (1856).
Eugenia calycosema O.Berg in C.F.P.von Martius & auct. suc. (eds.), Fl. Bras. 14(1): 276 (1857). *Luma calycosema* (O.Berg) Herter, Revista Sudamer. Bot. 7: 218 (1943).
Eugenia guabiju O.Berg in C.F.P.von Martius & auct. suc. (eds.), Fl. Bras. 14(1): 251 (1857).
Eugenia maschalantha O.Berg in C.F.P.von Martius & auct. suc. (eds.), Fl. Bras. 14(1): 278 (1857).
Eugenia opaca O.Berg in C.F.P.von Martius & auct. suc. (eds.), Fl. Bras. 14(1): 278 (1857), nom. illeg. *Luma opaca* (O.Berg) Herter, Revista Sudamer. Bot. 7: 219 (1943).
Eugenia opaca var. *brasiliensis* O.Berg in C.F.P.von Martius & auct. suc. (eds.), Fl. Bras. 14(1): 278 (1857).
Eugenia opaca var. *montevideensis* O.Berg in C.F.P. von Martius & auct. suc. (eds.), Fl. Bras. 14(1): 278 (1857).

Eugenia uxpanapensis P.E.Sánchez & L.M.Ortega, Phytologia 63: 404 (1987).
Mexico. 79 MXG. Phan.

Eugenia vacana Lundell, Phytologia 1: 308 (1939).
Belize. 80 BLZ. Phan.

Eugenia valvata McVaugh, Fieldiana, Bot. 29: 221 (1956).
Ecuador. 83 ECU. Nanophan. or phan.

Eugenia vanderveldei Urb. & Ekman, Ark. Bot. 21A(5): 33 (1927).
Haiti (Massif de la Hotte). 81 HAI. Nanophan. or phan.

Eugenia varia Britton & P.Wilson, Mem. Torrey Bot. Club 16: 89 (1920).
W. & WC. Cuba. 81 CUB. Nanophan.

Eugenia variabilis Baill., Hist. Pl. 6: 341 (1876).
Brazil. 84+. Nanophan. or phan.

Eugenia variareolata McVaugh, Fieldiana, Bot. 29: 222 (1956).
Colombia. 83 CLM. Nanophan. or phan.

Eugenia vatomandrensis H.Perrier, Mém. Inst. Sci. Madagascar, Sér. B, Biol. Vég. 4: 175 (1953).
NE. Madagascar. 29 MDG. Nanophan. or phan.

Eugenia vaughanii J.Guého & A.J.Scott, Kew Bull. 34: 482 (1980).
Mauritius. 29 MAU. Nanophan. or phan.

Eugenia venezuelensis O.Berg, Linnaea 27: 188 (1856).
Mexico to Venezuela. 79 MXE MXG MXT 80 BLZ COS ELS GUA NIC PAN 82 VEN 83 CLM. Nanophan. or phan.
Eugenia origanoides O.Berg, Linnaea 29: 229 (1858).
Eugenia conglobata Sessé & Moç., Pl. Nov. Hisp.: 83 (1888).
Eugenia banghamii Standl., J. Arnold Arbor. 11: 125 (1930).
Eugenia cobanensis Lundell, Wrightia 5: 69 (1974).

Eugenia verapazensis Lundell, Wrightia 5: 3 (1972).
Guatemala. 80 GUA. Nanophan. or phan.

Eugenia verdoorniae A.E.van Wyk, J. S. African Bot. 45: 273 (1979).
Cape Prov. to KwaZulu-Natal. 27 CPP NAT. Nanophan. or phan.

Eugenia vernicosa O.Berg in C.F.P.von Martius & auct. suc. (eds.), Fl. Bras. 14(1): 274 (1857).
Brazil (Bahia). 84 BZE. Nanophan. or phan.

Eugenia verruculata Barrie, Novon 15: 45 (2005).
Costa Rica. 80 COS. Nanophan. or phan.

Eugenia versicolor McVaugh, Fieldiana, Bot. 29: 223 (1956).
Colombia to N. Brazil. 83 CLM 84 BZN. Nanophan. or phan.

Eugenia verticillaris O.Berg in C.F.P.von Martius & auct. suc. (eds.), Fl. Bras. 14(1): 581 (1859).
Brazil (Minas Gerais). 84 BZL. Nanophan. or phan.

Eugenia verticillata (Vell.) Angely, Fl. Analit. Fitogeogr. São Paulo 3–4: 560 (1970).
SE. & S. Brazil. 84 BZL BZS. Nanophan. or phan.
**Myrtus verticillata* Vell., Fl. Flumin. 5: 215, t. 66 (1829). *Eugenia riedeliana* O.Berg in C.F.P.von Martius & auct. suc. (eds.), Fl. Bras. 14(1): 261 (1857), nom. illeg.
Eugenia riedeliana var. *ferruginea* O.Berg in C.F.P.von Martius & auct. suc. (eds.), Fl. Bras. 14(1): 261 (1857).
Eugenia riedeliana var. *ochracea* O.Berg in C.F.P.von Martius & auct. suc. (eds.), Fl. Bras. 14(1): 261 (1857).
Eugenia schuechiana O.Berg in C.F.P.von Martius & auct. suc. (eds.), Fl. Bras. 14(1): 258 (1857).
Eugenia schuchiana var. *grandifolia* O.Berg in C.F.P. von Martius & auct. suc. (eds.), Fl. Bras. 14(1): 576 (1859).
Eugenia pseudoverticillata Kiaersk., Enum. Myrt. Bras.: 140 (1893).
Eugenia brevipaniculata Merr., Philipp. J. Sci., C 11: 23 (1916).

Eugenia vesca Lundell, Wrightia 2: 212 (1961).
Guatemala. 80 GUA. Nanophan. or phan.
Eugenia cumbreana Lundell, Wrightia 44: 179 (1971).

Eugenia vetula DC., Prodr. 3: 271 (1828).
Brazil (Bahia to Minas Gerais). 84 BZE BZL. Nanophan. or phan.

Eugenia victoriana Cuatrec., Mutisia 32: 6 (1970).
Colombia. 83 CLM. Nanophan. or phan.

Eugenia victorinii Alain, Contr. Ocas. Mus. Hist. Nat. Colegio "De La Salle" 12: 11 (1953). *Pseudanamomis victorinii* (Alain) Bisse, Feddes Repert. 96: 511 (1985).
Cuba (I. de la Juventud). 81 CUB. Cham.

Eugenia vigiensis Urb., Ark. Bot. 20A(5): 25 (1926).
Haiti. 81 HAI. Nanophan.

Eugenia viguieriana H.Perrier, Mém. Inst. Sci. Madagascar, Sér. B, Biol. Vég. 4: 167 (1953).
C. Madagascar. 29 MDG. Nanophan. or phan.

Eugenia vilersii H.Perrier, Mém. Inst. Sci. Madagascar, Sér. B, Biol. Vég. 4: 165 (1953).
E. Madagascar. 29 MDG. Nanophan. or phan.

Eugenia villae-novae Kiaersk., Enum. Myrt. Bras.: 141 (1893).
Brazil (Rio de Janeiro). 84 BZL. Nanophan. or phan.

Eugenia violascens O.Berg in C.F.P.von Martius & auct. suc. (eds.), Fl. Bras. 14(1): 581 (1859).
Brazil (Rio de Janeiro). 84 BZL. Nanophan. or phan.

Eugenia viridiflora Cambess. in A.F.C.de Saint-Hilaire, Fl. Bras. Merid. 2: 344 (1832).
Brazil (N. Minas Gerais). 84 BZL. Nanophan.
Eugenia viridiflora var. *elliptica* O.Berg in C.F.P.von Martius & auct. suc. (eds.), Fl. Bras. 14(1): 265 (1857).
Eugenia viridiflora var. *oblongata* O.Berg in C.F.P.von Martius & auct. suc. (eds.), Fl. Bras. 14(1): 265 (1857).

Eugenia viridis (Vell.) O.Berg, Linnaea 27: 302 (1856).
Brazil (Rio de Janeiro). 84 BZL. Nanophan. or phan.
Myrtus viridis Vell., Fl. Flumin. 5: 217, t. 74 (1829).

Eugenia virotii Guillaumin, Mém. Mus. Natl. Hist. Nat., B, Bot. 4: 35 (1953).
New Caledonia. 60 NWC. Nanophan. or phan.

Eugenia warmingiana Kiaersk., Enum. Myrt. Bras.: 138 (1893).
NE. Brazil. Nanophan. or phan.
Eugenia bimarginata O.Berg in C.F.P.von Martius & auct. suc. (eds.), Fl. Bras. 14(1): 291 (1857), nom. illeg.
Eugenia cauliflora O.Berg in C.F.P.von Martius & auct. suc. (eds.), Fl. Bras. 14(1): 268 (1857), nom. illeg. *Eugenia ayacuchae* Steyerm., Fieldiana, Bot. 28: 1012 (1957). *Eugenia portomite* D.Legrand, Bradea 2: 6 (1975), nom. illeg.

Eugenia websteri Proctor, J. Arnold Arbor. 63: 276 (1982).
Jamaica. 81 JAM.

Eugenia wentii Amshoff, Recueil Trav. Bot. Néerl. 39: 158 (1942).
S. Trop. America. 82 FRG GUY SUR VEN 83 BOL ECU PER 84 BZN. Nanophan. or phan.
Calycorectes macrocalyx Rusby, Mem. New York Bot. Gard. 7: 313 (1927). *Eugenia macrocalyx* (Rusby) McVaugh, Fieldiana, Bot. 29: 212 (1956), nom. illeg.
Phyllocalyx wentii Amshoff, Recueil Trav. Bot. Néerl. 39: 158 (1942).

Eugenia whytei Sprague, J. Linn. Soc., Bot. 37: 98 (1905).
W. Trop. Africa. 22 IVO LBR SIE. Nanophan. or phan.

Eugenia widgrenii Sond. ex O.Berg in C.F.P.von Martius & auct. suc. (eds.), Fl. Bras. 14(1): 227 (1857).
Brazil (Minas Gerais). 84 BZL. Nanophan. or phan.

Eugenia wilsonella Fawc. & Rendle, J. Bot. 64: 15 (1926).
Jamaica. 81 JAM. Nanophan.

Eugenia winzerlingii Standl., Trop. Woods 11: 20 (1927).
Belize. 80 BLZ. Nanophan. or phan.

Eugenia woodburyana Alain, Phytologia 47: 185 (1980).
Puerto Rico. 81 PUE. Nanophan. or phan.

Eugenia woodfrediana Urb., Symb. Antill. 9: 489 (1928).
Cuba (Sierra de Nipe). 81 CUB. Nanophan.

Eugenia woodii Dummer, Gard. Chron., III, 52: 192 (1912).
Mozambique to S. Africa. 26 MOZ 27 CPP NAT SWZ TVL. Nanophan. or phan.

Eugenia wullschlaegeliana Amshoff, Natuurw. Stud. Suriname Curaçao 2: 20 (1948).
Suriname to French Guiana. 82 FRG SUR.

Eugenia xalapensis (Kunth) DC., Prodr. 3: 277 (1828).
Mexico to Belize. 79 MXE MXG MXS 80 BLZ. Nanophan. or phan.
Myrtus xalapensis Kunth in F.W.H.von Humboldt, A.J.A.Bonpland & C.S.Kunth, Nov. Gen. Sp. 6: 145 (1823).

Eugenia xanthoxyloides Cambess. in A.F.C.de Saint-Hilaire, Fl. Bras. Merid. 2: 366 (1833). *Stenocalyx xanthoxyloides* (Cambess.) O.Berg in C.F.P.von Martius & auct. suc. (eds.), Fl. Bras. 14(1): 346 (1857).
Brazil (Rio de Janeiro). 84 BZL. Nanophan. or phan.

Eugenia xerophytica Britton, Bull. Torrey Bot. Club 51: 11 (1924).
Puerto Rico. 81 PUE. Nanophan. or phan.

Eugenia xilitlensis McVaugh, Fieldiana, Bot. 29: 462 (1963).
NE. Mexico. 79 MXE. Nanophan. or phan.

Eugenia xiriricana Mattos, Loefgrenia 3: 1 (1961).
Brazil (São Paulo). 84 BZL. Nanophan. or phan.

Eugenia xystophylla O.Berg, Linnaea 27: 196 (1856).
W. Cuba. 81 CUB. Nanophan.

Eugenia yangambensis Amshoff, Acta Bot. Neerl. 9: 409 (1960).
Zaïre. 23 ZAI. Nanophan. or phan.

Eugenia yasuniana B.Holst & M.L.Kawas., Sida 22: 934 (2006).
Ecuador. 83 ECU.

Eugenia yatuae (McVaugh) B.Holst, Selbyana 23: 144 (2002).
S. Venezuela to N. Brazil. 82 VEN 84 BZN. Phan.
Calycorectes yatuae McVaugh, Mem. New York Bot. Gard. 18(2): 226 (1969).

Eugenia yautepecana Lundell, Wrightia 2: 107 (1960).
Mexico. 79 MXC MXG. Nanophan. or phan.

Eugenia yumana Alain, Phytologia 25: 268 (1973).
Dominican Rep. 81 DOM. Nanophan.

Eugenia yunckeri Standl., Publ. Field Mus. Nat. Hist., Bot. Ser. 17: 380 (1938).
Honduras. 80 HON. Nanophan. or phan.

Eugenia zelayensis P.E.Sánchez, Phytologia 63: 461 (1987).
Nicaragua. 80 NIC. Phan.

Eugenia zuccarinii O.Berg in C.F.P.von Martius & auct. suc. (eds.), Fl. Bras. 14(1): 257 (1857).
Brazil (Rio de Janeiro). 84 BZL. Nanophan. or phan.

Eugenia zuchowskiae Barrie, Novon 15: 47 (2005).
Costa Rica. 80 COS. Nanophan. or phan.

Eugenia zuluensis Dummer, Gard. Chron., III, 52: 152 (1912).
S. Africa. 27 CPP NAT TVL. Nanophan. or phan.

Eugenia zygophylla Govaerts, World Checklist Myrtaceae: 172 (2008).
N. & NE. Madagascar. 29 MDG. Nanophan. or phan.
**Eugenia oligantha* Baker, J. Linn. Soc., Bot. 22: 474 (1887), nom. illeg.

Synonyms:
Eugenia abbotiana Urb. = ***Myrcia abbotiana*** (Urb.) Alain
Eugenia abbreviata Elmer = ***Syzygium abbreviatum*** Merr.
Eugenia abeggii Urb. & Ekman = ***Plinia abeggii*** (Urb. & Ekman) Urb.
Eugenia aborensis Dunn = ***Syzygium aborense*** (Dunn) Rathakr. & N.C.Nair
Eugenia abortiva Gagnep. = ***Syzygium abortivum*** (Gagnep.) Merr. & L.M.Perry
Eugenia acetosans Poir. = ***Myrcia citrifolia*** (Aubl.) Urb.
Eugenia acka DC. = ***Acca macrostema*** (Ruiz & Pav. ex G.Don) McVaugh
Eugenia acris Wight & Arn. = ?
Eugenia acrophila C.B.Rob. = ***Syzygium acrophilum*** (C.B.Rob.) Merr.
Eugenia acrophylla O.Berg = ***Myrceugenia myrcioides*** (Cambess.) O.Berg
Eugenia acuminata Link = ***Pimenta pseudocaryophyllus*** (Gomes) Landrum var. ***pseudocaryophyllus***
Eugenia acuminata Roxb. = ***Syzygium acuminatum*** (Roxb.) Miq.
Eugenia acuminatissima O.Berg = ***Eugenia biflora*** (L.) DC.
Eugenia acuminatissima Miq. = ***Blepharocalyx salicifolius*** (Kunth) O.Berg
Eugenia acuminatissima (Blume) Kurz = ***Syzygium acuminatissimum*** (Blume) DC.
Eugenia acutangula K.Schum. = ***Syzygium acutangulum*** Nied.
Eugenia acutata Miq. = ***Calycorectes acutatus*** (Miq.) Toledo
Eugenia acutiflora Kiaersk. = ***Myrceugenia acutiflora*** (Kiaersk.) D.Legrand & Kausel
Eugenia acutiloba DC. = ***Eugenia biflora*** (L.) DC.
Eugenia acutisepala Proctor = ***Eugenia acrisepala*** Govaerts
Eugenia acutisepala Hayata = ***Syzygium formosanum*** (Hayata) Mori
Eugenia adamantium Cambess. = ***Blepharocalyx salicifolius*** (Kunth) O.Berg
Eugenia ademireana Mattos = ***Calycorectes cucullatus*** Mattos
Eugenia adenocalyx DC. = ***Eugenia citrifolia*** Poir.
Eugenia adenoclada Urb. = ***Pimenta adenoclada*** (Urb.) Alain
Eugenia adenodes Kiaersk. = ***Myrciaria pumila*** (Gardner) O.Berg
Eugenia adstringens Cambess. = ***Eugenia punicifolia*** (Kunth) DC.
Eugenia aegiceroides Korth. ex Miq. = ***Syzygium aegiceroides*** (Korth. ex Miq.) Korth.
Eugenia aemula Diels = [83 ECU]
Eugenia aemula (Blume) Koord. & Valeton = ***Syzygium aemulum*** (Blume) Amshoff
Eugenia affinis DC. = ***Eugenia glabrata*** (Sw.) DC.
Eugenia affinis Gillies ex Hook. & Arn. = ***Luma apiculata*** (DC.) Burret
Eugenia aggregata (Vell.) Kiaersk. = ***Eugenia cerasiflora*** Miq.
Eugenia aggregata Baker = ***Syzygium phillyreifolium*** (Baker) Labat & Schatz
Eugenia agusanensis Elmer = ***Syzygium rubrovenium*** (C.B.Rob.) Merr.
Eugenia alata Ridl. = ***Syzygium valdevenosum*** (Duthie) Merr. & L.M.Perry
Eugenia alaternifolia Benth. ex O.Berg = ***Eugenia pycnantha*** Benth.
Eugenia alba Roxb. = ***Syzygium aqueum*** (Burm.f.) Alston
Eugenia albanensis Sond. = ***Eugenia capensis*** subsp. ***albanensis*** (Sond.) F.White
Eugenia albida Humb. & Bonpl. = ***Eugenia biflora*** (L.) DC.
Eugenia albidiramea Merr. = ***Syzygium kunstleri*** (King) Bahadur & R.C.Gaur
Eugenia albiflora Duthie ex Kurz = ***Syzygium albiflorum*** (Duthie ex Kurz) Bahadur & R.C.Gaur
Eugenia albotomentosa Cambess. = ***Eugenia pyriformis*** Cambess.
Eugenia albotomentosa var. *goyazensis* Cambess. = ***Eugenia pyriformis*** Cambess.
Eugenia albotomentosa var. *urussangensis* O.Berg = ***Eugenia pyriformis*** Cambess.
Eugenia alcinae Merr. = ***Syzygium leucoxylon*** Korth.
Eugenia alegrensis Kiaersk. = ***Myrciaria cuspidata*** O.Berg
Eugenia alexandri Krug & Urb. = ***Eugenia biflora*** (L.) DC.
Eugenia alfaroana Standl. = ***Eugenia biflora*** (L.) DC.
Eugenia alpigena DC. = ***Myrceugenia alpigena*** (DC.) Landrum
Eugenia alternifolia Wight = ***Syzygium alternifolium*** (Wight) Walp.
Eugenia altipeta Greves = [43 NWG]
Eugenia alvarezii C.B.Rob. = ***Syzygium alvarezii*** (C.B.Rob.) Merr.
Eugenia alyxiifolia Ridl. = ***Syzygium alyxiifolium*** (Ridl.) I.M.Turner
Eugenia amanuensis Steyerm. = ***Eugenia biflora*** (L.) DC.
Eugenia amazonica O.Berg = ***Eugenia patens*** Poir.
Eugenia ambigua O.Berg = ***Eugenia punicifolia*** (Kunth) DC.
Eugenia amboinensis Greves = [42 MOL]
Eugenia amboinensis Nois. ex Link = ***Xanthostemon verus*** (Lindl.) Peter G.Wilson
Eugenia ambongensis Ridl. = ***Syzygium elopurae*** (Ridl.) Merr. & L.M.Perry
Eugenia ambongensis var. *havilandii* Ridl. = ***Syzygium leucocladum*** Merr. & L.M.Perry
Eugenia americana Makoy ex E.Morren = ***Eugenia macrocarpa*** Schltdl. & Cham.
Eugenia amicorum A.Gray = ***Syzygium amicorum*** (A.Gray) Müll.Stuttg.
Eugenia amoena var. *rotundata* Trimen = ***Eugenia rotundata*** (Trimen) Trimen
Eugenia amplectens Nied. = ***Eugenia sessilifolia*** DC.
Eugenia amplexicaulis Vell. = ***Myrcia amplexicaulis*** (Vell.) Hook.f.
Eugenia amplexicaulis (DC.) Roxb. = ***Syzygium amplexicaule*** (DC.) N.P.Balakr.

Eugenia ampliflora Koord. & Valeton = ***Syzygium ampliflorum*** (Koord. & Valeton) Amshoff
Eugenia ampullaria Stapf = ***Syzygium ampullarium*** (Stapf) Merr. & L.M.Perry
Eugenia anacardiifolia Craib = ***Syzygium anacardiifolium*** (Craib) Chantaran. & J.Parn.
Eugenia anceps O.Berg = ***Myrceugenia bracteosa*** (DC.) D.Legrand & Kausel
Eugenia andamanica King = ***Syzygium andamanicum*** (King) N.P.Balakr.
Eugenia andersonii Ridl. = ***Syzygium punctilimbum*** (Merr.) Merr. & L.M.Perry
Eugenia andina O.Berg = ?
Eugenia androsaemoides Bedd. = ***Syzygium cordifolium*** (Wight) Walp. subsp. ***cordifolium***
Eugenia androsaemoides (L.) DC. = ***Syzygium cordifolium*** subsp. ***spissum*** (Alston) P.S.Ashton
Eugenia androsiana Urb. = ***Psidium androsianum*** (Urb.) Correll
Eugenia angkae Craib = ***Syzygium angkae*** (Craib) Chantaran. & J.Parn.
Eugenia anglohondurensis Lundell = ***Chamguava schippii*** (Standl.) Landrum
Eugenia angolensis Engl. = ***Eugenia malangensis*** (O.Hoffm.) Nied.
Eugenia angophoroides F.Muell. = ***Syzygium angophoroides*** (F.Muell.) B.Hyland
Eugenia angularis Elmer = ***Syzygium angulare*** (Elmer) Merr.
Eugenia angulata C.B.Rob. = ***Syzygium angulatum*** (C.B.Rob.) Merr.
Eugenia angustibracteolata Baker f. = ***Austromyrtus vieillardii*** (Brongn. & Gris) Burret
Eugenia angustifolia O.Berg = ***Eugenia pilosula*** Krug & Urb.
Eugenia angustifolia Lam. = ***Eugenia pomifera*** (Aubl.) Urb.
Eugenia angustifolia Spreng. = ***Eugenia sprengelii*** DC.
Eugenia angustifolia Blume = ***Syzygium blumei*** (Steud.) Merr. & L.M.Perry
Eugenia angustifolia Roxb. = ***Syzygium polypetalum*** (Wall.) Merr. & L.M.Perry
Eugenia angustifolia var. *angustissima* (O.Berg) Krug & Urb. = ***Eugenia pomifera*** (Aubl.) Urb.
Eugenia angustifolia var. *latifolia* (O.Berg) Krug & Urb. = ***Eugenia pomifera*** (Aubl.) Urb.
Eugenia angustisepala Mattos = ***Eugenia subreticulata*** Glaz.
Eugenia anisopetala R.Parker = ***Syzygium anisopetalum*** (R.Parker) N.P.Balakr.
Eugenia anisosepala Duthie = ***Syzygium anisosepalum*** (Duthie) I.M.Turner
Eugenia anjouanensis H.Perrier = ***Eugenia humblotii*** Engl. & Brehmer
Eugenia anomala D.Legrand = ***Hexachlamys humilis*** O.Berg
Eugenia anomala var. *coriacea* D.Legrand = ***Hexachlamys humilis*** O.Berg
Eugenia anthera Small = ***Eugenia axillaris*** (Sw.) Willd. var. ***axillaris***
Eugenia anthica Ridl. = ***Syzygium anthicum*** (Ridl.) Merr. & L.M.Perry
Eugenia antiquae L.Riley = ***Eugenia acapulcensis*** Steud.
Eugenia antiseptica (Blume) Kuntze = ***Syzygium antisepticum*** (Blume) Merr. & L.M.Perry
Eugenia antoniana Elmer = ***Syzygium antonianum*** (Elmer) Merr.
Eugenia aphthosa Vieill. ex Brongn. & Gris = ***Austromyrtus aphthosa*** (Vieill. ex Brongn. & Gris) Burret
Eugenia apiculata DC. = ***Luma apiculata*** (DC.) Burret
Eugenia apiculata (O.Berg) Nied. = ***Myrcianthes cisplatensis*** (Cambess.) O.Berg
Eugenia apiculata var. *arnyan* Hook.f. = ***Luma apiculata*** (DC.) Burret
Eugenia apodantha Standl. = ***Chamguava gentlei*** var. ***apodantha*** (Standl.) Landrum
Eugenia apodophylla F.Muell. = ***Syzygium apodophyllum*** (F.Muell.) B.Hyland
Eugenia apoensis Elmer = ***Syzygium apoense*** (Elmer) Merr.
Eugenia aprica Trimen = ***Eugenia mabaeoides*** subsp. ***pedunculata*** (Trimen) P.S.Ashton
Eugenia aprica O.Berg = ***Myrceugenia euosma*** (O.Berg) D.Legrand
Eugenia aquea Burm.f. = ***Syzygium aqueum*** (Burm.f.) Alston
Eugenia araujoana O.Berg = ***Myrceugenia glaucescens*** (Cambess.) D.Legrand & Kausel var. ***glaucescens***
Eugenia arborea Baker f. = ***Syzygium arboreum*** (Baker f.) J.W.Dawson
Eugenia arbutifolia O.Berg = ***Eugenia punicifolia*** (Kunth) DC.
Eugenia arbutifolia var. *obscura* O.Berg = ***Eugenia punicifolia*** (Kunth) DC.
Eugenia arbutifolia var. *pallida* O.Berg = ***Eugenia punicifolia*** (Kunth) DC.
Eugenia arctostaphyloides O.Berg = ***Eugenia punicifolia*** (Kunth) DC.
Eugenia arctostaphyloides var. *ovalis* O.Berg = ***Eugenia punicifolia*** (Kunth) DC.
Eugenia arcuatinervia Merr. = ***Syzygium arcuatinervium*** (Merr.) Craven & Biffin
Eugenia arechavaletae Herter = ***Eugenia uniflora*** L.
Eugenia areolata (DC.) Duthie = ***Syzygium venosum*** DC.
Eugenia argutata Koord. & Valeton = ***Syzygium acutatum*** (Miq.) Amshoff
Eugenia argyrea Lundell = ***Eugenia galalonensis*** (C.Wright ex Griseb.) Krug & Urb.
Eugenia argyrocalyx Warb. = ***Syzygium argyrocalyx*** (Warb.) Merr. & L.M.Perry
Eugenia arivoa Aubl. = ***Eugenia undulata*** Aubl.
Eugenia armstrongii Benth. = ***Syzygium armstrongii*** (Benth.) B.Hyland
Eugenia arnoldsonii Urb. = ***Eugenia gibberosa*** Urb.
Eugenia arnottiana Wight = ***Syzygium densiflorum*** Wall. ex Wight & Arn.
Eugenia aromatica (O.Berg) Sond. ex B.D.Jacks. = ?
Eugenia aromatica O.Berg = ***Myrcianthes fragrans*** (Sw.) McVaugh
Eugenia aromatica (L.) Baill. = ***Syzygium aromaticum*** (L.) Merr. & L.M.Perry
Eugenia arthroopoda var. *ambalavensis* H.Perrier = ***Eugenia arthroopoda*** Baill. ex Drake
Eugenia arthroopoda var. *angustata* H.Perrier = ***Eugenia arthroopoda*** Baill. ex Drake
Eugenia arthroopoda var. *baroniana* H.Perrier = ***Eugenia arthroopoda*** Baill. ex Drake
Eugenia arthroopoda var. *elatipes* H.Perrier = ***Eugenia arthroopoda*** Baill. ex Drake
Eugenia arthroopoda var. *menabeensis* H.Perrier = ***Eugenia arthroopoda*** Baill. ex Drake
Eugenia asa-grayi Krug & Urb. = ***Myrciaria floribunda*** (H.West ex Willd.) O.Berg
Eugenia assamica Biswas & Purkay. = ***Syzygium assamicum*** (Biswas & Purkay.) Raizada

Eugenia assimilis (Thwaites) Bedd. = ***Syzygium assimile*** Thwaites
Eugenia astronioides C.B.Rob. = ***Syzygium astronioides*** (C.B.Rob.) Merr.
Eugenia atrata Mattos & D.Legrand = ***Eugenia suberosa*** Cambess.
Eugenia atronervia M.R.Hend. = ***Syzygium dyerianum*** (King) Chantaran. & J.Parn.
Eugenia atropunctata Steud. = ***Eugenia florida*** DC.
Eugenia atropunctata C.B.Rob. = ***Syzygium polyanthum*** (Wight) Walp.
Eugenia atropunctata var. *gracilis* O.Berg = ***Eugenia florida*** DC.
Eugenia atropunctata var. *robusta* O.Berg = ***Eugenia florida*** DC.
Eugenia attenuata (Miq.) Koord. & Valeton = ***Syzygium attenuatum*** (Miq.) Merr. & L.M.Perry
Eugenia attenuata var. *montana* M.R.Hend. = ***Syzygium attenuatum*** (Miq.) Merr. & L.M.Perry
Eugenia attenuata var. *ophirensis* M.R.Hend. = ***Syzygium attenuatum*** (Miq.) Merr. & L.M.Perry
Eugenia attenuatifolia Merr. = ***Syzygium acuminatissimum*** (Blume) DC.
Eugenia attopeuensis Gagnep. = ***Syzygium attopeuense*** (Gagnep.) Merr. & L.M.Perry
Eugenia aubletiana Mattos = ***Eugenia latifolia*** Aubl.
Eugenia aurantiaca H.Perrier = ***Syzygium aurantiacum*** (H.Perrier) Labat & Schatz
Eugenia aurea Elmer = ***Xanthomyrtus diplycosiifolia*** (C.B.Rob.) Merr.
Eugenia auriculata Ridl. = ***Syzygium hendersonii*** Merr.
Eugenia australis Colla = ***Eugenia biflora*** (L.) DC.
Eugenia australis J.C.Wendl. ex Link = ***Syzygium australe*** (J.C.Wendl. ex Link) B.Hyland
Eugenia austrocaledonica Seem. = ***Syzygium austrocaledonicum*** (Seem.) Guillaumin
Eugenia avenis (Miq.) M.R.Hend. = ***Syzygium avene*** Miq.
Eugenia axillaris Vell. = ***Eugenia supraaxillaris*** Spreng.
Eugenia axillaris G.Don = ?
Eugenia axillaris Koord. & Valeton = [42 JAW]
Eugenia axillaris var. *microcarpa* Krug & Urb. = ***Eugenia axillaris*** (Sw.) Willd. var. ***axillaris***
Eugenia ayacuchae Steyerm. = ***Eugenia warmingiana*** Kiaersk.
Eugenia baeuerlenii F.Muell. = ***Syzygium baeuerlenii*** (F.Muell.) Craven & Biffin
Eugenia bagensis O.Berg = ***Myrceugenia glaucescens*** (Cambess.) D.Legrand & Kausel var. ***glaucescens***
Eugenia bagensis var. *angustifolia* O.Berg = ***Myrceugenia glaucescens*** (Cambess.) D.Legrand & Kausel var. ***glaucescens***
Eugenia bagensis var. *avenia* O.Berg = ***Myrceugenia glaucescens*** (Cambess.) D.Legrand & Kausel var. ***glaucescens***
Eugenia bagensis var. *latifolia* O.Berg = ***Myrceugenia glaucescens*** (Cambess.) D.Legrand & Kausel var. ***glaucescens***
Eugenia bahamensis Kiaersk. = ***Mosiera longipes*** (O.Berg) Small
Eugenia bakeri Britton & P.Wilson = ***Eugenia rosariensis*** Borhidi
Eugenia bakeri Elmer = ***Syzygium astronioides*** (C.B.Rob.) Merr.
Eugenia balancanensis Lundell = ***Eugenia oerstediana*** O.Berg
Eugenia balbisiana (DC.) O.Berg = ***Myrcianthes fragrans*** (Sw.) McVaugh
Eugenia balbisiana var. *dives* O.Berg = ***Myrcianthes fragrans*** (Sw.) McVaugh
Eugenia balbisiana var. *triflora* O.Berg = ***Myrcianthes fragrans*** (Sw.) McVaugh
Eugenia balerensis C.B.Rob. = ***Syzygium balerense*** (C.B.Rob.) Merr.
Eugenia balfourii Baker = ***Syzygium balfourii*** (Baker) J.Guého & A.J.Scott
Eugenia balsamea Wight = ***Syzygium balsameum*** (Wight) Wall. ex Walp.
Eugenia balsamica Jacq. = ***Myrcianthes fragrans*** (Sw.) McVaugh
Eugenia banaba Elmer = ***Syzygium globosum*** (Elmer) Merr.
Eugenia banghamii Standl. = ***Eugenia venezuelensis*** O.Berg
Eugenia bankensis (Hassk.) Backer = ***Syzygium bankense*** (Hassk.) Merr. & L.M.Perry
Eugenia banksii Britten & S.Moore = ***Syzygium banksii*** (Britten & S.Moore) B.Hyland
Eugenia bantamensis Koord. & Valeton = [42 JAW]
Eugenia baramensis Merr. = ***Syzygium baramense*** (Merr.) Merr. & L.M.Perry
Eugenia barnesii (Merr.) Merr. = ***Syzygium barnesii*** (Merr.) Merr.
Eugenia barrerensis O.Berg = ***Eugenia complicata*** O.Berg
Eugenia barringtonii Hole ex Chalk & Chattaway = ***Syzygium circumscissum*** (Gagnep.) Craven & Biffin
Eugenia barringtonioides Ridl. = ***Syzygium barringtonioides*** (Ridl.) Masam.
Eugenia bartlettiana Lundell = ***Eugenia acapulcensis*** Steud.
Eugenia bartonii F.M.Bailey = ***Syzygium bartonii*** (F.M.Bailey) Merr. & L.M.Perry
Eugenia baruensis (Jacq.) Jacq. = ***Eugenia axillaris*** (Sw.) Willd. var. ***axillaris***
Eugenia baruensis Griseb. = ***Eugenia procera*** (Sw.) Poir.
Eugenia bataanensis (Merr.) Merr. = ***Syzygium bataanense*** (Merr.) Merr.
Eugenia batavorum (McVaugh) Mattos = ***Calycorectes batavorum*** McVaugh
Eugenia batucaryensis O.Berg = ***Eugenia uruguayensis*** Cambess.
Eugenia bauanguica Blanco = ***Syzygium laetum*** (Buch.-Ham.) Gandhi subsp. ***laetum***
Eugenia baviensis Gagnep. = ***Syzygium baviense*** (Gagnep.) Merr. & L.M.Perry
Eugenia beaurepairiana (Kiaersk.) D.Legrand = ***Eugenia ternatifolia*** Cambess.
Eugenia beccarii Ridl. = ***Syzygium beccarii*** (Ridl.) Merr. & L.M.Perry
Eugenia beddomei Duthie = ***Syzygium beddomei*** (Duthie) Chithra
Eugenia belemii (Mattos) Mattos = ***Calycorectes belemii*** Mattos
Eugenia belizensis Standl. = ***Calyptranthes hondurensis*** Standl.
Eugenia bella Cambess. = ***Eugenia modesta*** DC.
Eugenia bella Phil. = ***Luma chequen*** (Molina) A.Gray
Eugenia benguellensis Welw. ex Hiern = ***Syzygium benguellense*** (Welw. ex Hiern) Engl.
Eugenia benguetensis C.B.Rob. = ***Syzygium benguetense*** (C.B.Rob.) Merr.
Eugenia benjamina King = ***Syzygium perryae*** I.M.Turner
Eugenia benthamiana Wight ex Duthie = ***Syzygium densiflorum*** Wall. ex Wight & Arn.

Eugenia benthamii O.Berg = ***Eugenia punicifolia*** (Kunth) DC.
Eugenia benthamii A.Gray = ***Syzygium nitidum*** Benth.
Eugenia bergiana Griseb. = ***Eugenia lineata*** (Sw.) DC.
Eugenia berlynensis O.Berg = ***Eugenia procera*** (Sw.) Poir.
Eugenia bernardii King = ***Syzygium inophyllum*** DC.
Eugenia bernardoi Merr. = ***Syzygium bernardoi*** (Merr.) Merr.
Eugenia bernieri Baill. ex Drake = ***Syzygium bernieri*** (Baill. ex Drake) Labat & Schatz
Eugenia bernieri var. *lateriticolor* H.Perrier = ***Syzygium bernieri*** (Baill. ex Drake) Labat & Schatz
Eugenia bernieri var. *oblanceolata* H.Perrier = ***Syzygium bernieri*** (Baill. ex Drake) Labat & Schatz
Eugenia bernieri var. *tampinensis* H.Perrier = ***Syzygium bernieri*** (Baill. ex Drake) Labat & Schatz
Eugenia berteroana (Spreng.) B.D.Jacks. = ***Eugenia biflora*** (L.) DC.
Eugenia beruttii (Mattos) Mattos = ***Calycorectes beruttii*** Mattos
Eugenia besukiensis Merr. = ***Syzygium bankense*** (Hassk.) Merr. & L.M.Perry
Eugenia bibracteata Greves = ***Syzygium fastigiatum*** (Blume) Merr. & L.M.Perry
Eugenia bicincta Urb. & Ekman = ***Eugenia christii*** Urb.
Eugenia bicolor O.Berg = ***Eugenia hyemalis*** Cambess.
Eugenia bifaria Wall. = ***Syzygium laurifolium*** (DC.) N.P.Balakr.
Eugenia biflora var. *hoffmannseggii* (O.Berg) Amshoff = ***Eugenia biflora*** (L.) DC.
Eugenia biflora var. *lancea* (Poir.) Krug & Urb. = ***Eugenia biflora*** (L.) DC.
Eugenia biflora var. *ludibunda* (Bertero ex DC.) Krug & Urb. = ***Eugenia biflora*** (L.) DC.
Eugenia biflora var. *mini* (Aubl.) Amshoff = ***Eugenia biflora*** (L.) DC.
Eugenia biflora var. *myriostigma* (Sagot) Amshoff = ***Eugenia biflora*** (L.) DC.
Eugenia biflora var. *pallens* (Vahl) Krug & Urb. = ***Eugenia biflora*** (L.) DC.
Eugenia biflora var. *virgultosa* (Sw.) Krug & Urb. = ***Eugenia biflora*** (L.) DC.
Eugenia biflora var. *wallenii* (Macfad.) Krug & Urb. = ***Eugenia biflora*** (L.) DC.
Eugenia bifurcata McVaugh = ***Myrcianthes osteomeloides*** (Rusby) McVaugh
Eugenia bimarginata O.Berg = ***Eugenia warmingiana*** Kiaersk.
Eugenia bimarginata var. *impunctata* Cambess. = ***Eugenia bimarginata*** DC.
Eugenia bimarginata var. *rubrocincta* (O.Berg) Kiaersk. = ***Eugenia bimarginata*** DC.
Eugenia bimarginata var. *tomentosa* Cambess. = ***Eugenia bimarginata*** DC.
Eugenia bimarginata var. *umbellaris* (O.Berg) Kiaersk. = ***Eugenia bimarginata*** DC.
Eugenia binacag Elmer = ***Eucalyptus binacag*** (Elmer) Elmer
Eugenia biniflora Ridl. = [42 BOR SUM]
Eugenia biseriata Kiaersk. = ***Myrciaria disticha*** O.Berg
Eugenia blancoi Merr. = ***Syzygium blancoi*** (Merr.) Merr.
Eugenia bleeseri O.Schwarz = ***Syzygium eucalyptoides*** subsp. ***bleeseri*** (O.Schwarz) B.Hyland
Eugenia blepharidantha Urb. = ***Eugenia ehrenbergiana*** O.Berg
Eugenia blumeana Kuntze = ***Syzygium glabratum*** (DC.) Veldkamp
Eugenia blumei Steud. = ***Syzygium blumei*** (Steud.) Merr. & L.M.Perry
Eugenia boerlagei Merr. = ***Syzygium boerlagei*** (Merr.) Govaerts
Eugenia boisiana Gagnep. = ***Syzygium boisianum*** (Gagnep.) Merr. & L.M.Perry
Eugenia boliviensis O.Berg = ***Eugenia punicifolia*** (Kunth) DC.
Eugenia boliviensis Rusby = ***Myrcia mollis*** (Kunth) DC.
Eugenia bonii Gagnep. = ***Syzygium bonii*** (Gagnep.) Merr. & L.M.Perry
Eugenia boninensis Hayata ex Koidz. = ***Metrosideros boninensis*** (Hayata ex Koidz.) Tuyama
Eugenia bonplandiana O.Berg = ***Eugenia acapulcensis*** Steud.
Eugenia bordenii Merr. = ***Syzygium bordenii*** (Merr.) Merr.
Eugenia borinquensis Britton = ***Eugenia cordata*** var. ***sintenisii*** (Kiaersk.) Krug & Urb.
Eugenia borneensis Miq. = ***Syzygium borneense*** (Miq.) Miq.
Eugenia brachiata Roxb. = ***Syzygium brachiatum*** (Roxb.) Miq.
Eugenia brachyandra Maiden & Betche = ***Syzygium ingens*** (F.Muell. ex C.Moore) Craven & Biffin
Eugenia brachybotrya DC. = ***Mouriri guianensis*** (Melastomataceae)
Eugenia brachycalyx Baker f. = ***Syzygium brachycalyx*** (Baker f.) J.W.Dawson
Eugenia brachymischa Kiaersk. = ***Myrceugenia bracteosa*** (DC.) D.Legrand & Kausel
Eugenia brachymischa f. *pedunculata* Kiaersk. = ***Myrceugenia bracteosa*** (DC.) D.Legrand & Kausel
Eugenia brachypoda DC. = ***Eugenia ramiflora*** Desv.
Eugenia brachystachys O.Berg = ***Eugenia trinervia*** Vahl
Eugenia brackenridgei A.Gray = ***Syzygium brackenridgei*** (A.Gray) Müll.Stuttg.
Eugenia bracteata (Willd.) Raeusch. ex Roxb. = ***Eugenia roxburghii*** DC.
Eugenia bracteata Rich. = ***Myrcia guianensis*** (Aubl.) DC.
Eugenia bracteata var. *fasciculata* (Wall. ex Blume) Duthie = ***Eugenia roxburghii*** DC.
Eugenia bracteata var. *roxburghii* (DC.) Duthie = ***Eugenia roxburghii*** DC.
Eugenia bracteolaris Lam. ex DC. = ***Eugenia brasiliensis*** Lam.
Eugenia bracteolata Wight = ***Syzygium fastigiatum*** (Blume) Merr. & L.M.Perry
Eugenia bracteolosa Lundell = ***Eugenia acapulcensis*** Steud.
Eugenia bracteosa DC. = ***Myrceugenia bracteosa*** (DC.) D.Legrand & Kausel
Eugenia brantiana M.R.Hend. = ***Syzygium oblatum*** (Roxb.) Wall. ex A.M.Cowan & Cowan
Eugenia brasiliae Mattos & D.Legrand = ***Eugenia complicata*** O.Berg
Eugenia brassii Merr. & L.M.Perry = ***Pilidiostigma papuanum*** (Lauterb.) A.J.Scott
Eugenia bresolinii D.Legrand = ***Eugenia pluriflora*** DC.
Eugenia brevifolia A.Gray = ***Syzygium brevifolium*** (A.Gray) Müll.Stuttg.
Eugenia brevipaniculata Merr. = ***Eugenia verticillata*** (Vell.) Angely
Eugenia brevipetiolata Britton & P.Wilson = ***Eugenia gibberosa*** Urb.
Eugenia brevistipitata Lundell = ***Eugenia bumelioides*** Standl.
Eugenia brevistylis C.B.Rob. = ***Syzygium leucoxylon*** Korth.
Eugenia × *bridgesii* Hook. & Arn. = ***Myrceugenia* × *bridgesii*** (Hook. & Arn.) O.Berg

Eugenia brihondurensis Lundell = ***Eugenia gigas*** Lundell
Eugenia brittoniana C.B.Rob. = ***Syzygium brittonianum*** (C.B.Rob.) Merr.
Eugenia brownsbergii var. *glauca* Amshoff = ***Eugenia brownsbergii*** Amshoff
Eugenia brunnea Nied. = ***Psidium ovale*** (Spreng.) Burret
Eugenia brunnea C.B.Rob. = ***Syzygium balerense*** (C.B. Rob.) Merr.
Eugenia brunneoramea Merr. = ***Syzygium racemosum*** (Blume) DC.
Eugenia brunnescens Urb. = ***Eugenia cupuligera*** Urb.
Eugenia buettneriana K.Schum. = ***Syzygium buettnerianum*** (K.Schum.) Nied.
Eugenia bullockii Hance = ***Syzygium bullockii*** (Hance) Merr. & L.M.Perry
Eugenia bulusanensis Elmer = ***Syzygium vulcanicum*** Elmer ex Merr.
Eugenia bunchosiifolia Nied. = ***Eugenia santensis*** Kiaersk.
Eugenia bungadinnia F.M.Bailey = ***Syzygium bungadinnia*** (F.M.Bailey) B.Hyland
Eugenia burebidensis Elmer = ***Syzygium rosenbluthii*** (C.B.Rob.) Merr.
Eugenia burkilliana King = ***Syzygium burkillianum*** (King) I.M.Turner
Eugenia burkilliana var. *garcinifolioides* M.R.Hend. = ***Syzygium burkillianum*** (King) I.M.Turner
Eugenia buxifolia (Sw.) Willd. = ***Eugenia foetida*** Pers.
Eugenia buxifolia Phil. = ***Myrceugenia chrysocarpa*** (O.Berg) Kausel
Eugenia buxifolia var. *minor* Cordem. = ***Eugenia buxifolia*** Lam.
Eugenia buxoides Urb. = ***Mosiera cabanasensis*** (Britton & P.Wilson) Borhidi subsp. ***cabanasensis***
Eugenia cabanasensis Britton & P.Wilson = ***Mosiera cabanasensis*** (Britton & P.Wilson) Borhidi
Eugenia cabanisiana O.Berg = ***Eugenia axillaris*** (Sw.) Willd. var. ***axillaris***
Eugenia cabelludo Kiaersk. = ***Myrciaria glazioviana*** (Kiaersk.) G.M.Barroso ex Sobral
Eugenia cabelludo var. *glazioviana* Kiaersk. = ***Myrciaria glazioviana*** (Kiaersk.) G.M.Barroso ex Sobral
Eugenia cachoeirensis var. *sessilifolia* Kiaersk. = ***Eugenia cachoeirensis*** O.Berg
Eugenia cacuminis Alain = ***Eugenia cacuminoides*** Govaerts
Eugenia cacuminis Craib = ***Syzygium cacuminis*** (Craib) Chantaran. & J.Parn.
Eugenia cagayanensis Merr. = ***Syzygium cagayanense*** (Merr.) Merr.
Eugenia calcadensis Bedd. = ***Syzygium calcadense*** (Bedd.) Chandrash.
Eugenia calcicola Merr. = ***Syzygium calcicola*** (Merr.) Merr.
Eugenia calciphila Lundell = ***Eugenia laevis*** O.Berg
Eugenia caldensis Kiaersk. = ***Eugenia speciosa*** Cambess.
Eugenia calleryana C.B.Rob. = ***Syzygium calleryanum*** (C.B.Rob.) Merr.
Eugenia callophylla (Miq.) Reinw. ex de Vriese = ***Syzygium aqueum*** (Burm.f.) Alston
Eugenia calophyllifolia Wight = ***Syzygium calophyllifolium*** (Wight) Walp.
Eugenia calothyrsa Diels = ***Eugenia patens*** Poir.
Eugenia calubcob C.B.Rob. = ***Syzygium calubcob*** (C.B.Rob.) Merr.
Eugenia calvinii Elmer = ***Syzygium confertum*** (Korth.) Merr. & L.M.Perry
Eugenia calycina Benth. = ***Eugenia liberiana*** Amshoff
Eugenia calycolpoides Griseb. = ***Eugenia punicifolia*** (Kunth) DC.
Eugenia calycorectoides Guillaumin = [60 NWC]
Eugenia calycorectoides O.Berg = ***Calycorectes mexicanus*** O.Berg
Eugenia calycosema O.Berg = ***Eugenia uruguayensis*** Cambess.
Eugenia calyculata Bello = ***Eugenia domingensis*** O.Berg
Eugenia calyptrata Roxb. ex Wight & Arn. = ***Syzygium cumini*** (L.) Skeels
Eugenia cambessedeana O.Berg = ***Myrceugenia glaucescens*** (Cambess.) D.Legrand & Kausel var. ***glaucescens***
Eugenia cambodiana Gagnep. = ***Syzygium cambodianum*** (Gagnep.) Merr. & L.M.Perry
Eugenia cambucarana Kiaersk. = ***Eugenia supraaxillaris*** Spreng.
Eugenia camiguinensis Merr. = ***Syzygium incrassatum*** (Elmer) Merr.
Eugenia campechiana Lundell = ***Eugenia acapulcensis*** Steud.
Eugenia campestris Vell. = ***Eugenia arrabidae*** O.Berg
Eugenia campestris DC. = ***Myrceugenia campestris*** (DC.) D.Legrand & Kausel
Eugenia campestris var. *nervulosa* Spring = ***Myrceugenia campestris*** (DC.) D.Legrand & Kausel
Eugenia camporum Morong = ***Eugenia pitanga*** (O.Berg) Nied.
Eugenia camptophylla M.R.Hend. = ***Syzygium camptophyllum*** (M.R.Hend.) I.M.Turner
Eugenia campylocarpa Gagnep. = ***Syzygium campylocarpum*** (Gagnep.) Merr. & L.M.Perry
Eugenia candelabriformis C.B.Rob. = ***Syzygium candelabriforme*** (C.B.Rob.) Merr.
Eugenia canelonensis O.Berg = ***Myrceugenia glaucescens*** (Cambess.) D.Legrand & Kausel var. ***glaucescens***
Eugenia canescens O.Berg = ***Eugenia monticola*** (Sw.) DC.
Eugenia capensis subsp. *aschersoniana* (F.Hoffm.) F.White = ***Eugenia aschersoniana*** F.Hoffm.
Eugenia capensis var. *major* Sond. = ***Eugenia capensis*** (Eckl. & Zeyh.) Harv. subsp. ***capensis***
Eugenia capitata Merr. = ***Syzygium capitatum*** (Merr.) Merr. & L.M.Perry
Eugenia capitellata Arn. = (Melastomataceae)
Eugenia capizensis Merr. = ***Syzygium urdanetense*** (Elmer) Merr.
Eugenia capoasensis Merr. = ***Syzygium capoasense*** (Merr.) Merr.
Eugenia capuli var. *lindeniana* (O.Berg) Lundell = ***Eugenia capuli*** (Schltdl. & Cham.) Hook. & Arn.
Eugenia capuli var. *macroterantha* O.Berg = ***Eugenia capuli*** (Schltdl. & Cham.) Hook. & Arn.
Eugenia capuli var. *micrantha* O.Berg = ***Eugenia capuli*** (Schltdl. & Cham.) Hook. & Arn.
Eugenia capuli var. *rigida* O.Berg = ***Eugenia capuli*** (Schltdl. & Cham.) Hook. & Arn.
Eugenia cardiophylla Merr. = ***Syzygium cardiophyllum*** (Merr.) Merr.
Eugenia carissoides F.Muell. = ***Eugenia reinwardtiana*** (Blume) A.Cunn. ex DC.
Eugenia carolinensis Koidz. = ***Syzygium carolinense*** (Koidz.) Hosok.
Eugenia carthagenensis Jacq. = ***Eugenia acapulcensis*** Steud.

Eugenia carthagenensis var. *baruensis* Jacq. = ***Eugenia axillaris*** (Sw.) Willd. var. ***axillaris***
Eugenia cartilaginea McVaugh = ***Myrcianthes myrsinoides*** (Kunth) Grifo
Eugenia caryophylla St.-Lag. = ***Pimenta pseudocaryophyllus*** (Gomes) Landrum var. ***pseudocaryophyllus***
Eugenia caryophyllaea Wight = ***Syzygium caryophyllatum*** (L.) Alston
Eugenia caryophyllata Thunb. = ***Syzygium aromaticum*** (L.) Merr. & L.M.Perry
Eugenia caryophylliflora Ridl. = ***Syzygium caryophylliflorum*** (Ridl.) Merr. & L.M.Perry
Eugenia caryophyllifolia Lam. = ***Syzygium cumini*** (L.) Skeels
Eugenia caryophyllus (Spreng.) Bullock & S.G.Harrison = ***Syzygium aromaticum*** (L.) Merr. & L.M.Perry
Eugenia casaretteana O.Berg = ***Eugenia patens*** Poir.
Eugenia casiguranensis Quisumb. = ***Syzygium casiguranense*** (Quisumb.) Merr.
Eugenia cassapensis O.Berg = ***Eugenia limbosa*** O.Berg
Eugenia cassinoides O.Berg = [84 BZE]
Eugenia cassinoides var. *difformis* O.Berg = [84 BZE]
Eugenia cassinoides var. *gracilis* O.Berg = [84 BZE]
Eugenia cassinoides var. *robusta* O.Berg = [84 BZE]
Eugenia castanea Merr. = ***Syzygium castaneum*** (Merr.) Merr. & L.M.Perry
Eugenia cataractarum Guillaumin = ***Austromyrtus cataractarum*** (Guillaumin) Burret
Eugenia cati Britton & P.Wilson = ***Eugenia catingiflora*** Griseb.
Eugenia catinga Baill. = ***Eugenia moschata*** (Aubl.) Nied. ex T.Durand & B.D.Jacks.
Eugenia caudata King = ***Syzygium urophyllum*** Merr.
Eugenia caudatifolia Merr. = ***Syzygium urdanetense*** (Elmer) Merr.
Eugenia caudatilimba Merr. = ***Syzygium caudatilimbum*** (Merr.) Merr. & L.M.Perry
Eugenia cauliflora O.Berg = ***Eugenia warmingiana*** Kiaersk.
Eugenia cauliflora Ridl. = [42 MLY]
Eugenia cauliflora (Mart.) DC. = ***Plinia cauliflora*** (Mart.) Kausel
Eugenia cauliflora Miq. = ***Plinia peruviana*** (Poir.) Govaerts
Eugenia caurensis Steyerm. = ***Eugenia biflora*** (L.) DC.
Eugenia ceibensis Standl. = ***Eugenia farameoides*** A.Rich.
Eugenia celebica (Blume) Merr. = ?
Eugenia cephalophora Ridl. = ***Syzygium cephalophorum*** (Ridl.) Merr. & L.M.Perry
Eugenia cerasiflora Kurz = ***Syzygium kurzii*** (Duthie) N.P.Balakr.
Eugenia cerasiformis (Blume) DC. = ***Syzygium racemosum*** (Blume) DC.
Eugenia cerasoides Roxb. = ***Syzygium operculatum*** (Roxb.) Nied.
Eugenia cerina M.R.Hend. = ***Syzygium incarnatum*** (Elmer) Merr. & L.M.Perry
Eugenia cerina var. *montana* M.R.Hend. = ***Syzygium incarnatum*** (Elmer) Merr. & L.M.Perry
Eugenia cerina var. *turbinata* M.R.Hend. = ***Syzygium incarnatum*** (Elmer) Merr. & L.M.Perry
Eugenia chacteana Lundell = ***Eugenia galalonensis*** (C.Wright ex Griseb.) Krug & Urb.
Eugenia championii (Benth.) Hemsl. = ***Syzygium championii*** (Benth.) Merr. & L.M.Perry
Eugenia chanlos Gagnep. = ***Syzygium chanlos*** (Gagnep.) Merr. & L.M.Perry
Eugenia chapelieri Baill. ex Drake = ***Syzygium bernieri*** (Baill. ex Drake) Labat & Schatz
Eugenia chavaran Bourd. = ***Syzygium chavaran*** (Bourd.) Gamble
Eugenia chekan DC. = ***Luma chequen*** (Molina) A.Gray
Eugenia chequen Molina = ***Luma chequen*** (Molina) A.Gray
Eugenia chinensis Regel = ?
Eugenia chiquimulana Standl. & Steyerm. = ***Eugenia pachychlamys*** Donn.Sm.
Eugenia chiquitensis O.Berg = ***Eugenia orbignyana*** O.Berg
Eugenia chirindensis Baker f. = ***Eugenia capensis*** subsp. ***nyassensis*** (Engl.) F.White
Eugenia chlorantha Duthie = ***Syzygium chloranthum*** (Duthie) Merr. & L.M.Perry
Eugenia chloroleuca King = ***Syzygium chloroleucum*** (King) Masam.
Eugenia chnoosepala Kiaersk. = ***Siphoneugena densiflora*** O.Berg var. ***densiflora***
Eugenia chnoosepala var. *angustifolia* Kiaersk. = ***Siphoneugena kiaerskoviana*** (Burret) Kausel
Eugenia chnoosepala var. *elliptica* Kiaersk. = ***Siphoneugena kiaerskoviana*** (Burret) Kausel
Eugenia chnoosepala var. *latifolia* Kiaersk. = ***Siphoneugena densiflora*** O.Berg var. ***densiflora***
Eugenia chnoosepala var. *regnelliana* Kiaersk. = ***Siphoneugena densiflora*** O.Berg var. ***densiflora***
Eugenia christovana Kiaersk. = ***Eugenia candolleana*** DC.
Eugenia chrysantha O.Berg = ***Eugenia aurata*** O.Berg
Eugenia chrysocarpa O.Berg = ***Myrceugenia chrysocarpa*** (O.Berg) Kausel
Eugenia chrysophylloides DC. = ***Eugenia chrysophyllum*** Poir.
Eugenia ciarensis O.Berg = ***Eugenia punicifolia*** (Kunth) DC.
Eugenia ciliaris Ridl. = ***Decaspermum montanum*** Ridl.
Eugenia ciliatosetosa Merr. = ***Syzygium ciliatosetosum*** (Merr.) Merr.
Eugenia ciliolata Cambess. = ***Myrciaria floribunda*** (H.West ex Willd.) O.Berg
Eugenia cincta (Merr. & L.M.Perry) Whitmore = ***Syzygium cinctum*** Merr. & L.M.Perry
Eugenia cinerea Kurz = ***Syzygium cinereum*** (Kurz) Chantaran. & J.Parn.
Eugenia cinnamomea Vidal = ***Syzygium cinnamomeum*** (Vidal) Merr.
Eugenia circumscissa Gagnep. = ***Syzygium circumscissum*** (Gagnep.) Craven & Biffin
Eugenia cisplatensis Cambess. = ***Myrcianthes cisplatensis*** (Cambess.) O.Berg
Eugenia cisplatensis var. *angustifolia* Lillo = ***Myrcianthes cisplatensis*** (Cambess.) O.Berg
Eugenia cisplatensis var. *gigantea* D.Legrand = ***Myrcianthes gigantea*** (D.Legrand) D.Legrand
Eugenia clarkeana King = ***Syzygium dyerianum*** (King) Chantaran. & J.Parn.
Eugenia clausa C.B.Rob. = ***Syzygium operculatum*** (Roxb.) Nied.
Eugenia clavata (Korth.) Merr. = ***Syzygium claviflorum*** (Roxb.) Wall. ex A.M.Cowan & Cowan
Eugenia clavellata Merr. = ***Syzygium clavellatum*** (Merr.) Merr.
Eugenia claviflora Roxb. = ***Syzygium claviflorum*** (Roxb.) Wall. ex A.M.Cowan & Cowan
Eugenia claviflora var. *excavata* King = ***Syzygium claviflorum*** (Roxb.) Wall. ex A.M.Cowan & Cowan

Eugenia claviflora var. *glandulosa* King = ***Syzygium claviflorum*** (Roxb.) Wall. ex A.M.Cowan & Cowan
Eugenia claviflora var. *maingayi* (Duthie) King = ***Syzygium claviflorum*** (Roxb.) Wall. ex A.M.Cowan & Cowan
Eugenia claviflora var. *montana* M.R.Hend. = ***Syzygium claviflorum*** (Roxb.) Wall. ex A.M.Cowan & Cowan
Eugenia claviflora var. *oblongifolia* Hayata = ***Syzygium taiwanicum*** H.T.Chang & R.H.Miao
Eugenia claviflora var. *riparia* M.R.Hend. = ***Syzygium claviflorum*** (Roxb.) Wall. ex A.M.Cowan & Cowan
Eugenia clavimyrtus Koord. & Valeton = ***Syzygium glabratum*** (DC.) Veldkamp
Eugenia clavimyrtus var. *minor* Koord. & Valeton = ***Syzygium filiforme*** (Wall. ex Duthie) Chantaran. & J.Parn.
Eugenia cleistocalyx Merr. = ***Syzygium operculatum*** (Roxb.) Nied.
Eugenia clementis C.B.Rob. = ***Syzygium clementis*** (C.B.Rob.) Merr.
Eugenia cleyerifolia Yatabe = ***Syzygium cleyerifolium*** (Yatabe) Makino
Eugenia clinocarpa DC. = ***Eugenia punicifolia*** (Kunth) DC.
Eugenia clusiifolia A.Gray = ***Syzygium clusiifolium*** (A.Gray) Müll.Stuttg.
Eugenia clusioides Brongn. & Gris = ***Austromyrtus clusioides*** (Brongn. & Gris) Burret
Eugenia clypeolata Ridl. = ***Syzygium clypeolatum*** (Ridl.) I.M.Turner
Eugenia coalita Greves = ***Syzygium coalitum*** (Greves) T.G.Hartley & L.M.Perry
Eugenia coarensis DC. = ***Eugenia punicifolia*** (Kunth) DC.
Eugenia cobanensis Lundell = ***Eugenia venezuelensis*** O.Berg
Eugenia coccolobifolia (Kunth) Burret = ***Myrcianthes rhopaloides*** (Kunth) McVaugh
Eugenia cochinchinensis Gagnep. = ***Syzygium ripicola*** (Craib) Merr. & L.M.Perry
Eugenia cocquericotensis Lundell = ***Eugenia oerstediana*** O.Berg
Eugenia coffeifolia var. *grandifolia* O.Berg = ***Eugenia coffeifolia*** DC.
Eugenia coffeifolia var. *parvifolia* O.Berg = ***Eugenia coffeifolia*** DC.
Eugenia coffeoides Lundell = ***Eugenia letreroana*** Lundell
Eugenia colchaguensis Phil. = ***Myrceugenia colchaguensis*** (Phil.) Navas
Eugenia collina DC. = ***Myrcia pubiflora*** DC.
Eugenia collinsae Craib = ***Syzygium antisepticum*** (Blume) Merr. & L.M.Perry
Eugenia coloradoensis Standl. = ***Eugenia florida*** DC.
Eugenia colorata Duthie = ***Syzygium subdecussatum*** (Duthie) I.M.Turner
Eugenia comitanensis Lundell = ***Eugenia acapulcensis*** Steud.
Eugenia compactiflora Spring ex Mart. = ***Eugenia monosperma*** Vell.
Eugenia complanata Gardner = ***Myrceugenia myrcioides*** (Cambess.) O.Berg
Eugenia compongensis Gagnep. = ***Syzygium compongense*** (Gagnep.) Merr. & L.M.Perry
Eugenia compta A.Rich. ex O.Berg = ***Eugenia pseudopsidium*** Jacq.
Eugenia comptonii Baker f. = ***Syzygium neocaledonicum*** (Seem.) J.W.Dawson
Eugenia conceptionis (Kuntze) K.Schum. = ***Eugenia pyriformis*** Cambess.
Eugenia concinna Thwaites = ***Eugenia mooniana*** Wight
Eugenia concinna Phil. = ***Legrandia concinna*** (Phil.) Kausel
Eugenia condensata (Miq.) Ridl. = ?
Eugenia condensata Baker = ***Syzygium condensatum*** (Baker) Labat & Schatz
Eugenia condensata var. *thouvenotii* Danguy ex Lecomte = ***Syzygium emirnense*** (Baker) Labat & Schatz
Eugenia confertiflora A.Gray = ***Syzygium confertiflorum*** (A.Gray) Müll.Stuttg.
Eugenia confertiflora Koord. & Valeton = ***Syzygium fastigiatum*** (Blume) Merr. & L.M.Perry
Eugenia congesta Merr. = ***Syzygium congestum*** (Merr.) Merr.
Eugenia congestissima Diels = ***Eugenia dittocrepis*** O.Berg
Eugenia conglobata Sessé & Moç. = ***Eugenia venezuelensis*** O.Berg
Eugenia conglobata C.B.Rob. = ***Syzygium conglobatum*** Merr.
Eugenia conglomerata Duthie = ***Syzygium conglomeratum*** (Duthie) I.M.Turner
Eugenia conglomerata var. *paniculata* M.R.Hend. = ***Syzygium conglomeratum*** (Duthie) I.M.Turner
Eugenia consanguinea Merr. = ***Syzygium consanguineum*** (Merr.) Merr.
Eugenia contracta Poir. = ***Syzygium contractum*** (Poir.) J.Guého & A.J.Scott
Eugenia contrerasii Lundell = ***Eugenia capuli*** (Schltdl. & Cham.) Hook. & Arn.
Eugenia conzattii Standl. = ***Eugenia oerstediana*** O.Berg
Eugenia coolminiana C.Moore = ***Syzygium oleosum*** (F.Muell.) B.Hyland
Eugenia copelandii C.B.Rob. = ***Syzygium copelandii*** (C.B.Rob.) Merr.
Eugenia coralina Merr. = ***Syzygium curtisii*** (King) Merr. & L.M.Perry
Eugenia cordata (Hochst. ex Krauss) G.Lawson = ***Syzygium cordatum*** Hochst. ex Krauss
Eugenia cordata var. *ovata* Kiaersk. = ***Eugenia cordata*** (Sw.) DC. var. ***cordata***
Eugenia cordata var. *parvifolia* Schltdl. = ***Eugenia cordata*** (Sw.) DC. var. ***cordata***
Eugenia cordatilimba Merr. = ***Syzygium cordatilimbum*** (Merr.) Merr.
Eugenia cordifolia Wight = ***Syzygium cordifolium*** (Wight) Walp.
Eugenia cordifoliata Ridl. = ***Syzygium cordifoliatum*** (Ridl.) I.M.Turner
Eugenia coriacea F.Dietr. = ***Eugenia buxifolia*** Lam.
Eugenia coriacea O.Berg = ***Eugenia limbosa*** O.Berg
Eugenia cormiflora F.Muell. = ***Syzygium cormiflorum*** (F.Muell.) B.Hyland
Eugenia corralensis Phil. = ***Myrceugenia exsucca*** (DC.) O.Berg
Eugenia correae O.Berg = ***Eugenia lambertiana*** DC.
Eugenia correifolia Hook. & Arn. = ***Myrceugenia correifolia*** (Hook. & Arn.) O.Berg
Eugenia corrugata King = ***Syzygium dyerianum*** (King) Chantaran. & J.Parn.
Eugenia corticopapyracea Elmer = ***Syzygium corticopapyraceum*** (Elmer) Merr.
Eugenia corticosa Lour. = ***Syzygium corticosum*** (Lour.) Merr. & L.M.Perry
Eugenia corymbifera Koord. & Valeton = ***Syzygium corymbosum*** (Blume) DC.
Eugenia corymbosa O.Berg = ***Eugenia myrobalana*** DC.

Eugenia corymbosa Lam. = ***Syzygium caryophyllatum*** (L.) Alston
Eugenia corymbosa Roxb. = ***Syzygium pycnanthum*** Merr. & L.M.Perry
Eugenia corynantha F.Muell. = ***Syzygium corynanthum*** (F.Muell.) L.A.S.Johnson
Eugenia corynocarpa A.Gray = ***Syzygium corynocarpum*** (A.Gray) Müll.Stuttg.
Eugenia costata O.Berg = ***Eugenia moschata*** (Aubl.) Nied. ex T.Durand & B.D.Jacks.
Eugenia costata Cambess. = ***Eugenia uniflora*** L.
Eugenia costata Bello = [81 PUE]
Eugenia costata (Miq.) Koord. & Valeton = ?
Eugenia costata Brongn. & Gris = ***Syzygium brousmichei*** Govaerts
Eugenia costulata C.B.Rob. = ***Syzygium costulatum*** (C.B.Rob.) Merr.
Eugenia cotinifolia subsp. *codyensis* (Munro ex Wight) P.S.Ashton = ***Eugenia codyensis*** Munro ex Wight
Eugenia cotinifolia var. *elliptica* (Lam.) Baker ex Engl. & Nied. = ***Eugenia elliptica*** Lam.
Eugenia cotinifolia var. *gardneri* Baker = ***Eugenia sieberi*** J.Guého & A.J.Scott
Eugenia cotinifolia var. *gracilis* (Duthie) M.R.Almeida = ***Eugenia phillyraeoides*** Trimen
Eugenia cotinifolia subsp. *phyllyraeoides* (Trimen) P.S.Ashton = ***Eugenia phillyraeoides*** Trimen
Eugenia cotinifolioides H.Perrier = (Melastomataceae)
Eugenia coumeta Aubl. = ***Myrcia coumeta*** (Aubl.) DC.
Eugenia cozumelensis Lundell = ***Eugenia axillaris*** var. ***cozumelensis*** (Lundell) Lundell
Eugenia crassibracteata Merr. = ***Syzygium crassibracteatum*** (Merr.) Merr.
Eugenia crassiflora Kiaersk. = ***Eugenia umbrosa*** O.Berg
Eugenia crassifolia Vieill. ex Brongn. & Gris = ***Eugenia brongniartiana*** Guillaumin
Eugenia crassifolia Ant.Molina = ***Eugenia hondurensis*** Ant.Molina
Eugenia crassifolia Miq. = ***Myrcia obovata*** (O.Berg) Nied.
Eugenia crassilimba Merr. = ***Syzygium crassilimbum*** (Merr.) Merr.
Eugenia crassipes C.B.Rob. = ***Syzygium crassipes*** (C.B.Rob.) Merr.
Eugenia crassissima Merr. = ***Syzygium crassissimum*** (Merr.) Merr.
Eugenia creaghii Ridl. = ***Syzygium creaghii*** (Ridl.) Merr. & L.M.Perry
Eugenia crebrifolia Steyerm. = ***Myrcianthes crebrifolia*** (Steyerm.) McVaugh
Eugenia crebrinervis C.T.White = ***Syzygium crebrinerve*** (C.T.White) L.A.S.Johnson
Eugenia crenata O.Berg = ***Eugenia harrisii*** Krug & Urb.
Eugenia crenulata Duthie = ***Syzygium pseudocrenulatum*** (M.R.Hend.) I.M.Turner
Eugenia crenulata var. *cubensis* O.Berg = ***Eugenia crenulata*** (Sw.) Willd.
Eugenia crenulata var. *domingensis* O.Berg = ***Eugenia crenulata*** (Sw.) Willd.
Eugenia crosbyi Burkill = ***Syzygium inophylloides*** (A.Gray) Müll.Stuttg.
Eugenia cruckshanksii Hook. & Arn. = ***Blepharocalyx cruckshanksii*** (Hook. & Arn.) Nied.
Eugenia cruegeri Krug & Urb. = ***Eugenia limbosa*** O.Berg
Eugenia cryptadena Amshoff = ***Eugenia pseudopsidium*** Jacq.
Eugenia cryptadena var. *gracilis* Amshoff = ***Eugenia cowanii*** McVaugh
Eugenia cryptophlebia F.Muell. = ***Syzygium wilsonii*** subsp. ***cryptophlebium*** (F.Muell.) B.Hyland
Eugenia cumbreana Lundell = ***Eugenia vesca*** Lundell
Eugenia cumingiana Vidal = ***Syzygium acuminatissimum*** (Blume) DC.
Eugenia cumingii Hook. & Arn. = ***Myrceugenia ovata*** (Hook. & Arn.) O.Berg var. ***ovata***
Eugenia cumini (L.) Druce = ***Syzygium cumini*** (L.) Skeels
Eugenia cuneata B.Heyne ex Bedd. = ***Eugenia indica*** (Wight) Chithra
Eugenia cuneata Duthie = ***Syzygium cuneatum*** Brahmam & H.O.Saxena
Eugenia cuneifolia Baker = ***Syzygium emirnense*** (Baker) Labat & Schatz
Eugenia cuprea Koord. & Valeton = ***Syzygium antisepticum*** (Blume) Merr. & L.M.Perry
Eugenia cupulata var. *macrophylla* McVaugh = ***Eugenia cupulata*** Amshoff
Eugenia cupulifera H.Perrier = (Rhizophoraceae)
Eugenia curranii C.B.Rob. = ***Syzygium curranii*** (C.B. Rob.) Merr.
Eugenia curtiflora Elmer = ***Syzygium curtiflorum*** (Elmer) Merr.
Eugenia curtisii King = ***Syzygium curtisii*** (King) Merr. & L.M.Perry
Eugenia curtisii var. *holttumii* (Ridl.) M.R.Hend. = ***Syzygium curtisii*** (King) Merr. & L.M.Perry
Eugenia curtisii var. *minor* King = ***Syzygium curtisii*** (King) Merr. & L.M.Perry
Eugenia curvatopetiolata Kiaersk. = ***Eugenia dodonaeifolia*** Cambess.
Eugenia curvistyla Gillespie = ***Syzygium curvistylum*** (Gillespie) Merr. & L.M.Perry
Eugenia cuspidata O.Berg = ***Eugenia excelsa*** O.Berg
Eugenia cuspidata Phil. = ***Luma apiculata*** (DC.) Burret
Eugenia cuspidato-obovata Hayata = ***Syzygium acuminatissimum*** (Blume) DC.
Eugenia cyanocarpa (F.Muell.) Maiden & Betche = ***Syzygium oleosum*** (F.Muell.) B.Hyland
Eugenia cyanophylla P.C.Kanjilal & D.Das = ***Syzygium cyanophyllum*** (P.C.Kanjilal & D.Das) Raizada
Eugenia cycliantha D.Legrand = ***Eugenia hyemalis*** Cambess.
Eugenia cyclophylla Baker = ***Syzygium loiseleurioides*** (Baker) Govaerts
Eugenia cyclophylla Thwaites ex Duthie = ***Syzygium revolutum*** subsp. ***cyclophyllum*** (Thwaites ex Duthie) P.S.Ashton
Eugenia cylindrica Wight = ***Syzygium cylindricum*** (Wight) Alston
Eugenia cymosa Lam. = ***Syzygium cymosum*** (Lam.) DC.
Eugenia cymosa Roxb. = ***Syzygium grande*** (Wight) Walp.
Eugenia cyrtophylloides Ridl. = ***Syzygium cyrtophylloides*** (Ridl.) I.M.Turner
Eugenia daaouiensis var. *glabriflora* Guillaumin = ***Eugenia daaouiensis*** Guillaumin
Eugenia dallachyana F.Muell. ex Benth. = ***Gossia dallachyana*** (F.Muell. ex Benth.) N.Snow & Guymer
Eugenia danguyana H.Perrier = ***Syzygium danguyanum*** (H.Perrier) Labat & Schatz
Eugenia danguyana var. *rotranala* H.Perrier = ***Syzygium danguyanum*** (H.Perrier) Labat & Schatz
Eugenia daphne Ridl. = ***Syzygium daphne*** (Ridl.) Merr. & L.M.Perry
Eugenia daphnoides Greves = ***Syzygium longipes*** Merr. & L.M.Perry
Eugenia darwinii Hook.f. = ***Amomyrtus luma*** (Molina) D.Legrand & Kausel

Eugenia dasyantha O.Berg = ***Eugenia punicifolia*** (Kunth) DC.
Eugenia dasyblasta (O.Berg) Nied. = ***Eugenia uniflora*** L.
Eugenia davaoensis Elmer = ***Syzygium davaoense*** (Elmer) Merr.
Eugenia dawei Hutch. & Dalziel = ***Eugenia kalbreyeri*** Engl. & Brehmer
Eugenia dealbata Burkill = ***Syzygium dealbatum*** (Burkill) A.C.Sm.
Eugenia decidua Merr. = ***Eugenia uniflora*** L.
Eugenia decipiens Koord. & Valeton = ***Syzygium decipiens*** (Koord. & Valeton) Merr. & L.M.Perry
Eugenia deckeri Gagnep. = ***Syzygium odoratum*** (Lour.) DC.
Eugenia decora Thwaites = ***Eugenia glabra*** Alston
Eugenia decora Salisb. = ***Syzygium jambos*** (L.) Alston
Eugenia decorticans O.Berg = ***Eugenia punicifolia*** (Kunth) DC.
Eugenia decumbens Cambess. = ***Blepharocalyx salicifolius*** (Kunth) O.Berg
Eugenia deflexa Poir. = ***Myrcia deflexa*** (Poir.) DC.
Eugenia delicatula DC. = ***Myrciaria delicatula*** (DC.) O.Berg
Eugenia delicatula var. *conferta* Kiaersk. = ***Myrciaria delicatula*** (DC.) O.Berg
Eugenia deltoidea Standl. = ***Eugenia acapulcensis*** Steud.
Eugenia demeusei f. *lukolelaensis* De Wild. = ***Eugenia demeusei*** De Wild.
Eugenia dempoensis Greves = ***Syzygium dempoense*** (Greves) Govaerts
Eugenia densepunctata Koord. & Valeton = ***Syzygium confertum*** (Korth.) Merr. & L.M.Perry
Eugenia densiflora (Blume) DC. = ***Syzygium pycnanthum*** Merr. & L.M.Perry
Eugenia densiflora var. *angustifolia* Ridl. = ***Syzygium foxworthianum*** (Ridl.) Merr. & L.M.Perry
Eugenia densinervia Merr. = ***Syzygium densinervium*** (Merr.) Merr.
Eugenia depauperata Cambess. = ***Blepharocalyx salicifolius*** (Kunth) O.Berg
Eugenia depauperata Cordem. = ***Syzygium cymosum*** (Lam.) DC. var. ***cymosum***
Eugenia deserti Cambess. = ***Blepharocalyx salicifolius*** (Kunth) O.Berg
Eugenia desvauxiana O.Berg = ***Campomanesia aromatica*** (Aubl.) Griseb.
Eugenia diantha O.Berg = ***Eugenia punicifolia*** (Kunth) DC.
Eugenia diantha var. *ciliata* O.Berg = ***Eugenia punicifolia*** (Kunth) DC.
Eugenia diantha var. *glabra* O.Berg = ***Eugenia punicifolia*** (Kunth) DC.
Eugenia diaphana Kiaersk. = ***Eugenia bergii***
Eugenia dichotoma (Poir.) DC. = ***Myrcianthes fragrans*** (Sw.) McVaugh
Eugenia dichotoma var. *fragrans* (Sw.) Nutt. = ***Myrcianthes fragrans*** (Sw.) McVaugh
Eugenia dicrana O.Berg = ***Myrcianthes fragrans*** (Sw.) McVaugh
Eugenia dicrossa O.Berg = ***Eugenia bimarginata*** DC.
Eugenia dicrossa var. *latifolia* O.Berg = ***Eugenia bimarginata*** DC.
Eugenia dicrossa var. *longifolia* O.Berg = ***Eugenia bimarginata*** DC.
Eugenia dicrossa var. *parvifolia* O.Berg = ***Eugenia bimarginata*** DC.
Eugenia diffusa Merr. = ***Syzygium diffusiflorum*** Merr.
Eugenia diffusa Turrill = ***Syzygium diffusum*** (Turrill) Merr. & L.M.Perry
Eugenia dimorpha var. *australis* Mattos = ***Eugenia dimorpha*** O.Berg
Eugenia dimorpha var. *oeidocarpa* (O.Berg) Mattos = ***Eugenia dimorpha*** O.Berg
Eugenia diospyrifolia Merr. = ***Eugenia montalbanica*** Merr.
Eugenia diospyrifolia Wall. ex Duthie = ***Syzygium diospyrifolium*** (Wall. ex Duthie) S.N.Mitra
Eugenia diospyrifolia var. *lanceolata* Craib = ***Syzygium diospyrifolium*** (Wall. ex Duthie) S.N.Mitra
Eugenia diplycosiifolia C.B.Rob. = ***Xanthomyrtus diplycosiifolia*** (C.B.Rob.) Merr.
Eugenia dipoda DC. = ***Eugenia punicifolia*** (Kunth) DC.
Eugenia dipoda var. *brachypoda* DC. = ***Eugenia punicifolia*** (Kunth) DC.
Eugenia dipoda f. *grandifolia* Miq. = ***Eugenia punicifolia*** (Kunth) DC.
Eugenia discolor (Kunth) DC. = ***Myrcianthes discolor*** (Kunth) McVaugh
Eugenia discophora Koord. & Valeton = ***Syzygium discophorum*** (Koord. & Valeton) Amshoff
Eugenia dispansa Ridl. = ***Syzygium dispansum*** (Ridl.) Craven & Biffin
Eugenia dissitiflora Lundell = ***Eugenia flavoviridis*** Lundell
Eugenia distans O.Berg = ***Myrceugenia campestris*** (DC.) D.Legrand & Kausel
Eugenia disticha var. *galalonensis* C.Wright ex Griseb. = ***Eugenia galalonensis*** (C.Wright ex Griseb.) Krug & Urb.
Eugenia distoma O.Berg = ***Myrceugenia planipes*** (Hook. & Arn.) O.Berg
Eugenia divaricata Willd. ex O.Berg = ***Eugenia axillaris*** (Sw.) Willd. var. ***axillaris***
Eugenia divaricata O.Berg = ***Blepharocalyx cruckshanksii*** (Hook. & Arn.) Nied.
Eugenia divaricata Lam. = ***Myrcia splendens*** (Sw.) DC.
Eugenia divaricata Benth. = ***Myrciaria dubia*** (Kunth) McVaugh
Eugenia divaricata var. *obovata* O.Berg = ***Blepharocalyx cruckshanksii*** (Hook. & Arn.) Nied.
Eugenia divaricata var. *ovalis* O.Berg = ***Blepharocalyx cruckshanksii*** (Hook. & Arn.) Nied.
Eugenia divaricata var. *pauciflora* O.Berg = ***Blepharocalyx cruckshanksii*** (Hook. & Arn.) Nied.
Eugenia divaricatocymosa Hayata = ***Syzygium operculatum*** (Roxb.) Nied.
Eugenia diversiflora O.Berg = ***Eugenia punicifolia*** (Kunth) DC.
Eugenia diversifolia Brongn. & Gris = ***Austromyrtus diversifolia*** (Brongn. & Gris) Burret
Eugenia djalonensis A.Chev. = ***Eugenia elliotii*** Engl. & Brehmer
Eugenia djouat Perrier = [42 PHI]
Eugenia dodoneifolia var. *grandifolia* O.Berg = ***Eugenia dodonaeifolia*** Cambess.
Eugenia dolichophylla Kiaersk. = ***Eugenia pitanga*** (O.Berg) Nied.
Eugenia doligophylla Koord. & Valeton = ***Syzygium confusum*** (Blume) Bakh.f.
Eugenia dombeyana DC. = ***Myrceugenia lanceolata*** (Juss. ex J.St.-Hil.) Kausel
Eugenia dombeyi Skeels = ***Eugenia brasiliensis*** Lam.
Eugenia domestica Baill. = ***Syzygium malaccense*** (L.) Merr. & L.M.Perry
Eugenia domingensis var. *membranacea* O.Berg = ***Eugenia domingensis*** O.Berg
Eugenia doniana O.Berg = ***Eugenia stictopetala*** DC.

Eugenia dugandii Standl. = ***Myrcianthes fragrans*** (Sw.) McVaugh
Eugenia dumetorum DC. = ***Rhodamnia dumetorum*** (DC.) Merr. & L.M.Perry
Eugenia dumicola Barb.Rodr. = ***Eugenia pyriformis*** Cambess.
Eugenia duplomarginata Greves = ***Syzygium duplomarginatum*** (Greves) Merr. & L.M.Perry
Eugenia dupontii Baker = ***Syzygium dupontii*** (Baker) Govaerts
Eugenia dura Merr. = ***Syzygium durum*** (Merr.) Merr.
Eugenia durifolia A.C.Sm. = ***Syzygium phaeophyllum*** Merr. & L.M.Perry
Eugenia dussii Krug & Urb. = ?
Eugenia duthieana King = ***Syzygium duthieanum*** (King) Masam.
Eugenia dyeriana King = ***Syzygium dyerianum*** (King) Chantaran. & J.Parn.
Eugenia dysantha Benth. = ***Eugenia biflora*** (L.) DC.
Eugenia ebracteata Phil. = ***Luma apiculata*** (DC.) Burret
Eugenia eburnea Gagnep. = ***Syzygium eburneum*** (Gagnep.) Merr. & L.M.Perry
Eugenia ecostulata Elmer = ***Syzygium ecostulatum*** (Elmer) Merr.
Eugenia edulis (O.Berg) Kiaersk. = ***Eugenia selloi*** B.D.Jacks.
Eugenia edulis (O.Berg) Benth. & Hook.f. ex Griseb. = ***Hexachlamys edulis*** (O.Berg) Kausel & D.Legrand
Eugenia edulis Vell. = ***Plinia edulis*** (Vell.) Sobral
Eugenia effusa A.Gray = ***Syzygium effusum*** (A.Gray) Müll.Stuttg.
Eugenia egensis var. *angustifolia* O.Berg = ***Eugenia egensis*** DC.
Eugenia egensis var. *bimarginata* O.Berg = ***Eugenia egensis*** DC.
Eugenia egensis var. *grandifolia* O.Berg = ***Eugenia egensis*** DC.
Eugenia egensis var. *latifolia* O.Berg = ***Eugenia egensis*** DC.
Eugenia egensis var. *parvifolia* O.Berg = ***Eugenia egensis*** DC.
Eugenia egensis var. *tenuiramis* (Miq.) O.Berg = ***Syzygium zeylanicum*** (L.) DC.
Eugenia ekmanii (Urb.) Mattos = ***Hottea ekmanii*** (Urb.) Borhidi
Eugenia elegans O.Berg = ***Myrceugenia glaucescens*** (Cambess.) D.Legrand & Kausel var. ***glaucescens***
Eugenia elliptica Phil. = ***Blepharocalyx cruckshanksii*** (Hook. & Arn.) Nied.
Eugenia elliptica Hook. & Arn. = ***Myrcianthes cisplatensis*** (Cambess.) O.Berg
Eugenia elliptica Sm. = ***Syzygium smithii*** (Poir.) Nied.
Eugenia elliptica var. *levinervis* Fosberg = ***Eugenia levinervis*** (Fosberg) A.J.Scott
Eugenia elliptifolia Merr. = ***Syzygium elliptifolium*** (Merr.) Merr.
Eugenia elliptilimba Merr. = ***Syzygium elliptilimbum*** (Merr.) Merr. & L.M.Perry
Eugenia elmeri Merr. = ***Syzygium fastigiatum*** (Blume) Merr. & L.M.Perry
Eugenia elopurae Ridl. = ***Syzygium elopurae*** (Ridl.) Merr. & L.M.Perry
Eugenia emarginata Vahl = [81 LEE]
Eugenia embelioides Ridl. = ***Syzygium embelioides*** (Ridl.) Masam.
Eugenia emirnensis Baker = ***Syzygium emirnense*** (Baker) Labat & Schatz
Eugenia emirnensis f. *cuneifolia* (Baker) H.Perrier = ***Syzygium emirnense*** (Baker) Labat & Schatz
Eugenia emirnensis var. *elongata* Hochr. = ***Syzygium emirnense*** (Baker) Labat & Schatz
Eugenia emirnensis var. *podocarpifolia* H.Perrier = ***Syzygium emirnense*** (Baker) Labat & Schatz
Eugenia emirnensis f. *podocarpifolia* (H.Perrier) H.Perrier = ***Syzygium emirnense*** (Baker) Labat & Schatz
Eugenia emirnensis var. *submaritima* H.Perrier = ***Syzygium emirnense*** (Baker) Labat & Schatz
Eugenia emirnensis var. *subrotundifolia* H.Perrier = ***Syzygium emirnense*** (Baker) Labat & Schatz
Eugenia emirnensis f. *subrotundifolia* (H.Perrier) H.Perrier = ***Syzygium emirnense*** (Baker) Labat & Schatz
Eugenia enormis (McVaugh) Mattos = ***Calycorectes enormis*** McVaugh
Eugenia erythrocalyx C.T.White = ***Syzygium erythrocalyx*** (C.T.White) B.Hyland
Eugenia erythrodoxa S.Moore = ***Syzygium erythrodoxum*** (S.Moore) B.Hyland
Eugenia eschholtziana O.Berg = ***Eugenia stictopetala*** DC.
Eugenia eschholtziana var. *angustifolia* O.Berg = ***Eugenia stictopetala*** DC.
Eugenia eschholtziana var. *latifolia* O.Berg = ***Eugenia stictopetala*** DC.
Eugenia escuintlensis Lundell = ***Eugenia acapulcensis*** Steud.
Eugenia esculenta O.Berg = ***Pseudanamomis umbellulifera*** (Kunth) Kausel
Eugenia esnardiana Urb. & Ekman = ***Myrcianthes esnardiana*** (Urb. & Ekman) Alain
Eugenia esquirolii H.Lév. = ***Decaspermum gracilentum*** (Hance) Merr. & L.M.Perry
Eugenia essingtoniana S.Moore = ***Syzygium minutuliflorum*** (F.Muell.) B.Hyland
Eugenia estrellensis O.Berg = ***Myrceugenia myrcioides*** (Cambess.) O.Berg
Eugenia eucalyptoides (F.Muell.) F.Muell. = ***Syzygium eucalyptoides*** (F.Muell.) B.Hyland
Eugenia eucaudata Elmer ex Merr. = ***Syzygium acuminatissimum*** (Blume) DC.
Eugenia euneura (Miq.) Craib = ***Syzygium euneuron*** Miq.
Eugenia euonymifolia F.P.Metcalf = ***Syzygium euonymifolium*** (F.P.Metcalf) Merr. & L.M.Perry
Eugenia euosma O.Berg = ***Myrceugenia euosma*** (O.Berg) D.Legrand
Eugenia euosma var. *lutescens* O.Berg = ***Myrceugenia euosma*** (O.Berg) D.Legrand
Eugenia euosma var. *rufescens* O.Berg = ***Myrceugenia euosma*** (O.Berg) D.Legrand
Eugenia euphlebia Merr. = ***Syzygium escritorii*** Merr.
Eugenia euphlebia Hayata = ***Syzygium euphlebium*** (Hayata) Mori
Eugenia eurycheila var. *grandifolia* O.Berg = ***Eugenia eurycheila*** O.Berg
Eugenia eurycheila var. *parvifolia* O.Berg = ***Eugenia eurycheila*** O.Berg
Eugenia eurysepala Kiaersk. = ***Eugenia nutans*** O.Berg
Eugenia eustipitata Lundell = ***Eugenia savannarum*** Standl. & Steyerm.
Eugenia eutenuipes Lundell = ***Eugenia oerstediana*** O.Berg
Eugenia evansii Ridl. = ***Syzygium racemosum*** (Blume) DC.
Eugenia everettii C.B.Rob. = ***Syzygium everettii*** (C.B. Rob.) Merr.
Eugenia exechusa O.Berg = ***Eugenia arenaria*** Cambess.
Eugenia exechusa var. *uniflora* Kiaersk. = ***Eugenia arenaria*** Cambess.

Eugenia expallens O.Berg = ***Myrceugenia bracteosa*** (DC.) D.Legrand & Kausel
Eugenia expansa Duthie = ***Syzygium racemosum*** (Blume) DC.
Eugenia exsucca DC. = ***Myrceugenia exsucca*** (DC.) O.Berg
Eugenia exsucca var. *apiculata* O.Berg = ***Myrceugenia exsucca*** (DC.) O.Berg
Eugenia exsucca var. *patagua* O.Berg = ***Myrceugenia exsucca*** (DC.) O.Berg
Eugenia exsucca var. *peruviana* O.Berg = ***Myrceugenia exsucca*** (DC.) O.Berg
Eugenia exsucca var. *temu* (Hook. & Arn.) O.Berg = ***Myrceugenia exsucca*** (DC.) O.Berg
Eugenia ezechiasii Mattos = ***Calycorectes heringerianus*** Mattos
Eugenia fadyenii Krug & Urb. = ***Eugenia aeruginea*** DC.
Eugenia fadyenii var. *glabra* Krug & Urb. = ***Eugenia aeruginea*** DC.
Eugenia fajardensis (Krug & Urb.) Urb. = ***Myrcianthes fragrans*** (Sw.) McVaugh
Eugenia falkenbergiana Mattos = ***Calycorectes duarteanus*** D.Legrand
Eugenia fallax Rich. = ***Myrcia splendens*** (Sw.) DC.
Eugenia fasciculata J.Guého & A.J.Scott = ***Eugenia neofasciculata*** Bennet
Eugenia fasciculata Wall. ex Blume = ***Eugenia roxburghii*** DC.
Eugenia fasciculiflora O.Berg = ***Eugenia moschata*** (Aubl.) Nied. ex T.Durand & B.D.Jacks.
Eugenia fastigiata (Blume) Koord. & Valeton = ***Syzygium fastigiatum*** (Blume) Merr. & L.M.Perry
Eugenia feijoi O.Berg = ***Eugenia moschata*** (Aubl.) Nied. ex T.Durand & B.D.Jacks.
Eugenia felisbertii DC. = ***Myrcia felisbertii*** (DC.) O.Berg
Eugenia fenicis C.B.Rob. = ***Syzygium fenicis*** (C.B.Rob.) Merr.
Eugenia fenzliana O.Berg = ***Eugenia patens*** Poir.
Eugenia fergusonii Trimen = ***Syzygium fergusonii*** (Trimen) Gamble
Eugenia fergusonii var. *minor* Trimen = ***Syzygium montis-adam*** Kosterm.
Eugenia fernandeziana (Hook. & Arn.) Barnéoud = ***Myrceugenia fernandeziana*** (Hook. & Arn.) Johow
Eugenia ferreyrae McVaugh = ***Myrcianthes ferreyrae*** (McVaugh) McVaugh
Eugenia ferruginea A.Rich. = ***Calyptranthes cuprea*** O.Berg
Eugenia ferruginea Poir. = ***Marlierea ferruginea*** (Poir.) McVaugh
Eugenia ferruginea Hook. & Arn. = ***Myrceugenia rufa*** (Colla) Skottsb.
Eugenia ferruginea Sieber ex C.Presl = ***Psidium cattleianum*** Afzel. ex Sabine
Eugenia ferruginea Wight = ***Syzygium thumra*** (Roxb.) Merr. & L.M.Perry subsp. ***thumra***
Eugenia ferruginosa Mattos = ***Calycorectes ferrugineus*** Mattos
Eugenia fetida Poir. = ***Eugenia foetida*** Pers.
Eugenia fibrosa F.M.Bailey = ***Syzygium fibrosum*** (F.M.Bailey) T.G.Hartley & L.M.Perry
Eugenia fieldingii O.Berg = ***Eugenia biflora*** (L.) DC.
Eugenia filiformis Macfad. = ***Eugenia confusa***
Eugenia filiformis Wall. ex Duthie = ***Syzygium filiforme*** (Wall. ex Duthie) Chantaran. & J.Parn.
Eugenia filiformis var. *constricta* Kochummen = ***Syzygium filiforme*** (Wall. ex Duthie) Chantaran. & J.Parn.
Eugenia filiformis var. *parvifolia* Craib = ***Syzygium urophyllum*** Merr.
Eugenia filipes Baill. = ***Eugenia brasiliensis*** Lam.
Eugenia fimbriata (Kunth) DC. = ***Myrcianthes fimbriata*** (Kunth) McVaugh
Eugenia finetii Gagnep. = ***Syzygium finetii*** (Gagnep.) Merr. & L.M.Perry
Eugenia finisterrae Lauterb. = ***Syzygium finisterrae*** (Lauterb.) Merr. & L.M.Perry
Eugenia firma DC. = ***Eugenia capparidifolia*** DC.
Eugenia firma (Blume) Koord. & Valeton = ***Syzygium ampliflorum*** (Koord. & Valeton) Amshoff
Eugenia fiscalensis Donn.Sm. = ***Eugenia rhombea*** (O.Berg) Krug & Urb.
Eugenia fischeri Mattos = [84 BZC]
Eugenia fischeri Merr. = ***Syzygium fischeri*** (Merr.) Merr.
Eugenia fissurata Mattos = ***Calycorectes langsdorffii*** O.Berg
Eugenia fitzgeraldii F.M.Bailey = ***Syzygium hodgkinsoniae*** (F.Muell.) L.A.S.Johnson
Eugenia flava O.Berg = ***Eugenia punicifolia*** (Kunth) DC.
Eugenia flavescens Ridl. = ***Syzygium flavescens*** Merr. & L.M.Perry
Eugenia flavescens var. *guyanensis* O.Berg = ***Eugenia flavescens*** DC.
Eugenia flavescens var. *longifolia* O.Berg = ***Eugenia flavescens*** DC.
Eugenia flavescens var. *parvifolia* O.Berg = ***Eugenia flavescens*** DC.
Eugenia flavicans Urb. & Ekman = ***Mosiera cabanasensis*** subsp. ***flavicans*** (Urb. & Ekman) Borhidi
Eugenia flavida Lundell = ***Eugenia flavoviridis*** Lundell
Eugenia flavifolia Standl. = ***Eugenia farameoides*** A.Rich.
Eugenia flavonigra A.Rich. ex O.Berg = ***Eugenia lambertiana*** DC.
Eugenia flavonigra var. *guadalupensis* O.Berg = ***Eugenia lambertiana*** DC.
Eugenia flavonigra var. *martinicensis* O.Berg = ***Eugenia lambertiana*** DC.
Eugenia flavovirens O.Berg = ***Eugenia monticola*** (Sw.) DC.
Eugenia flavovirens var. *obscura* O.Berg = ***Eugenia monticola*** (Sw.) DC.
Eugenia flavovirens var. *pallida* O.Berg = ***Eugenia monticola*** (Sw.) DC.
Eugenia floccifera Thwaites = ***Eugenia fulva*** Thwaites
Eugenia floribunda H.West ex Willd. = ***Myrciaria floribunda*** (H.West ex Willd.) O.Berg
Eugenia flosculifera M.R.Hend. = ***Syzygium flosculiferum*** (M.R.Hend.) Sreek.
Eugenia fluviatilis Hemsl. = ***Syzygium fluviatile*** (Hemsl.) Merr. & L.M.Perry
Eugenia foetida var. *parvifolia* O.Berg = ***Eugenia monticola*** (Sw.) DC.
Eugenia foetida var. *rhombea* O.Berg = ***Eugenia rhombea*** (O.Berg) Krug & Urb.
Eugenia foliosa (Kunth) DC. = ***Myrcianthes leucoxyla*** (Ortega) McVaugh
Eugenia fontellae Mattos = ***Calycorectes fluminensis*** Mattos
Eugenia forbesii Greves = ***Syzygium gonatanthum*** (Diels) Merr. & L.M.Perry
Eugenia formonica Urb. & Ekman = ***Eugenia picardiae*** Krug & Urb.
Eugenia formosa Cambess. = ***Eugenia neoformosa*** Sobral

Eugenia formosa Wall. = ***Syzygium formosum*** (Wall.) Masam.
Eugenia formosana Hayata = ***Syzygium formosanum*** (Hayata) Mori
Eugenia forsteri O.Berg = ***Eugenia polystachya*** Rich.
Eugenia fortis F.Muell. = ***Syzygium forte*** (F.Muell.) B.Hyland
Eugenia fourcadei Dummer = ***Syzygium guineense*** (Willd.) DC. subsp. ***guineense***
Eugenia foveolata O.Berg = ***Myrceugenia leptocalyx*** D.Legrand
Eugenia foxworthiana Ridl. = ***Syzygium foxworthianum*** (Ridl.) Merr. & L.M.Perry
Eugenia foxworthyi Ridl. = ***Syzygium foxworthianum*** (Ridl.) Merr. & L.M.Perry
Eugenia foxworthyi Elmer = ***Syzygium foxworthyi*** (Elmer) Merr.
Eugenia fragrans (Sw.) Willd. = ***Myrcianthes fragrans*** (Sw.) McVaugh
Eugenia fragrans var. *brachyrhiza* Krug & Urb. = ***Myrcianthes fragrans*** (Sw.) McVaugh
Eugenia fragrans var. *fajardensis* Krug & Urb. = ***Myrcianthes fragrans*** (Sw.) McVaugh
Eugenia fragrans var. *obovata* O.Berg = ***Myrcianthes fragrans*** (Sw.) McVaugh
Eugenia franciscensis O.Berg = ***Myrceugenia franciscensis*** (O.Berg) Landrum
Eugenia francisii F.M.Bailey = ***Syzygium francisii*** (F.M.Bailey) L.A.S.Johnson
Eugenia fraseri Ridl. = ***Syzygium claviflorum*** (Roxb.) Wall. ex A.M.Cowan & Cowan
Eugenia freireana O.Berg = ***Eugenia biflora*** (L.) DC.
Eugenia freireana var. *angustifolia* O.Berg = ***Eugenia biflora*** (L.) DC.
Eugenia freireana var. *latifolia* O.Berg = ***Eugenia biflora*** (L.) DC.
Eugenia friburgensis Glaz. = ***Eugenia candolleana*** DC.
Eugenia friedrichsthalii (O.Berg) Hemsl. = ***Ugni myricoides*** (Kunth) O.Berg
Eugenia frondosa Wall. ex Duthie = ***Syzygium venosum*** DC.
Eugenia fruticosa (DC.) Roxb. = ***Syzygium fruticosum*** DC.
Eugenia fruticulosa DC. = ***Eugenia punicifolia*** (Kunth) DC.
Eugenia fuliginea O.Berg = ***Myrceugenia alpigena*** (DC.) Landrum var. ***alpigena***
Eugenia fuliginea var. *rufa* O.Berg = ***Myrceugenia alpigena*** (DC.) Landrum var. ***alpigena***
Eugenia fulvescens Mart. ex DC. = ***Pimenta pseudocaryophyllus*** var. ***fulvescens*** (Mart. ex DC.) Landrum
Eugenia fulvipes Sagot = ***Eugenia ramiflora*** Desv.
Eugenia furfuracea Craib = ***Syzygium putii*** Chantaran. & J.Parn.
Eugenia fuscescens Craib = ***Syzygium fuscescens*** (Craib) Chantaran. & J.Parn.
Eugenia fusiformis Duthie = ***Syzygium glabratum*** (DC.) Veldkamp
Eugenia fusticulifera Ridl. = ***Syzygium fusticuliferum*** (Ridl.) Merr. & L.M.Perry
Eugenia gageana King = ***Syzygium gageanum*** (King) I.M.Turner
Eugenia garberi Sarg. = ***Eugenia confusa*** DC.
Eugenia garciniifolia King = ***Syzygium garciniifolium*** (King) Merr. & L.M.Perry
Eugenia garcinioides Engl. & Brehmer = [23 CMN]
Eugenia garcinioides Ridl. = ***Syzygium garcinioides*** (Ridl.) Merr. & L.M.Perry
Eugenia gardneri Fern.-Vill. = ***Syzygium acuminatissimum*** (Blume) DC.
Eugenia gardneri (Thwaites) Bedd. = ***Syzygium gardneri*** Thwaites
Eugenia gardneriana O.Berg = ***Eugenia florida*** DC.
Eugenia gardneriana var. *depauperata* O.Berg = ***Eugenia florida*** DC.
Eugenia gardneriana var. *dives* O.Berg = ***Eugenia florida*** DC.
Eugenia gardneriana var. *ovata* O.Berg = ***Eugenia florida*** DC.
Eugenia gardneriana var. *rigida* O.Berg = ***Eugenia florida*** DC.
Eugenia gaultherioides Ridl. = ***Syzygium gaultherioides*** (Ridl.) Merr. & L.M.Perry
Eugenia gayana Barnéoud = ***Luma chequen*** (Molina) A.Gray
Eugenia gemmiflora O.Berg = ***Eugenia dysenterica*** DC.
Eugenia gentlei Lundell = ***Chamguava gentlei*** (Lundell) Landrum
Eugenia geraensis (D.Legrand & Mattos) Mattos = ***Hexachlamys geraensis*** D.Legrand & Mattos
Eugenia gerrardii (Harv. ex Hook.f.) Sim = ***Syzygium gerrardii*** (Harv. ex Hook.f.) Burtt Davy
Eugenia gigantea Ridl. = ***Syzygium operculatum*** (Roxb.) Nied.
Eugenia gigantifolia Merr. = ***Syzygium gigantifolium*** (Merr.) Merr.
Eugenia gilliesii Hook. & Arn. = ***Luma apiculata*** (DC.) Burret
Eugenia ginoriaefolia Britton & P.Wilson = ***Eugenia maestrensis*** Urb.
Eugenia gitingensis Elmer = ***Syzygium gitingense*** (Elmer) Merr.
Eugenia gladiata Ridl. = ***Syzygium gladiatum*** (Ridl.) Merr. & L.M.Perry
Eugenia glandulifera Roxb. = ***Syzygium zeylanicum*** (L.) DC.
Eugenia glanduliflora Kiaersk. = ***Myrciaria glanduliflora*** (Kiaersk.) Mattos & D.Legrand
Eugenia glanduligera Ridl. = ***Syzygium glanduligerum*** (Ridl.) Merr. & L.M.Perry
Eugenia glandulosa Blanco = [42 PHI]
Eugenia glandulosissima Kiaersk. = ***Eugenia candolleana*** DC.
Eugenia glareosa O.Berg = ***Eugenia punicifolia*** (Kunth) DC.
Eugenia glauca King = ***Syzygium glaucum*** (King) Chantaran. & J.Parn.
Eugenia glauca var. *pseudoglauca* King = ***Syzygium glaucum*** (King) Chantaran. & J.Parn.
Eugenia glaucescens Cambess. = ***Myrceugenia glaucescens*** (Cambess.) D.Legrand & Kausel
Eugenia glaucicalyx Merr. = ***Syzygium antisepticum*** (Blume) Merr. & L.M.Perry
Eugenia glaucissima Haines = ***Syzygium glaucissimum*** (Haines) Rathakr. & N.C.Nair
Eugenia glazioviana Kiaersk. = ***Eugenia francavilleana*** O.Berg
Eugenia glazioviana (Kiaersk.) D.Legrand = ?
Eugenia glazioviana var. *macrophylla* Kiaersk. = ***Eugenia francavilleana*** O.Berg
Eugenia glazioviana f. *parvifolia* Kiaersk. = ***Eugenia francavilleana*** O.Berg
Eugenia glazioviana var. *pauciflora* Kiaersk. = ***Eugenia francavilleana*** O.Berg
Eugenia globiflora Craib = ***Syzygium globiflorum*** (Craib) Chantaran. & J.Parn.

Eugenia globosa Elmer = ***Syzygium globosum*** (Elmer) Merr.
Eugenia glomerata Spring ex Mart. = ***Eugenia neoglomerata*** Sobral
Eugenia glomerata Sieber ex C.Presl = ***Calyptranthes zuzygium*** (L.) Sw.
Eugenia glomerata Lam. = ***Syzygium glomeratum*** (Lam.) DC.
Eugenia glomerata Koord. & Valeton = ***Syzygium glomeruliferum*** Amshoff
Eugenia glomerata Warb. = ***Syzygium warburgii*** Merr. & L.M.Perry
Eugenia glomerata (K.Schum. & Lauterb.) Diels = ***Syzygium warburgii*** Merr. & L.M.Perry
Eugenia glomerata var. *grandifolia* O.Berg = ***Eugenia neoglomerata*** Sobral
Eugenia glomerata var. *parvifolia* O.Berg = ***Eugenia neoglomerata*** Sobral
Eugenia glomerulata Gagnep. = ***Syzygium glomerulatum*** (Gagnep.) Merr. & L.M.Perry
Eugenia goniocalyx Ridl. = ***Syzygium cameronum*** I.M.Turner
Eugenia goodenovii King = ***Syzygium goodenovii*** (King) Masam.
Eugenia goudotiana H.Perrier = ***Syzygium zeylanicum*** (L.) DC.
Eugenia gracilenta Hance = ***Decaspermum gracilentum*** (Hance) Merr. & L.M.Perry
Eugenia gracilipes A.Gray = ***Syzygium gracilipes*** (A.Gray) Merr. & L.M.Perry
Eugenia gracilis Bedd. = ***Eugenia mooniana*** Wight
Eugenia gracilis O.Berg = [84 BZL]
Eugenia graeme-andersoniae Ridl. = ***Syzygium graeme-andersoniae*** (Ridl.) I.M.Turner
Eugenia grandiflora (O.Berg) Mattos = ***Eugenia megaflora*** Govaerts
Eugenia grandiglandulosa Kiaersk. = ***Myrciaria dubia*** (Kunth) McVaugh
Eugenia grandis Wight = ***Syzygium grande*** (Wight) Walp.
Eugenia grandisepala Mattos = ***Eugenia cuprea*** (O.Berg) Nied.
Eugenia granulata O.Berg = ***Myrcianthes fragrans*** (Sw.) McVaugh
Eugenia grata Wight = ***Syzygium antisepticum*** (Blume) Merr. & L.M.Perry
Eugenia grayi Seem. = ***Syzygium grayi*** (Seem.) Merr. & L.M.Perry
Eugenia grenadensis Urb. = ***Eugenia trinitatis*** DC.
Eugenia griffithii Duthie = ***Syzygium griffithii*** (Duthie) Merr. & L.M.Perry
Eugenia grijsii Hance = ***Syzygium grijsii*** (Hance) Merr. & L.M.Perry
Eugenia grisea Mattos = [84 BZC]
Eugenia grisea C.B.Rob. = ***Syzygium griseum*** (C.B.Rob.) Airy Shaw
Eugenia grisii Baker f. = ***Syzygium austrocaledonicum*** (Seem.) Guillaumin
Eugenia guabiju O.Berg = ***Eugenia uruguayensis*** Cambess.
Eugenia guadalupensis DC. = ***Eugenia axillaris*** (Sw.) Willd. var. ***axillaris***
Eugenia guanacastensis Standl. = ***Eugenia truncata*** O.Berg
Eugenia guapurium DC. = ***Plinia peruviana*** (Poir.) Govaerts
Eugenia guaquiea Kiaersk. = ***Myrciaria guaquiea*** (Kiaersk.) Mattos & D.Legrand
Eugenia guayabillo A.Rich. = ***Myrcianthes fragrans*** (Sw.) McVaugh
Eugenia guayavillo Benth. = ***Psidium salutare*** var. ***salutare***
Eugenia gudilla (Colla) Barnéoud = ***Myrceugenia lanceolata*** (Juss. ex J.St.-Hil.) Kausel
Eugenia gueinzii Sond. = ***Eugenia capensis*** subsp. ***gueinzii*** (Sond.) F.White
Eugenia guianensis Aubl. = ***Myrcia guianensis*** (Aubl.) DC.
Eugenia guilii Speg. = ***Amomyrtella guilii*** (Speg.) Kausel
Eugenia guineensis (Willd.) Baill. ex Laness. = ***Syzygium guineense*** (Willd.) DC.
Eugenia guineensis var. *huillensis* Hiern = ***Syzygium guineense*** subsp. ***huillense*** (Hiern) F.White
Eugenia gustavioides F.M.Bailey = ***Syzygium gustavioides*** (F.M.Bailey) B.Hyland
Eugenia haeckeliana Trimen = ***Eugenia fulva*** Thwaites
Eugenia hagelundii Mattos = ***Eugenia arenosa*** Mattos
Eugenia hagendorffii (O.Berg) Kiaersk. = ***Plinia rivularis*** (Cambess.) Rotman
Eugenia hainanensis Merr. = ***Decaspermum hainanense*** (Merr.) Merr.
Eugenia hallii O.Berg = ***Myrcianthes hallii*** (O.Berg) McVaugh
Eugenia halophila Merr. = ***Syzygium halophilum*** (Merr.) Masam.
Eugenia hamiltonii (Mattos) Mattos = ***Hexachlamys hamiltonii*** Mattos
Eugenia handroi (Mattos) Mattos = ***Hexachlamys handroi*** Mattos
Eugenia haniffii M.R.Hend. = ***Syzygium haniffii*** (M.R.Hend.) I.M.Turner
Eugenia hankeana H.J.P.Winkl. = ***Eugenia kameruniana*** Engl.
Eugenia harmandii Gagnep. = ***Syzygium harmandii*** (Gagnep.) Merr. & L.M.Perry
Eugenia harrisii var. *grandiflora* Krug & Urb. = ***Eugenia harrisii*** Krug & Urb.
Eugenia hartii Kiaersk. = ***Eugenia biflora*** (L.) DC.
Eugenia hassleriana Barb.Rodr. = ***Eugenia pyriformis*** Cambess.
Eugenia hauthalii (Kuntze) K.Schum. = ***Psidium guineense*** Sw.
Eugenia havanensis Britton & P.Wilson = ***Eugenia camarioca*** C.Wright
Eugenia havilandii Merr. = ***Syzygium havilandii*** (Merr.) Merr. & L.M.Perry
Eugenia hazompasika var. *extensa* H.Perrier = ***Eugenia hazompasika*** H.Perrier
Eugenia heckelii Pancher & Sebert = ***Austromyrtus vieillardii*** (Brongn. & Gris) Burret
Eugenia hedraiophylla F.Muell. = ***Syzygium hedraiophyllum*** (F.Muell.) Craven & Biffin
Eugenia helferi Duthie = ***Syzygium helferi*** (Duthie) Chantaran. & J.Parn.
Eugenia hemilampra F.Muell. = ***Syzygium hemilamprum*** (F.Muell.) Craven & Biffin
Eugenia hemispherica Wight = ***Syzygium hemisphericum*** (Wight) Alston
Eugenia hemsleyana King = ***Syzygium hemsleyanum*** (King) Chantaran. & J.Parn.
Eugenia henryi Hance = ***Syzygium championii*** (Benth.) Merr. & L.M.Perry
Eugenia herbacea A.Chev. = ***Eugenia subherbacea*** A.Chev.
Eugenia hermannii Cordem. = ***Eugenia mespiloides*** Lam.
Eugenia heteroclada Merr. = ***Syzygium scortechinii*** (King) Chantaran. & J.Parn.

Eugenia heteroclita Tussac = ***Myrcianthes fragrans*** (Sw.) McVaugh
Eugenia heteromorpha Cordem. = ***Eugenia buxifolia*** Lam.
Eugenia heterophylla Urb. = ***Eugenia ramoniana*** Urb.
Eugenia hexovulata McVaugh = ?
Eugenia heyneana Duthie = ***Syzygium salicifolium*** (Wight) J.Graham
Eugenia heyneana var. *alternans* Duthie = ***Syzygium salicifolium*** (Wight) J.Graham
Eugenia heynei Rathakr. & N.C.Nair = ***Eugenia roxburghii*** DC.
Eugenia hintonii Lundell = ***Eugenia crenularis*** Lundell
Eugenia hirsuta Ruiz & Pav. = ***Myrcia guianensis*** (Aubl.) DC.
Eugenia hislopii F.M.Bailey = ***Syzygium cormiflorum*** (F.Muell.) B.Hyland
Eugenia hodgkinsoniae F.Muell. = ***Syzygium hodgkinsoniae*** (F.Muell.) L.A.S.Johnson
Eugenia hoffmannseggii O.Berg = ***Eugenia biflora*** (L.) DC.
Eugenia hoffmannseggii var. *grandifolia* O.Berg = ***Eugenia biflora*** (L.) DC.
Eugenia hoffmannseggii var. *parvifolia* O.Berg = ***Eugenia biflora*** (L.) DC.
Eugenia holmanii Elmer = ***Syzygium polyanthum*** (Wight) Walp.
Eugenia holttumii Ridl. = ***Syzygium curtisii*** (King) Merr. & L.M.Perry
Eugenia holtzei F.Muell. = ***Syzygium operculatum*** (Roxb.) Nied.
Eugenia holtziana F.Muell. = ***Syzygium operculatum*** (Roxb.) Nied.
Eugenia homei Seem. = ***Eugenia gacognei*** Montrouz.
Eugenia hookeri Steud. = ***Luma apiculata*** (DC.) Burret
Eugenia horsfieldii (Miq.) Koord. & Valeton = ?
Eugenia hoseana King = ***Syzygium hoseanum*** (King) Merr. & L.M.Perry
Eugenia hulletiana King = ***Syzygium hulletianum*** (King) Chantaran. & J.Parn.
Eugenia humblotii H.Perrier = ***Syzygium humblotii*** Labat & Schatz
Eugenia humifusa Phil. = ***Myrteola nummularia*** (Lam.) O.Berg
Eugenia humilis (O.Berg) Benth. ex Nied. = ***Hexachlamys humilis*** O.Berg
Eugenia humilis Vell. = ***Myrceugenia myrcioides*** (Cambess.) O.Berg
Eugenia hutchinsonii C.B.Robinson = ***Syzygium hutchinsonii*** (C.B.Robinson) Merr.
Eugenia hydrophila Baker f. = ***Syzygium pancheri*** Brongn. & Gris
Eugenia hyemalis var. *marginata* (O.Berg) D.Legrand = ***Eugenia hyemalis*** Cambess.
Eugenia hypericifolia Gardner = ***Myrceugenia myrcioides*** (Cambess.) O.Berg
Eugenia hypericifolia (DC.) Koord. & Valeton = ***Syzygium blumei*** (Steud.) Merr. & L.M.Perry
Eugenia hypoleuca Thwaites ex Kosterm. = ***Eugenia codyensis*** Munro ex Wight
Eugenia hypospodia F.Muell. = ***Eugenia reinwardtiana*** (Blume) A.Cunn. ex DC.
Eugenia ibitensis Drake = ***Syzygium parkeri*** (Baker) Labat & Schatz
Eugenia ibitipocensis Cambess. = ***Myrceugenia bracteosa*** (DC.) D.Legrand & Kausel
Eugenia ilhensis O.Berg = ***Eugenia maritima*** DC.
Eugenia ilocana Merr. = ***Syzygium ilocanum*** (Merr.) Merr.
Eugenia inaequiloba DC. = ***Myrcia inaequiloba*** (DC.) Lemée
Eugenia inasensis King = ***Syzygium inasense*** (King) I.M.Turner
Eugenia incanescens var. *complicata* O.Berg = ***Eugenia incanescens*** Benth.
Eugenia incanescens var. *elliptica* O.Berg = ***Eugenia incanescens*** Benth.
Eugenia incanescens var. *parvifolia* O.Berg = ***Eugenia incanescens*** Benth.
Eugenia incarnata Elmer = ***Syzygium incarnatum*** (Elmer) Merr. & L.M.Perry
Eugenia incerta Dummer = [27 NAT]
Eugenia incertissima Sobral = [85 PAR]
Eugenia incrassata Elmer = ***Syzygium incrassatum*** (Elmer) Merr.
Eugenia indifferens McVaugh = ***Myrcianthes indifferens*** (McVaugh) McVaugh
Eugenia inocarpa DC. = ***Eugenia procera*** (Sw.) Poir.
Eugenia inophylla (DC.) Roxb. = ***Syzygium inophyllum*** DC.
Eugenia inophylloides A.Gray = ***Syzygium inophylloides*** (A.Gray) Müll.Stuttg.
Eugenia insignis Thwaites = ***Eugenia fulva*** Thwaites
Eugenia insipida Cambess. = ***Eugenia punicifolia*** (Kunth) DC.
Eugenia insipida Mart. = ***Eugenia punicifolia*** (Kunth) DC.
Eugenia insularis O.Berg = ***Eugenia monticola*** (Sw.) DC.
Eugenia intermedia Koord. & Valeton = [42 JAW]
Eugenia intumescens C.B.Rob. = ***Syzygium intumescens*** (C.B.Rob.) Merr.
Eugenia inundata var. *acutifolia* O.Berg = ***Eugenia biflora*** (L.) DC.
Eugenia inundata var. *coriacea* O.Berg = ***Eugenia inundata*** DC.
Eugenia inundata var. *membranacea* O.Berg = ***Eugenia inundata*** DC.
Eugenia involucrata var. *minutifolia* Mattos & D.Legrand = ***Eugenia involucrata*** DC.
Eugenia ipehuensis Barb.Rodr. ex Chodat & Hassl. = ***Blepharocalyx salicifolius*** (Kunth) O.Berg
Eugenia irosinensis Elmer = ***Syzygium everettii*** (C.B. Rob.) Merr.
Eugenia irregularis Craib = ***Syzygium borneense*** (Miq.) Miq.
Eugenia isabelensis Quisumb. = ***Syzygium isabelense*** (Quisumb.) Merr.
Eugenia itacolumensis O.Berg = ***Eugenia cerasiflora*** Miq.
Eugenia itararensis (Mattos) Mattos = ***Hexachlamys itararensis*** Mattos
Eugenia itatiaiensis (Mattos) Mattos = ***Hexachlamys handroi*** Mattos
Eugenia itatiaiensis Kiaersk. = ***Myrceugenia ovata*** var. ***regnelliana*** (O.Berg) Landrum
Eugenia itzana Lundell = ***Eugenia axillaris*** (Sw.) Willd. var. ***axillaris***
Eugenia iwahigensis Elmer = ***Syzygium iwahigense*** (Elmer) Merr.
Eugenia ixoroides Elmer = ***Syzygium cinereum*** (Kurz) Chantaran. & J.Parn.
Eugenia jaboticaba (Vell.) Kiaersk. = ***Plinia cauliflora*** (Mart.) Kausel
Eugenia jaguariaivensis Mattos = ***Eugenia calycina*** var. ***herbacea*** (O.Berg) Mattos

Eugenia jaguariaivensis var. *brevipedunculata* Mattos = ***Eugenia calycina*** var. ***herbacea*** (O.Berg) Mattos
Eugenia jamaicensis O.Berg = ***Eugenia biflora*** (L.) DC.
Eugenia jamboides Wender. = ***Syzygium jambos*** (L.) Alston
Eugenia jambolana Lam. = ***Syzygium cumini*** (L.) Skeels
Eugenia jambolana var. *caryophyllifolia* (Lam.) Duthie = ***Syzygium cumini*** (L.) Skeels
Eugenia jambolana var. *obtusifolia* Duthie = ***Syzygium cumini*** (L.) Skeels
Eugenia jambolifera Roxb. ex Wight & Arn. = ***Syzygium cumini*** (L.) Skeels
Eugenia jamboloides Koord. & Valeton = ***Syzygium racemosum*** (Blume) DC.
Eugenia jambos L. = ***Syzygium jambos*** (L.) Alston
Eugenia jambosa Crantz = ***Syzygium jambos*** (L.) Alston
Eugenia jambosoides (Lauterb.) O.Schwarz = ***Syzygium suborbiculare*** (Benth.) T.G.Hartley & L.M.Perry
Eugenia jasminifolia Ridl. = ***Syzygium jasminifolium*** (Ridl.) Chantaran. & J.Parn.
Eugenia javanica Lam. = ***Syzygium samarangense*** (Blume) Merr. & L.M.Perry
Eugenia javanica var. *parviflora* Craib = ***Syzygium samarangense*** (Blume) Merr. & L.M.Perry
Eugenia javensis Koord. & Valeton = ***Syzygium pyrifolium*** (Blume) DC.
Eugenia jequitinhonhensis Cambess. = ***Eugenia candolleana*** DC.
Eugenia jeremiensis Urb. & Ekman = ***Calyptrogenia jeremiensis*** (Urb. & Ekman) Burret
Eugenia johnsonii F.Muell. = ***Syzygium johnsonii*** (F.Muell.) B.Hyland
Eugenia johorensis Ridl. = ***Syzygium pulaiense*** I.M. Turner
Eugenia johorensis Ridl. = ***Syzygium rugosum*** Korth.
Eugenia jossinia Duthie = ***Eugenia indica*** (Wight) Chithra
Eugenia jugalis Ridl. = ***Syzygium oreophilum*** I.M.Turner
Eugenia jugorum Craib = ***Syzygium laetum*** subsp. ***jugorum*** (Craib) Chantaran. & J.Parn.
Eugenia junghuhniana Miq. = ***Syzygium polyanthum*** (Wight) Walp.
Eugenia jurujubensis Kiaersk. = ***Eugenia prasina*** O.Berg
Eugenia jutiapensis Standl. & Steyerm. = ***Eugenia oreinoma*** O.Berg
Eugenia kabaensis Greves = ***Syzygium kabaense*** (Greves) Govaerts
Eugenia kalahiensis Miq. = ***Syzygium kalahiense*** Korth.
Eugenia kanalaensis Hochr. = ***Austromyrtus kanalaensis*** (Hochr.) Burret
Eugenia kanarensis Talbot = ***Syzygium kanarense*** (Talbot) Raizada
Eugenia kangeanensis Valeton = ***Eugenia reinwardtiana*** (Blume) A.Cunn. ex DC.
Eugenia karsteniana O.Berg = ***Myrcianthes karsteniana*** (O.Berg) McVaugh
Eugenia karsteniana var. *albicans* O.Berg = ***Myrcianthes karsteniana*** (O.Berg) McVaugh
Eugenia karsteniana var. *virens* O.Berg = ***Myrcianthes karsteniana*** (O.Berg) McVaugh
Eugenia kashotoensis Hayata = ***Syzygium paucivenium*** (C.B.Rob.) Merr.
Eugenia kellermanii Lundell = ***Eugenia pittieri*** Standl.
Eugenia kemamanensis M.R.Hend. = ***Syzygium kemamanense*** (M.R.Hend.) I.M.Turner
Eugenia khasiana Duthie = ***Syzygium khasianum*** (Duthie) N.P.Balakr.
Eugenia kiaerskoviana Mattos & D.Legrand = ***Eugenia moschata*** (Aubl.) Nied. ex T.Durand & B.D.Jacks.
Eugenia kiahii M.R.Hend. = ***Syzygium kiahii*** (M.R.Hend.) I.M.Turner
Eugenia kiahii var. *angustifolia* M.R.Hend. = ***Syzygium kiahii*** (M.R.Hend.) I.M.Turner
Eugenia kiauensis Merr. = ***Syzygium kiauense*** (Merr.) Merr. & L.M.Perry
Eugenia kinabaluensis Stapf = ***Syzygium kinabaluense*** (Stapf) Merr. & L.M.Perry
Eugenia kingii Merr. = ***Syzygium scortechinii*** (King) Chantaran. & J.Parn.
Eugenia klampok (Miq.) Koord. & Valeton = ***Syzygium klampok*** (Miq.) Amshoff
Eugenia klappenbachiana Mattos & D.Legrand = ***Eugenia gracillima*** Kiaersk.
Eugenia klossii Ridl. = ***Syzygium klossii*** (Ridl.) Masam.
Eugenia klotzschiana var. *glabrata* O.Berg = ***Eugenia klotzschiana*** O.Berg
Eugenia kochiana DC. = ***Eugenia punicifolia*** (Kunth) DC.
Eugenia koikokoensis Greves = ?
Eugenia koolauensis var. *glabra* O.Deg. = ***Eugenia reinwardtiana*** (Blume) A.Cunn. ex DC.
Eugenia koordersiana King = ***Syzygium koordersianum*** (King) I.M.Turner
Eugenia koordersii Herter = [42 JAW]
Eugenia korthalsiana Miq. = ***Syzygium korthalsianum*** (Miq.) Miq.
Eugenia krugii Kiaersk. = ***Eugenia confusa*** DC.
Eugenia kuchingensis Merr. = ***Syzygium palembanicum*** Miq.
Eugenia kunstleri King = ***Syzygium kunstleri*** (King) Bahadur & R.C.Gaur
Eugenia kunthiana DC. = ***Eugenia punicifolia*** (Kunth) DC.
Eugenia kunthiana var. *opaca* O.Berg = ***Eugenia punicifolia*** (Kunth) DC.
Eugenia kunthiana var. *pellucida* O.Berg = ***Eugenia punicifolia*** (Kunth) DC.
Eugenia kuranda F.M.Bailey = ***Syzygium kuranda*** (F.M.Bailey) B.Hyland
Eugenia kurzii Duthie = ***Syzygium kurzii*** (Duthie) N.P.Balakr.
Eugenia kurzii var. *andamanica* King = ***Syzygium kurzii*** (Duthie) N.P.Balakr.
Eugenia kusukusensis Hayata = ***Syzygium kusukusuense*** (Hayata) Mori
Eugenia kwangtungensis Merr. = ***Syzygium kwangtungense*** (Merr.) Merr.
Eugenia lacustris C.B.Rob. = ***Syzygium lacustre*** (C.B.Rob.) Merr.
Eugenia laeta Buch.-Ham. = ***Syzygium laetum*** (Buch.-Ham.) Gandhi
Eugenia laeta var. *pauciflora* Duthie = ***Syzygium laetum*** (Buch.-Ham.) Gandhi subsp. ***laetum***
Eugenia laevicaulis Duthie = ***Syzygium oblatum*** (Roxb.) Wall. ex A.M.Cowan & Cowan
Eugenia laevifolia Ridl. = ***Syzygium acuminatissimum*** (Blume) DC.
Eugenia laevigata (O.Berg) D.Legrand = ***Eugenia involucrata*** DC.
Eugenia laevigata Miq. = [42 BOR]
Eugenia laevigata DC. = ***Marlierea laevigata*** (DC.) Kiaersk.
Eugenia laevis var. *gaumeri* (Standl.) McVaugh = ***Eugenia gaumeri*** Standl.
Eugenia lambertiana var. *hispidula* McVaugh = ***Eugenia lambertiana*** DC.
Eugenia lambii Elmer = ***Syzygium polyanthum*** (Wight) Walp.

Eugenia lancea Poir. = ***Eugenia biflora*** (L.) DC.
Eugenia lanceifolia Roxb. = ***Syzygium praecox*** (Roxb.) Rathakr. & N.C.Nair
Eugenia lanceolaria Roxb. = ***Syzygium lanceolarium*** (Roxb.) N.P.Balakr.
Eugenia lanceolata O.Berg = ***Eugenia fusca*** O.Berg
Eugenia lanceolata Lam. = ***Syzygium lanceolatum*** (Lam.) Wight & Arn.
Eugenia lancifolia Miq. = ***Syzygium insigne*** (Blume) Merr. & L.M.Perry
Eugenia lancilimba Merr. = ***Syzygium lancilimbum*** (Merr.) Merr.
Eugenia lanosa Mattos & D.Legrand = ***Eugenia joenssonii*** Kausel
Eugenia laosensis Gagnep. = ***Syzygium grande*** (Wight) Walp.
Eugenia laosensis var. *quocensis* Gagnep. = ***Syzygium grande*** (Wight) Walp.
Eugenia lasiopus Mart. ex DC. = ***Myrcia decorticans*** DC.
Eugenia laterifolia Pers. ex DC. = ***Eugenia sessiliflora*** Vahl
Eugenia latilimba Merr. = ***Syzygium megacarpum*** (Craib) Rathakr. & N.C.Nair
Eugenia laughlinii Lundell = ***Eugenia guatemalensis*** Donn.Sm.
Eugenia laurentii Engl. = ***Eugenia malangensis*** (O.Hoffm.) Nied.
Eugenia laurifolia Cambess. = ***Eugenia neolaurifolia*** Sobral
Eugenia laurifolia Spring ex Mart. = ***Calycorectes acutatus*** (Miq.) Toledo
Eugenia laurifolia (DC.) Roxb. = ***Syzygium laurifolium*** (DC.) N.P.Balakr.
Eugenia laurina Willd. = (Symplocaceae)
Eugenia laxiflora Poir. = ***Myrcia splendens*** (Sw.) DC.
Eugenia laxiflora (Blume) Koord. & Valeton = ***Syzygium laxiflorum*** (Blume) DC.
Eugenia laxiuscula Ridl. = ***Syzygium oblatum*** (Roxb.) Wall. ex A.M.Cowan & Cowan
Eugenia leandriana O.Berg = ***Pimenta pseudocaryophyllus*** (Gomes) Landrum var. ***pseudocaryophyllus***
Eugenia legrandii (Mattos) Mattos = ***Hexachlamys legrandii*** Mattos
Eugenia lehuntii F.M.Bailey = ***Syzygium lehuntii*** (F.M.Bailey) Merr. & L.M.Perry
Eugenia leiophloea Urb. = ***Mosiera bullata*** subsp. ***leiophloea*** (Urb.) Bisse
Eugenia leonorae Mattos = ***Calycorectes schottianus*** O.Berg
Eugenia lepida Mattos & D.Legrand = ***Eugenia handroana*** D.Legrand
Eugenia lepidocarpa Wall. ex Kurz = ***Syzygium palembanicum*** Miq.
Eugenia lepidota var. *corymbosa* O.Berg = ***Eugenia lepidota*** O.Berg
Eugenia lepidota var. *pauciflora* O.Berg = [82 GUY]
Eugenia leptalea Craib = ***Syzygium claviflorum*** (Roxb.) Wall. ex A.M.Cowan & Cowan
Eugenia leptantha Benth. = ***Eugenia inundata*** DC.
Eugenia leptantha Wight = ***Syzygium claviflorum*** (Roxb.) Wall. ex A.M.Cowan & Cowan
Eugenia leptantha var. *parvifolia* F.M.Bailey = ***Syzygium luehmannii*** (F.Muell.) L.A.S.Johnson
Eugenia leptogyna C.B.Rob. = ***Syzygium glabratum*** (DC.) Veldkamp
Eugenia leptomischa Kiaersk. = ***Eugenia gracillima*** Kiaersk.
Eugenia leptopa Lundell = ***Eugenia rhombea*** (O.Berg) Krug & Urb.
Eugenia leptophlebia Diels = ***Eugenia biflora*** (L.) DC.
Eugenia leptophylla Barb.Rodr. = ***Eugenia herbacea*** O.Berg
Eugenia leptosperma Roxb. = ?
Eugenia leptospermoides DC. = ***Myrceugenia leptospermoides*** (DC.) Kausel
Eugenia leptospermoides var. *latifolia* O.Berg = ***Myrceugenia leptospermoides*** (DC.) Kausel
Eugenia leptospermoides var. *longifolia* O.Berg = ***Myrceugenia leptospermoides*** (DC.) Kausel
Eugenia leptospermoides var. *microphylla* O.Berg = ***Myrceugenia leptospermoides*** (DC.) Kausel
Eugenia leptostemon (Korth.) Miq. = ***Syzygium leptostemon*** (Korth.) Merr. & L.M.Perry
Eugenia leptosticta Urb. = ***Eugenia camarioca*** C.Wright
Eugenia leriocarpa (C.Wright) Krug & Urb. = ***Pimenta ferruginea*** (Griseb.) Burret
Eugenia leriocarpa (C.Wright) Krug & Urb. = ***Pimenta ferruginea*** (Griseb.) Burret
Eugenia leucanthera O.Berg = ***Eugenia moschata*** (Aubl.) Nied. ex T.Durand & B.D.Jacks.
Eugenia leucocarpa Merr. = ***Syzygium mainitense*** (Elmer) Merr.
Eugenia leucocarpa Gagnep. = ***Syzygium tsoongii*** (Merr.) Merr. & L.M.Perry
Eugenia leucomyrtillus (Griseb.) Phil. = ***Myrteola nummularia*** (Lam.) O.Berg
Eugenia leucophloea (O.Berg) Kiaersk. = ***Myrciaria floribunda*** (H.West ex Willd.) O.Berg
Eugenia leucophloea var. *warmingiana* Kiaersk. = ***Myrciaria floribunda*** (H.West ex Willd.) O.Berg
Eugenia leucoxylon (Korth.) Miq. = ***Syzygium leucoxylon*** Korth.
Eugenia levinei Merr. = ***Syzygium levinei*** (Merr.) Merr.
Eugenia leytensis Elmer = ***Syzygium leytense*** (Elmer) Merr.
Eugenia ligustrina Miq. = ***Eugenia subreticulata*** Glaz.
Eugenia ligustrina var. *minensis* (O.Berg) Mattos = ***Eugenia ligustrina*** (Sw.) Willd. var. ***ligustrina***
Eugenia lilacina Merr. = ***Syzygium lilacinum*** (Merr.) Merr. & L.M.Perry
Eugenia limbata (Kunth) DC. = ***Myrcianthes fragrans*** (Sw.) McVaugh
Eugenia limnaea Ridl. = ***Syzygium oblatum*** (Roxb.) Wall. ex A.M.Cowan & Cowan
Eugenia lindeniana O.Berg = ***Eugenia capuli*** (Schltdl. & Cham.) Hook. & Arn.
Eugenia lindleyana (Kunth) DC. = ***Myrcianthes lindleyana*** (Kunth) McVaugh
Eugenia linearifolia O.Berg = ***Eugenia punicifolia*** (Kunth) DC.
Eugenia linearifolia var. *oppositifolia* O.Berg = ***Eugenia punicifolia*** (Kunth) DC.
Eugenia linearifolia var. *ternifolia* O.Berg = ***Eugenia punicifolia*** (Kunth) DC.
Eugenia linearis Duthie = ***Syzygium zeylanicum*** (L.) DC.
Eugenia linearis var. *longifolia* O.Berg = ***Eugenia linearis*** A.Rich. ex O.Berg
Eugenia linearis var. *parvifolia* O.Berg = ***Eugenia linearis*** A.Rich. ex O.Berg
Eugenia lineata (DC.) Duthie = ***Syzygium lineatum*** (DC.) Merr. & L.M.Perry
Eugenia lineata var. *bergiana* (Griseb.) Krug & Urb. = ***Eugenia lineata*** (Sw.) DC.
Eugenia lineata var. *racemosa* O.Berg = ***Eugenia lineata*** (Sw.) DC.
Eugenia lineatifolia Mattos = ***Eugenia bergii***

Eugenia lingua O.Berg = ***Eugenia cachoeirensis*** O.Berg
Eugenia linguiformis O.Berg = ***Eugenia monosperma*** Vell.
Eugenia linocieroidea King = ***Syzygium linocieroideum*** (King) I.M.Turner
Eugenia lissophylla (Thwaites) Bedd. = ***Syzygium rubicundum*** Wight & Arn.
Eugenia litseifolia Merr. = ***Syzygium borneense*** (Miq.) Miq.
Eugenia littoralis Mattos = ***Eugenia brunoi*** Mattos
Eugenia littoralis (Blume) Meijer Drees = ***Syzygium littorale*** (Blume) Amshoff
Eugenia littorea Engl. & Brehmer = ***Eugenia coronata*** Vahl ex DC.
Eugenia livida Elmer = ***Syzygium mananquil*** (Blanco) Merr.
Eugenia llanosii Merr. = ***Syzygium llanosii*** (Merr.) Merr.
Eugenia llewelynii Steyerm. = ***Eugenia chrysophyllum*** Poir.
Eugenia lobbii Ridl. = ***Syzygium multibracteolatum*** (Merr.) Merr. & L.M.Perry
Eugenia loiseleurioides Baker = ***Syzygium loiseleurioides*** (Baker) Govaerts
Eugenia lonchophylla Voigt = ***Syzygium lanceolarium*** (Roxb.) N.P.Balakr.
Eugenia longicalyx Ridl. = ***Syzygium lineatum*** (DC.) Merr. & L.M.Perry
Eugenia longicauda Ridl. = ***Syzygium zeylanicum*** (L.) DC.
Eugenia longiflora (C.Presl) Fern.-Vill. = ***Syzygium lineatum*** (DC.) Merr. & L.M.Perry
Eugenia longipedicellata (Merr.) C.B.Rob. = ***Syzygium longipedicellatum*** (Merr.) Merr.
Eugenia longipes O.Berg = ***Mosiera longipes*** (O.Berg) Small
Eugenia longipes Warb. = ***Syzygium longipes*** Merr. & L.M.Perry
Eugenia longissima Merr. = ***Syzygium longissimum*** (Merr.) Merr.
Eugenia longistyla Merr. = ***Syzygium longistylum*** (Merr.) Merr.
Eugenia lopeziana Ant.Molina = ***Myrcianthes fragrans*** (Sw.) McVaugh
Eugenia loretensis Diels = ***Eugenia biflora*** (L.) DC.
Eugenia lourteigiae (Mattos & D.Legrand) Mattos = ***Calycorectes lourteigiae*** Mattos & D.Legrand
Eugenia louvelii var. *suborbiculata* H.Perrier = ***Eugenia louvelii*** H.Perrier
Eugenia lucayana (Britton) Alain = ***Myrcianthes fragrans*** (Sw.) McVaugh
Eugenia lucescens Nied. = ***Eugenia luschnathiana*** (O.Berg) Klotzsch ex B.D.Jacks.
Eugenia lucida Cambess. = ***Eugenia cerasiflora*** Miq.
Eugenia lucida Banks ex Gaertn. = ***Gossia lucida*** (Gaertn.) N.Snow & Guymer
Eugenia lucidula Miq. = ***Syzygium polyanthum*** (Wight) Walp.
Eugenia ludibunda Bertero ex DC. = ***Eugenia biflora*** (L.) DC.
Eugenia luehmannii F.Muell. = ***Syzygium luehmannii*** (F.Muell.) L.A.S.Johnson
Eugenia lugens O.Berg = ***Eugenia patens*** Poir.
Eugenia lugubris H.Perrier = ***Syzygium lugubre*** (H.Perrier) Labat & Schatz
Eugenia luma O.Berg = ***Luma apiculata*** (DC.) Burret
Eugenia lumboy Elmer = ***Syzygium merrittianum*** (C.B. Rob.) Merr.
Eugenia lumilla Phil. = ***Myrceugenia fernandeziana*** (Hook. & Arn.) Johow
Eugenia lundellii Standl. = ***Eugenia gaumeri*** Standl.
Eugenia lundiana Kiaersk. = ***Eugenia goyazensis*** Nied.
Eugenia lunduensis Merr. = ***Syzygium lunduense*** (Merr.) Merr. & L.M.Perry
Eugenia lutea C.B.Rob. = ***Syzygium luteum*** (C.B.Rob.) Merr.
Eugenia luzonensis (Merr.) Merr. = ***Syzygium luzonense*** (Merr.) Merr.
Eugenia macancheana Lundell = ***Eugenia laevis*** O.Berg
Eugenia macarensis O.Berg = ***Eugenia biflora*** (L.) DC.
Eugenia macgregorii C.B.Rob. = ***Syzygium macgregorii*** (C.B.Rob.) Merr.
Eugenia maclurei Merr. = ***Syzygium championii*** (Benth.) Merr. & L.M.Perry
Eugenia macoorai F.M.Bailey = ***Syzygium wilsonii*** subsp. ***cryptophlebium*** (F.Muell.) B.Hyland
Eugenia macrantha var. *glomerata* O.Berg = ***Eugenia macrantha*** O.Berg
Eugenia macrocalyx (Rusby) McVaugh = ***Eugenia wentii*** Amshoff
Eugenia macrocarpa Cham. & Schltdl. = ***Eugenia macrocarpa*** Schltdl. & Cham.
Eugenia macrocarpa Roxb. = ***Syzygium megacarpum*** (Craib) Rathakr. & N.C.Nair
Eugenia macroclada O.Berg = ***Eugenia punicifolia*** (Kunth) DC.
Eugenia macrohila C.T.White & W.D.Francis = ***Eugenia reinwardtiana*** (Blume) A.Cunn. ex DC.
Eugenia macromyrtus Koord. & Valeton = ***Syzygium macromyrtus*** (Koord. & Valeton) Merr. & L.M.Perry
Eugenia macrophylla O.Berg = ***Eugenia cupulata*** var. ***macrophylla*** McVaugh
Eugenia macrophylla Lam. = ***Syzygium malaccense*** (L.) Merr. & L.M.Perry
Eugenia macrorhyncha Miq. = ***Syzygium zeylanicum*** (L.) DC.
Eugenia macrosepala (O.Berg) Mattos = ***Eugenia macrocalyx*** Mart. ex B.D.Jacks.
Eugenia macrosepala Duthie = ***Eugenia roxburghii*** DC.
Eugenia macrostemon O.Berg = ***Eugenia foetida*** Pers.
Eugenia maculata O.Berg = ***Eugenia egensis*** DC.
Eugenia magnifica Brongn. & Gris = [60 NWC]
Eugenia magnoliifolia (Blume) Koord. & Valeton = ***Syzygium magnoliifolium*** (Blume) DC.
Eugenia maingayi Duthie = ***Syzygium claviflorum*** (Roxb.) Wall. ex A.M.Cowan & Cowan
Eugenia mainitensis Elmer = ***Syzygium mainitense*** (Elmer) Merr.
Eugenia maire A.Cunn. = ***Syzygium maire*** (A.Cunn.) Sykes & Garn.-Jones
Eugenia malabarica Bedd. = ***Syzygium malabaricum*** (Bedd.) Gamble
Eugenia malaccensis Lour. = ***Syzygium aqueum*** (Burm.f.) Alston
Eugenia malaccensis Blanco = ***Syzygium jambos*** (L.) Alston
Eugenia malaccensis L. = ***Syzygium malaccense*** (L.) Merr. & L.M.Perry
Eugenia malaccensis f. *cericarpa* (O.Deg.) H.St.John = ***Syzygium jambos*** (L.) Alston
Eugenia malagsam Elmer = ***Syzygium malagsam*** (Elmer) Merr.
Eugenia malangensis Urb. & Ekman = ***Hottea malangensis*** Urb.
Eugenia malayana Gagnep. = ***Syzygium confusum*** (Blume) Bakh.f.
Eugenia maleolens Pers. = ***Eugenia monticola*** (Sw.) DC.
Eugenia mananquil Blanco = ***Syzygium mananquil*** (Blanco) Merr.

Eugenia manausensis Mattos = ***Calycorectes costatus*** Mattos & D.Legrand
Eugenia mangasiana Urb. = ***Eugenia plicatula*** C.Wright
Eugenia mangifolia Wall. ex Duthie = ***Syzygium reticulatum*** (Wight) Walp.
Eugenia manii King = ***Syzygium manii*** (King) N.P. Balakr.
Eugenia maquilingensis Elmer = ***Syzygium alvarezii*** (C.B.Rob.) Merr.
Eugenia maracayuensis Barb.Rodr. = ***Eugenia bimarginata*** DC.
Eugenia maranhensis (O.Berg) Kiaersk. = ***Myrciaria floribunda*** (H.West ex Willd.) O.Berg
Eugenia margarettae Alain = ***Myrcia margarettae*** (Alain) Alain
Eugenia marginata Poir. = ***Myrcia citrifolia***
Eugenia mariquitensis O.Berg = ***Myrcianthes rhopaloides*** (Kunth) McVaugh
Eugenia maritima (Kunth) DC. = ***Eugenia acapulcensis*** Steud.
Eugenia maritima Barnéoud = ***Myrceugenia correifolia*** (Hook. & Arn.) O.Berg
Eugenia maritima Merr. = ***Syzygium halophilum*** (Merr.) Masam.
Eugenia marivelesensis Merr. = ***Syzygium lineatum*** (DC.) Merr. & L.M.Perry
Eugenia maroana Aug.DC. = ***Syzygium bernieri*** (Baill. ex Drake) Labat & Schatz
Eugenia marquesii Engl. = ***Eugenia malangensis*** (O.Hoffm.) Nied.
Eugenia martelinoi Merr. = ***Syzygium martelinoi*** (Merr.) Merr.
Eugenia martiana (O.Berg) Mattos = ***Calycorectes martianus*** O.Berg
Eugenia martini O.Berg = ***Eugenia brownsbergii*** Amshoff
Eugenia martiusiana DC. = ***Eugenia stictopetala*** DC.
Eugenia mascarensis C.Presl = ***Syzygium cymosum*** (Lam.) DC. var. ***cymosum***
Eugenia mascarensis Cordem. = ***Syzygium cymosum*** (Lam.) DC. var. ***cymosum***
Eugenia maschalantha O.Berg = ***Eugenia uruguayensis*** Cambess.
Eugenia maschalantha Kiaersk. = ***Myrciaria delicatula*** (DC.) O.Berg
Eugenia masukuensis Baker = ***Syzygium masukuense*** (Baker) R.E.Fr.
Eugenia matanzensis O.Berg = ***Eugenia axillaris*** (Sw.) Willd. var. ***axillaris***
Eugenia mato Griseb. = ***Myrcianthes mato*** (Griseb.) McVaugh
Eugenia maxima (McVaugh) Mattos = ***Calycorectes maximus*** McVaugh
Eugenia maximiliana O.Berg = ***Eugenia punicifolia*** (Kunth) DC.
Eugenia maximiliana DC. = ***Myrciaria floribunda*** (H.West ex Willd.) O.Berg
Eugenia mayana Standl. = ***Eugenia foetida*** Pers.
Eugenia maynensis O.Berg = ***Eugenia limbosa*** O.Berg
Eugenia media (Sagot) Nied. ex T.Durand & B.D.Jacks. = ?
Eugenia megacarpa Craib = ***Syzygium megacarpum*** (Craib) Rathakr. & N.C.Nair
Eugenia megalantha C.B.Rob. = ***Syzygium megalanthum*** (C.B.Rob.) Merr.
Eugenia megalocarpa Urb. = ***Eugenia pseudopsidium*** Jacq.
Eugenia megalophylla Merr. = [42 PHI]
Eugenia mekongensis Gagnep. = ***Syzygium mekongense*** (Gagnep.) Merr. & L.M.Perry
Eugenia melanosticta Standl. = ***Eugenia florida*** DC.
Eugenia melanosticta (Miq.) Koord. & Valeton = ***Syzygium melanostictum*** (Miq.) Craven & Biffin
Eugenia melastomifolia (Blume) Merr. = ?
Eugenia melastomoides Elmer = ***Eugenia aherniana*** C.B.Rob.
Eugenia melinonis Sagot = ***Eugenia coffeifolia*** DC.
Eugenia melliodora C.B.Rob. = ***Syzygium melliodorum*** (C.B.Rob.) Merr.
Eugenia membranacea O.Berg = ***Eugenia florida*** DC.
Eugenia mendute Guillaumin = ***Austromyrtus mendute*** (Guillaumin) Burret
Eugenia meridionalis Mattos = ***Hexachlamys hamiltonii*** Mattos
Eugenia merokensis Greves = ***Syzygium merokense*** (Greves) Merr. & L.M.Perry
Eugenia merrillii C.B.Rob. = ***Syzygium mauritsii*** Govaerts
Eugenia merrittiana C.B.Rob. = ***Syzygium merrittianum*** (C.B.Rob.) Merr.
Eugenia metriosa Rojas = [85 AGE]
Eugenia mexiae Standl. = ***Eugenia salamensis*** Donn.Sm.
Eugenia mexicana Steud. = ***Eugenia macrocarpa*** Schltdl. & Cham.
Eugenia meyeriana O.Berg = ***Eugenia biflora*** (L.) DC.
Eugenia meyeriana var. *depauperata* O.Berg = ***Eugenia biflora*** (L.) DC.
Eugenia meyeriana var. *dives* O.Berg = ***Eugenia biflora*** (L.) DC.
Eugenia michelii Lam. = ***Eugenia uniflora*** L.
Eugenia michelii Parodi = [85 PAR]
Eugenia micrandra Ridl. = ***Syzygium micrandrum*** (Ridl.) Merr. & L.M.Perry
Eugenia micrantha (Kunth) DC. = ***Eugenia monticola*** (Sw.) DC.
Eugenia micrantha Bertol. = ***Pimenta dioica*** (L.) Merr.
Eugenia micrantha (Thwaites) Bedd. = ***Syzygium micranthum*** Thwaites
Eugenia microbotrya Miq. = ***Syzygium polyanthum*** (Wight) Walp.
Eugenia microcalyx Duthie = ***Syzygium borneense*** (Miq.) Miq.
Eugenia microcarpa O.Berg = ***Eugenia tinguyensis*** Cambess.
Eugenia microcarpos Lam. = ***Eugenia biflora*** (L.) DC.
Eugenia microcyma Koord. & Valeton = ***Syzygium microcymum*** (Koord. & Valeton) Amshoff
Eugenia micropetala Mart. = ***Myrcia micropetala*** (Mart.) Nied.
Eugenia microphylla A.Rich. = ***Eugenia asperifolia*** O.Berg
Eugenia microphylla A.Rich. ex O.Berg = ***Eugenia picardiae*** Krug & Urb.
Eugenia microphylla Abel = ***Syzygium buxifolium*** Hook. & Arn.
Eugenia microphylla Bedd. = ***Syzygium microphyllum*** Gamble
Eugenia micropoda Baker = ***Syzygium micropodum*** (Baker) Labat & Schatz
Eugenia micropoda var. *littoralis* H.Perrier = ***Syzygium micropodum*** (Baker) Labat & Schatz
Eugenia miegeana Aké Assi = ***Eugenia gabonensis*** Amshoff
Eugenia miersiana Gardner = ***Myrceugenia miersiana*** (Gardner) D.Legrand & Kausel
Eugenia miersiana var. *costata* O.Berg = ***Myrceugenia miersiana*** (Gardner) D.Legrand & Kausel
Eugenia miersiana var. *glomerata* O.Berg = ***Myrceugenia miersiana*** (Gardner) D.Legrand & Kausel

Eugenia miersiana var. *membranacea* O.Berg = ***Myrceugenia miersiana*** (Gardner) D.Legrand & Kausel
Eugenia miersiana var. *venosa* O.Berg = ***Myrceugenia miersiana*** (Gardner) D.Legrand & Kausel
Eugenia mikaniana O.Berg = ***Eugenia pruniformis*** Cambess.
Eugenia mikaniana DC. = ***Myrcia splendens*** (Sw.) DC.
Eugenia mikanioides O.Berg = ***Eugenia candolleana*** DC.
Eugenia milletiana Hemsl. = ***Syzygium odoratum*** (Lour.) DC.
Eugenia millsii M.R.Hend. = ***Syzygium millsii*** (M.R.Hend.) I.M.Turner
Eugenia mimica Merr. = ***Syzygium mimicum*** (Merr.) Merr.
Eugenia mindanaensis C.B.Rob. = ***Syzygium aqueum*** (Burm.f.) Alston
Eugenia mindorensis C.B.Rob. = ***Syzygium mindorense*** (C.B.Rob.) Masam.
Eugenia minensis (O.Berg) Mattos = ***Eugenia ligustrina*** (Sw.) Willd. var. ***ligustrina***
Eugenia minensis (O.Berg) Kiaersk. = ***Myrciaria cuspidata*** O.Berg
Eugenia mini Aubl. = ***Eugenia biflora*** (L.) DC.
Eugenia miniata S.Moore = ***Eugenia orbignyana*** O.Berg
Eugenia minima Blume = ***Syzygium minimum*** (Blume) Airy Shaw
Eugenia minimifolia McVaugh = ***Myrcianthes minimifolia*** (McVaugh) McVaugh
Eugenia minutiflora Hance = ***Syzygium hancei*** Merr. & L.M.Perry
Eugenia minutifolia (Mattos & D.Legrand) Mattos = ***Eugenia involucrata*** DC.
Eugenia minutifolia (Mattos) Mattos = ***Eugenia neibensis*** Mattos
Eugenia minutuliflora F.Muell. = ***Syzygium minutuliflorum*** (F.Muell.) B.Hyland
Eugenia miqueliana Greves = ?
Eugenia miquelii Elmer = ***Syzygium lineatum*** (DC.) Merr. & L.M.Perry
Eugenia mirabilis Merr. = ***Syzygium mirabile*** (Merr.) Merr.
Eugenia mirandae D.Ramírez Cantu = ***Eugenia cantuana*** Lundell
Eugenia mirandae Merr. = ***Syzygium mirandae*** (Merr.) Merr.
Eugenia mirtiflora Rojas = [85 AGE]
Eugenia moaensis R.A.Howard = ***Mosiera moaensis*** (R.A.Howard) Bisse
Eugenia modesta Phil. = ***Luma apiculata*** (DC.) Burret
Eugenia modesta var. *brasiliensis* O.Berg = ***Eugenia modesta*** DC.
Eugenia modesta var. *jamaicensis* O.Berg = ***Eugenia biflora*** (L.) DC.
Eugenia mogotensis Urb. & Ekman = ***Eugenia banderensis*** Urb.
Eugenia molliana O.Berg = [8+]
Eugenia mollis Willd. ex O.Berg = ***Myrcia splendens*** (Sw.) DC.
Eugenia mollis King = ***Syzygium pseudomolle*** (M.R.Hend.) I.M.Turner
Eugenia molokaiana K.A.Wilson & Rock = ***Eugenia koolauensis*** O.Deg.
Eugenia moluccana Merr. = ***Syzygium acuminatum*** (Roxb.) Miq.
Eugenia monantha Merr. = ***Syzygium jambos*** (L.) Alston
Eugenia monetaria Ridl. = ***Syzygium monetarium*** (Ridl.) Merr. & L.M.Perry
Eugenia montana Aubl. = ***Marlierea montana*** (Aubl.) Amshoff
Eugenia montana Cambess. = ***Myrceugenia bracteosa*** (DC.) D.Legrand & Kausel
Eugenia montana Cordem. = ***Syzygium cymosum*** var. ***montanum*** J.Guého & A.J.Scott
Eugenia montana Wight = ***Syzygium grande*** (Wight) Walp.
Eugenia montevidensis Mattos = ***Hexachlamys edulis*** (O.Berg) Kausel & D.Legrand
Eugenia monticola var. *flavovirens* (O.Berg) Stehlé & Quentin = ***Eugenia monticola*** (Sw.) DC.
Eugenia monticola var. *latifolia* Krug & Urb. = ***Eugenia monticola*** (Sw.) DC.
Eugenia monticola var. *racemosa* Amshoff = ***Eugenia monticola*** (Sw.) DC.
Eugenia montigena Barb.Rodr. = ***Eugenia pitanga*** (O.Berg) Nied.
Eugenia mooniana Gardner = ***Eugenia subundulata*** Kiaersk. var. ***subundulata***
Eugenia mooniana var. *gracilis* Duthie = ***Eugenia phillyraeoides*** Trimen
Eugenia moorei F.Muell. = ***Syzygium moorei*** (F.Muell.) L.A.S.Johnson
Eugenia moraviana O.Berg = ***Eugenia modesta*** DC.
Eugenia moraviana var. *gardneriana* (O.Berg) Mattos = ***Eugenia florida*** DC.
Eugenia moraviana var. *impunctata* O.Berg = ***Eugenia paracatuana*** O.Berg
Eugenia mornicola Urb. = ***Eugenia isabeliana*** Kiaersk.
Eugenia mosenii (Kausel) Sobral = ***Eugenia magnifica*** Spring ex Mart.
Eugenia mosquitensis O.Berg = ***Eugenia acapulcensis*** Steud.
Eugenia mossambicensis Engl. = ***Eugenia aschersoniana*** F.Hoffm.
Eugenia motleyi Ridl. = ***Syzygium rugosum*** Korth.
Eugenia moultonii Merr. = ***Syzygium moultonii*** (Merr.) Merr. & L.M.Perry
Eugenia mouririoides Lundell = ***Chamguava schippii*** (Standl.) Landrum
Eugenia mucronata Phil. = ***Luma apiculata*** (DC.) Burret
Eugenia muelleri Miq. = ***Syzygium muelleri*** (Miq.) Miq.
Eugenia mugiensis O.Berg = ***Eugenia punicifolia*** (Kunth) DC.
Eugenia multibracteolata Merr. = ***Syzygium multibracteolatum*** (Merr.) Merr. & L.M.Perry
Eugenia multicostata var. *octocostata* D.Legrand = ***Eugenia multicostata*** D.Legrand
Eugenia multiflora Cambess. = ***Eugenia hyemalis*** Cambess.
Eugenia multiflora Rich. = ***Myrcia multiflora***
Eugenia multiflora Hook. & Arn. = ***Myrceugenia exsucca*** (DC.) O.Berg
Eugenia multiflora Lam. = ***Myrcia multiflora*** (Lam.) DC.
Eugenia multiflora var. *glabra* O.Berg = ***Eugenia hyemalis*** Cambess.
Eugenia multiflora var. *lutescens* Cambess. = ***Eugenia hyemalis*** Cambess.
Eugenia multiflora var. *rubiginosa* Cambess. = ***Eugenia hyemalis*** Cambess.
Eugenia multinervis C.B.Rob. = ***Syzygium multinerve*** (C.B.Rob.) Merr.
Eugenia multiovulata Mattos & D.Legrand = ***Eugenia chlorophylla*** O.Berg
Eugenia multipetala (Pancher ex Brongn. & Gris) Baker f. = ***Syzygium multipetalum*** Pancher ex Brongn. & Gris
Eugenia multipunctata Mattos & D.Legrand = ***Eugenia gracillima*** Kiaersk.

Eugenia multipunctata Merr. = ***Decaspermum montanum*** Ridl.
Eugenia mundagam Bourd. = ***Syzygium mundagam*** (Bourd.) Chithra
Eugenia munroii Miq. = ***Syzygium munronii*** (Wight) N.P.Balakr.
Eugenia munronii Wight = ***Syzygium munronii*** (Wight) N.P.Balakr.
Eugenia muricata DC. = ***Eugenia patens*** Poir.
Eugenia muricata var. *guyanensis* O.Berg = ***Eugenia patens*** Poir.
Eugenia musarum Standl. & Steyerm. = ***Chamguava musarum*** (Standl. & Steyerm.) Landrum
Eugenia mutabilis (O.Berg) Nied. = ***Eugenia longipetiolata*** Mattos
Eugenia mutabilis O.Berg = ***Pimenta pseudocaryophyllus*** var. ***fulvescens*** (Mart. ex DC.) Landrum
Eugenia myhendrae Bedd. ex Brandis = ***Syzygium myhendrae*** (Bedd. ex Brandis) Gamble
Eugenia myrcianthes Nied. = ***Hexachlamys edulis*** (O.Berg) Kausel & D.Legrand
Eugenia myrcianthes var. *nana* D.Legrand = ***Hexachlamys humilis*** O.Berg
Eugenia myrcinifolia Hance = ?
Eugenia myrcioides Cambess. = ***Myrceugenia myrcioides*** (Cambess.) O.Berg
Eugenia myrciopsis (Kuntze) K.Schum. = ***Myrcianthes osteomeloides*** (Rusby) McVaugh
Eugenia myriadena (Merr. & L.M.Perry) Whitmore = ***Syzygium myriadenum*** Merr. & L.M.Perry
Eugenia myriantha King = ***Syzygium myrianthum*** (King) I.M.Turner
Eugenia myriophylla Casar. = ***Blepharocalyx myriophyllus*** Morais & Sobral
Eugenia myriostigma Sagot = ***Eugenia biflora*** (L.) DC.
Eugenia myrsinifolia Hance = ***Syzygium myrsinifolium*** (Hance) Merr. & L.M.Perry
Eugenia myrsinocarpa F.Muell. = ***Gossia myrsinocarpa*** (F.Muell.) N.Snow & Guymer
Eugenia myrsinoides (Kunth) Burret ex Diels = ***Myrcianthes myrsinoides*** (Kunth) Grifo
Eugenia myrtifolia Cambess. = ***Eugenia neomyrtifolia*** Sobral
Eugenia myrtifolia Salisb. = ***Eugenia uniflora*** L.
Eugenia myrtifolia Sims = ***Syzygium australe*** (J.C.Wendl. ex Link) B.Hyland
Eugenia myrtifolia Roxb. = ***Syzygium myrtifolium*** Walp.
Eugenia myrtillifolia DC. = ***Eugenia punicifolia*** (Kunth) DC.
Eugenia myrtillus Stapf = ***Syzygium myrtillus*** (Stapf) Merr. & L.M.Perry
Eugenia myrtoides G.Don = ***Eugenia coronata*** Vahl ex DC.
Eugenia myrtoides Poir. = ***Eugenia foetida*** Pers.
Eugenia myrtoides Brongn. & Gris = ***Uromyrtus neomyrtoides*** Burret
Eugenia myrtomimeta Diels = ***Luma chequen*** (Molina) A.Gray
Eugenia nana O.Berg = ***Myrceugenia euosma*** (O.Berg) D.Legrand
Eugenia nana var. *congesta* O.Berg = ***Myrceugenia euosma*** (O.Berg) D.Legrand
Eugenia nana var. *effusa* O.Berg = ***Myrceugenia euosma*** (O.Berg) D.Legrand
Eugenia nana var. *robusta* Glaz. = ***Myrceugenia euosma*** (O.Berg) D.Legrand
Eugenia nandarivatensis Gillespie = ***Syzygium nandarivatense*** (Gillespie) L.M.Perry
Eugenia napiformis Koord. & Valeton = ***Syzygium napiforme*** (Koord. & Valeton) Merr. & L.M.Perry
Eugenia natalitia Sond. = ***Eugenia capensis*** subsp. ***natalitia*** (Sond.) F.White
Eugenia neei Merr. = ***Syzygium striatulum*** (C.B.Rob.) Merr.
Eugenia neesiana (Arn.) Wight = ***Syzygium neesianum*** Arn.
Eugenia nemestrina M.R.Hend. = ***Syzygium nemestrinum*** (M.R.Hend.) I.M.Turner
Eugenia nemoralis DC. = ***Eugenia ferreiraeana*** O.Berg
Eugenia nemoricola Ridl. = ***Syzygium pseudoformosum*** (King) Merr. & L.M.Perry
Eugenia neoaustralis var. *impressovenosa* (D.Legrand) Mattos = ***Eugenia neoaustralis*** Sobral
Eugenia neocaledonica Baker f. = ***Syzygium toninense*** (Baker f.) J.W.Dawson
Eugenia neograndifolia Mattos = ***Calycorectes grandifolius*** O.Berg
Eugenia neolanceolata Sobral = ***Eugenia fusca*** O.Berg
Eugenia neomooniana Sobral = ***Eugenia subundulata*** Kiaersk. var. ***subundulata***
Eugenia neomultipunctata Sobral = ***Eugenia gracillima*** Kiaersk.
Eugenia nervosa (A.Cunn. ex DC.) Bedd. = ***Eugenia nervosa*** Lour.
Eugenia nettiana Kiaersk. = [84 BZL] *Myrcia* sp.
Eugenia neurocalyx A.Gray = ***Syzygium neurocalyx*** (A.Gray) Christoph.
Eugenia neurocalyx K.Schum. = ***Syzygium schumannianum*** (Nied.) Diels
Eugenia ngadimaniana M.R.Hend. = ***Syzygium ngadimanianum*** (M.R.Hend.) I.M.Turner
Eugenia ngoyensis Schltr. = ***Syzygium ngoyense*** (Schltr.) Guillaumin
Eugenia nhanica Cambess. = ***Eugenia punicifolia*** (Kunth) DC.
Eugenia nicaraguensis Amshoff = ***Eugenia hondurensis*** Ant.Molina
Eugenia nicholsii Fawc. & Rendle = ***Eugenia biflora*** (L.) DC.
Eugenia nicobarica King = ***Syzygium nicobaricum*** (King) Rathakr. & N.C.Nair
Eugenia nigra Pennant = [40 IND]
Eugenia nigra DC. = (Melastomataceae)
Eugenia nigrans Gagnep. = ***Syzygium nigrans*** (Gagnep.) Craven & Biffin
Eugenia nigrescens Poir. = ***Syzygium cymosum*** (Lam.) DC. var. ***cymosum***
Eugenia nigricans King = ***Syzygium nigricans*** (King) Merr. & L.M.Perry
Eugenia nitida Cambess. = ***Eugenia neonitida*** Sobral
Eugenia nitida Benth. = ***Myrcia inaequiloba*** (DC.) Lemée
Eugenia nitida Vell. = ***Myrcia subsericea*** A.Gray
Eugenia nitida Duthie = ***Syzygium polyanthum*** (Wight) Walp.
Eugenia nitidissima Merr. = ***Syzygium tenuirame*** (Miq.) Merr.
Eugenia nitidula Ridl. = ***Syzygium nitidulum*** (Ridl.) I.M.Turner
Eugenia nivifera Greves = ***Syzygium effusum*** (A.Gray) Müll.Stuttg.
Eugenia nodiflora Aubl. = ***Syzygium aqueum*** (Burm.f.) Alston
Eugenia nummularia C.Wright ex Griseb. = ***Psidium nummularia*** (C.Wright ex Griseb.) C.Wright
Eugenia nummularia Baker = ***Syzygium latifolium*** (Poir.) DC.

Eugenia nutans K.Schum. = ***Syzygium nutans*** Merr. & L.M.Perry
Eugenia nutans var. *pellucida* O.Berg = ***Eugenia nutans*** O.Berg
Eugenia nyassensis Engl. = ***Eugenia capensis*** subsp. ***nyassensis*** (Engl.) F.White
Eugenia oaxacana Standl. = ***Eugenia purpusii*** Standl.
Eugenia obcordata Raoul = ***Lophomyrtus obcordata*** (Raoul) Burret
Eugenia oblanceifolia Lundell = ***Eugenia ardisioides*** Lundell
Eugenia oblanceolata C.B.Rob. = ***Syzygium oblanceolatum*** (C.B.Rob.) Merr.
Eugenia oblata Roxb. = ***Syzygium oblatum*** (Roxb.) Wall. ex A.M.Cowan & Cowan
Eugenia obliquinervia Elmer = ***Syzygium obliquinervium*** (Elmer) Merr.
Eugenia oblongata Mattos & D.Legrand = ***Eugenia platysema*** O.Berg
Eugenia oblongata var. *glabrata* O.Berg = ? [84 BZL]
Eugenia oblongata var. *velutina* O.Berg = ***Eugenia oblongata*** O.Berg
Eugenia oblongifolia (O.Berg) Nied. = ***Eugenia uniflora*** L.
Eugenia oblongifolia (O.Berg) Arechav. = ***Eugenia uniflora*** L.
Eugenia oblongifolia (Sagot) Nied. ex T.Durand & B.D.Jacks. = ***Calycorectes grandifolius*** O.Berg
Eugenia oblongifolia Duthie = ***Syzygium claviflorum*** (Roxb.) Wall. ex A.M.Cowan & Cowan
Eugenia obovata O.Berg = ***Eugenia excelsa*** O.Berg
Eugenia obovata Poir. = ***Syzygium cumini*** (L.) Skeels
Eugenia obtusa Barb.Rodr. ex Chodat & Hassl. = ?
Eugenia obtusa DC. = ***Myrceugenia obtusa*** (DC.) O.Berg
Eugenia obtusata Willd. ex O.Berg = ***Eugenia monticola*** (Sw.) DC.
Eugenia obtusiflora Kiaersk. = ***Myrceugenia myrcioides*** (Cambess.) O.Berg
Eugenia obtusifolia Cambess. = ***Eugenia punicifolia*** (Kunth) DC.
Eugenia obtusifolia Roxb. = ***Syzygium cumini*** (L.) Skeels
Eugenia obversa O.Berg = ***Eugenia punicifolia*** (Kunth) DC.
Eugenia obversa Miq. = ***Syzygium aqueum*** (Burm.f.) Alston
Eugenia occidentalis Bourd. = ***Syzygium occidentale*** (Bourd.) D.N.Gandhi
Eugenia occlusa (Miq.) Kurz = ***Syzygium occlusum*** Miq.
Eugenia ochneocarpa Merr. = ***Syzygium oligomyrum*** Diels
Eugenia ochra O.Berg = ***Eugenia stictopetala*** DC.
Eugenia odorata O.Berg = ***Eugenia isabeliana*** Kiaersk.
Eugenia odorata Wight = ***Syzygium brachiatum*** (Roxb.) Miq.
Eugenia odorata var. *longifolia* O.Berg = ***Eugenia isabeliana*** Kiaersk.
Eugenia odorata var. *parvifolia* O.Berg = ***Eugenia isabeliana*** Kiaersk.
Eugenia oeidocarpa O.Berg = ***Eugenia dimorpha*** O.Berg
Eugenia oleifolia Thouars ex H.Perrier = ***Eugenia orbiculata*** Lam.
Eugenia oleifolia (Kunth) DC. = ***Eugenia punicifolia*** (Kunth) DC.
Eugenia oleina Wight = ***Syzygium myrtifolium*** Walp.
Eugenia oleioides H.Perrier = ***Eugenia orbiculata*** Lam.
Eugenia oleoides Planch. & Linden = ?
Eugenia oleosa F.Muell. = ***Syzygium oleosum*** (F.Muell.) B.Hyland
Eugenia oleosa var. *cyanocarpa* F.Muell. = ***Syzygium oleosum*** (F.Muell.) B.Hyland
Eugenia oligandra var. *macnabiana* Krug & Urb. = ***Eugenia macnabiana*** (Krug & Urb.) Urb.
Eugenia oligantha Baker = ***Eugenia zygophylla*** Govaerts
Eugenia oligantha (Thwaites) Bedd. = ***Syzygium oliganthum*** Thwaites
Eugenia oligoneura O.Berg = ***Eugenia florida*** DC.
Eugenia oligophylla A.Rich. ex O.Berg = ***Eugenia lambertiana*** DC.
Eugenia olivacea O.Berg = ***Eugenia pruniformis*** Cambess.
Eugenia olivifolia Duthie = ***Syzygium spathulatum*** Thwaites
Eugenia oneillii Lundell = ***Myrciaria floribunda*** (H.West ex Willd.) O.Berg
Eugenia onesima (Merr. & L.M.Perry) Whitmore = ***Syzygium onesimum*** Merr. & L.M.Perry
Eugenia onivensis H.Perrier = ***Syzygium onivense*** (H.Perrier) Labat & Schatz
Eugenia oonophylla Urb. = ***Mosiera oonophylla*** (Urb.) Bisse
Eugenia opaca O.Berg = ***Eugenia uruguayensis*** Cambess.
Eugenia opaca Koord. & Valeton = ?
Eugenia opaca var. *brasiliensis* O.Berg = ***Eugenia uruguayensis*** Cambess.
Eugenia opaca var. *montevideensis* O.Berg = ***Eugenia uruguayensis*** Cambess.
Eugenia operculata Roxb. = ***Syzygium operculatum*** (Roxb.) Nied.
Eugenia operculata var. *obovata* Kurz = ***Syzygium operculatum*** (Roxb.) Nied.
Eugenia operculata var. *orientalis* Craib = ***Syzygium cinereum*** (Kurz) Chantaran. & J.Parn.
Eugenia oraria Guillaumin = ***Eugenia littoralis*** Pancher ex Brongn. & Gris
Eugenia orbicularis O.Berg = ***Mosiera longipes*** (O.Berg) Small
Eugenia orbignyana var. *grandifolia* O.Berg = ***Eugenia orbignyana*** O.Berg
Eugenia oreophila Diels = ***Myrcianthes oreophila*** (Diels) McVaugh
Eugenia oreophila Rech. = ***Syzygium brevifolium*** (A.Gray) Müll.Stuttg.
Eugenia oreophila Ridl. = ***Syzygium oreophilum*** I.M.Turner
Eugenia organosia Urb. = ***Eugenia rigidifolia*** A.Rich. subsp. ***rigidifolia***
Eugenia origanoides O.Berg = ***Eugenia venezuelensis*** O.Berg
Eugenia orites Ridl. = ***Syzygium orites*** (Ridl.) I.M.Turner
Eugenia orlandoi (Mattos) Mattos = ***Calycorectes orlandoi*** Mattos
Eugenia orthioneura Urb. = ***Eugenia lindahlii*** Urb. & Ekman
Eugenia orthostemon O.Berg = ***Myrcianthes orthostemon*** (O.Berg) Grifo
Eugenia osoriana Mattos & D.Legrand = ***Eugenia pluriflora*** DC.
Eugenia ossaeana Urb. = ***Eugenia foetida*** Pers.
Eugenia osteomeloides (Rusby) McVaugh = ***Myrcianthes osteomeloides*** (Rusby) McVaugh
Eugenia ottonis O.Berg = ***Eugenia pubescens*** (Kunth) DC.

Eugenia ovalifolia Cambess. = ***Eugenia punicifolia*** (Kunth) DC.
Eugenia ovalifolia (Blume) Warb. = ***Syzygium ovalifolium*** (Blume) Merr. & L.M.Perry
Eugenia ovalifolia var. *chacoensis* D.Legrand = ***Eugenia chacoensis*** (D.Legrand) Kausel
Eugenia ovata Hook. & Arn. = ***Myrceugenia ovata*** (Hook. & Arn.) O.Berg
Eugenia ovatifolia Lundell = ***Eugenia acapulcensis*** Steud.
Eugenia ovigera Brongn. & Gris = ***Stereocaryum ovigerum*** (Brongn. & Gris) Burret
Eugenia owariensis P.Beauv. = ***Syzygium owariense*** (P.Beauv.) Benth.
Eugenia oxygona Koidz. = ***Psidium cattleianum*** Afzel. ex Sabine
Eugenia oxyoentophylla Kiaersk. = ***Eugenia repanda*** O.Berg
Eugenia oxyphylla O.Berg = ***Eugenia arvensis*** Vell.
Eugenia pachyphylla Kurz = ***Syzygium pachyphyllum*** (Kurz) Merr. & L.M.Perry
Eugenia pachysarca Gagnep. = ***Syzygium pachysarcum*** (Gagnep.) Merr. & L.M.Perry
Eugenia pacifica Elmer ex Merr. = ***Syzygium pallidum*** Merr.
Eugenia pahangensis Ridl. = ***Syzygium pahangense*** (Ridl.) I.M.Turner
Eugenia pahangensis var. *fraseri* M.R.Hend. = ***Syzygium pahangense*** (Ridl.) I.M.Turner
Eugenia palauensis Kaneh. = ***Syzygium palauense*** (Kaneh.) Hosok.
Eugenia palawanensis C.B.Rob. = ***Syzygium palawanense*** (C.B.Rob.) Merr. & L.M.Perry
Eugenia palembanica (Miq.) Merr. = ***Syzygium palembanicum*** Miq.
Eugenia palenae Phil. = ***Luma apiculata*** (DC.) Burret
Eugenia pallens (Vahl) DC. = ***Eugenia biflora*** (L.) DC.
Eugenia pallens Poir. = ***Calyptranthes pallens*** Griseb. var. ***pallens***
Eugenia pallida O.Berg = ***Myrceugenia glaucescens*** (Cambess.) D.Legrand & Kausel var. ***glaucescens***
Eugenia pallidifolia Merr. = ***Syzygium everettii*** (C.B.Rob.) Merr.
Eugenia pallidula Ridl. = ***Syzygium pallidulum*** (Ridl.) I.M.Turner
Eugenia pamatensis Miq. = ***Syzygium polyanthum*** (Wight) Walp.
Eugenia panayensis Merr. = ***Syzygium panayense*** (Merr.) Merr.
Eugenia pancheri Brongn. & Gris = ***Austromyrtus pancheri*** (Brongn. & Gris) Burret
Eugenia panduriformis Elmer = ***Syzygium panduriforme*** (Elmer) Merr.
Eugenia paniala Roxb. = ***Syzygium operculatum*** (Roxb.) Nied.
Eugenia paniculata Bello = [81 PUE]
Eugenia paniculata Cambess. = ***Marlierea paniculata*** (O.Berg) D.Legrand
Eugenia paniculata Jacq. = ***Myrcia citrifolia*** (Aubl.) Urb.
Eugenia paniculata Hils. & Sieber ex C.Presl = ***Myrcia ferruginea*** G.Don
Eugenia paniculata Lam. = ***Syzygium borbonicum*** J.Guého & A.J.Scott
Eugenia paniculata (Gaertn.) Britten = ***Syzygium paniculatum*** Gaertn.
Eugenia paniculiflora Steud. = ***Myrcia splendens*** (Sw.) DC.
Eugenia paniensis Baker f. = ***Syzygium paniense*** (Baker f.) J.W.Dawson
Eugenia pantagensis var. *elliptica* O.Berg = ***Eugenia pantagensis*** O.Berg
Eugenia papillosa Duthie = ***Syzygium papillosum*** (Duthie) Merr. & L.M.Perry
Eugenia paradoxa Merr. = ***Syzygium paradoxum*** (Merr.) Masam.
Eugenia paraensis O.Berg = ***Eugenia moschata*** (Aubl.) Nied. ex T.Durand & B.D.Jacks.
Eugenia paranaensis (Mattos) Mattos = ***Calycorectes paranaensis*** Mattos
Eugenia paranaguensis Mattos & D.Legrand = ***Eugenia cereja*** D.Legrand
Eugenia parkeri Baker = ***Syzygium parkeri*** (Baker) Labat & Schatz
Eugenia parkeri var. *ibitensis* (Drake) H.Perrier = ***Syzygium parkeri*** (Baker) Labat & Schatz
Eugenia parkeriana DC. = ***Eugenia procera*** (Sw.) Poir.
Eugenia parodiana Morong = ***Eugenia egensis*** DC.
Eugenia parva C.B.Rob. = ***Syzygium myrtifolium*** Walp.
Eugenia parviflora Lam. = ***Decaspermum parviflorum*** (Lam.) A.J.Scott
Eugenia parvifolia DC. = ***Myrceugenia parvifolia*** (DC.) Kausel
Eugenia parvifolia C.Moore = ***Syzygium luehmannii*** (F.Muell.) L.A.S.Johnson
Eugenia pascasioii Merr. = ***Syzygium pascasioii*** (Merr.) Merr.
Eugenia pastelillensis Urb. = ***Mosiera cabanasensis*** subsp. ***pastelillensis*** (Urb.) Borhidi
Eugenia patagonica Phil. = ***Myrceugenia chrysocarpa*** (O.Berg) Kausel
Eugenia patagonica var. *macrophylla* Phil. = ***Myrceugenia chrysocarpa*** (O.Berg) Kausel
Eugenia patagonica var. *microphylla* Phil. = ***Myrceugenia chrysocarpa*** (O.Berg) Kausel
Eugenia patalensis Standl. & Steyerm. = ***Eugenia guatemalensis*** Donn.Sm.
Eugenia patula DC. = ***Eugenia florida*** DC.
Eugenia pauciflora Wight = ***Syzygium laetum*** (Buch.-Ham.) Gandhi subsp. ***laetum***
Eugenia paucipunctata Koord. & Valeton = ***Syzygium paucipunctatum*** (Koord. & Valeton) Merr. & L.M.Perry
Eugenia paucipunctata Merr. = ***Syzygium pseudocalcicola*** Craven & Biffin
Eugenia paucivenia C.B.Rob. = ***Syzygium paucivenium*** (C.B.Rob.) Merr.
Eugenia pauper Guillaumin = ***Syzygium brousmichei*** Govaerts
Eugenia pauper Ridl. = ***Syzygium pauper*** (Ridl.) I.M.Turner
Eugenia pearcei McVaugh = ***Myrcianthes pearcei*** (McVaugh) McVaugh
Eugenia pearsoniana King = ***Syzygium praineanum*** (King) Chantaran. & J.Parn. subsp. ***praineanum***
Eugenia pedunculata Trimen = ***Eugenia mabaeoides*** subsp. ***pedunculata*** (Trimen) P.S.Ashton
Eugenia pellucida Fern.-Vill. = ***Syzygium cinnamomeum*** (Vidal) Merr.
Eugenia pellucida Duthie = ***Syzygium pellucidum*** (Duthie) N.P.Balakr.
Eugenia penangiana Duthie = ***Syzygium attenuatum*** (Miq.) Merr. & L.M.Perry
Eugenia penasii Merr. = ***Syzygium penasii*** (Merr.) Merr.
Eugenia pendens Duthie = ***Syzygium pendens*** (Duthie) I.M.Turner
Eugenia pendula (Blume) DC. = (Rubiaceae) ?
Eugenia peninsula Elmer = ***Syzygium tripinnatum*** (Blanco) Merr.

Eugenia perakensis King = ***Syzygium perakense*** (King) I.M.Turner
Eugenia perogrina McVaugh = ***Eugenia baileyi***
Eugenia perforata O.Berg = ***Eugenia egensis*** DC.
Eugenia pergamacea Greves = ***Syzygium pergamaceum*** (Greves) Merr. & L.M.Perry
Eugenia pergamentacea King = ***Syzygium pergamentaceum*** (King) Chantaran. & J.Parn.
Eugenia periplocifolia Jacq. = ***Myrcia splendens*** (Sw.) DC.
Eugenia peroblata Lundell = ***Plinia peroblata*** (Lundell) Lundell
Eugenia perorebi Parodi ex Speg. & Girola = ***Eugenia florida*** DC.
Eugenia perpallida Merr. = ***Syzygium pallidum*** Merr.
Eugenia perparvifolia Merr. = ***Syzygium perparvifolium*** (Merr.) Merr. & L.M.Perry
Eugenia perplexans R.O.Williams = ***Eugenia trinervia*** Vahl
Eugenia perpunctata Lundell = ***Eugenia savannarum*** Standl. & Steyerm.
Eugenia perpuncticulata Merr. = ***Syzygium pustulatum*** (Duthie) Merr.
Eugenia perratonii Proctor = ***Eugenia mandevillensis*** var. ***perratonii*** (Proctor) Proctor
Eugenia perspicuinervia Merr. = ***Syzygium perspicuinervium*** (Merr.) Masam.
Eugenia petenensis Lundell = ***Eugenia oerstediana*** O.Berg
Eugenia petiolata Phil. = ***Myrceugenia chrysocarpa*** (O.Berg) Kausel
Eugenia petriei C.T.White & W.D.Francis = ***Syzygium johnsonii*** (F.Muell.) B.Hyland
Eugenia phaea O.Berg = ***Eugenia disperma*** Vell.
Eugenia phanerophlebia C.B.Rob. = ***Syzygium phanerophlebium*** (C.B.Rob.) Merr.
Eugenia philippii O.Berg = ***Myrceugenia chrysocarpa*** (O.Berg) Kausel
Eugenia philippinensis C.B.Rob. = ***Syzygium philippinense*** (C.B.Rob.) Merr.
Eugenia philippinorum C.Nelson Sutherland = ***Syzygium conglobatum*** Merr.
Eugenia phillyraeoides Willd. ex O.Berg = ***Eugenia punicifolia*** (Kunth) DC.
Eugenia phillyreifolia Baker = ***Syzygium phillyreifolium*** (Baker) Labat & Schatz
Eugenia phillyreifolia f. *obscurifolia* H.Perrier = ***Syzygium phillyreifolium*** (Baker) Labat & Schatz
Eugenia phillyreifolia var. *obscurifolia* (H.Perrier) H.Perrier = ***Syzygium phillyreifolium*** (Baker) Labat & Schatz
Eugenia phitrantha Kiaersk. = ***Plinia phitrantha*** (Kiaersk.) Sobral
Eugenia phlebotomonides Kiaersk. = ***Eugenia pyriformis*** Cambess.
Eugenia phyllyreifolia A.Rich. = ***Eugenia rigidifolia*** A.Rich. subsp. ***rigidifolia***
Eugenia piauhiensis O.Berg = ***Eugenia stictopetala*** DC.
Eugenia piedadensis Kiaersk. = ***Blepharocalyx salicifolius*** (Kunth) O.Berg
Eugenia pierrei Gagnep. = ***Syzygium pierrei*** (Gagnep.) Merr. & L.M.Perry
Eugenia pilotantha Kiaersk. = ***Myrceugenia pilotantha*** (Kiaersk.) Landrum
Eugenia pimenta (L.) DC. = ***Pimenta dioica*** (L.) Merr.
Eugenia pimenta var. *longifolia* DC. = ***Pimenta dioica*** (L.) Merr.
Eugenia pimenta var. *ovalifolia* DC. = ***Pimenta dioica*** (L.) Merr.
Eugenia pinhaesensis Mattos = ***Eugenia platysema*** O.Berg
Eugenia pinifolia Phil. = ***Myrceugenia pinifolia*** (Phil.) Kausel
Eugenia pinnata (L.) Druce = ***Plinia pinnata*** L.
Eugenia pirataquinensis (Mattos) Mattos = ***Calycorectes pirataquinensis*** Mattos
Eugenia pisonis O.Berg = ***Eugenia moschata*** (Aubl.) Nied. ex T.Durand & B.D.Jacks.
Eugenia pitanga var. *camporum* (Morong) Mattos = ***Eugenia pitanga*** (O.Berg) Nied.
Eugenia pitanga var. *venosa* Mattos = ***Eugenia pitanga*** (O.Berg) Nied.
Eugenia pitoniana Urb. & Ekman = ***Eugenia picardiae*** Krug & Urb.
Eugenia pitra O.Berg = ***Myrceugenia exsucca*** (DC.) O.Berg
Eugenia planipes Hook. & Arn. = ***Myrceugenia planipes*** (Hook. & Arn.) O.Berg
Eugenia planipes var. *grandifolia* O.Berg = ***Myrceugenia planipes*** (Hook. & Arn.) O.Berg
Eugenia planiramea O.Berg = ***Myrceugenia rufescens*** (DC.) D.Legrand & Kausel
Eugenia platyclada O.Berg = ***Eugenia punicifolia*** (Kunth) DC.
Eugenia platyclada var. *cuneata* O.Berg = ***Eugenia punicifolia*** (Kunth) DC.
Eugenia platyclada var. *microphylla* O.Berg = ***Eugenia punicifolia*** (Kunth) DC.
Eugenia platyclada var. *multiflora* O.Berg = ***Eugenia punicifolia*** (Kunth) DC.
Eugenia platyclada var. *ovalis* O.Berg = ***Eugenia punicifolia*** (Kunth) DC.
Eugenia platyphylla Cordem. = ***Syzygium cordemoyi*** J.Bosser & Cadet
Eugenia pleiopetala F.Muell. = ***Octamyrtus pleiopetala*** Diels
Eugenia pleurosiphonea Diels = ***Eugenia moschata*** (Aubl.) Nied. ex T.Durand & B.D.Jacks.
Eugenia ploumensis Däniker = ***Austromyrtus ploumensis*** (Däniker) Burret
Eugenia plumbea King = ***Syzygium plumbeum*** (King) I.M.Turner
Eugenia plumea Ridl. = ***Syzygium plumeum*** (Ridl.) Merr. & L.M.Perry
Eugenia plumieri (O.Berg) Nied. = ***Plinia pinnata*** L.
Eugenia pluriflora Mart. = ?
Eugenia pluriflora var. *opaca* O.Berg = ***Eugenia pluriflora*** DC.
Eugenia pluriflora var. *perforata* O.Berg = ***Eugenia pluriflora*** DC.
Eugenia poeppigiana O.Berg = ***Myrceugenia rufescens*** (DC.) D.Legrand & Kausel
Eugenia poggei Engl. = ***Eugenia malangensis*** (O.Hoffm.) Nied.
Eugenia poiretii (Spreng.) DC. = ***Eugenia foetida*** Pers.
Eugenia polita King = ***Syzygium politum*** (King) I.M.Turner
Eugenia polyantha Phil. = ***Myrceugenia obtusa*** (DC.) O.Berg
Eugenia polyantha Miq. = ***Myrcia inaequiloba*** (DC.) Lemée
Eugenia polyantha Wight = ***Syzygium polyanthum*** (Wight) Walp.
Eugenia polyantha var. *sessilis* M.R.Hend. = ***Syzygium polyanthum*** (Wight) Walp.
Eugenia polycarpa O.Berg = ***Eugenia hyemalis*** Cambess.
Eugenia polycarpa var. *bimarginata* O.Berg = ***Eugenia hyemalis*** Cambess.

Eugenia polycarpa var. *marginata* O.Berg = ***Eugenia hyemalis*** Cambess.
Eugenia polycarpa var. *ovata* O.Berg = ***Eugenia hyemalis*** Cambess.
Eugenia polycephala Miq. = ***Syzygium polycephalum*** (Miq.) Merr. & L.M.Perry
Eugenia polycephaloides C.B.Rob. = ***Syzygium polycephaloides*** (C.B.Rob.) Merr.
Eugenia polygama Roxb. = ***Decaspermum parviflorum*** (Lam.) A.J.Scott subsp. ***parviflorum***
Eugenia polyneura (Miq.) Koord. & Valeton = ?
Eugenia polyneura Urb. = ***Myrciaria floribunda*** (H.West ex Willd.) O.Berg
Eugenia polypetala (Wall.) Wight = ***Syzygium polypetalum*** (Wall.) Merr. & L.M.Perry
Eugenia polyphylla O.Berg = ***Eugenia punicifolia*** (Kunth) DC.
Eugenia polyphylla var. *obovata* O.Berg = ***Eugenia punicifolia*** (Kunth) DC.
Eugenia polystachyoides Amshoff = ***Eugenia limbosa*** O.Berg
Eugenia pomatensis Miq. = [42 BOR]
Eugenia pomifera var. *angustissima* (O.Berg) Urb. = ***Eugenia pomifera*** (Aubl.) Urb.
Eugenia pomifera var. *latifolia* (O.Berg) Urb. = ***Eugenia pomifera*** (Aubl.) Urb.
Eugenia ponapensis Merr. & Kaneh. = ***Syzygium carolinense*** (Koidz.) Hosok.
Eugenia populifolia Baker = ***Syzygium populifolium*** (Baker) J.Guého & A.J.Scott
Eugenia porphyrantha Ridl. = ***Syzygium porphyranthum*** (Ridl.) I.M.Turner
Eugenia porphyrocarpa Greves = ***Syzygium porphyrocarpum*** (Greves) Merr. & L.M.Perry
Eugenia porphyroclada O.Berg = ***Myrcianthes rhopaloides*** (Kunth) McVaugh
Eugenia portomite D.Legrand = ***Eugenia warmingiana*** Kiaersk.
Eugenia portoricensis DC. = ***Eugenia pseudopsidium*** Jacq.
Eugenia portoricensis var. *brevipes* DC. = ***Eugenia pseudopsidium*** Jacq.
Eugenia praecox Roxb. = ***Syzygium praecox*** (Roxb.) Rathakr. & N.C.Nair
Eugenia praestigiosa M.R.Hend. = ***Syzygium praestigiosum*** (M.R.Hend.) I.M.Turner
Eugenia praetermissa Gage = ***Syzygium praetermissum*** (Gage) N.P.Balakr.
Eugenia praineana King = ***Syzygium praineanum*** (King) Chantaran. & J.Parn.
Eugenia prasina var. *grandifolia* Huber = ***Eugenia prasina*** O.Berg
Eugenia prasiniflora Ridl. = ***Syzygium prasiniflorum*** (Ridl.) Merr. & L.M.Perry
Eugenia prenleloupii Kiaersk. = ***Eugenia laevis*** O.Berg
Eugenia prieurii O.Berg = ***Eugenia pseudopsidium*** Jacq.
Eugenia prieurii var. *robusta* O.Berg = ***Eugenia pseudopsidium*** Jacq.
Eugenia prieurii var. *tenuiramis* O.Berg = ***Eugenia pseudopsidium*** Jacq.
Eugenia prismatica D.Legrand = ***Mosiera prismatica*** (D.Legrand) Landrum
Eugenia proba O.Berg = ***Luma apiculata*** (DC.) Burret
Eugenia prolixa var. *vestita* S.Moore = ***Eugenia prolixa*** S.Moore
Eugenia prominens O.Berg = ***Eugenia punicifolia*** (Kunth) DC.
Eugenia propinqua Merr. = ***Syzygium abulugense*** Merr.
Eugenia prora Burkill = ***Syzygium seemannii*** (A.Gray) Biffin & Craven
Eugenia prosoneura O.Berg = ***Eugenia gomesiana*** O.Berg
Eugenia protracta O.Berg = ***Eugenia protenta*** McVaugh
Eugenia protracta Steud. = ***Myrciaria floribunda*** (H.West ex Willd.) O.Berg
Eugenia psammophila Diels = ***Eugenia punicifolia*** (Kunth) DC.
Eugenia pseudocaryophyllus (Gomes) DC. = ***Pimenta pseudocaryophyllus*** (Gomes) Landrum
Eugenia pseudocaryophyllus var. *ocoteoides* DC. = ***Pimenta pseudocaryophyllus*** (Gomes) Landrum var. ***pseudocaryophyllus***
Eugenia pseudoclaviflora M.R.Hend. = ***Syzygium pseudoclaviflorum*** (M.R.Hend.) I.M.Turner
Eugenia pseudocrenulata M.R.Hend. = ***Syzygium pseudocrenulatum*** (M.R.Hend.) I.M.Turner
Eugenia pseudodichasiantha Kiaersk. = *Plinia* sp. ?
Eugenia pseudoformosa King = ***Syzygium pseudoformosum*** (King) Merr. & L.M.Perry
Eugenia pseudoglauca (King) Ridl. = ***Syzygium glaucum*** (King) Chantaran. & J.Parn.
Eugenia pseudomalaccensis Linden = ***Syzygium malaccense*** (L.) Merr. & L.M.Perry
Eugenia pseudomato D.Legrand = ***Myrcianthes pseudomato*** (D.Legrand) McVaugh
Eugenia pseudomollis M.R.Hend. = ***Syzygium pseudomolle*** (M.R.Hend.) I.M.Turner
Eugenia pseudopsidium var. *portoricensis* (DC.) Krug & Urb. = ***Eugenia pseudopsidium*** Jacq.
Eugenia pseudosubtilis King = ***Syzygium borneense*** (Miq.) Miq.
Eugenia pseudosubtilis var. *platyphylla* King = ***Syzygium borneense*** (Miq.) Miq.
Eugenia pseudosubtilis var. *subacuminata* King = ***Syzygium cinereum*** (Kurz) Chantaran. & J.Parn.
Eugenia pseudosyzygioides M.R.Hend. = ***Syzygium syzygioides*** (Miq.) Merr. & L.M.Perry
Eugenia pseudotetraptera King = ***Syzygium tetrapterum*** (Miq.) Chantaran. & J.Parn.
Eugenia pseudoverticillata S.Moore = ***Eugenia orbignyana*** O.Berg
Eugenia pseudoverticillata Kiaersk. = ***Eugenia verticillata*** (Vell.) Angely
Eugenia psidiiflora var. *riedelinana* (O.Berg) Mattos = ***Eugenia psidiiflora*** DC.
Eugenia psidiiflora var. *schultziana* (Mattos) Mattos = ***Eugenia psidiiflora*** DC.
Eugenia psidiiflora var. *triflora* (Mattos) Mattos = ***Eugenia psidiiflora*** DC.
Eugenia psidioides DC. = ***Eugenia pseudopsidium*** Jacq.
Eugenia psychotrioides Mart. ex O.Berg = ***Eugenia punicifolia*** (Kunth) DC.
Eugenia ptariensis Steyerm. = ***Eugenia anastomosans*** DC.
Eugenia pterocalyx Greves = ***Syzygium grevesianum*** Merr. & L.M.Perry
Eugenia pterocalyx (Brongn. & Gris) Baker f. = ***Syzygium pterocalyx*** Brongn. & Gris
Eugenia pterocaulis Miq. = (Melastomataceae)
Eugenia pulchella Roxb. = ***Syzygium pulchellum*** (Roxb.) Govaerts
Eugenia pulchra O.Berg = ***Luma chequen*** (Molina) A.Gray
Eugenia pulgarensis C.B.Rob. = ***Syzygium pulgarense*** (C.B.Rob.) Merr.
Eugenia pumila Gardner = ***Myrciaria pumila*** (Gardner) O.Berg

Eugenia punctata Vahl = ***Myrcianthes fragrans*** (Sw.) McVaugh
Eugenia punctata Baker f. = ***Syzygium micans*** Brongn. & Gris
Eugenia punctifolia Ridl. = ***Syzygium thumra*** subsp. ***punctifolium*** (Ridl.) Chantaran. & J.Parn.
Eugenia punctilimba Merr. = ***Syzygium punctilimbum*** (Merr.) Merr. & L.M.Perry
Eugenia punctulata F.M.Bailey = ***Syzygium corynanthum*** (F.Muell.) L.A.S.Johnson
Eugenia punctulata King = ***Syzygium incarnatum*** (Elmer) Merr. & L.M.Perry
Eugenia pungens O.Berg = ***Myrcianthes pungens*** (O.Berg) D.Legrand
Eugenia pungens var. *subalterna* (Benth.) Amshoff = ***Eugenia punicifolia*** (Kunth) DC.
Eugenia punicifolia var. *brachypoda* (DC.) Krug & Urb. = ***Eugenia punicifolia*** (Kunth) DC.
Eugenia purpurascens Baill. = ***Syzygium malaccense*** (L.) Merr. & L.M.Perry
Eugenia purpurea Roxb. = ***Syzygium malaccense*** (L.) Merr. & L.M.Perry
Eugenia purpuricarpa Elmer = ***Syzygium attenuatum*** (Miq.) Merr. & L.M.Perry
Eugenia purpuriflora Elmer = ***Syzygium purpuriflorum*** (Elmer) Merr.
Eugenia pusilana Lundell = ***Eugenia rhombea*** (O.Berg) Krug & Urb.
Eugenia pustulata Duthie = ***Syzygium pustulatum*** (Duthie) Merr.
Eugenia pycnoneura Urb. = ***Myrciaria floribunda*** (H.West ex Willd.) O.Berg
Eugenia pyramidalis O.Berg = ***Eugenia punicifolia*** (Kunth) DC.
Eugenia pyramidalis var. *angustifolia* O.Berg = ***Eugenia punicifolia*** (Kunth) DC.
Eugenia pyriflora var. *major* O.Berg = ***Eugenia pyriflora*** O.Berg
Eugenia pyriflora var. *minor* O.Berg = ***Eugenia pyriflora*** O.Berg
Eugenia pyrifolia Desv. = ***Myrcia pyrifolia*** (Desv.) Nied.
Eugenia pyrifolia (Blume) Duthie = ***Syzygium pyrifolium*** (Blume) DC.
Eugenia pyriformis var. *argentea* Mattos & D.Legrand = ***Eugenia pyriformis*** Cambess.
Eugenia pyriformis f. *ponhi* D.Legrand = ***Eugenia pyriformis*** Cambess.
Eugenia pyriformis var. *uvalha* (Cambess.) D.Legrand = ***Eugenia pyriformis*** Cambess.
Eugenia pyrocarpa Greves = ***Syzygium pyrocarpum*** (Greves) Merr. & L.M.Perry
Eugenia pyrroclada O.Berg = ***Eugenia punicifolia*** (Kunth) DC.
Eugenia pyrroclada var. *columbiensis* O.Berg = ***Eugenia punicifolia*** (Kunth) DC.
Eugenia pyrroclada var. *guianensis* O.Berg = ***Eugenia punicifolia*** (Kunth) DC.
Eugenia pyxophylla Hance = ***Syzygium buxifolium*** Hook. & Arn.
Eugenia quadrangulata A.Gray = ***Syzygium quadrangulatum*** (A.Gray) Merr. & L.M.Perry
Eugenia quadrata King = ***Syzygium quadratum*** (King) I.M.Turner
Eugenia quadrialata Craib = ***Syzygium aksornae*** Chantaran. & J.Parn.
Eugenia quadribracteata M.R.Hend. = ***Syzygium quadribracteatum*** (M.R.Hend.) I.M.Turner
Eugenia quaternifolia Guillaumin = ***Eugenia daenikeri*** Guillaumin
Eugenia quaternifolia Cambess. = ***Eugenia suberosa*** Cambess.
Eugenia quinqueloba McVaugh = ***Myrcianthes quinqueloba*** (McVaugh) McVaugh
Eugenia quitarensis Amshoff = ***Eugenia limbosa*** O.Berg
Eugenia quitarensis Benth. = ***Myrcia quitarensis*** (Benth.) Sagot
Eugenia rabeniana Kiaersk. = ***Plinia peruviana*** (Poir.) Govaerts
Eugenia racemifera Sagot = ***Eugenia florida*** DC.
Eugenia racemifera O.Berg = ***Eugenia patens*** Poir.
Eugenia racemoides Greves = ***Lindsayomyrtus racemoides*** (Greves) Craven
Eugenia racemosa DC. = ***Eugenia biflora*** (L.) DC.
Eugenia racemulosa O.Berg = ***Eugenia modesta*** DC.
Eugenia racemulosa var. *grandiflora* O.Berg = ***Eugenia modesta*** DC.
Eugenia racemulosa var. *parviflora* O.Berg = ***Eugenia modesta*** DC.
Eugenia rama-varmae Bourd. = ***Syzygium rama-varmae*** (Bourd.) Chithra
Eugenia ramiflora Miq. = ***Eugenia dodonaeifolia*** Cambess.
Eugenia ramiflora var. *montana* Amshoff = ***Eugenia ferreiraeana*** O.Berg
Eugenia ramosii C.B.Rob. = ***Syzygium ramosii*** (C.B.Rob.) Merr.
Eugenia ramosissima (Blume) Wall. ex Duthie = ***Syzygium ramosissimum*** (Blume) N.P.Balakr.
Eugenia rampans Baker = ***Syzygium rampans*** (Baker) J.Guého & A.J.Scott
Eugenia raran (Colla) Barnéoud = ***Myrceugenia obtusa*** (DC.) O.Berg
Eugenia rariflora Benth. = ***Eugenia reinwardtiana*** (Blume) A.Cunn. ex DC.
Eugenia rariflora var. *parvifolia* Hillebr. = ***Eugenia reinwardtiana*** (Blume) A.Cunn. ex DC.
Eugenia recurvata O.Berg = ***Eugenia candolleana*** DC.
Eugenia referta Craib = ***Syzygium refertum*** (Craib) Chantaran. & J.Parn.
Eugenia regeliana O.Berg = ***Eugenia stephanii*** O.Berg
Eugenia regnelliana O.Berg = ***Myrceugenia ovata*** var. ***regnelliana*** (O.Berg) Landrum
Eugenia reinwardtiana f. *lutea* H.St.John = ***Eugenia reinwardtiana*** (Blume) A.Cunn. ex DC.
Eugenia remotifolia Ridl. = ***Syzygium remotifolium*** (Ridl.) Merr. & L.M.Perry
Eugenia resinosa Gagnep. = ***Syzygium polyanthum*** (Wight) Walp.
Eugenia reticulata Trimen = [40 SRL]
Eugenia reticulata Wight = ***Syzygium reticulatum*** (Wight) Walp.
Eugenia retivenia C.Wright = ***Myrcia retivenia*** (C.Wright) Urb.
Eugenia retusa (O.Berg) Nied. = ***Eugenia speciosa*** Cambess.
Eugenia revoluta O.Berg = ***Eugenia oxysepala*** Urb.
Eugenia revoluta Willd. ex O.Berg = ***Eugenia tuberculata*** (Kunth) DC.
Eugenia revoluta Wight = ***Syzygium revolutum*** (Wight) Walp.
Eugenia rhadinantha S.Moore = ***Syzygium wilsonii*** subsp. ***cryptophlebium*** (F.Muell.) B.Hyland
Eugenia rhamphiphylla Craib = ***Syzygium circumscissum*** (Gagnep.) Craven & Biffin
Eugenia rheedioides Standl. & Steyerm. = ***Syzygium paniculatum*** Gaertn.

Eugenia rhizophora Boerl. & Koord.-Schum. = ***Syzygium rhizophorum*** (Boerl. & Koord.-Schum.) Govaerts
Eugenia rhododendrifolia Miq. = ***Syzygium claviflorum*** (Roxb.) Wall. ex A.M.Cowan & Cowan
Eugenia rhododendrifolia var. *forma* Miq. = ***Syzygium claviflorum*** (Roxb.) Wall. ex A.M.Cowan & Cowan
Eugenia rhodomelea Comm. ex DC. = ***Syzygium cymosum*** (Lam.) DC. var. ***cymosum***
Eugenia rhombocarpa O.Berg = ***Eugenia punicifolia*** (Kunth) DC.
Eugenia rhomboidea Ridl. = ***Syzygium rhomboideum*** (Ridl.) I.M.Turner
Eugenia rhopaloides (Kunth) DC. = ***Myrcianthes rhopaloides*** (Kunth) McVaugh
Eugenia rhynchophylla Merr. = ***Syzygium caudatum*** (Merr.) Airy Shaw
Eugenia ribeireana O.Berg = ***Myrceugenia glaucescens*** (Cambess.) D.Legrand & Kausel var. ***glaucescens***
Eugenia richardiana O.Berg = ***Eugenia biflora*** (L.) DC.
Eugenia richardiana Cordem. = ***Syzygium emirnense*** (Baker) Labat & Schatz
Eugenia richii A.Gray = ***Syzygium richii*** (A.Gray) Merr. & L.M.Perry
Eugenia ridleyi King = ***Syzygium ridleyi*** (King) Chantaran. & J.Parn.
Eugenia riedeliana O.Berg = ***Eugenia verticillata*** (Vell.) Angely
Eugenia riedeliana var. *ferruginea* O.Berg = ***Eugenia verticillata*** (Vell.) Angely
Eugenia riedeliana var. *ochracea* O.Berg = ***Eugenia verticillata*** (Vell.) Angely
Eugenia rigens Craib = ***Syzygium rigens*** (Craib) Chantaran. & J.Parn.
Eugenia rigida var. *ovalis* O.Berg = ***Eugenia rigida*** DC.
Eugenia rigida var. *ovata* O.Berg = ***Eugenia rigida*** DC.
Eugenia rigidifolia var. *angustifolia* O.Berg = ***Eugenia rigidifolia*** A.Rich. subsp. ***rigidifolia***
Eugenia rigidifolia var. *genuina* Krug & Urb. = ***Eugenia rigidifolia*** A.Rich.
Eugenia rigidifolia var. *latifolia* O.Berg = ***Eugenia rigidifolia*** A.Rich. subsp. ***rigidifolia***
Eugenia rigidifolia var. *phyllyreifolia* (A.Rich.) Krug & Urb. = ***Eugenia rigidifolia*** A.Rich. subsp. ***rigidifolia***
Eugenia rigidissima Cufod. = ***Myrcianthes storkii*** (Standl.) McVaugh
Eugenia rinconiensis Mattos = ***Calycorectes mexicanus*** O.Berg
Eugenia riparia DC. = ***Eugenia patens*** Poir.
Eugenia riparia Becc. = ***Syzygium odoardoi*** Merr. & L.M.Perry
Eugenia ripicola Craib = ***Syzygium ripicola*** (Craib) Merr. & L.M.Perry
Eugenia ripidocarpa Ruiz & Pav. = ***Eugenia patens*** Poir.
Eugenia rivularis Cambess. = ***Plinia rivularis*** (Cambess.) Rotman
Eugenia rivularis Seem. = ***Syzygium seemannianum*** Merr. & L.M.Perry
Eugenia rizalensis Merr. = ***Syzygium rizalense*** (Merr.) Merr.
Eugenia robertii Merr. = [42 PHI]
Eugenia robertoana Mattos = ***Hexachlamys handroi*** Mattos
Eugenia robinsoniana Ridl. = ***Syzygium racemosum*** (Blume) DC.
Eugenia robinsonii Elmer = ***Syzygium robinsonii*** (Elmer) Merr.
Eugenia robusta O.Berg = ***Eugenia capparidifolia*** DC.
Eugenia robusta var. *capparidifolia* (DC.) O.Berg = ***Eugenia capparidifolia*** DC.
Eugenia robusta var. *firma* (DC.) O.Berg = ***Eugenia capparidifolia*** DC.
Eugenia robustovenosa Kiaersk. = ***Eugenia umbrosa*** O.Berg
Eugenia rodriguesii (Mattos & D.Legrand) Mattos = ***Calycorectes rodriguesii*** Mattos & D.Legrand
Eugenia rojasiana (D.Legrand) Mattos = ***Hexachlamys rojasiana*** (D.Legrand) D.Legrand
Eugenia romana O.Berg = ***Eugenia punicifolia*** (Kunth) DC.
Eugenia romana var. *pickeli* Mattos = ***Eugenia punicifolia*** (Kunth) DC.
Eugenia rondonensis Steyerm. = ***Myrcianthes fragrans*** (Sw.) McVaugh
Eugenia roraimana O.Berg = ***Eugenia stictopetala*** DC.
Eugenia rosea Barb.Rodr. ex Chodat & Hassl. = [85 PAR]
Eugenia rosenbluthii C.B.Rob. = ***Syzygium rosenbluthii*** (C.B.Rob.) Merr.
Eugenia roseomarginata C.B.Rob. = ***Syzygium roseomarginatum*** (C.B.Rob.) Merr. & L.M.Perry
Eugenia rosmarinifolia Poir. = ***Eugenia pomifera*** (Aubl.) Urb.
Eugenia rosmarinifolia var. *angustissima* O.Berg = ***Eugenia pomifera*** (Aubl.) Urb.
Eugenia rosmarinifolia var. *latifolia* O.Berg = ***Eugenia pomifera*** (Aubl.) Urb.
Eugenia rosmarinifolia var. *longifolia* O.Berg = ***Eugenia pomifera*** (Aubl.) Urb.
Eugenia rostadonis Ridl. = ***Syzygium rostadonis*** (Ridl.) I.M.Turner
Eugenia rosulenta Ridl. = ***Syzygium rosulentum*** (Ridl.) Merr. & L.M.Perry
Eugenia rotata King ex Craib = ***Syzygium leptostemon*** (Korth.) Merr. & L.M.Perry
Eugenia rothii Panigrahi = ***Eugenia roxburghii*** DC.
Eugenia rothii var. *fasciculata* (Duthie) H.B.Naithani = ***Eugenia roxburghii*** DC.
Eugenia rotundifolia Casar. = [84 BZL]
Eugenia rotundifolia (Arn.) Wight = ***Syzygium rotundifolium*** Arn.
Eugenia rotundifolia var. *oblongata* O.Berg = [84 BZL]
Eugenia rubens Roxb. = ***Syzygium rubens*** (Roxb.) Walp.
Eugenia rubescens A.Gray = ***Syzygium rubescens*** (A.Gray) Müll.Stuttg.
Eugenia rubicunda (Wight & Arn.) Wight = ***Syzygium rubicundum*** Wight & Arn.
Eugenia rubida Ridl. = ***Syzygium siamense*** (Craib) Chantaran. & J.Parn.
Eugenia rubiginosa Cordem. = ***Eugenia mespiloides*** Lam.
Eugenia rubiginosa Cambess. = ***Marlierea rubiginosa*** (Cambess.) D.Legrand
Eugenia rubra (Blume) Reinw. ex de Vriese = ***Decaspermum bracteatum*** (Roxb.) A.J.Scott var. ***bracteatum***
Eugenia rubrescens Mattos & D.Legrand = ***Eugenia punicifolia*** (Kunth) DC.
Eugenia rubricaulis (Miq.) Duthie = ***Syzygium lineatum*** (DC.) Merr. & L.M.Perry
Eugenia rubrocincta O.Berg = ***Eugenia bimarginata*** DC.
Eugenia rubropunctata Ridl. = ***Syzygium rubropunctatum*** (Ridl.) Merr. & L.M.Perry
Eugenia rubropurpurea C.B.Rob. = ***Syzygium rubropurpureum*** (C.B.Rob.) Airy Shaw
Eugenia rubrovenia C.B.Rob. = ***Syzygium rubrovenium*** (C.B.Rob.) Merr.
Eugenia rudatisii Engl. & Brehmer = ***Eugenia capensis*** subsp. ***natalitia*** (Sond.) F.White

Eugenia rufa (Colla) Barnéoud = ***Myrceugenia rufa*** (Colla) Skottsb.
Eugenia rufescens DC. = ***Myrceugenia rufescens*** (DC.) D.Legrand & Kausel
Eugenia rugosa Ruiz & Pav. ex.DC. = ***Eugenia patens*** Poir.
Eugenia rugosa (Korth.) Merr. = ***Syzygium rugosum*** Korth.
Eugenia rugosa var. *cordata* M.R.Hend. = ***Syzygium rugosum*** Korth.
Eugenia rugosa var. *saxitana* (Ridl.) M.R.Hend. = ***Syzygium rugosum*** Korth.
Eugenia ruiziana O.Berg = [82 FRG]
Eugenia ruminata Koord. & Valeton = ***Syzygium claviflorum*** (Roxb.) Wall. ex A.M.Cowan & Cowan
Eugenia rumphii Merr. = ***Syzygium rumphii*** (Merr.) Govaerts
Eugenia rupestris Engl. & Brehmer = ***Eugenia leonensis*** Engl. & Brehmer
Eugenia rupestris O.Berg = ***Myrceugenia alpigena*** (DC.) Landrum var. ***alpigena***
Eugenia rutidocarpa Ruiz & Pav. ex G.Don = ***Eugenia patens*** Poir.
Eugenia sablanensis Elmer = ***Syzygium fastigiatum*** (Blume) Merr. & L.M.Perry
Eugenia sabulosa Cambess. = ***Eugenia punicifolia*** (Kunth) DC.
Eugenia sakalavarum H.Perrier = ***Syzygium sakalavarum*** (H.Perrier) Labat & Schatz
Eugenia salaccensis Koord. & Valeton = ***Syzygium pyrifolium*** (Blume) DC.
Eugenia salamancana Standl. = ***Plinia salamancana*** (Standl.) Barrie
Eugenia salamensis var. *hiraeifolia* (Standl.) McVaugh = ***Eugenia hiraeifolia*** Standl.
Eugenia salamensis var. *rensoniana* (Standl.) McVaugh = ***Eugenia salamensis*** Donn.Sm.
Eugenia salicifolia O.Berg = ***Eugenia biflora*** (L.) DC.
Eugenia salicifolia G.Lawson = ***Eugenia salacioides*** G.Lawson ex Hutch. & Dalziel
Eugenia salicifolia (Kunth) DC. = ***Blepharocalyx salicifolius*** (Kunth) O.Berg
Eugenia salicifolia Wight = ***Syzygium salicifolium*** (Wight) J.Graham
Eugenia salicina Ridl. = ***Syzygium salicinum*** (Ridl.) Merr. & L.M.Perry
Eugenia salictoides Ridl. = ***Syzygium salictoides*** (Ridl.) I.M.Turner
Eugenia saligna (Miq.) C.B.Rob. = ***Syzygium salignum*** (Miq.) Rathakr. & N.C.Nair
Eugenia salomonensis Hemsl. = ***Syzygium salomonense*** (Hemsl.) Merr. & L.M.Perry
Eugenia salpingantha Greves = ***Syzygium salpinganthum*** (Greves) Merr. & L.M.Perry
Eugenia salzmannii Benth. = ***Myrciaria floribunda*** (H.West ex Willd.) O.Berg
Eugenia samarangensis (Blume) O.Berg = ***Syzygium samarangense*** (Blume) Merr. & L.M.Perry
Eugenia samarensis Merr. = ***Syzygium leytense*** (Elmer) Merr.
Eugenia sambiranensis H.Perrier = ***Syzygium sambiranense*** (H.Perrier) Labat & Schatz
Eugenia samoensis Burkill = ***Syzygium samoense*** (Burkill) Whistler
Eugenia sancta DC. = ***Eugenia punicifolia*** (Kunth) DC.
Eugenia sandakanensis Merr. = ***Syzygium leptostemon*** (Korth.) Merr. & L.M.Perry
Eugenia sandwicensis A.Gray = ***Syzygium sandwicense*** (A.Gray) Müll.Stuttg.
Eugenia santosii Merr. = ***Syzygium santosii*** (Merr.) Merr.
Eugenia sarawacensis Merr. = ***Syzygium muelleri*** (Miq.) Miq.
Eugenia sarcocarpa Merr. = ***Syzygium leytense*** (Elmer) Merr.
Eugenia savaiiensis A.Gray = ***Syzygium savaiiense*** (A.Gray) Müll.Stuttg.
Eugenia saxitana Ridl. = ***Syzygium rugosum*** Korth.
Eugenia sayeri F.Muell. = ***Syzygium sayeri*** (F.Muell.) B.Hyland
Eugenia scalarinervis King = ***Syzygium scalarinerve*** (King) I.M.Turner
Eugenia scandens Baker = ***Syzygium latifolium*** (Poir.) DC.
Eugenia schaueriana Miq. = ***Myrcia graciliflora*** Sagot
Eugenia schiedeana Schltdl. = ***Eugenia capuli*** (Schltdl. & Cham.) Hook. & Arn.
Eugenia schippii Standl. = ***Chamguava schippii*** (Standl.) Landrum
Eugenia schlechtendaliana O.Berg = ***Eugenia polystachya*** Rich.
Eugenia schomburgkii Benth. = ***Eugenia lambertiana*** DC.
Eugenia schuchiana var. *grandifolia* O.Berg = ***Eugenia verticillata*** (Vell.) Angely
Eugenia schuechiana O.Berg = ***Eugenia verticillata*** (Vell.) Angely
Eugenia schumanniana (Nied.) Greves = ***Syzygium schumannianum*** (Nied.) Diels
Eugenia sclerocalyx var. *cambajuvensis* D.Legrand = ***Eugenia sclerocalyx*** D.Legrand
Eugenia sclerophylla (Thwaites) Bedd. = ***Syzygium sclerophyllum*** Thwaites
Eugenia scolopacina Ridl. = ***Xanthomyrtus scolopacina*** (Ridl.) Diels
Eugenia scolopophylla Ridl. = ***Syzygium scolopophyllum*** (Ridl.) Masam.
Eugenia scoparia Duthie = ***Syzygium avene*** Miq.
Eugenia scortechinii King = ***Syzygium scortechinii*** (King) Chantaran. & J.Parn.
Eugenia scortechinii var. *cuneata* M.R.Hend. = ***Syzygium scortechinii*** (King) Chantaran. & J.Parn.
Eugenia sechellarum Baker = ***Syzygium wrightii*** (Baker) A.J.Scott
Eugenia sehnemiana (Mattos) Mattos = ***Hexachlamys sehnemiana*** (Mattos) Mattos
Eugenia selangorensis Ridl. = ***Syzygium palembanicum*** Miq.
Eugenia selkirkii Hook. & Arn. = ***Ugni selkirkii*** (Hook. & Arn.) O.Berg
Eugenia sellowiana DC. = ***Psidium salutare*** var. ***sericeum*** (Cambess.) Landrum
Eugenia seriatopedunculata Kiaersk. = ***Myrceugenia alpigena*** (DC.) Landrum var. ***alpigena***
Eugenia seriatoracemosa Kiaersk. = ***Eugenia florida*** DC.
Eugenia seriatoramosa Kiaersk. = ***Myrceugenia seriatoramosa*** (Kiaersk.) D.Legrand & Kausel
Eugenia sericea O.Berg = ***Eugenia tinguyensis*** Cambess.
Eugenia sericea var. *angustifolia* O.Berg = ***Eugenia tinguyensis*** Cambess.
Eugenia sericea var. *robusta* O.Berg = ***Eugenia tinguyensis*** Cambess.
Eugenia sericiflora Benth. = ***Eugenia biflora*** (L.) DC.
Eugenia sessililimba Merr. = ***Syzygium sessililimbum*** (Merr.) Merr.
Eugenia setosa King = ***Syzygium setosum*** (King) I.M. Turner

Eugenia sexangulata (Miq.) Koord. & Valeton = ***Syzygium sexangulatum*** (Miq.) Amshoff
Eugenia shepherdii F.Muell. = ***Gossia shepherdii*** (F.Muell.) N.Snow & Guymer
Eugenia siamensis Craib = ***Syzygium siamense*** (Craib) Chantaran. & J.Parn.
Eugenia sibulanensis Elmer = ***Syzygium arcuatinervium*** (Merr.) Craven & Biffin
Eugenia sibunensis Lundell = ***Eugenia acapulcensis*** Steud.
Eugenia siderocola Merr. = ***Syzygium siderocola*** (Merr.) Merr.
Eugenia sieberiana DC. = ***Eugenia greggii*** (Sw.) Poir.
Eugenia sieberiana var. *crassifolia* O.Berg = ***Eugenia greggii*** (Sw.) Poir.
Eugenia silvestrei Elmer = ***Syzygium densinervium*** (Merr.) Merr. var. ***densinervium***
Eugenia silvestris (O.Berg) Mattos = ***Eugenia neosilvestris*** Sobral
Eugenia simii Dummer = ***Eugenia capensis*** subsp. ***simii*** (Dummer) F.White
Eugenia similis Merr. = ***Syzygium simile*** (Merr.) Merr.
Eugenia simmondsiae F.M.Bailey = ***Syzygium australe*** (J.C.Wendl. ex Link) B.Hyland
Eugenia simpsonii (Small) Sarg. = ***Myrcianthes fragrans*** (Sw.) McVaugh
Eugenia simulans King = [42 MLY]
Eugenia sinemariensis Aubl. = ***Eugenia coffeifolia*** DC.
Eugenia sinensis Hemsl. = ***Syzygium buxifolium*** Hook. & Arn.
Eugenia sintenisii Kiaersk. = ***Eugenia cordata*** var. ***sintenisii*** (Kiaersk.) Krug & Urb.
Eugenia sintenisii Krug & Urb. = ***Eugenia cordata*** var. ***sintenisii*** (Kiaersk.) Krug & Urb.
Eugenia sinubanensis Elmer = ***Syzygium myrtifolium*** Walp.
Eugenia siphonantha King ex Greves = ***Syzygium siphonanthum*** (King ex Greves) Amshoff
Eugenia skiophila Duthie = ***Syzygium skiophilum*** (Duthie) Airy Shaw
Eugenia skutchii C.V.Morton & Standl. = ***Eugenia pachychlamys*** Donn.Sm.
Eugenia smaliana Brandis = ***Syzygium smalianum*** (Brandis) D.G.Long
Eugenia smaragdina O.Berg = ***Eugenia lambertiana*** DC.
Eugenia smaragdina var. *angustifolia* O.Berg = ***Eugenia lambertiana*** DC.
Eugenia smaragdina var. *brevipes* O.Berg = ***Eugenia lambertiana*** DC.
Eugenia smaragdina var. *rigida* O.Berg = ***Eugenia lambertiana*** DC.
Eugenia smithii Poir. = ***Syzygium smithii*** (Poir.) Nied.
Eugenia smithii var. *coriacea* Domin = ***Syzygium smithii*** (Poir.) Nied.
Eugenia smithii var. *hemilampra* F.Muell. = ***Syzygium hemilamprum*** subsp. ***hemilamprum***
Eugenia smithii var. *minor* Maiden = ***Syzygium smithii*** (Poir.) Nied.
Eugenia sodiroi Diels = ***Eugenia pycnantha*** Benth.
Eugenia sogerensis Greves = ***Syzygium sogerense*** (Greves) Merr. & L.M.Perry
Eugenia somai Hayata = ***Syzygium buxifolium*** Hook. & Arn.
Eugenia somalensis Chiov. = ***Eugenia aschersoniana*** F.Hoffm.
Eugenia sordida F.M.Bailey = ***Syzygium wilsonii*** subsp. ***cryptophlebium*** (F.Muell.) B.Hyland
Eugenia sorsogonensis Merr. = ***Syzygium vidalianum*** (Elmer) Merr.
Eugenia soyauxii Engl. = ***Eugenia klaineana*** (Pierre) Engl.
Eugenia sparsiflora DC. = ***Campomanesia aromatica*** (Aubl.) Griseb.
Eugenia spathophylla O.Berg = ***Eugenia punicifolia*** (Kunth) DC.
Eugenia spathulata O.Berg = ***Eugenia punicifolia*** (Kunth) DC.
Eugenia spathulata (Thwaites) Bedd. = ***Syzygium spathulatum*** Thwaites
Eugenia speciosissima C.B.Rob. = ***Syzygium speciosissimum*** (C.B.Rob.) Merr.
Eugenia spectabilis Phil. = ***Luma apiculata*** (DC.) Burret
Eugenia sphaerantha Gagnep. = ***Syzygium sphaeranthum*** (Gagnep.) Merr. & L.M.Perry
Eugenia sphaerosperma DC. = ***Eugenia egensis*** DC.
Eugenia sphenophylla O.Berg = ***Eugenia dodonaeifolia*** Cambess.
Eugenia sphenophylla var. *coaetanea* O.Berg = ***Eugenia dodonaeifolia*** Cambess.
Eugenia sphenophylla var. *prasina* O.Berg = ***Eugenia dodonaeifolia*** Cambess.
Eugenia sphenophylla var. *tristis* O.Berg = ***Eugenia dodonaeifolia*** Cambess.
Eugenia spicata Lam. = ***Syzygium zeylanicum*** (L.) DC.
Eugenia spicata var. *cordata* Kochummen = ***Syzygium zeylanicum*** (L.) DC.
Eugenia spiciflora Nees & Mart. = ***Plinia spiciflora*** (Nees & Mart.) Sobral
Eugenia spissa Craib = ***Syzygium angkae*** subsp. ***spissum*** (Craib) Chantaran. & J.Parn.
Eugenia spissifolia Ridl. = ***Syzygium spissifolium*** (Ridl.) I.M.Turner
Eugenia springiana O.Berg = ***Calycorectes acutatus*** (Miq.) Toledo
Eugenia squamata Willd. ex O.Berg = ***Marlierea umbraticola*** (Kunth) O.Berg
Eugenia squamifera C.B.Rob. = ***Syzygium squamiferum*** (C.B.Rob.) Merr.
Eugenia squarrosa Urb. & Ekman = ***Eugenia anthacanthoides*** Urb. & Ekman
Eugenia stapfiana King = ***Syzygium stapfianum*** (King) I.M.Turner
Eugenia stelechantha (Diels) Kaneh. = ***Syzygium stelechanthum*** (Diels) Glassman
Eugenia stelechanthoides Kaneh. = ***Syzygium stelechanthum*** (Diels) Glassman
Eugenia stenophylla Hook. & Arn. = ***Myrceugenia lanceolata*** (Juss. ex J.St.-Hil.) Kausel
Eugenia stenophylla var. *angustifolia* O.Berg = ***Myrceugenia lanceolata*** (Juss. ex J.St.-Hil.) Kausel
Eugenia stephanii var. *angustifolia* O.Berg = ***Eugenia stephanii*** O.Berg
Eugenia stephanii var. *latifolia* O.Berg = ***Eugenia stephanii*** O.Berg
Eugenia steyermarkii Standl. = ***Myrcianthes fragrans*** (Sw.) McVaugh
Eugenia sticheromischa Kiaersk. = ***Myrceugenia alpigena*** (DC.) Landrum var. ***alpigena***
Eugenia stictosepala Kiaersk. = ***Eugenia prasina*** O.Berg
Eugenia stictosepala var. *melanogyna* D.Legrand = ***Eugenia melanogyna*** (D.Legrand) Sobral
Eugenia stigmatosa var. *acutata* O.Berg = ***Eugenia stigmatosa*** DC.
Eugenia stigmatosa var. *obtusata* O.Berg = ***Eugenia stigmatosa*** DC.
Eugenia stipitata subsp. *sororia* McVaugh = ***Eugenia stipitata*** McVaugh subsp. ***stipitata***

Eugenia stipularis (O.Berg) Mattos = ***Eugenia squamiflora*** Mattos
Eugenia stipularis (Blume) Miq. = ***Syzygium aqueum*** (Burm.f.) Alston
Eugenia stirpiflora (O.Berg) Krug & Urb. = ***Eugenia cordata*** (Sw.) DC. var. ***cordata***
Eugenia stocksii Duthie = ***Syzygium stocksii*** (Duthie) Gamble
Eugenia stokesii C.A.Gardner = ***Syzygium eucalyptoides*** subsp. ***bleeseri*** (O.Schwarz) B.Hyland
Eugenia stolzii Engl. & Brehmer = ***Eugenia malangensis*** (O.Hoffm.) Nied.
Eugenia storkii Standl. = ***Myrcianthes storkii*** (Standl.) McVaugh
Eugenia striata Mattos & D.Legrand = ***Eugenia oblongata*** O.Berg
Eugenia striata Koord. & Valeton = ***Syzygium pyrifolium*** (Blume) DC.
Eugenia striatula C.B.Rob. = ***Syzygium striatulum*** (C.B.Rob.) Merr.
Eugenia stricta (O.Berg) Kiaersk. = ***Eugenia strictissima*** Govaerts
Eugenia strigosa (O.Berg) Arechav. = ***Eugenia uniflora*** L.
Eugenia stuposa A.Rich. ex O.Berg = ***Calyptranthes cuprea*** O.Berg
Eugenia suaveolens Cambess. = ***Blepharocalyx salicifolius*** (Kunth) O.Berg
Eugenia suavis Ridl. = [42 MLY]
Eugenia subalata Ridl. = ***Syzygium subalatum*** (Ridl.) Merr. & L.M.Perry
Eugenia subalterna Benth. = ***Eugenia punicifolia*** (Kunth) DC.
Eugenia subavenis Duthie = ***Syzygium umbrosum*** Thwaites
Eugenia subcaudata Merr. = ***Syzygium subcaudatum*** (Merr.) Merr.
Eugenia subcordata O.Berg = [84 BZS]
Eugenia subcordata var. *australis* Mattos = [84 BZS]
Eugenia subcorymbosa O.Berg = ***Eugenia punicifolia*** (Kunth) DC.
Eugenia subdecurrens (Miq.) Merr. & Chun = ***Syzygium acuminatissimum*** (Blume) DC.
Eugenia subdecussata Duthie = ***Syzygium subdecussatum*** (Duthie) I.M.Turner
Eugenia subdecussata var. *montana* King = ***Syzygium subdecussatum*** (Duthie) I.M.Turner
Eugenia subfalcata C.B.Rob. = ***Syzygium subfalcatum*** (C.B.Rob.) Merr.
Eugenia subfoetida C.B.Rob. = ***Syzygium subfoetidum*** (C.B.Rob.) Merr.
Eugenia subglauca Koord. & Valeton = ***Syzygium littorale*** (Blume) Amshoff
Eugenia subhorizontalis King = ***Syzygium subhorizontale*** (King) Chantaran. & J.Parn.
Eugenia sublaeta Craib = ***Syzygium laetum*** subsp. ***sublaetum*** (Craib) Chantaran. & J.Parn.
Eugenia submimica Elmer = ***Syzygium mimicum*** (Merr.) Merr.
Eugenia subobliqua Benth. = [82 GUY]
Eugenia subopposita F.M.Bailey = ***Syzygium wilsonii*** (F.Muell.) B.Hyland subsp. ***wilsonii***
Eugenia suborbicularis Benth. = ***Syzygium suborbiculare*** (Benth.) T.G.Hartley & L.M.Perry
Eugenia subracemosa Merr. = ***Syzygium leptostemon*** (Korth.) Merr. & L.M.Perry
Eugenia subrigens Craib = ***Syzygium rigens*** (Craib) Chantaran. & J.Parn.
Eugenia subrotundifolia C.B.Rob. = ***Syzygium subrotundifolium*** (C.B.Rob.) Merr.
Eugenia subrufa King = ***Syzygium griffithii*** (Duthie) Merr. & L.M.Perry
Eugenia subsessiliflora Merr. = ***Syzygium subsessiliflorum*** (Merr.) Merr.
Eugenia subsessilifolia Merr. = ***Syzygium subsessilifolium*** (Merr.) Merr. & L.M.Perry
Eugenia subsessilis C.B.Rob. = ***Syzygium subsessile*** (C.B.Rob.) Merr.
Eugenia subsulcata Elmer = ***Syzygium conglobatum*** Merr.
Eugenia subverticillaris O.Berg = ***Eugenia laevis*** O.Berg
Eugenia subviridis Craib = ***Syzygium tetragonum*** (Wight) Wall. ex Walp.
Eugenia succulenta Elmer = ?
Eugenia suffrutescens Nied. = ***Eugenia calycina*** var. ***herbacea*** (O.Berg) Mattos
Eugenia suffrutescens var. *brevipedunculata* (Mattos) Mattos = ***Eugenia calycina*** var. ***herbacea*** (O.Berg) Mattos
Eugenia suffruticosa O.Berg = ***Eugenia punicifolia*** (Kunth) DC.
Eugenia suffruticosa var. *laeta* O.Berg = ***Eugenia punicifolia*** (Kunth) DC.
Eugenia suffruticosa var. *latifolia* O.Berg = ***Eugenia punicifolia*** (Kunth) DC.
Eugenia suffruticosa var. *opaca* O.Berg = ***Eugenia punicifolia*** (Kunth) DC.
Eugenia sulcata var. *latifolia* Kiaersk. = ***Eugenia sulcata*** Spring ex Mart.
Eugenia sulcata var. *pubescens* Mattos = ***Eugenia sulcata*** Spring ex Mart.
Eugenia sulcata var. *stricta* (O.Berg) Mattos = ***Eugenia sulcata*** Spring ex Mart.
Eugenia sulcistyla C.B.Rob. = ***Syzygium sulcistylum*** (C.B.Rob.) Merr.
Eugenia sulphurata Ridl. = ***Syzygium sulphuratum*** (Ridl.) Govaerts
Eugenia suluensis Merr. = ***Syzygium elliptilimbum*** (Merr.) Merr. & L.M.Perry
Eugenia surigaensis Merr. = ***Syzygium surigaense*** (Merr.) Merr.
Eugenia surinamensis Miq. = ***Eugenia punicifolia*** (Kunth) DC.
Eugenia surinamensis var. *impunctata* O.Berg = ***Eugenia punicifolia*** (Kunth) DC.
Eugenia surinamensis var. *punctata* O.Berg = ***Eugenia punicifolia*** (Kunth) DC.
Eugenia suringariana Koord. & Valeton = ***Syzygium suringarianum*** (Koord. & Valeton) Amshoff
Eugenia suzukii Kaneh. = [62 CRL]
Eugenia swettenhamiana King = ***Syzygium swettenhamianum*** (King) I.M.Turner
Eugenia sylvana Ridl. = ***Syzygium effusum*** (A.Gray) Müll.Stuttg.
Eugenia sylvatica Cambess. = ***Eugenia florida*** DC.
Eugenia sylvatica Gardner = ***Marlierea sylvatica*** (O.Berg) Kiaersk.
Eugenia sylvestris Moon ex Wight = ***Syzygium makul*** Gaertn.
Eugenia sylvicola C.Wright = ***Eugenia scaphephylla*** C.Wright
Eugenia symingtoniana M.R.Hend. = ***Syzygium symingtonianum*** (M.R.Hend.) I.M.Turner
Eugenia symphysipetala Koord. & Valeton = ***Syzygium occlusum*** Miq.
Eugenia syzygioides (Miq.) M.R.Hend. = ***Syzygium syzygioides*** (Miq.) Merr. & L.M.Perry
Eugenia tabascensis Lundell = ***Eugenia karwinskyana*** O.Berg

Eugenia tabasco (Willd. ex Schltdl. & Cham.) G.Don = ***Pimenta racemosa*** (Mill.) J.W.Moore var. ***racemosa***
Eugenia tahanensis Ridl. = ***Syzygium tahanense*** (Ridl.) I.M.Turner
Eugenia taipingensis M.R.Hend. = ***Syzygium taipingense*** (M.R.Hend.) I.M.Turner
Eugenia tanalensis Baker = [29 MDG] *Syzygium* sp. ?
Eugenia tapacumensis O.Berg = ***Eugenia stictopetala*** DC.
Eugenia tapacumensis var. *angustifolia* O.Berg = ***Eugenia stictopetala*** DC.
Eugenia tapacumensis var. *latifolia* O.Berg = ***Eugenia stictopetala*** DC.
Eugenia tapiaka H.Perrier = ***Syzygium tapiaka*** (H.Perrier) Labat & Schatz
Eugenia tapiraguayensis Barb.Rodr. ex Chodat & Hassl. = [85 PAR]
Eugenia tawaensis Merr. = ***Syzygium peregrinum*** (Blume) Merr. & L.M.Perry
Eugenia tayabensis Quisumb. & Merr. = ***Syzygium tayabense*** (Quisumb. & Merr.) Merr.
Eugenia taytayensis Merr. = ***Syzygium taytayense*** (Merr.) Merr.
Eugenia tecta King = ***Syzygium tectum*** (King) I.M.Turner
Eugenia teffensis O.Berg = ***Eugenia procera*** (Sw.) Poir.
Eugenia teffensis var. *subcordata* O.Berg = ***Eugenia procera*** (Sw.) Poir.
Eugenia teffensis var. *truncata* O.Berg = ***Eugenia procera*** (Sw.) Poir.
Eugenia teixeireana (Mattos) Mattos = ***Calycorectes teixeireanus*** Mattos
Eugenia tekuensis M.R.Hend. = ***Syzygium tekuense*** (M.R.Hend.) I.M.Turner
Eugenia temu Hook. & Arn. = ***Myrceugenia exsucca*** (DC.) O.Berg
Eugenia tenejapensis Lundell = ***Eugenia karwinskyana*** O.Berg
Eugenia tenella Miq. = ***Myrciaria cuspidata*** O.Berg
Eugenia tenella DC. = ***Myrciaria tenella*** (DC.) O.Berg
Eugenia tenella var. *elliptica* Kiaersk. = ***Myrciaria tenella*** (DC.) O.Berg
Eugenia tenella var. *glazioviana* Kiaersk. = ***Myrciaria tenella*** (DC.) O.Berg
Eugenia tenella var. *macrocarpa* Kiaersk. = ***Myrciaria tenella*** (DC.) O.Berg
Eugenia tenella var. *minor* Cambess. = ***Myrciaria tenella*** (DC.) O.Berg
Eugenia tenella var. *spathulata* Kiaersk. = ***Myrciaria tenella*** (DC.) O.Berg
Eugenia tenuicuspis Koord. & Valeton = ***Syzygium rostratum*** (Blume) DC.
Eugenia tenuifolia O.Berg = ***Eugenia decussata*** (Vell.) Mattos
Eugenia tenuifolia var. *axillaris* O.Berg = ***Eugenia decussata*** (Vell.) Mattos
Eugenia tenuifolia var. *laterifolia* O.Berg = ***Eugenia decussata*** (Vell.) Mattos
Eugenia tenuipes Merr. = ***Syzygium tenuipes*** (Merr.) Merr.
Eugenia tenuiramis Miq. = ***Syzygium zeylanicum*** (L.) DC.
Eugenia tenuis Duthie = ***Syzygium tenue*** (Duthie) N.P.Balakr.
Eugenia tephrodes Hance = ***Syzygium tephrodes*** (Hance) Merr. & L.M.Perry
Eugenia teretiflora Koord. & Valeton = ***Syzygium teretiflorum*** (Koord. & Valeton) Amshoff
Eugenia ternifolia O.Berg = ***Myrcianthes myrsinoides*** (Kunth) Grifo
Eugenia ternifolia Roxb. = ***Syzygium formosum*** (Wall.) Masam.
Eugenia tetraedra Duthie = ?
Eugenia tetragona Wight = ***Syzygium tetragonum*** (Wight) Wall. ex Walp.
Eugenia tetraptera (Miq.) M.R.Hend. = ***Syzygium tetrapterum*** (Miq.) Chantaran. & J.Parn.
Eugenia tetrasperma Bello = [81 PUE]
Eugenia teysmannii (Miq.) Koord. & Valeton = ***Syzygium lineatum*** (DC.) Merr. & L.M.Perry
Eugenia teyu-iba Parodi = [85 AGE]
Eugenia thalassaia O.Berg = ***Myrceugenia correifolia*** (Hook. & Arn.) O.Berg
Eugenia thalassaia var. *colchaguensis* (Phil.) Reiche = ***Myrceugenia colchaguensis*** (Phil.) Navas
Eugenia theodorae var. *brevipedunculata* Kiaersk. = ***Eugenia theodorae*** Kiaersk.
Eugenia theodori-wolfii Domin = ***Syzygium tierneyanum*** (F.Muell.) T.G.Hartley & L.M.Perry
Eugenia thorelii Gagnep. = ***Syzygium thorelii*** (Gagnep.) Merr. & L.M.Perry
Eugenia thumra Roxb. = ***Syzygium thumra*** (Roxb.) Merr. & L.M.Perry
Eugenia thumra var. *penangiana* King = ***Syzygium thumra*** var. ***penangianum*** (King) I.M.Turner
Eugenia thwaitesii Duthie = ***Eugenia mooniana*** Wight
Eugenia thymifolia Phil. ex Reiche = ***Myrceugenia leptospermoides*** (DC.) Kausel
Eugenia tiburona Urb. & Ekman = ***Mosiera tiburona*** (Urb. & Ekman) Borhidi
Eugenia tierneyana F.Muell. = ***Syzygium tierneyanum*** (F.Muell.) T.G.Hartley & L.M.Perry
Eugenia tinctoria Gagnep. = ***Syzygium tinctorium*** (Gagnep.) Merr. & L.M.Perry
Eugenia tinge-lingua S.Moore = ***Eugenia florida*** DC.
Eugenia tiumanensis Ridl. = ***Syzygium tiumanense*** (Ridl.) I.M.Turner
Eugenia tocaiana O.Berg = ***Eugenia itapemirimensis*** Cambess.
Eugenia toddalioides Wight = ***Syzygium toddalioides*** (Wight) Walp.
Eugenia toledoi (Mattos) Mattos = ***Hexachlamys toledoi*** Mattos
Eugenia tomentosa Cambess. = [84 BZL]
Eugenia tomentosa Aubl. = ***Myrcia tomentosa*** (Aubl.) DC.
Eugenia tomentulosa Standl. = ***Eugenia salamensis*** Donn.Sm.
Eugenia tomlinsonii Maiden & Betche = ***Syzygium francisii*** (F.M.Bailey) L.A.S.Johnson
Eugenia toninensis Baker f. = ***Syzygium toninense*** (Baker f.) J.W.Dawson
Eugenia tonkinensis Gagnep. = ***Syzygium tonkinense*** (Gagnep.) Merr. & L.M.Perry
Eugenia toppingii Elmer = ***Syzygium toppingii*** (Elmer) Merr.
Eugenia trachyphloia C.T.White = ***Syzygium trachyphloium*** (C.T.White) B.Hyland
Eugenia tramnion Gagnep. = ***Syzygium tramnion*** (Gagnep.) Merr. & L.M.Perry
Eugenia triantha Merr. = ***Syzygium trianthum*** (Merr.) Merr.
Eugenia trichocarpa Phil. = ***Myrceugenia ovata*** (Hook. & Arn.) O.Berg var. ***ovata***
Eugenia trichotoma Greves = ***Syzygium trichotomum*** (Greves) Merr. & L.M.Perry
Eugenia triflora Willd. ex O.Berg = ***Campomanesia dichotoma*** (O.Berg) Mattos
Eugenia triflora Sessé & Moç. = ***Myrcianthes fragrans*** (Sw.) McVaugh

Eugenia trinervia DC. = ***Rhodamnia rubescens*** (Benth.) Miq.
Eugenia triphylla O.Berg = ***Eugenia punicifolia*** (Kunth) DC.
Eugenia triphylla C.B.Rob. = ***Syzygium triphyllum*** Merr.
Eugenia tripinnata (Blanco) C.B.Rob. = ***Syzygium tripinnatum*** (Blanco) Merr.
Eugenia triplinervia O.Berg = ***Eugenia foetida*** Pers.
Eugenia triplinervia var. *angustifolia* O.Berg = ***Eugenia foetida*** Pers.
Eugenia triplinervia var. *buxifolia* (Sw.) O.Berg = ***Eugenia foetida*** Pers.
Eugenia triplinervia var. *laevigata* O.Berg = ***Eugenia foetida*** Pers.
Eugenia triplinervia var. *latissima* O.Berg = ***Eugenia foetida*** Pers.
Eugenia triplinervia var. *oblongata* O.Berg = ***Eugenia foetida*** Pers.
Eugenia triplinervia var. *obovata* O.Berg = ***Eugenia foetida*** Pers.
Eugenia triquetra O.Berg = ***Myrcianthes myrsinoides*** (Kunth) Grifo
Eugenia triquetra var. *aequatorialis* O.Berg = ***Myrcianthes myrsinoides*** (Kunth) Grifo
Eugenia tristis D.Legrand = ***Eugenia neotristis*** Sobral
Eugenia tristis Kurz = ***Syzygium triste*** (Kurz) N.P.Balakr.
Eugenia trivenis Ridl. = ***Syzygium trivene*** (Ridl.) Merr. & L.M.Perry
Eugenia tropophylla var. *majoriflora* H.Perrier = ***Eugenia tropophylla*** H.Perrier
Eugenia tropophylla var. *piriformis* H.Perrier = ***Eugenia tropophylla*** H.Perrier
Eugenia trukensis Hosok. = [62 CRL]
Eugenia trunciflora Ridl. = ***Syzygium tekuense*** (M.R.Hend.) I.M.Turner
Eugenia tsoi Merr. & Chun = ***Syzygium cumini*** (L.) Skeels
Eugenia tsoongii Merr. = ***Syzygium tsoongii*** (Merr.) Merr. & L.M.Perry
Eugenia tuberculata var. *uniflora* O.Berg = ***Eugenia tuberculata*** (Kunth) DC.
Eugenia tula Merr. = ***Syzygium tula*** (Merr.) Merr.
Eugenia tumida Duthie = ***Syzygium pyrifolium*** (Blume) DC.
Eugenia turbinata O.Berg = ***Eugenia pyriformis*** Cambess.
Eugenia turumiquirensis Steyerm. = ***Myrcianthes fragrans*** (Sw.) McVaugh
Eugenia tussacii Urb. & Ekman = ***Mosiera tussacii*** (Urb. & Ekman) Borhidi
Eugenia tutuilensis A.Gray = ***Syzygium savaiiense*** (A.Gray) Müll.Stuttg.
Eugenia tweediei Hook. & Arn. = ***Blepharocalyx salicifolius*** (Kunth) O.Berg
Eugenia ubensis Cambess. = ***Eugenia brasiliensis*** Lam.
Eugenia ugni (Molina) Hook. & Arn. = ***Ugni molinae*** Turcz.
Eugenia ugoensis C.B.Rob. = ***Syzygium myrtillus*** (Stapf) Merr. & L.M.Perry
Eugenia umbellaris O.Berg = ***Eugenia bimarginata*** DC.
Eugenia umbellata DC. = ***Eugenia bimarginata*** DC.
Eugenia umbellata Spreng. = ***Eugenia erythrocarpa*** (Kunth) DC.
Eugenia umbellulifera (Kunth) Krug & Urb. = ***Pseudanamomis umbellulifera*** (Kunth) Kausel
Eugenia umbilicata Koord. & Valeton = ***Syzygium umbilicatum*** (Koord. & Valeton) Amshoff
Eugenia umbrosa Bedd. = ***Syzygium umbrosum*** Thwaites
Eugenia undulatifolia Nied. = ***Eugenia speciosa*** Cambess.
Eugenia unedifolia Spreng. ex O.Berg = ***Eugenia foetida*** Pers.
Eugenia uniflora O.Berg = ***Eugenia bergii***
Eugenia uniflora var. *atropurpurea* Mattos = ***Eugenia uniflora*** L.
Eugenia urceolata Cordem. = ***Psidium cattleianum*** Afzel. ex Sabine
Eugenia urceolata (Korth. ex Miq.) King = ***Syzygium leptostemon*** (Korth.) Merr. & L.M.Perry
Eugenia urdanetensis Elmer = ***Syzygium urdanetense*** (Elmer) Merr.
Eugenia urschiana var. *subacuminata* H.Perrier = ***Eugenia urschiana*** H.Perrier
Eugenia utilis Talbot = ***Syzygium utilis*** (Talbot) Rathakr. & N.C.Nair
Eugenia uvalha Cambess. = ***Eugenia pyriformis*** Cambess.
Eugenia vacciniifolia Baker = ***Syzygium loiseleurioides*** (Baker) Govaerts
Eugenia vaccinioides Elmer = ***Syzygium vaccinifolium*** Merr.
Eugenia vaga O.Berg = ***Eugenia punicifolia*** (Kunth) DC.
Eugenia vaga var. *brachypoda* (DC.) O.Berg = ***Eugenia punicifolia*** (Kunth) DC.
Eugenia vaga var. *brasiliensis* O.Berg = ***Eugenia punicifolia*** (Kunth) DC.
Eugenia vaga var. *dipoda* (DC.) O.Berg = ***Eugenia punicifolia*** (Kunth) DC.
Eugenia vaga var. *pumila* O.Berg = ***Eugenia punicifolia*** (Kunth) DC.
Eugenia vaga var. *punicifolia* (Kunth) O.Berg = ***Eugenia punicifolia*** (Kunth) DC.
Eugenia vaga var. *rigida* O.Berg = ***Eugenia punicifolia*** (Kunth) DC.
Eugenia vahlii DC. = ***Eugenia trinervia*** Vahl
Eugenia valdevenosa Duthie = ***Syzygium valdevenosum*** (Duthie) Merr. & L.M.Perry
Eugenia valenzuelana A.Rich. = ***Myrcia valenzuelana*** (A.Rich.) Griseb.
Eugenia valerioi Standl. = ***Eugenia costaricensis*** O.Berg
Eugenia valetoniana King = [42 MLY]
Eugenia vallis Standl. = ***Myrcia tomentosa*** (Aubl.) DC.
Eugenia vallsiana Mattos = ***Hexachlamys humilis*** O.Berg
Eugenia vanderwateri Ridl. = ***Syzygium vanderwateri*** (Ridl.) Merr. & L.M.Perry
Eugenia varians Miq. = ***Syzygium zeylanicum*** (L.) DC.
Eugenia variolosa King = ***Syzygium variolosum*** (King) Chantaran. & J.Parn.
Eugenia vattimoana Mattos = ***Eugenia neoaustralis*** Sobral
Eugenia vauthiereana O.Berg = ***Eugenia pyriformis*** Cambess.
Eugenia velloziana O.Berg = ***Eugenia crenata*** Vell.
Eugenia vellozii O.Berg = ***Eugenia procera*** (Sw.) Poir.
Eugenia velutiflora Kiaersk. = ***Eugenia bahiensis*** DC.
Eugenia velutina O.Berg = ***Eugenia macrosperma*** DC.
Eugenia venosa Lam. = ***Syzygium dupontii*** (Baker) Govaerts
Eugenia ventenatii Benth. = ***Syzygium floribundum*** F.Muell.
Eugenia venulosa Duthie = ***Syzygium muelleri*** (Miq.) Miq.

Eugenia venusta Roxb. = ***Syzygium venustum*** (Roxb.) N.P.Balakr.
Eugenia verecunda Duthie = ***Syzygium leucoxylon*** Korth.
Eugenia vernonioides Elmer = ***Syzygium vernonioides*** (Elmer) Merr.
Eugenia verrucosa A.Rich. = ***Eugenia axillaris*** (Sw.) Willd. var. ***axillaris***
Eugenia verrucosa D.Legrand = ***Eugenia neoverrucosa*** Sobral
Eugenia verrucosa f. *glabrata* D.Legrand = ***Eugenia neoverrucosa*** Sobral
Eugenia verruculosa DC. = ***Eugenia polystachya*** Rich.
Eugenia verticilligera Ridl. = ***Syzygium caudatilimbum*** (Merr.) Merr. & L.M.Perry
Eugenia viburnifolia Ridl. = ***Syzygium muelleri*** (Miq.) Miq.
Eugenia vicozensis O.Berg = ***Eugenia tinguyensis*** Cambess.
Eugenia vicozensis var. *glabrata* O.Berg = ***Eugenia tinguyensis*** Cambess.
Eugenia vicozensis var. *sericea* O.Berg = ***Eugenia tinguyensis*** Cambess.
Eugenia vidaliana Elmer = ***Syzygium vidalianum*** (Elmer) Merr.
Eugenia vieillardii Brongn. & Gris = ***Austromyrtus vieillardii*** (Brongn. & Gris) Burret
Eugenia villamilii Merr. = ***Syzygium villamilii*** (Merr.) Merr. & L.M.Perry
Eugenia villifera Ridl. = ***Syzygium villiferum*** (Ridl.) Masam.
Eugenia villosa Poir. = (Monimiaceae)
Eugenia viminalis O.Berg = ***Eugenia pyriformis*** Cambess.
Eugenia vincentina Krug & Urb. = ***Eugenia oerstediana*** O.Berg
Eugenia vincifolia O.Berg = ***Eugenia mansoi*** O.Berg
Eugenia vinifera Rojas = [85 AGE]
Eugenia virens (Blume) Koord. & Valeton = ***Syzygium glabratum*** (DC.) Veldkamp
Eugenia virgata Macfad. = [81 JAM]
Eugenia virgata Gardner = ***Myrceugenia alpigena*** var. ***virgata*** (Gardner) Proença
Eugenia virgultosa (Sw.) DC. = ***Eugenia biflora*** (L.) DC.
Eugenia virgultosa var. *jamaicensis* (O.Berg) Proctor = ***Eugenia biflora*** (L.) DC.
Eugenia viridescens Ridl. = ***Syzygium viridescens*** (Ridl.) I.M.Turner
Eugenia viridiflora var. *elliptica* O.Berg = ***Eugenia viridiflora*** Cambess.
Eugenia viridiflora var. *oblongata* O.Berg = ***Eugenia viridiflora*** Cambess.
Eugenia viridifolia Elmer = ***Syzygium claviflorum*** (Roxb.) Wall. ex A.M.Cowan & Cowan
Eugenia vismeifolia Benth. = ***Myrciaria vismeifolia*** (Benth.) O.Berg
Eugenia vismioides DC. = ***Syzygium vismioides*** (DC.) Govaerts
Eugenia vitiensis Turrill = ***Syzygium gracilipes*** (A.Gray) Merr. & L.M.Perry
Eugenia vitis-idaea C.Wright = [81 CUB]
Eugenia vitis-idaea Raoul = ***Neomyrtus pedunculata*** (Hook.f.) Allan
Eugenia vrieseana (Miq.) Koord. & Valeton = ***Syzygium vrieseanum*** (Miq.) Amshoff
Eugenia vulgaris Baill. = ***Syzygium jambos*** (L.) Alston
Eugenia waianensis O.Deg. = ***Eugenia reinwardtiana*** (Blume) A.Cunn. ex DC.
Eugenia wallenii Macfad. = ***Eugenia biflora*** (L.) DC.
Eugenia wallichii Wight = ***Syzygium praecox*** (Roxb.) Rathakr. & N.C.Nair
Eugenia warmingiana var. *pubescens* Kiaersk. = ***Eugenia bimarginata*** DC.
Eugenia warszewiczii (O.Berg) Hemsl. = ***Ugni myricoides*** (Kunth) O.Berg
Eugenia watsoniana M.R.Hend. = ***Syzygium watsonianum*** (M.R.Hend.) I.M.Turner
Eugenia weberbaueri Diels = ***Myrcianthes myrsinoides*** (Kunth) Grifo
Eugenia wenzelii Merr. = ***Syzygium wenzelii*** (Merr.) Merr.
Eugenia whitfordii Merr. = ***Syzygium whitfordii*** (Merr.) Merr.
Eugenia widgreniana (O.Berg) Kiaersk. = ***Siphoneugena widgreniana*** O.Berg
Eugenia wightiana Wight = ***Syzygium lanceolatum*** (Lam.) Wight & Arn.
Eugenia wightii Bedd. = ***Syzygium laetum*** (Buch.-Ham.) Gandhi subsp. ***laetum***
Eugenia willdenowii Nied. = ***Eugenia pseudopsidium*** Jacq.
Eugenia willdenowii Wight = ***Eugenia roxburghii*** DC.
Eugenia willdenowii (Spreng.) DC. = ***Eugenia uniflora*** L.
Eugenia williamsii C.B.Rob. = ***Syzygium williamsii*** (C.B.Rob.) Merr.
Eugenia wilsonii F.Muell. = ***Syzygium wilsonii*** (F.Muell.) B.Hyland
Eugenia winitii Craib = ***Syzygium winitii*** (Craib) Merr. & L.M.Perry
Eugenia winitii var. *terminalis* Craib = ***Syzygium winitii*** (Craib) Merr. & L.M.Perry
Eugenia wolfii Gillespie = ***Syzygium wolfii*** (Gillespie) Merr. & L.M.Perry
Eugenia wollastonii Ridl. = ***Syzygium wollastonii*** (Ridl.) Merr. & L.M.Perry
Eugenia woodii Merr. = ***Syzygium creaghii*** (Ridl.) Merr. & L.M.Perry
Eugenia wrayi King = ***Syzygium wrayi*** (King) I.M.Turner
Eugenia wrightii Baker = ***Syzygium wrightii*** (Baker) A.J.Scott
Eugenia wurdackii (McVaugh) Mattos = ***Calycorectes wurdackii*** McVaugh
Eugenia wynaadensis Bedd. = ***Meteoromyrtus wynaadensis*** (Bedd.) Gamble
Eugenia xanthocarpa Thwaites = ***Eugenia rufofulva*** Thwaites
Eugenia xanthocarpa Mart. = ***Campomanesia xanthocarpa*** (Mart.) O.Berg
Eugenia xanthochlora O.Berg = ***Myrceugenia ovata*** var. ***regnelliana*** (O.Berg) Landrum
Eugenia xanthophylla C.B.Rob. = ***Syzygium xanthophyllum*** (C.B.Rob.) Merr.
Eugenia xiphophylla Merr. = ***Syzygium xiphophyllum*** (Merr.) Merr.
Eugenia xylantha Greves = ***Syzygium pyrocarpum*** (Greves) Merr. & L.M.Perry
Eugenia xylopifolia DC. = ***Eugenia biflora*** (L.) DC.
Eugenia xylopifolia var. *brevipes* O.Berg = ***Eugenia biflora*** (L.) DC.
Eugenia ybaviyu Parodi = ***Myrcianthes pungens*** (O.Berg) D.Legrand
Eugenia ypanemensis O.Berg = ***Eugenia repanda*** O.Berg
Eugenia yrirensis O.Berg = ***Eugenia irirensis*** O.Berg
Eugenia yucatanensis Standl. = ***Eugenia karwinskyana*** O.Berg

Eugenia yumuryensis O.Berg = ***Eugenia axillaris*** (Sw.) Willd. var. ***axillaris***
Eugenia zamboangensis C.B.Rob. = ***Syzygium zamboangense*** (C.B.Rob.) Merr.
Eugenia zenkeri Engl. = ?
Eugenia zetekiana Standl. = ***Myrcia zetekiana*** (Standl.) B.Holst
Eugenia zeyheri (Harv.) Harv. = ***Eugenia capensis*** subsp. ***zeyheri*** (Harv.) F.White
Eugenia zeylanica Roxb. = ***Eugenia roxburghii*** DC.
Eugenia zeylanica Willd. = ***Eugenia uniflora*** L.
Eugenia zeylanica (L.) Wight = ***Syzygium zeylanicum*** (L.) DC.
Eugenia zimmermannii Warb. ex Craib = ***Syzygium zimmermannii*** (Warb. ex Craib) Merr. & L.M.Perry
Eugenia zippeliana (Miq.) Koord. & Valeton = ***Syzygium lineatum*** (DC.) Merr. & L.M.Perry
Eugenia zollingeriana (Miq.) Koord. & Valeton = ***Syzygium zollingerianum*** (Miq.) Amshoff

Unplaced Names:
Eugenia acris Wight & Arn., Prodr. Fl. Ind. Orient.: 331 (1834). = ?
Eugenia aemula Diels, Bot. Jahrb. Syst. 40(91): 47 (1907), nom. illeg. = [83 ECU]
Eugenia altipeta Greves, J. Bot. 61(Suppl.): 14 (1923). = [43 NWG]
Eugenia amboinensis Greves, J. Bot. 62(Suppl.): 36 (1924), nom. illeg. = [42 MOL]
Eugenia andina O.Berg, Linnaea 27: 274 (1856). = ?
Eugenia axillaris G.Don, Gen. Hist. 2: 866 (1832), nom. illeg. = ?
Eugenia axillaris Koord. & Valeton, Bull. Inst. Bot. Buitenzorg 2: 6 (1899), nom. illeg. = [42 JAW]
Eugenia bantamensis Koord. & Valeton, Bull. Inst. Bot. Buitenzorg 2: 6 (1899). = [42 JAW]
Eugenia biniflora Ridl., Bull. Misc. Inform. Kew 1925: 80 (1925). = [42 BOR SUM]
Eugenia calycorectoides Guillaumin, Bull. Mus. Natl. Hist. Nat., II, 20: 364 (1948), nom. illeg. = [60 NWC]
Eugenia cassinoides O.Berg in C.F.P.von Martius & auct. suc. (eds.), Fl. Bras. 14(1): 286 (1857), nom. illeg. = [84 BZE]
Eugenia cassinoides var. *difformis* O.Berg in C.F.P.von Martius & auct. suc. (eds.), Fl. Bras. 14(1): 580 (1859). = [84 BZE]
Eugenia cassinoides var. *gracilis* O.Berg in C.F.P.von Martius & auct. suc. (eds.), Fl. Bras. 14(1): 286 (1857). = [84 BZE]
Eugenia cassinoides var. *robusta* O.Berg in C.F.P.von Martius & auct. suc. (eds.), Fl. Bras. 14(1): 286 (1857). = [84 BZE]
Eugenia cauliflora Ridl., Trans. Linn. Soc. London, Bot. 3: 299 (1893), nom. illeg. = [42 MLY]
Eugenia chinensis Regel, Cat. Pl. Hort. Aksakov.: 57 (1860). = ?
Eugenia costata Bello, Anales Soc. Esp. Hist. Nat. 10: 272 (1881), nom. illeg. = [81 PUE]
Eugenia costata (Miq.) Koord. & Valeton, Meded. Lands Plantentuin 40: 162 (1900), nom. illeg. = ?
Eugenia djouat Perrier, Mém. Soc. Linn. Paris 3: 116 (1825). = [42 PHI]
Eugenia emarginata Vahl in H.West, Bidr. Beskr. Ste Croix: 290 (1793), nom. inval. = [81 LEE]
Eugenia fischeri Mattos, Loefgrenia 119: 1 (2004), nom. illeg. = [84 BZC]
Eugenia garcinioides Engl. & Brehmer, Bot. Jahrb. Syst. 54: 334 (1917), nom. illeg. = [23 CMN]
Eugenia glandulosa Blanco, Fl. Filip.: 417 (1837), nom. illeg. = [42 PHI]
Eugenia gracilis O.Berg in C.F.P.von Martius & auct. suc. (eds.), Fl. Bras. 14(1): 222 (1857), nom. illeg. = [84 BZL]
Eugenia grisea Mattos, Loefgrenia 107: 2 (1995), nom. illeg. = [84 BZC]
Eugenia hexovulata McVaugh, Fieldiana, Bot. 29: 210 (1956). = ?
Eugenia incerta Dummer, Gard. Chron., III, 52: 179 (1912). = [27 NAT]
Eugenia incertissima Sobral, Candollea 42: 109 (1987), nom. inval. = [85 PAR]
Eugenia intermedia Koord. & Valeton, Bull. Inst. Bot. Buitenzorg 2: 7 (1899), nom. illeg. = [42 JAW]
Eugenia koikokoensis Greves, J. Bot. 61(Suppl.): 20 (1923). = ?
Eugenia koordersii Herter, Revista Sudamer. Bot. 7: 219 (1943). = [42 JAW]
Eugenia laevigata Miq., Anal. Bot. Ind. 1: 25 (1850), nom. illeg. = [42 BOR]
Eugenia lepidota var. *pauciflora* O.Berg, Linnaea 27: 227 (1856). = [82 GUY]
Eugenia leptosperma Roxb., Fl. Ind. ed. 1832, 2: 495 (1832). = ?
Eugenia magnifica Brongn. & Gris, Bull. Soc. Bot. France 12: 178 (1865), nom. illeg. = [60 NWC]
Eugenia marginata W.Hill, Cat. Nat. Indust. Prod. Queensland: 23 (1862), nom. illeg. = ?
Eugenia megalophylla Merr., Philipp. J. Sci. 27: 42 (1925). = [42 PHI]
Eugenia metriosa Rojas, Cat. Hist. Nat. Corrientes: 62 (1897). = [85 AGE]
Eugenia michelii Parodi, Contr. Fl. Paraguay 4: 122 (1879), nom. illeg. = [85 PAR]
Eugenia miqueliana Greves, J. Bot. 61(Suppl.): 16 (1923). = ?
Eugenia mirtiflora Rojas, Cat. Hist. Nat. Corrientes: 61 (1897). = [85 AGE]
Eugenia molliana O.Berg, Linnaea 27: 295 (1856). = [8+]
Eugenia multiflora Rich., Actes Soc. Hist. Nat. Paris 1: 110 (1792). = [82 FRG]
Eugenia myrcinifolia Hance, J. Bot. 23: 8 (1885). = ?
Eugenia nettiana Kiaersk., Enum. Myrt. Bras.: 179 (1893). = [84 BZL] *Myrcia* sp.
Eugenia nigra Pennant, Outlin. Globe 4: 277 (1800). = [40 IND]
Eugenia nigra DC., Prodr. 3: 268 (1828), nom. illeg. = (Melastomataceae)
Eugenia oleoides Planch. & Linden, Bot. Zeitung (Berlin) 12: 364 (1854). = ?
Eugenia paniculata Bello, Anales Soc. Esp. Hist. Nat. 10: 271 (1881), nom. illeg. = [81 PUE]
Eugenia pomatensis Miq., Anal. Bot. Ind. 1: 22 (1850). = [42 BOR]
Eugenia pseudodichasiantha Kiaersk., Enum. Myrt. Bras.: 177 (1893). = *Plinia* sp. ?
Eugenia reticulata Trimen, Handb. Fl. Ceylon 2: 185 (1894), nom. illeg. = [40 SRL]
Eugenia robertii Merr., Philipp. J. Sci. 1(Suppl.): 106 (1906). = [42 PHI]
Eugenia rosea Barb.Rodr. ex Chodat & Hassl., Bull. Herb. Boissier, II, 7: 805 (1907), nom. subnud. = [85 PAR]
Eugenia rotundifolia Casar., Nov. Stirp. Bras.: 40 (1842), nom. illeg. = [84 BZL]
Eugenia rotundifolia var. *oblongata* O.Berg in C.F.P.von Martius & auct. suc. (eds.), Fl. Bras. 14(1): 287 (1857). = [84 BZL]

Eugenia ruiziana O.Berg, Linnaea 27: 167 (1856). = [82 FRG]
Eugenia simulans King, J. Asiat. Soc. Bengal, Pt. 2, Nat. Hist. 70: 128 (1901). = [42 MLY]
Eugenia suavis Ridl., J. Fed. Malay States Mus. 5: 160 (1915). = [42 MLY]
Eugenia subcordata O.Berg in C.F.P.von Martius & auct. suc. (eds.), Fl. Bras. 14(1): 300 (1857), nom. illeg. = [84 BZS]
Eugenia subcordata var. *australis* Mattos, Loefgrenia 70: 4 (1976). = [84 BZS]
Eugenia subobliqua Benth., J. Bot. (Hooker) 2: 322 (1840). = [82 GUY]
Eugenia suzukii Kaneh., Bot. Mag. (Tokyo) 45: 335 (1931). = [62 CRL]
Eugenia tanalensis Baker, J. Bot. 20: 111 (1882). = [29 MDG] *Syzygium* sp. ?
Eugenia tapiraguayensis Barb.Rodr. ex Chodat & Hassl., Bull. Herb. Boissier, II, 7: 804 (1907), nom. inval. = [85 PAR]
Eugenia tetraedra Duthie in J.D.Hooker, Fl. Brit. India 2: 476 (1878). = ?
Eugenia tetrasperma Bello, Anales Soc. Esp. Hist. Nat. 10: 271 (1881). = [81 PUE]
Eugenia teyu-iba Parodi, Anales Soc. Ci. Argent. 7: 116 (1879). = [85 AGE]
Eugenia tomentosa Cambess. in A.F.C.de Saint-Hilaire, Fl. Bras. Merid. 2: 353 (1832), nom. illeg. = [84 BZL]
Eugenia trukensis Hosok., J. Jap. Bot. 13: 281 (1937). = [62 CRL]
Eugenia valetoniana King, J. Asiat. Soc. Bengal, Pt. 2, Nat. Hist. 70: 112 (1901). = [42 MLY]
Eugenia vinifera Rojas, Cat. Hist. Nat. Corrientes: 62 (1897). = [85 AGE]
Eugenia virgata Macfad., Fl. Jamaica 2: 121 (1850), nom. illeg. = [81 JAM]
Eugenia vitis-idaea C.Wright, Anales Acad. Ci. Méd. Habana 5: 431 (1868), nom. illeg. = [81 CUB]
Eugenia zenkeri Engl., Notizbl. Königl. Bot. Gart. Berlin 2: 291 (1899). = ?

Eugeniopsis

Eugeniopsis O.Berg = ***Marlierea*** Cambess.
Eugeniopsis acuminatissima O.Berg = ***Marlierea acuminatissima*** (O.Berg) D.Legrand
Eugeniopsis affinis O.Berg = ***Marlierea racemosa*** (Vell.) Kiaersk.
Eugeniopsis angustifolia O.Berg = ***Marlierea angustifolia*** (O.Berg) Mattos
Eugeniopsis cannifolia O.Berg = ***Marlierea racemosa*** (Vell.) Kiaersk.
Eugeniopsis clausseniana O.Berg = ***Marlierea clausseniana*** (O.Berg) Kiaersk.
Eugeniopsis clausseniana var. *glabrata* O.Berg = ***Marlierea clausseniana*** (O.Berg) Kiaersk.
Eugeniopsis clausseniana var. *rufa* O.Berg = ***Marlierea clausseniana*** (O.Berg) Kiaersk.
Eugeniopsis gardneriana O.Berg = ***Marlierea gardneriana*** (O.Berg) Nied.
Eugeniopsis gaudichaudiana O.Berg = ***Marlierea gaudichaudiana*** (O.Berg) Loefgr. & Everett
Eugeniopsis grandifolia O.Berg = ***Marlierea racemosa*** (Vell.) Kiaersk.
Eugeniopsis laevigata (DC.) O.Berg = ***Marlierea laevigata*** (DC.) Kiaersk.
Eugeniopsis luschnathiana O.Berg = ***Marlierea luschnathiana*** (O.Berg) D.Legrand
Eugeniopsis ovata O.Berg = ***Marlierea eugenioides*** (Cambess.) D.Legrand
Eugeniopsis paniculata O.Berg = ***Marlierea eugenioides*** (Cambess.) D.Legrand
Eugeniopsis polygama O.Berg = ***Marlierea polygama*** (O.Berg) D.Legrand
Eugeniopsis richardiana O.Berg = ***Marlierea ferruginea*** (Poir.) McVaugh
Eugeniopsis riedeliana O.Berg = ***Marlierea riedeliana*** (O.Berg) D.Legrand
Eugeniopsis rubiginosa (Cambess.) O.Berg = ***Marlierea rubiginosa*** (Cambess.) D.Legrand
Eugeniopsis schottiana O.Berg = ***Marlierea schottii*** D.Legrand
Eugeniopsis sylvatica O.Berg = ***Marlierea sylvatica*** (O.Berg) Kiaersk.
Eugeniopsis teuscheriana O.Berg = ***Marlierea teuscheriana*** (O.Berg) D.Legrand

Euryomyrtus

Euryomyrtus Schauer, Linnaea 17: 239 (1843).
Australia. 50 NSW SOA TAS VIC WAU.
7 Species

Euryomyrtus denticulata (Maiden & Betche) Trudgen, Nuytsia 13: 563 (2001).
SE. New South Wales. 50 NSW. Nanophan.
**Baeckea denticulata* Maiden & Betche, Proc. Linn. Soc. New South Wales 34: 361 (1909).

Euryomyrtus inflata Trudgen, Nuytsia 13: 553 (2001).
SW. Australia. 50 WAU. Nanophan.

Euryomyrtus leptospermoides (C.A.Gardner) Trudgen, Nuytsia 13: 550 (2001).
SW. Australia. 50 WAU. Nanophan.
**Baeckea leptospermoides* C.A.Gardner, J. Proc. Roy. Soc. W. Australia 27: 188 (1942).

Euryomyrtus maidenii (Ewart & Jean White) Trudgen, Nuytsia 13: 557 (2001).
SW. Australia. 50 WAU. Nanophan.
**Baeckea maidenii* Ewart & Jean White, J. Proc. Roy. Soc. New S. Wales 42: 184 (1908 publ. 1909).

Euryomyrtus patrickiae Trudgen, Nuytsia 13: 559 (2001).
SW. Australia. 50 WAU. Nanophan.

Euryomyrtus ramosissima (A.Cunn.) M.E.Trudgen, Nuytsia 13: 560 (2001).
SE. Australia. 50 NSW SOA TAS VIC. Cham. or nanophan.
**Baeckea ramosissima* A.Cunn., Field New South Wales: 249 (1825).

subsp. ***prostrata*** (Hook.f.) M.E.Trudgen, Nuytsia 13: 563 (2001).
SE. Australia. 50 NSW TAS VIC. Cham. or nanophan.
**Baeckea prostrata* Hook.f., Hooker's Icon. Pl. 3: 284b (1840). *Baeckea ramosissima* subsp. *prostrata* (Hook.f.) G.W.Carr, Telopea 1: 416 (1980).
Euryomyrtus parviflora F.Muell. ex Miq., Ned. Kruidk. Arch. 4: 149 (1856).

subsp. ***ramosissima***
SE. Australia. 50 NSW SOA TAS VIC. Cham. or nanophan.
Baeckea diffusa Sieber ex DC., Prodr. 3: 230 (1828). *Euryomyrtus diffusa* (Sieber ex DC.) Schauer, Linnaea 17: 239 (1843).
Baeckea diffusa var. *striata* DC., Prodr. 3: 230 (1828).
Baeckea alpina Lindl. in T.L.Mitchell, Three Exped. Australia 2: 177 (1838). *Euryomyrtus*

alpina (Lindl.) Schauer, Linnaea 17: 239 (1843).
Baeckea affinis Hook., Hooker's Icon. Pl. 3: 284c (1840), nom. illeg.
Baeckea thymifolia Hook.f., Hooker's Icon. Pl. 3: 284a (1840). *Euryomyrtus thymifolia* (Hook.f.) Schauer, Linnaea 17: 239 (1843).
Baeckea stuartiana F.Muell. ex Miq., Ned. Kruidk. Arch. 4: 149 (1856).
Euryomyrtus stuartiana F.Muell. ex Miq., Ned. Kruidk. Arch. 4: 149 (1856).

Euryomyrtus recurva Trudgen, Nuytsia 13: 555 (2001). SW. Australia. 50 WAU. Nanophan.

Synonyms:
Euryomyrtus alpina (Lindl.) Schauer = ***Euryomyrtus ramosissima*** (A.Cunn.) M.E.Trudgen subsp. ***ramosissima***
Euryomyrtus diffusa (Sieber ex DC.) Schauer = ***Euryomyrtus ramosissima*** (A.Cunn.) M.E.Trudgen subsp. ***ramosissima***
Euryomyrtus parviflora F.Muell. ex Miq. = ***Euryomyrtus ramosissima*** subsp. ***prostrata*** (Hook.f.) M.E.Trudgen
Euryomyrtus stuartiana F.Muell. ex Miq. = ***Euryomyrtus ramosissima*** (A.Cunn.) M.E.Trudgen subsp. ***ramosissima***
Euryomyrtus thymifolia (Hook.f.) Schauer = ***Euryomyrtus ramosissima*** (A.Cunn.) M.E.Trudgen subsp. ***ramosissima***

Evanesca

Evanesca Raf. = ***Pimenta*** Lindl.
Evanesca crassifolia Raf. = ***Pimenta dioica*** (L.) Merr.

Fabricia

Fabricia Gaertn. = ***Neofabricia*** Joy Thomps.
Fabricia angustifolia Otto ex DC. = ***Agonis marginata*** (Labill.) Sweet
Fabricia bracteata Roxb. = ***Decaspermum bracteatum*** (Roxb.) A.J.Scott
Fabricia coriacea F.Muell. ex Miq. = ***Leptospermum coriaceum*** (F.Muell. ex Miq.) Cheel
Fabricia incana Dum.Cours. = ***Leptospermum lanigerum*** (Aiton) Sm.
Fabricia laevigata Gaertn. = ***Leptospermum laevigatum*** (Gaertn.) F.Muell.
Fabricia minor Dum.Cours. = ***Leptospermum lanigerum*** (Aiton) Sm.
Fabricia myrtifolia Heynh. = ***Agonis marginata*** (Labill.) Sweet
Fabricia myrtifolia Sieber ex Benth. = ***Leptospermum laevigatum*** (Gaertn.) F.Muell.
Fabricia myrtifolia Gaertn. = ***Neofabricia myrtifolia*** (Gaertn.) Joy Thomps.
Fabricia pubescens (Lam.) Schauer = ***Leptospermum lanigerum*** (Aiton) Sm.
Fabricia sericea Sweet = ***Leptospermum lanigerum*** (Aiton) Sm.
Fabricia stricta Lodd. = ***Agonis marginata*** (Labill.) Sweet

Feijoa

Feijoa O.Berg = ***Acca*** O.Berg
Feijoa obovata (O.Berg) O.Berg = ***Acca sellowiana*** (O. Berg) Burret
Feijoa schenckiana Kiaersk. = ***Acca sellowiana*** (O.Berg) Burret
Feijoa sellowiana (O.Berg) O.Berg = ***Acca sellowiana*** (O.Berg) Burret
Feijoa sellowiana f. *elongata* Voronova = ***Acca sellowiana*** (O.Berg) Burret
Feijoa sellowiana var. *rugosa* Mattos = ***Acca sellowiana*** (O.Berg) Burret

Felicianea

Felicianea Cambess. = ***Myrrhinium*** Schott
Felicianea rubriflora Cambess. = ***Myrrhinium atropurpureum*** Schott var. ***atropurpureum***

Fenzlia

Fenzlia Endl. = ***Lithomyrtus*** F.Muell.
Fenzlia beccarii (F.Muell.) Burret = ***Myrtella beccarii*** F.Muell.
Fenzlia bennigseniana (Volkens) Burret = ***Myrtella bennigseniana*** (Volkens) Diels
Fenzlia hirsutula (F.Muell.) Burret = ***Kania hirsutula*** (F.Muell.) A.J.Scott
Fenzlia microphylla (Benth.) Domin = ***Lithomyrtus microphylla*** (Benth.) N.Snow & Guymer
Fenzlia obtusa Endl. = ***Lithomyrtus obtusa*** (Endl.) N.Snow & Guymer
Fenzlia obtusa var. *microphylla* Benth. = ***Lithomyrtus microphylla*** (Benth.) N.Snow & Guymer
Fenzlia phebalioides W.Fitzg. = ***Lithomyrtus retusa*** (Endl.) N.Snow & Guymer
Fenzlia retusa Endl. = ***Lithomyrtus retusa*** (Endl.) N.Snow & Guymer

Francisia

Francisia Endl. = ***Darwinia*** Rudge

Fremya

Fremya Brongn. & Gris = ***Xanthostemon*** F.Muell.
Fremya aurantiaca Brongn. & Gris = ***Xanthostemon aurantiacus*** (Brongn. & Gris) Schltr.
Fremya austrocaledonica Vieill. ex Guillaumin = ***Pleurocalyptus austrocaledonicus*** (Guillaumin) J.W.Dawson
Fremya baudouinii Bureau ex Guillaumin = ***Xanthostemon multiflorus*** (Montrouz.) Beauvis.
Fremya ciliata (J.R.Forst. & G.Forst.) Brongn. & Gris = ***Purpureostemon ciliatus*** (J.R.Forst. & G.Forst.) Gugerli
Fremya deplanchei Brongn. & Gris = ***Xanthostemon vieillardii*** (Brongn. & Gris) Nied.
Fremya elegans Brongn. & Gris = ***Xanthostemon multiflorus*** (Montrouz.) Beauvis.
Fremya flava Brongn. & Gris = ***Xanthostemon multiflorus*** (Montrouz.) Beauvis.
Fremya glauca Vieill. ex Pamp. & Pampal. = ***Xanthostemon glaucus*** Pamp.
Fremya integrifolia Brongn. & Gris ex Baker f. = ***Xanthostemon myrtifolius*** (Brongn. & Gris) Pamp. ex Pampal.
Fremya lateriflora Bureau ex Guillaumin = ***Xanthostemon lateriflorus*** Guillaumin
Fremya laurina Vieill. ex Pamp. & Pampal. = ***Xanthostemon laurinus*** (Pamp.) Guillaumin
Fremya lucens Bureau ex Guillaumin = ***Stereocaryum rubiginosum*** (Brongn. & Gris) Burret
Fremya myrtifolia Brongn. & Gris = ***Xanthostemon myrtifolius*** (Brongn. & Gris) Pamp. ex Pampal.
Fremya pancheri Brongn. & Gris = ***Pleurocalyptus pancheri*** (Brongn. & Gris) J.W.Dawson
Fremya pubescens Brongn. & Gris = ***Xanthostemon pubescens*** (Brongn. & Gris) Sebert & Pancher

Fremya rubra Brongn. & Gris = ***Xanthostemon ruber*** (Brongn. & Gris) Sebert & Pancher
Fremya speciosa Brongn. & Gris = ***Xanthostemon gugerlii*** Merr.
Fremya vieillardii Brongn. & Gris = ***Xanthostemon vieillardii*** (Brongn. & Gris) Nied.

Fropiera

Fropiera Bouton ex Hook.f. = ***Psiloxylon*** Thouars ex Tul.
Fropiera mauritiana Bouton ex Hook.f. = ***Psiloxylon mauritianum*** (Bouton ex Hook.f.) Baill.

Gaslondia

Gaslondia Vieill. = ***Syzygium*** Gaertn.
Gaslondia amphoricarpa Vieill. = ***Syzygium neocaledonicum*** (Seem.) J.W.Dawson

Gelpkea

Gelpkea Blume = ***Syzygium*** Gaertn.
Gelpkea pendula (Blume) Blume = (Rubiaceae) ?
Gelpkea stipularis Blume = ***Syzygium aqueum*** (Burm.f.) Alston

Genetyllis

Genetyllis DC. = ***Darwinia*** Rudge
Genetyllis affinis Turcz. = ***Darwinia diosmoides*** (DC.) Benth.
Genetyllis agathosmoides Turcz. = ***Darwinia vestita*** (Endl.) Benth.
Genetyllis alpestris Lindl. = ***Calytrix alpestris*** (Lindl.) Court
Genetyllis citriodora Endl. = ***Darwinia citriodora*** (Endl.) Benth.
Genetyllis diosmoides DC. = ***Darwinia diosmoides*** (DC.) Benth.
Genetyllis drummondii Turcz. = ***Chamelaucium drummondii*** (Turcz.) Meisn.
Genetyllis fimbriata Kippist = ***Darwinia squarrosa*** (Turcz.) Domin
Genetyllis helichrysoides Meisn. = ***Darwinia helichrysoides*** (Meisn.) Benth.
Genetyllis hookeriana Meisn. = ***Darwinia hypericifolia*** (Turcz.) Domin
Genetyllis hypericifolia Turcz. = ***Darwinia hypericifolia*** (Turcz.) Domin
Genetyllis leiostyla Turcz. = ***Darwinia leiostyla*** (Turcz.) Domin
Genetyllis macrostegia Hook. = ***Darwinia hypericifolia*** (Turcz.) Domin
Genetyllis macrostegia Turcz. = ***Darwinia macrostegia*** (Turcz.) Benth.
Genetyllis meisneri Kippist = ***Darwinia oxylepis*** (Turcz.) N.G.Marchant & Keighery
Genetyllis micropetala F.Muell. = ***Darwinia micropetala*** (F.Muell.) Benth.
Genetyllis oederoides Turcz. = ***Darwinia oederoides*** (Turcz.) Benth.
Genetyllis oxylepis Turcz. = ***Darwinia oxylepis*** (Turcz.) N.G.Marchant & Keighery
Genetyllis pauciflora Turcz. = ***Chamelaucium ciliatum*** Desf.
Genetyllis pimeloides F.Muell. = ***Darwinia citriodora*** (Endl.) Benth.
Genetyllis pinifolia (Lindl.) Schauer = ***Darwinia pinifolia*** (Lindl.) Benth.
Genetyllis purpurea (Endl.) Schauer = ***Darwinia purpurea*** (Endl.) Benth.
Genetyllis sanguinea Meisn. = ***Darwinia sanguinea*** (Meisn.) Benth.
Genetyllis schuermannii F.Muell. = ***Homoranthus homoranthoides*** (F.Muell.) Craven & S.R.Jones
Genetyllis speciosa Meisn. = ***Darwinia speciosa*** (Meisn.) Benth.
Genetyllis squarrosa Turcz. = ***Darwinia squarrosa*** (Turcz.) Domin
Genetyllis thymoides (Lindl.) Schauer = ***Darwinia thymoides*** (Lindl.) Benth.
Genetyllis tulipifera (Lindl.) Planch. = ***Darwinia macrostegia*** (Turcz.) Benth.
Genetyllis vestita Endl. = ***Darwinia vestita*** (Endl.) Benth.
Genetyllis virescens Meisn. = ***Darwinia virescens*** (Meisn.) Benth.

Glaphyria

Glaphyria Jack = ***Leptospermum*** J.R.Forst. & G.Forst.
Glaphyria annae Stein = ***Leptospermum javanicum*** Blume
Glaphyria nitida Jack = ***Leptospermum javanicum*** Blume
Glaphyria sericea Jack = ***Leptospermum polygalifolium*** Salisb. subsp. ***polygalifolium***

Gomidesia

Gomidesia O.Berg = ***Myrcia*** DC. ex Guill.
Gomidesia affinis (Cambess.) D.Legrand = ***Myrcia hebepetala*** DC.
Gomidesia affinis var. *catharinensis* D.Legrand = ***Myrcia cordiifolia*** DC.
Gomidesia affinis var. *pohliana* (O.Berg) D.Legrand = ***Myrcia cordiifolia*** DC.
Gomidesia amplexicaulis (Vell.) O.Berg = ***Myrcia amplexicaulis*** (Vell.) Hook.f.
Gomidesia anacardiifolia (Gardner) O.Berg = ***Myrcia anacardiifolia*** Gardner
Gomidesia anacardiifolia var. *oblongata* O.Berg = ***Myrcia anacardiifolia*** Gardner
Gomidesia anacardiifolia var. *opaca* O.Berg = ***Myrcia anacardiifolia*** Gardner
Gomidesia anacardiifolia var. *ovalis* O.Berg = ***Myrcia anacardiifolia*** Gardner
Gomidesia aubletiana O.Berg = ***Myrcia pubescens*** DC.
Gomidesia banisteriifolia (DC.) O.Berg = ***Myrcia palustris*** DC.
Gomidesia barituensis Legname = [83 BOL 85 AGE]
Gomidesia beyrichiana O.Berg = ***Myrcia eriocalyx*** DC.
Gomidesia beyrichiana var. *acutata* O.Berg = ***Myrcia eriocalyx*** DC.
Gomidesia blanchetiana O.Berg = [84 BZE]
Gomidesia bonnetiasylvestris Steyerm. = ***Myrcia bonnetiasylvestris*** (Steyerm.) Steyerm.
Gomidesia browniana (Gardner) O.Berg = ***Myrcia spectabilis*** DC.
Gomidesia brunnea (Cambess.) D.Legrand = ***Myrcia brunnea*** Cambess.
Gomidesia cambessedeana O.Berg = [84 BZL]
Gomidesia candolleana O.Berg = ***Myrcia hebepetala*** DC.
Gomidesia casaretteana O.Berg = ***Myrcia pubescens*** DC.
Gomidesia cerqueiria Nied. = ?
Gomidesia chamissoeana O.Berg = ***Myrcia subsericea*** A.Gray
Gomidesia clausseniana O.Berg = ***Myrcia ouropretoensis*** Kiaersk.
Gomidesia cordiifolia (DC.) NicLugh. = ***Myrcia cordiifolia*** DC.

Gomidesia crocea O.Berg = ***Myrcia amplexicaulis*** (Vell.) Hook.f.
Gomidesia crocea Nied. = ***Myrcia grandifolia*** Cambess.
Gomidesia eriocalyx (DC.) O.Berg = ***Myrcia eriocalyx*** DC.
Gomidesia eriocalyx var. *acuminata* D.Legrand = ***Myrcia eriocalyx*** DC.
Gomidesia fenzliana O.Berg = ***Myrcia ilheosensis*** Kiaersk.
Gomidesia fenzliana var. *obovata* O.Berg = ***Myrcia ilheosensis*** Kiaersk.
Gomidesia fenzliana var. *spathulata* O.Berg = ***Myrcia ilheosensis*** Kiaersk.
Gomidesia flagellaris D.Legrand = [84 BZS]
Gomidesia freyreissiana O.Berg = ***Myrcia freyreissiana*** (O.Berg) Kiaersk.
Gomidesia gardneriana O.Berg = ***Myrcia palustris*** DC.
Gomidesia garopabensis (Cambess.) D.Legrand = ***Myrcia palustris*** DC.
Gomidesia gaudichaudiana O.Berg = ***Myrcia eriocalyx*** DC.
Gomidesia gaudichaudiana var. *acutata* O.Berg = ***Myrcia eriocalyx*** DC.
Gomidesia gestasiana (Cambess.) D.Legrand = ***Myrcia gestasiana*** Cambess.
Gomidesia glazioviana (Kiaersk.) D.Legrand = ***Myrciaria glazioviana*** (Kiaersk.) G.M.Barroso ex Sobral
Gomidesia grandifolia (Cambess.) Mattos & D.Legrand = ***Myrcia grandifolia*** Cambess.
Gomidesia haenkeana O.Berg = [83 BOL]
Gomidesia hartwegiana O.Berg = ***Myrcia hartwegiana*** (O.Berg) Kiaersk.
Gomidesia hebepetala (DC.) O.Berg = ***Myrcia hebepetala*** DC.
Gomidesia hilariana O.Berg = ***Myrcia eriocalyx*** DC.
Gomidesia hookeriana O.Berg = ***Myrcia hebepetala*** DC.
Gomidesia innovans (Kiaersk.) D.Legrand = ***Myrcia innovans*** Kiaersk.
Gomidesia jacquiniana O.Berg = ***Myrcia grandifolia*** Cambess.
Gomidesia klotzschiana O.Berg = [84 BZL]
Gomidesia kunthiana O.Berg = [84 BZL]
Gomidesia kunthiana var. *angustifolia* O.Berg = ?
Gomidesia kunthiana var. *latifolia* O.Berg = ?
Gomidesia langsdorffii O.Berg = ***Myrcia amplexicaulis*** (Vell.) Hook.f.
Gomidesia lindeniana O.Berg = ***Myrcia fenzliana*** O.Berg
Gomidesia linkiana O.Berg = ***Myrcia cordiifolia*** DC.
Gomidesia lutescens (Cambess.) D.Legrand = ***Myrcia lutescens*** Cambess.
Gomidesia magnifolia O.Berg = ***Myrcia magnifolia*** (O.Berg) Kiaersk.
Gomidesia martiana O.Berg = ***Myrcia vittoriana*** Kiaersk.
Gomidesia mikaniana O.Berg = ***Myrcia pubescens*** DC.
Gomidesia minutiflora Mattos & D.Legrand = ***Myrcia amazonica*** DC.
Gomidesia miqueliana O.Berg = ***Myrcia eriocalyx*** DC.
Gomidesia miqueliana var. *angustifolia* O.Berg = ***Myrcia eriocalyx*** DC.
Gomidesia miqueliana var. *brunnea* O.Berg = ***Myrcia eriocalyx*** DC.
Gomidesia miqueliana var. *ferruginea* O.Berg = ***Myrcia eriocalyx*** DC.
Gomidesia miqueliana var. *obtusata* O.Berg = ***Myrcia eriocalyx*** DC.
Gomidesia myrcioides Mattos & D.Legrand = [83 BOL]
Gomidesia nitida (Vell.) Nied. = ***Myrcia subsericea*** A.Gray
Gomidesia palustris (DC.) Kausel = ***Myrcia palustris*** DC.
Gomidesia palustris var. *angustifolia* (O.Berg) Mattos = ***Myrcia palustris*** DC.
Gomidesia poeppigiana O.Berg = [84 BZL]
Gomidesia pohliana O.Berg = ***Myrcia hebepetala*** DC.
Gomidesia pubescens (DC.) D.Legrand = ***Myrcia pubescens*** DC.
Gomidesia pubescens var. *casaretteana* (O.Berg) D.Legrand = ***Myrcia pubescens*** DC.
Gomidesia pubescens var. *widgreniana* (O.Berg) D.Legrand = ***Myrcia pubescens*** DC.
Gomidesia raddiana O.Berg = ***Myrcia pubescens*** DC.
Gomidesia regeliana O.Berg = ***Myrcia eriocalyx*** DC.
Gomidesia reticulata (Cambess.) O.Berg = ***Myrcia reticulata*** Cambess.
Gomidesia reticulata var. *angustifolia* O.Berg = ***Myrcia reticulata*** Cambess.
Gomidesia reticulata var. *latifolia* O.Berg = ***Myrcia reticulata*** Cambess.
Gomidesia riedeliana O.Berg = ***Myrcia anacardiifolia*** Gardner
Gomidesia schaueriana O.Berg = ***Myrcia brasiliensis*** Kiaersk.
Gomidesia schaueriana var. *spathulata* Mattos = ***Myrcia brasiliensis*** Kiaersk.
Gomidesia sellowiana O.Berg = ***Myrcia hartwegiana*** (O.Berg) Kiaersk.
Gomidesia sonderiana O.Berg = ***Myrcia dolicopetala*** Kiaersk.
Gomidesia spectabilis (DC.) O.Berg = ***Myrcia spectabilis*** DC.
Gomidesia spectabilis var. *farinosa* O.Berg = ***Myrcia spectabilis*** DC.
Gomidesia spectabilis var. *ovata* O.Berg = ***Myrcia spectabilis*** DC.
Gomidesia sprengeliana O.Berg = ***Myrcia hartwegiana*** (O.Berg) Kiaersk.
Gomidesia springiana O.Berg = ***Myrcia springiana*** (O.Berg) Kiaersk.
Gomidesia springiana var. *coriacea* O.Berg = ***Myrcia springiana*** (O.Berg) Kiaersk.
Gomidesia springiana var. *membranacea* O.Berg = ***Myrcia springiana*** (O.Berg) Kiaersk.
Gomidesia spruceana O.Berg = [84 BZL]
Gomidesia spruceana var. *acutifolia* O.Berg = ?
Gomidesia spruceana var. *obtusifolia* O.Berg = ?
Gomidesia squamata Mattos & D.Legrand = [84 BZS]
Gomidesia ticuensis (Kiaersk.) D.Legrand = ***Myrcia grandifolia*** Cambess.
Gomidesia tijucensis (Kiaersk.) D.Legrand = ***Myrcia tijucensis*** Kiaersk.
Gomidesia tijucensis var. *flexuosa* Mattos & D.Legrand = ***Myrcia tijucensis*** Kiaersk.
Gomidesia velutiflora Mattos & D.Legrand = [84 BZC]
Gomidesia warmingiana (Kiaersk.) D.Legrand = ***Myrcia warmingiana*** Kiaersk.
Gomidesia widgreniana O.Berg = ***Myrcia pubescens*** DC.
Gomidesia willdenowiana O.Berg = [84 BZL]

Unplaced Names:
Gomidesia barituensis Legname, Lilloa 35: 79 (1978). = [83 BOL 85 AGE]
Gomidesia blanchetiana O.Berg in C.F.P.von Martius & auct. suc. (eds.), Fl. Bras. 14(1): 14 (1857). = [84 BZE]
Gomidesia cambessedeana O.Berg in C.F.P.von Martius & auct. suc. (eds.), Fl. Bras. 14(1): 24 (1857). = [84 BZL]
Gomidesia flagellaris D.Legrand, Sellowia 13: 279 (1961). = [84 BZS]

Gomidesia haenkeana O.Berg in C.F.P.von Martius & auct. suc. (eds.), Fl. Bras. 14(1): 515 (1858). = [83 BOL]
Gomidesia klotzschiana O.Berg in C.F.P.von Martius & auct. suc. (eds.), Fl. Bras. 14(1): 534 (1859). = [84 BZL]
Gomidesia kunthiana O.Berg in C.F.P.von Martius & auct. suc. (eds.), Fl. Bras. 14(1): 23 (1857). = [84 BZL]
Gomidesia myrcioides Mattos & D.Legrand, Loefgrenia 67: 14 (1975). = [83 BOL]
Gomidesia poeppigiana O.Berg in C.F.P.von Martius & auct. suc. (eds.), Fl. Bras. 14(1): 14 (1857). = [84 BZL]
Gomidesia spruceana O.Berg in C.F.P.von Martius & auct. suc. (eds.), Fl. Bras. 14(1): 534 (1859). = [84 BZL]
Gomidesia squamata Mattos & D.Legrand, Loefgrenia 67: 14 (1975). = [84 BZS]
Gomidesia velutiflora Mattos & D.Legrand, Loefgrenia 67: 15 (1975). = [84 BZC]
Gomidesia willdenowiana O.Berg in C.F.P.von Martius & auct. suc. (eds.), Fl. Bras. 14(1): 19 (1857). = [84 BZL]

Gomphotis

Gomphotis Raf. = ***Thryptomene*** Endl.
Gomphotis saxicola (A.Cunn. ex Hook.) Raf. = ***Thryptomene saxicola*** (A.Cunn. ex Hook.) Schauer

Gossia

Gossia N.Snow & Guymer, Syst. Bot. Monogr. 65: 31 (2003).
New Guinea to SW. Pacific and E. Australia. 43 NWG SOL 50 NSW QLD 60 VAN. [Myrtaceae]
27 Species

Gossia acmenoides (F.Muell.) N.Snow & Guymer, Syst. Bot. Monogr. 65: 45 (2003).
CE. Queensland to CE. New South Wales. 50 NSW QLD. Phan.
**Myrtus acmenoides* F.Muell., Fragm. 1: 77 (1858). *Austromyrtus acmenoides* (F.Muell.) Burret, Notizbl. Bot. Gart. Berlin-Dahlem 15: 501 (1941).
Austromyrtus acutiuscula Burret, Notizbl. Bot. Gart. Berlin-Dahlem 15: 502 (1941).

Gossia aneityensis (Guillaumin) N.Snow, Novon 15: 478 (2005).
Vanuatu. 60 VAN. Nanophan. or phan.
**Myrtus aneityensis* Guillaumin, J. Arnold Arbor. 12: 254 (1931). *Austromyrtus aneityensis* (Guillaumin) Burret, Notizbl. Bot. Gart. Berlin-Dahlem 15: 506 (1941).

Gossia bamagensis N.Snow & Guymer, Syst. Bot. Monogr. 65: 65 (2003).
N. & NE. Queensland. 50 QLD. Phan.

Gossia bidwillii (Benth.) N.Snow & Guymer, Syst. Bot. Monogr. 65: 39 (2003).
E. Queensland to NE. New South Wales. 50 NSW QLD. Phan.
**Myrtus bidwillii* Benth., Fl. Austral. 3: 275 (1867). *Austromyrtus bidwillii* (Benth.) Burret, Notizbl. Bot. Gart. Berlin-Dahlem 15: 501 (1941).
Myrtus racemulosa Benth., Fl. Austral. 3: 276 (1867). *Austromyrtus racemulosa* Burret, Notizbl. Bot. Gart. Berlin-Dahlem 15: 501 (1941).
Myrtus racemulosa var. *conferta* Benth., Fl. Austral. 3: 276 (1867).

Gossia byrnesii N.Snow & Guymer, Syst. Bot. Monogr. 65: 81 (2003).
N. Queensland. 50 QLD. Phan.

Gossia dallachyana (F.Muell. ex Benth.) N.Snow & Guymer, Syst. Bot. Monogr. 65: 58 (2003).
N. & NE. Queensland. 50 QLD. Phan.
**Eugenia dallachyana* F.Muell. ex Benth., Fl. Austral. 3: 287 (1867). *Austromyrtus dallachyana* (F.Muell. ex Benth.) L.S.Sm., Proc. Roy. Soc. Queensland 67: 35 (1956).

Gossia eugenioides (A.J.Scott) N.Snow, Novon 15: 477 (2005).
W. New Guinea. 43 NWG. Nanophan.
**Myrtus eugenioides* A.J.Scott, Kew Bull. 33: 514 (1979).

Gossia floribunda (A.J.Scott) N.Snow & Guymer, Syst. Bot. Monogr. 65: 72 (2003).
S. New Guinea to N. Queensland. 43 NWG 50 QLD. Nanophan. or phan.
**Backhousia floribunda* A.J.Scott, Kew Bull. 39: 659 (1984). *Austromyrtus floribunda* (A.J.Scott) Guymer, Austrobaileya 2: 569 (1988).

Gossia fragrantissima (F.Muell. ex Benth.) N.Snow & Guymer, Syst. Bot. Monogr. 65: 75 (2003).
SE. Queensland to NE. New South Wales. 50 NSW QLD. Phan.
**Myrtus fragrantissima* F.Muell. ex Benth., Fl. Austral. 3: 277 (1867). *Austromyrtus fragrantissima* (F.Muell. ex Benth.) Burret, Notizbl. Bot. Gart. Berlin-Dahlem 15: 501 (1941).

Gossia gonoclada (F.Muell. ex Benth.) N.Snow & Guymer, Syst. Bot. Monogr. 65: 47 (2003).
SE. Queensland. 50 QLD. Phan.
**Myrtus gonoclada* F.Muell. ex Benth., Fl. Austral. 3: 275 (1867). *Austromyrtus gonoclada* (F.Muell. ex Benth.) Burret, Notizbl. Bot. Gart. Berlin-Dahlem 15: 501 (1941).

Gossia grayi N.Snow & Guymer, Syst. Bot. Monogr. 65: 61 (2003).
NE. Queensland. 50 QLD. Phan.

Gossia hillii (Benth.) N.Snow & Guymer, Syst. Bot. Monogr. 65: 50 (2003).
Queensland. 50 QLD. Nanophan. or phan.
**Myrtus hillii* Benth., Fl. Austral. 3: 275 (1867). *Austromyrtus hillii* (Benth.) Burret, Notizbl. Bot. Gart. Berlin-Dahlem 15: 501 (1941).
Myrtus opaca C.T.White, Proc. Roy. Soc. Queensland 53: 218 (1942). *Austromyrtus opaca* (C.T.White) L.S.Sm., Proc. Roy. Soc. Queensland 67: 35 (1956).

Gossia inophloia (J.F.Bailey & C.T.White) N.Snow & Guymer, Syst. Bot. Monogr. 65: 35 (2003).
SE. Queensland. 50 QLD. Nanophan. or phan.
**Myrtus inophloia* J.F.Bailey & C.T.White, Bot. Bull. Dept. Agric. Queensland 19: 8 (1917). *Austromyrtus inophloia* (J.F.Bailey & C.T.White) Burret, Notizbl. Bot. Gart. Berlin-Dahlem 15: 502 (1941).
Myrtus decaspermoides Domin, Biblioth. Bot. 89: 475 (1928). *Austromyrtus decapermoides* (Domin) Burret, Notizbl. Bot. Gart. Berlin-Dahlem 15: 502 (1941).

Gossia lewisensis N.Snow & Guymer, Syst. Bot. Monogr. 65: 53 (2003).
NE. Queensland. 50 QLD. Phan.

Gossia longipetiolata N.Snow, Austrobaileya 7: 328 (2006).
Papua New Guinea. 43 NWG. Phan.

Gossia lucida (Gaertn.) N.Snow & Guymer, Syst. Bot. Monogr. 65: 70 (2003).
NE. Queensland. 50 QLD. Phan.
Eugenia lucida Banks ex Gaertn., Fruct. Sem. Pl. 1: 167 (1788), nom. inval.
**Syzygium lucidum* Gaertn., Fruct. Sem. Pl. 1: 167 (1788). *Austromyrtus lucida* (Gaertn.) L.S.Sm., Proc. Roy. Soc. Queensland 67: 35 (1956).
Myrtus nitida J.F.Gmel., Syst. Nat.: 792 (1791).
Calyptranthes nitida Raeusch., Nomencl. Bot., ed. 3: 144 (1797).
Myrtus monosperma F.Muell., Victorian Naturalist 9: 9 (1892).

Gossia macilwraithensis N.Snow & Guymer, Syst. Bot. Monogr. 65: 63 (2003).
NE. Queensland. 50 QLD. Phan.

Gossia myrsinocarpa (F.Muell.) N.Snow & Guymer, Syst. Bot. Monogr. 65: 73 (2003).
NE. & CE. Queensland. 50 QLD. Phan.
**Eugenia myrsinocarpa* F.Muell., Victoria Naturalist 8: 200 (1892).
Austromyrtus minutiflora Burret, Notizbl. Bot. Gart. Berlin-Dahlem 15: 502 (1941).

Gossia pubiflora (C.T.White) N.Snow & Guymer, Syst. Bot. Monogr. 65: 38 (2003).
E. Queensland. 50 QLD. Nanophan. or phan.
**Myrtus pubiflora* C.T.White, Proc. Roy. Soc. Queensland 50: 77 (1938 publ. 1939). *Austromyrtus pubiflora* (C.T.White) L.S.Sm., Proc. Roy. Soc. Queensland 67: 35 (1956).

Gossia punctata N.Snow & Guymer, Syst. Bot. Monogr. 65: 55 (2003).
SE. Queensland to NE. New South Wales. 50 NSW QLD. Phan.

Gossia randiana (Merr. & L.M.Perry) N.Snow, Novon 15: 477 (2005).
New Guinea. 43 NWG. Nanophan.
**Myrtus randiana* Merr. & L.M.Perry, J. Arnold Arbor. 23: 237 (1942).

Gossia retusa N.Snow & Guymer, Syst. Bot. Monogr. 65: 79 (2003).
NE. Queensland. 50 QLD. Nanophan. or phan.

Gossia salomonensis (A.J.Scott) N.Snow, Novon 15: 478 (2005).
Solomon Is. 43 SOL. Nanophan. or phan.
**Myrtus salomonensis* A.J.Scott, Kew Bull. 35: 412 (1980).

Gossia sankowskiorum N.Snow & Guymer, Syst. Bot. Monogr. 65: 77 (2003).
N. Queensland. 50 QLD. Phan.

Gossia scottiana N.Snow, Austrobaileya 7: 326 (2006).
Papua New Guinea. 43 NWG. Phan.

Gossia shepherdii (F.Muell.) N.Snow & Guymer, Syst. Bot. Monogr. 65: 68 (2003).
NE. Queensland. 50 QLD. Phan.
Eugenia shepherdii F.Muell., Fragm. 9: 148 (1875).
**Myrtus shepherdii* F.Muell., Fragm. 9: 148 (1875). *Austromyrtus shepherdii* (F.Muell.) L.S.Sm., Proc. Roy. Soc. Queensland 67: 35 (1956).

Gossia versteeghii (Merr. & L.M.Perry) N.Snow, Novon 15: 477 (2005).
New Guinea. 43 NWG. Nanophan.
**Myrtus versteeghii* Merr. & L.M.Perry, J. Arnold Arbor. 23: 238 (1942).

Greggia

Greggia Gaertn. = ***Eugenia*** P.Micheli ex L.
Greggia aromatica Gaertn. = ***Eugenia greggii*** (Sw.) Poir.

Guajava

Guajava Mill. = ***Psidium*** L.
Guajava acutangula (Mart. ex DC.) Kuntze = ***Psidium acutangulum*** Mart. ex DC.
Guajava aeruginea (O.Berg) Kuntze = ***Psidium laruotteanum*** Cambess.
Guajava alata (O.Berg) Kuntze = ***Psidium australe*** var. ***australe***
Guajava albida (Cambess.) Kuntze = ***Psidium guineense*** Sw.
Guajava amplexicaulis (Pers.) Kuntze = ***Psidium amplexicaule*** Pers.
Guajava anceps (O.Berg) Kuntze = ***Psidium australe*** var. ***australe***
Guajava anthomega (Vell.) Kuntze = ?
Guajava aquatica (Benth.) Kuntze = ***Psidium striatulum*** DC.
Guajava arborea (Vell.) Kuntze = ***Psidium arboreum*** Vell.
Guajava argenta (O.Berg) Kuntze = ***Psidium australe*** var. ***argenteum*** (O.Berg) Landrum
Guajava australis (Cambess.) Kuntze = ***Psidium australe*** Cambess.
Guajava basantha (O.Berg) Kuntze = ***Psidium laruotteanum*** Cambess.
Guajava benthamiana (O.Berg) Kuntze = ***Psidium guineense*** Sw.
Guajava berteroana (O.Berg) Kuntze = ?
Guajava buxifolia (Nutt.) Kuntze = ?
Guajava campomanesiodes (O.Berg) Kuntze = ***Calycolpus calophyllus*** (Kunth) O.Berg var. ***calophyllus***
Guajava cattleiana (Afzel. ex Sabine) Kuntze = ***Psidium cattleianum*** Afzel. ex Sabine
Guajava ciliata (Benth.) Kuntze = ***Psidium salutare*** var. ***salutare***
Guajava cinerea (Mart. ex DC.) Kuntze = ***Psidium grandifolium*** Mart. ex DC.
Guajava costa-ricensis (O.Berg) Kuntze = ***Psidium guineense*** Sw.
Guajava crenata (O.Berg) Kuntze = ***Psidium maribense*** Mart. ex DC.
Guajava cuneata (Cambess.) Kuntze = ***Psidium australe*** var. ***argenteum*** (O.Berg) Landrum
Guajava cuprea (O.Berg) Kuntze = ***Psidium rufum*** Mart. ex DC.
Guajava decussata (DC.) Kuntze = ***Psidium salutare*** var. ***decussatum*** (DC.) Landrum
Guajava densicoma (Mart. ex DC.) Kuntze = ***Psidium densicomum*** Mart. ex DC.
Guajava doniana (O.Berg) Kuntze = ***Psidium donianum*** O.Berg
Guajava elegans (DC.) Kuntze = ***Accara elegans*** (DC.) Landrum
Guajava firma (O.Berg) Kuntze = ***Psidium firmum*** O.Berg
Guajava fluviatilis Kuntze = ***Psidium guyanense*** Pers.
Guajava gardneriana (O.Berg) Kuntze = ***Psidium myrsinites*** DC.
Guajava glaucescens (O.Berg) Kuntze = ***Psidium laruotteanum*** Cambess.
Guajava grandifolia (Mart. ex DC.) Kuntze = ***Psidium grandifolium*** Mart. ex DC.
Guajava guajabita (A.Rich.) Kuntze = ***Psidium salutare*** var. ***salutare***

Guajava guineensis (Sw.) Kuntze = ***Psidium guineense*** Sw.
Guajava herbacea (O.Berg) Kuntze = ***Eugenia arenosa*** Mattos
Guajava hians (Mart. ex DC.) Kuntze = ***Campomanesia pubescens*** (Mart. ex DC.) O.Berg
Guajava humilis (Vell.) Kuntze = ?
Guajava inaequilatera (O.Berg) Kuntze = ***Psidium inaequilaterum*** O.Berg
Guajava incanescens (Mart. ex DC.) Kuntze = ***Psidium grandifolium*** Mart. ex DC.
Guajava indica (Raddi) Kuntze = ?
Guajava itanareensis (O.Berg) Kuntze = ***Psidium itanareense*** O.Berg
Guajava lactea (O.Berg) Kuntze = ***Psidium grandifolium*** Mart. ex DC.
Guajava lanceolata (O.Berg) Kuntze = ***Psidium salutare*** var. ***salutare***
Guajava langsdorffii (O.Berg) Kuntze = ***Psidium langsdorffii*** O.Berg
Guajava laruotteana (Cambess.) Kuntze = ***Psidium laruotteanum*** Cambess.
Guajava laurifolia (O.Berg) Kuntze = ***Psidium guineense*** Sw.
Guajava leptoclada (O.Berg) Kuntze = ***Psidium striatulum*** DC.
Guajava macrosperma (O.Berg) Kuntze = ***Psidium rufum*** Mart. ex DC.
Guajava maranhensis (O.Berg) Kuntze = ***Psidium riparium*** Mart. ex DC.
Guajava maribensis (Mart. ex DC.) Kuntze = ***Psidium maribense*** Mart. ex DC.
Guajava mengahiensis (Cambess.) Kuntze = ***Psidium riparium*** Mart. ex DC.
Guajava microcarpa (Cambess.) Kuntze = ***Psidium grandifolium*** Mart. ex DC.
Guajava mollis (Bertol.) Kuntze = ***Psidium guineense*** Sw.
Guajava montana (Sw.) Kuntze = ***Psidium montanum*** Sw.
Guajava moritziana (O.Berg) Kuntze = ***Psidium brownianum*** Mart. ex DC.
Guajava multiflora (Cambess.) Kuntze = ***Psidium guineense*** Sw.
Guajava myrsinites (DC.) Kuntze = ***Psidium myrsinites*** DC.
Guajava myrsinoides (O.Berg) Kuntze = ***Psidium myrtoides*** O.Berg
Guajava myrtoides (O.Berg) Kuntze = ***Psidium myrtoides*** O.Berg
Guajava nutans (O.Berg) Kuntze = ***Psidium nutans*** O.Berg
Guajava oblongata (O.Berg) Kuntze = ***Psidium oblongatum*** O.Berg
Guajava oblongifolia (O.Berg) Kuntze = ***Psidium oblongifolium*** O.Berg
Guajava obovata (Mart. ex DC.) Kuntze = ***Psidium cattleianum*** Afzel. ex Sabine
Guajava oerstediana (O.Berg) Kuntze = ***Psidium salutare*** var. ***salutare***
Guajava oligosperma (Mart. ex DC.) Kuntze = ***Psidium oligospermum*** Mart. ex DC.
Guajava ooidea (O.Berg) Kuntze = ***Psidium guineense*** Sw.
Guajava ovatifolia (O.Berg) Kuntze = ***Psidium densicomum*** Mart. ex DC.
Guajava paraensis (O.Berg) Kuntze = ***Psidium riparium*** Mart. ex DC.
Guajava paranensis (O.Berg) Kuntze = ***Psidium riparium*** Mart. ex DC.
Guajava parviflora (Benth.) Kuntze = ***Psidium striatulum*** DC.
Guajava persicifolia (O.Berg) Kuntze = ***Psidium striatulum*** DC.
Guajava pilosa (Vell.) Kuntze = ***Psidium rufum*** Mart. ex DC.
Guajava pohliana (O.Berg) Kuntze = ***Psidium salutare*** var. ***pohlianum*** (O.Berg) Landrum
Guajava polycarpa (Lamb.) Kuntze = ***Psidium guineense*** Sw.
Guajava pumila (Vahl) Kuntze = ***Psidium guajava*** L.
Guajava pyrifera (L.) Kuntze = ***Psidium guajava*** L.
Guajava radicans (O.Berg) Kuntze = ***Psidium grandifolium*** Mart. ex DC.
Guajava refracta (O.Berg) Kuntze = ***Psidium refractum*** O.Berg
Guajava rhombea (O.Berg) Kuntze = ***Psidium rhombeum*** O.Berg
Guajava richardiana (O.Berg) Kuntze = ***Psidium guyanense*** Pers.
Guajava riedeliana (O.Berg) Kuntze = ***Psidium grandifolium*** Mart. ex DC.
Guajava riparia (Mart. ex DC.) Kuntze = ***Psidium riparium*** Mart. ex DC.
Guajava robusta (O.Berg) Kuntze = ***Psidium robustum*** O.Berg
Guajava rubescens (O.Berg) Kuntze = ***Psidium grandifolium*** Mart. ex DC.
Guajava ruiziana (O.Berg) Kuntze = ***Psidium rutidocarpum*** Ruiz & Pav. ex G.Don
Guajava salutaris (Kunth) Kuntze = ***Psidium salutare*** (Kunth) O.Berg
Guajava schiedeana (O.Berg) Kuntze = ***Psidium guineense*** Sw.
Guajava sericea (O.Berg) Kuntze = ***Psidium grandifolium*** Mart. ex DC.
Guajava sieberiana (O.Berg) Kuntze = ***Psidium riparium*** Mart. ex DC.
Guajava sorocabensis (O.Berg) Kuntze = ***Psidium sorocabense*** O.Berg
Guajava sprucei (O.Berg) Kuntze = ***Psidium guyanense*** Pers.
Guajava striatula (DC.) Kuntze = ***Psidium striatulum*** DC.
Guajava suffruticosa (O.Berg) Kuntze = ***Psidium australe*** var. ***suffruticosum*** (O.Berg) Landrum
Guajava turbiniflora (Mart. ex DC.) Kuntze = ***Psidium striatulum*** DC.
Guajava umbrosa (O.Berg) Kuntze = ***Psidium guyanense*** Pers.
Guajava widgreniana (O.Berg) Kuntze = ***Psidium rufum*** Mart. ex DC.
Guajava ypanemensis (O.Berg) Kuntze = ***Psidium guineense*** Sw.

Guapurium

Guapurium Juss. = ***Plinia*** Plum. ex L.
Guapurium fruticosum Spreng. = ***Plinia peruviana*** (Poir.) Govaerts
Guapurium herbaceum Spreng. = (Loasaceae)
Guapurium peruvianum Poir. = ***Plinia peruviana*** (Poir.) Govaerts

Guapurum

Guapurum J.F.Gmel. = ***Plinia*** Plum. ex L.

Guayaba

Guayaba Noronha = ***Psidium*** L.

Gymnagathis

Gymnagathis Schauer = ***Melaleuca*** L.
Gymnagathis teretifolia (Endl.) Schauer = ***Melaleuca teretifolia*** Endl.

Harmogia

Harmogia Schauer = ***Babingtonia*** Lindl.
Harmogia baueriana Schauer = ***Babingtonia densifolia*** (Sm.) F.Muell.
Harmogia corynophylla F.Muell. = ***Baeckea corynophylla*** (F.Muell.) F.Muell.
Harmogia crenulata Miq. = ***Babingtonia crenulata*** (F.Muell.) A.R.Bean
Harmogia crispiflora F.Muell. = ***Baeckea crispiflora*** (F.Muell.) F.Muell.
Harmogia cunninghamii Schauer = ***Babingtonia cunninghamii*** (Schauer) A.R.Bean
Harmogia densifolia (Sm.) Schauer = ***Babingtonia densifolia*** (Sm.) F.Muell.
Harmogia leptophylla Turcz. = ***Baeckea leptophylla*** (Turcz.) Domin
Harmogia ovalifolia F.Muell. = ***Baeckea ovalifolia*** (F.Muell.) F.Muell.
Harmogia parviflora Turcz. = [50]
Harmogia parvula Schauer = ***Babingtonia virgata*** (J.R.Forst. & G.Forst.) F.Muell.
Harmogia pentandra F.Muell. = ***Baeckea pentandra*** (F.Muell.) F.Muell.
Harmogia pinifolia (Labill.) Schauer = ***Babingtonia pinifolia*** (Labill.) A.R.Bean
Harmogia propinqua Schauer = ***Babingtonia densifolia*** (Sm.) F.Muell.
Harmogia serpyllifolia Turcz. = ***Baeckea serpyllifolia*** (Turcz.) F.Muell.
Harmogia umbellata F.Muell. = ***Babingtonia virgata*** (J.R.Forst. & G.Forst.) F.Muell.
Harmogia virgata (J.R.Forst. & G.Forst.) Schauer = ***Babingtonia virgata*** (J.R.Forst. & G.Forst.) F.Muell.

***Unplaced Names*:**
Harmogia parviflora Turcz., Bull. Cl. Phys.-Math. Acad. Imp. Sci. Saint-Pétersbourg 10: 330 (1852). = [50]

Hedaroma

Hedaroma Lindl. = ***Darwinia*** Rudge
Hedaroma fimbriatum (Kippist) Baines = ***Darwinia squarrosa*** (Turcz.) Domin
Hedaroma hookeri Baines = ***Darwinia hypericifolia*** (Turcz.) Domin
Hedaroma latifolium Lindl. = ***Darwinia citriodora*** (Endl.) Benth.
Hedaroma pinifolium Lindl. = ***Darwinia pinifolia*** (Lindl.) Benth.
Hedaroma thymoides Lindl. = ***Darwinia thymoides*** (Lindl.) Benth.
Hedaroma tulipiferum Lindl. = ***Darwinia macrostegia*** (Turcz.) Benth.

Heteromyrtus

Heteromyrtus Blume = ***Blepharocalyx*** O.Berg
Heteromyrtus umbilicata (Cambess.) Blume = ***Blepharocalyx salicifolius*** (Kunth) O.Berg

Heteropyxis

Heteropyxis Harv., Thes. Cap. 2: 18 (1863).
S. Trop. & S. Africa. 26 MLW MOZ ZIM 27 NAT SWZ TVL.
3 Species

Heteropyxis canescens Oliv., Hooker's Icon. Pl. 25: t. 2407 (1895).
S. Africa. 27 SWZ TVL. Phan.
Heteropyxis transvaalensis Schinz, Bull. Herb. Boissier 4: 439 (1896).

Heteropyxis dehniae Suess., Trans. Rhodesia Sci. Assoc. 43: 27 (1951).
S. Trop. Africa to Northern Prov. 26 MLW MOZ ZIM 27 TVL. Phan.

Heteropyxis natalensis Harv., Thes. Cap. 2: 18 (1863).
S. Trop. & S. Africa. 26 MOZ ZIM 27 NAT SWZ TVL. Phan.

***Synonyms*:**
Heteropyxis transvaalensis Schinz = ***Heteropyxis canescens*** Oliv.

Hexachlamys

Hexachlamys O.Berg, Linnaea 27: 345 (1856).
Brazil to NE. Argentina. 83 BOL 84 BZC BZL BZS 85 AGE PAR URU.
13 Species
Calomyrtus Blume, Mus. Bot. 1: 76 (1850), nom. inval.

Hexachlamys edulis (O.Berg) Kausel & D.Legrand, Darwiniana 9: 302 (1950).
Brazil to NE. Argentina. 83 BOL 84 BZC BZL BZS 85 AGE PAR URU. Nanophan.
Myrtus excelsa Cambess. in A.F.C.de Saint-Hilaire, Fl. Bras. Merid. 2: 293 (1832). *Calomyrtus excelsa* (Cambess.) Blume, Mus. Bot. 1: 76 (1850). *Hexachlamys excelsa* (Cambess.) Mattos, Loefgrenia 94: 12 (1989). *Eugenia montevidensis* Mattos, Loefgrenia 105: 2 (1995).
Psidium amygdalinum Hook. & Arn., Bot. Misc. 3: 317 (1833), nom. rej.
Myrcia gemmiflora O.Berg in C.F.P.von Martius & auct. suc. (eds.), Fl. Bras. 14(1): 190 (1857).
**Myrcianthes edulis* O.Berg in C.F.P.von Martius & auct. suc. (eds.), Fl. Bras. 14(1): 353 (1857), nom. cons. *Eugenia edulis* (O.Berg) Benth. & Hook.f. ex Griseb., Abh. Königl. Ges. Wiss. Göttingen 24: 126 (1879), nom. illeg. *Eugenia myrcianthes* Nied. in H.G.A.Engler & K.A.E.Prantl, Nat. Pflanzenfam. 3(8): 79 (1893). *Luma myrcianthes* (Nied.) Herter, Revista Sudamer. Bot. 7: 219 (1943). *Luma grisebachii* Herter, Revista Sudamer. Bot. 7: 219 (1943).
Campomanesia cagaiteira Kiaersk., Enum. Myrt. Bras.: 12 (1893).
Myrcia sparsifolia Barb.Rodr., Myrt. Paraguay: 1 (1903).

Hexachlamys emerichii Mattos, Loefgrenia 78: 2 (1983).
Brazil (Rio Grande do Sul), Argentina (Corrientes). 84 BZS 85 AGE. Cham.

Hexachlamys geraensis D.Legrand & Mattos, Loefgrenia 62: 3 (1974). *Eugenia geraensis* (D.Legrand & Mattos) Mattos, Loefgrenia 105: 2 (1995).
SE. Brazil. 84 BZL. Nanophan.

Hexachlamys hamiltonii Mattos, Loefgrenia 11: 1 (1963). *Eugenia hamiltonii* (Mattos) Mattos, Loefgrenia 105: 2 (1995).
S. Brazil. 84 BZS. Nanophan.
Eugenia meridionalis Mattos, Loefgrenia 66: 6 (1975).

Hexachlamys handroi Mattos, Loefgrenia 1: 1 (1961). *Eugenia handroi* (Mattos) Mattos, Loefgrenia 105: 2 (1995).
SE. & S. Brazil. 84 BZL BZS. Nanophan.
Hexachlamys itatiaiensis Mattos, Loefgrenia 32: 2 (1969). *Eugenia itatiaiensis* (Mattos) Mattos, Loefgrenia 105: 2 (1995), nom. illeg.
Hexachlamys itatiaiensis var. *kleinii* D.Legrand ex Mattos, Loefgrenia 61: 1 (1974). *Hexachlamys kleinii* (D.Legrand ex Mattos) Mattos, Roessléria 5: 277 (1983). *Eugenia robertoana* Mattos, Loefgrenia 105: 2 (1995).

Hexachlamys humilis O.Berg, Linnaea 27: 345 (1856). *Eugenia humilis* (O.Berg) Benth. ex Nied. in H.G.A.Engler & K.A.E.Prantl, Nat. Pflanzenfam. 3(7): 82 (1893), nom. illeg. *Eugenia vallsiana* Mattos, Loefgrenia 105: 2 (1995).
Brazil to NE. Argentina. 84 BZC BZL BZS 85 AGE PAR URU. Cham.
Eugenia anomala D.Legrand, Anales Mus. Hist. Nat. Montevideo, II, 4(11): 61 (1936). *Hexachlamys anomala* (D.Legrand) D.Legrand, Darwiniana 9: 302 (1950).
Eugenia anomala var. *coriacea* D.Legrand, Anales Mus. Hist. Nat. Montevideo, II, 4(11): 61 (1936).
Eugenia myrcianthes var. *nana* D.Legrand, Darwiniana 5: 481 (1941).
Hexachlamys humilis var. *tomentosa* Loefgr. ex Mattos, Arq. Bot. Estado São Paulo, n.s., 3: 288 (1962).

Hexachlamys itararensis Mattos, Loefgrenia 53: 1 (1971). *Eugenia itararensis* (Mattos) Mattos, Loefgrenia 105: 3 (1995).
Brazil (São Paulo). 84 BZL. Cham.

Hexachlamys legrandii Mattos, Loefgrenia 62: 2 (1974). *Eugenia legrandii* (Mattos) Mattos, Loefgrenia 105: 2 (1995).
Brazil (São Paulo). 84 BZL. Phan.

Hexachlamys macedoi D.Legrand, Loefgrenia 55: 1 (1972).
Brazil (Minas Gerais). 84 BZL. Cham.
Hexachlamys macedoi var. *pubescens* Mattos, Loefgrenia 61: 2 (1974).

Hexachlamys minarum Mattos & D.Legrand, Loefgrenia 67: 13 (1975).
Brazil (Minas Gerais). 84 BZL. Nanophan.

Hexachlamys rojasiana (D.Legrand) D.Legrand, Loefgrenia 55: 2 (1972).
Paraguay. 85 PAR. Cham.
**Hexachlamys anomala* var. *rojasiana* D.Legrand, Bol. Soc. Argent. Bot. 10: 9 (1961). *Eugenia rojasiana* (D.Legrand) Mattos, Loefgrenia 105: 2 (1995).

Hexachlamys sehnemiana (Mattos) Mattos, Loefgrenia 78: 3 (1983).
Brazil (Rio Grande do Sul). 84 BZS. Cham.
**Hexachlamys hamiltonii* var. *sehnemiana* Mattos, Loefgrenia 61: 2 (1974). *Eugenia sehnemiana* (Mattos) Mattos, Loefgrenia 105: 2 (1995).

Hexachlamys toledoi Mattos, Arq. Bot. Estado São Paulo 4: 277 (1970). *Eugenia toledoi* (Mattos) Mattos, Loefgrenia 105: 3 (1995).
Brazil (São Paulo to Paraná). 84 BZL BZS. Nanophan.

Synonyms:
Hexachlamys anomala (D.Legrand) D.Legrand = ***Hexachlamys humilis*** O.Berg
Hexachlamys anomala var. *rojasiana* D.Legrand = ***Hexachlamys rojasiana*** (D.Legrand) D.Legrand
Hexachlamys boliviana D.Legrand = ***Eugenia boliviana*** (D.Legrand) Mattos
Hexachlamys boliviensis Kausel = ***Eugenia boliviana*** (D.Legrand) Mattos
Hexachlamys excelsa (Cambess.) Mattos = ***Hexachlamys edulis*** (O.Berg) Kausel & D.Legrand
Hexachlamys hamiltonii var. *sehnemiana* Mattos = ***Hexachlamys sehnemiana*** (Mattos) Mattos
Hexachlamys humilis var. *tomentosa* Loefgr. ex Mattos = ***Hexachlamys humilis*** O.Berg
Hexachlamys itatiaiensis Mattos = ***Hexachlamys handroi*** Mattos
Hexachlamys itatiaiensis var. *kleinii* D.Legrand ex Mattos = ***Hexachlamys handroi*** Mattos
Hexachlamys kleinii (D.Legrand ex Mattos) Mattos = ***Hexachlamys handroi*** Mattos
Hexachlamys macedoi var. *pubescens* Mattos = ***Hexachlamys macedoi*** D.Legrand
Hexachlamys tapiraguaiensis (Barb.Rodr. ex Chodat & Hassl.) Mattos = ?

Homalocalyx

Homalocalyx F.Muell., Hooker's J. Bot. Kew Gard. Misc. 9: 309 (1857).
Australia. 50 NTA QLD WAU.
11 Species
Wehlia F.Muell., Fragm. 10: 22 (1876).

Homalocalyx aureus (C.A.Gardner) Craven, Brunonia 10: 147 (1987).
W. Western Australia. 50 WAU. Nanophan.
**Wehlia aurea* C.A.Gardner, J. Roy. Soc. Western Australia 47: 61 (1964).

Homalocalyx chapmanii Craven, Brunonia 10: 147 (1987).
W. Western Australia. 50 WAU. Cham.

Homalocalyx coarctatus (F.Muell.) Craven, Brunonia 10: 148 (1987).
WSW. Western Australia. 50 WAU. Cham. or nanophan.
**Wehlia coarctata* F.Muell., Fragm. 10: 23 (1876).

Homalocalyx echinulatus Craven, Brunonia 10: 149 (1987).
C. Western Australia. 50 WAU. Nanophan.

Homalocalyx ericaeus F.Muell., Hooker's J. Bot. Kew Gard. Misc. 9: 309 (1857). *Thryptomene homalocalyx* F.Muell., Fragm. 4: 63 (1864).
N. Western Australia to Northern Territory. 50 NTA WAU. Nanophan.

Homalocalyx grandiflorus (C.A.Gardner) Craven, Brunonia 10: 151 (1987).
SC. Western Australia. 50 WAU. Nanophan.
**Wehlia grandiflora* C.A.Gardner, J. Roy. Soc. Western Australia 47: 61 (1964).

Homalocalyx inerrabundus Craven, Brunonia 10: 152 (1987).
W. Western Australia. 50 WAU. Cham.

Homalocalyx polyandrus (F.Muell.) Benth., Fl. Austral. 3: 56 (1867).
S. Queensland. 50 QLD. Nanophan.
**Thryptomene polyandra* F.Muell., Fragm. 4: 77 (1864).

Homalocalyx pulcherrimus (Ewart & B.Rees) Craven, Brunonia 10: 154 (1987).
S. Western Australia. 50 WAU. Cham. or nanophan.
Wehlia pedicellata Ewart & B.Rees, Proc. Roy. Soc. Victoria, n.s., 24: 266 (1912).
**Wehlia pulcherrima* Ewart & B.Rees, Proc. Roy. Soc. Victoria, n.s., 24: 267 (1912).

Homalocalyx staminosus (F.Muell.) Craven, Brunonia 10: 155 (1987).
WC. Western Australia. 50 WAU. Nanophan.
**Wehlia staminosa* F.Muell., S. Sci. Rec. 3: 282 (1884).

Homalocalyx thryptomenoides (F.Muell.) Craven, Brunonia 10: 156 (1987).
Western Australia. 50 WAU. Nanophan.
**Wehlia thryptomenoides* F.Muell., Fragm. 10: 22 (1876).
Wehlia thryptomenoides var. *microphylla* Ewart & B.Rees, Proc. Roy. Soc. Queensland, II, 24: 266 (1912).

Homalospermum

Homalospermum Schauer, Linnaea 17: 242 (1843).
SW. Australia. 50 WAU.
1 Species

Homalospermum firmum Schauer, Linnaea 17: 242 (1843). *Leptospermum firmum* (Schauer) Benth., Fl. Austral. 3: 104 (1867).
SW. Australia. 50 WAU. Nanophan.

Homoranthus

Homoranthus A.Cunn. ex Schauer, Linnaea 10: 310 (1836).
E. & SE. Australia. 50 NSW QLD SOA.
23 Species
Enosanthes A.Cunn. ex Schauer, Nov. Actorum Acad. Caes. Leop.-Carol. Nat. Cur. 19(Suppl. 2): 191 (1841), nom. inval.
Schuermannia F.Muell., Linnaea 25: 386 (1853).
Rylstonea R.T.Baker, Proc. Linn. Soc. New South Wales 23: 768 (1898).

Homoranthus biflorus Craven & S.R.Jones, Austral. Syst. Bot. 4: 516 (1991).
NE. New South Wales. 50 NSW. Nanophan.

Homoranthus binghiensis J.T.Hunter, Telopea 9: 431 (2001).
New South Wales. 50 NSW. Nanophan.

Homoranthus bornhardtiensis J.T.Hunter, Telopea 8: 37 (1998).
New South Wales. 50 NSW. Nanophan.

Homoranthus cernuus (R.T.Baker) Craven & S.R.Jones, Austral. Syst. Bot. 4: 518 (1991).
E. New South Wales. 50 NSW. Nanophan.
**Rylstonea cernua* R.T.Baker, Proc. Linn. Soc. New South Wales 23: 768 (1898).

Homoranthus coracinus A.R.Bean, Austrobaileya 5: 687 (2000).
Queensland. 50 QLD. Nanophan.

Homoranthus croftianus J.T.Hunter, Telopea 8: 39 (1998).
New South Wales. 50 NSW. Nanophan.

Homoranthus darwinioides (Maiden & Betche) Cheel, J. Proc. Roy. Soc. New S. Wales 56: 77 (1922).
New South Wales. 50 NSW. Nanophan.
**Verticordia darwinioides* Maiden & Betche, Proc. Linn. Soc. New South Wales 23: 17 (1898). *Rylstonea darwinioides* (Maiden & Betche) R.T.Baker, Proc. Linn. Soc. New South Wales 25: 665 (1900).

Homoranthus decasetus Byrnes, Austrobaileya 1: 374 (1981).
E. Queensland. 50 QLD. Nanophan.

Homoranthus decumbens (Byrnes) Craven & S.R.Jones, Austral. Syst. Bot. 4: 520 (1991).
SE. Queensland. 50 QLD. Nanophan.
**Darwinia decumbens* Byrnes, Austrobaileya 2: 15 (1984).

Homoranthus flavescens Schauer, Monogr. Myrt. Xerocarp.: 192 (1843).
NE. New South Wales. 50 NSW. Nanophan.
Darwinia flavescens A.Cunn. ex Schauer, Nov. Actorum Acad. Caes. Leop.-Carol. Nat. Cur. 19(Suppl. 2): 192 (1841), nom. inval.
Enosanthes flavescens A.Cunn. ex Schauer, Nov. Actorum Acad. Caes. Leop.-Carol. Nat. Cur. 19(Suppl. 2): 192 (1841), nom. inval.

Homoranthus floydii Craven & S.R.Jones, Austral. Syst. Bot. 4: 521 (1991).
NE. New South Wales. 50 NSW. Nanophan.

Homoranthus homoranthoides (F.Muell.) Craven & S.R.Jones, Austral. Syst. Bot. 4: 522 (1991).
South Australia (Eyre Pen.). 50 SOA. Nanophan.
**Schuermannia homoranthoides* F.Muell., Linnaea 25: 387 (1853). *Genetyllis schuermannii* F.Muell., Fragm. 1: 12 (1858), nom. superfl. *Chamelaucium schuermannii* (F.Muell.) F.Muell., Fragm. 4: 138 (1864), nom. illeg. *Darwinia schuermannii* (F.Muell.) Benth., J. Linn. Soc., Bot. 9: 181 (1867), nom. illeg. *Darwinia homoranthoides* (F.Muell.) J.M.Black, Fl. S. Austral.: 424 (1926).
Chamelaucium thomasii F.Muell., Fragm. 4: 137 (1864). *Darwinia thomasii* (F.Muell.) Benth., J. Linn. Soc., Bot. 9: 181 (1867).

Homoranthus lunatus Craven & S.R.Jones, Austral. Syst. Bot. 4: 522 (1991).
NE. New South Wales. 50 NSW. Nanophan.

Homoranthus melanostictus Craven & S.R.Jones, Austral. Syst. Bot. 4: 523 (1991).
SE. Queensland. 50 QLD. Nanophan.

Homoranthus montanus Craven & S.R.Jones, Austral. Syst. Bot. 4: 524 (1991).
SE. Queensland. 50 QLD. Nanophan.

Homoranthus papillatus Byrnes, Austrobaileya 1: 373 (1981).
SE. Queensland. 50 QLD. Nanophan.

Homoranthus porteri (C.T.White) Craven & S.R.Jones, Austral. Syst. Bot. 4: 525 (1991).
NE. Queensland. 50 QLD. Nanophan.
**Darwinia porteri* C.T.White, Proc. Roy. Soc. Queensland 43: 15 (1931 publ. 1932).

Homoranthus prolixus Craven & S.R.Jones, Austral. Syst. Bot. 4: 525 (1991).
NE. New South Wales. 50 NSW. Cham.

Homoranthus thomasii (F.Muell.) Craven & S.R.Jones,

Austral. Syst. Bot. 4: 526 (1991).
NC. Queensland. 50 QLD. Nanophan.

Homoranthus tropicus Byrnes, Austrobaileya 1: 375 (1981).
N. Queensland. 50 QLD. Nanophan.

Homoranthus virgatus A.Cunn. ex Schauer, Nov. Actorum Acad. Caes. Leop.-Carol. Nat. Cur. 19(Suppl. 2): 193 (1842 publ. 1843). *Darwinia virgata* (A.Cunn. ex Schauer) F.Muell., Fragm. 9: 176 (1875).
E. Queensland to NE. New South Wales. 50 NSW QLD. Nanophan.
Enosanthes virgata A.Cunn. ex Schauer, Nov. Actorum Acad. Caes. Leop.-Carol. Nat. Cur. 19(Suppl. 2): 193 (1841), nom. inval.

Homoranthus wilhelmii (F.Muell.) Cheel, J. Proc. Roy. Soc. New S. Wales 56: 77 (1922).
South Australia. 50 SOA. Nanophan.
**Verticordia wilhelmii* F.Muell., Trans. Philos. Soc. Victoria 1: 122 (1855). *Rylstonea wilhelmii* (F.Muell.) R.T.Baker, Proc. Linn. Soc. New South Wales 25: 665 (1900).

Homoranthus zeteticorum Craven & S.R.Jones, Austral. Syst. Bot. 4: 529 (1991).
EC. Queensland. 50 QLD. Nanophan.

Hottea

Hottea Urb., Ark. Bot. 22A(10): 40 (1929).
E. Cuba to Hispaniola. 81 CUB DOM HAI.
9 Species

Hottea crispula (Urb.) Urb., Ark. Bot. 22A(10): 42 (1929).
Haiti (Massif de la Hotte). 81 HAI. Nanophan. or phan.
**Psidium crispulum* Urb., Ark. Bot. 17(7): 44 (1922).

Hottea ekmanii (Urb.) Borhidi, Acta Bot. Hung. 37: 77 (1992).
E. Cuba. 81 CUB. Nanophan. or phan.
**Calycorectes ekmanii* Urb., Symb. Antill. 9: 110 (1923). *Eugenia ekmanii* (Urb.) Mattos, Loefgrenia 120: 5 (2005), with incorrect basionym ref.

Hottea goavensis Urb., Ark. Bot. 24A(4): 29 (1932).
Haiti (Massif de la Hotte). 81 HAI. Nanophan.

Hottea malangensis Urb., Ark. Bot. 24A(4): 31 (1932).
Haiti (Massif de la Selle). 81 HAI. Nanophan. or phan.
**Eugenia malangensis* Urb. & Ekman, Ark. Bot. 21A(5): 30 (1927), nom. illeg.

Hottea micrantha Urb. & Ekman, Ark. Bot. 24A(4): 30 (1932). *Hottea goavensis* var. *micrantha* (Urb. & Ekman) Borhidi, Acta Bot. Hung. 29: 186 (1983).
Haiti (Massif de la Hotte). 81 HAI. Nanophan.

Hottea miragoanae Urb., Ark. Bot. 22A(10): 40 (1929).
Haiti (Massif de la Hotte). 81 HAI. Nanophan. or phan.

Hottea moana (Borhidi & O.Muñiz) Borhidi, Acta Bot. Hung. 37: 78 (1992).
E. Cuba. 81 CUB. Nanophan. or phan.
**Calycorectes moana* Borhidi & O.Muñiz, Acta Bot. Acad. Sci. Hung. 21: 225 (1975 publ. 1976).

Hottea neibensis Alain, Phytologia 25: 269 (1973).
Dominican Rep. (Sierra de Neiba). 81 DOM. Nanophan.

Hottea torbeciana Urb. & Ekman, Ark. Bot. 22A(10): 41 (1929).
Haiti (Massif de la Hotte). 81 HAI. Nanophan.

Synonyms:
Hottea goavensis var. *micrantha* (Urb. & Ekman) Borhidi = ***Hottea micrantha*** Urb. & Ekman

Hypocalymma

Hypocalymma (Endl.) Endl., Gen. Pl.: 1230 (1840).
SW. Australia. 50 WAU.
22 Species

Hypocalymma angustifolium (Endl.) Schauer, Linnaea 17: 241 (1843).
SW. Australia. 50 WAU. Cham. or nanophan.
**Leptospermum angustifolium* Endl. in S.L.Endlicher & al., Enum. Pl.: 50 (1837).

subsp. ***angustifolium***
SW. Australia. 50 WAU. Nanophan. or phan.
Hypocalymma suavis Lindl., Edwards's Bot. Reg. 30(Misc.): 27 (1844).
Hypocalymma angustifolium var. *acerosa* Schauer in J.G.C.Lehmann, Pl. Preiss. 1: 112 (1845).
Hypocalymma angustifolium var. *linophylla* Schauer in J.G.C.Lehmann, Pl. Preiss. 1: 112 (1845).
Hypocalymma angustifolium var. *verrucosa* Schauer in J.G.C.Lehmann, Pl. Preiss. 1: 112 (1845).

subsp. ***longifolium*** (F.Muell.) Strid & Keighery, Nord. J. Bot. 22: 543 (2002 publ. 2003).
SW. Australia. 50 WAU. Cham. or nanophan.
**Hypocalymma longifolium* F.Muell., Fragm. 2: 28 (1860).

Hypocalymma asperum Schauer in J.G.C.Lehmann, Pl. Preiss. 1: 111 (1845).
SSW. Western Australia. 50 WAU. Cham.

Hypocalymma connatum Strid & Keighery, Nord. J. Bot. 22: 572 (2002 publ. 2003).
WSW. Western Australia. 50 WAU†. Cham.

Hypocalymma cordifolium Schauer in J.G.C.Lehmann, Pl. Preiss. 1: 112 (1844).
SW. Australia. 50 WAU. Cham. or nanophan.

subsp. ***cordifolium***
SW. Western Australia. 50 WAU. Cham. or nanophan.

subsp. ***minus*** Strid & Keighery, Nord. J. Bot. 22: 567 (2002 publ. 2003).
SW. Australia. 50 WAU. Cham. or nanophan.

Hypocalymma ericifolium Benth., Fl. Austral. 3: 94 (1867).
SW. Australia. 50 WAU. Cham. or nanophan.

Hypocalymma gardneri Strid & Keighery, Nord. J. Bot. 22: 556 (2002 publ. 2003).
WSW. Western Australia. 50 WAU. Cham.

Hypocalymma hirsutum Strid & Keighery, Nord. J. Bot. 22: 553 (2002 publ. 2003).
WSW. Western Australia. 50 WAU. Cham.

Hypocalymma jessicae Strid & Keighery, Nord. J. Bot. 22: 562 (2002 publ. 2003).
SSW. Western Australia. 50 WAU. Nanophan.

Hypocalymma melaleucoides Gardner ex Strid & Keighery, Nord. J. Bot. 22: 564 (2002 publ. 2003).
SSW. Western Australia. 50 WAU. Cham. or nanophan.

Hypocalymma myrtifolium Turcz., Bull. Cl. Phys.-Math. Acad. Imp. Sci. Saint-Pétersbourg 10: 333 (1952). *Hypocalymma hypericifolium* Benth., Fl. Austral. 3: 95 (1867), nom. illeg.
SW. Western Australia. 50 WAU. Cham. or nanophan.

Hypocalymma phillipsii Harv., Nat. Hist. Rev. 5(2): 2966 (1858).
SW. Western Australia. 50 WAU. Cham. or nanophan.

Hypocalymma puniceum C.A.Gardner, J. Proc. Roy. Soc. W. Australia 9(2): 103 (1923).
SW. Australia. 50 WAU. Nanophan.

Hypocalymma robustum (Endl.) Lindl., Edwards's Bot. Reg.: t. 8 (1843).
SW. Australia. 50 WAU. Nanophan.
**Leptospermum robustum* Endl. in S.L.Endlicher & al., Enum. Pl.: 50 (1837).

Hypocalymma scariosum Schauer in J.G.C.Lehmann, Pl. Preiss. 1: 111 (1844). *Hypocalymma angustifolium* var. *densiflorum* Benth., Fl. Austral. 3: 94 (1867).
SW. Western Australia. 50 WAU. Cham.

Hypocalymma serrulatum Strid & Keighery, Nord. J. Bot. 22: 550 (2002 publ. 2003).
WSW. Western Australia. 50 WAU. Nanophan.

Hypocalymma speciosum Turcz., Bull. Cl. Phys.-Math. Acad. Imp. Sci. Saint-Pétersbourg 10: 332 (1852).
SW. Western Australia. 50 WAU. Cham.
Hypocalymma boroniaceum F.Muell. ex Benth., Fl. Austral. 3: 95 (1867).

Hypocalymma strictum Schauer in J.G.C.Lehmann, Pl. Preiss. 1: 111 (1845).
SW. Western Australia. 50 WAU. Cham. or nanophan.

subsp. ***elongatum*** Strid & Keighery, Nord. J. Bot. 22: 560 (2002 publ. 2003).
SW. Australia. 50 WAU. Cham.

subsp. ***strictum***
SW. Western Australia. 50 WAU. Cham.
Baeckea speciosa A.Cunn., J. Bot. (Hooker) 4: 243 (1841).
Hypocalymma cunninghamii Schauer, Linnaea 17: 241 (1843).
Hypocalymma strictum var. *pedunculatum* Benth., Fl. Austral. 3: 93 (1867).

Hypocalymma sylvestris Strid & Keighery, Nord. J. Bot. 22: 558 (2002 publ. 2003).
SW. Australia. 50 WAU. Cham. or nanophan.

Hypocalymma tenuatum Strid & Keighery, Nord. J. Bot. 22: 545 (2002 publ. 2003).
WSW. Western Australia. 50 WAU. Cham.

Hypocalymma tetrapterum Turcz., Bull. Soc. Imp. Naturalistes Moscou 35(2): 325 (1862).
WSW. Western Australia. 50 WAU. Nanophan.
Hypocalymma linifolium Turcz., Bull. Soc. Imp. Naturalistes Moscou 35(2): 325 (1862).

Hypocalymma uncinatum Strid & Keighery, Nord. J. Bot. 22: 544 (2002 publ. 2003).
SW. Australia. 50 WAU. Nanophan.

Hypocalymma xanthopetalum F.Muell., Fragm. 2: 29 (1860).
WSW. Western Australia. 50 WAU. Nanophan.
Hypocalymma ciliatum Turcz., Bull. Soc. Imp. Naturalistes Moscou 35(2): 325 (1862).
Hypocalymma cuneatum Turcz., Bull. Soc. Imp. Naturalistes Moscou 35(2): 325 (1862).

Synonyms:
Hypocalymma angustifolium var. *acerosa* Schauer = ***Hypocalymma angustifolium*** subsp. ***angustifolium***
Hypocalymma angustifolium var. *densiflorum* Benth. = ***Hypocalymma scariosum*** Schauer
Hypocalymma angustifolium var. *linophylla* Schauer = ***Hypocalymma angustifolium*** subsp. ***angustifolium***
Hypocalymma angustifolium var. *verrucosa* Schauer = ***Hypocalymma angustifolium*** subsp. ***angustifolium***
Hypocalymma boroniaceum F.Muell. ex Benth. = ***Hypocalymma speciosum*** Turcz.
Hypocalymma ciliatum Turcz. = ***Hypocalymma xanthopetalum*** F.Muell.
Hypocalymma cuneatum Turcz. = ***Hypocalymma xanthopetalum*** F.Muell.
Hypocalymma cunninghamii Schauer = ***Hypocalymma strictum*** subsp. ***strictum***
Hypocalymma dimorphandrum (F.Muell. ex Benth.) Nied. = ***Rinzia dimorphandra*** (F.Muell. ex Benth.) Trudgen
Hypocalymma drummondii (Benth.) Nied. = ?
Hypocalymma fumanum (Schauer) Nied. = ***Rinzia fumana*** Schauer
Hypocalymma hypericifolium Benth. = ***Hypocalymma myrtifolium*** Turcz.
Hypocalymma linifolium Turcz. = ***Hypocalymma tetrapterum*** Turcz.
Hypocalymma longifolium F.Muell. = ***Hypocalymma angustifolium*** subsp. ***longifolium*** (F.Muell.) Strid & Keighery
Hypocalymma oxycoccoides (Turcz.) Nied. = ***Rinzia oxycoccoides*** Turcz.
Hypocalymma platystemonum (Benth.) Nied. = ***Rinzia crassifolia*** Turcz.
Hypocalymma schollerifolium (Lehm.) Nied. = ***Rinzia schollerifolia*** (Lehm.) Trudgen
Hypocalymma strictum var. *pedunculatum* Benth. = ***Hypocalymma strictum*** subsp. ***strictum***
Hypocalymma suavis Lindl. = ***Hypocalymma angustifolium*** subsp. ***angustifolium***

Imbricaria

Imbricaria Sm. = ***Baeckea*** L.
Imbricaria ciliata Sm. = ***Micromyrtus ciliata*** (Sm.) Druce
Imbricaria crenulata Sm. = ***Baeckea imbricata*** (Gaertn.) Druce
Imbricaria tenella (Gaertn.) Dryand. = ***Baeckea imbricata*** (Gaertn.) Druce

Jambolifera

Jambolifera Houtt. = ***Syzygium*** Gaertn.
Jambolifera chinensis Spreng. = ***Syzygium cumini*** (L.) Skeels
Jambolifera coromandelica Houtt. = ***Syzygium cumini*** (L.) Skeels
Jambolifera pedunculata Houtt. = ***Syzygium cumini*** (L.) Skeels

Jambos

Jambos Adans. = ***Syzygium*** Gaertn.

Jambosa

Jambosa DC. = ***Syzygium*** Gaertn.
Jambosa acida auct. = ?

Jambosa acris Pancher ex Guillaumin = ***Syzygium acre*** (Pancher ex Guillaumin) J.W.Dawson
Jambosa acuminata (Roxb.) Sweet = ***Syzygium acuminatum*** (Roxb.) Miq.
Jambosa acuminatissima (Blume) Hassk. = ***Syzygium acuminatissimum*** (Blume) DC.
Jambosa acutata Miq. = ***Syzygium acutatum*** (Miq.) Amshoff
Jambosa aemula Blume = ***Syzygium aemulum*** (Blume) Amshoff
Jambosa aeorantha Diels = ***Syzygium aeoranthum*** (Diels) Merr. & L.M.Perry
Jambosa alba (Roxb.) G.Don = ***Syzygium aqueum*** (Burm.f.) Alston
Jambosa alutacea Diels = ***Syzygium alutaceum*** (Diels) Merr. & L.M.Perry
Jambosa ambigua Blume = ***Syzygium aqueum*** (Burm.f.) Alston
Jambosa amplexicaulis DC. = ***Syzygium amplexicaule*** (DC.) N.P.Balakr.
Jambosa anastomosans Miq. = [42 SUM]
Jambosa appendiculata Blume = (Melastomataceae)
Jambosa aquea (Burm.f.) DC. = ***Syzygium aqueum*** (Burm.f.) Alston
Jambosa arfakensis Gibbs = ***Syzygium effusum*** (A.Gray) Müll.Stuttg.
Jambosa aromatica Miq. = ***Syzygium antisepticum*** (Blume) Merr. & L.M.Perry
Jambosa attenuata Miq. = ***Syzygium attenuatum*** (Miq.) Merr. & L.M.Perry
Jambosa auriculata Blume = ***Syzygium novoguineense*** Merr. & L.M.Perry
Jambosa australis (J.C.Wendl. ex Link) DC. = ***Syzygium australe*** (J.C.Wendl. ex Link) B.Hyland
Jambosa baladensis (Brongn. & Gris) Nied. = ***Syzygium baladense*** (Brongn. & Gris) J.W.Dawson
Jambosa barnesii Merr. = ***Syzygium barnesii*** (Merr.) Merr.
Jambosa bartonii (F.M.Bailey) Diels = ***Syzygium bartonii*** (F.M.Bailey) Merr. & L.M.Perry
Jambosa bataanensis Merr. = ***Syzygium bataanense*** (Merr.) Merr.
Jambosa beddomei (Duthie) Gamble = ***Syzygium beddomei*** (Duthie) Chithra
Jambosa bifaria (Wall.) Miq. = ***Syzygium laurifolium*** (DC.) N.P.Balakr.
Jambosa bisulca Miq. = [42 SUM]
Jambosa borneensis Miq. = ***Syzygium claviflorum*** (Roxb.) Wall. ex A.M.Cowan & Cowan
Jambosa bourdillonii Gamble = ***Syzygium bourdillonii*** (Gamble) Rathakr. & N.C.Nair
Jambosa brackenridgei Brongn. & Gris = [60 NWC]
Jambosa bracteata Miq. = ***Syzygium zeylanicum*** (L.) DC.
Jambosa brevicyma Diels = ***Syzygium brevicymum*** (Diels) Merr. & L.M.Perry
Jambosa bruynii Diels = ***Syzygium bruynii*** (Diels) Merr. & L.M.Perry
Jambosa buxifolia Miq. = ***Syzygium bankense*** (Hassk.) Merr. & L.M.Perry
Jambosa calophylla Miq. = ***Syzygium aqueum*** (Burm.f.) Alston
Jambosa canalensis Pancher ex Guillaumin = ***Eugenia mouensis*** Baker f.
Jambosa caryophyllata (L.) Bedevian = ***Syzygium caryophyllatum*** (L.) Alston
Jambosa caryophylloides Lauterb. = ***Syzygium caryophylloides*** (Lauterb.) Merr. & L.M.Perry
Jambosa caryophyllus (Thunb.) Nied. = ***Syzygium aromaticum*** (L.) Merr. & L.M.Perry
Jambosa cauliflora DC. = ***Syzygium polycephalum*** (Miq.) Merr. & L.M.Perry
Jambosa celebica Blume = [42 SUL]
Jambosa cerasiformis (Blume) Hassk. = ***Syzygium racemosum*** (Blume) DC.
Jambosa cladoptera Diels = ***Syzygium cladopterum*** (Diels) Merr. & L.M.Perry
Jambosa clavata Korth. = ***Syzygium claviflorum*** (Roxb.) Wall. ex A.M.Cowan & Cowan
Jambosa coarctata Blume = ***Syzygium megacarpum*** (Craib) Rathakr. & N.C.Nair
Jambosa coarctata var. *firma* Blume = [42 SUM]
Jambosa combretiflora Diels = ***Syzygium combretiflorum*** (Diels) Merr. & L.M.Perry
Jambosa commersonii Blume = ***Syzygium cymosum*** (Lam.) DC. var. ***cymosum***
Jambosa condensata Miq. = [42 JAW]
Jambosa conferta Korth. = ***Syzygium confertum*** (Korth.) Merr. & L.M.Perry
Jambosa confusa Blume = ***Syzygium confusum*** (Blume) Bakh.f.
Jambosa cornifolia Blume = ***Syzygium cornifolium*** (Blume) Merr. & L.M.Perry
Jambosa corymbosa (Blume) Miq. = ***Syzygium corymbosum*** (Blume) DC.
Jambosa costata Miq. = [42 JAW]
Jambosa courtallensis Gamble = [40 IND SRL]
Jambosa cuneata Blume = ***Syzygium leucophloium*** (Blume) Merr. & L.M.Perry
Jambosa cuspidata (Blume) Blume ex Miq. = ***Syzygium confertum*** (Korth.) Merr. & L.M.Perry
Jambosa cylindrica (Wight) Thwaites = ***Syzygium cylindricum*** (Wight) Alston
Jambosa cymifera E.Mey. = ***Syzygium cordatum*** Hochst. ex Krauss subsp. ***cordatum***
Jambosa decoriflora Diels = ***Syzygium decoriflorum*** (Diels) Merr. & L.M.Perry
Jambosa densiflora (Blume) DC. = ***Syzygium pycnanthum*** Merr. & L.M.Perry
Jambosa diospyrifolia (Wall. ex Duthie) C.E.C.Fisch. = ***Syzygium diospyrifolium*** (Wall. ex Duthie) S.N.Mitra
Jambosa dolichophylla K.Schum. & Lauterb. = ***Syzygium dolichophyllum*** (K.Schum. & Lauterb.) Merr. & L.M.Perry
Jambosa dolichostyla Diels = ***Syzygium dolichostylum*** (Diels) Merr. & L.M.Perry
Jambosa domestica DC. = ***Syzygium malaccense*** (L.) Merr. & L.M.Perry
Jambosa elegans (Brongn. & Gris) Nied. = ***Syzygium elegans*** (Brongn. & Gris) J.W.Dawson
Jambosa eucalyptoides F.Muell. = ***Syzygium eucalyptoides*** (F.Muell.) B.Hyland
Jambosa eximiiflora Diels = ***Syzygium eximiiflorum*** (Diels) Merr. & L.M.Perry
Jambosa firma (Blume) Miq. = ***Syzygium ampliflorum*** (Koord. & Valeton) Amshoff
Jambosa firma Blume = ***Syzygium grande*** (Wight) Walp.
Jambosa floribunda Diels = ***Syzygium tierneyanum*** (F.Muell.) T.G.Hartley & L.M.Perry
Jambosa formosa (Wall.) G.Don = ***Syzygium formosum*** (Wall.) Masam.
Jambosa garciae Merr. = ***Syzygium garciae*** (Merr.) Merr.
Jambosa glabra Zoll. ex Koord. & Valeton = ***Syzygium klampok*** (Miq.) Amshoff
Jambosa glabrata DC. = ***Syzygium glabratum*** (DC.) Veldkamp
Jambosa glandulifera (Roxb.) Miq. = ***Syzygium zeylanicum*** (L.) DC.
Jambosa glandulosa Korth. = [42 BOR]

Jambosa glomerata K.Schum. & Lauterb. = ***Syzygium warburgii*** Merr. & L.M.Perry
Jambosa gonatantha Diels = ***Syzygium gonatanthum*** (Diels) Merr. & L.M.Perry
Jambosa goniocalyx Lauterb. = ***Syzygium goniocalyx*** (Lauterb.) Merr. & L.M.Perry
Jambosa gonioptera Diels = ***Syzygium goniopterum*** (Diels) Merr. & L.M.Perry
Jambosa gracilipes (A.Gray) Müll.Stuttg. = ***Syzygium gracilipes*** (A.Gray) Merr. & L.M.Perry
Jambosa gracilis Korth. = ***Syzygium glabratum*** (DC.) Veldkamp
Jambosa grandis (Wight) Blume = ***Syzygium grande*** (Wight) Walp.
Jambosa hardwickia G.Don = [29 MAU]
Jambosa hemispherica Wight = ***Syzygium hemisphericum*** (Wight) Alston
Jambosa hirta Korth. = ***Syzygium hirtum*** (Korth.) Merr. & L.M.Perry
Jambosa horsfieldii Miq. = [42 JAW]
Jambosa hylocharis Diels = ***Syzygium hylochare*** (Diels) Merr. & L.M.Perry
Jambosa hylophila K.Schum. & Lauterb. = ***Syzygium hylophilum*** (K.Schum. & Lauterb.) Merr. & L.M.Perry
Jambosa hypericifolia DC. = ***Syzygium blumei*** (Steud.) Merr. & L.M.Perry
Jambosa inophylla (DC.) Miq. = ***Syzygium inophyllum*** DC.
Jambosa insignis Blume = ***Syzygium insigne*** (Blume) Merr. & L.M.Perry
Jambosa jambos (L.) Millsp. = ***Syzygium jambos*** (L.) Alston
Jambosa javanica (Lam.) K.Schum. & Lauterb. = ***Syzygium samarangense*** (Blume) Merr. & L.M.Perry
Jambosa keroantha Diels = ***Syzygium keroanthum*** (Diels) Merr. & L.M.Perry
Jambosa keysseri Schltr. ex Diels = ***Syzygium keysseri*** (Schltr. ex Diels) Merr. & L.M.Perry
Jambosa kigurrung Blume = [42 JAW]
Jambosa klampok Miq. = ***Syzygium klampok*** (Miq.) Amshoff
Jambosa koenigii Blume = ***Syzygium zeylanicum*** (L.) DC.
Jambosa korthalsii Blume = ***Syzygium confusum*** (Blume) Bakh.f.
Jambosa kurzii (Duthie) A.M.Cowan & Cowan = ***Syzygium kurzii*** (Duthie) N.P.Balakr.
Jambosa laeta (Buch.-Ham.) Blume = ***Syzygium laetum*** (Buch.-Ham.) Gandhi
Jambosa laevis Montrouz. = ***Syzygium laeve*** (Montrouz.) Govaerts
Jambosa lagynocalyx Diels = ***Syzygium trachyanthum*** (Diels) Merr. & L.M.Perry
Jambosa lanceolaria (Roxb.) Blume = ***Syzygium lanceolarium*** (Roxb.) N.P.Balakr.
Jambosa lanceolata Korth. ex Miq. = ***Syzygium confusum*** (Blume) Bakh.f.
Jambosa lancifolia Miq. = ***Syzygium insigne*** (Blume) Merr. & L.M.Perry
Jambosa latifolia (Blume) Miq. = ***Syzygium lineatum*** (DC.) Merr. & L.M.Perry
Jambosa laurifolia DC. = ***Syzygium laurifolium*** (DC.) N.P.Balakr.
Jambosa leonhardii Diels = ***Syzygium leonhardii*** (Diels) Merr. & L.M.Perry
Jambosa leptostachya Blume = ***Syzygium leptostachyum*** (Blume) Merr. & L.M.Perry
Jambosa leptostemon Korth. = ***Syzygium leptostemon*** (Korth.) Merr. & L.M.Perry
Jambosa linearis Korth. = ***Syzygium lineare*** (Korth.) Masam.
Jambosa lineata DC. = ***Syzygium lineatum*** (DC.) Merr. & L.M.Perry
Jambosa littoralis Blume = ***Syzygium littorale*** (Blume) Amshoff
Jambosa longifolia Brongn. & Gris = ***Syzygium longifolium*** (Brongn. & Gris) J.W.Dawson
Jambosa longipedicellata Merr. = ***Syzygium longipedicellatum*** (Merr.) Merr.
Jambosa luzonensis Merr. = ***Syzygium luzonense*** (Merr.) Merr.
Jambosa macrocarpa Sweet = ***Syzygium megacarpum*** (Craib) Rathakr. & N.C.Nair
Jambosa macrophylla (Lam.) DC. = ***Syzygium malaccense*** (L.) Merr. & L.M.Perry
Jambosa madagascariensis Blume = ***Syzygium aqueum*** (Burm.f.) Alston
Jambosa malaccensis (L.) DC. = ***Syzygium malaccense*** (L.) Merr. & L.M.Perry
Jambosa malaccensis f. *cericarpa* O.Deg. = ***Syzygium jambos*** (L.) Alston
Jambosa mappacea Korth. = ***Syzygium formosum*** (Wall.) Masam.
Jambosa marginata (Blume) Miq. = ***Syzygium glabratum*** (DC.) Veldkamp
Jambosa maritima Miq. = [42 JAW]
Jambosa media Korth. = ***Syzygium medium*** (Korth.) Merr. & L.M.Perry
Jambosa megalosperma K.Schum. & Lauterb. = ***Syzygium megalospermum*** (K.Schum. & Lauterb.) Merr. & L.M.Perry
Jambosa melanocarpa Miq. = ***Syzygium claviflorum*** (Roxb.) Wall. ex A.M.Cowan & Cowan
Jambosa melanosticta Miq. = ***Syzygium melanostictum*** (Miq.) Craven & Biffin
Jambosa melastomifolia Blume = ?
Jambosa micrantha Rech. = ***Syzygium rechingeri*** Merr. & L.M.Perry
Jambosa mundagam (Bourd.) Gamble = ***Syzygium mundagam*** (Bourd.) Chithra
Jambosa munronii (Wight) Walp. = ***Syzygium munronii*** (Wight) N.P.Balakr.
Jambosa myrtifolia Heynh. = ***Syzygium australe*** (J.C.Wendl. ex Link) B.Hyland
Jambosa naiadum Diels = ***Syzygium naiadum*** (Diels) Merr. & L.M.Perry
Jambosa neriifolia Brongn. & Gris = ***Syzygium schistaceum*** J.W.Dawson
Jambosa nervosa Vieill. ex Guillaumin = ***Syzygium guillauminii*** J.W.Dawson
Jambosa neurocalyx (A.Gray) Müll.Stuttg. = ***Syzygium neurocalyx*** (A.Gray) Christoph.
Jambosa nitida Korth. = ***Syzygium operculatum*** (Roxb.) Nied.
Jambosa nutans (K.Schum.) Nied. = ***Syzygium nutans*** Merr. & L.M.Perry
Jambosa obtusissima (Blume) DC. = ***Syzygium aqueum*** (Burm.f.) Alston
Jambosa occidentalis (Bourd.) Gamble = ***Syzygium occidentale*** (Bourd.) D.N.Gandhi
Jambosa ovalifolia Blume = ***Syzygium ovalifolium*** (Blume) Merr. & L.M.Perry
Jambosa owariensis (P.Beauv.) DC. = ***Syzygium owariense*** (P.Beauv.) Benth.
Jambosa pachyantha Diels = ***Syzygium pachyanthum*** (Diels) Merr. & L.M.Perry

Jambosa pachyclada K.Schum. & Lauterb. = ***Syzygium pachycladum*** (K.Schum. & Lauterb.) Merr. & L.M.Perry
Jambosa palembanica Blume = ***Syzygium jambos*** (L.) Alston
Jambosa papuana (Lauterb.) Diels = ***Pilidiostigma papuanum*** (Lauterb.) A.J.Scott
Jambosa pauciflora Walp. = ***Syzygium laetum*** (Buch.-Ham.) Gandhi subsp. ***laetum***
Jambosa peregrina Blume = ***Syzygium peregrinum*** (Blume) Merr. & L.M.Perry
Jambosa perforata Miq. = [42 SUM]
Jambosa phacelantha Diels = ***Syzygium phacelanthum*** (Diels) Merr. & L.M.Perry
Jambosa pilgeriana K.Schum. & Lauterb. = ***Syzygium pilgerianum*** (K.Schum. & Lauterb.) Merr. & L.M.Perry
Jambosa platycarpa Diels = ***Syzygium platycarpum*** (Diels) Merr. & L.M.Perry
Jambosa polycephala (Miq.) Miq. = ***Syzygium polycephalum*** (Miq.) Merr. & L.M.Perry
Jambosa polyneura Miq. = [42 JAW]
Jambosa polypetala Wall. = ***Syzygium polypetalum*** (Wall.) Merr. & L.M.Perry
Jambosa polyphlebia Diels = ***Syzygium polyphlebium*** (Diels) Merr. & L.M.Perry
Jambosa praecox (Roxb.) A.M.Cowan & Cowan = ***Syzygium praecox*** (Roxb.) Rathakr. & N.C.Nair
Jambosa pseudodensiflora Hochr. = [42 JAW]
Jambosa pseudolaeta C.E.C.Fisch. = ***Syzygium pseudolaetum*** (C.E.C.Fisch.) Merr. & L.M.Perry
Jambosa pseudomalaccensis Vieill. ex Brongn. & Gris = ***Syzygium pseudomalaccense*** (Vieill. ex Brongn. & Gris) Govaerts
Jambosa pterocarpa (Vieill. ex Pancher & Sebert) Nied. = ***Syzygium pterocarpum*** (Vieill. ex Pancher & Sebert) Govaerts
Jambosa pterocaulis Korth. = [42 BOR]
Jambosa pteropoda K.Schum. & Lauterb. = ***Syzygium pteropodum*** (K.Schum. & Lauterb.) Merr. & L.M.Perry
Jambosa pulchella Miq. = ***Syzygium oblatum*** (Roxb.) Wall. ex A.M.Cowan & Cowan
Jambosa puncticulata Miq. = [42 SUM]
Jambosa purpurascens DC. = ***Syzygium malaccense*** (L.) Merr. & L.M.Perry
Jambosa purpurea Korth. = [42 SUM]
Jambosa purpurea (Roxb.) Wight & Arn. = ***Syzygium malaccense*** (L.) Merr. & L.M.Perry
Jambosa pycnantha Diels = ***Syzygium dielsianum*** Merr. & L.M.Perry
Jambosa quadrangulata (A.Gray) Müll.Stuttg. = ***Syzygium quadrangulatum*** (A.Gray) Merr. & L.M.Perry
Jambosa rama-varmae (Bourd.) Gamble = ***Syzygium rama-varmae*** (Bourd.) Chithra
Jambosa ramosissima (Blume) A.M.Cowan & Cowan = ***Syzygium ramosissimum*** (Blume) N.P.Balakr.
Jambosa recurvovenosa Lauterb. = ***Syzygium recurvovenosum*** (Lauterb.) Diels
Jambosa reticulata (Wight) Blume = ***Syzygium reticulatum*** (Wight) Walp.
Jambosa rhyncophylla Miq. = [42 JAW]
Jambosa rhytidocarpa Zoll. = ?
Jambosa richii (A.Gray) Müll.Stuttg. = ***Syzygium richii*** (A.Gray) Merr. & L.M.Perry
Jambosa riparia Diels = ***Syzygium riparium*** (Diels) Merr. & L.M.Perry
Jambosa roemeri Lauterb. = ***Syzygium roemeri*** (Lauterb.) Merr. & L.M.Perry
Jambosa rostrata Miq. = [42 SUM]
Jambosa rotundifolia (Arn.) G.Don = ***Syzygium rotundifolium*** Arn.
Jambosa rubella Rech. = ***Syzygium roemeri*** (Lauterb.) Merr. & L.M.Perry
Jambosa rubens (Roxb.) Blume = ***Syzygium rubens*** (Roxb.) Walp.
Jambosa rubricaule Miq. = ***Syzygium lineatum*** (DC.) Merr. & L.M.Perry
Jambosa rufotomentosa Gibbs = ***Syzygium hirtum*** (Korth.) Merr. & L.M.Perry
Jambosa sabangensis Lauterb. = ***Syzygium sabangense*** (Lauterb.) Merr. & L.M.Perry
Jambosa salicina Diels = [43 NWG]
Jambosa saligna Miq. = ***Syzygium salignum*** (Miq.) Rathakr. & N.C.Nair
Jambosa samarangensis (Blume) DC. = ***Syzygium samarangense*** (Blume) Merr. & L.M.Perry
Jambosa sargentiana Diels = ***Syzygium coalitum*** (Greves) T.G.Hartley & L.M.Perry
Jambosa schumanniana Nied. = ***Syzygium schumannianum*** (Nied.) Diels
Jambosa sexangulata Miq. = ***Syzygium sexangulatum*** (Miq.) Amshoff
Jambosa soliflora Diels = ***Syzygium subalatum*** (Ridl.) Merr. & L.M.Perry
Jambosa splendens (Blume) Miq. = ***Syzygium splendens*** (Blume) Merr. & L.M.Perry
Jambosa stelechantha Diels = ***Syzygium stelechanthum*** (Diels) Glassman
Jambosa stelechanthoides (Kaneh.) Hosok. = ***Syzygium stelechanthum*** (Diels) Glassman
Jambosa subsessilis Miq. = ***Syzygium aqueum*** (Burm.f.) Alston
Jambosa sulcata Teijsm. & Binn. = ?
Jambosa sumatrana Miq. = [42 SUM]
Jambosa suzukii (Kaneh.) Hosok. = ?
Jambosa symphytocarpa (Blume) Korth. ex Miq. = ***Syzygium lineatum*** (DC.) Merr. & L.M.Perry
Jambosa synaptoneura K.Schum. & Lauterb. = ***Syzygium synaptoneurum*** (K.Schum. & Lauterb.) Merr. & L.M.Perry
Jambosa syzygioides Miq. = ***Syzygium syzygioides*** (Miq.) Merr. & L.M.Perry
Jambosa tawahensis Korth. = ***Syzygium tawahense*** (Korth.) Merr. & L.M.Perry
Jambosa tenuicuspis Miq. = ***Syzygium rostratum*** (Blume) DC.
Jambosa tenuifolia Sweet = ***Syzygium formosum*** (Wall.) Masam.
Jambosa tenuiramis Miq. = ***Syzygium tenuirame*** (Miq.) Merr.
Jambosa tetraedra (Duthie) Miq. = ?
Jambosa tetragona Blume = ***Syzygium umbilicatum*** (Koord. & Valeton) Amshoff
Jambosa tetraptera Miq. = ***Syzygium tetrapterum*** (Miq.) Chantaran. & J.Parn.
Jambosa tetraquetra Miq. = ***Syzygium minimum*** (Blume) Airy Shaw
Jambosa teysmannii Miq. = ***Syzygium lineatum*** (DC.) Merr. & L.M.Perry
Jambosa thomsenii Diels = ***Syzygium thomsenii*** (Diels) Merr. & L.M.Perry
Jambosa thozetiana F.Muell. = ***Syzygium australe*** (J.C.Wendl. ex Link) B.Hyland
Jambosa tierneyana (F.Muell.) Diels = ***Syzygium tierneyanum*** (F.Muell.) T.G.Hartley & L.M.Perry
Jambosa timorensis Blume = ***Syzygium aqueum*** (Burm.f.) Alston

Jambosa trachyantha Diels = ***Syzygium trachyanthum*** (Diels) Merr. & L.M.Perry
Jambosa tricolor Diels = ***Syzygium tricolor*** (Diels) Merr. & L.M.Perry
Jambosa trifida (Blume) Miq. = ***Syzygium confertum*** (Korth.) Merr. & L.M.Perry
Jambosa trukensis (Hosok.) Hosok. = ?
Jambosa tympanantha Diels = ***Syzygium tympananthum*** (Diels) Merr. & L.M.Perry
Jambosa urceolata Korth. ex Miq. = ***Syzygium leptostemon*** (Korth.) Merr. & L.M.Perry
Jambosa venosa (Lam.) DC. = ***Syzygium dupontii*** (Baker) Govaerts
Jambosa verniciflora Diels = ***Syzygium verniciflorum*** (Diels) Merr. & L.M.Perry
Jambosa versteegii Lauterb. = ***Syzygium versteegii*** (Lauterb.) Merr. & L.M.Perry
Jambosa virens (Blume) Miq. = ***Syzygium glabratum*** (DC.) Veldkamp
Jambosa vrieseana Miq. = ***Syzygium vrieseanum*** (Miq.) Amshoff
Jambosa vulgaris DC. = ***Syzygium jambos*** (L.) Alston
Jambosa weinlandii K.Schum. = ***Syzygium hylophilum*** (K.Schum. & Lauterb.) Merr. & L.M.Perry
Jambosa wightiana Blume = ***Syzygium rubens*** (Roxb.) Walp.
Jambosa xylopiacea Diels = ***Syzygium xylopiaceum*** (Diels) Merr. & L.M.Perry
Jambosa zollingeriana Miq. = ***Syzygium zollingerianum*** (Miq.) Amshoff

Unplaced Names:
Jambosa acida auct., Gard. Chron. 1876(1): 735 (1876). = ?
Jambosa anastomosans Miq., Fl. Ned. Ind., Eerste Bijv.: 309 (1861). = [42 SUM]
Jambosa bisulca Miq., Fl. Ned. Ind., Eerste Bijv.: 310 (1861). = [42 SUM]
Jambosa brackenridgei Brongn. & Gris, Bull. Soc. Bot. France 12: 181 (1865). = [60 NWC]
Jambosa celebica Blume, Mus. Bot. 1: 107 (1850). = [42 SUL]
Jambosa coarctata var. *firma* Blume, Mus. Bot. 1: 100 (1850). = [42 SUM]
Jambosa condensata Miq., Fl. Ned. Ind. 1(1): 419 (1855). = [42 JAW]
Jambosa costata Miq., Fl. Ned. Ind. 1(1): 415 (1855). = [42 JAW]
Jambosa courtallensis Gamble, Bull. Misc. Inform. Kew 1918: 239 (1918). = [40 IND SRL]
Jambosa glandulosa Korth., Ned. Kruidk. Arch. 1: 201 (1847). = [42 BOR]
Jambosa hardwickia G.Don, Gen. Hist. 2: 869 (1832). = [29 MAU]
Jambosa horsfieldii Miq., Fl. Ned. Ind. 1(1): 420 (1855). = [42 JAW]
Jambosa kigurrung Blume, Mus. Bot. 1: 107 (1850). = [42 JAW]
Jambosa maritima Miq., Fl. Ned. Ind. 1(1): 435 (1855). = [42 JAW]
Jambosa melastomifolia Blume, Mus. Bot. 1: 102 (1850). = ?
Jambosa perforata Miq., Fl. Ned. Ind., Eerste Bijv.: 309 (1861). = [42 SUM]
Jambosa polyneura Miq., Fl. Ned. Ind., Eerste Bijv.: 415 (1861). = [42 JAW]
Jambosa pseudodensiflora Hochr., Candollea 2: 452 (1925). = [42 JAW]
Jambosa pterocaulis Korth., Ned. Kruidk. Arch. 1: 200 (1847). = [42 BOR]
Jambosa puncticulata Miq., Fl. Ned. Ind., Eerste Bijv.: 310 (1861). = [42 SUM]
Jambosa purpurea Korth., Ned. Kruidk. Arch. 1: 201 (1847). = [42 SUM]
Jambosa rhyncophylla Miq., Fl. Ned. Ind. 1(1): 430 (1855). = [42 JAW]
Jambosa rhytidocarpa Zoll. in F.a.W.Miquel, Choix Pl. Buitenzorg: t. 4 (1863). = ?
Jambosa rostrata Miq., Fl. Ned. Ind. 1(1): 436 (1855). = [42 SUM]
Jambosa salicina Diels, Bot. Jahrb. Syst. 57: 390 (1922). = [43 NWG]
Jambosa sulcata Teijsm. & Binn., Natuurk. Tijdschr. Ned.-Indië 27: 54 (1864). = ?
Jambosa sumatrana Miq., Fl. Ned. Ind. 1(1): 419 (1855). = [42 SUM]

Jossinia

Jossinia Comm. ex DC. = ***Eugenia*** P.Micheli ex L.
Jossinia aherniana (C.B.Rob.) Merr. = ***Eugenia aherniana*** C.B.Rob.
Jossinia brachypoda Merr. = [42 PHI]
Jossinia bryanii (Kaneh.) Hosok. = ***Eugenia bryanii*** Kaneh.
Jossinia buxifolia (Lam.) DC. = ***Eugenia buxifolia*** Lam.
Jossinia buxifolia var. *microphylla* Blume = ***Eugenia buxifolia*** Lam.
Jossinia cassinoides (Lam.) DC. = ***Eugenia cassinoides*** Lam.
Jossinia celebica Blume = [42 SUL]
Jossinia cordifolia Bojer = ***Eugenia elliptica*** Lam.
Jossinia costenoblei (Merr.) Diels = ***Eugenia costenoblei*** Merr.
Jossinia cotinifolia (Jacq.) DC. = ***Eugenia cotinifolia*** Jacq.
Jossinia desmantha Diels = [43 NWG]
Jossinia elliptica DC. = ***Eugenia elliptica*** Lam.
Jossinia ferruginea Bojer = ***Eugenia tinifolia*** Lam.
Jossinia glomerata (Lam.) Blume = ***Syzygium glomeratum*** (Lam.) DC.
Jossinia hastilis Blume = ***Eugenia hastilis*** (Blume) J.Guého & A.J.Scott
Jossinia heterophylla (Merr.) Merr. = ?
Jossinia indica Wight = ***Eugenia indica*** (Wight) Chithra
Jossinia kamelii (Merr.) Merr. = ***Eugenia kamelii*** Merr.
Jossinia lamarckii Blume = ***Eugenia lucida*** Lam.
Jossinia lanceolata Bojer = ***Eugenia elliptica*** Lam.
Jossinia littoralis Blume = [43 NWG]
Jossinia loheri (C.B.Rob.) Merr. = ***Eugenia loheri*** C.B.Rob.
Jossinia lucida (Lam.) DC. = ***Eugenia lucida*** Lam.
Jossinia mespiloides (Lam.) DC. = ***Eugenia mespiloides*** Lam.
Jossinia montalbanica (Merr.) Merr. = ***Eugenia montalbanica*** Merr.
Jossinia orbiculata (Lam.) DC. = ***Eugenia orbiculata*** Lam.
Jossinia palumbis (Merr.) Diels = ***Eugenia palumbis*** Merr.
Jossinia pasacaensis (C.B.Rob.) Merr. = ***Eugenia pasacaensis*** C.B.Rob.
Jossinia pinnata (L.) Heynh. = ***Plinia pinnata*** L.
Jossinia reinwardtiana (Blume) Blume = ***Eugenia reinwardtiana*** (Blume) A.Cunn. ex DC.
Jossinia revoluta Bojer = ***Eugenia tinifolia*** Lam.
Jossinia richardii Blume = [29 MDG]
Jossinia sargentii (Merr.) Merr. = ***Eugenia sargentii*** Merr.
Jossinia schlechteri Diels = [43 NWG]

Jossinia sieberiana Blume = ***Eugenia sieberi*** J.Guého & A.J.Scott
Jossinia tahitensis Nadeaud = ***Eugenia reinwardtiana*** (Blume) A.Cunn. ex DC.
Jossinia terminalis Bojer = ***Eugenia orbiculata*** Lam.
Jossinia tinifolia (Lam.) DC. = ***Eugenia tinifolia*** Lam.
Jossinia tulanan (Merr.) Merr. = ***Eugenia tulanan*** Merr.

***Unplaced Names*:**
Jossinia brachypoda Merr., Philipp. J. Sci. 79: 3359 (1951). = [42 PHI]
Jossinia celebica Blume, Mus. Bot. 1: 124 (1850). = [42 SUL]
Jossinia desmantha Diels, J. Arnold Arbor. 10: 82 (1929). = [43 NWG]
Jossinia littoralis Blume, Mus. Bot. 1: 124 (1850). = [43 NWG]
Jossinia richardii Blume, Mus. Bot. 1: 123 (1850). = [29 MDG]
Jossinia schlechteri Diels, Bot. Jahrb. Syst. 57: 377 (1922). = [43 NWG]

Jungia

Jungia Gaertn. = ***Baeckea*** L.
Jungia imbricata Gaertn. = ***Baeckea imbricata*** (Gaertn.) Druce
Jungia tenella Gaertn. = ***Baeckea imbricata*** (Gaertn.) Druce

Kajuputi

Kajuputi Adans. = ***Melaleuca*** L.

Kamptzia

Kamptzia Nees = ***Syncarpia*** Ten.
Kamptzia albens (A.Cunn. ex DC.) Nees = ***Syncarpia glomulifera*** (Sm.) Nied. subsp. ***glomulifera***

Kania

Kania Schltr., Bot. Jahrb. Syst. 52: 118 (1914).
Philippines, New Guinea. 42 PHI 43 NWG.
6 Species

Kania eugenioides Schltr., Bot. Jahrb. Syst. 52: 120 (1914). *Metrosideros eugenioides* (Schltr.) Steenis, Blumea 16: 358 (1969).
New Guinea. 43 NWG. Nanophan. or phan.
Backhousia aurea Ridl., Trans. Linn. Soc. London, Bot. 9: 43 (1916). *Metrosideros aurea* (Ridl.) Diels, Bot. Jahrb. Syst. 57: 418 (1922).
Metrosideros pullei Diels, Bot. Jahrb. Syst. 57: 417 (1922).
Metrosideros parviflora C.T.White, J. Arnold Arbor. 23: 80 (1942).

Kania hirsutula (F.Muell.) A.J.Scott, Kew Bull. 45: 206 (1990).
New Guinea. 43 NWG. Nanophan. or phan.
**Myrtella hirsutula* F.Muell., Descr. Notes Papuan Pl. 1: 106 (1877). *Fenzlia hirsutula* (F.Muell.) Burret, Notizbl. Bot. Gart. Berlin-Dahlem 15: 500 (1941).
Backhousia arfakensis Gibbs, Fl. Arfak Mts.: 153 (1917). *Metrosideros gibbsiae* Diels, Bot. Jahrb. Syst. 57: 418 (1922).

Kania microphylla (Quisumb. & Merr.) Peter G.Wilson, Blumea 28: 179 (1982).
Philippines. 42 PHI. Nanophan. or phan.
**Tristania microphylla* Quisumb. & Merr., Philipp. J. Sci. 37: 176 (1928).

Kania nettotensis A.J.Scott, Kew Bull. 38: 309 (1983).
W. New Guinea. 43 NWG. Nanophan. or phan.

Kania platyphylla A.J.Scott, Kew Bull. 38: 309 (1983).
W. New Guinea. 43 NWG. Nanophan. or phan.

Kania urdanetensis (Elmer) Peter G.Wilson, Blumea 28: 179 (1982).
Philippines (Mindanao). 42 PHI. Nanophan. or phan.
Cloezia urdanetensis (Elmer) Merr., Philipp. J. Sci. 14: 429 (1919).
Mooria urdanetensis (Elmer) Merr., Enum. Philipp. Fl. Pl. 3: 182 (1923).

Kjellbergiodendron

Kjellbergiodendron Burret, Notizbl. Bot. Gart. Berlin-Dahlem 13: 101 (1936).
C. Malesia. 42 SUL.
1 Species

Kjellbergiodendron celebicum (Koord.) Merr., J. Arnold Arbor. 33: 162 (1952).
Sulawesi. 42 SUL. Nanophan. or phan.
**Xanthostemon celebicus* Koord., Meded. Lands Plantentuin 19: 637 (1898).
Tristania celebica Koord.-Schum., Syst. Verz. 3: 96 (1914), nom. nud.
Tristania anacardiifolia Ridl., J. Bot. 68: 39 (1930).
Kjellbergiodendron hylogeiton Burret, Notizbl. Bot. Gart. Berlin-Dahlem 13: 102 (1936).
Kjellbergiodendron limnogeiton Burret, Notizbl. Bot. Gart. Berlin-Dahlem 13: 102 (1936).

***Synonyms*:**
Kjellbergiodendron hylogeiton Burret = ***Kjellbergiodendron celebicum*** (Koord.) Merr.
Kjellbergiodendron limnogeiton Burret = ***Kjellbergiodendron celebicum*** (Koord.) Merr.

Krokia

Krokia Urb. = ***Pimenta*** Lindl.
Krokia adenoclada (Urb.) Borhidi & O.Muñiz = ***Pimenta adenoclada*** (Urb.) Alain
Krokia cainitoides (Urb.) Urb. = ***Pimenta cainitoides*** (Urb.) Burret
Krokia ferruginea (Griseb.) Urb. = ***Pimenta ferruginea*** (Griseb.) Burret
Krokia filipes (Urb.) Borhidi & O.Muñiz = ***Pimenta filipes*** (Urb.) Burret
Krokia leonis Borhidi & O.Muñiz = ***Pimenta filipes*** (Urb.) Burret
Krokia moaensis (Areces) Borhidi & O.Muñiz = ***Pimenta filipes*** (Urb.) Burret
Krokia nipensis Urb. = ***Pimenta cainitoides*** (Urb.) Burret
Krokia odiolens (Urb.) Borhidi & O.Muñiz = ***Pimenta odiolens*** (Urb.) Burret
Krokia oligantha (Urb.) Borhidi & O.Muñiz = ***Pimenta oligantha*** (Urb.) Burret
Krokia pilotoana (Urb.) Borhidi = ***Pimenta cainitoides*** (Urb.) Burret
Krokia podocarpoides (Areces) Borhidi & O.Muñiz = ***Pimenta podocarpoides*** (Areces) Landrum

Krugia

Krugia Urb. = ***Marlierea*** Cambess.
Krugia elliptica (Griseb.) Urb. = ***Marlierea ferruginea*** (Poir.) McVaugh
Krugia ferruginea (Poir.) Urb. = ***Marlierea ferruginea*** (Poir.) McVaugh

Kunzea

Kunzea Rchb., Consp. Regn. Veg.: 175 (1828).
Australia, New Zealand. 50 NSW QLD SOA TAS VIC WAU 51 NZN NZS.
38 Species
Tillospermum Salisb., Griffiths, Monthly Rev. 75: 74 (1814).
Pentagonaster Klotzsch, Allg. Gartenzeitung 4: 112 (1836).
Salisia Lindl., Sketch Veg. Swan R.: 10 (1839).
Stenospermum Sweet ex Heynh., Nom. Bot. Hort. 1: 787 (1840).

Kunzea acuminata Toelken, J. Adelaide Bot. Gard. 17: 88 (1996).
S. Western Australia. 50 WAU. Nanophan.

Kunzea affinis S.Moore, J. Linn. Soc., Bot. 45: 202 (1920).
S. Western Australia. 50 WAU. Nanophan.

Kunzea ambigua (Sm.) Druce, Rep. Bot. Exch. Club Brit. Isles 1916: 629 (1917).
SE. Australia. 50 NSW TAS VIC. Nanophan.
**Leptospermum ambiguum* Sm., Trans. Linn. Soc. London 3: 264 (1797).
Metrosideros corifolia Vent., Jard. Malmaison: 46 (1804). *Stenospermum corifolium* (Vent.) Sweet ex Heynh., Nom. Bot. Hort. 1: 787 (1840). *Kunzea corifolia* (Vent.) Schauer in J.G.C.Lehmann, Pl. Preiss. 1: 124 (1844).
Metrosideros abietina Hoffmanns., Verz. Pfl.-Kult.: 80 (1824).
Kunzea pelagia F.Muell. ex Miq., Ned. Kruidk. Arch. 4: 145 (1856).

Kunzea baxteri (Klotzsch) Schauer in J.G.C.Lehmann, Pl. Preiss. 1: 123 (1844).
S. Western Australia. 50 WAU. Nanophan.
Calothamnus baxteri Klotzsch, Allg. Gartenzeitung 4: 115 (1836).
**Pentagonaster baxteri* Klotzsch, Allg. Gartenzeitung 4: 115 (1836).
Callistemon microstachyus Lindl., Edwards's Bot. Reg. 24: t. 7 (1838).
Calothamnus spathulatus Steud., Nomencl. Bot., ed. 2, 1: 261 (1840), nom. inval.
Callistemon hainesii F.Muell., Fragm. 3: 153 (1863).

Kunzea bracteolata Maiden & Betche, Proc. Linn. Soc. New South Wales 30: 363 (1905).
SE. Queensland to NE. New South Wales. 50 NSW QLD. Nanophan.

Kunzea calida F.Muell., Fragm. 6: 23 (1868).
Queensland. 50 QLD. Nanophan.

Kunzea cambagei Maiden & Betche, Proc. Linn. Soc. New South Wales 38: 246 (1913).
EC. New South Wales. 50 NSW. Nanophan.

Kunzea capitata (Sm.) Heynh., Alph. Aufz. Gew.: 337 (1846).
SE. Queensland to New South Wales. 50 NSW QLD. Nanophan.
**Metrosideros capitata* Sm., Trans. Linn. Soc. London 3: 273 (1797). *Callistemon capitatus* (Sm.) Rchb., Iconogr. Bot. Exot. 1: 59 (1825). *Stenospermum capitatum* (Sm.) Sweet ex Heynh., Nom. Bot. Hort. 1: 787 (1840). *Kunzea schaueri* Schauer in J.G.C.Lehmann, Pl. Preiss. 1: 124 (1844), nom. illeg.
Melaleuca eriocephala Sieber ex Spreng., Syst. Veg. 3: 336 (1826).
Kunzea hirsuta Turcz., Bull. Soc. Imp. Naturalistes Moscou 35(2): 326 (1862).
Kunzea capitata var. *glabrescens* Benth., Fl. Austral. 3: 116 (1867).

Kunzea ciliata Toelken, J. Adelaide Bot. Gard. 17: 64 (1996).
SW. Australia. 50 WAU. Cham. or nanophan.

Kunzea cincinnata Toelken, J. Adelaide Bot. Gard. 17: 101 (1996).
S. Western Australia. 50 WAU. Nanophan.

Kunzea clavata Toelken, J. Adelaide Bot. Gard. 17: 50 (1996).
SW. Australia. 50 WAU. Nanophan. or phan.

Kunzea ericifolia (Sm.) Heynh., Alph. Aufz. Gew.: 338 (1846).
SW. Australia. 50 WAU. Nanophan.
**Metrosideros ericifolia* Sm. in A.Rees, Cycl. 23: 16 (1813). *Stenospermum ericifolium* (Sm.) Heynh., Nom. Bot. Hort. 1: 787 (1840).

subsp. ***ericifolia***
SW. Australia. 50 WAU. Nanophan.
Metrosideros propinqua Endl. in S.L.Endlicher & al., Enum. Pl.: 50 (1837), nom. illeg. *Kunzea propinqua* Schauer in J.G.C.Lehmann, Pl. Preiss. 1: 126 (1844).
Kunzea vestita Schauer in J.G.C.Lehmann, Pl. Preiss. 1: 126 (1844).

subsp. ***subulata*** Toelken, J. Adelaide Bot. Gard. 17: 47 (1996).
SW. Australia. 50 WAU. Nanophan.

Kunzea ericoides (A.Rich.) Joy Thomps., Telopea 2: 379 (1983).
E. & SE. Australia, New Zealand. 50 NSW QLD VIC 51 NZN NZS. Nanophan. or phan.
**Leptospermum ericoides* A.Rich. in J.S.C.Dumont d'Urville, Voy. Astrolabe 1: 338 (1832).
Baeckea phylicoides A.Cunn. ex Schauer in W.G.Walpers, Repert. Bot. Syst. 2: 921 (1843). *Kunzea phylicioides* (A.Cunn. ex Schauer) Druce, Rep. Bot. Exch. Club Brit. Isles 1916: 629 (1917). *Leptospermum phylicoideum* (A.Cunn. ex Schauer) Cheel, J. Proc. Roy. Soc. New S. Wales 76: 231 (1943).
Kunzea peduncularis F.Muell., Defin. Austral. Pl.: 44 (1855).
Kunzea leptospermoides F.Muell., Ned. Kruidk. Arch. 4: 146 (1856).
Kunzea peduncularis var. *brachyandra* Benth., Fl. Austral. 3: 115 (1867).
Leptospermum ericoides var. *lineare* Kirk, Forest Fl. New Zealand: 125 (1889). *Kunzea ericoides* var. *linearis* (Kirk) W.Harris, New Zealand J. Bot. 25: 134 (1987).
Baeckea virgata var. *polyandra* Maiden & Betche, Proc. Linn. Soc. New South Wales 23: 12 (1898).
Kunzea glabriuscula Gand., Bull. Soc. Bot. France 65: 26 (1918).
Leptospermum ericoides var. *microflorum* G.Simpson, Trans. & Proc. Roy. Soc. New Zealand 75: 189 (1945). *Kunzea ericoides* var. *microflora* (G. Simpson) W.Harris, New Zealand J. Bot. 25: 134 (1987).

Kunzea eriocalyx F.Muell., Fragm. 2: 28 (1860).
S. Western Australia. 50 WAU. Nanophan.

Kunzea flavescens C.T.White & W.D.Francis, Proc. Roy. Soc. Queensland 33: 155 (1921).
Queensland. 50 QLD. Nanophan.

Kunzea glabrescens Toelken, J. Adelaide Bot. Gard. 17: 41 (1996).
SW. Australia. 50 WAU. Nanophan. or phan.
Kunzea ericifolia var. *glabrior* Benth., Fl. Austral. 3: 113 (1867).

Kunzea graniticola Byrnes, Austrobaileya 1: 468 (1982 publ. 1983).
NE. Queensland. 50 QLD. Nanophan.

Kunzea jucunda Diels & E.Pritz., Bot. Jahrb. Syst. 35: 424 (1904).
S. Western Australia. 50 WAU. Nanophan.

Kunzea micrantha Schauer in J.G.C.Lehmann, Pl. Preiss. 1: 125 (1844).
SW. Australia. 50 WAU. Cham. or nanophan.

subsp. ***hirtiflora*** Toelken, J. Adelaide Bot. Gard. 17: 77 (1996).
SW. Australia. 50 WAU. Cham. or nanophan.

subsp. ***micrantha***
SW. Australia. 50 WAU. Cham. or nanophan.

subsp. ***oligandra*** (Turcz.) Toelken, J. Adelaide Bot. Gard. 17: 75 (1996).
SW. Australia. 50 WAU. Cham. or nanophan.
**Kunzea oligandra* Turcz., Bull. Cl. Phys.-Math. Acad. Imp. Sci. Saint-Pétersbourg 10: 336 (1852).

subsp. ***petiolata*** Toelken, J. Adelaide Bot. Gard. 17: 73 (1996).
WSW. Western Australia. 50 WAU. Cham. or nanophan.

Kunzea micromera Schauer in J.G.C.Lehmann, Pl. Preiss. 1: 223 (1845).
SW. Australia. 50 WAU. Cham. or nanophan.

Kunzea montana (Diels) Domin, Vestn. Král. Ceské Spolecn. Nauk, Tr. Mat.-Prír. 1921–1922(2): 87 (1923).
SW. Australia. 50 WAU. Nanophan.
**Kunzea recurva* var. *montana* Diels, Bot. Jahrb. Syst. 35: 424 (1904).

Kunzea muelleri Benth., Fl. Austral. 3: 113 (1867).
SE. New South Wales. 50 NSW. Nanophan.
**Kunzea ericifolia* F.Muell., Trans. Philos. Soc. Victoria 1: 123 (1855), nom. illeg.
Agonis ericoides F.M.Bailey, Bot. Bull. Dept. Agric. Queensland 2: 37 (1891).

Kunzea newbeyi Toelken, J. Adelaide Bot. Gard. 17: 60 (1996).
SW. Australia. 50 WAU. Nanophan.

Kunzea obovata Byrnes, Austrobaileya 1: 469 (1982 publ. 1983).
Queensland to New South Wales. 50 NSW QLD. Nanophan.

Kunzea opposita F.Muell., Fragm. 6: 24 (1868).
E. Australia. 50 NSW QLD. Nanophan.

var. ***leichhardtii*** Byrnes, Austrobaileya 1: 470 (1982 publ. 1983).
Queensland. 50 QLD. Nanophan.

var. ***opposita***
E. Australia. 50 NSW QLD. Nanophan.

Kunzea parvifolia Schauer in J.G.C.Lehmann, Pl. Preiss. 1: 123 (1844).
SE. Queensland to Victoria. 50 NSW QLD VIC. Nanophan.
Kunzea parvifolia var. *laba* Maiden & Betche, Proc. Linn. Soc. New South Wales 38: 246 (1913).

Kunzea pauciflora Schauer in J.G.C.Lehmann, Pl. Preiss. 1: 124 (1844).
S. Western Australia. 50 WAU. Cham. or nanophan.
Pericalymma teretifolium Turcz., Bull. Cl. Phys.-Math. Acad. Imp. Sci. Saint-Pétersbourg 10: 334 (1852).

Kunzea pomifera F.Muell., Trans. Philos. Soc. Victoria 1: 124 (1855).
South Australia to Victoria. 50 SOA VIC. Nanophan.

Kunzea praestans Schauer in J.G.C.Lehmann, Pl. Preiss. 1: 124 (1844). *Kunzea recurva* var. *praestans* (Schauer) Benth., Fl. Austral. 3: 114 (1867).
WSW. Western Australia. 50 WAU. Nanophan.

Kunzea preissiana Schauer in J.G.C.Lehmann, Pl. Preiss. 1: 125 (1844).
SW. Australia. 50 WAU. Nanophan.
Kunzea villiceps Schauer in J.G.C.Lehmann, Pl. Preiss. 1: 125 (1844). *Kunzea preissiana* var. *villiceps* (Schauer) Benth., Fl. Austral. 3: 114 (1867).
Kunzea villiceps var. *glabrior* Domin, Vestn. Král. Ceské Spolecn. Nauk, Tr. Mat.-Prír. 2(2): 87 (1923).

Kunzea pulchella (Lindl.) A.S.George, W. Austral. Naturalist 10: 32 (1966).
SW. Australia. 50 WAU. Nanophan.
**Salisia pulchella* Lindl., Sketch Veg. Swan R.: x (1839).
Kunzea sericea Turcz., Bull. Soc. Imp. Naturalistes Moscou 20(1): 162 (1847).
Kunzea sericea var. *albiflora* S.Moore, J. Linn. Soc., Bot. 34: 192 (1899).
Kunzea sericea var. *glabrata* C.A.Gardner, J. Proc. Roy. Soc. New S. Wales 9: 35 (1923).

Kunzea recurva Schauer in J.G.C.Lehmann, Pl. Preiss. 1: 125 (1844).
SW. Australia. 50 WAU. Nanophan.
Kunzea recurva var. *melaleucoides* Benth., Fl. Austral. 3: 114 (1867).
Kunzea spicata S.Moore, J. Linn. Soc., Bot. 45: 203 (1920).

Kunzea × rosea (Turcz.) Govaerts *K. juncea* × *K. preissiana*
S. Western Australia. 50 WAU. Nanophan.
**Pericalymma* × *roseum* Turcz., Bull. Cl. Phys.-Math. Acad. Imp. Sci. Saint-Pétersbourg 10: 334 (1852).

Kunzea rostrata Toelken, J. Adelaide Bot. Gard. 17: 62 (1996).
SW. Australia. 50 WAU. Nanophan.

Kunzea rupestris Blakely, Proc. Linn. Soc. New South Wales 54: 683 (1929).
E. New South Wales. 50 NSW. Nanophan.

Kunzea similis Toelken, J. Adelaide Bot. Gard. 17: 86 (1996).
S. Western Australia (near Hopetown). 50 WAU. Nanophan.

Kunzea sinclairii (Kirk) W.Harris, New Zealand J. Bot. 25: 134 (1987).

New Zealand North I. 51 NZN. Nanophan.
Leptospermum sinclairii Kirk, Stud. Fl. New Zealand: 158 (1899).

Kunzea spathulata Toelken, J. Adelaide Bot. Gard. 17: 48 (1996).
SW. Australia. 50 WAU. Nanophan. or phan.

Kunzea* × *squarrosa Turcz., Bull. Cl. Phys.-Math. Acad. Imp. Sci. Saint-Pétersbourg 10: 335 (1852). *K. montana* × *K. recurvata*.
SW. Australia. 50 WAU. Nanophan.

Kunzea sulphurea Tovey & P.Morris, Proc. Roy. Soc. Victoria, n.s., 35: 194 (1923).
SW. Australia. 50 WAU. Nanophan. or phan.

***Synonyms*:**
Kunzea brachyandra F.Muell. = ***Leptospermum brachyandrum*** (F.Muell.) Druce
Kunzea capitata var. *glabrescens* Benth. = ***Kunzea capitata*** (Sm.) Heynh.
Kunzea corifolia (Vent.) Schauer = ***Kunzea ambigua*** (Sm.) Druce
Kunzea ericifolia F.Muell. = ***Kunzea muelleri*** Benth.
Kunzea ericifolia var. *glabrior* Benth. = ***Kunzea glabrescens*** Toelken
Kunzea ericoides var. *linearis* (Kirk) W.Harris = ***Kunzea ericoides*** (A.Rich.) Joy Thomps.
Kunzea ericoides var. *microflora* (G.Simpson) W.Harris = ***Kunzea ericoides*** (A.Rich.) Joy Thomps.
Kunzea glabriuscula Gand. = ***Kunzea ericoides*** (A.Rich.) Joy Thomps.
Kunzea hirsuta Turcz. = ***Kunzea capitata*** (Sm.) Heynh.
Kunzea leptospermoides F.Muell. = ***Kunzea ericoides*** (A.Rich.) Joy Thomps.
Kunzea oligandra Turcz. = ***Kunzea micrantha*** subsp. ***oligandra*** (Turcz.) Toelken
Kunzea parvifolia var. *laba* Maiden & Betche = ***Kunzea parvifolia*** Schauer
Kunzea peduncularis F.Muell. = ***Kunzea ericoides*** (A.Rich.) Joy Thomps.
Kunzea peduncularis var. *brachyandra* Benth. = ***Kunzea ericoides*** (A.Rich.) Joy Thomps.
Kunzea pelagia F.Muell. ex Miq. = ***Kunzea ambigua*** (Sm.) Druce
Kunzea phylicioides (A.Cunn. ex Schauer) Druce = ***Kunzea ericoides*** (A.Rich.) Joy Thomps.
Kunzea podantha F.Muell. = ***Leptospermum oligandrum*** Turcz.
Kunzea preissiana var. *villiceps* (Schauer) Benth. = ***Kunzea preissiana*** Schauer
Kunzea propinqua Schauer = ***Kunzea ericifolia*** (Sm.) Heynh. subsp. ***ericifolia***
Kunzea recurva var. *melaleucoides* Benth. = ***Kunzea recurva*** Schauer
Kunzea recurva var. *montana* Diels = ***Kunzea montana*** (Diels) Domin
Kunzea recurva var. *praestans* (Schauer) Benth. = ***Kunzea praestans*** Schauer
Kunzea schaueri Schauer = ***Kunzea capitata*** (Sm.) Heynh.
Kunzea sericea Turcz. = ***Kunzea pulchella*** (Lindl.) A.S. George
Kunzea sericea var. *albiflora* S.Moore = ***Kunzea pulchella*** (Lindl.) A.S.George
Kunzea sericea var. *glabrata* C.A.Gardner = ***Kunzea pulchella*** (Lindl.) A.S.George
Kunzea spicata S.Moore = ***Kunzea recurva*** Schauer
Kunzea sprengelioides Turcz. = [50]
Kunzea vestita Schauer = ***Kunzea ericifolia*** (Sm.) Heynh. subsp. ***ericifolia***
Kunzea villiceps Schauer = ***Kunzea preissiana*** Schauer
Kunzea villiceps var. *glabrior* Domin = ***Kunzea preissiana*** Schauer

***Unplaced Names*:**
Kunzea sprengelioides Turcz., Bull. Cl. Phys.-Math. Acad. Imp. Sci. Saint-Pétersbourg 10: 336 (1852). = [50]

×*Kunzspermum*

×*Kunzspermum* W.Harris = *Kunzea* × *Leptospermum*
×*Kunzspermum hirakimata* W.Harris = *Kunzea sinclairii* × *Leptospermum scoparium*

***Unplaced Names*:**
×*Kunzspermum* W.Harris, Hort. New Zealand 4(2): 10 (1993). = *Kunzea* × *Leptospermum*
×*Kunzspermum hirakimata* W.Harris, Hort. New Zealand 4: 10 (1993). = *Kunzea sinclairii* × *Leptospermum scoparium*

Lacerdaea

Lacerdaea O.Berg = ***Campomanesia*** Ruiz & Pav.
Lacerdaea luschnathiana O.Berg = ***Campomanesia guazumifolia*** (Cambess.) O.Berg

Lamarchea

Lamarchea Gaudich., Voy. Uranie: 483 (1830).
Australia. 50 NTA WAU.
2 Species
Lamarkea Rchb., Consp. Regn. Veg.: 175 (1828), orth. var.

Lamarchea hakeifolia Gaudich., Voy. Uranie: 484 (1830).
Western Australia (Irwin). 50 WAU. Nanophan. or phan.

var. ***brevifolia*** A.S.George, Nuytsia 1: 272 (1972).
Western Australia (Irwin). 50 WAU. Nanophan.

var. ***hakeifolia***
Western Australia (Irwin). 50 WAU. Nanophan. or phan.

Lamarchea sulcata A.S.George, Nuytsia 1: 275 (1972).
Western Australia to Northern Territory. 50 NTA WAU. Nanophan.

Lamarkea

Lamarkea Rchb. = ***Lamarchea*** Gaudich.

Legrandia

Legrandia Kausel, Revista Argent. Agron. 11: 321 (1944).
Chile. 85 CLC.
1 Species

Legrandia concinna (Phil.) Kausel, Revista Argent. Agron. 11: 322 (1944).
C. Chile. 85 CLC. Nanophan. or phan.
**Eugenia concinna* Phil., Linnaea 28: 640 (1857). *Luma concinna* (Phil.) Herter, Revista Sudamer. Bot. 7: 218 (1943).

Lenwebbia

Lenwebbia N.Snow & Guymer, Syst. Bot. Monogr. 65: 25 (2003).
E. Australia. 50 NSW QLD.
2 Species

Lenwebbia lasioclada (F.Muell.) N.Snow & Guymer, Syst. Bot. Monogr. 65: 26 (2003).
NE. & SE. Queensland to NE. New South Wales. 50 NSW QLD. Nanophan. or phan.
**Myrtus lasioclada* F.Muell., Fragm. 9: 148 (1875). *Austromyrtus lasioclada* (F.Muell.) L.S.Sm., Proc. Roy. Soc. Queensland 67: 35 (1956).

Lenwebbia prominens N.Snow & Guymer, Syst. Bot. Monogr. 65: 30 (2003).
SE. Queensland to NE. New South Wales. 50 NSW QLD. Nanophan. or phan.

Leptomyrtus

Leptomyrtus (Miq.) O.Berg = ***Syzygium*** Gaertn.

Leptospermopsis

Leptospermopsis S.Moore = ***Leptospermum*** J.R.Forst. & G.Forst.
Leptospermopsis myrtifolia S.Moore = ***Leptospermum oligandrum*** Turcz.

Leptospermum

Leptospermum J.R.Forst. & G.Forst., Char. Gen. Pl.: 36 (1775).
Indo-China to New Zealand. (25) ken uga 41 MYA THA 42 BOR JAW LSI MLY MOL PHI SUL SUM 43 NWG 50 NFK NSW NTA QLD SOA TAS VIC WAU 51 CTM NZN NZS (63) haw (76) cal.
87 Species
Glaphyria Jack, Trans. Linn. Soc. London 14: 128 (1823).
Agonomyrtus Schauer ex Rchb., Handb. Nat. Pfl.-Syst.: 253 (1837).
Macklottia Korth., Ned. Kruidk. Arch. 1: 196 (1847).
Leptospermum floribundum Jungh., Java 1: 578 (1851).
Leptospermopsis S.Moore, J. Linn. Soc., Bot. 45: 202 (1920).

Leptospermum anfractum A.R.Bean, Telopea 10: 831 (2004).
Queensland. 50 QLD.

Leptospermum arachnoides Gaertn., Fruct. Sem. Pl. 1: 175 (1788).
SE. Queensland to New South Wales. 50 NSW QLD. Nanophan.
Leptospermum recurvifolium K.D.Koenig & Sims in R.A.Salisbury, Prodr. Stirp. Chap. Allerton: 350 (1796).
Leptospermum arachnoideum Sm., Trans. Linn. Soc. London 3: 263 (1797), nom. illeg.
Leptospermum baccatum Sm., Trans. Linn. Soc. London 3: 264 (1797).
Leptospermum juniperifolium Cav., Icon. 4: 18 (1797).
Melaleuca arachnoidea Raeusch., Nomencl. Bot., ed. 3: 143 (1797).
Leptospermum triloculare Vent., Jard. Malmaison: 88 (1804).

Leptospermum argenteum Joy Thomps., Telopea 3: 413 (1989).
E. New South Wales. 50 NSW. Nanophan.

Leptospermum barneyense A.R.Bean, Telopea 10: 836 (2004).
Queensland. 50 QLD.

Leptospermum benwellii A.R.Bean, Telopea 10: 832 (2004).
NE. New South Wales. 50 NSW.

Leptospermum blakelyi Joy Thomps., Telopea 3: 378 (1989).
EC. New South Wales. 50 NSW. Nanophan.

Leptospermum brachyandrum (F.Muell.) Druce, Rep. Bot. Exch. Club Brit. Isles 1916: 632 (1917).
Queensland to NE. New South Wales. 50 NSW QLD. Nanophan. or phan.
**Kunzea brachyandra* F.Muell., Fragm. 2: 27 (1860).
Leptospermum abnorme F.Muell. ex Benth., Fl. Austral. 3: 109 (1867). *Agonis abnormis* (F.Muell. ex Benth.) C.T.White, Bot. Bull. Dept. Agric. Queensland 22: 20 (1920).

Leptospermum brevipes F.Muell., Trans. Philos. Soc. Victoria 1: 125 (1855).
SE. Queensland to NE. Victoria. 50 NSW QLD VIC. Nanophan. or phan.

Leptospermum confertum Joy Thomps., Telopea 3: 362 (1989).
S. Western Australia. 50 WAU. Nanophan.

Leptospermum continentale Joy Thomps., Telopea 3: 417 (1989).
SE. Australia. (25) ken uga 50 NSW SOA VIC. Nanophan. or phan.
Leptospermum scoparium f. *angustifolium* Miq., Ned. Kruidk. Arch. 4: 143 (1856).

Leptospermum coriaceum (F.Muell. ex Miq.) Cheel, J. Proc. Roy. Soc. New S. Wales 57: 128 (1923).
SE. Australia. 50 NSW SOA VIC. Nanophan.
**Fabricia coriacea* F.Muell. ex Miq., Ned. Kruidk. Arch. 4: 147 (1856).
Leptospermum laevigatum var. *minus* Benth., Fl. Austral. 3: 103 (1867).

Leptospermum crassifolium Joy Thomps., Telopea 3: 438 (1989).
SE. New South Wales (Budawang Range). 50 NSW. Nanophan.

Leptospermum deanei Joy Thomps., Telopea 3: 364 (1989).
EC. New South Wales. 50 NSW. Nanophan. or phan.

Leptospermum deuense Joy Thomps., Telopea 3: 424 (1989).
SE. New South Wales. 50 NSW. Nanophan.

Leptospermum divaricatum Schauer in W.G.Walpers, Repert. Bot. Syst. 2: 923 (1843).
New South Wales. 50 NSW. Nanophan.
Leptospermum trivalve Cheel, J. Proc. Roy. Soc. New S. Wales 65: 201 (1931 publ. 1932).

Leptospermum emarginatum H.L.Wendl. ex Link, Enum. Hort. Berol. Alt. 2: 25 (1822).
New South Wales to Victoria. 50 NSW VIC. Nanophan.
Leptospermum odoratum Cheel, J. Proc. Roy. Soc. New S. Wales 53: 122 (1919).

Leptospermum epacridoideum Cheel, J. Proc. Roy. Soc. New S. Wales 53: 121 (1920).
SE. New South Wales. 50 NSW. Nanophan.

Leptospermum erubescens Schauer in J.G.C.Lehmann, Pl. Preiss. 1: 121 (1844).
SW. Australia. 50 WAU. Nanophan.

Leptospermum erubescens var. *strictum* Benth., Fl. Austral. 3: 109 (1867).

Leptospermum exsertum Joy Thomps., Telopea 3: 352 (1989).
Western Australia (NW. Avon). 50 WAU. Cham. or nanophan.

Leptospermum fastigiatum S.Moore, J. Linn. Soc., Bot. 45: 201 (1920).
S. Western Australia to SW. South Australia. 50 SOA WAU. Nanophan.

Leptospermum glabrescens N.A.Wakef., Victorian Naturalist 72: 43 (1955).
Victoria (East Gippsland Distr.). 50 VIC. Nanophan.

Leptospermum glaucescens Schauer, Linnaea 15: 421 (1841).
Tasmania. 50 TAS. Nanophan. or phan.
Leptospermum erythrocarpum Summerh. & H.F.Comber in H.F.Comber, Field Notes Tasman. Pl.: 34 (1930).

Leptospermum grandiflorum Lodd., Bot. Cab. 6: t. 514 (1821). *Leptospermum flavescens* var. *grandiflorum* (Lodd.) Benth., Fl. Austral. 3: 105 (1867). *Leptospermum flavescens* f. *grandiflorum* (Lodd.) Siebert & Voss, Vilm. Blumengärtn. ed. 3, 1: 310 (1896). *Leptospermum polygalifolium* var. *grandiflorum* (Lodd.) Domin, Biblioth. Bot. 89: 1006 (1928).
E. Tasmania. 50 TAS. Nanophan. or phan.
Leptospermum buxifolium Dehnh., Rivista Napol. 1(3): 172 (1839), nom. illeg.
Leptospermum tortuosum Dehnh., Rivista Napol. 1: 171 (1839).
Leptospermum grandiflorum var. *minus* S.Schauer, Linnaea 15: 438 (1841).
Leptospermum nobile F.Muell. ex Miq., Ned. Kruidk. Arch. 4: 145 (1856).
Leptospermum rodwayanum Summerh. & H.F.Comber in H.F.Comber, Field Notes Tasman. Pl.: 55 (1930).

Leptospermum grandifolium Sm., Trans. Linn. Soc. London 6: 299 (1802). *Leptospermum lanigerum* var. *grandifolium* (Sm.) Hook.f., Fl. Tasman.: 139 (1857). *Leptospermum pubescenes* var. *grandifolium* (Sm.) Domin, Biblioth. Bot. 89: 1007 (1928).
New South Wales to NE. Victoria. 50 NSW VIC. Nanophan.
Leptospermum grandifolium f. *angustum* S.Schauer, Linnaea 15: 413 (1841).
Leptospermum grandifolium f. *argyreum* S.Schauer, Linnaea 15: 413 (1841). *Leptospermum pubescens* f. *argyreum* (S.Schauer) Domin, Biblioth. Bot. 89: 1007 (1928).
Leptospermum subargenteum Gand., Bull. Soc. Bot. France 65: 26 (1918).

Leptospermum gregarium Joy Thomps., Telopea 3: 411 (1989).
S. Queensland to N. New South Wales. 50 NSW QLD. Nanophan.

Leptospermum incanum Turcz., Bull. Cl. Phys.-Math. Acad. Imp. Sci. Saint-Pétersbourg 10: 335 (1852).
S. & SC. Western Australia. 50 WAU. Nanophan.

Leptospermum inelegans Joy Thomps., Telopea 3: 376 (1989).
SW. Australia. 50 WAU. Nanophan.

Leptospermum javanicum Blume, Bijdr.: 1100 (1827). *Macklottia javanica* (Blume) Korth., Ned. Kruidk. Arch. 1: 196 (1847).
Indo-China to Malesia. 41 MYA THA 42 BOR JAW LSI MLY MOL PHI SUL SUM. Nanophan. or phan.
Glaphyria nitida Jack, Trans. Linn. Soc. London 14: 128 (1823).
Leptospermum alpestre Blume, Bijdr.: 1100 (1827).
Leptospermum amboinense Reinw. ex Blume, Bijdr.: 1100 (1827). *Macklottia amboinensis* (Reinw. ex Blume) Korth., Ned. Kruidk. Arch. 1: 196 (1847).
Glaphyria annae Stein, Gartenflora 34: 67 (1885).
Leptospermum annae Stein, Gartenflora 34: 66 (1885).

Leptospermum jingera Lyne & Crisp, Austral. Syst. Bot. 9: 301 (1996).
NE. Victoria. 50 VIC. Nanophan.

Leptospermum juniperinum Sm., Trans. Linn. Soc. London 3: 263 (1797). *Leptospermum scoparium* var. *juniperinum* (Sm.) Domin, Biblioth. Bot. 89: 1007 (1928).
E. Australia. 50 NSW QLD. Nanophan.
Leptospermum acerosum Schauer ex Otto & A.Dietr., Allg. Gartenzeitung 9: 250 (1841).
Leptospermum aciculare S.Schauer, Linnaea 15: 429 (1841). *Leptospermum scoparium* var. *aciculare* (S.Schauer) Domin, Biblioth. Bot. 89: 1007 (1928).
Leptospermum aciculare var. *majus* S.Schauer, Linnaea 15: 429 (1841).
Leptospermum aciculare var. *minor* S.Schauer, Linnaea 15: 429 (1841).

Leptospermum laevigatum (Gaertn.) F.Muell., Rep. Gov. Bot. Director Bot. Zool. Gard.: 22 (1858).
SE. Australia. (25) ken (27) cpp 50 NSW qld SOA TAS VIC wau (51) nzn (76) cal. Nanophan. or phan.
**Fabricia laevigata* Gaertn., Fruct. Sem. Pl. 1: 175 (1788).
Fabricia myrtifolia Sieber ex Benth., Fl. Austral. 3: 103 (1867).

Leptospermum lamellatum Joy Thomps., Telopea 3: 384 (1989).
Queensland (Leichhardt). 50 QLD. Nanophan. or phan.

Leptospermum lanigerum (Aiton) Sm., Trans. Linn. Soc. London 3: 263 (1797).
E. & SE. Australia. 50 NSW QLD SOA TAS VIC. Nanophan. or phan.
Leptospermum pubescens Lam., Encycl. 3: 466 (1792). *Fabricia pubescens* (Lam.) Schauer, Linnaea 15: 422 (1841).
Leptospermum australe K.D.Koenig & Sims in R.A.Salisbury, Prodr. Stirp. Chap. Allerton: 350 (1796), nom. illeg.
Leptospermum pubescens Muhl. ex Willd., Sp. Pl. 2: 950 (1799), nom. illeg. *Leptospermum lanigerum* var. *pubescens* (Muhl. ex Willd.) A.Cunn. ex DC., Prodr.: 227 (1828).
Fabricia incana Dum.Cours., Bot. Cult., ed. 2, 5: 387 (1811).
Fabricia minor Dum.Cours., Bot. Cult., ed. 2, 5: 387 (1811).
Leptospermum tomentosum Dum.Cours., Bot. Cult., ed. 2, 5: 384 (1811).
Leptospermum cuneatum Hoffmanns., Verz. Pfl.-Kult.: 72 (1824).
Leptospermum cuspidatum Hoffmanns., Verz. Pfl.-Kult.: 72 (1824).
Leptospermum microphyllum Hoffmanns., Verz. Pfl.-Kult.: 72 (1824).
Fabricia sericea Sweet, Hort. Brit.: 155 (1826).
Leptospermum ovalifolium Sweet, Hort. Brit.: 155 (1826).

Leptospermum candollei Schauer, Linnaea 15: 441 (1841).
Leptospermum grandifolium var. *vestitum* S.Schauer, Linnaea 15: 413 (1841). *Leptospermum pubescens* f. *vestitum* (S.Schauer) Domin, Biblioth. Bot. 89: 1007 (1928).
Leptospermum lanigerum var. *concolor* S.Schauer, Linnaea 15: 415 (1841).
Leptospermum lanigerum var. *discolor* S.Schauer, Linnaea 15: 415 (1841).
Leptospermum sericophyllum Schauer, Linnaea 15: 420 (1841).
Leptospermum tonsum Schauer, Linnaea 15: 422 (1841). *Leptospermum pubescens* var. *tonsum* (Schauer) Domin, Biblioth. Bot. 89: 1007 (1928).
Leptospermum villosum Otto & A.Dietr., Allg. Gartenzeitung 9: 242 (1841), nom. illeg.
Leptospermum pilosum Schauer in W.G.Walpers, Repert. Bot. Syst. 2: 923 (1843).
Leptospermum splendens Schauer in W.G.Walpers, Repert. Bot. Syst. 2: 923 (1843).
Leptospermum microphyllum F.Muell. ex Miq., Ned. Kruidk. Arch. 4: 142 (1856), nom. illeg.
Leptospermum microphyllum var. *glaucum* Miq., Ned. Kruidk. Arch. 4: 143 (1856).
Leptospermum microphyllum var. *viride* Miq., Ned. Kruidk. Arch. 4: 143 (1856).
Leptospermum pubescens f. *angustifolium* Miq., Ned. Kruidk. Arch. 4: 144 (1856).
Leptospermum pubescens f. *minor* Miq., Ned. Kruidk. Arch. 4: 144 (1856).
Leptospermum lanigerum var. *macrocarpum* Maiden & Betche, Proc. Linn. Soc. New South Wales 23: 12 (1898).
Leptospermum lanigerum var. *montanum* Rodway, Tasm. Fl.: 52 (1903).

Leptospermum liversidgei R.T.Baker & H.G.Sm., J. Proc. Roy. Soc. New S. Wales 39: 124 (1905 publ. 1906).
Queensland to NE. New South Wales. 50 NSW QLD. Nanophan.
Leptospermum flavescens var. *citriodorum* F.M.Bailey, Queensland Agric. J. 15: 781 (1905). *Leptospermum polygalifolium* var. *citriodorum* (F.M.Bailey) Domin, Biblioth. Bot. 89: 1006 (1928).

Leptospermum luehmannii F.M.Bailey, Queensl. Fl. 2: 592 (1900). *Agonis luehmannii* (F.M.Bailey) C.T. White & W.D.Francis, Bot. Bull. Dept. Agric. Queensland 22: 21 (1920).
SE. Queensland. 50 QLD. Nanophan. or phan.

Leptospermum macgillivrayi Joy Thomps., Telopea 3: 347 (1989).
SW. Australia (Coolgardie). 50 QLD. Nanophan.

Leptospermum macrocarpum (Maiden & Betche) Joy Thomps., Telopea 3: 432 (1989).
CE. New South Wales. 50 NSW. Nanophan.

Leptospermum madidum A.R.Bean, Austrobaileya 3: 645 (1992).
N. Australia. 50 NTA QLD WAU. Nanophan.

subsp. ***madidum***
Queensland. 50 QLD. Nanophan.
Agonis longifolia C.T.White & W.D.Francis, Bot. Bull. Dept. Agric. Queensland 22: 18 (1920).

subsp. ***sativum*** A.R.Bean, Austrobaileya 3: 646 (1992).
Western Australia to Northern Territory. 50 NTA WAU. Nanophan.

Leptospermum maxwellii S.Moore, J. Linn. Soc., Bot. 45: 201 (1920).
S. Western Australia. 50 WAU. Nanophan.

Leptospermum microcarpum Cheel, J. Proc. Roy. Soc. New S. Wales 57: 126 (1923).
SE. Queensland to NE. New South Wales. 50 NSW QLD. Nanophan.

Leptospermum micromyrtus Miq., Ned. Kruidk. Arch. 4: 145 (1856).
SE. New South Wales to NE. Victoria. 50 NSW VIC. Nanophan.

Leptospermum minutifolium (Benth.) C.T.White, Proc. Roy. Soc. Queensland 57: 26 (1946).
SE. Queensland to NE. New South Wales. 50 NSW QLD. Nanophan.
**Leptospermum flavescens* var. *minutifolium* Benth., Fl. Austral. 3: 105 (1867). *Leptospermum flavescens* f. *minutifolium* (Benth.) Siebert & Voss, Vilm. Blumengärtn. ed. 3, 1: 310 (1896). *Leptospermum polygalifolium* var. *minutifolium* (Benth.) Domin, Biblioth. Bot. 89: 1006 (1928).

Leptospermum morrisonii Joy Thomps., Telopea 3: 402 (1989).
New South Wales. 50 NSW. Nanophan. or phan.
Leptospermum virgatum S.Schauer, Linnaea 15: 410 (1841), nom. illeg.

Leptospermum multicaule A.Cunn., Field New South Wales: 349 (1825).
New South Wales to Victoria. 50 NSW VIC. Nanophan.

Leptospermum myrsinoides Schltdl., Linnaea 20: 653 (1847).
SE. Australia. 50 NSW SOA VIC. Nanophan.
Leptospermum myrsinoides var. *angustifolium* Miq., Ned. Kruidk. Arch. 4: 144 (1856).
Leptospermum myrsinoides var. *latifolium* Miq., Ned. Kruidk. Arch. 4: 144 (1856).

Leptospermum myrtifolium Sieber ex DC., Prodr. 3: 228 (1828).
New South Wales to NE. Victoria. 50 NSW VIC. Nanophan.
Leptospermum thymifolium Hoffmanns., Verz. Pfl.-Kult.: 22, 176 (1824).
Leptospermum thymifolium A.Cunn., Field New South Wales: 349 (1825), nom. illeg. *Leptospermum cunninghamii* Schauer, Linnaea 15: 420 (1841).
Leptospermum pubescens var. *parviflorum* Domin, Biblioth. Bot. 89: 1007 (1928).

Leptospermum namadgiensis Lyne, Telopea 5: 319 (1993).
New South Wales. 50 NSW. Nanophan.

Leptospermum neglectum Joy Thomps., Telopea 3: 383 (1989).
E. Queensland. 50 QLD. Nanophan.
Leptospermum attenuatum var. *subsessile* C.T.White, Proc. Roy. Soc. Queensland 50: 76 (1939).

Leptospermum nitens Turcz., Bull. Cl. Phys.-Math. Acad. Imp. Sci. Saint-Pétersbourg 10: 335 (1852).
SW. Australia. 50 WAU. Nanophan.

Leptospermum nitidum Hook.f., Fl. Tasman. 1: 139 (1856). *Leptospermum flavescens* var. *nitidum* (Hook.f.) Rodway, Tasm. Fl.: 53 (1903). *Leptospermum pubescens* var. *nitidum* (Hook.f.) Domin, Biblioth. Bot. 89: 1007 (1928).

Tasmania. 50 TAS. Nanophan.

Leptospermum novae-angliae Joy Thomps., Telopea 3: 405 (1989).
SE. Queensland to NE. New South Wales. 50 NSW QLD. Nanophan.

Leptospermum obovatum Sweet, Fl. Australas.: t. 36 (1827). *Leptospermum flavescens* var. *obovatum* (Sweet) Benth., Fl. Austral. 3: 105 (1867). *Leptospermum flavescens* f. *obovatum* (Sweet) Siebert & Voss, Vilm. Blumengärtn. ed. 3, 1: 310 (1896). *Leptospermum polygalifolium* var. *obovatum* (Sweet) Domin, Biblioth. Bot. 89: 1006 (1928).
New South Wales to Victoria. 50 NSW VIC. Nanophan.

Leptospermum oligandrum Turcz., Bull. Cl. Phys.-Math. Acad. Imp. Sci. Saint-Pétersbourg 10: 335 (1852).
SW. Australia. 50 WAU. Cham. or nanophan.
Kunzea podantha F.Muell., Fragm. 2: 28 (1860). *Leptospermum erubescens* var. *psilocalyx* Benth., Fl. Austral. 3: 109 (1867). *Leptospermum podanthum* (F.Muell.) Diels, Bot. Jahrb. Syst. 35: 423 (1904).
Leptospermopsis myrtifolia S.Moore, J. Linn. Soc., Bot. 45: 202 (1920).

Leptospermum oreophilum Joy Thomps., Telopea 3: 404 (1989).
SE. Queensland. 50 QLD. Nanophan.

Leptospermum pallidum A.R.Bean, Austrobaileya 3: 645 (1992).
Queensland. 50 QLD. Nanophan.

Leptospermum parviflorum Valeton, Bull. Dép. Agric. Indes Néerl. 10: 39 (1907).
New Guinea to N. Australia. 43 NWG 50 NTA QLD WAU. Nanophan. or phan.
Leptospermum longifolium A.Cunn., J. Bot. (Hooker) 4: 243 (1841).

Leptospermum parvifolium Sm., Trans. Linn. Soc. London 3: 263 (1797).
E. New South Wales. 50 NSW. Nanophan.
Leptospermum ovale Dum.Cours., Bot. Cult., ed. 2, 5: 385 (1811).
Leptospermum eriocalyx Sieber ex Spreng., Syst. Veg. 4(2): 194 (1827).

Leptospermum petersonii F.M.Bailey, Queensland Agric. J. 15: 781 (1905).
Queensland to New South Wales. (25) ken 50 NSW QLD. Nanophan. or phan.

subsp. ***lanceolatum*** Joy Thomps., Telopea 3: 394 (1989).
NE. Queensland to New South Wales. 50 NSW QLD. Nanophan. or phan.

subsp. ***petersonii***
Queensland to NE. New South Wales. 50 NSW QLD. Nanophan. or phan.
Leptospermum flavescens var. *citratum* J.F.Bailey & C.T.White, Queensland Agric. J., II, 18: 8 (1916). *Leptospermum citratum* (J.F.Bailey & C.T.White) Challinor, Cheel & A.R.Penfold, J. Proc. Roy. Soc. New S. Wales 52: 175 (1918 publ. 1919).

Leptospermum petraeum Joy Thomps., Telopea 3: 436 (1989).
EC. New South Wales. 50 NSW. Nanophan.

Leptospermum polyanthum Joy Thomps., Telopea 3: 381 (1989).
New South Wales. 50 NSW. Nanophan. or phan.

Leptospermum polygalifolium Salisb., Prodr. Stirp. Chap. Allerton: 350 (1796).
E. Australia to Lord Howe I. 50 NFK NSW QLD. Nanophan.

subsp. ***cismontanum*** Joy Thomps., Telopea 3: 400 (1989).
Queensland to New South Wales. 50 NSW QLD. Nanophan.
Leptospermum flavescens var. *microphyllum* Benth., Fl. Austral. 3: 105 (1867). *Leptospermum flavescens* f. *microphyllum* (Benth.) Siebert & Voss, Vilm. Blumengärtn. ed. 3, 1: 310 (1896).
Leptospermum flavescens var. *leptophyllum* Cheel, J. Proc. Roy. Soc. New S. Wales 56: 166 (1922).

subsp. ***howense*** Joy Thomps., Telopea 3: 401 (1989).
S. Lord Howe I. 50 NFK. Nanophan. or phan.

subsp. ***montanum*** Joy Thomps., Telopea 3: 399 (1989).
SE. Queensland to New South Wales. 50 NSW QLD. Nanophan.

subsp. ***polygalifolium***
New South Wales. 50 NSW. Nanophan.
Melaleuca thea Schrad. & J.C.Wendl., Sert. Hannov.: 24 (1796). *Leptospermum thea* (Schrad. & J.C.Wendl.) Muhl. ex Willd., Sp. Pl. 2: 949 (1799).
Leptospermum flavescens Sm., Trans. Linn. Soc. London 3: 262 (1797).
Leptospermum porophyllum Cav., Icon. 4: 17 (1797).
Melaleuca aromatica Dum.Cours., Bot. Cult. 5: 426 (1802).
Leptospermum tuberculatum Poir. in J.B.A.P.M.de Lamarck, Encycl., Suppl. 3: 338 (1813).
Glaphyria sericea Jack, Trans. Linn. Soc. London 14: 129 (1823).
Leptospermum tuberculatum var. *subenerve* A.Cunn. ex DC., Prodr. 3: 227 (1828).
Leptospermum obtusum Sweet ex G.Don in R.Sweet, Hort. Brit., ed. 2: 196 (1830).
Leptospermum buxifolium H.L.Wendl., Allg. Gartenzeitung 1: 186 (1833).
Leptospermum acutifolium Otto & Dietr., Allg. Gartenzeitung 9: 225 (1841).
Leptospermum aquaticum Otto & Dietr., Allg. Gartenzeitung 9: 243 (1841).
Leptospermum blumei Steud., Nomencl. Bot., ed. 2, 2: 31 (1841).
Leptospermum nervosum Otto & Dietr., Allg. Gartenzeitung 9: 260 (1841).
Leptospermum paludosum Schauer, Linnaea 15: 410 (1841), nom. inval.
Leptospermum retusum Otto & A.Dietr., Allg. Gartenzeitung 9: 243 (1841).
Leptospermum roseum Otto & A.Dietr., Allg. Gartenzeitung 9: 243 (1841).
Leptospermum flavescens var. *commune* Benth., Fl. Austral. 3: 104 (1867). *Leptospermum polygalifolium* var. *commune* (Benth.) Domin, Biblioth. Bot. 89: 1006 (1928).

Leptospermum flavescens var. *javanica* King, J. Asiat. Soc. Bengal, Pt. 2, Nat. Hist. 70(2): 69 (1901).
Leptospermum flavescens var. *angustifolia* Ridl., Fl. Malay Penins. 1: 713 (1922).

subsp. ***transmontanum*** Joy Thomps., Telopea 3: 401 (1989).
Queensland to New South Wales. 50 NSW QLD. Nanophan.

subsp. ***tropicum*** Joy Thomps., Telopea 3: 401 (1989).
N. & NE. Queensland. 50 QLD. Nanophan. or phan.

Leptospermum purpurascens Joy Thomps., Telopea 3: 355 (1989).
N. Queensland. 50 QLD. Nanophan. or phan.

Leptospermum recurvum Hook.f., Hooker's Icon. Pl. 9: t. 893 (1852).
Borneo (Gunung Kinabalu), Sulawesi. 42 BOR SUL. Nanophan. or phan.

Leptospermum riparium D.I.Morris, Rec. Queen Victoria Mus. 50: 2 (1974).
W. Tasmania. 50 TAS. Nanophan.

Leptospermum roei Benth., Fl. Austral. 3: 110 (1867).
SW. Australia. 50 WAU. Nanophan.

Leptospermum rotundifolium (Maiden & Betche) F.A.Rodway, J. Proc. Roy. Soc. New S. Wales 53: 122 (1919).
CE. New South Wales. 50 NSW. Nanophan.
Leptospermum rotundifolium var. *scoparium* Maiden & Betche, Proc. Linn. Soc. New South Wales 25: 101 (1900).
**Leptospermum scoparium* var. *rotundifolium* Maiden & Betche, Proc. Linn. Soc. New South Wales 25: 101 (1900). *Leptospermum polygalifolium* var. *rotundifolium* (Maiden & Betche) Domin, Biblioth. Bot. 89: 1006 (1928), nom. illeg.

Leptospermum rupestre Hook.f., Hooker's Icon. Pl. 4: t. 308 (1841).
Tasmania. 50 TAS. Nanophan.
Leptospermum scoparium var. *microphyllum* S.Schauer, Linnaea 15: 425 (1841).
Leptospermum grandifolium var. *compactum* Miq., Ned. Kruidk. Arch. 4: 144 (1856).

Leptospermum rupicola Joy Thomps., Telopea 3: 420 (1989).
CE. New South Wales. 50 NSW. Cham. or nanophan.

Leptospermum scoparium J.R.Forst. & G.Forst., Char. Gen. Pl.: 36 (1775). *Melaleuca scoparia* (J.R.Forst. & G.Forst.) L.f., Suppl. Pl.: 343 (1782). *Leptospermum scoparium* var. *vulgare* Domin, Biblioth. Bot. 89: 452 (1928), nom. inval.
SE. Australia, New Zealand, Chatham Is. 50 NSW TAS VIC 51 CTM NZN NZS (63) haw. Nanophan.
Leptospermum floribundum Salisb., Prodr. Stirp. Chap. Allerton: 349 (1796), nom. illeg.
Leptospermum multiflorum Cav., Icon. 4: 17 (1797).
Melaleuca tenuifolia J.C.Wendl., Bot. Beob.: 50 (1798).
Leptospermum linifolium (Sol.) Dum.Cours., Bot. Cult., ed. 2, 5: 381 (1811), nom. illeg.
Leptospermum scoparium var. *linifolium* (Sol.) Dum.Cours., Bot. Cult., ed. 2, 5: 384 (1811).
Leptospermum scoparium var. *myrtifolium* (Aiton) W.T.Aiton, Hortus Kew. 3: 181 (1811).
Leptospermum obliquum Colla, Hortus Ripul., App. 2: 351 (1825).
Leptospermum humifusum A.Cunn. ex Schauer, Linnaea 15: 425 (1841).
Leptospermum oxycedrus Schauer, Linnaea 15: 432 (1841).
Leptospermum pungens Otto & A.Dietr., Allg. Gartenzeitung 9: 260 (1841), nom. illeg.
Leptospermum scoparium var. *confertifolium* S.Schauer, Linnaea 15: 425 (1841).
Leptospermum scoparium var. *forsteri* S.Schauer, Linnaea 15: 425 (1841).
Leptospermum scoparium var. *linearifolium* Otto & A.Dietr., Allg. Gartenzeitung 9: 260 (1841).
Leptospermum scoparium var. *linifolium* Hook.f., Fl. Nov.-Zel. 1: 69 (1852), nom. illeg.
Leptospermum scoparium var. *myrtifolium* Hook.f., Fl. Nov.-Zel. 1: 69 (1852), nom. illeg.
Leptospermum scoparium var. *prostratum* Hook.f., Fl. Nov.-Zel. 1: 70 (1852).
Leptospermum scoparium var. *sericeum* Regel, Index Seminum (LE) 1863: 34 (1863).
Leptospermum scoparium var. *parvum* Kirk, Stud. Fl. New Zealand: 158 (1899).
Leptospermum nichollsii Dorr.Sm., Gard. Chron., III, 43: 399 (1908). *Leptospermum scoparium* var. *nichollsii* (Dorr.Sm.) Ewart, Fl. Victoria: 856 (1931).
Leptospermum scoparium var. *bullatum* Rehder, Gardening Ill. 30: 415 (1909).
Leptospermum bullatum Fitzh., Gard. Chron., III, 5: 100 (1912).
Leptospermum scoparium var. *incanum* Cackayne, Trans. & Proc. New Zealand Inst. 49: 58 (1917).
Leptospermum scoparium var. *eximea* B.L.Burtt, Bot. Mag. 162: t. 9582 (1939).

Leptospermum sejunctum Joy Thomps., Telopea 3: 407 (1989).
CE. New South Wales. 50 NSW. Nanophan.

Leptospermum semibaccatum Cheel, J. Proc. Roy. Soc. New S. Wales 65: 203 (1931 publ. 1932).
SE. Queensland to NE. New South Wales. 50 NSW QLD. Cham. or nanophan.

Leptospermum sericatum Lindl. in T.L.Mitchell, J. Exped. Trop. Australia: 298 (1848). *Leptospermum stellatum* f. *sericatum* (Lindl.) Domin, Biblioth. Bot. 89: 1008 (1928).
Queensland (Leichhardt). 50 QLD. Nanophan.

Leptospermum sericeum Labill., Nov. Holl. Pl. 2: 9 (1806).
S. Western Australia. 50 WAU. Nanophan.

Leptospermum speciosum Schauer in W.G.Walpers, Repert. Bot. Syst. 2: 923 (1843). *Melaleuca leucadendra* var. *speciosa* (Schauer) Domin, Biblioth. Bot. 89: 457 (1928). *Agonis speciosa* (Schauer) C.T.White, Proc. Roy. Soc. Queensland 53: 218 (1942).
Queensland to NE. New South Wales. 50 NSW QLD. Nanophan.
Agonis scortechiniana F.Muell., Fragm. 11: 118 (1881).

Leptospermum spectabile Joy Thomps., Telopea 3: 431 (1989).
CE. New South Wales. 50 NSW. Nanophan.

Leptospermum sphaerocarpum Cheel, J. Proc. Roy. Soc. New S. Wales 65: 204 (1931 publ. 1932).
CE. New South Wales. 50 NSW. Nanophan.

Leptospermum spinescens Endl. in S.L.Endlicher & al., Enum. Pl.: 50 (1837).
SW. Australia. 50 WAU. Nanophan.

Leptospermum squarrosum Gaertn., Fruct. Sem. Pl. 1: 174 (1788).
CE. New South Wales. 50 NSW. Nanophan.
Leptospermum rubricaule Cels ex Link, Enum. Hort. Berol. Alt. 2: 25 (1822). *Leptospermum scoparium* var. *rubricaule* (Cels ex Link) A.Cunn. ex DC., Prodr. 3: 227 (1828).
Leptospermum persiciflorum Rchb., Iconogr. Bot. Exot. 3: 8 (1830).
Leptospermum scoparium var. *grandiflorum* Hook., Bot. Mag. 62: t. 3419 (1835).
Leptospermum baccatum var. *roseum* S.Schauer, Linnaea 15: 429 (1841).

Leptospermum subglabratum Joy Thomps., Telopea 3: 368 (1989).
S. New South Wales. 50 NSW. Nanophan.

Leptospermum subtenue Joy Thomps., Telopea 3: 353 (1989).
SW. Australia. 50 WAU. Nanophan.

Leptospermum thompsonii Joy Thomps., Telopea 3: 429 (1989).
SE. New South Wales. 50 NSW. Nanophan.

Leptospermum trinervium (Sm.) Joy Thomps., Telopea 3: 366 (1989).
Queensland to Victoria. 50 NSW QLD VIC. Nanophan. or phan.
**Melaleuca trinervia* Sm., J. Voy. New South Wales: 229 (1790).
Leptospermum attenuatum Sm., Trans. Linn. Soc. London 3: 262 (1797).
Leptospermum stellatum Cav., Icon. 4: 16 (1797). *Leptospermum stellatum* var. *typicum* Hochr., Candollea 2: 465 (1925), nom. inval.
Leptospermum trinerve Sm. in A.Rees, Cycl. 20: no. 5 (1812).
Leptospermum acuminatum J.C.Wendl. ex Steud., Nomencl. Bot.: 473 (1821).
Leptospermum pendulum Spreng., Syst. Veg. 4(2): 194 (1827).
Leptospermum gnidiaefolium DC., Prodr. 3: 228 (1828).
Leptospermum lucidum Schauer, Linnaea 15: 435 (1841).
Leptospermum stellatum var. *grandiflorum* Benth., Fl. Austral. 3: 107 (1867).
Leptospermum stellatum f. *angustifolium* Domin, Biblioth. Bot. 89: 1008, (1928).
Leptospermum stellatum f. *fallax* Domin, Biblioth. Bot. 89: 1008 (1928).

Leptospermum turbinatum Joy Thomps., Telopea 3: 437 (1989).
W. Victoria. 50 VIC. Nanophan.

Leptospermum variabile Joy Thomps., Telopea 3: 403 (1989).
SE. Queensland to New South Wales. 50 NSW QLD. Nanophan. or phan.

Leptospermum venustum A.R.Bean, Austrobaileya 3: 649 (1992).
Queensland. 50 QLD. Nanophan.

Leptospermum whitei Cheel, J. Proc. Roy. Soc. New S. Wales 65: 199 (1931 publ. 1932).
Queensland to NE. New South Wales. 50 NSW QLD. Nanophan.
**Agonis elliptica* C.T.White & W.D.Francis, Bot. Bull. Dept. Agric. Queensland 22: 16 (1920).
Agonis elliptica var. *angustifolia* C.T.White & W.D. Francis, Bot. Bull. Dept. Agric. Queensland 22: 18 (1920).

Leptospermum wooroonooran F.M.Bailey in Meston, Rep. Gov. Sci. Exped. Bellenden-Ker Range, Bot.: 40 (1889).
N. Queensland. 50 QLD. Nanophan. or phan.

Synonyms:
Leptospermum abnorme F.Muell. ex Benth. = ***Leptospermum brachyandrum*** (F.Muell.) Druce
Leptospermum acerosum Schauer ex Otto & A.Dietr. = ***Leptospermum juniperinum*** Sm.
Leptospermum aciculare S.Schauer = ***Leptospermum juniperinum*** Sm.
Leptospermum aciculare var. *majus* S.Schauer = ***Leptospermum juniperinum*** Sm.
Leptospermum aciculare var. *minor* S.Schauer = ***Leptospermum juniperinum*** Sm.
Leptospermum acuminatum J.C.Wendl. ex Steud. = ***Leptospermum trinervium*** (Sm.) Joy Thomps.
Leptospermum acutifolium Otto & Dietr. = ***Leptospermum polygalifolium*** Salisb. subsp. ***polygalifolium***
Leptospermum alpestre Blume = ***Leptospermum javanicum*** Blume
Leptospermum ambiguum Sm. = ***Kunzea ambigua*** (Sm.) Druce
Leptospermum amboinense Reinw. ex Blume = ***Leptospermum javanicum*** Blume
Leptospermum angustifolium Endl. = ***Hypocalymma angustifolium*** (Endl.) Schauer
Leptospermum annae Stein = ***Leptospermum javanicum*** Blume
Leptospermum aquaticum Otto & Dietr. = ***Leptospermum polygalifolium*** Salisb. subsp. ***polygalifolium***
Leptospermum arachnoideum Sm. = ***Leptospermum arachnoides*** Gaertn.
Leptospermum attenuatum Sm. = ***Leptospermum trinervium*** (Sm.) Joy Thomps.
Leptospermum attenuatum var. *subsessile* C.T.White = ***Leptospermum neglectum*** Joy Thomps.
Leptospermum australe K.D.Koenig & Sims = ***Leptospermum lanigerum*** (Aiton) Sm.
Leptospermum baccatum Sm. = ***Leptospermum arachnoides*** Gaertn.
Leptospermum baccatum var. *roseum* S.Schauer = ***Leptospermum squarrosum*** Gaertn.
Leptospermum bennigsenianum Volkens = ***Myrtella bennigseniana*** (Volkens) Diels
Leptospermum blumei Steud. = ***Leptospermum polygalifolium*** Salisb. subsp. ***polygalifolium***
Leptospermum brevifolium Rudge = ***Baeckea brevifolia*** (Rudge) DC.
Leptospermum bullatum Fitzh. = ***Leptospermum scoparium*** J.R.Forst. & G.Forst.
Leptospermum buxifolium Dehnh. = ***Leptospermum grandiflorum*** Lodd.
Leptospermum buxifolium H.L.Wendl. = ***Leptospermum polygalifolium*** Salisb. subsp. ***polygalifolium***
Leptospermum candollei Schauer = ***Leptospermum lanigerum*** (Aiton) Sm.
Leptospermum celsianum Graeffer ex Ten. = ?
Leptospermum ciliatum J.R.Forst. & G.Forst. = ***Purpureostemon ciliatus*** (J.R.Forst. & G.Forst.) Gugerli
Leptospermum ciliolatum Otto & A.Dietr. = [50]
Leptospermum citratum (J.F.Bailey & C.T.White) Challinor, Cheel & A.R.Penfold = ***Leptospermum petersonii*** F.M.Bailey subsp. ***petersonii***

Leptospermum collinum J.R.Forst. & G.Forst. = ***Metrosideros collina*** (J.R.Forst. & G.Forst.) A.Gray
Leptospermum crassipes Lehm. = ***Pericalymma crassipes*** (Lehm.) Schauer
Leptospermum cuneatum Hoffmanns. = ***Leptospermum lanigerum*** (Aiton) Sm.
Leptospermum cuneiforme Otto & A.Dietr. = [50]
Leptospermum cunninghamii Schauer = ***Leptospermum myrtifolium*** Sieber ex DC.
Leptospermum cupressinum Otto & A.Dietr. = [50]
Leptospermum cuspidatum Hoffmanns. = ***Leptospermum lanigerum*** (Aiton) Sm.
Leptospermum dubium Spreng. = ***Astartea affinis*** (Endl.) Rye
Leptospermum ellipticum Endl. = ***Pericalymma ellipticum*** (Endl.) Schauer
Leptospermum ericoides A.Rich. = ***Kunzea ericoides*** (A.Rich.) Joy Thomps.
Leptospermum ericoides var. *lineare* Kirk = ***Kunzea ericoides*** (A.Rich.) Joy Thomps.
Leptospermum ericoides var. *microflorum* G.Simpson = ***Kunzea ericoides*** (A.Rich.) Joy Thomps.
Leptospermum eriocalyx Sieber ex Spreng. = ***Leptospermum parvifolium*** Sm.
Leptospermum erubescens var. *psilocalyx* Benth. = ***Leptospermum oligandrum*** Turcz.
Leptospermum erubescens var. *strictum* Benth. = ***Leptospermum erubescens*** Schauer
Leptospermum erythrocarpum Summerh. & H.F.Comber = ***Leptospermum glaucescens*** Schauer
Leptospermum fabricia Benth. = ***Neofabricia myrtifolia*** (Gaertn.) Joy Thomps.
Leptospermum firmum (Schauer) Benth. = ***Homalospermum firmum*** Schauer
Leptospermum flavescens Sm. = ***Leptospermum polygalifolium*** Salisb. subsp. ***polygalifolium***
Leptospermum flavescens var. *angustifolia* Ridl. = ***Leptospermum polygalifolium*** Salisb. subsp. ***polygalifolium***
Leptospermum flavescens var. *citratum* J.F.Bailey & C.T.White = ***Leptospermum petersonii*** F.M.Bailey subsp. ***petersonii***
Leptospermum flavescens var. *citriodorum* F.M.Bailey = ***Leptospermum liversidgei*** R.T.Baker & H.G.Sm.
Leptospermum flavescens var. *commune* Benth. = ***Leptospermum polygalifolium*** Salisb. subsp. ***polygalifolium***
Leptospermum flavescens var. *grandiflorum* (Lodd.) Benth. = ***Leptospermum grandiflorum*** Lodd.
Leptospermum flavescens f. *grandiflorum* (Lodd.) Siebert & Voss = ***Leptospermum grandiflorum*** Lodd.
Leptospermum flavescens var. *javanica* King = ***Leptospermum polygalifolium*** Salisb. subsp. ***polygalifolium***
Leptospermum flavescens var. *leptophyllum* Cheel = ***Leptospermum polygalifolium*** subsp. ***cismontanum*** Joy Thomps.
Leptospermum flavescens var. *microphyllum* Benth. = ***Leptospermum polygalifolium*** subsp. ***cismontanum*** Joy Thomps.
Leptospermum flavescens f. *microphyllum* (Benth.) Siebert & Voss = ***Leptospermum polygalifolium*** subsp. ***cismontanum*** Joy Thomps.
Leptospermum flavescens var. *minutifolium* Benth. = ***Leptospermum minutifolium*** (Benth.) C.T.White
Leptospermum flavescens f. *minutifolium* (Benth.) Siebert & Voss = ***Leptospermum minutifolium*** (Benth.) C.T.White
Leptospermum flavescens var. *nitidum* (Hook.f.) Rodway = ***Leptospermum nitidum*** Hook.f.
Leptospermum flavescens var. *obovatum* (Sweet) Benth. = ***Leptospermum obovatum*** Sweet
Leptospermum flavescens f. *obovatum* (Sweet) Siebert & Voss = ***Leptospermum obovatum*** Sweet
Leptospermum flexuosum (Muhl. ex Willd.) Spreng. = ***Agonis flexuosa*** (Muhl. ex Willd.) Sweet
Leptospermum floribundum Jungh. = ***Leptospermum*** J.R.Forst. & G.Forst.
Leptospermum floribundum Salisb. = ***Leptospermum scoparium*** J.R.Forst. & G.Forst.
Leptospermum floridum (Schauer) Benth. = ***Pericalymma ellipticum*** var. ***floridum*** (Schauer) Cranfield
Leptospermum glomeratum H.L.Wendl. = ***Agonis flexuosa*** (Muhl. ex Willd.) Sweet
Leptospermum gnidiaefolium DC. = ***Leptospermum trinervium*** (Sm.) Joy Thomps.
Leptospermum grandiflorum var. *minus* S.Schauer = ***Leptospermum grandiflorum*** Lodd.
Leptospermum grandifolium f. *angustum* S.Schauer = ***Leptospermum grandifolium*** Sm.
Leptospermum grandifolium f. *argyreum* S.Schauer = ***Leptospermum grandifolium*** Sm.
Leptospermum grandifolium var. *compactum* Miq. = ***Leptospermum rupestre*** Hook.f.
Leptospermum grandifolium var. *vestitum* S.Schauer = ***Leptospermum lanigerum*** (Aiton) Sm.
Leptospermum humifusum A.Cunn. ex Schauer = ***Leptospermum scoparium*** J.R.Forst. & G.Forst.
Leptospermum hypericifolium Otto & A.Dietr. = ***Agonis hypericifolia*** (Otto & A.Dietr.) Schauer
Leptospermum imbricatum Sm. = ***Triplarina imbricata*** (Sm.) A.R.Bean
Leptospermum juniperifolium Cav. = ***Leptospermum arachnoides*** Gaertn.
Leptospermum laevigatum var. *minus* Benth. = ***Leptospermum coriaceum*** (F.Muell. ex Miq.) Cheel
Leptospermum lanigerum var. *concolor* S.Schauer = ***Leptospermum lanigerum*** (Aiton) Sm.
Leptospermum lanigerum var. *discolor* S.Schauer = ***Leptospermum lanigerum*** (Aiton) Sm.
Leptospermum lanigerum var. *grandifolium* (Sm.) Hook.f. = ***Leptospermum grandifolium*** Sm.
Leptospermum lanigerum var. *macrocarpum* Maiden & Betche = ***Leptospermum lanigerum*** (Aiton) Sm.
Leptospermum lanigerum var. *montanum* Rodway = ***Leptospermum lanigerum*** (Aiton) Sm.
Leptospermum lanigerum var. *pubescens* (Muhl. ex Willd.) A.Cunn. ex DC. = ***Leptospermum lanigerum*** (Aiton) Sm.
Leptospermum laricifolium A.Cunn. ex Schauer = ***Astartea laricifolia*** Schauer
Leptospermum leucodendron (L.) J.R.Forst. & G.Forst. = ***Melaleuca leucadendra*** (L.) L.
Leptospermum linearifolium DC. = ***Agonis linearifolia*** (DC.) Sweet
Leptospermum linifolium (Sol.) Dum.Cours. = ***Leptospermum scoparium*** J.R.Forst. & G.Forst.
Leptospermum longifolium (C.T.White & W.D.Francis) S.T.Blake = ***Leptospermum madidum*** A.R.Bean
Leptospermum longifolium A.Cunn. = ***Leptospermum parviflorum*** Valeton
Leptospermum lucidum Schauer = ***Leptospermum trinervium*** (Sm.) Joy Thomps.
Leptospermum marginatum Labill. = ***Agonis marginata*** (Labill.) Sweet
Leptospermum marginatum var. *glabratum* A.Cunn. ex DC. = ***Agonis marginata*** (Labill.) Sweet

Leptospermum microphyllum Hoffmanns. = ***Leptospermum lanigerum*** (Aiton) Sm.
Leptospermum microphyllum F.Muell. ex Miq. = ***Leptospermum lanigerum*** (Aiton) Sm.
Leptospermum microphyllum var. *glaucum* Miq. = ***Leptospermum lanigerum*** (Aiton) Sm.
Leptospermum microphyllum var. *viride* Miq. = ***Leptospermum lanigerum*** (Aiton) Sm.
Leptospermum mjoebergii Cheel = ***Neofabricia mjoebergii*** (Cheel) Joy Thomps.
Leptospermum multiflorum Cav. = ***Leptospermum scoparium*** J.R.Forst. & G.Forst.
Leptospermum myrsinoides var. *angustifolium* Miq. = ***Leptospermum myrsinoides*** Schltdl.
Leptospermum myrsinoides var. *latifolium* Miq. = ***Leptospermum myrsinoides*** Schltdl.
Leptospermum nervosum Otto & Dietr. = ***Leptospermum polygalifolium*** Salisb. subsp. ***polygalifolium***
Leptospermum nichollsii Dorr.Sm. = ***Leptospermum scoparium*** J.R.Forst. & G.Forst.
Leptospermum nobile F.Muell. ex Miq. = ***Leptospermum grandiflorum*** Lodd.
Leptospermum obliquum Colla = ***Leptospermum scoparium*** J.R.Forst. & G.Forst.
Leptospermum obtusum Sweet ex G.Don = ***Leptospermum polygalifolium*** Salisb. subsp. ***polygalifolium***
Leptospermum odoratum Cheel = ***Leptospermum emarginatum*** H.L.Wendl. ex Link
Leptospermum ovale Dum.Cours. = ***Leptospermum parvifolium*** Sm.
Leptospermum ovalifolium Sweet = ***Leptospermum lanigerum*** (Aiton) Sm.
Leptospermum oxycedrus Schauer = ***Leptospermum scoparium*** J.R.Forst. & G.Forst.
Leptospermum paludosum Schauer = ***Leptospermum polygalifolium*** Salisb. subsp. ***polygalifolium***
Leptospermum parvulum Labill. = ***Babingtonia virgata*** (J.R.Forst. & G.Forst.) F.Muell.
Leptospermum pendulum Spreng. = ***Leptospermum trinervium*** (Sm.) Joy Thomps.
Leptospermum perforatum J.R.Forst. & G.Forst. = ***Metrosideros perforata*** (J.R.Forst. & G.Forst.) Druce
Leptospermum persiciflorum Rchb. = ***Leptospermum squarrosum*** Gaertn.
Leptospermum phylicoideum (A.Cunn. ex Schauer) Cheel = ***Kunzea ericoides*** (A.Rich.) Joy Thomps.
Leptospermum pilosum Schauer = ***Leptospermum lanigerum*** (Aiton) Sm.
Leptospermum pinifolium Labill. = ***Babingtonia pinifolia*** (Labill.) A.R.Bean
Leptospermum podanthum (F.Muell.) Diels = ***Leptospermum oligandrum*** Turcz.
Leptospermum polygalifolium var. *citriodorum* (F.M.Bailey) Domin = ***Leptospermum liversidgei*** R.T.Baker & H.G.Sm.
Leptospermum polygalifolium var. *commune* (Benth.) Domin = ***Leptospermum polygalifolium*** Salisb. subsp. ***polygalifolium***
Leptospermum polygalifolium var. *grandiflorum* (Lodd.) Domin = ***Leptospermum grandiflorum*** Lodd.
Leptospermum polygalifolium var. *minutifolium* (Benth.) Domin = ***Leptospermum minutifolium*** (Benth.) C.T.White
Leptospermum polygalifolium var. *obovatum* (Sweet) Domin = ***Leptospermum obovatum*** Sweet
Leptospermum polygalifolium var. *rotundifolium* (Maiden & Betche) Domin = ***Leptospermum rotundifolium*** (Maiden & Betche) F.A.Rodway
Leptospermum porophyllum Cav. = ***Leptospermum polygalifolium*** Salisb. subsp. ***polygalifolium***
Leptospermum pubescenes var. *grandifolium* (Sm.) Domin = ***Leptospermum grandifolium*** Sm.
Leptospermum pubescens Lam. = ***Leptospermum lanigerum*** (Aiton) Sm.
Leptospermum pubescens Muhl. ex Willd. = ***Leptospermum lanigerum*** (Aiton) Sm.
Leptospermum pubescens f. *angustifolium* Miq. = ***Leptospermum lanigerum*** (Aiton) Sm.
Leptospermum pubescens f. *argyreum* (S.Schauer) Domin = ***Leptospermum grandifolium*** Sm.
Leptospermum pubescens f. *minor* Miq. = ***Leptospermum lanigerum*** (Aiton) Sm.
Leptospermum pubescens var. *nitidum* (Hook.f.) Domin = ***Leptospermum nitidum*** Hook.f.
Leptospermum pubescens var. *parviflorum* Domin = ***Leptospermum myrtifolium*** Sieber ex DC.
Leptospermum pubescens var. *tonsum* (Schauer) Domin = ***Leptospermum lanigerum*** (Aiton) Sm.
Leptospermum pubescens f. *vestitum* (S.Schauer) Domin = ***Leptospermum lanigerum*** (Aiton) Sm.
Leptospermum pungens Otto & A.Dietr. = ***Leptospermum scoparium*** J.R.Forst. & G.Forst.
Leptospermum pungens Banks ex Dum.Cours. = ?
Leptospermum recurvifolium K.D.Koenig & Sims = ***Leptospermum arachnoides*** Gaertn.
Leptospermum resiniferum Bertol. = ***Agonis flexuosa*** (Muhl. ex Willd.) Sweet
Leptospermum retusum Otto & A.Dietr. = ***Leptospermum polygalifolium*** Salisb. subsp. ***polygalifolium***
Leptospermum robustum Endl. = ***Hypocalymma robustum*** (Endl.) Lindl.
Leptospermum rodwayanum Summerh. & H.F.Comber = ***Leptospermum grandiflorum*** Lodd.
Leptospermum roseum Otto & A.Dietr. = ***Leptospermum polygalifolium*** Salisb. subsp. ***polygalifolium***
Leptospermum rotundifolium Domin = ***Leptospermum rotundifolium*** (Maiden & Betche) F.A.Rodway
Leptospermum rotundifolium var. *scoparium* Maiden & Betche = ***Leptospermum rotundifolium*** (Maiden & Betche) F.A.Rodway
Leptospermum rubricaule Cels ex Link = ***Leptospermum squarrosum*** Gaertn.
Leptospermum salicifolium Lam. = [42 JAW]
Leptospermum scandens J.R.Forst. & G.Forst. = ***Metrosideros fulgens*** Sol. ex Gaertn.
Leptospermum scoparium var. *aciculare* (S.Schauer) Domin = ***Leptospermum juniperinum*** Sm.
Leptospermum scoparium f. *angustifolium* Miq. = ***Leptospermum continentale*** Joy Thomps.
Leptospermum scoparium var. *bullatum* Rehder = ***Leptospermum scoparium*** J.R.Forst. & G.Forst.
Leptospermum scoparium var. *confertifolium* S.Schauer = ***Leptospermum scoparium*** J.R.Forst. & G.Forst.
Leptospermum scoparium var. *eximea* B.L.Burtt = ***Leptospermum scoparium*** J.R.Forst. & G.Forst.
Leptospermum scoparium var. *forsteri* S.Schauer = ***Leptospermum scoparium*** J.R.Forst. & G.Forst.
Leptospermum scoparium var. *grandiflorum* Hook. = ***Leptospermum squarrosum*** Gaertn.
Leptospermum scoparium var. *incanum* Cackayne = ***Leptospermum scoparium*** J.R.Forst. & G.Forst.
Leptospermum scoparium var. *juniperinum* (Sm.) Domin = ***Leptospermum juniperinum*** Sm.
Leptospermum scoparium var. *linearifolium* Otto & A.Dietr. = ***Leptospermum scoparium*** J.R.Forst. & G.Forst.

Leptospermum scoparium var. *linifolium* (Sol.) Dum. Cours. = ***Leptospermum scoparium*** J.R.Forst. & G.Forst.
Leptospermum scoparium var. *linifolium* Hook.f. = ***Leptospermum scoparium*** J.R.Forst. & G.Forst.
Leptospermum scoparium var. *microphyllum* S.Schauer = ***Leptospermum rupestre*** Hook.f.
Leptospermum scoparium var. *myrtifolium* (Aiton) W.T.Aiton = ***Leptospermum scoparium*** J.R.Forst. & G.Forst.
Leptospermum scoparium var. *myrtifolium* Hook.f. = ***Leptospermum scoparium*** J.R.Forst. & G.Forst.
Leptospermum scoparium var. *nichollsii* (Dorr.Sm.) Ewart = ***Leptospermum scoparium*** J.R.Forst. & G.Forst.
Leptospermum scoparium var. *parvum* Kirk = ***Leptospermum scoparium*** J.R.Forst. & G.Forst.
Leptospermum scoparium var. *prostratum* Hook.f. = ***Leptospermum scoparium*** J.R.Forst. & G.Forst.
Leptospermum scoparium var. *rotundifolium* Maiden & Betche = ***Leptospermum rotundifolium*** (Maiden & Betche) F.A.Rodway
Leptospermum scoparium var. *rubricaule* (Cels ex Link) A.Cunn. ex DC. = ***Leptospermum squarrosum*** Gaertn.
Leptospermum scoparium var. *sericeum* Regel = ***Leptospermum scoparium*** J.R.Forst. & G.Forst.
Leptospermum scoparium var. *vulgare* Domin = ***Leptospermum scoparium*** J.R.Forst. & G.Forst.
Leptospermum sericophyllum Schauer = ***Leptospermum lanigerum*** (Aiton) Sm.
Leptospermum sinclairii Kirk = ***Kunzea sinclairii*** (Kirk) W.Harris
Leptospermum splendens Schauer = ***Leptospermum lanigerum*** (Aiton) Sm.
Leptospermum stellatum Cav. = ***Leptospermum trinervium*** (Sm.) Joy Thomps.
Leptospermum stellatum f. *angustifolium* Domin = ***Leptospermum trinervium*** (Sm.) Joy Thomps.
Leptospermum stellatum f. *fallax* Domin = ***Leptospermum trinervium*** (Sm.) Joy Thomps.
Leptospermum stellatum var. *grandiflorum* Benth. = ***Leptospermum trinervium*** (Sm.) Joy Thomps.
Leptospermum stellatum f. *sericatum* (Lindl.) Domin = ***Leptospermum sericatum*** Lindl.
Leptospermum stellatum var. *typicum* Hochr. = ***Leptospermum trinervium*** (Sm.) Joy Thomps.
Leptospermum styphelioides (Sm.) Schauer = ***Melaleuca styphelioides*** Sm.
Leptospermum subargenteum Gand. = ***Leptospermum grandifolium*** Sm.
Leptospermum thea (Schrad. & J.C.Wendl.) Muhl. ex Willd. = ***Leptospermum polygalifolium*** Salisb. subsp. ***polygalifolium***
Leptospermum theiforme A.Cunn. ex Schauer = ***Agonis hypericifolia*** (Otto & A.Dietr.) Schauer
Leptospermum thymifolium Hoffmanns. = ***Leptospermum myrtifolium*** Sieber ex DC.
Leptospermum thymifolium A.Cunn. = ***Leptospermum myrtifolium*** Sieber ex DC.
Leptospermum tomentosum Dum.Cours. = ***Leptospermum lanigerum*** (Aiton) Sm.
Leptospermum tonsum Schauer = ***Leptospermum lanigerum*** (Aiton) Sm.
Leptospermum tortuosum Dehnh. = ***Leptospermum grandiflorum*** Lodd.
Leptospermum triloculare Vent. = ***Leptospermum arachnoides*** Gaertn.
Leptospermum trinerve Sm. = ***Leptospermum trinervium*** (Sm.) Joy Thomps.
Leptospermum trivalve Cheel = ***Leptospermum divaricatum*** Schauer
Leptospermum tuberculatum Poir. = ***Leptospermum polygalifolium*** Salisb. subsp. ***polygalifolium***
Leptospermum tuberculatum var. *subenerve* A.Cunn. ex DC. = ***Leptospermum polygalifolium*** Salisb. subsp. ***polygalifolium***
Leptospermum umbellatum Gaertn. = ***Eucalyptus tereticornis*** Sm.
Leptospermum villosum Otto & A.Dietr. = ***Leptospermum lanigerum*** (Aiton) Sm.
Leptospermum villosum Fisch. = ?
Leptospermum × violipurpureum W.Harris, M.I.Dawson & Heenan = *L. rotundifolium × L. spectabile*
Leptospermum virgatum S.Schauer = ***Leptospermum morrisonii*** Joy Thomps.
Leptospermum virgatum J.R.Forst. & G.Forst. = ***Babingtonia virgata*** (J.R.Forst. & G.Forst.) F.Muell.

Unplaced Names:
Leptospermum celsianum Graeffer ex Ten., Cat. Pl. Neapol. 1813, app.: 11 (1815). = ?
Leptospermum ciliolatum Otto & A.Dietr., Allg. Gartenzeitung 9: 241 (1841). = [50]
Leptospermum cuneiforme Otto & A.Dietr., Allg. Gartenzeitung 9: 251 (1841). = [50]
Leptospermum cupressinum Otto & A.Dietr., Allg. Gartenzeitung 9: 250 (1841). = [50]
Leptospermum pungens Banks ex Dum.Cours., Bot. Cult., ed. 2, 5: 385 (1811). = ?
Leptospermum salicifolium Lam., Encycl. 3: 467 (1792). = [42 JAW]
Leptospermum villosum Fisch., Cat. Jard. Gorenki, ed. 2: 63 (1812). = ?
Leptospermum × violipurpureum W.Harris, M.I.Dawson & Heenan, Hort. New Zealand 6: 6 (1995). = *L. rotundifolium × L. spectabile*

Lhotskya

Lhotskya Schauer = ***Calytrix*** Labill.
Lhotskya acutifolia Lindl. = ***Calytrix acutifolia*** (Lindl.) Craven
Lhotskya alpestris (Lindl.) Druce = ***Calytrix alpestris*** (Lindl.) Court
Lhotskya alpestris var. *bracteosa* (Benth.) J.M.Black = ***Calytrix alpestris*** (Lindl.) Court
Lhotskya brevifolia Schauer = ***Calytrix sylvana*** Craven
Lhotskya ciliata F.Muell. ex Benth. = ***Calytrix nematoclada*** Craven
Lhotskya cuspidata F.Muell. = ***Calytrix achaeta*** (F.Muell.) Benth.
Lhotskya ericoides Schauer = ***Calytrix acutifolia*** (Lindl.) Craven
Lhotskya genethylloides F.Muell. = ***Calytrix alpestris*** (Lindl.) Court
Lhotskya genethylloides var. *bracteosa* Benth. = ***Calytrix alpestris*** (Lindl.) Court
Lhotskya genethylloides var. *glabra* F.Muell. = ***Calytrix alpestris*** (Lindl.) Court
Lhotskya glaberrima F.Muell. = ***Calytrix glaberrima*** (F.Muell.) Craven
Lhotskya glaberrima var. *magnisepala* J.M.Black = ***Calytrix glaberrima*** (F.Muell.) Craven
Lhotskya harvestiana F.Muell. = ***Calytrix harvestiana*** (F.Muell.) Craven
Lhotskya hirta Regel = ***Calytrix acutifolia*** (Lindl.) Craven
Lhotskya purpurea F.Muell. = ***Calytrix purpurea*** (F.Muell.) Craven

Lhotskya scabra Turcz. = ***Calytrix acutifolia*** (Lindl.) Craven
Lhotskya smeatoniana F.Muell. = ***Calytrix smeatoniana*** (F.Muell.) Craven
Lhotskya violacea Lindl. = ***Calytrix violacea*** (Lindl.) Craven
Lhotskya violacea var. *merrilliana* F.Muell. & Tate = ***Calytrix merrelliana*** (F.Muell. & Tate) Craven

Lindsayomyrtus

Lindsayomyrtus B.Hyland & Steenis, Blumea 21: 190 (1973).
Maluku to N. Queensland. 42 MOL 43 NWG 50 QLD.
1 Species

Lindsayomyrtus racemoides (Greves) Craven, Austral. Syst. Bot. 3: 731 (1990).
Maluku to N. Queensland. 42 MOL 43 NWG 50 QLD. Phan.
**Eugenia racemoides* Greves, J. Bot. 61(Suppl.): 20 (1923).
Xanthostemon brachyandrus C.T.White, Proc. Roy. Soc. Queensland 53: 219 (1942). *Lindsayomyrtus brachyandrus* (C.T.White) B.Hyland & Steenis, Blumea 21: 190 (1973).
Metrosideros nigroviridis Steenis, Acta Bot. Neerl. 2: 299 (1953).

Synonyms:
Lindsayomyrtus brachyandrus (C.T.White) B.Hyland & Steenis = ***Lindsayomyrtus racemoides*** (Greves) Craven

Lithomyrtus

Lithomyrtus F.Muell., Hooker's J. Bot. Kew Gard. Misc. 9: 228 (1857).
Australia. 50 NTA QLD WAU.
11 Species
**Fenzlia* Endl., Atakta Bot.: 19 (1833), nom. illeg.

Lithomyrtus cordata (A.J.Scott) N.Snow & Guymer, Austrobaileya 5: 184 (1999).
Northern Territory. 50 NTA. Nanophan. or phan.
**Myrtella cordata* A.J.Scott, Kew Bull. 33: 301 (1978).

Lithomyrtus densifolia N.Snow & Guymer, Austrobaileya 5: 185 (1999).
Northern Territory. 50 NTA. Cham. or nanophan.

Lithomyrtus dunlopii N.Snow & Guymer, Austrobaileya 5: 186 (1999).
Northern Territory. 50 NTA. Nanophan.

Lithomyrtus grandifolia N.Snow & Guymer, Austrobaileya 5: 189 (1999).
Northern Territory. 50 NTA. Nanophan.

Lithomyrtus hypoleuca F.Muell. ex N.Snow & Guymer, Austrobaileya 5: 191 (1999).
Northern Territory to Queensland. 50 NTA QLD. Nanophan.

Lithomyrtus kakaduensis N.Snow & Guymer, Austrobaileya 5: 192 (1999).
Northern Territory. 50 NTA. Nanophan.

Lithomyrtus linariifolia N.Snow & Guymer, Austrobaileya 5: 193 (1999).
Northern Territory. 50 NTA. Cham.

Lithomyrtus microphylla (Benth.) N.Snow & Guymer, Austrobaileya 5: 193 (1999).
Queensland. 50 QLD. Cham. or nanophan.
**Fenzlia obtusa* var. *microphylla* Benth., Fl. Austral. 3: 279 (1867). *Fenzlia microphylla* (Benth.) Domin, Biblioth. Bot. 89: 476 (1928). *Myrtella microphylla* (Benth.) A.J.Scott, Kew Bull. 33: 301 (1978).

Lithomyrtus obtusa (Endl.) N.Snow & Guymer, Austrobaileya 5: 194 (1999).
Queensland. 50 QLD. Cham. or nanophan.
**Fenzlia obtusa* Endl., Atakta Bot.: 19 (1834). *Myrtella obtusa* (Endl.) A.J.Scott, Kew Bull. 33: 300 (1978).

Lithomyrtus repens N.Snow & Guymer, Austrobaileya 5: 197 (1999).
Northern Territory. 50 NTA. Cham.

Lithomyrtus retusa (Endl.) N.Snow & Guymer, Austrobaileya 5: 198 (1999).
N. Australia. 50 NTA QLD WAU. Nanophan. or phan.
**Fenzlia retusa* Endl., Atakta Bot.: 20 (1834). *Myrtella retusa* (Endl.) A.J.Scott, Kew Bull. 33: 300 (1978).
Fenzlia phebalioides W.Fitzg., J. Proc. Roy. Soc. W. Australia 3: 189 (1918). *Myrtella phebalioides* (W.Fitzg.) A.J.Scott, Kew Bull. 33: 301 (1978).

Lomastelma

Lomastelma Raf. = ***Syzygium*** Gaertn.
Lomastelma elliptica Raf. = ***Syzygium smithii*** (Poir.) Nied.
Lomastelma smithii (Poir.) J.H.Willis = ***Syzygium smithii*** (Poir.) Nied.

Lophomyrtus

Lophomyrtus Burret, Notizbl. Bot. Gart. Berlin-Dahlem 15: 489 (1941).
New Zealand. 51 NZN NZS.
2 Species

Lophomyrtus bullata Burret, Notizbl. Bot. Gart. Berlin-Dahlem 15: 489 (1941).
New Zealand. 51 NZN NZS. Nanophan. or phan.
**Myrtus bullata* Sol. ex A.Cunn., Ann. Nat. Hist. 3: 115 (1839), nom. illeg. *Lophomyrtus aotearoana* E.C. Nelson, New Zealand J. Bot. 33: 558 (1995), nom. illeg. *Myrtus aotearoana* (E.C.Nelson) E.C.Nelson, New Zealand J. Bot. 33: 558 (1995), nom. inval.

Lophomyrtus obcordata (Raoul) Burret, Notizbl. Bot. Gart. Berlin-Dahlem 15: 490 (1941).
New Zealand. 51 NZN NZS. Nanophan. or phan.
**Eugenia obcordata* Raoul, Ann. Sci. Nat., Bot., III, 2: 122 (1844). *Myrtus obcordata* (Raoul) Hook.f., Fl. Nov.-Zel. 1: 71 (1852).

Lophomyrtus* × *ralphii (Hook.f.) Burret, Notizbl. Bot. Gart. Berlin-Dahlem 15: 490 (1941). *L. bullata* × *L. obcordata*.
New Zealand. 51 NZN NZS. Nanophan. or phan.
**Myrtus* × *ralphii* Hook.f., Fl. Nov.-Zel. 2: 329 (1855).
Myrtus × *bullobcordata* Cockayne & Allan, Nature 118: 624 (1926). *Lophomyrtus* × *bullobcordata* (Cockayne & Allan) Burret, Repert. Spec. Nov. Regni Veg. 50: 60 (1941).

Synonyms:
Lophomyrtus aotearoana E.C.Nelson = ***Lophomyrtus bullata*** Burret
Lophomyrtus × *bullobcordata* (Cockayne & Allan) Burret = ***Lophomyrtus* × *ralphii*** (Hook.f.) Burret

Lophostemon

Lophostemon Schott, Wiener Z. Kunst 3: 772 (1830).
New Guinea to Australia. (25) ken tan uga 43 NWG 50 NSW NTA QLD WAU.
4 Species

Lophostemon confertus (R.Br.) Peter G.Wilson & J.T.Waterh., Austral. J. Bot. 30: 424 (1982).
E. Australia. (25) ken tan uga 50 NSW QLD. Phan.
**Tristania conferta* R.Br. in W.T.Aiton, Hortus Kew. 4: 417 (1812). *Melaleuca conferta* (R.Br.) Steud., Nomencl. Bot.: 515 (1821). *Tristania conferta* var. *typica* Domin, Biblioth. Bot. 89: 471 (1928), nom. inval.
Lophostemon arborescens Schott, Wiener Z. Kunst 3: 772 (1830).
Tristania subverticillata H.Wendl., Allg. Gartenzeitung 1: 186 (1833).
Tristania depressa A.Cunn., Edwards's Bot. Reg. 22: t. 1839 (1836).
Tristania macrophylla A.Cunn., Edwards's Bot. Reg. 22: t. 1839 (1836).
Tristania conferta Griff., Not. Pl. Asiat. 4: 649 (1854), sensu auct. *Tristania griffithii* Kurz, Prelim. Rep. Forest Pegu, App. B: 50 (1875).
Tristania conferta var. *fibrosa* F.M.Bailey, Queensl. Fl. 2: 636 (1900).
Tristania conferta var. *microcarpa* Domin, Biblioth. Bot. 89: 1025 (1928).

Lophostemon grandiflorus (Benth.) Peter G.Wilson & J.T.Waterh., Austral. J. Bot. 30: 429 (1982).
N. Australia. 50 NTA QLD WAU. Phan.
**Tristania suaveolens* var. *grandiflora* Benth., Fl. Austral. 3: 263 (1867). *Tristania grandiflora* (Benth.) Cheel, Fl. N. Territory: 290 (1917).

subsp. ***grandiflorus***
N. Western Australia to Northern Territory. 50 NTA WAU. Phan.

subsp. ***riparius*** (Domin) Peter G.Wilson & J.T.Waterh., Austral. J. Bot. 30: 430 (1982).
N. Australia. 50 NTA QLD WAU. Phan.
**Tristania suaveolens* var. *riparia* Domin, Biblioth. Bot. 89: 1025 (1928).

Lophostemon lactifluus (F.Muell.) Peter G.Wilson & J.T.Waterh., Austral. J. Bot. 30: 428 (1982).
N. Western Australia to Northern Territory. 50 NTA WAU. Phan.
**Tristania lactiflua* F.Muell., Fragm. 1: 82 (1858).

Lophostemon suaveolens (Sol. ex Gaertn.) Peter G. Wilson & J.T.Waterh., Austral. J. Bot. 30: 425 (1982).
New Guinea to E. Australia. 43 NWG 50 NSW QLD. Phan.
**Melaleuca suaveolens* Sol. ex Gaertn., Fruct. Sem. Pl. 1: 173 (1788). *Tristania suaveolens* (Sol. ex Gaertn.) Sm. in A.Rees, Cycl. 36: 2 (1817). *Tristania suaveolens* var. *vulgaris* Domin, Biblioth. Bot. 89: 470 (1928), nom. inval.
Tristania depressa Link, Enum. Hort. Berol. Alt. 2: 273 (1822), nom. nud.
Lophostemon depressus Schott, Wiener Z. Kunst 3: 772 (1830).
Tristania salicifolia Link ex Steud., Nomencl. Bot., ed. 2, 2: 714 (1841).
Tristania rhytiphloia F.Muell., Fragm. 1: 81 (1858).
Tristania suaveolens var. *glabrescens* F.M.Bailey, Syn. Queensl. Fl.: 182 (1883).

Synonyms:
Lophostemon arborescens Schott = ***Lophostemon confertus*** (R.Br.) Peter G.Wilson & J.T.Waterh.
Lophostemon depressus Schott = ***Lophostemon suaveolens*** (Sol. ex Gaertn.) Peter G.Wilson & J.T.Waterh.

Luma

Luma A.Gray, U.S. Expl. Exped., Phan. 1: 535 (1854).
Chile to W. Argentina. (76) cal (83) bol per 85 AGS CLC CLS.
2 Species
Myrceugenella Kausel, Revista Argent. Agron. 9: 42 (1942).

Luma apiculata (DC.) Burret, Notizbl. Bot. Gart. Berlin-Dahlem 16: 523 (1941).
SC. & S. Chile to SW. Argentina. (76) cal 85 AGS CLC CLS. Nanophan. or phan.
**Eugenia apiculata* DC., Prodr. 3: 276 (1828). *Myrceugenia apiculata* (DC.) Nied. in H.G.A.Engler & K.A.E.Prantl, Nat. Pflanzenfam. 3(7): 74 (1893). *Myrtus chekenilla* Kuntze, Revis. Gen. Pl. 3(2): 90 (1898). *Myrceugenella apiculata* (DC.) Kausel, Revista Argent. Agron. 9: 47 (1942). *Myrceugenella apiculata* var. *genuina* Kausel, Revista Argent. Agron. 9: 48 (1942), nom. inval.
Eugenia affinis Gillies ex Hook. & Arn., Bot. Misc. 3: 321 (1833), nom. illeg. *Eugenia hookeri* Steud., Nomencl. Bot., ed. 2, 1: 603 (1840). *Luma hookeri* (Steud.) Burret, Notizbl. Bot. Gart. Berlin-Dahlem 15: 526 (1941).
Eugenia gilliesii Hook. & Arn., Bot. Misc. 3: 320 (1833). *Luma gilliesii* (Hook. & Arn.) Burret, Notizbl. Bot. Gart. Berlin-Dahlem 15: 523 (1941).
Eugenia apiculata var. *arnyan* Hook.f., Fl. Antarct. 2: 277 (1846).
Eugenia luma O.Berg, Linnaea 27: 251 (1856).
Eugenia spectabilis Phil., Linnaea 28: 639 (1857). *Luma spectabilis* (Phil.) Burret, Notizbl. Bot. Gart. Berlin-Dahlem 15: 523 (1941). *Myrceugenella apiculata* var. *spectabilis* (Phil.) Kausel, Revista Argent. Agron. 9: 49 (1942).
Eugenia modesta Phil., Linnaea 28: 688 (1858), nom. illeg. *Eugenia proba* O.Berg, Linnaea 29: 225 (1858).
Eugenia mucronata Phil., Anales Univ. Chile 1865(2): 327 (1865), nom. illeg.
Eugenia cuspidata Phil., Anales Univ. Chile 84: 755 (1893), nom. illeg.
Eugenia ebracteata Phil., Anales Univ. Chile 84: 758 (1893).
Eugenia palenae Phil., Anales Univ. Chile 84: 756 (1893).
Myrceugenella apiculata var. *australis* Kausel, Revista Argent. Agron. 9: 49 (1942).
Myrceugenella apiculata var. *nahulhuapensis* Kausel, Revista Argent. Agron. 9: 236 (1942).
Myrceugenella grandjotii Kausel, Revista Argent. Agron. 9: 50 (1942).
Myrceugenia apiculata var. *australis* Kausel, Revista Argent. Agron. 9: 49 (1942).

Luma chequen (Molina) A.Gray, U.S. Expl. Exped., Phan. 1: 536 (1854).
C. & SC. Chile. (83) bol per 85 CLC CLS. Nanophan. or phan.
**Eugenia chequen* Molina, Sag. Stor. Nat. Chili, ed. 2: 148 (1810). *Myrtus chequen* (Molina) Spreng., Syst.

Veg. 2: 485 (1825). *Myrceugenella chequen* (Molina) Kausel, Revista Argent. Agron. 9: 43 (1942).
Eugenia chekan DC., Prodr. 3: 278 (1828), orth. var.
Myrtus luma Schauer, Nov. Actorum Acad. Caes. Leop.-Carol. Nat. Cur. 19(Suppl. 1): 333 (1843), nom. illeg.
Eugenia gayana Barnéoud in C.Gay, Fl. Chil. 2: 390 (1847). *Myrtus gayana* (Barnéoud) O.Berg, Linnaea 27: 399 (1856). *Luma gayana* (Barnéoud) Burret, Notizbl. Bot. Gart. Berlin-Dahlem 15: 523 (1941). *Myrceugenella gayana* (Barnéoud) Kausel, Revista Argent. Agron. 9: 46 (1942).
Myrtus uliginosa Miq., Linnaea 25: 652 (1853).
Myrtus gayana var. *major* O.Berg, Bot. Zeitung (Berlin) 15: 859 (1857).
Eugenia bella Phil., Linnaea 28: 641 (1858), nom. illeg. *Eugenia pulchra* O.Berg, Linnaea 29: 224 (1858).
Eugenia myrtomimeta Diels, Bot. Jahrb. Syst. 37: 598 (1906). *Myrceugenella chequen* var. *myrtomimeta* (Diels) Kausel, Lilloa 33: 104 (1971 publ. 1972).
Myrceugenella langerfeldtii Kausel, Revista Argent. Agron. 11: 324 (1944).

***Synonyms*:**
Luma acrophylla (O.Berg) Burret = ***Myrceugenia myrcioides*** (Cambess.) O.Berg
Luma acutiflora (Kiaersk.) Burret = ***Myrceugenia acutiflora*** (Kiaersk.) D.Legrand & Kausel
Luma adenocarpa (O.Berg) Burret = ***Eugenia adenocarpa*** O.Berg
Luma alpigena (DC.) Burret = ***Myrceugenia alpigena*** (DC.) Landrum
Luma anceps (O.Berg) Burret = ***Myrceugenia bracteosa*** (DC.) D.Legrand & Kausel
Luma angustifolia Burret = ***Myrceugenia euosma*** (O.Berg) D.Legrand
Luma angustior Burret = ***Myrceugenia glaucescens*** (Cambess.) D.Legrand & Kausel var. ***glaucescens***
Luma aprica (O.Berg) Burret = ***Myrceugenia euosma*** (O.Berg) D.Legrand
Luma araujoana (O.Berg) Burret = ***Myrceugenia glaucescens*** (Cambess.) D.Legrand & Kausel var. ***glaucescens***
Luma arechavaletae (Herter) Herter = ***Eugenia uniflora*** L.
Luma baeckeoides Griseb. = ***Myrceugenia parvifolia*** (DC.) Kausel
Luma bagensis (O.Berg) Burret = ***Myrceugenia glaucescens*** (Cambess.) D.Legrand & Kausel var. ***glaucescens***
Luma bergii Herter = ***Eugenia moschata*** (Aubl.) Nied. ex T.Durand & B.D.Jacks.
Luma brachymischa (Kiaersk.) Burret = ***Myrceugenia bracteosa*** (DC.) D.Legrand & Kausel
Luma bracteosa (DC.) Burret = ***Myrceugenia bracteosa*** (DC.) D.Legrand & Kausel
Luma brevipedicellata Burret = ***Myrceugenia brevipedicellata*** (Burret) D.Legrand & Kausel
Luma brevipes Burret = ***Myrceugenia miersiana*** (Gardner) D.Legrand & Kausel
Luma × *bridgesii* (Hook. & Arn.) A.Gray ex F.Phil. = ***Myrceugenia* × *bridgesii*** (Hook. & Arn.) O.Berg
Luma calycosema (O.Berg) Herter = ***Eugenia uruguayensis*** Cambess.
Luma cambessedeana (O.Berg) Burret = ***Myrceugenia glaucescens*** (Cambess.) D.Legrand & Kausel var. ***glaucescens***
Luma campestris (DC.) Burret = ***Myrceugenia campestris*** (DC.) D.Legrand & Kausel
Luma canelonensis (O.Berg) Burret = ***Myrceugenia glaucescens*** (Cambess.) D.Legrand & Kausel var. ***glaucescens***
Luma chrysocarpa (O.Berg) Burret = ***Myrceugenia chrysocarpa*** (O.Berg) Kausel
Luma cinerea Burret = ***Myrceugenia euosma*** (O.Berg) D.Legrand
Luma cinnamomeotomentosa Burret = ***Myrceugenia myrtoides*** O.Berg
Luma cisplatenais (Cambess.) Herter = ***Myrcianthes cisplatensis*** (Cambess.) O.Berg
Luma concinna (Phil.) Herter = ***Legrandia concinna*** (Phil.) Kausel
Luma corralensis (Phil.) Burret = ***Myrceugenia exsucca*** (DC.) O.Berg
Luma correifolia (Hook. & Arn.) A.Gray = ***Myrceugenia correifolia*** (Hook. & Arn.) O.Berg
Luma costata (Cambess.) Herter = ***Eugenia uniflora*** L.
Luma cruckshanksii (Hook. & Arn.) A.Gray = ***Blepharocalyx cruckshanksii*** (Hook. & Arn.) Nied.
Luma cumingii (Hook. & Arn.) Burret = ***Myrceugenia ovata*** (Hook. & Arn.) O.Berg var. ***ovata***
Luma dasyblasta (O.Berg) Herter = ***Eugenia uniflora*** L.
Luma dimorpha (O.Berg) Herter = ***Eugenia dimorpha*** O.Berg
Luma distans (O.Berg) Burret = ***Myrceugenia campestris*** (DC.) D.Legrand & Kausel
Luma distoma (O.Berg) Burret = ***Myrceugenia planipes*** (Hook. & Arn.) O.Berg
Luma dombeyana (DC.) Burret = ***Myrceugenia lanceolata*** (Juss. ex J.St.-Hil.) Kausel
Luma elegans (O.Berg) Burret = ***Myrceugenia glaucescens*** (Cambess.) D.Legrand & Kausel var. ***glaucescens***
Luma ereophila Burret = ***Myrceugenia ovata*** var. ***regnelliana*** (O.Berg) Landrum
Luma estrellensis (O.Berg) Burret = ***Myrceugenia myrcioides*** (Cambess.) O.Berg
Luma euosma (O.Berg) Burret = ***Myrceugenia euosma*** (O.Berg) D.Legrand
Luma expallens (O.Berg) Burret = ***Myrceugenia bracteosa*** (DC.) D.Legrand & Kausel
Luma exsucca (DC.) Burret = ***Myrceugenia exsucca*** (DC.) O.Berg
Luma fernandeziana (Hook. & Arn.) Burret = ***Myrceugenia fernandeziana*** (Hook. & Arn.) Johow
Luma ferruginea A.Gray = ***Myrceugenia rufa*** (Colla) Skottsb.
Luma filibracteata Burret = ***Myrceugenia miersiana*** (Gardner) D.Legrand & Kausel
Luma filipes Burret = [84 BZL]
Luma foveolata (O.Berg) Burret = ***Myrceugenia foveolata*** (O.Berg) Sobral
Luma franciscensis (O.Berg) Burret = ***Myrceugenia franciscensis*** (O.Berg) Landrum
Luma fuliginea (O.Berg) Burret = ***Myrceugenia alpigena*** (DC.) Landrum var. ***alpigena***
Luma fuscovelutina Burret = ***Myrceugenia miersiana*** (Gardner) D.Legrand & Kausel
Luma gayana (Barnéoud) Burret = ***Luma chequen*** (Molina) A.Gray
Luma gilliesii (Hook. & Arn.) Burret = ***Luma apiculata*** (DC.) Burret
Luma glaucescens (Cambess.) Burret = ***Myrceugenia glaucescens*** (Cambess.) D.Legrand & Kausel
Luma gracilis Burret = ***Myrceugenia ovata*** var. ***regnelliana*** (O.Berg) Landrum
Luma grisebachii Herter = ***Hexachlamys edulis*** (O.Berg) Kausel & D.Legrand

Luma hoehnei Burret = ***Myrceugenia hoehnei*** (Burret) D.Legrand & Kausel
Luma hookeri (Steud.) Burret = ***Luma apiculata*** (DC.) Burret
Luma itatiaiensis (Kiaersk.) Burret = ***Myrceugenia ovata*** var. ***regnelliana*** (O.Berg) Landrum
Luma latior Burret = ***Myrceugenia glaucescens*** var. ***latior*** (Burret) Landrum
Luma leptospermoides (DC.) Burret = ***Myrceugenia leptospermoides*** (DC.) Kausel
Luma longifolia Burret = ***Myrceugenia alpigena*** var. ***longifolia*** (Burret) Landrum
Luma lucidissima Herter = ***Eugenia cerasiflora*** Miq.
Luma macahensis Burret = ***Myrceugenia pilotantha*** (Kiaersk.) Landrum var. ***pilotantha***
Luma macromischa Burret = ***Myrceugenia ovata*** var. ***regnelliana*** (O.Berg) Landrum
Luma macrosepala Burret = ***Myrceugenia oxysepala*** (Burret) D.Legrand & Kausel
Luma maschalantha (O.Berg) Herter = ***Myrciaria delicatula*** (DC.) O.Berg
Luma mesomischa Burret = ***Myrceugenia mesomischa*** (Burret) D.Legrand & Kausel
Luma miersiana (Gardner) Burret = ***Myrceugenia miersiana*** (Gardner) D.Legrand & Kausel
Luma montana Burret = ***Myrceugenia bracteosa*** (DC.) D.Legrand & Kausel
Luma montevidensis (O.Berg) Burret = ***Myrceugenia myrtoides*** O.Berg
Luma mucronata Burret = ***Myrceugenia franciscensis*** (O.Berg) Landrum
Luma multiflora Herter = ***Eugenia hyemalis*** Cambess.
Luma myrcianthes (Nied.) Herter = ***Hexachlamys edulis*** (O.Berg) Kausel & D.Legrand
Luma myrcioides (Cambess.) Burret = ***Myrceugenia myrcioides*** (Cambess.) O.Berg
Luma myrtoides (O.Berg) Burret = ***Myrceugenia myrtoides*** O.Berg
Luma nana (O.Berg) Burret = ***Myrceugenia euosma*** (O.Berg) D.Legrand
Luma nannophylla Burret = ***Myrceugenia ovata*** var. ***nannophylla*** (Burret) Landrum
Luma obtusa (DC.) A.Gray = ***Myrceugenia obtusa*** (DC.) O.Berg
Luma obtusiflora (Kiaersk.) Burret = ***Myrceugenia myrcioides*** (Cambess.) O.Berg
Luma opaca (O.Berg) Herter = ***Eugenia uruguayensis*** Cambess.
Luma oreophila Burret = ***Myrceugenia ovata*** var. ***regnelliana*** (O.Berg) Landrum
Luma ovata (Hook. & Arn.) A.Gray ex F.Phil. = ***Myrceugenia ovata*** (Hook. & Arn.) O.Berg
Luma oxysepala Burret = ***Myrceugenia oxysepala*** (Burret) D.Legrand & Kausel
Luma pallida (O.Berg) Burret = ***Myrceugenia glaucescens*** (Cambess.) D.Legrand & Kausel var. ***glaucescens***
Luma parvifolia (DC.) Burret = ***Myrceugenia parvifolia*** (DC.) Kausel
Luma philippii (O.Berg) Burret = ***Myrceugenia chrysocarpa*** (O.Berg) Kausel
Luma pilotantha (Kiaersk.) Burret = ***Myrceugenia pilotantha*** (Kiaersk.) Landrum
Luma pitanga (O.Berg) Herter = ***Eugenia pitanga*** (O.Berg) Nied.
Luma pitra (O.Berg) Burret = ***Myrceugenia exsucca*** (DC.) O.Berg
Luma planipes (Hook. & Arn.) A.Gray = ***Myrceugenia planipes*** (Hook. & Arn.) O.Berg
Luma planiramea (O.Berg) Burret = ***Myrceugenia rufescens*** (DC.) D.Legrand & Kausel
Luma poeppigiana (O.Berg) Burret = ***Myrceugenia rufescens*** (DC.) D.Legrand & Kausel
Luma pungens (O.Berg) Herter = ***Myrcianthes pungens*** (O.Berg) D.Legrand
Luma regnelliana (O.Berg) Burret = ***Myrceugenia ovata*** var. ***regnelliana*** (O.Berg) Landrum
Luma repanda (O.Berg) Herter = ***Eugenia repanda*** O.Berg
Luma retusa (O.Berg) Herter = ?
Luma ribeireana (O.Berg) Burret = ***Myrceugenia glaucescens*** (Cambess.) D.Legrand & Kausel var. ***glaucescens***
Luma rufa (Colla) Burret = ***Myrceugenia rufa*** (Colla) Skottsb.
Luma rufescens (DC.) Burret = ***Myrceugenia rufescens*** (DC.) D.Legrand & Kausel
Luma ruiziana (O.Berg) Burret = ?
Luma rupestris (O.Berg) Burret = ***Myrceugenia alpigena*** (DC.) Landrum var. ***alpigena***
Luma schulzii (Johow) Burret = ***Myrceugenia schulzii*** Johow
Luma sellowiana (O.Berg) Burret = ***Myrceugenia myrtoides*** O.Berg
Luma seriatopedunculata (Kiaersk.) Burret = ***Myrceugenia alpigena*** (DC.) Landrum var. ***alpigena***
Luma seriatoramosa (Kiaersk.) Burret = ***Myrceugenia seriatoramosa*** (Kiaersk.) D.Legrand & Kausel
Luma spectabilis (Phil.) Burret = ***Luma apiculata*** (DC.) Burret
Luma stenophylla (Hook. & Arn.) A.Gray = ***Myrceugenia lanceolata*** (Juss. ex J.St.-Hil.) Kausel
Luma sticheromischa (Kiaersk.) Burret = ***Myrceugenia alpigena*** (DC.) Landrum var. ***alpigena***
Luma strigosa (O.Berg) Herter = ***Eugenia uniflora*** L.
Luma temu (Hook. & Arn.) A.Gray = ***Myrceugenia exsucca*** (DC.) O.Berg
Luma thalassaia (O.Berg) Burret = ***Myrceugenia correifolia*** (Hook. & Arn.) O.Berg
Luma turbinata (O.Berg) Herter = ***Eugenia pyriformis*** Cambess.
Luma ulei Burret = ***Myrceugenia myrcioides*** (Cambess.) O.Berg
Luma uruguayensis (Cambess.) Herter = ***Eugenia uruguayensis*** Cambess.
Luma verticillaris (O.Berg) Herter = ***Myrcia verticillaris*** O.Berg
Luma xanthochlora (O.Berg) Burret = ***Myrceugenia ovata*** var. ***regnelliana*** (O.Berg) Landrum

***Unplaced Names*:**
Luma filipes Burret, Notizbl. Bot. Gart. Berlin-Dahlem 15: 529 (1941). = [84 BZL]

Lysicarpus

Lysicarpus F.Muell., Trans. & Proc. Philos. Inst. Victoria 2: 67 (1858).
Queensland. 50 QLD.
1 Species

Lysicarpus angustifolius (Hook.) Druce, Rep. Bot. Exch. Club Brit. Isles 1916: 634 (1917).
Queensland. 50 QLD. Nanophan.
**Tristania angustifolia* Hook. in T.L.Mitchell, J. Exped. Trop. Australia: 198 (1848).
Lysicarpus ternifolius F.Muell., Trans. & Proc. Philos. Inst. Victoria 2: 68 (1858). *Nania ternifolia* (F.Muell.) Kuntze, Revis. Gen. Pl. 1: 242 (1891).

Metrosideros ternifolia F.Muell., Docum. Intercol. Exhib. 5: 30 (1867), nom. inval.

Synonyms:
Lysicarpus ternifolius F.Muell. = ***Lysicarpus angustifolius*** (Hook.) Druce

Macklottia

Macklottia Korth. = ***Leptospermum*** J.R.Forst. & G.Forst.
Macklottia amboinensis (Reinw. ex Blume) Korth. = ***Leptospermum javanicum*** Blume
Macklottia javanica (Blume) Korth. = ***Leptospermum javanicum*** Blume

Macromyrtus

Macromyrtus Miq. = ***Syzygium*** Gaertn.
Macromyrtus javanica Miq. = ***Syzygium macromyrtus*** (Koord. & Valeton) Merr. & L.M.Perry

Macropsidium

Macropsidium Blume = ***Syzygium*** Gaertn.
Macropsidium elegans Blume = ***Rhodomyrtus elegans*** (Blume) A.J.Scott
Macropsidium rubrum (Lour.) Blume = ?

Malidra

Malidra Raf. = ***Syzygium*** Gaertn.
Malidra aquea (Burm.f.) Raf. = ***Syzygium aqueum*** (Burm.f.) Alston

Malleostemon

Malleostemon J.W.Green, Nuytsia 4: 296 (1983).
W. & SW. Australia. 50 WAU.
6 Species

Malleostemon hursthousei (W.Fitzg.) J.W.Green, Nuytsia 4: 297 (1983).
W. Western Australia. 50 WAU. Nanophan.
**Micromyrtus hursthousei* W.Fitzg., J. West Austral. Nat. Hist. Soc. 2: 22 (1905).

Malleostemon minilyaensis J.W.Green, Nuytsia 4: 299 (1983).
W. Western Australia. 50 WAU. Nanophan.

Malleostemon pedunculatus J.W.Green, Nuytsia 4: 301 (1983).
W. Western Australia. 50 WAU. Nanophan.

Malleostemon peltiger (S.Moore) J.W.Green, Nuytsia 4: 303 (1983).
SW. Australia. 50 WAU. Nanophan.
**Micromyrtus peltigera* S.Moore, J. Linn. Soc., Bot. 45: 200 (1920).

Malleostemon roseus (E.Pritz.) J.W.Green, Nuytsia 4: 305 (1983).
SW. Australia. 50 WAU. Nanophan.
**Thryptomene rosea* E.Pritz., Bot. Jahrb. Syst. 35: 413 (1904).
Micromyrtus rosea S.Moore, J. Linn. Soc., Bot. 45: 200 (1920).

Malleostemon tuberculatus (E.Pritz.) J.W.Green, Nuytsia 4: 308 (1983).
SW. Australia. 50 WAU. Nanophan.
**Thryptomene tuberculata* E.Pritz., Bot. Jahrb. Syst. 35: 411 (1904).

Manglesia

Manglesia Lindl. = ***Beaufortia*** R.Br.
Manglesia purpurea Lindl. = ***Beaufortia purpurea*** Lindl.

Marlierea

Marlierea Cambess. in A.F.C.de Saint-Hilaire, Fl. Bras. Merid. 2: 373 (1833).
Trop. America. 80 COS 81 PUE TRT WIN 82 FRG GUY SUR VEN 83 BOL CLM ECU PER 84 BZC BZE BZL BZN BZS.
90 Species
Eugeniopsis O.Berg, Linnaea 27: 80 (1855).
Rubachia O.Berg, Linnaea 27: 11 (1855).
Krugia Urb., Ber. Deutsch. Bot. Ges. 11: 375 (1893).

Marlierea acuminatissima (O.Berg) D.Legrand, Comun. Bot. Mus. Hist. Nat. Montevideo 3(40): 30 (1962).
NE. Brazil. 84 BZE. Nanophan. or phan.
**Eugeniopsis acuminatissima* O.Berg in C.F.P.von Martius & auct. suc. (eds.), Fl. Bras. 14(1): !44 (1857).

Marlierea angustifolia (O.Berg) Mattos, Ci. & Cult. 19: 333 (1967).
SE. Brazil. 84 BZL. Nanophan. or phan.
**Eugeniopsis angustifolia* O.Berg in C.F.P.von Martius & auct. suc. (eds.), Fl. Bras. 14(1): 143 (1857).
Myrcia lenheirensis Kiaersk., Enum. Myrt. Bras.: 98 (1893).

Marlierea antonia (O.Berg) D.Legrand, Comun. Bot. Mus. Hist. Nat. Montevideo 3(40): 29 (1962).
SE. Brazil. 84 BZL. Nanophan. or phan.
**Rubachia antonia* O.Berg in C.F.P.von Martius & auct. suc. (eds.), Fl. Bras. 14(1): 29 (1857).

Marlierea areolata McVaugh, Fieldiana, Bot. 29: 175 (1956).
N. Peru. 83 PER. Nanophan. or phan.

Marlierea bipennis (O.Berg) McVaugh, Mem. New York Bot. Gard. 10(1): 79 (1958).
Venezuela, N. Brazil, Peru. 82 VEN 83 PER 84 BZN. Nanophan.
**Myrciaria bipennis* O.Berg, Linnaea 31: 259 (1862).
Myrcia bipennis (O.Berg) McVaugh, Fieldiana, Bot. 29: 189 (1956).

Marlierea biptera Amshoff, Recueil Trav. Bot. Néerl. 42: 3 (1950).
Suriname. 82 SUR. Nanophan. or phan.

Marlierea buxifolia Amshoff, Bull. Torrey Bot. Club 75: 529 (1948).
Venezuela to Guyana. 82 GUY VEN. Nanophan. or phan.

Marlierea caesariata McVaugh, Mem. New York Bot. Gard. 10(1): 81 (1958).
Venezuela. 82 VEN. Nanophan. or phan.

Marlierea cana McVaugh, Mem. New York Bot. Gard. 10(1): 82 (1958).
Venezuela (Amazonas). 82 VEN. Nanophan. or phan.

Marlierea capitata O.Berg in C.F.P.von Martius & auct. suc. (eds.), Fl. Bras. 14(1): 33 (1857).
Brazil (Rio de Janeiro). 84 BZL. Nanophan. or phan.

Marlierea caudata McVaugh, Fieldiana, Bot. 29: 176 (1956).
S. Venezuela to Peru. 82 VEN 83 CLM PER 84 BZC BZN. Nanophan. or phan.

Marlierea choriophylla Kiaersk., Enum. Myrt. Bras.: 47 (1893).
SE. Brazil. 84 BZL. Nanophan. or phan.

Marlierea clausseniana (O.Berg) Kiaersk., Enum. Myrt. Bras.: 51 (1893).
Brazil. 84 BZC BZE BZL. Phan.
**Eugeniopsis clausseniana* O.Berg in C.F.P.von Martius & auct. suc. (eds.), Fl. Bras. 14(1): 145 (1857).
Eugeniopsis clausseniana var. *glabrata* O.Berg in C.F.P.von Martius & auct. suc. (eds.), Fl. Bras. 14(1): 146 (1857).
Eugeniopsis clausseniana var. *rufa* O.Berg in C.F.P. von Martius & auct. suc. (eds.), Fl. Bras. 14(1): 146 (1857).

Marlierea convexivenia B.Holst, Selbyana 23: 144 (2002).
S. Venezuela. 82 VEN. Nanophan. or phan.

Marlierea cordata Mattos, Loefgrenia 94: 7 (1989).
Brazil (Espírito Santo). 84 BZL. Nanophan. or phan.

Marlierea cuprea Amshoff, Bull. Torrey Bot. Club 75: 530 (1948).
S. Venezuela to Guyana. 82 GUY VEN. Nanophan. or phan.

Marlierea dimorpha O.Berg in C.F.P.von Martius & auct. suc. (eds.), Fl. Bras. 14(1): 538 (1859).
Brazil (Rio de Janeiro). 84 BZL. Nanophan. or phan.

Marlierea ensiformis McVaugh, Mem. New York Bot. Gard. 18(2): 64 (1969).
N. Brazil. 84 BZN. Phan.

Marlierea estrellensis O.Berg in C.F.P.von Martius & auct. suc. (eds.), Fl. Bras. 14(1): 34 (1857).
SE. Brazil. 84 BZL. Phan.

Marlierea eugenioides (Cambess.) D.Legrand, Notul. Syst. (Paris) 15: 265 (1958).
E. Brazil. 84 BZE BZL. Phan.
Eugenia paniculata Cambess. in A.F.C.de Saint-Hilaire, Fl. Bras. Merid. 2: 338 (1832), nom. illeg. *Eugeniopsis paniculata* O.Berg, Linnaea 27: 82 (1855). *Marlierea paniculata* (O.Berg) D.Legrand, Comun. Bot. Mus. Hist. Nat. Montevideo 3(40): 31 (1962).
**Myrcia eugenioides* Cambess. in A.F.C.de Saint-Hilaire, Fl. Bras. Merid. 2: 302 (1832). *Calyptromyrcia eugenioides* (Cambess.) O.Berg in C.F.P.von Martius & auct. suc. (eds.), Fl. Bras. 14(1): 56 (1857).
Calyptromyrcia eugenioides var. *glabra* O.Berg in C.F.P.von Martius & auct. suc. (eds.), Fl. Bras. 14(1): 56 (1857).
Calyptromyrcia eugenioides var. *puberula* O.Berg in C.F.P.von Martius & auct. suc. (eds.), Fl. Bras. 14(1): 56 (1857).
Eugeniopsis ovata O.Berg in C.F.P.von Martius & auct. suc. (eds.), Fl. Bras. 14(1): 146 (1857). *Marlierea ovata* (O.Berg) Nied. in H.G.A.Engler & K.A.E.Prantl, Nat. Pflanzenfam. 3(7): 76 (1893). *Marlierea eugenioides* var. *ovata* (O.Berg) D.Legrand, Comun. Bot. Mus. Hist. Nat. Montevideo 3(40): 33 (1962).

Marlierea eugeniopsoides (D.Legrand & Kausel) D.Legrand, Bradea 2: 7 (1975).
S. Brazil. 84 BZS. Nanophan. or phan.
**Calyptranthes eugeniopsoides* D.Legrand & Kausel, Lilloa 31: 194 (1962).

Marlierea excoriata Mart., Flora 20(2 Beibl.): 88 (1837).
E. & S. Brazil. 84 BZE BZL BZS. Phan.
Marlierea excoriata var. *angustifolia* O.Berg in C.F.P. von Martius & auct. suc. (eds.), Fl. Bras. 14(1): 36 (1857).
Marlierea excoriata var. *latifolia* O.Berg in C.F.P.von Martius & auct. suc. (eds.), Fl. Bras. 14(1): 35 (1857).
Marlierea parviflora O.Berg in C.F.P.von Martius & auct. suc. (eds.), Fl. Bras. 14(1): 537 (1859).

Marlierea ferruginea (Poir.) McVaugh, Mem. New York Bot. Gard. 10(1): 83 (1958).
Windward Is. to N. Brazil. 81 TRT WIN 82 FRG GUY SUR VEN 84 BZN. Nanophan. or phan.
**Eugenia ferruginea* Poir. in J.B.A.P.M.de Lamarck, Encycl., Suppl. 3: 124 (1813). *Myrtus ferruginea* (Poir.) Spreng., Syst. Veg. 2: 487 (1825). *Myrcia ferruginea* (Poir.) DC., Prodr. 3: 245 (1828). *Krugia ferruginea* (Poir.) Urb., Bot. Jahrb. Syst. 19: 604 (1895).
Marlierea acuminata O.Berg in A.F.C.de Saint-Hilaire, Fl. Bras. Merid. 2: 373 (1833).
Marlierea elliptica Griseb., Fl. Brit. W. I.: 233 (1860). *Krugia elliptica* (Griseb.) Urb., Ber. Deutsch. Bot. Ges. 11: 376 (1893).
Eugeniopsis richardiana O.Berg, Linnaea 30: 665 (1861). *Marlierea richardiana* (O.Berg) Nied. in H.G.A.Engler & K.A.E.Prantl, Nat. Pflanzenfam. 3(7): 76 (1892), nom. illeg. *Marlierea guyanensis* D.Legrand, Comun. Bot. Mus. Hist. Nat. Montevideo 3(40): 36 (1962).

Marlierea foveolata B.Holst, Selbyana 23: 145 (2002).
S. Venezuela. 82 VEN. Nanophan. or phan.

Marlierea gardneriana (O.Berg) Nied. in H.G.A.Engler & K.A.E.Prantl, Nat. Pflanzenfam. 3(7): 76 (1893).
SE. Brazil. 84 BZL. Nanophan. or phan.
**Eugeniopsis gardneriana* O.Berg in C.F.P.von Martius & auct. suc. (eds.), Fl. Bras. 14(1): 145 (1857).

Marlierea gaudichaudiana (O.Berg) Loefgr. & Everett, Syst. Analyt. S. Paulo: 219 (1919).
SE. Brazil. 84 BZL.
**Eugeniopsis gaudichaudiana* O.Berg in C.F.P.von Martius & auct. suc. (eds.), Fl. Bras. 14(1): 147 (1857).

Marlierea glabra Cambess. in A.F.C.de Saint-Hilaire, Fl. Bras. Merid. 2: 374 (1833).
SE. Brazil. 84 BZL. Nanophan.
Marlierea grandifolia O.Berg in C.F.P.von Martius & auct. suc. (eds.), Fl. Bras. 14(1): 37 (1857).
Marlierea schottiana O.Berg in C.F.P.von Martius & auct. suc. (eds.), Fl. Bras. 14(1): 36 (1857).
Marlierea glabra var. *gracilis* O.Berg in C.F.P.von Martius & auct. suc. (eds.), Fl. Bras. 14(1): 538 (1859).
Marlierea grandifolia var. *minor* O.Berg in C.F.P.von Martius & auct. suc. (eds.), Fl. Bras. 14(1): 537 (1859).

Marlierea glazioviana Kiaersk., Enum. Myrt. Bras.: 48 (1893).
SE. Brazil. 84 BZL. Nanophan. or phan.

Marlierea guanabarina Mattos & D.Legrand, Loefgrenia 67: 6 (1975).
Brazil (Rio de Janeiro). 84 BZL. Nanophan. or phan.

Marlierea guildingiana (Griseb.) Krug & Urb., Bot. Jahrb. Syst. 19: 590 (1895).

Trinidad to N. South America. 81 TRT WIN? 82 GUY VEN. Nanophan. or phan.

Marlierea imperfecta McVaugh, Fieldiana, Bot. 29: 176 (1956).
Peru. 83 PER. Phan.

Marlierea insignis McVaugh, Fieldiana, Bot. 29: 176 (1956).
S. Venezuela to N. Peru. 82 VEN 83 CLM PER. Nanophan. or phan.

Marlierea involucrata (O.Berg) Nied. in H.G.A.Engler & K.A.E.Prantl, Nat. Pflanzenfam. 3(7): 77 (1893).
Brazil (Rio de Janeiro). 84 BZL. Nanophan. or phan.
**Rubachia involucrata* O.Berg in C.F.P.von Martius & auct. suc. (eds.), Fl. Bras. 14(1): 29 (1857).

Marlierea karuaiensis (Steyerm.) McVaugh, Mem. New York Bot. Gard. 10(1): 85 (1958).
Venezuela to Guyana. 82 GUY VEN. Nanophan. or phan.
**Aulomyrcia karuaiensis* Steyerm., Fieldiana, Bot. 28: 1005 (1957).

Marlierea krapovickae D.Legrand, in Fl. Illustr. Catar. 1(Mirt.): 484 (1971).
S. Brazil. 84 BZS. Phan.

Marlierea laevigata (DC.) Kiaersk., Enum. Myrt. Bras.: 51 (1893).
E. Brazil. 84 BZE BZL. Nanophan. or phan.
**Eugenia laevigata* DC., Prodr. 3: 283 (1828). *Eugeniopsis laevigata* (DC.) O.Berg, Linnaea 27: 81 (1855).
Syzygium laevigatum Miq., Fl. Ned. Ind. 1(1): 457 (1855).
Marlierea laevigata var. *glazioviana* Kiaersk., Enum. Myrt. Bras.: 51 (1893).

Marlierea langsdorffii O.Berg in C.F.P.von Martius & auct. suc. (eds.), Fl. Bras. 14(1): 536 (1859).
SE. Brazil. 84 BZL. Nanophan. or phan.

Marlierea leal-costae G.M.Barroso & Peixoto, Revista Brasil. Bot. 18: 105 (1995).
Brazil (Bahia). 84 BZE. Nanophan. or phan.

Marlierea ligustrina McVaugh, Mem. New York Bot. Gard. 10(1): 86 (1958).
Venezuela. 82 GUY? VEN. Nanophan. or phan.

Marlierea lituatinervia (O.Berg) McVaugh, Mem. New York Bot. Gard. 18(2): 65 (1969).
Guyana to Brazil. 82 GUY 84 BZE?. Nanophan. or phan.
**Myrciaria lituatinervia* O.Berg, Linnaea 27: 322 (1856).

Marlierea luschnathiana (O.Berg) D.Legrand, Comun. Bot. Mus. Hist. Nat. Montevideo 3(40): 33 (1962).
NE. Brazil. 84 BZE. Nanophan. or phan.
**Eugeniopsis luschnathiana* O.Berg in C.F.P.von Martius & auct. suc. (eds.), Fl. Bras. 14(1): 147 (1857).

Marlierea macrophylla Amshoff, Recueil Trav. Bot. Néerl. 42: 2 (1950).
Guyana. 82 GUY. Nanophan. or phan.

Marlierea maguirei McVaugh, Mem. New York Bot. Gard. 10(1): 87 (1958).
Venezuela. 82 VEN. Nanophan. or phan.

Marlierea martinellii G.M.Barroso & Peixoto, Bol. Mus. Biol. Prof. Mello-Leitão, Nova Sér. 1: 86 (1992).
Brazil (Rio de Janeiro). 84 BZL. Nanophan. or phan.

Marlierea mcvaughii B.Holst, Selbyana 23: 147 (2002).
Colombia to S. Venezuela. 82 VEN 83 CLM 84 BZN. Phan.

Marlierea mesoamericana P.E.Sánchez, Brenesia 35: 117 (1991).
Costa Rica. 80 COS. Nanophan. or phan.

Marlierea montana (Aubl.) Amshoff, Recueil Trav. Bot. Néerl. 39: 147 (1942).
N. South America. 82 FRG GUY SUR VEN. Nanophan. or phan.
**Eugenia montana* Aubl., Hist. Pl. Guiane 1: 495 (1775). *Cumetea montana* (Aubl.) Raf., Sylva Tellur.: 106 (1838). *Myrcianthes montana* (Aubl.) C.Nelson Sutherland, Anales Jard. Bot. Madrid 57: 406 (2000).
Calyptranthes obtusa Miq., Linnaea 22: 172 (1849), nom. illeg.
Marlierea obtusa O.Berg, Linnaea 27: 15 (1855).
Marlierea suffruticosa O.Berg, Linnaea 27: 16 (1855).
Marlierea richardiana O.Berg, Linnaea 30: 650 (1861).

Marlierea multiglomerata Amshoff, Recueil Trav. Bot. Néerl. 42: 3 (1950).
N. South America. 82 FRG SUR VEN. Nanophan. or phan.

Marlierea neuwiedeana (O.Berg) Nied. in H.G.A.Engler & K.A.E.Prantl, Nat. Pflanzenfam. 3(7): 77 (1893).
Brazil (Rio de Janeiro). 84 BZL. Nanophan. or phan.
**Rubachia neuwiedeana* O.Berg in C.F.P.von Martius & auct. suc. (eds.), Fl. Bras. 14(1): 29 (1857).

Marlierea obscura O.Berg in C.F.P.von Martius & auct. suc. (eds.), Fl. Bras. 14(1): 36 (1857).
SE. & S. Brazil. 84 BZL BZS. Phan.
Marlierea warmingiana Kiaersk., Enum. Myrt. Bras.: 44 (1893).

Marlierea obversa D.Legrand, Comun. Bot. Mus. Hist. Nat. Montevideo 3(40): 28 (1962).
Brazil (Rio de Janeiro). 84 BZL. Nanophan. or phan.
**Rubachia spathulata* O.Berg in C.F.P.von Martius & auct. suc. (eds.), Fl. Bras. 14(1): 535 (1859).

Marlierea occhionii D.Legrand, Comun. Bot. Mus. Hist. Nat. Montevideo 3(40): 37 (1962).
Brazil (Rio de Janeiro). 84 BZL. Nanophan. or phan.

Marlierea parvifolia O.Berg in C.F.P.von Martius & auct. suc. (eds.), Fl. Bras. 14(1): 33 (1857).
NE. Brazil. 84 BZE. Nanophan. or phan.

Marlierea polygama (O.Berg) D.Legrand, Comun. Bot. Mus. Hist. Nat. Montevideo 3(40): 30 (1962).
SE. Brazil. 84 BZL. Nanophan. or phan.
**Eugeniopsis polygama* O.Berg in C.F.P.von Martius & auct. suc. (eds.), Fl. Bras. 14(1): 147 (1857).

Marlierea pudica McVaugh, Brittonia 33: 35 (1981).
Venezuela. 82 VEN. Nanophan. or phan.

Marlierea racemosa (Vell.) Kiaersk., Enum. Myrt. Bras.: 51 (1893).
E. & S. Brazil. 84 BZE BZL BZS. Nanophan. or phan.
**Myrtus racemosa* Vell., Fl. Flumin. 5: 213, t. 60 (1829). *Eugeniopsis cannifolia* O.Berg in C.F.P.von Martius & auct. suc. (eds.), Fl. Bras. 14(1): 148 (1857). *Marlierea cannifolia* Nied. in H.G.A.Engler & K.A.E.Prantl, Nat. Pflanzenfam. 3(7): 76 (1893).
Eugeniopsis affinis O.Berg in C.F.P.von Martius & auct. suc. (eds.), Fl. Bras. 14(1): 149 (1857). *Marlierea affinis* (O.Berg) D.Legrand, Comun. Bot. Mus. Hist. Nat. Montevideo 3(40): 30 (1962).

Eugeniopsis grandifolia O.Berg in C.F.P.von Martius & auct. suc. (eds.), Fl. Bras. 14(1): 149 (1857). *Marlierea bergiana* D.Legrand, Comun. Bot. Mus. Hist. Nat. Montevideo 3(40): 30 (1962).

Marlierea regeliana O.Berg in C.F.P.von Martius & auct. suc. (eds.), Fl. Bras. 14(1): 537 (1859).
SE. Brazil. 84 BZL. Nanophan. or phan.
Marlierea regeliana var. *parviflora* Kiaersk., Enum. Myrt. Bras.: 44 (1893).

Marlierea reitzii D.Legrand, Sellowia 13: 284 (1961).
Brazil (São Paulo to Santa Catarina). 84 BZL BZS. Phan.

Marlierea resupinata (Vell.) O.Berg in C.F.P.von Martius & auct. suc. (eds.), Fl. Bras. 14(1): 37 (1857).
Brazil (Rio de Janeiro). 84 BZL. Nanophan. or phan.
**Myrtus resupinata* Vell., Fl. Flumin. 5: 214, t. 61 (1829).

Marlierea riedeliana (O.Berg) D.Legrand, Comun. Bot. Mus. Hist. Nat. Montevideo 3(40): 31 (1962).
NE. Brazil. 84 BZE. Nanophan. or phan.
**Eugeniopsis riedeliana* O.Berg in C.F.P.von Martius & auct. suc. (eds.), Fl. Bras. 14(1): 561 (1859).

Marlierea rubiginosa (Cambess.) D.Legrand, Comun. Bot. Mus. Hist. Nat. Montevideo 3(40): 32 (1962).
Brazil (Minas Gerais). 84 BZL. Nanophan. or phan.
**Eugenia rubiginosa* Cambess. in A.F.C.de Saint-Hilaire, Fl. Bras. Merid. 2: 338 (1833). *Eugeniopsis rubiginosa* (Cambess.) O.Berg, Linnaea 27: 82 (1855). *Myrciaria rubiginosa* (Cambess.) O.Berg in C.F.P.von Martius & auct. suc. (eds.), Fl. Bras. 14(1): 359 (1857).

Marlierea rufa O.Berg in C.F.P.von Martius & auct. suc. (eds.), Fl. Bras. 14(1): 37 (1857).
Brazil (Rio de Janeiro). 84 BZL. Nanophan. or phan.
**Myrtus rufa* Vell., Fl. Flumin. 5: 217, t. 75 (1829).

Marlierea rugosior McVaugh, Mem. New York Bot. Gard. 18(2): 67 (1969).
Venezuela (Bolívar). 82 VEN. Nanophan. or phan.

Marlierea salticola Amshoff, Bull. Torrey Bot. Club 75: 529 (1948).
Guyana. 82 GUY. Nanophan. or phan.

Marlierea schomburgkiana O.Berg, Linnaea 29: 209 (1858).
N. South America to Peru. 82 GUY SUR VEN 83 ECU PER. Nanophan. or phan.

Marlierea schottii D.Legrand, Comun. Bot. Mus. Hist. Nat. Montevideo 3(40): 33 (1962).
Brazil (Rio de Janeiro). 84 BZL. Nanophan. or phan.
**Eugeniopsis schottiana* O.Berg in C.F.P.von Martius & auct. suc. (eds.), Fl. Bras. 14(1): 148 (1857).

Marlierea scytophylla Diels, Verh. Bot. Vereins Prov. Brandenburg 48: 187 (1906 publ. 1907).
S. Venezuela to Peru. 82 VEN 83 PER 84 BZN. Phan.

Marlierea sessiliflora O.Berg in C.F.P.von Martius & auct. suc. (eds.), Fl. Bras. 14(1): 34 (1857).
Brazil (Rio de Janeiro). 84 BZL. Nanophan. or phan.

Marlierea silvestris (Vell.) Mattos, Loefgrenia 102: 4 (1993).
SE. Brazil. 84 BZL. Nanophan. or phan.
**Myrtus silvestris* Vell., Fl. Flumin. 5: 215, t. 64 (1829).

Marlierea sintenisii Kiaersk., Bot. Tidsskr. 17: 252 (1889). *Plinia sintenisii* (Kiaersk.) Britton, Sci. Surv. Porto Rico & Virgin Islands 6: 30 (1925).
Puerto Rico (Luquillo Mts.). 81 PUE. Nanophan. or phan.

Marlierea skortzoviana Mattos, Loefgrenia 54: 1 (1971).
SE. Brazil. 84 BZL. Nanophan. or phan.
**Marlierea subulata* Mattos, Ci. & Cult. 19: 332 (1967), nom. illeg.

Marlierea spruceana O.Berg in C.F.P.von Martius & auct. suc. (eds.), Fl. Bras. 14(1): 34 (1857).
Venezuela to Peru. 82 VEN 83 CLM PER 84 BZC BZN. Phan.
Marlierea spruceana var. *angustifolia* O.Berg in C.F.P.von Martius & auct. suc. (eds.), Fl. Bras. 14(1): 515 (1858).
Marlierea spruceana var. *latifolia* O.Berg in C.F.P.von Martius & auct. suc. (eds.), Fl. Bras. 14(1): 515 (1858).
Marlierea uaupensis O.Berg in C.F.P.von Martius & auct. suc. (eds.), Fl. Bras. 14(1): 516 (1858).

Marlierea suaveolens Cambess. in A.F.C.de Saint-Hilaire, Fl. Bras. Merid. 2: 374 (1833).
SE. & S. Brazil. 84 BZL BZS. Phan.

Marlierea subacuminata Kiaersk., Enum. Myrt. Bras.: 50 (1893).
SE. Brazil. 84 BZL. Nanophan. or phan.

Marlierea subcordata B.Holst, Selbyana 23: 150 (2002).
S. Venezuela. 82 VEN. Phan.

Marlierea suborbicularis McVaugh, Mem. New York Bot. Gard. 10(1): 88 (1958).
Venezuela (Amazonas). 82 VEN. Nanophan. or phan.

Marlierea subulata McVaugh, Fieldiana, Bot. 29: 177 (1956).
S. Venezuela to Bolivia. 82 VEN 83 BOL PER 84 BZN. Nanophan. or phan.

Marlierea sucrei G.M.Barroso & Peixoto, Acta Bot. Brasil. 4: 15 (1990).
Brazil (Bahia, Espírito Santo). 84 BZE BZL. Nanophan. or phan.

Marlierea summa McVaugh, Mem. New York Bot. Gard. 10(1): 89 (1958).
Venezuela to Guyana and N. Brazil. 82 GUY VEN 84 BZN. Nanophan. or phan.
Marlierea summa var. *superior* McVaugh, Mem. New York Bot. Gard. 10: 91 (1958).
Marlierea summa var. *calva* McVaugh, Mem. New York Bot. Gard. 18(2): 69 (1969).

Marlierea sylvatica (O.Berg) Kiaersk., Enum. Myrt. Bras.: 51 (1893).
SE. & S. Brazil. 84 BZL BZS. Phan.
Eugenia sylvatica Gardner, London J. Bot. 2: 352 (1843), nom. illeg. **Eugeniopsis sylvatica* O.Berg, Linnaea 27: 81 (1855).

Marlierea teuscheriana (O.Berg) D.Legrand, Comun. Bot. Mus. Hist. Nat. Montevideo 3(40): 30 (1962).
SE. Brazil. 84 BZL. Nanophan. or phan.
**Eugeniopsis teuscheriana* O.Berg, Linnaea 31: 249 (1862).

Marlierea tomentosa Cambess. in A.F.C.de Saint-Hilaire, Fl. Bras. Merid. 2: 373 (1833).
E. & S. Brazil. 84 BZE BZL BZS. Phan.

Myrcia strigipes Mart., Flora 24(2 Beibl.): 108 (1841).
Marlierea strigipes (Mart.) O.Berg in C.F.P.von Martius & auct. suc. (eds.), Fl. Bras. 14(1): 32 (1857).
Marlierea spathulata O.Berg in C.F.P.von Martius & auct. suc. (eds.), Fl. Bras. 14(1): 32 (1857).

Marlierea tovarensis O.Berg, Linnaea 27: 13 (1855).
Venezuela. 82 VEN. Phan.

Marlierea umbraticola (Kunth) O.Berg in C.F.P.von Martius & auct. suc. (eds.), Fl. Bras. 14(1): 35 (1857).
S. Trop. America. 82 FRG VEN 83 BOL CLM ECU PER 84 BZN. Nanophan. or phan.
**Myrtus umbraticola* Kunth in F.W.H.von Humboldt, A.J.A.Bonpland & C.S.Kunth, Nov. Gen. Sp. 7: 258 (1825).
Myrcia nigrescens DC., Prodr. 3: 246 (1828).
Aulomyrcia nigrescens (DC.) O.Berg, Linnaea 27: 71 (1855).
Eugenia squamata Willd. ex O.Berg, Linnaea 27: 17 (1855).
Aulomyrcia nigrescens var. *angustifolia* O.Berg in C.F.P.von Martius & auct. suc. (eds.), Fl. Bras. 14(1): 119 (1857).
Aulomyrcia nigrescens var. *dives* O.Berg in C.F.P.von Martius & auct. suc. (eds.), Fl. Bras. 14(1): 120 (1857).
Aulomyrcia nigrescens var. *latifolia* O.Berg in C.F.P. von Martius & auct. suc. (eds.), Fl. Bras. 14(1): 120 (1857).
Marlierea insculpta Diels, Verh. Bot. Vereins Prov. Brandenburg 48: 188 (1906 publ. 1907).
Marlierea squarrosa McVaugh, Fieldiana, Bot. 29: 177 (1956).

Marlierea uniflora McVaugh, Mem. New York Bot. Gard. 18(2): 69 (1969).
S. Venezuela to N. Brazil. 82 VEN 84 BZN. Nanophan.

Marlierea velutina McVaugh, Fieldiana, Bot. 29: 178 (1956).
N. Brazil, Peru to Argentina (Buenos Aires). 83 BOL PER 84 BZN 85 AGE. Nanophan. or phan.

Marlierea ventuarensis B.Holst, Selbyana 23: 152 (2002).
S. Venezuela. 82 VEN. Nanophan. or phan.

Marlierea verticillaris O.Berg in C.F.P.von Martius & auct. suc. (eds.), Fl. Bras. 14(1): 538 (1859).
NE. Brazil. 84 BZE. Nanophan. or phan.

Marlierea vicina McVaugh, Mem. New York Bot. Gard. 18(2): 70 (1969).
Guyana. 82 GUY. Nanophan. or phan.

Marlierea villas-boasii Mattos, Loefgrenia 118: 1 (2003).
Brazil (São Paulo). 84 BZL. Nanophan.

Synonyms:
Marlierea acuminata O.Berg = ***Marlierea ferruginea*** (Poir.) McVaugh
Marlierea affinis (O.Berg) D.Legrand = ***Marlierea racemosa*** (Vell.) Kiaersk.
Marlierea antrocola Kiaersk. = ***Myrciaria glazioviana*** (Kiaersk.) G.M.Barroso ex Sobral
Marlierea bergiana D.Legrand = ***Marlierea racemosa*** (Vell.) Kiaersk.
Marlierea brachymischa Kiaersk. = ***Myrciaria floribunda*** (H.West ex Willd.) O.Berg
Marlierea cannifolia Nied. = ***Marlierea racemosa*** (Vell.) Kiaersk.
Marlierea cubensis (Griseb.) Krug & Urb. = ***Myrciaria floribunda*** (H.West ex Willd.) O.Berg
Marlierea dussii Krug & Urb. = ***Siphoneugena dussii*** (Krug & Urb.) Proença
Marlierea edulis Nied. = ***Plinia edulis*** (Vell.) Sobral
Marlierea elliptica Griseb. = ***Marlierea ferruginea*** (Poir.) McVaugh
Marlierea elliptica var. *cubensis* Griseb. = [81 CUB]
Marlierea eugenioides var. *ovata* (O.Berg) D.Legrand = ***Marlierea eugenioides*** (Cambess.) D.Legrand
Marlierea excoriata var. *angustifolia* O.Berg = ***Marlierea excoriata*** Mart.
Marlierea excoriata var. *latifolia* O.Berg = ***Marlierea excoriata*** Mart.
Marlierea glabra var. *gracilis* O.Berg = ***Marlierea glabra*** Cambess.
Marlierea gleasonii McVaugh = ***Myrcia decorticans*** DC.
Marlierea glomerata O.Berg = ***Plinia pinnata*** L.
Marlierea grandifolia O.Berg = ***Marlierea glabra*** Cambess.
Marlierea grandifolia var. *minor* O.Berg = ***Marlierea glabra*** Cambess.
Marlierea guyanensis D.Legrand = ***Marlierea ferruginea*** (Poir.) McVaugh
Marlierea insculpta Diels = ***Marlierea umbraticola*** (Kunth) O.Berg
Marlierea intonsa McVaugh = ***Myrcia intonsa*** (McVaugh) B.Holst
Marlierea laevigata var. *glazioviana* Kiaersk. = ***Marlierea laevigata*** (DC.) Kiaersk.
Marlierea lateriflora (DC.) D.Legrand = ***Plinia spiciflora*** (Nees & Mart.) Sobral
Marlierea macedoi D.Legrand = ***Myrciaria dubia*** (Kunth) McVaugh
Marlierea obtusa O.Berg = ***Marlierea montana*** (Aubl.) Amshoff
Marlierea obumbrans (O.Berg) Nied. = ***Myrcia obumbrans*** (O.Berg) McVaugh
Marlierea ovata (O.Berg) Nied. = ***Marlierea eugenioides*** (Cambess.) D.Legrand
Marlierea paniculata (O.Berg) D.Legrand = ***Marlierea eugenioides*** (Cambess.) D.Legrand
Marlierea parviflora O.Berg = ***Marlierea excoriata*** Mart.
Marlierea parviflora var. *brachybotrya* D.Legrand = ***Plinia brachybotrya*** (D.Legrand) Sobral
Marlierea pilodes (Kiaersk.) M.L.Kawas. = ***Myrcia mutabilis*** (O.Berg) N.Silveira
Marlierea regeliana var. *parviflora* Kiaersk. = ***Marlierea regeliana*** O.Berg
Marlierea richardiana (O.Berg) Nied. = ***Marlierea ferruginea*** (Poir.) McVaugh
Marlierea richardiana O.Berg = ***Marlierea montana*** (Aubl.) Amshoff
Marlierea schottiana O.Berg = ***Marlierea glabra*** Cambess.
Marlierea spathulata O.Berg = ***Marlierea tomentosa*** Cambess.
Marlierea spiciflora (Nees & Mart.) Nied. = ***Plinia spiciflora*** (Nees & Mart.) Sobral
Marlierea spruceana var. *angustifolia* O.Berg = ***Marlierea spruceana*** O.Berg
Marlierea spruceana var. *latifolia* O.Berg = ***Marlierea spruceana*** O.Berg
Marlierea squarrosa McVaugh = ***Marlierea umbraticola*** (Kunth) O.Berg
Marlierea strigipes (Mart.) O.Berg = ***Marlierea tomentosa*** Cambess.
Marlierea subulata Mattos = ***Marlierea skortzoviana*** Mattos

Marlierea suffruticosa O.Berg = ***Marlierea montana*** (Aubl.) Amshoff
Marlierea summa var. *calva* McVaugh = ***Marlierea summa*** McVaugh
Marlierea summa var. *superior* McVaugh = ***Marlierea summa*** McVaugh
Marlierea uaupensis O.Berg = ***Marlierea spruceana*** O.Berg
Marlierea warmingiana Kiaersk. = ***Marlierea obscura*** O.Berg

***Unplaced Names*:**
Marlierea elliptica var. *cubensis* Griseb., Cat. Pl. Cub.: 86 (1866). = [81 CUB]

Marlieriopsis

Marlieriopsis Kiaersk. = ***Blepharocalyx*** O.Berg
Marlieriopsis eggersii Kiaersk. = ***Blepharocalyx eggersii*** (Kiaersk.) Landrum

Mearnsia

Mearnsia Merr. = ***Metrosideros*** Banks ex Gaertn.
Mearnsia cordata C.T.White & W.D.Francis = ***Metrosideros cordata*** (C.T.White & W.D.Francis) J.W.Dawson
Mearnsia halconensis Merr. = ***Metrosideros halconensis*** (Merr.) J.W.Dawson
Mearnsia ovata C.T.White = ***Metrosideros ovata*** (C.T.White) J.W.Dawson
Mearnsia porphyrea (Schltr.) Diels = ***Metrosideros porphyrea*** Schltr.
Mearnsia ramiflora (Lauterb.) Diels = ***Metrosideros ramiflora*** Lauterb.
Mearnsia ramiflora var. *humilis* Diels = ***Metrosideros ramiflora*** var. ***humilis*** (Diels) J.W.Dawson
Mearnsia ramiflora var. *villosa* C.T.White = ***Metrosideros ramiflora*** var. ***villosa*** (C.T.White) J.W.Dawson
Mearnsia salomonensis C.T.White = ***Metrosideros tetragyna*** J.W.Dawson
Mearnsia scandens C.T.White = ***Metrosideros whiteana*** J.W.Dawson

Meladendron

Meladendron St.-Lag. = ***Melaleuca*** L.
Meladendron leucocladum St.-Lag. = ***Melaleuca leucadendra*** (L.) L.

Melaleuca

Melaleuca L., Mant. Pl. 1: 14 (1767). *Myrtoleucodendron* Burm. ex Kuntze, Revis. Gen. Pl. 1: 241 (1891), nom. illeg.
Indo-China to Australia, New Caledonia, Society Is. (22) ben (23) gab (40) srl 41 MYA THA VIE 42 BOR JAW LSI MLY MOL SUM 43 NWG 50 NFK NSW NTA QLD SOA TAS VIC WAU 60 NWC 61 SCI.
246 Species
Caju-puti Adans., Fam. Pl. 2: 84 (1763).
Kajuputi Adans., Fam. Pl. 2: 84 (1763).
Ozandra Raf., Autik. Bot.: 148 (1840).
Gymnagathis Schauer, Linnaea 17: 242 (1843).
Meladendron St.-Lag., Ann. Soc. Bot. Lyon 7: 64 (1880).
Melaleucon St.-Lag., Ann. Soc. Bot. Lyon 7: 64 (1880), orth. var.
Melanoleuce St.-Lag., Ann. Soc. Bot. Lyon 8: 203 (1881), orth. var.

Melaleuca acacioides F.Muell., Fragm. 3: 116 (1862). *Myrtoleucodendron acacioides* (F.Muell.) Kuntze, Revis. Gen. Pl. 1: 241 (1891).
S. New Guinea, Northern Territory to Queensland. 43 NWG 50 NTA QLD. Nanophan. or phan.
Melaleuca graminea S.Moore, J. Linn. Soc., Bot. 45: 204 (1920).

Melaleuca acuminata F.Muell., Fragm. 1: 15 (1858). *Myrtoleucodendron acuminatum* (F.Muell.) Kuntze, Revis. Gen. Pl. 1: 241 (1891).
S. & SE. Australia. 50 NSW SOA VIC WAU. Nanophan. or phan.

subsp. ***acuminata***
S. & SE. Australia. 50 NSW SOA VIC WAU. Nanophan. or phan.

subsp. ***websteri*** (S.Moore) Barlow ex Craven, Austral. Syst. Bot. 12: 858 (1999).
Western Australia (Avon). 50 WAU. Nanophan. or phan.
Melaleuca concava S.Moore, J. Linn. Soc., Bot. 45: 178 (1920).
**Melaleuca websteri* S.Moore, J. Linn. Soc., Bot. 45: 204 (1920).

Melaleuca adenostyla K.J.Cowley, Austral. Syst. Bot. 3: 171 (1990).
SW. Australia. 50 WAU. Nanophan.

Melaleuca adnata Turcz., Bull. Cl. Phys.-Math. Acad. Imp. Sci. Saint-Pétersbourg 10: 343 (1852). *Myrtoleucodendron adnatum* (Turcz.) Kuntze, Revis. Gen. Pl. 1: 241 (1891).
Western Australia. 50 WAU. Nanophan.
Melaleuca eleuterostachya var. *abietina* Benth., Fl. Austral. 3: 140 (1867).
Melaleuca adnata var. *abietina* F.Muell., J. Bot. 15: 280 (1877).

Melaleuca aestuosa G.Forst., Fl. Ins. Austr.: 38 (1786).
Society Is. (Tahiti). 61 SCI. Nanophan.

Melaleuca agathosmoides C.A.Gardner, Hooker's Icon. Pl. 34: t. 3381 (1939).
S. Western Australia. 50 WAU. Nanophan.

Melaleuca alsophila A.Cunn. ex Benth., Fl. Austral. 3: 137 (1867). *Myrtoleucodendron alsophilum* (A.Cunn. ex Benth.) Kuntze, Revis. Gen. Pl. 1: 241 (1891). *Melaleuca acacioides* subsp. *alsophila* (A.Cunn. ex Benth.) Barlow, Brunonia 9: 167 (1987).
N. Western Australia. 50 WAU. Nanophan.

Melaleuca alternifolia (Maiden & Betche) Cheel, J. Proc. Roy. Soc. New S. Wales 58: 195 (1924).
Queensland to NE. New South Wales. 50 NSW QLD. Nanophan.
**Melaleuca linariifolia* var. *alternifolia* Maiden & Betche, Proc. Linn. Soc. New South Wales 29: 742 (1905).

Melaleuca amydra Craven, Austral. Syst. Bot. 12: 859 (1999).
Western Australia. 50 WAU. Nanophan. or phan.

Melaleuca apodocephala Turcz., Bull. Cl. Phys.-Math. Acad. Imp. Sci. Saint-Pétersbourg 10: 340 (1852).
S. Western Australia. 50 WAU. Nanophan.

subsp. ***apodocephala***
S. Western Australia. 50 WAU. Nanophan.

subsp. ***calcicola*** Craven, Austral. Syst. Bot. 12: 860 (1999).
S. Western Australia. 50 WAU. Nanophan. or phan.

Melaleuca apostiba K.J.Cowley, Austral. Syst. Bot. 3: 182 (1990).
SC. Western Australia. 50 WAU. Nanophan.

Melaleuca araucarioides Barlow, Nuytsia 8: 334 (1992).
SSW. Western Australia. 50 WAU. Nanophan.

Melaleuca arcana S.T.Blake, Contr. Queensland Herb. 1: 54 (1968).
N. Queensland. 50 QLD. Phan.
Melaleuca leucadendra f. *ruscifolia* Cheel in A.J. Ewart & O.B.Davies, Fl. N. Territory: 302 (1917).
Melaleuca leucadendra f. *ruscifolia* Cheel, Fl. N. Territory: 302 (1917).

Melaleuca* × *arenicola S.Moore, J. Linn. Soc., Bot. 45: 205 (1920). *M. depauperata* × *M. laxiflora*.
SW. Australia. 50 WAU. Nanophan.

Melaleuca argentea W.Fitzg., J. Proc. Roy. Soc. W. Australia 3: 187 (1918).
New Guinea to N. Australia. 43 NWG 50 NTA QLD WAU. Phan.

Melaleuca armillaris (Sol. ex Gaertn.) Sm., Trans. Linn. Soc. London 3: 277 (1797).
SE. Australia. 50 NSW SOA TAS VIC. Nanophan. or phan.
**Metrosideros armillaris* Sol. ex Gaertn., Fruct. Sem. Pl. 1: 171 (1788). *Myrtoleucodendron armillare* (Sol. ex Gaertn.) Kuntze, Revis. Gen. Pl. 1: 241 (1891).

subsp. ***akineta*** F.C.Quinn, Austral. Syst. Bot. 3: 188 (1990).
SE. South Australia. 50 SOA. Nanophan. or phan.

subsp. ***armillaris***
SE. Australia. 50 NSW TAS VIC. Nanophan. or phan.
Melaleuca ericifolia Andrews, Bot. Repos. 3: t. 175 (1801), nom. illeg.
Melaleuca taxifolia Schltdl. ex Spreng., Syst. Veg. 3: 336 (1826).

Melaleuca aspalathoides Schauer in J.G.C.Lehmann, Pl. Preiss. 1: 140 (1844). *Myrtoleucodendron aspalathodes* (Schauer) Kuntze, Revis. Gen. Pl. 1: 241 (1891).
Western Australia. 50 WAU. Nanophan.

Melaleuca atroviridis Craven & Lepschi, Austral. Syst. Bot. 17: 259 (2004).
SW. Australia. 50 WAU. Phan.

Melaleuca barlowii Craven, Austral. Syst. Bot. 12: 861 (1999).
Western Australia. 50 WAU. Nanophan. or phan.

Melaleuca basicephala Benth., Fl. Austral. 3: 133 (1867). *Myrtoleucodendron basicephalum* (Benth.) Kuntze, Revis. Gen. Pl. 1: 241 (1891).
SW. Australia. 50 WAU. Nanophan.

Melaleuca beardii Craven, Austral. Syst. Bot. 12: 862 (1999).
Western Australia. 50 WAU. Nanophan. or phan.

Melaleuca biconvexa Byrnes, Austrobaileya 2: 74 (1984).
E. New South Wales. 50 NSW. Nanophan. or phan.

Melaleuca bisulcata F.Muell., Fragm. 3: 118 (1862). *Melaleuca acerosa* var. *bracteata* Benth., Fl. Austral. 3: 161 (1867).
Western Australia. 50 WAU. Nanophan.
Melaleuca acerosa var. *bracteosa* Benth., Fl. Austral. 3: 161 (1867).

Melaleuca blaeriifolia Turcz., Bull. Soc. Imp. Naturalistes Moscou 20(1): 165 (1847). *Myrtoleucodendron blaeriifolium* (Turcz.) Kuntze, Revis. Gen. Pl. 1: 241 (1891).
SW. Australia. 50 WAU. Nanophan.
Melaleuca eruciformis Turcz., Bull. Cl. Phys.-Math. Acad. Imp. Sci. Saint-Pétersbourg 10: 344 (1852).

Melaleuca boelophylla Craven, Austral. Syst. Bot. 12: 863 (1999).
Western Australia. 50 WAU. Nanophan. or phan.

Melaleuca boeophylla Craven, Austral. Syst. Bot. 12: 863 (1999).
SW. Australia. 50 WAU.

Melaleuca borealis Craven, Austral. Syst. Bot. 12: 863 (1999).
Queensland. 50 QLD. Nanophan. or phan.

Melaleuca bracteata F.Muell., Fragm. 1: 15 (1858).
Australia. 50 NSW NTA QLD SOA WAU. Phan.
Melaleuca glaucocalyx Gand., Bull. Soc. Bot. France 65: 26 (1918).
Melaleuca genistifolia var. *coriacea* Ewart, L.R.Kerr & Derrick, Proc. Roy. Soc. Queensland 38: 84 (1926).
Melaleuca monticola J.M.Black, Trans. & Proc. Roy. Soc. South Australia 58: 179 (1934).
Melaleuca daleana Blakely, Austral. Naturalist 11: 9 (1941).

Melaleuca bracteosa Turcz., Bull. Cl. Phys.-Math. Acad. Imp. Sci. Saint-Pétersbourg 10: 340 (1852).
S. Western Australia. 50 WAU. Nanophan.

Melaleuca brevifolia Turcz., Bull. Cl. Phys.-Math. Acad. Imp. Sci. Saint-Pétersbourg 10: 342 (1852).
S. Western Australia to W. Victoria. 50 SOA VIC WAU. Nanophan.
Melaleuca fasciculiflora Benth., Fl. Austral. 3: 137 (1867). *Myrtoleucodendron fasciculiflorum* (Benth.) Kuntze, Revis. Gen. Pl. 1: 241 (1891).
Melaleuca neglecta Ewart & B.Wood, Proc. Roy. Soc. Victoria, n.s., 23: 60 (1910).
Melaleuca oraria J.M.Black, Trans. Roy. Soc. South Australia 69: 309 (1945).

Melaleuca brevisepala (J.W.Dawson) Craven & J.W.Dawson, Adansonia, III, 20: 193 (1998).
WNW. New Caledonia. 60 NWC. Nanophan. or phan.
**Callistemon brevisepalus* J.W.Dawson, in Fl. Nouv.-Caléd. 18: 242 (1992).

Melaleuca bromelioides Barlow, Austral. Syst. Bot. 1: 112 (1988).
S. Western Australia. 50 WAU. Nanophan.

Melaleuca brongniartii Däniker, Vierteljahrsschr. Naturf. Ges. Zürich 79(19): 318 (1933).
SE. New Caledonia. 60 NWC. Cham. or nanophan.
**Melaleuca pungens* Brongn. & Gris, Ann. Sci. Nat., Bot., V, 2: 139 (1864), nom. illeg.

Melaleuca brophyi Craven, Austral. Syst. Bot. 12: 864 (1999).
Western Australia. 50 WAU. Nanophan. or phan.

Melaleuca buseana (Guillaumin) Craven & J.W.Dawson, Adansonia, III, 20: 192 (1998).
SE. New Caledonia. 60 NWC. Nanophan. or phan.
**Callistemon buseanus* Guillaumin, Bull. Mus. Natl. Hist. Nat., II, 11: 414 (1939).
Callistemon buseanus var. *longifolius* Guillaumin, Mém. Mus. Natl. Hist. Nat., B, Bot. 8: 276 (1962).

Melaleuca caeca Craven, Austral. Syst. Bot. 12: 865 (1999).
Western Australia. 50 WAU. Nanophan. or phan.

Melaleuca cajuputi Powell, Pharm. Roy. Coll. Phys. Transl.: 22 (1809).
Indo-China to N. Northern Territory. (40) srl 41 MYA THA VIE 42 BOR JAW LSI MLY MOL SUM 43 NWG 50 NTA QLD WAU. Phan.

subsp. ***cajuputi***
Lesser Sunda Is. to N. Northern Territory. 42 LSI MOL 50 NTA WAU. Phan.
Myrtus saligna J.F.Gmel., Syst. Nat.: 793 (1791). *Melaleuca trinervis* Buch.-Ham., Mem. Wern. Nat. Hist. Soc. 6: 302 (1832), nom. illeg. *Pimentus saligna* (J.F.Gmel.) Raf., Sylva Tellur.: 105 (1838). *Melaleuca saligna* Reinw. ex Blume, Mus. Bot. 1: 66 (1849), nom. illeg.
Melaleuca minor Sm. in A.Rees, Cycl. 23: 2 (1812). *Melaleuca leucadendra* var. *minor* (Sm.) Duthie in J.D.Hooker, Fl. Brit. India 2: 465 (1878).
Melaleuca viridiflora var. *angustifolia* Blume, Bijdr.: 1099 (1826). *Melaleuca angustifolia* (Blume) Blume, Mus. Bot. 1: 83 (1850), nom. illeg.
Metrosideros comosa Roxb., Fl. Ind. ed. 1832, 2: 478 (1832). *Nania comosa* (Roxb.) Kuntze, Revis. Gen. Pl. 1: 242 (1891).
Melaleuca lancifolia Turcz., Bull. Soc. Imp. Naturalistes Moscou 10: 164 (1847).
Melaleuca commutata Miq., Anal. Bot. Ind. 1: 14 (1850).
Melaleuca eriorhachis Gand., Bull. Soc. Bot. France 65: 26 (1918).

subsp. ***cumingiana*** (Turcz.) Barlow, Novon 7: 113 (1997).
Indo-China to W. Malesia. (40) srl 41 MYA THA VIE 42 BOR JAW MLY SUM. Phan.
**Melaleuca cumingiana* Turcz., Bull. Soc. Imp. Naturalistes Moscou 20: 164 (1847).

subsp. ***platyphylla*** Barlow, Novon 7: 113 (1997).
New Guinea to N. Queensland. 43 NWG 50 QLD. Phan.

Melaleuca calothamnoides F.Muell., Fragm. 3: 114 (1862). *Myrtoleucodendron calothamnoides* (F.Muell.) Kuntze, Revis. Gen. Pl. 1: 241 (1891).
WSW. Western Australia. 50 WAU. Nanophan.

Melaleuca calycina R.Br. in W.T.Aiton, Hortus Kew. 4: 416 (1812). *Myrtoleucodendron calycinum* (R.Br.) Kuntze, Revis. Gen. Pl. 1: 241 (1891).
S. Western Australia. 50 WAU. Nanophan.
Melaleuca carinata Turcz., Bull. Cl. Phys.-Math. Acad. Imp. Sci. Saint-Pétersbourg 10: 344 (1852).

Melaleuca calyptroides Craven, Austral. Syst. Bot. 12: 866 (1999).
Western Australia. 50 WAU. Nanophan. or phan.

Melaleuca campanae Craven, Austral. Syst. Bot. 12: 866 (1999).
Western Australia. 50 WAU. Nanophan. or phan.

Melaleuca camptoclada F.C.Quinn, Austral. Syst. Bot. 3: 199 (1990).
SW. Australia. 50 WAU. Nanophan.

Melaleuca capitata Cheel, J. Proc. Roy. Soc. New S. Wales 58: 194 (1924).
New South Wales. 50 NSW. Nanophan.

Melaleuca cardiophylla F.Muell., Fragm. 1: 225 (1859). *Myrtoleucodendron cardiophyllum* (F.Muell.) Kuntze, Revis. Gen. Pl. 1: 241 (1891).
Western Australia. 50 WAU. Nanophan.

Melaleuca carrii Craven, Austral. Syst. Bot. 12: 867 (1999).
Western Australia. 50 WAU. Nanophan. or phan.

Melaleuca cheelii C.T.White, Proc. Roy. Soc. Queensland 43: 16 (1931 publ. 1932).
SE. Queensland (Wide Bay). 50 QLD. Nanophan. or phan.

Melaleuca ciliosa Turcz., Bull. Soc. Imp. Naturalistes Moscou 35(2): 326 (1862). *Myrtoleucodendron ciliosum* (Turcz.) Kuntze, Revis. Gen. Pl. 1: 241 (1891).
SW. Australia. 50 WAU. Nanophan.

Melaleuca citrolens Barlow, Brunonia 9: 168 (1987).
NE. Northern Territory to N. Queensland. 50 NTA QLD. Phan.
Melaleuca acacioides var. *angustifolia* Domin, Biblioth. Bot. 89: 1009 (1928).

Melaleuca clarksonii Barlow, Novon 7: 114 (1997).
N. Queensland. 50 QLD. Phan.

Melaleuca clavifolia Craven, Austral. Syst. Bot. 12: 868 (1999).
Western Australia. 50 WAU. Nanophan. or phan.

Melaleuca cliffortioides Diels, Bot. Jahrb. Syst. 35: 427 (1904).
SW. Australia. 50 WAU. Nanophan.

Melaleuca coccinea A.S.George, W. Austral. Naturalist 10: 28 (1966).
S. Western Australia. 50 WAU. Nanophan.

Melaleuca concinna Turcz., Bull. Cl. Phys.-Math. Acad. Imp. Sci. Saint-Pétersbourg 10: 339 (1852).
Western Australia. 50 WAU. Nanophan.

Melaleuca concreta F.Muell., Fragm. 3: 118 (1862). *Myrtoleucodendron concretum* (F.Muell.) Kuntze, Revis. Gen. Pl. 1: 241 (1891).
W. Western Australia. 50 WAU. Nanophan. or phan.
Melaleuca concreta var. *brevifolia* Benth., Fl. Austral. 3: 151 (1867).

Melaleuca condylosa Craven, Austral. Syst. Bot. 12: 869 (1999).
Western Australia. 50 WAU. Nanophan. or phan.

Melaleuca conothamnoides C.A.Gardner, J. Roy. Soc. Western Australia 47: 61 (1964).
Western Australia. 50 WAU. Nanophan.
Melaleuca cordata var. *ovata* Ewart, Proc. Roy. Soc. Queensland, II, 19: 41 (1907).

Melaleuca cordata Turcz., Bull. Cl. Phys.-Math. Acad. Imp. Sci. Saint-Pétersbourg 10: 339 (1852). *Myrtoleucodendron cordatum* (Turcz.) Kuntze, Revis. Gen. Pl. 1: 241 (1891).
Western Australia. 50 WAU. Nanophan.
Melaleuca cordata Benth., Fl. Austral. 3: 149 (1867), nom. illeg.

Melaleuca cornucopiae Byrnes, Austrobaileya 2: 74 (1984).
N. Northern Territory. 50 NTA. Nanophan.

Melaleuca coronicarpa D.A.Herb., J. Proc. Roy. Soc. W. Australia 8: 35 (1922).
Western Australia. 50 WAU. Nanophan.

Melaleuca cardiophylla var. *parviflora* Benth., Fl. Austral. 3: 135 (1867).

Melaleuca croxfordiae Craven, Austral. Syst. Bot. 12: 870 (1999).
Western Australia. 50 WAU. Nanophan. or phan.

Melaleuca ctenoides F.C.Quinn, Austral. Syst. Bot. 3: 194 (1990).
SW. Australia. 50 WAU. Nanophan.

Melaleuca cucullata Turcz., Bull. Cl. Phys.-Math. Acad. Imp. Sci. Saint-Pétersbourg 10: 343 (1852).
S. Western Australia. 50 WAU. Nanophan.
Melaleuca deltoidea Benth., Fl. Austral. 3: 161 (1867). *Myrtoleucodendron deltoideum* (Benth.) Kuntze, Revis. Gen. Pl. 1: 241 (1891).

Melaleuca cuticularis Labill., Nov. Holl. Pl. 2: 30 (1806). *Myrtoleucodendron cuticulare* (Labill.) Kuntze, Revis. Gen. Pl. 1: 241 (1891). *Cajuputi cuticularis* (Labill.) Skeels, Bull. Bur. Pl. Industr. U.S.D.A. 282: 60 (1913).
SW. & S. Western Australia, South Australia (Kangaroo I.). 50 SOA WAU. Nanophan.
Melaleuca abietina Sm. in A.Rees, Cycl. 23: 15 (1812).

Melaleuca dawsonii Craven, Adansonia, III, 20: 192 (1998).
SE. New Caledonia. 60 NWC. Nanophan.
**Callistemon suberosus* Pancher ex Brongn. & Gris, Bull. Soc. Bot. France 11: 183 (1864).

Melaleuca dealbata S.T.Blake, Contr. Queensland Herb. 1: 41 (1968).
New Guinea to N. Australia. 43 NWG 50 NTA QLD WAU. Phan.

Melaleuca deanei F.Muell., Proc. Linn. Soc. New South Wales, II, 1: 1106 (1887). *Myrtoleucodendron deanei* (F.Muell.) Kuntze, Revis. Gen. Pl. 1: 241 (1891).
CE. New South Wales. 50 NSW. Nanophan. or phan.
Melaleuca lanceolata R.Br. ex R.T.Baker, Proc. Linn. Soc. New South Wales 38: 601 (1914), nom. illeg.

Melaleuca decora (Salisb.) Britten, J. Bot. 54: 62 (1916).
E. Australia. 50 NSW QLD. Nanophan. or phan.
**Metrosideros decora* Salisb., Prodr. Stirp. Chap. Allerton: 352 (1796).
Melaleuca genistifolia Sm., Trans. Linn. Soc. London 3: 277 (1797). *Myrtoleucodendron genistifolium* (Sm.) Kuntze, Revis. Gen. Pl. 1: 241 (1891).
Melaleuca tubulata Dum.Cours., Bot. Cult., ed. 2, 7: 275 (1814).
Melaleuca polygonoides Hoffmanns., Verz. Pfl.-Kult.: 78 (1824).
Melaleuca ferrea A.Cunn. ex A.Gray, U.S. Expl. Exped., Phan. 1: 566 (1854).

Melaleuca decussata R.Br. in W.T.Aiton, Hortus Kew. 4: 415 (1812). *Myrtoleucodendron decussatum* (R.Br.) Kuntze, Revis. Gen. Pl. 1: 241 (1891).
SE. South Australia to C. Victoria. 50 SOA VIC. Nanophan. or phan.
Melaleuca parviflora Rchb., Mag. Aesthetischen Bot. 1: t. VIII. (1821).
Melaleuca tetragona Lodd. ex Otto in J.H.F.Link & C.F.Otto, Icon. Pl. Select.: 37 (1821).
Melaleuca pumila A.Cunn. ex DC., Prodr. 3: 214 (1828).
Melaleuca elegans Hornsch., Allg. Gartenzeitung 2: 329 (1834).
Melaleuca affinis Regel & Körn., Index Seminum (LE) 1867: 56 (1867).
Melaleuca decussata var. *ovoidea* J.M.Black, Trans. & Proc. Roy. Soc. South Australia 49: 275 (1925).

Melaleuca delta Craven, Austral. Syst. Bot. 12: 872 (1999).
Western Australia. 50 WAU. Nanophan. or phan.

Melaleuca dempta (Barlow) Craven, Austral. Syst. Bot. 12: 872 (1999).
S. Western Australia. 50 WAU. Nanophan. or phan.
**Melaleuca calycina* subsp. *dempta* Barlow, Austral. Syst. Bot. 1: 104 (1988).

Melaleuca densa R.Br. in W.T.Aiton, Hortus Kew. 4: 416 (1812). *Myrtoleucodendron densum* (R.Br.) Kuntze, Revis. Gen. Pl. 1: 241 (1891).
SW. Australia. 50 WAU. Nanophan.
Melaleuca epacridioides Turcz., Bull. Soc. Imp. Naturalistes Moscou 20(1): 165 (1847).
Melaleuca dorrien-smithii Domin, Vestn. Král. Ceské Spolecn. Nauk, Tr. Mat.-Prír. 1921–1922(2): 88 (1923).

Melaleuca densispicata Byrnes, Austrobaileya 2: 74 (1984).
SE. Queensland to NE. New South Wales. 50 NSW QLD. Nanophan. or phan.

Melaleuca depauperata Turcz., Bull. Cl. Phys.-Math. Acad. Imp. Sci. Saint-Pétersbourg 10: 343 (1852). *Myrtoleucodendron depauperatum* (Turcz.) Kuntze, Revis. Gen. Pl. 1: 241 (1891).
SW. Australia. 50 WAU. Cham. or nanophan.
Melaleuca elachophylla F.Muell., Fragm. 3: 120 (1862). *Myrtoleucodendron elachophyllum* (F. Muell.) Kuntze, Revis. Gen. Pl. 1: 241 (1891).

Melaleuca depressa Diels, Bot. Jahrb. Syst. 35: 428 (1904).
SW. Australia. 50 WAU. Nanophan.
Melaleuca depressa var. *geraldtonensis* Hochr., Candollea 2: 466 (1925).

Melaleuca dichroma Craven & Lepschi, Austral. Syst. Bot. 14: 971 (2001).
Western Australia. 50 WAU.
**Melaleuca urceolaris* var. *virgata* Benth., Fl. Austral. 3: 155 (1867), nom. illeg. *Melaleuca virgata* (Benth.) Craven, Austral. Syst. Bot. 12: 914 (1999), nom. illeg.

Melaleuca diosmatifolia Dum.Cours., Bot. Cult., ed. 2, 5: 373 (1811).
SE. Queensland to New South Wales. 50 NSW QLD. Nanophan.
Melaleuca erubescens Otto in J.H.F.Link & C.F.Otto, Icon. Pl. Select.: 37 (1821), nom. illeg. *Melaleuca ericifolia* var. *erubescens* Benth., Fl. Austral. 3: 159 (1867).
Melaleuca fraseri Hook., Bot. Mag. 60: t. 3210 (1833).
Melaleuca armillaris var. *tenuifolia* Benth., Fl. Austral. 3: 146 (1867). *Melaleuca cylindrica* F.Muell., Syst. Census Austral. Pl. 2: 95 (1889). *Myrtoleucodendron cylindricum* (F.Muell.) Kuntze, Revis. Gen. Pl. 1: 241 (1891).

Melaleuca diosmifolia Andrews, Bot. Repos. 7: t. 476 (1806). *Myrtoleucodendron diosmifolium* (Andrews) Kuntze, Revis. Gen. Pl. 1: 241 (1891).
SW. Australia. 50 WAU. Nanophan.
Melaleuca foliosa Dum.Cours., Bot. Cult., ed. 2, 5: 373 (1811).
Melaleuca chlorantha Bonpl., Descr. Pl. Malmaison: 22 (1813).

Melaleuca dissitiflora F.Muell., Fragm. 3: 153 (1863). *Myrtoleucodendron dissitiflorum* (F.Muell.) Kuntze, Revis. Gen. Pl. 1: 241 (1891).
C. Australia. 50 NTA QLD SOA. Nanophan.

Melaleuca eleuterostachya F.Muell., Fragm. 3: 117 (1862). *Myrtoleucodendron eleutherostachyum* (F.Muell.) Kuntze, Revis. Gen. Pl. 1: 241 (1891).
C. & S. Western Australia to South Australia. 50 SOA WAU. Nanophan.

Melaleuca elliptica Labill., Nov. Holl. Pl. 2: 31 (1806). *Myrtoleucodendron ellipticum* (Labill.) Kuntze, Revis. Gen. Pl. 1: 241 (1891).
SW. & S. Western Australia. 50 WAU. Nanophan.
Melaleuca nummularia Turcz., Bull. Cl. Phys.-Math. Acad. Imp. Sci. Saint-Pétersbourg 10: 341 (1852).

Melaleuca ericifolia Sm., Trans. Linn. Soc. London 3: 276 (1797). *Myrtoleucodendron ericifolium* (Sm.) Kuntze, Revis. Gen. Pl. 1: 241 (1891). *Cajuputi ericifolia* (Sm.) A.Lyons, Plant Nam.: 74 (1900).
SE. Australia. 50 NSW TAS VIC. Nanophan. or phan.
Melaleuca pinifolia Colla, Hortus Ripul., App. 2: 352 (1825).
Melaleuca axillaris Steud., Nomencl. Bot., ed. 2, 2: 111 (1841).
Melaleuca gunniana Schauer in W.G.Walpers, Repert. Bot. Syst. 2: 928 (1843).
Melaleuca gunniana var. *capitata* Miq., Ned. Kruidk. Arch. 4: 120 (1856).
Melaleuca ternifolia F.Muell. ex Miq., Ned. Kruidk. Arch. 4: 123 (1856).
Melaleuca heliophila F.Muell. ex Miq., Ned. Kruidk. Arch. 4: 120 (1859).

Melaleuca erubescens Nees, Horae Phys. Berol.: 37 (1820).
SE. Queensland to New South Wales. 50 NSW QLD. Nanophan.

Melaleuca eulobata Craven, Austral. Syst. Bot. 12: 875 (1999).
Western Australia. 50 WAU. Nanophan. or phan.

Melaleuca eurystoma Craven, Austral. Syst. Bot. 12: 875 (1999).
Western Australia. 50 WAU. Nanophan. or phan.

Melaleuca eximia (K.J.Cowley) Craven, Austral. Syst. Bot. 12: 876 (1999).
S. Western Australia. 50 WAU. Nanophan. or phan.
**Melaleuca coccinea* subsp. *eximia* K.J.Cowley, Austral. Syst. Bot. 3: 179 (1990).

Melaleuca exuvia Craven & Lepschi, Austral. Syst. Bot. 17: 262 (2004).
S. Western Australia. 50 WAU.

Melaleuca fabri Craven, Austral. Syst. Bot. 12: 876 (1999).
Western Australia. 50 WAU. Nanophan. or phan.

Melaleuca filifolia F.Muell., Fragm. 3: 119 (1862). *Myrtoleucodendron filifolium* (F.Muell.) Kuntze, Revis. Gen. Pl. 1: 241 (1891).
Western Australia. 50 WAU. Nanophan.
Melaleuca nematophylla F.Muell., Fragm. 3: 119, (1862).

Melaleuca fissurata Barlow, Nuytsia 8: 336 (1992).
S. Western Australia. 50 WAU. Nanophan.

Melaleuca fluviatilis Barlow, Novon 7: 116 (1997).
N. Queensland. 50 QLD. Phan.
Melaleuca nervosa f. *pendulina* Byrnes, Austrobaileya 2: 74 (1984).

Melaleuca foliolosa A.Cunn. ex Benth., Fl. Austral. 3: 161 (1867). *Myrtoleucodendron foliolosum* (A.Cunn. ex Benth.) Kuntze, Revis. Gen. Pl. 1: 241 (1891).
N. Queensland. 50 QLD. Nanophan. or phan.

Melaleuca fulgens R.Br. in W.T.Aiton, Hortus Kew. 4: 415 (1812). *Myrtoleucodendron fulgens* (R.Br.) Kuntze, Revis. Gen. Pl. 1: 241 (1891).
SW. & C. Australia. 50 NTA SOA WAU. Nanophan.

subsp. ***corrugata*** (J.M.Black ex Eardley) K.J.Cowley, Austral. Syst. Bot. 3: 174 (1990).
C. Australia. 50 NTA SOA WAU. Nanophan.
**Melaleuca corrugata* J.M.Black ex Eardley, Fl. S. Austral., ed. 2: 945 (1957).

subsp. ***fulgens***
SW. Australia. 50 WAU. Nanophan.

subsp. ***steedmanii*** (C.A.Gardner) K.J.Cowley, Austral. Syst. Bot. 3: 172 (1990).
SW. Australia. 50 WAU. Nanophan.
**Melaleuca steedmanii* C.A.Gardner, J. Roy. Soc. Western Australia 14: 81 (1928).

Melaleuca gibbosa Labill., Nov. Holl. Pl. 2: 30 (1806). *Myrtoleucodendron gibbosum* (Labill.) Kuntze, Revis. Gen. Pl. 1: 241 (1891).
SE. South Australia to Tasmania. 50 SOA TAS VIC. Nanophan. or phan.

Melaleuca glaberrima F.Muell., Fragm. 3: 119 (1862). *Myrtoleucodendron glaberrimum* (F.Muell.) Kuntze, Revis. Gen. Pl. 1: 241 (1891).
SW. Australia. 50 WAU. Nanophan.

Melaleuca glauca Domin, Biblioth. Bot. 89: 1011 (1928).
S. Western Australia. 50 WAU. Nanophan.

Melaleuca glena Craven, Austral. Syst. Bot. 12: 878 (1999).
Western Australia. 50 WAU. Nanophan. or phan.

Melaleuca globifera R.Br. in W.T.Aiton, Hortus Kew. 4: 410 (1812). *Myrtoleucodendron globiferum* (R.Br.) Kuntze, Revis. Gen. Pl. 1: 241 (1891).
S. Western Australia. 50 WAU. Nanophan.

Melaleuca glomerata F.Muell., Rep. Pl. Babage's Exped.: 8 (1859). *Myrtoleucodendron glomeratum* (F.Muell.) Kuntze, Revis. Gen. Pl. 1: 241 (1891).
C. Australia. 50 NSW NTA SOA WAU. Nanophan. or phan.
Melaleuca hakeoides F.Muell. ex Benth., Fl. Austral. 3: 151 (1867). *Myrtoleucodendron hakeoides* (F.Muell. ex Benth.) Kuntze, Revis. Gen. Pl. 1: 241 (1891).

Melaleuca gnidioides Brongn. & Gris, Ann. Sci. Nat., Bot., V, 2: 139 (1864).
SE. New Caledonia. 60 NWC. Cham. or nanophan.
Melaleuca bonatiana Schltr., Bot. Jahrb. Syst. 40(92): 32 (1908).
Melaleuca acicularis Brongn. & Gris ex Guillaumin, Bull. Mus. Natl. Hist. Nat. 17: 456 (1911), pro syn.

Melaleuca grieveana Craven, Austral. Syst. Bot. 12: 879 (1999).
Western Australia. 50 WAU. Nanophan. or phan.

Melaleuca groveana Cheel & C.T.White, Proc. Roy. Soc. Queensland 36: 41 (1924).
SE. Queensland to E. New South Wales. 50 NSW QLD. Nanophan. or phan.

Melaleuca halmaturorum F.Muell. ex Miq., Ned. Kruidk. Arch. 4: 122 (1859).
SW. & SE. Australia. 50 SOA VIC WAU. Nanophan. or phan.

subsp. ***cymbifolia*** (Benth.) Barlow, Austral. Syst. Bot. 1: 102 (1988).
SW. Australia. 50 WAU. Nanophan. or phan.
**Melaleuca cymbifolia* Benth., Fl. Austral. 3: 148 (1867). *Myrtoleucodendron cymbifolium* (Benth.) Kuntze, Revis. Gen. Pl. 1: 241 (1891).
Melaleuca cuticularis var. *brachyphylla* Domin, Vestn. Král. Ceské Spolecn. Nauk, Tr. Mat.-Prír. 2(2): 89 (1923).

subsp. ***halmaturorum***
SE. South Australia to W. Victoria. 50 SOA VIC. Nanophan. or phan.
Melaleuca halmaturorum var. *enervis* Miq., Ned. Kruidk. Arch. 4: 122 (1856).
Melaleuca halmaturorum var. *tuberculifera* Miq., Ned. Kruidk. Arch. 4: 122 (1856).

Melaleuca halophila Craven, Austral. Syst. Bot. 12: 880 (1999).
Western Australia. 50 WAU. Nanophan. or phan.

Melaleuca hamata Field. & G.A.Gardner, Sert. Pl.: t. 74 (1843).
SW. Australia. 50 WAU. Nanophan. or phan.
Melaleuca drummondii Schauer in J.G.C.Lehmann, Pl. Preiss. 1: 138 (1844).

Melaleuca hamulosa Turcz., Bull. Soc. Imp. Naturalistes Moscou 20(1): 165 (1847). *Myrtoleucodendron hamulosum* (Turcz.) Kuntze, Revis. Gen. Pl. 1: 241 (1891).
SW. Australia. 50 WAU. Nanophan. or phan.

Melaleuca haplantha Barlow, Austral. Syst. Bot. 1: 105 (1988).
SW. Australia. 50 WAU. Nanophan.

Melaleuca hnatiukii Craven, Austral. Syst. Bot. 12: 880 (1999).
Western Australia. 50 WAU. Nanophan. or phan.

Melaleuca hollidayi Craven, Austral. Syst. Bot. 12: 881 (1999).
Western Australia. 50 WAU. Nanophan. or phan.

Melaleuca holosericea Schauer in J.G.C.Lehmann, Pl. Preiss. 1: 139 (1844). *Myrtoleucodendron holosericeum* (Schauer) Kuntze, Revis. Gen. Pl. 1: 241 (1891).
SW. Australia. 50 WAU. Nanophan.

Melaleuca howeana Cheel, J. Proc. Roy. Soc. New S. Wales 58: 192 (1924).
Lord Howe I. 50 NFK. Nanophan.

Melaleuca huegelii Endl. in S.L.Endlicher & al., Enum. Pl.: 48 (1837). *Myrtoleucodendron huegelii* (Endl.) Kuntze, Revis. Gen. Pl. 1: 241 (1891).
W. & SW. Australia. 50 WAU. Nanophan.

subsp. ***huegelii***
W. & SW. Australia. 50 WAU. Nanophan.

subsp. ***pristicensis*** Barlow, Nuytsia 8: 338 (1992).
W. Western Australia. 50 WAU. Nanophan.

Melaleuca huttensis Craven, Austral. Syst. Bot. 12: 881 (1999).
Western Australia. 50 WAU. Nanophan. or phan.

Melaleuca hypericifolia Sm., Trans. Linn. Soc. London 3: 279 (1797). *Metrosideros hypericifolia* (Sm.) Salisb., Prodr. Stirp. Chap. Allerton: 351 (1796). *Myrtoleucodendron hypericifolium* (Sm.) Kuntze, Revis. Gen. Pl. 1: 241 (1891). *Cajuputi hypericifolia* (Sm.) Skeels, Bull. Bur. Pl. Industr. U.S.D.A. 242: 39 (1912).
New South Wales. 50 NSW. Nanophan. or phan.

Melaleuca idana Craven, Austral. Syst. Bot. 12: 882 (1999).
Western Australia. 50 WAU. Nanophan. or phan.

Melaleuca incana R.Br., Bot. Reg. 5: t. 410 (1819). *Myrtoleucodendron incanum* (R.Br.) Kuntze, Revis. Gen. Pl. 1: 241 (1891).
SW. Australia. 50 WAU. Nanophan.

subsp. ***incana***
SW. Australia. 50 WAU. Nanophan.
Melaleuca canescens Otto in J.H.F.Link & C.F.Otto, Icon. Pl. Select.: 37 (1821).
Melaleuca lanata Nois. ex Steud., Nomencl. Bot.: 515 (1821).
Melaleuca tomentosa Colla, Hortus Ripul.: 87 (1824).
Melaleuca lanigera J.C.Wendl. ex DC., Prodr. 3: 215 (1828).
Melaleuca hypochondriaca Dehnh., Rivista Napol. 1(3): 177 (1839).
Melaleuca polygaloides Schauer in J.G.C.Lehmann, Pl. Preiss. 1: 142 (1844). *Myrtoleucodendron polygaloides* (Schauer) Kuntze, Revis. Gen. Pl. 1: 241 (1891).

subsp. ***tenella*** (Benth.) Barlow, Nuytsia 8: 340 (1992).
S. Western Australia. 50 WAU. Nanophan.
**Melaleuca tenella* Benth., Fl. Austral. 3: 160 (1867). *Myrtoleucodendron tenellum* (Benth.) Kuntze, Revis. Gen. Pl. 1: 241 (1891).

Melaleuca interioris Craven & Lepschi, Austral. Syst. Bot. 17: 263 (2004).
Australia. 50 NSW NTA QLD WAU.

Melaleuca irbyana F.Muell. ex R.T.Baker, Proc. Linn. Soc. New South Wales 37: 587 (1912 publ. 1913). *Melaleuca tamariscina* subsp. *irbyana* (F.Muell. ex R.T.Baker) Barlow, Brunonia 9: 173 (1987).
SE. Queensland to NE. New South Wales. 50 NSW QLD. Nanophan.

Melaleuca johnsonii Craven, Austral. Syst. Bot. 12: 883 (1999).
Western Australia. 50 WAU. Nanophan. or phan.

Melaleuca keigheryi Craven, Austral. Syst. Bot. 12: 883 (1999).
Western Australia. 50 WAU. Nanophan. or phan.

Melaleuca kunzeoides Byrnes, Austrobaileya 2: 75 (1984).
SW. Queensland. 50 QLD. Nanophan.

Melaleuca laetifica Craven, Austral. Syst. Bot. 12: 884 (1999).
Western Australia. 50 WAU. Nanophan. or phan.

Melaleuca lanceolata Otto in J.H.F.Link & C.F.Otto, Icon. Pl. Select.: 36 (1821).
E. & S. Australia. 50 NSW QLD SOA VIC WAU. Nanophan. or phan.

subsp. ***lanceolata***
E. & S. Australia. 50 NSW QLD SOA VIC WAU. Nanophan. or phan.

Melaleuca pubescens Schauer in W.G.Walpers, Repert. Bot. Syst. 2: 928 (1843). *Melaleuca parviflora* var. *pubescens* (Schauer) Domin, Biblioth. Bot. 89: 1011 (1928). *Cajuputi pubescens* (Schauer) Skeels, Bull. Bur. Pl. Industr. U.S.D.A. 242: 41 (1912).
Melaleuca curvifolia Schltdl., Linnaea 20: 654 (1847).

subsp. ***occidentalis*** Barlow, Austral. Syst. Bot. 1: 121 (1988).
W. & SW. Western Australia. 50 WAU. Nanophan. or phan.

subsp. ***planifolia*** Barlow, Austral. Syst. Bot. 1: 119 (1988).
S. Western Australia. 50 WAU. Nanophan. or phan.
Melaleuca leiostachya var. *preissiana* Benth., Fl. Austral. 3: 145 (1867).
Melaleuca preissiana var. *leiostachya* Benth., Fl. Austral. 3: 145 (1867).
Melaleuca parviflora var. *leiostachya* (Benth.) Domin, Biblioth. Bot. 89: 1012 (1928).

subsp. ***thaeroides*** Barlow, Austral. Syst. Bot. 1: 121 (1988).
SW. & S. Western Australia to SW. South Australia. 50 SOA WAU. Nanophan. or phan.
Melaleuca seorsiflora F.Muell., Australas. J. Pharm. 1: 278 (1886). *Myrtoleucodendron seorsiflorum* (F.Muell.) Kuntze, Revis. Gen. Pl. 1: 241 (1891).

Melaleuca lara Craven, Austral. Syst. Bot. 12: 885 (1999).
Western Australia. 50 WAU. Nanophan. or phan.

Melaleuca lasiandra F.Muell., Fragm. 3: 115 (1862). *Myrtoleucodendron lasiandrum* (F.Muell.) Kuntze, Revis. Gen. Pl. 1: 241 (1891).
N. Australia. 50 NTA QLD WAU. Nanophan. or phan.
Melaleuca loguei W.Fitzg., J. Proc. Roy. Soc. W. Australia 3: 188 (1918).

Melaleuca lateralis Turcz., Bull. Cl. Phys.-Math. Acad. Imp. Sci. Saint-Pétersbourg 10: 339 (1852).
SW. Australia. 50 WAU. Nanophan.

Melaleuca lateriflora Benth., Fl. Austral. 3: 136 (1867). *Myrtoleucodendron lateriflorum* (Benth.) Kuntze, Revis. Gen. Pl. 1: 241 (1891).
SW. Australia. 50 WAU. Nanophan.

subsp. ***acutifolia*** (Benth.) Craven, Austral. Syst. Bot. 12: 885 (1999).
SW. Western Australia. 50 WAU. Nanophan. or phan.
**Melaleuca lateriflora* var. *acutifolia* Benth., Fl. Austral. 3: 136 (1867).

subsp. ***lateriflora***
SW. Australia. 50 WAU. Nanophan.
Melaleuca lateriflora var. *elliptica* Benth., Fl. Austral. 3: 136 (1867).

Melaleuca lateritia A.Dietr., Allg. Gartenzeitung 2: 257 (1834). *Myrtoleucodendron lateritium* (A.Dietr.) Kuntze, Revis. Gen. Pl. 1: 241 (1891).
W. & SW. Western Australia. 50 WAU. Nanophan.
Melaleuca callistemonea Lindl., Sketch Veg. Swan R.: viii (1839).

Melaleuca laxiflora Turcz., Bull. Cl. Phys.-Math. Acad. Imp. Sci. Saint-Pétersbourg 10: 341 (1852). *Myrtoleucodendron laxiflorum* (Turcz.) Kuntze, Revis. Gen. Pl. 1: 241 (1891).
SW. Australia. 50 WAU. Nanophan.
Melaleuca parviflora Lindl., Sketch Veg. Swan R.: viii (1839), nom. illeg. *Myrtoleucodendron parviflorum* Kuntze, Revis. Gen. Pl. 1: 241 (1891).
Melaleuca crassifolia Benth., Fl. Austral. 3: 145 (1867). *Myrtoleucodendron crassifolium* (Benth.) Kuntze, Revis. Gen. Pl. 1: 241 (1891).

Melaleuca lecanantha Barlow, Austral. Syst. Bot. 1: 113 (1988).
SW. Australia. 50 WAU. Nanophan.

Melaleuca leiocarpa F.Muell., Fragm. 10: 55 (1877). *Myrtoleucodendron leiocarpum* (F.Muell.) Kuntze, Revis. Gen. Pl. 1: 241 (1891).
SE. Western Australia to S. South Australia. 50 SOA WAU. Nanophan.

Melaleuca leiopyxis F.Muell. ex Benth., Fl. Austral. 3: 160 (1867). *Myrtoleucodendron leiopyxe* (F.Muell. ex Benth.) Kuntze, Revis. Gen. Pl. 1: 241 (1891).
W. Western Australia. 50 WAU. Nanophan.

Melaleuca leptospermoides Schauer in J.G.C.Lehmann, Pl. Preiss. 1: 139 (1844).
W. & SW. Western Australia. 50 WAU. Nanophan.
Melaleuca cuneata Turcz., Bull. Cl. Phys.-Math. Acad. Imp. Sci. Saint-Pétersbourg 10: 339 (1852).
Melaleuca eriantha Benth., Fl. Austral. 3: 153 (1867). *Myrtoleucodendron erianthum* (Benth.) Kuntze, Revis. Gen. Pl. 1: 241 (1891).

Melaleuca leucadendra (L.) L., Mant. Pl. 1: 105 (1767).
Maluku to N. Australia. (22) ben (23) gab 42 MOL 43 NWG 50 NTA QLD WAU. Phan.
**Myrtus leucadendra* L., Herb. Amb.: 9 (1754). *Leptospermum leucodendron* (L.) J.R.Forst. & G.Forst., Char. Gen. Pl.: 48 (1775). *Meladendron leucocladum* St.-Lag., Ann. Soc. Bot. Lyon 7: 64 (1880). *Cajuputi leucadendron* (L.) A.Lyons, Plant Nam.: 479 (1900).
Myrtus saligna Burm.f., Fl. Indica: 116 (1768).
Myrtus alba Noronha, Verh. Batav. Genootsch. Kunsten 5(4): 20 (1790).
Metrosideros coriacea K.D.Koenig & Sims in R.A. Salisbury, Prodr. Stirp. Chap. Allerton: 352 (1796).
Melaleuca rigida Roxb., Fl. Ind. ed. 1832, 3: 399 (1832).
Melaleuca mimosoides A.Cunn. ex Schauer in W.G. Walpers, Repert. Bot. Syst. 2: 927 (1843). *Melaleuca leucadendra* var. *mimosoides* (A.Cunn. ex Schauer) Cheel in A.J.Ewart & O.B.Davies, Fl. N. Territory: 295 (1917).
Melaleuca conferta Benth., Fl. Austral. 3: 142 (1867), nom. illeg.
Melaleuca leucadendra var. *angusta* C.Rivière, Bull. Soc. Natl. Acclim. France 3: 537 (1882).
Melaleuca leucadendra var. *cunninghamii* F.M.Bailey, Syn. Queensl. Fl.: 171 (1883).
Melaleuca leucadendra var. *lancifolia* F.M.Bailey, Syn. Queensl. Fl.: 170 (1883).
Melaleuca leucadendra var. *albida* Cheel in A.J.Ewart & O.B.Davies, Fl. N. Territory: 301 (1917).
Melaleuca amboinensis Gand., Bull. Soc. Bot. France 65: 26 (1918).

Melaleuca leuropoma Craven, Austral. Syst. Bot. 12: 888 (1999).
Western Australia. 50 WAU. Nanophan. or phan.

Melaleuca linariifolia Sm., Trans. Linn. Soc. London 3: 278 (1797). *Myrtoleucodendron linariifolium* (Sm.) Kuntze, Revis. Gen. Pl. 1: 241 (1891). *Melaleuca linariifolia* var. *typica* Domin, Biblioth. Bot. 89: 456 (1928), nom. inval.

E. Australia. 50 NSW QLD. Nanophan. or phan.
Metrosideros hyssopifolia Cav., Icon. 4: 20 (1797), nom. illeg. *Melaleuca hyssopifolia* (Cav.) Dum.Cours., Bot. Cult., ed. 2, 5: 375 (1811). *Ozandra hyssopifolia* (Cav.) Raf., Autik. Bot.: 148 (1840).
Melaleuca stricta Dum.Cours., Bot. Cult., ed. 2, 5: 375 (1811), nom. inval.

Melaleuca linguiformis Craven, Austral. Syst. Bot. 12: 888 (1999).
Western Australia. 50 WAU. Nanophan. or phan.

Melaleuca linophylla F.Muell., Fragm. 3: 115 (1862). *Myrtoleucodendron linophyllum* (F.Muell.) Kuntze, Revis. Gen. Pl. 1: 241 (1891).
Western Australia (Pilbara area). 50 WAU. Nanophan. or phan.

Melaleuca longistaminea (F.Muell.) Barlow ex Craven, Austral. Syst. Bot. 12: 889 (1999).
Western Australia. 50 WAU. Nanophan. or phan.
**Melaleuca cardiophylla* var. *longistaminea* F.Muell., Fragm. 3: 164 (1862).

subsp. ***longistaminea***
Western Australia. 50 WAU. Nanophan. or phan.

subsp. ***spectabilis*** Craven, Austral. Syst. Bot. 12: 889 (1999).
Western Australia. 50 WAU. Nanophan. or phan.

Melaleuca macronychia Turcz., Bull. Cl. Phys.-Math. Acad. Imp. Sci. Saint-Pétersbourg 10: 340 (1852). *Myrtoleucodendron macronychium* (Turcz.) Kuntze, Revis. Gen. Pl. 1: 241 (1891).
SW. Australia. 50 WAU. Nanophan.

subsp. ***macronychia***
SW. Australia. 50 WAU. Nanophan.
Melaleuca longicoma Benth., Fl. Austral. 3: 129 (1867). *Myrtoleucodendron longicomum* (Benth.) Kuntze, Revis. Gen. Pl. 1: 241 (1891).

subsp. ***trygonoides*** K.J.Cowley, Austral. Syst. Bot. 3: 180 (1990).
SW. Australia. 50 WAU. Nanophan.

Melaleuca manglesii Schauer in J.G.C.Lehmann, Pl. Preiss. 1: 135 (1844).
Western Australia. 50 WAU. Nanophan.

Melaleuca megacephala F.Muell., Fragm. 3: 117 (1862). *Myrtoleucodendron megacephalum* (F.Muell.) Kuntze, Revis. Gen. Pl. 1: 241 (1891).
WSW. Western Australia. 50 WAU. Nanophan.

Melaleuca micromera Schauer in J.G.C.Lehmann, Pl. Preiss. 1: 146 (1844). *Myrtoleucodendron micromerum* (Schauer) Kuntze, Revis. Gen. Pl. 1: 241 (1891).
SW. Australia. 50 WAU. Nanophan.

Melaleuca microphylla Sm. in A.Rees, Cycl. 23: 9 (1812). *Myrtoleucodendron microphyllum* (Sm.) Kuntze, Revis. Gen. Pl. 1: 241 (1891).
SW. Australia. 50 WAU. Nanophan.
Melaleuca brachyphylla Schauer, Allg. Gartenzeitung 3: 162 (1835).
Melaleuca tenuissima Tausch, Flora 19: 411 (1836).
Melaleuca brevifolia F.Muell., Fragm. 1: 116 (1859), nom. illeg.

Melaleuca minutifolia F.Muell., Trans. & Proc. Philos. Inst. Victoria 3: 45 (1859). *Myrtoleucodendron minutifolium* (F.Muell.) Kuntze, Revis. Gen. Pl. 1: 241 (1891).
N. Western Australia to Northern Territory. 50 NTA WAU. Nanophan. or phan.

Melaleuca monantha (Barlow) Craven, Austral. Syst. Bot. 12: 891 (1999).
NE. Queensland. 50 QLD. Nanophan. or phan.
**Melaleuca minutifolia* subsp. *monantha* Barlow, Brunonia 9: 176 (1987).

Melaleuca nanophylla Carrick, J. Adelaide Bot. Gard. 1: 300 (1979).
C. Australia. 50 SOA WAU. Nanophan.

Melaleuca nematophylla F.Muell. ex Craven, Austral. Syst. Bot. 12: 892 (1999).
Western Australia. 50 WAU. Nanophan. or phan.

Melaleuca nervosa (Lindl.) Cheel, J. Proc. Roy. Soc. New S. Wales 78: 665 (1945).
Papua New Guinea to N. Australia. 43 NWG 50 NTA QLD WAU. Phan.
**Callistemon nervosus* Lindl. in T.L.Mitchell, J. Exped. Trop. Australia: 235 (1848). *Melaleuca leucadendra* var. *parvifolia* Benth., Fl. Austral. 3: 143 (1867). *Melaleuca leucadendra* var. *nervosa* (Lindl.) Domin, Biblioth. Bot. 89: 457 (1928), nom. illeg.

subsp. ***crosslandiana*** (W.Fitzg.) Barlow ex Craven, Austral. Syst. Bot. 12: 892 (1999).
N. Australia. 50 NTA QLD WAU. Nanophan. or phan.
**Melaleuca crosslandiana* W.Fitzg., J. Proc. Roy. Soc. W. Australia 3: 186 (1918).

subsp. ***nervosa***
Papua New Guinea to N. Australia. 43 NWG 50 NTA QLD WAU. Phan.
Melaleuca nervosa f. *latifolia* Byrnes, Austrobaileya 2: 74 (1984).

Melaleuca nesophila F.Muell., Fragm. 3: 113 (1862). *Myrtoleucodendron nesophyllum* (F.Muell.) Kuntze, Revis. Gen. Pl. 1: 241 (1891).
W. & S. Western Australia. 50 WAU. Nanophan.

Melaleuca nodosa (Sol. ex Gaertn.) Sm., Trans. Linn. Soc. London 3: 276 (1797).
E. Australia. 50 NSW QLD. Nanophan. or phan.
**Metrosideros nodosa* Sol. ex Gaertn., Fruct. Sem. Pl. 1: 172 (1788). *Metrosideros gracilis* K.D.Koenig & Sims in R.A.Salisbury, Prodr. Stirp. Chap. Allerton: 352 (1796), nom. superfl. *Myrtoleucodendron nodosum* (Sol. ex Gaertn.) Kuntze, Revis. Gen. Pl. 1: 241 (1891).
Metrosideros juniperina Rchb. ex Spreng., Syst. Veg. 4(2): 194 (1827). *Melaleuca juniperina* (Rchb. ex Spreng.) Sieber ex Rchb., Iconogr. Bot. Exot. 2: 5 (1829). *Callistemon juniperinus* (Rchb. ex Spreng.) Heynh., Nom. Bot. Hort. 1: 148 (1840).
Metrosideros pungens Rchb., Syst. Veg. 4(2): 194 (1827).
Melaleuca juniperoides DC., Prodr. 3: 213 (1828).
Melaleuca tenuifolia DC., Prodr. 3: 213 (1828). *Melaleuca nodosa* var. *tenuifolia* (DC.) Cheel ex A.R.Penfold, J. Proc. Roy. Soc. New S. Wales 63: 107 (1930).
Metrosideros juniperoides Rchb. ex DC., Prodr. 3: 213 (1828).

Melaleuca oldfieldii F.Muell., Fragm. 3: 118 (1862). *Myrtoleucodendron oldfieldii* (F.Muell.) Kuntze, Revis. Gen. Pl. 1: 241 (1891).
W. Western Australia. 50 WAU. Nanophan.

Melaleuca orbicularis Craven, Austral. Syst. Bot. 12: 893 (1999).
Western Australia. 50 WAU. Nanophan. or phan.

Melaleuca ordinifolia Barlow, Nuytsia 8: 342 (1992).
SW. Australia. 50 WAU. Cham. or nanophan.

Melaleuca osullivanii Craven & Lepschi, Austral. Syst. Bot. 17: 265 (2004).
WSW. Western Australia. 50 WAU.

Melaleuca oxyphylla Carrick, J. Adelaide Bot. Gard. 1: 304 (1979).
South Australia. 50 SOA. Nanophan.

Melaleuca pallescens Byrnes, Austrobaileya 2: 72 (1984). *Melaleuca tamariscina* subsp. *pallescens* (Byrnes) Barlow, Brunonia 9: 173 (1987).
SE. Queensland. 50 QLD. Nanophan.

Melaleuca pancheri (Brongn. & Gris) Craven & J.W.Dawson, Adansonia, III, 20: 192 (1998).
SE. New Caledonia. 60 NWC. Nanophan. or phan.
**Callistemon pancheri* Brongn. & Gris, Bull. Soc. Bot. France 11: 183 (1864).
Callistemon suberosus var. *microphyllus* Guillaumin, Mém. Mus. Natl. Hist. Nat., B, Bot. 8: 277 (1962).

Melaleuca papillosa Craven, Austral. Syst. Bot. 12: 894 (1999).
Western Australia. 50 WAU. Nanophan. or phan.

Melaleuca parviceps Lindl., Sketch Veg. Swan R.: viii (1839).
Western Australia. 50 WAU. Nanophan.

Melaleuca parvistaminea Byrnes, Austrobaileya 2: 75 (1984).
New South Wales to Victoria. 50 NSW VIC. Nanophan. or phan.

Melaleuca pauciflora Turcz., Bull. Soc. Imp. Naturalistes Moscou 20(1): 166 (1847). *Myrtoleucodendron pauciflorum* (Turcz.) Kuntze, Revis. Gen. Pl. 1: 241 (1891).
Western Australia. 50 WAU. Nanophan.
Melaleuca leptoclada Benth., Fl. Austral. 3: 132 (1867). *Myrtoleucodendron leptocladum* (Benth.) Kuntze, Revis. Gen. Pl. 1: 241 (1891).

Melaleuca pauperiflora F.Muell., Fragm. 3: 116 (1862). *Myrtoleucodendron pauperiflorum* (F.Muell.) Kuntze, Revis. Gen. Pl. 1: 241 (1891).
SW. & S Australia. 50 SOA WAU. Nanophan.

subsp. ***fastigiata*** Barlow, Austral. Syst. Bot. 1: 109 (1988).
SW. & S. Western Australia. 50 WAU. Nanophan.

subsp. ***mutica*** Barlow, Austral. Syst. Bot. 1: 109 (1988).
S. South Australia. 50 SOA. Nanophan.

subsp. ***pauperiflora***
S. Western Australia. 50 WAU. Nanophan.

Melaleuca penicula (K.J.Cowley) Craven, Austral. Syst. Bot. 12: 896 (1999).
S. Western Australia. 50 WAU. Nanophan. or phan.
**Melaleuca coccinea* subsp. *penicula* K.J.Cowley, Austral. Syst. Bot. 3: 178 (1990).

Melaleuca pentagona Labill., Nov. Holl. Pl. 2: 27 (1806). *Myrtoleucodendron pentagonum* (Labill.) Kuntze, Revis. Gen. Pl. 1: 241 (1891).
Western Australia. 50 WAU. Nanophan.

var. ***latifolia*** Benth., Fl. Austral. 3: 152 (1867).
Western Australia. 50 WAU. Nanophan.

var. ***pentagona***
Western Australia. 50 WAU. Nanophan.
Melaleuca waeberi Rchb. ex Schauer, Allg. Gartenzeitung 3: 167 (1835).
Melaleuca pentagona var. *planifolia* Schauer in J.G.C.Lehmann, Pl. Preiss. 1: 136 (1845).
Melaleuca pentagona var. *subulifolia* Schauer in J.G.C.Lehmann, Pl. Preiss. 1: 136 (1845).
Melaleuca canaliculata Turcz., Bull. Cl. Phys.-Math. Acad. Imp. Sci. Saint-Pétersbourg 10: 432 (1852).

var. ***raggedensis*** Craven, Austral. Syst. Bot. 12: 897 (1999).
Western Australia. 50 WAU. Nanophan. or phan.

Melaleuca phoidophylla Barlow ex Craven, Austral. Syst. Bot. 12: 897 (1999).
Western Australia. 50 WAU. Nanophan. or phan.

Melaleuca platycalyx Diels, Bot. Jahrb. Syst. 35: 426 (1904).
SW. Australia. 50 WAU. Nanophan.
Melaleuca violacea var. *petiolata* Benth., Fl. Austral. 3: 135 (1867).

Melaleuca plumea Craven, Austral. Syst. Bot. 12: 897 (1999).
Western Australia. 50 WAU. Nanophan. or phan.

Melaleuca podiocarpa Barlow ex Craven, Austral. Syst. Bot. 12: 898 (1999).
Western Australia. 50 WAU. Nanophan. or phan.

Melaleuca polycephala Benth., Fl. Austral. 3: 152 (1867). *Myrtoleucodendron polycephalum* (Benth.) Kuntze, Revis. Gen. Pl. 1: 241 (1891).
S. Western Australia. 50 WAU. Nanophan.
Melaleuca serpyllifolia Turcz., Bull. Cl. Phys.-Math. Acad. Imp. Sci. Saint-Pétersbourg 10: 339 (1852), nom. illeg.

Melaleuca pomphostoma Barlow, Nuytsia 8: 343 (1992).
SW. Australia. 50 WAU. Nanophan.

Melaleuca preissiana Schauer in J.G.C.Lehmann, Pl. Preiss. 1: 143 (1844).
SW. Australia. 50 WAU. Nanophan.

Melaleuca pritzelii (Domin) Barlow, Nuytsia 8: 345 (1992).
SW. Australia. 50 WAU. Nanophan.
**Melaleuca densa* var. *pritzelii* Domin, Vestn. Král. Ceské Spolecn. Nauk, Tr. Mat.-Prír. 2(2): 90 (1923).

Melaleuca procera Craven, Austral. Syst. Bot. 12: 899 (1999).
Western Australia. 50 WAU. Nanophan. or phan.

Melaleuca propinqua Schauer, Allg. Gartenzeitung 3: 166 (1835).
S. Western Australia. 50 WAU. Nanophan.

Melaleuca psammophila Diels, Bot. Jahrb. Syst. 35: 429 (1904).
W. Western Australia. 50 WAU. Nanophan.

Melaleuca pulchella R.Br. in W.T.Aiton, Hortus Kew. 4: 414 (1812). *Myrtoleucodendron pulchellum* (R.Br.) Kuntze, Revis. Gen. Pl. 1: 241 (1891). *Cajuputi pulchella* (R.Br.) Skeels, Bull. Bur. Pl. Industr. U.S.D.A. 181: 60 (1913).
S. Western Australia. 50 WAU. Nanophan.

Melaleuca pungens Schauer in J.G.C.Lehmann, Pl. Preiss. 1: 138 (1844). *Myrtoleucodendron pungens* (Schauer) Kuntze, Revis. Gen. Pl. 1: 241 (1891).
W. & S. Western Australia. 50 WAU. Nanophan. Provisionally accepted.
Melaleuca pungens var. *obtusifolia* Benth., Fl. Austral. 3: 159 (1867).

Melaleuca pustulata Hook.f., London J. Bot. 6: 476 (1847). *Myrtoleucodendron pustulatum* (Hook.f.) Kuntze, Revis. Gen. Pl. 1: 241 (1891).
CE. Tasmania. 50 TAS. Nanophan. or phan.

Melaleuca quadrifaria F.Muell., S. Sci. Rec., n.s., 2: (Apr. 1886). *Myrtoleucodendron quadrifarium* (F.Muell.) Kuntze, Revis. Gen. Pl. 1: 241 (1891).
S. Western Australia to SW. South Australia. 50 SOA WAU. Nanophan.

Melaleuca quinquenervia (Cav.) S.T.Blake, Proc. Roy. Soc. Queensland 69: 76 (1958).
New Guinea, New Caledonia, E. Australia. 43 NWG 50 NSW QLD 60 NWC (61) sci. Nanophan. or phan.
Melaleuca leucadendra var. *angustifolia* L.f., Suppl. Pl.: 342 (1781). *Melaleuca viridiflora* var. *angustifolia* (L.f.) Byrnes, Austrobaileya 2: 74 (1984), nom. illeg.
**Metrosideros quinquenervia* Cav., Icon. 4: 19 (1797).
Metrosideros coriacea Poir. in J.B.A.P.M.de Lamarck, Encycl., Suppl. 3: 685 (1813), nom. illeg. *Melaleuca leucadendra* var. *coriacea* (Poir.) Cheel in A.J.Ewart & O.B.Davies, Fl. N. Territory: 297 (1917).
Metrosideros albida Sieber ex DC., Prodr. 3: 212 (1828), nom. inval.
Melaleuca viridiflora var. *rubriflora* Pancher ex Brongn. & Gris, Bull. Soc. Bot. France 11: 183 (1864).
Melaleuca maidenii R.T.Baker, Proc. Linn. Soc. New South Wales 38: 598 (1913 publ. 1914).
Melaleuca smithii R.T.Baker, Proc. Linn. Soc. New South Wales 38: 599 (1913 publ. 1914). *Melaleuca leucadendra* var. *albida* Cheel, Fl. N. Terr.: 301 (1917).

Melaleuca radula Lindl., Sketch Veg. Swan R.: viii (1839). *Myrtoleucodendron radula* (Lindl.) Kuntze, Revis. Gen. Pl. 1: 241 (1891).
SW. Australia. 50 WAU. Nanophan.

Melaleuca rhaphiophylla Schauer in J.G.C.Lehmann, Pl. Preiss. 1: 143 (1844). *Myrtoleucodendron rhaphiophyllum* (Schauer) Kuntze, Revis. Gen. Pl. 1: 241 (1891).
Western Australia. 50 WAU. Nanophan.

Melaleuca rigidifolia Turcz., Bull. Cl. Phys.-Math. Acad. Imp. Sci. Saint-Pétersbourg 10: 342 (1852).
SW. Australia. 50 WAU. Nanophan.

Melaleuca ringens Barlow, Nuytsia 8: 346 (1992).
Western Australia. 50 WAU. Nanophan.

Melaleuca ryeae Craven, Austral. Syst. Bot. 12: 901 (1999).
Western Australia. 50 WAU. Nanophan. or phan.

Melaleuca saligna Schauer in W.G.Walpers, Repert. Bot. Syst. 2: 927 (1843). *Melaleuca leucadendra* var. *saligna* (Schauer) F.M.Bailey, Syn. Queensl. Fl.: 170 (1883).
N. Queensland. 50 QLD. Nanophan. or phan.
Melaleuca stenostachya var. *pendula* Byrnes, Austrobaileya 2: 74 (1984).

Melaleuca sapientes Craven, Austral. Syst. Bot. 12: 901 (1999).
Western Australia. 50 WAU. Nanophan. or phan.

Melaleuca scabra R.Br. in W.T.Aiton, Hortus Kew. 4: 414 (1812). *Myrtoleucodendron scabrum* (R.Br.) Kuntze, Revis. Gen. Pl. 1: 241 (1891).
SW. Australia. 50 WAU. Nanophan.
Melaleuca muricata A.Cunn. ex DC., Prodr. 3: 213 (1828).

Melaleuca scalena Craven & Lepschi, Austral. Syst. Bot. 17: 266 (2004).
SW. Australia. 50 WAU.

Melaleuca sciotostyla Barlow, Austral. Syst. Bot. 1: 102 (1988).
SW. Australia. 50 WAU. Nanophan.

Melaleuca sclerophylla Diels, Bot. Jahrb. Syst. 35: 428 (1904).
SW. Australia. 50 WAU. Nanophan.

Melaleuca sculponeata Barlow, Nuytsia 8: 347 (1992).
Western Australia (near Ravensthorpe). 50 WAU. Cham.

Melaleuca seriata Lindl., Sketch Veg. Swan R.: viii (1839). *Myrtoleucodendron seriatum* (Lindl.) Kuntze, Revis. Gen. Pl. 1: 241 (1891).
W. & S. Western Australia. 50 WAU. Nanophan.
Metrosideros sororia Endl. in S.L.Endlicher & al., Enum. Pl.: 49 (1837). *Melaleuca endlicheriana* Schauer in J.G.C.Lehmann, Pl. Preiss. 1: 134 (1844), nom. illeg.
Melaleuca ornata Schauer in J.G.C.Lehmann, Pl. Preiss. 1: 135 (1844).

Melaleuca sericea Byrnes, Austrobaileya 2: 74 (1984).
N. Western Australia to Northern Territory. 50 NTA WAU. Nanophan. or phan.

Melaleuca sheathiana W.Fitzg., J. Proc. Mueller Bot. Soc. Western Australia 9: 16 (1902).
S. Western Australia. 50 WAU. Nanophan.

Melaleuca sieberi Schauer in W.G.Walpers, Repert. Bot. Syst. 2: 928 (1843).
SE. Queensland to NE. New South Wales. 50 NSW QLD. Nanophan. or phan.
Melaleuca parviflora var. *latifolia* Maiden & Betche, Census N.S.W. Pl.: 156 (1916).

Melaleuca similis Craven, Austral. Syst. Bot. 12: 903 (1999).
Western Australia. 50 WAU. Nanophan. or phan.

Melaleuca societatis Craven, Austral. Syst. Bot. 12: 904 (1999).
Western Australia. 50 WAU. Nanophan. or phan.

Melaleuca sparsiflora Turcz., Bull. Soc. Imp. Naturalistes Moscou 20(1): 167 (1847). *Myrtoleucodendron sparsiflorum* (Turcz.) Kuntze, Revis. Gen. Pl. 1: 241 (1891).
SW. Australia. 50 WAU. Nanophan.

Melaleuca spathulata Schauer in J.G.C.Lehmann, Pl. Preiss. 1: 134 (1844). *Myrtoleucodendron spathulatum* (Schauer) Kuntze, Revis. Gen. Pl. 1: 241 (1891).
SW. Australia. 50 WAU. Nanophan.

Melaleuca sphaerodendra Craven & J.W.Dawson, Adansonia, III, 20: 192 (1998).
New Caledonia. 60 NWC. Nanophan. or phan.
**Callistemon gnidioides* Guillaumin, Bull. Soc. Bot. France 81: 6 (1934).

var. ***microphylla*** (Virot) Craven & J.W.Dawson, Adansonia, III, 20: 193 (1998).
NW. & WC. New Caledonia. 60 NWC. Nanophan. or phan.
**Callistemon gnidioides* var. *microphyllus* Virot, Mém. Mus. Natl. Hist. Nat., B, Bot. 4: 30 (1953).

var. ***sphaerodendra***
New Caledonia. 60 NWC. Nanophan. or phan.

Melaleuca spicigera S.Moore, J. Bot. 40: 25 (1902).
SW. Australia. 50 WAU. Nanophan.

Melaleuca squamea Labill., Nov. Holl. Pl. 2: 28 (1806). *Myrtoleucodendron squameum* (Labill.) Kuntze, Revis. Gen. Pl. 1: 241 (1891).
SE. Australia. 50 NSW SOA TAS VIC. Nanophan.
Melaleuca ottonis Schauer, Allg. Gartenzeitung 3: 167 (1835).
Melaleuca squamea var. *glabra* Cheel, J. Proc. Roy. Soc. New S. Wales 58: 193 (1924).

Melaleuca squamophloia (Byrnes) Craven, Novon 7: 118 (1997).
N. Queensland. 50 QLD. Phan.
**Melaleuca styphelioides* var. *squamophloia* Byrnes, Austrobaileya 2: 74 (1984).

Melaleuca squarrosa Sm., Trans. Linn. Soc. London 6: 300 (1802). *Myrtoleucodendron squarrosum* (Sm.) Kuntze, Revis. Gen. Pl. 1: 241 (1891).
SE. Australia. 50 NSW SOA TAS VIC. Nanophan. or phan.
Melaleuca myrtifolia Vent., Jard. Malmaison: 47 (1804).
Melaleuca caja-putti DC., Prodr. 3: 215 (1828).
Melaleuca squarrosa var. *glabrata* Miq., Ned. Kruidk. Arch. 4: 122 (1856).

Melaleuca stenostachya S.T.Blake, Contr. Queensland Herb. 1: 50 (1968).
Northern Territory to N. Queensland. 50 NTA QLD. Phan.

Melaleuca stereophloia Craven, Austral. Syst. Bot. 12: 905 (1999).
W. Western Australia. 50 WAU. Nanophan.

Melaleuca stipitata Craven, Novon 7: 118 (1997).
N. Northern Territory. 50 NTA. Nanophan. or phan.

Melaleuca stramentosa Craven, Austral. Syst. Bot. 12: 906 (1999).
Western Australia. 50 WAU. Nanophan. or phan.

Melaleuca striata Labill., Nov. Holl. Pl. 2: 26 (1806). *Myrtoleucodendron striatum* (Labill.) Kuntze, Revis. Gen. Pl. 1: 241 (1891).
SW. Australia. 50 WAU. Nanophan. Provisionally accepted.

Melaleuca strobophylla Barlow, Austral. Syst. Bot. 1: 116 (1988).
SW. Australia. 50 WAU. Nanophan. or phan.

Melaleuca styphelioides Sm., Trans. Linn. Soc. London 3: 275 (1797). *Leptospermum styphelioides* (Sm.) Schauer, Linnaea 15: 423 (1841). *Myrtoleucodendron styphelioides* (Sm.) Kuntze, Revis. Gen. Pl. 1: 241 (1891).
SE. Queensland to New South Wales. 50 NSW QLD. Phan.
Melaleuca obliqua Büse ex de Vriese, Hort. Spaarn-Berg.: 119 (1839).

Melaleuca subalaris Barlow, Austral. Syst. Bot. 1: 106 (1988).
S. Western Australia. 50 WAU. Nanophan.

Melaleuca suberosa (Schauer) C.A.Gardner, Enum. Pl. Austr. Occ.: 91 (1931).
SW. Australia. 50 WAU. Nanophan.
**Calothamnus suberosus* Schauer in J.G.C.Lehmann, Pl. Preiss. 1: 156 (1844).
Melaleuca angulata Turcz., Bull. Cl. Phys.-Math. Acad. Imp. Sci. Saint-Pétersbourg 10: 342 (1852).
Melaleuca exarata F.Muell., Fragm. 3: 114 (1862). *Myrtoleucodendron exaratum* (F.Muell.) Kuntze, Revis. Gen. Pl. 1: 241 (1891).

Melaleuca subfalcata Turcz., Bull. Cl. Phys.-Math. Acad. Imp. Sci. Saint-Pétersbourg 10: 341 (1852). *Myrtoleucodendron subfalcatum* (Turcz.) Kuntze, Revis. Gen. Pl. 1: 241 (1891).
S. Western Australia. 50 WAU. Nanophan.
Melaleuca brachystachya F.Muell., Fragm. 1: 119 (1859). *Myrtoleucodendron brachystachyum* (F.Muell.) Kuntze, Revis. Gen. Pl. 1: 241 (1891).

Melaleuca subtrigona Schauer in J.G.C.Lehmann, Pl. Preiss. 1: 139 (1844). *Myrtoleucodendron subtrigonum* (Schauer) Kuntze, Revis. Gen. Pl. 1: 241 (1891).
SW. Australia. 50 WAU. Nanophan.

Melaleuca sylvana Craven & A.J.Ford, Muelleria 20: 3 (2004).
Queensland. 50 QLD.

Melaleuca systena Craven, Austral. Syst. Bot. 12: 907 (1999).
Western Australia. 50 WAU. Nanophan. or phan.
**Melaleuca acerosa* Schauer in J.G.C.Lehmann, Pl. Preiss. 1: 137 (1844), nom. illeg. *Myrtoleucodendron acerosum* Kuntze, Revis. Gen. Pl. 1: 241 (1891).

Melaleuca tamariscina Hook. in T.L.Mitchell, J. Exped. Trop. Australia: 262 (1848). *Myrtoleucodendron tamariscinum* (Hook.) Kuntze, Revis. Gen. Pl. 1: 241 (1891).
Queensland. 50 QLD. Nanophan. or phan.

Melaleuca teretifolia Endl. in S.L.Endlicher & al., Enum. Pl.: 49 (1837). *Gymnagathis teretifolia* (Endl.) Schauer, Linnaea 17: 243 (1843). *Myrtoleucodendron teretifolium* (Endl.) Kuntze, Revis. Gen. Pl. 1: 241 (1891).
Western Australia. 50 WAU. Nanophan.
Melaleuca semiteres Schauer in J.G.C.Lehmann, Pl. Preiss. 1: 143 (1844).
Melaleuca pinifolia Turcz., Bull. Soc. Imp. Naturalistes Moscou 20(1): 166 (1847), nom. illeg.
Melaleuca hakeacea F.Muell., Fragm. 3: 117 (1862).

Melaleuca teuthidoides Barlow, Austral. Syst. Bot. 1: 110 (1988).
S. Western Australia. 50 WAU. Nanophan.

Melaleuca thapsina Craven, Austral. Syst. Bot. 12: 908 (1999).
Western Australia. 50 WAU. Nanophan. or phan.

Melaleuca thymifolia Sm., Trans. Linn. Soc. London 3: 278 (1797). *Myrtoleucodendron thymifolium* (Sm.) Kuntze, Revis. Gen. Pl. 1: 241 (1891).
SE. Queensland to E. New South Wales. 50 NSW QLD. Nanophan. or phan.
Metrosideros calycina Cav., Icon. 4: 20 (1797).
Melaleuca coronata Andrews, Bot. Repos. 4: t. 278 (1803).

Melaleuca gnidiifolia Vent., Jard. Malmaison: 4 (1803).
Melaleuca fimbriata Donn, Hortus Cantabrig., ed. 6: 207 (1811).
Melaleuca serpyllifolia Dum.Cours., Bot. Cult., ed. 2, 5: 374 (1811).
Melaleuca parvifolia Otto in J.H.F.Link & C.F.Otto, Icon. Pl. Select.: 37 (1821).
Melaleuca discolor Rchb. ex Spreng., Syst. Veg. 3: 337 (1826).
Ozandra granulata Raf., Autik. Bot.: 148 (1840).
Melaleuca oligantha F.Muell. ex Miq., Ned. Kruidk. Arch. 4: 123 (1856).

Melaleuca thymoides Labill., Nov. Holl. Pl. 2: 27 (1806). *Myrtoleucodendron thymoides* (Labill.) Kuntze, Revis. Gen. Pl. 1: 241 (1891).
SW. Australia. 50 WAU. Nanophan. Provisionally accepted.
Melaleuca spinosa Lindl., Sketch Veg. Swan R.: viii (1839), provisional synonym.

Melaleuca thyoides Turcz., Bull. Soc. Imp. Naturalistes Moscou 20(1): 167 (1847). *Myrtoleucodendron thyoides* (Turcz.) Kuntze, Revis. Gen. Pl. 1: 241 (1891).
SW. Australia. 50 WAU. Nanophan.
Melaleuca cupressina F.Muell., Fragm. 3: 114 (1862).

Melaleuca tinkeri Craven, Austral. Syst. Bot. 12: 909 (1999).
Western Australia. 50 WAU. Nanophan. or phan.

Melaleuca torquata Barlow, Austral. Syst. Bot. 1: 110 (1988).
SW. & S. Western Australia. 50 WAU. Nanophan.

Melaleuca tortifolia Byrnes, Austrobaileya 2: 74 (1984).
NE. New South Wales. 50 NSW. Nanophan. or phan.

Melaleuca trichophylla Lindl., Sketch Veg. Swan R.: viii (1839). *Myrtoleucodendron trichophyllum* (Lindl.) Kuntze, Revis. Gen. Pl. 1: 241 (1891).
WSW. Western Australia. 50 WAU. Nanophan.
Melaleuca eremaea F.Muell., Fragm. 3: 114 (1862).

Melaleuca trichostachya Lindl. in T.L.Mitchell, J. Exped. Trop. Australia: 277 (1848). *Melaleuca linariifolia* var. *trichostachya* (Lindl.) Benth., Fl. Austral. 3: 141 (1867). *Myrtoleucodendron trichostachyum* (Lindl.) Kuntze, Revis. Gen. Pl. 1: 241 (1891).
C. Australia. 50 NSW NTA QLD SOA. Nanophan.

Melaleuca triumphalis Craven, Muelleria 11: 1 (1998).
Northern Territory. 50 NTA. Nanophan.

Melaleuca tuberculata Schauer in J.G.C.Lehmann, Pl. Preiss. 1: 139 (1844).
Western Australia. 50 WAU. Nanophan.

var. ***arenaria*** (C.A.Gardner) Craven, Austral. Syst. Bot. 12: 911 (1999).
Western Australia. 50 WAU. Nanophan. or phan.
**Melaleuca arenaria* C.A.Gardner, J. Proc. Roy. Soc. W. Australia 9(2): 104 (1923).

var. ***macrophylla*** Craven, Austral. Syst. Bot. 12: 911 (1999).
Western Australia. 50 WAU. Nanophan. or phan.

var. ***tuberculata***
Western Australia. 50 WAU. Nanophan.

Melaleuca uncinata R.Br. in W.T.Aiton, Hortus Kew. 4: 414 (1812). *Myrtoleucodendron uncinatum* (R.Br.) Kuntze, Revis. Gen. Pl. 1: 242 (1891).
S. Western Australia. 50 NSW NTA QLD SOA VIC WAU. Nanophan.
Melaleuca nodosa var. *stenostoma* Domin, Biblioth. Bot. 89: 1012 (1928).

Melaleuca undulata Benth., Fl. Austral. 3: 135 (1867). *Myrtoleucodendron undulatum* (Benth.) Kuntze, Revis. Gen. Pl. 1: 242 (1891).
SW. Australia. 50 WAU. Nanophan.
Melaleuca undulata var. *minor* Benth., Fl. Austral. 3: 136 (1867).

Melaleuca urceolaris F.Muell. ex Benth., Fl. Austral. 3: 154 (1867). *Myrtoleucodendron urceolare* (F.Muell. ex Benth.) Kuntze, Revis. Gen. Pl. 1: 242 (1891).
SW. Australia. 50 WAU. Nanophan.

Melaleuca uxorum Craven, G.Holmes & Sankowsky, Muelleria 18: 3 (2003).
Queensland. 50 QLD.

Melaleuca venusta Craven, Austral. Syst. Bot. 12: 912 (1999).
Western Australia. 50 WAU. Nanophan. or phan.

Melaleuca villosisepala Craven, Austral. Syst. Bot. 12: 913 (1999).
Western Australia. 50 WAU. Nanophan. or phan.

Melaleuca viminea Lindl., Sketch Veg. Swan R.: viii (1839). *Myrtoleucodendron vimineum* (Lindl.) Kuntze, Revis. Gen. Pl. 1: 242 (1891).
SW. Australia. 50 WAU. Nanophan.

subsp. ***appressa*** Barlow, Nuytsia 8: 349 (1992).
SW. Australia. 50 WAU. Nanophan.

subsp. ***demissa*** Craven, Austral. Syst. Bot. 12: 913 (1999).
SW. Australia. 50 WAU. Nanophan. or phan.

subsp. ***viminea***
SW. Australia. 50 WAU. Nanophan.
Melaleuca lehmannii Schauer in J.G.C.Lehmann, Pl. Preiss. 1: 142 (1844).
Melaleuca viminea var. *major* Benth., Fl. Austral. 3: 159 (1867).

Melaleuca vinnula Craven & Lepschi, Austral. Syst. Bot. 17: 269 (2004).
SW. Australia. 50 WAU.

Melaleuca violacea Schauer in J.G.C.Lehmann, Pl. Preiss. 1: 146 (1844). *Myrtoleucodendron violaceum* (Schauer) Kuntze, Revis. Gen. Pl. 1: 242 (1891).
SW. Australia. 50 WAU. Nanophan.
Melaleuca divaricata Turcz., Bull. Cl. Phys.-Math. Acad. Imp. Sci. Saint-Pétersbourg 10: 344 (1852).

Melaleuca viridiflora Sol. ex Gaertn., Fruct. Sem. Pl. 1: 175 (1788). *Myrtoleucodendron viridiflorum* (Sol. ex Gaertn.) Kuntze, Revis. Gen. Pl. 1: 241 (1891). *Cajuputi viridiflora* (Sol. ex Gaertn.) A.Lyons, Plant Nam.: 74 (1900). *Melaleuca leucadendra* var. *viridiflora* (Sol. ex Gaertn.) Cheel in A.J.Ewart & O.B.Davies, Fl. N. Territory: 299 (1917).
New Guinea to N. Australia. 43 NWG 50 NTA QLD WAU. Phan.
Melaleuca leucadendra var. *latifolia* Raeusch., Nomencl. Bot., ed. 3: 142 (1797).
Melaleuca cunninghamii Schauer in W.G.Walpers, Repert. Bot. Syst. 2: 927 (1843).
Melaleuca leucadendra var. *sanguinea* Cheel in A.J.Ewart & O.B.Davies, Fl. N. Territory: 296 (1917), nom. illeg.

Melaleuca leucadendra var. *sanguinea* Cheel, Fl. N. Territory: 296 (1917).
Melaleuca sanguinea Cheel, Fl. N. Territory: 296 (1917), nom. inval.
Melaleuca cunninghamii var. *glabra* C.T.White, J. Arnold Arbor. 23: 87 (1942). *Melaleuca viridiflora* var. *glabra* (C.T.White) Byrnes, Austrobaileya 2: 74 (1984).
Melaleuca viridiflora var. *attenuata* Byrnes, Austrobaileya 2: 74 (1984).
Melaleuca viridiflora var. *canescens* Byrnes, Austrobaileya 2: 74 (1984).

Melaleuca wilsonii F.Muell., Fragm. 2: 124 (1860). *Myrtoleucodendron wilsonii* (F.Muell.) Kuntze, Revis. Gen. Pl. 1: 242 (1891). *Cajuputi wilsonii* (F.Muell.) Skeels, Bull. Bur. Pl. Industr. U.S.D.A. 282: 60 (1913).
SE. South Australia to W. Victoria. 50 SOA VIC. Nanophan.

Melaleuca wonganensis Craven, Austral. Syst. Bot. 12: 915 (1999).
Western Australia (Wongan Hills). 50 WAU. Nanophan. or phan.

Melaleuca xerophila Barlow, Austral. Syst. Bot. 1: 122 (1988).
Western Australia to South Australia. 50 SOA WAU. Nanophan. or phan.

Melaleuca zeteticorum Craven & Lepschi, Austral. Syst. Bot. 17: 269 (2004).
WSW. Western Australia. 50 WAU.

Melaleuca zonalis Craven, Austral. Syst. Bot. 12: 916 (1999).
Western Australia. 50 WAU. Nanophan. or phan.

Synonyms:
Melaleuca abietina Sm. = ***Melaleuca cuticularis*** Labill.
Melaleuca acacioides subsp. *alsophila* (A.Cunn. ex Benth.) Barlow = ***Melaleuca alsophila*** A.Cunn. ex Benth.
Melaleuca acacioides var. *angustifolia* Domin = ***Melaleuca citrolens*** Barlow
Melaleuca acerosa Schauer = ***Melaleuca systena*** Craven
Melaleuca acerosa (Colla) Sweet ex G.Don = ***Calothamnus quadrifidus*** R.Br. ex W.T.Aiton
Melaleuca acerosa var. *bracteata* Benth. = ***Melaleuca bisulcata*** F.Muell.
Melaleuca acerosa var. *bracteosa* Benth. = ***Melaleuca bisulcata*** F.Muell.
Melaleuca acicularis Brongn. & Gris ex Guillaumin = ***Melaleuca gnidioides*** Brongn. & Gris
Melaleuca adnata var. *abietina* F.Muell. = ***Melaleuca adnata*** Turcz.
Melaleuca affinis Regel & Körn. = ***Melaleuca decussata*** R.Br.
Melaleuca albiflora (Sol. ex Gaertn.) Raeusch. = ***Metrosideros albiflora*** Sol. ex Gaertn.
Melaleuca amboinensis Gand. = ***Melaleuca leucadendra*** (L.) L.
Melaleuca angulata Turcz. = ***Melaleuca suberosa*** (Schauer) C.A.Gardner
Melaleuca angustifolia (Blume) Blume = ***Melaleuca cajuputi*** Powell subsp. ***cajuputi***
Melaleuca angustifolia Gaertn. = ***Asteromyrtus angustifolia*** (Gaertn.) Craven
Melaleuca arachnoidea Raeusch. = ***Leptospermum arachnoides*** Gaertn.
Melaleuca arenaria C.A.Gardner = ***Melaleuca tuberculata*** var. ***arenaria*** (C.A.Gardner) Craven
Melaleuca armillaris var. *tenuifolia* Benth. = ***Melaleuca diosmatifolia*** Dum.Cours.
Melaleuca arnhemica Byrnes = ***Asteromyrtus arnhemica*** (Byrnes) Craven
Melaleuca aromatica Dum.Cours. = ***Leptospermum polygalifolium*** Salisb. subsp. ***polygalifolium***
Melaleuca axillaris Steud. = ***Melaleuca ericifolia*** Sm.
Melaleuca baxteri Benth. = *Agonis* sp.
Melaleuca bicolor Poir. = (Melastomataceae)
Melaleuca bonatiana Schltr. = ***Melaleuca gnidioides*** Brongn. & Gris
Melaleuca brachyandra (Lindl.) Craven = ***Callistemon brachyandrus*** Lindl.
Melaleuca brachyphylla Schauer = ***Melaleuca microphylla*** Sm.
Melaleuca brachystachya F.Muell. = ***Melaleuca subfalcata*** Turcz.
Melaleuca brassii Byrnes = ***Asteromyrtus brassii*** (Byrnes) Craven
Melaleuca brevifolia F.Muell. = ***Melaleuca microphylla*** Sm.
Melaleuca caja-putti DC. = ***Melaleuca squarrosa*** Sm.
Melaleuca callistemonea Lindl. = ***Melaleuca lateritia*** A.Dietr.
Melaleuca calycina subsp. *dempta* Barlow = ***Melaleuca dempta*** (Barlow) Craven
Melaleuca canaliculata Turcz. = ***Melaleuca pentagona*** Labill. var. ***pentagona***
Melaleuca canescens Otto = ***Melaleuca incana*** R.Br. subsp. ***incana***
Melaleuca cardiophylla var. *longistaminea* F.Muell. = ***Melaleuca longistaminea*** (F.Muell.) Barlow ex Craven
Melaleuca cardiophylla var. *parviflora* Benth. = ***Melaleuca coronicarpa*** D.A.Herb.
Melaleuca carinata Turcz. = ***Melaleuca calycina*** R.Br.
Melaleuca chisholmii (Cheel) Craven = ***Callistemon chisholmii*** Cheel
Melaleuca chlorantha Bonpl. = ***Melaleuca diosmifolia*** Andrews
Melaleuca ciliata (J.R.Forst. & G.Forst.) G.Forst. = ***Purpureostemon ciliatus*** (J.R.Forst. & G.Forst.) Gugerli
Melaleuca citrina Turcz. = [50]
Melaleuca citrina (Curtis) Dum.Cours. = ***Callistemon citrinus*** (Curtis) Skeels
Melaleuca coccinea subsp. *eximia* K.J.Cowley = ***Melaleuca eximia*** (K.J.Cowley) Craven
Melaleuca coccinea subsp. *penicula* K.J.Cowley = ***Melaleuca penicula*** (K.J.Cowley) Craven
Melaleuca comboynensis (Cheel) Craven = ***Callistemon comboynensis*** Cheel
Melaleuca commutata Miq. = ***Melaleuca cajuputi*** Powell subsp. ***cajuputi***
Melaleuca concava S.Moore = ***Melaleuca acuminata*** subsp. ***websteri*** (S.Moore) Barlow ex Craven
Melaleuca concreta var. *brevifolia* Benth. = ***Melaleuca concreta*** F.Muell.
Melaleuca conferta Benth. = ***Melaleuca leucadendra*** (L.) L.
Melaleuca conferta (R.Br.) Steud. = ***Lophostemon confertus*** (R.Br.) Peter G.Wilson & J.T.Waterh.
Melaleuca cordata Benth. = ***Melaleuca cordata*** Turcz.
Melaleuca cordata var. *ovata* Ewart = ***Melaleuca conothamnoides*** C.A.Gardner
Melaleuca coronata Andrews = ***Melaleuca thymifolia*** Sm.
Melaleuca corrugata J.M.Black ex Eardley = ***Melaleuca fulgens*** subsp. ***corrugata*** (J.M.Black ex Eardley) K.J.Cowley

Melaleuca costata (Gaertn.) Raeusch. = ***Angophora costata*** (Gaertn.) Hochr. ex Britten
Melaleuca crassifolia Benth. = ***Melaleuca laxiflora*** Turcz.
Melaleuca crosslandiana W.Fitzg. = ***Melaleuca nervosa*** subsp. ***crosslandiana*** (W.Fitzg.) Barlow ex Craven
Melaleuca cumingiana Turcz. = ***Melaleuca cajuputi*** subsp. ***cumingiana*** (Turcz.) Barlow
Melaleuca cuneata Turcz. = ***Melaleuca leptospermoides*** Schauer
Melaleuca cunninghamii Schauer = ***Melaleuca viridiflora*** Sol. ex Gaertn.
Melaleuca cunninghamii var. *glabra* C.T.White = ***Melaleuca viridiflora*** Sol. ex Gaertn.
Melaleuca cupressina F.Muell. = ***Melaleuca thyoides*** Turcz.
Melaleuca curvifolia Schltdl. = ***Melaleuca lanceolata*** Otto subsp. ***lanceolata***
Melaleuca cuspidata Turcz. = ***Conothamnus trinervis*** Lindl.
Melaleuca cuticularis var. *brachyphylla* Domin = ***Melaleuca halmaturorum*** subsp. ***cymbifolia*** (Benth.) Barlow
Melaleuca cylindrica F.Muell. = ***Melaleuca diosmatifolia*** Dum.Cours.
Melaleuca cymbifolia Benth. = ***Melaleuca halmaturorum*** subsp. ***cymbifolia*** (Benth.) Barlow
Melaleuca cyrtodonta Turcz. = ***Beaufortia cyrtodonta*** (Turcz.) Benth.
Melaleuca daleana Blakely = ***Melaleuca bracteata*** F.Muell.
Melaleuca decussata var. *ovoidea* J.M.Black = ***Melaleuca decussata*** R.Br.
Melaleuca deltoidea Benth. = ***Melaleuca cucullata*** Turcz.
Melaleuca densa var. *pritzelii* Domin = ***Melaleuca pritzelii*** (Domin) Barlow
Melaleuca depressa var. *geraldtonensis* Hochr. = ***Melaleuca depressa*** Diels
Melaleuca diffusa G.Forst. = ***Metrosideros diffusa*** (G.Forst.) Sm.
Melaleuca discolor Rchb. ex Spreng. = ***Melaleuca thymifolia*** Sm.
Melaleuca divaricata Turcz. = ***Melaleuca violacea*** Schauer
Melaleuca dorrien-smithii Domin = ***Melaleuca densa*** R.Br.
Melaleuca drummondii Schauer = ***Melaleuca hamata*** Field. & G.A.Gardner
Melaleuca elachophylla F.Muell. = ***Melaleuca depauperata*** Turcz.
Melaleuca elegans Hornsch. = ***Melaleuca decussata*** R.Br.
Melaleuca eleuterostachya var. *abietina* Benth. = ***Melaleuca adnata*** Turcz.
Melaleuca empetrifolia Rchb. = ***Beaufortia empetrifolia*** (Rchb.) Schauer
Melaleuca endlicheriana Schauer = ***Melaleuca seriata*** Lindl.
Melaleuca epacridioides Turcz. = ***Melaleuca densa*** R.Br.
Melaleuca eremaea F.Muell. = ***Melaleuca trichophylla*** Lindl.
Melaleuca eriantha Benth. = ***Melaleuca leptospermoides*** Schauer
Melaleuca ericifolia Andrews = ***Melaleuca armillaris*** (Sol. ex Gaertn.) Sm. subsp. ***armillaris***
Melaleuca ericifolia var. *erubescens* Benth. = ***Melaleuca diosmatifolia*** Dum.Cours.
Melaleuca eriocephala Sieber ex Spreng. = ***Kunzea capitata*** (Sm.) Heynh.
Melaleuca eriorhachis Gand. = ***Melaleuca cajuputi*** Powell subsp. ***cajuputi***
Melaleuca erubescens Otto = ***Melaleuca diosmatifolia*** Dum.Cours.
Melaleuca eruciformis Turcz. = ***Melaleuca blaeriifolia*** Turcz.
Melaleuca exarata F.Muell. = ***Melaleuca suberosa*** (Schauer) C.A.Gardner
Melaleuca fascicularis Labill. = ***Astartea fascicularis*** (Labill.) A.Cunn. ex DC.
Melaleuca fasciculiflora Benth. = ***Melaleuca brevifolia*** Turcz.
Melaleuca faucicola Craven = ***Callistemon pauciflorus*** R.D.Spencer & Lumley
Melaleuca ferrea A.Cunn. ex A.Gray = ***Melaleuca decora*** (Salisb.) Britten
Melaleuca fimbriata Donn = ***Melaleuca thymifolia*** Sm.
Melaleuca flammea Craven = ***Callistemon acuminatus*** Cheel
Melaleuca flavovirens (Cheel) Craven = ***Callistemon flavovirens*** (Cheel) Cheel
Melaleuca florida G.Forst. = ***Metrosideros diffusa*** (G.Forst.) Sm.
Melaleuca foliosa Dum.Cours. = ***Melaleuca diosmifolia*** Andrews
Melaleuca formosa (S.T.Blake) Craven = ***Callistemon formosus*** S.T.Blake
Melaleuca fraseri Hook. = ***Melaleuca diosmatifolia*** Dum.Cours.
Melaleuca genistifolia Sm. = ***Melaleuca decora*** (Salisb.) Britten
Melaleuca genistifolia var. *coriacea* Ewart, L.R.Kerr & Derrick = ***Melaleuca bracteata*** F.Muell.
Melaleuca glaucocalyx Gand. = ***Melaleuca bracteata*** F.Muell.
Melaleuca gnidiifolia Vent. = ***Melaleuca thymifolia*** Sm.
Melaleuca graminea S.Moore = ***Melaleuca acacioides*** F.Muell.
Melaleuca grandiflora Blanco = (Bombaceae)
Melaleuca grandis Fisch. ex Steud. = ?
Melaleuca gummifera Steud. = ***Eucalyptus resinifera*** Sm.
Melaleuca gunniana Schauer = ***Melaleuca ericifolia*** Sm.
Melaleuca gunniana var. *capitata* Miq. = ***Melaleuca ericifolia*** Sm.
Melaleuca hakeacea F.Muell. = ***Melaleuca teretifolia*** Endl.
Melaleuca hakeoides F.Muell. ex Benth. = ***Melaleuca glomerata*** F.Muell.
Melaleuca halmaturorum var. *enervis* Miq. = ***Melaleuca halmaturorum*** F.Muell. ex Miq. subsp. ***halmaturorum***
Melaleuca halmaturorum var. *tuberculifera* Miq. = ***Melaleuca halmaturorum*** F.Muell. ex Miq. subsp. ***halmaturorum***
Melaleuca heliophila F.Muell. ex Miq. = ***Melaleuca ericifolia*** Sm.
Melaleuca hispida J.C.Wendl. ex DC. = ?
Melaleuca hypochondriaca Dehnh. = ***Melaleuca incana*** R.Br. subsp. ***incana***
Melaleuca hyssopifolia (Cav.) Dum.Cours. = ***Melaleuca linariifolia*** Sm.
Melaleuca imbricata Link = ?
Melaleuca juniperina (Rchb. ex Spreng.) Sieber ex Rchb. = ***Melaleuca nodosa*** (Sol. ex Gaertn.) Sm.
Melaleuca juniperoides DC. = ***Melaleuca nodosa*** (Sol. ex Gaertn.) Sm.
Melaleuca lanata Nois. ex Steud. = ***Melaleuca incana*** R.Br. subsp. ***incana***

Melaleuca lanceolata R.Br. ex R.T.Baker = ***Melaleuca deanei*** F.Muell.
Melaleuca lancifolia Turcz. = ***Melaleuca cajuputi*** Powell subsp. ***cajuputi***
Melaleuca lanigera J.C.Wendl. ex DC. = ***Melaleuca incana*** R.Br. subsp. ***incana***
Melaleuca lateriflora var. *acutifolia* Benth. = ***Melaleuca lateriflora*** subsp. ***acutifolia*** (Benth.) Craven
Melaleuca lateriflora var. *elliptica* Benth. = ***Melaleuca lateriflora*** Benth. subsp. ***lateriflora***
Melaleuca laurina Sm. = ***Tristaniopsis laurina*** (Sm.) Peter G.Wilson & J.T.Waterh.
Melaleuca lehmannii Schauer = ***Melaleuca viminea*** Lindl. subsp. ***viminea***
Melaleuca leiostachya var. *preissiana* Benth. = ***Melaleuca lanceolata*** subsp. ***planifolia*** Barlow
Melaleuca leptoclada Benth. = ***Melaleuca pauciflora*** Turcz.
Melaleuca leucadendra var. *albida* Cheel = ***Melaleuca leucadendra*** (L.) L.
Melaleuca leucadendra var. *albida* Cheel = ***Melaleuca quinquenervia*** (Cav.) S.T.Blake
Melaleuca leucadendra var. *angusta* C.Rivière = ***Melaleuca leucadendra*** (L.) L.
Melaleuca leucadendra var. *angustifolia* L.f. = ***Melaleuca quinquenervia*** (Cav.) S.T.Blake
Melaleuca leucadendra var. *coriacea* (Poir.) Cheel = ***Melaleuca quinquenervia*** (Cav.) S.T.Blake
Melaleuca leucadendra f. *crosslandiana* (W.Fitzg.) Cheel = ***Melaleuca nervosa*** subsp. ***crosslandiana*** (W.Fitzg.) Barlow ex Craven
Melaleuca leucadendra var. *cunninghamii* F.M.Bailey = ***Melaleuca leucadendra*** (L.) L.
Melaleuca leucadendra var. *lancifolia* F.M.Bailey = ***Melaleuca leucadendra*** (L.) L.
Melaleuca leucadendra var. *latifolia* Raeusch. = ***Melaleuca viridiflora*** Sol. ex Gaertn.
Melaleuca leucadendra var. *mimosoides* (A.Cunn. ex Schauer) Cheel = ***Melaleuca leucadendra*** (L.) L.
Melaleuca leucadendra var. *minor* (Sm.) Duthie = ***Melaleuca cajuputi*** Powell subsp. ***cajuputi***
Melaleuca leucadendra var. *nervosa* (Lindl.) Domin = ***Melaleuca nervosa*** (Lindl.) Cheel
Melaleuca leucadendra var. *parvifolia* Benth. = ***Melaleuca nervosa*** (Lindl.) Cheel
Melaleuca leucadendra f. *ruscifolia* Cheel = ***Melaleuca arcana*** S.T.Blake
Melaleuca leucadendra f. *ruscifolia* Cheel = ***Melaleuca arcana*** S.T.Blake
Melaleuca leucadendra var. *saligna* (Schauer) F.M.Bailey = ***Melaleuca saligna*** Schauer
Melaleuca leucadendra var. *sanguinea* Cheel = ***Melaleuca viridiflora*** Sol. ex Gaertn.
Melaleuca leucadendra var. *sanguinea* Cheel = ***Melaleuca viridiflora*** Sol. ex Gaertn.
Melaleuca leucadendra var. *speciosa* (Schauer) Domin = ***Leptospermum speciosum*** Schauer
Melaleuca leucadendra var. *viridiflora* (Sol. ex Gaertn.) Cheel = ***Melaleuca viridiflora*** Sol. ex Gaertn.
Melaleuca linariifolia var. *alternifolia* Maiden & Betche = ***Melaleuca alternifolia*** (Maiden & Betche) Cheel
Melaleuca linariifolia var. *trichostachya* (Lindl.) Benth. = ***Melaleuca trichostachya*** Lindl.
Melaleuca linariifolia var. *typica* Domin = ***Melaleuca linariifolia*** Sm.
Melaleuca linearifolia (Link) Craven = ***Callistemon linearifolius*** (Link) DC.
Melaleuca linearis Schrad. & J.C.Wendl. = ***Callistemon linearis*** (Schrad. & J.C.Wendl.) Colv. ex Sweet
Melaleuca linearis var. *pinifolia* (J.C.Wendl.) Craven = ***Callistemon pinifolius*** (J.C.Wendl.) Sweet
Melaleuca loguei W.Fitzg. = ***Melaleuca lasiandra*** F.Muell.
Melaleuca longicoma Benth. = ***Melaleuca macronychia*** Turcz. subsp. ***macronychia***
Melaleuca lucida L.f. = ***Metrosideros diffusa*** (G.Forst.) Sm.
Melaleuca lucida G.Forst. = ***Metrosideros umbellata*** Cav.
Melaleuca lysicephala F.Muell. & F.M.Bailey = ***Asteromyrtus lysicephala*** (F.Muell. & F.M.Bailey) Craven
Melaleuca magnifica Specht = ***Asteromyrtus magnifica*** (Specht) Craven
Melaleuca maidenii R.T.Baker = ***Melaleuca quinquenervia*** (Cav.) S.T.Blake
Melaleuca mimosoides A.Cunn. ex Schauer = ***Melaleuca leucadendra*** (L.) L.
Melaleuca minor Sm. = ***Melaleuca cajuputi*** Powell subsp. ***cajuputi***
Melaleuca minutifolia subsp. *monantha* Barlow = ***Melaleuca monantha*** (Barlow) Craven
Melaleuca montana (C.T.White ex S.T.Blake) Craven = ***Callistemon montanus*** C.T.White ex S.T.Blake
Melaleuca monticola J.M.Black = ***Melaleuca bracteata*** F.Muell.
Melaleuca muricata A.Cunn. ex DC. = ***Melaleuca scabra*** R.Br.
Melaleuca myrtifolia Vent. = ***Melaleuca squarrosa*** Sm.
Melaleuca neglecta Ewart & B.Wood = ***Melaleuca brevifolia*** Turcz.
Melaleuca nematophylla F.Muell. = ***Melaleuca filifolia*** F.Muell.
Melaleuca neriifolia Sieber ex Sims = ***Tristania neriifolia*** (Sieber ex Sims) R.Br.
Melaleuca nervosa f. *latifolia* Byrnes = ***Melaleuca nervosa*** (Lindl.) Cheel subsp. ***nervosa***
Melaleuca nervosa f. *pendulina* Byrnes = ***Melaleuca fluviatilis*** Barlow
Melaleuca nodosa var. *stenostoma* Domin = ***Melaleuca uncinata*** R.Br.
Melaleuca nodosa var. *tenuifolia* (DC.) Cheel ex A.R. Penfold = ***Melaleuca nodosa*** (Sol. ex Gaertn.) Sm.
Melaleuca nummularia Turcz. = ***Melaleuca elliptica*** Labill.
Melaleuca obliqua Büse ex de Vriese = ***Melaleuca styphelioides*** Sm.
Melaleuca oligantha F.Muell. ex Miq. = ***Melaleuca thymifolia*** Sm.
Melaleuca oraria J.M.Black = ***Melaleuca brevifolia*** Turcz.
Melaleuca ornata Schauer = ***Melaleuca seriata*** Lindl.
Melaleuca orophila Craven = ***Callistemon teretifolius*** F.Muell.
Melaleuca ottonis Schauer = ***Melaleuca squamea*** Labill.
Melaleuca ovatifolia Poir. = (Melastomataceae)
Melaleuca pachyphylla (Cheel) Craven = ***Callistemon pachyphyllus*** Cheel
Melaleuca pallida (Bonpl.) Craven = ***Callistemon pallidus*** (Bonpl.) DC.
Melaleuca paludicola Craven = ***Callistemon sieberi*** DC.
Melaleuca paludosa Schltdl. = ***Callistemon salignus*** (Sm.) Colv. ex Sweet var. ***salignus***
Melaleuca paludosa R.Br. = ***Callistemon speciosus*** (Sims) Sweet
Melaleuca parviflora Rchb. = ***Melaleuca decussata*** R.Br.
Melaleuca parviflora Lindl. = ***Melaleuca laxiflora*** Turcz.
Melaleuca parviflora var. *latifolia* Maiden & Betche = ***Melaleuca sieberi*** Schauer

Melaleuca parviflora var. *leiostachya* (Benth.) Domin = ***Melaleuca lanceolata*** subsp. ***planifolia*** Barlow
Melaleuca parviflora var. *pubescens* (Schauer) Domin = ***Melaleuca lanceolata*** Otto subsp. ***lanceolata***
Melaleuca parvifolia Otto = ***Melaleuca thymifolia*** Sm.
Melaleuca pearsonii (R.D.Spencer & Lumley) Craven = ***Callistemon pearsonii*** R.D.Spencer & Lumley
Melaleuca pendulina G.Don ex Loudon = ?
Melaleuca pentagona var. *planifolia* Schauer = ***Melaleuca pentagona*** Labill. var. ***pentagona***
Melaleuca pentagona var. *subulifolia* Schauer = ***Melaleuca pentagona*** Labill. var. ***pentagona***
Melaleuca perforata (J.R.Forst. & G.Forst.) G.Forst. = ***Metrosideros perforata*** (J.R.Forst. & G.Forst.) Druce
Melaleuca phoenicea (Lindl.) Craven = ***Callistemon phoeniceus*** Lindl.
Melaleuca pinifolia Colla = ***Melaleuca ericifolia*** Sm.
Melaleuca pinifolia Turcz. = ***Melaleuca teretifolia*** Endl.
Melaleuca pithyoides F.Muell. ex Benth. = ***Callistemon salignus*** (Sm.) Colv. ex Sweet var. ***salignus***
Melaleuca pityoides (F.Muell.) Craven = ***Callistemon pityoides*** F.Muell.
Melaleuca polandii (F.M.Bailey) Craven = ***Callistemon polandii*** F.M.Bailey
Melaleuca polygaloides Schauer = ***Melaleuca incana*** R.Br. subsp. ***incana***
Melaleuca polygonoides Hoffmanns. = ***Melaleuca decora*** (Salisb.) Britten
Melaleuca preissiana var. *leiostachya* Benth. = ***Melaleuca lanceolata*** subsp. ***planifolia*** Barlow
Melaleuca pubescens Schauer = ***Melaleuca lanceolata*** Otto subsp. ***lanceolata***
Melaleuca pumila A.Cunn. ex DC. = ***Melaleuca decussata*** R.Br.
Melaleuca pungens Brongn. & Gris = ***Melaleuca brongniartii*** Däniker
Melaleuca pungens var. *obtusifolia* Benth. = ***Melaleuca pungens*** Schauer
Melaleuca punicea Byrnes = ***Petraeomyrtus punicea*** (Byrnes) Craven
Melaleuca recurva (R.D.Spencer & Lumley) Craven = ***Callistemon recurvus*** R.D.Spencer & Lumley
Melaleuca regelii Planch. = ***Beaufortia micrantha*** Schauer var. ***micrantha***
Melaleuca rigida Roxb. = ***Melaleuca leucadendra*** (L.) L.
Melaleuca rotundifolia Sweet = ***Regelia ciliata*** Schauer
Melaleuca rugulosa (Link) Craven = ***Callistemon macropunctatus*** (Dum.Cours.) Court var. ***macropunctatus***
Melaleuca salicifolia Andrews = ***Tristania neriifolia*** (Sieber ex Sims) R.Br.
Melaleuca salicina Craven = ***Callistemon salignus*** (Sm.) Colv. ex Sweet
Melaleuca saligna Reinw. ex Blume = ***Melaleuca cajuputi*** Powell subsp. ***cajuputi***
Melaleuca sanguinea Cheel = ***Melaleuca viridiflora*** Sol. ex Gaertn.
Melaleuca scandens Raeusch. = ***Metrosideros fulgens*** Sol. ex Gaertn.
Melaleuca scoparia (J.R.Forst. & G.Forst.) L.f. = ***Leptospermum scoparium*** J.R.Forst. & G.Forst.
Melaleuca semiteres Schauer = ***Melaleuca teretifolia*** Endl.
Melaleuca seorsiflora F.Muell. = ***Melaleuca lanceolata*** subsp. ***thaeroides*** Barlow
Melaleuca serpyllifolia Turcz. = ***Melaleuca polycephala*** Benth.
Melaleuca serpyllifolia Dum.Cours. = ***Melaleuca thymifolia*** Sm.
Melaleuca shiressii (Blakely) Craven = ***Callistemon shiressii*** Blakely
Melaleuca smithii R.T.Baker = ***Melaleuca quinquenervia*** (Cav.) S.T.Blake
Melaleuca spectabilis Raeusch. = ***Metrosideros fulgens*** Sol. ex Gaertn.
Melaleuca spinosa Lindl. = ***Melaleuca thymoides*** Labill.
Melaleuca splendens J.Lee ex J.Kern. = ***Metrosideros fulgens*** Sol. ex Gaertn.
Melaleuca sprengelioides DC. = ***Beaufortia sprengelioides*** (DC.) Craven
Melaleuca squamea var. *glabra* Cheel = ***Melaleuca squamea*** Labill.
Melaleuca squarrosa var. *glabrata* Miq. = ***Melaleuca squarrosa*** Sm.
Melaleuca steedmanii C.A.Gardner = ***Melaleuca fulgens*** subsp. ***steedmanii*** (C.A.Gardner) K.J.Cowley
Melaleuca stenostachya var. *pendula* Byrnes = ***Melaleuca saligna*** Schauer
Melaleuca stricta Dum.Cours. = ***Melaleuca linariifolia*** Sm.
Melaleuca styphelioides var. *squamophloia* Byrnes = ***Melaleuca squamophloia*** (Byrnes) Craven
Melaleuca suaveolens Sol. ex Gaertn. = ***Lophostemon suaveolens*** (Sol. ex Gaertn.) Peter G.Wilson & J.T.Waterh.
Melaleuca subulata (Cheel) Craven = ***Callistemon subulatus*** Cheel
Melaleuca symphyocarpa F.Muell. = ***Asteromyrtus symphyocarpa*** (F.Muell.) Craven
Melaleuca tamariscina subsp. *irbyana* (F.Muell. ex R.T.Baker) Barlow = ***Melaleuca irbyana*** F.Muell. ex R.T.Baker
Melaleuca tamariscina subsp. *pallescens* (Byrnes) Barlow = ***Melaleuca pallescens*** Byrnes
Melaleuca taxifolia Schltdl. ex Spreng. = ***Melaleuca armillaris*** (Sol. ex Gaertn.) Sm. subsp. ***armillaris***
Melaleuca tenella Benth. = ***Melaleuca incana*** subsp. ***tenella*** (Benth.) Barlow
Melaleuca tenuifolia DC. = ***Melaleuca nodosa*** (Sol. ex Gaertn.) Sm.
Melaleuca tenuifolia J.C.Wendl. = ***Leptospermum scoparium*** J.R.Forst. & G.Forst.
Melaleuca tenuissima Tausch = ***Melaleuca microphylla*** Sm.
Melaleuca ternifolia F.Muell. ex Miq. = ***Melaleuca ericifolia*** Sm.
Melaleuca tetragona Lodd. ex Otto = ***Melaleuca decussata*** R.Br.
Melaleuca thea Schrad. & J.C.Wendl. = ***Leptospermum polygalifolium*** Salisb. subsp. ***polygalifolium***
Melaleuca tomentosa Colla = ***Melaleuca incana*** R.Br. subsp. ***incana***
Melaleuca trinervia Dum.Cours. = ?
Melaleuca trinervia Sm. = ***Leptospermum trinervium*** (Sm.) Joy Thomps.
Melaleuca trinervis Buch.-Ham. = ***Melaleuca cajuputi*** Powell subsp. ***cajuputi***
Melaleuca tubulata Dum.Cours. = ***Melaleuca decora*** (Salisb.) Britten
Melaleuca undulata var. *minor* Benth. = ***Melaleuca undulata*** Benth.
Melaleuca uniflora Spreng. = ?
Melaleuca urceolaris var. *virgata* Benth. = ***Melaleuca dichroma*** Craven & Lepschi
Melaleuca villosa L.f. = ***Metrosideros collina*** var. ***villosa*** (L.f.) A.Gray
Melaleuca viminalis (Sol. ex Gaertn.) Byrnes = ***Callistemon viminalis*** (Sol. ex Gaertn.) G.Don ex Loudon

Melaleuca viminalis var. *minor* Byrnes = ***Callistemon viminalis*** (Sol. ex Gaertn.) G.Don ex Loudon
Melaleuca viminalis var. *minor* Byrnes = ***Callistemon viminalis*** (Sol. ex Gaertn.) G.Don ex Loudon
Melaleuca viminea var. *major* Benth. = ***Melaleuca viminea*** Lindl. subsp. ***viminea***
Melaleuca violacea var. *petiolata* Benth. = ***Melaleuca platycalyx*** Diels
Melaleuca virens Craven = ***Callistemon viridiflorus*** (Sieber ex Sims) Sweet
Melaleuca virgata (Benth.) Craven = ***Melaleuca dichroma*** Craven & Lepschi
Melaleuca virgata (J.R.Forst. & G.Forst.) L.f. = ***Babingtonia virgata*** (J.R.Forst. & G.Forst.) F.Muell.
Melaleuca viridiflora var. *angustifolia* Blume = ***Melaleuca cajuputi*** Powell subsp. ***cajuputi***
Melaleuca viridiflora var. *angustifolia* (L.f.) Byrnes = ***Melaleuca quinquenervia*** (Cav.) S.T.Blake
Melaleuca viridiflora var. *attenuata* Byrnes = ***Melaleuca viridiflora*** Sol. ex Gaertn.
Melaleuca viridiflora var. *canescens* Byrnes = ***Melaleuca viridiflora*** Sol. ex Gaertn.
Melaleuca viridiflora var. *glabra* (C.T.White) Byrnes = ***Melaleuca viridiflora*** Sol. ex Gaertn.
Melaleuca viridiflora var. *rubriflora* Pancher ex Brongn. & Gris = ***Melaleuca quinquenervia*** (Cav.) S.T.Blake
Melaleuca waeberi Rchb. ex Schauer = ***Melaleuca pentagona*** Labill. var. ***pentagona***
Melaleuca websteri S.Moore = ***Melaleuca acuminata*** subsp. ***websteri*** (S.Moore) Barlow ex Craven
Melaleuca williamsii Craven = ***Callistemon pungens*** Lumley & R.D.Spencer

***Unplaced Names*:**
Melaleuca baxteri Benth., Fl. Austral. 3: 138 (1867). = *Agonis* sp.
Melaleuca citrina Turcz., Bull. Cl. Phys.-Math. Acad. Imp. Sci. Saint-Pétersbourg 10: 341 (1852), nom. illeg. = [50]
Melaleuca grandis Fisch. ex Steud., Nomencl. Bot., ed. 2, 2: 112 (1841). = ?
Melaleuca hispida J.C.Wendl. ex DC., Prodr. 3: 215 (1828). = ?
Melaleuca imbricata Link, Enum. Hort. Berol. Alt. 2: 272 (1822). = ?
Melaleuca pendulina G.Don ex Loudon, Hort. Brit.: 319 (1830). = ?
Melaleuca trinervia Dum.Cours., Bot. Cult., ed. 2: 276 (1814), nom. illeg. = ?
Melaleuca uniflora Spreng., Syst. Veg. 4(2): 194 (1827). = ?

Melaleucon

Melaleucon St.-Lag. = ***Melaleuca*** L.

Melanoleuce

Melanoleuce St.-Lag. = ***Melaleuca*** L.

Mentodendron

Mentodendron Lundell = ***Pimenta*** Lindl.
Mentodendron guatamalense (Lundell) Lundell = ***Pimenta guatemalensis*** (Lundell) Lundell

Meteoromyrtus

Meteoromyrtus Gamble, Bull. Misc. Inform. Kew 1918: 241 (1918).
SW. India. 40 IND.
1 Species

Meteoromyrtus wynaadensis (Bedd.) Gamble, Bull. Misc. Inform. Kew 1918: 241 (1918).
SW. India. 40 IND. Nanophan.
**Eugenia wynaadensis* Bedd., Madras J. Lit. Sci., III, 1: 47 (1864).

Metrosideros

Metrosideros Banks ex Gaertn., Fruct. Sem. Pl. 1: 170 (1788).
Cape Prov, Ogasawara-shoto to Malesia and Pacific. 27 CPP 38 OGA 42 PHI 43 NWG SOL 50 NFK QLD 51 ATP ctm KER NZN NZS 60 FIJ NWC SAM TON TUA VAN 61 MRQ SCI TUB 63 HAW.
54 Species
Agalmanthus (Endl.) Hombr. & Jacquinot in J.S.C. Dumont d'Urville, Voy. Pôle Sud, Atlas: 78 (1845).
Ballardia Montrouz., Mém. Acad. Roy. Sci. Lyon, Sect. Sci. 10: 204 (1860).
Mearnsia Merr., Philipp. J. Sci., C 2: 283 (1907).
Carpolepis (J.W.Dawson) J.W.Dawson, Bull. Mus. Natl. Hist. Nat., B, Adansonia 1984: 466 (1984 publ. 1985).
Microsideros Baum.-Bod., Syst. Fl. Neu-Caledonien 5: 77 (1989), nom. inval.

Metrosideros albiflora Sol. ex Gaertn., Fruct. Sem. Pl. 1: 172 (1788). *Melaleuca albiflora* (Sol. ex Gaertn.) Raeusch., Nomencl. Bot., ed. 3: 142 (1797). *Nania albiflora* (Sol. ex Gaertn.) Kuntze, Revis. Gen. Pl. 1: 242 (1891).
New Zealand North I. 51 NZN. Cl.
Metrosideros diffusa A.Cunn., Ann. Nat. Hist. 3: 114 (1839), nom. illeg.

Metrosideros angustifolia (L.) Sm., Trans. Linn. Soc. London 3: 270 (1797).
Cape Prov. 27 CPP. Nanophan.
**Myrtus angustifolia* L., Mant. Pl. 1: 74 (1767). *Nania angustifolia* (L.) Kuntze, Revis. Gen. Pl. 1: 242 (1891).

Metrosideros arfakensis Gibbs, Fl. Arfak Mts.: 154 (1917).
New Guinea. 43 NWG. Nanophan. or phan.

Metrosideros bartlettii J.W.Dawson, New Zealand J. Bot. 23: 607 (1985 publ. 1986).
NW. New Zealand North I. 51 NZN. Phan.

Metrosideros boninensis (Hayata ex Koidz.) Tuyama, Bot. Mag. (Tokyo) 52: 569 (1938).
Ogasawara-shoto (Chichi-jima). 38 OGA. Nanophan. or phan.
**Eugenia boninensis* Hayata ex Koidz., Bot. Mag. (Tokyo) 32: 135 (1918).

Metrosideros brevistylis J.W.Dawson, Bull. Mus. Natl. Hist. Nat., B, Adansonia 1984: 475 (1984 publ. 1985).
New Caledonia. 60 NWC. Nanophan. or phan.

Metrosideros cacuminum J.W.Dawson, Bull. Mus. Natl. Hist. Nat., B, Adansonia 1984: 485 (1984 publ. 1985).
NE. & C. New Caledonia. 60 NWC. Nanophan. or phan.

Metrosideros carminea W.R.B.Oliv., Trans. & Proc. New Zealand Inst. 59: 420 (1928).
New Zealand North I. 51 NZN. Cl.

Metrosideros cherrieri J.W.Dawson, Bull. Mus. Natl. Hist. Nat., B, Adansonia 1984: 473 (1984 publ. 1985).
NE. New Caledonia. 60 NWC. Nanophan.

Metrosideros colensoi Hook.f., Fl. Nov.-Zel. 1: 68 (1852). *Nania colensoi* (Hook.f.) Kuntze, Revis. Gen. Pl. 1: 242 (1891).
New Zealand. 51 NZN NZS. Cl.

var. ***colensoi***
New Zealand. 51 NZN NZS. Cl.

var. ***pendens*** (Colenso) Kirk, Stud. Fl. New Zealand: 162 (1899).
New Zealand North I. 51 NZN. Cl.
**Metrosideros pendens* Colenso, Trans. & Proc. New Zealand Inst. 12: 360 (1880).

Metrosideros collina (J.R.Forst. & G.Forst.) A.Gray, U.S. Expl. Exped., Phan. 1: 558 (1854).
Queensland to S. Pacific. 50 QLD 60 FIJ TON TUA VAN 61 MRQ SCI TUB. Nanophan. or phan.
**Leptospermum collinum* J.R.Forst. & G.Forst., Char. Gen. Pl.: 36 (1775). *Nania collina* (J.R.Forst. & G.Forst.) Kuntze, Revis. Gen. Pl. 1: 242 (1891).

var. ***collina***
S. Pacific. 60 FIJ TON VAN 61 SCI. Nanophan. or phan.
Metrosideros collina var. *vitiensis* A.Gray, U.S. Expl. Exped., Phan. 1: 559 (1854).
Metrosideros collina var. *temehaniensis* J.W.Moore, Bernice P. Bishop Mus. Bull. 226: 27 (1963).

var. ***fruticosa*** J.W.Moore, Bernice P. Bishop Mus. Bull. 226: 24 (1963).
S. Pacific. 60 FIJ VAN 61 SCI. Nanophan.
Tristania vitiensis A.C.Sm., Bernice P. Bishop Mus. Bull. 141: 110 (1936).

var. ***villosa*** (L.f.) A.Gray, U.S. Expl. Exped., Phan. 1: 558 (1854).
Queensland to S. Pacific. 50 QLD 60 FIJ TUA VAN 61 MRQ SCI TUB. Phan.
**Melaleuca villosa* L.f., Suppl. Pl.: 342 (1782). *Metrosideros villosa* (L.f.) Sm., Trans. Linn. Soc. London 3: 268 (1797).
Metrosideros spectabilis Sol. ex Gaertn., Fruct. Sem. Pl. 1: 172 (1788).
Metrosideros obovata Hook. & Arn., Bot. Beechey Voy.: 63 (1832). *Nania obovata* (Hook. & Arn.) Kuntze, Revis. Gen. Pl. 1: 242 (1891).
Metrosideros villosa var. *glaberrima* Bertero ex Guillaumin, Ann. Sci. Nat., Bot., II, 7: 358 (1837). *Metrosideros collina* var. *glaberrima* (Bertero ex Guillaumin) A.Gray, U.S. Expl. Exped., Phan. 1: 558 (1854).
Metrosideros acuminata Decne. in J.S.C.Dumont d'Urville, Voy. Pôle Sud 2: 29 (1853).
Metrosideros taitensis Decne. in J.S.C.Dumont d'Urville, Voy. Pôle Sud 2: 30 (1853).
Metrosideros lutea A.Gray, U.S. Expl. Exped., Phan. 1: 560 (1854).
Nania lutea A.Heller, Minnesota Bot. Stud. 1: 867 (1894).

Metrosideros cordata (C.T.White & W.D.Francis) J.W.Dawson, Blumea 23: 11 (1976).
New Guinea. 43 NWG.
**Mearnsia cordata* C.T.White & W.D.Francis, Proc. Roy. Soc. Queensland 39: 67 (1927 publ. 1928).

Metrosideros diffusa (G.Forst.) Sm., Trans. Linn. Soc. London 3: 268 (1797).
New Zealand. 51 NZN NZS. Cl.
Melaleuca lucida L.f., Suppl. Pl.: 342 (1782), nom. illeg.
**Melaleuca diffusa* G.Forst., Fl. Ins. Austr.: 37 (1786). *Nania diffusa* (G.Forst.) Kuntze, Revis. Gen. Pl. 1: 242 (1891).
Melaleuca florida G.Forst., Fl. Ins. Austr.: 37 (1786).
Metrosideros myrtifolia Sol. ex Gaertn., Fruct. Sem. Pl. 1: 172 (1788).
Metrosideros hypericifolia A.Cunn., Ann. Nat. Hist. 3: 114 (1839), nom. illeg. *Nania hypericifolia* (A.Cunn.) Kuntze, Revis. Gen. Pl. 1: 242 (1891).
Metrosideros homeana Turcz. ex Hook.f., Handb. New Zeal. Fl.: 71 (1864).
Metrosideros subsimilis Colenso, Trans. & Proc. New Zealand Inst. 12: 361 (1880).

Metrosideros dolichandra Schltr. ex Guillaumin, Bull. Soc. Bot. France 81: 10 (1934).
SE. New Caledonia. 60 NWC. Nanophan. or phan.

Metrosideros elegans (Montrouz.) Beauvis., Ann. Soc. Bot. Lyon 26: 39 (1901).
New Caledonia. 60 NWC. Nanophan. or phan.
**Ballardia elegans* Montrouz., Mém. Acad. Roy. Sci. Lyon, Sect. Sci. 10: 205 (1860). *Carpolepis elegans* (Montrouz.) J.W.Dawson, Bull. Mus. Natl. Hist. Nat., B, Adansonia 1984: 466 (1984 publ. 1985).

Metrosideros engleriana Schltr., Bot. Jahrb. Syst. 39: 205 (1906).
New Caledonia. 60 NWC. Nanophan. or phan.

Metrosideros excelsa Sol. ex Gaertn., Fruct. Sem. Pl. 1: 172 (1788).
New Zealand North I. (50) nfk 51 NZN. Phan.
Metrosideros tomentosa A.Rich. in J.S.C.Dumontd' Urville, Voy. Astrolabe 1: 336 (1832). *Nania tomentosa* (A.Rich.) Kuntze, Revis. Gen. Pl. 1: 242 (1891).

Metrosideros fulgens Sol. ex Gaertn., Fruct. Sem. Pl. 1: 172 (1788).
New Zealand. 51 NZN NZS. Cl.
Leptospermum scandens J.R.Forst. & G.Forst., Char. Gen. Pl.: 36 (1775). *Nania scandens* (J.R.Forst. & G.Forst.) Kuntze, Revis. Gen. Pl. 1: 242 (1891). *Metrosideros scandens* (J.R.Forst. & G.Forst.) Druce, Rep. Bot. Exch. Club Brit. Isles 1916: 635 (1917), nom. illeg.
Melaleuca scandens Raeusch., Nomencl. Bot., ed. 3: 142 (1797).
Melaleuca spectabilis Raeusch., Nomencl. Bot., ed. 3: 142 (1797).
Metrosideros florida Sm., Trans. Linn. Soc. London 3: 269 (1797). *Nania florida* (Sm.) Kuntze, Revis. Gen. Pl. 1: 242 (1891).
Metrosideros speciosa Colenso, Trans. & Proc. New Zealand Inst. 22: 463 (1889 publ. 1890), nom. illeg.
Metrosideros aurata Colenso, Trans. & Proc. New Zealand Inst. 23: 385 (1890 publ. 1891).

Metrosideros gregoryi Christoph., Bernice P. Bishop Mus. Bull. 154: 28 (1938).
Samoa. 60 SAM. Nanophan. or phan.

Metrosideros halconensis (Merr.) J.W.Dawson, Blumea 23: 11 (1976).
Philippines. 42 PHI. Nanophan. or phan.
**Mearnsia halconensis* Merr., Philipp. J. Sci., C 2: 284 (1907).

Metrosideros humboldtiana Guillaumin, Bull. Soc. Bot. France 85: 627 (1938 publ. 1939).
SE. New Caledonia. 60 NWC. Nanophan. or phan.

Metrosideros kermadecensis W.R.B.Oliv., Trans. & Proc. New Zealand Inst. 59: 422 (1928).
Kermadec Is. (51) ctm KER. Nanophan. or phan.
**Metrosideros polymorpha* J.R.Forst. ex Hook.f., Handb. N. Zeal. Fl.: 73 (1864).

Metrosideros laurifolia Brongn. & Gris, Bull. Soc. Bot. France 12: 300 (1865). *Carpolepis laurifolia* (Brongn. & Gris) J.W.Dawson, Bull. Mus. Natl. Hist. Nat., B, Adansonia 1984: 468 (1984 publ. 1985).
New Caledonia. 60 NWC. Nanophan. or phan.
Metrosideros demonstrans Tison, Compt. Rend. Assoc. Franç. Avancem. Sci. 5: 462 (1877). *Carpolepis laurifolia* var. *demonstrans* (Tison) J.W.Dawson, Bull. Mus. Natl. Hist. Nat., B, Adansonia 1984: 468 (1984 publ. 1985).
Carpolepis tardiflora J.W.Dawson, Bull. Mus. Natl. Hist. Nat., B, Adansonia 1984: 468 (1984 publ. 1985).

Metrosideros longipetiolata J.W.Dawson, Bull. Mus. Natl. Hist. Nat., B, Adansonia 1984: 478 (1984 publ. 1985).
NE. New Caledonia. 60 NWC. Nanophan.

Metrosideros macropus Hook. & Arn., Bot. Beechey Voy.: 83 (1832). *Nania macropus* (Hook. & Arn.) Kuntze, Revis. Gen. Pl. 1: 242 (1891).
Hawaiian Is. (Oahu). 63 HAW. Phan.
Metrosideros macropus f. *ruber* H.St.John, Occas. Pap. Bernice Pauahi Bishop Mus. 11(13): 5 (1935).

Metrosideros microphylla (Schltr.) J.W.Dawson, Bull. Mus. Natl. Hist. Nat., B, Adansonia 1984: 472 (1984 publ. 1985).
SE. New Caledonia. 60 NWC. Nanophan. or phan.
**Metrosideros engleriana* var. *microphylla* Schltr., Bot. Jahrb. Syst. 39: 206 (1907).
Metrosideros balansae Brongn. & Gris ex Guillaumin, Ann. Inst. Bot.-Géol. Colon. Marseille, II, 9: 147 (1911), pro syn.

Metrosideros nervulosa C.Moore & F.Muell. in F.J.h.von Mueller, Fragm. 8: 15 (1874). *Nania nervulosa* (C.Moore & F.Muell.) Kuntze, Revis. Gen. Pl. 1: 242 (1891).
Lord Howe I. 50 NFK. Nanophan. or phan.

Metrosideros nitida Brongn. & Gris, Bull. Soc. Bot. France 11: 182 (1864). *Nania nitida* (Brongn. & Gris) Kuntze, Revis. Gen. Pl. 1: 242 (1891).
New Caledonia. 60 NWC. Nanophan. or phan.

Metrosideros ochrantha A.C.Sm., Amer. J. Bot. 60: 490 (1973).
Fiji (Vanua Levu: Mt. Kasi). 60 FIJ. Nanophan. or phan.

Metrosideros operculata Labill., Sert. Austro-Caledon.: 61 (1825). *Nania operculata* (Labill.) Kuntze, Revis. Gen. Pl. 1: 242 (1891).
New Caledonia. 60 NWC. Nanophan.

var. ***francii*** J.W.Dawson, Bull. Mus. Natl. Hist. Nat., B, Adansonia 1984: 480 (1984 publ. 1985).
New Caledonia. 60 NWC. Nanophan.

var. ***operculata***
New Caledonia. 60 NWC. Nanophan.
Metrosideros operculata var. *longifolia* Brongn. & Gris, Bull. Soc. Bot. France 11: 182 (1864).
Metrosideros operculata var. *myrtifolia* Brongn. & Gris, Bull. Soc. Bot. France 11: 182 (1864).
Metrosideros francii Schltr. ex Guillaumin, Notul. Syst. (Paris) 1: 110 (1909).

Metrosideros oreomyrtus Däniker, Vierteljahrsschr. Naturf. Ges. Zürich 78(19): 308 (1933).
New Caledonia. 60 NWC. Phan.

Metrosideros ovata (C.T.White) J.W.Dawson, Blumea 23: 11 (1976).
New Guinea. 43 NWG. Nanophan. or phan.
**Mearnsia ovata* C.T.White, J. Arnold Arbor. 23: 81 (1942).

Metrosideros paniensis J.W.Dawson, Bull. Mus. Natl. Hist. Nat., B, Adansonia 1984: 484 (1984 publ. 1985).
SE. New Caledonia. 60 NWC. Nanophan. or phan.

Metrosideros parallelinervis C.T.White, J. Arnold Arbor. 23: 79 (1942).
New Guinea. 43 NWG. Nanophan. or phan.

Metrosideros parkinsonii Buchanan, Trans. & Proc. New Zealand Inst. 15: 339 (1883).
New Zealand. 51 NZN NZS. Nanophan. or phan.

Metrosideros patens J.W.Dawson, Bull. Mus. Natl. Hist. Nat., B, Adansonia 1984: 482 (1984 publ. 1985).
SE. New Caledonia. 60 NWC. Nanophan.

Metrosideros perforata (J.R.Forst. & G.Forst.) Druce, Rep. Bot. Exch. Club Brit. Isles 1916: 635 (1917).
New Zealand. 51 NZN NZS. Cl.
**Leptospermum perforatum* J.R.Forst. & G.Forst., Char. Gen. Pl.: 36 (1775). *Melaleuca perforata* (J.R.Forst. & G.Forst.) G.Forst., Fl. Ins. Austr.: 37 (1786).
Metrosideros scandens Sol. ex Gaertn., Fruct. Sem. Pl. 1: 172 (1788).
Metrosideros myrtifolia Dum.Cours., Bot. Cult., ed. 2, 5: 379 (1811), nom. illeg.
Metrosideros buxifolia A.Cunn., Ann. Nat. Hist. 3: 111 (1839), nom. illeg.
Metrosideros vesiculata Colenso, Trans. & Proc. New Zealand Inst. 16: 327 (1884).
Metrosideros tenuifolia Colenso, Trans. & Proc. New Zealand Inst. 24: 387 (1891 publ. 1892).

Metrosideros polymorpha Gaudich., Voy. Uranie: 482 (1830). *Nania polymorpha* (Gaudich.) A.Heller, Minnesota Bot. Stud. 1: 864 (1894). *Metrosideros collina* subsp. *polymorpha* (Gaudich.) Rock, Bot. Bull. Hawaii Board Agric. For. 4: 47 (1917).
Hawaiian Is. 63 HAW. Nanophan. or phan.

var. ***dieteri*** J.W.Dawson & Stemmerm., Bishop Mus. Occas. Pap. 29: 108 (1989).
Hawaiian Is. (Kauai). 63 HAW. Nanophan. or phan.

var. ***imbricata*** (Rock) H.St.John, Phytologia 42: 216 (1979).
Hawaiian Is. 63 HAW. Nanophan. or phan.
Nania polymorpha var. *incana* H.Lév., Repert. Spec. Nov. Regni Veg. 10: 149 (1911). *Metrosideros collina* var. *incana* (H.Lév.) Rock, Bot. Bull. Hawaii Board Agric. For. 4: 51 (1917). *Metrosideros polymorpha* subsp. *incana* (H.Lév.) Skottsb., Acta Horti Gothob. 10: 147 (1935).
Nania polymorpha var. *nummularifolia* H.Lév., Repert. Spec. Nov. Regni Veg. 10: 149 (1911).
**Metrosideros collina* var. *imbricata* Rock, Bot. Bull. Hawaii Board Agric. For. 4: 49 (1917). *Metrosideros polymorpha* subsp. *imbricata* (Rock) Skottsb., Acta Horti Gothob. 10: 147 (1935).
Metrosideros collina f. *lurida* Rock, Bot. Bull. Hawaii Board Agric. For. 4: 54 (1917). *Metrosideros polymorpha* f. *lurida* (Rock) H.St.John, Phytologia 42: 217 (1979).

Metrosideros polymorpha var. *subimbricaria* Skottsb., Acta Horti Gothob. 10: 147 (1935).
Metrosideros polymorpha f. *psilophylla* Skottsb., Acta Horti Gothob. 15: 406 (1944).

var. ***lutea*** H.Mann, Proc. Amer. Acad. Arts 7: 166 (1866).
Hawaiian Is. 63 HAW. Nanophan. or phan.
Nania glabrifolia A.Heller, Minnesota Bot. Stud. 1: 866 (1894). *Metrosideros collina* var. *glabrifolia* (A.Heller) Rock, Bot. Bull. Hawaii Board Agric. For. 4: 67 (1917). *Metrosideros polymorpha* subsp. *glabrifolia* (A.Heller) Skottsb., Acta Horti Gothob. 10: 148 (1935). *Metrosideros polymorpha* var. *glabrifolia* (A.Heller) H.St.John, Phytologia 42: 216 (1979).
Nania polymorpha var. *glaberrima* H.Lév., Repert. Spec. Nov. Regni Veg. 10: 149 (1911). *Metrosideros collina* var. *glaberrima* (H.Lév.) Rock, Bot. Bull. Hawaii Board Agric. For. 4: 69 (1917), nom. illeg. *Metrosideros polymorpha* subsp. *glaberrima* (H.Lév.) Skottsb., Acta Horti Gothob. 15: 408 (1944). *Metrosideros polymorpha* var. *glaberrima* (H.Lév.) H.St.John, Phytologia 42: 216 (1979).
Metrosideros collina var. *haleakalesis* Rock, Bot. Bull. Hawaii Board Agric. For. 4: 56 (1917). *Metrosideros polymorpha* var. *haleakalensis* (Rock) H.St.John, Phytologia 42: 216 (1979).
Metrosideros collina f. *sericea* Rock, Bot. Bull. Hawaii Board Agric. For. 4: 71 (1917).
Metrosideros collina var. *hemilanata* Hochr., Candollea 2: 456 (1925). *Metrosideros polymorpha* var. *hemilanata* (Hochr.) H.St.John, Phytologia 42: 216 (1979).
Metrosideros polymorpha var. *nuda* Skottsb., Acta Horti Gothob. 10: 147 (1935).
Metrosideros polymorpha f. *calva* Skottsb., Acta Horti Gothob. 15: 405 (1944).
Metrosideros polymorpha f. *obovata* Skottsb., Acta Horti Gothob. 15: 404 (1944).
Metrosideros polymorpha var. *parviflora* Skottsb., Acta Horti Gothob. 15: 405 (1944).

var. ***macrophylla*** (Rock) H.St.John, Phytologia 42: 217 (1979).
Hawaiian Is. (Hawaii). 63 HAW. Nanophan. or phan.
**Metrosideros collina* var. *macrophylla* Rock, Bot. Bull. Hawaii Board Agric. For. 4: 58 (1917).
Metrosideros polymorpha subsp. *micrantha* Skottsb., Acta Horti Gothob. 15: 407 (1944). *Metrosideros polymorpha* var. *micrantha* (Skottsb.) H.St.John, Phytologia 42: 217 (1979).

var. ***newellii*** (Rock) H.St.John, Phytologia 42: 217 (1979).
Hawaiian Is. (Hawaii). 63 HAW. Nanophan. or phan.
**Metrosideros collina* var. *newellii* Rock, Bot. Bull. Hawaii Board Agric. For. 4: 58 (1917).

var. ***polymorpha***
Hawaiian Is. 63 HAW. Nanophan. or phan.
Metrosideros hillebrandii H.Lév. & Vaniot, Repert. Spec. Nov. Regni Veg. 13: 357 (1914).
Metrosideros polymorpha f. *humilis* Skottsb., Acta Horti Gothob. 15: 404 (1944).
Metrosideros polymorpha var. *macrostemon* Skottsb., Acta Horti Gothob. 10: 147 (1944).

var. ***pseudorugosa*** Skottsb., Acta Horti Gothob. 10: 147 (1935). *Metrosideros polymorpha* f. *pseudorugosa* (Skottsb.) Skottsb., Acta Horti Gothob. 15: 404 (1944).
Hawaiian Is. (W. Maui). 63 HAW. Cham. or nanophan.

var. ***pumila*** (A.Heller) Skottsb., Acta Horti Gothob. 15: 406 (1944).
Hawaiian Is. (Kauai, Molokai, Maui). 63 HAW. Nanophan.
**Nania pumila* A.Heller, Minnesota Bot. Stud. 1: 864 (1894). *Metrosideros collina* var. *pumila* (A.Heller) Rock, Bot. Bull. Hawaii Board Agric. For. 4: 61 (1917). *Metrosideros pumila* (A.Heller) Hochr., Candollea 2: 459 (1925).
Metrosideros collina var. *prostrata* Rock, Bot. Bull. Hawaii Board Agric. For. 4: 61 (1917). *Metrosideros polymorpha* var. *prostrata* (Rock) H.St.John, Phytologia 42: 217 (1979).
Metrosideros collina f. *strigosa* Rock, Bot. Bull. Hawaii Board Agric. For. 4: 64 (1917). *Metrosideros polymorpha* f. *strigosa* (Rock) H.St.John, Phytologia 42: 218 (1979).
Metrosideros polymorpha f. *perglabra* Skottsb., Acta Horti Gothob. 15: 407 (1944).

Metrosideros porphyrea Schltr., Bot. Jahrb. Syst. 39: 206 (1906). *Mearnsia porphyrea* (Schltr.) Diels, Bot. Jahrb. Syst. 57: 419 (1922).
New Caledonia. 60 NWC. Nanophan. or phan.
Metrosideros porphyrea var. *lucida* Bonati & Petitm., Bull. Herb. Boissier, II, 7: 652 (1907).

Metrosideros punctata J.W.Dawson, Bull. Mus. Natl. Hist. Nat., B, Adansonia 1984: 470 (1984 publ. 1985).
New Caledonia. 60 NWC. Nanophan. or phan.

Metrosideros ramiflora Lauterb., Nova Guinea 8: 853 (1912). *Mearnsia ramiflora* (Lauterb.) Diels, Bot. Jahrb. Syst. 57: 419 (1922).
New Guinea. 43 NWG. Nanophan. or phan.

var. ***humilis*** (Diels) J.W.Dawson, Blumea 23: 11 (1976).
New Guinea. 43 NWG.
**Mearnsia ramiflora* var. *humilis* Diels, Bot. Jahrb. Syst. 57: 420 (1922).

var. ***ramiflora***
New Guinea. 43 NWG. Nanophan. or phan.

var. ***villosa*** (C.T.White) J.W.Dawson, Blumea 23: 11 (1976).
New Guinea. 43 NWG. Nanophan. or phan.
**Mearnsia ramiflora* var. *villosa* C.T.White, J. Arnold Arbor. 32: 142 (1951).

Metrosideros regelii F.Muell., Trans. Roy. Soc. Victoria, n.s., 1(2): 6 (1889).
New Guinea. 43 NWG. Nanophan. or phan.

Metrosideros robusta A.Cunn., Ann. Nat. Hist. 3: 112 (1839). *Nania robusta* (A.Cunn.) Kuntze, Revis. Gen. Pl. 1: 242 (1891).
New Zealand. 51 NZN NZS. Phan.
Metrosideros florida Hook.f., Bot. Mag. 75: t. 4471 (1849).

Metrosideros rotundifolia J.W.Dawson, Blumea 45: 437 (2000).
New Caledonia. 60 NWC.

Metrosideros rugosa A.Gray, U.S. Expl. Exped., Phan. 1: 561 (1854). *Nania rugosa* (A.Gray) Kuntze, Revis. Gen. Pl. 1: 242 (1891).
Hawaiian Is. (Oahu). 63 HAW. Nanophan. or phan.

Metrosideros salomonensis C.T.White, J. Arnold Arbor. 32: 143 (1951).
Solomon Is. 43 SOL. Nanophan. or phan.

Metrosideros sclerocarpa J.W.Dawson, Kew Bull. 45: 244 (1990).
Lord Howe I. 50 NFK. Phan.

Metrosideros* × *subtomentosa Carse, Trans. & Proc. New Zealand Inst. 57: 92 (1927). *M. excelsa* × *M. robusta*.
New Zealand North I. 51 NZN. Nanophan. or phan.
Metrosideros robusta var. *intermedia* Kirk, Stud. Fl. New Zealand: 162 (1899).

Metrosideros tetragyna J.W.Dawson, Blumea 23: 11 (1976).
Solomon Is. 43 SOL. Nanophan. or phan.
**Mearnsia salomonensis* C.T.White, J. Arnold Arbor. 32: 142 (1951).

Metrosideros tetrasticha Guillaumin, Mém. Mus. Natl. Hist. Nat., B, Bot. 4: 32 (1953).
SE. New Caledonia. 60 NWC. Nanophan.

Metrosideros tremuloides (A.Heller) Rock, Indig. Trees Haw. Isl.: 333 (1913).
Hawaiian Is. (Oahu). 63 HAW. Nanophan. or phan.
**Nania tremuloides* A.Heller, Minnesota Bot. Stud. 1: 866 (1894). *Metrosideros polymorpha* var. *tremuloides* (A.Heller) Skottsb., Acta Horti Gothob. 15: 408 (1944).
Nania feddei H.Lév., Repert. Spec. Nov. Regni Veg. 10: 150 (1911).

Metrosideros umbellata Cav., Icon. 4: 20 (1797). *Agalmanthus umbellata* (Cav.) Hombr. & Jacquinot in J.S.C.Dumont d'Urville, Voy. Pôle Sud, Atlas: 78 (1845).
New Zealand, Antipodean Is. 51 ATP NZN NZS. Phan.
Melaleuca lucida G.Forst., Fl. Ins. Austr.: 38 (1786), nom. illeg. *Metrosideros lucida* A.Rich. in J.S.C. Dumont d'Urville, Voy. Astrolabe 1: 333 (1832). *Nania lucida* (A.Rich.) Kuntze, Revis. Gen. Pl. 1: 242 (1891).

Metrosideros waialealae (Rock) Rock, Board Agric. For., Bull. 4: 43 (1917).
Hawaiian Is. 63 HAW. Nanophan. or phan.
**Metrosideros tremuloides* var. *waialealae* Rock, Indig. Trees Haw. Isl.: 335 (1913).

var. ***fauriei*** (H.Lév.) J.W.Dawson & Stemmerm., Bishop Mus. Occas. Pap. 29: 109 (1989).
Hawaiian Is. (Lanai, Molokai). 63 HAW. Nanophan. or phan.
**Nania fauriei* H.Lév., Repert. Spec. Nov. Regni Veg. 10: 150 (1911). *Metrosideros collina* var. *fauriei* (H.Lév.) Rock, Bot. Bull. Hawaii Board Agric. For. 4: 67 (1917). *Metrosideros polymorpha* var. *fauriei* (H.Lév.) Skottsb., Acta Horti Gothob. 15: 406 (1944).

var. ***waialealae***
Hawaiian Is. (Kauai). 63 HAW. Nanophan. or phan.

Metrosideros whitakeri J.W.Dawson, Blumea 45: 435 (2000).
New Caledonia. 60 NWC.

Metrosideros whiteana J.W.Dawson, Blumea 23: 11 (1976).
New Guinea. 43 NWG. Cl.
**Mearnsia scandens* C.T.White, J. Arnold Arbor. 23: 81 (1942).

Synonyms:
Metrosideros abietina Hoffmanns. = ***Kunzea ambigua*** (Sm.) Druce
Metrosideros acuminata Decne. = ***Metrosideros collina*** var. ***villosa*** (L.f.) A.Gray
Metrosideros albida Sieber ex DC. = ***Melaleuca quinquenervia*** (Cav.) S.T.Blake
Metrosideros angustifolia Dum.Cours. = ***Callistemon lanceolatus*** (Sm.) Sweet
Metrosideros anomala Vent. = ***Angophora hispida*** (Sm.) Blaxell
Metrosideros apocynifolia Salisb. = ***Angophora costata*** (Gaertn.) Hochr. ex Britten
Metrosideros armillaris Sol. ex Gaertn. = ***Melaleuca armillaris*** (Sol. ex Gaertn.) Sm.
Metrosideros aromatica K.D.Koenig & Sims = ***Eucalyptus piperita*** Sm. subsp. ***piperita***
Metrosideros aspera Dum.Cours. = ?
Metrosideros aspera Hoffmanns. = ***Callistemon rigidus*** (Muhl. ex Willd.) R.Br.
Metrosideros aurata Colenso = ***Metrosideros fulgens*** Sol. ex Gaertn.
Metrosideros aurea (Ridl.) Diels = ***Kania eugenioides*** Schltr.
Metrosideros australis R.Br. ex Bonpl. = ***Callistemon salignus*** (Sm.) Colv. ex Sweet var. ***salignus***
Metrosideros balansae Brongn. & Gris ex Guillaumin = ***Metrosideros microphylla*** (Schltr.) J.W.Dawson
Metrosideros brachyanthera Diels = ***Thaleropia iteophylla*** (Diels) Peter G.Wilson
Metrosideros buxifolia A.Cunn. = ***Metrosideros perforata*** (J.R.Forst. & G.Forst.) Druce
Metrosideros buxifolia Dum.Cours. = ***Purpureostemon ciliatus*** (J.R.Forst. & G.Forst.) Gugerli
Metrosideros calycina Cav. = ***Melaleuca thymifolia*** Sm.
Metrosideros calyculatus Sieber ex DC. = ***Callistemon linearis*** (Schrad. & J.C.Wendl.) Colv. ex Sweet
Metrosideros canaliculata Dum.Cours. = ?
Metrosideros canaliculata Colla = ?
Metrosideros capitata Sm. = ***Kunzea capitata*** (Sm.) Heynh.
Metrosideros chrysantha F.Muell. = ***Xanthostemon chrysanthus*** (F.Muell.) Benth.
Metrosideros ciliata (J.R.Forst. & G.Forst.) Sm. = ***Purpureostemon ciliatus*** (J.R.Forst. & G.Forst.) Gugerli
Metrosideros citrina Curtis = ***Callistemon citrinus*** (Curtis) Skeels
Metrosideros collina var. *fauriei* (H.Lév.) Rock = ***Metrosideros waialealae*** var. ***fauriei*** (H.Lév.) J.W.Dawson & Stemmerm.
Metrosideros collina var. *glaberrima* (Bertero ex Guillaumin) A.Gray = ***Metrosideros collina*** var. ***villosa*** (L.f.) A.Gray
Metrosideros collina var. *glaberrima* (H.Lév.) Rock = ***Metrosideros polymorpha*** var. ***lutea*** H.Mann
Metrosideros collina var. *glabrifolia* (A.Heller) Rock = ***Metrosideros polymorpha*** var. ***lutea*** H.Mann
Metrosideros collina var. *haleakalesis* Rock = ***Metrosideros polymorpha*** var. ***lutea*** H.Mann
Metrosideros collina var. *hemilanata* Hochr. = ***Metrosideros polymorpha*** var. ***lutea*** H.Mann
Metrosideros collina var. *imbricata* Rock = ***Metrosideros polymorpha*** var. ***imbricata*** (Rock) H.St.John
Metrosideros collina var. *incana* (H.Lév.) Rock = ***Metrosideros polymorpha*** var. ***imbricata*** (Rock) H.St.John
Metrosideros collina f. *lurida* Rock = ***Metrosideros polymorpha*** var. ***imbricata*** (Rock) H.St.John

Metrosideros collina var. *macrophylla* Rock = ***Metrosideros polymorpha*** var. ***macrophylla*** (Rock) H.St.John
Metrosideros collina var. *newellii* Rock = ***Metrosideros polymorpha*** var. ***newellii*** (Rock) H.St.John
Metrosideros collina subsp. *polymorpha* (Gaudich.) Rock = ***Metrosideros polymorpha*** Gaudich.
Metrosideros collina var. *prostrata* Rock = ***Metrosideros polymorpha*** var. ***pumila*** (A.Heller) Skottsb.
Metrosideros collina var. *pumila* (A.Heller) Rock = ***Metrosideros polymorpha*** var. ***pumila*** (A.Heller) Skottsb.
Metrosideros collina f. *sericea* Rock = ***Metrosideros polymorpha*** var. ***lutea*** H.Mann
Metrosideros collina f. *strigosa* Rock = ***Metrosideros polymorpha*** var. ***pumila*** (A.Heller) Skottsb.
Metrosideros collina var. *temehaniensis* J.W.Moore = ***Metrosideros collina*** (J.R.Forst. & G.Forst.) A.Gray var. ***collina***
Metrosideros collina var. *vitiensis* A.Gray = ***Metrosideros collina*** (J.R.Forst. & G.Forst.) A.Gray var. ***collina***
Metrosideros comosa Roxb. = ***Melaleuca cajuputi*** Powell subsp. ***cajuputi***
Metrosideros connata Desf. = ***Syzygium floribundum*** F.Muell.
Metrosideros cordifolia (Cav.) Pers. = ***Angophora hispida*** (Sm.) Blaxell
Metrosideros coriacea K.D.Koenig & Sims = ***Melaleuca leucadendra*** (L.) L.
Metrosideros coriacea Poir. = ***Melaleuca quinquenervia*** (Cav.) S.T.Blake
Metrosideros corifolia Vent. = ***Kunzea ambigua*** (Sm.) Druce
Metrosideros costata Gaertn. = ***Angophora costata*** (Gaertn.) Hochr. ex Britten
Metrosideros crassifolia Dum.Cours. = ***Callistemon speciosus*** (Sims) Sweet
Metrosideros decora Salisb. = ***Melaleuca decora*** (Salisb.) Britten
Metrosideros decurrens Steud. = ?
Metrosideros demonstrans Tison = ***Metrosideros laurifolia*** Brongn. & Gris
Metrosideros diffusa A.Cunn. = ***Metrosideros albiflora*** Sol. ex Gaertn.
Metrosideros engleriana var. *microphylla* Schltr. = ***Metrosideros microphylla*** (Schltr.) J.W.Dawson
Metrosideros ericifolia Sm. = ***Kunzea ericifolia*** (Sm.) Heynh.
Metrosideros eucalyptoides (F.Muell.) F.Muell. = ***Xanthostemon eucalyptoides*** F.Muell.
Metrosideros eugenioides (Schltr.) Steenis = ***Kania eugenioides*** Schltr.
Metrosideros falcata Dehnh. = ?
Metrosideros falcata Dum.Cours. = ***Callistemon lanceolatus*** (Sm.) Sweet
Metrosideros flexuosa Muhl. ex Willd. = ***Agonis flexuosa*** (Muhl. ex Willd.) Sweet
Metrosideros floribunda Sm. = ***Angophora floribunda*** (Sm.) Sweet
Metrosideros floribunda Vent. = ***Syzygium floribundum*** F.Muell.
Metrosideros florida Sm. = ***Metrosideros fulgens*** Sol. ex Gaertn.
Metrosideros florida Hook.f. = ***Metrosideros robusta*** A.Cunn.
Metrosideros francii Schltr. ex Guillaumin = ***Metrosideros operculata*** Labill. var. ***operculata***
Metrosideros gibbsiae Diels = ***Kania hirsutula*** (F.Muell.) A.J.Scott
Metrosideros glandulosa Desf. = ***Callistemon rigidus*** (Muhl. ex Willd.) R.Br.
Metrosideros glauca Dum.Cours. = ***Callistemon speciosus*** (Sims) Sweet
Metrosideros glauca Bonpl. = ***Callistemon speciosus*** (Sims) Sweet
Metrosideros glomulifera Sm. = ***Syncarpia glomulifera*** (Sm.) Nied.
Metrosideros glomulifera var. *glabra* (Benth.) C.Moore & Betche = ***Syncarpia glomulifera*** subsp. ***glabra*** (Benth.) A.R.Bean
Metrosideros gracilis K.D.Koenig & Sims = ***Melaleuca nodosa*** (Sol. ex Gaertn.) Sm.
Metrosideros gummifera Gaertn. = ***Corymbia gummifera*** (Gaertn.) K.D.Hill & L.A.S.Johnson
Metrosideros hillebrandii H.Lév. & Vaniot = ***Metrosideros polymorpha*** Gaudich. var. ***polymorpha***
Metrosideros hirsuta Andrews = ***Angophora hispida*** (Sm.) Blaxell
Metrosideros hispida Sm. = ***Angophora hispida*** (Sm.) Blaxell
Metrosideros homeana Turcz. ex Hook.f. = ***Metrosideros diffusa*** (G.Forst.) Sm.
Metrosideros hypargyrea Diels = ***Thaleropia hypargyrea*** (Diels) Peter G.Wilson
Metrosideros hypericifolia A.Cunn. = ***Metrosideros diffusa*** (G.Forst.) Sm.
Metrosideros hypericifolia (Sm.) Salisb. = ***Melaleuca hypericifolia*** Sm.
Metrosideros hyssopifolia Cav. = ***Melaleuca linariifolia*** Sm.
Metrosideros iteophylla Diels = ***Thaleropia iteophylla*** (Diels) Peter G.Wilson
Metrosideros juniperina Rchb. ex Spreng. = ***Melaleuca nodosa*** (Sol. ex Gaertn.) Sm.
Metrosideros juniperoides Rchb. ex DC. = ***Melaleuca nodosa*** (Sol. ex Gaertn.) Sm.
Metrosideros lanceolata Pers. = ***Angophora costata*** (Gaertn.) Hochr. ex Britten
Metrosideros lanceolata Sm. = ***Callistemon lanceolatus*** (Sm.) Sweet
Metrosideros latifolia Dum.Cours. = ***Callistemon lanceolatus*** (Sm.) Sweet
Metrosideros laurifolia Dum.Cours. = ***Syzygium floribundum*** F.Muell.
Metrosideros leptopetala (F.Muell.) F.Muell. = ***Choricarpia leptopetala*** (F.Muell.) Domin
Metrosideros linearifolia Link = ***Callistemon linearifolius*** (Link) DC.
Metrosideros linearis (Schrad. & J.C.Wendl.) Sm. = ***Callistemon linearis*** (Schrad. & J.C.Wendl.) Colv. ex Sweet
Metrosideros linearis Muhl. ex Willd. = ***Callistemon rigidus*** (Muhl. ex Willd.) R.Br.
Metrosideros linifolia Dum.Cours. = ***Callistemon rigidus*** (Muhl. ex Willd.) R.Br.
Metrosideros longifolia Dum.Cours. = ***Callistemon lanceolatus*** (Sm.) Sweet
Metrosideros lophantha Vent. = ***Callistemon salignus*** (Sm.) Colv. ex Sweet var. ***salignus***
Metrosideros lucida A.Rich. = ***Metrosideros umbellata*** Cav.
Metrosideros lutea A.Gray = ***Metrosideros collina*** var. ***villosa*** (L.f.) A.Gray
Metrosideros macrophylla Lam. = [29 MDG] (?) not a Myrtaceae.
Metrosideros macropunctata Dum.Cours. = ***Callistemon macropunctatus*** (Dum.Cours.) Court
Metrosideros macropus f. *ruber* H.St.John = ***Metrosideros***

macropus Hook. & Arn.
Metrosideros marginata Cav. = ***Callistemon lanceolatus*** (Sm.) Sweet
Metrosideros myrtifolia Sol. ex Gaertn. = ***Metrosideros diffusa*** (G.Forst.) Sm.
Metrosideros myrtifolia Dum.Cours. = ***Metrosideros perforata*** (J.R.Forst. & G.Forst.) Druce
Metrosideros myrtifolia Hoffmanns. = ***Callistemon lanceolatus*** (Sm.) Sweet
Metrosideros nana C.C.Gmel. ex DC. = ?
Metrosideros nigroviridis Steenis = ***Lindsayomyrtus racemoides*** (Greves) Craven
Metrosideros nodosa Sol. ex Gaertn. = ***Melaleuca nodosa*** (Sol. ex Gaertn.) Sm.
Metrosideros obovata Hook. & Arn. = ***Metrosideros collina*** var. ***villosa*** (L.f.) A.Gray
Metrosideros operculata var. *longifolia* Brongn. & Gris = ***Metrosideros operculata*** Labill. var. ***operculata***
Metrosideros operculata var. *myrtifolia* Brongn. & Gris = ***Metrosideros operculata*** Labill. var. ***operculata***
Metrosideros ornata C.T.White = ***Thaleropia hypargyrea*** (Diels) Peter G.Wilson
Metrosideros pachysperma (F.Muell. & F.M.Bailey) F.Muell. = ***Ristantia pachysperma*** (F.Muell. & F.M.Bailey) Peter G.Wilson & J.T.Waterh.
Metrosideros pallida Bonpl. = ***Callistemon pallidus*** (Bonpl.) DC.
Metrosideros paradoxa (F.Muell.) F.Muell. = ***Xanthostemon paradoxus*** F.Muell.
Metrosideros parviflora C.T.White = ***Kania eugenioides*** Schltr.
Metrosideros pauciflora Endl. = ***Eremaea pauciflora*** (Endl.) Druce
Metrosideros pendens Colenso = ***Metrosideros colensoi*** var. ***pendens*** (Colenso) Kirk
Metrosideros petiolata (Valeton) Koord. = ***Xanthostemon petiolatus*** (Valeton) Peter G.Wilson
Metrosideros pictipetala Blanco = ***Xanthostemon verus*** (Lindl.) Peter G.Wilson
Metrosideros pinifolia J.C.Wendl. = ***Callistemon pinifolius*** (J.C.Wendl.) Sweet
Metrosideros pleurocalyptus Baill. = ***Pleurocalyptus pancheri*** (Brongn. & Gris) J.W.Dawson
Metrosideros polymorpha J.R.Forst. ex Hook.f. = ***Metrosideros kermadecensis*** W.R.B.Oliv.
Metrosideros polymorpha f. *calva* Skottsb. = ***Metrosideros polymorpha*** var. ***lutea*** H.Mann
Metrosideros polymorpha var. *fauriei* (H.Lév.) Skottsb. = ***Metrosideros waialealae*** var. ***fauriei*** (H.Lév.) J.W.Dawson & Stemmerm.
Metrosideros polymorpha subsp. *glaberrima* (H.Lév.) Skottsb. = ***Metrosideros polymorpha*** var. ***lutea*** H.Mann
Metrosideros polymorpha var. *glaberrima* (H.Lév.) H.St. John = ***Metrosideros polymorpha*** var. ***lutea*** H.Mann
Metrosideros polymorpha subsp. *glabrifolia* (A.Heller) Skottsb. = ***Metrosideros polymorpha*** var. ***lutea*** H.Mann
Metrosideros polymorpha var. *glabrifolia* (A.Heller) H.St.John = ***Metrosideros polymorpha*** var. ***lutea*** H.Mann
Metrosideros polymorpha var. *haleakalensis* (Rock) H.St.John = ***Metrosideros polymorpha*** var. ***lutea*** H.Mann
Metrosideros polymorpha var. *hemilanata* (Hochr.) H.St. John = ***Metrosideros polymorpha*** var. ***lutea*** H.Mann
Metrosideros polymorpha f. *humilis* Skottsb. = ***Metrosideros polymorpha*** Gaudich. var. ***polymorpha***
Metrosideros polymorpha subsp. *imbricata* (Rock) Skottsb. = ***Metrosideros polymorpha*** var. ***imbricata*** (Rock) H.St.John
Metrosideros polymorpha subsp. *incana* (H.Lév.) Skottsb. = ***Metrosideros polymorpha*** var. ***imbricata*** (Rock) H.St.John
Metrosideros polymorpha f. *lurida* (Rock) H.St.John = ***Metrosideros polymorpha*** var. ***imbricata*** (Rock) H.St.John
Metrosideros polymorpha var. *macrostemon* Skottsb. = ***Metrosideros polymorpha*** Gaudich. var. ***polymorpha***
Metrosideros polymorpha subsp. *micrantha* Skottsb. = ***Metrosideros polymorpha*** var. ***macrophylla*** (Rock) H.St.John
Metrosideros polymorpha var. *micrantha* (Skottsb.) H.St.John = ***Metrosideros polymorpha*** var. ***macrophylla*** (Rock) H.St.John
Metrosideros polymorpha var. *nuda* Skottsb. = ***Metrosideros polymorpha*** var. ***lutea*** H.Mann
Metrosideros polymorpha f. *obovata* Skottsb. = ***Metrosideros polymorpha*** var. ***lutea*** H.Mann
Metrosideros polymorpha var. *parviflora* Skottsb. = ***Metrosideros polymorpha*** var. ***lutea*** H.Mann
Metrosideros polymorpha f. *perglabra* Skottsb. = ***Metrosideros polymorpha*** var. ***pumila*** (A.Heller) Skottsb.
Metrosideros polymorpha var. *prostrata* (Rock) H.St.John = ***Metrosideros polymorpha*** var. ***pumila*** (A.Heller) Skottsb.
Metrosideros polymorpha f. *pseudorugosa* (Skottsb.) Skottsb. = ***Metrosideros polymorpha*** var. ***pseudorugosa*** Skottsb.
Metrosideros polymorpha f. *psilophylla* Skottsb. = ***Metrosideros polymorpha*** var. ***imbricata*** (Rock) H.St.John
Metrosideros polymorpha f. *strigosa* (Rock) H.St.John = ***Metrosideros polymorpha*** var. ***pumila*** (A.Heller) Skottsb.
Metrosideros polymorpha var. *subimbricaria* Skottsb. = ***Metrosideros polymorpha*** var. ***imbricata*** (Rock) H.St.John
Metrosideros polymorpha var. *tremuloides* (A.Heller) Skottsb. = ***Metrosideros tremuloides*** (A.Heller) Rock
Metrosideros porphyrea var. *lucida* Bonati & Petitm. = ***Metrosideros porphyrea*** Schltr.
Metrosideros procera K.D.Koenig & Sims = ***Syncarpia glomulifera*** (Sm.) Nied. subsp. ***glomulifera***
Metrosideros propinqua Endl. = ***Kunzea ericifolia*** (Sm.) Heynh. subsp. ***ericifolia***
Metrosideros propinqua K.D.Koenig & Sims = ***Syncarpia glomulifera*** (Sm.) Nied. subsp. ***glomulifera***
Metrosideros pubescens (Brongn. & Gris) Baill. = ***Xanthostemon pubescens*** (Brongn. & Gris) Sebert & Pancher
Metrosideros pullei Diels = ***Kania eugenioides*** Schltr.
Metrosideros pumila (A.Heller) Hochr. = ***Metrosideros polymorpha*** var. ***pumila*** (A.Heller) Skottsb.
Metrosideros pungens Rchb. = ***Melaleuca nodosa*** (Sol. ex Gaertn.) Sm.
Metrosideros queenslandica L.S.Sm. = ***Thaleropia queenslandica*** (L.S.Sm.) Peter G.Wilson
Metrosideros quinquenervia Cav. = ***Melaleuca quinquenervia*** (Cav.) S.T.Blake
Metrosideros rigida (R.Br.) Dum.Cours. = ***Callistemon rigidus*** (Muhl. ex Willd.) R.Br.
Metrosideros rigidifolia Hoffmanns. = ***Callistemon rigidus*** (Muhl. ex Willd.) R.Br.
Metrosideros robusta var. *intermedia* Kirk = ***Metrosideros*** × ***subtomentosa*** Carse

Metrosideros rubra (Brongn. & Gris) Baill. = ***Xanthostemon ruber*** (Brongn. & Gris) Sebert & Pancher
Metrosideros rugulosa Link = ***Callistemon macropunctatus*** (Dum.Cours.) Court var. ***macropunctatus***
Metrosideros ruscifolia Pépin = ?
Metrosideros salicifolia A.Cunn. = (Oleaceae)
Metrosideros salicifolia Sol. ex Gaertn. = ***Eucalyptus crebra*** F.Muell.
Metrosideros saligna Sm. = ***Callistemon salignus*** (Sm.) Colv. ex Sweet
Metrosideros scabra Colla = ***Callistemon macropunctatus*** (Dum.Cours.) Court var. ***macropunctatus***
Metrosideros scandens (J.R.Forst. & G.Forst.) Druce = ***Metrosideros fulgens*** Sol. ex Gaertn.
Metrosideros scandens Sol. ex Gaertn. = ***Metrosideros perforata*** (J.R.Forst. & G.Forst.) Druce
Metrosideros scariosa J.C.Wendl. ex Hornem. = ?
Metrosideros semperflorens Lodd. = ***Callistemon lanceolatus*** (Sm.) Sweet
Metrosideros sororia Endl. = ***Melaleuca seriata*** Lindl.
Metrosideros speciosa Colenso = ***Metrosideros fulgens*** Sol. ex Gaertn.
Metrosideros speciosa Sims = ***Callistemon speciosus*** (Sims) Sweet
Metrosideros spectabilis Sol. ex Gaertn. = ***Metrosideros collina*** var. ***villosa*** (L.f.) A.Gray
Metrosideros stipularis (Hook. & Arn.) Hook.f. = ***Tepualia stipularis*** (Hook. & Arn.) Griseb.
Metrosideros suberosa Roxb. = ***Xanthostemon verus*** (Lindl.) Peter G.Wilson
Metrosideros subsimilis Colenso = ***Metrosideros diffusa*** (G.Forst.) Sm.
Metrosideros taitensis Decne. = ***Metrosideros collina*** var. ***villosa*** (L.f.) A.Gray
Metrosideros tenuifolia Colenso = ***Metrosideros perforata*** (J.R.Forst. & G.Forst.) Druce
Metrosideros ternifolia F.Muell. = ***Lysicarpus angustifolius*** (Hook.) Druce
Metrosideros tetrapetala F.Muell. = ***Xanthostemon umbrosus*** (A.Cunn. ex Lindl.) Peter G.Wilson & J.T.Waterh.
Metrosideros tomentosa A.Rich. = ***Metrosideros excelsa*** Sol. ex Gaertn.
Metrosideros tremuloides var. *waialealae* Rock = ***Metrosideros waialealae*** (Rock) Rock
Metrosideros vera Lindl. = ***Xanthostemon verus*** (Lindl.) Peter G.Wilson
Metrosideros verticillata C.T.White & W.D.Francis = ***Xanthostemon verticillatus*** (C.T.White & W.D.Francis) L.S.Sm.
Metrosideros vesiculata Colenso = ***Metrosideros perforata*** (J.R.Forst. & G.Forst.) Druce
Metrosideros villosa (L.f.) Sm. = ***Metrosideros collina*** var. ***villosa*** (L.f.) A.Gray
Metrosideros villosa var. *glaberrima* Bertero ex Guillaumin = ***Metrosideros collina*** var. ***villosa*** (L.f.) A.Gray
Metrosideros viminalis Sol. ex Gaertn. = ***Callistemon viminalis*** (Sol. ex Gaertn.) G.Don ex Loudon
Metrosideros viridiflora Sieber ex Sims = ***Callistemon viridiflorus*** (Sieber ex Sims) Sweet

***Unplaced Names*:**
Metrosideros aspera Dum.Cours., Bot. Cult., ed. 2, 5: 381 (1811). = ?
Metrosideros canaliculata Dum.Cours., Bot. Cult., ed. 2, 5: 381 (1811). = ?
Metrosideros canaliculata Colla, Hortus Ripul., App. 3: 155 (1826). = ?
Metrosideros decurrens Steud., Nomencl. Bot., ed. 2, 2: 137 (1841). = ?
Metrosideros falcata Dehnh., Rivista Napol. 1(3): 172 (1839). = ?
Metrosideros macrophylla Lam., Tabl. Encycl. 2: 538 (1819). = [29 MDG] (?) not a Myrtaceae.
Metrosideros nana C.C.Gmel. ex DC., Prodr. 3: 226 (1828). = ?
Metrosideros ruscifolia Pépin, Ann. Fl. Pomone 1838–1839: 380 (1839). = ?
Metrosideros scariosa J.C.Wendl. ex Hornem., Hort. Bot. Hafn., Suppl.: 139 (1819). = ?

Microjambosa

Microjambosa Blume = ***Syzygium*** Gaertn.
Microjambosa bankensis Hassk. = ***Syzygium bankense*** (Hassk.) Merr. & L.M.Perry
Microjambosa besukiensis Hassk. ex Miq. = ***Syzygium bankense*** (Hassk.) Merr. & L.M.Perry
Microjambosa conferta (Korth.) Blume = ***Syzygium confertum*** (Korth.) Merr. & L.M.Perry
Microjambosa cuspidata Blume = ***Syzygium confertum*** (Korth.) Merr. & L.M.Perry
Microjambosa ferruginea Blume = ***Syzygium thumra*** (Roxb.) Merr. & L.M.Perry subsp. ***thumra***
Microjambosa reticulata Blume = ***Syzygium reticulatum*** (Wight) Walp.
Microjambosa splendens Blume = ***Syzygium splendens*** (Blume) Merr. & L.M.Perry
Microjambosa trifida Blume = ***Syzygium confertum*** (Korth.) Merr. & L.M.Perry

Micromyrtus

Micromyrtus Benth. in G.Bentham & J.D.Hooker, Gen. Pl. 1: 700 (1865).
Australia. 50 NSW NTA QLD SOA VIC WAU.
43 Species

Micromyrtus acuta Rye, Nuytsia 16: 126 (2006).
WC. Western Australia. 50 WAU. Nanophan.

Micromyrtus albicans A.R.Bean, Austrobaileya 4: 463 (1997).
SE. Queensland. 50 QLD. Nanophan.

Micromyrtus barbata J.W.Green, Nuytsia 3: 203 (1980).
C. Western Australia. 50 WAU. Cham. or nanophan.

Micromyrtus blakelyi J.W.Green, Nuytsia 4: 327 (1983).
NE. New South Wales. 50 NSW. Nanophan.

Micromyrtus capricornia A.R.Bean, Austrobaileya 4: 464 (1997).
E. Queensland. 50 QLD. Nanophan.

Micromyrtus carinata A.R.Bean, Austrobaileya 4: 464 (1997).
SE. Queensland. 50 QLD. Nanophan.

Micromyrtus chrysodema Rye, Nuytsia 16: 127 (2006).
SC. Western Australia. 50 WAU. Nanophan.

Micromyrtus ciliata (Sm.) Druce, Rep. Bot. Exch. Club Brit. Isles 1916: 636 (1917).
SE. Australia. 50 NSW SOA VIC. Nanophan.
**Imbricaria ciliata* Sm., Trans. Linn. Soc. London 3: 257 (1797). *Thryptomene ciliata* (Sm.) Woolls, Pl. Sydney: 23 (1880).
Baeckea gracilis A.Cunn., Field New South Wales: 349 (1825).

Baeckea microphylla Sieber ex Spreng., Syst. Veg. 4(2): 149 (1827). *Micromyrtus microphylla* (Sieber ex Spreng.) Benth., Fl. Austral. 3: 65 (1867).
Schidiomyrtus tenella Schauer, Linnaea 17: 237 (1843).
Schidiomyrtus ericacea F.Muell. ex Miq., Ned. Kruidk. Arch. 4: 148 (1856).
Schidiomyrtus tenella var. *ciliata* Miq., Ned. Kruidk. Arch. 4: 148 (1856).
Baeckea plicata F.Muell., Fragm. 1: 30 (1858). *Thryptomene plicata* (F.Muell.) F.Muell., Fragm. 4: 63 (1864).

Micromyrtus clavata J.W.Green ex Rye, Nuytsia 16: 128 (2006).
WC. & SC. Western Australia. 50 WAU. Cham. or nanophan.

Micromyrtus delicata A.R.Bean, Austrobaileya 4: 457 (1997).
N. Queensland. 50 QLD. Nanophan.

Micromyrtus elobata (F.Muell.) Benth., Fl. Austral. 3: 64 (1867).
S. Western Australia. 50 WAU. Nanophan.
**Thryptomene elobata* F.Muell., Fragm. 4: 63 (1864).

subsp. ***elobata***
S. Western Australia. 50 WAU. Cham. or nanophan.

subsp. ***scopula*** Rye, Nuytsia 16: 132 (2006).
S. Western Australia. 50 WAU. Cham. or nanophan.

Micromyrtus erichsenii Hemsl., Hooker's Icon. Pl. 28: t. 2780 (1905).
SW. Australia. 50 WAU. Nanophan.

Micromyrtus fimbrisepala J.W.Green, Nuytsia 3: 198 (1980).
C. Western Australia to W. South Australia. 50 SOA WAU. Cham. or nanophan.

Micromyrtus flaviflora (F.Muell.) J.M.Black, Fl. S. Austral. 3: 424 (1926).
C. Australia. 50 NTA SOA WAU. Cham. or nanophan.
**Thryptomene flaviflora* F.Muell., Fragm. 8: 13 (1874).
Thryptomene trachycalyx F.Muell., Fragm. 10: 25 (1876). *Micromyrtus trachycalyx* (F.Muell.) C.A. Gardner, Enum. Pl. Austr. Occ.: 96 (1931).

Micromyrtus forsteri A.R.Bean, Austrobaileya 4: 471 (1997).
N. Queensland. 50 QLD. Nanophan.

Micromyrtus gracilis A.R.Bean, Austrobaileya 4: 466 (1997).
EC. Queensland. 50 QLD. Nanophan.

Micromyrtus grandis J.T.Hunter, Telopea 7: 77 (1996).
New South Wales (NW. Slopes). 50 NSW. Nanophan.

Micromyrtus helmsii (F.Muell. & Tate) J.W.Green, Nuytsia 3: 200 (1980).
E. Western Australia. 50 WAU. Nanophan.
**Thryptomene helmsii* F.Muell. & Tate, Trans. & Proc. Roy. Soc. South Australia 16: 356 (1896).

Micromyrtus hexamera (Maiden & Betche) Maiden & Betche, Census N.S.W. Pl.: 157 (1916).
S. Queensland to New South Wales. 50 NSW QLD. Nanophan.
**Thryptomene hexamera* Maiden & Betche, Proc. Linn. Soc. New South Wales 26: 82 (1901).

Micromyrtus hymenonema (F.Muell.) C.A.Gardner, Enum. Pl. Austr. Occ.: 96 (1931).
C. Western Australia. 50 WAU. Cham. or nanophan.
**Thryptomene hymenonema* F.Muell., Fragm. 10: 26 (1876).

Micromyrtus imbricata Benth., Fl. Austral. 3: 64 (1867).
Thryptomene imbricata (Benth.) F.Muell., Fragm. 8: 13 (1873).
SSW. Western Australia. 50 WAU. Cham. or nanophan.

Micromyrtus leptocalyx (F.Muell.) Benth., Fl. Austral. 3: 65 (1867).
EC. Queensland. 50 QLD. Nanophan.
**Baeckea leptocalyx* F.Muell., Fragm. 1: 30 (1858). *Thryptomene leptocalyx* (F.Muell.) F.Muell., Syst. Census Austral. Pl. 1: 53 (1882).

Micromyrtus littoralis A.R.Bean, Austrobaileya 4: 467 (1997).
SE. Queensland. 50 QLD. Nanophan.

Micromyrtus minutiflora Benth., Fl. Austral. 3: 65 (1867). *Thryptomene minutiflora* (Benth.) F.Muell. ex Woolls, Pl. Sydney: 23 (1880).
New South Wales. 50 NSW. Nanophan.

Micromyrtus monotaxis Rye, Nuytsia 15: 111 (2002).
SW. Australia. 50 WAU. Cham. or nanophan.

Micromyrtus navicularis Rye, Nuytsia 16: 133 (2006).
S. Western Australia. 50 WAU. Nanophan.

Micromyrtus ninghanensis Rye, Nuytsia 15: 113 (2002).
WSW. Western Australia. 50 WAU. Cham.

Micromyrtus obovata (Turcz.) J.W.Green, Census Vasc. Pl. W. Australia, ed. 2: 6 (1985).
SW. Australia. 50 WAU. Nanophan.
**Thryptomene obovata* Turcz., Bull. Cl. Phys.-Math. Acad. Imp. Sci. Saint-Pétersbourg 10: 322 (1852).
Micromyrtus drummondii Benth., Fl. Austral. 3: 64 (1867), nom. illeg. *Thryptomene drummondii* (Benth.) F.Muell., Syst. Census Austral. Pl. 1: 53 (1882).

Micromyrtus papillosa J.W.Green ex Rye, Nuytsia 15: 116 (2002).
SSW. Western Australia. 50 WAU. Cham. or nanophan.

Micromyrtus patula A.R.Bean, Austrobaileya 4: 458 (1997).
SE. Queensland. 50 QLD. Nanophan.

Micromyrtus placoides Rye, Nuytsia 16: 134 (2006).
WC. Western Australia. 50 WAU. Nanophan.

Micromyrtus racemosa Benth., Fl. Austral. 3: 64 (1867).
Thryptomene racemosa (Benth.) F.Muell., Syst. Census Austral. Pl. 1: 53 (1882).
SW. Australia. 50 WAU. Nanophan.

Micromyrtus redita Rye, Nuytsia 16: 136 (2006).
WSW. Western Australia. 50 WAU. Nanophan.

Micromyrtus rogeri J.W.Green ex Rye, Nuytsia 15: 117 (2002).
WSW. Western Australia. 50 WAU. Cham.

Micromyrtus rotundifolia A.R.Bean, Austrobaileya 4: 460 (1997).
EC. Queensland. 50 QLD. Nanophan.

Micromyrtus serrulata J.W.Green, Nuytsia 3: 200 (1980).
SC. Western Australia. 50 WAU. Cham. or nanophan.

Micromyrtus sessilis J.W.Green, Nuytsia 4: 322 (1983).
SE. Queensland to New South Wales. 50 NSW QLD. Nanophan.

Micromyrtus stenocalyx (F.Muell.) J.W.Green, Nuytsia 3: 201 (1980).
SC. Western Australia. 50 WAU. Nanophan.
**Thryptomene stenocalyx* F.Muell., Fragm. 10: 23 (1876).

Micromyrtus striata J.W.Green, Nuytsia 4: 324 (1983).
S. Queensland to New South Wales. 50 NSW QLD. Nanophan.

Micromyrtus sulphurea W.Fitzg., J. West Austral. Nat. Hist. Soc. 1: 19 (1904).
WC. Western Australia. 50 WAU. Cham.

Micromyrtus triptycha Rye, Nuytsia 16: 143 (2006).
SW. Western Australia. 50 WAU. Cham.

Micromyrtus uniovula Rye, Nuytsia 15: 118 (2002).
WSW. Western Australia. 50 WAU. Cham.

Micromyrtus vernicosa A.R.Bean, Austrobaileya 4: 469 (1997).
Queensland. 50 QLD. Nanophan.

***Synonyms*:**
Micromyrtus drummondii Benth. = ***Micromyrtus obovata*** (Turcz.) J.W.Green
Micromyrtus hursthousei W.Fitzg. = ***Malleostemon hursthousei*** (W.Fitzg.) J.W.Green
Micromyrtus microphylla (Sieber ex Spreng.) Benth. = ***Micromyrtus ciliata*** (Sm.) Druce
Micromyrtus peltigera S.Moore = ***Malleostemon peltiger*** (S.Moore) J.W.Green
Micromyrtus rosea S.Moore = ***Malleostemon roseus*** (E.Pritz.) J.W.Green
Micromyrtus trachycalyx (F.Muell.) C.A.Gardner = ***Micromyrtus flaviflora*** (F.Muell.) J.M.Black

Microsideros

Microsideros Baum.-Bod. = ***Metrosideros*** Banks ex Gaertn.

Mitranthes

Mitranthes O.Berg, Linnaea 27: 316 (1856).
Cuba, Jamaica. 81 CUB JAM.
7 Species

Mitranthes clarendonensis (Proctor) Proctor, J. Arnold Arbor. 63: 280 (1982).
C. Jamaica. 81 JAM. Nanophan.
**Calyptranthes clarendonensis* Proctor, Rhodora 60: 323 (1959).

Mitranthes glabra Proctor, J. Arnold Arbor. 63: 280 (1982).
Jamaica. 81 JAM. Nanophan.

Mitranthes macrophylla Proctor, J. Arnold Arbor. 63: 280 (1982).
Jamaica. 81 JAM. Nanophan. or phan.

Mitranthes maxonii (Britton & Urb.) Proctor, J. Arnold Arbor. 63: 278 (1982).
NW. Jamaica. 81 JAM. Nanophan.
**Calyptranthes maxonii* Britton & Urb. in I.Urban, Symb. Antill. 7: 296 (1912).

Mitranthes nivea Proctor, J. Arnold Arbor. 63: 278 (1982).
Jamaica. 81 JAM. Nanophan.

Mitranthes ottonis O.Berg, Linnaea 27: 316 (1856). *Calyptranthes ottonis* (O.Berg) C.Wright, Anales Acad. Ci. Méd. Habana 5: 429 (1868). *Chytraculia ottonis* (O.Berg) Kuntze, Revis. Gen. Pl. 1: 238 (1891).
W. & C. Cuba. 81 CUB. Nanophan.

Mitranthes urbaniana Burret, Notizbl. Bot. Gart. Berlin-Dahlem 15: 537 (1941).
Jamaica. 81 JAM. Nanophan.

***Synonyms*:**
Mitranthes amblymitra Burret = ***Neomitranthes amblymitra*** (Burret) Mattos
Mitranthes apiculata Burret = ***Neomitranthes apiculata*** (Burret) Mattos
Mitranthes browniana (Mart. ex DC.) O.Berg = ***Psidium brownianum*** Mart. ex DC.
Mitranthes capivariensis Mattos = ***Neomitranthes capivariensis*** (Mattos) Mattos
Mitranthes castellanosii Mattos & D.Legrand = ***Neomitranthes castellanosii*** (Mattos & D.Legrand) Mattos
Mitranthes cordifolia D.Legrand = ***Neomitranthes cordifolia*** (D.Legrand) D.Legrand
Mitranthes dusenii Kausel = ***Neomitranthes hoehnei*** (Burret) Mattos
Mitranthes eggersii (Kiaersk.) Nied. = ***Blepharocalyx eggersii*** (Kiaersk.) Landrum
Mitranthes eugenioides (Cambess.) O.Berg = ***Calyptranthes eugenioides*** Cambess.
Mitranthes eugenioides var. *oblongifolia* O.Berg = ***Calyptranthes eugenioides*** Cambess.
Mitranthes eugenioides var. *ovata* O.Berg = ***Calyptranthes eugenioides*** Cambess.
Mitranthes gardneriana O.Berg = [84 BZE]
Mitranthes gemballae D.Legrand = ***Neomitranthes gemballae*** (D.Legrand) D.Legrand
Mitranthes glomerata D.Legrand = ***Neomitranthes glomerata*** (D.Legrand) Govaerts
Mitranthes gracilis Burret = ***Neomitranthes gracilis*** (Burret) N.Silveira
Mitranthes hoehnei Burret = ***Neomitranthes hoehnei*** (Burret) Mattos
Mitranthes langsdorffii O.Berg = ***Neomitranthes langsdorffii*** (O.Berg) Mattos
Mitranthes maria-aemiliae D.Legrand = ***Myrceugenia ovalifolia*** (O.Berg) Landrum
Mitranthes obscura (DC.) D.Legrand = ***Neomitranthes obscura*** (DC.) N.Silveira
Mitranthes ovalifolia O.Berg = ***Myrceugenia ovalifolia*** (O.Berg) Landrum
Mitranthes pedicellata Burret = ***Neomitranthes pedicellata*** (Burret) Mattos
Mitranthes pilosa Burret = ***Siphoneugena kiaerskoviana*** (Burret) Kausel
Mitranthes pubescens Burret = ***Siphoneugena widgreniana*** O.Berg
Mitranthes riedeliana O.Berg = ***Neomitranthes riedeliana*** (O.Berg) Mattos
Mitranthes sartoriana O.Berg = ***Psidium sartorianum*** (O.Berg) Nied.
Mitranthes warmingiana (Kiaersk.) Burret = ***Neomitranthes warmingiana*** (Kiaersk.) Mattos
Mitranthes widgreniana (O.Berg) Burret = ***Siphoneugena widgreniana*** O.Berg

***Unplaced Names*:**
Mitranthes gardneriana O.Berg in C.F.P.von Martius & auct. suc. (eds.), Fl. Bras. 14(1): 354 (1857). = [84 BZE]

Mitrantia

Mitrantia Peter G.Wilson & B.Hyland, Telopea 3: 264 (1988).
NE. Australia. 50 QLD.
1 Species

Mitrantia bilocularis Peter G.Wilson & B.Hyland, Telopea 3: 265 (1988).
N. Queensland. 50 QLD. Phan.

Mitropsidium [*Myrtaceae*]

Mitropsidium Burret = ***Psidium*** L.
Mitropsidium brownianum (Mart. ex DC.) Burret = ***Psidium brownianum*** Mart. ex DC.
Mitropsidium eugenioides (Cambess.) Burret = ***Calyptranthes eugenioides*** Cambess.
Mitropsidium gardnerianum (O.Berg) Burret = ?
Mitropsidium oblanceolatum Burret = ***Psidium sartorianum*** (O.Berg) Nied.
Mitropsidium oligospermum (Mart. ex DC.) Burret = ***Psidium oligospermum*** Mart. ex DC.
Mitropsidium ovalifolium (O.Berg) Burret = ***Myrceugenia ovalifolia*** (O.Berg) Landrum
Mitropsidium pittieri Burret = ***Psidium sartorianum*** (O.Berg) Nied.
Mitropsidium sartorianum (O.Berg) Burret = ***Psidium sartorianum*** (O.Berg) Nied.
Mitropsidium sintenisii (Kiaersk.) Burret = ***Psidium sintenisii*** (Kiaersk.) Alain

Mollia

Mollia J.F.Gmel. = ***Baeckea*** L.
Mollia imbricata (Gaertn.) J.F.Gmel. = ***Baeckea imbricata*** (Gaertn.) Druce

Monimiastrum

Monimiastrum J.Guého & A.J.Scott, Kew Bull. 34: 483 (1980).
Mauritius. 29 MAU.
5 Species

Monimiastrum acutisepalum J.Guého & A.J.Scott, Kew Bull. 34: 485 (1980).
Mauritius. 29 MAU. Nanophan.

Monimiastrum fasciculatum J.Guého & A.J.Scott, Kew Bull. 34: 486 (1980).
Mauritius. 29 MAU†. Nanophan. or phan. Last coll. in 1864

Monimiastrum globosum J.Guého & A.J.Scott, Kew Bull. 34: 486 (1980).
Mauritius. 29 MAU. Nanophan. or phan.

Monimiastrum psidiodeum J.Guého & A.J.Scott, Kew Bull. 34: 487 (1980).
Mauritius. 29 MAU. Nanophan.

Monimiastrum pyxidatum J.Guého & A.J.Scott, Kew Bull. 34: 488 (1980).
Mauritius. 29 MAU. Nanophan. or phan.

Monoxora

Monoxora Wight = ***Rhodamnia*** Jack
Monoxora latifolia Benth. = ***Rhodamnia latifolia*** (Benth.) Miq.
Monoxora rubescens Benth. = ***Rhodamnia rubescens*** (Benth.) Miq.
Monoxora spectabilis (Blume) Wight = ***Rhodamnia cinerea*** Jack

Mooria

Mooria Montrouz. = ***Cloezia*** Brongn. & Gris
Mooria angustifolia (Baker f.) Guillaumin = ***Cloezia artensis*** (Montrouz.) P.S.Green var. ***artensis***
Mooria aquarum Guillaumin = ***Cloezia aquarum*** (Guillaumin) J.W.Dawson
Mooria artensis Montrouz. = ***Cloezia artensis*** (Montrouz.) P.S.Green
Mooria artensis var. *angustifolia* (Brongn. & Gris) Guillaumin = ***Cloezia floribunda*** Brongn. & Gris
Mooria artensis var. *latifolia* (Brongn. & Gris) Guillaumin = ***Cloezia floribunda*** Brongn. & Gris
Mooria buxifolia (Brongn. & Gris) Guillaumin = ***Cloezia buxifolia*** Brongn. & Gris
Mooria canescens (Brongn. & Gris) Beauvis. ex Guillaumin = ***Cloezia artensis*** (Montrouz.) P.S.Green var. ***artensis***
Mooria canescens f. *parvifolia* Däniker = ***Cloezia artensis*** (Montrouz.) P.S.Green var. ***artensis***
Mooria deplanchei (Brongn. & Gris) Guillaumin = ***Cloezia deplanchei*** Brongn. & Gris
Mooria floribunda (Brongn. & Gris) Guillaumin = ***Cloezia floribunda*** Brongn. & Gris
Mooria × *glaberrima* Guillaumin = ***Cloezia*** × ***glaberrima*** (Guillaumin) J.W.Dawson
Mooria microphylla A.C.Sm. = ***Decaspermum cryptanthum*** A.J.Scott
Mooria sessilifolia (Brongn. & Gris) Guillaumin = ***Cloezia floribunda*** Brongn. & Gris
Mooria streptophylla Guillaumin = ***Cloezia aquarum*** (Guillaumin) J.W.Dawson
Mooria urdanetensis (Elmer) Merr. = ***Kania urdanetensis*** (Elmer) Peter G.Wilson

Mosiera

Mosiera Small, Man. S.E. Fl.: 936 (1933).
S. Florida, Mexico to Guatemala, Caribbean, S. Brasil. 78 FLA 79 MXE MXS? MXT 80 GUA 81 BAH CUB DOM HAI LEE NLA PUE TCI 84 BZS.
24 Species

Mosiera acunae (Borhidi & O.Muñiz) Bisse, Revista Jard. Bot. Nac. Univ. Habana 6(3): 5 (1985 publ. 1986).
Cuba. 81 CUB. Nanophan.
**Myrtus acunae* Borhidi & O.Muñiz, Acta Bot. Acad. Sci. Hung. 21: 227 (1975 publ. 1976).

Mosiera araneosa (Urb.) Bisse, Revista Jard. Bot. Nac. Univ. Habana 6(3): 5 (1985 publ. 1986).
SE. Cuba. 81 CUB. Nanophan.
**Psidium araneosum* Urb., Repert. Spec. Nov. Regni Veg. 19: 304 (1924). *Myrtus araneosa* (Urb.) Bisse, Ciencias (Havana), Ser. 10, 12: 10 (1976 publ. 1977).

Mosiera bullata (Britton & P.Wilson) Bisse, Revista Jard. Bot. Nac. Univ. Habana 6(3): 5 (1985 publ. 1986).
C. & E. Cuba. 81 CUB. Nanophan. or phan.
**Psidium bullatum* Britton & P.Wilson, Mem. Torrey Bot. Club 16: 85 (1920). *Myrtus anomala* (Britton & P.Wilson) Burret, Notizbl. Bot. Gart. Berlin-Dahlem 15: 480 (1941). *Myrtus bullata* (Britton & P.Wilson) Bisse, Ciencias (Havana), Ser. 10, 12: 10 (1976 publ. 1977), nom. illeg.

subsp. ***bullata***
C. & E. Cuba. 81 CUB. Nanophan. or phan.

subsp. ***leiophloea*** (Urb.) Bisse, Revista Jard. Bot. Nac. Univ. Habana 6(3): 5 (1985 publ. 1986).
Cuba (Sierra de Nipe). 81 CUB. Nanophan.
**Eugenia leiophloea* Urb., Symb. Antill. 9: 108 (1923). *Psidium leiophloeum* (Urb.) Urb., Symb.

Antill. 9: 459 (1928). *Myrtus leiophloea* (Urb.) Bisse, Ciencias (Havana), Ser. 10, 12: 10 (1976 publ. 1977).

Mosiera cabanasensis (Britton & P.Wilson) Borhidi, Acta Bot. Hung. 37: 78 (1992).
EC. & E. Cuba. 81 CUB. Nanophan. or phan.
**Eugenia cabanasensis* Britton & P.Wilson, Mem. Torrey Bot. Club 16: 88 (1920). *Myrtus cabanasensis* (Britton & P.Wilson) Alain, Brittonia 20: 159 (1968). *Mosiera flavicans* subsp. *cabanasensis* (Britton & P.Wilson) Bisse, Revista Jard. Bot. Nac. Univ. Habana 6(3): 5 (1985 publ. 1986).

subsp. ***cabanasensis***
EC. & E. Cuba. 81 CUB. Nanophan. or phan.
Eugenia buxoides Urb., Symb. Antill. 9: 99 (1923). *Myrtus buxoides* (Urb.) Burret, Notizbl. Bot. Gart. Berlin-Dahlem 15: 483 (1941). *Mosiera buxoides* (Urb.) Bisse, Revista Jard. Bot. Nac. Univ. Habana 6(3): 4 (1985 publ. 1986).

subsp. ***flavicans*** (Urb. & Ekman) Borhidi, Acta Bot. Hung. 37: 78 (1992).
Cuba (Sierra de Nipe). 81 CUB. Nanophan. or phan.
**Eugenia flavicans* Urb. & Ekman in I.Urban, Symb. Antill. 9: 488 (1928). *Mosiera flavicans* (Urb. & Ekman) Bisse, Revista Jard. Bot. Nac. Univ. Habana 6(3): 5 (1985 publ. 1986).

subsp. ***pastelillensis*** (Urb.) Borhidi, Acta Bot. Hung. 37: 78 (1992).
E. Cuba. 81 CUB. Nanophan. or phan.
**Eugenia pastelillensis* Urb., Symb. Antill. 9: 510 (1928). *Mosiera flavicans* subsp. *pastelillensis* (Urb.) Bisse, Revista Jard. Bot. Nac. Univ. Habana 6(3): 5 (1985 publ. 1986).

Mosiera calycolpoides (Griseb.) Borhidi, Acta Bot. Hung. 37: 78 (1992).
E. Cuba. 81 CUB. Nanophan.
**Psidium calycolpoides* Griseb., Mem. Amer. Acad. Arts, n.s., 8: 183 (1861). *Myrtus calycolpoides* (Griseb.) Burret, Notizbl. Bot. Gart. Berlin-Dahlem 15: 481 (1941).
Psidium versicolor Urb., Symb. Antill. 9: 85 (1923).

Mosiera contrerasii (Lundell) Landrum, Novon 2: 27 (1992).
SE. Mexico to Guatemala. 79 MXT 80 GUA. Phan.
**Psidium contrerasii* Lundell, Wrightia 5: 84 (1975).

Mosiera crenulata (Urb. & Ekman) Borhidi, Acta Bot. Hung. 37: 78 (1992).
EC. Cuba. 81 CUB. Nanophan.
**Psidium crenulatum* Urb. & Ekman in I.Urban, Symb. Antill. 9: 460 (1928). *Myrtus crenulata* (Urb. & Ekman) Bisse, Ciencias (Havana), Ser. 10, 12: 10 (1976 publ. 1977), nom. illeg.

Mosiera del-riscoi (Borhidi & O.Muñiz) Borhidi, Acta Bot. Hung. 37: 79 (1992).
Cuba. 81 CUB. Nanophan.
**Myrtus del-riscoi* Borhidi & O.Muñiz, Bot. Közlem. 64: 219 (1977 publ. 1978).

Mosiera ehrenbergii (O.Berg) Landrum, Novon 2: 28 (1992).
Mexico. 79 MXE MXS?. Nanophan. or phan.
**Myrtus ehrenbergii* O.Berg, Linnaea 27: 404 (1856). *Psidium ehrenbergii* (O.Berg) Burret, Notizbl. Bot. Gart. Berlin-Dahlem 15: 483 (1941).

Mosiera ekmanii (Urb.) Bisse, Revista Jard. Bot. Nac. Univ. Habana 6(3): 4 (1985 publ. 1986).
Cuba (Sierra de Nipe). 81 CUB. Nanophan.
**Myrtus ekmanii* Urb., Symb. Antill. 9: 79 (1923).

Mosiera elliptica (C.Wright) Bisse, Revista Jard. Bot. Nac. Univ. Habana 6(3): 5 (1985 publ. 1986).
C. & E. Cuba. 81 CUB. Nanophan.
**Myrtus elliptica* C.Wright, Anales Acad. Ci. Méd. Habana 5: 410 (1868), nom. illeg.

subsp. ***elliptica***
C. & E. Cuba. 81 CUB. Nanophan.
Psidium saxicola Britton & P.Wilson, Mem. Torrey Bot. Club 16: 86 (1920).

subsp. ***matanzasia*** (Urb.) Bisse, Revista Jard. Bot. Nac. Univ. Habana 6(3): 5 (1985 publ. 1986).
WC. Cuba. 81 CUB. Nanophan.
**Myrtus matanzasia* Urb., Symb. Antill. 9: 459 (1928).

Mosiera havanensis (Urb.) Bisse, Revista Jard. Bot. Nac. Univ. Habana 6(3): 4 (1985 publ. 1986).
W. Cuba. 81 CUB. Nanophan. or phan.
**Psidium havanense* Urb., Symb. Antill. 9: 461 (1928).

Mosiera jackii (Urb.) Bisse, Revista Jard. Bot. Nac. Univ. Habana 6(3): 5 (1985 publ. 1986).
SE. Cuba. 81 CUB. Nanophan. or phan.
**Psidium jackii* Urb., Symb. Antill. 9: 467 (1928). *Myrtus calycolpoides* subsp. *jackii* (Urb.) Borhidi, Bot. Közlem. 64: 218 (1977 publ. 1978).

Mosiera longipes (O.Berg) Small, Man. S.E. Fl.: 937 (1933).
S. Florida, Caribbean. 78 FLA 81 BAH HAI LEE NLA PUE TCI. Nanophan. or phan.
**Eugenia longipes* O.Berg, Linnaea 27: 150 (1856). *Anamomis longipes* (O.Berg) Britton ex Small, Fl. Miami: 132 (1913). *Psidium longipes* (O.Berg) McVaugh, J. Arnold Arbor. 54: 312 (1973).
Myrtus verrucosa O.Berg, Linnaea 27: 405 (1856).
Eugenia orbicularis O.Berg, Linnaea 30: 678 (1861). *Myrtus orbicularis* (O.Berg) Burret, Notizbl. Bot. Gart. Berlin-Dahlem 15: 482 (1941). *Psidium longipes* var. *orbiculare* (O.Berg) McVaugh, J. Arnold Arbor. 54: 314 (1973).
Eugenia bahamensis Kiaersk., Bot. Tidsskr. 17: 266 (1890). *Anamomis bahamensis* (Kiaersk.) Britton ex Small, Fl. Florida Keys: 104 (1913). *Myrtus bahamensis* (Kiaersk.) Urb., Ark. Bot. 21A(5): 18 (1927). *Mosiera bahamensis* (Kiaersk.) Small, Man. S.E. Fl.: 937 (1933).
Psidium insulanum Alain, Phytologia 47: 187 (1980).

Mosiera × miraflorensis (Borhidi & O.Muñiz) Borhidi, Acta Bot. Hung. 37: 79 (1992). *M. ekmannii* × *M. ophiticola*.
E. Cuba. 81 CUB. Nanophan. or phan.
**Myrtus* × *miraflorensis* Borhidi & O.Muñiz, Acta Bot. Acad. Sci. Hung. 21: 227 (1975 publ. 1976).

Mosiera moaensis (R.A.Howard) Bisse, Revista Jard. Bot. Nac. Univ. Habana 6(3): 4 (1985 publ. 1986).
Cuba (Sierra de Moa). 81 CUB. Nanophan.
**Eugenia moaensis* R.A.Howard, J. Arnold Arbor. 28: 122 (1947).

Mosiera munizii (Borhidi) Bisse, Revista Jard. Bot. Nac. Univ. Habana 6(3): 5 (1985 publ. 1986).
Cuba. 81 CUB. Nanophan. or phan.
**Myrtus munizii* Borhidi, Bot. Közlem. 64: 218 (1977 publ. 1978).

Mosiera nummularioides (Britton & P.Wilson) Bisse, Revista Jard. Bot. Nac. Univ. Habana 6(3): 5 (1985 publ. 1986).
E. Cuba. 81 CUB. Nanophan.
**Psidium nummularioides* Britton & P.Wilson, Mem. Torrey Bot. Club 16: 85 (1920). *Myrtus nummularioides* (Britton & P.Wilson) Urb., Ark. Bot. 24A(4): 16 (1932).

Mosiera oonophylla (Urb.) Bisse, Revista Jard. Bot. Nac. Univ. Habana 6(3): 4 (1985 publ. 1986).
SE. Cuba. 81 CUB. Nanophan.
**Eugenia oonophylla* Urb., Symb. Antill. 9: 105 (1923). *Myrtus oonophylla* (Urb.) Burret, Notizbl. Bot. Gart. Berlin-Dahlem 15: 483 (1941).

Mosiera ophiticola (Britton & P.Wilson) Bisse, Revista Jard. Bot. Nac. Univ. Habana 6(3): 5 (1985 publ. 1986).
Cuba (Sierra de Moa). 81 CUB. Nanophan.
**Psidium ophiticola* Britton & P.Wilson, Mem. Torrey Bot. Club 16: 86 (1920). *Myrtus ophiticola* (Britton & P.Wilson) Alain, Contr. Ocas. Mus. Hist. Nat. Colegio "De La Salle" 12: 10 (1953).
Myrtus tripliphylla Urb., Symb. Antill. 9: 80 (1923).
Psidium confertum R.A.Howard, J. Arnold Arbor. 28: 123 (1947).

Mosiera prismatica (D.Legrand) Landrum, Brittonia 49: 522 (1997).
S. Brazil. 84 BZS. Nanophan. or phan.
**Eugenia prismatica* D.Legrand, in Fl. Ilustr. Catar. 1(Mirt.): 205 (1969).

Mosiera tiburona (Urb. & Ekman) Borhidi, Acta Bot. Hung. 37: 79 (1992).
Haiti (Massif de la Hotte). 81 HAI. Nanophan. or phan.
**Eugenia tiburona* Urb. & Ekman, Ark. Bot. 14A(4): 26 (1932). *Myrtus tiburona* (Urb. & Ekman) Borhidi, Acta Bot. Hung. 29: 186 (1983).

Mosiera tussacii (Urb. & Ekman) Borhidi, Acta Bot. Hung. 37: 79 (1992).
Haiti. 81 HAI. Nanophan. or phan.
**Eugenia tussacii* Urb. & Ekman, Ark. Bot. 21A(5): 37 (1927). *Myrtus tussacii* (Urb. & Ekman) Burret, Notizbl. Bot. Gart. Berlin-Dahlem 15: 483 (1941).

Mosiera urbaniana Borhidi, Acta Bot. Hung. 37: 79 (1992).
Dominican Rep. 81 DOM. Nanophan.
**Myrtus flavicans* Urb. & Ekman, Ark. Bot. 24A(4): 15 (1932).

Mosiera wrightii (Krug & Urb.) Borhidi, Acta Bot. Hung. 37: 79 (1992).
E. Cuba. 81 CUB. Nanophan.
**Psidium wrightii* Krug & Urb., Bot. Jahrb. Syst. 19: 570 (1894), nom. illeg.

***Synonyms*:**
Mosiera bahamensis (Kiaersk.) Small = ***Mosiera longipes*** (O.Berg) Small
Mosiera buxoides (Urb.) Bisse = ***Mosiera cabanasensis*** (Britton & P.Wilson) Borhidi subsp. ***cabanasensis***
Mosiera flavicans (Urb. & Ekman) Bisse = ***Mosiera cabanasensis*** subsp. ***flavicans*** (Urb. & Ekman) Borhidi
Mosiera flavicans subsp. *cabanasensis* (Britton & P. Wilson) Bisse = ***Mosiera cabanasensis*** (Britton & P.Wilson) Borhidi
Mosiera flavicans subsp. *pastelillensis* (Urb.) Bisse = ***Mosiera cabanasensis*** subsp. ***pastelillensis*** (Urb.) Borhidi
Mosiera guineensis (Sw.) Bisse = ***Psidium guineense*** Sw.
Mosiera sagraea (O.Berg) Bisse = ***Psidium salutare*** var. ***salutare***

Mozartia

Mozartia Urb. = ***Myrcia*** DC. ex Guill.
Mozartia abbottiana (Urb.) Urb. = ***Myrcia abbotiana*** (Urb.) Alain
Mozartia albescens Alain = ***Myrcia albescens*** (Alain) Alain
Mozartia emarginata Moldenke = ***Pimenta oligantha*** (Urb.) Burret
Mozartia gundlachii (Krug & Urb.) Urb. = ***Myrcia gundlachii*** Krug & Urb.
Mozartia maestrensis Urb. = ***Myrcia maestrensis*** (Urb.) Alain
Mozartia manacalensis (Urb.) Urb. = ***Myrcia manacalensis*** Urb.
Mozartia oligostemon Urb. = ***Myrcia oligostemon*** (Urb.) Alain

Murrinea

Murrinea Raf. = ***Baeckea*** L.

Myrceugenella

Myrceugenella Kausel = ***Luma*** A.Gray
Myrceugenella apiculata (DC.) Kausel = ***Luma apiculata*** (DC.) Burret
Myrceugenella apiculata var. *australis* Kausel = ***Luma apiculata*** (DC.) Burret
Myrceugenella apiculata var. *genuina* Kausel = ***Luma apiculata*** (DC.) Burret
Myrceugenella apiculata var. *nahulhuapensis* Kausel = ***Luma apiculata*** (DC.) Burret
Myrceugenella apiculata var. *spectabilis* (Phil.) Kausel = ***Luma apiculata*** (DC.) Burret
Myrceugenella chequen (Molina) Kausel = ***Luma chequen*** (Molina) A.Gray
Myrceugenella chequen var. *myrtomimeta* (Diels) Kausel = ***Luma chequen*** (Molina) A.Gray
Myrceugenella gayana (Barnéoud) Kausel = ***Luma chequen*** (Molina) A.Gray
Myrceugenella grandjotii Kausel = ***Luma apiculata*** (DC.) Burret
Myrceugenella langerfeldtii Kausel = ***Luma chequen*** (Molina) A.Gray

Myrceugenia

Myrceugenia O.Berg, Linnaea 27: 131 (1856).
Brazil to S. South America. 84 BZC BZE BZL BZS 85 AGE AGS CLC CLS JNF PAR URU.
44 Species
Nothomyrcia Kausel, Lilloa 13: 147 (1947).

Myrceugenia acutiflora (Kiaersk.) D.Legrand & Kausel, Comun. Bot. Mus. Hist. Nat. Montevideo 2(28): 5 (1953).
SE. & S. Brazil. 84 BZL BZS. Phan.
**Eugenia acutiflora* Kiaersk., Enum. Myrt. Bras.: 164 (1893). *Luma acutiflora* (Kiaersk.) Burret, Notizbl. Bot. Gart. Berlin-Dahlem 15: 518 (1941).

Myrceugenia alpigena (DC.) Landrum, Brittonia 32: 372 (1980).
Brazil. 84 BZC BZE BZL BZS. Nanophan. or phan.
**Eugenia alpigena* DC., Prodr. 3: 265 (1828). *Myrtus alpigena* (DC.) Mart., Prodr. 3: 265 (1828). *Luma alpigena* (DC.) Burret, Notizbl. Bot. Gart. Berlin-

Dahlem 15: 535 (1941). *Myrceugenia bracteosa* var. *alpigena* (DC.) D.Legrand, Comun. Bot. Mus. Hist. Nat. Montevideo 2(28): 7 (1953).

var. ***alpigena***
Brazil. 84 BZC BZE BZL BZS. Nanophan. or phan.
Eugenia fuliginea O.Berg in C.F.P.von Martius & auct. suc. (eds.), Fl. Bras. 14(1): 233 (1857). *Luma fuliginea* (O.Berg) Burret, Notizbl. Bot. Gart. Berlin-Dahlem 15: 528 (1941). *Myrceugenia bracteosa* var. *fuliginea* (O.Berg) D.Legrand, Comun. Bot. Mus. Hist. Nat. Montevideo 2(28): 7 (1953). *Myrceugenia alpigena* var. *fuliginea* (O.Berg) Landrum, Brittonia 43: 199 (1991).
Eugenia fuliginea var. *rufa* O.Berg in C.F.P.von Martius & auct. suc. (eds.), Fl. Bras. 14(1): 571 (1859). *Myrceugenia alpigena* var. *rufa* (O.Berg) Landrum, Brittonia 32: 372 (1980).
Eugenia rupestris O.Berg in C.F.P.von Martius & auct. suc. (eds.), Fl. Bras. 14(1): 570 (1859). *Luma rupestris* (O.Berg) Burret, Notizbl. Bot. Gart. Berlin-Dahlem 15: 531 (1941).
Psidium itatiaiae Wawra, Itin. Princ. S. Coburgi 1: 20 (1883).
Eugenia seriatopedunculata Kiaersk., Enum. Myrt. Bras.: 170 (1893). *Luma seriatopedunculata* (Kiaersk.) Burret, Notizbl. Bot. Gart. Berlin-Dahlem 15: 527 (1941). *Myrceugenia bracteosa* var. *seriatopedunculata* (Kiaersk.) D.Legrand, Comun. Bot. Mus. Hist. Nat. Montevideo 2(28): 7 (1953).
Eugenia sticheromischa Kiaersk., Enum. Myrt. Bras.: 172 (1893). *Luma sticheromischa* (Kiaersk.) Burret, Notizbl. Bot. Gart. Berlin-Dahlem 15: 526 (1941).
Myrceugenia leptorhyncha D.Legrand, in Fl. Ilustr. Catar. 1(Mirt.): 392 (1970).
Myrceugenia bracteosa var. *guaratubensis* Mattos, Loefgrenia 66: 6 (1975).
Myrceugenia catharinae D.Legrand, in Fl. Ilustr. Catar. 1(Mirt., Suppl. 1): 30 (1977).

var. ***longifolia*** (Burret) Landrum, Brittonia 32: 372 (1980).
Brazil (Goiás). 84 BZC. Nanophan. or phan.
**Luma longifolia* Burret, Notizbl. Bot. Gart. Berlin-Dahlem 15: 532 (1941). *Myrceugenia longifolia* (Burret) D.Legrand & Kausel, Comun. Bot. Mus. Hist. Nat. Montevideo 2(28): 5 (1953).

var. ***virgata*** (Gardner) Proença, Bradea 8: 312 (2002).
Brazil (Rio de Janeiro). 84 BZL. Nanophan. or phan.
**Eugenia virgata* Gardner, London J. Bot. 4: 102 (1845).

Myrceugenia bocaiuvensis Mattos, Loefgrenia 66: 4 (1975).
Brazil (Paraná). 84 BZS. Nanophan. or phan. Provisionally accepted.

Myrceugenia bracteosa (DC.) D.Legrand & Kausel, Comun. Bot. Mus. Hist. Nat. Montevideo 2(28) 6 (1953).
SE. Brazil. 84 BZL. Nanophan. or phan.
**Eugenia bracteosa* DC., Prodr. 3: 276 (1828). *Luma bracteosa* (DC.) Burret, Notizbl. Bot. Gart. Berlin-Dahlem 15: 528 (1941).
Eugenia ibitipocensis Cambess. in A.F.C.de Saint-Hilaire, Fl. Bras. Merid. 2: 365 (1833). *Myrceugenia bracteosa* var. *ibitipocensis* (Cambess.) D.Legrand, Comun. Bot. Mus. Hist. Nat. Montevideo 2(28): 7 (1953).
Eugenia montana Cambess. in A.F.C.de Saint-Hilaire, Fl. Bras. Merid. 2: 365 (1833), nom. illeg. *Luma montana* Burret, Notizbl. Bot. Gart. Berlin-Dahlem 15: 529 (1941). *Myrceugenia cambessedeana* D.Legrand & Kausel, Comun. Bot. Mus. Hist. Nat. Montevideo 2(28): 4 (1953). *Myrceugenia monticola* D.Legrand & Kausel, Darwiniana 11: 321 (1957), nom. illeg.
Eugenia anceps O.Berg in C.F.P.von Martius & auct. suc. (eds.), Fl. Bras. 14(1): 260 (1857). *Luma anceps* (O.Berg) Burret, Notizbl. Bot. Gart. Berlin-Dahlem 15: 528 (1941).
Eugenia expallens O.Berg in C.F.P.von Martius & auct. suc. (eds.), Fl. Bras. 14(1): 260 (1857). *Luma expallens* (O.Berg) Burret, Notizbl. Bot. Gart. Berlin-Dahlem 15: 528 (1941). *Myrceugenia bracteosa* var. *expallens* (O.Berg) D.Legrand, Comun. Bot. Mus. Hist. Nat. Montevideo 2(28): 6 (1953).
Eugenia brachymischa f. *pedunculata* Kiaersk., Enum. Myrt. Bras.: 166 (1893).
Eugenia brachymischa Kiaersk., Enum. Myrt. Bras.: 165 (1895). *Luma brachymischa* (Kiaersk.) Burret, Notizbl. Bot. Gart. Berlin-Dahlem 15: 526 (1941).
Myrceugenia bracteosa var. *australis* D.Legrand, Darwiniana 11: 335 (1957).

Myrceugenia brevipedicellata (Burret) D.Legrand & Kausel, Comun. Bot. Mus. Hist. Nat. Montevideo 2(28): 4 (1953).
Brazil (S. Minas Gerais, São Paulo). 84 BZL. Nanophan.
**Luma brevipedicellata* Burret, Notizbl. Bot. Gart. Berlin-Dahlem 15: 527 (1941).

Myrceugenia* × *bridgesii (Hook. & Arn.) O.Berg, Linnaea 30: 671 (1861). *M. exsucca* × *M. lanceolata*.
C. Chile. 85 CLC. Nanophan. or phan.
**Eugenia* × *bridgesii* Hook. & Arn., Bot. Misc. 3: 322 (1833). *Luma* × *bridgesii* (Hook. & Arn.) A.Gray ex F.Phil., Cat. Pl. Vasc. Chil.: 76 (1881). *Myrceugenia exsucca* var. *bridgessi* (Hook. & Arn.) Kausel, Lilloa 13: 142 (1947).

Myrceugenia camargoana Mattos, Loefgrenia 78: 1 (1983).
Brazil (Rio Grande do Sul). 84 BZS. Nanophan. or phan.

Myrceugenia campestris (DC.) D.Legrand & Kausel, Comun. Bot. Mus. Hist. Nat. Montevideo 2(28): 12 (1953).
SE. & S. Brazil. 84 BZL BZS. Nanophan. or phan.
**Eugenia campestris* DC., Prodr. 3: 274 (1828). *Luma campestris* (DC.) Burret, Notizbl. Bot. Gart. Berlin-Dahlem 15: 529 (1941).
Eugenia campestris var. *nervulosa* Spring, Flora 20(2 Beibl.): 87 (1837).
Eugenia distans O.Berg in C.F.P.von Martius & auct. suc. (eds.), Fl. Bras. 14(1): 263 (1857). *Luma distans* (O.Berg) Burret, Notizbl. Bot. Gart. Berlin-Dahlem 15: 529 (1941). *Myrceugenia distans* (O.Berg) D.Legrand & Kausel, Comun. Bot. Mus. Hist. Nat. Montevideo 2(28): 12 (1953). *Myrceugenia campestris* var. *distans* (O.Berg) D.Legrand, in Fl. Ilustr. Catar., Mirt.: 447 (1970).
Psidium paraibicum Wawra, Itin. Princ. S. Coburgi 1: 19 (1883).

Myrceugenia chrysocarpa (O.Berg) Kausel, Revista Argent. Agron. 9: 58 (1942).
SC. & S. Chile to SW. Argentina. 85 AGS CLC CLS. Nanophan.
**Eugenia chrysocarpa* O.Berg, Linnaea 27: 168 (1856). *Luma chrysocarpa* (O.Berg) Burret, Notizbl. Bot. Gart. Berlin-Dahlem 15: 524 (1941).
Eugenia philippii O.Berg, Linnaea 27: 145 (1856). *Luma philippii* (O.Berg) Burret, Notizbl. Bot. Gart. Berlin-Dahlem 15: 523 (1941).
Myrtus chrysocarpa Poepp. ex O.Berg, Linnaea 27: 169 (1856).
Eugenia buxifolia Phil., Linnaea 28: 640 (1857), nom. illeg. *Myrceugenia buxifolia* Reiche, Anales Univ. Chile 98: 712 (1897).
Eugenia patagonica Phil., Linnaea 33: 22 (1864).
Eugenia patagonica var. *macrophylla* Phil., Linnaea 33: 73 (1864).
Eugenia patagonica var. *microphylla* Phil., Linnaea 33: 73 (1864).
Eugenia petiolata Phil., Anales Univ. Chile 84: 757 (1893).

Myrceugenia colchaguensis (Phil.) Navas, Bol. Mus. Nac. Hist. Nat., Santiago de Chile 29: 230 (1970).
C. Chile. 85 CLC. Nanophan.
**Eugenia colchaguensis* Phil., Linnaea 33: 72 (1864). *Eugenia thalassaia* var. *colchaguensis* (Phil.) Reiche, Anales Univ. Chile 98: 720 (1897).
Myrceugenia malvillana Kausel, Lilloa 17: 51 (1949).

Myrceugenia correifolia (Hook. & Arn.) O.Berg, Linnaea 30: 670 (1861).
NC. Chile. 85 CLC. Nanophan. or phan.
**Eugenia correifolia* Hook. & Arn., Bot. Misc. 3: 819 (1833). *Luma correifolia* (Hook. & Arn.) A.Gray, U.S. Expl. Exped., Phan. 1: 542 (1854).
Eugenia maritima Barnéoud in C.Gay, Fl. Chil. 2: 390 (1847), nom. illeg. *Eugenia thalassaia* O.Berg, Linnaea 27: 179 (1856). *Myrceugenia thalassaia* (O.Berg) Gusinde ex Fuentes, Bol. Mus. Nac. Hist. Nat., Santiago de Chile 13: 108 (1930). *Luma thalassaia* (O.Berg) Burret, Notizbl. Bot. Gart. Berlin-Dahlem 15: 525 (1941). *Myrceugenia maritima* Johow, Revista Chilena Hist. Nat. 49–50: 205 (1948).
Myrceugenia johowii Gusinde, Descr. Chil. Jen. Myrceugenia: 4 (1917).

Myrceugenia cucullata D.Legrand, Darwiniana 11: 347 (1957).
S. Brazil. 84 BZS. Nanophan. or phan.

Myrceugenia decussata Mattos, Loefgrenia 94: 8 (1989).
Brazil (São Paulo). 84 BZL. Nanophan. or phan.

Myrceugenia × diemii Kausel, Revista Argent. Agron. 9: 239 (1942). *M. exsucca* × *M. ovata* var. *nannophylla*.
C. Chile, SW. Argentina. 85 AGS CLC. Nanophan. or phan.
Myrceugenia exsucca var. *quetrihuensis* Kausel, Lilloa 17: 53 (1949).

Myrceugenia euosma (O.Berg) D.Legrand, Anales Mus. Hist. Nat. Montevideo, II, 4(11): 40 (1936).
SE. & S. Brazil to Argentina (Misiones). 84 BZL BZS 85 AGE PAR URU. Nanophan. or phan.
Eugenia aprica O.Berg in C.F.P.von Martius & auct. suc. (eds.), Fl. Bras. 14(1): 218 (1857). *Myrceugenia euosma* var. *aprica* (O.Berg) D.Legrand, Anales Mus. Hist. Nat. Montevideo, II, 4(11): 42 (1936). *Luma aprica* (O.Berg) Burret, Notizbl. Bot. Gart. Berlin-Dahlem 15: 535 (1941).
**Eugenia euosma* O.Berg in C.F.P.von Martius & auct. suc. (eds.), Fl. Bras. 14(1): 233 (1857). *Luma euosma* (O.Berg) Burret, Notizbl. Bot. Gart. Berlin-Dahlem 15: 534 (1941).
Eugenia euosma var. *lutescens* O.Berg in C.F.P.von Martius & auct. suc. (eds.), Fl. Bras. 14(1): 233 (1857).
Eugenia euosma var. *rufescens* O.Berg in C.F.P.von Martius & auct. suc. (eds.), Fl. Bras. 14(1): 233 (1857).
Eugenia nana O.Berg in C.F.P.von Martius & auct. suc. (eds.), Fl. Bras. 14(1): 244 (1857). *Myrceugenia euosma* var. *nana* (O.Berg) D.Legrand, Anales Mus. Hist. Nat. Montevideo, II, 4(11): 42 (1936). *Luma nana* (O.Berg) Burret, Notizbl. Bot. Gart. Berlin-Dahlem 15: 535 (1941).
Eugenia nana var. *congesta* O.Berg in C.F.P.von Martius & auct. suc. (eds.), Fl. Bras. 14(1): 244 (1857).
Eugenia nana var. *effusa* O.Berg in C.F.P.von Martius & auct. suc. (eds.), Fl. Bras. 14(1): 244 (1857).
Eugenia nana var. *robusta* Glaz., Bull. Soc. Bot. France 54(3c): 235 (1908).
Luma angustifolia Burret, Repert. Spec. Nov. Regni Veg. 50: 53 (1941).
Luma cinerea Burret, Notizbl. Bot. Gart. Berlin-Dahlem 15: 534 (1941).
Myrceugenia euosma var. *oblongata* Mattos, Loefgrenia 66: 5 (1975).

Myrceugenia exsucca (DC.) O.Berg, Linnaea 30: 671 (1861).
C. & S. Chile to SW. Argentina. 85 AGS CLC CLS. Phan.
**Eugenia exsucca* DC., Prodr. 3: 278 (1828). *Luma exsucca* (DC.) Burret, Notizbl. Bot. Gart. Berlin-Dahlem 15: 525 (1941).
Eugenia temu Hook. & Arn., Bot. Beechey Voy.: 56 (1832). *Luma temu* (Hook. & Arn.) A.Gray, U.S. Expl. Exped., Phan. 1: 539 (1854). *Eugenia exsucca* var. *temu* (Hook. & Arn.) O.Berg, Linnaea 27: 256 (1856).
Eugenia multiflora Hook. & Arn., Bot. Misc. 3: 322 (1833), nom. illeg. *Myrceugenia pitra* var. *angustifolia* Reiche, Anales Univ. Chile 98: 715 (1897). *Myrceugenia multiflora* (Hook. & Arn.) Kausel, Revista Argent. Agron. 9: 63 (1942).
Eugenia exsucca var. *apiculata* O.Berg, Linnaea 27: 257 (1856). *Myrceugenia exsucca* var. *apiculata* (O.Berg) Reiche, Anales Univ. Chile 98: 715 (1897).
Eugenia exsucca var. *patagua* O.Berg, Linnaea 27: 256 (1856). *Myrceugenia exsucca* var. *patagua* (O.Berg) Reiche, Anales Univ. Chile 98: 714 (1897).
Eugenia exsucca var. *peruviana* O.Berg, Linnaea 27: 255 (1856).
Eugenia pitra O.Berg, Linnaea 27: 264 (1856). *Myrceugenia pitra* (O.Berg) O.Berg, Linnaea 30: 671 (1861). *Luma pitra* (O.Berg) Burret, Notizbl. Bot. Gart. Berlin-Dahlem 15: 525 (1941).
Myrceugenia camphorata O.Berg, Linnaea 27: 134 (1856). *Myrtus camphorata* (O.Berg) Baill., Hist. Pl. 6: 340 (1876).
Myrceugenia lechleriana O.Berg, Linnaea 27: 133 (1856).
Eugenia corralensis Phil., Linnaea 33: 72 (1864). *Luma corralensis* (Phil.) Burret, Notizbl. Bot. Gart. Berlin-Dahlem 15: 525 (1941).

Myrceugenia exsucca var. *temu* Reiche, Anales Univ. Chile 98: 714 (1897).

Myrceugenia fernandeziana (Hook. & Arn.) Johow, Estud. I. Juan Fernandez: 94 (1896).
Juan Fernández Is. (I. Robinson Crusoe). 85 JNF. Phan.
Myrtus maxima Molina, Sag. Stor. Nat. Chili: 173 (1782), provisional synonym. *Nothomyrcia maxima* (Molina) Gunckel, Notas Mens. Mus. Nac. Hist. Nat. (Chile) 17: 11 (1972).
**Myrtus fernandeziana* Hook. & Arn., Bot. Misc. 3: 316 (1833). *Eugenia fernandeziana* (Hook. & Arn.) Barnéoud in C.Gay, Fl. Chil. 2: 392 (1847). *Myrceugenia luma* O.Berg, Linnaea 30: 671 (1861), nom. illeg. *Luma fernandeziana* (Hook. & Arn.) Burret, Notizbl. Bot. Gart. Berlin-Dahlem 15: 526 (1941). *Nothomyrcia fernandeziana* (Hook. & Arn.) Kausel, Lilloa 13: 148 (1947).
Eugenia lumilla Phil., Bot. Zeitung (Berlin) 14: 643 (1856).

Myrceugenia foveolata (O.Berg) Sobral, Roessléria 8: 43 (1986).
SE. & S. Brazil. 84 BZL BZS. Nanophan. or phan.
**Eugenia foveolata* O.Berg in C.F.P.von Martius & auct. suc. (eds.), Fl. Bras. 14(1): 261 (1857). *Luma foveolata* (O.Berg) Burret, Notizbl. Bot. Gart. Berlin-Dahlem 15: 529 (1941).
Myrceugenia leptocalyx D.Legrand, Darwiniana 11: 313 (1957).
Myrceugenia leptocalyx var. *coriacea* Mattos, Loefgrenia 66: 5 (1975).

Myrceugenia franciscensis (O.Berg) Landrum, Brittonia 32: 372 (1980).
Brazil (São Paulo to Paraná). 84 BZS. Nanophan.
**Eugenia franciscensis* O.Berg in C.F.P.von Martius & auct. suc. (eds.), Fl. Bras. 14(1): 233 (1857). *Luma franciscensis* (O.Berg) Burret, Notizbl. Bot. Gart. Berlin-Dahlem 15: 528 (1941). *Myrceugenia bracteosa* var. *franciscensis* (O.Berg) D.Legrand, Comun. Bot. Mus. Hist. Nat. Montevideo 2(28): 7 (1953).
Luma mucronata Burret, Notizbl. Bot. Gart. Berlin-Dahlem 15: 533 (1941). *Myrceugenia mucronata* (Burret) D.Legrand & Kausel, Comun. Bot. Mus. Hist. Nat. Montevideo 2(28): 10 (1953).
Myrceugenia longipedunculata Mattos & D.Legrand, Loefgrenia 67: 18 (1975).

Myrceugenia gertii Landrum, Brittonia 36: 163 (1984).
S. Brazil. 84 BZS. Nanophan. or phan.
**Calyptrogenia hatschbachii* D.Legrand, Comun. Bot. Mus. Hist. Nat. Montevideo 3(36): 8 (1958). *Neomitranthes hatschbachii* (D.Legrand) Mattos, Loefgrenia 99: 6 (1990).

Myrceugenia glaucescens (Cambess.) D.Legrand & Kausel, Comun. Bot. Mus. Hist. Nat. Montevideo 1(7): 7 (1943).
Brazil to Paraguay and Uruguay. 84 BZL BZS 85 AGE PAR URU. Nanophan. or phan.
**Eugenia glaucescens* Cambess. in A.F.C.de Saint-Hilaire, Fl. Bras. Merid. 2: 68 (1829). *Luma glaucescens* (Cambess.) Burret, Notizbl. Bot. Gart. Berlin-Dahlem 15: 531 (1941).

var. ***glaucescens***
Paraguay to Uruguay. 84 BZL BZS 85 AGE PAR URU. Nanophan. or phan.
Eugenia araujoana O.Berg in C.F.P.von Martius & auct. suc. (eds.), Fl. Bras. 14(1): 219 (1857). *Luma araujoana* (O.Berg) Burret, Notizbl. Bot. Gart. Berlin-Dahlem 15: 532 (1941).
Eugenia bagensis O.Berg in C.F.P.von Martius & auct. suc. (eds.), Fl. Bras. 14(1): 231 (1857). *Luma bagensis* (O.Berg) Burret, Notizbl. Bot. Gart. Berlin-Dahlem 15: 532 (1941).
Eugenia bagensis var. *angustifolia* O.Berg in C.F.P.von Martius & auct. suc. (eds.), Fl. Bras. 14(1): 231 (1857).
Eugenia bagensis var. *avenia* O.Berg in C.F.P.von Martius & auct. suc. (eds.), Fl. Bras. 14(1): 231 (1857).
Eugenia bagensis var. *latifolia* O.Berg in C.F.P.von Martius & auct. suc. (eds.), Fl. Bras. 14(1): 231 (1857).
Eugenia cambessedeana O.Berg in C.F.P.von Martius & auct. suc. (eds.), Fl. Bras. 14(1): 230 (1857). *Luma cambessedeana* (O.Berg) Burret, Notizbl. Bot. Gart. Berlin-Dahlem 15: 532 (1941).
Eugenia canelonensis O.Berg in C.F.P.von Martius & auct. suc. (eds.), Fl. Bras. 14(1): 232 (1857). *Luma canelonensis* (O.Berg) Burret, Notizbl. Bot. Gart. Berlin-Dahlem 15: 532 (1941).
Eugenia elegans O.Berg in C.F.P.von Martius & auct. suc. (eds.), Fl. Bras. 14(1): 232 (1857). *Luma elegans* (O.Berg) Burret, Notizbl. Bot. Gart. Berlin-Dahlem 15: 533 (1941).
Eugenia pallida O.Berg in C.F.P.von Martius & auct. suc. (eds.), Fl. Bras. 14(1): 231 (1857). *Luma pallida* (O.Berg) Burret, Notizbl. Bot. Gart. Berlin-Dahlem 15: 533 (1941). *Myrceugenia pallida* (O.Berg) D.Legrand & Kausel, Comun. Bot. Mus. Hist. Nat. Montevideo 1(7): 8 (1943). *Myrceugenia glaucescens* f. *pallida* (O.Berg) D.Legrand, Comun. Bot. Mus. Hist. Nat. Montevideo 2(28): 11 (1953). *Myrceugenia glaucescens* var. *pallida* (O.Berg) Kausel, Lilloa 32: 352 (1967).
Eugenia ribeireana O.Berg in C.F.P.von Martius & auct. suc. (eds.), Fl. Bras. 14(1): 307 (1857). *Luma ribeireana* (O.Berg) Burret, Notizbl. Bot. Gart. Berlin-Dahlem 15: 532 (1941). *Myrceugenia ribeireana* (O.Berg) D.Legrand & Kausel, Comun. Bot. Mus. Hist. Nat. Montevideo 2(28): 11 (1953). *Myrceugenia glaucescens* var. *ribeireana* (O.Berg) D.Legrand, Sellowia 13: 304 (1961).
Luma angustior Burret, Notizbl. Bot. Gart. Berlin-Dahlem 15: 530 (1941).
Myrceugenia glaucescens f. *catharinensis* D.Legrand & R.M.Klein, in Fl. Ilustr. Catar., Mirt.: 384 (1970).
Myrceugenia glaucescens f. *debilis* D.Legrand & R.M.Klein, in Fl. Ilustr. Catar., Mirt.: 383 (1970).
Myrceugenia grisea D.Legrand, in Fl. Ilustr. Catar. 1(Mirt.): 387 (1970).

var. ***latior*** (Burret) Landrum, Brittonia 32: 372 (1980).
SE. & S. Brazil. 84 BZL BZS. Nanophan.
**Luma latior* Burret, Notizbl. Bot. Gart. Berlin-Dahlem 15: 530 (1941). *Myrceugenia latior* (Burret) D.Legrand & Kausel, Comun. Bot. Mus. Hist. Nat. Montevideo 2(28): 11 (1953).

Myrceugenia hatschbachii Landrum, Brittonia 32: 372 (1980).
Brazil (Paraná). 84 BZS. Nanophan.

Myrceugenia hoehnei (Burret) D.Legrand & Kausel, Comun. Bot. Mus. Hist. Nat. Montevideo 2(28): 5 (1953).

Brazil (São Paulo to Santa Catarina). 84 BZL BZS. Nanophan.
Luma hoehnei Burret, Repert. Spec. Nov. Regni Veg. 50: 54 (1941).

Myrceugenia kleinii D.Legrand & Kausel, Sellowia 13: 306 (1961).
Brazil (São Paulo to Santa Catarina). 84 BZL BZS. Phan.

Myrceugenia lanceolata (Juss. ex J.St.-Hil.) Kausel, Lilloa 13: 135 (1947).
C. Chile. 85 CLC. Nanophan.
**Myrtus lanceolata* Juss. ex J.St.-Hil. in H.L.Duhamel du Monceau, Traité Arbr. Arbust. 1: 208 (1803). *Eugenia dombeyana* DC., Prodr. 3: 276 (1828). *Luma dombeyana* (DC.) Burret, Notizbl. Bot. Gart. Berlin-Dahlem 15: 525 (1941).
Eugenia stenophylla Hook. & Arn., Bot. Misc. 3: 322 (1833). *Luma stenophylla* (Hook. & Arn.) A.Gray, U.S. Expl. Exped., Phan. 1: 540 (1854). *Myrceugenia stenophylla* (Hook. & Arn.) O.Berg, Linnaea 30: 670 (1861). *Myrceugenia stenophylla* var. *angustifolia* Reiche, Anales Univ. Chile 98: 713 (1897).
Myrtus gudilla Colla, Mem. Reale Accad. Sci. Torino 37: 66 (1834). *Eugenia gudilla* (Colla) Barnéoud in C.Gay, Fl. Chil. 2: 396 (1847).
Eugenia stenophylla var. *angustifolia* O.Berg in C.F.P.von Martius & auct. suc. (eds.), Fl. Bras. 14(1): 288 (1857).
Myrceugenia stenophylla var. *latifolia* Reiche, Anales Univ. Chile 98: 713 (1897).

Myrceugenia leptospermoides (DC.) Kausel, Revista Argent. Agron. 9: 52 (1942).
SC. Chile. 85 CLC. Nanophan.
**Eugenia leptospermoides* DC., Prodr. 3: 266 (1828). *Luma leptospermoides* (DC.) Burret, Notizbl. Bot. Gart. Berlin-Dahlem 15: 524 (1941).
Eugenia leptospermoides var. *latifolia* O.Berg, Linnaea 27: 143 (1856).
Eugenia leptospermoides var. *longifolia* O.Berg, Linnaea 27: 143 (1856).
Eugenia leptospermoides var. *microphylla* O.Berg, Linnaea 27: 143 (1856).
Myrtus brachyphylla Kunze ex O.Berg, Linnaea 27: 144 (1856).
Eugenia thymifolia Phil. ex Reiche, Fl. Chile 2: 303 (1897).

Myrceugenia mesomischa (Burret) D.Legrand & Kausel, Comun. Bot. Mus. Hist. Nat. Montevideo 2(28): 10 (1953).
Brazil (Rio Grande do Sul). 84 BZS. Nanophan. or phan. Provisionally accepted.
**Luma mesomischa* Burret, Notizbl. Bot. Gart. Berlin-Dahlem 15: 529 (1941).

Myrceugenia miersiana (Gardner) D.Legrand & Kausel, Comun. Bot. Mus. Hist. Nat. Montevideo 2(28): 8 (1953).
SE. & S. Brazil. 84 BZL BZS. Nanophan. or phan.
**Eugenia miersiana* Gardner, London J. Bot. 4: 103 (1845). *Luma miersiana* (Gardner) Burret, Notizbl. Bot. Gart. Berlin-Dahlem 15: 527 (1941).
Eugenia miersiana var. *costata* O.Berg in C.F.P.von Martius & auct. suc. (eds.), Fl. Bras. 14(1): 574 (1859).
Eugenia miersiana var. *glomerata* O.Berg in C.F.P.von Martius & auct. suc. (eds.), Fl. Bras. 14(1): 574 (1859).
Eugenia miersiana var. *membranacea* O.Berg in C.F.P.von Martius & auct. suc. (eds.), Fl. Bras. 14(1): 574 (1859).
Eugenia miersiana var. *venosa* O.Berg in C.F.P.von Martius & auct. suc. (eds.), Fl. Bras. 14(1): 574 (1859).
Luma brevipes Burret, Repert. Spec. Nov. Regni Veg. 50: 52 (1941). *Myrceugenia brevipes* (Burret) D.Legrand & Kausel, Comun. Bot. Mus. Hist. Nat. Montevideo 2(28): 8 (1953).
Luma filibracteata Burret, Repert. Spec. Nov. Regni Veg. 50: 52 (1941). *Myrceugenia filibracteata* (Burret) D.Legrand, Sellowia 13: 301 (1961).
Luma fuscovelutina Burret, Repert. Spec. Nov. Regni Veg. 50: 51 (1941). *Myrceugenia fuscovelutina* (Burret) D.Legrand & Kausel, Comun. Bot. Mus. Hist. Nat. Montevideo 2(28): 13 (1953).
Myrceugenia nothorufa var. *venosa* D.Legrand, in Fl. Ilustr. Catar. 1(Mirt.): 413 (1970).

Myrceugenia myrcioides (Cambess.) O.Berg, Linnaea 27: 134 (1856).
E. & S. Brazil. 84 BZE BZL BZS. Nanophan. or phan.
Eugenia humilis Vell., Fl. Flumin. 5: 210, t. 43 (1829), provisional synonym.
**Eugenia myrcioides* Cambess. in A.F.C.de Saint-Hilaire, Fl. Bras. Merid. 2: 353 (1832). *Luma myrcioides* (Cambess.) Burret, Notizbl. Bot. Gart. Berlin-Dahlem 15: 531 (1941).
Eugenia complanata Gardner, London J. Bot. 2: 353 (1843).
Eugenia hypericifolia Gardner, London J. Bot. 2: 354 (1843). *Myrceugenia myrcioides* var. *hypericifolia* (Gardner) D.Legrand, in Fl. Ilustr. Catar., Mirt.: 407 (1970).
Eugenia estrellensis O.Berg in C.F.P.von Martius & auct. suc. (eds.), Fl. Bras. 14(1): 250 (1857). *Luma estrellensis* (O.Berg) Burret, Notizbl. Bot. Gart. Berlin-Dahlem 15: 529 (1941). *Myrceugenia estrellensis* (O.Berg) D.Legrand & Kausel, Comun. Bot. Mus. Hist. Nat. Montevideo 2(28): 9 (1953).
Eugenia acrophylla O.Berg in C.F.P.von Martius & auct. suc. (eds.), Fl. Bras. 14(1): 575 (1859). *Luma acrophylla* (O.Berg) Burret, Notizbl. Bot. Gart. Berlin-Dahlem 15: 531 (1941). *Myrceugenia myrcioides* var. *acrophylla* (O.Berg) D.Legrand, Comun. Bot. Mus. Hist. Nat. Montevideo 2(28): 9 (1953). *Myrceugenia acrophylla* (O.Berg) D.Legrand, in Fl. Ilustr. Catar. 1(Mirt.): 342 (1970).
Eugenia obtusiflora Kiaersk., Enum. Myrt. Bras.: 168 (1893). *Luma obtusiflora* (Kiaersk.) Burret, Notizbl. Bot. Gart. Berlin-Dahlem 15: 531 (1941).
Luma ulei Burret, Notizbl. Bot. Gart. Berlin-Dahlem 15: 531 (1941). *Myrceugenia myrcioides* var. *ulei* (Burret) D.Legrand, Comun. Bot. Mus. Hist. Nat. Montevideo 2(28): 9 (1953). *Myrceugenia acrophylla* var. *ulei* (Burret) D.Legrand, in Fl. Ilustr. Catar., Mirt.: 347 (1970).
Myrceugenia myrcioides var. *paranensis* D.Legrand, Darwiniana 11: 341 (1957).
Myrceugenia ferreira-limana R.M.Klein & D.Legrand, in Fl. Ilustr. Catar. 1(Mirt.): 374 (1970).
Myrceugenia paranaguensis Mattos, Loefgrenia 66: 4 (1975).
Myrceugenia myrcioides var. *loefgreniana* Mattos, Loefgrenia 71: 1 (1977).

Myrceugenia myrtoides O.Berg in C.F.P.von Martius & auct. suc. (eds.), Fl. Bras. 14(1): 211 (1857). *Myrtus myrtoides* (O.Berg) Arechav., Anales Mus. Nac.

Montevideo 5: 35 (1905). *Luma myrtoides* (O.Berg) Burret, Notizbl. Bot. Gart. Berlin-Dahlem 15: 528 (1941).
S. Brazil to Uruguay. 84 BZS 85 URU. Nanophan. or phan.
Myrceugenia montevidensis O.Berg in C.F.P.von Martius & auct. suc. (eds.), Fl. Bras. 14(1): 211 (1857). *Myrtus montevidensis* (O.Berg) Arechav., Anales Mus. Nac. Montevideo 5: 87 (1905). *Luma montevidensis* (O.Berg) Burret, Notizbl. Bot. Gart. Berlin-Dahlem 15: 533 (1941).
Myrceugenia myrtoides var. *conferta* O.Berg in C.F.P.von Martius & auct. suc. (eds.), Fl. Bras. 14(1): 211 (1857).
Myrceugenia myrtoides var. *stricta* O.Berg in C.F.P.von Martius & auct. suc. (eds.), Fl. Bras. 14(1): 211 (1857).
Myrceugenia sellowiana O.Berg in C.F.P.von Martius & auct. suc. (eds.), Fl. Bras. 14(1): 212 (1857). *Luma sellowiana* (O.Berg) Burret, Notizbl. Bot. Gart. Berlin-Dahlem 15: 533 (1941). *Myrceugenia montevidensis* var. *sellowiana* (O.Berg) Mattos, Loefgrenia 78: 2 (1983). *Myrceugenia myrtoides* var. *sellowiana* (O.Berg) Mattos, Loefgrenia 85: 2 (1984).
Luma cinnamomeotomentosa Burret, Repert. Spec. Nov. Regni Veg. 50: 51 (1941).

Myrceugenia obtusa (DC.) O.Berg, Linnaea 30: 669 (1861).
C. Chile. 85 CLC. Nanophan. or phan.
**Eugenia obtusa* DC., Prodr. 3: 266 (1828). *Luma obtusa* (DC.) A.Gray, U.S. Expl. Exped., Phan. 1: 541 (1854). *Myrceugenia chilensis* O.Berg, Linnaea 27: 132 (1856), nom. illeg.
Myrtus raran Colla, Mem. Reale Accad. Sci. Torino 37: 66 (1834). *Eugenia raran* (Colla) Barnéoud in C.Gay, Fl. Chil. 2: 388 (1847). *Myrceugenia obtusa* var. *raran* (Colla) O.Berg, Linnaea 30: 669 (1861).
Eugenia polyantha Phil., Linnaea 28: 639 (1857). *Myrceugenia obtusa* var. *polyantha* (Phil.) O.Berg, Linnaea 30: 669 (1861).
Myrceugenia obtusa var. *berteroana* O.Berg, Linnaea 30: 669 (1861).

Myrceugenia ovalifolia (O.Berg) Landrum, Brittonia 36: 163 (1984).
Brazil (São Paulo to Paraná). 84 BZL BZS. Nanophan. or phan.
**Mitranthes ovalifolia* O.Berg in C.F.P.von Martius & auct. suc. (eds.), Fl. Bras. 14(1): 356 (1857). *Chytraculia selloana* Kuntze, Revis. Gen. Pl. 1: 238 (1891). *Psidium ovalifolium* (O.Berg) Nied. in H.G.A.Engler & K.A.E.Prantl, Nat. Pflanzenfam. 3(7): 69 (1893). *Mitropsidium ovalifolium* (O.Berg) Burret, Notizbl. Bot. Gart. Berlin-Dahlem 15: 486 (1941).
Mitranthes maria-aemiliae D.Legrand, Comun. Bot. Mus. Hist. Nat. Montevideo 4(49): 1 (1969). *Neomitranthes maria-emiliae* (D.Legrand) Mattos, Loefgrenia 76: 3 (1981).

Myrceugenia ovata (Hook. & Arn.) O.Berg, Linnaea 30: 670 (1861).
Brazil to Chile. 84 BZL BZS 85 AGS CLC CLS. Nanophan. or phan.
**Eugenia ovata* Hook. & Arn., Bot. Misc. 3: 319 (1833). *Luma ovata* (Hook. & Arn.) A.Gray ex F.Phil., Cat. Pl. Vasc. Chil.: 76 (1881).

var. ***acutata*** (D.Legrand) Landrum, Brittonia 32: 374 (1980).
SE. & S. Brazil. 84 BZL BZS. Nanophan. or phan.
**Myrceugenia acutata* D.Legrand, Darwiniana 11: 351 (1957). *Myrceugenia cucullata* var. *acutata* (D.Legrand) Mattos, Loefgrenia 85: 1 (1984).

var. ***nannophylla*** (Burret) Landrum, Brittonia 32: 374 (1980).
SC. & S. Chile to SW. Argentina (Neuquén). 85 AGS CLC CLS. Nanophan. or phan.
**Luma nannophylla* Burret, Repert. Spec. Nov. Regni Veg. 50: 50 (1941). *Myrceugenia nannophylla* (Burret) Kausel, Revista Argent. Agron. 18: 233 (1951).
Myrceugenia montana Kausel, Revista Argent. Agron. 9: 55 (1942).
Myrceugenia valientei Kausel, Revista Argent. Agron. 9: 323 (1944).

var. ***ovata***
SC. Chile (I. Mocha, Los Lagos). 85 CLS. Nanophan. or phan.
Eugenia cumingii Hook. & Arn., Bot. Misc. 3: 319 (1833). *Luma cumingii* (Hook. & Arn.) Burret, Notizbl. Bot. Gart. Berlin-Dahlem 15: 524 (1941). *Myrceugenia cumingii* (Hook. & Arn.) Kausel, Revista Argent. Agron. 9: 239 (1942).
Eugenia trichocarpa Phil., Linnaea 28: 688 (1858).

var. ***regnelliana*** (O.Berg) Landrum, Brittonia 43: 200 (1991).
SE. & S. Brazil. 84 BZL BZS. Nanophan. or phan.
**Eugenia regnelliana* O.Berg in C.F.P.von Martius & auct. suc. (eds.), Fl. Bras. 14(1): 245 (1857). *Luma regnelliana* (O.Berg) Burret, Notizbl. Bot. Gart. Berlin-Dahlem 15: 535 (1941). *Myrceugenia regnelliana* (O.Berg) D.Legrand & Kausel, Comun. Bot. Mus. Hist. Nat. Montevideo 2(28): 11 (1953).
Eugenia xanthochlora O.Berg in C.F.P.von Martius & auct. suc. (eds.), Fl. Bras. 14(1): 216 (1857). *Luma xanthochlora* (O.Berg) Burret, Notizbl. Bot. Gart. Berlin-Dahlem 15: 533 (1941). *Myrceugenia regnelliana* f. *xanthochlora* (O.Berg) D.Legrand, in Fl. Ilustr. Catar., Mirt.: 425 (1970).
Eugenia itatiaiensis Kiaersk., Enum. Myrt. Bras.: 166 (1893). *Luma itatiaiensis* (Kiaersk.) Burret, Notizbl. Bot. Gart. Berlin-Dahlem 15: 534 (1941). *Myrceugenia itatiaiensis* (Kiaersk.) D.Legrand & Kausel, Comun. Bot. Mus. Hist. Nat. Montevideo 2(28): 12 (1953). *Myrceugenia regnelliana* f. *itatiaiensis* (Kiaersk.) D.Legrand, in Fl. Ilustr. Catar., Mirt.: 426 (1970).
Luma ereophila Burret, Repert. Spec. Nov. Regni Veg. 50: 54 (1941).
Luma gracilis Burret, Notizbl. Bot. Gart. Berlin-Dahlem 15: 534 (1941). *Myrceugenia regnelliana* var. *gracilis* (Burret) D.Legrand, Darwiniana 11: 354 (1957). *Myrceugenia ovata* var. *gracilis* (Burret) Landrum, Brittonia 32: 374 (1980).
Luma macromischa Burret, Repert. Spec. Nov. Regni Veg. 50: 53 (1941).
Luma oreophila Burret, Repert. Spec. Nov. Regni Veg. 50: 54 (1941).
Myrceugenia regnelliana var. *leptophylla* D.Legrand, Darwiniana 11: 354 (1957).
Myrceugenia regnelliana var. *capillaris* D.Legrand, Sellowia 13: 303 (1961).
Myrceugenia regnelliana var. *coriacea* D.Legrand, Sellowia 13: 303 (1961).

Myrceugenia oxysepala (Burret) D.Legrand & Kausel, Comun. Bot. Mus. Hist. Nat. Montevideo 2(28): 5 (1953).
SE. & S. Brazil. 84 BZL BZS. Nanophan. or phan.
Luma macrosepala Burret, Repert. Spec. Nov. Regni Veg. 50: 53 (1941). *Myrceugenia macrosepala* (Burret) D.Legrand & Kausel, Comun. Bot. Mus. Hist. Nat. Montevideo 2(28): 12 (1953).
**Luma oxysepala* Burret, Repert. Spec. Nov. Regni Veg. 50: 51 (1941).
Myrceugenia macrosepala var. *obovata* Mattos, Loefgrenia 66: 5 (1975).

Myrceugenia parvifolia (DC.) Kausel, Revista Argent. Agron. 9: 53 (1942).
C. & S. Chile. 85 CLC CLS. Nanophan. or phan.
**Eugenia parvifolia* DC., Prodr. 3: 266 (1828). *Luma parvifolia* (DC.) Burret, Notizbl. Bot. Gart. Berlin-Dahlem 15: 524 (1941).
Luma baeckeoides Griseb., Abh. Königl. Ges. Wiss. Göttingen 6: 120 (1854).

Myrceugenia pilotantha (Kiaersk.) Landrum, Brittonia 32: 374 (1980).
SE. & S. Brazil. 84 BZL BZS. Nanophan. or phan.
**Eugenia pilotantha* Kiaersk., Enum. Myrt. Bras.: 169 (1893). *Luma pilotantha* (Kiaersk.) Burret, Notizbl. Bot. Gart. Berlin-Dahlem 15: 527 (1941).

var. **nothorufa** (D.Legrand) Landrum, Brittonia 43: 200 (1991).
S. Brazil. 84 BZS. Nanophan. or phan.
Myrceugenia ramboi D.Legrand, Darwiniana 11: 320 (1957).
**Myrceugenia nothorufa* D.Legrand, in Fl. Ilustr. Catar. 1(Mirt.): 409 (1970).
Myrceugenia nothorufa var. *major* D.Legrand, in Fl. Ilustr. Catar., Mirt.: 414 (1970).
Myrceugenia pilotantha var. *major* (D.Legrand) Landrum, Brittonia 32: 374 (1980).

var. ***pilotantha***
SE. & S. Brazil. 84 BZL BZS. Nanophan. or phan.
Luma macahensis Burret, Notizbl. Bot. Gart. Berlin-Dahlem 15: 527 (1941). *Myrceugenia miersiana* var. *macahensis* (Burret) D.Legrand, Comun. Bot. Mus. Hist. Nat. Montevideo 2(28): 8 (1953).

Myrceugenia pinifolia (Phil.) Kausel, Revista Argent. Agron. 9: 54 (1942).
C. Chile. 85 CLC. Nanophan.
**Eugenia pinifolia* Phil., Anales Univ. Chile 84: 758 (1893). *Myrceugenia stenophylla* var. *pinifolia* (Phil.) Reiche, Anales Univ. Chile 98: 713 (1897).

Myrceugenia planipes (Hook. & Arn.) O.Berg, Bot. Zeitung (Berlin) 16: 250 (1858).
SC. & S. Chile to Argentina (Neuquén). 85 AGS CLC CLS. Nanophan. or phan.
**Eugenia planipes* Hook. & Arn., Bot. Misc. 3: 323 (1833). *Luma planipes* (Hook. & Arn.) A.Gray, U.S. Expl. Exped., Phan. 1: 540 (1854). *Myrcia planipes* (Hook. & Arn.) Kiaersk., Enum. Myrt. Bras.: 115 (1893). *Myrceugenia planipes* var. *genuina* Reiche, Anales Univ. Chile 98: 712 (1897), nom. inval.
Eugenia distoma O.Berg, Linnaea 27: 155 (1856). *Luma distoma* (O.Berg) Burret, Notizbl. Bot. Gart. Berlin-Dahlem 15: 526 (1941). *Myrceugenia distoma* (O.Berg) Kausel, Revista Argent. Agron. 9: 238 (1942).
Eugenia planipes var. *grandifolia* O.Berg, Linnaea 27: 162 (1856).
Myrtus elegantula Kunze ex O.Berg, Linnaea 27: 155 (1856).
Myrceugenia planipes var. *grandifolia* Reiche, Anales Univ. Chile 98: 712 (1897).

Myrceugenia reitzii D.Legrand & Kausel, Sellowia 13: 305 (1961).
S. Brazil. 84 BZS. Phan.

Myrceugenia rufa (Colla) Skottsb., Lilloa 13: 134 (1947).
Coastal C. Chile. 85 CLC. Nanophan.
Myrtus chrysophyllum Spreng., Syst. Veg. 2: 482 (1825), provisional synonym.
Eugenia ferruginea Hook. & Arn., Bot. Misc. 3: 319 (1833), nom. illeg. *Luma ferruginea* A.Gray, U.S. Expl. Exped., Phan. 1: 542 (1854). *Myrceugenia ferruginea* (Hook. & Arn.) Reiche, Anales Univ. Chile 98: 710 (1897).
**Myrtus rufa* Colla, Mem. Reale Accad. Sci. Torino 37: 66 (1834). *Eugenia rufa* (Colla) Barnéoud in C.Gay, Fl. Chil. 2: 387 (1847). *Luma rufa* (Colla) Burret, Notizbl. Bot. Gart. Berlin-Dahlem 15: 524 (1941).

Myrceugenia rufescens (DC.) D.Legrand & Kausel, Comun. Bot. Mus. Hist. Nat. Montevideo 2(28): 8 (1953).
SE. & S. Brazil. 84 BZL BZS. Nanophan. or phan.
**Eugenia rufescens* DC., Prodr. 3: 279 (1828). *Luma rufescens* (DC.) Burret, Notizbl. Bot. Gart. Berlin-Dahlem 15: 528 (1941).
Eugenia planiramea O.Berg in C.F.P.von Martius & auct. suc. (eds.), Fl. Bras. 14(1): 234 (1857). *Luma planiramea* (O.Berg) Burret, Notizbl. Bot. Gart. Berlin-Dahlem 15: 527 (1941).
Eugenia poeppigiana O.Berg in C.F.P.von Martius & auct. suc. (eds.), Fl. Bras. 14(1): 232 (1857). *Myrtus ugni* var. *poeppigii* Kuntze, Revis. Gen. Pl. 3(2): 93 (1898). *Luma poeppigiana* (O.Berg) Burret, Notizbl. Bot. Gart. Berlin-Dahlem 15: 533 (1941).
Myrceugenia rufescens var. *alegrensis* D.Legrand, in Fl. Ilustr. Catar., Mirt.: 433 (1970).
Myrceugenia jonssonii Kausel, Lilloa 33: 106 (1971 publ. 1972).

Myrceugenia schulzii Johow, Estud. I. Juan Fernandez: 96 (1896). *Luma schulzii* (Johow) Burret, Notizbl. Bot. Gart. Berlin-Dahlem 15: 526 (1941).
Juan Fernández Is. (I. Alejandro Selkirk). 85 JNF. Phan.

Myrceugenia scutellata C.D.Legrand, in Fl. Ilustr. Catar. 1(Mirt.): 436 (1970).
SE. & S. Brazil. 84 BZL BZS. Phan.

Myrceugenia seriatoramosa (Kiaersk.) D.Legrand & Kausel, Comun. Bot. Mus. Hist. Nat. Montevideo 2(28): 5 (1953).
Brazil (Rio de Janeiro to Paraná). 84 BZL BZS. Nanophan. or phan.
**Eugenia seriatoramosa* Kiaersk., Enum. Myrt. Bras.: 170 (1893). *Luma seriatoramosa* (Kiaersk.) Burret, Notizbl. Bot. Gart. Berlin-Dahlem 15: 528 (1941).

Myrceugenia smithii Landrum, Brittonia 32: 374 (1980).
Brazil (Santa Catarina). 84 BZS. Nanophan.

Myrceugenia venosa D.Legrand, Darwiniana 11: 323 (1957).
Brazil (Santa Catarina). 84 BZS. Nanophan. or phan.
Myrceugenia miersiana var. *lanceolata* D.Legrand, in Fl. Ilustr. Catar. 1(Mirt.): 401 (1970).

***Synonyms*:**
Myrceugenia acrophylla (O.Berg) D.Legrand = ***Myrceugenia myrcioides*** (Cambess.) O.Berg
Myrceugenia acrophylla var. *ulei* (Burret) D.Legrand = ***Myrceugenia myrcioides*** (Cambess.) O.Berg
Myrceugenia acutata D.Legrand = ***Myrceugenia ovata*** var. ***acutata*** (D.Legrand) Landrum
Myrceugenia alpigena var. *fuliginea* (O.Berg) Landrum = ***Myrceugenia alpigena*** (DC.) Landrum var. ***alpigena***
Myrceugenia alpigena var. *rufa* (O.Berg) Landrum = ***Myrceugenia alpigena*** (DC.) Landrum var. ***alpigena***
Myrceugenia apiculata (DC.) Nied. = ***Luma apiculata*** (DC.) Burret
Myrceugenia apiculata var. *australis* Kausel = ***Luma apiculata*** (DC.) Burret
Myrceugenia bracteosa var. *alpigena* (DC.) D.Legrand = ***Myrceugenia alpigena*** (DC.) Landrum var. ***alpigena***
Myrceugenia bracteosa var. *australis* D.Legrand = ***Myrceugenia bracteosa*** (DC.) D.Legrand & Kausel
Myrceugenia bracteosa var. *expallens* (O.Berg) D.Legrand = ***Myrceugenia bracteosa*** (DC.) D.Legrand & Kausel
Myrceugenia bracteosa var. *franciscensis* (O.Berg) D.Legrand = ***Myrceugenia franciscensis*** (O.Berg) Landrum
Myrceugenia bracteosa var. *fuliginea* (O.Berg) D.Legrand = ***Myrceugenia alpigena*** (DC.) Landrum var. ***alpigena***
Myrceugenia bracteosa var. *guaratubensis* Mattos = ***Myrceugenia alpigena*** (DC.) Landrum var. ***alpigena***
Myrceugenia bracteosa var. *ibitipocensis* (Cambess.) D.Legrand = ***Myrceugenia bracteosa*** (DC.) D.Legrand & Kausel
Myrceugenia bracteosa var. *seriatopedunculata* (Kiaersk.) D.Legrand = ***Myrceugenia alpigena*** (DC.) Landrum var. ***alpigena***
Myrceugenia brevipes (Burret) D.Legrand & Kausel = ***Myrceugenia miersiana*** (Gardner) D.Legrand & Kausel
Myrceugenia buxifolia Reiche = ***Myrceugenia chrysocarpa*** (O.Berg) Kausel
Myrceugenia cambessedeana D.Legrand & Kausel = ***Myrceugenia bracteosa*** (DC.) D.Legrand & Kausel
Myrceugenia campestris var. *distans* (O.Berg) D.Legrand = ***Myrceugenia campestris*** (DC.) D.Legrand & Kausel
Myrceugenia camphorata O.Berg = ***Myrceugenia exsucca*** (DC.) O.Berg
Myrceugenia candolleana (DC.) D.Legrand = ***Psidium salutare*** var. ***sericeum*** (Cambess.) Landrum
Myrceugenia catharinae D.Legrand = ***Myrceugenia alpigena*** (DC.) Landrum var. ***alpigena***
Myrceugenia chilensis O.Berg = ***Myrceugenia obtusa*** (DC.) O.Berg
Myrceugenia cucullata var. *acutata* (D.Legrand) Mattos = ***Myrceugenia ovata*** var. ***acutata*** (D.Legrand) Landrum
Myrceugenia cumingii (Hook. & Arn.) Kausel = ***Myrceugenia ovata*** (Hook. & Arn.) O.Berg var. ***ovata***
Myrceugenia distans (O.Berg) D.Legrand & Kausel = ***Myrceugenia campestris*** (DC.) D.Legrand & Kausel
Myrceugenia distoma (O.Berg) Kausel = ***Myrceugenia planipes*** (Hook. & Arn.) O.Berg
Myrceugenia estrellensis (O.Berg) D.Legrand & Kausel = ***Myrceugenia myrcioides*** (Cambess.) O.Berg
Myrceugenia euosma var. *aprica* (O.Berg) D.Legrand = ***Myrceugenia euosma*** (O.Berg) D.Legrand
Myrceugenia euosma var. *nana* (O.Berg) D.Legrand = ***Myrceugenia euosma*** (O.Berg) D.Legrand
Myrceugenia euosma var. *oblongata* Mattos = ***Myrceugenia euosma*** (O.Berg) D.Legrand
Myrceugenia exsucca var. *apiculata* (O.Berg) Reiche = ***Myrceugenia exsucca*** (DC.) O.Berg
Myrceugenia exsucca var. *bridgessi* (Hook. & Arn.) Kausel = ***Myrceugenia*** × ***bridgesii*** (Hook. & Arn.) O.Berg
Myrceugenia exsucca var. *patagua* (O.Berg) Reiche = ***Myrceugenia exsucca*** (DC.) O.Berg
Myrceugenia exsucca var. *quetrihuensis* Kausel = ***Myrceugenia*** × ***diemii*** Kausel
Myrceugenia exsucca var. *temu* Reiche = ***Myrceugenia exsucca*** (DC.) O.Berg
Myrceugenia ferreira-limana R.M.Klein & D.Legrand = ***Myrceugenia myrcioides*** (Cambess.) O.Berg
Myrceugenia ferruginea (Hook. & Arn.) Reiche = ***Myrceugenia rufa*** (Colla) Skottsb.
Myrceugenia filibracteata (Burret) D.Legrand = ***Myrceugenia miersiana*** (Gardner) D.Legrand & Kausel
Myrceugenia fuscovelutina (Burret) D.Legrand & Kausel = ***Myrceugenia miersiana*** (Gardner) D.Legrand & Kausel
Myrceugenia glaucescens f. *catharinensis* D.Legrand & R.M.Klein = ***Myrceugenia glaucescens*** (Cambess.) D.Legrand & Kausel var. ***glaucescens***
Myrceugenia glaucescens f. *debilis* D.Legrand & R.M.Klein = ***Myrceugenia glaucescens*** (Cambess.) D.Legrand & Kausel var. ***glaucescens***
Myrceugenia glaucescens f. *pallida* (O.Berg) D.Legrand = ***Myrceugenia glaucescens*** (Cambess.) D.Legrand & Kausel var. ***glaucescens***
Myrceugenia glaucescens var. *pallida* (O.Berg) Kausel = ***Myrceugenia glaucescens*** (Cambess.) D.Legrand & Kausel var. ***glaucescens***
Myrceugenia glaucescens var. *ribeireana* (O.Berg) D.Legrand = ***Myrceugenia glaucescens*** (Cambess.) D.Legrand & Kausel var. ***glaucescens***
Myrceugenia grisea D.Legrand = ***Myrceugenia glaucescens*** (Cambess.) D.Legrand & Kausel var. ***glaucescens***
Myrceugenia itatiaiensis (Kiaersk.) D.Legrand & Kausel = ***Myrceugenia ovata*** var. ***regnelliana*** (O.Berg) Landrum
Myrceugenia johowii Gusinde = ***Myrceugenia correifolia*** (Hook. & Arn.) O.Berg
Myrceugenia jonssonii Kausel = ***Myrceugenia rufescens*** (DC.) D.Legrand & Kausel
Myrceugenia latior (Burret) D.Legrand & Kausel = ***Myrceugenia glaucescens*** var. ***latior*** (Burret) Landrum
Myrceugenia lechleriana O.Berg = ***Myrceugenia exsucca*** (DC.) O.Berg
Myrceugenia leptocalyx D.Legrand = ***Myrceugenia foveolata*** (O.Berg) Sobral
Myrceugenia leptocalyx var. *coriacea* Mattos = ***Myrceugenia foveolata*** (O.Berg) Sobral
Myrceugenia leptorhyncha D.Legrand = ***Myrceugenia alpigena*** (DC.) Landrum var. ***alpigena***
Myrceugenia longifolia (Burret) D.Legrand & Kausel = ***Myrceugenia alpigena*** var. ***longifolia*** (Burret) Landrum
Myrceugenia longipedicellata Barb.Rodr. = [85 PAR]
Myrceugenia longipedunculata Mattos & D.Legrand = ***Myrceugenia franciscensis*** (O.Berg) Landrum
Myrceugenia luma O.Berg = ***Myrceugenia fernandeziana*** (Hook. & Arn.) Johow
Myrceugenia luma (Molina) I.M.Johnst. = ***Amomyrtus luma*** (Molina) D.Legrand & Kausel

Myrceugenia macrosepala (Burret) D.Legrand & Kausel = ***Myrceugenia oxysepala*** (Burret) D.Legrand & Kausel
Myrceugenia macrosepala var. *obovata* Mattos = ***Myrceugenia oxysepala*** (Burret) D.Legrand & Kausel
Myrceugenia malvillana Kausel = ***Myrceugenia colchaguensis*** (Phil.) Navas
Myrceugenia maritima Johow = ***Myrceugenia correifolia*** (Hook. & Arn.) O.Berg
Myrceugenia miersiana var. *lanceolata* D.Legrand = ***Myrceugenia venosa*** D.Legrand
Myrceugenia miersiana var. *macahensis* (Burret) D.Legrand = ***Myrceugenia pilotantha*** (Kiaersk.) Landrum var. ***pilotantha***
Myrceugenia montana Kausel = ***Myrceugenia ovata*** var. ***nannophylla*** (Burret) Landrum
Myrceugenia montevidensis O.Berg = ***Myrceugenia myrtoides*** O.Berg
Myrceugenia montevidensis var. *sellowiana* (O.Berg) Mattos = ***Myrceugenia myrtoides*** O.Berg
Myrceugenia monticola D.Legrand & Kausel = ***Myrceugenia bracteosa*** (DC.) D.Legrand & Kausel
Myrceugenia mosenii Kausel = ***Eugenia magnifica*** Spring ex Mart.
Myrceugenia mucronata (Burret) D.Legrand & Kausel = ***Myrceugenia franciscensis*** (O.Berg) Landrum
Myrceugenia multiflora (Hook. & Arn.) Kausel = ***Myrceugenia exsucca*** (DC.) O.Berg
Myrceugenia myrcioides var. *acrophylla* (O.Berg) D.Legrand = ***Myrceugenia myrcioides*** (Cambess.) O.Berg
Myrceugenia myrcioides var. *hypericifolia* (Gardner) D.Legrand = ***Myrceugenia myrcioides*** (Cambess.) O.Berg
Myrceugenia myrcioides var. *loefgreniana* Mattos = ***Myrceugenia myrcioides*** (Cambess.) O.Berg
Myrceugenia myrcioides var. *paranensis* D.Legrand = ***Myrceugenia myrcioides*** (Cambess.) O.Berg
Myrceugenia myrcioides var. *ulei* (Burret) D.Legrand = ***Myrceugenia myrcioides*** (Cambess.) O.Berg
Myrceugenia myrtoides var. *conferta* O.Berg = ***Myrceugenia myrtoides*** O.Berg
Myrceugenia myrtoides var. *sellowiana* (O.Berg) Mattos = ***Myrceugenia myrtoides*** O.Berg
Myrceugenia myrtoides var. *stricta* O.Berg = ***Myrceugenia myrtoides*** O.Berg
Myrceugenia nannophylla (Burret) Kausel = ***Myrceugenia ovata*** var. ***nannophylla*** (Burret) Landrum
Myrceugenia nothorufa D.Legrand = ***Myrceugenia pilotantha*** var. ***nothorufa*** (D.Legrand) Landrum
Myrceugenia nothorufa var. *major* D.Legrand = ***Myrceugenia pilotantha*** var. ***nothorufa*** (D.Legrand) Landrum
Myrceugenia nothorufa var. *venosa* D.Legrand = ***Myrceugenia miersiana*** (Gardner) D.Legrand & Kausel
Myrceugenia obtusa var. *berteroana* O.Berg = ***Myrceugenia obtusa*** (DC.) O.Berg
Myrceugenia obtusa var. *polyantha* (Phil.) O.Berg = ***Myrceugenia obtusa*** (DC.) O.Berg
Myrceugenia obtusa var. *raran* (Colla) O.Berg = ***Myrceugenia obtusa*** (DC.) O.Berg
Myrceugenia ovata var. *gracilis* (Burret) Landrum = ***Myrceugenia ovata*** var. ***regnelliana*** (O.Berg) Landrum
Myrceugenia pallida (O.Berg) D.Legrand & Kausel = ***Myrceugenia glaucescens*** (Cambess.) D.Legrand & Kausel var. ***glaucescens***
Myrceugenia paranaguensis Mattos = ***Myrceugenia myrcioides*** (Cambess.) O.Berg
Myrceugenia pilotantha var. *major* (D.Legrand) Landrum = ***Myrceugenia pilotantha*** var. ***nothorufa*** (D.Legrand) Landrum
Myrceugenia pitra (O.Berg) O.Berg = ***Myrceugenia exsucca*** (DC.) O.Berg
Myrceugenia pitra var. *angustifolia* Reiche = ***Myrceugenia exsucca*** (DC.) O.Berg
Myrceugenia planipes var. *genuina* Reiche = ***Myrceugenia planipes*** (Hook. & Arn.) O.Berg
Myrceugenia planipes var. *grandifolia* Reiche = ***Myrceugenia planipes*** (Hook. & Arn.) O.Berg
Myrceugenia ramboi D.Legrand = ***Myrceugenia pilotantha*** var. ***nothorufa*** (D.Legrand) Landrum
Myrceugenia regnelliana (O.Berg) D.Legrand & Kausel = ***Myrceugenia ovata*** var. ***regnelliana*** (O.Berg) Landrum
Myrceugenia regnelliana var. *capillaris* D.Legrand = ***Myrceugenia ovata*** var. ***regnelliana*** (O.Berg) Landrum
Myrceugenia regnelliana var. *coriacea* D.Legrand = ***Myrceugenia ovata*** var. ***regnelliana*** (O.Berg) Landrum
Myrceugenia regnelliana var. *dubia* D.Legrand = *Myrceugenia euosma* × *Myrceugenia glaucescens*
Myrceugenia regnelliana var. *gracilis* (Burret) D.Legrand = ***Myrceugenia ovata*** var. ***regnelliana*** (O.Berg) Landrum
Myrceugenia regnelliana f. *itatiaiensis* (Kiaersk.) D.Legrand = ***Myrceugenia ovata*** var. ***regnelliana*** (O.Berg) Landrum
Myrceugenia regnelliana var. *leptophylla* D.Legrand = ***Myrceugenia ovata*** var. ***regnelliana*** (O.Berg) Landrum
Myrceugenia regnelliana f. *xanthochlora* (O.Berg) D.Legrand = ***Myrceugenia ovata*** var. ***regnelliana*** (O.Berg) Landrum
Myrceugenia ribeireana (O.Berg) D.Legrand & Kausel = ***Myrceugenia glaucescens*** (Cambess.) D.Legrand & Kausel var. ***glaucescens***
Myrceugenia rufescens var. *alegrensis* D.Legrand = ***Myrceugenia rufescens*** (DC.) D.Legrand & Kausel
Myrceugenia sellowiana O.Berg = ***Myrceugenia myrtoides*** O.Berg
Myrceugenia stenophylla (Hook. & Arn.) O.Berg = ***Myrceugenia lanceolata*** (Juss. ex J.St.-Hil.) Kausel
Myrceugenia stenophylla var. *angustifolia* Reiche = ***Myrceugenia lanceolata*** (Juss. ex J.St.-Hil.) Kausel
Myrceugenia stenophylla var. *latifolia* Reiche = ***Myrceugenia lanceolata*** (Juss. ex J.St.-Hil.) Kausel
Myrceugenia stenophylla var. *pinifolia* (Phil.) Reiche = ***Myrceugenia pinifolia*** (Phil.) Kausel
Myrceugenia thalassaia (O.Berg) Gusinde ex Fuentes = ***Myrceugenia correifolia*** (Hook. & Arn.) O.Berg
Myrceugenia valientei Kausel = ***Myrceugenia ovata*** var. ***nannophylla*** (Burret) Landrum

***Unplaced Names*:**
Myrceugenia longipedicellata Barb.Rodr., Myrt. Paraguay: 2 (1903). = [85 PAR]
Myrceugenia regnelliana var. *dubia* D.Legrand, in Fl. Ilustr. Catar. 1(Mirt.): 429 (1970). = *Myrceugenia euosma* × *Myrceugenia glaucescens*

Myrcia

Myrcia DC. ex Guill. in J.B.G. Bory de Saint-Vincent, Dict. Class. Hist. Nat. 11: 401 (1827).
Mexico to Trop. America. 79 MXG MXS MXT 80

BLZ COS GUA HON NIC PAN 81 CUB DOM HAI JAM LEE NLA PUE TRT WIN 82 FRG GUY SUR VEN 83 BOL CLM ECU PER 84 BZC BZE BZL BZN BZS 85 AGE PAR URU.
374 Species
Aguava Raf., Sylva Tellur.: 107 (1838).
Cumetea Raf., Sylva Tellur.: 106 (1838).
Aulomyrcia O.Berg, Linnaea 27: 35 (1855).
Calyptromyrcia O.Berg, Linnaea 27: 34 (1855).
Cerqueiria O.Berg, Linnaea 27: 5 (1855).
Gomidesia O.Berg, Linnaea 27: 6 (1855).
Calycampe O.Berg, Linnaea 27: 129 (1856).
Mozartia Urb., Symb. Antill. 9: 87 (1923).

Myrcia abbotiana (Urb.) Alain, Mem. New York Bot. Gard. 21(2): 138 (1971).
Dominican Rep. 81 DOM. Nanophan. or phan.
**Eugenia abbotiana* Urb., Repert. Spec. Nov. Regni Veg. 20: 341 (1924). *Mozartia abbottiana* (Urb.) Urb., Ark. Bot. 22A(10): 24 (1929).

Myrcia acunae Borhidi, Bot. Közlem. 64: 20 (1977).
Cuba. 81 CUB. Nanophan. or phan.

Myrcia aegiphylloides Mattos, Loefgrenia 66: 2 (1975).
SE. Brazil. 84 BZL. Nanophan. or phan.
**Aulomyrcia anomala* O.Berg in C.F.P.von Martius & auct. suc. (eds.), Fl. Bras. 14(1): 96 (1857).

Myrcia aethusa (O.Berg) N.Silveira, Roessléria 7: 67 (1975).
SE. & S. Brazil. 84 BZL BZS. Nanophan. or phan.
**Aulomyrcia aethusa* O.Berg in C.F.P.von Martius & auct. suc. (eds.), Fl. Bras. 14(1): 112 (1857).
Aulomyrcia richardiana O.Berg, Linnaea 30: 664 (1861). *Myrcia richardiana* (O.Berg) Kiaersk., Enum. Myrt. Bras.: 97 (1893).

Myrcia albescens (Alain) Alain, Mem. New York Bot. Gard. 21(2): 138 (1971).
Cuba (Sierra de Moa). 81 CUB. Nanophan.
**Mozartia albescens* Alain, Contr. Ocas. Mus. Hist. Nat. Colegio "De La Salle" 12: 10 (1953).

Myrcia albidotomentosa (Amshoff) McVaugh, Mem. New York Bot. Gard. 18(2): 79 (1969).
N. South America to N. Brazil. 82 GUY SUR VEN 84 BZN. Phan.
**Aulomyrcia albidotomentosa* Amshoff, Bull. Torrey Bot. Club 75: 532 (1948).
Aulomyrcia amshoffiana Steyerm., Fieldiana, Bot. 28: 1003 (1957).

Myrcia albobrunnea McVaugh, Fieldiana, Bot. 29: 187 (1956).
Peru. 83 PER. Phan.

Myrcia albotomentosa DC., Prodr. 3: 254 (1828). *Aulomyrcia albotomentosa* (DC.) O.Berg, Linnaea 27: 62 (1855).
WC. & SE. Brazil. 84 BZC BZL. Nanophan. or phan.
Myrcia albotomentosa var. *humilis* Cambess. in A.F.C.P.de Saint-Hilaire & al., Fl. Bras. Merid. 2: 330 (1832).
Myrcia albotomentosa var. *nivea* Cambess. in A.F.C.P. de Saint-Hilaire & al., Fl. Bras. Merid. 2: 330 (1832).
Aulomyrcia albotomentosa var. *lutescens* O.Berg in C.F.P.von Martius & auct. suc. (eds.), Fl. Bras. 14(1): 105 (1857).
Aulomyrcia albotomentosa var. *rubrescens* O.Berg in C.F.P.von Martius & auct. suc. (eds.), Fl. Bras. 14(1): 105 (1857).
Aulomyrcia albotomentosa var. *subcordata* O.Berg in C.F.P.von Martius & auct. suc. (eds.), Fl. Bras. 14(1): 549 (1859).

Myrcia aliena McVaugh, Publ. Field Mus. Nat. Hist., Bot. Ser. 13(4): 627 (1958).
S. Trop. America. 82 VEN 83 BOL ECU PER 84 BZN. Phan.
**Aulomyrcia chilensis* O.Berg, Linnaea 27: 38 (1855).

Myrcia almasensis NicLugh., Kew Bull. 49: 322 (1994).
Brazil (Bahia). 84 BZE. Nanophan. or phan.

Myrcia amapensis McVaugh, Mem. New York Bot. Gard. 18(2): 80 (1969).
Brazil (Amapá). 84 BZN. Nanophan. or phan.

Myrcia amazonica DC., Prodr. 3: 250 (1828). *Aulomyrcia amazonica* (DC.) O.Berg, Linnaea 27: 41 (1855).
Trop. America. 80 BLZ GUA HON NIC 81 DOM HAI JAM LEE PUE TRT WIN 82 FRG GUY SUR VEN 83 BOL 84 BZC BZE BZL BZN BZS. Phan.
Myrcia corymbosa DC., Prodr. 3: 252 (1828). *Aulomyrcia corymbosa* (DC.) O.Berg, Linnaea 27: 41 (1855).
Myrcia leptoclada DC., Prodr. 3: 244 (1828). *Aulomyrcia leptoclada* (DC.) O.Berg, Linnaea 27: 40 (1855).
Myrcia detergens Miq., Linnaea 22: 795 (1850). *Aulomyrcia detergens* (Miq.) O.Berg, Linnaea 27: 46 (1855).
Aulomyrcia hostmanniana var. *gracilior* O.Berg, Linnaea 27: 43 (1855).
Aulomyrcia hostmanniana O.Berg, Linnaea 27: 442 (1856). *Myrcia hostmanniana* (O.Berg) Kiaersk., Enum. Myrt. Bras.: 83 (1893).
Aulomyrcia detergens var. *depauperata* O.Berg in C.F.P.von Martius & auct. suc. (eds.), Fl. Bras. 14(1): 77 (1857).
Aulomyrcia detergens var. *dives* O.Berg in C.F.P.von Martius & auct. suc. (eds.), Fl. Bras. 14(1): 77 (1857).
Aulomyrcia paraensis O.Berg in C.F.P.von Martius & auct. suc. (eds.), Fl. Bras. 14(1): 76 (1857). *Myrcia paraensis* (O.Berg) Kiaersk., Enum. Myrt. Bras.: 78 (1893).
Aulomyrcia spruceana O.Berg in C.F.P.von Martius & auct. suc. (eds.), Fl. Bras. 14(1): 76 (1857).
Myrcia oitchi O.Berg in C.F.P.von Martius & auct. suc. (eds.), Fl. Bras. 14(1): 563 (1859).
Myrcia hostmanniana var. *robustior* Kiaersk., Enum. Myrt. Bras.: 83 (1893).
Myrcia leptoclada var. *glazioviana* Kiaersk., Enum. Myrt. Bras.: 78 (1893).
Myrcia gentlei Lundell, Wrightia 2: 214 (1961).
Gomidesia minutiflora Mattos & D.Legrand, Loefgrenia 67: 15 (1975).

Myrcia ambivalens McVaugh, Fieldiana, Bot. 29: 188 (1956).
N. Peru. 83 PER. Nanophan.

Myrcia amblyophylla Kiaersk., Enum. Myrt. Bras.: 76 (1893).
SE. Brazil. 84 BZL. Nanophan. or phan.

Myrcia amethystina (O.Berg) Kiaersk., Enum. Myrt. Bras.: 90 (1893).
SE. Brazil. 84 BZL. Nanophan. or phan.
**Aulomyrcia amethystina* O.Berg in C.F.P.von Martius & auct. suc. (eds.), Fl. Bras. 14(1): 108 (1857).

Aulomyrcia amethystina var. *dealbata* O.Berg in C.F.P.von Martius & auct. suc. (eds.), Fl. Bras. 14(1): 108 (1857).
Aulomyrcia amethystina var. *pulchra* O.Berg in C.F.P.von Martius & auct. suc. (eds.), Fl. Bras. 14(1): 108 (1857).
Myrcia amethystina var. *dealbata* Kiaersk., Enum. Myrt. Bras.: 90 (1893).
Myrcia amethystina var. *pulchra* Kiaersk., Enum. Myrt. Bras.: 90 (1893).

Myrcia amplexicaulis (Vell.) Hook.f., Bot. Mag. 95: t. 5790 (1869).
E. Brazil. 84 BZE BZL. Nanophan. or phan.
**Eugenia amplexicaulis* Vell., Fl. Flumin. 5: 210, t. 44 (1829). *Gomidesia amplexicaulis* (Vell.) O.Berg in C.F.P.von Martius & auct. suc. (eds.), Fl. Bras. 14(1): 13 (1857).
Gomidesia crocea O.Berg in C.F.P.von Martius & auct. suc. (eds.), Fl. Bras. 14(1): 533 (1859).
Gomidesia langsdorffii O.Berg in C.F.P.von Martius & auct. suc. (eds.), Fl. Bras. 14(1): 531 (1859).

Myrcia anacardiifolia Gardner, London J. Bot. 2: 354 (1843). *Gomidesia anacardiifolia* (Gardner) O.Berg in C.F.P.von Martius & auct. suc. (eds.), Fl. Bras. 14(1): 15 (1857).
SE. & S. Brazil. 84 BZL BZS. Nanophan. or phan.
Gomidesia anacardiifolia var. *oblongata* O.Berg in C.F.P.von Martius & auct. suc. (eds.), Fl. Bras. 14(1): 15 (1857).
Gomidesia anacardiifolia var. *opaca* O.Berg in C.F.P.von Martius & auct. suc. (eds.), Fl. Bras. 14(1): 15 (1857).
Gomidesia anacardiifolia var. *ovalis* O.Berg in C.F.P.von Martius & auct. suc. (eds.), Fl. Bras. 14(1): 15 (1857).
Gomidesia riedeliana O.Berg in C.F.P.von Martius & auct. suc. (eds.), Fl. Bras. 14(1): 532 (1859). *Myrcia estrellensis* Kiaersk., Enum. Myrt. Bras.: 107 (1893).

Myrcia anceps (Spreng.) O.Berg in C.F.P.von Martius & auct. suc. (eds.), Fl. Bras. 14(1): 186 (1857).
SE. Brazil. 84 BZL. Nanophan. or phan.
**Myrtus anceps* Spreng., Neue Entd. 2: 170 (1821).
Myrcia anceps var. *brevipes* O.Berg in C.F.P.von Martius & auct. suc. (eds.), Fl. Bras. 14(1): 186 (1857).
Myrcia anceps var. *depauperata* O.Berg in C.F.P.von Martius & auct. suc. (eds.), Fl. Bras. 14(1): 186 (1857).
Myrcia anceps var. *dives* O.Berg in C.F.P.von Martius & auct. suc. (eds.), Fl. Bras. 14(1): 186 (1857).

Myrcia angusta G.Don, Gen. Hist. 2: 845 (1832).
NE. Brazil. 84 BZE. Nanophan. or phan.

Myrcia angustifolia (O.Berg) Nied. in H.G.A.Engler & K.A.E.Prantl, Nat. Pflanzenfam. 3(7): 76 (1893).
SE. Brazil. 84 BZL. Nanophan. or phan.
**Aulomyrcia angustifolia* O.Berg in C.F.P.von Martius & auct. suc. (eds.), Fl. Bras. 14(1): 135 (1857).

Myrcia anomala Cambess. in A.F.C.de Saint-Hilaire, Fl. Bras. Merid. 2: 328 (1832).
Bolivia to Brazil. 83 BOL 84 BZC BZL BZS 85 AGE PAR. Cham.
Myrcia anomala var. *ramosa* Cambess. in A.F.C.de Saint-Hilaire, Fl. Bras. Merid. 2: 328 (1832).
Myrcia alpestris Barb.Rodr. ex Chodat & Hassl., Bull. Herb. Boissier, II, 7: 803 (1907), nom. nud.
Myrcia cotonosa Barb.Rodr. ex Chodat & Hassl., Bull. Herb. Boissier, II, 7: 802 (1907), nom. nud.

Myrcia antillana McVaugh, J. Arnold Arbor. 54: 311 (1973).
Lesser Antilles. 81 LEE WIN. Nanophan. or phan.
Myrcia edulis var. *dominicana* Krug & Urb., Bot. Jahrb. Syst. 19: 582 (1893).

Myrcia apiocarpa (O.Berg) N.Silveira, Roessléria 7: 65 (1985).
SE. Brazil. 84 BZL. Nanophan. or phan.
**Aulomyrcia apiocarpa* O.Berg in C.F.P.von Martius & auct. suc. (eds.), Fl. Bras. 14(1): 64 (1857).

Myrcia apodocarpa Urb., Symb. Antill. 9: 87 (1923).
SE. Cuba. 81 CUB. Nanophan.

Myrcia atramentifera Barb.Rodr., Vellosia, sec. ed.: 31 (1891).
N. Brazil. 84 BZN. Phan.

Myrcia atropilosa (O.Berg) N.Silveira, Roessléria 7: 65 (1985).
S. Brazil. 84 BZS. Nanophan. or phan.
**Aulomyrcia atropilosa* O.Berg in C.F.P.von Martius & auct. suc. (eds.), Fl. Bras. 14(1): 101 (1857).

Myrcia atropunctata Kiaersk., Enum. Myrt. Bras.: 68 (1893).
SE. Brazil. 84 BZL. Nanophan. or phan.

Myrcia atrorufa McVaugh, Fieldiana, Bot. 29: 188 (1956).
Peru. 83 PER. Nanophan.

Myrcia ayabambensis Hieron., Bot. Jahrb. Syst. 20(49): 64 (1895).
Ecuador. 83 ECU. Phan.

Myrcia badia (O.Berg) N.Silveira, Roessléria 7: 66 (1985).
NE. Brazil. 84 BZE. Nanophan. or phan.
**Aulomyrcia badia* O.Berg in C.F.P.von Martius & auct. suc. (eds.), Fl. Bras. 14(1): 547 (1859).

Myrcia bella Cambess. in A.F.C.de Saint-Hilaire, Fl. Bras. Merid. 2: 322 (1832). *Aulomyrcia bella* (Cambess.) O.Berg in C.F.P.von Martius & auct. suc. (eds.), Fl. Bras. 14(1): 71 (1857).
Brazil. 84 BZC BZL. Nanophan. or phan.
Myrcia dasyblasta O.Berg in C.F.P.von Martius & auct. suc. (eds.), Fl. Bras. 14(1): 207 (1857).
Myrcia dasyblasta var. *cordata* O.Berg in C.F.P.von Martius & auct. suc. (eds.), Fl. Bras. 14(1): 568 (1859).
Myrcia dasyblasta var. *ovata* O.Berg in C.F.P.von Martius & auct. suc. (eds.), Fl. Bras. 14(1): 568 (1859).

Myrcia bergiana O.Berg in C.F.P.von Martius & auct. suc. (eds.), Fl. Bras. 14(1): 194 (1857).
Brazil (Bahia). 84 BZE. Nanophan. or phan.
Myrcia bergiana var. *angustifolia* O.Berg in C.F.P.von Martius & auct. suc. (eds.), Fl. Bras. 14(1): 567 (1859).

Myrcia bicarinata (O.Berg) D.Legrand, Sellowia 13: 298 (1961).
SE. & S. Brazil. 84 BZL BZS. Nanophan. or phan.
**Aulomyrcia bicarinata* O.Berg in C.F.P.von Martius & auct. suc. (eds.), Fl. Bras. 14(1): 118 (1857).

Myrcia bicolor Kiaersk., Enum. Myrt. Bras.: 65 (1893).
SE. Brazil. 84 BZL. Nanophan. or phan.

Myrcia billardiana (Kunth) DC., Prodr. 3: 255 (1828).
Colombia to NE. Venezuela. 82 VEN 83 CLM.

Nanophan. or phan.
Myrtus billardiana Kunth in F.W.H.von Humboldt, A.J.A.Bonpland & C.S.Kunth, Nov. Gen. Sp. 6: 139 (1823). *Opanea billardiana* (Kunth) Raf., Sylva Tellur.: 106 (1838). *Aulomyrcia billardiana* (Kunth) O.Berg, Linnaea 27: 51 (1855).
Aulomyrcia billardiana var. *grandifolia* O.Berg, Linnaea 27: 51 (1855).
Aulomyrcia billardiana var. *parvifolia* O.Berg, Linnaea 27: 51 (1855).

Myrcia blanchetiana (O.Berg) Mattos, Arq. Bot. Estado São Paulo, n.s., f.m., 4: 59 (1966).
Brazil (Bahia, Minas Gerais). 84 BZE BZL. Nanophan. or phan.
Aulomyrcia blanchetiana O.Berg in C.F.P.von Martius & auct. suc. (eds.), Fl. Bras. 14(1): 65 (1857).

Myrcia bolivarensis (Steyerm.) McVaugh, Mem. New York Bot. Gard. 18(2): 81 (1969).
Venezuela to Guyana. 82 GUY VEN. Nanophan. or phan.
Aulomyrcia bolivarensis Steyerm., Fieldiana, Bot. 28: 1004 (1957).

Myrcia bombycina (O.Berg) Nied. in H.G.A.Engler & K.A.E.Prantl, Nat. Pflanzenfam. 3(7): 75 (1893).
Brazil to NE. Argentina. 84 BZC BZL BZS 85 AGE PAR. Nanophan. or phan.
Aulomyrcia bombycina O.Berg in C.F.P.von Martius & auct. suc. (eds.), Fl. Bras. 14(1): 66 (1857).

Myrcia bonnetiasylvestris (Steyerm.) Steyerm., Ann. Missouri Bot. Gard. 71: 330 (1984).
Venezuela (Bolívar). 82 VEN. Nanophan. or phan.
Gomidesia bonnetiasylvestris Steyerm., Fieldiana, Bot. 28: 1016 (1957).

Myrcia borhidii O.Muñiz, Bot. Közlem. 64: 20 (1977).
Cuba. 81 CUB. Nanophan.

Myrcia bracteata (Rich.) DC., Prodr. 3: 245 (1828).
S. Trop. America. 82 FRG GUY SUR VEN 83 ECU PER 84 BZC BZL BZN. Nanophan. or phan.
Eugenia bracteata Rich., Actes Soc. Hist. Nat. Paris 1: 110 (1792).
Myrcia lanceolata Cambess. in A.F.C.de Saint-Hilaire, Fl. Bras. Merid. 2: 329 (1832).
Myrcia lanceolata var. *angustifolia* Cambess. in A.F.C.P.de Saint-Hilaire & al., Fl. Bras. Merid. 2: 329 (1832).
Myrcia lanceolata var. *avenia* O.Berg in C.F.P.von Martius & auct. suc. (eds.), Fl. Bras. 14(1): 156 (1857).
Myrcia lanceolata var. *grandifolia* O.Berg in C.F.P.von Martius & auct. suc. (eds.), Fl. Bras. 14(1): 155 (1857).
Myrcia lanceolata var. *latifolia* O.Berg in C.F.P.von Martius & auct. suc. (eds.), Fl. Bras. 14(1): 155 (1857).
Myrcia lanceolata var. *racemosa* O.Berg in C.F.P.von Martius & auct. suc. (eds.), Fl. Bras. 14(1): 156 (1857).
Myrcia hirtellifolia Gleason, Bull. Torrey Bot. Club 58: 411 (1931).
Eugenia hirsuta Ruiz & Pav., Anales Inst. Bot. Cavanilles 15: 186 (1957).

Myrcia brasiliae Mattos & D.Legrand, Loefgrenia 67: 4 (1975).
Brazil (Brasília D.F.). 84 BZC. Nanophan. or phan.

Myrcia brasiliensis Kiaersk., Enum. Myrt. Bras.: 102 (1893).
SE. & S. Brazil. 84 BZL BZS. Phan.
Gomidesia schaueriana O.Berg in C.F.P.von Martius & auct. suc. (eds.), Fl. Bras. 14(1): 18 (1857).
Gomidesia schaueriana var. *spathulata* Mattos, Loefgrenia 66: 6 (1975).

Myrcia bredemeyeriana O.Berg, Linnaea 27: 85 (1855).
Venezuela. 82 VEN. Nanophan. or phan.

Myrcia brunnea Cambess. in A.F.C.de Saint-Hilaire, Fl. Bras. Merid. 2: 306 (1832). *Gomidesia brunnea* (Cambess.) D.Legrand, Notul. Syst. (Paris) 15: 261 (1958).
Brazil (Minas Gerais). 84 BZL. Nanophan.
Myrcia brunnea var. *opaca* O.Berg in C.F.P.von Martius & auct. suc. (eds.), Fl. Bras. 14(1): 567 (1859).

Myrcia bullata O.Berg in C.F.P.von Martius & auct. suc. (eds.), Fl. Bras. 14(1): 153 (1857).
E. Brazil. 84 BZE BZL. Nanophan. or phan.

Myrcia buxifolia Gardner, London J. Bot. 4: 101 (1845).
Brazil (Rio de Janeiro). 84 BZL. Nanophan. or phan.
Myrcia buxifolia var. *glazioviana* Kiaersk., Enum. Myrt. Bras.: 61 (1893).

Myrcia calcicola Proctor, J. Arnold Arbor. 63: 281 (1982).
Jamaica. 81 JAM. Nanophan. or phan.

Myrcia calycampa Amshoff, Recueil Trav. Bot. Néerl. 39: 153 (1943).
N. South America to N. Brazil. 82 FRG? GUY SUR VEN 84 BZN. Phan.
Calycampe angustifolia O.Berg, Linnaea 27: 131 (1856).
Calycampe latifolia O.Berg, Linnaea 27: 130 (1856). *Myrtus latifolia* (O.Berg) V.M.Badillo in H.F.Pittier & al., Cat. Fl. Venez. 2: 196 (1947), nom. illeg.

Myrcia calyptranthoides (O.Berg) Mattos, Arq. Bot. Estado São Paulo, n.s., f.m., 4: 60 (1966).
NE. Brazil. 84 BZE. Nanophan. or phan.
Aulomyrcia calyptranthoides O.Berg in C.F.P.von Martius & auct. suc. (eds.), Fl. Bras. 14(1): 67 (1857).

Myrcia camapuanensis N.Silveira, Loefgrenia 86: 2 (1985).
WC. Brazil. 84 BZC. Phan.
Aulomyrcia capitata O.Berg in C.F.P.von Martius & auct. suc. (eds.), Fl. Bras. 14(1): 554 (1859). *Myrcia capitata* (O.Berg) Nied. in H.G.A.Engler & K.A.E.Prantl, Nat. Pflanzenfam. 3(7): 76 (1893), nom. illeg.

Myrcia cambessedesiana O.Berg in C.F.P.von Martius & auct. suc. (eds.), Fl. Bras. 14(1): 202 (1857).
Brazil (Minas Gerais). 84 BZL. Nanophan. or phan.

Myrcia campestris DC., Prodr. 3: 247 (1828). *Aulomyrcia campestris* (DC.) O.Berg, Linnaea 27: 74 (1855).
SE. Brazil. 84 BZL. Nanophan. or phan.
Aulomyrcia campestris var. *brunnea* O.Berg in C.F.P.von Martius & auct. suc. (eds.), Fl. Bras. 14(1): 128 (1857).
Aulomyrcia campestris var. *rufa* O.Berg in C.F.P.von Martius & auct. suc. (eds.), Fl. Bras. 14(1): 128 (1857).

Myrcia cancellata O.Berg in C.F.P.von Martius & auct. suc. (eds.), Fl. Bras. 14(1): 154 (1857).
SE. Brazil. 84 BZL. Nanophan. or phan.

Myrcia canescens O.Berg in C.F.P.von Martius & auct. suc. (eds.), Fl. Bras. 14(1): 206 (1857).
Brazil (Mato Grosso, Goiás, Minas Gerais). 84 BZC BZL. Nanophan. or phan.
Myrcia canescens var. *reticulata* O.Berg in C.F.P.von Martius & auct. suc. (eds.), Fl. Bras. 14(1): 567 (1859).

Myrcia capitata O.Berg in C.F.P.von Martius & auct. suc. (eds.), Fl. Bras. 14(1): 154 (1857).
SE. Brazil. 84 BZL. Nanophan. or phan.

Myrcia caracasana (O.Berg) Klotzsch ex O.Berg, Linnaea 27: 92 (1855).
Venezuela. 82 VEN. Nanophan. or phan.
**Aulomyrcia caracasana* O.Berg, Linnaea 27: 53 (1855).

Myrcia cardiaca O.Berg in C.F.P.von Martius & auct. suc. (eds.), Fl. Bras. 14(1): 204 (1857).
WC. Brazil. 84 BZC. Nanophan. or phan.

Myrcia cardiophylla (O.Berg) Reichardt, Verh. K. K. Zool.-Bot. Ges. Wien 33: 324 (1884).
SE. Brazil. 84 BZL. Nanophan. or phan.
**Aulomyrcia cardiophylla* O.Berg in C.F.P.von Martius & auct. suc. (eds.), Fl. Bras. 14(1): 103 (1857).
Aulomyrcia cardiophylla var. *densiflora* O.Berg in C.F.P.von Martius & auct. suc. (eds.), Fl. Bras. 14(1): 104 (1857).
Aulomyrcia cardiophylla var. *laxiflora* O.Berg in C.F.P.von Martius & auct. suc. (eds.), Fl. Bras. 14(1): 104 (1857).

Myrcia chapadensis S.Moore, Trans. Linn. Soc. London, Bot. 4: 355 (1895).
WC. Brazil. 84 BZC. Phan.

Myrcia chapadinhaeana Glaz. ex Mattos & D.Legrand, Loefgrenia 67: 3 (1975).
WC. Brazil. 84 BZC. Nanophan. or phan.

var. ***chapadinhaeana***
Brazil (Brasília D.F.). 84 BZC. Nanophan. or phan.

var. ***meiapotensis*** Glaz. ex Mattos & D.Legrand, Loefgrenia 67: 4 (1975).
Brazil (Goiás). 84 BZC. Nanophan. or phan.

Myrcia citrifolia (Aubl.) Urb., Repert. Spec. Nov. Regni Veg. 16: 150 (1919).
Caribbean, S. Trop. America. 81 CUB DOM HAI LEE PUE WIN 82 FRG GUY SUR VEN 84 BZC BZL BZN BZS. Phan.
**Myrtus citrifolia* Aubl., Hist. Pl. Guiane 1: 20 (1775). *Myrtus coriacea* Vahl, Symb. Bot. 2: 59 (1791), nom. illeg. *Myrcia coriacea* DC., Prodr. 3: 243 (1828), nom. illeg. *Aulomyrcia coriacea* (DC.) O.Berg, Linnaea 27: 70 (1855). *Myrtus cotinifolia* J.F.Gmel., Syst. Nat. 2: 792 (1791), nom. illeg. *Aulomyrcia citrifolia* (Aubl.) Amshoff, Bull. Torrey Bot. Club 75: 531 (1948).
Eugenia paniculata Jacq., Collectanea 2: 108 (1789). *Aulomyrcia jacquiniana* O.Berg, Linnaea 27: 69 (1855), nom. illeg. *Myrcia coriacea* var. *jacquiniana* Griseb., Fl. Brit. W. I.: 234 (1860). *Myrcia citrifolia* var. *jacquiniana* Stehlé & Quentin, Fl. Guadeloupe 2(3): 57 (1949). *Myrcia paniculata* (Jacq.) Krug & Urb., Bot. Jahrb. Syst. 19: 577 (1895).
Eugenia marginata Pers., Syn. Pl. 2: 28 (1806). *Myrtus marginata* (Pers.) Spreng., Syst. Veg. 2: 438 (1825).
Eugenia acetosans Poir. in J.B.A.P.M.de Lamarck, Encycl., Suppl. 3: 125 (1813). *Myrtus acetosans* (Poir.) Spreng., Syst. Veg. 2: 488 (1825). *Aulomyrcia acetosans* (Poir.) O.Berg, Linnaea 30: 662 (1861).
Myrcia vernicosa DC., Prodr. 3: 256 (1828).
Myrcia coriacea var. *imrayana* Griseb., Fl. Brit. W. I.: 234 (1860). *Myrcia paniculata* var. *imrayana* (Griseb.) Duss, Ann. Inst. Bot.-Géol. Colon. Marseille 3: 264 (1897). *Myrcia citrifolia* var. *imrayana* (Griseb.) Stehlé & Quentin, Fl. Guadeloupe 2(3): 57 (1949).
Myrcia coriacea var. *swartziana* Griseb., Fl. Brit. W. I.: 234 (1860).
Aulomyrcia coriacea var. *parvifolia* O.Berg, Linnaea 30: 662 (1861).
Myrcia coriacea var. *reticulata* Griseb., Mem. Amer. Acad. Arts, n.s., 8: 182 (1861).
Myrcia coriacea var. *acutifolia* Griseb., Cat. Pl. Cub.: 86 (1866).

Myrcia clausa McVaugh, Mem. New York Bot. Gard. 18(2): 128 (1969).
Venezuela (Bolívar). 82 GUY? VEN. Nanophan. or phan.

Myrcia clavija Sobral, Novon 16: 520 (2006).
Brazil (Minas Gerais). 84 BZL.

Myrcia clusiifolia (Kunth) DC., Prodr. 3: 255 (1828).
SE. Colombia to N. Brazil. 82 VEN 83 CLM 84 BZN. Nanophan. or phan.
**Myrtus clusiifolia* Kunth in F.W.H.von Humboldt, A.J.A.Bonpland & C.S.Kunth, Nov. Gen. Sp. 6: 138 (1823).

Myrcia coelosepala Kiaersk., Enum. Myrt. Bras.: 81 (1893).
SE. Brazil. 84 BZL. Nanophan. or phan.

Myrcia collina S.Moore, Trans. Linn. Soc. London, Bot. 4: 356 (1895).
WC. Brazil. 84 BZC. Nanophan. or phan.

Myrcia colpodes Kiaersk., Enum. Myrt. Bras.: 80 (1893).
SE. Brazil. 84 BZL. Nanophan. or phan.

Myrcia compta McVaugh, Mem. New York Bot. Gard. 18(2): 82 (1969).
Venezuela (Amazonas). 82 VEN. Nanophan. or phan.

Myrcia concava McVaugh, Fieldiana, Bot. 29: 189 (1956).
N. Peru. 83 PER. Phan.

Myrcia concinna B.Holst & M.L.Kawas., Selbyana 25: 93 (2004).
Panama. 80 PAN. Phan.

Myrcia connata McVaugh, Fieldiana, Bot. 29: 189 (1956).
Bolivia. 83 BOL. Nanophan. or phan.

Myrcia cordiifolia DC., Prodr. 3: 248 (1828). *Gomidesia linkiana* O.Berg in C.F.P.von Martius & auct. suc. (eds.), Fl. Bras. 14(1): 16 (1857), nom. illeg. *Gomidesia cordiifolia* (DC.) NicLugh., Kew Bull. 53: 243 (1998).
SE. Brazil. 84 BZL. Nanophan. or phan.
Myrcia linkiana DC., Prodr. 3: 248 (1828).
Gomidesia affinis var. *catharinensis* D.Legrand, in Fl. Ilustr. Catar. 1(Mirt.): 13 (1967).

Myrcia coumete (Aubl.) DC., Prodr. 3: 245 (1828).
C. America to N. South America. 80 COS PAN 82 FRG SUR. Nanophan. or phan.

Eugenia coumete Aubl., Hist. Pl. Guiane: 497 (1775). *Myrtus coumete* (Aubl.) Spreng., Syst. Veg. 2: 488 (1825). *Aulomyrcia coumete* (Aubl.) O.Berg, Linnaea 27: 60 (1855).
Myrtus carnea G.Mey., Prim. Fl. Esseq.: 191 (1818). *Myrcia carnea* (G.Mey.) DC., Prodr. 3: 245 (1828).
Cumetea alba Raf., Sylva Tellur.: 106 (1838).

Myrcia crassimarginata McVaugh, Fieldiana, Bot. 29: 190 (1956).
N. Peru. 83 PER. Phan.

Myrcia crispa McVaugh, Mem. New York Bot. Gard. 18(2): 130 (1969).
Venezuela (Amazonas). 82 VEN. Nanophan. or phan.

Myrcia cristalensis Borhidi & O.Muñiz, Bot. Közlem. 64: 20 (1977).
E. Cuba. 81 CUB. Nanophan. or phan.

Myrcia cujabensis O.Berg in C.F.P.von Martius & auct. suc. (eds.), Fl. Bras. 14(1): 161 (1857).
Brazil (Mato Grosso). 84 BZC. Nanophan. or phan.
Myrcia cujabensis var. *latifolia* O.Berg in C.F.P.von Martius & auct. suc. (eds.), Fl. Bras. 14(1): 162 (1857).
Myrcia cujabensis var. *longifolia* O.Berg in C.F.P.von Martius & auct. suc. (eds.), Fl. Bras. 14(1): 161 (1857).

Myrcia cuprea (O.Berg) Kiaersk., Enum. Myrt. Bras.: 95 (1893).
N. South America to N. Brazil. 82 FRG SUR 84 BZN. Nanophan. or phan.
Aulomyrcia chrysophylla O.Berg in C.F.P.von Martius & auct. suc. (eds.), Fl. Bras. 14(1): 125 (1857).
Aulomyrcia cuprea O.Berg in C.F.P.von Martius & auct. suc. (eds.), Fl. Bras. 14(1): 77 (1857).

Myrcia curassavica (Amshoff) Stoffers, Proc. Kon. Ned. Akad. Wetensch., C 84: 349 (1981).
Bonaire, Curaçao. 81 NLA. Nanophan. or phan.
Aulomyrcia curassavica Amshoff, Recueil Trav. Bot. Néerl. 42: 10 (1950).
Aulomyrcia curassavica var. *acutata* Amshoff, Recueil Trav. Bot. Néerl. 42: 10 (1950).

Myrcia cymosa (O.Berg) Nied. in H.G.A.Engler & K.A.E.Prantl, Nat. Pflanzenfam. 3(7): 76 (1893).
SE. Brazil. 84 BZL. Nanophan. or phan.
Aulomyrcia cymosa O.Berg in C.F.P.von Martius & auct. suc. (eds.), Fl. Bras. 14(1): 552 (1859).

Myrcia dealbata DC., Prodr. 3: 254 (1828). *Aulomyrcia dealbata* (DC.) O.Berg, Linnaea 27: 61 (1855).
SE. Brazil. 84 BZL. Nanophan. or phan.
Aulomyrcia dealbata var. *glaucescens* O.Berg in C.F.P.von Martius & auct. suc. (eds.), Fl. Bras. 14(1): 102 (1857).
Aulomyrcia dealbata var. *pallida* O.Berg in C.F.P.von Martius & auct. suc. (eds.), Fl. Bras. 14(1): 103 (1857).

Myrcia debilis Cambess. in A.F.C.de Saint-Hilaire, Fl. Bras. Merid. 2: 326 (1832). *Aulomyrcia debilis* (Cambess.) O.Berg, Linnaea 27: 62 (1855).
SE. & S. Brazil. 84 BZL BZS. Nanophan. or phan.

Myrcia decorticans DC., Prodr. 3: 252 (1828). *Aulomyrcia polymorpha* var. *decorticans* (DC.) O.Berg in C.F.P. von Martius & auct. suc. (eds.), Fl. Bras. 14(1): 78 (1857).
Tobago to S. Trop. America. 81 TRT 82 FRG GUY SUR VEN 84 BZC BZE BZL BZN. Nanophan. or phan.
Eugenia lasiopus Mart. ex DC., Prodr. 3: 253 (1828), nom. inval.
Myrcia duriuscula Mart. ex DC., Prodr. 3: 253 (1828). *Aulomyrcia polymorpha* var. *duriuscula* (Mart. ex DC.) O.Berg in C.F.P.von Martius & auct. suc. (eds.), Fl. Bras. 14(1): 78 (1857).
Myrcia lasiopus DC., Prodr. 3: 253 (1828). *Aulomyrcia polymorpha* var. *lasiopus* (DC.) O.Berg in C.F.P. von Martius & auct. suc. (eds.), Fl. Bras. 14(1): 78 (1857).
Myrcia leucophloea DC., Prodr. 3: 247 (1828). *Aulomyrcia polymorpha* var. *leucophloea* (DC.) O.Berg in C.F.P.von Martius & auct. suc. (eds.), Fl. Bras. 14(1): 78 (1857).
Aulomyrcia polymorpha O.Berg, Linnaea 27: 46 (1855).
Calyptranthes tobagensis Krug & Urb., Bot. Jahrb. Syst. 19: 593 (1895). *Myrcia tobagensis* (Krug & Urb.) Urb., Repert. Spec. Nov. Regni Veg. 14: 336 (1916). *Aulomyrcia tobagensis* (Krug & Urb.) Amshoff, Recueil Trav. Bot. Néerl. 39: 155 (1942).
Aulomyrcia tetramera Amshoff, Recueil Trav. Bot. Néerl. 42: 9 (1950). *Myrcia tetramera* (Amshoff) Lemée, in Fl. Guyane Franç. 3: 150 (1954).
Marlierea gleasonii McVaugh, Mem. New York Bot. Gard. 10(1): 83 (1958).

Myrcia deflexa (Poir.) DC., Prodr. 3: 244 (1828).
Caribbean, S. Trop. America. 81 CUB DOM JAM LEE PUE TRT WIN 82 FRG GUY SUR VEN 83 CLM PER 84 BZC BZN. Nanophan. or phan.
Eugenia deflexa Poir. in J.B.A.P.M.de Lamarck, Encycl., Suppl. 3: 124 (1813). *Myrcia ferruginea* var. *martinicensis* O.Berg, Linnaea 27: 90 (1855), nom. inval.
Myrcia crassinervia DC., Prodr. 3: 245 (1828).
Myrcia duchassaingiana O.Berg, Linnaea 27: 88 (1855).
Myrcia ferruginea var. *domingensis* O.Berg, Linnaea 27: 90 (1855).
Myrcia sulcata O.Berg, Linnaea 30: 667 (1861).
Myrcia deflexa var. *dussii* Krug & Urb., Bot. Jahrb. Syst. 19: 588 (1895).

Myrcia densa (DC.) Sobral, Novon 16: 136 (2006).
Brazil (Bahia, Minas Gerais). 84 BZE BZL. Nanophan.
Calyptranthes densa DC., Prodr. 3: 257 (1828). *Aulomyrcia densa* (DC.) O.Berg, Linnaea 27: 37 (1855).

Myrcia dermatophylla Kiaersk., Enum. Myrt. Bras.: 97 (1893).
SE. Brazil. 84 BZL. Nanophan. or phan.

Myrcia desertorum (O.Berg) N.Silveira, Loefgrenia 88: 1 (1985).
Brazil (Minas Gerais). 84 BZL. Nanophan. or phan.
Aulomyrcia desertorum O.Berg in C.F.P.von Martius & auct. suc. (eds.), Fl. Bras. 14(1): 556 (1859).

Myrcia diaphana (O.Berg) N.Silveira, Roessléria 7: 66 (1985).
SE. Brazil. 84 BZL. Nanophan. or phan.
Aulomyrcia diaphana O.Berg in C.F.P.von Martius & auct. suc. (eds.), Fl. Bras. 14(1): 82 (1857).

Myrcia dichasialis McVaugh, Fieldiana, Bot. 29: 190 (1956).
Venezuela to Peru. 82 VEN 83 PER. Nanophan. or phan.

Myrcia dichrophylla D.Legrand, Sellowia 13: 294 (1961).

Brazil (São Paulo to Santa Catarina). 84 BZL BZS. Nanophan. or phan.

Myrcia dictyophylla (O.Berg) Mattos & D.Legrand, Loefgrenia 67: 5 (1975).
SE. Brazil. 84 BZL. Nanophan. or phan.
**Aulomyrcia dictyophylla* O.Berg in C.F.P.von Martius & auct. suc. (eds.), Fl. Bras. 14(1): 72 (1857).
Aulomyrcia dictyophylla var. *robustior* O.Berg in C.F.P.von Martius & auct. suc. (eds.), Fl. Bras. 14(1): 545 (1859). *Myrcia dictyophylla* var. *robustior* (O.Berg) Mattos, Loefgrenia 66: 2 (1975).
Aulomyrcia dictyophylla var. *subcordata* O.Berg in C.F.P.von Martius & auct. suc. (eds.), Fl. Bras. 14(1): 545 (1859). *Myrcia dictyophylla* var. *subcordata* (O.Berg) N.Silveira, Loefgrenia 86: 2 (1985).

Myrcia didrichseniana Kiaersk., Enum. Myrt. Bras.: 82 (1893).
Brazil (Rio de Janeiro). 84 BZL. Phan.
**Aulomyrcia bracteata* O.Berg in C.F.P.von Martius & auct. suc. (eds.), Fl. Bras. 14(1): 554 (1859).

Myrcia dilucida G.M.Barroso, Bradea 5: 358 (1990).
SE. Brazil. 84 BZL. Nanophan. or phan.
**Aulomyrcia lucida* O.Berg in C.F.P.von Martius & auct. suc. (eds.), Fl. Bras. 14(1): 118 (1857).
Aulomyrcia lucida var. *grandifolia* O.Berg in C.F.P.von Martius & auct. suc. (eds.), Fl. Bras. 14(1): 119 (1857).
Aulomyrcia lucida var. *parvifolia* O.Berg in C.F.P.von Martius & auct. suc. (eds.), Fl. Bras. 14(1): 118 (1857).

Myrcia dimorpha (O.Berg) N.Silveira, Loefgrenia 88: 1 (1985).
SE. Brazil. 84 BZL. Nanophan. or phan.
**Aulomyrcia dimorpha* O.Berg in C.F.P.von Martius & auct. suc. (eds.), Fl. Bras. 14(1): 101 (1857).

Myrcia directa McVaugh, Publ. Field Mus. Nat. Hist., Bot. Ser. 13(4): 636 (1958).
Peru. 83 PER. Phan.

Myrcia dispar McVaugh, Fieldiana, Bot. 29: 191 (1956).
N. Brazil. 84 BZN. Nanophan. or phan.

Myrcia dolicopetala Kiaersk., Enum. Myrt. Bras.: 106 (1893).
Brazil (Rio de Janeiro). 84 BZL. Nanophan.
**Gomidesia sonderiana* O.Berg in C.F.P.von Martius & auct. suc. (eds.), Fl. Bras. 14(1): 533 (1859).

Myrcia doloresensis Hieron., Bot. Jahrb. Syst. 20(49): 63 (1895).
Colombia. 83 CLM. Nanophan. or phan.

Myrcia doniana O.Berg in C.F.P.von Martius & auct. suc. (eds.), Fl. Bras. 14(1): 518 (1858).
Brazil (Maranhão). 84 BZE. Nanophan. or phan.
Aulomyrcia doniana O.Berg in C.F.P.von Martius & auct. suc. (eds.), Fl. Bras. 14(1): 516 (1858).

Myrcia egensis (O.Berg) McVaugh, Fieldiana, Bot. 29: 191 (1956).
N. Brazil to Peru. 83 PER 84 BZN. Phan.
**Aulomyrcia egensis* O.Berg in C.F.P.von Martius & auct. suc. (eds.), Fl. Bras. 14(1): 99 (1857).
Aulomyrcia macrophylla O.Berg in C.F.P.von Martius & auct. suc. (eds.), Fl. Bras. 14(1): 99 (1857).

Myrcia ehrenbergiana (O.Berg) McVaugh, Mem. New York Bot. Gard. 18(2): 85 (1969).
N. South America to N. Brazil. 82 GUY VEN 84 BZN. Nanophan. or phan.
**Myrciaria ehrenbergiana* O.Berg, Linnaea 27: 321 (1856). *Aulomyrcia ehrenbergiana* (O.Berg) Amshoff, Recueil Trav. Bot. Néerl. 42: 8 (1950).

Myrcia elattophylla Diels, Bot. Jahrb. Syst. 37: 595 (1906).
Peru. 83 PER. Nanophan. or phan. Provisionally accepted.

Myrcia emarginata (O.Berg) Nied. in H.G.A.Engler & K.A.E.Prantl, Nat. Pflanzenfam. 3(7): 76 (1893).
Brazil. 84+. Nanophan. or phan.
**Aulomyrcia emarginata* O.Berg in C.F.P.von Martius & auct. suc. (eds.), Fl. Bras. 14(1): 134 (1857).

Myrcia eriocalyx DC., Prodr. 3: 247 (1828). *Gomidesia eriocalyx* (DC.) O.Berg in C.F.P.von Martius & auct. suc. (eds.), Fl. Bras. 14(1): 25 (1857).
Brazil (Goiás to Rio de Janeiro). 84 BZC BZL. Nanophan. or phan.
Myrcia pauciflora Cambess. in A.F.C.de Saint-Hilaire, Fl. Bras. Merid. 2: 321 (1832). *Gomidesia eriocalyx* var. *acuminata* D.Legrand, Notul. Syst. (Paris) 15: 262 (1958).
Myrcia elliptica Gardner, London J. Bot. 2: 352 (1843).
Gomidesia beyrichiana O.Berg in C.F.P.von Martius & auct. suc. (eds.), Fl. Bras. 14(1): 25 (1857). *Myrcia eriocalyx* var. *beyrichiana* (O.Berg) Kiaersk., Enum. Myrt. Bras.: 110 (1893).
Gomidesia gaudichaudiana O.Berg in C.F.P.von Martius & auct. suc. (eds.), Fl. Bras. 14(1): 27 (1857). *Myrcia alpina* Kiaersk., Enum. Myrt. Bras.: 113 (1893).
Gomidesia hilariana O.Berg in C.F.P.von Martius & auct. suc. (eds.), Fl. Bras. 14(1): 26 (1857).
Gomidesia miqueliana O.Berg in C.F.P.von Martius & auct. suc. (eds.), Fl. Bras. 14(1): 24 (1857). *Myrcia eriocalyx* var. *miqueliana* (O.Berg) Kiaersk., Enum. Myrt. Bras.: 111 (1893).
Gomidesia miqueliana var. *brunnea* O.Berg in C.F.P.von Martius & auct. suc. (eds.), Fl. Bras. 14(1): 25 (1857).
Gomidesia miqueliana var. *ferruginea* O.Berg in C.F.P.von Martius & auct. suc. (eds.), Fl. Bras. 14(1): 24 (1857).
Myrcia pauciflora var. *brunnea* O.Berg in C.F.P.von Martius & auct. suc. (eds.), Fl. Bras. 14(1): 201 (1857).
Myrcia pauciflora var. *rubiginosa* O.Berg in C.F.P.von Martius & auct. suc. (eds.), Fl. Bras. 14(1): 201 (1857).
Gomidesia beyrichiana var. *acutata* O.Berg in C.F.P.von Martius & auct. suc. (eds.), Fl. Bras. 14(1): 534 (1859).
Gomidesia gaudichaudiana var. *acutata* O.Berg in C.F.P.von Martius & auct. suc. (eds.), Fl. Bras. 14(1): 535 (1859).
Gomidesia miqueliana var. *angustifolia* O.Berg in C.F.P.von Martius & auct. suc. (eds.), Fl. Bras. 14(1): 534 (1859).
Gomidesia miqueliana var. *obtusata* O.Berg in C.F.P.von Martius & auct. suc. (eds.), Fl. Bras. 14(1): 534 (1859).
Gomidesia regeliana O.Berg in C.F.P.von Martius & auct. suc. (eds.), Fl. Bras. 14(1): 535 (1859). *Myrcia sessilifolia* Kiaersk., Enum. Myrt. Bras.: 115 (1893).
Myrcia eriocalyx f. *brunnea* Kiaersk., Enum. Myrt. Bras.: 111 (1893).

Myrcia eriocalyx f. *obtusata* Kiaersk., Enum. Myrt. Bras.: 111 (1893).

Myrcia eriopus DC., Prodr. 3: 255 (1828).
WC. & SE. Brazil. 84 BZC BZL. Nanophan.
Myrcia eriopus var. *grandifolia* O.Berg in C.F.P.von Martius & auct. suc. (eds.), Fl. Bras. 14(1): 152 (1857).
Myrcia eriopus var. *parvifolia* O.Berg in C.F.P.von Martius & auct. suc. (eds.), Fl. Bras. 14(1): 152 (1857).

Myrcia eumecephylla (O.Berg) Nied. in H.G.A.Engler & K.A.E.Prantl, Nat. Pflanzenfam. 3(7): 76 (1893).
SE. Brazil. 84 BZL. Nanophan. or phan.
**Aulomyrcia eumecephylla* O.Berg in C.F.P.von Martius & auct. suc. (eds.), Fl. Bras. 14(1): 98 (1857).

Myrcia eximia DC., Prodr. 3: 248 (1828).
Brazil (Minas Gerais). 84 BZL. Nanophan. or phan.

Myrcia exploratoris McVaugh, Mem. New York Bot. Gard. 18(2): 86 (1969).
N. South America. 82 GUY SUR VEN. Nanophan. or phan.

Myrcia extranea McVaugh, Mem. New York Bot. Gard. 18(2): 87 (1969).
Guyana. 82 GUY. Phan.

Myrcia fasciata McVaugh, Fieldiana, Bot. 29: 192 (1956).
Ecuador. 83 ECU. Phan.

Myrcia fascicularis O.Berg, Linnaea 27: 87 (1855).
Peru to Bolivia. 83 BOL PER. Nanophan. or phan.
Myrcia bangii Rusby, Mem. Torrey Bot. Club 6: 36 (1895).

Myrcia felisbertii (DC.) O.Berg in C.F.P.von Martius & auct. suc. (eds.), Fl. Bras. 14(1): 562 (1859).
N. Brazil. 84 BZN. Nanophan. or phan.
**Eugenia felisbertii* DC., Prodr. 3: 266 (1828). *Myrcia involucrata* O.Berg in C.F.P.von Martius & auct. suc. (eds.), Fl. Bras. 14(1): 156 (1857), nom. illeg.

Myrcia fenestrata DC., Prodr. 3: 251 (1828).
French Guiana to Bolivia. 82 FRG 83 BOL 84 BZN. Nanophan. or phan.

Myrcia fenzliana O.Berg in C.F.P.von Martius & auct. suc. (eds.), Fl. Bras. 14(1): 196 (1857).
Caribbean to Brazil. 81 CUB DOM HAI JAM LEE PUE WIN 82 GUY VEN 83 BOL CLM 84 BZC BZE BZL. Nanophan. or phan.
Gomidesia lindeniana O.Berg, Linnaea 29: 208 (1858). *Myrcia lindeniana* (O.Berg) C.Wright, Anales Acad. Ci. Méd. Habana 5: 429 (1868), nom. illeg. *Myrcia sintenisii* Kiaersk., Bot. Tidsskr. 17: 257 (1890).

Myrcia ferruginea G.Don, Gen. Hist. 2: 845 (1832).
NE. Brazil. 84 BZE. Nanophan. or phan.
Eugenia paniculata Hils. & Sieber ex C.Presl, Isis (Oken) 21: 274 (1828).

Myrcia filibracteata Mattos & D.Legrand, Loefgrenia 67: 2 (1975).
Brazil (Goiás). 84 BZC. Nanophan. or phan.

Myrcia florida Lem., Jard. Fleur. 3(Misc.): 100 (1853).
Brazil. 84+. Nanophan. or phan.

Myrcia follii G.M.Barroso & Peixoto, Acta Bot. Brasil. 4(2): 4 (1990).
E. Brazil. 84 BZE BZL. Phan.

Myrcia fosteri Croat, Ann. Missouri Bot. Gard. 61: 886 (1974).
Panama. 80 PAN. Nanophan. or phan.

Myrcia freyreissiana (O.Berg) Kiaersk., Enum. Myrt. Bras.: 102 (1893).
NE. Brazil. 84 BZE. Nanophan. or phan.
**Gomidesia freyreissiana* O.Berg in C.F.P.von Martius & auct. suc. (eds.), Fl. Bras. 14(1): 19 (1857).

Myrcia fusca B.Holst & M.L.Kawas., Selbyana 25: 95 (2004).
Panama. 80 PAN. Phan.

Myrcia gatunensis Standl., Publ. Field Mus. Nat. Hist., Bot. Ser. 4: 154 (1929).
Panama. 80 PAN. Nanophan. or phan.

Myrcia gentryi B.Holst, BioLlania 10: 4 (1994).
Venezuela. 82 VEN. Nanophan. or phan.

Myrcia gestasiana Cambess. in A.F.C.de Saint-Hilaire, Fl. Bras. Merid. 2: 303 (1832). *Gomidesia gestasiana* (Cambess.) D.Legrand, Notul. Syst. (Paris) 15: 261 (1958).
Brazil (Rio de Janeiro). 84 BZL. Nanophan. or phan.

Myrcia gigantea (O.Berg) Nied. in H.G.A.Engler & K.A.E.Prantl, Nat. Pflanzenfam. 3(7): 76 (1893).
NE. Brazil. 84 BZE. Nanophan. or phan.
**Aulomyrcia gigantea* O.Berg in C.F.P.von Martius & auct. suc. (eds.), Fl. Bras. 14(1): 548 (1859).

Myrcia gigas McVaugh, Mem. New York Bot. Gard. 18(2): 88 (1969).
French Guiana to Brazil (Amapá). 82 FRG 84 BZN. Phan.

Myrcia gilsoniana G.M.Barroso & Peixoto, Acta Bot. Brasil. 4(2): 7 (1990).
Brazil (Espírito Santo). 84 BZL. Nanophan. or phan.

Myrcia glabra (O.Berg) D.Legrand, Sellowia 13: 298 (1961).
Brazil. 84 BZC BZL BZN BZS. Phan.
**Aulomyrcia glabra* O.Berg in C.F.P.von Martius & auct. suc. (eds.), Fl. Bras. 14(1): 119 (1857).
Myrcia citrifolia D.Legrand, Sellowia 13: 294 (1961), nom. illeg.

Myrcia glauca Cambess. in A.F.C.de Saint-Hilaire, Fl. Bras. Merid. 2: 318 (1832). *Aulomyrcia glauca* (Cambess.) O.Berg, Linnaea 27: 50 (1855).
SE. Brazil. 84 BZL. Nanophan. or phan.

Myrcia glaziovii Mattos & D.Legrand, Loefgrenia 67: 1 (1975).
Brazil (Minas Gerais). 84 BZL. Cham. or nanophan.

Myrcia gollmeriana O.Berg, Linnaea 29: 221 (1858).
Venezuela. 82 VEN. Nanophan. or phan.

Myrcia gomidesioides Kiaersk., Enum. Myrt. Bras.: 87 (1893).
Brazil (Minas Gerais). 84 BZL. Nanophan. or phan.

Myrcia gonini McVaugh, Mem. New York Bot. Gard. 18(2): 89 (1969).
Suriname. 82 SUR. Phan.

Myrcia govinha S.Moore, Trans. Linn. Soc. London, Bot. 4: 354 (1895).
WC. Brazil. 84 BZC. Nanophan. or phan.

Myrcia goyazensis Cambess. in A.F.C.de Saint-Hilaire, Fl. Bras. Merid. 2: 305 (1832).
WC. Brazil. 84 BZC. Nanophan. or phan.
Myrcia goyazensis var. *angustifolia* O.Berg in C.F.P. von Martius & auct. suc. (eds.), Fl. Bras. 14(1): 188 (1857).
Myrcia goyazensis var. *latifolia* O.Berg in C.F.P.von Martius & auct. suc. (eds.), Fl. Bras. 14(1): 188 (1857).

Myrcia graciliflora Sagot, Ann. Sci. Nat., Bot., VI, 20: 185 (1885).
N. South America. 82 FRG GUY SUR. Nanophan. or phan.
**Eugenia schaueriana* Miq., Stirp. Surinam. Select.: 41 (1851). *Myrciaria schaueriana* (Miq.) O.Berg, Linnaea 27: 323 (1856). *Aulomyrcia schaueriana* (Miq.) Amshoff, Recueil Trav. Bot. Néerl. 39: 155 (1942).
Aulomyrcia triflora O.Berg, Linnaea 27: 79 (1855).

Myrcia grandifolia Cambess. in A.F.C.de Saint-Hilaire, Fl. Bras. Merid. 2: 298 (1832). *Gomidesia grandifolia* (Cambess.) Mattos & D.Legrand, Loefgrenia 67: 17 (1975).
SE. Brazil. 84 BZL. Nanophan. or phan.
Gomidesia jacquiniana O.Berg in C.F.P.von Martius & auct. suc. (eds.), Fl. Bras. 14(1): 17 (1857).
Gomidesia crocea Nied. in H.G.A.Engler & K.A.E. Prantl, Nat. Pflanzenfam. 3(7): 77 (1893), nom. illeg.
Myrcia crocea var. *blanchetiana* Kiaersk., Enum. Myrt. Bras.: 106 (1893).
Myrcia ticuensis Kiaersk., Enum. Myrt. Bras.: 107 (1893). *Gomidesia ticuensis* (Kiaersk.) D.Legrand, Comun. Bot. Mus. Hist. Nat. Montevideo 3(37): 15 (1958).

Myrcia grandiglandulosa Kiaersk., Enum. Myrt. Bras.: 83 (1893).
SE. Brazil. 84 BZL. Nanophan. or phan.

Myrcia grandis McVaugh, Mem. New York Bot. Gard. 18(2): 114 (1969).
Panama to N. Brazil. 80 PAN 82 GUY VEN 83 CLM 84 BZN. Nanophan. or phan.

Myrcia guavira Parodi, Anales Soc. Ci. Argent. 7: 182 (1878).
Paraguay. 85 PAR. Nanophan. or phan.

Myrcia guavira-mi Parodi, Anales Soc. Ci. Argent. 7: 140 (1879).
Paraguay. 85 PAR. Nanophan. or phan.

Myrcia guianensis (Aubl.) DC., Prodr. 3: 245 (1828).
Trinidad to S. Trop. America. 81 TRT 82 FRG GUY SUR VEN 83 BOL CLM ECU PER 84 BZC BZE BZL BZN BZS. Nanophan. or phan.
**Eugenia guianensis* Aubl., Hist. Pl. Guiane 1: 506 (1775). *Myrtus guianensis* (Aubl.) Ham., Prodr. Pl. Ind. Occid.: 45 (1825). *Aguava guianensis* (Aubl.) Raf., Sylva Tellur.: 107 (1838).
Myrtus pyrifolia J.St.-Hil. in H.L.Duhamel du Monceau, Traité Arbr. Arbust., ed. 2, 1: 208 (1803).
Myrcia cassinioides DC., Prodr. 3: 249 (1828). *Aulomyrcia cassinioides* (DC.) O.Berg, Linnaea 27: 74 (1855).
Myrcia daphnoides DC., Prodr. 3: 246 (1828). *Aulomyrcia daphnoides* (DC.) O.Berg, Linnaea 27: 77 (1855).
Myrcia elaeodendra DC., Prodr. 3: 250 (1828). *Aulomyrcia elaeodendra* (DC.) O.Berg, Linnaea 27: 75 (1855).
Myrcia elegans DC., Prodr. 3: 251 (1828). *Calyptromyrcia elegans* (DC.) O.Berg in C.F.P.von Martius & auct. suc. (eds.), Fl. Bras. 14(1): 58 (1857).
Myrcia exsucca DC., Prodr. 3: 247 (1828). *Aulomyrcia exsucca* (DC.) O.Berg, Linnaea 27: 79 (1855).
Myrcia lauriflora DC., Prodr. 3: 252 (1828). *Aulomyrcia lauriflora* (DC.) O.Berg, Linnaea 27: 64 (1855).
Myrcia pallens DC., Prodr. 3: 252 (1828). *Aulomyrcia pallens* (DC.) O.Berg, Linnaea 27: 73 (1855).
Myrcia schrankiana DC., Prodr. 3: 247 (1828). *Aulomyrcia schrankiana* (DC.) O.Berg, Linnaea 27: 79 (1855).
Myrcia spixiana DC., Prodr. 3: 251 (1828). *Calyptromyrcia spixiana* (DC.) O.Berg in C.F.P.von Martius & auct. suc. (eds.), Fl. Bras. 14(1): 59 (1857).
Myrtus elegans Mart. ex DC., Prodr. 3: 251 (1828).
Myrtus exsucca Mart. ex DC., Prodr. 3: 247 (1828), nom. inval.
Myrcia crassicaulis Cambess. in A.F.C.de Saint-Hilaire, Fl. Bras. Merid. 2: 311 (1832). *Aulomyrcia crassicaulis* (Cambess.) O.Berg, Linnaea 27: 74 (1855).
Myrcia microcarpa Cambess. in A.F.C.de Saint-Hilaire, Fl. Bras. Merid. 2: 324 (1832). *Aulomyrcia microcarpa* (Cambess.) O.Berg in C.F.P.von Martius & auct. suc. (eds.), Fl. Bras. 14(1): 81 (1857).
Myrcia suaveolens Cambess. in A.F.C.de Saint-Hilaire, Fl. Bras. Merid. 2: 315 (1832). *Aulomyrcia suaveolens* (Cambess.) O.Berg, Linnaea 27: 78 (1855).
Myrcia obtusa Schauer, Linnaea 21: 273 (1848). *Aulomyrcia obtusa* (Schauer) O.Berg, Linnaea 27: 66 (1855).
Myrcia alternifolia Miq., Linnaea 22: 534 (1849). *Aulomyrcia alternifolia* (Miq.) O.Berg, Linnaea 27: 72 (1855). *Myrcia obtecta* var. *alternifolia* (Miq.) D.Legrand, in Fl. Ilustr. Catar. 1(Mirt.): 283 (1969).
Myrcia surinamensis Miq., Linnaea 22: 170 (1849). *Aulomyrcia surinamensis* (Miq.) O.Berg, Linnaea 27: 64 (1855). *Aulomyrcia obtusa* var. *surinamensis* (Miq.) Amshoff, Recueil Trav. Bot. Néerl. 39: 154 (1942).
Aulomyrcia conduplicata O.Berg, Linnaea 27: 76 (1855).
Aulomyrcia cuneata O.Berg, Linnaea 27: 72 (1855). *Myrcia cuneata* (O.Berg) Nied. in H.G.A.Engler & K.A.E.Prantl, Nat. Pflanzenfam. 3(7): 76 (1893). *Myrcia guianensis* var. *cuneata* (O.Berg) McVaugh, Mem. New York Bot. Gard. 18(2): 93 (1969).
Aulomyrcia dichroma O.Berg, Linnaea 27: 65 (1855).
Aulomyrcia glandulosa O.Berg, Linnaea 27: 79 (1855). *Myrcia glandulosa* (O.Berg) Kiaersk., Enum. Myrt. Bras.: 83 (1893).
Aulomyrcia obtusa var. *grandifolia* O.Berg, Linnaea 27: 66 (1855).
Aulomyrcia obtusa var. *longipes* O.Berg, Linnaea 27: 67 (1855).
Aulomyrcia obtusa var. *panicularis* O.Berg, Linnaea 27: 67 (1855).
Aulomyrcia obtusa var. *pauciflora* O.Berg, Linnaea 27: 67 (1855).
Aulomyrcia obtusa var. *tenuifolia* O.Berg, Linnaea 27: 67 (1855).
Aulomyrcia poeppigiana O.Berg, Linnaea 27: 73 (1855).

Aulomyrcia roraimensis O.Berg, Linnaea 27: 68 (1855).
Aulomyrcia schomburgkiana O.Berg, Linnaea 27: 75 (1855). *Aulomyrcia obtusa* var. *schomburgkiana* (O.Berg) Amshoff in A.A.Pulle, Fl. Suriname 3(2):79 (1951).
Aulomyrcia lingua O.Berg, Linnaea 27: 177 (1856). *Myrcia lingua* (O.Berg) Mattos, Dusenia 8: 161 (1968).
Aulomyrcia botrys O.Berg in C.F.P.von Martius & auct. suc. (eds.), Fl. Bras. 14(1): 116 (1857). *Myrcia botrys* (O.Berg) N.Silveira, Loefgrenia 91: 1 (1987).
Aulomyrcia buxifolia O.Berg in C.F.P.von Martius & auct. suc. (eds.), Fl. Bras. 14(1): 80 (1857). *Myrcia taubatensis* Kiaersk., Enum. Myrt. Bras.: 79 (1893).
Aulomyrcia buxifolia var. *elliptica* O.Berg in C.F.P.von Martius & auct. suc. (eds.), Fl. Bras. 14(1): 80 (1857).
Aulomyrcia buxifolia var. *ovalis* O.Berg in C.F.P.von Martius & auct. suc. (eds.), Fl. Bras. 14(1): 80 (1857). *Myrcia taubatensis* var. *ovalis* (O.Berg) Kiaersk., Enum. Myrt. Bras.: 79 (1893).
Aulomyrcia cassinioides var. *glabrata* O.Berg in C.F.P.von Martius & auct. suc. (eds.), Fl. Bras. 14(1): 129 (1857).
Aulomyrcia cassinioides var. *velutina* O.Berg in C.F.P.von Martius & auct. suc. (eds.), Fl. Bras. 14(1): 129 (1857).
Aulomyrcia fragilis O.Berg in C.F.P.von Martius & auct. suc. (eds.), Fl. Bras. 14(1): 117 (1857).
Aulomyrcia gardneriana O.Berg in C.F.P.von Martius & auct. suc. (eds.), Fl. Bras. 14(1): 129 (1857). *Myrcia renatoana* Mattos, Arq. Bot. Estado São Paulo, n.s., f.m., 4: 62 (1966).
Aulomyrcia gardneriana var. *caerulescens* O.Berg in C.F.P.von Martius & auct. suc. (eds.), Fl. Bras. 14(1): 130 (1857).
Aulomyrcia gardneriana var. *virescens* O.Berg in C.F.P.von Martius & auct. suc. (eds.), Fl. Bras. 14(1): 130 (1857).
Aulomyrcia glandulosa var. *elliptica* O.Berg in C.F.P.von Martius & auct. suc. (eds.), Fl. Bras. 14(1): 139 (1857).
Aulomyrcia glandulosa var. *longifolia* O.Berg in C.F.P.von Martius & auct. suc. (eds.), Fl. Bras. 14(1): 139 (1857).
Aulomyrcia glandulosa var. *obovata* O.Berg in C.F.P.von Martius & auct. suc. (eds.), Fl. Bras. 14(1): 139 (1857). *Myrcia glandulosa* var. *obovata* (O.Berg) N.Silveira, Roessléria 7: 67 (1985).
Aulomyrcia hepatica O.Berg in C.F.P.von Martius & auct. suc. (eds.), Fl. Bras. 14(1): 132 (1857). *Myrcia hepatica* (O.Berg) Kiaersk., Enum. Myrt. Bras.: 86 (1893).
Aulomyrcia intermedia O.Berg in C.F.P.von Martius & auct. suc. (eds.), Fl. Bras. 14(1): 107 (1857). *Myrcia intermedia* (O.Berg) Kiaersk., Enum. Myrt. Bras.: 90 (1893).
Aulomyrcia lingua var. *glabrata* O.Berg in C.F.P.von Martius & auct. suc. (eds.), Fl. Bras. 14(1): 130 (1857).
Aulomyrcia lingua var. *rufa* O.Berg in C.F.P.von Martius & auct. suc. (eds.), Fl. Bras. 14(1): 130 (1857). *Myrcia lingua* var. *rufa* (O.Berg) Mattos, Loefgrenia 70: 5 (1976).
Aulomyrcia obscura O.Berg in C.F.P.von Martius & auct. suc. (eds.), Fl. Bras. 14(1): 132 (1857). *Myrcia obscura* (O.Berg) N.Silveira, Roessléria 7: 66 (1985).
Aulomyrcia obtecta O.Berg in C.F.P.von Martius & auct. suc. (eds.), Fl. Bras. 14(1): 117 (1857). *Myrcia obtecta* (O.Berg) Kiaersk., Enum. Myrt. Bras.: 89 (1893).
Aulomyrcia pallens var. *ovalis* O.Berg in C.F.P.von Martius & auct. suc. (eds.), Fl. Bras. 14(1): 122 (1857). *Myrcia pallens* var. *ovalis* (O.Berg) Kiaersk., Enum. Myrt. Bras.: 83 (1893).
Aulomyrcia pallens var. *ovata* O.Berg in C.F.P.von Martius & auct. suc. (eds.), Fl. Bras. 14(1): 123 (1857).
Aulomyrcia pallens var. *petiolaris* O.Berg in C.F.P.von Martius & auct. suc. (eds.), Fl. Bras. 14(1): 123 (1857).
Aulomyrcia pallida O.Berg in C.F.P.von Martius & auct. suc. (eds.), Fl. Bras. 14(1): 87 (1857). *Myrcia pallida* (O.Berg) N.Silveira, Loefgrenia 88: 2 (1985).
Aulomyrcia pruinosa O.Berg in C.F.P.von Martius & auct. suc. (eds.), Fl. Bras. 14(1): 114 (1857).
Aulomyrcia retusa O.Berg in C.F.P.von Martius & auct. suc. (eds.), Fl. Bras. 14(1): 142 (1857). *Myrcia retusa* (O.Berg) Nied. in H.G.A.Engler & K.A.E.Prantl, Nat. Pflanzenfam. 3(7): 76 (1893).
Myrcia alagoensis O.Berg in C.F.P.von Martius & auct. suc. (eds.), Fl. Bras. 14(1): 165 (1857).
Myrcia terebinthacea Poepp. ex O.Berg in C.F.P.von Martius & auct. suc. (eds.), Fl. Bras. 14(1): 155 (1857).
Aulomyrcia uaupensis O.Berg in C.F.P.von Martius & auct. suc. (eds.), Fl. Bras. 14(1): 518 (1858).
Aulomyrcia maritima O.Berg in C.F.P.von Martius & auct. suc. (eds.), Fl. Bras. 14(1): 553 (1859).
Aulomyrcia obscura var. *longipes* O.Berg in C.F.P.von Martius & auct. suc. (eds.), Fl. Bras. 14(1): 556 (1859). *Myrcia obscura* var. *longipes* (O.Berg) N.Silveira, Roessléria 7: 66 (1985).
Aulomyrcia pallens var. *subcordata* O.Berg in C.F.P.von Martius & auct. suc. (eds.), Fl. Bras. 14(1): 553 (1859). *Myrcia pallens* var. *subcordata* (O.Berg) Kiaersk., Enum. Myrt. Bras.: 83 (1893).
Aulomyrcia rorida O.Berg in C.F.P.von Martius & auct. suc. (eds.), Fl. Bras. 14(1): 522 (1859). *Myrcia rorida* (O.Berg) Kiaersk., Enum. Myrt. Bras.: 79 (1893).
Aulomyrcia androsaemoides O.Berg, Linnaea 30: 661 (1861). *Myrcia androsaemoides* (O.Berg) Krug & Urb., Bot. Jahrb. Syst. 19: 579 (1895).
Aulomyrcia buxizans O.Berg, Linnaea 30: 664 (1861).
Aulomyrcia daphnoides var. *ochracea* O.Berg, Linnaea 30: 663 (1861).
Myrcia roraimae Oliv., Trans. Linn. Soc. London, Bot. 2: 273 (1887). *Aulomyrcia roraimae* (Oliv.) Steyerm., Fieldiana, Bot. 28: 1007 (1957).
Myrcia adpressepilosa Kiaersk., Enum. Myrt. Bras.: 75 (1893).
Myrcia cymosopaniculata Kiaersk., Enum. Myrt. Bras.: 90 (1893).
Myrcia diaphanosticta Kiaersk., Enum. Myrt. Bras.: 91 (1893).
Myrcia fastigiata Kiaersk., Enum. Myrt. Bras.: 92 (1893).
Myrcia fastigiata var. *coriacea* Kiaersk., Enum. Myrt. Bras.: 93 (1893).
Myrcia rhabdoides Kiaersk., Enum. Myrt. Bras.: 99 (1893).
Myrcia yungasensis Rusby, Mem. Torrey Bot. Club 3(3): 27 (1893).
Myrcia androsaemoides var. *parvifolia* Krug & Urb., Bot. Jahrb. Syst. 19: 579 (1895).

Myrcia daphnoides var. *nervosa* Kiaersk., Bull. Soc. Bot. France 54(3c): 218 (1908).
Myrcia arimensis Britton, Bull. Torrey Bot. Club 48: 334 (1921 publ. 1922).
Myrcia incisa D.Legrand, Sellowia 13: 290 (1961).
Myrcia obcordata Mattos, Loefgrenia 19: 1 (1964).
Myrcia rorida var. *microphylla* Glaz. ex Mattos, Loefgrenia 66: 2 (1975).
Myrcia stemmeriana D.Legrand, in Fl. Ilustr. Catar. 1(Mirt., Suppl. 1): 4 (1977).

Myrcia guildingiana (Griseb.) Krug & Urb., Bot. Jahrb. Syst. 19: 57 (1895).
St. Vincent. 81 WIN. Nanophan. or phan.
**Psidium guildingianum* Griseb., Fl. Brit. W. I.: 242 (1860).

Myrcia gundlachii Krug & Urb., Bot. Jahrb. Syst. 19: 581 (1895). *Mozartia gundlachii* (Krug & Urb.) Urb., Symb. Antill. 9: 88 (1923).
E. Cuba. 81 CUB. Nanophan. or phan.

Myrcia hartwegiana (O.Berg) Kiaersk., Enum. Myrt. Bras.: 109 (1893).
SE. & S. Brazil. 84 BZL BZS. Nanophan. or phan.
**Gomidesia hartwegiana* O.Berg in C.F.P.von Martius & auct. suc. (eds.), Fl. Bras. 14(1): 22 (1857).
Gomidesia sellowiana O.Berg in C.F.P.von Martius & auct. suc. (eds.), Fl. Bras. 14(1): 21 (1857). *Myrcia sellowiana* (O.Berg) Arechav., Anales Mus. Nac. Montevideo 5: 45 (1905), nom. illeg.
Gomidesia sprengeliana O.Berg in C.F.P.von Martius & auct. suc. (eds.), Fl. Bras. 14(1): 21 (1857).

Myrcia hatschbachii D.Legrand, Sellowia 13: 293 (1961).
S. Brazil to Uruguay. 84 BZS 85 URU. Phan.

Myrcia hebepetala DC., Prodr. 3: 246 (1828). *Gomidesia hebepetala* (DC.) O.Berg in C.F.P.von Martius & auct. suc. (eds.), Fl. Bras. 14(1): 18 (1857).
Brazil. 84 BZC BZL BZS. Phan.
Myrcia cordiifolia var. *minor* DC., Prodr. 3: 248 (1828). *Gomidesia pohliana* O.Berg in C.F.P.von Martius & auct. suc. (eds.), Fl. Bras. 14(1): 16 (1857).
Myrcia affinis Cambess. in A.F.C.de Saint-Hilaire, Fl. Bras. Merid. 2: 307 (1832). *Gomidesia affinis* (Cambess.) D.Legrand, Notul. Syst. (Paris) 15: 260 (1958).
Myrcia itajuruensis Cambess. in A.F.C.de Saint-Hilaire, Fl. Bras. Merid. 2: 307 (1832).
Gomidesia candolleana O.Berg in C.F.P.von Martius & auct. suc. (eds.), Fl. Bras. 14(1): 17 (1857). *Myrcia candolleana* (O.Berg) Kiaersk., Enum. Myrt. Bras.: 105 (1893).
Gomidesia hookeriana O.Berg in C.F.P.von Martius & auct. suc. (eds.), Fl. Bras. 14(1): 18 (1857). *Myrcia hookeriana* (O.Berg) Kiaersk., Enum. Myrt. Bras.: 107 (1893).

Myrcia heliandina Diels, Bot. Jahrb. Syst. 37: 594 (1906).
Peru. 83 PER. Nanophan. or phan. Provisionally accepted.

Myrcia heringii D.Legrand, Sellowia 13: 298 (1961).
Brazil (São Paulo to Santa Catarina). 84 BZL BZS. Nanophan. or phan.

Myrcia hernandezii Parra-Os., Caldasia 24: 99 (2002).
Colombia. 83 CLM.

Myrcia hexasticha Kiaersk., Enum. Myrt. Bras.: 72 (1893).
SE. Brazil. 84 BZL. Nanophan. or phan.

Myrcia hiemalis Cambess. in A.F.C.de Saint-Hilaire, Fl. Bras. Merid. 2: 332 (1832).
Brazil (Mato Grosso, Minas Gerais). 84 BZC BZL. Nanophan. or phan.

Myrcia hilariana O.Berg in C.F.P.von Martius & auct. suc. (eds.), Fl. Bras. 14(1): 199 (1857).
Brazil (Minas Gerais). 84 BZL. Nanophan. or phan.
Myrcia formosiana Cambess. in A.F.C.de Saint-Hilaire, Fl. Bras. Merid. 2: 299 (1832), nom. illeg.

Myrcia hirsuta O.Berg in C.F.P.von Martius & auct. suc. (eds.), Fl. Bras. 14(1): 151 (1857).
SE. Brazil. 84 BZL. Nanophan. or phan.

Myrcia hispida O.Berg in C.F.P.von Martius & auct. suc. (eds.), Fl. Bras. 14(1): 152 (1857).
SE. Brazil. 84 BZL. Nanophan. or phan.
Myrcia hispida var. *angustifolia* O.Berg in C.F.P.von Martius & auct. suc. (eds.), Fl. Bras. 14(1): 153 (1857).
Myrcia hispida var. *latifolia* O.Berg in C.F.P.von Martius & auct. suc. (eds.), Fl. Bras. 14(1): 153 (1857).
Myrcia hispida var. *panicularis* O.Berg in C.F.P.von Martius & auct. suc. (eds.), Fl. Bras. 14(1): 152 (1857).

Myrcia hoffmannseggii O.Berg in C.F.P.von Martius & auct. suc. (eds.), Fl. Bras. 14(1): 157 (1857).
N. Brazil. 84 BZN. Nanophan. or phan.

Myrcia hotteana Urb. & Ekman, Ark. Bot. 21A(5): 22 (1927).
Haiti. 81 HAI. Nanophan. or phan.

Myrcia huallagae McVaugh, Fieldiana, Bot. 29: 192 (1956).
Peru. 83 PER. (Cl.) nanophan.

Myrcia hypericoides Cambess. in A.F.C.de Saint-Hilaire, Fl. Bras. Merid. 2: 317 (1832). *Aulomyrcia hypericoides* (Cambess.) O.Berg, Linnaea 27: 62 (1855).
SE. Brazil. 84 BZL. Nanophan. or phan.

Myrcia hypoleuca Spring ex Mart., Flora 20(2 Beibl.): 80 (1837).
Brazil (Minas Gerais). 84 BZL. Nanophan. or phan.

Myrcia ilheosensis Kiaersk., Enum. Myrt. Bras.: 109 (1893).
S. Trop. America. 82 FRG GUY SUR VEN 84 BZE BZL BZN BZS. Nanophan. or phan.
**Gomidesia fenzliana* O.Berg in C.F.P.von Martius & auct. suc. (eds.), Fl. Bras. 14(1): 20 (1857).
Gomidesia fenzliana var. *obovata* O.Berg in C.F.P.von Martius & auct. suc. (eds.), Fl. Bras. 14(1): 20 (1857).
Gomidesia fenzliana var. *spathulata* O.Berg in C.F.P. von Martius & auct. suc. (eds.), Fl. Bras. 14(1): 20 (1857).

Myrcia imperatoris-maximiliani Wawra, Oesterr. Bot. Z. 13: 11 (1863).
Brazil. 84+. Nanophan. or phan.

Myrcia imperfecta McVaugh, Mem. New York Bot. Gard. 18(2): 95 (1969).
Guyana. 82 GUY. Phan.

Myrcia inaequiloba (DC.) Lemée, in Fl. Guyane Franç. 3: 150 (1954).
Panama to N. Brazil. 80 PAN 82 FRG GUY SUR VEN 83 CLM 84 BZN. Phan.
**Eugenia inaequiloba* DC., Prodr. 3: 282 (1828). *Aulomyrcia inaequiloba* (DC.) Amshoff, Recueil Trav. Bot. Néerl. 42: 7 (1950).

Eugenia nitida Benth., J. Bot. (Hooker) 22: 322 (1840). *Myrciaria nitida* (Benth.) O.Berg, Linnaea 27: 324 (1856). *Aulomyrcia inaequiloba* var. *nitida* (Benth.) Amshoff, Recueil Trav. Bot. Néerl. 42: 8 (1950).
Eugenia polyantha Miq., Linnaea 18: 741 (1845). *Myrciaria polyantha* (Miq.) O.Berg, Linnaea 27: 322 (1856).
Aulomyrcia paniculata O.Berg, Linnaea 27: 50 (1855). *Aulomyrcia inaequiloba* var. *paniculata* (O.Berg) Amshoff, Recueil Trav. Bot. Néerl. 42: 7 (1950).
Aulomyrcia pirarensis O.Berg, Linnaea 27: 41 (1855).
Myrciaria nitida var. *chartacea* O.Berg, Linnaea 27: 324 (1856).
Myrciaria nitida var. *coriacea* O.Berg, Linnaea 27: 325 (1856).
Myrciaria nitida var. *dives* O.Berg, Linnaea 27: 325 (1856).
Aulomyrcia lancifolia O.Berg, Linnaea 30: 658 (1861).

Myrcia inconspicua L.Kollmann & Sobral, Novon 16: 501 (2006).
Brazil (Espírito Santo). 84 BZL.

Myrcia induta McVaugh, Mem. New York Bot. Gard. 18(2): 99 (1969).
Venezuela (Amazonas). 82 VEN. Nanophan. or phan.

Myrcia innovans Kiaersk., Enum. Myrt. Bras.: 100 (1893). *Gomidesia innovans* (Kiaersk.) D.Legrand, Comun. Bot. Mus. Hist. Nat. Montevideo 3(37): 23 (1958).
SE. Brazil. 84 BZL. Nanophan. or phan.

Myrcia insularis Gardner, London J. Bot. 1: 536 (1842). *Aulomyrcia insularis* (Gardner) O.Berg in C.F.P.von Martius & auct. suc. (eds.), Fl. Bras. 14(1): 98 (1857).
SE. Brazil. 84 BZL. Nanophan. or phan.
Aulomyrcia insularis var. *opaca* O.Berg in C.F.P.von Martius & auct. suc. (eds.), Fl. Bras. 14(1): 98 (1857).
Aulomyrcia insularis var. *punctata* O.Berg in C.F.P.von Martius & auct. suc. (eds.), Fl. Bras. 14(1): 98 (1857).

Myrcia intonsa (McVaugh) B.Holst, Selbyana 23: 152 (2002).
S. Venezuela to N. Brazil. 82 VEN 84 BZN.
**Marlierea intonsa* McVaugh, Mem. New York Bot. Gard. 10(1): 85 (1958).

Myrcia irwinii Mattos & D.Legrand, Loefgrenia 67: 4 (1975).
SE. Brazil. 84 BZL.
**Aulomyrcia crassifolia* O.Berg in C.F.P.von Martius & auct. suc. (eds.), Fl. Bras. 14(1): 128 (1857).

Myrcia isaiana G.M.Barroso & Peixoto, Acta Bot. Brasil. 4(2): 8 (1990).
Brazil (Espírito Santo). 84 BZL. Nanophan. or phan.

Myrcia jacobinensis Mattos, Arq. Bot. Estado São Paulo, n.s., f.m., 4: 60 (1966).
NE. Brazil. 84 BZE. Nanophan. or phan.
**Aulomyrcia supraaxillaris* O.Berg in C.F.P.von Martius & auct. suc. (eds.), Fl. Bras. 14(1): 69 (1857).

Myrcia kylistophylla B.Holst, Selbyana 23: 154 (2002).
S. Venezuela. 82 VEN. Nanophan.

Myrcia lacerdaeana O.Berg in C.F.P.von Martius & auct. suc. (eds.), Fl. Bras. 14(1): 193 (1857).
NE. Brazil. 84 BZE. Nanophan. or phan.

Myrcia lacunosa (O.Berg) N.Silveira, Loefgrenia 88: 1 (1985).
SE. Brazil. 84 BZL. Nanophan. or phan.
**Aulomyrcia lacunosa* O.Berg in C.F.P.von Martius & auct. suc. (eds.), Fl. Bras. 14(1): 545 (1859).

Myrcia laevis G.Don, Gen. Hist. 2: 845 (1832).
NE. Brazil. 84 BZE. Nanophan. or phan.

Myrcia lajeana D.Legrand, Sellowia 13: 291 (1961).
S. Brazil. 84 BZS. Nanophan. or phan.
**Aulomyrcia undulata* O.Berg in C.F.P.von Martius & auct. suc. (eds.), Fl. Bras. 14(1): 89 (1857).

Myrcia lanuginosa O.Berg in C.F.P.von Martius & auct. suc. (eds.), Fl. Bras. 14(1): 205 (1857).
Brazil (Goiás). 84 BZC. Nanophan. or phan.

Myrcia lapensis N.Silveira, Roessléria 7: 67 (1985).
Brazil (Minas Gerais). 84 BZL. Nanophan.
**Aulomyrcia ferruginea* O.Berg in C.F.P.von Martius & auct. suc. (eds.), Fl. Bras. 14(1): 552 (1859).

Myrcia lapidulosa B.Holst & M.L.Kawas., Selbyana 25: 97 (2004).
Panama. 80 PAN. Phan.

Myrcia laricina (O.Berg) Burret ex Luetzelb., Estud. Bot. Nordéste 3: 201 (1926).
NE. Brazil. 84 BZE. Nanophan. or phan.
**Aulomyrcia laricina* O.Berg in C.F.P.von Martius & auct. suc. (eds.), Fl. Bras. 14(1): 61 (1857).

Myrcia laruotteana Cambess. in A.F.C.de Saint-Hilaire, Fl. Bras. Merid. 2: 311 (1832). *Aulomyrcia laruotteana* (Cambess.) O.Berg, Linnaea 27: 53 (1855).
Brazil to Paraguay. 84 BZC BZE BZL BZS 85 AGE PAR. Nanophan. or phan.

var. ***laruotteana***
Brazil. 84 BZC BZE BZL BZS. Nanophan. or phan.
Myrcia laruotteana var. *glabriuscula* Cambess. in A.F.C. de Saint-Hilaire, Fl. Bras. Merid. 2: 311 (1832).
Myrcia laruotteana var. *impunctata* Cambess. in A.F.C.P.de Saint-Hilaire, Fl. Bras. Merid. 2: 311 (1832).
Aulomyrcia acutifolia O.Berg in C.F.P.von Martius & auct. suc. (eds.), Fl. Bras. 14(1): 89 (1857).
Aulomyrcia acutifolia var. *pubescens* O.Berg in C.F.P.von Martius & auct. suc. (eds.), Fl. Bras. 14(1): 90 (1857).
Aulomyrcia acutifolia var. *villosa* O.Berg in C.F.P.von Martius & auct. suc. (eds.), Fl. Bras. 14(1): 89 (1857).
Aulomyrcia laruotteana var. *membranacea* O.Berg in C.F.P.von Martius & auct. suc. (eds.), Fl. Bras. 14(1): 90 (1857).
Aulomyrcia laruotteana var. *opaca* O.Berg in C.F.P.von Martius & auct. suc. (eds.), Fl. Bras. 14(1): 90 (1857).
Aulomyrcia laruotteana var. *punctata* O.Berg in C.F.P.von Martius & auct. suc. (eds.), Fl. Bras. 14(1): 90 (1857).
Myrcia ochracea O.Berg in C.F.P.von Martius & auct. suc. (eds.), Fl. Bras. 14(1): 190 (1857).
Aulomyrcia laruotteana var. *paraguayensis* O.Berg, Linnaea 30: 661 (1861).
Myrcia assumptionis Morong, Ann. New York Acad. Sci. 7: 106 (1893). *Aulomyrcia assumptionis* (Morong) Kausel, Lilloa 32: 350 (1967).

Myrcia laruotteana var. *macrophylla* Kiaersk., Enum. Myrt. Bras.: 75 (1893).
Myrcia corrientinensis Barb.Rodr. ex Chodat & Hassl., Bull. Herb. Boissier, II, 7: 802 (1907), nom. nud.
Myrcia perorebimi Barb.Rodr. ex Chodat & Hassl., Bull. Herb. Boissier, II, 7: 802 (1907), nom. nud.
Myrcia laruotteana var. *australis* D.Legrand, in Fl. Ilustr. Catar. 1(Mirt.): 317 (1969).

var. ***paraguayensis*** (O.Berg) D.Legrand, in Fl. Ilustr. Catar. 1(Mirt.): 321 (1969).
SE. & S. Brazil to NE. Argentina. 84 BZL BZS 85 AGE PAR. Nanophan. or phan.

Myrcia lasiantha DC., Prodr. 3: 254 (1828).
WC. & SE. Brazil. 84 BZC BZL. Nanophan. or phan.
Myrcia cordifolia O.Berg in C.F.P.von Martius & auct. suc. (eds.), Fl. Bras. 14(1): 205 (1857), nom. illeg.
Myrcia cordifolia var. *acuminata* O.Berg in C.F.P.von Martius & auct. suc. (eds.), Fl. Bras. 14(1): 205 (1857).
Myrcia cordifolia var. *alba* O.Berg in C.F.P.von Martius & auct. suc. (eds.), Fl. Bras. 14(1): 205 (1857).
Myrcia cordifolia var. *fuscescens* O.Berg in C.F.P.von Martius & auct. suc. (eds.), Fl. Bras. 14(1): 205 (1857).
Myrcia cordifolia var. *nivea* O.Berg in C.F.P.von Martius & auct. suc. (eds.), Fl. Bras. 14(1): 205 (1857).
Myrcia lasiantha var. *multiflora* O.Berg in C.F.P.von Martius & auct. suc. (eds.), Fl. Bras. 14(1): 203 (1857).
Myrcia lasiantha var. *pauciflora* O.Berg in C.F.P.von Martius & auct. suc. (eds.), Fl. Bras. 14(1): 204 (1857).
Myrcia cordifolia var. *glabrescens* Kiaersk., Enum. Myrt. Bras.: 63 (1893).

Myrcia lateriflora Kiaersk., Enum. Myrt. Bras.: 77 (1893).
SE. Brazil. 84 BZL. Nanophan. or phan.

Myrcia laurifolia Cambess. in A.F.C.de Saint-Hilaire, Fl. Bras. Merid. 2: 301 (1832).
SE. Brazil. 84 BZL. Nanophan. or phan.
Aulomyrcia laurifolia O.Berg, Linnaea 27: 64 (1855).

Myrcia laxiflora Cambess. in A.F.C.de Saint-Hilaire, Fl. Bras. Merid. 2: 319 (1832). *Aulomyrcia laxiflora* (Cambess.) O.Berg in C.F.P.von Martius & auct. suc. (eds.), Fl. Bras. 14(1): 114 (1857).
SE. Brazil. 84 BZL. Nanophan. or phan.
Aulomyrcia laxiflora var. *angustifolia* O.Berg in C.F.P.von Martius & auct. suc. (eds.), Fl. Bras. 14(1): 114 (1857).
Aulomyrcia laxiflora var. *latifolia* O.Berg in C.F.P.von Martius & auct. suc. (eds.), Fl. Bras. 14(1): 114 (1857).

Myrcia lehmannii Hieron., Bot. Jahrb. Syst. 20(49): 65 (1895).
Colombia. 83 CLM. Nanophan. or phan.

Myrcia leucadendron DC., Prodr. 3: 251 (1828). *Aulomyrcia leucadendron* (DC.) O.Berg, Linnaea 27: 64 (1855).
SE. Brazil. 84 BZL. Phan.

Myrcia liesneri B.Holst, Selbyana 23: 154 (2002).
S. Venezuela. 82 VEN. Nanophan. or phan.

Myrcia limae G.M.Barroso & Peixoto, Acta Bot. Brasil. 4(2): 11 (1990).
Brazil (Espírito Santo). 84 BZL. Nanophan. or phan.

Myrcia linearifolia Cambess. in A.F.C.de Saint-Hilaire, Fl. Bras. Merid. 2: 334 (1832). *Aulomyrcia linearifolia* (Cambess.) O.Berg, Linnaea 27: 35 (1855).
WC. Brazil. 84 BZC. Nanophan. or phan.

Myrcia lineata (O.Berg) Nied. in H.G.A.Engler & K.A.E.Prantl, Nat. Pflanzenfam. 3(7): 76 (1893).
E. Brazil. 84 BZE BZL. Nanophan. or phan.
**Aulomyrcia lineata* O.Berg in C.F.P.von Martius & auct. suc. (eds.), Fl. Bras. 14(1): 68 (1857).

Myrcia littoralis DC., Prodr. 3: 249 (1828).
NE. Brazil. 84 BZE. Nanophan. or phan.

Myrcia lucida McVaugh, Mem. New York Bot. Gard. 18(2): 100 (1969).
Venezuela to Bolivia. 82 VEN 83 BOL 84 BZN. Nanophan. or phan.
**Myrcia laevis* O.Berg, Linnaea 31: 252 (1862), nom. illeg.

var. ***attenuata*** McVaugh, Mem. New York Bot. Gard. 18(2): 100 (1969).
Brazil (Amazonas). 84 BZN. Nanophan.

var. ***lucida***
Venezuela, Bolivia. 82 VEN 83 BOL. Nanophan. or phan.

Myrcia lundiana Kiaersk., Enum. Myrt. Bras.: 78 (1893).
SE. Brazil. 84 BZL. Nanophan. or phan.
**Aulomyrcia vautheriana* O.Berg, Linnaea 30: 655 (1861).

Myrcia luschnathiana O.Berg in C.F.P.von Martius & auct. suc. (eds.), Fl. Bras. 14(1): 194 (1857).
Brazil (Bahia). 84 BZE. Nanophan. or phan.

Myrcia lutescens Cambess. in A.F.C.de Saint-Hilaire, Fl. Bras. Merid. 2: 301 (1832). *Gomidesia lutescens* (Cambess.) D.Legrand, Notul. Syst. (Paris) 15: 262 (1958).
Brazil (Minas Gerais). 84 BZL. Phan.

Myrcia macrocarpa DC., Prodr. 3: 249 (1828). *Aulomyrcia macrocarpa* (DC.) O.Berg, Linnaea 27: 41 (1855).
Brazil (Rio de Janeiro). 84 BZL. Nanophan. or phan.

Myrcia madida McVaugh, Fieldiana, Bot. 29: 192 (1956).
N. Peru. 83 PER. Phan.

Myrcia maestrensis (Urb.) Alain, Mem. New York Bot. Gard. 21(2): 138 (1971).
SE. Cuba (Sierra Maestra). 81 CUB. Phan.
**Mozartia maestrensis* Urb., Symb. Antill. 9: 472 (1928).

Myrcia magnifolia (O.Berg) Kiaersk., Enum. Myrt. Bras.: 107 (1893).
Brazil (Rio de Janeiro). Nanophan. or phan.
**Gomidesia magnifolia* O.Berg in C.F.P.von Martius & auct. suc. (eds.), Fl. Bras. 14(1): 531 (1859).

Myrcia majaguitana Alain & M.M.Mejia, Moscosoa 9: 18 (1997).
Dominican Rep. 81 DOM. Nanophan. or phan.

Myrcia manacalensis Urb., Symb. Antill. 9: 85 (1923). *Mozartia manacalensis* (Urb.) Urb., Symb. Antill. 9: 473 (1928).

SE. Cuba (Sierra Maestra). 81 CUB. Phan.
Anamomis reticulata Britton & P.Wilson, Bull. Torrey Bot. Club 50: 45 (1923).

Myrcia mansoniana O.Berg in C.F.P.von Martius & auct. suc. (eds.), Fl. Bras. 14(1): 163 (1857).
Brazil (Mato Grosso). 84 BZC. Nanophan. or phan.

Myrcia mansonii (O.Berg) N.Silveira, Loefgrenia 88: 1 (1985).
WC. Brazil. 84 BZC. Nanophan. or phan.
**Aulomyrcia mansonii* O.Berg in C.F.P.von Martius & auct. suc. (eds.), Fl. Bras. 14(1): 121 (1857).

Myrcia maraguana DC., Prodr. 3: 249 (1828). *Aulomyrcia maraguana* (DC.) O.Berg in C.F.P.von Martius & auct. suc. (eds.), Fl. Bras. 14(1): 120 (1857).
NE. Brazil. 84 BZE. Nanophan. or phan.

Myrcia margarettae (Alain) Alain, Phytologia 58: 328 (1985).
C. Puerto Rico. 81 PUE. Nanophan. or phan.
**Eugenia margarettae* Alain, Bull. Torrey Bot. Club 90: 190 (1963).

Myrcia marginata O.Berg in C.F.P.von Martius & auct. suc. (eds.), Fl. Bras. 14(1): 565 (1859).
Brazil (Goiás). 84 BZC. Cham.

Myrcia maricaensis Alain, Phytologia 58: 327 (1985).
Puerto Rico. 81 PUE. Phan.

Myrcia mathewsiana (O.Berg) McVaugh, Publ. Field Mus. Nat. Hist., Bot. Ser. 13(4): 646 (1958).
Peru to Bolivia. 83 BOL PER. Nanophan. or phan.
**Aulomyrcia mathewsiana* O.Berg, Linnaea 27: 45 (1855).

Myrcia maximiliana O.Berg, Linnaea 27: 89 (1855).
SE. Brazil. 84 BZL. Nanophan. or phan.
Myrtus greggii Nees & Mart., Nova Acta Phys.-Med. Acad. Caes. Leop.-Carol. Nat. Cur. 12(1): 51 (1824), nom. illeg.

Myrcia melastomoides DC., Prodr. 3: 256 (1828).
Tobago. 81 TRT. Nanophan. or phan.

Myrcia micropetala (Mart.) Nied. in H.G.A.Engler & K.A.E.Prantl, Nat. Pflanzenfam. 3(7): 76 (1893).
NE. Brazil. 84 BZE. Nanophan. or phan.
**Eugenia micropetala* Mart., Flora 24(2 Beibl.): 108 (1841). *Aulomyrcia micropetala* (Mart.) O.Berg, Linnaea 27: 56 (1855).

Myrcia microphylla O.Berg in C.F.P.von Martius & auct. suc. (eds.), Fl. Bras. 14(1): 566 (1859).
Brazil (Goiás). 84 BZC. Nanophan. or phan.

Myrcia microstachya Krug & Urb., Bot. Jahrb. Syst. 15: 361 (1892).
Jamaica. 81 JAM. Nanophan. or phan.

Myrcia minutiflora Sagot, Ann. Sci. Nat., Bot., VI, 20: 185 (1885). *Aulomyrcia minutiflora* (Sagot) Amshoff, Bull. Torrey Bot. Club 75: 532 (1948).
S. Trop. America. 82 FRG GUY SUR VEN 83 PER 84 BZE BZN. Nanophan. or phan.

Myrcia mischophylla Kiaersk., Enum. Myrt. Bras.: 61 (1893).
Brazil (Bahia to Minas Gerais). 84 BZE BZL. Nanophan. or phan.

Myrcia mollis (Kunth) DC., Prodr. 3: 256 (1828).
W. South America. 83 BOL CLM ECU PER. Nanophan. or phan.
**Myrtus mollis* Kunth in F.W.H.von Humboldt, A.J.A. Bonpland & C.S.Kunth, Nov. Gen. Sp. 6: 141 (1823).
Myrcia huanocensis O.Berg, Linnaea 31: 250 (1862).
Eugenia boliviensis Rusby, Mem. Torrey Bot. Club 3(3): 28 (1893), nom. illeg.
Myrcia huanocensis var. *glazioviana* Kiaersk., Enum. Myrt. Bras.: 55 (1893).

Myrcia montana Cambess. in A.F.C.de Saint-Hilaire, Fl. Bras. Merid. 2: 325 (1832). *Aulomyrcia montana* (Cambess.) O.Berg, Linnaea 27: 36 (1855).
Brazil (Espírito Santo, Minas Gerais). 84 BZL. Nanophan. or phan.

Myrcia morroqueimadensis Kiaersk., Enum. Myrt. Bras.: 73 (1893).
SE. Brazil. 84 BZL. Nanophan. or phan.

Myrcia multiflora (Lam.) DC., Prodr. 3: 244 (1828).
Trinidad to S. Trop. America. 81 TRT 82 FRG GUY SUR VEN 83 BOL PER 84 BZC BZE BZL BZN BZS 85 PAR URU. Nanophan. or phan.
**Eugenia multiflora* Lam., Encycl. 3: 202 (1789). *Myrtus multiflora* (Lam.) Spreng., Syst. Veg. 2: 485 (1825). *Cumetea multiflora* (Lam.) Raf., Sylva Tellur.: 106 (1838). *Aulomyrcia multiflora* (Lam.) O.Berg, Linnaea 27: 47 (1855).
Eugenia multiflora Rich., Actes Soc. Hist. Nat. Paris 1: 110 (1792), nom. illeg.
Myrcia camaraeana DC., Prodr. 3: 251 (1828). *Aulomyrcia camareana* (DC.) O.Berg in C.F.P.von Martius & auct. suc. (eds.), Fl. Bras. 14(1): 83 (1857).
Myrcia sphaerocarpa DC., Prodr. 3: 251 (1828). *Aulomyrcia sphaerocarpa* (DC.) O.Berg, Linnaea 27: 51 (1855).
Myrcia ellipticifolia Cambess. in A.F.C.de Saint-Hilaire, Fl. Bras. Merid. 2: 312 (1832).
Aulomyrcia caerulescens O.Berg in C.F.P.von Martius & auct. suc. (eds.), Fl. Bras. 14(1): 80 (1857). *Myrcia caerulescens* (O.Berg) Kiaersk., Enum. Myrt. Bras.: 79 (1893).
Aulomyrcia caesia O.Berg in C.F.P.von Martius & auct. suc. (eds.), Fl. Bras. 14(1): 83 (1857).
Aulomyrcia glaucescens O.Berg in C.F.P.von Martius & auct. suc. (eds.), Fl. Bras. 14(1): 81 (1857). *Myrcia glaucescens* (O.Berg) Kiaersk., Enum. Myrt. Bras.: 83 (1893). *Myrcia multiflora* f. *glaucescens* (O.Berg) D.Legrand, in Fl. Ilustr. Catar. 1(Mirt.): 305 (1969). *Myrcia multiflora* var. *glaucescens* (O.Berg) D.Legrand, in Fl. Ilustr. Catar. 1(Mirt.): 305 (1969).
Aulomyrcia glaucescens var. *grandifolia* O.Berg in C.F.P.von Martius & auct. suc. (eds.), Fl. Bras. 14(1): 81 (1857).
Aulomyrcia glaucescens var. *parvifolia* O.Berg in C.F.P.von Martius & auct. suc. (eds.), Fl. Bras. 14(1): 82 (1857).
Aulomyrcia laruotteana var. *peruviana* O.Berg in C.F.P.von Martius & auct. suc. (eds.), Fl. Bras. 14(1): 91 (1857).
Aulomyrcia ovalifolia O.Berg in C.F.P.von Martius & auct. suc. (eds.), Fl. Bras. 14(1): 81 (1857). *Myrcia ovalifolia* (O.Berg) Kiaersk., Enum. Myrt. Bras.: 84 (1893). *Myrcia multiflora* f. *ovalifolia* (O.Berg) D.Legrand, in Fl. Ilustr. Catar. 1(Mirt.): 308 (1969).
Aulomyrcia ovalis O.Berg in C.F.P.von Martius & auct. suc. (eds.), Fl. Bras. 14(1): 107 (1857).
Aulomyrcia perforata O.Berg in C.F.P.von Martius & auct. suc. (eds.), Fl. Bras. 14(1): 83 (1857).

Aulomyrcia sphaerocarpa var. *arborescens* O.Berg in C.F.P.von Martius & auct. suc. (eds.), Fl. Bras. 14(1): 85 (1857).
Aulomyrcia sphaerocarpa var. *complicata* O.Berg in C.F.P.von Martius & auct. suc. (eds.), Fl. Bras. 14(1): 86 (1857).
Aulomyrcia sphaerocarpa var. *gracilis* O.Berg in C.F.P.von Martius & auct. suc. (eds.), Fl. Bras. 14(1): 85 (1857).
Aulomyrcia sphaerocarpa var. *intermedia* O.Berg in C.F.P.von Martius & auct. suc. (eds.), Fl. Bras. 14(1): 85 (1857).
Aulomyrcia sphaerocarpa var. *obtusata* O.Berg in C.F.P.von Martius & auct. suc. (eds.), Fl. Bras. 14(1): 86 (1857).
Aulomyrcia sphaerocarpa var. *ovata* O.Berg in C.F.P.von Martius & auct. suc. (eds.), Fl. Bras. 14(1): 86 (1857).
Aulomyrcia sphaerocarpa var. *pauciflora* O.Berg in C.F.P.von Martius & auct. suc. (eds.), Fl. Bras. 14(1): 86 (1857).
Aulomyrcia multiflora var. *grandifolia* O.Berg, Linnaea 30: 660 (1861).
Myrcia glaberrima Barb.Rodr. ex Chodat & Hassl., Bull. Herb. Boissier, II, 7: 803 (1907), nom. nud.
Aulomyrcia vinacea Steyerm., Fieldiana, Bot. 28: 1008 (1957).
Myrcia multiflora var. *ramulosa* D.Legrand, in Fl. Ilustr. Catar. 1(Mirt.): 309 (1969).

Myrcia mutabilis (O.Berg) N.Silveira, Loefgrenia 88: 1 (1985).
Brazil. 84 BZC BZE BZL. Nanophan. or phan.
**Aulomyrcia mutabilis* O.Berg in C.F.P.von Martius & auct. suc. (eds.), Fl. Bras. 14(1): 70 (1857).
Myrcia pilodes Kiaersk., Enum. Myrt. Bras.: 67 (1893). *Marlierea pilodes* (Kiaersk.) M.L.Kawas., Bol. Bot. (São Paulo) 11: 126 (1989).

Myrcia myoporina DC., Prodr. 3: 246 (1828).
Brazil. 84+. Nanophan. or phan.

Myrcia myriantha McVaugh, Mem. New York Bot. Gard. 18(2): 116 (1969).
Guyana. 82 GUY. Phan.

Myrcia myrtillifolia DC., Prodr. 3: 250 (1828). *Aulomyrcia myrtillifolia* (DC.) O.Berg, Linnaea 27: 36 (1855).
E. Brazil. 84 BZE BZL. Nanophan. or phan.

Myrcia nandu-apysa Parodi, Anales Soc. Ci. Argent. 7: 183 (1879).
Paraguay. 85 PAR. Nanophan. or phan.

Myrcia neesiana DC., Prodr. 3: 249 (1828). *Aulomyrcia neesiana* (DC.) O.Berg, Linnaea 27: 36 (1855).
Peru to Bolivia and N. Brazil. 83 BOL PER 84 BZN. Phan.
Aulomyrcia holosericea O.Berg in C.F.P.von Martius & auct. suc. (eds.), Fl. Bras. 14(1): 99 (1857).

Myrcia neorostrata Sobral, Novon 16: 136 (2006).
Brazil (Bahia, Espírito Santo). 84 BZE BZL. Nanophan. or phan.
**Aulomyrcia rostrata* O.Berg in C.F.P.von Martius & auct. suc. (eds.), Fl. Bras. 14(1): 544 (1859).

Myrcia nigricans (O.Berg) N.Silveira, Loefgrenia 88: 1 (1985).
SE. Brazil. 84 BZL. Nanophan. or phan.
**Aulomyrcia nigricans* O.Berg in C.F.P.von Martius & auct. suc. (eds.), Fl. Bras. 14(1): 101 (1857).

Myrcia nigropunctata (O.Berg) N.Silveira, Loefgrenia 88: 1 (1985).
Brazil (Rio de Janeiro). 84 BZL. Nanophan. or phan.
**Aulomyrcia nigropunctata* O.Berg in C.F.P.von Martius & auct. suc. (eds.), Fl. Bras. 14(1): 116 (1857).

Myrcia nitida Cambess. in A.F.C.de Saint-Hilaire, Fl. Bras. Merid. 2: 309 (1832).
Brazil (Minas Gerais). 84 BZL. Nanophan.

Myrcia nivea Cambess. in A.F.C.de Saint-Hilaire, Fl. Bras. Merid. 2: 332 (1832). *Aulomyrcia nivea* (Cambess.) O.Berg, Linnaea 27: 61 (1855).
SE. Brazil. 84 BZL. Nanophan. or phan.
Aulomyrcia nivea var. *andromedifolia* O.Berg in C.F.P. von Martius & auct. suc. (eds.), Fl. Bras. 14(1): 103 (1857).
Aulomyrcia nivea var. *rosmarinifolia* O.Berg in C.F.P. von Martius & auct. suc. (eds.), Fl. Bras. 14(1): 103 (1857). *Myrcia nivea* var. *rosmarinifolia* (O.Berg) Mattos, Loefgrenia 66: 3 (1975).

Myrcia nobilis O.Berg in C.F.P.von Martius & auct. suc. (eds.), Fl. Bras. 14(1): 195 (1857).
Brazil (Minas Gerais). 84 BZL. Nanophan. or phan.
Myrtus firma Spreng., Syst. Veg. 2: 487 (1825), provisional synonym.

Myrcia nubicola McVaugh, Mem. New York Bot. Gard. 18(2): 116 (1969).
S. Venezuela to Guyana. 82 GUY VEN. Nanophan. or phan.

Myrcia oblongata DC., Prodr. 3: 251 (1828). *Aulomyrcia oblongata* (DC.) O.Berg, Linnaea 27: 37 (1855).
Brazil (Minas Gerais). 84 BZL. Nanophan. or phan.

Myrcia obovata (O.Berg) Nied. in H.G.A.Engler & K.A.E.Prantl, Nat. Pflanzenfam. 3(7): 761 (1893).
SE. Brazil. 84 BZL. Nanophan. or phan.
Eugenia crassifolia Miq., Linnaea 19: 439 (1846), nom. illeg. *Myrcia crassifolia* Kiaersk., Enum. Myrt. Bras.: 89 (1893).
**Aulomyrcia obovata* O.Berg, Linnaea 27: 72 (1855).
Aulomyrcia atrovirens O.Berg in C.F.P.von Martius & auct. suc. (eds.), Fl. Bras. 14(1): 121 (1857).

Myrcia obumbrans (O.Berg) McVaugh, Fieldiana, Bot. 29: 193 (1956).
Ecuador to Peru. 83 ECU PER 84 BZN. Nanophan. or phan.
**Rubachia obumbrans* O.Berg in C.F.P.von Martius & auct. suc. (eds.), Fl. Bras. 14(1): 28 (1857). *Marlierea obumbrans* (O.Berg) Nied. in H.G.A. Engler & K.A.E.Prantl, Nat. Pflanzenfam. 3(7): 76 (1893).

Myrcia ochroides O.Berg in C.F.P.von Martius & auct. suc. (eds.), Fl. Bras. 14(1): 208 (1857).
NE. Brazil. 84 BZE. Nanophan. or phan.

Myrcia oligantha O.Berg in C.F.P.von Martius & auct. suc. (eds.), Fl. Bras. 14(1): 184 (1857).
SE. & S. Brazil. 84 BZL BZS. Nanophan. or phan.
Aulomyrcia fenzliana O.Berg, Linnaea 27: 63 (1855). *Myrcia kauseliana* D.Legrand, Sellowia 13: 296 (1961).

Myrcia oligostemon (Urb.) Alain, Mem. New York Bot. Gard. 21(2): 138 (1971).
C. Cuba. 81 CUB. Nanophan.
**Mozartia oligostemon* Urb., Symb. Antill. 9: 473 (1928).

Myrcia oreioeca Kiaersk., Enum. Myrt. Bras.: 94 (1893).
SE. Brazil. 84 BZL. Nanophan. or phan.

Myrcia orthophylla (O.Berg) Kiaersk., Enum. Myrt. Bras.: 91 (1893).
WC. Brazil. 84 BZC. Nanophan. or phan.
**Aulomyrcia orthophylla* O.Berg in C.F.P.von Martius & auct. suc. (eds.), Fl. Bras. 14(1): 549 (1859).

Myrcia ouropretoensis Kiaersk., Enum. Myrt. Bras.: 113 (1893).
Brazil (Minas Gerais). 84 BZL. Nanophan.
**Gomidesia clausseniana* O.Berg in C.F.P.von Martius & auct. suc. (eds.), Fl. Bras. 14(1): 23 (1857).

Myrcia ovalis (O.Berg) N.Silveira, Loefgrenia 88: 2 (1985).
SE. Brazil. 84 BZL. Nanophan. or phan.

Myrcia ovata Cambess. in A.F.C.de Saint-Hilaire, Fl. Bras. Merid. 2: 319 (1832).
SE. Brazil. 84 BZL. Nanophan. or phan.
Myrcia ovata var. *subcordata* O.Berg in C.F.P.von Martius & auct. suc. (eds.), Fl. Bras. 14(1): 167 (1857).

Myrcia paganii Krug & Urb., Bot. Jahrb. Syst. 19: 587 (1895).
Puerto Rico. 81 PUE. Phan.

Myrcia paivae O.Berg in C.F.P.von Martius & auct. suc. (eds.), Fl. Bras. 14(1): 179 (1857).
Costa Rica ti S. Trop. America. 80 COS PAN 82 GUY SUR VEN 83 BOL CLM PER 84 BZN. Nanophan. or phan.
Myrcia frontinensis Hieron., Bot. Jahrb. Syst. 20(49): 63 (1895).
Myrcia frontinensis var. *gamaeana* Glaz., Bull. Soc. Bot. France 54(3c): 210 (1908).
Myrcia paivae var. *gracilis* Lingelsh., Repert. Spec. Nov. Regni Veg. 7: 243 (1909).

Myrcia palustris DC., Prodr. 3: 246 (1828). *Gomidesia palustris* (DC.) Kausel, Lilloa 32: 348 (1966 publ. 1967).
Brazil to NE. Argentina. 84 BZC BZL BZS 85 AGE PAR URU. Nanophan. or phan.
Myrcia banisteriifolia DC., Prodr. 3: 246 (1828). *Gomidesia banisteriifolia* (DC.) O.Berg in C.F.P.von Martius & auct. suc. (eds.), Fl. Bras. 14(1): 23 (1857).
Myrcia garopabensis Cambess. in A.F.C.de Saint-Hilaire, Fl. Bras. Merid. 2: 324 (1832). *Aulomyrcia garopabensis* (Cambess.) O.Berg, Linnaea 27: 47 (1855). *Gomidesia garopabensis* (Cambess.) D.Legrand, Sellowia 13: 281 (1961).
Gomidesia gardneriana O.Berg in C.F.P.von Martius & auct. suc. (eds.), Fl. Bras. 14(1): 22 (1857).
Myrcia anomala var. *multiceps* O.Berg in C.F.P.von Martius & auct. suc. (eds.), Fl. Bras. 14(1): 191 (1857).
Myrcia palustris var. *acutata* O.Berg in C.F.P.von Martius & auct. suc. (eds.), Fl. Bras. 14(1): 191 (1857).
Myrcia palustris var. *angustifolia* O.Berg in C.F.P.von Martius & auct. suc. (eds.), Fl. Bras. 14(1): 191 (1857). *Gomidesia palustris* var. *angustifolia* (O.Berg) Mattos, Loefgrenia 61: 3 (1974).
Myrcia palustris var. *bracteata* O.Berg in C.F.P.von Martius & auct. suc. (eds.), Fl. Bras. 14(1): 191 (1857).
Myrcia palustris var. *stictophylla* O.Berg in C.F.P.von Martius & auct. suc. (eds.), Fl. Bras. 14(1): 191 (1857).
Myrcia hartwegiana D.Legrand, Anales Mus. Hist. Nat. Montevideo, IV, 11: 37 (1936), nom. illeg.

Myrcia panamensis B.Holst & M.L.Kawas., Selbyana 25: 97 (2004).
Panama. 80 PAN. Phan.

Myrcia panicularis (O.Berg) N.Silveira, Roessléria 7: 66 (1985).
Brazil (Rio de Janeiro). 84 BZL. Nanophan. or phan.
**Aulomyrcia panicularis* O.Berg in C.F.P.von Martius & auct. suc. (eds.), Fl. Bras. 14(1): 84 (1857).

Myrcia paracatuensis Kiaersk., Enum. Myrt. Bras.: 99 (1893).
WC. Brazil. 84 BZC. Nanophan. or phan.
**Aulomyrcia suffruticosa* O.Berg in C.F.P.von Martius & auct. suc. (eds.), Fl. Bras. 14(1): 136 (1857).
Aulomyrcia suffruticosa var. *arenaria* O.Berg in C.F.P.von Martius & auct. suc. (eds.), Fl. Bras. 14(1): 559 (1859). *Myrcia paracatuensis* var. *arenaria* (O.Berg) Kiaersk., Enum. Myrt. Bras.: 99 (1893).
Aulomyrcia suffruticosa var. *venosa* O.Berg in C.F.P.von Martius & auct. suc. (eds.), Fl. Bras. 14(1): 559 (1859). *Myrcia paracatuensis* var. *venosa* (O.Berg) Mattos, Loefgrenia 66: 1 (1975).
Myrcia paracatuensis var. *pumila* Glaz., Bull. Soc. Bot. France 54(3c): 220 (1908).
Myrcia paracatuensis var. *linearis* Mattos, Loefgrenia 66: 2 (1975).

Myrcia parnahibensis (O.Berg) Kiaersk., Enum. Myrt. Bras.: 89 (1893).
WC. Brazil. 84 BZC. Nanophan. or phan.
**Aulomyrcia parnahibensis* O.Berg in C.F.P.von Martius & auct. suc. (eds.), Fl. Bras. 14(1): 135 (1857).
Aulomyrcia parnahibensis var. *acutifolia* O.Berg in C.F.P.von Martius & auct. suc. (eds.), Fl. Bras. 14(1): 558 (1859). *Myrcia parnahibensis* var. *acutifolia* (O.Berg) Kiaersk., Enum. Myrt. Bras.: 89 (1893). *Myrcia parnahibensis* var. *acutifolia* (O.Berg) Mattos, Loefgrenia 66: 2 (1975).
Aulomyrcia parnahibensis var. *imbricata* O.Berg in C.F.P.von Martius & auct. suc. (eds.), Fl. Bras. 14(1): 559 (1859). *Myrcia parnahibensis* var. *imbricata* (O.Berg) Kiaersk., Enum. Myrt. Bras.: 89 (1893).
Aulomyrcia parnahibensis var. *verticillata* O.Berg in C.F.P.von Martius & auct. suc. (eds.), Fl. Bras. 14(1): 559 (1859).

Myrcia pentagona McVaugh, Fieldiana, Bot. 29: 193 (1956).
N. Peru. 83 PER. Phan.

Myrcia perforata O.Berg in C.F.P.von Martius & auct. suc. (eds.), Fl. Bras. 14(1): 197 (1857).
Brazil (Minas Gerais). 84 BZL. Nanophan. or phan.

Myrcia pernambucensis O.Berg in C.F.P.von Martius & auct. suc. (eds.), Fl. Bras. 14(1): 194 (1857).
NE. Brazil. 84 BZE. Nanophan. or phan.

Myrcia pertusa DC., Prodr. 3: 251 (1828). *Aulomyrcia pertusa* (DC.) O.Berg, Linnaea 27: 47 (1855).
N. Brazil to Peru. 83 PER 84 BZN. Nanophan. or phan.

Myrcia phaeoclada O.Berg in C.F.P.von Martius & auct. suc. (eds.), Fl. Bras. 14(1): 167 (1857).
Brazil to Bolivia. 83 BOL 84 BZE. Nanophan. or phan.

Myrcia phaeoclada var. *alagoensis* O.Berg in C.F.P. von Martius & auct. suc. (eds.), Fl. Bras. 14(1): 167 (1857).

Myrcia picardiae Krug & Urb., Bot. Jahrb. Syst. 15: 359 (1892).
Haiti. 81 HAI. Nanophan. or phan.

Myrcia pineticola Borhidi & O.Muñiz, Acta Bot. Acad. Sci. Hung. 21: 226 (1975 publ. 1976).
Cuba. 81 CUB. Nanophan. or phan.

Myrcia pinifolia Cambess. in A.F.C.de Saint-Hilaire, Fl. Bras. Merid. 2: 333 (1832). *Aulomyrcia pinifolia* (Cambess.) O.Berg, Linnaea 27: 35 (1855).
WC. Brazil to Bolivia. 83 BOL 84 BZC. Cham.

Myrcia pistrinalis McVaugh, Mem. New York Bot. Gard. 18(2): 117 (1969).
SE. Venezuela, Suriname. 82 SUR VEN. Nanophan. or phan.

Myrcia platycaula Diels, Bot. Jahrb. Syst. 37: 595 (1906).
Peru. 83 PER. Nanophan. or phan. Provisionally accepted.

Myrcia platyclada DC., Prodr. 3: 244 (1828). *Aulomyrcia platyclada* (DC.) Amshoff, Bull. Torrey Bot. Club 75: 531 (1948), nom. illeg.
Lesser Antilles, Panama to N. Brazil. 80 PAN 81 LEE TRT WIN 82 FRG GUY SUR VEN 84 BZN. Nanophan. or phan.
Aulomyrcia dumosa O.Berg, Linnaea 30: 656 (1861). *Myrcia dumosa* (O.Berg) Krug & Urb., Bot. Jahrb. Syst. 19: 580 (1895).
Aulomyrcia edulis O.Berg, Linnaea 30: 657 (1861). *Myrcia edulis* (O.Berg) Krug & Urb., Bot. Jahrb. Syst. 19: 581 (1895).
Myrcia dussii Krug & Urb., Bot. Jahrb. Syst. 19: 580 (1895).
Aulomyrcia platyclada var. *kaieteurensis* Amshoff, Bull. Torrey Bot. Club 75: 532 (1948).

Myrcia plusiantha Kiaersk., Enum. Myrt. Bras.: 66 (1893).
Brazil (Rio de Janeiro). 84 BZL. Nanophan. or phan.

Myrcia poeppigiana O.Berg in C.F.P.von Martius & auct. suc. (eds.), Fl. Bras. 14(1): 157 (1857).
N. Brazil to Peru. 83 PER 84 BZN. Nanophan. or phan.

Myrcia polyantha DC., Prodr. 3: 252 (1828). *Aulomyrcia polyantha* (DC.) O.Berg in C.F.P.von Martius & auct. suc. (eds.), Fl. Bras. 14(1): 82 (1857).
NE. Brazil. 84 BZE. Nanophan. or phan.
Aulomyrcia polyantha var. *coriacea* O.Berg in C.F.P.von Martius & auct. suc. (eds.), Fl. Bras. 14(1): 82 (1857).
Aulomyrcia polyantha var. *membranacea* O.Berg in C.F.P.von Martius & auct. suc. (eds.), Fl. Bras. 14(1): 82 (1857).
Aulomyrcia polyantha var. *parviflora* O.Berg in C.F.P.von Martius & auct. suc. (eds.), Fl. Bras. 14(1): 82 (1857).
Myrcia polyantha var. *brasiliensis* O.Berg in C.F.P.von Martius & auct. suc. (eds.), Fl. Bras. 14(1): 185 (1857).
Myrcia polyantha var. *orinocensis* O.Berg in C.F.P.von Martius & auct. suc. (eds.), Fl. Bras. 14(1): 185 (1857).

Myrcia polyneura (Urb.) Borhidi, Acta Bot. Hung. 29: 186 (1983).
Cuba (Sierra de Nipe). 81 CUB. Nanophan. or phan.
**Calyptranthes polyneura* Urb., Symb. Antill. 9: 94 (1923).

Myrcia popayanensis Hieron., Bot. Jahrb. Syst. 20(49): 62 (1895).
Colombia. 83 CLM. Nanophan. or phan.

Myrcia porphyrea McVaugh, Mem. New York Bot. Gard. 18(2): 119 (1969).
Guyana. 82 GUY. Nanophan.

Myrcia ptariensis (Steyerm.) McVaugh, Mem. New York Bot. Gard. 18(2): 103 (1969).
Venezuela (Bolívar). 82 VEN. Nanophan. or phan.
**Aulomyrcia ptariensis* Steyerm., Fieldiana, Bot. 28: 1006 (1957).

Myrcia pubescens DC., Prodr. 3: 247 (1828). *Gomidesia pubescens* (DC.) D.Legrand, Comun. Bot. Mus. Hist. Nat. Montevideo 3(37): 20 (1958).
Brazil to Bolivia. 83 BOL 84 BZC BZE BZL BZS. Phan.
Gomidesia aubletiana O.Berg in C.F.P.von Martius & auct. suc. (eds.), Fl. Bras. 14(1): 27 (1857).
Gomidesia casaretteana O.Berg in C.F.P.von Martius & auct. suc. (eds.), Fl. Bras. 14(1): 20 (1857). *Gomidesia pubescens* var. *casaretteana* (O.Berg) D.Legrand, Comun. Bot. Mus. Hist. Nat. Montevideo 3(37): 20 (1959).
Gomidesia mikaniana O.Berg in C.F.P.von Martius & auct. suc. (eds.), Fl. Bras. 14(1): 21 (1857).
Gomidesia raddiana O.Berg in C.F.P.von Martius & auct. suc. (eds.), Fl. Bras. 14(1): 26 (1857).
Gomidesia widgreniana O.Berg in C.F.P.von Martius & auct. suc. (eds.), Fl. Bras. 14(1): 26 (1857). *Myrcia minensis* Kiaersk., Enum. Myrt. Bras.: 110 (1893). *Gomidesia pubescens* var. *widgreniana* (O.Berg) D.Legrand, Comun. Bot. Mus. Hist. Nat. Montevideo 3(37): 21 (1959).
Myrcia minensis var. *subcordata* Kiaersk., Enum. Myrt. Bras.: 110 (1893).
Myrcia schenckiana Kiaersk., Enum. Myrt. Bras.: 114 (1893).

Myrcia pubiflora DC., Prodr. 3: 249 (1828). *Aulomyrcia pubiflora* (DC.) O.Berg, Linnaea 27: 40 (1855).
Brazil to Bolivia. 83 BOL 84 BZE BZL BZS 85 PAR. Nanophan. or phan.
Eugenia collina DC., Prodr. 3: 281 (1828). *Aulomyrcia collina* (DC.) O.Berg, Linnaea 27: 40 (1855). *Myrcia collina* (DC.) J.F.Macbr., Candollea 5: 394 (1934), nom. illeg.
Aulomyrcia pohliana O.Berg in C.F.P.von Martius & auct. suc. (eds.), Fl. Bras. 14(1): 72 (1857). *Myrcia calumbaensis* Kiaersk., Enum. Myrt. Bras.: 77 (1893).
Myrcia pubiflora var. *glazioviana* Kiaersk., Enum. Myrt. Bras.: 19 (1893).

Myrcia pubipetala Miq., Linnaea 19: 441 (1846).
Brazil to Bolivia. 83 BOL 84 BZC BZE BZL BZS. Phan.
Aulomyrcia grandiflora O.Berg in C.F.P.von Martius & auct. suc. (eds.), Fl. Bras. 14(1): 113 (1857). *Myrcia grandiflora* (O.Berg) Nied. in H.G.A.Engler & K.A.E.Prantl, Nat. Pflanzenfam. 3(7): 77 (1893).
Myrcia pubipetala var. *magnifolia* D.Legrand, Sellowia 13: 293 (1961).

Myrcia pulchra (O.Berg) Kiaersk., Enum. Myrt. Bras.: 65 (1893).
SE. & S. Brazil. 84 BZL BZS. Nanophan. or phan.

Aulomyrcia breviramis O.Berg in C.F.P.von Martius & auct. suc. (eds.), Fl. Bras. 14(1): 66 (1857). *Myrcia breviramis* (O.Berg) D.Legrand, Sellowia 13: 294 (1961).
**Aulomyrcia pulchra* O.Berg in C.F.P.von Martius & auct. suc. (eds.), Fl. Bras. 14(1): 68 (1857).
Aulomyrcia rufa O.Berg in C.F.P.von Martius & auct. suc. (eds.), Fl. Bras. 14(1): 65 (1857). *Myrcia rufa* (O.Berg) N.Silveira, Roesslёria 7: 66 (1985).
Aulomyrcia widgreniana O.Berg in C.F.P.von Martius & auct. suc. (eds.), Fl. Bras. 14(1): 70 (1857). *Myrcia widgreniana* (O.Berg) Mattos, Dusenia 8: 161 (1968).
Myrcia pilotantha Kiaersk., Enum. Myrt. Bras.: 64 (1893).
Myrcia jaguariaivensis Mattos & D.Legrand, Loefgrenia 67: 2 (1975).

Myrcia pyrifolia (Desv.) Nied. in H.G.A.Engler & K.A.E.Prantl, Nat. Pflanzenfam. 3(7): 76 (1893).
S. Trop. America. 82 FRG GUY SUR VEN 83 CLM PER 84 BZN. Nanophan. or phan.
**Eugenia pyrifolia* Desv. in W.Hamilton, Prodr. Pl. Ind. Occid.: 44 (1825). *Aulomyrcia pyrifolia* (Desv.) O.Berg, Linnaea 27: 44 (1855).
Myrcia divergens DC., Prodr. 3: 245 (1828).
Aulomyrcia pyrifolia var. *gracilis* O.Berg, Linnaea 27: 45 (1855).
Aulomyrcia pyrifolia var. *robusta* O.Berg, Linnaea 27: 44 (1855).
Aulomyrcia ovata O.Berg, Linnaea 27: 467 (1856).

Myrcia quitarensis (Benth.) Sagot, Ann. Sci. Nat., Bot., VI, 20: 184 (1885).
SE. Venezuela to Guyana. 82 GUY VEN. Nanophan. or phan.
**Eugenia quitarensis* Benth., J. Bot. (Hooker) 2: 322 (1840). *Myrciaria quitarensis* (Benth.) O.Berg, Linnaea 27: 323 (1856).

Myrcia racemosa (O.Berg) Kiaersk., Enum. Myrt. Bras.: 72 (1893).
SE. & S. Brazil. 84 BZL BZS. Phan.
**Aulomyrcia racemosa* O.Berg, Linnaea 27: 52 (1855).
Aulomyrcia gaudichaudiana O.Berg in C.F.P.von Martius & auct. suc. (eds.), Fl. Bras. 14(1): 88 (1857).
Myrcia acuminatissima O.Berg in C.F.P.von Martius & auct. suc. (eds.), Fl. Bras. 14(1): 167 (1857).

Myrcia racemulosa DC., Prodr. 3: 254 (1828).
WC. & SE. Brazil. 84 BZC BZL. Nanophan. or phan.
Myrcia scutulifera DC., Prodr. 3: 254 (1828). *Aulomyrcia imbricata* var. *scutulifera* (DC.) O.Berg in C.F.P.von Martius & auct. suc. (eds.), Fl. Bras. 14(1): 100 (1857).
Myrcia imbricata Gardner in H.B.Fielding & G.Gardner, Sert. Pl.: t. 75 (1844). *Aulomyrcia imbricata* (Gardner) O.Berg, Linnaea 27: 61 (1855).
Aulomyrcia imbricata var. *intermedia* O.Berg in C.F.P.von Martius & auct. suc. (eds.), Fl. Bras. 14(1): 100 (1857).
Aulomyrcia imbricata var. *microphylla* O.Berg in C.F.P.von Martius & auct. suc. (eds.), Fl. Bras. 14(1): 100 (1857).
Aulomyrcia imbricata var. *rotundifolia* O.Berg in C.F.P.von Martius & auct. suc. (eds.), Fl. Bras. 14(1): 100 (1857).

Myrcia ramageana Krug & Urb., Bot. Jahrb. Syst. 19: 586 (1895).
Windward Is. 81 WIN. Nanophan. or phan.

Myrcia ramuliflora (O.Berg) N.Silveira, Roesslёria 7: 66 (1985).
NE. Brazil. 84 BZE. Nanophan. or phan.
**Aulomyrcia ramuliflora* O.Berg in C.F.P.von Martius & auct. suc. (eds.), Fl. Bras. 14(1): 64 (1857).
Aulomyrcia sphenoides O.Berg in C.F.P.von Martius & auct. suc. (eds.), Fl. Bras. 14(1): 63 (1857).

Myrcia recurvata O.Berg in C.F.P.von Martius & auct. suc. (eds.), Fl. Bras. 14(1): 166 (1857).
SE. Brazil. 84 BZL. Nanophan. or phan.
Myrcia recurvata var. *grandifolia* O.Berg in C.F.P.von Martius & auct. suc. (eds.), Fl. Bras. 14(1): 166 (1857).
Myrcia recurvata var. *parvifolia* O.Berg in C.F.P.von Martius & auct. suc. (eds.), Fl. Bras. 14(1): 166 (1857).

Myrcia regeliana O.Berg in C.F.P.von Martius & auct. suc. (eds.), Fl. Bras. 14(1): 562 (1859).
Brazil (Goiás). 84 BZC. Nanophan. or phan.

Myrcia regnelliana O.Berg in C.F.P.von Martius & auct. suc. (eds.), Fl. Bras. 14(1): 174 (1857).
WC. & SE. Brazil. 84 BZC BZL. Nanophan. or phan.
Myrcia regnelliana var. *leptocalama* O.Berg in C.F.P. von Martius & auct. suc. (eds.), Fl. Bras. 14(1): 174 (1857).
Myrcia regnelliana var. *robustior* O.Berg in C.F.P.von Martius & auct. suc. (eds.), Fl. Bras. 14(1): 174 (1857).

Myrcia reticulata Cambess. in A.F.C.de Saint-Hilaire, Fl. Bras. Merid. 2: 304 (1832). *Gomidesia reticulata* (Cambess.) O.Berg in C.F.P.von Martius & auct. suc. (eds.), Fl. Bras. 14(1): 15 (1857).
SE. Brazil. 84 BZL. Nanophan. or phan.
Gomidesia reticulata var. *angustifolia* O.Berg in C.F.P. von Martius & auct. suc. (eds.), Fl. Bras. 14(1): 16 (1857).
Gomidesia reticulata var. *latifolia* O.Berg in C.F.P.von Martius & auct. suc. (eds.), Fl. Bras. 14(1): 15 (1857).

Myrcia reticulosa Miq., Linnaea 22: 794 (1850). *Aulomyrcia reticulosa* (Miq.) O.Berg, Linnaea 27: 62 (1855).
Brazil (Bahia, Minas Gerais). 84 BZE BZL. Nanophan.

Myrcia retivenia (C.Wright) Urb., Symb. Antill. 9: 86 (1923).
E. Cuba. 81 CUB. Nanophan.
**Eugenia retivenia* C.Wright, Anales Acad. Ci. Méd. Habana 5: 433 (1868).
Myrcia pungens Urb., Symb. Antill. 9: 86 (1923).

Myrcia retorta Cambess. in A.F.C.de Saint-Hilaire, Fl. Bras. Merid. 2: 322 (1832).
SE. & S. Brazil. 84 BZL BZS. Nanophan. or phan.
Myrcia arborescens O.Berg in C.F.P.von Martius & auct. suc. (eds.), Fl. Bras. 14(1): 200 (1857).
Myrcia itambensis O.Berg in C.F.P.von Martius & auct. suc. (eds.), Fl. Bras. 14(1): 190 (1857).

Myrcia revolutifolia McVaugh, Mem. New York Bot. Gard. 18(2): 121 (1969).
SE. Colombia to N. Brazil. 82 VEN 83 CLM 84 BZN. Nanophan. or phan.
Myrcia planifolia McVaugh, Mem. New York Bot. Gard. 18(2): 118 (1969).

Myrcia rimosa Cambess. in A.F.C.de Saint-Hilaire, Fl. Bras. Merid. 2: 333 (1832).

Brazil (Mato Grosso, Minas Gerais). 84 BZC BZL. Nanophan. or phan.

Myrcia riodocensis G.M.Barroso & Peixoto, Acta Bot. Brasil. 4(2): 13 (1990).
Brazil (Espírito Santo). 84 BZL. Nanophan. or phan.

Myrcia robusta Sobral, Bol. Mus. Biol. Mello Leitão. Nova Sér. 20: 75 (2006 publ. 2007).
Brazil (Espírito Santo). 84 BZL. Phan.

Myrcia rotundata (Amshoff) McVaugh, Mem. New York Bot. Gard. 18(2): 122 (1969).
Venezuela to Guyana. 82 GUY VEN. Nanophan. or phan.
**Aulomyrcia rotundata* Amshoff, Recueil Trav. Bot. Néerl. 42: 8 (1950).
Aulomyrcia parvifolia Steyerm., Fieldiana, Bot. 28: 1006 (1957).
Myrcia rotundata var. *atrans* McVaugh, Mem. New York Bot. Gard. 18(2): 122 (1969).

Myrcia rotundifolia (O.Berg) Kiaersk., Enum. Myrt. Bras.: 89 (1893).
NE. Brazil. 84 BZE. Nanophan. or phan.
**Aulomyrcia rotundifolia* O.Berg in C.F.P.von Martius & auct. suc. (eds.), Fl. Bras. 14(1): 123 (1857).

Myrcia rubella Cambess. in A.F.C.de Saint-Hilaire, Fl. Bras. Merid. 2: 317 (1832). *Aulomyrcia rubella* (Cambess.) O.Berg, Linnaea 27: 40 (1855).
WC. Brazil. 84 BZC. Nanophan. or phan.
Myrcia rubella var. *puberula* Cambess. in A.F.C.P.de Saint-Hilaire & al., Fl. Bras. Merid. 2: 317 (1832).
Aulomyrcia rubella var. *glabra* O.Berg in C.F.P.von Martius & auct. suc. (eds.), Fl. Bras. 14(1): 74 (1857).
Aulomyrcia rubella var. *puberula* O.Berg in C.F.P.von Martius & auct. suc. (eds.), Fl. Bras. 14(1): 74 (1857).

Myrcia rubiginosa Cambess. in A.F.C.de Saint-Hilaire, Fl. Bras. Merid. 2: 300 (1832).
Brazil (Rio de Janeiro). 84 BZL. Nanophan. or phan.
Myrcia pyramidata O.Berg in C.F.P.von Martius & auct. suc. (eds.), Fl. Bras. 14(1): 193 (1857).

Myrcia rufipes DC., Prodr. 3: 247 (1828). *Aulomyrcia rufipes* (DC.) O.Berg, Linnaea 27: 77 (1855). *Aulomyrcia rufipes* var. *bracteata* O.Berg in C.F.P.von Martius & auct. suc. (eds.), Fl. Bras. 14(1): 131 (1857), nom. inval.
SE. Brazil. 84 BZL. Nanophan.
Aulomyrcia rufipes var. *angustifolia* O.Berg in C.F.P.von Martius & auct. suc. (eds.), Fl. Bras. 14(1): 131 (1857).
Aulomyrcia rufipes var. *dives* O.Berg in C.F.P.von Martius & auct. suc. (eds.), Fl. Bras. 14(1): 131 (1857). *Myrcia rufipes* var. *dives* (O.Berg) N.Silveira, Loefgrenia 91: 2 (1987).
Aulomyrcia rufipes var. *grandiflora* O.Berg in C.F.P.von Martius & auct. suc. (eds.), Fl. Bras. 14(1): 131 (1857).
Aulomyrcia rufipes var. *latifolia* O.Berg in C.F.P.von Martius & auct. suc. (eds.), Fl. Bras. 14(1): 131 (1857).
Aulomyrcia pilantha O.Berg in C.F.P.von Martius & auct. suc. (eds.), Fl. Bras. 14(1): 555 (1859). *Myrcia rufipes* var. *pilantha* (O.Berg) Kiaersk., Enum. Myrt. Bras.: 95 (1893).
Aulomyrcia pilantha var. *latifolia* O.Berg in C.F.P.von Martius & auct. suc. (eds.), Fl. Bras. 14(1): 555 (1859).
Aulomyrcia pilantha var. *longifolia* O.Berg in C.F.P.von Martius & auct. suc. (eds.), Fl. Bras. 14(1): 555 (1859).
Aulomyrcia pilantha var. *parvifolia* O.Berg in C.F.P.von Martius & auct. suc. (eds.), Fl. Bras. 14(1): 555 (1859).
Myrcia rufipes var. *grandiflora* Kiaersk., Enum. Myrt. Bras.: 95 (1893).

Myrcia rufipila McVaugh, Mem. New York Bot. Gard. 18(2): 104 (1969).
N. South America to N. Brazil. 82 FRG GUY SUR 84 BZN. Nanophan. or phan.
**Aulomyrcia divaricata* O.Berg, Linnaea 27: 58 (1855). *Myrcia divaricata* (O.Berg) Lemée, in Fl. Guyane Franç. 3: 146 (1954), nom. illeg.
Myrcia rufipila var. *piresiana* Mattos, Loefgrenia 62: 4 (1974).

Myrcia rupicola D.Legrand, Sellowia 13: 289 (1961).
S. Brazil. 84 BZS. Nanophan. or phan.

Myrcia rupta M.L.Kawas. & B.K.Holst, Brittonia 46: 141 (1994).
French Guiana. 82 FRG. Phan.

Myrcia saliana Alain, Moscosoa 1: 29 (1976).
Dominican Rep. 81 DOM. Nanophan. or phan.

Myrcia salicifolia DC., Prodr. 3: 246 (1828). *Aulomyrcia salicifolia* (DC.) O.Berg, Linnaea 27: 78 (1855).
N. Brazil to Peru. 83 PER 84 BZN. Nanophan. or phan.

Myrcia salticola (Steyerm.) McVaugh, Mem. New York Bot. Gard. 18(2): 123 (1969).
Venezuela (Bolívar). 82 VEN. Nanophan. or phan.
**Aulomyrcia salticola* Steyerm., Fieldiana, Bot. 28: 1007 (1957).

Myrcia salzmannii O.Berg in C.F.P.von Martius & auct. suc. (eds.), Fl. Bras. 14(1): 207 (1857).
NE. Brazil. 84 BZE. Nanophan. or phan.

Myrcia sanisidrensis Steyerm., Fieldiana, Bot. 28: 1018 (1957).
Venezuela. 82 VEN. Nanophan. or phan.

Myrcia saxatilis (Amshoff) McVaugh, Mem. New York Bot. Gard. 18(2): 105 (1969).
N. South America to N. Brazil. 82 FRG SUR 84 BZN. Nanophan. or phan.
**Aulomyrcia saxatilis* Amshoff, Recueil Trav. Bot. Néerl. 39: 154 (1942).

Myrcia schaueriana O.Berg in C.F.P.von Martius & auct. suc. (eds.), Fl. Bras. 14(1): 163 (1857).
SE. Brazil. 84 BZL. Nanophan. or phan.
**Myrtus dioica* Spreng., Syst. Veg. 2: 486 (1825), nom. illeg.

Myrcia schottiana O.Berg in C.F.P.von Martius & auct. suc. (eds.), Fl. Bras. 14(1): 188 (1857).
WC. & SE. Brazil. 84 BZC BZL. Nanophan. or phan.
Myrcia schottiana var. *sericea* O.Berg in C.F.P.von Martius & auct. suc. (eds.), Fl. Bras. 14(1): 566 (1859).

Myrcia scrobiculata (O.Berg) O.Berg, Linnaea 30: 668 (1861).
S. Brazil. 84 BZS. Nanophan. or phan.
**Aulomyrcia scrobiculata* O.Berg in C.F.P.von Martius & auct. suc. (eds.), Fl. Bras. 14(1): 137 (1857).

Myrcia selloi (Spreng.) N.Silveira, Loefgrenia 89: 5 (1986).

Bolivia to Brazil and N. Argentina. 83 BOL 84 BZC BZL BZS 85 AGE PAR URU. Nanophan. or phan.
Myrtus selloi Spreng., Syst. Veg. 2: 482 (1825). *Aulomyrcia selloi* (Spreng.) Kausel, Lilloa 32: 350 (19667).
Myrcia ramulosa DC., Prodr. 3: 250 (1828). *Aulomyrcia ramulosa* (DC.) O.Berg, Linnaea 27: 36 (1855).
Myrcia ramulosa var. *multiflora* DC., Prodr. 3: 250 (1828).
Aulomyrcia ramulosa var. *acutata* O.Berg in C.F.P.von Martius & auct. suc. (eds.), Fl. Bras. 14(1): 62 (1857). *Myrcia ramulosa* var. *acutata* (O.Berg) Mattos, Loefgrenia 61: 3 (1974). *Myrcia selloi* var. *acutata* (O.Berg) N.Silveira, Loefgrenia 89: 6 (1986).
Aulomyrcia ramulosa var. *colorata* O.Berg in C.F.P. von Martius & auct. suc. (eds.), Fl. Bras. 14(1): 62 (1857).
Aulomyrcia ramulosa var. *panicularis* O.Berg in C.F.P. von Martius & auct. suc. (eds.), Fl. Bras. 14(1): 63 (1857).
Aulomyrcia ramulosa var. *pauciflora* O.Berg in C.F.P. von Martius & auct. suc. (eds.), Fl. Bras. 14(1): 62 (1857).
Aulomyrcia ramulosa var. *subcordata* O.Berg in C.F.P. von Martius & auct. suc. (eds.), Fl. Bras. 14(1): 62 (1857).
Aulomyrcia ramulosa var. *australis* O.Berg in C.F.P. von Martius & auct. suc. (eds.), Fl. Bras. 14(1): 544 (1859).
Myrcia ramulosa var. *leptophylla* Kiaersk., Enum. Myrt. Bras.: 80 (1893).
Myrcia ramulosa var. *pauciflora* Kiaersk., Enum. Myrt. Bras.: 80 (1893).
Myrcia hassleriana Barb.Rodr., Myrt. Paraguay: 2 (1903).
Myrcia microsiphonata D.Legrand, Sellowia 13: 295 (1961). *Myrcia ramulosa* var. *microsiphonata* (D.Legrand) D.Legrand, in Fl. Ilustr. Catar., Mirt.: 312 (1969). *Myrcia selloi* var. *microsiphonata* (D.Legrand) N.Silveira, Loefgrenia 89: 7 (1986).
Myrcia smithii D.Legrand & Kausel, Sellowia 13: 290 (1961).
Myrcia ramulosa var. *megapotamica* D.Legrand, Bol. Fac. Agron. Univ. Montevideo 101: 22 (1968).
Myrcia selloi var. *australis* (O.Berg) N.Silveira, Loefgrenia 89: 7 (1986).
Myrcia selloi var. *megapotamica* (Legrand) N.Silveira, Loefgrenia 92: 3 (1988).

Myrcia sericea G.Don, Gen. Hist. 2: 844 (1832).
NE. Brazil. 84 BZE. Nanophan. or phan.

Myrcia servata McVaugh, Mem. New York Bot. Gard. 18(2): 139 (1969).
N. South America. 82 FRG GUY SUR VEN. Nanophan. or phan.

Myrcia sessiliflora McVaugh, Mem. New York Bot. Gard. 18(2): 105 (1969).
Venezuela (Amazonas). 82 VEN. Nanophan.

Myrcia sipapensis McVaugh, Mem. New York Bot. Gard. 18(2): 140 (1969).
Venezuela (Amazonas). 82 VEN. Phan.

Myrcia skeldingii Proctor, Rhodora 60: 325 (1959).
Jamaica. 81 JAM. Nanophan. or phan.

Myrcia sororopanensis Steyerm., Fieldiana, Bot. 28: 1019 (1957).
Venezuela to Guyana. 82 GUY VEN. Nanophan. or phan.

Myrcia sosias D.Legrand, in Fl. Ilustr. Catar. 1(Mirt.): 244 (1969).
S. Brazil. 84 BZS. Phan.

Myrcia spathulata (O.Berg) Kiaersk., Enum. Myrt. Bras.: 72 (1893).
SE. Brazil. 84 BZL. Nanophan. or phan.
**Aulomyrcia spathulata* O.Berg in C.F.P.von Martius & auct. suc. (eds.), Fl. Bras. 14(1): 91 (1857).

Myrcia speciosa (Amshoff) McVaugh, Mem. New York Bot. Gard. 18(2): 106 (1969).
Guyana. 82 GUY. Nanophan. or phan.
**Aulomyrcia speciosa* Amshoff, Recueil Trav. Bot. Néerl. 42: 5 (1950).

Myrcia spectabilis DC., Prodr. 3: 248 (1828). *Gomidesia spectabilis* (DC.) O.Berg in C.F.P.von Martius & auct. suc. (eds.), Fl. Bras. 14(1): 12 (1857).
E. & S. Brazil. 84 BZE BZL BZS. Nanophan. or phan.
Myrcia browniana Gardner, London J. Bot. 2: 354 (1843). *Gomidesia browniana* (Gardner) O.Berg in C.F.P.von Martius & auct. suc. (eds.), Fl. Bras. 14(1): 13 (1857).
Gomidesia spectabilis var. *farinosa* O.Berg in C.F.P.von Martius & auct. suc. (eds.), Fl. Bras. 14(1): 12 (1857). *Myrcia spectabilis* var. *farinosa* (O.Berg) Kiaersk., Enum. Myrt. Bras.: 101 (1893).
Gomidesia spectabilis var. *ovata* O.Berg in C.F.P.von Martius & auct. suc. (eds.), Fl. Bras. 14(1): 13 (1857).

Myrcia spinifolia Borhidi & O.Muñiz, Bot. Közlem. 64: 215 (1977 publ. 1978).
Cuba. 81 CUB. Nanophan. or phan.

Myrcia splendens (Sw.) DC., Prodr. 3: 244 (1828).
Mexico to Trop. America. 79 MXG MXS MXT 80 BLZ COS GUA HON NIC PAN 81 CUB DOM HAI JAM LEE PUE TRT WIN 82 FRG GUY SUR VEN 83 BOL CLM ECU PER 84 BZC BZE BZL BZN BZS 85 AGE PAR. Nanophan. or phan.
**Myrtus splendens* Sw., Prodr.: 79 (1788).
Eugenia divaricata Lam., Encycl. 3: 202 (1789). *Myrcia divaricata* (Lam.) DC., Prodr. 3: 243 (1828). *Cumetea divaricata* (Lam.) Raf., Sylva Tellur.: 106 (1838).
Eugenia periplocifolia Jacq., Collectanea 2: 108 (1789).
Eugenia fallax Rich., Actes Soc. Hist. Nat. Paris 1: 110 (1792). *Myrcia fallax* (Rich.) DC., Prodr. 3: 244 (1828).
Myrtus bracteolaris Poir. in J.B.A.M.de Lamarck, Encycl. 4: 411 (1798). *Myrcia bracteolaris* (Poir.) DC., Prodr. 3: 245 (1828).
Eugenia laxiflora Poir. in J.B.A.P.M.de Lamarck, Encycl., Suppl. 3: 123 (1813).
Myrtus acuminata Kunth in F.W.H.von Humboldt, A.J.A.Bonpland & C.S.Kunth, Nov. Gen. Sp. 6: 141 (1823). *Myrcia acuminata* (Kunth) DC., Prodr. 3: 256 (1828).
Myrtus complicata Kunth in F.W.H.von Humboldt, A.J.A.Bonpland & C.S.Kunth, Nov. Gen. Sp. 6: 140 (1823). *Myrcia complicata* (Kunth) DC., Prodr. 3: 255 (1828).
Myrtus deflexa Kunth in F.W.H.von Humboldt, A.J.A. Bonpland & C.S.Kunth, Nov. Gen. Sp. 6: 142 (1823).
Myrtus polyantha Kunth in F.W.H.von Humboldt, A.J.A.Bonpland & C.S.Kunth, Nov. Gen. Sp. 6: 140

(1823). *Myrcia humboldtiana* DC., Prodr. 3: 256 (1828), nom. illeg. *Myrcia kunthiana* Steud., Nomencl. Bot., ed, 2, 2: 172 (1841).
Myrtus stoupii Spreng., Syst. Veg. 2: 484 (1825).
Eugenia mikaniana DC., Prodr. 3: 283 (1828). *Myrcia mikaniana* (DC.) O.Berg in C.F.P.von Martius & auct. suc. (eds.), Fl. Bras. 14(1): 182 (1857).
Myrcia berberis DC., Prodr. 3: 254 (1828).
Myrcia costata DC., Prodr. 3: 252 (1828). *Aulomyrcia costata* (DC.) O.Berg in C.F.P.von Martius & auct. suc. (eds.), Fl. Bras. 14(1): 79 (1857). *Calyptromyrcia costata* (DC.) O.Berg in C.F.P.von Martius & auct. suc. (eds.), Fl. Bras. 14(1): 56 (1857).
Myrcia formosiana DC., Prodr. 3: 255 (1828).
Myrcia hayneana DC., Prodr. 3: 246 (1828).
Myrcia macrophylla DC., Prodr. 3: 249 (1828).
Myrcia magnoliifolia DC., Prodr. 3: 248 (1828).
Myrcia pseudomini DC., Prodr. 3: 252 (1828). *Myrcia rostrata* f. *pseudomini* (DC.) D.Legrand, in Fl. Ilustr. Catar. 1(Mirt.): 243 (1969).
Myrcia rostrata DC., Prodr. 3: 255 (1828).
Myrcia sepiaria DC., Prodr. 3: 249 (1828).
Myrcia sororia DC., Prodr. 3: 243 (1828).
Myrcia oocarpa Cambess. in A.F.C.de Saint-Hilaire, Fl. Bras. Merid. 2: 298 (1832).
Myrcia rostrata var. *brunea* Cambess. in A.F.C.P.de Saint-Hilaire & al., Fl. Bras. Merid. 2: 321 (1832).
Myrcia rufidula Schltdl., Linnaea 13: 416 (1839).
Eugenia paniculiflora Steud., Flora 26: 762 (1843).
Myrcia rufula Miq., Linnaea 19: 440 (1846).
Myrcia berberis Schauer, Linnaea 21: 273 (1848), nom. illeg.
Aulomyrcia wullschlaegeliana O.Berg, Linnaea 27: 48 (1855).
Eugenia mollis Willd. ex O.Berg, Linnaea 27: 121 (1855).
Myrcia acuminata var. *bullata* O.Berg, Linnaea 27: 94 (1855).
Myrcia acuminata var. *meridensis* O.Berg, Linnaea 27: 94 (1855).
Myrcia acuminata var. *peruviana* O.Berg, Linnaea 27: 94 (1855).
Myrcia acuminata var. *tovarensis* O.Berg, Linnaea 27: 95 (1855).
Myrcia ayresiana O.Berg, Linnaea 27: 100 (1855).
Myrcia chilensis O.Berg, Linnaea 27: 99 (1855).
Myrcia costa-ricensis O.Berg, Linnaea 27: 104 (1855).
Myrcia coumetoides O.Berg, Linnaea 27: 102 (1855).
Myrcia cucullata O.Berg, Linnaea 27: 97 (1855).
Myrcia discolor O.Berg, Linnaea 27: 111 (1855).
Myrcia humboldtiana var. *caribaea* O.Berg, Linnaea 27: 121 (1855).
Myrcia humboldtiana var. *orinocensis* O.Berg, Linnaea 27: 120 (1855).
Myrcia kegeliana O.Berg, Linnaea 27: 99 (1855).
Myrcia lindeniana O.Berg, Linnaea 27: 86 (1855).
Myrcia melanoclada O.Berg, Linnaea 27: 113 (1855).
Myrcia oerstediana O.Berg, Linnaea 27: 112 (1855).
Myrcia plicatocostata O.Berg, Linnaea 27: 114 (1855).
Myrcia reticulata O.Berg, Linnaea 27: 101 (1855).
Myrcia saxicola O.Berg, Linnaea 27: 92 (1855).
Myrcia sericea O.Berg, Linnaea 27: 114 (1855), nom. illeg.
Myrcia splendens var. *micropora* O.Berg, Linnaea 27: 105 (1855).
Myrcia splendens var. *obscura* O.Berg, Linnaea 27: 105 (1855).
Myrcia venezuelensis O.Berg, Linnaea 27: 96 (1855).
Myrcia splendens var. *chrysocoma* McVaugh, Fieldiana, Bot. 29: 193 (1856).
Myrcia acutata O.Berg in C.F.P.von Martius & auct. suc. (eds.), Fl. Bras. 14(1): 175 (1857).
Myrcia acutiloba O.Berg in C.F.P.von Martius & auct. suc. (eds.), Fl. Bras. 14(1): 189 (1857).
Myrcia alagoensis var. *intermedia* O.Berg in C.F.P.von Martius & auct. suc. (eds.), Fl. Bras. 14(1): 165 (1857).
Myrcia alagoensis var. *oblongata* O.Berg in C.F.P.von Martius & auct. suc. (eds.), Fl. Bras. 14(1): 165 (1857).
Myrcia alagoensis var. *ovata* O.Berg in C.F.P.von Martius & auct. suc. (eds.), Fl. Bras. 14(1): 165 (1857).
Myrcia barrensis O.Berg in C.F.P.von Martius & auct. suc. (eds.), Fl. Bras. 14(1): 187 (1857).
Myrcia berberis var. *angustifolia* O.Berg in C.F.P.von Martius & auct. suc. (eds.), Fl. Bras. 14(1): 170 (1857).
Myrcia berberis var. *latifolia* O.Berg in C.F.P.von Martius & auct. suc. (eds.), Fl. Bras. 14(1): 170 (1857).
Myrcia brandamii O.Berg in C.F.P.von Martius & auct. suc. (eds.), Fl. Bras. 14(1): 164 (1857).
Myrcia catharinae O.Berg in C.F.P.von Martius & auct. suc. (eds.), Fl. Bras. 14(1): 176 (1857).
Myrcia ciarensis O.Berg in C.F.P.von Martius & auct. suc. (eds.), Fl. Bras. 14(1): 178 (1857).
Myrcia communis O.Berg in C.F.P.von Martius & auct. suc. (eds.), Fl. Bras. 14(1): 183 (1857). *Myrcia rostrata* f. *communis* (O.Berg) D.Legrand, in Fl. Ilustr. Catar. 1(Mirt.): 243 (1969).
Myrcia communis var. *glabrata* O.Berg in C.F.P.von Martius & auct. suc. (eds.), Fl. Bras. 14(1): 183 (1857).
Myrcia communis var. *latifolia* O.Berg in C.F.P.von Martius & auct. suc. (eds.), Fl. Bras. 14(1): 183 (1857).
Myrcia corcovadensis O.Berg in C.F.P.von Martius & auct. suc. (eds.), Fl. Bras. 14(1): 177 (1857).
Myrcia costata var. *bahiensis* O.Berg in C.F.P.von Martius & auct. suc. (eds.), Fl. Bras. 14(1): 157 (1857).
Myrcia costata var. *minensis* O.Berg in C.F.P.von Martius & auct. suc. (eds.), Fl. Bras. 14(1): 157 (1857).
Myrcia elongata O.Berg in C.F.P.von Martius & auct. suc. (eds.), Fl. Bras. 14(1): 159 (1857).
Myrcia elongata var. *brunnea* O.Berg in C.F.P.von Martius & auct. suc. (eds.), Fl. Bras. 14(1): 160 (1857).
Myrcia elongata var. *grandifolia* O.Berg in C.F.P.von Martius & auct. suc. (eds.), Fl. Bras. 14(1): 160 (1857).
Myrcia elongata var. *ochracea* O.Berg in C.F.P.von Martius & auct. suc. (eds.), Fl. Bras. 14(1): 160 (1857).
Myrcia erythroxylon O.Berg in C.F.P.von Martius & auct. suc. (eds.), Fl. Bras. 14(1): 173 (1857).
Myrcia erythroxylon var. *caerulescens* O.Berg in C.F.P.von Martius & auct. suc. (eds.), Fl. Bras. 14(1): 173 (1857).
Myrcia friburgensis O.Berg in C.F.P.von Martius & auct. suc. (eds.), Fl. Bras. 14(1): 161 (1857).
Myrcia gardneriana O.Berg in C.F.P.von Martius & auct. suc. (eds.), Fl. Bras. 14(1): 184 (1857).
Myrcia gracilis O.Berg in C.F.P.von Martius & auct. suc. (eds.), Fl. Bras. 14(1): 174 (1857). *Myrcia*

rostrata f. *gracilis* (O.Berg) D.Legrand, in Fl. Ilustr. Catar. 1(Mirt.): 240 (1969).
Myrcia gracilis var. *opaca* O.Berg in C.F.P.von Martius & auct. suc. (eds.), Fl. Bras. 14(1): 165 (1857).
Myrcia gracilis var. *prasina* O.Berg in C.F.P.von Martius & auct. suc. (eds.), Fl. Bras. 14(1): 174 (1857).
Myrcia guajavifolia O.Berg in C.F.P.von Martius & auct. suc. (eds.), Fl. Bras. 14(1): 160 (1857).
Myrcia guajavifolia var. *bullata* O.Berg in C.F.P.von Martius & auct. suc. (eds.), Fl. Bras. 14(1): 161 (1857).
Myrcia guajavifolia var. *impunctata* O.Berg in C.F.P. von Martius & auct. suc. (eds.), Fl. Bras. 14(1): 160 (1857).
Myrcia guajavifolia var. *perforata* O.Berg in C.F.P.von Martius & auct. suc. (eds.), Fl. Bras. 14(1): 160 (1857).
Myrcia hayneana var. *paraensis* O.Berg in C.F.P.von Martius & auct. suc. (eds.), Fl. Bras. 14(1): 187 (1857).
Myrcia impressa O.Berg in C.F.P.von Martius & auct. suc. (eds.), Fl. Bras. 14(1): 184 (1857).
Myrcia kegeliana var. *angustifolia* O.Berg in C.F.P. von Martius & auct. suc. (eds.), Fl. Bras. 14(1): 168 (1857).
Myrcia kegeliana var. *latifolia* O.Berg in C.F.P.von Martius & auct. suc. (eds.), Fl. Bras. 14(1): 168 (1857).
Myrcia kegeliana var. *longifolia* O.Berg in C.F.P.von Martius & auct. suc. (eds.), Fl. Bras. 14(1): 168 (1857).
Myrcia kegeliana var. *pendula* O.Berg in C.F.P.von Martius & auct. suc. (eds.), Fl. Bras. 14(1): 168 (1857).
Myrcia kegeliana var. *vulgaris* O.Berg in C.F.P.von Martius & auct. suc. (eds.), Fl. Bras. 14(1): 168 (1857).
Myrcia klotzschiana O.Berg in C.F.P.von Martius & auct. suc. (eds.), Fl. Bras. 14(1): 169 (1857).
Myrcia laevigata O.Berg in C.F.P.von Martius & auct. suc. (eds.), Fl. Bras. 14(1): 175 (1857).
Myrcia laevigata var. *brunnea* O.Berg in C.F.P.von Martius & auct. suc. (eds.), Fl. Bras. 14(1): 175 (1857).
Myrcia laevigata var. *canescens* O.Berg in C.F.P.von Martius & auct. suc. (eds.), Fl. Bras. 14(1): 175 (1857).
Myrcia latifolia O.Berg in C.F.P.von Martius & auct. suc. (eds.), Fl. Bras. 14(1): 170 (1857).
Myrcia magnoliifolia var. *angustifolia* O.Berg in C.F.P.von Martius & auct. suc. (eds.), Fl. Bras. 14(1): 162 (1857).
Myrcia magnoliifolia var. *latifolia* O.Berg in C.F.P.von Martius & auct. suc. (eds.), Fl. Bras. 14(1): 162 (1857).
Myrcia magnoliifolia var. *parvifolia* O.Berg in C.F.P.von Martius & auct. suc. (eds.), Fl. Bras. 14(1): 162 (1857).
Myrcia martiana O.Berg in C.F.P.von Martius & auct. suc. (eds.), Fl. Bras. 14(1): 159 (1857). *Myrcia rufula* var. *martiana* (O.Berg) Kiaersk., Enum. Myrt. Bras.: 55 (1893).
Myrcia micrantha O.Berg in C.F.P.von Martius & auct. suc. (eds.), Fl. Bras. 14(1): 169 (1857).
Myrcia mikaniana var. *angustifolia* O.Berg in C.F.P.von Martius & auct. suc. (eds.), Fl. Bras. 14(1): 182 (1857).
Myrcia mikaniana var. *latifolia* O.Berg in C.F.P.von Martius & auct. suc. (eds.), Fl. Bras. 14(1): 182 (1857).
Myrcia negrensis O.Berg in C.F.P.von Martius & auct. suc. (eds.), Fl. Bras. 14(1): 187 (1857).
Myrcia nitens O.Berg in C.F.P.von Martius & auct. suc. (eds.), Fl. Bras. 14(1): 178 (1857).
Myrcia opaca O.Berg in C.F.P.von Martius & auct. suc. (eds.), Fl. Bras. 14(1): 177 (1857).
Myrcia opaca var. *angustifolia* O.Berg in C.F.P.von Martius & auct. suc. (eds.), Fl. Bras. 14(1): 177 (1857).
Myrcia opaca var. *latifolia* O.Berg in C.F.P.von Martius & auct. suc. (eds.), Fl. Bras. 14(1): 177 (1857).
Myrcia pellucida O.Berg in C.F.P.von Martius & auct. suc. (eds.), Fl. Bras. 14(1): 173 (1857).
Myrcia phaeoclada var. *guyanensis* O.Berg in C.F.P. von Martius & auct. suc. (eds.), Fl. Bras. 14(1): 167 (1857).
Myrcia pohliana O.Berg in C.F.P.von Martius & auct. suc. (eds.), Fl. Bras. 14(1): 199 (1857).
Myrcia riparia O.Berg in C.F.P.von Martius & auct. suc. (eds.), Fl. Bras. 14(1): 179 (1857).
Myrcia schuechiana O.Berg in C.F.P.von Martius & auct. suc. (eds.), Fl. Bras. 14(1): 181 (1857).
Myrcia sellowiana O.Berg in C.F.P.von Martius & auct. suc. (eds.), Fl. Bras. 14(1): 197 (1857).
Myrcia sellowiana var. *bullata* O.Berg in C.F.P.von Martius & auct. suc. (eds.), Fl. Bras. 14(1): 197 (1857).
Myrcia sellowiana var. *costata* O.Berg in C.F.P.von Martius & auct. suc. (eds.), Fl. Bras. 14(1): 197 (1857).
Myrcia sericiflora O.Berg in C.F.P.von Martius & auct. suc. (eds.), Fl. Bras. 14(1): 178 (1857). *Myrcia rostrata* f. *sericiflora* (O.Berg) D.Legrand, in Fl. Ilustr. Catar. 1(Mirt.): 243 (1969).
Myrcia spruceana O.Berg in C.F.P.von Martius & auct. suc. (eds.), Fl. Bras. 14(1): 165 (1857).
Myrcia superba O.Berg in C.F.P.von Martius & auct. suc. (eds.), Fl. Bras. 14(1): 198 (1857).
Myrcia velutina O.Berg in C.F.P.von Martius & auct. suc. (eds.), Fl. Bras. 14(1): 182 (1857).
Myrcia velutina var. *canescens* O.Berg in C.F.P.von Martius & auct. suc. (eds.), Fl. Bras. 14(1): 183 (1857).
Myrcia velutina var. *ochracea* O.Berg in C.F.P.von Martius & auct. suc. (eds.), Fl. Bras. 14(1): 183 (1857).
Myrcia ypanemensis O.Berg in C.F.P.von Martius & auct. suc. (eds.), Fl. Bras. 14(1): 186 (1857).
Myrcia sartoriana O.Berg, Linnaea 29: 220 (1858).
Myrcia saxicola var. *grandifolia* O.Berg, Linnaea 29: 219 (1858).
Myrcia gracilis var. *sessiliflora* O.Berg in C.F.P.von Martius & auct. suc. (eds.), Fl. Bras. 14(1): 564 (1859).
Myrcia klotzschiana var. *impellucida* O.Berg in C.F.P.von Martius & auct. suc. (eds.), Fl. Bras. 14(1): 563 (1859).
Myrcia langsdorffii O.Berg in C.F.P.von Martius & auct. suc. (eds.), Fl. Bras. 14(1): 562 (1859).
Myrcia riedeliana O.Berg in C.F.P.von Martius & auct. suc. (eds.), Fl. Bras. 14(1): 565 (1859).
Myrcia tingens O.Berg in C.F.P.von Martius & auct. suc. (eds.), Fl. Bras. 14(1): 564 (1859).
Myrcia splendens var. *robustior* Kuntze, Revis. Gen. Pl. 1: 241 (1891).
Myrcia augustana Kiaersk., Enum. Myrt. Bras.: 52 (1893).

Myrcia guajavifolia f. *grandifolia* Kiaersk., Enum. Myrt. Bras.: 54 (1893).
Myrcia melanosticta Kiaersk., Enum. Myrt. Bras.: 54 (1893).
Myrcia oxyoentophylla Kiaersk., Enum. Myrt. Bras.: 57 (1893).
Myrcia martinicensis Krug & Urb., Bot. Jahrb. Syst. 19: 586 (1895).
Myrcia brachylopadia Diels, Bot. Jahrb. Syst. 37: 595 (1906).
Myrcia dictyoneura Diels, Bot. Jahrb. Syst. 37: 594 (1906).
Myrcia lamprosericea Diels, Bot. Jahrb. Syst. 37: 596 (1906).
Myrcia coroicensis Rusby, Bull. New York Bot. Gard. 4: 354 (1907).
Myrcia luetzelburgii Burret ex Luetzelb., Estud. Bot. Nordéste 3: 201 (1923).
Myrcia aguitensis Gleason, Bull. Torrey Bot. Club 58: 409 (1931).
Myrcia compressa Gleason, Bull. Torrey Bot. Club 58: 410 (1931).
Myrcia longicaudata Lundell, Amer. Midl. Naturalist 29: 481 (1943).
Myrcia schippii Lundell, Amer. Midl. Naturalist 29: 482 (1943).
Myrcia belizensis Lundell, Wrightia 2: 213 (1961).
Myrcia splendens var. *guantanamana* Borhidi & O.Muñiz, Bot. Közlem. 64: 20 (1977).
Myrcia rostrata f. *flexuosa* Soares-Silva, Bradea 8: 323 (2002).

Myrcia sporadosticta Kiaersk., Enum. Myrt. Bras.: 59 (1893).
SE. Brazil. 84 BZL. Nanophan. or phan.

Myrcia sprengeliana O.Berg, Linnaea 27: 109 (1855).
Virgin Is. 81 LEE. Nanophan. or phan.

Myrcia springiana (O.Berg) Kiaersk., Enum. Myrt. Bras.: 102 (1893).
E. Brazil. 84 BZE BZL. Nanophan.
**Gomidesia springiana* O.Berg in C.F.P.von Martius & auct. suc. (eds.), Fl. Bras. 14(1): 13 (1857).
Gomidesia springiana var. *coriacea* O.Berg in C.F.P. von Martius & auct. suc. (eds.), Fl. Bras. 14(1): 13 (1857).
Gomidesia springiana var. *membranacea* O.Berg in C.F.P.von Martius & auct. suc. (eds.), Fl. Bras. 14(1): 14 (1857). *Myrcia springiana* var. *membranacea* (O.Berg) Kiaersk., Enum. Myrt. Bras.: 102 (1893).

Myrcia stenocarpa Krug & Urb., Bot. Jahrb. Syst. 19: 584 (1895).
Trinidad. 81 TRT. Nanophan. or phan.

Myrcia stenocymbia Diels, Bot. Jahrb. Syst. 37: 596 (1906).
Peru. 83 PER. Nanophan. or phan. Provisionally accepted.

Myrcia stewartiana O.Berg, Linnaea 29: 219 (1858).
Brazil. 84+. Nanophan. or phan.

Myrcia stictophylla (O.Berg) N.Silveira, Roessléria 7: 66 (1985).
Brazil (Rio de Janeiro). 84 BZL. Nanophan. or phan.
**Aulomyrcia stictophylla* O.Berg in C.F.P.von Martius & auct. suc. (eds.), Fl. Bras. 14(1): 67 (1857).

Myrcia stigmatosa O.Berg in C.F.P.von Martius & auct. suc. (eds.), Fl. Bras. 14(1): 564 (1859).
NE. Brazil. 84 BZE. Nanophan. or phan.

Myrcia stricta (O.Berg) Kiaersk., Enum. Myrt. Bras.: 99 (1893).
WC. Brazil. 84 BZC. Nanophan. or phan.
**Aulomyrcia stricta* O.Berg in C.F.P.von Martius & auct. suc. (eds.), Fl. Bras. 14(1): 548 (1859).

Myrcia subalpestris DC., Prodr. 3: 250 (1828). *Aulomyrcia subalpestris* (DC.) O.Berg, Linnaea 27: 73 (1855).
SE. Brazil. 84 BZL. Nanophan. or phan.

Myrcia subavenia (O.Berg) N.Silveira, Roessléria 7: 66 (1985).
SE. Brazil. 84 BZL. Nanophan. or phan.
**Aulomyrcia subavenia* O.Berg in C.F.P.von Martius & auct. suc. (eds.), Fl. Bras. 14(1): 69 (1857).

Myrcia subcordata DC., Prodr. 3: 253 (1828). *Aulomyrcia subcordata* (DC.) O.Berg, Linnaea 27: 62 (1855).
Brazil (Minas Gerais). 84 BZL. Nanophan. or phan.
Calyptranthes cordata O.Berg in C.F.P.von Martius & auct. suc. (eds.), Fl. Bras. 14(1): 48 (1857). *Chytraculia cordata* (O.Berg) Kuntze, Revis. Gen. Pl. 1: 238 (1891).

Myrcia subglabra McVaugh, Fieldiana, Bot. 29: 194 (1956).
Bolivia. 83 BOL. Nanophan. or phan.

Myrcia subobliqua (O.Berg) Nied. in H.G.A.Engler & K.A.E.Prantl, Nat. Pflanzenfam. 3(7): 76 (1893).
Guianas. 82 FRG GUY SUR. Nanophan. or phan.
**Aulomyrcia subobliqua* O.Berg, Linnaea 27: 57 (1855).

Myrcia subrugosa Kiaersk., Enum. Myrt. Bras.: 86 (1893).
SE. Brazil. 84 BZL. Nanophan. or phan.

Myrcia subsericea A.Gray, U.S. Expl. Exped., Phan. 1: 533 (1854).
SE. Brazil. 84 BZL. Nanophan. or phan.
Eugenia nitida Vell., Fl. Flumin. 5: 208, t. 35 (1829). *Gomidesia chamissoeana* O.Berg in C.F.P.von Martius & auct. suc. (eds.), Fl. Bras. 14(1): 11 (1857), nom. illeg. *Gomidesia nitida* (Vell.) Nied. in H.G.A.Engler & K.A.E.Prantl, Nat. Pflanzenfam. 3(7): 77 (1893). *Myrcia nitida* (Vell.) Kiaersk., Enum. Myrt. Bras.: 102 (1893), nom. illeg.

Myrcia subsessilis O.Berg, Linnaea 31: 251 (1862).
N. & W. South America. 82 FRG VEN 83 BOL CLM. Nanophan. or phan.
Myrcia subsessilis var. *ovalis* O.Berg, Linnaea 31: 252 (1862).
Myrcia subsessilis var. *subcordata* O.Berg, Linnaea 31: 251 (1862).

Myrcia subverticillaris (O.Berg) Kiaersk., Enum. Myrt. Bras.: 88 (1893).
WC. & E. Brazil. 84 BZC BZE BZL. Nanophan. or phan.
**Aulomyrcia subverticillaris* O.Berg, Linnaea 27: 73 (1855).
Aulomyrcia subverticillaris var. *angustifolia* O.Berg in C.F.P.von Martius & auct. suc. (eds.), Fl. Bras. 14(1): 124 (1857).
Aulomyrcia subverticillaris var. *incanescens* O.Berg in C.F.P.von Martius & auct. suc. (eds.), Fl. Bras. 14(1): 124 (1857).
Aulomyrcia subverticillaris var. *rufa* O.Berg in C.F.P.von Martius & auct. suc. (eds.), Fl. Bras. 14(1): 124 (1857).

Myrcia suffruticosa O.Berg in C.F.P.von Martius & auct. suc. (eds.), Fl. Bras. 14(1): 189 (1857).
WC. Brazil. 84 BZC. Cham.

Myrcia susannae Borhidi, Bot. Közlem. 64: 215 (1977 publ. 1978).
Cuba. 81 CUB. Nanophan. or phan.

Myrcia sylvatica (G.Mey.) DC., Prodr. 3: 244 (1828).
Costa Rica to S. Trop. America. 80 COS PAN 82 FRG GUY SUR VEN 83 CLM PER 84 BZC BZE BZN. Nanophan. or phan.
Myrtus lucida L., Syst. Nat. ed. 10, 2: 1056 (1759).
**Myrtus sylvatica* G.Mey., Prim. Fl. Esseq.: 191 (1818). *Myrcia ambigua* var. *sylvatica* (G.Mey.) O.Berg in C.F.P.von Martius & auct. suc. (eds.), Fl. Bras. 14(1): 180 (1857).
Myrcia ambigua DC., Prodr. 3: 252 (1828).
Myrcia ambigua var. *pauciflora* DC., Prodr. 3: 252 (1828).
Myrcia ambigua var. *dives* O.Berg in C.F.P.von Martius & auct. suc. (eds.), Fl. Bras. 14(1): 180 (1857).
Myrcia ambigua var. *latifolia* O.Berg in C.F.P.von Martius & auct. suc. (eds.), Fl. Bras. 14(1): 181 (1857).
Myrcia ambigua var. *multiflora* O.Berg in C.F.P.von Martius & auct. suc. (eds.), Fl. Bras. 14(1): 180 (1857).
Myrcia ambigua var. *rostrata* O.Berg in C.F.P.von Martius & auct. suc. (eds.), Fl. Bras. 14(1): 181 (1857).

Myrcia tafelbergica Amshoff, Bull. Torrey Bot. Club 75: 533 (1948).
Guyana to Suriname. 82 GUY SUR. Nanophan. or phan.

Myrcia tenuifolia (O.Berg) Sobral, Novon 16: 136 (2006).
Brazil (Bahia, Minas Gerais). 84 BZE BZL. Nanophan. or phan.
**Aulomyrcia tenuifolia* O.Berg in C.F.P.von Martius & auct. suc. (eds.), Fl. Bras. 14(1): 68 (1857).

Myrcia tenuivenosa Kiaersk., Enum. Myrt. Bras.: 84 (1893).
SE. & S. Brazil. 84 BZL BZS. Nanophan. or phan.

Myrcia tepuiensis Steyerm., Fieldiana, Bot. 28: 1019 (1957).
Venezuela (Bolívar). 82 VEN. Nanophan. or phan.

Myrcia thomasiana DC., Prodr. 3: 244 (1828).
Virgin Is. 81 LEE. Nanophan. or phan.

Myrcia thyrsoidea O.Berg in C.F.P.von Martius & auct. suc. (eds.), Fl. Bras. 14(1): 192 (1857).
NE. Brazil. 84 BZL. Nanophan. or phan.

Myrcia tiburoniana Urb. & Ekman, Ark. Bot. 24A(4): 18 (1932).
Haiti (Massif de la Hotte). 81 HAI. Nanophan. or phan.

Myrcia tijucensis Kiaersk., Enum. Myrt. Bras.: 102 (1893). *Gomidesia tijucensis* (Kiaersk.) D.Legrand, Comun. Bot. Mus. Hist. Nat. Montevideo 3(37): 23 (1958).
SE. & S. Brazil. 84 BZL BZS. Phan.
Gomidesia tijucensis var. *flexuosa* Mattos & D.Legrand, Loefgrenia 67: 17 (1975).

Myrcia toaensis Borhidi & O.Muñiz, Bot. Közlem. 64: 215 (1977 publ. 1978).
Cuba. 81 CUB. Nanophan. or phan.

Myrcia tomentosa (Aubl.) DC., Prodr. 3: 245 (1828).
Trinidad, Panama to S. Trop. America. 80 PAN 81 TRT 82 FRG GUY SUR VEN 83 BOL ECU 84 BZC BZE BZL BZN BZS. Nanophan. or phan.
**Eugenia tomentosa* Aubl., Hist. Pl. Guiane 1: 504 (1775). *Myrtus aubletii* Spreng., Syst. Veg. 2: 486 (1825). *Aguava tomentosa* (Aubl.) Raf., Sylva Tellur.: 107 (1838). *Cumetea tomentosa* (Aubl.) Raf., Sylva Tellur.: 106 (1838). *Aulomyrcia tomentosa* (Aubl.) Amshoff, Recueil Trav. Bot. Néerl. 39: 153 (1942).
Myrcia curatellifolia DC., Prodr. 3: 253 (1828). *Aulomyrcia curatellifolia* (DC.) O.Berg, Linnaea 27: 55 (1855).
Myrcia hirtiflora DC., Prodr. 3: 249 (1828). *Aulomyrcia hirtiflora* (DC.) O.Berg, Linnaea 27: 40 (1855).
Myrcia pilosa DC., Prodr. 3: 253 (1828).
Myrcia prunifolia DC., Prodr. 3: 253 (1828). *Aulomyrcia prunifolia* (DC.) O.Berg, Linnaea 27: 55 (1855).
Myrcia prunifolia var. *angustior* DC., Prodr. 3: 253 (1828).
Myrcia prunifolia var. *obovata* DC., Prodr. 3: 253 (1828).
Myrcia prunifolia var. *ovata* DC., Prodr. 3: 253 (1828).
Myrcia capivarhyensis Cambess. in A.F.C.de Saint-Hilaire, Fl. Bras. Merid. 2: 328 (1832). *Aulomyrcia capivarhyensis* (Cambess.) O.Berg, Linnaea 27: 36 (1855).
Myrcia puberula Cambess. in A.F.C.de Saint-Hilaire, Fl. Bras. Merid. 2: 316 (1832). *Aulomyrcia puberula* (Cambess.) O.Berg, Linnaea 27: 54 (1855). *Calyptromyrcia puberula* (Cambess.) O.Berg in C.F.P.von Martius & auct. suc. (eds.), Fl. Bras. 14(1): 57 (1857).
Myrcia floribunda Miq., Linnaea 22: 534 (1849).
Aulomyrcia alloiota O.Berg, Linnaea 27: 54 (1855). *Myrcia alloiota* (O.Berg) Kiaersk., Enum. Myrt. Bras.: 70 (1893).
Aulomyrcia aureolanata O.Berg, Linnaea 27: 52 (1855).
Aulomyrcia confusa O.Berg, Linnaea 27: 56 (1855).
Aulomyrcia longipes O.Berg, Linnaea 27: 55 (1855). *Myrcia longipes* (O.Berg) Kiaersk., Enum. Myrt. Bras.: 70 (1893).
Aulomyrcia ottonis O.Berg, Linnaea 27: 55 (1855).
Aulomyrcia rosulans O.Berg, Linnaea 27: 54 (1855). *Myrcia rosulans* (O.Berg) Kiaersk., Enum. Myrt. Bras.: 72 (1893).
Aulomyrcia alloiota var. *cuneata* O.Berg in C.F.P.von Martius & auct. suc. (eds.), Fl. Bras. 14(1): 91 (1857).
Aulomyrcia alloiota var. *obovata* O.Berg in C.F.P.von Martius & auct. suc. (eds.), Fl. Bras. 14(1): 91 (1857).
Aulomyrcia alloiota var. *ovalis* O.Berg in C.F.P.von Martius & auct. suc. (eds.), Fl. Bras. 14(1): 91 (1857).
Aulomyrcia alloiota var. *pyramidalis* O.Berg in C.F.P.von Martius & auct. suc. (eds.), Fl. Bras. 14(1): 91 (1857).
Aulomyrcia alloiota var. *subcordata* O.Berg in C.F.P.von Martius & auct. suc. (eds.), Fl. Bras. 14(1): 91 (1857). *Myrcia alloiota* var. *subcordata* (O.Berg) N.Silveira, Loefgrenia 86: 1 (1985).
Aulomyrcia curatellifolia var. *grandifolia* O.Berg in C.F.P.von Martius & auct. suc. (eds.), Fl. Bras. 14(1): 95 (1857).
Aulomyrcia curatellifolia var. *parvifolia* O.Berg in C.F.P.von Martius & auct. suc. (eds.), Fl. Bras. 14(1): 95 (1857).

Aulomyrcia lancea O.Berg in C.F.P.von Martius & auct. suc. (eds.), Fl. Bras. 14(1): 88 (1857). *Myrcia lancea* (O.Berg) Mattos, Arq. Bot. Estado São Paulo, n.s., f.m., 4: 62 (1966).

Aulomyrcia lanuginosa O.Berg in C.F.P.von Martius & auct. suc. (eds.), Fl. Bras. 14(1): 102 (1857). *Myrcia lanuginosa* (O.Berg) Nied. in H.G.A.Engler & K.A.E.Prantl, Nat. Pflanzenfam. 3(7): 76 (1893), nom. illeg. *Myrcia rhodeosepala* Kiaersk., Enum. Myrt. Bras.: 75 (1893).

Aulomyrcia leucantha O.Berg in C.F.P.von Martius & auct. suc. (eds.), Fl. Bras. 14(1): 93 (1857). *Myrcia leucantha* (O.Berg) N.Silveira, Roessléria 7: 66 (1985).

Aulomyrcia longipes var. *latifolia* O.Berg in C.F.P.von Martius & auct. suc. (eds.), Fl. Bras. 14(1): 94 (1857).

Aulomyrcia longipes var. *obovata* O.Berg in C.F.P.von Martius & auct. suc. (eds.), Fl. Bras. 14(1): 94 (1857).

Aulomyrcia longipes var. *spathulata* O.Berg in C.F.P.von Martius & auct. suc. (eds.), Fl. Bras. 14(1): 94 (1857).

Aulomyrcia prunifolia var. *brevipes* O.Berg in C.F.P.von Martius & auct. suc. (eds.), Fl. Bras. 14(1): 95 (1857).

Aulomyrcia prunifolia var. *longipes* O.Berg in C.F.P.von Martius & auct. suc. (eds.), Fl. Bras. 14(1): 96 (1857).

Calyptromyrcia venosa O.Berg in C.F.P.von Martius & auct. suc. (eds.), Fl. Bras. 14(1): 57 (1857).

Myrcia piauhiensis O.Berg in C.F.P.von Martius & auct. suc. (eds.), Fl. Bras. 14(1): 196 (1857).

Aulomyrcia micrantha O.Berg in C.F.P.von Martius & auct. suc. (eds.), Fl. Bras. 14(1): 516 (1858). *Myrcia marahanensis* Kiaersk., Enum. Myrt. Bras.: 71 (1893).

Aulomyrcia curatellifolia var. *australis* O.Berg in C.F.P.von Martius & auct. suc. (eds.), Fl. Bras. 14(1): 547 (1859).

Aulomyrcia lanuginosa var. *pyramidata* O.Berg in C.F.P.von Martius & auct. suc. (eds.), Fl. Bras. 14(1): 548 (1859).

Myrcia alloiota var. *obovata* Kiaersk., Enum. Myrt. Bras.: 70 (1893).

Myrcia curatellifolia var. *grandifolia* Kiaersk., Enum. Myrt. Bras.: 70 (1893).

Myrcia curatellifolia var. *parvifolia* Kiaersk., Enum. Myrt. Bras.: 70 (1893).

Myrcia longipes f. *obovata* Kiaersk., Enum. Myrt. Bras.: 70 (1893).

Myrcia membranacea Kiaersk., Enum. Myrt. Bras.: 71 (1893).

Eugenia vallis Standl., Ann. Missouri Bot. Gard. 37: 323 (1940).

Myrcia torta DC., Prodr. 3: 250 (1828). *Aulomyrcia torta* (DC.) O.Berg, Linnaea 27: 78 (1855).
WC. & E. Brazil. 84 BZC BZE BZL. Cham. or nanophan.

Aulomyrcia jequitinhonhensis O.Berg in C.F.P.von Martius & auct. suc. (eds.), Fl. Bras. 14(1): 137 (1857).

Aulomyrcia martiana O.Berg in C.F.P.von Martius & auct. suc. (eds.), Fl. Bras. 14(1): 138 (1857).

Aulomyrcia jequitinhonhensis var. *glauca* O.Berg in C.F.P.von Martius & auct. suc. (eds.), Fl. Bras. 14(1): 560 (1859).

Aulomyrcia jequitinhonhensis var. *grandifolia* O.Berg in C.F.P.von Martius & auct. suc. (eds.), Fl. Bras. 14(1): 559 (1859).

Aulomyrcia jequitinhonhensis var. *parvifolia* O.Berg in C.F.P.von Martius & auct. suc. (eds.), Fl. Bras. 14(1): 559 (1859).

Myrcia torta f. *glauca* Kiaersk., Enum. Myrt. Bras.: 79 (1893).

Myrcia torta f. *grandifolia* Kiaersk., Enum. Myrt. Bras.: 79 (1893).

Myrcia torta f. *parvifolia* Kiaersk., Enum. Myrt. Bras.: 79 (1893).

Myrcia tortuosa (O.Berg) N.Silveira, Roessléria 7: 67 (1985).
WC. Brazil. 84 BZC. Nanophan. or phan.

**Aulomyrcia tortuosa* O.Berg in C.F.P.von Martius & auct. suc. (eds.), Fl. Bras. 14(1): 558 (1859).

Myrcia tovarensis O.Berg, Linnaea 27: 118 (1855). *Calyptranthes tovarensis* (O.Berg) Steyerm., Fieldiana, Bot. 28: 1010 (1957).
Venezuela. 82 VEN. Nanophan. or phan.

Myrcia uberavensis O.Berg in C.F.P.von Martius & auct. suc. (eds.), Fl. Bras. 14(1): 568 (1859).
Brazil. 84 BZC BZL. Nanophan. or phan.

Myrcia uberavensis var. *ovata* O.Berg in C.F.P.von Martius & auct. suc. (eds.), Fl. Bras. 14(1): 568 (1859).

Myrcia uberavensis var. *rotundifolia* O.Berg in C.F.P. von Martius & auct. suc. (eds.), Fl. Bras. 14(1): 568 (1859).

Myrcia undulata O.Berg in C.F.P.von Martius & auct. suc. (eds.), Fl. Bras. 14(1): 185 (1857).
Brazil (Minas Gerais). 84 BZL. Nanophan. or phan.

Myrcia vacciniifolia (O.Berg) Nied. in H.G.A.Engler & K.A.E.Prantl, Nat. Pflanzenfam. 3(7): 76 (1893).
Brazil (Minas Gerais). 84 BZL. Nanophan. or phan.

**Aulomyrcia vacciniifolia* O.Berg in C.F.P.von Martius & auct. suc. (eds.), Fl. Bras. 14(1): 140 (1857).

Myrcia valenzuelana (A.Rich.) Griseb., Cat. Pl. Cub.: 86 (1866).
W. Cuba. 81 CUB. Nanophan.

**Eugenia valenzuelana* A.Rich., Hist. Phys. Cuba, Pl. Vasc.: 593 (1846).

Aulomyrcia sagraea O.Berg, Linnaea 30: 655 (1861).

Myrcia variabilis Mart. ex DC., Prodr. 3: 254 (1828). *Aulomyrcia variabilis* (Mart. ex DC.) O.Berg, Linnaea 27: 62 (1855).
Brazil. 84 BZC BZE BZL BZN. Cham. or nanophan.

Myrcia variabilis var. *intermedia* Mart. ex DC., Prodr. 3: 254 (1828).

Myrcia variabilis var. *nummularia* Mart. ex DC., Prodr. 3: 254 (1828).

Myrcia variabilis var. *ovatilofia* Mart. ex DC., Prodr. 3: 254 (1828).

Myrcia cordata Cambess. in A.F.C.de Saint-Hilaire, Fl. Bras. Merid. 2: 330 (1832).

Aulomyrcia variabilis var. *intermedia* Mart. ex O.Berg in C.F.P.von Martius & auct. suc. (eds.), Fl. Bras. 14(1): 106 (1857).

Aulomyrcia variabilis var. *nummulaira* Mart. ex O.Berg in C.F.P.von Martius & auct. suc. (eds.), Fl. Bras. 14(1): 106 (1857).

Aulomyrcia variabilis var. *ovatifolia* Mart. ex O.Berg in C.F.P.von Martius & auct. suc. (eds.), Fl. Bras. 14(1): 106 (1857).

Aulomyrcia variabilis var. *suffruticosa* O.Berg in C.F.P. von Martius & auct. suc. (eds.), Fl. Bras. 14(1): 550 (1859).

Myrcia vauthiereana O.Berg in C.F.P.von Martius & auct. suc. (eds.), Fl. Bras. 14(1): 154 (1857).
SE. Brazil. 84 BZL. Nanophan. or phan.

Myrcia velhensis (O.Berg) N.Silveira, Loefgrenia 91: 1 (1987).
Brazil (Goiás). 84 BZC. Nanophan. or phan.
**Aulomyrcia velhensis* O.Berg in C.F.P.von Martius & auct. suc. (eds.), Fl. Bras. 14(1): 560 (1859).

Myrcia velloziana O.Berg in C.F.P.von Martius & auct. suc. (eds.), Fl. Bras. 14(1): 171 (1857).
SE. Brazil. 84 BZL. Nanophan. or phan.

Myrcia venulosa DC., Prodr. 3: 250 (1828). *Aulomyrcia venulosa* (DC.) O.Berg, Linnaea 27: 63 (1855).
Brazil. 84 BZC BZE BZL BZS. Phan.

var. ***dives*** (O.Berg) Soares-Silva, Bradea 8: 325 (2002).
S. Brazil. 84 BZS. Phan.
Aulomyrcia reticulata O.Berg in C.F.P.von Martius & auct. suc. (eds.), Fl. Bras. 14(1): 108 (1857). *Myrcia shirleyana* Mattos, Loefgrenia 66: 2 (1975).
**Aulomyrcia reticulata* var. *dives* O.Berg in C.F.P. von Martius & auct. suc. (eds.), Fl. Bras. 14(1): 109 (1857).
Aulomyrcia reticulata var. *pauciflora* O.Berg in C.F.P.von Martius & auct. suc. (eds.), Fl. Bras. 14(1): 109 (1857).

var. ***venulosa***
Brazil. 84 BZC BZE BZL BZS. Phan.
Myrcia venulosa var. *capoeirensis* DC., Prodr. 3: 250 (1828). *Aulomyrcia venulosa* var. *capoeirensis* (DC.) O.Berg in C.F.P.von Martius & auct. suc. (eds.), Fl. Bras. 14(1): 109 (1857).
Myrcia andromedoides Cambess. in A.F.C.de Saint-Hilaire, Fl. Bras. Merid. 2: 323 (1832). *Aulomyrcia andromedoides* (Cambess.) O.Berg, Linnaea 27: 40 (1855).
Myrcia oleifolia Cambess. in A.F.C.de Saint-Hilaire, Fl. Bras. Merid. 2: 313 (1832).
Aulomyrcia castrensis O.Berg in C.F.P.von Martius & auct. suc. (eds.), Fl. Bras. 14(1): 111 (1857). *Myrcia castrensis* (O.Berg) D.Legrand, Sellowia 13: 297 (1961).
Aulomyrcia laureola O.Berg in C.F.P.von Martius & auct. suc. (eds.), Fl. Bras. 14(1): 110 (1857). *Myrcia laureola* (O.Berg) Kiaersk., Enum. Myrt. Bras.: 86 (1893).
Aulomyrcia rugosa O.Berg in C.F.P.von Martius & auct. suc. (eds.), Fl. Bras. 14(1): 110 (1857). *Myrcia rugosa* (O.Berg) Kiaersk., Enum. Myrt. Bras.: 96 (1893).
Aulomyrcia sonderiana O.Berg in C.F.P.von Martius & auct. suc. (eds.), Fl. Bras. 14(1): 110 (1857). *Myrcia rabeniana* Kiaersk., Enum. Myrt. Bras.: 86 (1893). *Myrcia sonderiana* (O.Berg) Mattos, Ci. & Cult. 19: 333 (1967).
Aulomyrcia venulosa var. *parvifolia* O.Berg in C.F.P.von Martius & auct. suc. (eds.), Fl. Bras. 14(1): 109 (1857).
Aulomyrcia venulosa var. *rufa* O.Berg in C.F.P.von Martius & auct. suc. (eds.), Fl. Bras. 14(1): 109 (1857). *Myrcia venulosa* var. *rufa* (O.Berg) N.Silveira, Roessléria 7: 67 (1985).
Aulomyrcia dictyophleba O.Berg in C.F.P.von Martius & auct. suc. (eds.), Fl. Bras. 14(1): 560 (1859). *Myrcia dictyophleba* (O.Berg) D.Legrand, Sellowia 13: 297 (1961).
Aulomyrcia riedeliana O.Berg in C.F.P.von Martius & auct. suc. (eds.), Fl. Bras. 14(1): 551 (1859).
Aulomyrcia venulosa var. *ochracea* O.Berg in C.F.P.von Martius & auct. suc. (eds.), Fl. Bras. 14(1): 550 (1859).
Myrcia melanosepala Kiaersk., Enum. Myrt. Bras.: 60 (1893).
Myrcia pyrrhopilodes Kiaersk., Enum. Myrt. Bras.: 96 (1893).
Myrcia carassana Glaz., Bull. Soc. Bot. France 54(Mem. 3): 222 (1908), nom. inval.

Myrcia verrucosa Sobral, Bol. Mus. Biol. Mello Leitão. Nova Sér. 20: 77 (2006 publ. 2007).
Brazil (Espírito Santo). 84 BZL. Phan.

Myrcia verruculata S.Moore, Trans. Linn. Soc. London, Bot. 4: 355 (1895).
WC. Brazil. 84 BZC. Nanophan. or phan.

Myrcia verticillaris O.Berg in C.F.P.von Martius & auct. suc. (eds.), Fl. Bras. 14(1): 206 (1857). *Luma verticillaris* (O.Berg) Herter, Revista Sudamer. Bot. 7: 220 (1943).
S. Brazil to N. Uruguay. 84 BZS 85 URU. Cham.
Myrcia verticillaris var. *glomerata* O.Berg in C.F.P.von Martius & auct. suc. (eds.), Fl. Bras. 14(1): 206 (1857).
Myrcia verticillaris var. *laxa* O.Berg in C.F.P.von Martius & auct. suc. (eds.), Fl. Bras. 14(1): 207 (1857).
Myrcia verticillaris var. *multicaulis* O.Berg in C.F.P. von Martius & auct. suc. (eds.), Fl. Bras. 14(1): 206 (1857).
Myrcia verticillaris var. *paniculata* O.Berg in C.F.P. von Martius & auct. suc. (eds.), Fl. Bras. 14(1): 206 (1857).
Myrcia verticillaris var. *pygmaea* O.Berg in C.F.P.von Martius & auct. suc. (eds.), Fl. Bras. 14(1): 207 (1857).

Myrcia vestita DC., Prodr. 3: 248 (1828). *Aulomyrcia vestita* (DC.) O.Berg, Linnaea 27: 74 (1855).
Brazil. 84 BZC BZE BZL. Nanophan. or phan.
Myrcia vestita var. *obtusifolia* DC., Prodr. 3: 248 (1828). *Aulomyrcia thyrsiflora* var. *obtusifolia* (DC.) O.Berg in C.F.P.von Martius & auct. suc. (eds.), Fl. Bras. 14(1): 126 (1857).
Aulomyrcia corymbiflora O.Berg, Linnaea 27: 74 (1855).
Aulomyrcia thyrsiflora O.Berg, Linnaea 27: 74 (1855).
Aulomyrcia linguiformis O.Berg in C.F.P.von Martius & auct. suc. (eds.), Fl. Bras. 14(1): 125 (1857). *Myrcia linguiformis* (O.Berg) N.Silveira, Roessléria 7: 66 (1985).
Aulomyrcia pachyclada O.Berg in C.F.P.von Martius & auct. suc. (eds.), Fl. Bras. 14(1): 133 (1857). *Myrcia pachyclada* (O.Berg) N.Silveira, Loefgrenia 88: 2 (1985).
Aulomyrcia pachyclada var. *elliptica* O.Berg in C.F.P.von Martius & auct. suc. (eds.), Fl. Bras. 14(1): 134 (1857).
Aulomyrcia pachyclada var. *spathulata* O.Berg in C.F.P.von Martius & auct. suc. (eds.), Fl. Bras. 14(1): 133 (1857). *Myrcia pachyclada* var. *spathulata* (O.Berg) N.Silveira, Loefgrenia 88: 2 (1985).
Aulomyrcia thyrsiflora var. *lateriflora* O.Berg in C.F.P.von Martius & auct. suc. (eds.), Fl. Bras. 14(1): 126 (1857).

Aulomyrcia thyrsiflora var. *petiolaris* O.Berg in C.F.P.von Martius & auct. suc. (eds.), Fl. Bras. 14(1): 126 (1857).
Aulomyrcia vestita var. *grandifolia* O.Berg in C.F.P.von Martius & auct. suc. (eds.), Fl. Bras. 14(1): 127 (1857). *Myrcia vestita* var. *grandifolia* (O.Berg) Kiaersk., Enum. Myrt. Bras.: 97 (1893).
Aulomyrcia vestita var. *parvifolia* O.Berg in C.F.P.von Martius & auct. suc. (eds.), Fl. Bras. 14(1): 127 (1857). *Myrcia vestita* var. *parvifolia* (O.Berg) Kiaersk., Enum. Myrt. Bras.: 97 (1893).
Aulomyrcia pachyclada var. *prolifera* O.Berg in C.F.P.von Martius & auct. suc. (eds.), Fl. Bras. 14(1): 558 (1859). *Myrcia pachyclada* var. *prolifera* (O.Berg) N.Silveira, Loefgrenia 88: 2 (1985).

Myrcia virgata Cambess. in A.F.C.de Saint-Hilaire, Fl. Bras. Merid. 2: 320 (1832). *Aulomyrcia virgata* (Cambess.) O.Berg in C.F.P.von Martius & auct. suc. (eds.), Fl. Bras. 14(1): 136 (1857).
WC. Brazil. 84 BZC. Nanophan. or phan.

Myrcia vittoriana Kiaersk., Enum. Myrt. Bras.: 102 (1893).
Brazil (Rio de Janeiro). 84 BZL. Nanophan. or phan.
**Gomidesia martiana* O.Berg in C.F.P.von Martius & auct. suc. (eds.), Fl. Bras. 14(1): 12 (1857).
Myrcia vittoriana var. *piratiningensis* Kiaersk., Enum. Myrt. Bras.: 102 (1893).

Myrcia warmingiana Kiaersk., Enum. Myrt. Bras.: 104 (1893). *Gomidesia warmingiana* (Kiaersk.) D.Legrand, Comun. Bot. Mus. Hist. Nat. Montevideo 3(37): 23 (1958).
SE. Brazil. 84 BZL. Nanophan. or phan.

Myrcia xylopioides (Kunth) DC., Prodr. 3: 256 (1828).
Colombia. 83 CLM. Nanophan. or phan.
**Myrtus xylopioides* Kunth in F.W.H.von Humboldt, A.J.A.Bonpland & C.S.Kunth, Nov. Gen. Sp. 7: 257 (1825).

Myrcia zetekiana (Standl.) B.Holst, Novon 15: 296 (2005).
Panama to Colombia. 80 PAN 83 CLM. Nanophan. or phan.
**Eugenia zetekiana* Standl., J. Wash. Acad. Sci. 15: 286 (1925). *Aulomyrcia zetekiana* (Standl.) Amshoff, Ann. Missouri Bot. Gard. 45: 170 (1958).

Synonyms:
Myrcia acris (Sw.) DC. = ***Pimenta racemosa*** (Mill.) J.W.Moore var. ***racemosa***
Myrcia acuminata (Kunth) DC. = ***Myrcia splendens*** (Sw.) DC.
Myrcia acuminata var. *bullata* O.Berg = ***Myrcia splendens*** (Sw.) DC.
Myrcia acuminata var. *meridensis* O.Berg = ***Myrcia splendens*** (Sw.) DC.
Myrcia acuminata var. *peruviana* O.Berg = ***Myrcia splendens*** (Sw.) DC.
Myrcia acuminata var. *tovarensis* O.Berg = ***Myrcia splendens*** (Sw.) DC.
Myrcia acuminatissima O.Berg = ***Myrcia racemosa*** (O.Berg) Kiaersk.
Myrcia acuminatissima Hieron. = [84 BZL]
Myrcia acutata O.Berg = ***Myrcia splendens*** (Sw.) DC.
Myrcia acutiloba O.Berg = ***Myrcia splendens*** (Sw.) DC.
Myrcia adpressepilosa Kiaersk. = ***Myrcia guianensis*** (Aubl.) DC.
Myrcia affinis Cambess. = ***Myrcia hebepetala*** DC.
Myrcia aguitensis Gleason = ***Myrcia splendens*** (Sw.) DC.
Myrcia alagoensis O.Berg = ***Myrcia guianensis*** (Aubl.) DC.
Myrcia alagoensis var. *intermedia* O.Berg = ***Myrcia splendens*** (Sw.) DC.
Myrcia alagoensis var. *oblongata* O.Berg = ***Myrcia splendens*** (Sw.) DC.
Myrcia alagoensis var. *ovata* O.Berg = ***Myrcia splendens*** (Sw.) DC.
Myrcia albotomentosa var. *humilis* Cambess. = ***Myrcia albotomentosa*** DC.
Myrcia albotomentosa var. *nivea* Cambess. = ***Myrcia albotomentosa*** DC.
Myrcia alloiota (O.Berg) Kiaersk. = ***Myrcia tomentosa*** (Aubl.) DC.
Myrcia alloiota var. *obovata* Kiaersk. = ***Myrcia tomentosa*** (Aubl.) DC.
Myrcia alloiota var. *subcordata* (O.Berg) N.Silveira = ***Myrcia tomentosa*** (Aubl.) DC.
Myrcia alpestris Barb.Rodr. ex Chodat & Hassl. = ***Myrcia anomala*** Cambess.
Myrcia alpina Kiaersk. = ***Myrcia eriocalyx*** DC.
Myrcia alternifolia Miq. = ***Myrcia guianensis*** (Aubl.) DC.
Myrcia ambigua DC. = ***Myrcia sylvatica*** (G.Mey.) DC.
Myrcia ambigua var. *dives* O.Berg = ***Myrcia sylvatica*** (G.Mey.) DC.
Myrcia ambigua var. *latifolia* O.Berg = ***Myrcia sylvatica*** (G.Mey.) DC.
Myrcia ambigua var. *multiflora* O.Berg = ***Myrcia sylvatica*** (G.Mey.) DC.
Myrcia ambigua var. *pauciflora* DC. = ***Myrcia sylvatica*** (G.Mey.) DC.
Myrcia ambigua var. *rostrata* O.Berg = ***Myrcia sylvatica*** (G.Mey.) DC.
Myrcia ambigua var. *sylvatica* (G.Mey.) O.Berg = ***Myrcia sylvatica*** (G.Mey.) DC.
Myrcia amethystina var. *dealbata* Kiaersk. = ***Myrcia amethystina*** (O.Berg) Kiaersk.
Myrcia amethystina var. *pulchra* Kiaersk. = ***Myrcia amethystina*** (O.Berg) Kiaersk.
Myrcia anceps var. *brevipes* O.Berg = ***Myrcia anceps*** (Spreng.) O.Berg
Myrcia anceps var. *depauperata* O.Berg = ***Myrcia anceps*** (Spreng.) O.Berg
Myrcia anceps var. *dives* O.Berg = ***Myrcia anceps*** (Spreng.) O.Berg
Myrcia andromedoides Cambess. = ***Myrcia venulosa*** var. ***venulosa***
Myrcia androsaemoides (O.Berg) Krug & Urb. = ***Myrcia guianensis*** (Aubl.) DC.
Myrcia androsaemoides var. *parvifolia* Krug & Urb. = ***Myrcia guianensis*** (Aubl.) DC.
Myrcia anomala var. *multiceps* O.Berg = ***Myrcia palustris*** DC.
Myrcia anomala var. *ramosa* Cambess. = ***Myrcia anomala*** Cambess.
Myrcia arborescens O.Berg = ***Myrcia retorta*** Cambess.
Myrcia arimensis Britton = ***Myrcia guianensis*** (Aubl.) DC.
Myrcia aromatica Schltdl. = ***Calyptranthes schiedeana*** O.Berg
Myrcia assumptionis Morong = ***Myrcia laruotteana*** var. ***laruotteana***
Myrcia augustana Kiaersk. = ***Myrcia splendens*** (Sw.) DC.
Myrcia australasiae F.Muell. = (Melastomataceae)
Myrcia ayresiana O.Berg = ***Myrcia splendens*** (Sw.) DC.
Myrcia balbisiana DC. = ***Myrcianthes fragrans*** (Sw.) McVaugh

Myrcia bangii Rusby = ***Myrcia fascicularis*** O.Berg
Myrcia banisteriifolia DC. = ***Myrcia palustris*** DC.
Myrcia barrensis O.Berg = ***Myrcia splendens*** (Sw.) DC.
Myrcia belizensis Lundell = ***Myrcia splendens*** (Sw.) DC.
Myrcia berberis DC. = ***Myrcia splendens*** (Sw.) DC.
Myrcia berberis Schauer = ***Myrcia splendens*** (Sw.) DC.
Myrcia berberis var. *angustifolia* O.Berg = ***Myrcia splendens*** (Sw.) DC.
Myrcia berberis var. *latifolia* O.Berg = ***Myrcia splendens*** (Sw.) DC.
Myrcia bergiana var. *angustifolia* O.Berg = ***Myrcia bergiana*** O.Berg
Myrcia bipennis (O.Berg) McVaugh = ***Marlierea bipennis*** (O.Berg) McVaugh
Myrcia botrys (O.Berg) N.Silveira = ***Myrcia guianensis*** (Aubl.) DC.
Myrcia brachylopadia Diels = ***Myrcia splendens*** (Sw.) DC.
Myrcia bracteolaris (Poir.) DC. = ***Myrcia splendens*** (Sw.) DC.
Myrcia brandamii O.Berg = ***Myrcia splendens*** (Sw.) DC.
Myrcia breviramis (O.Berg) D.Legrand = ***Myrcia pulchra*** (O.Berg) Kiaersk.
Myrcia browniana Gardner = ***Myrcia spectabilis*** DC.
Myrcia brunnea var. *opaca* O.Berg = ***Myrcia brunnea*** Cambess.
Myrcia buxifolia var. *glazioviana* Kiaersk. = ***Myrcia buxifolia*** Gardner
Myrcia caerulescens (O.Berg) Kiaersk. = ***Myrcia multiflora*** (Lam.) DC.
Myrcia calumbaensis Kiaersk. = ***Myrcia pubiflora*** DC.
Myrcia camaraeana DC. = ***Myrcia multiflora*** (Lam.) DC.
Myrcia candolleana (O.Berg) Kiaersk. = ***Myrcia hebepetala*** DC.
Myrcia canescens var. *reticulata* O.Berg = ***Myrcia canescens*** O.Berg
Myrcia capitata (O.Berg) Nied. = ***Myrcia camapuanensis*** N.Silveira
Myrcia capivarhyensis Cambess. = ***Myrcia tomentosa*** (Aubl.) DC.
Myrcia carassana Glaz. = ***Myrcia venulosa*** var. ***venulosa***
Myrcia carnea (G.Mey.) DC. = ***Myrcia coumete*** (Aubl.) DC.
Myrcia cassinioides DC. = ***Myrcia guianensis*** (Aubl.) DC.
Myrcia castrensis (O.Berg) D.Legrand = ***Myrcia venulosa*** var. ***venulosa***
Myrcia catharinae O.Berg = ***Myrcia splendens*** (Sw.) DC.
Myrcia chilensis O.Berg = ***Myrcia splendens*** (Sw.) DC.
Myrcia ciarensis O.Berg = ***Myrcia splendens*** (Sw.) DC.
Myrcia citrifolia D.Legrand = ***Myrcia glabra*** (O.Berg) D.Legrand
Myrcia citrifolia var. *imrayana* (Griseb.) Stehlé & Quentin = ***Myrcia citrifolia*** (Aubl.) Urb.
Myrcia citrifolia var. *jacquiniana* Stehlé & Quentin = ***Myrcia citrifolia*** (Aubl.) Urb.
Myrcia clusiifolia Miq. = ***Calyptranthes clusiifolia*** O.Berg
Myrcia coccolobifolia (Kunth) DC. = ***Myrcianthes rhopaloides*** (Kunth) McVaugh
Myrcia collina (DC.) J.F.Macbr. = ***Myrcia pubiflora*** DC.
Myrcia communis O.Berg = ***Myrcia splendens*** (Sw.) DC.
Myrcia communis var. *glabrata* O.Berg = ***Myrcia splendens*** (Sw.) DC.
Myrcia communis var. *latifolia* O.Berg = ***Myrcia splendens*** (Sw.) DC.
Myrcia complicata (Kunth) DC. = ***Myrcia splendens*** (Sw.) DC.
Myrcia compressa Gleason = ***Myrcia splendens*** (Sw.) DC.
Myrcia corcovadensis O.Berg = ***Myrcia splendens*** (Sw.) DC.
Myrcia cordata Cambess. = ***Myrcia variabilis*** Mart. ex DC.
Myrcia cordifolia O.Berg = ***Myrcia lasiantha*** DC.
Myrcia cordifolia var. *acuminata* O.Berg = ***Myrcia lasiantha*** DC.
Myrcia cordifolia var. *alba* O.Berg = ***Myrcia lasiantha*** DC.
Myrcia cordifolia var. *fuscescens* O.Berg = ***Myrcia lasiantha*** DC.
Myrcia cordifolia var. *glabrescens* Kiaersk. = ***Myrcia lasiantha*** DC.
Myrcia cordifolia var. *nivea* O.Berg = ***Myrcia lasiantha*** DC.
Myrcia cordiifolia var. *minor* DC. = ***Myrcia hebepetala*** DC.
Myrcia coriacea DC. = ***Myrcia citrifolia*** (Aubl.) Urb.
Myrcia coriacea var. *acutifolia* Griseb. = ***Myrcia citrifolia*** (Aubl.) Urb.
Myrcia coriacea var. *imrayana* Griseb. = ***Myrcia citrifolia*** (Aubl.) Urb.
Myrcia coriacea var. *jacquiniana* Griseb. = ***Myrcia citrifolia*** (Aubl.) Urb.
Myrcia coriacea var. *reticulata* Griseb. = ***Myrcia citrifolia*** (Aubl.) Urb.
Myrcia coriacea var. *swartziana* Griseb. = ***Myrcia citrifolia*** (Aubl.) Urb.
Myrcia coroicensis Rusby = ***Myrcia splendens*** (Sw.) DC.
Myrcia corrientinensis Barb.Rodr. ex Chodat & Hassl. = ***Myrcia laruotteana*** var. ***laruotteana***
Myrcia corymbosa DC. = ***Myrcia amazonica*** DC.
Myrcia costa-ricensis O.Berg = ***Myrcia splendens*** (Sw.) DC.
Myrcia costata DC. = ***Myrcia splendens*** (Sw.) DC.
Myrcia costata var. *bahiensis* O.Berg = ***Myrcia splendens*** (Sw.) DC.
Myrcia costata var. *minensis* O.Berg = ***Myrcia splendens*** (Sw.) DC.
Myrcia cotonosa Barb.Rodr. ex Chodat & Hassl. = ***Myrcia anomala*** Cambess.
Myrcia coumetoides O.Berg = ***Myrcia splendens*** (Sw.) DC.
Myrcia crassicaulis Cambess. = ***Myrcia guianensis*** (Aubl.) DC.
Myrcia crassifolia Kiaersk. = ***Myrcia obovata*** (O.Berg) Nied.
Myrcia crassinervia DC. = ***Myrcia deflexa*** (Poir.) DC.
Myrcia crocea Kiaersk. = ***Myrcia amplexicaulis*** (Vell.) Hook.f.
Myrcia crocea var. *blanchetiana* Kiaersk. = ***Myrcia grandifolia*** Cambess.
Myrcia cubensis (Griseb.) Krug & Urb. = ***Myrciaria floribunda*** (H.West ex Willd.) O.Berg
Myrcia cucullata O.Berg = ***Myrcia splendens*** (Sw.) DC.
Myrcia cujabensis var. *latifolia* O.Berg = ***Myrcia cujabensis*** O.Berg
Myrcia cujabensis var. *longifolia* O.Berg = ***Myrcia cujabensis*** O.Berg
Myrcia cuneata (O.Berg) Nied. = ***Myrcia guianensis*** (Aubl.) DC.
Myrcia curatellifolia DC. = ***Myrcia tomentosa*** (Aubl.) DC.
Myrcia curatellifolia var. *grandifolia* Kiaersk. = ***Myrcia tomentosa*** (Aubl.) DC.

Myrcia curatellifolia var. *parvifolia* Kiaersk. = ***Myrcia tomentosa*** (Aubl.) DC.
Myrcia cymosa (O.Berg) Nied. = ***Myrcia guianensis*** (Aubl.) DC.
Myrcia cymosopaniculata Kiaersk. = ***Myrcia guianensis*** (Aubl.) DC.
Myrcia daphnoides DC. = ***Myrcia guianensis*** (Aubl.) DC.
Myrcia daphnoides var. *nervosa* Kiaersk. = ***Myrcia guianensis*** (Aubl.) DC.
Myrcia dasyblasta O.Berg = ***Myrcia bella*** Cambess.
Myrcia dasyblasta var. *cordata* O.Berg = ***Myrcia bella*** Cambess.
Myrcia dasyblasta var. *ovata* O.Berg = ***Myrcia bella*** Cambess.
Myrcia deflexa var. *dussii* Krug & Urb. = ***Myrcia deflexa*** (Poir.) DC.
Myrcia detergens Miq. = ***Myrcia amazonica*** DC.
Myrcia diaphanosticta Kiaersk. = ***Myrcia guianensis*** (Aubl.) DC.
Myrcia dictyoneura Diels = ***Myrcia splendens*** (Sw.) DC.
Myrcia dictyophleba (O.Berg) D.Legrand = ***Myrcia venulosa*** var. ***venulosa***
Myrcia dictyophylla var. *robustior* (O.Berg) Mattos = ***Myrcia dictyophylla*** (O.Berg) Mattos & D.Legrand
Myrcia dictyophylla var. *subcordata* (O.Berg) N.Silveira = ***Myrcia dictyophylla*** (O.Berg) Mattos & D.Legrand
Myrcia discolor O.Berg = ***Myrcia splendens*** (Sw.) DC.
Myrcia divaricata (O.Berg) Lemée = ***Myrcia rufipila*** McVaugh
Myrcia divaricata (Lam.) DC. = ***Myrcia splendens*** (Sw.) DC.
Myrcia divergens DC. = ***Myrcia pyrifolia*** (Desv.) Nied.
Myrcia duchassaingiana O.Berg = ***Myrcia deflexa*** (Poir.) DC.
Myrcia dumosa (O.Berg) Krug & Urb. = ***Myrcia platyclada*** DC.
Myrcia duriuscula Mart. ex DC. = ***Myrcia decorticans*** DC.
Myrcia dussii Krug & Urb. = ***Myrcia platyclada*** DC.
Myrcia edulis (O.Berg) Krug & Urb. = ***Myrcia platyclada*** DC.
Myrcia edulis var. *dominicana* Krug & Urb. = ***Myrcia antillana*** McVaugh
Myrcia elaeodendra DC. = ***Myrcia guianensis*** (Aubl.) DC.
Myrcia elegans DC. = ***Myrcia guianensis*** (Aubl.) DC.
Myrcia elliptica Gardner = ***Myrcia eriocalyx*** DC.
Myrcia ellipticifolia Cambess. = ***Myrcia multiflora*** (Lam.) DC.
Myrcia elongata O.Berg = ***Myrcia splendens*** (Sw.) DC.
Myrcia elongata var. *brunnea* O.Berg = ***Myrcia splendens*** (Sw.) DC.
Myrcia elongata var. *grandifolia* O.Berg = ***Myrcia splendens*** (Sw.) DC.
Myrcia elongata var. *ochracea* O.Berg = ***Myrcia splendens*** (Sw.) DC.
Myrcia emarginata (Moldenke) Alain = ***Pimenta oligantha*** (Urb.) Burret
Myrcia eriocalyx var. *beyrichiana* (O.Berg) Kiaersk. = ***Myrcia eriocalyx*** DC.
Myrcia eriocalyx f. *brunnea* Kiaersk. = ***Myrcia eriocalyx*** DC.
Myrcia eriocalyx var. *miqueliana* (O.Berg) Kiaersk. = ***Myrcia eriocalyx*** DC.
Myrcia eriocalyx f. *obtusata* Kiaersk. = ***Myrcia eriocalyx*** DC.
Myrcia eriopus var. *grandifolia* O.Berg = ***Myrcia eriopus*** DC.
Myrcia eriopus var. *parvifolia* O.Berg = ***Myrcia eriopus*** DC.
Myrcia erythroxylon O.Berg = ***Myrcia splendens*** (Sw.) DC.
Myrcia erythroxylon var. *caerulescens* O.Berg = ***Myrcia splendens*** (Sw.) DC.
Myrcia erythroxylon var. *virescens* O.Berg = ***Eugenia biflora*** (L.) DC.
Myrcia estrellensis Kiaersk. = ***Myrcia anacardiifolia*** Gardner
Myrcia eugenioides Cambess. = ***Marlierea eugenioides*** (Cambess.) D.Legrand
Myrcia exsucca DC. = ***Myrcia guianensis*** (Aubl.) DC.
Myrcia fallax (Rich.) DC. = ***Myrcia splendens*** (Sw.) DC.
Myrcia fastigiata Kiaersk. = ***Myrcia guianensis*** (Aubl.) DC.
Myrcia fastigiata var. *coriacea* Kiaersk. = ***Myrcia guianensis*** (Aubl.) DC.
Myrcia ferruginea (Poir.) DC. = ***Marlierea ferruginea*** (Poir.) McVaugh
Myrcia ferruginea var. *domingensis* O.Berg = ***Myrcia deflexa*** (Poir.) DC.
Myrcia ferruginea var. *martinicensis* O.Berg = ***Myrcia deflexa*** (Poir.) DC.
Myrcia floribunda Miq. = ***Myrcia tomentosa*** (Aubl.) DC.
Myrcia formosiana Cambess. = ***Myrcia hilariana*** O.Berg
Myrcia formosiana DC. = ***Myrcia splendens*** (Sw.) DC.
Myrcia friburgensis O.Berg = ***Myrcia splendens*** (Sw.) DC.
Myrcia frontinensis Hieron. = ***Myrcia paivae*** O.Berg
Myrcia frontinensis var. *gamaeana* Glaz. = ***Myrcia paivae*** O.Berg
Myrcia gardneriana O.Berg = ***Myrcia splendens*** (Sw.) DC.
Myrcia garopabensis Cambess. = ***Myrcia palustris*** DC.
Myrcia gemmiflora O.Berg = ***Hexachlamys edulis*** (O.Berg) Kausel & D.Legrand
Myrcia gentlei Lundell = ***Myrcia amazonica*** DC.
Myrcia glaberrima Barb.Rodr. ex Chodat & Hassl. = ***Myrcia multiflora*** (Lam.) DC.
Myrcia glandulosa (O.Berg) Kiaersk. = ***Myrcia guianensis*** (Aubl.) DC.
Myrcia glandulosa var. *obovata* (O.Berg) N.Silveira = ***Myrcia guianensis*** (Aubl.) DC.
Myrcia glaucescens (O.Berg) Kiaersk. = ***Myrcia multiflora*** (Lam.) DC.
Myrcia glazioviana Kiaersk. = ***Myrciaria glazioviana*** (Kiaersk.) G.M.Barroso ex Sobral
Myrcia glazioviana var. *villosa* Kiaersk. = ***Myrciaria glazioviana*** (Kiaersk.) G.M.Barroso ex Sobral
Myrcia goyazensis var. *angustifolia* O.Berg = ***Myrcia goyazensis*** Cambess.
Myrcia goyazensis var. *latifolia* O.Berg = ***Myrcia goyazensis*** Cambess.
Myrcia gracilis O.Berg = ***Myrcia splendens*** (Sw.) DC.
Myrcia gracilis var. *opaca* O.Berg = ***Myrcia splendens*** (Sw.) DC.
Myrcia gracilis var. *prasina* O.Berg = ***Myrcia splendens*** (Sw.) DC.
Myrcia gracilis var. *sessiliflora* O.Berg = ***Myrcia splendens*** (Sw.) DC.
Myrcia grandiflora (O.Berg) Nied. = ***Myrcia pubipetala*** Miq.
Myrcia granulata R.O.Williams = ***Plinia rivularis*** (Cambess.) Rotman
Myrcia guajavifolia O.Berg = ***Myrcia splendens*** (Sw.) DC.

Myrcia guajavifolia var. *bullata* O.Berg = ***Myrcia splendens*** (Sw.) DC.
Myrcia guajavifolia f. *grandifolia* Kiaersk. = ***Myrcia splendens*** (Sw.) DC.
Myrcia guajavifolia var. *impunctata* O.Berg = ***Myrcia splendens*** (Sw.) DC.
Myrcia guajavifolia var. *perforata* O.Berg = ***Myrcia splendens*** (Sw.) DC.
Myrcia guatemalensis Lundell = ***Pimenta guatemalensis*** (Lundell) Lundell
Myrcia guianensis var. *cuneata* (O.Berg) McVaugh = ***Myrcia guianensis*** (Aubl.) DC.
Myrcia hartwegiana D.Legrand = ***Myrcia palustris*** DC.
Myrcia hassleriana Barb.Rodr. = ***Myrcia selloi*** (Spreng.) N.Silveira
Myrcia hayneana DC. = ***Myrcia splendens*** (Sw.) DC.
Myrcia hayneana var. *paraensis* O.Berg = ***Myrcia splendens*** (Sw.) DC.
Myrcia hepatica (O.Berg) Kiaersk. = ***Myrcia guianensis*** (Aubl.) DC.
Myrcia hirtellifolia Gleason = ***Myrcia bracteata*** (Rich.) DC.
Myrcia hirtiflora DC. = ***Myrcia tomentosa*** (Aubl.) DC.
Myrcia hispida var. *angustifolia* O.Berg = ***Myrcia hispida*** O.Berg
Myrcia hispida var. *latifolia* O.Berg = ***Myrcia hispida*** O.Berg
Myrcia hispida var. *panicularis* O.Berg = ***Myrcia hispida*** O.Berg
Myrcia hookeriana (O.Berg) Kiaersk. = ***Myrcia hebepetala*** DC.
Myrcia hostmanniana (O.Berg) Kiaersk. = ***Myrcia amazonica*** DC.
Myrcia hostmanniana var. *robustior* Kiaersk. = ***Myrcia amazonica*** DC.
Myrcia huanocensis O.Berg = ***Myrcia mollis*** (Kunth) DC.
Myrcia huanocensis var. *glazioviana* Kiaersk. = ***Myrcia mollis*** (Kunth) DC.
Myrcia humboldtiana DC. = ***Myrcia splendens*** (Sw.) DC.
Myrcia humboldtiana var. *caribaea* O.Berg = ***Myrcia splendens*** (Sw.) DC.
Myrcia humboldtiana var. *orinocensis* O.Berg = ***Myrcia splendens*** (Sw.) DC.
Myrcia imbricata Gardner = ***Myrcia racemulosa*** DC.
Myrcia impressa O.Berg = ***Myrcia splendens*** (Sw.) DC.
Myrcia incisa D.Legrand = ***Myrcia guianensis*** (Aubl.) DC.
Myrcia intermedia (O.Berg) Kiaersk. = ***Myrcia guianensis*** (Aubl.) DC.
Myrcia involucrata O.Berg = ***Myrcia felisbertii*** (DC.) O.Berg
Myrcia itajuruensis Cambess. = ***Myrcia hebepetala*** DC.
Myrcia itambensis O.Berg = ***Myrcia retorta*** Cambess.
Myrcia jaboticaba (Vell.) Baill. = ***Plinia cauliflora*** (Mart.) Kausel
Myrcia jaguariaivensis Mattos & D.Legrand = ***Myrcia pulchra*** (O.Berg) Kiaersk.
Myrcia karsteniana (O.Berg) Steyerm. = ***Myrcianthes karsteniana*** (O.Berg) McVaugh
Myrcia kauseliana D.Legrand = ***Myrcia oligantha*** O.Berg
Myrcia kegeliana O.Berg = ***Myrcia splendens*** (Sw.) DC.
Myrcia kegeliana var. *angustifolia* O.Berg = ***Myrcia splendens*** (Sw.) DC.
Myrcia kegeliana var. *latifolia* O.Berg = ***Myrcia splendens*** (Sw.) DC.
Myrcia kegeliana var. *longifolia* O.Berg = ***Myrcia splendens*** (Sw.) DC.
Myrcia kegeliana var. *pendula* O.Berg = ***Myrcia splendens*** (Sw.) DC.
Myrcia kegeliana var. *vulgaris* O.Berg = ***Myrcia splendens*** (Sw.) DC.
Myrcia klotzschiana O.Berg = ***Myrcia splendens*** (Sw.) DC.
Myrcia klotzschiana var. *impellucida* O.Berg = ***Myrcia splendens*** (Sw.) DC.
Myrcia kunthiana Steud. = ***Myrcia splendens*** (Sw.) DC.
Myrcia kunthiana (O.Berg) Kiaersk. = ?
Myrcia kunthiana var. *latifolia* Kiaersk. = ?
Myrcia kunthiana var. *latissima* Kiaersk. = ?
Myrcia kunthiana var. *microphylla* Kiaersk. = ?
Myrcia laevigata O.Berg = ***Myrcia splendens*** (Sw.) DC.
Myrcia laevigata var. *brunnea* O.Berg = ***Myrcia splendens*** (Sw.) DC.
Myrcia laevigata var. *canescens* O.Berg = ***Myrcia splendens*** (Sw.) DC.
Myrcia laevis O.Berg = ***Myrcia lucida*** McVaugh
Myrcia lamprosericea Diels = ***Myrcia splendens*** (Sw.) DC.
Myrcia lancea (O.Berg) Mattos = ***Myrcia tomentosa*** (Aubl.) DC.
Myrcia lanceolata Cambess. = ***Myrcia bracteata*** (Rich.) DC.
Myrcia lanceolata var. *angustifolia* Cambess. = ***Myrcia bracteata*** (Rich.) DC.
Myrcia lanceolata var. *avenia* O.Berg = ***Myrcia bracteata*** (Rich.) DC.
Myrcia lanceolata var. *grandifolia* O.Berg = ***Myrcia bracteata*** (Rich.) DC.
Myrcia lanceolata var. *latifolia* O.Berg = ***Myrcia bracteata*** (Rich.) DC.
Myrcia lanceolata var. *racemosa* O.Berg = ***Myrcia bracteata*** (Rich.) DC.
Myrcia langsdorffii O.Berg = ***Myrcia splendens*** (Sw.) DC.
Myrcia lanuginosa (O.Berg) Nied. = ***Myrcia tomentosa*** (Aubl.) DC.
Myrcia laruotteana var. *australis* D.Legrand = ***Myrcia laruotteana*** var. ***laruotteana***
Myrcia laruotteana var. *glabriuscula* Cambess. = ***Myrcia laruotteana*** var. ***laruotteana***
Myrcia laruotteana var. *impunctata* Cambess. = ***Myrcia laruotteana*** var. ***laruotteana***
Myrcia laruotteana var. *macrophylla* Kiaersk. = ***Myrcia laruotteana*** var. ***laruotteana***
Myrcia lasiantha var. *multiflora* O.Berg = ***Myrcia lasiantha*** DC.
Myrcia lasiantha var. *pauciflora* O.Berg = ***Myrcia lasiantha*** DC.
Myrcia lasiopus DC. = ***Myrcia decorticans*** DC.
Myrcia latifolia O.Berg = ***Myrcia splendens*** (Sw.) DC.
Myrcia laureola (O.Berg) Kiaersk. = ***Myrcia venulosa*** var. ***venulosa***
Myrcia lauriflora DC. = ***Myrcia guianensis*** (Aubl.) DC.
Myrcia lechleriana Miq. = ***Amomyrtus luma*** (Molina) D.Legrand & Kausel
Myrcia lenheirensis Kiaersk. = ***Marlierea angustifolia*** (O.Berg) Mattos
Myrcia leptoclada DC. = ***Myrcia amazonica*** DC.
Myrcia leptoclada var. *glazioviana* Kiaersk. = ***Myrcia amazonica*** DC.
Myrcia leucantha (O.Berg) N.Silveira = ***Myrcia tomentosa*** (Aubl.) DC.
Myrcia leucophloea DC. = ***Myrcia decorticans*** DC.
Myrcia lindeniana (O.Berg) C.Wright = ***Myrcia fenzliana*** O.Berg

Myrcia lindeniana O.Berg = ***Myrcia splendens*** (Sw.) DC.
Myrcia lingua (O.Berg) Mattos = ***Myrcia guianensis*** (Aubl.) DC.
Myrcia lingua var. *rufa* (O.Berg) Mattos = ***Myrcia guianensis*** (Aubl.) DC.
Myrcia linguiformis (O.Berg) N.Silveira = ***Myrcia vestita*** DC.
Myrcia linkiana DC. = ***Myrcia cordiifolia*** DC.
Myrcia longicaudata Lundell = ***Myrcia splendens*** (Sw.) DC.
Myrcia longifolia (Kunth) DC. = (Melastomataceae)
Myrcia longipes (O.Berg) Kiaersk. = ***Myrcia tomentosa*** (Aubl.) DC.
Myrcia longipes f. *obovata* Kiaersk. = ***Myrcia tomentosa*** (Aubl.) DC.
Myrcia luetzelburgii Burret ex Luetzelb. = ***Myrcia splendens*** (Sw.) DC.
Myrcia macrocarpa Barb.Rodr. ex Chodat & Hassl. = ?
Myrcia macrochlamys DC. = ***Algrizea macrochlamys*** (DC.) Proença & NicLugh.
Myrcia macrophylla DC. = ***Myrcia splendens*** (Sw.) DC.
Myrcia magnoliifolia DC. = ***Myrcia splendens*** (Sw.) DC.
Myrcia magnoliifolia var. *angustifolia* O.Berg = ***Myrcia splendens*** (Sw.) DC.
Myrcia magnoliifolia var. *latifolia* O.Berg = ***Myrcia splendens*** (Sw.) DC.
Myrcia magnoliifolia var. *parvifolia* O.Berg = ***Myrcia splendens*** (Sw.) DC.
Myrcia marahanensis Kiaersk. = ***Myrcia tomentosa*** (Aubl.) DC.
Myrcia martiana O.Berg = ***Myrcia splendens*** (Sw.) DC.
Myrcia martinicensis Krug & Urb. = ***Myrcia splendens*** (Sw.) DC.
Myrcia melanoclada O.Berg = ***Myrcia splendens*** (Sw.) DC.
Myrcia melanosepala Kiaersk. = ***Myrcia venulosa*** var. ***venulosa***
Myrcia melanosticta Kiaersk. = ***Myrcia splendens*** (Sw.) DC.
Myrcia membranacea Kiaersk. = ***Myrcia tomentosa*** (Aubl.) DC.
Myrcia micrantha O.Berg = ***Myrcia splendens*** (Sw.) DC.
Myrcia microcarpa Cambess. = ***Myrcia guianensis*** (Aubl.) DC.
Myrcia microsiphonata D.Legrand = ***Myrcia selloi*** (Spreng.) N.Silveira
Myrcia mikaniana (DC.) O.Berg = ***Myrcia splendens*** (Sw.) DC.
Myrcia mikaniana var. *angustifolia* O.Berg = ***Myrcia splendens*** (Sw.) DC.
Myrcia mikaniana var. *latifolia* O.Berg = ***Myrcia splendens*** (Sw.) DC.
Myrcia minensis Kiaersk. = ***Myrcia pubescens*** DC.
Myrcia minensis var. *subcordata* Kiaersk. = ***Myrcia pubescens*** DC.
Myrcia mini (Aubl.) Sweet = ***Eugenia biflora*** (L.) DC.
Myrcia mugiensis Cambess. = ***Blepharocalyx salicifolius*** (Kunth) O.Berg
Myrcia multiflora f. *glaucescens* (O.Berg) D.Legrand = ***Myrcia multiflora*** (Lam.) DC.
Myrcia multiflora var. *glaucescens* (O.Berg) D.Legrand = ***Myrcia multiflora*** (Lam.) DC.
Myrcia multiflora var. *microsiphonata* (D.Legrand) D.Legrand = ***Myrcia multiflora*** (Lam.) DC.
Myrcia multiflora f. *ovalifolia* (O.Berg) D.Legrand = ***Myrcia multiflora*** (Lam.) DC.
Myrcia multiflora var. *ramulosa* D.Legrand = ***Myrcia multiflora*** (Lam.) DC.
Myrcia negrensis O.Berg = ***Myrcia splendens*** (Sw.) DC.
Myrcia nigrescens DC. = ***Marlierea umbraticola*** (Kunth) O.Berg
Myrcia nitens O.Berg = ***Myrcia splendens*** (Sw.) DC.
Myrcia nitida (Vell.) Kiaersk. = ***Myrcia subsericea*** A.Gray
Myrcia nivea var. *rosmarinifolia* (O.Berg) Mattos = ***Myrcia nivea*** Cambess.
Myrcia obcordata Mattos = ***Myrcia guianensis*** (Aubl.) DC.
Myrcia obscura (O.Berg) N.Silveira = ***Myrcia guianensis*** (Aubl.) DC.
Myrcia obscura var. *longipes* (O.Berg) N.Silveira = ***Myrcia guianensis*** (Aubl.) DC.
Myrcia obtecta (O.Berg) Kiaersk. = ***Myrcia guianensis*** (Aubl.) DC.
Myrcia obtecta var. *alternifolia* (Miq.) D.Legrand = ***Myrcia guianensis*** (Aubl.) DC.
Myrcia obtusa Schauer = ***Myrcia guianensis*** (Aubl.) DC.
Myrcia ochracea O.Berg = ***Myrcia laruotteana*** var. ***laruotteana***
Myrcia oerstediana O.Berg = ***Myrcia splendens*** (Sw.) DC.
Myrcia oitchi O.Berg = ***Myrcia amazonica*** DC.
Myrcia oleifolia Cambess. = ***Myrcia venulosa*** var. ***venulosa***
Myrcia oocarpa Cambess. = ***Myrcia splendens*** (Sw.) DC.
Myrcia opaca O.Berg = ***Myrcia splendens*** (Sw.) DC.
Myrcia opaca var. *angustifolia* O.Berg = ***Myrcia splendens*** (Sw.) DC.
Myrcia opaca var. *latifolia* O.Berg = ***Myrcia splendens*** (Sw.) DC.
Myrcia ovalifolia (O.Berg) Kiaersk. = ***Myrcia multiflora*** (Lam.) DC.
Myrcia ovata var. *subcordata* O.Berg = ***Myrcia ovata*** Cambess.
Myrcia oxyoentophylla Kiaersk. = ***Myrcia splendens*** (Sw.) DC.
Myrcia pachyclada (O.Berg) N.Silveira = ***Myrcia vestita*** DC.
Myrcia pachyclada var. *prolifera* (O.Berg) N.Silveira = ***Myrcia vestita*** DC.
Myrcia pachyclada var. *spathulata* (O.Berg) N.Silveira = ***Myrcia vestita*** DC.
Myrcia paivae var. *gracilis* Lingelsh. = ***Myrcia paivae*** O.Berg
Myrcia pallens DC. = ***Myrcia guianensis*** (Aubl.) DC.
Myrcia pallens var. *ovalis* (O.Berg) Kiaersk. = ***Myrcia guianensis*** (Aubl.) DC.
Myrcia pallens var. *subcordata* (O.Berg) Kiaersk. = ***Myrcia guianensis*** (Aubl.) DC.
Myrcia pallida (O.Berg) N.Silveira = ***Myrcia guianensis*** (Aubl.) DC.
Myrcia palustris var. *acutata* O.Berg = ***Myrcia palustris*** DC.
Myrcia palustris var. *angustifolia* O.Berg = ***Myrcia palustris*** DC.
Myrcia palustris var. *bracteata* O.Berg = ***Myrcia palustris*** DC.
Myrcia palustris var. *stictophylla* O.Berg = ***Myrcia palustris*** DC.
Myrcia paniculata (Jacq.) Krug & Urb. = ***Myrcia citrifolia*** (Aubl.) Urb.
Myrcia paniculata var. *imrayana* (Griseb.) Duss = ***Myrcia citrifolia*** (Aubl.) Urb.
Myrcia paracatuensis var. *arenaria* (O.Berg) Kiaersk. = ***Myrcia paracatuensis*** Kiaersk.
Myrcia paracatuensis var. *linearis* Mattos = ***Myrcia paracatuensis*** Kiaersk.

Myrcia paracatuensis var. *pumila* Glaz. = ***Myrcia paracatuensis*** Kiaersk.
Myrcia paracatuensis var. *venosa* (O.Berg) Mattos = ***Myrcia paracatuensis*** Kiaersk.
Myrcia paraensis (O.Berg) Kiaersk. = ***Myrcia amazonica*** DC.
Myrcia parnahibensis var. *acutifolia* (O.Berg) Kiaersk. = ***Myrcia parnahibensis*** (O.Berg) Kiaersk.
Myrcia parnahibensis var. *acutifolia* (O.Berg) Mattos = ***Myrcia parnahibensis*** (O.Berg) Kiaersk.
Myrcia parnahibensis var. *imbricata* (O.Berg) Kiaersk. = ***Myrcia parnahibensis*** (O.Berg) Kiaersk.
Myrcia pauciflora Cambess. = ***Myrcia eriocalyx*** DC.
Myrcia pauciflora (O.Berg) Nied. = ?
Myrcia pauciflora var. *brunnea* O.Berg = ***Myrcia eriocalyx*** DC.
Myrcia pauciflora var. *rubiginosa* O.Berg = ***Myrcia eriocalyx*** DC.
Myrcia pellucida O.Berg = ***Myrcia splendens*** (Sw.) DC.
Myrcia perorebimi Barb.Rodr. ex Chodat & Hassl. = ***Myrcia laruotteana*** var. ***laruotteana***
Myrcia phaeoclada var. *alagoensis* O.Berg = ***Myrcia phaeoclada*** O.Berg
Myrcia phaeoclada var. *guyanensis* O.Berg = ***Myrcia splendens*** (Sw.) DC.
Myrcia piauhiensis O.Berg = ***Myrcia tomentosa*** (Aubl.) DC.
Myrcia pilodes Kiaersk. = ***Myrcia mutabilis*** (O.Berg) N.Silveira
Myrcia pilosa DC. = ***Myrcia tomentosa*** (Aubl.) DC.
Myrcia pilotantha Kiaersk. = ***Myrcia pulchra*** (O.Berg) Kiaersk.
Myrcia pimentoides DC. = ***Pimenta racemosa*** (Mill.) J.W.Moore var. ***racemosa***
Myrcia pinaster Mart. ex O.Berg = ***Blepharocalyx myriophyllus*** Morais & Sobral
Myrcia planifolia McVaugh = ***Myrcia revolutifolia*** McVaugh
Myrcia planipes (Hook. & Arn.) Kiaersk. = ***Myrceugenia planipes*** (Hook. & Arn.) O.Berg
Myrcia plicatocostata O.Berg = ***Myrcia splendens*** (Sw.) DC.
Myrcia pohliana O.Berg = ***Myrcia splendens*** (Sw.) DC.
Myrcia polyantha var. *brasiliensis* O.Berg = ***Myrcia polyantha*** DC.
Myrcia polyantha var. *orinocensis* O.Berg = ***Myrcia polyantha*** DC.
Myrcia prunifolia DC. = ***Myrcia tomentosa*** (Aubl.) DC.
Myrcia prunifolia var. *angustior* DC. = ***Myrcia tomentosa*** (Aubl.) DC.
Myrcia prunifolia var. *obovata* DC. = ***Myrcia tomentosa*** (Aubl.) DC.
Myrcia prunifolia var. *ovata* DC. = ***Myrcia tomentosa*** (Aubl.) DC.
Myrcia pseudomini DC. = ***Myrcia splendens*** (Sw.) DC.
Myrcia puberula Cambess. = ***Myrcia tomentosa*** (Aubl.) DC.
Myrcia pubescens G.Don = ?
Myrcia pubiflora var. *glazioviana* Kiaersk. = ***Myrcia pubiflora*** DC.
Myrcia pubipetala var. *magnifolia* D.Legrand = ***Myrcia pubipetala*** Miq.
Myrcia punctata (Vahl) DC. = ***Myrcianthes fragrans*** (Sw.) McVaugh
Myrcia pungens Urb. = ***Myrcia retivenia*** (C.Wright) Urb.
Myrcia pycnantha (Benth.) Steyerm. = ***Eugenia pycnantha*** Benth.
Myrcia pyramidata O.Berg = ***Myrcia rubiginosa*** Cambess.
Myrcia pyrrhopilodes Kiaersk. = ***Myrcia venulosa*** var. ***venulosa***
Myrcia rabeniana Kiaersk. = ***Myrcia venulosa*** var. ***venulosa***
Myrcia ramulosa DC. = ***Myrcia selloi*** (Spreng.) N.Silveira
Myrcia ramulosa var. *acutata* (O.Berg) Mattos = ***Myrcia selloi*** (Spreng.) N.Silveira
Myrcia ramulosa var. *leptophylla* Kiaersk. = ***Myrcia selloi*** (Spreng.) N.Silveira
Myrcia ramulosa var. *megapotamica* D.Legrand = ***Myrcia selloi*** (Spreng.) N.Silveira
Myrcia ramulosa var. *microsiphonata* (D.Legrand) D.Legrand = ***Myrcia selloi*** (Spreng.) N.Silveira
Myrcia ramulosa var. *multiflora* DC. = ***Myrcia selloi*** (Spreng.) N.Silveira
Myrcia ramulosa var. *pauciflora* Kiaersk. = ***Myrcia selloi*** (Spreng.) N.Silveira
Myrcia recurvata var. *grandifolia* O.Berg = ***Myrcia recurvata*** O.Berg
Myrcia recurvata var. *parvifolia* O.Berg = ***Myrcia recurvata*** O.Berg
Myrcia regnelliana var. *leptocalama* O.Berg = ***Myrcia regnelliana*** O.Berg
Myrcia regnelliana var. *robustior* O.Berg = ***Myrcia regnelliana*** O.Berg
Myrcia renatoana Mattos = ***Myrcia guianensis*** (Aubl.) DC.
Myrcia reticulata O.Berg = ***Myrcia splendens*** (Sw.) DC.
Myrcia reticulata Krug & Urb. = [81 DOM]
Myrcia retusa (O.Berg) Nied. = ***Myrcia guianensis*** (Aubl.) DC.
Myrcia rhabdoides Kiaersk. = ***Myrcia guianensis*** (Aubl.) DC.
Myrcia rhodeosepala Kiaersk. = ***Myrcia tomentosa*** (Aubl.) DC.
Myrcia richardiana (O.Berg) Kiaersk. = ***Myrcia aethusa*** (O.Berg) N.Silveira
Myrcia riedeliana O.Berg = ***Myrcia splendens*** (Sw.) DC.
Myrcia riparia O.Berg = ***Myrcia splendens*** (Sw.) DC.
Myrcia roraimae Oliv. = ***Myrcia guianensis*** (Aubl.) DC.
Myrcia rorida (O.Berg) Kiaersk. = ***Myrcia guianensis*** (Aubl.) DC.
Myrcia rorida var. *microphylla* Glaz. ex Mattos = ***Myrcia guianensis*** (Aubl.) DC.
Myrcia rostrata DC. = ***Myrcia splendens*** (Sw.) DC.
Myrcia rostrata var. *brunea* Cambess. = ***Myrcia splendens*** (Sw.) DC.
Myrcia rostrata f. *communis* (O.Berg) D.Legrand = ***Myrcia splendens*** (Sw.) DC.
Myrcia rostrata f. *flexuosa* Soares-Silva = ***Myrcia splendens*** (Sw.) DC.
Myrcia rostrata f. *gracilis* (O.Berg) D.Legrand = ***Myrcia splendens*** (Sw.) DC.
Myrcia rostrata f. *pseudomini* (DC.) D.Legrand = ***Myrcia splendens*** (Sw.) DC.
Myrcia rostrata f. *sericiflora* (O.Berg) D.Legrand = ***Myrcia splendens*** (Sw.) DC.
Myrcia rosulans (O.Berg) Kiaersk. = ***Myrcia tomentosa*** (Aubl.) DC.
Myrcia rotundata var. *atrans* McVaugh = ***Myrcia rotundata*** (Amshoff) McVaugh
Myrcia rubella var. *puberula* Cambess. = ***Myrcia rubella*** Cambess.
Myrcia rufa (O.Berg) N.Silveira = ***Myrcia pulchra*** (O.Berg) Kiaersk.
Myrcia rufidula Schltdl. = ***Myrcia splendens*** (Sw.) DC.
Myrcia rufipes var. *dives* (O.Berg) N.Silveira = ***Myrcia rufipes*** DC.

Myrcia rufipes var. *grandiflora* Kiaersk. = ***Myrcia rufipes*** DC.
Myrcia rufipes var. *pilantha* (O.Berg) Kiaersk. = ***Myrcia rufipes*** DC.
Myrcia rufipila var. *piresiana* Mattos = ***Myrcia rufipila*** McVaugh
Myrcia rufula Miq. = ***Myrcia splendens*** (Sw.) DC.
Myrcia rufula var. *martiana* (O.Berg) Kiaersk. = ***Myrcia splendens*** (Sw.) DC.
Myrcia rugosa (O.Berg) Kiaersk. = ***Myrcia venulosa*** var. ***venulosa***
Myrcia sartoriana O.Berg = ***Myrcia splendens*** (Sw.) DC.
Myrcia saxicola O.Berg = ***Myrcia splendens*** (Sw.) DC.
Myrcia saxicola var. *grandifolia* O.Berg = ***Myrcia splendens*** (Sw.) DC.
Myrcia schenckiana Kiaersk. = ***Myrcia pubescens*** DC.
Myrcia schippii Lundell = ***Myrcia splendens*** (Sw.) DC.
Myrcia schomburgkiana O.Berg = ***Eugenia biflora*** (L.) DC.
Myrcia schottiana var. *sericea* O.Berg = ***Myrcia schottiana*** O.Berg
Myrcia schrankiana DC. = ***Myrcia guianensis*** (Aubl.) DC.
Myrcia schuechiana O.Berg = ***Myrcia splendens*** (Sw.) DC.
Myrcia scutulifera DC. = ***Myrcia racemulosa*** DC.
Myrcia seleriana Donn.Sm. = ***Myrcianthes fragrans*** (Sw.) McVaugh
Myrcia selloi var. *acutata* (O.Berg) N.Silveira = ***Myrcia selloi*** (Spreng.) N.Silveira
Myrcia selloi var. *australis* (O.Berg) N.Silveira = ***Myrcia selloi*** (Spreng.) N.Silveira
Myrcia selloi var. *megapotamica* (Legrand) N.Silveira = ***Myrcia selloi*** (Spreng.) N.Silveira
Myrcia selloi var. *microsiphonata* (D.Legrand) N.Silveira = ***Myrcia selloi*** (Spreng.) N.Silveira
Myrcia sellowiana (O.Berg) Arechav. = ***Myrcia hartwegiana*** (O.Berg) Kiaersk.
Myrcia sellowiana O.Berg = ***Myrcia splendens*** (Sw.) DC.
Myrcia sellowiana var. *bullata* O.Berg = ***Myrcia splendens*** (Sw.) DC.
Myrcia sellowiana var. *costata* O.Berg = ***Myrcia splendens*** (Sw.) DC.
Myrcia sepiaria DC. = ***Myrcia splendens*** (Sw.) DC.
Myrcia sericea O.Berg = ***Myrcia splendens*** (Sw.) DC.
Myrcia sericiflora O.Berg = ***Myrcia splendens*** (Sw.) DC.
Myrcia sessilifolia Kiaersk. = ***Myrcia eriocalyx*** DC.
Myrcia shirleyana Mattos = ***Myrcia venulosa*** var. ***dives*** (O.Berg) Soares-Silva
Myrcia silvatica Barb.Rodr. ex Chodat & Hassl. = ***Plinia rivularis*** (Cambess.) Rotman
Myrcia sintenisii Kiaersk. = ***Myrcia fenzliana*** O.Berg
Myrcia smithii D.Legrand & Kausel = ***Myrcia selloi*** (Spreng.) N.Silveira
Myrcia sonderiana (O.Berg) Mattos = ***Myrcia venulosa*** var. ***venulosa***
Myrcia sororia DC. = ***Myrcia splendens*** (Sw.) DC.
Myrcia sparsifolia Barb.Rodr. = ***Hexachlamys edulis*** (O.Berg) Kausel & D.Legrand
Myrcia spectabilis var. *farinosa* (O.Berg) Kiaersk. = ***Myrcia spectabilis*** DC.
Myrcia sphaerocarpa DC. = ***Myrcia multiflora*** (Lam.) DC.
Myrcia spixiana DC. = ***Myrcia guianensis*** (Aubl.) DC.
Myrcia splendens var. *chrysocoma* McVaugh = ***Myrcia splendens*** (Sw.) DC.
Myrcia splendens var. *guantanamana* Borhidi & O.Muñiz = ***Myrcia splendens*** (Sw.) DC.
Myrcia splendens var. *micropora* O.Berg = ***Myrcia splendens*** (Sw.) DC.
Myrcia splendens var. *mini* (Aubl.) DC. = ***Eugenia biflora*** (L.) DC.
Myrcia splendens var. *obscura* O.Berg = ***Myrcia splendens*** (Sw.) DC.
Myrcia splendens var. *robustior* Kuntze = ***Myrcia splendens*** (Sw.) DC.
Myrcia springiana var. *membranacea* (O.Berg) Kiaersk. = ***Myrcia springiana*** (O.Berg) Kiaersk.
Myrcia spruceana O.Berg = ***Myrcia splendens*** (Sw.) DC.
Myrcia stemmeriana D.Legrand = ***Myrcia guianensis*** (Aubl.) DC.
Myrcia strigipes Mart. = ***Marlierea tomentosa*** Cambess.
Myrcia suaveolens Cambess. = ***Myrcia guianensis*** (Aubl.) DC.
Myrcia subsessilis var. *ovalis* O.Berg = ***Myrcia subsessilis*** O.Berg
Myrcia subsessilis var. *subcordata* O.Berg = ***Myrcia subsessilis*** O.Berg
Myrcia sulcata O.Berg = ***Myrcia deflexa*** (Poir.) DC.
Myrcia superba O.Berg = ***Myrcia splendens*** (Sw.) DC.
Myrcia surinamensis Miq. = ***Myrcia guianensis*** (Aubl.) DC.
Myrcia taubatensis Kiaersk. = ***Myrcia guianensis*** (Aubl.) DC.
Myrcia taubatensis var. *ovalis* (O.Berg) Kiaersk. = ***Myrcia guianensis*** (Aubl.) DC.
Myrcia terebinthacea Poepp. ex O.Berg = ***Myrcia guianensis*** (Aubl.) DC.
Myrcia tetramera (Amshoff) Lemée = ***Myrcia decorticans*** DC.
Myrcia ticuensis Kiaersk. = ***Myrcia grandifolia*** Cambess.
Myrcia tingens O.Berg = ***Myrcia splendens*** (Sw.) DC.
Myrcia tobagensis (Krug & Urb.) Urb. = ***Myrcia decorticans*** DC.
Myrcia torta f. *glauca* Kiaersk. = ***Myrcia torta*** DC.
Myrcia torta f. *grandifolia* Kiaersk. = ***Myrcia torta*** DC.
Myrcia torta f. *parvifolia* Kiaersk. = ***Myrcia torta*** DC.
Myrcia triantha DC. = ?
Myrcia triflora (O.Berg) Cambess. = ***Myrcia graciliflora*** Sagot
Myrcia uberavensis var. *ovata* O.Berg = ***Myrcia uberavensis*** O.Berg
Myrcia uberavensis var. *rotundifolia* O.Berg = ***Myrcia uberavensis*** O.Berg
Myrcia umbellulifera (Kunth) DC. = ***Pseudanamomis umbellulifera*** (Kunth) Kausel
Myrcia variabilis var. *intermedia* Mart. ex DC. = ***Myrcia variabilis*** Mart. ex DC.
Myrcia variabilis var. *nummularia* Mart. ex DC. = ***Myrcia variabilis*** Mart. ex DC.
Myrcia variabilis var. *ovatilofia* Mart. ex DC. = ***Myrcia variabilis*** Mart. ex DC.
Myrcia velutina O.Berg = ***Myrcia splendens*** (Sw.) DC.
Myrcia velutina var. *canescens* O.Berg = ***Myrcia splendens*** (Sw.) DC.
Myrcia velutina var. *ochracea* O.Berg = ***Myrcia splendens*** (Sw.) DC.
Myrcia venezuelensis O.Berg = ***Myrcia splendens*** (Sw.) DC.
Myrcia venulosa var. *capoeirensis* DC. = ***Myrcia venulosa*** var. ***venulosa***
Myrcia venulosa var. *rufa* (O.Berg) N.Silveira = ***Myrcia venulosa*** var. ***venulosa***
Myrcia vernicosa DC. = ***Myrcia citrifolia*** (Aubl.) Urb.
Myrcia verticillaris var. *glomerata* O.Berg = ***Myrcia verticillaris*** O.Berg

Myrcia verticillaris var. *laxa* O.Berg = ***Myrcia verticillaris*** O.Berg
Myrcia verticillaris var. *multicaulis* O.Berg = ***Myrcia verticillaris*** O.Berg
Myrcia verticillaris var. *paniculata* O.Berg = ***Myrcia verticillaris*** O.Berg
Myrcia verticillaris var. *pygmaea* O.Berg = ***Myrcia verticillaris*** O.Berg
Myrcia vestita var. *grandifolia* (O.Berg) Kiaersk. = ***Myrcia vestita*** DC.
Myrcia vestita var. *obtusifolia* DC. = ***Myrcia vestita*** DC.
Myrcia vestita var. *parvifolia* (O.Berg) Kiaersk. = ***Myrcia vestita*** DC.
Myrcia vittoriana var. *piratiningensis* Kiaersk. = ***Myrcia vittoriana*** Kiaersk.
Myrcia widgreniana (O.Berg) Mattos = ***Myrcia pulchra*** (O.Berg) Kiaersk.
Myrcia ypanemensis O.Berg = ***Myrcia splendens*** (Sw.) DC.
Myrcia yungasensis Rusby = ***Myrcia guianensis*** (Aubl.) DC.

***Unplaced Names*:**
Myrcia acuminatissima Hieron., Bot. Jahrb. Syst. 20(49): 64 (1895), nom. illeg. = [84 BZL]
Myrcia pubescens G.Don, Gen. Hist. 2: 845 (1832). = ?
Myrcia reticulata Krug & Urb., Bot. Jahrb. Syst. 15: 306 (1892), nom. illeg. = [81 DOM]
Myrcia triantha DC., Prodr. 3: 256 (1828). = ?

Myrcialeucus

Myrcialeucus Rojas = ***Eugenia*** P.Micheli ex L.
Myrcialeucus odorifolius Rojas = [85 AGE]

***Unplaced Names*:**
Myrcialeucus odorifolius Rojas, Bull. Acad. Int. Géogr. Bot. 24: 217 (1914). = [85 AGE]

Myrcianthes

Myrcianthes O.Berg, Linnaea 27: 315 (1856).
S. Florida to Trop. America. 78 FLA 79 MXC MXE MXG MXS MXT 80 BLZ COS ELS GUA HON NIC PAN 81 BAH CAY CUB DOM HAI JAM LEE PUE TCS WIN 82 GUY SUR VEN 83 BOL CLM ECU PER 84 BZL BZN BZS 85 AGE AGW CLC PAR URU.
35 Species
Anamomis Griseb., Fl. Brit. W. I.: 240 (1860).
Reichea Kausel, Revista Argent. Agron. 7: 364 (1940).
Aspidogenia Burret, Notizbl. Bot. Gart. Berlin-Dahlem 15: 521 (1941).
Reicheia Kausel, Revista Argent. Agron. 9: 41 (1942), orth. var.
Acreugenia Kausel, Ark. Bot., a.s., 3: 510 (1956).
Pseudomyrcianthes Kausel, Ark. Bot., a.s., 3: 504 (1956).

Myrcianthes borealis McVaugh, Fieldiana, Bot. 29: 481 (1963).
Colombia to Venezuela. 82 VEN 83 CLM. Nanophan. or phan.

Myrcianthes bradeana Mattos & D.Legrand, Bradea 2: 5 (1975).
Brazil (Minas Gerais). 84 BZL. Nanophan.

Myrcianthes callicoma McVaugh, Fieldiana, Bot. 29: 482 (1963).
Bolivia to Argentina (Tucumán). 83 BOL 85 AGW. Phan.

Myrcianthes cavalcantei Mattos, Loefgrenia 115: 1 (2000).
Brazil (Roraima). 84 BZN. Phan.

Myrcianthes cisplatensis (Cambess.) O.Berg, Linnaea 27: 315 (1856).
S. Brazil to N. Argentina. 84 BZS 85 AGE AGW PAR URU. Nanophan. or phan.
**Eugenia cisplatensis* Cambess. in A.F.C.de Saint-Hilaire, Fl. Bras. Merid. 2: 342 (1832). *Luma cisplatenais* (Cambess.) Herter, Revista Sudamer. Bot. 7: 218 (1943).
Eugenia elliptica Hook. & Arn., Bot. Misc. 3: 323 (1833). *Blepharocalyx ellipticus* (Hook. & Arn.) O.Berg, Linnaea 27: 415 (1856). *Myrtus elliptica* Arechav., Anales Mus. Nac. Montevideo 5: 35 (1905), nom. illeg.
Myrcianthes apiculata O.Berg in C.F.P.von Martius & auct. suc. (eds.), Fl. Bras. 14(1): 352 (1857). *Eugenia apiculata* (O.Berg) Nied. in H.G.A.Engler & K.A.E.Prantl, Nat. Pflanzenfam. 3(7): 81 (1893), nom. illeg.
Myrcianthes cisplatensis var. *latifolia* O.Berg in C.F.P.von Martius & auct. suc. (eds.), Fl. Bras. 14(1): 353 (1857).
Eugenia cisplatensis var. *angustifolia* Lillo, Seg. Contr. Arb. Argent.: 38 (1917).

Myrcianthes coquimbensis (Barnéoud) Landrum & Grifo, Brittonia 40: 290 (1988).
NC. Chile (coastal Coquimbo). 85 CLC. Nanophan.
**Myrtus coquimbensis* Barnéoud in C.Gay, Fl. Chil. 2: 382 (1847). *Reichea coquimbensis* (Barnéoud) Kausel, Revista Argent. Agron. 7: 364 (1940). *Aspidogenia coquimbensis* (Barnéoud) Burret, Notizbl. Bot. Gart. Berlin-Dahlem 15: 522 (1941).
Myrtus coquimbensis var. *rotundifolia* O.Berg, Linnaea 29: 255 (1858).

Myrcianthes crebrifolia (Steyerm.) McVaugh, Fieldiana, Bot. 29: 484 (1963).
Colombia to Venezuela. 82 VEN 83 CLM. Nanophan. or phan.
**Eugenia crebrifolia* Steyerm., Fieldiana, Bot. 28: 101 (1957).

Myrcianthes cymosa (O.Berg) Mattos, Loefgrenia 115: 1 (2000).
SE. Brazil. 84 BZL. Phan.
**Calyptromyrcia cymosa* O.Berg in C.F.P.von Martius & auct. suc. (eds.), Fl. Bras. 14(1): 58 (1857).
Calyptromyrcia cymosa var. *major* O.Berg in C.F.P.von Martius & auct. suc. (eds.), Fl. Bras. 14(1): 58 (1857).
Calyptromyrcia cymosa var. *minor* O.Berg in C.F.P.von Martius & auct. suc. (eds.), Fl. Bras. 14(1): 58 (1857).

Myrcianthes discolor (Kunth) McVaugh, Publ. Field Mus. Nat. Hist., Bot. Ser. 13(4): 754 (1958).
S. Ecuador to Bolivia. 83 BOL ECU PER. Nanophan. or phan.
**Myrtus discolor* Kunth in F.W.H.von Humboldt, A.J.A.Bonpland & C.S.Kunth, Nov. Gen. Sp. 6: 134 (1823). *Eugenia discolor* (Kunth) DC., Prodr. 3: 277 (1828). *Amyrsia discolor* (Kunth) Raf., Sylva Tellur.: 106 (1838).
Myrtus bicolor Kunth in F.W.H.von Humboldt, A.J.A.Bonpland & C.S.Kunth, Nov. Gen. Sp. 6: t. 540 (1824).
Myrcianthes gracilipes Kausel, Lilloa 33: 107 (1971 publ. 1972).

Myrcianthes esnardiana (Urb. & Ekman) Alain, Brittonia 20: 159 (1968).
Haiti (Massif de la Hotte). 81 HAI. Nanophan. or phan.
**Eugenia esnardiana* Urb. & Ekman, Ark. Bot. 21A(5): 36 (1927).

Myrcianthes ferreyrae (McVaugh) McVaugh, Publ. Field Mus. Nat. Hist., Bot. Ser. 13(4): 756 (1958).
Peru. 83 PER. Nanophan. or phan.
**Eugenia ferreyrae* McVaugh, Fieldiana, Bot. 29: 209 (1956).

Myrcianthes fimbriata (Kunth) McVaugh, Publ. Field Mus. Nat. Hist., Bot. Ser. 13(4): 757 (1958).
Ecuador to Peru. 83 ECU PER. Nanophan. or phan. Generic placement uncertain.
**Myrtus fimbriata* Kunth in F.W.H.von Humboldt, A.J.A.Bonpland & C.S.Kunth, Nov. Gen. Sp. 6: 137 (1823). *Eugenia fimbriata* (Kunth) DC., Prodr. 3: 278 (1828).

Myrcianthes fragrans (Sw.) McVaugh, Fieldiana, Bot. 29: 485 (1963).
S. Florida to Trop. America. 78 FLA 79 MXC MXE MXG MXS MXT 80 BLZ COS ELS GUA HON NIC PAN 81 BAH CAY CUB DOM HAI JAM LEE PUE TCS WIN 82 VEN 83 CLM ECU PER. Nanophan. or phan.
**Myrtus fragrans* Sw., Prodr.: 79 (1788), nom. cons. *Eugenia fragrans* (Sw.) Willd., Sp. Pl. 2: 964 (1799). *Cumetea fragrans* (Sw.) Raf., Sylva Tellur.: 106 (1838). *Eugenia dichotoma* var. *fragrans* (Sw.) Nutt., N. Amer. Sylv. 1: 103 (1842). *Anamomis fragrans* (Sw.) Griseb., Fl. Brit. W. I.: 240 (1860).
Eugenia punctata Vahl in H.West, Bidr. Beskr. Ste Croix: 289 (1793). *Myrtus punctata* (Vahl) Spreng., Syst. Veg. 2: 482 (1825). *Myrcia punctata* (Vahl) DC., Prodr. 3: 243 (1828).
Eugenia balsamica Jacq., Fragm. Bot.: 40 (1805). *Myrtus balsamica* (Jacq.) Spreng., Syst. Veg. 2: 481 (1825).
Eugenia heteroclita Tussac, Fl. Antill. 1: 98 (1808). *Pseudocaryophyllus heteroclitus* (Tussac) O.Berg, Linnaea 27: 416 (1856). *Myrtus heteroclita* (Tussac) Nied. in H.G.A.Engler & K.A.E.Prantl, Nat. Pflanzenfam. 3(7): 67 (1893).
Myrtus dichotoma Poir. in J.B.A.P.M.de Lamarck, Encycl., Suppl. 4: 53 (1816). *Eugenia dichotoma* (Poir.) DC., Prodr. 3: 278 (1828). *Anamomis punctata* var. *dichotoma* (Poir.) Kiaersk., Bot. Tidsskr. 17: 279 (1890). *Anamomis dichotoma* (Poir.) Sarg., Gard. & Forest 6: 130 (1893).
Myrtus compressa Kunth in F.W.H.von Humboldt, A.J.A.Bonpland & C.S.Kunth, Nov. Gen. Sp. 6: 135 (1823). *Amyrsia compressa* (Kunth) Raf., Sylva Tellur.: 106 (1838). *Myrcianthes compressa* (Kunth) McVaugh, Publ. Field Mus. Nat. Hist., Bot. Ser. 13(4): 754 (1958).
Myrtus limbata Kunth in F.W.H.von Humboldt, A.J.A.Bonpland & C.S.Kunth, Nov. Gen. Sp. 6: 136 (1823). *Eugenia limbata* (Kunth) DC., Prodr. 3: 278 (1828). *Amyrsia limbata* (Kunth) Kausel, Ark. Bot., a.s., 3: 513 (1956). *Myrcianthes limbata* (Kunth) McVaugh, Publ. Field Mus. Nat. Hist., Bot. Ser. 13(4): 762 (1958).
Myrcia balbisiana DC., Prodr. 3: 243 (1828). *Eugenia balbisiana* (DC.) O.Berg, Linnaea 27: 268 (1856).
Eugenia guayabillo A.Rich., Hist. Phys. Cuba, Pl. Vasc.: 590 (1846). *Anamomis guayabillo* (A.Rich.) Griseb., Cat. Pl. Cub.: 90 (1866). *Myrtus guayabillo* (A.Rich.) C.Wright, Anales Acad. Ci. Méd. Habana 5: 410 (1868).
Eugenia aromatica O.Berg, Linnaea 27: 263 (1856).
Eugenia balbisiana var. *dives* O.Berg, Linnaea 27: 269 (1856).
Eugenia balbisiana var. *triflora* O.Berg, Linnaea 27: 269 (1856).
Eugenia dicrana O.Berg, Linnaea 27: 259 (1856). *Anamomis dicrana* (O.Berg) Britton, N. Amer. Trees: 728 (1908). *Myrcianthes dicrana* (O.Berg) K.A.Wilson, J. Arnold Arbor. 41: 276 (1960).
Eugenia fragrans var. *obovata* O.Berg, Linnaea 27: 272 (1856).
Eugenia granulata O.Berg, Linnaea 27: 260 (1856).
Anamomis fragrans var. *cuneata* Griseb., Fl. Brit. W. I.: 240 (1860).
Anamomis punctata Griseb., Fl. Brit. W. I.: 240 (1860).
Eugenia triflora Sessé & Moç., Pl. Nov. Hisp.: 83 (1888), nom. illeg.
Eugenia fragrans var. *brachyrhiza* Krug & Urb., Bot. Jahrb. Syst. 19: 665 (1895).
Eugenia fragrans var. *fajardensis* Krug & Urb., Bot. Jahrb. Syst. 19: 665 (1895). *Eugenia fajardensis* (Krug & Urb.) Urb., Symb. Antill. 9: 109 (1923).
Myrcia seleriana Donn.Sm., Bot. Gaz. 27: 332 (1899).
Anamomis grandis Britton, Bull. Torrey Bot. Club 37: 355 (1910).
Anamomis simpsonii Small, Torreya 17: 222 (1917 publ. 1918). *Eugenia simpsonii* (Small) Sarg., Man. Trees, ed. 2: 775 (1922). *Myrcianthes simpsonii* (Small) K.A.Wilson, J. Arnold Arbor. 41: 276 (1960). *Myrcianthes fragrans* var. *simpsonii* (Small) R.W.Long, Rhodora 72: 23 (1970). *Myrcianthes fragrans* subsp. *simpsonii* (Small) E.Murray, Kalmia 13: 9 (1983).
Anamomis lucayana Britton in N.L.Britton & C.F.Millspaugh, Bahama Fl.: 306 (1920). *Eugenia lucayana* (Britton) Alain, Contr. Ocas. Mus. Hist. Nat. Colegio "De La Salle" 12: 11 (1953).
Myrtus guayabo Larrañaga, Escritos D. A. Larrañaga 3: 85 (1924).
Eugenia dugandii Standl., Trop. Woods 52: 28 (1937). *Myrcianthes dugandii* (Standl.) McVaugh, Fieldiana, Bot. 29: 484 (1963).
Eugenia steyermarkii Standl., Publ. Field Mus. Nat. Hist., Bot. Ser. 22: 360 (1940).
Eugenia lopeziana Ant.Molina, Ceiba 3: 170 (1953).
Eugenia rondonensis Steyerm., Fieldiana, Bot. 28: 1013 (1957).
Eugenia turumiquirensis Steyerm., Fieldiana, Bot. 28: 1015 (1957).
Myrcianthes fragrans var. *hispidula* McVaugh, Fieldiana, Bot. 29: 489 (1963).

Myrcianthes gigantea (D.Legrand) D.Legrand, Darwiniana 9: 300 (1950).
Brazil (S. Minas Gerais) to Argentina (Misiones). 84 BZL BZS 85 AGE URU. Phan.
**Eugenia cisplatensis* var. *gigantea* D.Legrand, Anales Mus. Hist. Nat. Montevideo, IV, 11: 56 (1936).

Myrcianthes hallii (O.Berg) McVaugh, Publ. Field Mus. Nat. Hist., Bot. Ser. 13(4): 759 (1958).
Colombia to Ecuador. 83 CLM ECU. Phan.
**Eugenia hallii* O.Berg, Linnaea 27: 250 (1856). *Amyrsia hallii* (O.Berg) Kausel, Ark. Bot., a.s., 3: 513 (1956).

Myrcianthes indifferens (McVaugh) McVaugh, Publ. Field Mus. Nat. Hist., Bot. Ser. 13(4): 760 (1958).
Peru. 83 PER. Nanophan. or phan.

**Eugenia indifferens* McVaugh, Fieldiana, Bot. 29: 211 (1956).

Myrcianthes karsteniana (O.Berg) McVaugh, Fieldiana, Bot. 29: 290 (1963).
Colombia to Venezuela. 82 VEN 83 CLM. Phan.
**Eugenia karsteniana* O.Berg, Linnaea 27: 277 (1856). *Myrcia karsteniana* (O.Berg) Steyerm., Fieldiana, Bot. 28: 1017 (1957).
Eugenia karsteniana var. *albicans* O.Berg, Linnaea 27: 277 (1856).
Eugenia karsteniana var. *virens* O.Berg, Linnaea 27: 278 (1856).

Myrcianthes lanosa McVaugh, Publ. Field Mus. Nat. Hist., Bot. Ser. 13(4): 761 (1958).
Peru (Cajamarca). 83 PER. Phan.

Myrcianthes leucoxyla (Ortega) McVaugh, Fieldiana, Bot. 29: 491 (1963).
Colombia to Venezuela. 82 VEN 83 CLM. Nanophan. or phan.
**Myrtus leucoxyla* Ortega, Nov. Pl. Descr. Dec.: 129 (1800).
Myrtus foliosa Kunth in F.W.H.von Humboldt, A.J.A. Bonpland & C.S.Kunth, Nov. Gen. Sp. 6: 134 (1823). *Eugenia foliosa* (Kunth) DC., Prodr. 3: 277 (1828). *Amyrsia foliosa* (Kunth) Raf., Sylva Tellur.: 106 (1838). *Myrcianthes foliosa* (Kunth) McVaugh, Publ. Field Mus. Nat. Hist., Bot. Ser. 13(4): 758 (1958).

Myrcianthes lindleyana (Kunth) McVaugh, Publ. Field Mus. Nat. Hist., Bot. Ser. 13(4): 763 (1958).
W. South America to Venezuela. 82 VEN 83 CLM ECU PER. Nanophan. or phan.
**Myrtus lindleyana* Kunth in F.W.H.von Humboldt, A.J.A.Bonpland & C.S.Kunth, Nov. Gen. Sp. 6: 138 (1823). *Eugenia lindleyana* (Kunth) DC., Prodr. 3: 278 (1828).

Myrcianthes mato (Griseb.) McVaugh, Fieldiana, Bot. 29: 491 (1963).
Bolivia to NW. Argentina. 83 BOL 85 AGW. Nanophan. or phan.
**Eugenia mato* Griseb., Abh. Königl. Ges. Wiss. Göttingen 24: 125 (1879). *Acreugenia mato* (Griseb.) Kausel, Lilloa 32: 343 (1967).

Myrcianthes minimifolia (McVaugh) McVaugh, Fieldiana, Bot. 29: 492 (1963).
Peru to Argentina (Jujuy). 83 PER 85 AGW. Nanophan. or phan.
**Eugenia minimifolia* McVaugh, Fieldiana, Bot. 19: 213 (1956).

Myrcianthes myrsinoides (Kunth) Grifo, Monogr. Syst. Bot. Missouri Bot. Gard. 45: 1256 (1993).
W. South America to Venezuela. 82 VEN 83 BOL CLM ECU PER. Nanophan. or phan.
**Myrtus myrsinoides* Kunth in F.W.H.von Humboldt, A.J.A.Bonpland & C.S.Kunth, Nov. Gen. Sp. 6: 132 (1823). *Cluacena myrsinoides* (Kunth) Raf., Sylva Tellur.: 104 (1838). *Myrteola myrsinoides* (Kunth) O.Berg, Linnaea 27: 396 (1856). *Eugenia myrsinoides* (Kunth) Burret ex Diels, Biblioth. Bot. 29(116): 115 (1937).
Eugenia ternifolia O.Berg, Linnaea 27: 246 (1856). *Myrcianthes ternifolia* (O.Berg) McVaugh, Fieldiana, Bot. 29: 497 (1963).
Eugenia triquetra O.Berg, Linnaea 27: 141 (1856).
Eugenia triquetra var. *aequatorialis* O.Berg, Linnaea 31: 253 (1862).
Eugenia weberbaueri Diels, Bot. Jahrb. Syst. 37: 598 (1906).
Eugenia cartilaginea McVaugh, Fieldiana, Bot. 29: 205 (1956).

Myrcianthes oreophila (Diels) McVaugh, Publ. Field Mus. Nat. Hist., Bot. Ser. 13(4): 767 (1958).
S. Peru to Bolivia. 83 BOL PER. Phan.
**Eugenia oreophila* Diels, Bot. Jahrb. Syst. 37: 597 (1906).

Myrcianthes orthostemon (O.Berg) Grifo, Monogr. Syst. Bot. Missouri Bot. Gard. 45: 1257 (1993).
S. Colombia to Peru and Venezuela. 82 VEN 83 CLM ECU PER. Nanophan. or phan.
**Eugenia orthostemon* O.Berg, Linnaea 27: 179 (1856).

Myrcianthes osteomeloides (Rusby) McVaugh, Publ. Field Mus. Nat. Hist., Bot. Ser. 13(4): 768 (1958).
S. Peru to Bolivia. 83 BOL PER. Nanophan. or phan.
**Myrtus osteomeloides* Rusby, Mem. Torrey Bot. Club 6: 36 (1895). *Eugenia osteomeloides* (Rusby) McVaugh, Fieldiana, Bot. 29: 214 (1956).
Myrtus myrciopsis Kuntze, Revis. Gen. Pl. 3(2): 92 (1898).
Eugenia bifurcata McVaugh, Fieldiana, Bot. 29: 205 (1956). *Myrcianthes bifurcata* (McVaugh) McVaugh, Publ. Field Mus. Nat. Hist., Bot. Ser. 13(4): 752 (1958).

Myrcianthes pearcei (McVaugh) McVaugh, Publ. Field Mus. Nat. Hist., Bot. Ser. 13(4): 769 (1958).
Bolivia (La Paz). 83 BOL. Nanophan.
**Eugenia pearcei* McVaugh, Fieldiana, Bot. 29: 214 (1956).

Myrcianthes pedersenii D.Legrand, Bol. Soc. Argent. Bot. 10: 1 (1962).
Paraguay. 85 PAR. Nanophan.

Myrcianthes prodigiosa McVaugh, Fieldiana, Bot. 29: 492 (1963).
N. South America, Ecuador. 82 GUY SUR VEN 83 ECU. Phan.

Myrcianthes pseudomato (D.Legrand) McVaugh, Fieldiana, Bot. 29: 493 (1963).
Bolivia to NW. Argentina. 83 BOL 85 AGW. Phan.
**Eugenia pseudomato* D.Legrand, Lilloa 10: 477 (1944). *Pseudomyrcianthes pseudomato* (D.Legrand) Kausel, Ark. Bot., a.s., 3: 505 (1956).

Myrcianthes pungens (O.Berg) D.Legrand, Bol. Fac. Agron. Univ. Montevideo 101: 52 (1968).
Brazil to N. Argentina. 83 BOL 84 BZC BZL BZS 85 AGE AGW PAR URU. Phan.
**Eugenia pungens* O.Berg in C.F.P.von Martius & auct. suc. (eds.), Fl. Bras. 14(1): 224 (1857). *Luma pungens* (O.Berg) Herter, Revista Sudamer. Bot. 7: 220 (1943). *Acreugenia pungens* (O.Berg) Kausel, Ark. Bot., a.s., 3: 510 (1956).
Eugenia ybaviyu Parodi, Anales Soc. Ci. Argent. 7: 69 (1879).

Myrcianthes quinqueloba (McVaugh) McVaugh, Publ. Field Mus. Nat. Hist., Bot. Ser. 13(4): 770 (1958).
Peru. 83 PER. Nanophan. or phan.
**Eugenia quinqueloba* McVaugh, Fieldiana, Bot. 29: 217 (1956).

Myrcianthes rhopaloides (Kunth) McVaugh, Publ. Field Mus. Nat. Hist., Bot. Ser. 13(4): 771 (1958).
Venezuela to W. South America. 82 VEN 83 BOL CLM ECU PER. Nanophan. or phan.

Myrtus coccolobifolia Kunth in F.W.H.von Humboldt, A.J.A.Bonpland & C.S.Kunth, Nov. Gen. Sp. 6: 139 (1823). *Myrcia coccolobifolia* (Kunth) DC., Prodr. 3: 255 (1828). *Pseudocaryophyllus coccolobifolius* (Kunth) O.Berg, Linnaea 27: 416 (1856). *Eugenia coccolobifolia* (Kunth) Burret, Notizbl. Bot. Gart. Berlin-Dahlem 15: 521 (1941).
**Myrtus rhopaloides* Kunth in F.W.H.von Humboldt, A.J.A.Bonpland & C.S.Kunth, Nov. Gen. Sp. 6: 137 (1823). *Eugenia rhopaloides* (Kunth) DC., Prodr. 3: 278 (1828).
Eugenia mariquitensis O.Berg, Linnaea 27: 267 (1856).
Eugenia porphyroclada O.Berg, Linnaea 27: 266 (1856).

Myrcianthes sessilis McVaugh, Fieldiana, Bot. 29: 495 (1963).
Colombia. 83 CLM. Phan.

Myrcianthes storkii (Standl.) McVaugh, Fieldiana, Bot. 29: 496 (1963).
Costa Rica to Panama. 80 COS PAN. Nanophan. or phan.
**Eugenia storkii* Standl., Publ. Field Mus. Nat. Hist., Bot. Ser. 8: 143 (1930).
Eugenia rigidissima Cufod., Arch. Bot. (Forlì) 9: 198 (1933). *Myrcianthes rigidissima* (Cufod.) W.D. Stevens, Phytologia 64: 333 (1988).

***Synonyms*:**
Myrcianthes apiculata O.Berg = ***Myrcianthes cisplatensis*** (Cambess.) O.Berg
Myrcianthes bifurcata (McVaugh) McVaugh = ***Myrcianthes osteomeloides*** (Rusby) McVaugh
Myrcianthes brunnea O.Berg = ***Psidium ovale*** (Spreng.) Burret
Myrcianthes brunnea var. *grandifolia* O.Berg = ***Psidium ovale*** (Spreng.) Burret
Myrcianthes brunnea var. *parvifolia* O.Berg = ***Psidium ovale*** (Spreng.) Burret
Myrcianthes campomanesioides Mattos & D.Legrand = ***Campomanesia eugenioides*** (Cambess.) D.Legrand
Myrcianthes cionei Mattos = ?
Myrcianthes cisplatensis var. *angustifolia* O.Berg = ***Blepharocalyx salicifolius*** (Kunth) O.Berg
Myrcianthes cisplatensis var. *brevipedunculata* O.Berg = ***Blepharocalyx salicifolius*** (Kunth) O.Berg
Myrcianthes cisplatensis var. *latifolia* O.Berg = ***Myrcianthes cisplatensis*** (Cambess.) O.Berg
Myrcianthes compressa (Kunth) McVaugh = ***Myrcianthes fragrans*** (Sw.) McVaugh
Myrcianthes dicrana (O.Berg) K.A.Wilson = ***Myrcianthes fragrans*** (Sw.) McVaugh
Myrcianthes dugandii (Standl.) McVaugh = ***Myrcianthes fragrans*** (Sw.) McVaugh
Myrcianthes edulis O.Berg = ***Hexachlamys edulis*** (O.Berg) Kausel & D.Legrand
Myrcianthes foliosa (Kunth) McVaugh = ***Myrcianthes leucoxyla*** (Ortega) McVaugh
Myrcianthes fragrans var. *hispidula* McVaugh = ***Myrcianthes fragrans*** (Sw.) McVaugh
Myrcianthes fragrans var. *simpsonii* (Small) R.W.Long = ***Myrcianthes fragrans*** (Sw.) McVaugh
Myrcianthes fragrans subsp. *simpsonii* (Small) E.Murray = ***Myrcianthes fragrans*** (Sw.) McVaugh
Myrcianthes gracilipes Kausel = ***Myrcianthes discolor*** (Kunth) McVaugh
Myrcianthes irregularis McVaugh = ***Psidium guineense*** Sw.
Myrcianthes krugii Kiaersk. = ***Pseudanamomis umbellulifera*** (Kunth) Kausel
Myrcianthes limbata (Kunth) McVaugh = ***Myrcianthes fragrans*** (Sw.) McVaugh
Myrcianthes montana (Aubl.) C.Nelson Sutherland = ***Marlierea montana*** (Aubl.) Amshoff
Myrcianthes reptans D.Legrand = ***Psidium reptans*** (D.Legrand) Soares-Silva & Proença
Myrcianthes rigidissima (Cufod.) W.D.Stevens = ***Myrcianthes storkii*** (Standl.) McVaugh
Myrcianthes simpsonii (Small) K.A.Wilson = ***Myrcianthes fragrans*** (Sw.) McVaugh
Myrcianthes terminalis Mattos & D.Legrand = [84 BZN]
Myrcianthes ternifolia (O.Berg) McVaugh = ***Myrcianthes myrsinoides*** (Kunth) Grifo
Myrcianthes umbellulifera (Kunth) Alain = ***Pseudanamomis umbellulifera*** (Kunth) Kausel

***Unplaced Names*:**
Myrcianthes cionei Mattos, Loefgrenia 54: 2 (1971). = ?
Myrcianthes terminalis Mattos & D.Legrand, Loefgrenia 67: 12 (1975). = [84 BZN]

Myrciaria

Myrciaria O.Berg, Linnaea 27: 320 (1856).
Trop. America. (25) tan 79 MXT 80 BLZ COS ELS GUA NIC PAN 81 CUB DOM HAI JAM LEE PUE TRT WIN 82 FRG GUY SUR VEN 83 BOL CLM ECU PER 84 BZC BZE BZL BZN BZS 85 AGE AGW PAR URU.
22 Species
Myrciariopsis Kausel, Ark. Bot., a.s., 3: 509 (1956).
Paramyrciaria Kausel, Lilloa 32: 345 (1966 publ. 1967).

Myrciaria borinquena Alain, Phytologia 47: 187 (1980).
Puerto Rico. 81 PUE. Nanophan. or phan.

Myrciaria cordata O.Berg, Linnaea 27: 337 (1856).
SE. Venezuela to N. Brazil. 82 GUY VEN 84 BZN. Nanophan. or phan.

Myrciaria cuspidata O.Berg in C.F.P.von Martius & auct. suc. (eds.), Fl. Bras. 14(1): 367 (1857).
Brazil to Paraguay. 83 BOL 84 BZC BZE BZL BZS 85 PAR. Cham. or nanophan.
**Eugenia tenella* Miq., Linnaea 22: 532 (1849), nom. illeg.
Myrciaria cuspidata var. *acuminatissima* O.Berg in C.F.P.von Martius & auct. suc. (eds.), Fl. Bras. 14(1): 367 (1857).
Myrciaria cuspidata var. *diffusa* O.Berg in C.F.P.von Martius & auct. suc. (eds.), Fl. Bras. 14(1): 367 (1857).
Myrciaria cuspidata var. *humilis* O.Berg in C.F.P.von Martius & auct. suc. (eds.), Fl. Bras. 14(1): 367 (1857).
Myrciaria cuspidata var. *latifolia* O.Berg in C.F.P.von Martius & auct. suc. (eds.), Fl. Bras. 14(1): 367 (1857).
Myrciaria cuspidata var. *stricta* O.Berg in C.F.P.von Martius & auct. suc. (eds.), Fl. Bras. 14(1): 367 (1857).
Myrciaria herbacea O.Berg in C.F.P.von Martius & auct. suc. (eds.), Fl. Bras. 14(1): 369 (1857).
Myrciaria minensis O.Berg in C.F.P.von Martius & auct. suc. (eds.), Fl. Bras. 14(1): 369 (1857). *Eugenia minensis* (O.Berg) Kiaersk., Enum. Myrt. Bras.: 184 (1893).
Eugenia alegrensis Kiaersk., Enum. Myrt. Bras.: 180 (1893).
Myrciaria apiculata Barb.Rodr. ex Chodat & Hassl., Bull. Herb. Boissier, II, 7: 807 (1907), nom. nud.

Myrciaria recurvipetala Barb.Rodr. ex Chodat & Hassl., Bull. Herb. Boissier, II, 7: 808 (1907), nom. nud.

Myrciaria delicatula (DC.) O.Berg in C.F.P.von Martius & auct. suc. (eds.), Fl. Bras. 14(1): 362 (1857).
Brazil to Bolivia. 83 BOL 84 BZC BZL BZS 85 AGE PAR. Nanophan. or phan.
**Eugenia delicatula* DC., Prodr. 3: 273 (1828). *Paramyrciaria delicatula* (DC.) Kausel, Lilloa 32: 345 (1966 publ. 1967).
Myrciaria delicatula var. *acutifolia* O.Berg in C.F.P.von Martius & auct. suc. (eds.), Fl. Bras. 14(1): 363 (1857).
Myrciaria delicatula var. *angustifolia* O.Berg in C.F.P.von Martius & auct. suc. (eds.), Fl. Bras. 14(1): 363 (1857).
Myrciaria delicatula var. *conferta* O.Berg in C.F.P.von Martius & auct. suc. (eds.), Fl. Bras. 14(1): 363 (1857).
Myrciaria delicatula var. *latifolia* O.Berg in C.F.P.von Martius & auct. suc. (eds.), Fl. Bras. 14(1): 363 (1857).
Myrciaria linearifolia O.Berg in C.F.P.von Martius & auct. suc. (eds.), Fl. Bras. 14(1): 362 (1857). *Paramyrciaria delicatula* var. *linearifolia* (O.Berg) O.Berg, Lilloa 32: 346 (1967).
Eugenia delicatula var. *conferta* Kiaersk., Enum. Myrt. Bras.: 180 (1893).
Eugenia maschalantha Kiaersk., Enum. Myrt. Bras.: 183 (1893). *Myrciaria maschalantha* (Kiaersk.) Mattos & D.Legrand, Loefgrenia 67: 6 (1975).
Myrciaria micrantha Barb.Rodr. ex Chodat & Hassl., Bull. Herb. Boissier, II, 7: 808 (1907), nom. nud.
Myrciaria macrocarpa Usteri, Fl. Umgeb. Stadt S. Paulo: 213 (1911).
Paramyrciaria delicatula var. *argentinensis* Kausel, Lilloa 32: 346 (1966 publ. 1967).

Myrciaria disticha O.Berg in C.F.P.von Martius & auct. suc. (eds.), Fl. Bras. 14(1): 366 (1857). *Eugenia biseriata* Kiaersk., Enum. Myrt. Bras.: 180 (1893).
E. Brazil. 84 BZE BZL. Nanophan. or phan.

var. ***bahiensis*** O.Berg in C.F.P.von Martius & auct. suc. (eds.), Fl. Bras. 14(1): 367 (1857).
Brazil (Bahia). 84 BZE. Nanophan.

var. ***disticha***
SE. Brazil. 84 BZL. Nanophan. or phan.
Myrciaria disticha var. *fluminensis* O.Berg in C.F.P.von Martius & auct. suc. (eds.), Fl. Bras. 14(1): 366 (1857).

Myrciaria dubia (Kunth) McVaugh, Fieldiana, Bot. 29: 501 (1963).
S. Trop. America. 82 GUY VEN 83 CLM ECU PER 84 BZC BZN. Nanophan. or phan.
**Psidium dubium* Kunth in F.W.H.von Humboldt, A.J.A. Bonpland & C.S.Kunth, Nov. Gen. Sp. 6: 152 (1823).
Eugenia divaricata Benth., J. Bot. (Hooker) 2: 319 (1840). *Myrciaria divaricata* (Benth.) O.Berg, Linnaea 27: 334 (1856).
Myrciaria phillyraeoides O.Berg, Linnaea 27: 326 (1856). *Myrtus phillyraeoides* (O.Berg) Willd. ex O.Berg, Linnaea 27: 326 (1856).
Myrciaria lanceolata O.Berg in C.F.P.von Martius & auct. suc. (eds.), Fl. Bras. 14(1): 363 (1857).
Myrciaria lanceolata var. *angustifolia* O.Berg in C.F.P.von Martius & auct. suc. (eds.), Fl. Bras. 14(1): 363 (1857).
Myrciaria lanceolata var. *glomerata* O.Berg in C.F.P.von Martius & auct. suc. (eds.), Fl. Bras. 14(1): 363 (1857).
Myrciaria lanceolata var. *laxa* O.Berg in C.F.P.von Martius & auct. suc. (eds.), Fl. Bras. 14(1): 363 (1857).
Myrciaria obscura O.Berg in C.F.P.von Martius & auct. suc. (eds.), Fl. Bras. 14(1): 374 (1857).
Myrciaria paraensis O.Berg in C.F.P.von Martius & auct. suc. (eds.), Fl. Bras. 14(1): 364 (1857).
Myrciaria spruceana O.Berg in C.F.P.von Martius & auct. suc. (eds.), Fl. Bras. 14(1): 365 (1857). *Eugenia grandiglandulosa* Kiaersk., Enum. Myrt. Bras.: 181 (1893).
Myrciaria riedeliana O.Berg in C.F.P.von Martius & auct. suc. (eds.), Fl. Bras. 14(1): 598 (1859).
Myrciaria caurensis Steyerm., Fieldiana, Bot. 28: 1020 (1957).
Marlierea macedoi D.Legrand, Comun. Bot. Mus. Hist. Nat. Montevideo 3(40): 27 (1962), no latin descr.

Myrciaria floribunda (H.West ex Willd.) O.Berg, Linnaea 27: 330 (1856).
Trop. America. (25) tan 79 MXT 80 BLZ COS ELS GUA NIC PAN 81 CUB DOM HAI JAM LEE PUE TRT WIN 82 FRG GUY SUR VEN 83 BOL CLM ECU PER 84 BZC BZE BZL BZN BZS 85 AGW PAR. Nanophan. or phan. Widely cultivated for its edible fruits.
**Eugenia floribunda* H.West ex Willd., Sp. Pl. 2: 960 (1799). *Myrtus floribunda* (H.West ex Willd.) Spreng., Syst. Veg. 2: 484 (1825). *Calyptranthes floribunda* (H.West ex Willd.) Blume, Bijdr.: 1091 (1826). *Caryophyllus floribundus* (H.West ex Willd.) Blume in A.P.de Candolle, Prodr. 3: 262 (1828).
Myrtus micrantha Nees & Mart., Nova Acta Phys.-Med. Acad. Caes. Leop.-Carol. Nat. Cur. 12(1): 51 (1824).
Eugenia maximiliana DC., Prodr. 3: 270 (1828). *Myrciaria maximiliana* (DC.) O.Berg in C.F.P.von Martius & auct. suc. (eds.), Fl. Bras. 14(1): 370 (1857).
Eugenia ciliolata Cambess. in A.F.C.de Saint-Hilaire, Fl. Bras. Merid. 2: 344 (1832). *Myrciaria ciliolata* (Cambess.) O.Berg in C.F.P.von Martius & auct. suc. (eds.), Fl. Bras. 14(1): 366 (1857). *Paramyrciaria ciliolata* (Cambess.) Rotman, Bol. Soc. Argent. Bot. 24: 418 (1986).
Eugenia salzmannii Benth., J. Bot. (Hooker) 2: 319 (1840). *Myrciaria salzmannii* (Benth.) O.Berg, Linnaea 27: 331 (1856).
Eugenia protracta Steud., Flora 26: 762 (1843). *Myrciaria protracta* (Steud.) O.Berg, Linnaea 27: 330 (1856).
Myrciaria uliginosa O.Berg, Linnaea 27: 329 (1856).
Myrciaria verticillata O.Berg, Linnaea 27: 332 (1856).
Myrciaria amazonica O.Berg in C.F.P.von Martius & auct. suc. (eds.), Fl. Bras. 14(1): 374 (1857).
Myrciaria axillaris O.Berg in C.F.P.von Martius & auct. suc. (eds.), Fl. Bras. 14(1): 373 (1857).
Myrciaria leucadendron O.Berg in C.F.P.von Martius & auct. suc. (eds.), Fl. Bras. 14(1): 364 (1857).
Myrciaria leucophloea O.Berg in C.F.P.von Martius & auct. suc. (eds.), Fl. Bras. 14(1): 370 (1857). *Eugenia leucophloea* (O.Berg) Kiaersk., Enum. Myrt. Bras.: 183 (1893).
Myrciaria leucophloea var. *conferta* O.Berg in C.F.P.von Martius & auct. suc. (eds.), Fl. Bras. 14(1): 370 (1857).

Myrciaria leucophloea var. *laxa* O.Berg in C.F.P.von Martius & auct. suc. (eds.), Fl. Bras. 14(1): 370 (1857).
Myrciaria maragnanensis O.Berg in C.F.P.von Martius & auct. suc. (eds.), Fl. Bras. 14(1): 372 (1857).
Myrciaria prasina O.Berg in C.F.P.von Martius & auct. suc. (eds.), Fl. Bras. 14(1): 373 (1857).
Myrciaria schuechiana O.Berg in C.F.P.von Martius & auct. suc. (eds.), Fl. Bras. 14(1): 373 (1857).
Myrciaria sellowiana O.Berg in C.F.P.von Martius & auct. suc. (eds.), Fl. Bras. 14(1): 374 (1857).
Myrciaria splendens O.Berg in C.F.P.von Martius & auct. suc. (eds.), Fl. Bras. 14(1): 371 (1857).
Myrciaria tenuiramis O.Berg in C.F.P.von Martius & auct. suc. (eds.), Fl. Bras. 14(1): 371 (1857).
Myrciaria tolypantha O.Berg in C.F.P.von Martius & auct. suc. (eds.), Fl. Bras. 14(1): 372 (1857).
Myrciaria tolypantha var. *angustifolia* O.Berg in C.F.P.von Martius & auct. suc. (eds.), Fl. Bras. 14(1): 372 (1857).
Myrciaria tolypantha var. *latifolia* O.Berg in C.F.P.von Martius & auct. suc. (eds.), Fl. Bras. 14(1): 372 (1857).
Myrciaria ferruginea O.Berg in C.F.P.von Martius & auct. suc. (eds.), Fl. Bras. 14(1): 597 (1859).
Myrciaria longipes O.Berg in C.F.P.von Martius & auct. suc. (eds.), Fl. Bras. 14(1): 599 (1859).
Myrciaria longipes var. *opaca* O.Berg in C.F.P.von Martius & auct. suc. (eds.), Fl. Bras. 14(1): 600 (1859).
Myrciaria longipes var. *pellucida* O.Berg in C.F.P.von Martius & auct. suc. (eds.), Fl. Bras. 14(1): 600 (1859).
Myrciaria maranhensis O.Berg in C.F.P.von Martius & auct. suc. (eds.), Fl. Bras. 14(1): 652 (1859). *Eugenia maranhensis* (O.Berg) Kiaersk., Enum. Myrt. Bras.: 183 (1893).
Myrciaria schuechiana var. *deflexa* O.Berg in C.F.P.von Martius & auct. suc. (eds.), Fl. Bras. 14(1): 599 (1859).
Myrciaria schuechiana var. *latifolia* O.Berg in C.F.P.von Martius & auct. suc. (eds.), Fl. Bras. 14(1): 599 (1859).
Calycorectes cubensis Griseb., Cat. Pl. Cub.: 90 (1866). *Marlierea cubensis* (Griseb.) Krug & Urb., Bot. Jahrb. Syst. 19: 589 (1895). *Myrcia cubensis* (Griseb.) Krug & Urb., Bot. Jahrb. Syst. 19: 589 (1895). *Plinia cubensis* (Griseb.) Urb., Repert. Spec. Nov. Regni Veg. 15: 413 (1919).
Eugenia leucophloea var. *warmingiana* Kiaersk., Enum. Myrt. Bras.: 183 (1893). *Myrciaria leucophloea* var. *warmingiana* (Kiaersk.) Mattos, Loefgrenia 64: 3 (1975). *Myrciaria ciliolata* var. *warmingiana* (Kiaersk.) Mattos, Loefgrenia 78: 3 (1983).
Marlierea brachymischa Kiaersk., Enum. Myrt. Bras.: 47 (1893).
Eugenia asa-grayi Krug & Urb., Bot. Jahrb. Syst. 19: 658 (1895). *Plinia asa-grayi* (Krug & Urb.) Urb., Symb. Antill. 9: 474 (1928).
Eugenia polyneura Urb., Symb. Antill. 5: 446 (1908), nom. illeg. *Eugenia pycnoneura* Urb., Symb. Antill. 6: 25 (1909).
Plinia formosa Urb., Symb. Antill. 9: 89 (1923).
Plinia rubrinervis Urb., Symb. Antill. 9: 474 (1928).
Plinia acutissima Urb., Ark. Bot. 22A(10): 25 (1929).
Eugenia oneillii Lundell, Bull. Torrey Bot. Club 64: 555 (1937). *Myrciaria oneillii* (Lundell) I.M.Johnst., Sargentia 8: 228 (1949).
Myrciaria arborea D.Legrand, Sellowia 13: 328 (1961).
Siphoneugena cantareirae Mattos, Ci. & Cult. 19: 332 (1967).
Myrciaria longicaudata Lundell, Wrightia 4: 144 (1970).
Siphoneugena micrantha Kausel, Lilloa 33: 111 (1971 publ. 1972).
Myrciaria mexicana Lundell, Wrightia 5: 44 (1974).
Myrciaria arborea var. *rostrata* Mattos, Loefgrenia 66: 3 (1975).

Myrciaria glanduliflora (Kiaersk.) Mattos & D.Legrand, Loefgrenia 67: 6 (1975).
Brazil (Minas Gerais). 84 BZL. Nanophan.
**Eugenia glanduliflora* Kiaersk., Enum. Myrt. Bras.: 180 (1893).

Myrciaria glazioviana (Kiaersk.) G.M.Barroso ex Sobral, Novon 16: 137 (2006).
SE. Brazil. (84) bze BZL bzs. Nanophan. or phan.
Myrciaria glomerata O.Berg in C.F.P.von Martius & auct. suc. (eds.), Fl. Bras. 14(1): 365 (1857). *Eugenia cabelludo* Kiaersk., Enum. Myrt. Bras.: 180 (1893). *Plinia glomerata* (O.Berg) Amshoff, Recueil Trav. Bot. Néerl. 42: 11 (1950). *Paramyrciaria glomerata* (O.Berg) Sobral, Candollea 46: 515 (1991).
**Eugenia cabelludo* var. *glazioviana* Kiaersk., Enum. Myrt. Bras.: 180 (1893). *Paramyrciaria glazioviana* (Kiaersk.) Sobral, Candollea 46: 515 (1991).
Marlierea antrocola Kiaersk., Enum. Myrt. Bras.: 45 (1893).
Myrcia glazioviana Kiaersk., Enum. Myrt. Bras.: 111 (1893). *Gomidesia glazioviana* (Kiaersk.) D. Legrand, Comun. Bot. Mus. Hist. Nat. Montevideo 3(37): 22 (1958).
Myrcia glazioviana var. *villosa* Kiaersk., Enum. Myrt. Bras.: 112 (1893).

Myrciaria guaquiea (Kiaersk.) Mattos & D.Legrand, Loefgrenia 67: 6 (1975).
E. Brazil. 84 BZE BZL. Nanophan. or phan.
**Eugenia guaquiea* Kiaersk., Enum. Myrt. Bras.: 181 (1893). *Paramyrciaria guaquiea* (Kiaersk.) Sobral, Candollea 46: 515 (1991).

Myrciaria ibarrae Lundell, Wrightia 2: 213 (1961).
SE. Mexico to Guatemala. 79 MXT 80 GUA. Nanophan. or phan.

Myrciaria myrtifolia Alain, Phytologia 54: 109 (1983).
Puerto Rico. 81 PUE. Nanophan. or phan.

Myrciaria pallida O.Berg in C.F.P.von Martius & auct. suc. (eds.), Fl. Bras. 14(1): 368 (1857).
SE. Brazil. 84 BZL. Nanophan. or phan.
Myrciaria sulcata Mattos, Loefgrenia 90: 2 (1986).

Myrciaria pilosa Sobral & Couto, Novon 16: 523 (2006).
Brazil (Bahia, Minas Gerais). 84 BZE BZL.

Myrciaria plinioides D.Legrand, Sellowia 13: 329 (1961).
S. Brazil. 84 BZS. Nanophan. or phan.

Myrciaria puberulenta B.Holst, Selbyana 23: 156 (2002).
S. Venezuela. 82 VEN. Phan.

Myrciaria pumila (Gardner) O.Berg in C.F.P.von Martius & auct. suc. (eds.), Fl. Bras. 14(1): 366 (1857).
Brazil (Rio de Janeiro). 84 BZL.
**Eugenia pumila* Gardner, London J. Bot. 4: 102 (1845).

Eugenia adenodes Kiaersk., Enum. Myrt. Bras.: 176 (1893). *Myrciaria adenodes* (Kiaersk.) Mattos & D.Legrand, Loefgrenia 67: 6 (1975).

Myrciaria rojasii D.Legrand, Bol. Soc. Argent. Bot. 10: 3 (1963).
WC. Brazil to Paraguay. 84 BZC 85 PAR. Nanophan. or phan.
Myrciaria tapiraguayensis Barb.Rodr. ex Chodat & Hassl., Bull. Herb. Boissier, II, 7: 807 (1907), nom. nud. *Paramyrciaria tapiraguayensis* (Barb.Rodr. ex Chodat & Hassl.) Sobral, Candollea 46: 521 (1991), nom. inval.

Myrciaria strigipes O.Berg in C.F.P.von Martius & auct. suc. (eds.), Fl. Bras. 14(1): 364 (1857). *Plinia strigipes* (O.Berg) Sobral, Bol. Mus. Bot. Munic. 63: 3 (1985). *Paramyrciaria strigipes* (O.Berg) Sobral, Candollea 46: 515 (1991).
E.Brazil. 84 BZE BZL. Nanophan. or phan.
Myrciaria strigipes var. *longifolia* O.Berg in C.F.P.von Martius & auct. suc. (eds.), Fl. Bras. 14(1): 597 (1859).

Myrciaria tenella (DC.) O.Berg in C.F.P.von Martius & auct. suc. (eds.), Fl. Bras. 14(1): 368 (1857).
Hispaniola, S. Trop. America. 81 DOM HAI 82 FRG VEN 83 BOL PER 84 BZC BZE BZL BZN BZS 85 AGE AGW PAR URU. Nanophan. or phan.
**Eugenia tenella* DC., Prodr. 3: 272 (1828).
Eugenia tenella var. *minor* Cambess. in A.F.C.P.de Saint-Hilaire & al., Fl. Bras. Merid. 2: 346 (1832).
Myrciaria tenella var. *elliptica* O.Berg in C.F.P.von Martius & auct. suc. (eds.), Fl. Bras. 14(1): 368 (1857).
Myrciaria tenella var. *spathulata* O.Berg in C.F.P.von Martius & auct. suc. (eds.), Fl. Bras. 14(1): 368 (1857).
Myrciaria undulata O.Berg in C.F.P.von Martius & auct. suc. (eds.), Fl. Bras. 14(1): 368 (1857).
Eugenia tenella var. *elliptica* Kiaersk., Enum. Myrt. Bras.: 184 (1893).
Eugenia tenella var. *glazioviana* Kiaersk., Enum. Myrt. Bras.: 184 (1893).
Eugenia tenella var. *macrocarpa* Kiaersk., Enum. Myrt. Bras.: 184 (1893).
Eugenia tenella var. *spathulata* Kiaersk., Enum. Myrt. Bras.: 184 (1893).
Plinia haitiensis Urb. & Ekman, Ark. Bot. 20A(5): 22 (1926).
Plinia montecristina Urb. & Ekman, Ark. Bot. 24A(4): 19 (1932).

Myrciaria vexator McVaugh, Fieldiana, Bot. 29: 503 (1963).
Costa Rica to Venezuela and Ecuador. 80 COS PAN 82 VEN 83 ECU. Phan.

Myrciaria vismeifolia (Benth.) O.Berg, Linnaea 27: 336 (1856).
N. South America to N. Brazil. 82 FRG GUY SUR VEN 84 BZN. Nanophan. or phan.
**Eugenia vismeifolia* Benth., J. Bot. (Hooker) 2: 320 (1840).

Synonyms:
Myrciaria adenodes (Kiaersk.) Mattos & D.Legrand = ***Myrciaria pumila*** (Gardner) O.Berg
Myrciaria amazonica O.Berg = ***Myrciaria floribunda*** (H.West ex Willd.) O.Berg
Myrciaria angustifolia (O.Berg) Mattos = ***Blepharocalyx salicifolius*** (Kunth) O.Berg
Myrciaria apiculata Barb.Rodr. ex Chodat & Hassl. = ***Myrciaria cuspidata*** O.Berg
Myrciaria arborea D.Legrand = ***Myrciaria floribunda*** (H.West ex Willd.) O.Berg
Myrciaria arborea var. *rostrata* Mattos = ***Myrciaria floribunda*** (H.West ex Willd.) O.Berg
Myrciaria atiraensis Barb.Rodr. ex Chodat & Hassl. = ***Eugenia bimarginata*** DC.
Myrciaria aureana Mattos = ***Plinia phitrantha*** (Kiaersk.) Sobral
Myrciaria axillaris O.Berg = ***Myrciaria floribunda*** (H.West ex Willd.) O.Berg
Myrciaria baporeti D.Legrand = ***Plinia rivularis*** (Cambess.) Rotman
Myrciaria bipennis O.Berg = ***Marlierea bipennis*** (O.Berg) McVaugh
Myrciaria brevipedunculata (O.Berg) Mattos = ***Blepharocalyx salicifolius*** (Kunth) O.Berg
Myrciaria cauliflora (Mart.) O.Berg = ***Plinia cauliflora*** (Mart.) Kausel
Myrciaria caurensis Steyerm. = ***Myrciaria dubia*** (Kunth) McVaugh
Myrciaria chartacea O.Berg = [84 BZL]
Myrciaria ciliolata (Cambess.) O.Berg = ***Myrciaria floribunda*** (H.West ex Willd.) O.Berg
Myrciaria ciliolata var. *warmingiana* (Kiaersk.) Mattos = ***Myrciaria floribunda*** (H.West ex Willd.) O.Berg
Myrciaria cordifolia D.Legrand = ***Plinia cordifolia*** (D.Legrand) Sobral
Myrciaria coronata Mattos = ***Plinia coronata*** (Mattos) Mattos
Myrciaria cuspidata var. *acuminatissima* O.Berg = ***Myrciaria cuspidata*** O.Berg
Myrciaria cuspidata var. *diffusa* O.Berg = ***Myrciaria cuspidata*** O.Berg
Myrciaria cuspidata var. *humilis* O.Berg = ***Myrciaria cuspidata*** O.Berg
Myrciaria cuspidata var. *latifolia* O.Berg = ***Myrciaria cuspidata*** O.Berg
Myrciaria cuspidata var. *stricta* O.Berg = ***Myrciaria cuspidata*** O.Berg
Myrciaria delicatula var. *acutifolia* O.Berg = ***Myrciaria delicatula*** (DC.) O.Berg
Myrciaria delicatula var. *angustifolia* O.Berg = ***Myrciaria delicatula*** (DC.) O.Berg
Myrciaria delicatula var. *conferta* O.Berg = ***Myrciaria delicatula*** (DC.) O.Berg
Myrciaria delicatula var. *latifolia* O.Berg = ***Myrciaria delicatula*** (DC.) O.Berg
Myrciaria deserti (Cambess.) O.Berg = ***Blepharocalyx salicifolius*** (Kunth) O.Berg
Myrciaria dichotoma D.Legrand = ***Blepharocalyx salicifolius*** (Kunth) O.Berg
Myrciaria disticha var. *fluminensis* O.Berg = ***Myrciaria disticha*** O.Berg var. ***disticha***
Myrciaria divaricata (Benth.) O.Berg = ***Myrciaria dubia*** (Kunth) McVaugh
Myrciaria dumicola (Barb.Rodr.) Chodat & Hassl. = ***Eugenia pyriformis*** Cambess.
Myrciaria edulis (Vell.) Skeels = ***Plinia edulis*** (Vell.) Sobral
Myrciaria ehrenbergiana O.Berg = ***Myrcia ehrenbergiana*** (O.Berg) McVaugh
Myrciaria ferruginea O.Berg = ***Myrciaria floribunda*** (H.West ex Willd.) O.Berg
Myrciaria glomerata O.Berg = ***Myrciaria glazioviana*** (Kiaersk.) G.M.Barroso ex Sobral
Myrciaria grandifolia Mattos = ***Plinia grandifolia*** (Mattos) Sobral

Myrciaria guapurium (DC.) O.Berg = ***Plinia peruviana*** (Poir.) Govaerts
Myrciaria hagendorffii O.Berg = ***Plinia rivularis*** (Cambess.) Rotman
Myrciaria hatschbachii Mattos = ***Plinia hatschbachii*** (Mattos) Sobral
Myrciaria herbacea O.Berg = ***Myrciaria cuspidata*** O.Berg
Myrciaria involucrata O.Berg = ***Plinia involucrata*** (O.Berg) McVaugh
Myrciaria itacurubiensis Barb.Rodr. ex Chodat & Hassl. = ***Eugenia hyemalis*** Cambess.
Myrciaria jaboticaba (Vell.) O.Berg = ***Plinia cauliflora*** (Mart.) Kausel
Myrciaria lanceolata O.Berg = ***Myrciaria dubia*** (Kunth) McVaugh
Myrciaria lanceolata var. *angustifolia* O.Berg = ***Myrciaria dubia*** (Kunth) McVaugh
Myrciaria lanceolata var. *glomerata* O.Berg = ***Myrciaria dubia*** (Kunth) McVaugh
Myrciaria lanceolata var. *laxa* O.Berg = ***Myrciaria dubia*** (Kunth) McVaugh
Myrciaria leptophylla (Barb.Rodr.) Chodat & Hassl. = ***Eugenia herbacea*** O.Berg
Myrciaria leucadendron O.Berg = ***Myrciaria floribunda*** (H.West ex Willd.) O.Berg
Myrciaria leucophloea O.Berg = ***Myrciaria floribunda*** (H.West ex Willd.) O.Berg
Myrciaria leucophloea var. *conferta* O.Berg = ***Myrciaria floribunda*** (H.West ex Willd.) O.Berg
Myrciaria leucophloea var. *laxa* O.Berg = ***Myrciaria floribunda*** (H.West ex Willd.) O.Berg
Myrciaria leucophloea var. *warmingiana* (Kiaersk.) Mattos = ***Myrciaria floribunda*** (H.West ex Willd.) O.Berg
Myrciaria linearifolia O.Berg = ***Myrciaria delicatula*** (DC.) O.Berg
Myrciaria lituatinervia O.Berg = ***Marlierea lituatinervia*** (O.Berg) McVaugh
Myrciaria longicaudata Lundell = ***Myrciaria floribunda*** (H.West ex Willd.) O.Berg
Myrciaria longipes O.Berg = ***Myrciaria floribunda*** (H.West ex Willd.) O.Berg
Myrciaria longipes var. *opaca* O.Berg = ***Myrciaria floribunda*** (H.West ex Willd.) O.Berg
Myrciaria longipes var. *pellucida* O.Berg = ***Myrciaria floribunda*** (H.West ex Willd.) O.Berg
Myrciaria macrocarpa Usteri = ***Myrciaria delicatula*** (DC.) O.Berg
Myrciaria maragnanensis O.Berg = ***Myrciaria floribunda*** (H.West ex Willd.) O.Berg
Myrciaria maranhensis O.Berg = ***Myrciaria floribunda*** (H.West ex Willd.) O.Berg
Myrciaria marowynensis (Miq.) O.Berg = ***Eugenia marowynensis*** Miq.
Myrciaria maschalantha (Kiaersk.) Mattos & D.Legrand = ***Myrciaria delicatula*** (DC.) O.Berg
Myrciaria maximiliana (DC.) O.Berg = ***Myrciaria floribunda*** (H.West ex Willd.) O.Berg
Myrciaria mexicana Lundell = ***Myrciaria floribunda*** (H.West ex Willd.) O.Berg
Myrciaria micrantha Barb.Rodr. ex Chodat & Hassl. = ***Myrciaria delicatula*** (DC.) O.Berg
Myrciaria micrantha O.Berg = ?
Myrciaria minensis O.Berg = ***Myrciaria cuspidata*** O.Berg
Myrciaria myriophylla (Casar.) O.Berg = ***Blepharocalyx myriophyllus*** Morais & Sobral
Myrciaria nettiana (Kiaersk.) Mattos & D.Legrand = ?
Myrciaria nitida (Benth.) O.Berg = ***Myrcia inaequiloba*** (DC.) Lemée
Myrciaria nitida var. *chartacea* O.Berg = ***Myrcia inaequiloba*** (DC.) Lemée
Myrciaria nitida var. *coriacea* O.Berg = ***Myrcia inaequiloba*** (DC.) Lemée
Myrciaria nitida var. *dives* O.Berg = ***Myrcia inaequiloba*** (DC.) Lemée
Myrciaria oblongata Mattos = ***Plinia oblongata*** (Mattos) Mattos
Myrciaria obscura O.Berg = ***Myrciaria dubia*** (Kunth) McVaugh
Myrciaria oneillii (Lundell) I.M.Johnst. = ***Myrciaria floribunda*** (H.West ex Willd.) O.Berg
Myrciaria paraensis O.Berg = ***Myrciaria dubia*** (Kunth) McVaugh
Myrciaria perforata O.Berg = ***Neomitranthes obscura*** (DC.) N.Silveira
Myrciaria peruviana (Poir.) Mattos = ***Plinia peruviana*** (Poir.) Govaerts
Myrciaria peruviana var. *trunciflora* (O.Berg) Mattos = ***Plinia peruviana*** (Poir.) Govaerts
Myrciaria phillyraeoides O.Berg = ***Myrciaria dubia*** (Kunth) McVaugh
Myrciaria phitrantha (Kiaersk.) Mattos = ***Plinia phitrantha*** (Kiaersk.) Sobral
Myrciaria piedadensis (Kiaersk.) Mattos & D.Legrand = ***Blepharocalyx salicifolius*** (Kunth) O.Berg
Myrciaria plicatocostata O.Berg = ***Plinia edulis*** (Vell.) Sobral
Myrciaria polyantha (Miq.) O.Berg = ***Myrcia inaequiloba*** (DC.) Lemée
Myrciaria prasina O.Berg = ***Myrciaria floribunda*** (H.West ex Willd.) O.Berg
Myrciaria protracta (Steud.) O.Berg = ***Myrciaria floribunda*** (H.West ex Willd.) O.Berg
Myrciaria pseudodichasiantha (Kiaersk.) Mattos & D.Legrand = ?
Myrciaria quitarensis (Benth.) O.Berg = ***Myrcia quitarensis*** (Benth.) Sagot
Myrciaria ramiflora O.Berg = ***Eugenia coffeifolia*** DC.
Myrciaria recurvipetala Barb.Rodr. ex Chodat & Hassl. = ***Myrciaria cuspidata*** O.Berg
Myrciaria riedeliana O.Berg = ***Myrciaria dubia*** (Kunth) McVaugh
Myrciaria rivularis (Cambess.) O.Berg = ***Plinia rivularis*** (Cambess.) Rotman
Myrciaria rivularis var. *baporeti* (D.Legrand) D.Legrand = ***Plinia rivularis*** (Cambess.) Rotman
Myrciaria rubiginosa (Cambess.) O.Berg = ***Marlierea rubiginosa*** (Cambess.) D.Legrand
Myrciaria salzmannii (Benth.) O.Berg = ***Myrciaria floribunda*** (H.West ex Willd.) O.Berg
Myrciaria schaueriana (Miq.) O.Berg = ***Myrcia graciliflora*** Sagot
Myrciaria schuechiana O.Berg = ***Myrciaria floribunda*** (H.West ex Willd.) O.Berg
Myrciaria schuechiana var. *deflexa* O.Berg = ***Myrciaria floribunda*** (H.West ex Willd.) O.Berg
Myrciaria schuechiana var. *latifolia* O.Berg = ***Myrciaria floribunda*** (H.West ex Willd.) O.Berg
Myrciaria sellowiana O.Berg = ***Myrciaria floribunda*** (H.West ex Willd.) O.Berg
Myrciaria sericea O.Berg = ***Eugenia neosericea*** Morais & Sobral
Myrciaria silveirana D.Legrand = [84 BZS]
Myrciaria spirito-santensis Mattos = ***Plinia spirito-santensis*** (Mattos) Mattos
Myrciaria splendens O.Berg = ***Myrciaria floribunda***

(H.West ex Willd.) O.Berg
Myrciaria spruceana O.Berg = ***Myrciaria dubia*** (Kunth) McVaugh
Myrciaria stirpiflora O.Berg = ***Eugenia cordata*** (Sw.) DC. var. ***cordata***
Myrciaria strigipes var. *longifolia* O.Berg = ***Myrciaria strigipes*** O.Berg
Myrciaria sulcata Mattos = ***Myrciaria pallida*** O.Berg
Myrciaria tapiraguayensis Barb.Rodr. ex Chodat & Hassl. = ***Myrciaria rojasii*** D.Legrand
Myrciaria tenella var. *elliptica* O.Berg = ***Myrciaria tenella*** (DC.) O.Berg
Myrciaria tenella var. *spathulata* O.Berg = ***Myrciaria tenella*** (DC.) O.Berg
Myrciaria tenuiramis O.Berg = ***Myrciaria floribunda*** (H.West ex Willd.) O.Berg
Myrciaria tolypantha O.Berg = ***Myrciaria floribunda*** (H.West ex Willd.) O.Berg
Myrciaria tolypantha var. *angustifolia* O.Berg = ***Myrciaria floribunda*** (H.West ex Willd.) O.Berg
Myrciaria tolypantha var. *latifolia* O.Berg = ***Myrciaria floribunda*** (H.West ex Willd.) O.Berg
Myrciaria tolypantha var. *pubescens* O.Berg = ***Neomitranthes obscura*** (DC.) N.Silveira
Myrciaria trinitatis O.Berg = ***Plinia pinnata*** L.
Myrciaria trunciflora O.Berg = ***Plinia peruviana*** (Poir.) Govaerts
Myrciaria uliginosa O.Berg = ***Myrciaria floribunda*** (H.West ex Willd.) O.Berg
Myrciaria undulata O.Berg = ***Myrciaria tenella*** (DC.) O.Berg
Myrciaria verticillata O.Berg = ***Myrciaria floribunda*** (H.West ex Willd.) O.Berg

Unplaced Names:
Myrciaria chartacea O.Berg in C.F.P.von Martius & auct. suc. (eds.), Fl. Bras. 14(1): 371 (1857). = [84 BZL]
Myrciaria micrantha O.Berg in C.F.P.von Martius & auct. suc. (eds.), Fl. Bras. 14(1): 360 (1857). = ?
Myrciaria silveirana D.Legrand, in Fl. Ilustr. Catar. 1(Mirt., Suppl. 1): 7 (1977). = [84 BZS]

Myrciariopsis

Myrciariopsis Kausel = ***Myrciaria*** O.Berg
Myrciariopsis baporeti (D.Legrand) Kausel = ***Plinia rivularis*** (Cambess.) Rotman

Myrrhinium

Myrrhinium Schott in K.P.J.Sprengel, Syst. Veg. 4(2): 404 (1827).
SE. Brazil. 83 CLM ECU PER 84 BZL BZS 85 AGE AGW URU.
1 Species
Felicianea Cambess. in A.F.C.de Saint-Hilaire, Fl. Bras. Merid. 2: 375 (1833).
Tetrastemon Hook. & Arn., Bot. Misc. 3: 317 (1833).

Myrrhinium atropurpureum Schott in K.P.J.Sprengel, Syst. Veg. 4(2): 404 (1827).
Andes to SE. & S. Brazil. 83 CLM ECU PER 84 BZL BZS 85 AGE AGW URU. Nanophan. or phan.

var. ***atropurpureum***
SE. Brazil. 84 BZL. Nanophan. or phan.
Felicianea rubriflora Cambess. in A.F.C.de Saint-Hilaire, Fl. Bras. Merid. 2: 375 (1833).

var. ***octandrum*** Benth., Pl. Hartw.: 131 (1844). *Myrrhinium octandrum* (Benth.) Mattos, Loefgrenia 78: 2 (1983).
Colombia to Uruguay. 83 CLM ECU PER 84 BZS 85 AGE AGW URU. Nanophan.
Tetrastemon loranthoides Hook. & Arn., Bot. Misc. 3: 318 (1833). *Myrrhinium rubriflorum* O.Berg in C.F.P.von Martius & auct. suc. (eds.), Fl. Bras. 14(1): 466 (1857), nom. illeg. *Myrrhinium loranthoides* (Hook. & Arn.) Burret, Notizbl. Bot. Gart. Berlin-Dahlem 15: 508 (1941).
Myrrhinium peruvianum O.Berg, Linnaea 26: 438 (1854).
Myrrhinium sarcopetalum Lem., Rev. Hort., IV, 7: 444 (1858).
Myrrhinium salicinum Gand., Bull. Soc. Bot. France 65: 26 (1918).
Myrrhinium lanceolatum Burret, Notizbl. Bot. Gart. Berlin-Dahlem 15: 508 (1941).

Synonyms:
Myrrhinium lanceolatum Burret = ***Myrrhinium atropurpureum*** var. ***octandrum*** Benth.
Myrrhinium loranthoides (Hook. & Arn.) Burret = ***Myrrhinium atropurpureum*** var. ***octandrum*** Benth.
Myrrhinium octandrum (Benth.) Mattos = ***Myrrhinium atropurpureum*** var. ***octandrum*** Benth.
Myrrhinium peruvianum O.Berg = ***Myrrhinium atropurpureum*** var. ***octandrum*** Benth.
Myrrhinium rubriflorum O.Berg = ***Myrrhinium atropurpureum*** var. ***octandrum*** Benth.
Myrrhinium salicinum Gand. = ***Myrrhinium atropurpureum*** var. ***octandrum*** Benth.
Myrrhinium sarcopetalum Lem. = ***Myrrhinium atropurpureum*** var. ***octandrum*** Benth.

Myrtastrum

Myrtastrum Burret, Notizbl. Bot. Gart. Berlin-Dahlem 15: 494 (1941).
New Caledonia. 60 NWC.
1 Species

Myrtastrum rufopunctatum (Pancher ex Brongn. & Gris) Burret, Notizbl. Bot. Gart. Berlin-Dahlem 15: 494 (1941).
New Caledonia. 60 NWC. Nanophan. or phan.
**Myrtus rufopunctata* Pancher ex Brongn. & Gris, Bull. Soc. Bot. France 12: 176 (1865).

Myrtekmania

Myrtekmania Urb. = ***Pimenta*** Lindl.
Myrtekmania adenoclada (Urb.) Urb. = ***Pimenta adenoclada*** (Urb.) Alain
Myrtekmania clementis Alain = ***Pimenta odiolens*** (Urb.) Burret
Myrtekmania filipes Urb. = ***Pimenta filipes*** (Urb.) Burret
Myrtekmania intermedia Bisse = [81 CUB]
Myrtekmania moaensis Areces = ***Pimenta filipes*** (Urb.) Burret
Myrtekmania odiolens (Urb.) Alain = ***Pimenta odiolens*** (Urb.) Burret
Myrtekmania podocarpoides Areces = ***Pimenta podocarpoides*** (Areces) Landrum

Unplaced Names:
Myrtekmania intermedia Bisse, Revista Jard. Bot. Nac. Univ. Habana 4(2): 6 (1983). = [81 CUB]

Myrtella

Myrtella F.Muell., Descr. Notes Papuan Pl. 1: 105 (1877).

New Guinea to NE. Pacific. 43 NWG SOL 62 CRL MRN.
2 Species
Saffordiella Merr., Philipp. J. Sci., C 9: 124 (1914).

Myrtella beccarii F.Muell., Descr. Notes Papuan Pl. 1: 106 (1877). *Fenzlia beccarii* (F.Muell.) Burret, Notizbl. Bot. Gart. Berlin-Dahlem 15: 500 (1941).
New Guinea to Solomon Is. 43 NWG SOL. Nanophan. or phan.

Myrtella bennigseniana (Volkens) Diels, Bot. Jahrb. Syst. 56: 529 (1921).
New Guinea to NE. Pacific. 43 NWG 62 CRL MRN. Nanophan. or phan.
**Leptospermum bennigsenianum* Volkens, Bot. Jahrb. Syst. 31: 470 (1901). *Fenzlia bennigseniana* (Volkens) Burret, Notizbl. Bot. Gart. Berlin-Dahlem 15: 500 (1941).
Saffordiella bennigseniana Merr., Philipp. J. Sci., C 9: 124 (1914).

Synonyms:
Myrtella cordata A.J.Scott = ***Lithomyrtus cordata*** (A.J.Scott) N.Snow & Guymer
Myrtella hirsutula F.Muell. = ***Kania hirsutula*** (F.Muell.) A.J.Scott
Myrtella microphylla (Benth.) A.J.Scott = ***Lithomyrtus microphylla*** (Benth.) N.Snow & Guymer
Myrtella obtusa (Endl.) A.J.Scott = ***Lithomyrtus obtusa*** (Endl.) N.Snow & Guymer
Myrtella phebalioides (W.Fitzg.) A.J.Scott = ***Lithomyrtus retusa*** (Endl.) N.Snow & Guymer
Myrtella retusa (Endl.) A.J.Scott = ***Lithomyrtus retusa*** (Endl.) N.Snow & Guymer
Myrtella rostrata Lauterb. = ***Uromyrtus rostrata*** (Lauterb.) N.Snow & Guymer

Myrteola

Myrteola O.Berg, Linnaea 27: 393 (1856).
W. South America to Falkland Is. 83 BOL CLM ECU PER 85 AGS CLC CLS 90 FAL.
3 Species
Amyrsia Raf., Sylva Tellur.: 106 (1838).
Cluacena Raf., Sylva Tellur.: 104 (1838).
Orestion Kunze ex O.Berg, Linnaea 27: 394 (1856).

Myrteola acerosa (O.Berg) Burret, Notizbl. Bot. Gart. Berlin-Dahlem 15: 495 (1941).
Peru. 83 PER. Nanophan.
**Myrtus acerosa* O.Berg, Linnaea 29: 253 (1858).

Myrteola nummularia (Lam.) O.Berg, Linnaea 27: 396 (1856).
W. South America to Falkland Is. 83 BOL CLM ECU PER 85 AGS CLC CLS 90 FAL. Cham. or nanophan.
**Myrtus nummularia* Lam., Encycl. 4: 407 (1798).
Myrtus vaccinioides Kunth in F.W.H.von Humboldt, A.J.A.Bonpland & C.S.Kunth, Nov. Gen. Sp. 6: 130 (1823). *Cluacena vaccinioides* (Kunth) Raf., Sylva Tellur.: 104 (1838). *Myrteola vaccinioides* (Kunth) O.Berg, Linnaea 27: 395 (1856).
Myrtus oxycoccoides Benth., Pl. Hartw.: 174 (1845). *Myrteola oxycoccoides* (Benth.) O.Berg, Linnaea 27: 396 (1856).
Myrtus nummularia var. *major* Hook.f., Fl. Antarct. 2: 276 (1846).
Myrtus nummularia Barnéoud in C.Gay, Fl. Chil. 2: 379 (1847), nom. illeg.
Myrtus leucomyrtillus Griseb., Abh. Königl. Ges. Wiss. Göttingen 6: 120 (1854). *Eugenia leucomyrtillus* (Griseb.) Phil., Anales Univ. Chile 84: 761 (1893). *Myrteola leucomyrtillus* (Griseb.) Reiche, Fl. Chile 2: 286 (1897).
Myrteola bullata O.Berg, Linnaea 27: 394 (1856).
Myrtus barneoudii O.Berg, Linnaea 27: 398 (1856). *Myrteola barneoudii* (O.Berg) O.Berg, Linnaea 29: 252 (1858). *Myrteola nummularia* var. *barneoudii* (O.Berg) Kausel, Lilloa 32: 360 (1966).
Orestion vaccinioides Kunze ex O.Berg, Linnaea 27: 294 (1856).
Eugenia humifusa Phil., Linnaea 28: 639 (1857). *Myrteola humifusa* (Phil.) O.Berg, Linnaea 29: 252 (1858). *Myrteola barneoudii* var. *humifusa* (Phil.) Reiche, Anales Univ. Chile 98: 701 (1897).
Myrteola bullata var. *pentamera* O.Berg, Bot. Zeitung (Berlin) 15: 858 (1857).
Myrteola bullata var. *tetramera* O.Berg, Bot. Zeitung (Berlin) 15: 858 (1857).
Myrtus leucomyrtus Griseb. ex Phil., Linnaea 28: 637 (1857).
Myrtus repens Phil., Bot. Zeitung (Berlin) 15: 400 (1857). *Myrteola repens* (Phil.) O.Berg, Bot. Zeitung (Berlin) 16: 250 (1858). *Myrteola nummularia* var. *repens* (Phil.) Reiche, Anales Univ. Chile 98: 701 (1897).
Myrteola philippii O.Berg, Linnaea 29: 252 (1858).
Myrteola nannophylla Burret, Notizbl. Bot. Gart. Berlin-Dahlem 15: 494 (1941).
Myrteola vaccinioides var. *carabaya* McVaugh, Fieldiana, Bot. 29: 228 (1956).

Myrteola phylicoides (Benth.) Landrum, Brittonia 43: 199 (1991).
Ecuador to Bolivia. 83 BOL ECU PER. Nanophan.
**Myrtus phylicoides* Benth., Pl. Hartw.: 131 (1844).

var. ***glabrata*** (O.Berg) Landrum, Brittonia 43: 199 (1991).
Peru to Bolivia. 83 BOL PER. Nanophan.
Myrteola weberbaueri Diels, Bot. Jahrb. Syst. 37: 593 (1906).

var. ***phylicoides***
Ecuador. 83 ECU. Nanophan.
Myrtus microphylla Humb. & Bonpl., Pl. Aequinoct. 1: 19 (1805), nom. illeg. *Amyrsia microphyla* Raf., Sylva Tellur.: 106 (1838). *Myrteola microphylla* (Humb. & Bonpl.) O.Berg, Linnaea 27: 393 (1856).
Myrteola microphylla var. *angustifolia* O.Berg, Linnaea 30: 709 (1861).
Myrteola microphylla var. *glabrata* O.Berg, Linnaea 30: 709 (1861).
Myrteola microphylla var. *latifolia* O.Berg, Linnaea 30: 709 (1861).
Myrteola microphylla var. *australis* Diels, Bot. Jahrb. Syst. 37: 593 (1906).

Synonyms:
Myrteola barneoudii (O.Berg) O.Berg = ***Myrteola nummularia*** (Lam.) O.Berg
Myrteola barneoudii var. *humifusa* (Phil.) Reiche = ***Myrteola nummularia*** (Lam.) O.Berg
Myrteola bullata O.Berg = ***Myrteola nummularia*** (Lam.) O.Berg
Myrteola bullata var. *pentamera* O.Berg = ***Myrteola nummularia*** (Lam.) O.Berg
Myrteola bullata var. *tetramera* O.Berg = ***Myrteola nummularia*** (Lam.) O.Berg
Myrteola humifusa (Phil.) O.Berg = ***Myrteola nummularia*** (Lam.) O.Berg

Myrteola leucomyrtillus (Griseb.) Reiche = ***Myrteola nummularia*** (Lam.) O.Berg
Myrteola microphylla (Humb. & Bonpl.) O.Berg = ***Myrteola phylicoides*** (Benth.) Landrum var. ***phylicoides***
Myrteola microphylla var. *angustifolia* O.Berg = ***Myrteola phylicoides*** (Benth.) Landrum var. ***phylicoides***
Myrteola microphylla var. *australis* Diels = ***Myrteola phylicoides*** (Benth.) Landrum var. ***phylicoides***
Myrteola microphylla var. *glabrata* O.Berg = ***Myrteola phylicoides*** (Benth.) Landrum var. ***phylicoides***
Myrteola microphylla var. *latifolia* O.Berg = ***Myrteola phylicoides*** (Benth.) Landrum var. ***phylicoides***
Myrteola myrsinoides (Kunth) O.Berg = ***Myrcianthes myrsinoides*** (Kunth) Grifo
Myrteola nannophylla Burret = ***Myrteola nummularia*** (Lam.) O.Berg
Myrteola nummularia var. *barneoudii* (O.Berg) Kausel = ***Myrteola nummularia*** (Lam.) O.Berg
Myrteola nummularia var. *repens* (Phil.) Reiche = ***Myrteola nummularia*** (Lam.) O.Berg
Myrteola oxycoccoides (Benth.) O.Berg = ***Myrteola nummularia*** (Lam.) O.Berg
Myrteola philippii O.Berg = ***Myrteola nummularia*** (Lam.) O.Berg
Myrteola repens (Phil.) O.Berg = ***Myrteola nummularia*** (Lam.) O.Berg
Myrteola vaccinioides (Kunth) O.Berg = ***Myrteola nummularia*** (Lam.) O.Berg
Myrteola vaccinioides var. *carabaya* McVaugh = ***Myrteola nummularia*** (Lam.) O.Berg
Myrteola weberbaueri Diels = ***Myrteola phylicoides*** var. ***glabrata*** (O.Berg) Landrum

Myrthoides

Myrthoides Wolf = ***Syzygium*** Gaertn.

Myrthus

Myrthus Scop. = ***Myrtus*** Tourn. ex L.

Myrtoleucodendron

Myrtoleucodendron Burm. ex Kuntze = ***Melaleuca*** L.
Myrtoleucodendron seorsiflorum (F.Muell.) Kuntze = ***Melaleuca lanceolata*** subsp. ***thaeroides*** Barlow
Myrtoleucodendron acacioides (F.Muell.) Kuntze = ***Melaleuca acacioides*** F.Muell.
Myrtoleucodendron acerosum Kuntze = ***Melaleuca systena*** Craven
Myrtoleucodendron acuminatum (F.Muell.) Kuntze = ***Melaleuca acuminata*** F.Muell.
Myrtoleucodendron adnatum (Turcz.) Kuntze = ***Melaleuca adnata*** Turcz.
Myrtoleucodendron alsophilum (A.Cunn. ex Benth.) Kuntze = ***Melaleuca alsophila*** A.Cunn. ex Benth.
Myrtoleucodendron angustifolium (Gaertn.) Kuntze = ***Asteromyrtus angustifolia*** (Gaertn.) Craven
Myrtoleucodendron armillare (Sol. ex Gaertn.) Kuntze = ***Melaleuca armillaris*** (Sol. ex Gaertn.) Sm.
Myrtoleucodendron aspalathodes (Schauer) Kuntze = ***Melaleuca aspalathoides*** Schauer
Myrtoleucodendron basicephalum (Benth.) Kuntze = ***Melaleuca basicephala*** Benth.
Myrtoleucodendron baxteri (Benth.) Kuntze = ?
Myrtoleucodendron blaeriifolium (Turcz.) Kuntze = ***Melaleuca blaeriifolia*** Turcz.
Myrtoleucodendron brachystachyum (F.Muell.) Kuntze = ***Melaleuca subfalcata*** Turcz.
Myrtoleucodendron calothamnoides (F.Muell.) Kuntze = ***Melaleuca calothamnoides*** F.Muell.
Myrtoleucodendron calycinum (R.Br.) Kuntze = ***Melaleuca calycina*** R.Br.
Myrtoleucodendron cardiophyllum (F.Muell.) Kuntze = ***Melaleuca cardiophylla*** F.Muell.
Myrtoleucodendron ciliosum (Turcz.) Kuntze = ***Melaleuca ciliosa*** Turcz.
Myrtoleucodendron concretum (F.Muell.) Kuntze = ***Melaleuca concreta*** F.Muell.
Myrtoleucodendron confertum Kuntze = ***Melaleuca lecanantha*** Barlow
Myrtoleucodendron cordatum (Turcz.) Kuntze = ***Melaleuca cordata*** Turcz.
Myrtoleucodendron crassifolium (Benth.) Kuntze = ***Melaleuca laxiflora*** Turcz.
Myrtoleucodendron cuticulare (Labill.) Kuntze = ***Melaleuca cuticularis*** Labill.
Myrtoleucodendron cylindricum (F.Muell.) Kuntze = ***Melaleuca diosmatifolia*** Dum.Cours.
Myrtoleucodendron cymbifolium (Benth.) Kuntze = ***Melaleuca halmaturorum*** subsp. ***cymbifolia*** (Benth.) Barlow
Myrtoleucodendron deanei (F.Muell.) Kuntze = ***Melaleuca deanei*** F.Muell.
Myrtoleucodendron decussatum (R.Br.) Kuntze = ***Melaleuca decussata*** R.Br.
Myrtoleucodendron deltoideum (Benth.) Kuntze = ***Melaleuca cucullata*** Turcz.
Myrtoleucodendron densum (R.Br.) Kuntze = ***Melaleuca densa*** R.Br.
Myrtoleucodendron depauperatum (Turcz.) Kuntze = ***Melaleuca depauperata*** Turcz.
Myrtoleucodendron diosmifolium (Andrews) Kuntze = ***Melaleuca diosmifolia*** Andrews
Myrtoleucodendron dissitiflorum (F.Muell.) Kuntze = ***Melaleuca dissitiflora*** F.Muell.
Myrtoleucodendron elachophyllum (F.Muell.) Kuntze = ***Melaleuca depauperata*** Turcz.
Myrtoleucodendron eleutherostachyum (F.Muell.) Kuntze = ***Melaleuca eleuterostachya*** F.Muell.
Myrtoleucodendron ellipticum (Labill.) Kuntze = ***Melaleuca elliptica*** Labill.
Myrtoleucodendron erianthum (Benth.) Kuntze = ***Melaleuca leptospermoides*** Schauer
Myrtoleucodendron ericifolium (Sm.) Kuntze = ***Melaleuca ericifolia*** Sm.
Myrtoleucodendron exaratum (F.Muell.) Kuntze = ***Melaleuca suberosa*** (Schauer) C.A.Gardner
Myrtoleucodendron fasciculiflorum (Benth.) Kuntze = ***Melaleuca brevifolia*** Turcz.
Myrtoleucodendron filifolium (F.Muell.) Kuntze = ***Melaleuca filifolia*** F.Muell.
Myrtoleucodendron foliolosum (A.Cunn. ex Benth.) Kuntze = ***Melaleuca foliolosa*** A.Cunn. ex Benth.
Myrtoleucodendron fulgens (R.Br.) Kuntze = ***Melaleuca fulgens*** R.Br.
Myrtoleucodendron genistifolium (Sm.) Kuntze = ***Melaleuca decora*** (Salisb.) Britten
Myrtoleucodendron gibbosum (Labill.) Kuntze = ***Melaleuca gibbosa*** Labill.
Myrtoleucodendron glaberrimum (F.Muell.) Kuntze = ***Melaleuca glaberrima*** F.Muell.
Myrtoleucodendron globiferum (R.Br.) Kuntze = ***Melaleuca globifera*** R.Br.
Myrtoleucodendron glomeratum (F.Muell.) Kuntze = ***Melaleuca glomerata*** F.Muell.
Myrtoleucodendron hakeoides (F.Muell. ex Benth.) Kuntze = ***Melaleuca glomerata*** F.Muell.

Myrtoleucodendron hamulosum (Turcz.) Kuntze = ***Melaleuca hamulosa*** Turcz.
Myrtoleucodendron holosericeum (Schauer) Kuntze = ***Melaleuca holosericea*** Schauer
Myrtoleucodendron huegelii (Endl.) Kuntze = ***Melaleuca huegelii*** Endl.
Myrtoleucodendron hypericifolium (Sm.) Kuntze = ***Melaleuca hypericifolia*** Sm.
Myrtoleucodendron incanum (R.Br.) Kuntze = ***Melaleuca incana*** R.Br.
Myrtoleucodendron lasiandrum (F.Muell.) Kuntze = ***Melaleuca lasiandra*** F.Muell.
Myrtoleucodendron lateriflorum (Benth.) Kuntze = ***Melaleuca lateriflora*** Benth.
Myrtoleucodendron lateritium (A.Dietr.) Kuntze = ***Melaleuca lateritia*** A.Dietr.
Myrtoleucodendron laxiflorum (Turcz.) Kuntze = ***Melaleuca laxiflora*** Turcz.
Myrtoleucodendron leiocarpum (F.Muell.) Kuntze = ***Melaleuca leiocarpa*** F.Muell.
Myrtoleucodendron leiopyxe (F.Muell. ex Benth.) Kuntze = ***Melaleuca leiopyxis*** F.Muell. ex Benth.
Myrtoleucodendron leptocladum (Benth.) Kuntze = ***Melaleuca pauciflora*** Turcz.
Myrtoleucodendron linariifolium (Sm.) Kuntze = ***Melaleuca linariifolia*** Sm.
Myrtoleucodendron linophyllum (F.Muell.) Kuntze = ***Melaleuca linophylla*** F.Muell.
Myrtoleucodendron longicomum (Benth.) Kuntze = ***Melaleuca macronychia*** Turcz. subsp. ***macronychia***
Myrtoleucodendron macronychium (Turcz.) Kuntze = ***Melaleuca macronychia*** Turcz.
Myrtoleucodendron megacephalum (F.Muell.) Kuntze = ***Melaleuca megacephala*** F.Muell.
Myrtoleucodendron micromerum (Schauer) Kuntze = ***Melaleuca micromera*** Schauer
Myrtoleucodendron microphyllum (Sm.) Kuntze = ***Melaleuca microphylla*** Sm.
Myrtoleucodendron minutifolium (F.Muell.) Kuntze = ***Melaleuca minutifolia*** F.Muell.
Myrtoleucodendron nesophyllum (F.Muell.) Kuntze = ***Melaleuca nesophila*** F.Muell.
Myrtoleucodendron nodosum (Sol. ex Gaertn.) Kuntze = ***Melaleuca nodosa*** (Sol. ex Gaertn.) Sm.
Myrtoleucodendron oldfieldii (F.Muell.) Kuntze = ***Melaleuca oldfieldii*** F.Muell.
Myrtoleucodendron parviflorum Kuntze = ***Melaleuca laxiflora*** Turcz.
Myrtoleucodendron pauciflorum (Turcz.) Kuntze = ***Melaleuca pauciflora*** Turcz.
Myrtoleucodendron pauperiflorum (F.Muell.) Kuntze = ***Melaleuca pauperiflora*** F.Muell.
Myrtoleucodendron pentagonum (Labill.) Kuntze = ***Melaleuca pentagona*** Labill.
Myrtoleucodendron polycephalum (Benth.) Kuntze = ***Melaleuca polycephala*** Benth.
Myrtoleucodendron polygaloides (Schauer) Kuntze = ***Melaleuca incana*** R.Br. subsp. ***incana***
Myrtoleucodendron pulchellum (R.Br.) Kuntze = ***Melaleuca pulchella*** R.Br.
Myrtoleucodendron pungens (Schauer) Kuntze = ***Melaleuca pungens*** Schauer
Myrtoleucodendron pustulatum (Hook.f.) Kuntze = ***Melaleuca pustulata*** Hook.f.
Myrtoleucodendron quadrifarium (F.Muell.) Kuntze = ***Melaleuca quadrifaria*** F.Muell.
Myrtoleucodendron radula (Lindl.) Kuntze = ***Melaleuca radula*** Lindl.
Myrtoleucodendron rhaphiophyllum (Schauer) Kuntze = ***Melaleuca rhaphiophylla*** Schauer
Myrtoleucodendron scabrum (R.Br.) Kuntze = ***Melaleuca scabra*** R.Br.
Myrtoleucodendron seriatum (Lindl.) Kuntze = ***Melaleuca seriata*** Lindl.
Myrtoleucodendron sparsiflorum (Turcz.) Kuntze = ***Melaleuca sparsiflora*** Turcz.
Myrtoleucodendron spathulatum (Schauer) Kuntze = ***Melaleuca spathulata*** Schauer
Myrtoleucodendron squameum (Labill.) Kuntze = ***Melaleuca squamea*** Labill.
Myrtoleucodendron squarrosum (Sm.) Kuntze = ***Melaleuca squarrosa*** Sm.
Myrtoleucodendron striatum (Labill.) Kuntze = ***Melaleuca striata*** Labill.
Myrtoleucodendron styphelioides (Sm.) Kuntze = ***Melaleuca styphelioides*** Sm.
Myrtoleucodendron subfalcatum (Turcz.) Kuntze = ***Melaleuca subfalcata*** Turcz.
Myrtoleucodendron subtrigonum (Schauer) Kuntze = ***Melaleuca subtrigona*** Schauer
Myrtoleucodendron symphyocarpum (F.Muell.) Kuntze = ***Asteromyrtus symphyocarpa*** (F.Muell.) Craven
Myrtoleucodendron tamariscinum (Hook.) Kuntze = ***Melaleuca tamariscina*** Hook.
Myrtoleucodendron tenellum (Benth.) Kuntze = ***Melaleuca incana*** subsp. ***tenella*** (Benth.) Barlow
Myrtoleucodendron teretifolium (Endl.) Kuntze = ***Melaleuca teretifolia*** Endl.
Myrtoleucodendron thymifolium (Sm.) Kuntze = ***Melaleuca thymifolia*** Sm.
Myrtoleucodendron thymoides (Labill.) Kuntze = ***Melaleuca thymoides*** Labill.
Myrtoleucodendron thyoides (Turcz.) Kuntze = ***Melaleuca thyoides*** Turcz.
Myrtoleucodendron trichophyllum (Lindl.) Kuntze = ***Melaleuca trichophylla*** Lindl.
Myrtoleucodendron trichostachyum (Lindl.) Kuntze = ***Melaleuca trichostachya*** Lindl.
Myrtoleucodendron uncinatum (R.Br.) Kuntze = ***Melaleuca uncinata*** R.Br.
Myrtoleucodendron undulatum (Benth.) Kuntze = ***Melaleuca undulata*** Benth.
Myrtoleucodendron urceolare (F.Muell. ex Benth.) Kuntze = ***Melaleuca urceolaris*** F.Muell. ex Benth.
Myrtoleucodendron vimineum (Lindl.) Kuntze = ***Melaleuca viminea*** Lindl.
Myrtoleucodendron violaceum (Schauer) Kuntze = ***Melaleuca violacea*** Schauer
Myrtoleucodendron viridiflorum (Sol. ex Gaertn.) Kuntze = ***Melaleuca viridiflora*** Sol. ex Gaertn.
Myrtoleucodendron wilsonii (F.Muell.) Kuntze = ***Melaleuca wilsonii*** F.Muell.

Myrtomera

Myrtomera B.C.Stone = ***Arillastrum*** Pancher ex Baill.
Myrtomera gummifera (Brongn. & Gris) B.C.Stone = ***Arillastrum gummiferum*** (Brongn. & Gris) Pancher ex Baill.
Myrtomera rubiginosa (Brongn. & Gris) B.C.Stone = ***Stereocaryum rubiginosum*** (Brongn. & Gris) Burret

Myrtopsis

Myrtopsis O.Hoffm. = ***Eugenia*** P.Micheli ex L.
Myrtopsis malangensis O.Hoffm. = ***Eugenia malangensis*** (O.Hoffm.) Nied.

Myrtus

Myrtus Tourn. ex L., Sp. Pl.: 471 (1753).
Macaronesia to Pakistan. 12 BAL COR FRA POR SAR SPA 13 ALB GRC ITA KRI SIC YUG 20 ALG LBY MOR TUN 21 AZO CNY MDR 24 CHA ERI ETH 27 CPP 34 AFG CYP EAI IRN IRQ LBS PAL TUR 35 YEM 40 IND? PAK.
2 Species
Myrthus Scop., Intr. Hist. Nat.: 218 (1777).

Myrtus communis L., Sp. Pl.: 471 (1753).
Macaronesia to Pakistan. 12 BAL COR FRA POR SAR SPA 13 ALB GRC ITA KRI SIC YUG 20 ALG LBY MOR TUN 21 AZO CNY MDR 24 ERI ETH (27) cpp 34 AFG CYP EAI IRN IRQ LBS PAL TUR 35 YEM 40 IND? PAK. Nanophan. or phan.

subsp. ***communis***
Macaronesia to Pakistan. 12 BAL COR FRA POR SAR SPA 13 ALB GRC ITA KRI SIC YUG 20 ALG LBY MOR TUN 21 AZO CNY MDR 24 ERI ETH (27) cpp 34 AFG CYP EAI IRN IRQ LBS PAL TUR 35 YEM 40 IND? PAK. Nanophan. or phan.
Myrtus communis var. *acutifolia* L., Sp. Pl.: 471 (1753). *Myrtus acutifolia* (L.) Sennen & Teodoro, Bull. Soc. Dendrol. France 68: 14 (1929).
Myrtus communis var. *angustifolia* L., Sp. Pl.: 471 (1753).
Myrtus communis var. *baetica* L., Sp. Pl.: 471 (1753). *Myrtus baetica* (L.) Mill., Gard. Dict. ed. 8: 4 (1768).
Myrtus communis var. *belgica* L., Sp. Pl.: 471 (1753). *Myrtus belgica* (L.) Mill., Gard. Dict. ed. 8: 2 (1768). *Myrtus communis* var. *acuminata* Rouy & E.G.Camus in G.Rouy & J.Foucaud, Fl. France 7: 155 (1901), nom. superfl.
Myrtus communis var. *mucronata* L., Sp. Pl.: 471 (1753).
Myrtus communis var. *romana* L., Sp. Pl.: 471 (1753). *Myrtus romana* (L.) Hoffmanns., Verz. Pfl.-Kult.: 82 (1824).
Myrtus major Garsault, Fig. Pl. Méd.: t. 396 (1764), opus utique oppr.
Myrtus minor Garsault, Fig. Pl. Méd.: t. 396 (1764), opus utique oppr.
Myrtus acuta Mill., Gard. Dict. ed. 8: 3 (1768). *Myrtus communis* var. *lusitanica* Rouy in G.Rouy & J.Foucaud, Fl. France 7: 155 (1901).
Myrtus italica Mill., Gard. Dict. ed. 8: 5 (1768). *Myrtus communis* var. *italica* (Mill.) Rouy & E.G.Camus in G.Rouy & J.Foucaud, Fl. France 7: 155 (1901).
Myrtus minima Mill., Gard. Dict. ed. 8: 7 (1768).
Myrtus littoralis Salisb., Prodr. Stirp. Chap. Allerton: 353 (1796).
Myrtus macrophylla J.St.-Hil. in H.L.Duhamel du Monceau, Traité Arbr. Arbust., ed. 2, 1: 208 (1803).
Myrtus microphylla J.St.-Hil. in H.L.Duhamel du Monceau, Traité Arbr. Arbust., ed. 2, 1: 207 (1803).
Myrtus romanifolia J.St.-Hil. in H.L.Duhamel du Monceau, Traité Arbr. Arbust., ed. 2, 1: 207 (1803).
Myrtus communis subsp. *mucronata* Pers., Syn. Pl. 2: 30 (1806).
Myrtus media Hoffmanns., Verz. Pfl.-Kult.: 82 (1824).
Myrtus angustifolia Raf., Sylva Tellur.: 105 (1838), nom. illeg.
Myrtus buxifolia Raf., Sylva Tellur.: 104 (1838), nom. illeg.
Myrtus lanceolata Raf., Sylva Tellur.: 105 (1838), nom. illeg.
Myrtus latifolia Raf., Sylva Tellur.: 104 (1838), nom. illeg.
Myrtus oerstedeana O.Berg, Linnaea 27: 405 (1856).
Myrtus sparsifolia O.Berg, Linnaea 27: 402 (1856).
Myrtus veneris Bubani, Fl. Pyren. 2: 639 (1899).
Myrtus borbonis Sennen, Ann. Soc. Linn. Lyon, n.s., 79: 63 (1923 publ. 1924).
Myrtus baetica var. *vidalii* Sennen & Teodoro, Exsicc. (Pl. Esp.): 6479 (1928). *Myrtus vidalii* (Sennen & Teodoro) Sennen & Teodoro, Bull. Soc. Dendrol. France 68: 15 (1929).
Myrtus communis var. *christinae* Sennen & Teodoro, Exsicc. (Pl. Esp.): 6480 (1928). *Myrtus christinae* (Sennen & Teodoro) Sennen & Teodoro, Bull. Soc. Dendrol. France 68: 15 (1929).
Myrtus communis var. *eusebii* Sennen & Teodoro, Exsicc. (Pl. Esp.): 6478 (1928). *Myrtus eusebii* (Sennen & Teodoro) Sennen & Teodoro, Bull. Soc. Dendrol. France 68: 11 (1929).
Myrtus communis var. *gervasii* Sennen & Teodoro, Exsicc. (Pl. Esp.): 6759 (1928). *Myrtus gervasii* (Sennen & Teodoro) Sennen & Teodoro, Bull. Soc. Dendrol. France 68: 12 (1929).
Myrtus italica var. *briquetii* Sennen & Teodoro, Exsicc. (Pl. Esp.): 6350 (1928). *Myrtus briquetii* (Sennen & Teodoro) Sennen & Teodoro, Bull. Soc. Dendrol. France 68: 12 (1929).
Myrtus italica var. *petri-ludovici* Sennen & Teodoro, Exsicc. (Pl. Esp.): 6482 (1928). *Myrtus petri-ludovici* (Sennen & Teodoro) Sennen & Teodoro, Bull. Soc. Dendrol. France 68: 15 (1929).
Myrtus augustinii Sennen & Teodoro, Bull. Soc. Dendrol. France 68: 16 (1929).
Myrtus baui Sennen & Teodoro, Bull. Soc. Dendrol. France 68: 13 (1929).
Myrtus communis var. *balearica* Sennen & Teodoro, Bull. Soc. Dendrol. France 68: 17 (1929).
Myrtus communis var. *foucaudii* Sennen & Teodoro, Bull. Soc. Dendrol. France 68: 17 (1929).
Myrtus communis var. *grandifolia* Sennen & Teodoro, Bull. Soc. Dendrol. France 68: 17 (1929).
Myrtus communis var. *joussetii* Sennen & Teodoro, Bull. Soc. Dendrol. France 68: 17 (1929).
Myrtus communis var. *neapolitana* Sennen & Teodoro, Bull. Soc. Dendrol. France 68: 17 (1929).
Myrtus josephi Sennen & Teodoro, Bull. Soc. Dendrol. France 68: 14 (1929).
Myrtus mirifolia Sennen & Teodoro, Bull. Soc. Dendrol. France 68: 13 (1929).
Myrtus rodesi Sennen & Teodoro, Bull. Soc. Dendrol. France 68: 14 (1929).
Myrtus theodori Sennen, Bull. Soc. Dendrol. France 68: 16 (1929).

subsp. ***tarentina*** (L.) Nyman, Consp. Fl. Eur. 2: 245 (1879).
S. Europe. 12 COR FRA por SAR SPA 13 ITA KRI YUG. Nanophan.
**Myrtus communis* var. *tarentina* L., Sp. Pl.: 471 (1753). *Myrtus tarentina* (L.) Mill., Gard. Dict. ed. 8: 6 (1768).

Myrtus nivelii Batt. & Trab., Bull. Soc. Bot. France 58: 671 (1911 publ. 1912).
S. Algeria, SW. Libya ?, Chad (Tibesti). 20 ALG LBY 24 CHA. Nanophan.

subsp. ***nivelii***
S. Algeria, SW. Libya ?. 20 ALG LBY. Nanophan.

subsp. ***tibesticus*** Quézel, Miss. Bot. Tibesti: 155 (1958).
Chad (Tibesti). 24 CHA. Nanophan.

***Synonyms*:**
Myrtus acerosa O.Berg = ***Myrteola acerosa*** (O.Berg) Burret
Myrtus acetosans (Poir.) Spreng. = ***Myrcia citrifolia*** (Aubl.) Urb.
Myrtus acka Juss. ex DC. = ***Eugenia foetida*** Pers.
Myrtus acmenoides F.Muell. = ***Gossia acmenoides*** (F.Muell.) N.Snow & Guymer
Myrtus acris Sw. = ***Pimenta racemosa*** (Mill.) J.W. Moore var. ***racemosa***
Myrtus acuminata Sessé & Moç. = [81 PUE]
Myrtus acuminata Kunth = ***Myrcia splendens*** (Sw.) DC.
Myrtus acuminatissima Blume = ***Syzygium acuminatissimum*** (Blume) DC.
Myrtus acunae Borhidi & O.Muñiz = ***Mosiera acunae*** (Borhidi & O.Muñiz) Bisse
Myrtus acuta Mill. = ***Myrtus communis*** L. subsp. ***communis***
Myrtus acutata O.Berg = ***Psidium salutare*** var. ***mucronatum*** (Cambess.) Landrum
Myrtus acutifolia (L.) Sennen & Teodoro = ***Myrtus communis*** L. subsp. ***communis***
Myrtus aemulans Schltr. = [60 NWC]
Myrtus aequalis Vell. = [84 BZL]
Myrtus aeruginosa Griseb. = [85 AGE]
Myrtus afzelii (Engl.) Kuntze = ***Eugenia afzelii*** Engl.
Myrtus aggregata Vell. = ***Eugenia cerasiflora*** Miq.
Myrtus alata Zipp. ex Span. = (Lecythidaceae)
Myrtus alaternoides Brongn. & Gris = ***Austromyrtus alaternoides*** (Brongn. & Gris) Burret
Myrtus alba Noronha = ***Melaleuca leucadendra*** (L.) L.
Myrtus albida Kunth = ?
Myrtus albida Vell. = ***Eugenia bimarginata*** DC.
Myrtus alpigena (DC.) Mart. = ***Myrceugenia alpigena*** (DC.) Landrum
Myrtus alpina Sw. = ***Eugenia alpina*** (Sw.) Willd.
Myrtus alternifolia Gleason = ***Calycolpus alternifolius*** (Gleason) Landrum
Myrtus amara (O.Berg) Arechav. = ***Blepharocalyx salicifolius*** (Kunth) O.Berg
Myrtus ambrosiaca Moritzi ex O.Berg = ***Calycolpus moritzianus*** (O.Berg) Burret
Myrtus anceps Spreng. = ***Myrcia anceps*** (Spreng.) O.Berg
Myrtus androsaemoides L. = ***Syzygium cordifolium*** subsp. ***spissum*** (Alston) P.S.Ashton
Myrtus aneityensis Guillaumin = ***Gossia aneityensis*** (Guillaumin) N.Snow
Myrtus angolensis (Engl.) Kuntze = ***Eugenia malangensis*** (O.Hoffm.) Nied.
Myrtus anguillensis Urb. = [81 LEE]
Myrtus angustifolia Raf. = ***Myrtus communis*** L. subsp. ***communis***
Myrtus angustifolia Parodi = [85 PAR]
Myrtus angustifolia (O.Berg) Arechav. = ***Blepharocalyx salicifolius*** (Kunth) O.Berg
Myrtus angustifolia (Lam.) Spreng. = ***Eugenia pomifera*** (Aubl.) Urb.
Myrtus angustifolia L. = ***Metrosideros angustifolia*** (L.) Sm.
Myrtus angustissima (O.Berg) Arechav. = ***Blepharocalyx salicifolius*** (Kunth) O.Berg
Myrtus ankarensis H.Perrier = ***Eugenia ankarensis*** (H.Perrier) A.J.Scott
Myrtus anomala (Britton & P.Wilson) Burret = ***Mosiera bullata*** (Britton & P.Wilson) Bisse
Myrtus aotearoana (E.C.Nelson) E.C.Nelson = ***Lophomyrtus bullata*** Burret
Myrtus apiculata (O.Berg) Kiaersk. = ***Blepharocalyx salicifolius*** (Kunth) O.Berg
Myrtus apiculata var. *microphylla* Kiaersk. = ***Blepharocalyx salicifolius*** (Kunth) O.Berg
Myrtus araneosa (Urb.) Bisse = ***Mosiera araneosa*** (Urb.) Bisse
Myrtus arayan Kunth = ***Psidium salutare*** var. ***salutare***
Myrtus arbutifolia (O.Berg) Kuntze = ***Eugenia punicifolia*** (Kunth) DC.
Myrtus archboldiana Merr. & L.M.Perry = ***Uromyrtus archboldiana*** (Merr. & L.M.Perry) A.J.Scott
Myrtus arfakensis Gibbs = ***Xanthomyrtus arfakensis*** (Gibbs) Diels
Myrtus aromatica Salisb. = ***Pimenta dioica*** (L.) Merr.
Myrtus aromatica Poir. = ***Pimenta dioica*** (L.) Merr.
Myrtus artensis (Montrouz.) Guillaumin & Beauvis. = ***Uromyrtus artensis*** (Montrouz.) Burret
Myrtus aubletii Spreng. = ***Myrcia tomentosa*** (Aubl.) DC.
Myrtus augustinii Sennen & Teodoro = ***Myrtus communis*** L. subsp. ***communis***
Myrtus auriculata (Blume) Zipp. ex Blume = ***Syzygium novoguineense*** Merr. & L.M.Perry
Myrtus australis (J.C.Wendl. ex Link) Spreng. = ***Syzygium australe*** (J.C.Wendl. ex Link) B.Hyland
Myrtus axillaris Sw. = ***Eugenia axillaris*** (Sw.) Willd.
Myrtus axillaris Poir. = ***Eugenia foetida*** Pers.
Myrtus baetica (L.) Mill. = ***Myrtus communis*** L. subsp. ***communis***
Myrtus baetica var. *vidalii* Sennen & Teodoro = ***Myrtus communis*** L. subsp. ***communis***
Myrtus bahamensis (Kiaersk.) Urb. = ***Mosiera longipes*** (O.Berg) Small
Myrtus baladensis Brongn. & Gris = ***Archirhodomyrtus baladensis*** (Brongn. & Gris) Burret
Myrtus balsamica (Jacq.) Spreng. = ***Myrcianthes fragrans*** (Sw.) McVaugh
Myrtus barneoudii O.Berg = ***Myrteola nummularia*** (Lam.) O.Berg
Myrtus baruensis (Jacq.) Spreng. = ***Eugenia axillaris*** (Sw.) Willd. var. ***axillaris***
Myrtus baui Sennen & Teodoro = ***Myrtus communis*** L. subsp. ***communis***
Myrtus baumanii Guillaumin = ***Uromyrtus baumanii*** (Guillaumin) N.Snow & Guymer
Myrtus beaurepairiana Kiaersk. = ***Eugenia ternatifolia*** Cambess.
Myrtus beckleri F.Muell. = ***Archirhodomyrtus beckleri*** (F.Muell.) A.J.Scott
Myrtus belgica (L.) Mill. = ***Myrtus communis*** L. subsp. ***communis***
Myrtus bellonis (Krug & Urb.) Burret = ***Eugenia bellonis*** Krug & Urb.
Myrtus bergiana Nied. = ***Psidium laruotteanum*** Cambess.
Myrtus berlandiereana O.Berg = [79 MXE MXG]
Myrtus berteroana Spreng. = ***Eugenia biflora*** (L.) DC.
Myrtus berteroi Phil. = ***Ugni selkirkii*** (Hook. & Arn.) O.Berg
Myrtus bicolor Kunth = ***Myrcianthes discolor*** (Kunth) McVaugh
Myrtus bidwillii Benth. = ***Gossia bidwillii*** (Benth.) N.Snow & Guymer

Myrtus biflora L. = ***Eugenia biflora*** (L.) DC.
Myrtus biflora var. *salicifolia* (O.Berg) Kuntze = ***Eugenia biflora*** (L.) DC.
Myrtus biflora f. *subsericea* Kuntze = ***Eugenia biflora*** (L.) DC.
Myrtus biflora var. *yapacani* Kuntze = ***Eugenia biflora*** (L.) DC.
Myrtus billardiana Kunth = ***Myrcia billardiana*** (Kunth) DC.
Myrtus blanchetiana O.Berg = ***Psidium salutare*** var. ***salutare***
Myrtus borbonica Spreng. = ***Eugenia buxifolia*** Lam.
Myrtus borbonis Sennen = ***Myrtus communis*** L. subsp. ***communis***
Myrtus brabantica Garsault = (Myricaceae)
Myrtus brachyphylla Kunze ex O.Berg = ***Myrceugenia leptospermoides*** (DC.) Kausel
Myrtus brachystemon DC. = ***Eugenia procera*** (Sw.) Poir.
Myrtus bracteata Willd. = ***Eugenia roxburghii*** DC.
Myrtus bracteifolia Sessé & Moç. = [81 PUE]
Myrtus bracteolaris Poir. = ***Myrcia splendens*** (Sw.) DC.
Myrtus brasiliana L. = ***Eugenia uniflora*** L.
Myrtus brasiliana var. *diversifolia* Kuntze = ***Eugenia uniflora*** L.
Myrtus brasiliana var. *lanceolata* Kuntze = ***Eugenia uniflora*** L.
Myrtus brasiliana var. *lucida* (O.Berg) Kuntze = ***Eugenia uniflora*** L.
Myrtus brasiliana var. *normalis* Kuntze = ***Eugenia uniflora*** L.
Myrtus brasiliana var. *parkeriana* (DC.) Kuntze = ***Eugenia procera*** (Sw.) Poir.
Myrtus brassii Merr. & L.M.Perry = ***Uromyrtus brassii*** (Merr. & L.M.Perry) A.J.Scott
Myrtus briquetii (Sennen & Teodoro) Sennen & Teodoro = ***Myrtus communis*** L. subsp. ***communis***
Myrtus brunnea (O.Berg) Kiaersk. = ***Blepharocalyx salicifolius*** (Kunth) O.Berg
Myrtus buchholzii (Engl.) Kuntze = ***Eugenia buchholzii*** Engl.
Myrtus bukobensis (Engl.) Kuntze = ***Eugenia capensis*** subsp. ***nyassensis*** (Engl.) F.White
Myrtus bullata Salisb. = ***Calyptranthes bullata*** (Salisb.) DC.
Myrtus bullata Sol. ex A.Cunn. = ***Lophomyrtus bullata*** Burret
Myrtus bullata (Britton & P.Wilson) Bisse = ***Mosiera bullata*** (Britton & P.Wilson) Bisse
Myrtus × *bullobcordata* Cockayne & Allan = ***Lophomyrtus* × *ralphii*** (Hook.f.) Burret
Myrtus buxifolia Raf. = ***Myrtus communis*** L. subsp. ***communis***
Myrtus buxifolia (Lam.) Poir. = ***Eugenia buxifolia*** Lam.
Myrtus buxifolia Sw. = ***Eugenia foetida*** Pers.
Myrtus buxoides (Urb.) Burret = ***Mosiera cabanasensis*** (Britton & P.Wilson) Borhidi subsp. ***cabanasensis***
Myrtus cabanasensis (Britton & P.Wilson) Alain = ***Mosiera cabanasensis*** (Britton & P.Wilson) Borhidi
Myrtus cainitoides Urb. = ***Pimenta cainitoides*** (Urb.) Burret
Myrtus caledoniae Spreng. = ***Syzygium deplanchei*** (Guillaumin) J.W.Dawson
Myrtus calophylla Kunth = ***Calycolpus calophyllus*** (Kunth) O.Berg
Myrtus calycolpoides (Griseb.) Burret = ***Mosiera calycolpoides*** (Griseb.) Borhidi
Myrtus calycolpoides subsp. *jackii* (Urb.) Borhidi = ***Mosiera jackii*** (Urb.) Bisse
Myrtus camphorata (O.Berg) Baill. = ***Myrceugenia exsucca*** (DC.) O.Berg
Myrtus candollei Barnéoud = ***Ugni candollei*** (Barnéoud) O.Berg
Myrtus canescens Lour. = ***Rhodomyrtus tomentosa*** (Aiton) Hassk. var. ***tomentosa***
Myrtus capensis Burm.f. = [27 CPP]
Myrtus capensis (Eckl. & Zeyh.) Harv. = ***Eugenia capensis*** (Eckl. & Zeyh.) Harv.
Myrtus capuli Schltdl. & Cham. = ***Eugenia capuli*** (Schltdl. & Cham.) Hook. & Arn.
Myrtus carnea G.Mey. = ***Myrcia coumete*** (Aubl.) DC.
Myrtus caryophyllata Vell. = ***Pimenta pseudocaryophyllus*** (Gomes) Landrum
Myrtus caryophyllata Jacq. = ***Pimenta racemosa*** (Mill.) J.W.Moore var. ***racemosa***
Myrtus caryophyllata L. = ***Syzygium caryophyllatum*** (L.) Alston
Myrtus caryophyllus Spreng. = ***Syzygium aromaticum*** (L.) Merr. & L.M.Perry
Myrtus casearioides Kunth = ***Eugenia casearioides*** (Kunth) DC.
Myrtus cassinoides Zipp. ex Blume = ?
Myrtus cassinoides (Lam.) Spreng. = ***Eugenia cassinoides*** Lam.
Myrtus caudata Wall. = ***Syzygium singaporense*** (King) Airy Shaw
Myrtus cauliflora Mart. = ***Plinia cauliflora*** (Mart.) Kausel
Myrtus cauliflora Blume = ***Syzygium cauliflorum*** T.G.Hartley & L.M.Perry
Myrtus cayennensis Spreng. = ***Eugenia citrifolia*** Poir.
Myrtus cerasiformis Blume = ***Syzygium racemosum*** (Blume) DC.
Myrtus cerasina Vahl = ***Eugenia ligustrina*** (Sw.) Willd. var. ***ligustrina***
Myrtus chekenilla Kuntze = ***Luma apiculata*** (DC.) Burret
Myrtus chequen (Molina) Spreng. = ***Luma chequen*** (Molina) A.Gray
Myrtus chinensis Lour. = (Symplocaceae)
Myrtus christinae (Sennen & Teodoro) Sennen & Teodoro = ***Myrtus communis*** L. subsp. ***communis***
Myrtus chrysocarpa Poepp. ex O.Berg = ***Myrceugenia chrysocarpa*** (O.Berg) Kausel
Myrtus chrysophyllum Spreng. = ***Myrceugenia rufa*** (Colla) Skottsb.
Myrtus chytraculia L. = ***Calyptranthes chytraculia*** (L.) Sw.
Myrtus citrifolia Willd. ex O.Berg = ***Calycolpus calophyllus*** (Kunth) O.Berg var. ***calophyllus***
Myrtus citrifolia Aubl. = ***Myrcia citrifolia*** (Aubl.) Urb.
Myrtus citrifolia Poir. = ***Pimenta racemosa*** (Mill.) J.W.Moore var. ***racemosa***
Myrtus claraensis (Urb.) Bisse = ***Psidium claraense*** Urb.
Myrtus clusiifolia Kunth = ***Myrcia clusiifolia*** (Kunth) DC.
Myrtus coaetanea (O.Berg) Kuntze = ***Eugenia coaetanea*** O.Berg
Myrtus coccolobifolia Kunth = ***Myrcianthes rhopaloides*** (Kunth) McVaugh
Myrtus commersonii Spreng. = ***Eugenia lucida*** Lam.
Myrtus communis Blanco = ***Decaspermum blancoi*** Vidal
Myrtus communis var. *acuminata* Rouy & E.G.Camus = ***Myrtus communis*** L. subsp. ***communis***
Myrtus communis var. *acutifolia* L. = ***Myrtus communis*** L. subsp. ***communis***
Myrtus communis var. *angustifolia* L. = ***Myrtus communis*** L. subsp. ***communis***

Myrtus communis var. *baetica* L. = ***Myrtus communis*** L. subsp. ***communis***
Myrtus communis var. *balearica* Sennen & Teodoro = ***Myrtus communis*** L. subsp. ***communis***
Myrtus communis var. *belgica* L. = ***Myrtus communis*** L. subsp. ***communis***
Myrtus communis var. *christinae* Sennen & Teodoro = ***Myrtus communis*** L. subsp. ***communis***
Myrtus communis var. *eusebii* Sennen & Teodoro = ***Myrtus communis*** L. subsp. ***communis***
Myrtus communis var. *foucaudii* Sennen & Teodoro = ***Myrtus communis*** L. subsp. ***communis***
Myrtus communis var. *gervasii* Sennen & Teodoro = ***Myrtus communis*** L. subsp. ***communis***
Myrtus communis var. *grandifolia* Sennen & Teodoro = ***Myrtus communis*** L. subsp. ***communis***
Myrtus communis var. *italica* (Mill.) Rouy & E.G.Camus = ***Myrtus communis*** L. subsp. ***communis***
Myrtus communis var. *joussetii* Sennen & Teodoro = ***Myrtus communis*** L. subsp. ***communis***
Myrtus communis var. *lusitanica* Rouy = ***Myrtus communis*** L. subsp. ***communis***
Myrtus communis var. *mucronata* L. = ***Myrtus communis*** L. subsp. ***communis***
Myrtus communis subsp. *mucronata* Pers. = ***Myrtus communis*** L. subsp. ***communis***
Myrtus communis var. *neapolitana* Sennen & Teodoro = ***Myrtus communis*** L. subsp. ***communis***
Myrtus communis var. *romana* L. = ***Myrtus communis*** L. subsp. ***communis***
Myrtus communis var. *tarentina* L. = ***Myrtus communis*** subsp. ***tarentina*** (L.) Nyman
Myrtus compacta Ridl. = ***Xanthomyrtus compacta*** (Ridl.) Diels
Myrtus complicata Kunth = ***Myrcia splendens*** (Sw.) DC.
Myrtus compressa Kunth = ***Myrcianthes fragrans*** (Sw.) McVaugh
Myrtus conceptionis Kuntze = ***Eugenia pyriformis*** Cambess.
Myrtus conferta Sessé & Moç. = [79]
Myrtus conspicua Vieill. ex Guillaumin = ***Austromyrtus conspicua*** (Vieill. ex Guillaumin) Burret
Myrtus coquimbensis Barnéoud = ***Myrcianthes coquimbensis*** (Barnéoud) Landrum & Grifo
Myrtus coquimbensis var. *rotundifolia* O.Berg = ***Myrcianthes coquimbensis*** (Barnéoud) Landrum & Grifo
Myrtus cordata Sw. = ***Eugenia cordata*** (Sw.) DC.
Myrtus coriacea Vahl = ***Myrcia citrifolia*** (Aubl.) Urb.
Myrtus coriandriformis Zipp. ex Blume = ***Decaspermum bracteatum*** (Roxb.) A.J.Scott var. ***bracteatum***
Myrtus coromandelina J.König ex Roxb. = ***Eugenia roxburghii*** DC.
Myrtus corticosa Spreng. = ***Syzygium cumini*** (L.) Skeels
Myrtus corynantha Kiaersk. = [84 BZL] *Psidium* sp. ?
Myrtus cotinifolia (Jacq.) Spreng. = ***Eugenia cotinifolia*** Jacq.
Myrtus cotinifolia J.F.Gmel. = ***Myrcia citrifolia*** (Aubl.) Urb.
Myrtus coumete (Aubl.) Spreng. = ***Myrcia coumete*** (Aubl.) DC.
Myrtus crenulata Sw. = ***Eugenia crenulata*** (Sw.) Willd.
Myrtus crenulata (Urb. & Ekman) Bisse = ***Mosiera crenulata*** (Urb. & Ekman) Borhidi
Myrtus cruckshanksii (Hook. & Arn.) Kuntze = ***Blepharocalyx cruckshanksii*** (Hook. & Arn.) Nied.
Myrtus cumini L. = ***Syzygium cumini*** (L.) Skeels
Myrtus curvipes Gand. = ***Uromyrtus curvipes*** (Gand.) Burret
Myrtus cuspidata O.Berg = ***Psidium salutare*** var. ***mucronatum*** (Cambess.) Landrum
Myrtus cuspidata var. *pentamera* O.Berg. = ***Psidium salutare*** var. ***mucronatum*** (Cambess.) Landrum
Myrtus cuspidata var. *tetramera* O.Berg. = ***Psidium salutare*** var. ***mucronatum*** (Cambess.) Landrum
Myrtus cymiflora F.Muell. = ***Archirhodomyrtus beckleri*** (F.Muell.) A.J.Scott
Myrtus cymosa (Lam.) Spreng. = ***Syzygium cymosum*** (Lam.) DC.
Myrtus cymosa Blume = ***Syzygium polyanthum*** (Wight) Walp.
Myrtus daphnoides Spreng. = ***Eugenia ruscifolia*** Poir.
Myrtus darwinii (Hook.f.) Barnéoud = ***Amomyrtus luma*** (Molina) D.Legrand & Kausel
Myrtus decaspermoides Domin = ***Gossia inophloia*** (J.F.Bailey & C.T.White) N.Snow & Guymer
Myrtus decussata Vell. = ***Eugenia decussata*** (Vell.) Mattos
Myrtus deflexa Kunth = ***Myrcia splendens*** (Sw.) DC.
Myrtus del-riscoi Borhidi & O.Muñiz = ***Mosiera del-riscoi*** (Borhidi & O.Muñiz) Borhidi
Myrtus densiflora Blume = ***Syzygium pycnanthum*** Merr. & L.M.Perry
Myrtus dichotoma Salisb. = ***Calyptranthes zuzygium*** (L.) Sw.
Myrtus dichotoma Poir. = ***Myrcianthes fragrans*** (Sw.) McVaugh
Myrtus dioica Spreng. = ***Myrcia schaueriana*** O.Berg
Myrtus dioica L. = ***Pimenta dioica*** (L.) Merr.
Myrtus discolor Kunth = ***Myrcianthes discolor*** (Kunth) McVaugh
Myrtus disperma Sessé & Moç. = [79]
Myrtus disticha Sw. = ***Eugenia disticha*** (Sw.) DC.
Myrtus divaricata Ham. = ***Eugenia axillaris*** (Sw.) Willd. var. ***axillaris***
Myrtus diversifolia (Brongn. & Gris) Guillaumin = ***Austromyrtus diversifolia*** (Brongn. & Gris) Burret
Myrtus dombeyi Spreng. = ***Eugenia brasiliensis*** Lam.
Myrtus dulcis C.T.White = ***Austromyrtus dulcis*** (C.T.White) L.S.Sm.
Myrtus dumetorum Poir. = ***Rhodamnia dumetorum*** (DC.) Merr. & L.M.Perry
Myrtus dumosa Spreng. = ***Blepharocalyx salicifolius*** (Kunth) O.Berg
Myrtus dumosa Vahl = ***Eugenia dumosa*** (Vahl) DC.
Myrtus dumosa L'Hér. ex DC. = ***Eugenia lheritieriana*** DC.
Myrtus dusenii (Engl.) Kuntze = ***Eugenia dusenii*** Engl.
Myrtus dysenterica Mart. = ***Eugenia dysenterica*** DC.
Myrtus ehrenbergii O.Berg = ***Mosiera ehrenbergii*** (O.Berg) Landrum
Myrtus ekmanii Urb. = ***Mosiera ekmanii*** (Urb.) Bisse
Myrtus elachantha F.Muell. = ***Decaspermum parviflorum*** (Lam.) A.J.Scott subsp. ***parviflorum***
Myrtus elegans DC. = ***Accara elegans*** (DC.) Landrum
Myrtus elegans Mart. ex DC. = ***Myrcia guianensis*** (Aubl.) DC.
Myrtus elegantula Kunze ex O.Berg = ***Myrceugenia planipes*** (Hook. & Arn.) O.Berg
Myrtus elliptica (Lam.) Spreng. = ***Eugenia elliptica*** Lam.
Myrtus elliptica C.Wright = ***Mosiera elliptica*** (C.Wright) Bisse
Myrtus elliptica Arechav. = ***Myrcianthes cisplatensis*** (Cambess.) O.Berg
Myrtus emarginata Sessé & Moç. = [81 PUE]
Myrtus emarginata Kunth = ***Eugenia emarginata*** (Kunth) DC.

Myrtus emarginata Pancher ex Brongn. & Gris = ***Uromyrtus emarginata*** (Pancher ex Baker f.) Burret
Myrtus engleriana Schltr. = [60 NWC]
Myrtus erythrocarpa Kunth = ***Eugenia erythrocarpa*** (Kunth) DC.
Myrtus erythroxyloides Kunth = ***Pseudanamomis umbellulifera*** (Kunth) Kausel
Myrtus eugeniiflora Kunth ex O.Berg = ***Acca macrostema*** (Ruiz & Pav. ex G.Don) McVaugh
Myrtus eugenioides A.J.Scott = ***Gossia eugenioides*** (A.J.Scott) N.Snow
Myrtus eusebii (Sennen & Teodoro) Sennen & Teodoro = ***Myrtus communis*** L. subsp. ***communis***
Myrtus exaltata F.M.Bailey = ***Syzygium luehmannii*** (F.Muell.) L.A.S.Johnson
Myrtus excelsa Cambess. = ***Hexachlamys edulis*** (O.Berg) Kausel & D.Legrand
Myrtus exsucca Mart. ex DC. = ***Myrcia guianensis*** (Aubl.) DC.
Myrtus fascicularis DC. = ***Campomanesia aromatica*** (Aubl.) Griseb.
Myrtus fasciculata Vell. = [84 BZL]
Myrtus fernandeziana Hook. & Arn. = ***Myrceugenia fernandeziana*** (Hook. & Arn.) Johow
Myrtus ferruginea (Poir.) Spreng. = ***Marlierea ferruginea*** (Poir.) McVaugh
Myrtus fimbriata Kunth = ***Myrcianthes fimbriata*** (Kunth) McVaugh
Myrtus firma Spreng. = ***Myrcia nobilis*** O.Berg
Myrtus flavicans Urb. & Ekman = ***Mosiera urbaniana*** Borhidi
Myrtus flavida Schltr. = [60 NWC]
Myrtus flavida Stapf = ***Xanthomyrtus flavida*** (Stapf) Diels
Myrtus flavida var. *glabrescens* Gibbs = ***Xanthomyrtus scolopacina*** (Ridl.) Diels
Myrtus floribunda (H.West ex Willd.) Spreng. = ***Myrciaria floribunda*** (H.West ex Willd.) O.Berg
Myrtus foetida (Pers.) Spreng. = ***Eugenia foetida*** Pers.
Myrtus foliosa Kunth = ***Myrcianthes leucoxyla*** (Ortega) McVaugh
Myrtus formosa Barb.Rodr. = ***Psidium laruotteanum*** Cambess.
Myrtus fragrans Sw. = ***Myrcianthes fragrans*** (Sw.) McVaugh
Myrtus fragrantissima F.Muell. ex Benth. = ***Gossia fragrantissima*** (F.Muell. ex Benth.) N.Snow & Guymer
Myrtus friedrichsthalii (O.Berg) Donn.Sm. & Standl. = ***Ugni myricoides*** (Kunth) O.Berg
Myrtus friedrichsthalii var. *brevipes* Donn.Sm. = ***Ugni myricoides*** (Kunth) O.Berg
Myrtus fulva Spreng. = [84]
Myrtus fulvescens (Mart. ex DC.) Kiaersk. = ***Pimenta pseudocaryophyllus*** var. ***fulvescens*** (Mart. ex DC.) Landrum
Myrtus fulvescens f. *glazioviana* Kiaersk. = ***Pimenta pseudocaryophyllus*** (Gomes) Landrum var. ***pseudocaryophyllus***
Myrtus gayana (Barnéoud) O.Berg = ***Luma chequen*** (Molina) A.Gray
Myrtus gayana var. *major* O.Berg = ***Luma chequen*** (Molina) A.Gray
Myrtus gervasii (Sennen & Teodoro) Sennen & Teodoro = ***Myrtus communis*** L. subsp. ***communis***
Myrtus glabra Vell. = ***Eugenia badia*** O.Berg
Myrtus glabrata Sw. = ***Eugenia glabrata*** (Sw.) DC.
Myrtus glabrata Blume = ***Syzygium glabratum*** (DC.) Veldkamp
Myrtus glazioviana Kiaersk. = [84 BZL] *Eugenia* sp. ?
Myrtus globosa Korth. = ***Rhodamnia cinerea*** Jack
Myrtus glomerata (Lam.) Spreng. = ***Syzygium glomeratum*** (Lam.) DC.
Myrtus goetheana Mart. ex DC. = ***Calycolpus goetheanus*** (Mart. ex DC.) O.Berg
Myrtus gomonenensis Guillaumin = ***Uromyrtus gomonenensis*** (Guillaumin) Burret
Myrtus gonoclada F.Muell. ex Benth. = ***Gossia gonoclada*** (F.Muell. ex Benth.) N.Snow & Guymer
Myrtus grammica Spreng. = ***Calyptranthes grammica*** (Spreng.) D.Legrand
Myrtus grandifolia O.Berg = ***Psidium firmum*** O.Berg
Myrtus greggii Sw. = ***Eugenia greggii*** (Sw.) Poir.
Myrtus greggii Nees & Mart. = ***Myrcia maximiliana*** O.Berg
Myrtus grumixama Vell. = ***Eugenia brasiliensis*** Lam.
Myrtus guajava (L.) Kuntze = ***Psidium guajava*** L.
Myrtus guayabillo (A.Rich.) C.Wright = ***Myrcianthes fragrans*** (Sw.) McVaugh
Myrtus guayabo Larrañaga = ***Myrcianthes fragrans*** (Sw.) McVaugh
Myrtus guayaquilensis Kunth = ***Eugenia guayaquilensis*** (Kunth) DC.
Myrtus gudilla Colla = ***Myrceugenia lanceolata*** (Juss. ex J.St.-Hil.) Kausel
Myrtus guianensis (Aubl.) Ham. = ***Myrcia guianensis*** (Aubl.) DC.
Myrtus guilii (Speg.) D.Legrand = ***Amomyrtella guilii*** (Speg.) Kausel
Myrtus guineensis (Sw.) Kuntze = ***Psidium guineense*** Sw.
Myrtus hassleriana Barb.Rodr. = ***Psidium salutare*** var. ***salutare***
Myrtus hauthalii Kuntze = ***Psidium guineense*** Sw.
Myrtus heteroclita (Tussac) Nied. = ***Myrcianthes fragrans*** (Sw.) McVaugh
Myrtus heynei Spreng. = ***Eugenia roxburghii*** DC.
Myrtus hillii Benth. = ***Gossia hillii*** (Benth.) N.Snow & Guymer
Myrtus horizontalis Vent. = ***Eugenia disticha*** (Sw.) DC.
Myrtus hypericifolia Salisb. = [29 REU]
Myrtus hypericifolia Blume = ***Syzygium blumei*** (Steud.) Merr. & L.M.Perry
Myrtus incana O.Berg = ***Psidium salutare*** var. ***sericeum*** (Cambess.) Landrum
Myrtus indica Walp. = ***Eugenia indica*** (Wight) Chithra
Myrtus inophloia J.F.Bailey & C.T.White = ***Gossia inophloia*** (J.F.Bailey & C.T.White) N.Snow & Guymer
Myrtus italica Mill. = ***Myrtus communis*** L. subsp. ***communis***
Myrtus italica var. *briquetii* Sennen & Teodoro = ***Myrtus communis*** L. subsp. ***communis***
Myrtus italica var. *petri-ludovici* Sennen & Teodoro = ***Myrtus communis*** L. subsp. ***communis***
Myrtus jaboticaba Vell. = ***Plinia cauliflora*** (Mart.) Kausel
Myrtus jacquiniana O.Berg = ***Psidium jacquinianum*** (O.Berg) Mattos
Myrtus jambos (L.) Kunth = ***Syzygium jambos*** (L.) Alston
Myrtus javanica Spreng. = ***Syzygium laurifolium*** (DC.) N.P.Balakr.
Myrtus javanica (Lam.) Blume = ***Syzygium samarangense*** (Blume) Merr. & L.M.Perry
Myrtus josephi Sennen & Teodoro = ***Myrtus communis*** L. subsp. ***communis***
Myrtus kalah Miq. = ***Syzygium kalahiense*** Korth.

Myrtus kameruniana (Engl.) Kuntze = ***Eugenia kameruniana*** Engl.
Myrtus klossii Ridl. = ***Xanthomyrtus compacta*** (Ridl.) Diels
Myrtus klossii var. *brevipedunculata* Diels = ***Xanthomyrtus arfakensis*** (Gibbs) Diels
Myrtus koebrensis Gibbs = ***Xanthomyrtus koebrensis*** (Gibbs) Diels
Myrtus krausei Phil. = ***Ugni candollei*** (Barnéoud) O.Berg
Myrtus kuma Barnéoud = [85 CLC]
Myrtus laevis Thunb. = [38 JAP]
Myrtus lancea (Poir.) Spreng. = ***Eugenia biflora*** (L.) DC.
Myrtus lanceolata Raf. = ***Myrtus communis*** L. subsp. ***communis***
Myrtus lanceolata (O.Berg) Arechav. = ***Blepharocalyx salicifolius*** (Kunth) O.Berg
Myrtus lanceolata Juss. ex J.St.-Hil. = ***Myrceugenia lanceolata*** (Juss. ex J.St.-Hil.) Kausel
Myrtus langsdorffii (O.Berg) Kuntze = ***Eugenia langsdorffii*** O.Berg
Myrtus lasioclada F.Muell. = ***Lenwebbia lasioclada*** (F.Muell.) N.Snow & Guymer
Myrtus latifolia Raf. = ***Myrtus communis*** L. subsp. ***communis***
Myrtus latifolia (Aubl.) Spreng. = ***Eugenia latifolia*** Aubl.
Myrtus latifolia B.Heyne ex Roth = ***Eugenia roxburghii*** DC.
Myrtus latifolia (O.Berg) V.M.Badillo = ***Myrcia calycampa*** Amshoff
Myrtus laurentii (Engl.) Kuntze = ***Eugenia malangensis*** (O.Hoffm.) Nied.
Myrtus laurina Retz. = (Symplocaceae)
Myrtus lechleriana (Miq.) Sealy = ***Amomyrtus luma*** (Molina) D.Legrand & Kausel
Myrtus ledophylla Standl. = ***Eugenia ledophylla*** (Standl.) McVaugh
Myrtus leiophloea (Urb.) Bisse = ***Mosiera bullata*** subsp. ***leiophloea*** (Urb.) Bisse
Myrtus leriocarpa C.Wright = ***Pimenta ferruginea*** (Griseb.) Burret
Myrtus leucadendra L. = ***Melaleuca leucadendra*** (L.) L.
Myrtus leucomyrtillus Griseb. = ***Myrteola nummularia*** (Lam.) O.Berg
Myrtus leucomyrtus Griseb. ex Phil. = ***Myrteola nummularia*** (Lam.) O.Berg
Myrtus leucoxyla Ortega = ***Myrcianthes leucoxyla*** (Ortega) McVaugh
Myrtus leucoxylon Korth. ex Miq. = ***Syzygium leucoxylon*** Korth.
Myrtus ligustrina Sw. = ***Eugenia ligustrina*** (Sw.) Willd.
Myrtus limbata Kunth = ***Myrcianthes fragrans*** (Sw.) McVaugh
Myrtus limbosa Ruiz ex O.Berg = ***Eugenia limbosa*** O.Berg
Myrtus lindleyana Kunth = ***Myrcianthes lindleyana*** (Kunth) McVaugh
Myrtus linearis J.F.Gmel. = ***Eugenia pomifera*** (Aubl.) Urb.
Myrtus lineata Sw. = ***Eugenia lineata*** (Sw.) DC.
Myrtus lineata Blume = ***Syzygium lineatum*** (DC.) Merr. & L.M.Perry
Myrtus linifolia Spreng. = ***Eugenia sprengelii*** DC.
Myrtus littoralis Salisb. = ***Myrtus communis*** L. subsp. ***communis***
Myrtus littoralis Roxb. ex Wight & Arn. = ***Eugenia roxburghii*** DC.
Myrtus longifolia Kunth = (Melastomataceae)
Myrtus longifolia Reinw. ex Blume = ***Decaspermum parviflorum*** (Lam.) A.J.Scott subsp. ***parviflorum***
Myrtus longipes (O.Berg) Kiaersk. = ***Blepharocalyx salicifolius*** (Kunth) O.Berg
Myrtus lotoides Guillaumin = ***Austromyrtus lotoides*** (Guillaumin) Burret
Myrtus loureiroi Spreng. = ***Eugenia nervosa*** Lour.
Myrtus lucida L. = ***Myrcia sylvatica*** (G.Mey.) DC.
Myrtus luma Molina = ***Amomyrtus luma*** (Molina) D.Legrand & Kausel
Myrtus luma Barnéoud = ***Amomyrtus luma*** (Molina) D.Legrand & Kausel
Myrtus luma Schauer = ***Luma chequen*** (Molina) A.Gray
Myrtus lurida Spreng. = ***Psidium salutare*** var. ***mucronatum*** (Cambess.) Landrum
Myrtus luteoviridis Baker f. = ***Austromyrtus luteoviridis*** (Baker f.) Burret
Myrtus macrochlamys (DC.) Mart. ex O.Berg = ***Algrizea macrochlamys*** (DC.) Proença & NicLugh.
Myrtus macrophylla J.St.-Hil. = ***Myrtus communis*** L. subsp. ***communis***
Myrtus macrophylla (Lam.) Spreng. = ***Syzygium malaccense*** (L.) Merr. & L.M.Perry
Myrtus madagascariensis H.Perrier = ***Eugenia madagascariensis*** (H.Perrier) A.J.Scott
Myrtus magnoliifolia Blume = ***Syzygium magnoliifolium*** (Blume) DC.
Myrtus major Garsault = ***Myrtus communis*** L. subsp. ***communis***
Myrtus major Kuntze = ***Eugenia capensis*** (Eckl. & Zeyh.) Harv. subsp. ***capensis***
Myrtus makul (Gaertn.) J.F.Gmel. = ***Syzygium makul*** Gaertn.
Myrtus malaccensis (L.) Spreng. = ***Syzygium malaccense*** (L.) Merr. & L.M.Perry
Myrtus malpighioides Kunth = ***Eugenia malpighioides*** (Kunth) DC.
Myrtus mapirensis Rusby = [83 BOL]
Myrtus marginata (Pers.) Spreng. = ***Myrcia citrifolia*** (Aubl.) Urb.
Myrtus maritima Kunth = ***Eugenia acapulcensis*** Steud.
Myrtus marquesii (Engl.) Kuntze = ***Eugenia malangensis*** (O.Hoffm.) Nied.
Myrtus matanzasia Urb. = ***Mosiera elliptica*** subsp. ***matanzasia*** (Urb.) Bisse
Myrtus matudai Lundell = ***Ugni myricoides*** (Kunth) O.Berg
Myrtus maxima Molina = ***Myrceugenia fernandeziana*** (Hook. & Arn.) Johow
Myrtus media Hoffmanns. = ***Myrtus communis*** L. subsp. ***communis***
Myrtus megapotamica Spreng. = [84]
Myrtus melastomoides F.Muell. = ***Rhodamnia rubescens*** (Benth.) Miq.
Myrtus meli Phil. = ***Amomyrtus meli*** (Phil.) D.Legrand & Kausel
Myrtus mespiloides (Lam.) Spreng. = ***Eugenia mespiloides*** Lam.
Myrtus metrosideros F.M.Bailey = ***Uromyrtus metrosideros*** (F.M.Bailey) A.J.Scott
Myrtus micarensis Urb. = [81 CUB]
Myrtus micrantha Kunth = ***Eugenia monticola*** (Sw.) DC.
Myrtus micrantha Nees & Mart. = ***Myrciaria floribunda*** (H.West ex Willd.) O.Berg
Myrtus microphylla J.St.-Hil. = ***Myrtus communis*** L. subsp. ***communis***
Myrtus microphylla Humb. & Bonpl. = ***Myrteola phylicoides*** (Benth.) Landrum var. ***phylicoides***

Myrtus mini (Aubl.) Spreng. = ***Eugenia biflora*** (L.) DC.
Myrtus minima Mill. = ***Myrtus communis*** L. subsp. ***communis***
Myrtus minor Garsault = ***Myrtus communis*** L. subsp. ***communis***
Myrtus × *miraflorensis* Borhidi & O.Muñiz = ***Mosiera*** × ***miraflorensis*** (Borhidi & O.Muñiz) Borhidi
Myrtus mirifolia Sennen & Teodoro = ***Myrtus communis*** L. subsp. ***communis***
Myrtus moana Urb. = [81 CUB]
Myrtus molinae (Turcz.) Barnéoud = ***Ugni molinae*** Turcz.
Myrtus mollis Kunth = ***Myrcia mollis*** (Kunth) DC.
Myrtus monosperma F.Muell. = ***Gossia lucida*** (Gaertn.) N.Snow & Guymer
Myrtus montana Benth. = ***Ugni myricoides*** (Kunth) O.Berg
Myrtus montevidensis Kuntze = [85 URU]
Myrtus montevidensis (O.Berg) Arechav. = ***Myrceugenia myrtoides*** O.Berg
Myrtus monticola Sw. = ***Eugenia monticola*** (Sw.) DC.
Myrtus moraviana (O.Berg) Kuntze = ***Eugenia moraviana*** O.Berg
Myrtus mossambicensis (Engl.) Kuntze = ***Eugenia aschersoniana*** F.Hoffm.
Myrtus moultonii Merr. = ***Xanthomyrtus moultonii*** (Merr.) Merr.
Myrtus mucronata Cambess. = ***Psidium salutare*** var. ***mucronatum*** (Cambess.) Landrum
Myrtus mucronata var. *opaca* O.Berg. = ***Psidium salutare*** var. ***mucronatum*** (Cambess.) Landrum
Myrtus mucronata var. *perforata* O.Berg. = ***Psidium salutare*** var. ***mucronatum*** (Cambess.) Landrum
Myrtus mucronata var. *thea* (Griseb.) Griseb. = ***Psidium salutare*** var. ***mucronatum*** (Cambess.) Landrum
Myrtus mulleri Korth. = ***Rhodamnia mulleri*** (Korth.) Blume
Myrtus multiflora Juss. ex J.St.-Hil. = ***Amomyrtus luma*** (Molina) D.Legrand & Kausel
Myrtus multiflora (Lam.) Spreng. = ***Myrcia multiflora*** (Lam.) DC.
Myrtus muniziana (Borhidi) Borhidi = ***Psidium munizianum*** Borhidi
Myrtus munizii Borhidi = ***Mosiera munizii*** (Borhidi) Bisse
Myrtus myrciopsis Kuntze = ***Myrcianthes osteomeloides*** (Rusby) McVaugh
Myrtus myricoides Kunth = ***Ugni myricoides*** (Kunth) O.Berg
Myrtus myricoides var. *roraimensis* (N.E.Br.) Steyerm. = ***Ugni myricoides*** (Kunth) O.Berg
Myrtus myricoides var. *stenophylla* (Oliv. ex Thurn) Steyerm. = ***Ugni myricoides*** (Kunth) O.Berg
Myrtus myricoides var. *turumiquirensis* Steyerm. = ***Ugni myricoides*** (Kunth) O.Berg
Myrtus myrsinoides Kunth = ***Myrcianthes myrsinoides*** (Kunth) Grifo
Myrtus myrtoides (O.Berg) Arechav. = ***Myrceugenia myrtoides*** O.Berg
Myrtus nekouana Guillaumin = ***Uromyrtus nekouana*** (Guillaumin) Burret
Myrtus neocaledonica Nied. = [60 NWC]
Myrtus ngoyensis Schltr. = ***Uromyrtus ngoyensis*** (Schltr.) Burret
Myrtus nigripes Guillaumin = ***Austromyrtus nigripes*** (Guillaumin) Burret
Myrtus nitida Vell. = ***Eugenia candolleana*** DC.
Myrtus nitida J.F.Gmel. = ***Gossia lucida*** (Gaertn.) N.Snow & Guymer
Myrtus nivalis Ridl. = ***Decaspermum nivale*** (Ridl.) Merr. & L.M.Perry
Myrtus nivea O.Berg = ***Psidium salutare*** var. ***sericeum*** (Cambess.) Landrum
Myrtus nobilis Larrañaga = [85 URU]
Myrtus nodosa (Engl.) Kuntze = ***Eugenia nodosa*** Engl.
Myrtus nummularia Lam. = ***Myrteola nummularia*** (Lam.) O.Berg
Myrtus nummularia Barnéoud = ***Myrteola nummularia*** (Lam.) O.Berg
Myrtus nummularia var. *major* Hook.f. = ***Myrteola nummularia*** (Lam.) O.Berg
Myrtus nummularioides (Britton & P.Wilson) Urb. = ***Mosiera nummularioides*** (Britton & P.Wilson) Bisse
Myrtus nyassensis (Engl.) Kuntze = ***Eugenia capensis*** subsp. ***nyassensis*** (Engl.) F.White
Myrtus oaxacana Standl. = ***Eugenia standleyi*** McVaugh
Myrtus obcordata (Raoul) Hook.f. = ***Lophomyrtus obcordata*** (Raoul) Burret
Myrtus obovata (Poir.) Spreng. = ***Syzygium cumini*** (L.) Skeels
Myrtus obovata Korth. ex Miq. = ***Syzygium muelleri*** (Miq.) Miq.
Myrtus obscura Lindl. = ***Eugenia obscura*** (Lindl.) DC.
Myrtus obtusissima Blume = ***Syzygium aqueum*** (Burm.f.) Alston
Myrtus oerstedeana O.Berg = ***Myrtus communis*** L. subsp. ***communis***
Myrtus oerstedii (O.Berg) Hemsl. = ***Ugni myricoides*** (Kunth) O.Berg
Myrtus oleifolia Kunth = ***Eugenia punicifolia*** (Kunth) DC.
Myrtus oligantha Urb. = ***Pimenta oligantha*** (Urb.) Burret
Myrtus oligoneura Korth. ex Blume = (Melastomataceae)
Myrtus oonophylla (Urb.) Burret = ***Mosiera oonophylla*** (Urb.) Bisse
Myrtus opaca C.T.White = ***Gossia hillii*** (Benth.) N.Snow & Guymer
Myrtus ophiticola (Britton & P.Wilson) Alain = ***Mosiera ophiticola*** (Britton & P.Wilson) Bisse
Myrtus orbicularis (O.Berg) Burret = ***Mosiera longipes*** (O.Berg) Small
Myrtus orbiculata (Lam.) Spreng. = ***Eugenia orbiculata*** Lam.
Myrtus oreogena Schltr. = [60 NWC]
Myrtus osteomeloides Rusby = ***Myrcianthes osteomeloides*** (Rusby) McVaugh
Myrtus ovalifolia Cambess. = [84+] Psidium ?
Myrtus ovalis O.Berg = ***Eugenia ovalis*** O.Berg
Myrtus ovalis Spreng. = ***Psidium ovale*** (Spreng.) Burret
Myrtus ovalis O.Berg = ***Psidium salutare*** var. ***mucronatum*** (Cambess.) Landrum
Myrtus oxycoccoides Benth. = ***Myrteola nummularia*** (Lam.) O.Berg
Myrtus paitensis Schltr. = ***Archirhodomyrtus paitensis*** (Schltr.) Burret
Myrtus pallens Vahl = ***Eugenia biflora*** (L.) DC.
Myrtus parviflora Sessé & Moç. = [81 PUE]
Myrtus parviflora (Lam.) Spreng. = ***Decaspermum parviflorum*** (Lam.) A.J.Scott
Myrtus patrisii (Vahl) Spreng. = ***Eugenia patrisii*** Vahl
Myrtus pauciflora Cambess. = ***Psidium salutare*** var. ***mucronatum*** (Cambess.) Landrum
Myrtus paulotchensis Guillaumin = ***Uromyrtus paulotchensis*** (Guillaumin) Burret
Myrtus pavonii Poepp. ex O.Berg = [83 PER]
Myrtus pedunculata Hook.f. = ***Neomyrtus pedunculata*** (Hook.f.) Allan
Myrtus pendula Blume = [42 JAW] (Rubiaceae) ?

Myrtus petri-ludovici (Sennen & Teodoro) Sennen & Teodoro = ***Myrtus communis*** L. subsp. ***communis***
Myrtus phillyraeoides (O.Berg) Willd. ex O.Berg = ***Myrciaria dubia*** (Kunth) McVaugh
Myrtus phillyreifolia (Baker) Kuntze = ***Syzygium phillyreifolium*** (Baker) Labat & Schatz
Myrtus phylicoides Benth. = ***Myrteola phylicoides*** (Benth.) Landrum
Myrtus pimenta L. = ***Pimenta dioica*** (L.) Merr.
Myrtus pimenta var. *breviflora* Hayne = ***Pimenta dioica*** (L.) Merr.
Myrtus pimenta var. *longifolia* Sims = ***Pimenta dioica*** (L.) Merr.
Myrtus pimentoides (DC.) T.Nees = ***Pimenta dioica*** (L.) Merr.
Myrtus piperita Sessé & Moç. = ***Pimenta dioica*** (L.) Merr.
Myrtus pitanga (O.Berg) Kuntze = ***Eugenia pitanga*** (O.Berg) Nied.
Myrtus pitanga var. *angustifolia* Kuntze = ***Eugenia pitanga*** (O.Berg) Nied.
Myrtus pitanga f. *fasciculata* Kuntze = ***Eugenia pitanga*** (O.Berg) Nied.
Myrtus pitanga f. *subsolitaria* Kuntze = ***Eugenia pitanga*** (O.Berg) Nied.
Myrtus poggei (Engl.) Kuntze = ***Eugenia malangensis*** (O.Hoffm.) Nied.
Myrtus poimbailensis Guillaumin = ***Austromyrtus poimbailensis*** (Guillaumin) Burret
Myrtus poiretii Spreng. = ***Eugenia foetida*** Pers.
Myrtus polyantha Kunth = ***Myrcia splendens*** (Sw.) DC.
Myrtus pomifera Aubl. = ***Eugenia pomifera*** (Aubl.) Urb.
Myrtus procera Sw. = ***Eugenia procera*** (Sw.) Poir.
Myrtus prolixa Baker f. = ***Austromyrtus prolixa*** (Baker f.) Burret
Myrtus prostrata Gibbs = ***Xanthomyrtus compacta*** (Ridl.) Diels
Myrtus pseudocaryophyllus Gomes = ***Pimenta pseudocaryophyllus*** (Gomes) Landrum
Myrtus pseudopsidium (Jacq.) Spreng. = ***Eugenia pseudopsidium*** Jacq.
Myrtus psidioides Desv. = ***Campomanesia aromatica*** (Aubl.) Griseb.
Myrtus psychotrioides Colla = ***Eugenia punicifolia*** (Kunth) DC.
Myrtus pubescens Kunth = ***Eugenia pubescens*** (Kunth) DC.
Myrtus pubescens O.Berg = ***Psidium salutare*** var. ***sericeum*** (Cambess.) Landrum
Myrtus pubiflora C.T.White = ***Gossia pubiflora*** (C.T.White) N.Snow & Guymer
Myrtus pulchella Regel = ?
Myrtus pulchrifolius Guillaumin = [60 NWC]
Myrtus punctata (Vahl) Spreng. = ***Myrcianthes fragrans*** (Sw.) McVaugh
Myrtus punicifolia Kunth = ***Eugenia punicifolia*** (Kunth) DC.
Myrtus pyrifolia J.St.-Hil. = ***Myrcia guianensis*** (Aubl.) DC.
Myrtus pyriformis (Cambess.) Parodi = ***Eugenia pyriformis*** Cambess.
Myrtus quadrangularis Buch.-Ham. ex Duthie = ***Syzygium antisepticum*** (Blume) Merr. & L.M.Perry
Myrtus quadriflora Vell. = ***Syzygium lanceolatum*** (Lam.) Wight & Arn.
Myrtus quadripartita Royen ex Blume = ***Eugenia roxburghii*** DC.
Myrtus quadrisperma Vell. = ***Eugenia pruniformis*** Cambess.
Myrtus racemosa Sessé & Moç. = ?
Myrtus racemosa Sessé & Moç. = [79]
Myrtus racemosa Vell. = ***Marlierea racemosa*** (Vell.) Kiaersk.
Myrtus racemulosa Benth. = ***Gossia bidwillii*** (Benth.) N.Snow & Guymer
Myrtus racemulosa var. *conferta* Benth. = ***Gossia bidwillii*** (Benth.) N.Snow & Guymer
Myrtus × ralphii Hook.f. = ***Lophomyrtus × ralphii*** (Hook.f.) Burret
Myrtus ramiflora Vahl = [81 LEE]
Myrtus randiana Merr. & L.M.Perry = ***Gossia randiana*** (Merr. & L.M.Perry) N.Snow
Myrtus raran Colla = ***Myrceugenia obtusa*** (DC.) O.Berg
Myrtus reinhardtiana Kiaersk. = ***Blepharocalyx salicifolius*** (Kunth) O.Berg
Myrtus reinhardtiana f. *subcordata* Kiaersk. = ***Blepharocalyx salicifolius*** (Kunth) O.Berg
Myrtus reinwardtiana Blume = ***Eugenia reinwardtiana*** (Blume) A.Cunn. ex DC.
Myrtus reloncavi Barnéoud ex F.Phil. = ***Amomyrtus luma*** (Molina) D.Legrand & Kausel
Myrtus repens Phil. = ***Myrteola nummularia*** (Lam.) O.Berg
Myrtus resupinata Vell. = ***Marlierea resupinata*** (Vell.) O.Berg
Myrtus reticulata Kunze ex C.Gay = [85]
Myrtus revoluta Schauer = ***Calycolpus revolutus*** (Schauer) O.Berg
Myrtus rhopaloides Kunth = ***Myrcianthes rhopaloides*** (Kunth) McVaugh
Myrtus rhytisperma F.Muell. = ***Pilidiostigma rhytispermum*** (F.Muell.) Burret
Myrtus rhytisperma var. *grandifolia* Benth. = ***Pilidiostigma glabrum*** Burret
Myrtus rigida (Sw.) J.F.Gmel. = ***Calyptranthes rigida*** Sw.
Myrtus rigida O.Berg = ***Psidium salutare*** var. ***salutare***
Myrtus rigidifolia (A.Rich.) Kuntze = ***Eugenia rigidifolia*** A.Rich.
Myrtus rodesi Sennen & Teodoro = ***Myrtus communis*** L. subsp. ***communis***
Myrtus romana (L.) Hoffmanns. = ***Myrtus communis*** L. subsp. ***communis***
Myrtus romanifolia J.St.-Hil. = ***Myrtus communis*** L. subsp. ***communis***
Myrtus roraimensis N.E.Br. = ***Ugni myricoides*** (Kunth) O.Berg
Myrtus rosmarinifolia (Poir.) Pers. = ***Eugenia pomifera*** (Aubl.) Urb.
Myrtus rufa Vell. = ***Marlierea rufa*** O.Berg
Myrtus rufa Colla = ***Myrceugenia rufa*** (Colla) Skottsb.
Myrtus rufescens (DC.) Spreng. = ***Myrceugenia rufescens*** (DC.) D.Legrand & Kausel
Myrtus rufopunctata Pancher ex Brongn. & Gris = ***Myrtastrum rufopunctatum*** (Pancher ex Brongn. & Gris) Burret
Myrtus ruscifolia Willd. = ***Eugenia roxburghii*** DC.
Myrtus sagraea O.Berg = ***Psidium salutare*** var. ***salutare***
Myrtus salicifolia Kunth = ***Blepharocalyx salicifolius*** (Kunth) O.Berg
Myrtus salicifolia Willd. ex O.Berg = ***Psidium maribense*** Mart. ex DC.
Myrtus saligna J.F.Gmel. = ***Melaleuca cajuputi*** Powell subsp. ***cajuputi***
Myrtus saligna Burm.f. = ***Melaleuca leucadendra*** (L.) L.
Myrtus salomonensis A.J.Scott = ***Gossia salomonensis*** (A.J.Scott) N.Snow

Myrtus salutaris Kunth = ***Psidium salutare*** (Kunth) O.Berg
Myrtus samarangensis Blume = ***Syzygium samarangense*** (Blume) Merr. & L.M.Perry
Myrtus sarandi Larrañaga = [85 URU]
Myrtus scabra Arruda = [84]
Myrtus selkirkii (Hook. & Arn.) Hemsl. = ***Ugni selkirkii*** (Hook. & Arn.) O.Berg
Myrtus selloi Spreng. = ***Myrcia selloi*** (Spreng.) N.Silveira
Myrtus sellowiana O.Berg = ***Psidium salutare*** var. ***mucronatum*** (Cambess.) Landrum
Myrtus sericea Cambess. = ***Psidium salutare*** var. ***sericeum*** (Cambess.) Landrum
Myrtus sericea var. *fruticosa* O.Berg. = ***Psidium salutare*** var. ***sericeum*** (Cambess.) Landrum
Myrtus sericea var. *sufffruticosa* O.Berg. = ***Psidium salutare*** var. ***sericeum*** (Cambess.) Landrum
Myrtus sericocalyx C.T.White = ***Decaspermum humile*** (Sweet ex G.Don) A.J.Scott
Myrtus sessiliflora (Vahl) Spreng. = ***Eugenia sessiliflora*** Vahl
Myrtus sessilis Korth. ex Miq. = ***Syzygium korthalsianum*** (Miq.) Miq.
Myrtus shepherdii F.Muell. = ***Gossia shepherdii*** (F.Muell.) N.Snow & Guymer
Myrtus silvestris Vell. = ***Marlierea silvestris*** (Vell.) Mattos
Myrtus sinemariensis (Aubl.) Spreng. = ***Eugenia coffeifolia*** DC.
Myrtus sintenisii (Krug & Urb.) Kiaersk. = ***Eugenia cordata*** var. ***sintenisii*** (Kiaersk.) Krug & Urb.
Myrtus smithii (Poir.) Spreng. = ***Syzygium smithii*** (Poir.) Nied.
Myrtus sonneratii Spreng. = ***Syzygium lanceolatum*** (Lam.) Wight & Arn.
Myrtus soyauxii (Engl.) Kuntze = ***Eugenia klaineana*** (Pierre) Engl.
Myrtus sparsifolia O.Berg = ***Myrtus communis*** L. subsp. ***communis***
Myrtus spectabilis Blume = ***Rhodamnia cinerea*** Jack
Myrtus spiciflora Spreng. = ***Eugenia sprengelii*** DC.
Myrtus splendens Sw. = ***Myrcia splendens*** (Sw.) DC.
Myrtus stahlii Kiaersk. = ***Eugenia stahlii*** (Kiaersk.) Krug & Urb.
Myrtus stenophylla Oliv. ex Thurn = ***Ugni myricoides*** (Kunth) O.Berg
Myrtus stictophylla Kiaersk. = ***Accara elegans*** (DC.) Landrum
Myrtus stipularis Hook. & Arn. = ***Tepualia stipularis*** (Hook. & Arn.) Griseb.
Myrtus stoupii Spreng. = ***Myrcia splendens*** (Sw.) DC.
Myrtus striatula (DC.) Kuntze = ***Psidium striatulum*** DC.
Myrtus stricta (O.Berg) Arechav. = ***Blepharocalyx salicifolius*** (Kunth) O.Berg
Myrtus styphelioides Schltr. = ***Austromyrtus styphelioides*** (Schltr.) Burret
Myrtus suaveolens Parodi = [85 PAR]
Myrtus subglomerata (Spring ex Mart.) Kuntze = ***Eugenia neoglomerata*** Sobral
Myrtus subrubens Blanco = ***Syzygium tripinnatum*** (Blanco) Merr.
Myrtus suffruticosa O.Berg = ***Psidium salutare*** var. ***mucronatum*** (Cambess.) Landrum
Myrtus suffruticosa var. *angustifolia* O.Berg. = ***Psidium salutare*** var. ***mucronatum*** (Cambess.) Landrum
Myrtus suffruticosa var. *latifolia* O.Berg. = ***Psidium salutare*** var. ***mucronatum*** (Cambess.) Landrum
Myrtus sunshinensis Guillaumin = ***Uromyrtus sunshinensis*** (Guillaumin) N.Snow & Guymer
Myrtus supraaxillaris Guillaumin = ***Uromyrtus supraaxillaris*** (Guillaumin) Burret
Myrtus sylvatica G.Mey. = ***Myrcia sylvatica*** (G.Mey.) DC.
Myrtus syzygium L. = ***Calyptranthes zuzygium*** (L.) Sw.
Myrtus tabasco Willd. ex Schltdl. & Cham. = ***Pimenta dioica*** (L.) Merr.
Myrtus tarentina (L.) Mill. = ***Myrtus communis*** subsp. ***tarentina*** (L.) Nyman
Myrtus taxifolia Ridl. = ***Xanthomyrtus taxifolia*** (Ridl.) Merr.
Myrtus tenuifolia Sm. = ***Austromyrtus tenuifolia*** (Sm.) Burret
Myrtus tenuifolia var. *latifolia* Maiden & Betche = ***Austromyrtus dulcis*** (C.T.White) L.S.Sm.
Myrtus theodori Sennen = ***Myrtus communis*** L. subsp. ***communis***
Myrtus thymifolia (Planch. ex Guillaumin) Guillaumin = ***Uromyrtus thymifolia*** (Planch. ex Guillaumin) Burret
Myrtus thyrsoidea Kuntze = ***Psidium riparium*** Mart. ex DC.
Myrtus tiburona (Urb. & Ekman) Borhidi = ***Mosiera tiburona*** (Urb. & Ekman) Borhidi
Myrtus timorensis Zipp. ex Span. = ***Syzygium aqueum*** (Burm.f.) Alston
Myrtus tinifolia (Lam.) Spreng. = ***Eugenia tinifolia*** Lam.
Myrtus togoensis (Engl.) Kuntze = ***Eugenia togoensis*** Engl.
Myrtus tomentosa Aiton = ***Rhodomyrtus tomentosa*** (Aiton) Hassk.
Myrtus tozerii F.Muell. = ***Rhodomyrtus psidioides*** (G.Don) Benth.
Myrtus trifida Sessé & Moç. = [81 PUE]
Myrtus triflora Spreng. = [85 AGE]
Myrtus triflora Jacq. = ***Eugenia triflora*** (Jacq.) Ham.
Myrtus trinervia Lour. = ***Rhodamnia dumetorum*** (DC.) Merr. & L.M.Perry
Myrtus trinervia Sm. = ***Rhodamnia rubescens*** (Benth.) Miq.
Myrtus trineura F.Muell. = ***Rhodomyrtus trineura*** (F.Muell.) Benth.
Myrtus tripinnata Blanco = ***Syzygium tripinnatum*** (Blanco) Merr.
Myrtus tripliphylla Urb. = ***Mosiera ophiticola*** (Britton & P.Wilson) Bisse
Myrtus trunciflora Schltdl. & Cham. = ***Eugenia trunciflora*** (Schltdl. & Cham.) G.Don
Myrtus tuberculata Kunth = ***Eugenia tuberculata*** (Kunth) DC.
Myrtus turbinata Schltr. = ***Archirhodomyrtus turbinata*** (Schltr.) Burret
Myrtus tussacii (Urb. & Ekman) Burret = ***Mosiera tussacii*** (Urb. & Ekman) Borhidi
Myrtus ugni Molina = ***Ugni molinae*** Turcz.
Myrtus ugni var. *angustifolia* Phil. = ***Ugni molinae*** Turcz.
Myrtus ugni var. *latifolia* Kuntze = ***Ugni molinae*** Turcz.
Myrtus ugni var. *poeppigii* Kuntze = ***Myrceugenia rufescens*** (DC.) D.Legrand & Kausel
Myrtus uliginosa Miq. = ***Luma chequen*** (Molina) A.Gray
Myrtus umbellata Desv. = (Melastomataceae)
Myrtus umbellulifera Kunth = ***Pseudanamomis umbellulifera*** (Kunth) Kausel
Myrtus umbilicata Cambess. = ***Blepharocalyx salicifolius*** (Kunth) O.Berg
Myrtus umbraticola Kunth = ***Marlierea umbraticola*** (Kunth) O.Berg
Myrtus undulata (Aubl.) Spreng. = ***Eugenia undulata*** Aubl.

Myrtus vaccinioides Pancher ex Brongn. & Gris = ?
Myrtus vaccinioides Kunth = ***Myrteola nummularia*** (Lam.) O.Berg
Myrtus vahlii Spreng. = ***Eugenia trinervia*** Vahl
Myrtus valdiviana Phil. = ***Amomyrtus luma*** (Molina) D.Legrand & Kausel
Myrtus variegata Blume = ***Syzygium minimum*** (Blume) Airy Shaw
Myrtus velutina (O.Berg) Kiaersk. = ***Pimenta pseudocaryophyllus*** var. ***fulvescens*** (Mart. ex DC.) Landrum
Myrtus velutina f. *macrophylla* Kiaersk. = ***Pimenta pseudocaryophyllus*** (Gomes) Landrum var. ***pseudocaryophyllus***
Myrtus veneris Bubani = ***Myrtus communis*** L. subsp. ***communis***
Myrtus venosa (Lam.) Spreng. = ***Syzygium dupontii*** (Baker) Govaerts
Myrtus verrucosa O.Berg = ***Mosiera longipes*** (O.Berg) Small
Myrtus versteeghii Merr. & L.M.Perry = ***Gossia versteeghii*** (Merr. & L.M.Perry) N.Snow
Myrtus verticillata Vell. = ***Eugenia verticillata*** (Vell.) Angely
Myrtus verticillata Salzm. ex O.Berg = ***Myrciaria floribunda*** (H.West ex Willd.) O.Berg
Myrtus vestita Spreng. = [84]
Myrtus vidalii (Sennen & Teodoro) Sennen & Teodoro = ***Myrtus communis*** L. subsp. ***communis***
Myrtus vieillardii Brongn. & Gris = ***Archirhodomyrtus vieillardii*** (Brongn. & Gris) Burret
Myrtus villosa Spreng. = (Monimiaceae)
Myrtus virgultosa Sw. = ***Eugenia biflora*** (L.) DC.
Myrtus viridis Vell. = ***Eugenia viridis*** (Vell.) O.Berg
Myrtus virotii Guillaumin = [60 NWC]
Myrtus vitiensis (A.Gray) F.Muell. = ***Decaspermum vitiense*** (A.Gray) Nied.
Myrtus vitis-idaea (Raoul) Druce = ***Neomyrtus pedunculata*** (Hook.f.) Allan
Myrtus vulcani Korth. = [42 SUM]
Myrtus warmingiana Kiaersk. = ***Eugenia rizzoana*** Mattos
Myrtus widgrenii (O.Berg) Kiaersk. = ***Blepharocalyx salicifolius*** (Kunth) O.Berg
Myrtus willdenowii Spreng. = ***Eugenia uniflora*** L.
Myrtus xalapensis Kunth = ***Eugenia xalapensis*** (Kunth) DC.
Myrtus xylopioides Kunth = ***Myrcia xylopioides*** (Kunth) DC.
Myrtus yapacani Kuntze = [84]
Myrtus zenkeri (Engl.) Kuntze = ?
Myrtus zeyheri Harv. = ***Eugenia capensis*** subsp. ***zeyheri*** (Harv.) F.White
Myrtus zeylanica L. = ***Syzygium zeylanicum*** (L.) DC.
Myrtus zuzygium L. = ***Calyptranthes zuzygium*** (L.) Sw.

***Unplaced Names*:**
Myrtus acuminata Sessé & Moç., Fl. Mexic., ed. 2: 124 (1894). = [81 PUE]
Myrtus aemulans Schltr., Bot. Jahrb. Syst. 40(29): 29 (1908). = [60 NWC]
Myrtus aequalis Vell., Fl. Flumin. 5: 217, t. 76 (1829). = [84 BZL]
Myrtus aeruginosa Griseb., Abh. Königl. Ges. Wiss. Göttingen 24: 127 (1879). = [85 AGE]
Myrtus anguillensis Urb., Symb. Antill. 6: 21 (1909). = [81 LEE]
Myrtus angustifolia Parodi, Anales Soc. Ci. Argent. 7: 186 (1879), nom. illeg. = [85 PAR]
Myrtus berlandiereana O.Berg, Linnaea 27: 403 (1856). = [79 MXE MXG]
Myrtus bracteifolia Sessé & Moç., Fl. Mexic., ed. 2: 124 (1894). = [81 PUE]
Myrtus capensis Burm.f., Fl. Indica, Prodr. Fl. Cap.: 14 (1768). = [27 CPP]
Myrtus conferta Sessé & Moç., Fl. Mexic., ed. 2: 125 (1894). = [79]
Myrtus corynantha Kiaersk., Enum. Myrt. Bras.: 18 (1893). = [84 BZL] *Psidium* sp. ?
Myrtus disperma Sessé & Moç., Fl. Mexic., ed. 2: 125 (1894). = [79]
Myrtus emarginata Sessé & Moç., Fl. Mexic., ed. 2: 124 (1894), nom. illeg. = [81 PUE]
Myrtus engleriana Schltr., Bot. Jahrb. Syst. 40(92): 29 (1908). = [60 NWC]
Myrtus fasciculata Vell., Fl. Flumin. 5: 210, t. 69 (1829). = [84 BZL]
Myrtus flavida Schltr., Bot. Jahrb. Syst. 40(91): 30 (1908), nom. illeg. = [60 NWC]
Myrtus fulva Spreng., Syst. Veg. 2: 487 (1825). = [84]
Myrtus glazioviana Kiaersk., Enum. Myrt. Bras.: 23 (1893). = [84 BZL] *Eugenia* sp. ?
Myrtus hypericifolia Salisb., Prodr. Stirp. Chap. Allerton: 354 (1796). = [29 REU]
Myrtus kuma Barnéoud in C.Gay, Fl. Chil. 2: 384 (1847). = [85 CLC]
Myrtus laevis Thunb. in J.A.Murray, Syst. Veg. ed. 14: 461 (1784). = [38 JAP]
Myrtus longifolia Kunth in F.W.H.von Humboldt, A.J.A.Bonpland & C.S.Kunth, Nov. Gen. Sp. 6: 258 (1824). = (Melastomataceae)
Myrtus mapirensis Rusby, Bull. New York Bot. Gard. 8: 108 (1912). = [83 BOL]
Myrtus megapotamica Spreng., Syst. Veg. 4(2): 193 (1827). = [84]
Myrtus micarensis Urb., Symb. Antill. 9: 81 (1923). = [81 CUB]
Myrtus moana Urb., Symb. Antill. 9: 81 (1923). = [81 CUB]
Myrtus montevidensis Kuntze, Revis. Gen. Pl. 3(2): 92 (1898). = [85 URU]
Myrtus neocaledonica Nied. in H.G.A.Engler & K.A.E.Prantl, Nat. Pflanzenfam. 3(7): 67 (1893). = [60 NWC]
Myrtus nobilis Larrañaga, Escritos D. A. Larrañaga 2: 169 (1923). = [85 URU]
Myrtus oreogena Schltr., Bot. Jahrb. Syst. 40(92): 30 (1908). = [60 NWC]
Myrtus ovalifolia Cambess. in A.F.C.de Saint-Hilaire, Fl. Bras. Merid. 2: 381 (1833). = [84+] Psidium ?
Myrtus parviflora Sessé & Moç., Fl. Mexic., ed. 2: 124 (1894). = [81 PUE]
Myrtus pavonii Poepp. ex O.Berg, Linnaea 27: 38 (1855). = [83 PER]
Myrtus pendula Blume, Bijdr.: 1085 (1826). = [42 JAW] (Rubiaceae) ?
Myrtus pulchella Regel, Index Seminum (LE) 1857: 56 (1857). = ?
Myrtus pulchrifolius Guillaumin, Mém. Mus. Natl. Hist. Nat., B, Bot. 8: 289 (1962). = [60 NWC]
Myrtus racemosa Sessé & Moç., Fl. Mexic., ed. 2: 125, col. 1 (1894). = ?
Myrtus racemosa Sessé & Moç., Fl. Mexic., ed. 2: 125, col. 2 (1894). = [79]
Myrtus ramiflora Vahl in H.West, Bidr. Beskr. Ste Croix: 290 (1793). = [81 LEE]
Myrtus reticulata Kunze ex C.Gay, Fl. Chil. 2: 380 (1847). = [85]
Myrtus sarandi Larrañaga, Escritos D. A. Larrañaga 3: 85 (1924). = [85 URU]

Myrtus scabra Arruda in H.Koster, Trav. Brazil: 491 (1816). = [84]
Myrtus suaveolens Parodi, Anales Soc. Ci. Argent. 7: 220 (1879). = [85 PAR]
Myrtus trifida Sessé & Moç., Fl. Mexic., ed. 2: 124 (1894). = [81 PUE]
Myrtus triflora Spreng., Syst. Veg. 2: 482 (1825). = [85 AGE]
Myrtus vestita Spreng., Syst. Veg. 4(2): 193 (1827). = [84]
Myrtus virotii Guillaumin, Mém. Mus. Natl. Hist. Nat., B, Bot. 4: 33 (1953). = [60 NWC]
Myrtus vulcani Korth., Ned. Kruidk. Arch. 1: 197 (1847). = [42 SUM]
Myrtus yapacani Kuntze, Revis. Gen. Pl. 3(2): 89 (1898). = [84]

Nani

Nani Adans. = ***Xanthostemon*** F.Muell.

Nania

Nania Miq. = ***Xanthostemon*** F.Muell.
Nania albiflora (Sol. ex Gaertn.) Kuntze = ***Metrosideros albiflora*** Sol. ex Gaertn.
Nania angustifolia (L.) Kuntze = ***Metrosideros angustifolia*** (L.) Sm.
Nania buxifolia (Brongn. & Gris) Kuntze = ***Cloezia buxifolia*** Brongn. & Gris
Nania canescens (Brongn. & Gris) Kuntze = ***Cloezia artensis*** (Montrouz.) P.S.Green var. ***artensis***
Nania chrysantha (F.Muell.) Kuntze = ***Xanthostemon chrysanthus*** (F.Muell.) Benth.
Nania ciliata (J.R.Forst. & G.Forst.) Kuntze = ***Purpureostemon ciliatus*** (J.R.Forst. & G.Forst.) Gugerli
Nania colensoi (Hook.f.) Kuntze = ***Metrosideros colensoi*** Hook.f.
Nania collina (J.R.Forst. & G.Forst.) Kuntze = ***Metrosideros collina*** (J.R.Forst. & G.Forst.) A.Gray
Nania comosa (Roxb.) Kuntze = ***Melaleuca cajuputi*** Powell subsp. ***cajuputi***
Nania deplanchei (Brongn. & Gris) Kuntze = ***Xanthostemon vieillardii*** (Brongn. & Gris) Nied.
Nania diffusa (G.Forst.) Kuntze = ***Metrosideros diffusa*** (G.Forst.) Sm.
Nania elegans (Brongn. & Gris) Kuntze = ***Xanthostemon multiflorus*** (Montrouz.) Beauvis.
Nania eucalyptoides (F.Muell.) Kuntze = ***Xanthostemon eucalyptoides*** F.Muell.
Nania fauriei H.Lév. = ***Metrosideros waialealae*** var. ***fauriei*** (H.Lév.) J.W.Dawson & Stemmerm.
Nania feddei H.Lév. = ***Metrosideros tremuloides*** (A.Heller) Rock
Nania flava (Brongn. & Gris) Kuntze = ***Xanthostemon multiflorus*** (Montrouz.) Beauvis.
Nania floribunda (Brongn. & Gris) Kuntze = ***Cloezia floribunda*** Brongn. & Gris
Nania florida (Sm.) Kuntze = ***Metrosideros fulgens*** Sol. ex Gaertn.
Nania glabrifolia A.Heller = ***Metrosideros polymorpha*** var. ***lutea*** H.Mann
Nania glomulifera (Sm.) Kuntze = ***Syncarpia glomulifera*** (Sm.) Nied.
Nania grisii Kuntze = ***Cloezia deplanchei*** Brongn. & Gris
Nania gummifera (Brongn. & Gris) Kuntze = ***Arillastrum gummiferum*** (Brongn. & Gris) Pancher ex Baill.
Nania hypericifolia (A.Cunn.) Kuntze = ***Metrosideros diffusa*** (G.Forst.) Sm.
Nania leptopetala (F.Muell.) Kuntze = ***Choricarpia leptopetala*** (F.Muell.) Domin
Nania ligustrina (Brongn. & Gris) Kuntze = ***Cloezia floribunda*** Brongn. & Gris
Nania lucida (A.Rich.) Kuntze = ***Metrosideros umbellata*** Cav.
Nania lutea A.Heller = ***Metrosideros collina*** var. ***villosa*** (L.f.) A.Gray
Nania macropus (Hook. & Arn.) Kuntze = ***Metrosideros macropus*** Hook. & Arn.
Nania nervulosa (C.Moore & F.Muell.) Kuntze = ***Metrosideros nervulosa*** C.Moore & F.Muell.
Nania nitida (Brongn. & Gris) Kuntze = ***Metrosideros nitida*** Brongn. & Gris
Nania obovata (Hook. & Arn.) Kuntze = ***Metrosideros collina*** var. ***villosa*** (L.f.) A.Gray
Nania operculata (Labill.) Kuntze = ***Metrosideros operculata*** Labill.
Nania paradoxa (F.Muell.) Kuntze = ***Xanthostemon paradoxus*** F.Muell.
Nania petiolata Valeton = ***Xanthostemon petiolatus*** (Valeton) Peter G.Wilson
Nania polymorpha (Gaudich.) A.Heller = ***Metrosideros polymorpha*** Gaudich.
Nania polymorpha var. *glaberrima* H.Lév. = ***Metrosideros polymorpha*** var. ***lutea*** H.Mann
Nania polymorpha var. *incana* H.Lév. = ***Metrosideros polymorpha*** var. ***imbricata*** (Rock) H.St.John
Nania polymorpha var. *nummularifolia* H.Lév. = ***Metrosideros polymorpha*** var. ***imbricata*** (Rock) H.St.John
Nania pumila A.Heller = ***Metrosideros polymorpha*** var. ***pumila*** (A.Heller) Skottsb.
Nania robusta (A.Cunn.) Kuntze = ***Metrosideros robusta*** A.Cunn.
Nania rubiginosa (Brongn. & Gris) Kuntze = ***Stereocaryum rubiginosum*** (Brongn. & Gris) Burret
Nania rubra (Brongn. & Gris) Kuntze = ***Xanthostemon ruber*** (Brongn. & Gris) Sebert & Pancher
Nania rugosa (A.Gray) Kuntze = ***Metrosideros rugosa*** A.Gray
Nania scandens (J.R.Forst. & G.Forst.) Kuntze = ***Metrosideros fulgens*** Sol. ex Gaertn.
Nania sessilifolia (Brongn. & Gris) Kuntze = ***Cloezia floribunda*** Brongn. & Gris
Nania stipularis (Hook. & Arn.) Kuntze = ***Tepualia stipularis*** (Hook. & Arn.) Griseb.
Nania suberosa (Roxb.) Kuntze = ***Xanthostemon verus*** (Lindl.) Peter G.Wilson
Nania ternifolia (F.Muell.) Kuntze = ***Lysicarpus angustifolius*** (Hook.) Druce
Nania tetrapetala (F.Muell.) Kuntze = ***Xanthostemon umbrosus*** (A.Cunn. ex Lindl.) Peter G.Wilson & J.T.Waterh.
Nania tomentosa (A.Rich.) Kuntze = ***Metrosideros excelsa*** Sol. ex Gaertn.
Nania tremuloides A.Heller = ***Metrosideros tremuloides*** (A.Heller) Rock
Nania vera (Lindl.) Miq. = ***Xanthostemon verus*** (Lindl.) Peter G.Wilson
Nania vieillardii (Brongn. & Gris) Kuntze = ***Xanthostemon vieillardii*** (Brongn. & Gris) Nied.

Nelitris

Nelitris Gaertn. = ***Decaspermum*** J.R.Forst. & G.Forst.
Nelitris alba Blume = ***Decaspermum parviflorum*** (Lam.) A.J.Scott subsp. ***parviflorum***
Nelitris alternifolia Miq. = ***Decaspermum parviflorum*** (Lam.) A.J.Scott subsp. ***parviflorum***

Nelitris billardierei Seem. = ***Uromyrtus billardierei*** (Seem.) A.J.Scott
Nelitris bracteata Blume = ***Decaspermum parviflorum*** (Lam.) A.J.Scott subsp. ***parviflorum***
Nelitris confinis Blume = ***Decaspermum parviflorum*** (Lam.) A.J.Scott subsp. ***parviflorum***
Nelitris coriandri Blume = ***Decaspermum bracteatum*** (Roxb.) A.J.Scott var. ***bracteatum***
Nelitris forsteri Seem. = ***Decaspermum fruticosum*** J.R.Forst. & G.Forst.
Nelitris fruticosa (J.R.Forst. & G.Forst.) A.Gray = ***Decaspermum fruticosum*** J.R.Forst. & G.Forst.
Nelitris humilis Sweet ex G.Don = ***Decaspermum humile*** (Sweet ex G.Don) A.J.Scott
Nelitris ingens F.Muell. ex C.Moore = ***Syzygium ingens*** (F.Muell. ex C.Moore) Craven & Biffin
Nelitris jambosella Gaertn. = ***Decaspermum fruticosum*** J.R.Forst. & G.Forst.
Nelitris laxiflora Blume = ***Decaspermum bracteatum*** (Roxb.) A.J.Scott var. ***bracteatum***
Nelitris leucocoma Miq. = ***Decaspermum parviflorum*** (Lam.) A.J.Scott subsp. ***parviflorum***
Nelitris lucida Blume = ***Decaspermum parviflorum*** (Lam.) A.J.Scott subsp. ***parviflorum***
Nelitris myrsinoides Blume = ***Decaspermum parviflorum*** (Lam.) A.J.Scott subsp. ***parviflorum***
Nelitris pallescens Miq. = ***Decaspermum parviflorum*** (Lam.) A.J.Scott subsp. ***parviflorum***
Nelitris paniculata Lindl. = ***Decaspermum parviflorum*** (Lam.) A.J.Scott subsp. ***parviflorum***
Nelitris paniculata var. *laxiflora* Benth. = ***Decaspermum humile*** (Sweet ex G.Don) A.J.Scott
Nelitris parviflora (Lam.) Blume = ***Decaspermum parviflorum*** (Lam.) A.J.Scott
Nelitris polygama Spreng. = ***Decaspermum parviflorum*** (Lam.) A.J.Scott subsp. ***parviflorum***
Nelitris polymorpha Blume = ***Decaspermum parviflorum*** (Lam.) A.J.Scott subsp. ***parviflorum***
Nelitris psidioides G.Don = ***Rhodomyrtus psidioides*** (G.Don) Benth.
Nelitris pubescens Miq. = ***Decaspermum parviflorum*** (Lam.) A.J.Scott subsp. ***parviflorum***
Nelitris pyrifolia Miq. = ***Decaspermum parviflorum*** (Lam.) A.J.Scott subsp. ***parviflorum***
Nelitris rubra Vidal = ***Decaspermum blancoi*** Vidal
Nelitris rubra Blume = ***Decaspermum bracteatum*** (Roxb.) A.J.Scott var. ***bracteatum***
Nelitris trinervia (Lour.) Spreng. = ***Rhodamnia dumetorum*** (DC.) Merr. & L.M.Perry
Nelitris urvillei DC. = ***Decaspermum urvillei*** (DC.) A.J.Scott
Nelitris vitiensis A.Gray = ***Decaspermum vitiense*** (A.Gray) Nied.

Neofabricia

Neofabricia Joy Thomps., Telopea 2: 380 (1983).
N. Queensland. 50 QLD.
3 Species
Fabricia Gaertn., Fruct. Sem. Pl. 1: 175 (1788), nom. illeg.

Neofabricia mjoebergii (Cheel) Joy Thomps., Telopea 2: 381 (1983).
N. Queensland. 50 QLD. Nanophan. or phan.
**Leptospermum mjoebergii* Cheel, J. Proc. Roy. Soc. New S. Wales 53: 120 (1919).

Neofabricia myrtifolia (Gaertn.) Joy Thomps., Telopea 2: 380 (1983).
N. Queensland. 50 QLD. Nanophan. or phan.
**Fabricia myrtifolia* Gaertn., Fruct. Sem. Pl. 1: 175 (1788).
Leptospermum fabricia Benth., Fl. Austral. 3: 102 (1867).

Neofabricia sericisepala J.R.Clarkson & Joy Thomps., Telopea 3: 294 (1989).
N. Queensland. 50 QLD. Nanophan. or phan.

Neomitranthes

Neomitranthes D.Legrand, in Fl. Ilustr. Catar. 1(Mirt.): 671 (1977).
Brazil. 84 BZC BZE BZL BZS.
17 Species

Neomitranthes amblymitra (Burret) Mattos, Loefgrenia 76: 2 (1981).
Brazil (São Paulo). 84 BZL. Phan.
**Mitranthes amblymitra* Burret, Notizbl. Bot. Gart. Berlin-Dahlem 15: 539 (1941).

Neomitranthes apiculata (Burret) Mattos, Loefgrenia 76: 2 (1981).
Brazil (Rio de Janeiro). 84 BZL. Phan.
**Mitranthes apiculata* Burret, Notizbl. Bot. Gart. Berlin-Dahlem 15: 539 (1941).

Neomitranthes capivariensis (Mattos) Mattos, Loefgrenia 76: 2 (1981).
Brazil (São Paulo). 84 BZL. Phan.
**Mitranthes capivariensis* Mattos, Arq. Bot. Estado São Paulo, n.s., f.m., 4: 53 (1966).

Neomitranthes castellanosii (Mattos & D.Legrand) Mattos, Loefgrenia 76: 2 (1981).
SE. Brazil. 84 BZL. Phan.
**Mitranthes castellanosii* Mattos & D.Legrand, Loefgrenia 67: 7 (1975).

Neomitranthes cordifolia (D.Legrand) D.Legrand, Fl. Ill. Rio Grande do Sul, Mirtac.: 141 (1989).
S. Brazil. 84 BZS. Nanophan. or phan.
**Mitranthes cordifolia* D.Legrand, Comun. Bot. Mus. Hist. Nat. Montevideo 3(36): 9 (1958).

Neomitranthes gemballae (D.Legrand) D.Legrand, in Fl. Ilustr. Catar. 1(Mirt.): 676 (1977).
Brazil (Paraná, Santa Catarina). 84 BZS. Phan.
**Mitranthes gemballae* D.Legrand, Sellowia 13: 332 (1961).

Neomitranthes glomerata (D.Legrand) Govaerts
S. Brazil (to São Paulo). 84 BZL BZS. Phan.
**Mitranthes glomerata* D.Legrand, Comun. Bot. Mus. Hist. Nat. Montevideo 3(36): 14 (1958).
Calyptranthes ranulphii D.Legrand, in Fl. Ilustr. Catar. 1(Mirtac.): 495 (1971).

Neomitranthes gracilis (Burret) N.Silveira, Roessléria 4: 123 (1981).
Brazil (São Paulo). 84 BZL. Phan.
**Mitranthes gracilis* Burret, Notizbl. Bot. Gart. Berlin-Dahlem 15: 538 (1941).

Neomitranthes hoehnei (Burret) Mattos, Loefgrenia 76: 2 (1981).
Brazil (São Paulo to Paraná). 84 BZL BZS. Phan.
**Mitranthes hoehnei* Burret, Notizbl. Bot. Gart. Berlin-Dahlem 15: 538 (1941).
Mitranthes dusenii Kausel, Lilloa 33: 104 (1971 publ. 1972).

Neomitranthes langsdorffii (O.Berg) Mattos, Loefgrenia 76: 2 (1981).

NE. Brazil. 84 BZE. Nanophan. or phan.
Mitranthes langsdorffii O.Berg in C.F.P.von Martius & auct. suc. (eds.), Fl. Bras. 14(1): 595 (1859). *Chytraculia bergiana* Kuntze, Revis. Gen. Pl. 1: 238 (1891).

Neomitranthes nitida Mattos, Loefgrenia 110: 2 (1997).
Brazil (São Paulo). 84 BZL. Nanophan. or phan.

Neomitranthes obscura (DC.) N.Silveira, Roesslèria 4: 123 (1981).
Brazil. 84 BZC BZL BZS. Nanophan. or phan.
Calyptranthes obscura DC., Prodr. 3: 257 (1828). *Chytraculia obscura* (DC.) Kuntze, Revis. Gen. Pl. 1: 238 (1891). *Mitranthes obscura* (DC.) D.Legrand, Comun. Bot. Mus. Hist. Nat. Montevideo 3(36): 16 (1958).
Calyptranthes obscura Mart., Flora 20(2): 87 (1837), nom. illeg.
Calyptranthes maschalantha O.Berg in C.F.P.von Martius & auct. suc. (eds.), Fl. Bras. 14(1): 53 (1857). *Chytraculia maschalantha* (O.Berg) Kuntze, Revis. Gen. Pl. 1: 238 (1891).
Calyptranthes tuberculata O.Berg in C.F.P.von Martius & auct. suc. (eds.), Fl. Bras. 14(1): 52 (1857). *Chytraculia tuberculata* (O.Berg) Kuntze, Revis. Gen. Pl. 1: 238 (1891).
Myrciaria tolypantha var. *pubescens* O.Berg in C.F.P.von Martius & auct. suc. (eds.), Fl. Bras. 14(1): 372 (1857). *Myrciaria perforata* O.Berg in C.F.P.von Martius & auct. suc. (eds.), Fl. Bras. 14(1): 598 (1859).
Calyptranthes obscura var. *fluminensis* O.Berg in C.F.P.von Martius & auct. suc. (eds.), Fl. Bras. 14(1): 542 (1859).
Calyptranthes maschalantha var. *rotundifolia* O.Berg, Linnaea 30: 652 (1861).

Neomitranthes obtusa Sobral & Zambom, Novon 12: 112 (2002).
Brazil (Espírito Santo). 84 BZL.

Neomitranthes pedicellata (Burret) Mattos, Loefgrenia 76: 2 (1981).
Brazil (São Paulo). 84 BZL. Phan.
Mitranthes pedicellata Burret, Repert. Spec. Nov. Regni Veg. 50: 55 (1941).

Neomitranthes riedeliana (O.Berg) Mattos, Loefgrenia 99: 6 (1990).
SE. Brazil. 84 BZL. Nanophan. or phan.
Mitranthes riedeliana O.Berg in C.F.P.von Martius & auct. suc. (eds.), Fl. Bras. 14(1): 595 (1859). *Chytraculia riedeliana* (O.Berg) Kuntze, Revis. Gen. Pl. 1: 238 (1891). *Calyptrogenia riedeliana* (O.Berg) Burret, Notizbl. Bot. Gart. Berlin-Dahlem 15: 546 (1941).

Neomitranthes warmingiana (Kiaersk.) Mattos, Loefgrenia 76: 2 (1981).
SE. Brazil. 84 BZL. Nanophan. or phan.
Calyptranthes warmingiana Kiaersk., Enum. Myrt. Bras.: 40 (1893). *Mitranthes warmingiana* (Kiaersk.) Burret, Notizbl. Bot. Gart. Berlin-Dahlem 15: 537 (1941).

Neomitranthes wilsoniana Mattos, Loefgrenia 94: 7 (1989).
Brazil (Rio de Janeiro). 84 BZL. Nanophan.

Synonyms:
Neomitranthes ekmanii (Urb.) Mattos = ***Calyptrogenia grandiflora*** Burret
Neomitranthes hatschbachii (D.Legrand) Mattos = ***Myrceugenia gertii*** Landrum
Neomitranthes maria-emiliae (D.Legrand) Mattos = ***Myrceugenia ovalifolia*** (O.Berg) Landrum

Neomyrtus

Neomyrtus Burret, Notizbl. Bot. Gart. Berlin-Dahlem 15: 493 (1941).
New Zealand. 51 NZN NZS.
1 Species

Neomyrtus pedunculata (Hook.f.) Allan, Fl. N. Zeal. 1: 330 (1961).
New Zealand. 51 NZN NZS. Nanophan. or phan.
Eugenia vitis-idaea Raoul, Ann. Sci. Nat., Bot., III, 2: 122 (1844). *Myrtus vitis-idaea* (Raoul) Druce, Rep. Bot. Exch. Club Brit. Isles 1916: 637 (1917). *Neomyrtus vitis-idaea* (Raoul) Burret, Notizbl. Bot. Gart. Berlin-Dahlem 15: 494 (1941).
Myrtus pedunculata Hook.f., Hooker's Icon. Pl. 7: t. 629 (1844).

Synonyms:
Neomyrtus vitis-idaea (Raoul) Burret = ***Neomyrtus pedunculata*** (Hook.f.) Allan

Neuhofia

Neuhofia Stokes = ***Baeckea*** L.
Neuhofia rosmarinifolia Stokes = ***Baeckea frutescens*** L.

Nothomyrcia

Nothomyrcia Kausel = ***Myrceugenia*** O.Berg
Nothomyrcia fernandeziana (Hook. & Arn.) Kausel = ***Myrceugenia fernandeziana*** (Hook. & Arn.) Johow
Nothomyrcia maxima (Molina) Gunckel = ***Myrceugenia fernandeziana*** (Hook. & Arn.) Johow

Ochrosperma

Ochrosperma Trudgen, Nuytsia 6: 11 (1987).
E. Australia. 50 NSW NTA QLD.
6 Species

Ochrosperma adpressum A.R.Bean, Austrobaileya 4: 387 (1995).
NE. Queensland. 50 QLD. Nanophan.

Ochrosperma citriodorum (Penfold & J.L.Willis) Trudgen, Nuytsia 6: 14 (1987).
NE. New South Wales. 50 NSW. Nanophan.
Baeckea citriodora Penfold & J.L.Willis, J. Proc. Roy. Soc. New S. Wales 89: 186 (1956).

Ochrosperma lineare (C.T.White) Trudgen, Nuytsia 6: 12 (1987).
Queensland to NE. New South Wales. 50 NSW QLD. Nanophan.
Baeckea linearis C.T.White, Proc. Roy. Soc. Queensland 55: 65 (1944).

Ochrosperma obovatum A.R.Bean, Austrobaileya 5: 499 (1999).
SE. Queensland. 50 QLD. Nanophan.

Ochrosperma oligomerum (Radlk.) A.R.Bean, Austrobaileya 4: 389 (1995).
EC. New South Wales. 50 NSW. Nanophan.
Baeckea oligomera Radlk., Ber. Deutsch. Bot. Ges. 2: 264 (1884).
Ochrosperma monticola Trudgen, Nuytsia 6: 15 (1987).

Ochrosperma sulcatum A.R.Bean, Austrobaileya 4: 648 (1997).
Northern Territory. 50 NTA. Nanophan.

***Synonyms*:**
Ochrosperma monticola Trudgen = ***Ochrosperma oligomerum*** (Radlk.) A.R.Bean

Octamyrtus

Octamyrtus Diels, Bot. Jahrb. Syst. 57: 373 (1922).
Maluku to New Guinea. 42 MOL 43 NWG.
6 Species

Octamyrtus arfakensis Kaneh. & Hatus. ex C.T.White, J. Arnold Arbor. 32: 144 (1951). *Octamyrtus pleiopetala* var. *arfakensis* (Kaneh. & Hatus. ex C.T.White) A.J.Scott, Kew Bull. 33: 305 (1978).
W. New Guinea. 43 NWG. Phan.

Octamyrtus behrmannii Diels, Bot. Jahrb. Syst. 57: 376 (1922).
New Guinea. 43 NWG. Phan.

Octamyrtus glomerata Kaneh. & Hatus. ex C.T.White, J. Arnold Arbor. 32: 145 (1951).
New Guinea. 43 NWG. Phan.

Octamyrtus halmaherensis Craven & Sunarti, Gard. Bull. Singapore 56: 150 (2004).
Maluku. 42 MOL. Phan.

Octamyrtus insignis Diels, Bot. Jahrb. Syst. 57: 374 (1922).
New Guinea. 43 NWG. Phan.

Octamyrtus pleiopetala Diels, Bot. Jahrb. Syst. 57: 373 (1922).
New Guinea (incl. Kep. Aru). 43 NWG. Phan.
**Eugenia pleiopetala* F.Muell., Descr. Notes Papuan Pl. 1: 106 (1877), nom. provis.
Octamyrtus lanceolata C.T.White, J. Arnold Arbor. 32: 145 (1951).

***Synonyms*:**
Octamyrtus elegans (Blume) Steenis = ***Rhodomyrtus elegans*** (Blume) A.J.Scott
Octamyrtus lanceolata C.T.White = ***Octamyrtus pleiopetala*** Diels
Octamyrtus pleiopetala var. *arfakensis* (Kaneh. & Hatus. ex C.T.White) A.J.Scott = ***Octamyrtus arfakensis*** Kaneh. & Hatus. ex C.T.White

Olynthia

Olynthia Lindl. = ***Eugenia*** P.Micheli ex L.
Olynthia disticha (Sw.) Lindl. = ***Eugenia disticha*** (Sw.) DC.

Opa

Opa Lour. = ***Syzygium*** Gaertn.
Opa integerrima Seem. = (Rosaceae)
Opa japonica Seem. = (Rosaceae)
Opa mertensii Seem. = (Rosaceae)
Opa metrosideros Lour. = (Rosaceae)
Opa odorata Lour. = ***Syzygium odoratum*** (Lour.) DC.
Opa spiralis Seem. = (Rosaceae)

Opanea

Opanea Raf. = ***Rhodamnia*** Jack
Opanea billardiana (Kunth) Raf. = ***Myrcia billardiana*** (Kunth) DC.
Opanea trinervia (Lour.) Raf. = ***Rhodamnia dumetorum*** (DC.) Merr. & L.M.Perry

Orestion

Orestion Kunze ex O.Berg = ***Myrteola*** O.Berg
Orestion vaccinioides Kunze ex O.Berg = ***Myrteola nummularia*** (Lam.) O.Berg

Orthostemon

Orthostemon O.Berg = ***Acca*** O.Berg
Orthostemon obovatus O.Berg = ***Acca sellowiana*** (O.Berg) Burret
Orthostemon sellowianus O.Berg = ***Acca sellowiana*** (O.Berg) Burret

Osbornia

Osbornia F.Muell., Fragm. 3: 30 (1862).
Philippines, N. Australia. 42 JAW? PHI 50 NTA QLD WAU.
1 Species

Osbornia octodonta F.Muell., Fragm. 3: 31 (1862).
Philippines, N. Australia. 42 JAW? PHI 50 NTA QLD WAU. Nanophan. or phan.

Oxymyrrhine

Oxymyrrhine Schauer = ***Baeckea*** L.
Oxymyrrhine gracilis Schauer = ***Baeckea polyandra*** F.Muell.

Ozandra

Ozandra Raf. = ***Melaleuca*** L.
Ozandra granulata Raf. = ***Melaleuca thymifolia*** Sm.
Ozandra hyssopifolia (Cav.) Raf. = ***Melaleuca linariifolia*** Sm.

Paivaea

Paivaea O.Berg = ***Campomanesia*** Ruiz & Pav.
Paivaea langsdorffii O.Berg = ***Campomanesia phaea*** (O.Berg) Landrum
Paivaea langsdorffii var. *lauroana* Mattos = ***Campomanesia phaea*** (O.Berg) Landrum
Paivaea phaea (O.Berg) Mattos = ***Campomanesia phaea*** (O.Berg) Landrum
Paivaea phaea var. *lauroana* (Mattos) Mattos = ***Campomanesia phaea*** (O.Berg) Landrum

Paramitranthes

Paramitranthes Burret = ***Siphoneugena*** O.Berg
Paramitranthes bracteata Burret = ***Siphoneugena densiflora*** O.Berg var. ***densiflora***
Paramitranthes chnoosepala (Kiaersk.) Burret = ***Siphoneugena densiflora*** O.Berg var. ***densiflora***
Paramitranthes densiflora (O.Berg) Burret = ***Siphoneugena densiflora*** O.Berg
Paramitranthes kiaerskoviana Burret = ***Siphoneugena kiaerskoviana*** (Burret) Kausel
Paramitranthes macrophylla Burret = ***Siphoneugena densiflora*** O.Berg var. ***densiflora***
Paramitranthes regnelliana (Kiaersk.) Burret = ***Siphoneugena densiflora*** O.Berg var. ***densiflora***
Paramitranthes sulcata Burret = [84 BZL] *Plinia* sp. ?

***Unplaced Names*:**
Paramitranthes sulcata Burret, Notizbl. Bot. Gart. Berlin-Dahlem 15: 542 (1941). = [84 BZL] *Plinia* sp. ?

Paramyrciaria

Paramyrciaria Kausel = ***Myrciaria*** O.Berg
Paramyrciaria ciliolata (Cambess.) Rotman = ***Myrciaria floribunda*** (H.West ex Willd.) O.Berg
Paramyrciaria delicatula (DC.) Kausel = ***Myrciaria delicatula*** (DC.) O.Berg
Paramyrciaria delicatula var. *argentinensis* Kausel = ***Myrciaria delicatula*** (DC.) O.Berg
Paramyrciaria delicatula var. *linearifolia* (O.Berg) O.Berg = ***Myrciaria delicatula*** (DC.) O.Berg
Paramyrciaria glazioviana (Kiaersk.) Sobral = ***Myrciaria glazioviana*** (Kiaersk.) G.M.Barroso ex Sobral
Paramyrciaria glomerata (O.Berg) Sobral = ***Myrciaria glazioviana*** (Kiaersk.) G.M.Barroso ex Sobral
Paramyrciaria guaquiea (Kiaersk.) Sobral = ***Myrciaria guaquiea*** (Kiaersk.) Mattos & D.Legrand
Paramyrciaria strigipes (O.Berg) Sobral = ***Myrciaria strigipes*** O.Berg
Paramyrciaria tapiraguayensis (Barb.Rodr. ex Chodat & Hassl.) Sobral = ***Myrciaria rojasii*** D.Legrand

Pareugenia

Pareugenia Turrill = ***Syzygium*** Gaertn.
Pareugenia brackenridgei (A.Gray) A.C.Sm. = ***Syzygium brackenridgei*** (A.Gray) Müll.Stuttg.
Pareugenia imthurnii Turrill = ***Syzygium brackenridgei*** (A.Gray) Müll.Stuttg.
Pareugenia oblongifolia Gillespie = ***Syzygium oblongifolium*** (Gillespie) Merr. & L.M.Perry
Pareugenia oligadelpha Christoph. = ***Syzygium oligadelphum*** (Christoph.) Merr. & L.M.Perry

Paryphantha

Paryphantha Schauer = ***Thryptomene*** Endl.
Paryphantha camphorata F.Muell. ex Miq. = ***Thryptomene saxicola*** (A.Cunn. ex Hook.) Schauer
Paryphantha cuspidata Turcz. = ***Thryptomene cuspidata*** (Turcz.) J.W.Green
Paryphantha mitchelliana Schauer = ***Thryptomene calycina*** (Lindl.) Stapf

Pentagonaster

Pentagonaster Klotzsch = ***Kunzea*** Rchb.
Pentagonaster baxteri Klotzsch = ***Kunzea baxteri*** (Klotzsch) Schauer

Pericalymma

Pericalymma (Endl.) Endl., Gen. Pl.: 1230 (1840).
SW. Australia. 50 WAU.
4 Species
Pericalymna Meisn., Pl. Vasc. Gen.: 354 (1843), orth. var.

Pericalymma crassipes (Lehm.) Schauer in J.G.C.Lehmann, Pl. Preiss. 1: 120 (1844).
SW. Australia. 50 WAU. Cham.
**Leptospermum crassipes* Lehm., Index Seminum (HBG) 1842: 5 (1842).

Pericalymma ellipticum (Endl.) Schauer in J.G.C.Lehmann, Pl. Preiss. 1: 120 (1844).
SW. & S. Western Australia. 50 WAU. Nanophan.
**Leptospermum ellipticum* Endl. in S.L.Endlicher & al., Enum. Pl.: 51 (1837). *Billottia elliptica* (Endl.) Sweet, Hort. Brit., ed. 3: 249 (1839).

var. ***ellipticum***
SW. & S. Western Australia. 50 WAU. Nanophan.

var. ***floridum*** (Schauer) Cranfield, Nuytsia 13: 15 (1999).
SW. Australia. 50 WAU. Nanophan.
**Pericalymma floridum* Schauer in J.G.C.Lehmann, Pl. Preiss. 1: 121 (1844). *Leptospermum floridum* (Schauer) Benth., Fl. Austral. 3: 110 (1867).

Pericalymma megaphyllum Cranfield, Nuytsia 13: 18 (1999).
SW. Australia. 50 WAU. Cham.

Pericalymma spongiocaule Cranfield, Nuytsia 13: 20 (1999).
SW. Australia. 50 WAU. Nanophan.

Synonyms:
Pericalymma floridum Schauer = ***Pericalymma ellipticum*** var. ***floridum*** (Schauer) Cranfield
Pericalymma × *roseum* Turcz. = ***Kunzea*** × ***rosea*** (Turcz.) Govaerts
Pericalymma teretifolium Turcz. = ***Kunzea pauciflora*** Schauer

Pericalymna

Pericalymna Meisn. = ***Pericalymma*** (Endl.) Endl.

Petraeomyrtus

Petraeomyrtus Craven, Austral. Syst. Bot. 12: 678 (1999).
N. Australia. 50 NTA.
1 Species

Petraeomyrtus punicea (Byrnes) Craven, Austral. Syst. Bot. 12: 678 (1999).
Northern Territory (Alligator Rivers Reg.). 50 NTA. Nanophan.
**Melaleuca punicea* Byrnes, Austrobaileya 2: 74 (1984). *Regelia punicea* (Byrnes) Barlow, Brunonia 9: 95 (1987).

Phyllocalyx

Phyllocalyx O.Berg = ***Eugenia*** P.Micheli ex L.
Phyllocalyx calycinus (Cambess.) O.Berg = ***Eugenia calycina*** Cambess.
Phyllocalyx calystegius O.Berg = ***Eugenia calystegia*** (O.Berg) Nied.
Phyllocalyx cerasiflorus O.Berg = ***Eugenia cerasiflora*** Miq.
Phyllocalyx cupreus O.Berg = ***Eugenia cuprea*** (O.Berg) Nied.
Phyllocalyx edulis O.Berg = ***Eugenia selloi*** B.D.Jacks.
Phyllocalyx edulis var. *depauperata* O.Berg = ***Eugenia selloi*** B.D.Jacks.
Phyllocalyx edulis var. *dives* O.Berg = ***Eugenia selloi*** B.D.Jacks.
Phyllocalyx formosus (Cambess.) O.Berg = ***Eugenia neoformosa*** Sobral
Phyllocalyx glandulosus (Cambess.) D.Legrand = ***Eugenia glandulosa*** Cambess.
Phyllocalyx grandifolius O.Berg = ***Eugenia santensis*** Kiaersk.
Phyllocalyx grandifolius var. *pyriformis* O.Berg = ***Eugenia santensis*** Kiaersk.
Phyllocalyx herbaceus O.Berg = ***Eugenia calycina*** var. ***herbacea*** (O.Berg) Mattos
Phyllocalyx involucratus (DC.) O.Berg = ***Eugenia involucrata*** DC.
Phyllocalyx laevigatus O.Berg = ***Eugenia involucrata*** DC.
Phyllocalyx ligustrinus (Sw.) O.Berg = ***Eugenia ligustrina*** (Sw.) Willd.

Phyllocalyx limbatus O.Berg = ***Eugenia speciosa*** Cambess.
Phyllocalyx luschnathianus O.Berg = ***Eugenia luschnathiana*** (O.Berg) Klotzsch ex B.D.Jacks.
Phyllocalyx macrosepalus O.Berg = ***Eugenia macrocalyx*** Mart. ex B.D.Jacks.
Phyllocalyx marginatus O.Berg = ***Eugenia speciosa*** Cambess.
Phyllocalyx membranaceus O.Berg = ***Eugenia membranifolia*** Nied.
Phyllocalyx pubescens O.Berg = ***Eugenia puberula*** Nied.
Phyllocalyx racemosus O.Berg = ***Eugenia elongata*** Nied.
Phyllocalyx regelianus O.Berg = ***Eugenia goyazensis*** Nied.
Phyllocalyx retusus O.Berg = ***Eugenia speciosa*** Cambess.
Phyllocalyx riedelianus O.Berg = ***Eugenia cavalcanteana*** Mattos
Phyllocalyx speciosus (Cambess.) O.Berg = ***Eugenia speciosa*** Cambess.
Phyllocalyx strictus O.Berg = ***Eugenia strictissima*** Govaerts
Phyllocalyx tomentosus O.Berg = ?
Phyllocalyx wentii Amshoff = ***Eugenia wentii*** Amshoff

Phymatocarpus

Phymatocarpus F.Muell., Fragm. 3: 120 (1862).
SW. Australia. 50 WAU.
3 Species

Phymatocarpus interioris Craven, Muelleria 12: 133 (1999).
S. Western Australia. 50 WAU. Nanophan.

Phymatocarpus maxwellii F.Muell., Fragm. 9: 45 (1862).
S. Western Australia. 50 WAU. Nanophan.
Regelia sparsiflora W.Fitzg., J. Bot. 50: 21 (1912).

Phymatocarpus porphyrocephalus F.Muell., Fragm. 3: 121 (1862).
W. Western Australia. 50 WAU. Nanophan.

Pileanthus

Pileanthus Labill., Nov. Holl. Pl. 2: 11 (1806).
SW. Australia. 50 WAU.
8 Species

Pileanthus aurantiacus Keighery, Nuytsia 15: 44 (2002).
W. Western Australia. 50 WAU. Nanophan.

Pileanthus bellus Keighery, Nuytsia 15: 46 (2002).
W. Western Australia. 50 WAU. Nanophan.

Pileanthus filifolius Meisn., J. Proc. Linn. Soc., Bot. 1: 45 (1857).
WSW. Western Australia. 50 WAU. Nanophan.

Pileanthus limacis Labill., Nov. Holl. Pl. 2: 11 (1806).
W. Western Australia. 50 WAU. Cham. or nanophan.

Pileanthus peduncularis Endl., Stirp. Herb. Hügel.: 3 (1838).
WSW. Western Australia. 50 WAU. Nanophan.

subsp. ***peduncularis***
W. & WSW. Western Australia. 50 WAU. Nanophan.
Chamelaucium dilatata J.Drumm., J. Bot. (Hooker) 5: 402 (1853).

subsp. ***pilifer*** Keighery, Nuytsia 15: 50 (2002).
W. Western Australia. 50 WAU. Nanophan.

Pileanthus rubronitidus Keighery, Nuytsia 15: 48 (2002).
W. Western Australia. 50 WAU. Nanophan.

Pileanthus septentrionalis Keighery, Nuytsia 15: 43 (2002).
W. Western Australia. 50 WAU. Nanophan.

Pileanthus vernicosus F.Muell., Fragm. 1: 255 (1859).
W. Western Australia. 50 WAU. Nanophan.

Pilidiostigma

Pilidiostigma Burret, Notizbl. Bot. Gart. Berlin-Dahlem 15: 547 (1941).
New Guinea to NE. New South Wales. 43 NWG 50 NSW QLD.
6 Species

Pilidiostigma glabrum Burret, Notizbl. Bot. Gart. Berlin-Dahlem 15: 548 (1941).
SE. Queensland to NE. New South Wales. 50 NSW QLD. Nanophan. or phan.
Myrtus rhytisperma var. *grandifolia* Benth., Fl. Austral. 3: 274 (1867).

Pilidiostigma papuanum (Lauterb.) A.J.Scott, Kew Bull. 33: 327 (1978).
S. New Guinea to N. Queensland. 43 NWG 50 QLD. Nanophan. or phan.
**Decaspermum papuanum* Lauterb., Nova Guinea 8: 319 (1910). *Jambosa papuana* (Lauterb.) Diels, Bot. Jahrb. Syst. 57: 384 (1922).
Rhodomyrtus recurva C.T.White, Proc. Roy. Soc. Queensland 47: 63 (1935 publ. 1936). *Pilidiostigma recurvum* (C.T.White) A.J.Scott, Kew Bull. 33: 326 (1978).
Eugenia brassii Merr. & L.M.Perry, J. Arnold Arbor. 23: 247 (1942).

Pilidiostigma rhytispermum (F.Muell.) Burret, Notizbl. Bot. Gart. Berlin-Dahlem 15: 550 (1941).
SE. Queensland to NE. New South Wales. 50 NSW QLD. Nanophan. or phan.
**Myrtus rhytisperma* F.Muell., Fragm. 1: 77 (1858).

Pilidiostigma sessile N.Snow, Syst. Bot. 29: 395 (2004).
NE. Queensland. 50 QLD.

Pilidiostigma tetramerum L.S.Sm., Proc. Roy. Soc. Queensland 70: 30 (1959).
N. Queensland. 50 QLD. Nanophan. or phan.

Pilidiostigma tropicum L.S.Sm., Proc. Roy. Soc. Queensland 67: 36 (1956).
NE. Queensland. 50 QLD. Nanophan. or phan.

Synonyms:
Pilidiostigma cuneatum Burret = ***Archirhodomyrtus beckleri*** (F.Muell.) A.J.Scott
Pilidiostigma parviflorum Burret = ***Archirhodomyrtus beckleri*** (F.Muell.) A.J.Scott
Pilidiostigma recurvum (C.T.White) A.J.Scott = ***Pilidiostigma papuanum*** (Lauterb.) A.J.Scott

Piliocalyx

Piliocalyx Brongn. & Gris, Bull. Soc. Bot. France 12: 185 (1865).
New Caledonia. 60 NWC.
8 Species

Piliocalyx boudoninii Brongn. & Gris, Bull. Soc. Bot. France 12: 186 (1865).
New Caledonia. 60 NWC. Nanophan. or phan.

Piliocalyx bullatus Brongn. & Gris, Bull. Soc. Bot. France 13: 470 (1866).
New Caledonia. 60 NWC. Nanophan. or phan.

Piliocalyx eugenioides Guillaumin, Bull. Soc. Bot. France 85: 651 (1938 publ. 1939).
New Caledonia. 60 NWC. Nanophan. or phan.

Piliocalyx francii Guillaumin, Bull. Mus. Natl. Hist. Nat., II, 10: 625 (1938 publ. 1939).
New Caledonia. 60 NWC. Nanophan. or phan.

Piliocalyx laurifolius Brongn. & Gris, Bull. Soc. Bot. France 12: 186 (1865).
New Caledonia. 60 NWC. Nanophan. or phan.

Piliocalyx micranthus Brongn. & Gris, Bull. Soc. Bot. France 12: 186 (1865).
New Caledonia. 60 NWC. Nanophan. or phan.

Piliocalyx robustus Brongn. & Gris, Bull. Soc. Bot. France 12: 185 (1865).
New Caledonia. 60 NWC. Nanophan. or phan.

Piliocalyx wagapensis Brongn. & Gris, Bull. Soc. Bot. France 13: 471 (1866).
New Caledonia. 60 NWC. Nanophan. or phan.

Synonyms:
Piliocalyx concinnus A.C.Sm. = ***Syzygium concinnum*** (A.C.Sm.) Craven & Biffin

Pilothecium

Pilothecium (Kiaersk.) Kausel = ***Eugenia*** P.Micheli ex L.
Pilothecium beaurepairianum (Kiaersk.) Kausel = ***Eugenia ternatifolia*** Cambess.
Pilothecium glazíovianum (Kiaersk.) Kausel = ?
Pilothecium itajurense (Cambess.) Kausel = ***Eugenia itajurensis*** Cambess.
Pilothecium lutescens (Cambess.) Kausel = ***Eugenia lutescens*** Cambess.
Pilothecium ternatifolium (Cambess.) Kausel = ***Eugenia ternatifolia*** Cambess.

Pimenta

Pimenta Lindl., Coll. Bot.: t. 19 (1821).
Mexico Trop. America. (61) sci 79 MXG MXS MXT 80 BLZ COS ELS GUA HON NIC PAN 81 CUB DOM HAI JAM LEE PUE TRT WIN 82 VEN 83 BOL 84 BZC BZE BZL BZS.
15 Species
Evanesca Raf., Sylva Tellur.: 105 (1838).
Pimentus Raf., Sylva Tellur.: 105 (1838).
Amomis O.Berg, Linnaea 27: 416 (1856).
Pseudocaryophyllus O.Berg, Linnaea 27: 415 (1856).
Cryptorhiza Urb., Repert. Spec. Nov. Regni Veg. 17: 403 (1921).
Krokia Urb., Symb. Antill. 9: 468 (1928).
Myrtekmania Urb., Symb. Antill. 9: 484 (1928).
Mentodendron Lundell, Wrightia 4: 180 (1971).

Pimenta adenoclada (Urb.) Alain, Mem. New York Bot. Gard. 21(2): 138 (1971).
Cuba. 81 CUB. Nanophan. or phan.
**Eugenia adenoclada* Urb., Symb. Antill. 9: 109 (1923). *Myrtekmania adenoclada* (Urb.) Urb., Symb. Antill. 9: 485 (1928). *Krokia adenoclada* (Urb.) Borhidi & O.Muñiz, Bot. Közlem. 64: 213 (1977 publ. 1978).

Pimenta cainitoides (Urb.) Burret, Notizbl. Bot. Gart. Berlin-Dahlem 15: 514 (1941).
E. Cuba to Dominican Rep. 81 CUB DOM. Phan.
**Myrtus cainitoides* Urb., Repert. Spec. Nov. Regni Veg. 19: 301 (1924). *Krokia cainitoides* (Urb.) Urb., Symb. Antill. 9: 470 (1928).
Amomis pilotoana Urb., Symb. Antill. 9: 471 (1928). *Pimenta pilotoana* (Urb.) Borhidi, Bot. Közlem. 62: 26 (1975). *Krokia pilotoana* (Urb.) Borhidi, Bot. Közlem. 64: 213 (1977 publ. 1978).
Krokia nipensis Urb., Symb. Antill. 9: 468 (1928). *Pimenta nipensis* (Urb.) Burret, Notizbl. Bot. Gart. Berlin-Dahlem 15: 512 (1941).

Pimenta dioica (L.) Merr., Contr. Gray Herb., n.s., 165: 337 (1947).
Mexico to C. America, Caribbean. (61) sci 79 MXG MXS MXT 80 BLZ COS ELS GUA HON NIC 81 CUB DOM JAM pue (82) ven. Phan. Dried unripe fruit are known as allspice.
Myrtus pimenta L., Sp. Pl.: 472 (1753). *Caryophyllus pimenta* (L.) Mill., Gard. Dict. ed. 8: 2 (1768). *Myrtus aromatica* Salisb., Prodr. Stirp. Chap. Allerton: 354 (1796), nom. illeg. *Pimentus aromatica* Raf., Sylva Tellur.: 158 (1838). *Pimenta officinalis* Lindl., Coll. Bot.: t. 19 (1821). *Eugenia pimenta* (L.) DC., Prodr. 3: 285 (1828). *Pimenta aromatica* Kostel., Allg. Med.-Pharm. Fl. 4: 1525 (1835), nom. illeg. *Pimentus vera* Raf., Sylva Tellur.: 105 (1838), nom. illeg. *Pimenta pimenta* (L.) H.Karst., Deut. Fl.: 790 (1882), nom. inval.
**Myrtus dioica* L., Syst. Nat. ed. 10, 2: 1056 (1759). *Evanesca crassifolia* Raf., Sylva Tellur.: 105 (1838), nom. illeg.
Myrtus aromatica Poir. in J.B.A.P.M.de Lamarck, Encycl. 4: 410 (1798), nom. illeg.
Myrtus pimenta var. *longifolia* Sims, Bot. Mag. 30: t. 1236 (1809). *Pimenta officinalis* var. *longifolia* (Sims) O.Berg, Linnaea 27: 423 (1856).
Myrtus pimenta var. *breviflora* Hayne, Getreue Darstell. Gew. 9: t. 37 (1825).
Eugenia pimenta var. *longifolia* DC., Prodr. 3: 285 (1828).
Eugenia pimenta var. *ovalifolia* DC., Prodr. 3: 285 (1828). *Pimenta officinalis* var. *ovalifolia* (DC.) O.Berg, Linnaea 27: 424 (1856).
Pimenta vulgaris Lindl. in J.C. Loudon, Encycl. Pl.: 418 (1829).
Myrtus tabasco Willd. ex Schltdl. & Cham., Linnaea 5: 559 (1830). *Pimenta officinalis* var. *tabasco* (Willd. ex Schltdl. & Cham.) O.Berg, Linnaea 27: 425 (1856). *Pimenta dioica* var. *tabasco* (Willd. ex Schltdl. & Cham.) Standl., Ceiba 3: 172 (1953). *Eugenia tabasco* (Willd. ex Schltdl. & Cham.) G.Don, Gen. Hist. 2: 866 (1832). *Pimenta tabasco* (Willd. ex Schltdl. & Cham.) Lundell, Wrightia 2: 58 (1960).
Myrtus pimentoides (DC.) T.Nees in M.F.Weihe & al., Pl. Offic., Suppl.: t. 89 (1833).
Pimentus geminata Raf., Sylva Tellur.: 105 (1838).
Eugenia micrantha Bertol., Fl. Guatimal.: 22 (1840).
Pimenta officinalis var. *cumanensis* Schiede & Deppe, Linnaea 27: 424 (1856).
Pimenta communis Benth. & Hook.f., Gen. Pl. 1: 717 (1865).
Pimenta vulgaris Bello, Anales Soc. Esp. Hist. Nat. 10: 270 (1881).
Myrtus piperita Sessé & Moç., Fl. Mexic., ed. 2: 124 (1894).

Pimenta ferruginea (Griseb.) Burret, Notizbl. Bot. Gart. Berlin-Dahlem 15: 513 (1941).
NW. Cuba. 81 CUB. Nanophan. or phan.

Anamomis ferruginea Griseb., Cat. Pl. Cub.: 91 (1866). *Myrtus leriocarpa* C.Wright, Anales Acad. Ci. Méd. Habana 5: 410 (1868). *Eugenia leriocarpa* (C.Wright) Krug & Urb., Bot. Jahrb. Syst. 19: 662 (1895). *Krokia ferruginea* (Griseb.) Urb., Symb. Antill. 9: 469 (1928).

Eugenia leriocarpa (C.Wright) Krug & Urb., Bot. Jahrb. Syst. 19: 662 (1895).

Pimenta filipes (Urb.) Burret, Notizbl. Bot. Gart. Berlin-Dahlem 15: 513 (1941).
EC. & NE. Cuba. 81 CUB. Nanophan. or phan.
Myrtekmania filipes Urb., Symb. Antill. 9: 486 (1928). *Krokia filipes* (Urb.) Borhidi & O.Muñiz, Bot. Közlem. 64: 214 (1977 publ. 1978).
Myrtekmania moaensis Areces, Ciencias (Havana), Ser. 10, 3: 6 (1975 publ. 1976). *Krokia moaensis* (Areces) Borhidi & O.Muñiz, Bot. Közlem. 64: 214 (1977 publ. 1978).
Pimenta moaensis Borhidi & O.Muñiz, Acta Bot. Acad. Sci. Hung. 21: 226 (1975 publ. 1976). *Krokia leonis* Borhidi & O.Muñiz, Bot. Közlem. 64: 213 (1977 publ. 1978).

Pimenta guatemalensis (Lundell) Lundell, Wrightia 4: 159 (1971).
C. America. 80 COS GUA PAN. Phan.
Myrcia guatemalensis Lundell, Wrightia 4: 42 (1968). *Mentodendron guatamalense* (Lundell) Lundell, Wrightia 4: 180 (1971).

Pimenta haitiensis (Urb.) Landrum, Brittonia 36: 242 (1984).
S. Hispaniola. 81 DOM HAI. Nanophan. or phan.
Cryptorhiza haitiensis Urb., Repert. Spec. Nov. Regni Veg. 17: 404 (1921).

Pimenta jamaicensis (Britton & Harris) Proctor, Bull. Inst. Jamaica, Sci. Ser. 16: 34 (1967).
C. Jamaica. 81 JAM. Phan.
Amomis jamaicensis Britton & Harris, J. New York Bot. Gard. 21: 39 (1920).

Pimenta obscura Proctor, J. Arnold Arbor. 63: 281 (1982).
C. Jamaica. 81 JAM. Nanophan. or phan.

Pimenta odiolens (Urb.) Burret, Notizbl. Bot. Gart. Berlin-Dahlem 15: 514 (1941).
E. Cuba. 81 CUB. Nanophan. or phan.
Amomis odiolens Urb., Symb. Antill. 9: 471 (1928). *Myrtekmania odiolens* (Urb.) Alain, Phytologia 8: 370 (1962). *Krokia odiolens* (Urb.) Borhidi & O.Muñiz, Bot. Közlem. 64: 213 (1977 publ. 1978).
Myrtekmania clementis Alain, Contr. Ocas. Mus. Hist. Nat. Colegio "De La Salle" 12: 10 (1953).

Pimenta oligantha (Urb.) Burret, Notizbl. Bot. Gart. Berlin-Dahlem 15: 513 (1941).
E. Cuba. 81 CUB. Nanophan. or phan.
Myrtus oligantha Urb., Symb. Antill. 9: 79 (1923). *Krokia oligantha* (Urb.) Borhidi & O.Muñiz, Bot. Közlem. 64: 213 (1977 publ. 1978).
Pimenta cubensis Urb., Symb. Antill. 9: 470 (1928).
Mozartia emarginata Moldenke, Phytologia 2: 226 (1947). *Myrcia emarginata* (Moldenke) Alain, Mem. New York Bot. Gard. 21(2): 138 (1971), nom. illeg.
Pimenta myrciifolia Borhidi & O.Muñiz, Bot. Közlem. 64: 214 (1977 publ. 1978).

Pimenta podocarpoides (Areces) Landrum, Fl. Neotrop. Monogr. 45: 99 (1986).
E. Cuba. 81 CUB. Nanophan. or phan.
Myrtekmania podocarpoides Areces, Ciencias (Havana), Ser. 10, 3: 4 (1975 publ. 1976). *Krokia podocarpoides* (Areces) Borhidi & O.Muñiz, Bot. Közlem. 64: 213 (1977 publ. 1978).

Pimenta pseudocaryophyllus (Gomes) Landrum, Brittonia 36: 242 (1984).
Bolivia to Brazil. 83 BOL 84 BZC BZE BZL BZS. Nanophan. or phan.
Myrtus caryophyllata Vell., Fl. Flumin. 5: 216, t. 70 (1829), nom. illeg.

var. ***fulvescens*** (Mart. ex DC.) Landrum, Fl. Neotrop. Monogr. 45: 104 (1986).
Bolivia to Brazil. 83 BOL 84 BZC BZE BZL. Nanophan. or phan.
Eugenia fulvescens Mart. ex DC., Prodr. 3: 283 (1828). *Pimenta fulvescens* (Mart. ex DC.) Blume, Mus. Bot. 1: 80 (1850). *Pseudocaryophyllus fulvescens* (Mart. ex DC.) O.Berg, Linnaea 27: 416 (1856). *Myrtus fulvescens* (Mart. ex DC.) Kiaersk., Enum. Myrt. Bras.: 24 (1893).
Eugenia mutabilis O.Berg in C.F.P.von Martius & auct. suc. (eds.), Fl. Bras. 14(1): 311 (1857). *Pseudocaryophyllus mutabilis* (O.Berg) Burret, Notizbl. Bot. Gart. Berlin-Dahlem 15: 518 (1941).
Pseudocaryophyllus velutinus O.Berg in C.F.P.von Martius & auct. suc. (eds.), Fl. Bras. 14(1): 607 (1859). *Myrtus velutina* (O.Berg) Kiaersk., Enum. Myrt. Bras.: 25 (1893).
Pseudocaryophyllus platyphyllus Burret ex Luetzelb., Estud. Bot. Nordéste 3: 202 (1923).

var. ***hoehnei*** (Burret) Landrum, Fl. Neotrop. Monogr. 45: 104 (1986).
SE. & S. Brazil. 84 BZL BZS. Nanophan. or phan.
Pseudocaryophyllus emarginatus Burret, Notizbl. Bot. Gart. Berlin-Dahlem 15: 516 (1941).
Pseudocaryophyllus hoehnei Burret, Notizbl. Bot. Gart. Berlin-Dahlem 15: 516 (1941).
Pseudocaryophyllus theifer Toledo, Arq. Bot. Estado São Paulo, n.s., f.m., 3: 64 (1953).

var. ***pseudocaryophyllus***
SE. & S. Brazil. 84 BZL BZS. Nanophan. or phan.
Eugenia acuminata Link, Enum. Hort. Berol. Alt. 2: 28 (1822). *Pseudocaryophyllus acuminatus* (Link) Burret, Notizbl. Bot. Gart. Berlin-Dahlem 15: 520 (1941).
Eugenia pseudocaryophyllus var. *ocoteoides* DC., Prodr. 3: 282 (1828).
Eugenia leandriana O.Berg in C.F.P.von Martius & auct. suc. (eds.), Fl. Bras. 14(1): 311 (1857). *Pseudocaryophyllus leandrianus* (O.Berg) Burret, Notizbl. Bot. Gart. Berlin-Dahlem 15: 517 (1941).
Pseudocaryophyllus costatus O.Berg in C.F.P.von Martius & auct. suc. (eds.), Fl. Bras. 14(1): 607 (1859).
Eugenia caryophylla St.-Lag., Ann. Soc. Bot. Lyon 7: 126 (1880).
Myrtus fulvescens f. *glazioviana* Kiaersk., Enum. Myrt. Bras.: 24 (1893). *Pseudocaryophyllus glazriovianus* (Kiaersk.) Burret, Notizbl. Bot. Gart. Berlin-Dahlem 15: 518 (1941).
Myrtus velutina f. *macrophylla* Kiaersk., Enum. Myrt. Bras.: 25 (1893).
Pseudocaryophyllus chrysophyllus Burret, Notizbl. Bot. Gart. Berlin-Dahlem 15: 519 (1941).
Pseudocaryophyllus organensis Burret, Notizbl. Bot. Gart. Berlin-Dahlem 15: 520 (1941).

Pseudocaryophyllus crenatus D.Legrand, Sellowia 13: 342 (1961).
Pseudocaryophyllus jaccoudii Mattos, Loefgrenia 2: 1 (1961).

Pimenta racemosa (Mill.) J.W.Moore, Bernice P. Bishop Mus. Bull. 102: 33 (1933).
Caribbean to Venezuela. (61) sci 81 CUB DOM HAI jam LEE PUE TRT WIN 82 VEN. Phan.
**Caryophyllus racemosus* Mill., Gard. Dict. ed. 8: 5 (1768).

var. ***grisea*** (Kiaersk.) Fosberg, Amer. Midl. Naturalist 27: 762 (1942).
Dominican Rep. to Virgin Is. 81 DOM LEE PUE. Nanophan. or phan.
**Pimenta acris* var. *grisea* Kiaersk., Bot. Tidsskr. 17: 289 (1890). *Amomis caryophyllata* var. *grisea* (Kiaersk.) Kiaersk., Bot. Jahrb. Syst. 19: 575 (1894). *Amomis grisea* (Kiaersk.) Britton, Sci. Surv. Porto Rico & Virgin Islands 6: 28 (1925).

var. ***hispaniolensis*** (Urb.) Landrum, Brittonia 36: 242 (1984).
Hispaniola. 81 DOM HAI. Nanophan. or phan.
**Amomis hispaniolensis* Urb., Ark. Bot. 20A(5): 21 (1926). *Pimenta hispaniolensis* (Urb.) Burret, Notizbl. Bot. Gart. Berlin-Dahlem 15: 511 (1941).
Amomis pauciflora Urb., Ark. Bot. 21A(5): 21 (1927). *Pimenta pauciflora* (Urb.) Burret, Notizbl. Bot. Gart. Berlin-Dahlem 15: 511 (1941).
Pimenta crenulata Alain, Moscosoa 1: 29 (1976).

var. ***ozua*** (Urb. & Ekman) Landrum, Brittonia 36: 242 (1984).
NC. Hispaniola. 81 DOM HAI. Nanophan. or phan.
Amomis anisomera Urb. & Ekman, Ark. Bot. 22A(10): 23 (1929). *Pimenta anisomera* (Urb. & Ekman) Burret, Notizbl. Bot. Gart. Berlin-Dahlem 15: 511 (1941).
**Amomis ozua* Urb. & Ekman, Ark. Bot. 22A(10): 22 (1929). *Pimenta ozua* (Urb. & Ekman) Burret, Notizbl. Bot. Gart. Berlin-Dahlem 15: 511 (1941).

var. ***racemosa***
Caribbean to Venezuela. 81 CUB DOM HAI jam LEE PUE TRT WIN 82 VEN. Phan.
Myrtus caryophyllata Jacq., Observ. Bot. 2: 1 (1767), nom. illeg. *Amomis caryophyllata* Krug & Urb., Bot. Jahrb. Syst. 19: 573 (1894).
Myrtus acris Sw., Prodr.: 79 (1788). *Myrcia acris* (Sw.) DC., Prodr. 3: 243 (1828). *Pimenta acris* (Sw.) Kostel., Allg. Med.-Pharm. Fl. 4: 1526 (1835). *Amomis acris* (Sw.) O.Berg, Linnaea 27: 417 (1856).
Myrtus citrifolia Poir. in J.B.A.P.M.de Lamarck, Encycl. 4: 410 (1798). *Myrcia pimentoides* DC., Prodr. 3: 243 (1828). *Pimenta acris* var. *pimentoides* (DC.) Griseb., Fl. Brit. W. I.: 241 (1860). *Pimenta citrifolia* (Poir.) Kostel., Allg. Med.-Pharm. Fl. 4: 1525 (1835).
Pimentus cotinifolia Raf., Sylva Tellur.: 105 (1838).
Amomis oblongata var. *occidentalis* O.Berg, Handb. Pharm. Bot.: 340 (1855).
Amomis acris var. *grandifolia* O.Berg, Linnaea 27: 418 (1856).
Amomis acris var. *obtusata* O.Berg, Linnaea 27: 418 (1856).
Amomis acris var. *parvifolia* O.Berg, Linnaea 27: 418 (1856).
Amomis oblongata O.Berg, Linnaea 27: 421 (1856).
Amomis pimento O.Berg, Linnaea 27: 418 (1856).
Amomis pimento var. *jamaicensis* O.Berg, Linnaea 27: 419 (1856).
Amomis pimento var. *surinamensis* O.Berg, Linnaea 27: 419 (1856).
Amomis pimentoides O.Berg, Linnaea 27: 420 (1856).
Pimenta pimento Griseb., Fl. Brit. W. I.: 241 (1860).
Pimenta acuminata Bello, Anales Soc. Esp. Hist. Nat. 10: 270 (1881).

var. ***terebinthina*** (Burret) Landrum, Brittonia 36: 243 (1984).
NE. Dominican Rep. 81 DOM. Nanophan. or phan.
**Pimenta terebinthina* Burret, Notizbl. Bot. Gart. Berlin-Dahlem 15: 511 (1941).

Pimenta richardii Proctor, J. Arnold Arbor. 63: 282 (1982).
Jamaica (St. Ann). 81 JAM. Phan.

Synonyms:

Pimenta acris (Sw.) Kostel. = ***Pimenta racemosa*** (Mill.) J.W.Moore var. ***racemosa***
Pimenta acris var. *grisea* Kiaersk. = ***Pimenta racemosa*** var. ***grisea*** (Kiaersk.) Fosberg
Pimenta acris var. *pimentoides* (DC.) Griseb. = ***Pimenta racemosa*** (Mill.) J.W.Moore var. ***racemosa***
Pimenta acuminata Bello = ***Pimenta racemosa*** (Mill.) J.W.Moore var. ***racemosa***
Pimenta anisomera (Urb. & Ekman) Burret = ***Pimenta racemosa*** var. ***ozua*** (Urb. & Ekman) Landrum
Pimenta aromatica Kostel. = ***Pimenta dioica*** (L.) Merr.
Pimenta citrifolia (Poir.) Kostel. = ***Pimenta racemosa*** (Mill.) J.W.Moore var. ***racemosa***
Pimenta communis Benth. & Hook.f. = ***Pimenta dioica*** (L.) Merr.
Pimenta crenulata Alain = ***Pimenta racemosa*** var. ***hispaniolensis*** (Urb.) Landrum
Pimenta cubensis Urb. = ***Pimenta oligantha*** (Urb.) Burret
Pimenta dioica var. *tabasco* (Willd. ex Schltdl. & Cham.) Standl. = ***Pimenta dioica*** (L.) Merr.
Pimenta fulvescens (Mart. ex DC.) Blume = ***Pimenta pseudocaryophyllus*** var. ***fulvescens*** (Mart. ex DC.) Landrum
Pimenta hispaniolensis (Urb.) Burret = ***Pimenta racemosa*** var. ***hispaniolensis*** (Urb.) Landrum
Pimenta moaensis Borhidi & O.Muñiz = ***Pimenta filipes*** (Urb.) Burret
Pimenta myrciifolia Borhidi & O.Muñiz = ***Pimenta oligantha*** (Urb.) Burret
Pimenta nipensis (Urb.) Burret = ***Pimenta cainitoides*** (Urb.) Burret
Pimenta officinalis Lindl. = ***Pimenta dioica*** (L.) Merr.
Pimenta officinalis var. *cumanensis* Schiede & Deppe = ***Pimenta dioica*** (L.) Merr.
Pimenta officinalis var. *longifolia* (Sims) O.Berg = ***Pimenta dioica*** (L.) Merr.
Pimenta officinalis var. *ovalifolia* (DC.) O.Berg = ***Pimenta dioica*** (L.) Merr.
Pimenta officinalis var. *tabasco* (Willd. ex Schltdl. & Cham.) O.Berg = ***Pimenta dioica*** (L.) Merr.
Pimenta ozua (Urb. & Ekman) Burret = ***Pimenta racemosa*** var. ***ozua*** (Urb. & Ekman) Landrum
Pimenta pauciflora (Urb.) Burret = ***Pimenta racemosa*** var. ***hispaniolensis*** (Urb.) Landrum
Pimenta pilotoana (Urb.) Borhidi = ***Pimenta cainitoides*** (Urb.) Burret
Pimenta pimenta (L.) H.Karst. = ***Pimenta dioica*** (L.) Merr.

Pimenta pimento Griseb. = ***Pimenta racemosa*** (Mill.) J.W.Moore var. ***racemosa***
Pimenta tabasco (Willd. ex Schltdl. & Cham.) Lundell = ***Pimenta racemosa*** (Mill.) J.W.Moore var. ***racemosa***
Pimenta terebinthina Burret = ***Pimenta racemosa*** var. ***terebinthina*** (Burret) Landrum
Pimenta vulgaris Lindl. = ***Pimenta dioica*** (L.) Merr.
Pimenta vulgaris Bello = ***Pimenta dioica*** (L.) Merr.

Pimentus

Pimentus Raf. = ***Pimenta*** Lindl.
Pimentus aromatica Raf. = ***Pimenta dioica*** (L.) Merr.
Pimentus cotinifolia Raf. = ***Pimenta racemosa*** (Mill.) J.W.Moore var. ***racemosa***
Pimentus geminata Raf. = ***Pimenta dioica*** (L.) Merr.
Pimentus greggii (Sw.) Raf. = ***Eugenia greggii*** (Sw.) Poir.
Pimentus laurinus Raf. = (Symplocaceae)
Pimentus saligna (J.F.Gmel.) Raf. = ***Melaleuca cajuputi*** Powell subsp. ***cajuputi***
Pimentus triflora (Jacq.) Raf. = ***Eugenia triflora*** (Jacq.) Ham.
Pimentus vera Raf. = ***Pimenta dioica*** (L.) Merr.

Piptandra

Piptandra Turcz. = ***Scholtzia*** Schauer
Piptandra spathulata Turcz. = ***Scholtzia spathulata*** (Turcz.) Benth.

Pleurocalyptus

Pleurocalyptus Brongn. & Gris, Bull. Soc. Bot. France 14: 264 (1867).
New Caledonia. 60 NWC.
2 Species

Pleurocalyptus austrocaledonicus (Guillaumin) J.W.Dawson, in Fl. Nouv.-Caléd. 118: 155 (1992).
C. New Caledonia. 60 NWC. Nanophan. or phan.
Fremya austrocaledonica Vieill. ex Guillaumin, Ann. Inst. Bot.-Géol. Colon. Marseille, II, 9: 148 (1911), nom. nud.
**Xanthostemon austrocaledonicus* Guillaumin, Bull. Soc. Bot. France 81: 12 (1934).

Pleurocalyptus pancheri (Brongn. & Gris) J.W.Dawson, in Fl. Nouv.-Caléd. 18: 152 (1992).
C. & SE. New Caledonia. 60 NWC. Nanophan. or phan.
**Fremya pancheri* Brongn. & Gris, Bull. Soc. Bot. France 10: 373 (1863). *Xanthostemon pancheri* (Brongn. & Gris) Schltr., Bot. Jahrb. Syst. 36: 24 (1905).
Salisia rugosa Pancher, Bull. Soc. Bot. France 10: 374 (1863).
Pleurocalyptus deplanchei Brongn. & Gris, Bull. Soc. Bot. France 14: 264 (1867). *Metrosideros pleurocalyptus* Baill., Hist. Pl. 6: 337 (1876).

***Synonyms*:**
Pleurocalyptus deplanchei Brongn. & Gris = ***Pleurocalyptus pancheri*** (Brongn. & Gris) J.W.Dawson

Plinia

Plinia Plum. ex L., Sp. Pl.: 516 (1753).
Trop. America. 80 BLZ COS GUA NIC PAN 81 CUB DOM HAI LEE TRT WIN 82 GUY SUR VEN 83 BOL ECU? PER 84 BZC BZE BZL BZN BZS 85 AGE PAR URU.
67 Species
Guapurium Juss., Gen. Pl.: 324 (1789).
Guapurum J.F.Gmel., Syst. Nat.: 793 (1791).

Plinia abeggii (Urb. & Ekman) Urb., Ark. Bot. 22A(10): 25 (1929).
Hispaniola. 81 DOM HAI. Nanophan. or phan.
**Eugenia abeggii* Urb. & Ekman, Ark. Bot. 21A(5): 35 (1927).

Plinia acunae Borhidi & O.Muñiz, Bot. Közlem. 64: 19 (1977).
Cuba. 81 CUB. Nanophan. or phan.

Plinia anonyma Sobral, Bol. Mus. Bot. Munic. 63: 1 (1985).
SE. Brazil. 84 BZL. Nanophan. or phan.

Plinia baileyi R.O.Williams ex R.S.Marsh., Trees Trinidad Tobago: 60 (1934).
Trinidad-Tobago. 81 TRT. Nanophan. or phan.

Plinia baracoensis Borhidi, Bot. Közlem. 64: 19 (1977).
SE. Cuba. 81 CUB. Nanophan. or phan.

Plinia brachybotrya (D.Legrand) Sobral, Hoehnea 21: 202 (1994 publ. 1995).
S. Brazil. 84 BZS. Nanophan. or phan.
**Marlierea parviflora* var. *brachybotrya* D.Legrand, in Fl. Ilustr. Catar. 1(Mirt.): 475 (1971).

Plinia callosa Sobral, Napaea 4: 12 (1988).
Brazil (Bahia). 84 BZE. Nanophan. or phan.

Plinia caricensis Urb., Ark. Bot. 22A(10): 26 (1929).
NW. & C. Hispaniola. 81 DOM HAI. Nanophan. or phan.

Plinia cauliflora (Mart.) Kausel, Ark. Bot., a.s., 3: 508 (1956).
Bolivia, SE. & S. Brazil. 83 BOL 84 BZL BZS. Phan. Widely cultivated for its edible fruits (jaboticaba).
**Myrtus cauliflora* Mart., Reise Bras. 1: 285 (1823). *Eugenia cauliflora* (Mart.) DC., Prodr. 3: 273 (1828). *Myrciaria cauliflora* (Mart.) O.Berg in C.F.P.von Martius & auct. suc. (eds.), Fl. Bras. 14(1): 361 (1857).
Myrtus jaboticaba Vell., Fl. Flumin. 5: 214, t. 62 (1829). *Myrciaria jaboticaba* (Vell.) O.Berg in C.F.P.von Martius & auct. suc. (eds.), Fl. Bras. 14(1): 361 (1857). *Myrcia jaboticaba* (Vell.) Baill., Hist. Pl. 6: 345 (1876). *Eugenia jaboticaba* (Vell.) Kiaersk., Enum. Myrt. Bras.: 185 (1893). *Plinia jaboticaba* (Vell.) Kausel, Ark. Bot., a.s., 3: 508 (1956).

Plinia cerrocampanensis Barrie, Novon 14: 382 (2004).
Panama. 80 PAN. Nanophan. or phan.

Plinia cidrensis Urb., Ark. Bot. 22A(10): 27 (1929).
Plinia acutissima var. *cidrensis* (Urb.) Borhidi, Acta Bot. Hung. 29: 186 (1983).
Haiti (Massif du Nord). 81 HAI. Nanophan. or phan.

Plinia clausa McVaugh, Fieldiana, Bot. 29: 224 (1956).
N. Peru. 83 PER. Nanophan. or phan.

Plinia coclensis Barrie, Novon 14: 384 (2004).
Panama. 80 PAN. Nanophan. or phan.

Plinia complanata M.L.Kawas. & B.Holst, Brittonia 54: 94 (2002).
Brazil (São Paulo). 84 BZL.

Plinia cordifolia (D.Legrand) Sobral, Hoehnea 21: 202 (1994 publ. 1995).
S. Brazil. 84 BZS. Nanophan. or phan.

Myrciaria cordifolia D.Legrand, Sellowia 13: 330 (1961).

Plinia coronata (Mattos) Mattos, Loefgrenia 112: 5 (1998).
SE. Brazil. 84 BZL. Nanophan. or phan.
**Myrciaria coronata* Mattos, Loefgrenia 51: 1 (1976).

Plinia costata Amshoff, Recueil Trav. Bot. Néerl. 42: 11 (1950).
Guyana to Suriname. 82 GUY SUR. Nanophan. or phan.

Plinia cuspidata Gómez-Laur. & Valverde, Lankesteriana 3: 11 (2002).
Costa Rica to Panama. 80 COS PAN.

Plinia darienensis Barrie, Novon 14: 386 (2004).
Panama. 80 PAN. Nanophan. or phan.

Plinia dermatodes Urb., Symb. Antill. 9: 476 (1928).
W. Cuba. 81 CUB. Nanophan.

Plinia duplipilosa McVaugh, Fieldiana, Bot. 29: 224 (1956).
Peru (Loreto). 83 PER. Nanophan. or phan.

Plinia edulis (Vell.) Sobral, Bol. Mus. Bot. Munic. 63: 2 (1985).
SE. & S. Brazil to NE. Argentina. 84 BZL BZS 85 AGE. Nanophan. or phan.
**Eugenia edulis* Vell., Fl. Flumin. 5: 208, t. 34 (1829). *Myrciaria plicatocostata* O.Berg in C.F.P.von Martius & auct. suc. (eds.), Fl. Bras. 14(1): 366 (1857), nom. illeg. *Plinia plicatocostata* (O.Berg) Amshoff, Recueil Trav. Bot. Néerl. 42: 11 (1950), nom. illeg. *Myrciaria edulis* (Vell.) Skeels, Bull. Bur. Pl. Industr. U.S.D.A. 148: 14 (1909).
Rubachia glomerata O.Berg in C.F.P.von Martius & auct. suc. (eds.), Fl. Bras. 14(1): 31 (1857). *Marlierea edulis* Nied. in H.G.A.Engler & K.A.E.Prantl, Nat. Pflanzenfam. 3(7): 77 (1893).

Plinia ekmaniana Urb., Ark. Bot. 21A(5): 23 (1927).
Haiti (Massif du Nord). 81 HAI. Nanophan. or phan.

Plinia gentryi Barrie, Novon 14: 387 (2004).
Panama. 80 PAN. Nanophan. or phan.

Plinia grandifolia (Mattos) Sobral, Hoehnea 21: 202 (1994 publ. 1995).
SE. Brazil. 84 BZL. Nanophan. or phan.
**Myrciaria grandifolia* Mattos, Loefgrenia 51: 4 (1976).

Plinia guanacastensis Barrie, Novon 14: 389 (2004).
Costa Rica. 80 COS. Nanophan. or phan.

Plinia hatschbachii (Mattos) Sobral, Hoehnea 21: 202 (1994 publ. 1995).
Brazil (Paraná). 84 BZS. Nanophan. or phan.
**Myrciaria hatschbachii* Mattos, Loefgrenia 15: 1 (1964).

Plinia icardiana Urb., Ark. Bot. 22A(10): 27 (1929).
Haiti (Massif de la Hotte). 81 HAI. Nanophan.

Plinia ilhensis G.M.Barroso, Napaea 10: 2 (1994).
Brazil (Rio de Janeiro). 84 BZL. Nanophan. or phan.

Plinia inflata McVaugh, Fieldiana, Bot. 29: 225 (1956).
N. Brazil. 84 BZN. Nanophan. or phan.

Plinia involucrata (O.Berg) McVaugh, Mem. New York Bot. Gard. 18(2): 228 (1969).
Venezuela to N. Brazil. 82 VEN 84 BZN. Nanophan. or phan.
**Myrciaria involucrata* O.Berg in C.F.P.von Martius & auct. suc. (eds.), Fl. Bras. 14(1): 375 (1857).

Plinia longiacuminata Sobral, Novon 16: 525 (2006).
Brazil (Bahia). 84 BZE. Nanophan. or phan.

Plinia marqueteana G.M.Barroso, Napaea 10: 2 (1994).
Brazil (Rio de Janeiro). 84 BZL. Nanophan. or phan.

Plinia martinellii G.M.Barroso & Peron, Reserva Ecol. Macaé de Cima 1: 289 (1994).
Brazil (Rio de Janeiro). 84 BZL. Nanophan. or phan.

Plinia microcycla Urb., Ark. Bot. 22A(10): 26 (1929).
SW. Dominican Rep. 81 DOM. Nanophan. or phan.

Plinia moaensis Borhidi, Bot. Közlem. 64: 18 (1977).
E. Cuba. 81 CUB. Nanophan. or phan.

Plinia moralesii Barrie, Novon 14: 390 (2004).
Costa Rica. 80 COS. Nanophan. or phan.

Plinia muricata Sobral, Hoehnea 21: 200 (1994 publ. 1995).
Brazil (Bahia). 84 BZE. Phan.

Plinia nana Sobral, Novon 15: 587 (2005).
Brazil (Minas Gerais). 84 BZL. Epiphyte.

Plinia nicaraguensis Barrie, Novon 14: 392 (2004).
Nicaragua. 80 NIC. Nanophan. or phan.

Plinia oblongata (Mattos) Mattos, Loefgrenia 112: 5 (1998).
Brazil (São Paulo). 84 BZL. Nanophan. or phan.
**Myrciaria oblongata* Mattos, Loefgrenia 51: 2 (1976).

Plinia orthoclada Urb., Symb. Antill. 9: 476 (1928).
W. Cuba. 81 CUB. Nanophan.

Plinia panamensis Barrie, Novon 14: 393 (2004).
Panama. 80 PAN. Nanophan. or phan.

Plinia pauciflora M.L.Kawas. & B.Holst, Brittonia 54: 97 (2002).
Brazil (São Paulo). 84 BZL.

Plinia peroblata (Lundell) Lundell, Wrightia 3: 124 (1965).
Belize. 80 BLZ. Nanophan. or phan.
**Eugenia peroblata* Lundell, Wrightia 2: 124 (1961).

Plinia peruviana (Poir.) Govaerts, World Checklist Myrtaceae: 344 (2008).
Peru to Brazil and NE. Argentina. 83 BOL PER 84 BZC BZL BZS 85 AGE PAR. Nanophan. or phan.
**Guapurium peruvianum* Poir. in G.-F.Cuvier, Dict. Sci. Nat., ed. 2, 20: 11 (1821). *Eugenia guapurium* DC., Prodr. 3: 273 (1828), nom. superfl. *Myrciaria guapurium* (DC.) O.Berg, Linnaea 27: 325 (1856), nom. superfl. *Myrciaria peruviana* (Poir.) Mattos, Loefgrenia 78: 3 (1983).
Guapurium fruticosum Spreng., Syst. Veg. 2: 479 (1825).
Eugenia cauliflora Miq., Linnaea 22: 537 (1849), nom. illeg.
Myrciaria trunciflora O.Berg in C.F.P.von Martius & auct. suc. (eds.), Fl. Bras. 14(1): 361 (1857). *Eugenia rabeniana* Kiaersk., Enum. Myrt. Bras.: 186 (1893). *Plinia trunciflora* (O.Berg) Kausel, Ark. Bot., a.s., 3: 507 (1956). *Myrciaria peruviana* var. *trunciflora* (O.Berg) Mattos, Loefgrenia 78: 3 (1983).

Plinia phitrantha (Kiaersk.) Sobral, Hoehnea 21: 202 (1994 publ. 1995).
Brazil (S. Minas Gerais, São Paulo). 84 BZL.

Nanophan. or phan.
Eugenia phitrantha Kiaersk., Enum. Myrt. Bras.: 185 (1893). *Myrciaria phitrantha* (Kiaersk.) Mattos, Arq. Bot. Estado São Paulo, n.s., f.m., 3: 293 (1962).
Myrciaria aureana Mattos, Loefgrenia 51: 5 (1976). *Plinia aureana* (Mattos) Mattos, Loefgrenia 112: 5 (1998).

Plinia pinnata L., Sp. Pl.: 516 (1753). *Jossinia pinnata* (L.) Heynh., Nom. Bot. Hort. 1: 431 (1840). *Eugenia pinnata* (L.) Druce, Rep. Bot. Exch. Club Brit. Isles 3: 418 (1913 publ. 1914).
S. Caribbean. 81 LEE TRT WIN 82 GUY? SUR? 83 ECU?. Nanophan. or phan.
Plinia crocea L., Mant. Pl. 2: 244 (1771).
Plinia pentapetala L., Mant. Pl. 2: 402 (1771).
Marlierea glomerata O.Berg, Linnaea 27: 14 (1855).
Myrciaria trinitatis O.Berg, Linnaea 27: 336 (1856).
Stenocalyx plumieri O.Berg, Linnaea 30: 698 (1861). *Eugenia plumieri* (O.Berg) Nied. in H.G.A. Engler & K.A.E.Prantl, Nat. Pflanzenfam. 3(7): 82 (1893).

Plinia povedae P.E.Sánchez, Brenesia 25–26: 313 (1986 publ. 1988).
E. Costa Rica to W. Panama. 80 COS GUA NIC. Phan.

Plinia punctata Urb., Symb. Antill. 9: 89 (1923).
E. Cuba. 81 CUB. Nanophan.

Plinia puriscalensis P.E.Sánchez & Q.Jiménez, Brenesia 32: 113 (1989 publ. 1990).
Costa Rica. 80 COS. Nanophan. or phan.

Plinia ramosissima (Urb.) Urb., Symb. Antill. 9: 475 (1928).
E. Cuba. 81 CUB. Nanophan.
Calyptranthes ramosissima Urb., Symb. Antill. 9: 92 (1923).

Plinia rara Sobral, Hoehnea 21: 200 (1994 publ. 1995).
Brazil (Bahia). 84 BZE. Nanophan. or phan.

Plinia recurvata Urb., Symb. Antill. 9: 477 (1928).
W. Cuba. 81 CUB. Nanophan.

Plinia renatiana G.M.Barroso & Peixoto, Atas Soc. Bot. Brasil, Secç. Rio de Janeiro 3: 98 (1991).
Brazil (Espírito Santo). 84 BZL. Phan.

Plinia rivularis (Cambess.) Rotman, Bol. Soc. Argent. Bot. 24: 195 (1985).
Trinidad to NE. Argentina. 81 TRT 82 VEN 84 BZC BZE BZL BZN BZS 85 AGE PAR URU. Nanophan. or phan.
Eugenia rivularis Cambess. in A.F.C.de Saint-Hilaire, Fl. Bras. Merid. 2: 337 (1832). *Myrciaria rivularis* (Cambess.) O.Berg in C.F.P.von Martius & auct. suc. (eds.), Fl. Bras. 14(1): 360 (1857).
Myrciaria hagendorffii O.Berg in C.F.P.von Martius & auct. suc. (eds.), Fl. Bras. 14(1): 360 (1857). *Eugenia hagendorffii* (O.Berg) Kiaersk., Enum. Myrt. Bras.: 179 (1893).
Eugenia variifolia Barb.Rodr. ex Chodat & Hassl., Bull. Herb. Boissier, II, 7: 806 (1907), nom. nud.
Myrcia silvatica Barb.Rodr. ex Chodat & Hassl., Bull. Herb. Boissier, II, 7: 803 (1907), nom. inval.
Myrcia granulata R.O.Williams, Fl. Trinidad 1: 340 (1934).
Myrciaria baporeti D.Legrand, Anales Mus. Hist. Nat. Montevideo, II, 4(11): 63 (1936). *Myrciariopsis baporeti* (D.Legrand) Kausel, Ark. Bot., a.s., 3: 509 (1956). *Siphoneugena baporeti* (D.Legrand) Kausel, Lilloa 32: 345 (1967). *Myrciaria rivularis* var. *baporeti* (D.Legrand) D.Legrand, in Fl. Ilustr. Catar. 1(Mirt.): 768 (1978). *Plinia baporeti* (D.Legrand) Rotman, Darwiniana 24: 169 (1982).
Siphoneugena legrandii Mattos & N.Silveira, Loefgrenia 87: 2 (1985).

Plinia rogersiana Mattos, Loefgrenia 106: 3 (1995).
Brazil (São Paulo). 84 BZL. Nanophan. or phan.

Plinia rupestris Ekman & Urb. in I.Urban, Symb. Antill. 9: 474 (1928).
W. Cuba. 81 CUB. Nanophan. or phan.

Plinia salamancana (Standl.) Barrie, Novon 14: 397 (2004).
C. Panama. 80 PAN. Nanophan. or phan.
Eugenia salamancana Standl., Ann. Missouri Bot. Gard. 26: 295 (1939).

Plinia salticola McVaugh, Fieldiana, Bot. 29: 505 (1963).
Costa Rica to Panama. 80 COS PAN. Nanophan. or phan.

Plinia sebastianopolitana G.M.Barroso, Napaea 10: 2 (1994).
Brazil (Rio de Janeiro). 84 BZL. Nanophan. or phan.

Plinia spiciflora (Nees & Mart.) Sobral, Hoehnea 21: 202 (1994 publ. 1995).
E. Brazil. 84 BZE BZL. Phan.
Eugenia spiciflora Nees & Mart., Nova Acta Phys.-Med. Acad. Caes. Leop.-Carol. Nat. Cur. 12: 52 (1824). *Rubachia spiciflora* (Nees & Mart.) O.Berg, Linnaea 27: 12 (1855). *Marlierea spiciflora* (Nees & Mart.) Nied. in H.G.A.Engler & K.A.E.Prantl, Nat. Pflanzenfam. 3(7): 77 (1893).
Calyptranthes lateriflora DC., Prodr. 3: 258 (1828). *Rubachia lateriflora* (DC.) O.Berg, Linnaea 27: 12 (1855). *Marlierea lateriflora* (DC.) D.Legrand, Comun. Bot. Mus. Hist. Nat. Montevideo 3(40): 28 (1962).
Rubachia spiciflora var. *grandifolia* O.Berg in C.F.P. von Martius & auct. suc. (eds.), Fl. Bras. 14(1): 536 (1859).

Plinia spirito-santensis (Mattos) Mattos, Loefgrenia 112: 5 (1998).
SE. Brazil. 84 BZL. Nanophan. or phan.
Myrciaria spirito-santensis Mattos, Loefgrenia 51: 3 (1976).

Plinia stenophylla Urb., Symb. Antill. 9: 90 (1923).
Cuba (Sierra de Nipe). 81 CUB. Nanophan.

Plinia stictophylla G.M.Barroso & Peixoto, Atas Soc. Bot. Brasil, Secç. Rio de Janeiro 3: 100 (1991).
Brazil (Espírito Santo). 84 BZL. Phan.

Plinia subavenia Sobral, Novon 16: 526 (2006).
Brazil (Bahia, Espírito Santo). 84 BZE BZL. Nanophan. or phan.

Plinia toscanosia Urb., Symb. Antill. 9: 477 (1928).
NW. Cuba. 81 CUB. Nanophan.

Synonyms:
Plinia acutissima Urb. = ***Myrciaria floribunda*** (H.West ex Willd.) O.Berg
Plinia acutissima var. *cidrensis* (Urb.) Borhidi = ***Plinia cidrensis*** Urb.

Plinia asa-grayi (Krug & Urb.) Urb. = ***Myrciaria floribunda*** (H.West ex Willd.) O.Berg
Plinia aureana (Mattos) Mattos = ***Plinia phitrantha*** (Kiaersk.) Sobral
Plinia baporeti (D.Legrand) Rotman = ***Plinia rivularis*** (Cambess.) Rotman
Plinia crocea L. = ***Plinia pinnata*** L.
Plinia cubensis (Griseb.) Urb. = ***Myrciaria floribunda*** (H.West ex Willd.) O.Berg
Plinia dussii (Krug & Urb.) Urb. = ***Siphoneugena dussii*** (Krug & Urb.) Proença
Plinia formosa Urb. = ***Myrciaria floribunda*** (H.West ex Willd.) O.Berg
Plinia fruticosa Steyerm. = ***Siphoneugena dussii*** (Krug & Urb.) Proença var. ***dussii***
Plinia glomerata (O.Berg) Amshoff = ***Myrciaria glazioviana*** (Kiaersk.) G.M.Barroso ex Sobral
Plinia guildingiana (Griseb.) Urb. = ***Marlierea guildingiana*** (Griseb.) Krug & Urb.
Plinia haitiensis Urb. & Ekman = ***Myrciaria tenella*** (DC.) O.Berg
Plinia jaboticaba (Vell.) Kausel = ***Plinia cauliflora*** (Mart.) Kausel
Plinia jambos (L.) M.Gómez = ***Syzygium jambos*** (L.) Alston
Plinia montecristina Urb. & Ekman = ***Myrciaria tenella*** (DC.) O.Berg
Plinia pedunculata L.f. = ***Eugenia uniflora*** L.
Plinia pentapetala L. = ***Plinia pinnata*** L.
Plinia petiolata L. = ***Eugenia uniflora*** L.
Plinia plicatocostata (O.Berg) Amshoff = ***Plinia edulis*** (Vell.) Sobral
Plinia rubra L. = ***Eugenia uniflora*** L.
Plinia rubrinervis Urb. = ***Myrciaria floribunda*** (H.West ex Willd.) O.Berg
Plinia sintenisii (Kiaersk.) Britton = ***Marlierea sintenisii*** Kiaersk.
Plinia strigipes (O.Berg) Sobral = ***Myrciaria strigipes*** O.Berg
Plinia tetrapetala L. = ***Eugenia uniflora*** L.
Plinia trunciflora (O.Berg) Kausel = ***Plinia peruviana*** (Poir.) Govaerts

Polyzone

Polyzone Endl. = ***Darwinia*** Rudge
Polyzone purpurea Endl. = ***Darwinia purpurea*** (Endl.) Benth.

Pritzelia

Pritzelia Schauer = ***Scholtzia*** Schauer
Pritzelia obovata (DC.) Schauer = ***Scholtzia obovata*** (DC.) Schauer

Pseudanamomis

Pseudanamomis Kausel, Ark. Bot., a.s., 3: 511 (1956).
Caribbean, Colombia to N. South America. 81 DOM NLA PUE TRT 82 GUY VEN 83 CLM.
1 Species

Pseudanamomis umbellulifera (Kunth) Kausel, Ark. Bot., a.s., 3: 512 (1956).
Caribbean, Colombia to N. South America. 81 DOM NLA PUE TRT 82 GUY VEN 83 CLM. Nanophan. or phan.
Myrtus erythroxyloides Kunth in F.W.H.von Humboldt, A.J.A.Bonpland & C.S.Kunth, Nov. Gen. Sp. 6: 149 (1823).
**Myrtus umbellulifera* Kunth in F.W.H.von Humboldt, A.J.A.Bonpland & C.S.Kunth, Nov. Gen. Sp. 6: 135 (1823). *Myrcia umbellulifera* (Kunth) DC., Prodr. 3: 255 (1828). *Eugenia umbellulifera* (Kunth) Krug & Urb., Bot. Jahrb. Syst. 19: 665 (1895). *Anamomis umbellulifera* (Kunth) Britton, Sci. Surv. Porto Rico & Virgin Islands 6: 42 (1925). *Myrcianthes umbellulifera* (Kunth) Alain, Brittonia 20: 195 (1968).
Eugenia esculenta O.Berg, Linnaea 27: 273 (1856). *Anamomis esculenta* (O.Berg) Griseb., Fl. Brit. W. I.: 240 (1860).

Synonyms:
Pseudanamomis cati (Britton & P.Wilson) Bisse = ***Eugenia catingiflora*** Griseb.
Pseudanamomis catingiflora (Griseb.) Bisse = ***Eugenia catingiflora*** Griseb.
Pseudanamomis cordata (Sw.) Bisse = ***Eugenia cordata*** (Sw.) DC.
Pseudanamomis cupuligera (Urb.) Bisse = ***Eugenia cupuligera*** Urb.
Pseudanamomis gibberosa (Urb.) Bisse = ***Eugenia gibberosa*** Urb.
Pseudanamomis jambosoides (C.Wright ex Griseb.) Bisse = ***Eugenia jambosoides*** C.Wright ex Griseb.
Pseudanamomis macrocarpa (Schltdl. & Cham.) Bisse = ***Eugenia macrocarpa*** Schltdl. & Cham.
Pseudanamomis maestrensis (Urb.) Bisse = ***Eugenia maestrensis*** Urb.
Pseudanamomis nipensis Bisse = [81 CUB]
Pseudanamomis samuelssonii (Ekman & Urb.) Bisse = ***Eugenia samuelssonii*** Ekman & Urb.
Pseudanamomis trunciflora (Schltdl. & Cham.) Bisse = ***Eugenia trunciflora*** (Schltdl. & Cham.) G.Don
Pseudanamomis victorinii (Alain) Bisse = ***Eugenia victorinii*** Alain

Unplaced Names:
Pseudanamomis nipensis Bisse, Feddes Repert. 96: 512 (1985). = [81 CUB]

Pseudeugenia

Pseudeugenia D.Legrand & Mattos = ***Eugenia*** P.Micheli ex L.
Pseudeugenia stolonifera D.Legrand & Mattos = [84 BZL] *Eugenia* sp. ?

Unplaced Names:
Pseudeugenia stolonifera D.Legrand & Mattos, Arq. Bot. Estado São Paulo, n.s., f.m., 4: 63 (1966). = [84 BZL] *Eugenia* sp. ?

Pseudocaryophyllus

Pseudocaryophyllus O.Berg = ***Pimenta*** Lindl.
Pseudocaryophyllus acuminatus (Link) Burret = ***Pimenta pseudocaryophyllus*** (Gomes) Landrum var. ***pseudocaryophyllus***
Pseudocaryophyllus burkartianus D.Legrand = ***Eugenia burkartiana*** (D.Legrand) D.Legrand
Pseudocaryophyllus chrysophyllus Burret = ***Pimenta pseudocaryophyllus*** (Gomes) Landrum var. ***pseudocaryophyllus***
Pseudocaryophyllus coccolobifolius (Kunth) O.Berg = ***Myrcianthes rhopaloides*** (Kunth) McVaugh
Pseudocaryophyllus costatus O.Berg = ***Pimenta pseudocaryophyllus*** (Gomes) Landrum var. ***pseudocaryophyllus***
Pseudocaryophyllus crenatus D.Legrand = ***Pimenta pseudocaryophyllus*** (Gomes) Landrum var. ***pseudocaryophyllus***

Pseudocaryophyllus darwinii (Hook.f.) Burret = ***Amomyrtus luma*** (Molina) D.Legrand & Kausel
Pseudocaryophyllus emarginatus Burret = ***Pimenta pseudocaryophyllus*** var. ***hoehnei*** (Burret) Landrum
Pseudocaryophyllus fulvescens (Mart. ex DC.) O.Berg = ***Pimenta pseudocaryophyllus*** var. ***fulvescens*** (Mart. ex DC.) Landrum
Pseudocaryophyllus glaziovianus (Kiaersk.) Burret = ***Pimenta pseudocaryophyllus*** (Gomes) Landrum var. ***pseudocaryophyllus***
Pseudocaryophyllus guilii (Speg.) Burret = ***Amomyrtella guilii*** (Speg.) Kausel
Pseudocaryophyllus heteroclitus (Tussac) O.Berg = ***Myrcianthes fragrans*** (Sw.) McVaugh
Pseudocaryophyllus hoehnei Burret = ***Pimenta pseudocaryophyllus*** var. ***hoehnei*** (Burret) Landrum
Pseudocaryophyllus jaccoudii Mattos = ***Pimenta pseudocaryophyllus*** (Gomes) Landrum var. ***pseudocaryophyllus***
Pseudocaryophyllus leandrianus (O.Berg) Burret = ***Pimenta pseudocaryophyllus*** (Gomes) Landrum var. ***pseudocaryophyllus***
Pseudocaryophyllus meli (Phil.) Burret = ***Amomyrtus meli*** (Phil.) D.Legrand & Kausel
Pseudocaryophyllus multiflorus (Juss. ex J.St.-Hil.) Burret = ***Amomyrtus luma*** (Molina) D.Legrand & Kausel
Pseudocaryophyllus mutabilis (O.Berg) Burret = ***Pimenta pseudocaryophyllus*** var. ***fulvescens*** (Mart. ex DC.) Landrum
Pseudocaryophyllus organensis Burret = ***Pimenta pseudocaryophyllus*** (Gomes) Landrum var. ***pseudocaryophyllus***
Pseudocaryophyllus pabstianus D.Legrand = [84 BZL]
Pseudocaryophyllus platyphyllus Burret ex Luetzelb. = ***Pimenta pseudocaryophyllus*** var. ***fulvescens*** (Mart. ex DC.) Landrum
Pseudocaryophyllus seemannii Triana ex Hemsl. = ***Psidium salutare*** var. ***salutare***
Pseudocaryophyllus sericeus O.Berg = ***Pimenta pseudocaryophyllus*** (Gomes) Landrum var. ***pseudocaryophyllus***
Pseudocaryophyllus theifer Toledo = ***Pimenta pseudocaryophyllus*** var. ***hoehnei*** (Burret) Landrum
Pseudocaryophyllus uniflorus Burret = [84 BZL]
Pseudocaryophyllus velutinus O.Berg = ***Pimenta pseudocaryophyllus*** var. ***fulvescens*** (Mart. ex DC.) Landrum

***Unplaced Names*:**
Pseudocaryophyllus pabstianus D.Legrand, Bradea 1: 153 (1972). = [84 BZL]
Pseudocaryophyllus uniflorus Burret, Notizbl. Bot. Gart. Berlin-Dahlem 15: 515 (1941). = [84 BZL]

Pseudoeugenia

Pseudoeugenia Scort. = ***Syzygium*** Gaertn.
Pseudoeugenia perakiana Scort. = ***Syzygium skiophilum*** (Duthie) Airy Shaw
Pseudoeugenia singaporensis King = ***Syzygium singaporense*** (King) Airy Shaw
Pseudoeugenia tenuifolia Ridl. = ***Syzygium tenuifolium*** (Ridl.) Airy Shaw

Pseudomyrcianthes

Pseudomyrcianthes Kausel = ***Myrcianthes*** O.Berg
Pseudomyrcianthes adamantium (Cambess.) Kausel = ***Blepharocalyx salicifolius*** (Kunth) O.Berg
Pseudomyrcianthes kochiana (DC.) Kausel = ***Eugenia punicifolia*** (Kunth) DC.
Pseudomyrcianthes pseudomato (D.Legrand) Kausel = ***Myrcianthes pseudomato*** (D.Legrand) McVaugh
Pseudomyrcianthes pyriformis (Cambess.) Kausel = ***Eugenia pyriformis*** Cambess.
Pseudomyrcianthes rosea (DC.) Kausel = ***Eugenia rosea*** DC.

Psidiastrum

Psidiastrum Bello = ***Eugenia*** P.Micheli ex L.
Psidiastrum dubium Bello = ***Eugenia axillaris*** (Sw.) Willd. var. ***axillaris***

Psidiomyrtus

Psidiomyrtus Guillaumin = ***Rhodomyrtus*** (DC.) Rchb.
Psidiomyrtus locellata Guillaumin = ***Rhodomyrtus locellata*** (Guillaumin) Burret

Psidiopsis

Psidiopsis O.Berg = ***Calycolpus*** O.Berg
Psidiopsis moritzianum O.Berg = ***Calycolpus moritzianus*** (O.Berg) Burret

Psidium

Psidium L., Sp. Pl.: 470 (1753).
Mexico to Trop. America. (21) cvi mdr (22) ivo mtn (23) gab (25) ken tan uga (26) ang mlw moz zam zim (27) cpp nat tvl (29) mdg (38) kzn oga (40) ind (42) bor mly sul (43) bis (50) nfk (51) ker nzn (60) gil wal (61) lin sci (63) haw 78 FLA 79 MXE MXG MXI MXN MXS MXT 80 BLZ COS ELS GUA HON NIC PAN 81 BAH CAY CUB DOM HAI JAM LEE NLA PUE TRT WIN 82 FRG GUY SUR VEN 83 BOL CLM ECU GAL PER 84 BZC BZE BZL BZN BZS 85 AGE AGW jnf PAR URU.
93 Species
Cuiavus Trew, Pl. Select. 4: 12 (1754).
Guajava Mill., Gard. Dict. Abr. ed. 4 (1754).
Guayaba Noronha, Verh. Batav. Genootsch. Kunsten 5(4): 16 (1790), nom. inval.
Calyptropsidium O.Berg, Linnaea 27: 349 (1856).
Mitropsidium Burret, Notizbl. Bot. Gart. Berlin-Dahlem 15: 486 (1941).
Corynemyrtus (Kiaersk.) Mattos, Loefgrenia 10: 1 (1963).

Psidium acranthum Urb., Repert. Spec. Nov. Regni Veg. 18: 367 (1922).
Dominican Rep. 81 DOM. Nanophan. or phan. Provisionally accepted.

Psidium acunae Borhidi, Acta Bot. Acad. Sci. Hung. 17: 17 (1971 publ. 1972).
Cuba. 81 CUB. Nanophan. or phan. Provisionally accepted.

Psidium acutangulum Mart. ex DC., Prodr. 3: 233 (1828). *Guajava acutangula* (Mart. ex DC.) Kuntze, Revis. Gen. Pl. 1: 239 (1891).
S. Trop. America. 82 FRG GUY SUR VEN 83 BOL ECU PER 84 BZC BZE BZN. Nanophan. or phan.
Psidium acutangulum var. *acidum* DC., Prodr. 3: 233 (1828). *Britoa acida* (DC.) O.Berg, Linnaea 27: 436 (1856).
Psidium acidum Mart. ex O.Berg, Linnaea 27: 436 (1856), nom. inval.

Psidium acutangulum var. *crassirame* O.Berg in C.F.P.von Martius & auct. suc. (eds.), Fl. Bras. 14(1): 409 (1857).
Psidium acutangulum var. *tenuirame* O.Berg in C.F.P. von Martius & auct. suc. (eds.), Fl. Bras. 14(1): 409 (1857).
Psidium grandiflorum Ruiz & Pav., Anales Inst. Bot. Cavanilles 15: 194 (1957), nom. illeg.
Psidium persoonii McVaugh, Mem. New York Bot. Gard. 18(2): 255 (1969).
Psidium acutangulum var. *oblongatum* Mattos, Loefgrenia 94: 12 (1989).

Psidium albescens Urb., Symb. Antill. 5: 441 (1908).
Jamaica. 81 JAM. Nanophan. or phan.

Psidium amplexicaule Pers., Syn. Pl. 2: 27 (1806). *Guajava amplexicaulis* (Pers.) Kuntze, Revis. Gen. Pl. 1: 240 (1891).
Puerto Rico to Leeward Is. 81 LEE PUE. Nanophan. or phan.
Psidium cordatum Sims, Bot. Mag. 43: t. 1779 (1815).
Psidium cordatum var. *parvifolium* Griseb., Cat. Pl. Cub.: 91 (1866). *Psidium parvifolium* (Griseb.) Griseb., Cat. Pl. Cub.: 285 (1866).

Psidium androsianum (Urb.) Correll, J. Arnold Arbor. 58: 41 (1977).
Bahamas. 81 BAH. Nanophan. or phan.
**Eugenia androsiana* Urb., Repert. Spec. Nov. Regni Veg. 13: 467 (1915).

Psidium apiculatum Mattos, Anais Congr. Soc. Bot. Brasil 14: 29 (1964).
Brazil (Bahia). 84 BZE. Nanophan. or phan. Provisionally accepted.

Psidium appendiculatum Kiaersk., Enum. Myrt. Bras.: 32 (1893).
E. Brazil. 84 BZE BZL. Nanophan. or phan.

Psidium arboreum Vell., Fl. Flumin. 5: 211, t. 50 (1829). *Psidium sellowianum* O.Berg in C.F.P.von Martius & auct. suc. (eds.), Fl. Bras. 14(1): 400 (1857), nom. illeg. *Guajava arborea* (Vell.) Kuntze, Revis. Gen. Pl. 1: 239 (1891).
Brazil (Rio de Janeiro). 84 BZL. Phan. Provisionally accepted.

Psidium australe Cambess. in A.F.C.de Saint-Hilaire, Fl. Bras. Merid. 2: 283 (1832). *Guajava australis* (Cambess.) Kuntze, Revis. Gen. Pl. 1: 239 (1891).
Venezuela to Paraguay. 82 GUY VEN 83 BOL 84 BZC BZE BZL BZN BZS 85 PAR. Cham. or nanophan.

var. ***argenteum*** (O.Berg) Landrum, Sida 21: 1342 (2005).
Brazil. 84 BZC BZL BZS. Nanophan.
Psidium cuneatum Cambess. in A.F.C.de Saint-Hilaire, Fl. Bras. Merid. 2: 285 (1832). *Guajava cuneata* (Cambess.) Kuntze, Revis. Gen. Pl. 1: 239 (1891). *Psidium albidum* var. *cuneatum* (Cambess.) Mattos, Loefgrenia 116: 2 (2001).
**Psidium argenteum* O.Berg in C.F.P.von Martius & auct. suc. (eds.), Fl. Bras. 14(1): 388 (1857). *Guajava argenta* (O.Berg) Kuntze, Revis. Gen. Pl. 1: 239 (1891).
Psidium argenteum var. *angustifolium* O.Berg in C.F.P.von Martius & auct. suc. (eds.), Fl. Bras. 14(1): 388 (1857).
Psidium argenteum var. *grandifolium* O.Berg in C.F.P.von Martius & auct. suc. (eds.), Fl. Bras. 14(1): 388 (1857).
Psidium argenteum var. *pumilum* O.Berg in C.F.P. von Martius & auct. suc. (eds.), Fl. Bras. 14(1): 388 (1857).
Psidium argenteum var. *purpureum* O.Berg in C.F.P.von Martius & auct. suc. (eds.), Fl. Bras. 14(1): 388 (1857).
Psidium cuneatum var. *incanescens* O.Berg in C.F.P.von Martius & auct. suc. (eds.), Fl. Bras. 14(1): 405 (1857).
Psidium cuneatum var. *niveum* O.Berg in C.F.P.von Martius & auct. suc. (eds.), Fl. Bras. 14(1): 405 (1857).

var. ***australe***
Venezuela to Brazil. 82 GUY VEN 84 BZC BZE BZL BZN BZS 85 PAR. Cham. or nanophan.
Psidium anceps O.Berg in C.F.P.von Martius & auct. suc. (eds.), Fl. Bras. 14(1): 395 (1857). *Guajava anceps* (O.Berg) Kuntze, Revis. Gen. Pl. 1: 239 (1891).
Psidium triphyllum Barb.Rodr., Myrt. Paraguay: 12 (1903).
Psidium mucronatum Barb.Rodr. ex Chodat & Hassl., Bull. Herb. Boissier 7: 798 (1907), nom. nud.
Psidium piribebuiense Barb.Rodr. ex Chodat & Hassl., Bull. Herb. Boissier 7: 797 (1907), nom. nud.
Psidium submetrale McVaugh, Mem. New York Bot. Gard. 18(2): 261 (1969).

var. ***suffruticosum*** (O.Berg) Landrum, Sida 21: 1344 (2005).
Bolivia to Brazil. 83 BOL 84 BZC BZL BZS 85 PAR. Cham.
**Psidium suffruticosum* O.Berg in C.F.P.von Martius & auct. suc. (eds.), Fl. Bras. 14(1): 387 (1857). *Guajava suffruticosa* (O.Berg) Kuntze, Revis. Gen. Pl. 1: 239 (1891).
Psidium alatum O.Berg in C.F.P.von Martius & auct. suc. (eds.), Fl. Bras. 14(1): 604 (1859). *Psidium suffruticosum* var. *alatum* (O.Berg) Kiaersk., Enum. Myrt. Bras.: 27 (1893).

Psidium balium Urb., Symb. Antill. 9: 84 (1923).
E. Cuba. 81 CUB. Nanophan. Provisionally accepted.

Psidium brevifolium Alain, Moscosoa 1: 33 (1976).
Dominican Rep. 81 DOM. Nanophan. Provisionally accepted.

Psidium brownianum Mart. ex DC., Prodr. 3: 236 (1828). *Mitranthes browniana* (Mart. ex DC.) O.Berg in C.F.P.von Martius & auct. suc. (eds.), Fl. Bras. 14(1): 355 (1857). *Chytraculia browniana* (Mart. ex DC.) Kuntze, Revis. Gen. Pl. 1: 238 (1891). *Mitropsidium brownianum* (Mart. ex DC.) Burret, Notizbl. Bot. Gart. Berlin-Dahlem 15: 486 (1941).
Venezuela to E. Brazil. 82 VEN 84 BZE BZL.
Psidium moritzianum O.Berg, Linnaea 27: 359 (1856). *Guajava moritziana* (O.Berg) Kuntze, Revis. Gen. Pl. 1: 240 (1891).
Psidium macahense O.Berg in C.F.P.von Martius & auct. suc. (eds.), Fl. Bras. 14(1): 605 (1859).

Psidium calyptranthoides Alain, Phytologia 54: 109 (1983).
Puerto Rico. 81 PUE. Phan. Provisionally accepted.

Psidium canum Mattos, Loefgrenia 94: 11 (1989).
WC. Brazil. 84 BZC. Provisionally accepted.

Psidium cattleianum Afzel. ex Sabine, Trans. Hort. Soc. London 4: 317 (1821). *Guajava cattleiana* (Afzel. ex Sabine) Kuntze, Revis. Gen. Pl. 1: 239 (1891).
E. & S. Brazil to NE. Uruguay, widely introduced elsewere. (23) gab (25) ken tan uga (26) mlw moz zam zim (27) cpp nat (29) mdg (38) kzn oga (40) ind (42) mly (50) nfk (51) ker nzn (61) sci (63) haw (80) els (81) jam lee win 84 BZE BZL BZS (85) jnf URU. Phan. Widely cultivated for its edible fruits (strawberry guava).
Psidium littorale Raddi, Opusc. Sci. 4: 254 (1823). *Psidium cattleianum* var. *littorale* (Raddi) Fosberg, Occas. Pap. Bernice Pauahi Bishop Mus. 23: 37 (1962). *Psidium cattleianum* var. *littorale* (Raddi) Mattos, Loefgrenia 85: 1 (1984), nom. illeg.
Eugenia ferruginea Sieber ex C.Presl, Isis (Oken) 21: 274 (1828).
Psidium ferrugineum C.Presl, Isis (Oken) 21: 274 (1828).
Psidium obovatum Mart. ex DC., Prodr. 3: 236 (1828). *Guajava obovata* (Mart. ex DC.) Kuntze, Revis. Gen. Pl. 1: 239 (1891).
Psidium indicum Bojer, Hortus Maurit.: 139 (1837), nom. inval.
Psidium coriaceum var. *grandifolium* O.Berg in C.F.P. von Martius & auct. suc. (eds.), Fl. Bras. 14(1): 461 (1857).
Psidium coriaceum var. *longipes* O.Berg in C.F.P.von Martius & auct. suc. (eds.), Fl. Bras. 14(1): 402 (1857). *Psidium littorale* var. *longipes* (O.Berg) Fosberg, Proc. Biol. Soc. Wash. 54: 180 (1941).
Psidium coriaceum var. *obovatum* O.Berg in C.F.P.von Martius & auct. suc. (eds.), Fl. Bras. 14(1): 461 (1857).
Psidium variabile O.Berg in C.F.P.von Martius & auct. suc. (eds.), Fl. Bras. 14(1): 400 (1857).
Eugenia urceolata Cordem., Fl. Réunion: 426 (1895).
Eugenia oxygona Koidz., Bot. Mag. (Tokyo) 32: 258 (1918).
Psidium cattleianum f. *lucidum* O.Deg., Fl. Hawaiiensis 273: s.p. (1939).
Psidium cattleianum var. *pyriformis* Mattos, Loefgrenia 76: 1 (1981).

Psidium cauliflorum Landrum & Sobral, Sida 22: 927 (2006).
Brazil (Bahia). 84 BZE.

Psidium celastroides Urb., Symb. Antill. 9: 463 (1928).
C. Cuba. 81 CUB. Nanophan. Provisionally accepted.

Psidium claraense Urb., Symb. Antill. 9: 466 (1928). *Myrtus claraensis* (Urb.) Bisse, Ciencias (Havana), Ser. 10, 12: 10 (1976 publ. 1977).
EC. Cuba. 81 CUB. Nanophan. or phan.

Psidium cymosum Urb., Symb. Antill. 9: 464 (1928).
W. Cuba. 81 CUB. Nanophan.

Psidium densicomum Mart. ex DC., Prodr. 3: 235 (1828). *Guajava densicoma* (Mart. ex DC.) Kuntze, Revis. Gen. Pl. 1: 239 (1891).
S. Trop. America. 82 GUY VEN 83 CLM PER 84 BZN. Nanophan. or phan.
Psidium ovatifolium O.Berg in C.F.P.von Martius & auct. suc. (eds.), Fl. Bras. 14(1): 385 (1857). *Guajava ovatifolia* (O.Berg) Kuntze, Revis. Gen. Pl. 1: 239 (1891).
Psidium ovatifolium var. *glabrum* Amshoff, Bull. Torrey Bot. Club 75: 538 (1948).

Psidium dictyophyllum Urb. & Ekman, Ark. Bot. 21A(5): 19 (1927).
Hispaniola. 81 DOM HAI. Nanophan. or phan. Provisionally accepted.

Psidium donianum O.Berg in C.F.P.von Martius & auct. suc. (eds.), Fl. Bras. 14(1): 521 (1858). *Guajava doniana* (O.Berg) Kuntze, Revis. Gen. Pl. 1: 239 (1891).
Brazil (Maranhão). 84 BZE.

Psidium dumetorum Proctor, Bull. Inst. Jamaica, Sci. Ser. 16: 37 (1967).
C. Jamaica. 81 JAM. Nanophan. Provisionally accepted.

Psidium eugenii Kiaersk., Enum. Myrt. Bras.: 26 (1893).
SE. Brazil. 84 BZL. Nanophan. Provisionally accepted.

Psidium firmum O.Berg in C.F.P.von Martius & auct. suc. (eds.), Fl. Bras. 14(1): 390 (1857). *Guajava firma* (O.Berg) Kuntze, Revis. Gen. Pl. 1: 239 (1891).
WC. & SE. Brazil. 84 BZC BZL. Cham. or nanophan.
Myrtus grandifolia O.Berg in C.F.P.von Martius & auct. suc. (eds.), Fl. Bras. 14(1): 419 (1857). *Psidium grandifolium* (O.Berg) Burret, Notizbl. Bot. Gart. Berlin-Dahlem 15: 484 (1941), nom. illeg. *Psidium minense* Mattos, Loefgrenia 42: 2 (1970).
Psidium firmum var. *subcordatum* O.Berg in C.F.P.von Martius & auct. suc. (eds.), Fl. Bras. 14(1): 601 (1859).
Psidium macedoi Kausel, Lilloa 33: 108 (1971 publ. 1972).

Psidium friedrichsthalianum (O.Berg) Nied. in H.G.A.Engler & K.A.E.Prantl, Nat. Pflanzenfam. 3(7): 69 (1893).
S. Mexico to Colombia. 79 MXS 80 COS ELS GUA NIC PAN 83 CLM. Nanophan. or phan.
**Calyptropsidium friedrichsthalianum* O.Berg, Linnaea 27: 350 (1856).

Psidium fulvum McVaugh, Fieldiana, Bot. 29: 226 (1956).
Peru. 83 PER. Nanophan. or phan.

Psidium galapagaeum Hook.f., Trans. Linn. Soc. London 20: 224 (1847).
Galápagos. 83 GAL. Nanophan. or phan.
Psidium galapagaeum var. *howellii* D.M.Porter, Ann. Missouri Bot. Gard. 55: 370 (1969).

Psidium giganteum Mattos, Loefgrenia 28: 2 (1969).
Brazil (S. Minas Gerais to São Paulo). 84 BZL.

Psidium glaziovianum Kiaersk., Enum. Myrt. Bras.: 33 (1893).
SE. Brazil. 84 BZL.

Psidium globosum Larrañaga, Escritos D. A. Larrañaga 2: 168 (1923).
Uruguay. 85 URU. Provisionally accepted.

Psidium gracilipes Alain, Phytologia 25: 269 (1973).
Dominican Rep. 81 DOM. Nanophan. or phan.

Psidium grandifolium Mart. ex DC., Prodr. 3: 234 (1828). *Guajava grandifolia* (Mart. ex DC.) Kuntze, Revis. Gen. Pl. 1: 239 (1891).
Brazil to NE. Argentina. 83 BOL 84 BZC BZE BZL BZS 85 AGE PAR. Nanophan.
Psidium cinereum Mart. ex DC., Prodr. 3: 234 (1828). *Guajava cinerea* (Mart. ex DC.) Kuntze, Revis. Gen. Pl. 1: 239 (1891).

Psidium incanescens Mart. ex DC., Prodr. 3: 234 (1828). *Guajava incanescens* (Mart. ex DC.) Kuntze, Revis. Gen. Pl. 1: 239 (1891). *Psidium cinereum* var. *incanescens* (Mart. ex DC.) D.Legrand, in Fl. Ilustr. Catar. 1(Mirt.): 692 (1977).
Psidium microcarpum Cambess. in A.F.C.de Saint-Hilaire, Fl. Bras. Merid. 2: 284 (1832). *Guajava microcarpa* (Cambess.) Kuntze, Revis. Gen. Pl. 1: 239 (1891).
Psidium ternatifolium Cambess. in A.F.C.de Saint-Hilaire, Fl. Bras. Merid. 2: 278 (1832). *Psidium grandifolium* var. *ternatifolium* (Cambess.) O.Berg in C.F.P.von Martius & auct. suc. (eds.), Fl. Bras. 14(1): 407 (1857).
Psidium albidum Miq., Linnaea 22: 532 (1849).
Psidium cinereum var. *angustifolium* O.Berg in C.F.P. von Martius & auct. suc. (eds.), Fl. Bras. 14(1): 405 (1857).
Psidium cinereum var. *brevipes* O.Berg in C.F.P.von Martius & auct. suc. (eds.), Fl. Bras. 14(1): 405 (1857).
Psidium cinereum var. *grandifolium* O.Berg in C.F.P. von Martius & auct. suc. (eds.), Fl. Bras. 14(1): 405 (1857).
Psidium cinereum var. *intermedium* O.Berg in C.F.P. von Martius & auct. suc. (eds.), Fl. Bras. 14(1): 405 (1857).
Psidium grandifolium var. *heterophyllum* O.Berg in C.F.P.von Martius & auct. suc. (eds.), Fl. Bras. 14(1): 407 (1857).
Psidium grandifolium var. *intermedium* O.Berg in C.F.P.von Martius & auct. suc. (eds.), Fl. Bras. 14(1): 407 (1857).
Psidium grandifolium var. *parvifolium* O.Berg in C.F.P.von Martius & auct. suc. (eds.), Fl. Bras. 14(1): 407 (1857).
Psidium grandifolium var. *tenuinerve* O.Berg in C.F.P. von Martius & auct. suc. (eds.), Fl. Bras. 14(1): 407 (1857).
Psidium incanescens var. *cuneatum* O.Berg in C.F.P. von Martius & auct. suc. (eds.), Fl. Bras. 14(1): 403 (1857).
Psidium incanescens var. *parvifolium* O.Berg in C.F.P. von Martius & auct. suc. (eds.), Fl. Bras. 14(1): 403 (1857).
Psidium incanescens var. *rotundifolium* O.Berg in C.F.P.von Martius & auct. suc. (eds.), Fl. Bras. 14(1): 403 (1857).
Psidium lacteum O.Berg in C.F.P.von Martius & auct. suc. (eds.), Fl. Bras. 14(1): 403 (1857). *Guajava lactea* (O.Berg) Kuntze, Revis. Gen. Pl. 1: 239 (1891).
Psidium radicans O.Berg in C.F.P.von Martius & auct. suc. (eds.), Fl. Bras. 14(1): 402 (1857). *Guajava radicans* (O.Berg) Kuntze, Revis. Gen. Pl. 1: 239 (1891).
Psidium rubescens O.Berg in C.F.P.von Martius & auct. suc. (eds.), Fl. Bras. 14(1): 395 (1857). *Guajava rubescens* (O.Berg) Kuntze, Revis. Gen. Pl. 1: 239 (1891).
Psidium sericeum O.Berg in C.F.P.von Martius & auct. suc. (eds.), Fl. Bras. 14(1): 389 (1857). *Guajava sericea* (O.Berg) Kuntze, Revis. Gen. Pl. 1: 239 (1891).
Psidium grandifolium var. *albidum* O.Berg in C.F.P.von Martius & auct. suc. (eds.), Fl. Bras. 14(1): 603 (1859).
Psidium grandifolium var. *incanescens* O.Berg in C.F.P.von Martius & auct. suc. (eds.), Fl. Bras. 14(1): 603 (1859).
Psidium riedelianum O.Berg in C.F.P.von Martius & auct. suc. (eds.), Fl. Bras. 14(1): 603 (1859). *Guajava riedeliana* (O.Berg) Kuntze, Revis. Gen. Pl. 1: 239 (1891).
Psidium eriophyllum Barb.Rodr., Myrt. Paraguay: 12 (1903).
Psidium lanatum Barb.Rodr., Myrt. Paraguay: 13 (1903).
Psidium spodophyllum Barb.Rodr., Myrt. Paraguay: 14 (1903).
Psidium apaense Barb.Rodr. ex Chodat & Hassl., Bull. Herb. Boissier 7: 798 (1907), nom. inval.
Psidium paraguayense Barb.Rodr. ex Chodat & Hassl., Bull. Herb. Boissier 7: 798 (1907), nom. inval.
Psidium psychrophyllum Barb.Rodr. ex Chodat & Hassl., Bull. Herb. Boissier 7: 797 (1907), nom. inval.
Psidium yacaense Barb.Rodr. ex Chodat & Hassl., Bull. Herb. Boissier 7: 797 (1907), nom. inval.
Psidium cinereum var. *paraguariae* D.Legrand, in Fl. Ilustr. Catar. 1(Mirt.): 694 (1977).

Psidium guajava L., Sp. Pl.: 470 (1753). *Myrtus guajava* (L.) Kuntze, Revis. Gen. Pl. 3(2): 91 (1898).
Trop. & Subtrop. America, widely introduced elsewere. (21) cvi mdr (22) mtn (23) gab (25) ken tan uga (26) ang mlw moz zam zim (27) cpp nat tvl (38) oga (40) ind (42) bor mly sul (43) bis (51) ker (60) gil wal (61) lin sci (63) haw 78 FLA 79 MXT 80 BLZ COS ELS GUA HON NIC PAN 81 CAY CUB DOM HAI JAM LEE TRT WIN (82) frg guy sur VEN 83 BOL CLM 84 BZC BZE BZL BZS 85 AGE AGW PAR. Phan. Widely cultivated for its edible fruits (common guava).
Psidium cujavus L., Herb. Amb.: 7 (1754).
Psidium pomiferum L., Sp. Pl. ed. 2: 672 (1762).
Psidium pyriferum L., Sp. Pl. ed. 2: 672 (1762). *Guajava pyrifera* (L.) Kuntze, Revis. Gen. Pl. 1: 239 (1891).
Psidium cujavillus Burm.f., Fl. Indica: 114 (1768). *Psidium guajava* var. *cujavillum* (Burm.f.) Krug & Urb., Bot. Jahrb. Syst. 19: 566 (1894).
Psidium angustifolium Lam., Encycl. 3: 16 (1789).
Psidium pumilum Vahl, Symb. Bot. 2: 56 (1791). *Guajava pumila* (Vahl) Kuntze, Revis. Gen. Pl. 1: 240 (1891).
Psidium vulgare Rich., Actes Soc. Hist. Nat. Paris 1: 110 (1792).
Psidium sapidissimum Jacq., Pl. Hort. Schoenbr. 3: 62 (1798).
Psidium pumilum var. *guadalupense* DC., Prodr. 3: 233 (1828).
Psidium aromaticum Blanco, Fl. Filip.: 417 (1837), nom. illeg.
Psidium pyriferum var. *glabrum* Benth., J. Bot. (Hooker) 2: 318 (1840).
Psidium fragrans Macfad., Fl. Jamaica 2: 108 (1850).
Psidium intermedium Zipp. ex Blume, Mus. Bot. 1: 72 (1850).
Psidium prostratum O.Berg, Linnaea 27: 364 (1856).
Syzygium ellipticum K.Schum. & Lauterb., Fl. Schutzgeb. Südsee: 476 (1900).
Psidium igatemyense Barb.Rodr., Myrt. Paraguay: 10 (1903).
Psidium guajava var. *minor* Mattos, Loefgrenia 70: 5 (1976).

Psidium guineense Sw., Prodr.: 77 (1788). *Guajava guineensis* (Sw.) Kuntze, Revis. Gen. Pl. 1: 239 (1891). *Myrtus guineensis* (Sw.) Kuntze, Revis. Gen. Pl. 3(2):

91 (1898). *Mosiera guineensis* (Sw.) Bisse, Revista Jard. Bot. Nac. Univ. Habana 6(3): 4 (1985 publ. 1986).

Mexico to Trop. America. (22) ivo (23) gab (40) ind (61) sci 79 MXG MXN MXT 80 BLZ COS ELS GUA HON NIC PAN (81) cub hai jam lee TRT WIN 82 FRG GUY SUR VEN 83 BOL CLM ECU PER 84 BZC BZE BZL BZN BZS 85 AGE PAR. Nanophan. Widely cultivated for its edible fruits (guisaro).

Psidium polycarpon Lamb., Trans. Linn. Soc. London 11: 231 (1813). *Guajava polycarpa* (Lamb.) Kuntze, Revis. Gen. Pl. 1: 239 (1891).

Campomanesia tomentosa Kunth in F.W.H.von Humboldt, A.J.A.Bonpland & C.S.Kunth, Nov. Gen. Sp. 6: 151 (1823).

Psidium araca Raddi, Opusc. Sci. 4: 252 (1823).

Psidium dichotomum Weinm., Syll. Pl. Nov. 2: 166 (1828).

Psidium minus Mart. ex DC., Prodr. 3: 235 (1828), nom. inval.

Psidium albidum Cambess. in A.F.C.de Saint-Hilaire, Fl. Bras. Merid. 2: 283 (1832). *Guajava albida* (Cambess.) Kuntze, Revis. Gen. Pl. 1: 239 (1891).

Psidium multiflorum Cambess. in A.F.C. de Saint-Hilaire, Fl. Bras. 2: 281 (1832). *Campomanesia multiflora* (Cambess.) O.Berg in C.F.P.von Martius & auct. suc. (eds.), Fl. Bras. 14(1): 443 (1857). *Guajava multiflora* (Cambess.) Kuntze, Revis. Gen. Pl. 1: 239 (1891).

Psidium molle Bertol., Fl. Guatimal.: 22 (1840). *Guajava mollis* (Bertol.) Kuntze, Revis. Gen. Pl. 1: 240 (1891).

Psidium sericiflorum Benth., Pl. Hartw.: 176 (1845).

Psidium monticola O.Berg, Vidensk. Meddel. Naturhist. Foren. Kjøbenhavn 1855: 11 (1855).

Psidium benthamianum O.Berg, Linnaea 27: 362 (1856). *Guajava benthamiana* (O.Berg) Kuntze, Revis. Gen. Pl. 1: 240 (1891).

Psidium costa-ricense O.Berg, Linnaea 27: 368 (1856). *Guajava costa-ricensis* (O.Berg) Kuntze, Revis. Gen. Pl. 1: 240 (1891).

Psidium laurifolium O.Berg, Linnaea 27: 364 (1856). *Guajava laurifolia* (O.Berg) Kuntze, Revis. Gen. Pl. 1: 240 (1891).

Psidium molle var. *gracile* O.Berg, Linnaea 27: 370 (1856).

Psidium molle var. *robustum* O.Berg, Linnaea 27: 370 (1856).

Psidium monticola var. *gracile* O.Berg, Vidensk. Meddel. Dansk Naturhist. Foren. Kjøbenhavn 1855: 11 (1856).

Psidium monticola var. *robustum* O.Berg, Vidensk. Meddel. Dansk Naturhist. Foren. Kjøbenhavn 1855: 11 (1856).

Psidium schiedeanum O.Berg, Linnaea 27: 368 (1856). *Guajava schiedeana* (O.Berg) Kuntze, Revis. Gen. Pl. 1: 240 (1891).

Psidium ooideum O.Berg in C.F.P.von Martius & auct. suc. (eds.), Fl. Bras. 14(1): 398 (1857). *Guajava ooidea* (O.Berg) Kuntze, Revis. Gen. Pl. 1: 239 (1891).

Psidium ypanemense O.Berg in C.F.P.von Martius & auct. suc. (eds.), Fl. Bras. 14(1): 395 (1857). *Guajava ypanemensis* (O.Berg) Kuntze, Revis. Gen. Pl. 1: 239 (1891).

Psidium ooideum var. *grandifolium* O.Berg in C.F.P.von Martius & auct. suc. (eds.), Fl. Bras. 14(1): 602 (1859).

Psidium ooideum var. *intermedium* O.Berg in C.F.P.von Martius & auct. suc. (eds.), Fl. Bras. 14(1): 602 (1859).

Psidium ooideum var. *parvifolium* O.Berg in C.F.P.von Martius & auct. suc. (eds.), Fl. Bras. 14(1): 602 (1859).

Psidium ooideum var. *longipedunculatum* Rusby, Mem. Torrey Bot. Club 3(3): 27 (1893).

Myrtus hauthalii Kuntze, Revis. Gen. Pl. 3(2): 91 (1898).

Psidium campicolum Barb.Rodr., Myrt. Paraguay: 11 (1903).

Psidium rufinervum Barb.Rodr., Myrt. Paraguay: 15 (1903).

Psidium lehmannii Diels, Bot. Jahrb. Syst. 37: 594 (1906).

Psidium araca var. *sampaionis* Herter, Arq. Mus. Nac. Rio de Janeiro 18: 12 (1916).

Psidium chrysobalanoides Standl., Publ. Field Mus. Nat. Hist., Bot. Ser. 8: 319 (1931).

Psidium rotundifolium Standl., Publ. Field Mus. Nat. Hist., Bot. Ser. 8: 318 (1931).

Psidium schippii Standl., Publ. Field Mus. Nat. Hist., Bot. Ser. 8: 319 (1931).

Psidium popenoei Standl., Ceiba 1: 41 (1950).

Myrcianthes irregularis McVaugh, Fieldiana, Bot. 29: 489 (1963).

Psidium guyanense Pers., Syn. Pl. 2: 27 (1806). *Psidium fluviatile* Rich. ex DC., Prodr. 3: 235 (1828), nom. illeg. *Guajava fluviatilis* Kuntze, Revis. Gen. Pl. 1: 240 (1891).

N. South America to N. Brazil. 82 FRG VEN 84 BZN. Phan.

Psidium sprucei O.Berg in C.F.P.von Martius & auct. suc. (eds.), Fl. Bras. 14(1): 396 (1857). *Guajava sprucei* (O.Berg) Kuntze, Revis. Gen. Pl. 1: 239 (1891).

Psidium umbrosum O.Berg in C.F.P.von Martius & auct. suc. (eds.), Fl. Bras. 14(1): 599 (1859). *Guajava umbrosa* (O.Berg) Kuntze, Revis. Gen. Pl. 1: 239 (1891).

Psidium richardianum O.Berg, Linnaea 30: 705 (1861). *Guajava richardiana* (O.Berg) Kuntze, Revis. Gen. Pl. 1: 240 (1891).

Psidium haitiense Alain, Brittonia 20: 159 (1969).

Haiti. 81 HAI. Nanophan. Provisionally accepted.

Psidium harrisianum Urb., Symb. Antill. 7: 294 (1912).

Jamaica. 81 JAM. Nanophan. or phan. Provisionally accepted.

Psidium* × *hasslerianum Barb.Rodr., Myrt. Paraguay: 9 (1903). *P. guajava* × *P. guineense*.

Trop. America. 80 BLZ GUA HON 85 PAR. Phan.

Psidium × *hypoglaucum* Standl., Publ. Field Mus. Nat. Hist., Bot. Ser. 8: 320 (1931).

Psidium hotteanum Urb. & Ekman, Ark. Bot. 22A(10): 21 (1929).

Haiti (Massif de la Hotte). 81 HAI. Nanophan. or phan. Provisionally accepted.

Psidium huanucoense Landrum, Novon 15: 444 (2005).

Peru (Huánuco). 83 PER. Phan.

Psidium inaequilaterum O.Berg in C.F.P.von Martius & auct. suc. (eds.), Fl. Bras. 14(1): 399 (1857). *Guajava inaequilatera* (O.Berg) Kuntze, Revis. Gen. Pl. 1: 239 (1891).

SE. Brazil. 84 BZL. Provisionally accepted.

Psidium itanareense O.Berg in C.F.P.von Martius & auct. suc. (eds.), Fl. Bras. 14(1): 402 (1857). *Guajava itanareensis* (O.Berg) Kuntze, Revis. Gen. Pl. 1: 239 (1891).
Brazil (São Paulo). 84 BZL.

Psidium jacquinianum (O.Berg) Mattos, Loefgrenia 116: 2 (2001).
S. America (?). 8+.
**Myrtus jacquiniana* O.Berg, Linnaea 27: 406 (1856).

Psidium jakuscianum Borhidi, Bot. Közlem. 64: 214 (1977 publ. 1978).
Cuba. 81 CUB. Provisionally accepted.

Psidium kennedyanum Morong, Ann. New York Acad. Sci. 7: 104 (1892).
Paraguay to NE. Argentina. 84 BZC 85 AGE PAR. Nanophan. or phan.
Psidium tripartitum S.Moore, Trans. Linn. Soc. London, Bot. 4: 353 (1895).

Psidium langsdorffii O.Berg in C.F.P.von Martius & auct. suc. (eds.), Fl. Bras. 14(1): 599 (1859). *Guajava langsdorffii* (O.Berg) Kuntze, Revis. Gen. Pl. 1: 239 (1891).
Brazil (Minas Gerais). 84 BZL. Provisionally accepted.

Psidium laruotteanum Cambess. in A.F.C.de Saint-Hilaire, Fl. Bras. Merid. 2: 282 (1832). *Guajava laruotteana* (Cambess.) Kuntze, Revis. Gen. Pl. 1: 239 (1891).
Costa Rica to Paraguay. 80 COS 82 GUY SUR VEN 83 BOL 84 BZC BZL BZS 85 PAR. Nanophan.
Campomanesia suffruticosa O.Berg in C.F.P.von Martius & auct. suc. (eds.), Fl. Bras. 14(1): 448 (1857). *Myrtus bergiana* Nied. in H.G.A.Engler & K.A.E.Prantl, Nat. Pflanzenfam. 3(7): 66 (1893). *Psidium bergianum* (Nied.) Burret, Notizbl. Bot. Gart. Berlin-Dahlem 15: 485 (1941). *Psidium warmingianum* Kiaersk., Enum. Myrt. Bras.: 28 (1893), nom. illeg.
Psidium aerugineum O.Berg in C.F.P.von Martius & auct. suc. (eds.), Fl. Bras. 14(1): 391 (1857). *Guajava aeruginea* (O.Berg) Kuntze, Revis. Gen. Pl. 1: 239 (1891).
Psidium aerugineum var. *angustifolium* O.Berg in C.F.P.von Martius & auct. suc. (eds.), Fl. Bras. 14(1): 601 (1859).
Psidium basanthum O.Berg in C.F.P.von Martius & auct. suc. (eds.), Fl. Bras. 14(1): 601 (1859). *Guajava basantha* (O.Berg) Kuntze, Revis. Gen. Pl. 1: 239 (1891).
Psidium glaucescens O.Berg in C.F.P.von Martius & auct. suc. (eds.), Fl. Bras. 14(1): 600 (1859). *Guajava glaucescens* (O.Berg) Kuntze, Revis. Gen. Pl. 1: 239 (1891).
Psidium warmingianum var. *verticillatum* Kiaersk., Enum. Myrt. Bras.: 28 (1893).
Psidium savannarum Donn.Sm., Bot. Gaz. 23: 244 (1897).
Myrtus formosa Barb.Rodr., Myrt. Paraguay: 16 (1903). *Psidium formosum* (Barb.Rodr.) Burret, Notizbl. Bot. Gart. Berlin-Dahlem 15: 485 (1941).
Psidium capibaryense Barb.Rodr. ex Chodat & Hassl., Bull. Herb. Boissier, II, 7: 797 (1907).
Psidium quinquedentatum Amshoff, Recueil Trav. Bot. Néerl. 39: 164 (1942).

Psidium longipetiolatum D.Legrand, Sellowia 13: 341 (1961).
S. Brazil. 84 BZS. Phan.

Psidium lourteigiae D.Legrand, Bradea 1: 157 (1972).
WC. Brazil. 84 BZC. Nanophan. Provisionally accepted.

Psidium loustalotii Britton & P.Wilson, Bull. Torrey Bot. Club 48: 342 (1921 publ. 1922).
Cuba. 81 CUB. Provisionally accepted.

Psidium maribense Mart. ex DC., Prodr. 3: 233 (1828). *Guajava maribensis* (Mart. ex DC.) Kuntze, Revis. Gen. Pl. 1: 239 (1891).
Colombia to N. Brazil. 82 VEN 83 CLM 84 BZN. Nanophan. or phan.
Myrtus salicifolia Willd. ex O.Berg, Linnaea 27: 374 (1856), nom. inval.
Psidium crenatum O.Berg, Linnaea 27: 373 (1856). *Guajava crenata* (O.Berg) Kuntze, Revis. Gen. Pl. 1: 240 (1891).

Psidium minutifolium Krug & Urb., Bot. Jahrb. Syst. 19: 569 (1894).
NE. Cuba. 81 CUB. Nanophan. or phan. Provisionally accepted.

Psidium misionum D.Legrand, Darwiniana 9: 284 (1950).
Paraguay to NE. Argentina (Misiones). 85 AGE PAR. Nanophan.

Psidium montanum Sw., Prodr.: 77 (1788). *Guajava montana* (Sw.) Kuntze, Revis. Gen. Pl. 1: 240 (1891).
Jamaica. 81 JAM. Phan.
Psidium wrightii D.Don ex W.Wright in T.L.Mitchell, Mem. W. Wright: 278 (1828).

Psidium munizianum Borhidi, Acta Bot. Acad. Sci. Hung. 21: 225 (1975 publ. 1976). *Myrtus muniziana* (Borhidi) Borhidi, Acta Bot. Acad. Sci. Hung. 25: 27 (1979).
Cuba. 81 CUB. Provisionally accepted.

Psidium myrsinites DC., Prodr. 3: 236 (1828). *Guajava myrsinites* (DC.) Kuntze, Revis. Gen. Pl. 1: 239 (1891).
Brazil. 84 BZC BZE BZL. Nanophan.
Psidium gardnerianum O.Berg in C.F.P.von Martius & auct. suc. (eds.), Fl. Bras. 14(1): 389 (1857). *Guajava gardneriana* (O.Berg) Kuntze, Revis. Gen. Pl. 1: 239 (1891).
Psidium malmei Kausel, Lilloa 33: 108 (1971 publ. 1972).

Psidium myrtoides O.Berg in C.F.P.von Martius & auct. suc. (eds.), Fl. Bras. 14(1): 384 (1857). *Guajava myrtoides* (O.Berg) Kuntze, Revis. Gen. Pl. 1: 239 (1891).
E. Brazil. 84 BZE BZL. Nanophan. or phan.
Psidium myrsinoides O.Berg in C.F.P.von Martius & auct. suc. (eds.), Fl. Bras. 14(1): 384 (1857). *Guajava myrsinoides* (O.Berg) Kuntze, Revis. Gen. Pl. 1: 239 (1891).

Psidium nannophyllum Alain, Phytologia 25: 270 (1973).
Dominican Rep. 81 DOM. Cham. or nanophan. Provisionally accepted.

Psidium nummularia (C.Wright ex Griseb.) C.Wright, Anales Acad. Ci. Méd. Habana 5: 433 (1868).
Cuba. 81 CUB. Nanophan.
**Eugenia nummularia* C.Wright ex Griseb., Cat. Pl. Cub.: 86 (1866).

Psidium nutans O.Berg in C.F.P.von Martius & auct. suc. (eds.), Fl. Bras. 14(1): 394 (1857). *Guajava nutans*

(O.Berg) Kuntze, Revis. Gen. Pl. 1: 239 (1891).
Brazil to NE. Argentina. 84 BZE 85 AGE. Phan. Provisionally accepted.

Psidium oblongatum O.Berg in C.F.P.von Martius & auct. suc. (eds.), Fl. Bras. 14(1): 392 (1857). *Guajava oblongata* (O.Berg) Kuntze, Revis. Gen. Pl. 1: 239 (1891).
Brazil (Espírito Santo, Minas Gerais). 84 BZL. Phan.

Psidium oblongifolium O.Berg in C.F.P.von Martius & auct. suc. (eds.), Fl. Bras. 14(1): 602 (1859). *Guajava oblongifolia* (O.Berg) Kuntze, Revis. Gen. Pl. 1: 239 (1891).
SE. Brazil. 84 BZL. Provisionally accepted.

Psidium oligospermum Mart. ex DC., Prodr. 3: 236 (1828). *Guajava oligosperma* (Mart. ex DC.) Kuntze, Revis. Gen. Pl. 1: 239 (1891). *Mitropsidium oligospermum* (Mart. ex DC.) Burret, Notizbl. Bot. Gart. Berlin-Dahlem 15: 486 (1941).
Brazil (Bahia). 84 BZE.

Psidium oncocalyx Burret, Repert. Spec. Nov. Regni Veg. 50: 56 (1941).
Brazil (Bahia). 84 BZE.

Psidium orbifolium Urb., Symb. Antill. 9: 462 (1928).
SE. Cuba. 81 CUB. Nanophan. or phan. Provisionally accepted.

Psidium ovale (Spreng.) Burret, Notizbl. Bot. Gart. Berlin-Dahlem 15: 485 (1941).
Brazil (S. Minas Gerais to E. Santa Catarina). 84 BZL BZS. Nanophan. or phan.
**Myrtus ovalis* Spreng., Syst. Veg. 2: 479 (1825). *Myrcianthes brunnea* O.Berg, Linnaea 27: 315 (1856), nom. superfl. *Eugenia brunnea* Nied. in H.G.A.Engler & K.A.E.Prantl, Nat. Pflanzenfam. 3(7): 81 (1893). *Myrcianthes brunnea* var. *parvifolia* O.Berg in C.F.P.von Martius & auct. suc. (eds.), Fl. Bras. 14(1): 352 (1857), nom. inval.
Myrcianthes brunnea var. *grandifolia* O.Berg in C.F.P.von Martius & auct. suc. (eds.), Fl. Bras. 14(1): 352 (1857).
Psidium spathulatum Mattos, Loefgrenia 22: 1 (1965).
Psidium hatschbachii D.Legrand, Bol. Univ. Paraná, Bot. 25: 1 (1971).
Psidium imaruinense Mattos, Loefgrenia 66: 1 (1975).
Psidium hagelundianum Mattos, Loefgrenia 94: 11 (1989).

Psidium parvifolium Griseb., Cat. Pl. Cub.: 91 (1866).
Cuba. 81 CUB. Nanophan.
Psidium nitidum C.Wright, Anales Acad. Ci. Méd. Habana 5: 433 (1868).
Psidium paucinerve Urb., Symb. Antill. 9: 82 (1923).

Psidium pedicellatum McVaugh, Fieldiana, Bot. 29: 227 (1956).
Ecuador. 83 ECU. Phan.

Psidium pigmeum Arruda in H.Koster, Trav. Brazil: 492 (1816).
SE. Brazil. 84 BZL. Provisionally accepted.

Psidium pulverulentum Krug & Urb., Bot. Jahrb. Syst. 19: 567 (1894).
Jamaica. 81 JAM.

Psidium raimondii Burret, Repert. Spec. Nov. Regni Veg. 50: 56 (1941).
Peru. 83 PER. Provisionally accepted.

Psidium ramboanum Mattos, Loefgrenia 116: 2 (2001).
Brazil (Mato Grosso). 84 BZC. Nanophan.
**Psidium nigrum* Mattos & D.Legrand, Loefgrenia 67: 10 (1975), nom. illeg.

Psidium refractum O.Berg in C.F.P.von Martius & auct. suc. (eds.), Fl. Bras. 14(1): 394 (1857). *Guajava refracta* (O.Berg) Kuntze, Revis. Gen. Pl. 1: 239 (1891).
Brazil (Goiás). 84 BZC. Provisionally accepted.

Psidium reptans (D.Legrand) Soares-Silva & Proença, Kew Bull. 61: 203 (2006).
Brazil (Paraná). 84 BZS. Cham.
**Myrcianthes reptans* D.Legrand, Bol. Univ. Paraná, Bot. 27: 1 (1971).

Psidium reversum Urb., Symb. Antill. 9: 84 (1923). *Calycolpus reversus* (Urb.) Bisse, Revista Jard. Bot. Nac. Univ. Habana 4: 6 (1983).
E. Cuba (Sierra Sagua Baracoa). 81 CUB. Nanophan. or phan.

Psidium rhombeum O.Berg in C.F.P.von Martius & auct. suc. (eds.), Fl. Bras. 14(1): 383 (1857). *Guajava rhombea* (O.Berg) Kuntze, Revis. Gen. Pl. 1: 239 (1891).
Brazil (Bahia). 84 BZE.

Psidium riparium Mart. ex DC., Prodr. 3: 235 (1828). *Guajava riparia* (Mart. ex DC.) Kuntze, Revis. Gen. Pl. 1: 239 (1891).
Brazil. 84 BZC BZE BZL BZN. Nanophan.
Psidium mengahiense Cambess. in A.F.C.de Saint-Hilaire, Fl. Bras. Merid. 2: 286 (1832). *Guajava mengahiensis* (Cambess.) Kuntze, Revis. Gen. Pl. 1: 239 (1891).
Psidium maranhense O.Berg in C.F.P.von Martius & auct. suc. (eds.), Fl. Bras. 14(1): 386 (1857). *Guajava maranhensis* (O.Berg) Kuntze, Revis. Gen. Pl. 1: 239 (1891).
Psidium paraense O.Berg in C.F.P.von Martius & auct. suc. (eds.), Fl. Bras. 14(1): 386 (1857). *Guajava paraensis* (O.Berg) Kuntze, Revis. Gen. Pl. 1: 239 (1891).
Psidium sieberianum O.Berg in C.F.P.von Martius & auct. suc. (eds.), Fl. Bras. 14(1): 387 (1857). *Guajava sieberiana* (O.Berg) Kuntze, Revis. Gen. Pl. 1: 239 (1891).
Psidium sieberianum var. *gracile* O.Berg in C.F.P.von Martius & auct. suc. (eds.), Fl. Bras. 14(1): 387 (1857).
Psidium sieberianum var. *robustum* O.Berg in C.F.P.von Martius & auct. suc. (eds.), Fl. Bras. 14(1): 387 (1857).
Psidium paranense O.Berg in C.F.P.von Martius & auct. suc. (eds.), Fl. Bras. 14(1): 604 (1859). *Guajava paranensis* (O.Berg) Kuntze, Revis. Gen. Pl. 1: 239 (1891).
Psidium insulincola S.Moore, Trans. Linn. Soc. London, Bot. 4: 353 (1895).
Myrtus thyrsoidea Kuntze, Revis. Gen. Pl. 3(2): 92 (1898). *Psidium thyrsoideum* (Kuntze) K.Schum., Just's Bot. Jahresber. 26(1): 359 (1898).

Psidium robustum O.Berg in C.F.P.von Martius & auct. suc. (eds.), Fl. Bras. 14(1): 400 (1857). *Guajava robusta* (O.Berg) Kuntze, Revis. Gen. Pl. 1: 239 (1891).
Brazil (Minas Gerais). 84 BZL.

Psidium rostratum McVaugh, Fieldiana, Bot. 29: 227 (1956).
Peru. 83 PER. Nanophan. or phan.

Psidium rotundatum Griseb., Cat. Pl. Cub.: 92 (1866).
Cuba. 81 CUB. Nanophan. Provisionally accepted.
Psidium rotundatum var. *triflorum* Griseb., Cat. Pl. Cub.: 92 (1866).

Psidium rufum Mart. ex DC., Prodr. 3: 234 (1828).
Brazil. 84 BZC BZE BZL BZS. Nanophan.
Psidium pilosum Vell., Fl. Flumin. 5: 212, t. 54 (1829). *Guajava pilosa* (Vell.) Kuntze, Revis. Gen. Pl. 1: 239 (1891).
Abbevillea recurvata O.Berg in C.F.P.von Martius & auct. suc. (eds.), Fl. Bras. 14(1): 438 (1857). *Campomanesia recurvata* (O.Berg) Nied. in H.G.A.Engler & K.A.E.Prantl, Nat. Pflanzenfam. 3(7): 73 (1893).
Psidium cupreum O.Berg in C.F.P.von Martius & auct. suc. (eds.), Fl. Bras. 14(1): 393 (1857). *Guajava cuprea* (O.Berg) Kuntze, Revis. Gen. Pl. 1: 239 (1891).
Psidium macrospermum O.Berg in C.F.P.von Martius & auct. suc. (eds.), Fl. Bras. 14(1): 392 (1857). *Guajava macrosperma* (O.Berg) Kuntze, Revis. Gen. Pl. 1: 239 (1891).
Psidium widgrenianum O.Berg in C.F.P.von Martius & auct. suc. (eds.), Fl. Bras. 14(1): 392 (1857). *Guajava widgreniana* (O.Berg) Kuntze, Revis. Gen. Pl. 1: 239 (1891).
Abbevillea regeliana O.Berg in C.F.P.von Martius & auct. suc. (eds.), Fl. Bras. 14(1): 608 (1859). *Campomanesia regeliana* (O.Berg) Kiaersk., Enum. Myrt. Bras.: 8 (1893).
Campomanesia martiana O.Berg in C.F.P.von Martius & auct. suc. (eds.), Fl. Bras. 14(1): 610 (1859).
Psidium cupreum var. *glabratum* Kiaersk., Enum. Myrt. Bras.: 29 (1893).
Psidium lagoense Kiaersk., Enum. Myrt. Bras.: 30 (1893).
Psidium rufum var. *rotundifolium* Kiaersk., Enum. Myrt. Bras.: 31 (1893).

Psidium rutidocarpum Ruiz & Pav. ex G.Don, Gen. Hist. 2: 833 (1832).
Peru. 83 PER. Nanophan.
Psidium pratense Poepp. ex O.Berg, Linnaea 27: 365 (1856), nom. inval.
Psidium ruizianum O.Berg, Linnaea 27: 365 (1856). *Guajava ruiziana* (O.Berg) Kuntze, Revis. Gen. Pl. 1: 240 (1891).
Psidium xidocarpum Ruiz ex O.Berg, Linnaea 27: 365 (1856), nom. inval.

Psidium salutare (Kunth) O.Berg, Linnaea 27: 356 (1856).
Mexico to Trop. America. 79 MXG MXS MXT 80 BLZ COS ELS GUA HON NIC PAN 81 CUB DOM 82 GUY SUR VEN 83 BOL CLM ECU 84 BZC BZE BZL BZN BZS 85 AGE AGW PAR URU. Cham. or nanophan.
**Myrtus salutaris* Kunth in F.W.H.von Humboldt, A.J.A.Bonpland & C.S.Kunth, Nov. Gen. Sp. 6: 132 (1823). *Psidium salutare* var. *strictum* O.Berg, Linnaea 27: 356 (1856), nom. inval. *Guajava salutaris* (Kunth) Kuntze, Revis. Gen. Pl. 1: 240 (1891).

var. ***decussatum*** (DC.) Landrum, Sida 20: 1463 (2003).
Brazil. 84 BZC BZL BZS. Cham.
**Psidium decussatum* DC., Prodr. 3: 235 (1828). *Guajava decussata* (DC.) Kuntze, Revis. Gen. Pl. 1: 239 (1891).

var. ***mucronatum*** (Cambess.) Landrum, Sida 20: 1463 (2003).
Brazil to N. Argentina. 84 BZE BZL BZS 85 AGE AGW PAR URU. Nanophan. or phan.
Myrtus lurida Spreng., Syst. Veg. 2: 480 (1825). *Psidium luridum* (Spreng.) Burret, Notizbl. Bot. Gart. Berlin-Dahlem 15: 484 (1941).
**Myrtus mucronata* Cambess. in A.F.C.de Saint-Hilaire, Fl. Bras. Merid. 2: 294 (1832). *Myrtus mucronata* var. *perforata* O.Berg. in C.F.P.von Martius & auct. suc. (eds.), Fl. Bras. 14(1): 416 (1857), nom. inval. *Psidium mucronatum* (Cambess.) Burret, Notizbl. Bot. Gart. Berlin-Dahlem 15: 483 (1941).
Myrtus pauciflora Cambess. in A.F.C.de Saint-Hilaire, Fl. Bras. Merid. 2: 296 (1832). *Psidium pauciflorum* (Cambess.) Burret, Notizbl. Bot. Gart. Berlin-Dahlem 15: 483 (1951). *Psidium luridum* var. *pauciflora* (Cambess.) Mattos, Loefgrenia 71: 2 (1977).
Myrtus acutata O.Berg in C.F.P.von Martius & auct. suc. (eds.), Fl. Bras. 14(1): 415 (1857). *Psidium acutatum* (O.Berg) Burret, Notizbl. Bot. Gart. Berlin-Dahlem 15: 484 (1941).
Myrtus cuspidata O.Berg in C.F.P.von Martius & auct. suc. (eds.), Fl. Bras. 14(1): 415 (1857). *Psidium cuspidatum* (O.Berg) Burret, Notizbl. Bot. Gart. Berlin-Dahlem 15: 483 (1941).
Myrtus cuspidata var. *pentamera* O.Berg. in C.F.P. von Martius & auct. suc. (eds.), Fl. Bras. 14(1): 415 (1857).
Myrtus cuspidata var. *tetramera* O.Berg. in C.F.P.von Martius & auct. suc. (eds.), Fl. Bras. 14(1): 415 (1857).
Myrtus mucronata var. *opaca* O.Berg. in C.F.P.von Martius & auct. suc. (eds.), Fl. Bras. 14(1): 416 (1857).
Myrtus ovalis O.Berg in C.F.P.von Martius & auct. suc. (eds.), Fl. Bras. 14(1): 417 (1857), nom. illeg. *Psidium pubifolium* Burret, Notizbl. Bot. Gart. Berlin-Dahlem 15: 484 (1941).
Myrtus sellowiana O.Berg in C.F.P.von Martius & auct. suc. (eds.), Fl. Bras. 14(1): 413 (1857). *Psidium affine* Burret, Notizbl. Bot. Gart. Berlin-Dahlem 15: 484 (1941).
Myrtus suffruticosa O.Berg in C.F.P.von Martius & auct. suc. (eds.), Fl. Bras. 14(1): 418 (1857). *Psidium latius* Burret, Notizbl. Bot. Gart. Berlin-Dahlem 15: 484 (1941).
Myrtus suffruticosa var. *angustifolia* O.Berg. in C.F.P.von Martius & auct. suc. (eds.), Fl. Bras. 14(1): 419 (1857).
Myrtus suffruticosa var. *latifolia* O.Berg. in C.F.P. von Martius & auct. suc. (eds.), Fl. Bras. 14(1): 418 (1857).
Psidium thea Griseb., Abh. Königl. Ges. Wiss. Göttingen 19: 139 (1874). *Myrtus mucronata* var. *thea* (Griseb.) Griseb., Abh. Königl. Ges. Wiss. Göttingen 24: 127 (1879).
Psidium luridum var. *pubescens* Mattos, Loefgrenia 61: 3 (1974).
Psidium luridum var. *cinereum* Mattos, Loefgrenia 64: 2 (1975).
Psidium pubifolium f. *nanum* Rotman, Darwiniana 20: 433 (1976).

var. ***pohlianum*** (O.Berg) Landrum, Sida 20: 1466 (2003).
S. Venezuela to Brazil. 82 VEN 83 BOL 84 BZC BZE BZL. Nanophan. or phan.
**Psidium pohlianum* O.Berg in C.F.P.von Martius & auct. suc. (eds.), Fl. Bras. 14(1): 390 (1857).

Guajava pohliana (O.Berg) Kuntze, Revis. Gen. Pl. 1: 239 (1891).
Psidium pohlianum var. *brevipes* O.Berg. in C.F.P. von Martius & auct. suc. (eds.), Fl. Bras. 14(1): 601 (1857).

var. ***salutare***

Mexico to Trop. America. 79 MXG MXS MXT 80 BLZ COS ELS GUA HON NIC PAN 81 CUB DOM 82 GUY SUR VEN 83 BOL CLM ECU 84 BZC BZE BZL BZN BZS 85 PAR. Nanophan.

Myrtus arayan Kunth in F.W.H.von Humboldt, A.J.A.Bonpland & C.S.Kunth, Nov. Gen. Sp. 6: 133 (1823). *Pseudocaryophyllus seemannii* Triana ex Hemsl., Biol. Cent.-Amer., Bot. 1: 407 (1880). *Psidium arayan* (Kunth) Burret, Notizbl. Bot. Gart. Berlin-Dahlem 15: 484 (1941).
Psidium ciliatum Benth., J. Bot. (Hooker) 2: 318 (1840). *Guajava ciliata* (Benth.) Kuntze, Revis. Gen. Pl. 1: 240 (1891).
Eugenia guayavillo Benth., Pl. Hartw.: 174 (1845).
Psidium guayabita A.Rich., Hist. Phys. Cuba, Pl. Vasc.: 581 (1846). *Guajava guajabita* (A.Rich.) Kuntze, Revis. Gen. Pl. 1: 240 (1891).
Psidium oerstedianum O.Berg, Linnaea 27: 360 (1856). *Guajava oerstediana* (O.Berg) Kuntze, Revis. Gen. Pl. 1: 240 (1891).
Psidium salutare var. *laxum* O.Berg, Linnaea 27: 357 (1856).
Psidium salutare var. *subalternum* O.Berg, Linnaea 27: 357 (1856).
Myrtus blanchetiana O.Berg in C.F.P.von Martius & auct. suc. (eds.), Fl. Bras. 14(1): 418 (1857). *Psidium blanchetianum* (O.Berg) Burret, Notizbl. Bot. Gart. Berlin-Dahlem 15: 483 (1941).
Myrtus rigida O.Berg in C.F.P.von Martius & auct. suc. (eds.), Fl. Bras. 14(1): 417 (1857), nom. illeg. *Psidium rigidum* Burret, Notizbl. Bot. Gart. Berlin-Dahlem 15: 484 (1941).
Myrtus sagraea O.Berg, Linnaea 30: 710 (1861). *Mosiera sagraea* (O.Berg) Bisse, Revista Jard. Bot. Nac. Univ. Habana 6(3): 4 (1985 publ. 1986).
Psidium lanceolatum O.Berg, Linnaea 30: 704 (1861). *Guajava lanceolata* (O.Berg) Kuntze, Revis. Gen. Pl. 1: 240 (1891).
Psidium lanceolatum var. *grandifolium* O.Berg, Linnaea 30: 705 (1861).
Psidium guayabita var. *angustifolium* Griseb., Cat. Pl. Cub.: 91 (1866).
Psidium guayabita var. *oblongatum* Griseb., Cat. Pl. Cub.: 91 (1866).
Calycolpus parviflorus Sagot, Ann. Sci. Nat., Bot., VI, 20: 181 (1885).
Myrtus hassleriana Barb.Rodr., Myrt. Paraguay: 16 (1903). *Psidium barbosianum* Burret, Notizbl. Bot. Gart. Berlin-Dahlem 15: 485 (1941).
Psidium deltosepalum Barb.Rodr. ex Chodat & Hassl., Bull. Herb. Boissier, II, 7: 799 (1907).
Psidium valenzuelense Barb.Rodr. ex Chodat & Hassl., Bull. Herb. Boissier, II, 7: 798 (1907).
Psidium gentlei Lundell, Amer. Midl. Naturalist 29: 483 (1943).
Psidium chiapasense Lundell, Wrightia 2: 204 (1961).

var. ***sericeum*** (Cambess.) Landrum, Sida 20: 1467 (2003).

Bolivia to Uruguay. 83 BOL 84 BZS 85 AGE AGW PAR URU. Cham.

Eugenia sellowiana DC., Prodr. 3: 262 (1828). *Myrceugenia candolleana* (DC.) D.Legrand, Darwiniana 11: 314 (1957).
**Myrtus sericea* Cambess. in A.F.C.de Saint-Hilaire, Fl. Bras. Merid. 2: 295 (1832). *Psidium tomentellum* Burret, Notizbl. Bot. Gart. Berlin-Dahlem 15: 485 (1941).
Myrtus incana O.Berg in C.F.P.von Martius & auct. suc. (eds.), Fl. Bras. 14(1): 416 (1857). *Psidium incanum* (O.Berg) Burret, Notizbl. Bot. Gart. Berlin-Dahlem 15: 485 (1941).
Myrtus nivea O.Berg in C.F.P.von Martius & auct. suc. (eds.), Fl. Bras. 14(1): 414 (1857). *Psidium niveum* (O.Berg) Herter, Revista Sudamer. Bot. 7: 221 (1943).
Myrtus pubescens O.Berg in C.F.P.von Martius & auct. suc. (eds.), Fl. Bras. 14(1): 415 (1857), nom. illeg. *Psidium pubigerum* Burret, Notizbl. Bot. Gart. Berlin-Dahlem 15: 485 (1941). *Psidium incanum* var. *pubescens* Mattos, Loefgrenia 70: 4 (1976).
Myrtus sericea var. *fruticosa* O.Berg. in C.F.P.von Martius & auct. suc. (eds.), Fl. Bras. 14(1): 414 (1857).
Myrtus sericea var. *sufffruticosa* O.Berg. in C.F.P. von Martius & auct. suc. (eds.), Fl. Bras. 14(1): 414 (1857).

Psidium sartorianum (O.Berg) Nied. in H.G.A.Engler & K.A.E.Prantl, Nat. Pflanzenfam. 3(7): 69 (1893).

Mexico to Trop. America. 79 MXE MXG MXI MXS MXT 80 BLZ COS ELS GUA HON PAN 81 CUB NLA 82 FRG GUY SUR VEN 83 CLM ECU 84 BZC BZE BZL BZN. Nanophan. or phan.

Psidium ciliatum O.Berg, Linnaea 27: 353 (1856), nom. illeg. *Psidium minutiflorum* Amshoff, Recueil Trav. Bot. Néerl. 42: 19 (1950).
**Mitranthes sartoriana* O.Berg, Linnaea 29: 248 (1858). *Chytraculia sartoriana* (O.Berg) Kuntze, Revis. Gen. Pl. 1: 238 (1891). *Calyptropsidium sartorianum* (O.Berg) Krug & Urb., Bot. Jahrb. Syst. 19: 581 (1894). *Mitropsidium sartorianum* (O.Berg) Burret, Notizbl. Bot. Gart. Berlin-Dahlem 15: 487 (1941).
Calycorectes protractus Griseb., Cat. Pl. Cub.: 284 (1866). *Psidium protractum* (Griseb.) Lundell, Wrightia 5: 70 (1974).
Calyptranthes tonduzii Donn.Sm., Bot. Gaz. 23: 245 (1896).
Psidium socorrense I.M.Johnst., Proc. Calif. Acad. Sci., IV, 20: 81 (1931).
Mitropsidium oblanceolatum Burret, Notizbl. Bot. Gart. Berlin-Dahlem 15: 487 (1941).
Mitropsidium pittieri Burret, Notizbl. Bot. Gart. Berlin-Dahlem 15: 488 (1941).
Psidium yucatanense Lundell, Contr. Univ. Michigan Herb. 7: 35 (1942).
Psidium solisii Standl., Publ. Field Mus. Nat. Hist., Bot. Ser. 23: 133 (1944).
Psidium molinae Amshoff, Acta Bot. Neerl. 5: 277 (1956).
Psidium sartorianum var. *yucatanense* McVaugh, Fieldiana, Bot. 29: 527 (1963).

Psidium schenckianum Kiaersk., Enum. Myrt. Bras.: 34 (1893).

NE. Brazil. 84 BZE. Provisionally accepted.

Psidium scopulorum Ekman & Urb. in I.Urban, Symb. Antill. 9: 465 (1928).

W. Cuba. 81 CUB. Nanophan. or phan. Provisionally accepted.
Psidium tomasianum Urb. & Ekman in I.Urban, Symb. Antill. 9: 465 (1928).

Psidium sessilifolium Alain, Phytologia 25: 270 (1973).
Dominican Rep. 81 DOM. Nanophan. or phan. Provisionally accepted.

Psidium sintenisii (Kiaersk.) Alain, Mem. New York Bot. Gard. 21(2): 138 (1971).
Puerto Rico. 81 PUE. Nanophan. or phan. Provisionally accepted.
**Calyptropsidium sintenisii* Kiaersk., Bot. Tidsskr. 17: 280 (1890). *Mitropsidium sintenisii* (Kiaersk.) Burret, Notizbl. Bot. Gart. Berlin-Dahlem 15: 489 (1941).

Psidium sorocabense O.Berg in C.F.P.von Martius & auct. suc. (eds.), Fl. Bras. 14(1): 398 (1857). *Guajava sorocabensis* (O.Berg) Kuntze, Revis. Gen. Pl. 1: 239 (1891).
SE. Brazil. 84 BZL. Provisionally accepted.

Psidium striatulum DC., Prodr. 2: 233 (1828). *Guajava striatula* (DC.) Kuntze, Revis. Gen. Pl. 1: 239 (1891). *Myrtus striatula* (DC.) Kuntze, Revis. Gen. Pl. 3(2): 92 (1898).
N. South America to Brazil. 82 GUY SUR VEN 84 BZC BZE BZN. Nanophan. or phan.
Psidium turbiniflorum Mart. ex DC., Prodr. 3: 234 (1828). *Guajava turbiniflora* (Mart. ex DC.) Kuntze, Revis. Gen. Pl. 1: 239 (1891).
Psidium aquaticum Benth., J. Bot. (Hooker) 2: 318 (1840). *Guajava aquatica* (Benth.) Kuntze, Revis. Gen. Pl. 1: 240 (1891).
Psidium parviflorum Benth., J. Bot. (Hooker) 2: 318 (1840). *Guajava parviflora* (Benth.) Kuntze, Revis. Gen. Pl. 1: 240 (1891).
Psidium aquaticum var. *triflorum* O.Berg, Linnaea 27: 355 (1856).
Psidium aquaticum var. *uniflorum* O.Berg, Linnaea 27: 354 (1856).
Psidium leptocladum O.Berg in C.F.P.von Martius & auct. suc. (eds.), Fl. Bras. 14(1): 409 (1857). *Guajava leptoclada* (O.Berg) Kuntze, Revis. Gen. Pl. 1: 239 (1891).
Psidium persicifolium O.Berg in C.F.P.von Martius & auct. suc. (eds.), Fl. Bras. 14(1): 407 (1857). *Guajava persicifolia* (O.Berg) Kuntze, Revis. Gen. Pl. 1: 239 (1891).
Psidium striatulum var. *australe* O.Berg in C.F.P.von Martius & auct. suc. (eds.), Fl. Bras. 14(1): 604 (1859).
Psidium striatulum var. *paranense* O.Berg in C.F.P.von Martius & auct. suc. (eds.), Fl. Bras. 14(1): 603 (1859).
Psidium parvifolium var. *planifolium* Krug & Urb., Bot. Jahrb. Syst. 19: 569 (1894).
Psidium parviflorum var. *saramaccense* Amshoff, Bull. Torrey Bot. Club 75: 537 (1948).
Psidium parviflorum var. *coppenamense* Amshoff in A.A.Pulle, Fl. Suriname 3(2): 153 (1951).

Psidium tenuirame Urb., Symb. Antill. 9: 83 (1923).
E. Cuba. 81 CUB. Nanophan. or phan. Provisionally accepted.
Psidium leonis Urb., Symb. Antill. 9: 464 (1928).

Psidium trilobum Urb. & Ekman, Ark. Bot. 22A(10): 20 (1929).
Haiti (Massif du Nord). 81 HAI. Nanophan. Provisionally accepted.

Synonyms:
Psidium acidum Mart. ex O.Berg = ***Psidium acutangulum*** Mart. ex DC.
Psidium acre Ten. = ?
Psidium acutangulum var. *acidum* DC. = ***Psidium acutangulum*** Mart. ex DC.
Psidium acutangulum var. *crassirame* O.Berg = ***Psidium acutangulum*** Mart. ex DC.
Psidium acutangulum var. *oblongatum* Mattos = ***Psidium acutangulum*** Mart. ex DC.
Psidium acutangulum var. *tenuirame* O.Berg = ***Psidium acutangulum*** Mart. ex DC.
Psidium acutatum (O.Berg) Burret = ***Psidium salutare*** var. ***mucronatum*** (Cambess.) Landrum
Psidium adamantium Cambess. = ***Campomanesia adamantium*** (Cambess.) O.Berg
Psidium aerugineum O.Berg = ***Psidium laruotteanum*** Cambess.
Psidium aerugineum var. *angustifolium* O.Berg = ***Psidium laruotteanum*** Cambess.
Psidium affine Burret = ***Psidium salutare*** var. ***mucronatum*** (Cambess.) Landrum
Psidium alatum O.Berg = ***Psidium australe*** var. ***suffruticosum*** (O.Berg) Landrum
Psidium albidum Miq. = ***Psidium grandifolium*** Mart. ex DC.
Psidium albidum Cambess. = ***Psidium guineense*** Sw.
Psidium albidum var. *cuneatum* (Cambess.) Mattos = ***Psidium australe*** var. ***argenteum*** (O.Berg) Landrum
Psidium amygdalinum Hook. & Arn. = ***Hexachlamys edulis*** (O.Berg) Kausel & D.Legrand
Psidium anceps O.Berg = ***Psidium australe*** var. ***australe***
Psidium anglohondurense (Lundell) McVaugh = ***Chamguava schippii*** (Standl.) Landrum
Psidium angustifolium Lam. = ***Psidium guajava*** L.
Psidium anthomega Vell. = [84 BZL]
Psidium apaense Barb.Rodr. ex Chodat & Hassl. = ***Psidium grandifolium*** Mart. ex DC.
Psidium apodanthum (Standl.) McVaugh = ***Chamguava gentlei*** var. ***apodantha*** (Standl.) Landrum
Psidium apricum Vell. = ***Campomanesia aprica*** (Vell.) O.Berg
Psidium apysa Parodi = [85 AGE]
Psidium aquaticum Benth. = ***Psidium striatulum*** DC.
Psidium aquaticum var. *triflorum* O.Berg = ***Psidium striatulum*** DC.
Psidium aquaticum var. *uniflorum* O.Berg = ***Psidium striatulum*** DC.
Psidium araca Raddi = ***Psidium guineense*** Sw.
Psidium araca var. *sampaionis* Herter = ***Psidium guineense*** Sw.
Psidium araneosum Urb. = ***Mosiera araneosa*** (Urb.) Bisse
Psidium arasa-hu Parodi = [85 AGE]
Psidium arasa-pe Parodi = [85 AGE]
Psidium arasope-mi Parodi = [85 AGE]
Psidium arayan (Kunth) Burret = ***Psidium salutare*** var. ***salutare***
Psidium argenteum O.Berg = ***Psidium australe*** var. ***argenteum*** (O.Berg) Landrum
Psidium argenteum var. *angustifolium* O.Berg = ***Psidium australe*** var. ***argenteum*** (O.Berg) Landrum
Psidium argenteum var. *grandifolium* O.Berg = ***Psidium australe*** var. ***argenteum*** (O.Berg) Landrum
Psidium argenteum var. *pumilum* O.Berg = ***Psidium australe*** var. ***argenteum*** (O.Berg) Landrum
Psidium argenteum var. *purpureum* O.Berg = ***Psidium australe*** var. ***argenteum*** (O.Berg) Landrum
Psidium aromaticum Blanco = ***Psidium guajava*** L.

Psidium aromaticum Aubl. = ***Campomanesia aromatica*** (Aubl.) Griseb.
Psidium bahorucanum Alain & Ricardo García = [81 DOM]
Psidium barbosianum Burret = ***Psidium salutare*** var. ***salutare***
Psidium basanthum O.Berg = ***Psidium laruotteanum*** Cambess.
Psidium benthamianum O.Berg = ***Psidium guineense*** Sw.
Psidium bergianum (Nied.) Burret = ***Psidium laruotteanum*** Cambess.
Psidium berteroanum O.Berg = [81 PUE]
Psidium biloculare McVaugh = ***Chamguava gentlei*** (Lundell) Landrum
Psidium blanchetianum (O.Berg) Burret = ***Psidium salutare*** var. ***salutare***
Psidium bullatum Britton & P.Wilson = ***Mosiera bullata*** (Britton & P.Wilson) Bisse
Psidium buxifolium Nutt. = ?
Psidium cacuminis Britton & P.Wilson = ***Eugenia maestrensis*** Urb.
Psidium calycolpoides Griseb. = ***Mosiera calycolpoides*** (Griseb.) Borhidi
Psidium campestre Cambess. = ***Campomanesia adamantium*** (Cambess.) O.Berg
Psidium campicolum Barb.Rodr. = ***Psidium guineense*** Sw.
Psidium campomanesioides O.Berg = ***Calycolpus calophyllus*** (Kunth) O.Berg var. ***calophyllus***
Psidium caninum Lour. = (Rosaceae)
Psidium capibaryense Barb.Rodr. ex Chodat & Hassl. = ***Psidium laruotteanum*** Cambess.
Psidium cattleianum var. *coriaceum* (O.Berg) Kiaersk. = ?
Psidium cattleianum var. *littorale* (Raddi) Fosberg = ***Psidium cattleianum*** Afzel. ex Sabine
Psidium cattleianum var. *littorale* (Raddi) Mattos = ***Psidium cattleianum*** Afzel. ex Sabine
Psidium cattleianum f. *lucidum* O.Deg. = ***Psidium cattleianum*** Afzel. ex Sabine
Psidium cattleianum var. *pyriformis* Mattos = ***Psidium cattleianum*** Afzel. ex Sabine
Psidium caudatum McVaugh = ***Calycolpus moritzianus*** (O.Berg) Burret
Psidium cerasoides Cambess. = ***Campomanesia guaviroba*** (DC.) Kiaersk.
Psidium chiapasense Lundell = ***Psidium salutare*** var. ***salutare***
Psidium chrysobalanoides Standl. = ***Psidium guineense*** Sw.
Psidium chrysophyllum (O.Berg) F.Muell. = ***Campomanesia eugenioides*** (Cambess.) D.Legrand
Psidium ciliatum O.Berg = ***Psidium sartorianum*** (O.Berg) Nied.
Psidium ciliatum Benth. = ***Psidium salutare*** var. ***salutare***
Psidium cinereum Mart. ex DC. = ***Psidium grandifolium*** Mart. ex DC.
Psidium cinereum var. *angustifolium* O.Berg = ***Psidium grandifolium*** Mart. ex DC.
Psidium cinereum var. *brevipes* O.Berg = ***Psidium grandifolium*** Mart. ex DC.
Psidium cinereum var. *grandifolium* O.Berg = ***Psidium grandifolium*** Mart. ex DC.
Psidium cinereum var. *incanescens* (Mart. ex DC.) D.Legrand = ***Psidium grandifolium*** Mart. ex DC.
Psidium cinereum var. *intermedium* O.Berg = ***Psidium grandifolium*** Mart. ex DC.
Psidium cinereum var. *paraguariae* D.Legrand = ***Psidium grandifolium*** Mart. ex DC.
Psidium confertum R.A.Howard = ***Mosiera ophiticola*** (Britton & P.Wilson) Bisse
Psidium contrerasii Lundell = ***Mosiera contrerasii*** (Lundell) Landrum
Psidium cordatum Sims = ***Psidium amplexicaule*** Pers.
Psidium cordatum var. *parvifolium* Griseb. = ***Psidium amplexicaule*** Pers.
Psidium coriaceum O.Berg = ?
Psidium coriaceum var. *grandifolium* O.Berg = ***Psidium cattleianum*** Afzel. ex Sabine
Psidium coriaceum var. *longipes* O.Berg = ***Psidium cattleianum*** Afzel. ex Sabine
Psidium coriaceum var. *obovatum* O.Berg = ***Psidium cattleianum*** Afzel. ex Sabine
Psidium corymbosum Cambess. = ***Campomanesia pubescens*** (Mart. ex DC.) O.Berg
Psidium corymbosum f. *angustifolium* Miq. = ***Campomanesia pubescens*** (Mart. ex DC.) O.Berg
Psidium corynanthum (Kiaersk.) Burret = ?
Psidium costa-ricense O.Berg = ***Psidium guineense*** Sw.
Psidium crenatum O.Berg = ***Psidium maribense*** Mart. ex DC.
Psidium crenulatum Urb. & Ekman = ***Mosiera crenulata*** (Urb. & Ekman) Borhidi
Psidium crispulum Urb. = ***Hottea crispula*** (Urb.) Urb.
Psidium cujavillus Burm.f. = ***Psidium guajava*** L.
Psidium cujavus L. = ***Psidium guajava*** L.
Psidium cuneatum Cambess. = ***Psidium australe*** var. ***argenteum*** (O.Berg) Landrum
Psidium cuneatum var. *incanescens* O.Berg = ***Psidium australe*** var. ***argenteum*** (O.Berg) Landrum
Psidium cuneatum var. *niveum* O.Berg = ***Psidium australe*** var. ***argenteum*** (O.Berg) Landrum
Psidium cuneifolium Ten. = ?
Psidium cupreum O.Berg = ***Psidium rufum*** Mart. ex DC.
Psidium cupreum var. *glabratum* Kiaersk. = ***Psidium rufum*** Mart. ex DC.
Psidium cuspidatum (O.Berg) Burret = ***Psidium salutare*** var. ***mucronatum*** (Cambess.) Landrum
Psidium cuspidatum Alain = [81 DOM]
Psidium decaspermum L.f. = ***Decaspermum fruticosum*** J.R.Forst. & G.Forst.
Psidium decussatum DC. = ***Psidium salutare*** var. ***decussatum*** (DC.) Landrum
Psidium deltosepalum Barb.Rodr. ex Chodat & Hassl. = ***Psidium salutare*** var. ***salutare***
Psidium desertorum Mart. ex DC. = ***Campomanesia eugenioides*** (Cambess.) D.Legrand
Psidium dichotomum Weinm. = ***Psidium guineense*** Sw.
Psidium dubium Kunth = ***Myrciaria dubia*** (Kunth) McVaugh
Psidium dulce Vell. = ***Campomanesia guaviroba*** (DC.) Kiaersk.
Psidium ehrenbergii (O.Berg) Burret = ***Mosiera ehrenbergii*** (O.Berg) Landrum
Psidium elegans Miq. = [42 MOL]
Psidium elegans (DC.) Mart. ex O.Berg = ***Accara elegans*** (DC.) Landrum
Psidium elegans var. *angustifolium* O.Berg = ***Accara elegans*** (DC.) Landrum
Psidium elegans var. *latifolium* O.Berg = ***Accara elegans*** (DC.) Landrum
Psidium emarginatum Ruiz & Pav. ex G.Don = ***Eugenia racemiflora*** O.Berg
Psidium erianthum Cambess. = ***Campomanesia pubescens*** (Mart. ex DC.) O.Berg

Psidium eriophyllum Barb.Rodr. = ***Psidium grandifolium*** Mart. ex DC.
Psidium erosum Miq. = ***Campomanesia pubescens*** (Mart. ex DC.) O.Berg
Psidium eugenioides Cambess. = ***Campomanesia eugenioides*** (Cambess.) D.Legrand
Psidium ferrugineum C.Presl = ***Psidium cattleianum*** Afzel. ex Sabine
Psidium firmum var. *subcordatum* O.Berg = ***Psidium firmum*** O.Berg
Psidium fluviatile Rich. ex DC. = ***Psidium guyanense*** Pers.
Psidium formosum (Barb.Rodr.) Burret = ***Psidium laruotteanum*** Cambess.
Psidium fragrans Macfad. = ***Psidium guajava*** L.
Psidium fruticosum Vell. = ***Campomanesia fruticosa*** (Vell.) O.Berg
Psidium galapagaeum var. *howellii* D.M.Porter = ***Psidium galapagaeum*** Hook.f.
Psidium gardnerianum O.Berg = ***Psidium myrsinites*** DC.
Psidium gentlei Lundell = ***Psidium salutare*** var. ***salutare***
Psidium glaucescens O.Berg = ***Psidium laruotteanum*** Cambess.
Psidium grandiflorum Ruiz & Pav. = ***Psidium acutangulum*** Mart. ex DC.
Psidium grandiflorum Aubl. = ***Campomanesia grandiflora*** (Aubl.) Sagot
Psidium grandifolium (O.Berg) Burret = ***Psidium firmum*** O.Berg
Psidium grandifolium var. *albidum* O.Berg = ***Psidium grandifolium*** Mart. ex DC.
Psidium grandifolium var. *heterophyllum* O.Berg = ***Psidium grandifolium*** Mart. ex DC.
Psidium grandifolium var. *incanescens* O.Berg = ***Psidium grandifolium*** Mart. ex DC.
Psidium grandifolium var. *intermedium* O.Berg = ***Psidium grandifolium*** Mart. ex DC.
Psidium grandifolium var. *parvifolium* O.Berg = ***Psidium grandifolium*** Mart. ex DC.
Psidium grandifolium var. *tenuinerve* O.Berg = ***Psidium grandifolium*** Mart. ex DC.
Psidium grandifolium var. *ternatifolium* (Cambess.) O.Berg = ***Psidium grandifolium*** Mart. ex DC.
Psidium guajava var. *cujavillum* (Burm.f.) Krug & Urb. = ***Psidium guajava*** L.
Psidium guajava var. *minor* Mattos = ***Psidium guajava*** L.
Psidium guaviroba DC. = ***Campomanesia guaviroba*** (DC.) Kiaersk.
Psidium guayabita A.Rich. = ***Psidium salutare*** var. ***salutare***
Psidium guayabita var. *angustifolium* Griseb. = ***Psidium salutare*** var. ***salutare***
Psidium guayabita var. *oblongatum* Griseb. = ***Psidium salutare*** var. ***salutare***
Psidium guazumifolium Cambess. = ***Campomanesia guazumifolia*** (Cambess.) O.Berg
Psidium guazumifolium var. *griseum* Cambess. = ***Campomanesia guazumifolia*** (Cambess.) O.Berg
Psidium guildingianum Griseb. = ***Myrcia guildingiana*** (Griseb.) Krug & Urb.
Psidium hagelundianum Mattos = ***Psidium ovale*** (Spreng.) Burret
Psidium hatschbachii D.Legrand = ***Psidium ovale*** (Spreng.) Burret
Psidium havanense Urb. = ***Mosiera havanensis*** (Urb.) Bisse
Psidium herbaceum O.Berg = ***Eugenia arenosa*** Mattos
Psidium hians Mart. ex DC. = ***Campomanesia pubescens*** (Mart. ex DC.) O.Berg
Psidium hians var. *cuneatum* O.Berg = ***Campomanesia pubescens*** (Mart. ex DC.) O.Berg
Psidium hians var. *truncatum* O.Berg = ***Campomanesia pubescens*** (Mart. ex DC.) O.Berg
Psidium humile Vell. = [84 BZL]
Psidium × *hypoglaucum* Standl. = ***Psidium* × *hasslerianum*** Barb.Rodr.
Psidium igatemyense Barb.Rodr. = ***Psidium guajava*** L.
Psidium imaruinense Mattos = ***Psidium ovale*** (Spreng.) Burret
Psidium incanescens Mart. ex DC. = ***Psidium grandifolium*** Mart. ex DC.
Psidium incanescens var. *cuneatum* O.Berg = ***Psidium grandifolium*** Mart. ex DC.
Psidium incanescens var. *parvifolium* O.Berg = ***Psidium grandifolium*** Mart. ex DC.
Psidium incanescens var. *rotundifolium* O.Berg = ***Psidium grandifolium*** Mart. ex DC.
Psidium incanum (O.Berg) Burret = ***Psidium salutare*** var. ***sericeum*** (Cambess.) Landrum
Psidium incanum var. *pubescens* Mattos = ***Psidium salutare*** var. ***sericeum*** (Cambess.) Landrum
Psidium indicum Bojer = ***Psidium cattleianum*** Afzel. ex Sabine
Psidium indicum Raddi = ?
Psidium insulanum Alain = ***Mosiera longipes*** (O.Berg) Small
Psidium insulincola S.Moore = ***Psidium riparium*** Mart. ex DC.
Psidium intermedium Zipp. ex Blume = ***Psidium guajava*** L.
Psidium itatiaiae Wawra = ***Myrceugenia alpigena*** (DC.) Landrum var. ***alpigena***
Psidium jackii Urb. = ***Mosiera jackii*** (Urb.) Bisse
Psidium jollyanum A.Chev. = [22 IVO]
Psidium kuakuense Baker f. = ***Austromyrtus kuakuensis*** (Baker f.) Burret
Psidium lacteum O.Berg = ***Psidium grandifolium*** Mart. ex DC.
Psidium lagoense Kiaersk. = ***Psidium rufum*** Mart. ex DC.
Psidium lanatum Barb.Rodr. = ***Psidium grandifolium*** Mart. ex DC.
Psidium lanceolatum O.Berg = ***Psidium salutare*** var. ***salutare***
Psidium lanceolatum var. *grandifolium* O.Berg = ***Psidium salutare*** var. ***salutare***
Psidium lanuginosum Ruiz & Pav. ex G.Don = ***Acca lanuginosa*** (Ruiz & Pav. ex G.Don) McVaugh
Psidium latifolium Link = ?
Psidium latius Burret = ***Psidium salutare*** var. ***mucronatum*** (Cambess.) Landrum
Psidium laurifolium O.Berg = ***Psidium guineense*** Sw.
Psidium lehmannii Diels = ***Psidium guineense*** Sw.
Psidium leiophloeum (Urb.) Urb. = ***Mosiera bullata*** subsp. ***leiophloea*** (Urb.) Bisse
Psidium leonis Urb. = ***Psidium tenuirame*** Urb.
Psidium leptocladum O.Berg = ***Psidium striatulum*** DC.
Psidium lineatifolium (Ruiz & Pav.) Pers. = ***Campomanesia lineatifolia*** Ruiz & Pav.
Psidium littorale Raddi = ***Psidium cattleianum*** Afzel. ex Sabine
Psidium littorale var. *longipes* (O.Berg) Fosberg = ***Psidium cattleianum*** Afzel. ex Sabine
Psidium longifolium Schumach. = [22]
Psidium longipes (O.Berg) McVaugh = ***Mosiera longipes*** (O.Berg) Small
Psidium longipes var. *orbiculare* (O.Berg) McVaugh = ***Mosiera longipes*** (O.Berg) Small

Psidium luridum (Spreng.) Burret = ***Psidium salutare*** var. ***mucronatum*** (Cambess.) Landrum
Psidium luridum var. *cinereum* Mattos = ***Psidium salutare*** var. ***mucronatum*** (Cambess.) Landrum
Psidium luridum var. *pauciflora* (Cambess.) Mattos = ***Psidium salutare*** var. ***mucronatum*** (Cambess.) Landrum
Psidium luridum var. *pubescens* Mattos = ***Psidium salutare*** var. ***mucronatum*** (Cambess.) Landrum
Psidium macahense O.Berg = ***Psidium brownianum*** Mart. ex DC.
Psidium macedoi Kausel = ***Psidium firmum*** O.Berg
Psidium macrochlamys (DC.) Mattos = ***Algrizea macrochlamys*** (DC.) Proença & NicLugh.
Psidium macrospermum O.Berg = ***Psidium rufum*** Mart. ex DC.
Psidium macrostemum Ruiz & Pav. ex G.Don = ***Acca macrostema*** (Ruiz & Pav. ex G.Don) McVaugh
Psidium malifolium F.Muell. = ***Campomanesia xanthocarpa*** (Mart.) O.Berg var. ***xanthocarpa***
Psidium malmei Kausel = ***Psidium myrsinites*** DC.
Psidium maranhense O.Berg = ***Psidium riparium*** Mart. ex DC.
Psidium mediterraneum Vell. = ***Campomanesia mediterranea*** (Vell.) O.Berg
Psidium mengahiense Cambess. = ***Psidium riparium*** Mart. ex DC.
Psidium microcarpum Cambess. = ***Psidium grandifolium*** Mart. ex DC.
Psidium minense Mattos = ***Psidium firmum*** O.Berg
Psidium minus Mart. ex DC. = ***Psidium guineense*** Sw.
Psidium minutiflorum Amshoff = ***Psidium sartorianum*** (O.Berg) Nied.
Psidium molinae Amshoff = ***Psidium sartorianum*** (O.Berg) Nied.
Psidium molle Bertol. = ***Psidium guineense*** Sw.
Psidium molle var. *gracile* O.Berg = ***Psidium guineense*** Sw.
Psidium molle var. *robustum* O.Berg = ***Psidium guineense*** Sw.
Psidium monticola O.Berg = ***Psidium guineense*** Sw.
Psidium monticola var. *gracile* O.Berg = ***Psidium guineense*** Sw.
Psidium monticola var. *robustum* O.Berg = ***Psidium guineense*** Sw.
Psidium moritzianum O.Berg = ***Psidium brownianum*** Mart. ex DC.
Psidium mouririoides (Lundell) McVaugh = ***Chamguava schippii*** (Standl.) Landrum
Psidium mucronatum (Cambess.) Burret = ***Psidium salutare*** var. ***mucronatum*** (Cambess.) Landrum
Psidium mucronatum Barb.Rodr. ex Chodat & Hassl. = ***Psidium australe*** var. ***australe***
Psidium multiflorum Cambess. = ***Psidium guineense*** Sw.
Psidium musarum (Standl. & Steyerm.) McVaugh = ***Chamguava musarum*** (Standl. & Steyerm.) Landrum
Psidium myrsinoides O.Berg = ***Psidium myrtoides*** O.Berg
Psidium navasense Britton & P.Wilson = [81 CUB] Eugenia ?
Psidium nigrum Mattos & D.Legrand = ***Psidium ramboanum*** Mattos
Psidium nigrum Lour. = (Rosaceae)
Psidium nitidum C.Wright = ***Psidium parvifolium*** Griseb.
Psidium niveum (O.Berg) Herter = ***Psidium salutare*** var. ***sericeum*** (Cambess.) Landrum
Psidium nummularioides Britton & P.Wilson = ***Mosiera nummularioides*** (Britton & P.Wilson) Bisse
Psidium obovatum Mart. ex DC. = ***Psidium cattleianum*** Afzel. ex Sabine
Psidium obversum Miq. = ***Campomanesia pubescens*** (Mart. ex DC.) O.Berg
Psidium oerstedianum O.Berg = ***Psidium salutare*** var. ***salutare***
Psidium ooideum O.Berg = ***Psidium guineense*** Sw.
Psidium ooideum var. *grandifolium* O.Berg = ***Psidium guineense*** Sw.
Psidium ooideum var. *intermedium* O.Berg = ***Psidium guineense*** Sw.
Psidium ooideum var. *longipedunculatum* Rusby = ***Psidium guineense*** Sw.
Psidium ooideum var. *parvifolium* O.Berg = ***Psidium guineense*** Sw.
Psidium ophiticola Britton & P.Wilson = ***Mosiera ophiticola*** (Britton & P.Wilson) Bisse
Psidium ovalifolium (O.Berg) Nied. = ***Myrceugenia ovalifolia*** (O.Berg) Landrum
Psidium ovatifolium O.Berg = ***Psidium densicomum*** Mart. ex DC.
Psidium ovatifolium var. *glabrum* Amshoff = ***Psidium densicomum*** Mart. ex DC.
Psidium paraense O.Berg = ***Psidium riparium*** Mart. ex DC.
Psidium paraguayense Barb.Rodr. ex Chodat & Hassl. = ***Psidium grandifolium*** Mart. ex DC.
Psidium paraibicum Wawra = ***Myrceugenia campestris*** (DC.) D.Legrand & Kausel
Psidium paranense O.Berg = ***Psidium riparium*** Mart. ex DC.
Psidium parviflorum Benth. = ***Psidium striatulum*** DC.
Psidium parviflorum var. *coppenamense* Amshoff = ***Psidium striatulum*** DC.
Psidium parviflorum var. *saramaccense* Amshoff = ***Psidium striatulum*** DC.
Psidium parvifolium (Griseb.) Griseb. = ***Psidium amplexicaule*** Pers.
Psidium parvifolium var. *planifolium* Krug & Urb. = ***Psidium striatulum*** DC.
Psidium passeanum André = ?
Psidium pauciflorum (Cambess.) Burret = ***Psidium salutare*** var. ***mucronatum*** (Cambess.) Landrum
Psidium paucinerve Urb. = ***Psidium parvifolium*** Griseb.
Psidium persicifolium O.Berg = ***Psidium striatulum*** DC.
Psidium persoonii McVaugh = ***Psidium acutangulum*** Mart. ex DC.
Psidium pilosum Vell. = ***Psidium rufum*** Mart. ex DC.
Psidium piribebuiense Barb.Rodr. ex Chodat & Hassl. = ***Psidium australe*** var. ***australe***
Psidium pohlianum O.Berg = ***Psidium salutare*** var. ***pohlianum*** (O.Berg) Landrum
Psidium pohlianum var. *brevipes* O.Berg. = ***Psidium salutare*** var. ***pohlianum*** (O.Berg) Landrum
Psidium polycarpon Lamb. = ***Psidium guineense*** Sw.
Psidium pomiferum L. = ***Psidium guajava*** L.
Psidium popenoei Standl. = ***Psidium guineense*** Sw.
Psidium pratense Poepp. ex O.Berg = ***Psidium rutidocarpum*** Ruiz & Pav. ex G.Don
Psidium prostratum O.Berg = ***Psidium guajava*** L.
Psidium protractum (Griseb.) Lundell = ***Psidium sartorianum*** (O.Berg) Nied.
Psidium psychrophyllum Barb.Rodr. ex Chodat & Hassl. = ***Psidium grandifolium*** Mart. ex DC.
Psidium pubescens Mart. ex DC. = ***Campomanesia pubescens*** (Mart. ex DC.) O.Berg
Psidium pubifolium Burret = ***Psidium salutare*** var. ***mucronatum*** (Cambess.) Landrum

Psidium pubifolium f. *nanum* Rotman = ***Psidium salutare*** var. ***mucronatum*** (Cambess.) Landrum
Psidium pubigerum Burret = ***Psidium salutare*** var. ***sericeum*** (Cambess.) Landrum
Psidium pumilum Vahl = ***Psidium guajava*** L.
Psidium pumilum var. *guadalupense* DC. = ***Psidium guajava*** L.
Psidium punctulatum DC. = ***Campomanesia guaviroba*** (DC.) Kiaersk.
Psidium punctulatum Miq. = ***Campomanesia xanthocarpa*** (Mart.) O.Berg var. ***xanthocarpa***
Psidium pyriferum L. = ***Psidium guajava*** L.
Psidium pyriferum var. *glabrum* Benth. = ***Psidium guajava*** L.
Psidium quinquedentatum Amshoff = ***Psidium laruotteanum*** Cambess.
Psidium racemosum Vell. = ***Campomanesia racemosa*** (Vell.) O.Berg
Psidium radicans O.Berg = ***Psidium grandifolium*** Mart. ex DC.
Psidium rensonianum Standl. = ***Eugenia salamensis*** Donn.Sm.
Psidium richardianum O.Berg = ***Psidium guyanense*** Pers.
Psidium riedelianum O.Berg = ***Psidium grandifolium*** Mart. ex DC.
Psidium rigidum Burret = ***Psidium salutare*** var. ***salutare***
Psidium rivulare Mart. ex DC. = ***Campomanesia lineatifolia*** Ruiz & Pav.
Psidium roraimense Mattos = ***Calycolpus roraimensis*** Steyerm.
Psidium rotundatum var. *triflorum* Griseb. = ***Psidium rotundatum*** Griseb.
Psidium rotundifolium Standl. = ***Psidium guineense*** Sw.
Psidium rubescens O.Berg = ***Psidium grandifolium*** Mart. ex DC.
Psidium rubrum Lour. = [41 VIE]
Psidium rufinervum Barb.Rodr. = ***Psidium guineense*** Sw.
Psidium rufum var. *rotundifolium* Kiaersk. = ***Psidium rufum*** Mart. ex DC.
Psidium ruizianum O.Berg = ***Psidium rutidocarpum*** Ruiz & Pav. ex G.Don
Psidium salutare var. *laxum* O.Berg = ***Psidium salutare*** var. ***salutare***
Psidium salutare var. *strictum* O.Berg = ***Psidium salutare*** (Kunth) O.Berg
Psidium salutare var. *subalternum* O.Berg = ***Psidium salutare*** var. ***salutare***
Psidium sapidissimum Jacq. = ***Psidium guajava*** L.
Psidium sartorianum var. *yucatanense* McVaugh = ***Psidium sartorianum*** (O.Berg) Nied.
Psidium savannarum Donn.Sm. = ***Psidium laruotteanum*** Cambess.
Psidium saxicola Britton & P.Wilson = ***Mosiera elliptica*** (C.Wright) Bisse subsp. ***elliptica***
Psidium schiedeanum O.Berg = ***Psidium guineense*** Sw.
Psidium schippii Standl. = ***Psidium guineense*** Sw.
Psidium sellowianum O.Berg = ***Psidium arboreum*** Vell.
Psidium sericeum O.Berg = ***Psidium grandifolium*** Mart. ex DC.
Psidium sericiflorum Benth. = ***Psidium guineense*** Sw.
Psidium sieberianum O.Berg = ***Psidium riparium*** Mart. ex DC.
Psidium sieberianum var. *gracile* O.Berg = ***Psidium riparium*** Mart. ex DC.
Psidium sieberianum var. *robustum* O.Berg = ***Psidium riparium*** Mart. ex DC.
Psidium socorrense I.M.Johnst. = ***Psidium sartorianum*** (O.Berg) Nied.
Psidium solisii Standl. = ***Psidium sartorianum*** (O.Berg) Nied.
Psidium spathulatum Mattos = ***Psidium ovale*** (Spreng.) Burret
Psidium speciosum Diels = ***Campomanesia speciosa*** (Diels) McVaugh
Psidium spodophyllum Barb.Rodr. = ***Psidium grandifolium*** Mart. ex DC.
Psidium sprucei O.Berg = ***Psidium guyanense*** Pers.
Psidium stictophyllum (Kiaersk.) Mattos = ***Accara elegans*** (DC.) Landrum
Psidium striatulum var. *australe* O.Berg = ***Psidium striatulum*** DC.
Psidium striatulum var. *paranense* O.Berg = ***Psidium striatulum*** DC.
Psidium suaveolens Cambess. = ***Campomanesia pubescens*** (Mart. ex DC.) O.Berg
Psidium submetrale McVaugh = ***Psidium australe*** var. ***australe***
Psidium subrostrifolium Mattos = [84 BZL]
Psidium suffruticosum O.Berg = ***Psidium australe*** var. ***suffruticosum*** (O.Berg) Landrum
Psidium suffruticosum var. *alatum* (O.Berg) Kiaersk. = ***Psidium australe*** var. ***suffruticosum*** (O.Berg) Landrum
Psidium tenuifolium Mart. ex DC. = ***Campomanesia aromatica*** (Aubl.) Griseb.
Psidium terminale Vell. = ***Campomanesia terminalis*** (Vell.) Mattos
Psidium ternatifolium Cambess. = ***Psidium grandifolium*** Mart. ex DC.
Psidium thea Griseb. = ***Psidium salutare*** var. ***mucronatum*** (Cambess.) Landrum
Psidium thyrsoideum (Kuntze) K.Schum. = ***Psidium riparium*** Mart. ex DC.
Psidium tomasianum Urb. & Ekman = ***Psidium scopulorum*** Ekman & Urb.
Psidium tomentellum Burret = ***Psidium salutare*** var. ***sericeum*** (Cambess.) Landrum
Psidium tomentosum Parodi = [85 AGE]
Psidium transalpinum Vell. = ***Campomanesia transalpina*** (Vell.) O.Berg
Psidium tripartitum S.Moore = ***Psidium kennedyanum*** Morong
Psidium triphyllum Barb.Rodr. = ***Psidium australe*** var. ***australe***
Psidium turbinatum Mattos = [84 BZC]
Psidium turbiniflorum Mart. ex DC. = ***Psidium striatulum*** DC.
Psidium ubatubense Mattos = [84 BZL]
Psidium ulei Diels = ***Eugenia ulei*** (Diels) McVaugh
Psidium umbrosum O.Berg = ***Psidium guyanense*** Pers.
Psidium valenzuelense Barb.Rodr. ex Chodat & Hassl. = ***Psidium salutare*** var. ***salutare***
Psidium variabile O.Berg = ***Psidium cattleianum*** Afzel. ex Sabine
Psidium velutinum Cambess. = ***Campomanesia velutina*** (Cambess.) O.Berg
Psidium versicolor Urb. = ***Mosiera calycolpoides*** (Griseb.) Borhidi
Psidium vicentinum Urb. = ***Eugenia cristata*** C.Wright
Psidium vulgare Rich. = ***Psidium guajava*** L.
Psidium warmingianum Kiaersk. = ***Psidium laruotteanum*** Cambess.
Psidium warmingianum var. *verticillatum* Kiaersk. = ***Psidium laruotteanum*** Cambess.
Psidium widgrenianum O.Berg = ***Psidium rufum*** Mart. ex DC.
Psidium willemetianum DC. = [29 MAU]

Psidium wrightii D.Don ex W.Wright = ***Psidium montanum*** Sw.
Psidium wrightii Krug & Urb. = ***Mosiera wrightii*** (Krug & Urb.) Borhidi
Psidium xidocarpum Ruiz ex O.Berg = ***Psidium rutidocarpum*** Ruiz & Pav. ex G.Don
Psidium yacaense Barb.Rodr. ex Chodat & Hassl. = ***Psidium grandifolium*** Mart. ex DC.
Psidium ypanemense O.Berg = ***Psidium guineense*** Sw.
Psidium yucatanense Lundell = ***Psidium sartorianum*** (O.Berg) Nied.

Unplaced Names:
Psidium acre Ten., Index Seminum (NAP) 1829: 17 (1829). = ?
Psidium anthomega Vell., Fl. Flumin. 5: 212, t. 51 (1829). = [84 BZL]
Psidium apysa Parodi, Anales Soc. Ci. Argent. 7: 66 (1879). = [85 AGE]
Psidium arasa-hu Parodi, Anales Soc. Ci. Argent. 7: 65 (1879). = [85 AGE]
Psidium arasa-pe Parodi, Anales Soc. Ci. Argent. 7: 65 (1879). = [85 AGE]
Psidium arasope-mi Parodi, Anales Soc. Ci. Argent. 7: 65 (1879). = [85 AGE]
Psidium bahorucanum Alain & Ricardo García, Moscosoa 9: 22 (1997). = [81 DOM]
Psidium berteroanum O.Berg, Linnaea 27: 374 (1856). = [81 PUE]
Psidium buxifolium Nutt., N. Amer. Sylv. 1: 115 (1842). = ?
Psidium caninum Lour., Fl. Cochinch.: 310 (1790). = (Rosaceae)
Psidium cuneifolium Ten., Index Seminum (NAP) 1833: 14 (1833). = ?
Psidium cuspidatum Alain, Brittonia 20: 159 (1968), nom. illeg. = [81 DOM]
Psidium elegans Miq., Fl. Ned. Ind. 1(1): 470 (1855). = [42 MOL]
Psidium humile Vell., Fl. Flumin. 5: 211, t. 49 (1829). = [84 BZL]
Psidium indicum Raddi, Alc. Sp. Pero: 6 (1821). = ?
Psidium jollyanum A.Chev., Explor. Bot. Afrique Occ. Franç. 1: 266 (1920). = [22 IVO]
Psidium latifolium Link, Enum. Hort. Berol. Alt. 2: 27 (1822). = ?
Psidium longifolium Schumach., Beskr. Guin. Pl.: 229 (1827). = [22]
Psidium navasense Britton & P.Wilson, Mem. Torrey Bot. Club 16: 85 (1920). = [81 CUB] Eugenia ?
Psidium nigrum Lour., Fl. Cochinch.: 311 (1790). = (Rosaceae)
Psidium passeanum André, Rev. Hort. 68: 233 (1890). = ?
Psidium rubrum Lour., Fl. Cochinch.: 311 (1790). = [41 VIE]
Psidium subrostrifolium Mattos, Loefgrenia 90: 3 (1986). = [84 BZL]
Psidium tomentosum Parodi, Anales Soc. Ci. Argent. 7: 64 (1879). = [85 AGE]
Psidium turbinatum Mattos, Loefgrenia 94: 12 (1989). = [84 BZC]
Psidium ubatubense Mattos, Ci. & Cult. 19: 332 (1967). = [84 BZL]
Psidium willemetianum DC., Prodr. 3: 237 (1828). = [29 MAU]

Psiloxylon

Psiloxylon Thouars ex Tul., Ann. Sci. Nat., Bot., IV, 6: 138 (1856).
Mascarenes. 29 MAU REU.
1 Species
Fropiera Bouton ex Hook.f., J. Proc. Linn. Soc., Bot. 5: 2 (1860).

Psiloxylon mauritianum (Bouton ex Hook.f.) Baill., Adansonia 10: 41 (1871).
Mascarenes. 29 MAU REU. Phan.
**Fropiera mauritiana* Bouton ex Hook.f., J. Proc. Linn. Soc., Bot. 5: 2 (1860).

Punicella

Punicella Turcz. = ***Balaustion*** Hook.
Punicella carinata Turcz. = ***Balaustion pulcherrimum*** Hook.

Purpureostemon

Purpureostemon Gugerli, Repert. Spec. Nov. Regni Veg. 46: 299 (1939).
New Caledonia. 60 NWC.
1 Species

Purpureostemon ciliatus (J.R.Forst. & G.Forst.) Gugerli, Repert. Spec. Nov. Regni Veg. 46: 230 (1939).
NW. & C. New Caledonia. 60 NWC. Nanophan.
**Leptospermum ciliatum* J.R.Forst. & G.Forst., Char. Gen. Pl.: 36 (1775). *Melaleuca ciliata* (J.R.Forst. & G.Forst.) G.Forst., Fl. Ins. Austr.: 38 (1786). *Metrosideros ciliata* (J.R.Forst. & G.Forst.) Sm., Trans. Linn. Soc. London 3: 271 (1797). *Stenospermum ciliatum* (J.R.Forst. & G.Forst.) Heynh., Nom. Bot. Hort. 1: 787 (1840). *Fremya ciliata* (J.R.Forst. & G.Forst.) Brongn. & Gris, Bull. Soc. Bot. France 10: 374 (1863). *Nania ciliata* (J.R.Forst. & G.Forst.) Kuntze, Revis. Gen. Pl. 1: 242 (1891). *Xanthostemon ciliatum* (J.R.Forst. & G.Forst.) Nied. in H.G.A.Engler & K.A.E.Prantl, Nat. Pflanzenfam. 3(7): 88 (1893).
Metrosideros buxifolia Dum.Cours., Bot. Cult., ed. 2, 5: 379 (1811).

Pyrenocarpa

Pyrenocarpa H.T.Chang & R.H.Miao = ***Decaspermum*** J.R.Forst. & G.Forst.
Pyrenocarpa hainanensis (Merr.) H.T.Chang & R.H.Miao = ***Decaspermum hainanense*** (Merr.) Merr.
Pyrenocarpa teretis H.T.Chang & R.H.Miao = ***Decaspermum hainanense*** (Merr.) Merr.

Regelia

Regelia Schauer, Linnaea 17: 243 (1843).
SW. Australia. 50 WAU.
5 Species

Regelia ciliata Schauer, Linnaea 17: 244 (1843).
SW. Australia. 50 WAU.
Melaleuca rotundifolia Sweet, Hort. Brit.: 156 (1826).

Regelia cymbifolia (Diels) C.A.Gardner, J. Roy. Soc. Western Australia 47: 60 (1964).
SW. Australia. 50 WAU. Nanophan.
**Beaufortia cymbifolia* Diels, Bot. Jahrb. Syst. 35: 431 (1905).

Regelia inops (Schauer) Schauer in J.G.C.Lehmann, Pl. Preiss. 2: 224 (1848).
SW. Australia. 50 WAU. Nanophan.
**Beaufortia inops* Schauer, Nov. Actorum Acad. Caes. Leop.-Carol. Nat. Cur. 21: 17 (1844).
Regelia gibbosa Turcz., Bull. Soc. Imp. Naturalistes Moscou 20(1): 168 (1847).

Regelia megacephala C.A.Gardner, J. Roy. Soc. Western Australia 47: 60 (1964).
SW. Australia. 50 WAU. Nanophan.

Regelia velutina (Turcz.) C.A.Gardner, J. Roy. Soc. Western Australia 47: 60 (1964).
S. Western Australia. 50 WAU. Nanophan.
**Beaufortia velutina* Turcz., Bull. Cl. Phys.-Math. Acad. Imp. Sci. Saint-Pétersbourg 10: 345 (1852).
Regelia grandiflora Benth., Fl. Austral. 3: 170 (1867).

Synonyms:
Regelia adpressa Turcz. = ***Beaufortia micrantha*** Schauer var. ***micrantha***
Regelia gibbosa Turcz. = ***Regelia inops*** (Schauer) Schauer
Regelia grandiflora Benth. = ***Regelia velutina*** (Turcz.) C.A.Gardner
Regelia punicea (Byrnes) Barlow = ***Petraeomyrtus punicea*** (Byrnes) Craven
Regelia sparsiflora W.Fitzg. = ***Phymatocarpus maxwellii*** F.Muell.

Reichea

Reichea Kausel = ***Myrcianthes*** O.Berg
Reichea coquimbensis (Barnéoud) Kausel = ***Myrcianthes coquimbensis*** (Barnéoud) Landrum & Grifo

Reicheia

Reicheia Kausel = ***Myrcianthes*** O.Berg

Rhodamnia

Rhodamnia Jack, Malayan Misc. 2(7): 48 (1822).
Hainan, Indo-China to SW. Pacific. 36 CHH 41 AND CBD LAO MYA NCB THA VIE 42 BOR JAW LSI MLY MOL SUL SUM 43 BIS NWG SOL 50 NSW NTA QLD 60 NWC.
34 Species
Opanea Raf., Sylva Tellur.: 106 (1838).
Monoxora Wight, Ill. Ind. Bot. 2: 12 (1841).

Rhodamnia acuminata C.T.White, Blumea, Suppl. 1: 217 (1937).
SE. Queensland. 50 QLD. Nanophan. or phan.

Rhodamnia andromedoides Guillaumin, Bull. Soc. Bot. France 85: 629 (1938 publ. 1939).
New Caledonia. 60 NWC. Phan.

Rhodamnia angustifolia N.Snow & Guymer, Austrobaileya 5: 421 (1999).
Queensland (Port Curtis). 50 QLD. Phan.

Rhodamnia arenaria N.Snow, Syst. Bot. Monogr. 82: 22 (2007).
N. Queensland. 50 QLD. Nanophan. or phan.

Rhodamnia argentea Benth., Fl. Austral. 3: 278 (1867).
Queensland to NE. New South Wales. 50 NSW QLD. Phan.

Rhodamnia australis A.J.Scott, Kew Bull. 33: 447 (1979).
Northern Territory to N. Queensland. 50 NTA QLD. Nanophan. or phan.

Rhodamnia blairiana F.Muell., Fragm. 9: 141 (1875).
Papua New Guinea to Queensland. 43 NWG 50 QLD. Phan.

var. ***blairiana***
Papua New Guinea to Queensland. 43 NWG 50 QLD. Phan.

var. ***propinqua*** (C.T.White) A.J.Scott, Kew Bull. 33: 455 (1979).
Papua New Guinea. 43 NWG. Phan.
**Rhodamnia propinqua* C.T.White, J. Arnold Arbor. 32: !47 (1951).

Rhodamnia cinerea Jack, Malayan Misc. 2(7): 48 (1822).
Indo-China to W. Malesia. 41 MYA NCB THA 42 BOR JAW MLY SUM. Nanophan. or phan.
Myrtus spectabilis Blume, Bijdr.: 1083 (1826). *Monoxora spectabilis* (Blume) Wight, Ill. Ind. Bot. 2: 12 (1841). *Rhodamnia spectabilis* (Blume) Blume, Mus. Bot. 1: 78 (1850).
Myrtus globosa Korth., Ned. Kruidk. Arch. 1: 197 (1847). *Rhodamnia globosa* (Korth.) Blume, Mus. Bot. 1: 79 (1850).
Rhodamnia subtriflora Blume, Mus. Bot. 1: 79 (1850).
Rhodamnia nageli Miq., Fl. Ned. Ind. 1(1): 478 (1855).
Rhodamnia concolor Miq., Fl. Ned. Ind., Eerste Bijv.: 315 (1861). *Rhodamnia cinerea* var. *concolor* (Miq.) Blume, Ann. Mus. Bot. Lugduno-Batavi 1: 78 (1863).
Rhodamnia cinerea var. *laxiflora* Blume, Ann. Mus. Bot. Lugduno-Batavi 1: 78 (1863).
Rhodamnia cinerea var. *macrophylla* Blume, Ann. Mus. Bot. Lugduno-Batavi 1: 78 (1863).

Rhodamnia costata A.J.Scott, Kew Bull. 33: 453 (1979).
NE. Queensland. 50 QLD. Phan.

Rhodamnia dumetorum (DC.) Merr. & L.M.Perry, J. Arnold Arbor. 19: 195 (1938).
Hainan, Indo-China to Pen. Malaysia (Langkawi Is.). 36 CHH 41 AND CBD LAO THA VIE 42 MLY. Nanophan. or phan.
Myrtus trinervia Lour., Fl. Cochinch.: 312 (1790). *Myrtus dumetorum* Poir. in J.B.A.P.M.de Lamarck, Encycl., Suppl. 4: 52 (1816), nom. illeg. **Eugenia dumetorum* DC., Prodr. 3: 284 (1828). *Nelitris trinervia* (Lour.) Spreng., Syst. Veg. 2: 488 (1825). *Opanea trinervia* (Lour.) Raf., Sylva Tellur.: 106 (1838).
Rhodamnia siamensis Craib, Bull. Misc. Inform. Kew 1926: 167 (1926).
Rhodamnia dumetorum var. *hainanensis* Merr. & L.M.Perry, J. Arnold Arbor. 19: 196 (1938).

Rhodamnia dumicola Guymer & Jessup, Austrobaileya 2: 228 (1986).
Queensland. 50 QLD. Nanophan. or phan.

Rhodamnia fordii N.Snow, Syst. Bot. Monogr. 82: 47 (2007).
N. Queensland. 50 QLD. Nanophan. or phan.

Rhodamnia glabrescens Guymer & Jessup, Austrobaileya 2: 231 (1986).
E. Queensland. 50 QLD. Nanophan. or phan.

Rhodamnia glauca Blume, Mus. Bot. 1: 79 (1850).
Papua New Guinea to N. & NE. Queensland. 43 NWG 50 QLD. Nanophan. or phan.

Rhodamnia trinervia var. *spongiosa* F.M.Bailey, Queensl. Fl. 2: 652 (1900). *Rhodamnia spongiosa* (F.M.Bailey) Domin, Biblioth. Bot. 89: 476 (1928).

Rhodamnia hylandii N.Snow, Syst. Bot. Monogr. 82: 50 (2007).
N. Queensland. 50 QLD. Nanophan. or phan.

Rhodamnia kerrii J.Parn. & NicLugh., Kew Bull. 47: 705 (1992).
NE. Thailand. 41 THA. Nanophan.
**Rhodomyrtus parvifolia* Craib, Bull. Misc. Inform. Kew 1928: 70 (1928).

Rhodamnia lancifolia A.J.Scott, Kew Bull. 33: 442 (1979).
Papua New Guinea. 43 NWG. Phan.

Rhodamnia latifolia (Benth.) Miq., Fl. Ned. Ind. 1(1): 480 (1855).
Maluku to Bismarck Arch. 42 MOL 43 BIS NWG. Phan.
**Monoxora latifolia* Benth., London J. Bot. 2: 218 (1843).
Rhodamnia lamprophylla Diels, Bot. Jahrb. Syst. 57: 360 (1922).
Rhodamnia polyantha Diels, Bot. Jahrb. Syst. 57: 360 (1922).

Rhodamnia longisepala N.Snow & A.J.Ford, Novon 11: 479 (2001).
NE. Queensland. 50 QLD. Nanophan. or phan.

Rhodamnia maideniana C.T.White, Blumea, Suppl. 1: 215 (1937).
SE. Queensland to NE. New South Wales. 50 NSW QLD. Nanophan. or phan.
Rhodamnia trinervia var. *glabra* Maiden & Betche, Proc. Linn. Soc. New South Wales 24: 146 (1900).

Rhodamnia moluccana Burret, Notizbl. Bot. Gart. Berlin-Dahlem 15: 498 (1941).
Jawa to New Guinea. 42 JAW LSI MOL SUL 43 NWG. Nanophan. or phan.

Rhodamnia mulleri (Korth.) Blume, Mus. Bot. 1: 79 (1850).
Borneo. 42 BOR. Nanophan. or phan.
**Myrtus mulleri* Korth., Ned. Kruidk. Arch. 1: 197 (1847).

Rhodamnia novoguineensis A.J.Scott, Kew Bull. 33: 444 (1979).
New Guinea to N. Queensland. 43 NWG 50 QLD. Phan.

Rhodamnia pachyloba A.J.Scott, Kew Bull. 33: 447 (1979).
Maluku to W. New Guinea. 42 MOL 43 NWG. Phan.

Rhodamnia parviflora A.J.Scott, Kew Bull. 33: 451 (1979).
W. New Guinea. 43 NWG. Nanophan.

Rhodamnia pauciovulata Guymer, Austrobaileya 2: 515 (1988).
Queensland. 50 QLD. Nanophan. or phan.

Rhodamnia reticulata A.J.Scott, Kew Bull. 33: 444 (1979).
W. New Guinea. 43 NWG. Phan.

Rhodamnia rubescens (Benth.) Miq., Fl. Ned. Ind. 1(1): 480 (1855).
E. Australia. 50 NSW QLD. Phan.
Myrtus trinervia Sm., Trans. Linn. Soc. London 3: 280 (1797), nom. illeg. *Eugenia trinervia* DC., Prodr. 3: 279 (1828), nom. illeg. **Monoxora rubescens* Benth., London J. Bot. 2: 219 (1843). *Rhodamnia trinervia* Reinw. ex Blume, Mus. Bot. 1: 79 (1850). *Myrtus melastomoides* F.Muell., Fragm. 1: 76 (1858), nom. illeg.

Rhodamnia sepicana Diels, Bot. Jahrb. Syst. 57: 359 (1922).
Maluku to Solomon Is. 42 MOL 43 NWG SOL. Nanophan. or phan.
Rhodamnia ledermannii Diels, Bot. Jahrb. Syst. 57: 359 (1922).

Rhodamnia sessiliflora Benth., Fl. Austral. 3: 277 (1867).
N. & NE. Queensland. 50 QLD. Nanophan. or phan.

Rhodamnia sharpeana N.Snow, Syst. Bot. Monogr. 82: 41 (2007).
N. Queensland. 50 QLD. Nanophan. or phan.

Rhodamnia tessellata Steenis ex A.J.Scott, Kew Bull. 33: 441 (1979).
Sumatera. 42 SUM. Nanophan. or phan.

Rhodamnia uniflora (Ridl.) Burkill, Gard. Bull. Straits Settlem. 3: 42 (1923).
Pen. Malaysia to Borneo. 42 BOR MLY. Nanophan. or phan.
**Rhodamnia trinervia* var. *uniflora* Ridl., J. Fed. Malay States Mus. 6: 146 (1915).

Rhodamnia whiteana Guymer & Jessup, Austrobaileya 2: 229 (1986).
Queensland to NE. New South Wales. 50 NSW QLD. Nanophan. or phan.

Synonyms:
Rhodamnia cinerea var. *concolor* (Miq.) Blume = ***Rhodamnia cinerea*** Jack
Rhodamnia cinerea var. *laxiflora* Blume = ***Rhodamnia cinerea*** Jack
Rhodamnia cinerea var. *macrophylla* Blume = ***Rhodamnia cinerea*** Jack
Rhodamnia concolor Miq. = ***Rhodamnia cinerea*** Jack
Rhodamnia dumetorum var. *hainanensis* Merr. & L.M. Perry = ***Rhodamnia dumetorum*** (DC.) Merr. & L.M.Perry
Rhodamnia glabra Vidal = (Melastomataceae)
Rhodamnia globosa (Korth.) Blume = ***Rhodamnia cinerea*** Jack
Rhodamnia lamprophylla Diels = ***Rhodamnia latifolia*** (Benth.) Miq.
Rhodamnia ledermannii Diels = ***Rhodamnia sepicana*** Diels
Rhodamnia nageli Miq. = ***Rhodamnia cinerea*** Jack
Rhodamnia parvifolia Ridl. = ***Decaspermum parvifolium*** (Ridl.) A.J.Scott
Rhodamnia polyantha Diels = ***Rhodamnia latifolia*** (Benth.) Miq.
Rhodamnia propinqua C.T.White = ***Rhodamnia blairiana*** var. ***propinqua*** (C.T.White) A.J.Scott
Rhodamnia salomonensis C.T.White = ***Rhodomyrtus salomonensis*** (C.T.White) A.J.Scott
Rhodamnia siamensis Craib = ***Rhodamnia dumetorum*** (DC.) Merr. & L.M.Perry
Rhodamnia spectabilis (Blume) Blume = ***Rhodamnia cinerea*** Jack
Rhodamnia spongiosa (F.M.Bailey) Domin = ***Rhodamnia glauca*** Blume
Rhodamnia subtriflora Blume = ***Rhodamnia cinerea*** Jack

Rhodamnia trinervia Reinw. ex Blume = ***Rhodamnia rubescens*** (Benth.) Miq.
Rhodamnia trinervia var. *glabra* Maiden & Betche = ***Rhodamnia maideniana*** C.T.White
Rhodamnia trinervia var. *macrophylla* (Domin) C.T.White = ***Rhodomyrtus trineura*** var. ***macrophylla*** Domin
Rhodamnia trinervia var. *spongiosa* F.M.Bailey = ***Rhodamnia glauca*** Blume
Rhodamnia trinervia var. *uniflora* Ridl. = ***Rhodamnia uniflora*** (Ridl.) Burkill

Rhodomyrtus

Rhodomyrtus (DC.) Rchb., Deut. Bot. Herb.-Buch: 117 (1841).
Trop. & Subtrop. Asia to SW. Pacific. 36 CHH CHS 38 NNS TAI 40 IND SRL 41 CBD LAO MYA THA VIE 42 BOR JAW LSI MLY MOL PHI SUL 43 BIS NWG SOL 50 NSW QLD 60 NWC (61) sci (63) haw (78) fla.
17 Species
Cynomyrtus Scriv., J. Fed. Malay States Mus. 6: 253 (1916).
Psidiomyrtus Guillaumin, Bull. Mus. Natl. Hist. Nat., II, 4: 696 (1932).

Rhodomyrtus effusa Guymer, Austrobaileya 3: 382 (1991).
Queensland. 50 QLD. Nanophan. or phan.

Rhodomyrtus elegans (Blume) A.J.Scott, Kew Bull. 33: 320 (1978).
Maluku to New Guinea. 42 MOL 43 NWG. Nanophan. or phan.
**Macropsidium elegans* Blume, Mus. Bot. 1: 85 (1850). *Octamyrtus elegans* (Blume) Steenis, Blumea 19: 146 (1971).
Rhodomyrtus calophlebia C.T.White, J. Arnold Arbor. 23: 91 (1942).

Rhodomyrtus kaweaensis N.Snow, Austrobaileya 7: 331 (2006).
Papua New Guinea. 43 NWG. Nanophan. or phan.

Rhodomyrtus lanata Guymer, Austrobaileya 3: 384 (1991).
Papua New Guinea. 43 NWG. Nanophan. or phan.

Rhodomyrtus locellata (Guillaumin) Burret, Notizbl. Bot. Gart. Berlin-Dahlem 15: 496 (1941).
New Caledonia. 60 NWC. Nanophan. or phan.
**Psidiomyrtus locellata* Guillaumin, Bull. Mus. Natl. Hist. Nat., II, 4: 693 (1932).

Rhodomyrtus macrocarpa Benth., Fl. Austral. 3: 273 (1867).
New Guinea (incl. Kep. Aru) to N. & NE. Queensland. 43 NWG 50 QLD. Phan.

Rhodomyrtus mengenensis N.Snow, Austrobaileya 7: 333 (2006).
Papua New Guinea. 43 NWG. Nanophan. or phan.

Rhodomyrtus montana Guymer, Austrobaileya 3: 386 (1991).
W. New Guinea. 43 NWG. Nanophan. or phan.

Rhodomyrtus obovata C.T.White, J. Arnold Arbor. 32: 148 (1951).
Papua New Guinea. 43 NWG. Phan.

Rhodomyrtus pervagata Guymer, Austrobaileya 3: 380 (1991).
Queensland. 50 QLD. Nanophan. or phan.

Rhodomyrtus pinnatinervis C.T.White, J. Arnold Arbor. 23: 90 (1942).
New Guinea to Bismarck Arch. 43 BIS NWG. Nanophan. or phan.

Rhodomyrtus psidioides (G.Don) Benth., Fl. Austral. 3: 272 (1867).
SE. Queensland to New South Wales. 50 NSW QLD. Nanophan. or phan.
**Nelitris psidioides* G.Don, Gen. Hist. 2: 829 (1832).
Myrtus tozerii F.Muell., Fragm. 2: 86 (1860).

Rhodomyrtus salomonensis (C.T.White) A.J.Scott, Kew Bull. 33: 321 (1978).
Solomon Is. 43 SOL. Phan.
**Rhodamnia salomonensis* C.T.White, J. Arnold Arbor. 32: 147 (1951).

Rhodomyrtus sericea Burret, Notizbl. Bot. Gart. Berlin-Dahlem 15: 497 (1941).
N. Queensland. 50 QLD. Nanophan. or phan.

Rhodomyrtus surigaoensis Elmer, Leafl. Philipp. Bot. 7: 2344 (1914).
Philippines (Mindanao). 42 PHI. Nanophan. or phan.

Rhodomyrtus tomentosa (Aiton) Hassk., Flora 25(Beibl.): 35 (1842).
Trop. & Subtrop. Asia. 36 CHH CHS 38 NNS TAI 40 IND SRL 41 CBD LAO MYA THA VIE 42 BOR JAW LSI MLY MOL PHI SUL (61) sci (63) haw (78) fla. Nanophan. or phan.
**Myrtus tomentosa* Aiton, Hort. Kew. 2: 159 (1789).
Cynomyrtus tomentosa (Aiton) Scriv., J. Fed. Malay States Mus. 6: 253 (1916).

var. ***parviflora*** (Alston) A.J.Scott, Kew Bull. 33: 315 (1978).
S. India, Sri Lanka. 40 IND SRL. Nanophan. or phan.
**Rhodomyrtus parviflora* Alston in H.Trimen, Handb. Fl. Ceylon 6(Suppl.): 111 (1931).

var. ***tomentosa***
Trop. & Subtrop. Asia. 36 CHH CHS 38 NNS TAI 40 IND SRL 41 CBD LAO MYA THA VIE 42 BOR JAW LSI MLY MOL PHI SUL (61) sci (63) haw (78) fla. Nanophan. or phan.
Myrtus canescens Lour., Fl. Cochinch.: 311 (1790).

Rhodomyrtus trineura (F.Muell.) Benth., Fl. Austral. 3: 272 (1867).
Maliku to Queensland. 42 MOL 43 BIS NWG 50 QLD. Nanophan. or phan.
**Myrtus trineura* F.Muell., Fragm. 4: 117 (1864).

var. ***canescens*** (C.T.White) A.J.Scott, Kew Bull. 33: 325 (1978).
NE. Queensland. 50 QLD. Nanophan. or phan.
**Rhodomyrtus canescens* C.T.White, Bot. Bull. Dept. Agric. Queensland 22: 26 (1920).

subsp. ***capensis*** Guymer, Austrobaileya 3: 378 (1991).
N. Queensland. 50 QLD. Nanophan. or phan.

var. ***macrophylla*** Domin, Biblioth. Bot. 89: 1028 (1928). *Rhodamnia trinervia* var. *macrophylla* (Domin) C.T.White, Contr. Arnold Arbor. 4: 76 (1933).
NE. Queensland. 50 QLD. Nanophan. or phan.

var. ***novoguineensis*** (Diels) A.J.Scott, Kew Bull. 33: 324 (1978).
Maluku to Bismarck Arch. 42 MOL 43 BIS NWG. Nanophan. or phan.

Rhodomyrtus novoguineensis Diels, Bot. Jahrb. Syst. 57: 378 (1922).

subsp. ***trineura***
N. & NE. Queensland. 50 QLD. Nanophan. or phan.

Synonyms:
Rhodomyrtus baladensis (Brongn. & Gris) Nied. = ***Archirhodomyrtus baladensis*** (Brongn. & Gris) Burret
Rhodomyrtus beckleri (F.Muell.) L.S.Sm. = ***Archirhodomyrtus beckleri*** (F.Muell.) A.J.Scott
Rhodomyrtus calophlebia C.T.White = ***Rhodomyrtus elegans*** (Blume) A.J.Scott
Rhodomyrtus canescens C.T.White = ***Rhodomyrtus trineura*** var. ***canescens*** (C.T.White) A.J.Scott
Rhodomyrtus cymiflora (F.Muell.) Benth. = ***Archirhodomyrtus beckleri*** (F.Muell.) A.J.Scott
Rhodomyrtus emarginata Pancher ex Baker f. = ***Uromyrtus emarginata*** (Pancher ex Baker f.) Burret
Rhodomyrtus novoguineensis Diels = ***Rhodomyrtus trineura*** var. ***novoguineensis*** (Diels) A.J.Scott
Rhodomyrtus paitensis (Schltr.) Burret = ***Archirhodomyrtus paitensis*** (Schltr.) Burret
Rhodomyrtus parviflora Alston = ***Rhodomyrtus tomentosa*** var. ***parviflora*** (Alston) A.J.Scott
Rhodomyrtus parvifolia Craib = ***Rhodamnia kerrii*** J.Parn. & NicLugh.
Rhodomyrtus recurva C.T.White = ***Pilidiostigma papuanum*** (Lauterb.) A.J.Scott
Rhodomyrtus thymifolia Planch. ex Guillaumin = ***Uromyrtus thymifolia*** (Planch. ex Guillaumin) Burret
Rhodomyrtus turbinata (Schltr.) Burret = ***Archirhodomyrtus turbinata*** (Schltr.) Burret
Rhodomyrtus vieillardii (Brongn. & Gris) Burret = ***Archirhodomyrtus vieillardii*** (Brongn. & Gris) Burret

Rinzia

Rinzia Schauer, Linnaea 17: 239 (1843).
SW. Australia. 50 WAU.
12 Species

Rinzia affinis Trudgen, Nuytsia 5: 431 (1986).
S. Western Australia. 50 WAU. Cham.

Rinzia carnosa (S.Moore) Trudgen, Nuytsia 5: 426 (1986).
SW. Australia. 50 WAU. Cham.
**Baeckea carnosa* S.Moore, J. Linn. Soc., Bot. 45: 175 (1920).
Baeckea minutifolia Cheel, J. Proc. Roy. Soc. W. Australia 10: 5 (1923).

Rinzia communis Trudgen, Nuytsia 5: 435 (1986).
S. Western Australia. 50 WAU. Cham.

Rinzia crassifolia Turcz., Bull. Cl. Phys.-Math. Acad. Imp. Sci. Saint-Pétersbourg 10: 3331 (1852).
SW. Australia. 50 WAU. Cham.
Baeckea platystemona Benth., Fl. Austral. 3: 74 (1867). *Hypocalymma platystemonum* (Benth.) Nied. in H.G.A.Engler & K.A.E.Prantl, Nat. Pflanzenfam. 3(7): 99 (1893).

Rinzia dimorphandra (F.Muell. ex Benth.) Trudgen, Nuytsia 5: 430 (1986).
S. Western Australia. 50 WAU. Cham.
**Baeckea dimorphandra* F.Muell. ex Benth., Fl. Austral. 3: 74 (1867). *Hypocalymma dimorphandrum* (F. Muell. ex Benth.) Nied. in H.G.A.Engler & K.A.E. Prantl, Nat. Pflanzenfam. 3(7): 99 (1893).

Rinzia fumana Schauer, Linnaea 17: 239 (1843). *Baeckea fumana* (Schauer) F.Muell., Fragm. 4: 68 (1864). *Hypocalymma fumanum* (Schauer) Nied. in H.G.A.Engler & K.A.E.Prantl, Nat. Pflanzenfam. 3(7): 99 (1893).
SW. Australia. 50 WAU. Cham.

Rinzia longifolia Turcz., Bull. Cl. Phys.-Math. Acad. Imp. Sci. Saint-Pétersbourg 10: 331 (1852).
SW. Australia. 50 WAU. Cham.

Rinzia morrisonii Trudgen, Nuytsia 5: 423 (1986).
S. Western Australia. 50 WAU. Cham.

Rinzia oxycoccoides Turcz., Bull. Cl. Phys.-Math. Acad. Imp. Sci. Saint-Pétersbourg 10: 331 (1852). *Baeckea oxycoccoides* (Turcz.) Benth., Fl. Austral. 3: 75 (1867). *Hypocalymma oxycoccoides* (Turcz.) Nied. in H.G.A.Engler & K.A.E.Prantl, Nat. Pflanzenfam. 3(7): 99 (1893).
S. Western Australia. 50 WAU. Cham.

Rinzia rubra Trudgen, Nuytsia 5: 427 (1986).
SW. Australia. 50 WAU. Cham.

Rinzia schollerifolia (Lehm.) Trudgen, Nuytsia 5: 422 (1986).
SW. Australia. 50 WAU. Cham.
**Baeckea schollerifolia* Lehm., Pl. Preiss. 2: 369 (1848). *Hypocalymma schollerifolium* (Lehm.) Nied. in H.G.A.Engler & K.A.E.Prantl, Nat. Pflanzenfam. 3(7): 99 (1893).

Rinzia sessilis Trudgen, Nuytsia 5: 436 (1986).
Western Australia (Roe). 50 WAU. Cham.

Ristantia

Ristantia Peter G.Wilson & J.T.Waterh., Austral. J. Bot. 30: 442 (1982).
Queensland. 50 QLD.
3 Species

Ristantia gouldii Peter G.Wilson & B.Hyland, Telopea 3: 267 (1988).
N. Queensland. 50 QLD. Phan.

Ristantia pachysperma (F.Muell. & F.M.Bailey) Peter G.Wilson & J.T.Waterh., Austral. J. Bot. 30: 443 (1982).
N. & NE. Queensland. 50 QLD. Phan.
**Xanthostemon pachyspermus* F.Muell. & F.M.Bailey, Occas. Pap. Queensland Fl. 1: 4 (1886). *Metrosideros pachysperma* (F.Muell. & F.M.Bailey) F.Muell., Syst. Census Austral. Pl. 2: 100 (1889). *Tristania pachysperma* (F.Muell. & F.M.Bailey) W.D.Francis, Queensland Naturalist 14: 56 (1951).
Tristania odorata C.T.White & W.D.Francis, Bot. Bull. Dept. Agric. Queensland 22: 22 (1920).

Ristantia waterhousei Peter G.Wilson & B.Hyland, Telopea 3: 268 (1988).
NE. Queensland. 50 QLD. Phan.

Rubachia

Rubachia O.Berg = ***Marlierea*** Cambess.
Rubachia antonia O.Berg = ***Marlierea antonia*** (O.Berg) D.Legrand
Rubachia glomerata O.Berg = ***Plinia edulis*** (Vell.) Sobral
Rubachia involucrata O.Berg = ***Marlierea involucrata*** (O.Berg) Nied.
Rubachia lateriflora (DC.) O.Berg = ***Plinia spiciflora*** (Nees & Mart.) Sobral

Rubachia neuwiedeana O.Berg = ***Marlierea neuwiedeana*** (O.Berg) Nied.
Rubachia obumbrans O.Berg = ***Myrcia obumbrans*** (O.Berg) McVaugh
Rubachia spathulata O.Berg = ***Marlierea obversa*** D.Legrand
Rubachia spiciflora (Nees & Mart.) O.Berg = ***Plinia spiciflora*** (Nees & Mart.) Sobral
Rubachia spiciflora var. *grandifolia* O.Berg = ***Plinia spiciflora*** (Nees & Mart.) Sobral

Rylstonea

Rylstonea R.T.Baker = ***Homoranthus*** A.Cunn. ex Schauer
Rylstonea cernua R.T.Baker = ***Homoranthus cernuus*** (R.T.Baker) Craven & S.R.Jones
Rylstonea darwinioides (Maiden & Betche) R.T.Baker = ***Homoranthus darwinioides*** (Maiden & Betche) Cheel
Rylstonea wilhelmii (F.Muell.) R.T.Baker = ***Homoranthus wilhelmii*** (F.Muell.) Cheel

Saffordiella

Saffordiella Merr. = ***Myrtella*** F.Muell.
Saffordiella bennigseniana Merr. = ***Myrtella bennigseniana*** (Volkens) Diels

Salisia

Salisia Lindl. = ***Kunzea*** Rchb.
Salisia Pancher ex Brongn. & Gris = ***Xanthostemon*** F.Muell.
Salisia aurantiaca (Brongn. & Gris) Pancher = ***Xanthostemon aurantiacus*** (Brongn. & Gris) Schltr.
Salisia flava Pancher = ***Xanthostemon multiflorus*** (Montrouz.) Beauvis.
Salisia pulchella Lindl. = ***Kunzea pulchella*** (Lindl.) A.S.George
Salisia rubra Pancher = ***Xanthostemon ruber*** (Brongn. & Gris) Sebert & Pancher
Salisia rugosa Pancher = ***Pleurocalyptus pancheri*** (Brongn. & Gris) J.W.Dawson

Schidiomyrtus

Schidiomyrtus Schauer = ***Baeckea*** L.
Schidiomyrtus crenulata (Sm.) Schauer = ***Baeckea imbricata*** (Gaertn.) Druce
Schidiomyrtus diosmifolia (Rudge) Schauer = ***Baeckea diosmifolia*** Rudge
Schidiomyrtus ericacea F.Muell. ex Miq. = ***Micromyrtus ciliata*** (Sm.) Druce
Schidiomyrtus micrantha (DC.) Schauer = ***Thryptomene baeckeacea*** F.Muell.
Schidiomyrtus sieberi Schauer = ***Baeckea imbricata*** (Gaertn.) Druce
Schidiomyrtus tenella Schauer = ***Micromyrtus ciliata*** (Sm.) Druce
Schidiomyrtus tenella var. *ciliata* Miq. = ***Micromyrtus ciliata*** (Sm.) Druce

Schizocalomyrtus

Schizocalomyrtus Kausel = ***Calycorectes*** O.Berg
Schizocalomyrtus pohliana (O.Berg) Kausel = ***Calycorectes pohlianus*** (O.Berg) Kiaersk.

Schizocalyx

Schizocalyx O.Berg = ***Calycorectes*** O.Berg
Schizocalyx neocaledonica Brongn. & Gris = ***Stereocaryum neocaledonicum*** (Brongn. & Gris) A.J.Scott
Schizocalyx pohliana O.Berg = ***Calycorectes pohlianus*** (O.Berg) Kiaersk.
Schizocalyx pohliana var. *panicularis* O.Berg = ***Calycorectes pohlianus*** (O.Berg) Kiaersk.
Schizocalyx pohliana var. *triflora* O.Berg = ***Calycorectes pohlianus*** (O.Berg) Kiaersk.
Schizocalyx rubiginosa (Brongn. & Gris) Brongn. & Gris = ***Stereocaryum rubiginosum*** (Brongn. & Gris) Burret

Schizopleura

Schizopleura Endl. = ***Beaufortia*** R.Br.
Schizopleura dampieri (A.Cunn.) Walp. = ***Beaufortia sprengelioides*** (DC.) Craven
Schizopleura macrostemon (Lindl.) Walp. = ***Beaufortia macrostemon*** Lindl.
Schizopleura purpurea (Lindl.) Walp. = ***Beaufortia purpurea*** Lindl.

Scholtzia

Scholtzia Schauer, Linnaea 17: 241 (1843).
SW. Australia. 50 WAU.
13 Species
Pritzelia Schauer, Flora 26: 407 (1843).
Piptandra Turcz., Bull. Soc. Imp. Naturalistes Moscou 35(2): 323 (1862).

Scholtzia capitata F.Muell. ex Benth., Fl. Austral. 3: 69 (1867). *Baeckea capitata* (F.Muell. ex Benth.) F.Muell., Syst. Census Austral. Pl. 1: 54 (1882).
SW. Australia. 50 WAU. Nanophan.

Scholtzia ciliata F.Muell., Fragm. 4: 76 (1864). *Baeckea ciliata* (F.Muell.) F.Muell., Syst. Census Austral. Pl. 1: 54 (1882).
W. Western Australia. 50 WAU. Nanophan.

Scholtzia drummondii Benth., Fl. Austral. 3: 70 (1867).
SW. Australia. 50 WAU. Nanophan.

Scholtzia eatoniana (Ewart & Jean White) C.A.Gardner, Enum. Pl. Austr. Occ.: 95 (1931).
SW. Australia. 50 WAU. Nanophan.
**Baeckea eatoniana* Ewart & Jean White, Proc. Roy. Soc. Victoria, n.s. 21: 540 (1909).

Scholtzia laxiflora Benth., Fl. Austral. 3: 69 (1867). *Baeckea laxiflora* (Benth.) F.Muell., Syst. Census Austral. Pl. 1: 54 (1882).
SW. Australia. 50 WAU. Nanophan.

Scholtzia leptantha Benth., Fl. Austral. 3: 69 (1867). *Baeckea leptantha* (Benth.) F.Muell., Syst. Census Austral. Pl. 1: 54 (1882).
W. Western Australia. 50 WAU. Nanophan.

Scholtzia obovata (DC.) Schauer, Linnaea 17: 241 (1843).
SW. Australia. 50 WAU. Nanophan.
**Baeckea obovata* DC., Prodr. 3: 230 (1828). *Pritzelia obovata* (DC.) Schauer in W.G.Walpers, Repert. Bot. Syst. 2: 922 (1843).
Baeckea involucrata Endl. in S.L.Endlicher & al., Enum. Pl.: 51 (1837). *Scholtzia involucrata* (Endl.) Druce, Rep. Bot. Exch. Club Brit. Isles 1916: 645 (1917).

Scholtzia oligandra F.Muell. ex Benth., Fl. Austral. 3: 70 (1867). *Baeckea oligandra* (F.Muell. ex Benth.) F.Muell., Syst. Census Austral. Pl. 1: 54 (1882).
SW. Australia. 50 WAU. Nanophan.

Scholtzia parviflora F.Muell., Fragm. 4: 76 (1864). *Baeckea parviflora* (F.Muell.) F.Muell., Syst. Census Austral. Pl. 1: 54 (1882).
SW. Australia. 50 WAU. Nanophan.

Scholtzia spathulata (Turcz.) Benth., Fl. Austral. 3: 68 (1867).
SW. Australia. 50 WAU. Nanophan.
**Piptandra spathulata* Turcz., Bull. Soc. Imp. Naturalistes Moscou 35(2): 324 (1862). *Baeckea spatulata* (Turcz.) F.Muell., Syst. Census Austral. Pl. 1: 54 (1882).

Scholtzia teretifolia Benth., Fl. Austral. 3: 70 (1867). *Baeckea teretifolia* (Benth.) F.Muell., Syst. Census Austral. Pl. 1: 54 (1882).
WSW. Western Australia. 50 WAU. Nanophan.

Scholtzia uberiflora F.Muell., Fragm. 4: 74 (1864). *Baeckea uberiflora* (F.Muell.) F.Muell., Syst. Census Austral. Pl. 1: 53 (1882).
W. Western Australia. 50 WAU. Nanophan.

Scholtzia umbellifera F.Muell., Fragm. 4: 75 (1864). *Baeckea umbellifera* (F.Muell.) F.Muell., Syst. Census Austral. Pl. 1: 54 (1882).
W. Western Australia. 50 WAU. Nanophan.

***Synonyms*:**
Scholtzia decandra F.Muell. = ***Thryptomene saxicola*** (A.Cunn. ex Hook.) Schauer
Scholtzia decussata W.Fitzg. = ***Thryptomene decussata*** (W.Fitzg.) J.W.Green
Scholtzia denticulata F.Muell. = ***Thryptomene denticulata*** (F.Muell.) Benth.
Scholtzia involucrata (Endl.) Druce = ***Scholtzia obovata*** (DC.) Schauer

Schuermannia

Schuermannia F.Muell. = ***Homoranthus*** A.Cunn. ex Schauer
Schuermannia homoranthoides F.Muell. = ***Homoranthus homoranthoides*** (F.Muell.) Craven & S.R.Jones

Sinoga

Sinoga S.T.Blake = ***Asteromyrtus*** Schauer
Sinoga lysicephala (F.Muell. & F.M.Bailey) S.T.Blake = ***Asteromyrtus lysicephala*** (F.Muell. & F.M.Bailey) Craven

Siphoneugena

Siphoneugena O.Berg, Linnaea 27: 344 (1856).
Caribbean, S. Trop. America. 81 LEE PUE WIN 82 GUY SUR VEN 83 BOL ECU PER 84 BZC BZE BZL BZN BZS 85 AGW PAR.
9 Species
Paramitranthes Burret, Notizbl. Bot. Gart. Berlin-Dahlem 15: 541 (1941).

Siphoneugena delicata Sobral & Proença, Novon 16: 530 (2006).
SE. Brazil. 84 BZL. Phan.

Siphoneugena densiflora O.Berg in C.F.P.von Martius & auct. suc. (eds.), Fl. Bras. 14(1): 379 (1857). *Calycorectes densiflorus* (O.Berg) Nied. in H.G.A. Engler & K.A.E.Prantl, Nat. Pflanzenfam. 3(7): 82 (1893). *Paramitranthes densiflora* (O.Berg) Burret, Notizbl. Bot. Gart. Berlin-Dahlem 15: 541 (1941).
Brazil. 84 BZC BZE BZL BZN. Phan.

var. ***cipoensis*** Proença, Edinburgh J. Bot. 47: 250 (1990).
Brazil (Minas Gerais). 84 BZL. Phan.

var. ***densiflora***
Brazil. 84 BZC BZE BZL BZN. Phan.
Eugenia chnoosepala Kiaersk., Enum. Myrt. Bras.: 174 (1893). *Paramitranthes chnoosepala* (Kiaersk.) Burret, Notizbl. Bot. Gart. Berlin-Dahlem 15: 543 (1941). *Siphoneugena chnoosepala* (Kiaersk.) Kausel, Lilloa 32: 367 (1967).
Eugenia chnoosepala var. *regnelliana* Kiaersk., Enum. Myrt. Bras.: 195 (1893). *Paramitranthes regnelliana* (Kiaersk.) Burret, Notizbl. Bot. Gart. Berlin-Dahlem 15: 543 (1941). *Siphoneugena regnelliana* (Kiaersk.) Kausel, Lilloa 32: 367 (1967).
Eugenia chnoosepala var. *latifolia* Kiaersk., Bull. Soc. Bot. France 54(3c): 236 (1908).
Paramitranthes bracteata Burret, Notizbl. Bot. Gart. Berlin-Dahlem 15: 544 (1941). *Siphoneugena bracteata* (Burret) Kausel, Lilloa 32: 367 (1967).
Paramitranthes macrophylla Burret, Notizbl. Bot. Gart. Berlin-Dahlem 15: 544 (1941). *Siphoneugena macrophylla* (Burret) Kausel, Lilloa 32: 367 (1967). *Siphoneugena chnoosepala* var. *macrophylla* (Burret) Mattos & N.Silveira, Loefgrenia 87: 1 (1985).
Siphoneugena macrophylla var. *brasiliae* Mattos, Loefgrenia 71: 1 (1977).
Siphoneugena chnoosepala var. *pilosa* Mattos & N.Silveira, Loefgrenia 87: 1 (1985).

Siphoneugena dussii (Krug & Urb.) Proença, Edinburgh J. Bot. 47: 251 (1990).
Caribbean, S. Trop. America. 81 LEE PUE WIN 82 GUY SUR VEN 83 ECU PER 84 BZC BZL BZN. Phan.
**Marlierea dussii* Krug & Urb., Bot. Jahrb. Syst. 19: 590 (1895). *Plinia dussii* (Krug & Urb.) Urb., Repert. Spec. Nov. Regni Veg. 15: 413 (1919).

var. ***dussii***
Caribbean, S. Trop. America. 81 LEE PUE WIN 82 GUY SUR VEN 83 ECU PER 84 BZC BZL BZN. Phan.
Plinia fruticosa Steyerm., Fieldiana, Bot. 28: 1023 (1957).
Siphoneugena densiflora var. *tepuiensis* Steyerm., Fieldiana, Bot. 28: 1024 (1957).

var. ***salicifolia*** (McVaugh) Proença, Edinburgh J. Bot. 47: 254 (1990).
S. Venezuela. 82 GUY? VEN. Phan.
**Siphoneugena densiflora* var. *salicifolia* McVaugh, Mem. New York Bot. Gard. 18(2): 231 (1969).

Siphoneugena guilfoyleiana Proença, Edinburgh J. Bot. 47: 254 (1990).
Brazil (São Paulo). 84 BZL. Phan.

Siphoneugena kiaerskoviana (Burret) Kausel, Lilloa 32: 367 (1967).
SE. Brazil. 84 BZL. Nanophan. or phan.
Eugenia chnoosepala var. *angustifolia* Kiaersk., Enum. Myrt. Bras.: 175 (1893). *Siphoneugena angustifolia* (Kiaersk.) Mattos & N.Silveira, Loefgrenia 87: 1 (1985).
Eugenia chnoosepala var. *elliptica* Kiaersk., Enum. Myrt. Bras.: 175 (1893). *Mitranthes pilosa* Burret, Notizbl. Bot. Gart. Berlin-Dahlem 15: 540 (1941).

Siphoneugena angustifolia var. *elliptica* (Kiaersk.) Mattos & N.Silveira, Loefgrenia 87: 1 (1985).
**Paramitranthes kiaerskoviana* Burret, Notizbl. Bot. Gart. Berlin-Dahlem 15: 542 (1941).

Siphoneugena kuhlmannii Mattos, Ci. & Cult. 19: 334 (1967).
E. Brazil. 84 BZE BZL. Phan.
Siphoneugena rubensiana Mattos & N.Silveira, Loefgrenia 87: 2 (1985).

Siphoneugena occidentalis D.Legrand, Bol. Soc. Argent. Bot. 10: 5 (1962).
WC. Brazil, S. Bolivia to NW. Argentina (Salta). 83 BOL 84 BZC 85 AGW PAR. Phan.
Siphoneugena parviflora Kausel, Lilloa 33: 112 (1971 publ. 1972).

Siphoneugena reitzii D.Legrand, Sellowia 8: 78 (1957).
SE. & S. Brazil. 84 BZL BZS. Phan.
Siphoneugena gomesiana Mattos, Loefgrenia 19: 1 (1964).
Siphoneugena dusenii Kausel, Lilloa 33: 109 (1971 publ. 1972).
Siphoneugena boraceiensis Mattos & N.Silveira, Loefgrenia 87: 1 (1985).

Siphoneugena widgreniana O.Berg in C.F.P.von Martius & auct. suc. (eds.), Fl. Bras. 14(1): 379 (1857). *Calycorectes widgrenianus* (O.Berg) Nied. in H.G.A. Engler & K.A.E.Prantl, Nat. Pflanzenfam. 3(7): 82 (1893). *Eugenia widgreniana* (O.Berg) Kiaersk., Enum. Myrt. Bras.: 175 (1893). *Mitranthes widgreniana* (O.Berg) Burret, Notizbl. Bot. Gart. Berlin-Dahlem 15: 540 (1941).
Brazil (Minas Gerais to Paraná). 84 BZL BZS. Nanophan.
Mitranthes pubescens Burret, Repert. Spec. Nov. Regni Veg. 50: 55 (1941).

Synonyms:
Siphoneugena angustifolia (Kiaersk.) Mattos & N.Silveira = ***Siphoneugena kiaerskoviana*** (Burret) Kausel
Siphoneugena angustifolia var. *elliptica* (Kiaersk.) Mattos & N.Silveira = ***Siphoneugena kiaerskoviana*** (Burret) Kausel
Siphoneugena aromatica O.Berg = [84+] *Syzygium* sp. ?
Siphoneugena baporeti (D.Legrand) Kausel = ***Plinia rivularis*** (Cambess.) Rotman
Siphoneugena boraceiensis Mattos & N.Silveira = ***Siphoneugena reitzii*** D.Legrand
Siphoneugena bracteata (Burret) Kausel = ***Siphoneugena densiflora*** O.Berg var. ***densiflora***
Siphoneugena cantareirae Mattos = ***Myrciaria floribunda*** (H.West ex Willd.) O.Berg
Siphoneugena chnoosepala (Kiaersk.) Kausel = ***Siphoneugena densiflora*** O.Berg var. ***densiflora***
Siphoneugena chnoosepala var. *macrophylla* (Burret) Mattos & N.Silveira = ***Siphoneugena densiflora*** O.Berg var. ***densiflora***
Siphoneugena chnoosepala var. *pilosa* Mattos & N. Silveira = ***Siphoneugena densiflora*** O.Berg var. ***densiflora***
Siphoneugena densiflora var. *salicifolia* McVaugh = ***Siphoneugena dussii*** var. ***salicifolia*** (McVaugh) Proença
Siphoneugena densiflora var. *tepuiensis* Steyerm. = ***Siphoneugena dussii*** (Krug & Urb.) Proença var. ***dussii***
Siphoneugena dusenii Kausel = ***Siphoneugena reitzii*** D.Legrand
Siphoneugena gomesiana Mattos = ***Siphoneugena reitzii*** D.Legrand
Siphoneugena krukoffiana Kausel = [83 BOL]
Siphoneugena legrandii Mattos & N.Silveira = ***Plinia rivularis*** (Cambess.) Rotman
Siphoneugena macrophylla (Burret) Kausel = ***Siphoneugena densiflora*** O.Berg var. ***densiflora***
Siphoneugena macrophylla var. *brasiliae* Mattos = ***Siphoneugena densiflora*** O.Berg var. ***densiflora***
Siphoneugena micrantha Kausel = ***Myrciaria floribunda*** (H.West ex Willd.) O.Berg
Siphoneugena parviflora Kausel = ***Siphoneugena occidentalis*** D.Legrand
Siphoneugena regnelliana (Kiaersk.) Kausel = ***Siphoneugena densiflora*** O.Berg var. ***densiflora***
Siphoneugena rubensiana Mattos & N.Silveira = ***Siphoneugena kuhlmannii*** Mattos
Siphoneugena sulcata (Burret) Kausel = ?

Unplaced Names:
Siphoneugena aromatica O.Berg, Linnaea 27: 345 (1856). = [84+] *Syzygium* sp. ?
Siphoneugena krukoffiana Kausel, Lilloa 33: 110 (1971 publ. 1972). = [83 BOL]

Spermolepis

Spermolepis Brongn. & Gris = ***Arillastrum*** Pancher ex Baill.
Spermolepis gummifera Brongn. & Gris = ***Arillastrum gummiferum*** (Brongn. & Gris) Pancher ex Baill.
Spermolepis rubiginosa Brongn. & Gris = ***Stereocaryum rubiginosum*** (Brongn. & Gris) Burret
Spermolepis rubra Vieill. ex Guillanmin = ***Arillastrum gummiferum*** (Brongn. & Gris) Pancher ex Baill.
Spermolepis tannifera Heckel = ***Arillastrum gummiferum*** (Brongn. & Gris) Pancher ex Baill.

Sphaerantia

Sphaerantia Peter G.Wilson & B.Hyland, Telopea 3: 260 (1988).
N. Queensland. 50 QLD.
2 Species

Sphaerantia chartacea Peter G.Wilson & B.Hyland, Telopea 3: 262 (1988).
N. Queensland. 50 QLD.

Sphaerantia discolor Peter G.Wilson & B.Hyland, Telopea 3: 262 (1988).
N. Queensland. 50 QLD.

Stenocalyx

Stenocalyx O.Berg = ***Eugenia*** P.Micheli ex L.
Stenocalyx affinis O.Berg = ***Eugenia uniflora*** L.
Stenocalyx alagoensis O.Berg = ***Eugenia alagoensis*** (O.Berg) Mattos
Stenocalyx albicans O.Berg = ***Eugenia albicans*** (O.Berg) Urb.
Stenocalyx bahiensis O.Berg = ***Eugenia bahiana*** Mattos
Stenocalyx blastanthus O.Berg = ***Eugenia blastantha*** (O.Berg) D.Legrand
Stenocalyx brasiliensis (Lam.) O.Berg = ***Eugenia brasiliensis*** Lam.
Stenocalyx brasiliensis var. *erythrocarpa* (Cambess.) O.Berg = ***Eugenia brasiliensis*** Lam.
Stenocalyx brasiliensis var. *iocarpa* O.Berg = ***Eugenia brasiliensis*** Lam.
Stenocalyx brasiliensis var. *leucocarpa* (Cambess.) O.Berg = ***Eugenia brasiliensis*** Lam.

Stenocalyx brasiliensis var. *silvestris* O.Berg = ***Eugenia brasiliensis*** Lam.
Stenocalyx brunneus O.Berg = ***Eugenia uniflora*** L.
Stenocalyx costatus (Cambess.) O.Berg = ***Eugenia uniflora*** L.
Stenocalyx dasyblastus O.Berg = ***Eugenia uniflora*** L.
Stenocalyx dentatus O.Berg = ***Eugenia dentata*** (O.Berg) Nied.
Stenocalyx duarteanus (Cambess.) O.Berg = ***Eugenia duarteana*** Cambess.
Stenocalyx dysentericus (DC.) O.Berg = ***Eugenia dysenterica*** DC.
Stenocalyx gardnerianus O.Berg = ***Eugenia piresii*** Mattos
Stenocalyx gemmiflorus O.Berg = ***Eugenia ophthalmantha*** Kiaersk.
Stenocalyx glaber O.Berg = ***Eugenia uniflora*** L.
Stenocalyx grandiflorus O.Berg = ***Eugenia megaflora*** Govaerts
Stenocalyx grandifolius O.Berg = ***Eugenia fuscopunctata*** Kiaersk.
Stenocalyx impressus O.Berg = ***Eugenia impressa*** (O.Berg) Mattos
Stenocalyx impunctatus O.Berg = ***Eugenia uniflora*** L.
Stenocalyx involucratus (DC.) Kausel = ***Eugenia involucrata*** DC.
Stenocalyx lanceolatus O.Berg = ***Eugenia pyriformis*** Cambess.
Stenocalyx langsdorffii O.Berg = ***Eugenia itaguahiensis*** Nied.
Stenocalyx laxus (DC.) O.Berg = ***Eugenia laxa*** DC.
Stenocalyx ligustrina (Sw.) O.Berg = ***Eugenia ligustrina*** (Sw.) Willd.
Stenocalyx ligustrina var. *caribaea* O.Berg = ***Eugenia ligustrina*** (Sw.) Willd. var. ***ligustrina***
Stenocalyx ligustrinus var. *fluminensis* O.Berg = ***Eugenia ligustrina*** (Sw.) Willd. var. ***ligustrina***
Stenocalyx ligustrinus var. *minensis* O.Berg = ***Eugenia ligustrina*** (Sw.) Willd. var. ***ligustrina***
Stenocalyx longipes O.Berg = ***Eugenia longipedunculata*** Nied.
Stenocalyx lucidus O.Berg = ***Eugenia uniflora*** L.
Stenocalyx michelii (Lam.) O.Berg = ***Eugenia uniflora*** L.
Stenocalyx michelii var. *membranacea* O.Berg = ***Eugenia uniflora*** L.
Stenocalyx michelii var. *rigida* O.Berg = ***Eugenia uniflora*** L.
Stenocalyx mollicoma O.Berg = ***Eugenia mollicoma*** Mart. ex O.Berg
Stenocalyx mutabilis O.Berg = ***Eugenia longipetiolata*** Mattos
Stenocalyx myrtifolius O.Berg = ***Eugenia neomyrtifolia*** Sobral
Stenocalyx nanus Barb.Rodr. = ***Eugenia barbosae*** Barb.Rodr. ex Chodat & Hassl.
Stenocalyx oblongifolius O.Berg = ***Eugenia uniflora*** L.
Stenocalyx patrisii (Vahl) O.Berg = ***Eugenia patrisii*** Vahl
Stenocalyx patrisii var. *grandifolius* O.Berg = ***Eugenia patrisii*** Vahl
Stenocalyx patrisii var. *parvifolius* O.Berg = ***Eugenia patrisii*** Vahl
Stenocalyx pistaciifolius (DC.) O.Berg = ***Eugenia pistaciifolia*** DC.
Stenocalyx pitanga O.Berg = ***Eugenia pitanga*** (O.Berg) Nied.
Stenocalyx pitanga var. *nana* Mattos = ***Eugenia pitanga*** (O.Berg) Nied.
Stenocalyx pius (DC.) O.Berg = ***Eugenia pia*** DC.
Stenocalyx plumieri O.Berg = ***Plinia pinnata*** L.
Stenocalyx portoricensis (DC.) O.Berg = ***Eugenia pseudopsidium*** Jacq.
Stenocalyx protractus O.Berg = ***Eugenia prostrata*** Mattos
Stenocalyx pseudopsidium (Jacq.) O.Berg = ***Eugenia pseudopsidium*** Jacq.
Stenocalyx reticulatus O.Berg = ***Eugenia subreticulata*** Glaz.
Stenocalyx rhampiri Barb.Rodr. = ***Eugenia uniflora*** L.
Stenocalyx riedelianus O.Berg = ***Eugenia plicata*** Nied.
Stenocalyx ruber (L.) Kausel = ***Eugenia uniflora*** L.
Stenocalyx rufescens Kausel = ***Eugenia psidiiflora*** O.Berg
Stenocalyx sessilifolius (DC.) O.Berg = ***Eugenia sessilifolia*** DC.
Stenocalyx silvestris O.Berg = ***Eugenia neosilvestris*** Sobral
Stenocalyx squamiflorus O.Berg = ***Eugenia ligustrina*** (Sw.) Willd. var. ***ligustrina***
Stenocalyx stipularis O.Berg = ***Eugenia squamiflora*** Mattos
Stenocalyx strigosus O.Berg = ***Eugenia uniflora*** L.
Stenocalyx sulcatus (Spring ex Mart.) O.Berg = ***Eugenia sulcata*** Spring ex Mart.
Stenocalyx sulcatus var. *latifolius* O.Berg = ***Eugenia sulcata*** Spring ex Mart.
Stenocalyx sulcatus var. *ligustrinus* O.Berg = ***Eugenia sulcata*** Spring ex Mart.
Stenocalyx sulcatus var. *longifolius* O.Berg = ***Eugenia sulcata*** Spring ex Mart.
Stenocalyx sulcatus var. *strictus* O.Berg = ***Eugenia sulcata*** Spring ex Mart.
Stenocalyx ubensis (Cambess.) O.Berg = ***Eugenia brasiliensis*** Lam.
Stenocalyx uniflorus (L.) Kausel = ***Eugenia uniflora*** L.
Stenocalyx xanthoxyloides (Cambess.) O.Berg = ***Eugenia xanthoxyloides*** Cambess.

Stenospermum

Stenospermum Sweet ex Heynh. = ***Kunzea*** Rchb.
Stenospermum capitatum (Sm.) Sweet ex Heynh. = ***Kunzea capitata*** (Sm.) Heynh.
Stenospermum ciliatum (J.R.Forst. & G.Forst.) Heynh. = ***Purpureostemon ciliatus*** (J.R.Forst. & G.Forst.) Gugerli
Stenospermum corifolium (Vent.) Sweet ex Heynh. = ***Kunzea ambigua*** (Sm.) Druce
Stenospermum ericifolium (Sm.) Heynh. = ***Kunzea ericifolia*** (Sm.) Heynh.

Stenostegia

Stenostegia A.R.Bean, Muelleria 11: 127 (1998).
N. Australia. 50 NTA.
1 Species

Stenostegia congesta A.R.Bean, Muelleria 11: 130 (1998).
NW. Northern Territory. 50 NTA. Nanophan. or phan.
Baeckea umbellata F.Muell., Fragm. 4: 69 (1864), nom. inval.

Stereocaryum

Stereocaryum Burret, Notizbl. Bot. Gart. Berlin-Dahlem 15: 546 (1941).
New Caledonia. 60 NWC.
3 Species

Stereocaryum neocaledonicum (Brongn. & Gris) A.J.Scott, Kew Bull. 34: 496 (1980).

New Caledonia. 60 NWC. Nanophan. or phan.
**Schizocalyx neocaledonica* Brongn. & Gris, Ann. Sci. Nat., Bot., V, 13: 382 (1871).

Stereocaryum ovigerum (Brongn. & Gris) Burret, Notizbl. Bot. Gart. Berlin-Dahlem 15: 547 (1941).
New Caledonia. 60 NWC. Nanophan. or phan.
**Eugenia ovigera* Brongn. & Gris, Bull. Soc. Bot. France 12: 179 (1865). *Calycorectes ovigerus* (Brongn. & Gris) Guillaumin, Bull. Mus. Natl. Hist. Nat. 25: 503 (1919).

Stereocaryum rubiginosum (Brongn. & Gris) Burret, Notizbl. Bot. Gart. Berlin-Dahlem 15: 546 (1941).
New Caledonia. 60 NWC. Nanophan. or phan.
**Spermolepis rubiginosa* Brongn. & Gris, Bull. Soc. Bot. France 10: 578 (1863). *Schizocalyx rubiginosa* (Brongn. & Gris) Brongn. & Gris, Ann. Sci. Nat., Bot., V, 13: 380 (1871). *Nania rubiginosa* (Brongn. & Gris) Kuntze, Revis. Gen. Pl. 1: 242 (1891). *Calycorectes rubiginosa* (Brongn. & Gris) Guillaumin, Notul. Syst. (Paris) 2: 131 (1911). *Fremya lucens* Bureau ex Guillaumin, Notul. Syst. (Paris) 2: 130 (1911). *Myrtomera rubiginosa* (Brongn. & Gris) B.C.Stone, Pacific Sci. 16: 241 (1962).

Synonyms:
Stereocaryum symingtonianum (M.R.Hend.) A.J.Scott = ***Syzygium symingtonianum*** (M.R.Hend.) I.M.Turner
Stereocaryum watsonianum (M.R.Hend.) A.J.Scott = ***Syzygium watsonianum*** (M.R.Hend.) I.M.Turner

Stockwellia

Stockwellia D.J.Carr, S.G.M.Carr & B.Hyland, Bot. J. Linn. Soc. 139: 416 (2002).
Queensland. 50 QLD.
1 Species

Stockwellia quadrifida D.J.Carr, S.G.M.Carr & B.Hyland, Bot. J. Linn. Soc. 139: 416 (2002).
Queensland. 50 QLD.

Strongylocalyx

Strongylocalyx Blume = ***Syzygium*** Gaertn.
Strongylocalyx hemisphaericus (Wight) Blume = ***Syzygium hemisphericum*** (Wight) Alston
Strongylocalyx lanceifolius (Roxb.) Blume = ***Syzygium praecox*** (Roxb.) Rathakr. & N.C.Nair
Strongylocalyx leptostachyus Blume ex Miq. = ***Syzygium leptostemon*** (Korth.) Merr. & L.M.Perry
Strongylocalyx leptostemon (Korth.) Blume = ***Syzygium leptostemon*** (Korth.) Merr. & L.M.Perry
Strongylocalyx oblatus (Roxb.) Blume = ***Syzygium oblatum*** (Roxb.) Wall. ex A.M.Cowan & Cowan
Strongylocalyx praecox (Roxb.) Blume = ***Syzygium praecox*** (Roxb.) Rathakr. & N.C.Nair

Suzygium

Suzygium P.Browne = ***Calyptranthes*** Sw.

Syllisium

Syllisium Endl. = ***Syzygium*** Gaertn.

Syllysium

Syllysium Meyen & Schauer = ***Syzygium*** Gaertn.
Syllysium buxifolium (Hook. & Arn.) Meyen & Schauer = ***Syzygium buxifolium*** Hook. & Arn.

Symphyomyrtus

Symphyomyrtus Schauer = ***Eucalyptus*** L'Hér.
Symphyomyrtus lehmannii Schauer = ***Eucalyptus lehmannii*** (Schauer) Benth.

Syncarpia

Syncarpia Ten., Index Seminum (NAP) 1839: 12 (1839).
E. Australia. (25) ken tan uga (27) cpp tvl 50 NSW QLD (63) haw.
3 Species
Kamptzia Nees, Kamptziae: 5 (1840).

Syncarpia glomulifera (Sm.) Nied. in H.G.A.Engler & K.A.E.Prantl, Nat. Pflanzenfam. 3(7): 88 (1893).
E. Australia. (25) ken tan uga (27) cpp tvl 50 NSW QLD (63) haw. Phan.
**Metrosideros glomulifera* Sm., Trans. Linn. Soc. London 3: 269 (1797). *Nania glomulifera* (Sm.) Kuntze, Revis. Gen. Pl. 1: 242 (1891).

subsp. ***glabra*** (Benth.) A.R.Bean, Austrobaileya 4: 339 (1995).
CE. New South Wales. 50 NSW. Phan.
**Syncarpia laurifolia* var. *glabra* Benth., Fl. Austral. 3: 266 (1867). *Metrosideros glomulifera* var. *glabra* (Benth.) C.Moore & Betche, Handb. Fl. N.S.W.: 205 (1893). *Syncarpia procera* var. *glabra* (Benth.) Domin, Biblioth. Bot. 89: 1026 (1928).

subsp. ***glomulifera***
E. Australia. (25) ken tan uga (27) cpp 50 NSW QLD (63) haw. Phan.
Metrosideros procera K.D.Koenig & Sims in R.A.Salisbury, Prodr. Stirp. Chap. Allerton: 351 (1796), provisional synonym. *Syncarpia procera* (K.D.Koenig & Sims) Domin, Biblioth. Bot. 89: 472 (1928).
Metrosideros propinqua K.D.Koenig & Sims in R.A.Salisbury, Prodr. Stirp. Chap. Allerton: 351 (1796).
Tristania albens A.Cunn. ex DC., Prodr. 3: 210 (1828). *Kamptzia albens* (A.Cunn. ex DC.) Nees, Kamptziae: 6 (1840).
Syncarpia laurifolia Graeffer ex Ten., Index Seminum (NAP) 1839(App.): 12 (1839).

Syncarpia hillii F.M.Bailey, Proc. Roy. Soc. Queensland 1: 86 (1884).
SE. Queensland to NE. New South Wales. 50 NSW QLD. Phan.

Syncarpia verecunda A.R.Bean, Austrobaileya 4: 340 (1995).
SE. Queensland. 50 QLD. Phan.

Synonyms:
Syncarpia laurifolia Graeffer ex Ten. = ***Syncarpia glomulifera*** (Sm.) Nied. subsp. ***glomulifera***
Syncarpia laurifolia var. *glabra* Benth. = ***Syncarpia glomulifera*** subsp. ***glabra*** (Benth.) A.R.Bean
Syncarpia leptopetala F.Muell. = ***Choricarpia leptopetala*** (F.Muell.) Domin
Syncarpia procera (K.D.Koenig & Sims) Domin = ***Syncarpia glomulifera*** (Sm.) Nied. subsp. ***glomulifera***
Syncarpia procera var. *glabra* (Benth.) Domin = ***Syncarpia glomulifera*** subsp. ***glabra*** (Benth.) A.R.Bean
Syncarpia subargentea C.T.White = ***Choricarpia subargentea*** (C.T.White) L.A.S.Johnson

Syncarpia subargentea var. *latifolia* C.T.White = ***Choricarpia subargentea*** (C.T.White) L.A.S.Johnson

Syncarpia vertholenii Teijsm. & Binn. = ***Xanthostemon verus*** (Lindl.) Peter G.Wilson

Syzygium

Syzygium Gaertn., Fruct. Sem. Pl. 1: 166 (1788), nom. cons.

Trop. & Subtrop. Old World to Pacific. (21) mdr 22 BEN GAM GHA GNB GUI IVO LBR MLI NGA SEN SIE TOG 23 BUR CAF CMN CON GAB GGI RWA ZAI 24 ERI ETH SOM SUD 25 KEN TAN UGA 26 ANG MLW MOZ TAN UGA ZAM ZIM 27 BOT CPP CPV NAM NAT SWZ TVL 29 COM MAU MDG REU ROD† SEY 35 SAU YEM 36 CHC CHH CHS CHT 38 JAP NNS OGA TAI 40 ASS BAN EHM IND NEP SRL 41 AND CBD LAO MYA NCB THA VIE 42 BOR JAW LSI MLY MOL PHI SUL SUM XMS 43 BIS NWG SOL 50 NFK NSW NTA QLD VIC? WAU 51 NZN NZS 60 FIJ NUE NWC SAM TON VAN WAL (61) mrq sci 62 CRL 63 HAW (78) fla (80) blz els gua (81) lee win (84) bze bzl. [Myrtaceae]

1119 Species

Caryophyllus L., Sp. Pl.: 515 (1753).
Jambos Adans., Fam. Pl. 2: 88 (1763).
Jambolifera Houtt., Handl. Pl.-Kruidk. 2: 272 (1774).
Myrthoides Wolf, Gen. Pl.: 73 (1776).
Cetra Noronha, Verh. Batav. Genootsch. Kunsten 5(4): 2 (1790).
Opa Lour., Fl. Cochinch.: 308 (1790).
Calyptranthus Blume, Bijdr.: 1089 (1826).
Acmena DC. in J.B.G. Bory de Saint-Vincent, Dict. Class. Hist. Nat. 11: 401 (1827).
Jambosa DC., Prodr. 3: 286 (1828), nom. illeg.
Lomastelma Raf., Sylva Tellur.: 107 (1838).
Malidra Raf., Sylva Tellur.: 107 (1838).
Cerocarpus Colebr. ex Hassk., Flora 25(2): 36 (1842).
Syllisium Endl., Gen. Pl., Suppl. 3: 102 (1843), orth. var.
Syllysium Meyen & Schauer, Nov. Actorum Acad. Caes. Leop.-Carol. Nat. Cur. 19(Suppl. 1): 334 (1843).
Clavimyrtus Blume, Mus. Bot. 1: 113 (1850).
Cleistocalyx Blume, Mus. Bot. 1: 84 (1850).
Gelpkea Blume, Mus. Bot. 1: 88 (1850).
Macropsidium Blume, Mus. Bot. 1: 85 (1850).
Microjambosa Blume, Mus. Bot. 1: 117 (1850).
Strongylocalyx Blume, Mus. Bot. 1: 89 (1850).
Acicalyptus A.Gray, U.S. Expl. Exped., Phan. 1: 551 (1854).
Macromyrtus Miq., Fl. Ned. Ind. 1(1): 439 (1855).
Bostrychode (Miq.) O.Berg in C.F.P.von Martius & auct. suc. (eds.), Fl. Bras. 14(1): 634 (1859).
Leptomyrtus (Miq.) O.Berg in C.F.P.von Martius & auct. suc. (eds.), Fl. Bras. 14(1): 635 (1859).
Cupheanthus Seem., Fl. Vit.: 76 (1865).
Gaslondia Vieill., Bull. Soc. Linn. Normandie 10: 96 (1866).
Aphanomyrtus Miq., Fl. Ned. Ind. 1: 480 (1885).
Pseudoeugenia Scort., J. Bot. 23: 153 (1885).
Xenodendron K.Schum. & Lauterb., Fl. Schutzgeb. Südsee: 461 (1900).
Pareugenia Turrill, Hooker's Icon. Pl. 31: t. 3001 (1915).
Tetraeugenia Merr., J. Straits Branch Roy. Asiat. Soc. 77: 230 (1917).
Acmenosperma Kausel, Ark. Bot., a.s., 3: 609 (1957).
Waterhousea B.Hyland, Austral. J. Bot., Suppl. Ser. 9: 138 (1983).
Anetholea Peter G.Wilson, Austral. Syst. Bot. 13: 434 (2000).

Syzygium abbreviatum Merr., Philipp. J. Sci. 79: 372 (1951).
Philippines. 42 PHI. Phan.
**Eugenia abbreviata* Elmer, Leafl. Philipp. Bot. 8: 2776 (1915), nom. illeg.

Syzygium aborense (Dunn) Rathakr. & N.C.Nair, J. Econ. Taxon. Bot. 4: 287 (1983).
Arunachal Pradesh. 40 EHM. Nanophan. or phan.
**Eugenia aborensis* Dunn, Bull. Misc. Inform. Kew 1920: 109 (1920).

Syzygium abortivum (Gagnep.) Merr. & L.M.Perry, J. Arnold Arbor. 19: 101 (1938).
Indo-China. 41 LAO MYA THA VIE. Nanophan. or phan.
**Eugenia abortiva* Gagnep., Notul. Syst. (Paris) 3: 316 (1917).

Syzygium abulugense Merr., Philipp. J. Sci. 89: 373 (1951).
Philippines. 42 PHI. Phan.
**Eugenia propinqua* Merr., Philipp. J. Sci., C 7: 315 (1912).

Syzygium aciculinum Merr. & L.M.Perry, Brittonia 4: 126 (1941).
N. Myanmar. 41 MYA. Phan.

Syzygium acre (Pancher ex Guillaumin) J.W.Dawson, in Fl. Nouv.-Caléd. 23: 44 (1999).
New Caledonia. 60 NWC. Phan.
**Jambosa acris* Pancher ex Guillaumin, Bull. Soc. Bot. France 85: 642 (1938 publ. 1939).

Syzygium acrophilum (C.B.Rob.) Merr., Philipp. J. Sci. 79: 373 (1951).
Philippines. 42 PHI. Phan.
**Eugenia acrophila* C.B.Rob., Philipp. J. Sci., C 4: 389 (1909).

Syzygium acuminatissimum (Blume) DC., Prodr. 3: 261 (1828).
S. China to Solomon Is. 36 CHS 38 TAI 41 AND MYA NCB THA VIE 42 BOR JAW LSI MLY PHI SUL SUM 43 NWG SOL. Nanophan. or phan.
**Myrtus acuminatissima* Blume, Bijdr.: 1088 (1826). *Jambosa acuminatissima* (Blume) Hassk., Cat. Hort. Bot. Bogor.: 362 (1844). *Eugenia acuminatissima* (Blume) Kurz, Prelim. Rep. Forest Pegu, App. A: lxiii (1875), nom. illeg. *Acmena acuminatissima* (Blume) Merr. & L.M.Perry, J. Arnold Arbor. 19: 12 (1938).
Syzygium subdecurrens Miq., Fl. Ned. Ind. 1(1): 449 (1855). *Eugenia subdecurrens* (Miq.) Merr. & Chun, Sunyatsenia 2: 289 (1935).
Eugenia gardneri Fern.-Vill. in F.M.Blanco, Fl. Filip., ed. 3, 4(13A): 85 (1880).
Eugenia cumingiana Vidal, Phan. Cuming. Philipp.: 173 (1885). *Syzygium cumingianum* (Vidal) Gibbs, J. Linn. Soc., Bot. 42: 73 (1914).
Xenodendron polyanthum K.Schum. & Lauterb., Fl. Schutzgeb. Südsee: 461 (1900). *Acmena polyantha* (K.Schum. & Lauterb.) Merr. & L.M.Perry, J. Arnold Arbor. 19: 11 (1938).
Eugenia cuspidato-obovata Hayata, Icon. Pl. Formosan. 3: 116 (1913). *Syzygium cuspidato-obovatum* (Hayata) Mori, Trans. Nat. Hist. Soc. Taiwan 28: 439 (1938).
Eugenia laevifolia Ridl., Trans. Linn. Soc. London, Bot. 9: 48 (1916). *Acmena laevifolia* (Ridl.) Merr. & L.M.Perry, J. Arnold Arbor. 19: 18 (1938).
Eugenia attenuatifolia Merr., Philipp. J. Sci. 18: 299 (1921).

Eugenia eucaudata Elmer ex Merr., Enum. Philipp. Fl. Pl. 3: 176 (1923), pro syn.
Acmena dielsii Merr. & L.M.Perry, J. Arnold Arbor. 19: 18 (1938).

Syzygium acuminatum (Roxb.) Miq., Fl. Ned. Ind. 1(1): 451 (1855).
Maluku. 42 MOL. Phan.
**Eugenia acuminata* Roxb., Fl. Ind. ed. 1832, 2: 492 (1832), nom. illeg. *Eugenia moluccana* Merr., Philipp. J. Sci., C 11: 296 (1916).

Syzygium acutangulum Nied. in H.G.A.Engler & K.A.E.Prantl, Nat. Pflanzenfam. 3(7): 85 (1893).
Papuasia. 43 BIS NWG SOL. Phan.
**Eugenia acutangula* K.Schum. in K.M.Schumann & U.M.Hollrung, Fl. Kais. Wilh. Land: 89 (1889), nom. illeg.
Syzygium papuasicum Merr. & L.M.Perry, J. Arnold Arbor. 23: 251 (1942).

Syzygium acutatum (Miq.) Amshoff in C.A.Backer & R.C.Bakhuizen van der Brink, Bekn. Fl. Java 4b(98): 17 (1944).
Jawa. 42 JAW. Phan.
**Jambosa acutata* Miq., Fl. Ned. Ind. 1(1): 432 (1855).
Eugenia argutata Koord. & Valeton, Meded. Lands Plantentuin 40: 146 (1900).

Syzygium adelphicum Diels, Nova Guinea 14: 93 (1924).
New Guinea. 43 NWG. Phan.
Syzygium adelphicum var. *adenanthum* Merr. & L.M. Perry, J. Arnold Arbor. 23: 292 (1942).

Syzygium adenophyllum Merr. & L.M.Perry, Mem. Amer. Acad. Arts 18: 184 (1939).
Borneo. 42 BOR. Phan.

Syzygium aegiceroides (Korth. ex Miq.) Korth., Ned. Kruidk. Arch. 1: 203 (1848).
Borneo. 42 BOR. Phan.
**Eugenia aegiceroides* Korth. ex Miq., Anal. Bot. Ind. 1: 22 (1850).

Syzygium aemulum (Blume) Amshoff in C.A.Backer & R.C.Bakhuizen van der Brink, Bekn. Fl. Java 4b(98): 28 (1944).
Jawa. 42 JAW. Phan.
**Jambosa aemula* Blume, Mus. Bot. 1: 99 (1850).
Eugenia aemula (Blume) Koord. & Valeton, Meded. Lands Plantentuin 40: 76 (1900).

Syzygium aeoranthum (Diels) Merr. & L.M.Perry, J. Arnold Arbor. 23: 249 (1942).
New Guinea. 43 NWG. Phan.
**Jambosa aeorantha* Diels, Nova Guinea 14: 90 (1924).

Syzygium affine Merr., Philipp. J. Sci. 79: 373 (1951).
Philippines (Luzon). 42 PHI. Phan.

Syzygium aggregatum J.W.Dawson, in Fl. Nouv.-Caléd. 23: 19 (1999).
New Caledonia. 60 NWC. Phan.

Syzygium aksornae Chantaran. & J.Parn., Kew Bull. 48: 592 (1993).
Pen. Thailand. 41 THA. Phan.
**Eugenia quadrialata* Craib, Bull. Misc. Inform. Kew 1929: 115 (1929).

Syzygium alatoramulum B.Hyland, Austral. J. Bot., Suppl. Ser. 9: 42 (1983).
N. Queensland. 50 QLD. Phan.

Syzygium alatum (Lauterb.) Diels, Bot. Jahrb. Syst. 57: 411 (1922).
New Guinea. 43 NWG. Phan.
**Aphanomyrtus alata* Lauterb., Nova Guinea 8: 854 (1912).
Syzygium vaccinioides Merr. & L.M.Perry, J. Arnold Arbor. 23: 256 (1942).

Syzygium albayense Merr., Philipp. J. Sci. 79: 374 (1951).
Philippines (Luzon). 42 PHI. Phan.

Syzygium albiflorum (Duthie ex Kurz) Bahadur & R.C.Gaur, Indian J. Forest. 1: 349 (1978).
W. Indo-China. 41 MYA THA. Phan.
**Eugenia albiflora* Duthie ex Kurz, Forest Fl. Burma 1: 491 (1877).

Syzygium album Q.F.Zheng, in Fl. Fujianica 4: 633 (1989 publ. 1990).
SE. China. 36 CHS. Phan.

Syzygium alliiligneum B.Hyland, Austral. J. Bot., Suppl. Ser. 9: 44 (1983).
N. Queensland. 50 QLD. Phan.

Syzygium alternifolium (Wight) Walp., Repert. Bot. Syst. 2: 179 (1843).
India to Assam. 40 ASS IND. Phan.
**Eugenia alternifolia* Wight, Icon. Pl. Ind. Orient. 2: t. 537 (1842).

Syzygium alubo Kosterm., Quart. J. Taiwan Mus. 34: 127 (1981).
Sri Lanka. 40 SRL. Phan.

Syzygium alutaceum (Diels) Merr. & L.M.Perry, J. Arnold Arbor. 23: 270 (1942).
New Guinea. 43 NWG. Phan.
**Jambosa alutacea* Diels, Bot. Jahrb. Syst. 57: 386 (1922).

Syzygium alvarezii (C.B.Rob.) Merr., Philipp. J. Sci. 79: 375 (1951).
Philippines. 42 PHI. Phan.
**Eugenia alvarezii* C.B.Rob., Philipp. J. Sci., C 4: 390 (1909).
Eugenia maquilingensis Elmer, Leafl. Philipp. Bot. 8: 3096 (1919).

Syzygium alyxiifolium (Ridl.) I.M.Turner, J. Singapore Natl. Acad. Sci. 22–24: 16 (1997).
Pen. Malaysia. 42 MLY.
**Eugenia alyxiifolia* Ridl., J. Bot. 62: 296 (1924).

Syzygium amicorum (A.Gray) Müll.Stuttg. in W.G.Walpers, Ann. Bot. Syst. 4: 839 (1858).
Fiji. 60 FIJ. Nanophan. or phan.
**Eugenia amicorum* A.Gray, U.S. Expl. Exped., Phan. 1: 524 (1854).

Syzygium amieuense (Guillaumin) J.W.Dawson, in Fl. Nouv.-Caléd. 23: 30 (1999).
NW. & C. New Caledonia. 60 NWC. Phan.
**Caryophyllus amieuensis* Guillaumin, Mém. Mus. Natl. Hist. Nat., B, Bot. 8: 147 (1959).

Syzygium amphoraecarpus Kosterm., Quart. J. Taiwan Mus. 34: 129 (1981).
Sri Lanka. 40 SRL. Nanophan. or phan.

Syzygium amplexicaule (DC.) N.P.Balakr., Bull. Bot. Surv. India 22: 173 (1980 publ. 1982).
Bangladesh. 40 BAN. Phan.

Jambosa amplexicaulis DC., Prodr. 3: 287 (1828). *Eugenia amplexicaulis* (DC.) Roxb., Fl. Ind. ed. 1832, 2: 483 (1832), nom. illeg.

Syzygium ampliflorum (Koord. & Valeton) Amshoff in C.A.Backer & R.C.Bakhuizen van der Brink, Bekn. Fl. Java 4b(98): 21 (1944).
Jawa. 42 JAW. Nanophan. or phan.
Clavimyrtus firma Blume, Mus. Bot. 1: 113 (1850). *Jambosa firma* (Blume) Miq., Fl. Ned. Ind. 1(1): 429 (1855). *Eugenia firma* (Blume) Koord. & Valeton, Meded. Lands Plantentuin 40: 163 (1900), nom. illeg.
**Eugenia ampliflora* Koord. & Valeton, Bull. Inst. Bot. Buitenzorg 2: 5 (1899).

Syzygium amplifolium L.M.Perry, J. Arnold Arbor. 31: 365 (1950).
Fiji (S. Viti Levu). 60 FIJ. Nanophan. or phan.

Syzygium amplum T.G.Hartley & L.M.Perry, J. Arnold Arbor. 54: 192 (1973).
New Guinea to N. Queensland. 43 NWG 50 QLD. Phan.

Syzygium ampullarium (Stapf) Merr. & L.M.Perry, Mem. Amer. Acad. Arts 18: 164 (1939).
Borneo. 42 BOR. Phan.
**Eugenia ampullaria* Stapf, Trans. Linn. Soc. London, Bot. 4: 153 (1894).

Syzygium anacardiifolium (Craib) Chantaran. & J.Parn., Kew Bull. 48: 592 (1993).
Thailand to Pen. Malaysia. 41 THA 42 MLY. Phan.
**Eugenia anacardiifolia* Craib, Bull. Misc. Inform. Kew 1929: 114 (1929).

Syzygium andamanicum (King) N.P.Balakr., Bull. Bot. Surv. India 22: 174 (1980 publ. 1982).
Andaman Is. 41 AND. Phan.
**Eugenia andamanica* King, J. Asiat. Soc. Bengal, Pt. 2, Nat. Hist. 70: 106 (1901).

Syzygium aneityense Guillaumin, J. Arnold Arbor. 12: 256 (1931).
Vanuatu. 60 VAN. Phan.

Syzygium angkae (Craib) Chantaran. & J.Parn., Kew Bull. 48: 592 (1993).
Indo-China. 41 LAO MYA THA. Phan.
**Eugenia angkae* Craib, Bull. Misc. Inform. Kew 1929: 115 (1929).

subsp. ***angkae***
Indo-China. 41 LAO MYA THA. Phan.

subsp. ***spissum*** (Craib) Chantaran. & J.Parn., Kew Bull. 48: 592 (1993).
Indo-China. 41 MYA THA. Phan.
**Eugenia spissa* Craib, Bull. Misc. Inform. Kew 1930: 169 (1930).

Syzygium angophoroides (F.Muell.) B.Hyland, Austral. J. Bot., Suppl. Ser. 9: 47 (1983).
N. Australia. 50 NTA QLD WAU. Phan.
**Eugenia angophoroides* F.Muell., Fragm. 5: 33 (1865).

Syzygium angulare (Elmer) Merr., Philipp. J. Sci. 79: 375 (1951).
Philippines. 42 PHI. Phan.
**Eugenia angularis* Elmer, Leafl. Philipp. Bot. 4: 1434 (1912).

Syzygium angulatum (C.B.Rob.) Merr., Philipp. J. Sci. 79: 375 (1951).
Philippines. 42 PHI. Phan.
**Eugenia angulata* C.B.Rob., Philipp. J. Sci., C 4: 354 (1909).

Syzygium anisatum (Vickery) Craven & Biffin, Blumea 50: 159 (2005).
NE. New South Wales. 50 NSW. Nanophan.
**Backhousia anisata* Vickery, Contr. New South Wales Natl. Herb. 1: 129 (1941). *Anetholea anisata* (Vickery) Peter G.Wilson, Austral. Syst. Bot. 13: 434 (2000).

Syzygium anisopetalum (R.Parker) N.P.Balakr., Bull. Bot. Surv. India 22: 174 (1980 publ. 1982).
Assam. 40 ASS. Nanophan. or phan.
**Eugenia anisopetala* R.Parker, Repert. Spec. Nov. Regni Veg. 31: 125 (1932).

Syzygium anisosepalum (Duthie) I.M.Turner, J. Singapore Natl. Acad. Sci. 22–24: 16 (1997).
Pen. Malaysia. 42 MLY. Phan.
**Eugenia anisosepala* Duthie in J.D.Hooker, Fl. Brit. India 2: 481 (1878).

Syzygium anomalum Lauterb., Nova Guinea 8: 853 (1912).
New Guinea. 43 NWG. Phan.

Syzygium anthicoides P.S.Ashton, Kew Bull. 61: 108 (2006).
Borneo (Sarawak). 42 BOR. Phan.

Syzygium anthicum (Ridl.) Merr. & L.M.Perry, Mem. Amer. Acad. Arts 18: 165 (1939).
Borneo. 42 BOR. Phan.
**Eugenia anthica* Ridl., J. Bot. 68: 11 (1930).

Syzygium antisepticum (Blume) Merr. & L.M.Perry, Mem. Amer. Acad. Arts 18: 159 (1939).
Assam to Philippines. 40 ASS 41 AND MYA NCB THA 42 BOR JAW MLY PHI SUM. Phan.
Calyptranthes aromatica Blume, Bijdr.: 1092 (1826), nom. illeg. **Caryophyllus antisepticus* Blume in A.P.de Candolle, Prodr. 3: 262 (1828). *Eugenia antiseptica* (Blume) Kuntze, Revis. Gen. Pl. 1: 238 (1891). *Jambosa aromatica* Miq., Fl. Ned. Ind. 1(1): 436 (1855).
Eugenia grata Wight, Ill. Ind. Bot. 2: 15 (1841). *Acmena grata* (Wight) Walp., Repert. Bot. Syst. 2: 181 (1843). *Syzygium gratum* (Wight) S.N.Mitra, Indian Forester 99: 100 (1973).
Myrtus quadrangularis Buch.-Ham. ex Duthie in J.D.Hooker, Fl. Brit. India 2: 486 (1878).
Eugenia cuprea Koord. & Valeton, Bull. Inst. Bot. Buitenzorg 2: 6 (1899), nom. illeg.
Eugenia glaucicalyx Merr., Publ. Bur. Sci. Gov. Lab. 35: 50 (1905). *Syzygium glaucicalyx* (Merr.) Merr., Philipp. J. Sci. 79: 391 (1951).
Eugenia collinsae Craib, Bull. Misc. Inform. Kew 1928: 237 (1928).
Syzygium ovatifolium Merr. & L.M.Perry, Mem. Amer. Acad. Arts 18: 161 (1939).
Syzygium gratum var. *confertum* Chantaran. & J.Parn., Kew Bull. 48: 599 (1993).

Syzygium antonianum (Elmer) Merr., Philipp. J. Sci. 79: 375 (1951).
Philippines. 42 PHI. Phan.
**Eugenia antoniana* Elmer, Leafl. Philipp. Bot. 4: 1425 (1912).

Syzygium aoupinianum J.W.Dawson, in Fl. Nouv.-Caléd. 23: 127 (1999).
WC. New Caledonia. 60 NWC. Nanophan. or phan.

Syzygium apetiolatum J.W.Dawson, in Fl. Nouv.-Caléd. 23: 28 (1999).
C. New Caledonia. 60 NWC. Nanophan. or phan.

Syzygium apiarii P.S.Ashton, Kew Bull. 61: 108 (2006).
Borneo (NE. Sarawak). 42 BOR. Nanophan. or phan.

Syzygium apodophyllum (F.Muell.) B.Hyland, Austral. J. Bot., Suppl. Ser. 9: 49 (1983).
N. Queensland. 50 QLD. Phan.
**Eugenia apodophylla* F.Muell., Victorian Naturalist 8: 197 (1892).

Syzygium apodum Miq., Fl. Ned. Ind., Eerste Bijv.: 312 (1861).
Sumatera. 42 SUM. Phan.

Syzygium apoense (Elmer) Merr., Philipp. J. Sci. 79: 376 (1951).
Philippines (Mindanao). 42 PHI. Phan.
**Eugenia apoensis* Elmer, Leafl. Philipp. Bot. 4: 1401 (1912).

Syzygium aqueum (Burm.f.) Alston, Ann. Roy. Bot. Gard. (Peradeniya) 11: 204 (1929).
Trop. Asia to N. Queensland. 40 BAN SRL 41 MYA NCB tha 42 BOR JAW MLY MOL SUL SUM 43 NWG 50 QLD. Phan. Widely cultivated for its edible fruits (water rose-apple).
**Eugenia aquea* Burm.f., Fl. Indica: 114 (1768). *Jambosa aquea* (Burm.f.) DC., Prodr. 3: 288 (1828). *Malidra aquea* (Burm.f.) Raf., Sylva Tellur.: 107 (1838). *Cerocarpus aqueus* (Burm.f.) Hassk., Flora 25(Beibl. 2): 36 (1842).
Eugenia nodiflora Aubl., Hist. Pl. Guiane 2: 140 (1775).
Eugenia malaccensis Lour., Fl. Cochinch.: 306 (1790), nom. illeg.
Myrtus obtusissima Blume, Bijdr.: 1086 (1826). *Jambosa obtusissima* (Blume) DC., Prodr. 3: 287 (1828).
Eugenia alba Roxb., Fl. Ind. ed. 1832, 2: 493 (1832). *Jambosa alba* (Roxb.) G.Don, Gen. Hist. 2: 868 (1832).
Myrtus timorensis Zipp. ex Span., Linnaea 15: 204 (1841).
Eugenia obversa Miq., Anal. Bot. Ind. 1: 18 (1850). *Syzygium obversum* (Miq.) Masam., Enum. Phan. Born.: 535 (1942).
Gelpkea stipularis Blume, Mus. Bot. 1: 88 (1850). *Eugenia stipularis* (Blume) Miq., Fl. Ned. Ind. 1(1): 441 (1855).
Jambosa ambigua Blume, Mus. Bot. 1: 96 (1850).
Jambosa madagascariensis Blume, Mus. Bot. 1: 103 (1850).
Jambosa timorensis Blume, Mus. Bot. 1: 97 (1850).
Jambosa calophylla Miq., Fl. Ned. Ind. 1(1): 423 (1855). *Eugenia callophylla* (Miq.) Reinw. ex de Vriese, Pl. Ind. Bat. Orient.: 69 (1856).
Jambosa subsessilis Miq., Fl. Ned. Ind., Eerste Bijv.: 309 (1861).
Eugenia mindanaensis C.B.Rob., Philipp. J. Sci., C 4: 363 (1909).

Syzygium araiocladum Merr. & L.M.Perry, J. Arnold Arbor. 19: 225 (1938).
SE. China. 36 CHS. Nanophan. or phan.

Syzygium arboreum (Baker f.) J.W.Dawson, in Fl. Nouv.-Caléd. 23: 30 (1999).
New Caledonia. 60 NWC. Phan.
**Eugenia arborea* Baker f., J. Linn. Soc., Bot. 45: 315 (1921).
Caryophyllus undulatus Guillaumin, Bull. Soc. Bot. France 85: 649 (1938 publ. 1939).

Syzygium arcanum P.S.Ashton, Kew Bull. 61: 108 (2006).
Borneo (Sarawak). 42 BOR. Phan.

Syzygium arcuatinervium (Merr.) Craven & Biffin, Blumea 51: 135 (2006).
Philippines. 42 PHI. Phan.
**Eugenia arcuatinervia* Merr., Philipp. J. Sci. 1(Suppl.): 104 (1906). *Cleistocalyx arcuatinervius* (Merr.) Merr. & L.M.Perry, J. Arnold Arbor. 18: 333 (1937).
Eugenia sibulanensis Elmer, Leafl. Philipp. Bot. 10: 3808 (1939), pro syn.

Syzygium arenitense Craven, Beagle 19: 87 (2003).
Northern Territory. 50 NTA. Phan.

Syzygium argyrocalyx (Warb.) Merr. & L.M.Perry, J. Arnold Arbor. 23: 264 (1942).
New Guinea (Kep. Aru). 43 NWG. Phan.
**Eugenia argyrocalyx* Warb., Bot. Jahrb. Syst. 13: 390 (1891).

Syzygium argyropedicum B.Hyland, Austral. J. Bot., Suppl. Ser. 9: 52 (1983).
N. Queensland. 50 QLD. Phan.

Syzygium armstrongii (Benth.) B.Hyland, Austral. J. Bot., Suppl. Ser. 9: 54 (1983).
N. Western Australia to Northern Territory. 50 NTA WAU. Phan.
**Eugenia armstrongii* Benth., Fl. Austral. 3: 286 (1867).

Syzygium aromaticum (L.) Merr. & L.M.Perry, Mem. Amer. Acad. Arts 18: 196 (1939).
Maluku. (25) tan (29) sey 42 MOL. Nanophan. or phan. Widely cultivated for cloves.
**Caryophyllus aromaticus* L., Sp. Pl.: 515 (1753). *Eugenia aromatica* (L.) Baill., Hist. Pl. 6: 311 (1876), nom. illeg.
Eugenia caryophyllata Thunb., Caryoph. Arom.: 1 (1788). *Jambosa caryophyllus* (Thunb.) Nied. in H.G.A.Engler & K.A.E.Prantl, Nat. Pflanzenfam. 3(7): 85 (1893).
Caryophyllus hortensis Noronha, Verh. Batav. Genootsch. Kunsten 5(4): 11 (1790).
Myrtus caryophyllus Spreng., Syst. Veg. 2: 485 (1825). *Eugenia caryophyllus* (Spreng.) Bullock & S.G.Harrison, Kew Bull. 13: 52 (1958).
Caryophyllus silvestris Teijsm. ex Hassk., Abh. Naturf. Ges. Halle 9: 167 (1866).

Syzygium assamicum (Biswas & Purkay.) Raizada, Indian Forester 74: 336 (1948).
Assam. 40 ASS. Nanophan. or phan.
**Eugenia assamica* Biswas & Purkay., Bull. Misc. Inform. Kew 1938: 262 (1938).

Syzygium assimile Thwaites, Enum. Pl. Zeyl.: 116 (1859). *Eugenia assimilis* (Thwaites) Bedd., Fl. Sylv. S. India: 107 (1872).
S. India, Sri Lanka. 40 IND SRL. Phan.

Syzygium astronioides (C.B.Rob.) Merr., Philipp. J. Sci. 79: 376 (1951).
Philippines. 42 PHI. Phan.
**Eugenia astronioides* C.B.Rob., Philipp. J. Sci., C 4: 393 (1909).
Eugenia bakeri Elmer, Leafl. Philipp. Bot. 7: 2355 (1914).

Syzygium attenuatum (Miq.) Merr. & L.M.Perry, Mem. Amer. Acad. Arts 18: 185 (1939).
Thailand to Philippines. 41 THA 42 BOR JAW MLY PHI. Phan.
**Jambosa attenuata* Miq., Fl. Ned. Ind. 1(1): 437 (1855). *Eugenia attenuata* (Miq.) Koord. & Valeton, Meded. Lands Plantentuin 40: 121 (1900).
Eugenia penangiana Duthie in J.D.Hooker, Fl. Brit. India 2: 486 (1878).
Eugenia purpuricarpa Elmer, Leafl. Philipp. Bot. 4: 1435 (1912).
Eugenia attenuata var. *montana* M.R.Hend., Gard. Bull. Singapore 12: 242 (1949). *Syzygium attenuatum* var. *montanum* (M.R.Hend.) Chantaran. & J.Parn., Kew Bull. 48: 593 (1993).
Eugenia attenuata var. *ophirensis* M.R.Hend., Gard. Bull. Singapore 12: 241 (1949). *Syzygium attenuatum* var. *ophirense* (M.R.Hend.) I.M.Turner, J. Singapore Natl. Acad. Sci. 22–24: 16 (1997).

Syzygium attopeuense (Gagnep.) Merr. & L.M.Perry, J. Arnold Arbor. 19: 107 (1938).
Laos. 41 LAO. Phan.
**Eugenia attopeuensis* Gagnep., Notul. Syst. (Paris) 3: 316 (1917).

Syzygium aurantiacum (H.Perrier) Labat & Schatz, Novon 12: 201 (2002).
NE. Madagascar. 29 MDG. Nanophan.
**Eugenia aurantiaca* H.Perrier, Mém. Inst. Sci. Madagascar, Sér. B, Biol. Vég. 4: 190 (1953).

Syzygium auriculatum Brongn. & Gris, Bull. Soc. Bot. France 12: 184 (1865).
NW. & C. New Caledonia. 60 NWC. Nanophan. or phan.

Syzygium australe (J.C.Wendl. ex Link) B.Hyland, Austral. J. Bot., Suppl. Ser. 9: 55 (1983).
E. Australia. 50 NSW QLD (51) nzn. Nanophan. or phan.
Eugenia myrtifolia Sims, Bot. Mag. 48: t. 2230 (1821), nom. illeg. *Jambosa myrtifolia* Heynh., Nom. Bot. Hort. 1: 428 (1840).
**Eugenia australis* J.C.Wendl. ex Link, Enum. Hort. Berol. Alt. 2: 28 (1822). *Myrtus australis* (J.C.Wendl. ex Link) Spreng., Syst. Veg. 2: 482 (1825). *Jambosa australis* (J.C.Wendl. ex Link) DC., Prodr. 3: 287 (1828).
Jambosa thozetiana F.Muell., Fragm. 1: 255 (1859).
Eugenia simmondsiae F.M.Bailey, Queensland Agric. J. 23: 297 (1909).

Syzygium austrocaledonicum (Seem.) Guillaumin, Bull. Soc. Bot. France 85: 643 (1938 publ. 1939).
New Caledonia. 60 NWC. Nanophan. or phan.
**Eugenia austrocaledonica* Seem., Fl. Vit.: 76 (1865).
Syzygium nitidum Brongn. & Gris, Bull. Soc. Bot. France 12: 183 (1865), nom. illeg. *Syzygium artense* Montrouz. ex Guillaumin & Beauvis., Ann. Soc. Bot. Lyon 38: 92 (1913 publ. 1914). *Eugenia grisii* Baker f., J. Linn. Soc., Bot. 45: 316 (1921).

Syzygium austrosinense (Merr. & L.M.Perry) H.T. Chang & R.H.Miao, Acta Bot. Yunnan. 4: 24 (1982).
SE. China. 36 CHS.
**Syzygium buxifolium* var. *austrosinense* Merr. & L.M.Perry, J. Arnold Arbor. 19: 236 (1938).

Syzygium austroyunnanense H.T.Chang & R.H.Miao, Acta Bot. Yunnan. 4: 17 (1982).
SC. China. 36 CHC.

Syzygium avene Miq., Fl. Ned. Ind., Eerste Bijv.: 312 (1861). *Eugenia avenis* (Miq.) M.R.Hend., Gard. Bull. Singapore 12: 171 (1949).
S. India, Pen. Malaysia to Sumatera. 40 IND 42 MLY SUM. Phan.
Eugenia scoparia Duthie in J.D.Hooker, Fl. Brit. India 2: 489 (1878). *Syzygium scoparium* (Duthie) Wall. ex Murugan & Manickam, J. Econ. Taxon. Bot. 28: 523 (2004), nom. nud.

Syzygium badescens P.S.Ashton, Kew Bull. 61: 111 (2006).
Borneo (Sabah). 42 BOR. Phan.

Syzygium badium Merr. & L.M.Perry, J. Arnold Arbor. 23: 281 (1942).
New Guinea. 43 NWG. Phan.

Syzygium baeuerlenii (F.Muell.) Craven & Biffin, Blumea 51: 135 (2006).
New Guinea. 43 NWG. Phan.
**Eugenia baeuerlenii* F.Muell., Australas. J. Pharm., June 1886: 19 (1886). *Cleistocalyx baeuerlenii* (F.Muell.) Merr. & L.M.Perry, J. Arnold Arbor. 18: 331 (1937).

Syzygium bakoense P.S.Ashton, Kew Bull. 61: 111 (2006).
Borneo (Sarawak). 42 BOR. Phan.

Syzygium baladense (Brongn. & Gris) J.W.Dawson, in Fl. Nouv.-Caléd. 23: 88 (1999).
New Caledonia. 60 NWC. Nanophan. or phan.
**Caryophyllus baladensis* Brongn. & Gris, Bull. Soc. Bot. France 12: 185 (1865). *Jambosa baladensis* (Brongn. & Gris) Nied. in H.G.A.Engler & K.A.E.Prantl, Nat. Pflanzenfam. 3(7): 84 (1893).
Caryophyllus vieillardii Lenorm. ex Guillaumin, Bull. Soc. Bot. France 85: 649 (1938 publ. 1939).

Syzygium balansae (Guillaumin) J.W.Dawson, in Fl. Nouv.-Caléd. 23: 24 (1999).
New Caledonia. 60 NWC. Phan.
**Caryophyllus balansae* Guillaumin, Bull. Soc. Bot. France 85: 648 (1938 publ. (1939).

Syzygium balerense (C.B.Rob.) Merr., Philipp. J. Sci. 79: 377 (1951).
Philippines. 42 PHI. Phan.
Eugenia brunnea C.B.Rob., Philipp. J. Sci., C 4: 372 (1909), nom. illeg. **Eugenia balerensis* C.B.Rob., Philipp. J. Sci., C 6: 346 (1911).

Syzygium balfourii (Baker) J.Guého & A.J.Scott, Kew Bull. 34: 488 (1980).
Rodrigues. 29 ROD†. Nanophan. or phan.
**Eugenia balfourii* Baker, Fl. Mauritius: 116 (1877).

Syzygium balsameum (Wight) Wall. ex Walp., Repert. Bot. Syst. 2: 179 (1843).
SC. China to Pen. Malaysia. 36 CHC 40 ASS BAN EHM 41 MYA THA VIE 42 MLY. Phan.
**Eugenia balsamea* Wight, Ill. Ind. Bot. 2: 16 (1841).

Syzygium bamagense B.Hyland, Austral. J. Bot., Suppl. Ser. 9: 56 (1983).
N. Queensland. 50 QLD. Phan.

Syzygium bankense (Hassk.) Merr. & L.M.Perry, Mem. Amer. Acad. Arts 18: 160 (1939).
Sumatera to Philippines. 42 BOR jaw PHI SUM. Phan.
Jambosa buxifolia Miq., Fl. Ned. Ind. 1(1): 1086 (1858). *Microjambosa besukiensis* Hassk. ex Miq., Fl. Ned. Ind., Eerste Bijv.: 311 (1861).

**Microjambosa bankensis* Hassk., Hort. Bogor. Descr.: 276 (1858). *Eugenia bankensis* (Hassk.) Backer, Schoolfl. Java: 508 (1911).
Eugenia besukiensis Merr., J. Straits Branch Roy. Asiat. Soc. 77: 226 (1917). *Syzygium besukiense* (Merr.) Masam., Enum. Phan. Born.: 524 (1942).

Syzygium banksii (Britten & S.Moore) B.Hyland, Austral. J. Bot., Suppl. Ser. 9: 58 (1983).
N. Queensland. 50 QLD. Phan.
**Eugenia banksii* Britten & S.Moore, J. Bot. 40: 26 (1902).

Syzygium baramense (Merr.) Merr. & L.M.Perry, Mem. Amer. Acad. Arts 18: 189 (1939).
Borneo. 42 BOR. Phan.
**Eugenia baramensis* Merr., J. Straits Branch Roy. Asiat. Soc. 77: 218 (1917).

Syzygium barnesii (Merr.) Merr., Philipp. J. Sci. 79: 377 (1951).
Philippines. 42 PHI. Phan.
**Jambosa barnesii* Merr., Publ. Bur. Sci. Gov. Lab. 17: 37 (1904). *Eugenia barnesii* (Merr.) Merr., Philipp. J. Sci. 1(Suppl.): 104 (1906).

Syzygium barringtonioides (Ridl.) Masam., Enum. Phan. Born.: 524 (1942).
Borneo. 42 BOR. Phan.
**Eugenia barringtonioides* Ridl., J. Bot. 68: 12 (1930). *Cleistocalyx barringtonioides* (Ridl.) Merr. & L.M.Perry, J. Arnold Arbor. 18: 332 (1937).

var. ***barringtonioides***
Borneo. 42 BOR. Phan.

var. ***quadrisepalum*** P.S.Ashton, Kew Bull. 61: 113 (2006).
Borneo. 42 BOR. Phan.

Syzygium bartonii (F.M.Bailey) Merr. & L.M.Perry, J. Arnold Arbor. 23: 250 (1942).
New Guinea. 43 NWG. Phan.
**Eugenia bartonii* F.M.Bailey, Proc. Roy. Soc. Queensland 18: 2 (1902). *Jambosa bartonii* (F.M.Bailey) Diels, Bot. Jahrb. Syst. 57: 383 (1922).

Syzygium bataanense (Merr.) Merr., Philipp. J. Sci. 79: 377 (1951).
Philippines. 42 PHI. Phan.
**Jambosa bataanensis* Merr., Publ. Bur. Sci. Gov. Lab. 17: 36 (1904). *Eugenia bataanensis* (Merr.) Merr., Philipp. J. Sci. 1(Suppl.): 104 (1906).

Syzygium batadamba Kosterm., Quart. J. Taiwan Mus. 34: 132 (1981).
Sri Lanka. 40 SRL.

Syzygium baviense (Gagnep.) Merr. & L.M.Perry, J. Arnold Arbor. 19: 102 (1938).
SC. China to Vietnam. 36 CHC 41 VIE.
**Eugenia baviensis* Gagnep., Notul. Syst. (Paris) 3: 317 (1917).

Syzygium beccarii (Ridl.) Merr. & L.M.Perry, Mem. Amer. Acad. Arts 18: 170 (1938).
Borneo, S. Sulawesi. 42 BOR SUL. Phan.
**Eugenia beccarii* Ridl., J. Bot. 68: 12 (1930).

Syzygium beddomei (Duthie) Chithra in N.C.Nair & A.N.Henry, Fl. Tamil Nadu 1: 155 (1983).
S. India. 40 IND. Phan.
**Eugenia beddomei* Duthie in J.D.Hooker, Fl. Brit. India 2: 476 (1878). *Jambosa beddomei* (Duthie) Gamble, Fl. Madras: 475 (1919).

Syzygium benguellense (Welw. ex Hiern) Engl. in H.G.A.Engler & C.G.O.Drude, Veg. Erde 9(3; 2): 737 (1921).
Angola. 23 ZAI? 26 ANG. Phan.
**Eugenia benguellensis* Welw. ex Hiern, Cat. Afr. Pl. 1: 360 (1898).

Syzygium benguetense (C.B.Rob.) Merr., Philipp. J. Sci. 79: 377 (1951).
Philippines. 42 PHI. Phan.
**Eugenia benguetensis* C.B.Rob., Philipp. J. Sci., C 4: 374 (1909).

Syzygium benjaminum Diels, Bot. Jahrb. Syst. 57: 411 (1922).
New Guinea. 43 NWG. Phan.
Syzygium micropetalum Merr. & L.M.Perry, J. Arnold Arbor. 23: 293 (1942).

Syzygium bernardoi (Merr.) Merr., Philipp. J. Sci. 79: 378 (1951).
Philippines. 42 PHI. Phan.
**Eugenia bernardoi* Merr., Philipp. J. Sci. 18: 304 (1921).

Syzygium bernieri (Baill. ex Drake) Labat & Schatz, Novon 12: 202 (2002).
E. Madagascar. 29 MDG. Phan.
**Eugenia bernieri* Baill. ex Drake, Bull. Mens. Soc. Linn. Paris 2: 1221 (1896).
Eugenia chapelieri Baill. ex Drake, Bull. Mens. Soc. Linn. Paris 2: 1222 (1896).
Eugenia maroana Aug.DC., Bull. Herb. Boissier, II, 1: 573 (1901).
Eugenia bernieri var. *lateriticolor* H.Perrier, Mém. Inst. Sci. Madagascar, Sér. B, Biol. Vég. 4: 191 (1952).
Eugenia bernieri var. *oblanceolata* H.Perrier, Mém. Inst. Sci. Madagascar, Sér. B, Biol. Vég. 4: 191 (1952).
Eugenia bernieri var. *tampinensis* H.Perrier, Mém. Inst. Sci. Madagascar, Sér. B, Biol. Vég. 4: 192 (1952).

Syzygium bicolor Merr. & L.M.Perry, J. Arnold Arbor. 23: 286 (1942).
New Guinea. 43 NWG.

Syzygium bicostatum P.S.Ashton, Kew Bull. 61: 113 (2006).
Borneo (Sarawak). 42 BOR. Phan.

Syzygium bijouxii J.Guého & A.J.Scott, Kew Bull. 34: 491 (1980).
Mauritius. 29 MAU. Nanophan.

Syzygium blancoi (Merr.) Merr., Philipp. J. Sci. 79: 378 (1951).
Philippines. 42 PHI.
**Eugenia blancoi* Merr., Philipp. J. Sci., C 10: 208 (1915).

Syzygium blumei (Steud.) Merr. & L.M.Perry, Mem. Amer. Acad. Arts 18: 164 (1939).
W. Jawa. 42 JAW.
Eugenia angustifolia Blume, Flora 7: 291 (1824), nom. illeg.
Myrtus hypericifolia Blume, Bijdr.: 1082 (1826), nom. illeg. *Jambosa hypericifolia* DC., Prodr. 3: 287 (1828). *Eugenia hypericifolia* (DC.) Koord. & Valeton, Meded. Lands Plantentuin 40: 69 (1900), nom. illeg. **Eugenia blumei* Steud., Nomencl. Bot., ed. 2, 1: 601 (1840).

Syzygium boerlagei (Merr.) Govaerts, World Checklist Myrtaceae: 377 (2008).
Maluku. 42 MOL. Nanophan. or phan.
**Eugenia boerlagei* Merr., Philipp. J. Sci., C 11: 296 (1916).

Syzygium boisianum (Gagnep.) Merr. & L.M.Perry, J. Arnold Arbor. 19: 115 (1938).
Hainan to Indo-China. 36 CHH 41 THA VIE.
**Eugenia boisiana* Gagnep., Notul. Syst. (Paris) 3: 318 (1917).

subsp. ***boisianum***
Hainan, Vietnam. 36 CHH 41 VIE.

subsp. ***longifolium*** Chantaran. & J.Parn., Kew Bull. 48: 593 (1993).
Pen. Thailand. 41 THA.

Syzygium bonii (Gagnep.) Merr. & L.M.Perry, J. Arnold Arbor. 19: 102 (1938).
Vietnam. 41 VIE.
**Eugenia bonii* Gagnep., Notul. Syst. (Paris) 3: 318 (1917).

Syzygium boonjee B.Hyland, Austral. J. Bot., Suppl. Ser. 9: 59 (1983).
N. Queensland. 50 QLD. Phan.

Syzygium borbonicum J.Guého & A.J.Scott, Kew Bull. 34: 489 (1980).
Réunion. 29 REU. Phan.
**Eugenia paniculata* Lam., Encycl. 3: 199 (1789), nom. illeg. *Calyptranthes paniculata* (Lam.) Raeusch., Nomencl. Bot., ed. 3: 144 (1797). *Syzygium paniculatum* DC., Prodr. 3: 259 (1828), nom. illeg.

Syzygium bordenii (Merr.) Merr., Philipp. J. Sci. 79: 378 (1951).
Philippines. 42 PHI. Phan.
**Eugenia bordenii* Merr., Publ. Bur. Sci. Gov. Lab. 35: 47 (1905).

Syzygium borneense (Miq.) Miq., Fl. Ned. Ind. 1(1): 453 (1855).
Indo-China to W. Malesia. 41 THA VIE 42 BOR MLY. Phan.
**Eugenia borneensis* Miq., Anal. Bot. Ind. 1: 24 (1850). *Syzygium myrtillus* var. *borneense* (Miq.) Chantaran. & J.Parn., Kew Bull. 48: 605 (1993).
Syzygium glaucescens Miq., Anal. Bot. Ind. 1: 24 (1850), nom. inval.
Eugenia microcalyx Duthie in J.D.Hooker, Fl. Brit. India 2: 493 (1878).
Eugenia pseudosubtilis King, J. Asiat. Soc. Bengal, Pt. 2, Nat. Hist. 70: 123 (1901).
Eugenia pseudosubtilis var. *platyphylla* King, J. Asiat. Soc. Bengal, Pt. 2, Nat. Hist. 70(2): 124 (1901).
Eugenia litseifolia Merr., J. Straits Branch Roy. Asiat. Soc. 77: 215 (1917). *Syzygium litseifolium* (Merr.) Merr. & L.M.Perry, Mem. Amer. Acad. Arts 18: 191 (1939).
Syzygium hackenbergii Diels, Bot. Jahrb. Syst. 60: 312 (1926).
Eugenia irregularis Craib, Bull. Misc. Inform. Kew 1930: 167 (1930). *Syzygium irregulare* (Craib) Merr. & L.M.Perry, J. Arnold Arbor. 19: 107 (1938).

Syzygium boulindaense J.W.Dawson, in Fl. Nouv.-Caléd. 23: 91 (1999).
WC. New Caledonia (Mt. Boulinda). 60 NWC. Cham.

Syzygium bourdillonii (Gamble) Rathakr. & N.C.Nair, J. Econ. Taxon. Bot. 4: 287 (1983).
SW. India. 40 IND. Phan.
**Jambosa bourdillonii* Gamble, Bull. Misc. Inform. Kew 1918: 239 (1918).

Syzygium brachiatum (Roxb.) Miq., Fl. Ned. Ind. 1(1): 460 (1855).
Maluku. 42 MOL. Phan.
**Eugenia brachiata* Roxb., Fl. Ind. ed. 1832, 2: 488 (1832).
Eugenia odorata Wight, Ill. Ind. Bot. 2: 16 (1841).

Syzygium brachyanthelium Diels, Bot. Jahrb. Syst. 57: 412 (1922).
New Guinea. 43 NWG. Phan.

Syzygium brachybotryum Miq., Fl. Ned. Ind. 1(1): 452 (1855).
Sumatera. 42 SUM. Phan.

Syzygium brachycalyx (Baker f.) J.W.Dawson, in Fl. Nouv.-Caléd. 23: 133 (1999).
WC. & C. New Caledonia. 60 NWC. Nanophan. or phan.
**Eugenia brachycalyx* Baker f., J. Linn. Soc., Bot. 45: 314 (1921).

Syzygium brachypodum Merr. & L.M.Perry, Mem. Amer. Acad. Arts 18: 179 (1939).
Borneo. 42 BOR. Phan.

Syzygium brachyrachis Merr. & L.M.Perry, Mem. Amer. Acad. Arts 18: 181 (1939).
Borneo. 42 BOR. Phan.

Syzygium brachythyrsum Merr. & L.M.Perry, J. Arnold Arbor. 19: 239 (1938).
SC. China. 36 CHC.

Syzygium brachyurum Merr., Philipp. J. Sci. 79: 379 (1951).
Philippines (Luzon). 42 PHI.

Syzygium brackenridgei (A.Gray) Müll.Stuttg. in W.G.Walpers, Ann. Bot. Syst. 4: 838 (1858).
Fiji, Tonga. 60 FIJ TON. Phan.
**Eugenia brackenridgei* A.Gray, U.S. Expl. Exped., Phan. 1: 521 (1854). *Pareugenia brackenridgei* (A.Gray) A.C.Sm., Bernice P. Bishop Mus. Bull. 141: 109 (1936).
Pareugenia imthurnii Turrill, Hooker's Icon. Pl. 31: t. 3001 (1915). *Syzygium imthurnii* (Turrill) Merr. & L.M.Perry, Sargentia 1: 75 (1942).

Syzygium bracteosum Merr. & L.M.Perry, J. Arnold Arbor. 23: 250 (1942).
New Guinea. 43 NWG.

Syzygium branderhorstii Lauterb., Nova Guinea 8: 322 (1910).
New Guinea to N. Queensland. 43 NWG 50 QLD. Phan.
Syzygium acetosum Merr. & L.M.Perry, J. Arnold Arbor. 23: 280 (1942).
Syzygium leptophlebioides Merr. & L.M.Perry, J. Arnold Arbor. 23: 282 (1942).

Syzygium brassii Merr. & L.M.Perry, J. Arnold Arbor. 23: 277 (1942).
Papua New Guinea. 43 NWG.

Syzygium brazzavillense Aubrév. & Pellegr., Notul. Syst. (Paris) 14: 61 (1950).
Congo. 23 CON.

Syzygium brevicymum (Diels) Merr. & L.M.Perry, J. Arnold Arbor. 23: 261 (1942).
New Guinea to Bismarck Arch. 43 BIS NWG. Phan.
**Jambosa brevicyma* Diels, Bot. Jahrb. Syst. 57: 389 (1922).
Syzygium caudiferum Merr. & L.M.Perry, J. Arnold Arbor. 23: 272 (1942).

Syzygium brevifolium (A.Gray) Müll.Stuttg. in W.G. Walpers, Ann. Bot. Syst. 4: 839 (1858).
Samoa. 60 SAM. Phan.
**Eugenia brevifolia* A.Gray, U.S. Expl. Exped., Phan. 1: 531 (1854).
Eugenia oreophila Rech., Denkschr. Kaiserl. Akad. Wiss., Wien. Math.-Naturwiss. Kl. 85: 145 (1910), nom. illeg.

Syzygium brevioperculatum J.W.Dawson, in Fl. Nouv.-Caléd. 23: 18 (1999).
C. New Caledonia. 60 NWC. Nanophan. or phan.

Syzygium brevipaniculatum (Merr.) Merr., Philipp. J. Sci. 79: 380 (1951).
Philippines. 42 PHI.

Syzygium brevipes (Brongn. & Gris) J.W.Dawson, in Fl. Nouv.-Caléd. 23: 101 (1999).
NW. New Caledonia. 60 NWC. Nanophan. or phan.
**Syzygium tenuiflorum* var. *brevipes* Brongn. & Gris, Bull. Soc. Bot. France 12: 183 (1865).

Syzygium brittonianum (C.B.Rob.) Merr., Philipp. J. Sci. 79: 380 (1951).
Philippines. 42 PHI.
**Eugenia brittoniana* C.B.Rob., Philipp. J. Sci., C 4: 398 (1909).

Syzygium brongniartii (Merr. & L.M.Perry) J.W. Dawson, in Fl. Nouv.-Caléd. 23: 16 (1999).
New Caledonia. 60 NWC. Phan.
Acicalyptus nitida Brongn. & Gris, Bull. Soc. Bot. France 12: 186 (1865). **Cleistocalyx brongniartii* Merr. & L.M.Perry, J. Arnold Arbor. 18: 337 (1937).

Syzygium brousmichei Govaerts, World Checklist Myrtaceae: 378 (2008).
New Caledonia. 60 NWC. Nanophan. or phan.
**Eugenia costata* Brongn. & Gris, Bull. Soc. Bot. France 12: 178 (1865), nom. illeg. *Eugenia pauper* Guillaumin, Bull. Mus. Natl. Hist. Nat., II, 5: 245 (1933), nom. illeg.

Syzygium bruynii (Diels) Merr. & L.M.Perry, J. Arnold Arbor. 23: 249 (1942).
New Guinea. 43 NWG.
**Jambosa bruynii* Diels, Nova Guinea 14: 92 (1924).

Syzygium bubengense C.Chen, Harvard Pap. Bot. 11: 25 (2006).
China (Yunnan). 36 CHC.

Syzygium buettnerianum (K.Schum.) Nied. in H.G.A. Engler & K.A.E.Prantl, Nat. Pflanzenfam. 3(7): 85 (1893).
Papuasia to Vanuatu. 43 NWG SOL 50 QLD 60 VAN. Phan.
**Eugenia buettneriana* K.Schum. in K.M.Schumann & U.M.Hollrung, Fl. Kais. Wilh. Land: 89 (1889).

Syzygium bujangii P.S.Ashton, Kew Bull. 61: 113 (2006).
Borneo (Brunei, Sarawak). 42 BOR. Phan.

Syzygium bullockii (Hance) Merr. & L.M.Perry, J. Arnold Arbor. 19: 107 (1938).
SE. China to Indo-China. 36 CHH CHS 41 LAO VIE.
**Eugenia bullockii* Hance, J. Bot. 16: 227 (1878).

Syzygium bungadinnia (F.M.Bailey) B.Hyland, Austral. J. Bot., Suppl. Ser. 9: 64 (1983).
N. Queensland. 50 QLD. Phan.
**Eugenia bungadinnia* F.M.Bailey, Queensland Agric. J. 2: 285 (1898).

Syzygium burepense T.G.Hartley & L.M.Perry, J. Arnold Arbor. 54: 184 (1973).
Papua New Guinea. 43 NWG. Phan.

Syzygium burkillianum (King) I.M.Turner, J. Singapore Natl. Acad. Sci. 22–24: 16 (1997).
Pen. Malaysia. 42 MLY. Phan.
**Eugenia burkilliana* King, J. Asiat. Soc. Bengal, Pt. 2, Nat. Hist. 70: 94 (1901).
Eugenia burkilliana var. *garcinifolioides* M.R.Hend., Gard. Bull. Singapore 12: 118 (1949). *Syzygium burkillianum* var. *garcinifolioides* (M.R.Hend.) I.M.Turner, J. Singapore Natl. Acad. Sci. 22–24: 16 (1997).

Syzygium busuense T.G.Hartley & L.M.Perry, J. Arnold Arbor. 54: 206 (1973).
Papua New Guinea. 43 NWG. Phan.

Syzygium buxifolioideum H.T.Chang & R.H.Miao, Acta Bot. Yunnan. 4: 20 (1982).
Hainan. 36 CHH.

Syzygium buxifolium Hook. & Arn., Bot. Beechey Voy.: 187 (1833). *Syllysium buxifolium* (Hook. & Arn.) Meyen & Schauer, Nov. Actorum Acad. Caes. Leop.-Carol. Nat. Cur. 19(Suppl. 1): 334 (1843). *Eugenia sinensis* Hemsl., J. Linn. Soc., Bot. 23: 298 (1887).
Japan (S. Kyushu) to Vietnam. 36 CHS 38 JAP NNS TAI 41 VIE. Nanophan. or phan.
Eugenia microphylla Abel, Narr. Journey China: 364 (1818).
Eugenia pyxophylla Hance, J. Bot. 9: 6 (1871).
Eugenia somai Hayata in R.Kanehira, Formos. Trees: 255 (1918). *Syzygium somai* (Hayata) Mori, Trans. Nat. Hist. Soc. Taiwan 28: 440 (1938).
Syzygium buxifolium var. *verticillatum* C.Chen, Harvard Pap. Bot. 11: 25 (2006).

Syzygium cacuminis (Craib) Chantaran. & J.Parn., Kew Bull. 48: 595 (1993).
Thailand. 41 THA. Phan.
**Eugenia cacuminis* Craib, Bull. Misc. Inform. Kew 11930: 165 (1930).

subsp. ***cacuminis***
Pen. Thailand. 41 THA. Phan.

subsp. ***inthanonense*** Chantaran. & J.Parn., Kew Bull. 48: 595 (1993).
N. Thailand. 41 THA. Phan.

Syzygium cagayanense (Merr.) Merr., Philipp. J. Sci. 79: 380 (1951).
Philippines. 42 PHI. Phan.
**Eugenia cagayanensis* Merr., Philipp. J. Sci., C 10: 214 (1915).

Syzygium calcadense (Bedd.) Chandrash., Biol. Membr. Abstr. 2: 57 (1977).
India. 40 IND. Nanophan. or phan.
**Eugenia calcadensis* Bedd., Fl. Sylv. S. India: 110 (1872).

Syzygium calcicola (Merr.) Merr., Philipp. J. Sci. 79: 380 (1951).

Philippines. 42 PHI. Phan.
Eugenia calcicola Merr., Philipp. J. Sci., C 10: 209 (1915).

Syzygium calcimontanum P.S.Ashton, Kew Bull. 61: 114 (2006).
Borneo (Sarawak). 42 BOR. Phan.

Syzygium calleryanum (C.B.Rob.) Merr., Philipp. J. Sci. 79: 380 (1951).
Philippines. 42 PHI. Phan.
**Eugenia calleryana* C.B.Rob., Philipp. J. Sci., C 6: 348 (1911).

Syzygium callianthum Merr. & L.M.Perry, J. Arnold Arbor. 23: 252 (1942).
New Guinea. 43 NWG. Phan.

Syzygium calophyllifolium (Wight) Walp., Repert. Bot. Syst. 2: 180 (1843).
S. India, Sri Lanka. 40 IND SRL. Phan.
**Eugenia calophyllifolia* Wight, Icon. Pl. Ind. Orient. 3: t. 1000 (1845).

Syzygium calubcob (C.B.Rob.) Merr., Philipp. J. Sci. 79: 380 (1951).
Philippines to N. Sulawesi. 42 PHI SUL. Phan.
**Eugenia calubcob* C.B.Rob., Philipp. J. Sci., C 4: 364 (1909).

Syzygium calyptrocalyx P.S.Ashton, Blumea 51: 136 (2006).
NW. Borneo. 42 BOR. Phan.
**Cleistocalyx leucocladus* Merr. & L.M.Perry, J. Arnold Arbor. 18: 336 (1937).

Syzygium cambodianum (Gagnep.) Merr. & L.M.Perry, J. Arnold Arbor. 19: 105 (1938).
Cambodia. 41 CBD. Phan.
**Eugenia cambodiana* Gagnep., Notul. Syst. (Paris) 3: 319 (1917).

Syzygium cameronum I.M.Turner, J. Singapore Natl. Acad. Sci. 22–24: 16 (1997).
Pen. Malaysia (Pahang). 42 MLY. Phan.
**Eugenia goniocalyx* Ridl., Fl. Malay Penins. 5: 309 (1925).

Syzygium camptodromum Merr. & L.M.Perry, J. Arnold Arbor. 23: 285 (1942).
Solomon Is. 43 SOL. Phan.

Syzygium camptophyllum (M.R.Hend.) I.M.Turner, J. Singapore Natl. Acad. Sci. 22–24: 16 (1997).
Pen. Malaysia. 42 MLY. Phan.
**Eugenia camptophylla* M.R.Hend., Gard. Bull. Singapore 11: 311 (1947).

Syzygium campylocarpum (Gagnep.) Merr. & L.M. Perry, J. Arnold Arbor. 19: 116 (1938).
Laos. 41 LAO. Phan.
**Eugenia campylocarpa* Gagnep., Notul. Syst. (Paris) 3: 320 (1917).

Syzygium candelabriforme (C.B.Rob.) Merr., Philipp. J. Sci. 79: 380 (1951).
Philippines. 42 PHI. Phan.
**Eugenia candelabriformis* C.B.Rob., Philipp. J. Sci., C 4: 375 (1909).

Syzygium canicortex B.Hyland, Austral. J. Bot., Suppl. Ser. 9: 66 (1983).
N. Queensland. 50 QLD. Phan.

Syzygium capillaceum (Brongn. & Gris) J.W.Dawson, in Fl. Nouv.-Caléd. 23: 121 (1999).
New Caledonia. 60 NWC. Nanophan. or phan.
**Syzygium tenuiflorum* var. *capillaceum* Brongn. & Gris, Bull. Soc. Bot. France 13: 469 (1866).

Syzygium capitatum (Merr.) Merr. & L.M.Perry, Mem. Amer. Acad. Arts 18: 150 (1939).
Borneo. 42 BOR. Phan.
**Eugenia capitata* Merr., J. Straits Branch Roy. Asiat. Soc. 77: 208 (1917).

Syzygium capituliferum Merr. & L.M.Perry, J. Arnold Arbor. 23: 287 (1942).
New Guinea. 43 NWG. Phan.

Syzygium capoasense (Merr.) Merr., Philipp. J. Sci. 79: 381 (1951).
Philippines. 42 PHI. Phan.
**Eugenia capoasensis* Merr., Philipp. J. Sci., C 10: 209 (1915).

Syzygium cardiophyllum (Merr.) Merr., Philipp. J. Sci. 79: 381 (1951).
Philippines. 42 PHI. Phan.
**Eugenia cardiophylla* Merr., Philipp. J. Sci. 18: 305 (1921).

Syzygium caroli Diels, Bot. Jahrb. Syst. 57: 404 (1922).
New Guinea. 43 NWG. Phan.

Syzygium carolinense (Koidz.) Hosok., J. Jap. Bot. 16: 542 (1940).
Caroline Is. (Pohnpei), Samoa. 60 SAM 62 CRL. Phan.
**Eugenia carolinensis* Koidz., Bot. Mag. (Tokyo) 30: 402 (1916).
Eugenia ponapensis Merr. & Kaneh., Trans. Nat. Hist. Soc. Taiwan 6: 43 (1916). *Syzygium ponapense* (Merr. & Kaneh.) Diels, Bot. Jahrb. Syst. 56: 533 (1921).

Syzygium carrii T.G.Hartley & L.M.Perry, J. Arnold Arbor. 54: 209 (1973).
W. New Guinea. 43 NWG. Phan.

Syzygium cartilagineum Merr. & L.M.Perry, J. Arnold Arbor. 23: 287 (1942).
Solomon Is. 43 SOL. Phan.

Syzygium caryophyllatum (L.) Alston in H.Trimen, Handb. Fl. Ceylon 6(Suppl.): 116 (1931).
S. India, Sri Lanka. 40 IND SRL. Nanophan. or phan.
**Myrtus caryophyllata* L., Sp. Pl.: 472 (1753). *Calyptranthes caryophyllata* (L.) Pers., Syn. Pl. 2: 32 (1806). *Jambosa caryophyllata* (L.) Bedevian, Ill. Polyglot. Dict.: 266 (1936), nom. inval.
Syzygium caryophyllaeum Gaertn., Fruct. Sem. Pl. 1: 166 (1788).
Eugenia corymbosa Lam., Encycl. 3: 199 (1789).
Eugenia caryophyllaea Wight, Icon. Pl. Ind. Orient. 2: t. 540 (1842).

Syzygium caryophylliflorum (Ridl.) Merr. & L.M.Perry, Mem. Amer. Acad. Arts 18: 173 (1939).
Borneo. 42 BOR. Phan.
**Eugenia caryophylliflora* Ridl., J. Bot. 68: 15 (1930).

Syzygium caryophylloides (Lauterb.) Merr. & L.M.Perry, J. Arnold Arbor. 23: 264 (1942).
Bismarck Arch. 43 BIS. Phan.
**Jambosa caryophylloides* Lauterb., Bot. Jahrb. Syst. 45: 363 (1911).

Syzygium casiguranense (Quisumb.) Merr., Philipp. J. Sci. 79: 381 (1951).
Philippines. 42 PHI. Phan.
**Eugenia casiguranensis* Quisumb., Philipp. J. Sci. 41: 337 (1930).

Syzygium castaneum (Merr.) Merr. & L.M.Perry, Mem. Amer. Acad. Arts 18: 156 (1939).
Pen. Malaysia, Borneo. 42 BOR MLY. Phan.
**Eugenia castanea* Merr., J. Straits Branch Roy. Asiat. Soc. 77: 212 (1917).

subsp. ***altecastaneum*** P.S.Ashton, Kew Bull. 61: 116 (2006).
Borneo (Sarawak). 42 BOR. Phan.

subsp. ***castaneum***
Pen. Malaysia, Borneo. 42 BOR MLY. Phan.

Syzygium cathayense Merr. & L.M.Perry, J. Arnold Arbor. 19: 232 (1938).
SE. China. 36 CHS. Phan.

Syzygium caudatilimbum (Merr.) Merr. & L.M.Perry, Mem. Amer. Acad. Arts 18: 181 (1939).
Borneo. 42 BOR. Phan.
**Eugenia caudatilimba* Merr., J. Straits Branch Roy. Asiat. Soc. 77: 216 (1917).
Eugenia verticilligera Ridl., J. Bot. 68: 12 (1930). *Syzygium verticilligerum* (Ridl.) Masam., Enum. Phan. Born.: 541 (1942).

Syzygium caudatum (Merr.) Airy Shaw, Kew Bull. 4: 122 (1949).
Pen. Malaysia, Borneo. 42 BOR MLY. Phan.
**Tetraeugenia caudata* Merr., J. Straits Branch Roy. Asiat. Soc. 77: 320 (1917).
Eugenia rhynchophylla Merr., J. Straits Branch Roy. Asiat. Soc. 79: 26 (1918). *Syzygium rhynchophyllum* (Merr.) Merr. & L.M.Perry, Mem. Amer. Acad. Arts 18: 189 (1939).
Syzygium aphanomyrtoides Merr. & L.M.Perry, Mem. Amer. Acad. Arts 18: 190 (1939).

Syzygium cauliflorum T.G.Hartley & L.M.Perry, J. Arnold Arbor. 54: 182 (1973).
Papua New Guinea. 43 NWG. Phan.
Myrtus cauliflora Blume, Bijdr.: 1086 (1826), nom. illeg. *Jambosa cauliflora* DC., Prodr. 3: 287 (1828). *Syzygium cauliflorum* (DC.) Bennet, Indian J. Forest. 1: 22 (1978), nom. illeg.

Syzygium cavitense Merr., Philipp. J. Sci. 79: 381 (1951).
Philippines (Luzon). 42 PHI. Phan.

Syzygium cephalophorum (Ridl.) Merr. & L.M.Perry, Mem. Amer. Acad. Arts 18: 150 (1939).
Borneo. 42 BOR. Phan.
**Eugenia cephalophora* Ridl., J. Bot. 68: 13 (1930).

Syzygium chaii P.S.Ashton, Kew Bull. 61: 116 (2006).
Borneo (Sarawak). 42 BOR. Phan.

Syzygium chamaebuxus Diels, Nova Guinea 14: 93 (1924).
New Guinea. 43 NWG.

Syzygium championii (Benth.) Merr. & L.M.Perry, J. Arnold Arbor. 19: 219 (1938).
SE. China to Hainan. 36 CHH CHS.
**Acmena championii* Benth., Hooker's J. Bot. Kew Gard. Misc. 4: 118 (1852). *Eugenia championii* (Benth.) Hemsl., J. Linn. Soc., Bot. 23: 296 (1887).
Eugenia henryi Hance, J. Bot. 23: 7 (1885).
Eugenia maclurei Merr., Philipp. J. Sci. 21: 350 (1922).

Syzygium chandrasekharanii Chandrab. & V.Chandras., J. Bombay Nat. Hist. Soc. 78: 354 (1981).
S. India. 40 IND. Phan.

Syzygium chanlos (Gagnep.) Merr. & L.M.Perry, J. Arnold Arbor. 19: 109 (1938).
Cambodia. 41 CBD. Phan.
**Eugenia chanlos* Gagnep., Notul. Syst. (Paris) 3: 320 (1917).

Syzygium chavaran (Bourd.) Gamble, Fl. Madras: 480 (1919).
S. India, Pen. Thailand. 40 IND 41 THA. Phan.
**Eugenia chavaran* Bourd., Forest Trees Travancore: 188 (1908).

Syzygium chloranthum (Duthie) Merr. & L.M.Perry, Mem. Amer. Acad. Arts 18: 173 (1939).
W. Malesia. 42 BOR MLY SUM. Phan.
**Eugenia chlorantha* Duthie in J.D.Hooker, Fl. Brit. India 2: 487 (1878).

Syzygium chloroleucum (King) Masam., Enum. Phan. Born.: 526 (1942).
Pen. Malaysia. 42 MLY. Phan.
**Eugenia chloroleuca* King, J. Asiat. Soc. Bengal, Pt. 2, Nat. Hist. 70: 113 (1901).

Syzygium christmannii Merr. & L.M.Perry, Mem. Amer. Acad. Arts 18: 185 (1939).
Borneo. 42 BOR. Phan.

Syzygium christophersenii Whistler, J. Arnold Arbor. 69: 186 (1988).
Samoa. 60 SAM. Phan.

Syzygium chunianum Merr. & L.M.Perry, J. Arnold Arbor. 19: 240 (1938).
Hainan. 36 CHH. Phan.

Syzygium ciliatosetosum (Merr.) Merr., Philipp. J. Sci. 79: 382 (1951).
Philippines. 42 PHI. Phan.
**Eugenia ciliatosetosa* Merr., Philipp. J. Sci., C 7: 315 (1912).

Syzygium cinctum Merr. & L.M.Perry, J. Arnold Arbor. 23: 272 (1942). *Eugenia cincta* (Merr. & L.M.Perry) Whitmore, Gard. Bull. Singapore 22: 16 (1967), nom. illeg.
Solomon Is. 43 SOL. Phan.

Syzygium cinereum (Kurz) Chantaran. & J.Parn., Kew Bull. 48: 596 (1993).
SE. China, Indo-China to Philippines. 36 CHS 41 MYA THA VIE 42 MLY PHI. Phan.
Syzygium cinerascens C.Presl, Abh. Königl. Böhm. Ges. Wiss., V, 3: 500 (1845), provisional synonym.
**Eugenia cinerea* Kurz, Prelim. Rep. Forest Pegu, App. B: 50 (1875).
Eugenia pseudosubtilis var. *subacuminata* King, J. Asiat. Soc. Bengal, Pt. 2, Nat. Hist. 70(2): 124 (1901).
Eugenia ixoroides Elmer, Leafl. Philipp. Bot. 4: 1426 (1912).
Eugenia operculata var. *orientalis* Craib, Fl. Siam. 1: 654 (1931).

Syzygium cinnamomeum (Vidal) Merr., Philipp. J. Sci. 79: 382 (1951).
Philippines. 42 PHI. Phan.
Eugenia pellucida Fern.-Vill. in F.M.Blanco, Fl. Filip., ed. 3, 4(13A): 85 (1880).
**Eugenia cinnamomea* Vidal, Phan. Cuming. Philipp.: 173 (1885).

Syzygium circumscissum (Gagnep.) Craven & Biffin, Blumea 51: 136 (2006).

Indo-China to Pen. Malaysia. 41 MYA THA VIE 42 MLY. Phan.

Eugenia circumscissa Gagnep., Notul. Syst. (Paris) 3: 321 (1918). *Cleistocalyx circumscissa* (Gagnep.) P.H.Hô, Cayco Vietnam 2(1): 63 (1992). *Syzygium attenuatum* subsp. *circumscissum* (Gagnep.) Chantaran. & J.Parn., Kew Bull. 48: 593 (1993).

Eugenia rhamphiphylla Craib, Bull. Misc. Inform. Kew 1930: 168 (1930). *Syzygium rhamphiphyllum* (Craib) C.E.C.Fisch., Bull. Misc. Inform. Kew 1937: 438 (1937).

Eugenia barringtonii Hole ex Chalk & Chattaway, Bull. Misc. Inform. Kew 1937: 438 (1937), pro syn. *Syzygium holei* Bahadur & R.C.Gaur, Indian J. Forest. 1: 349 (1978).

Syzygium cladopterum (Diels) Merr. & L.M.Perry, J. Arnold Arbor. 23: 249 (1942).
New Guinea. 43 NWG. Phan.

Jambosa cladoptera Diels, Bot. Jahrb. Syst. 57: 391 (1922).

Syzygium clavellatum (Merr.) Merr., Philipp. J. Sci. 79: 382 (1951).
Philippines. 42 PHI. Phan.

Eugenia clavellata Merr., Philipp. J. Sci. 1(Suppl.): 104 (1906).

Syzygium claviflorum (Roxb.) Wall. ex A.M.Cowan & Cowan, Trees N. Bengal: 67 (1929).
SC. China to N. Australia. 36 CHC CHH 40 ASS BAN EHM 41 AND MYA NCB THA VIE 42 BOR JAW MLY PHI SUL 43 NWG 50 NTA QLD. Phan.

Eugenia claviflora Roxb., Fl. Ind. ed. 1832, 2: 488 (1832). *Acmena claviflora* (Roxb.) Walp., Repert. Bot. Syst. 2: 181 (1843). *Clavimyrtus claviflora* (Roxb.) Blume, Mus. Bot. 1: 113 (1850). *Acmenosperma claviflorum* (Roxb.) Kausel, Ark. Bot., a.s., 3: 609 (1957).

Jambosa clavata Korth., Ned. Kruidk. Arch. 1: 201 (1847). *Eugenia clavata* (Korth.) Merr., J. Straits Branch Roy. Asiat. Soc. 77: 225 (1917). *Syzygium clavatum* (Korth.) Merr. & L.M.Perry, Mem. Amer. Acad. Arts 18: 180 (1939).

Eugenia leptantha Wight, Ill. Ind. Bot. 2: 15 (1850), nom. illeg. *Eugenia leptalea* Craib, Fl. Siam. 1: 649 (1931).

Eugenia rhododendrifolia Miq., Anal. Bot. Ind. 1: 19 (1850). *Syzygium rhododendrifolium* (Miq.) Masam., Enum. Phan. Born.: 538 (1942).

Eugenia rhododendrifolia var. *forma* Miq., Anal. Bot. Ind. 1: 19 (1850).

Jambosa borneensis Miq., Fl. Ned. Ind. 1(1): 434 (1855).

Jambosa melanocarpa Miq., Fl. Ned. Ind. 1(1): 430 (1855).

Eugenia maingayi Duthie in J.D.Hooker, Fl. Brit. India 2: 484 (1878). *Eugenia claviflora* var. *maingayi* (Duthie) King, J. Asiat. Soc. Bengal, Pt. 2, Nat. Hist. 70(2): 108 (1901). *Syzygium claviflorum* var. *maingayi* (Duthie) Chantaran. & J.Parn., Kew Bull. 48: 590 (1993).

Eugenia oblongifolia Duthie in J.D.Hooker, Fl. Brit. India 2: 491 (1878). *Syzygium maingayi* Chantaran. & J.Parn., Kew Bull. 48: 605 (1993).

Eugenia ruminata Koord. & Valeton, Bull. Inst. Bot. Buitenzorg 2: 8 (1899). *Syzygium ruminatum* (Koord. & Valeton) Amshoff in C.A.Backer & R.C.Bakhuizen van der Brink, Bekn. Fl. Java 4b(98): 11 (1944).

Eugenia claviflora var. *excavata* King, J. Asiat. Soc. Bengal, Pt. 2, Nat. Hist. 70(2): 108 (1901). *Syzygium claviflorum* var. *excavatum* (King) I.M.Turner, J. Singapore Natl. Acad. Sci. 22–24: 17 (1997).

Eugenia claviflora var. *glandulosa* King, J. Asiat. Soc. Bengal, Pt. 2, Nat. Hist. 70(2): 108 (1901). *Syzygium claviflorum* var. *glandulosum* (King) Chantaran. & J.Parn., Kew Bull. 48: 590 (1993).

Eugenia viridifolia Elmer, Leafl. Philipp. Bot. 4: 1420 (1912). *Syzygium viridifolium* (Elmer) Merr. & L.M.Perry, Mem. Amer. Acad. Arts 18: 183 (1939).

Eugenia fraseri Ridl., J. Bot. 68: 33 (1930). *Syzygium fraseri* (Ridl.) Masam., Enum. Phan. Born.: 528 (1942).

Eugenia claviflora var. *montana* M.R.Hend., Gard. Bull. Singapore 12: 260 (1949). *Syzygium claviflorum* var. *montanum* (M.R.Hend.) I.M.Turner, J. Singapore Natl. Acad. Sci. 22–24: 17 (1997).

Eugenia claviflora var. *riparia* M.R.Hend., Gard. Bull. Singapore 12: 257 (1949). *Syzygium claviflorum* var. *riparium* (M.R.Hend.) I.M.Turner, J. Singapore Natl. Acad. Sci. 22–24: 18 (1997).

Syzygium clementis (C.B.Rob.) Merr., Philipp. J. Sci. 79: 383 (1951).
Philippines. 42 PHI. Phan.

Eugenia clementis C.B.Rob., Philipp. J. Sci., C 4: 383 (1909).

Syzygium cleyerifolium (Yatabe) Makino, Bot. Mag. (Tokyo) 16: 15 (1902).
Ogasawara-shoto. 38 OGA. Nanophan.

Eugenia cleyerifolia Yatabe, Bot. Mag. (Tokyo) 6: 405 (1892). *Syzygium buxifolium* var. *cleyerifolium* (Yatabe) Y.Tateishi, in Fl. Japan 2c: 212 (1999).

Syzygium clusiifolium (A.Gray) Müll.Stuttg. in W.G.Walpers, Ann. Bot. Syst. 4: 839 (1858).
SW. Pacific. 60 SAM TON VAN WAL. Phan.

Eugenia clusiifolia A.Gray, U.S. Expl. Exped., Phan. 1: 528 (1854).

Syzygium clypeolatum (Ridl.) I.M.Turner, J. Singapore Natl. Acad. Sci. 22–24: 18 (1997).
Pen. Malaysia. 42 MLY. Nanophan.

Eugenia clypeolata Ridl., J. Straits Branch Roy. Asiat. Soc. 82: 185 (1920).

Syzygium coalitum (Greves) T.G.Hartley & L.M.Perry, J. Arnold Arbor. 54: 196 (1973).
New Guinea. 43 NWG. Phan.

Eugenia coalita Greves, J. Bot. 61(Suppl.): 15 (1923).

Jambosa sargentiana Diels, J. Arnold Arbor. 10: 83 (1929). *Syzygium sargentianum* (Diels) Merr. & L.M.Perry, J. Arnold Arbor. 23: 249 (1942).

Syzygium coccineum J.W.Dawson, in Fl. Nouv.-Caléd. 23: 140 (1999).
C. & SE. New Caledonia. 60 NWC. Nanophan. or phan.

Syzygium combretiflorum (Diels) Merr. & L.M.Perry, J. Arnold Arbor. 23: 286 (1942).
New Guinea. 43 NWG. Phan.

Jambosa combretiflora Diels, Bot. Jahrb. Syst. 57: 392 (1922).

Syzygium commersonii J.Guého & A.J.Scott, Kew Bull. 34: 491 (1980).
Mauritius. 29 MAU. Nanophan. or phan.

Syzygium compongense (Gagnep.) Merr. & L.M.Perry, J. Arnold Arbor. 19: 112 (1938).
Cambodia. 41 CBD. Phan.

Eugenia compongensis Gagnep., Notul. Syst. (Paris) 3: 323 (1918).

Syzygium conceptionis Guillaumin, Bull. Soc. Bot. France 85: 643 (1938 publ. 1939).
WC. & SE. New Caledonia. 60 NWC. Phan.

Syzygium concinnum (A.C.Sm.) Craven & Biffin, Blumea 51: 136 (2006).
Fiji (Viti Levu). 60 FIJ. Phan.
**Piliocalyx concinnus* A.C.Sm., Pacific Sci. 25: 496 (1971).

Syzygium condensatum (Baker) Labat & Schatz, Novon 12: 202 (2002).
C. Madagascar. 29 MDG. Nanophan. or phan.
**Eugenia condensata* Baker, J. Bot. 20: 112 (1882).

Syzygium confertiflorum (A.Gray) Müll.Stuttg. in W.G.Walpers, Ann. Bot. Syst. 4: 838 (1858).
Fiji. 60 FIJ. Phan.
**Eugenia confertiflora* A.Gray, U.S. Expl. Exped., Phan. 1: 523 (1854).

Syzygium confertum (Korth.) Merr. & L.M.Perry, Mem. Amer. Acad. Arts 18: 177 (1939).
W. Malesia to W. Philippines. 42 BOR JAW PHI SUM. Phan.
**Jambosa conferta* Korth., Ned. Kruidk. Arch. 1: 202 (1847). *Microjambosa conferta* (Korth.) Blume, Mus. Bot. 1: 118 (1850).
Microjambosa cuspidata Blume, Mus. Bot. 1: 119 (1850). *Jambosa cuspidata* (Blume) Blume ex Miq., Fl. Ned. Ind. 1(1): 435 (1855).
Microjambosa trifida Blume, Mus. Bot. 1: 118 (1850). *Jambosa trifida* (Blume) Miq., Fl. Ned. Ind. 1(1): 435 (1855).
Eugenia densipunctata Koord. & Valeton, Bull. Inst. Bot. Buitenzorg 2: 6 (1899).
Eugenia calvinii Elmer, Leafl. Philipp. Bot. 4: 1419 (1912). *Syzygium calvinii* (Elmer) Masam., Enum. Phan. Born.: 525 (1942).

Syzygium confusum (Blume) Bakh.f., Blumea 12: 61 (1963).
W. Malesia. 42 JAW MLY SUM. Nanophan. or phan.
**Jambosa confusa* Blume, Mus. Bot. 1: 101 (1850). *Eugenia doligophylla* Koord. & Valeton, Meded. Lands Plantentuin 40: 78 (1900), nom. illeg. *Syzygium amshoffianum* Merr., Philipp. J. Sci. 79: 366 (1951), nom. superfl. *Eugenia malayana* Gagnep. in H.Lecomte, Fl. Indo-Chine 2: 838 (1921). *Syzygium malayanum* (Gagnep.) I.M. Turner, J. Singapore Natl. Acad. Sci. 22–24: 21 (1997), nom. superfl.
Jambosa korthalsii Blume, Mus. Bot. 1: 101 (1850).
Jambosa lanceolata Korth. ex Miq., Fl. Ned. Ind. 1(1): 426 (1855).

Syzygium congestiflorum H.T.Chang & R.H.Miao, Acta Bot. Yunnan. 4: 19 (1982).
SC. China. 36 CHC.

Syzygium congestum (Merr.) Merr., Philipp. J. Sci. 79: 383 (1951).
Philippines. 42 PHI. Phan.
**Eugenia congesta* Merr., Publ. Bur. Sci. Gov. Lab. 35: 49 (1905).

Syzygium conglobatum Merr., Philipp. J. Sci. 79: 383 (1951).
Borneo, Philippines. 42 BOR PHI. Phan.
**Eugenia conglobata* C.B.Rob., Philipp. J. Sci., C 4: 359 (1909), nom. illeg. *Eugenia philippinorum* C.Nelson Sutherland, Anales Jard. Bot. Madrid 56: 162 (1998).
Eugenia subsulcata Elmer, Leafl. Philipp. Bot. 8: 3095 (1919).

Syzygium conglomeratum (Duthie) I.M.Turner, J. Singapore Natl. Acad. Sci. 22–24: 18 (1997).
Pen. Malaysia. 42 MLY. Phan.
**Eugenia conglomerata* Duthie in J.D.Hooker, Fl. Brit. India 2: 497 (1879).
Eugenia conglomerata var. *paniculata* M.R.Hend., Gard. Bull. Singapore 12: 201 (1949). *Syzygium conglomeratum* var. *paniculatum* (M.R.Hend.) I.M.Turner, J. Singapore Natl. Acad. Sci. 22–24: 18 (1997).

Syzygium congolense Vermoesen, Acta Bot. Neerl. 9: 406 (1960).
WC. & E. Trop. Africa. 23 BUR CAF CMN GAB ZAI 26 TAN UGA. Phan.

Syzygium conicum Korth., Ned. Kruidk. Arch. 1: 204 (1848).
Borneo. 42 BOR. Phan.

Syzygium consanguineum (Merr.) Merr., Philipp. J. Sci. 79: 384 (1951).
Philippines. 42 PHI. Phan.
**Eugenia consanguinea* Merr., Philipp. J. Sci. 18: 300 (1921).

Syzygium consimile Merr., Philipp. J. Sci. 79: 384 (1951).
Philippines (Samar). 42 PHI. Phan.

Syzygium conspersipunctatum (Merr. & L.M.Perry) Craven & Biffin, Blumea 51: 136 (2006).
Hainan. 36 CHH. Phan.
**Cleistocalyx conspersipunctatus* Merr. & L.M.Perry, J. Arnold Arbor. 18: 335 (1937).

Syzygium contractum (Poir.) J.Guého & A.J.Scott, Kew Bull. 34: 489 (1980).
Mauritius. 29 MAU. Nanophan. or phan.
**Eugenia contracta* Poir. in J.B.A.P.M.de Lamarck, Encycl., Suppl. 3: 125 (1813).

Syzygium copelandii (C.B.Rob.) Merr., Philipp. J. Sci. 79: 385 (1951).
Philippines. 42 PHI. Phan.
**Eugenia copelandii* C.B.Rob., Philipp. J. Sci., C 4: 352 (1909).

Syzygium cordatilimbum (Merr.) Merr., Philipp. J. Sci. 79: 385 (1951).
Philippines. 42 PHI. Phan.
**Eugenia cordatilimba* Merr., Philipp. J. Sci. 27: 40 (1925).

Syzygium cordatum Hochst. ex Krauss, Flora 27: 425 (1844). *Eugenia cordata* (Hochst. ex Krauss) G.Lawson in D.Oliver & auct. suc. (eds.), Fl. Trop. Afr. 2: 438 (1871), nom. illeg.
Uganda to S. Africa. 23 BUR GAB RWA ZAI 25 KEN TAN UGA 26 ANG MLW MOZ ZAM ZIM 27 BOT CPP CPV NAT SWZ TVL. Nanophan. or phan.

subsp. ***cordatum***
Uganda to S. Africa. 23 BUR GAB RWA ZAI 25 KEN TAN UGA 26 ANG MLW MOZ ZAM ZIM 27 BOT CPP NAT SWZ TVL. Nanophan. or phan.
Jambosa cymifera E.Mey. in J.F.Drège, Zwei Pflanzengeogr. Dokum.: 194 (1843). *Syzygium cymiferum* (E.Mey.) C.Presl, Abh. Königl. Böhm. Ges. Wiss., V, 3: 500 (1845).
Syzygium cordifolium Klotzsch in W.C.H.Peters, Naturw. Reise Mossambique 6: 63 (1861).

subsp. ***shimbaense*** Verdc., in Fl. Trop. E. Afr., Myrtac.: 74 (2001).
E. Trop. Africa. 25 KEN TAN.

Syzygium cordemoyi J.Bosser & Cadet, Bull. Mus. Natl. Hist. Nat., B, Adansonia 9: 30 (1987).
Réunion. 29 REU. Nanophan. or phan.
Eugenia platyphylla Cordem., Fl. Réunion: 432 (1895), nom. illeg.

Syzygium cordifoliatum (Ridl.) I.M.Turner, J. Singapore Natl. Acad. Sci. 22–24: 18 (1997).
Pen. Malaysia. 42 MLY. Phan.
**Eugenia cordifoliata* Ridl., J. Straits Branch Roy. Asiat. Soc. 79: 66 (1918).

Syzygium cordifolium (Wight) Walp., Repert. Bot. Syst. 2: 179 (1843).
Sri Lanka. 40 SRL. Phan.
**Eugenia cordifolia* Wight, Icon. Pl. Ind. Orient. 2: t. 544 (1842).

subsp. ***cordifolium***
Sri Lanka. 40 SRL. Phan.
Calyptranthes cordifolia Moon, Cat. Pl. Ceylon: 39 (1824), nom. nud.
Eugenia androsaemoides Bedd., Fl. Sylv. S. India: cvii (1872).

subsp. ***spissum*** (Alston) P.S.Ashton, in Revised Handb. Fl. Ceyl. 2: 441 (1981).
Sri Lanka. 40 SRL. Phan.
Myrtus androsaemoides L., Sp. Pl.: 472 (1753). *Eugenia androsaemoides* (L.) DC., Prodr. 3: 284 (1828). *Syzygium androsaemoides* (L.) Walp., Repert. Bot. Syst. 2: 179 (1843). **Syzygium spissum* Alston in H.Trimen, Handb. Fl. Ceylon 6(Suppl.): 117 (1931).

Syzygium coriaceum Bosser & J.Guého, Bull. Mus. Natl. Hist. Nat., B, Adansonia 9: 34 (1987).
Mauritius. 29 MAU. Nanophan.

Syzygium cormiflorum (F.Muell.) B.Hyland, Austral. J. Bot., Suppl. Ser. 9: 68 (1983).
N. & NE. Queensland. 50 QLD. Phan.
**Eugenia cormiflora* F.Muell., Fragm. 5: 23 (1865).
Eugenia hislopii F.M.Bailey, Queensland Agric. J. 5: 483 (1899).

Syzygium cornifolium (Blume) Merr. & L.M.Perry, J. Arnold Arbor. 23: 278 (1942).
Sulawesi. 42 SUL. Phan.
**Jambosa cornifolia* Blume, Mus. Bot. 1: 92 (1850).

Syzygium cornuflorum P.S.Ashton, Kew Bull. 61: 116 (2006).
Borneo (Sabah). 42 BOR. Phan.

Syzygium corticopapyraceum (Elmer) Merr., Philipp. J. Sci. 79: 385 (1951).
Philippines. 42 PHI. Phan.
**Eugenia corticopapyracea* Elmer, Leafl. Philipp. Bot. 4: 1405 (1912).

Syzygium corticosum (Lour.) Merr. & L.M.Perry, J. Arnold Arbor. 19: 105 (1938).
Vietnam. 41 VIE. Phan.
**Eugenia corticosa* Lour., Fl. Cochinch.: 308 (1790).

Syzygium corymbosum (Blume) DC., Prodr. 3: 261 (1828).
Jawa. 42 JAW. Phan.
**Calyptranthes corymbosa* Blume, Flora 7(1; 19): 291 (1824). *Jambosa corymbosa* (Blume) Miq., Fl. Ned. Ind. 1(1): 420 (1855). *Eugenia corymbifera* Koord. & Valeton, Meded. Lands Plantentuin 40: 98 (1900).

Syzygium corynanthum (F.Muell.) L.A.S.Johnson, Contr. New South Wales Natl. Herb. 3: 99 (1962).
SE. Queensland to NE. New South Wales. 50 NSW QLD. Nanophan. or phan.
**Eugenia corynantha* F.Muell., Fragm. 9: 145 (1875).
Eugenia punctulata F.M.Bailey, Bot. Bull. Dept. Agric. Queensland 13: 10 (1896).

Syzygium corynocarpum (A.Gray) Müll.Stuttg. in W.G.Walpers, Ann. Bot. Syst. 4: 839 (1858).
SW. Pacific. 60 FIJ NUE SAM TON WAL. Phan.
**Eugenia corynocarpa* A.Gray, U.S. Expl. Exped., Phan. 1: 526 (1854).

Syzygium costulatum (C.B.Rob.) Merr., Philipp. J. Sci. 79: 385 (1951).
Philippines. 42 PHI. Phan.
**Eugenia costulata* C.B.Rob., Philipp. J. Sci., C 4: 393 (1909).

Syzygium craibii Chantaran. & J.Parn., Kew Bull. 48: 596 (1993).
Pen. Thailand. 41 THA. Phan.

Syzygium crassibracteatum (Merr.) Merr., Philipp. J. Sci. 79: 385 (1951).
Philippines. 42 PHI. Phan.
**Eugenia crassibracteata* Merr., Philipp. J. Sci., C 10: 210 (1915).

Syzygium crassiflorum Merr. & L.M.Perry, J. Arnold Arbor. 19: 114 (1938).
S. Vietnam. 41 VIE. Phan.

Syzygium crassilimbum (Merr.) Merr., Philipp. J. Sci. 79: 385 (1951).
Philippines. 42 PHI. Phan.
**Eugenia crassilimba* Merr., Philipp. J. Sci. 27: 41 (1925).

Syzygium crassipes (C.B.Rob.) Merr., Philipp. J. Sci. 79: 385 (1951).
Philippines. 42 PHI. Phan.
**Eugenia crassipes* C.B.Rob., Philipp. J. Sci., C 4: 361 (1909).

Syzygium crassissimum (Merr.) Merr., Philipp. J. Sci. 79: 386 (1951).
Philippines. 42 PHI. Phan.
**Eugenia crassissima* Merr., Philipp. J. Sci., C 9: 211 (1914).

Syzygium cratermontensis W.N.Takeuchi, Edinburgh J. Bot. 59: 260 (2002).
E. New Guinea. 43 NWG.

Syzygium creaghii (Ridl.) Merr. & L.M.Perry, Mem. Amer. Acad. Arts 18: 164 (1939).
Borneo. 42 BOR. Phan.
Eugenia woodii Merr., J. Straits Branch Roy. Asiat. Soc. 86: 336 (1922), nom. illeg. *Syzygium woodii* Masam., Enum. Phan. Born.: 541 (1942).
**Eugenia creaghii* Ridl., J. Bot. 68: 14 (1930).

Syzygium crebrinerve (C.T.White) L.A.S.Johnson, Contr. New South Wales Natl. Herb. 3: 99 (1962).
SE. Queensland to NE. New South Wales. 50 NSW QLD. Nanophan. or phan.
**Eugenia crebrinervis* C.T.White, Proc. Roy. Soc. Queensland 57: 25 (1947).

Syzygium cruriflorum Diels, Bot. Jahrb. Syst. 57: 402 (1922).
New Guinea. 43 NWG. Phan.

Syzygium crypteronioides P.S.Ashton, Kew Bull. 61: 118 (2006).
Borneo (Sarawak). 42 BOR. Phan.

Syzygium cumini (L.) Skeels, Bull. Bur. Pl. Industr. U.S.D.A. 248: 25 (1912).
Trop. & Subtrop. Asia to N. Queensland. (25) ken tan uga 36 CHH CHS 40 ASS BAN EHM IND NEP SRL 41 AND MYA THA VIE 42 JAW MLY phi SUL 50 QLD (61) mrq sci (63) haw (78) fla (80) blz. Nanophan. or phan. Widely cultivated for its edible fruits (Java plum).
**Myrtus cumini* L., Sp. Pl.: 471 (1753). *Calyptranthes cumini* (L.) Pers., Syn. Pl. 2: 32 (1806). *Eugenia cumini* (L.) Druce, Rep. Bot. Exch. Club Brit. Isles 3: 418 (1913 publ. 1914).
Jambolifera coromandelica Houtt., Handl. Pl.-Kruidk. 2: 275 (1774).
Jambolifera pedunculata Houtt., Handl. Pl.-Kruidk. 2: 273 (1774).
Eugenia caryophyllifolia Lam., Encycl. 3: 198 (1789). *Syzygium caryophyllifolium* (Lam.) DC., Prodr. 3: 260 (1828). *Eugenia jambolana* var. *caryophyllifolia* (Lam.) Duthie in J.D.Hooker, Fl. Brit. India 2: 499 (1879). *Syzygium cumini* var. *caryophyllifolium* (Lam.) K.K.Khanna, Fl. Bihar, Analysis: 199 (2001).
Eugenia jambolana Lam., Encycl. 3: 198 (1789). *Calyptranthes jambolana* (Lam.) Willd., Ann. Bot. (Usteri) 17: 23 (1796). *Syzygium jambolanum* (Lam.) DC., Prodr. 3: 259 (1828).
Calyptranthes caryophyllifolia Willd., Ann. Bot. (Usteri) 17: 22 (1796).
Calyptranthes cuminodora Stokes, Bot. Mat. Med. 3: 67 (1812).
Calyptranthes jambolifera Stokes, Bot. Mat. Med. 3: 68 (1812).
Caryophyllus corticosus Stokes, Bot. Mat. Med. 3: 75 (1812).
Caryophyllus jambos Stokes, Bot. Mat. Med. 3: 73 (1812).
Eugenia obovata Poir. in J.B.A.P.M.de Lamarck, Encycl., Suppl. 3: 124 (1813). *Myrtus obovata* (Poir.) Spreng., Syst. Veg. 2: 486 (1825). *Syzygium obovatum* (Poir.) DC., Prodr. 3: 259 (1828).
Jambolifera chinensis Spreng., Syst. Veg. 2: 216 (1825).
Myrtus corticosa Spreng., Syst. Veg. 2: 488 (1825).
Calyptranthes capitellata Buch.-Ham. ex Wall., Numer. List: 3560B (1831), nom. nud.
Eugenia obtusifolia Roxb., Fl. Ind. ed. 1832, 2: 485 (1832). *Syzygium obtusifolium* (Roxb.) Kostel., Allg. Med.-Pharm. Fl. 4: 1532 (1835). *Syzygium cumini* var. *obtusifolium* (Roxb.) K.K.Khanna, Fl. Bihar, Analysis: 202 (2001).
Eugenia calyptrata Roxb. ex Wight & Arn., Prodr. Fl. Ind. Orient.: 329 (1834).
Eugenia jambolifera Roxb. ex Wight & Arn., Prodr. Fl. Ind. Orient.: 329 (1834).
Eugenia jambolana var. *obtusifolia* Duthie in J.D.Hooker, Fl. Brit. India 2: 500 (1879).
Eugenia tsoi Merr. & Chun, Sunyatsenia 2: 291 (1935). *Syzygium cumini* var. *tsoi* (Merr. & Chun) H.T. Chang & R.H.Miao, Acta Bot. Yunnan. 4: 22 (1982).
Calyptranthes oneillii Lundell, Bull. Torrey Bot. Club 64: 554 (1937).

Syzygium cuneatum Brahmam & H.O.Saxena, J. Bombay Nat. Hist. Soc. 78: 417 (1981).
Assam to Bangladesh. 40 ASS BAN. Phan.
**Eugenia cuneata* Duthie in J.D.Hooker, Fl. Brit. India 2: 495 (1878). *Syzygium schmidii* Rathakr. & N.C.Nair, J. Econ. Taxon. Bot. 4: 288 (1983).

Syzygium cuneiforme Merr. & L.M.Perry, Mem. Amer. Acad. Arts 18: 152 (1939).
Borneo. 42 BOR. Phan.

Syzygium curranii (C.B.Rob.) Merr., Philipp. J. Sci. 79: 386 (1951).
Philippines. 42 PHI. Phan.
**Eugenia curranii* C.B.Rob., Philipp. J. Sci., C 4: 351 (1909).

Syzygium curtiflorum (Elmer) Merr., Philipp. J. Sci. 79: 386 (1951).
Philippines. 42 PHI. Phan.
**Eugenia curtiflora* Elmer, Leafl. Philipp. Bot. 1: 328 (1908).

Syzygium curtisii (King) Merr. & L.M.Perry, Mem. Amer. Acad. Arts 18: 182 (1939).
Pen. Thailand to W. Malesia. 41 THA 42 BOR MLY SUM. Phan.
**Eugenia curtisii* King, J. Asiat. Soc. Bengal, Pt. 2, Nat. Hist. 70: 129 (1901).
Eugenia curtisii var. *minor* King, J. Asiat. Soc. Bengal, Pt. 2, Nat. Hist. 70(2): 129 (1901). *Syzygium curtisii* var. *minus* (King) I.M.Turner, J. Singapore Natl. Acad. Sci. 22–24: 18 (1997).
Eugenia coralina Merr., J. Straits Branch Roy. Asiat. Soc. 77: 207 (1917). *Syzygium coralinum* (Merr.) Masam., Enum. Phan. Born.: 526 (1942).
Eugenia holttumii Ridl., J. Bot. 62: 296 (1924). *Eugenia curtisii* var. *holttumii* (Ridl.) M.R.Hend., Gard. Bull. Singapore 12: 176 (1949). *Syzygium curtisii* var. *holttumii* (Ridl.) I.M.Turner, J. Singapore Natl. Acad. Sci. 22–24: 18 (1997).

Syzygium curvistylum (Gillespie) Merr. & L.M.Perry, Sargentia 1: 75 (1942).
Fiji, Samoa. 60 FIJ SAM. Phan.
**Eugenia curvistyla* Gillespie, Bernice P. Bishop Mus. Bull. 83: 21 (1931).
Syzygium curvistylum var. *parvifolium* L.M.Perry, J. Arnold Arbor. 31: 360 (1950).

Syzygium cuttingii Merr. & L.M.Perry, Brittonia 4: 126 (1941).
N. Myanmar. 41 MYA. Phan.

Syzygium cyanophyllum (P.C.Kanjilal & D.Das) Raizada, Indian Forester 74: 336 (1948).
Assam. 40 ASS. Phan.
**Eugenia cyanophylla* P.C.Kanjilal & D.Das, Assam Forest Rec., Bot. 2: 12 (1937).

Syzygium cylindricum (Wight) Alston in H.Trimen, Handb. Fl. Ceylon 6(Suppl.): 115 (1931).
Sri Lanka. 40 SRL. Nanophan. or phan.
**Eugenia cylindrica* Wight, Icon. Pl. Ind. Orient. 2: t. 527 (1842). *Clavimyrtus cylindrica* (Wight) Blume, Mus. Bot. 1: 113 (1850). *Jambosa cylindrica* (Wight) Thwaites, Enum. Pl. Zeyl.: 115 (1859).

Syzygium cymosum (Lam.) DC., Prodr. 3: 259 (1828).
Mascarenes. 29 MAU REU. Nanophan. or phan.
**Eugenia cymosa* Lam., Encycl. 3: 199 (1789). *Myrtus cymosa* (Lam.) Spreng., Syst. Veg. 2: 486 (1825).

var. ***cymosum***
Mascarenes. 29 MAU REU. Nanophan. or phan.
Calyptranthes pollicina Willemet, Ann. Bot. (Usteri) 18: 39 (1796).
Eugenia nigrescens Poir. in J.B.A.P.M.de Lamarck, Encycl., Suppl. 3: 123 (1813).
Eugenia mascarensis C.Presl, Isis (Oken) 21: 274 (1828).
Eugenia rhodomelea Comm. ex DC., Prodr. 3: 259 (1828).
Jambosa commersonii Blume, Mus. Bot. 1: 105 (1850).
Eugenia depauperata Cordem., Fl. Réunion: 431 (1895), nom. illeg.
Eugenia mascarensis Cordem., Fl. Réunion: 429 (1895), nom. illeg.

var. ***montanum*** J.Guého & A.J.Scott, Kew Bull. 34: 490 (1980).
Réunion. 29 REU. Nanophan. or phan.
Eugenia montana Cordem., Fl. Réunion: 432 (1895), nom. illeg.

Syzygium cyrtophylloides (Ridl.) I.M.Turner, J. Singapore Natl. Acad. Sci. 22–24: 18 (1997).
Pen. Malaysia. 42 MLY. Phan.
**Eugenia cyrtophylloides* Ridl., J. Straits Branch Roy. Asiat. Soc. 79: 65 (1918).

Syzygium danguyanum (H.Perrier) Labat & Schatz, Novon 12: 202 (2002).
E. Madagascar. 29 MDG. Phan.
Eugenia danguyana var. *rotranala* H.Perrier, Mém. Inst. Sci. Madagascar, Sér. B, Biol. Vég. 4: 195 (1952).
**Eugenia danguyana* H.Perrier, Mém. Inst. Sci. Madagascar, Sér. B, Biol. Vég. 4: 195 (1953).

Syzygium dansiei B.Hyland, Austral. J. Bot., Suppl. Ser. 9: 75 (1983).
N. Queensland. 50 QLD. Phan.

Syzygium daphne (Ridl.) Merr. & L.M.Perry, J. Arnold Arbor. 23: 254 (1942).
New Guinea. 43 NWG. Phan.
**Eugenia daphne* Ridl., Trans. Linn. Soc. London, Bot. 9: 45 (1916).

Syzygium dasyphyllum Merr. & L.M.Perry, Mem. Amer. Acad. Arts 18: 153 (1939).
Borneo. 42 BOR. Phan.

Syzygium davaoense (Elmer) Merr., Philipp. J. Sci. 79: 386 (1951).
Philippines. 42 PHI. Phan.
**Eugenia davaoensis* Elmer, Leafl. Philipp. Bot. 4: 1439 (1912).

Syzygium dealbatum (Burkill) A.C.Sm., Bernice P. Bishop Mus. Bull. 220: 203 (1959).
SW. Pacific. 60 NUE SAM TON WAL. Phan.
**Eugenia dealbata* Burkill, J. Linn. Soc., Bot. 35: 37 (1901).

Syzygium decipiens (Koord. & Valeton) Merr. & L.M.Perry, J. Arnold Arbor. 23: 281 (1942).
Jawa to Solomon Is. 42 JAW PHI 43 NWG SOL. Phan.
**Eugenia decipiens* Koord. & Valeton, Bull. Inst. Bot. Buitenzorg 2: 6 (1899).
Syzygium megalanthelium Diels, Nova Guinea 14: 93 (1924).
Syzygium rectangulare Merr. & L.M.Perry, J. Arnold Arbor. 23: 282 (1942).

Syzygium decoriflorum (Diels) Merr. & L.M.Perry, J. Arnold Arbor. 23: 249 (1942).
New Guinea. 43 NWG. Phan.
**Jambosa decoriflora* Diels, Bot. Jahrb. Syst. 57: 396 (1922).

Syzygium decussatum (A.C.Sm.) Biffin & Craven, Blumea 50(2): 385 (2005).
Fiji (Vanua Levu, Viti Levu). 60 FIJ. Phan.
**Cleistocalyx decussatus* A.C.Sm., Pacific Sci. 25: 495 (1971).

Syzygium delicatulum Merr. & L.M.Perry, J. Arnold Arbor. 23: 273 (1942).
Solomon Is. 43 SOL. Phan.

Syzygium dempoense (Greves) Govaerts, World Checklist Myrtaceae: 385 (2008).
Sumatera. 42 SUM. Nanophan. or phan.
**Eugenia dempoensis* Greves, J. Bot. 62(Suppl.): 36 (1924).

Syzygium densiflorum Wall. ex Wight & Arn., Prodr. Fl. Ind. Orient.: 329 (1834). *Eugenia arnottiana* Wight, Ill. Ind. Bot. 2: 17 (1841). *Syzygium arnottianum* (Wight) Walp., Repert. Bot. Syst. 2: 180 (1843), nom. superfl.
S. India. 40 IND. Phan.
Eugenia benthamiana Wight ex Duthie in J.D.Hooker, Fl. Brit. India 2: 484 (1878).
Syzygium benthamianum Gamble, Fl. Madras: 478 (1919).

Syzygium densinervium (Merr.) Merr., Philipp. J. Sci. 79: 387 (1951).
Taiwan (Lan Yü, P'ingtung) to Philippines. 38 TAI 42 PHI. Phan.
**Eugenia densinervia* Merr., Philipp. J. Sci. 1(Suppl.): 105 (1905).

var. ***densinervium***
Philippines. 42 PHI. Phan.
Eugenia silvestrei Elmer, Leafl. Philipp. Bot. 8: 3095 (1919).

var. ***insulare*** C.E.Chang, Bull. Taiwan Prov. Inst. Agric. Pintung 5: 52 (1964).
Taiwan (Lan Yü, P'ingtung). 38 TAI. Nanophan. or phan.

Syzygium deplanchei (Guillaumin) J.W.Dawson, Fl. Nouv.-Caléd. 23: 83 (1999).
New Caledonia (incl. Î. des Pins). 60 NWC. Nanophan. or phan.
Caryophyllus ellipticus Labill., Sert. Austro-Caledon.: 64 (1825). *Myrtus caledoniae* Spreng., Syst. Veg. 4(2): 1993 (1827).
**Caryophyllus deplanchei* Guillaumin, Bull. Soc. Bot. France 85: 648 (1938 publ. (1939).

Syzygium dictyoneurum Diels, Bot. Jahrb. Syst. 57: 404 (1922).
New Guinea. 43 NWG. Phan.

Syzygium dielsianum Merr. & L.M.Perry, J. Arnold Arbor. 23: 259 (1942).
New Guinea. 43 NWG. Phan.
**Jambosa pycnantha* Diels, Bot. Jahrb. Syst. 57: 394 (1922).

Syzygium diffusiflorum Merr., Philipp. J. Sci. 79: 387 (1951).
Philippines (Luzon). 42 PHI. Phan.
**Eugenia diffusa* Merr., Philipp. J. Sci. 18: 301 (1921), nom. illeg.

Syzygium diffusum (Turrill) Merr. & L.M.Perry, Sargentia 1: 76 (1942).
Fiji. 60 FIJ. Phan.
**Eugenia diffusa* Turrill, J. Linn. Soc., Bot. 43: 20 (1915).

Syzygium diospyrifolium (Wall. ex Duthie) S.N.Mitra, Indian Forester 99: 100 (1973).
Assam to Pen. Malaysia. 40 ASS 41 MYA THA 42 MLY. Phan.
**Eugenia diospyrifolia* Wall. ex Duthie in J.D.Hooker, Fl. Brit. India 2: 472 (1878). *Jambosa diospyrifolia* (Wall. ex Duthie) C.E.C.Fisch., Rec. Bot. Surv. India 12: 95 (1938).
Eugenia diospyrifolia var. *lanceolata* Craib, Fl. Siam. 1: 639 (1931).

Syzygium discophorum (Koord. & Valeton) Amshoff in C.A.Backer & R.C.Bakhuizen van der Brink, Bekn. Fl. Java 4b(98): 25 (1944).
Jawa. 42 JAW. Phan.
**Eugenia discophora* Koord. & Valeton, Bull. Inst. Bot. Buitenzorg 2: 6 (1899).

Syzygium dispansum (Ridl.) Craven & Biffin, Blumea 51: 136 (2006).
W. New Guinea. 43 NWG. Phan.
**Eugenia dispansa* Ridl., Trans. Linn. Soc. London, Bot. 9: 47 (1916). *Acmena dispansa* (Ridl.) Merr. & L.M.Perry, J. Arnold Arbor. 19: 18 (1938).

Syzygium divaricatum (Merr. & L.M.Perry) Craven & Biffin, Blumea 51: 137 (2006).
NE. Queensland. 50 QLD. Phan.
**Acmena divaricata* Merr. & L.M.Perry, J. Arnold Arbor. 19: 17 (1938).

Syzygium dolichophyllum (K.Schum. & Lauterb.) Merr. & L.M.Perry, J. Arnold Arbor. 23: 249 (1942).
New Guinea. 43 NWG. Phan.
**Jambosa dolichophylla* K.Schum. & Lauterb., Fl. Schutzgeb. Südsee: 471 (1900).

Syzygium dolichorhynchum Diels, Bot. Jahrb. Syst. 57: 413 (1922).
New Guinea. 43 NWG. Phan.

Syzygium dolichostylum (Diels) Merr. & L.M.Perry, J. Arnold Arbor. 23: 272 (1942).
New Guinea. 43 NWG. Phan.
**Jambosa dolichostyla* Diels, Nova Guinea 14: 91 (1924).

Syzygium dubium (L.M.Perry) A.C.Sm., Fl. Vitiensis Nova 3: 324 (1985).
Fiji (Vanua Levu). 60 FIJ. Phan.
**Syzygium brackenridgei* var. *dubium* L.M.Perry, J. Arnold Arbor. 31: 354 (1950).

Syzygium duplomarginatum (Greves) Merr. & L.M.Perry, J. Arnold Arbor. 23: 249 (1942).
New Guinea. 43 NWG. Phan.
**Eugenia duplomarginata* Greves, J. Bot. 61(Suppl.): 15 (1923).

Syzygium dupontii (Baker) Govaerts, World Checklist Myrtaceae: 386 (2008).
Mauritius. 29 MAU. Nanophan. or phan.
Eugenia venosa Lam., Encycl. 3: 200 (1789). *Myrtus venosa* (Lam.) Spreng., Syst. Veg. 2: 485 (1825). *Jambosa venosa* (Lam.) DC., Prodr. 3: 286 (1828). *Syzygium venosum* (Lam.) J.Guého & A.J.Scott, Kew Bull. 34: 495 (1980), nom. illeg. *Syzygium guehoscottii* Bennet & Raizada, Indian Forester 109: 220 (1983).
**Eugenia dupontii* Baker, Fl. Mauritius: 116 (1877).

Syzygium durifolium Merr. & L.M.Perry, Mem. Amer. Acad. Arts 18: 176 (1939).
Borneo. 42 BOR. Phan.
Syzygium hallieri Merr. & L.M.Perry, Mem. Amer. Acad. Arts 18: 176 (1939).

Syzygium durum (Merr.) Merr., Philipp. J. Sci. 79: 387 (1951).
Philippines. 42 PHI. Phan.
**Eugenia dura* Merr., Philipp. J. Sci., C 11: 24 (1916).

Syzygium duthieanum (King) Masam., Enum. Phan. Born.: 527 (1942).
Pen. Thailand to Pen. Malaysia. 41 THA 42 MLY. Phan.
**Eugenia duthieana* King, J. Asiat. Soc. Bengal, Pt. 2, Nat. Hist. 70: 103 (1901).

Syzygium dyerianum (King) Chantaran. & J.Parn., Kew Bull. 48: 596 (1993).
Pen. Thailand to Pen. Malaysia. 41 THA 42 MLY. Phan.
Eugenia clarkeana King, J. Asiat. Soc. Bengal, Pt. 2, Nat. Hist. 70: 93 (1901).
Eugenia corrugata King, J. Asiat. Soc. Bengal, Pt. 2, Nat. Hist. 70: 93 (1901).
**Eugenia dyeriana* King, J. Asiat. Soc. Bengal, Pt. 2, Nat. Hist. 70: 88 (1901).
Eugenia atronervia M.R.Hend., Gard. Bull. Singapore 11: 299 (1947).

Syzygium ebaloii Merr., Philipp. J. Sci. 79: 387 (1951).
Philippines (Basilan). 42 PHI. Phan.

Syzygium eburneum (Gagnep.) Merr. & L.M.Perry, J. Arnold Arbor. 19: 107 (1938).
Cambodia. 41 CBD. Phan.
**Eugenia eburnea* Gagnep., Notul. Syst. (Paris) 3: 324 (1918).

Syzygium ecostulatum (Elmer) Merr., Philipp. J. Sci. 79: 388 (1951).
Philippines. 42 PHI. Phan.
**Eugenia ecostulata* Elmer, Leafl. Philipp. Bot. 4: 1428 (1912).

Syzygium effusum (A.Gray) Müll.Stuttg. in W.G.Walpers, Ann. Bot. Syst. 4: 838 (1858).
New Guinea to SW. Pacific. 43 NWG SOL 60 FIJ SAM. Phan.
**Eugenia effusa* A.Gray, U.S. Expl. Exped., Phan. 1: 524 (1854).
Eugenia sylvana Ridl., Trans. Linn. Soc. London, Bot. 9: 48 (1916). *Syzygium sylvanum* (Ridl.) Merr. & L.M.Perry, J. Arnold Arbor. 23: 295 (1942).
Jambosa arfakensis Gibbs, Fl. Arfak Mts.: 153 (1917).
Syzygium leucoderme Diels, Bot. Jahrb. Syst. 57: 409 (1922).
Eugenia nivifera Greves, J. Bot. 61(Suppl.): 19 (1923). *Syzygium niviferum* (Greves) Merr. & L.M.Perry, J. Arnold Arbor. 23: 250 (1942).
Syzygium doctersii Merr. & L.M.Perry, J. Arnold Arbor. 23: 283 (1942).
Syzygium obtusum Merr. & L.M.Perry, J. Arnold Arbor. 23: 294 (1942).

Syzygium elegans (Brongn. & Gris) J.W.Dawson, Fl. Nouv.-Caléd. 23: 82 (1999).
NW. New Caledonia. 60 NWC. Nanophan.
**Caryophyllus elegans* Brongn. & Gris, Bull. Soc. Bot. France 12: 184 (1865). *Jambosa elegans* (Brongn. & Gris) Nied. in H.G.A.Engler & K.A.E.Prantl, Nat. Pflanzenfam. 3(7): 84 (1893).

Syzygium elliptifolium (Merr.) Merr., Philipp. J. Sci. 79: 388 (1951).
Philippines. 42 PHI. Phan.
**Eugenia elliptifolia* Merr., Philipp. J. Sci. 18: 291 (1921).

Syzygium elliptilimbum (Merr.) Merr. & L.M.Perry, Mem. Amer. Acad. Arts 18: 187 (1939).
Borneo to Philippines. 42 BOR PHI. Phan.
**Eugenia elliptilimba* Merr., J. Straits Branch Roy. Asiat. Soc. 77: 211 (1917).
Eugenia suluensis Merr., Philipp. J. Sci. 30: 416 (1926).

Syzygium elopurae (Ridl.) Merr. & L.M.Perry, Mem. Amer. Acad. Arts 18: 177 (1939).
Borneo. 42 BOR. Phan.
Eugenia ambongensis Ridl., J. Bot. 68: 16 (1930). *Syzygium ambongense* (Ridl.) Masam., Enum. Phan. Born.: 523 (1942).
**Eugenia elopurae* Ridl., J. Bot. 68: 15 (1930).

Syzygium embelioides (Ridl.) Masam., Enum. Phan. Born.: 528 (1942).
Borneo. 42 BOR. Phan.
**Eugenia embelioides* Ridl., J. Bot. 68: 36 (1930).

Syzygium emirnense (Baker) Labat & Schatz, Novon 12: 202 (2002).
Madagascar. 29 MDG. Nanophan. or phan.
Eugenia cuneifolia Baker, J. Linn. Soc., Bot. 20: 144 (1883). *Eugenia emirnensis* f. *cuneifolia* (Baker) H.Perrier, in Fl. Madag. 152: 54 (1953).
**Eugenia emirnensis* Baker, J. Linn. Soc., Bot. 20: 145 (1883).
Syzygium cuneifolium Bojer ex Baker, J. Linn. Soc., Bot. 20: 144 (1883).
Eugenia richardiana Cordem., Fl. Réunion: 430 (1895), nom. illeg. *Syzygium richardianum* J.Guého & A.J.Scott, Kew Bull. 34: 494 (1980).
Eugenia emirnensis var. *elongata* Hochr., Annuaire Conserv. Jard. Bot. Genève 11–12: 176 (1908).
Eugenia condensata var. *thouvenotii* Danguy ex Lecomte, Bois For. Analamazoatra, Ser. B: 106 (1922).
Eugenia emirnensis var. *podocarpifolia* H.Perrier, Mém. Inst. Sci. Madagascar, Sér. B, Biol. Vég. 4: 187 (1952). *Eugenia emirnensis* f. *podocarpifolia* (H.Perrier) H.Perrier, in Fl. Madag. 152: 54 (1953).
Eugenia emirnensis var. *submaritima* H.Perrier, Mém. Inst. Sci. Madagascar, Sér. B, Biol. Vég. 4: 187 (1952).
Eugenia emirnensis var. *subrotundifolia* H.Perrier, Mém. Inst. Sci. Madagascar, Sér. B, Biol. Vég. 4: 186 (1952). *Eugenia emirnensis* f. *subrotundifolia* (H.Perrier) H.Perrier, in Fl. Madag. 152: 54 (1953).

Syzygium endertii Merr. & L.M.Perry, Mem. Amer. Acad. Arts 18: 167 (1939).
Borneo. 42 BOR. Phan.

Syzygium endophloium B.Hyland, Austral. J. Bot., Suppl. Ser. 9: 77 (1983).
Queensland. 50 QLD. Nanophan. or phan.

Syzygium erythranthum Merr. & L.M.Perry, Mem. Amer. Acad. Arts 18: 166 (1939).
Borneo. 42 BOR. Phan.

Syzygium erythrocalyx (C.T.White) B.Hyland, Austral. J. Bot., Suppl. Ser. 9: 79 (1983).
N. Queensland. 50 QLD. Phan.
**Eugenia erythrocalyx* C.T.White, Contr. Arnold Arbor. 4: 78 (1933).

Syzygium erythrodoxum (S.Moore) B.Hyland, Austral. J. Bot., Suppl. Ser. 9: 81 (1983).
Queensland. 50 QLD. Phan.
**Eugenia erythrodoxa* S.Moore, J. Bot. 55: 304 (1917).

Syzygium erythropetalum T.G.Hartley & L.M.Perry, J. Arnold Arbor. 54: 186 (1973).
Papua New Guinea. 43 NWG. Phan.

Syzygium escritorii Merr., Philipp. J. Sci. 79: 389 (1951).
Philippines (Luzon). 42 PHI. Phan.
**Eugenia euphlebia* Merr., Philipp. J. Sci., C 10: 217 (1915), nom. illeg.

Syzygium eucalyptoides (F.Muell.) B.Hyland, Austral. J. Bot., Suppl. Ser. 9: 82 (1983).
N. Australia. 50 NTA QLD WAU. Phan.
**Jambosa eucalyptoides* F.Muell., Fragm. 1: 226 (1859). *Eugenia eucalyptoides* (F.Muell.) F.Muell., Fragm. 4: 55 (1864).

subsp. ***bleeseri*** (O.Schwarz) B.Hyland, Austral. J. Bot., Suppl. Ser. 9: 84 (1983).
NW. Australia. 50 NTA WAU. Phan.
**Eugenia bleeseri* O.Schwarz, Repert. Spec. Nov. Regni Veg. 24: 90 (1927).
Eugenia stokesii C.A.Gardner, J. Roy. Soc. Western Australia 27: 184 (1942).

subsp. ***eucalyptoides***
N. Australia. 50 NTA QLD WAU. Phan.

Syzygium eugeniiforme P.S.Ashton, Kew Bull. 61: 118 (2006).
Borneo (Sarawak). 42 BOR. Phan.

Syzygium eugenioides (Merr. & L.M.Perry) Biffin & Craven, Blumea 50: 385 (2005).
Fiji. 60 FIJ. Phan.
Calyptranthes eugenioides Seem., Fl. Vit.: 81 (1866), nom. illeg. *Acicalyptus eugenioides* F.Muell., Fragm. 8: 17 (1873). **Cleistocalyx eugenioides* (F.Muell.) Merr. & L.M.Perry, J. Arnold Arbor. 18: 330 (1937).

Syzygium euneuron Miq., Fl. Ned. Ind., Eerste Bijv.: 314 (1861). *Eugenia euneura* (Miq.) Craib, Fl. Siam. 1: 640 (1931).
Sumatera. 42 SUM. Phan.

Syzygium euonymifolium (F.P.Metcalf) Merr. & L.M.Perry, J. Arnold Arbor. 19: 242 (1938).
SE. China. 36 CHS. Nanophan. or phan.
Eugenia euonymifolia F.P.Metcalf, Lingnan Sci. J. 11: 22 (1932).

Syzygium euphlebium (Hayata) Mori, Trans. Nat. Hist. Soc. Taiwan 28: 439 (1938).
Taiwan (Hengch'un Pen.). 38 TAI. Phan.
**Eugenia euphlebia* Hayata, Icon. Pl. Formosan. 3: 119 (1913).

Syzygium evenulosum Merr. & L.M.Perry, J. Arnold Arbor. 23: 261 (1942).
New Guinea. 43 NWG. Phan.

Syzygium everettii (C.B.Rob.) Merr., Philipp. J. Sci. 79: 389 (1951).
Philippines. 42 PHI. Phan.
**Eugenia everettii* C.B.Rob., Philipp. J. Sci., C 4: 371 (1909).
Eugenia pallidifolia Merr., Philipp. J. Sci., C 10: 222 (1915).
Eugenia irosinensis Elmer, Leafl. Philipp. Bot. 10: 3767 (1939), no latin descr. *Syzygium irosinense* Merr., Philipp. J. Sci. 79: 396 (1951).

Syzygium exiguifolium Merr. & L.M.Perry, Mem. Amer. Acad. Arts 18: 161 (1939).
Borneo. 42 BOR. Phan.

Syzygium eximiiflorum (Diels) Merr. & L.M.Perry, J. Arnold Arbor. 23: 252 (1942).
New Guinea. 43 NWG. Phan.
**Jambosa eximiiflora* Diels, Nova Guinea 14: 92 (1924).

Syzygium faciflorum P.S.Ashton, Kew Bull. 61: 118 (2006).
Borneo. 42 BOR. Phan.

Syzygium fastigiatum (Blume) Merr. & L.M.Perry, Mem. Amer. Acad. Arts 18: 152 (1939).
Indo-China to Papuasia. 41 MYA THA VIE 42 BOR JAW MLY PHI SUM 43 BIS NWG. Phan.
**Calyptranthes fastigiata* Blume, Bijdr.: 1090 (1826). *Caryophyllus fastigiatus* (Blume) Blume in A.P.de Candolle, Prodr. 3: 262 (1828). *Eugenia fastigiata* (Blume) Koord. & Valeton, Meded. Lands Plantentuin 40: 104 (1900).
Eugenia bracteolata Wight, Icon. Pl. Ind. Orient. 2: t. 531 (1842). *Acmena bracteolata* (Wight) Walp., Repert. Bot. Syst. 2: 181 (1843). *Syzygium bracteolatum* (Wight) Masam., Enum. Phan. Born.: 524 (1942).
Eugenia confertiflora Koord. & Valeton, Meded. Lands Plantentuin 40: 106 (1900), nom. illeg.
Eugenia sablanensis Elmer, Leafl. Philipp. Bot. 1: 328 (1908).
Eugenia bibracteata Greves, J. Bot. 61(Suppl.): 18 (1923). *Syzygium bibracteatum* (Greves) Merr. & L.M.Perry, J. Arnold Arbor. 23: 249 (1942).
Eugenia elmeri Merr., Univ. Calif. Publ. Bot. 15: 218 (1929). *Syzygium elmeri* (Merr.) Masam., Enum. Phan. Born.: 528 (1942).

Syzygium fenicis (C.B.Rob.) Merr., Philipp. J. Sci. 79: 390 (1951).
Philippines. 42 PHI. Phan.
**Eugenia fenicis* C.B.Rob., Philipp. J. Sci., C 4: 355 (1909).

Syzygium fergusonii (Trimen) Gamble, Bull. Misc. Inform. Kew 1920: 52 (1920).
S. India, Sri Lanka. 40 IND SRL. Nanophan. or phan.
**Eugenia fergusonii* Trimen, Handb. Fl. Ceylon 2: 172 (1894).

Syzygium fibrosum (F.M.Bailey) T.G.Hartley & L.M.Perry, J. Arnold Arbor. 54: 201 (1973).
New Guinea to N. Australia. 43 NWG 50 NTA QLD. Phan.
**Eugenia fibrosa* F.M.Bailey, Queensl. Fl.: 662 (1900).

Syzygium fijiense L.M.Perry, J. Arnold Arbor. 31: 361 (1950).
Fiji. 60 FIJ. Phan.
Syzygium rubescens var. *koroense* L.M.Perry, J. Arnold Arbor. 31: 363 (1950).

Syzygium filicaudum Merr. & L.M.Perry, Mem. Amer. Acad. Arts 18: 189 (1939).
Borneo. 42 BOR. Phan.

Syzygium filiflorum J.W.Dawson, in Fl. Nouv.-Caléd. 23: 107 (1999).
N. & NW. New Caledonia. 60 NWC. Nanophan. or phan.

Syzygium filiforme Chantaran. & J.Parn., Kew Bull. 48: 598 (1993).
SW. Thailand to W. Malesia. 41 THA 42 JAW MLY. Phan.
**Eugenia filiformis* Wall. ex Duthie in J.D.Hooker, Fl. Brit. India 2: 478 (1878), nom. illeg.
Eugenia clavimyrtus var. *minor* Koord. & Valeton, Meded. Lands Plantentuin 40: 112 (1900).
Eugenia filiformis var. *constricta* Kochummen, Gard. Bull. Singapore 28: 227 (1976). *Syzygium filiforme* var. *constrictum* (Kochummen) I.M.Turner, J. Singapore Natl. Acad. Sci. 22–24: 18 (1997).

Syzygium filipes Merr., Philipp. J. Sci. 79: 390 (1951).
Philippines (Mindanao). 42 PHI. Phan.

Syzygium finetii (Gagnep.) Merr. & L.M.Perry, J. Arnold Arbor. 19: 116 (1938).
Vietnam. 41 VIE. Phan.
**Eugenia finetii* Gagnep., Notul. Syst. (Paris) 3: 324 (1918).

Syzygium finisterrae (Lauterb.) Merr. & L.M.Perry, J. Arnold Arbor. 23: 295 (1942).
New Guinea. 43 NWG. Phan.
**Eugenia finisterrae* Lauterb., Repert. Spec. Nov. Regni Veg. 13: 240 (1914).

Syzygium fischeri (Merr.) Merr., Philipp. J. Sci. 79: 391 (1951).
Philippines. 42 PHI. Phan.
**Eugenia fischeri* Merr., Philipp. J. Sci., C 10: 218 (1915).

Syzygium flavescens Merr. & L.M.Perry, J. Arnold Arbor. 23: 278 (1942).
New Guinea. 43 NWG. Phan.
**Eugenia flavescens* Ridl., Trans. Linn. Soc. London, Bot. 9: 46 (1916), nom. illeg.

Syzygium flavidum T.G.Hartley & L.M.Perry, J. Arnold Arbor. 54: 183 (1973).
Papua New Guinea. 43 NWG. Phan.

Syzygium floribundum F.Muell., Fragm. 4: 58 (1864). *Waterhousea floribunda* (F.Muell.) B.Hyland, Austral. J. Bot., Suppl. Ser. 9: 139 (1983).
E. Australia. 50 NSW QLD. Nanophan. or phan.
Metrosideros connata Desf., Tabl. École Bot.: 171 (1804), nom. nud.
**Metrosideros floribunda* Vent., Jard. Malmaison: t. 75 (1804), nom. illeg. *Eugenia ventenatii* Benth., Fl. Austral. 3: 283 (1867), nom. illeg.
Metrosideros laurifolia Dum.Cours., Bot. Cult., ed. 2, 5: 376 (1811), nom. inval.

Syzygium flosculiferum (M.R.Hend.) Sreek., J. Econ. Taxon. Bot. 17: 454 (1993).
Pen. Malaysia. 42 MLY. Phan.
**Eugenia flosculifera* M.R.Hend., Gard. Bull. Singapore 11: 329 (1947).

Syzygium fluviatile (Hemsl.) Merr. & L.M.Perry, J. Arnold Arbor. 19: 241 (1938).
SE. China. 36 CHS.
**Eugenia fluviatilis* Hemsl., J. Linn. Soc., Bot. 23: 296 (1887).

Syzygium fluvicola (T.G.Hartley & Craven) Craven & Biffin, Blumea 51: 137 (2006).
Papua New Guinea. 43 NWG. Phan.
**Acmena fluvicola* T.G.Hartley & Craven, J. Arnold Arbor. 58: 336 (1977).

Syzygium formosanum (Hayata) Mori, Trans. Nat. Hist. Soc. Taiwan 28: 439 (1938).

Taiwan. 38 TAI. Phan.
Eugenia acutisepala Hayata, J. Coll. Sci. Imp. Univ. Tokyo 30(1): 112 (1911). *Syzygium acutisepalum* (Hayata) Mori, Trans. Nat. Hist. Soc. Taiwan 28: 438 (1938).
**Eugenia formosana* Hayata, J. Coll. Sci. Imp. Univ. Tokyo 30(1): 113 (1911).

Syzygium formosum (Wall.) Masam., Enum. Phan. Born.: 528 (1942).
C. Himalaya to Indo-China. 40 ASS BAN EHM NEP 41 MYA THA VIE. Phan.
**Eugenia formosa* Wall., Pl. Asiat. Rar. 2: 6 (1830). *Jambosa formosa* (Wall.) G.Don, Gen. Hist. 2: 969 (1832).
Jambosa tenuifolia Sweet, Hort. Brit., ed. 2: 212 (1830).
Eugenia ternifolia Roxb., Fl. Ind. ed. 1832, 2: 489 (1832).
Jambosa mappacea Korth., Ned. Kruidk. Arch. 1: 200 (1847). *Syzygium mappaceum* (Korth.) Merr. & L.M.Perry, Mem. Amer. Acad. Arts 18: 164 (1939).

Syzygium forrestii Merr. & L.M.Perry, J. Arnold Arbor. 19: 238 (1938).
SC. China. 36 CHC.

Syzygium forte (F.Muell.) B.Hyland, Austral. J. Bot., Suppl. Ser. 9: 89 (1983).
New Guinea to N. Australia. 43 NWG 50 NTA QLD WAU. Phan.
**Eugenia fortis* F.Muell., Fragm. 5: 13 (1865).

subsp. ***forte***
New Guinea to N. Australia. 43 NWG 50 NTA QLD WAU. Phan.
Syzygium rubiginosum Merr. & L.M.Perry, J. Arnold Arbor. 23: 289 (1942).

subsp. ***potamophilum*** B.Hyland, Austral. J. Bot., Suppl. Ser. 9: 90 (1983).
N. Australia. 50 NTA QLD WAU. Phan.

Syzygium fossiramulosum P.S.Ashton, Kew Bull. 61: 121 (2006).
Borneo (E. Sabah). 42 BOR. Phan.

Syzygium foxworthianum (Ridl.) Merr. & L.M.Perry, Mem. Amer. Acad. Arts 18: 168 (1939).
Pen. Thailand to W. Malesia. 41 THA 42 BOR MLY SUM. Phan.
Eugenia densiflora var. *angustifolia* Ridl., Fl. Malay Penins. 1: 729 (1922).
Eugenia foxworthyi Ridl., Fl. Malay Penins. 1: 728 (1922), nom. illeg. **Eugenia foxworthiana* Ridl., Fl. Malay Penins. 5: 308 (1925).

Syzygium foxworthyi (Elmer) Merr., Philipp. J. Sci. 79: 391 (1951).
Philippines. 42 PHI. Phan.
**Eugenia foxworthyi* Elmer, Leafl. Philipp. Bot. 4: 1414 (1912).

Syzygium francisii (F.M.Bailey) L.A.S.Johnson, Contr. New South Wales Natl. Herb. 3: 99 (1962).
Queensland to NE. New South Wales. 50 NSW QLD. Nanophan. or phan.
**Eugenia francisii* F.M.Bailey, Queensland Agric. J. 26: 315 (1911).
Eugenia tomlinsonii Maiden & Betche, Proc. Linn. Soc. New South Wales 38: 247 (1913).

Syzygium fraternum Miq., Fl. Ned. Ind. 1(1): 458 (1855).
Jawa. 42 JAW. Phan.

Syzygium fratris Craven, Blumea 48: 480 (2003).
Queensland. 50 QLD. Phan.

Syzygium frutescens Brongn. & Gris, Bull. Soc. Bot. France 12: 184 (1865).
New Caledonia. 60 NWC. Nanophan. or phan.
Syzygium koghiense Guillaumin, Bull. Soc. Bot. France 89: 17 (1934).

Syzygium fruticosum DC., Prodr. 3: 260 (1828). *Eugenia fruticosa* (DC.) Roxb., Fl. Ind. ed. 1832, 2: 487 (1832).
SC. China to Indo-China. 36 CHC 40 BAN 41 LAO MYA THA. Phan.

Syzygium fullagarii (F.Muell.) Craven, Muelleria 11: 95 (1998).
Lord Howe I. 50 NFK. Phan.
**Acicalyptus fullagarii* F.Muell., Fragm. 8: 15 (1874). *Cleistocalyx fullagarii* (F.Muell.) Merr. & L.M. Perry, J. Arnold Arbor. 18: 331 (1937).

Syzygium fulvotomentosum P.S.Ashton, Kew Bull. 61: 121 (2006).
Borneo. 42 BOR. Phan.

Syzygium furfuraceum Merr. & L.M.Perry, J. Arnold Arbor. 23: 276 (1942).
New Guinea to Bismarck Arch. 43 BIS NWG. Phan.
Syzygium folidorhachis Merr. & L.M.Perry, J. Arnold Arbor. 23: 276 (1942).

Syzygium fuscescens (Craib) Chantaran. & J.Parn., Kew Bull. 49: 598 (1993).
Pen. Thailand. 41 THA. Phan.
**Eugenia fuscescens* Craib, Bull. Misc. Inform. Kew 1930: 166 (1930).

Syzygium fusticuliferum (Ridl.) Merr. & L.M.Perry, Mem. Amer. Acad. Arts 18: 184 (1939).
Borneo. 42 BOR. Phan.
**Eugenia fusticulifera* Ridl., J. Bot. 68: 33 (1930).

Syzygium gageanum (King) I.M.Turner, J. Singapore Natl. Acad. Sci. 22–24: 19 (1997).
Pen. Malaysia. 42 MLY. Phan.
**Eugenia gageana* King, J. Asiat. Soc. Bengal, Pt. 2, Nat. Hist. 70: 96 (1901).

Syzygium ganophyllum Diels, Bot. Jahrb. Syst. 57: 408 (1922).
New Guinea. 43 NWG. Phan.

Syzygium garciae (Merr.) Merr., Philipp. J. Sci. 79: 391 (1951).
Philippines. 42 PHI. Phan.
**Jambosa garciae* Merr., Publ. Bur. Sci. Gov. Lab. 17: 36 (1904).

Syzygium garciniifolium (King) Merr. & L.M.Perry, Mem. Amer. Acad. Arts 18: 167 (1939).
W. Malesia. 42 BOR MLY SUM. Phan.
**Eugenia garciniifolia* King, J. Asiat. Soc. Bengal, Pt. 2, Nat. Hist. 70: 90 (1901).

var. ***garciniifolium***
W. Malesia. 42 BOR MLY SUM. Phan.

var. ***patentinervium*** P.S.Ashton, Kew Bull. 61(1): 123 (2006).
Sumatera, Borneo. 42 BOR SUM. Phan.

Syzygium garcinioides (Ridl.) Merr. & L.M.Perry, J. Arnold Arbor. 23: 249 (1942).
New Guinea. 43 NWG. Phan.

Eugenia garcinioides Ridl., Trans. Linn. Soc. London, Bot. 9: 44 (1916).

Syzygium gardneri Thwaites, Enum. Pl. Zeyl.: 117 (1859). *Eugenia gardneri* (Thwaites) Bedd., Fl. Sylv. S. India: 108 (1872).
S. India, Sri Lanka. 40 IND SRL. Phan.

Syzygium gaultherioides (Ridl.) Merr. & L.M.Perry, Mem. Amer. Acad. Arts 18: 161 (1939).
Borneo. 42 BOR.
Eugenia gaultherioides Ridl., J. Bot. 68: 16 (1930).

Syzygium germainii Amshoff, Acta Bot. Neerl. 8: 54 (1959).
Zaïre. 23 ZAI. Phan.

Syzygium gerrardii (Harv. ex Hook.f.) Burtt Davy, Ann. Transvaal Mus. 3: 122 (1912).
S. Africa. 27 CPP NAT SWZ TVL. Nanophan. or phan.
Acmena gerrardii Harv. ex Hook.f., Gen. S. Afr. Pl., ed. 2: 112 (1868). *Eugenia gerrardii* (Harv. ex Hook.f.) Sim, Forest Fl. Cape: 226 (1907). *Syzygium guineense* subsp. *gerrardii* (Harv. ex Hook.f.) F.White, Kirkia 10: 404 (1977).

Syzygium gigantifolium (Merr.) Merr., Philipp. J. Sci. 79: 391 (1951).
Philippines. 42 PHI. Phan.
Eugenia gigantifolia Merr., Philipp. J. Sci., C 4: 349 (1909).

Syzygium gillespiei Merr. & L.M.Perry, Sargentia 1: 78 (1942).
Fiji. 60 FIJ. Nanophan. or phan.

Syzygium gilletii De Wild., Bull. Jard. Bot. État 4: 375 (1914).
WC. Trop. Africa. 23 CAF CON GAB ZAI. Phan.

Syzygium giorgii De Wild., Bull. Jard. Bot. État 4: 376 (1914).
Zaïre. 23 ZAI. Nanophan. or phan.

Syzygium gitingense (Elmer) Merr., Philipp. J. Sci. 79: 391 (1951).
Philippines. 42 PHI. Phan.
Eugenia gitingensis Elmer, Leafl. Philipp. Bot. 4: 1409 (1912).

Syzygium gjellerupii Lauterb., Nova Guinea 8: 852 (1912).
New Guinea. 43 NWG. Phan.

Syzygium glabratum (DC.) Veldkamp, Blumea 48: 489 (2003).
W. Malesia to Philippines. 42 BOR JAW LSI MLY PHI SUM. Phan.
Myrtus glabrata Blume, Bijdr.: 1088 (1826), nom. illeg. **Jambosa glabrata* DC., Prodr. 3: 287 (1828). *Clavimyrtus glabrata* (DC.) Blume, Mus. Bot. 1: 113 (1850). *Eugenia blumeana* Kuntze, Revis. Gen. Pl. 1: 239 (1891).
Jambosa gracilis Korth., Ned. Kruidk. Arch. 1: 202 (1847). *Eugenia clavimyrtus* Koord. & Valeton, Meded. Lands Plantentuin 40: 110 (1900). *Syzygium gracile* (Korth.) Amshoff in C.A.Backer & R.C.Bakhuizen van der Brink, Bekn. Fl. Java 4b(98): 22 (1944).
Clavimyrtus marginata Blume, Mus. Bot. 1: 113 (1850). *Jambosa marginata* (Blume) Miq., Fl. Ned. Ind. 1(1) 428 (1855).
Clavimyrtus virens Blume, Mus. Bot. 1: 113 (1850). *Jambosa virens* (Blume) Miq., Fl. Ned. Ind. 1(1): 428 (1855). *Eugenia virens* (Blume) Koord. & Valeton, Meded. Lands Plantentuin 40: 113 (1900).
Eugenia fusiformis Duthie in J.D.Hooker, Fl. Brit. India 2: 479 (1878). *Syzygium fusiforme* (Duthie) Merr. & L.M.Perry, Mem. Amer. Acad. Arts 18: 176 (1939).
Eugenia leptogyna C.B.Rob., Philipp. J. Sci., C 4: 368 (1909).

Syzygium gladiatum (Ridl.) Merr. & L.M.Perry, Mem. Amer. Acad. Arts 18: 181 (1939).
Borneo. 42 BOR. Phan.
Eugenia gladiata Ridl., J. Bot. 68: 35 (1930).

Syzygium glanduligerum (Ridl.) Merr. & L.M.Perry, Mem. Amer. Acad. Arts 18: 169 (1939).
Borneo. 42 BOR. Phan.
Eugenia glanduligera Ridl., J. Bot. 68: 14 (1930).

Syzygium glaucissimum (Haines) Rathakr. & N.C.Nair, J. Econ. Taxon. Bot. 4: 287 (1983).
India. 40 IND. Nanophan. or phan.
Eugenia glaucissima Haines, Bot. Bihar Orissa 3: 362 (1922).

Syzygium glaucum (King) Chantaran. & J.Parn., Kew Bull. 48: 598 (1993).
Indo-China to Pen. Malaysia. 41 MYA THA 42 MLY. Phan.
Eugenia glauca King, J. Asiat. Soc. Bengal, Pt. 2, Nat. Hist. 70: 102 (1901).
Eugenia glauca var. *pseudoglauca* King, J. Asiat. Soc. Bengal, Pt. 2, Nat. Hist. 70(2): 102 (1901). *Eugenia pseudoglauca* (King) Ridl., Fl. Malay Penins. 1: 737 (1922).

Syzygium glenum Craven, Blumea 48: 482 (2003).
Queensland. 50 QLD. Phan.

Syzygium globiflorum (Craib) Chantaran. & J.Parn., Kew Bull. 48: 598 (1993).
Hainan, SC. China to N. Thailand. 36 CHC CHH 41 THA. Phan.
Eugenia globiflora Craib, Bull. Misc. Inform. Kew 1930: 167 (1930).
Syzygium brachyantherum Merr. & L.M.Perry, J. Arnold Arbor. 19: 218 (1938).

Syzygium globosum (Elmer) Merr., Philipp. J. Sci. 79: 392 (1951).
Philippines. 42 PHI. Phan.
Eugenia banaba Elmer, Leafl. Philipp. Bot. 4: 1443 (1912).
Eugenia globosa Elmer, Leafl. Philipp. Bot. 4: 1404 (1912).

Syzygium glomeratum (Lam.) DC., Prodr. 3: 259 (1828).
Mauritius. 29 MAU. Nanophan. or phan.
Eugenia glomerata Lam., Encycl. 3: 199 (1789). *Myrtus glomerata* (Lam.) Spreng., Syst. Veg. 2: 483 (1825). *Jossinia glomerata* (Lam.) Blume, Mus. Bot. 1: 122 (1850).

Syzygium glomerulatum (Gagnep.) Merr. & L.M.Perry, J. Arnold Arbor. 19: 112 (1938).
Cambodia. 41 CBD. Phan.
Eugenia glomerulata Gagnep., Notul. Syst. (Paris) 3: 325 (1918).

Syzygium glomeruliferum Amshoff in C.A.Backer & R.C.Bakhuizen van der Brink, Bekn. Fl. Java 4b(98): 14 (1944).
Jawa. 42 JAW. Phan.
Eugenia glomerata Koord. & Valeton, Bull. Inst. Bot. Buitenzorg 2: 7 (1899), nom. illeg.

Syzygium gonatanthum (Diels) Merr. & L.M.Perry, J. Arnold Arbor. 23: 256 (1942).
New Guinea. 43 NWG. Phan.
**Jambosa gonatantha* Diels, Bot. Jahrb. Syst. 57: 384 (1922).
Eugenia forbesii Greves, J. Bot. 61(Suppl.): 16 (1923). *Syzygium forbesii* (Greves) Merr. & L.M.Perry, J. Arnold Arbor. 23: 261 (1942).

Syzygium goniocalyx (Lauterb.) Merr. & L.M.Perry, J. Arnold Arbor. 23: 262 (1942).
New Guinea. 43 NWG. Phan.
**Jambosa goniocalyx* Lauterb., Nova Guinea 8: 851 (1912).

Syzygium goniopterum (Diels) Merr. & L.M.Perry, J. Arnold Arbor. 23: 259 (1942).
New Guinea. 43 NWG. Phan.
**Jambosa gonioptera* Diels, Bot. Jahrb. Syst. 57: 391 (1922).

Syzygium gonshanense P.Y.Bai, Acta Bot. Yunnan., Suppl. 5: 26 (1992).
China (Yunnan). 36 CHC. Phan.

Syzygium goodenovii (King) Masam., Enum. Phan. Born.: 529 (1942).
Pen. Malaysia to Sumatera. 42 MLY SUM. Phan.
**Eugenia goodenovii* King, J. Asiat. Soc. Bengal, Pt. 2, Nat. Hist. 70: 117 (1901).

Syzygium gracilipaniculum P.S.Ashton, Kew Bull. 61: 123 (2006).
Borneo. 42 BOR. Phan.

Syzygium gracilipes (A.Gray) Merr. & L.M.Perry, Sargentia 1: 78 (1942).
Fiji. 60 FIJ. Nanophan. or phan.
**Eugenia gracilipes* A.Gray, U.S. Expl. Exped., Phan. 1: 513 (1854). *Jambosa gracilipes* (A.Gray) Müll. Stuttg. in W.G.Walpers, Ann. Bot. Syst. 4: 849 (1858).
Eugenia vitiensis Turrill, J. Linn. Soc., Bot. 43: 21 (1915). *Syzygium vitiense* (Turrill) Merr. & L.M. Perry, Sargentia 1: 78 (1942).

Syzygium graeffei Whistler, J. Arnold Arbor. 69: 175 (1988).
Samoa. 60 SAM. Phan.

Syzygium graeme-andersoniae (Ridl.) I.M.Turner, J. Singapore Natl. Acad. Sci. 22–24: 19 (1997).
Pen. Malaysia. 42 MLY. Phan.
**Eugenia graeme-andersoniae* Ridl., J. Fed. Malay States Mus. 10: 134 (1920).

Syzygium grande (Wight) Walp., Repert. Bot. Syst. 2: 180 (1843).
Seychelles to W. Malesia. 29 SEY 36 CHC 40 ASS BAN IND SRL 41 AND LAO MYA THA VIE 42 BOR MLY. Phan.
Eugenia cymosa Roxb., Fl. Ind. ed. 1832, 2: 492 (1832).
**Eugenia grandis* Wight, Ill. Ind. Bot. 2: 17 (1841). *Jambosa grandis* (Wight) Blume, Mus. Bot. 1: 108 (1850).
Eugenia montana Wight, Icon. Pl. Ind. Orient. 3: t. 1060 (1846), nom. illeg. *Syzygium montanum* Thwaites & Hook.f. in G.H.K.Thwaites, Enum. Pl. Zeyl.: 116 (1859). *Syzygium tamilnadensis* Rathakr. & V.Chithra in N.C.Nair & A.N.Henry, Fl. Tamil Nadu 1: 158 (1983), nom. superfl. *Syzygium gadgilii* M.R.Almeida, Fl. Maharashtra 2: 273 (1998), nom. superfl.
Jambosa firma Blume, Mus. Bot. 1: 108 (1850). *Syzygium firmum* (Blume) Thwaites, Enum. Pl. Zeyl.: 417 (1864).
Eugenia laosensis Gagnep., Notul. Syst. (Paris) 3: 326 (1918). *Syzygium laosense* (Gagnep.) Merr. & L.M.Perry, J. Arnold Arbor. 19: 113 (1938).
Eugenia laosensis var. *quocensis* Gagnep., Notul. Syst. (Paris) 3: 327 (1918). *Syzygium laosense* var. *quocense* (Gagnep.) H.T.Chang & R.H.Miao, in Fl. Reipubl. Popul. Sin. 53(1): 70 (1984).
Syzygium megalophyllum Merr. & L.M.Perry, Mem. Amer. Acad. Arts 18: 179 (1939).
Syzygium grande var. *parviflorum* Chantaran. & J.Parn., Kew Bull. 48: 598 (1993).

Syzygium graveolens (F.M.Bailey) Craven & Biffin, Blumea 51: 137 (2006).
NE. Queensland. 50 QLD. Phan.
Acmena macrocarpa C.T.White, Proc. Roy. Soc. Queensland 53: 217 (1942).
Acmena graveolens (F.M.Bailey) L.S.Sm., Proc. Roy. Soc. Queensland 67: 34 (1956).

Syzygium grayi (Seem.) Merr. & L.M.Perry, Sargentia 1: 76 (1942).
Fiji. 60 FIJ. Phan.
**Eugenia grayi* Seem., Fl. Vit.: 79 (1865).

Syzygium grevesianum Merr. & L.M.Perry, J. Arnold Arbor. 23: 253 (1942).
New Guinea. 43 NWG. Phan.
**Eugenia pterocalyx* Greves, J. Bot. 61(Suppl.): 17 (1923), nom. illeg.

Syzygium griffithii (Duthie) Merr. & L.M.Perry, Mem. Amer. Acad. Arts 18: 174 (1939).
Pen. Malaysia to Borneo. 42 BOR MLY. Phan.
**Eugenia griffithii* Duthie in J.D.Hooker, Fl. Brit. India 2: 481 (1878).
Eugenia subrufa King, J. Asiat. Soc. Bengal, Pt. 2, Nat. Hist. 70: 102 (1901). *Syzygium subrufum* (King) Masam., Enum. Phan. Born.: 539 (1942).

Syzygium grijsii (Hance) Merr. & L.M.Perry, J. Arnold Arbor. 19: 233 (1938).
SE. China. 36 CHS. Phan.
**Eugenia grijsii* Hance, J. Bot. 9: 5 (1871).

Syzygium griseum (C.B.Rob.) Airy Shaw, Kew Bull. 4: 124 (1949).
Philippines. 42 PHI. Phan.
**Eugenia grisea* C.B.Rob., Philipp. J. Sci. 55: 395 (1909).

Syzygium guangxiense H.T.Chang & R.H.Miao, Acta Bot. Yunnan. 4: 22 (1982).
China (Guangxi). 36 CHS. Phan.

Syzygium guehoi Bosser & Florens, Adansonia, III, 22: 193 (2000).
Mauritius. 29 MAU.

Syzygium guillauminii J.W.Dawson, in Fl. Nouv.-Caléd. 23: 49 (1999).
New Caledonia. 60 NWC. Nanophan. or phan.
**Jambosa nervosa* Vieill. ex Guillaumin, Bull. Soc. Bot. France 85: 642 (1938 publ. 1939).

Syzygium guineense (Willd.) DC., Prodr. 3: 259 (1828).
Trop. & S. Africa, SW. Arabian Pen. 22 BEN GAM GHA GNB GUI IVO LBR MLI NGA SEN SIE TOG 23 BUR CAF CMN GAB GGI RWA ZAI 24 ERI ETH SOM SUD 25 KEN TAN UGA 26 ANG

MLW MOZ ZAM ZIM 27 BOT CPV NAM NAT SWZ TVL 35 SAU YEM. Phan.
**Calyptranthes guineensis* Willd., Sp. Pl. 2: 974 (1799). *Eugenia guineensis* (Willd.) Baill. ex Laness., Pl. Util. Col. Franç.: 822 (1886).

subsp. ***afromontana*** F.White, Forest Fl. N. Rhodes.: 455 (1962).
Sudan to S. Trop. Africa. 23 ZAI 24 SUD 25 KEN TAN UGA 26 ANG MLW MOZ ZAM ZIM. Phan.

subsp. ***barotsense*** F.White, Forest Fl. N. Rhodes.: 455 (1962).
Zaïre to S. Africa. 23 ZAI 26 ANG ZAM ZIM 27 BOT CPV. Phan.

subsp. ***guineense***
Trop. & S. Africa, SW. Arabian Pen. 22 BEN GAM GHA GNB GUI IVO LBR MLI NGA SEN SIE TOG 23 BUR CAF CMN GAB RWA ZAI 24 ERI ETH SOM SUD 25 KEN TAN UGA 26 MLW MOZ ZAM ZIM 27 BOT NAM NAT SWZ TVL 35 SAU YEM. Phan.
Eugenia fourcadei Dummer, Gard. Chron., III, 52: 152 (1912). *Syzygium fourcadei* (Dummer) Burtt Davy, Man. Pl. Transvaal 1: 50 (1926).
Syzygium guineense var. *macrocarpum* Engl. in H.G.A.Engler & C.G.O.Drude, Veg. Erde 3(2): 738 (1921). *Syzygium guineense* subsp. *macrocarpum* (Engl.) F.White, Forest Fl. N. Rhodes.: 455 (1962).

subsp. ***huillense*** (Hiern) F.White, Forest Fl. N. Rhodes.: 455 (1962).
Zaïre to S. Trop. Africa. 23 ZAI 25 TAN 26 ANG MLW ZAM ZIM. Cham.
**Eugenia guineensis* var. *huillensis* Hiern, Cat. Afr. Pl. 1: 359 (1898). *Syzygium huillense* (Hiern) Engl., Bot. Jahrb. Syst. 54: 339 (1917).
Syzygium mumbwaense Greenway, Bull. Misc. Inform. Kew 1928: 196 (1928).

var. ***littorale*** Keay, Kew Bull. 8: 289 (1953). *Syzygium guineense* subsp. *littorale* (Keay) Boutique, in Fl. Congo Belge, Myrt.: 18 (1968).
W. & WC. Trop. Africa. 22 BEN GHA IVO NGA TOG 23 CMN GAB GGI ZAI. Phan.
Syzygium littorale Aubrév., Fl. Forest. Côte d'Ivoire 3: 70 (1936), no latin descr.

Syzygium gustavioides (F.M.Bailey) B.Hyland, Austral. J. Bot., Suppl. Ser. 9: 93 (1983).
N. & NE. Queensland. 50 QLD. Phan.
**Eugenia gustavioides* F.M.Bailey, Queensland Agric. J. 5: 389 (1899). *Cleistocalyx gustavioides* (F.M. Bailey) Merr. & L.M.Perry, J. Arnold Arbor. 18: 334 (1937).

Syzygium gyrostemoneum Diels, Bot. Jahrb. Syst. 57: 412 (1922).
New Guinea. 43 NWG. Phan.

Syzygium hainanense H.T.Chang & R.H.Miao, Acta Bot. Yunnan. 4: 20 (1982).
Hainan. 36 CHH. Phan.

Syzygium halophilum (Merr.) Masam., Enum. Phan. Born.: 529 (1942).
Borneo (Banggi) to Philippines. 42 BOR PHI. Phan.
Eugenia maritima Merr., Philipp. J. Sci., C 10: 212 (1915), nom. illeg. **Eugenia halophila* Merr., Enum. Philipp. Fl. Pl. 3: 167 (1923).

Syzygium hancei Merr. & L.M.Perry, J. Arnold Arbor. 19: 242 (1938).
SE. China. 36 CHS. Phan.
**Eugenia minutiflora* Hance, J. Bot. 9: 5 (1871).

Syzygium handelii Merr. & L.M.Perry, J. Arnold Arbor. 19: 233 (1938).
S. China. 36 CHC CHS. Nanophan. or phan.

Syzygium haniffii (M.R.Hend.) I.M.Turner, J. Singapore Natl. Acad. Sci. 22–24: 19 (1997).
Pen. Malaysia. 42 MLY. Phan.
**Eugenia haniffii* M.R.Hend., Gard. Bull. Singapore 11: 309 (1947).

Syzygium harmandii (Gagnep.) Merr. & L.M.Perry, J. Arnold Arbor. 19: 115 (1938).
Laos. 41 LAO. Phan.
**Eugenia harmandii* Gagnep., Notul. Syst. (Paris) 3: 325 (1918).

Syzygium havilandii (Merr.) Merr. & L.M.Perry, Mem. Amer. Acad. Arts 18: 193 (1939).
Borneo. 42 BOR. Phan.
**Eugenia havilandii* Merr., J. Straits Branch Roy. Asiat. Soc. 77: 222 (1917).

Syzygium hebephyllum Melville, Kew Bull. 10: 293 (1955).
Samoa. 60 SAM. Phan.

Syzygium hedraiophyllum (F.Muell.) Craven & Biffin, Blumea 51: 137 (2006).
N. & NE. Queensland. 50 QLD. Nanophan. or phan.
**Eugenia hedraiophylla* F.Muell., Victorian Naturalist 8: 198 (1892). *Waterhousea hedraiophylla* (F.Muell.) B.Hyland, Austral. J. Bot., Suppl. Ser. 9: 141 (1983).

Syzygium helferi (Duthie) Chantaran. & J.Parn., Kew Bull. 48: 599 (1993).
Indo-China to W. Malesia. 41 MYA THA 42 BOR MLY. Nanophan. or phan.
**Eugenia helferi* Duthie in J.D.Hooker, Fl. Brit. India 2: 480 (1878).

Syzygium heloanthum Diels, Bot. Jahrb. Syst. 57: 403 (1922).
New Guinea. 43 NWG. Phan.

Syzygium hemilamprum (F.Muell.) Craven & Biffin, Blumea 51: 137 (2006).
Papua New Guinea, Northern Territory (Melvill I.), E. Australia. 43 NWG 50 NSW NTA QLD. Phan.
**Eugenia hemilampra* F.Muell., Fragm. 9: 45 (1875). *Acmena hemilampra* (F.Muell.) Merr. & L.M.Perry, J. Arnold Arbor. 19: 15 (1938).

subsp. ***hemilamprum***
Papua New Guinea, Northern Territory (Melvill I.), E. Australia. 43 NWG 50 NSW NTA QLD. Phan.
Eugenia smithii var. *hemilampra* F.Muell., Fragm. 9: 145 (1875).
Syzygium acmenoides Merr. & L.M.Perry, J. Arnold Arbor. 23: 291 (1942).

subsp. ***orophilum*** (B.Hyland) Craven & Biffin, Blumea 51: 137 (2006).
E. Queensland. 50 QLD. Phan.
**Acmena hemilampra* subsp. *orophila* B.Hyland, Austral. J. Bot., Suppl. Ser. 9: 13 (1983).

Syzygium hemisphericum (Wight) Alston in H.Trimen, Handb. Fl. Ceylon 6(Suppl.): 115 (1931).
S. India, Sri Lanka, Thailand. 40 IND SRL 41 THA. Phan.

Eugenia hemispherica Wight, Ill. Ind. Bot. 2: 14 (1841). *Strongylocalyx hemisphaericus* (Wight) Blume, Mus. Bot. 1: 89 (1850).
Jambosa hemispherica Wight, Ill. Ind. Bot. 2: 14 (1841), pro syn.

Syzygium hemsleyanum (King) Chantaran. & J.Parn., Kew Bull. 48: 599 (1993).
Pen. Thailand to Pen. Malaysia. 41 THA 42 MLY. Phan.
Eugenia hemsleyana King, J. Asiat. Soc. Bengal, Pt. 2, Nat. Hist. 70: 88 (1901).

subsp. ***hemsleyanum***
Pen. Malaysia. 42 MLY. Phan.

subsp. ***paucinervium*** Chantaran. & J.Parn., Kew Bull. 48: 599 (1993).
Pen. Thailand. 41 THA. Phan.

Syzygium hendersonii Merr., Philipp. J. Sci. 79: 369 (1951).
Pen. Malaysia. 42 MLY. Phan.
Eugenia auriculata Ridl., J. Straits Branch Roy. Asiat. Soc. 61: 7 (1912).

Syzygium heterobotrys Merr. & L.M.Perry, J. Arnold Arbor. 23: 257 (1942).
New Guinea. 43 NWG. Phan.

Syzygium hirtum (Korth.) Merr. & L.M.Perry, Mem. Amer. Acad. Arts 18: 157 (1939).
Sumatera to Borneo. 42 BOR SUM. Phan.
Jambosa hirta Korth., Ned. Kruidk. Arch. 1: 200 (1847).
Jambosa rufotomentosa Gibbs, J. Linn. Soc., Bot. 42: 77 (1914). *Syzygium rufotomentosum* (Gibbs) Masam., Enum. Phan. Born.: 538 (1942).

Syzygium hodgkinsoniae (F.Muell.) L.A.S.Johnson, Contr. New South Wales Natl. Herb. 3: 99 (1962).
SE. Queensland to NE. New South Wales. 50 NSW QLD. Nanophan. or phan.
Eugenia hodgkinsoniae F.Muell., Fragm. 9: 145 (1875).
Eugenia fitzgeraldii F.M.Bailey, Bot. Bull. Dept. Agric. Queensland 3: 11 (1891).

Syzygium homichlophilum Diels, Bot. Jahrb. Syst. 57: 409 (1922).
New Guinea. 43 NWG. Phan.
Syzygium retivenium Merr. & L.M.Perry, J. Arnold Arbor. 23: 294 (1942).

Syzygium hoseanum (King) Merr. & L.M.Perry, Mem. Amer. Acad. Arts 18: 150 (1939).
Pen. Malaysia. 42 MLY. Phan.
Eugenia hoseana King, J. Asiat. Soc. Bengal, Pt. 2, Nat. Hist. 70: 106 (1901).

Syzygium houttuynii Merr. & L.M.Perry, Mem. Amer. Acad. Arts 18: 168 (1939).
Borneo. 42 BOR. Phan.

Syzygium houttuyniifolia P.S.Ashton, Kew Bull. 61: 125 (2006).
Borneo (Sarawak). 42 BOR. Phan.

Syzygium howii Merr. & L.M.Perry, J. Arnold Arbor. 19: 243 (1938).
Hainan. 36 CHH. Phan.

Syzygium hughcumingii Merr., Philipp. J. Sci. 79: 394 (1951).
Philippines (Luzon). 42 PHI. Phan.

Syzygium hulletianum (King) Chantaran. & J.Parn., Kew Bull. 48: 601 (1993).
Pen. Thailand to Sumatera. 41 THA 42 MLY SUM. Phan.
Eugenia hulletiana King, J. Asiat. Soc. Bengal, Pt. 2, Nat. Hist. 70: 97 (1901).

Syzygium humblotii Labat & Schatz, Novon 12: 203 (2002).
Comoros (Mayotte). 29 COM. Nanophan. or phan.
Eugenia humblotii H.Perrier, Mém. Inst. Sci. Madagascar, Sér. B, Biol. Vég. 4: 193 (1953), nom. illeg.

Syzygium hutchinsonii (C.B.Robinson) Merr., Philipp. J. Sci. 79: 395 (1951).
Philippines. 42 PHI. Phan.
Eugenia hutchinsonii C.B.Robinson, Philipp. J. Sci., C 4: 376 (1909).

Syzygium hylochare (Diels) Merr. & L.M.Perry, J. Arnold Arbor. 23: 249 (1942).
New Guinea. 43 NWG. Phan.
Jambosa hylocharis Diels, J. Arnold Arbor. 10: 83 (1929).

Syzygium hylophilum (K.Schum. & Lauterb.) Merr. & L.M.Perry, J. Arnold Arbor. 23: 273 (1942).
New Guinea. 43 NWG. Phan.
Jambosa hylophila K.Schum. & Lauterb., Fl. Schutzgeb. Südsee: 471 (1900).
Jambosa weinlandii K.Schum. in K.M.Schumann & C.A.G.Lauterbach, Fl. Schutzgeb. Südsee, Nachtr. : 326 (1905).

Syzygium hypsipetes Airy Shaw, Kew Bull. 4: 120 (1949).
Borneo. 42 BOR. Phan.

Syzygium idrisii P.S.Ashton, Kew Bull. 61: 125 (2006).
Borneo. 42 BOR. Phan.

Syzygium iliasii P.S.Ashton, Kew Bull. 61: 125 (2006).
Borneo (Sarawak). 42 BOR. Phan.

Syzygium ilocanum (Merr.) Merr., Philipp. J. Sci. 79: 395 (1951).
Philippines. 42 PHI. Phan.
Eugenia ilocana Merr., Philipp. J. Sci. 18: 291 (1921).

Syzygium imitans Merr. & L.M.Perry, J. Arnold Arbor. 19: 113 (1938).
SE. China to Vietnam. 36 CHS 41 VIE. Phan.

Syzygium imperiale P.S.Ashton, Kew Bull. 61: 127 (2006).
Borneo. 42 BOR. Phan.

Syzygium inasense (King) I.M.Turner, J. Singapore Natl. Acad. Sci. 22–24: 20 (1997).
Pen. Malaysia. 42 MLY. Phan.
Eugenia inasensis King, J. Asiat. Soc. Bengal, Pt. 2, Nat. Hist. 70: 120 (1901).

Syzygium incarnatum (Elmer) Merr. & L.M.Perry, Mem. Amer. Acad. Arts 18: 195 (1939).
W. Malesia to W. Philippines. 42 BOR MLY PHI SUM. Phan.
Syzygium punctulatum Wall., Numer. List: 3583 (1831), nom. nud.
Eugenia punctulata King, J. Asiat. Soc. Bengal, Pt. 2, Nat. Hist. 70: 122 (1901), nom. illeg.
Eugenia incarnata Elmer, Leafl. Philipp. Bot. 4: 1416 (1912).

Eugenia cerina M.R.Hend., Gard. Bull. Singapore 11: 322 (1947). *Syzygium cerinum* (M.R.Hend.) I.M. Turner, J. Singapore Natl. Acad. Sci. 22–24: 17 (1997).
Eugenia cerina var. *montana* M.R.Hend., Gard. Bull. Singapore 12: 171 (1949). *Syzygium cerinum* var. *montanum* (M.R.Hend.) I.M.Turner, J. Singapore Natl. Acad. Sci. 22–24: 17 (1997).
Eugenia cerina var. *turbinata* M.R.Hend., Gard. Bull. Singapore 12: 170 (1949). *Syzygium cerinum* var. *turbinatum* (M.R.Hend.) I.M.Turner, J. Singapore Natl. Acad. Sci. 22–24: 17 (1997).

Syzygium incrassatum (Elmer) Merr., Philipp. J. Sci. 79: 395 (1951).
Philippines. 42 PHI. Phan.
**Eugenia incrassata* Elmer, Leafl. Philipp. Bot. 2: 581 (1909).
Eugenia camiguinensis Merr., Philipp. J. Sci., C 7: 314 (1912).

Syzygium infrarubiginosum H.T.Chang & R.H.Miao, Acta Bot. Yunnan. 4: 23 (1982).
Hainan. 36 CHH. Phan.

Syzygium ingens (F.Muell. ex C.Moore) Craven & Biffin, Blumea 51: 137 (2006).
SE. Queensland to NE. New South Wales. 50 NSW QLD. Phan.
**Nelitris ingens* F.Muell. ex C.Moore, Cat. Nat. Indust. Prod. New South Wales:. *Acmena ingens* (F.Muell. ex C.Moore) Guymer & B.Hyland, Muelleria 6: 437 (1988).
Eugenia brachyandra Maiden & Betche, Proc. Linn. Soc. New South Wales 23: 15 (1898). *Acmena brachyandra* (Maiden & Betche) Merr. & L.M.Perry, J. Arnold Arbor. 19: 17 (1938).
Acmena australis L.A.S.Johnson, Contr. New South Wales Natl. Herb. 3: 100 (1962), nom. illeg.

Syzygium inophylloides (A.Gray) Müll.Stuttg. in W.G.Walpers, Ann. Bot. Syst. 4: 838 (1858).
SW. Pacific. 60 NUE SAM TON WAL. Phan.
**Eugenia inophylloides* A.Gray, U.S. Expl. Exped., Phan. 1: 521 (1854).
Eugenia crosbyi Burkill, J. Linn. Soc., Bot. 35: 38 (1901).

Syzygium inophyllum DC., Prodr. 3: 260 (1828). *Eugenia inophylla* (DC.) Roxb., Fl. Ind. ed. 1832, 2: 496 (1832). *Acmena inophylla* (DC.) Wight, Icon. Pl. Ind. Orient. 2: 623 (1842). *Jambosa inophylla* (DC.) Miq., Fl. Ned. Ind. 1(1): 433 (1855).
Bangladesh, W. Malesia. 40 BAN 42 BOR MLY. Phan.
Calyptranthes obtusifolia Buch.-Ham. ex Wall., Numer. List: 3600 B (1831), nom. nud.
Eugenia bernardii King, J. Asiat. Soc. Bengal, Pt. 2, Nat. Hist. 70: 115 (1901). *Syzygium inophyllum* var. *bernardii* (King) I.M.Turner, J. Singapore Natl. Acad. Sci. 22–24: 20 (1997).

Syzygium inopinatum Amshoff, Blumea 5: 501 (1945).
Jawa. 42 JAW. Nanophan. or phan.

Syzygium insigne (Blume) Merr. & L.M.Perry, Mem. Amer. Acad. Arts 18: 163 (1939).
Borneo. 42 BOR. Phan.
Eugenia lancifolia Miq., Anal. Bot. Ind. 1: 17 (1850).
**Jambosa insignis* Blume, Mus. Bot. 1: 100 (1850).
Jambosa lancifolia Miq., Fl. Ned. Ind. 1(1): 427 (1855). *Syzygium lancifolium* (Miq.) Merr. & L.M. Perry, Mem. Amer. Acad. Arts 18: 196 (1939).

Syzygium insulare T.G.Hartley & L.M.Perry, J. Arnold Arbor. 54: 183 (1973).
Bismarck Arch. to Solomon Is. 43 BIS SOL. Phan.

Syzygium* × *intermedium Engl. & Brehmer, Bot. Jahrb. Syst. 54: 339 (1917). *S. cordatum* × *S. guineense*.
Trop. & S. Africa. 23 ZAI 25 KEN TAN UGA 26 ANG MLW MOZ ZAM ZIM 27 BOT TVL. Nanophan. or phan.
Syzygium × *deiningeri* Engl., Bot. Jahrb. Syst. 54: 340 (1917).

Syzygium intumescens (C.B.Rob.) Merr., Philipp. J. Sci. 79: 396 (1951).
Philippines. 42 PHI. Phan.
**Eugenia intumescens* C.B.Rob., Philipp. J. Sci., C 4: 401 (1909).

Syzygium isabelense (Quisumb.) Merr., Philipp. J. Sci. 79: 397 (1951).
Philippines. 42 PHI. Phan.
**Eugenia isabelensis* Quisumb., Philipp. J. Sci. 41: 339 (1930).

Syzygium iteophyllum Diels, Bot. Jahrb. Syst. 57: 414 (1922).
New Guinea. 43 NWG. Phan.

Syzygium iwahigense (Elmer) Merr., Philipp. J. Sci. 79: 397 (1951).
Philippines. 42 PHI. Phan.
**Eugenia iwahigensis* Elmer, Leafl. Philipp. Bot. 4: 1417 (1912).

Syzygium ixoroides Chantaran. & J.Parn., Kew Bull. 48: 601 (1993).
Pen. Thailand. 41 THA. Phan.

Syzygium jaffrei J.W.Dawson, in Fl. Nouv.-Caléd. 23: 112 (1999).
C. New Caledonia. 60 NWC. Nanophan. or phan.

Syzygium jaheri Merr. & L.M.Perry, Mem. Amer. Acad. Arts 18: 156 (1939).
Borneo. 42 BOR. Phan.

Syzygium jainii Harid. & R.R.Rao, Forest Fl. Meghalaya: 398 (1985).
Assam. 40 ASS. Phan.

Syzygium jambos (L.) Alston in H.Trimen, Handb. Fl. Ceylon 6(Suppl.): 115 (1931).
Himalaya to W. Malesia. (21) mdr (25) ken tan uga 36 CHC chs 40 EHM NEP srl 41 MYA THA VIE 42 BOR MLY phi (60) wal (61) sci (63) haw (80) blz els (81) lee win (84) bze bzl. Phan. Widely cultivated for its edible fruits (rose apple).
**Eugenia jambos* L., Sp. Pl.: 470 (1753). *Myrtus jambos* (L.) Kunth in F.W.H.von Humboldt, A.J.A. Bonpland & C.S.Kunth, Nov. Gen. Sp. 6: 144 (1823). *Jambosa vulgaris* DC., Prodr. 3: 286 (1828), nom. illeg. *Jambosa jambos* (L.) Millsp., Publ. Field Columb. Mus., Bot. Ser. 2: 80 (1896). *Plinia jambos* (L.) M.Gómez, Fl. Haban.: 292 (1914).
Eugenia jambosa Crantz, Inst. Rei Herb. 2: 201 (1766).
Eugenia decora Salisb., Prodr. Stirp. Chap. Allerton: 353 (1796).
Eugenia jamboides Wender., Schriften Ges. Beförd. Gesammten Naturwiss. Marburg 2: 254 (1831).
Eugenia malaccensis Blanco, Fl. Filip.: 415 (1837), nom. illeg.
Jambosa palembanica Blume, Mus. Bot. 1: 93 (1850).
Eugenia vulgaris Baill., Hist. Pl. 6: 345 (1876).
Eugenia monantha Merr., J. Straits Branch Roy. Asiat.

Soc. 79: 22 (1918). *Syzygium monanthum* (Merr.) Merr. & L.M.Perry, Mem. Amer. Acad. Arts 18: 163 (1939). *Syzygium merrillii* Masam., Enum. Phan. Born.: 534 (1942), nom. superfl.

Jambosa malaccensis f. *cericarpa* O.Deg., Fl. Hawaiiensis 273: s.p. (1933). *Eugenia malaccensis* f. *cericarpa* (O.Deg.) H.St.John, Mem. Pacific Trop. Bot. Gard. 1: 252 (1973).

Syzygium jambos var. *linearilimbum* H.T.Chang & R.H.Miao, Acta Bot. Yunnan. 4: 17 (1982).

Syzygium jasminifolium (Ridl.) Chantaran. & J.Parn., Kew Bull. 48: 602 (1993).
Pen. Thailand to Pen. Malaysia. 41 THA 42 MLY. Nanophan.
**Eugenia jasminifolia* Ridl., J. Fed. Malay States Mus. 10: 133 (1920).

Syzygium jienfunicum H.T.Chang & R.H.Miao, Acta Bot. Yunnan. 4: 19 (1982).
Hainan. 36 CHH. Phan.

Syzygium johnsonii (F.Muell.) B.Hyland, Austral. J. Bot., Suppl. Ser. 9: 95 (1983).
Queensland. 50 QLD. Phan.
**Eugenia johnsonii* F.Muell., Victorian Naturalist 8: 199 (1892).
Eugenia petriei C.T.White & W.D.Francis, Proc. Roy. Soc. Queensland 35: 71 (1923).

Syzygium kabaense (Greves) Govaerts, World Checklist Myrtaceae: 395 (2008).
Sumatera. 42 SUM. Nanophan. or phan.
**Eugenia kabaensis* Greves, J. Bot. 62(Suppl.): 38 (1924).

Syzygium kajewskii Guillaumin, J. Arnold Arbor. 12: 256 (1931).
Vanuatu. 60 VAN. Phan.

Syzygium kalahiense Korth., Ned. Kruidk. Arch. 1: 205 (1848). *Eugenia kalahiensis* (Korth.) Miq., Anal. Bot. Ind. 1: 23 (1850).
Borneo. 42 BOR. Phan.
Myrtus kalah Miq., Anal. Bot. Ind. 1: 23 (1850), pro syn.

Syzygium kanarense (Talbot) Raizada, Indian Forester 74: 336 (1948).
India. 40 IND. Nanophan. or phan.
**Eugenia kanarensis* Talbot, J. Bombay Nat. Hist. Soc. 11: 236 (1897).

Syzygium kanneliyensis Kosterm., Quart. J. Taiwan Mus. 34: 142 (1981).
Sri Lanka. 40 SRL. Nanophan. or phan.

Syzygium karimatense Merr. & L.M.Perry, Mem. Amer. Acad. Arts 18: 198 (1939).
Borneo. 42 BOR. Phan.

Syzygium kemamanense (M.R.Hend.) I.M.Turner, J. Singapore Natl. Acad. Sci. 22–24: 20 (1997).
Pen. Malaysia. 42 MLY. Phan.
**Eugenia kemamanensis* M.R.Hend., Gard. Bull. Singapore 11: 303 (1947).

Syzygium keroanthum (Diels) Merr. & L.M.Perry, J. Arnold Arbor. 23: 271 (1942).
New Guinea. 43 NWG. Phan.
**Jambosa keroantha* Diels, Bot. Jahrb. Syst. 57: 385 (1922).

Syzygium kerrii Chantaran. & J.Parn., Kew Bull. 48: 602 (1993).
SE. Thailand. 41 THA. Phan.

Syzygium kerstingii Engl., Bot. Jahrb. Syst. 54: 340 (1917).
Togo. 22 TOG. Phan.

Syzygium keysseri (Schltr. ex Diels) Merr. & L.M.Perry, J. Arnold Arbor. 23: 253 (1942).
New Guinea. 43 NWG. Phan.
**Jambosa keysseri* Schltr. ex Diels, Bot. Jahrb. Syst. 62: 485 (1929).

Syzygium khaoyaiense (Chantaranothai & J.Parn.) Craven & Biffin, Blumea 51: 137 (2006).
SE. Thailand. 41 THA. Phan.
**Cleistocalyx khaoyaiensis* Chantaranothai & J.Parn., Kew Bull. 48: 590 (1993).

Syzygium khasianum (Duthie) N.P.Balakr., Bull. Bot. Surv. India 22: 174 (1980 publ. 1982).
Assam. 40 ASS. Phan.
**Eugenia khasiana* Duthie in J.D.Hooker, Fl. Brit. India 2: 491 (1878).

Syzygium khoonmengianum P.S.Ashton, Kew Bull. 61: 127 (2006).
Borneo (Sabah). 42 BOR. Phan.

Syzygium kiahii (M.R.Hend.) I.M.Turner, J. Singapore Natl. Acad. Sci. 22–24: 20 (1997).
Pen. Malaysia. 42 MLY. Phan.
**Eugenia kiahii* M.R.Hend., Gard. Bull. Singapore 11: 307 (1947).
Eugenia kiahii var. *angustifolia* M.R.Hend., Gard. Bull. Singapore 12: 116 (1949). *Syzygium kiahii* var. *angustifolium* (M.R.Hend.) I.M.Turner, J. Singapore Natl. Acad. Sci. 22–24: 20 (1997).

Syzygium kiauense (Merr.) Merr. & L.M.Perry, Mem. Amer. Acad. Arts 18: 162 (1939).
Borneo. 42 BOR. Phan.
**Eugenia kiauensis* Merr., J. Straits Branch Roy. Asiat. Soc. 77: 209 (1917).

Syzygium kietanum Rech., Repert. Spec. Nov. Regni Veg. 11: 183 (1912).
Solomon Is. 43 SOL. Phan.

Syzygium kinabaluense (Stapf) Merr. & L.M.Perry, Mem. Amer. Acad. Arts 18: 160 (1939).
Borneo. 42 BOR. Phan.
**Eugenia kinabaluensis* Stapf, Trans. Linn. Soc. London, Bot. 4: 152 (1894).

Syzygium kipidamasii W.N.Takeuchi, Edinburgh J. Bot. 59: 261 (2002).
E. New Guinea. 43 NWG.

Syzygium klampok (Miq.) Amshoff in C.A.Backer & R.C.Bakhuizen van der Brink, Bekn. Fl. Java 4b(98): 25 (1944).
E. Jawa. 42 JAW. Phan.
**Jambosa klampok* Miq., Fl. Ned. Ind. 1(1): 421 (1855). *Eugenia klampok* (Miq.) Koord. & Valeton, Meded. Lands Plantentuin 40: 68 (1900).
Jambosa glabra Zoll. ex Koord. & Valeton, Meded. Lands Plantentuin 40: 68 (1900).

Syzygium klossii (Ridl.) Masam., Enum. Phan. Born.: 531 (1942).
Pen. Malaysia. 42 MLY. Phan.
**Eugenia klossii* Ridl., J. Straits Branch Roy. Asiat. Soc. 79: 65 (1918).

Syzygium koghianum Petitm. & Bonati, Bull. Herb. Boissier, II, 7: 651 (1907).
New Caledonia (incl. Îs. Loyauté). 60 NWC. Phan.

Syzygium densiflorum Brongn. & Gris, Bull. Soc. Bot. France 12: 182 (1865), nom. illeg.

Syzygium koniamboense J.W.Dawson, in Fl. Nouv.-Caléd. 23: 63 (1999).
NW. & WC. New Caledonia. 60 NWC. Nanophan. or phan.

Syzygium koordersianum (King) I.M.Turner, J. Singapore Natl. Acad. Sci. 22–24: 20 (1997).
Pen. Malaysia to Sumatera. 42 MLY SUM. Phan.
**Eugenia koordersiana* King, J. Asiat. Soc. Bengal, Pt. 2, Nat. Hist. 70: 128 (1901).

Syzygium korthalsianum (Miq.) Miq., Fl. Ned. Ind. 1(1): 454 (1855).
Sumatera, Borneo. 42 BOR SUM. Phan.
**Eugenia korthalsiana* Miq., Anal. Bot. Ind. 1: 25 (1850).
Myrtus sessilis Korth. ex Miq., Anal. Bot. Ind. 1: 25 (1850).
Syzygium sessile Blume ex Miq., Anal. Bot. Ind. 1: 25 (1850).

Syzygium koumacense J.W.Dawson, in Fl. Nouv.-Caléd. 23: 126 (1999).
NW. New Caledonia. 60 NWC. Phan.

Syzygium kriegeri (Guillaumin) J.W.Dawson, in Fl. Nouv.-Caléd. 34: 58 (1999).
NW. & WC. New Caledonia. 60 NWC. Nanophan.
**Caryophyllus kriegeri* Guillaumin, Bull. Soc. Bot. France 85: 649 (1938 publ. 1939).

Syzygium kudatense P.S.Ashton, Kew Bull. 61: 130 (2006).
Borneo (Sabah). 42 BOR. Phan.

Syzygium kuebiniense J.W.Dawson, in Fl. Nouv.-Caléd. 23: 13 (1999).
SE. New Caledonia. 60 NWC. Nanophan.

Syzygium kunstleri (King) Bahadur & R.C.Gaur, Indian J. Forest. 1: 349 (1978).
Pen. Malaysia, Borneo. 42 BOR MLY. Phan.
**Eugenia kunstleri* King, J. Asiat. Soc. Bengal, Pt. 2, Nat. Hist. 70: 127 (1901).
Eugenia albidiramea Merr., Univ. Calif. Publ. Bot. 15: 221 (1929). *Syzygium albidirameum* (Merr.) Merr. & L.M.Perry, Mem. Amer. Acad. Arts 18: 190 (1939).
Syzygium chrysanthum Merr. & L.M.Perry, Mem. Amer. Acad. Arts 18: 192 (1939).
Syzygium stictophyllum Merr. & L.M.Perry, Mem. Amer. Acad. Arts 18: 192 (1939).

Syzygium kuranda (F.M.Bailey) B.Hyland, Austral. J. Bot., Suppl. Ser. 9: 96 (1983).
N. & NE. Queensland. 50 QLD. Phan.
**Eugenia kuranda* F.M.Bailey, Queensl. Fl.: 658 (1900).

Syzygium kurzii (Duthie) N.P.Balakr., Bull. Bot. Surv. India 22: 174 (1980 publ. 1982).
C. Himalaya to Indo-China. 40 ASS EHM NEP 41 AND MYA THA. Phan.
Eugenia cerasiflora Kurz, J. Asiat. Soc. Bengal, Pt. 2, Nat. Hist. 42(2): 233 (1873), nom. illeg. **Eugenia kurzii* Duthie, J. Asiat. Soc. Bengal, Pt. 2, Nat. Hist. 46(2): 68 (1877). *Jambosa kurzii* (Duthie) A.M. Cowan & Cowan, Trees N. Bengal: 67 (1929).
Eugenia kurzii var. *andamanica* King, J. Asiat. Soc. Bengal, Pt. 2, Nat. Hist. 70: 105 (1901). *Syzygium kurzii* var. *andamanica* (King) N.P.Balakr., Bull. Bot. Surv. India 22: 174 (1980 publ. 1982).

Syzygium kusukusuense (Hayata) Mori, Trans. Nat. Hist. Soc. Taiwan 28: 439 (1938).
Taiwan (Hengch'un Pen.). 38 TAI. Phan.
**Eugenia kusukusensis* Hayata, Icon. Pl. Formosan. 3: 119 (1913).

Syzygium kwangtungense (Merr.) Merr., J. Arnold Arbor. 19: 241 (1938).
SE. China. 36 CHS.
**Eugenia kwangtungensis* Merr., Sunyatsenia 1: 202 (1934).

Syzygium lacustre (C.B.Rob.) Merr., Philipp. J. Sci. 79: 397 (1951).
Philippines. 42 PHI. Phan.
**Eugenia lacustris* C.B.Rob., Philipp. J. Sci., C 4: 377 (1909).

Syzygium laetum (Buch.-Ham.) Gandhi, Fl. Hassan Distr. Karnataka: 282 (1976).
SW. India, Pen. Thailand. 40 IND 41 THA. Nanophan.
**Eugenia laeta* Buch.-Ham., Mem. Wern. Nat. Hist. Soc. 5: 338 (1826). *Jambosa laeta* (Buch.-Ham.) Blume, Mus. Bot. 1: 101 (1850).

subsp. ***jugorum*** (Craib) Chantaran. & J.Parn., Kew Bull. 48: 602 (1993).
Pen. Thailand. 41 THA. Nanophan.
**Eugenia jugorum* Craib, Bull. Misc. Inform. Kew 1930: 168 (1930).

subsp. ***laetum***
SW. India. 40 IND. Nanophan.
Eugenia bauanguica Blanco, Fl. Filip.: 418 (1837).
Eugenia pauciflora Wight, Ill. Ind. Bot. 2: 14 (1841), nom. illeg. *Jambosa pauciflora* Walp., Repert. Bot. Syst. 2: 192 (1843). *Clavimyrtus pauciflora* (Walp.) Blume, Mus. Bot. 1: 113 (1850).
Eugenia wightii Bedd., Fl. Sylv. S. India: 109 (1872).
Eugenia laeta var. *pauciflora* Duthie in J.D.Hooker, Fl. Brit. India 2: 479 (1878). *Syzygium laetum* var. *pauciflorum* (Duthie) M.R.Almeida, Fl. Maharashtra 2: 272 (1998).

subsp. ***sublaetum*** (Craib) Chantaran. & J.Parn., Kew Bull. 48: 602 (1993).
Pen. Thailand. 41 THA. Nanophan.
**Eugenia sublaeta* Craib, Bull. Misc. Inform. Kew 1930: 169 (1930).

Syzygium laeve (Montrouz.) Govaerts, World Checklist Myrtaceae: 396 (2008).
New Caledonia. 60 NWC. Nanophan. or phan.
**Jambosa laevis* Montrouz., Mém. Acad. Roy. Sci. Lyon, Sect. Sci. 10: 208 (1860).

Syzygium lagerstmemioides Merr. & L.M.Perry, J. Arnold Arbor. 23: 262 (1942).
New Guinea. 43 NWG. Phan.

Syzygium lakshnakarae Chantaran. & J.Parn., Kew Bull. 48: 605 (1993).
NE. Thailand. 41 THA. Phan.

Syzygium lambirense P.S.Ashton, Kew Bull. 61: 130 (2006).
Borneo. 42 BOR. Nanophan. or phan.

Syzygium lamii Merr. & L.M.Perry, Mem. Amer. Acad. Arts 18: 173 (1939).
Borneo. 42 BOR. Phan.

Syzygium lamprophyllum Diels, Bot. Jahrb. Syst. 57: 411 (1922).
New Guinea. 43 NWG. Phan.

Syzygium lanceolarium (Roxb.) N.P.Balakr., Bull. Bot. Surv. India 22: 174 (1980 publ. 1982).
Bangladesh. 40 BAN. Nanophan. or phan.
**Eugenia lanceolaria* Roxb., Fl. Ind. ed. 1832, 2: 494 (1832). *Jambosa lanceolaria* (Roxb.) Blume, Mus. Bot. 1: 101 (1850).
Eugenia lonchophylla Voigt, Hort. Suburb. Calcutt.: 49 (1845).

Syzygium lanceolatum (Lam.) Wight & Arn., Prodr. Fl. Ind. Orient.: 330 (1834).
S. India, Sri Lanka. 40 IND SRL. Phan.
**Eugenia lanceolata* Lam., Encycl. 3: 200 (1789). *Acmena lanceolata* (Lam.) Thwaites, Enum. Pl. Zeyl.: 119 (1859).
Myrtus sonneratii Spreng., Syst. Veg. 2: 485 (1825).
Myrtus quadriflora Vell., Fl. Flumin. 5: 217, t. 77 (1829).
Syzygium wightianum Wall. ex Wight & Arn., Prodr. Fl. Ind. Orient. 1: 330 (1834).
Eugenia wightiana Wight, Icon. Pl. Ind. Orient. 2: t. 529 (1842). *Acmena wightiana* (Wight) Walp., Repert. Bot. Syst. 2: 181 (1843).

Syzygium lancilimbum (Merr.) Merr., Philipp. J. Sci. 79: 397 (1951).
Philippines. 42 PHI. Phan.
**Eugenia lancilimba* Merr., Philipp. J. Sci. 20: 413 (1922).

Syzygium laqueatum Merr. & L.M.Perry, J. Arnold Arbor. 23: 257 (1942).
New Guinea. 43 NWG. Phan.

Syzygium lasianthifolium H.T.Chang & R.H.Miao, Acta Bot. Yunnan. 4: 18 (1982).
China (Guangdong). 36 CHS. Phan.

Syzygium lateriflorum Brongn. & Gris, Bull. Soc. Bot. France 12: 183 (1865).
New Caledonia. 60 NWC. Nanophan. or phan.

Syzygium latifolium (Poir.) DC., Prodr. 3: 259 (1828).
Mauritius. 29 MAU. Nanophan. or phan.
**Calyptranthes latifolia* Poir. in J.B.A.M.de Lamarck, Encycl., Suppl. 2: 43 (1811).
Syzygium scandens Bojer, Hortus Maurit.: 143 (1837), nom. nud.
Eugenia nummularia Baker, Fl. Mauritius: 118 (1877).
Eugenia scandens Baker, Fl. Mauritius: 118 (1877). *Syzygium scandens* (Baker) J.Guého & A.J.Scott, Kew Bull. 34: 494 (1980).

Syzygium laurifolium (DC.) N.P.Balakr., Bull. Bot. Surv. India 22: 174 (1980 publ. 1982).
Bangladesh. 40 BAN. Phan.
Myrtus javanica Spreng., Syst. Veg. 2: 484 (1825).
**Jambosa laurifolia* DC., Prodr. 3: 287 (1828). *Eugenia laurifolia* (DC.) Roxb., Fl. Ind. ed. 1832, 2: 489 (1832).
Eugenia bifaria Wall., Pl. Asiat. Rar. 2: 47 (1831). *Jambosa bifaria* (Wall.) Miq., Fl. Ned. Ind. 1(1): 422 (1855).

Syzygium laxeracemosum (Guillaumin) J.W.Dawson, in Fl. Nouv.-Caléd. 23: 43 (1999).
SE. New Caledonia. 60 NWC. Nanophan.
**Caryophyllus laxeracemosus* Guillaumin, Bull. Mus. Natl. Hist. Nat., II, 28: 312 (1956).

Syzygium laxiflorum (Blume) DC., Prodr. 3: 261 (1828).
Jawa. 42 JAW. Phan.
**Calyptranthes laxiflora* Blume, Bijdr.: 1090 (1826). *Eugenia laxiflora* (Blume) Koord. & Valeton, Meded. Lands Plantentuin 40: 139 (1900), nom. illeg.

Syzygium lecardii Guillaumin, Bull. Soc. Bot. France 85: 643 (1938 publ. 1939).
New Caledonia. 60 NWC. Nanophan. or phan.

Syzygium legatii Burtt Davy & Greenway in J.B.Davy, Man. Pl. Transvaal 1: 240 (1926). *Syzygium guineense* subsp. *legatii* (Burtt Davy & Greenway) F.White, Kirkia 10: 404 (1977).
Northern Prov. 27 TVL. Phan.

Syzygium lehuntii (F.M.Bailey) Merr. & L.M.Perry, J. Arnold Arbor. 23: 250 (1942).
New Guinea. 43 NWG. Phan.
**Eugenia lehuntii* F.M.Bailey, Queensland Agric. J. 9: 411 (1905).

Syzygium lenbrassii Craven & Biffin, Blumea 51: 137 (2006).
Papua New Guinea. 43 NWG. Nanophan. or phan.
**Acmena brassii* Craven, Austral. Syst. Bot. 3: 727 (1990).

Syzygium leonhardii (Diels) Merr. & L.M.Perry, J. Arnold Arbor. 23: 258 (1942).
New Guinea. 43 NWG. Phan.
**Jambosa leonhardii* Diels, Bot. Jahrb. Syst. 57: 384 (1922).

Syzygium leptoneurum Diels, Bot. Jahrb. Syst. 57: 407 (1922).
New Guinea. 43 NWG. Phan.

Syzygium leptophlebium Diels, Bot. Jahrb. Syst. 57: 406 (1922).
New Guinea. 43 NWG. Phan.

Syzygium leptopodium Merr. & L.M.Perry, J. Arnold Arbor. 23: 284 (1942).
New Guinea. 43 NWG. Phan.

Syzygium leptostachyum (Blume) Merr. & L.M.Perry, Mem. Amer. Acad. Arts 18: 175 (1939).
Borneo. 42 BOR. Phan.
**Jambosa leptostachya* Blume, Mus. Bot. 1: 99 (1850).

Syzygium leptostemon (Korth.) Merr. & L.M.Perry, Mem. Amer. Acad. Arts 18: 156 (1939).
Pen. Thailand to W. Malesia. 41 THA 42 BOR MLY. Phan.
**Jambosa leptostemon* Korth., Ned. Kruidk. Arch. 1: 201 (1847). *Strongylocalyx leptostemon* (Korth.) Blume, Mus. Bot. 1: 89 (1850). *Eugenia leptostemon* (Korth.) Miq., Fl. Ned. Ind. 1(1): 442 (1855).
Jambosa urceolata Korth. ex Miq., Fl. Ned. Ind. 1(1): 418 (1855). *Eugenia urceolata* (Korth. ex Miq.) King, J. Asiat. Soc. Bengal, Pt. 2, Nat. Hist. 70: 101 (1901). *Eugenia rotata* King ex Craib, Fl. Siam. 1: 660 (1931). *Syzygium urceolatum* (Korth. ex Miq.) Merr. & L.M.Perry, Mem. Amer. Acad. Arts 18: 174 (1939).
Strongylocalyx leptostachyus Blume ex Miq., Fl. Ned. Ind. 1(1): 443 (1855).
Eugenia subracemosa Merr., J. Straits Branch Roy. Asiat. Soc. 79: 23 (1918). *Syzygium subracemosum* (Merr.) Masam., Enum. Phan. Born.: 539 (1942).
Eugenia sandakanensis Merr., J. Straits Branch Roy. Asiat. Soc. 86: 335 (1922). *Syzygium sandakanense* (Merr.) Merr. & L.M.Perry, Mem. Amer. Acad. Arts 18: 155 (1939).

Syzygium leucanthum L.M.Perry, J. Arnold Arbor. 31: 3557 (1950).
Fiji. 60 FIJ. Phan.

Syzygium leucocladum Merr. & L.M.Perry, Mem. Amer. Acad. Arts 18: 178 (1939).
Borneo. 42 BOR. Phan.
**Eugenia ambongensis* var. *havilandii* Ridl., J. Bot. 68: 16 (1930). *Syzygium ambongense* var. *havilandii* (Ridl.) Masam., Enum. Phan. Born.: 523 (1942).

Syzygium leucophloium (Blume) Merr. & L.M.Perry, Mem. Amer. Acad. Arts 18: 174 (1939).
Jawa. 42 JAW. Phan.
**Jambosa cuneata* Blume, Mus. Bot. 1: 105 (1850).

Syzygium leucoxylon Korth., Ned. Kruidk. Arch. 1: 203 (1848). *Eugenia leucoxylon* (Korth.) Miq., Anal. Bot. Ind. 1: 26 (1850).
Malaya to Philippines. 42 BOR MLY PHI. Phan.
Myrtus leucoxylon Korth. ex Miq., Anal. Bot. Ind. 1: 26 (1850).
Eugenia verecunda Duthie in J.D.Hooker, Fl. Brit. India 2: 496 (1878).
Eugenia brevistylis C.B.Rob., Philipp. J. Sci., C 6: 347 (1911).
Eugenia alcinae Merr., Philipp. J. Sci., C 10: 216 (1915). *Syzygium alcinae* (Merr.) Merr. & L.M. Perry, Mem. Amer. Acad. Arts 18: 194 (1939).

Syzygium levinei (Merr.) Merr., J. Arnold Arbor. 19: 110 (1938).
SE. China to Vietnam. 36 CHH CHS 41 VIE. Phan.
**Eugenia levinei* Merr., Lingnan Sci. J. 13: 39 (1934).

Syzygium lewisii Alston in H.Trimen, Handb. Fl. Ceylon 6(Suppl.): 117 (1931).
Sri Lanka. 40 SRL. Phan.

Syzygium leytense (Elmer) Merr., Philipp. J. Sci. 79: 398 (1951).
Philippines. 42 PHI. Phan.
**Eugenia leytensis* Elmer, Leafl. Philipp. Bot. 1: 329 (1908).
Eugenia samarensis Merr., Philipp. J. Sci., C 10: 223 (1915).
Eugenia sarcocarpa Merr., Philipp. J. Sci. 18: 295 (1921).

Syzygium lifuanum Däniker, Vierteljahrsschr. Naturf. Ges. Zürich 78(19): 302 (1933).
New Caledonia (Îs. Loyauté). 60 NWC. Phan.

Syzygium lilacinum (Merr.) Merr. & L.M.Perry, Mem. Amer. Acad. Arts 18: 165 (1939).
Borneo. 42 BOR. Phan.
**Eugenia lilacina* Merr., Univ. Calif. Publ. Bot. 15: 219 (1929).

Syzygium lineare (Korth.) Masam., Enum. Phan. Born.: 532 (1942).
Borneo. 42 BOR. Phan.
**Jambosa linearis* Korth., Ned. Kruidk. Arch. 1: 199 (1847). *Syzygium pauciflorum* Merr. & L.M.Perry, Mem. Amer. Acad. Arts 18: 170 (1939).

Syzygium lineatum (DC.) Merr. & L.M.Perry, J. Arnold Arbor. 19: 109 (1938).
SE. China, Indo-China to W. Malesia and Philippines. 36 CHS 41 MYA THA VIE 42 BOR JAW MLY PHI SUM. Phan.
Myrtus lineata Blume, Bijdr.: 1087 (1826), nom. illeg. **Jambosa lineata* DC., Prodr. 3: 287 (1828). *Clavimyrtus lineata* (DC.) Blume, Mus. Bot. 1: 113 (1850). *Eugenia lineata* (DC.) Duthie in J.D.Hooker, Fl. Brit. India 2: 487 (1878), nom. illeg.
Syzygium longiflorum C.Presl, Abh. Königl. Böhm. Ges. Wiss., V, 3: 500 (1845). *Eugenia longiflora* (C.Presl) Fern.-Vill. in F.M.Blanco, Fl. Filip., ed. 3, 2(13A): 86 (1880).
Clavimyrtus latifolia Blume, Mus. Bot. 1: 113 (1850). *Jambosa latifolia* (Blume) Miq., Fl. Ned. Ind. 1(1): 429 (1855).
Clavimyrtus symphytocarpa Blume, Mus. Bot. 1: 113 (1850). *Jambosa symphytocarpa* (Blume) Korth. ex Miq., Fl. Ned. Ind. 1(1): 430 (1855).
Jambosa rubricaule Miq., Fl. Ned. Ind. 1(1): 432 (1855). *Eugenia rubricaulis* (Miq.) Duthie in J.D.Hooker, Fl. Brit. India 2: 487 (1878).
Syzygium zippelianum Miq., Fl. Ned. Ind. 1(1): 449 (1855). *Eugenia zippeliana* (Miq.) Koord. & Valeton, Meded. Lands Plantentuin 40: 142 (1900).
Jambosa teysmannii Miq., Fl. Ned. Ind., Eerste Bijv.: 429 (1861). *Eugenia teysmannii* (Miq.) Koord. & Valeton, Meded. Lands Plantentuin 40: 164 (1900). *Syzygium teysmannii* (Miq.) Masam., Enum. Phan. Born.: 540 (1942).
Eugenia marivelesensis Merr., Philipp. J. Sci. 1(Suppl.): 106 (1906).
Eugenia miquelii Elmer, Leafl. Philipp. Bot. 4: 1441 (1912).
Eugenia longicalyx Ridl., J. Bot. 68: 11 (1930). *Syzygium longicalyx* (Ridl.) Masam., Enum. Phan. Born.: 533 (1942).

Syzygium linocieroideum (King) I.M.Turner, J. Singapore Natl. Acad. Sci. 22–24: 21 (1997).
Pen. Malaysia. 42 MLY. Phan.
**Eugenia linocieroidea* King, J. Asiat. Soc. Bengal, Pt. 2, Nat. Hist. 70: 118 (1901).

Syzygium littorale (Blume) Amshoff in C.A.Backer & R.C.Bakhuizen van der Brink, Bekn. Fl. Java 4b(98): 26 (1944).
Jawa. 42 JAW. Phan.
**Jambosa littoralis* Blume, Mus. Bot. 1: 102 (1850). *Eugenia littoralis* (Blume) Meijer Drees, Commun. Forest Res. Inst., Bogor 33: 89 (1951), nom. illeg.
Eugenia subglauca Koord. & Valeton, Bull. Inst. Bot. Buitenzorg 2: 8 (1899).

Syzygium llanosii (Merr.) Merr., Philipp. J. Sci. 79: 398 (1951).
Philippines. 42 PHI. Phan.
**Eugenia llanosii* Merr., Philipp. J. Sci., C 10: 220 (1915).

Syzygium loiseleurioides (Baker) Govaerts, World Checklist Myrtaceae: 398 (2008).
C. Madagascar. 29 MDG. Nanophan.
Eugenia cyclophylla Baker, J. Bot. 20: 111 (1882), nom. illeg. *Syzygium baronii* Labat & G.E.Schatz, Novon 12: 202 (2002).
Eugenia vacciniifolia Baker, J. Linn. Soc., Bot. 20: 145 (1883).
**Eugenia loiseleurioides* Baker, J. Linn. Soc., Bot. 21: 341 (1884).

Syzygium longifolium (Brongn. & Gris) J.W.Dawson, in Fl. Nouv.-Caléd. 23: 46 (1999).
New Caledonia. 60 NWC. Nanophan. or phan.
**Jambosa longifolia* Brongn. & Gris, Bull. Soc. Bot. France 12: 181 (1865).

Syzygium longipedicellatum (Merr.) Merr., Philipp. J. Sci. 79: 399 (1951).
Philippines. 42 PHI. Phan.
**Jambosa longipedicellata* Merr., Publ. Bur. Sci. Gov. Lab. 17: 37 (1904). *Eugenia longipedicellata* (Merr.) C.B.Rob., Philipp. J. Sci., C 4: 349 (1909).

Syzygium longipes Merr. & L.M.Perry, J. Arnold Arbor. 23: 274 (1942).
New Guinea. 43 NWG. Phan.
**Eugenia longipes* Warb., Bot. Jahrb. Syst. 13: 391 (1891), nom. illeg.
Eugenia daphnoides Greves, J. Bot. 61(Suppl.): 15 (1923). *Syzygium daphnoides* (Greves) Merr. & L.M.Perry, J. Arnold Arbor. 23: 270 (1942).

Syzygium longissimum (Merr.) Merr., Philipp. J. Sci. 79: 399 (1951).
Philippines. 42 PHI. Phan.
**Eugenia longissima* Merr., Publ. Bur. Sci. Gov. Lab. 35: 50 (1905).

Syzygium longistylum (Merr.) Merr., Philipp. J. Sci. 79: 399 (1951).
Philippines. 42 PHI. Phan.
**Eugenia longistyla* Merr., Philipp. J. Sci., C 10: 220 (1915).

Syzygium lorentzianum Lauterb., Nova Guinea 8: 852 (1912).
New Guinea. 43 NWG. Phan.

Syzygium lorofolium Merr., Philipp. J. Sci. 79: 399 (1951).
Philippines (Luzon). 42 PHI. Phan.

Syzygium luehmannii (F.Muell.) L.A.S.Johnson, Contr. New South Wales Natl. Herb. 3: 99 (1962).
New Guinea (Goodenough I.), Queensland to NE. New South Wales. 43 NWG 50 NSW QLD. Phan.
**Eugenia luehmannii* F.Muell., Victorian Naturalist 9: 10 (1892).
Eugenia parvifolia C.Moore, J. Proc. Roy. Soc. New S. Wales 27: 85 (1893), nom. illeg.
Myrtus exaltata F.M.Bailey, Bot. Bull. Dept. Agric. Queensland 8: 77 (1893). *Austromyrtus exaltata* (F.M.Bailey) Burret, Notizbl. Bot. Gart. Berlin-Dahlem 15: 501 (1941).
Eugenia leptantha var. *parvifolia* F.M.Bailey, Queensl. Fl. 2: 659 (1900).

Syzygium lugubre (H.Perrier) Labat & Schatz, Novon 12: 203 (2002).
CE. Madagascar. 29 MDG. Phan.
**Eugenia lugubris* H.Perrier, Mém. Inst. Sci. Madagascar, Sér. B, Biol. Vég. 4: 187 (1953).

Syzygium lunduense (Merr.) Merr. & L.M.Perry, Mem. Amer. Acad. Arts 18: 180 (1939).
Borneo. 42 BOR. Phan.
**Eugenia lunduensis* Merr., J. Straits Branch Roy. Asiat. Soc. 79: 25 (1918).

Syzygium luteum (C.B.Rob.) Merr., Philipp. J. Sci. 79: 401 (1951).
Philippines. 42 PHI. Phan.
**Eugenia lutea* C.B.Rob., Philipp. J. Sci., C 6: 350 (1911).

Syzygium luzonense (Merr.) Merr., Philipp. J. Sci. 79: 401 (1951).
Philippines. 42 PHI. Phan.
**Jambosa luzonensis* Merr., Publ. Bur. Sci. Gov. Lab. 17: 37 (1904). *Eugenia luzonensis* (Merr.) Merr., Philipp. J. Sci. 1(Suppl.): 105 (1906).

Syzygium macgregorii (C.B.Rob.) Merr., Philipp. J. Sci. 79: 401 (1951).
Philippines. 42 PHI. Phan.
**Eugenia macgregorii* C.B.Rob., Philipp. J. Sci., C 4: 367 (1909).

Syzygium macilwraithianum B.Hyland, Austral. J. Bot., Suppl. Ser. 9: 99 (1983).
N. Queensland. 50 QLD. Phan.

Syzygium mackinnonianum (B.Hyland) Craven & Biffin, Blumea 51: 138 (2006).
Queensland (E. Cape York Pen.). 50 QLD. Phan.
**Acmena mackinnoniana* B.Hyland, Austral. J. Bot., Suppl. Ser. 9: 14 (1983).

Syzygium macranthum Brongn. & Gris, Bull. Soc. Bot. France 12: 182 (1865).
New Caledonia (incl. Î. des Pins). 60 NWC. Phan.

Syzygium macrocalyx Merr. & L.M.Perry, J. Arnold Arbor. 23: 253 (1942).
New Guinea. 43 NWG. Phan.

Syzygium macromyrtus (Koord. & Valeton) Merr. & L.M.Perry, Mem. Amer. Acad. Arts 18: 169 (1939).
Jawa. 42 JAW. Phan.
Macromyrtus javanica Miq., Fl. Ned. Ind. 1(1): 440 (1855). **Eugenia macromyrtus* Koord. & Valeton, Meded. Lands Plantentuin 40: 109 (1900).

Syzygium madangense T.G.Hartley & L.M.Perry, J. Arnold Arbor. 54: 188 (1973).
Papua New Guinea. 43 NWG. Phan.

Syzygium magnoliifolium (Blume) DC., Prodr. 3: 261 (1828).
Jawa. 42 JAW. Phan.
**Myrtus magnoliifolia* Blume, Bijdr.: 1088 (1826). *Eugenia magnoliifolia* (Blume) Koord. & Valeton, Meded. Lands Plantentuin 40: 153 (1900).

Syzygium mainitense (Elmer) Merr., Philipp. J. Sci. 79: 401 (1951).
Philippines. 42 PHI. Phan.
**Eugenia mainitensis* Elmer, Leafl. Philipp. Bot. 4: 1415 (1912).
Eugenia leucocarpa Merr., Philipp. J. Sci., C 11: 23 (1916).

Syzygium maire (A.Cunn.) Sykes & Garn.-Jones, J. Arnold Arbor. 60: 400 (1979).
New Zealand. 51 NZN NZS. Phan.
**Eugenia maire* A.Cunn., Ann. Nat. Hist. 3: 114 (1839).

Syzygium makul Gaertn., Fruct. Sem. Pl. 1: 166 (1788). *Myrtus makul* (Gaertn.) J.F.Gmel., Syst. Nat.: 792 (1791). *Calyptranthes makul* (Gaertn.) Raeusch., Nomencl. Bot., ed. 3: 144 (1797).
Sri Lanka. 40 SRL. Phan.
Eugenia sylvestris Moon ex Wight, Icon. Pl. Ind. Orient. 2: t. 532 (1842). *Syzygium sylvestre* (Moon ex Wight) Walp., Repert. Bot. Syst. 2: 178 (1843).

Syzygium malabaricum (Bedd.) Gamble, Fl. Madras: 481 (1919).
SW. India. 40 IND. Phan.
**Eugenia malabarica* Bedd., Fl. Sylv. S. India 1: t. 199 (1872).

Syzygium malaccense (L.) Merr. & L.M.Perry, J. Arnold Arbor. 19: 215 (1938).
Indo-China to Vanuatu. (40) srl 41 MYA THA (42) bor jaw MLY phi 43 BIS NWG SOL 50 QLD (60) fij VAN (61) mrq sci (63) haw (80) blz (81) lee win. Nanophan. or phan. Widely cultivated for its edible fruits (Malay apple).
**Eugenia malaccensis* L., Sp. Pl.: 470 (1753). *Caryophyllus malaccensis* (L.) Stokes, Bot. Mat.

Med. 3: 72 (1812). *Myrtus malaccensis* (L.) Spreng., Syst. Veg. 2: 484 (1825). *Jambosa malaccensis* (L.) DC., Prodr. 3: 286 (1828).
Eugenia macrophylla Lam., Encycl. 3: 196 (1789). *Myrtus macrophylla* (Lam.) Spreng., Syst. Veg. 2: 483 (1825), nom. illeg. *Jambosa macrophylla* (Lam.) DC., Prodr. 3: 286 (1828).
Jambosa purpurascens DC., Prodr. 2: 286 (1825).
Jambosa domestica DC., Prodr. 3: 288 (1828).
Eugenia purpurea Roxb., Fl. Ind. ed. 1832, 2: 483 (1832). *Jambosa purpurea* (Roxb.) Wight & Arn., Prodr. Fl. Ind. Orient.: 333 (1834).
Eugenia pseudomalaccensis Linden, Fl. Serres Jard. Eur. 5: t. 429 (1849).
Eugenia domestica Baill., Hist. Pl. 6: 345 (1876).
Eugenia purpurascens Baill., Hist. Pl. 6: 345 (1876).

Syzygium malagsam (Elmer) Merr., Philipp. J. Sci. 79: 402 (1951).
Philippines. 42 PHI. Phan.
**Eugenia malagsam* Elmer, Leafl. Philipp. Bot. 4: 1403 (1912).

Syzygium mamillatum Bosser & J.Guého, Bull. Mus. Natl. Hist. Nat., B, Adansonia 9: 36 (1987).
Mauritius. 29 MAU. Nanophan. or phan.

Syzygium mananquil (Blanco) Merr., Philipp. J. Sci. 79: 402 (1951).
Philippines. 42 PHI. Phan.
**Eugenia mananquil* Blanco, Fl. Filip., ed. 2: 290 (1845).
Eugenia livida Elmer, Leafl. Philipp. Bot. 7: 2349 (1914), nom. illeg.

Syzygium manii (King) N.P.Balakr., Bull. Bot. Surv. India 22: 174 (1980 publ. 1982).
Andaman Is. 41 AND. Phan.
**Eugenia manii* King, J. Asiat. Soc. Bengal, Pt. 2, Nat. Hist. 70: 104 (1901).

Syzygium maraca Craven & Biffin, Blumea 50: 160 (2005).
Queensland. 50 QLD.

Syzygium marginatum Korth., Ned. Kruidk. Arch. 1: 203 (1848).
Sumatera. 42 SUM. Phan.

Syzygium martelinoi (Merr.) Merr., Philipp. J. Sci. 79: 402 (1951).
Philippines. 42 PHI. Phan.
**Eugenia martelinoi* Merr., Philipp. J. Sci. 18: 306 (1921).

Syzygium masukuense (Baker) R.E.Fr., Wiss. Erg. Schwed. Rhod.-Kongo Exped. 1: 77 (1914).
Tanzania to S. Trop. Africa. 25 TAN 26 MLW MOZ ZIM. Phan.
**Eugenia masukuensis* Baker, Bull. Misc. Inform. Kew 1897: 267 (1897).

subsp. ***masukuense***
Tanzania to S. Trop. Africa. 25 TAN 26 MLW ZIM. Phan.

subsp. ***pachyphyllum*** F.White, Kirkia 10: 404 (1977).
S. Trop. Africa. 26 MLW MOZ ZIM. Phan.

Syzygium mauritianum J.Guého & A.J.Scott, Kew Bull. 34: 493 (1980).
Mauritius. 29 MAU. Phan.

Syzygium mauritsii Govaerts, World Checklist Myrtaceae: 400 (2008). Named after Maurits Nachtergaele.
Philippines. 42 PHI. Phan.
**Eugenia merrillii* C.B.Rob., Philipp. J. Sci., C 4: 349 (1909). *Syzygium merrillii* (C.B.Rob.) Merr., Philipp. J. Sci. 79: 402 (1951), nom. illeg.

Syzygium medium (Korth.) Merr. & L.M.Perry, Mem. Amer. Acad. Arts 18: 166 (1939).
Borneo. 42 BOR. Phan.
**Jambosa media* Korth., Ned. Kruidk. Arch. 1: 199 (1847).

Syzygium megacarpum (Craib) Rathakr. & N.C.Nair, J. Econ. Taxon. Bot. 4: 287 (1983).
Assam to Thailand, Hainan. 36 CHH 40 ASS BAN 41 MYA THA. Phan.
Eugenia macrocarpa Roxb., Fl. Ind. ed. 1832, 2: 497 (1832), nom. illeg. **Eugenia megacarpa* Craib, Fl. Siam. 1: 652 (1931). *Syzygium macrocarpum* Bahadur & R.C.Gaur, Indian J. Forest. 1: 349 (1978).
Jambosa coarctata Blume, Mus. Bot. 1: 99 (1850).
Eugenia latilimba Merr., Lingnan Sci. J. 13: 64 (1934). *Syzygium latilimbum* (Merr.) Merr. & L.M.Perry, J. Arnold Arbor. 19: 216 (1938).

Syzygium megalanthum (C.B.Rob.) Merr., Philipp. J. Sci. 79: 402 (1951).
Philippines. 42 PHI. Phan.
**Eugenia megalantha* C.B.Rob., Philipp. J. Sci., C 4: 374 (1909).

Syzygium megalospermum (K.Schum. & Lauterb.) Merr. & L.M.Perry, J. Arnold Arbor. 23: 264 (1942).
New Guinea. 43 NWG. Phan.
**Jambosa megalosperma* K.Schum. & Lauterb., Fl. Schutzgeb. Südsee: 472 (1900).

Syzygium megistophyllum Merr. & L.M.Perry, J. Arnold Arbor. 23: 279 (1942).
New Guinea. 43 NWG. Phan.

Syzygium mekongense (Gagnep.) Merr. & L.M.Perry, J. Arnold Arbor. 19: 102 (1938).
Indo-China. 41 CBD THA VIE. Phan.
**Eugenia mekongensis* Gagnep., Notul. Syst. (Paris) 3: 328 (1918).

Syzygium melanophilum H.T.Chang & R.H.Miao, Acta Bot. Yunnan. 4: 22 (1982).
SC. China. 36 CHC. Nanophan. or phan.

Syzygium melanostictum (Miq.) Craven & Biffin, Blumea 51: 138 (2006).
Jawa, Philippines. 42 JAW PHI. Phan.
**Jambosa melanosticta* Miq., Fl. Ned. Ind. 1(1): 432 (1855). *Eugenia melanosticta* (Miq.) Koord. & Valeton, Meded. Lands Plantentuin 40: 159 (1900). *Acmena melanosticta* (Miq.) Merr. & L.M.Perry, J. Arnold Arbor. 19: 12 (1938).

Syzygium melliodorum (C.B.Rob.) Merr., Philipp. J. Sci. 79: 402 (1951).
Philippines. 42 PHI. Phan.
**Eugenia melliodora* C.B.Rob., Philipp. J. Sci., C 4: 401 (1909).

Syzygium meorianum J.W.Dawson, in Fl. Nouv.-Caléd. 23: 132 (1999).
C. New Caledonia. 60 NWC. Nanophan. or phan.

Syzygium merokense (Greves) Merr. & L.M.Perry, J. Arnold Arbor. 23: 249 (1942).
New Guinea. 43 NWG. Phan.
**Eugenia merokensis* Greves, J. Bot. 61(Suppl.): 16 (1923).

Syzygium merrittianum (C.B.Rob.) Merr., Philipp. J. Sci. 79: 403 (1951).
Philippines. 42 PHI. Phan.
**Eugenia merrittiana* C.B.Rob., Philipp. J. Sci., C 4: 369 (1909).
Eugenia lumboy Elmer, Leafl. Philipp. Bot. 4: 1431 (1912).

Syzygium metrosideros DC., Prodr. 3: 261 (1828).
Vietnam. 41 VIE. Phan.

Syzygium micans Brongn. & Gris, Bull. Soc. Bot. France 13: 468 (1866).
New Caledonia. 60 NWC. Phan.
Syzygium punctatum Vieill. ex Guillanmin, Ann. Inst. Bot.- Géol. Colon. Marseille, II, 9: 154 (1911), nom. nud.
Eugenia punctata Baker f., J. Linn. Soc., Bot. 45: 315 (1921), nom. illeg.

Syzygium micklethwaitii Verdc., Kew Bull. 52: 682 (1997).
E. Trop. Africa. 25 KEN TAN. Phan.
**Syzygium sclerophyllum* Brenan, Kew Bull. 4: 79 (1949), nom. illeg.

var. ***dryas*** Verdc., in Fl. Trop. E. Afr., Myrtac.: 84 (2001).
Tanzania (W. Usambara Mts.). 25 TAN. Phan.

subsp. ***micklethwaitii***
E. Trop. Africa. 25 KEN TAN. Phan.

subsp. ***subcordatum*** Verdc., in Fl. Trop. E. Afr., Myrtac.: 84 (2001).
Tanzania (Uluguru Mts.). 25 TAN. Phan.
Syzygium micklethwaitii var. *subcordatum* Verdc., in Fl. Trop. E. Afr., Myrtac.: 84 (2001).

Syzygium micrandrum (Ridl.) Merr. & L.M.Perry, J. Arnold Arbor. 23: 294 (1942).
New Guinea to Solomon Is. 43 NWG SOL. Phan.
**Eugenia micrandra* Ridl., Trans. Linn. Soc. London, Bot. 9: 48 (1916).

Syzygium micranthum Thwaites, Enum. Pl. Zeyl.: 117 (1859). *Eugenia micrantha* (Thwaites) Bedd., Fl. Sylv. S. India: 108 (1872), nom. illeg.
Sri Lanka. 40 SRL. Phan.

Syzygium microcymum (Koord. & Valeton) Amshoff in C.A.Backer & R.C.Bakhuizen van der Brink, Bekn. Fl. Java 4b(98): 15 (1944).
Jawa. 42 JAW. Phan.
**Eugenia microcyma* Koord. & Valeton, Bull. Inst. Bot. Buitenzorg 2: 7 (1899).

Syzygium microphyllum Gamble, Fl. Madras: 479 (1919).
S. India. 40 IND. Phan.
**Eugenia microphylla* Bedd., Fl. Sylv. S. India: 110 (1872), nom. illeg. *Syzygium gambleanum* Rathakr. & V.Chithra in N.C.Nair & A.N.Henry, Fl. Tamil Nadu 1: 156 (1983), nom. illeg.

Syzygium micropodum (Baker) Labat & Schatz, Novon 12: 203 (2002).
Madagascar. 29 MDG. Phan.
**Eugenia micropoda* Baker, J. Linn. Soc., Bot. 20: 143 (1883).
Eugenia micropoda var. *littoralis* H.Perrier, Mém. Inst. Sci. Madagascar, Sér. B, Biol. Vég. 4: 186 (1952).

Syzygium millsii (M.R.Hend.) I.M.Turner, J. Singapore Natl. Acad. Sci. 22–24: 21 (1997).
Pen. Malaysia. 42 MLY. Phan.
**Eugenia millsii* M.R.Hend., Gard. Bull. Singapore 11: 301 (1947).

Syzygium mimicum (Merr.) Merr., Philipp. J. Sci. 79: 403 (1951).
Philippines. 42 PHI. Phan.
**Eugenia mimica* Merr., Philipp. J. Sci. 1(Suppl.): 212 (1906).
Eugenia submimica Elmer, Leafl. Philipp. Bot. 4: 1438 (1912).

Syzygium mindorense (C.B.Rob.) Masam., Enum. Phan. Born.: 534 (1942).
Philippines. 42 PHI. Phan.
**Eugenia mindorensis* C.B.Rob., Philipp. J. Sci., C 4: 399 (1909).

Syzygium minimum (Blume) Airy Shaw, Kew Bull. 4: 124 (1949).
Sumatera to Jawa. 42 JAW SUM.
**Eugenia minima* Blume, Catalogus: 75 (1823).
Aphanomyrtus minima (Blume) Merr., Blumea, Suppl. 1: 108 (1937).
Myrtus variegata Blume, Bijdr.: 1082 (1826).
Jambosa tetraquetra Miq., Fl. Ned. Ind., Eerste Bijv.: 433 (1861). *Aphanomyrtus tetraquetra* (Miq.) Valeton, Repert. Spec. Nov. Regni Veg. 4: 235 (1907).
Aphanomyrtus octandra Koord. & Valeton, Bull. Inst. Bot. Buitenzorg 2: 5 (1899).
Aphanomyrtus forbesii Greves, J. Bot. 62(Suppl.): 36 (1924).

Syzygium minus A.C.Sm., Fl. Vitiensis Nova 3: 336 (1985).
Fiji (Viti Levu). 60 FIJ. Phan.

Syzygium minutiflorum Miq., Fl. Ned. Ind., Eerste Bijv.: 311 (1861).
Sumatera. 42 SUM. Phan.

Syzygium minutuliflorum (F.Muell.) B.Hyland, Austral. J. Bot., Suppl. Ser. 9: 104 (1983).
Northern Territory. 50 NTA. Phan.
**Eugenia minutuliflora* F.Muell., Victorian Naturalist 8: 197 (1892).
Eugenia essingtoniana S.Moore, J. Linn. Soc., Bot. 45: 206 (1920).

Syzygium mirabile (Merr.) Merr., Philipp. J. Sci. 79: 403 (1951).
Philippines. 42 PHI. Phan.
**Eugenia mirabilis* Merr., Philipp. J. Sci. 20: 412 (1922).

Syzygium mirandae (Merr.) Merr., Philipp. J. Sci. 79: 403 (1951).
Philippines. 42 PHI. Phan.
**Eugenia mirandae* Merr., Philipp. J. Sci., C 10: 221 (1915).

Syzygium mishmiense Chatterjee, Kew Bull. 3: 61 (1948).
Arunachal Pradesh. 40 EHM. Phan.

Syzygium monetarium (Ridl.) Merr. & L.M.Perry, J. Arnold Arbor. 23: 249 (1942).
New Guinea. 43 NWG. Phan.
**Eugenia monetaria* Ridl., Trans. Linn. Soc. London, Bot. 9: 49 (1916).

Syzygium monimioides Craven, Blumea 48: 485 (2003).
Queensland. 50 QLD. Phan.

Syzygium monospermum Craven, Blumea 48: 486 (2003).
Queensland. 50 QLD. Phan.

Syzygium monticola Merr. & L.M.Perry, Mem. Amer. Acad. Arts 18: 173 (1939).
Borneo. 42 BOR. Nanophan. or phan.

Syzygium montis-adam Kosterm., Quart. J. Taiwan Mus. 34: 145 (1981).
C. Sri Lanka. 40 SRL. Phan.
**Eugenia fergusonii* var. *minor* Trimen, Handb. Fl. Ceyl. 2: 173 (1894). *Syzygium fergusonii* subsp. *minor* (Trimen) Alston, in Revised Handb. Fl. Ceyl. 2: 433 (1981).

Syzygium moorei (F.Muell.) L.A.S.Johnson, Contr. New South Wales Natl. Herb. 3: 99 (1962).
SE. Queensland to NE. New South Wales. 50 NSW QLD. Nanophan. or phan.
**Eugenia moorei* F.Muell., Fragm. 5: 33 (1865).

Syzygium mouanum Guillaumin, Bull. Soc. Bot. France 85: 644 (1938 publ. 1939).
New Caledonia. 60 NWC. Nanophan. or phan.

Syzygium moultonii (Merr.) Merr. & L.M.Perry, Mem. Amer. Acad. Arts 18: 151 (1939).
Borneo. 42 BOR. Phan.
**Eugenia moultonii* Merr., J. Straits Branch Roy. Asiat. Soc. 77: 221 (1917).

Syzygium muelleri (Miq.) Miq., Fl. Ned. Ind. 1(1): 453 (1855).
Pen. Thailand to W. Malesia. 41 THA 42 BOR MLY. Phan.
Syzygium obovatum Korth., Ned. Kruidk. Arch. 1: 205 (1848), nom. illeg.
**Eugenia muelleri* Miq., Anal. Bot. Ind. 1: 23 (1850).
Syzygium furcatum Blume ex Miq., Anal. Bot. Ind. 1: 23 (1850).
Myrtus obovata Korth. ex Miq., Fl. Ned. Ind. 1(1): 453 (1855).
Eugenia venulosa Duthie in J.D.Hooker, Fl. Brit. India 2: 490 (1878).
Eugenia sarawacensis Merr., J. Straits Branch Roy. Asiat. Soc. 77: 214 (1917). *Syzygium sarawacense* (Merr.) Merr. & L.M.Perry, Mem. Amer. Acad. Arts 18: 186 (1939).
Eugenia viburnifolia Ridl., J. Bot. 68: 15 (1930). *Syzygium viburnifolium* (Ridl.) Masam., Enum. Phan. Born.: 541 (1942).

Syzygium mulgraveanum (B.Hyland) Craven & Biffin, Blumea 51: 138 (2006).
N. Queensland. 50 QLD. Phan.
**Waterhousea mulgraveana* B.Hyland, Austral. J. Bot., Suppl. Ser. 9: 143 (1983).

Syzygium multibracteolatum (Merr.) Merr. & L.M.Perry, Mem. Amer. Acad. Arts 18: 158 (1939).
Borneo. 42 BOR. Phan.
**Eugenia multibracteolata* Merr., J. Straits Branch Roy. Asiat. Soc. 77: 219 (1917).
Eugenia lobbii Ridl., J. Bot. 68: 17 (1930). *Syzygium lobbii* (Ridl.) Masam., Enum. Phan. Born.: 533 (1942).

Syzygium multiglandulosum Merr. & L.M.Perry, J. Arnold Arbor. 23: 269 (1942).
New Guinea. 43 NWG. Phan.

Syzygium multinerve (C.B.Rob.) Merr., Philipp. J. Sci. 79: 403 (1951).
Philippines. 42 PHI. Phan.
**Eugenia multinervis* C.B.Rob., Philipp. J. Sci., C 4: 352 (1909).

Syzygium multipetalum Pancher ex Brongn. & Gris, Bull. Soc. Bot. France 12: !82 (1865). *Eugenia multipetala* (Pancher ex Brongn. & Gris) Baker f., J. Linn. Soc., Bot. 45: 316 (1921). *Caryophyllus multipetalus* (Pancher ex Brongn. & Gris) Guillaumin, Bull. Soc. Bot. France 85: 648 (1938 publ. 1939).
EC. & SE. New Caledonia. 60 NWC. Nanophan. or phan.

Syzygium multipuncticulatum Merr., Philipp. J. Sci. 79: 404 (1951).
Philippines (Samar). 42 PHI. Phan.

Syzygium mundagam (Bourd.) Chithra in N.C.Nair & A.N.Henry, Fl. Tamil Nadu 1: 157 (1983).
S. India. 40 IND. Phan.
**Eugenia mundagam* Bourd., Forest Trees Travancore: 182 (1908). *Jambosa mundagam* (Bourd.) Gamble, Fl. Madras: 473 (1919).

Syzygium munronii (Wight) N.P.Balakr., Bull. Bot. Surv. India 22: 174 (1980 publ. 1982).
S. India, Assam. 40 ASS IND. Phan.
**Eugenia munronii* Wight, Ill. Ind. Bot. 2: 14 (1841). *Jambosa munronii* (Wight) Walp., Repert. Bot. Syst. 2: 191 (1843).
Eugenia munroii Miq., Anal. Bot. Ind. 1: 18 (1850), orth. var.

Syzygium myhendrae (Bedd. ex Brandis) Gamble, Fl. Madras: 478 (1919).
India. 40 IND. Phan.
**Eugenia myhendrae* Bedd. ex Brandis, Indian Trees: 325 (1906).

Syzygium myriadenum Merr. & L.M.Perry, J. Arnold Arbor. 23: 293 (1942). *Eugenia myriadena* (Merr. & L.M.Perry) Whitmore, Gard. Bull. Singapore 22: 16 (1967).
Solomon Is. 43 SOL. Phan.

Syzygium myrianthum (King) I.M.Turner, J. Singapore Natl. Acad. Sci. 22–24: 21 (1997).
Pen. Malaysia. 42 MLY. Phan.
**Eugenia myriantha* King, J. Asiat. Soc. Bengal, Pt. 2, Nat. Hist. 70: 125 (1901).

Syzygium myrsinifolium (Hance) Merr. & L.M.Perry, J. Arnold Arbor. 19: 226 (1938).
Hainan. 36 CHH. Phan.
**Eugenia myrsinifolia* Hance, J. Bot. 23: 8 (1885).
Syzygium myrsinifolium var. *grandiflorum* H.T.Chang & R.H.Miao, Acta Bot. Yunnan. 4: 21 (1982).

Syzygium myrtifolium Walp., Repert. Bot. Syst. (Walpers) 2: 178 (1843).
Indo-China to W. & C. Malesia. 41 MYA THA 42 BOR JAW MLY PHI SUM. Nanophan. or phan.
**Eugenia myrtifolia* Roxb., Fl. Ind. ed. 1832, 2: 490 (1832), nom. illeg.
Eugenia oleina Wight, Ill. Ind. Bot. 2: 15 (1841), nom. nud.
Syzygium campanulatum Korth., Ned. Kruidk. Arch. 1: 203 (1848).
Syzygium campanellum Miq., Fl. Ned. Ind. 1(1): 451 (1855).
Eugenia parva C.B.Rob., Philipp. J. Sci., C 4: 391 (1909).
Eugenia sinubanensis Elmer, Leafl. Philipp. Bot. 4: 1424 (1912). *Syzygium sinubanense* (Elmer) Diels, Bot. Jahrb. Syst. 57: 411 (1922).
Syzygium campanulatum var. *longistylum* Chantaran. & J.Parn., Kew Bull. 48: 595 (1993).

Syzygium myrtilloides Merr. & L.M.Perry, Mem. Amer. Acad. Arts 18: 172 (1939).
Borneo. 42 BOR. Phan.

Syzygium myrtillus (Stapf) Merr. & L.M.Perry, Mem. Amer. Acad. Arts 18: 171 (1939).
Borneo to Philippines. 42 BOR PHI. Phan.
**Eugenia myrtillus* Stapf, Trans. Linn. Soc. London, Bot. 4: 153 (1894). *Syzygium borneense* var. *myrtillus* (Stapf) Chantaran. & J.Parn., Thai Forest Bull., Bot. 21: 45 (1994).
Eugenia ugoensis C.B.Rob., Philipp. J. Sci., C 4: 389 (1909). *Syzygium ugoense* (C.B.Rob.) Masam., Enum. Phan. Born.: 540 (1942).

Syzygium myrtoides (A.Gray) R.Schmid, Bot. Jahrb. Syst. 92: 478 (1972).
Fiji (Viti Levu). 60 FIJ. Phan.
**Acicalyptus myrtoides* A.Gray, U.S. Expl. Exped., Phan. 1: 551 (1854). *Calyptranthes myrtoides* (A.Gray) Seem., Fl. Vit.: 81 (1866). *Cleistocalyx myrtoides* (A.Gray) Merr. & L.M.Perry, J. Arnold Arbor. 18: 329 (1937).

Syzygium naiadum (Diels) Merr. & L.M.Perry, J. Arnold Arbor. 23: 249 (1942).
New Guinea. 43 NWG. Phan.
**Jambosa naiadum* Diels, J. Arnold Arbor. 10: 82 (1929).

Syzygium nandarivatense (Gillespie) L.M.Perry, J. Arnold Arbor. 31: 367 (1950).
Fiji (Viti Levu). 60 FIJ. Phan.
**Eugenia nandarivatensis* Gillespie, Bernice P. Bishop Mus. Bull. 83: 22 (1931).

Syzygium nanpingense Y.Y.Qian, Guihaia 11: 210 (1991).
China (Yunnan). 36 CHC. Nanophan. or phan.

Syzygium nanum J.W.Dawson, in Fl. Nouv.-Caléd. 23: 92 (1999).
SE. New Caledonia. 60 NWC. Cham.

Syzygium napiforme (Koord. & Valeton) Merr. & L.M.Perry, Mem. Amer. Acad. Arts 18: 183 (1939).
W. Malesia. 42 BOR JAW MLY. Phan.
**Eugenia napiformis* Koord. & Valeton, Bull. Inst. Bot. Buitenzorg 2: 7 (1899).

Syzygium neepau Guillaumin, J. Arnold Arbor. 12: 257 (1931).
Vanuatu. 60 VAN. Phan.

Syzygium neesianum Arn., Nova Acta Phys.-Med. Acad. Caes. Leop.-Carol. Nat. Cur. 18(1): 335 (1836). *Eugenia neesiana* (Arn.) Wight, Ill. Ind. Bot. 2: 15 (1841).
Sri Lanka. 40 SRL. Phan.

Syzygium nemestrinum (M.R.Hend.) I.M.Turner, J. Singapore Natl. Acad. Sci. 22–24: 21 (1997).
Pen. Malaysia. 42 MLY. Phan.
**Eugenia nemestrina* M.R.Hend., Gard. Bull. Singapore 11: 324 (1947).

Syzygium nemorale Merr. & L.M.Perry, J. Arnold Arbor. 23: 280 (1942).
Bismarck Arch. to Solomon Is. 43 BIS SOL. Phan.

Syzygium neocaledonicum (Seem.) J.W.Dawson, in Fl. Nouv.-Caléd. 23: 32 (1999).
NW. & WC. New Caledonia. 60 NWC. Phan.
**Cupheanthus neocaledonicus* Seem., Fl. Vit.: 76 (1865).
Gaslondia amphoricarpa Vieill., Bull. Soc. Linn. Normandie 10: 97 (1866).
Cupheanthus austrocaledonicus Guillaumin, Ann. Inst. Bot.-Géol. Colon. Marseille, II, 9: 155 (1911), sphalm.
Eugenia comptonii Baker f., J. Linn. Soc., Bot. 45: 317 (1921). *Cupheanthus comptonii* (Baker f.) Guillaumin, Bull. Soc. Bot. France 85: 651 (1938 publ. 1939).

Syzygium neriifolium Becc. ex Merr. & L.M.Perry, Mem. Amer. Acad. Arts 18: 151 (1939).
Borneo. 42 BOR. Phan.

Syzygium neurocalyx (A.Gray) Christoph., Bernice P. Bishop Mus. Bull. 154: 27 (1938).
SW. Pacific. 60 FIJ SAM TON WAL. Nanophan. or phan.
**Eugenia neurocalyx* A.Gray, U.S. Expl. Exped., Phan. 1: 512 (1854). *Jambosa neurocalyx* (A.Gray) Müll.Stuttg. in W.G.Walpers, Ann. Bot. Syst. 4: 849 (1858).

Syzygium ngadimanianum (M.R.Hend.) I.M.Turner, J. Singapore Natl. Acad. Sci. 22–24: 21 (1997).
Pen. Malaysia. 42 MLY. Phan.
**Eugenia ngadimaniana* M.R.Hend., Gard. Bull. Singapore 11: 305 (1947).

Syzygium ngoyense (Schltr.) Guillaumin, Bull. Soc. Bot. France 85: 644 (1938 publ. 1939).
SE. New Caledonia. 60 NWC. Cham. or nanophan.
Syzygium patens Pancher ex Brongn. & Gris, Bull. Soc. Bot. France 12: 184 (1865), nom. illeg.
**Eugenia ngoyensis* Schltr., Bot. Jahrb. Syst. 39: 204 (1906).

Syzygium nicobaricum (King) Rathakr. & N.C.Nair, J. Econ. Taxon. Bot. 4: 287 (1983).
Nicobar Is. 41 NCB. Phan.
**Eugenia nicobarica* King, J. Asiat. Soc. Bengal, Pt. 2, Nat. Hist. 70: 130 (1901). *Cleistocalyx nicobaricus* (King) Merr. & L.M.Perry, J. Arnold Arbor. 18: 341 (1937).

Syzygium nidie Guillaumin, J. Arnold Arbor. 12: 257 (1931).
SW. Pacific. 60 FIJ VAN. Phan.

Syzygium nigrans (Gagnep.) Craven & Biffin, Blumea 51: 138 (2006).
Indo-China. 41 CBD THA VIE. Phan.
**Eugenia nigrans* Gagnep., Notul. Syst. (Paris) 3: 329 (1918). *Cleistocalyx nigrans* (Gagnep.) Merr. & L.M.Perry, J. Arnold Arbor. 18: 336 (1937).

Syzygium nigricans (King) Merr. & L.M.Perry, Mem. Amer. Acad. Arts 18: 194 (1939).
Pen. Malaysia, Borneo, Sulawesi. 42 BOR MLY SUL. Phan.
**Eugenia nigricans* King, J. Asiat. Soc. Bengal, Pt. 2, Nat. Hist. 70: 114 (1901).

subsp. ***nigricans***
Pen. Malaysia, Borneo. 42 BOR MLY. Phan.

subsp. ***phaeophyllum*** (Merr. & L.M.Perry) P.S.Ashton, Kew Bull. 61: 132 (2006).
NE. Borneo, Sulawesi. 42 BOR SUL. Phan.
**Syzygium leucoxylon* var. *phaeophyllum* Merr. & L.M.Perry, Mem. Amer. Acad. Arts 18: 194 (1939).

Syzygium nigropunctatum Merr. & L.M.Perry, Mem. Amer. Acad. Arts 18: 195 (1939).
Borneo. 42 BOR. Phan.

Syzygium nitens J.W.Dawson, in Fl. Nouv.-Caléd. 23: 26 (1999).
SE. New Caledonia. 60 NWC. Nanophan. or phan.

Syzygium nitidulum (Ridl.) I.M.Turner, J. Singapore Natl. Acad. Sci. 22–24: 21 (1997).
Pen. Malaysia. 42 MLY. Phan.
**Eugenia nitidula* Ridl., Fl. Malay Penins. 5: 308 (1925).

Syzygium nitidum Benth., London J. Bot. 2: 221 (1843). *Eugenia benthamii* A.Gray, U.S. Expl. Exped., Phan. 1: 520 (1854).
Philippines, Caroline Is. (Tobi), New Guinea (Yapen). 42 PHI 43 NWG 62 CRL. Phan.
Syzygium japenense Merr. & L.M.Perry, J. Arnold Arbor. 23: 288 (1942).

Syzygium nitrasirirakii Chantaran. & J.Parn., Kew Bull. 48: 607 (1993).
Pen. Thailand. 41 THA. Phan.

Syzygium nomoa Guillaumin, J. Arnold Arbor. 12: 258 (1931).
Vanuatu. 60 VAN. Phan.

Syzygium normanbiensc T.G.Hartley & L.M.Perry, J. Arnold Arbor. 54: 189 (1973).
Papua New Guinea. 43 NWG. Phan.

Syzygium novoguineense Merr. & L.M.Perry, J. Arnold Arbor. 23: 286 (1942).
New Guinea. 43 NWG. Phan.
**Jambosa auriculata* Blume, Mus. Bot. 1: 104 (1850). *Myrtus auriculata* (Blume) Zipp. ex Blume, Mus. Bot. 1: 98 (1850).

Syzygium nummularium Airy Shaw, Kew Bull. 4: 119 (1949).
Borneo. 42 BOR. Phan.

Syzygium nutans Merr. & L.M.Perry, J. Arnold Arbor. 23: 263 (1942).
New Guinea. 43 NWG. Phan.
**Eugenia nutans* K.Schum. in K.M.Schumann & U.M.Hollrung, Fl. Kais. Wilh. Land: 90 (1889), nom. illeg. *Jambosa nutans* (K.Schum.) Nied. in H.G.A.Engler & K.A.E.Prantl, Nat. Pflanzenfam. 3(7): 84 (1893).

Syzygium oblanceolatum (C.B.Rob.) Merr., Philipp. J. Sci. 79: 405 (1951).
Borneo to Philippines. 42 BOR PHI. Phan.
**Eugenia oblanceolata* C.B.Rob., Philipp. J. Sci., C 4: 400 (1909).
Syzygium kihamense Merr. & L.M.Perry, Mem. Amer. Acad. Arts 18: 150 (1939).

Syzygium oblancilimbum H.T.Chang & R.H.Miao, Acta Bot. Yunnan. 4: 23 (1982).
SC. China. 36 CHC. Nanophan. or phan.

Syzygium oblatum (Roxb.) Wall. ex A.M.Cowan & Cowan, Trees N. Bengal: 68 (1929).
Assam to SC. China and W. Malesia. 36 CHC 40 ASS BAN 41 CBD MYA NCB THA VIE 42 BOR MLY. Phan.
**Eugenia oblata* Roxb., Fl. Ind. ed. 1832, 2: 493 (1832). *Acmena oblata* (Roxb.) Walp., Repert. Bot. Syst. 2: 181 (1843). *Strongylocalyx oblatus* (Roxb.) Blume, Mus. Bot. 1: 89 (1850).
Jambosa pulchella Miq., Fl. Ned. Ind. 1(1): 422 (1855).
Eugenia laevicaulis Duthie in J.D.Hooker, Fl. Brit. India 2: 492 (1878). *Syzygium laevicaule* (Duthie) Masam., Enum. Phan. Born.: 531 (1942). *Syzygium oblatum* var. *laevicaule* (Duthie) Chantaran. & J.Parn., Kew Bull. 48: 607 (1993).
Eugenia limnaea Ridl., J. Straits Branch Roy. Asiat. Soc. 79: 64 (1918).
Eugenia laxiuscula Ridl., J. Fed. Malay States Mus. 10: 133 (1920).
Eugenia brantiana M.R.Hend., Gard. Bull. Singapore 11: 313 (1947).

Syzygium obliquinervium (Elmer) Merr., Philipp. J. Sci. 79: 405 (1951).
Philippines. 42 PHI. Phan.
**Eugenia obliquinervia* Elmer, Leafl. Philipp. Bot. 10: 3768 (1939), no latin descr.

Syzygium oblongifolium (Gillespie) Merr. & L.M.Perry, Sargentia 1: 75 (1942).
Fiji (Viti Levu). 60 FIJ. Phan.
**Pareugenia oblongifolia* Gillespie, Bernice P. Bishop Mus. Bull. 83: 23 (1931).

Syzygium occidentale (Bourd.) D.N.Gandhi in C.J. Saldanha & D.H. Nicolson, Fl. Hassan Distr. Karnataka: 282 (1976).
SW. India. 40 IND. Phan.
**Eugenia occidentalis* Bourd., Indian Forester 30: 195 (1904). *Jambosa occidentalis* (Bourd.) Gamble, Fl. Madras: 474 (1919).

Syzygium occlusum Miq., Fl. Ned. Ind. 1(1): 460 (1855). *Eugenia occlusa* (Miq.) Kurz, J. Asiat. Soc. Bengal, Pt. 2, Nat. Hist. 45(2): 130 (1876). *Eugenia symphysipetala* Koord. & Valeton, Meded. Lands Plantentuin 40: 161 (1900), nom. illeg.
Nicobar Is., Jawa. 41 NCB 42 JAW. Phan.

Syzygium odoardoi Merr. & L.M.Perry, Mem. Amer. Acad. Arts 18: 151 (1939).
Borneo. 42 BOR. Phan.
**Eugenia riparia* Becc., For. Borneo: 524 (1902), nom. illeg. *Syzygium riparium* (Becc.) Masam., Enum. Phan. Born.: 538 (1942), nom. illeg.

Syzygium odoratum (Lour.) DC., Prodr. 3: 260 (1828).
SE. China to Vietnam. 36 CHH CHS 41 VIE. Phan.
**Opa odorata* Lour., Fl. Cochinch.: 309 (1790). *Eugenia milletiana* Hemsl., J. Linn. Soc., Bot. 23: 297 (1887).
Eugenia deckeri Gagnep., Notul. Syst. (Paris) 3: 323 (1918).

Syzygium oleosum (F.Muell.) B.Hyland, Austral. J. Bot., Suppl. Ser. 9: 107 (1983).
E. Australia. 50 NSW QLD. Nanophan. or phan.
**Eugenia oleosa* F.Muell., Fragm. 5: 15 (1865).
Eugenia oleosa var. *cyanocarpa* F.Muell., Fragm. 9: 146 (1875). *Eugenia cyanocarpa* (F.Muell.) Maiden & Betche, Proc. Linn. Soc. New South Wales 29: 740 (1904 publ. 1905).
Eugenia coolminiana C.Moore, Handb. Fl. N.S.W.: 207 (1893). *Syzygium coolminianum* (C.Moore) L.A.S. Johnson, Contr. New South Wales Natl. Herb. 3: 98 (1962).

Syzygium oligadelphum (Christoph.) Merr. & L.M.Perry, Sargentia 1: 75 (1942).
Samoa (Upolu). 60 SAM. Phan.
**Pareugenia oligadelpha* Christoph., Bernice P. Bishop Mus. Bull. 154: 20 (1938).

Syzygium oliganthum Thwaites, Enum. Pl. Zeyl.: 118 (1859). *Eugenia oligantha* (Thwaites) Bedd., Fl. Sylv. S. India: 108 (1872).
Sri Lanka. 40 SRL. Nanophan. or phan.

Syzygium oligomyrum Diels, Bot. Jahrb. Syst. 60: 313 (1926).
Borneo. 42 BOR. Phan.
Eugenia ochneocarpa Merr., Univ. Calif. Publ. Bot. 15: 217 (1929). *Syzygium ochneocarpum* (Merr.) Merr. & L.M.Perry, Mem. Amer. Acad. Arts 18: 184 (1939).

Syzygium onesimum Merr. & L.M.Perry, J. Arnold Arbor. 23: 296 (1942). *Eugenia onesima* (Merr. & L.M.Perry) Whitmore, Gard. Bull. Singapore 22: 17 (1967).
Solomon Is. 43 SOL. Phan.

Syzygium onivense (H.Perrier) Labat & Schatz, Novon 12: 204 (2002).
E. Madagascar. 29 MDG. Nanophan. or phan.
**Eugenia onivensis* H.Perrier, Mém. Inst. Sci. Madagascar, Sér. B, Biol. Vég. 4: 188 (1953).

Syzygium operculatum (Roxb.) Nied. in H.G.A.Engler & K.A.E.Prantl, Nat. Pflanzenfam. 3(7): 85 (1893).
Nepal to N. Australia. 36 CHC CHH CHS 40 ASS BAN EHM NEP SRL 41 MYA THA VIE 42 BOR JAW MLY MOL PHI SUM XMS 50 NTA WAU. Phan.
Syzygium nervosum A.Cunn. ex DC., Prodr. 3: 260 (1828). *Cleistocalyx nervosus* (A.Cunn. ex DC.) Kosterm., Bull. Bot. Surv. India 29: 17 (1987 publ. 1989), nom. illeg.
Calyptranthes costata Buch.-Ham. in N.Wallich, Numer. List: 3556 (1831), nom. nud.
Calyptranthes cuneata Buch.-Ham. ex Wall., Numer. List: 3557 (1831), nom. nud.
Calyptranthes grandis Buch.-Ham. ex Wall., Numer. List: 3554 (1831), nom. nud.
Calyptranthes tatna Buch.-Ham. ex Wall., Numer. List: 3555 (1831), nom. nud.
Eugenia cerasoides Roxb., Fl. Ind. ed. 1832, 2: 488 (1832). *Syzygium cerasoides* (Roxb.) Raizada, Indian Forester 84: 478 (1958). *Cleistocalyx cerasoides* (Roxb.) I.M.Turner, Gard. Bull. Singapore 57: 26 (2005).
**Eugenia operculata* Roxb., Fl. Ind. ed. 1832, 2: 486 (1832). *Cleistocalyx operculatus* (Roxb.) Merr. & L.M.Perry, J. Arnold Arbor. 18: 337 (1937).
Eugenia paniala Roxb., Fl. Ind. ed. 1832, 2: 489 (1832). *Cleistocalyx operculatus* var. *paniala* (Roxb.) Chantaranothai & J.Parn., Kew Bull. 48: 591 (1993). *Cleistocalyx nervosus* var. *paniala* (Roxb.) J.Parn. & Chantaranothai, Novon 6: 201 (1996). *Cleistocalyx cerasoides* var. *paniala* (Roxb.) I.M.Turner, Gard. Bull. Singapore 57: 26 (2005). *Syzygium nervosum* var. *paniala* (Roxb.) Craven & Biffin, Blumea 51: 138 (2006).
Calyptranthes makul Blanco, Fl. Filip.: 419 (1837).
Calyptranthes zuzygium Blanco, Fl. Filip., ed. 2: 293 (1845), nom. illeg.
Syzygium wallichianum C.Presl, Abh. Königl. Böhm. Ges. Wiss., V, 3: 500 (1845).
Jambosa nitida Korth., Ned. Kruidk. Arch. 1: 202 (1847). *Cleistocalyx nitidus* (Korth.) Blume, Mus. Bot. 1: 84 (1850). *Eugenia cleistocalyx* Merr., Philipp. J. Sci., C 13: 98 (1918). *Syzygium cleistocalyx* (Merr.) P.S.Ashton, Blumea 51: 136 (2006).
Calyptranthes mangiferifolia Hance ex Walp., Ann. Bot. Syst. 2: 629 (1852).
Syzygium angkolanum Miq., Fl. Ned. Ind. 1(1): 448 (1855).
Syzygium nodosum Miq., Fl. Ned. Ind. 1(1): 447 (1855).
Syzygium polyanthum Thwaites, Enum. Pl. Zeyl.: 116 (1859), nom. illeg.
Eugenia operculata var. *obovata* Kurz, Forest Fl. Burma 1: 482 (1877). *Syzygium nervosum* var. *obovatum* (Kurz) A.Kumar, J. Econ. Taxon. Bot. 7: 664 (1985 publ. 1986).
Eugenia holtzei F.Muell., Australas. J. Pharm. 1: 199 (1886).
Eugenia holtziana F.Muell., Syst. Census Austral. Pl. 2: 101 (1889).
Eugenia gigantea Ridl., J. Straits Branch Roy. Asiat. Soc. 45: 192 (1906).
Eugenia clausa C.B.Rob., Philipp. J. Sci., C 4: 380 (1909).
Eugenia divaricatocymosa Hayata, Icon. Pl. Formosan. 3: 118 (1913).
Syzygium cerasoides Raizada, Indian Forester 93: 755 (1967), nom. illeg.

Syzygium oreophilum I.M.Turner, J. Singapore Natl. Acad. Sci. 22–24: 22 (1997).
Pen. Malaysia. 42 MLY. Nanophan.
**Eugenia oreophila* Ridl., J. Straits Branch Roy. Asiat. Soc. 61: 9 (1912), nom. illeg.
Eugenia jugalis Ridl., J. Fed. Malay States Mus. 6: 47 (1915).

Syzygium orites (Ridl.) I.M.Turner, J. Singapore Natl. Acad. Sci. 22–24: 22 (1997).
Pen. Malaysia. 42 MLY. Phan.
**Eugenia orites* Ridl., Fl. Malay Penins. 5: 308 (1925).

Syzygium orthoneurum Diels, Bot. Jahrb. Syst. 57: 400 (1922).
New Guinea. 43 NWG. Phan.

Syzygium ovale Korth., Ned. Kruidk. Arch. 1: 205 (1848).
Borneo. 42 BOR. Phan.

Syzygium ovalifolium (Blume) Merr. & L.M.Perry, J. Arnold Arbor. 23: 260 (1942).
New Guinea. 43 NWG. Phan.
**Jambosa ovalifolia* Blume, Mus. Bot. 1: 98 (1850). *Eugenia ovalifolia* (Blume) Warb., Bot. Jahrb. Syst. 13: 390 (1891), nom. illeg.

Syzygium owariense (P.Beauv.) Benth. in W.J.Hooker, Niger Fl.: 359 (1849).
Trop. Africa. 22 BEN IVO LBR NGA SEN SIE 23 CMN GAB GGI ZAI 25 TAN UGA 26 ANG MLW MOZ ZAM ZIM. Phan.
**Eugenia owariensis* P.Beauv., Fl. Oware 2: 20 (1810). *Jambosa owariensis* (P.Beauv.) DC., Prodr. 3: 287 (1828).

Syzygium oxyphyllum Diels, Bot. Jahrb. Syst. 57: 407 (1922).
New Guinea. 43 NWG. Phan.

Syzygium pachyanthum (Diels) Merr. & L.M.Perry, J. Arnold Arbor. 23: 249 (1942).
New Guinea. 43 NWG. Phan.
**Jambosa pachyantha* Diels, Bot. Jahrb. Syst. 57: 395 (1922).

Syzygium pachycladum (K.Schum. & Lauterb.) Merr. & L.M.Perry, J. Arnold Arbor. 23: 258 (1942).
New Guinea. 43 NWG. Phan.
**Jambosa pachyclada* K.Schum. & Lauterb., Fl. Schutzgeb. Südsee: 474 (1900).

Syzygium pachyphyllum (Kurz) Merr. & L.M.Perry, Mem. Amer. Acad. Arts 18: 168 (1939).

Indo-China to W. Malesia. 41 MYA THA 42 BOR MLY. Phan.
**Eugenia pachyphylla* Kurz, J. Asiat. Soc. Bengal, Pt. 2, Nat. Hist. 42(2): 232 (1873).

Syzygium pachyrrachis Amshoff, Blumea 5: 497 (1945).
W. Jawa. 42 JAW. Phan.

Syzygium pachysarcum (Gagnep.) Merr. & L.M.Perry, J. Arnold Arbor. 19: 111 (1938).
Indo-China. 41 CBD THA VIE. Nanophan. or phan.
**Eugenia pachysarca* Gagnep., Notul. Syst. (Paris) 3: 329 (1918).

Syzygium pachysepalum Merr. & L.M.Perry, Mem. Amer. Acad. Arts 18: 193 (1939).
Borneo. 42 BOR. Phan.

Syzygium pahangense (Ridl.) I.M.Turner, J. Singapore Natl. Acad. Sci. 22–24: 22 (1997).
Pen. Malaysia. 42 MLY. Phan.
**Eugenia pahangensis* Ridl., J. Linn. Soc., Bot. 38: 307 (1908).
Eugenia pahangensis var. *fraseri* M.R.Hend., Gard. Bull. Singapore 12: 102 (1949). *Syzygium pahangense* var. *fraseri* (M.R.Hend.) I.M.Turner, J. Singapore Natl. Acad. Sci. 22–24: 22 (1997).

Syzygium palauense (Kaneh.) Hosok., J. Jap. Bot. 16: 543 (1940).
Caroline Is. (Palau). 62 CRL. Phan.
**Eugenia palauensis* Kaneh., Bot. Mag. (Tokyo) 45: 335 (1931).

Syzygium palawanense (C.B.Rob.) Merr. & L.M.Perry, Mem. Amer. Acad. Arts 18: 177 (1939).
Borneo to W. Philippines. 42 BOR PHI. Phan.
**Eugenia palawanensis* C.B.Rob., Philipp. J. Sci., C 4: 377 (1909).

Syzygium palembanicum Miq., Fl. Ned. Ind., Eerste Bijv.: 313 (1861). *Eugenia palembanica* (Miq.) Merr., J. Straits Branch Roy. Asiat. Soc. 77: 225 (1917).
Myanmar to W. Malesia. 41 MYA 42 BOR MLY SUM. Phan.
Eugenia lepidocarpa Wall. ex Kurz, J. Asiat. Soc. Bengal, Pt. 2, Nat. Hist. 46(2): 68 (1877).
Eugenia selangorensis Ridl., J. Fed. Malay States Mus. 5: 32 (1914).
Eugenia kuchingensis Merr., J. Straits Branch Roy. Asiat. Soc. 77: 213 (1917). *Syzygium kuchingense* (Merr.) Merr. & L.M.Perry, Mem. Amer. Acad. Arts 18: 175 (1939).

Syzygium palghatense Gamble, Bull. Misc. Inform. Kew 1918: 240 (1918).
SW. India. 40 IND. Phan.

Syzygium pallens Merr. & L.M.Perry, J. Arnold Arbor. 23: 251 (1942).
New Guinea. 43 NWG. Phan.

Syzygium pallidilimbum Merr. & L.M.Perry, Mem. Amer. Acad. Arts 18: 183 (1939).
Borneo. 42 BOR. Phan.

Syzygium pallidulum (Ridl.) I.M.Turner, J. Singapore Natl. Acad. Sci. 22–24: 22 (1997).
Pen. Malaysia. 42 MLY. Phan.
**Eugenia pallidula* Ridl., Fl. Malay Penins. 1: 748 (1922).

Syzygium pallidum Merr., Publ. Bur. Sci. Gov. Lab. 17: 38 (1904).
Philippines. 42 PHI. Phan.
Eugenia perpallida Merr., Philipp. J. Sci. 1(Suppl.): 106 (1906).
Eugenia pacifica Elmer ex Merr., Enum. Philipp. Fl. Pl. 3: 174 (1923), pro syn.

Syzygium paludosum P.S.Ashton, Kew Bull. 61: 132 (2006).
Borneo (Brunei, Sarawak). 42 BOR. Phan.

Syzygium panayense (Merr.) Merr., Philipp. J. Sci. 79: 407 (1951).
Philippines. 42 PHI. Phan.
**Eugenia panayensis* Merr., Philipp. J. Sci. 18: 292 (1921).

Syzygium pancheri Brongn. & Gris, Bull. Soc. Bot. France 12: 183 (1865).
New Caledonia. 60 NWC. Nanophan. or phan.
Syzygium neglectum Brongn. & Gris, Bull. Soc. Bot. France 12: 184 (1865).
Syzygium micranthum Montrouz. ex Guillaumin, Ann. Inst. Bot.-Géol. Colon. Marseille, II, 9: 153 (1911), nom. nud.
Eugenia hydrophila Baker f., J. Linn. Soc., Bot. 45: 315 (1921).

Syzygium panduriforme (Elmer) Merr., Philipp. J. Sci. 79: 408 (1951).
Philippines. 42 PHI. Phan.
**Eugenia panduriformis* Elmer, Leafl. Philipp. Bot. 4: 1412 (1912).

Syzygium paniculatum Gaertn., Fruct. Sem. Pl. 1: 167 (1788). *Eugenia paniculata* (Gaertn.) Britten, J. Bot. 37: 247 (1899), nom. illeg.
E. Australia. 50 NSW QLD (80) gua. Nanophan. or phan.
Eugenia rheedioides Standl. & Steyerm., Publ. Field Mus. Nat. Hist., Bot. Ser. 23: 131 (1944).

Syzygium paniense (Baker f.) J.W.Dawson, in Fl. Nouv.-Caléd. 23: 37 (1999).
NW. & WC. New Caledonia. 60 NWC. Phan.
**Eugenia paniensis* Baker f., J. Linn. Soc., Bot. 45: 318 (1921). *Cupheanthus paniensis* (Baker f.) Guillaumin, Bull. Soc. Bot. France 85: 651 (1938 publ. 1939).
Caryophyllus garciniifolius Guillaumin, Bull. Mus. Natl. Hist. Nat., II, 31: 176 (1959).

Syzygium panzeri Merr. & L.M.Perry, Mem. Amer. Acad. Arts 18: 162 (1939).
Borneo. 42 BOR. Phan.

Syzygium papillosum (Duthie) Merr. & L.M.Perry, Mem. Amer. Acad. Arts 18: 157 (1939).
Pen. Thailand to W. Malesia. 41 THA 42 BOR MLY. Phan.
**Eugenia papillosa* Duthie in J.D.Hooker, Fl. Brit. India 2: 495 (1878).

Syzygium papyraceum B.Hyland, Austral. J. Bot., Suppl. Ser. 9: 111 (1983).
Queensland. 50 QLD. Phan.

Syzygium paradoxum (Merr.) Masam., Enum. Phan. Born.: 536 (1942).
NW. Borneo. 42 BOR. Phan.
**Eugenia paradoxa* Merr., J. Straits Branch Roy. Asiat. Soc. 77: 210 (1917). *Cleistocalyx paradoxus* (Merr.) Merr. & L.M.Perry, J. Arnold Arbor. 18: 331 (1937).

Syzygium paraiense Merr. & L.M.Perry, Mem. Amer. Acad. Arts 18: 169 (1939).
Borneo. 42 BOR. Phan.

Syzygium parameswaranii M.Mohanan & A.N.Henry, J. Bombay Nat. Hist. Soc. 84: 409 (1987 publ. 1988).
SW. India. 40 IND. Phan.

Syzygium parkeri (Baker) Labat & Schatz, Novon 12: 204 (2002).
Madagascar. 29 MDG. Phan.
**Eugenia parkeri* Baker, J. Linn. Soc., Bot. 20: 144 (1883).
Eugenia ibitensis Drake, Bull. Mus. Hist. Nat. (Paris) 1903: 42 (1903). *Eugenia parkeri* var. *ibitensis* (Drake) H.Perrier, in Fl. Madag. 152: 72 (1953).

Syzygium parvicarpum J.W.Dawson, in Fl. Nouv.-Caléd. 23: 113 (1999).
EC. New Caledonia. 60 NWC. Nanophan.

Syzygium parvifolium (Engl.) Mildbr., Wiss. Erg. Deut. Zentr.-Afr. Exped., Bot. 2: 623 (1914).
C. & E. Trop. Africa. 23 BUR RWA ZAI 25 KEN TAN UGA. Phan.
**Syzygium guineense* var. *parvifolium* Engl. in G.W.J. Mildbraed, Wiss. Erg. Deut. Zentr.-Afr. Exped., Bot. 2: 582 (1913). *Syzygium guineense* subsp. *parvifolium* (Engl.) F.White, Kirkia 10: 404 (1977).

Syzygium parvulum Mildbr. ex Amshoff, Acta Bot. Neerl. 8: 54 (1959).
Tanzania (Uluguru Mts.). 25 TAN. Nanophan. or phan.

Syzygium pascasioii (Merr.) Merr., Philipp. J. Sci. 79: 408 (1951).
Philippines (Bucas Grande). 42 PHI. Phan.
**Eugenia pascasioii* Merr., Philipp. J. Sci. 18: 307 (1921).

Syzygium patens Korth., Ned. Kruidk. Arch. 1: 203 (1848).
Sumatera. 42 SUM. Phan.

Syzygium patentinerve Christoph., Bernice P. Bishop Mus. Bull. 154: 24 (1938).
Samoa. 60 SAM. Phan.

Syzygium paucipunctatum (Koord. & Valeton) Merr. & L.M.Perry, Mem. Amer. Acad. Arts 18: 169 (1939).
Jawa, Borneo. 42 BOR JAW. Phan.
**Eugenia paucipunctata* Koord. & Valeton, Bull. Inst. Bot. Buitenzorg 2: 8 (1899).

Syzygium paucivenium (C.B.Rob.) Merr., Philipp. J. Sci. 79: 408 (1951).
Taiwan (Lan Yü, Lü Tao) to Philippines. 38 TAI 42 PHI. Phan.
**Eugenia paucivenia* C.B.Rob., Philipp. J. Sci., C 4: 382 (1909).
Eugenia kashotoensis Hayata, J. Coll. Sci. Imp. Univ. Tokyo 30(1): 113 (1911). *Syzygium kashotoense* (Hayata) Mori, Trans. Nat. Hist. Soc. Taiwan 28: 439 (1938).

Syzygium pauper (Ridl.) I.M.Turner, J. Singapore Natl. Acad. Sci. 22–24: 22 (1997).
Pen. Malaysia. 42 MLY. Nanophan. or phan.
**Eugenia pauper* Ridl., J. Straits Branch Roy. Asiat. Soc. 79: 65 (1918).

Syzygium peekelii Diels, Bot. Jahrb. Syst. 57: 414 (1922).
Bismarck Arch. 43 BIS. Phan.

Syzygium pellucidum (Duthie) N.P.Balakr., Bull. Bot. Surv. India 22: 174 (1980 publ. 1982).
Myanmar. 41 MYA. Phan.
**Eugenia pellucida* Duthie in J.D.Hooker, Fl. Brit. India 2: 485 (1878).

Syzygium penasii (Merr.) Merr., Philipp. J. Sci. 79: 408 (1951).
Philippines. 42 PHI. Phan.
**Eugenia penasii* Merr., Philipp. J. Sci. 18: 293 (1921).

Syzygium pendens (Duthie) I.M.Turner, J. Singapore Natl. Acad. Sci. 22–24: 22 (1997).
Pen. Malaysia to Borneo. 42 BOR MLY. Phan.
**Eugenia pendens* Duthie in J.D.Hooker, Fl. Brit. India 2: 475 (1878).

Syzygium pendulinum J.W.Dawson, in Fl. Nouv.-Caléd. 23: 117 (1999).
C. & SE. New Caledonia. 60 NWC. Nanophan. or phan.

Syzygium penibukanense Merr. & L.M.Perry, Mem. Amer. Acad. Arts 18: 153 (1939).
Borneo. 42 BOR. Phan.

Syzygium pennelii (Guillaumin) J.W.Dawson, in Fl. Nouv.-Caléd. 23: 14 (1999).
C. & WC. New Caledonia. 60 NWC. Phan.
**Acicalyptus pennelii* Guillaumin, Bull. Soc. Bot. France 85: 653 (1938 publ. 1939). *Cleistocalyx pennelii* (Guillaumin) Merr., Bull. Soc. Bot. France 86: 379 (1940).

Syzygium perakense (King) I.M.Turner, J. Singapore Natl. Acad. Sci. 22–24: 22 (1997).
Pen. Thailand to Pen. Malaysia. 41 THA 42 MLY. Phan.
**Eugenia perakensis* King, J. Asiat. Soc. Bengal, Pt. 2, Nat. Hist. 70: 81 (1901).

Syzygium peregrinum (Blume) Merr. & L.M.Perry, Mem. Amer. Acad. Arts 18: 154 (1939).
Borneo to Philippines. 42 BOR PHI. Phan.
**Jambosa peregrina* Blume, Mus. Bot. 1: 92 (1850).
Eugenia tawaensis Merr., Univ. Calif. Publ. Bot. 15: 220 (1929). *Syzygium tawaense* (Merr.) Masam., Enum. Phan. Born.: 539 (1942).

Syzygium pergamaceum (Greves) Merr. & L.M.Perry, J. Arnold Arbor. 23: 250 (1942).
New Guinea. 43 NWG. Phan.
**Eugenia pergamacea* Greves, J. Bot. 61(Suppl.): 16 (1923).
Syzygium archboldianum Merr. & L.M.Perry, J. Arnold Arbor. 23: 271 (1942).
Syzygium discolor Merr. & L.M.Perry, J. Arnold Arbor. 23: 274 (1942).

Syzygium pergamentaceum (King) Chantaran. & J.Parn., Kew Bull. 48: 607 (1993).
Pen. Thailand to Pen. Malaysia. 41 THA 42 MLY. Phan.
**Eugenia pergamentacea* King, J. Asiat. Soc. Bengal, Pt. 2, Nat. Hist. 70: 87 (1901).

Syzygium periyarensis Augustine & Sasidh., Rheedea 9: 1555 (1999).
SW. India. 40 IND. Phan.

Syzygium perparvifolium (Merr.) Merr. & L.M.Perry, Mem. Amer. Acad. Arts 18: 162 (1939).
Borneo. 42 BOR. Phan.
**Eugenia perparvifolia* Merr., J. Straits Branch Roy. Asiat. Soc. 77: 220 (1917).

Syzygium perryae I.M.Turner, J. Singapore Natl. Acad. Sci. 22–24: 22 (1997).
Pen. Malaysia to Sumatera. 42 MLY SUM. Nanophan.

Eugenia benjamina King, J. Asiat. Soc. Bengal, Pt. 2, Nat. Hist. 70: 106 (1901).

Syzygium perspicuinervium (Merr.) Masam., Enum. Phan. Born.: 537 (1942).
N. Borneo. 42 BOR. Phan.
**Eugenia perspicuinervia* Merr., Univ. Calif. Publ. Bot. 15: 218 (1929). *Cleistocalyx perspicuinervius* (Merr.) Merr. & L.M.Perry, J. Arnold Arbor. 18: 332 (1937).

Syzygium petakense Merr. & L.M.Perry, Mem. Amer. Acad. Arts 18: 150 (1939).
Borneo. 42 BOR. Phan.

Syzygium petelotii Merr. & L.M.Perry, J. Arnold Arbor. 19: 103 (1938).
N. Vietnam. 41 VIE. Nanophan. or phan.

Syzygium petraeum Diels, Bot. Jahrb. Syst. 57: 410 (1922).
New Guinea. 43 NWG. Phan.

Syzygium petrinense Bosser & J.Guého, Bull. Mus. Natl. Hist. Nat., B, Adansonia 9: 36 (1987).
Mauritius. 29 MAU. Nanophan.

Syzygium petrophilum Merr. & L.M.Perry, Mem. Amer. Acad. Arts 18: 195 (1939).
Borneo. 42 BOR. Phan.

Syzygium phacelanthum (Diels) Merr. & L.M.Perry, J. Arnold Arbor. 23: 249 (1942).
New Guinea. 43 NWG. Phan.
**Jambosa phacelantha* Diels, Bot. Jahrb. Syst. 57: 390 (1922).

Syzygium phaeophyllum Merr. & L.M.Perry, J. Arnold Arbor. 26: 103 (1945).
Fiji (Taveuni). 60 FIJ. Phan.
**Eugenia durifolia* A.C.Sm., Bernice P. Bishop Mus. Bull. 141: 105 (1936). *Syzygium durifolium* (A.C. Sm.) Merr. & L.M.Perry, Sargentia 1: 76 (1942), nom. illeg.

Syzygium phaeostictum Merr. & L.M.Perry, J. Arnold Arbor. 23: 270 (1942).
New Guinea. 43 NWG. Phan.

Syzygium phanerophlebium (C.B.Rob.) Merr., Philipp. J. Sci. 79: 408 (1951).
Philippines. 42 PHI. Phan.
**Eugenia phanerophlebia* C.B.Rob., Philipp. J. Sci., C 4: 353 (1909).

Syzygium phengklaii (Chantar. & J.Parn.) Craven & Biffin, Blumea 51: 138 (2006).
SW. Thailand. 41 THA. Phan.
**Cleistocalyx phengklaii* Chantar. & J.Parn., Kew Bull. 48: 591 (1993).

Syzygium philippinense (C.B.Rob.) Merr., Philipp. J. Sci. 79: 409 (1951).
Philippines. 42 PHI. Phan.
**Eugenia philippinensis* C.B.Rob., Philipp. J. Sci., C 4: 378 (1909).

Syzygium phillyreifolium (Baker) Labat & Schatz, Novon 12: 204 (2002).
Madagascar. 29 MDG. Phan.
**Eugenia phillyreifolia* Baker, J. Linn. Soc., Bot. 20: 145 (1883). *Myrtus phillyreifolia* (Baker) Kuntze, Revis. Gen. Pl. 3(2): 92 (1898).
Eugenia aggregata Baker, J. Linn. Soc., Bot. 22: 475 (1887).
Eugenia phillyreifolia f. *obscurifolia* H.Perrier, Mém. Inst. Sci. Madagascar, Sér. B, Biol. Vég. 4: 188 (1952). *Eugenia phillyreifolia* var. *obscurifolia* (H.Perrier) H.Perrier, in Fl. Madag. 152: 58 (1953).

Syzygium phryganodes Merr. & L.M.Perry, Mem. Amer. Acad. Arts 18: 178 (1939).
Borneo. 42 BOR. Phan.

Syzygium pierrei (Gagnep.) Merr. & L.M.Perry, J. Arnold Arbor. 19: 114 (1938).
Vietnam. 41 VIE. Phan.
**Eugenia pierrei* Gagnep., Notul. Syst. (Paris) 3: 330 (1918).

Syzygium pilgerianum (K.Schum. & Lauterb.) Merr. & L.M.Perry, J. Arnold Arbor. 23: 255 (1942).
New Guinea. 43 NWG. Phan.
**Jambosa pilgeriana* K.Schum. & Lauterb., Fl. Schutzgeb. Südsee: 473 (1900).

Syzygium piluliferum Craven & Biffin, Blumea 51: 138 (2006).
Borneo. 42 BOR. Nanophan. or phan.
**Acmena caudata* Merr. & L.M.Perry, J. Arnold Arbor. 19: 11 (1938).

Syzygium platycarpum (Diels) Merr. & L.M.Perry, J. Arnold Arbor. 23: 278 (1942).
New Guinea. 43 NWG. Phan.
**Jambosa platycarpa* Diels, Bot. Jahrb. Syst. 57: 385 (1922).

Syzygium platypodum Diels, Bot. Jahrb. Syst. 57: 406 (1922).
New Guinea. 43 NWG. Phan.

Syzygium plumbeum (King) I.M.Turner, J. Singapore Natl. Acad. Sci. 22–24: 22 (1997).
Pen. Malaysia. 42 MLY. Phan.
**Eugenia plumbea* King, J. Asiat. Soc. Bengal, Pt. 2, Nat. Hist. 70: 85 (1901).

Syzygium plumeum (Ridl.) Merr. & L.M.Perry, J. Arnold Arbor. 23: 296 (1942).
New Guinea. 43 NWG. Phan.
**Eugenia plumea* Ridl., Trans. Linn. Soc. London, Bot. 9: 46 (1916).
Syzygium leptanthelium Diels, Bot. Jahrb. Syst. 57: 410 (1922).

Syzygium pluviatile T.G.Hartley & L.M.Perry, J. Arnold Arbor. 54: 199 (1973).
New Guinea. 43 NWG. Phan.

Syzygium polisense Merr., Philipp. J. Sci. 79: 409 (1951).
Philippines (Luzon). 42 PHI. Phan.

Syzygium politum (King) I.M.Turner, J. Singapore Natl. Acad. Sci. 22–24: 23 (1997).
Pen. Malaysia. 42 MLY. Nanophan. or phan.
**Eugenia polita* King, J. Asiat. Soc. Bengal, Pt. 2, Nat. Hist. 70: 110 (1901).

Syzygium polyanthum (Wight) Walp., Repert. Bot. Syst. 2: 180 (1843).
Indo-China to Philippines. 41 AND MYA NCB THA VIE 42 BOR JAW MLY PHI SUM. Phan.
Myrtus cymosa Blume, Bijdr.: 1086 (1826), nom. illeg.
**Eugenia polyantha* Wight, Ill. Ind. Bot. 2: 17 (1841).
Syzygium cymosum Korth., Ned. Kruidk. Arch. 1: 202 (1847), nom. illeg.
Eugenia microbotrya Miq., Anal. Bot. Ind. 1: 27 (1850). *Syzygium microbotryum* (Miq.) Masam., Enum. Phan. Born.: 534 (1942).

Eugenia pamatensis Miq., Anal. Bot. Ind. 1: 22 (1850). *Syzygium pamatense* (Miq.) Masam., Enum. Phan. Born.: 536 (1942).
Syzygium micranthum Blume ex Miq., Anal. Bot. Ind. 1: 22 (1850).
Eugenia junghuhniana Miq., Fl. Ned. Ind. 1(1): 441 (1855).
Eugenia lucidula Miq., Fl. Ned. Ind. 1(1): 444 (1855).
Eugenia nitida Duthie in J.D.Hooker, Fl. Brit. India 2: 496 (1878), nom. illeg.
Eugenia atropunctata C.B.Rob., Philipp. J. Sci., C 4: 385 (1909), nom. illeg.
Eugenia lambii Elmer, Leafl. Philipp. Bot. 4: 1430 (1912).
Eugenia holmanii Elmer, Leafl. Philipp. Bot. 7: 2354 (1914).
Eugenia resinosa Gagnep., Notul. Syst. (Paris) 3: 331 (1918).
Eugenia polyantha var. *sessilis* M.R.Hend., Gard. Bull. Singapore 12: 212 (1949). *Syzygium polyanthum* var. *sessile* (M.R.Hend.) I.M.Turner, J. Singapore Natl. Acad. Sci. 22–24: 23 (1997).

Syzygium polycephaloides (C.B.Rob.) Merr., Philipp. J. Sci. 79: 410 (1951).
Philippines, SE. Sulawesi. 42 PHI SUL. Phan.
**Eugenia polycephaloides* C.B.Rob., Philipp. J. Sci., C 4: 399 (1909).

Syzygium polycephalum (Miq.) Merr. & L.M.Perry, Mem. Amer. Acad. Arts 18: 153 (1939).
W. & C. Malesia. 42 BOR JAW SUL. Phan.
**Eugenia polycephala* Miq., Anal. Bot. Ind. 1: 19 (1850). *Jambosa polycephala* (Miq.) Miq., Fl. Ned. Ind., Eerste Bijv.: 439 (1861).

Syzygium polycladum Merr. & L.M.Perry, Mem. Amer. Acad. Arts 18: 161 (1939).
Borneo. 42 BOR. Phan.

Syzygium polypetaloideum Merr. & L.M.Perry, J. Arnold Arbor. 19: 217 (1938).
S. China. 36 CHC CHS. Phan.

Syzygium polypetalum (Wall.) Merr. & L.M.Perry, Brittonia 4: 125 (1941).
Assam to Bangladesh. 40 ASS BAN. Phan.
Eugenia angustifolia Roxb., Fl. Ind. ed. 1832, 2: 490 (1832), nom. illeg. **Jambosa polypetala* Wall., Rep. Calcutta Bot. Gard.: 27 (1840). *Eugenia polypetala* (Wall.) Wight, Icon. Pl. Ind. Orient. 2(3): 8 (1842).

Syzygium polyphlebium (Diels) Merr. & L.M.Perry, J. Arnold Arbor. 23: 249 (1942).
New Guinea. 43 NWG. Phan.
**Jambosa polyphlebia* Diels, Bot. Jahrb. Syst. 57: 391 (1922).

Syzygium pondoense Engl., Bot. Jahrb. Syst. 54: 241 (1917).
Cape Prov. to KwaZulu-Natal. 27 CPP NAT. Nanophan. or phan.

Syzygium pontianakense Merr. & L.M.Perry, Mem. Amer. Acad. Arts 18: 178 (1939).
Borneo. 42 BOR. Phan.

Syzygium populifolium (Baker) J.Guého & A.J.Scott, Kew Bull. 34: 493 (1980).
Mauritius. 29 MAU. Nanophan. or phan.
**Eugenia populifolia* Baker, Fl. Mauritius: 118 (1877).

Syzygium porphyranthum (Ridl.) I.M.Turner, J. Singapore Natl. Acad. Sci. 22–24: 23 (1997).
Pen. Malaysia. 42 MLY. Phan.
**Eugenia porphyrantha* Ridl., J. Straits Branch Roy. Asiat. Soc. 61: 8 (1912).

Syzygium porphyrocarpum (Greves) Merr. & L.M.Perry, J. Arnold Arbor. 23: 250 (1942).
New Guinea. 43 NWG. Phan.
**Eugenia porphyrocarpa* Greves, J. Bot. 61(Suppl.): 17 (1923).

Syzygium potamicum Kosterm., Quart. J. Taiwan Mus. 34: 147 (1981).
Sri Lanka. 40 SRL. Phan.

Syzygium poyanum J.W.Dawson, in Fl. Nouv.-Caléd. 23: 124 (1999).
C. New Caledonia. 60 NWC. Nanophan. or phan.

Syzygium praecox (Roxb.) Rathakr. & N.C.Nair, J. Econ. Taxon. Bot. 4: 288 (1983).
E. Nepal to Thailand. 40 ASS BAN EHM NEP 41 MYA THA. Phan.
Eugenia lanceifolia Roxb., Fl. Ind. ed. 1832, 2: 494 (1832). *Strongylocalyx lanceifolius* (Roxb.) Blume, Mus. Bot. 1: 89 (1850). *Syzygium roxburghianum* Raizada, Indian Forester 74: 336 (1948).
**Eugenia praecox* Roxb., Fl. Ind. ed. 1832, 2: 488 (1832). *Strongylocalyx praecox* (Roxb.) Blume, Mus. Bot. 1: 89 (1850). *Jambosa praecox* (Roxb.) A.M.Cowan & Cowan, Trees N. Bengal: 67 (1929).
Eugenia wallichii Wight, Icon. Pl. Ind. Orient. 2: t. 536 (1842). *Syzygium wallichii* (Wight) Walp., Repert. Bot. Syst. 2: 180 (1843).

Syzygium praestigiosum (M.R.Hend.) I.M.Turner, J. Singapore Natl. Acad. Sci. 22–24: 23 (1997).
Pen. Malaysia. 42 MLY. Phan.
**Eugenia praestigiosa* M.R.Hend., Gard. Bull. Singapore 11: 318 (1947).

Syzygium praetermissum (Gage) N.P.Balakr., Bull. Bot. Surv. India 22: 174 (1980 publ. 1982).
Myanmar. 41 MYA. Phan.
**Eugenia praetermissa* Gage, Indian Forester 32: 6 (1906).

Syzygium praineanum (King) Chantaran. & J.Parn., Kew Bull. 48: 608 (1993).
Pen. Thailand to Pen. Malaysia. 41 THA 42 MLY. Phan.
**Eugenia praineana* King, J. Asiat. Soc. Bengal, Pt. 2, Nat. Hist. 70: 116 (1901).

subsp. ***minor*** Chantaran. & J.Parn., Kew Bull. 48: 608 (1993).
Pen. Thailand. 41 THA. Phan.

subsp. ***praineanum***
Pen. Malaysia to Borneo. 42 BOR MLY. Phan.
Eugenia pearsoniana King, J. Asiat. Soc. Bengal, Pt. 2, Nat. Hist. 70: 116 (1901). *Syzygium pearsonianum* (King) I.M.Turner, J. Singapore Natl. Acad. Sci. 22–24: 22 (1997).

Syzygium prasiniflorum (Ridl.) Merr. & L.M.Perry, Mem. Amer. Acad. Arts 18: 194 (1939).
Borneo. 42 BOR. Phan.
**Eugenia prasiniflora* Ridl., J. Bot. 68: 35 (1930).

Syzygium pringlei (B.Hyland) Craven & Biffin, Blumea 51: 139 (2006).
E. Queensland. 50 QLD. Phan.
**Acmenosperma pringlei* B.Hyland, Austral. J. Bot., Suppl. Ser. 9: 26 (1983).

Syzygium propinquum (Guillaumin) J.W.Dawson, Fl. Nouv.-Caléd. 23: 86 (1999).
C. New Caledonia. 60 NWC. Nanophan. or phan.
**Caryophyllus propinquus* Guillaumin, Bull. Soc. Bot. France 85: 649 (1938 publ. 1939).

Syzygium pseudocalcicola Craven & Biffin, Blumea 51: 139 (2006).
Philippines (Luzon). 42 PHI. Phan.
**Eugenia paucipunctata* Merr., Philipp. J. Sci., C 10: 215 (1915), nom. illeg. *Cleistocalyx paucipunctatus* Merr. & L.M.Perry, J. Arnold Arbor. 18: 336 (1937).

Syzygium pseudoclaviflorum (M.R.Hend.) I.M.Turner, J. Singapore Natl. Acad. Sci. 22–24: 23 (1997).
Pen. Malaysia. 42 MLY. Phan.
**Eugenia pseudoclaviflora* M.R.Hend., Gard. Bull. Singapore 11: 331 (1947).

Syzygium pseudocrenulatum (M.R.Hend.) I.M.Turner, J. Singapore Natl. Acad. Sci. 22–24: 23 (1997).
Pen. Malaysia to Borneo. 42 BOR MLY. Phan.
Eugenia crenulata Duthie in J.D.Hooker, Fl. Brit. India 2: 490 (1878), nom. illeg. **Eugenia pseudocrenulata* M.R.Hend., Gard. Bull. Singapore 12: 216 (1949).

Syzygium pseudofastigiatum B.Hyland, Austral. J. Bot., Suppl. Ser. 9: 113 (1983).
N. Queensland. 50 QLD. Phan.

Syzygium pseudoformosum (King) Merr. & L.M.Perry, Mem. Amer. Acad. Arts 18: 165 (1939).
Pen. Thailand to W. Malesia. 41 THA 42 BOR JAW MLY SUM. Nanophan. or phan.
**Eugenia pseudoformosa* King, J. Asiat. Soc. Bengal, Pt. 2, Nat. Hist. 70: 83 (1901).
Eugenia nemoricola Ridl., J. Straits Branch Roy. Asiat. Soc. 61: 9 (1912).

Syzygium pseudojambolana Miq., Fl. Ned. Ind. 1(1): 458 (1855).
Jawa. 42 JAW. Phan.

Syzygium pseudolaetum (C.E.C.Fisch.) Merr. & L.M.Perry, Brittonia 4: 125 (1941).
Myanmar. 41 MYA. Phan.
**Jambosa pseudolaeta* C.E.C.Fisch., Bull. Misc. Inform. Kew 1927: 312 (1927).

Syzygium pseudomalaccense (Vieill. ex Brongn. & Gris) Govaerts, World Checklist Myrtaceae: 410 (2008).
New Caledonia. 60 NWC. Nanophan. or phan.
**Jambosa pseudomalaccensis* Vieill. ex Brongn. & Gris, Bull. Soc. Bot. France 12: 182 (1865).

Syzygium pseudomegistophyllum W.N.Takeuchi, Edinburgh J. Bot. 59: 266 (2002).
E. New Guinea. 43 NWG.

Syzygium pseudomolle (M.R.Hend.) I.M.Turner, J. Singapore Natl. Acad. Sci. 22–24: 23 (1997).
Pen. Malaysia. 42 MLY. Phan.
Eugenia mollis King, J. Asiat. Soc. Bengal, Pt. 2, Nat. Hist. 70: 86 (1901), nom. illeg. **Eugenia pseudomollis* M.R.Hend., Gard. Bull. Singapore 12: 75 (1949).

Syzygium pseudopinnatum Däniker, Vierteljahrsschr. Naturf. Ges. Zürich 78(19): 304 (1933).
SE. New Caledonia (incl. Îs. Loyauté). 60 NWC. Phan.

Syzygium pterocalyx Brongn. & Gris, Ann. Sci. Nat., Bot., V, 13: 386 (1871). *Eugenia pterocalyx* (Brongn. & Gris) Baker f., J. Linn. Soc., Bot. 45: 316 (1921). *Caryophyllus pterocalyx* (Brongn. & Gris) Guillaumin, Bull. Soc. Bot. France 85: 648 (1938 publ. 1939).
New Caledonia. 60 NWC. Phan.

Syzygium pterocarpum (Vieill. ex Pancher & Sebert) Govaerts, World Checklist Myrtaceae: 410 (2008).
New Caledonia. 60 NWC. Nanophan. or phan.
**Caryophyllus pterocarpus* Vieill. ex Pancher & Sebert, Not. Bois Nouv. Caléd.: 256 (1874). *Jambosa pterocarpa* (Vieill. ex Pancher & Sebert) Nied. in H.G.A.Engler & K.A.E.Prantl, Nat. Pflanzenfam. 3(7): 85 (1893).

Syzygium pterophorum Merr. & L.M.Perry, Mem. Amer. Acad. Arts 18: 158 (1939).
Borneo. 42 BOR. Phan.

Syzygium pteropodum (K.Schum. & Lauterb.) Merr. & L.M.Perry, J. Arnold Arbor. 23: 263 (1942).
New Guinea. 43 NWG. Phan.
**Jambosa pteropoda* K.Schum. & Lauterb., Fl. Schutzgeb. Südsee: 473 (1900).

Syzygium puberulum Merr. & L.M.Perry, J. Arnold Arbor. 23: 263 (1942).
New Guinea to N. Queensland. 43 NWG 50 QLD. Phan.

Syzygium pulaiense I.M.Turner, J. Singapore Natl. Acad. Sci. 22–24: 23 (1997).
Pen. Malaysia (Johor). 42 MLY. Phan.
**Eugenia johorensis* Ridl., Fl. Malay Penins. 5: 308 (1925), nom. illeg. *Syzygium johorense* Masam., Enum. Phan. Born.: 530 (1942).

Syzygium pulchellum (Roxb.) Govaerts, World Checklist Myrtaceae: 410 (2008).
Maluku. 42 MOL. Nanophan. or phan.
**Eugenia pulchella* Roxb., Fl. Ind. ed. 1832, 2: 496 (1832).

Syzygium pulgarense (C.B.Rob.) Merr., Philipp. J. Sci. 79: 410 (1951).
Philippines. 42 PHI. Phan.
**Eugenia pulgarensis* C.B.Rob., Philipp. J. Sci., C 4: 380 (1909).

Syzygium pullei Diels, Bot. Jahrb. Syst. 57: 401 (1922).
New Guinea. 43 NWG. Phan.

Syzygium punctilimbum (Merr.) Merr. & L.M.Perry, Mem. Amer. Acad. Arts 18: 194 (1939).
Borneo. 42 BOR. Phan.
**Eugenia punctilimba* Merr., J. Straits Branch Roy. Asiat. Soc. 77: 217 (1917).
Eugenia andersonii Ridl., J. Bot. 68: 36 (1930). *Syzygium andersonii* (Ridl.) Masam., Enum. Phan. Born.: 523 (1942).

Syzygium purpureum (L.M.Perry) A.C.Sm., Fl. Vitiensis Nova 3: 332 (1985).
Fiji. 60 FIJ. Phan.
**Syzygium diffusum* var. *purpureum* L.M.Perry, J. Arnold Arbor. 31: 356 (1950).

Syzygium purpuriflorum (Elmer) Merr., Philipp. J. Sci. 79: 411 (1951).
Philippines (Palawan). 42 PHI. Phan.
**Eugenia purpuriflora* Elmer, Leafl. Philipp. Bot. 4: 1432 (1912).

Syzygium pustulatum (Duthie) Merr., Philipp. J. Sci. 79: 421 (1951).
Pen. Thailand to Pen. Malaysia, Borneo. 41 THA 42 BOR MLY. Nanophan. or phan.

Eugenia pustulata Duthie in J.D.Hooker, Fl. Brit. India 2: 495 (1878).
Eugenia perpuncticulata Merr., Univ. Calif. Publ. Bot. 15: 220 (1929). *Syzygium perpuncticulatum* (Merr.) Merr. & L.M.Perry, Mem. Amer. Acad. Arts 18: 179 (1939).

Syzygium putii Chantaran. & J.Parn., Kew Bull. 48: 608 (1993).
Pen. Thailand. 41 THA. Phan.
Eugenia furfuracea Craib, Bull. Misc. Inform. Kew 1930: 165 (1930).

Syzygium pycnanthum Merr. & L.M.Perry, Mem. Amer. Acad. Arts 18: 168 (1939).
Pen. Thailand to C. Malesia. 41 THA 42 BOR JAW MLY SUL SUM. Phan.
Myrtus densiflora Blume, Bijdr.: 1087 (1826). *Eugenia densiflora* (Blume) DC., Prodr. 3: 287 (1828). *Jambosa densiflora* (Blume) DC., Prodr. 3: 287 (1828).
Eugenia corymbosa Roxb., Fl. Ind. ed. 1832, 2: 497 (1832), nom. illeg.

Syzygium pyrifolium (Blume) DC., Prodr. 3: 261 (1828).
Pen. Thailand to W. Malesia. 41 THA 42 JAW MLY. Phan.
Calyptranthus pyrifolia Blume, Bijdr.: 1090 (1826). *Eugenia pyrifolia* (Blume) Duthie in J.D.Hooker, Fl. Brit. India 2: 487 (1878), nom. illeg.
Syzygium truncatum Miq., Fl. Ned. Ind. 1(1): 455 (1855). *Eugenia striata* Koord. & Valeton, Meded. Lands Plantentuin 40: 145 (1900).
Eugenia tumida Duthie in J.D.Hooker, Fl. Brit. India 2: 487 (1878).
Eugenia javensis Koord. & Valeton, Bull. Inst. Bot. Buitenzorg 2: 7 (1899).
Eugenia salaccensis Koord. & Valeton, Meded. Lands Plantentuin 40: 144 (1900).

Syzygium pyriforme Merr. & L.M.Perry, J. Arnold Arbor. 23: 259 (1942).
New Guinea. 43 NWG. Phan.

Syzygium pyrocarpum (Greves) Merr. & L.M.Perry, J. Arnold Arbor. 23: 280 (1942).
New Guinea. 43 NWG. Phan.
Eugenia pyrocarpa Greves, J. Bot. 61(Suppl.): 17 (1923).
Eugenia xylantha Greves, J. Bot. 61(Suppl.): 18 (1923). *Syzygium xylanthum* (Greves) Merr. & L.M. Perry, J. Arnold Arbor. 23: 280 (1942).

Syzygium pyrrophloeum Diels, Bot. Jahrb. Syst. 57: 413 (1922).
Bismarck Arch. 43 BIS. Phan.

Syzygium quadrangulare Guillaumin, Bull. Soc. Bot. France 85: 645 (1938 publ. 1939).
New Caledonia. 60 NWC. Nanophan. or phan.

var. ***microsemmifolium*** (Guillaumin) J.W.Dawson, in Fl. Nouv.-Caléd. 23: 98 (1999).
New Caledonia. 60 NWC. Nanophan. or phan.
Syzygium microsemmifolium Guillaumin, Acta Horti Gothob. 18: 254 (1950).

var. ***quadrangulare***
New Caledonia. 60 NWC. Nanophan. or phan.

Syzygium quadrangulatum (A.Gray) Merr. & L.M.Perry, Sargentia 1: 77 (1942).
Fiji, Tonga. 60 FIJ TON. Phan.
Eugenia quadrangulata A.Gray, U.S. Expl. Exped., Phan. 1: 511 (1854). *Jambosa quadrangulata* (A.Gray) Müll.Stuttg. in W.G.Walpers, Ann. Bot. Syst. 4: 849 (1858).

Syzygium quadratum (King) I.M.Turner, J. Singapore Natl. Acad. Sci. 22–24: 24 (1997).
Pen. Malaysia. 42 MLY. Phan.
Eugenia quadrata King, J. Asiat. Soc. Bengal, Pt. 2, Nat. Hist. 70: 86 (1901).

Syzygium quadrialatum Teijsm. & Binn., Cat. Hort. Bot. Bogor.: 246 (1866).
Maluku. 42 MOL. Phan.

Syzygium quadribracteatum (M.R.Hend.) I.M.Turner, J. Singapore Natl. Acad. Sci. 22–24: 24 (1997).
Pen. Malaysia. 42 MLY. Phan.
Eugenia quadribracteata M.R.Hend., Gard. Bull. Singapore 11: 320 (1947).

Syzygium quadricostatum P.S.Ashton, Kew Bull. 61: 134 (2006).
Borneo. 42 BOR. Phan.

Syzygium racemosum (Blume) DC., Prodr. 3: 261 (1828).
Thailand to W. Malesia. 41 THA 42 BOR JAW MLY. Phan.
Calyptranthus racemosa Blume, Bijdr.: 1089 (1826). *Eugenia jamboloides* Koord. & Valeton, Meded. Lands Plantentuin 40: 136 (1900).
Myrtus cerasiformis Blume, Bijdr.: 1087 (1826). *Eugenia cerasiformis* (Blume) DC., Prodr. 3: 274 (1828). *Jambosa cerasiformis* (Blume) Hassk., Cat. Hort. Bot. Bogor.: 262 (1844). *Syzygium cerasiforme* (Blume) Merr. & L.M.Perry, Mem. Amer. Acad. Arts 18: 187 (1939).
Syzygium costatum Miq., Fl. Ned. Ind. 1(1): 451 (1855).
Syzygium javanicum Miq., Fl. Ned. Ind. 1(1): 461 (1855).
Eugenia expansa Duthie in J.D.Hooker, Fl. Brit. India 2: 491 (1878), nom. illeg.
Eugenia robinsoniana Ridl., J. Fed. Malay States Mus. 4: 13 (1909).
Eugenia evansii Ridl., J. Fed. Malay States Mus. 10: 134 (1920).
Eugenia brunneoramea Merr., Univ. Calif. Publ. Bot. 15: 217 (1929). *Syzygium brunneorameum* (Merr.) Masam., Enum. Phan. Born.: 524 (1942).

Syzygium rama-varmae (Bourd.) Chithra in N.C.Nair & A.N.Henry, Fl. Tamil Nadu 1: 157 (1983).
S. India. 40 IND. Phan.
Eugenia rama-varmae Bourd., Indian Forester 30: 147 (1904). *Jambosa rama-varmae* (Bourd.) Gamble, Fl. Madras: 474 (1919).

Syzygium ramiflorum Airy Shaw, Kew Bull. 4: 118 (1949).
Borneo. 42 BOR. Phan.

Syzygium ramilepis J.W.Dawson, in Fl. Nouv.-Caléd. 23: 110 (1999).
C. New Caledonia. 60 NWC. Nanophan. or phan.

Syzygium ramosii (C.B.Rob.) Merr., Philipp. J. Sci. 79: 411 (1951).
Philippines. 42 PHI. Phan.
Eugenia ramosii C.B.Rob., Philipp. J. Sci., C 4: 349 (1909).

Syzygium ramosissimum (Blume) N.P.Balakr., Fl. Jowai 1: 200 (1981).
C. & E. Himalaya to Bangladesh. 40 ASS BAN EHM NEP. Phan.

Clavimyrtus ramosissima Blume, Mus. Bot. 1: 113 (1850). *Eugenia ramosissima* (Blume) Wall. ex Duthie in J.D.Hooker, Fl. Brit. India 2: 480 (1878). *Jambosa ramosissima* (Blume) A.M.Cowan & Cowan, Trees N. Bengal: 67 (1929).

Syzygium rampans (Baker) J.Guého & A.J.Scott, Kew Bull. 34: 493 (1980).
Mauritius. 29 MAU. Nanophan. or phan.
Eugenia rampans Baker, Fl. Mauritius: 116 (1877).

Syzygium randianum Merr. & L.M.Perry, J. Arnold Arbor. 23: 264 (1942).
New Guinea. 43 NWG. Phan.

Syzygium rechingeri Merr. & L.M.Perry, J. Arnold Arbor. 23: 249 (1942).
New Guinea. 43 NWG. Phan.
Jambosa micrantha Rech., Repert. Spec. Nov. Regni Veg. 11: 183 (1912).

Syzygium recurvovenosum (Lauterb.) Diels, Bot. Jahrb. Syst. 57: 403 (1922).
New Guinea. 43 NWG. Phan.
Jambosa recurvovenosa Lauterb., Nova Guinea 8: 851 (1912).

Syzygium refertum (Craib) Chantaran. & J.Parn., Kew Bull. 48: 608 (1993).
NE. Thailand. 41 THA. Phan.
Eugenia referta Craib, Bull. Misc. Inform. Kew 1929: 116 (1929).

Syzygium rehderianum Merr. & L.M.Perry, J. Arnold Arbor. 19: 243 (1938).
S. China. 36 CHC CHS. Nanophan. or phan.

Syzygium rejangense Merr. & L.M.Perry, Mem. Amer. Acad. Arts 18: 162 (1939).
Borneo. 42 BOR. Phan.

Syzygium tetragonocladum Merr. & L.M.Perry, Mem. Amer. Acad. Arts 18: 153 (1939).

Syzygium remotifolium (Ridl.) Merr. & L.M.Perry, Mem. Amer. Acad. Arts 18: 188 (1939).
Borneo. 42 BOR. Phan.
Eugenia remotifolia Ridl., J. Bot. 68: 36 (1930).

Syzygium resa (B.Hyland) Craven & Biffin, Blumea 51: 139 (2006).
E. Queensland. 50 QLD. Phan.
Acmena resa B.Hyland, Austral. J. Bot., Suppl. Ser. 9: 16 (1983).

Syzygium reticulatum (Wight) Walp., Repert. Bot. Syst. 2: 179 (1843).
Assam to Bangladesh. 40 ASS BAN. Phan.
Eugenia reticulata Wight, Icon. Pl. Ind. Orient. 2: t. 541 (1842). *Jambosa reticulata* (Wight) Blume, Mus. Bot. 1: 92 (1850).
Microjambosa reticulata Blume, Mus. Bot. 1: 118 (1850).
Eugenia mangifolia Wall. ex Duthie in J.D.Hooker, Fl. Brit. India 2: 480 (1878).

Syzygium retinervium (Merr. & L.M.Perry) Craven & Biffin, Blumea 51: 139 (2006).
S. Vietnam. 41 VIE. Phan.
Cleistocalyx retinervius Merr. & L.M.Perry, J. Arnold Arbor. 18: 334 (1937).

Syzygium revolutum (Wight) Walp., Repert. Bot. Syst. 2: 180 (1843).
S. India, Sri Lanka. 40 IND SRL. Phan.
Eugenia revoluta Wight, Icon. Pl. Ind. Orient. 2: t. 534 (1842).

subsp. ***cyclophyllum*** (Thwaites ex Duthie) P.S.Ashton, in Revised Handb. Fl. Ceyl. 2: 450 (1981).
Sri Lanka. 40 SRL. Phan.
Eugenia cyclophylla Thwaites ex Duthie in J.D.Hooker, Fl. Brit. India 2: 494 (1878). *Syzygium cyclophyllum* (Thwaites ex Duthie) Alston in H.Trimen, Handb. Fl. Ceylon 6(Suppl.): 117 (1931).

subsp. ***revolutum***
S. India, Sri Lanka. 40 IND SRL. Phan.

Syzygium rheophyticum P.S.Ashton, Kew Bull. 61: 134 (2006).
Borneo. 42 BOR. Nanophan.

Syzygium rhizophorum (Boerl. & Koord.-Schum.) Govaerts, World Checklist Myrtaceae: 412 (2008).
Sumatera. 42 SUM. Nanophan. or phan.
Eugenia rhizophora Boerl. & Koord.-Schum. in A.Koorders-Schumacher, Syst. Verz. 2: 44 (1911).

Syzygium rhomboideum (Ridl.) I.M.Turner, J. Singapore Natl. Acad. Sci. 22–24: 24 (1997).
Pen. Malaysia. 42 MLY. Nanophan.
Eugenia rhomboidea Ridl., J. Fed. Malay States Mus. 5: 33 (1914).

Syzygium rhopalanthum Schltr., Bot. Jahrb. Syst. 39: 205 (1906). *Caryophyllus rhopalanthus* (Schltr.) Guillaumin, Bull. Soc. Bot. France 85: 648 (1938 publ. 1939).
New Caledonia. 60 NWC. Nanophan. or phan.

Syzygium richardsonianum Merr. & L.M.Perry, J. Arnold Arbor. 23: 274 (1942).
New Guinea. 43 NWG. Phan.

Syzygium richii (A.Gray) Merr. & L.M.Perry, Sargentia 1: 77 (1942).
SW. Pacific. 60 FIJ NUE SAM TON. Nanophan. or phan.
Eugenia richii A.Gray, U.S. Expl. Exped., Phan. 1: 510 (1854). *Jambosa richii* (A.Gray) Müll.Stuttg. in W.G.Walpers, Ann. Bot. Syst. 4: 849 (1858).

Syzygium ridleyi (King) Chantaran. & J.Parn., Kew Bull. 48: 608 (1993).
Pen. Thailand to Pen. Malaysia. 41 THA 42 MLY. Phan.
Eugenia ridleyi King, J. Asiat. Soc. Bengal, Pt. 2, Nat. Hist. 70: 98 (1901).

Syzygium rigens (Craib) Chantaran. & J.Parn., Kew Bull. 48: 609 (1993).
Pen. Thailand. 41 THA. Phan.
Eugenia rigens Craib, Bull. Misc. Inform. Kew 1929: 116 (1929).
Eugenia subrigens Craib, Bull. Misc. Inform. Kew 1929: 117 (1929).

Syzygium rigidifolium Merr., Philipp. J. Sci. 79: 411 (1951).
Philippines (Luzon). 42 PHI. Phan.

Syzygium riparium (Diels) Merr. & L.M.Perry, J. Arnold Arbor. 23: 249 (1942).
New Guinea. 43 NWG. Phan.
Jambosa riparia Diels, Bot. Jahrb. Syst. 57: 389 (1922).

Syzygium ripicola (Craib) Merr. & L.M.Perry, Brittonia 4: 127 (1941).
Indo-China. 41 CBD LAO MYA THA VIE. Phan.
**Eugenia ripicola* Craib, Bull. Misc. Inform. Kew 1915: 428 (1915).
Eugenia cochinchinensis Gagnep., Notul. Syst. (Paris) 3: 322 (1918). *Syzygium cochinchinense* (Gagnep.) Merr. & L.M.Perry, J. Arnold Arbor. 19: 107 (1938).

Syzygium rivulare Vieill. ex Guillaumin, Bull. Soc. Bot. France 85: 645 (1938 publ. 1939).
WC. New Caledonia. 60 NWC. Nanophan.

Syzygium rizalense (Merr.) Merr., Philipp. J. Sci. 79: 412 (1951).
Philippines. 42 PHI. Phan.
**Eugenia rizalensis* Merr., Philipp. J. Sci. 18: 302 (1921).

Syzygium robbinsii T.G.Hartley & L.M.Perry, J. Arnold Arbor. 54: 209 (1973).
Papua New Guinea. 43 NWG. Phan.

Syzygium robinsonii (Elmer) Merr., Philipp. J. Sci. 79: 412 (1951).
Philippines. 42 PHI. Phan.
**Eugenia robinsonii* Elmer, Leafl. Philipp. Bot. 2: 358 (1909).

Syzygium robustum Miq., Fl. Ned. Ind. 1(1): 1086 (1858).
Sumatera (Bangka). 42 SUM. Phan.

Syzygium rockii Merr. & L.M.Perry, J. Arnold Arbor. 19: 223 (1938).
China (Yunnan). 36 CHC. Nanophan. or phan.

Syzygium roemeri (Lauterb.) Merr. & L.M.Perry, J. Arnold Arbor. 23: 275 (1942).
New Guinea. 43 NWG. Phan.
**Jambosa roemeri* Lauterb., Nova Guinea 8: 851 (1912).
Jambosa rubella Rech., Repert. Spec. Nov. Regni Veg. 11: 183 (1912). *Syzygium rubellum* (Rech.) Merr. & L.M.Perry, J. Arnold Arbor. 23: 275 (1942).

Syzygium rolfei Merr., Philipp. J. Sci. 79: 412 (1951).
Philippines (Panay). 42 PHI. Phan.

Syzygium rosaceum Diels, Bot. Jahrb. Syst. 57: 406 (1922).
New Guinea. 43 NWG. Phan.

Syzygium rosenbluthii (C.B.Rob.) Merr., Philipp. J. Sci. 79: 413 (1951).
Philippines. 42 PHI. Phan.
**Eugenia rosenbluthii* C.B.Rob., Philipp. J. Sci., C 4: 384 (1909).
Eugenia burebidensis Elmer, Leafl. Philipp. Bot. 4: 1436 (1912).

Syzygium roseomarginatum (C.B.Rob.) Merr. & L.M.Perry, Mem. Amer. Acad. Arts 18: 191 (1939).
N. Borneo to Philippines. 42 BOR PHI. Phan.
**Eugenia roseomarginata* C.B.Rob., Philipp. J. Sci., C 4: 390 (1909).

Syzygium roseum Merr. & L.M.Perry, J. Arnold Arbor. 23: 270 (1942).
New Guinea. 43 NWG. Phan.

Syzygium rostadonis (Ridl.) I.M.Turner, J. Singapore Natl. Acad. Sci. 22–24: 24 (1997).
Pen. Malaysia. 42 MLY. Phan.
**Eugenia rostadonis* Ridl., J. Straits Branch Roy. Asiat. Soc. 61: 8 (1912).

Syzygium rostratum (Blume) DC., Prodr. 3: 261 (1828).
Malesia to New Guinea. 42 BOR JAW SUM 43 NWG. Phan.
**Calyptranthus rostrata* Blume, Bijdr.: 1092 (1826).
Jambosa tenuicuspis Miq., Fl. Ned. Ind. 1(1): 431 (1855).
Eugenia tenuicuspis Koord. & Valeton, Meded. Lands Plantentuin 40: 129 (1900).

Syzygium rosulentum (Ridl.) Merr. & L.M.Perry, Mem. Amer. Acad. Arts 18: 150 (1939).
Borneo. 42 BOR. Phan.
**Eugenia rosulenta* Ridl., J. Bot. 68: 34 (1930).

Syzygium rotundifolium Arn., Nova Acta Phys.-Med. Acad. Caes. Leop.-Carol. Nat. Cur. 18(1): 335 (1836). *Jambosa rotundifolia* (Arn.) G.Don, Gen. Hist. 2: 869 (1832). *Eugenia rotundifolia* (Arn.) Wight, Ill. Ind. Bot. 2: 17 (1841).
Sri Lanka. 40 SRL. Phan.

Syzygium rowlandii Sprague, Bull. Herb. Boissier, II, 5: 1170 (1905).
W. & WC. Trop. Africa. 22 IVO LBR NGA SIE 23 CMN RWA ZAI. Phan.
Syzygium abidjaneose Aubrév. & Pellegr. in A.Aubréville, Fl. Forest. Côte d'Ivoire 3: 68 (1936), nom. inval.

Syzygium rubens (Roxb.) Walp., Repert. Bot. Syst. 2: 180 (1843).
Bangladesh to Myanmar. 40 BAN 41 MYA. Phan.
**Eugenia rubens* Roxb., Fl. Ind. ed. 1832, 2: 496 (1832). *Jambosa rubens* (Roxb.) Blume, Mus. Bot. 1: 106 (1850).
Jambosa wightiana Blume, Mus. Bot. 1: 106 (1850).

Syzygium rubescens (A.Gray) Müll.Stuttg. in W.G.Walpers, Ann. Bot. Syst. 4: 839 (1858).
Fiji. 60 FIJ. Phan.
**Eugenia rubescens* A.Gray, U.S. Expl. Exped., Phan. 1: 525 (1854).

Syzygium rubicundum Wight & Arn., Prodr. Fl. Ind. Orient.: 330 (1834). *Eugenia rubicunda* (Wight & Arn.) Wight, Icon. Pl. Ind. Orient. 2: t. 538 (1842).
SW. India, Sri Lanka. 40 IND SRL. Phan.
Syzygium lissophyllum Thwaites, Enum. Pl. Zeyl.: 117 (1859). *Eugenia lissophylla* (Thwaites) Bedd., Fl. Sylv. S. India: 108 (1872).

Syzygium rubrimolle B.Hyland, Austral. J. Bot., Suppl. Ser. 9: 117 (1983).
N. Queensland. 50 QLD. Phan.

Syzygium rubropunctatum (Ridl.) Merr. & L.M.Perry, J. Arnold Arbor. 23: 268 (1942).
New Guinea. 43 NWG. Phan.
**Eugenia rubropunctata* Ridl., Trans. Linn. Soc. London, Bot. 9: 46 (1916).
Syzygium insculptum Merr. & L.M.Perry, J. Arnold Arbor. 23: 267 (1942).

Syzygium rubropurpureum (C.B.Rob.) Airy Shaw, Kew Bull. 4: 118 (1949).
Philippines. 42 PHI. Phan.
**Eugenia rubropurpurea* C.B.Rob., Philipp. J. Sci., C 4: 358 (1909).

Syzygium rubrovenium (C.B.Rob.) Merr., Philipp. J. Sci. 79: 413 (1951).
Philippines. 42 PHI. Phan.
**Eugenia rubrovenia* C.B.Rob., Philipp. J. Sci., C 4: 358 (1909).

Eugenia agusanensis Elmer, Leafl. Philipp. Bot. 7: 2357 (1914).

Syzygium rugosum Korth., Ned. Kruidk. Arch. 1: 204 (1848). *Eugenia rugosa* (Korth.) Merr., J. Straits Branch Roy. Asiat. Soc. 77: 224 (1917).
W. Malesia. 42 BOR JAW MLY. Phan.
Eugenia johorensis Ridl., J. Straits Branch Roy. Asiat. Soc. 61: 8 (1912).
Eugenia saxitana Ridl., Bull. Misc. Inform. Kew 1928: 74 (1928). *Eugenia rugosa* var. *saxitana* (Ridl.) M.R.Hend., Gard. Bull. Singapore 12: 246 (1949).
Eugenia motleyi Ridl., J. Bot. 68: 33 (1930). *Syzygium motleyi* (Ridl.) Masam., Enum. Phan. Born.: 534 (1942).
Eugenia rugosa var. *cordata* M.R.Hend., Gard. Bull. Singapore 12: 246 (1949). *Syzygium rugosum* var. *cordatum* (M.R.Hend.) I.M.Turner, J. Singapore Natl. Acad. Sci. 22–24: 24 (1997).

Syzygium rumphii (Merr.) Govaerts, World Checklist Myrtaceae: 414 (2008).
Maluku. 42 MOL. Nanophan. or phan.
**Eugenia rumphii* Merr., Interpr. Herb. Amboin.: 396 (1917).

Syzygium rysopodum Merr. & L.M.Perry, J. Arnold Arbor. 19: 211 (1938).
Hainan. 36 CHH. Phan.

Syzygium sabangense (Lauterb.) Merr. & L.M.Perry, J. Arnold Arbor. 23: 249 (1942).
New Guinea. 43 NWG. Phan.
**Jambosa sabangensis* Lauterb., Nova Guinea 8: 320 (1910).

Syzygium sakalavarum (H.Perrier) Labat & Schatz, Novon 12: 204 (2002).
W. Madagascar. 29 MDG. Phan.
**Eugenia sakalavarum* H.Perrier, Mém. Inst. Sci. Madagascar, Sér. B, Biol. Vég. 4: 189 (1953).

Syzygium salicifolium (Wight) J.Graham, Cat. Pl. Bombay: 73 (1839).
SW. India. 40 IND. Nanophan. or phan.
Calyptranthes danca Buch.-Ham. ex Wall., Numer. List: 3560 (1831), nom. nud.
**Eugenia salicifolia* Wight, Ill. Ind. Bot. 2: 16 (1841).
Eugenia heyneana Duthie in J.D.Hooker, Fl. Brit. India 2: 500 (1879). *Syzygium heyneanum* (Duthie) Gamble, Fl. Madras: 482 (1919).
Eugenia heyneana var. *alternans* Duthie in J.D.Hooker, Fl. Brit. India 2: 500 (1879). *Syzygium heyneanum* var. *alternans* (Duthie) B.G.Kulk. & Lakshmin., Fl. Maharashtra State, Dicot. 2: 11 (2001).
Syzygium alternans Miq. ex Duthie in J.D.Hooker, Fl. Brit. India 2: 500 (1879).

Syzygium saliciforme Merr. & L.M.Perry, J. Arnold Arbor. 23: 253 (1942).
New Guinea. 43 NWG. Phan.

Syzygium salicinum (Ridl.) Merr. & L.M.Perry, J. Arnold Arbor. 23: 254 (1942).
New Guinea. 43 NWG. Phan.
**Eugenia salicina* Ridl., Trans. Linn. Soc. London, Bot. 9: 45 (1916).

Syzygium salictoides (Ridl.) I.M.Turner, J. Singapore Natl. Acad. Sci. 22–24: 24 (1997).
Pen. Malaysia. 42 MLY. Nanophan. or phan.
**Eugenia salictoides* Ridl., J. Straits Branch Roy. Asiat. Soc. 68: 12 (1915).

Syzygium salignum (Miq.) Rathakr. & N.C.Nair, J. Econ. Taxon. Bot. 4: 288 (1983).
Jawa. 42 JAW. Phan.
**Jambosa saligna* Miq., Fl. Ned. Ind. 1(1): 432 (1855). *Eugenia saligna* (Miq.) C.B.Rob., Philipp. J. Sci., C 4: 392 (1909).

Syzygium salomonense (Hemsl.) Merr. & L.M.Perry, J. Arnold Arbor. 23: 273 (1942).
Solomon Is. 43 SOL. Phan.
**Eugenia salomonensis* Hemsl., J. Linn. Soc., Bot. 30: 212 (1894).

Syzygium salpinganthum (Greves) Merr. & L.M.Perry, J. Arnold Arbor. 23: 250 (1942).
New Guinea. 43 NWG. Phan.
**Eugenia salpingantha* Greves, J. Bot. 61(Suppl.): 19 (1923).

Syzygium salwinense Merr. & L.M.Perry, J. Arnold Arbor. 19: 237 (1938).
China (Yunnan). 36 CHC. Nanophan. or phan.

Syzygium samarangense (Blume) Merr. & L.M.Perry, J. Arnold Arbor. 19: 115 (1938).
Bangladesh to Solomon Is., widely introduced elsewere. (29) mau reu sey (36) chs 40 BAN srl 41 AND MYA NCB THA 42 BOR JAW MLY PHI SUL 43 NWG SOL (60) fij nue sam wal (81) lee win. Phan. Widely cultivated for its edible fruits (wax jambu).
Eugenia javanica Lam., Encycl. 3: 200 (1789). *Myrtus javanica* (Lam.) Blume, Bijdr.: 1084 (1826). *Jambosa javanica* (Lam.) K.Schum. & Lauterb., Fl. Schutzgeb. Südsee: 470 (1900).
**Myrtus samarangensis* Blume, Bijdr.: 1084 (1826). *Jambosa samarangensis* (Blume) DC., Prodr. 3: 286 (1828). *Eugenia samarangensis* (Blume) O.Berg in C.F.P.von Martius & auct. suc. (eds.), Fl. Bras. 14(1): 646 (1859).
Eugenia javanica var. *parviflora* Craib, Fl. Siam. 1: 647 (1931). *Syzygium samarangense* var. *parviflorum* (Craib) Chantaran. & J.Parn., Kew Bull. 48: 609 (1993).

Syzygium sambiranense (H.Perrier) Labat & Schatz, Novon 12: 204 (2002).
N. Madagascar. 29 MDG. Phan.
**Eugenia sambiranensis* H.Perrier, Mém. Inst. Sci. Madagascar, Sér. B, Biol. Vég. 4: 193 (1953).

Syzygium sambogense T.G.Hartley & L.M.Perry, J. Arnold Arbor. 54: 207 (1973).
Papua New Guinea. 43 NWG. Phan.

Syzygium samoense (Burkill) Whistler, Phytologia 38: 410 (1978).
Wallis, Samoa. 60 SAM WAL. Phan.
**Eugenia samoensis* Burkill, J. Linn. Soc., Bot. 35: 38 (1901).

Syzygium sandwicense (A.Gray) Müll.Stuttg. in W.G.Walpers, Ann. Bot. Syst. 4: 838 (1864).
Hawaiian Is. 63 HAW. Nanophan. or phan.
**Eugenia sandwicensis* A.Gray, U.S. Expl. Exped., Phan. 1: 519 (1854).
Syzygium oahuense O.Deg & Ludwig, Mitt. Bot. Staatssamml. München 4: 113 (1952).

Syzygium santosii (Merr.) Merr., Philipp. J. Sci. 79: 414 (1951).
Philippines. 42 PHI. Phan.
**Eugenia santosii* Merr., Philipp. J. Sci. 18: 294 (1921).

Syzygium sarmentosum J.W.Dawson, in Fl. Nouv.-Caléd. 23: 116 (1999).
NW. New Caledonia. 60 NWC. Nanophan. or phan.

Syzygium savaiiense (A.Gray) Müll.Stuttg. in W.G.Walpers, Ann. Bot. Syst. 4: 839 (1858).
Samoa. 60 SAM. Phan.
Eugenia savaiiensis A.Gray, U.S. Expl. Exped., Phan. 1: 530 (1854).
Eugenia tutuilensis A.Gray, U.S. Expl. Exped., Phan. 1: 529 (1854). *Syzygium tutuilense* (A.Gray) Müll. Stuttg. in W.G.Walpers, Ann. Bot. Syst. 4: 839 (1858).

Syzygium saxatile H.T.Chang & R.H.Miao, Acta Bot. Yunnan. 4: 22 (1982).
SC. China. 36 CHC. Nanophan. or phan.

Syzygium sayeri (F.Muell.) B.Hyland, Austral. J. Bot., Suppl. Ser. 9: 119 (1983).
New Guinea to N. & NE. Queensland. 43 NWG 50 QLD. Phan.
Eugenia sayeri F.Muell., Victorian Naturalist 8: 199 (1891).
Syzygium dictyophlebium Merr. & L.M.Perry, J. Arnold Arbor. 23: 267 (1942).

Syzygium scalarinerve (King) I.M.Turner, J. Singapore Natl. Acad. Sci. 22–24: 24 (1997).
Pen. Malaysia. 42 MLY. Phan.
Eugenia scalarinervis King, J. Asiat. Soc. Bengal, Pt. 2, Nat. Hist. 70: 87 (1901).

Syzygium schistaceum J.W.Dawson, in Fl. Nouv.-Caléd. 23: 48 (1999).
New Caledonia. 60 NWC. Nanophan.
Jambosa neriifolia Brongn. & Gris, Bull. Soc. Bot. France 12: 181 (1865).

Syzygium schlechteri Diels, Bot. Jahrb. Syst. 57: 402 (1922).
New Guinea. 43 NWG. Phan.

Syzygium schlechterianum Hochr., Bull. New York Bot. Gard. 6: 279 (1910).
NW. & SC. New Caledonia. 60 NWC. Nanophan. or phan.

Syzygium schumannianum (Nied.) Diels, Bot. Jahrb. Syst. 57: 402 (1922).
New Guinea. 43 NWG. Phan.
Eugenia neurocalyx K.Schum. in K.M.Schumann & U.M.Hollrung, Fl. Kais. Wilh. Land: 90 (1889), nom. illeg. *Jambosa schumanniana* Nied. in H.G.A.Engler & K.A.E.Prantl, Nat. Pflanzenfam. 3(7): 85 (1893). *Eugenia schumanniana* (Nied.) Greves, J. Bot. 61(Suppl.): 18 (1923).

Syzygium schwenckii Teijsm. & Binn., Cat. Hort. Bot. Bogor.: 246 (1866).
Sumatera. 42 SUM. Phan.

Syzygium sclerophyllum Thwaites, Enum. Pl. Zeyl.: 118 (1859). *Eugenia sclerophylla* (Thwaites) Bedd., Fl. Sylv. S. India: 108 (1872).
Sri Lanka. 40 SRL. Phan.

Syzygium scolopophyllum (Ridl.) Masam., Enum. Phan. Born.: 539 (1942).
Borneo. 42 BOR. Phan.
Eugenia scolopophylla Ridl., J. Bot. 68: 17 (1930).

Syzygium scortechinii (King) Chantaran. & J.Parn., Kew Bull. 48: 609 (1993).
Pen. Thailand to Borneo. 41 THA 42 BOR MLY. Phan.
Eugenia scortechinii King, J. Asiat. Soc. Bengal, Pt. 2, Nat. Hist. 70: 85 (1901).
Eugenia heteroclada Merr., J. Straits Branch Roy. Asiat. Soc. 77: 218 (1917). *Syzygium heterocladum* (Merr.) Merr. & L.M.Perry, Mem. Amer. Acad. Arts 18: 163 (1939).
Eugenia kingii Merr., J. Straits Branch Roy. Asiat. Soc. 79: 22 (1918). *Syzygium kingii* (Merr.) Merr. & L.M.Perry, Mem. Amer. Acad. Arts 18: 163 (1939).
Eugenia scortechinii var. *cuneata* M.R.Hend., Gard. Bull. Singapore 12: 67 (1949). *Syzygium scortechinii* var. *cuneatum* (M.R.Hend.) I.M.Turner, J. Singapore Natl. Acad. Sci. 22–24: 24 (1997).

Syzygium scytophyllum Diels, Bot. Jahrb. Syst. 57: 407 (1922).
New Guinea. 43 NWG. Phan.

Syzygium seemannianum Merr. & L.M.Perry, Sargentia 1: 76 (1942).
Fiji. 60 FIJ. Nanophan. or phan.
Eugenia rivularis Seem., Fl. Vit.: 80 (1865), nom. illeg.

Syzygium seemannii (A.Gray) Biffin & Craven, Blumea 50: 386 (2005).
Vanuatu, Fiji. 60 FIJ VAN. Phan.
Acicalyptus myrtoides Seem., Bonplandia (Hannover) 9: 206 (1861), nom. illeg. *Acicalyptus seemannii* A.Gray, Bonplandia (Hannover) 10: 35 (1862). *Calyptranthes seemannii* (A.Gray) Seem., Fl. Vit.: 81 (1866). *Cleistocalyx seemannii* (A.Gray) Merr. & L.M.Perry, J. Arnold Arbor. 18: 330 (1937).
Eugenia prora Burkill, Bull. Misc. Inform. Kew 1906: 4 (1906).
Acicalyptus elliptica A.C.Sm., Bernice P. Bishop Mus. Bull. 141: 107 (1936). *Cleistocalyx ellipticus* (A.C.Sm.) Merr. & L.M.Perry, J. Arnold Arbor. 18: 330 (1937).
Acicalyptus longiflora A.C.Sm., Bernice P. Bishop Mus. Bull. 141: 109 (1936). *Cleistocalyx longiflorus* (A.C.Sm.) Merr. & L.M.Perry, J. Arnold Arbor. 18: 329 (1937).
Cleistocalyx seemannii var. *punctatus* Merr. & L.M. Perry, J. Arnold Arbor. 18: 330 (1937).
Cleistocalyx kasiensis A.C.Sm., Pacific Sci. 25: 494 (1971).

Syzygium selukaifolium P.S.Ashton, Kew Bull. 61: 136 (2006).
Borneo (NE. Sarawak). 42 BOR. Phan.

Syzygium sessililimbum (Merr.) Merr., Philipp. J. Sci. 79: 414 (1951).
Philippines. 42 PHI. Phan.
Eugenia sessililimba Merr., Philipp. J. Sci. 18: 296 (1921).

Syzygium setosum (King) I.M.Turner, J. Singapore Natl. Acad. Sci. 22–24: 24 (1997).
Pen. Malaysia. 42 MLY. (Cl.) nanophan.
Eugenia setosa King, J. Asiat. Soc. Bengal, Pt. 2, Nat. Hist. 70: 120 (1901).

Syzygium sexangulatum (Miq.) Amshoff in C.A.Backer & R.C.Bakhuizen van der Brink, Bekn. Fl. Java 4b(98): 28 (1944).
Jawa. 42 JAW. Phan.
Jambosa sexangulata Miq., Fl. Ned. Ind. 1(1): 423 (1855). *Eugenia sexangulata* (Miq.) Koord. & Valeton, Meded. Lands Plantentuin 60: 79 (1902).

Syzygium sharonae B.Hyland, Austral. J. Bot., Suppl. Ser. 9: 122 (1983).
N. Queensland. 50 QLD. Phan.

Syzygium siamense (Craib) Chantaran. & J.Parn., Kew Bull. 48: 609 (1993).
Myanmar to Pen. Malaysia. 41 MYA THA 42 MLY. Phan.
**Eugenia siamensis* Craib, Bull. Misc. Inform. Kew 1912: 153 (1912).
Eugenia rubida Ridl., J. Fed. Malay States Mus. 10: 90 (1920).

Syzygium sichuanense H.T.Chang & R.H.Miao, Acta Bot. Yunnan. 4: 21 (1982).
SC. China. 36 CHC. Nanophan. or phan.

Syzygium siderocola (Merr.) Merr., Philipp. J. Sci. 79: 414 (1951).
Philippines. 42 PHI. Phan.
**Eugenia siderocola* Merr., Philipp. J. Sci. 18: 303 (1921).

Syzygium silamense P.S.Ashton, Kew Bull. 61: 136 (2006).
Borneo (Sabah). 42 BOR. Phan.

Syzygium simile (Merr.) Merr., Philipp. J. Sci. 79: 414 (1951).
Taiwan (Lan Yü, Lü Tao) to Philippines. 38 TAI 42 PHI. Phan.
Calyptranthes ramiflora Blanco, Fl. Filip.: 420 (1837).
**Eugenia similis* Merr., Philipp. J. Sci. 1(Suppl.): 106 (1906).
Syzygium lanyuense C.E.Chang, Bull. Taiwan Prov. Inst. Agric. Pintung 5: 51 (1964).

Syzygium simillimum Merr. & L.M.Perry, Sargentia 1: 76 (1942).
Fiji. 60 FIJ. Phan.

Syzygium singaporense (King) Airy Shaw, Kew Bull. 4: 124 (1949).
Pen. Malaysia. 42 MLY. Phan.
Myrtus caudata Wall., Numer. List: 3631 (1831), nom. nud.
Aphanomyrtus rostrata Miq., Fl. Ned. Ind. 1: 481 (1885).
**Pseudoeugenia singaporensis* King, J. Asiat. Soc. Bengal, Pt. 2, Nat. Hist. 70: 133 (1901).

Syzygium siphonanthum (King ex Greves) Amshoff, Blumea 5: 496 (1945).
Sumatera, Borneo. 42 BOR SUM. Phan.
**Eugenia siphonantha* King ex Greves, J. Bot. 62(Suppl.): 38 (1924).

Syzygium skiophilum (Duthie) Airy Shaw, Kew Bull. 4: 124 (1949).
Pen. Thailand to Pen. Malaysia. 41 THA 42 MLY. Phan.
**Eugenia skiophila* Duthie in J.D.Hooker, Fl. Brit. India 2: 486 (1878). *Aphanomyrtus skiophila* (Duthie) Valeton, Repert. Spec. Nov. Regni Veg. 4: 235 (1907).
Pseudoeugenia perakiana Scort., J. Bot. 23: 153 (1885).
Aphanomyrtus camphorata Valeton, Ann. Jard. Bot. Buitenzorg, Suppl. 2: 149 (1898).

Syzygium slootenii Merr. & L.M.Perry, Mem. Amer. Acad. Arts 18: 193 (1939).
Sumatera, Borneo. 42 BOR SUM. Phan.

Syzygium smalianum (Brandis) D.G.Long, Edinburgh J. Bot. 47: 357 (1990).
N. Myanmar to N. Thailand. 41 MYA THA. Phan.
**Eugenia smaliana* Brandis, Indian Trees: 320 (1906). *Syzygium subrufum* var. *smalianum* (Brandis) Chantaran. & J.Parn., Kew Bull. 48: 609 (1993).

Syzygium smithii (Poir.) Nied. in H.G.A.Engler & K.A.E.Prantl, Nat. Pflanzenfam. 3(7): 85 (1893).
E. Australia. 50 NSW QLD VIC? (51) nzn. Phan.
Eugenia elliptica Sm., Trans. Linn. Soc. London 3: 281 (1797), nom. illeg. **Eugenia smithii* Poir. in J.B.A.P.M.de Lamarck, Encycl., Suppl. 3: 126 (1813). *Myrtus smithii* (Poir.) Spreng., Syst. Veg. 2: 487 (1825). *Acmena smithii* (Poir.) Merr. & L.M.Perry, J. Arnold Arbor. 19: 16 (1938). *Lomastelma smithii* (Poir.) J.H.Willis, Victorian Naturalist 73: 197 (1957).
Acmena floribunda DC. in J.B.G. Bory de Saint-Vincent, Dict. Class. Hist. Nat. 11: 401 (1827).
Acmena floribunda var. *elliptica* A.Cunn. ex DC., Prodr.: 262 (1828).
Acmena kingii G.Don, Gen. Hist. 2: 851 (1832).
Lomastelma elliptica Raf., Sylva Tellur.: 107 (1838).
Acmena elliptica G.Don ex Steud., Nomencl. Bot., ed. 2, 1: 7 (1840).
Acmena elliptica G.Benn., Gatherings Natural. Austral.: 327 (1860).
Acmena pendula G.Benn., Gatherings Natural. Austral.: 327 (1860).
Syzygium brachynemum F.Muell., Fragm. 4: 59 (1864).
Eugenia smithii var. *minor* Maiden, Agric. Gaz. New South Wales 9: 581 (1898). *Acmena smithii* var. *minor* (Maiden) Merr. & L.M.Perry, J. Arnold Arbor. 19: 16 (1938).
Eugenia smithii var. *coriacea* Domin, Biblioth. Bot. 89: 1031 (1928).

Syzygium soepadmoi P.S.Ashton, Kew Bull. 61: 136 (2006).
Borneo (Sabah). 42 BOR. Phan.

Syzygium sogerense (Greves) Merr. & L.M.Perry, J. Arnold Arbor. 23: 250 (1942).
New Guinea. 43 NWG. Phan.
**Eugenia sogerensis* Greves, J. Bot. 61(Suppl.): 18 (1923).

Syzygium sorongense (T.G.Hartley & Craven) Craven & Biffin, Blumea 51: 139 (2006).
W. New Guinea. 43 NWG. Phan.
**Acmena sorongensis* T.G.Hartley & Craven, J. Arnold Arbor. 58: 339 (1977).

Syzygium spathulatum Thwaites, Enum. Pl. Zeyl.: 118 (1859). *Eugenia spathulata* (Thwaites) Bedd., Fl. Sylv. S. India: 108 (1872), nom. illeg. *Eugenia olivifolia* Duthie in J.D.Hooker, Fl. Brit. India 2: 495 (1878). *Syzygium olivifolium* (Duthie) Gamble, Bull. Misc. Inform. Kew 1920: 52 (1920), nom. superfl.
Sri Lanka. 40 SRL. Phan.

Syzygium speciosissimum (C.B.Rob.) Merr., Philipp. J. Sci. 79: 414 (1951).
Philippines. 42 PHI. Phan.
**Eugenia speciosissima* C.B.Rob., Philipp. J. Sci., C 4: 348 (1909).

Syzygium spectabile Merr. & L.M.Perry, J. Arnold Arbor. 23: 265 (1942).
New Guinea. 43 NWG. Phan.

Syzygium sphaeranthum (Gagnep.) Merr. & L.M.Perry, J. Arnold Arbor. 19: 113 (1938).
Laos. 41 LAO. Phan.
**Eugenia sphaerantha* Gagnep., Notul. Syst. (Paris) 3: 333 (1918).

Syzygium spissifolium (Ridl.) I.M.Turner, J. Singapore Natl. Acad. Sci. 22–24: 24 (1997).
Pen. Malaysia. 42 MLY. Nanophan.
**Eugenia spissifolia* Ridl., J. Fed. Malay States Mus. 5: 32 (1914).

Syzygium splendens (Blume) Merr. & L.M.Perry, Mem. Amer. Acad. Arts 18: 180 (1939).
Jawa. 42 JAW. Phan.
**Microjambosa splendens* Blume, Mus. Bot. 1: 119 (1850). *Jambosa splendens* (Blume) Miq., Fl. Ned. Ind. 1(1): 435 (1855).

Syzygium squamatum Merr. & L.M.Perry, J. Arnold Arbor. 23: 277 (1942).
New Guinea. 43 NWG. Phan.

Syzygium squamiferum (C.B.Rob.) Merr., Philipp. J. Sci. 79: 415 (1951).
Philippines. 42 PHI. Phan.
**Eugenia squamifera* C.B.Rob., Philipp. J. Sci., C 4: 373 (1909).

Syzygium sriganesanii K.Ravik. & V.Lakshm., Rheedea 9: 59 (1999).
India (Tamil Nadu). 40 IND. Phan.

Syzygium stapfianum (King) I.M.Turner, J. Singapore Natl. Acad. Sci. 22–24: 25 (1997).
Philippines. 42 PHI. Phan.
**Eugenia stapfiana* King, J. Asiat. Soc. Bengal, Pt. 2, Nat. Hist. 70: 119 (1901).

Syzygium staudtii (Engl.) Mildbr., Wiss. Erg. Deut. Zentr.-Afr. Exped., Bot. 2: 188 (1922).
W. & WC. Trop. Africa. 22 GUI IVO LBR 23 BUR CAF CMN CON GAB GGI ZAI. Phan.
**Syzygium guineense* var. *staudtii* Engl. in G.W.J. Mildbraed, Wiss. Erg. Deut. Zentr.-Afr. Exped., Bot. 2: 582 (1913).
Syzygium marounzense Pellegr., Bull. Mus. Natl. Hist. Nat. 29: 269 (1923).
Syzygium montanum Aubrév., Fl. Forest. Côte d'Ivoire 3: 70 (1936), nom. nud.
Syzygium guineense subsp. *bamendae* F.White, Kirkia 10: 404 (1977).
Syzygium guineense subsp. *obovatum* F.White, Kirkia 10: 404 (1977).
Syzygium guineense subsp. *occidentale* F.White, Kirkia 10: 403 (1977).

Syzygium steenisii Merr. & L.M.Perry, Mem. Amer. Acad. Arts 18: 180 (1939).
Borneo. 42 BOR. Phan.

Syzygium stelechanthum (Diels) Glassman, Bernice P. Bishop Mus. Bull. 209: 62 (1952).
Caroline Is. 62 CRL. Phan.
**Jambosa stelechantha* Diels, Bot. Jahrb. Syst. 56: 533 (1921). *Eugenia stelechantha* (Diels) Kaneh., Bot. Mag. (Tokyo) 45: 334 (1931).
Eugenia stelechanthoides Kaneh., Bot. Mag. (Tokyo) 46: 669 (1932). *Jambosa stelechanthoides* (Kaneh.) Hosok., J. Jap. Bot. 16: 544 (1940).

Syzygium stenocladum Merr. & L.M.Perry, J. Arnold Arbor. 19: 220 (1938).
Hainan. 36 CHH. Phan.

Syzygium stenurum Merr. & L.M.Perry, Brittonia 4: 125 (1941).
N. Myanmar. 41 MYA. Phan.

Syzygium sterrophyllum Merr. & L.M.Perry, J. Arnold Arbor. 19: 103 (1938).
SE. China to N. Vietnam. 36 CHS 41 VIE. Nanophan.

Syzygium stictanthum Merr. & L.M.Perry, J. Arnold Arbor. 19: 116 (1938).
Laos. 41 LAO. Phan.

Syzygium stipitatum P.S.Ashton, Kew Bull. 61: 138 (2006).
Borneo (Sabah). 42 BOR. Phan.

Syzygium stocksii (Duthie) Gamble, Fl. Madras: 481 (1919).
India. 40 IND. Phan.
**Eugenia stocksii* Duthie in J.D.Hooker, Fl. Brit. India 2: 498 (1879).

Syzygium striatulum (C.B.Rob.) Merr., Philipp. J. Sci. 79: 415 (1951).
Philippines. 42 PHI. Phan.
**Eugenia striatula* C.B.Rob., Philipp. J. Sci., C 4: 397 (1909).
Eugenia neei Merr., Philipp. J. Sci. 18: 301 (1921).

Syzygium subalatum (Ridl.) Merr. & L.M.Perry, J. Arnold Arbor. 23: 249 (1942).
New Guinea. 43 NWG. Phan.
**Eugenia subalata* Ridl., Trans. Linn. Soc. London, Bot. 9: 44 (1916).
Jambosa soliflora Diels, Bot. Jahrb. Syst. 57: 393 (1922). *Syzygium soliflorum* (Diels) Merr. & L.M. Perry, J. Arnold Arbor. 23: 254 (1942).

Syzygium subamplexicaule Merr. & L.M.Perry, J. Arnold Arbor. 23: 285 (1942).
New Guinea. 43 NWG. Phan.

Syzygium subcapitulatum Miq., Fl. Ned. Ind. 1(1): 450 (1855).
Jawa. 42 JAW. Phan.

Syzygium subcaudatum (Merr.) Merr., Philipp. J. Sci. 79: 415 (1951).
Philippines. 42 PHI. Phan.
**Eugenia subcaudata* Merr., Philipp. J. Sci., C 11: 21 (1916).

Syzygium subcorymbosum Merr. & L.M.Perry, J. Arnold Arbor. 23: 297 (1942).
Papuasia. 43 BIS NWG SOL. Phan.

Syzygium subcrenatum Merr. & L.M.Perry, Mem. Amer. Acad. Arts 18: 190 (1939).
Borneo. 42 BOR. Phan.

Syzygium subdecussatum (Duthie) I.M.Turner, J. Singapore Natl. Acad. Sci. 22–24: 25 (1997).
Pen. Malaysia. 42 MLY. Phan.
Eugenia colorata Duthie in J.D.Hooker, Fl. Brit. India 2: 492 (1878).
**Eugenia subdecussata* Duthie in J.D.Hooker, Fl. Brit. India 2: 491 (1878).
Eugenia subdecussata var. *montana* King, J. Asiat. Soc. Bengal, Pt. 2, Nat. Hist. 70(2): 121 (1901). *Syzygium subdecussatum* var. *montanum* (King) I.M.Turner, J. Singapore Natl. Acad. Sci. 22–24: 25 (1997).

Syzygium subfalcatum (C.B.Rob.) Merr., Philipp. J. Sci. 79: 415 (1951).
Philippines. 42 PHI. Phan.

Eugenia subfalcata C.B.Rob., Philipp. J. Sci., C 4: 382 (1909).

Syzygium subfoetidum (C.B.Rob.) Merr., Philipp. J. Sci. 79: 416 (1951).
Philippines. 42 PHI. Phan.
**Eugenia subfoetida* C.B.Rob., Philipp. J. Sci., C 4: 360 (1909).

Syzygium subglobosum Merr. & L.M.Perry, J. Arnold Arbor. 23: 290 (1942).
New Guinea. 43 NWG. Phan.

Syzygium subhorizontale (King) Chantaran. & J.Parn., Kew Bull. 48: 590 (1993).
Pen. Malaysia to Sumatera. 42 MLY SUM. Phan.
**Eugenia subhorizontalis* King, J. Asiat. Soc. Bengal, Pt. 2, Nat. Hist. 70: 112 (1901).

Syzygium subisense P.S.Ashton, Kew Bull. 61: 138 (2006).
Borneo (Sarawak). 42 BOR. Phan.

Syzygium subnodosum Miq., Fl. Ned. Ind., Eerste Bijv.: 313 (1861).
Sumatera. 42 SUM. Phan.

Syzygium suborbiculare (Benth.) T.G.Hartley & L.M.Perry, J. Arnold Arbor. 54: 189 (1973).
New Guinea to N. Australia. 43 NWG 50 NTA QLD WAU. Phan.
**Eugenia suborbicularis* Benth., Fl. Austral. 2: 285 (1864).
Careya jambosoides Lauterb., Nova Guinea 8: 313 (1910). *Eugenia jambosoides* (Lauterb.) O.Schwarz, Repert. Spec. Nov. Regni Veg. 24: 90 (1927), nom. illeg. *Syzygium jambosoides* (Lauterb.) Merr. & L.M.Perry, J. Arnold Arbor. 23: 262 (1942).

Syzygium subrotundifolium (C.B.Rob.) Merr., Philipp. J. Sci. 79: 416 (1951).
Philippines. 42 PHI.
**Eugenia subrotundifolia* C.B.Rob., Philipp. J. Sci., C 4: 362 (1909).

Syzygium subsessile (C.B.Rob.) Merr., Philipp. J. Sci. 79: 416 (1951).
Philippines. 42 PHI. Phan.
**Eugenia subsessilis* C.B.Rob., Philipp. J. Sci., C 4: 360 (1909).

Syzygium subsessiliflorum (Merr.) Merr., Philipp. J. Sci. 79: 416 (1951).
Philippines. 42 PHI. Phan.
**Eugenia subsessiliflora* Merr., Philipp. J. Sci., C 10: 216 (1915).

Syzygium subsessilifolium (Merr.) Merr. & L.M.Perry, Mem. Amer. Acad. Arts 18: 180 (1939).
Borneo. 42 BOR. Phan.
**Eugenia subsessilifolia* Merr., J. Straits Branch Roy. Asiat. Soc. 79: 24 (1918).

Syzygium subsimile Diels, Bot. Jahrb. Syst. 57: 401 (1922).
New Guinea. 43 NWG. Phan.

Syzygium subtile Miq., Fl. Ned. Ind., Eerste Bijv.: 313 (1861).
Sumatera. 42 SUM. Phan.

Syzygium sulcistylum (C.B.Rob.) Merr., Philipp. J. Sci. 79: 416 (1951).
Philippines. 42 PHI. Phan.
**Eugenia sulcistyla* C.B.Rob., Philipp. J. Sci., C 4: 368 (1909).

Syzygium sulitii Merr., Philipp. J. Sci. 79: 416 (1951).
Philippines (Luzon). 42 PHI. Phan.

Syzygium sulphuratum (Ridl.) Govaerts, World Checklist Myrtaceae: 418 (2008).
Sumatera. 42 SUM. Nanophan. or phan.
**Eugenia sulphurata* Ridl., J. Malayan Branch Roy. Asiat. Soc. 1: 59 (1923).

Syzygium surigaense (Merr.) Merr., Philipp. J. Sci. 79: 417 (1951).
Philippines. 42 PHI. Phan.
**Eugenia surigaensis* Merr., Philipp. J. Sci. 18: 297 (1921).

Syzygium suringarianum (Koord. & Valeton) Amshoff in C.A.Backer & R.C.Bakhuizen van der Brink, Bekn. Fl. Java 4b(98): 22 (1944).
Jawa. 42 JAW. Phan.
**Eugenia suringariana* Koord. & Valeton, Bull. Inst. Bot. Buitenzorg 2: 8 (1899).

Syzygium swettenhamianum (King) I.M.Turner, J. Singapore Natl. Acad. Sci. 22–24: 25 (1997).
Pen. Malaysia. 42 MLY. Phan.
**Eugenia swettenhamiana* King, J. Asiat. Soc. Bengal, Pt. 2, Nat. Hist. 70: 126 (1901).

Syzygium sylvicola T.G.Hartley & L.M.Perry, J. Arnold Arbor. 54: 198 (1973).
New Guinea. 43 NWG. Phan.

Syzygium symingtonianum (M.R.Hend.) I.M.Turner, J. Singapore Natl. Acad. Sci. 22–24: 25 (1997).
Pen. Malaysia. 42 MLY. Phan.
**Eugenia symingtoniana* M.R.Hend., Gard. Bull. Singapore 11: 333 (1947). *Stereocaryum symingtonianum* (M.R.Hend.) A.J.Scott, Kew Bull. 34: 496 (1980).

Syzygium synaptoneurum (K.Schum. & Lauterb.) Merr. & L.M.Perry, J. Arnold Arbor. 23: 249 (1942).
New Guinea. 43 NWG. Phan.
**Jambosa synaptoneura* K.Schum. & Lauterb., Fl. Schutzgeb. Südsee: 475 (1900).

Syzygium syzygioides (Miq.) Merr. & L.M.Perry, J. Arnold Arbor. 19: 109 (1938).
Assam to W. Malesia. 40 ASS BAN 41 AND MYA NCB THA VIE 42 BOR JAW MLY SUM. Nanophan. or phan.
Syzygium caudatum Wall., Numer. List: 3591 (1831), nom. nud.
**Jambosa syzygioides* Miq., Fl. Ned. Ind. 1(1): 431 (1855). *Eugenia syzygioides* (Miq.) M.R.Hend., Gard. Bull. Singapore 12: 154 (1949).
Syzygium nelitricarpum Teijsm. & Binn., Natuurk. Tijdschr. Ned.-Indië 27: 53 (1864).
Eugenia pseudosyzygioides M.R.Hend., Gard. Bull. Singapore 11: 315 (1947).

Syzygium szemaoense Merr. & L.M.Perry, J. Arnold Arbor. 19: 105 (1938).
SC. China to N. Vietnam. 36 CHC 41 VIE. Nanophan.

Syzygium taeniatum Diels, Bot. Jahrb. Syst. 57: 411 (1922).
New Guinea. 43 NWG. Phan.
Syzygium maschalocladum Merr. & L.M.Perry, J. Arnold Arbor. 23: 284 (1942).

Syzygium tahanense (Ridl.) I.M.Turner, J. Singapore Natl. Acad. Sci. 22–24: 25 (1997).
Pen. Malaysia. 42 MLY. Phan.

**Eugenia tahanensis* Ridl., J. Fed. Malay States Mus. 6: 146 (1915).

Syzygium taipingense (M.R.Hend.) I.M.Turner, J. Singapore Natl. Acad. Sci. 22–24: 25 (1997).
Pen. Malaysia. 42 MLY. Phan.
**Eugenia taipingensis* M.R.Hend., Gard. Bull. Singapore 11: 327 (1947).

Syzygium taiwanicum H.T.Chang & R.H.Miao, Acta Bot. Yunnan. 4: 18 (1982).
Taiwan, SE. China to Pen. Malaysia. 36 CHS 38 TAI 41 THA VIE 42 MLY. Phan.
**Eugenia claviflora* var. *oblongifolia* Hayata, Icon. Pl. Formosan. 3: 116 (1913). *Syzygium claviflorum* var. *oblongifolium* (Hayata) Mori, Trans. Nat. Hist. Soc. Taiwan 28: 438 (1938).

Syzygium tapiaka (H.Perrier) Labat & Schatz, Novon 12: 204 (2002).
W. Madagascar. 29 MDG. Phan.
**Eugenia tapiaka* H.Perrier, Mém. Inst. Sci. Madagascar, Sér. B, Biol. Vég. 4: 190 (1953).

Syzygium tawahense (Korth.) Merr. & L.M.Perry, Mem. Amer. Acad. Arts 18: 174 (1939).
Borneo. 42 BOR. Phan.
**Jambosa tawahensis* Korth., Ned. Kruidk. Arch. 1: 202 (1847).

Syzygium tayabense (Quisumb. & Merr.) Merr., Philipp. J. Sci. 79: 417 (1951).
Philippines. 42 PHI. Phan.
**Eugenia tayabensis* Quisumb. & Merr., Philipp. J. Sci. 37: 175 (1928).

Syzygium taytayense (Merr.) Merr., Philipp. J. Sci. 79: 417 (1951).
Borneo to Philippines (Palawan). 42 BOR PHI. Phan.
**Eugenia taytayensis* Merr., Philipp. J. Sci., C 10: 223 (1915).

Syzygium tchambaense J.W.Dawson, in Fl. Nouv.-Caléd. 23: 38 (1999).
WC. New Caledonia. 60 NWC. Nanophan. or phan.

Syzygium tectum (King) I.M.Turner, J. Singapore Natl. Acad. Sci. 22–24: 25 (1997).
Pen. Malaysia. 42 MLY. Phan.
**Eugenia tecta* King, J. Asiat. Soc. Bengal, Pt. 2, Nat. Hist. 70: 109 (1901).

Syzygium tekuense (M.R.Hend.) I.M.Turner, J. Singapore Natl. Acad. Sci. 22–24: 25 (1997).
Pen. Malaysia. 42 MLY. Phan.
Eugenia trunciflora Ridl., Fl. Malay Penins. 1: 724 (1922), nom. illeg. **Eugenia tekuensis* M.R.Hend., Gard. Bull. Singapore 12: 48 (1949). *Syzygium trunciflorum* Airy Shaw, Kew Bull. 4: 118 (1949).

Syzygium tenellum Blume ex Miq., Anal. Bot. Ind. 1: 27 (1850).
Borneo. 42 BOR. Phan.

Syzygium tenue (Duthie) N.P.Balakr., Bull. Bot. Surv. India 22: 175 (1980 publ. 1982).
NW. India. 40 IND. Nanophan. or phan.
Calyptranthes tenuis Buch.-Ham. ex Wall., Numer. List: 3570 (1831), nom. nud.
**Eugenia tenuis* Duthie in J.D.Hooker, Fl. Brit. India 2: 500 (1879).

Syzygium tenuicaudatum Merr. & L.M.Perry, Mem. Amer. Acad. Arts 18: 175 (1939).
Borneo. 42 BOR. Phan.

Syzygium tenuicorticum P.S.Ashton, Kew Bull. 61: 140 (2006).
Borneo (Sarawak). 42 BOR. Phan.

Syzygium tenuiflorum Brongn. & Gris, Bull. Soc. Bot. France 12: 183 (1865).
New Caledonia. 60 NWC. Nanophan.

Syzygium tenuifolium (Ridl.) Airy Shaw, Kew Bull. 4: 124 (1949).
Pen. Malaysia. 42 MLY. Phan.
**Pseudoeugenia tenuifolia* Ridl., Bull. Misc. Inform. Kew 1929: 123 (1929). *Aphanomyrtus tenuifiolia* (Ridl.) Merr., Blumea, Suppl. 1: 110 (1937).

Syzygium tenuilimbum P.S.Ashton, Kew Bull. 61: 140 (2006).
Borneo (Brunei, NE. Sarawak). 42 BOR. Phan.

Syzygium tenuipes (Merr.) Merr., Philipp. J. Sci. 79: 417 (1951).
Philippines. 42 PHI. Phan.
**Eugenia tenuipes* Merr., Philipp. J. Sci., C 7: 316 (1912).

Syzygium tenuirame (Miq.) Merr., Philipp. J. Sci. 79: 418 (1951).
Sumatera to Philippines. 42 PHI SUM. Phan.
**Jambosa tenuiramis* Miq., Fl. Ned. Ind. 1(1): 437 (1855).
Eugenia nitidissima Merr., Philipp. J. Sci., C 10: 213 (1915).

Syzygium tenuirhachis H.T.Chang & R.H.Miao, Acta Bot. Yunnan. 4: 19 (1982).
SE. China. 36 CHS. Nanophan. or phan.

Syzygium tephrodes (Hance) Merr. & L.M.Perry, J. Arnold Arbor. 19: 223 (1938).
Hainan. 36 CHH. Phan.
**Eugenia tephrodes* Hance, J. Bot. 23: 7 (1885).

Syzygium teretiflorum (Koord. & Valeton) Amshoff in C.A.Backer & R.C.Bakhuizen van der Brink, Bekn. Fl. Java 4b(98): 11 (1944).
Jawa. 42 JAW. Phan.
**Eugenia teretiflora* Koord. & Valeton, Bull. Inst. Bot. Buitenzorg 2: 9 (1899).

Syzygium tesselatum Korth., Ned. Kruidk. Arch. 1: 203 (1848).
Borneo. 42 BOR. Phan.

Syzygium tetragonum (Wight) Wall. ex Walp., Repert. Bot. Syst. 2: 179 (1843).
Nepal to S. China. 36 CHC CHC CHS 40 ASS BAN EHM NEP 41 MYA THA. Nanophan. or phan.
**Eugenia tetragona* Wight, Ill. Ind. Bot. 2: 16 (1841).
Eugenia subviridis Craib, Bull. Misc. Inform. Kew 1929: 117 (1929).
Syzygium nienkui Merr. & L.M.Perry, J. Arnold Arbor. 19: 228 (1938).

Syzygium tetrapleurum L.M.Perry, J. Arnold Arbor. 31: 369 (1950).
Fiji (Viti Levu). 60 FIJ. Nanophan. or phan.

Syzygium tetrapterum (Miq.) Chantaran. & J.Parn., Kew Bull. 48: 590 (1993).
Pen. Malaysia to Sumatera. 42 MLY SUM. Nanophan.
**Jambosa tetraptera* Miq., Fl. Ned. Ind., Eerste Bijv.: 311 (1861). *Eugenia tetraptera* (Miq.) M.R.Hend., Gard. Bull. Singapore 12: 218 (1949).
Eugenia pseudotetraptera King, J. Asiat. Soc. Bengal, Pt. 2, Nat. Hist. 70: 109 (1901). *Syzygium*

pseudotetrapterum (King) Chantaran. & J.Parn., Kew Bull. 48: 590 (1993). *Syzygium tetrapterum* var. *pseudotetrapterum* (King) I.M.Turner, J. Singapore Natl. Acad. Sci. 22–24: 25 (1997).

Syzygium thalassicum Merr. & L.M.Perry, J. Arnold Arbor. 23: 258 (1942).
Solomon Is. 43 SOL. Phan.

Syzygium thomsenii (Diels) Merr. & L.M.Perry, J. Arnold Arbor. 23: 249 (1942).
New Guinea. 43 NWG. Phan.
**Jambosa thomsenii* Diels, Nova Guinea 14: 91 (1924).

Syzygium thorelii (Gagnep.) Merr. & L.M.Perry, J. Arnold Arbor. 19: 107 (1938).
Indo-China. 41 LAO THA. Nanophan.
**Eugenia thorelii* Gagnep., Notul. Syst. (Paris) 3: 333 (1918).

Syzygium thornei T.G.Hartley & L.M.Perry, J. Arnold Arbor. 54: 206 (1973).
New Guinea. 43 NWG. Phan.

Syzygium thumra (Roxb.) Merr. & L.M.Perry, J. Arnold Arbor. 20: 103 (1939).
SC. China to Pen. Malaysia. 36 CHC 41 LAO MYA THA 42 MLY. Nanophan. or phan.
**Eugenia thumra* Roxb., Fl. Ind. ed. 1832, 2: 495 (1832).

var. ***penangianum*** (King) I.M.Turner, J. Singapore Natl. Acad. Sci. 22–24: 25 (1997).
Pen. Malaysia. 42 MLY. Phan.
**Eugenia thumra* var. *penangiana* King, J. Asiat. Soc. Bengal, Pt. 2, Nat. Hist. 70(2): 92 (1901).

subsp. ***punctifolium*** (Ridl.) Chantaran. & J.Parn., Kew Bull. 48: 610 (1993).
Pen. Thailand. 41 THA. Phan.
**Eugenia punctifolia* Ridl., J. Fed. Malay States Mus. 10: 91 (1920).

subsp. ***thumra***
SC. China to Indo-China. 36 CHC 41 LAO MYA THA. Nanophan. or phan.
Eugenia ferruginea Wight, Icon. Pl. Ind. Orient. 2: t. 554 (1842), nom. illeg. *Microjambosa ferruginea* Blume, Mus. Bot. 1: 118 (1850).

Syzygium tierneyanum (F.Muell.) T.G.Hartley & L.M.Perry, J. Arnold Arbor. 54: 200 (1973).
Papuasia to N. & NE. Queensland. 43 NWG SOL 50 QLD. Phan.
**Eugenia tierneyana* F.Muell., Fragm. 4: 14 (1863). *Jambosa tierneyana* (F.Muell.) Diels, Bot. Jahrb. Syst. 57: 389 (1922).
Syzygium floribundum K.Schum. & Lauterb., Fl. Schutzgeb. Südsee: 476 (1900), nom. illeg. *Jambosa floribunda* Diels, Bot. Jahrb. Syst. 57: 388 (1922). *Syzygium lauterbachianum* Merr. & L.M. Perry, J. Arnold Arbor. 23: 268 (1942).
Eugenia theodori-wolfii Domin, Biblioth. Bot. 89: 477 (1928).

Syzygium timorianum Decne., Nouv. Ann. Mus. Hist. Nat. 3: 455 (1834).
Lesser Sunda Is. 42 LSI. Phan.

Syzygium tinctorium (Gagnep.) Merr. & L.M.Perry, J. Arnold Arbor. 19: 112 (1938).
Cambodia. 41 CBD. Phan.
**Eugenia tinctoria* Gagnep., Notul. Syst. (Paris) 3: 334 (1918).

Syzygium tiumanense (Ridl.) I.M.Turner, J. Singapore Natl. Acad. Sci. 22–24: 26 (1997).
Pen. Malaysia. 42 MLY. Phan.
**Eugenia tiumanensis* Ridl., Trans. Linn. Soc. London, Bot. 3: 299 (1893).

Syzygium toddalioides (Wight) Walp., Repert. Bot. Syst. 2: 179 (1843).
E. Himalaya to Indo-China. 36 CHC 40 ASS EHM 41 MYA THA VIE. Phan.
**Eugenia toddalioides* Wight, Icon. Pl. Ind. Orient. 2: t. 542 (1842).
Syzygium augustinii Merr. & L.M.Perry, J. Arnold Arbor. 19: 231 (1938).

Syzygium tolypanthum Diels, Bot. Jahrb. Syst. 57: 408 (1922).
New Guinea. 43 NWG. Phan.

Syzygium toninense (Baker f.) J.W.Dawson, in Fl. Nouv.-Caléd. 23: 34 (1999).
NW. & C. New Caledonia. 60 NWC. Phan.
Eugenia neocaledonica Baker f., J. Linn. Soc., Bot. 45: 317 (1921). *Cupheanthus neocaledonicus* (Baker f.) Guillaumin, Bull. Soc. Bot. France 85: 651 (1938 publ. 1939), nom. illeg. *Cupheanthus serpentinus* Airy Shaw, Kew Bull. 15: 336 (1961).
**Eugenia toninensis* Baker f., J. Linn. Soc., Bot. 45: 317 (1921). *Cupheanthus toninensis* (Baker f.) Guillaumin, Bull. Soc. Bot. France 85: 651 (1938 publ. 1939).

Syzygium tonkinense (Gagnep.) Merr. & L.M.Perry, J. Arnold Arbor. 19: 105 (1938).
N. Vietnam. 41 VIE. Phan.
**Eugenia tonkinensis* Gagnep., Notul. Syst. (Paris) 3: 334 (1918).

Syzygium tontoutaense J.W.Dawson, in Fl. Nouv.-Caléd. 23: 100 (1999).
SE. New Caledonia. 60 NWC. Nanophan. or phan.

Syzygium toppingii (Elmer) Merr., Philipp. J. Sci. 79: 418 (1951).
Philippines. 42 PHI. Phan.
**Eugenia toppingii* Elmer, Leafl. Philipp. Bot. 4: 1407 (1912).

Syzygium torricellianum Diels, Bot. Jahrb. Syst. 57: 405 (1922).
New Guinea. 43 NWG. Phan.

Syzygium touranense Merr. & L.M.Perry, J. Arnold Arbor. 19: 111 (1938).
S. Vietnam. 41 VIE. Phan.

Syzygium trachyanthum (Diels) Merr. & L.M.Perry, J. Arnold Arbor. 23: 249 (1942).
New Guinea. 43 NWG. Phan.
Jambosa lagynocalyx Diels, Bot. Jahrb. Syst. 57: 394 (1922). *Syzygium lagynocalyx* (Diels) Merr. & L.M. Perry, J. Arnold Arbor. 23: 249 (1942).
**Jambosa trachyantha* Diels, Bot. Jahrb. Syst. 57: 394 (1922).

Syzygium trachyphloium (C.T.White) B.Hyland, Austral. J. Bot., Suppl. Ser. 9: 128 (1983).
N. & NE. Queensland. 50 QLD. Phan.
**Eugenia trachyphloia* C.T.White, Contr. Arnold Arbor. 55: 80 (1933).

Syzygium tramnion (Gagnep.) Merr. & L.M.Perry, J. Arnold Arbor. 19: 111 (1938).
Vietnam. 41 VIE. Phan.

Eugenia tramnion Gagnep., Notul. Syst. (Paris) 3: 332 (1918).

Syzygium travancoricum Gamble, Bull. Misc. Inform. Kew 1918: 240 (1918).
S. India. 40 IND. Phan.

Syzygium treubii Merr. & L.M.Perry, Mem. Amer. Acad. Arts 18: 192 (1939).
Borneo. 42 BOR. Phan.

Syzygium trianthum (Merr.) Merr., Philipp. J. Sci. 79: 418 (1951).
Philippines. 42 PHI. Phan.
Eugenia triantha Merr., Philipp. J. Sci., C 10: 224 (1915).

Syzygium trichotomum (Greves) Merr. & L.M.Perry, J. Arnold Arbor. 23: 250 (1942).
New Guinea. 43 NWG. Phan.
Eugenia trichotoma Greves, J. Bot. 61(Suppl.): 19 (1923).

Syzygium tricolor (Diels) Merr. & L.M.Perry, J. Arnold Arbor. 23: 255 (1942).
New Guinea. 43 NWG. Phan.
Jambosa tricolor Diels, Bot. Jahrb. Syst. 57: 393 (1922).

Syzygium tripetalum Guillaumin, Bull. Soc. Bot. France 85: 645 (1938 publ. 1939).
New Caledonia. 60 NWC. Nanophan. or phan.

Syzygium triphlebium Diels, Bot. Jahrb. Syst. 57: 400 (1922). *Acmena triphlebia* (Diels) T.G.Hartley & Craven, J. Arnold Arbor. 58: 340 (1977).
New Guinea. 43 NWG. Phan.

Syzygium triphyllum Merr., Philipp. J. Sci. 79: 419 (1951).
Philippines. 42 PHI. Phan.
Eugenia triphylla C.B.Rob., Philipp. J. Sci., C 4: 371 (1909), nom. illeg.

Syzygium tripinnatum (Blanco) Merr., Philipp. J. Sci. 79: 419 (1951).
Taiwan (Lan Yü) to Philippines. 38 TAI 42 PHI. Phan.
Myrtus tripinnata Blanco, Fl. Filip.: 421 (1837). *Eugenia tripinnata* (Blanco) C.B.Rob., Philipp. J. Sci., C 4: 357 (1909). *Syzygium jambos* var. *tripinnatum* (Blanco) C.Chen, Harvard Pap. Bot. 11: 27 (2006).
Myrtus subrubens Blanco, Fl. Filip., ed. 2: 294 (1845).
Syzygium okudai Mori, Trans. Nat. Hist. Soc. Taiwan 28: 439 (1938).
Eugenia peninsula Elmer, Leafl. Philipp. Bot. 10: 3769 (1939), no latin descr.

Syzygium triplinervium Teijsm. & Binn., Cat. Hort. Bot. Bogor.: 246 (1866).
Jawa. 42 JAW. Phan.
Syzygium modestum Diels, Bot. Jahrb. Syst. 57: 400 (1922).

Syzygium triste (Kurz) N.P.Balakr., Bull. Bot. Surv. India 22: 175 (1980 publ. 1982).
Myanmar. 41 MYA. Phan.
Eugenia tristis Kurz, J. Asiat. Soc. Bengal, Pt. 2, Nat. Hist. 42(2): 233 (1873).

Syzygium trivene (Ridl.) Merr. & L.M.Perry, J. Arnold Arbor. 23: 288 (1942).
Papuasia. 43 NWG SOL. Phan.
Eugenia trivenis Ridl., Trans. Linn. Soc. London, Bot. 9: 47 (1916).

Syzygium tsoongii (Merr.) Merr. & L.M.Perry, J. Arnold Arbor. 19: 112 (1938).
SE. China to Vietnam. 36 CHH CHS 41 VIE. Nanophan. or phan.
Eugenia leucocarpa Gagnep., Notul. Syst. (Paris) 3: 327 (1918), nom. illeg.
Eugenia tsoongii Merr., Philipp. J. Sci. 21: 504 (1922).

Syzygium tula (Merr.) Merr., Philipp. J. Sci. 79: 419 (1951).
Philippines. 42 PHI. Phan.
Eugenia tula Merr., Philipp. J. Sci. 18: 297 (1921).

Syzygium turbinatum Alston in H.Trimen, Handb. Fl. Ceylon 6(Suppl.): 114 (1931).
Sri Lanka. 40 SRL. Phan.

Syzygium tympananthum (Diels) Merr. & L.M.Perry, J. Arnold Arbor. 23: 255 (1942).
New Guinea. 43 NWG. Phan.
Jambosa tympanantha Diels, Bot. Jahrb. Syst. 57: 393 (1922).

Syzygium ubogoensis W.N.Takeuchi, Edinburgh J. Bot. 59: 268 (2002).
Papua New Guinea. 43 NWG.

Syzygium ultramaficum P.S.Ashton, Kew Bull. 61: 140 (2006).
Borneo (Sabah). 42 BOR. Phan.

Syzygium umbellatum Korth., Ned. Kruidk. Arch. 1: 205 (1848).
Sumatera. 42 SUM. Phan.

Syzygium umbilicatum (Koord. & Valeton) Amshoff in C.A.Backer & R.C.Bakhuizen van der Brink, Bekn. Fl. Java 4b(98): 24 (1944).
Jawa. 42 JAW. Phan.
Jambosa tetragona Blume, Mus. Bot. 1: 106 (1850). *Eugenia umbilicata* Koord. & Valeton, Meded. Lands Plantentuin 40: 62 (1900).

Syzygium umbrosum Thwaites, Enum. Pl. Zeyl.: 118 (1859). *Eugenia subavenis* Duthie in J.D.Hooker, Fl. Brit. India 2: 489 (1878).
Sri Lanka. 40 SRL. Phan.
Eugenia umbrosa Bedd., Fl. Sylv. S. India: 108 (1872), nom. illeg.

Syzygium uniflorum Merr. & L.M.Perry, J. Arnold Arbor. 23: 255 (1942).
New Guinea. 43 NWG. Phan.

Syzygium unipunctatum (B.Hyland) Craven & Biffin, Blumea 51: 139 (2006).
N. & NE. Queensland. 50 QLD. Nanophan. or phan.
Waterhousea unipunctata B.Hyland, Austral. J. Bot., Suppl. Ser. 9: 145 (1983).

Syzygium urdanetense (Elmer) Merr., Philipp. J. Sci. 79: 420 (1951).
Philippines. 42 PHI. Phan.
Eugenia urdanetensis Elmer, Leafl. Philipp. Bot. 7: 2356 (1914).
Eugenia caudatifolia Merr., Philipp. J. Sci., C 10: 211 (1915).
Eugenia capizensis Merr., Philipp. J. Sci. 18: 305 (1921).

Syzygium urophyllum Merr., Philipp. J. Sci. 79: 369 (1951).
Pen. Thailand to Pen. Malaysia. 41 THA 42 MLY. Phan.
Eugenia caudata King, J. Asiat. Soc. Bengal, Pt. 2, Nat. Hist. 70: 105 (1901). *Syzygium enervosum* Chantaran. & J.Parn., Kew Bull. 48: 596 (1993), nom. illeg.

Eugenia filiformis var. *parvifolia* Craib, Fl. Siam. 1: 641 (1931).

Syzygium utilis (Talbot) Rathakr. & N.C.Nair, J. Econ. Taxon. Bot. 4: 288 (1983).
India. 40 IND. Phan.
**Eugenia utilis* Talbot, J. Bombay Nat. Hist. Soc. 11: 235 (1897).

Syzygium vaccinifolium Merr., Philipp. J. Sci. 79: 420 (1951).
Philippines. 42 PHI.
**Eugenia vaccinioides* Elmer, Leafl. Philipp. Bot. 7: 2350 (1914).

Syzygium valdecoriaceum P.S.Ashton, Kew Bull. 61: 142 (2006).
Borneo. 42 BOR. Phan.

Syzygium valdepunctatum Merr., Philipp. J. Sci. 79: 420 (1951).
Philippines (Mindanao). 42 PHI. Phan.

Syzygium valdevenosum (Duthie) Merr. & L.M.Perry, Mem. Amer. Acad. Arts 18: 182 (1939).
Pen. Malaysia to Borneo. 42 BOR MLY. Phan.
**Eugenia valdevenosa* Duthie in J.D.Hooker, Fl. Brit. India 2: 489 (1878).
Eugenia alata Ridl., J. Straits Branch Roy. Asiat. Soc. 86: 293 (1922).

Syzygium validinerve T.G.Hartley & L.M.Perry, J. Arnold Arbor. 54: 193 (1973).
Papua New Guinea. 43 NWG. Phan.

Syzygium validum Korth., Ned. Kruidk. Arch. 1: 204 (1848).
Sumatera. 42 SUM. Phan.

Syzygium vanderwateri (Ridl.) Merr. & L.M.Perry, J. Arnold Arbor. 23: 249 (1942).
New Guinea. 43 NWG. Phan.
**Eugenia vanderwateri* Ridl., Trans. Linn. Soc. London, Bot. 9: 45 (1916).

Syzygium variabile T.G.Hartley & L.M.Perry, J. Arnold Arbor. 54: 195 (1973).
Papua New Guinea. 43 NWG. Phan.

Syzygium variifolium Miq., Fl. Ned. Ind., Eerste Bijv.: 311 (1861).
Sumatera (Bangka). 42 SUM. Phan.

Syzygium variolosum (King) Chantaran. & J.Parn., Kew Bull. 48: 590 (1993).
Pen. Malaysia. 42 MLY. Phan.
**Eugenia variolosa* King, J. Asiat. Soc. Bengal, Pt. 2, Nat. Hist. 70: 107 (1901).

Syzygium vaughanii J.Guého & A.J.Scott, Kew Bull. 34: 494 (1980).
Mauritius. 29 MAU. Nanophan. or phan.

Syzygium vaupelii Whistler, J. Arnold Arbor. 69: 172 (1988).
Samoa. 60 SAM. Phan.

Syzygium veillonii J.W.Dawson, in Fl. Nouv.-Caléd. 23: 108 (1999).
C. New Caledonia. 60 NWC. Nanophan. or phan.

Syzygium velae B.Hyland, Austral. J. Bot., Suppl. Ser. 9: 130 (1983).
N. Queensland. 50 QLD. Phan.

Syzygium velutinum A.P.Davis, Kew Bull. 52: 720 (1997).
Borneo. 42 BOR. Phan.

Syzygium venosum DC., Prodr. 3: 260 (1828).
Nepal to Bhutan, S. India. 40 EHM IND NEP. Phan.
Syzygium areolatum DC., Prodr. 3: 260 (1828).
Eugenia areolata (DC.) Duthie in J.D.Hooker, Fl. Brit. India 2: 490 (1878).
Eugenia frondosa Wall. ex Duthie in J.D.Hooker, Fl. Brit. India 2: 490 (1878).

Syzygium venustum (Roxb.) N.P.Balakr., Bull. Bot. Surv. India 22: 175 (1980 publ. 1982).
Myanmar. 41 MYA. Phan.
**Eugenia venusta* Roxb., Fl. Ind. ed. 1832, 2: 491 (1832).

Syzygium verniciflorum (Diels) Merr. & L.M.Perry, J. Arnold Arbor. 23: 249 (1942).
New Guinea. 43 NWG. Phan.
**Jambosa verniciflora* Diels, Bot. Jahrb. Syst. 57: 387 (1922).

Syzygium vernicosum Merr. & L.M.Perry, J. Arnold Arbor. 23: 260 (1942).
New Guinea. 43 NWG. Phan.

Syzygium vernonioides (Elmer) Merr., Philipp. J. Sci. 79: 421 (1951).
Philippines. 42 PHI. Phan.
**Eugenia vernonioides* Elmer, Leafl. Philipp. Bot. 7: 2352 (1914).

Syzygium verrucosum Däniker, Vierteljahrsschr. Naturf. Ges. Zürich 78(19): 305 (1933).
New Caledonia (Îs. Loyauté). 60 NWC. Phan.

Syzygium versteegii (Lauterb.) Merr. & L.M.Perry, J. Arnold Arbor. 23: 256 (1942).
New Guinea. 43 NWG. Phan.
**Jambosa versteegii* Lauterb., Nova Guinea 8: 321 (1910).

Syzygium vestitum Merr. & L.M.Perry, J. Arnold Arbor. 19: 110 (1938).
SC. China to Vietnam. 36 CHC 41 VIE. Phan.

Syzygium viburnoides Diels, Bot. Jahrb. Syst. 57: 405 (1922).
New Guinea. 43 NWG. Phan.

Syzygium vidalianum (Elmer) Merr., Philipp. J. Sci. 79: 421 (1951).
Philippines. 42 PHI. Phan.
**Eugenia vidaliana* Elmer, Leafl. Philipp. Bot. 2: 584 (1909).
Eugenia sorsogonensis Merr., Philipp. J. Sci., C 11: 22 (1916).

Syzygium villamilii (Merr.) Merr. & L.M.Perry, Mem. Amer. Acad. Arts 18: 187 (1939).
Borneo. 42 BOR. Phan.
**Eugenia villamilii* Merr., Philipp. J. Sci., C 13: 98 (1918).
Syzygium richardsii Airy Shaw, Kew Bull. 4: 122 (1949).

Syzygium villiferum (Ridl.) Masam., Enum. Phan. Born.: 541 (1942).
Borneo. 42 BOR. Phan.
**Eugenia villifera* Ridl., J. Bot. 68: 13 (1930).

Syzygium virescens Merr. & L.M.Perry, J. Arnold Arbor. 23: 260 (1942).
New Guinea. 43 NWG. Phan.

Syzygium viridescens (Ridl.) I.M.Turner, J. Singapore Natl. Acad. Sci. 22–24: 26 (1997).
Pen. Malaysia. 42 MLY. Nanophan.

Eugenia viridescens Ridl., J. Linn. Soc., Bot. 38: 308 (1908).

Syzygium virotii J.W.Dawson, in Fl. Nouv.-Caléd. 23: 40 (1999).
SE. New Caledonia. 60 NWC. Cham. or nanophan.
Cupheanthus microphyllus Guillaumin, Mém. Mus. Natl. Hist. Nat., B, Bot. 4: 36 (1953).

Syzygium vismioides (DC.) Govaerts, World Checklist Myrtaceae: 423 (2008).
Maluku. 42 MOL. Nanophan. or phan.
Eugenia vismioides DC., Prodr. 3: 267 (1828).

Syzygium vrieseanum (Miq.) Amshoff in C.A.Backer & R.C.Bakhuizen van der Brink, Bekn. Fl. Java 4b(98): 27 (1944).
Jawa. 42 JAW. Phan.
Jambosa vrieseana Miq., Fl. Ned. Ind. 1(1): 424 (1855). *Eugenia vrieseana* (Miq.) Koord. & Valeton, Meded. Lands Plantentuin 40: 75 (1900).

Syzygium vulcanicum Elmer ex Merr., Philipp. J. Sci. 79: 422 (1951).
Philippines (Luzon). 42 PHI. Phan.
Eugenia bulusanensis Elmer, Leafl. Philipp. Bot. 9: 3135 (1925), nom. nud.

Syzygium wagapense Brongn. & Gris, Bull. Soc. Bot. France 13: 468 (1866).
New Caledonia. 60 NWC. Nanophan. or phan.

Syzygium waikaiunense T.G.Hartley & L.M.Perry, J. Arnold Arbor. 54: 194 (1973).
Papua New Guinea. 43 NWG. Phan.

Syzygium walkeri Merr. & L.M.Perry ex C.T.White, J. Arnold Arbor. 31: 101 (1950).
Solomon Is. 43 SOL. Phan.

Syzygium warburgii Merr. & L.M.Perry, J. Arnold Arbor. 23: 250 (1942).
New Guinea. 43 NWG. Phan.
Eugenia glomerata Warb., Bot. Jahrb. Syst. 13: 390 (1891), nom. illeg. *Jambosa glomerata* K.Schum. & Lauterb., Fl. Schutzgeb. Südsee: 470 (1900). *Eugenia glomerata* (K.Schum. & Lauterb.) Diels, Nova Guinea 19: 90 (1924), nom. illeg.

Syzygium waterhousei Merr. & L.M.Perry, J. Arnold Arbor. 23: 290 (1942).
Bismarck Arch. to Solomon Is. 43 BIS SOL. Phan.

Syzygium watsonianum (M.R.Hend.) I.M.Turner, J. Singapore Natl. Acad. Sci. 22–24: 26 (1997).
Pen. Malaysia. 42 MLY. Phan.
Eugenia watsoniana M.R.Hend., Gard. Bull. Singapore 11: 336 (1947). *Stereocaryum watsonianum* (M.R.Hend.) A.J.Scott, Kew Bull. 34: 496 (1980).

Syzygium wenshanense H.T.Chang & R.H.Miao, Acta Bot. Yunnan. 4: 21 (1982).
SC. China. 36 CHC. Phan.

Syzygium wenzelii (Merr.) Merr., Philipp. J. Sci. 79: 423 (1951).
Philippines. 42 PHI. Phan.
Eugenia wenzelii Merr., Philipp. J. Sci., C 9: 380 (1914).

Syzygium wesa B.Hyland, Austral. J. Bot., Suppl. Ser. 9: 132 (1983).
Queensland. 50 QLD. Phan.

Syzygium whitfordii (Merr.) Merr., Philipp. J. Sci. 79: 423 (1951).
Philippines. 42 PHI. Phan.
Eugenia whitfordii Merr., Publ. Bur. Sci. Gov. Lab. 35: 49 (1905).

Syzygium williamsii (C.B.Rob.) Merr., Philipp. J. Sci. 79: 423 (1951).
Philippines. 42 PHI. Phan.
Eugenia williamsii C.B.Rob., Philipp. J. Sci., C 4: 365 (1909).

Syzygium wilsonii (F.Muell.) B.Hyland, Austral. J. Bot., Suppl. Ser. 9: 134 (1983).
Queensland. 50 QLD. Phan.
Eugenia wilsonii F.Muell., Fragm. 5: 12 (1865).

subsp. ***cryptophlebium*** (F.Muell.) B.Hyland, Austral. J. Bot., Suppl. Ser. 9: 135 (1983).
Queensland. 50 QLD. Phan.
Eugenia cryptophlebia F.Muell., Fragm. 9: 144 (1875). *Syzygium cryptophlebium* (F.Muell.) Craven & Biffin, Blumea 50: 159 (2005).
Eugenia macoorai F.M.Bailey, Bot. Bull. Dept. Agric. Queensland 5: 15 (1892).
Eugenia sordida F.M.Bailey, Bot. Bull. Dept. Agric. Queensland 5: 15 (1892).
Eugenia rhadinantha S.Moore, J. Bot. 55: 303 (1917).

subsp. ***epigaeum*** Craven & Biffin, Blumea 50: 159 (2005).
Queensland. 50 QLD. Phan.

subsp. ***wilsonii***
Queensland. 50 QLD. Phan.
Eugenia subopposita F.M.Bailey, Queensl. Fl.: 2004 (1902).

Syzygium winckelii Amshoff, Blumea 5: 498 (1945).
W. Jawa. 42 JAW. Phan.

Syzygium winitii (Craib) Merr. & L.M.Perry, Brittonia 4: 127 (1941).
Indo-China. 41 MYA THA. Phan.
Eugenia winitii Craib, Bull. Misc. Inform. Kew 1929: 118 (1929).
Eugenia winitii var. *terminalis* Craib, Fl. Siam. 1: 666 (1931).

Syzygium wolfii (Gillespie) Merr. & L.M.Perry, Sargentia 1: 75 (1942).
Fiji. 60 FIJ. Phan.
Eugenia wolfii Gillespie, Bernice P. Bishop Mus. Bull. 83: 22 (1931).

Syzygium wollastonii (Ridl.) Merr. & L.M.Perry, J. Arnold Arbor. 23: 249 (1942).
New Guinea. 43 NWG. Phan.
Eugenia wollastonii Ridl., Trans. Linn. Soc. London, Bot. 9: 47 (1916).

Syzygium womersleyi T.G.Hartley & L.M.Perry, J. Arnold Arbor. 54: 193 (1973).
Papua New Guinea. 43 NWG. Phan.

Syzygium wrayi (King) I.M.Turner, J. Singapore Natl. Acad. Sci. 22–24: 26 (1997).
Pen. Malaysia. 42 MLY. Nanophan. or phan.
Eugenia wrayi King, J. Asiat. Soc. Bengal, Pt. 2, Nat. Hist. 70: 119 (1901).

Syzygium wrightii (Baker) A.J.Scott, Kew Bull. 34: 496 (1980).
Seychelles. 29 SEY. Nanophan. or phan.
Eugenia sechellarum Baker, Proc. Roy. Irish Acad., II, 2: 160 (1876).

Eugenia wrightii Baker, Proc. Roy. Irish Acad., II, 2: 160 (1876).

Syzygium xanthophyllum (C.B.Rob.) Merr., Philipp. J. Sci. 79: 424 (1951).
Philippines. 42 PHI. Phan.
Eugenia xanthophylla C.B.Rob., Philipp. J. Sci., C 4: 370 (1909).

Syzygium xanthostemifolium (Guillaumin) J.W.Dawson, Fl. Nouv.-Caléd. 23: 66 (1999).
SE. New Caledonia. 60 NWC. Nanophan. or phan.
Caryophyllus xanthostemifolius Guillaumin, Bull. Soc. Bot. France 85: 650 (1938 publ. 1939).

Syzygium xerampelinum B.Hyland, Austral. J. Bot., Suppl. Ser. 9: 136 (1983).
N. Queensland. 50 QLD. Phan.

Syzygium xiphophyllum (Merr.) Merr., Philipp. J. Sci. 79: 424 (1951).
Philippines. 42 PHI. Phan.
Eugenia xiphophylla Merr., Philipp. J. Sci. 18: 298 (1921).

Syzygium xizangense H.T.Chang & R.H.Miao, in Fl. Xizangica 3: 343 (1986).
SE. Tibet. 36 CHT. Phan.

Syzygium xylopiaceum (Diels) Merr. & L.M.Perry, J. Arnold Arbor. 23: 249 (1942).
New Guinea. 43 NWG. Phan.
Jambosa xylopiacea Diels, Bot. Jahrb. Syst. 57: 392 (1922).

Syzygium yunnanense Merr. & L.M.Perry, J. Arnold Arbor. 19: 227 (1938).
SC. China. 36 CHC. Phan.

Syzygium zamboangense (C.B.Rob.) Merr., Philipp. J. Sci. 79: 424 (1951).
Philippines. 42 PHI. Phan.
Eugenia zamboangensis C.B.Rob., Philipp. J. Sci., C 4: 379 (1909).

Syzygium zeylanicum (L.) DC., Prodr. 3: 260 (1828).
E. Madagascar, India to SE. China and Malesia. 29 MDG 36 CHS 40 BAN IND SRL 41 AND MYA THA VIE 42 BOR JAW MLY SUL SUM. Phan.
Myrtus zeylanica L., Sp. Pl.: 472 (1753). *Eugenia zeylanica* (L.) Wight, Icon. Pl. Ind. Orient. 1: 73 (1838), nom. illeg. *Acmena zeylanica* (L.) Thwaites, Enum. Pl. Zeyl.: 118 (1859).
Eugenia spicata Lam., Encycl. 3: 201 (1789). *Syzygium spicatum* (Lam.) DC., Prodr. 3: 260 (1828).
Calyptranthes malabarica Dennst., Schlüssel Hortus Malab.: 31 (1818).
Acmena parviflora DC., Prodr. 3: 262 (1828).
Syzygium bellutta DC., Prodr. 3: 261 (1828).
Eugenia glandulifera Roxb., Fl. Ind. ed. 1832, 2: 496 (1832). *Jambosa glandulifera* (Roxb.) Miq., Fl. Ned. Ind. 1(1): 438 (1855).
Eugenia macrorhyncha Miq., Anal. Bot. Ind. 1: 21 (1850).
Eugenia varians Miq., Anal. Bot. Ind. 1: 21 (1850).
Jambosa koenigii Blume, Mus. Bot. 1: 100 (1850).
Eugenia tenuiramis Miq., Stirp. Surinam. Select.: 39 (1851). *Eugenia egensis* var. *tenuiramis* (Miq.) O.Berg in C.F.P.von Martius & auct. suc. (eds.), Fl. Bras. 14(1): 296 (1857).
Caryophyllus rugosus Blume ex Miq., Fl. Ned. Ind. 1(1): 437 (1855).
Jambosa bracteata Miq., Fl. Ned. Ind. 1(1): 437 (1855).
Syzygium coarctatum Blume ex Miq., Fl. Ned. Ind. 1(1): 437 (1855).
Eugenia linearis Duthie in J.D.Hooker, Fl. Brit. India 2: 486 (1878), nom. illeg. *Syzygium zeylanicum* var. *lineare* Alston in H.Trimen, Handb. Fl. Ceylon 6: 115 (1931).
Eugenia longicauda Ridl., J. Straits Branch Roy. Asiat. Soc. 61: 7 (1912).
Eugenia goudotiana H.Perrier, Mém. Inst. Sci. Madagascar, Sér. B, Biol. Vég. 4: 193 (1953).
Eugenia spicata var. *cordata* Kochummen, Gard. Bull. Singapore 28: 227 (1976).
Syzygium zeylanicum var. *ellipticum* A.N.Henry, Chandrab. & N.C.Nair in K.S. Manilal, Bot. Hist. Hortus Malabaricus: 161 (1980).
Syzygium zeylanicum var. *magamalayanum* K.Ravik. & V.Lakshm., Rheedea 9: 61 (1999).

Syzygium zhenghei Craven & Biffin, Blumea 51: 139 (2006).
Papua New Guinea. 43 NWG. Phan.
Acmena montana T.G.Hartley & Craven, J. Arnold Arbor. 58: 338 (1977).

Syzygium zimmermannii (Warb. ex Craib) Merr. & L.M.Perry, J. Arnold Arbor. 19: 114 (1938).
Indo-China. 41 LAO MYA THA VIE. Phan.
Eugenia zimmermannii Warb. ex Craib, Bull. Misc. Inform. Kew 1914: 124 (1914).

Syzygium zollingerianum (Miq.) Amshoff in C.A.Backer & R.C.Bakhuizen van der Brink, Bekn. Fl. Java 4b(98): 27 (1944).
Jawa. 42 JAW. Phan.
Jambosa zollingeriana Miq., Fl. Ned. Ind. 1(1): 413 (1855). *Eugenia zollingeriana* (Miq.) Koord. & Valeton, Meded. Lands Plantentuin 40: 71 (1900).

Synonyms:

Syzygium abidjaneose Aubrév. & Pellegr. = ***Syzygium rowlandii*** Sprague
Syzygium acetosum Merr. & L.M.Perry = ***Syzygium branderhorstii*** Lauterb.
Syzygium acmenoides Merr. & L.M.Perry = ***Syzygium hemilamprum*** subsp. ***hemilamprum***
Syzygium acutisepalum (Hayata) Mori = ***Syzygium formosanum*** (Hayata) Mori
Syzygium adelphicum var. *adenanthum* Merr. & L.M. Perry = ***Syzygium adelphicum*** Diels
Syzygium albidirameum (Merr.) Merr. & L.M.Perry = ***Syzygium kunstleri*** (King) Bahadur & R.C.Gaur
Syzygium alcinae (Merr.) Merr. & L.M.Perry = ***Syzygium leucoxylon*** Korth.
Syzygium alternans Miq. ex Duthie = ***Syzygium salicifolium*** (Wight) J.Graham
Syzygium ambongense (Ridl.) Masam. = ***Syzygium elopurae*** (Ridl.) Merr. & L.M.Perry
Syzygium ambongense var. *havilandii* (Ridl.) Masam. = ***Syzygium leucocladum*** Merr. & L.M.Perry
Syzygium amshoffianum Merr. = ***Syzygium confusum*** (Blume) Bakh.f.
Syzygium andersonii (Ridl.) Masam. = ***Syzygium punctilimbum*** (Merr.) Merr. & L.M.Perry
Syzygium androsaemoides (L.) Walp. = ***Syzygium cordifolium*** subsp. ***spissum*** (Alston) P.S.Ashton
Syzygium angkolanum Miq. = ***Syzygium operculatum*** (Roxb.) Nied.
Syzygium aphanomyrtoides Merr. & L.M.Perry = ***Syzygium caudatum*** (Merr.) Airy Shaw
Syzygium archboldianum Merr. & L.M.Perry = ***Syzygium pergamaceum*** (Greves) Merr. & L.M.Perry

Syzygium areolatum DC. = ***Syzygium venosum*** DC.
Syzygium arnottianum (Wight) Walp. = ***Syzygium densiflorum*** Wall. ex Wight & Arn.
Syzygium artense Montrouz. ex Guillaumin & Beauvis. = ***Syzygium austrocaledonicum*** (Seem.) Guillaumin
Syzygium attenuatum subsp. *circumscissum* (Gagnep.) Chantaran. & J.Parn. = ***Syzygium circumscissum*** (Gagnep.) Craven & Biffin
Syzygium attenuatum var. *montanum* (M.R.Hend.) Chantaran. & J.Parn. = ***Syzygium attenuatum*** (Miq.) Merr. & L.M.Perry
Syzygium attenuatum var. *ophirense* (M.R.Hend.) I.M.Turner = ***Syzygium attenuatum*** (Miq.) Merr. & L.M.Perry
Syzygium augustinii Merr. & L.M.Perry = ***Syzygium toddalioides*** (Wight) Walp.
Syzygium baronii Labat & G.E.Schatz = ***Syzygium loiseleurioides*** (Baker) Govaerts
Syzygium bellutta DC. = ***Syzygium zeylanicum*** (L.) DC.
Syzygium benthamianum Gamble = ***Syzygium densiflorum*** Wall. ex Wight & Arn.
Syzygium besukiense (Merr.) Masam. = ***Syzygium bankense*** (Hassk.) Merr. & L.M.Perry
Syzygium bibracteatum (Greves) Merr. & L.M.Perry = ***Syzygium fastigiatum*** (Blume) Merr. & L.M.Perry
Syzygium borneense var. *myrtillus* (Stapf) Chantaran. & J.Parn. = ***Syzygium myrtillus*** (Stapf) Merr. & L.M.Perry
Syzygium brachyantherum Merr. & L.M.Perry = ***Syzygium globiflorum*** (Craib) Chantaran. & J.Parn.
Syzygium brachynemum F.Muell. = ***Syzygium smithii*** (Poir.) Nied.
Syzygium brackenridgei var. *dubium* L.M.Perry = ***Syzygium dubium*** (L.M.Perry) A.C.Sm.
Syzygium bracteatum (Willd.) Raizada = ***Eugenia roxburghii*** DC.
Syzygium bracteolatum (Wight) Masam. = ***Syzygium fastigiatum*** (Blume) Merr. & L.M.Perry
Syzygium brunneorameum (Merr.) Masam. = ***Syzygium racemosum*** (Blume) DC.
Syzygium burkillianum var. *garcinifolioides* (M.R.Hend.) I.M.Turner = ***Syzygium burkillianum*** (King) I.M.Turner
Syzygium buxifolium var. *austrosinense* Merr. & L.M.Perry = ***Syzygium austrosinense*** (Merr. & L.M.Perry) H.T.Chang & R.H.Miao
Syzygium buxifolium var. *cleyerifolium* (Yatabe) Y.Tateishi = ***Syzygium cleyerifolium*** (Yatabe) Makino
Syzygium buxifolium var. *verticillatum* C.Chen = ***Syzygium buxifolium*** Hook. & Arn.
Syzygium calvinii (Elmer) Masam. = ***Syzygium confertum*** (Korth.) Merr. & L.M.Perry
Syzygium campanellum Miq. = ***Syzygium myrtifolium*** Walp.
Syzygium campanulatum Korth. = ***Syzygium myrtifolium*** Walp.
Syzygium campanulatum var. *longistylum* Chantaran. & J.Parn. = ***Syzygium myrtifolium*** Walp.
Syzygium caryophyllaeum Gaertn. = ***Syzygium caryophyllatum*** (L.) Alston
Syzygium caryophyllifolium (Lam.) DC. = ***Syzygium cumini*** (L.) Skeels
Syzygium caudatum Wall. = ***Syzygium syzygioides*** (Miq.) Merr. & L.M.Perry
Syzygium caudiferum Merr. & L.M.Perry = ***Syzygium brevicymum*** (Diels) Merr. & L.M.Perry
Syzygium cauliflorum (DC.) Bennet = ***Syzygium polycephalum*** (Miq.) Merr. & L.M.Perry
Syzygium cerasiforme (Blume) Merr. & L.M.Perry = ***Syzygium racemosum*** (Blume) DC.
Syzygium cerasoides (Roxb.) Raizada = ***Syzygium operculatum*** (Roxb.) Nied.
Syzygium cerasoides Raizada = ***Syzygium operculatum*** (Roxb.) Nied.
Syzygium cerinum (M.R.Hend.) I.M.Turner = ***Syzygium incarnatum*** (Elmer) Merr. & L.M.Perry
Syzygium cerinum var. *montanum* (M.R.Hend.) I.M. Turner = ***Syzygium incarnatum*** (Elmer) Merr. & L.M. Perry
Syzygium cerinum var. *turbinatum* (M.R.Hend.) I.M. Turner = ***Syzygium incarnatum*** (Elmer) Merr. & L.M.Perry
Syzygium chrysanthum Merr. & L.M.Perry = ***Syzygium kunstleri*** (King) Bahadur & R.C.Gaur
Syzygium cinerascens C.Presl = ***Syzygium cinereum*** (Kurz) Chantaran. & J.Parn.
Syzygium clavatum (Korth.) Merr. & L.M.Perry = ***Syzygium claviflorum*** (Roxb.) Wall. ex A.M.Cowan & Cowan
Syzygium claviflorum var. *excavatum* (King) I.M.Turner = ***Syzygium claviflorum*** (Roxb.) Wall. ex A.M.Cowan & Cowan
Syzygium claviflorum var. *glandulosum* (King) Chantaran. & J.Parn. = ***Syzygium claviflorum*** (Roxb.) Wall. ex A.M.Cowan & Cowan
Syzygium claviflorum var. *maingayi* (Duthie) Chantaran. & J.Parn. = ***Syzygium claviflorum*** (Roxb.) Wall. ex A.M.Cowan & Cowan
Syzygium claviflorum var. *montanum* (M.R.Hend.) I.M.Turner = ***Syzygium claviflorum*** (Roxb.) Wall. ex A.M.Cowan & Cowan
Syzygium claviflorum var. *oblongifolium* (Hayata) Mori = ***Syzygium taiwanicum*** H.T.Chang & R.H. Miao
Syzygium claviflorum var. *riparium* (M.R.Hend.) I.M.Turner = ***Syzygium claviflorum*** (Roxb.) Wall. ex A.M.Cowan & Cowan
Syzygium cleistocalyx (Merr.) P.S.Ashton = ***Syzygium operculatum*** (Roxb.) Nied.
Syzygium coarctatum Blume ex Miq. = ***Syzygium zeylanicum*** (L.) DC.
Syzygium cochinchinense (Gagnep.) Merr. & L.M.Perry = ***Syzygium ripicola*** (Craib) Merr. & L.M.Perry
Syzygium codyensis (Munro ex Wight) Chandrash. = ***Eugenia codyensis*** Munro ex Wight
Syzygium conglomeratum var. *paniculatum* (M.R.Hend.) I.M.Turner = ***Syzygium conglomeratum*** (Duthie) I.M.Turner
Syzygium coolminianum (C.Moore) L.A.S.Johnson = ***Syzygium oleosum*** (F.Muell.) B.Hyland
Syzygium coralinum (Merr.) Masam. = ***Syzygium curtisii*** (King) Merr. & L.M.Perry
Syzygium cordifolium Klotzsch = ***Syzygium cordatum*** Hochst. ex Krauss subsp. ***cordatum***
Syzygium costatum Miq. = ***Syzygium racemosum*** (Blume) DC.
Syzygium cryptophlebium (F.Muell.) Craven & Biffin = ***Syzygium wilsonii*** subsp. ***cryptophlebium*** (F.Muell.) B.Hyland
Syzygium cumingianum (Vidal) Gibbs = ***Syzygium acuminatissimum*** (Blume) DC.
Syzygium cumini var. *caryophyllifolium* (Lam.) K.K.Khanna = ***Syzygium cumini*** (L.) Skeels
Syzygium cumini var. *obtusifolium* (Roxb.) K.K.Khanna = ***Syzygium cumini*** (L.) Skeels
Syzygium cumini var. *tsoi* (Merr. & Chun) H.T.Chang & R.H.Miao = ***Syzygium cumini*** (L.) Skeels
Syzygium cuneifolium Bojer ex Baker = ***Syzygium emirnense*** (Baker) Labat & Schatz

Syzygium curtisii var. *holttumii* (Ridl.) I.M.Turner = ***Syzygium curtisii*** (King) Merr. & L.M.Perry
Syzygium curtisii var. *minus* (King) I.M.Turner = ***Syzygium curtisii*** (King) Merr. & L.M.Perry
Syzygium curvistylum var. *parvifolium* L.M.Perry = ***Syzygium curvistylum*** (Gillespie) Merr. & L.M.Perry
Syzygium cuspidato-obovatum (Hayata) Mori = ***Syzygium acuminatissimum*** (Blume) DC.
Syzygium cyclophyllum (Thwaites ex Duthie) Alston = ***Syzygium revolutum*** subsp. ***cyclophyllum*** (Thwaites ex Duthie) P.S.Ashton
Syzygium cymiferum (E.Mey.) C.Presl = ***Syzygium cordatum*** Hochst. ex Krauss subsp. ***cordatum***
Syzygium cymosum Korth. = ***Syzygium polyanthum*** (Wight) Walp.
Syzygium daphnoides (Greves) Merr. & L.M.Perry = ***Syzygium longipes*** Merr. & L.M.Perry
Syzygium × *deiningeri* Engl. = ***Syzygium* × *intermedium*** Engl. & Brehmer
Syzygium densiflorum Brongn. & Gris = ***Syzygium koghianum*** Petitm. & Bonati
Syzygium dictyophlebium Merr. & L.M.Perry = ***Syzygium sayeri*** (F.Muell.) B.Hyland
Syzygium diffusum var. *purpureum* L.M.Perry = ***Syzygium purpureum*** (L.M.Perry) A.C.Sm.
Syzygium discolor Merr. & L.M.Perry = ***Syzygium pergamaceum*** (Greves) Merr. & L.M.Perry
Syzygium doctersii Merr. & L.M.Perry = ***Syzygium effusum*** (A.Gray) Müll.Stuttg.
Syzygium durifolium (A.C.Sm.) Merr. & L.M.Perry = ***Syzygium phaeophyllum*** Merr. & L.M.Perry
Syzygium ellipticum K.Schum. & Lauterb. = ***Psidium guajava*** L.
Syzygium elmeri (Merr.) Masam. = ***Syzygium fastigiatum*** (Blume) Merr. & L.M.Perry
Syzygium enervosum Chantaran. & J.Parn. = ***Syzygium urophyllum*** Merr.
Syzygium fergusonii subsp. *minor* (Trimen) Alston = ***Syzygium montis-adam*** Kosterm.
Syzygium filiforme var. *clavimyrtus* (Koord. & Valeton) I.M.Turner = ***Syzygium filiforme*** (Wall. ex Duthie) Chantaran. & J.Parn.
Syzygium filiforme var. *constrictum* (Kochummen) I.M.Turner = ***Syzygium filiforme*** (Wall. ex Duthie) Chantaran. & J.Parn.
Syzygium firmum (Blume) Thwaites = ***Syzygium grande*** (Wight) Walp.
Syzygium floribundum K.Schum. & Lauterb. = ***Syzygium tierneyanum*** (F.Muell.) T.G.Hartley & L.M.Perry
Syzygium folidorhachis Merr. & L.M.Perry = ***Syzygium furfuraceum*** Merr. & L.M.Perry
Syzygium forbesii (Greves) Merr. & L.M.Perry = ***Syzygium gonatanthum*** (Diels) Merr. & L.M.Perry
Syzygium fourcadei (Dummer) Burtt Davy = ***Syzygium guineense*** (Willd.) DC. subsp. ***guineense***
Syzygium fraseri (Ridl.) Masam. = ***Syzygium claviflorum*** (Roxb.) Wall. ex A.M.Cowan & Cowan
Syzygium furcatum Blume ex Miq. = ***Syzygium muelleri*** (Miq.) Miq.
Syzygium fusiforme (Duthie) Merr. & L.M.Perry = ***Syzygium glabratum*** (DC.) Veldkamp
Syzygium gadgilii M.R.Almeida = ***Syzygium grande*** (Wight) Walp.
Syzygium gambleanum Rathakr. & V.Chithra = ***Syzygium microphyllum*** Gamble
Syzygium glaucescens Miq. = ***Syzygium borneense*** (Miq.) Miq.
Syzygium glaucicalyx (Merr.) Merr. = ***Syzygium antisepticum*** (Blume) Merr. & L.M.Perry
Syzygium gracile (Korth.) Amshoff = ***Syzygium glabratum*** (DC.) Veldkamp
Syzygium gracilentum (Hance) Hu = ***Decaspermum gracilentum*** (Hance) Merr. & L.M.Perry
Syzygium grande var. *parviflorum* Chantaran. & J.Parn. = ***Syzygium grande*** (Wight) Walp.
Syzygium gratum (Wight) S.N.Mitra = ***Syzygium antisepticum*** (Blume) Merr. & L.M.Perry
Syzygium gratum var. *confertum* Chantaran. & J.Parn. = ***Syzygium antisepticum*** (Blume) Merr. & L.M.Perry
Syzygium guehoscottii Bennet & Raizada = ***Syzygium dupontii*** (Baker) Govaerts
Syzygium guineense subsp. *bamendae* F.White = ***Syzygium staudtii*** (Engl.) Mildbr.
Syzygium guineense subsp. *gerrardii* (Harv. ex Hook.f.) F.White = ***Syzygium gerrardii*** (Harv. ex Hook.f.) Burtt Davy
Syzygium guineense subsp. *legatii* (Burtt Davy & Greenway) F.White = ***Syzygium legatii*** Burtt Davy & Greenway
Syzygium guineense subsp. *littorale* (Keay) Boutique = ***Syzygium guineense*** var. ***littorale*** Keay
Syzygium guineense var. *macrocarpum* Engl. = ***Syzygium guineense*** (Willd.) DC. subsp. ***guineense***
Syzygium guineense subsp. *macrocarpum* (Engl.) F.White = ***Syzygium guineense*** (Willd.) DC. subsp. ***guineense***
Syzygium guineense subsp. *obovatum* F.White = ***Syzygium staudtii*** (Engl.) Mildbr.
Syzygium guineense subsp. *occidentale* F.White = ***Syzygium staudtii*** (Engl.) Mildbr.
Syzygium guineense var. *parvifolium* Engl. = ***Syzygium parvifolium*** (Engl.) Mildbr.
Syzygium guineense subsp. *parvifolium* (Engl.) F.White = ***Syzygium parvifolium*** (Engl.) Mildbr.
Syzygium guineense var. *staudtii* Engl. = ***Syzygium staudtii*** (Engl.) Mildbr.
Syzygium hackenbergii Diels = ***Syzygium borneense*** (Miq.) Miq.
Syzygium hallieri Merr. & L.M.Perry = ***Syzygium durifolium*** Merr. & L.M.Perry
Syzygium heterocladum (Merr.) Merr. & L.M.Perry = ***Syzygium scortechinii*** (King) Chantaran. & J.Parn.
Syzygium heyneanum (Duthie) Gamble = ***Syzygium salicifolium*** (Wight) J.Graham
Syzygium heyneanum var. *alternans* (Duthie) B.G.Kulk. & Lakshmin. = ***Syzygium salicifolium*** (Wight) J.Graham
Syzygium holei Bahadur & R.C.Gaur = ***Syzygium circumscissum*** (Gagnep.) Craven & Biffin
Syzygium huillense (Hiern) Engl. = ***Syzygium guineense*** subsp. ***huillense*** (Hiern) F.White
Syzygium hypericifolium (Salisb.) DC. = ?
Syzygium imthurnii (Turrill) Merr. & L.M.Perry = ***Syzygium brackenridgei*** (A.Gray) Müll.Stuttg.
Syzygium inophyllum var. *bernardii* (King) I.M.Turner = ***Syzygium inophyllum*** DC.
Syzygium insculptum Merr. & L.M.Perry = ***Syzygium rubropunctatum*** (Ridl.) Merr. & L.M.Perry
Syzygium irosinense Merr. = ***Syzygium everettii*** (C.B.Rob.) Merr.
Syzygium irregulare (Craib) Merr. & L.M.Perry = ***Syzygium borneense*** (Miq.) Miq.
Syzygium jambolanum (Lam.) DC. = ***Syzygium cumini*** (L.) Skeels
Syzygium jambos var. *linearilimbum* H.T.Chang & R.H.Miao = ***Syzygium jambos*** (L.) Alston
Syzygium jambos var. *tripinnatum* (Blanco) C.Chen = ***Syzygium tripinnatum*** (Blanco) Merr.

Syzygium jambosoides (Lauterb.) Merr. & L.M.Perry = ***Syzygium suborbiculare*** (Benth.) T.G.Hartley & L.M.Perry
Syzygium japenense Merr. & L.M.Perry = ***Syzygium nitidum*** Benth.
Syzygium javanicum Miq. = ***Syzygium racemosum*** (Blume) DC.
Syzygium johorense Masam. = ***Syzygium rugosum*** Korth.
Syzygium kashotoense (Hayata) Mori = ***Syzygium paucivenium*** (C.B.Rob.) Merr.
Syzygium kiahii var. *angustifolium* (M.R.Hend.) I.M.Turner = ***Syzygium kiahii*** (M.R.Hend.) I.M.Turner
Syzygium kihamense Merr. & L.M.Perry = ***Syzygium oblanceolatum*** (C.B.Rob.) Merr.
Syzygium kingii (Merr.) Merr. & L.M.Perry = ***Syzygium scortechinii*** (King) Chantaran. & J.Parn.
Syzygium koghiense Guillaumin = ***Syzygium frutescens*** Brongn. & Gris
Syzygium kuchingense (Merr.) Merr. & L.M.Perry = ***Syzygium palembanicum*** Miq.
Syzygium kurzii var. *andamanica* (King) N.P.Balakr. = ***Syzygium kurzii*** (Duthie) N.P.Balakr.
Syzygium laetum var. *pauciflorum* (Duthie) M.R.Almeida = ***Syzygium laetum*** (Buch.-Ham.) Gandhi subsp. ***laetum***
Syzygium laevicaule (Duthie) Masam. = ***Syzygium oblatum*** (Roxb.) Wall. ex A.M.Cowan & Cowan
Syzygium laevigatum Miq. = ***Marlierea laevigata*** (DC.) Kiaersk.
Syzygium lagynocalyx (Diels) Merr. & L.M.Perry = ***Syzygium trachyanthum*** (Diels) Merr. & L.M.Perry
Syzygium lancifolium (Miq.) Merr. & L.M.Perry = ***Syzygium insigne*** (Blume) Merr. & L.M.Perry
Syzygium lanyuense C.E.Chang = ***Syzygium simile*** (Merr.) Merr.
Syzygium laosense (Gagnep.) Merr. & L.M.Perry = ***Syzygium grande*** (Wight) Walp.
Syzygium laosense var. *quocense* (Gagnep.) H.T.Chang & R.H.Miao = ***Syzygium grande*** (Wight) Walp.
Syzygium latifolium Blanco = ***Eugenia roxburghii*** DC.
Syzygium latilimbum (Merr.) Merr. & L.M.Perry = ***Syzygium megacarpum*** (Craib) Rathakr. & N.C.Nair
Syzygium lauterbachianum Merr. & L.M.Perry = ***Syzygium tierneyanum*** (F.Muell.) T.G.Hartley & L.M.Perry
Syzygium leptanthelium Diels = ***Syzygium plumeum*** (Ridl.) Merr. & L.M.Perry
Syzygium leptanthum (Walp.) Nied. = ***Syzygium claviflorum*** (Roxb.) Wall. ex A.M.Cowan & Cowan
Syzygium leptophlebioides Merr. & L.M.Perry = ***Syzygium branderhorstii*** Lauterb.
Syzygium leucoderme Diels = ***Syzygium effusum*** (A.Gray) Müll.Stuttg.
Syzygium leucoxylon var. *phaeophyllum* Merr. & L.M. Perry = ***Syzygium nigricans*** subsp. ***phaeophyllum*** (Merr. & L.M.Perry) P.S.Ashton
Syzygium lissophyllum Thwaites = ***Syzygium rubicundum*** Wight & Arn.
Syzygium litseifolium (Merr.) Merr. & L.M.Perry = ***Syzygium borneense*** (Miq.) Miq.
Syzygium littorale Aubrév. = ***Syzygium guineense*** var. ***littorale*** Keay
Syzygium lobbii (Ridl.) Masam. = ***Syzygium multibracteolatum*** (Merr.) Merr. & L.M.Perry
Syzygium longicalyx (Ridl.) Masam. = ***Syzygium lineatum*** (DC.) Merr. & L.M.Perry
Syzygium longiflorum C.Presl = ***Syzygium lineatum*** (DC.) Merr. & L.M.Perry
Syzygium lucidum Gaertn. = ***Gossia lucida*** (Gaertn.) N.Snow & Guymer
Syzygium macrocarpum Bahadur & R.C.Gaur = ***Syzygium megacarpum*** (Craib) Rathakr. & N.C.Nair
Syzygium maingayi Chantaran. & J.Parn. = ***Syzygium claviflorum*** (Roxb.) Wall. ex A.M.Cowan & Cowan
Syzygium malayanum (Gagnep.) I.M.Turner = ***Syzygium confusum*** (Blume) Bakh.f.
Syzygium mappaceum (Korth.) Merr. & L.M.Perry = ***Syzygium formosum*** (Wall.) Masam.
Syzygium marounzense Pellegr. = ***Syzygium staudtii*** (Engl.) Mildbr.
Syzygium maschalocladum Merr. & L.M.Perry = ***Syzygium taeniatum*** Diels
Syzygium megalanthelium Diels = ***Syzygium decipiens*** (Koord. & Valeton) Merr. & L.M.Perry
Syzygium megalophyllum Merr. & L.M.Perry = ***Syzygium grande*** (Wight) Walp.
Syzygium megistophyllum Merr. = ?
Syzygium merrillii Masam. = ***Syzygium jambos*** (L.) Alston
Syzygium merrillii (C.B.Rob.) Merr. = ***Syzygium mauritsii*** Govaerts
Syzygium michelii (Lam.) Duthie = ***Eugenia uniflora*** L.
Syzygium micklethwaitii var. *subcordatum* Verdc. = ***Syzygium micklethwaitii*** subsp. ***subcordatum*** Verdc.
Syzygium micranthum Montrouz. ex Guillaumin = ***Syzygium pancheri*** Brongn. & Gris
Syzygium micranthum Blume ex Miq. = ***Syzygium polyanthum*** (Wight) Walp.
Syzygium microbotryum (Miq.) Masam. = ***Syzygium polyanthum*** (Wight) Walp.
Syzygium micropetalum Merr. & L.M.Perry = ***Syzygium benjaminum*** Diels
Syzygium microsemmifolium Guillaumin = ***Syzygium quadrangulare*** var. ***microsemmifolium*** (Guillaumin) J.W.Dawson
Syzygium modestum Diels = ***Syzygium triplinervium*** Teijsm. & Binn.
Syzygium monanthum (Merr.) Merr. & L.M.Perry = ***Syzygium jambos*** (L.) Alston
Syzygium montanum Thwaites & Hook.f. = ***Syzygium grande*** (Wight) Walp.
Syzygium montanum Aubrév. = ***Syzygium staudtii*** (Engl.) Mildbr.
Syzygium motleyi (Ridl.) Masam. = ***Syzygium rugosum*** Korth.
Syzygium mumbwaense Greenway = ***Syzygium guineense*** subsp. ***huillense*** (Hiern) F.White
Syzygium myrsinifolium var. *grandiflorum* H.T.Chang & R.H.Miao = ***Syzygium myrsinifolium*** (Hance) Merr. & L.M.Perry
Syzygium myrtillus var. *borneense* (Miq.) Chantaran. & J.Parn. = ***Syzygium borneense*** (Miq.) Miq.
Syzygium neglectum Brongn. & Gris = ***Syzygium pancheri*** Brongn. & Gris
Syzygium nelitricarpum Teijsm. & Binn. = ***Syzygium syzygioides*** (Miq.) Merr. & L.M.Perry
Syzygium nervosum A.Cunn. ex DC. = ***Syzygium operculatum*** (Roxb.) Nied.
Syzygium nervosum var. *obovatum* (Kurz) A.Kumar = ***Syzygium operculatum*** (Roxb.) Nied.
Syzygium nervosum var. *paniala* (Roxb.) Craven & Biffin = ***Syzygium operculatum*** (Roxb.) Nied.
Syzygium nienkui Merr. & L.M.Perry = ***Syzygium tetragonum*** (Wight) Wall. ex Walp.
Syzygium nitidum Brongn. & Gris = ***Syzygium austrocaledonicum*** (Seem.) Guillaumin
Syzygium niviferum (Greves) Merr. & L.M.Perry = ***Syzygium effusum*** (A.Gray) Müll.Stuttg.

Syzygium nodosum Miq. = ***Syzygium operculatum*** (Roxb.) Nied.
Syzygium oahuense O.Deg & Ludwig = ***Syzygium sandwicense*** (A.Gray) Müll.Stuttg.
Syzygium oblatum var. *laevicaule* (Duthie) Chantaran. & J.Parn. = ***Syzygium oblatum*** (Roxb.) Wall. ex A.M.Cowan & Cowan
Syzygium obovatum (Poir.) DC. = ***Syzygium cumini*** (L.) Skeels
Syzygium obovatum Korth. = ***Syzygium muelleri*** (Miq.) Miq.
Syzygium obtusifolium (Roxb.) Kostel. = ***Syzygium cumini*** (L.) Skeels
Syzygium obtusum Merr. & L.M.Perry = ***Syzygium effusum*** (A.Gray) Müll.Stuttg.
Syzygium obversum (Miq.) Masam. = ***Syzygium aqueum*** (Burm.f.) Alston
Syzygium ochneocarpum (Merr.) Merr. & L.M.Perry = ***Syzygium oligomyrum*** Diels
Syzygium okudai Mori = ***Syzygium tripinnatum*** (Blanco) Merr.
Syzygium olivifolium (Duthie) Gamble = ***Syzygium spathulatum*** Thwaites
Syzygium ovatifolium Merr. & L.M.Perry = ***Syzygium antisepticum*** (Blume) Merr. & L.M.Perry
Syzygium pahangense var. *fraseri* (M.R.Hend.) I.M. Turner = ***Syzygium pahangense*** (Ridl.) I.M.Turner
Syzygium pamatense (Miq.) Masam. = ***Syzygium polyanthum*** (Wight) Walp.
Syzygium paniculatum DC. = ***Syzygium borbonicum*** J.Guého & A.J.Scott
Syzygium papuasicum Merr. & L.M.Perry = ***Syzygium acutangulum*** Nied.
Syzygium patens Pancher ex Brongn. & Gris = ***Syzygium ngoyense*** (Schltr.) Guillaumin
Syzygium pauciflorum Merr. & L.M.Perry = ***Syzygium lineare*** (Korth.) Masam.
Syzygium pearsonianum (King) I.M.Turner = ***Syzygium praineanum*** (King) Chantaran. & J.Parn. subsp. ***praineanum***
Syzygium perpuncticulatum (Merr.) Merr. & L.M.Perry = ***Syzygium pustulatum*** (Duthie) Merr.
Syzygium phillyraeoides (Trimen) Santapau = ***Eugenia phillyraeoides*** Trimen
Syzygium polyanthum Thwaites = ***Syzygium operculatum*** (Roxb.) Nied.
Syzygium polyanthum var. *sessile* (M.R.Hend.) I.M.Turner = ***Syzygium polyanthum*** (Wight) Walp.
Syzygium ponapense (Merr. & Kaneh.) Diels = ***Syzygium carolinense*** (Koidz.) Hosok.
Syzygium pseudotetrapterum (King) Chantaran. & J.Parn. = ***Syzygium tetrapterum*** (Miq.) Chantaran. & J.Parn.
Syzygium punctatum Vieill. ex Guillanmin = ***Syzygium micans*** Brongn. & Gris
Syzygium punctulatum Wall. = ***Syzygium incarnatum*** (Elmer) Merr. & L.M.Perry
Syzygium rectangulare Merr. & L.M.Perry = ***Syzygium decipiens*** (Koord. & Valeton) Merr. & L.M.Perry
Syzygium retivenium Merr. & L.M.Perry = ***Syzygium homichlophilum*** Diels
Syzygium rhamphiphyllum (Craib) C.E.C.Fisch. = ***Syzygium circumscissum*** (Gagnep.) Craven & Biffin
Syzygium rhododendrifolium (Miq.) Masam. = ***Syzygium claviflorum*** (Roxb.) Wall. ex A.M.Cowan & Cowan
Syzygium rhynchophyllum (Merr.) Merr. & L.M.Perry = ***Syzygium caudatum*** (Merr.) Airy Shaw
Syzygium richardianum J.Guého & A.J.Scott = ***Syzygium emirnense*** (Baker) Labat & Schatz
Syzygium richardsii Airy Shaw = ***Syzygium villamilii*** (Merr.) Merr. & L.M.Perry
Syzygium riparium (Becc.) Masam. = ***Syzygium odoardoi*** Merr. & L.M.Perry
Syzygium roxburghianum Raizada = ***Syzygium praecox*** (Roxb.) Rathakr. & N.C.Nair
Syzygium rubellum (Rech.) Merr. & L.M.Perry = ***Syzygium roemeri*** (Lauterb.) Merr. & L.M.Perry
Syzygium rubescens var. *koroense* L.M.Perry = ***Syzygium fijiense*** L.M.Perry
Syzygium rubiginosum Merr. & L.M.Perry = ***Syzygium forte*** (F.Muell.) B.Hyland subsp. ***forte***
Syzygium rufotomentosum (Gibbs) Masam. = ***Syzygium hirtum*** (Korth.) Merr. & L.M.Perry
Syzygium rugosum var. *cordatum* (M.R.Hend.) I.M.Turner = ***Syzygium rugosum*** Korth.
Syzygium ruminatum (Koord. & Valeton) Amshoff = ***Syzygium claviflorum*** (Roxb.) Wall. ex A.M.Cowan & Cowan
Syzygium ruscifolium (Willd.) Santapau & Wagh = ***Eugenia roxburghii*** DC.
Syzygium samarangense var. *parviflorum* (Craib) Chantaran. & J.Parn. = ***Syzygium samarangense*** (Blume) Merr. & L.M.Perry
Syzygium sandakanense (Merr.) Merr. & L.M.Perry = ***Syzygium leptostemon*** (Korth.) Merr. & L.M.Perry
Syzygium sarawacense (Merr.) Merr. & L.M.Perry = ***Syzygium muelleri*** (Miq.) Miq.
Syzygium sargentianum (Diels) Merr. & L.M.Perry = ***Syzygium coalitum*** (Greves) T.G.Hartley & L.M.Perry
Syzygium scandens Bojer = ***Syzygium latifolium*** (Poir.) DC.
Syzygium scandens (Baker) J.Guého & A.J.Scott = ***Syzygium latifolium*** (Poir.) DC.
Syzygium schmidii Rathakr. & N.C.Nair = ***Syzygium cuneatum*** Brahmam & H.O.Saxena
Syzygium sclerophyllum Brenan = ***Syzygium micklethwaitii*** Verdc.
Syzygium scoparium (Duthie) Wall. ex Murugan & Manickam = ***Syzygium avene*** Miq.
Syzygium scortechinii var. *cuneatum* (M.R.Hend.) I.M.Turner = ***Syzygium scortechinii*** (King) Chantaran. & J.Parn.
Syzygium sessile Blume ex Miq. = ***Syzygium korthalsianum*** (Miq.) Miq.
Syzygium sieberianum C.Presl = [29 MAU]
Syzygium sinubanense (Elmer) Diels = ***Syzygium myrtifolium*** Walp.
Syzygium soliflorum (Diels) Merr. & L.M.Perry = ***Syzygium subalatum*** (Ridl.) Merr. & L.M.Perry
Syzygium somai (Hayata) Mori = ***Syzygium buxifolium*** Hook. & Arn.
Syzygium spicatum (Lam.) DC. = ***Syzygium zeylanicum*** (L.) DC.
Syzygium spissum Alston = ***Syzygium cordifolium*** subsp. ***spissum*** (Alston) P.S.Ashton
Syzygium stictophyllum Merr. & L.M.Perry = ***Syzygium kunstleri*** (King) Bahadur & R.C.Gaur
Syzygium subdecurrens Miq. = ***Syzygium acuminatissimum*** (Blume) DC.
Syzygium subdecussatum var. *montanum* (King) I.M. Turner = ***Syzygium subdecussatum*** (Duthie) I.M.Turner
Syzygium subracemosum (Merr.) Masam. = ***Syzygium leptostemon*** (Korth.) Merr. & L.M.Perry
Syzygium subrufum (King) Masam. = ***Syzygium griffithii*** (Duthie) Merr. & L.M.Perry
Syzygium subrufum var. *smalianum* (Brandis) Chantaran. & J.Parn. = ***Syzygium smalianum*** (Brandis) D.G. Long

Syzygium sylvanum (Ridl.) Merr. & L.M.Perry = ***Syzygium effusum*** (A.Gray) Müll.Stuttg.
Syzygium sylvestre (Moon ex Wight) Walp. = ***Syzygium makul*** Gaertn.
Syzygium tamilnadensis Rathakr. & V.Chithra = ***Syzygium grande*** (Wight) Walp.
Syzygium tawaense (Merr.) Masam. = ***Syzygium peregrinum*** (Blume) Merr. & L.M.Perry
Syzygium tenuiflorum var. *brevipes* Brongn. & Gris = ***Syzygium brevipes*** (Brongn. & Gris) J.W.Dawson
Syzygium tenuiflorum var. *capillaceum* Brongn. & Gris = ***Syzygium capillaceum*** (Brongn. & Gris) J.W. Dawson
Syzygium tetragonocladum Merr. & L.M.Perry = ***Syzygium rejangense*** Merr. & L.M.Perry
Syzygium tetrapterum var. *pseudotetrapterum* (King) I.M.Turner = ***Syzygium tetrapterum*** (Miq.) Chantaran. & J.Parn.
Syzygium teysmannii (Miq.) Masam. = ***Syzygium lineatum*** (DC.) Merr. & L.M.Perry
Syzygium truncatum Miq. = ***Syzygium pyrifolium*** (Blume) DC.
Syzygium trunciflorum Airy Shaw = ***Syzygium tekuense*** (M.R.Hend.) I.M.Turner
Syzygium tutuilense (A.Gray) Müll.Stuttg. = ***Syzygium savaiiense*** (A.Gray) Müll.Stuttg.
Syzygium ugoense (C.B.Rob.) Masam. = ***Syzygium myrtillus*** (Stapf) Merr. & L.M.Perry
Syzygium urceolatum (Korth. ex Miq.) Merr. & L.M. Perry = ***Syzygium leptostemon*** (Korth.) Merr. & L.M.Perry
Syzygium vaccinioides Merr. & L.M.Perry = ***Syzygium alatum*** (Lauterb.) Diels
Syzygium venosum (Lam.) J.Guého & A.J.Scott = ***Syzygium dupontii*** (Baker) Govaerts
Syzygium verticilligerum (Ridl.) Masam. = ***Syzygium caudatilimbum*** (Merr.) Merr. & L.M.Perry
Syzygium viburnifolium (Ridl.) Masam. = ***Syzygium muelleri*** (Miq.) Miq.
Syzygium viridifolium (Elmer) Merr. & L.M.Perry = ***Syzygium claviflorum*** (Roxb.) Wall. ex A.M.Cowan & Cowan
Syzygium vitiense (Turrill) Merr. & L.M.Perry = ***Syzygium gracilipes*** (A.Gray) Merr. & L.M.Perry
Syzygium wallichianum C.Presl = ***Syzygium operculatum*** (Roxb.) Nied.
Syzygium wallichii (Wight) Walp. = ***Syzygium praecox*** (Roxb.) Rathakr. & N.C.Nair
Syzygium wightianum Wall. ex Wight & Arn. = ***Syzygium lanceolatum*** (Lam.) Wight & Arn.
Syzygium woodii Masam. = ***Syzygium creaghii*** (Ridl.) Merr. & L.M.Perry
Syzygium xylanthum (Greves) Merr. & L.M.Perry = ***Syzygium pyrocarpum*** (Greves) Merr. & L.M.Perry
Syzygium zeylanicum var. *ellipticum* A.N.Henry, Chandrab. & N.C.Nair = ***Syzygium zeylanicum*** (L.) DC.
Syzygium zeylanicum var. *lineare* Alston = ***Syzygium zeylanicum*** (L.) DC.
Syzygium zeylanicum var. *magamalayanum* K.Ravik. & V.Lakshm. = ***Syzygium zeylanicum*** (L.) DC.
Syzygium zippelianum Miq. = ***Syzygium lineatum*** (DC.) Merr. & L.M.Perry

***Unplaced Names*:**
Syzygium sieberianum C.Presl, Abh. Königl. Böhm. Ges. Wiss., V, 3: 499 (1845). = [29 MAU]

Temu

Temu O.Berg = ***Blepharocalyx*** O.Berg
Temu cruckshanksii (Hook. & Arn.) O.Berg = ***Blepharocalyx cruckshanksii*** (Hook. & Arn.) Nied.
Temu divaricatum O.Berg = ***Blepharocalyx cruckshanksii*** (Hook. & Arn.) Nied.

Tepualia

Tepualia Griseb., Abh. Königl. Ges. Wiss. Göttingen 6: 119 (1854).
S. South America. 85 AGS CLC CLS.
1 Species

Tepualia stipularis (Hook. & Arn.) Griseb., Abh. Königl. Ges. Wiss. Göttingen 6: 119 (1854).
C. & S. Chile to SW. Argentina. 85 AGS CLC CLS. Nanophan. or phan.
**Myrtus stipularis* Hook. & Arn., Bot. Misc. 3: 316 (1833). *Metrosideros stipularis* (Hook. & Arn.) Hook.f., Fl. Antarct.: 275 (1846). *Nania stipularis* (Hook. & Arn.) Kuntze, Revis. Gen. Pl. 1: 242 (1891).
Tepualia philippiana Griseb., Abh. Königl. Ges. Wiss. Göttingen 6: 120 (1854). *Tepualia stipularis* var. *philippiana* (Griseb.) Speg., Anales Mus. Nac. Buenos Aires 7: 284 (1902).
Tepualia philippii Griseb. ex Phil., Linnaea 28: 637 (1857).
Tepualia patagonica Phil., Anales Univ. Chile 84: 755 (1893). *Tepualia stipularis* var. *patagonica* (Phil.) Reiche, Anales Univ. Chile 98: 726 (1897).

***Synonyms*:**
Tepualia patagonica Phil. = ***Tepualia stipularis*** (Hook. & Arn.) Griseb.
Tepualia philippiana Griseb. = ***Tepualia stipularis*** (Hook. & Arn.) Griseb.
Tepualia philippii Griseb. ex Phil. = ***Tepualia stipularis*** (Hook. & Arn.) Griseb.
Tepualia stipularis var. *patagonica* (Phil.) Reiche = ***Tepualia stipularis*** (Hook. & Arn.) Griseb.
Tepualia stipularis var. *philippiana* (Griseb.) Speg. = ***Tepualia stipularis*** (Hook. & Arn.) Griseb.

Tetraeugenia

Tetraeugenia Merr. = ***Syzygium*** Gaertn.
Tetraeugenia caudata Merr. = ***Syzygium caudatum*** (Merr.) Airy Shaw

Tetrapora

Tetrapora Schauer = ***Babingtonia*** Lindl.
Tetrapora glomerata Turcz. = ***Baeckea pentandra*** (F.Muell.) F.Muell.
Tetrapora gunniana (Schauer ex Walp.) Miq. = ***Baeckea gunniana*** Schauer ex Walp.
Tetrapora preissiana Schauer = ***Baeckea preissiana*** (Schauer) Druce
Tetrapora verrucosa Turcz. = ***Baeckea preissiana*** (Schauer) Druce

Tetraspora

Tetraspora Miq. = ***Baeckea*** L.
Tetraspora gunniana (Schauer ex Walp.) Miq. = ***Baeckea gunniana*** Schauer ex Walp.

Tetrastemon

Tetrastemon Hook. & Arn. = ***Myrrhinium*** Schott
Tetrastemon loranthoides Hook. & Arn. = ***Myrrhinium atropurpureum*** var. ***octandrum*** Benth.

Thaleropia

Thaleropia Peter G.Wilson, Austral. Syst. Bot. 6: 255 (1993).
New Guinea to N. Queensland. 43 NWG 50 QLD.
3 Species

Thaleropia hypargyrea (Diels) Peter G.Wilson, Austral. Syst. Bot. 6: 257 (1993).
New Guinea. 43 NWG. Phan.
**Metrosideros hypargyrea* Diels, Bot. Jahrb. Syst. 57: 417 (1922).
Metrosideros ornata C.T.White, J. Arnold Arbor. 23: 79 (1942).

Thaleropia iteophylla (Diels) Peter G.Wilson, Austral. Syst. Bot. 6: 256 (1993).
New Guinea. 43 NWG. Phan.
Metrosideros brachyanthera Diels, Bot. Jahrb. Syst. 57: 416 (1922).
**Metrosideros iteophylla* Diels, Bot. Jahrb. Syst. 57: 416 (1922).

Thaleropia queenslandica (L.S.Sm.) Peter G.Wilson, Austral. Syst. Bot. 6: 258 (1993).
N. Queensland. 50 QLD. Phan.
**Metrosideros queenslandica* L.S.Sm., Proc. Roy. Soc. Queensland 69: 50 (1958).

Thryptomene

Thryptomene Endl., Stirp. Herb. Hügel.: 4 (1838), nom. cons.
Australia. 50 NSW NTA QLD SOA TAS VIC WAU.
32 Species
Gomphotis Raf., Sylva Tellur.: 103 (1838).
Astraea Schauer, Linnaea 17: 238 (1843), nom. illeg.
Paryphantha Schauer, Linnaea 17: 235 (1843).
Tryptomene Walp., Repert. Bot. Syst. 2: 157 (1843), orth. var.
Bucheria Heynh., Alph. Aufz. Gew. 2: 80 (1846).

Thryptomene australis Endl., Stirp. Herb. Hügel.: 4 (1838).
SW. Australia. 50 WAU. Cham. or nanophan.

subsp. ***australis***
SW. Australia. 50 WAU. Cham. or nanophan.

subsp. ***brachyandra*** Rye & Trudgen, Nuytsia 13: 516 (2001).
SW. Australia. 50 WAU.

Thryptomene baeckeacea F.Muell., Fragm. 4: 65 (1864).
W. Western Australia. 50 WAU. Cham. or nanophan.
Baeckea micrantha DC., Prodr. 3: 230 (1828). *Schidiomyrtus micrantha* (DC.) Schauer, Linnaea 17: 237 (1843). *Thryptomene micrantha* (DC.) C.A.Gardner, Enum. Pl. Austr. Occ.: 97 (1931), nom. illeg.

Thryptomene biseriata J.W.Green, Fl. S. Austral. 2: 948 (1986).
SE. Western Australia to S. South Australia. 50 SOA WAU. Cham. or nanophan.

Thryptomene calycina (Lindl.) Stapf, Bot. Mag. 149: t. 8995 (1924).
SE. South Australia. 50 SOA. Cham. or nanophan.
**Baeckea calycina* Lindl. in T.L.Mitchell, Three Exped. Australia 2: 189 (1838). *Paryphantha mitchelliana* Schauer, Linnaea 17: 235 (1843), nom. illeg. *Thryptomene mitchelliana* (Schauer) F.Muell., Fragm. 1: 11 (1858), nom. illeg.

Thryptomene costata Rye & Trudgen, Nuytsia 13: 518 (2001).
SW. Australia. 50 WAU.

Thryptomene cuspidata (Turcz.) J.W.Green, Census Vasc. Pl. W. Australia, ed. 2: 6 (1985).
SW. Australia. 50 WAU. Cham. or nanophan.
**Paryphantha cuspidata* Turcz., Bull. Cl. Phys.-Math. Acad. Imp. Sci. Saint-Pétersbourg 10: 321 (1852).
Thryptomene tenella Benth., Fl. Austral. 3: 59 (1867).
Thryptomene thymifolia Stapf, Bot. Mag. 149: t. 8995 (1924).

Thryptomene decussata (W.Fitzg.) J.W.Green, Census Vasc. Pl. W. Australia, ed. 2: 6 (1985).
SW. Australia. 50 WAU. Cham. or nanophan.
**Scholtzia decussata* W.Fitzg., J. West Austral. Nat. Hist. Soc. 1: 19 (1904).

Thryptomene denticulata (F.Muell.) Benth., Fl. Austral. 3: 60 (1867).
W. Western Australia. 50 WAU. Cham. or nanophan.
**Scholtzia denticulata* F.Muell., Fragm. 4: 75 (1864).

Thryptomene duplicata Rye & Trudgen, Nuytsia 13: 520 (2001).
SW. Australia. 50 WAU.

Thryptomene elliottii F.Muell., Fragm. 9: 62 (1875).
SE. Western Australia to S. South Australia. 50 SOA WAU. Cham. or nanophan.
Thryptomene whiteae J.M.Black, Trans. & Proc. Roy. Soc. South Australia 41: 384 (1917).

Thryptomene eremaea Rye & Trudgen, Nuytsia 13: 521 (2001).
SW. Australia. 50 WAU.

Thryptomene ericaea F.Muell., Fragm. 1: 12 (1858).
South Australia. 50 SOA. Cham. or nanophan.

Thryptomene hexandra C.T.White, Proc. Roy. Soc. Queensland 55: 67 (1944).
S. Queensland to New South Wales. 50 NSW QLD. Nanophan.

Thryptomene hyporhytis Turcz., Bull. Soc. Imp. Naturalistes Moscou 35(2): 324 (1862).
W. Western Australia. 50 WAU. Cham. or nanophan.

Thryptomene johnsonii F.Muell., Fragm. 4: 77 (1864).
SW. Australia. 50 WAU. Nanophan.

Thryptomene kochii E.Pritz., Repert. Spec. Nov. Regni Veg. 10: 133 (1911).
Western Australia. 50 WAU. Cham. or nanophan.

Thryptomene longifolia J.W.Green, Fl. S. Austral. 2: 950 (1986).
South Australia. 50 SOA. Cham. or nanophan.

Thryptomene micrantha Hook.f., Hooker's J. Bot. Kew Gard. Misc. 5: 299 (1853).
SE. Australia. 50 SOA TAS VIC. Cham. or nanophan.

Thryptomene mucronulata Turcz., Bull. Soc. Imp. Naturalistes Moscou 20(1): 156 (1847).
SW. Australia. 50 WAU. Cham. or nanophan.
Thryptomene prolifera Turcz., Bull. Soc. Imp. Naturalistes Moscou 35(2): 324 (1862).
Thryptomene dielsiana E.Pritz., Bot. Jahrb. Syst. 35: 412 (1904).
Thryptomene davisiae Diels, Biol. Meddel. Kongel. Danske Vidensk. Selsk. 3(2): 95 (1921).

Thryptomene naviculata J.W.Green, Nuytsia 3: 188 (1980).
C. Western Australia. 50 WAU. Cham. or nanophan.

Thryptomene nealensis J.W.Green, Nuytsia 3: 190 (1980).
SE. Western Australia. 50 WAU. Cham. or nanophan.

Thryptomene oligandra F.Muell., Fragm. 1: 11 (1858).
Queensland. 50 QLD. Cham. or nanophan.

Thryptomene parviflora (Benth.) Domin, Biblioth. Bot. 89: 449 (1928).
Queensland. 50 QLD. Cham. or nanophan.
**Thryptomene oligandra* var. *parviflora* Benth., Fl. Austral. 3: 63 (1867).

Thryptomene racemulosa Turcz., Bull. Soc. Imp. Naturalistes Moscou 20(1): 156 (1847).
S. Western Australia. 50 WAU. Cham. or nanophan.

Thryptomene remota A.R.Bean, Austrobaileya 4: 647 (1997).
Northern Territory. 50 NTA. Cham. or nanophan.

Thryptomene salina Rye & Trudgen, Nuytsia 13: 525 (2001).
SW. Australia. 50 WAU.

Thryptomene saxicola (A.Cunn. ex Hook.) Schauer in J.G.C.Lehmann, Pl. Preiss. 1: 102 (1844).
SW. Australia. 50 WAU. Cham. or nanophan.
**Baeckea saxicola* A.Cunn. ex Hook., Bot. Mag. 59: t. 3160 (1832). *Gomphotis saxicola* (A.Cunn. ex Hook.) Raf., Sylva Tellur.: 103 (1838). *Astraea saxicola* (A.Cunn. ex Hook.) Schauer, Linnaea 17: 239 (1843). *Bucheria saxicola* (A.Cunn. ex Hook.) Heynh., Alph. Aufz. Gew. 2: 80 (1846).
Paryphantha camphorata F.Muell. ex Miq., Ned. Kruidk. Arch. 4: 116 (1856).
Thryptomene miqueliana F.Muell., Fragm. 1: 11 (1858).
Scholtzia decandra F.Muell., Fragm. 4: 75 (1864).

Thryptomene stenophylla E.Pritz., Bot. Jahrb. Syst. 35: 412 (1904).
W. Western Australia. 50 WAU. Cham. or nanophan.

Thryptomene striata Rye & Trudgen, Nuytsia 13: 526 (2001).
SW. Australia. 50 WAU.

Thryptomene strongylophylla F.Muell. ex Benth., Fl. Austral. 3: 61 (1867).
W. Western Australia. 50 WAU. Cham. or nanophan.

Thryptomene urceolaris F.Muell., Fragm. 10: 25 (1876).
SW. Australia. 50 WAU. Cham. or nanophan.

Thryptomene wittweri J.W.Green, Nuytsia 3: 190 (1980).
Western Australia to Northern Territory. 50 NTA WAU. Nanophan.

***Synonyms*:**
Thryptomene appressa C.R.P.Andrews = ***Aluta appressa*** (C.R.P.Andrews) Rye & Trudgen
Thryptomene aspera E.Pritz. = ***Aluta aspera*** (E.Pritz.) Rye & Trudgen
Thryptomene auriculata F.Muell. = ***Aluta maisonneuvei*** subsp. ***auriculata*** (F.Muell.) Rye & Trudgen
Thryptomene ciliata (Sm.) Woolls = ***Micromyrtus ciliata*** (Sm.) Druce
Thryptomene davisiae Diels = ***Thryptomene mucronulata*** Turcz.
Thryptomene dielsiana E.Pritz. = ***Thryptomene mucronulata*** Turcz.
Thryptomene drummondii (Benth.) F.Muell. = ***Micromyrtus obovata*** (Turcz.) J.W.Green
Thryptomene elobata F.Muell. = ***Micromyrtus elobata*** (F.Muell.) Benth.
Thryptomene fimbriata D.A.Herb. = [50 WAU]
Thryptomene flaviflora F.Muell. = ***Micromyrtus flaviflora*** (F.Muell.) J.M.Black
Thryptomene helmsii F.Muell. & Tate = ***Micromyrtus helmsii*** (F.Muell. & Tate) J.W.Green
Thryptomene hexamera Maiden & Betche = ***Micromyrtus hexamera*** (Maiden & Betche) Maiden & Betche
Thryptomene homalocalyx F.Muell. = ***Homalocalyx ericaeus*** F.Muell.
Thryptomene hymenonema F.Muell. = ***Micromyrtus hymenonema*** (F.Muell.) C.A.Gardner
Thryptomene imbricata (Benth.) F.Muell. = ***Micromyrtus imbricata*** Benth.
Thryptomene leptocalyx (F.Muell.) F.Muell. = ***Micromyrtus leptocalyx*** (F.Muell.) Benth.
Thryptomene maisonneuvei F.Muell. = ***Aluta maisonneuvei*** (F.Muell.) Rye & Trudgen
Thryptomene micrantha (DC.) C.A.Gardner = ***Thryptomene baeckeacea*** F.Muell.
Thryptomene minutiflora (Benth.) F.Muell. ex Woolls = ***Micromyrtus minutiflora*** Benth.
Thryptomene miqueliana F.Muell. = ***Thryptomene saxicola*** (A.Cunn. ex Hook.) Schauer
Thryptomene mitchelliana (Schauer) F.Muell. = ***Thryptomene calycina*** (Lindl.) Stapf
Thryptomene obovata Turcz. = ***Micromyrtus obovata*** (Turcz.) J.W.Green
Thryptomene oligandra var. *parviflora* Benth. = ***Thryptomene parviflora*** (Benth.) Domin
Thryptomene plicata (F.Muell.) F.Muell. = ***Micromyrtus ciliata*** (Sm.) Druce
Thryptomene polyandra F.Muell. = ***Homalocalyx polyandrus*** (F.Muell.) Benth.
Thryptomene prolifera Turcz. = ***Thryptomene mucronulata*** Turcz.
Thryptomene racemosa (Benth.) F.Muell. = ***Micromyrtus racemosa*** Benth.
Thryptomene rosea E.Pritz. = ***Malleostemon roseus*** (E.Pritz.) J.W.Green
Thryptomene stenocalyx F.Muell. = ***Micromyrtus stenocalyx*** (F.Muell.) J.W.Green
Thryptomene tenella Benth. = ***Thryptomene cuspidata*** (Turcz.) J.W.Green
Thryptomene thymifolia Stapf = ***Thryptomene cuspidata*** (Turcz.) J.W.Green
Thryptomene trachycalyx F.Muell. = ***Micromyrtus flaviflora*** (F.Muell.) J.M.Black
Thryptomene tuberculata E.Pritz. = ***Malleostemon tuberculatus*** (E.Pritz.) J.W.Green
Thryptomene whiteae J.M.Black = ***Thryptomene elliottii*** F.Muell.

***Unplaced Names*:**
Thryptomene fimbriata D.A.Herb., J. Proc. Roy. Soc. W. Australia 7: 88 (1921). = [50 WAU]

Tillospermum

Tillospermum Salisb. = ***Kunzea*** Rchb.

Tjongina

Tjongina Adans. = ***Baeckea*** L.

Trichobasis

Trichobasis Turcz. = ***Conothamnus*** Lindl.
Trichobasis aurea Turcz. = ***Conothamnus aureus*** (Turcz.) Domin

Trichocalyx

Trichocalyx Schauer = ***Calytrix*** Labill.

Triphelia

Triphelia R.Br. ex Endl. = ***Actinodium*** Schauer ex Schltdl.
Triphelia brunioides R.Br. ex Endl. = ***Actinodium cunninghamii*** Schauer ex Lindl.

Triplarina

Triplarina Raf., Sylva Tellur.: 104 (1838).
E. Australia. 50 NSW QLD.
7 Species
Eremopyxis Baill., Adansonia 2: 328 (1862).

Triplarina bancroftii A.R.Bean, Austrobaileya 4: 360 (1995).
Queensland (Burnett). 50 QLD. Nanophan.

Triplarina calophylla A.R.Bean, Austrobaileya 4: 357 (1995).
NE. Queensland. 50 QLD. Nanophan.

Triplarina imbricata (Sm.) A.R.Bean, Austrobaileya 4: 362 (1995).
E. New South Wales. 50 NSW. Nanophan.
**Leptospermum imbricatum* Sm., Trans. Linn. Soc. London 6: 300 (1802).
Baeckea camphorata R.Br. ex Sims, Bot. Mag. 53: t. 2694 (1826). *Triplarina camphorata* (R.Br. ex Sims) Raf., Sylva Tellur.: 104 (1838). *Camphoromyrtus brownii* Schauer, Linnaea 17: 240 (1843), nom. illeg. *Eremopyxis camphorata* (R.Br. ex Sims) Baill., Adansonia 2: 329 (1862).

Triplarina nitchaga A.R.Bean, Austrobaileya 4: 357 (1995).
NE. Queensland. 50 QLD. Nanophan.

Triplarina nowraensis A.R.Bean, Austrobaileya 4: 364 (1995).
New South Wales (SC. Coast). 50 NSW. Nanophan.

Triplarina paludosa A.R.Bean, Austrobaileya 4: 358 (1995).
Queensland (Blackdown Tableland). 50 QLD. Nanophan.

Triplarina volcanica A.R.Bean, Austrobaileya 4: 361 (1995).
SE. Queensland. 50 QLD. Nanophan.

subsp. ***borealis*** A.R.Bean, Austrobaileya 4: 362 (1995).
SE. Queensland. 50 QLD. Nanophan.

subsp. ***volcanica***
SE. Queensland. 50 QLD. Nanophan.

***Synonyms*:**
Triplarina camphorata (R.Br. ex Sims) Raf. = ***Triplarina imbricata*** (Sm.) A.R.Bean

Tristania

Tristania R.Br. in W.T.Aiton, Hortus Kew. 4: 417 (1812).
E. Australia. 50 NSW.
1 Species
Callobuxus Pancher ex Brongn. & Gris, Bull. Soc. Bot. France 10: 372 (1863).

Tristania neriifolia (Sieber ex Sims) R.Br. in W.T.Aiton, Hortus Kew. 4: 417 (1812).
E. New South Wales. 50 NSW. Nanophan. or phan.
**Melaleuca neriifolia* Sieber ex Sims, Bot. Mag. 26: t. 1058 (1807).
Melaleuca salicifolia Andrews, Bot. Repos. 7: t. 485 (1807).
Tristania persicifolia A.Cunn., Field New South Wales: 350 (1825).
Tristania salicina A.Cunn., Edwards's Bot. Reg. 22: t. 1839 (1836).

***Synonyms*:**
Tristania albens A.Cunn. ex DC. = ***Syncarpia glomulifera*** (Sm.) Nied. subsp. ***glomulifera***
Tristania albicans G.Benn. = ?
Tristania anacardiifolia Ridl. = ***Kjellbergiodendron celebicum*** (Koord.) Merr.
Tristania angustifolia Hook. = ***Lysicarpus angustifolius*** (Hook.) Druce
Tristania anomala Merr. = ***Tristaniopsis anomala*** (Merr.) Peter G.Wilson & J.T.Waterh.
Tristania bakeriana Gand. = ***Tristaniopsis laurina*** (Sm.) Peter G.Wilson & J.T.Waterh.
Tristania bakhuizenii Backer = ***Tristaniopsis merguensis*** subsp. ***merguensis***
Tristania beccarii Ridl. = ***Tristaniopsis beccarii*** (Ridl.) Peter G.Wilson & J.T.Waterh.
Tristania bilocularis Stapf = ***Tristaniopsis bilocularis*** (Stapf) Peter G.Wilson & J.T.Waterh.
Tristania brownii S.Moore = ***Welchiodendron longivalve*** (F.Muell.) Peter G.Wilson & J.T.Waterh.
Tristania burmanica Griff. = ***Tristaniopsis burmanica*** (Griff.) Peter G.Wilson & J.T.Waterh.
Tristania callobuxus (Brongn. & Gris) Nied. = ***Tristaniopsis callobuxus*** Brongn. & Gris
Tristania callobuxus var. *lanceolata* Guillaumin = ***Tristaniopsis callobuxus*** Brongn. & Gris
Tristania capitellata Baill. ex Laness. = ***Tristaniopsis capitulata*** Brongn. & Gris
Tristania capitulata (Brongn. & Gris) Nied. = ***Tristaniopsis capitulata*** Brongn. & Gris
Tristania celebica Koord.-Schum. = ***Kjellbergiodendron celebicum*** (Koord.) Merr.
Tristania clementis Merr. = ***Tristaniopsis clementis*** (Merr.) Peter G.Wilson & J.T.Waterh.
Tristania conferta R.Br. = ***Lophostemon confertus*** (R.Br.) Peter G.Wilson & J.T.Waterh.
Tristania conferta Griff. = ***Lophostemon confertus*** (R.Br.) Peter G.Wilson & J.T.Waterh.
Tristania conferta var. *fibrosa* F.M.Bailey = ***Lophostemon confertus*** (R.Br.) Peter G.Wilson & J.T.Waterh.
Tristania conferta var. *microcarpa* Domin = ***Lophostemon confertus*** (R.Br.) Peter G.Wilson & J.T.Waterh.
Tristania conferta var. *typica* Domin = ***Lophostemon confertus*** (R.Br.) Peter G.Wilson & J.T.Waterh.
Tristania decorticata Merr. = ***Tristaniopsis decorticata*** (Merr.) Peter G.Wilson & J.T.Waterh.
Tristania densiflora Carr = ?
Tristania depressa A.Cunn. = ***Lophostemon confertus*** (R.Br.) Peter G.Wilson & J.T.Waterh.
Tristania depressa Link = ***Lophostemon suaveolens*** (Sol. ex Gaertn.) Peter G.Wilson & J.T.Waterh.
Tristania elliptica Stapf = ***Tristaniopsis elliptica*** (Stapf) Peter G.Wilson & J.T.Waterh.
Tristania exiliflora F.Muell. = ***Tristaniopsis exiliflora*** (F.Muell.) Peter G.Wilson & J.T.Waterh.

Tristania ferruginea C.T.White = ***Tristaniopsis ferruginea*** (C.T.White) Peter G.Wilson & J.T.Waterh.
Tristania floribunda Schltr. = ***Tristaniopsis vieillardii*** Brongn. & Gris
Tristania fruticosa Ridl. = ***Tristaniopsis fruticosa*** (Ridl.) Peter G.Wilson & J.T.Waterh.
Tristania glauca (Brongn. & Gris) Nied. = ***Tristaniopsis glauca*** Brongn. & Gris
Tristania grandiflora (Benth.) Cheel = ***Lophostemon grandiflorus*** (Benth.) Peter G.Wilson & J.T.Waterh.
Tristania grandifolia Ridl. = ***Tristaniopsis merguensis*** subsp. ***merguensis***
Tristania griffithii Kurz = ***Lophostemon confertus*** (R.Br.) Peter G.Wilson & J.T.Waterh.
Tristania guillainii (Vieill. ex Brongn. & Gris) Tison = ***Tristaniopsis guillainii*** Vieill. ex Brongn. & Gris
Tristania guillainii var. *balansana* Tison = ***Tristaniopsis guillainii*** var. ***balansana*** (Tison) J.W.Dawson
Tristania insularis Vieill. ex Brongn. & Gris = ***Tristaniopsis vieillardii*** Brongn. & Gris
Tristania lactiflua F.Muell. = ***Lophostemon lactifluus*** (F.Muell.) Peter G.Wilson & J.T.Waterh.
Tristania laurina (Sm.) R.Br. = ***Tristaniopsis laurina*** (Sm.) Peter G.Wilson & J.T.Waterh.
Tristania littoralis Merr. = ***Tristaniopsis littoralis*** (Merr.) Peter G.Wilson & J.T.Waterh.
Tristania longivalvis F.Muell. = ***Welchiodendron longivalve*** (F.Muell.) Peter G.Wilson & J.T.Waterh.
Tristania macrophylla A.Cunn. = ***Lophostemon confertus*** (R.Br.) Peter G.Wilson & J.T.Waterh.
Tristania macrosperma F.Muell. = ***Tristaniopsis macrosperma*** (F.Muell.) Peter G.Wilson & J.T. Waterh.
Tristania maingayi Duthie = ***Tristaniopsis merguensis*** subsp. ***merguensis***
Tristania merguensis Griff. = ***Tristaniopsis merguensis*** (Griff.) Peter G.Wilson & J.T.Waterh.
Tristania micrantha Merr. = ***Tristaniopsis micrantha*** (Merr.) Peter G.Wilson & J.T.Waterh.
Tristania microphylla Quisumb. & Merr. = ***Kania microphylla*** (Quisumb. & Merr.) Peter G.Wilson
Tristania motleyi Ridl. = ***Tristaniopsis whiteana*** subsp. ***whiteana***
Tristania moultoniana W.W.Sm. = ***Whiteodendron moultonianum*** (W.W.Sm.) Steenis
Tristania oblongifolia Merr. = ***Tristaniopsis oblongifolia*** (Merr.) Peter G.Wilson & J.T.Waterh.
Tristania obovata Benn. = ***Tristaniopsis obovata*** (Benn.) Peter G.Wilson & J.T.Waterh.
Tristania odorata C.T.White & W.D.Francis = ***Ristantia pachysperma*** (F.Muell. & F.M.Bailey) Peter G.Wilson & J.T.Waterh.
Tristania oreophila Diels = ***Tristaniopsis oreophila*** (Diels) Peter G.Wilson & J.T.Waterh.
Tristania pachysperma (F.Muell. & F.M.Bailey) W.D.Francis = ***Ristantia pachysperma*** (F.Muell. & F.M.Bailey) Peter G.Wilson & J.T.Waterh.
Tristania pentandra Merr. = ***Tristaniopsis pentandra*** (Merr.) Peter G.Wilson & J.T.Waterh.
Tristania persicifolia A.Cunn. = ***Tristania neriifolia*** (Sieber ex Sims) R.Br.
Tristania polyandra Guillaumin = ***Tristaniopsis polyandra*** (Guillaumin) Peter G.Wilson & J.T.Waterh.
Tristania pontianensis M.R.Hend. = ***Tristaniopsis pontianensis*** (M.R.Hend.) Peter G.Wilson & J.T.Waterh.
Tristania psidioides A.Cunn. ex Lindl. = ***Xanthostemon psidioides*** (A.Cunn. ex Lindl.) Peter G.Wilson & J.T.Waterh.
Tristania razakiana Kochummen = ***Tristaniopsis razakiana*** (Kochummen) Peter G.Wilson & J.T.Waterh.
Tristania rhytiphloia F.Muell. = ***Lophostemon suaveolens*** (Sol. ex Gaertn.) Peter G.Wilson & J.T.Waterh.
Tristania rufescens Hance = ***Tristaniopsis burmanica*** var. ***rufescens*** (Hance) J.Parn. & NicLugh.
Tristania salicifolia Link ex Steud. = ***Lophostemon suaveolens*** (Sol. ex Gaertn.) Peter G.Wilson & J.T.Waterh.
Tristania salicina A.Cunn. = ***Tristania neriifolia*** (Sieber ex Sims) R.Br.
Tristania spathulata Ridl. = ***Tristaniopsis obovata*** (Benn.) Peter G.Wilson & J.T.Waterh.
Tristania stellata Ridl. = ***Tristaniopsis merguensis*** subsp. ***merguensis***
Tristania suaveolens (Sol. ex Gaertn.) Sm. = ***Lophostemon suaveolens*** (Sol. ex Gaertn.) Peter G.Wilson & J.T.Waterh.
Tristania suaveolens var. *glabrescens* F.M.Bailey = ***Lophostemon suaveolens*** (Sol. ex Gaertn.) Peter G.Wilson & J.T.Waterh.
Tristania suaveolens var. *grandiflora* Benth. = ***Lophostemon grandiflorus*** (Benth.) Peter G.Wilson & J.T.Waterh.
Tristania suaveolens var. *riparia* Domin = ***Lophostemon grandiflorus*** subsp. ***riparius*** (Domin) Peter G.Wilson & J.T.Waterh.
Tristania suaveolens var. *vulgaris* Domin = ***Lophostemon suaveolens*** (Sol. ex Gaertn.) Peter G.Wilson & J.T.Waterh.
Tristania subauriculata King = ***Tristaniopsis merguensis*** subsp. ***merguensis***
Tristania subverticillata H.Wendl. = ***Lophostemon confertus*** (R.Br.) Peter G.Wilson & J.T.Waterh.
Tristania sumatrana Miq. = ***Tristaniopsis whiteana*** subsp. ***whiteana***
Tristania umbrosa A.Cunn. ex Lindl. = ***Xanthostemon umbrosus*** (A.Cunn. ex Lindl.) Peter G.Wilson & J.T.Waterh.
Tristania undulata Pancher ex Guillaumin = ***Tristaniopsis capitulata*** Brongn. & Gris
Tristania vieillardii (Brongn. & Gris) Nied. = ***Tristaniopsis vieillardii*** Brongn. & Gris
Tristania vieillardii var. *grandiflora* Bonati & Petitm. = ***Tristaniopsis polyandra*** (Guillaumin) Peter G.Wilson & J.T.Waterh.
Tristania vitiensis A.C.Sm. = ***Metrosideros collina*** var. ***fruticosa*** J.W.Moore
Tristania whitiana Griff. = ***Tristaniopsis whiteana*** (Griff.) Peter G.Wilson & J.T.Waterh.
Tristania wightiana Duthie = ***Tristaniopsis whiteana*** subsp. ***whiteana***

***Unplaced Names*:**
Tristania albicans G.Benn., Gatherings Natural. Austral.: 285 (1860). = ?
Tristania densiflora Carr, Rev. Hort. 1881: 420 (1881). = ?

Tristaniopsis

Tristaniopsis Brongn. & Gris, Bull. Soc. Bot. France 10: 371 (1863).
Indo-China to SW. Pacific. 41 CBD MYA THA 42 BOR JAW MLY PHI SUM 43 NWG 50 NSW QLD VIC 60 NWC.
40 Species

Tristaniopsis anomala (Merr.) Peter G.Wilson & J.T.Waterh., Austral. J. Bot. 30: 439 (1982).

Borneo. 42 BOR. Phan.
Tristania anomala Merr., J. Straits Branch Roy. Asiat. Soc. 77: 227 (1917).

Tristaniopsis beccarii (Ridl.) Peter G.Wilson & J.T.Waterh., Austral. J. Bot. 30: 439 (1982).
Borneo. 42 BOR. Phan.
**Tristania beccarii* Ridl., J. Bot. 68: 37 (1930).

Tristaniopsis bilocularis (Stapf) Peter G.Wilson & J.T.Waterh., Austral. J. Bot. 30: 439 (1982).
Borneo. 42 BOR. Phan.
**Tristania bilocularis* Stapf, Trans. Linn. Soc. London, Bot. 4: 152 (1894).

Tristaniopsis burmanica (Griff.) Peter G.Wilson & J.T.Waterh., Austral. J. Bot. 30: 439 (1982).
Indo-China. 41 CBD MYA THA. Nanophan. or phan.
**Tristania burmanica* Griff., Pl. Cantor: 17 (1837).

var. ***burmanica***
Indo-China. 41 MYA THA. Nanophan. or phan.

var. ***rufescens*** (Hance) J.Parn. & NicLugh., Kew Bull. 47: 705 (1992).
Indo-China. 41 CBD THA. Nanophan. or phan.
**Tristania rufescens* Hance, J. Bot. 14: 259 (1876). *Tristaniopsis rufescens* (Hance) Peter G.Wilson & J.T.Waterh., Austral. J. Bot. 30: 440 (1982).

Tristaniopsis callobuxus Brongn. & Gris, Bull. Soc. Bot. France 10: 372 (1863). *Tristania callobuxus* (Brongn. & Gris) Nied. in H.G.A.Engler & K.A.E.Prantl, Nat. Pflanzenfam. 3(7): 89 (1893).
New Caledonia. 60 NWC. Nanophan. or phan.
Tristania callobuxus var. *lanceolata* Guillaumin, Mém. Mus. Natl. Hist. Nat., B, Bot. 8: 283 (1962).

Tristaniopsis capitulata Brongn. & Gris, Bull. Soc. Bot. France 10: 372 (1863). *Tristania capitulata* (Brongn. & Gris) Nied. in H.G.A.Engler & K.A.E.Prantl, Nat. Pflanzenfam. 3(7): 89 (1893).
C. & SE. New Caledonia. 60 NWC. Phan.
Tristania capitellata Baill. ex Laness., Pl. Util. Col. Franç.: 258 (1886), orth. var.
Tristania undulata Pancher ex Guillaumin, Ann. Inst. Bot.-Géol. Colon. Marseille, II, 9: 146 (1911), nom. nud.

Tristaniopsis clementis (Merr.) Peter G.Wilson & J.T.Waterh., Austral. J. Bot. 30: 439 (1982).
Borneo. 42 BOR. Phan.
**Tristania clementis* Merr., J. Straits Branch Roy. Asiat. Soc. 77: 229 (1917).

Tristaniopsis collina Peter G.Wilson & J.T.Waterh., Austral. J. Bot. 30: 436 (1982).
SE. Queensland to New South Wales. 50 NSW QLD. Phan.

Tristaniopsis decorticata (Merr.) Peter G.Wilson & J.T.Waterh., Austral. J. Bot. 30: 439 (1982).
Philippines. 42 PHI. Phan.
**Tristania decorticata* Merr., Publ. Bur. Sci. Gov. Lab. 35: 51 (1905).

Tristaniopsis elliptica (Stapf) Peter G.Wilson & J.T.Waterh., Austral. J. Bot. 30: 439 (1982).
Borneo. 42 BOR. Phan.
**Tristania elliptica* Stapf, Trans. Linn. Soc. London, Bot. 4: 151 (1894).

Tristaniopsis exiliflora (F.Muell.) Peter G.Wilson & J.T.Waterh., Austral. J. Bot. 30: 434 (1982).
N. & NE. Queensland. 50 QLD. Phan.
**Tristania exiliflora* F.Muell., Fragm. 5: 11 (1865).

Tristaniopsis ferruginea (C.T.White) Peter G.Wilson & J.T.Waterh., Austral. J. Bot. 30: 439 (1983).
New Guinea. 43 NWG. Phan.
**Tristania ferruginea* C.T.White, J. Arnold Arbor. 23: 83 (1942).

Tristaniopsis fruticosa (Ridl.) Peter G.Wilson & J.T.Waterh., Austral. J. Bot. 30: 439 (1982).
Pen. Malaysia. 42 MLY. Phan.
**Tristania fruticosa* Ridl., J. Fed. Malay States Mus. 6: 147 (1915).

Tristaniopsis glauca Brongn. & Gris, Bull. Soc. Bot. France 13: 471 (1866). *Tristania glauca* (Brongn. & Gris) Nied. in H.G.A.Engler & K.A.E.Prantl, Nat. Pflanzenfam. 33(7): 89 (1893).
SE. New Caledonia. 60 NWC. Nanophan. or phan.

Tristaniopsis guillainii Vieill. ex Brongn. & Gris, Ann. Sci. Nat., Bot., V, 13: 348 (1871). *Tristania guillainii* (Vieill. ex Brongn. & Gris) Tison, Rech. Placent. Myrtac.: 55 (1876).
New Caledonia. 60 NWC. Nanophan. or phan.

var. ***balansana*** (Tison) J.W.Dawson, Bull. Mus. Natl. Hist. Nat., B, Adansonia 1985: 180 (1985).
SE. New Caledonia. 60 NWC. Nanophan. or phan.
**Tristania guillainii* var. *balansana* Tison, Rech. Placent. Myrtac.: 55 (1876).

var. ***guillainii***
New Caledonia. 60 NWC. Nanophan. or phan.

Tristaniopsis jaffrei J.W.Dawson, Bull. Mus. Natl. Hist. Nat., B, Adansonia 1985: 178 (1985).
NW. New Caledonia. 60 NWC. Nanophan.

Tristaniopsis kinabaluensis P.S.Ashton, Gard. Bull. Singapore 57: 269 (2005).
Borneo (Sabah). 42 BOR. Phan.

subsp. ***kinabaluensis***
Borneo (Sabah). 42 BOR. Phan.

subsp. ***silamensis*** P.S.Ashton, Gard. Bull. Singapore 57: 271 (2005).
Borneo (Sabah). 42 BOR. Phan.

Tristaniopsis laurina (Sm.) Peter G.Wilson & J.T.Waterh., Austral. J. Bot. 30: 435 (1982).
Queensland to Victoria. 50 NSW QLD VIC. Phan.
**Melaleuca laurina* Sm., Trans. Linn. Soc. London 3: 275 (1797). *Tristania laurina* (Sm.) R.Br. in W.T.Aiton, Hortus Kew. 4: 417 (1812).
Tristania bakeriana Gand., Bull. Soc. Bot. France 65: 27 (1918).

Tristaniopsis littoralis (Merr.) Peter G.Wilson & J.T.Waterh., Austral. J. Bot. 30: 439 (1982).
Philippines. 42 PHI. Phan.
**Tristania littoralis* Merr., Philipp. J. Sci., C 7: 317 (1912).

Tristaniopsis lucida J.W.Dawson, Bull. Mus. Natl. Hist. Nat., B, Adansonia 1985: 186 (1985).
SE. New Caledonia. 60 NWC. Phan.

Tristaniopsis macphersonii J.W.Dawson, Bull. Mus. Natl. Hist. Nat., B, Adansonia 1985: 182 (1985).
C. & SE. New Caledonia. 60 NWC. Nanophan. or phan.

Tristaniopsis macrosperma (F.Muell.) Peter G.Wilson & J.T.Waterh., Austral. J. Bot. 30: 439 (1982).

New Guinea. 43 NWG. Phan.
**Tristania macrosperma* F.Muell., Descr. Notes Papuan Pl. 1: 104 (1877).

Tristaniopsis merguensis (Griff.) Peter G.Wilson & J.T.Waterh., Austral. J. Bot. 30: 439 (1982).
S. Myanmar to W. Malesia. 41 MYA 42 BOR JAW MLY SUM. Phan.
**Tristania merguensis* Griff., Pl. Cantor: 18 (1837).

subsp. ***merguensis***
S. Myanmar to W. Malesia. 41 MYA 42 BOR JAW MLY SUM. Phan.
Tristania maingayi Duthie in J.D.Hooker, Fl. Brit. India 2: 467 (1878).
Tristania subauriculata King, J. Asiat. Soc. Bengal, Pt. 2, Nat. Hist. 70: 72 (1901).
Tristania grandifolia Ridl., J. Bot. 68: 38 (1930). *Tristaniopsis grandifolia* (Ridl.) Peter G.Wilson & J.T.Waterh., Austral. J. Bot. 30: 439 (1982).
Tristania stellata Ridl., J. Bot. 68: 38 (1930). *Tristaniopsis stellata* (Ridl.) Peter G.Wilson & J.T.Waterh., Austral. J. Bot. 30: 440 (1982).
Tristania bakhuizenii Backer, Blumea 5: 502 (1945).

subsp. ***tavaiensis*** P.S.Ashton, Gard. Bull. Singapore 57: 273 (2005).
Borneo (Sabah). 42 BOR. Phan.

Tristaniopsis micrantha (Merr.) Peter G.Wilson & J.T.Waterh., Austral. J. Bot. 30: 439 (1982).
Philippines. 42 PHI. Phan.
**Tristania micrantha* Merr., Philipp. J. Sci., C 12: 288 (1917).

Tristaniopsis microcarpa P.S.Ashton, Gard. Bull. Singapore 57: 273 (2005).
Borneo. 42 BOR. Phan.

Tristaniopsis minutiflora J.W.Dawson, Bull. Mus. Natl. Hist. Nat., B, Adansonia 1985: 188 (1985).
NW. New Caledonia. 60 NWC. Phan.

Tristaniopsis ninndoensis J.W.Dawson, Bull. Mus. Natl. Hist. Nat., B, Adansonia 1985: 192 (1985).
NW. New Caledonia. 60 NWC. Nanophan.

Tristaniopsis oblongifolia (Merr.) Peter G.Wilson & J.T.Waterh., Austral. J. Bot. 30: 439 (1982).
Philippines. 42 PHI. Phan.
**Tristania oblongifolia* Merr., Philipp. J. Sci. 14: 430 (1919).

Tristaniopsis obovata (Benn.) Peter G.Wilson & J.T.Waterh., Austral. J. Bot. 30: 439 (1982).
W. Malesia. 42 BOR MLY SUM. Phan.
**Tristania obovata* Benn., Pl. Jav. Rar.: 127 (1840).
Tristania spathulata Ridl., J. Straits Branch Roy. Asiat. Soc. 82: 184 (1920).

Tristaniopsis oreophila (Diels) Peter G.Wilson & J.T.Waterh., Austral. J. Bot. 30: 439 (1982).
New Guinea. 43 NWG. Phan.
**Tristania oreophila* Diels, Bot. Jahrb. Syst. 57: 421 (1922).

Tristaniopsis parvifolia A.J.Scott, Kew Bull. 39: 660 (1984).
W. New Guinea. 43 NWG. Phan.

Tristaniopsis pentandra (Merr.) Peter G.Wilson & J.T.Waterh., Austral. J. Bot. 30: 439 (1982).
Borneo. 42 BOR. Phan.
**Tristania pentandra* Merr., J. Straits Branch Roy. Asiat. Soc. 77: 228 (1917).

Tristaniopsis polyandra (Guillaumin) Peter G.Wilson & J.T.Waterh., Austral. J. Bot. 30: 440 (1982).
SE. New Caledonia. 60 NWC. Nanophan. or phan.
Tristania vieillardii var. *grandiflora* Bonati & Petitm., Bull. Herb. Boissier, II, 7: 652 (1907).
**Tristania polyandra* Guillaumin, Bull. Soc. Bot. France 81: 5 (1934).

Tristaniopsis pontianensis (M.R.Hend.) Peter G.Wilson & J.T.Waterh., Austral. J. Bot. 30: 440 (1982).
Pen. Malaysia. 42 MLY. Phan.
**Tristania pontianensis* M.R.Hend., Gard. Bull. Singapore 14: 1 (1953).

Tristaniopsis razakiana (Kochummen) Peter G.Wilson & J.T.Waterh., Austral. J. Bot. 30: 440 (1982).
Pen. Malaysia. 42 MLY. Phan.
**Tristania razakiana* Kochummen, Gard. Bull. Singapore 28: 227 (1976).

Tristaniopsis reticulata J.W.Dawson, Bull. Mus. Natl. Hist. Nat., B, Adansonia 1985: 184 (1985).
SE. New Caledonia. 60 NWC. Phan.

Tristaniopsis rubiginosa Teo ex P.S.Ashton, Gard. Bull. Singapore 57: 274 (2005).
Borneo. 42 BOR. Phan.

Tristaniopsis vieillardii Brongn. & Gris, Bull. Soc. Bot. France 12: 300 (1865). *Tristania vieillardii* (Brongn. & Gris) Nied. in H.G.A.Engler & K.A.E.Prantl, Nat. Pflanzenfam. 3(7): 89 (1893).
NC. & SE. New Caledonia. 60 NWC. Phan.
Tristania insularis Vieill. ex Brongn. & Gris, Bull. Soc. Bot. France 12: 300 (1865), nom. inval.
Tristania floribunda Schltr., Bot. Jahrb. Syst. 40(92): 31 (1908).

Tristaniopsis whiteana (Griff.) Peter G.Wilson & J.T.Waterh., Austral. J. Bot. 30: 440 (1982).
W. Malesia. 42 BOR MLY SUM. Phan.
**Tristania whitiana* Griff., Pl. Cantor: 18 (1837).

subsp. ***monostemon*** P.S.Ashton, Gard. Bull. Singapore 57: 277 (2005).
Borneo. 42 BOR. Phan.

subsp. ***whiteana***
W. Malesia. 42 BOR MLY SUM. Phan.
Tristania sumatrana Miq., Fl. Ned. Ind., Eerste Bijv.: 308 (1861).
Tristania wightiana Duthie in J.D.Hooker, Fl. Brit. India 2: 466 (1878), orth. var.
Tristania motleyi Ridl., J. Bot. 68: 37 (1930).

Tristaniopsis yateensis J.W.Dawson, Bull. Mus. Natl. Hist. Nat., B, Adansonia 1985: 190 (1985).
SE. New Caledonia. 60 NWC. Nanophan.

Synonyms:
Tristaniopsis grandifolia (Ridl.) Peter G.Wilson & J.T.Waterh. = ***Tristaniopsis merguensis*** subsp. ***merguensis***
Tristaniopsis rufescens (Hance) Peter G.Wilson & J.T.Waterh. = ***Tristaniopsis burmanica*** var. ***rufescens*** (Hance) J.Parn. & NicLugh.
Tristaniopsis stellata (Ridl.) Peter G.Wilson & J.T.Waterh. = ***Tristaniopsis merguensis*** subsp. ***merguensis***

Tryptomene

Tryptomene Walp. = ***Thryptomene*** Endl.

Ugni

Ugni Turcz., Bull. Soc. Imp. Naturalistes Moscou 21(1): 579 (1848).
S. Mexico to Chile. (51) nzn 79 MXS 80 COS ELS GUA HON NIC PAN 82 GUY VEN 83 CLM ECU PER 85 AGS CLC CLS JNF.
4 Species

Ugni candollei (Barnéoud) O.Berg, Linnaea 27: 388 (1856).
SC. Chile. 85 CLC CLS. Cham. or nanophan.
**Myrtus candollei* Barnéoud in C.Gay, Fl. Chil. 2: 382 (1847).
Myrtus krausei Phil., Linnaea 28: 638 (1857). *Ugni krausei* (Phil.) Phil. ex O.Berg, Bot. Zeitung (Berlin) 16: 250 (1858).
Ugni candollei f. *litoralis* Kausel, Lilloa 17: 54 (1949).
Ugni candollei f. *monticola* Kausel, Lilloa 17: 54 (1949).

Ugni molinae Turcz., Bull. Soc. Imp. Naturalistes Moscou 21(1): 579 (1848). *Myrtus molinae* (Turcz.) Barnéoud in C.Gay, Fl. Chil. 2: 381 (1847).
C. & S. Chile to SW. Argentina. (51) nzn 85 AGS CLC CLS jnf. Cham. or nanophan. Widely cultivated for its edible fruits (strawberry myrtle).
**Myrtus ugni* Molina, Sag. Stor. Nat. Chili: 161 (1782). *Eugenia ugni* (Molina) Hook. & Arn., Bot. Misc. 3: 318 (1833). *Ugni ugni* (Molina) Voss, Vilm. Blumengärtn. ed. 3, 1: 315 (1894), nom. inval. *Ugni myrtus* Macloskie, Torreya 5: 198 (1905), nom. illeg.
Ugni bridgesii O.Berg, Linnaea 27: 389 (1856).
Ugni philippii O.Berg, Linnaea 27: 387 (1856).
Ugni poeppigii O.Berg, Linnaea 27: 386 (1856). *Myrtus ugni* var. *angustifolia* Phil., Anales Univ. Chile 98: 698 (1897).
Ugni lanceolata O.Berg, Bot. Zeitung (Berlin) 15: 857 (1857).
Myrtus ugni var. *latifolia* Kuntze, Revis. Gen. Pl. 3(2): 93 (1898).

Ugni myricoides (Kunth) O.Berg, Linnaea 27: 391 (1856).
S. Mexico to W. & N. South America. 79 MXS 80 COS ELS GUA HON NIC PAN 82 GUY VEN 83 CLM ECU PER. Nanophan. or phan.
**Myrtus myricoides* Kunth in F.W.H.von Humboldt, A.J.A.Bonpland & C.S.Kunth, Nov. Gen. Sp. 6: 131 (1823).
Myrtus montana Benth., Pl. Hartw.: 61 (1840). *Ugni montana* (Benth.) O.Berg, Linnaea 27: 392 (1856).
Ugni friedrichsthalii O.Berg, Linnaea 27: 388 (1856). *Eugenia friedrichsthalii* (O.Berg) Hemsl., Biol. Cent.-Amer., Bot. 1: 410 (1880). *Myrtus friedrichsthalii* (O.Berg) Donn.Sm. & Standl., Contr. U. S. Natl. Herb. 23: 1039 (1924).
Ugni friedrichsthalii var. *brevipes* O.Berg, Linnaea 27: 389 (1856).
Ugni friedrichsthalii var. *longipes* O.Berg, Linnaea 27: 389 (1856). *Ugni myricoides* var. *longipes* (O.Berg) McVaugh, Mem. New York Bot. Gard. 18(2): 264 (1969).
Ugni oerstedii O.Berg, Linnaea 27: 389 (1856). *Myrtus oerstedii* (O.Berg) Hemsl., Biol. Cent.-Amer., Bot. 1: 407 (1880). *Ugni myricoides* var. *oerstedii* (O.Berg) McVaugh, Mem. New York Bot. Gard. 18(2): 265 (1969).
Ugni warscewiczii O.Berg, Linnaea 27: 390 (1856). *Eugenia warszewiczii* (O.Berg) Hemsl., Biol. Cent.-Amer., Bot. 1: 411 (1880).
Myrtus stenophylla Oliv. ex Thurn, Timehri 2(5): 273 (1886). *Ugni stenophylla* (Oliv. ex Thurn) Burret, Notizbl. Bot. Gart. Berlin-Dahlem 15: 507 (1941). *Myrtus myricoides* var. *stenophylla* (Oliv. ex Thurn) Steyerm., Fieldiana, Bot. 28: 1022 (1957). *Ugni myricoides* f. *stenophylla* (Oliv. ex Thurn) McVaugh, Mem. New York Bot. Gard. 18(2): 266 (1969).
Myrtus friedrichsthalii var. *brevipes* Donn.Sm., Enum. Pl. Guatem. 3: 25 (1893).
Myrtus roraimensis N.E.Br., Trans. Linn. Soc. London, Bot. 6: 26 (1901). *Ugni roraimensis* (N.E.Br.) Burret, Notizbl. Bot. Gart. Berlin-Dahlem 15: 507 (1941). *Myrtus myricoides* var. *roraimensis* (N.E.Br.) Steyerm., Fieldiana, Bot. 28: 1022 (1957). *Ugni myricoides* var. *roraimensis* (N.E.Br.) McVaugh, Mem. New York Bot. Gard. 18(2): 265 (1969).
Myrtus matudai Lundell, Phytologia 1: 246 (1937).
Ugni angustifolia Burret, Notizbl. Bot. Gart. Berlin-Dahlem 15: 507 (1941).
Ugni vaccinioides Burret, Notizbl. Bot. Gart. Berlin-Dahlem 15: 506 (1941).
Myrtus myricoides var. *turumiquirensis* Steyerm., Fieldiana, Bot. 28: 1022 (1957).
Ugni disterigmoides Ant.Molina, Ceiba 11(1): 69 (1965).
Ugni myricoides f. *bifaria* McVaugh, Mem. New York Bot. Gard. 18(2): 267 (1969).
Ugni myricoides f. *grossa* McVaugh, Mem. New York Bot. Gard. 18(2): 266 (1969).
Ugni myricoides f. *oligandra* McVaugh, Mem. New York Bot. Gard. 18(2): 266 (1969).

Ugni selkirkii (Hook. & Arn.) O.Berg, Linnaea 27: 392 (1856).
Juan Fernández Is. (I. Robinson Crusoe). 85 JNF. Cham.
**Eugenia selkirkii* Hook. & Arn., Bot. Misc. 3: 318 (1833). *Myrtus selkirkii* (Hook. & Arn.) Hemsl., Rep. Challenger, Bot. 1(3): 36 (1884).
Myrtus berteroi Phil., Bot. Zeitung (Berlin) 14: 644 (1856). *Ugni berteroi* (Phil.) Phil.f., Cat. Pl. Vasc. Chil.: 79 (1881).

Synonyms:
Ugni angustifolia Burret = ***Ugni myricoides*** (Kunth) O.Berg
Ugni berteroi (Phil.) Phil.f. = ***Ugni selkirkii*** (Hook. & Arn.) O.Berg
Ugni bridgesii O.Berg = ***Ugni molinae*** Turcz.
Ugni candollei f. *litoralis* Kausel = ***Ugni candollei*** (Barnéoud) O.Berg
Ugni candollei f. *monticola* Kausel = ***Ugni candollei*** (Barnéoud) O.Berg
Ugni disterigmoides Ant.Molina = ***Ugni myricoides*** (Kunth) O.Berg
Ugni friedrichsthalii O.Berg = ***Ugni myricoides*** (Kunth) O.Berg
Ugni friedrichsthalii var. *brevipes* O.Berg = ***Ugni myricoides*** (Kunth) O.Berg
Ugni friedrichsthalii var. *longipes* O.Berg = ***Ugni myricoides*** (Kunth) O.Berg
Ugni krausei (Phil.) Phil. ex O.Berg = ***Ugni candollei*** (Barnéoud) O.Berg
Ugni lanceolata O.Berg = ***Ugni molinae*** Turcz.
Ugni montana (Benth.) O.Berg = ***Ugni myricoides*** (Kunth) O.Berg

Ugni myricoides f. *bifaria* McVaugh = ***Ugni myricoides*** (Kunth) O.Berg
Ugni myricoides f. *grossa* McVaugh = ***Ugni myricoides*** (Kunth) O.Berg
Ugni myricoides var. *longipes* (O.Berg) McVaugh = ***Ugni myricoides*** (Kunth) O.Berg
Ugni myricoides var. *oerstedii* (O.Berg) McVaugh = ***Ugni myricoides*** (Kunth) O.Berg
Ugni myricoides f. *oligandra* McVaugh = ***Ugni myricoides*** (Kunth) O.Berg
Ugni myricoides var. *roraimensis* (N.E.Br.) McVaugh = ***Ugni myricoides*** (Kunth) O.Berg
Ugni myricoides f. *stenophylla* (Oliv. ex Thurn) McVaugh = ***Ugni myricoides*** (Kunth) O.Berg
Ugni myrtus Macloskie = ***Ugni molinae*** Turcz.
Ugni oerstedii O.Berg = ***Ugni myricoides*** (Kunth) O.Berg
Ugni philippii O.Berg = ***Ugni molinae*** Turcz.
Ugni poeppigii O.Berg = ***Ugni molinae*** Turcz.
Ugni roraimensis (N.E.Br.) Burret = ***Ugni myricoides*** (Kunth) O.Berg
Ugni stenophylla (Oliv. ex Thurn) Burret = ***Ugni myricoides*** (Kunth) O.Berg
Ugni ugni (Molina) Voss = ***Ugni molinae*** Turcz.
Ugni vaccinioides Burret = ***Ugni myricoides*** (Kunth) O.Berg
Ugni warscewiczii O.Berg = ***Ugni myricoides*** (Kunth) O.Berg

Uromyrtus

Uromyrtus Burret, Notizbl. Bot. Gart. Berlin-Dahlem 15: 490 (1941).
Borneo, New Guinea to E. Australia and New Caledonia. 42 BOR 43 NWG 50 NSW QLD 60 NWC.
22 Species

Uromyrtus archboldiana (Merr. & L.M.Perry) A.J.Scott, Kew Bull. 33: 511 (1979).
New Guinea. 43 NWG. Nanophan. or phan.
**Myrtus archboldiana* Merr. & L.M.Perry, J. Arnold Arbor. 23: 239 (1942).

Uromyrtus artensis (Montrouz.) Burret, Notizbl. Bot. Gart. Berlin-Dahlem 15: 490 (1941).
New Caledonia. 60 NWC. Nanophan. or phan.
Myrtus artensis (Montrouz.) Guillaumin & Beauvis., Ann. Soc. Bot. Lyon 38: 91 (1913 publ. 1914).

Uromyrtus australis A.J.Scott, Kew Bull. 33: 512 (1979).
NE. New South Wales. 50 NSW. Nanophan. or phan.

Uromyrtus baumanii (Guillaumin) N.Snow & Guymer, Syst. Bot. 26: 742 (2001).
New Caledonia. 60 NWC.
**Myrtus baumanii* Guillaumin, Mém. Mus. Natl. Hist. Nat., B, Bot. 8: 287 (1962).

Uromyrtus billardierei (Seem.) A.J.Scott, Kew Bull. 33: 512 (1979).
New Caledonia. 60 NWC. Nanophan. or phan.
**Nelitris billardierei* Seem., Fl. Vit.: 81 (1866).

Uromyrtus brassii (Merr. & L.M.Perry) A.J.Scott, Kew Bull. 33: 511 (1979).
New Guinea. 43 NWG. Nanophan. or phan.
**Myrtus brassii* Merr. & L.M.Perry, J. Arnold Arbor. 23: 239 (1942).

Uromyrtus curvipes (Gand.) Burret, Notizbl. Bot. Gart. Berlin-Dahlem 15: 491 (1941).
New Caledonia. 60 NWC. Nanophan. or phan.
**Myrtus curvipes* Gand., Bull. Soc. Bot. France 65: 26 (1918).

Uromyrtus emarginata (Pancher ex Baker f.) Burret, Notizbl. Bot. Gart. Berlin-Dahlem 15: 492 (1941).
New Caledonia. 60 NWC. Nanophan. or phan.
Myrtus emarginata Pancher ex Brongn. & Gris, Bull. Soc. Bot. France 12: 177 (1865), nom. illeg. **Rhodomyrtus emarginata* Pancher ex Baker f., J. Linn. Soc., Bot. 45: 311 (1921).

Uromyrtus gomonenensis (Guillaumin) Burret, Notizbl. Bot. Gart. Berlin-Dahlem 15: 493 (1941).
New Caledonia. 60 NWC. Nanophan. or phan.
**Myrtus gomonenensis* Guillaumin, Bull. Soc. Bot. France 85: 631 (1938 publ. 1939). *Austromyrtus gomonenensis* (Guillaumin) Burret, Notizbl. Bot. Gart. Berlin-Dahlem 15: 493 (1941).

Uromyrtus lamingtonensis N.Snow & Guymer, Syst. Bot. 26: 734 (2001).
E. Australia. 50 NSW QLD.

Uromyrtus metrosideros (F.M.Bailey) A.J.Scott, Kew Bull. 41: 286 (1986).
N. Queensland. 50 QLD. Nanophan. or phan.
**Myrtus metrosideros* F.M.Bailey, Syn. Queensl. Fl., Suppl. 3: 27 (1890). *Austromyrtus metrosideros* (F.M.Bailey) Burret, Notizbl. Bot. Gart. Berlin-Dahlem 15: 501 (1941).

Uromyrtus nekouana (Guillaumin) Burret, Notizbl. Bot. Gart. Berlin-Dahlem 15: 493 (1941).
New Caledonia. 60 NWC. Nanophan. or phan.
**Myrtus nekouana* Guillaumin, Bull. Soc. Bot. France 85: 632 (1938 publ. 1939).

Uromyrtus neomyrtoides Burret, Notizbl. Bot. Gart. Berlin-Dahlem 15: 491 (1941).
New Caledonia. 60 NWC. Nanophan. or phan.
**Eugenia myrtoides* Brongn. & Gris, Bull. Soc. Bot. France 12: 180 (1865), nom. illeg.

Uromyrtus ngoyensis (Schltr.) Burret, Notizbl. Bot. Gart. Berlin-Dahlem 15: 491 (1941).
New Caledonia. 60 NWC. Nanophan. or phan.
**Myrtus ngoyensis* Schltr., Bot. Jahrb. Syst. 39: 202 (1906).

Uromyrtus novoguineensis A.J.Scott, Kew Bull. 33: 512 (1979).
Papua New Guinea. 43 NWG. Nanophan. or phan.

Uromyrtus paulotchensis (Guillaumin) Burret, Notizbl. Bot. Gart. Berlin-Dahlem 15: 493 (1941).
New Caledonia. 60 NWC. Nanophan. or phan.
**Myrtus paulotchensis* Guillaumin, Bull. Soc. Bot. France 85: 633 (1938 publ. 1939).

Uromyrtus rostrata (Lauterb.) N.Snow & Guymer, Austrobaileya 5: 181 (1999).
New Guinea. 43 NWG. Nanophan. or phan.
**Myrtella rostrata* Lauterb., Nova Guinea 8: 855 (1912).

Uromyrtus sarawakensis A.J.Scott, Kew Bull. 33: 513 (1979).
Borneo (Sarawak). 42 BOR. Nanophan. or phan.

Uromyrtus sunshinensis (Guillaumin) N.Snow & Guymer, Syst. Bot. 26: 742 (2001).
New Caledonia. 60 NWC. Nanophan.
**Myrtus sunshinensis* Guillaumin, Mém. Mus. Natl. Hist. Nat., B, Bot. 8: 290 (1962).

Uromyrtus supraaxillaris (Guillaumin) Burret, Notizbl. Bot. Gart. Berlin-Dahlem 15: 492 (1941).
New Caledonia. 60 NWC. Nanophan. or phan.
**Myrtus supraaxillaris* Guillaumin, Bull. Mus. Natl. Hist. Nat. 26: 177 (1920).

Uromyrtus tenella N.Snow & Guymer, Syst. Bot. 26: 741 (2001).
Queensland. 50 QLD.

Uromyrtus thymifolia (Planch. ex Guillaumin) Burret, Notizbl. Bot. Gart. Berlin-Dahlem 15: 492 (1941).
New Caledonia. 60 NWC. Nanophan. or phan.
**Rhodomyrtus thymifolia* Planch. ex Guillaumin, Bull. Mus. Natl. Hist. Nat. 17: 458 (1911).

Verticordia

Verticordia DC., Prodr. 3: 208 (1828), nom. cons.
Australia. 50 NTA QLD WAU.
100 Species
Diplachne R.Br. ex. Desf., Mém. Mus. Hist. Nat. 5: 272 (1819), nom. illeg.
Chrysorhoe Lindl., Companion Bot. Mag. 2: 357 (1836).

Verticordia acerosa Lindl., Sketch Veg. Swan R.: vi (1839).
SW. Australia. 50 WAU. Tuber nanophan.

var. ***acerosa***
SW. Australia. 50 WAU. Tuber nanophan.

var. ***preissii*** (Schauer) A.S.George, Nuytsia 7: 284 (1991).
SW. Australia. 50 WAU. Tuber nanophan.
**Verticordia preissii* Schauer in J.G.C.Lehmann, Pl. Preiss. 1: 101 (1844).

Verticordia aereiflora Eliz.A.George & A.S.George, Nuytsia 9: 333 (1994).
WSW. Western Australia. 50 WAU. Tuber cham. or nanophan.

Verticordia albida A.S.George, Nuytsia 7: 285 (1991).
SW. Australia. 50 WAU. Nanophan.

Verticordia amphigia A.S.George, Nuytsia 7: 285 (1991).
WSW. Western Australia. 50 WAU. Nanophan.

Verticordia apecta Eliz.A.George & A.S.George, Nuytsia 9: 335 (1994).
SW. Australia (SW. Mt. Barker). 50 WAU. Cham.

Verticordia argentea A.S.George, Nuytsia 7: 286 (1991).
WSW. Western Australia. 50 WAU. Nanophan.

Verticordia attenuata A.S.George, Nuytsia 7: 286 (1991).
SW. Western Australia. 50 WAU. Cham. or nanophan.

Verticordia aurea A.S.George, Nuytsia 7: 287 (1991).
WSW. Western Australia. 50 WAU. Nanophan.

Verticordia auriculata A.S.George, Nuytsia 7: 288 (1991).
WSW. Western Australia. 50 WAU. Cham. or nanophan.

Verticordia bifimbriata A.S.George, Nuytsia 7: 288 (1991).
WSW. Western Australia. 50 WAU. Cham. or nanophan.

Verticordia blepharophylla A.S.George, Nuytsia 7: 290 (1991).
WSW. Western Australia. 50 WAU. Cham. or nanophan.

Verticordia brachypoda Turcz., Bull. Soc. Imp. Naturalistes Moscou 20(1): 158 (1847).
SW. Australia. 50 WAU. Nanophan.
Verticordia stylotricha Diels, Bot. Jahrb. Syst. 35: 403 (1904).

Verticordia brevifolia A.S.George, Nuytsia 7: 291 (1991).
SW. Australia. 50 WAU. Tuber cham. or nanophan.

subsp. ***brevifolia***
SW. Australia. 50 WAU. Tuber cham. or nanophan.

subsp. ***stirlingensis*** A.S.George, Nuytsia 7: 293 (1991).
SW. Australia. 50 WAU. Tuber cham. or nanophan.

Verticordia brownii (Desf.) A.Cunn. ex DC., Prodr. 3: 209 (1828).
S. Western Australia. 50 WAU. Cham. or nanophan.
**Chamelaucium brownii* Desf., Mém. Mus. Hist. Nat. 5: 271 (1819).

Verticordia capillaris A.S.George, Nuytsia 7: 294 (1991).
W. Western Australia. 50 WAU. Nanophan.

Verticordia carinata Turcz., Bull. Soc. Imp. Naturalistes Moscou 22(1): 19 (1849).
SW. Australia. 50 WAU. Nanophan.

Verticordia centipeda A.S.George, Nuytsia 7: 295 (1991).
WSW. Western Australia. 50 WAU. Cham. or nanophan.

Verticordia chrysantha Endl., Stirp. Herb. Hügel.: 7 (1838). *Verticordia chrysantha* var. *roei* Schauer in J.G.C.Lehmann, Pl. Preiss. 1: 102 (1845), nom. inval.
SW. Australia. 50 WAU. Nanophan.
Verticordia gilbertii Turcz., Bull. Soc. Imp. Naturalistes Moscou 20(1): 160 (1847).

Verticordia chrysanthella A.S.George, Nuytsia 7: 297 (1991).
SW. Australia. 50 WAU. Nanophan.
Verticordia chrysantha var. *preissii* Schauer in J.G.C.Lehmann, Pl. Preiss. 1: 102 (1845).

Verticordia chrysostachys Meisn., J. Proc. Linn. Soc., Bot. 1: 41 (1856).
W. Western Australia. 50 WAU. Nanophan.

var. ***chrysostachys***
W. Western Australia. 50 WAU. Nanophan.

var. ***pallida*** A.S.George, Nuytsia 7: 300 (1991).
W. Western Australia. 50 WAU. Nanophan.

Verticordia citrella A.S.George, Nuytsia 7: 301 (1991).
SW. Australia (NE. of Noble Falls). 50 WAU. Tuber nanophan.

Verticordia comosa A.S.George, Nuytsia 7: 301 (1991).
WSW. Western Australia. 50 WAU. Nanophan.

Verticordia cooloomia A.S.George, Nuytsia 7: 303 (1991).
W. Western Australia (N. of lowre Murchison River). 50 WAU. Nanophan.

Verticordia coronata A.S.George, Nuytsia 7: 304 (1991).
SW. Australia. 50 WAU. Tuber cham.

Verticordia crebra A.S.George, Nuytsia 7: 304 (1991).
SW. Australia. 50 WAU. Cham. or nanophan.

Verticordia cunninghamii Schauer, Nov. Actorum Acad. Caes. Leop.-Carol. Nat. Cur. 19(Suppl. 2): 207 (1841).
N. Australia. 50 NTA QLD WAU. Nanophan. or phan.

Verticordia cunninghamii var. *longistyla* C.A.Gardner, W. Austral. Forests Dept. Bull., Bot. Notes 32: 74 (1923).

Verticordia dasystylis A.S.George, Nuytsia 7: 306 (1991).
SW. Australia. 50 WAU. Cham.

subsp. ***dasystylis***
Western Australia (Yellowdine area). 50 WAU. Cham.

subsp. ***kalbarriensis*** A.S.George, Nuytsia 7: 308 (1991).
W. Western Australia. 50 WAU. Cham.

subsp. ***oestopoia*** A.S.George, Nuytsia 7: 309 (1991).
WSW. Western Australia. 50 WAU. Cham.

Verticordia decussata S.T.Blake ex Byrnes, Austrobaileya 1: 47 (1977).
N. Northern Territory. 50 NTA. Nanophan.

Verticordia densiflora Lindl., Sketch Veg. Swan R.: vi (1839).
SW. Australia. 50 WAU. Cham. or nanophan.

var. ***cespitosa*** (Turcz.) A.S.George, Nuytsia 7: 310 (1991).
SW. Australia. 50 WAU. Cham. or nanophan.
**Verticordia cespitosa* Turcz., Bull. Soc. Imp. Naturalistes Moscou 20(1): 157 (1847).

var. ***densiflora***
SW. Australia. 50 WAU. Cham. or nanophan.

var. ***pedunculata*** A.S.George, Nuytsia 7: 311 (1991).
SW. Australia. 50 WAU. Cham. or nanophan.

var. ***roseostella*** A.S.George, Nuytsia 7: 312 (1991).
WSW. Western Australia. 50 WAU. Cham. or nanophan.

var. ***stelluligera*** (Meisn.) A.S.George, Nuytsia 7: 312 (1991).
W. Western Australia. 50 WAU. Cham. or nanophan.
**Verticordia stelluligera* Meisn., J. Proc. Linn. Soc., Bot. 1: 38 (1857).

Verticordia dichroma A.S.George, Nuytsia 7: 313 (1991).
W. Western Australia. 50 WAU. Cham. or nanophan.

var. ***dichroma***
W. Western Australia. 50 WAU. Cham. or nanophan.

var. ***syntoma*** A.S.George, Nuytsia 7: 314 (1991).
W. Western Australia. 50 WAU. Cham. or nanophan.

Verticordia drummondii Schauer, Nov. Actorum Acad. Caes. Leop.-Carol. Nat. Cur. 19(Suppl. 2): 208 (1841).
WSW. Western Australia. 50 WAU. Cham. or nanophan.

Verticordia endlicheriana Schauer in J.G.C.Lehmann, Pl. Preiss. 1: 101 (1844).
SW. Australia. 50 WAU. Cham. or nanophan.

var. ***angustifolia*** A.S.George, Nuytsia 7: 318 (1991).
SW. Western Australia. 50 WAU. Cham. or nanophan.

var. ***compacta*** A.S.George, Nuytsia 7: 317 (1991).
WSW. Western Australia. 50 WAU. Cham. or nanophan.

var. ***endlicheriana***
SW. Australia. 50 WAU. Cham. or nanophan.
Verticordia hirta Turcz., Bull. Cl. Phys.-Math. Acad. Imp. Sci. Saint-Pétersbourg 10: 327 (1852).

var. ***major*** A.S.George, Nuytsia 7: 318 (1991).
SW. Australia. 50 WAU. Cham. or nanophan.

var. ***manicula*** A.S.George, Nuytsia 7: 317 (1991).
WSW. Western Australia. 50 WAU. Cham. or nanophan.

Verticordia eriocephala A.S.George, Nuytsia 7: 319 (1991).
SW. Australia. 50 WAU. Nanophan.

Verticordia etheliana C.A.Gardner, J. Roy. Soc. Western Australia 27: 190 (1942).
W. Western Australia. 50 WAU. Nanophan.

var. ***etheliana***
W. Western Australia. 50 WAU. Nanophan.

var. ***formosa*** A.S.George, Nuytsia 7: 321 (1991).
W. Western Australia. 50 WAU. Nanophan.

Verticordia × eurardyensis Eliz.A.George & A.S.George, Nuytsia 9: 337 (1994). *V. dichroma × V. spicata*.
W. Western Australia. 50 WAU. Nanophan.

Verticordia fastigiata Turcz., Bull. Cl. Phys.-Math. Acad. Imp. Sci. Saint-Pétersbourg 10: 327 (1852).
S. Western Australia. 50 WAU. Cham.
Verticordia conferta Benth., Fl. Austral. 3: 22 (1867).

Verticordia fimbrilepis Turcz., Bull. Soc. Imp. Naturalistes Moscou 20(1): 158 (1847).
SW. Australia. 50 WAU. Nanophan.

subsp. ***australis*** A.S.George, Nuytsia 7: 322 (1991).
SW. Australia. 50 WAU. Nanophan.

subsp. ***fimbrilepis***
SW. Australia. 50 WAU. Nanophan.

Verticordia forrestii F.Muell., S. Sci. Rec. 3: 67 (1883).
NW. Western Australia. 50 WAU. Nanophan.

Verticordia fragrans A.S.George, Nuytsia 7: 323 (1991).
WSW. Western Australia. 50 WAU. Nanophan.

Verticordia galeata A.S.George, Nuytsia 7: 325 (1991).
WSW. Western Australia (lower Murchison River). 50 WAU. Nanophan.

Verticordia gracilis A.S.George, Nuytsia 7: 325 (1991).
SW. Australia. 50 WAU. Cham.

Verticordia grandiflora Endl., Stirp. Herb. Hügel.: 7 (1838).
SW. Australia. 50 WAU. Cham. or nanophan.
Verticordia heliantha Lindl., Sketch Veg. Swan R.: vi (1839).

Verticordia grandis J.Drumm., Hooker's J. Bot. Kew Gard. Misc. 5: 119 (1853).
WSW. Western Australia. 50 WAU. Tuber nanophan.

Verticordia habrantha Schauer in J.G.C.Lehmann, Pl. Preiss. 1: 100 (1844).
SW. Australia. 50 WAU. (Tuber) cham. or nanophan.
Verticordia umbellata Turcz., Bull. Soc. Imp. Naturalistes Moscou 20(1): 159 (1847).
Verticordia brachystylis F.Muell., Fragm. 1: 164 (1859).

Verticordia halophila A.S.George, Nuytsia 7: 327 (1991).
WSW. Western Australia. 50 WAU. Cham. or nanophan.

Verticordia harveyi Benth., Fl. Austral. 3: 22 (1867).
SW. Australia. 50 WAU. Tuber nanophan.
Verticordia harveyi var. *nudipetala* Benth., Fl. Austral. 3: 22 (1867).

Verticordia helichrysantha F.Muell. ex Benth., Fl. Austral. 3: 21 (1867).
SW. Australia. 50 WAU. Cham.

Verticordia helmsii S.Moore, J. Linn. Soc., Bot. 34: 190 (1899).
SC. & SW. Australia. 50 WAU. Tuber nanophan.
Verticordia adenocalyx Diels, Bot. Jahrb. Syst. 35: 405 (1904).

Verticordia huegelii Endl. in S.L.Endlicher & al., Enum. Pl.: 46 (1837).
SW. Australia. 50 WAU. Cham. or nanophan.

var. ***decumbens*** A.S.George, Nuytsia 7: 331 (1991).
WSW. Australia. 50 WAU. Cham.

var. ***huegelii***
WSW. Australia. 50 WAU. Cham.
Verticordia platystigma A.Cunn. ex Schauer, Nov. Actorum Acad. Caes. Leop.-Carol. Nat. Cur. 19(Suppl. 2): 214 (1841 publ. 1843).
Verticordia fimbripetala Turcz., Bull. Soc. Imp. Naturalistes Moscou 22(2): 19 (1849).

var. ***stylosa*** (Turcz.) A.S.George, Nuytsia 7: 331 (1991).
SW. Australia. 50 WAU. Cham. or nanophan.
**Verticordia stylosa* Turcz., Bull. Soc. Imp. Naturalistes Moscou 20(1): 160 (1847).

var. ***tridens*** A.S.George, Nuytsia 7: 332 (1991).
SW. Australia. 50 WAU. Cham. or nanophan.

Verticordia hughanii F.Muell., Fragm. 11: 10 (1879).
WSW. Australia. 50 WAU. Cham.

Verticordia humilis Benth., Fl. Austral. 3: 26 (1867).
S. Western Australia. 50 WAU. Cham.

Verticordia inclusa A.S.George, Nuytsia 7: 332 (1991).
SC. & S. Western Australia. 50 WAU. Cham. or nanophan.

Verticordia insignis Endl. in S.L.Endlicher & al., Enum. Pl.: 47 (1837).
SW. Australia. 50 WAU. Nanophan.

subsp. ***compta*** (Endl.) A.S.George, Nuytsia 7: 335 (1991).
SW. Australia. 50 WAU. Nanophan.
**Verticordia compta* Endl., Stirp. Herb. Hügel.: 6 (1838).

subsp. ***eomagis*** A.S.George, Nuytsia 7: 334 (1991).
WSW. Western Australia. 50 WAU. Nanophan.

subsp. ***insignis***
WSW. Western Australia. 50 WAU. Nanophan.

Verticordia integra A.S.George, Nuytsia 7: 335 (1991).
SW. Australia. 50 WAU. Cham. or nanophan.

Verticordia interioris C.A.Gardner ex A.S.George, Nuytsia 7: 336 (1991).
Western Australia. 50 WAU. Cham. or nanophan.

Verticordia jamiesonii F.Muell., S. Sci. Rec. 3: 68 (1883).
Western Australia (Austin). 50 WAU. Cham. or nanophan.

Verticordia laciniata A.S.George, Nuytsia 7: 336 (1991).
WSW. Western Australia. 50 WAU. Nanophan.

Verticordia lehmannii Schauer in J.G.C.Lehmann, Pl. Preiss. 1: 99 (1844).
SW. Western Australia. 50 WAU. Cham. or nanophan.

Verticordia lepidophylla F.Muell., Fragm. 1: 228 (1859).
W. Western Australia. 50 WAU. Nanophan.

var. ***lepidophylla***
W. Western Australia. 50 WAU. Nanophan.

var. ***quantula*** A.S.George, Nuytsia 7: 338 (1991).
W. Western Australia. 50 WAU. Nanophan.

Verticordia lindleyi Schauer, Nov. Actorum Acad. Caes. Leop.-Carol. Nat. Cur. 19(Suppl. 2): 210 (1841 publ. 1943). *Verticordia drummondii* var. *lindleyi* (Schauer) Benth., Fl. Austral. 3: 31 (1867).
SW. Australia. 50 WAU. Nanophan.

subsp. ***lindleyi***
WSW. Western Australia. 50 WAU. Nanophan.

subsp. ***purpurea*** A.S.George, Nuytsia 7: 339 (1991).
SW. Australia. 50 WAU. Nanophan.

Verticordia longistylis A.S.George, Nuytsia 7: 339 (1991).
SSW. Australia (lower Fritzgerald River). 50 WAU. Cham. or nanophan.

Verticordia luteola A.S.George, Nuytsia 7: 341 (1991).
WSW. Western Australia. 50 WAU. Nanophan.

var. ***luteola***
WSW. Western Australia. 50 WAU. Nanophan.

var. ***rosea*** Eliz.A.George & A.S.George, Nuytsia 9: 339 (1994).
WSW. Western Australia. 50 WAU. Nanophan.

Verticordia minutiflora F.Muell., Fragm. 4: 58 (1864).
S. Western Australia. 50 WAU. Cham. or nanophan.
Verticordia fontanesii var. *parviflora* Benth., Fl. Austral. 3: 21 (1867).

Verticordia mirabilis Eliz.George & A.S.George, Nuytsia 13: 465 (2001).
C. Western Australia (S. Gibson Desert). 50 WAU. Cham. or nanophan.

Verticordia mitchelliana C.A.Gardner, J. Roy. Soc. Western Australia 19: 89 (1932–1933 publ. 1934).
SW. Australia. 50 WAU. Cham. or nanophan.

Verticordia mitodes A.S.George, Nuytsia 7: 343 (1991).
SW. Australia. 50 WAU. Cham.

Verticordia monadelpha Turcz., Bull. Soc. Imp. Naturalistes Moscou 20(1): 158 (1847).
SW. Australia. 50 WAU. Cham. or nanophan.

var. ***callitricha*** (Meisn.) A.S.George, Nuytsia 7: 345 (1991).
WSW. Australia. 50 WAU. Cham. or nanophan.
**Verticordia callitricha* Meisn., J. Proc. Linn. Soc., Bot. 1: 39 (1856).

var. ***monadelpha***
SW. Australia. 50 WAU. Cham. or nanophan.

Verticordia muelleriana E.Pritz., Bot. Jahrb. Syst. 35: 407 (1904).
WSW. Australia. 50 WAU. Nanophan.

subsp. ***minor*** A.S.George, Nuytsia 7: 346 (1991).
WSW. Western Australia (E. of Geraldton). 50 WAU. Nanophan.

subsp. ***muelleriana***
WSW. Australia. 50 WAU. Nanophan.

Verticordia multiflora Turcz., Bull. Soc. Imp. Naturalistes Moscou 20(1): 159 (1847).
SW. Australia. 50 WAU. Cham. or nanophan.

subsp. ***multiflora***
SW. Australia. 50 WAU. Cham. or nanophan.

subsp. ***solox*** A.S.George, Nuytsia 7: 347 (1991).
SW. Australia. 50 WAU. Cham. or nanophan.

Verticordia nitens (Lindl.) Endl., Stirp. Herb. Hügel.: 194 (1838).
WSW. Western Australia. 50 WAU. Nanophan.
**Chrysorhoe nitens* Lindl., Companion Bot. Mag. 2: 357 (1836).

Verticordia nobilis Meisn., J. Proc. Linn. Soc., Bot. 1: 39 (1857).
WSW. Western Australia. 50 WAU. Nanophan.

Verticordia oculata Meisn., J. Proc. Linn. Soc., Bot. 1: 41 (1857).
W. Western Australia. 50 WAU. Tuber cham. or nanophan.

Verticordia ovalifolia Meisn., J. Proc. Linn. Soc., Bot. 1: 40 (1857).
SW. Australia. 50 WAU. Cham. or nanophan.

Verticordia oxylepis Turcz., Bull. Cl. Phys.-Math. Acad. Imp. Sci. Saint-Pétersbourg 10: 327 (1852).
S. Western Australia. 50 WAU. Cham.
Verticordia demissa F.Muell. ex Benth., Fl. Austral. 3: 25 (1867).

Verticordia paludosa A.S.George, Nuytsia 7: 349 (1991).
WSW. Western Australia. 50 WAU. Cham. or nanophan.

Verticordia patens A.S.George, W. Austral. Naturalist 10: 30 (1966).
WSW. Western Australia. 50 WAU. Nanophan.

Verticordia penicillaris F.Muell., Fragm. 1: 226 (1859).
W. Western Australia. 50 WAU. Cham.

Verticordia pennigera Endl. in S.L.Endlicher & al., Enum. Pl.: 46 (1837).
SW. Australia. 50 WAU. Tuber cham. or nanophan.
Verticordia setigera Lindl., Sketch Veg. Swan R.: vii (1839).
Verticordia cuspidata A.Cunn. ex Schauer, Nov. Actorum Acad. Caes. Leop.-Carol. Nat. Cur. 19(Suppl. 2): 212 (1841).

Verticordia pholidophylla F.Muell., Fragm. 1: 227 (1859).
W. Western Australia. 50 WAU. Cham. or nanophan.

Verticordia picta Endl., Stirp. Herb. Hügel.: 6 (1838).
SW. Australia. 50 WAU. Nanophan.
Verticordia pentandra Turcz., Bull. Soc. Imp. Naturalistes Moscou 20(1): !57 (1847).

Verticordia pityrhops A.S.George, Nuytsia 7: 351 (1991).
SSW. Western Australia (E. Mt. Barren). 50 WAU. Cham.

Verticordia plumosa (Desf.) Druce, Rep. Bot. Exch. Club Brit. Isles 1916: 651 (1917).
SW. Australia. 50 WAU. Cham. or nanophan.
**Chamelaucium plumosum* Desf., Mém. Mus. Hist. Nat. 5: 42 (1819). *Verticordia fontanesii* DC., Prodr. 3: 209 (1828), nom. illeg.

var. ***ananeotes*** A.S.George, Nuytsia 7: 355 (1991).
SW. Australia. 50 WAU. Tuber cham. or nanophan.

var. ***brachyphylla*** (Diels) A.S.George, Nuytsia 7: 356 (1991).
SW. Australia. 50 WAU. Cham. or nanophan.
**Verticordia fontanesii* var. *brachyphylla* Diels, Bot. Jahrb. Syst. 35: 403 (1904). *Verticordia brachyphylla* (Diels) A.S.George, Nuytsia 7: 356 (1991).

var. ***grandiflora*** (Benth.) A.S.George, Nuytsia 7: 358 (1991).
SSW. Australia. 50 WAU. Cham. or nanophan.
Verticordia pectinata Turcz., Bull. Acad. Imp. Sci. Saint-Pétersbourg 10: 327 (1862).
**Verticordia fontanesii* var. *grandiflora* Benth., Fl. Austral. 3: 21 (1867).

var. ***incrassata*** A.S.George, Nuytsia 7: 357 (1991).
SW. Australia. 50 WAU. Cham. or nanophan.

var. ***pleiobotrya*** A.S.George, Nuytsia 7: 354 (1991).
WSW. Australia. 50 WAU. Cham. or nanophan.

var. ***plumosa***
SW. Australia. 50 WAU. Cham. or nanophan.
Diplachne baueri R.Br. ex Desf., Mém. Mus. Hist. Nat. 5: 272 (1819).

var. ***vassensis*** A.S.George, Nuytsia 7: 356 (1991).
SW. Australia. 50 WAU. Cham. or nanophan.

Verticordia polytricha Benth., Fl. Austral. 3: 25 (1867).
W. Western Australia. 50 WAU. Nanophan.

Verticordia pritzelii Diels, Bot. Jahrb. Syst. 35: 404 (1904).
SW. Australia. 50 WAU. Cham. or nanophan.

Verticordia pulchella A.S.George, Nuytsia 7: 359 (1991).
SW. Australia. 50 WAU. Cham.

Verticordia rennieana F.Muell. & Tate, Trans. & Proc. Roy. Soc. South Australia 16: 354 (1895).
SW. Australia. 50 WAU. Nanophan.

Verticordia roei Endl., Stirp. Herb. Hügel.: 6 (1838).
SW. Australia. 50 WAU. Cham. or nanophan.

subsp. ***meiogona*** A.S.George, Nuytsia 7: 361 (1991).
WSW. Western Australia. 50 WAU. Cham. or nanophan.

subsp. ***roei***
SW. Australia. 50 WAU. Cham. or nanophan.

Verticordia rutilastra A.S.George, Nuytsia 7: 362 (1991).
WSW. Western Australia. 50 WAU. Cham. or nanophan.

Verticordia serotina A.S.George, Nuytsia 7: 363 (1991).
W. Western Australia (Cape Range). 50 WAU. Nanophan.

Verticordia serrata (Lindl.) Schauer, Nov. Actorum Acad. Caes. Leop.-Carol. Nat. Cur. 19(Suppl. 2): 222 (1841).
SW. Australia. 50 WAU. Nanophan.
**Chrysorhoe serrata* Lindl., Sketch Veg. Swan R.: vi (1839).

var. ***ciliata*** A.S.George, Nuytsia 7: 364 (1991).
SW. Australia. 50 WAU. Nanophan.

var. ***linearis*** A.S.George, Nuytsia 7: 365 (1991).
WSW. Australia. 50 WAU. Nanophan.

var. ***serrata***
SW. Australia. 50 WAU. Nanophan.

Verticordia sieberi Diesing ex Schauer, Nov. Actorum Acad. Caes. Leop.-Carol. Nat. Cur. 19(Suppl. 2): 201 (1841).
S. Western Australia. 50 WAU. Nanophan.

var. ***curta*** A.S.George, Nuytsia 7: 367 (1991).
SSW. Australia. 50 WAU. Nanophan.

var. ***lomata*** A.S.George, Nuytsia 7: 366 (1991).
S. Western Australia. 50 WAU. Nanophan.

var. ***pachyphylla*** A.S.George, Nuytsia 7: 367 (1991).
S. Western Australia (NE. of Lake King). 50 WAU. Nanophan.

var. ***sieberi***
SSW. Australia. 50 WAU. Nanophan.

Verticordia spicata F.Muell., Fragm. 1: 226 (1859).
W. Western Australia. 50 WAU. Nanophan.

subsp. ***spicata***
W. Western Australia. 50 WAU. Nanophan.

subsp. ***squamosa*** A.S.George, Nuytsia 7: 368 (1991).
WSW. Western Australia. 50 WAU. Nanophan.

Verticordia staminosa C.A.Gardner & A.S.George, J. Roy. Soc. Western Australia 46: 132 (1963).
SW. Australia. 50 WAU. Cham. or nanophan.

subsp. ***cylindracea*** A.S.George, Nuytsia 7: 371 (1991).
SW. Australia. 50 WAU. Cham. or nanophan.
Verticordia staminosa var. *cylindracea* A.S.George, Nuytsia 7: 371 (1991).

var. ***erecta*** A.S.George, Nuytsia 7: 372 (1991).
SW. Australia (NW. of Newdegate). 50 WAU. Cham. or nanophan.

subsp. ***staminosa***
WSW. Australia (Wongan Hills). 50 WAU. Cham. or nanophan.

Verticordia stenopetala Diels, Bot. Jahrb. Syst. 35: 402 (1904).
SW. Australia. 50 WAU. Cham.

Verticordia subulata A.S.George, Nuytsia 7: 372 (1991).
SW. Australia. 50 WAU. Cham. or nanophan.

Verticordia tumida A.S.George, Nuytsia 7: 373 (1991).
SW. Australia. 50 WAU. Tuber cham. or nanophan.

subsp. ***therogana*** A.S.George, Nuytsia 7: 375 (1991).
SW. Australia. 50 WAU. Tuber cham. or nanophan.

subsp. ***tumida***
SW. Australia. 50 WAU. Tuber cham. or nanophan.

Verticordia venusta A.S.George, Nuytsia 7: 376 (1991).
WSW. Western Australia. 50 WAU. Nanophan.

Verticordia verticillata Byrnes, Austrobaileya 1: 48 (1977).
N. Western Australia to Northern Territory. 50 NTA WAU. Nanophan. or phan.

Verticordia verticordina (F.Muell.) A.S.George, Nuytsia 7: 377 (1991).
S. Western Australia. 50 WAU. Cham.
**Chamelaucium verticordinum* F.Muell., Fragm. 4: 57 (1864). *Darwinia verticordina* (F.Muell.) Benth., J. Linn. Soc., Bot. 9: 181 (1867).
Verticordia integrisepala F.Muell., Fragm. 4: 57 (1864).

Verticordia vicinella A.S.George, Nuytsia 7: 377 (1991).
S. Western Australia. 50 WAU. Nanophan.

Verticordia wonganensis A.S.George, Nuytsia 7: 379 (1991).
WSW. Western Australia (Wongan Hills). 50 WAU. Cham. or nanophan.

***Synonyms*:**

Verticordia adenocalyx Diels = ***Verticordia helmsii*** S.Moore
Verticordia brachyphylla (Diels) A.S.George = ***Verticordia plumosa*** var. ***brachyphylla*** (Diels) A.S.George
Verticordia brachystylis F.Muell. = ***Verticordia habrantha*** Schauer
Verticordia callitricha Meisn. = ***Verticordia monadelpha*** var. ***callitricha*** (Meisn.) A.S.George
Verticordia cespitosa Turcz. = ***Verticordia densiflora*** var. ***cespitosa*** (Turcz.) A.S.George
Verticordia chrysantha var. *preissii* Schauer = ***Verticordia chrysanthella*** A.S.George
Verticordia chrysantha var. *roei* Schauer = ***Verticordia chrysantha*** Endl.
Verticordia compta Endl. = ***Verticordia insignis*** subsp. ***compta*** (Endl.) A.S.George
Verticordia conferta Benth. = ***Verticordia fastigiata*** Turcz.
Verticordia cunninghamii var. *longistyla* C.A.Gardner = ***Verticordia cunninghamii*** Schauer
Verticordia cuspidata A.Cunn. ex Schauer = ***Verticordia pennigera*** Endl.
Verticordia darwinioides Maiden & Betche = ***Homoranthus darwinioides*** (Maiden & Betche) Cheel
Verticordia demissa F.Muell. ex Benth. = ***Verticordia oxylepis*** Turcz.
Verticordia drummondii var. *lindleyi* (Schauer) Benth. = ***Verticordia lindleyi*** Schauer
Verticordia fimbripetala Turcz. = ***Verticordia huegelii*** Endl. var. ***huegelii***
Verticordia fontanesii DC. = ***Verticordia plumosa*** (Desf.) Druce
Verticordia fontanesii var. *brachyphylla* Diels = ***Verticordia plumosa*** var. ***brachyphylla*** (Diels) A.S.George
Verticordia fontanesii var. *grandiflora* Benth. = ***Verticordia plumosa*** var. ***grandiflora*** (Benth.) A.S.George
Verticordia fontanesii var. *parviflora* Benth. = ***Verticordia minutiflora*** F.Muell.
Verticordia gilbertii Turcz. = ***Verticordia chrysantha*** Endl.
Verticordia harveyi var. *nudipetala* Benth. = ***Verticordia harveyi*** Benth.
Verticordia heliantha Lindl. = ***Verticordia grandiflora*** Endl.
Verticordia hirta Turcz. = ***Verticordia endlicheriana*** Schauer var. ***endlicheriana***
Verticordia integrisepala F.Muell. = ***Verticordia verticordina*** (F.Muell.) A.S.George
Verticordia pectinata Turcz. = ***Verticordia plumosa*** var. ***grandiflora*** (Benth.) A.S.George
Verticordia pentandra Turcz. = ***Verticordia picta*** Endl.
Verticordia platystigma A.Cunn. ex Schauer = ***Verticordia huegelii*** Endl. var. ***huegelii***
Verticordia preissii Schauer = ***Verticordia acerosa*** var. ***preissii*** (Schauer) A.S.George
Verticordia setigera Lindl. = ***Verticordia pennigera*** Endl.
Verticordia staminosa var. *cylindracea* A.S.George = ***Verticordia staminosa*** subsp. ***cylindracea*** A.S.George
Verticordia stelluligera Meisn. = ***Verticordia densiflora*** var. ***stelluligera*** (Meisn.) A.S.George
Verticordia stylosa Turcz. = ***Verticordia huegelii*** var. ***stylosa*** (Turcz.) A.S.George
Verticordia stylotricha Diels = ***Verticordia brachypoda*** Turcz.

Verticordia umbellata Turcz. = ***Verticordia habrantha*** Schauer
Verticordia wilhelmii F.Muell. = ***Homoranthus wilhelmii*** (F.Muell.) Cheel

Waterhousea

Waterhousea B.Hyland = ***Syzygium*** Gaertn.
Waterhousea floribunda (F.Muell.) B.Hyland = ***Syzygium floribundum*** F.Muell.
Waterhousea hedraiophylla (F.Muell.) B.Hyland = ***Syzygium hedraiophyllum*** (F.Muell.) Craven & Biffin
Waterhousea mulgraveana B.Hyland = ***Syzygium mulgraveanum*** (B.Hyland) Craven & Biffin
Waterhousea unipunctata B.Hyland = ***Syzygium unipunctatum*** (B.Hyland) Craven & Biffin

Wehlia

Wehlia F.Muell. = ***Homalocalyx*** F.Muell.
Wehlia aurea C.A.Gardner = ***Homalocalyx aureus*** (C.A.Gardner) Craven
Wehlia coarctata F.Muell. = ***Homalocalyx coarctatus*** (F.Muell.) Craven
Wehlia grandiflora C.A.Gardner = ***Homalocalyx grandiflorus*** (C.A.Gardner) Craven
Wehlia pedicellata Ewart & B.Rees = ***Homalocalyx pulcherrimus*** (Ewart & B.Rees) Craven
Wehlia pulcherrima Ewart & B.Rees = ***Homalocalyx pulcherrimus*** (Ewart & B.Rees) Craven
Wehlia staminosa F.Muell. = ***Homalocalyx staminosus*** (F.Muell.) Craven
Wehlia thryptomenoides F.Muell. = ***Homalocalyx thryptomenoides*** (F.Muell.) Craven
Wehlia thryptomenoides var. *microphylla* Ewart & B.Rees = ***Homalocalyx thryptomenoides*** (F.Muell.) Craven

Welchiodendron

Welchiodendron Peter G.Wilson & J.T.Waterh., Austral. J. Bot. 30: 440 (1982).
New Guinea to N. Queensland. 43 NWG 50 QLD.
1 Species

Welchiodendron longivalve (F.Muell.) Peter G.Wilson & J.T.Waterh., Austral. J. Bot. 30: 411 (1982).
New Guinea to N. Queensland. 43 NWG 50 QLD. Nanophan. or phan.
**Tristania longivalvis* F.Muell., S. Sci. Rec., n.s., 2: (1886).
Tristania brownii S.Moore, J. Bot. 40: 25 (1902).

Whiteodendron

Whiteodendron Steenis, Acta Bot. Neerl. 1: 436 (1952).
Borneo. 42 BOR.
1 Species

Whiteodendron moultonianum (W.W.Sm.) Steenis, Acta Bot. Neerl. 1: 439 (1952).
Borneo. 42 BOR. Phan.
**Tristania moultoniana* W.W.Sm., Notes Roy. Bot. Gard. Edinburgh 8: 328 (1915).

Xanthomyrtus

Xanthomyrtus Diels, Bot. Jahrb. Syst. 57: 362 (1922).
Malesia to New Caledonia. 42 BOR MOL PHI SUL 43 BIS NWG 60 NWC.
23 Species

Xanthomyrtus angustifolia A.J.Scott, Kew Bull. 33: 466 (1979).
Sulawesi to New Guinea. 42 SUL 43 NWG. Phan.

Xanthomyrtus arfakensis (Gibbs) Diels, Bot. Jahrb. Syst. 57: 366 (1922).
W. New Guinea. 43 NWG. Nanophan. or phan.
**Myrtus arfakensis* Gibbs, Fl. Arfak Mts.: 152 (1917).
Myrtus klossii var. *brevipedunculata* Diels, Nova Guinea 14: 86 (1924).

Xanthomyrtus bryophila Diels, Nova Guinea 14: 86 (1924).
W. New Guinea (Snow Mts.). 43 NWG. Nanophan.

Xanthomyrtus cardiophylla Merr. & L.M.Perry, J. Arnold Arbor. 23: 242 (1942).
W. New Guinea (Snow Mts.). 43 NWG. Nanophan.
Xanthomyrtus cardiophylla var. *parvifolia* Merr. & L.M.Perry, J. Arnold Arbor. 23: 243 (1942).

Xanthomyrtus compacta (Ridl.) Diels, Bot. Jahrb. Syst. 57: 367 (1922).
Maluku to New Guinea. 42 MOL 43 NWG. Nanophan. or phan.
**Myrtus compacta* Ridl., Trans. Linn. Soc. London, Bot. 9: 42 (1916).
Myrtus klossii Ridl., Trans. Linn. Soc. London, Bot. 9: 42 (1916). *Xanthomyrtus klossii* (Ridl.) Diels, Bot. Jahrb. Syst. 57: 366 (1922).
Myrtus prostrata Gibbs, Fl. Arfak Mts.: 151 (1917). *Xanthomyrtus prostrata* (Gibbs) Diels, Bot. Jahrb. Syst. 57: 367 (1922).
Xanthomyrtus calytrichoides Diels, Bot. Jahrb. Syst. 57: 367 (1922).
Xanthomyrtus linnaeifolia Diels, Bot. Jahrb. Syst. 57: 366 (1922).
Xanthomyrtus dielsiana Merr. & L.M.Perry, J. Arnold Arbor. 23: 244 (1942).

Xanthomyrtus diplycosiifolia (C.B.Rob.) Merr., Philipp. J. Sci. 30: 415 (1926).
Philippines to Maluku. 42 MOL PHI. Phan.
**Eugenia diplycosiifolia* C.B.Rob., Philipp. J. Sci., C 4: 347 (1909).
Eugenia aurea Elmer, Leafl. Philipp. Bot. 4: 1400 (1912). *Xanthomyrtus aurea* (Elmer) Merr., Philipp. J. Sci. 30: 415 (1926).

Xanthomyrtus flavida (Stapf) Diels, Bot. Jahrb. Syst. 57: 366 (1922).
Borneo to Sulawesi. 42 BOR SUL. Nanophan. or phan.
**Myrtus flavida* Stapf, Hooker's Icon. Pl. 23: t. 2290 (1894).

var. ***flavida***
Borneo to Sulawesi. 42 BOR SUL. Nanophan. or phan.

var. ***latifolia*** Airy Shaw, Kew Bull. 4: 117 (1949).
NW. Borneo. 42 BOR. Nanophan. or phan.

Xanthomyrtus grandiflora A.J.Scott, Kew Bull. 33: 471 (1979).
W. New Guinea. 43 NWG. Nanophan.

Xanthomyrtus hienghenensis Guillaumin, Bull. Soc. Bot. France 81: 16 (1934).
New Caledonia. 60 NWC. Nanophan. or phan.
Xanthomyrtus hienghenensis var. *latifolia* Guillaumin, Bull. Soc. Bot. France 81: 17 (1934).

Xanthomyrtus humilis Merr. & L.M.Perry, J. Arnold Arbor. 23: 243 (1942).
New Guinea. 43 NWG. Nanophan. or phan.

Xanthomyrtus koebrensis (Gibbs) Diels, Bot. Jahrb. Syst. 57: 367 (1922).
W. New Guinea. 43 NWG. Nanophan. or phan.
**Myrtus koebrensis* Gibbs, Fl. Arfak Mts.: 152 (1917).
Xanthomyrtus exigua Merr. & L.M.Perry, J. Arnold Arbor. 23: 246 (1942).

Xanthomyrtus lanceolata Merr. & L.M.Perry, J. Arnold Arbor. 23: 241 (1942).
New Guinea. 43 NWG. Phan.

Xanthomyrtus leeuwenii Diels ex A.J.Scott, Kew Bull. 33: 478 (1979).
New Guinea. 43 NWG. Nanophan. or phan.

Xanthomyrtus montis-sucklingii A.J.Scott, Kew Bull. 33: 472 (1979).
Papua New Guinea. 43 NWG. Phan.

Xanthomyrtus montivaga A.J.Scott, Kew Bull. 33: 471 (1979).
New Guinea. 43 NWG. Nanophan. or phan.

Xanthomyrtus moultonii (Merr.) Merr., Sarawak Mus. J. 3: 533 (1928).
Borneo (Sarawak). 42 BOR. Nanophan.
**Myrtus moultonii* Merr., J. Straits Branch Roy. Asiat. Soc. 86: 337 (1922).

Xanthomyrtus oreophila A.J.Scott, Kew Bull. 33: 470 (1979).
W. New Guinea. 43 NWG. Nanophan. or phan.

Xanthomyrtus ovata A.J.Scott, Kew Bull. 33: 481 (1979).
New Guinea. 43 NWG. Nanophan. or phan.

Xanthomyrtus papuana Merr. & L.M.Perry, J. Arnold Arbor. 23: 244 (1942).
New Guinea. 43 NWG. Nanophan. or phan.
Xanthomyrtus papuana var. *parviflora* Merr. & L.M.Perry, J. Arnold Arbor. 23: 244 (1942).

Xanthomyrtus polyclada Diels, Bot. Jahrb. Syst. 57: 365 (1922).
New Guinea. 43 NWG. Phan.

Xanthomyrtus schlechteri Diels, Bot. Jahrb. Syst. 57: 364 (1922).
New Guinea to Bismarck Arch. 43 BIS NWG. Phan.
Xanthomyrtus fasciculata Diels, Bot. Jahrb. Syst. 57: 363 (1922).
Xanthomyrtus rostrata Merr. & L.M.Perry, J. Arnold Arbor. 23: 240 (1942).

Xanthomyrtus scolopacina (Ridl.) Diels, Bot. Jahrb. Syst. 57: 365 (1922).
New Guinea. 43 NWG. Phan.
**Eugenia scolopacina* Ridl., Trans. Linn. Soc. London, Bot. 9: 49 (1916).
Myrtus flavida var. *glabrescens* Gibbs, Fl. Arfak Mts.: 150 (1917).
Xanthomyrtus longicuspis Diels, Bot. Jahrb. Syst. 57: 364 (1922).
Xanthomyrtus longicuspis var. *fruticosa* Diels, Bot. Jahrb. Syst. 57: 364 (1922).
Xanthomyrtus pullei Diels, Bot. Jahrb. Syst. 57: 365 (1922).

Xanthomyrtus taxifolia (Ridl.) Merr., Sarawak Mus. J. 3: 532 (1928).
Borneo (Sarawak). 42 BOR. Nanophan. or phan.
**Myrtus taxifolia* Ridl., Bull. Misc. Inform. Kew 1914: 209 (1914).

Synonyms:
Xanthomyrtus aurea (Elmer) Merr. = ***Xanthomyrtus diplycosiifolia*** (C.B.Rob.) Merr.
Xanthomyrtus calytrichoides Diels = ***Xanthomyrtus compacta*** (Ridl.) Diels
Xanthomyrtus cardiophylla var. *parvifolia* Merr. & L.M.Perry = ***Xanthomyrtus cardiophylla*** Merr. & L.M.Perry
Xanthomyrtus dielsiana Merr. & L.M.Perry = ***Xanthomyrtus compacta*** (Ridl.) Diels
Xanthomyrtus exigua Merr. & L.M.Perry = ***Xanthomyrtus koebrensis*** (Gibbs) Diels
Xanthomyrtus fasciculata Diels = ***Xanthomyrtus schlechteri*** Diels
Xanthomyrtus hienghenensis var. *latifolia* Guillaumin = ***Xanthomyrtus hienghenensis*** Guillaumin
Xanthomyrtus klossii (Ridl.) Diels = ***Xanthomyrtus compacta*** (Ridl.) Diels
Xanthomyrtus linnaeifolia Diels = ***Xanthomyrtus compacta*** (Ridl.) Diels
Xanthomyrtus longicuspis Diels = ***Xanthomyrtus scolopacina*** (Ridl.) Diels
Xanthomyrtus longicuspis var. *fruticosa* Diels = ***Xanthomyrtus scolopacina*** (Ridl.) Diels
Xanthomyrtus papuana var. *parviflora* Merr. & L.M. Perry = ***Xanthomyrtus papuana*** Merr. & L.M.Perry
Xanthomyrtus pergracilis Diels = ***Eugenia sarasinii*** Guillaumin
Xanthomyrtus prostrata (Gibbs) Diels = ***Xanthomyrtus compacta*** (Ridl.) Diels
Xanthomyrtus pullei Diels = ***Xanthomyrtus scolopacina*** (Ridl.) Diels
Xanthomyrtus rostrata Merr. & L.M.Perry = ***Xanthomyrtus schlechteri*** Diels
Xanthomyrtus sarasinii (Guillaumin) Guillaumin = ***Eugenia sarasinii*** Guillaumin

Xanthostemon

Xanthostemon F.Muell., Hooker's J. Bot. Kew Gard. Misc. 9: 17 (1857), nom. cons.
C. Malesia to New Caledonia. 42 JAW? MOL PHI SUL 43 NWG SOL 50 NTA QLD WAU 60 NWC.
47 Species
Nani Adans., Fam. Pl. 2: 88 (1763).
Nania Miq., Fl. Ned. Ind. 1(1): 399 (1855).
Draparnaudia Montrouz., Mém. Acad. Roy. Sci. Lyon, Sect. Sci., II, 10: 205 (1860).
Fremya Brongn. & Gris, Bull. Soc. Bot. France 10: 372 (1863).
Salisia Pancher ex Brongn. & Gris, Bull. Soc. Bot. France 10: 373 (1863), nom. illeg.

Xanthostemon arenarius Peter G.Wilson, Telopea 5: 305 (1993).
N. Queensland. 50 QLD. Nanophan. or phan.

Xanthostemon aurantiacus (Brongn. & Gris) Schltr., Bot. Jahrb. Syst. 36: 26 (1905).
SE. New Caledonia. 60 NWC. Nanophan.
**Fremya aurantiaca* Brongn. & Gris, Bull. Soc. Bot. France 10: 373 (1863). *Salisia aurantiaca* (Brongn. & Gris) Pancher, Bull. Soc. Bot. France 10: 373 (1863).

Xanthostemon bracteatus Merr., Philipp. J. Sci., C 12: 289 (1917).
Philippines (Luzon). 42 PHI. Phan.

Xanthostemon brassii Merr., J. Arnold Arbor. 33: 154 (1952).
New Guinea. 43 NWG. Phan.

Xanthostemon carlii J.W.Dawson, in Fl. Nouv.-Caléd. 18: 200 (1992).
WNW. New Caledonia. 60 NWC. Nanophan. or phan.

Xanthostemon chrysanthus (F.Muell.) Benth., Fl. Austral. 3: 268 (1867).
N. & NE. Queensland. 50 QLD. Phan.
**Metrosideros chrysantha* F.Muell., Fragm. 4: 159 (1864). *Nania chrysantha* (F.Muell.) Kuntze, Revis. Gen. Pl. 1: 242 (1891).

Xanthostemon confertiflorus Merr., J. Arnold Arbor. 33: 158 (1952).
Sulawesi. 42 SUL. Phan.

Xanthostemon crenulatus C.T.White, J. Arnold Arbor. 23: 82 (1942).
S. New Guinea to N. Queensland. 43 NWG 50 QLD. Phan.

Xanthostemon eucalyptoides F.Muell., Fragm. 1: 81 (1858). *Metrosideros eucalyptoides* (F.Muell.) F.Muell., Fragm. 1: 243 (1859). *Nania eucalyptoides* (F.Muell.) Kuntze, Revis. Gen. Pl. 1: 242 (1891).
N. Western Australia to N. Northern Territory. 50 NTA WAU. Phan.

Xanthostemon ferrugineus J.W.Dawson, in Fl. Nouv.-Caléd. 18: 184 (1992).
SC. New Caledonia. 60 NWC. Nanophan. or phan.

Xanthostemon formosus Peter G.Wilson, Telopea 3: 462 (1990).
N. Queensland. 50 QLD. Phan.

Xanthostemon francii Guillaumin, Bull. Mus. Natl. Hist. Nat., II, 4: 691 (1932).
SE. New Caledonia. 60 NWC. Nanophan.
Xanthostemon sulphureus var. *tomentosus* Däniker, Vierteljahrsschr. Naturf. Ges. Zürich 78(19): 315 (1933).

Xanthostemon fruticosus Peter G.Wilson & Co, Sida 18: 283 (1998).
Philippines. 42 PHI. Phan.

Xanthostemon glaucus Pamp., Nuovo Giorn. Bot. Ital., n.s., 12: 682 (1904).
WNW. New Caledonia. 60 NWC. Nanophan.
Fremya glauca Vieill. ex Pamp. & Pampal., Nuovo Giorn. Bot. Ital., n.s., 12: 682 (1904).

Xanthostemon grandiflorus Gugerli, Repert. Spec. Nov. Regni Veg. Beih. 120: 102 (1940).
New Caledonia. 60 NWC. Nanophan.

Xanthostemon graniticus Peter G.Wilson, Telopea 3: 470 (1990).
N. Queensland. 50 QLD. Phan.

Xanthostemon grisii Guillaumin, Bull. Soc. Bot. France 81: 13 (1934).
NE. & NC. New Caledonia. 60 NWC. Phan.
Xanthostemon rugosus Gugerli, Repert. Spec. Nov. Regni Veg. Beih. 120: 72 (1940).

Xanthostemon gugerlii Merr., J. Arnold Arbor. 33: 151 (1952).
C. & SE. New Caledonia. 60 NWC. Nanophan.
**Fremya speciosa* Brongn. & Gris, Bull. Soc. Bot. France 10: 299 (1863). *Xanthostemon multiflorus* var. *speciosus* (Brongn. & Gris) Pamp., Nuovo Giorn. Bot. Ital., n.s., 12: 678 (1905). *Xanthostemon speciosus* (Brongn. & Gris) Pamp., Nuovo Giorn. Bot. Ital., n.s., 12: 688 (1905), nom. illeg.
Xanthostemon speciosus var. *lanceolatus* Gugerli, Repert. Spec. Nov. Regni Veg. Beih. 120: 99 (1940).

Xanthostemon* × *intermedius Gugerli, Repert. Spec. Nov. Regni Veg. Beih. 120: 66 (1940). *X. aurantiacus* × *X. myrtifolius*.
SE. New Caledonia. 60 NWC. Nanophan.

Xanthostemon lateriflorus Guillaumin, Bull. Soc. Bot. France 81: 13 (1934).
New Caledonia (î. Art). 60 NWC. Nanophan.
Fremya lateriflora Bureau ex Guillaumin, Ann. Inst. Bot.-Géol. Colon. Marseille, II, 9: 148 (1911), nom. nud.

Xanthostemon laurinus (Pamp.) Guillaumin, Bull. Soc. Bot. France 81: 15 (1934).
NW. & WC. New Caledonia. 60 NWC. Nanophan. or phan.
Fremya laurina Vieill. ex Pamp. & Pampal., Nuovo Giorn. Bot. Ital., n.s., 12: 678 (1904), nom. inval.
**Xanthostemon multiflorus* f. *laurinus* Pamp., Nuovo Giorn. Bot. Ital., n.s., 12: 678 (1905).

Xanthostemon longipes Guillaumin, Bull. Mus. Natl. Hist. Nat., II, 4: 691 (1932).
SE. New Caledonia. 60 NWC. Nanophan.

Xanthostemon macrophyllus Pamp., Nuovo Giorn. Bot. Ital., n.s., 12: 683 (1904).
EC. New Caledonia. 60 NWC. Nanophan. or phan.

Xanthostemon melanoxylon Peter G.Wilson & Pitisopa, Telopea 11: 399 (2007).
Solomon Is. 43 SOL. Phan.

Xanthostemon multiflorus (Montrouz.) Beauvis., Ann. Soc. Bot. Lyon 26: 46 (1901).
New Caledonia. 60 NWC. Nanophan. or phan.
**Draparnaudia multiflora* Montrouz., Mém. Acad. Roy. Sci. Lyon, Sect. Sci., II, 10: 206 (1860).
Fremya elegans Brongn. & Gris, Bull. Soc. Bot. France 10: 374 (1863). *Nania elegans* (Brongn. & Gris) Kuntze, Revis. Gen. Pl. 1: 242 (1891). *Xanthostemon elegans* (Brongn. & Gris) Nied. in H.G.A.Engler & K.A.E.Prantl, Nat. Pflanzenfam. 3(7): 88 (1893). *Xanthostemon multiflorus* f. *elegans* (Brongn. & Gris) Pamp., Nuovo Giorn. Bot. Ital., n.s., 12: 675 (1905).
Fremya flava Brongn. & Gris, Bull. Soc. Bot. France 10: 373 (1863). *Nania flava* (Brongn. & Gris) Kuntze, Revis. Gen. Pl. 1: 242 (1891). *Xanthostemon flavus* (Brongn. & Gris) Schltr., Nuovo Giorn. Bot. Ital., n.s., 12: 684 (1904). *Xanthostemon multiflorus* f. *flavus* (Brongn. & Gris) Pamp., Nuovo Giorn. Bot. Ital., n.s., 12: 677 (1905).
Salisia flava Pancher, Bull. Soc. Bot. France 10: 373 (1863), nom. inval.
Xanthostemon montrouzieri Pamp., Nuovo Giorn. Bot. Ital., n.s., 12: 687 (1904).
Xanthostemon multiflorus f. *latifolius* Pamp., Nuovo Giorn. Bot. Ital., n.s., 12: 676 (1905). *Xanthostemon elegans* subsp. *latifolius* (Pamp.) Gugerli, Repert. Spec. Nov. Regni Veg. Beih. 120: 116 (1940).
Fremya baudouinii Bureau ex Guillaumin, Ann. Inst. Bot.-Géol. Colon. Marseille, II, 9: 148 (1911), nom. nud.
Xanthostemon baudouinii Guillaumin, Bull. Soc. Bot. France 81: 12 (1934).
Xanthostemon acutifolius Gugerli, Repert. Spec. Nov. Regni Veg. Beih. 120: 92 (1940).
Xanthostemon elegans var. *microphyllus* Gugerli, Repert. Spec. Nov. Regni Veg. Beih. 120: 115 (1940).

Xanthostemon elegans subsp. *pampaninii* Gugerli, Repert. Spec. Nov. Regni Veg. Beih. 120: 115 (1940).
Xanthostemon guillauminii Gugerli, Repert. Spec. Nov. Regni Veg. Beih. 120: 119 (1940).
Xanthostemon longifolius Gugerli, Repert. Spec. Nov. Regni Veg. Beih. 120: 122 (1940).
Xanthostemon ouenensis Gugerli, Repert. Spec. Nov. Regni Veg. Beih. 120: 128 (1940).
Xanthostemon pininsularis Gugerli, Repert. Spec. Nov. Regni Veg. Beih. 120: 109 (1940).

Xanthostemon myrtifolius (Brongn. & Gris) Pamp. ex Pampal., Nuovo Giorn. Bot. Ital., n.s., 13: 135 (1906).
SE. New Caledonia. 60 NWC. Nanophan.
**Fremya myrtifolia* Brongn. & Gris, Bull. Soc. Bot. France 10: 299 (1863).
Xanthostemon integrifolium Baker f., J. Linn. Soc., Bot. 45: 311 (1921).
Fremya integrifolia Brongn. & Gris ex Baker f., J. Linn. Soc., Bot. 45: 311 (1931), nom. inval.
Xanthostemon ramosus Gugerli, Repert. Spec. Nov. Regni Veg. Beih. 120: 69 (1940).

Xanthostemon novaguineensis Valeton, Icon. Bogor.: t. 239 (1907).
New Guinea. 43 NWG. Phan.
Xanthostemon papuanus Lauterb., Nova Guinea 8: 854 (1912).

Xanthostemon oppositifolius F.M.Bailey, Cat. Indig. Natur. Pl. Queensl. Add.: 109 (1890).
SE. Queensland. 50 QLD. Phan.

Xanthostemon paabaensis Gugerli, Repert. Spec. Nov. Regni Veg. Beih. 120: 103 (1940).
New Caledonia. 60 NWC. Nanophan. or phan.

Xanthostemon paradoxus F.Muell., Hooker's J. Bot. Kew Gard. Misc. 9: 18 (1857). *Metrosideros paradoxa* (F.Muell.) F.Muell., Fragm. 1: 243 (1859). *Nania paradoxa* (F.Muell.) Kuntze, Revis. Gen. Pl. 1: 242 (1891).
N. Western Australia to Northern Territory. 50 NTA WAU. Phan.

Xanthostemon petiolatus (Valeton) Peter G.Wilson, Telopea 3: 459 (1990).
Sulawesi. 42 JAW? SUL. Phan.
**Nania petiolata* Valeton, Meded. Lands Plantentuin 40: 172 (1900). *Metrosideros petiolata* (Valeton) Koord., Exkurs.-Fl. Java 2: 686 (1912).

Xanthostemon philippinensis Merr., Philipp. J. Sci., C 12: 289 (1917).
Philippines (Luzon). 42 PHI. Phan.

Xanthostemon psidioides (A.Cunn. ex Lindl.) Peter G.Wilson & J.T.Waterh., Austral. J. Bot. 30: 444 (1982).
N. Western Australia to Northern Territory. 50 NTA WAU. Phan.
**Tristania psidioides* A.Cunn. ex Lindl., Edwards's Bot. Reg. 22: t. 1839 (1836).

Xanthostemon pubescens (Brongn. & Gris) Sebert & Pancher, Rev. Marit. Colon. 41: 227 (1874).
New Caledonia. 60 NWC. Nanophan. or phan.
**Fremya pubescens* Brongn. & Gris, Bull. Soc. Bot. France 10: 373 (1863). *Metrosideros pubescens* (Brongn. & Gris) Baill., Hist. Pl. 6: 337 (1876). *Xanthostemon multiflorus* f. *pubescens* (Brongn. & Gris) Pamp., Nuovo Giorn. Bot. Ital., n.s., 12: 675 (1905).

Xanthostemon retusus Gugerli, Repert. Spec. Nov. Regni Veg. Beih. 120: 120 (1940).
New Caledonia. 60 NWC. Nanophan. or phan.

Xanthostemon ruber (Brongn. & Gris) Sebert & Pancher, Rev. Marit. Colon. 41: 227 (1874).
SE. New Caledonia. 60 NWC. Nanophan. or phan.
**Fremya rubra* Brongn. & Gris, Bull. Soc. Bot. France 10: 372 (1863). *Metrosideros rubra* (Brongn. & Gris) Baill., Hist. Pl. 6: 337 (1876). *Nania rubra* (Brongn. & Gris) Kuntze, Revis. Gen. Pl. 1: 242 (1891).
Salisia rubra Pancher, Bull. Soc. Bot. France 10: 373 (1863).

Xanthostemon sebertii Guillaumin, Bull. Soc. Bot. France 81: 14 (1934).
SE. New Caledonia. 60 NWC. Phan.

Xanthostemon speciosus Merr., Publ. Bur. Sci. Gov. Lab. 6: 10 (1904).
W. & WC. Philippines. 42 PHI. Phan.
Xanthostemon merrillii Pamp., Nuovo Giorn. Bot. Ital., n.s., 12: 688 (1904).
Xanthostemon purpureus Gugerli, Repert. Spec. Nov. Regni Veg. Beih. 120: 53 (1940).

Xanthostemon sulfureus Guillaumin in C.F.Sarasin & J.Roux, Nova Caledonia, Bot. 1: 194 (1921).
SE. New Caledonia. 60 NWC. Nanophan. or phan.

Xanthostemon umbrosus (A.Cunn. ex Lindl.) Peter G.Wilson & J.T.Waterh., Austral. J. Bot. 30: 444 (1982).
N. Australia. 50 NTA QLD WAU. Phan.
**Tristania umbrosa* A.Cunn. ex Lindl., Edwards's Bot. Reg. 22: t. 1839 (1836).
Metrosideros tetrapetala F.Muell., Fragm. 7: 41 (1870). *Nania tetrapetala* (F.Muell.) Kuntze, Revis. Gen. Pl. 1: 242 (1891).

Xanthostemon velutinus (Gugerli) J.W.Dawson, in Fl. Nouv.-Caléd. 18: 183 (1992).
SE. New Caledonia. 60 NWC. Phan.
**Xanthostemon flavus* var. *velutinus* Gugerli, Repert. Spec. Nov. Regni Veg. Beih. 120: 109 (1940).

Xanthostemon verdugonianus Naves ex Fern.-Vill. in F.M.Blanco, Fl. Filip., ed. 3, 4(13A): 82 (1880).
C. & S. Philippines. 42 PHI. Phan.

Xanthostemon verticillatus (C.T.White & W.D.Francis) L.S.Sm., Proc. Roy. Soc. Queensland 67: 38 (1956).
N. Queensland. 50 QLD. Phan.
**Metrosideros verticillata* C.T.White & W.D.Francis, Bot. Bull. Dept. Agric. Queensland 22: 24 (1920).

Xanthostemon verus (Lindl.) Peter G.Wilson, Telopea 3: 459 (1990).
Sulawesi to Maluku. 42 MOL SUL. Phan.
**Metrosideros vera* Lindl., Coll. Bot.: t. 18 (1821). *Nania vera* (Lindl.) Miq., Fl. Ned. Ind. 1(1): 400 (1855).
Eugenia amboinensis Nois. ex Link, Enum. Hort. Berol. Alt. 2: 28 (1822).
Metrosideros suberosa Roxb., Fl. Ind. ed. 1832, 2: 478 (1832). *Nania suberosa* (Roxb.) Kuntze, Revis. Gen. Pl. 1: 242 (1891).
Metrosideros pictipetala Blanco, Fl. Filip., ed. 2: 295 (1845).
Syncarpia vertholenii Teijsm. & Binn., Ned. Kruidk. Arch. 3: 411 (1855).

Xanthostemon vieillardii (Brongn. & Gris) Nied. in H.G.A.Engler & K.A.E.Prantl, Nat. Pflanzenfam. 3(7): 88 (1893).
SE. New Caledonia. 60 NWC. Nanophan. or phan.

Fremya deplanchei Brongn. & Gris, Bull. Soc. Bot. France 10: 378 (1863). *Nania deplanchei* (Brongn. & Gris) Kuntze, Revis. Gen. Pl. 1: 242 (1891). *Xanthostemon deplanchei* (Brongn. & Gris) Guillaumin, Bull. Soc. Bot. France 81: 152: 675 (1934).
**Fremya vieillardii* Brongn. & Gris, Bull. Soc. Bot. France 10: 378 (1863). *Nania vieillardii* (Brongn. & Gris) Kuntze, Revis. Gen. Pl. 1: 242 (1891). *Xanthostemon multiflorus* f. *vieillardii* (Brongn. & Gris) Pamp., Nuovo Giorn. Bot. Ital., n.s., 12: 677 (1905).
Xanthostemon beauvisagei Pamp., Nuovo Giorn. Bot. Ital., n.s., 12: 684 (1904).

Xanthostemon whitei Gugerli, Bull. Soc. Bot. France 81: 83 (1934).
N. Queensland. 50 QLD. Phan.
**Xanthostemon pubescens* C.T.White, Proc. Roy. Soc. Queensland 29: 57 (1917), nom. illeg.

Xanthostemon xerophilus Peter G.Wilson, Telopea 3: 465 (1990).
N. Queensland. 50 QLD. Phan.

Xanthostemon youngii C.T.White & W.D.Francis, Proc. Roy. Soc. Queensland 37: 159 (1925 publ. 1926).
N. Queensland. 50 QLD. Nanophan. or phan.

***Synonyms*:**
Xanthostemon acutifolius Gugerli = ***Xanthostemon multiflorus*** (Montrouz.) Beauvis.
Xanthostemon austrocaledonicus Guillaumin = ***Pleurocalyptus austrocaledonicus*** (Guillaumin) J.W.Dawson
Xanthostemon baudouinii Guillaumin = ***Xanthostemon multiflorus*** (Montrouz.) Beauvis.
Xanthostemon beauvisagei Pamp. = ***Xanthostemon vieillardii*** (Brongn. & Gris) Nied.
Xanthostemon brachyandrus C.T.White = ***Lindsayomyrtus racemoides*** (Greves) Craven
Xanthostemon celebicus Koord. = ***Kjellbergiodendron celebicum*** (Koord.) Merr.
Xanthostemon ciliatum (J.R.Forst. & G.Forst.) Nied. = ***Purpureostemon ciliatus*** (J.R.Forst. & G.Forst.) Gugerli
Xanthostemon deplanchei (Brongn. & Gris) Guillaumin = ***Xanthostemon vieillardii*** (Brongn. & Gris) Nied.
Xanthostemon elegans (Brongn. & Gris) Nied. = ***Xanthostemon multiflorus*** (Montrouz.) Beauvis.
Xanthostemon elegans subsp. *latifolius* (Pamp.) Gugerli = ***Xanthostemon multiflorus*** (Montrouz.) Beauvis.
Xanthostemon elegans var. *microphyllus* Gugerli = ***Xanthostemon multiflorus*** (Montrouz.) Beauvis.
Xanthostemon elegans subsp. *pampaninii* Gugerli = ***Xanthostemon multiflorus*** (Montrouz.) Beauvis.
Xanthostemon flavus (Brongn. & Gris) Schltr. = ***Xanthostemon multiflorus*** (Montrouz.) Beauvis.
Xanthostemon flavus var. *velutinus* Gugerli = ***Xanthostemon velutinus*** (Gugerli) J.W.Dawson
Xanthostemon guillauminii Gugerli = ***Xanthostemon multiflorus*** (Montrouz.) Beauvis.
Xanthostemon integrifolium Baker f. = ***Xanthostemon myrtifolius*** (Brongn. & Gris) Pamp. ex Pampal.
Xanthostemon longifolius Gugerli = ***Xanthostemon multiflorus*** (Montrouz.) Beauvis.
Xanthostemon merrillii Pamp. = ***Xanthostemon speciosus*** Merr.
Xanthostemon montrouzieri Pamp. = ***Xanthostemon multiflorus*** (Montrouz.) Beauvis.
Xanthostemon multiflorus f. *elegans* (Brongn. & Gris) Pamp. = ***Xanthostemon multiflorus*** (Montrouz.) Beauvis.
Xanthostemon multiflorus f. *flavus* (Brongn. & Gris) Pamp. = ***Xanthostemon multiflorus*** (Montrouz.) Beauvis.
Xanthostemon multiflorus f. *latifolius* Pamp. = ***Xanthostemon multiflorus*** (Montrouz.) Beauvis.
Xanthostemon multiflorus f. *laurinus* Pamp. = ***Xanthostemon laurinus*** (Pamp.) Guillaumin
Xanthostemon multiflorus f. *pubescens* (Brongn. & Gris) Pamp. = ***Xanthostemon pubescens*** (Brongn. & Gris) Sebert & Pancher
Xanthostemon multiflorus var. *speciosus* (Brongn. & Gris) Pamp. = ***Xanthostemon gugerlii*** Merr.
Xanthostemon multiflorus f. *vieillardii* (Brongn. & Gris) Pamp. = ***Xanthostemon vieillardii*** (Brongn. & Gris) Nied.
Xanthostemon ouenensis Gugerli = ***Xanthostemon multiflorus*** (Montrouz.) Beauvis.
Xanthostemon pachyspermus F.Muell. & F.M.Bailey = ***Ristantia pachysperma*** (F.Muell. & F.M.Bailey) Peter G.Wilson & J.T.Waterh.
Xanthostemon pancheri (Brongn. & Gris) Schltr. = ***Pleurocalyptus pancheri*** (Brongn. & Gris) J.W.Dawson
Xanthostemon papuanus Lauterb. = ***Xanthostemon novaguineensis*** Valeton
Xanthostemon pininsularis Gugerli = ***Xanthostemon multiflorus*** (Montrouz.) Beauvis.
Xanthostemon pubescens C.T.White = ***Xanthostemon whitei*** Gugerli
Xanthostemon purpureus Gugerli = ***Xanthostemon speciosus*** Merr.
Xanthostemon ramosus Gugerli = ***Xanthostemon myrtifolius*** (Brongn. & Gris) Pamp. ex Pampal.
Xanthostemon rugosus Gugerli = ***Xanthostemon grisii*** Guillaumin
Xanthostemon speciosus (Brongn. & Gris) Pamp. = ***Xanthostemon gugerlii*** Merr.
Xanthostemon speciosus var. *lanceolatus* Gugerli = ***Xanthostemon gugerlii*** Merr.
Xanthostemon sulphureus var. *tomentosus* Däniker = ***Xanthostemon francii*** Guillaumin

Xenodendron

Xenodendron K.Schum. & Lauterb. = ***Syzygium*** Gaertn.
Xenodendron polyanthum K.Schum. & Lauterb. = ***Syzygium acuminatissimum*** (Blume) DC.

Unplaced Names

Acmena chinensis Planch., Hort. Donat.: 84 (1858). = [36 CHS]
Aulomyrcia abrantea O.Berg, Linnaea 30: 659 (1861). = [84]
Aulomyrcia acrantha O.Berg in C.F.P.von Martius & auct. suc. (eds.), Fl. Bras. 14(1): 71 (1857). = [84 BZS]
Aulomyrcia acutata O.Berg in C.F.P.von Martius & auct. suc. (eds.), Fl. Bras. 14(1): 71 (1857). = [84 BZL]
Aulomyrcia alagoensis O.Berg in C.F.P.von Martius & auct. suc. (eds.), Fl. Bras. 14(1): 120 (1857). = [84 BZE]
Aulomyrcia bahiensis O.Berg in C.F.P.von Martius & auct. suc. (eds.), Fl. Bras. 14(1): 113 (1857). = [84 BZE]
Aulomyrcia bicudoensis O.Berg in C.F.P.von Martius & auct. suc. (eds.), Fl. Bras. 14(1): 557 (1859). = [84 BZC]
Aulomyrcia biformis O.Berg in C.F.P.von Martius & auct. suc. (eds.), Fl. Bras. 14(1): 141 (1857). = [84 BZL]
Aulomyrcia bimarginata O.Berg in C.F.P.von Martius & auct. suc. (eds.), Fl. Bras. 14(1): 115 (1857). = [84 BZL]
Aulomyrcia bullata O.Berg in C.F.P.von Martius & auct. suc. (eds.), Fl. Bras. 14(1): 96 (1857). = [84 BZE]
Aulomyrcia caesia O.Berg in C.F.P.von Martius & auct. suc. (eds.), Fl. Bras. 14(1): 83 (1857). = [84 BZE]
Aulomyrcia cambessedeana O.Berg, Linnaea 27: 40 (1855). = [84 BZL]
Aulomyrcia chapadensis O.Berg in C.F.P.von Martius & auct. suc. (eds.), Fl. Bras. 14(1): 554 (1859). = [84 BZC]
Aulomyrcia clausseniana O.Berg in C.F.P.von Martius & auct. suc. (eds.), Fl. Bras. 14(1): 118 (1857). = [84 BZL]
Aulomyrcia comosa O.Berg in C.F.P.von Martius & auct. suc. (eds.), Fl. Bras. 14(1): 133 (1857). = [84 BZL]
Aulomyrcia crenulata O.Berg in C.F.P.von Martius & auct. suc. (eds.), Fl. Bras. 14(1): 141 (1857). = [84 BZL]
Aulomyrcia decrescens O.Berg in C.F.P.von Martius & auct. suc. (eds.), Fl. Bras. 14(1): 135 (1857). = [84 BZC BZL]
Aulomyrcia goyazensis O.Berg in C.F.P.von Martius & auct. suc. (eds.), Fl. Bras. 14(1): 85 (1857). = [84 BZC]
Aulomyrcia grandifolia O.Berg in C.F.P.von Martius & auct. suc. (eds.), Fl. Bras. 14(1): 97 (1857). = [84 BZN]
Aulomyrcia klotzschiana O.Berg in C.F.P.von Martius & auct. suc. (eds.), Fl. Bras. 14(1): 111 (1857). = [84 BZL]
Aulomyrcia langsdorffii O.Berg in C.F.P.von Martius & auct. suc. (eds.), Fl. Bras. 14(1): 550 (1859). = [84 BZL]
Aulomyrcia obtusata O.Berg, Linnaea 27: 39 (1855). = [82 GUY]
Aulomyrcia parvifolia Steyerm., Fieldiana, Bot. 28: 1006 (1957). = [82 VEN]
Aulomyrcia pauciflora O.Berg in C.F.P.von Martius & auct. suc. (eds.), Fl. Bras. 14(1): 560 (1859). = [84 BZL]
Aulomyrcia piauhiensis O.Berg in C.F.P.von Martius & auct. suc. (eds.), Fl. Bras. 14(1): 87 (1857). = [84 BZE]
Aulomyrcia platyclada O.Berg in C.F.P.von Martius & auct. suc. (eds.), Fl. Bras. 14(1): 545 (1859). = [84 BZL]
Aulomyrcia plumbea O.Berg in C.F.P.von Martius & auct. suc. (eds.), Fl. Bras. 14(1): 142 (1857). = [84 BZL]
Aulomyrcia punctata O.Berg in C.F.P.von Martius & auct. suc. (eds.), Fl. Bras. 14(1): 546 (1859). = [84 BZL]
Aulomyrcia pusilla O.Berg in C.F.P.von Martius & auct. suc. (eds.), Fl. Bras. 14(1): 140 (1857). = [84 BZL]
Aulomyrcia regeliana O.Berg in C.F.P.von Martius & auct. suc. (eds.), Fl. Bras. 14(1): 557 (1859). = [84 BZC BZL]
Aulomyrcia riedeliana O.Berg in C.F.P.von Martius & auct. suc. (eds.), Fl. Bras. 14(1): 551 (1859). = [84 BZL]
Aulomyrcia salzmannii O.Berg in C.F.P.von Martius & auct. suc. (eds.), Fl. Bras. 14(1): 116 (1857). = [84 BZE]
Aulomyrcia stenophylla O.Berg, Linnaea 30: 654 (1861). = [85 PAR]
Aulomyrcia ternifolia O.Berg in C.F.P.von Martius & auct. suc. (eds.), Fl. Bras. 14(1): 134 (1857). = [84 BZL]
Aulomyrcia trichantha Wawra, Oesterr. Bot. Z. 29: 215 (1879). = [84]
Aulomyrcia trifolia O.Berg in C.F.P.von Martius & auct. suc. (eds.), Fl. Bras. 14(1): 107 (1857). = [84 BZL]
Baeckea drummondii Benth., Fl. Austral. 3: 75 (1867). = *Rinzia* sp. ?
Baeckea imbricata S.Moore, J. Linn. Soc., Bot. 45: 178 (1920), nom. illeg. = ?
Baeckea spinosa Sieber ex Spreng., Syst. Veg. 4(2): 149 (1827). = [50+]
Baeckea tetragona F.Muell. ex Benth., Fl. Austral. 3: 77 (1867). = [50 WAU]
Beaufortia puberula Turcz., Bull. Cl. Phys.-Math. Acad. Imp. Sci. Saint-Pétersbourg 10: 345 (1852). = [50 WAU]
Blepharocalyx spirioides Stapf, Bot. Mag. 133: t. 8123 (1907). = [84]
Callistemon amoenus Lem., Ill. Hort. 7: t. 247 (1860). = ?
Callistemon cunninghamii K.Koch, Wochenschr. Gärtnerei Pflanzenk. 6: 294 (1863). = ?
Callistemon hortensis Cheel, Proc. Linn. Soc. New South Wales 68: 184 (1943). = ?
Callistemon hybridus DC., Prodr. 3: 224 (1828). = ?
Calyptranthes fastigiata O.Berg in C.F.P.von Martius & auct. suc. (eds.), Fl. Bras. 14(1): 50 (1857), nom. illeg. = [84 BZL]
Calyptranthes rostrata Griseb., Mem. Amer. Acad. Arts, n.s., 8: 181 (1861), nom. illeg. = [81 CUB]
Calyptromyrcia cordata O.Berg in C.F.P.von Martius & auct. suc. (eds.), Fl. Bras. 14(1): 56 (1857). = [84 BZE]
Calyptromyrcia paniculata O.Berg in C.F.P.von Martius & auct. suc. (eds.), Fl. Bras. 14(1): 57 (1857). = [84 BZL]
Campomanesia arenaria O.Berg in C.F.P.von Martius & auct. suc. (eds.), Fl. Bras. 14(1): 448 (1857). = [85 URU]

Campomanesia dentata O.Berg in C.F.P.von Martius & auct. suc. (eds.), Fl. Bras. 14(1): 457 (1857). = [84 BZS]
Caryophyllus adenolepis Miq., Fl. Ned. Ind. 1: 466 (1855). = [42 JAW]
Catinga media Sagot in ?. = ?
Cerqueiria sellowiana O.Berg, Linnaea 27: 6 (1855). = [84 BZE]
Darwinia forrestii F.Muell., Fragm. 11: 9 (1878). = *Chamelaucium* sp. ?
Eucalyptus alata Sweet, Hort. Brit.: 157 (1826). = ?
Eucalyptus albicans F.Muell., Fragm. 7: 42 (1869). = ?
Eucalyptus albicaulis Sweet, Hort. Brit.: 157 (1826). = ?
Eucalyptus × *algeriensis* A.Vilm. ex Trab., Rev. Hort. Algérie Tunisie Maroc 8: 146 (1904). = *E. camaldulensis* × *E. rudris*
Eucalyptus amygdalina var. *hypericifolia* Benth., Fl. Austral. 3: 203 (1867). = [50 TAS] *E. amygdalina* × *E. risdonii*
Eucalyptus androsaemifolia Hoffmanns., Verz. Pfl.-Kult., Nachtr. 2: 113 (1841). = ?
Eucalyptus × *antipolitensis* Trab. ex Maiden, Crit. Revis. Eucalyptus 6: 70 (1922). = ?
Eucalyptus apodophylla var. *brachyphylla* Blakely & Jacobs, Key Eucalypts: 166 (1934). = *E. alba* × *E. apodophylla*
Eucalyptus bauerlesii F.Muell., Victorian Naturalist 7: 76 (1890). = ?
Eucalyptus beauchampiana Elwes & A.Henry, Trees Great Britain 6: 1615 (1912). = ?
Eucalyptus × *biangularis* Simmonds in J.H.Maiden, Crit. Revis. Eucalyptus 7: 382 (1927). = *E. globulus* × *E. urnigera*
Eucalyptus botryoides var. *lynei* Blakely, Key Eucalypts: 97 (1934). = [50 NSW] *E. botryoides* × *E. robusta*
Eucalyptus × *bourlieri* Trab., Rev. Hort. Algérie Tunisie Maroc 1903: 327 (1903). = ?
Eucalyptus calcicultrix var. *obscura* Blakely, Key Eucalypts: 224 (1934). = [50 SOA] *E. albens* × *E. odorata*
Eucalyptus × *californica* Kinney, Eucalyptus: 191 (1895). = ?
Eucalyptus calophylla var. *hawkeyi* Blakely, Key Eucalypts: 85 (1934). = [50 WAU] *Corymbia calophylla* × *Corymbia ficifolia*
Eucalyptus carnea C.A.Gardner, J. Roy. Soc. Western Australia 14: 82 (1928), nom. illeg. = ?
Eucalyptus × *coniophloia* D.J.Carr & S.G.M.Carr, Eucalyptus 1: 97 (1985). = [50 NTA WAU] *Corymbia capricornia* × *Corymbia drysdalensis*
Eucalyptus connata Dum.Cours., Bot. Cult., ed. 2, 7: 280 (1814). = ?
Eucalyptus cotinifolia Colla, Herb. Pedem. 2: 423 (1834). = ?
Eucalyptus × *cultrifolia* Naudin, Descr. Emploi Eucalypt.: 64 (1891). = ?
Eucalyptus curvula Sieber ex Spreng., Syst. Veg. 4(2): 195 (1827). = ?
Eucalyptus ficifolia var. *alba* Blakely, Key Eucalypts: 86 (1934), nom. illeg. = *E. calophylla* × *E. ficifolia*
Eucalyptus ficifolia var. *guilfoylei* F.M.Bailey, Proc. Roy. Soc. Queensland 10: 18 (1894). = *Corymbia calophylla* × *Corymbia ficifolia*
Eucalyptus firma F.Muell. ex Miq., Ned. Kruidk. Arch. 4: 133 (1856). = ?
Eucalyptus fissilis Guilf., Rep. (Annual) Director Bot. Gard. Melbourne 1877: 25 (1877). = ?
Eucalyptus flavieri Añon, Rev. Int. Bot. Appl. Agric. Colon. 26: 234 (1946). = ?
Eucalyptus flexilis Regel, Gartenflora 7 7: 284 (1858). = ?
Eucalyptus gillii var. *petiolaris* Maiden, J. Proc. Roy. Soc. New S. Wales 53: 69 (1919). = [50 SOA] *E. gillii* × *E. socialis*
Eucalyptus globoidea var. *largifructa* Blakely, Key Eucalypts: 189 (1934). = ?
Eucalyptus globulus var. *compacta* Maiden, Crit. Revis. Eucalyptus 8: 23 (1929). = ?
Eucalyptus × *gomphocornuta* A.Vilm. ex Trab., Rev. Hort., II, 3: 326 (1903). = *E. cornuta* × *E. gomphocephala*
Eucalyptus grandis var. *grandiflora* Maiden, J. Proc. Roy. Soc. New S. Wales 52: 502 (1919). = *E. grandis* × *E. robusta*
Eucalyptus inophloea Brogli, Beitr. Anat. Myrtac.-Rinden: 20 (1926). = [50]
Eucalyptus insingiana Maiden, Crit. Revis. Eucalyptus 6: 353 (1923). = ?
Eucalyptus × *insizwaensis* Maiden, Crit. Revis. Eucalyptus 6: 82 (1922). = *E. globulus* × *E. robusta*
Eucalyptus irbyi R.T.Baker & H.F.Sm., Res. Eucalypts, ed. 2: 242 (1920). = [50 TAS]
Eucalyptus × *jugalis* Naudin, Descr. Emploi Eucalypt.: 37 (1891). = ?
Eucalyptus kitsonii H.Deane in ?. = Fossil species.
Eucalyptus laevopinea var. *turbinata* Blakely, Key Eucalypts: 183 (1934). = ?
Eucalyptus × *lamprocalyx* Blakely, Key Eucalypts: 87 (1934). = [50 WAU] *Corymbia cadophora* × *Corymbia polycarpa*
Eucalyptus laseronii var. *doleiformis* Blakely, Key Eucalypts: 193 (1934). = ?
Eucalyptus laseronii var. *maxima* Blakely, Key Eucalypts: 193 (1934). = ?
Eucalyptus leucoxylon var. *pluriflora* Miq., Ned. Kruidk. Arch. 4: 127 (1856). = ?
Eucalyptus longifolia var. *multiflora* Maiden, Crit. Revis. Eucalyptus 6: 432 (1923). = *E. longifolia* × *E. robusta*
Eucalyptus longifolia var. *turbinata* Blakely, Key Eucalypts: 102 (1934). = [50 NSW] *E. longifolia* × *E. tereticornis*
Eucalyptus maidenii var. *williamsonii* Blakely, Key Eucalypts: 157 (1934). = [50 VIC] *E. botruoides* × *E. globulus* subsp. *pseudoglobulus*
Eucalyptus mazeliana Naudin, Descr. Emploi Eucalypt.: 41 (1891). = ?
Eucalyptus × *mcclatchiei* Kinney, Eucalyptus: 188 (1895). = *E. globulus* × *E. ovata*
Eucalyptus × *media* Link, Enum. Hort. Berol. Alt. 2: 30 (1822). = [50+]
Eucalyptus melanophloia var. *senta* Blakely, Key Eucalypts: 253 (1934). = ?
Eucalyptus microphylla Muhl. ex Willd., Enum. Pl.: 515 (1809). = not Eucalyptus
Eucalyptus microtheca var. *cymbaliformis* Blakely & Jacobs in W.F.Blakely, Key Eucalypts: 245 (1934). = [50 NTA] *E. cyanoclada* × *E. microtheca*
Eucalyptus mitchellii Ettingsh., Blatt-Skel. Dikot.: 203 (1861). = Fossil species.
Eucalyptus moorei var. *arborea* Blakely, Key Eucalypts: 207 (1934). = *E. moorei* × *E. piperita*
Eucalyptus × *mortoniana* Kinney, Eucalyptus: 193 (1895). = *E. globulus* × *E. viminalis*
Eucalyptus mucronata Link, Enum. Hort. Berol. Alt. 2: 30 (1822). = ?
Eucalyptus multiflora Rich. ex A.Gray, U.S. Expl. Exped., Phan. 1: 554 (1854), nom. illeg. = [42 PHI]

Eucalyptus myrtifolia Link, Enum. Hort. Berol. Alt. 2: 30 (1822). = ?
Eucalyptus × *nervosa* Hoffmanns., Verz. Pfl.-Kult. 1: 134 (1824). = ?
Eucalyptus × *nowraensis* Maiden, Crit. Revis. Eucalyptus 7: 68 (1924). = [50 NSW] *Corymbia gummifera* × *Corymbia maculata*
Eucalyptus obliqua var. *discocarpa* Blakely, Key Eucalypts: 194 (1934). = [50 VIC] *E. muelleriana* × *E. obliqua*
Eucalyptus obliqua var. *microstoma* Blakely, Key Eucalypts: 194 (1934). = ?
Eucalyptus obliqua var. *pilula* Blakely, Key Eucalypts: 194 (1934). = ?
Eucalyptus occidentalis var. *californica* Kinney, Eucalyptus: 92 (1895). = ?
Eucalyptus odorata var. *macrocarpa* Blakely, Key Eucalypts: 226 (1934). = [50 SOA] *E. leucoxyla* × *E. odorata*
Eucalyptus pachypoda Benth., Fl. Austral. 3: 233 (1867). = ?
Eucalyptus × *pachypoda* F.Muell., Fragm. 7: 41 (1870), nom. illeg. = ?
Eucalyptus pauciflora var. *cylindrocarpa* Blakely, Key Eucalypts: 205 (1934). = [50 NSW]
Eucalyptus pauciflora var. *densiflora* Blakely & McKie, Proc. Linn. Soc. New South Wales 63: 66 (1938). = [50 NSW]
Eucalyptus pauciflora var. *rusticata* Blakely, Key Eucalypts: 205 (1934). = ?
Eucalyptus piperita var. *orophila* Blakely, Key Eucalypts: 339 (1934). = [50 NSW] *E. oreodes* × *E. piperita*
Eucalyptus × *plurilocularis* F.Muell., Fragm. 2: 70 (1860). = ?
Eucalyptus × *purpurascens* Link, Enum. Hort. Berol. Alt. 2: 31 (1822). = ?
Eucalyptus × *rameliana* A.Vilm. ex Trab., Assoc. Franç. 'Avancem. Sci. Marseille 1891: 463 (1892), nom. illeg. = ?
Eucalyptus × *rameliana* var. *trabutii* Vilm. ex Trab., Bull. Stat. Recherch. Forest. Nord Afr. 1: 141 (1917). = *E. botryoides* × *E. camaldulensis*
Eucalyptus × *reticulata* Link, Enum. Hort. Berol. Alt. 2: 29 (1822). = ?
Eucalyptus × *rigida* Hoffmanns., Verz. Pfl.-Kult., Nachtr. 2: 114 (1841). = ?
Eucalyptus × *rigida* var. *luehmanniana* F.Muell., Eucalyptographia 4: t. 6 (1879). = ?
Eucalyptus × *rubricaulis* Desf., Tabl. École Bot., ed. 3: 408 (1829). = ?
Eucalyptus scyphoidea Maiden, Proc. Linn. Soc. New South Wales 28: 899 (1904). = ?
Eucalyptus stenophylla Link, Enum. Hort. Berol. Alt. 2: 30 (1822). = ?
Eucalyptus stricta var. *pyrifera* Blakely, Key Eucalypts: 202 (1934). = [50 NSW] *E. sieberi* × *E. stricta*
Eucalyptus stricta var. *rigida* H.Deane & Maiden, Proc. Linn. Soc. New South Wales 22: 710 (1898). = ?
Eucalyptus tereticornis var. *latifolia* Benth., Fl. Austral. 3: 242 (1867). = [50 QLD] *E. platyphylla* × *E. tereticornis*
Eucalyptus teretiuscula Ettingsh., Blatt-Skel. Dikot.: 201 (1861). = [50]
Eucalyptus × *tokwa* D.J.Carr & S.G.M.Carr, Eucalyptus 2: 152 (1987). = [43 NWG 50 NTA QLD WAU] *Corymbia latifolia* × *Corymbia novoguinensis*
Eucalyptus triartha Domin, Biblioth. Bot. 89: 461 (1928). = [50]
Eucalyptus trinervis Ettingsh., Blatt-Skel. Dikot.: 204 (1861). = [50]
Eucalyptus tuberculata Parm. ex DC., Prodr. 3: 221 (1828). = ?
Eucalyptus umbellata Dum.Cours., Bot. Cult., ed. 2, 7: 279 (1814). = [50]
Eucalyptus umbellata var. *media* Blakely, Key Eucalypts: 130 (1934). = [50 VIC] *E. camaldulensis* × *E. tereticornis*
Eucalyptus × *urnularis* D.J.Carr & S.G.M.Carr, Eucalyptus 1: 87 (1985). = [50 NTA] *Corymbia dichromophloia* × *Corymbia latifolia*
Eucalyptus verrucosa Sweet, Hort. Brit.: 157 (1826). = ?
Eucalyptus verrucosa Colla, Herb. Pedem. 2: 422 (1834), nom. illeg. = ?
Eucalyptus viridis var. *latiuscula* Blakely, Key Eucalypts: 229 (1934). = [50 QLD] *E. microcarpa* × *E. viridis*
Eucalyptus × *vittellina* Naudin, Descr. Emploi Eucalypt.: 65 (1891). = ?
Eucalyptus whittingehamensis G.Nicholson ex Elwes & Henry, Trees Great Britain 6: 1642 (1912). = ?
Eucalyptus yumbarrana subsp. *striata* Boomsma, J. Adelaide Bot. Gard. 2: 298 (1980). = [50 SOA] *E. canescens* × *E. volkesensis*
Eugenia acris Wight & Arn., Prodr. Fl. Ind. Orient.: 331 (1834). = ?
Eugenia aemula Diels, Bot. Jahrb. Syst. 40(91): 47 (1907), nom. illeg. = [83 ECU]
Eugenia altipeta Greves, J. Bot. 61(Suppl.): 14 (1923). = [43 NWG]
Eugenia amboinensis Greves, J. Bot. 62(Suppl.): 36 (1924), nom. illeg. = [42 MOL]
Eugenia andina O.Berg, Linnaea 27: 274 (1856). = ?
Eugenia axillaris G.Don, Gen. Hist. 2: 866 (1832), nom. illeg. = ?
Eugenia axillaris Koord. & Valeton, Bull. Inst. Bot. Buitenzorg 2: 6 (1899), nom. illeg. = [42 JAW]
Eugenia bantamensis Koord. & Valeton, Bull. Inst. Bot. Buitenzorg 2: 6 (1899). = [42 JAW]
Eugenia biniflora Ridl., Bull. Misc. Inform. Kew 1925: 80 (1925). = [42 BOR SUM]
Eugenia calycorectoides Guillaumin, Bull. Mus. Natl. Hist. Nat., II, 20: 364 (1948), nom. illeg. = [60 NWC]
Eugenia cassinoides O.Berg in C.F.P.von Martius & auct. suc. (eds.), Fl. Bras. 14(1): 286 (1857), nom. illeg. = [84 BZE]
Eugenia cassinoides var. *difformis* O.Berg in C.F.P.von Martius & auct. suc. (eds.), Fl. Bras. 14(1): 580 (1859). = [84 BZE]
Eugenia cassinoides var. *gracilis* O.Berg in C.F.P.von Martius & auct. suc. (eds.), Fl. Bras. 14(1): 286 (1857). = [84 BZE]
Eugenia cassinoides var. *robusta* O.Berg in C.F.P.von Martius & auct. suc. (eds.), Fl. Bras. 14(1): 286 (1857). = [84 BZE]
Eugenia cauliflora Ridl., Trans. Linn. Soc. London, Bot. 3: 299 (1893), nom. illeg. = [42 MLY]
Eugenia chinensis Regel, Cat. Pl. Hort. Aksakov.: 57 (1860). = ?
Eugenia costata Bello, Anales Soc. Esp. Hist. Nat. 10: 272 (1881), nom. illeg. = [81 PUE]
Eugenia costata (Miq.) Koord. & Valeton, Meded. Lands Plantentuin 40: 162 (1900), nom. illeg. = ?
Eugenia djouat Perrier, Mém. Soc. Linn. Paris 3: 116 (1825). = [42 PHI]
Eugenia emarginata Vahl in H.West, Bidr. Beskr. Ste Croix: 290 (1793), nom. inval. = [81 LEE]

Eugenia fischeri Mattos, Loefgrenia 119: 1 (2004), nom. illeg. = [84 BZC]
Eugenia garcinioides Engl. & Brehmer, Bot. Jahrb. Syst. 54: 334 (1917), nom. illeg. = [23 CMN]
Eugenia glandulosa Blanco, Fl. Filip.: 417 (1837), nom. illeg. = [42 PHI]
Eugenia gracilis O.Berg in C.F.P.von Martius & auct. suc. (eds.), Fl. Bras. 14(1): 222 (1857), nom. illeg. = [84 BZL]
Eugenia grisea Mattos, Loefgrenia 107: 2 (1995), nom. illeg. = [84 BZC]
Eugenia hexovulata McVaugh, Fieldiana, Bot. 29: 210 (1956). = ?
Eugenia incerta Dummer, Gard. Chron., III, 52: 179 (1912). = [27 NAT]
Eugenia incertissima Sobral, Candollea 42: 109 (1987), nom. inval. = [85 PAR]
Eugenia intermedia Koord. & Valeton, Bull. Inst. Bot. Buitenzorg 2: 7 (1899), nom. illeg. = [42 JAW]
Eugenia koikokoensis Greves, J. Bot. 61(Suppl.): 20 (1923). = ?
Eugenia koordersii Herter, Revista Sudamer. Bot. 7: 219 (1943). = [42 JAW]
Eugenia laevigata Miq., Anal. Bot. Ind. 1: 25 (1850), nom. illeg. = [42 BOR]
Eugenia lepidota var. *pauciflora* O.Berg, Linnaea 27: 227 (1856). = [82 GUY]
Eugenia leptosperma Roxb., Fl. Ind. ed. 1832, 2: 495 (1832). = ?
Eugenia magnifica Brongn. & Gris, Bull. Soc. Bot. France 12: 178 (1865), nom. illeg. = [60 NWC]
Eugenia marginata W.Hill, Cat. Nat. Indust. Prod. Queensland: 23 (1862), nom. illeg. = ?
Eugenia megalophylla Merr., Philipp. J. Sci. 27: 42 (1925). = [42 PHI]
Eugenia metriosa Rojas, Cat. Hist. Nat. Corrientes: 62 (1897). = [85 AGE]
Eugenia michelii Parodi, Contr. Fl. Paraguay 4: 122 (1879), nom. illeg. = [85 PAR]
Eugenia miqueliana Greves, J. Bot. 61(Suppl.): 16 (1923). = ?
Eugenia mirtiflora Rojas, Cat. Hist. Nat. Corrientes: 61 (1897). = [85 AGE]
Eugenia molliana O.Berg, Linnaea 27: 295 (1856). = [8+]
Eugenia multiflora Rich., Actes Soc. Hist. Nat. Paris 1: 110 (1792). = [82 FRG]
Eugenia myrcinifolia Hance, J. Bot. 23: 8 (1885). = ?
Eugenia nettiana Kiaersk., Enum. Myrt. Bras.: 179 (1893). = [84 BZL] *Myrcia* sp.
Eugenia nigra Pennant, Outlin. Globe 4: 277 (1800). = [40 IND]
Eugenia nigra DC., Prodr. 3: 268 (1828), nom. illeg. = (Melastomataceae)
Eugenia oleoides Planch. & Linden, Bot. Zeitung (Berlin) 12: 364 (1854). = ?
Eugenia paniculata Bello, Anales Soc. Esp. Hist. Nat. 10: 271 (1881), nom. illeg. = [81 PUE]
Eugenia pomatensis Miq., Anal. Bot. Ind. 1: 22 (1850). = [42 BOR]
Eugenia pseudodichasiantha Kiaersk., Enum. Myrt. Bras.: 177 (1893). = *Plinia* sp. ?
Eugenia reticulata Trimen, Handb. Fl. Ceylon 2: 185 (1894), nom. illeg. = [40 SRL]
Eugenia robertii Merr., Philipp. J. Sci. 1(Suppl.): 106 (1906). = [42 PHI]
Eugenia rosea Barb.Rodr. ex Chodat & Hassl., Bull. Herb. Boissier, II, 7: 805 (1907), nom. subnud. = [85 PAR]
Eugenia rotundifolia Casar., Nov. Stirp. Bras.: 40 (1842), nom. illeg. = [84 BZL]
Eugenia rotundifolia var. *oblongata* O.Berg in C.F.P.von Martius & auct. suc. (eds.), Fl. Bras. 14(1): 287 (1857). = [84 BZL]
Eugenia ruiziana O.Berg, Linnaea 27: 167 (1856). = [82 FRG]
Eugenia simulans King, J. Asiat. Soc. Bengal, Pt. 2, Nat. Hist. 70: 128 (1901). = [42 MLY]
Eugenia suavis Ridl., J. Fed. Malay States Mus. 5: 160 (1915). = [42 MLY]
Eugenia subcordata O.Berg in C.F.P.von Martius & auct. suc. (eds.), Fl. Bras. 14(1): 300 (1857), nom. illeg. = [84 BZS]
Eugenia subcordata var. *australis* Mattos, Loefgrenia 70: 4 (1976). = [84 BZS]
Eugenia subobliqua Benth., J. Bot. (Hooker) 2: 322 (1840). = [82 GUY]
Eugenia suzukii Kaneh., Bot. Mag. (Tokyo) 45: 335 (1931). = [62 CRL]
Eugenia tanalensis Baker, J. Bot. 20: 111 (1882). = [29 MDG] *Syzygium* sp. ?
Eugenia tapiraguayensis Barb.Rodr. ex Chodat & Hassl., Bull. Herb. Boissier, II, 7: 804 (1907), nom. inval. = [85 PAR]
Eugenia tetraedra Duthie in J.D.Hooker, Fl. Brit. India 2: 476 (1878). = ?
Eugenia tetrasperma Bello, Anales Soc. Esp. Hist. Nat. 10: 271 (1881). = [81 PUE]
Eugenia teyu-iba Parodi, Anales Soc. Ci. Argent. 7: 116 (1879). = [85 AGE]
Eugenia tomentosa Cambess. in A.F.C.de Saint-Hilaire, Fl. Bras. Merid. 2: 353 (1832), nom. illeg. = [84 BZL]
Eugenia trukensis Hosok., J. Jap. Bot. 13: 281 (1937). = [62 CRL]
Eugenia valetoniana King, J. Asiat. Soc. Bengal, Pt. 2, Nat. Hist. 70: 112 (1901). = [42 MLY]
Eugenia vinifera Rojas, Cat. Hist. Nat. Corrientes: 62 (1897). = [85 AGE]
Eugenia virgata Macfad., Fl. Jamaica 2: 121 (1850), nom. illeg. = [81 JAM]
Eugenia vitis-idaea C.Wright, Anales Acad. Ci. Méd. Habana 5: 431 (1868), nom. illeg. = [81 CUB]
Eugenia zenkeri Engl., Notizbl. Königl. Bot. Gart. Berlin 2: 291 (1899). = ?
Gomidesia barituensis Legname, Lilloa 35: 79 (1978). = [83 BOL 85 AGE]
Gomidesia blanchetiana O.Berg in C.F.P.von Martius & auct. suc. (eds.), Fl. Bras. 14(1): 14 (1857). = [84 BZE]
Gomidesia cambessedeana O.Berg in C.F.P.von Martius & auct. suc. (eds.), Fl. Bras. 14(1): 24 (1857). = [84 BZL]
Gomidesia flagellaris D.Legrand, Sellowia 13: 279 (1961). = [84 BZS]
Gomidesia haenkeana O.Berg in C.F.P.von Martius & auct. suc. (eds.), Fl. Bras. 14(1): 515 (1858). = [83 BOL]
Gomidesia klotzschiana O.Berg in C.F.P.von Martius & auct. suc. (eds.), Fl. Bras. 14(1): 534 (1859). = [84 BZL]
Gomidesia kunthiana O.Berg in C.F.P.von Martius & auct. suc. (eds.), Fl. Bras. 14(1): 23 (1857). = [84 BZL]
Gomidesia myrcioides Mattos & D.Legrand, Loefgrenia 67: 14 (1975). = [83 BOL]
Gomidesia poeppigiana O.Berg in C.F.P.von Martius & auct. suc. (eds.), Fl. Bras. 14(1): 14 (1857). = [84 BZL]

Gomidesia spruceana O.Berg in C.F.P.von Martius & auct. suc. (eds.), Fl. Bras. 14(1): 534 (1859). = [84 BZL]
Gomidesia squamata Mattos & D.Legrand, Loefgrenia 67: 14 (1975). = [84 BZS]
Gomidesia velutiflora Mattos & D.Legrand, Loefgrenia 67: 15 (1975). = [84 BZC]
Gomidesia willdenowiana O.Berg in C.F.P.von Martius & auct. suc. (eds.), Fl. Bras. 14(1): 19 (1857). = [84 BZL]
Harmogia parviflora Turcz., Bull. Cl. Phys.-Math. Acad. Imp. Sci. Saint-Pétersbourg 10: 330 (1852). = [50]
Jambosa acida auct., Gard. Chron. 1876(1): 735 (1876). = ?
Jambosa anastomosans Miq., Fl. Ned. Ind., Eerste Bijv.: 309 (1861). = [42 SUM]
Jambosa bisulca Miq., Fl. Ned. Ind., Eerste Bijv.: 310 (1861). = [42 SUM]
Jambosa brackenridgei Brongn. & Gris, Bull. Soc. Bot. France 12: 181 (1865). = [60 NWC]
Jambosa celebica Blume, Mus. Bot. 1: 107 (1850). = [42 SUL]
Jambosa coarctata var. *firma* Blume, Mus. Bot. 1: 100 (1850). = [42 SUM]
Jambosa condensata Miq., Fl. Ned. Ind. 1(1): 419 (1855). = [42 JAW]
Jambosa costata Miq., Fl. Ned. Ind. 1(1): 415 (1855). = [42 JAW]
Jambosa courtallensis Gamble, Bull. Misc. Inform. Kew 1918: 239 (1918). = [40 IND SRL]
Jambosa glandulosa Korth., Ned. Kruidk. Arch. 1: 201 (1847). = [42 BOR]
Jambosa hardwickia G.Don, Gen. Hist. 2: 869 (1832). = [29 MAU]
Jambosa horsfieldii Miq., Fl. Ned. Ind. 1(1): 420 (1855). = [42 JAW]
Jambosa kigurrung Blume, Mus. Bot. 1: 107 (1850). = [42 JAW]
Jambosa maritima Miq., Fl. Ned. Ind. 1(1): 435 (1855). = [42 JAW]
Jambosa melastomifolia Blume, Mus. Bot. 1: 102 (1850). = ?
Jambosa perforata Miq., Fl. Ned. Ind., Eerste Bijv.: 309 (1861). = [42 SUM]
Jambosa polyneura Miq., Fl. Ned. Ind., Eerste Bijv.: 415 (1861). = [42 JAW]
Jambosa pseudodensiflora Hochr., Candollea 2: 452 (1925). = [42 JAW]
Jambosa pterocaulis Korth., Ned. Kruidk. Arch. 1: 200 (1847). = [42 BOR]
Jambosa puncticulata Miq., Fl. Ned. Ind., Eerste Bijv.: 310 (1861). = [42 SUM]
Jambosa purpurea Korth., Ned. Kruidk. Arch. 1: 201 (1847). = [42 SUM]
Jambosa rhyncophylla Miq., Fl. Ned. Ind. 1(1): 430 (1855). = [42 JAW]
Jambosa rhytidocarpa Zoll. in F.a.W.Miquel, Choix Pl. Buitenzorg: t. 4 (1863). = ?
Jambosa rostrata Miq., Fl. Ned. Ind. 1(1): 436 (1855). = [42 SUM]
Jambosa salicina Diels, Bot. Jahrb. Syst. 57: 390 (1922). = [43 NWG]
Jambosa sulcata Teijsm. & Binn., Natuurk. Tijdschr. Ned.-Indië 27: 54 (1864). = ?
Jambosa sumatrana Miq., Fl. Ned. Ind. 1(1): 419 (1855). = [42 SUM]
Jossinia brachypoda Merr., Philipp. J. Sci. 79: 3359 (1951). = [42 PHI]
Jossinia celebica Blume, Mus. Bot. 1: 124 (1850). = [42 SUL]
Jossinia desmantha Diels, J. Arnold Arbor. 10: 82 (1929). = [43 NWG]
Jossinia littoralis Blume, Mus. Bot. 1: 124 (1850). = [43 NWG]
Jossinia richardii Blume, Mus. Bot. 1: 123 (1850). = [29 MDG]
Jossinia schlechteri Diels, Bot. Jahrb. Syst. 57: 377 (1922). = [43 NWG]
Kunzea sprengelioides Turcz., Bull. Cl. Phys.-Math. Acad. Imp. Sci. Saint-Pétersbourg 10: 336 (1852). = [50]
?*Kunzspermum* W.Harris, Hort. New Zealand 4(2): 10 (1993). = *Kunzea* × *Leptospermum*
?*Kunzspermum hirakimata* W.Harris, Hort. New Zealand 4: 10 (1993). = *Kunzea sinclairii* × *Leptospermum scoparium*
Leptospermum celsianum Graeffer ex Ten., Cat. Pl. Neapol. 1813, app.: 11 (1815). = ?
Leptospermum ciliolatum Otto & A.Dietr., Allg. Gartenzeitung 9: 241 (1841). = [50]
Leptospermum cuneiforme Otto & A.Dietr., Allg. Gartenzeitung 9: 251 (1841). = [50]
Leptospermum cupressinum Otto & A.Dietr., Allg. Gartenzeitung 9: 250 (1841). = [50]
Leptospermum pungens Banks ex Dum.Cours., Bot. Cult., ed. 2, 5: 385 (1811). = ?
Leptospermum salicifolium Lam., Encycl. 3: 467 (1792). = [42 JAW]
Leptospermum villosum Fisch., Cat. Jard. Gorenki, ed. 2: 63 (1812). = ?
Leptospermum × *violipurpureum* W.Harris, M.I.Dawson & Heenan, Hort. New Zealand 6: 6 (1995). = *L. rotundifolium* × *L. spectabile*
Luma filipes Burret, Notizbl. Bot. Gart. Berlin-Dahlem 15: 529 (1941). = [84 BZL]
Marlierea elliptica var. *cubensis* Griseb., Cat. Pl. Cub.: 86 (1866). = [81 CUB]
Melaleuca baxteri Benth., Fl. Austral. 3: 138 (1867). = *Agonis* sp.
Melaleuca citrina Turcz., Bull. Cl. Phys.-Math. Acad. Imp. Sci. Saint-Pétersbourg 10: 341 (1852), nom. illeg. = [50]
Melaleuca grandis Fisch. ex Steud., Nomencl. Bot., ed. 2, 2: 112 (1841). = ?
Melaleuca hispida J.C.Wendl. ex DC., Prodr. 3: 215 (1828). = ?
Melaleuca imbricata Link, Enum. Hort. Berol. Alt. 2: 272 (1822). = ?
Melaleuca pendulina G.Don ex Loudon, Hort. Brit.: 319 (1830). = ?
Melaleuca trinervia Dum.Cours., Bot. Cult., ed. 2: 276 (1814), nom. illeg. = ?
Melaleuca uniflora Spreng., Syst. Veg. 4(2): 194 (1827). = ?
Metrosideros aspera Dum.Cours., Bot. Cult., ed. 2, 5: 381 (1811). = ?
Metrosideros canaliculata Dum.Cours., Bot. Cult., ed. 2, 5: 381 (1811). = ?
Metrosideros canaliculata Colla, Hortus Ripul., App. 3: 155 (1826). = ?
Metrosideros decurrens Steud., Nomencl. Bot., ed. 2, 2: 137 (1841). = ?
Metrosideros falcata Dehnh., Rivista Napol. 1(3): 172 (1839). = ?
Metrosideros macrophylla Lam., Tabl. Encycl. 2: 538 (1819). = [29 MDG] (?) not a Myrtaceae.
Metrosideros nana C.C.Gmel. ex DC., Prodr. 3: 226 (1828). = ?
Metrosideros ruscifolia Pépin, Ann. Fl. Pomone 1838–1839: 380 (1839). = ?

Metrosideros scariosa J.C.Wendl. ex Hornem., Hort. Bot. Hafn., Suppl.: 139 (1819). = ?
Mitranthes gardneriana O.Berg in C.F.P.von Martius & auct. suc. (eds.), Fl. Bras. 14(1): 354 (1857). = [84 BZE]
Myrceugenia longipedicellata Barb.Rodr., Myrt. Paraguay: 2 (1903). = [85 PAR]
Myrceugenia regnelliana var. *dubia* D.Legrand, in Fl. Ilustr. Catar. 1(Mirt.): 429 (1970). = *Myrceugenia euosma* × *Myrceugenia glaucescens*
Myrcia acuminatissima Hieron., Bot. Jahrb. Syst. 20(49): 64 (1895), nom. illeg. = [84 BZL]
Myrcia pubescens G.Don, Gen. Hist. 2: 845 (1832). = ?
Myrcia reticulata Krug & Urb., Bot. Jahrb. Syst. 15: 306 (1893), nom. illeg. = [81 DOM]
Myrcia triantha DC., Prodr. 3: 256 (1828). = ?
Myrcialeucus odorifolius Rojas, Bull. Acad. Int. Géogr. Bot. 24: 217 (1914). = [85 AGE]
Myrcianthes cionei Mattos, Loefgrenia 54: 2 (1971). = ?
Myrcianthes terminalis Mattos & D.Legrand, Loefgrenia 67: 12 (1975). = [82 GUY]
Myrciaria chartacea O.Berg in C.F.P.von Martius & auct. suc. (eds.), Fl. Bras. 14(1): 371 (1857). = [84 BZL]
Myrciaria micrantha O.Berg in C.F.P.von Martius & auct. suc. (eds.), Fl. Bras. 14(1): 360 (1857). = ?
Myrciaria silveirana D.Legrand, in Fl. Ilustr. Catar. 1(Mirt., Suppl. 1): 7 (1977). = [84 BZS]
Myrtekmania intermedia Bisse, Revista Jard. Bot. Nac. Univ. Habana 4(2): 6 (1983). = [81 CUB]
Myrtus acuminata Sessé & Moç., Fl. Mexic., ed. 2: 124 (1894). = [81 PUE]
Myrtus aemulans Schltr., Bot. Jahrb. Syst. 40(29): 29 (1908). = [60 NWC]
Myrtus aequalis Vell., Fl. Flumin. 5: 217, t. 76 (1829). = [84 BZL]
Myrtus aeruginosa Griseb., Abh. Königl. Ges. Wiss. Göttingen 24: 127 (1879). = [85 AGE]
Myrtus anguillensis Urb., Symb. Antill. 6: 21 (1909). = [81 LEE]
Myrtus angustifolia Parodi, Anales Soc. Ci. Argent. 7: 186 (1879), nom. illeg. = [85 PAR]
Myrtus berlandiereana O.Berg, Linnaea 27: 403 (1856). = [79 MXE MXG]
Myrtus bracteifolia Sessé & Moç., Fl. Mexic., ed. 2: 124 (1894). = [81 PUE]
Myrtus capensis Burm.f., Fl. Indica, Prodr. Fl. Cap.: 14 (1768). = [27 CPP]
Myrtus conferta Sessé & Moç., Fl. Mexic., ed. 2: 125 (1894). = [79]
Myrtus corynantha Kiaersk., Enum. Myrt. Bras.: 18 (1893). = [84 BZL] *Psidium* sp. ?
Myrtus disperma Sessé & Moç., Fl. Mexic., ed. 2: 125 (1894). = [79]
Myrtus emarginata Sessé & Moç., Fl. Mexic., ed. 2: 124 (1894), nom. illeg. = [81 PUE]
Myrtus engleriana Schltr., Bot. Jahrb. Syst. 40(92): 29 (1908). = [60 NWC]
Myrtus fasciculata Vell., Fl. Flumin. 5: 210, t. 69 (1829). = [84 BZL]
Myrtus flavida Schltr., Bot. Jahrb. Syst. 40(91): 30 (1908), nom. illeg. = [60 NWC]
Myrtus fulva Spreng., Syst. Veg. 2: 487 (1825). = [84]
Myrtus glazioviana Kiaersk., Enum. Myrt. Bras.: 23 (1893). = [84 BZL] *Eugenia* sp. ?
Myrtus hypericifolia Salisb., Prodr. Stirp. Chap. Allerton: 354 (1796). = [29 REU]
Myrtus kuma Barnéoud in C.Gay, Fl. Chil. 2: 384 (1847). = [85 CLC]
Myrtus laevis Thunb. in J.A.Murray, Syst. Veg. ed. 14: 461 (1784). = [38 JAP]
Myrtus longifolia Kunth in F.W.H.von Humboldt, A.J.A.Bonpland & C.S.Kunth, Nov. Gen. Sp. 6: 258 (1824). = (Melastomataceae)
Myrtus mapirensis Rusby, Bull. New York Bot. Gard. 8: 108 (1912). = [83 BOL]
Myrtus megapotamica Spreng., Syst. Veg. 4(2): 193 (1827). = [84]
Myrtus micarensis Urb., Symb. Antill. 9: 81 (1923). = [81 CUB]
Myrtus moana Urb., Symb. Antill. 9: 81 (1923). = [81 CUB]
Myrtus montevidensis Kuntze, Revis. Gen. Pl. 3(2): 92 (1898). = [85 URU]
Myrtus neocaledonica Nied. in H.G.A.Engler & K.A.E.Prantl, Nat. Pflanzenfam. 3(7): 67 (1893). = [60 NWC]
Myrtus nobilis Larrañaga, Escritos D. A. Larrañaga 2: 169 (1923). = [85 URU]
Myrtus oreogena Schltr., Bot. Jahrb. Syst. 40(92): 30 (1908). = [60 NWC]
Myrtus ovalifolia Cambess. in A.F.C.de Saint-Hilaire, Fl. Bras. Merid. 2: 381 (1833). = [84+] Psidium ?
Myrtus parviflora Sessé & Moç., Fl. Mexic., ed. 2: 124 (1894). = [81 PUE]
Myrtus pavonii Poepp. ex O.Berg, Linnaea 27: 38 (1855). = [83 PER]
Myrtus pendula Blume, Bijdr.: 1085 (1826). = [42 JAW] (Rubiaceae) ?
Myrtus pulchella Regel, Index Seminum (LE) 1857: 56 (1857). = ?
Myrtus pulchrifolius Guillaumin, Mém. Mus. Natl. Hist. Nat., B, Bot. 8: 289 (1962). = [60 NWC]
Myrtus racemosa Sessé & Moç., Fl. Mexic., ed. 2: 125, col. 1 (1894). = ?
Myrtus racemosa Sessé & Moç., Fl. Mexic., ed. 2: 125, col. 2 (1894). = [79]
Myrtus ramiflora Vahl in H.West, Bidr. Beskr. Ste Croix: 290 (1793). = [81 LEE]
Myrtus reticulata Kunze ex C.Gay, Fl. Chil. 2: 380 (1847). = [85]
Myrtus sarandi Larrañaga, Escritos D. A. Larrañaga 3: 85 (1924). = [85 URU]
Myrtus scabra Arruda in H.Koster, Trav. Brazil: 491 (1816). = [84]
Myrtus suaveolens Parodi, Anales Soc. Ci. Argent. 7: 220 (1879). = [85 PAR]
Myrtus trifida Sessé & Moç., Fl. Mexic., ed. 2: 124 (1894). = [81 PUE]
Myrtus triflora Spreng., Syst. Veg. 2: 482 (1825). = [85 AGE]
Myrtus vestita Spreng., Syst. Veg. 4(2): 193 (1827). = [84]
Myrtus virotii Guillaumin, Mém. Mus. Natl. Hist. Nat., B, Bot. 4: 33 (1953). = [60 NWC]
Myrtus vulcani Korth., Ned. Kruidk. Arch. 1: 197 (1847). = [42 SUM]
Myrtus yapacani Kuntze, Revis. Gen. Pl. 3(2): 89 (1898). = [84]
Paramitranthes sulcata Burret, Notizbl. Bot. Gart. Berlin-Dahlem 15: 542 (1941). = [84 BZL] *Plinia* sp. ?
Pseudanamomis nipensis Bisse, Feddes Repert. 96: 512 (1985). = [81 CUB]
Pseudeugenia stolonifera D.Legrand & Mattos, Arq. Bot. Estado São Paulo, n.s., f.m., 4: 63 (1966). = [84 BZL] *Eugenia* sp. ?
Pseudocaryophyllus pabstianus D.Legrand, Bradea 1: 153 (1972). = [84 BZL]

Pseudocaryophyllus uniflorus Burret, Notizbl. Bot. Gart. Berlin-Dahlem 15: 515 (1941). = [84 BZL]
Psidium acre Ten., Index Seminum (NAP) 1829: 17 (1829). = ?
Psidium anthomega Vell., Fl. Flumin. 5: 212, t. 51 (1829). = [84 BZL]
Psidium apysa Parodi, Anales Soc. Ci. Argent. 7: 66 (1879). = [85 AGE]
Psidium arasa-hu Parodi, Anales Soc. Ci. Argent. 7: 65 (1879). = [85 AGE]
Psidium arasa-pe Parodi, Anales Soc. Ci. Argent. 7: 65 (1879). = [85 AGE]
Psidium arasope-mi Parodi, Anales Soc. Ci. Argent. 7: 65 (1879). = [85 AGE]
Psidium bahorucanum Alain & Ricardo García, Moscosoa 9: 22 (1997). = [81 DOM]
Psidium berteroanum O.Berg, Linnaea 27: 374 (1856). = [81 PUE]
Psidium buxifolium Nutt., N. Amer. Sylv. 1: 115 (1842). = ?
Psidium caninum Lour., Fl. Cochinch.: 310 (1790). = (Rosaceae)
Psidium cuneifolium Ten., Index Seminum (NAP) 1833: 14 (1833). = ?
Psidium cuspidatum Alain, Brittonia 20: 159 (1968), nom. illeg. = [81 DOM]
Psidium elegans Miq., Fl. Ned. Ind. 1(1): 470 (1855). = [42 MOL]
Psidium humile Vell., Fl. Flumin. 5: 211, t. 49 (1829). = [84 BZL]
Psidium indicum Raddi, Alc. Sp. Pero: 6 (1821). = ?
Psidium jollyanum A.Chev., Explor. Bot. Afrique Occ. Franç. 1: 266 (1920). = [22 IVO]
Psidium latifolium Link, Enum. Hort. Berol. Alt. 2: 27 (1822). = ?
Psidium longifolium Schumach., Beskr. Guin. Pl.: 229 (1827). = [22]
Psidium navasense Britton & P.Wilson, Mem. Torrey Bot. Club 16: 85 (1920). = [81 CUB] Eugenia ?
Psidium nigrum Lour., Fl. Cochinch.: 311 (1790). = (Rosaceae)
Psidium passeanum André, Rev. Hort. 68: 233 (1890). = ?
Psidium rubrum Lour., Fl. Cochinch.: 311 (1790). = [41 VIE]
Psidium subrostrifolium Mattos, Loefgrenia 90: 3 (1986). = [84 BZL]
Psidium tomentosum Parodi, Anales Soc. Ci. Argent. 7: 64 (1879). = [85 AGE]
Psidium turbinatum Mattos, Loefgrenia 94: 12 (1989). = [84 BZC]
Psidium ubatubense Mattos, Ci. & Cult. 19: 332 (1967). = [84 BZL]
Psidium willemetianum DC., Prodr. 3: 237 (1828). = [29 MAU]
Siphoneugena aromatica O.Berg, Linnaea 27: 345 (1856). = [84+] *Syzygium* sp. ?
Siphoneugena krukoffiana Kausel, Lilloa 33: 110 (1971 publ. 1972). = [83 BOL]
Syzygium sieberianum C.Presl, Abh. Königl. Böhm. Ges. Wiss., V, 3: 499 (1845). = [29 MAU]
Thryptomene fimbriata D.A.Herb., J. Proc. Roy. Soc. W. Australia 7: 88 (1921). = [50 WAU]
Tristania albicans G.Benn., Gatherings Natural. Austral.: 285 (1860). = ?
Tristania densiflora Carr, Rev. Hort. 1881: 420 (1881). = ?

www.ingramcontent.com/pod-product-compliance
Ingram Content Group UK Ltd.
Pitfield, Milton Keynes, MK11 3LW, UK
UKHW012145240726
13966UKWH00001B/162